CW00938083

International Handbook of
Earthquake and Engineering Seismology

To:

Susan Francis

Best wishes,

Willie Lee

This is Volume 81B in the
INTERNATIONAL GEOPHYSICS SERIES
A series of monographs and textbooks
Edited by RENATA DMOWSKA, JAMES R. HOLTON AND H. THOMAS ROSSBY

A complete list of books in this series appears at the end of this volume

International Handbook of Earthquake and Engineering Seismology

PART B

Edited by

William H. K. Lee, Hiroo Kanamori, Paul C. Jennings, and Carl Kisslinger

A project of the Committee on Education
International Association of Seismology and Physics of the Earth's Interior

in collaboration with
International Association for Earthquake Engineering

Published by Academic Press for
International Association of Seismology
and Physics of the Earth's Interior

Amsterdam Boston Heidelberg London New York Oxford
Paris San Diego San Francisco Singapore Sydney Tokyo

Senior Editor, Earth Sciences	Frank Cynar
Editorial Coordinator	Jennifer Helé
Senior Project Manager	Angela Dooley
Senior Marketing Manager	Linda Beattie
Project Management	Matrix Productions
Composition	International Typesetting & Composition
Printer	Maple-Vail

This book is printed on acid-free paper.

Academic Press
An Imprint of Elsevier Science
84 Theobald's Road, London WC1X 8RR, UK
http://www.academicpress.com

Academic Press
An Imprint of Elsevier Science
525 B Street, Suite 1900, San Diego, California 92101-4495, USA
http://www.academicpress.com

Academic Press
200 Wheeler Road, Burlington, Massachusetts 01803, USA
http://www.academicpressbooks.com

ISBN: 0-12-440658-0
CD ISBNs: CD#2 0-12-440654-8
CD#3 0-12-440659-9

Library of Congress Control Number: 2002103787

A catalogue record for this book is available from the British Library

03 04 05 06 07 9 8 7 6 5 4 3 2 1

Contents of Part B

VII. Strong-Motion Seismology

VIII. Selected Topics in Earthquake Engineering

IX. Earthquake Prediction and Hazards Mitigation

X. National and International Reports to IASPEI

XI. General Information and Miscellaneous Data

Contributors to Part B

Abrahamson, Norman A.
Geosciences Department, Mail Code N4C
Pacific Gas and Electric Co.
245 Market Street
San Francisco, CA 94177, USA

Ahern, Timothy K.
IRIS
1408 NE 45th Street #201
Seattle, WA 98105, USA

Aki, Keiiti
Observatoire du Piton de la Fournaise
14 RN3 27 Eme Km
97418 La Plaine Des Cafres
La Réunion, France

Anderson, John G.
Seismological Laboratory
Mackay School of Mines
University of Nevada at Reno
Reno, NV 89557, USA

Baer, Manfred
Institute of Geophysics & Swiss Seismological Service
Eidgenössische Technische Hochschule Zürich
CH-8093 Zürich, Switzerland

Bardet, Jean-Pierre
Department of Civil Engineering
University of Southern California
Los Angeles, CA 90089
USA

Ben-Zion, Yehuda
Department of Earth Sciences
University of Southern California
Los Angeles, CA 90089-0740, USA

Bolt, Bruce
1491 Greenwood Terrace
Berkeley, CA 94708, USA

Boore, David M.
U.S. Geological Survey, MS 977
345 Middlefield Road
Menlo Park, CA 94025 USA

Borcherdt, Roger D.
United States Geological Survey, MS 977
345 Middlefield Road
Menlo Park, CA 94025, USA

Campbell, Kenneth W.
ABS Consulting and EQECAT Inc.
1030 NW 161st Place
Beaverton, OR 97006, USA

Cherry, Sheldon
Department of Civil Engineering
University of British Columbia
Vancouver B.C. V6T 1Z4, Canada

Christensen, Nikolas I.
Department of Geology and Geophysics
University of Wisconsin
1215 W. Dayton Street
Madison, WI 53706, USA

Clévédé, Éric
Institut de Physique du Globe
de Paris
4, Place Jussieu
75252 Paris cedex 05, France

Diggles, Michael F.
U.S. Geological Survey. MS 951
345 Middlefield Road
Menlo Park, CA 94025 USA

Dodge, Doug
L-205
Lawrence Livermore National Laboratory
Livermore, CA 94550, USA

Dohrenwend, John C.
223 South State Street
Teasdale, UT 84773, USA.

Dost, Bernard
ORFEUS & Seismology Division
Royal Netherlands Meteorological
Institute (KNMI)
P.O. Box 201
3730 AE De Bilt, The Netherlands

Dreger, Douglas S.
Berkeley Seismological Laboratory
University of California
Berkeley, CA 94720, USA

Espinosa-Aranda, J. M.
Centro de Instrumentación y Registro Sísmico A.C.
Anaxágoras 814, Narvarte
México D. F. 03020, México

Faccioli, Ezio
Department of Structural Engineering
Politecnico
Piazza Leonardo da Vinci 32
20133 Milano, Italy

Firpo, Mike
L-205
Lawrence Livermore National Laboratory
Livermore, CA 94550, USA

Garcia-Fernandez, Mariano
Institute of Earth Sciences "Jaume Almera" / C.S.I.C.
Llúis Soĺei Sabaŕis, s/n
E-08028 Barcelona, Spain

Gee, Lind
Berkeley Seismological Laboratory
University of California
Berkeley, CA 94720, USA

Giardini, Domenico
Institute of Geophysics
ETH-Hoenggerberg
Zurich, CH-8093, Switzerland

Goldstein, Peter
L-205
Lawrence Livermore National Laboratory
Livermore, CA 94550, USA

Graizer, V. M.
California Strong Motion Instr. Program
California Geological Survey
801 K St., MS 13-35
Sacramento, CA 95814, USA

Gruenthal, Gottfried
GeoForschungsZentrum
Telegrafenberg E428
Potsdam D-14473, Germany

Hall, John
Department of Civil Engineering, MC 104-44
California Institute of Technology
Pasadena, CA 91125, USA

Hamburger, Ronald O.
Principal, Simpson Gumpertz & Heger Inc.
San Francisco, CA 94108, USA

Harris, Ruth A.
U.S. Geological Survey, Mail Stop 977
345 Middlefield Road
Menlo Park, CA 94025, USA

Hauksson, Egill
Seismology Lab 252-21
California Institute of Technology
Pasadena, CA, 91125, USA

Havskov, Jens
Department of Geoscience
University of Bergen
Allegt. 41
5007 Bergen, Norway

Howell, Benjamin Franklin, Jr.
Penn State University
406 Deike Building
University Park, PA 16802, USA

Huang, M. J.
California Strong Motion Instr. Program
California Geological Survey
801 K St., MS 13-35
Sacramento, CA 95814, USA

Jennings, Paul C.
Department Civil Engineering and
Applied Mechanics
MC 104-44
California Institute of Technology
Pasadena, CA 91125, USA

Jones, L. M.
U.S. Geological Survey
525 S. Wilson Avenue
Pasadena, CA, 91106, USA

Kanamori, Hiroo
Seismo Lab, MS 252-21
California Institute of Technology
Pasadena, CA 91125, USA

Kawase, Hiroshi
Graduate School of Human
Environment Studies
Kyushu University
6-10-1 Hakozaki, Higashi-ku
Fukuoka 812-8581, Japan

Kennett, Brian L. N.
Research School of Earth Sciences
Australian National University
Canberra ACT 0200, Australia

Kinoshita, Shigeo
Faculty of Science, Yokohama City University
Seto22-2, Kanazawa-ku
Yokohama 236-0027, Japan

Kircher, Charles A.
Principal, Kircher & Associates
1121 San Antonio Road Suite D-202
Palo Alto, CA 94305, USA

Kisslinger, Carl
CIRES/Campus Box 216
University of Colorado
Boulder, CO 80309, USA

Klein, Fred
U.S. Geological Survey, MS 977
345 Middlefield Road
Menlo Park, CA 94025 USA

Lahr, John C.
U.S. Geological Survey, MS 966
P. O. Box 25046
Denver, CO 80225, USA

Lee, William H. K.
U.S. Geological Survey, MS 977
345 Middlefield Road
Menlo Park, CA 94025 USA

Liu, Chun-Chi
Institute of Earth Sciences
Academia Sinica
128, Section 2, Yen Chiu Yuan Road
Nankang, Taipei 115, Taiwan

Lognonné, Philippe
Institut de Physique du Globe de Paris
Departement des Etudes Spatiales
4 avenue de Neptune
Saint Maur des Fosses 94107, France

Mileti, Dennis S.
Natural Hazards Research & Applications Information Center
University of Colorado
Campus Box 482
Boulder, CO 8039-0482, USA

Minner, Lee
Lawrence Livermore National Laboratory, L-205
Livermore, CA 94550, USA

Moriwaki, Yoshiharu
GeoPentech
601 North Parkcenter Drive, Suite 110
Santa Ana, CA 92705, USA

Neuhauser, Douglas
Berkeley Seismological Laboratory
University of California
Berkeley, CA 94720, USA

North, Robert
Science Applications International Corporation
Center for Monitoring Research
1300 N. 17th Street, Suite 1450
Arlington, VA 22209, USA

O'Brien, Paul W.
Department of Sociology/Criminal Justice
California State University, Stanislaus
801 W. Monte Vista
Turlock, CA 95382, USA

Ogata, Yosihiko
Institute of Statistical Mathematics
Minami-Azabu 4-6-7
Minato-Ku, Tokyo, Japan

Ordaz, M.
Instituto de Ingenieria
UNAM, Ciudad Universita
04150 Mexico DF, Mexico

Ottemöller, Lars
British Geological Survey
Murchison House
West Mains Road
Edinburgh EH9 3LA, Scotland, UK

Pacheco, J. F.
Instituto de Geofisica
UNAM, Ciudad Universita
04150 Mexico DF, Mexico

Pasyanos, Michael
L-205
Lawrence Livermore National Laboratory
Livermore, CA 94550, USA

Pessina, Vera
INGV — Instituto Nazionale di Geofisica e Vulcanologia
Sezione "Pericolosita'e Rischio Sismico"
via Bassini 15
20133 Milano, Italy

Pujol, Jose
Center for Earthquake Research and Information
The University of Memphis
Memphis, TN 38152, USA

Richards, Paul G.
Lamont-Doherty Earth Observatory
Columbia University
61 Route 9W
Palisades, NY 10964, USA

Rodríquez-Cayeros, F. H.
RO Consultores Especializados S. C.
Providencia 1206-2, Col del Valle
México D. F. 03100, México

Romanowicz, Barbara
Berkeley Seismological Laboratory
University of California
Berkeley, CA 94720, USA

Romney, Carl F.
4105 Sulgrave Drive
Alexandria, VA 22309, USA

Rymer, Michael J.
U.S. Geological Survey, MS 977
345 Middlefield Road
Menlo Park, CA 94025 USA

Sambridge, Malcolm
Research School of Earth Sciences
Australian National University
Canberra ACT 0200, Australia

Schweitzer, Johannes
NORSAR
P.O. Box 53
N-2027 Kjeller, Norway

Shakal, Anthony F.
California Strong Motion Instr. Program
California Geological Survey
801 K St., MS 13-35
Sacramento, CA 95814, USA

Shedlock, Kaye
U.S. Geological Survey, MS 966
Box 25046, Denver Federal Center
Denver, CO 80225, USA

Shin, Tzay Chyn
Central Weather Bureau
64 Kung Yuan Road
Taipei, Taiwan

Singh, S. K.
Instituto de Geofisica
UNAM, Ciudad Universita
04150 Mexico DF, Mexico

Snoke, J. Arthur
Department of Geological Sciences (0420)
Virginia Tech and State University
Blacksburg, VA 24061, USA

Somerville, Paul G.
URS Corporation
566 El Dorado Street
Pasadena, CA 91101, USA

Spudich, Paul
U.S. Geological Survey. MS 977
345 Middlefield Road
Menlo Park, CA 94025 USA

Stanley, Darrell
Department of Geology and Geophysics
University of Wisconsin
1215 W. Dayton Street
Madison, WI 53706, USA

Tsai, Yi-Ben
Department of Geophysics
National Central University
Chung-Li 32054, Taiwan

Uhrhammer, Robert
Berkeley Seismological Laboratory
University of California
Berkeley, CA 94720, USA

Valdes, C. M.
Instituto de Geofisica
UNAM, Ciudad Universita
04150 Mexico DF, Mexico

van Eck, Torild
ORFEUS & Seismology Division
Royal Netherlands Meteorological Institute (KNMI)
P.O. Box 201
3730 AE De Bilt, The Netherlands

Vere-Jones, David
Victoria University and Statistical
Research Associates
P. O. Box 600 Wellington, New Zealand

Wielandt, Erhard
University of Stuttgart
Institute of Geophysics
Richard-Wagner-Str. 44
Stuttgart, D-70184, Germany

Wieprecht, David E.
US Geological Survey
Cascades Volcano Observatory
1300 S.E. Cardinal Ct., Bldg 10, Suite 100
Vancouver, WA 98683, USA

Wu, Yih-Min.
Seismology Center
Central Weather Bureau
64 Kung Yuan Road
Taipei, Taiwan

Xu, Lisheng
Institute of Geophysics
China Seismological Bureau
Beijing 100081, China

Yang, Xiaoping
Science Applications International Corporation
Center for Monitoring Research
1300 N. 17th Street, Suite 1450
Arlington, VA 22209, USA

Yeh, Yeong Tein
Digital Earth and Disaster Reduction Research Center
Ching Yun Institute of Technology
229 Chien-Hsin Road
Jung-Li 320, Taiwan

Youd, T. Leslie
Department of Civil and Environmental Engineering
368 Clyde Building
Brigham Young University
Provo, UT 84602-4009, USA

Zhang, Peizhen
Institute of Geology
China Seismological Bureau
Beijing 100029, China

Preface to Part B

This volume completes the two-volume set of the *International Handbook of Earthquake and Engineering Seismology*, with 34 chapters, numerous appendices, and two additional CD-ROMs.

Section VII features eight chapters on strong motion seismology. Section VIII includes seven chapters on earthquake engineering. Section IX presents seven chapters on earthquake prediction and hazards mitigation. Section X contains three chapters (consisting of 84 sub-chapters) of summarized national, institutional, and international reports to the International Association of Seismology and Physics of Earth's Interior (IASPEI) and the International Association for Earthquake Engineering (IAEE). Section XI includes detailed descriptions of earthquake seismology software. The four appendices feature a glossary of seismological terms, key formulae, biographical sketches, and a user's manual for each of the three CD-ROMs. The two compact discs included with Part B include supplementary materials to chapters in this volume, as well as unabridged national and international reports. Each CD includes a convenient "Start_Here" file listing the complete contents, and tips to facilitate their utility.

Together, the three CDs included in the *Handbook* constitute a digital library of selected earthquake literature and data, the equivalent of approximately 300,000 pages. The CDs include scanned images of more than 100 hard-to-find and out-of-print publications.

The accomplishments and history of 56 countries are detailed within the national reports. Collectively, they complement the topical reviews printed in the first 78 chapters of the *Handbook*. The *Handbook's* digital library also features a unique collection of more than 2,100 biographical sketches of earthquake scientists and engineers.

Additional features of the *Handbook* include a global inventory of seismographic networks, a collection of useful earthquake seismology software, an extensive glossary of seismological and earthquake engineering terms, key formulae employed in earthquake seismology, detailed biographies of selected scientists and engineers, and an extensive collection of seismic bulletins and reports published prior to 1920.

The *Handbook* was completed with the efforts of some two thousand volunteers over an eight-year period. We hope that this long-awaited *Handbook* presents a useful summary of the status and accomplishments in earthquake and engineering seismology as we go forward into the 21st century.

William H. K. Lee, Hiroo Kanamori, Paul C. Jennings,
and Carl Kisslinger, Editors
April 17, 2003

VII

Strong-Motion Seismology

57

Strong-Motion Seismology

John G. Anderson
University of Nevada, Reno, Nevada, USA

1. Introduction

Strong-motion seismology is concerned with the measurement, interpretation, and estimation of strong shaking generated by potentially damaging earthquakes. Measurements of strong shaking generated by large earthquakes provide a principal research tool. First, these data are essential to understand the high-frequency nature of crustal seismogenic failure processes, the nature of seismic radiation from the source, and the nature of crustal wave-propagation phenomena near the source, all of which have a first order effect on the seismic loads applied to the physical environment. Second, these measurements are a principal tool used to develop empirical descriptions of the character of strong shaking. Principal goals of strong-motion seismology are to improve the scientific understanding of the physical processes that control strong shaking and to develop reliable estimates of seismic hazards for the reduction of loss of life and property during future earthquakes through improved earthquake-resistant design and retrofit.

The strongest earthquake motions that have been recorded to date have peak accelerations between $1g$ and $3g$, where $1g$ (=980 cm/sec^2) is the acceleration of the Earth's gravity field, although records with such large peak values are rare. It is less clear what threshold of ground motion needs to be exceeded to be considered "strong motion." Probably a logical level to choose would be about 10 cm/sec^2, as the older strong-motion instruments (accelerographs) that traditionally defined the field are not able to resolve ground accelerations with amplitudes smaller than this. Modern digital accelerographs are much more sensitive, able to resolve peak accelerations to 0.1 cm/sec^2 or smaller. People at rest are able to feel motions as small as 1 cm/sec^2. In moderate magnitude earthquakes, damage to structures that are not designed for earthquake resistance appears at accelerations of about 100 cm/sec^2.

Earthquakes with magnitude less than 5 are of minor concern for strong-motion seismology. They are not known to damage structures of modern construction. Only a tiny fraction of such small events has caused deaths, and the number of deaths is small except in extraordinary circumstances of extremely poor construction and faulting directly under a population center. As the magnitude grows, both the destructive capability and average number of deaths per event also grows. Events with magnitudes between 6 and 7.5 are most commonly responsible for significant disasters. Globally, on average, there are 130 earthquakes per year with magnitude in the range 6.0–6.9, and 15 earthquakes per year with magnitude in the range 7.0–7.9 (see Chapter 41 by Engdahl and Villasenor). Many of these, of course, are in remote locations and have little impact. Events with magnitude of 8 or above, with their immense destructive potential, fortunately occur at an average rate of only about 0.7 per year, with some in remote locations, so disasters caused by such events are less frequent.

Observation of strong motion is more difficult than observation in other fields of seismology due to the infrequency of large earthquakes and the difficulty of anticipating areas of strong shaking for instrumentation. Weak motions from an earthquake with magnitude greater than 6 can be recorded worldwide. Thus, a seismologist who studies teleseisms can record over 100 earthquakes per year, for interpretations of earth structure or tectonics. Similarly, local networks are generally set to the most sensitive level possible to detect and locate the smallest earthquakes, which are much more abundant. The sensitive instruments used for these two branches of earthquake studies are driven off scale by strong shaking, and thus their records cannot be used. A specialized instrument, the strong-motion accelerograph, was developed in the 1930s to record strong motion. A network of these specialized instruments must have the good fortune to be located close to the earthquake, and must be maintained, often for decades, in a state of readiness to record the rare strong shaking. In the 1990s, the situation changed somewhat with improvements in digital recording technology. A strong-motion accelerograph manufactured in the second half of the 1990s has

INTERNATIONAL HANDBOOK OF EARTHQUAKE AND ENGINEERING SEISMOLOGY, VOLUME 81B ISBN: 0-12-440658-0

an analog-to-digital converter with resolution of at least 16, and typically 19, bits, allowing it to record nearby earthquakes with magnitude as small as 2–3 with a useful signal-to-noise ratio, without sacrificing the ability to faithfully record the strongest shaking. With that capability, it can make a meaningful contribution to network seismology and, consequently, is worth including as part of the real-time networks in seismic areas. This overlap in instrumental capabilities promises a significant increase in the amount and quality of data that is useful for the objectives of strong-motion seismology.

Useful reviews of strong-motion seismology can be found in Kramer (1996) and Campbell (2002).

2. Strong-Motion Measurements

2.1 Instruments

The founding father of the strong-motion instrumentation program in the United States is John R. Freeman. After the Tokyo, Japan, earthquake of 1923, and the Santa Barbara, California, and Montreal, Quebec, earthquakes of 1925, he stimulated important early interactions between US and Japanese institutions on earthquake engineering and wrote the first significant book in the English language on earthquake engineering, *Earthquake Damage and Earthquake Insurance* (Freeman, 1932). He particularly recognized the urgent need for an instrument to record the strong shaking during earthquakes, and the result of his lobbying efforts was that the Coast and Geodetic Survey was authorized to develop and install such instruments in 1932. Nine months after the first instruments were installed, the first significant strong-motion records were obtained from the March 10, 1933, Long Beach, California, earthquake. Chapter 2 (G. Housner) gives a more complete review of the early history of strong-motion seismology.

Strong-motion seismology requires a specialized instrument called the strong-motion accelerograph. The instrument is a self-contained unit in as compact a container as possible, e.g., a box smaller than one foot ($\sim$30 cm) on each side. The sensor (accelerometer) is typically a damped spring-mass system. The spring is stiff, giving the system a high natural frequency (typically 25–50 Hz), with the parameters selected so that accelerations of 1–2g (typically) will cause a deflection corresponding to full scale of the recording system. Some of the stiffness may be introduced electronically in a force-balance feedback loop that seeks to keep the seismic mass motionless relative to the frame of the instrument. The instrument is called an accelerometer because when the ground vibrates with frequencies less than the natural frequency of the sensor (and these frequencies usually predominate), the deflection of the sensor is proportional to the acceleration. Every accelerograph contains three accelerometers to measure the one vertical and two perpendicular horizontal components of acceleration.

In accelerographs designed in the United States through the late 1970s, the recording medium was usually photographic film. The enclosure was thus made into a light-proof box. A Japanese design used a pen to mechanically scratch a waxed paper. Newer designs convert an electrical signal to a digital format, which is recorded in a digital memory within the unit. However, probably half of the accelerographs in operation in the year 2000 still used the film recording. To record frequencies of ground motion above 20 Hz, the film must move through the camera at a fairly high rate, 1 cm/sec being typical. Digital accelerographs record 100 to 200 samples per second. To conserve recording medium, the accelerographs are built to activate the recorder when strong motion above some threshold is detected, and turn it off when the strong shaking has stopped. Trigger mechanisms in old instruments were mechanical or electromechanical sensors installed separately from the accelerometers. Digital systems monitor the signal amplitude coming from the sensors themselves. Usually, internal batteries operate the system, so that power failure will not cause a loss of data, and the batteries are kept fully charged by a trickle charging system to maintain readiness. Timing systems were not used in many of the early analog accelerographs, but they became more important as the analysis became more sophisticated. High-precision time, usually from satellite systems, is now essential with the digital accelerographs that provide records of aftershocks and other small events in large numbers. Precision time together with pre-event memory allow digital accelerographs to pinpoint earthquake locations beyond what is possible using only the traditional high-gain seismic networks.

2.2 Networks

The first network of strong-motion instruments, totaling 51 instruments, was installed by the US Coast and Geodetic Survey by the end of 1935. The total number of instruments deployed worldwide remained low until the mid-1960s, when commercially produced accelerographs became available and building ordinances began to require their installation in tall structures. In 2000, there were between 10,000 and 20,000 strong-motion instruments operating worldwide.

Any effort to identify all of the active accelerograph networks in the world is certain to be incomplete. However, an attempt, which succeeds in including at least most of the major networks, is made in Chapter 87 by Lahr and van Eck. Most seismically active countries have at least a few accelerographs in operation. Industrial nations with high seismic hazards have extensive programs. Countries with the most extensive strong-motion networks are Japan (see Chapter 63 by Kinoshita for a review of the largest Japanese network) and Taiwan (see Chapter 64 by Shin *et al.*). The United States has networks operated by the US Geological Survey (Borcherdt, 1997), the California Division of Mines and Geology (Shakal, 1997), the US Bureau of Reclamation (Viksne *et al.*, 1997), the US Army Corps of Engineers (Franklin, 1997), and other organizations, several of which are identified in the US Geological Survey (1999). Other countries whose networks have produced data from major earthquakes in recent years are Mexico (Anderson *et al.*, 1994; Quaas, 1997) and Turkey (Inan *et al.*, 1996). Ambraseys (1997) estimated that there are 2000 strong-motion instruments operating in Europe, and that these have produced over 2500 recordings. Unfortunately, it is still possible for large earthquakes to occur in populated regions of the world but not be recorded by any local strong-motion instruments (e.g., Gujarat, India, January 26, 2001, $M_w = 7.6$). Every such earthquake is both a tragedy and a lost opportunity, because there is still a critical shortage of recordings of ground motions from close distances to large earthquakes. Confident knowledge of ground motions in those circumstances can only be obtained from empirical observations.

2.3 Processing Data

Significant accelerograms recorded on film formats are eventually digitized and distributed. A great deal of effort has been devoted to this process because of the importance of accurate information for earthquake-resistant design. Although the instruments record acceleration, estimates of the ground velocity and displacement are also important for both engineering and geophysical research. Thus numerical integration of the records is important. The challenge in doing this accurately is determining exactly where the true zero level of acceleration is located on the accelerograms, both on the film record and on digital recordings (since an accelerograph is sensitive to a static field, e.g., a tilt). From film recordings, ground motions can generally be recovered for frequencies greater than about 0.2 Hz; this lower limit depends strongly on the size of the earthquake and the type of instrument. At high frequencies, data is reliable to 10 to 25 Hz on analog accelerographs and to 50 Hz or more on modern digital accelerographs. The history and current methods for data processing are discussed in Chapter 58 by Shakal *et al.*

2.4 Archiving of Accelerograms

With so many different organizations operating strong-motion accelerograph networks, a single global system of access to strong-motion data has never been completely achieved. Various systems for data recovery have been put in place. Perhaps the most thorough for the late 1900s was the World Data Center, which sought to compile and reproduce various tapes or computer disks. Seekins *et al.* (1992) prepared a CD-ROM with much of the data gathered in North America through 1986. Other CD-ROM data sets have been prepared by Alcantara *et al.* (1997), Cousins (1998), Ambraseys *et al.* (2000), Lee (2001), and Celebi *et al.* (2001). Distribution over the Internet also began in the late 1990s. In 2000, many strong-motion programs operated active Web sites for data distribution. A particularly useful US program for distribution of strong-motion data was established by the Southern California Earthquake Center (SCEC) for southern California data, and expanded with support from the Consortium of Organizations for Strong-Motion Observation Systems (COSMOS) to other data sets. The database could be accessed through the World Wide Web operated COSMOS (*http://db.cosmos-eq.org*). Data from large and small earthquakes recorded on the extensive Kyoshin Net and Kiban-Kyoshin Net in Japan can be accessed through their excellent Web sites (*http://www.k-net.bosai.go.jp/* and *http://www.kik.bosai.go.jp/*, respectively). The content of the European data is also available online (*http://www.isesd.cv.ic.ac.uk/esd/frameset.htm*, Ambraseys *et al.*, 2001).

3. Empirical Descriptions of Strong Ground Motion

Empirical descriptions of strong motion invariably incorporate a dependence on the earthquake size and distance to the active faulting. The challenge of developing empirical descriptions is discussed by both Bolt and Abrahamson in Chapter 59 and by Campbell in Chapter 60, and thus the current chapter is confined to discussion of elementary material and definitions that are fundamental to those detailed discussions.

Earthquake size is measured using a variety of magnitude scales. Important magnitudes in the United States include the moment magnitude (discussed below), the local magnitude (M_L) originally defined by Richter using an instrument with a natural period of 0.8 seconds (see Richter, 1958), the body wave magnitude (m_b) determined from short-period teleseismic P waves measured on instruments with a natural period of 1.0 seconds, the surface wave magnitude (M_S) determined from teleseismic 20-second Rayleigh waves, and the coda duration magnitude (determined by the duration of the S-wave coda on short-period seismic networks using instruments with natural periods of a fraction of a second). While all of these scales are calibrated to give similar numbers at the magnitudes of overlap, it is impossible, due to the different methods of their determination, for them to all agree at every magnitude. The average relationship among these scales and the reasons for their differences are discussed by Heaton *et al.* (1986), and in Chapter 44 by Utsu (see also Chapter 60 by Campbell). The moment magnitude is at present the preferred scale for correlation with ground-motion characteristics, even though it is the last one to have been developed.

3.1 Parameters to Describe Strong Ground Motion

Because of the importance of strong-motion recordings for earthquake engineering, a number of different parameters have come into use to represent various characteristics of strong-motion recordings. To illustrate these, it is convenient to consider an example drawn from the collection of significant strong-motion records. Figure 1 shows four of the strong-motion accelerograms that were obtained in the September 19, 1985, Michoacan, Mexico, earthquake ($M_w = 8.0$). A report on these data is given by Anderson *et al.* (1986). Figure 2 shows the velocity at one of those stations, Caleta de Campos, integrated from the

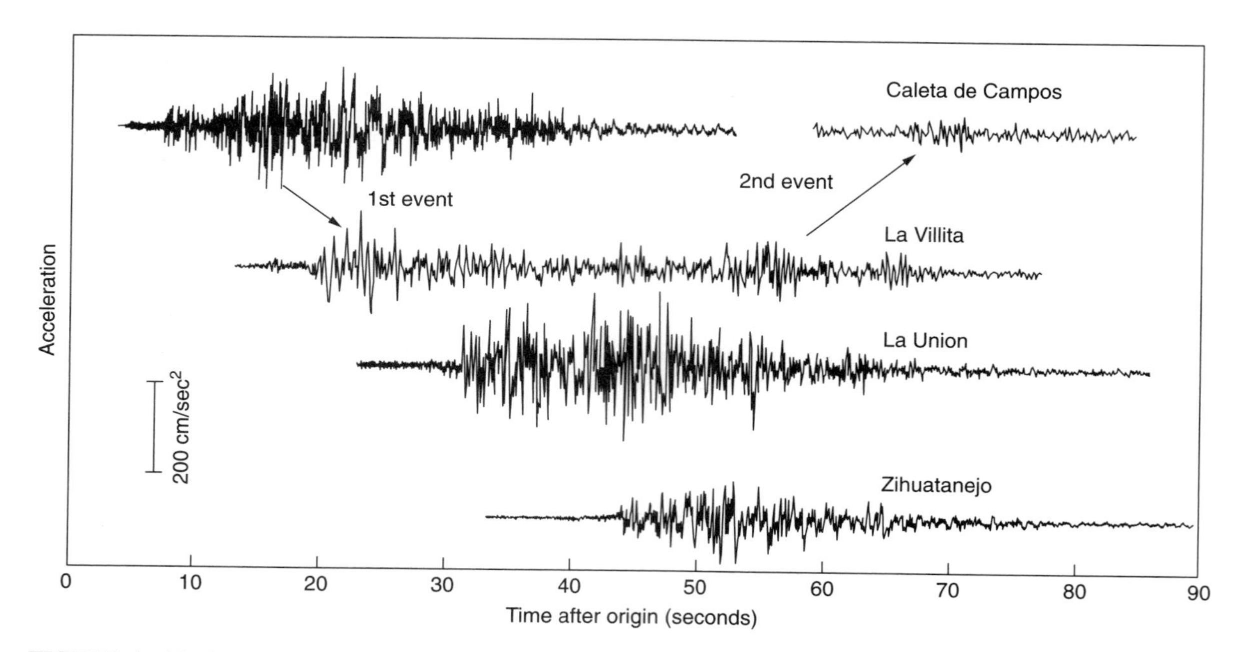

FIGURE 1 North–south component of acceleration at four strong-motion stations that recorded the September 19, 1985, Michoacan, Mexico, earthquake ($M_w = 8.0$). The Caleta de Campos station is almost directly above the hypocenter, the Zihuatanejo station is near the end of the rupture, 146 km east-southeast of Caleta de Campos, and the other stations are approximately along a line between Caleta de Campos and Zihuatanejo. The vertical separation of the traces is proportional to the separation of the stations on a projection along the strike of the causative fault. All records are aligned in absolute time after the origin of the earthquake (from Anderson *et al.*, 1986).

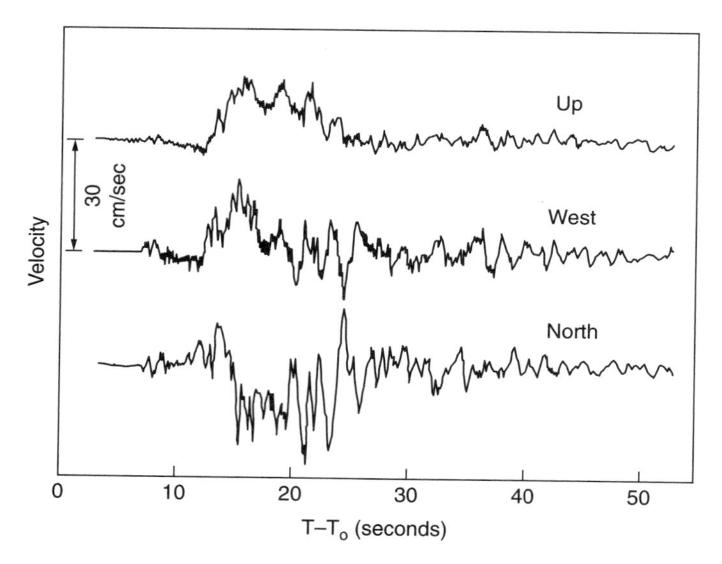

FIGURE 2 Three components of velocity at Caleta de Campos during the September 19, 1985, Michoacan earthquake (from Anderson *et al.*, 1986).

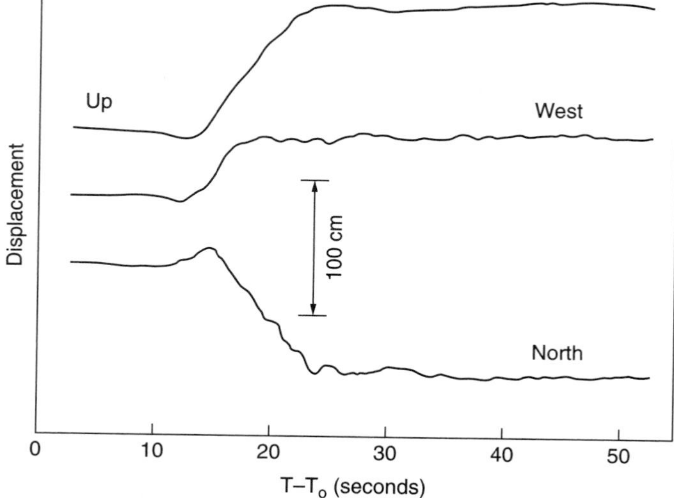

FIGURE 3 Three components of displacement at Caleta de Campos during the September 19, 1985, Michoacan earthquake (from Anderson *et al.*, 1986).

accelerograms. Figure 3 shows the corresponding displacement, integrated from the velocities given in Figure 2. The Caleta de Campos station is sited on rock above the fault plane; the fault is between 15 and 25 km below the station, dipping at about 15°. Figure 4 shows the integral of acceleration-squared from the record. Figure 5 shows the Fourier amplitude spectrum for the north component of the Caleta de Campos accelerogram, and Figure 6 shows the corresponding response spectrum.

Figures 1–6 illustrate the set of parameters typically used to characterize strong-motion records. The simplest of these is peak acceleration, which is easily obtained from the unprocessed accelerogram (Figure 1). After records are digitized, it is common

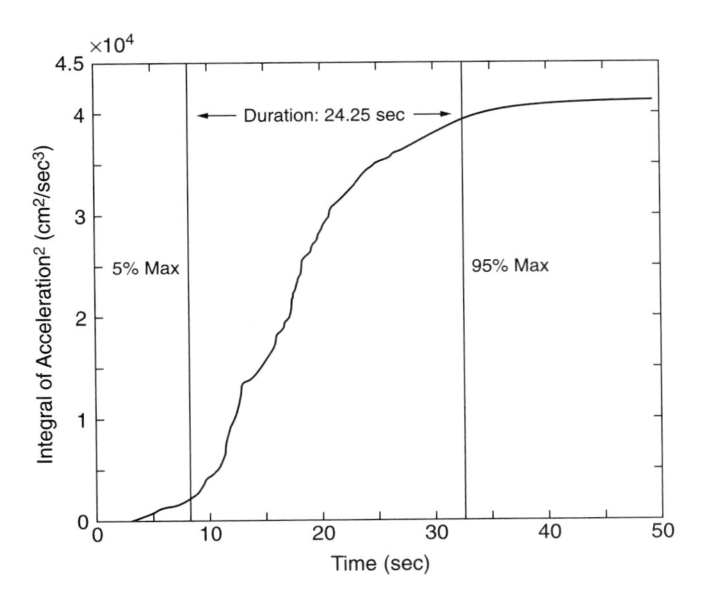

FIGURE 4 Integral of acceleration-squared determined from the north–south component of acceleration recorded at Caleta de Campos during the September 19, 1985, Michoacan earthquake. This integration is used as one method to measure the duration of the strong ground motions.

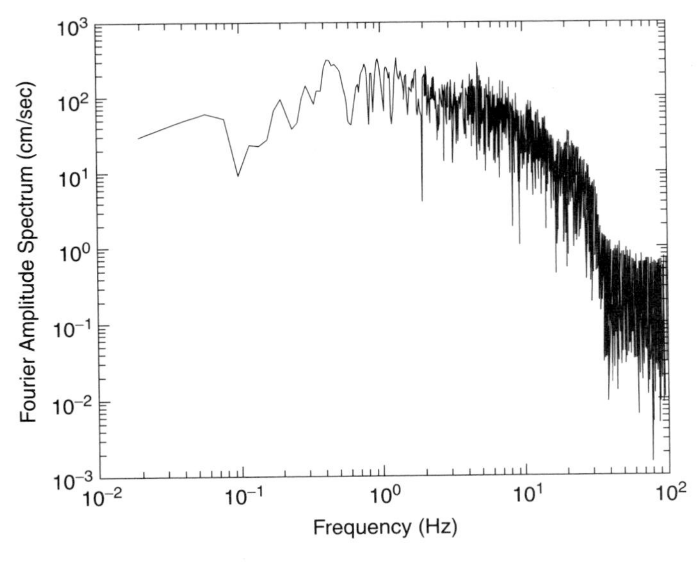

FIGURE 5 Fourier spectrum of the north–south component of acceleration recorded at Caleta de Campos during the September 19, 1985, Michoacan earthquake.

to obtain several additional parameters. Time-domain parameters often include peak velocity (Figure 2) and peak displacement (Figure 3, although it should be noted that records from which a static offset is recovered are extremely rare). Duration of the strong shaking is generally considered important, but there is not a unique definition of duration. One simple approach is to measure the interval between the times when the peak acceleration first and last exceeds some threshold, usually $0.05g$ (Bolt, 1969; Kramer, 1996). Figure 4 shows the alternative approach that was used by Trifunac and Brady (1975), namely to define the amount of time in which 90% of the integral of the acceleration-squared takes place. These two definitions lead to opposite results as distance increases. At large distances the peak ground motions decrease so that the interval duration goes to zero even though the ground was moving. On the other hand, the energy becomes dispersed, resulting in an increase in the time interval over which 90% of the total energy in the seismogram arrives. In the frequency domain, the Fourier amplitude spectrum (Figure 5) and a class of spectra known as response spectra (one of which is shown in Figure 6) are generally determined.

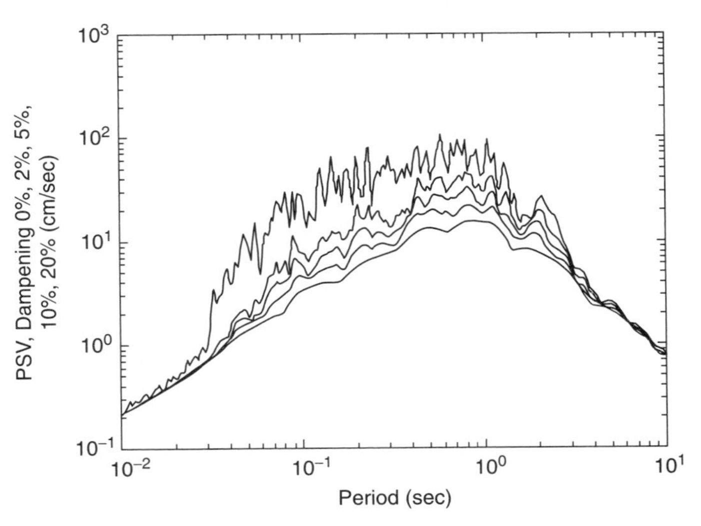

FIGURE 6 Pseudo-relative velocity response spectrum determined from the north–south component of acceleration recorded at Caleta de Campos during the September 19, 1985, Michoacan earthquake.

Response spectra describe peak time–domain response of a suite of single degree of freedom oscillators to the seismic excitation. Five types of response spectra are defined: relative displacement (*Sd*), relative velocity (*Sv*), absolute acceleration (*Sa*), pseudo-relative velocity (*PSV*), and pseudo-relative acceleration (*PSA*) (e.g., Hudson, 1979). Response spectra play an important role in the development of engineering designs. Consider a damped oscillator, which operates on the same principle as an inertial seismograph. Let the undamped natural period of this oscillator be T_o and let the fraction of critical damping be h. When the base of this oscillator is subjected to an accelerogram, there is relative motion between the seismic mass and the base. The maximum value of the relative displacement is the value of the relative displacement spectrum (*Sd*) at the period and damping of the oscillator. Thus to calculate a relative displacement spectrum, it is necessary to calculate the response to the accelerogram of a suite of oscillators with a range of T_o. The relative velocity response (*Sv*) has the value of the peak relative velocity between the seismic mass and the base. The absolute acceleration response (*Sa*) is the maximum acceleration of the seismic mass in an inertial reference frame. The pseudo-relative velocity, *PSV*, is obtained from *Sd* by $PSV = (2\pi/T_o)\,Sd$. The pseudo-relative acceleration, *PSA*, is obtained from *Sd* by $PSA = (2\pi/T_o)^2 Sd$. In general, $PSA \approx Sa$ and $PSV \approx Sv$, although these different spectra can have different asymptotic properties at high and low frequencies. The spectra are usually computed for a range of damping, from $h = 0\%$ (undamped) to $h = 20\%$ of critical. This range is used because most manmade structures are similarly lightly damped. A damping of $h = 5\%$ is the most likely to be reported. Newmark and Hall (1982), Kramer (1996), and Jennings in Chapter 67 present a review of response spectra and their applications.

Figure 6 gives *PSV* for a 5% damped oscillator for one of the accelerograms shown in Figure 1. Comparison of Figures 5 and 6 shows that in the central frequency ranges, *PSV* takes values that are comparable to a smoothed Fourier spectrum of acceleration. This is expected, as the Fourier spectral integral is in fact closely related to the Duhamel integral, which is one method used to compute the response spectra (Udwadia and Trifunac, 1973).

3.2 Ground-Motion Prediction Equations

Based on past recordings of strong earthquake motion, consisting of data such as those in Figures 1–6, the peak acceleration and other peak parameters can be described by a "ground-motion prediction equation" as a function of the earthquake magnitude, distance from the fault to the site, general site condition parameter, and sometimes other parameters. Figure 7 shows an example of one such relationship, developed by Abrahamson and Silva (1997) for crustal earthquakes in tectonically active areas. Naturally the ground-motion prediction equations do not exactly describe peaks of past earthquake ground motions. The deviation of a datum from the predicted mean peak acceleration (the "residual") is treated as a random variable, and it is consistent with a lognormal distribution function out to two standard deviations at least. The standard deviation for Figure 7 depends on magnitude. Abrahamson and Silva estimate that it is 0.7 natural (i.e., base e) logarithm units for $M < 5$, 0.43 for $M > 7$, and a linearly decreasing function of magnitude in between. For reference, a standard deviation of 0.5 corresponds to a multiplicative factor of 1.6 (times the mean value) to obtain the value that exceeds 84% of the data. Peak acceleration is a commonly employed parameter for these regressions, but peak velocity, peak displacement, spectral amplitudes, and duration have also been modeled in this manner. Recent ground-motion prediction equations are reviewed thoroughly by Campbell in Chapter 60, and also discussed by Bolt and Abrahamson in Chapter 59, and Faccioli and Pessina in Chapter 62.

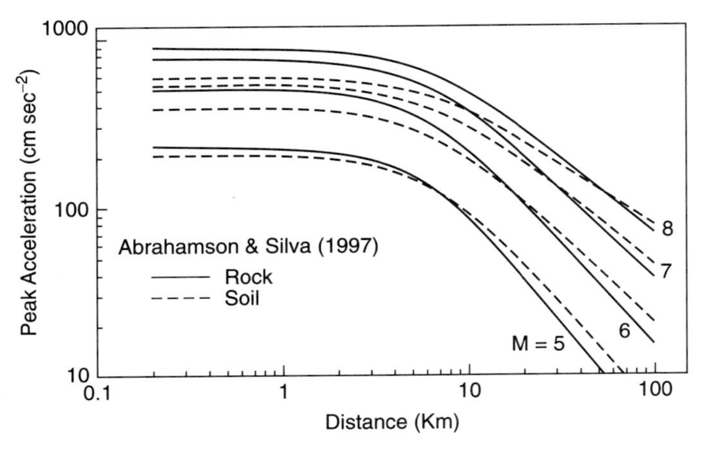

FIGURE 7 Peak acceleration as a function of magnitude and distance from the fault, as given by the ground-motion prediction equation of Abrahamson and Silva (1997).

Because ground-motion prediction equations are a key component of probabilistic seismic hazard analyses, it is clear that their development will be an important part of seismological research for some time to come. Until 1999, the estimates from these equations for large magnitudes and short distances were based more on extrapolation than on extensive amounts of data. In 1999, there were two significant earthquakes that contributed very important data at short distances. One was the Izmit, Turkey, earthquake (August 17, 1999, $M_w = 7.6$), for which Anderson *et al.* (2000a) found the peak accelerations on rock at near-fault distances to be significantly below four different ground-motion prediction equations. The other was the Chi-Chi, Taiwan, earthquake (September 20, 1999, $M_w = 7.6$), for which peak accelerations were significantly below three different recent ground-motion prediction equations developed primarily from North American data but presumed to apply to the same tectonic environment (Tsai and Huang, 2000). The discrepancy could be caused by any of several factors. Faults with a large total slip tend to be smooth and devoid of asperities, thus radiating less high-frequency energy than a fault with less total slip (Anderson *et al.*, 2000a). Faults with a high slip rate have less time to heal between earthquakes, and thus earthquakes on these faults may have lower stress drops and less high-frequency energy. Differences could arise from differing lithology in the rocks cut by the fault. The nature of the regional stress field (e.g., extensional vs. compressional) could affect the high-frequency generation (Spudich *et al.*, 1999). Somerville (2000) suggested that the discrepancy could be related to whether or not the earthquake ruptured the surface. Finally, the discrepancy could be due to the dynamics of rupture, which under some models can change when the slip becomes large enough. This issue illustrates that considerably more data, research, and basic physical understanding will be needed to reduce the uncertainties inherent in the present generation of ground-motion prediction equations.

4. Physics Contributing to the Nature of Strong-Motion Seismograms

A major goal of strong-motion seismology is to be able to synthesize strong-motion seismograms suitable for use in engineering analyses (Aki, 1980). The ability to synthesize observations depends on understanding of the physical phenomena affecting the ground motions at a site of interest when a fault identified by the geologists breaks.

In the last part of the 20th century, there was major progress toward this goal. A full understanding requires input from two disciplines. One is earthquake source physics, where the minimum requirement is a kinematical description of the slip function on the fault. The other is wave propagation, across the entire frequency band, from the fault to the station. The following sections discuss each of these fields.

4.1 Mathematical Framework for Ground-Motion Modeling

Earthquakes are caused when the rocks on opposite sides of a fault slip suddenly. The characteristics of strong ground motions that result from this instability are strongly affected by the geometrical and dynamic characteristics of the faulting. The geometrical characteristics include the size, shape, depth, and orientation of the fault area that slipped during the earthquake and the amount and direction of slip. The orientation of the fault and the amount and direction of offset may be variable over the fault surface. The important dynamic parameters include where on the fault the rupture initiated (hypocenter), how rapidly it spread over the fault (rupture velocity), how quickly (rise time) and how smoothly the slip took place at each point on the fault, and how coherently adjacent points on the fault moved. These dynamic parameters are functions of the stresses acting on the fault, the physical properties of the rock surrounding the fault, and the strength of the fault itself, and are variables over the dimension of a large fault during major earthquakes.

The process is generally expressed mathematically using a representation theorem. Using the notation of Aki and Richards (1980, p. 39), one form for the representation theorem is

$$u_n(\mathbf{x}, t) = \int_{-\infty}^{\infty} d\tau \iint_{\Sigma} [u_i(\boldsymbol{\xi}, \tau)] c_{ijpq} \nu_j \frac{\partial G_{np}(\mathbf{x}, t-\tau; \boldsymbol{\xi}, 0)}{\partial \xi_q} d\Sigma \tag{1}$$

In this equation, $u_n(\mathbf{x}, t)$ gives the nth component of the displacement of the ground at an arbitrary location $\mathbf{x}$ and at time t. The vector $\mathbf{v}$ is normal to the fault, and the positive direction of the normal defines the positive side of the fault for defining the slip discontinuity. The ith component of the discontinuity in the slip across the fault is given by

$$[u_i(\boldsymbol{\xi}, \tau)] = u_i^+(\boldsymbol{\xi}, \tau) - u_i^-(\boldsymbol{\xi}, \tau) \tag{2}$$

where ξ represents a location on the fault surface Σ and τ is the time that this displacement occurs. Because the fault is represented by the surface Σ, $d\Sigma$ represents two spatial dimensions. The Green's function is given by $G_{np}(\mathbf{x}, t; \xi, \tau)$. This gives the motion in the n direction at location $\mathbf{x}$ and time t caused by a point force acting in the p direction at location ξ and time τ. Finally, c_{ijpq} gives the elastic constants of the medium. For an isotropic medium,

$$c_{ijpq} = \lambda\delta_{ij}\delta_{kl} + \mu(\delta_{ik}\delta_{jl} + \delta_{il}\delta_{jk}), \tag{3}$$

where λ and μ are the Lame constants. The constant μ is the familiar shear modulus; there is no similarly simple experiment by which λ can be measured. By convention, in Eq. (1) summation takes place over repeated indices.

Equation (1) represents the ground motion at the site as the linear combination, through the integral over space, of the contributions from each point on the fault surface. The convolution over time incorporates the effect of the rupture at each point taking a finite amount of time to reach its final value. Through the representation theorem, the problem of predicting ground motions requires specification of the offset on the fault as a function of location and time, which incorporates earthquake source physics, and a specification of the Green's function, which incorporates wave propagation. Spudich and Archuleta (1987) discuss several techniques that are used for evaluation of Eq. (1).

Equation (1) has certain limitations. For large amplitudes of seismic waves, the stress–strain relationship in the Earth, especially near the surface, becomes nonlinear. In this case, the assumed linear superposition of waves from different parts of the fault does not apply. A common approximation for this case is to predict the motion in "rock" at some depth beneath the surface, where linearity is assumed to apply, and then treat the wave propagation from that depth to the surface as a vertical propagation problem through the nonlinear medium. On the earthquake source side, writing the time dependence in the Green's function as $(t - \tau)$ assumes that the Green's function is independent of time. This assumption can break down if the faulting process affects the propagation of the seismic waves from the source to the station. For instance, if seismic shear waves are not transmitted through the fault where it is slipping, but they are transmitted through the fault where it is not slipping, then Eq. (1) would not hold exactly. In spite of these limitations, Eq. (1) forms the mathematical basis for nearly all model-based ground-motion predictions and has enabled the inversion of strong ground-motion records to obtain models of the slip function for numerous earthquakes.

4.2 Effects of the Earthquake Source

4.2.1 Source Effects in the Far-Field

Some understanding of seismic radiation in the far-field is important for strong-motion seismology. The far-field spectrum is important because most places shaken by an earthquake are in the far-field region. For the eastern United States, models of strong motion have focused primarily on estimation of far-field ground motions. In addition, nearly all determinations of the seismic moment are made in the far-field region. In the near-field region, we expect that we cannot disregard the contributions of the far-field Green's function terms to the ground-motion spectrum, nor can we disregard the effects of spatial variation of the far-field radiation pattern for waves originating at different locations along the fault. Source effects on the far-field seismogram are discussed by Aki and Richards (1980, Chapter 14).

The seismic moment, M_0, is generally regarded as the best available single number to describe the size of an earthquake. A rigorous definition starts with the seismic moment density, which represents the internal stresses necessary to cancel the strain produced by the internal nonlinear processes that cause the earthquake. These stresses can be interpreted as a double-couple force, and the seismic moment is the magnitude of one of these couples integrated over the fault (e.g., Aki and Richards, 1980; Madariaga, 1983). The magnitude of the forces controls the amplitudes of radiated seismic waves, and thus seismic moment can be determined from seismograms. In an infinite, homogeneous, isotropic medium, seismic moment is also given by the equation

$$M_0 = \mu A D \tag{4}$$

where A is the fault area (e.g., length, L, times width, W), D is the average slip, and μ is the shear modulus. This links seismic observations with the magnitude of geological deformations. The moment magnitude, usually designated as M_w, is essentially a change of variable from seismic moment,

$$M_w = \frac{2}{3} \log M_0 - 10.73, \tag{5}$$

when the units of M_0 are dyne cm. Notation and terminology are a little confusing, as common usage differs slightly from the original literature (see Chapter 44 by Utsu). Kanamori (1977) defined an energy magnitude M_w such that

$$\log(M_0) = 1.5 M_w + 16.1, \tag{6}$$

and Hanks and Kanamori (1979) defined a moment magnitude $\mathbf{M}$ such that

$$\log(M_0) = 1.5\,\mathbf{M} + 16.1. \tag{7}$$

Harvard University, in their project to determine seismic moments for all large earthquakes worldwide, uses Eq. (5), which differs slightly from both of these definitions. The Harvard catalog is available online at *http://www.seismology.harvard.edu/projects/CMT/*.

A couple of examples of the way seismic moment can be used are in order. According to the Harvard CMT solution, the September 19, 1985, Michoacan, Mexico, earthquake had a seismic moment of 1.1×10^{28} dyne cm and $M_w = 8.0$. Based on the aftershock area (Anderson *et al.*, 1986) the fault was about $L = 170$ km long and $W = 50$ km wide. Assuming a shear modulus of 4.0×10^{11} dyne/cm^2, Eq. (4) implies $D = 320$ cm. The fault could not be examined in this case, but the static offsets inferred from the accelerograms (e.g., Figure 3) are consistent with this estimate of the slip to within a factor of less than two. A second example is the Northridge, California, earthquake of January 17, 1994. This event had a Harvard moment of 1.18×10^{26} dyne cm and $M_w = 6.7$. The fault dimension, according to Wald *et al.* (1996) was 15 km along strike and 20 km along the dip. The source was shallower than the Michoacan earthquake, so the shear modulus can be taken to be 3.3×10^{11} dyne/cm^2, implying an average slip $D = 120$ cm. Because slip of more than $\sim$10 m has practically never been observed, it becomes clear that a large seismic moment and consequently a large moment magnitude requires a very large fault area. Increasing fault dimensions have a fundamental impact on the characteristics of strong motions, as will be described following.

Because of the relationship of the moment to the size of the fault that caused the earthquake (length, width, dip, and the slip distribution), and because the fault size and slip may in some cases be estimated from geological studies, these geometrical characteristics can be used to estimate the seismic moment of anticipated earthquakes. Alternatively, when the geometry is not completely known, regression equations, such as those given by Wells and Coppersmith (1994), can relate moment magnitudes to the fault dimensions. Scatter of data relative to regression equations arises because the ratios of fault length to width to slip vary, perhaps, in part, correlated with the slip rate and/or repeat time on the fault (Kanamori and Allen, 1986; Scholz *et al.*, 1986; Anderson *et al.*, 1996).

Theoretically, the moment has a fundamental effect on strong ground motion in determining the spectral amplitudes of low-frequency ground motions. To see this, we begin with an expression for the complete vector displacement field from a point source in an infinite, homogeneous, isotropic solid. This expression, derivable from Eq. (1) (Aki and Richards, 1980, p. 81), is

$$\begin{aligned}
\mathbf{u}(\mathbf{x}, t) = {} & \frac{1}{4\pi\rho}\mathbf{A}^N \frac{1}{r^4}\int_{r/\alpha}^{r/\beta} \tau M_0(t-\tau)\,d\tau \\
& + \frac{1}{4\pi\rho\alpha^2}\mathbf{A}^{IP}\frac{1}{r^2} M_0\left(t - \frac{r}{\alpha}\right) \\
& + \frac{1}{4\pi\rho\beta^2}\mathbf{A}^{IS}\frac{1}{r^2} M_0\left(t - \frac{r}{\beta}\right) \\
& + \frac{1}{4\pi\rho\alpha^3}\mathbf{A}^{FP}\frac{1}{r}\dot{M}_0\left(t - \frac{r}{\alpha}\right) \\
& + \frac{1}{4\pi\rho\beta^3}\mathbf{A}^{FS}\frac{1}{r}\dot{M}_0\left(t - \frac{r}{\beta}\right),
\end{aligned} \tag{8}$$

where the radiation pattern terms are

$$\begin{aligned}
\mathbf{A}^N &= 9\sin 2\theta\,\cos\varphi\hat{\mathbf{r}} - 6(\cos 2\theta\,\cos\varphi\hat{\boldsymbol{\theta}} - \cos\theta\,\sin\varphi\hat{\boldsymbol{\varphi}}) \\
\mathbf{A}^{IP} &= 4\sin 2\theta\,\cos\varphi\hat{\mathbf{r}} - 2(\cos 2\theta\,\cos\varphi\hat{\boldsymbol{\theta}} - \cos\theta\,\sin\varphi\hat{\boldsymbol{\varphi}}) \\
\mathbf{A}^{IS} &= -3\sin 2\theta\,\cos\varphi\hat{\mathbf{r}} + 3(\cos 2\theta\,\cos\varphi\hat{\boldsymbol{\theta}} - \cos\theta\,\sin\varphi\hat{\boldsymbol{\varphi}}) \\
\mathbf{A}^{FP} &= \sin 2\theta\,\cos\varphi\hat{\mathbf{r}} \\
\mathbf{A}^{FS} &= (\cos 2\theta\,\cos\varphi\hat{\boldsymbol{\theta}} - \cos\theta\,\sin\varphi\hat{\boldsymbol{\varphi}}).
\end{aligned} \tag{9}$$

In Eq. (8), **u** is the displacement due to the earthquake. The earthquake is assumed to be a point source located at the origin, and $\mathbf{x} = (r, \theta, \varphi)$ is the location of the observer in spherical coordinates. The observation point is a distance r from the earthquake. $\hat{\mathbf{r}}$, $\hat{\theta}$, and $\hat{\varphi}$ are unit vectors in the direction of r, θ, and φ, increasing (respectively), so that the radial direction is along $\hat{\mathbf{r}}$ away from the source toward the site, and $\hat{\theta}$ and $\hat{\varphi}$ are both transverse directions. The angle θ is a co-latitudinal polar angle, measured between the direction that is normal to the fault and the direction to the station. The angle φ is measured in the plane containing the fault, between the direction of slip on the fault and the normal projection of the observation point into the plane. The vectors **A** are radiation patterns, which give the relative amplitudes of the different components of motion at location **x**. The superscripts on the components of **A** correspond to the types of waves: N to near-field waves, *IP* to intermediate-field P-waves, *IS* to intermediate-field S-waves, *FP* to far-field P-waves, and *FS* to far-field S-waves. $\dot{M}_0(t)$ is the moment rate. The constants in the equation are ρ, the material density, and α and β, the compressional and shear velocities in the medium.

From Eq. (8), the P-waves arrive at the location **x** after a delay of duration $\frac{r}{\alpha}$. The S-waves arrive at the location **x** after a delay of duration $\frac{r}{\beta}$. The first three terms in Eq. (8), the near- and intermediate-field motions, carry a permanent offset of the ground. The intermediate-field offsets arrive with the P-wave and the S-wave velocities, and the near-field offset arrives continuously between the P- and the S-waves. The amplitude of the near-field term decreases as r^{-4} while the intermediate-field terms decrease as r^{-2}. Thus these first three terms diminish rapidly and do not carry any energy into the far-field, but they are seen on some strong-motion records after integration to displacement. The far-field radiation terms decrease in amplitude as r^{-1}. Because the energy content of the wave is proportional to the amplitude squared, and the surface of a sphere with radius r increases as r^2, the energy carried by these waves is the same through a sphere of any radius (neglecting anelastic losses), and these terms do carry energy into the far-field.

The moment rate, $\dot{M}_0(t)$, reflects the fact that the slip on the fault is not instantaneous. Rather, it takes some time for rupture to spread from the hypocenter to the entire extent of the fault, and also it takes some time for the slip at each point to be completed. This equation treats the fault as a point source, in which the effects of these phenomena must be concentrated into a single effective moment rate. Thus it is a far-field approximation when it is applied to the interpretation of seismograms. Aki and Richards (1980, p. 805) suggest that a conservative criterion under which this approximation is strictly valid is when the wavelength λ satisfies the criteria $\lambda \gg \frac{2L^2}{r}$. For a station 30 km from a fault with length $L = 10$ km, this condition is met for frequencies substantially less than 0.5 Hz. At higher frequencies, the apparent moment rate that one would infer from the seismogram would be affected by the source finiteness, and show an azimuthal dependence. When fault finiteness is a factor, it may be possible to treat the fault as a set of several point sources, with each source radiating as in Eq. (8). Several studies have shown that the far-field moment-rate function is the Radon transform of the slip-rate function as a function of space and time (e.g., Kostrov, 1970; Ruff, 1984), and some have inverted seismograms to recover source behavior. Good examples of the source pulses seen for small earthquakes at differing azimuths are Frankel *et al.* (1986) and Fletcher and Spudich (1998).

Much can be learned from the relationship between seismic moment and the far-field P- and S-waves. We focus on the far-field S wave in Eq. (8), given by

$$\mathbf{u}^{FS}(\mathbf{x}, t) = \frac{\mathbf{A}^{FS}(\theta, \varphi)}{4\pi\rho\beta^3 r}\dot{M}_0\left(t - \frac{r}{\beta}\right), \quad (10)$$

recognizing that the application to the far-field P-wave is largely the same. The amplitude of the displacement pulse of S-waves in an infinite homogeneous medium is proportional to the moment rate. The integral of that displacement pulse over time is proportional to M_0, because

$$M_0 = \int_0^\infty \dot{M}_0(t)\, dt. \quad (11)$$

The limit of integration is written here as infinity, but the intent is that it be only long enough to include the duration of actual slip on the fault, and thus typically totals a few seconds. Consequently, moment can be estimated from the low frequencies of ground motion observations at large distances from the fault, with appropriate corrections for geometrical spreading, radiation pattern, and complexities of wave propagation.

As recognized by Aki (1967), spectral properties of the source spectrum follow from properties of the Fourier transform and from the far-field terms in Eq. (8). The far-field displacement pulse predicted by Eq. (8) is, in general, a one-sided pulse. A property of the amplitude of the Fourier transform of any one-sided pulse is that at low frequencies, it is asymptotic to the area under the pulse. Thus, M_0 can be estimated from the low-frequency asymptote of the Fourier transform of the displacement seismogram.

The Fourier transform of any one-sided pulse also has a corner frequency, f_C, say, that is inversely proportional to the duration of the pulse, with the amplitude of the transform decreasing for frequencies above the corner frequency. For example, for a boxcar function [$b(t) = 1$ for $0 \le t \le T_b$ and 0 otherwise] with a duration of T_b, the corner frequency is $f_C = 1/(2T_b)$. At the fault, the duration of faulting is controlled by the time it takes rupture to cross the fault, usually using a rupture velocity of about 90% of the shear velocity. Thus by measuring the corner frequency from the Fourier amplitude spectrum, it is possible to estimate the apparent duration of faulting at the source and the fault dimension. A relationship between corner frequency and fault dimension (radius, r_f) that is commonly used for small earthquakes was proposed by Brune (1970, 1971): $r_f = 2.34\,\beta/(2\pi f_C)$. Geometrical uncertainties (e.g., is rupture unilateral or bilateral, etc.) contribute about a factor of two to overall uncertainty.

Aki and Richards (1980) discuss additional relations and considerations for relating the corner frequency to the fault dimension. For very small events, the corner frequency can be obscured by attenuation. For larger earthquakes, strong-motion accelerograms cannot generally be assumed to be in the far-field, and the finite fault size and source-station geometry have a significant impact on the records. Nonetheless, the general prediction is that a larger earthquake, with a larger fault dimension, will generally have a smaller corner frequency. It was recognized by Aki (1967) that if $D \propto L \propto W$, implying that the stress drop is constant (e.g., Kanamori and Anderson, 1975), then $M_0 \propto L^3$, $f_C \propto L^{-1}$ and $f_C \propto M_0^{-1/3}$.

Figures 8–10 illustrate the way accelerograms recorded 25 km from the epicenter depend on magnitude, and can be used to illustrate several of these predictions. Figure 8 shows accelerograms from several earthquakes, of different magnitudes, plotted on a common scale. As the magnitude of the earthquake increases, the amplitudes of ground acceleration generally increase, and the duration increases dramatically. The increase in duration follows from the increase in the dimension of the fault. Figure 9 shows Fourier amplitude spectra corresponding to each of the accelerograms in Figure 8, and Figure 10 shows the *PSV* response spectra. The response spectra, in this case, are shown on tripartite axes. This is a common type of presentation that allows one to read the values of *PSA* and *Sd* also directly from the figure. The shape of the Fourier spectrum of acceleration can be generalized as in Figure 11. Figures 9 and 10 show that as

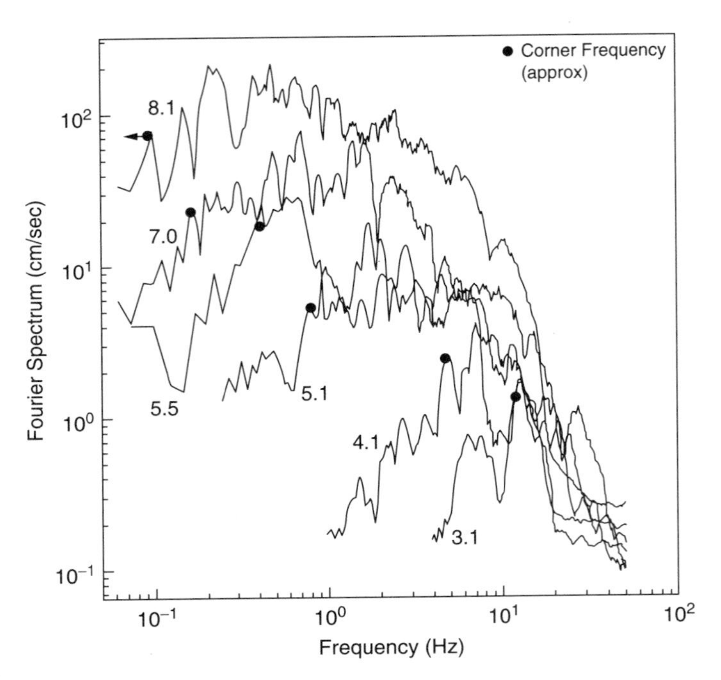

FIGURE 9 Smoothed Fourier amplitude spectra for the accelerograms shown in Figure 8 (modified from Anderson and Quaas, 1988). Circles identify approximate corner frequency on each spectrum.

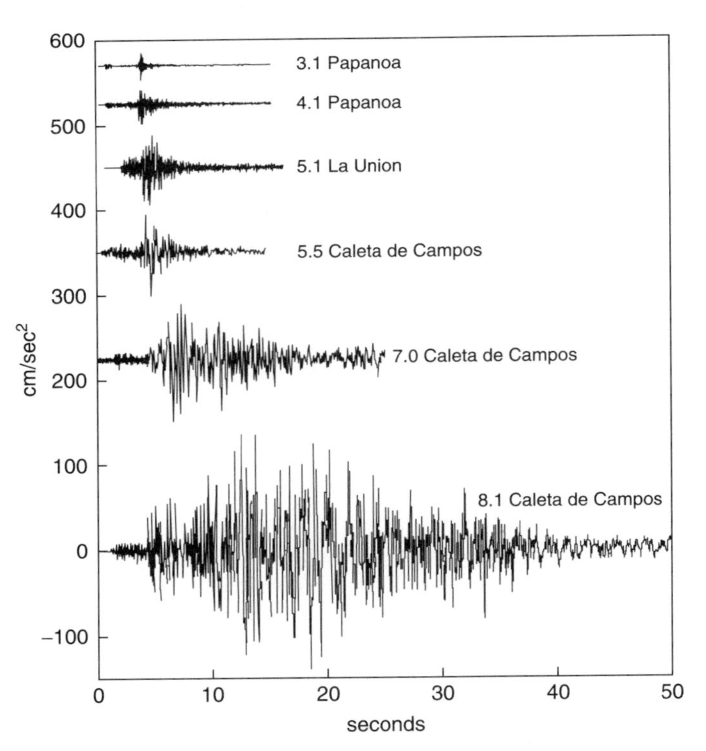

FIGURE 8 Accelerograms recorded in Guerrero, Mexico, from earthquakes of several magnitudes (from Anderson and Quaas, 1988).

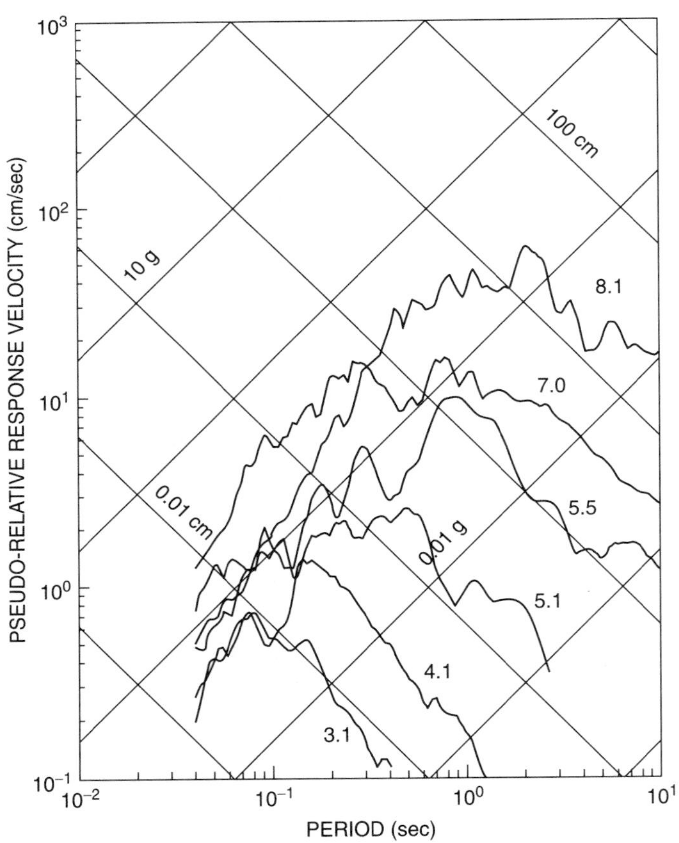

FIGURE 10 Pseudo-relative velocity response spectra of the accelerograms shown in Figure 8 (from Anderson and Quaas, 1988).

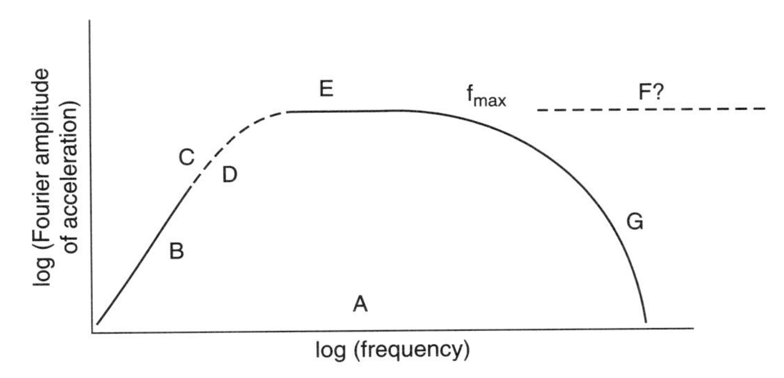

FIGURE 11 Qualitative illustration of the far-field Fourier amplitude spectrum of a strong-motion accelerogram. The overall level (A) is controlled by the distance from the seismic source. At low frequencies (B), the spectrum increases proportional to f^2, where f is the frequency. The corner frequency (C) is controlled by the fault dimension and directivity, and occurs at the frequency where this asymptotic behavior is replaced with a lower slope. The shape of the spectrum immediately above the corner frequency (D) is controlled by large-scale complexity of the fault slip as a function of space and time. The flat portion (E) is characteristic of an ω–*square* source model, meaning that the source displacement spectrum has a high-frequency rolloff of ω^{-2}. The spectrum rolls off at high frequencies (G), i.e., frequencies greater than $f_{\max}$. It is not known to what extent the spectrum at the fault is flat (F) with the rolloff due to attenuation, and to what extent the rolloff occurs due to processes at the earthquake source.

the magnitude increases, the amplitudes of the low-frequency waves increase dramatically, while the amplitudes of the high frequencies increase slowly. This change in spectral shape as the moment of the earthquake increases is the reason different magnitude scales cannot be calibrated to be identical: Different magnitudes are measured at different frequencies depending on the characteristics of the instruments, and spectral amplitudes at different frequencies do not change with the same ratio as earthquake size increases.

The spectra in Figures 5, 9, and 11 are Fourier spectra of the acceleration of the ground, not of the displacement. The relationship of the two at each frequency is

$$|\tilde{a}(f)| = (2\pi f)^2 |\tilde{u}(f)|, \tag{12}$$

where $\tilde{u}(f)$ is the Fourier transform of the displacement pulse. Thus, where the displacement spectrum is predicted to be constant at low frequencies, the acceleration spectrum $[\tilde{a}(f)]$ is predicted to increase proportional to f^2. The corner frequency is recognized at the transition from this rapid increase to a more nearly flat spectrum in the higher frequency band (Figure 11); it clearly decreases as magnitude increases (Figure 9). The magnitude dependence of the corner frequency introduces the magnitude dependence into the frequency band within which the accelerograph signal is above the instrumental noise level.

Above the corner frequency, a spectrum that falls off proportional to f^{-2} is called an ω–*square* (or ω^{-2}) model, and one that falls off proportional to f^{-3} is an ω–*cube* (or ω^{-3}) model ($\omega = 2\pi f$). In a frequency band where the displacement spectrum is proportional to f^{-2}, the acceleration spectrum is independent of frequency. In Figure 5, as is typical with strong-motion accelerograms (Hanks, 1979, 1982), this is the approximate spectral behavior above the corner frequency and up to a frequency of 5–10 Hz ($f_{\max}$, Figure 11). This approximately flat acceleration spectrum over a sizeable frequency band in strong-motion records for larger magnitude earthquakes has motivated models of strong motion as band-limited white noise (e.g., Hanks, 1979). The amplitude of this flat part increases gradually with magnitude (Figure 9). Below the corner frequency, the spectrum increases by a factor of 10–30 for each increase of a unit magnitude. These overall trends of the acceleration spectra at low to intermediate frequencies (e.g., below several Hz) are consistent with the ω–*square* model for the earthquake spectrum.

From a signal processing perspective, the rate of falloff of the spectrum of a general one-sided pulse is controlled by the sharpness of discontinuities in the displacement time series. The spectrum of the boxcar falls off proportional to f^{-1}. If the steps at the start and end of the boxcar are replaced with ramps, the falloff is proportional to f^{-2} instead. In the asymptotic limit at high frequencies, it is necessary for the seismic spectrum to fall off at least as fast as $f^{-1.5}$, because the radiated energy is finite. The more interesting interpretations, however, are in terms of the earthquake source time functions, summarized by Aki and Richards (1980). They suggest that the ω–*square* spectral shape is the result of dominance of stopping phases (i.e., signals from the edge of the expanding fault as it is arrested) in the seismic spectrum, as models for the spectrum from the start of rupture are predicted to have an ω–*cube* behavior. Indeed, some studies of strong motions from well-recorded earthquakes find evidence that high-frequency radiation is released preferentially from the margins of the areas with the largest slip, rather than from the areas that slipped the most (e.g., Zeng *et al.*, 1993b; Kakehi and Irikura, 1996, 1997; Kakehi *et al.*, 1996; Nakahara, 1998; Nakahara *et al.*, 1999).

The fine structure on most spectra make it difficult to determine shapes definitively. Many spectra seem to show a spectral fall-off closer to f^{-1} in a limited frequency band immediately above the corner frequency (e.g., Tucker and Brune, 1973); consider also the frequencies above the corner frequency in Figure 9 for the $M \sim$ 7–8 earthquakes. Brune (1970) proposed a physical interpretation of the high-frequency falloff: that f^{-2} follows from an earthquake with complete stress drop, whereas f^{-1} is an indication of partial stress drop (i.e., rupture is arrested before all of the stress driving the earthquake has been relieved).

4.2.2 Finite Source Models

While the characteristics of ground motions in the far-field are informative for strong-motion seismology, several phenomena observed in strong motion can only be understood in the context of finite source models. These include directivity, near-source pulse motions, and static offsets.

Description of the spatial distribution of slip is feasible for earthquakes that rupture a sufficiently large fault surface. Models can be developed using teleseismic data, but models that are also constrained by strong-motion data are most interesting as the strong-motion data is significantly less affected by attenuation and thus includes a much broader frequency band than teleseismic waves. These models are generally kinematic models, in that they describe a slip function as a function of location and time on the fault without assuring that this function satisfies all known physics of earthquake rupture. Some dynamic models that are consistent with known earthquake physics have also been developed recently (e.g., Mikumo and Miyatake, 1995; Olsen *et al.*, 1997, Belardinelli *et al.*, 1999). Dynamic rupture of earthquakes is reviewed by Madariaga and Olsen in Chapter 12.

Haskell (1964) proposed that a propagating ramp was a useful description of an earthquake slip function. In this model, rupture starts at one edge of a rectangular fault and propagates to the opposite edge with a constant velocity, the rupture velocity. When the rupture front reaches a point on the fault, slip begins and one side of the fault slips past the other with a constant slip velocity until a final offset is reached. The time required to reach this final offset is the rise time. Current models make use of the same concepts, but generally allow the rupture velocity, rise time, slip direction, and final offset all to be variables over the fault surface. These additional variables can be thought of as adding roughness to the propagating ramp model. Many of the earthquakes for which slip functions are available are listed in Table 1. A representative slip function, for the Landers earthquake, is given in Chapter 12 by Madariaga and Olsen. Most of these events have been modeled with a rupture velocity less than the shear velocity, in the range of 2.4–3.0 km/sec, with no evidence of a dependence of rupture velocity on seismic moment (Somerville *et al.*, 1999). Supershear rupture velocity has been observed in three crustal earthquakes; the 1979 Imperial Valley earthquake (Olson and Apsel, 1982; Archuleta, 1984; Spudich and Cranswick, 1984), the 1999 Izmit/Kocaeli, Turkey, earthquake (Ellsworth and Celebi, 1999; Anderson *et al.*, 2000b; Bouchon *et al.*, 2000; Sekiguchi and Iwata, 2002), and the 1999 Duzce, Turkey, earthquake (Bouchon *et al.*, 2000).

Heaton (1990) attempted to generalize the results of these inversions. His important conclusion, which has been supported in subsequent studies (e.g., Somerville *et al.*, 1999), is that the rise time on the fault appears to be short, implying that the rupture can be approximated as a slip pulse that is propagating along the fault (e.g., Figure 13 of Chapter 12 by Madariaga and Olsen), like the propagating ramp of Haskell (1964). Horton (1996) concluded that in the 1989 Loma Prieta earthquake, the duration of the slip pulse was approximately as expected if it is controlled by the fault width. If the duration of the rise time is shorter than this, as Heaton (1990) suggests, there are significant implications for earthquake rupture mechanics. From the practical viewpoint, a short rise time introduces a constraint on the approach to developing scenario source functions for possible future earthquakes, for the purposes of simulating plausible future ground motions.

At the same distance from the fault, directivity has a first order effect on the ground motion, in addition to its effect on the radiation pattern [Eq. (8)]. Directivity is the effect on the seismograms of rupture propagation along the fault. It impacts both high-frequency and low-frequency seismograms. To illustrate

TABLE 1 Earthquake Slip Models

Earthquake	Date	Reference
Parkfield ($M = 5.5$)	June 28, 1966	Aki (1968); Trifunac and Udwadia (1974)
San Fernando ($M_w = 6.5$)	Feb 9, 1971	Heaton (1982)
Coyote Lake ($M_w = 5.7$)	June 8, 1979	Liu and Helmberger (1983)
Tabas ($M_w = 7.1$)	Sep 16, 1978	Hartzell and Mendoza (1991)
Imperial Valley ($M_w = 6.4$)	Oct 15, 1979	Olson and Apsel (1982); Hartzell and Heaton (1983); Archuleta (1984)
Borah Peak ($M_w = 6.9$)	Oct 28, 1983	Mendoza and Hartzell (1988)
Morgan Hill ($M_w = 6.2$)	April 24, 1984	Hartzell and Heaton (1986); Beroza and Spudich (1988)
Nahanni ($M_w = 6.8$)	Oct 5, 1985	Hartzell *et al.* (1994)
Michoacan, Mexico ($M_w = 8.0$)	Sep 19, 1985	Mendoza and Hartzell (1988, 1989); Mendez and Anderson (1991); Wald *et al.* (1991)
North Palm Springs ($M_w = 6.1$)	Aug 7, 1986	Mendoza and Hartzell (1988); Hartzell (1989)
Whittier Narrows ($M_w = 6.0$)	Oct 1, 1987	Hartzell and Iida (1990); Zeng *et al.* (1993a)
Superstition Hills ($M_w = 6.3$)	Nov 24, 1987	Wald *et al.* (1990)
Loma Prieta ($M_w = 7.0$)	Oct 17, 1989	Beroza (1991); Steidl *et al.* (1991); Wald *et al.* (1991); Horton *et al.* (1994); Horton (1996)
Landers ($M_w = 7.3$)	June 28, 1992	Campillo and Archuleta (1993); Wald and Heaton (1994); Cohee and Beroza (1994); Cotton and Campillo (1995); Wald *et al.* (1996); Olsen *et al.* (1997); Hernandez *et al.* (1999)
Northridge ($M_w = 6.7$)	Jan 17, 1994	Hartzell *et al.* (1996); Wald *et al.* (1996); Zeng and Anderson (1996)
Hyogo-ken Nanbu (Kobe) ($M_w = 6.9$)	Jan 17, 1995	Sekiguchi *et al.* (1996); Horikawa *et al.* (1996); Kakehi *et al.* (1996); Wald (1996); Ide *et al.* (1996); Yoshida *et al.* (1996); Kamae and Irikura (1998); Cho and Nakanishi (2000)
Kocaeli, Turkey ($M_w = 7.6$)	Aug 17, 1999	Yagi and Kikuchi (2000)
Chi-Chi, Taiwan ($M_w = 7.6$)	Sep 20, 1999	Ma *et al.* (2001); Zeng and Chen (2001); Oglesby and Day (2001)

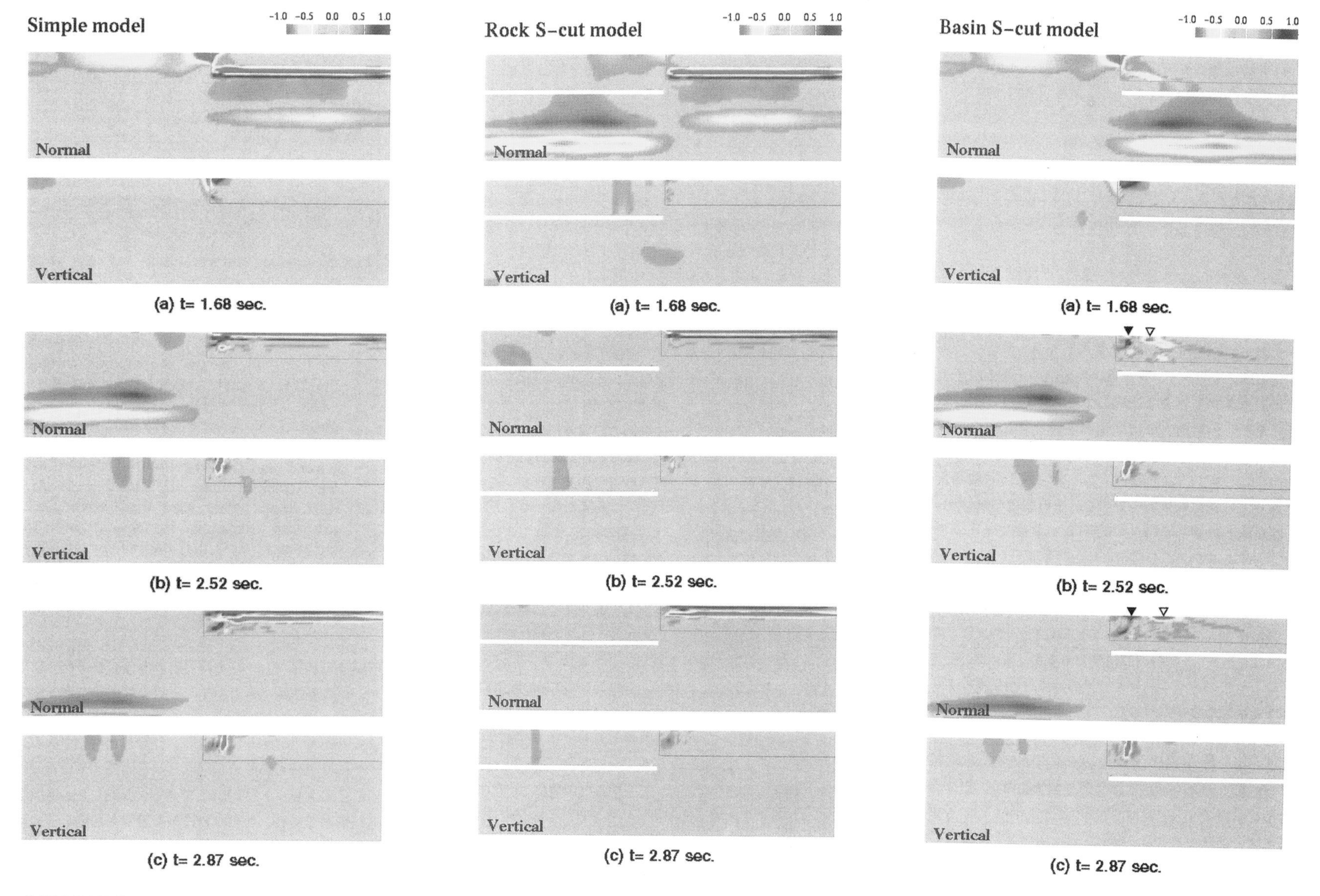

COLOR PLATE 25 Three snapshots each of the velocity responses on the vertical section of: (1) a simple 2-D model (Left), (2) the Rock-S-cut model (Center), and (3) the Basin-S-cut model (Right). Open and solid triangles show approximate locations of the edge-induced waves of two different types, and t is a relative time from the start of analysis.

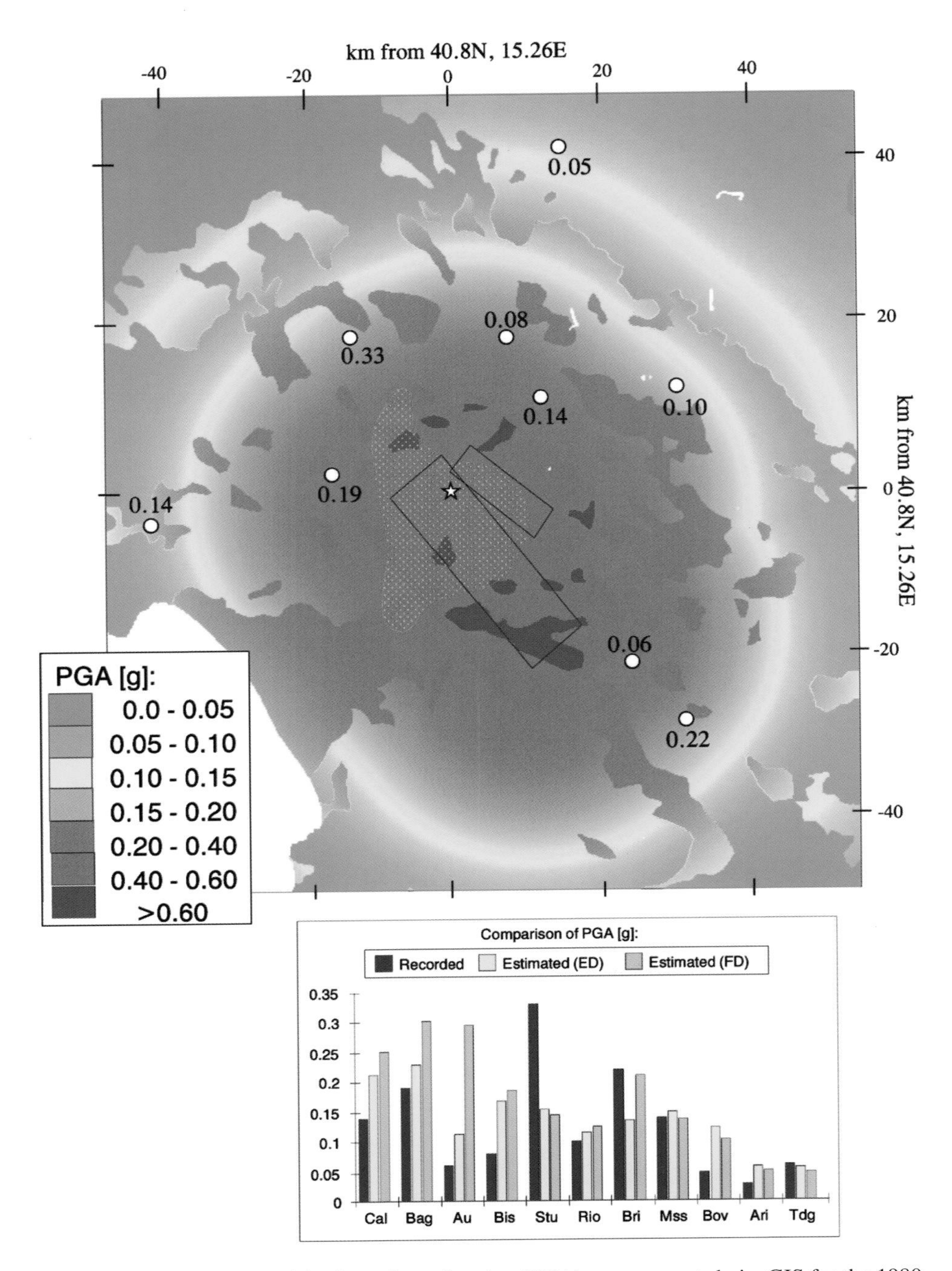

COLOR PLATE 26 Peak horizontal acceleration (PGA) map generated via GIS for the 1980 Irpinia, Italy earthquake using the surface projection of the rupture area represented by the two rectangles, the attenuation relation of Sabetta and Pugliese (1996) and a digital geological map of Italy at 1:500.000 scale. Also shown are the instrumental epicenter (empty star), the location of accelerograph stations (empty circles) with the recorded PGA value (in g), and the area with IX MCS felt intensity (dotted shading). In the bar diagram the observed PGA values are compared with the predictions using both the fault distance (FD) and epicentral distance (ED).

COLOR PLATE 27 Example of GIS case history database of liquefaction-induced lateral spreads showing locations of soil boreholes in Niigata city (Bardet *et al.*, 1999).

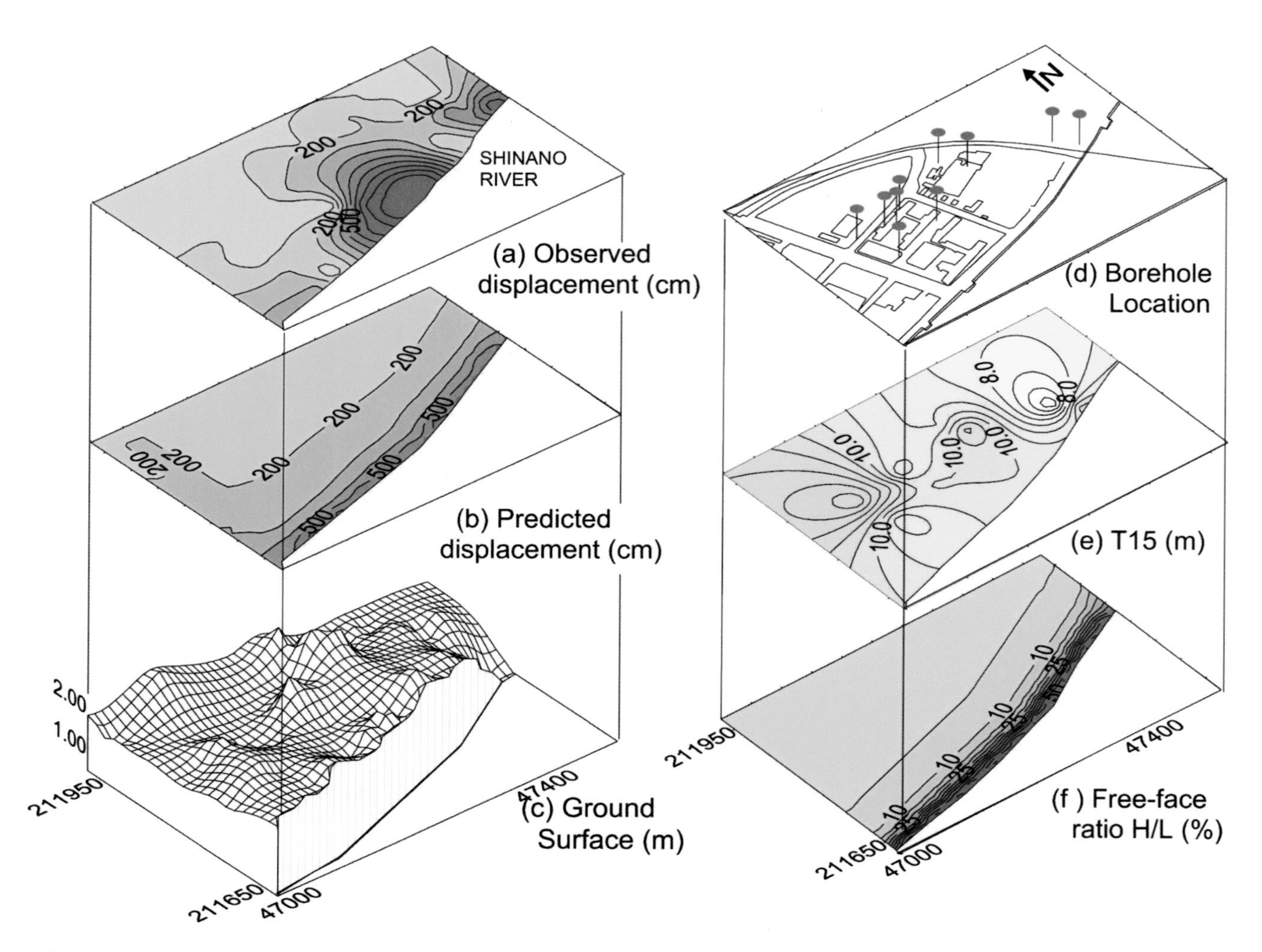

COLOR PLATE 28 Representation of (a) observed and (b) predicted (Bardet *et al.*, 1999) amplitude of liquefaction-induced lateral ground deformation, (c) ground surface, (d) borehole location, spatial distribution of (e) thickness T15, and (f) free-face ratio for slide G10-FF′ (Kawagishi-cho) in Niigata during the 1964 Niigata, Japan, earthquake (data after Hamada, 1999; coordinates are in meters; displacements are in centimeters).

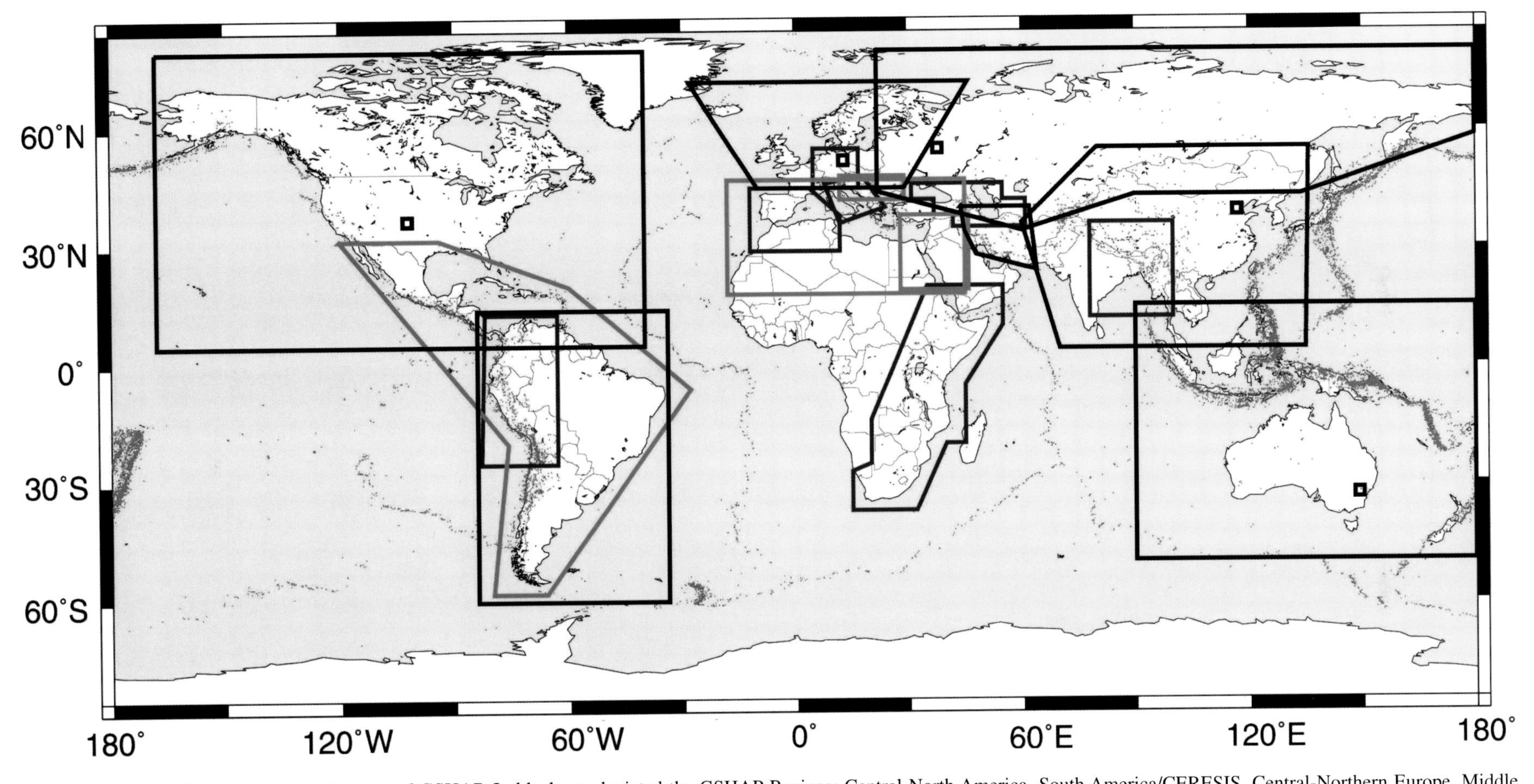

COLOR PLATE 29 The regional structure of GSHAP. In black are depicted the GSHAP Regions: Central-North America, South America/CERESIS, Central-Northern Europe, Middle East/Iran, Northern Eurasia, Eastern Asia, South-West Pacific. In blue, the GSHAP Test Areas: PILOTO, CAUCAS, ADRIA, East African Rift, India-China-Tibet, Ibero-Maghreb, DACH. In green, cooperating projects: PAIGH-IDRC, Circum Pannonian Basin, Eastern Mediterranean, SESAME, Turkey.

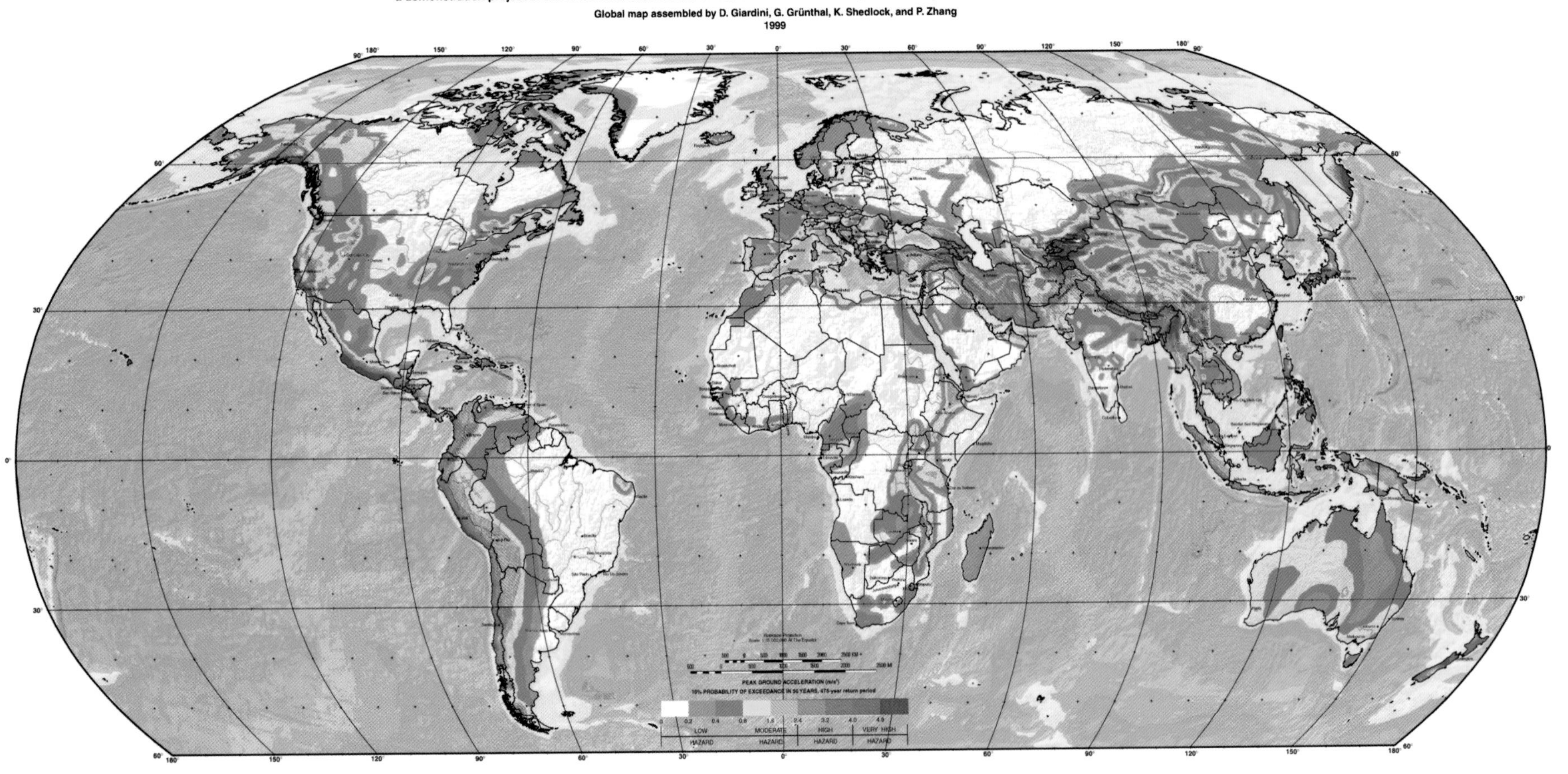

COLOR PLATE 30 The GSHAP Global Seismic Hazard Map, depicting Peak-Ground-Acceleration expected with 10% exceedance probability in 50 years, corresponding to a 475 yr average return period. The pga values are combined with a shaded relief base map using ARC/Info 7.2.1 geographic information system (GIS) software. The cell size of the pga and relief base grids is 0.0833 degrees.

COLOR PLATE 31 (a) A satellite image of Death Valley and associated faults, southeastern California. (b) Eruption column rising over the summit of Augustine volcano, Alaska. (c) Mount Taranaki, the second largest volcano in New Zealand. (d) Scarp along the Johnson Valley segment of rupture associated with the 1992 Landers, California, earthquake. For further explanations, *see* Chapter 90 compiled by Rymer *et al*.

COLOR PLATE 32 (a) Surface rupture on the Nojima fault, Awaji Island, associated with the 1995 Hyogo-ken Nambu (Kobe), Japan, earthquake. (b) Left-lateral offsets preserved in rice patties by the 1990 Luzon, Philippines, earthquake. (c) An oblique aerial view of the Pearce scarp at mouth of Miller Basin associated with the 1915 Pleasant Valley, Nevada, earthquake. (d) Tall, faceted spurs and preserved mole track associated with the 1932 Chang Ma earthquake, Gansu Province, China. (e) Rupture along the Bogd fault associated with the 1957 Gobi Altai earthquake, Mongolia. For further explanations, *see* Chapter 90 compiled by Rymer *et al.*

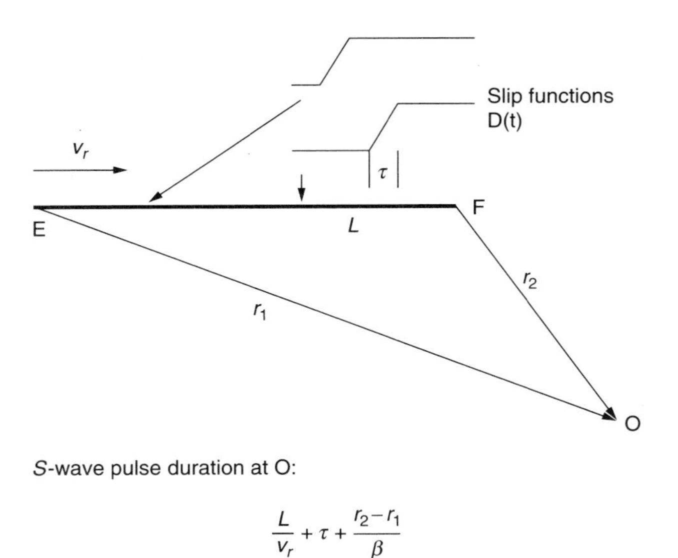

FIGURE 12 Geometry for consideration of the effect of directivity. This illustrates a vertical, strike-slip fault in map view, recorded at observation point O. The rupture is a Haskell propagating ramp with rupture velocity v_r and rise time τ. The shear velocity is β. The duration of direct arrivals at O is given by the equation in the figure. Rupture is assumed to originate below an epicenter (point E), at one end of the fault, and propagate to the far end (point F).

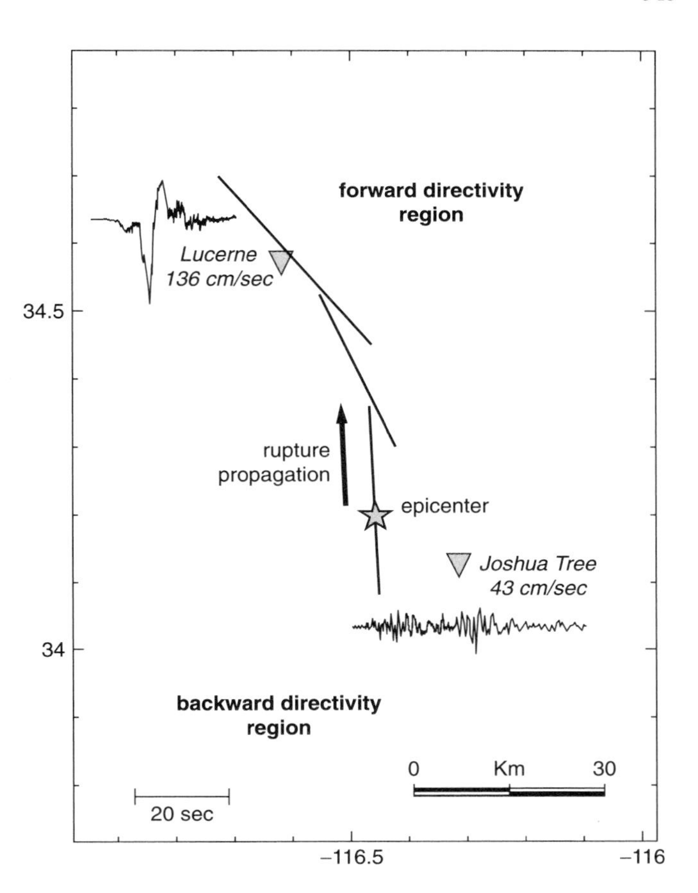

FIGURE 13 Illustration of directivity in velocity pulses recorded in the 1992 Landers, California, earthquake. Rupture propagated north from the epicenter, away from the station at Joshua Tree (with the long, low-amplitude velocity trace) and toward the Lucerne station (with the stronger, compact velocity pulse). From Somerville *et al.*, 1997.

the effect, consider first Figure 12. If the observation point O is located near the origin of the rupture (E), and if τ is small and $v_r \approx 0.9\beta$, then r_1 is small and r_2 is comparable to L, so the time interval over which direct arrivals of S waves will be arriving is on the order of $2L/v_r$. On the other hand, for a station located near the far end of the rupture (F), the duration terms in L and in r_1 approximately cancel and the duration of the seismogram is short. At high frequencies, directivity shows up as a short, intense accelerogram at the far end of the fault, in contrast with a lower-amplitude, long-duration accelerogram near the origin of rupture. This is illustrated in Figure 1, where the Caleta de Campos record is near the epicenter while the La Union and Zihuatanejo stations are located in the forward directivity direction.

At low frequencies, directivity associated with propagating rupture causes constructive interference of long period waves and high-velocity pulses. Recently, seismic engineers have become very concerned about these effects. Such pulses have time–function characteristics that are correlated with their polarizations, specifically, the fault-perpendicular component of ground velocity pulses observed in the region toward which the rupture is propagating is typically a two-sided pulse (little net displacement), whereas the fault-parallel component of ground velocity observed near the fault as the rupture passes by is a one-sided pulse having significant net displacement. This was shown by Archuleta and Hartzell (1981), Luco and Anderson (1983), Anderson and Luco (1983a, b), and recently Somerville *et al.* (1997), who have developed empirical ground-motion prediction relations that include this effect. Heaton *et al.* (1995) summarized such pulses observed up to that time, which had peak velocities of up to 170 cm/sec, and they showed that such pulses could be especially dangerous to buildings 10–25 stories tall. Figure 13, from Somerville *et al.* (1997), illustrates these phenomena. A feature of the intense velocity pulse, as seen in Figure 13, is that it tends to be strongest on the component perpendicular to the fault. This result, which may initially be surprising, is predicted by dislocation modeling. It was first recognized and explained in the interpretation of accelerograms from the 1966 Parkfield, California, earthquake (Aki, 1967; Haskell, 1969).

At locations sufficiently near a rupturing fault, the ground undergoes a static offset. Digital accelerograms have succeeded in capturing this static offset in a few cases (e.g., Mexico: Figure 3; Turkey: Rathje *et al.*, 2000; Safak and Erdik, 2000; Anderson *et al.*, 2000b; Taiwan: Shin *et al.*, 2000). The rise time of the static displacement in the Mexico earthquake is ~10 seconds (Figure 3). The rise time of the static displacements inferred in Turkey is ~3–6 seconds. These observations all are much shorter than the total duration of the earthquake, and are consistent with the hypothesis that rise times are generally short (Heaton, 1990). The 1999 Chi-Chi, Taiwan, earthquake has confirmed that the hanging walls of thrust faults move much more, and have greater high-frequency ground motions (peak accelerations) than the foot walls during earthquakes (Shin *et al.*, 2000). The low-frequency effect has been seen in laboratory models (Brune, 1996a), and it has been predicted in numerical models (Mikumo and Miyatake, 1993; Shi *et al.*, 1998; Oglesby *et al.*, 1998). The high-frequency effect has been detected in the development of some ground-motion prediction equations reviewed by Campbell in Chapter 60.

Isochrone theory is useful for understanding the relationship of ground motions in the near-source region (where the Green's function is not well approximated by a plane wave leaving the source). An isochrone is the locus of points on a fault from which the seismic waves all arrive at a selected station at the same time. The basic theory is given in Bernard and Madariaga (1984) and Spudich and Frazer (1984, 1987). In the basic theory, far-field ray theory Green's functions are used in the evaluation of the source radiation, but the assumption of plane waves is not made. The directivity of the source is manifested in a quantity, the isochrone velocity, which contains and is more general than the usual Ben-Menahem directivity function and which can be calculated as a function of position on the fault surface, showing exactly how source kinematics affect the radiation from each point of the fault. With the implicit assumption that the slip velocity at a point on the fault is discontinuous when the rupture starts and stops, Spudich and Fraser show that the high-frequency radiation reaching a station is large when the isochrone velocity for waves reaching that station is large or when the slip velocity of the dislocation on the fault is large. Archuleta and Hartzell (1981) and Spudich and Oppenheimer (1986) present several plots that show how near-source ground motions are affected by variation of radiation patterns and/or isochrone velocity over the fault surface. These effects are seen only in the near-source region. Joyner and Spudich (1994) extended the isochrone theory to include near-field terms of Green's functions, so that static offsets can be predicted using isochrone theory. Isochrone theory software will be included in the Handbook CD in the software package ISOSYN by Spudich and Xu in Chapter 85.14.

4.3 Effects of Wave Propagation

Complexities introduced into the strong ground motions by effects of wave propagation through the crust of the Earth are at least equal in importance to the complexities that are introduced at the source. These crustal complexities can be divided into four parts. The first is propagation of seismic waves through a flat-layered, attenuating Earth. The second is the effect of basins and other major systematic deviations from a flat layered model. The third is the effect of random variations in the velocity of the Earth on all scales. The fourth is the effect of nonlinearity. The success in incorporating these effects depends on how thoroughly the seismic properties of the Earth between the earthquake source and the site are known. Depending on the complexity of the Earth, success may also be limited by computer capabilities to calculate the response of the Earth. Altogether, the literature on this range of subjects is enormous. Relevant topics are covered, in part, in other chapters in this volume. This review is confined to selection of a few topics that, in the author's judgment, have particularly interesting impacts on strong ground motions.

4.3.1 Flat-Layered Attenuating Earth

The approximation of a flat- (or radially) layered Earth is widely used in all aspects of seismology, and is usually quite successful to first order for tasks such as locating earthquakes, describing arrival times of the various types of seismic waves on a global and regional scale, and modeling surface-wave dispersion at low frequencies. There are several methods to calculate the Green's functions [i.e., the response of a flat-layered, attenuating medium to seismic excitation applied at a point, including methods by Bouchon (1981), Luco and Apsel (1983), or Kennett (1983)]. Software to make this computation for a non-attenuating flat-layered structure is included in Chapter 85.14 by Spudich and Xu.

It is a little difficult to estimate the frequency band in which this model is accurate. Flat-layered models have been used to determine focal mechanisms, using moment tensor inversion, from regional records on broadband seismograms. For these inversions, records are preferably low-pass filtered to model only frequencies below 0.05 Hz (e.g., Ichinose *et al.*, 1998; Dreger and Kaverina, 2000; Ichinose, 2000). At these frequencies, with their corresponding long wavelengths, the presence of mountain ranges or basins can be argued to have a minimal effect, so the methods do not need to adjust the Green's functions for the specific source-station path, and may successfully model the complete seismogram. However, the direct S-wave part of the Green's function, and for smaller scales the complete Green's functions, may be reliable to much higher frequencies (e.g., 1–2 Hz) in some places.

These synthetic Green's functions have been used in inversions to find models for the slip on the causative fault, which is as consistent as possible with all observed ground motions. Observed seismic ground motions in the direct S waves at low frequencies (i.e., less than 1 Hz) have been reproduced with these methods, as demonstrated by several of the studies, cited in Table 1, that model the strong motion to infer a source time function. The success at low frequencies in determining the source time function may be a little misleading. Basins have a strong effect on the nature of seismic waves over a broad frequency range, starting at frequencies below 1 Hz. Inaccuracy in modeling the path can be mapped back into spurious source effects when flat-layered structure is assumed for the calculation of the

Green's function (e.g., Cormier and Su, 1994; Graves *et al.*, 1999; Graves and Wald, 2001; Wald and Graves, 2001).

4.3.2 Basin Response

Many major urban areas, particularly in regions with high seismic activity, are built in geological basins. The reasons are obvious. The tectonic activity that causes earthquakes also tends to build mountains. Flat-lying areas that develop where eroded sediments are deposited (i.e., basins) are far easier to build upon than are mountains. It is therefore of the utmost importance to understand the effects of basins on strong ground motions. The modeling of wave propagation in basins, using finite-difference or finite-element techniques, is an active field of research. The effects of basins are reviewed in Chapter 61 by Kawase, so selection of an interesting, state-of-the-art example is sufficient for this review.

The Kanto, Japan, basin (which includes Tokyo and adjacent cities) has been studied extensively (e.g., Phillips *et al.*, 1993; Sato *et al.*, 1998a, b, 1999; Koketsu and Kikuchi, 2000). Figures 14 and 15, from Sato *et al.* (1999), illustrate results of a finite-difference simulation of waveforms from a magnitude 5.1 earthquake. These figures illustrate several important consequences of basin response. The seismograms on rock near the basin edge (station ASK) are very simple, representing the moment rate function with probably relatively little distortion. Stations in the basin (KWS, HNG, FUT, CHB) have larger amplitudes in this frequency band, and much longer duration of strong shaking. Stations in the basin have strong arrivals much later than the *S* wave, caused by surface waves generated at the edge of the basin. In this frequency band, the surface waves dominate the seismograms in the basin. Studies in Los Angeles show all of these phenomena (e.g., Liu and Heaton, 1984; Vidale and Helmberger, 1988; Olsen *et al.*, 1995; Olsen and Archuleta, 1996; Wald and Graves, 1998; and Olsen, 2000). All of these effects are probably quite general properties of basins.

The effects of basin response are not restricted to low frequencies. For example, in the Kobe earthquake, the zone with the highest damage is not directly above the causative fault, but rather it is a linear band that is offset to the southeast. An explanation for this, supported by modeling studies, is that the damage zone occurs at the location of constructive interference of waves taking two distinctly different paths: one directly from the fault through the low-velocity basin structure, and the other through higher-velocity rock on the opposite side of the fault, which is then refracted into the basin from near the top of the fault (e.g., Kawase, 1996; Kawase *et al.*, 1998; Matsushima and Kawase, 1998; Pitarka *et al.*, 1997, 1998). A second good example comes from the Northridge earthquake, where there was an isolated zone of high damage in Santa Monica. Davis *et al.* (2000) find, from studies of large numbers of aftershocks, that this is the result of focusing of seismic energy by a 3-dimensional lens-like high-velocity structure within the basin.

4.3.3 Scattering and Attenuation

Scattering and attenuation have an important effect on the strong-motion spectrum, particularly at high frequencies. Sato *et al.* in Chapter 13 discuss this issue in more detail. The ω–*square* spectral model predicts that the Fourier acceleration spectra should be flat at high frequencies, but in Figures 4 and 9 they fall off quite rapidly above about 5 Hz. Hanks (1982) referred to the frequency above which the acceleration spectrum falls off as f_{max} (Figure 11). He pointed out that this could be either a source or a site effect.

In favor of the site-effect model, Anderson and Hough (1984) proposed that a good asymptotic approximation to the spectral shape at these high frequencies is exponential decay [$\sim\exp(-\pi\kappa f)$] where κ is the decay parameter. In their data, κ is a function of distance and the site conditions of the station. Figure 16 shows measured values of κ at two stations, one on slightly weathered granite (Pinyon Flat in the southern California batholith) and the other on deep alluvium (Imperial Valley in the Colorado River delta of southern California, 130 km southeast of Pinyon Flat). Both stations show a systematic increase in κ with distance, but the station on alluvium shows values offset to higher κ (more rapid decay) than the station on granite. Anderson (1986) explains this behavior as an effect of attenuation. In this model, the intercept of κ (i.e., at zero distance) is controlled by attenuation near the surface, while the increase of κ with distance is an effect of lateral propagation. It should be noted that κ is at most weakly dependent on distance in Mexico (Humphrey and Anderson, 1992). Based on this model, κ has been used in some models to simulate strong motions for engineering applications (e.g., Schneider *et al.*, 1993; Beresnev and Atkinson, 1998a, b). There is, however, a considerable amount of scatter in the individual measurements of κ in this figure, some of which comes from variability in the high-frequency spectrum as radiated at the source (e.g., Tsai and Chen, 2000) or highly variable attenuation in the source region (Castro *et al.*, 2000).

4.3.4 Site Effects Including Nonlinearity

The term *site effects* is generally used to refer to wave propagation in the immediate vicinity of the site, as opposed to the propagation effects, which refer to the complete path from the source to receiver. The boundary between a site effect and a propagation effect is not always clear, but it is useful to discuss them separately. Site effects can include modification of seismic waves by the local sedimentary cover, particularly where this local cover is not representative of the total path from the epicenter, the effect of alluvial valleys or basins, effect of local topography, and effects of the water table. Soil–structure interaction also might be considered a site effect, but it is not considered here because it represents modification of the ground motion by artificial structures rather than by the Earth in the absence of such

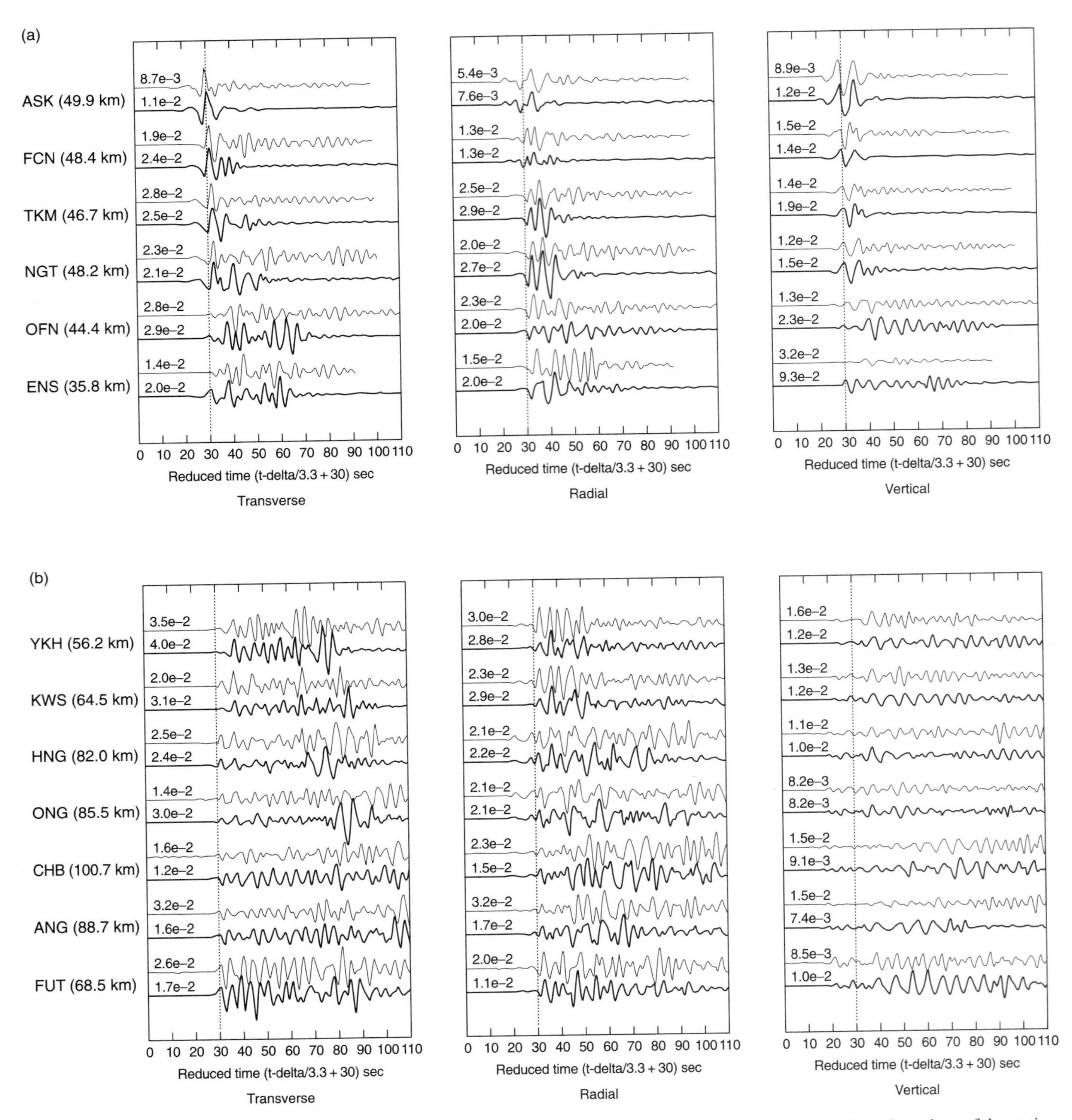

FIGURE 14 Observed (thin lines) and computed (bold lines) ground motions in the Kanto basin for various stations. Locations of the stations are shown in Figure 15 (some stations are not identified on this figure). All seismograms are aligned by the arrival time of the direct *S*-wave. From Sato *et al.*, 1999.

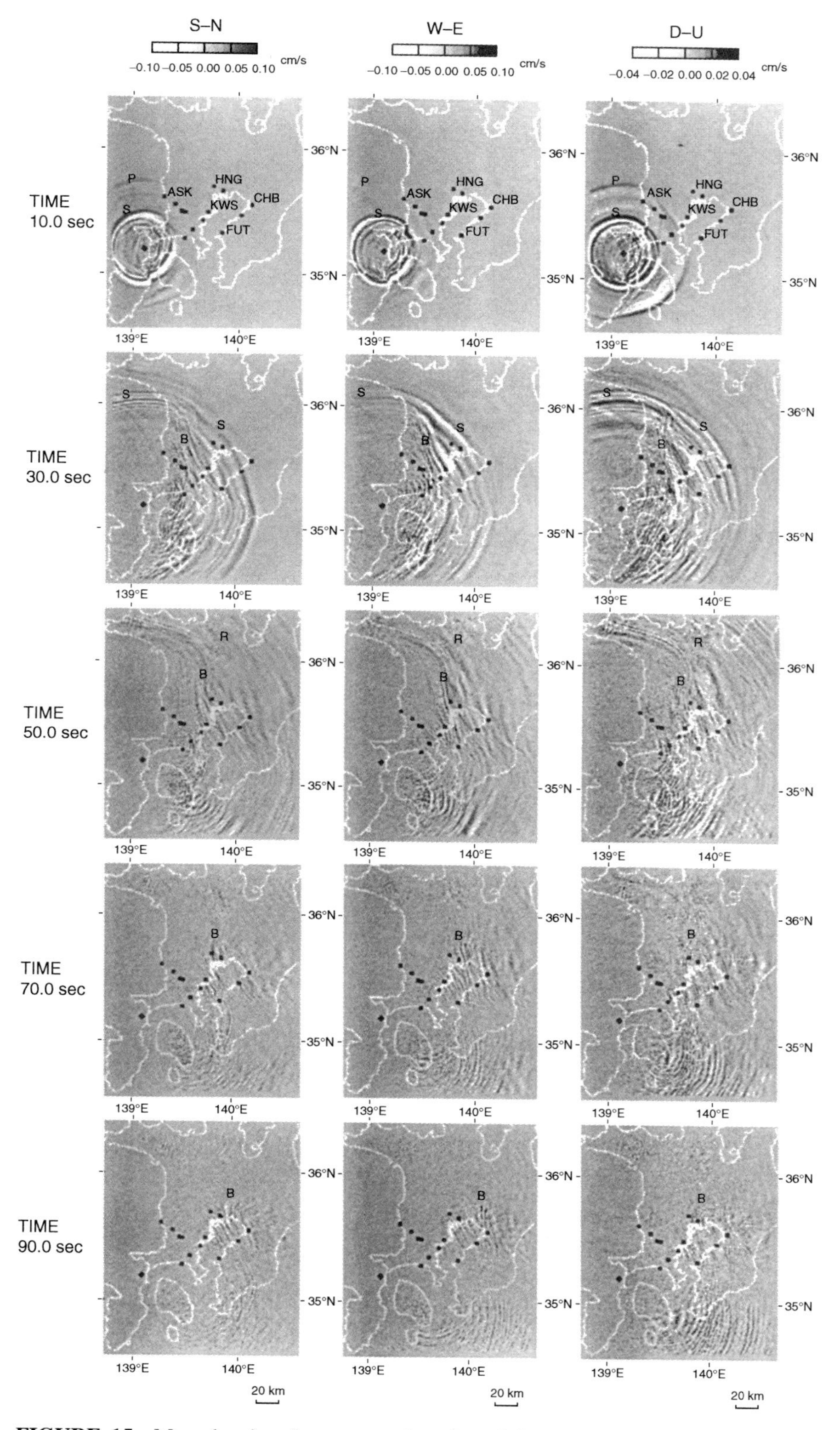

FIGURE 15 Map showing the computed surface deformation generated by a finite-difference simulation of the 1990 Odawara earthquake at five different times. From Sato *et al.* (1999). Data and calculations are bandpass filtered (0.1–0.3 Hz). On the map, P and S denote the direct *P*- and *S*-waves, and B denotes surface waves generated at the boundary of the basin.

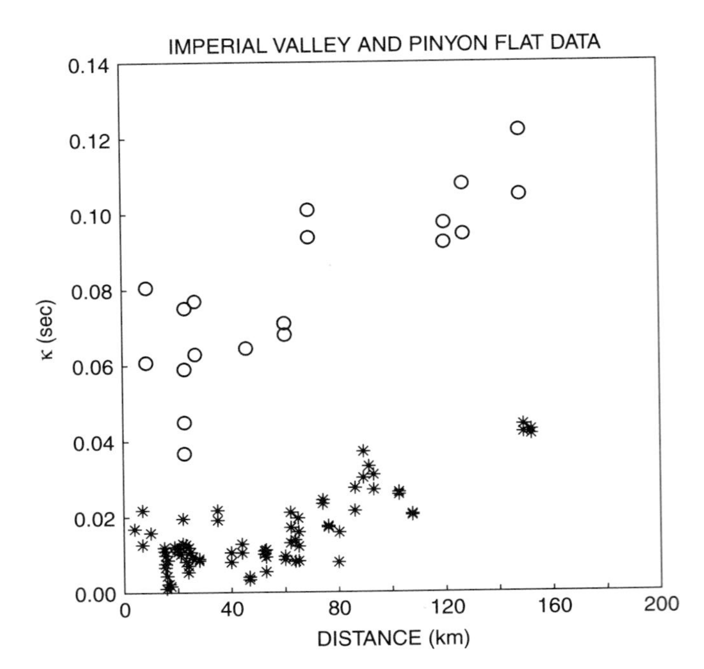

FIGURE 16 The high-frequency parameter κ as a function of distance for two stations in southern California (from Anderson, 1986). Open circles are measurements from strong-motion accelerograms recorded on deep sediments in Imperial Valley, and asterisks are from seismograms recorded on a rock station in the San Jacinto mountains west of Imperial Valley.

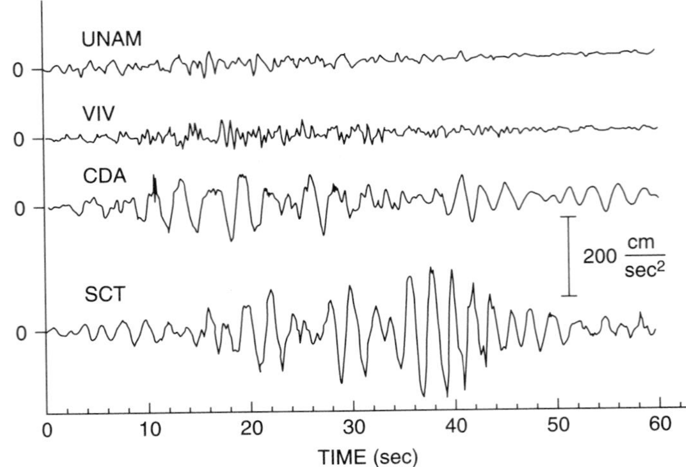

FIGURE 17 Most significant one-minute segment of strong-motion accelerograms recorded at four stations in Mexico City, in the September 19, 1985, earthquake ($M = 8.0$). Stations are 350 km from the epicenter, so differences can only be attributed to differences in geological conditions at the site. No time correlation exists among these traces (from Anderson *et al.*, 1986).

structures. Site effects are discussed in more detail in Chapter 61 by Kawase.

Figure 17 shows accelerograms recorded from four locations in Mexico City in the September 19, 1985, earthquake to illustrate a classic example of site effects. The stations UNAM and VIV are least affected, while CDA and SCT have been strongly amplified at a period of two to three seconds to cause the large accelerations as shown, and cause the terrible toll of death and destruction from that event (Anderson *et al.*, 1986). The main factor that causes such amplification is soft soils that form low-velocity layers near the Earth's surface. These low-velocity layers trap energy, amplify all frequencies due to the decrease in seismic impedance, and preferentially amplify resonant frequencies.

Effects of local topography, including valleys, hills, and ridges, may be grouped together and have been reviewed by Geli *et al.* (1988). The main results are that the topographic amplification is maximum at the top of the hill, and is maximum at the frequency at which one wavelength of the *S* wave equals the width of the hill base. Motions on the hill sides are not amplified much, and motions around the base of the hill are usually deamplified with respect to motions far from the hill. Geli *et al.* (1988) found that theoretical models of the topographic effects almost always underpredict the observed topographic effects. That discrepancy has not yet been resolved. For example, the Bouchon and Barker (1996) simulation of response at the Tarzana site, which recorded unexpectedly high accelerations in the 1994 Northridge, California, earthquake, underpredicts the observed amplification of Spudich *et al.* (1996).

It has been commonly assumed that the effects of wave propagation on strong and on weak ground motion are the same. However, nonlinear effects (i.e., a nonlinear stress–strain relationship) resulting in slower velocities and more rapid attenuation of large amplitude waves are predicted for soft materials. Field observational evidence has suggested this is a common phenomenon for many accelerograms (e.g., Chin and Aki, 1991; Field *et al.*, 1998; Su *et al.*, 1998; Irikura *et al.*, 1998). Su *et al.* (1998) concluded that in the Northridge earthquake, nonlinearity is recognizable in differences between site response (defined by spectral ratios) when peak accelerations exceed about 0.3g, peak velocity exceeds 20 cm/sec, or peak strain exceeds 0.06% (Figure 18). The main effects of nonlinearity are to decrease the effective shear velocity of the sediments and to increase the damping, thus shifting resonant peaks to lower frequencies and lower amplitudes, and generally reducing amplitudes overall. Understanding of nonlinearity is an important current issue for strong-motion seismology. Chapter 61 by Kawase reviews the topic in more detail.

4.4 Theoretical Models for Estimation of Strong Shaking

All methods of generating synthetic seismograms need to be calibrated. Complete calibration includes demonstration that the method produces realistic-appearing seismograms, for which measures of ground motion match the observations. They should be able to match the spectra of "calibration events"

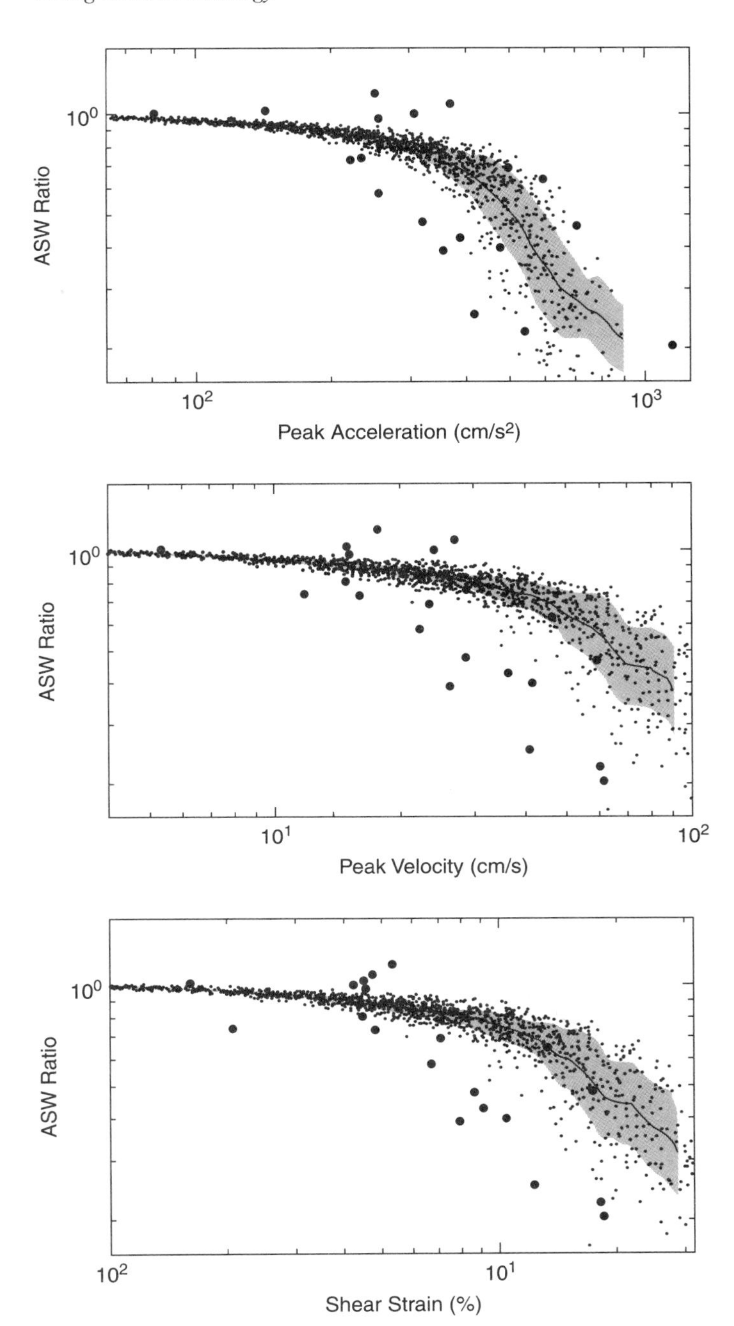

FIGURE 18 Effect of nonlinearity on the site-response contribution to the Fourier spectral amplification of strong motion. Large dots give the observed ASW ratio, from Su *et al.* (1999), defined as the average over frequency of the Fourier spectral amplification in the main shock (strong motion) to the average of the Fourier spectral amplification determined from several small aftershocks. The points are plotted as a function of the peak acceleration of the strong-motion record during the main shock (top), the peak velocity of the strong-motion record during the main shock (middle), and the peak strain related to the strong-motion record during the main shock (bottom). Peak strain is the ratio of peak velocity to average shear velocity estimated for the upper 30 meters below the station. Small points show the effect of numerical experiments based on the nonlinear model for average site conditions using the approach of Ni *et al.* (2000).

(e.g., Atkinson and Somerville, 1994; Anderson and Yu, 1996; Beresnev and Atkinson, 1998a, b) as well as the ground-motion prediction models (see Chapter 60 by Campbell) where these models are well constrained by data.

4.4.1 Stochastic Models

The broad flat portion of the Fourier spectrum motivates a model of strong motion as band-limited white noise, where the spectrum for white noise is proportional to f^0. Housner (1947) first modeled strong ground motion as a random process, and earthquake engineers employed band-limited white noise and both stationary and non-stationary Gaussian processes as models of strong-motion acceleration as early as the 1960's (see for example, Jennings *et al.*, 1969). More recent seismological models that use this approach include Hanks and McGuire (1981), Boore (1983), Boore and Atkinson (1987), Ou and Herrmann (1990), Chin and Aki (1991), Atkinson and Boore (1997), Toro *et al.* (1997), and Sokolov (1998). The process of constructing a seismogram with this approach begins with generating a white-noise time series, applying a shaping taper in the time domain to match the envelope of the expected strong motion, and then applying a bandpass filter in the frequency domain to mold the Fourier spectrum to the expected spectral shape. The frequency domain shaping can be controlled by assumptions about source spectral shape, Q, κ, and site response at high frequencies. In eastern North America, where there is little experience with large earthquakes, there is considerable uncertainty about the source spectral shape (e.g., Atkinson, 1993; Haddon, 1996b). For near-field calculations, Silva *et al.* (1990), Schneider *et al.* (1993), Beresnev and Atkinson (1998a, b) treat the total fault as a set of subfaults, which are summed to obtain seismograms with the statistical characteristics of strong motions. Software developed by D. M. Boore is given on the Handbook CD (under the directory \85\8513) and described in Chapter 85.13.

4.4.2 Synthetic Green's Function Models

A second general approach to generating synthetic strong ground motions follows the representation theorem [Eq. (1)] more closely than the stochastic methods, thus attempting to use a more complete description of the physics at every step on the process. These methods generate synthetic Green's functions based on Earth structure models of varying complexity, and combine these through Eq. (1) with a range of models of the earthquake source. Two examples of this approach are those that use a "composite source model" (Zeng *et al.*, 1994; Yu *et al.*, 1995; Anderson and Yu, 1996; Zeng and Anderson, 1996) and those that use an empirical source time function (e.g., Hadley and Helmberger, 1980; Somerville *et al.*, 1991; Cohee *et al.*, 1991; Atkinson and Somerville, 1994; Sato *et al.*, 1998b). Both methods need to develop representations for both the source and the wave propagation. In the simulations by Somerville and his associates, the source is represented by an empirical source time function. At high frequencies, ray theory in a layered medium is used to find the Green's function, and at low frequency, the Green's function has preferably been determined using a finite-element solution for ground motion in a realistic model for the regional geology, usually including a basin (e.g., Sato *et al.*, 1998b). Models reported so far using the composite source model use a Green's function developed for a flat-lying layered, attenuating half space. Scattering is added to the Green's functions for more realism at high frequencies (Zeng *et al.*, 1995).

The source component of the composite source model, motivated by a model by Frankel (1991), is interesting. This model describes the source as the superposition of overlapping circular cracks (subevents) with a power-law distribution of radii, $N(r) \sim r^{-p}$, where p is the fractal dimension. These subevents are distributed at random on the fault and each is triggered to radiate when a rupture front passes. Each subevent radiates a pulse of the type predicted by some model for radiation from a circular crack (e.g., Brune, 1970; Sato and Hirasawa, 1973). The source displacement spectrum of the subevents is proportional to f^{-2} at high frequencies, implying that the source acceleration spectrum is proportional to f^{0}. The source is kinematic, but this source description has the capability to generate realistic accelerograms with the proper frequency content (e.g., Zeng *et al.*, 1994; Yu *et al.*, 1995), and has a capability to predict ground motions (e.g., Anderson and Yu, 1996). Also, it is possible to find specific composite sources that are consistent with both the statistics and the phase of observed records (e.g., Zeng and Anderson, 1996). An example of seismograms generated by this approach is shown in Figure 19. Anderson (1997b) found a relationship between parameters of the composite source model and several physical earthquake source parameters (radiated energy, stress drop).

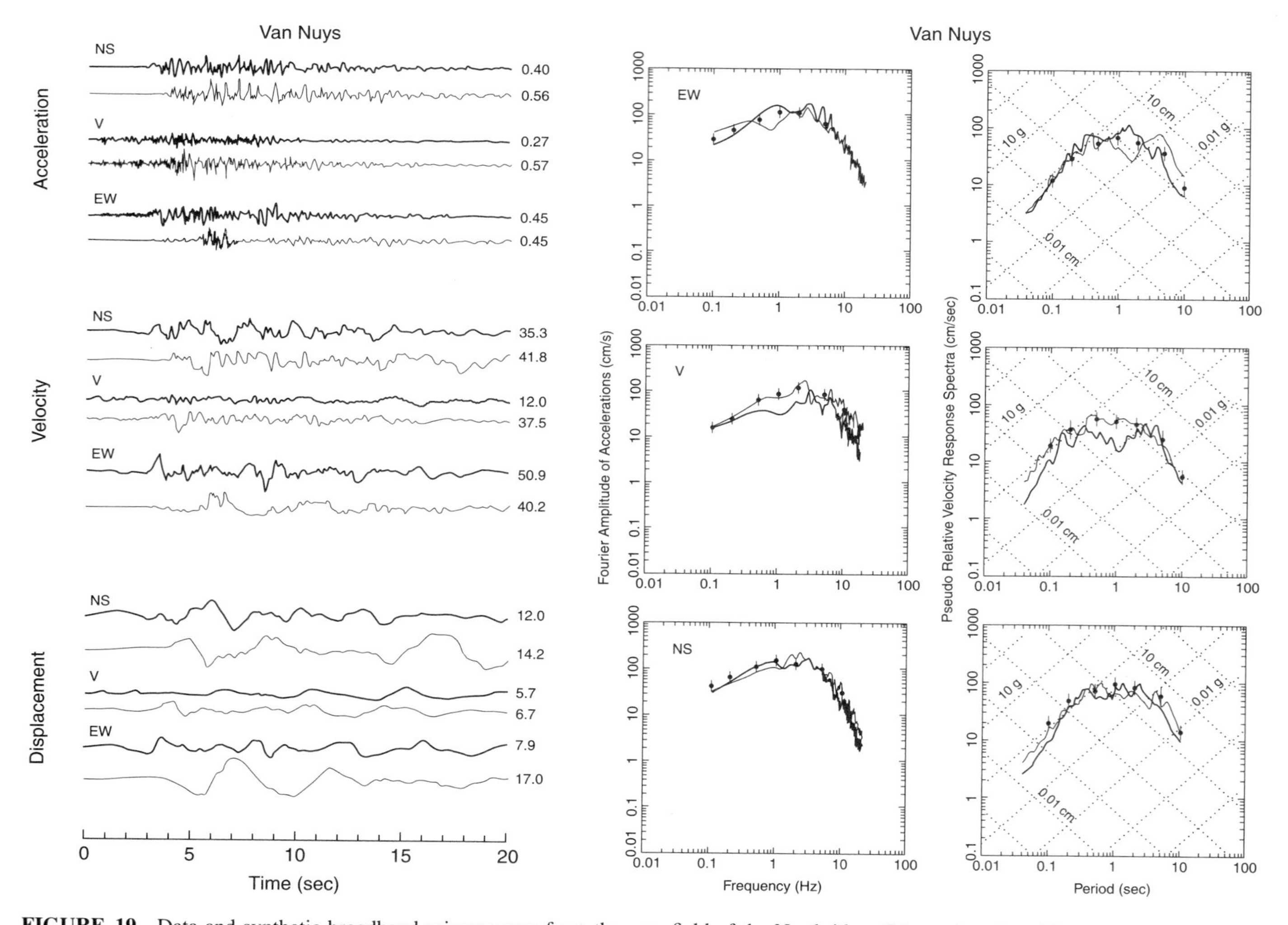

FIGURE 19 Data and synthetic broadband seismograms from the near-field of the Northridge, CA, earthquake of January 17, 1994 (from Anderson and Yu, 1996). Numbers at the end of each trace are peak acceleration (in fraction of g, the acceleration of gravity), peak velocity (in cm/s), and peak displacement (in cm). Upper trace in each pair (bold line) is data, and lower line is the synthetic. On the right, bold lines are data, and light lines are realization of one synthetic, and points with error bars are mean values of 50 realizations.

4.4.3 Empirical Green's Function Models

A third approach for simulating strong ground motions uses small earthquakes as empirical Green's functions. This method has been applied by Hartzell (1978), Wu (1978), Kanamori (1979), Irikura (1983, 1986), Joyner and Boore (1986), Takemura and Ikeura (1987), Dan *et al.*, (1990), Hutchings (1991, 1994), Hutchings and Wu (1990), Wennerberg (1990), Haddon (1996a), Jarpe and Kasameyer (1996), and others. Empirical Green's functions incorporate the effects of wave propagation by the use of small earthquake seismograms recorded in the same area. With proper care in scaling the seismograms and adding them with appropriate delays to represent propagation of rupture across the fault, it is possible to generate realistic synthetic seismograms. The advantage of this approach is that the seismograms used as empirical Green's functions incorporate all the complexities of wave propagation. The limitation is that the empirical Green's functions may not have an adequate signal-to-noise ratio at all the frequencies of interest, may not be available for the desired source–station pairs, and may originate from sources with a focal mechanism different from the desired mechanism. Care must be exercised in the superposition of multiple empirical Green's functions, in order to obtain an ω*–square* spectrum in the synthetic motions (e.g., Irikura, 1986; Takemura and Ikeura, 1987; Dan *et al.*, 1990; Wennerberg, 1990). Kamae *et al.* (1998) proposed a hybrid technique in which they synthesize the seismogram from a small event using 3-D finite difference at low frequencies and the stochastic method at high frequencies.

4.5 Seismic Hazard Analysis

Seismic hazard analysis transmits information on strong motions to allow for informed decisions on earthquake-resistant designs, governmental response to the hazards, and other societal impacts of earthquakes. (This topic is also reviewed by Somerville and Moriwaki in Chapter 65.) In its simplest form, a seismic hazard analysis might identify a scenario earthquake that might affect a region (see Anderson, 1997a, for a review), and estimate the ground motions expected from that earthquake, using techniques described previously. This approach does not necessarily convey information about the frequency with which this ground motion occurs. The alternative form, called a probabilistic seismic hazard analysis (PSHA), calculates a hazard curve (Fig. 20) giving the average annual probability that a ground-motion parameter, say Y, is equaled or exceeded. Thus, hazard curves predict the result of an experiment where an instrument at the site records ground motions for, say, 10^5 or 10^6 years, and the frequency of excedence of Y is tabulated. Milne and Davenport (1969) presented a method to estimate the hazard curve at relatively high occurrence rates directly from a seismic catalog and extrapolated to low probabilities using extreme value statistics, but most estimates synthesize earthquake sources and attenuation relations using an approach that can be traced to Cornell (1968).

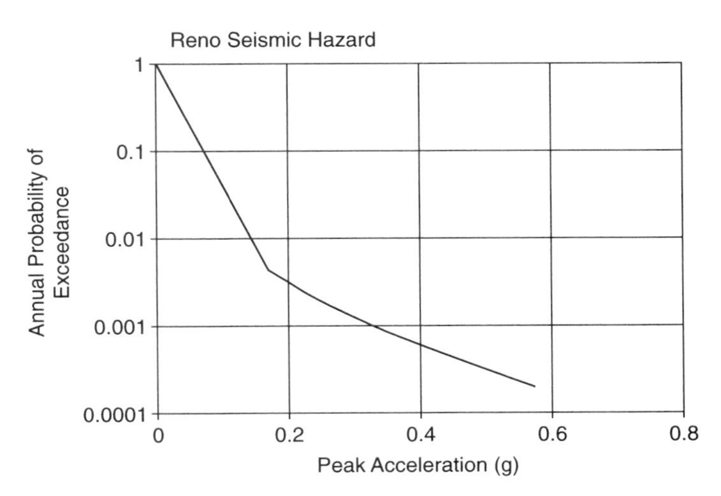

FIGURE 20 A seismic hazard curve for Reno, Nevada, based on the seismic hazard model of Frankel *et al.* (1996).

PSHA calculates the rate $r_C(Y)$ that ground motion Y is equaled or exceeded through the synthesis of the seismicity (see Chapter 41 by Engdahl and Villasenor) and ground motion prediction equations. Assuming that the earthquakes are random, uncorrelated events, then the Poisson model should hold, and we expect that the probability of exceedance in a time interval of duration T is

$$P_C(Y, T) = 1 - \exp[-r_C(Y)T]. \tag{13}$$

The curve $P_C(Y, T)$ is the *hazard curve*. The input seismicity describes the spatial distribution of earthquakes and their rates of occurrence. This is a function of both magnitude and location, $n(M, \mathbf{x})$, where M is the earthquake magnitude, $\mathbf{x}$ is the location vector of the source, and $n(M, \mathbf{x})\,dM\,d\mathbf{x}$ gives the number of earthquakes per year in a magnitude range of width dM and an area (or volume) of size $d\mathbf{x}$. Recognizing that earthquakes radiate energy from a finite fault, $n(M, \mathbf{x})$ must be consistent with fault geometry and the way the distance R from the earthquake to the site is defined.

A ground-motion prediction equation $\hat{Y}(M, R)$ estimates the mean ground motion as a function of the magnitude and distance (and sometimes additional parameters). As reviewed by Campbell in Chapter 60, $\hat{Y}(M, R)$ and its standard deviation σ_T are developed by regression using existing strong-motion data. The equation $\hat{Y}(M, R)$ can be inverted to find the cumulative probability Φ that a single realization of an earthquake with magnitude M at distance R will cause ground motion y in excess of Y, as follows:

$$\Phi(y \geq Y|\hat{Y}, \sigma_T) = \int_Y^{\infty} \frac{1}{\sigma_T\sqrt{2\pi}} \exp\left(-\frac{(y - \hat{Y}(M, R))^2}{2\sigma_T^2}\right) dy. \tag{14}$$

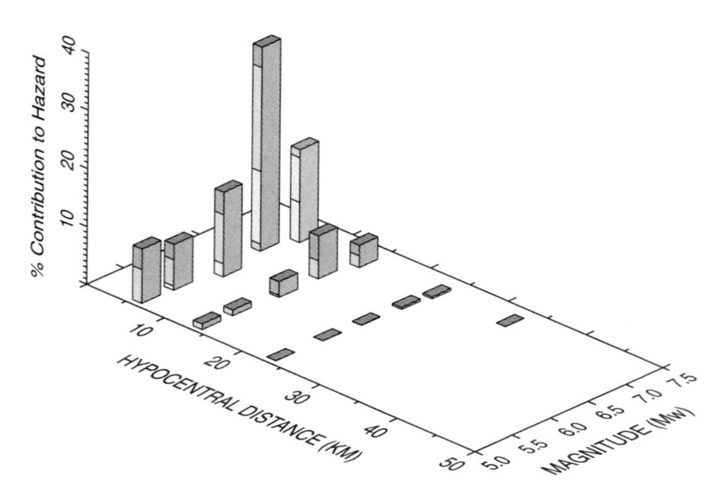

FIGURE 21 Deaggregation of the seismic hazard for Reno, Nevada (39.5°N, 119.8°W), based on the seismic hazard model developed by Frankel *et al.* (1996). This deaggregation is for a peak acceleration of 0.58 g, which is estimated to have a mean occurrence rate of 4.04×10^{-3} yr^{-1}, corresponding to a mean return time of 2475 years, or a probability of exceedance of 2% in 50 years. Differing shades of gray in each bar tell how much above (mostly in this case) or below average the ground motion must be to reach this peak acceleration from various earthquakes contributing to the hazard (modified from the USGS Web site, 2001).

The basic probabilistic seismic hazard analysis model (e.g., McGuire, 1995; SSHAC, 1997) finds

$$r_C(Y) = \iint n(M, R(\mathbf{x}))\, \Phi(y \geq Y | \hat{Y}(M, R(\mathbf{x})), \sigma_T)\, dM\, d\mathbf{x} \tag{15}$$

A more complete analysis distinguishes between aleatory and epistemic uncertainty. Aleatory uncertainty is introduced by true randomness in nature, while epistemic uncertainty is due to lack of knowledge (SSHAC, 1997). An extension of PSHA is the deaggregation or disaggregation of the seismic hazard (e.g., McGuire, 1995; Harmsen *et al.*, 1999; Bazzurro and Cornell, 1999). From the integrand in Equation (15), deaggregation identifies the magnitudes and distances that contribute the most to that hazard (e.g., Fig. 21). From this, specific important earthquakes can be identified, thus providing additional information to the engineers.

In principal, every specific site has its own seismic hazard curve. Spatial variation can be presented on a probabilistic map, which might contour, for instance, estimates of peak ground accelerations that occur with a probability of 10% in 50 years (e.g., Algermissen *et al.*, 1982; Frankel, 1995; Frankel *et al.*, 1996; Frankel *et al.*, 2000; Chapter 74 by Giardini *et al.*). Probabilistic maps thus communicate estimates of the relative hazards of different locations accounting for both severity of the potential ground motions and their frequency of occurrence (e.g., Cornell, 1995).

The difficulty in building the integrand for Eq. (15) should not be underestimated (e.g., Frankel, 1995; SSHAC, 1997; Ebel and Kafka, 1999). PSHA-generated hazard curves are thus a scientific prediction to be tested. However, it is very difficult to test because the hazard curve describes rates of events that are very rare compared to the history of rigorous strong-motion observations. At high probabilities ($\sim 10^0$–10^{-2}), the models can be tested by comparison with estimated ground motions from the historical seismicity (e.g., Ward, 1995). At somewhat lower probabilities ($\sim 10^{-2}$–10^{-3}), historical records for some old civilizations (Japan, Middle East) might be able to provide a test (Ordaz and Reyes, 1999). At even smaller probabilities, the only available recourse is to use geological indicators, in particular precarious rocks (e.g., Brune, 1996b; Brune, 1999; Anderson and Brune, 1999b). These studies find precarious rocks whose existence seems to contradict the results of PSHA. Anderson and Brune (1999a) propose that this discrepancy may be in part caused by treating spatial variability of ground motions, where site effects contribute to a large uncertainty, as a variability in time in Eq. (15). Another possibility, emphasized by the Turkey and Taiwan earthquakes (Sec. 3.2 above) is that the mean ground-motion curves are overestimating the ground motions (Brune, 1999; Anderson *et al.*, 2000a). The issue remains to be resolved by future data collection and research.

5. Future Directions for Strong-Motion Seismology

This article has attempted to give an overview of the field of strong-motion seismology. Coverage of several topics has necessarily been too brief, but it is believed that the information presented here is sufficient to provide a reasonable introduction and indication of the present state of research in this exciting field of seismology. This final section identifies some future challenges.

The first is in the field of data collection. The limited number of accelerograms collected to date for major earthquakes within the zones of major damage (<20 km on rock and 100 km on soft soil) remains a severe limitation of our science. For instance, the consensus of the Monterey Strong-Motion Workshop (Stepp, 1997) is that there is a need to install on the order of 20,000 instruments in densely urbanized areas of the United States in order to ensure that the next major damaging earthquake in the US is well documented. Worldwide, a stable mechanism for funding the collection and distribution of strong-motion data has not been developed in spite of the enormous value of the data to the engineering community.

Research in strong-motion seismology is proceeding toward a capability to take any hypothetical earthquake and to generate seismograms and corresponding spectra that incorporate an understanding of all the physical phenomena as they apply in the region of interest. There is reason for confidence that this physical approach is the best way to minimize the uncertainties in anticipating ground motion for future earthquakes. Models of this type can be expected to play a role in filling in gaps in data at critical magnitudes and distances, such as predicting the characteristics of the near-field directivity pulses. It can be argued that the modeling approach is the most reliable way of extrapolating from the present data set. However, models have sufficient flexibility in source parameters that must be extrapolated to large magnitudes that this modeling will not totally replace the important role that gathering more data must continue to play.

At low frequencies, modeling will require continued emphasis on the use of finite element/finite difference methods to predict ground motions in sedimentary basins beneath urban areas. Advances in computer codes, increasing computer speeds, and continual improvement in velocity models, will be needed to push the applicability of these methods to higher frequencies. Methods will be needed to utilize whole waveforms from multiple events to improve the urban velocity models, as exploration in these areas is most difficult.

There is little physical understanding of the role that various physical characteristics and processes play in controlling the high-frequency end of the spectrum. Factors that could contribute include the tectonic environment, fault characteristics such as total slip, time intervals between earthquakes, and lithology, issues related to the dynamics of rupture and the physics that control those dynamics, and nonlinear wave propagation along the path, especially near the source and in the near-surface geology. Collection of more data is essential, as the number of variables that are potentially involved outnumber the number of large earthquakes that have been recorded at short distances.

Because strong-motion seismology has an important application in the development of ground motions for use in engineering design, it is obvious that empirical characterization of strong-motion data will also need to continue. For the largest earthquakes, both the mean values and the uncertainty around the mean, together with the factors controlling those parameters, need to be understood much better in order to reduce uncertainties in seismic hazard analyses. Together with this, emphasis on prediction of strong motions through physical modeling is appropriate. By that approach, scientific understanding and improved prediction will go hand in hand, and the greatest progress will be made for description of seismic hazards.

Acknowledgments

The author thanks Willie Lee for the encouragement to write this article. The paper benefited enormously from several very thoughtful and extensive reviews. Five reviewers in particular, Roger Borcherdt, Kenneth Campbell, Hiroshi Kawase, Dan O'Connell, and Paul Spudich, contributed major ideas and even text and references to improve various sections of the article, and the quality of the result is enhanced enormously by their efforts. Carl Stepp made several critical recommendations. Aasha Pancha, Paul Caruso, and Jose A. Zepeda helped the process of reviewing and clarifying the manuscript. Figures 4, 5, and 6 were prepared by Matt Purvance. Figure 18 is an unpublished figure by Feng Su, compiling work from the paper by Su *et al.* (1998). The preparation of this paper was supported in part by the Southern California Earthquake Center. SCEC is funded by NSF Cooperative Agreement EAR-8920136 and USGS Cooperative Agreements 14-08-0001-A0899 and 1434-HQ-97AG01718. The SCEC contribution number for this paper is 640. The preparation of this paper was also supported in part by the National Science Foundation under Grant No. 0000050.

References

Abrahamson, N. A., and W. J. Silva (1997). Empirical response spectral attenuation relations for shallow crustal earthquakes. *Seismol. Res. Lett.* **68**, 94–127.

Aki, K. (1967). Scaling law of seismic spectrum. *J. Geophys. Res.* **72**, 1217–1231.

Aki, K. (1968). Seismic displacements near a fault. *J. Geophys. Res.* **73**, 5359–5376.

Aki, K. (1980). Possibilities of Seismology in the 1980's (Presidential Address). *Bull. Seismol. Soc. Am.* **70**, 1969–1976.

Aki, K., and P. G. Richards (1980). *Quantitative Seismology, Theory and Methods.* W. H. Freeman and Co., San Francisco.

Alcantara, L. E., Andrade, J. M. Espinosa, J. A. Flores, F. Gonzalez, C. Javier, B. Lopez, M. A. Macias, S. Medina, L. Munguia, J. A. Otero, C. Perez, R. Quaas, J. A. Roldan, H. Sandoval, R. and Vazquez (1997). Base Mexicana de datos de sismos fuertes, Disco Compacto Volumen 1, 1997, Sociedad Mexicana de Ingenieria Sismica, A.C.

Algermissen, S. T., D. M. Perkins, P. C. Thenhaus, and B. L. Bender (1982). Probabilistic estimates of maximum acceleration and velocity in rock in the contiguous United States, *US Geol. Surv. Open-File Report 82-1033*, 107 pp.

Ambraseys, N. N. (1997). Strong motion development and research in Europe. In: "Vision 2005: An Action Plan for Strong Motion Programs to Mitigate Earthquake Losses in Urbanized Areas" (J. C. Stepp, Ed.), US Committee for Advancement of Strong Motion Programs, Austin, Texas, 66–74.

Ambraseys, N., P. Smit, R. Berardi, D. Rinaldis, F. Cotton, and C. Berge-Thierry (2000). Dissemination of European Strong-Motion Data. CD-ROM collection. European Commission, Directorate General XII, Science, Research and Development, Environment and Climate Programme, Bruxelles.

Ambraseys, N., P. Smit, R. Sigbjornsson, P. Suhadolc, and B. Margaris (2001). Internet-Site for European Strong-Motion Data. *http://www.isesd.cv.ic.ac.uk*, European Commission, Research-Directorate General, Environment and Climate Programme.

Anderson, J. G. (1986). Implication of attenuation for studies of the earthquake source. In: "Earthquake Source Mechanics" (S. Das, J. Boatwright, and C. Scholz, Eds.), *Am. Geophys. Un. Geophysical Monograph* **37**, 311–318.

Anderson, J. G. (1997a). Benefits of scenario ground motion maps. *Engineering Geol.* **48**, 43–57.

Anderson, J. G. (1997b). Seismic energy and stress drop parameters for a composite source model. *Bull. Seismol. Soc. Am.* **87**, 85–96.

Anderson, J. G., and J. N. Brune (1999a). Probabilistic seismic hazard analysis without the ergodic assumption. *Seismol. Res. Lett.* **70**, 19–28.

Anderson, J. G., and J. N. Brune (1999b). Methodology for using precarious rocks in Nevada to test seismic hazard models. *Bull. Seismol. Soc. Am.* **89**, 456–467.

Anderson, J. G., and S. Hough (1984). A model for the shape of the Fourier amplitude spectrum of acceleration at high frequencies. *Bull. Seismol. Soc. Am.* **74**, 1969–1994.

Anderson, J. G., and J. E. Luco (1983a). Parametric study of near-field ground motion for a strike slip dislocation model. *Bull. Seismol. Soc. Am.* **73**, 23–43.

Anderson, J. G., and J. E. Luco (1983b). Parametric study of near-field ground motions for oblique-slip and dip-slip dislocation models. *Bull. Seismol. Soc. Am.* **73**, 45–57.

Anderson, J. G., and R. Quaas (1988). Effect of magnitude on the character of strong ground motion: an example from the Guerrero, Mexico, strong motion network. *Earthquake Spectra* **4**, 635–646.

Anderson, J. G., and G. Yu (1996). Predictability of strong motions from the Northridge, California, earthquake. *Bull. Seismol. Soc. Am.* **86**, S100–S114.

Anderson, J. G., P. Bodin, J. Brune, J. Prince, and S. K. Singh (1986). Strong ground motion and source mechanism of the Mexico earthquake of September 19, 1985. *Science* **233**, 1043–1049.

Anderson, J. G., J. N. Brune, J. Prince, R. Quaas, S. K. Singh, D. Almora, P. Bodin, M. Onate, R. Vazquez, and J. M. Velasco (1994). The Guerrero accelerograph network. *Geofisica Int.* **33**, 341–371.

Anderson, J. G., S. G. Wesnousky, and M. W. Stirling (1996). Earthquake size as a function of fault slip rate. *Bull. Seismol. Soc. Am.* **86**, 683–690.

Anderson, J. G., J. N. Brune, R. Anooshehpoor, and S.-D. Ni (2000a). New ground motion data and concepts in seismic hazard analysis. *Current Science* **79**, 1278–1290.

Anderson, J. G., H. Sucuoglu, A. Erberik, T. Yilmaz, E. Inan, E. Durukal, M. Erdik, R. Anooshehpoor, J. N. Brune, and S.-D. Ni (2000b). Implications for seismic hazard analysis. In: "Kocaeli, Turkey, earthquake of August 17, 1999 Reconnaissance Report" (T. L. Youd, J.-P. Bardet, and J. D. Bray, Eds.), *Earthquake Spectra*, Supplement A to Volume 16, 113–137.

Archuleta, R. J. (1984). A faulting model for the 1979 Imperial Valley earthquake. *J. Geophys. Res.* **89**, 4559–4585.

Archuleta, R. J., and S. H. Hartzell (1981). Effects on fault finiteness on near-source ground motion. *Bull. Seismol. Soc. Am.* **71**, 939–957.

Atkinson, G. (1993). Source spectra for earthquakes in eastern North America. *Bull. Seismol. Soc. Am.* **83**, 1778–1798.

Atkinson, G., and D. M. Boore (1997). Stochastic point-source modeling of strong motions in the Cascadia region. *Seismol. Res. Lett.* **68**, 74–85.

Atkinson, G. M., and P. G. Somerville (1994). Calibration of time-history simulation methods. *Bull. Seismol. Soc. Am.* **84**, 400–414.

Bazzurro, P., and C. A. Cornell (1999). Disaggregation of seismic hazard. *Bull. Seismol. Soc. Am.* **89**, 501–520.

Belardinelli, M. E., M. Cocco, O. Coutant, and F. Cotton (1999). Redistribution of dynamic stress during coseismic ruptures: evidence for fault interaction and earthquake triggering. *J. Geophys. Res.* **104**, 14925–14945.

Beresnev, I. A., and G. M. Atkinson (1998a). Stochastic finite-fault modeling of ground motions from the 1994 Northridge, California, earthquake, I. Validation on rock sites. *Bull. Seismol. Soc. Am.* **88**, 1392–1401.

Beresnev, I. A., and G. M. Atkinson (1998b). Stochastic finite-fault modeling of ground motions from the 1994 Northridge, California, earthquake, II. Widespread nonlinear response at soil sites. *Bull. Seismol. Soc. Am.* **88**, 1402–1410.

Bernard, P., and R. Madariaga (1984). A new asymptotic method for the modeling of near-field accelerograms. *Bull. Seismol. Soc. Am.* **74**, 539–557.

Beroza, G. C. (1991). Near-source modeling of the Loma Prieta earthquake: evidence for heterogeneous slip and implications for the earthquake hazards. *Bull. Seismol. Soc. Am.* **81**, 1603–1621.

Beroza, G. C., and P. Spudich (1988). Linearized inversion for fault rupture behavior: application to the 1984 Morgan Hill, California, earthquake. *J. Geophys. Res.* **93**, 6275–6296.

Bolt, B. (1969). Duration of strong motion. *Proc. 4th World Conf. Earthq. Engineering, Santiago, Chile*, 1304–1315.

Boore, D. M. (1983). Stochastic simulation of high-frequency ground motions based on seismological models of the radiated spectra. *Bull. Seismol. Soc. Am.* **73**, 1865–1894.

Boore, D. M., and G. M. Atkinson (1987). Stochastic prediction of ground motion and spectral response at hard-rock sites in eastern North America. *Bull. Seismol. Soc. Am.* **77**, 440–467.

Borcherdt, R. D. (1997). US National Strong Motion Program. In: "Vision 2005: An Action Plan for Strong Motion Programs to Mitigate Earthquake Losses in Urbanized Areas" (J. C. Stepp, Ed.), US Committee for Advancement of Strong Motion Programs, Austin, Texas, 81–85.

Bouchon, M. (1981). A simple method to calculate Green's functions for layered media. *Bull. Seismol. Soc. Am.* **71**, 959–971.

Bouchon, M., and J. S. Barker (1996). Seismic response of a hill: the example of Tarzana, California. *Bull. Seismol. Soc. Am.* **86**, 66–72.

Bouchon, M., M.-P. Bouin, and H. Karabulut (2000). How fast did rupture propagate during the Izmit and the Duzce earthquakes (abstract). *EOS, Trans. Am. Geophys. Un.* **81**, F836.

Brune, J. N. (1970). Tectonic stress and the spectra of seismic shear waves from earthquake. *J. Geophys. Res.* **75**, 4997–5009.

Brune, J. N. (1971). Correction. *J. Geophys. Res.* **76**, 5002.

Brune, J. N. (1996a). Particle motion in a physical model of shallow angle thrust faulting. *Proc. Indian Acad. Sci.* **105**, 197–206.

Brune, J. N. (1996b). Precariously balanced rocks and ground-motion maps for southern California. *Bull. Seismol. Soc. Am.* **86**, 43–54.

Brune, J. N. (1999). Precarious rocks along the Mojave section of the San Andreas fault, California: constraints on ground motion from great earthquakes. *Seismol. Res. Lett.* **70**, 29–33.

Campbell, K. W. (2002). Seismology, engineering. In: "Encyclopedia of Physical Science and Technology" (R. A. Meyers, Ed.), 3rd ed., Vol. 14, pp. 531–545, Academic Press, San Diego.

Campillo, M., and R. J. Archuleta (1993). A rupture model for the 28 June 1992 Landers, California, earthquake. *Geophys. Res. Lett.* **20**, 647–650.

Castro, R. R., L. Trojani, G. Monachesi, M. Mucciarelli, and M. Cattaneo (2000). The spectral decay parameter κ in the region of Umbria-Marche, Italy. *J. Geophys. Res.* **105**, 23811–23823.

Celebi, M., S. Akkar, U. Gulerce, A. Sanli, H. Bundock, and A. Salkin (2001). Main shock and aftershock records of the 1999 Izmit and Duzce, Turkey, earthquakes. *US Geol. Surv. Open File Report 01-163*.

Chin, B. H., and K Aki (1991). Simultaneous study of the source, path, and site effects on strong ground motion during the 1989 Loma Prieta earthquake: a preliminary result on pervasive nonlinear site effects. *Bull. Seismol. Soc. Am.* **81**, 1859–1884.

Cho, I., and I. Nakanishi (2000). Investigation of the three-dimensional fault geometry ruptured by the 1995 Hyogo-ken Nanbu earthquake using strong-motion and geodetic data. *Bull. Seismol. Soc. Am.* **90**, 450–467.

Cohee, B. P., and G. C. Beroza (1994). Slip distribution of the 1992 Landers earthquake and its implications for earthquake source mechanism. *Bull. Seismol. Soc. Am.* **84**, 692–712.

Cohee, B. P., P. G. Somerville, and N. A. Abrahamson (1991). Simulated ground motions for hypothesized $M_w = 8$ subduction earthquakes in Washington and Oregon. *Bull. Seismol. Soc. Am.* **81**, 28–56.

Cormier, V. F., and W.-J. Su (1994). Effects of three-dimensional crustal structure on the estimated history and ground motion of the Loma Prieta earthquake. *Bull. Seismol. Soc. Am.* **84**, 284–294.

Cornell, C. A. (1968). Engineering seismic risk analysis. *Bull. Seismol. Soc. Am.* **58**, 1583–1606.

Cornell, C. A. (1995). An advocation: map probabilistically derived quantities, ACT 35-2, National Earthquake Ground Motion Mapping Workshop, September 1995.

Cotton, F., and M. Campillo (1995). Frequency domain inversion of strong motions: application to the 1992 Landers earthquake. *J. Geophys. Res.* **100**, 3961–3975.

Cousins, W. J. (1998). New Zealand strong motion data 1965–1997. *Institute of Geological and Nuclear Sciences. Report 98/4*, Lower Hutt, New Zealand.

Dan, K., T. Watanabe, T. Tanaka, and R. Sato (1990). Stability of earthquake ground motion synthesized using different small-event records as empirical Green's functions. *Bull. Seismol. Soc. Am.* **80**, 1433–1455.

Davis, P. M., J. L. Rubenstein, K. H. Liu, S. S. Gao, and L. Knopoff (2000). Northridge earthquake damage caused by geologic focusing of seismic waves. *Science* **289**, 1746–1750.

Dreger, D. S., and A. Kaverina (2000). Seismic remote sensing for the earthquake source process and near-source shaking: a case study of the October 16, 1999, Hector Mine earthquake. *Geophys. Res. Lett.* **27**, 1941–1944.

Ebel, J. E., and A. L. Kafka (1999). A Monte Carlo approach to seismic hazard analysis. *Bull. Seismol. Soc. Am.* **89**, 854–866.

Ellsworth, W. L., and M. Celebi (1999). Near field displacement time histories of the M7.4 Kocaeli (Izmit), Turkey, earthquake of August 17, 1999 (abstract). *EOS, Trans. Am. Geophys. Un.* **80**(46), F648.

Field, E. H., S. Kramer, A. W. Elgemal, J. D. Bray, N. Matasovic, P. A. Johnson, C. Cramer, C. Roblee, D. Wald, L. F. Bonilla, P. P. Dimitriu, and J. G. Anderson (1998). Nonlinear site response: where we're at (a report from a SCEC/PEER seminar and workshop). *Seismol. Res. Lett.* **69**, 230–234.

Fletcher, J. B., and P. Spudich (1998). Rupture characteristics of the three M$\sim$4.7 (1992–1994) Parkfield earthquakes. *J. Geophys. Res.* **103**, 835–854.

Frankel, A. (1991). High-frequency spectral falloff of earthquakes, fractal dimension of complex rupture, b-value, and the scaling of strength of faults. *J. Geophys. Res.* **96**, 6291–6302.

Frankel, A. (1995). Mapping Seismic Hazard in the Central and Eastern United States. *Seismol. Res. Lett.* **66**(4), 8–21.

Frankel, A., J. Fletcher, F. Vernon, L. Haar, J. Berger, T. Hanks, and J. Brune (1986). Rupture characteristics and tomographic source imaging of $M_L \sim 3$ earthquakes near Anza, southern California. *J. Geophys. Res.* **91**, 12633–12650.

Frankel, A., C. Mueller, T. Barnhard, D. Perkins, E. V. Leyendecker, N. Dickman, S. Hanson, and M. Hopper (1996). National Seismic Hazard Maps, June 1996, U.S. Department of the Interior, U.S. Geological Survey, MS 966, Box 25046, Denver Federal Center, Denver, CO 80255.

Frankel, A., C. Mueller, S. Harmsen, R. Wesson, E. Leyendecker, F. Klein, T. Barnhard, D. Perkins, N. Dickman, S. Hanson, and M. Hopper (2000). USGS national seismic hazard maps: methodology and analysis. *Earthquake Spectra* **16**, 1–19.

Franklin, A. G. (1997). U. S. Army Corps of Engineers strong motion instrumentation program. In: "Vision 2005: An Action Plan for Strong Motion Programs to Mitigate Earthquake Losses in Urbanized Areas" (J. C. Stepp, Ed.), U.S. Committee for Advancement of Strong Motion Programs, Austin, Texas, 91.

Freeman, J. R. (1932). *Earthquake Damage and Earthquake Insurance*. McGraw-Hill, New York.

Geli, L., P. Bard, and B. Jullien (1988). The effect of topography on earthquake ground motion: a review and new results. *Bull. Seismol. Soc. Am.* **78**, 42–63.

Graves, R. W., and D. J. Wald (2001). Resolution analysis of finite fault source inversion using one- and three-dimensional Green's functions; 1, Strong motions, *J. Geophys. Res.* **106**, 8745–8766.

Graves, R. W., D. J. Wald, H. Kawase, and T. Sato (1999). Finite fault source inversion using 3D Green's functions. *Seismol. Res. Lett.* **70**, 251.

Haddon, R. A. W. (1996a). Use of empirical Green's functions, spectral ratios, and kinematic source models for simulating strong ground motion. *Bull. Seismol. Soc. Am.* **86**, 597–615.

Haddon, R. A. W (1996b). Earthquake source spectra in eastern North America. *Bull. Seismol. Soc. Am.* **86**, 1300–1313.

Hadley, D. M., and D. V. Helmberger (1980). Simulation of strong ground motions. *Bull. Seismol. Soc. Am.* **70**, 617–630.

Hanks, T. C. (1979). b-values and $\omega^{-\gamma}$ seismic source models: implications for tectonic stress variations along active crustal fault zones and estimation of high-frequency strong ground motion. *J. Geophys. Res.* **84**, 2235–2242.

Hanks, T. C. (1982). F_{max}. *Bull. Seismol. Soc. Am.* **74**, 1867–1880.

Hanks, T., and H. Kanamori (1979). A moment magnitude scale. *J. Geophys. Res.* **84**, 2348–2350.

Hanks, T. C., and R. K. McGuire (1981). The character of high-frequency strong ground motion. *Bull. Seismol. Soc. Am.* **71**, 2071–2095.

Harmsen, S., D. Perkins, and A. Frankel (1999). Deaggregation of probabilistic ground motions in the central and eastern United States. *Bull. Seismol. Soc. Am.* **89**, 1–13.

Hartzell, S. H. (1978). Earthquake aftershocks as Green's functions. *Geophys. Res. Lett.* **5**, 1–4.

Hartzell, S. H. (1989). Comparison of seismic waveform techniques for the rupture history of a finite fault: application to the 1986 North Palm Springs, California, earthquake. *J. Geophys. Res.* **94**, 7515–7534.

Hartzell, S. H., and T. H. Heaton (1983). Inversion of strong ground motion and teleseismic waveform data for the fault rupture history of the 1979 Imperial Valley, California, earthquake. *Bull. Seismol. Soc. Am.* **73**, 1553–1583.

Hartzell, S. H., and T. H. Heaton (1986). Rupture history of the 1984 Morgan Hill, California, earthquake from the inversion of strong motion records. *Bull. Seismol. Soc. Am.* **76**, 649–674.

Hartzell, S. H., and M. Iida (1990). Source complexity of the 1987 Whittier Narrows, California, earthquake from inversion of strong motion records. *J. Geophys. Res.* **95**, 12475–12485.

Hartzell, S. H., and C. Mendoza (1991). Application of an iterative least-square waveform inversion of strong-motion and teleseismic records to the 1978 Tabas, Iran, earthquake. *Bull. Seismol. Soc. Am.* **81**, 305–331.

Hartzell, S. H., C. Langer, and C. Mendoza (1994). Rupture histories of eastern North America earthquakes. *Bull. Seismol. Soc. Am.* **84**, 1703–1724.

Hartzell, S. H., P. Liu, and C. Mendoza (1996). The 1994 Northridge, California, earthquake: investigation of rupture velocity, rise time, and high-frequency radiation. *J. Geophys. Res.* **101**, 20091–20108.

Haskell, N. A. (1964). Total energy and energy spectral density of elastic wave radiation from propagating faults. *Bull. Seismol. Soc. Am.* **54**, 1811–1841.

Haskell, N. A. (1969). Elastic displacements in the near-field of a propagating fault. *Bull. Seismol. Soc. Am.* **59**, 865–908.

Heaton, T. H. (1982). The 1971 San Fernando earthquake: a double event? *Bull. Seismol. Soc. Am.* **72**, 2037–2062.

Heaton, T. H. (1990). Evidence for and implications of self-healing pulses of slip in earthquake rupture. *Phys. Earth Planet Inter.* **64**, 1–20.

Heaton, T. H., F. Tajima, and A. W. Mori (1986). Estimating ground motions using recorded accelerograms. *Surveys in Geophys.* **8**, 25–83.

Heaton, T. H., J. F. Hall, D. J. Wald, and M. W. Halling (1995). Response of high-rise and base-isolated buildings to a hypothetical M_w 7.0 blind thrust earthquake. *Science* **267**, 206–211.

Hernandez, B., F. Cotton, and M. Campillo (1999). Contribution of radar interferometry to a two-step inversion of the kinematic process of the 1992 Landers earthquake. *J. Geophys. Res.* **104**, 13083–13099.

Horikawa, H., H. Hirahara, Y. Umeda, M. Hashimoto, and F. Kasano (1996). Simultaneous inversion of geodetic and strong-motion data for the source process of the Hyogo-ken Nanbu, Japan, earthquake. *J. Phys. Earth* **44**, 455–471.

Horton, S. (1996). A fault model with variable slip duration for the 1989 Loma Prieta, California, earthquake determined from strong-ground-motion data. *Bull. Seismol. Soc. Am.* **86**, 122–132.

Horton, S. P., J. G. Anderson, and A. Mendez (1994). Frequency domain inversion for the character of rupture during the 1989 Loma Prieta, California, earthquake using strong motion and geodetic observations. In: "The Loma Prieta, California, Earthquake of October 17, 1989—Main-Shock Characteristics" (P. Spudich, Ed.), *US Geol. Surv. Prof. Paper 1550-A*.

Housner, G. W. (1947). Characteristics of strong-motion earthquakes. *Bull. Seismol. Soc. Am.* **37**, 19–31.

Hudson, D. E. (1979). *Reading and Interpreting Strong Motion Accelerograms*. Earthquake Engineering Research Institute, Berkeley, California, 112 pp.

Humphrey, J. R., and J. G. Anderson (1992). Shear wave attenuation and site response in Guerrero, Mexico. *Bull. Seismol. Soc. Am.* **82**, 1622–1645.

Hutchings, L. (1991). "Prediction" of strong ground motion for the 1989 Loma Prieta earthquake using empirical Green's functions. *Bull. Seismol. Soc. Am.* **81**, 88–121.

Hutchings, L. (1994). Kinematic earthquake models and synthesized ground motion using empirical Green's functions. *Bull. Seismol. Soc. Am.* **84**, 1028–1050.

Hutchings, L., and F. Wu (1990). Empirical Green's functions from small earthquakes—a waveform study of locally recorded aftershocks of the San Fernando earthquake. *J. Geophys. Res.* **95**, 1187–1214.

Ichinose, G. A. (2000). Seismicity and stress transfer studies in eastern California and Nevada: implications for earthquake sources and tectonics, Ph.D. Thesis, University of Nevada, Reno, Nevada, 356 pp.

Ichinose, G. A., K. D. Smith, and J. G. Anderson (1998). Moment tensor solutions of the 1994 to 1996 Double Spring Flat, Nevada, earthquake sequence and implications for local tectonic models. *Bull. Seismol. Soc. Am.* **88**, 1363–1378.

Ide, S., M. Takeo, and Y. Yoshida (1996). Source process of the 1995 Kobe earthquake: determination of spatiotemporal slip distribution by Bayesian modeling. *Bull. Seismol. Soc. Am.* **86**, 547–566.

Inan, E., Z. Colakoglu, N. Koc, N. Bayulke, and E. Coruh (1996). *Catalog of earthquakes between 1976–1996 acceleration records, Republic of Turkey*. Ministry of Public Works and Settlements, Directorate of Disaster Affairs, Earthquake Research Department, July 1996 (in Turkish).

Irikura, K. (1983). Semi-empirical estimation of strong ground motions during large earthquakes. *Bull. Disaster Prevention Res. Inst., Kyoto Univ.* **33**, 63–104.

Irikura, K. (1986). Prediction of strong acceleration motions using empirical Green's functions. *Proc. 7th Jap. Earthq. Eng. Symp.* 151–156.

Irikura, K., K. Kudo, H. Okada, and T. Sasatani (Eds.) (1998). "The Effects of Surface Geology on Seismic Motion." (Proc. of the Second International Symposium on Effects of Surface Geology on Strong Motions, Yokohama, Japan), Bulkema, Rotterdam.

Jarpe, S. P., and P. W. Kasameyer (1996). Validation of a procedure for calculating broadband strong-motion time histories with empirical Green's functions. *Bull. Seismol. Soc. Am.* **86**, 1116–1129.

Jennings, P. C., G. W. Housner, and N. C. Tsai (1969). Stimulated earthquake motions for design purposes. *Proc. 9th World Conf. Earthq. Eng., Santiage, Chile*, 145–160.

Joyner, W. B., and D. M. Boore (1986). On simulating large earthquakes by Green's function addition of smaller earthquakes. In: "Earthquake Source Mechanics" (S. Das, J. Boatwright, and C. Scholz, Eds.), *Am. Geophys. Un. Geophys. Monograph* **37**, 269–274.

Joyner, W. B., and P. Spudich (1994). Including near-field terms in the isochrone integration method for application to finite-fault or Kirchhoff boundary integral problems. *Bull. Seismol. Soc. Am.* **84**, 1260–1265.

Kakehi, Y., and K. Irikura (1996). Estimation of high-frequency wave radiation areas on the fault plane by the envelope inversion of acceleration seismograms. *Geophys. J. Int.* **125**, 892–900.

Kakehi, Y., and K. Irikura (1997). High-frequency radiation process during earthquake faulting—envelope inversion of acceleration seismograms from the 1993 Hokkaido-Nansei-Oki, Japan, earthquake. *Bull. Seismol. Soc. Am.* **87**, 904–917.

Kakehi, Y., K. Irikura, and M. Hoshiba (1996). Estimation of high-frequency wave radiation areas on the fault plane of the 1995 Hyogo-ken Nanbu earthquake by the envelope inversion of seismograms. *J. Phys. Earth* **44**, 505–517.

Kamae, K., and K. Irikura (1998). Source model of the 1995 Hyogo-ken Nanbu earthquake and simulation of near-source ground motion. *Bull. Seismol. Soc. Am.* **88**, 400–412.

Kamae, K., K. Irikura, and A. Pitarka (1998). A technique for simulating strong ground motion using hybrid Green's functions. *Bull. Seismol. Soc. Am.* **88**, 357–367.

Kanamori, H. (1977). The energy release in great earthquakes. *J. Geophys. Res.* **82**, 2981–2987.

Kanamori, H. (1979). A semi-empirical approach to prediction of long-period ground motions from great earthquakes. *Bull. Seismol. Soc. Am.* **69**, 1645–1670.

Kanamori, H., and C. R. Allen (1986). Earthquake repeat time and average stress drop. In: "Earthquake Source Mechanics" (S. Das,

J. Boatwright, and C. Scholz, Eds.), *Am. Geophys. Un. Geophys. Monograph* **37**, 227–235.

Kanamori, H., and D. L. Anderson (1975). Theoretical basis of some empirical relations in seismology. *Bull. Seismol. Soc. Am.* **65**, 1073–1095.

Kawase, H. (1996). The cause of the damage belt in Kobe: "The basin edge effect," constructive interference of the direct S-wave with the basin-induced diffracted Rayleigh waves. *Seismol. Res. Lett.* **67**(5), 25–34.

Kawase, H., S. Matsushima, R. W. Graves, and P. G. Somerville (1998). Three-dimensional wave propagation analysis of simple two-dimensional basin structures with special reference to: "The Basin Edge Effect." *Zisin, Series-2* **50**, 431–449 (in Japanese with English abstract).

Kennett, B. L. N. (1983). *Seismic Wave Propagation in Stratified Media*. Cambridge University Press, Cambridge, 342 pp.

Kramer, S. L. (1996). *Geotechnical Earthquake Engineering*. Prentice Hall, Upper Saddle River, New Jersey, 653 pp.

Koketsu, K., and M. Kikuchi (2000). Propagation of seismic ground motion in the Kanto basin, Japan. *Science* **288**, 1237–1239.

Kostrov, B. V. (1970). The theory of the focus for tectonic earthquakes. *Izv. Acad. Sci. USSR Phys. Solid Earth, Engl. Transl.*, No. 4, 84–101.

Lee, W. H. K. (Coordinator) (2001). CD-ROM Supplement of Seismic Data, Special Issue on the 1999 Chi-Chi, Taiwan, Earthquake. *Bull. Seismol. Soc. Am.* **91**, CD-ROM Supplement.

Liu, H.-L., and T. Heaton (1984). Array analysis of the ground velocities and accelerations from the 1971 San Fernando, California, earthquake. *Bull. Seismol. Soc. Am.* **74**, 1951–1968.

Liu, H. L., and D. V. Helmberger (1983). The near-source ground motion of the 6 August 1979 Coyote Lake, California, earthquake. *Bull. Seismol. Soc. Am.* **73**, 201–218.

Luco, J. E., and J. G. Anderson (1983). Steady state response of an elastic half-space to a moving dislocation of finite width. *Bull. Seismol. Soc. Am.* **73**, 1–22.

Luco, J. E., and R. J. Apsel (1983). On the Green's functions for a layered half space; Part I., *Bull. Seismol. Soc. Am.* **73**, 909–929.

Ma, K.-F., J. Mori, S.-J. Lee, and S. B. Yu (2001). Spatial and temporal distribution of slip for the 1999 Chi-Chi, Taiwan, earthquake. *Bull. Seismol. Soc. Am.* **91**, 1069–1087.

Madariaga, R. (1983). Earthquake source theory: a review. In: "Earthquakes: Observation, Theory and Interpretation" (H. Kanamori and E. Boschi, Eds.), North Holland Publishing Company, Amsterdam, 1–44.

Matsushima, S., and H. Kawase (1998). 3-D wave propagation analysis in Kobe referring to "The basin-edge effect." In: "The Effects of Surface Geology on Seismic Motions" (K. Irikura, K. Kudo, H. Okada, and T. Sasatani, Eds.), Bulkema, Rotterdam, 1377–1384.

McGuire, R. K. (1995). Probabilistic seismic hazard analysis and design earthquakes: closing the loop. *Bull. Seismol. Soc. Am.* **85**, 1275–1284.

Mendez, A. J., and J. G. Anderson (1991). The temporal and spatial evolution of the 19 September 1985 Michoacan earthquake as inferred from near-source ground-motion records. *Bull. Seismol. Soc. Am.* **81**, 844–861.

Mendoza, C., and S. H. Hartzell (1988). Inversion for slip distribution using GDSN P waves: North Palm Springs, Borah Peak, and Michoacan earthquakes. *Bull. Seismol. Soc. Am.* **78**, 1092–1111.

Mendoza, C., and S. H. Hartzell (1989). Slip distribution of the 19 September 1985 Michoacan Mexico earthquake: near source and teleseismic constraints. *Bull. Seismol. Soc. Am.* **79**, 655–669.

Mikumo, T., and T. Miyatake (1993). Dynamic rupture process on a dipping fault, and estimates of stress drop and strength excess from the results of waveform inversion. *Geophys. J. Int.* **112**, 481–496.

Mikumo, T., and T. Miyatake (1995). Heterogeneous distribution of dynamic stress drop and relative fault strength recovered from the results of waveform inversion: the 1984 Morgan Hill, California, earthquake. *Bull. Seismol. Soc. Am.* **85**, 178–193.

Milne, W. G., and A. G. Davenport (1969). Distribution of earthquake risk in Canada. *Bull. Seismol. Soc. Am.* **59**, 729–754.

Nakahara, H. (1998). Seismogram envelope inversion for the spatial distribution of high-frequency energy radiation from the earthquake fault; application to the 1994 far east off Sanriku earthquake, Japan. *J. Geophys. Res.* **103**, 855–867.

Nakahara, H., H. Sato, M. Ohtake, and T. Nishimura (1999). Spatial distribution of high-frequency energy radiation on the fault of the 1995 Hyogo-Ken Nanbu, Japan, earthquake (M_w 6.9) on the basis of the seismogram envelope Inversion. *Bull. Seismol. Soc. Am.* **89**, 22–35.

Newmark, N. M., and W. J. Hall (1982). *Earthquake Spectra and Design*. Earthquake Engineering Research Institute Monograph, Berkeley, Calif., 103 pp.

Ni, S.-D., J. G. Anderson, Y. Zeng, and R. V. Siddharthan (2000). Expected signature of nonlinearity on regression for strong ground motion parameters. *Bull. Seismol. Soc. Am.* **90**(6B), S53–S64.

Oglesby, D. D., and S. M. Day (2001). Fault geometry and the dynamics of the 1999 Chi-Chi (Taiwan) earthquake. *Bull. Seismol. Soc. Am.* **91**, 1099–1111.

Oglesby, D. D., R. J. Archuleta, and S. B. Nielsen (1998). Earthquakes on dipping faults: the effects of broken symmetry. *Science* **280**, 1055–1059.

Olsen, K. B. (2000). Site amplification in the Los Angeles basin from 3D modeling of ground motion. *Bull. Seismol. Soc. Am.* **90**(6B), S77–S94.

Olsen, K. B., and R. J. Archuleta (1996). Three-dimensional simulation of earthquakes on the Los Angeles fault system. *Bull. Seismol. Soc. Am.* **86**, 575–596.

Olsen, K. B., R. Madariaga, and R. J. Archuleta (1997). Three-dimensional dynamic simulation of the 1992 Landers earthquake. *Science* **278**, 834–838.

Olsen, K. B., R. J. Archuleta, and J. R. Matarese (1995). Three-dimensional simulation of a magnitude 7.75 earthquake on the San Andreas fault. *Science* **270**, 1628–1632.

Olsen, K. B., R. Madariaga, and R. J. Archuleta (1997). Three-dimensional dynamic simulation of the 1992 Landers earthquake. *Science* **278**, 834–838.

Olson, A. H., and R. J. Apsel (1982). Finite fault and inverse theory with applications to the 1979 Imperial Valley earthquake. *Bull. Seismol. Soc. Am.* **72**, 1969–2001.

Ordaz, M., and C. Reyes (1999). Earthquake hazard in Mexico City; observations versus computations. *Bull. Seismol. Soc. Am.* **89**, 1379–1383.

Ou, G. B., and R. B. Herrmann (1990). A statistical model for ground motion produced by earthquakes at local and regional distances. *Bull. Seismol. Soc. Am.* **80**, 1397–1417.

Phillips, W. S., S. Kinoshita, and H. Fujiwara (1993). Basin-induced Love waves observed using the strong-motion array at Fuchu, Japan. *Bull. Seismol. Soc. Am.* **83**, 65–84.

Pitarka, A., K. Irikura, and T. Iwata (1997). Modeling of ground motion in the Higashinada (Kobe) area for an aftershock of the 1995

January 17 Hyogo-ken Nanbu, Japan earthquake. *Geophys. J. Int.* **131**, 231–239.

Pitarka, A., K. Irikura, T. Iwata, and H. Sekiguchi (1998). Local geological structure effects on ground motion from earthquakes on basin-edge faults. In: "The Effects of Surface Geology on Seismic Motions" (K. Irikura, K. Kudo, H. Okada, and T. Sasatani, Eds.), Bulkema, Rotterdam, 901–906.

Quaas, R. (1997). Mexican strong motion program. In: "Vision 2005: An Action Plan for Strong Motion Programs to Mitigate Earthquake Losses in Urbanized Areas" (J. C. Stepp, Ed.), US Committee for Advancement of Strong Motion Programs, Austin, Texas, 53–65.

Rathje, E. (Coordinator) (2000). Strong ground motions and site effects. In *Earthquake Spectra*, Supplement A to Volume 16, 65–96.

Richter, C. F. (1958). *Elementary Seismolgy*. W. H. Freeman, San Francisco, 768 pp.

Ruff, L. J. (1984). Tomographic imaging of the earthquake rupture process. *Geophys. Res. Lett.* **11**, 629–632.

Safak, E., and M. Erdik (Coordinators) (2000). Recorded main shock and aftershock motions. *Earthquake Spectra*, Supplement A to Volume 16, 97–112.

Sato, T., and T. Hirasawa (1973). Body wave spectra from propagating shear cracks. *J. Phys. Earth* **21**, 415–431.

Sato, T., D. V. Helmberger, P. G. Somerville, R. W. Graves, and C. K. Saikia (1998a). Estimates of regional and local strong motions during the Great 1923 Kanto, Japan, earthquake (M_S 8.2). Part 1: source estimation of a calibration event and modeling of wave propagation paths. *Bull. Seismol. Soc. Am.* **88**, 183–205.

Sato, T., R. W. Graves, P. G. Somerville, and S. Kataoka (1998b). Estimates of regional and local strong motions during the Great 1923 Kanto, Japan, earthquake (M_S 8.2). Part 2: forward simulation of seismograms using variable-slip rupture models and estimation of near-fault long-period ground motions. *Bull. Seismol. Soc. Am.* **88**, 206–227.

Sato, T., R. W. Graves, and P. G. Somerville (1999). Three-dimensional finite-difference simulations of long-period strong motions in the Tokyo metropolitan area during the 1990 Odawara earthquake (M_J 5.1) and the Great 1923 Kanto earthquake (M_S 8.2) in Japan. *Bull. Seismol. Soc. Am.* **89**, 579–607.

Scholz, C. H., C. A. Aviles, and S. G. Wesnousky (1986). Scaling differences between large intraplate and interplate earthquakes. *Bull. Seismol. Soc. Am.* **76**, 65–70.

Schneider, J. F., W. J. Silva, and C. Stark (1993). Ground motion model for the 1989 M 6.9 Loma Prieta earthquake including effects of source, path, and site. *Earthquake Spectra* **9**, 251–287.

Seekins, L. C., A. G. Brady, C. Carpenter, and N. Brown (1992). Digitized strong-motion accelerograms of North and Central American earthquakes 1933–1986. *US Geol. Surv. Digital Data Series DDS-7.*

Sekiguchi, H., and T. Iwata (2002). Rupture process of the 1999 Kocaeli, Turkey, earthquake estimated from strong motion waveforms. *Bull. Seismol. Soc. Am.* **92**, 300–311.

Sekiguchi, H., K. Irikura, T. Iwata, Y. Kakehi, and M. Hoshiba (1996). Minute locating of faulting beneath Kobe and the waveform inversion of the source process during the 1995 Hyogo-ken Nanbu, Japan, earthquake using strong ground motion records. *J. Phys. Earth* **44**, 473–487.

Shakal, A. F. (1997). California strong motion instrumentation program. In: "Vision 2005: An Action Plan for Strong Motion Programs to Mitigate Earthquake Losses in Urbanized Areas" (J. C. Stepp, Ed.), US Committee for Advancement of Strong Motion Programs, Austin, Texas, 75–80.

Shi, B., A. Anooshehpoor, J. N. Brune, and Y. Zeng (1998). Dynamics of thrust faulting: 2D lattice model. *Bull. Seismol. Soc. Am.* **88**, 1484–1494.

Shin, T. C., K. W. Kuo, W. H. K. Lee, T. L. Teng, and Y. B. Tsai (2000). A preliminary report on the 1999 Chi-Chi (Taiwan) earthquake. *Seismol. Res. Lett.* **71**, 24–30.

Silva, W. J., R. B. Darragh, C. Stark, I. Wong, J. C. Stepp, J. Schneider, and S. J. Chiou (1990). A methodology to estimate design response spectra in the near-source region of large earthquakes using the band-limited-white-noise ground motion model. *Proc. 4th Int. Conf. Earthq. Engineering, Palm Springs, CA*. Earthquake Engineering Research Institute, Berkeley, 487–494.

Sokolov, V. Y. (1998). Spectral parameters of the ground motions in Caucasian seismogenic zones. *Bull. Seismol. Soc. Am.* **88**, 1438–1444.

Somerville, P. (2000). Magnitude scaling of near fault ground motions (abstract). *EOS, Trans. Am. Geophys. Un.* **81**, F822.

Somerville, P., M. Sen, and B. Cohee (1991). Simulation of strong ground motions recorded during the 1985 Michoacan, Mexico, and Valparaiso, Chile earthquakes. *Bull. Seismol. Soc. Am.* **81**, 1–27.

Somerville, P. G., N. F. Smith, R. W. Graves, and N. A. Abrahamson (1997). Modification of empirical strong ground motion attenuation relations to include the amplitude and duration effects of rupture directivity. *Seismol. Res. Lett.* **68**, 199–222.

Somerville, P., K. Irikura, R. W. Graves, S. Sawada, D. Wald, N. Abrahamson, Y. Iwasaki, T. Kagawa, N. Smith, and A. Kowada (1999). Characterizing crustal earthquake slip models for the prediction of strong ground motion. *Seismol. Res. Lett.* **70**, 59–80.

Spudich, P., and R. J. Archuleta (1987). Techniques for earthquake ground-motion calculations with applications to source parameterization of finite faults. In: "Seismic Strong Motion Synthetics" (B. A. Bolt, Ed.), Academic Press, Orlando, Florida, 205–265.

Spudich, P., and E. Cranswick (1984). Direct observation of rupture propagation during the 1979 Imperial Valley earthquake using a short baseline accelerometer array. *Bull. Seismol. Soc. Am.* **74**, 2083–2114.

Spudich, P., and L. N. Frazer (1984). Use of ray theory to calculate high-frequency radiation from earthquake sources having spatially variable rupture velocity and stress drop. *Bull. Seismol. Soc. Am.* **74**, 2061–2082.

Spudich, P., and L. N. Frazer (1987). Erratum. *Bull. Seismol. Soc. Am.* **77**, 2245.

Spudich, P., and D. Oppenheimer (1986). Dense seismograph array observations of earthquake rupture dynamics. In: "Earthquake Source Mechanics" (S. Das, J. Boatwright, and C. Scholz, Eds.), Am. Geophys. Un. *Geophys. Monograph,* **37**, 285–296.

Spudich, P., M. Hellweg, and W. H. K. Lee (1996). Directional topographic site response observed in aftershocks of the 1994 Northridge, California, earthquake: implications for main shock motions. *Bull. Seismol. Soc. Am.* **86**, S193–S208.

Spudich, P., W. B. Joyner, A. G. Lindh, D. M. Boore, B. M. Bargaris, and J. B. Fletcher (1999). SEA99: a revised ground motion prediction relation for use in extensional tectonic regimes. *Bull. Seismol. Soc. Am.* **89**, 1156–1170.

SSHAC (Senior Seismic Hazard Analysis Committee) (1997). Recommendations for Probabilistic Seismic Hazard Analysis: Guidance on Uncertainty and Use of Experts, NUREG/CR-6372, U.S. Nuclear Regulatory Commission.

Steidl, J. H., R. J. Archuleta, and S. H. Hartzell (1991). Rupture history of the 1989 Loma Prieta, California, earthquake. *Bull. Seismol. Soc. Am.* **81**, 1573–1602.

Stepp, J. C. (Ed.) (1997). *Vision 2005: An Action Plan for Strong Motion Programs to Mitigate Earthquake Losses in Urbanized Areas*, US Committee for Advancement of Strong Motion Programs, Austin, Texas. [This report is reproduced on the attached Handbook CD #2, under the directory of \57Anderson.]

Su, F., J. G. Anderson, and Y. Zeng (1998). Study of weak and strong ground motion including Nonlinearity from the Northridge, California, earthquake sequence. *Bull. Seismol. Soc. Am.* **88**, 1411–1425.

Takemura, M., and I. Ikeura (1987). Semi-empirical synthesis of strong ground motions for the description of inhomogeneous faulting. *J. Seismol. Soc. Japan* **40**, 77–88.

Trifunac, M. D., and A. G. Brady (1975). A study of the duration of strong earthquake ground motion. *Bull. Seismol. Soc. Am.* **65**, 581–626.

Trifunac, M. D., and F. E. Udwadia (1974). Parkfield, California, earthquake of June 27, 1966: a three-dimensional moving dislocation. *Bull. Seismol. Soc. Am.* **64**, 511–533.

Tsai, C.-C. P., and K.-C. Chen (2000). A model for the high-cut process of strong-motion accelerations in terms of distance, magnitude, and site condition: an example from the SMART 1 Array, Lotung, Taiwan. *Bull. Seismol. Soc. Am.* **90**, 1535–1542.

Tsai, Y.-B., and M.-W. Huang (2000). Strong ground motion characteristics of the Chi-Chi, Taiwan earthquake of September 21, 1999. *Earthq. Engin. and Engin. Seismol.* **2**, 1–21.

Toro, G. R., N. A. Abrahamson, and J. F. Schneider (1997). Model of strong ground motions from earthquakes in central and eastern North America: best estimates and uncertainties. *Seismol. Res. Lett.* **68**, 41–57.

Tucker, B. E., and Brune, J. N. (1973). Seismograms, S-wave spectra, and source parameters for aftershocks of San Fernando earthquake. In: "San Fernando, California, Earthquake of February 9, 1971" (N. A. Benfer, J. L. Coffman, and J. R. Bernick, Eds.), Volume III, US Dept. Commer., Natl. Oceanic Atmos. Adm., Environ. Res. Lab., Washington, D.C., 69–121.

Udwadia, F. E., and M. D. Trifunac (1973). Damped Fourier spectrum and response spectra. *Bull. Seismol. Soc. Am.* **63**, 1775–1783.

U.S. Geological Survey (1999). An Assessment of Seismic Monitoring in the United States. Requirement for an Advanced National Seismic System. *US Geol. Surv. Circular 1188*, US Department of Interior, Denver, Colorado.

USGS Web site (2001). *http://geohazards.cr.usgs.gov/eq/index.html*, accessed January 2001, National Seismic Hazard Mapping Project, United States Geological Survey, Golden, Colorado.

Vidale, J. E., and D. V. Helmberger (1988). Elastic finite-difference modeling of the 1971 San Fernando, California, earthquake. *Bull. Seismol. Soc. Am.* **78**, 122–141.

Viksne, A., C. Wood, and D. Copeland (1997). Bureau of Reclamation seismic monitoring/strong motion program and notification system. In: "Vision 2005: An Action Plan for Strong Motion Programs to Mitigate Earthquake Losses in Urbanized Areas" (J. C. Stepp, Ed.), US Committee for Advancement of Strong Motion Programs, Austin, Texas, 86–90.

Wald, D. J. (1996). Slip history of the 1995 Kobe, Japan, earthquake determined from strong-motion, teleseismic, and geodetic data. *J. Phys. Earth* **44**, 489–503.

Wald, D. J., and R. W. Graves (1998). The seismic response of the Los Angeles basin, California. *Bull. Seismol. Soc. Am.* **88**, 337–356.

Wald, D. J., and R. W. Graves (2001). Resolution analysis of finite fault source inversion using one- and three-dimensional Green's functions; 2, Combining seismic and geodetic data. *J. Geophys. Res.* **106**, 8767–8788.

Wald, D. J., and T. H. Heaton (1994). Spatial and temporal distribution of slip for the 1992 Landers, California, earthquake. *Bull. Seismol. Soc. Am.* **84**, 668–691.

Wald, D. J., D. V. Helmberger, and S. H. Hartzell (1990). Rupture process of the 1987 Superstition Hills earthquake from the inversion of strong motion data. *Bull. Seismol. Soc. Am.* **80**, 1079–1098.

Wald, D. J., D. V. Helmberger, and T. H. Heaton (1991). Rupture model of the 1989 Loma Prieta earthquake from the inversion of strong-motion and broadband teleseismic data. *Bull. Seismol. Soc. Am.* **81**, 1540–1572.

Wald, D. J., T. H. Heaton, and K. W. Hudnut (1996). The slip history of the 1994 Northridge, California, earthquake determined from strong-motion, teleseismic, GPS, and leveling data. *Bull. Seismol. Soc. Am.* **86**, S49–S70.

Ward, S. N. (1995). A multidisciplinary approach to seismic hazards in southern California. *Bull. Seismol. Soc. Am.* **85**, 1293–1309.

Wells, D. L., and K. J. Coppersmith (1994). New empirical relationships among magnitude, rupture length, rupture width, rupture area, and surface displacement. *Bull. Seismol. Soc. Am.* **84**, 974–1002.

Wennerberg, L. (1990). Stochastic summation of empirical Green's functions. *Bull. Seismol. Soc. Am.* **80**, 1418–1432.

Wu, F. T. (1978). Prediction of strong ground motion using small earthquakes. *Proc. 2nd Int. Microzonation Conf., San Francisco*, 701–704.

Yagi, Y., and M. Kikuchi (2000). Source rupture process of the Kocaeli, Turkey, earthquake of August 17, 1999, obtained by joint inversion of near-field data and teleseismic data. *Geophys. Res. Lett.* **27**, 1969–1972.

Yoshida, S., K. Koketsu, T. Sagiya, and T. Kato (1996). Joint inversion of near- and far-field waveforms and geodetic data for the rupture process of the 1995 Kobe earthquake. *J. Phys. Earth* **44**, 437–454.

Yu, G., K. N. Khattri, J. G. Anderson, J. N. Brune, and Y. Zeng (1995). Strong ground motion from the Uttarkashi, Himalaya, India earthquake: comparison of observations with synthetics using the composite source model. *Bull. Seismol. Soc. Am.* **85**, 31–50.

Zeng, Y., and J. G. Anderson (1996). A composite source model of the 1994 Northridge earthquake using genetic algorithms. *Bull. Seismol. Soc. Am.* **86**, S71–S83.

Zeng, Y., and C.-H. Chen (2001). Fault rupture process of the 20 September 1999 Chi-Chi, Taiwan, earthquake. *Bull. Seismol. Soc. Am.* **91**, 1088–1098.

Zeng, Y., K. Aki, and T. L. Teng (1993a). Source inversion of the 1987 Whittier Narrows earthquake, California, using the isochron method. *Bull. Seismol. Soc. Am.* **83**, 358–377.

Zeng, Y., K. Aki, and T. L. Teng (1993b). Mapping of the high-frequency source radiation for the Loma Prieta earthquake, California. *J. Geophys. Res.* **98**, 11981–11993.

Zeng, Y., J. G. Anderson, and G. Yu (1994). A composite source model for computing realistic synthetic strong ground motions. *Geophys. Res. Lett.* **21**, 725–728.

Zeng, Y., J. G. Anderson, and F. Su (1995). Subevent rake and random scattering effects in realistic strong ground motion simulation. *Geophys. Res. Lett.* **22**, 17–20.

58

Strong-Motion Data Processing

A. F. Shakal, M. J. Huang, and V. M. Graizer
California Geological Survey, Sacramento, CA, USA

1. Introduction

Initially very challenging, the digitizing and processing of recorded strong motion has seen great advances in the last 30 years. The evolution from the first digitization attempts, which defined problems still being dealt with today, through subsequent developments in computer digitizing and modern instrumentation has allowed significant progress in understanding near-source strong motion.

Although traditional seismological instruments were developed in the 1800s and perfected in the 1900s, instruments that could measure strong motion were not developed until the 1930s. The U.S. Coast and Geodetic Survey began the development of an accelerograph (Wenner, 1932) in response to a need expressed by earthquake engineers following an earthquake conference in Tokyo. Background aspects of the early development are reviewed by Housner (Scott, 1997; see also Chapter 2 by Housner). The first instruments were deployed in the Los Angeles area during the March 10, 1933, Long Beach earthquake, and very important records were obtained. Once earthquake recordings were obtained, processing of the records to extract crucial strong-motion information became important, and that forms the topic of this chapter.

2. Early Digitizing and Processing

2.1 Early Mechanical Analysis

The record from the Los Angeles Subway site from the 1933 Long Beach earthquake was one of the first strong-motion records. It was not exceeded in amplitude until the 1940 Imperial Valley earthquake, and it received extensive digitization and processing efforts.

Early analysis and digitization was difficult, partly because of the equipment limitations of the time, and partly because much was being learned about pitfalls in digitization and processing. The initial efforts involved cooperative work by the U.S. Coast and Geodetic Survey (USCGS), the National Bureau of Standards, and the Massachusetts Institute of Technology. The earliest analysis of a record was performed using a mechanical analyzer system at the USCGS (Neumann, 1943) by means of which operators could manually generate an approximate displacement record from an acceleration recording. A second approach utilized a mechanical differential analyzer already in use at MIT for studying other problems. These approaches using analyzing machines were paralleled by a manual numerical integration approach. The early work led to a symposium focused on "Determination of True Ground Motion by Integration of Strong Motion Records." Part of the Bulletin of the Seismological Society of America for January 1943 is devoted to the symposium, and it includes analysis of the differential analyzer and the importance of accuracy results from an engineering standpoint (Ruge, 1943a, b) and an appraisal of the numerical integration efforts (Neumann, 1943). The most important element from a historical perspective was the numerical methods approach, which became the basis of the progress that occurred in the succeeding decades.

2.2 Early Numerical Processing

The early procedures and difficulties in numerical integration of accelerograms are described in Neumann (1943), and many carry over in some form to the present. Digitization consisted of manually measuring and recording the position of the acceleration trace at fixed intervals along the length of the accelerogram. The accelerograms were photographically enlarged to make the measurement process more accurate and less difficult.

The ground motion could be calculated from these measurements using the fundamental equation of motion of a damped pendulum when subjected to an external acceleration,

$$-\mathrm{d}^2x(t)/\mathrm{d}t^2 = \mathrm{d}^2y(t)/\mathrm{d}t^2 + 2\mathrm{h}(2\pi/\mathrm{T_o})\mathrm{d}y(t)/\mathrm{d}t + (2\pi/\mathrm{T_o})^2y(t), \quad (1)$$

INTERNATIONAL HANDBOOK OF EARTHQUAKE AND ENGINEERING SEISMOLOGY, VOLUME 81B ISBN: 0-12-440658-0

where $x(t)$ is the instantaneous displacement of the instrument housing (i.e., normally the ground), $y(t)$ is the instantaneous displacement of the pendulum relative to the instantaneous position of the housing (measured from the record), h is a damping constant, and T_o is the instrument period. With integration twice, the housing or ground displacement is obtained as,

$$-x(t) = y(t) + 2h(2\pi/T_o)\int_o^t y(t)\,dt + (2\pi/T_o)^2\int_o^t\int_o^t y(t)\,dt\,dt + c_1 + c_2 t. \quad (2)$$

where c_1 and c_2 are the integration constants. Once the amplitude values have been scaled from the record, the housing (or ground) displacement can be approximated using the numerical equation (with C_1 and C_2 being the re-defined integration constants):

$$-x_i = y_i + 2h(2\pi/T_o)\left(\sum_0^t y_i\Delta t + C_1\right) + (2\pi/T_o)^2\left[\sum_0^t\left(\sum_0^t y_i\Delta t + C_1\right)\Delta t + C_2\right]. \quad (3)$$

The practical steps involve measuring, or scaling, the y_i ordinate values of the recorded accelerogram at constant time steps t_i. The effort and time required to accurately scale peak values on analog records even today underscores the extensive efforts required by the USCGS to obtain reasonable accuracy during this period. Once the measured values from the record were available, the process of numerical integration could begin, performed using an adding machine with which an operator tediously obtained the necessary partial sums to calculate the displacement.

The constants of integration, C_1 and C_2, had to be estimated once initial sums had been obtained. Note that regardless of advances in technology, the initial conditions remain an issue. However, most modern digital accelerographs have adequate pre-event memory so that the pre-event zero-acceleration values can be estimated.

If the ground is at rest during the time interval immediately prior to the event, then C_1 is the zero-acceleration value (or pre-event y mean value), and C_2 is zero. However, many analog or optical recorders are still deployed; in these units, recording is initiated by the shaking itself, and so some of the motion is necessarily lost. In general, if the threshold level necessary to initiate recording is set low enough, the effect will not be severe, especially for recordings with high-amplitude acceleration. However, if the shaking at the site is of low amplitude, an important part of the total motion will be missed. For example, a distant earthquake with only long-period motion at a site illustrates the problem. In the Mexico City earthquake of 1957, only $0.04g$ was recorded. Because the motion was at long period, near the period of many structures, significant damage occurred (Duke and Leeds, 1959), though many film recorders in that situation would trigger late in the motion (or not trigger). For late-triggered records, accurate determination of displacements is still problematic.

The determination of displacement by manual numeric integration was performed as an interactive process, each time obtaining better estimates of C_1 and C_2. Of course, the accuracy problems associated with summing many numbers in a machine with finite register size are now well known. The early work was groundbreaking, grappling with many of the problems that strong-motion processing continued to address for the rest of the century.

In a parallel instrumentation effort during this period, the USCGS deployed instruments designed to have a long-period response, called displacement meters. These units typically had a 10-second pendulum period so that the recording would more closely correspond to the displacement motion of the ground. Although not many of these instruments were installed because of difficulty and cost, some records were obtained. It was an important step to compare displacement records calculated by numeric integration of the accelerometer records with these displacement meter records, to help understand the source and amplitude of the various errors. These comparisons have been made for several records, generating confidence in the processing procedures (e.g., Trifunac and Lee, 1974).

2.3 Standardized Digitizing and Processing—The Caltech Bluebook Project

Until 1970, the USCGS continued to obtain strong-motion recordings, but a need developed for standardized processing of the records obtained so that investigators would be analyzing the same time histories and spectral inputs for engineering design and structural dynamics studies. Some studies highlighted the uncertainty associated with calculating long-period displacements (Hershberger, 1955; Berg and Housner, 1961). The records from El Centro in 1940, Kern County in 1952, and Parkfield in 1966 underscored the need for uniform processing, particularly because of the unexpected amplitudes and periods recorded.

A project initiated in 1968 at Caltech focused on the computer processing of all records available at the time in a standardized approach. The San Fernando earthquake that occurred in 1971 caused a large increase in the number of records while the project was underway. The San Fernando records included the largest motions recorded up to that time, which increased the importance of standard processing.

The Caltech project moved forward from the USCGS numerical integration effort in several aspects. The most important was the use of large, fast digital computers, which had become available at major universities and research centers. The project was very productive, and the series of reports produced, all in blue cover, gave the project its unofficial name. By releasing the complete results by means of printed reports and files on digital computer tapes, the project allowed major progress by many investigators in analytical studies of the data, and the records processed became the foundation of many research advances in subsequent years.

The second major difference between this project and earlier work was the use of a mechanical–electrical digitizing table incorporating a hand cursor with crosshairs. The cursor would be positioned over the center of the trace, and the (x, y) coordinate pair was output onto punched cards or magnetic tape when a button was pushed.

In one manual-digitizing approach, points on the waveform were digitized at equal time intervals on the record. In the second, the peaks, troughs, and points of inflection in the waveform were digitized, and the waveform was assumed to be comprised of straight-line segments between these points. The goal of this approach was to efficiently preserve most of the full-frequency content of the record while digitizing a reduced number of points. However, the effects of a varying sample rate are not well documented in signal theory. Current digitizing technology has made this approach unnecessary, and equal-interval sampling is now standard.

The use of a digital computer in the processing steps allowed much more complicated procedures to replace the adding machine in order to improve speed, accuracy, and precision. The resulting process became quite complicated, as reflected in Figure 1 (EERL, 1971), which illustrates that while there was

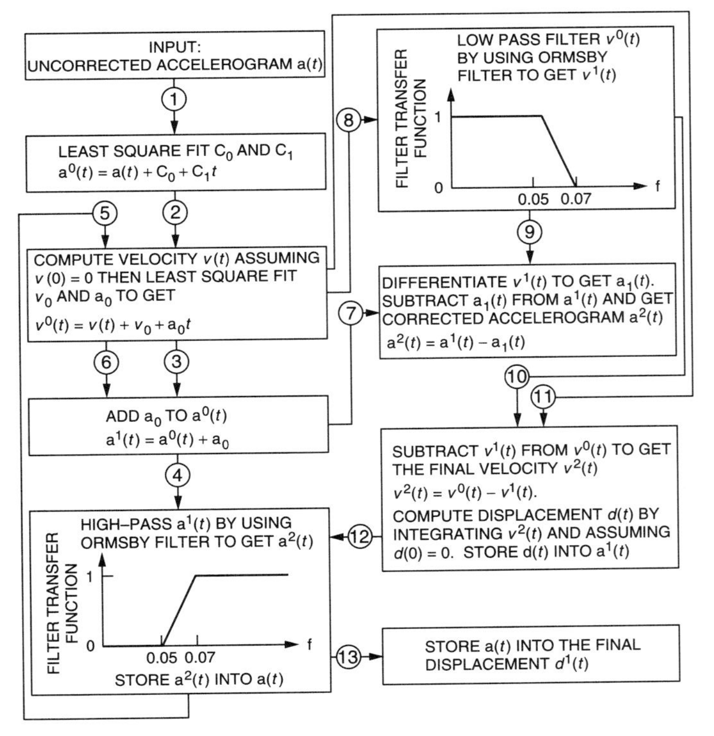

FIGURE 1 Flow chart summarizing the steps originally used in processing strong-motion accelerograms in the Caltech Bluebook Project (from EERL, 1971). Though simplified from the previous approaches, the process was still highly complex and iterative.

a significant advance upon the Neumann USCGS effort, it was still a highly complex, iterative process. The technical documentation and computer program listings for the Caltech project are included in Trifunac and Lee (1973).

2.3.1 Standard Data Products

Results were released at several stages in the Caltech processing, and the names for these have become traditional: (1) Volume 1: Raw acceleration, as digitized, usually given as acceleration–time pairs, and expressed in units of acceleration; no instrument correction or filtering applied. (2) Volume 2: "Corrected" acceleration, velocity, and displacement; basically the final time-history product, with constant time steps. The raw acceleration has been bandpass filtered, and various other processing steps, including correcting for the instrument response, applied to obtain the best estimate of the acceleration. This acceleration has been numerically integrated, with follow-up filtering as necessary, once to obtain velocity, and once again to obtain displacement. (It was unfortunate that the word "corrected" became attached to this product, because it implied to some engineers that the data needed to be corrected because of errors. However, this misunderstanding has become less common over the years.) (3) Volume 3: Response spectrum values, for five damping values (0 through 20%) and 91 periods (from 0.04 through 15 seconds), including spectral acceleration and the pseudovelocity and displacement spectra.

The preceding three products are sometimes denoted as Phase 1, 2, etc., rather than Volume 1, 2 (e.g., Brady *et al.*, 1980), and this usage avoids the confusion often associated with the word volume. Several additional numbered products were generated during the Caltech project, but these are not commonly used today. Nearly all strong-motion networks and projects, worldwide, generate these three primary data products.

3. Modern Digitization

3.1 Semi-automated Accelerogram Scanning and Digitization

Advances in computer hardware resulted in the commercial availability of computer-controlled scanning systems in the mid-1970s. Early scanning machines were expensive and required careful handling. The first semi-automatic, computer-controlled digitization system was introduced by Trifunac and Lee (1979) at the University of Southern California. This system replaced the cursor-and-crosshairs of the Caltech project with an optical-density sensor that moved across the film image under the control of a computer, yielding an x–y grid of optical density values. Curve-following software was applied to the grid of optical densities, ultimately yielding acceleration–time pairs like that obtained in manual digitization.

This system yielded a major increase in digitizing speed, though extensive work was still sometimes required at graphic workstations to construct accurate acceleration–time functions. The difficulty depended on the quality of the film accelerogram and the amplitude of the records. High-amplitude records with acceleration traces that repeatedly crossed traces from adjacent channels were particularly difficult.

A system like that of Trifunac and Lee has been used for many years at the California Strong Motion Instrumentation Program of the California Geological Survey (formerly the Div. of Mines and Geology) to digitize and process many records. As part of that effort, the system was further studied in terms of noise and long period response and certain improvements made to lower noise levels (Shakal and Ragsdale, 1984).

Another computer digitization system used during this time was a specialized commercial laser scanning and curve-following system that had some of the same difficulties and benefits (IOM/Towill; described in Converse, 1984). Many records were digitized by the U.S. Geological Survey through this company. Scanning systems were also developed in Japan (Iai *et al.*, 1978) and Italy (Berardi *et al.*, 1991) during this period.

3.2 PC-Based Desktop Digitization

As personal computers became more widespread and more powerful, it became practical to use them in strong-motion processing. In addition, development of scanning devices continued and economical desktop units became available that had marginally adequate resolution for use in accelerogram digitization. The first system of this type was developed by Lee and Trifunac (1990), for which they converted their earlier operator-intensive mainframe system (Trifunac and Lee, 1979) to use a desktop scanner and PC.

Another strategy was to take a fresh approach using the full capacity of a PC and the higher resolution desktop scanners that become available around 1990. Cao *et al.* (1994) describe a high-resolution desktop system that is almost fully automated except for the most difficult records. Many records have now been digitized using this system, including over 200 Northridge records from privately owned code-required instruments, which produced film records of lower optical quality than regular network records.

Other PC-based digitization systems have also been developed, but some have higher noise levels either because of inadequate step size or because the trace is approximated as a black-and-white image. The subtleties that gray-scale imaging allows yield results that more smoothly match the accelerogram. Of course quality must be carefully monitored, and some of these systems have encountered problems, as noted by Trifunac *et al.* (1998).

3.3 Digital Accelerographs

A transition in the 1980s significantly changed the strong-motion digitization and processing problem. As technology progressed, economical and reliable digitally recording accelerographs became available. For these units the onerous laboratory task of digitizing the recording is performed in the field by an internal analog-to-digital converter. The earliest digital accelerographs recorded data on magnetic tape but these did not approach the reliability of analog units (due to the effects of temperature cycling and aging, the magnetic tapes were often difficult, if not impossible, to decode). With the introduction of solid-state memory, the potential of digital accelerographs was fully realized. These units have now begun to be the source of new strong-motion records, and are supplanting analog accelerographs, worldwide.

Most accelerograms recorded by modern digital accelerographs are of high quality, so that processing of records is much less difficult. Certain serious problems can occur, however, and these are discussed subsequently. Excepting these cases, the traditional issues of base-line correction and high-frequency problems are often not as serious today. Recovering the maximum amount of information at long periods is still a challenge, as discussed further shortly. For most cases, however, the period band of most importance in earthquake engineering, up to 3 or 4 seconds, can be provided with confidence from digital accelerographs.

3.3.1 Accelerograph Dynamic Range and Noise Levels

The range of ground-motion amplitudes that can be recorded by an accelerograph is very important in controlling final processed signal quality and long-period noise levels. The largest acceleration that common analog accelerographs can record is nominally $1g$, and the smallest value that can be read from a film record is about $0.005g$. The ratio of the largest to smallest accelerations, the dynamic range, is about 200, or in the logarithmic decibel scale, about 50 dB. Recordings have shown that accelerographs need to be able to record accelerations as high as $2g$. Until about 1995, common digital accelerographs had a 12-bit analog-to-digital converter (A/D), so the smallest value they could record was the range of $\pm 2g$ divided into $2^{12} = 4096$ parts, or about $0.001g$. The dynamic range of these units, 72 dB, is significantly better than the analog film recorders. A few digital accelerographs currently use a 10-bit A/D. The dynamic range of these units is not much better than the analog units; although they have value in simple monitoring and alarm systems, noise levels make it very difficult to obtain processed data with quality similar to that of 12-bit accelerographs.

Most contemporary digital accelerographs have 18-bit or better resolution. For 18 bits, the least acceleration readable from a $2g$ sensor is about $\pm 2g/2^{18} = 0.015$ mg (108 dB dynamic range). The noise level at many urban sites where strong-motion instruments are installed is higher than 0.015 mg, so this resolution may be more than needed at these sites. The Northridge earthquake showed again that accelerations during earthquake shaking could exceed $2g$, at least in the near-field zone and in structures. Sensors with the next higher range, $4g$, are generally not more costly and the impact of raising the maximum (increasing the minimum recordable motion to 0.03 mg) has little negative effect at urban sites because of the noise level. This range yields a record unlikely to be clipped and with resolution low enough for performing accurate processing of the recorded accelerogram. It is noted that some accelerographs now have 24-bit A/D capability. These units, with a minimum recordable motion of 0.24 μg with a 2g sensor, have their most effective application in low-noise, non-urban environments.

Digital accelerographs have advanced so much in the last decade that the importance of the recorder itself has receded for strong-motion processing, and other noise sources have become the major concerns. The primary noise sources that may compromise processed strong-motion data include sensor noise and electronic/communication noise, discussed further following.

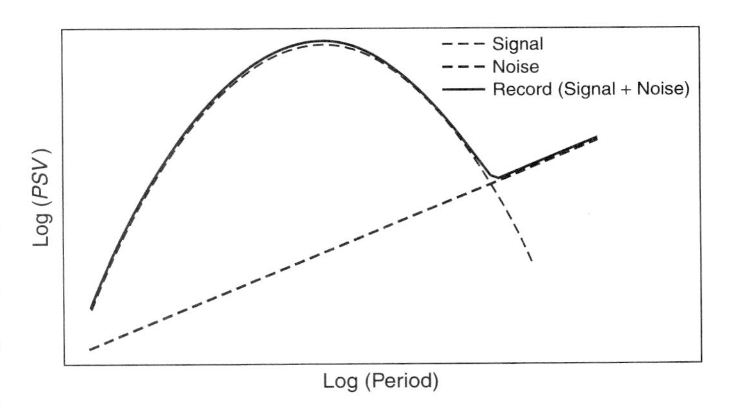

FIGURE 2 Schematic shape of the response spectrum of a typical digitized earthquake record (solid line). It is a combination of the actual earthquake spectrum (thin dash) and the noise from digitization and other sources (wide dash). The objective in data processing is to determine the maximum period band for which the final result preserves as much signal as possible but retains as little noise as possible.

4. Strong-Motion Processing Procedures

Whether a digital record has been recorded digitally or obtained through laboratory digitization of an analog record, procedures for processing the digital record are similar, and are described next.

4.1 Uniform Processing—Long-Period Filtering Dependent on Noise Spectra

A major lesson of the Caltech project was that the calculation of displacements for long periods was problematic. This problem was noted by Neumann (1943), but it became particularly important as more records were processed. Certain specific long-period problems were identified, for example, a problem at a spectral period near 14 seconds due to the practical aspect of digitizing table length (e.g., Hanks, 1973). More experience with the long-period problem was gained in the Caltech project, and during this period the method of choosing the most appropriate long-period filter was not yet standardized.

A realization developed that displacements computed for the longest periods were heavily influenced by noise intrinsic to the digitization and processing steps, and that this noise was largely understandable and could be handled in a uniform way (e.g., Trifunac, 1977). In general, the velocity response spectrum for records from intermediate-magnitude events increases from a low value at short periods until it reaches a maximum at intermediate periods, beyond which it decreases once again at longer periods. In contrast, the noise spectrum obtained by digitizing and processing a straight-line record increases in the long-period range, where the data spectrum is decreasing. As a result, the spectrum of a digitized accelerogram plotted against period, before filtering, often has a characteristic shape, which is shown schematically in Figure 2. The noise spectrum is present for both analog and digital records, though it is generally higher for optically digitized analog records than for records from digital recorders.

The principle of uniform processing is the utilization of this characteristic form as the basis for selecting the filter period. This approach was introduced in the reprocessing of the Caltech records by Trifunac and Lee (1978). They selected the long-period filter location uniformly based on where the accelerogram spectrum intersected the average spectrum obtained for a set of straight-line accelerograms (e.g., Figure 2). This was an important step in handling long-period noise in a coherent manner.

This approach coincides with signal-to-noise ratio approaches used in signal processing and is used by several strong-motion networks. A signal-to-noise ratio (SNR) of 2 to 3 may be used to obtain an initial estimate of the filter period. Specifically, a period is chosen to the left (toward shorter periods) of the minimum that occurs where the signal and noise spectra intersect. A period is then selected at which the spectrum has increased to 2–3 times the minimum.

A practical aspect remaining, even in uniform processing, is the choice of a single filter corner for all components of a given record. This means that some components (e.g., horizontal) are filtered more severely because the filter is set to control noise in the components with the lowest amplitudes (e.g., vertical). This practice continues at many networks, though the practice is arguable.

4.2 Procedural Steps in Strong-Motion Processing

The post-digitization processing procedures and steps described next have their origin in the procedures developed in the Caltech project (EERL, 1969, 1971, 1972) and modified by Trifunac and Lee (1978) and others. There are variations in the steps, and Converse and Brady (1992) use a procedure that is somewhat simpler, but the basic principles are shared in common. The seven steps are:

Step 1. Baseline Correction

The raw data are interpolated to obtain equal-interval sampling, if necessary (e.g., 200 points/sec, for a 100 Hz folding frequency), and converted to acceleration units using the sensitivity constant of the accelerometer. At least a first-order baseline operation is performed, to make the data zero-mean. More involved baseline correction may also be performed on certain records, though this is less common in the past, because there is normally no physical basis for removing a parabola or other high-level function from an accelerogram. In contrast, when baseline correction is performed via long-period filtering (Step 4) it has well-defined properties in the frequency domain; these are largely independent of the record length (e.g., Trifunac, 1971). Simple baseline correction using a constant or linear trend, where appropriate, is most effective in projects handling a large number of records. The results of this step are usually denoted as Volume (or Phase) 1, and released as the raw-data product.

Step 2. Instrument Correction

The baseline corrected data are corrected for instrument response using a simple finite-difference operator to obtain $dy(t)/dt$ and $d^2y(t)/dt^2$ in Equation (1). Note that the sensitivity constant has already been applied in Step 1. In frequency domain processing, discussed further following, the finite-difference process is replaced by simply dividing the spectrum by the instrument response spectrum.

Step 3. High-Frequency Filtering

After instrument correction, a filter is generally applied to remove high-frequency noise. In the Caltech and USC processing, an Ormsby filter (Ormsby, 1961) with a corner frequency at 23 Hz and a termination frequency at 25 Hz was applied. Current approaches typically use a more common filter (such as a Butterworth filter) with a corner frequency near 40% of the final sampling frequency (i.e., 80% of the Nyquist frequency) and a 3rd or 4th order decay. After filtering, the data are decimated to the final sample rate at which the acceleration data will be distributed to the user community (50 points/sec in the Caltech/USC data; 100 points/sec is common today for data recorded digitally). Note that if time-domain processing is used, the instrument correction should be performed prior to decimation, rather than after, in order to improve the high-frequency accuracy (Sunder and Connor, 1982; Shakal and Ragsdale, 1984). Some networks currently release processed data with the same sample rate as the original data; in this case, decimation is not performed.

Step 4. Initial Integration and Long-Period Filtering

An initial long-period filter is applied to the instrument-corrected acceleration data with a cutoff corner near 15 seconds period (longer periods are filtered out), which is the maximum period used in Caltech/USC processing. Recent data has shown that a longer period cutoff may be appropriate for digital records from large earthquakes. Velocity and displacement are obtained by numerically integrating the acceleration and filtered using the same low-frequency filter. In time-domain processing, if decimation is performed prior to the long-period filtering, prevention of spurious long-period energy requires use of an Ormsby or similar filter, rather than a running mean filter, because of its side-lobe leakage (Shakal and Ragsdale, 1984).

Step 5. Computation of Maximum-Bandwidth Response Spectra

The pseudovelocity (*PSV*) response spectra are calculated in the time domain for the full set of 91 periods defined in the Caltech project, ranging from 0.04 seconds to 15 seconds, using the method of Nigam and Jennings (1969) or an equivalent. The spectra are computed for damping values of 0, 2, 5, 10, and 20% of critical, using the acceleration obtained in Step 4, and plotted for 0.04 to 15 seconds (the full bandwidth) for use in comparative analyses to select the best filter.

Step 6. Long-Period Filter Selection

A suite of long-period filters is applied to the data obtained in Step 3 using corner periods near the long-period minimum of the spectrum obtained in Step 5. The long-period intersection of the maximum-bandwidth spectrum and the average noise spectrum determined for the system (e.g., Trifunac, 1977; Shakal and Ragsdale, 1984) is used to indicate the long-period limit of useful information in the record, as discussed previously. The final value of the filter corner is selected after studying the resulting suite of displacement plots, comparing them to one another, to displacement plots computed from noise records, and to records obtained for nearby stations, if available (e.g., Huang *et al.*, 1989). Choosing a filter period at which the signal-to-noise ratio (SNR) is 2 to 3 or greater usually gives a result with low noise, though this may be more conservative than some desire.

Selection of the optimal long-period filter remains the most difficult part of strong-motion processing. The effect of earthquake magnitude is to raise the response spectrum at long periods, such that the crossing with the noise spectrum may not occur in the usual strong-motion processing band. An approach has been suggested in which the long-period filter is selected based on a theoretical model of the spectrum radiated by a simplified source (e.g., Basili and Brady, 1978). However, as more data have become available, especially from larger events, it appears best to choose the long-period filter based on the signal

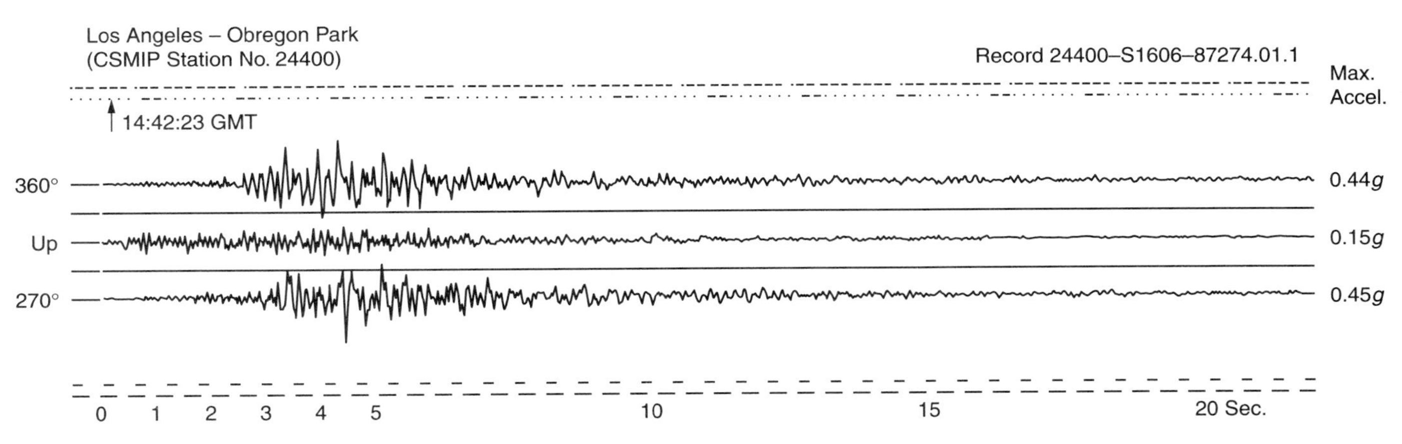

FIGURE 3 Copy of the analog (film) record obtained from the Obregon Park station in Los Angeles during the 1987 Whittier earthquake.

as recorded, regardless of theoretical models, especially since theoretical models will be revised as more data from large earthquakes becomes available.

Step 7. Final Output Preparations

The acceleration, velocity, and displacement time histories obtained using the final filter are plotted for presentation in reports and saved as files for distribution. The response spectra may be plotted with tripartite logarithmic scaling or the linear scaling most commonly used in building codes.

4.3 Illustrative Example

To illustrate the process just described, a copy of a raw accelerogram is shown in Figure 3. This analog record contains three acceleration traces corresponding to the three component directions, two fixed reference traces, a half-second time mark pattern at the bottom, and a time mark trace containing the IRIG-encoded time at the top. This record was digitized (not necessary for a digital record) and scaled to produce the Vol. 1 (Phase 1) data plotted in Figure 4. Steps 2 through 4 were then performed, and the full-bandwidth spectrum of Step 5 was computed. For the first component, the velocity spectrum of the full-bandwidth output is compared with the noise spectrum of the digitization system in Figure 5. As part of Step 6, a suite of displacement time histories, computed using a range of long-period filter corners, are compared in Figure 6 to guide the selection of the optimal filter. The final "corrected" acceleration, velocity, and displacement (Phase 2 data), and response spectra (Phase 3 data) of Step 7 are shown in Figures 7 and 8, respectively. Note that the final filter bandwidth is indicated in both plots.

4.4 Usable Data Bandwidth

Processed data has a bandwidth, extending from the period of the high-frequency (short-period) filter of Step 3 to the long-period filter of Step 6, within which the final spectrum corresponds to the original record. Outside of this band, filters have removed as much as possible of the information because of noise contamination, and this band may be called the Usable Data Bandwidth (Figure 9). The UDB defines the range within which the results can be used in modeling structure response, etc. Comparisons of model results to data outside this period band are not valid, of course, and may be misleading. Note that the spectra of the final result in Step 7 should only be plotted to the period of the long-period filter—beyond that the spectra are smoothly decreasing, but this is because of the asymptotic nature of the response spectrum at long periods (e.g., Hudson, 1979), and does not reflect the shape of the data spectrum.

4.5 Time Domain vs. Frequency Domain

Some operations in strong-motion processing may be most effectively performed in the frequency domain. The early Caltech processing was all performed in the time domain because it occurred before the Fast Fourier Transform (Cooley and Tukey, 1965) was widely available. The Ormsby filter was used because it is a relatively stable recursive filter. Today the FFT is very common, and filters such as the Butterworth are easily applied in the frequency domain, and integration to obtain velocity and displacement is as simple as dividing the spectrum by $-i\omega$. With fast computers and cheap memory, even the large transforms that may be needed are within the capacity of most personal computers.

Care must be taken in frequency-domain processing regarding certain aspects. Tapering the ends of the acceleration time series with a raised cosine bell, or Tukey interim window (e.g., Bergland, 1969), once the mean has been removed, prevents spurious side-lobe leakage in the spectrum. If there is an adequate pre-event data segment, this does not affect the data at the beginning of the data. In general, the time-history length to be transformed to the frequency domain must have a power-of-two

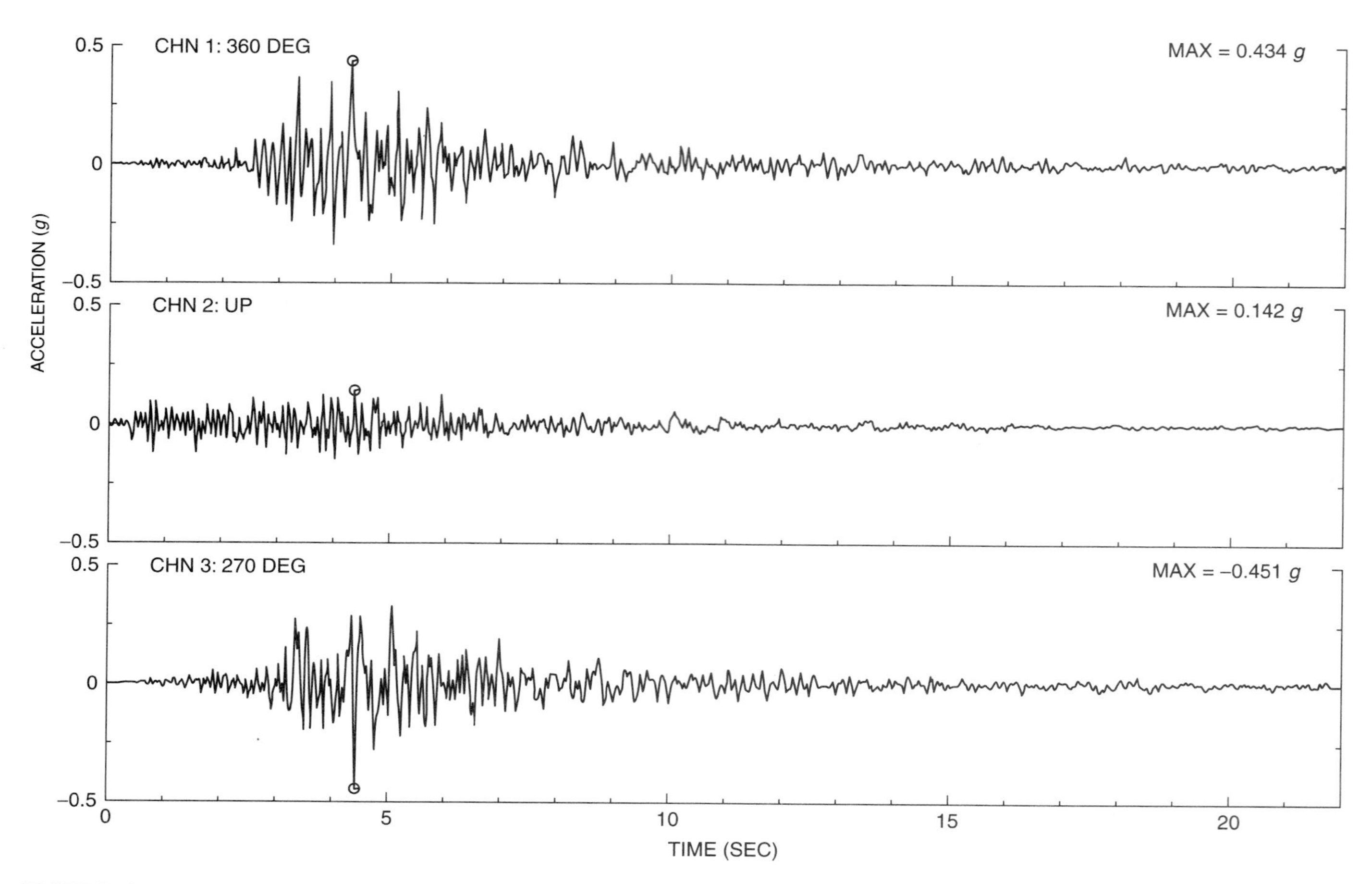

FIGURE 4 Acceleration records (Phase 1 data) after the data traces in the film record of Figure 3 have been digitized and converted to acceleration using the sensitivity constants of the accelerometers, and the means have been removed (Step 1).

number of points, and a zero-filled section should be included at the end of adequate length (perhaps 20% of the record length or more) to prevent problems with periodicity.

Some operations, such as complex baseline corrections, may be most effectively done in the time domain. For example, if there is an a priori reason that certain time functions should be used to compensate for a baseline shift that occurs at a certain time in an accelerogram (e.g., Iwan *et al.*, 1985), a time domain application is most effective.

In general, frequency domain methods are fast, understandable, stable, and give replicatable results in situations where many records are to be processed. They are also well understood in signal analysis and processing theory (e.g., Rabiner and Gold, 1975; Kanasewich, 1973). Response spectra calculations need to be at least evaluated in the time domain because they involve obtaining the time-domain maximum for each spectral period.

4.6 Processing by Other Networks in the World

Several networks in countries of Europe and the Pacific also process and release strong-motion data. For example, the Port and Harbor Research Institute in Japan has been processing and releasing data for many years using a system described by Iai *et al.* (1978) that shares many features of the approach developed in the Caltech project. They also use a semi-automated analog record digitization system for analog records. Of course, the recently instrumented Kyoshin network in Japan is a forward evolution with no analog instruments, like the TriNet network in the United States, both described elsewhere in this volume.

The New Zealand and Italian strong-motion networks have also digitized and released records from important events in the area, and key parts of their processing are also similar to, and grew out of, the Caltech system (e.g., Hodder, 1983; Berardi *et al.*, 1991).

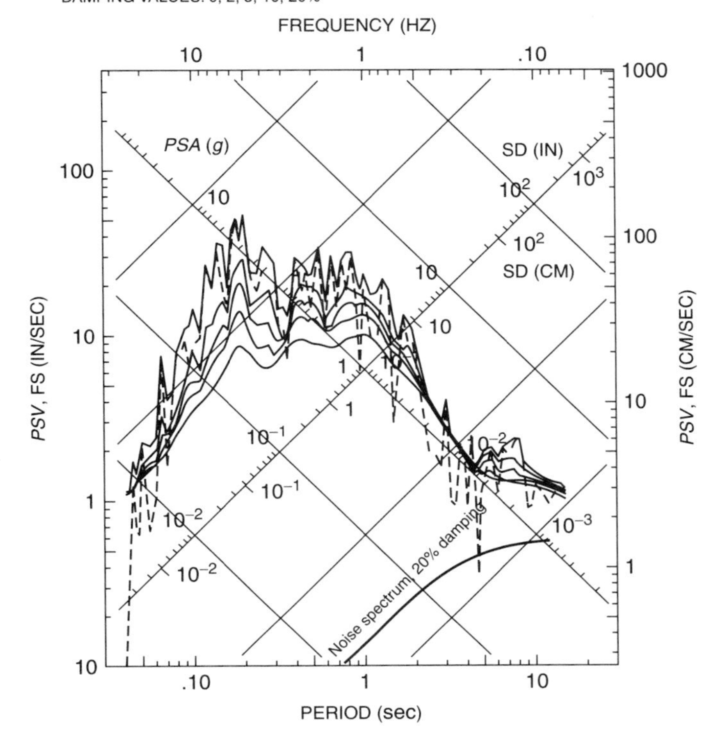

FIGURE 5 Spectrum computed (with a wide-bandwidth filter) for the first component of the record in Figure 4 is compared with the noise-level spectrum (*PSV*, 20% damping) of the digitization system to guide the selection of the long-period filter.

5. Specific Issues in Processing

Several issues are of particular importance in strong-motion processing. Some, such as natural frequency, have seen beneficial developments. Others, including DC offsets, can be as problematic today as they were years ago.

5.1 Instrument Response—Accelerometer Natural Frequency

Changes in accelerometer natural frequency over the last decades have had significant positive impacts on strong-motion processing. The earliest USCGS accelerometers of 1933 had a natural frequency of about 10 Hz (0.1 sec period); this increased to around 20 Hz with the commercial film recorders (e.g., SMA-1). The natural frequency increased again to 50 Hz for accelerometers used with the digital recorders. Some of the latest sensors have a natural frequency near 80 Hz.

The result of this increase in natural frequency is to make correction for the response of the sensor much less critical in the frequency range below 20 Hz, which is of most importance in strong motion. As the sensor natural frequency increases, the last term in the response equation (Eq. 1) dominates. The principal effect of instrument correction then becomes multiplication of the observed record by the static sensitivity. As a result, instrument correction has become much less critical than in the earlier years of strong-motion processing, when the natural frequency was 10 to 20 Hz, and errors in natural frequency or damping could significantly affect the results.

5.2 Accelerometer Offsets and Tilts

Calculated long-period displacements are very sensitive to the stability of the DC or zero level of the accelerometer. Small errors in the zero level can lead to significant errors in displacements and long-period spectrum levels. This problem is more important for modern digital recorders, largely because more information is being extracted from records due to their greater dynamic range. An illustrative example occurred in the Northridge earthquake records from a six-story office building in Los Angeles (Station 24652; Shakal *et al.*, 1994a, p. 153). The recording indicates that a shift of several percent *g* occurred around the time of the strongest shaking, implying a significant permanent tilt or adjustment at the 3rd-floor level of the building. However, the building was subsequently inspected and no structural changes were observed. Routine processing of a record like this would indicate significant results for velocity and especially displacement level, as well as spectral levels at long periods. In this case, the source of the implied structural change was actually a change in the zero of the sensor.

The ability of sensors to move to a new zero value during or after strong, or even weak, motion is an important caution to bear in mind in strong-motion processing. This phenomenon had been observed before, though never with such implications. For example, Iwan *et al.* (1985) described this in laboratory tests and proposed a numerical approach to correct for this effect. This behavior was also observed in early small-amplitude field recordings, especially in the early digital accelerographs that recorded on magnetic tape. It has also been observed in the records from other arrays and instrument types, such as those from the SMART array in Taiwan, and most recently the 1999 Taiwan earthquake (W. H. K. Lee, personal communication, 1999).

Correction for a changing zero level remains problematic, because it requires an assumption of the nature and timing of any changes. If the change can be approximated by a simple step, it is easy to introduce a correction in the processing. If a change can be approximated as a constant during an interval of unknown duration, followed by a new constant value, an approach suggested by Iwan *et al.* (1985) can be used. However, if the change occurs over a period of time as a ramp, curve, oscillatory function, or a series of steps (as slippage in the sensor might cause), the appropriate correction function is uncertain. Boore (2001) applied the Iwan approach in working with certain Taiwan earthquake data and found varying results.

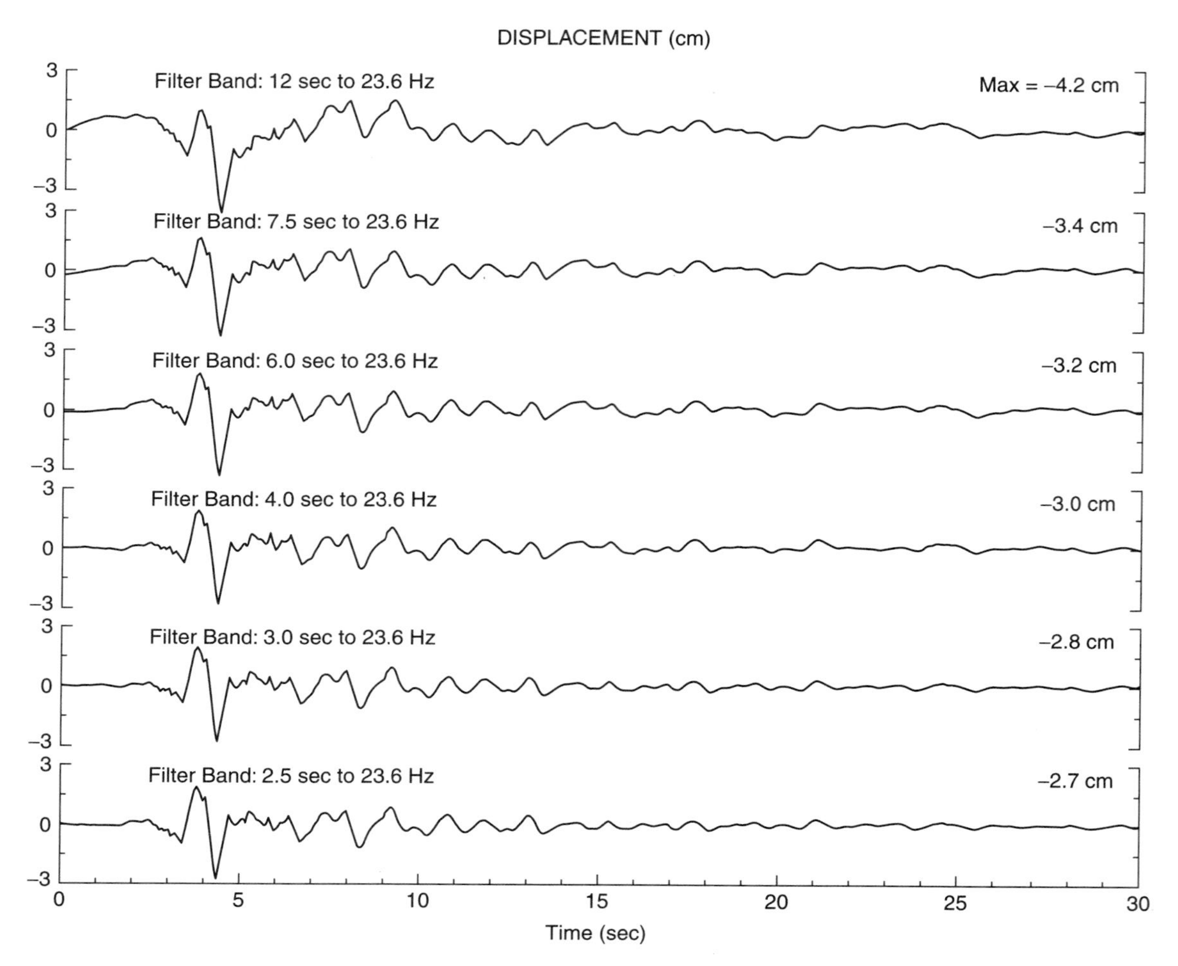

FIGURE 6 Displacement plots for the first component in Figure 4 processed with a suite of long-period filters. These plots are compared with each other, and the corresponding spectra are compared with the noise spectra in Figure 5, to select the long-period filter. The wandering of the uppermost trace is clearly due to noise, corresponding to the long-period noise in Figure 5. In contrast, very little noise is apparent in the bottom trace, also reflected in Figure 5 near 2 seconds, where the signal is well above the noise. In this case, the filter with a 3-second corner period (corresponding to a 0.2 to 0.4 Hz ramp) was selected (Huang *et al.*, 1989).

Of course, a zero-level change due to an actual tilt of the instrument is indistinguishable in a record from a sensor zero-offset problem. For example, the Pacoima Dam upper left abutment record in the Northridge earthquake showed offsets found in processing that were able to be substantiated in the field as actual tilts (Shakal *et al.*, 1994b).

In summary, zero-level offsets or tilts remain an important difficulty in strong-motion processing, especially in the recovery of the longest period motions. However, in the period band of most importance in engineering, this effect is not as important as it is in seismological source model research. For example, Shakal *et al.* (1994b) found little effects at periods less than 10 seconds, and Boore (2001) found no important effects below 20 seconds in the examples he studied.

5.3 Determination of Permanent Displacement

Permanent displacement is generally not important in earthquake engineering unless a structure straddles a fault. However, it is of scientific interest, and can help constrain the longest-period part of the source spectrum. The determination of permanent displacement has been attempted from the earliest tests (Neumann, 1943), and it has been fraught with controversy. In some cases it is possible to make estimates of the

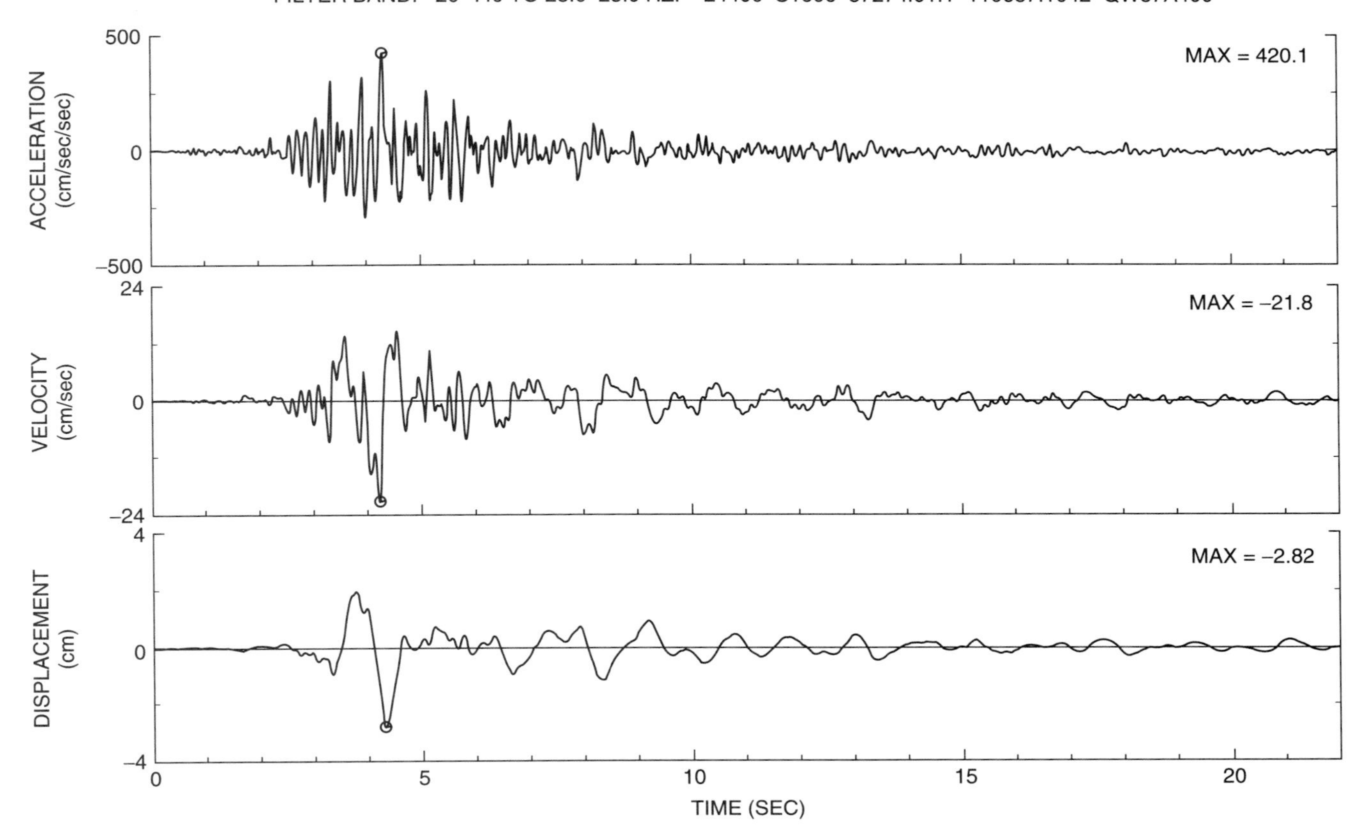

FIGURE 7 Instrument and baseline-corrected acceleration, velocity, and displacement (Phase 2 data) for the first component of the record in Figure 4. The filter bandwidth selected during the processing and analysis is indicated on the plot.

permanent displacement, but in general independent knowledge of the actual ground displacement is important to guide the calculations. One example is the record from the Lucerne station in the 1992 Landers earthquake (Hawkins *et al.*, 1993), from which Iwan and Chen (1994) estimated permanent displacements, and several possibilities occur in the data from the Taiwan earthquake of September 21, 1999, and the Hector Mine, California, earthquake of October 16, 1999. The potential presence of baseline offsets in the accelerogram challenge the determination of permanent displacements, though several analytic means have been suggested (e.g., Graizer, 1979; Iwan *et al.*, 1985).

In summary, except for records from sites at which the permanent displacement during an earthquake was relatively large, the determination of permanent displacements from acceleration records may not generally be practical, largely because of sensor limitations. Though this is an important scientific issue, it is not critical for most earthquake engineering applications.

5.4 Cross-Axis Sensitivity and Channel Cross Talk

A concern of long standing in strong motion is the sensitivity of an accelerometer to motion perpendicular to its sensing direction (e.g., Skinner and Stephenson, 1973; Wong and Trifunac, 1977). Mechanical limits in manufacturing and mechanical coupling effects limit how small this effect can be; common technical specifications list 3% or less. If a cross-axis signal is as large as this, it can be a significant noise (i.e., false signal) source. In general, it should be held to significantly less than this, and this and other aspects should be verified by static tilt tests (e.g., Lee, 1993) or other means.

In structural or geotechnical array configurations, where sensors may be located remotely from recorders, channel-to-channel electronic signal leaking, or cross talk, can be as important as, and indistinguishable from, cross-axis sensitivity of sensors. Cross talk must be prevented, before an earthquake, through signal tests and static tilt tests of individual sensors. Data analysis

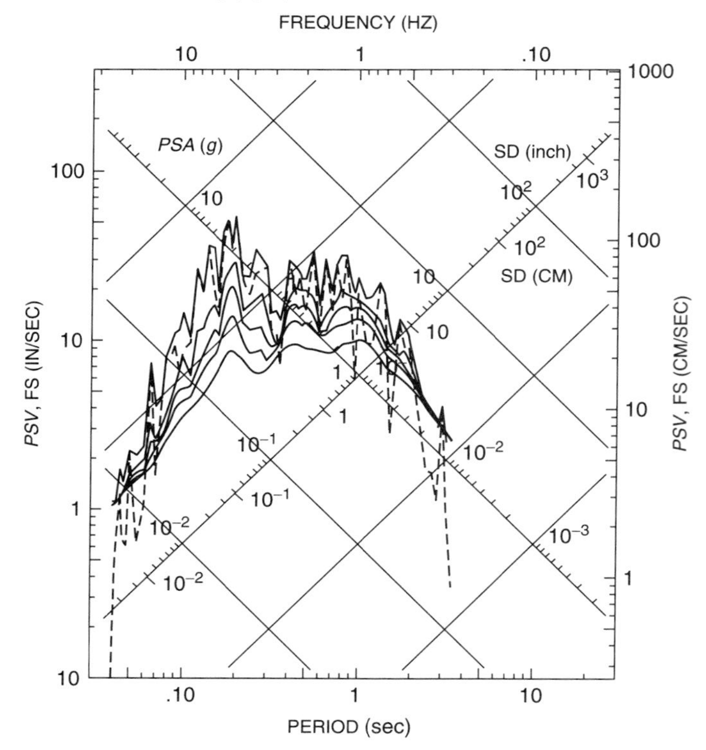

FIGURE 8 Response spectra (Phase 3 data) for the first component of the record in Figure 4. The spectra are plotted for periods within the filter bandwidth determined in the Phase 2 processing.

methods such as cross correlation and coherency are commonly used to study and characterize earthquake motion. Even low cross-axis sensitivity or channel cross talk will significantly contaminate these estimates.

5.5 Processing High-Amplitude Records

One continuing lesson of strong-motion recording in the last half of the century is that the largest recorded accelerations have continued to exceed expectations. These range from about 1.25g in the San Fernando earthquake of 1971 to nearly 2g in the 1994 Northridge earthquake. Some of these peak acceleration values are not of engineering significance because they are of short duration.

For strong-motion processing purposes the key issue is that the motion occurring be recorded and processed correctly. Recent data show that instruments need to be capable of recording more than 1g on the ground and 2g in certain structures. (Nearly all of the analog film recorders, and even some of the digital recorders, are not capable of that.) Assumptions are necessary to process

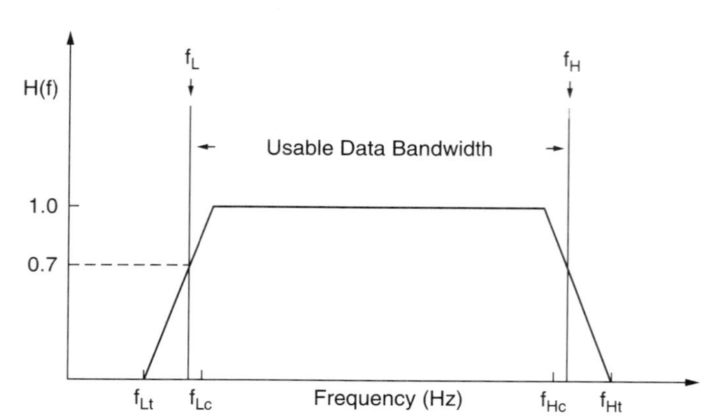

FIGURE 9 The "Usable Data Bandwidth" for a processed strong-motion record is the band between frequencies f_H and f_L, the high and low -3 dB corner frequencies. The data contains little useful information for analyses or modeling outside this band. The amplitude of the filter transfer function H(f) is reduced to approximately 0.7 at -3 dB and the transmitted power is reduced to 0.5. [The 3 dB corners are well defined for filters like the Butterworth. For Ormsby filters, they can be determined from the ramp corner (f_{Hc}, f_{Lc}) and termination frequencies (f_{Ht}, f_{Lt}) using $f_H = f_{Hc} + 0.3(f_{Ht} - f_{Hc})$ and $f_L = f_{Lc} - 0.3(f_{Lc} - f_{Lt})$. For example, the UDB for data filtered with Ormsby ramps from 0.3 to 0.6 Hz and 23 to 25 Hz is 0.51 Hz to 23.6 Hz, or 0.042 to 2.0 seconds period.]

any record that exceeds the instrument maximum, even for a short time. These assumptions, which can be challenged, can taint the perception of users regarding the quality of processing for the many other high-quality, lower amplitude records. The most effective way to manage this problem is to field instruments with adequate recording range at the outset.

5.6 Electronic Noise Issues

As technological equipment becomes more widespread in the environment it can present a challenge for strong-motion accelerograph stations, especially in urban situations. It has been observed that some strong-motion sensors are susceptible to the effects of radio frequency (RF) electromagnetic radiation. This can cause false or phantom events, or distortion during the recording of actual events. In most cases, the nature of RF-caused records has allowed them to be recognizable, and the occurrence of the events can be decreased with good field practices. Recent years have seen an increasing amount of RF transmission, and it may be expected that this problem will increase as communication channels continue to expand.

6. Automatic Preliminary Processing

> "Civilization advances by extending the number of operations which we can perform without thinking about them. . . ."
>
> A. N. Whitehead

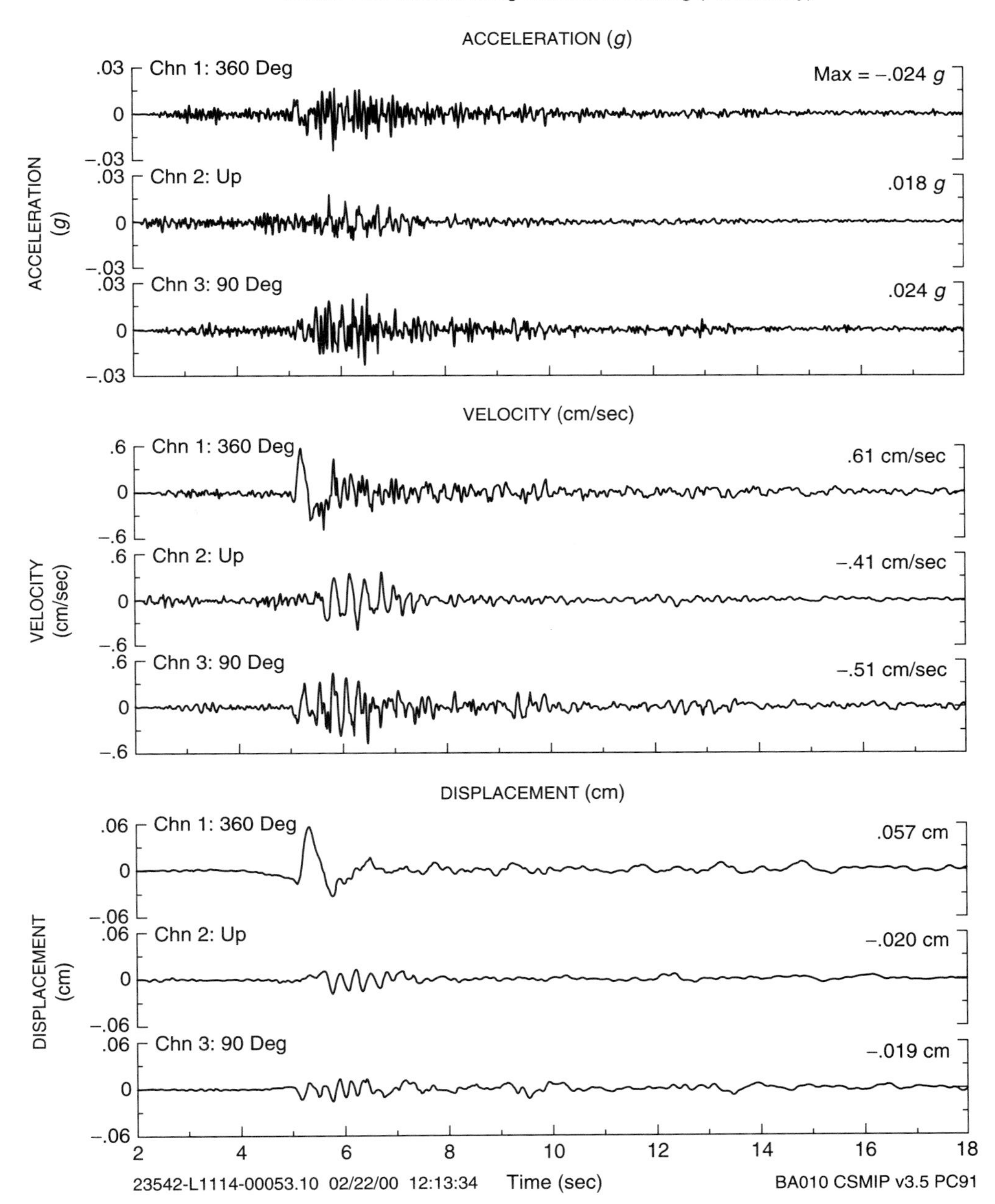

FIGURE 10 Automatic preliminary processing results—three components of acceleration, velocity, and displacement at San Bernardino from a M4.4 earthquake that occurred on February 21, 2000.

Advances in recording technology have opened the possibility of automated preliminary processing, and initial steps in this direction show promise for the future (e.g., Figure 10). Automated processing will never approach the quality level that human analysis can yield, but the rapid results obtained are useful if treated as preliminary. To get the best possible results, the preliminary processing must be followed up by professional analysis of noise levels and possible filter settings, just as has been performed for years in strong-motion processing.

A recent development that has opened new possibilities in rapid use of strong-motion recordings is communication with recorders at remote sites via telephone lines and modems.

With additional control at the field sites, it is possible to go to the next step of having field instruments initiate communication with a central site upon recording an event, and transmitting the recorded data. This has been done successfully for several years in the California Strong Motion Instrumentation Program (Shakal *et al.*, 1995) and the U.S. Bureau of Reclamation (Viksne *et al.*, 1995) and recently in the U.S. Geological Survey and other networks. This approach has shown such promise that accelerograph manufacturers have begun to include the necessary auxiliary field components within the accelerograph itself.

For automated, preliminary processing, a process similar to that described above may be used. For all but large-earthquake records, a long-period filter near 3 or 4 seconds can be used as a reasonable first estimate. Because sensors have a high natural frequency, as noted, an approximate instrument correction as simple as multiplication by a sensitivity factor can often be used with little penalty. Of course, this preliminary processing must be augmented by careful processing and filter selection by analysts at a later time to obtain optimal quality results.

7. Summary

The processing of strong-motion recordings has undergone great progress from the first recording in 1933 through the modern instruments available at the end of the century. Processing went from tedious, manual calculations to almost completely computerized processing. Some problems and issues associated with sensors, offsets and tilts, and electronics remain, despite the great progress. However, these problems are relatively small. The evolution of automated strong-motion processing in the 1990s will make strong-motion data increasingly useful for emergency response and rapid post-earthquake investigation of damage as well as for long-term research.

References

Basili, M., and A. G. Brady (1978). Low frequency filtering and the selection of limits for accelerogram corrections. *Proc. Sixth European Conf. on Earthq. Engineering*, Yugoslavia.

Berardi, R., G. Longhi, and D. Rinaldis (1991). Qualification of the European strong-motion databank: influence of the accelerometric station response and pre-processing techniques. *J. European Earthquake Engineering* **2**, 38–53.

Berg, G. V., and G. W. Housner (1961). Integrated velocity and displacement of strong earthquake ground motion. *Bull. Seism. Soc. Amer.* **51**, 175–189.

Bergland, G. D. (1969). A guided tour of the fast Fourier transform. *IEEE Spectrum* **6**, 51–52. In: "Digital Signal Processing" (L. R. Rabiner and C. M. Rader, Eds.), IEE Press, New York.

Boore, D. M. (2001). Effect of baseline corrections on response spectra for several recordings of the 1999 Chi-Chi, Taiwan, earthquake. *Bull. Seismol. Soc. Am.* **91**, 1199–1211.

Brady, A., V. Perez, and P. Mork (1980). The Imperial Valley earthquake, October 15, 1979, Digitization and processing of accelerograph records. *US Geol. Surv. Open-File Report 80-703*, Menlo Park.

Cao, T. Q., R. Darragh, and A. Shakal (1994). Digitization of strong motion accelerograms using a personal computer and flatbed scanner (abstract). *Seismol. Res. Lett.* **65**(1), 42.

Converse, A. (1984). AGRAM: a series of computer programs for processing digitized strong-motion accelerograms. *Open-File Report 84-525, U.S. Geological Survey*, Menlo Park.

Converse, A., and A. G. Brady (1992). BAP: basic strong motion accelerogram processing software. *Open-File Report, 92-296A, U.S. Geological Survey*, Menlo Park.

Cooley, J. W., and J. W. Tukey (1965). An algorithm for the machine calculation of complex Fourier series. *Math. Computation* **19**, 297–301.

Duke, C. M., and D. J. Leeds (1959). Soil conditions and damage in the Mexico earthquake of July 28, 1957. *Bull. Seismol. Soc. Amer.* **49**, 179–191.

Earthquake Engineering Research Laboratory (1969). Strong motion earthquake accelerograms, digitized and plotted data, Vol. I – Uncorrected accelerograms; Part A, *EERL Report 70-20*, California Institute of Technology, Pasadena.

Earthquake Engineering Research Laboratory (1971). Strong motion earthquake accelerograms, digitized and plotted data, Vol. II – Corrected accelerograms and integrated ground velocity and displacement curves; Part A, *EERL Report 71-50*, California Institute of Technology, Pasadena.

Earthquake Engineering Research Laboratory (1972). Analysis of strong motion earthquake accelerograms, Vol. III – Response spectra; Part A, *EERL Report 72-80*, California Institute of Technology, Pasadena.

Graizer, V. M. (1979). Determination of true ground displacement by using strong motion records. *Izv. Acad. Sci. USSR* (translated by Amer. Geophys. Un.) **15**, 875–884.

Hanks, T. (1973). Current assessment of long-period errors. In: "Strong-motion Earthquake Accelerograms, Digitized and Plotted Data." *Vol. II, Earthquake Engineering Research Laboratory, EERL 73-52*, California Institute of Technology, Pasadena.

Hawkins, H. G., D. K. Ostrom, and T. A. Kelley (1993). Comparison of strong motion records, Landers earthquake 28 June 1992, near Lucerne Valley, California. *Unpublished manuscript.*

Hershberger, J. (1955). Recent developments in strong motion analysis. *Bull. Seismol. Soc. Amer.* **45**, 11–22.

Hodder, S. B. (1983). Computer processing of New Zealand strong-motion accelerograms. *Bull. New Zealand Nat. Soc. Earthq. Eng.* **16**(3), 234–245.

Huang, M. J., T. Q. Cao, D. L. Parke, and A. F. Shakal (1989). Processed strong-motion data from the Whittier, California earthquake of 1 October 1987. *Report OSMS 89-03*, Calif. *Strong Motion Instrumentation Program, Calif. Div. Mines and Geology, Sacramento.*

Hudson, D. E. (1979). *Reading and Interpreting Strong Motion Accelerograms*. Earthquake Engineering Research Institute, Berkeley.

Iai, S., E. Kurata, and H. Tsuchida (1978). Digitization and correction of strong-motion accelerograms. *Technical Note 286, Port and Harbor Research Institute*, Yokosuka, Japan, 286 pp.

Iwan, W. D., and X. D. Chen (1994). Important near-field ground motion data from the Landers earthquake. *Proc. 10th European Conf. Earthq. Eng.*, Vienna, Austria.

Iwan, W. D., M. A. Moser, and C.-Y. Peng (1985). Some observations on strong-motion earthquake measurement using a digital accelerograph. *Bull. Seismol. Soc. Am.* **75**, 1225–1246.

Kanasewich, E. R. (1973). *Time Sequence Analysis in Geophysics*. Univ. Alberta Press, Canada, 352 pp.

Lee, V. W., and M. D. Trifunac (1979). Automatic digitization and processing of strong motion accelerograms, Part II: Computer processing of accelerograms. *Report CE 79-15 II*, Dept. Civil Engineering, Univ. Southern California, Los Angeles.

Lee, V. W., and M. D. Trifunac (1990). Automatic digitization and processing of accelerograms using PC. *Report CE 90-03*, Dept. Civil Engineering, Univ. Southern Calif. Los Angeles, 150 pp.

Lee, W. H. K. (1993). 1993 tilt test results. *Appendix 2 of the Central Weather Bureau Research Report No. 446*, Seismology Center, Central Weather Bureau, Taipei, Taiwan.

Neumann, F. (1943). An appraisal of numerical integration methods as applied to strong-motion data. *Bull. Seismol. Soc. Am.* **33**, 13–20.

Nigam, N. C., and P. C. Jennings (1969). Calculation of response spectra from strong-motion earthquake records. *Bull. Seismol. Soc. Am.* **59**, 909–922.

Ormsby, J. F. A. (1961). Design of numerical filters with application to missile data processing. *J. Assoc. Comput. Machinery* **8**, 440–466.

Rabiner, L. R., and B. Gold (1975). *Theory and Application of Digital Signal Processing*. Prentice Hall, New Jersey, 762 pp.

Ruge, A. C. (1943a). Analysis of accelerograms by means of the M.I.T. differential analyzer. *Bull. Seismol. Soc. Am.* **33**, 61–63.

Ruge, A. C. (1943b). Discussion of principal results from the engineering standpoint. *Bull. Seismol. Soc. Am.* **33**, 13–20.

Scott, S. (1997). "George W. Housner," *Connections, EERI Oral History Series No. 4*, Earthquake Engineering Research Institute, Oakland, CA, 275 pp.

Shakal, A. F., and J. T. Ragsdale (1984). Acceleration, velocity and displacement noise analysis for the CSMIP accelerogram digitization system. *Proc. 8th World Conf. Earthquake Engineering* **2**, 111–118.

Shakal, A., M. Huang, R. Darragh, T. Cao, R. Sherburne, P. Malhotra, C. Cramer, R. Sydnor, V. Graizer, G. Maldonado, C. Petersen, and J. Wampole (1994a). CSMIP strong-motion records from the Northridge, California earthquake of 17 January, 1994. *Calif. Strong Motion Instrumentation Program, Report 94-07*, Calif. Div. Mines and Geology, Sacramento, 308 pp.

Shakal, A., T. Cao, and R. Darragh (1994b). Processing of the upper left abutment record from Pacoima Dam for the Northridge earthquake. *Calif. Strong Motion Instrumentation Program, Report OSMS 94-13*, Calif. Div. Mines and Geology, Sacramento.

Shakal, A. F., C. D. Petersen, A. B. Cramlet, and R. B. Darragh (1995). CSMIP near-real-time strong motion monitoring system: Rapid data recovery and processing for event response. In: "Proceedings SMIP95 Seminar on Seismological and Engineering Implications of Recent Strong-Motion Data" (M. J. Huang, Ed.), pp. 1–10, Calif. Div. Mines and Geol., Sacramento.

Skinner, R. I., and W. R. Stephenson (1973). Accelerograph calibration and accelerogram correction. *Earthq. Eng. Struct. Dynamics* **2**, 71–86.

Sunder, S. S., and J. J. Connor (1982). A new procedure for processing strong-motion earthquake signals. *Bull. Seismol. Soc. Amer.* **72**, 643–661.

Trifunac, M. D. (1971). Zero baseline correction of strong-motion accelerograms. *Bull. Seismol. Soc. Amer.* **61**, 1201–1211.

Trifunac, M. D. (1977). Uniformly processed strong earthquake ground accelerations in the western United States of America for the period from 1933 to 1971: pseudo relative velocity spectra and processing noise. *Report CE 77-04*, Univ. Southern Calif., Los Angeles.

Trifunac, M. D., and V. Lee (1973). Routine computer processing of strong-motion accelerograms. *Earthquake Engineering Research Laboratory, Report EERL 73-03*, Pasadena.

Trifunac, M. D., and V. W. Lee (1974). A note on the accuracy of computed ground displacements from strong-motion accelerographs. *Bull. Seismol. Soc. Am.* **64**, 1209–1219.

Trifunac, M. D., and V. W. Lee (1978). Uniformly processed strong earthquake ground accelerations in the Western United States of America for the period from 1933 to 1971: corrected acceleration, velocity and displacement curves. *Report CE 78-01*, Dept. Civil Engineering, Univ. Southern California, Los Angeles.

Trifunac, M. D., and V. W. Lee (1979). Automatic digitization and processing of strong motion accelerograms, Part I: automatic Digitization. *Report CE 79-15 I, Dept. Civil Engineering*, Univ. Southern California, Los Angeles.

Trifunac, M. D., M. I. Todorovska, and V. W. Lee (1998). The Rinaldi strong motion accelerogram of the Northridge, California earthquake of 17 January 1994. *Earthquake Spectra* **14**, 225–240.

Viksne, A., C. Wood, and D. Copeland (1995). Seismic monitoring/strong motion program and notification system. In: "Water Operation and Maintenance Bulletin," No. 171, pp. 161–168, Bureau of Reclamation, Denver.

Wenner, F. (1932). Development of seismological instruments at the Bureau of Standards. *Bull. Seismol. Soc. Am.* **22**, 60–67.

Wong, H. L., and M. D. Trifunac (1977). Effects of cross-axis sensitivity and misalignment on the response of mechanical-optical accelerographs. *Bull. Seismol. Soc. Am.* **67**, 929–956.

59

Estimation of Strong Seismic Ground Motions

Bruce A. Bolt
University of California, Berkeley, CA, USA
Norman A. Abrahamson
Pacific Gas and Electric Co., San Francisco, CA, USA

1. Special Problems of Strong-Motion Seismology

At the beginning of instrumental seismology, seismologists such as John Milne and his colleagues in Japan endeavored to design seismographs that could record both weak and strong wave motion from earthquakes (see Chapter 1 by Agnew; Bolt, 1996). As instrumental technology evolved, the paths of seismologists recording teleseisms and of those who were concerned with measuring large amplitude seismic waves near to an earthquake source diverged. By the 1960s, strong-motion seismometers for the latter purpose were largely the domain of earthquake engineers, while seismographic stations operated sensitive seismographs that could record waves of order micrometer amplitude from small local or distant earthquakes. (A 1 μm displacement of a 1 Hz frequency seismic wave from a teleseism corresponds to a ground acceleration of only about $4 \times 10^{-6}g$.) In more recent decades, the problem of wide dynamic seismograph range in both amplitude (about 180 dB) and frequency (about 6 decades) was solved with the advent of digital recording with 18 bits or greater (Bullen and Bolt, 1985; see also Chapter 18 by Wielandt). Consequently, modern seismographic stations in earthquake country are now able to record both weak and strong motions over a very large frequency band. As a rough guide we define strong ground motion as seismic waves with accelerations in excess of about $0.05g$ to over $1g$. Similar ranges are of order 10 to 200 centimeters per second in velocity, and 1 to 100 centimeters in displacement. Examples of large strong ground motions from crustal earthquakes are given in Table 1.

Seismologists working on strong ground motion have a number of goals that are not shared by those interpreting records from highly sensitive seismographs (e.g., Heaton *et al.*, 1995; Vidale and Helmberger, 1987). They are concerned with accurate measurement of high-amplitude (even nonlinear) waves but are not concerned with microseisms. Strong-motion seismographs are often deliberately sited not only on hard rock (normal for sensitive instruments) but on soil of various types. Of particular interest is the progression of elastic wave motion from linear to nonlinear behavior (e.g., Darragh and Shakal, 1991); for example, in extreme cases as soil layers "liquefy." Finally, in order to understand the effect of soil variability, strong-motion instrumentation is needed throughout urban areas, both close to, and some distance from, engineered structures and across geological boundaries and topographic relief.

The preceding description makes clear that strong-motion seismology presents a number of specific scientific problems, many of which remain unsolved. For example, seismograms from teleseisms usually show a progression of different seismic phases, such as P, PP, SSS, ScS, PKP, surface-wave modes, etc. (e.g., see Bullen and Bolt, 1985). In contrast, strong-motion records near to an extended rupturing fault source often have much less easily identified wave patterns; in many cases, it is not possible to separate clearly the elastic wave types. Typically, the strong motions arrive at a recording site as a mix of source and path and site responses (e.g., Mikumo and Miyatake, 1987). The result is wave incoherency of various degrees. In estimation attempts, there is also the problem of differences of strong ground motion near to faults of different types (e.g., thrust-type sources versus strike-slip sources; see Archuleta, 1984; Spudich *et al.*, 1996) and the geological complexities that produce marked contrasts between horizontal and vertical components of ground motion (Amirbekian and Bolt, 1998).

INTERNATIONAL HANDBOOK OF EARTHQUAKE AND ENGINEERING SEISMOLOGY, VOLUME 81B ISBN: 0-12-440658-0

TABLE 1 Examples of Near-Fault Strong-Motion Recordings from Crustal Earthquakes with Large Peak Horizontal Ground Motions

Earthquake	Magnitude M_W	Source Mechanism	Distance km*	Acceleration (g)	Velocity (cm/sec)	Displacement (cm)
1940 Imperial Valley (El Centro, 270)	7.0	Strike-Slip	8	0.22	30	24
1971 San Fernando (Pacoima, 164)	6.7	Thrust	3	1.23	113	36
1979 Imperial Valley (EC #8, 140)	6.5	Strike-Slip	8	0.60	54	32
Erizican (Erizican, 000)	6.9	Strike-Slip	2	0.52	84	27
1989 Loma Prieta (Los Gatos, 000)	6.9	Oblique	5	0.56	95	41
1992 Landers (Lucerne, 260)	7.2	Strike-Slip	1	0.73	147	63
1992 Cape Mendocino (Cape Mendocino, 000)	7.1	Thrust	9	1.50	127	41
1994 Northridge (Rinaldi, 228)	6.7	Thrust	3	0.84	166	29
1995 Kobe (Takatori, 000)	6.9	Strike-Slip	1	0.61	127	36
1999 Kocaeli (SKR, 090)	7.4	Strike-Slip	3	0.41	80	205
1999 Chi-Chi (TCU068, 000)	7.6	Thrust	1	0.38	306	940

* rupture distance (see Figure 1).

In this chapter, we discuss recording, interpretations, and prediction of strong motions in large earthquakes. The approach is largely historical and concentrates on the direct use of recordings of strong ground motion for estimation of seismic ground shaking for engineering purposes. We address the modern success in acquiring and analyzing strong ground motion by seismologists, particularly those working on seismic hazard maps and geological site response. Our emphasis is on the presently available strong-motion recordings and a description of ground-motion estimation methods. We do not attempt to describe in detail the major allied seismological field of theoretical modeling of seismic wave motion.

2. Growth of Significant Recordings of Intense Ground Motion

Until the 1970s, only a small sample of ground-motion recordings near to sources generating large magnitude earthquakes was available for either seismological research or engineering use. The situation, while not yet completely satisfactory, has recently improved. Recording programs in several countries have produced many important records; a helpful summary has been given by Ambraseys (1988). A further major augmentation occurred in the 1990s: We note here the ground-motion recordings from the 1989 Loma Prieta, California, earthquake (M6.9); the 1994 Northridge, California, earthquake (M6.7); the 1992 Landers, California, earthquake (M7.2), and the 1995 Kobe, Japan, earthquake (M6.8). Particularly valuable recordings in the near field were obtained in the August 1999 Kocaeli, Turkey, earthquake (M7.4), the November 1999 Duzce, Turkey, earthquake (M7.2), and (an unprecedented number, over 400 digital recordings) in the September 1999 Chi-Chi, Taiwan, earthquake (M7.6).

As this chapter was being written it became clear that studies of the major 1999 Taiwan earthquake and its aftershocks are likely to throw light on many seminal seismological and engineering questions (see the Dedicated Issue on the Chi-Chi, Taiwan Earthquake of 20 September, 1999, edited by Teng *et al.*, 2001). The Central Weather Bureau (CWB) was operating at the time the densest network of digital strong-motion instruments in the world as described by Shin *et al.* in Chapter 64. For comparison, station spacing of the free-field accelerographs in Taiwan was about 3 km in the metropolitan areas versus a 25-km uniform spacing of a comparable digital system (the Kyoshin Net) in Japan. For a description of the Kyoshin Net, please see Chapter 63 by Kinoshita.

3. Strong Ground-Motion Characteristics

The ground-motion characteristics depend on the seismic source, wave propagation (attenuation), and site response. The most commonly used method for characterizing the ground motion is through attenuation relations, but more complex numerical simulations are also used. As mentioned earlier, in this chapter we do not address the large topic of numerical simulation procedures. In attenuation relations, the earthquake source, wave propagation, and site response are typically parameterized by the magnitude, fault-type, source-to-site distance, and site condition. Recently, additional parameters describing the rupture direction have also been used to parameterize the source (discussed next).

3.1 Ground-Motion Parameters

Beginning about the 1960s, the most used ground-motion parameter for quantification of ground motion was the peak ground acceleration. The peak acceleration was picked from accelerograms irrespective of seismic phase, wave type, or frequency band. In practice, the peak horizontal acceleration occurs in most recordings in the *S*-wave portion of the seismogram with predominant frequencies typically between 3 and 8 Hz.

The peak vertical acceleration occurs sometimes in the *P* wave and sometimes in the *S* waves, with predominant frequencies typically between 5 and 20 Hz. More recently in engineering risk assessment, more emphasis is given to values of maximum velocity and displacement (see Bolt, 1996; Gregor and Bolt, 1997). A sample of some of the largest ground motions (in terms of peak acceleration, velocity, or displacement) is given in Table 1.

In addition to peak acceleration, velocity, and displacement, a descriptive construct special to the field of strong-motion seismology is that of seismic response spectrum, first proposed by Benioff (1934) and developed by Housner (1959). This spectrum is defined as the maximum response of a damped harmonic system to input motions. While the ground motion may be represented fully by Fourier spectra, the response of the structure is better represented by a response spectrum. It is common for modern attenuation relations to include relations for response spectral accelerations as well as peak acceleration.

Another commonly used ground-motion parameter is the duration of shaking. While it is generally agreed that duration can impact the response of soils, foundations, and structures, particularly when strength or stiffness degradation is encountered, there is not general agreement as to how the duration of shaking should be parameterized. This is evidenced by the 30 different definitions of strong-motion duration reviewed by Bommer and Martinez-Pereira (1998). A thorough description of the ground motion for engineering applications should include an estimation of the duration of strong motion, but given the wide range of duration measures currently being used, it is important to be clear as to how the specified duration is defined.

3.2 Independent Parameters Used in Attenuation Relations

The most common source, ray path, and site parameters are magnitude, distance, style-of-fault, and site classification. In some studies, additional parameters are used: hanging wall flag, rupture directivity parameters, focal depth, and soil depth. These independent parameters are discussed subsequently.

Different measures of earthquake magnitude are used in different regions of the world. At the International Workshop on Strong Motion Data held in 1993 in Menlo Park, California, various researchers working on ground-motion attenuation problems agreed to the recommendation that moment magnitude (M_w) should be universally adopted as the measure of earthquake size. Moment magnitude has been adopted in most recent attenuation relations developed in the United States and New Zealand, but this is not the case worldwide. In many regions, region-specific attenuation relations are still developed for magnitudes other than M_w, making it difficult to compare relations from different regions.

Whereas agreement was reached on the preferred magnitude, no such consensus was developed for the definition of distance used in attenuation relations, but there was a recommendation from the 1993 workshop that some form of "closest" distance to the rupture be used rather than epicentral or hypocentral distance. A variety of measures of the closest distance from the rupture to the site are used in various attenuation relations. Some examples of closest distance measures used in attenuation relations are listed in Table 2. These different distance measures are shown graphically in Figure 1 for a vertical fault and for

TABLE 2 Alternative Definitions of Distance That Are Used in Attenuation Relations

Distance Definition	Distance Measure	Examples of Attenuation Relations Using the Distance Measure
Shortest horizontal distance to the vertical projection of the rupture	"Joyner-Boore" distance	Joyner and Boore (1981), Boore *et al.* (1997), Spudich *et al.* (1996)
Closest distance to the rupture surface	Rupture distance	Sadigh *et al.* (1997), Abrahamson and Silva (1997), Idriss (1995)
Closest distance to the seismogenic part of the rupture	Seismogenic distance	Campbell (1997)
Closest distance to the hypocenter	Hypocentral distance	Atkinson and Boore (1995)
Closest distance to the centroid	Centroid distance	Crouse (1991)

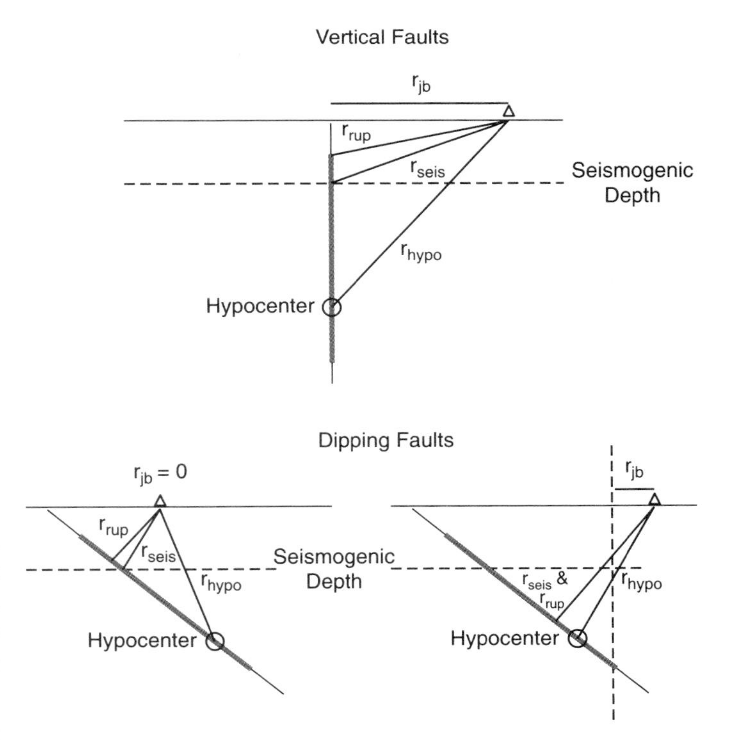

FIGURE 1 Source-to-site distance measures for ground-motion attenuation models (after Abrahamson and Shedlock, 1997).

a dipping fault. It is important to use the appropriate distance measure for a given attenuation relation.

Differences in site classification schemes in different regions also make comparison of ground motions and attenuation relations difficult. Most attenuation relations use broad site categories such as "rock," "stiff-soil," and "soft-soil." Recently, there has been a move toward using quantitative site classifications based on the shear-wave velocity measured at the strong-motion site, which would provide a standard site classification. The most commonly used parameter is the average shear-wave velocity over the top 30 m (e.g., Boore *et al.*, 1997). The difficulty with using quantitative site descriptions is that the information is not available for the majority of strong-motion sites that have recorded strong motions. This situation is improving as studies are now being conducted to systematically collect and compile shear-wave velocities at strong-motion sites.

The style-of-faulting parameter is used to distinguish between different source types. For crustal earthquakes, ground motions systematically differ when generated by strike-slip, thrust, or normal mechanisms (e.g., Somerville and Abrahamson, 1995; Spudich *et al.*, 1996). Given the same earthquake magnitude, distance to the site, and site condition, the ground motions from thrust earthquakes tend to be larger than the ground motions from strike-slip earthquakes (about 20–30% larger), and the ground motions from normal faulting earthquakes tend to be smaller than the ground motions from strike-slip earthquakes (about 20% smaller). For subduction earthquakes, the ground motions systematically differ when generated by interface or intraslab earthquakes (e.g., Youngs *et al.*, 1997). Again, for the same magnitude, distance, and site condition, the ground motions from intraslab earthquakes tend to be larger than the ground motions from interface earthquakes (about 40% larger).

Other independent parameters used in some attenuation relations include hanging wall, rupture directivity, and soil depth. Rupture directivity is discussed in detail in Section 4 and is not discussed here.

For thrust faults, high-frequency ground motions on the upthrown block (hanging wall side of a thrust fault) tend to be larger than on the downdropped block (footwall) (e.g., Somerville and Abrahamson, 1995). This increase in ground motions on the hanging wall side is in part an artifact of using a rupture distance measure. If a site on the hanging wall and footwall are at the same rupture distance (see Figure 1), the site on the hanging wall side is closer to more of the fault than the site on the footwall side. Using the Joyner-Boore distance measure implicitly accounts for this geometric difference because all sites over the hanging wall of the rupture are at a Joyner-Boore distance of zero. This distance definition effect does not account for all of the differences between the ground motions on the hanging wall and footwall.

Most attenuation relations simply use a site category such as "deep soil"; however, this broad category covers a wide range of soil depths from less than 100 m to several km of sediments. Some attenuation relations (e.g., Campbell, 1997) use an additional site parameter to describe the depth of the soil. This additional parameter improves the model predictions of long-period ($T > 1$ sec) ground motions.

3.3 Development of Attenuation Relations

For a selected ground-motion parameter and set of independent parameters (magnitude, distance, etc.), attenuation relations are typically developed using regression analyses. The first issue is the selection of the assumed functional form of the attenuation. There are in general two types of magnitude scaling that are used in attenuation relations (Figure 2). In the first case, the shape of the attenuation with distance is independent of magnitude. A typical form of this type of model for rock site conditions is given in Eq. (1).

$$\text{Ln } Y(M, R, F) = c_1 + c_2 M + c_3 M^2 + c_4 \ln(R + c_5) + c_6 F \quad (1)$$

where Y is the ground-motion parameter (peak acceleration or response spectral value), M is the magnitude, R is the distance measure, and F is a flag for the style-of-faulting (reverse, strike-slip, normal). This is one of the simplest forms used for attenuation relations. An example of an attenuation model that uses this form is Boore *et al.* (1997).

An alternative form allows the shape of the attenuation relation to depend on magnitude with the curves pinching together at short distances (Figure 2). This saturation of the ground motion implies that at short distances, moderate magnitude earthquakes produce similar levels of shaking as large magnitude earthquakes. Some attenuation relations use a combination of

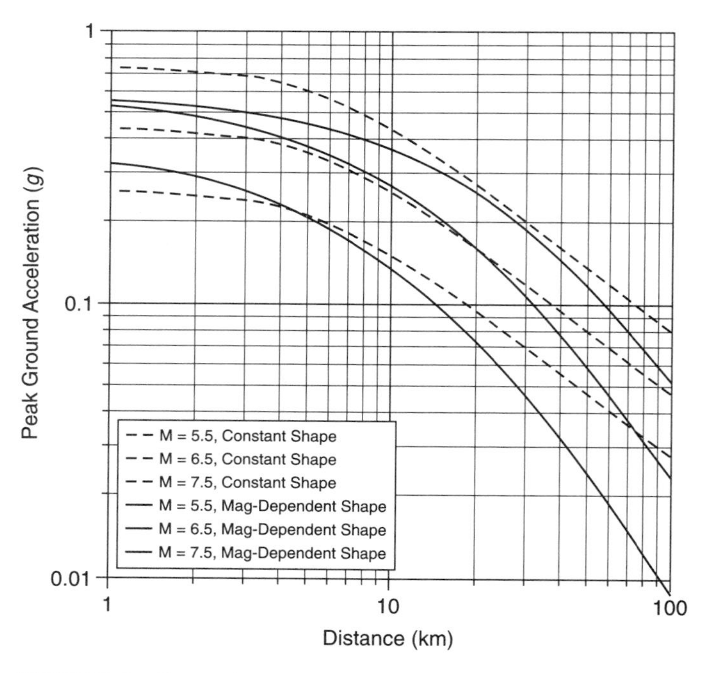

FIGURE 2 Comparison of constant and magnitude-dependent shapes of attenuation relations.

these two models with constant attenuation shapes for moderate magnitudes (e.g., $M < 6.5$) and saturation for large magnitudes (e.g., $M \geq 6.5$).

This type of saturation of the ground motion at short distances has been parameterized in two ways as shown in Eq. (2) and Eq. (3): replacing c_5 in Eq. (1) with a magnitude-dependent term, $f_1(M)$; or replacing the distance slope, c_4, in Eq. (1) with a magnitude-dependent slope, $f_2(M)$. For rock ground site conditions, two commonly used forms of the attenuation relation are

$$\mathrm{Ln}\,Y(M, R, F) = c_1 + c_2M + c_3M^2 + c_4 \ln[R + f_1(M)] + c_6F \quad (2)$$

and

$$\mathrm{Ln}\,Y(M, R, F) = c_1 + c_2M + c_3M^2 + (c_4 + f_2(M)) \ln(R + c_5) + c_6F \quad (3)$$

An example of an attenuation relation based on the model in Eq. (2) is Sadigh *et al.* (1997). A common form of the function $f_1(M)$ is

$$f_1(M) = c_7 \exp(c_8M) \quad (4)$$

If $c_3 = 0$ (which is common for high frequencies) and $c_8 = -c_2/c_4$, then the ground motion at zero distance is independent of magnitude and the model is said to have 100% magnitude saturation. If $c_8 > -c_2/c_4$, then the model is said to be oversaturated. In that case, at short distances, the median ground motion is reduced as the magnitude increases. An example of an attenuation relation based on the model in Eq. (3) is Abrahamson and Silva (1997). A common form of the function $f_2(M)$ is $f_2(M) = c_9M$. In this case, the model has 100% saturation if $c_8 = -c_2/\ln(c_5)$. If $c_8 < -c_2/\ln(c_5)$, then the model is oversaturated.

At distances less than about 50 km, these two different functional forms of the attenuation relation lead to similar curves, but at large distances, they become different (Figure 3). As most of the focus of attenuation relations for shallow crustal earthquakes in active tectonic regions has been on distances less than 50 km, these differences in slopes at large distances have not been subject to detailed studies. Part of the reason for this is that distant stations from moderate magnitude earthquakes were not available because the instruments did not trigger or the ground motions were too small to digitize. With the new digital recorders with wide dynamic range, we should be able to distinguish between these two alternatives.

In the eastern United States, it is common to incorporate a variation in the distance slope of the attenuation relation, which accommodates the increase in ground motions due to supercritical reflections from the base of the crust (e.g., Atkinson and Boore, 1995; Saikia and Somerville, 1997). This is usually done by incorporating a multi-linear form of the attenuation relation with different c_4 terms for different distance ranges. This typically leads to a flattening of the attenuation curve at distances of about 100 km (Figure 4). This is most significant for regions in which the high-activity sources are at a large distance from the site.

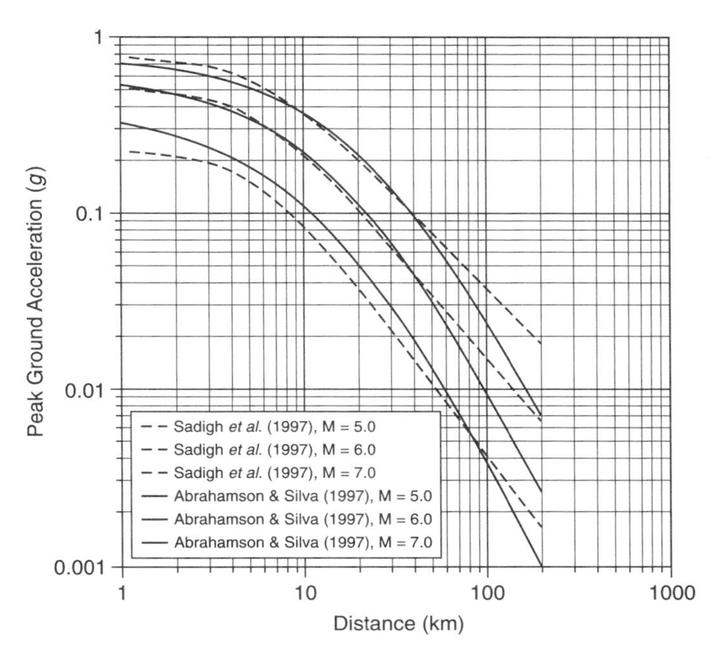

FIGURE 3 Comparison of predicted ground motions using alternative functional forms for saturation at short distances.

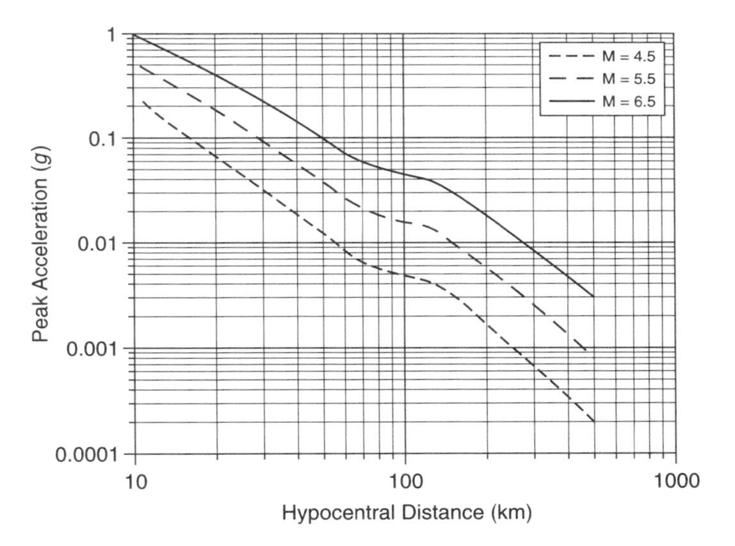

FIGURE 4 Form of attenuation relations in the eastern United States showing flattening of the attenuation relation due to crustal structure.

An important statistical issue in developing attenuation relations is the uneven sampling of the data from different earthquakes. For example, in some cases, an earthquake may have only one or two recordings (e.g., 1940 El Centro), whereas some recent earthquakes have hundreds of recordings (e.g., 1999 Chi-Chi earthquake). Should the well-recorded earthquakes overwhelm the poorly recorded earthquakes? Should the poorly recorded earthquakes be given equal weight to the well-recorded earthquakes? Weighting schemes are typically used as a way to account for this uneven sampling problem. There are two extremes: Give equal weight to each data point or give equal weight to each earthquake. We do not attempt to cover all of the various weighting procedures here, but two commonly used methods are the two-step procedure and the random effects method. The two-step regression procedure first used by Joyner and Boore (1981) gives equal weight to each earthquake. The random effect model (Brillinger and Preisler, 1984) uses a weighting scheme that is between equal weight to each earthquake and equal weight to each data point depending on the distribution of the data. As the number of data points from each earthquake becomes large, these two methods give the same result.

In addition to the median ground motion, the standard deviation of the ground motion is also important for either deterministic or probabilistic hazard analyses. Worldwide, it is most common to use a constant standard deviation, but recently, several attenuation relations have incorporated magnitude or amplitude dependence to the standard deviation. For example, Abrahamson and Silva (1997) allow the standard deviation to vary as a function of the magnitude of the earthquake. The result is that the standard deviation is smaller for large magnitude earthquakes (Youngs *et al.*, 1995). Campell (1997) allows the standard deviation to vary as a function of the amplitude of the ground motion. The result is that the standard deviation is smaller for larger amplitudes of ground motion. Both of these models can significantly affect the development of design ground motions.

3.4 Regional Differences in Attenuation

Many published studies have found significant differences in attenuation between various tectonic regions and also for various geologic conditions and seismic sources (e.g., Ambraseys and Bommer, 1991, 1995; Raoof *et al.*, 1999). Results for North America were recently summarized (Abrahamson and Shedlock, 1997). In the latter summary paper it was found useful to group the attenuation relations into three main tectonic categories: shallow crustal earthquakes in active tectonic regions (e.g., California), shallow crustal earthquakes in stable continental regions (e.g., eastern U.S.), and subduction zone earthquakes.

In some regions of the world, there are recordings from both crustal earthquakes and subduction earthquakes (e.g., Japan, New Zealand, Taiwan). It has been common to lump these data together in developing attenuation relations. Because peak ground motions from subduction zone earthquakes generally attenuate more slowly than those from shallow crustal earthquakes in tectonically active regions, ground motions from these different types of sources should be modeled separately.

As the number of recordings of strong ground motion increase, there has been a trend toward developing region-specific attenuation relations rather than just using the global average models developed for the broad tectonic categories. Often, there is a tendency to overemphasize region-specific data in developing region-specific attenuation. Typically, there is not enough data in a specific region to completely determine the attenuation relation. In particular, there is often not enough data close to the fault to constrain the behavior of the attenuation relation at short distances, or not enough earthquakes to constrain the magnitude scaling.

One way to address regionalization of attenuation relations is to only update parts of the global attenuation relations. For example, the simplest update is to estimate a constant scale factor to use to adjust a global attenuation model to a specific region. (This can reflect differences in the earthquake source or differences in the site categories.) If there is enough data over a range of distances, the slope of the attenuation could be updated while maintaining the magnitude scaling of the global model. As more strong-motion data become available, additional parameters can be made region specific. An example of this approach to regionalizing attenuation relations is McVerry and Zhao (1999).

3.5 Types of Uncertainties

In developing ground motions for design, random variability and scientific uncertainty in ground motions are treated differently (Toro *et al.*, 1997). Variability is the randomness in the ground motions that a model predicts will occur in future earthquakes. For example, variability in ground motion is often quantified by the standard deviation of an attenuation relation. In contrast, uncertainty represents the scientific uncertainty in the ground-motion model due to limited data. For example, the uncertainty in attenuation relations is often characterized by alternative attenuation relations. That is, uncertainty is captured by considering alternative models. In seismic hazard analyses, the terms "aleatory" and "epistemic" are used for variability and uncertainty, respectively. To keep the notation clear, in this chapter we will use the terms aleatory variability and epistemic uncertainty.

The distinction between aleatory variability and epistemic uncertainty is useful for seismic hazard analysis; however, a further subdivision is needed for the practical estimation of these two factors, which is shown in Table 3. There are two ways that the variability of ground-motion models is evaluated. First the models can be evaluated against recordings from past earthquakes. This is called the modeling term. The second way that a model is evaluated is by varying the free parameters of the model. This is called the parametric term.

The modeling component is a measure of the inability of the model to predict ground motions from past earthquakes. In general, the cause of the modeling variability is not understood. It is assumed to be random. If the cause of the modeling variability were understood, then the model could be improved to fit the observations. The parametric component is a measure of the variability of the ground motion from causes that are understood. That is, there is randomness in the earthquake source process that is understood and its effect on ground motion is part of the model.

In general, both the modeling and parametric terms need to be considered in a ground-motion model. As our understanding and

TABLE 3 Decomposition of Variability and Uncertainty

	Aleatory Variability	Epistemic Uncertainty
Modeling	Variability between model predictions of ground motions and observations of ground-motion data.	*Select (and weight) alternative ground-motion models.* Results in differences in both the median ground motion and the variability of the ground motion.
Parameteric	Variability of ground motions due to variability of *additional* earthquake source parameters. Additional source parameters are those that are not included in the specification of the design earthquake.	*Select (and weight) alternative models for the distributions of the additional parameters.* Results in differences in both the median ground motion and the variability of the ground motion.

modeling of earthquakes improves, there will be a trend of reducing the modeling variability (unexplainable variability), but this will likely be offset by an increase in the parametric variability (explainable variability). While the total variability (combination of the modeling variability and the parametric variability) may not be reduced significantly with improved models, there is an advantage to understanding the causes of the variability, particularly if the model is being extrapolated beyond the empirical data on which it was evaluated.

In empirical attenuation models, the standard deviation given for the model is modeling variability. In most attenuation models, there is no parametric variability component to the aleatory variability, but this is not necessarily the case. As new attenuation models are developed, they may begin to include a parametric variability component. For example, if an attenuation relation used the static stress-drop as an additional source parameter, then the variability of static stress-drop would be treated as parametric variability. Typically, for empirical attenuation relations, epistemic uncertainty is addressed by considering alternative attenuation relations.

4. The Special Case of Near-Fault Ground Motions

Near-fault ground motions often contain large long-period pulses in the ground motion. There are two causes of long-period pulses in near-fault ground motions. One is constructive interference of the dynamic shaking due to rupture directivity effects. The other is due to the movement of the ground associated with the permanent offset of the ground. This is the velocity of the ground due to the elastic rebound proposed by H. F. Reid after the 1906 earthquake. These two causes of long-period pulses attenuate very differently from one another. To keep these two effects separate, the terms "directivity pulse" and "fling-step" are used for the rupture directivity and elastic rebound effects, respectively.

Rupture directivity effects occur when the rupture is toward the site and the slip direction (on the fault plane) is aligned with the rupture direction (Somerville *et al.*, 1997). As described in the next section, it is strongest on the component of motion perpendicular to the strike of the fault (fault-normal component). Fling-step effects occur when the site is located close to a fault with significant surface rupture. It is polarized onto the component parallel to the slip direction. For strike-slip earthquakes, rupture directivity is observed on the fault-normal component and static displacement effects are observed on the fault-parallel component. Thus, for strike-slip earthquakes, the rupture directivity pulse and the fling-step pulse will naturally separate themselves on the two horizontal components. For dip-slip earthquakes, it is more complicated. The rupture directivity effect will be strongest on the fault-normal component at a location direct updip from the hypocenter. The fling-step will also be observed on the horizontal component perpendicular to the strike of the fault. Thus for dip-slip faults, directivity-pulse effects and fling-step effects occur on the same component.

The horizontal recordings of stations in the 1966 Parkfield, California, earthquake and the Pacoima station in the 1971 San Fernando, California (Bolt, 1971) earthquake were the first to be discussed in the literature as showing near-fault velocity pulses. These cases, with maximum amplitudes of 78 and 113 cm/sec, respectively, consisted predominantly of horizontally polarized SH wave motion and were relatively long period (about 2–3 sec). These velocity pulses were due to rupture directivity.

Additional recordings in the near field of large sources have confirmed the presence of energetic pulses of this type, and they are now included routinely in synthetic ground motions for seismic design purposes. Most recently, the availability of instrumented measured ground motion close to the sources of the 1994 Northridge earthquake (Heaton *et al.*, 1995), the 1995 Kobe earthquake (Nakamura, 1995), and the 1999 Chi-Chi earthquake (Lee *et al.*, 2001) provided important recordings of the velocity pulse.

4.1 Directivity

In the case of a fault rupture toward a site at a more or less constant velocity (almost as large as the shear-wave velocity), most of the seismic energy from the extended fault rupture arrives in a short time interval, resulting in a single large long-period pulse of motion, which occurs near the beginning of the record. This wave pulse represents the cumulative effect of almost all of the seismic radiation from the moving dislocation. In addition, the radiation pattern of the shear dislocation causes this large pulse of motion to be oriented mostly in the direction perpendicular to the fault. Coincidence of the radiation pattern maximum for tangential motion and the wave focusing due to the rupture propagation direction toward the site produces a large displacement pulse normal to the fault strike (see Bullen and Bolt, 1985, p. 443).

The directivity of the fault source rupture causes spatial variations in ground-motion amplitude and duration around faults and produces systematic differences between the strike-normal and strike-parallel components of horizontal ground-motion amplitudes (Somerville *et al.*, 1997). These variations become significant at a period of 0.6 seconds and generally grow in size with increasing period. Modifications to empirical strong ground-motion attenuation relations have been developed (Somerville *et al.*, 1997) to account for the effects of rupture directivity on strong-motion amplitudes and durations based on an empirical analysis of near-fault recordings. The ground-motion parameters that are modified include the average horizontal response spectral acceleration, the duration of the acceleration time history, and the ratio of strike-normal to strike-parallel spectral acceleration.

The results are that when rupture propagates toward a site, the spectral acceleration is larger for periods longer than 0.6 seconds, and the duration is smaller. That is, the duration and long-period amplitude are inversely correlated. For sites located close to faults, the strike-normal spectral acceleration is larger than the strike-parallel spectral acceleration at periods longer than 0.6 seconds in a manner that depends on magnitude, distance, and azimuth.

As in acoustics, the amplitude and frequency of the directivity pulse have a geometrical focusing factor, which depends on the angle between the direction of wave propagation from the source and the direction of the source velocity. Instrumental measurements show that such directivity focusing can modify the amplitude velocity pulses by a factor of up to 10, while reducing the duration by a factor of 2. The pulse may be single or multiple, with variations in the *impetus* nature of its onset and in its half-width period. A widely accepted illustration is the recorded ground velocity of the October 15, 1979, Imperial Valley, California, earthquake generated by a strike-slip fault source (see Figure 5). The main rupture front moved toward El Centro and away from Bonds Corner.

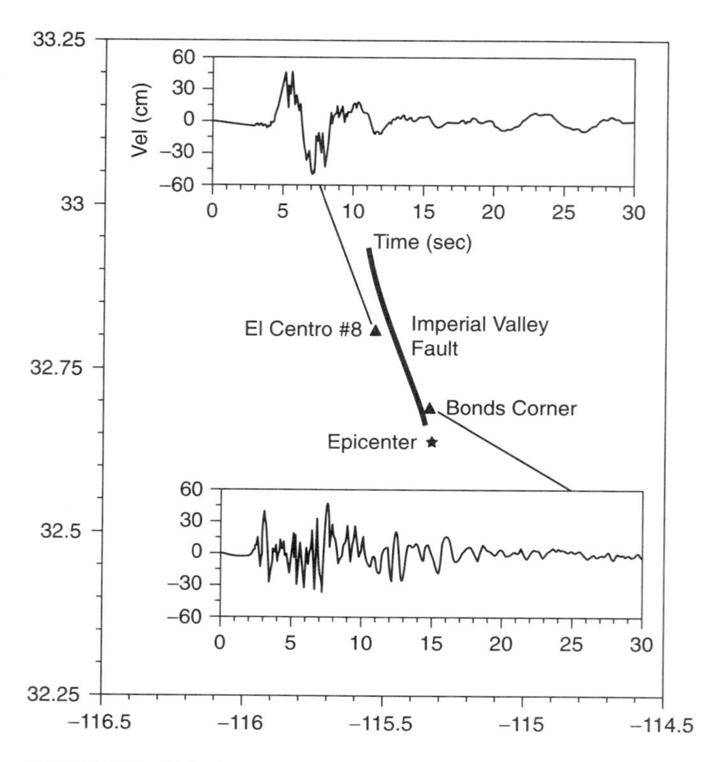

FIGURE 5 Velocity–time histories (230 Comp) from the 1979 Imperial Valley, California, earthquake at the Bonds Corner and El Centro Differential Array strong ground-motion recording sites.

4.2 Fling-Step

Prior to the 1999 Turkey and Taiwan earthquakes, nearly all of the observed large long-period pulses in near-fault ground motions were caused by rupture directivity effects. The Lucerne recording from the 1992 Landers earthquake contained a directivity pulse on the fault-normal component and a very long-period fling-step pulse on the fault-parallel component. The ground-motion data from the 1999 Turkey and Taiwan earthquakes contain large long-period velocity pulses due to the fling-step. As an example, the east–west components of velocity from selected recordings of the Chi-Chi earthquake are shown in Figure 6. The ground motions at station TCU068, located on the hanging wall near the northern end of the rupture, had the largest peak velocities ever recorded (300 cm/sec). The velocity pulse from the fling-step effect is one-sided. Most of the very large velocity at TCU068 is associated with the fling-step. If the fling-step is separated from the dynamic shaking, the peak

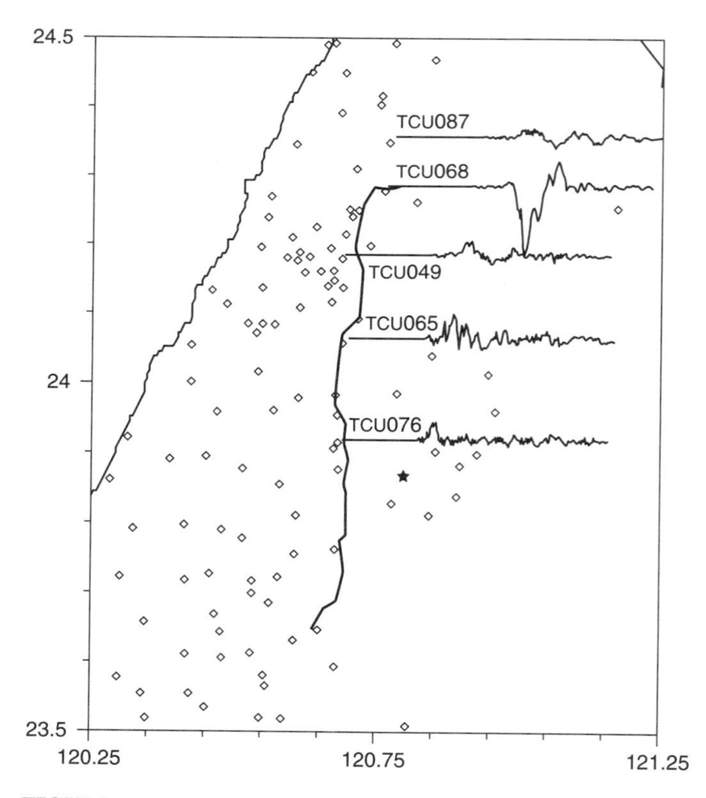

FIGURE 6 Selected velocity–time histories (east–west component) from the 1999 Chi-Chi earthquake.

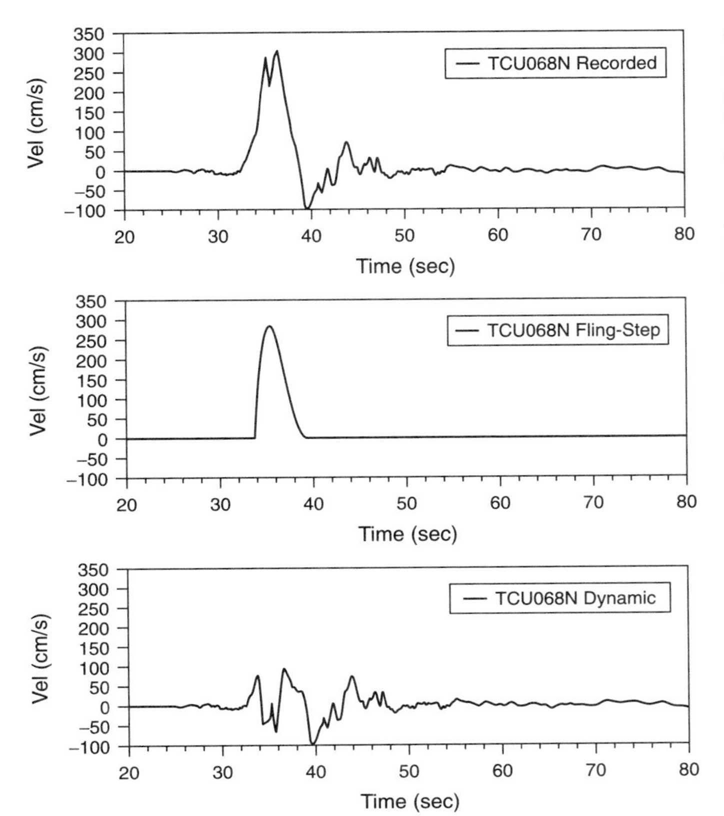

FIGURE 7 Velocity–time histories for the north–south component of station TCU068 (see Figure 6) separated into dynamic shaking and fling-step components.

velocity of the dynamic component of shaking is reduced to about 100 cm/sec (Figure 7). The very large fling-step velocity pulses do not attenuate at the same rate as the directivity pulses such as seen in Northridge and Kobe. Fling-step ground motion should not be simply thrown into the strong motion data set consisting of dynamic shaking. Separate attenuation relations will be needed for the fling-step and dynamic parts of the ground motion.

Models for the peak velocity from the fling-step have not been developed at this time. This represents an area of strong ground motion that can benefit from theoretical seismology. In displacement, the fling-step can be parameterized simply by the amplitude of the tectonic deformation and the rise time (time it takes for the fault to slip at a point).

5. Numerical Computation of Strong Ground Motion

Because of the lack of a complete library of strong ground motions and spectra from appropriate earthquakes, extrapolation from the available records is needed for estimation of large and unrepresented seismic motions. Confidence in such numerical predictions relies on the theoretical side of strong ground-motion seismology and from calibrating numerical techniques against empirical data. Important developments have occurred in this area in the last decades. In this section we briefly summarize some aspects of numerical computations of ground motions. This field is a major topic unto itself, which we only briefly cover here.

Numerous methods and examples of synthesizing seismograms numerically from source and wave theory in the near field of a large seismic source have been published (see, for example, Bolt, 1987; Irikura, 1983). Because the computation of site-dependent and phased-time histories is not unique, a number of alternative methods are available (see, e.g., Joyner and Boore, 1988; Vidale and Helmberger, 1987; Gusev, 1983; Mikumo and Miyatake, 1987; Panza and Suhadolc, 1987).

A more empirical approach is often adopted for engineering design practice (e.g., Bolt, 1994). The first step is to define, from geological and seismological information, the appropriate earthquake sources for the site of interest. The source selection may be deterministic or probabilistic and may be decided on the grounds of acceptable risk (Bolt, 1999a). Next, specification of the propagation path distance is needed, as well as the P-, S-, and surface-wave velocities in the zone. These speeds allow calculation of the appropriate wave propagation attenuation and angles of approach of the incident waves.

The construction of realistic motions consists of a series of iterations from the most appropriate observed strong-motion record already available to a set of more specific time histories that incorporate the seismologically defined wave patterns. Where feasible, strong-motion accelerograms are chosen that satisfy the seismic source (dip-slip, etc.) and path specifications for the seismic zone in question.

The iterative synthesis process is controlled by applying seismological (e.g., Kanamori, 1971; Anderson, 1997; Zeng *et al.*, 1994) and engineering constraints. For example, the response amplitude spectra might be required to fall within one standard error of a specified target spectrum obtained, say, from previous data analysis or an earthquake building code. Similarly, each seismogram is restricted to prespecified statistical bounds of peak-ground accelerations, velocities, and displacements. The durations of each wave section (mainly, P-, S-, and surface-wave portions) of each time history should also satisfy prescribed source, path, and site conditions. At the conclusion, experience with observed seismograms and knowledge of seismic wave theory are needed as a check on the results.

Arguably, currently, the most reliable estimation is based on a stochastic model of earthquake ground motions. This model uses simplified, yet physically based, representations of seismic energy release and wave propagation to obtain predictions of ground-motion amplitude for given values of earthquake size and depth, distance to the site, and model parameters. Tests are needed to confirm that any adopted stochastic model is consistent with recorded ground motions for the frequency, distance, and magnitude range of interest (i.e., the bias in the model predictions should be essentially zero for most frequencies, considering statistical uncertainty). The key parameters for representative stochastic models are stress-drop, crustal velocity structure, crustal anelastic attenuation (Q), near-site anelastic attenuation (κ or kappa), and focal depth. Adopted values for these parameters have considerable uncertainties in most cases.

Finally, we would like to stress that much of the modeling in the seismological literature depends on assumptions on a range of key model parameters (see Boore *et al.*, 1997; Brune and Anooshehpoor, 1999) and has until recently been restricted to two-dimensional structures (see cases in Bolt, 1987; Boore and Joyner, 1997; Takei and Kanamori, 1997). Significant surface topography has been shown to affect ground motions (e.g., Spudich *et al.*, 1995; Bard, 1982). Now, recent work has begun to produce synthetic ground motions based on realistic three-dimensional geological and source mechanics models (e.g., Olsen and Archuleta, 1996; Olsen *et al.*, 1995; Stidham *et al.*, 1999). There is clear observational evidence of the importance of three-dimensional structural variation in the surface motions, as observed, for example, in the isoseismal maps of the 1906 San Francisco earthquake (see Bolt, 1999b) and in the irregular patterns of seismic intensity in alluvial basins. O'Connell (1999) has addressed the key tradeoffs between nonlinear wave amplifications and linear scattering from crustal and basin velocity variations. A case has been published (Lomax and Bolt, 1992) that illustrates a significant lateral diffraction effect from the varied geology of the San Francisco Bay Area. We reproduce in Figure 8 the recordings of strong-motion displacements (transverse wave component) along three profiles that demonstrate the basin effect; the comparison is with synthetic motions calculated using the regional geological structures. In a follow-up study, Stidham *et al.* (1999) used full seismic-wave theory and three-dimensional modeling also to simulate the 1989 Loma Prieta earthquake. Their results demonstrated the significant variability in the radiated wave pattern due to different rock types across the San Andreas fault, the directivity focusing effect, and the azimuthally dependent amplification due to the Tertiary basin. Such physical predictions are likely to be a vigorous part of strong-motion research in seismology's future.

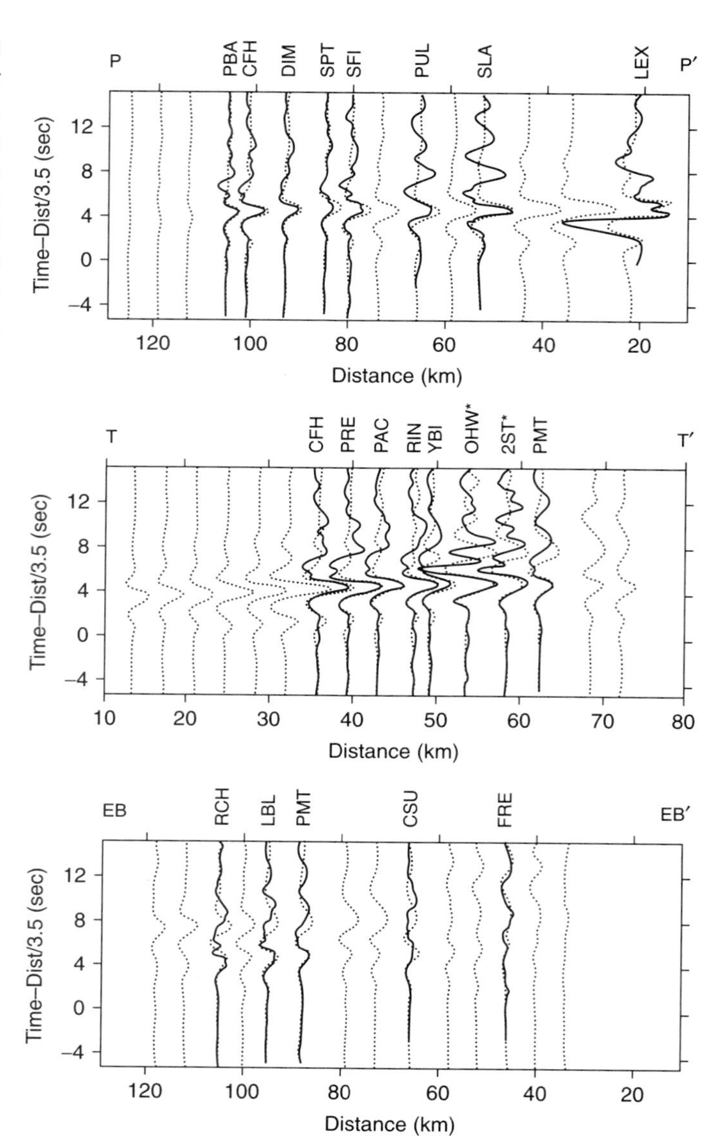

FIGURE 8 Observed strong-motion transverse displacement recordings of the 1989 Loma Prieta, California, earthquake (solid) and synthetics (dotted) shown along three lines: San Francisco peninsula (P–P′), Bay transverse (T–T′), and East Bay (EB–EB′). Traces are plotted as a function of distance along section; travel time is reduced with respect to epicentral distance. Trace distance on section P–P′ only is identical to epicentral distance. All amplitudes on each section are plotted to the same scale; the observed traces are bandpass filtered from 1 to 8 sec and shifted in time to align the interpreted direction SH arrival on the synthetics. (* indicates soft-soil sites.)

6. Seismic Strong-Motion Hazard and Design Ground Motions

It is now usual to use the word *hazard* to describe the actual danger itself (that is, in this discussion, the earthquake), and *risk* for the result of this hazard applied to vulnerable structures. The most commonly mapped ground-motion parameters are horizontal and vertical peak ground accelerations (PGA), peak ground velocity (PGV), and 5% damped spectral acceleration (SA) for a given site classification. The 1996 U.S. National Hazard maps include PGA and 0.2, 0.3, and 1.0 sec period SA with a 10%, 5%, and 2% chance of exceedance in 50 years (Frankel *et al.*, 2000), assuming a "firm rock" site (see *http://geohazards.cr.usgs.gov/eq/*).

There are two basic approaches to developing design spectra: deterministic and probabilistic. The deterministic approach uses selected individual earthquake scenarios (magnitude, distance, directivity, etc). The ground motion is then computed using appropriate attenuation relations with a specified probability of not being exceeded given that the specified scenario earthquake occurs. Typically, a probability of nonexceedance of either 0.5 (median) or 0.84 (median plus one standard deviation) is used.

The engineering seismic risk analysis (Cornell, 1968) differs from the probabilistic approach in that it considers the rate of occurrence of each scenario and variability of the ground motion (number of standard deviations above or below the median) and its associated probability distribution. The hazard curve gives the probability that any of the scenarios will produce a ground motion exceeding the ground-motion test value (ground motion on the x-axis). For probabilistic analyses, the design ground motion is typically given by an equal hazard spectrum. A description of probabilistic hazard analysis is given by Reiter (1990).

An equal hazard spectrum gives at each spectral period the response spectral value that has the specified return period. Equal hazard spectra are constructed by first computing the hazard at each spectral period independently (using response spectral attenuation relations, see Section 3). Because the hazard analysis is done independently at each period, an equal hazard spectrum may not correspond to the same earthquakes at both short and long periods. The equal hazard spectrum may not be physically realizable in a single event, but is meant to represent reasonable design criteria.

7. Design Time Histories

The construction of strong-motion seismograms is now an essential part of the definition of hazard for the design and testing of critical structures. There are two main approaches used to develop design ground motions: (a) scaling ground motions and (b) adjusting ground motions to match a design spectrum.

Scaling refers to multiplying a recorded or simulated time history by a constant factor at all time points. This approach has the advantage that the time histories maintain the natural phasing of the recorded motion and realistic peaks and troughs in the spectral shape; the disadvantage is that a large number of time histories need to be used to get a reliable estimate of the average response of the structure (e.g., 10 sets of time histories may be needed). The suite of time histories should be scaled such that their ensemble average is close to the design spectra over the critical period range of the structure.

Spectrum-compatible time histories refers to time histories that are modified in terms of their frequency content to match the entire design spectrum. Various methods have been developed to modify a reference time history so that its response spectrum is compatible with a specified target spectrum. A review of spectral-matching methods is given by Preumont (1984). A commonly used method adjusts the Fourier amplitude spectrum based on the ratio of the target response spectrum to the time history response spectrum while keeping the Fourier phase of the reference time history fixed. While this approach is straightforward, it has two drawbacks. First, it generally does not have good convergence properties, particularly for multiple damping spectra. Second, it can alter the nonstationary character of the time history if the shape of the Fourier amplitude spectrum is changed significantly.

There are two alternative approaches for spectral matching. The first is a frequency domain approach in which the Fourier amplitude spectrum of the initial motion is replaced by a Fourier amplitude spectrum that is consistent with the target spectrum based on random vibration theory. The second alternative approach for spectral matching adjusts the time history in the time domain by adding wavelets to the reference time history. A formal optimization procedure for this type of time-domain spectral matching was first proposed by Kaul (1978) and was extended to simultaneously match spectra at multiple damping values by Lilhanand and Tseng (1988). While this procedure is more complicated than the frequency-domain approach, it has good convergence properties and in most cases preserves the nonstationary character of the reference time history.

If spectrum-compatible motions are used, then a small number of sets of time histories can be used and still provide a reliable estimate of the average response of the structure (e.g., 3 sets may be adequate).

7.1 Selection of Initial Time Histories

The selection of the starting time histories for use in either scaling or spectral matching is important due to nonlinear response of the soil and structure. Potential motions should be based on their duration, site characteristics, event magnitude and recording distance, and general character of the displacement history. In particular, for near-fault time histories, the character of the displacement pulse as either one-sided, two-sided, or multi-sided should be considered so that the selected motions will have distinctly different time-signatures to thoroughly test the structural design. If the scaling approach is used, then selecting records with a similar site condition to the project site is also an important factor. However, if the spectral matching approach is used, then the site condition of the initial time history is not critical because the spectral matching process will adjust the frequency content to match the project site conditions to the extent that they are captured by the design spectrum.

8. Special Case for Extended Structures

For most engineering applications, the ground motion is defined at a single point. In reality, for viaducts, large bridges, and dams, out-of-phase wave motions over inter-support distances cause differential ground accelerations and differential rotations along the base of the structure.

The spatial variation of seismic ground motion results from several effects: non-vertical wave propagation (wave passage effect), scattering and complex 3–D wave propagation (spatial coherency effect), variable distance to the fault rupture (attenuation effect), and variable site conditions (local site effect). The variable site condition can be considered in the site response calculations, which are not discussed in this chapter. Studies of the spatial variation of strong ground motions from array observations have been published (see Abrahamson *et al.*, 1987), and the results have been incorporated into structural response analyses for some large critical structures.

8.1 Wave-Passage Effect

The wave-passage effect is due to non-vertical wave propagation that produces systematic time shifts in the arrival of the seismic waves at the support locations. The wave-passage effect depends on the apparent velocity of the waves along the axis of the bridge. This requires information on both the apparent velocity of the waves in the direction of the wave propagation and the orientation of the structure with respect to the earthquake. Studies of strong- and weak-motion array data have shown that the apparent velocity of S waves is typically in the 2.0 to 3.5 km/sec range (Abrahamson, 1992b). In other studies, numerical simulations have been used to estimate the horizontal wave speed. These numerical simulations typically lead to higher horizontal wave speeds than measured from array data. The discrepancy may be due to the use of simplified crustal models in the simulations. The array measurements of the wave speeds are considered to be more reliable than the numerical simulations.

The wave-passage effect is included by applying systematic time shifts to the time histories. Although the P, S, and surface wave will have different phase velocities, the S-wave phase velocity is applied to the entire time history. The rationale for this approach is that the S waves generally contain the strongest shaking. Therefore, for an engineering analysis, it is adequate to simply model the S-wave wave-passage effect. More complex wave propagation methods can be used if the differential motion caused by the P wave and surface waves are considered to be important for the structural response.

8.2 Spatial Coherency Effect

The scattering and complex wave propagation result in variations in both Fourier amplitude and Fourier phase of the ground motions (Hao *et al.*, 1989). The phase variations can be included through a spatial coherency function. The coherency is a complex number. The absolute value of the complex coherency is called the "lagged coherency" because it is equivalent to lining up (lagging) the ground motion at the two locations so that the wave passage effect is removed. The lagged coherency does not restrict the alignment of the ground motions to be consistent from frequency to frequency; that is, the apparent wave speeds can be different at each frequency. At high frequencies, the wave speed implied by the complex coherency becomes more random. However, in the application of the wave-passage effect, a constant wave speed is typically used for all frequencies. To be consistent with such an application, the coherency function needs to correspond to the coherency that would be computed by aligning the ground motion to a single wave speed. This coherency is called the "plane-wave coherency." Generic empirical models for the plane-wave coherency were developed from worldwide dense array recordings. An example of the coherency model for the horizontal component is given in Figure 9. Additional details of the coherency model are given in Fugro *Earth Mechanics* (1998).

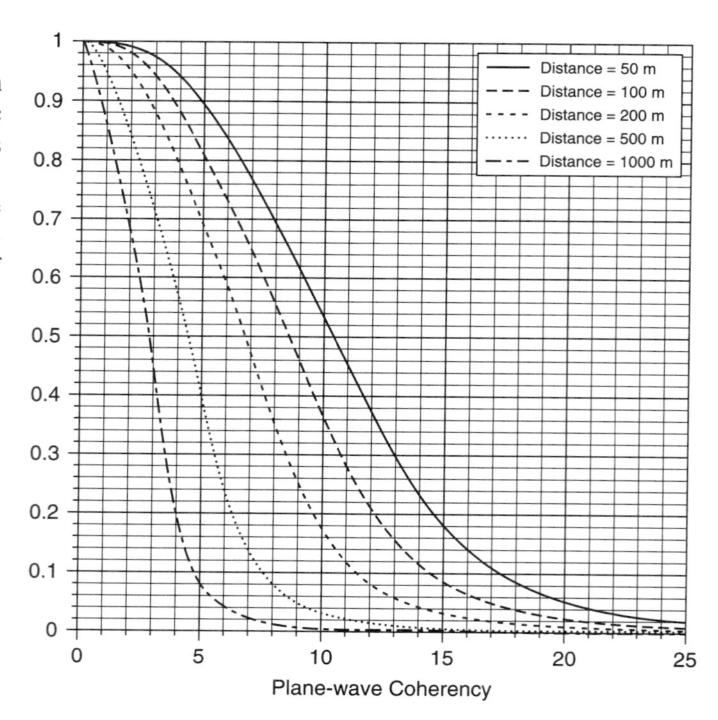

FIGURE 9 Plane-wave coherency model for the horizontal component.

8.3 Attenuation

For sources near to the structure, the different distances to the various support locations can result in different ground motions due to attenuation of the ground motion. The attenuation effect can be estimated simply using the attenuation relations with different distances to the controlling source for different parts of the structure.

8.4 Site Response

The effect of variable site conditions along extended structures is accommodated by conducting site-specific estimates of the site response using the appropriate rock motion for each location along the structure.

9. Example Application of Strong Ground-Motion Estimation to Critical Engineered Structural Design

In this concluding section, we briefly describe a recent application of the preceding methods to the estimation of the strong-motion parameters for a major bridge over San Francisco Bay, California. The details of the application are given in a report to the California Department of Transportation (Caltrans, Office of Structure Foundations) (Fugro Earth Mechanics, 1998). The San Francisco–Oakland Bay Bridge consists of two bridges: an East Crossing and a West Crossing. This example is for the East Crossing. The bridge lies between two major active faults: the Hayward fault 8 km to the east and the San Andreas fault 15 km to the west.

9.1 Design Spectrum

A design spectrum was derived based on a probabilistic seismic hazard. A response spectrum with a return period of 1500 years was selected as a "Safety Evaluation Earthquake." The dynamics of near-source effects, discussed in Section 4, are crucial in this application. Therefore, the hazard analysis must include the effects of rupture directivity. This was done by modifying the attenuation relations to include directivity focusing and modifying the hazard computation to include variability of rupture direction. The equal hazard spectrum for fault normal motion is given in Figure 10.

The hazard for the 1500-year return period was deaggregated to determine the controlling earthquakes (Bazzurro and Cornell, 1999). In this case, the deaggregation indicated that the seismic hazard at the bridge is dominated by a M7.8 earthquake at about 20 km distance on the San Andreas fault and a M7.0 earthquake at about 10 km distance on the Hayward fault. The deaggregation also showed that at a spectral period of 3 seconds (a critical period for the bridge foundations), the 1500-year return period ground motion is dominated by forward rupture directivity.

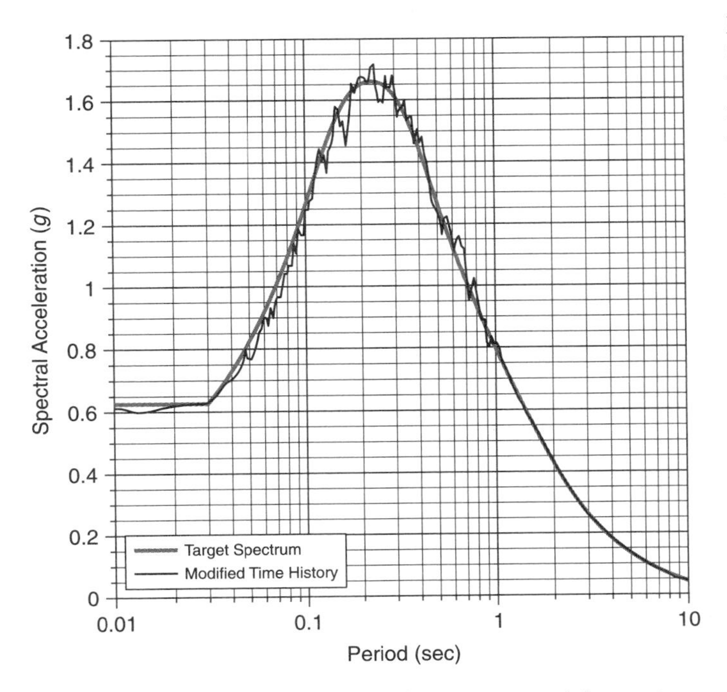

FIGURE 10 Comparison of the design spectrum and the spectrum of the modified time history (5% damping).

9.2 Design Time Histories for Rock Site Conditions

In order to allow for aleatory variability in the time histories associated with the design spectrum (see Section 3.5), three sets of three-component seismograms (time histories) were computed for each of the dominant sources (three sets for San Andreas fault sources and three sets for Hayward fault sources). Because forward directivity dominated the long-period hazard for the design return period, forward rupture direction was assumed for all six sets of ground-motion time histories.

Next, available recorded accelerograms from earthquakes with magnitudes greater than 6.5 and recorded at distances less than 15 km were evaluated (Tables 4a, 4b, 4c). Both rock and soil time histories were considered for selection because the response spectral matching adjusts to a large extent for the wave frequency differences at soil and rock sites (Aki, 1988). These candidate time histories were rotated to their principle directions (in displacement) and then classified in terms of the character of the displacement waveform: one-sided pulse, two-sided pulse, multiple pulses. Because the bridge will undergo significant nonlinear response under the design ground motions, it is important to get a range of nonstationary characteristics of the near-fault ground motions to represent a range of possible ground motions from future earthquakes in order to adequately test the performance of the bridge.

For the Hayward fault sources, suitable empirical time histories could be found and used directly as initial time histories; however, for the San Andreas source scenarios, there are as yet no recorded time histories that are directly applicable to the required magnitude and distance range. As an alternative, two procedures were used to develop the initial ground motions: splicing together empirical time histories and numerical simulations. As an example, the initial displacement time history for the fault-normal component derived by splicing the El Centro #6 recording from the 1979 earthquake onto the end of the 1940 El Centro recording (to increase the duration) is shown in Figure 11.

The selected initial time histories were modified to match the design response spectra using a time domain spectral-matching procedure (e.g., Lihanand and Tseng, 1988). The spectrum-compatible time histories are shown in Figure 12 and the fit to the spectrum is shown in Figure 10. Comparing the time histories in Figures 11 and 12 shows that this spectral matching procedure preserves the gross nonstationary characteristics of the acceleration, velocity, and displacement time histories.

9.3 Coherency and Multiple Support Time Histories

After development of the reference rock motions that are spectrum compatible, multiple-support rock ground motions were generated at all support points for the bridge structure to characterize the spatial variation of the rock ground motion

TABLE 4a Candidate Time Histories Considered for the Bay Bridge: One-Sided Displacement Pulses

Earthquake	Station	Rupture Distance (km)	Site	Major Axis (degrees)
1940 Imperial Valley, CA	El Centro #9	8.3	Soil	255
1985 Nahanni, Canada	Site 1	6.0	Rock	310
1989 Loma Prieta, CA	Corralitos	5.1	Rock	055
1989 Loma Prieta, CA	Gilroy–Historic Bldg	12.7	Soil	260
1989 Loma Prieta, CA	Gilroy Array #3	14.4	Soil	105
1992 Cape Mendocino, CA	Cape Mendocino	8.5	Rock	005
1992 Landers, CA	Lucerne	1.1	Rock	275
1994 Northridge, CA	Arleta	9.2	Soil	055
1994 Northridge, CA	W. Lost Canyon	13.0	Soil	075
1994 Northridge, CA	W. Pico Canyon	7.1	Soil	051
1994 Northridge, CA	Sylmar Converter Stn (east)	6.2	Soil	092
1994 Northridge, CA	Sylmar Converter Stn	6.1	Soil	333

TABLE 4b Candidate Time Histories Considered for the Bay Bridge: Two-Sided Displacement Pulses

Earthquake	Station	Rupture Distance (km)	Site	Major Axis (degrees)
1978 Tabas, Iran	Tabas	3.0	Soil	045
1989 Loma Prieta, CA	Gilroy–Gavilan College	11.6	Rock	097
1989 Loma Prieta, CA	Gilroy #1	11.2	Rock	045
1989 Loma Prieta, CA	Saratoga, Aloha Ave.	13.0	Soil	070
1992 Erzinkan,	Erzinkan	2.0	Soil	045
1992 Cape Mendocino, CA	Petrolia	9.5	Soil	100
1994 Northridge, CA	Jensen Filter Plant	6.2	Soil	012
1994 Northridge, CA	Newhall	7.1	Soil	160
1994 Northridge, CA	Pacoima Dam, Upper Left	8.0	Rock	199

TABLE 4c Candidate Time Histories Considered for the Bay Bridge: Multiple Displacement Pulses

Earthquake	Station	Rupture Distance (km)	Site	Major Axis (degrees)
1996 Gazli, USSR	Gazli	3.0	Rock	035
1986 Superstition Hills, CA	El Centro, ICC	13.9	Soil	045
1986 Superstition Hills, CA	Westmoreland Fire Station	13.3	Soil	240
1989 Loma Prieta, CA	Bran	10.3	Rock	085
1989 Loma Prieta, CA	Capitola	14.5	Soil	070
1989 Loma Prieta, CA	Gilroy #2	12.7	Soil	145
1989 Loma Prieta, CA	Los Gatos Pres. Center	6.1	Rock	100
1992 Landers, CA	Joshua Tree	11.6	Soil	005
1994 Northridge, CA	Montebello, Bluff Rd	12.3	Soil	211
1994 Northridge, CA	Cold Water Canyon	14.6	Soil	195
1994 Northridge, CA	Pacoima Dam Downstream	8.0	Rock	305
1994 Northridge, CA	Kagel Canyon	8.2	Rock	360
1994 Northridge, CA	Rinaldi Receiving Station	7.1	Soil	298
1994 Northridge, CA	Sepulveda VA	8.9	Soil	290
1994 Northridge, CA	Simi Valley, Katherine Rd.	14.6	Soil	115
1994 Northridge, CA	Sun Valley, Roscoe Blvd	12.3	Soil	045
1994 Northridge, CA	Sylmar, Olive View Med.	6.4	Soil	055
1995 Kobe, Japan	KJMA	0.6	Rock	050
1995 Kobe, Japan	Kobe Univ.	0.2	Rock	075
1995 Kobe, Japan	Amagasaki	10.2	Soil	030

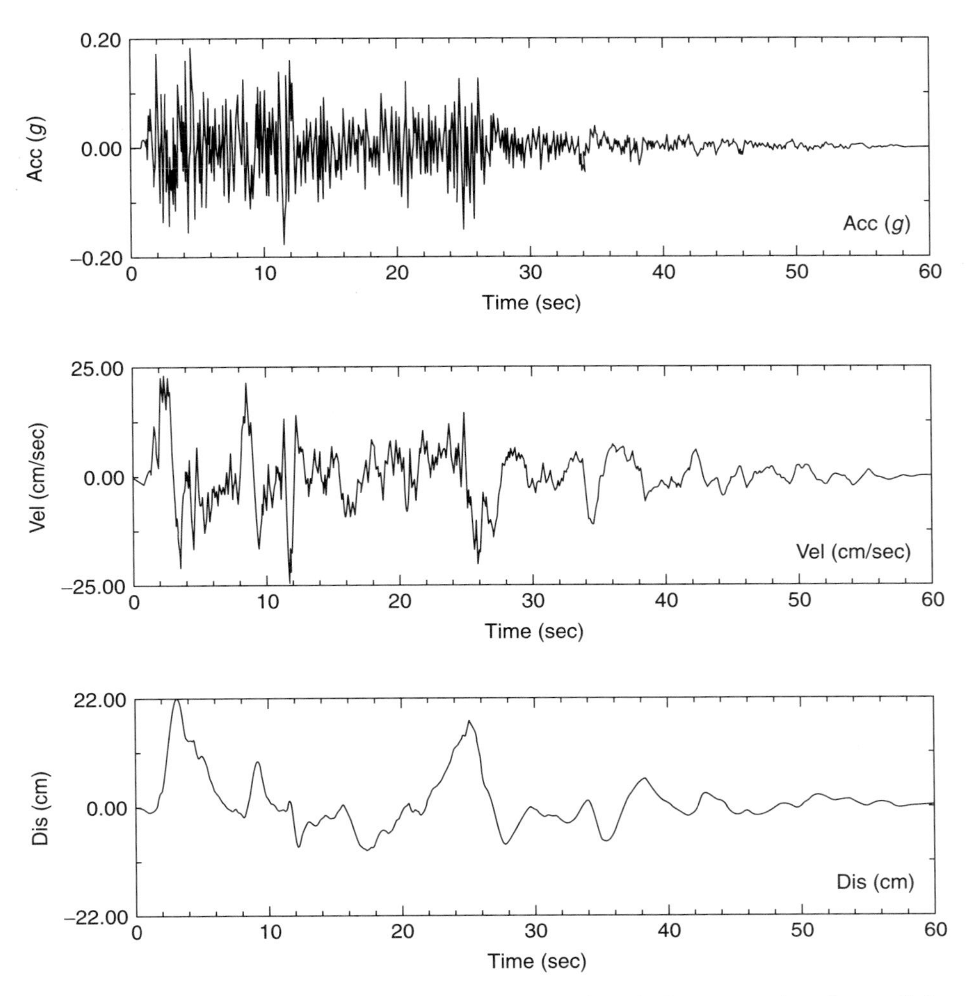

FIGURE 11 Initial fault-normal time history for the San Andreas event based on splicing together empirical recordings.

(see Section 8). The spatial variation of the rock motion resulted from several effects: non-vertical wave propagation (wave-passage effect) scattering and complex 3–D wave propagation (spatial coherency effect), and variable distance to the fault rupture (attenuation effect). Because most of the Bay Bridge supports are founded on piles driven into deep mud, the actual input motions also include spatial variability due to variability of the soil profile along the bridge. (Such site response variability in this project was incorporated using 1D site response calculations. In this chapter we only discuss the rock ground motions.) The steps in developing the incoherent rock ground motions were fourfold:

Step 1: Generate coherency compatible time histories at each support location.
Step 2: Modify the incoherent time histories to be compatible with the target spectrum using a frequency-domain approach.
Step 3: Apply a baseline correction to the incoherent time histories.
Step. 4: Apply the wave passage and attenuation effect.

9.3.1 Generation of Multiple Support Ground Motions

Multiple support time histories were generated at a total of 30 pier locations along the bridge using the procedure by Abrahamson (1992a). This procedure modifies the Fourier phase angles at each support location, using a relation between the spatial coherency and the degree of randomness in the phase angle.

9.3.2 Wave-Passage Effect

Based on the array data measurements, an apparent wave-speed of 2.5 km/sec for *S* waves was used. This velocity was selected because it is toward the lower end of the range of measured speeds and leads to somewhat conservative estimates of the differential motions. The wave propagation was projected along the axis of the bridge with apparent velocity along the bridge, V_{bridge}, given by

$$V_{bridge} = V_a / \sin(\theta) \tag{5}$$

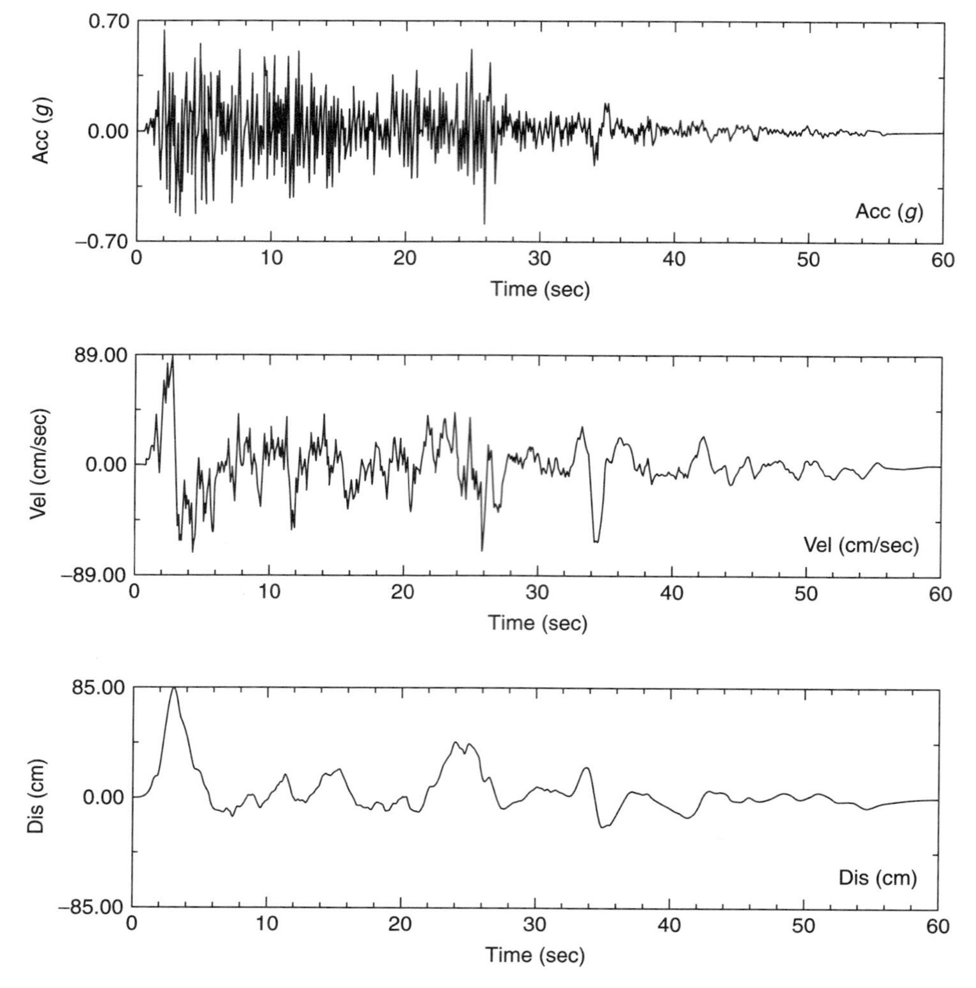

FIGURE 12 Spectrum compatible fault-normal time history.

where V_a is the apparent wave velocity and θ is the angle between the line perpendicular to the bridge axis and the line connecting the closest point of the fault and the center of the bridge. (Of course, waves arrive from all fault rupture segments in addition to waves arriving from the closest point on the fault, but here it is approximated by a single wave direction.)

9.3.3 Attenuation across the Structure

For fault sources near to long structures, such as the San Francisco Bay Bridge, the different distances to the various support locations result in different ground motions due to attenuation of the ground motion. To account for this attenuation, the ground motion at each bridge pier was scaled by the ratio of the predicted T = 1 second spectral acceleration to the target T = 1 second spectral acceleration using the empirical attenuation relations. While the attenuation could have been computed separately for each frequency, the T = 1 second spectral value was used for all periods to allow simple scaling of the ground motions. Because the structure is a long-period structure, this was considered reasonable.

9.3.4 Results

An example of the resulting incoherent rock ground motions for the two ends of the bridge is shown in Figure 13. The differential displacements are shown in Figure 14 as a function of the separation distance. These multiple supports are then propagated through site-specific soil profiles to incorporate the local site response.

10. Overview

An important point in summarizing the present status of assessment of seismic strong ground motions is that in a number of countries digital strong-motion systems linked to computer

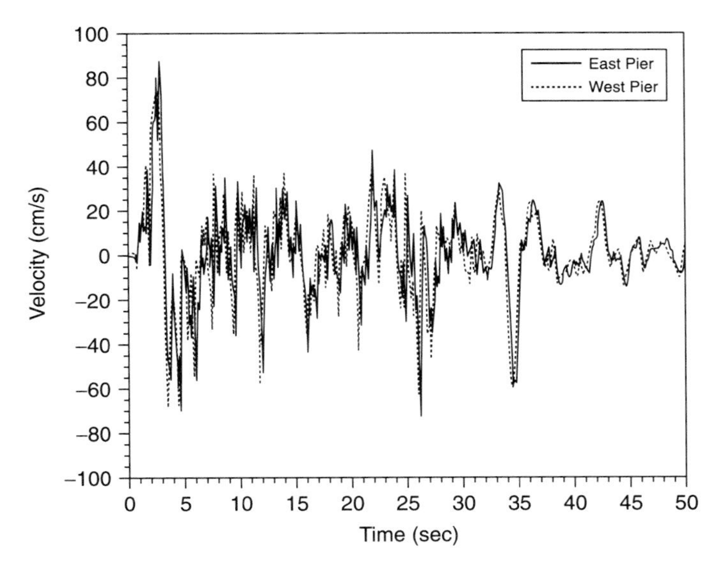

FIGURE 13 Incoherent rock time histories at the two ends of the Bay Bridge.

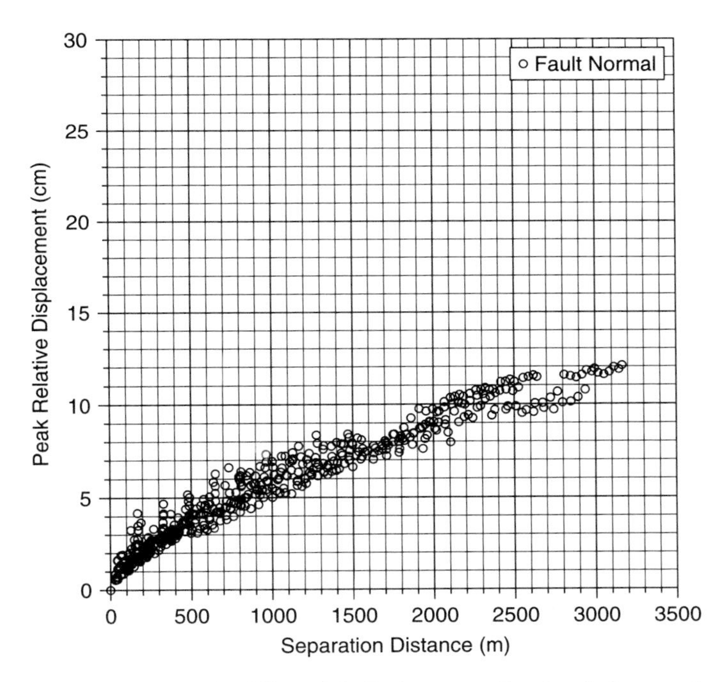

FIGURE 14 Peak differential displacement for the fault-normal component of motion from a set of incoherent spectrum-compatible time histories for the Oakland crossing of the San Francisco–Oakland Bay Bridge.

centers have now been installed. They provide processed observational data within a few minutes after shaking occurs. These near real-time data can be a basis for postearthquake response and evaluation of damage to structures, and for urgent risk decisions. By the end of 1999, computer-generated maps of intensity were being generated in California from the network of digital stations installed in the Los Angeles region. These displays, called "shake maps", available on the Internet for public use (*http://www.trinet.org*). One format of the maps contours peak ground velocity and spectral acceleration at 0.3, 1.0, and 3.0 seconds and displays in color a ground-motion strength-parameter called "instrumental intensity." Similar maps were provided immediately after the 1999 Chi-Chi earthquake in Taiwan. The maps have a variety of applications in postearthquake valuation of structures, damage and loss estimation, and in guiding emergency response but lack estimation detail.

The seismological problems dealt with in this chapter will no doubt be much extended in subsequent years. First, greater sampling of strong ground motions at all distances from fault sources of various mechanisms and magnitudes will inevitably become available. An excellent example is the wide recording of the 1999 Chi-Chi, Taiwan, earthquake (see Shin *et al.*, 2000; Lee *et al.*, 2001). Second, more realistic 3–D numerical models will solve the problem of the sequential development of the wave mixtures as the waves pass through different geological structures. Two difficulties may persist: the lack of knowledge of the roughness distribution along the dislocated fault and, in many places, quantitative knowledge of the soil, alluvium, and crustal rock variations in the region.

A new significant ingredient in general motion measurement is correlation with precisely mapped coseismic ground deformations. Networks of GPS instruments will help greatly in understanding of the source problem and the correct adjustment to strong-motion displacement records. A broad collection of standardized strong-motion time histories represented by both amplitude spectra and phase spectra is now being accumulated in virtual libraries for easy access on the Internet. Such records will provide greater confidence in seismologically sound selection of ground-motion estimates.

Additional instrumentation to record strong ground motion remains a crucial need in earthquake countries around the world (see e.g., Bolt, 1999b; Franke1, 1999). Such basic systems should measure not only free-field surface motions, but downhole motions to record the wave changes as they emerge at the Earth's surface.

Many contemporary attenuation estimates will no doubt be updated as more recorded measurements are included, rendering earlier models obsolete. To keep abreast of changes, ground-motion attenuation model information may be found at *http://www.geohazards.cr.usgs.gov/earthquake.html*. Click on Engineering Seismology, then on Ground Motion Information.

Acknowledgments

We thank the three referees for their comments on the manuscript draft.

References

Abrahamson, N. A. (1992a). Generation of spatially incoherent strong motion time histories. *Proc. 10th World Conf. Earthq. Eng.*, Madrid, Spain, 845–850.

Abrahamson, N. A. (1992b). Spatial variation of earthquake ground motion for application to soil–structure interaction. *Electric Power Research Institute, report TR-100463*.

Abrahamson, N. A., and B. A. Bolt (1987). Array analysis and synthesis mapping of strong ground motion. In: "Strong Motion Synthetics" (B. A. Bolt, Ed.), Academic Press, Orlando, Florida, 55–90.

Abrahamson, N. A., and K. M. Shedlock (1997). Overview. *Seism. Res. Lett.* **68**, 9–23.

Abrahamson, N. A., and W. Silva (1997). Empirical response spectral attenuation relations for shallow crustal earthquakes. *Seismol. Res. Lett.* **68**, 94–127.

Abrahamson, N. A., B. A. Bolt, R. D. Darragh, J. Penzien, and Y. B. Tsai (1987). The SMART1 accelerograph array (1980–1987): a review. *Earthq. Spectra* **3**, 263–288.

Aki, K. (1988). Local site effects on strong ground motion. In: *Proc. Earthq. Eng. and Soil Dyn. II*, Park City, Utah, 103–155.

Ambraseys, N. N. (1988). Engineering seismology, First Mallet Milne Lecture. *Earthq. Eng. Struc. Dyn.* **17**, 1–105.

Ambraseys, N. N., and J. J. Bommer (1995). Attenuation relations for use in Europe: An overview. *Proc. Fifth Society for Earthq. and Civil Eng. Dyn. Conf. on European Seismic Design Practice,* Chester, England, 67–74.

Ambraseys, N. N., and J. J. Bommer (1991). The attenuation of ground accelerations in Europe. *Earthq. Eng. Struc. Dyn.* **20**, 1179–1202.

Amirbekian, R. W., and B. A. Bolt (1998). Spectral comparison of vertical and horizontal seismic strong ground motions in alluvial basins. *Earthq. Spectra* **14**, 573–595.

Anderson, J. G. (1997). Seismic energy and stress-drop parameters for a composite source model. *Bull. Seismol. Soc. Am.* **87**, 85–96.

Archuleta, R. (1984). A faulting model for the 1969 Imperial Valley earthquake. *J. Geophys. Res.* **89**, 4559–4585.

Atkinson, G. M., and D. M. Boore (1995). New ground motion relations for eastern North America. *Bull. Seismol. Soc. Am.* **85**, 17–30.

Bard, P. Y. (1982). Diffracted waves and displacement field over two-dimensional elevated topographics. *Geophys. J. Roy. Astr. Soc.* **71**, 731–760.

Bazzurro, P., and C. A. Cornell (1999). Disaggregation of seismic hazard. *Bull. Seismol. Soc. Am.* **89**, 335–341.

Benioff, H. (1934). The physical evaluation of seismic destructiveness. *Bull. Seismol. Soc. Am.* **24**, 9–48.

Bolt, B. A. (1971). The San Fernando Valley, California, earthquake of February 9, 1971: data on seismic hazards. *Bull. Seismol. Soc. Am.* **61**, 501–510.

Bolt, B. A. (Ed.) (1987). *Seismic Strong Motion Synthetics*. Academic Press, Orlando, Florida, 328 pp.

Bolt, B. A. (1996). From earthquake acceleration to seismic displacement, Fifth Mallet-Milne Lecture. *Society for Earthq. and Civil Eng. Dyn.*, London, 50 pp.

Bolt, B. A. (1999a). Modern recording of seismic strong motion for hazard reduction. In: "Vrancea Earthquakes: Tectonic and Risk Mitigation" (F. Wenzel, Ed.), Kluwer, Netherlands, 1–14.

Bolt, B. A. (1999b). *Earthquakes*, 4th ed. W. H. Freeman, New York.

Bommer, J. J., and A. Martinez-Pereira (1998). The effective duration of earthquake strong motion. *J. Earthq. Eng.* **3**(2), 127–172.

Boore, D. M., and W. B. Joyner (1997). Site amplifications for generic rock sites. *Bull. Seismol. Soc. Am.* **87**, 327–341.

Boore, D. M., W. B. Joyner, and T. Fumal (1997). Equations for estimating horizontal response spectra and peak acceleration from western North American earthquakes: a summary of recent work. *Seismol. Res. Lett.* **68**, 128–153.

Brillinger, D. R., and H. K. Preisler (1984). Further analysis of the Joyner-Boore attenuation data. *Bull. Seismol. Soc. Am.* **75**, 611–614.

Brune, J., and A. Anooshehpoor (1999). Dynamic geometrical effects on strong ground motion in normal fault model. *J. Geophy. Res.* **104**, 809–815.

Bullen, K. E., and B. A. Bolt (1985). *An Introduction to the Theory of Seismology*. Cambridge University Press, Cambridge, 499 pp.

Campbell, K. W. (1997). Empirical near-source attenuation relationships for horizontal and vertical components of peak ground acceleration, peak ground velocity, and pseudo absolute acceleration response spectra. *Seismol. Res. Lett.* **68**, 154–179.

Cornell, C. A. (1968). Engineering seismic risk analysis. *Bull. Seismol. Soc. Am.* **58**, 1583–1606.

Crouse, C. B. (1991). Ground-motion attenuation equations for earthquakes on the Cascadia subduction zone. *Earthquake Spectra* **7**, 201–235; and 506.

Darragh, R. B., and A. F. Shakal (1991). The site response of two rock and soil station pairs to strong and weak ground motion. *Bull. Seismol. Soc. Am.* **81**, 1884–1899.

Frankel, A. D. (1999). How does the ground shake? *Science* **283**, 2032–2033.

Frankel, A. D., C. Mueller, T. Barnhard, D. Perkins, E. Leyendecker, N. Dickman, S. Hanson, and M. Hopper (2000). USGS national seismic hazard maps. *Earthquake Spectra* **16**, 1–19.

Fugro/Earth Mechanics (1998). Seismic ground motion report for San Francisco–Oakland Bay Bridge, East Span Seismc Safety Project, Report to Caltrans, Contract No. 59A0053. [A copy of this report may be accessed at both the Earthquake Engineering Research Center Library at the University of California, Berkeley, and at the Earthquake Engineering Research Library at Caltech, Pasadena.]

Gregor, N. J., and B. A. Bolt (1997). Peak strong motion attenuation relations for horizontal and vertical ground displacements. *J. Earthq. Eng.* **1**(2), 275–292.

Gusev, A. (1983). Descriptive statistical model of earthquake source radiation and its application to an estimate of short-period strong motion. *Geophys. J. R. Astr. Soc.* **74**, 787–808.

Hao, H., C. S. Oliveira, and J. Penzien (1989). Multiple-station ground motion processing and simulation based on SMART-1 array data. *Nuclear Eng. and Design* **11**, 293–310.

Heaton, T. H., J. F. Hall, D. J. Wald, and M. W. Halling (1995). Response of high-rise and base-isolated buildings to a hypothetical M_W 7.0 blind thrust earthquake. *Science* **267**, 206–211.

Housner, G. W. (1959). Behavior of structures during earthquakes. *Proc. ASCE EM4* **85**, 109–129.

Idriss, I. M. (1995). An overview of earthquake ground motions pertinent to seismic zonation. In: *Proc. 5th International Conference On Seismic Zonation*, Oct. 17–19, 1995, Nice, Vol. III, 2111–2126. Ouest Editions - Presses Academiques, Nantes Cedex, France.

Irikura, K. (1983). Semi-empirical estimation of strong ground motions during large earthquakes. *Bull. Disaster Prevention Res. Inst. (Kyoto Univ.)* **33**, 63–104.

Joyner, W. B., and D. M. Boore (1981). Peak horizontal acceleration and velocity from strong-motion records including records from the 1979 Imperial Valley, California, earthquake. *Bull. Seismol. Soc. Am.* **71**, 2021–2038.

Joyner, W. B., and D. M. Boore (1988). Measurement, characteristics and prediction of strong ground motion. *Proc. Specialty Conf. on Earthq. Eng. and Soil Dyn. II*, ASCE Park City, Utah, 43–102.

Kanamori, H. (1971). Faulting of the great Kanto earthquake of 1923 as revealed by seismological data. *Bull. Earthq. Res. Inst. Univ. Tokyo* **49**, 13–18.

Kaul, M. K. (1978). Spectrum-consistent time-history generation. *ASCE J. Eng. Mech., EM4*, 781–788.

Lee, W. H. K., T. C. Shin, K. W. Kuo, K. C. Chen, and C. F. Wu (2001). CWB free-field strong-motion data from the 21 September Chi-Chi, Taiwan, earthquake, *Bull. Seismol. Soc. Am.* **91**, 1370–1376.

Lilhanand, K., and W. S. Tseng (1988). Development and application of realistic earthquake time histories compatible with multiple damping response spectra. *Ninth World Conf. Earth. Eng.*, Tokyo, Japan, Vol II, 819–824.

Lomax, A., and B. A. Bolt (1992). Broadband waveform modelling of anomalous strong ground motion in the 1989 Loma Prieta earthquake using three-dimensional geological structures. *Geophys. Res. Lett.* **19**, 1963–1966.

McVerry, G. H., and J. X. Zhao (1999). A response spectrum model for New Zealand including high attenuation in the volcanic zone. *Institute of Geological and Nuclear Science, Ltd.*

Mikumo, T., and T, Miyatake (1987). Numerical modeling of realistic fault rupture processes. In: "Seismic Strong Motion Synthetics" (B. A. Bolt, Ed.), Academic Press, New York, 91–151.

O'Connell, D. R. (1999). Replication of apparent nonlinear seismic response with linear wave propagation model. *Science* **283**, 2045–2050.

Olsen, K. B., R. J. Archuleta, and J. R. Matarese (1995). Three-dimensional simulation of a magnitude 7.75 earthquake on the San Andreas fault. *Science* **270**, 1628–1632.

Olsen, K. B., and R. J. Archuleta (1996). Three-dimensional simulation of earthquakes on the Los Angeles fault system. *Bull. Seismol. Soc. Am.* **86**, 575–596.

Panza, G. F., and P. Suhadolc (1987). Complete strong motion synthetics. In: "Seismic strong motion synthetics" (B. A. Bolt, Ed.), Academic Press, New York, 153–204.

Preumont, A. (1984). The generation of spectrum compatible accelerograms for the design of nuclear power plants. *Earthq. Eng. Struct. Dyn.* **12**, 481–497.

Raoof, M., R. B. Herrmann, and L. Malagnini (1999). Attenuation and excitation of three-component ground motion in Southern California. *Bull. Seismol. Soc. Am.* **89**, 888–902.

Reiter, L. (1990). *Earthquake Hazard Analysis: Issues and Insights.* Columbia University Press, New York, 254 pp.

Sadigh, K., C.-Y. Chang, J. A. Egan, F. Makdisi, and R. R. Youngs (1997). Attenuation relationships for shallow crustal earthquakes based on California strong motion data. *Seismol. Res. Lett.* **68**, 180–189.

Saikia, C. K., and P. G. Somerville (1997). Simulated hard-rock motions in Saint Louis, Missouri, from large New Madrid earthquakes. *Bull. Seismol. Soc. Am.* **87**, 123–139.

Shin, T. C., K. W. Kuo, W. H. K. Lee, T. L. Teng, and Y. B. Tsai (2000). A preliminary report on the 1999 Chi-Chi (Taiwan) earthquake. *Seismol. Res. Lett.* **71**, 23–29.

Somerville, P. G., and N. A. Abrahamson (1995). Ground motion prediction for thrust earthquakes. *Proc. SMIP95 Seminar*, Cal. Div. of Mines and Geology, San Francisco, Calif., 11–23.

Somerville, P. G., N. F. Smith, R. W. Graves, and N. A. Abrahamson (1997). Modification of empirical strong ground motion attenuation relations to include the amplitude and duration effects of rupture directivity. *Seismol. Res. Lett.* **68**, 199–222.

Spudich, P., J. Fletcher, M. Hellweg, J. Boatwright, C. Sullivan, W. Joyner, T. Hanks, D. Boore, A. McGarr, L. Baker, and A. Lindh (1996). Earthquake ground motions in extensional tectonic regimes. *Open-File Report 96-292, U.S. Geological Survey,* 351 pp.

Spudich, P., M. Hellweg, and W. H. K. Lee (1995). Directional topographic site response at Tarzana observed in aftershocks of the 1994 Northridge, California, earthquake: implications for main shock motions. *Bull. Seismol. Soc. Am.* **85**, S193–S208.

Stidham, C., M. Antolik, D. Dreger, S. Larsen, and B. Romanowicz (1999). Three-dimensional structure influences on the strong-motion wavefield of the 1989 Loma Prieta earthquake. *Bull. Seismol. Soc. Am.* **89**, 1184–1202.

Takei, M., and H. Kanamori (1997). Simulation of long-period ground motions near a large earthquake. *Bull. Seismol. Soc. Am.* **87**, 140–156.

Teng, T. L., Y. B. Tsai, and W. H. K. Lee, Eds. (2001). Dedicated Issue on the Chi-Chi, Taiwan Earthquake of 20 September, 1999. *Bull. Seismol. Soc. Am.* **91**, 893–1395, with attached CD-ROM.

Toro, G. R., N. A. Abrahamson, and J. F. Schneider (1997). Model of strong ground motions from earthquakes in central and eastern North America: best estimates and uncertainties. *Seismol. Res. Lett.* **68**, 41–57.

Vidale, J., and D. V. Helmberger (1987). Path effects in strong ground motion seismology. In: "Seismic Strong Motion Synthetics" (B. A. Bolt, Ed.), Academic Press, New York, 267–319.

Youngs, R. R., N. A. Abrahamson, and K. Sadigh (1995). Magnitude-dependent variance of peak ground acceleration. *Bull. Seismol. Soc. Am.* **85**, 1161–1176.

Zeng, Y., J. G. Anderson, and G. Yu (1994). A composite source model for computing realistic synthetic strong ground motions. *Geophys. Res. Lett.* **21**, 725–728.

Editor's Note

Please see also Chapter 57, Strong-Motion Seismology, by Anderson; Chapter 58, Strong-Motion Data Processing, by Shakal *et al.*; Chapter 60, Strong-Motion Attenuation Relations, by Campbell; and Chapter 61, Site Effects of Strong Ground Motions, by Kawase.

60

Strong-Motion Attenuation Relations

Kenneth W. Campbell
ABS Consulting and EQECAT Inc., Beaverton, Oregon, USA

1. Introduction

An evaluation of seismic hazards, whether deterministic (scenario based) or probabilistic, requires an estimate of the expected ground motion at the site of interest. The most common means of estimating this ground motion in engineering practice, including probabilistic seismic hazard analysis (PSHA), is the use of an attenuation relation. An attenuation relation, or ground-motion model as seismologists prefer to call it, is a mathematical-based expression that relates a specific strong-motion parameter of ground shaking to one or more seismological parameters of an earthquake. These seismological parameters quantitatively characterize the earthquake source, the wave propagation path between the source and the site, and the soil and geological profile beneath the site.

In its most fundamental form, an attenuation relation can be described by an expression of the form

$$Y = c_1 e^{c_2 M} R^{-c_3} e^{-c_4 r} e^{c_5 F} e^{c_6 S} e^{\varepsilon} \tag{1}$$

or by its more common logarithmic form

$$\ln Y = c_1 + c_2 M - c_3 \ln R - c_4 r + c_5 F + c_6 S + \varepsilon \tag{2}$$

where "ln" represents the natural logarithm (log to the base e), Y is the strong-motion parameter of interest, M is earthquake magnitude, r is a measure of source-to-site distance, F is a parameter characterizing the type of faulting, S is a parameter characterizing the type of local site conditions, ε is a random error term with zero mean and standard deviation equal to the standard error of estimate of $\ln Y$ ($\sigma_{\ln Y}$), and R is a distance term often given by one of the alternative forms

$$R = \begin{cases} r + c_7 \exp(c_8 M) & \text{or} \\ \sqrt{r^2 + [c_7 + \exp(c_8 M)]^2} \end{cases} \tag{3}$$

Alternatively, the regression coefficients c_3, c_6, and c_7 can be defined as functions of M and R.

The mathematical relationships in Eqs. (1) to (3) have their roots in earthquake seismology (e.g., see Lay and Wallace, 1995, and contributions in this volume). The relationships $Y \propto e^{c_2 M}$ and $\ln Y \propto c_2 M$ are consistent with the definition of earthquake magnitude, which Richter (1935) originally defined as the logarithm of the amplitude of ground motion. The relationships $Y \propto R^{-c_3}$ and $\ln Y \propto -c_3 \ln R$ are consistent with the geometric attenuation of the seismic wave front as it propagates away from the earthquake source. Some attenuation relations set $c_3 = 1$, which is the theoretical value for the diminution of amplitude with distance from a point source in a half space, referred to as spherical spreading. If not constrained in the regression analysis, c_3 will be typically larger than unity. Sometimes c_3 is varied as a function of distance to accommodate differences in the attenuation rate of different wave types (e.g., direct waves versus surface waves) and critical reflections off the base of the crust (e.g., the Moho). The relationships $Y \propto e^{-c_4 r}$ and $\ln Y \propto -c_4 r$ are consistent with the anelastic attenuation caused by material damping and scattering as the seismic waves propagate through the crust. The relationship between Y and the remaining parameters has been established over the years from ground-motion observations and theoretical ground-motion studies.

The mathematical relationships defined in Eq. (3) are used to incorporate the widely held belief that short-period ground motion should become less dependent on magnitude close to the causative fault. Schnabel and Seed (1973) were the first to model this behavior theoretically using simple geometrical considerations. Since then similar results have been obtained using theoretical finite-source models (e.g., see Anderson, 2000). Campbell (1981) was one of the first investigators to identify and model this behavior empirically. Since then many other investigators have adopted similar models.

The remainder of this chapter defines the strong-motion and seismological parameters commonly found in an attenuation relation and provides a brief discussion of the various factors that can affect these parameters. The discussion specifically

INTERNATIONAL HANDBOOK OF EARTHQUAKE AND ENGINEERING SEISMOLOGY, VOLUME 81B   ISBN: 0-12-440658-0

addresses issues regarding the use of attenuation relations to estimate strong ground motion in engineering practice, including PSHA. Particular emphasis is placed on the estimation of peak time-domain and peak frequency-domain parameters. A more general discussion of strong-motion seismology is given in Chapter 57 by Anderson. The development of earthquake time histories and other related engineering estimates of ground motion is the topic of Chapter 59 by Bolt and Abrahamson. The estimation of seismic hazards, including PSHA, is the topic of Chapter 65 by Somerville and Moriwaki. Due to length limitations, this chapter can only provide a brief introduction to the attenuation relation and its constitutive strong-motion and seismological parameters. A contemporary guide to some recent attenuation relations and their use in engineering practice is included on the attached Handbook CD (hereafter referred to as the *Contemporary Guide*). This *Contemporary Guide* includes a comprehensive list of references to previous discussions on the empirical and theoretical prediction of strong ground motion as well as a selected compilation of recently published attenuation relations and guidance on their use in engineering practice. Most notable amongst these, because of their recency and comprehensiveness, are those of Douglas (2002), which cover relations developed from empirical methods, and Boore (2002), which cover relations developed from seismological models. Also included on the attached Handbook CD is an interactive electronic workbook that allows the user to evaluate the attenuation relations presented in the *Contemporary Guide* for a variety of seismological parameters.

Although this chapter is not intended to give a complete history of the attenuation relation, it would be remiss not to mention some of the pioneers whose early and continued development of attenuation relations first led to their widespread acceptance in engineering, particularly in the then-fledgling field of PSHA. These pioneers include (listed alphabetically) Neville Donovan, Luis Esteva, George Housner, Ed Idriss, Bill Joyner, Kiyoshi Kanai, Robin McGuire, H. Bolton Seed, and Mihailo Trifunac. Many others have followed in their footsteps, many whose contributions are significant and noted in the remainder of this chapter.

2. Strong-Motion Parameters

Strong-motion parameters represent in simple terms a specific characteristic of an earthquake time history (peak time-domain parameters) or its frequency-domain equivalent (peak frequency-domain parameters). A brief description of these two types of strong-motion parameters is given next.

2.1 Peak Time-Domain Parameters

Historically, peak ground acceleration (PGA) and to a lesser extent peak ground velocity (PGV) and peak ground displacement (PGD) have been the most common peak time-domain parameters used to describe strong ground motion in engineering practice. They represent the maximum absolute amplitude of ground motion scaled from a recorded or synthetic acceleration, velocity, or displacement time series. They serve as a simple representation of the short-, mid-, and long-period components of ground motion, respectively.

2.2 Peak Frequency-Domain Parameters

As seismic design procedures have become more sophisticated, engineers have begun to incorporate the natural period of a structure into its design through the use of a response spectrum (Chapter 68 by Borcherdt *et al.*). A response spectrum is a plot versus undamped natural period of the maximum response of a viscously damped, single-degree-of-freedom (SDOF) system subjected to ground-motion input at its base. (For a more complete description see Chapter 67 by Jennings.) The most common response-spectral parameters are pseudoacceleration (known variously as PSSA, PSA, or SA) and pseudovelocity (known variously as PSRV, PSV, or SV). The terms PSA and PSV are used throughout this chapter to represent pseudoacceleration and pseudovelocity in order to avoid confusion, noting that Gupta (1993) prefers the terms SA and SV. However, the latter notation has been used in the past to define absolute acceleration and relative velocity, which are similar, but not identical, to PSA and PSV (see Chapter 67 by Jennings or any text on earthquake engineering for a description of these parameters). PSA and PSV are related to one another and to relative displacement (SD) by the expression

$$\mathrm{PSA} = \frac{2\pi}{T_n}\,\mathrm{PSV} = \left(\frac{2\pi}{T_n}\right)^2 \mathrm{SD} \tag{4}$$

where T_n is the undamped natural period of a SDOF system. Relative displacement is the maximum absolute displacement of a SDOF system relative to its base.

Response spectra can be developed from PGA alone or from PGA, PGV, and PGD (Newmark and Hall, 1982; Campbell, 2002; see also Chapter 67 by Jennings), or they can be developed from one or two key response-spectral ordinates, such as PSA at $T_n = 0.2$ and $T_n = 1$ sec (Leyendecker *et al.*, 2000; see also Chapter 68 by Borcherdt *et al.*). Such procedures are typically used in building codes and other seismic regulations where a simple prescribed method for estimating a seismic design spectrum is required. However, it has become more common in engineering practice to develop a design response spectrum directly from attenuation relations representing individual spectral ordinates of PSA or PSV.

3. Earthquake Source Parameters

Earthquake source parameters describe a characteristic of an earthquake's source, such as its size, mechanism (type of faulting), stress drop, source directivity, or radiation pattern. Some common source parameters are discussed next.

3.1 Magnitude

Magnitude is used to define the "size" of an earthquake. There are many different scales that can be used to define magnitude. Some of the more common magnitude scales used in attenuation relations are moment magnitude (**M** or M_W), surface-wave magnitude (M_S), short-period body-wave magnitude (m_b), local magnitude (M_L), Lg-wave magnitude (m_{Lg} or m_N), and JMA magnitude (M_J). The symbol M_W used by engineers and engineering seismologists is used to denote moment magnitude throughout this chapter, noting that many seismologists prefer the symbol **M**. Network operators in different parts of the world will use one of the just-mentioned magnitude scales as a standard for quantifying magnitude in their region. However, it should be noted that even the use of the same magnitude scale can lead to different regional estimates of magnitude due to the way magnitude is locally defined and calculated. The mathematical definition of all peak time-domain magnitude scales can be given by the expression (e.g., Lay and Wallace, 1995)

$$M = \log(A/T) + f(R, h_{\text{hypo}}) + C_s + C_r \tag{5}$$

where "log" represents the common logarithm (log to the base 10), A is the recorded ground displacement of the seismic phase on which the amplitude scale is based, T is the period of the signal, $f(R, h_{\text{hypo}})$ is a correction for the distance from the earthquake source to the instrument (R) and the focal depth (h_{hypo}), C_s is a correction for the siting of the instrument, and C_r is a correction for the source region. Figure 1 gives a comparison of the different magnitude scales. A more thorough discussion of magnitude scales is given in Chapter 44 by Utsu.

There is an increasing tendency to adopt M_W as the worldwide standard for quantifying magnitude because of its strong physical and seismological basis (Bolt, 1993). By definition, M_W is related to seismic moment M_0, which is a measure of the energy radiated by an earthquake (Kanamori, 1978; Hanks and Kanamori, 1979). Seismic moment can be defined as

$$M_0 = \mu A_f \bar{D} = 2\mu E_S/\Delta\sigma \tag{6}$$

where μ is the shear modulus of the crust, A_f is the rupture area of the fault, $\bar{D}$ is the coseismic displacement averaged over the rupture area, $\Delta\sigma$ is the static stress drop averaged over the rupture area, and E_S is the radiated seismic energy. The definition based on $A_f\bar{D}$ allows M_0 to be derived from geological faulting parameters that can be measured in the field. The definition based on $E_S/\Delta\sigma$ allows M_0 to be derived from seismological data.

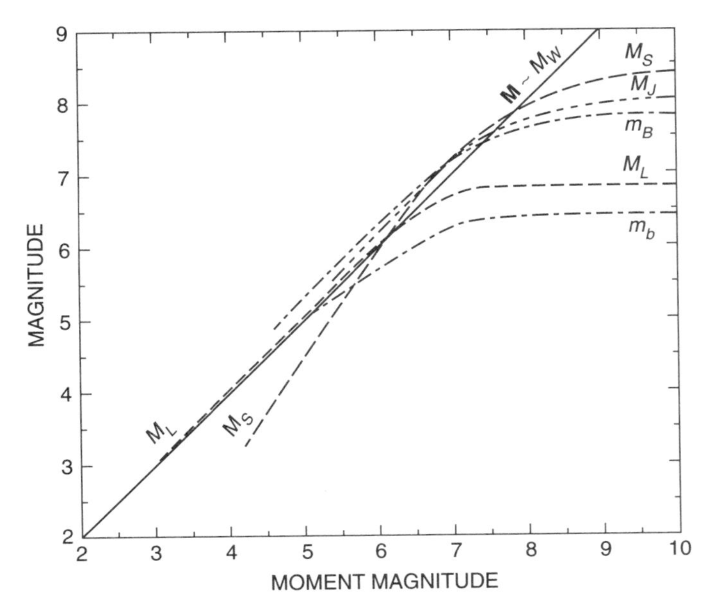

FIGURE 1 Comparison of magnitude scales (after Heaton *et al.*, 1986).

3.2 Type of Faulting and Focal Mechanism

The type of faulting, or focal mechanism, characterizes the orientation of slip on the fault plane, known as the rake, and the dip of the fault plane. Although rake is a continuous variable representing the angle between the direction of slip and the strike of the fault (the orientation of the fault on the Earth's surface), the type of faulting is typically classified into two or more categories for convenience. These categories will typically include strike slip (horizontal slip), reverse (dip slip with hanging-wall side up), thrust (same as reverse but with shallow dip) and normal (dip slip with hanging-wall side down). The values of rake corresponding to the pure form of these mechanisms are 0° (left-lateral slip) and 180° (right-lateral slip) for strike-slip faulting, 90° for reverse faulting, and 270° for normal faulting (Lay and Wallace, 1995).

Campbell (1981) was the first to empirically demonstrate that reverse and thrust-faulting earthquakes have relatively higher ground motions than strike-slip or normal-faulting earthquakes. All subsequent empirical and theoretical studies have found similar results (e.g., see the *Contemporary Guide* and Chapter 57 by Anderson). It has been common practice in the past to lump strike-slip events and normal-faulting events in the same category. However, a recent empirical study by Spudich *et al.* (1999) suggests that normal-faulting events, or for that matter strike-slip events in an extensional stress regime, might have lower ground motion than other types of shallow crustal events, an attribute first suggested by McGarr (1984). A preliminary interpretation of precarious rock observations by Brune (2000) suggests the presence of relatively low near-fault ground shaking on the footwall of normal faults, possibly due to a combination of faulting mechanism and footwall effects (see Section 4) that is not currently incorporated in any known attenuation relation (see the *Contemporary Guide*).

There has been a great deal of interest in blind-thrust faults after unusually large ground motions were observed during the 1987 Whittier Narrows, California, earthquake, the 1988 Saguenay, Canada, earthquake, and 1994 Northridge, California, earthquake. Whether similarly high ground motions can be expected from all future blind-thrust earthquakes is speculative. However, considering the current limited observational database, it cannot be ruled out. The higher ground motions observed during these earthquakes have been found to correspond to higher-than-average dynamic stress drops (see Section 3.3), which have been speculatively attributed to a lack of surface rupture (Somerville, 2000) or a relatively low total slip on the causative faults (see Chapter 57 by Anderson). More theoretical and empirical studies will be needed before there is a clear understanding of why these earthquakes produced high stress drops and how such events might be identified in the future.

3.3 Stress Drop

Stress drop, or more correctly dynamic stress drop, is the amount of stress that is relieved at the rupture front during an earthquake. Theoretical studies have shown that higher stress drops correspond to higher ground motion. It has been shown theoretically (e.g., Boore and Atkinson, 1987) and implied empirically (Campbell and Bozorgnia, 2000) that stress drop has a larger effect on short-period ground motion, including PGA, than on long-period ground motion. The author is not aware of any current attenuation relation that explicitly includes stress drop as a parameter (see the *Contemporary Guide*). However, stress drop is one of the parameters that must be included either explicitly or implicitly in the theoretical calculation of ground motion. A more thorough discussion of this parameter is given in the *Contemporary Guide*.

As mentioned in Section 3.2, high stress drops are the likely cause of the relatively high ground motions observed during some recent blind-thrust earthquakes. On the other hand, low stress drops might be the cause of the relatively low short-period ground motions observed during the 1999 Chi-Chi (M_W 7.6), Taiwan, earthquake (Tsai and Huang, 2000; Boore, 2001) and the 1999 Kocaeli (M_W 7.4), Turkey, earthquake (Anderson, 2000). The observation of relatively low ground motions during the Chi-Chi earthquake is particularly significant, because it is a large thrust earthquake, which had been expected from previous empirical and theoretical studies to have relatively large ground motions. The relatively low stress drops implied for the Taiwan and Turkey earthquakes could be caused by the large total slip on the causative faults (see Chapter 57 by Anderson) or because they ruptured the Earth's surface (Somerville, 2000). More study will be needed to better understand the phenomena that might have contributed to these low ground motions. If these earthquakes are found to be typical of similar large earthquakes worldwide, then the implication is that the current attenuation relations overpredict short-period ground motions from large earthquakes, an attribute suggested from observations of precarious rocks near great earthquakes on the San Andreas fault (Brune, 1999).

3.4 Source Directivity and Radiation Pattern

Radiation pattern is the geographic asymmetry of the ground motion caused by the faulting process. It is closely related to the focal mechanism of the earthquake. The radiation pattern can be perturbed by source directivity, which is an increase or decrease in the ground motion caused by the propagation of the rupture along the fault. Ground-motion amplitudes in the forward direction of rupture propagation will be increased while those in the backward direction will be decreased due to source directivity. This effect is particularly important during unilateral faulting for sites located close to the fault (Somerville *et al.*, 1997). Source directivity has its largest effect at long periods (periods greater than about 1 sec) on the horizontal component oriented perpendicular or normal to the fault plane.

Rupture directivity is a well-known seismological principle (Lay and Wallace, 1995). It has been observed or proposed as a factor in controlling the azimuthal dependence of ground motion during the 1979 Imperial Valley earthquake (Singh, 1985), the 1980 Livermore earthquakes (Boatwright and Boore, 1982), the 1989 Loma Prieta earthquake (Campbell, 1998), and the 1992 Landers earthquake (Campbell and Bozorgnia, 1994). Somerville *et al.* (1997) and Abrahamson (2000) present general empirical models for estimating the effects of source directivity and radiation pattern on the prediction of the fault-normal and fault-parallel components of the response spectrum. Somerville *et al.* also provide a list of near-source time histories that they believe to incorporate these effects. These models, along with a graphical description of the concept of radiation pattern and source directivity, are given in the *Contemporary Guide*.

Somerville (2000) suggests that the empirical models proposed by Abrahamson and Somerville *et al.* might be too simplistic. He has found that the near-fault directivity effects observed in several subsequent large earthquakes appear to manifest themselves as narrow-band pulses whose period markedly increases with increasing magnitude. This increase in period can actually lead to lower response-spectral ordinates at mid periods ($T_n \approx 1$ sec) as earthquakes exceed a magnitude threshold of around $7^1/_4$. This observation is inconsistent with the assumptions made in the Abrahamson and Somerville *et al.* directivity models that imply that directivity effects increase monotonically with magnitude at all periods. However, the directivity pulse model needs more development before it is ready to be used in engineering practice. Until then, the empirical models of Somerville *et al.* (1997) and Abrahamson (2000) remain the state of the art in the engineering characterization of source directivity effects.

4. Hanging-Wall and Footwall Effects

Generally speaking, the hanging wall is that portion of the crust that lies above the rupture plane of a dipping fault and the footwall is that portion of the crust that lies below this plane. The exact definitions of these two crustal regimes differ depending on the particular application. The definitions used to characterize hanging-wall and footwall effects on strong ground motion are given in the *Contemporary Guide*. Somerville and Abrahamson (1995, 2000), Abrahamson and Somerville (1996), and Abrahamson and Silva (1997) found empirical evidence to indicate that sites located on the hanging wall of a reverse or thrust fault generally exhibit higher-than-average ground motion. Furthermore, Somerville and Abrahamson (1995) found that sites located on the footwall generally have lower-than-average ground motion. Although based on limited data, both hanging-wall and footwall effects were clearly demonstrated in the 1999 Chi-Chi, Taiwan, thrust earthquake (Shin and Teng, 2001). The hanging-wall effect is probably caused by a combination of radiation pattern, source directivity, and the entrapment of seismic waves within the hanging-wall wedge of the fault. The authors present a generalized empirical model for estimating the effects of the hanging wall and footwall on the estimation of strong ground motion. These models, together with a graphical description of the concept of hanging-wall and footwall effects, are given in the *Contemporary Guide*.

5. Source-to-Site Distance

Source-to-site distance, or simply distance, is used to characterize the decrease in ground motion as it propagates away from the earthquake source. Distance measures can be grouped into two broad categories depending on whether they treat the source of an earthquake as a single point (point-source measures) or as a finite rupture (finite-source measures). A brief description of the distance measures commonly used in attenuation relations is given next.

5.1 Point-Source Distance Measures

Point-source distance measures include epicentral distance (r_{epi}) and hypocentral distance (r_{hypo}). Hypocentral distance is the distance from the site to the point within the Earth where the earthquake rupture initiated (the hypocenter). Epicentral distance is the distance from the site to the point on the Earth's surface directly above the hypocenter (the epicenter). Generally speaking, r_{epi} and r_{hypo} are poor measures of distance for large earthquakes with extended ruptures. They are primarily used for characterizing small earthquakes that can be reasonably represented by a point source or for characterizing large earthquakes when the fault-rupture plane cannot be identified for past or future (design) earthquakes. Experience shows that attenuation relations that use point-source measures should not be used to estimate ground motion from large earthquakes unless there is absolutely no other means available.

5.2 Finite-Source Distance Measures

There are three finite-source distance measures that are commonly used in attenuation relations: r_{jb}, the closest horizontal distance to the vertical projection of the rupture plane, introduced by Joyner and Boore (1981); r_{rup}, the closest distance to the rupture plane, introduced by Schnabel and Seed (1973); and r_{seis}, the closest distance to the seismogenic part of the rupture plane, introduced by Campbell (1987, 2000a). The distance measure r_{seis} assumes that fault rupture within the near-surface sediments or the shallow portions of fault gouge is non-seismogenic, as suggested by Marone and Scholz (1988). These different distance measures are compared in Figure 1 of Chapter 59 by Bolt and Abrahamson.

The distance measure r_{jb} is reasonably easy to estimate for a future (design) earthquake. On the other hand, r_{rup} and r_{seis} are not as easily determined, particularly when the earthquake is not expected to rupture the entire seismogenic width of the crust. In such cases, the average depth to the top of the inferred rupture plane (d_{rup}) or to the seismogenic part of this rupture plane (d_{seis}) for an earthquake of moment magnitude M_W can be calculated from the equation (generalized from Campbell, 1997, 2000b)

$$d_i = \begin{cases} \frac{1}{2}[H_{top} + H_{bot} - W\sin(\gamma)] & \text{for } d_i \geq H_i \\ H_i & \text{otherwise} \end{cases} \quad (7)$$

where i = rup or seis, H_{bot} is the depth to the bottom of the seismogenic part of the crust, H_{top} is the depth to the top of the fault, H_{seis} is the depth to the top of the seismogenic part of the fault (to be used with the distance measure r_{seis}), γ is the fault dip; and W is the down-dip width of the fault-rupture surface. The down-dip width of the rupture plane can be calculated from the expression (Wells and Coppersmith, 1994)

$$\log W = -1.01 + 0.32M_W \quad (8)$$

where W is in km and the standard deviation of log W is 0.15.

Campbell (1997) recommends $d_{seis} \geq 3$ km, even when the fault ruptures the Earth's surface. This recommendation is based on the following factors: (1) observations of aftershock distributions and background seismicity, (2) slip distributions from earthquake modeling studies, and (3) an independent assessment of seismogenic rupture by Marone and Scholz (1988). Representative values of d_{rup} and d_{seis} are given in the *Contemporary Guide*.

6. Local Site Conditions

Local site conditions describe the materials that lie directly beneath the site. They are usually defined in terms of surface and near-surface geology, shear-wave velocity, and sediment depth. The latter two descriptions are preferred because they represent parameters that can be related directly to the dynamic response of the site profile from vertically propagating body waves or horizontally propagating surface waves. Different classifications of local site conditions used in recently published attenuation relations are defined in the *Contemporary Guide*. A general description of these classifications is presented next. A more complete description of site effects is given in Chapter 61 by Kawase.

6.1 Surface and Near-Surface Geology

Traditionally, local site conditions have been classified as either soil or rock. Many attenuation relations still use this simple classification. Campbell (1981) proposed that soil should be further subdivided into shallow soil, soft soil, Holocene or firm soil, and Pleistocene or very firm soil, and that rock should be further subdivided into soft or primarily sedimentary rock and hard or primarily crystalline rock. Although this more refined geological classification has generally not been used, Campbell and Bozorgnia (2000) empirically demonstrate the importance of this more refined classification scheme in the estimation of near-source ground motion. Park and Elrick (1998) and Wills and Silva (1998) have also shown that a more refined geological classification might be warranted based on measurements of shear-wave velocity for different geologic units in California.

6.2 Shear-Wave Velocity

There are typically two methods for classifying a site in terms of shear-wave velocity (V_S). The first is the average value of V_S in the top 30 m (100 ft) of a site profile, known as 30-meter velocity. The second is the average value of V_S over a depth equal to a quarter-wavelength of the ground-motion frequency or period of interest, known as effective velocity. A brief description of these two velocity-based site parameters is given next.

6.2.1 30-Meter Velocity

The 1997 *Uniform Building Code* (UBC), the 1997 *NEHRP* (National Earthquake Hazard Reduction Program) *Recommended Provisions for Seismic Regulations for New Buildings and Other Structures*, and the 2000 *International Building Code* (IBC) have all adopted the 30-meter velocity ($\bar{v}_s$) as the primary basis for classifying a site for purposes of incorporating local site conditions in the estimation of strong ground motion (see Chapter 68 by Borcherdt *et al.*). Seven site classes, designated S_A or A (Hard Rock) through S_F or F (Soft Soil Profile), are defined in terms of $\bar{v}_s$ (Table 1). The value of $\bar{v}_s$ is determined from the formula

$$\bar{v}_s = \frac{\sum_{i=1}^{n} d_i}{\sum_{i=1}^{n} d_i / v_{si}} \tag{9}$$

where d_i is the thickness and v_{si} is the shear-wave velocity of site layer i. The summation in the numerator must equal 100 ft (30 m).

Boore *et al.* (1993) were the first to use site categories based on $\bar{v}_s$ in the development of an attenuation relation. In 1994, these same authors were the first to use $\bar{v}_s$ directly as a parameter in an attenuation relation (this same attenuation relation was later published in Boore *et al.*, 1997). Because of its adoption by the building codes, $\bar{v}_s$ has become the preferred site parameter to use in engineering practice. According to Boore and Joyner (1997), $\bar{v}_s = 310$ m/sec and $\bar{v}_s = 620$ m/sec are reasonable estimates of 30-meter velocity for generic soil and generic rock sites in western North America. An approximate correspondence among values of $\bar{v}_s$, their related site classes, and the site parameters used in recently published attenuation relations are given in the *Contemporary Guide*.

TABLE 1 Site Classes Defined in the U.S. Building Codes

Site Class	Soil Profile Name	30-Meter Velocity, $\bar{v}_s$ m/sec (ft/sec)
S_A or A	Hard Rock	>1500 (>5000)
S_B or B	Rock	760–1500 (2500–5000)
S_C or C	Very Dense Soil and Soft Rock	360–760 (1200–2500)
S_D or D	Stiff Soil Profile	180–360 (600–1200)
S_E or E	Soft Soil Profile	<180 (<600)
S_F or F	Soil Requiring Site-Specific Evaluation	

6.2.2 Effective Velocity

Joyner *et al.* (1981) proposed a V_S-based site parameter that is related to the nonresonant amplification produced as a result of the energy conservation of seismic waves propagating vertically upward through a site profile of gradually changing velocity. This parameter, which was later referred to as effective velocity by Boore and Joyner (1991), is defined as the average velocity from the surface to a depth corresponding to a quarter-wavelength of the ground-motion period or frequency of interest. Effective velocity can be calculated from Eq. (9) by summing to a depth corresponding to a quarter-wavelength rather than to 30 m. This depth is given by the expression

$$D_{1/4}(f) = \frac{\sum_{i=1}^{n} d_i}{4f \sum_{i=1}^{n} d_i / v_{si}} \tag{10}$$

where $T = 1/f$ is the period of interest. Progressively deeper soil layers are included in the summations until the equality $\sum_{i=1}^{n} 1/v_{si} = 1/(4f)$ is met.

Effective velocity is expected to represent site response better than 30-meter velocity because of its direct relationship to the period of the ground motion. Joyner and Fumal (1984) are the only investigators to include it as a parameter in an empirical attenuation relation (this same attenuation relation was later published in Joyner and Boore, 1988). Effective velocity has found widespread use in the calculation of site response in the stochastic simulation of ground motion (see the *Contemporary Guide*).

6.3 Sediment Depth

Sediment depth, also referred to as basin depth, is the depth to basement rock beneath the site. Basement rock is a geological term that is used to describe the more resistant, generally crystalline rock beneath layers or irregular deposits of younger, relatively deformed sedimentary rock. It was introduced as a site parameter by Trifunac and Lee (1979) and later used by Campbell (1987, 2000a) to quantify the response of long-period response-spectral ordinates. All of these investigators have continued to use this parameter in their more recent studies (e.g., Trifunac and Lee, 1989, 1992; Campbell, 1997). Recently, its importance has been recognized by several seismologists (e.g., Field, 2000; Lee and Anderson, 2000). Based on empirical and theoretical considerations, Joyner (2000) found that sediment depth appeared to be a reasonable proxy for the effects of traveling surface waves generated at the edge of a sedimentary basin. Field and the SCEC Phase III Working Group (2000) found that sediment depth could be used as a proxy for the three-dimensional response of the Los Angeles basin.

7. Site Location

Attenuation relations are intended to provide estimates of strong ground motion on level ground in the free field. This means that the strong-motion recordings used to develop these relations should not be located on or near a large structure, in an area of strong topographic relief, or below the ground surface. All of these situations have been shown to significantly modify free-field ground motion in some situations (e.g., Campbell, 1986, 1987, 2000a; Stewart, 2000). To avoid these effects, most investigators exclude certain recordings from their database. However, not all of these investigators agree on which types of recordings should be excluded. Also, because the majority of the available strong-motion recordings were obtained in or near a structure, it is impossible to restrict the database to truly free-field recordings without decimating it to the point where there are an insufficient number of recordings for a robust statistical analysis.

It has been shown both empirically and theoretically that recordings obtained in a large building, especially when located on soil, or on an embedded instrument, reduce the amplitude of ground motion at high frequencies. Therefore, it is clear that such recordings should be excluded. The effects of other site locations are not so easily quantified. Stewart (2000) has proposed some quantitative criteria that will help to identify which recordings should be excluded in the future. Different data selection criteria used to develop some recently published attenuation relations are described in the references to the attenuation relations provided in the *Contemporary Guide*.

8. Tectonic Environment

The tectonic environment has a significant impact on the amplitude and attenuation of strong ground motion. Tectonic environment can be classified into four basic types as follows: (1) shallow-crustal earthquakes in active tectonic regions, (2) shallow-crustal earthquakes in stable tectonic regions, (3) intermediate-depth earthquakes (also known as Wadati-Benioff or intraslab earthquakes) within a subducting plate, and (4) earthquakes on the interface of two subducting plates. The shallow-crustal environment can be further divided into compressional and extensional stress regimes. Subduction interface earthquakes occur on the seismogenic interface between two tectonic plates, where one plate (usually oceanic crust) thrusts, or is subducted, beneath another (usually continental crust). Depending on the age of the subducting plate, this interface can occur at depths ranging anywhere from 20 to 50 km. Wadati-Benioff earthquakes occur within the subducting plate as it descends within the Earth's mantle below the subduction interface zone. Recently published attenuation relations representing these different tectonic environments are given in the *Contemporary Guide*. Moores and Twiss (1995) provide a general discussion of the different tectonic environments that are found throughout the world. Johnston (1996) gives a map that shows the location of stable continental regions throughout the world. Zoback (1992) and Chapter 34 by Zoback and Zoback provide a map that shows the distribution of compressional and extensional stress regimes throughout the world.

9. Random Error

Random error is the scatter of the observed values of ground motion about their predicted values. It is derived from Eq. (2) by performing a regression analysis and calculating the standard error of estimate of $\ln Y$ from the resulting residuals. The basic assumption in this process, which is rarely tested statistically, is that $\ln Y$ can be represented by a normal, or Gaussian, distribution (i.e., Y is lognormally distributed).

9.1 Regression Analysis

Whether developed from empirical observations or theoretical "data," all attenuation relations are derived from a statistical fitting procedure known as regression analysis. This analysis is used to determine the best estimate of the coefficients c_1 through c_8 in Eqs. (2) and (3) using a fitting procedure such as least squares or maximum likelihood. Traditionally, there have been three different methods used to perform this analysis: (1) weighted nonlinear least-squares regression, introduced by Campbell (1981); (2) two-step regression, introduced by Joyner and Boore (1981) and later refined by Joyner and Boore (1994); and (3) random-effects regression, introduced by Brillinger and Preisler (1984). Each of these methods has its strengths and weaknesses, but they all have the same goal of reducing the bias introduced by the uneven distribution of recordings with respect to magnitude, distance, and other seismological parameters. The advantage of the latter two methods is that they provide a direct estimate of the intra-earthquake and inter-earthquake components of randomness (see Section 9.2).

9.2 Standard Deviation

The difference between a ground-motion observation and its predicted value is known as a residual. The standard deviation of the residuals is called the standard error of estimate of regression and represents a measure of aleatory variability (randomness) in the estimate of the ground motion. The standard deviation of $\ln Y$ is calculated from the expression

$$\sigma_{\ln Y} = \sqrt{\frac{1}{n-p}\sum_{i=1}^{n}(\ln Y_i - \overline{\ln Y_i})^2} \tag{11}$$

where n is the number of observations, p is the number of regression coefficients, $\ln Y_i$ is the ith observation, and $\overline{\ln Y_i}$ is the predicted value of the ith observation. A plot of the residuals versus a seismological parameter can be used to identify a trend or bias in the regression model. A bias that is observed either visually or statistically using a hypothesis test can indicate a problem with the functional form of the regression model or the need to include another parameter in the model.

It is often convenient to segregate $\sigma_{\ln Y}$ into its intra-earthquake and inter-earthquake components, traditionally designated σ and τ. Abrahamson and Youngs (1992) and Joyner and Boore (1993, 1994) give specific algorithms that can be used to estimate intra-earthquake and inter-earthquake standard deviations from random-effects and two-step regression analyses, respectively. Furthermore, if the regression analysis is performed on the geometric mean of the two horizontal components of a ground-motion parameter, equivalent to the average of the logarithms of these two components, and the variability between these components is desired, an additional component-to-component random error term is needed. Referring to the three random error terms as σ_{intra}, σ_{inter} and σ_{comp}, the total standard deviation of $\ln Y$ can be given by the expression

$$\sigma_{\ln Y} = \sqrt{\sigma^2_{\text{inter}} + \sigma^2_{\text{intra}} + \sigma^2_{\text{comp}}} \tag{12}$$

The standard deviation of $\ln Y$ has been found to be a function of magnitude (e.g., Youngs *et al.*, 1995) and the amplitude of ground motion (introduced by Donovan and Bornstein, 1978, and later statistically defined by Campbell, 1997). However, not all attenuation relations take these dependencies into account in defining the value of $\sigma_{\ln Y}$. These dependencies can be significant and will always result in relatively lower estimates of $\sigma_{\ln Y}$ at larger magnitudes or higher ground-motion amplitudes. Lower estimates of standard deviation can lead to a significant reduction in deterministic estimates of ground motion at the upper percentiles of the ground-motion distribution and in probabilistic estimates of ground motion at the longer return periods of the seismic hazard curve.

9.3 Predicted Value

Because the predicted value of ground motion from Eq. (2) is the logarithm of Y, it represents the mean of $\ln Y$ or equivalently the 50th-percentile (median value) of Y. The median is the value of Y that is exceeded by 50% of the observations and is, therefore, the value of Y that is expected to have a 50–50 chance of being exceeded by a future recording. The $100(1-\alpha)$-percentile estimate of the mean of n_0 future observations of $\ln Y$ is statistically defined by the expression (Draper and Smith, 1981)

$$\ln Y_{1-\alpha} = \ln Y + t_\nu(\alpha)\sqrt{\frac{\sigma^2_{\ln Y}}{n_0} + \sigma^2_{\overline{\ln Y}}} \tag{13}$$

where $t_\nu(\alpha)$ is the Student's t-statistic for an exceedance probability of α and for $\nu = n - p$ degrees of freedom (this statistic can be found in any statistics book) and $\sigma_{\overline{\ln Y}}$ is the standard error of the mean value of $\ln Y$ excluding random error (Draper and Smith, 1981). The $100(1-\alpha)$-percentile estimate of a single future observation of $\ln Y$, the most common application of Eq. (13), is calculated by setting $n_0 = 1$.

Equation (13) is most commonly used to calculate the $100(1-\alpha)$-percentile estimate of a single future value of $\ln Y$ by setting $\sigma_{\overline{\ln Y}} = 0$ and replacing the t-statistic with the standard normal variable, giving

$$\ln Y_{1-\alpha} = \ln Y + z_\alpha \sigma_{\ln Y} \tag{14}$$

where z_α is the standard normal variable for an exceedance probability of α (this variable can be found in any statistics book). Although statistically incorrect, the use of Eq. (14) does not result in significantly different results, except when the regression model is based on very few recordings, in which case the t-statistic should be used, or when the predicted value is based on an extrapolation of the regression equation, in which case the value of $\sigma_{\overline{\ln Y}}$ cannot be neglected.

The most common application of Eq. (14) is to estimate the median or 50th-percentile value of $Y(\alpha = 0.5)$ by setting $z_\alpha = 0$, or to estimate the 84th-percentile value of $Y(\alpha = 0.16)$ by setting $z_\alpha = 1$. The 84th-percentile is often used as a conservative estimate of Y to use for the design of important facilities. Epistemic uncertainty in these estimates is usually incorporated by using more than one attenuation relation to estimate Y.

References

Abrahamson, N. A. (2000). Effects of rupture directivity on probabilistic seismic hazard analysis. In: "Proc., 6th International Conference on Seismic Zonation," Nov. 12–15, 2000, Palm Springs, CA, Proc. CD-ROM, 6 pp. Earthquake Engineering Research Institute, Oakland, CA.

Abrahamson, N. A., and K. M. Shedlock (1997). Overview. *Seismol. Res. Lett.* **68**, 9–23.

Abrahamson, N. A., and W. J. Silva (1997). Empirical response spectral attenuation relations for shallow crustal earthquakes. *Seismol. Res. Lett.* **68**, 94–127.

Abrahamson, N. A., and P. C. Somerville (1996). Effects of the hanging wall and footwall on ground motions recorded during the Northridge earthquake. *Bull. Seismol. Soc. Am.* **86**, S93–S99.

Abrahamson, N. A., and R. R. Youngs (1992). A stable algorithm for regression analyses using the random effects model. *Bull. Seismol. Soc. Am.* **82**, 505–510.

Anderson, J. G. (2000). Expected shape of regressions for ground motion parameters on rock. *Bull. Seismol. Soc. Am.* **90**, S43–S52.

Boatwright, J. A., and D. M. Boore (1982). Analysis of the ground accelerations radiated by the 1980 Livermore Valley earthquakes for directivity and dynamic source characteristics. *Bull. Seismol. Soc. Am.* **72**, 1843–1865.

Bolt, B. A. (1993). *Earthquakes and Geological Discovery*. Scientific American Library, New York.

Boore, D. M. (2001). Comparisons of ground motions from the 1999 Chi-Chi earthquake with empirical predictions largely based on data from California. *Bull. Seismol. Soc. Am.* **91**, 1212–1217.

Boore, D. M. (2002). Prediction of ground motion using the stochastic method. *Pure Appl. Geophys.* (in press).

Boore, D. M., and G. M. Atkinson (1987). Stochastic prediction of ground motion and spectral response parameters at hard-rock sites in eastern North America. *Bull. Seismol. Soc. Am.* **77**, 440–467.

Boore, D. M., and W. B. Joyner (1991). Estimation of ground motion at deep-soil sites in eastern North America. *Bull. Seismol. Soc. Am.* **81**, 2167–2185.

Boore, D. M., and W. B. Joyner (1997). Site amplification for generic rock sites. *Bull. Seismol. Soc. Am.* **87**, 327–341.

Boore, D. M., W. B. Joyner, and T. E. Fumal (1993). Estimation of response spectra and peak accelerations from western North American earthquakes: An interim report. *Open-File Report 93-509, U.S. Geol. Surv.*, Menlo Park, California.

Boore, D. M., W. B. Joyner, and T. E. Fumal (1994). Estimation of response spectra and peak accelerations from western North American earthquakes: An interim report, part 2. *Open-File Report 94-127, U.S. Geol. Surv.*, Menlo Park, California.

Boore, D. M., W. B. Joyner, and T. E. Fumal (1997). Equations for estimating horizontal response spectra and peak acceleration from western North American earthquakes: a summary of recent work. *Seismol. Res. Lett.* **68**, 128–153.

Brillinger, D. R., and H. K. Preisler (1984). An exploratory analysis of the Joyner-Boore attenuation data. *Bull. Seismol. Soc. Am.* **74**, 1441–1450.

Brune, J. N. (1999). Precarious rocks along the Mojave section of the San Andreas fault, California: constraints on ground motion from great earthquakes. *Seismol. Res. Lett.* **70**, 29–33.

Brune, J. N. (2000). Precarious rock evidence for low ground shaking on the footwall of major normal faults. *Bull. Seismol. Soc. Am.* **90**, 1107–1112.

Campbell, K. W. (1981). Near-source attenuation of peak horizontal acceleration. *Bull. Seismol. Soc. Am.* **71**, 2039–2070.

Campbell, K. W. (1986). Empirical prediction of free-field ground motion using statistical regression models. In: "Proc., Soil Structure Interaction Workshop," NUREG/CR-0054, pp. 72–115. U.S. Nuclear Regulatory Commission, Washington, D.C.

Campbell, K. W. (1987). Predicting strong ground motion in Utah. In: "Assessment of Regional Earthquake Hazards and Risk Along the Wasatch Front, Utah" (P. L. Gori and W. W. Hays, Eds.), *U.S. Geol. Surv. Open-File Report 87-585*, Vol. II, pp. L1–L90, Reston, Virginia.

Campbell, K. W. (1997). Empirical near-source attenuation relationships for horizontal and vertical components of peak ground acceleration, peak ground velocity, and pseudo-absolute acceleration response spectra. *Seismol. Res. Lett.* **68**, 154–179.

Campbell, K. W. (1998). Empirical analysis of peak horizontal acceleration, peak horizontal velocity, and modified Mercalli intensity. In: "The Loma Prieta, California, Earthquake of October 17, 1989—Earth Structures and Engineering Characterization of Ground Motion" (T. L. Holzer, Ed.), *U.S. Geol. Surv., Profess. Paper 1552-D*, pp. D47–D68.

Campbell, K. W. (2000a). Predicting strong ground motion in Utah. In: "Assessment of Regional Earthquake Hazards and Risk Along the Wasatch Front, Utah" (P. L. Gori and W. W. Hays, Eds.), *U.S. Geol. Surv., Profess. Paper 1500-L*, pp. L1–L31.

Campbell, K. W. (2000b). Erratum: empirical near-source attenuation relationships for horizontal and vertical components of peak ground acceleration, peak ground velocity, and pseudo-absolute acceleration response spectra. *Seismol. Res. Lett.* **71**, 353–355.

Campbell, K. W. (2002). Engineering seismology. In: "Encyclopedia of Physical Science and Technology," 3rd ed., Academic Press, San Diego, Vol. 14, pp. 531–545.

Campbell, K. W., and Y. Bozorgnia (1994). Empirical analysis of strong ground motion from the 1992 Landers, California, earthquake. *Bull. Seismol. Soc. Am.* **84**, 573–588.

Campbell, K. W., and Y. Bozorgnia (2000). New empirical models for predicting near-source horizontal, vertical, and V/H response spectra: Implications for design. In: "Proc., 6th International Conference on Seismic Zonation," Nov. 12–15, 2000, Palm Springs, CA, Proc. CD-ROM, 6 pp. Earthquake Engineering Research Institute, Oakland, CA.

Donovan, N. C., and A. E. Bornstein (1978). Uncertainties in seismic risk procedures. *J. Geotech. Eng. Div.*, ASCE **104**, 869–887.

Douglas, J. (2002). Earthquake ground motion estimation using strong-motion records: a review of equations for the estimation of peak ground acceleration and response spectral ordinates. *Earth-Science Reviews* (submitted).

Draper, N. R., and H. Smith (1981). *Applied Regression Analysis*, 2nd ed. John Wiley and Sons, New York.

Field, E. H. (2000). A modified ground-motion attenuation relationship for southern California that accounts for detailed site classification and a basin-depth effect. *Bull. Seismol. Soc. Am.* **90**, S209–S221.

Field, E. H. and the SCEC Phase III Working Group (2000). Accounting for site-effects in probabilistic seismic hazard analyses of Southern California: Overview of the SCEC Phase III Report. *Bull. Seismo. Soc. Am.* **90**, S1–S31.

Gupta, A. K. (1993). *Response Spectrum Method in Seismic Analysis and Design.* CRC Press, Boca Raton, FL.

Hanks, T. C., and H. Kanamori (1979). A moment-magnitude scale. *J. Geophys. Res.* **84**, 2348–2350.

Heaton, T. H., F. Tajima, and A. W. Mori (1986). Estimating ground motions using recorded accelerograms. *Surv. in Geophys.* **8**, 25–83.

Johnston, A. C. (1996). Seismic moment assessment of earthquakes in stable continental regions, I. Instrumental seismicity. *Geophys. J. Int.* **124**, 381–414.

Joyner, W. B. (2000). Strong motion from surface waves in deep sedimentary basins. *Bull. Seismol. Soc. Am.* **90**, S95–S112.

Joyner, W. B., and D. M. Boore (1981). Peak horizontal acceleration and velocity from strong-motion records including records from the 1979 Imperial Valley, California, earthquake. *Bull. Seismol. Soc. Am.* **71**, 2011–2038.

Joyner, W. B., and D. M. Boore (1988). Measurement, characterization, and prediction of strong ground motion. In: "Proc., Conference on Earthquake Engineering and Soil Dynamics II—Recent Advances in Ground-Motion Evaluation" (J. L. Von Thun, Ed.), June 27–30, 1988, Park City, UT, Geot. Spec. Pub. No. 20, pp. 43–102. American Society of Civil Engineers, New York.

Joyner, W. B., and D. M. Boore (1993). Methods for regression analysis of strong-motion data. *Bull. Seismol. Soc. Am.* **83**, 469–487.

Joyner, W. B., and D. M. Boore (1994). Errata: methods for regression analysis of strong-motion data. *Bull. Seismol. Soc. Am.* **84**, 955–956.

Joyner, W. B., and T. E. Fumal (1984). Use of measured shear-wave velocity for predicting geologic site effects on strong ground motion. In: "Proc., 8th World Conference on Earthquake Engineering," Vol. 2, pp. 777–783.

Joyner, W. B., R. E. Warrick, and T. E. Fumal (1981). The effect of Quaternary alluvium on strong ground motion in the Coyote Lake, California, earthquake of 1979. *Bull. Seismol. Soc. Am.* **71**, 1333–1349.

Kanamori, H. (1978). Quantification of earthquakes. *Nature* **271**, 411–414.

Lay, T., and T. C. Wallace (1995). *Modern Global Seismology*. Academic Press, San Diego, CA.

Lee, Y., and J. G. Anderson (2000). Potential for improving ground-motion relations in Southern California by incorporating various site parameters. *Bull. Seismol. Soc. Am.* **90**, S170–S186.

Leyendecker, E. V., J. R. Hunt, A. D. Frankel, and, K. S. Rukstales (2000). Development of maximum considered earthquake ground motion maps. *Earthquake Spectra* **16**, 21–40.

Marone, C., and C. H. Scholz (1988). The depth of seismic faulting and the upper transition from stable to unstable slip regimes. *Geophys. Res. Lett.* **15**, 621–624.

McGarr, A. (1984). Scaling of ground motion parameters, state of stress, and focal depth. *J. Geophys. Res.* **89**, 6969–6979.

Moores, E. M., and R. J. Twiss (1995). *Tectonics*. W.H. Freeman and Company, New York.

Newmark, N. M., and W. J. Hall (1982). *Earthquake Spectra and Design*. Earthquake Engineering Research Institute, Berkeley, CA.

Park, S., and S. Elrick (1988). Predictions of shear-wave velocities in southern California using surface geology. *Bull. Seismol. Soc. Am.* **88**, 677–685.

Richter, C. F. (1935). An instrumental earthquake magnitude scale. *Bull. Seismol. Soc. Am.* **25**, 1–32.

Schnabel, P. B., and H. B. Seed (1973). Accelerations in rock for earthquakes in the western United States. *Bull. Seismol. Soc. Am.* **63**, 501–516.

Shin, T. C., and T. L. Teng (2001). An overview of the 1999 Chi-Chi, Taiwan, earthquake. *Bull. Seismol. Soc. Am.* **91**, 895–913.

Singh, J. P. (1985). Earthquake ground motions: implications for designing structures and reconciling structural damage. *Earthquake Spectra* **1**, 239–270.

Somerville, P. (2000). New developments in seismic hazard estimation. In: "Proc., 6th International Conference on Seismic Zonation," Nov. 12–15, 2000, Palm Springs, CA, Proc. CD-ROM, 25 pp. Earthquake Engineering Research Institute, Oakland, CA.

Somerville, P., and N. Abrahamson (1995). Ground motion prediction for thrust earthquakes. In: "Proc., SMIP95 Seminar on Seismological and Engineering Implications of Recent Strong-Motion Data," May 16, 1995, San Francisco, pp. 11–23. California Strong Motion Instrumentation Program, Sacramento, CA.

Somerville, P., and N. Abrahamson (2000). Prediction of ground motions for thrust earthquakes. *Data Utilization Rept. No. CSMIP/00-01 (OSMS 00-03)*, California Strong Motion Instrumentation Program, Sacramento, CA.

Somerville, P. G., N. F. Smith, R. W. Graves, and N. A. Abrahamson (1997). Modification of empirical strong ground motion attenuation relations to include the amplitude and duration effects of rupture directivity. *Seismol. Res. Lett.* **68**, 199–222.

Spudich, P., W. B. Joyner, A. G. Lindh, D. M. Boore, B. M. Margaris, and J. B. Fletcher (1999). SEA99: a revised ground motion prediction relation for use in extensional tectonic regimes. *Bull. Seismol. Soc. Am.* **89**, 1156–1170.

Stewart, J. P. (2000). Variations between foundation-level and free-field earthquake ground motions. *Earthquake Spectra* **16**, 511–532.

Trifunac, M. D., and V. W. Lee (1979). Dependence of pseudo-relative velocity spectra of strong motion acceleration on the depth of sedimentary deposits. *Report No. CE79-02*, Dept. of Civil Engineering, University of Southern California, Los Angeles.

Trifunac, M. D., and V. W. Lee (1989). Empirical models for scaling pseudo relative velocity spectra of strong earthquake accelerations in terms of magnitude, distance, site intensity and recording site conditions. *Soil Dyn. and Earthq. Eng.* **8**, 126–144.

Trifunac, M. D., and V. W. Lee (1992). A note on scaling peak acceleration, velocity and displacement of strong earthquake shaking by Modified Mercalli Intensity (MMI) and site soil and geologic conditions. *Soil Dyn. and Earthq. Eng.* **11**, 101–110.

Tsai, Y. B., and M. W. Huang (2000). Strong ground motion characteristics of the Chi-Chi, Taiwan earthquake of September 21, 1999. *Earthq. Eng. and Eng. Seismol.* **2**, 1–21.

Wells, D. L., and K. J. Coppersmith (1994). New empirical relationships among magnitude, rupture length, rupture width, rupture area, and surface displacement. *Bull. Seismol. Soc. Am.* **84**, 974–1002.

Wills, C. J., and W. Silva (1998). Shear-wave velocity characteristics of geologic units in California. *Earthquake Spectra* **14**, 533–556.

Youngs, R. R., N. A. Abrahamson, F. Makdisi, and K. Sadigh (1995). Magnitude dependent dispersion in peak ground acceleration. *Bull. Seismol. Soc. Am.* **85**, 1161–1176.

Zoback, M. L. (1992). First- and second-order patterns of stress in the lithosphere: the world stress map project. *J. Geophys. Res.* **97**, 11703–11728.

Editor's Note:

Due to space limitations, a supplemental contemporary guide to strong-motion attenuation relations is included as a computer-readable file on the attached Handbook CD under the directory \60Campbell. Also included in this directory is a supplemental electronic workbook that can be used to evaluate the attenuation relations and other engineering models that are presented in this *Contemporary Guide*.

61

Site Effects on Strong Ground Motions

Hiroshi Kawase

Graduate School of Human Environment Studies, Kyushu University, Fukuoka, Japan

1. Introduction

Site effects play a very important role in characterizing seismic ground motions because they may strongly amplify (or deamplify) seismic motions at the last moment just before reaching the surface of the ground or the basement of man-made structures. Because of the high level of amplification caused by site effects, which can be almost two orders of magnitude, we cannot neglect them in engineering practice. The purpose of this chapter is to review the essential aspects of site effects on strong ground motions.

For much of the history of seismological research, site effects or the effects of surface geology have received much less attention than they should, with the exception of Japan, where they have been well recognized through pioneering works by Sezawa and Ishimoto as early as the 1930s (Kawase and Aki, 1989). A typical example can be found in a paper by Hudson (1972), who found no strong correlation between the strength of observed ground motion and the surface geology of the stations for the San Fernando earthquake dataset. The situation was drastically changed by the catastrophic disaster in Mexico City during the Michoacan, Mexico earthquake of 1985, in which strong amplification due to extremely soft clay layers caused many high-rise buildings to collapse despite their long distance from the source. The real cause of the observed long duration of shaking during the earthquake is not well resolved yet even though considerable research has been conducted since then (e.g., Bard *et al.*, 1988; Kawase and Aki, 1989; Singh and Ordaz, 1993; Chavez-Garcia and Bard, 1994; Furumura and Kennett, 1998). However, there is no room for doubt that the primary cause of the large amplitude of strong motions in the soft soil (lakebed) zone relative to those in the hill zone is a simple one-dimensional (1-D) site effect of these soft layers.

The time was right when the joint working group of IASPEI and IAEE on the effects of surface geology (ESG) on strong motions was formed in 1985 as an international task force for promoting and coordinating research in this field (Kudo *et al.*, 1992). As a result of their activity, international symposia have been held twice, both in Japan, in 1992 and 1998, where many ongoing activities have been reported. Interested readers should refer to the proceedings of these symposia (JWG-ESG, 1992; Irikura *et al*, 1998).

There are plenty of ways to estimate site effects. The simplest way is to characterize them in terms of soil-type classification. Such soil-type specific amplification factors were implemented in the first version of the Japanese building code as early as the 1950s. Problems associated with such an idea are discussed in detail by Aki (1988), who concluded that the conventional broad classification of soil types is not effective for characterizing site effects. Because site amplification factors are strongly frequency and site dependent, any averaged values for different sites with the same site category yield relatively small and flat frequency characteristics, which is far from the reality at any sites in that category. Unfortunately, the soil-type approach continues to be a favored way to characterize site effects. Because our space is limited and further encouragement of the soil-type approach is not appropriate, we skip a review of studies for that approach.

Another empirical approach is to obtain site amplification factors in the frequency domain directly from the observed records. The observed ground motion itself is the final product of source, path, and site effects, so we need a way to extract only the site effects from the data. A basic but effective approach is to take spectral ratios of two adjacent records with different soil conditions. An ideal case is a free-field site on an intact hard rock outcrop next to a site on soft sedimentary layers. In reality, it is hard to find a site-effect-free reference site either because

INTERNATIONAL HANDBOOK OF EARTHQUAKE AND ENGINEERING SEISMOLOGY, VOLUME 81B ISBN: 0-12-440658-0

topographic irregularities around the site or weathered layers below the site yield strong site effects or because such a reference site is situated far away from the target site.

If we install borehole stations just beneath the surface station, i.e., conduct downhole measurements, then we can directly observe wave propagation phenomena between these stations. We should note, however, that a borehole seismogram inside the bedrock is not the same as an outcrop rock motion. In the borehole station we observe both incident waves coming up from below and reflected waves going back from the surface. Spectral ratios of surface records with respect to borehole records have predominant peaks either due to the amplification of surface soils or due to the interference (i.e., cancellation) of incident and reflected waves. Another drawback of borehole measurements is their high cost compared to a surface implementation.

Another recent innovation for finding site characteristics is to take ratios of horizontal to vertical components (H/V). We only need one three-component station on the surface. However, the physical meaning of the ratio is not so easy to interpret. And neither borehole ratios nor H/V ratios directly represent the real soil amplification relative to the unit input, so we also need to simulate these observed spectral ratios through physical modeling.

Physical modeling of the ground as a medium of seismic wave propagation from the source to the receiver or from the reference position to the receiver is the most rigorous approach to characterizing site effects. By modeling the ground theoretically, we can simulate ground motions for any arbitrary source of vibration. If the incident wave is so intense that soft-soil sediments go to the nonlinear regime, then we can introduce a nonlinear constitutive relationship in the physical model. If the layered structure is very complicated and has lens-like structures here and there, then we can construct irregular interfaces between layers to reproduce wave propagation and scattering through these interfaces. Once we establish an appropriate physical model around a target site, any kind of strong-motion prediction becomes possible.

The down side of this approach is that we must invest time and effort to characterize and model the ground, to develop theoretical or numerical methodologies, and to collect observed data. Once we have an initial model of the ground, an initial technique to solve the problem, and an initial database of observed data, we can calibrate our model and method to the observed data. If the initial matching with data is not satisfactory, then we will upgrade either our model or our method, or both. We may need additional data to constrain the model more strongly. The importance of such ongoing activity in the model-oriented approach is emphasized by Kawase (1993), who characterized the ground modeling, numerical technique development, and strong-motion observation as a triadic structure of the site-effect studies. We would like to emphasize here the importance of the calibration phase in the physical modeling approach because any results from modeling are just hypothesized ones if not calibrated. In this chapter we devote ourselves mainly to this modeling approach because this is the most efficient way to quantitatively characterize site effects in the long run.

2. Physical Modeling of Site Effects

When we model physical behavior of the actual ground by a seismic disturbance, we must introduce some kind of simplification because the actual ground is extremely complex in a three-dimensional manner. The simplest model of the ground is a homogeneous full-space, in which there is no site effect because there are no boundaries and no interfaces. The second simplest model is a homogeneous half-space, in which interaction with reflected waves may give very conspicuous amplification in a critical incidence (Aki, 1988) and Rayleigh waves may transport a significant part of wave energy for a shallow source. Although theoretically interesting, these models are too simple to be meaningful models for site-effect studies. We must at least take into account the impedance contrast due to younger sediments overlying older bedrock. When we assume a horizontally homogeneous but *vertically* varying medium, we can call it a one-dimensional (1-D) medium. Literally speaking, we could also consider a 1-D medium varying only in one *horizontal* axis, as is the case of a fault gouge deep inside the Earth. However, for most of the cases in site-effect studies a 1-D medium means a vertically varying structure.

If we introduce another variation in one horizontal axis, then it will be a two-dimensional (2-D) medium. Naturally, if we have variations in both horizontal directions then it will be a three-dimensional (3-D) medium. As the number of varying dimensions increases, the degree of complexity in wave propagation increases and so does the difficulty of modeling and calculation. We will start from 1-D modeling of site effects, including theoretical and observational aspects and a brief look at nonlinear problems. Then we review 2-D/3-D modeling in which we see several interesting phenomena due to constructive (and destructive) interference by different types of waves.

If the spatial variation is only on the surface of a homogeneous medium, the effects of such a surface irregularity, often called a topographic irregularity, should be much simpler than the variation inside the medium. In the last section we briefly describe topographic effects.

2.1 One-D Model

2.1.1 Amplification Factor

A very simple yet quite meaningful physical model of the ground for characterizing site effects is a single layer over a half-space extending infinitely in the horizontal direction. The frequency characteristics of the amplification factor for this 1-D two-layered model can be obtained by solving a simple wave equation of sinusoidal ($e^{i\omega t}$) input. In the case of vertical incidence of body waves, either S waves or P waves with propagation speeds of β_0 in the half-space and β_1 in the layer, to the layer with the thickness of h, we have the well-known amplification factor as

$$|U_1(\omega)| = 2.0\{\cos^2(\kappa_1 h) + \gamma^2 \sin^2(\kappa_1 h)\}^{-1/2} \quad (1)$$

where $\kappa_1 = \omega/\beta_1$ is the wavenumber in the surface layer and $\gamma = \rho_1\beta_1/\rho_0\beta_0$ is the impedance contrast of two media in which ρ_0 and ρ_1 are the densities of half-space and the surface layer. If the angular frequency $\omega = 2\pi f$ becomes zero, then Eq. (1) yields a numerical value of 2.0, the amplification due to the free surface of a half-space. When we compare the observed spectral ratios with respect to the outcrop motion, then we need to normalize $|U_1(w)|$ by 2.0. We should note that at the resonant frequencies where $\kappa_1 h = \pi(2n+1)/2\ (n = 0, 1, 2, \ldots)$ the maximum amplification factor is equal to twice the inverse of the impedance contrast, $2\gamma^{-1}$.

If we calculate the amplification factor of the surface with respect to the ground motion at the bottom of the surface layer, i.e., borehole at the depth of h, then

$$|U_1(\omega)/U_0(\omega)| = \{\cos(\kappa_1 h)\}^{-1} \quad (2)$$

The amplification factor is independent of the impedance contrast and becomes infinite at the same resonant frequencies, i.e., $\kappa_1 h = \pi(2n+1)/2\ (n = 0, 1, 2, \ldots)$. This is due to perfect cancellation of incident and reflected waves at the borehole level. Because it does not depend on the impedance contrast of the surface layer, even a half-space gives exactly the same amplification factor as any two-layered structure. We should also note that if the impedance contrast is zero, Eq. (1) becomes Eq. (2) except for the factor of 2.0. This means that the amplification factor relative to the borehole motion corresponds to the amplification factor of a structure with a rigid basement. In reality, we have intrinsic as well as scattering attenuation in the media so that the amplification factor always remains finite. Also, layering within surface sediments makes the amplification factor a function of impedance and thickness of each layer even for the surface/borehole spectral ratios. A complete calculation for any multiple-layered structures for arbitrary incidence of body wave is quite easy using the Thomson-Haskell matrix (e.g., Haskell, 1960) or the Propagator matrix (e.g., Aki and Richards, 1980).

In the case of a borehole measurement, we usually have an advantage for modeling as a by-product. That is, we usually conduct geotechnical as well as geophysical exploration when we excavate a hole before installing the instrument. This information is invaluable because it is an in-situ measurement independent from seismic observation. In particular, an *S*-wave velocity profile obtained by either *P-S* logging or suspension logging is important because the *S* wave usually has larger amplitude as input, stronger contrast between layers, and therefore, greater effect on the amplification factor. Once we obtain a 1-D structure by such an exploration, we can calculate site amplification factors using it as an initial model. Usually a 1-D model can give quite satisfactory results on average, but it is difficult to have exact matching on all the peak frequencies observed. Before introducing more complicated 2-D/3-D modeling, we should refine and improve our 1-D model as much as possible. Damping factors, or Q values as an inverse of damping, of surface layers are especially difficult to measure independently so that the calibration procedure is almost the only way to get their realistic values. Here we would like to show an example of such a procedure to get realistic 1-D models for several borehole stations.

2.1.2 Simulation

In the paper by Satoh *et al.* (1995a), the primary goal of the analysis is to obtain the so-called "engineering bedrock waves" from observed ground motions recorded in the borehole with various types of soil conditions. The definition of the engineering bedrock can depend on the purpose of the study, but usually it is Tertiary rock or hard Pleistocene commonly found in the studied region with some constraints on its geotechnical and geophysical properties. Satoh *et al.* (1995a) defined their engineering bedrock as the layer of sedimentary rock of Pliocene or earlier age with an *S*-wave velocity of 500 m/sec or larger and a SPT (Standard Penetration Test) blow count of 50 or more. From an engineering point of view, it is desirable to predict ground motions directly on the engineering bedrock on which foundation systems of important buildings and civil engineering structures will be constructed. Satoh *et al.*'s procedure for obtaining the engineering bedrock waves is as follows. First they identified a 1-D soil structure for each site using spectral ratios between surface and borehole records observed at each site among eighteen small- to moderate-sized ($3.4 \leq Mjma \leq 7.1$) earthquakes. Then they calculated the theoretical 1-D transfer function of the ground motion at the borehole level to the input motion at the engineering bedrock level and deconvolved it from the observed borehole record.

Data used were recorded by twelve borehole stations deployed in Sendai City, a cultural center of the Tohoku district, Japan, as shown in Figure 1. Among the twelve stations, two are situated on outcrops of the engineering bedrock so that the remaining ten stations are the targets of deconvolution. The initial *S*-wave velocity model for each site was developed from the *P-S* logging profile. The initial damping factors are determined to be $0.1f^{-0.5}$ for softer sites or $0.3f^{-0.5}$ for harder sites through several preliminary tests. Depths of the ten boreholes range from 27 to 81 meters.

The scheme of inversion for an optimal *S*-wave velocity and damping structure is a modified quasi-Newton method for the residual between a theoretical spectral amplification factor and an observed one. Layer properties between surface and borehole are identified and the remaining structure is assumed to be the same as the *P-S* logging. In their paper, the authors performed the inversion analyses for each horizontal component for each earthquake independently. This yields the best structure for each observed ratio and, hence, the most site-effect-free engineering bedrock wave. In most of the previous studies based

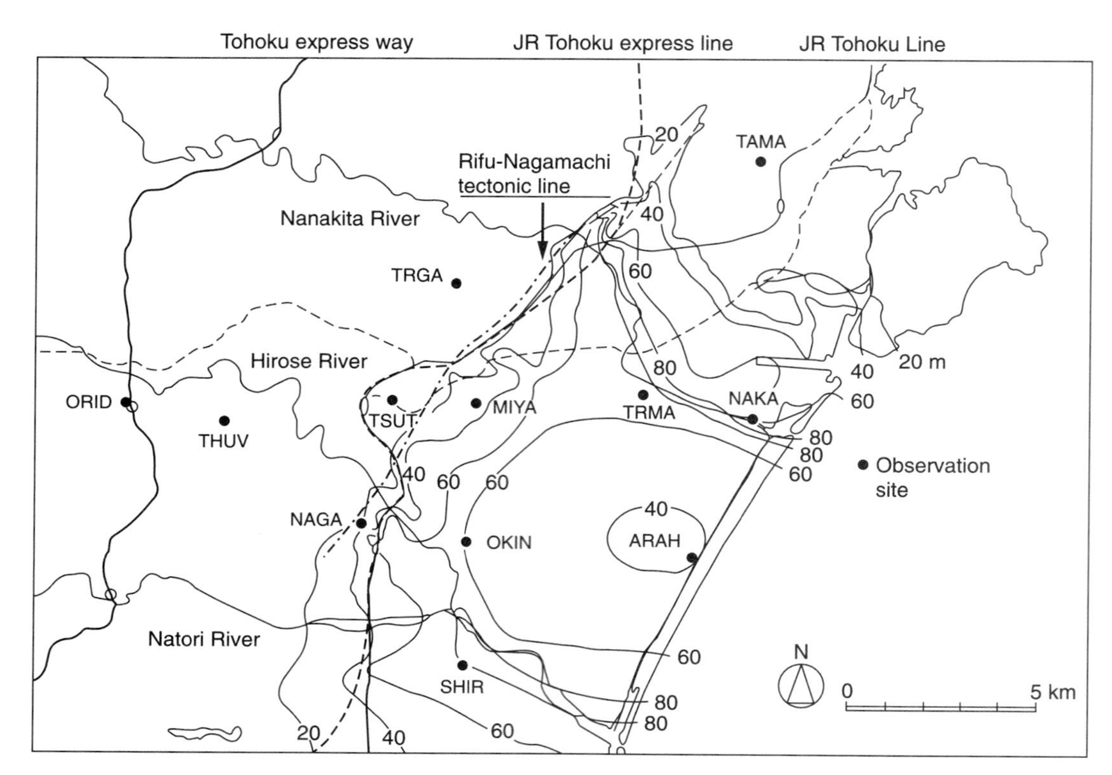

FIGURE 1 Twelve borehole stations deployed in Sendai City, a cultural center of the Tohoku district, Japan, used in Satoh *et al.* (1995a).

on the borehole data, the average of the observed ratios was determined first and then used as a target of inversion. If we use an averaged amplification ratio, the inverted structure may be not an averaged structure of the real ground but a smoothed one because of the stochastic nature of observed spectral ratios. This is especially true for the damping factor, which is very sensitive to peak values. Once we average observed ratios for different components for different earthquakes, the peak values are then smeared out so that the damping tends to be overestimated. If spectral ratios at a site are not stable and need to be averaged, then it means that the site effect there is not simple enough to be characterized by a 1-D model.

Figure 2 shows examples of the spectral ratios of all ten stations by one earthquake, EQ9101, for the EW component. A vertical line in each figure is the upper frequency limit for which the optimization is enforced. The matching between data and theory is quite good up to this high-frequency limit, especially at softer sites such as NAKA, OKIN, SHIR, and ARAH. Note that the peak amplification factors reach more than 10 primarily because these factors are of surface/borehole factors as in Eq. (2). In Figure 3, the authors compare the observed spectral ratios at NAKA and SHIR with two theoretical ones, namely, the optimal structures determined as previously mentioned and the predefined ones using the *P*-*S* logging velocities and a frequency-independent (2%) damping factor. At SHIR peak frequency shift between the observed ratio and that of the predefined model is apparent. Although the difference is small, this prompts us to refine the *S*-wave velocity structure to get the best matching. The differences of *S*-wave velocities between the initial models (=*P*-*S* logging values) and the optimal ones are about 10% in softer (≤200 m/sec) layers and less than that in harder layers. It is interesting to note that for most of the layers, the optimal values are smaller than the *P*-*S* logging values. Figure 3 also shows the importance of the frequency dependency in damping factors to explain spectral ratios in a wider frequency range. The obtained frequency dependency in damping factors (as power of frequency) range from 0.46 to 1.15, mostly centering around 0.8. The frequency dependency is not negligible because the spectral ratio between surface and borehole is controlled by the damping factor of soil layers between them.

Figure 4 shows the final spectral ratios at twelve stations with respect to the engineering bedrock waves at one station, ORID, averaged for all the data analyzed. Thick solid lines represent ratios of engineering bedrock waves, while thin solid lines represent those of surface records, and thin dotted lines represent the borehole records. The engineering bedrock waves are running between these observed data and are close to unity. We should note that the borehole records show quite deep troughs, which is exactly as expected from Eq. (2). Amplification higher than the lowest frequency of such troughs, shown by solid triangles, should be real amplification due to shallow soil layers, while amplification or deamplification lower than that, such as those shown by small arrows, should be due to a deep basin structure below the engineering bedrock.

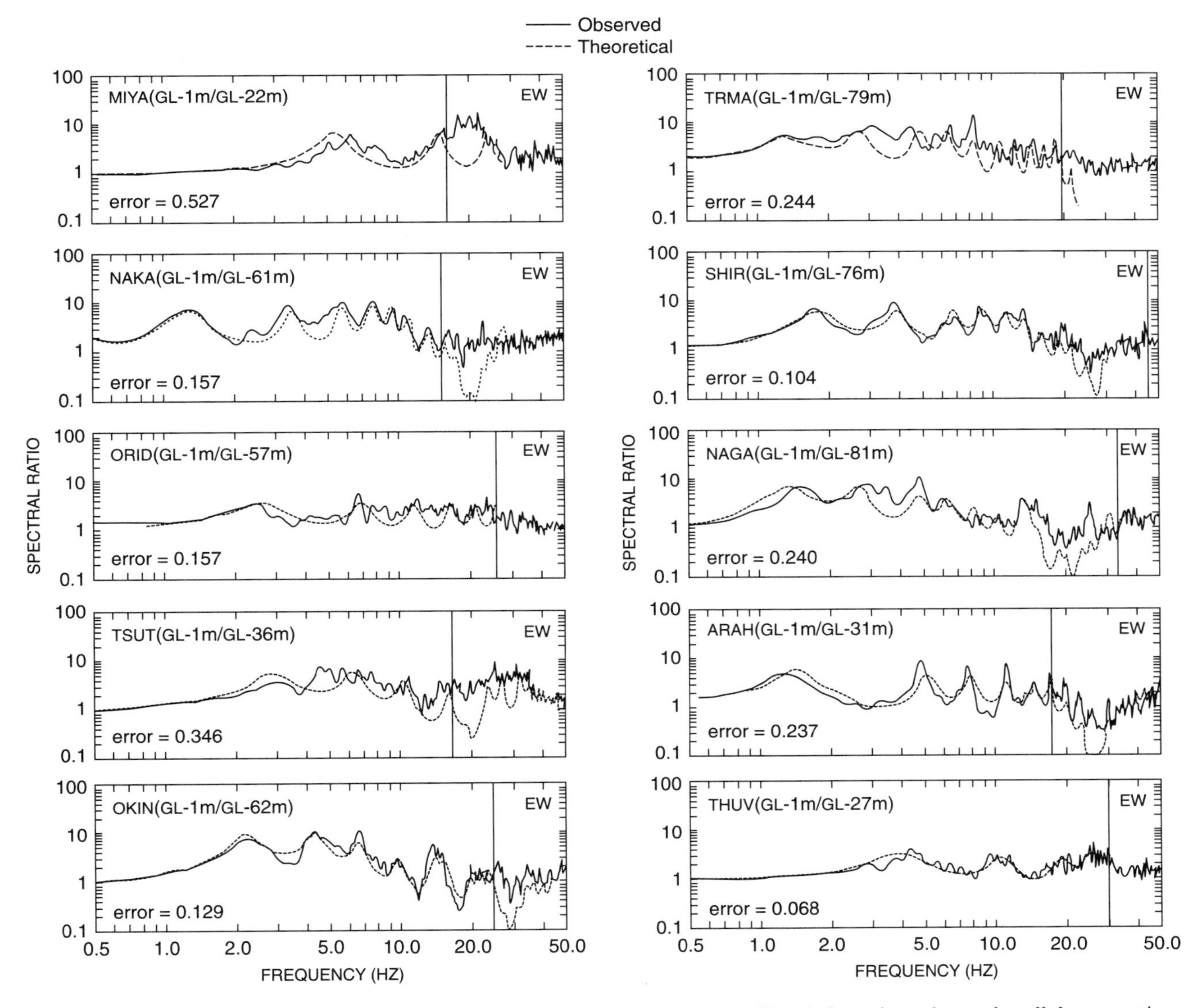

FIGURE 2 Examples of the spectral ratios of EW component between surface and borehole stations observed at all the ten stations during an earthquake of M_{JMA}4.7, EQ9101 (Satoh *et al.*, 1995a).

In conclusion, Satoh *et al.* (1995a) show that 1-D structures are capable of representing the basic characteristics of the site amplification due to shallow sedimentary layers if *S*-wave velocities and damping factors are properly assigned. We should note, however, that the assumed physical models in this study are considered to be equivalent ones, which are determined to reproduce the observed spectral ratios. For example, damping factors obtained here range from 10% to 50% at 1 Hz, which is quite large as a value of intrinsic attenuation of normal soft soil. These values should be the result of many plausible factors not considered in the simple 1-D model used, such as, 2-D/3-D basin effects, stochastic nature of incident waves, incidence angle rotation, different types of incident waves, and so on.

The effectiveness of such 1-D modeling in reproducing spectral ratios between observed surface and borehole records has also been reported by others such as Takahashi *et al.* (1988), Seale and Archuleta (1989), Kinoshita (1992), Kobayashi *et al.* (1992), Satoh *et al.* (1995b), and numbers of papers collected in the proceedings of the second International Symposium on the Effects of Surface Geology (Irikura *et al.*, 1998). In these proceedings, a good review by Archuleta (1998) on the recent borehole studies in the United States can be found.

2.1.3 Nonlinearity

If an input motion to soft-surface layers becomes so intense that the shear stain built up inside the layer reaches a certain threshold, then soil behaves nonlinearly. Soil nonlinearity is characterized by reduction of shear rigidity and, hence, reduction of shear wave velocity, and increase of damping factor. In terms of

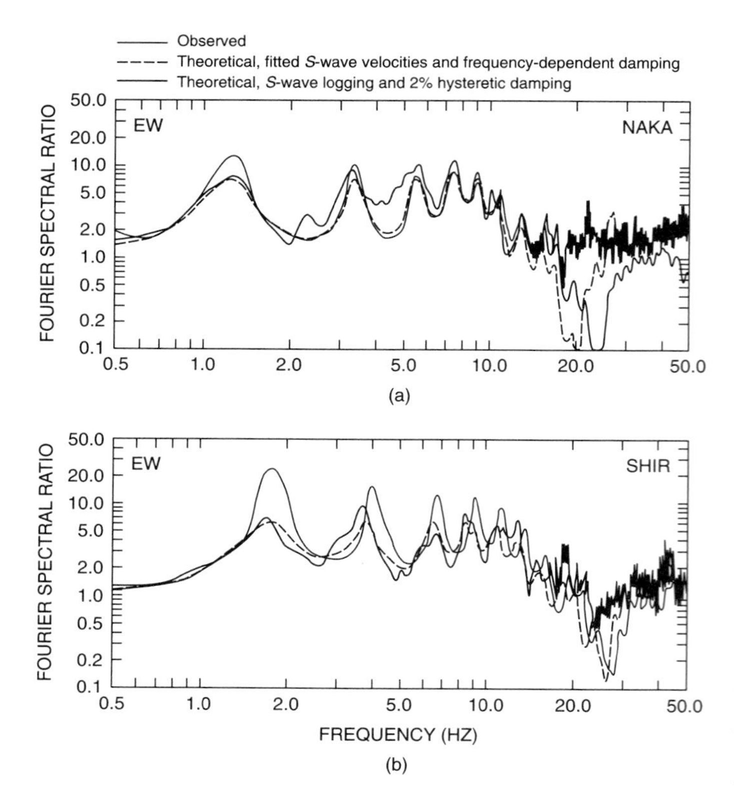

FIGURE 3 Comparisons of the observed spectral ratios at NAKA and SHIR (thick solid lines) with two theoretical ones, namely, the optimal structures determined by the inversion (thick broken lines) and the predefined ones using the *P-S* logging velocities and a frequency-independent damping factor of 2% (thin solid lines). Reproduced from Satoh *et al.* (1995a).

site effects, this results in prolongation of the predominant periods and reduction of amplification factors. In the geotechnical engineering field, nonlinearity in the soil as a microscopic (on the order of centimeters) level has been real and obvious based on laboratory experiments. Using a shear rigidity reduction factor and a damping increasing function in relation to the shear strain established by such laboratory experiments, Schnabel *et al.* (1972) proposed and distributed the famous SHAKE computer program. SHAKE combines a multiple 1-D wave propagation theory with these relationships as an equivalent linear method. In the equivalent linear method, iterative calculations are performed until convergence of the material properties used in the former calculation and the material properties in relation to the calculated strain is achieved. In essence, the equivalent linear method works if we focus our attention to only the major part of the seismogram with maximum strain.

On the other hand, the seismological community had been reluctant to accept the concept that nonlinearity was happening in a pervasive manner until Chin and Aki (1992) proposed a seismological method for finding evidence of nonlinearity in the observed strong-motion data. In similar work, Field *et al.* (1998) compared relative site responses for the mainshock and aftershocks of the Northridge earthquake sequence and reported a large difference between them. Although the data in these studies are extensive, they are indirect evidence of nonlinearity.

In Japan, borehole measurements are quite common and much strong-motion data has been observed in boreholes embedded in soft sediments. One example is a paper by Satoh *et al.* (1995b) in which the authors inverted soil properties between borehole and surface sensors and found quite good matching between the relationships of the calculated shear rigidity and damping factors to the shear strain and those of laboratory testing. Shear rigidities obtained by the equivalent linear method and those obtained by the inversion also coincide with each other, and both methods explain observed surface seismograms much better than the linear model determined by the inversion for a low strain input. Even for quite a strong level of input, such a scheme seems to work as Satoh *et al.* (1998) recently showed for borehole data of the Hyogo-ken Nanbu earthquake.

In the case of loose sand saturated by water, soil under strong shaking becomes liquefied because pore water pressure built up by shear strain destroys the grain structure. After much effort by the engineering community, we can now simulate ground motions on the surface of liquefied soil more or less quantitatively. A good example can be found in Kawase *et al.* (1996), who succeeded in reproducing Port Island borehole records during the Hyogo-ken Nanbu earthquake based on the 1-D effective stress analysis code developed by Fukutake *et al.* (1990). Many interesting papers, including one review paper on nonlinearity, are found in the proceedings of Irikura *et al.* (1998).

2.2 Two-D/three-D Models

2.2.1 Basin-induced Surface Wave

In 1-D modeling of soil layers, we assume that the soil layers are flat and extend infinitely in the horizontal directions. In reality, soil and sedimentary rock layers are confined by the surrounding intact rock to form sediment-filled basins. At the edge of the basin, strong diffraction takes place due to a large velocity contrast. In case of vertically or near vertically incident body waves, such diffraction at the edge creates basin-induced diffracted waves, which are transformed into surface waves very quickly. The basin-induced surface waves will propagate in the horizontal direction inside the basin, back and forth. Under normal circumstances these basin-induced waves will arrive later than the direct body waves on the surface of the basin simply because aspect ratios (horizontal extent/vertical extent) of normal basins are quite large and so surface waves have a longer distance to travel. A site close to the edge of the basin is an exception to this condition and if the shape of the basin edge is sharp (i.e., if the slope is steep), the edge-induced waves arrive at the same time as body waves, so that a strong constructive interference takes place. Kawase (1996) called such a phenomenon "the Edge Effect" and attributed the damage concentration formed during the Hyogo-ken Nanbu (Kobe) earthquake to it. Some details on the edge effect will follow later.

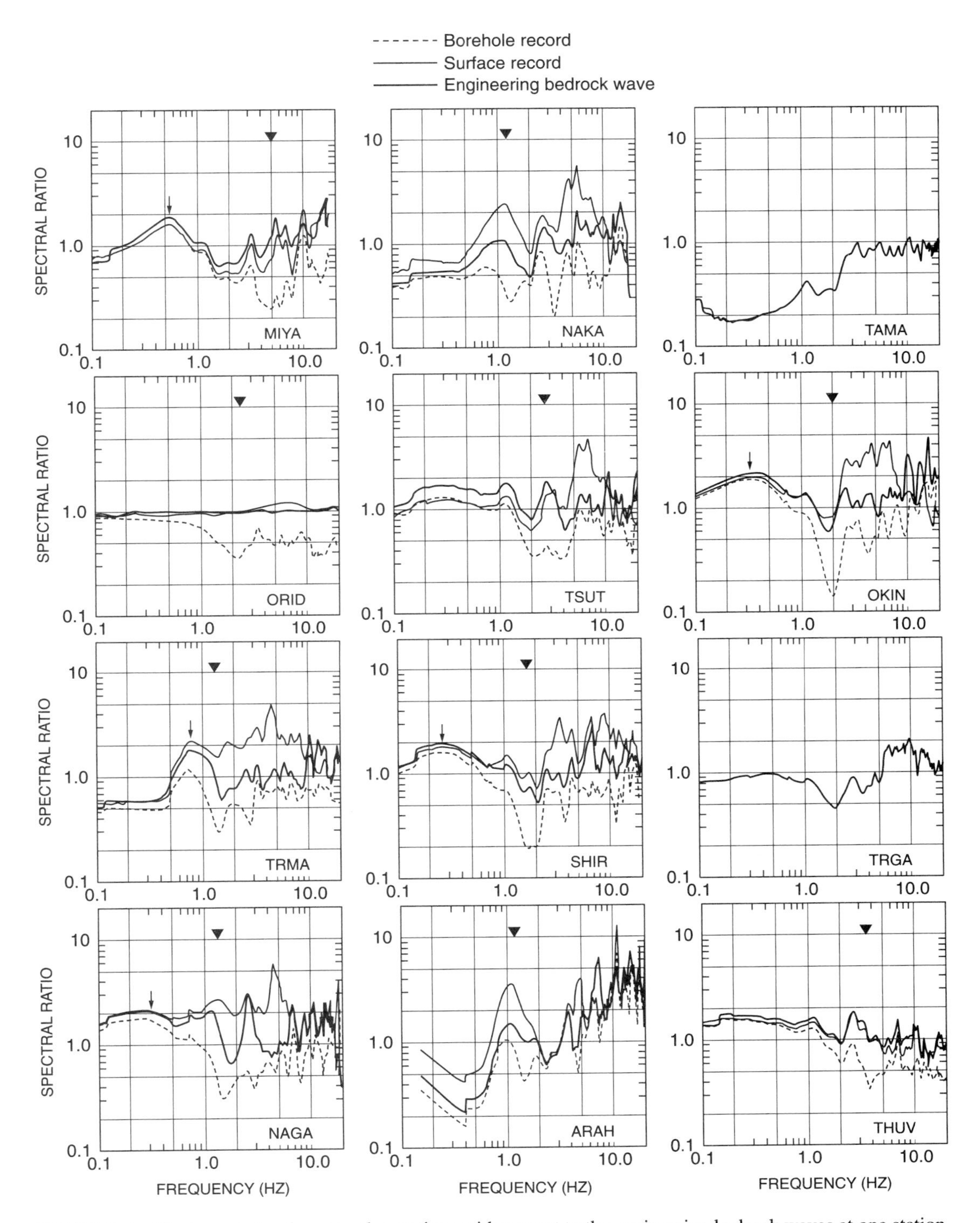

FIGURE 4 Final spectral ratios at twelve stations with respect to the engineering bedrock waves at one station, ORID, averaged for all the data analyzed. Thick solid lines represent ratios of engineering bedrock waves. Thin solid lines represent ratios of surface records. Thin dotted lines represent those of borehole records. Solid triangles correspond to the first resonant frequency of each surface–borehole pair, while arrows indicate low-frequency peak, probably due to the deeper basin amplification (Satoh *et al.*, 1995a).

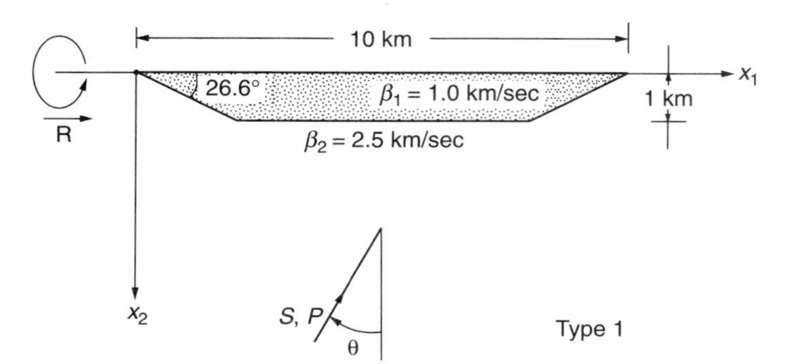

FIGURE 5 A simple hypothetical model of a trapezoidal 2-D basin used in Kawase and Aki (1989).

Let us go back to the basin-induced surface waves. Observational evidence of the basin-induced surface waves was reported for the first time by Toriumi (1975), who found distinctive later arrivals in the ground motions observed in the Osaka basin. At that time, no clear interpretation of the physical entity of the wave was given, but later it was confirmed to be basin-induced surface waves (Toriumi, 1984; Kagawa *et al.*, 1992). Comprehensive theoretical demonstration of this type of wave was made by Bard and Bouchon (1980) using many synthetic seismograms along the surface of a basin, calculated by the Aki-Larner method (Aki and Larner, 1970). Synthetic waveforms at several points for the same basin had been presented before (e.g., Boore 1972; Hong and Helmberger, 1978), but we cannot see clear wave propagation phenomenon without dense spatial sampling. Basic wave generation and propagation phenomena for a basin subject to incident body and surface waves can be better understood in the follow-up work by Kawase and Aki (1989), who used the so-called discrete wavenumber boundary element method (DWBEM) to obtain responses of a trapezoidal 2-D basin as shown in Figure 5. The responses of the basin for vertically incident SH waves with the time-domain shapes of Ricker wavelets with two different characteristic frequencies, namely, 0.25Hz and 0.5Hz, are plotted in Figure 6. We can clearly see that surface waves, Love waves in this case, are generated at the edges and propagate toward the opposite side. At the other side of the edge, a part of the surface wave energy is reflected back to the basin and the rest passes through to the surrounding rock. This symmetric generation and propagation of surface waves forms x-shaped wave patterns on the basin surface, their amplitude gradually decreasing as time goes by. As for the direct S-wave part, the amplitudes are the same on the surface of surrounding rock and in the flat part of the basin, but they are largest near the edges because of the constructive interference of the direct S-wave and the basin-induced surface waves—the edge effect.

Development of the theoretical techniques for irregular underground structures, including the Aki-Larner method and DWBEM, has been reviewed by Sanchez-Sesma (1987), Aki (1988), Kawase (1993), and Takenaka *et al.* (1998), among others. Through the development of such techniques and rapid increase of computational power, we have made it possible to simulate observed ground motions using 2-D/3-D models of realistic underground structures. Two examples of such simulation studies that try to reproduce observed strong motions by physical modeling of the ground are described following.

Kawase and Sato (1992) used strong-motion data and geological data distributed for a blind prediction experiment conducted by the Japanese Working Group on Effects of Surface Geology (Kudo *et al.*, 1992). First they analyzed observed ground motions at two stations on the soft soil inside the Ashigara basin. Figure 7 shows observed accelerograms and their nonstationary spectra at the station S8 during the East off Chiba earthquake. We can see a very isolated later phase at about 40 seconds in the N30°W component whose predominant period ranges from 1.0 to 1.5 second. Based on the polarization and an apparent group velocity between two stations, it is interpreted to be the basin-induced Love wave. We cannot see any similar later phases either in the N60°E component or in the horizontal components of the record on the surrounding rock. Kawase and Sato (1992) constructed a 2-D model, shown in Figure 8, and obtained its response by using their own finite element code similar to those used by Lysmer and Waas (1972). The convolution with a nearby rock record yielded the Fourier spectra indicated by a dashed line in Figure 9. For comparison, synthetic spectra using a 1-D model (dotted lines) as well as the observed spectra (solid lines) are shown. It is clear that the 2-D model gives additional amplification in the N30°W component at around 1.5 Hz, which is a little higher than the observed spectral peak at around 1 Hz; the amount of additional amplification is enough to fill the gap between the 1-D model and the observed Fourier spectra. Note, however, that spectral shapes of both 1-D and 2-D models coincide with the observed shapes in the frequency range higher than 2 Hz (for 2-D up to 5 Hz, maximum frequency for the finite element mesh). We should also note that there is no need to introduce a 2-D model for the N60°E component. The discrepancy of the 2-D amplification frequency between the model and reality is attributable to inappropriate modeling of soil layers between the edge of the basin and the site. Because the major part of the spectra is well reproduced by the 1-D model, we need to assign much softer properties for the layers in the traveling path of the surface wave. Unfortunately, the authors did not have any information on the properties of layers other than the P-S logging data at two borehole sites, so the simulation remained a qualitative one. However, the important lesson here is that a soil column just below the site controls the 1-D response, while the whole path from the edge to the site controls the 2-D response. Thus, we need much wider structural information for 2-D models than for 1-D models. This study also suggests that 1-D response should work for most of the cases and we should introduce 2-D/3-D models only if data show such a necessity.

Another simulation was provided by Hatayama *et al.* (1995) for an array measurement in the eastern part of the Osaka basin.

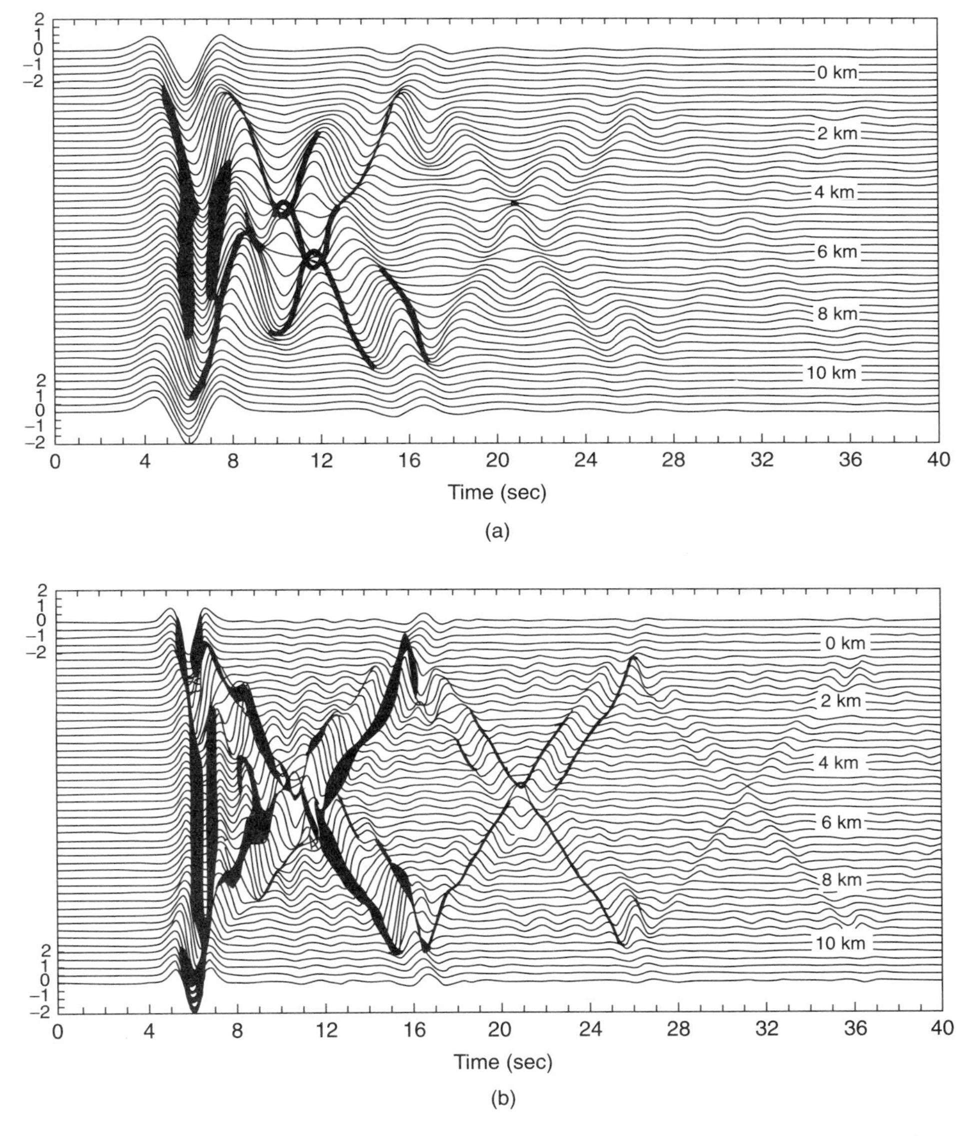

FIGURE 6 Responses of a trapezoidal basin for vertically incident SH waves with the time-domain shapes of Ricker wavelets with two different characteristic frequencies, namely, (a) 0.25Hz and (b) 0.5Hz.

In Figure 10, they compared the synthetic responses at four stations inside the basin calculated by their own 2-D boundary element code with the observed records. The topmost traces are for the rock motion used as a reference. The observed records inside the basin show clear propagation of the basin-induced surface waves, which is labeled as SL1 at the trace of OSA. They confirmed that this later phase was Love wave propagating from east to west based on a small array measurement near the OSA station. They succeeded in reproducing the basin-induced Love wave at OSA, but its peak amplitude was only half of the observed. They used relatively high Q, 100 for sediments, so it may be difficult to increase the amplitude by adjusting Q values. Similar amplitude deficiency for basin-induced surface waves in 2-D/3-D models has been commonly reported in the literature (e.g., Yamanaka *et al.*, 1989; Graves, 1998). This suggests that further study is necessary for the soil properties between the basin edge and the site to better simulate the basin-induced surface waves.

2.2.2 Edge Effect

Near the edge of a basin, generation and propagation of basin-induced diffracted and surface waves and incidence of body waves from the bottom of the basin are taking place simultaneously. If they meet in phase at some point, then constructive interference happens and amplitude of ground motion there becomes

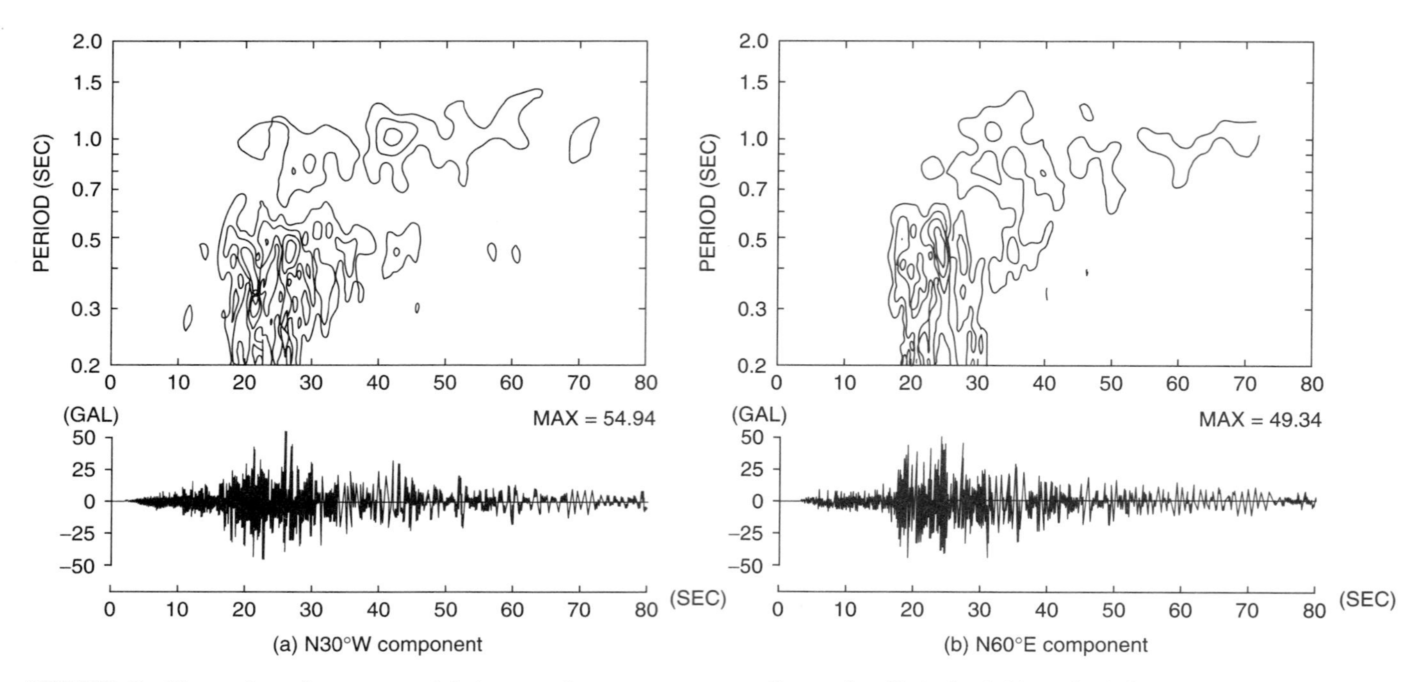

FIGURE 7 Observed accelerograms and their nonstationary spectra at a sediment site, S8, in the Ashigara basin in Kanto area, Japan, during the East off Chiba earthquake (Kawase and Sato, 1992).

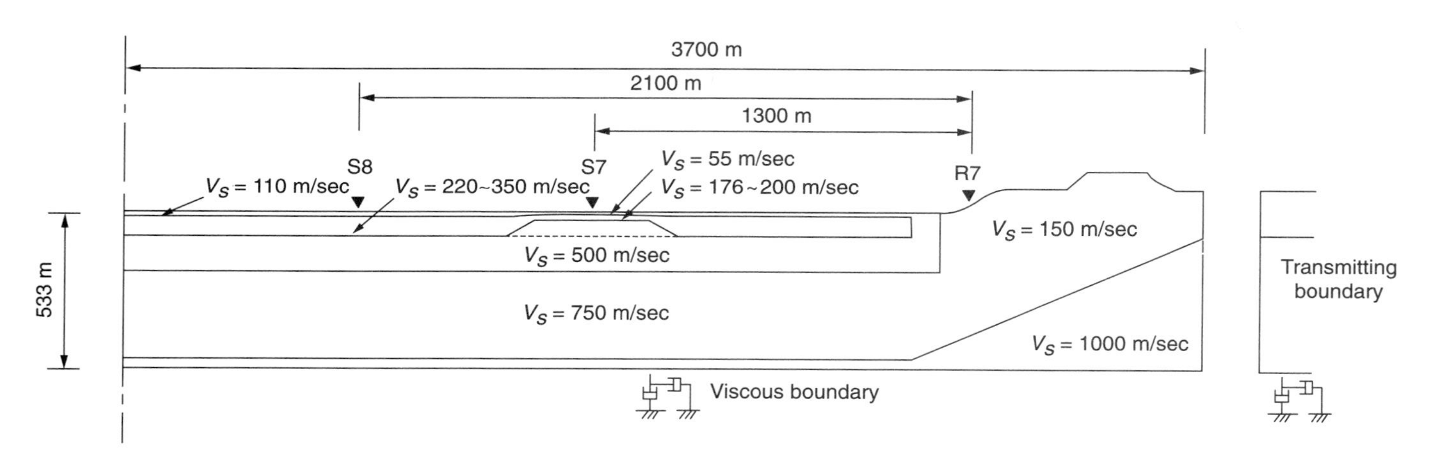

FIGURE 8 A two-dimensional model used by Kawase and Sato (1992) for the simulation.

much larger than a simple 1-D response. This amplification effect near the edge of the basin was named "the (Basin) Edge Effect" by Kawase (1996). The damage concentration found in Kobe during the Hyogo-ken Nanbu (Kobe) earthquake, often called the damage belt because of its large length (~20 km) compared to its small width (~1 km), was created as a consequence of both source extension along the strike of the damage belt and the edge effect along the northwestern edge of the Osaka basin. Kawase (1996) used a 2-D rectangular basin to simulate the observed strong motions in Sannomiya, downtown Kobe, where a heavy damage concentration was formed near the JR Sannomiya station. He showed snapshots for 1 Hz Ricker wavelet input in the vertical cross section to delineate the mechanisms of constructive interference happening near the edge of the basin. After his paper, Inoue and Miyatake (1997) showed snapshots of surface velocity response for a moving dislocation (strike-slip) source. They attributed the cause of the damage belt primarily to the source effect, rather than the edge effect, because it looks as if a high-amplitude zone is moving as rupture propagates. However, in defending Kawase's hypothesis, Kawase *et al.* (1998a) showed that the edge effect is present whatever the source of incident wave, and a 1-D response of the basin without the constructive interference with the edge-induced waves is not sufficient to cause the damage concentration seen in Kobe.

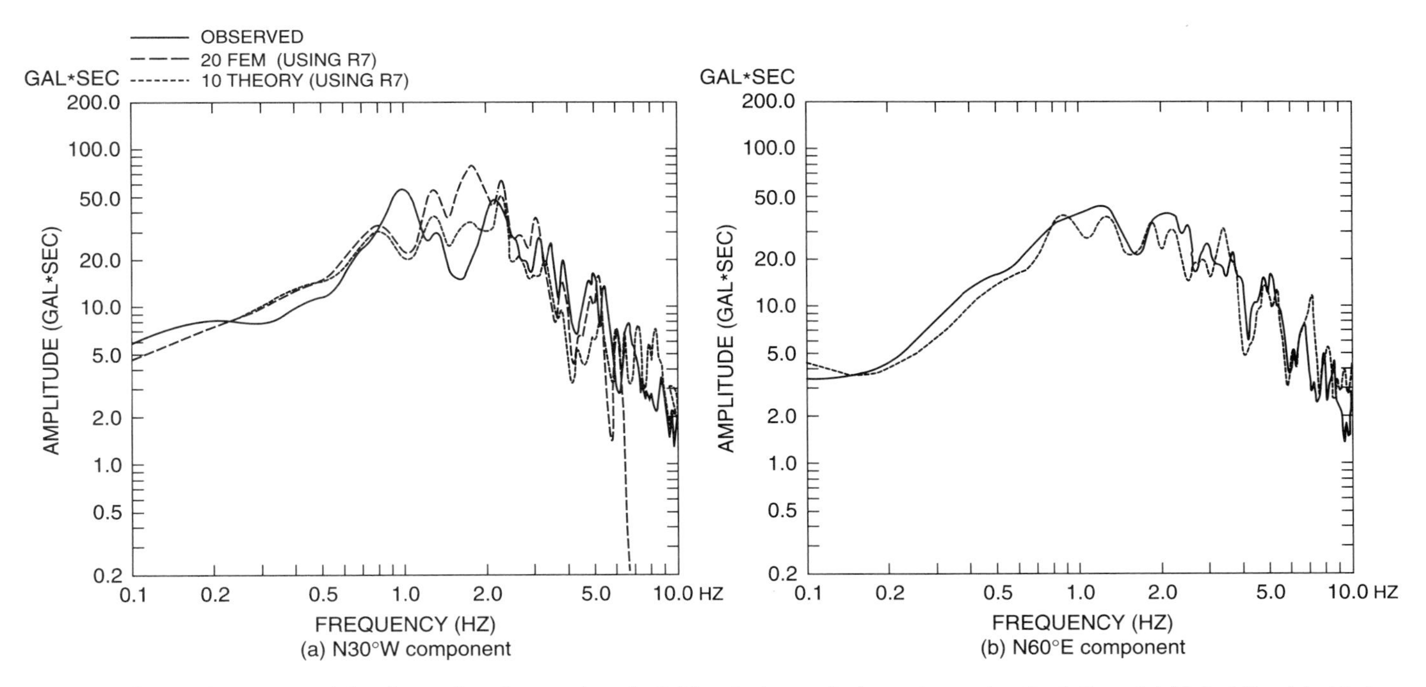

FIGURE 9 Fourier spectra of the observed surface motions (solid lines), the synthetic motions using the 1-D model (dotted lines) just below the site, and the synthetic motions using the 2-D model in Figure 8 (a broken line, only in the N30°W component).

More quantitative simulation for Kobe can be found in Matsushima and Kawase (1998), where a complex rupture process and a 3-D basin model were used in their finite difference calculation. Pitarka *et al.* (1997, 1998) support the idea of the basin-edge effect in Kobe. An introduction of the essential part of the edge effect study of Kawase *et al.* (1998a) is presented following.

Figure 11 shows a simple 2-D rectangular basin model used in the analysis. Kawase *et al.* considered three different models: 1) a simple 2-D model, 2) a 2-D model with a reflection layer in the rock side, and 3) a 2-D model with a reflection layer in the basin side. The reason for embedding a reflection layer in only one side of the model space is to separate incident and reverberated S waves in the basin (=1-D response) and the basin-induced diffracted/surface waves (=2-D additional amplification). An assumed incident wave was a vertically incident S wave with a bell-shaped function of about 1 second of the predominant period. The left panel of Color Plate 25 shows the whole response of a simple 2-D model, while the center panel and the right panel show responses of a 2-D model with rock-side reflection (called Rock-S-cut model), and a 2-D model with basin-side reflection (called Basin-S-cut model), respectively. In the center panel of Color Plate 25, clear vertical propagation and reflection of the S wave can be seen only in the basin side. This is a normal 1-D response of the basin. In the right panel of Color Plate 25, when an incident S wave in the rock side hits the surface, strong energy concentration appears at the edge and it radiates energy to the basin quite efficiently. The edge-induced waves, which are transforming themselves into surface waves quite rapidly as they propagate, have two distinctive phases that are indicated by open and solid triangles in the right panel of Color Plate 25. The faster phase consists primarily of diffracted P wave and a higher mode of Rayleigh wave, while the slower phase consists primarily of diffracted S wave and the fundamental mode of Rayleigh wave. The constructive interference mentioned in Kawase (1996) was only with the faster phase. Although the lobe of the faster phase is larger, the conspicuous amplification near the edge found for the realistic input (Figure 6 in Kawase, 1996) is mainly caused by the constructive interference with the slower phase. Note that these two phases of the edge-induced waves have opposite signs, the faster one sharing the same sign as the input. Thus for quantitative evaluation of the interference with the slower one, the waveform of the incident S wave is very important.

The basin-edge effect is not an extraordinary phenomenon found only in Kobe. Any kind of a basin edge would have the edge effect, although the degree of interference depends on the edge shape as well as the input waveform. An example for the northern edge of the great Los Angeles Basin near Santa Monica was reported by Graves *et al.* (1998), who succeeded in explaining velocity waveforms of the observed ground motion record at Santa Monica City Hall during the Northridge, California, earthquake of 1994.

2.2.3 Basin-transduced Surface Wave

If a source is distant and shallow, then the incident wave to a basin will consist mainly of surface waves. When they reach the basin, a part of the incident wave energy is reflected back

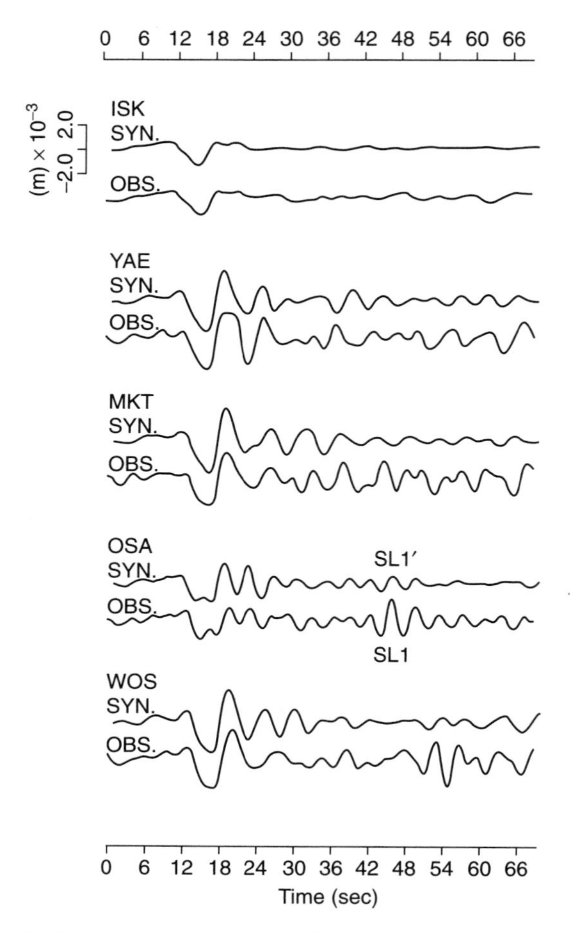

FIGURE 10 Comparisons of synthetic responses at five stations calculated by a 2-D model of the eastern part of the Osaka basin with the observed records (Hatayama *et al.*, 1995). The topmost traces are for the rock motion used as a reference.

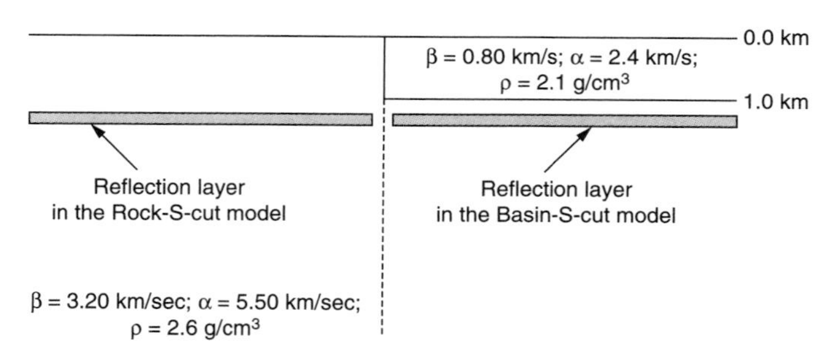

FIGURE 11 A simple 2-D rectangular basin model used in the analysis of the basin-edge effect (Kawase *et al.*, 1998a). In addition to the normal basin, two variations are considered. One has a reflection layer only below the rock surface (Rock-S-cut model) and the other has the same reflection layer only below the basin (Basin-S-cut model).

but the rest is impinging into the basin. At the edge, complex transformation from one mode of surface waves for the surrounding rock to different modes for the basin sediments, i.e., mode conversion between two different media, is taking place.

Kawase (1993) called these types of surface waves inside the basin basin-transduced surface waves. In this case, we do not see two distinctive phases as seen in the case of the basin-induced surface wave. Instead, we see continuous arrivals of long-period waves from the beginning because mainly basin-transduced surface waves exist.

The existence of basin-transduced surface wave is reported by Hanks (1975), who showed a series of displacement seismograms recorded during the San Fernando, California, earthquake of 1971. For the rock sites near the source, the duration of displacement records is short and the waveform is simple, while those inside the Los Angeles basin are quite long and dispersed. Hanks (1975) noted that despite the relatively short distance from the fault, the observed waveforms inside the basin have the characteristics of surface waves. Continuous arrivals of relatively short period (~3 seconds) waves are a clear indication of basin-transduced surface waves. Later, Vidale and Helmberger (1988) succeeded in simulating these observed velocity seismograms in the San Fernando and Los Angeles basins. Figure 12 shows the comparison of filtered transverse

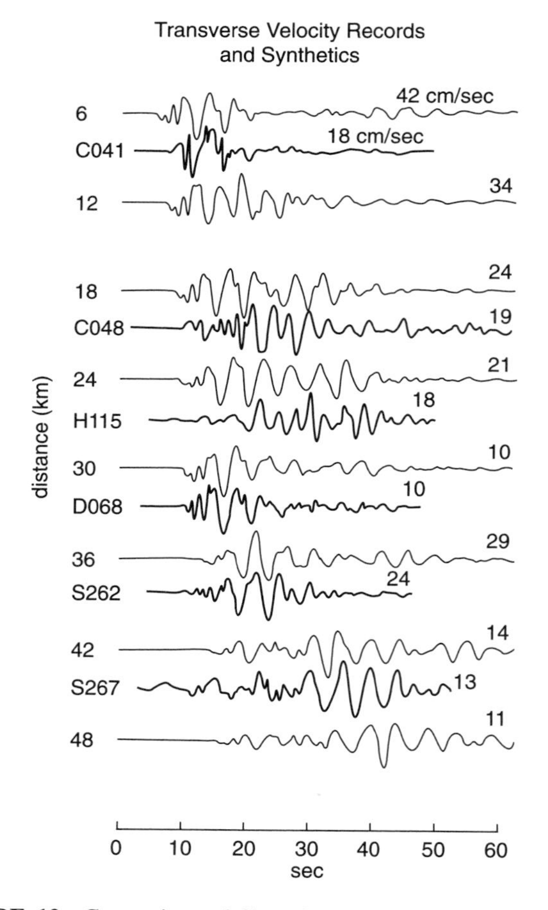

FIGURE 12 Comparison of filtered transverse components of the observed velocity records in the San Fernando and Los Angeles basins during the San Fernando earthquake of 1971 with those of the synthetics by Vidale and Helmberger (1988).

components of the observed velocity records with those of the synthetics. The matching of the synthetics with the data is generally very good. But we should note that source properties are controlled so as to match the synthetics with the observed at D068 and therefore the synthetics are not fully theoretical but similar to the convoluted ones with a reference record. Complete physical modeling of source, path, and site could be possible for simulation with basin-transduced surface waves because their predominant period tends to be longer than a few seconds. One pioneering work of this kind was done by Sato (1990, reproduced in Kawase, 1993), who succeeded in simulating long-period basin response of the Osaka basin during Kita-Mino, Japan, earthquake of 1961 by using the so-called thin-layer element method (TLEM) combined with reciprocity. Recently, Sato *et al.* (1999) show remarkably well-reproduced results of the displacement records observed at Tokyo during the Kanto earthquake of 1923, which is shown in Figure 13. They used a small recent earthquake that occurred near the hypocenter of the Kanto earthquake as a calibration event to refine the basin structure used in their 3-D finite difference calculation. 3-D basin effects can be clearly seen in Figure 14, in which they compare the observed records with 1-D and 3-D calculations. Note that before using the 3-D basin structure, they calibrated source, path, and site effects very carefully using 1-D models (Sato *et al.*, 1998a, 1998b), without which the good matching seen in Figures 13 and 14 could not be achieved.

2.2.4 Focusing Effects

In case of a basin with an irregular bottom shape subject to a body wave incidence below, we will see focusing effects in which seismic rays propagating in different paths meet together at certain points on the surface. At these points, there will be amplification or deamplification depending on their phases. Because such a ray concept is a high-frequency approximation, the focusing effects should emerge in the high-frequency amplification characteristics. Hartzell *et al.* (1996) attributed conspicuous amplification and large damage difference along the southern edge of the San Fernando Valley to such focusing effects created by irregular layer boundaries supposedly underlying the observation sites. We should note, however, that under normal circumstances the frequency range of our interest remains relatively low, a few Hz for most of the cases, and so the diffraction plays a major role in wave propagation and a simple ray concept would not always work.

2.3 Topographic Effects

Topographic effects have been reviewed in some detail by Geli *et al.* (1988), Aki (1988), and Kawase (1993). Because there are so many different observations on the strength of the topographic effect, we cannot draw clear conclusions. However, many studies in the past suggest that even a rock site has strong

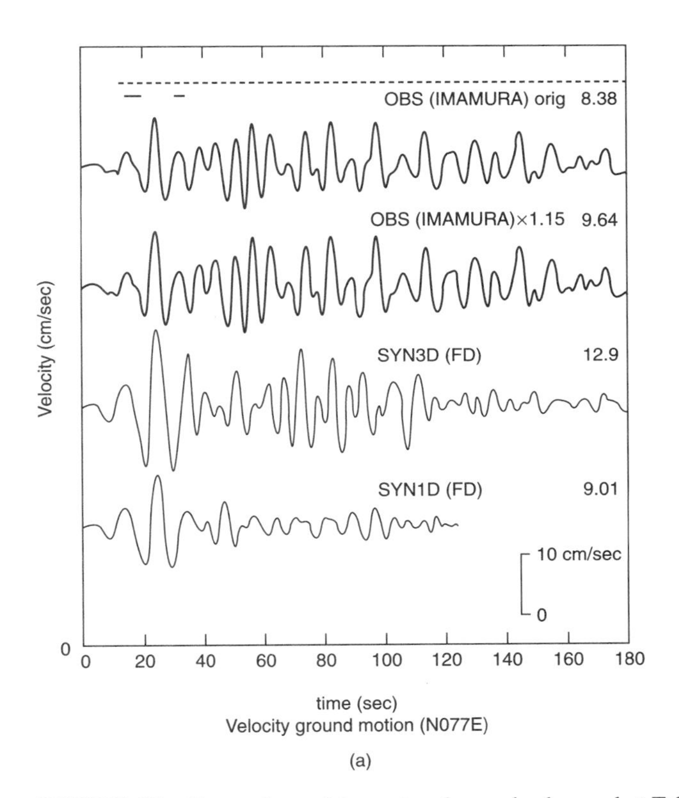

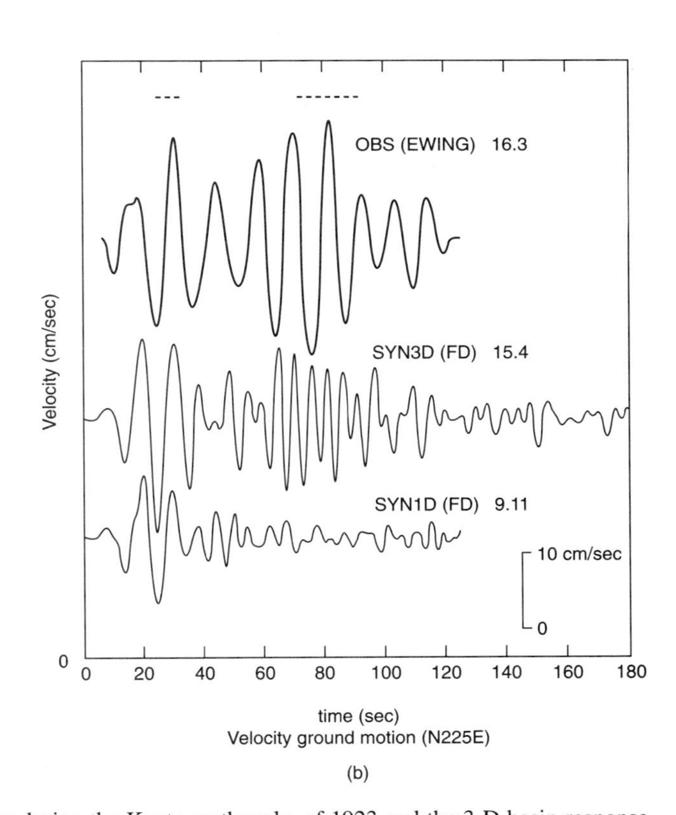

FIGURE 13 Comparison of the restored records observed at Tokyo during the Kanto earthquake of 1923 and the 3-D basin response calculated by Sato *et al.* (1999).

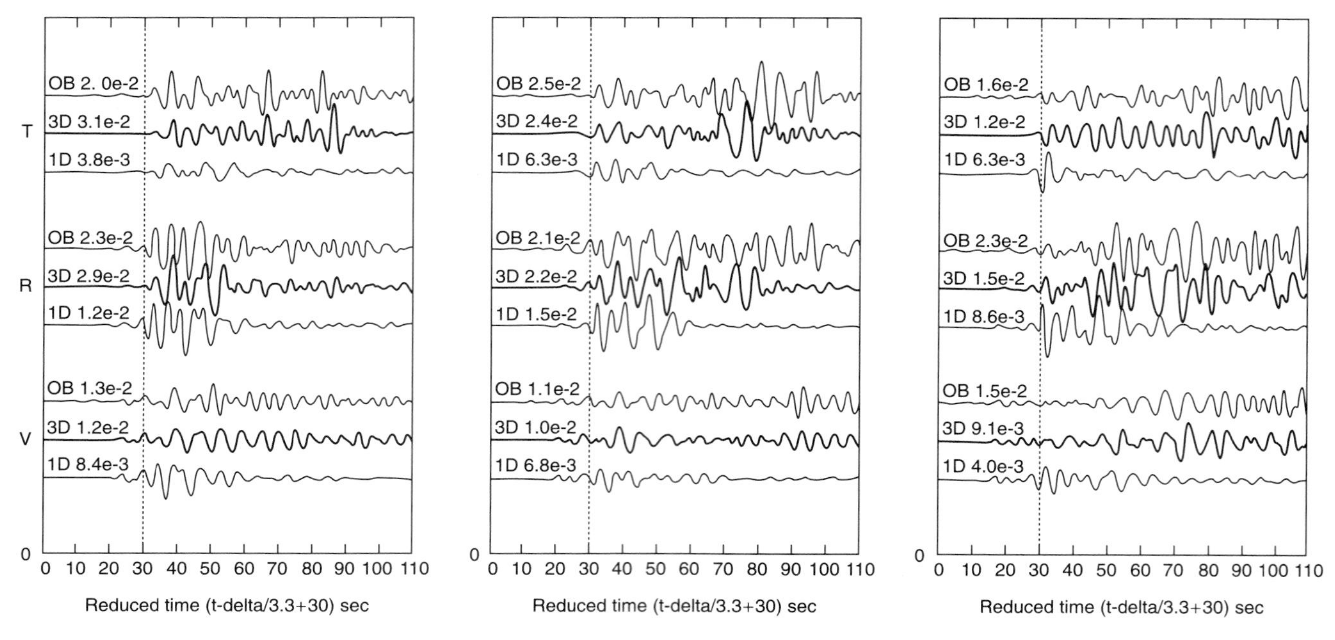

FIGURE 14 Comparison of the observed records used as a calibration event with 1-D and 3-D theoretical calculations by Sato *et al.* (1999).

site effects due to the subsurface structure below (e.g., Steidl *et al.*, 1996; Spudich *et al.*, 1996) so it may be difficult to distinguish the effects of the subsurface structure and those of the topography from observed site effects, unless we know the subsurface velocity structure of the site in detail. Theoretically speaking, the pure topographic amplification factor will be 2 at most, because constructive interference between two arrivals of different waves, either diffracted waves, surface waves, or body waves, is likely to happen but more than two arrivals with the same phase rarely meet at the same location. Besides, strong topography can be found only in the mountain area, where amplification itself is basically low compared to the basin surface and so practically speaking pure topographic effects are of secondary importance.

3. Empirical Modeling of Site Effects

If we obtain site effects directly from observed data, such an approach is called an empirical modeling approach. In the empirical modeling of site effects we must observe strong or weak ground motions due to one or more earthquakes and analyze data to extract site effects. It seems straightforward to extract site effects from data. It is, only if we have a very good reference site near the target site, as mentioned in the Introduction section. Here, we briefly review two empirical approaches for site-effects studies, namely, a separation method (sometimes called an inversion method) and a method based on coda and microtremors.

3.1 Separation of Source, Path, and Site

Empirical approaches to characterizing site effects have had a long history if we include studies of soil-type-specific amplification factors. Such an approach has been very common for engineering purposes. Over the last fifteen years we have been extracting site-specific empirical amplification factors from many observed data for different sites and different earthquakes. In a pioneering work, Andrews (1986) proposed a method to separate source effect, $S_i(f)$, path effect, $P_{ij}(f)$, and site effect, $G_j(f)$, by a singular value decomposition of the whole data space of observed Fourier spectra, $A_{ij}(f)$. The basic equation of the method is quite simple, as follows:

$$A_{ij}(f) = S_i(f)\, P_{ij}(f)\, G_j(f) \quad (3)$$

where i is a suffix for earthquake and j for site, while f is a frequency. If we assume *S*-wave propagation, $P_{ij}(f)$ could be expressed in terms of a geometrical spreading factor with respect to R_{ij}, the representative distance between a source and a site, and attenuation due to the intrinsic and scattering Q(f) as

$$P_{ij}(f) = R_{ij}^{-1} \exp(-\pi f R_{ij}/\beta Q(f)) \quad (4)$$

where β is the representative *S*-wave velocity of the whole path.

This technique has an important problem we need to note; it requires at least one independent constraint either on source, path, or site factors. Andrews (1986) assumed the logarithm of the sum of all the site factors to be zero, which physically means the average site factor should be a reference. Following Andrews, Iwata and Irikura (1988) used the same technique but a different constraint, in which they assumed any site factors to be more than 2.0. This physically means that a site with the least site factor among others should be a reference. Since these pioneering studies, many similar studies have followed as strong-motion databases have been expanding in different areas of the world (e.g., Kato *et al.*, 1992; Satoh *et al.*, 1997; Bonilla *et al.*, 1997). We should note that the scheme of the singular value decomposition is equivalent to the first step of the two-step decomposition used in the attenuation analysis for peak ground values (Joyner and Boore, 1981; Fukushima and Tanaka, 1990). Because decomposed source or site effects depend on the constraint, we should always be careful of its appropriateness. In an idealistic situation we can assume one station among all to be site-effect free so that the Fourier spectra at the reference site should consist of only source and path factors. If it is not the case, resultant source factors are directly biased by the site effect at the reference site. Such an example was reported in Kato *et al.* (1992). They used a hard-rock site, which was thought to be a good reference site from its surface *S*-wave velocity of 2.2 km/sec. After the separation, however, they found significant site effect present due to the V-shaped canyon topography around the site. This paper teaches us one important lesson; if we have a common frequency variation in the source (or site) factors for different earthquakes (or sites) we should check the validity of the assumption used as a constraint.

3.2 Microtremor and Coda

There has been a long history of using microtremors to delineate site characteristics. After the pioneering work of Kanai and his colleagues in the 1950s and 1960s in Japan, based on the microtremor data in Tokyo (e.g., Kanai *et al.*, 1954), different techniques have been proposed. The latest addition is to take a spectral ratio of the horizontal component with respect to the vertical component (H/V), which was proposed by Nakamura (1989) and is sometimes called Nakamura's technique. Because H/V ratios of microtremors often show similar site characteristics obtained by other independent methods, the H/V method has gained popularity quite rapidly (Lermo and Chavez-Garcia, 1993; Field and Jacob, 1995). As the method becomes widely used, skepticism on its validity as a method for directly obtaining *S*-wave amplification factors has arisen (e.g., Lachet and Bard, 1994; Zhao *et al.*, 1997). From array measurements of microtremors, the dominant wave type in the vertical component is found to be a Rayleigh wave (e.g., Horike, 1985; Kawase *et al.*, 1998b). A numbers of papers on this topic are found in the proceedings of the second International Symposium on Effects of Surface Geology (Irikura *et al.*, 1998); interested readers should refer to them.

Coda, the later part of the seismograms decreasing exponentially, has also been used for a long time for site correction. Physically, coda is considered to be scattered *S* waves in the whole volume between a source and a site, and is used to characterize the scattering property of the volume (e.g., Aki and Chouet, 1975; Tujiura, 1978; Sato, 1984). If the origin of coda is scattered *S* waves, then the local site amplification of coda should be an average of *S*-wave amplification for incident *S* waves with different azimuths and incidence angles and, therefore it could be more stable than that of a direct *S* wave (Chin and Aki, 1991; Kato *et al.*, 1995). From the results of studies reviewed here it seems too simplistic to assume that coda consists mainly of *S* waves scattered in the whole path for sites inside a soft sedimentary basin. Most of the previous studies on the site factors from coda draw the frequency limit in the lower frequency end at 1 Hz or higher so that contamination of basin-induced and/or basin-transduced surface waves may not be significant. Yet, physically, we should always expect to have some surface-wave contamination in the later part of the ground motion.

Recently Satoh *et al.* (2001) analyzed the data observed in Sendai, Japan, quite extensively and found that both *P*-coda and *S*-coda at soft soil sites, especially *P*-coda, are strongly contaminated by the basin-induced surface waves. They checked the characteristics of H/V ratios for initial parts of *P* and *S* waves, early coda, late coda, and microtremors for both soft-soil sites and rock or hard-soil sites. H/V ratios of coda are not the same as the direct *P* or *S* wave and are converging rapidly to those of microtremors. They also calculated H/H ratios with respect to the hardest rock site and found that they are varying with time for soft-soil sites but are stable for rock or hard-soil sites. Theoretical simulation based on the *S*-wave structures determined by array measurements of microtremors reveals that H/H ratios of direct *S* wave can be explained well by 1-D responses of the structures for the soft-soil sites and that H/V ratios of coda and microtremors can be explained well by the Rayleigh wave H/V ratios of the same structures. These results suggest that coda and microtremors on soft sediments consist mainly of surface waves so that they have amplification characteristics different from *S* waves.

4. Conclusions

The essential aspects of the site-effect studies are reviewed by focusing mainly on the physical modeling scheme to reproduce wave propagation phenomena in the shallower part of the Earth. The *S*-wave amplification can be characterized by 1-D modeling of soil sediments as a first-order approximation. In addition to that, we sometimes need to consider nonlinear effect and 2-D/3-D basin effect including the basin-induced surface-wave effect and the edge effect. In lower frequency range we need to consider the effect of the basin-transduced surface waves and in higher frequency range we do the focusing effect, as well.

However, no matter what is the most important effect for a specific site of target, the physical modeling scheme for site effect needs a precise model of the actual ground. If the model is strongly biased, then the prediction should also be strongly biased. Thus success of this approach depends on the information of the physical model that we can collect. The better the model, the better the prediction. To what precision and extent we must explore the underground structure for desired accuracy of prediction is an open question. A simple rule of thumb is that any variation of a structure with the same order of one-quarter wavelength may have non-negligible effect as Eq. (1) implies.

Physical properties of the actual complex structures of the Earth can be obtained by various geological, geophysical, and geotechnical methods, which can be used in the physical modeling scheme. Once a physical model of the whole area of interest is calibrated to actual observation, then such a model of the ground will be a common property of people, on which we can depend forever. From materials we have shown here, it seems appropriate to conclude that physical modeling of the ground is now a realistic and effective approach for practical evaluation/prediction of site effects. Needless to say, any physical models that we create in this manner must be validated with the observed data before we use them for prediction. Otherwise it will be a "castle on the sand." The advantage of the physical modeling approach is that it can predict site amplification for any hypothesized sources that have not yet happened but will happen in the near future.

From an engineering point of view, it seems too rigorous and cumbersome to include site-specific, spectral, or waveform representation of site effects as described here. However, a soil-type-specific approach commonly found in the current seismic codes does not represent the essential part of the site effects and, hence, the resulting engineered structures will be vulnerable to the actual input generated by a strong earthquake close to the site. This is exactly the case of the Hyogo-ken Nanbu, Japan, earthquake of 1995. Most of the damage-concentrated areas in Kobe belong to the moderate-to-stiff soil category in the Japanese code and it was not necessary to consider much amplification. But the damage concentration was created by the edge effect of the deeper structure near the edge of the Osaka basin, as reviewed here. Thus we may need to develop a way to translate site effects evaluated by a physical modeling approach into simple but effective engineering representations for better seismic design of structures.

References

Aki, K. (1988). Local site effects on strong ground motion. In: "Earthquake Engineering and Soil Dynamics II - Recent Advances in Ground Motion Evaluation," pp. 103–155. ASCE.

Aki, K., and B. Chouet (1975). Origin of coda waves: source, attenuation and scattering effects. *J. Geophys. Res.* **80**, 3322–3342.

Aki, K., and K. L. Larner (1970). Surface motion of a layered medium having an irregular interface due to incident plane SH waves. *J. Geophys. Res.* **75**, 933–954.

Aki, K., and P. G. Richards (1980). *Quantitative Seismology*. W. H. Freeman, San Francisco.

Andrews, D. J. (1986). Objective determination of source parameters and similarity of earthquakes of different size. In: "Earthquake Source Mechanics," *AGU Geophys. Monograph* **37**, 259–268.

Archuleta, R. (1998). ESG studies in the United States: Results from borehole arrays. In: "The Effects of Surface Geology on Seismic Motion" (K. Irikura, K. Kudo, H. Okada, and T. Sasatani, Eds.), pp. 3–14. Balkema, Rotterdam.

Bard, P.-Y., and M. Bouchon (1980). Seismic response of sediment-filled valleys, Part 1: The case of incident SH waves. *Bull. Seismol. Soc. Am.* **70**, 1263–1286.

Bard, P-Y., M. Campillo, F. J. Chavez-Garcia, and F. J. Sanchez-Sesma (1988). The Mexico earthquake of September 19, 1985—A theoretical investigation of large- and small-scale amplification effects in the Mexico City Valley. *Earthq. Spectra* **4**, 609–633.

Bonilla, L. F., J. H. Steidl, G. T. Lindley, A. G. Tumarkin, and R. Archuleta (1997). Site amplification in the San Fernando Valley, California: Validity of site effect estimation using the S-wave, coda, and H/V methods. *Bull. Seismol. Soc. Am.* **87**, 710–730.

Boore, D. M. (1972). Finite-difference methods for seismic waves. In: "Methods in Computational Physics" (B.A. Bolt, ed.), Vol. 11, pp. 1–37. Academic Press.

Chavez-Garcia, F. J., and P.-Y. Bard (1994). Site effects in Mexico City eight years after the September 1985 Michoacan earthquake. *Soil Dyn. Earthq. Eng.* **13**, 229–247.

Chin, B. H., and K. Aki (1991). Simultaneous study of the source, path, and site effects on strong motion during the 1989 Loma Prieta earthquake: A preliminary result on pervasive nonlinear site effects. *Bull. Seismol. Soc. Am.* **81**, 1859–1884.

Field, E. H., and K. H. Jacob (1995). A comparison and test of various site response estimation techniques, including three that are not reference-site-dependent. *Bull. Seismol. Soc. Am.* **85**, 1127–1143.

Field, E. H., Y. Zeng, J. G. Anderson, and I. A. Beresnev (1998). Pervasive nonlinear sediment response during the 1994 Northridge earthquake. In: "The Effects of Surface Geology on Seismic Motion" (K. Irikura, K. Kudo, H. Okada, and T. Sasatani, Eds.), pp. 773–778. Balkema, Rotterdam.

Fukushima, Y., and T. Tanaka (1990). A new attenuation relation for peak horizontal acceleration of strong earthquake ground motion in Japan. *Bull. Seismol. Soc. Am.* **80**, 757–783.

Fukutake, K., A. Ohtsuki, M. Sato, and Y. Shamoto (1990). Analysis of saturated dense sand-structure system and comparison with results from shaking table test. *Earthq. Eng. Struct. Dyn.* **19**, 977–992.

Furumura T., and B. L. N. Kennett (1998). On the nature of regional seismic phases-III. The influence of crustal heterogeneity on the wavefield for subduction earthquakes: the 1985 Michoacan and 1995 Copala, Guerrero, Mexico earthquakes. *Geophys. J. Int.* **135**, 1060–1084.

Geli, L., P.-Y. Bard, and B. Jullien (1988). The effect of topography on earthquake ground motion: A review and new results. *Bull. Seismol. Soc. Am.* **78**, 42–63.

Graves, R. W. (1998). Long period 3D finite difference modeling of the Kobe mainshock. In: "The Effects of Surface Geology on Seismic Motion" (K. Irikura, K. Kudo, H. Okada, and T. Sasatani, Eds.), pp. 1339–1345. Balkema, Rotterdam.

Graves, R. W., A. Pitarka, and P. G. Somerville (1998). Ground-motion amplification in the Santa Monica area: effect of shallow basin-edge structure. *Bull. Seismol. Soc. Am.* **88**, 1224–1242.

Hanks, T. C. (1975). Strong ground motion of the San Fernando, California, earthquake: ground displacements. *Bull. Seismol. Soc. Am.* **65**, 193–225.

Hartzell, S., A. Leeds, A. Frankel, and J. Michael (1996). Site response for urban Los Angeles using aftershocks of the Northridge earthquake. *Bull. Seismol. Soc. Am.* **86**, S168–S192.

Haskell, N. A. (1960). Crustal reflection of plane SH waves. *J. Geophys. Res.* **65**, 4147–4150.

Hatayama, K., K. Matsunami, T. Iwata, and K. Irikura (1995). Basin-induced Love waves in the eastern part of the Osaka basin. *J. Phys. Earth* **43**, 131–155.

Hong, T. L., and D. V. Helmberger (1978). Glorified optics and wave propagation in non planar structures. *Bull. Seismol. Soc. Am.* **68**, 1313–1330.

Horike, M. (1985). Inversion of phase velocity of long period microtremors to the S-wave velocity structure down to the basement in urbanized areas. *J. Phys. Earth* **33**, 59–96.

Hudson, D. E. (1972). Local distribution of strong earthquake ground motions. *Bull. Seismol. Soc. Am.* **62**, 1765–1786.

Inoue, T., and T. Miyatake (1997). 3-D simulation of near-field strong ground motion: Basin edge effect derived from rupture directivity. *Geophys. Res. Lett.* **24**, 905–908.

Irikura, K., K. Kudo, H. Okada, and T. Sasatani (Eds.) (1998). The Effects of Surface Geology on Seismic Motion, *Proc. of the Second International Symposium on Effects of Surface Geology on Strong Motions*. Yokohama, Japan, Balkema, Rotterdam.

Iwata T., and K. Irikura (1988). Source parameters of the 1983 Japan Sea earthquake sequence. *J. Phys. Earth* **36**, 155–184.

Joyner, W. D., and D. M. Boore (1981). Peak horizontal acceleration and velocity from strong-motion records including records from the 1979 Imperial Valley, California, earthquake. *Bull. Seismol. Soc. Am.* **71**, 2011–2038.

JWG-ESG (IASPEI/IAEE Joint Working Group on Effects of Surface Geology on Strong Motions) (1992). *Proc. of the First International Symposium on Effects of Surface Geology on Strong Motions*. Odawara, Japan.

Kagawa, T., S. Sawada, and Y. Iwasaki (1992). On the relationship between azimuth dependency of earthquake ground motion and deep basin structure beneath the Osaka plain. *J. Phys. Earth* **40**, 73–83.

Kanai, K., T. Osada, and T. Tanaka (1954). An investigation into the nature of microtremors. *Bull. Earthq. Res. Inst.* **32**, 199–209.

Kato, K., M. Takamura, T. Ikeura, K. Urano, and T. Uetake (1992). Preliminary analysis for evaluation of local site effects from strong motion spectra by an inversion method. *J. Phys. Earth* **40**, 175–191.

Kato, K., K. Aki, and M. Takemura (1995). Site amplification from coda waves: Validation and application to S wave site response. *Bull. Seismol. Soc. Am.* **85**, 467–477.

Kawase, H. (1993). Effects of surface and subsurface irregularities. In: "Earthquakes and Ground Motions", Pert I, Ch.3, Section 3.3, pp. 118–155. Architectural Institute of Japan.

Kawase, H. (1996). The cause of the damage belt in Kobe: "The basin-edge effect," Constructive interference of the direct S-wave with the basin-induced diffracted/Rayleigh waves. *Seism. Res. Lett.* **67**(5), 25–34.

Kawase, H., and K. Aki (1989). A study on the response of a soft basin for incident S, P, and Rayleigh waves with special reference to the long duration observed in Mexico City. *Bull. Seismol. Soc. Am.* **79**, 1361–1382.

Kawase, H., and T. Sato (1992). Simulation analysis of strong motions in the Ashigara Valley considering one- and two-dimensional geological structures. *J. Phys. Earth* **40**, 27–56.

Kawase, H., T. Satoh, and K. Fukutake (1996). Simulation of the borehole records observed at the Port Island in Kobe, Japan, during the Hyogo-ken Nanbu earthquake of 1995. *Proc. of 11th World Conf. on Earthq. Eng.*, Paper No.140.

Kawase, H., S. Matsushima, R. W. Graves, and P. G. Somerville (1998a). Three-dimensional wave propagation analysis of simple two-dimensional basin structures with special reference to "the Basin-Edge Effect." *Zisin (Series-2)* **50**, 431–449 (in Japanese with English abstract).

Kawase, H., T. Satoh, T. Iwata, and K. Irikura (1998b). S-wave velocity structures in the San Fernando and Santa Monica areas. In: "The Effects of Surface Geology on Seismic Motion" (K. Irikura, K. Kudo, H. Okada, and T. Sasatani, Eds.), pp. 733–740. Balkema, Rotterdam.

Kinoshita, S. (1992). Attenuation of shear waves in a sediment. *Proc. of 10th World Conf. on Earthq. Eng.*, pp. 685–690.

Kobayashi, K., F. Amaike, and Y. Abe (1992). Attenuation characteristics of soil deposits and its formulation, *Proc. of International Symposium on the Effects of Surface Geology on Seismic Motion*, **1**, pp. 269–274. Association for Earthquake Disaster Prevention, Japan.

Kudo, K., K. Irikura, and H. Kawase (1992). Effects of surface geology on seismic motion: Introduction to the special issue. *J. Phys. Earth* **40**, 1–4.

Lachet, C., and P.-Y. Bard (1994). Numerical and theoretical investigation on the possibilities and limitations of the Nakamura's technique. *J. Phys. Earth* **84**, 1574–1594.

Lermo, J., and F. J. Chavez-Garcia (1993). Site effect evaluation using spectral ratios with only one station. *Bull. Seismol. Soc. Am.* **83**, 1574–1594.

Lysmer, J., and G. Waas (1972). Shear waves in plain infinite structures. *J. Eng. Mech. Div.* **98**, 85–105. ASCE.

Matsushima, S., and H. Kawase (1998). 3-D wave propagation analysis in Kobe referring to "The basin-edge effect." In: "The Effects of Surface Geology on Seismic Motion" (K. Irikura, K. Kudo, H. Okada, and T. Sasatani, Eds.), pp. 1377–1384. Balkema, Rotterdam.

Nakamura, Y. (1989). A method for dynamic characteristics estimation of subsurface using microtremor on the ground surface. *QR of RTRI* **30**, 25–33.

Pitarka, A., K. Irikura, and T. Iwata (1997). Modeling of ground motion in the Higashinada (Kobe) area for an aftershock of the 1995 January 17 Hyogo-ken Nanbu, Japan earthquake. *Geophys. J. Int.* **131**, 231–239.

Pitarka, A., K. Irikura, T. Iwata, and H. Sekiguchi (1998). Local geological structure effects on ground motion from earthquakes on basin-edge faults. In: "The Effects of Surface Geology on Seismic Motion" (K. Irikura, K. Kudo, H. Okada, and T. Sasatani, Eds.), pp. 901–906. Balkema, Rotterdam.

Sanchez-Sesma, F. J. (1987). Site effects on strong ground motion. *Soil Dyn. Earthq. Eng.* **6**, 124–132.

Sato, H. (1984). Attenuation and envelope formulation of three component seismograms of small local earthquakes in randomly inhomogeneous lithosphere. *J. Geophys. Res.* **89**, 1221–1241.

Sato, T. (1990). Simulation of observed seismograms on sedimentary basin using theoretical seismograms in the period range from 2 sec to 20 sec—Synthesis of seismograms at Kobe and Osaka stations for the 1961 Kita-Mino earthquake, *Proc. of 8th Japan Earthq. Eng. Sym.*, pp. 193–198. Tokyo, Japan (in Japanese).

Sato, T., D. V. Helmberger, P. G. Somerville, R. W. Graves, and C. K. Saikia (1998a). Estimates of regional and local strong motions during the Great 1923 Kanto, Japan, earthquake (M_S8.2), Part 1: Source estimation of a calibration event and modeling of wave propagation paths. *Bull. Seismol. Soc. Am.* **88**, 183–205.

Sato, T., D. V. Helmberger, P. G. Somerville, R. W. Graves, and C. K. Saikia (1998b). Estimates of regional and local strong motions during the Great 1923 Kanto, Japan, earthquake (M_S 8.2), Part 2: Forward simulation of seismograms using variable-slip rupture models and estimation of near-fault long period ground motions. *Bull. Seismol. Soc. Am.* **88**, 206–227.

Sato, T., R. W. Graves, and P. G. Somerville (1999). 3-D Finite-Difference simulations of long-period strong motions in the Tokyo Metropolitan Area during the 1990 Odawara earthquake (M_J 5.1) and the Great 1923 Kanto earthquake (M_S8.2) in Japan. *Bull. Seismol. Soc. Am.* **89**, 579–607.

Satoh, T., T. Sato, and H. Kawase (1995a). Evaluation of local site effects and their removal from borehole records observed in the Sendai Region, Japan. *Bull. Seismol. Soc. Am.* **85**, 1770–1789.

Satoh, T., T. Sato, and H. Kawase (1995b). Nonlinear behavior of soil sediments identified by using borehole records observed at the Ashigara Valley, Japan. *Bull. Seismol. Soc. Am.* **85**, 1821–1834.

Satoh, T., T. Sato, and H. Kawase (1997). Statistical spectral model of earthquakes in the eastern Tohoku district, Japan, based on the surface and borehole records observed in Sendai. *Bull. Seismol. Soc. Am.* **87**, 446–462.

Satoh, T., M. Fushimi, and Y. Tatsumi (1998). Inversion of nonlinearity of soil sediments using borehole records in Amagasaki. In: "The Effects of Surface Geology on Seismic Motion" (K. Irikura, K. Kudo, H. Okada, and T. Sasatani, Eds.), pp. 823–830. Balkema, Rotterdam.

Satoh, T., H. Kawase, and S. Matsushima (2001). Differences between site characteristics obtained from microtremors, S-wave, P-wave, and codas, *Bull. Seismol. Soc. Am.* **91**, 313–334.

Schnabel, P. B., J. Lysmer, and H. B. Seed (1972). A computer program for earthquake response analysis of horizontally layered sites. *EERC Report 72-12*.

Seale, S. H., and R. Archuleta (1989). Site amplification and attenuation of strong ground motion. *Bull. Seismol. Soc. Am.* **79**, 1673–1696.

Singh, S. K., and M. Ordaz (1993). On the origin of long coda observed in the lake-bed strong-motion records of Mexico City. *Bull. Seismol. Soc. Am.* **83**, 1298–1306.

Spudich, P., M. Hellweg, and W. H. K. Lee (1996). Directional topographic site response at Tarzana observed in aftershocks of the 1994 Northridge, California, earthquake: Implications for mainshock motions. *Bull. Seismol. Soc. Am.* **86**, 192–208.

Steidl, J. H., A. G. Tumarkin, and R. Archuleta (1996). What is a reference site? *Bull. Seismol. Soc. Am.* **86**, 1733–1748.

Takahashi, K., S. Omote, T. Ohta, T. Ikeura, and S. Noda (1988). Observation of earthquake strong-motion with deep borehole—Comparison of seismic motions in the base rock and those on the rock outcrop, *Proc. of 8th World Conf. on Earthq. Eng.* **3**, 181–186.

Takenaka, H., T. Furumura, and H. Fujiwara (1998). Recent developments in numerical methods for ground motion simulation. In: "The Effects of Surface Geology on Seismic Motion" (K. Irikura, K. Kudo, H. Okada, and T. Sasatani, Eds.). Balkema, Rotterdam.

Toriumi, I. (1975). Earthquake characteristics in the Osaka plain. *Proc. of the 4th Japan Earthq. Eng. Sym.* pp. 129–136 (in Japanese).

Toriumi, I., S. Ohba, and N. Murai (1984). Earthquake motion characteristics of Osaka plain. *Proc. of 8th World Conf. on Earthq. Eng.* **2**, 761–768.

Tujiura, M. (1978). Spectral analysis of the coda waves from local earthquakes. *Bull. Earth. Res. Inst.* **53**, 1–48.

Vidale, J. E., and D. V. Helmberger (1988). Elastic finite difference modeling of the 1971 San Fernando, California, earthquake. *Bull. Seismol. Soc. Am.* **78**, 122–141.

Yamanaka, H., K. Seo, and T. Samano (1989). Effects of sedimentary layers on surface-wave propagation. *Bull. Seismol. Soc. Am.* **79**, 631–644.

Zhao, B., M. Horike, Y. Takeuchi, and H. Kawase (1997). Comparison of site-specific response characteristics inferred from seismic motions and microtremors, *Zisin (Series-2)* **50**, 67–87 (in Japanese with English abstract).

62

Use of Engineering Seismology Tools in Ground Shaking Scenarios

Ezio Faccioli
Department of Structural Engineering, Politecnico, Milano, Italy
Vera Pessina
INGV - Istituto Nazionale di Geofisica e Volcanologia, Milan, Italy

1. Introduction: Approaches to the Selection of Scenario Earthquakes

Damage scenario and loss estimation studies for destructive events represent a potentially powerful tool for enhancing the level of earthquake preparedness and for improving disaster prevention policies for densely populated areas in seismic regions. Especially for large and complex urban environments, a fundamental prerequisite for such a task is the realistic estimation of the spatial distribution of the ground-motion severity generated by scenario earthquakes. The recent cases of the 1995 Kobe (Hyogo-ken Nanbu) and 1999 Kocaeli and Duzce (Turkey) events stand out to signal the overwhelming influence that the spatial distribution of ground motions and its close dependence on source characteristics, local geological structure, and soil properties can have on the resulting damage.

A preliminary issue of critical relevance for the construction of damage scenarios is the choice between a deterministic and a probabilistic approach to the representation of ground shaking and to damage estimation. For ease of reference, the two approaches are briefly outlined and discussed below.

The deterministic approach develops a particular earthquake scenario upon which a ground-motion hazard evaluation is based. The scenario consists of the postulated occurrence of an earthquake of a specified size occurring at a specified location, typically a seismically active fault. The standard deterministic approach consists of the following steps:

1. Identification of the locations and characteristics of all significant earthquake sources that might affect the zone of interest, typically a city. The seismic potential is quantified by assigning to each source a significant earthquake that can be the maximum historical event known from that source (Maximum Probable Earthquake, or MPE), or the maximum earthquake that appears capable of occurring under the known tectonic framework (Maximum Credible Earthquake, or MCE). The shortest distance from the source to the sites in the zone of interest is generally assumed, in the absence of other information.
2. Selection of the attenuation relations enabling one to estimate the ground shaking (e.g., in terms of intensity, response spectral ordinates, or Arias intensity) within the zone of interest as a function of earthquake magnitude, source-to-site distance, and local ground conditions.
3. Definition of the controlling earthquake (i.e., the earthquake that is expected to produce the strongest level of shaking). The selection is made comparing, through the attenuation relation, the effects produced at the site by the combination of earthquake distance and magnitude identified at step 1. The resultant earthquake will define the seismic scenario.

However, depending on the historical seismicity of the region, variations with respect to this standard procedure are possible. Thus, it may be desirable to adopt more than one scenario earthquake, for instance, a destructive event (such as the MPE) and a less severe, but damaging, one with a higher likelihood of occurrence.

At the end of the process, the deterministic analysis gives as a result one or more scenarios that represent the worst situations expected, but it does not provide any information about their occurrence in time. In fact, this approach does not predict the likelihood of such occurrences during the lifetime of the

INTERNATIONAL HANDBOOK OF EARTHQUAKE AND ENGINEERING SEISMOLOGY, VOLUME 81B *ISBN: 0-12-440658-0*

TABLE 1 Occurrence Probabilities of Earthquakes Commonly Referred to in Earthquake Engineering

Type of occurrence	Return period[a]	Probability in 50 years
Frequent	43	0.70
Occasional	72	0.50
Rare	475	0.10
Very rare	970	0.05

[a] Based on Poisson occurrence process.

structures that make up the building and infrastructure stock in the zone of interest.

The probabilistic approach is a typical application of Probabilistic Seismic Hazard Analysis (PSHA), whose basic features have been established since the late 1960s (Cornell, 1968). The method has since become a worldwide standard and need not be recapitulated here in detail. It consists of four basic steps, some of which are partly coincident with those of the deterministic approach, namely:

- Seismic source zone identification
- Probabilistic characterization of source zone activity
- Selection of attenuation relation
- Integration over the whole range of magnitudes and distances for each source zone in order to obtain, for each site of interest, the probabilistic hazard values in the form of a cumulative distribution for the ground-motion parameter(s) of interest

A certain degree of combination of the deterministic and probabilistic approaches is also conceivable. For instance, in a zone with a well-documented history of damaging earthquakes one may choose the deterministic scenario, paying some consideration to the relative frequency of past events rather than to the MPE or MCE events. Table 1 may provide some guidance in this respect.

The 50 years reference term in Table 1 can be roughly regarded as the expected engineering lifetime of an ordinary building (this lifetime would increase to about 100 years for a bridge or tunnel, and to 100–150 years for a dam).

At a more sophisticated level, since probabilistic risk studies cover all possible damaging events in a region and estimate the cumulative losses from all of them (e.g., in the form of the annual loss rate), they can help in making an objective preselection of the basic features of the scenario earthquake. For instance, the latter could be the event that occurs on the seismogenic structure (fault or area source) that contributes most to the risk at the chosen return period, and has the magnitude and location contributing most to the loss (McGuire, 1995). Probabilistic ground-motion maps are normally developed on a regional basis for a single "reference" class of ground conditions, e.g., soft rock; one can then multiply such motions by frequency-dependent factors depending on the site conditions. Some of the difficulties inherent in this approach, which is not ideally suited for scenario studies in urban areas with complex geology, have been discussed by Frankel and Safak (1998).

While the two approaches generally point to different requirements on the side of risk analysts and managers (see, e.g., PELEM 1989), the deterministic approach is more intuitive and may be more appealing to local administrators because it assumes the occurrence of a well-defined earthquake (in terms of magnitude and source-to-site distance) that is used as input for all the successive steps of the risk analysis. However, it conceals to some extent the problem that the choice of the appropriate earthquake is a judgmental and not a quantitative issue.

Since the emphasis in this chapter is mainly on the quantification and representation of the effects that potentially control the ground shaking scenario within a limited area of interest, and especially those related to the type and severity of site effects and to the source process, such as the fault rupture geometry and directivity, we assume the deterministic approach as the natural underlying reference.

The practical feasibility of the deterministic approach may in itself be a nonnegligible problem. In many earthquake-prone areas of the world, and typically in most of the Mediterranean region, the seismicity is governed by earthquakes with maximum magnitudes in the 6.5–7.0 range, originating on faults that seldom generate unambiguous co-seismic surface ruptures, and whose geometry is ill-defined. It is consequently difficult to associate a scenario earthquake in such zones with a precisely defined tectonic lineament, as is often done in much of the western United States and Japan. This difficulty, as well as the need to provide ground-motion estimations with tools basically accessible to engineers, supports the use of simple methods. Therefore, in this chapter we shall mostly consider ground-motion representations obtained by combining appropriate attenuation relations of strong-motion parameters with digital geological or geotechnical maps of the study area. The combination is typically performed through Geographic Information System (GIS) technology in raster version. Since the attenuation relations play a central role, the following section is devoted entirely to an overview of them from the viewpoint of the user interested in scenario development.

Concerning the role of the different factors that control scenario ground motions, we devote somewhat more attention to site effects than to source effects, because the former are considered capable of causing stronger deviations with respect to average empirical predictions. Situations in which these predictions are particularly inadequate because of the occurrence of complex site effects, such as on sediment-filled valleys, are dealt with in Section 4.

2. Attenuation Relations for Ground-Motion Parameters: An Overview

The attenuation relations for ground-motion parameters are statistical regressions, either linear or nonlinear, on appropriate sets of recorded data. Since we are not interested here in the

TABLE 2 Distinctive Factors of Attenuation Relations

Ground-motion descriptor	Tectonic regime	Origin of data
• Response spectral ordinates • Macroseismic intensity • Arias intensity • Other	• Shallow earthquakes in active tectonic regions • Stable continental regions • Subduction zones	• Europe • North America • Japan • Worldwide

generation aspects of these relations but rather in their use as engineering tools, the emphasis in this section is mainly on practical application aspects.

Table 2 highlights three general criteria that can be used as a guide in the initial choice of an attenuation relation. The selection of the ground-motion parameter is linked to the scope of the scenario study and is discussed in more detail in the next section. The two other criteria—tectonics and geographical region—should be considered with great care, usually side by side. Geographical origin of data usually entails not only differences of tectonic regime affecting earthquake generation but also different prevailing geological features, different crustal attenuation properties, and to some extent, different characteristics of recording instruments.

A sufficiently general representation of attenuation relations, showing the specific independent physical variables that concur to the ground shaking estimation, has the following additive structure:

$$\log y = f_1(M, r, SD) + f_2(FT) + f_3(S) + \varepsilon \qquad (1)$$

where

y = ground-motion parameter, e.g.: peak ground acceleration (PGA) or velocity; response spectral ordinate (horizontal or vertical) for a prescribed fundamental period T of the oscillator and damping factor; Arias intensity; macroseismic intensity.

M = preferably moment magnitude M_w, or equivalent in appropriate range (e.g., M_S, M_L).

r = source distance; this can be epicentral, hypocentral (r_{hypo}), minimum distance from the fault rupture (r_{rup}), or closest distance to the vertical projection of rupture on the Earth surface; see Abrahamson and Shedlock (1997, their figure 1).

SD = source directivity factor.

FT = fault-type factor.

S = site conditions factor.

ε = random variable, introduced to account for the uncertainty of the prediction, usually assumed to have a normal distribution, with zero mean and standard deviation $\sigma_{\log y}$.

The term $f_1(M, r, SD)$ is the basic form describing the dependence on magnitude, source-to-site distance, and source rupture directivity (SD). For rock sites and y = larger horizontal PGA value (in units of g), two of the most commonly used among such basic forms are

$$f_1(M, r) = -0.038 + 0.216(M_w - 6) - 0.777 \log R$$
$$R = (r^2 + 5.48^2)^{1/2} \qquad \sigma_{\log y} = 0.205 \qquad (2)$$

(Boore *et al.*, 1993) and

$$f_1(M, r) = -1.48 + 0.266 M_S - 0.922 \log R$$
$$R = (r^2 + 3.5^2)^{1/2} \qquad \sigma_{\log y} = 0.25 \qquad (3)$$

(Ambraseys *et al.*, 1996). In both cases, R and r are in kilometers. Equation (2) is based on data from western North America, with moment magnitude M_w ranging between 5.5 and 7.0, and is valid up to 100-km distance, while Eq. (3) is applicable in Europe and the Middle East, with M_S in the 4.0–7.9 range. For $M_w \cong M_S = 7$ the values predicted by Eq. (2) are 1.2 to 2.0 times larger than those obtained with Eq. (3), depending on distance, while for $M_w = 6$ the ratio is 0.4 to 0.6. In Eqs. (2) and (3) attenuation with distance is independent of magnitude. In other relations (e.g., Abrahamson and Silva, 1997), a nonlinear magnitude-dependent factor is introduced in the distance attenuation term that produces faster attenuation of ground-motion amplitude at the smaller magnitudes.

For subduction zone earthquakes, where the distances are generally larger because of the offshore position or the depth of hypocenter, the distance dependence is r_{hypo}^{-1} (Atkinson and Boore, 1997). Other authors (e.g., Anderson, 1997; Youngs *et al.*, 1997) use instead r_{rup}, but r_{hypo} is indeed a good approximation for r_{rup} when the rupture surface is not defined.

The directivity term SD in Eq. (1) contains typically a cosine function of the angle between the direction of rupture propagation and that of waves traveling from the fault to the site. More specifically, in Somerville *et al.* (1997) the SD term is $\cos\theta$ for strike-slip faults and $\cos\phi$ for dip-slip faults, where θ and ϕ are defined as shown in Figure 1. Rupture directivity causes spatial variation in both amplitude and duration of the ground motion. Somerville *et al.* (1997) introduce the directivity term as an additive factor in existing attenuation relations, which account for the magnitude and distance dependence. Thus, using

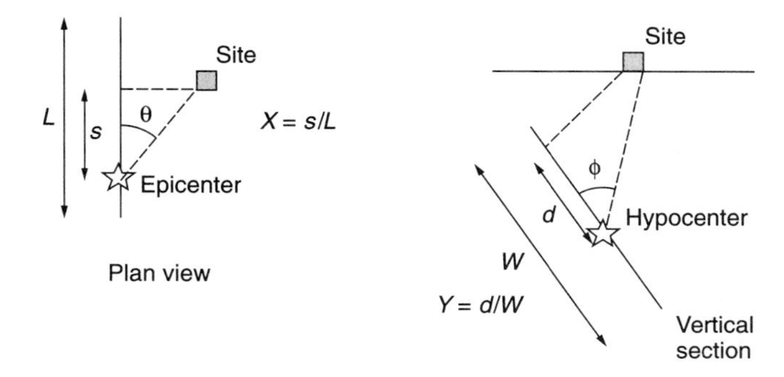

FIGURE 1 Definition of rupture directivity parameters θ and X for strike-slip fault, and ϕ and Y for dip-slip fault (from Somerville *et al.*, 1997).

the Abrahamson and Silva (1997) relation as a reference, the additional directivity term is

$$y = C_1 + C_2 X \cos\theta \qquad \text{for strike-slip faults and } M > 6.5$$

or

$$y = C_1 + C_2 Y \cos\phi \qquad \text{for dip-slip faults and } M > 6.5 \tag{4}$$

where

y = prediction residual of ln(spectral acceleration) at a given oscillator period.
X, Y = along strike and updip distance ratios, shown in Figure 1.
θ, ϕ = angles previously defined.
C_1, C_2 = period-dependent regression coefficients.

The influence of the directivity terms is significant at periods larger than 0.6 sec and increases with period: for a strike-slip fault, at $T = 2$ sec, the directivity effect on amplitude ranges between 0.6 and 1.8 times the average value neglecting directivity (for dip-slip fault the range is 0.8–1.2).

A somewhat different, analytical approach has been adopted to introduce directivity in an attenuation relation for the Arias intensity I_a, with parameters calibrated on strong-motion data from the Mediterranean region (Faccioli, 1983). This relation basically applies for a point source, and for horizontal motion has the form

$$\log I_a = 1.065 M_w - 2 \log r_{\text{hypo}} + \log D_a(\theta) - 4.63 \tag{5}$$

where I_a is in m/sec, and r_{hypo} is limited within the 10- to 50-km range. Source directivity is described by the function

$$D_a(\theta) = \frac{3 - m\cos\theta}{(1 - m\cos\theta)(2 - m\cos\theta)^2} \tag{6}$$

where θ is the angle between the direction of rupture propagation and the hypocenter-to-receiver direction, and m (<1) is the ratio between the rupture propagation and the S-wave propagation speeds, generally assumed between 0.7 and 0.85. The effects of directivity are best taken into account by the previous expression in the case of vertical (strike-slip) or steeply inclined (dip-slip) faults and rupture propagation along the strike of the fault.

The fault-type factor on the r.h.s. of Eq. (1), written out more explicitly, is

$$f_2(FT) = F_{\psi 1}(M, T) + HW_{\psi 2}(M, r_{\text{rup}}) \tag{7}$$

where $F_{\psi 1}(M, T)$ accounts for the fault style: reverse and thrust faulting generate larger high-frequency spectral ordinates than strike-slip and normal earthquakes. F is the fault-type index: it is 1 for reverse slip, 0.5 for reverse/oblique, and 0 otherwise in Abrahamson and Silva (1997). Boore *et al.* (1997) use instead a single coefficient to discriminate strike-slip earthquakes, reverse-slip ones, and the case of unspecified mechanism; $HW_{\psi 2}(M, r_{\text{rup}})$ describes the "overhanging wall" effect in reverse-fault ruptures. One attenuation relation showing this factor is due to Abrahamson and Silva (1997), where HW is equal to 1 for sites on the overhanging-wall side of the fault and 0 otherwise.

The third factor in Eq. (1), $f_3(S)$, quantifies the influence of the soil profile at the site. As a starting point, it is useful to recall the site classification of Boore *et al.* (1993) based on V_{S30}, i.e., the shear-wave velocity averaged in the upper 30 m of soil, reproduced in Table 3. A number of other commonly used descriptions, summarized in Table 4, refer to this classification.

The simplest option one can use in the estimation of response spectral ordinates, discriminating simply between rock and soil, is case I of Table 4, where C_S is a constant regression coefficient

TABLE 3 Site Classification (from Boore *et al.*, 1993)

Soil profile class	V_{S30} [m/sec]
A	>750
B	360–750
C	180–360
D	<180

TABLE 4 Available Options for the Site Factor $f_3(S)$ in Eq. (1)

Case	$f_3(S)$	Parameter values/note		Reference
I	$C_S S_S$	$S_S = 0$ for rock, $S_S = 1$ for soil C_S = regression coefficient (const.)		Spudich *et al.* (1999)
		C_S = a + b ln(PGA$_{\text{rock}}$ + c)		Abrahamson and Silva (1997)
II	$C_a S_a + C_S S_S$	Rock	$S_a = 0, S_S = 0$	Ambraseys *et al.* (1996)
	where	Stiff soil	$S_a = 1, S_S = 0$	Boore *et al.* (1993)
	S_a and S_S, or S_H,	Soft soil	$S_a = 0, S_S = 1$	
	describe the soil	Firm/stiff Quat. deposits > 10 m	$S_S = 0, S_H = 0$	Campbell (1997)
	classification and/or	Soft rock	$S_S = 1, S_H = 0$	
	deposit depth	Hard rock	$S_S = 0, S_H = 1$	
	C_a, C_S are	Rock	$S_a = 0, S_S = 0$	Sabetta and Pugliese (1996)
	regression coeff.	Shallow alluvium	$S_a = 1, S_S = 0$	
		Deep alluvium	$S_a = 0, S_S = 1$	
III	$C_S \log(V_{S30})$	V_{S30} values		Bommer *et al.* (1998) Boore *et al.* (1997)

depending on oscillator period (as in Spudich *et al.*, 1999). A considerable refinement, allowing to take into account nonlinearity in soil response, represents C_S as a function of expected PGA on rock (Abrahamson and Silva, 1997). Case II of Table 4 shows three different possibilities of seismic soil classification inspired by the four classes of Table 3; the S_a and S_s factors in this case allow also to take into account the influence of the depth of the soil deposit.

As a step toward a finer spatial resolution in the representation of soil response, one can also describe the site effects as a continuous function of the shear-wave velocity V_{S30} (case III), if known.

Especially for soil category C in Table 3, the magnitude of site effects estimated by the attenuation relations on response spectra strongly depends on the period; this is illustrated in Figure 2, which portrays the variation of the soil amplification factor $y(T;\text{soil})/y(T;\text{rock})$ for the different cases of Table 4. For $T = 0$ amplification is about 1.3, while the largest soil amplification, about 1.7, occurs at T around 1 sec. The amplification factors of the relation by Boore *et al.* (1997) actually coincide with those by Spudich *et al.* (1999). Note the significantly different amplification behavior of shallow soil deposits with respect to deep ones, consistent with the expectations.

Finally, concerning the uncertainty of the predictions, the standard deviation $\sigma_{\log y}$ is the commonly used measure of the data dispersion about the regression mean value. The typical range of $\sigma_{\log y}$ values between 0.2 and 0.3 found in most attenuation relations (for spectral ordinates it may be somewhat larger) means that the standard error band extends between about 0.6 and 1.8 times the predicted mean value.

In reality, in an attenuation relation the standard error depends on two different sources: (1) the statistical uncertainties due to the random scatter of data, and (2) the so-called epistemic uncertainties linked to the model chosen to represent attenuation.

The epistemic aspects of the standard error are difficult to define; we illustrate in Section 3.3 one example comparing

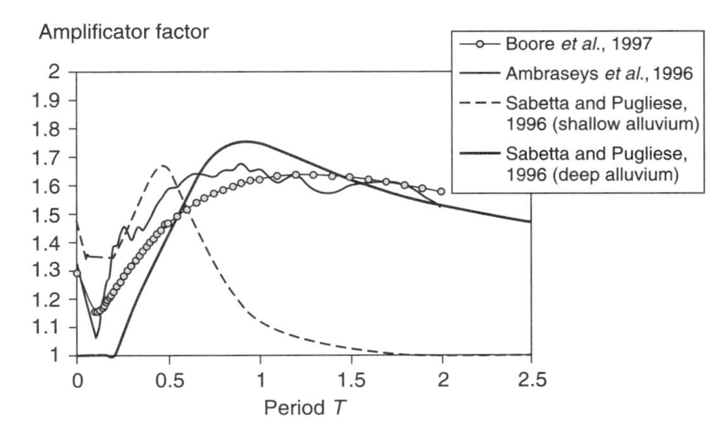

FIGURE 2 Variation with oscillator period of response spectral amplification factors of soft (category C in Table 3) and deep/shallow soil sites with respect to rock sites for different cases in Table 4.

alternative methods for the prediction of earthquake ground motion that can provide some quantitative appreciation of the variability to be expected from the use of different models.

About the random variability of the ground motion, a number of authors (mainly those who use a two-stage regression method) recognize a part of the standard error as due to the random variability of the records of the same event (Boore *et al.*, 1993), and a part collecting all the other (unspecified) components of variability. Other authors, using a random effect model analysis (Abrahamson and Silva, 1997) or a joint regression analysis (Youngs *et al.*, 1997), recognize the same intraevent variability term and, in addition, an earthquake-to-earthquake component of the variability. The total standard error is the square root of the overall variance of the regression.

It was also found (Youngs *et al.*, 1997) that the standard error depends on the magnitude of the earthquake or on the level of shaking (expressed by PGA), especially for PGA (Boore *et al.*, 1997).

3. Selection of Ground-Motion Descriptors

3.1 Ground Motions and Damage Estimation

A basic criterion for the selection of the parameters describing the shaking hazard in earthquake scenario studies is that the higher the damage state, the greater is generally the need for more detailed ground-motion information. Seismic risk scenarios do not usually produce damage predictions at the structure-by-structure level, which may require the knowledge of the full-time history of motion, but make use of some surrogates of it such as peak acceleration, or response spectral ordinates (related to global strength or displacement demand), or even macroseismic intensity. In addition, close to the seismic source, different descriptors of the ground motion may be useful. Some peculiar damage features observed, e.g., in Northridge 1994, especially for steel structures, have been linked to the distinct large-amplitude velocity pulses occurring in the near field. Because it has been found that the response spectrum cannot properly account for the destructiveness of these motions, a new measure of damage potential has been proposed, in the form of the drift demand spectrum (Iwan, 1994). On the other hand, integral measures that directly account for the ground-motion duration, such as the Arias intensity, appear as more suitable indicators of the shaking when geological damage, such as landslides and rockfalls, is of interest. Earthquake scenario maps using Arias intensity are discussed in more detail in Section 3.3.

A detailed discussion on the "best" ground-motion descriptor for earthquake scenarios is beyond the scope of this chapter since, as previously mentioned, the choice also depends on the parameters and methods used to estimate the damage to structures. Traditionally, because of scarcity of strong-motion recordings, macroseismic intensity has been used to quantify ground motion, and damage estimation methodologies were

TABLE 5 Ground Motion Descriptors and Damage Estimation Methods

Type of ground-motion parameter	Vulnerability (V)/damage (d) estimation	Object of estimation
Peak Ground Acceleration (PGA)	Vulnerability from score assignment, damage through empirical correlation e.g., $d = f(V, \text{PGA})$	Building and other structures
Acceleration or Displacement Response Spectrum	ATC–13 methodology, displacement limit state approach (Calvi, 1999; Faccioli *et al.*, 1999)	Building structures
Intensity (I_{MM}, $I_{\text{MCS}}, \ldots$)	Methods based on Damage Probability Matrices (DPM)	Building and other structures
Arias Intensity (I_a)	Empirical correlations, e.g., between I_a and failure displacements (Wilson and Keefer, 1985)	Natural/artificial slopes, ground deformation

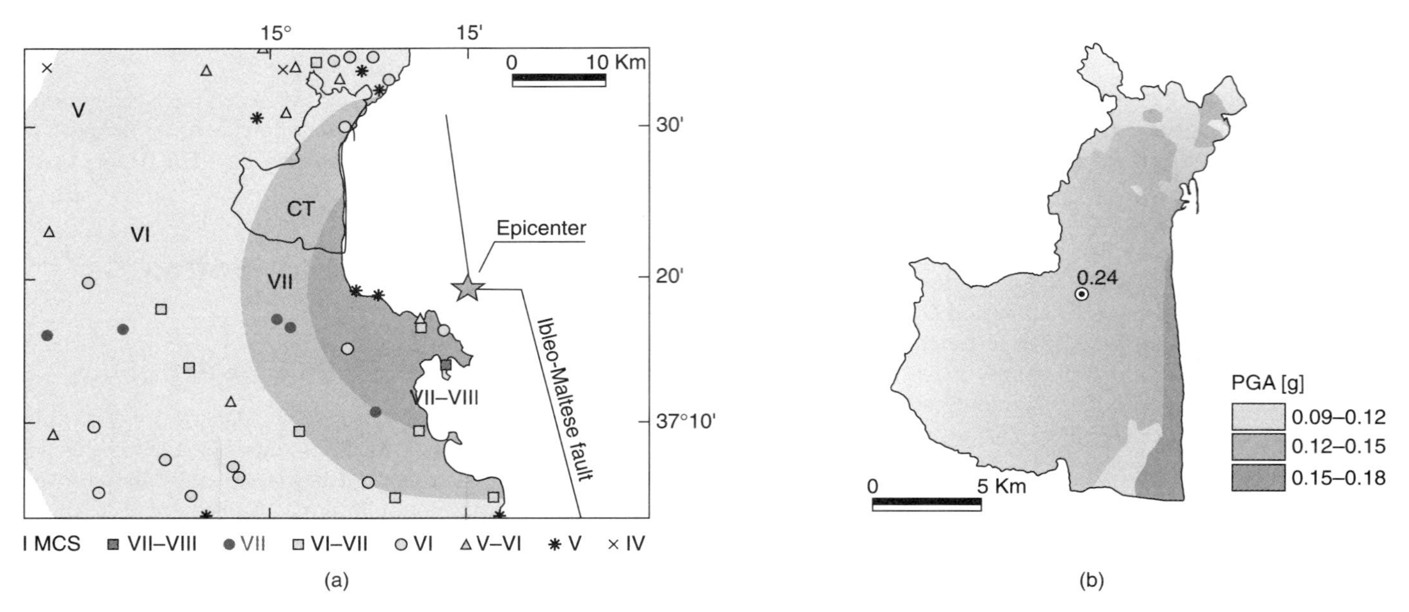

FIGURE 3 (a) Predicted and observed MCS intensities for the M_s 5.8, 1990 earthquake in eastern Sicily, including the Catania municipal area (denoted by CT). Symbols in legend are for observed intensities, while predicted intensities appear in different shades of gray and Latin numerals. (b) PGA values, in g, over the Catania municipal area predicted for the same earthquake using a geotechnical zonation map (see text); shown also is the only recorded PGA value (0.24 g).

predominantly based on this parameter. In recent years, following the vast increase of useful recorded data and the development of reliable numerical methods for simulating ground motions, instrumental measures have been increasingly used as ground-motion descriptors. The damage estimation methods have been updated accordingly, as in the US tool HAZUS (Whitman *et al.*, 1997) and in other approaches developed in Europe (e.g., Calvi, 1999; Faccioli *et al.*, 1999). In practice, in addition to intensity, peak ground acceleration (PGA) and response spectral ordinates, both in acceleration and displacement, are the parameters presently used in most earthquake scenario studies. For geological damage in hilly areas, as already pointed out, the Arias intensity is a convenient descriptor. Table 5 gives a summary of how these different descriptors of ground-motion severity can be linked to different methods of vulnerability and damage estimation.

Sensitivity considerations linked to the consequences on predicted damage resulting from the assumptions made in the creation of ground shaking maps are discussed in Section 3.4. We illustrate in the following section aspects related to the resolution of ground shaking maps resulting from different descriptors.

3.2 Resolution of Maps Using Different Descriptors: Intensity and PGA

To grasp some implications of using different hazard descriptors in terms of their spatial resolution, the comparison of predictive maps using macroseismic intensity (MCS scale) and PGA is instructive. An example is provided in Figures 3 and 4 for two different events affecting the municipal area of Catania, in eastern Sicily (Italy), which has a present population of about 500,000 and a historical record of destructive earthquakes. The maps in Figure 3 are for a moderate (M_S 5.8) earthquake that occurred in 1990, with maximum MCS intensity I_0 = VII–VIII. The epicenter, shown by a star in Figure 3a, is probably associated with a strike-slip E–W segment of a major N–S trending tectonic

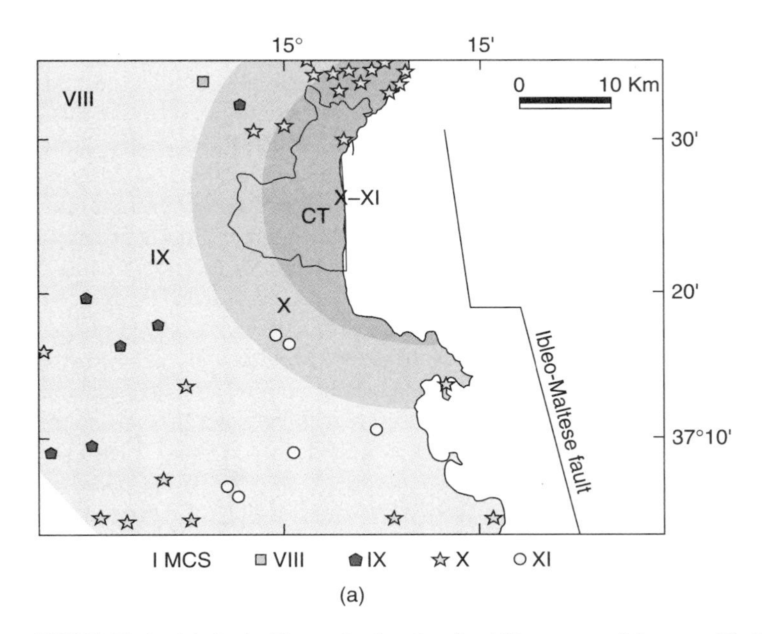

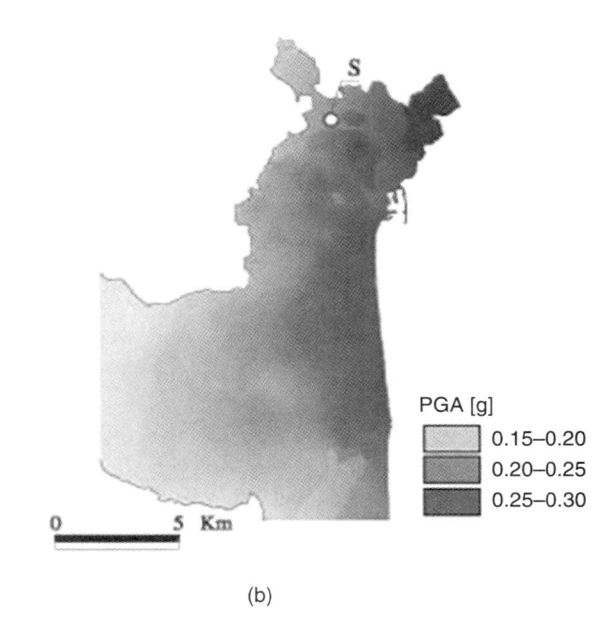

FIGURE 4 (a) As in Figure 3a, but for the $M7+$ event of January 11, 1693. (b) PGA map for Catania, generated assuming a 50-km-long rupture along the Ibleo-Maltese fault shown in Figure 3a. Site S in (b) is studied in the text, Section 4.3.

lineament, known as the Ibleo-Maltese fault (or escarpment). The relatively small size of this event allows us, for present purposes, to represent its source as a point.

The map of Figure 3a has been generated with a "circular" attenuation relation of the form $I = f(I_0, R_0, \text{distance})$ whereby I is predicted to decrease with distance only beyond the radius R_0 of earthquake effects on the built environment, and as such does not explicitly account for site conditions, as in the map of Figure 3a. The map area in this figure extends beyond the actual Catania municipality (labeled CT), to show both predicted and felt intensities in the surrounding region.

Note that the predicted variation across the Catania municipal area exceeds one intensity degree, while it is of one degree (from VII to VI) in the city area proper, which occupies the northern portion of the municipality. The felt intensity in 1990 in Catania was VI.

In contrast to the intensity map, the PGA map of Figure 3b (limited to the Catania municipal area) makes use of a geotechnical zonation in the form of a digital map of V_{S30}. The PGA map itself is generated from an attenuation relation that relies on a large set of European strong-motion data, including most of the significant records from Italian earthquakes, and is of type III in Table 4 as regards quantification of site conditions (Ambraseys, 1995). Thus, the map reflects the difference in near-surface soil properties between the city area proper, where massive and thin lava flows (from the Mt. Etna volcano) prevail, and the central and southern parts of the municipality, where the sediments of the so-called Catania plain predominate. The resolution (pixel size) of the latter map is 40 m, chosen on the basis of the density of available geotechnical borings in some city sections. The PGA values in Figure 3b correspond to the 81-percentile level; an observed value of PGA is shown for comparison in Figure 3b at the only accelerograph recording site in Catania.

Using a geotechnical zonation, as in Figure 3b, evidently introduces small-scale spatial variability in the ground-motion maps. However, in this case the distance attenuation effect predominates and the two maps in Figure 3 can be regarded as roughly equivalent in terms of input to a damage scenario for ordinary buildings, because of the moderate level of the ground motions involved, and also because of the position of the earthquake source with respect to the city.

The difference between the two types of maps, and the advantage of using an instrumental parameter, becomes much more evident if a substantially larger earthquake is considered. In the present example, the obvious candidate as a scenario earthquake is the 1693 event, with an estimated M_s between 7.0 and 7.5. This earthquake claimed over 10,000 victims (out of a total city population of 18,000) in Catania alone (Boschi *et al.*, 1995) and may have ruptured some 70 km of fault length on the Ibleo-Maltese escarpment (Sirovich and Pettenati, 1999). Intensity and PGA maps for this earthquake were constructed using the same attenuation relations as in Figure 3; the results, illustrated in Figure 4, indicate that the two maps can hardly be considered equivalent as an input for scenario damage prediction. The small-scale variability in ground motion introduced by the geotechnical zonation is now such that the local amplification of PGA values could control in certain areas the distribution of heavy damage to some classes of buildings. On the other hand, a hazard map such as that of Figure 4b could only be used as a rough preliminary scenario for emergency operations.

The same comments made on the PGA maps would qualitatively apply to maps of predicted spectral ordinates at periods $T \neq 0$.

3.3 Maps of Arias Intensity

The Arias intensity I_a, being proportional to the integral of the squared acceleration values from a strong-motion record, gives a measure of the energy released at a site by ground shaking. Hence, I_a should be a physically more meaningful parameter than peak acceleration for evaluating the susceptibility to the seismically induced failure of slopes in an area (see, e.g., Harp and Wilson, 1995). Because I_a varies in the far field as the inverse of the source distance squared, one would expect the highest concentration of slope failures to be predicted at close distances from the source. Figure 5 provides an example of the correlation observable in the M 6.9, 1995 Hyogo-ken Nanbu earthquake between I_a values (both predicted and observed) and the occurrence of slides and rockfalls on the slopes of the Rokko Mountains, steeply rising behind the city of Kobe. The predicted I_a values were generated with Eqs. (5) and (6), which account for directivity effects keeping the point source representation. In Figure 5, only the moment release on the Kobe side of the fault rupture (Irikura *et al.*, 1996) has been taken into account for the prediction, and the fault geometry proposed by Wald (1996) has been used.

The predictions of I_a in the near field appear reasonable, despite the simplification in the source description and the fact that the effects of surface geology are disregarded. The distribution of real slides is primarily controlled by slope steepness and surface geology, but it is also interesting to note that it clusters in a rather narrow strip aligned with the surface projection of the fault rupture, in agreement with previous studies (Faccioli, 1995).

3.4 Sensitivity Considerations

Sensitivity analyses on the consequences in terms of predicted damage can provide very helpful guidance to the choice of the attenuation relation for a hazard scenario. As a matter of fact, the implications of adopting one attenuation relation in preference to another—giving apparently similar ground-motion predictions—may not be obvious at first sight. We illustrate an example (Pessina, 1999) in which the scenario earthquake is the same considered in Figure 4a and two different attenuation relations were considered for PGA after a preliminary selection. The first ground-motion prediction is precisely the map of Figure 4b, while the second was generated using a relation based on a worldwide set of data recorded in extensional tectonic environments (Spudich *et al.*, 1997).

Shown in Figure 6 (top) are the two predicted frequency distributions of PGA across the Catania municipal area, using pixels of 40 × 40 m. Significant differences between the two distributions occur only for PGA > 0.25 g; only a small fraction of area is affected by values exceeding 0.30 g. However, based on fragility curves for masonry buildings in Italy, the limited differences in the PGA distributions are strongly amplified in the

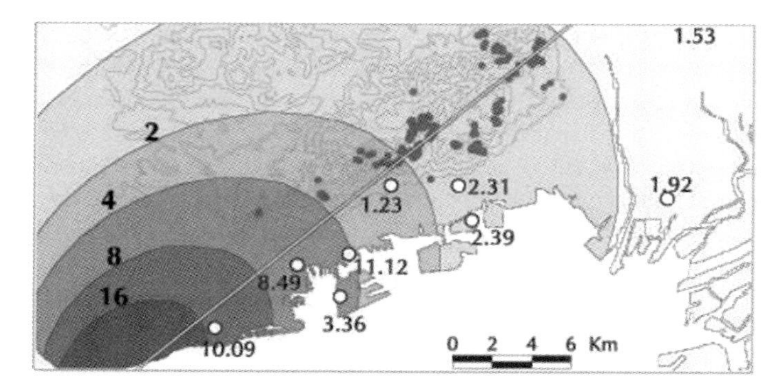

FIGURE 5 Predicted Arias intensities (I_a), shown by different shades of gray (in m/sec), in the Kobe area for the 1995 Hyogo-ken Nanbu earthquake; also shown are observed I_a values at selected strong-motion stations, the sites of landslide and rockfall occurrence (Okimura, 1995), the assumed surface projection of the ruptured fault, and the elevation contours at 100-m intervals.

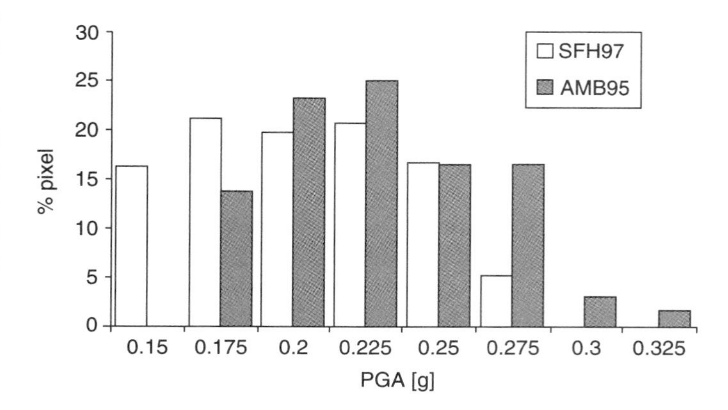

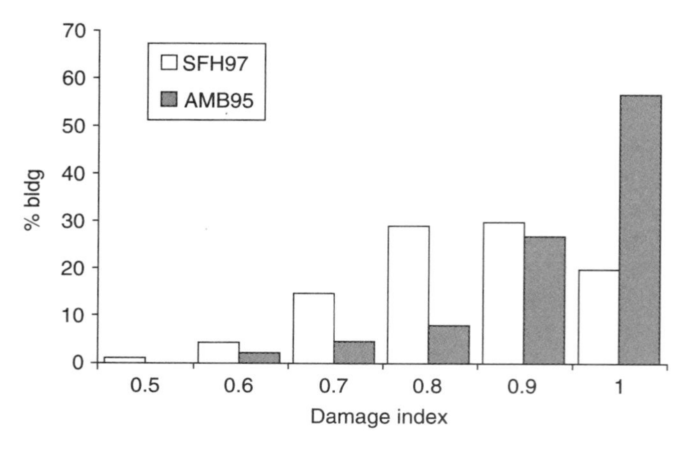

FIGURE 6 Sensitivity of damage predictions for masonry buildings in Catania to the ground motions for the destructive scenario earthquake. *Top*: frequency distributions of PGA in terms of percentage of elementary land cells (pixels) having the same level of ground shaking, predicted with the attenuation relation AMB95 (Ambraseys, 1995) and SFH97 (Spudich *et al.*, 1997). *Bottom*: predicted damage distributions, in percentage of buildings with a given damage index, for the previous two PGA distributions (from Pessina, 1999).

damage predictions. The results, portrayed in Figure 6 (bottom) in the form of mean damage distribution as a function of damage index, show a sharp increase of collapse or high damage incidence when the PGA map of Figure 4b is adopted. This is simply because the latter contains more pixels in the densely built city center where the collapse accelerations assumed for the typical masonry structures in Catania are attained.

4. Source-Related Factors

4.1 Representation of the Source Geometry

If the scenario ground shaking is to be predicted in the near-source range of events with $M > 6.0$, a finite source representation is obviously preferable even if rupture directivity effects may not be important (as is often the case of normal fault earthquakes). However, at larger distances an acceptable ground-motion scenario for $M \sim 7$ events can be estimated with a "circular" attenuation relation, e.g., from the epicenter, provided an appropriate site classification is used to account for average amplification effects. An example is presented in Color Plate 26 for the destructive $M = 6.8$, 1980 Irpinia earthquake in southern Italy.

The PGA map in Color Plate 26 was generated with an attenuation relation widely used in Italy, which separates rock/stiff sites from shallow alluvium sites (Sabetta and Pugliese, 1996; see case II in Table 4 and Fig. 2). A zonation into these two broad soil categories for the region of interest was derived from a digital geological map of Italy (originally at 1:500,000 scale), ignoring the detailed geotechnical characterization available for the accelerograph sites, and the acceleration map was then generated via GIS in raster format. The instrumental epicenter, the surface projection of the rupture area in the form of two adjacent rectangles (slightly simplified from Cocco and Pacor, 1993), and the area enclosed within the intensity IX (MCS) isoseism are also shown in Color Plate 26, together with the accelerograph site locations and the recorded horizontal PGA values. In the bar diagram of Color Plate 26, the observed peak accelerations are compared to the values predicted using both the finite source representation of Color Plate 26 and the distance from the epicenter.

At most accelerograph stations the latter assumption yields slightly better predictions. However, for the earthquake in question none of the acceleration records was obtained near the source, where the differences arising from the different source representation are expected to be large.

A point worth stressing here is that if a finite source is to be introduced for producing ground-motion maps of future earthquakes, its geometry must be assumed *a priori*, and this can prove especially difficult in the case of blind faults. In countries with a long and well-documented historical seismicity record, one can, however, rely on intensity maps of the strongest earthquakes in the region. In the last example, the intensity IX (MCS) isoseism appears to be a reasonable indicator of the size and position of the ruptured area, although some of its features seem well correlated with the presence of sedimentary deposits causing amplification. The determination of the geometry of earthquake sources from macroseismic intensity data is supported by a recent study on Italian earthquakes (Gasperini *et al.*, 1999).

4.2 Source Directivity

The 1994 Northridge and 1995 Hyogo-ken Nanbu events have provided, in different ways, strong evidence that the damage distribution occurring in the near-source region of a strong earthquake depends on the energy-focusing effects produced by the directivity of rupture on the rupturing fault. The impact of this evidence has been such that the directivity effects can now be accounted for through appropriate attenuation relations for response spectral ordinates; see Eqs. (1) and (4)–(6). On the other hand, as regards engineering design, the latest versions of building codes have introduced for the highest seismicity zones a classification of earthquake sources (in terms of magnitude and seismic rate) and "near-source factors" that can increase by up to a factor of 2 the ordinates of the design spectrum (ICBO, 1997). For these reasons, we are not dwelling here any further on the influence of directivity on the response spectra.

Unlike the response spectrum, which is rather insensitive to the significant duration of ground motions, the Arias intensity I_a, should be affected to a greater extent by directivity effects because it more closely reflects the influence of duration. A predictive I_a map, which takes directivity into account, generated with Eqs. (5) and (6), has already been shown in Figure 5; this prediction has been compared with that obtained for the same earthquake by an attenuation relation for I_a that disregards directivity (Wilson and Keefer, 1985). The comparison, not shown here, indicates that the residuals at close azimuths from the fault strike are in most cases decreased when Eqs. (5) and (6) are used.

4.3 Empirical vs Synthetic Ground-Motion Predictions

The preferred alternative to the use of empirical relations is obviously represented by numerical simulations based on realistic source representations (including finite dimensions, rupture propagation, etc.), which can be performed by a variety of seismological models and computational methods for wave propagation, and can produce ground-motion synthetics at any site within an area of interest. A recent overview of such methods and their performance has been provided in the context of a recent international prediction experiment (see Irikura *et al.*, 1999). Direct modeling is especially useful in parametric studies for the quantification of the uncertainties in the predicted spatial distribution of ground shaking, and of their dependence on the physical parameters of source and propagation path. Because of computational burden and lack of resolution in the crustal models, these calculations are typically limited to frequencies around 1 Hz, which leaves out a vast range of primary interest for engineering structures. Extension to higher frequencies can be achieved by adding an appropriate random component to the deterministically simulated time histories.

In the aforementioned scenario studies for Catania, in addition to the empirical ground-motion predictions, three different approaches were used to generate acceleration synthetics from the fault geometry and earthquake magnitude assigned to the destructive 1693 earthquake (Fig. 4). The variability of the results yielded by the different approaches at a given point in studies of this nature is an empirical measure of the model-to-model, or epistemic, uncertainty discussed in Section 2. As an illustration, we compare in Figure 7 the acceleration response spectra predicted at a site (S in Fig. 4) located on massive lava flows, at about 15 km source distance. The empirically predicted spectrum, shown in Figure 7 with its standard deviation band, has been generated with attenuation relations based on European data, already mentioned as an example of case II in Table 4 (Ambraseys *et al.*, 1996, labeled ASB96). The curve labeled 2D is the spectrum of a representative synthetic computed at point S by a 2D high-resolution spectral element analysis along a cross section perpendicular to the coastline (Priolo, 1999). This curve roughly matches the mean plus one standard error bound of the empirical prediction up to about 1 sec period. The bump of the 2D synthetic spectrum at longer periods is related to a strong phase reflected at a deep interface; this is a typical example of a feature that would be difficult to predict by empirical relations.

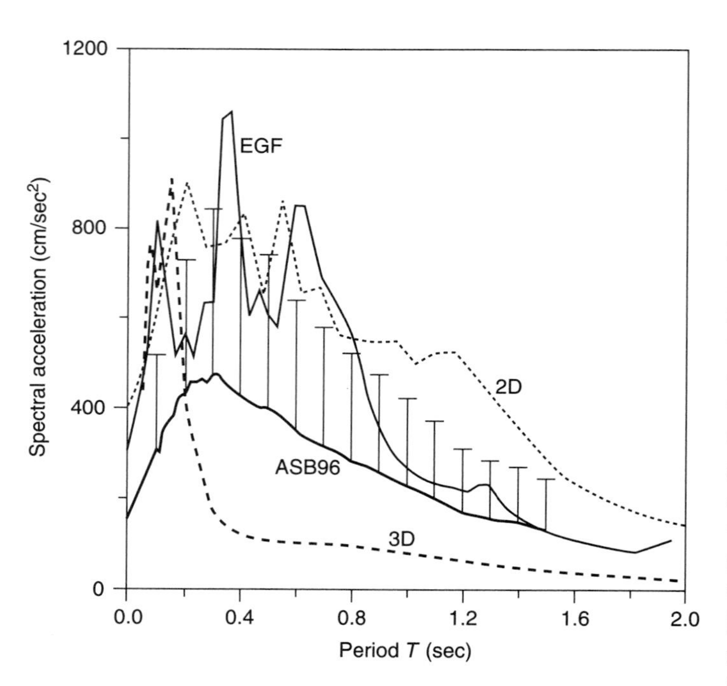

FIGURE 7 Comparison of acceleration response spectra generated at site S of Figure 4, on massive lava flows, by using empirical attenuation relations (ASB96, shown with plus one standard error bars), synthetic from 2D spectral analysis (labeled 2D), synthetic from 3D high-frequency method (labeled 3D), and empirical Green's function summation (labeled EGF).

The curve labeled 3D in Figure 7 is generated by a high-frequency approximation approach for the far-field analytical Green's functions combined with randomly generated models of the rupture process on the fault (Zollo *et al.*, 1997). Because of its computational efficiency, this approach is especially useful for obtaining estimates of the uncertainty in the spatial distribution of ground motions attached to the complexity of the source process, but the lack in intermediate and long period components in the associated spectrum of Figure 7 does not seem compatible with a $M7+$ earthquake at relatively close distance. Because of the high-frequency approximation, the model in question is more suitable for predicting peak ground acceleration than the full response spectra.

An independent assessment of the previous results has been obtained by the empirical Green's function (EGF) method, using a horizontal acceleration component recorded during the previously mentioned 1990 earthquake at a site denoted GRR (PGA = 0.04 g), located on the coast some 50 km north of the epicenter, on thick lava flows from Mt. Etna. GRR is to some extent representative of the ground conditions prevailing in several older sections of Catania, including site S under consideration. However, since S is only about 30 km from the 1990 epicenter, a simple distance geometrical spreading correction was applied to the amplitude of the original record used as EGF.

Taking a record from the 1990 earthquake as EGF for the large 1693 event seems acceptable in view of the probable location of the 1990 epicenter on the Ibleo-Maltese fault. A random summation scheme has been adopted to synthesize the ground motion for the large event, of which one only needs to specify the stress drop (Ordaz *et al.*, 1995). The extended target source area is approximated by a point source, whose rupture duration, however, is in accordance with its dimension. This method was selected in view of the uncertainty on the exact position and dimensions of the 1693 source and of the good results reported for the Mexican earthquakes by Ordaz *et al.* (1995). Also, strong directivity effects are not expected in Catania given its position with respect to the most probable seismogenic fault (Fig. 4). By taking a target seismic moment corresponding to M 7.0, and a stress drop of 100 bar for both the small and the large events, 22 different synthetic acceleration histories were generated, from which an average PGA of 0.28 g is obtained at the Catania site S, with a standard deviation of 0.07 g, in rather good agreement with the value shown on the map of Figure 4. One simulated time history with peak value equal to the sample average was then used to compute the EGF spectrum in Figure 7. Similar comparisons performed at soil sites confirm the indication of Figure 7, i.e., that in first approximation, in the absence of strong directivity effects, the standard error band associated with the empirical prediction is also a realistic measure of the spread of the synthetics generated by different methods, i.e., of the model-to-model uncertainty.

Layer	Material	V_S (m/sec)
1a	Alluvium	150
1b		250
1c		320
2	Marl	500
3	Limestone	1000–1500

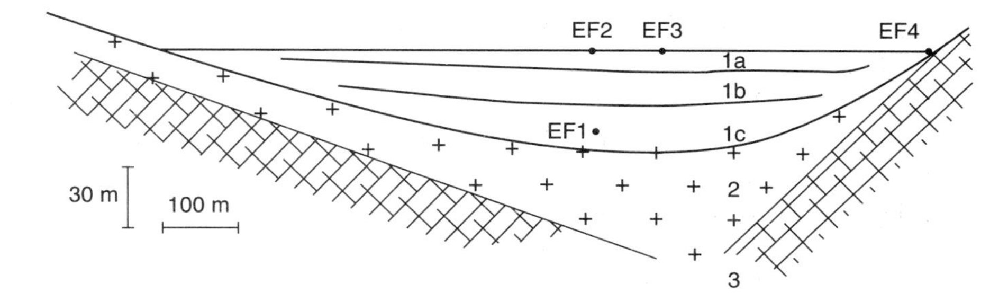

FIGURE 8 SW–NE cross section of alluvium valley at IONIANET test site in Kefalonia Island (Greece); EF1, EF2, EF3, and EF4 are the locations of digital accelerographs, while V_S denotes shear-wave velocity. Right-hand part of figure modified after Psarropoulos and Gazetas (1998).

5. Alternative Tools for the Treatment of Complex Site Effects

The inadequacy of zoning at a purely geotechnical level in the presence of markedly 2D or 3D near-surface configurations, e.g., alluvial valleys or sedimentary basins, has already been mentioned. The quantification of seismic site effects on such configurations can be a challenging task in a scenario investigation, especially if time-consuming numerical simulations are to be avoided for lack of sufficient geological and geophysical data. Of particular concern is the occurrence of 2D/3D resonant response of sediment valleys, which—although thoroughly understood from a theoretical standpoint (see Bard and Bouchon, 1985) and documented by remarkable data for weak-motion earthquakes (e.g., Paolucci *et al.*, 2000)—remains still scarcely supported by strong-motion observations. On the other hand, the main reason for the concern is the possible occurrence of soil response spectral amplification well in excess of the empirical values illustrated in Figure 2 as well as of those provided by 1D response analyses. We illustrate by means of two real case histories the applicability of alternative tools for quantifying 2D site effects: simplified analytical approaches and seismic noise spectra.

5.1 Use of Simplified Analytical Models: The IONIANET Test Site

The first case history stems from strong-motion observations obtained by the IONIANET digital accelerograph network, located on a small sediment valley in Kefalonia Island, in the most seismically active region of Greece. A cross section of the valley with the accelerograph location is shown in Figure 8. The valley, about 0.8 km wide at the test site and 6 km long, is filled with soft-to-medium dense Quaternary sediments (interbedded sand and clay layers) and surrounded by hills of limestone and marl. Geotechnical investigations were limited to two boreholes, one drilled near the rock outcrops at the edges and one in the middle of the valley, which disclosed about 46 m of soil overlying marl.

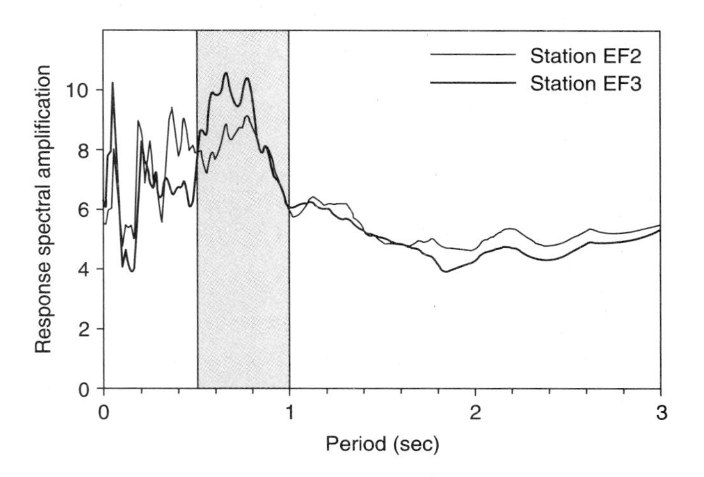

FIGURE 9 Response spectra ratios EF2/EF4 and EF3/EF4 for horizontal motion in the direction normal to the plane of Figure 5, averaged over three earthquakes. Shaded band indicates range of vibration periods where 2D resonance has probably occurred.

The shear-wave velocity values in Figure 8 were estimated from N_{SPT} measurements using empirical correlations.

In 1997 and 1998 IONIANET recorded four moderate earthquakes; three of them, with $4.2 < M < 5.2$, occurred during the same seismic sequence, at 20- to 45-km epicentral distance from the network, with focal depth of 5–7 km. The largest peak accelerations at EF4, on rock, were about 0.05 g, while on soil (at EF2) they reached about 0.22 g during the same event.

Previous analyses of IONIANET data had shown that 2D wave propagation and layering of the valley sediments must be accounted for to explain the amplitudes and duration of the accelerations recorded at EF3 and EF4 (Psarropoulos and Gazetas, 1998). We measure site amplification by the ratio between the response spectra of the soil recordings at EF2 and EF3 and those on rock at EF4, averaged over the three earthquakes. Figure 9 displays the ratios for the horizontal motion parallel to the valley axis. The six- to tenfold spectral amplification occurring up to 1 sec period should be noted in the first place. Particularly at EF3, a broad peak may be noted for vibration periods between about

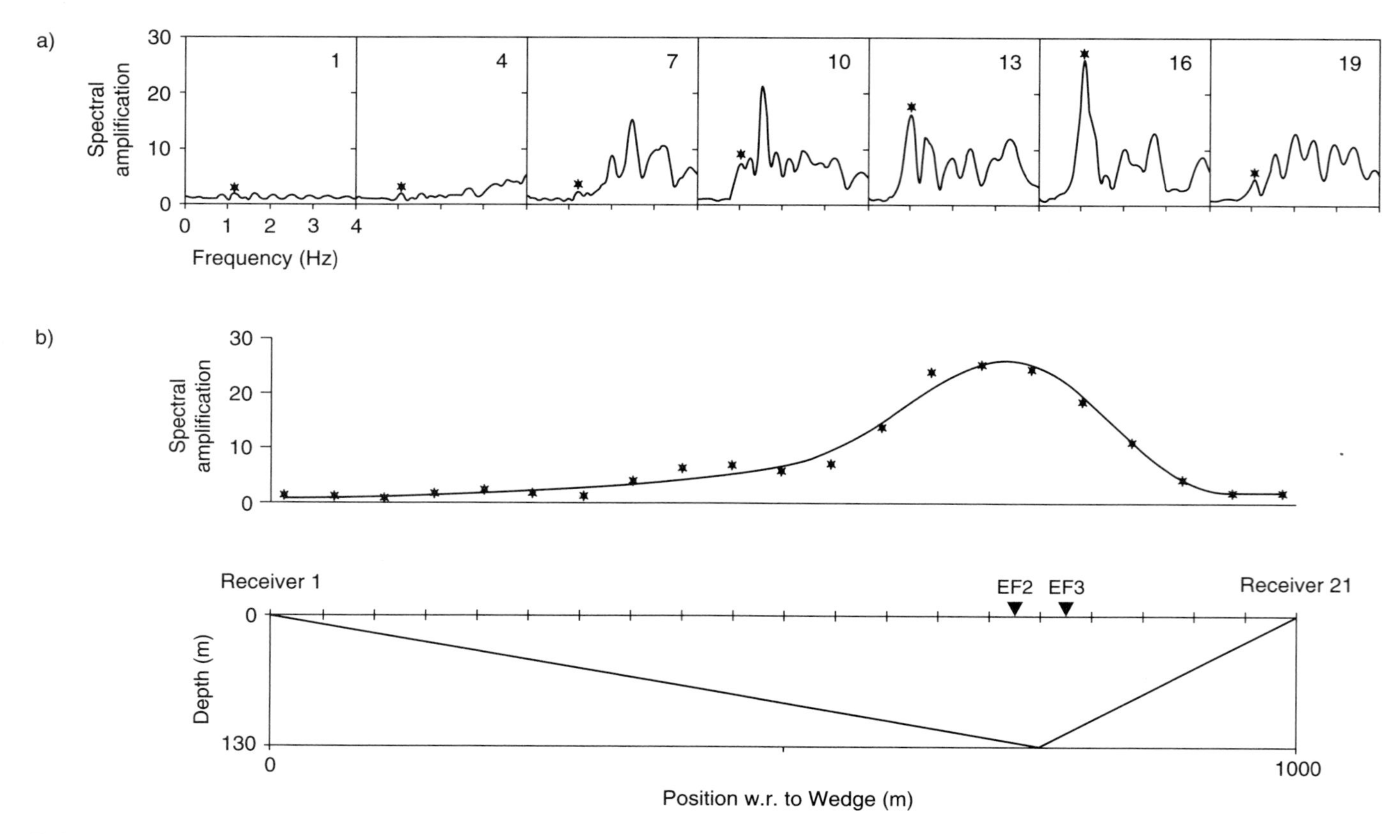

FIGURE 10 (Top) IONIANET site: amplification functions at selected surface receivers for antiplane (*SH*) motion calculated with the simple model of triangular homogeneous valley, shown at bottom. Receivers are numbered 1 to 21, from left to right, and equally spaced. Solid star in amplification plots indicates 2D resonance peak at 1.1 Hz. Receiver 16, exhibiting the highest peak, is between stations EF2 and EF3. Graph in the middle depicts variation of resonant peak amplitude across the valley.

0.5 and 1.0 sec (shaded band in Fig. 9). The observed amplification is substantially higher than the average empirical estimates shown in Figure 2.

To investigate 2D resonance effects as a possible cause of the high amplification observed, a very simple two-dimensional model of the valley was analyzed, consisting of homogeneous soil (with $V_S = 400$ m/sec) and having a triangular shape, shown in Figure 10. The simplicity stems from the fact that for specific values of the edge angles, a nearly exact solution of the problem can be quickly obtained for a vertically propagating plane *SH* wave (Paolucci *et al.*, 1992). The analytical amplification functions calculated with this method at different points across the valley, reproduced in Figure 10, clearly display a resonance peak common to all locations, which becomes very prominent in the zone where the valley sediments are deepest. The resonance peaks are indicated by a star in the figure. The common frequency of this peak is 1.1 Hz, while its amplitude variation across the valley is depicted in the middle graph in Figure 10. Based on well-known studies on the seismic response of valleys (Bard and Bouchon, 1985), the analytical results in Figure 10 are consistent with the fundamental resonant mode for *SH* waves at the frequency of 1.1 Hz (period 0.9 sec). Making due allowance for the approximations in the valley geometry and material properties of the simplified model, we conclude that the amplification peak in the shaded band of Figure 9 is caused by the fact that the three recorded earthquakes have also excited the 2D resonant response of the valley in its fundamental *SH* mode. As a rule of thumb, this example indicates that the 2D amplification effects would require, for the most critical sites, an extra multiplicative factor of 2 to 3 with respect to the predictions of attenuation relations (similar to those of current seismic codes such as UBC97).

In reality, due to the moderate earthquake magnitudes involved, the strongest ground response at the IONIANET site occurred between about 0.1 and 0.3 sec period, and little energy was present in the 0.6–0.9 interval where 2D resonance effects are identified. However, the 2D resonant motion could be strongly excited by higher-magnitude (say $M > 6$) future events, such as have repeatedly occurred in the past in the Ionian Islands. Obviously, nonlinearity of soil response should be considered in this case.

In conclusion, the IONIANET data interpretation indicates that relatively simple analytical or numerical models allowing

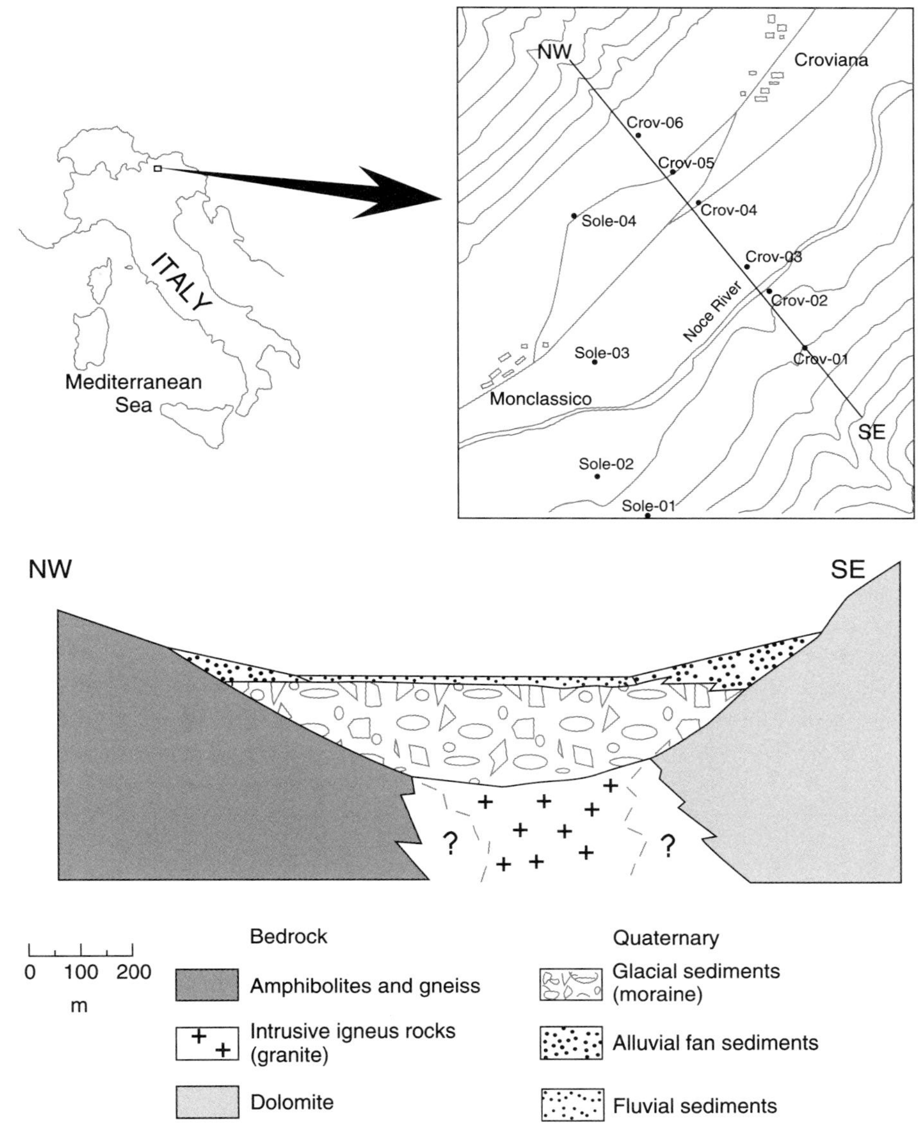

FIGURE 11 Location of the Val di Sole site and indicative vertical cross section of the valley along the NW–SE trace shown on the upper right map.

performance of quick parametric studies are essential for understanding complex site effects and determining their magnitude.

5.2 Use of Seismic Noise Spectral Ratios: The Val di Sole Site

A powerful alternative approach to the identification of resonant response on sedimentary valleys and basins relies on measurements of seismic noise and associated use of single-station, H/V spectral ratios (Nakamura, 1989; Field *et al.*, 1995). In fact, several studies show that this method can provide reliable estimates of dominant frequencies of motion, although it tends to underestimate the amplitude of the response when compared with the technique of the classical spectral ratios, it being generally accepted that the H/V ratios give a lower bound of the soil response amplitude for this frequency (see, e.g., Lachet and Bard, 1994).

Thus, to assess the capability of the H/V ratio technique in identifying the basic features of 2D response in a real valley, a second case history follows. The case study refers to a valley (Val di Sole) located in the Italian Alps (Fig. 11). The valley is filled by three types of Quaternary sediments. The oldest are glacial sediments (moraine), while alluvial fans are present on the hillsides of the valley, and the surface materials consist of fine detritic sediments accumulated by the river Noce. A sketch of the structure of the valley, based on observations made close to the study zone, is given in Figure 11. The sediment thickness is estimated at more than 200 m at the valley center. The alluvial fans have a thickness probably exceeding 50 m at some points, while the river sediments are thinner, probably less than 20 m. The shape ratio (thickness/half-width) of the valley is approximately 0.3.

Ambient noise was measured at 10 sites on the valley (Fig. 11), using a three-component seismometer of 1 sec period. After performing the FFT of the digitized noise recordings, the resulting spectra were smoothed with a sliding window and the spectral H/V ratio for each window and horizontal component was calculated. Finally, the average spectral ratio on both horizontal components was obtained. The resulting H/V spectral ratios are presented in Figure 12.

Apart from site Sole-01, which is located over very thin soil close to the sediment-rock (dolomites) contact, most of the remaining sites show an amplification peak between 1.16 and 1.77 Hz. This peak is maximum at the center of the valley (Crov-04 and Sole-03) and decreases toward the edges, where higher dominant frequencies exist with peaks of lesser amplitude (Crov-06 and Sole-04). These observations are in agreement with findings by Tucker and King (1984) and by Bard and Bouchon (1985) about the characteristics of the 2D response of valleys. Since no *in situ* data are available on the material properties of the sediments, they were estimated from correlations applicable in the same geological context (e.g., Frischknecht *et al.*, 1998); Table 6 summarizes the properties used in the models. A 1D dynamic response analysis at valley sites was performed first, based on the cross section of Figure 11. The results (not shown here) indicated that a 1D model does not fit the observations: computed resonance frequencies decrease from the edge toward the center of the valley, where they are lower than the observed one. Additionally, computed peak amplifications decrease from the edges toward the center of the valley, while the observed ones vary in just the opposite sense.

These preliminary analyses indicate that the dynamic ground response in the valley is more complex than in a simple 1D case, and it is possibly related to a 2D resonance condition. Due to the lack of data on the structure of the valley, the occurrence

TABLE 6 Mechanical Properties of the Val di Sole Materials (estimated)

Material	(kg/m^3)	V_P (m/sec)	V_S (m/sec)	Q
Fluvial	2000	1400	200	50
Alluvial fan	2100	1500	250	50
Moraine	2100	2000	700	100
Bedrock	2500	4000	2500	200

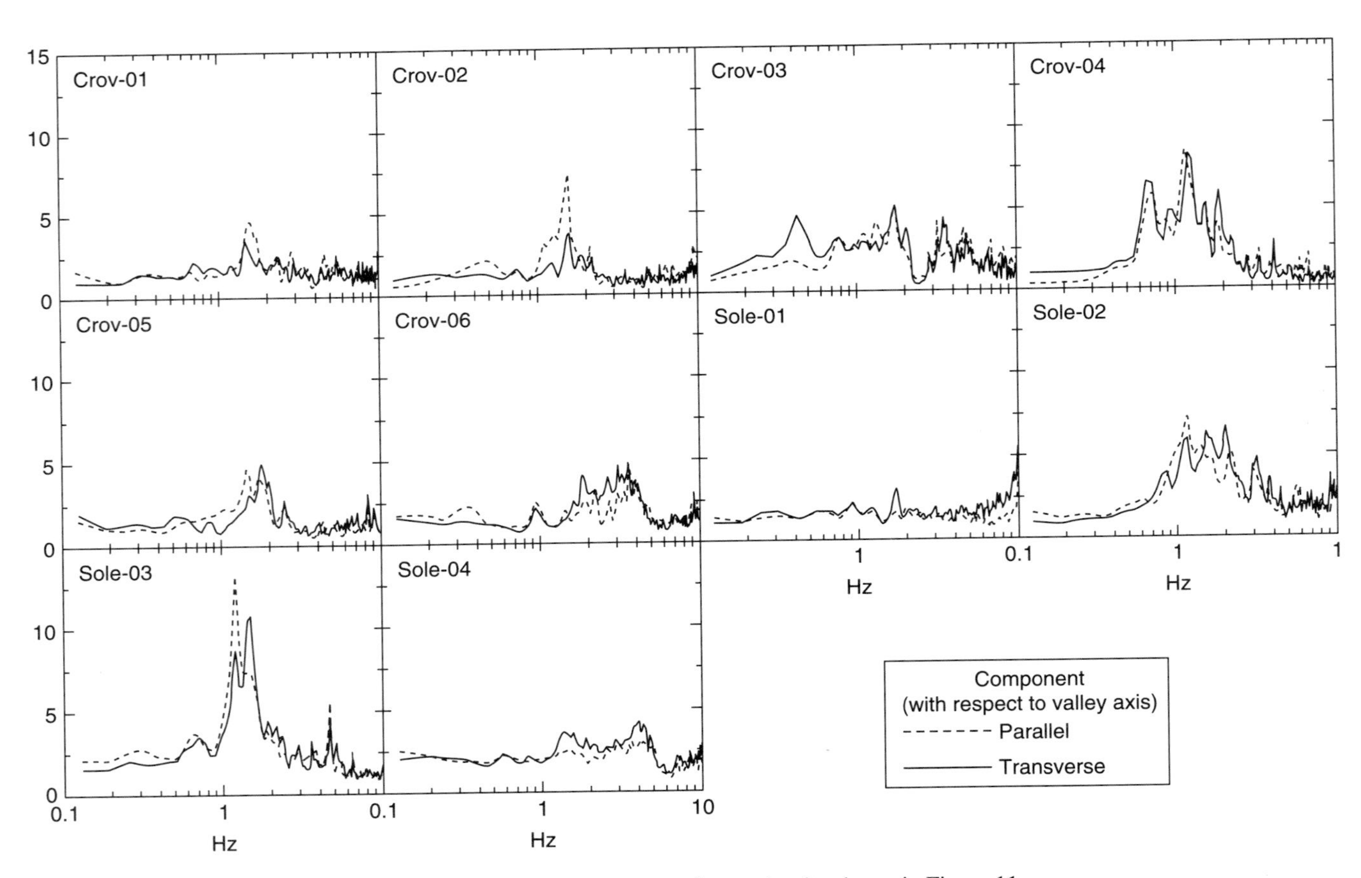

FIGURE 12 H/V spectral ratios of ambient noise measurements taken at the site shown in Figure 11.

of a 2D resonance was investigated first with a simplified method that allows estimation of the fundamental 2D resonance frequency, but not its amplification, through Rayleigh's approximation (Paolucci, 1999). A 2D valley model 1100 m wide was used, consisting of two layers (20 and 190 m thick) embedded in a half-space, with shear velocities equal to those for fluvial, moraine, and bedrock materials, respectively. The resulting 2D resonant frequencies vary between 0.95 and 1.10 Hz, depending on the shape of the model (elliptical, sinusoidal) and type of wave (*SH* or *SV*) used in the analysis. Such values are only slightly lower than those observed in the H/V ratios of noise at the center of the valley (1.16 Hz).

Based on this evidence, a full time-domain 2D analysis was performed next with the hybrid computer code AHNSE (Casadei and Gabellini, 1997), and transfer functions for the surface receivers were obtained. The code can combine spectral elements (Faccioli *et al.*, 1997) and—in the more irregular near-surface domains—finite elements; in 2D, it performs a *P–SV* propagation analysis. High-velocity materials (bedrock and moraine) were modeled with spectral elements while the low-velocity sediments were discretized by finite elements. The grid size was designed to propagate a maximum frequency of 9 Hz. As input motion, a plane Ricker wavelet with 3-Hz peak frequency was used.

To investigate the effect of the shape of the valley, the sediments were considered homogeneous, with properties identical to the moraine. In the frequency domain, Figure 13 compares the numerically computed amplification functions at different positions in the valley with the observed H/V ratios. For this comparison, sites Sole-01 to Sole-04 were projected onto the profile used in the simulation, shown in Figure 11.

The calculated fundamental frequency of resonance of the valley at 1.12 Hz is very evident in Figure 13. At this frequency, the maximum amplification occurs at the center of the valley and is about 2.5 times greater than the sediment/rock seismic impedance ratio (1D amplification); this level decreases toward the edges of the model. Additional evidence of 2D response is the good agreement between H/V ratios and numerical amplification functions at stations Sole-03 and Crov-04 near the valley center, as well as at Sole-02. This rather interesting result suggests that the H/V ratios, in addition to being a reliable tool for determining dominant frequencies of motion, could also be used for amplification estimates, at least for moderate excitation levels.

We may conclude from this second case history that inexpensive methods based on noise measurements are among the tools that can reliably be used to assess the effects of some basic features of ground response on complex geologic configurations.

As an appropriate complement to the previous case histories, in a recent study aimed at estimating a simplified "basin amplification factor" for alluvial valleys, Chávez-García and Faccioli (2000) show that, unlike the 1D site effects considered in current engineering practice, the additional amplification caused by 2D effects may not be safely evaluated using the soil properties in the near-surface layers, such as V_{S30}. As a minimum, some information is necessary also on the bedrock properties, to estimate the velocity contrast with respect to the underlying soft sediments. According to Chávez-García and Faccioli, this contrast is the most significant parameter controlling the severity of 2D effects.

To make complex site effects amenable to relatively simple empirical predictions for scenario studies, the most practical

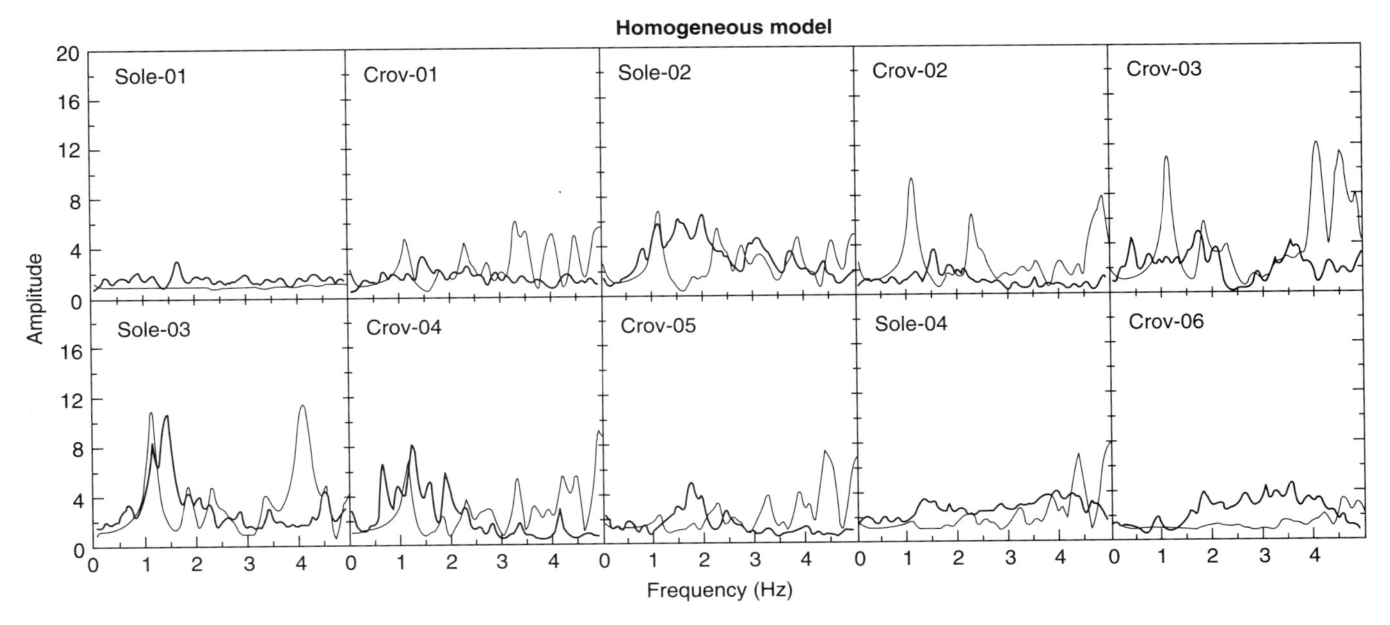

FIGURE 13 Comparison between H/V ratios of ambient noise (thick line) and computed 2D transfer functions (thin line). For this comparison, sites Sole-01 to Sole-04 were projected to profile used in the simulation. *P–SV* components transverse to valley axis are used for H/V ratios.

way is probably to perform regression analyses (incorporating a few simple geometrical and mechanical parameters) on the data already recorded by the dense accelerograph arrays installed on 2D/3D geological configurations around the world. While the use of average basin amplification factors derived from statistical analyses of many numerical simulations and observations could only be used as a rough first approximation in a scenario study where the source–receivers geometry is assigned, some of the general indications contained in the study by Chávez-García and Faccioli may provide useful guidelines, such as the following:

1. 2D effects in alluvial valleys are significant only at periods smaller than the 1D resonance period evaluated at the point where the valley sediments are deepest.
2. For realistic values of the sediment/bedrock velocity ratio (2.5 to 3.5), the basin amplification factor, i.e., the additional amplification factor by which one should multiply the 1D factor, is between 1.5 and 2.5.

6. Closing Remarks

After reviewing, as a basic preliminary step, the issue of a deterministic vs a probabilistic definition of a scenario earthquake, we have summarized the features of the main tool used in the engineering approach to the creation of a ground shaking scenario, namely attenuation relations for ground-motion parameters. Although such relations are discussed also in other chapters of this handbook, we have deemed it indispensable to recall here, from the user's standpoint, some quantitative aspects of the predictions obtained thereby, including the measures of statistical uncertainty.

We have subsequently discussed the selection of hazard descriptors for earthquake scenarios and highlighted some features of typical ground-motion maps generated through GIS technology, using for illustration a few examples from recent studies, notably those for the southern Italian city of Catania. Sensitivity analyses in terms of predicted damage are certainly apt to provide significant guidance on the choice of the tools for predicting ground motion. Also, sophisticated geotechnical zonation maps can be conveniently incorporated to produce ground shaking maps for an area, although they cannot bring into the picture the influence of sedimentary basins and of other nonsurficial geologic structures that are likely to generate complex site amplification effects.

One critical question in scenario studies concerns the extent to which physical factors that are likely to have a strong influence on the spatial distribution of ground motion can be realistically accounted for by the simple approach based on attenuation laws. For the source-related aspects of the problem (such as source finiteness and rupture directivity), the prediction tools made available after the Northridge and Kobe earthquakes can be considered adequate. However, the answer is mostly conditioned by the *a priori* knowledge of the geometry and orientation of the seismogenic structures, highly uncertain in many populated and highly seismic zones of the world. The rational exploitation of historical intensity data in regions with a sufficiently long historical record may be the best way to tackle this problem.

The situation is quite different for soil amplification arising on sediment-filled valleys and basins. Alternative methods for identifying complex site effects and determining their magnitude have been extensively illustrated in a specific section with the support of two real case histories. The quantitative results indicate that, although amplification factors in specific frequency bands can be well beyond the reach of current empirical predictions, some clue is already available on the values of the "basin amplification factors" coming into play in complex geological configurations of this type.

Regarding the relative performance of empirical predictions, and of advanced numerical methods for generating synthetic time histories at any desired location, the available results suggest that the former are probably adequate for most scenario studies using response spectra, in the sense that the model-to-model variability is basically within the standard error band of the empirical predictions.

Acknowledgments

We are grateful to L. Veronese and O. Groaz (Geological survey of the Trento Province administration) who collected the Val di Sole field data, and to J. Delgado (University of Alicante, Spain) who processed the data and performed the associated 2D numerical analyses. The Geological Survey (Servizio Geologico) of Italy provided the 1:500,000 digital geological map of Italy.

References

Abrahamson, N. A., and K. M. Shedlock (1997). Overview. *Seism. Res. Lett.* **68**, 9–23.

Abrahamson, N. A., and W. J. Silva (1997). Empirical response spectral attenuation relations for shallow crustal earthquakes. *Seism. Res. Lett.* **68**, 94–109.

Ambraseys, N. N. (1995). The prediction of earthquake peak ground acceleration in Europe. *Int. J. Earthquake Eng. Struct. Dyn.* **24**, 467–490.

Ambraseys, N. N., K. A. Simpson, and J. J. Bommer (1996). Prediction of horizontal response spectra in Europe. *Int. J. Earthquake Eng. Struct. Dyn.* **25**, 371–400.

Anderson, J. G. (1997). Non parametric description of peak acceleration above a subduction thrust. *Seism. Res. Lett.* **68**, 86–93.

Atkinson, G. M., and D. M. Boore (1997). Some comparison between recent ground-motion relations. *Seism. Res. Lett.* **68**, 24–40.

Bard, P. Y., and M. Bouchon (1985). The two-dimensional resonance of sediment-filled valleys. *Bull. Seism. Soc. Am.* **75**, 519–541.

Bommer, J. J., A. S. Elnashai, G. O. Chlimintzas, and D. Lee (1998). Review and development of response spectra for displacement-based seismic design. *ESEE Research Report N. 98-3*, Civ. Engng. Dept., Imperial College, London.

Boore, D. M., W. B. Joyner, and T. E. Fumal (1993). Estimation of response spectra and peak acceleration for Western North America earthquakes: an internal report. *Open-File Report 93-509*, US Geological Survey, Menlo Park, California.

Boore, D. M., W. B. Joyner, and T. E. Fumal (1997). Equations for estimating horizontal response spectra and peak acceleration for Western North America earthquakes: a summary of recent work. *Seism. Res. Lett.* **68**, 128–153.

Borcherdt, R., L. Lawson, V. Pessina, J. Bouabid, and H. Shah (1995). Applications of Geographic Information System Technology to Seismic Zonation and Earthquake Loss Estimation, State-of-art Paper. *Proc. Fifth Int. Conf. on Seismic Zonation*, Nice, France, III, 1933–1973.

Boschi, E., G. Ferrai, P. Gasperini, E. Guidoboni, G. Smriglio, and G. Valensise (1995). "Catalogo dei forti terremoti in Italia dal 461 a.C. al 1980." ING-SGA, Bologna (Italy), 973 pp. and a CD-ROM.

Calvi, G. M. (1999). A displacement-based approach for vulnerability evaluation of classes of buildings. *J. Earthquake Eng.* **3**, 411–438.

Campbell, K. W. (1997). Empirical near-source attenuation relationships for horizontal and vertical components of peak ground acceleration, peak ground velocity, and pseudo-absolute acceleration response spectra. *Seism. Res. Lett.* **68**, 154–179.

Casadei, F., and E. Gabellini (1997). Implementation of a 3D coupled spectral element/finite element solver for wave propagation and soil structure interaction simulations. *European Commission JRC Institute for Systems, Informatics and Safety Rep. EUR 17730 EN*, Ispra, Italy.

Chávez-García, F., and E. Faccioli (2000). Complex site effects and building codes: making the leap. *J. Seismology* **4**, 23–40.

Cocco, M., and F. Pacor (1993). Space-time evolution of the rupture process from the inversion of strong motion waveforms. *Ann. Geophys.* **36**, 109–130.

Cornell, C. A. (1968). Engineering seismic risk analysis. *Bull. Seism. Soc. Am.* **58**, 1583–1601.

Faccioli, E. (1983). Measures of strong ground motion derived from a stochastic source model. *Soil Dyn. Earthquake Eng.* **2**, 135–149.

Faccioli, E. (1995). Induced hazards: earthquake triggered landslides. State-of-art Paper. *Proc. Fifth Int. Conf. on Seismic Zonation*, Nice, France, III, 1908–1931.

Faccioli, E., F. Maggio, R. Paolucci, and A. Quarteroni (1997). 2D and 3D elastic wave propagation by a pseudo-spectral domain decomposition method. *J. Seismology* **1**, 237–251.

Faccioli, E., V. Pessina, G. M. Calvi, and B. Borzi (1999). A study on damage scenarios for residential buildings in Catania city. *J. Seismology* **3**, 327–343.

Field, E. H., A. C. Clement, V. Aharonian, P. A. Friberg, L. Carroll, T. O. Babaian, S. S. Karapetian, S. M. Hovanessian, and H. A. Abramian (1995). Earthquake site response study in Giumri (formerly Leninakan), Armenia, using ambient noise observations. *Bull. Seism. Soc. Am.* **85**, 349–353.

Frankel, A., and E. Safak (1998). Recent trends and future prospects in seismic hazard analysis. *Proc. Specialty Conf. on Geotechnical Earthq. Engng. and Soil Dynamics, ASCE Spec. Technical Publ. No. 75* **I**, 91–115.

Frischknecht, C., M. Gonzenbach, Ph. Rosset, and J. J.Wagner (1998). Estimation of site effects in an alpine valley. A comparison between ground ambient noise response and 2D modeling. *Proc. XI European Conf. on Earthquake Engineering*. Paris, September.

Gasperini, P., F. Bernardini, G. Valensise, and E. Boschi (1999). Defining seismic sources from historical earthquakes felt reports. *Bull. Seism. Soc. Am.* **89**, 94–110.

Harp, E. L., and R. C. Wilson (1995). Shaking intensity thresholds for rock falls and slides: examples from 1987 Whittier Narrows and Superstition Hills earthquake strong-motion records. *Bull. Seism. Soc. Am.* **85**, 1739–1757.

International Conference of Building Officials (ICBO) (1997). Uniform Building Code, Vol. 2, Structural Engineering Design Provisions, Chap. 16, Div. IV. Whittier, California.

Irikura, K., T. Iwata, H. Sekiguchi, and A. Pitarka (1996). Lesson from the 1995 Hyogo-ken Nanbu earthquake: why were such destructive motions generated to buildings? *J. Natural Disaster Sci.* **17**, 99–127.

Irikura, K., K. Kudo, H. Okada, and T. Sasatani, Editors (1999). Simultaneous simulation for Kobe. In: "The Effect of Surface Geology on Seimic Motion," *Proc. 2nd Int. Symp. on the Effects of Surface Geology on Seismic Motion*, Yokohama, December 1998, **3**, 1283–1514.

Iwan, W. (1994). Near-field considerations in specification of seismic design motions for structures. *Proc. 10th European Conf. on Earthq. Engin.*, August 28–September 2, Vienna, Austria, **1**, 257–267.

Lachet, C., and P. Y. Bard (1994). Numerical and theoretical investigations on the possibilities and limitations of Nakamura's technique. *J. Phys. Earth* **42**, 377–397.

McGuire, R. (1995). Scenario earthquakes for loss studies based on risk analysis. *Proc. 5th Int. Conf. on Seismic Zonation,* October 17–19, 1995, Nice, France, II, 1325–1333.

Nakamura, Y. (1989). A method for dynamic characteristics estimation of subsurface using microtremor on ground surface. *QR Railway Tech. Res. Inst.* **30**, 25–33.

Okimura, T. (1995). Damage to mountain slopes and embankments in residential areas. Preliminary Report on the Great Hanshin earthquake, Japan. Soc. of Civil Eng., 340–346.

Ordaz, M., J. Arboleda, and S. K. Singh (1995). A scheme of random summation of an empirical Green's function to estimate ground motions from future large earthquakes. *Bull. Seism. Soc. Am.* **85**, 1635–1647.

Paolucci, R., M. M. Suárez, and F. J. Sánchez Sesma (1992). Fast computation of SH seismic response for a class of alluvial valleys. *Bull. Seism. Soc. Am.* **82**, 2075–2086.

Paolucci, R. (1999). Shear resonance frequencies of alluvial valleys by Rayleigh's method. *Earthquake Spectra* **3**, 503–521.

Paolucci, R., E. Faccioli, F. Chiesa, and R. Cotignola (2000). Searching for 2D/3D site response patterns in weak and strong motion array data from different regions. Proc. 6th Int. Conf. on Seismic Zonation, Palm Springs, November 12–15, 2000, CD-Rom edited by EERI, Oakland, California, 6 pp.

PELEM (Panel on Earthquake Loss Estimation Methodology) (1989). Estimating losses from future earthquakes. *Panel Report*, National Academy Press, Washington, D.C.

Perkins, J., and J. Boatwright (1995). On shaky ground. *Association of Bay Area Governments Publication No.* P95001EQK, Oakland, California.

Pessina, V. (1999). Empirical prediction of the ground shaking scenario for the Catania area. *J. Seismology* **3**, 265–277.

Priolo, E. (1999). 2-D spectral element simulation of destructive ground shaking in Catania. *J. Seismology* **3**, 289–309.

Psarropoulos, P., and G. Gazetas (1998). Surface and borehole accelerograms versus spectral element simulation of alluvial basin in Cefalonia. *Proc. 2nd Int. Conf. on Earthquake Geotechnical Engineering,* Lisbon, 157–162.

Sabetta, F., and A. Pugliese (1996). Estimation of response spectra and simulation of nonstationary earthquake ground motions. *Bull. Seism. Soc. Am.* **86**, 337–352.

Sirovich, L., and F. Pettenati (1999). Seismotectonic outline of Eastern Sicily; an evaluation of the available options for the earthquake fault rupture scenario. *J. Seismology* **3**, 213–233.

Somerville, P. G., N. F. Smith, R. W. Graves, and N. A. Abrahamson (1997). Modification of empirical strong ground motion attenuation relations to include the amplitude and duration effects of directivity. *Seism. Res. Lett.* **68**, 199–222.

Spudich, P., J. Fletcher, M. Hellweg, J. Boatwright, C. Sullivan, W. Joyner, T. Hanks, D. Boore, A. McGarr, L. Baker, and A. Lindh (1997). SEA96—A new predictive relation for earthquake ground motion in extensional tectonic regimes. *Seism. Res. Lett.* **68**, 190–198.

Spudich, P., W. B. Joyner, A. G. Lindh, D. M. Boore, B. M. Margaris, and J. B. Fletcher (1999). SEA99—A revised ground motion prediction relation for use in extensional tectonic regimes. *Bull. Seism. Soc. Am.* **89**, 1156–1170.

Tucker, B. E., and J. L. King (1984) Dependence of sediment-filled valley response on the input amplitude and valley properties. *Bull. Seism. Soc. Am.* **74**, 153–165.

Wald, D. (1996). Slip history of the 1995 Kobe, Japan, earthquake, determined from strong motion, teleseismic, and geodetic data. *J. Phys. Earth* **44**, 489–503.

Whitman, R. V., T. Anagnos, C. A. Kircher, H. J. Lagorio, R. S. Lawson, and P. Schneider (1997). Development of a national earthquake loss estimation methodology. *Earthquake Spectra* **13**, 643–661.

Wilson, R., and D. Keefer (1985). Predicting areal limits of earthquake induced landsliding. In: "Evaluating earthquake hazards in the Los Angeles region" (J. Ziony, Ed.), *U.S. Geological Survey Professional Paper 1360*, 317–493.

Youngs, R. R., S. J. Chiou, W. J. Silva, and J. R. Humphrey (1997). Strong ground motion attenuation relationships for subduction zone earthquakes. *Seism. Res. Lett.* **68**, 58–73.

Zollo, A., A. Bobbio, A. Emolo, A. Herrero, and G. De Natale (1997). Modelling of ground acceleration in the near source range: the case of 1976, Friuli earthquake. *J. Seismology* **1**, 305–319.

Editor's Note

Please see Chapter 57 by Anderson for a general discussion of strong-motion seismology.

63

Kyoshin Net (K-NET), Japan

Shigeo Kinoshita
Yokohama City University, Yokohama, Japan

1. Introduction

After the Kobe (Hyogoken-nanbu) earthquake in 1995, the Japanese government decided to increase the number of strong-motion observation stations, to upgrade the observation network, and to release future strong-motion records as soon as possible. The National Research Institute for Earth Science and Disaster Prevention (NIED) of the Science and Technology Agency was given the responsibility to implement the program.

Kyoshin Net (K-NET) is one product resulting from this one-year program. The word *Kyoshin* stands for "Strong" (*Kyo*) "quake" (*shin*). The K-NET is a system that transmits strong-motion data on the Internet, data that are obtained from 1000 observatories deployed all over Japan. The K-NET was constructed on the basis of three policies. The first is to carry out systematic observation. All of the K-NET stations use the same seismograph, K-NET95, and they are all installed at free-field sites. The second is to create a network of uniform spacing. The K-NET consists of observatories having an almost equal station-to-station distance. The third policy is the release of all strong-motion records on the Internet. These easily available data files include the soil structures of all sites, obtained by downhole measurement.

We started to release the K-NET information, including strong-motion data, on June 3, 1996, through our graphical user interface on the World Wide Web. The text home page version of the K-NET Internet site has been available for quicker distribution of K-NET data since April 1, 1997. The Internet capacity in Japan is so poor that users found it impossible to connect to the K-NET Internet site immediately after the occurrence of moderate earthquakes. To resolve this problem, we constructed two new mirror sites and started to release K-NET information through the two new sites beginning on April 1, 1997.

In this article, we report on the hardware and software configurations and the development of the K-NET. The many uses to which K-NET data has been put during the last 6 years indicate that the K-NET information, without any access restrictions, is important not only before and long after a shock, but also immediately after a shock, and that the data provide an effective contribution to earthquake disaster countermeasures.

2. K-NET Instrumentation

2.1 Observation Network

The K-NET consists of 1000 strong-motion observation stations, a control center, and two mirror sites of the control center. Figure 1 shows the distribution of K-NET stations. The average station-to-station distance is about 25 km. This spacing of observation sites is designed to sample the epicentral region of an earthquake with a magnitude of 7 or larger occurring anywhere in Japan.

Each station is installed on a site 3 meters square. It commonly consists of a housing that is made of fiber-reinforced plastic (FRP), a concrete base on which a strong-motion seismograph is installed, facilities for electric power, a telephone line with lightning arresters, and a fence. The housing is designed to withstand snow of a depth of 4 m. Figure 2 shows the layout of a typical observatory. The effects of the concrete base and installation on the recorded motion appear in frequencies exceeding 10 Hz (Ikeura *et al.*, 1999). At sites where the temperature falls to below −20°C, the base is constructed about 80 cm below the ground's surface, as shown in Figure 2. Thus, the K-NET effectively offers a uniform, free-field, strong-motion data set throughout Japan.

The headquarters of the K-NET is the control center in Tsukuba city. The control center fully monitors the seismographs, acquires strong-motion data from the K-NET95 seismograph by telemetry, prepares the database for the strong-motion data files and earthquake catalogs, and releases the strong-motion files on the Internet. The procedure of the retrieval and

INTERNATIONAL HANDBOOK OF EARTHQUAKE AND ENGINEERING SEISMOLOGY, VOLUME 81B ISBN: 0-12-440658-0

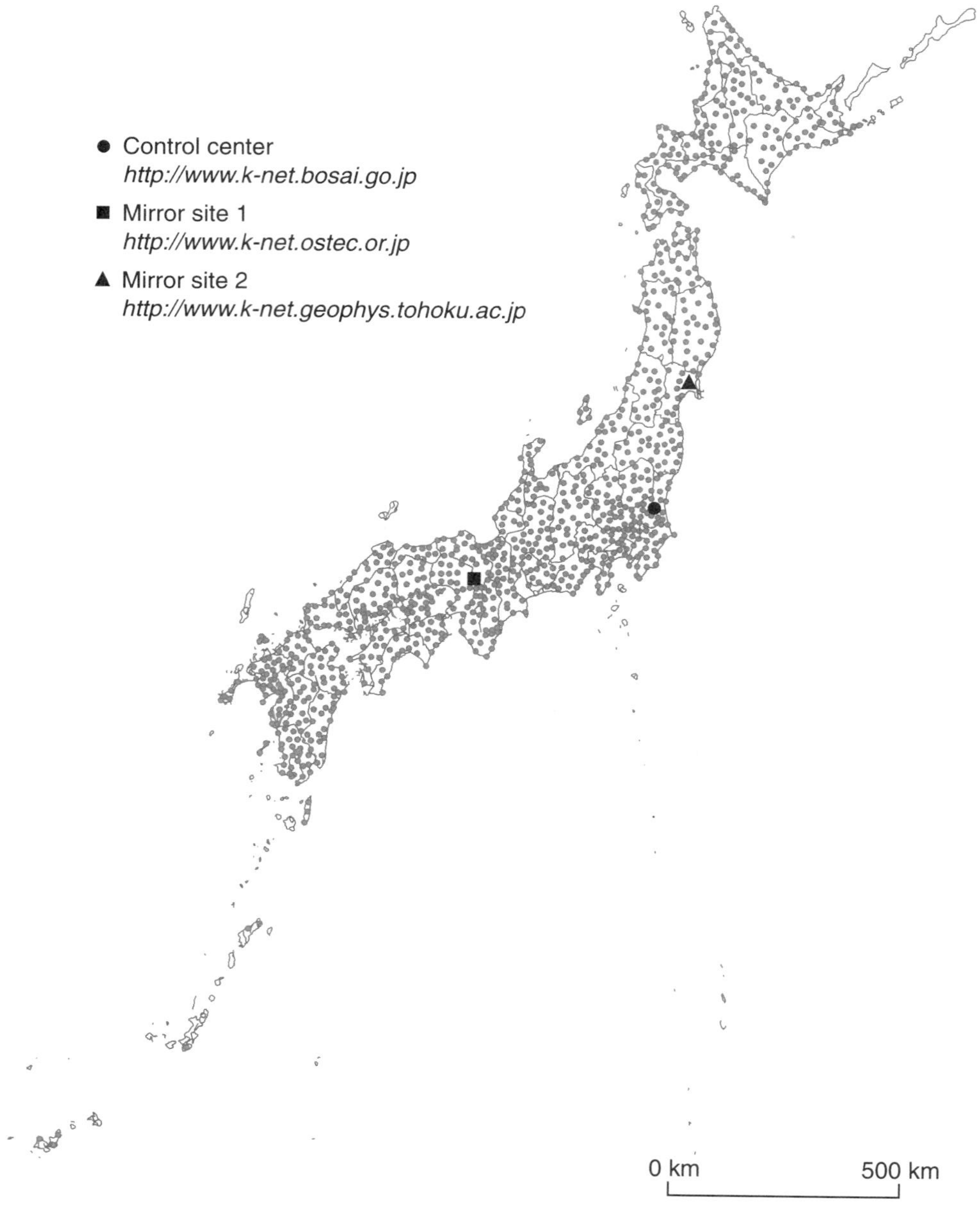

FIGURE 1 Distribution map of the K-NET strong-motion observatories. These 1000 stations, the control center, and two mirror sites comprise the K-NET. Reprinted from Kinoshita (1998) with permission from the Seismological Society of America.

release of data is as follows: When an earthquake occurs, the Japan Meteorological Agency (JMA) promptly determines and releases the source parameters (i.e., the origin time, epicenter location, depth, and magnitude) through a JMA weather satellite, provided that JMA observatories report the JMA intensity to be more than 3. The first task of the control center is to receive the JMA source parameters through an emergency information receiver installed at the center. After receiving this information, the control center automatically estimates the distribution of maximum acceleration by using an empirical function relating maximum acceleration to magnitude and distance (Fukushima and Tanaka, 1990) and then starts to retrieve the strong-motion records on the basis of the estimated maximum acceleration map.

The K-NET95 seismograph has two RS232C ports. The control center usually occupies one of the two ports and communicates with the K-NET95 by a dial-up procedure. The communication rate is automatically determined by the line condition. The maximum data transmission rate is 9600 bits/sec. The

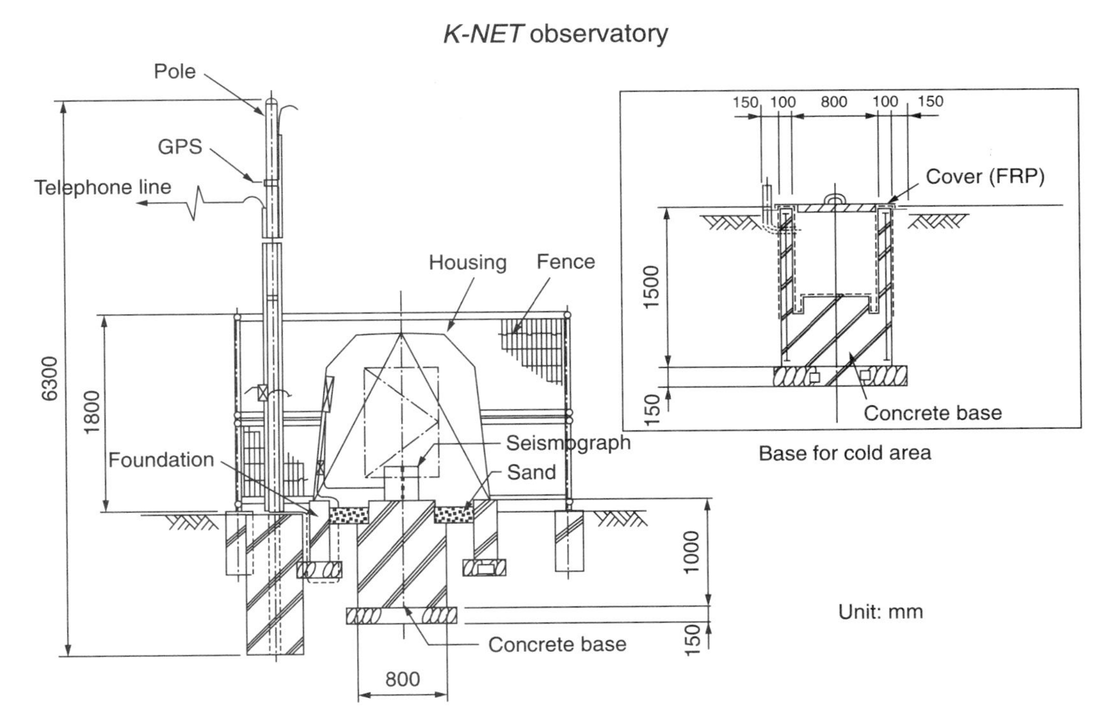

FIGURE 2 Layout of a typical observatory. In cold areas, the seismograph base is constructed beneath the ground's surface. Reprinted from Kinoshita (1998) with permission from the Seismological Society of America.

communication protocol is based on AT commands. After retrieving the strong-motion records obtained by the K-NET95 seismographs, the control center creates strong-motion data files with a common header that includes the JMA prompt source parameters and then creates a database of strong-motion files. Simultaneously, the earthquake catalog database is updated. These databases are maintained on a database server with a memory of 120 GB, which is installed in the control center. Finally, the control center simultaneously sends the strong-motion files to the three WWW servers installed at the control center and the two mirror sites. The servers make strong-motion data files available on the Internet. Usually, the release of K-NET data on the Internet is made within several hours of the occurrence of an earthquake in Japan. For example, in the case of the earthquake of March 16, 1997 ($M = 5.4$), we obtained 209 three-component seismograms and released these data after about 3 hours.

Another important job of the control center is to keep the seismographs in an operational condition. The K-NET95 seismograph continuously maintains a monitor file that has information on the state-of-health of the seismograph. The control center regularly checks these monitor files in order to maintain the K-NET. When the control center finds abnormal data in a monitor file, it orders a specified maintenance company to repair the seismograph.

The other RS232C port of the K-NET95 seismograph is connected to a local government office. The local government may be able to obtain the strong-motion data more quickly than the control center and to retrieve the data from the K-NET95 by visiting the site if AC power is lost or the telephone line is disconnected. This compensates for the control center's dependence on possibly unreliable telephone telemetry.

2.2 Type K-NET95 Seismograph

The strong-motion seismograph, type K-NET95 manufactured by Akashi Co., is an accelerograph with a wide frequency band and dynamic range. This seismograph is similar to a Kinemetrics K2. The K-NET95 has the following specifications.

2.2.1 Sensor

The sensor, type V403BT, is a triaxial, force-balance accelerometer with a natural frequency of 450 Hz and a damping factor of 0.707 (standard values). The output is 3 V/g and the resolution is more than 0.1 mGal.

2.2.2 Recording System

The recording system of the K-NET95 consists of six parts as shown in Figure 3. The amplification of the strong-motion records stored on IC memory is adjusted by the gain of a DC amplifier $G(f)$, where f is frequency in Hz. The analog low-pass filter $F(f)$ is an antialias filter for the A/D converter $C(f)$. This filter consists of a two-stage RC filter with the corner frequencies of 800 kHz and 160 kHz. The A/D converter $C(f)$ is a 24-bit type with a converter clock frequency of 1.6384 MHz. Strictly speaking, this A/D converter consists of a 1-bit sigma-delta modulator and the following digital decimation filter. This digital filter is a linear phase finite impulse response (FIR) filter with

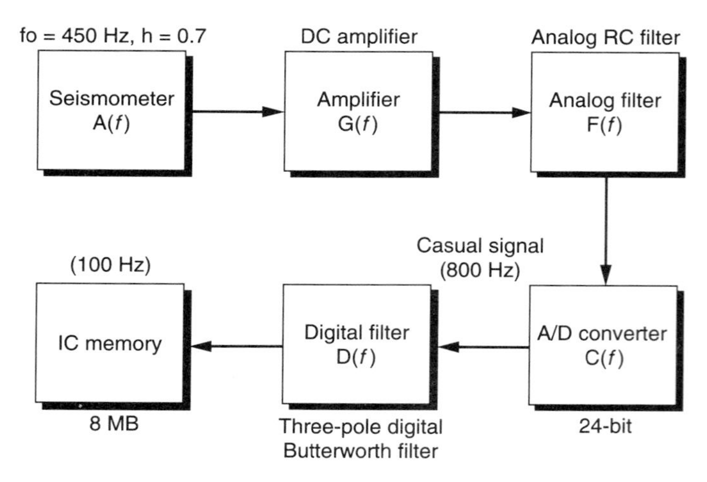

FIGURE 3 Block diagram of the K-NET recording system. The output of the A/D converter retains the causality of ground motion, and the sampling frequency is 800 Hz. The digital decimation filter D(f) converts the sampling frequency from 800 Hz to 100 Hz.

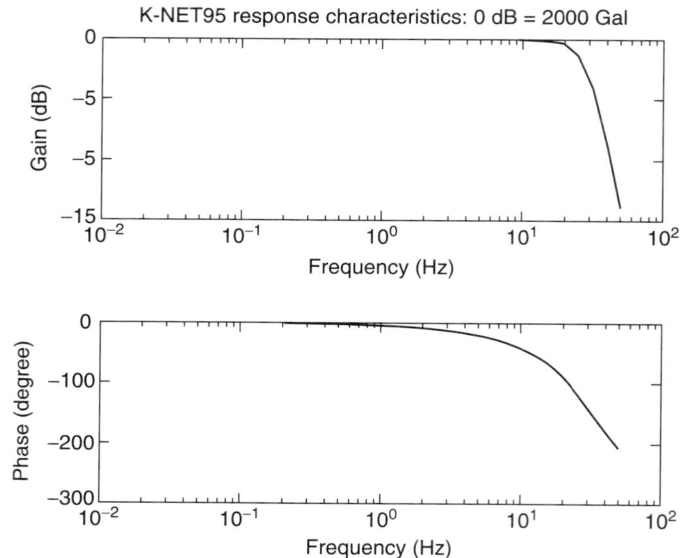

FIGURE 4 Overall frequency characteristics of the K-NET95 strong-motion seismograph.

a tap length of 1.25 ms, and the over sampling ratio is 2048. Thus, the output of $C(f)$ is a signal with a sampling frequency of 800 Hz that retains the causality of the original signal. The digital filter $D(f)$ is also a decimation filter. This filter is a three-pole Butterworth filter with a corner frequency of 30 Hz, and it also decreases the sampling frequency from 800 Hz to 100 Hz. This filter is designed by applying the bilinear transform to an analog Butterworth filter. Finally, the signal with a sampling frequency of 100 Hz is stored in IC memory. A flash memory card with a recording capacity of 8 MB is used for data storage. The maximum available time for recordings is 2.5 h. The maximum measurable acceleration is 2000 Gal. Strong-motion signals stored in the K-NET95 are deleted only by a command from the control center. This means that the K-NET95 keeps the main shock data in case of an earthquake swarm.

Figure 4 shows the overall frequency response characteristics of the K-NET95. The response characteristics are approximately equal to those of a three-pole digital Butterworth filter with a sampling frequency of 800 Hz and a filter corner at 30 Hz. Thus, instrument correction may be easily performed in the Fourier domain. However, it is difficult to execute instrument correction in the time domain because the digital Butterworth filter with a sampling frequency of 100 Hz does not preserve the digital Butterworth characteristics with a sampling frequency of 800 Hz in the frequency range from 30 Hz to 50 Hz. The reason is due to the decimation filter used in the K-NET95 recording system, in which the sampling frequency decreases from 800 Hz to 100 Hz.

2.2.3 Time Code

Timing in the K-NET95 is controlled by using Global Positioning System (GPS) satellite systems. A GPS antenna is installed at each site, and the K-NET95 has a GPS receiver.

2.2.4 Communication

The K-NET95 has two RS-232C communication ports. These ports are connected to modems, and the maximum speed of communication is 9600 bits/sec. Communication is executed using AT commands.

2.2.5 AC and DC Power

The main power for the K-NET95 is 100V AC. The K-NET95 has a battery with a capacity of 24 Ah. This battery is charged by a trickle-charger with a maximum current of 0.5 A at 12 V. Thus, the K-NET95 can function for approximately 24 hours using a fully trickle-charged battery if AC power is lost, because the power consumption of the K-NET95 is approximately 11 VA.

2.2.6 Packaging Case

The packaging case of the K-NET95 is made of FRP and is waterproof to a depth of 3 m. The dimensions are as follows: 180 mm (H), 380 mm (L), and 280 mm (W). The weight of the K-NET95 is approximately 7 kg.

2.2.7 Resolution and Dynamic Range of the K-NET95

The actual performance of the K-NET95 was tested at the factory and at a field test site, station TKN. This test yielded the following results (Kinoshita, 1998).

Figure 5 shows the maximum acceleration amplitude at each frequency, which is calculated from the acceleration power spectral density. The two dotted lines, LNM and HNM, show the USGS low- and high-noise models (Peterson, 1993), respectively. Also, Figure 5 shows the maximum acceleration

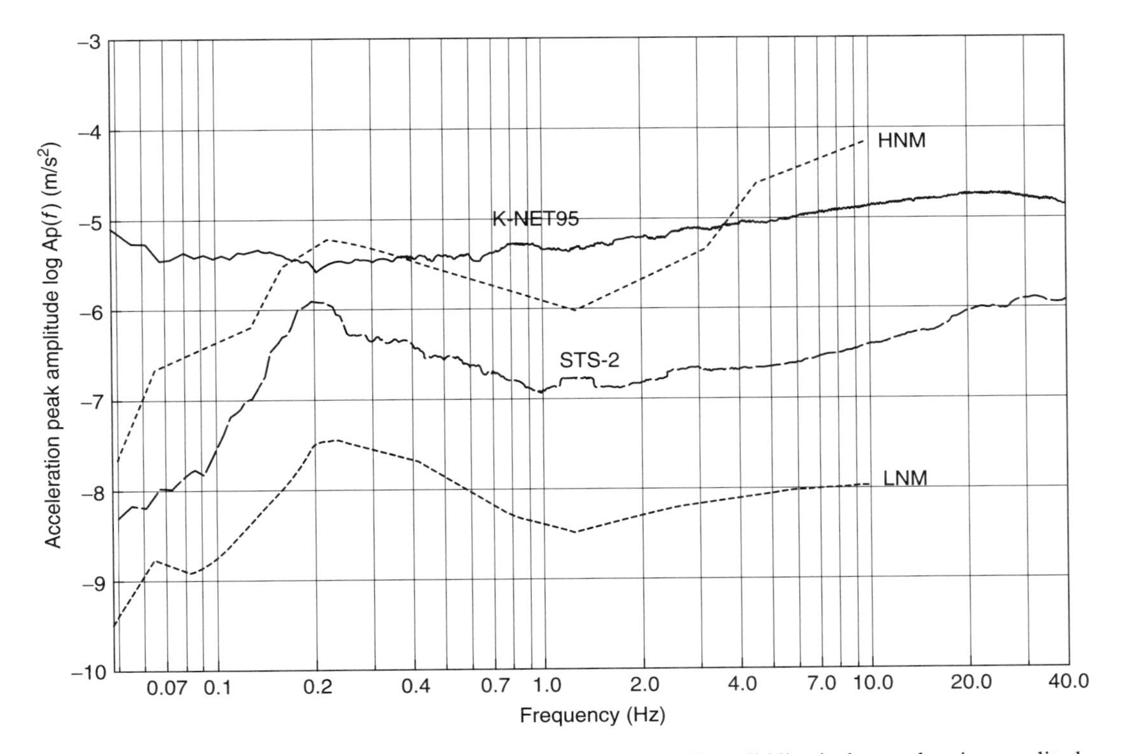

FIGURE 5 Amplitude spectra calculated from the power spectra. The solid line is the acceleration amplitude of a K-NET95. The two dotted lines, LNM and HNM, are the USGS low- and high-noise models, respectively, proposed by Peterson (1993).

amplitude of a UD component microtremor observed from the STS-2 seismometer installed on the same concrete base simultaneously. The K-NET95 and the STS-2 seismometers were installed on the same base. The result obtained from the STS-2 probably represents the actual microtremor at this site, and thus the result obtained by the K-NET95 probably consists almost entirely of instrument noise, due to the resolution amplitude of the K-NET95. This figure shows that the K-NET95 amplitude resolution is approximately 1 mGal throughout the entire frequency band. Thus, in this case, the dynamic range of the K-NET95 is 126 dB because the maximum measurable acceleration is 2000 Gal.

The V403BT negative feedback accelerometer has a test coil for calibrating the K-NET95 seismograph. This calibration is implemented by a command from the control center, and the calibration data are stored in the K-NET95. By retrieving the calibration file by a dial-up procedure, we can check the dynamic range of the K-NET system including transmission lines. This return test is performed every month.

2.2.8 Temperature Sensitivity and Temperature Dependence of the K-NET95 Responses

The K-NET95 functions within a nominal temperature range of -5 to 50°C, according to the specifications. However, in practice, the K-NET95 functions in a temperature range from -20 to 60°C as follows. Temperature sensitivity was measured by using an environment controller throughout a temperature range of from -10 to 50°C. The results showed that the temperature sensitivity of the K-NET95 is about 0.1%. The strong-motion data files available on the Internet include the temperature information at the time of recording, which is written under the heading of memorandum in the header part of the file.

2.2.9 DC Offset

There are two kinds of DC offset in the K-NET95. One is an initial drift, which continues for about 2 days until the internal temperature of the K-NET95 reaches a steady condition. The other is due to environmental temperature changes. After the initial offset, the K-NET95 has a hysteresis loop in the DC offset. This characteristic depends on the temperature transmission of the FRP case of the K-NET95. The average gradient of this loop is about 1 Gal/°C and the K-NET95 kept this value under a forced temperature change test within a temperature range of from -10 to 50°C.

2.2.10 Cross-Axis Sensitivity

The cross-axis sensitivity between the three components of the K-NET95 is measured by using a shake table. During the test, shaking within a frequency range of from 1 to 10 Hz, the cross-axis sensitivity was less than 1%.

3. K-NET Information

The K-NET provides four kinds of data and one software package. These include site information, strong-motion data, a maximum acceleration map, K-NET95 instrument parameters, and utility programs. The control center and the two mirror sites send these data and software across the Internet according to the user's request. The Internet addresses of the control center and the two mirror sites are as follows:

http://www.k-net.bosai.go.jp (control center in Tsukuba city)
http://www.k-net.ostec.co.jp (mirror site 1 in Osaka city)
http://www.k-net.geophys.tohoku.ac.jp (mirror site 2 in Sendai city)

3.1 Site Information

At each K-NET station, the velocity structure beneath the site to a depth of 10 or 20 meters was investigated by downhole measurement. The control center and the two mirror sites provide the location, elevation, address, and soil structure including N-values, bulk density, P- and S-wave structures, and soil profile for each site as part of the station information. Figure 6 shows an example of the site information sent across the Internet.

3.2 Strong-Motion Data

The control center makes three kinds of strong-motion data files available (UNIX, DOS, and ASCII) with a common header that includes the quick source parameters determined by JMA for each event. Users can select from among the three file types on the Internet. The UNIX and DOS files are compressed by gzip and lha compression commands, respectively.

A separate data file is made for each component of each seismogram. The numerical data in each file are raw accelerograms recovered from the K-NET95, without any corrections. The source parameters, origin time, epicenter location, focal depth, and magnitude, written in the header of the data file, are from the preliminary JMA report on the earthquake source. The control center does not release the JMA final source parameters on the Internet.

Since April 1, 1997, users have been able to download all strong-motion data files obtained from a specific event at one time. Such data retrieval is also possible for a set of events selected according to a user's request.

3.3 Maximum Acceleration Map

The map of maximum acceleration with equi-acceleration contour lines will be made after an event when the control center has recovered enough data, usually more than 50 three-component

Soil investigation
by downhole method (10–20 m)

KGS005 MIYANOJOH

Address: KAGOSHIMAKEN SATSUMAGUN MIYANOJOHCHO YACHI 887

Latitude: 31.8972N Longitude: 130.4536E Altitude: 42.50 m

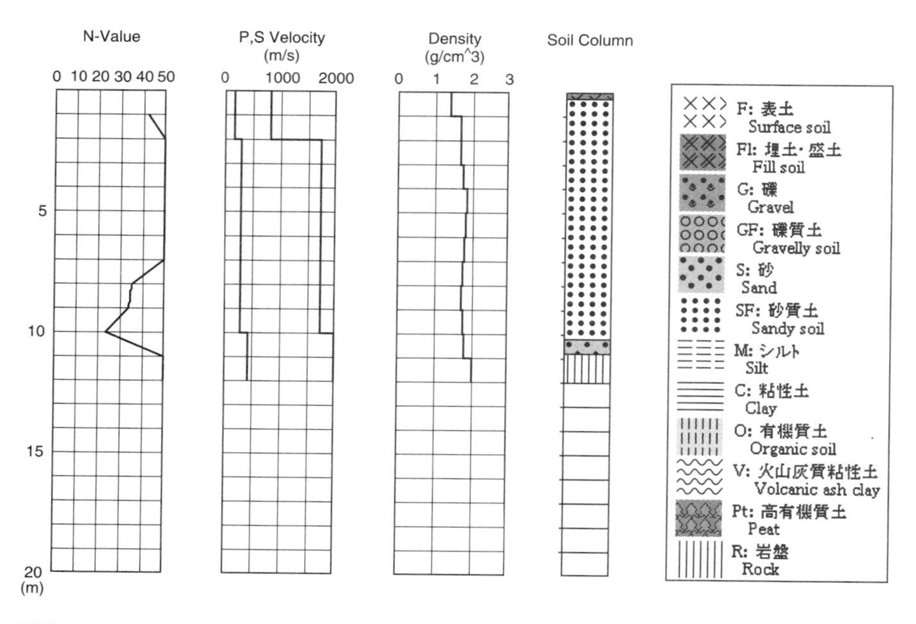

FIGURE 6 An example of site data sent out on the Internet.

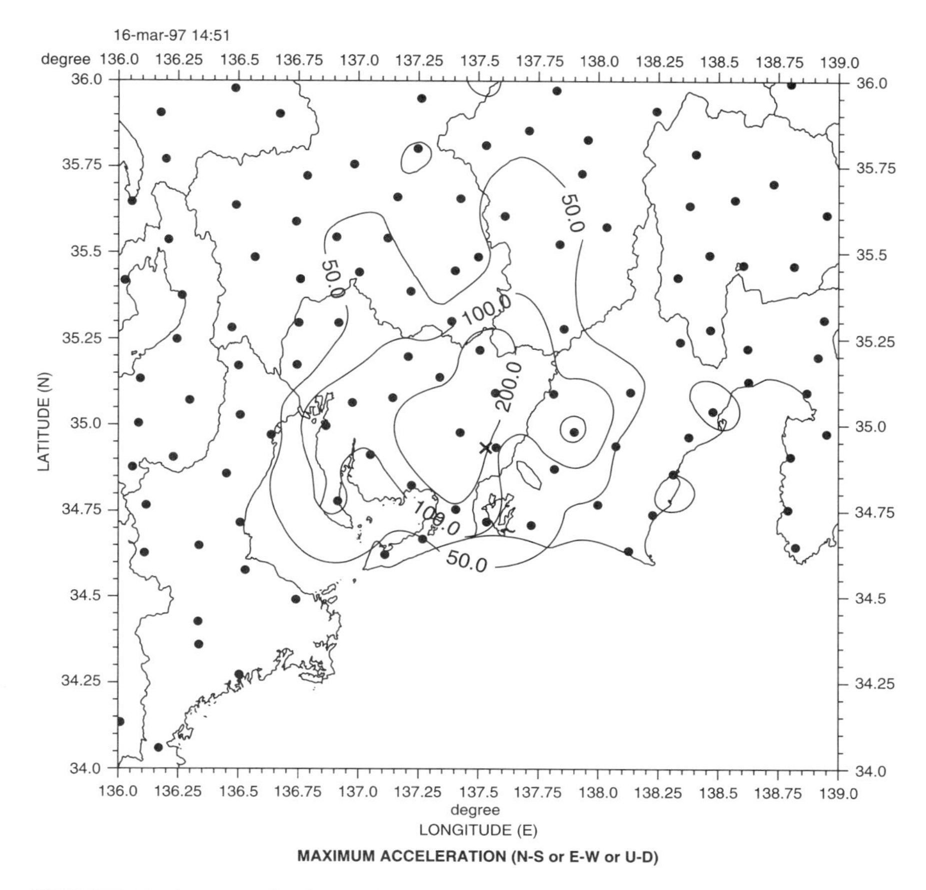

FIGURE 7 An example of a maximum acceleration map sent out on the Internet.

seismograms. The contour lines are smoothed by using Spline interpolation and indicate the center point of strong motion, which may be different from the epicenter of the earthquake. If the control center has not recovered enough data to construct such a map, only the table of maximum acceleration is released.

Figure 7 shows an example of a maximum acceleration map sent out on the Internet. This map was obtained from the Eastern Aichi prefecture earthquake of March 16, 1997 ($M = 5.4$). From such a map, it is possible to interpret the local characteristics of maximum acceleration.

3.4 The K-NET95 Instrument Parameters

The control center provides the information on the K-NET95 seismograph. These are the block diagram, overall frequency characteristics, and specifications, of the seismograph. The detailed information on the K-NET95 seismograph is provided by Kinoshita (1998) and Kinoshita *et al.* (1997).

3.5 Utility Software

In 1997, K-NET released utility software, which runs on Windows 95 or NT, for preliminary studies on the K-NET95 seismograms. The released application programs are applicable to DOS seismogram files, which are retrieved from our Web sites and are able to calculate and plot velocity and displacement seismograms in addition to the original accelerograms. Also, the utility programs can calculate Fourier, power, and response spectra and can implement various filtering functions. Plotting programs for the calculated spectra are included in the utility software. When we plot the three-component K-NET95 seismogram by using the utility program, the program simultaneously provides the value of the JMA intensity at the site.

3.6 Off-line Release of K-NET Information

Since April 1, 1997, individual users have been able to copy the K-NET strong-motion data and other information on to their MO and/or DAT media at the control center and mirror site 2. This is a self-service. We also distributed the strong-motion data obtained in 1996 and 2001 on ten CD-ROMs. Such a service is performed every 6 months.

4. Closing Remarks

From the Kobe earthquake of 1995, we learned that more information on strong motion in a hypocentral area is necessary to investigate the generation of strong earthquake shaking and to assess the degree of a disaster for quick mitigation response.

The K-NET deployed in 1995 to respond to these demands. This network was constructed on the basis of previous seismic networks installed in the United States, Japan, Mexico, the Republic of China, and other countries. The construction cost was US$40,000,000, and the maintenance cost for each year is 5% of the construction cost. Three persons including a computer engineer have been operating this system.

After about 1 year of operation, the K-NET has transmitted one million data files to users on the Internet. This means that the K-NET is responding very well to users' requests. In the future, the K-NET will provide even more useful information as the instruments and network are upgraded. The K-NET will continue to provide even more useful information as the instruments and network are upgraded. For example, K-NET upgraded the Web pages in 2001, in which a concise text of strong-motion entitled, "Fundamentals of strong-motion," was released.

Acknowledgments

The author thanks Makoto Miyamoto and Takao Sakai for the site negotiation of K-NET stations. Dr. Aketo Suzuki administrated the construction of the K-NET observatory. Masayosi Uehara, Toshio Tozawa, and Yasuji Wada built the network system of the K-NET including seismograph installation. I offer them my sincere gratitude for their invaluable assistance.

References

Fukushima, Y., and T. Tanaka (1990). A new attenuation relation for peak horizontal acceleration of strong earthquake ground motion in Japan. *Bull. Seism. Soc. Am.* **80**, 757–783.

Ikeura, T., S. Uchiyama, N. Adachi, T. Yamashita, T. Uetake, and M. Kikuchi (1999). Effects of K-NET seismometer installation conditions on its strong motion records. *Programme and abstracts, PO18, 1999 SSJ fall meeting* (in Japanese).

Kinoshita, S. (1998). Kyoshin net (K-NET). *Seism. Res. Lett.* **69**, 309–332.

Kinoshita, S., M. Uehara, T. Tozawa, Y. Wada, and Y. Ogue (1997). Recording characteristics of the K-NET95 strong-motion accelerograph. *Zishin (Series-2)* **49**, 467–481 (in Japanese with English abstract).

Peterson, J. (1993). Observation and modeling of seismic background noise. *Open-File Report 93-322*, US Geological Survey, Albuquerque, New Mexico.

64

Strong-Motion Instrumentation Programs in Taiwan

T. C. Shin
Central Weather Bureau, Taipei, Taiwan
Y. B. Tsai
National Central University, Chung-li, Taiwan
Y. T. Yeh
Ching Yun Institute of Technology, Jung-Li, Taiwan
C. C. Liu
Institute of Earth Sciences, Academia Sinica, Taipei, Taiwan
Y. M. Wu
Central Weather Bureau, Taipei, Taiwan

1. Introduction

Taiwan is located on the Circum-Pacific seismic belt. On the east side of Taiwan, the Philippine Sea plate subducts beneath the Eurasian plate at the Ryukyu trench, while at the south end of Taiwan, the South China Sea lithosphere subducts eastward under the Philippine Sea plate. The active convergent margin, connecting these two subduction zones, is characterized by rapid crustal deformation, regional-scale crustal faulting, and high seismicity. The densely populated western Taiwan, with high-rise buildings as a consequence of developing economy, is vulnerable to increasing earthquake hazard. Therefore, earthquake research has a high priority in Taiwan and considerable amounts of resources have been devoted to seismic instrumentation in general, and strong-motion instrumentation in particular.

Many disastrous earthquakes have occurred in the past, the most recent one being the 1999 Chi-Chi earthquake (Teng *et al.*, 2001). About 2500 people died and 300,000 were left homeless. The importance of strong-motion instrumentation has long been recognized, and we will summarize here the history of the strong-motion instrumentation programs in Taiwan. The earlier instrumentation programs were conducted primarily by the Institute of Earth Sciences, Academia Sinica, and have been mostly research oriented (Tsai, 1997; see also Report of the Institute of Earth Sciences under China (Taipei) in Chapter 79). The later efforts, involving an order-of-magnitude increase in the number of instruments, were conducted primarily by the Central Weather Bureau.

2. Strong-Motion Instrumentation Program by IES

2.1 Strong-Motion Accelerographs Network (SMA)

An islandwide strong-motion network was deployed by the Institute of Earth Sciences (IES), Academia Sinica, beginning in 1974, and by 1983, this network consisted of 72 stations. The instruments used were the standards at that time, i.e., the SMA-1s. The first strong-motion record obtained by this network was in April 1976. The purpose of this network is mainly to study earthquake source, structure responses, attenuation of ground motions, and risk analysis. By 1990, accelerographs of this network increased to 79, as a mix of analog and digital recording units. Most of them were installed on free-field sites, while some were on man-made structures. Most of those free-field stations were installed on the populous plain areas. After 1990, all stations of this network have been continuously upgraded to force-balance accelerometers with 16-bit resolution. Numbers of accelerograph stations on plain areas were reduced, and new stations were installed in the Central Range Mountain

INTERNATIONAL HANDBOOK OF EARTHQUAKE AND ENGINEERING SEISMOLOGY, VOLUME 81B ISBN: 0-12-440658-0

of Taiwan. The total number of stations in this network is currently 74. The purpose of the new installation is to study the topographic effects and attenuation behavior of strong motion in the mountainous area (Huang, 2000). Leaders of this project include Y. B. Tsai (1974–1978), C. S. Wang (1979–1980), Y. T. Yeh (1981–1992), and B. S. Huang (1992–present).

2.2 Strong-Motion Accelerograph Array in Taiwan, Phase 1 (SMART-1 Array)

SMART-1 Array was set up in Lotung in 1980 and closed at the end of 1990. This was a cooperative project between the Institute of Earth Sciences, Academia Sinica and University of California, Berkeley. The SMART-1 Array consisted of a central site and accelerographs in three concentric circles, with radii of 200 m, 1 km, and 2 km, respectively. Each circle had 12 evenly spaced sensors. All 43 accelerographs were tied to a common time base, with timing to better than ± 0.01 sec. Each accelerograph consisted of a triaxial force-balance accelerometer, capable of recording ± 2 g, connected to a digital event recorder that uses a magnetic tape cassette for recording. The accelerographs were triggered on either vertical or horizontal acceleration at an adjustable preset threshold. Signals were digitized with a 13-bit resolution at 100 samples per second. Each recorder had a digital preevent memory that stored the output signals from the force-balance accelerometer for approximately 2.5 sec before trigger. Such accelerographs had an obvious advantage of providing synchronous time history of the ground-motion acceleration. Hence, we could perform spatial and temporal correlation across the whole array. The recorded data on digital cassettes were played back at the central laboratory and transferred onto a regular 9-track magnetic tape in ASCII format. During the playback, a seismologist scanned the digital signals displayed on a minicomputer console and made corrections for glitches, gaps, time code errors, and offsets in DC level. A regular magnetic tape containing edited data was available for further analysis only hours after the recording, whereas the analog recording/processing commonly used at that time would take days to digitize and process (Tsai and Bolt, 1983). Many research papers were published using the SMART-1 data (e.g., Loh *et al.*, 1982; Abrahamson, 1988).

2.3 Lotung Large Scale Seismic Test Array (LSST)

The LSST program was set up for evaluating soil-structure interaction effects and backfill effects. These effects are important in seismic design of nuclear reactor facilities. A quarter-scale and a 1/12-scale model of the nuclear reactor containment structure were constructed inside the SMART-1 Array on October 1985. It was closed at the end of 1990, the same time that SMART-1 Array had completed its mission. The LSST program was a joint project between the Taiwan Power Company (Taipower) and the Electric Power Research Institute of USA (EPRI), under the management of H. T. Tang. Under a contract with Taipower, IES installed and maintained the instruments, as well as carried out data collection, reduction, and analysis. In the initial phase, four types of sensors were installed in the fields for data acquisition: the surface accelerometer, the downhole accelerometer, the structural response accelerometer, and the interfacial pressure transducer. These sensors were triaxial type, except the pressure transducer. The output of all accelerometers and pressure gauges was transmitted by hard wire to the central recording unit and was recorded on cassette tapes. These tapes were then processed and transcribed onto 9-track tapes using the ASCII format.

2.4 SMART-2 Array

The SMART-2 strong-motion array was deployed by IES in the northern part of the Longitudinal Valley in Hualien in December 1990, and was fully operational in 1992 (Chiu *et al.*, 1994). It consists of 45 Kinemetrics SSR-1 instruments as surface stations and two sets of downhole subarrays. All sensors used in this array are force-balance accelerometers. This array is designed to study the rupture process of seismic faults and the characteristics of near-source ground motions. Furthermore, the high-quality data from SMART-2 may be used for research in seismology and earthquake engineering (e.g., Huang and Chiu, 1996).

Chiu *et al.* (1995) studied the coherency of ground motions based on the SMART-2 data and compared it with the results of the SMART-1. A comparison of coherency functions for both vertical and horizontal motions from a magnitude 5.5 earthquake recorded by the SMART-2 indicated no significant difference in the range of 1 to 10 Hz for separation distance of 400, 800, and 1500 m.

2.5 Hualien Large Scale Seismic Test Array (HLSST)

Since 1993, EPRI and Taipower have sponsored a dense multiple-element array, the HLSST network, located at the Veteran's Marble Plant of Hualien within the SMART-2 deployment area of northeastern Taiwan. This is an international joint project operated by IES with the objectives of investigating soil-structure interaction behavior during severe earthquakes and verifying the validity of various analysis methods using the strong-motion records. To serve this purpose, a one-quarter-scale cylindrical reactor model and a cylindrical liquid-storage-tank model were constructed in Hualien, a high-seismicity region. The cylindrical liquid-storage-tank model was closed in July 1998.

2.6 Downhole Accelerometer Arrays in the Taipei Basin (DART)

A research project, "Integrated Survey of Subsurface Geology and Engineering Environment of the Taipei Basin," was proposed in early 1990s to collect data for the purposes of engineering construction, groundwater management, ground subsidence prediction, study of the basin effects of seismic waves, and geological sciences. This project has been sponsored by the Central Geological Survey (CGS), Ministry of Economic Affairs since August 1991. CGS contracted the study of the basin effects on seismic waves to IES, which proposed the DART program, in which one site was installed per year to analyze the variation of seismic waves propagating from the basement to ground surface. Each site includes one free-surface accelerometer and some downhole sensors. These force-balance accelerometers are connected to a PC-based central recording system or a K2 digital recording system with GPS timing and position information.

Wen *et al.* (1995) studied basin effects using a dense strong-motion array in Taipei Basin. Their results showed that site amplification is frequency dependent. They also indicated that both horizontal peak ground acceleration and the spectral ratio in low-frequency band are closely correlated with the geological structure of Taipei Basin.

3. Strong-Motion Instrumentation Program by CWB

In the late 1980s, Y. B. Tsai proposed an extensive strong-motion instrumentation program for the urban areas in Taiwan. Since the Central Weather Bureau has the official responsibility to monitor earthquakes in the Taiwan region, the Taiwan Strong-Motion Instrument Program (TSMIP; see Shin, 1993) was successfully implemented during 1991–1996.

The main goal of this program is to collect high-quality instrumental recordings of strong ground shaking from earthquakes, both at free-field sites and in buildings and bridges. These data are crucial for improving earthquake-resistant design of buildings and bridges and for understanding the earthquake source mechanisms, as well as seismic wave propagation from the source to the site of interest, including local site effects.

Two types of digital strong-motion instruments were deployed throughout Taiwan in this program, with special emphasis in nine metropolitan areas. One type is a digital triaxial accelerograph for recording free-field ground shakings (Liu *et al.*, 1999). The other type is a multichannel (32 or 64 channels), central-recording, accelerograph array system for monitoring shaking caused by earthquakes in buildings and other structures (Lee and Shin, 1997). By the end of 2000, a total of 640 free-field accelerographs and 56 structural arrays had been deployed. Locations of the free-field accelerographs are shown in Figure 1, and locations for the building arrays are shown in Figure 2.

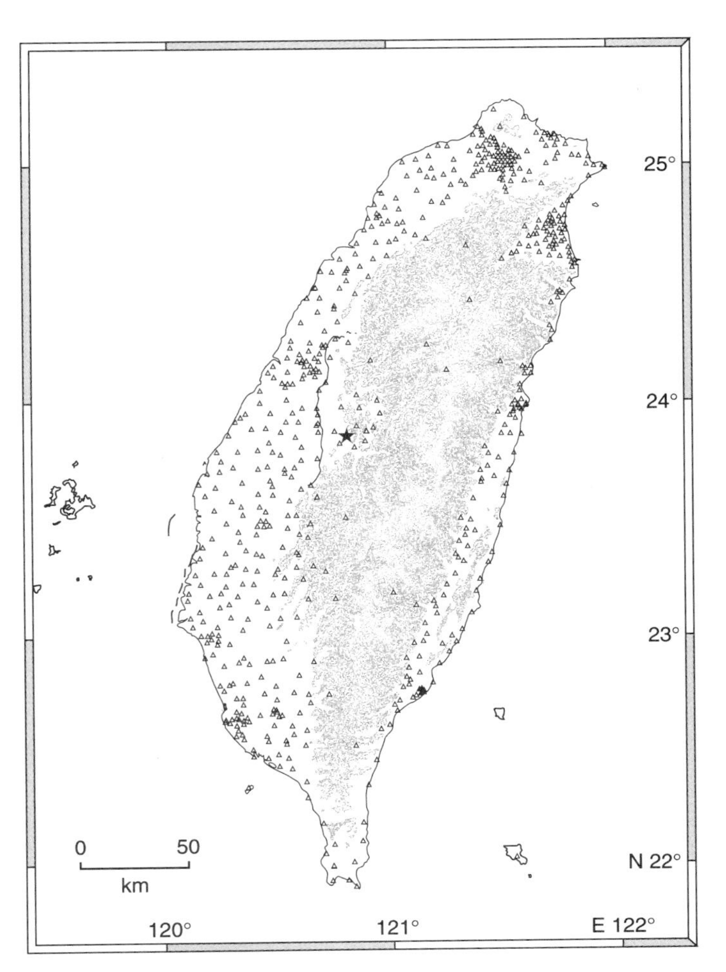

FIGURE 1 Locations of the CWB free-field, three-component, digital accelerograph stations. The star indicates the location of the Chi-Chi earthquake. Surface ruptures extending about 80 km north–south are shown to the left of the epicenter.

4. The Taiwan Rapid Earthquake Information Release System

The desire for seismological observation in real time has long been recognized, and significant advances have been made during the past decade in many countries (Kanamori *et al.*, 1997). The idea of an islandwide early earthquake warning system using the existing telemetry in Taiwan was first proposed by T. L. Teng in the early 1990s. The Taiwan Rapid Earthquake Information Release System (RTD) is based on a simple hardware/software design first introduced by Lee *et al.* (1989), and was subsequently improved and refined (Lee, 1994; Lee *et al.*, 1996; Shin *et al.*, 1996; Teng *et al.*, 1997; Wu *et al.*, 1997, 1998, 1999). The RTD system consists of 80 telemetered strong-motion accelerographs in Taiwan (Fig. 3). Digital signals are continuously telemetered to the headquarters of the Central Weather Bureau (CWB) in Taipei via 4800-baud leased telephone lines. Each telemetered signal contains three-component seismic data digitized at 50 samples per second and at 16-bit resolution. The full recording range is ± 2 g. The incoming digital data streams are processed by a computer program called XRTPDB (Tottingham and Mayle, 1994). Whenever the prespecified trigger criteria are met, the digital waveforms are stored in memory and are automatically analyzed by a series of programs (Wu *et al.*, 1998).

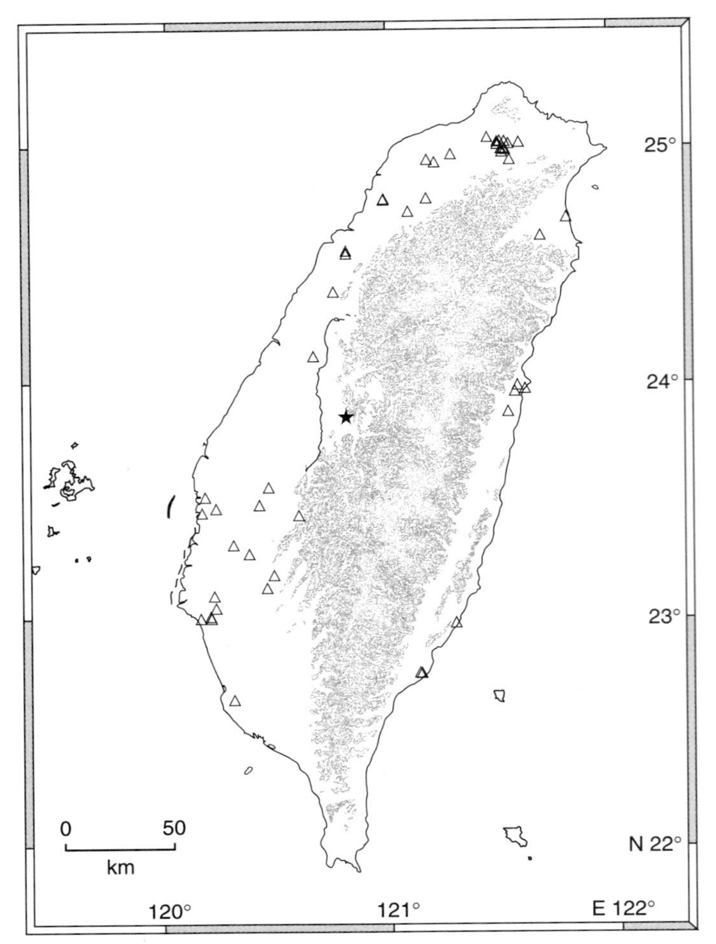

FIGURE 2 Locations of the CWB structural strong-motion arrays in buildings and bridges. The star indicates the location of the Chi-Chi earthquake. Surface ruptures extending about 80 km north–south are shown to the left of the epicenter.

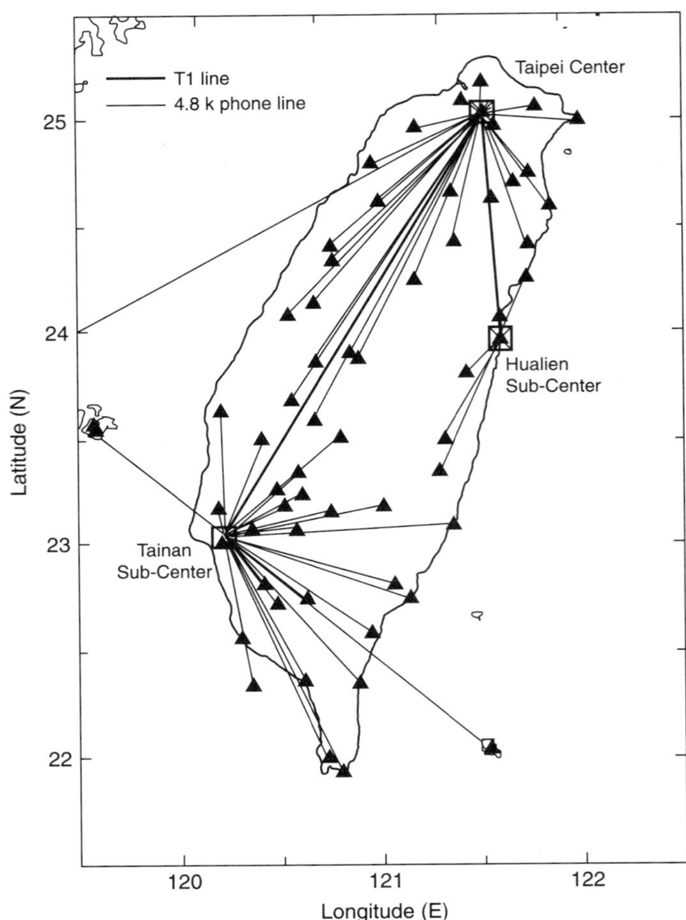

FIGURE 3 Map showing the telemetered stations of the Taiwan Earthquake Rapid Information Release System (RTD).

The results are immediately disseminated to emergency response agencies electronically in four ways, namely, by e-mail, World Wide Web, fax, and a pager system (Fig. 4).

5. The Chi-Chi Earthquake of September 21, 1999

The Chi-Chi earthquake occurred at 1:47 on September 21, 1999 (Taiwan local time) or at 17:47 on September 20, 1999 UTC. It was the largest ($M_w = 7.6$) earthquake to have occurred on land in Taiwan in the 20th century. For the main shock, 441 digital three-component, strong-motion records were successfully retrieved by the Taiwan Central Weather Bureau (CWB) from about 640 accelerographs deployed at the free-field sites. These preliminary strong-motion data from the Chi-Chi main shock were released on December 13, 1999, in the form of a prepublication data CD (Lee *et al.*, 1999). During the first 6 hours after the main shock, about 10,000 strong-motion records were recovered, and since then another 20,000 records were obtained. This is by far the best-recorded major earthquake in the world. There are over 60 three-component strong-motion records from within 20 km of the fault ruptures.

A preliminary report of this earthquake is given in Shin *et al.* (2000), and a detailed report of the processed free-field acceleration data is given in Lee *et al.* (2001a). These data are also archived in Lee *et al.* (2001b). Characteristics of the strong ground motion are given by Tsai and Huang (2000). At the time of the earthquake, the Taiwan Rapid Earthquake Information Release System (RTD) automatically determined the location and magnitude for the main shock and prepared a shake map within 102 seconds after the earthquake's origin time. This information was then sent out by a pager-telephone system, by an e-mail server, and by fax. Its performance during the Chi-Chi earthquake and numerous strong aftershocks has been documented in Wu *et al.* (2000).

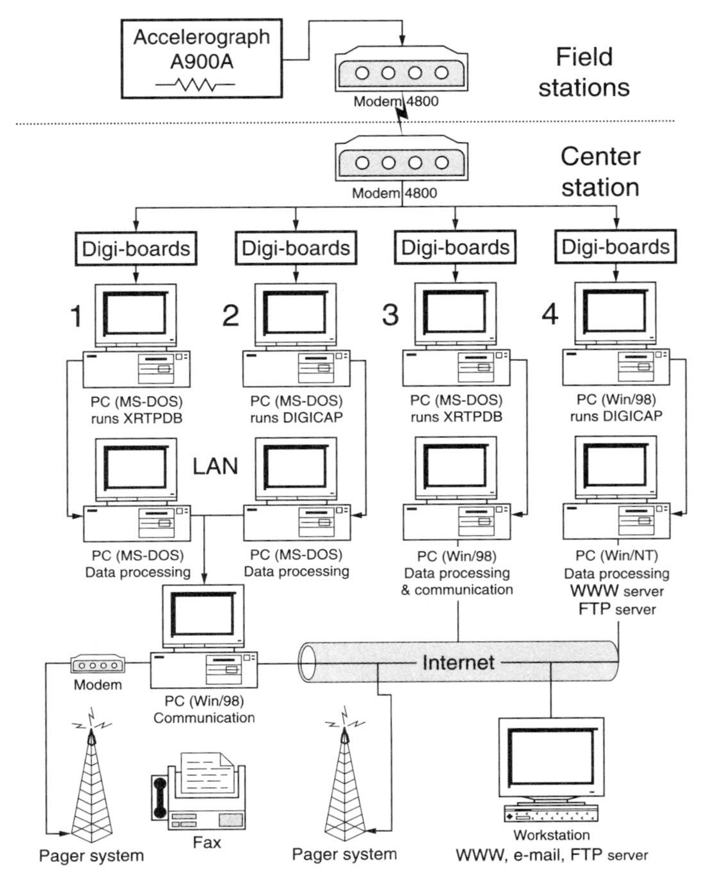

FIGURE 4 A block diagram showing the hardware of the RTD system.

6. Concluding Remarks

The Chi-Chi earthquake clearly demonstrated the usefulness of the extensive strong-motion instrumentation in Taiwan for purposes of emergency response and research in seismology and earthquake engineering. Readers are referred to several special issues and proceedings on the Chi-Chi earthquake (e.g., BERI, 2000; Loh and Liao, 2000; Wang *et al.*, 2000; Teng *et al.*, 2001).

Acknowledgments

The strong-motion instrumentation programs in Taiwan would not be possible without the labor of many people. We wish to thank many advisors and collaborators: B. A. Bolt, K. C. Chen, H. C. Chiu, B. S. Huang, G. C. Lee, W. H. K. Lee, K. S. Liu, S. C. Liu, C. H. Loh, S. T. Mau, G. B. Ou, M. S. Sheu, H. T. Tang, T. L. Teng, C. Y. Wang, K. L. Wen, C. F. Wu, F. T. Wu, Y. H. Yeh, and G. K. Yu.

References

Abrahamson, N. A. (1988). Statistical properties of peak ground acceleration recorded by the SMART 1 array. *Bull. Seism. Soc. Am.* **78**, 26–41.

BERI (2000). "Special Issue: The 1999 Chi-Chi, Taiwan earthquake." *Bull. Earthquake Res. Inst. Tokyo Univ.* **75**, Paret 1.

Chiu, H. C., Y. T. Yeh, S. D. Ni, L. Lee, W. S. Liu, C. F. Wen, and C. C. Liu (1994). A new strong-motion array in Taiwan—SMART-2. *Terres. Atmos. Ocean. Sci.* **5**, 443–455.

Chiu, H. C., R. V. Amirbekian, and B. A. Bolt (1995). Transferability of strong ground motion coherency between the SMART1 and SMART2 arrays. *Bull. Seism. Soc. Am.* **85**, 342–348.

Huang, B. S. (2000). Reconstruction of 2-D ground motions. *Geophys. Res. Lett.* **27**, 3025–3028.

Huang, H. C., and H. C. Chiu (1996). Estimation of site amplification from Dahan Downhole recording. *Int. J. Earthquake Eng. Struct. Dyn.* **25**, 319–332.

Kanamori, H., E. Hauksson, and T. Heaton (1997). Real-time seismology and earthquake hazard mitigation. *Nature* **390**, 461–464.

Lee, W. H. K. (Editor) (1994). "Realtime Seismic Data Acquisition and Processing," IASPEI Software Library, Volume 1 (2nd Edition), Seismological Society of America, El Cerrito, CA, (285 pp. and 3 diskettes).

Lee, W. H. K., D. M. Tottingham, and J. O. Ellis (1989). Design and implementation of a PC-based seismic data acquisition, processing, and analysis system. *IASPEI Software Library* **1**, 21–46.

Lee, W. H. K., T. C. Shin, and T. L. Teng (1996). Design and implementation of earthquake early warning systems in Taiwan. *Proc. 11th World Conf. Earthq. Eng., Paper No. 2133.*

Lee, W. H. K., and T. C. Shin (1997). Realtime seismic monitoring of buildings and bridges in Taiwan. In: *Structural Health Monitoring* (F. K. Chang, Ed.), Technomic Pub. Co., Lancaster, PA, 777–787.

Lee, W. H. K., T. C. Shin, K. W. Kuo, and K. C. Chen (1999). CWB Free-Field Strong-Motion Data from the 921 Chi-Chi Earthquake: Volume 1. Digital Acceleration Data, Pre-Publication CD, Central Weather Bureau, Taipei, Taiwan.

Lee, W. H. K., T. C. Shin, K. W. Kuo, K. C. Chen, and C. F. Wu (2001a). CWB Free-Field Strong-Motion Data from the 921 Chi-Chi Earthquake: Processed Acceleration Files on CD-ROM, *Strong-Motion Data Series CD-001*, Central Weather Bureau, Taipei, Taiwan.

Lee, W. H. K., T. C. Shin, K. W. Kuo, K. C. Chen, and C. F. Wu (2001b). Data Files from "CWB Free-Field Strong-Motion Data from the 21 September Chi-Chi, Taiwan, Earthquake," *Bull. Seism. Soc. Am.* **91**, 1390 and CD-ROM.

Liu, K. S., T. C. Shin, and Y. B. Tsai (1999). A free-field strong motion network in Taiwan: TSMIP. *Terres. Atmos. Ocean. Sci.* **10**, 377–396.

Loh, C. H., J. Penzien, and Y. B. Tsai (1982). Engineering analysis of SMART 1 array accelerograms. *Int. J. Earthquake Eng. Struct. Dyn.* **10**, 575–591.

Loh, C. H., and W. I. Liao (Editors) (2000). "Proceedings of International Workshop on Annual Commemoration of Chi-Chi Earthquake," 4 volumes, National Center for Research on Earthquake Engineering, Taipei, Taiwan.

Shin, T. C. (1993). Progress summary of the Taiwan strong-motion instrumentation program. In: "Symposium on the Taiwan Strong-Motion Program," Central Weather Bureau, 1–10.

Shin, T. C. (2000). Some seismological aspects of the 1999 Chi-Chi earthquakes in Taiwan. *Terres. Atmos. Ocean. Sci.* **11**, 555–566.

Shin, T. C., Y. B. Tsai, and Y. M. Wu (1996). Rapid response of large earthquake in Taiwan using a realtime telemetered network of digital accelerographs. *Proc. 11th World Conf. Earthq. Eng., Paper No. 2137.*

Shin, T. C., K. W. Kuo, W. H. K. Lee, T. L. Teng, and Y. B. Tsai (2000). A preliminary report on the 1999 Chi-Chi (Taiwan) earthquake. *Seism. Res. Lett.* **71**, 24–30.

Teng, T. L., Y. M. Wu, T. C. Shin, Y. B. Tsai, and W. H. K. Lee (1997). One minute after: strong motion map, effective epicenter, and effective magnitude. *Bull. Seism. Soc. Am.* **87**, 1209–1219.

Teng, T. L., Y. B. Tsai, and W. H. K. Lee (Editors) (2001). Dedicated Issue on the Chi-Chi (Taiwan) Earthquake of September 20, 1999. *Bull. Seism. Soc. Am.* **91**, 893–1395.

Tottingham, D. M., and A. J. Mayle (1994). User manual for XRTPDB. *IASPEI Software Library* **1** (2nd Edition), 255–263.

Tsai, Y. B. (1997). A brief review of strong motion instrumentation in Taiwan. In: "Vision 2005: An Action Plan for Strong Motion Programs to Mitigate Earthquake Losses in Urbanized Areas," A Workshop held in Monterey, California on April 2–4, 1997.

Tsai, Y. B., and B. A. Bolt (1983). An analysis of horizontal peak ground acceleration and velocity from SMART 1 array date. *Bull. Inst. Earth Sci., Acad. Sinica* **3**, 105–126.

Tsai, Y. B., and M. W. Huang (2000). Strong ground motion characteristics of the Chi-Chi, Taiwan earthquake of September 21, 1999. *Earthquake Eng. & Eng. Seism.* **2**, 1–21.

Wang, C. S., S. K. Hsu, H. Kao, and C. Y. Wang (Editors) (2000). Special Issue on the 1999 Chi-Chi Earthquake in Taiwan. *Terres. Atmos. Ocean. Sci.* **11**, 555–752.

Wen, K. L., H. Y. Peng, L. F. Liu, and T. C. Shin (1995). Basin effects analysis from a dense strong motion observation network. *Int. J. Earthquake Eng. Struct. Dyn.* **24**, 1069–1083

Wu, Y. M., T. C. Shin, C. C. Chen, Y. B. Tsai, W. H. K. Lee, and T. L. Teng (1997). Taiwan rapid earthquake information release system. *Seism. Res. Lett.* **68**, 931–943.

Wu, Y. M., C. C. Chen, J. K. Chung, and T. C. Shin (1998). An automatic phase picker of the real-time acceleration seismic network. *Meteorol. Bull., Central Weather Bureau, Taipei* **42**, 103–117 (in Chinese).

Wu, Y. M., J. K. Chung, T. C. Shin, N. C. Hsia, Y. B. Tsai, W. H. K. Lee, and T. L. Teng (1999). Development of an integrated seismic early warning system in Taiwan—case for Hualien area earthquakes. *Terres. Atmos. Ocean. Sci.* **10**, 719–736.

Wu, Y. M., W. H. K. Lee, C. C. Chen, T. C. Shin, T. L. Teng, and Y. B. Tsai (2000). Performance of the Taiwan Rapid Earthquake Information Release System (RTD) during the 1999 Chi-Chi (Taiwan) earthquake. *Seism. Res. Lett.* **71**, 338–343.

VIII

Selected Topics in Earthquake Engineering

65

Seismic Hazards and Risk Assessment in Engineering Practice

Paul Somerville and Yoshi Moriwaki*
URS Corporation, Pasadena, California, USA

1. Introduction

As a result of the plate tectonics revolution in the 1960s and the subsequent decades of intensive research, the long-term earthquake potential of most parts of the world, especially near plate boundaries, is in general fairly well understood. Earth scientists have a partial understanding of the long-term seismic behavior of some of the more active faults. However, with few exceptions, large earthquakes do not appear to occur at uniform time intervals, or to rupture the same segment of a fault from one earthquake to the next. Consequently, even for sites near the best-understood faults, earth scientists are not able to predict where and when the next large earthquake is going to occur. Even greater uncertainty exists in most regions of the Earth, where in most cases the locations of potentially active faults are unknown, the seismic potential is poorly understood, and earthquakes are assumed to occur in areal source zones because specific faults have not been identified.

Early approaches to the seismic design of critical facilities characterized seismic hazards using deterministic approaches. Deterministic estimates of ground motions are typically made for a single large scenario earthquake whose magnitude (and possibly other source parameters) and closest distance are specified. However, given the uncertainty in the timing, location, and magnitude of future earthquakes, and the objectives of performance-based seismic engineering described below, it is often more meaningful to use a probabilistic approach in characterizing the ground motion that a given site will experience in the future. A probabilistic seismic hazard analysis (PSHA) takes into account the ground motions from the full range of earthquake magnitudes that can occur on each fault or source zone that can affect the site. The PSHA numerically integrates this information using probability theory to produce the annual frequency of exceedance of each different ground-motion level for each ground-motion parameter of interest.

The probabilistic approach to seismic hazard characterization is very compatible with current trends in earthquake engineering and the development of building codes, which have embraced the concept of performance-based design. Examples of conceptual frameworks for performance-based design include SEAOC Vision 2000 (Structural Engineers' Association of California, 1996; Fig. 1), FEMA 273, and EERI (Earthquake Engineering Research Institute, 1998). In contrast to the traditional approach, performance-based design requires an explicit prediction of the structure's performance at each of several postulated ground-motion levels corresponding to a set of performance objectives. The performance objectives may range from continued function of the building during relatively small, frequent ground motions; to limiting damage below the life safety threshold in severe, less frequent ground motions; to prevention of collapse for very severe, infrequent ground motions. Each performance objective is associated with an annual probability of occurrence, with increasingly undesirable performance characteristics caused by increasing levels of strong ground motion having decreasing annual probability of occurrence.

The objective of this chapter is to provide an overview of how seismic hazards are evaluated and how seismic risk is assessed in current engineering practice in the United States. This is done within the context of the rapidly evolving framework of performance-based seismic engineering, which we have outlined above. For more detailed treatment of risk assessment, loss estimation, and building codes, the reader is referred to recent collections of papers on these topics published in *Earthquake Spectra*, volume 13, number 4 (1997), and volume 16, number 1 (Hamburger and Kircher, 2000). The focus throughout

*Present Affiliation: GeoPentech, Santa Ana, California, USA.

INTERNATIONAL HANDBOOK OF EARTHQUAKE AND ENGINEERING SEISMOLOGY, VOLUME 81B ISBN: 0-12-440658-0

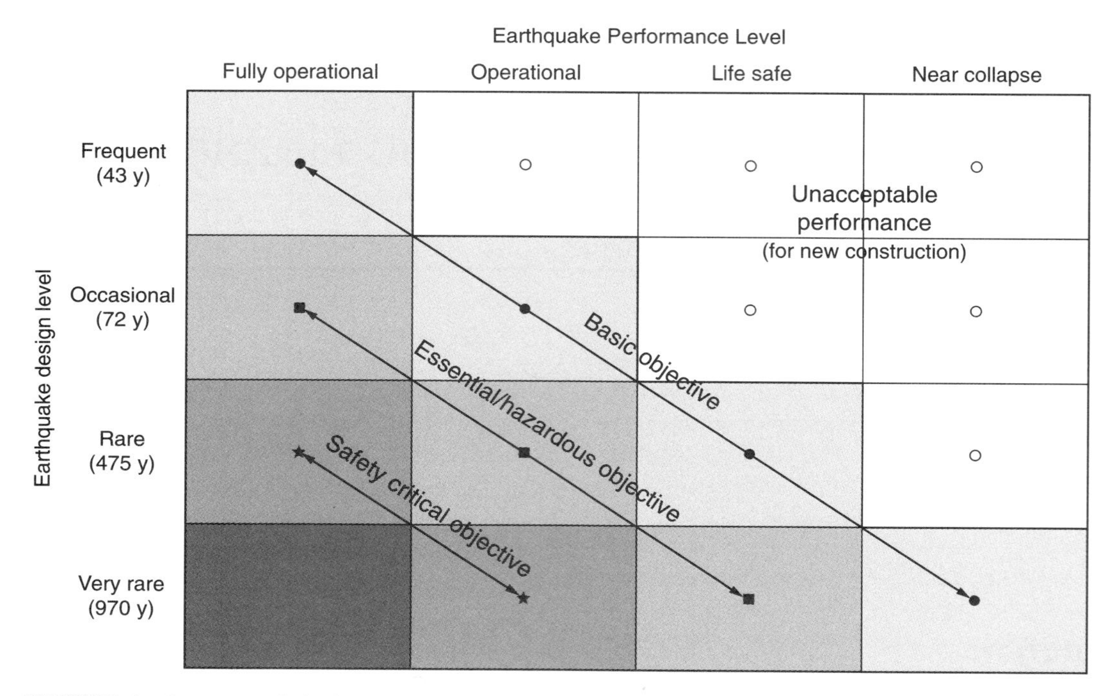

FIGURE 1 Recommended seismic performance objectives for buildings in SEAOC's Vision 2000 (SEAOC, 1996), showing increasingly undesirable performance characteristics from left to right on the horizontal axis and increasing level of ground motion from top to bottom on the vertical axis. Performance objectives for three categories of structures are shown by the diagonal lines. (From "Near-Source Ground Motion and its Effects on Flexible Buildings" by J. F. Hall *et al.*; *Earthquake Spectra*, Vol. 11, 1995. Earthquake Engineering Research Institute)

the chapter is on earthquake engineering practice in the United States. The important developments in seismic hazard and risk assessment that have occurred in parallel in other parts of the world such as Europe, Japan, and New Zealand are beyond the scope of this chapter.

2. Earthquake Source Characterization

In a seismic hazard evaluation for a given site, it is necessary to identify the seismic sources on which future earthquakes are likely to occur, to estimate the magnitudes and frequency of occurrence of earthquakes on each seismic source, and to identify the distance and orientation of each seismic source in relation to the site. If a deterministic approach is used to characterize the ground motions, then a single scenario earthquake is usually used to represent the seismic hazard, and its frequency of occurrence does not directly influence the level of the hazard. If a probabilistic approach is used, then the ground motions from a large number of possible earthquakes are considered, and their frequencies of occurrence are key parameters in the analysis.

Modern ground-motion models recognize differences in ground motions for different tectonic categories of earthquakes. These categories include crustal earthquakes in tectonically active regions, crustal earthquakes in tectonically stable regions, earthquakes on subduction plate interfaces, and earthquakes occurring within subducted slabs. Earthquake sources may be characterized as discrete faults, or as zones having uniform seismic potential. In tectonically active regions, it is usually possible to identify discrete earthquake faults. The regions surrounding these faults typically have some level of seismic activity, and are characterized by distributed source zones having uniform seismic potential. In moderately active regions, these distributed seismic zones often dominate the probabilistic seismic hazard even when identified faults are also present. In tectonically stable regions, all of the seismic sources may be characterized by distributed seismic zones.

The two basic parameters often used to estimate the ground motions at a particular site from an earthquake are the size of the earthquake and the distance of the earthquake from the site. However, as discussed further in the section on ground-motion estimation models, there are numerous other parameters that can also influence strong ground motions.

The size of the earthquake is characterized by its magnitude or seismic moment. Earthquake size is most conveniently characterized by seismic moment (or moment magnitude), because the seismic moment is directly related to the product of the fault rupture area and average fault displacement. Thus, the seismic moment of an earthquake on a given fault can be directly estimated from fault-specific estimates of the fault length, fault width, and average displacement. Alternatively, the seismic moment can be estimated from empirical relationships between seismic moment and fault parameters such as fault area, fault length, and average displacement that are based on the average properties of a large number of earthquakes (Wells and Coppersmith, 1994; Somerville *et al.*, 1999). The most reliable method of estimating seismic moment using empirical relationships is from fault area.

In order to identify the distance between the source and the site, the dimensions of the fault for a given magnitude need to be specified. Line source representation of distant sources may be adequate, but the fault dimensions along both strike and downdip may need to be specified for measuring the distance to the site when the source is close to the site. In some ground-motion models, additional source parameters specifying the dimensions and geometry of the source, the location of the site in relation to the source, and the location of the hypocenter are required. This is true for representing near-fault effects such as rupture directivity effects and hanging-wall effects, which are described later.

If numerical simulation methods are used to estimate the ground motions, then the spatial distribution of slip on the fault and the time function of slip on the fault also need to be characterized. Detailed studies of the spatial distribution of slip on the fault plane for earthquakes in tectonically active regions, derived from strong-motion recordings and other data, have shown that the slip distribution is highly variable, characterized by asperities (regions of large slip) surrounded by regions of low slip. These slip models have been used to develop relationships between seismic moment and a set of fault parameters that are needed for predicting strong ground motions from future earthquakes (Somerville *et al.*, 1999). These parameters include fault length, fault width, rise time (duration of slip at a point on the fault) and the size, slip contrast, and location of asperities. Predicting the locations of asperities in future earthquakes is a topic of ongoing research.

When the ground motions for evaluation and design are characterized by a scenario earthquake, the primary earthquake source parameter is the magnitude or seismic moment of the scenario earthquake. In a deterministic analysis, the scenario earthquake is typically the largest earthquake that is expected to occur on the source that controls the seismic hazard. The evaluation of the largest earthquake is complicated by the fact that fault systems are often composed of segments. While these segments usually rupture individually with characteristic magnitude earthquakes (Schwartz and Coppersmith, 1984), they may also rupture together in cascades, producing larger earthquakes than the characteristic events, as occurred in the 1992 Landers, California, earthquake (Wald and Heaton, 1994). The maximum earthquake magnitude thus has uncertainty associated with it, as do the other parameters used in seismic hazard analysis, and is best handled using a logic tree approach in a probabilistic seismic hazard analysis.

When the ground motions for evaluation and design are characterized probabilistically, the source characterization described above is repeated for earthquakes of all magnitudes on each potential seismic source. The uncertainties in the source parameters are addressed formally through the use of logic trees (Kulkarni *et al.*, 1984). In principle, the probabilistic analysis can make use of all of the detailed aspects of source characterization that may be used in a deterministic analysis, although in practice this may not be done.

Probabilistic analysis requires an important category of additional information: the expected frequencies of occurrence of earthquakes of various magnitudes on each potential seismic source. This information can be represented by the total seismic moment release rate on the source, together with an earthquake recurrence model that describes the partition of this seismic moment release rate into earthquakes of different seismic moment. The seismic moment release rate on a fault is given by the product of the fault area, the slip rate, and the shear modulus of the seismogenic zone.

The seismic moment release rate can be derived from one or more of three distinct categories of data. The first category, historical seismicity, is most reliable in tectonically active regions that have a relatively long historical earthquake record, and may be applicable both to individual faults and to a region. Historical seismicity can also be used to constrain the distribution of earthquake magnitudes, especially in the low-magnitude range. The second category, slip rates of active faults, is usually only available in tectonically active regions and is applicable to individual faults. Additional geological data, such as slip per event and average recurrence interval of events, may be used to constrain the distribution of earthquake magnitudes, especially in the high-magnitude range. The third category, geodetic strain rates, is in principle available in all tectonic environments and can be used to constrain the seismic moment rate over an extended region. Additional information is needed to partition the geodetic strain rate over individual faults and to constrain the distribution of earthquake magnitudes. Methods of integrating historical seismicity, geological slip rates, and geodetic measurements for characterizing earthquake recurrence are described by Ward (1994).

Earthquake recurrence is usually represented by one of two models in seismic hazard analyses (Youngs and Coppersmith, 1985). In the modified or truncated Gutenberg-Richter relationship (Gutenberg and Richter, 1964), the number of events is inversely proportional to the seismic moment of the event and is truncated at the maximum magnitude. In the characteristic recurrence relationship (Schwartz and Coppersmith, 1984), a truncated Gutenberg-Richter relationship, representing the occurrence of small and moderate earthquakes, is combined with a characteristic earthquake relationship, representing the occurrence of large earthquakes, which has a specified magnitude range and rate of occurrence. This rate of occurrence is higher than the extrapolated value of the truncated Gutenberg-Richter relationship. The truncated Gutenberg-Richter relationship is usually an appropriate representation of the seismicity of regions containing faults of various sizes. The characteristic recurrence model is usually an appropriate representation of the seismicity of individual faults.

If the seismicity of a source is based solely on historical seismicity, then the seismic moment rate of the source depends on the maximum earthquake magnitude in the earthquake recurrence model. In this case, the larger the maximum magnitude, the higher the calculated hazard. However, if the seismicity of the source is constrained by geologic slip rates or geodetic strain rates, the maximum earthquake in the earthquake recurrence model affects the calculated seismic hazard in a different way. Since there is a fixed seismic moment rate budget, selection of a larger maximum magnitude results in lower rates of occurrence of smaller events, and the calculated seismic hazard may be reduced (Youngs and Coppersmith, 1985).

3. Strong Ground Motion Estimation Models

3.1 Empirical Ground-Motion Models

The simplest ground-motion attenuation relations predict ground-motion parameters using a simplified model in which the effects of the earthquake source are represented by earthquake magnitude or moment; the effects of wave propagation from the earthquake source to the site region are specified by a distance; and the effects of the site are specified by a site category (Table 1).

Moment magnitude is the most appropriate measure of earthquake size because it is directly related to the seismic moment of the earthquake, and thus to the fault area and the average displacement on the fault. When earthquake occurrence rates are instead based on historical seismicity in seismic hazard analyses, other magnitude measures are sometimes used because the seismic moments of the historical earthquakes may be poorly known.

The distance between the source and the site is measured in a variety of ways, including closest distance to the rupture surface, closest distance to the seismogenic rupture surface, closest distance to the vertical projection of the rupture to the surface, and hypocentral distance, as summarized by Abrahamson and Shedlock (1997). The lack of a standard definition of distance hinders the comparison and use of multiple ground-motion attenuation relations, which is necessary to fully account for uncertainty in estimated ground motions.

Site categories are usually based either on geological criteria or on shear-wave velocity of the surficial materials. The use of shear-wave velocity (e.g., Boore *et al.*, 1997) has the advantage of being based on an objective measure that affects ground motions in a way that can be modeled. However, it cannot be directly applied to sites that lack shear-wave velocity measurements. Also, deeper geological structure such as sedimentary basins and laterally varying structure may have an equally strong or even stronger effect on site response under certain conditions (e.g., Campbell *et al.*, 1997; Trifunac, 1990).

A major shortcoming of existing attenuation relations is the lack of a standard definition of site categories. Consequently, a large source of difference in the ground-motion models of different investigators lies in the different criteria that are used for site conditions, and in the assignment of site categories to individual recording sites. The use of a standard site classification system, and the assignment of these categories to individual recording stations, has the potential to reduce these unnecessary discrepancies between different model estimates.

The important effect of nonlinear soil behavior on site response is recognized in the site response factors that are embodied in current building codes and provisions in the United States such as the 1997 UBC (SEAOC, 1996) and 1997 NEHRP (Federal Emergency Management Agency, 1998). Until recently, seismologists modeled the seismic response of near-surface soils on the assumption that they have a linear stress-strain relationship. However, cyclic testing of soils in geotechnical laboratories has been showing that soils have nonlinear behavior at the strain levels that correspond to strong ground shaking. These measurements show a progressive decrease in shear modulus and an increase in damping with increasing strain level. These strain-dependent changes in material properties tend to reduce the amplitudes of short-period ground motions and to move the period of the resonance of the soil layer to longer periods. In part based on such laboratory test results, geotechnical engineers have been addressing the effects of soil nonlinearity on ground motions for some years (e.g., Seed and Idriss, 1969).

However, these laboratory measurements may not fully represent the nonlinear behavior of soil *in situ*. For this reason, there has been a strong emphasis on using recorded strong-motion data to test the predictions of nonlinear soil response based on laboratory data. In particular, there is a need to know the degree of nonlinearity that is present, the amplitude level and frequency range over which it occurs, and the kinds of materials in which it occurs. There is a growing recognition that it may occur in soft rock as well as in soil.

TABLE 1 Ground Motion Estimation Methods

Method	Source	Path	Site
EMPIRICAL	Seismic Moment or Magnitude	Distance	Geological category
STOCHASTIC	Source spectrum, e.g. 1 corner (Brune) or 2 corners (Atkinson)	Attenuation function, e.g. $1/R - 1/R^{1/2}$, or empirical. Duration varies with R. Anelastic Q. No distinct body waves or surface waves.	Kappa or fmax
BROADBAND GREEN'S FUNCTION	Shear dislocation, slip time function specified on fault	Green's functions including body waves, surface waves, anelastic Q.	Kappa; empirical or theoretical receiver function

The recorded ground-motion data available for evaluating nonlinear soil behavior at moderate and high strain levels in North America were quite sparse until recently. Important sets of data became available from the 1985 Michoacan, 1989 Loma Prieta, and 1994 Northridge earthquakes. The data from the Michoacan and Loma Prieta earthquakes were recorded at relatively large distance and moderate strain levels, and were characterized by large amplifications due to large impedance contrast effects and relatively small nonlinear effects.

The effect of nonlinear soil response in strong ground motions was recognized in the adoption of amplitude-dependent site coefficients in the 1994 and 1997 NEHRP provisions, in the 1997 UBC, and in the 2000 IBC. These coefficients were based on work by Borcherdt (1994), Crouse and McGuire (1996), Martin (1994), Martin and Dobry (1994), and Seed (1994) and were developed in part in response to the large site effects observed in strong-motion recordings of the 1989 Loma Prieta earthquake. These recordings were used to develop the site coefficients at low levels of motion. For soft soil sites, the coefficients at high levels of motion were developed by analytical modeling based on laboratory data. For stiff soil sites, the coefficients at high levels of motion were obtained by interpolating between soft sites and rock sites, assuming the latter to have linear behavior. However, the data from the 1994 Northridge earthquake suggest that these coefficients overestimate the effects of nonlinearity at stiff soil sites.

During the past several decades, large sets of strong-motion recordings were obtained from numerous earthquakes, significantly expanding the database of strong-motion recordings available for the derivation of empirical ground-motion models. A collection of recent models based on recorded data and in some cases on seismological models, accompanied by an overview (Abrahamson and Shedlock, 1997), was published in *Seismological Research Letters*. These ground-motion models are for distinct tectonic categories of earthquakes: shallow crustal earthquakes in tectonically active regions; shallow crustal earthquakes in tectonically stable regions; and subduction zone earthquakes, which are subdivided into those that occur on the shallow plate interface, and those that occur at greater depths within the subducting plate. Strong ground motions from subduction earthquakes generally attenuate more gradually with distance than those from crustal earthquakes, and earthquakes occurring within the subducting slab produce stronger ground motions than those occurring on the subduction interface. Mixing data from different earthquake categories gives rise to attenuation relations that have large dispersion and that underpredict the ground motions from some tectonic categories of earthquakes and overpredict those from others, not to mention the impact on dispersion.

Current attenuation relations of peak acceleration are characterized by a number of features. There is a "distance saturation" in which the slope of the attenuation function decreases at close distances, reflecting the fact that the earthquake is a distributed source, not a point source. For related reasons, there is a "magnitude saturation" in which the ground motions increase more gradually with magnitude for large magnitudes. At large distances, the ground motions are typically larger on soil sites than on rock sites, reflecting the effect of amplification due to impedance contrast. However, at close distances where ground-motion levels are high, the nonlinear behavior of soils tends to offset the impedance contrast effects, to the extent that high-frequency ground motions on rock may exceed those on some soils at very close distances to large earthquakes.

3.2 Variability in Ground Motions

There is a large amount of variability in ground-motion characteristics due to effects that are more complex than the simple parameterization described above based on magnitude, distance, and site category. New methods for analyzing the origins of this variability have provided important insight into the nature of strong ground motions, and indicate the directions in which further research may be able to reduce the uncertainty in the estimation of ground motions for engineering application. Specifically, the random effects approach (Abrahamson and Youngs, 1992) has been applied to the strong-motion database to separately quantify two sources of variability: the variability in the average ground motions from one earthquake to the next (interevent variability) and the variability in ground motions from one site to another at the same closest distance from a given earthquake (intraevent variability). For earthquakes of a given tectonic category larger than about magnitude 6, the interevent or event-to-event variability is found to be insignificant compared with the intraevent variability (Youngs *et al.*, 1995). The overall variability thus decreases significantly for the larger magnitudes. The decrease in the variability of ground-motion amplitudes with increasing magnitude can have a significant effect on the estimation of ground motions for engineering analysis and design.

This finding indicates that while the average ground motions from one large earthquake are very similar to those of another, there are conditions that cause the ground motions to vary significantly from one location to another at the same distance from a given event. The factors that cause this variability are related to aspects of the earthquake source process, the propagation of seismic waves from the source region to the site region, and the site response that are not contained in the simple magnitude-distance-site category parameterization of standard attenuation relations. In these simple models, these other sources of variability are treated as randomness, whereas they potentially could be treated as resulting from specific effects, which may be predictable. While those predictable effects may have uncertainties of their own, in some cases it should be possible to reduce these uncertainties.

For example, current empirical ground-motion models are derived from recordings at many sites that fall within a given site category. For a given site category, these models treat site-to-site variability as random. For a given site, strong-motion recordings may exist that describe the site response, or the results of geotechnical investigation could potentially be used to characterize the response of the site. Although the site response would still have some uncertainty in these cases, that uncertainty should be less than the standard error of empirical attenuation relations that reflect large site-to-site variations in site response. In this situation, the variability among different sites in the strong motion data set is irrelevant to (and overestimates) the uncertainty in ground motion at the specific site. This is especially important for ground motions having low annual probabilities of exceedance, for which the variability in ground motion is the principal cause of the increase in calculated ground-motion level as the annual probability of occurrence decreases. These large calculated ground-motion levels in many cases are largely a reflection of our ignorance of the actual ground-motion conditions that exist at a given site rather than an accurate estimate of the potential for large ground motions at the site.

3.3 Enhanced Ground-Motion Models

The simple ground-motion models described above have a large degree of uncertainty because other conditions that are known to have an important influence on strong ground motions, such as near-fault rupture directivity effects, crustal waveguide effects, basin response effects, and site effects, are not treated as parameters in these simple models. In order to reduce the uncertainty in predicting the ground motions at a given site, we need to augment these ground-motion estimation models to include more realistic representations of source, path, and site effects.

One approach to accounting for these effects in ground-motion models is to include them in empirical models by using a larger number of predictive parameters related to source, path, and site conditions. Examples of this approach are the empirical models of hanging-wall effects and near-fault rupture directivity effects described further below.

Another approach is to use seismologically based ground-motion models that take account of the specific source, path, and site conditions. These methods can be used to augment the recorded data used to generate empirical models, or to generate suites of ground-motion estimates that can be used to develop independent ground-motion models. Ground-motion models based on synthetic seismograms can then be used to complement available empirical models. In some regions, the database of strong-motion recordings is too sparse to allow the development of empirical ground-motion models. In this case, ground-motion attenuation relations are based primarily on seismological models that are described next. Also, considerable progress has been made in understanding and predicting ground-motion variability, especially at periods longer than about 1 sec, through the modeling of specific effects such as rupture directivity and basin response. By incorporating these effects into ground-motion models, in addition to the standard parameters of magnitude, distance, and site category, we should be able to reduce the uncertainty in the ground-motion estimates for a specified site.

3.4 Numerical Ground-Motion Models

Three alternative procedures for estimating strong ground motion are summarized in Table 1. Empirical ground-motion attenuation relations of the kind described above are available for predicting the strong motions from earthquakes in tectonically active regions. However, the database of strong-motion recordings in tectonically stable regions is too sparse to permit the development of empirical attenuation relations based purely on recorded data. This necessitates the use of seismologically based methods to generate ground-motion attenuation relations.

The simplest seismologically based method is the Band-Limited White Noise–Random Vibration Theory method (Boore, 1983), often referred to as the stochastic method. This method models ground motion as a time sequence of band-limited white noise. A Fourier spectral model of the ground motion is constructed, starting with a model of the source spectrum (Brune, 1970) and modifying its shape by factors to represent wave propagation effects. In early models, simple half-space models represented the attenuation due to geometrical spreading. However, current models based on the stochastic method, including Atkinson and Boore (1997) and Toro *et al.* (1997), use empirical and simplified numerical procedures, respectively, to account for the effect of the crustal waveguide on ground-motion attenuation. Site effects are represented by a simple anelastic attenuation function (Anderson and Hough, 1984) that represents the absorption of seismic energy in the shallow subsurface below the site. There is currently debate among proponents of the stochastic method as to whether the source spectrum is best represented by a Brune spectrum with a single corner frequency (Brune, 1970; Frankel *et al.*, 1996) or by a model having two corner frequencies (Atkinson, 1993; Boatwright and Choy, 1992; Atkinson and Silva, 1997). Although the stochastic model may be somewhat oversimplified, it affords the opportunity to perform extensive analysis of the effects of source parameters on strong ground motions. Based on successful validation against the response spectra of recorded strong-motion data (e.g., Silva *et al.*, 1997), the stochastic procedure has been widely applied in engineering practice (e.g., Electric Power Research Institute, 1993).

The broadband Green's function method is a more rigorous procedure that contains fewer simplifications than the stochastic model. By using scaling relations for earthquake source parameters in conjunction with the elastodynamic representation theorem, the procedure can be used to construct ground-motion time histories without resorting to assumptions about the shape of the source spectrum. The Green's functions that are used in this procedure can be calculated from known crustal structure models, facilitating the use of the procedure in regions where recorded data are sparse or absent.

The broadband Green's function method (e.g., Somerville *et al.*, 1996) has a rigorous basis in theoretical and computational seismology (Helmberger, 1983; Graves, 1996). The earthquake source is represented as a shear dislocation on an extended fault plane, whose radiation pattern, and its tendency to become subdued at periods shorter than about 0.5 sec, are represented. The spatial and temporal variation of slip of the fault surface is represented using methods described by Somerville *et al.* (1999). Wave propagation is represented by Green's functions computed for the seismic velocity structure that contains the fault and the site, or by empirical Green's functions derived from strong-motion recordings of small earthquakes. These Green's functions contain both body waves and surface waves. Site effects are represented by a simple anelastic attenuation function (Anderson and Hough, 1984) when calculated Green's functions are used, and are included empirically when empirical Green's functions are used. The ground-motion time history is calculated in the time domain using the elastodynamic representation theorem. This involves integration over the fault surface of the convolution of the slip time function on the fault with the Green's function for the appropriate depth and distance.

This summation process is based on linear superposition and thus does not directly accommodate nonlinear soil effects. In practice, the simulation procedure is used to simulate the ground motion on "engineering bedrock," which is defined as not having significant nonlinear soil effects. This ground motion is then used as input into a nonlinear soil response analysis to obtain the ground motion at the soil surface.

Based on successful validation against the time histories and response spectra of recorded strong-motion data, the broadband simulation procedure has been widely applied in engineering practice. An example is the FEMA/SAC Steel Project, whose objective was to understand the causes of brittle failures in the moment frame connections of steel buildings during the 1994 Northridge earthquake. Before being applied to estimate ground motion time histories and response spectra at the sites of buildings damaged by the earthquake, the procedure was validated against strong-motion recordings of the Northridge earthquake (Somerville *et al.*, 1995). The ground-motion simulation procedure can reproduce the duration, peak accelerations, and short-period (less than 1 sec) response spectral accelerations within a factor of about 1.5, which is comparable to the uncertainty in empirical ground-motion estimation models. For peak velocities and response spectral accelerations at periods longer than 1 sec, the error in these simulation methods may be lower than that of empirical ground-motion models, especially if rupture directivity and basin response affect the data and are included in the simulations. This kind of validation against recorded data constitutes an important criterion for the selection of ground-motion simulation procedures for use in earthquake engineering.

Through the use of these seismological models of strong ground motion, the number of parameters used in the estimation of strong ground motion using ground-motion attenuation relations has grown beyond the simple set of magnitude, distance, and site category. For example, in recent ground-motion models, the source and path parameters allow for differences in the ground motions between the hanging wall and footwall of dipping faults (Abrahamson and Somerville, 1996). Recent models also address the spatial variations in ground motions around faults due to rupture directivity effects (Somerville *et al.*, 1997). The propagation of fault rupture toward a site at a velocity close to the shear-wave velocity causes most of the seismic energy from the rupture to arrive in a single large pulse of motion that occurs at the beginning of the record. This pulse of motion represents the cumulative effect of almost all of the seismic radiation from the fault. The radiation pattern of the shear dislocation on the fault causes this large pulse of motion to be oriented in the direction perpendicular to the fault, causing the strike-normal peak velocity to be larger than the strike-parallel peak velocity. The influence of the rupture directivity pulse on the response of structures has been evaluated by Bertero *et al.* (1978) and Hall *et al.* (1995). Hanging-wall effects and rupture directivity effects have been incorporated into the 1997 Uniform Building Code in the form of the near-source factor (International Council of Building Officials, 1997), which amplifies the design spectrum at locations within 10 to 15 km of major faults.

In addition to providing more realistic representations of the earthquake source, seismological models have also been used to enhance the representation of wave propagation effects in ground-motion models. In place of using simple monotonic models for the attenuation of ground motion with distance, more complex models that take account of seismic wave propagation in a layered crust have been developed (e.g., Cohee *et al.*, 1991; Electric Power Research Institute, 1993). In place of simple geological descriptors of site conditions or ranges of seismic velocity, the complexity of the interaction of waves arriving at the site with the strongly heterogeneous conditions that characterize the shallow geology of most sites has been recognized and analyzed (e.g., Kawase, 1996; Graves *et al.*, 1998; Sato *et al.*, 1999; Pitarka *et al.*, 1998).

Seismological methods are increasingly being used to constrain those aspects of ground-motion estimation that are poorly constrained by recorded strong-motion data. This is especially important in stable continental regions where recorded strong-motion data are sparse. Even in the most seismically active regions, there are few strong-motion recordings close to the large earthquakes that control the seismic design of most structures. The reliability of empirical ground-motion attenuation models can be enhanced by augmenting the strong-motion database with strong ground motion time histories simulated for these conditions.

4. Deterministic (Scenario) Ground-Motion Hazard Evaluations

The deterministic approach typically involves evaluating values of ground-motion parameters at a site for a specific earthquake event using empirical attenuation relationships or, less frequently, using more sophisticated analyses discussed in the section on ground-motion prediction models. Although the probabilistic characterization of ground motions, described below, provides a more comprehensive representation of the seismic hazard at a site, a deterministic approach may be required to satisfy some specific regulation or role.

Deterministic ground-motion estimates play an important role in planning, preparedness, and loss estimation for earthquakes in urban regions. Many of these activities are more readily related to a specific earthquake scenario than to a probabilistic representation of the seismic hazard. In many seismically active areas, it is convenient to characterize the seismic hazard as that due to the repeat of a significant historical earthquake, or to the occurrence of an earthquake that dominates the seismic hazard due to its potential size or proximity to an urban region. These earthquakes are best viewed as scenario earthquakes, because they focus on the occurrence of just one significant event, whereas the hazard at the site consists of many different potential earthquakes.

Deterministic ground-motion hazard estimates are sometimes made for use in evaluating seismic design at sites potentially affected by active fault(s) capable of producing significant earthquakes. Under such conditions, a large earthquake (e.g., characteristic earthquake) is postulated to occur on the active fault. Such an event may have a sufficiently high frequency of occurrence or high public visibility to warrant being addressed in seismic design or seismic retrofit in some manner. The site ground motions from such a "scenario" earthquake are estimated using a deterministic approach.

One prominent use of deterministic ground-motion hazard estimates, which is related to scenario earthquakes, is in the Maximum Considered Earthquake (MCE) in FEMA 273/302. The MCE shaking in most areas corresponds to a probabilistic ground motion having a 2% probability of exceedance in 50 years. However, for regions near identified active faults having sufficiently high slip rates, and having characteristic earthquakes with magnitude higher than about 6, the MCE is deterministically defined. The ground motions for the MCE then correspond to those of the characteristic earthquake at a level that is 50% above the median value estimated using ground-motion attenuation relations, corresponding approximately to the 84th percentile level.

Another role of deterministic ground-motion hazard estimates arises when they are required by a regulatory agency. One such example is the deterministic ground-motion requirements imposed by the California Division of Safety of Dams (DSOD) on seismic evaluations of dams under their jurisdiction. However, some agencies that formerly required deterministic ground-motion hazards are now moving toward more probabilistic requirements. One such example is the procedures prescribed by the US Nuclear Regulatory Commission for the seismic evaluation of nuclear power generating stations.

A probabilistic response spectrum represents the aggregated contribution of a range of earthquake magnitudes occurring at various rates on each of several discrete faults or seismic source zones located at various distances from the site, and includes the effect of random variability in the ground motions for a given magnitude and distance. However, in order to provide ground-motion time histories that represent the response spectrum, we must choose one or more discrete combinations of magnitude and distance to represent the probabilistic ground motion. Thus, another need for deterministic ground motions arises from the need to represent a probabilistic response spectrum by one or more time histories. The use of time histories in seismic evaluations may increase as the practice moves more toward performance-based seismic design and retrofit.

In many of the situations discussed above, extracting deterministic equivalents from the results of probabilistic seismic hazard analysis (PSHA) through the process of deaggregation may be desirable. This is because such a deterministic evaluation supplements the probabilistic approach in developing information (i.e., acceleration time histories) that is not probabilistic in nature but is consistent with the probabilistic approach. In the description of probabilistic seismic hazard analysis that follows, we describe the process of deaggregating the hazard to estimate the magnitude–distance combinations that make the most important contributions to the seismic hazard.

When ground-motion parameters are deterministically evaluated using empirical attenuation relations, for example, it is important to ensure that the effects of all the important factors are reflected in the results. Some of these factors, such as rupture directivity effects and effects of hanging wall and footwall, are discussed in the section on ground-motion estimation models. Further, each empirical attenuation relation has different features, giving rise to different results. These differences are usually controlled by the database used in developing a particular relation, the parameters used to estimate the ground motions, and the functional form of the equation that relates the source parameters to the ground-motion parameters. Thus, some empirical attenuation relations may be more applicable than others for a given site depending on the conditions associated with that site. Usually, it is desirable to use a weighted average of multiple attenuation relations in order to avoid unintended bias.

When a number of active faults are located close to a site, different faults may control different period ranges of the response spectrum estimated for the site using attenuation relations. For example, a nearby fault capable of producing magnitude 6 events may control the short-period range of the response spectrum, but a relatively distant fault capable of producing magnitude 8 events may control the long-period range. One way of addressing this situation is to envelope the response spectra associated with the various faults. However, when such enveloping is used, the results no longer correspond to a scenario earthquake or purely deterministic ground-motion hazard, which is conceptually associated with a single event.

The deterministic ground-motion hazard evaluated using empirical attenuation relations is not really deterministic in the sense that the results are determinate. In using a simple combination of magnitude–distance–site condition to estimate a ground-motion parameter, there is usually uncertainty associated with the earthquake magnitude, which may represent the characteristic earthquake for a given fault. Uncertainty also exists in the site conditions for a given site, as well as in the site conditions reflected in empirical attenuation relations that use recorded data from a variety of site conditions. Equally important is the probabilistic implication of the usual practice of selecting ground-motion parameter values at a median (50th percentile) or median plus one standard deviation (84th percentile level) using empirical attenuation relations. In all of these respects, the use of the deterministic approach to estimating ground-motion hazard really implies a choice of probability levels for a single event.

A related issue in using deterministic ground-motion hazard estimates is the potentially unintended inconsistency in the risk level selected for two similar structures that may be located relatively close to each other. For example, two structures may both be located the same distance from faults that have the same maximum magnitude earthquake, but one fault may have a slip rate that is twice as large as the other. Under these conditions, the two deterministically estimated ground-motion parameters should be the same if the same attenuation relations and the same percentile values are used. Thus, the resulting designs should be similar, but the implied seismic risk associated with these two structures would be significantly different.

One reasonable approach in evaluating ground-motion hazards is to use probabilistic results as the primary criteria for seismic design and retrofit. However, this could be supplemented by using ground-motion hazard estimates from one or more deterministic (scenario) earthquakes, which may be derived from the probabilistic seismic hazard analysis through deaggregation, as a check, and to provide ground-motion time histories when non-linear structural response analyses are needed. Such an approach would be consistent with trends in seismic design practice in recent years and would address many of the issues discussed above.

5. Probabilistic Ground-Motion Hazard Evaluations

Given the uncertainty in the timing, location, and magnitude of future earthquakes, and the uncertainty in the level of the ground motion that a specified earthquake will generate at a particular site, it is often appropriate to use a probabilistic approach to characterizing the ground motion that a given site will experience in the future. A PSHA takes into account the ground motions from the full range of earthquake magnitudes that can occur on each fault or source zone that can affect the site (Cornell, 1968; Kulkarni *et al.*, 1979). The PSHA numerically integrates this information using probability theory to produce the annual frequency of exceedance of each different ground-motion level for each ground-motion parameter of interest, reflecting the effects of all the postulated seismic sources in the region. More explicitly, the PSHA incorporates for each fault or source zone the following three factors:

- Mean frequency per year of occurrence of each different magnitude level of earthquakes
- Mean frequency per event of each possible source-to-site distance
- Mean frequency per event of each different level of ground motion from each possible magnitude–distance pair

The PSHA yields the annual frequency of exceedance of each different ground-motion level for each ground-motion parameter of interest. This relationship between ground-motion level and annual frequency of exceedance is called a ground-motion hazard curve. The products of a PSHA are ideally suited for performance-based design, because they specify the ground motions that are expected to occur for a range of different annual probabilities (or return periods).

Probabilistic seismic hazard analysis can formally take into account uncertainties in the input parameters. Recent trends in PSHA identify two distinct kinds of uncertainty, which are treated in different ways (Senior Seismic Hazard Analysis Committee, 1996; Toro *et al.*, 1997). Epistemic uncertainty, due to (potentially reducible) uncertainty in the true state of nature (e.g., the slip rate of an active fault), is addressed using a discrete set of probability-weighted values using logic trees (Kulkarni *et al.*, 1984). Aleatory uncertainty, due to (potentially irreducible) randomness in physical processes (such as the hypocentral location of a future earthquake on a specified fault, or the ground-motion level at a given distance from an earthquake of a given magnitude), is addressed by treating that parameter as a random variable in the analysis.

An example of a seismic hazard curve, showing the relationship between the ground-motion level and its annual probability of occurrence, is shown for Boston in Figure 2. This example is from the National Seismic Hazard Maps produced by the USGS

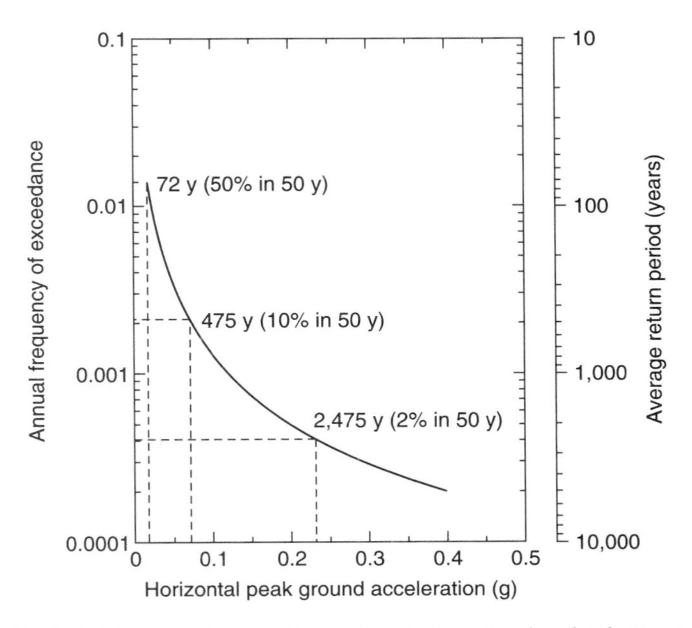

FIGURE 2 Seismic hazard curve for Boston, showing the decrease in annual frequency of exceedance (or increase in average return period) as the peak acceleration increases. The hazard curve is from the USGS National Seismic Hazard Maps, modified for soil site conditions.

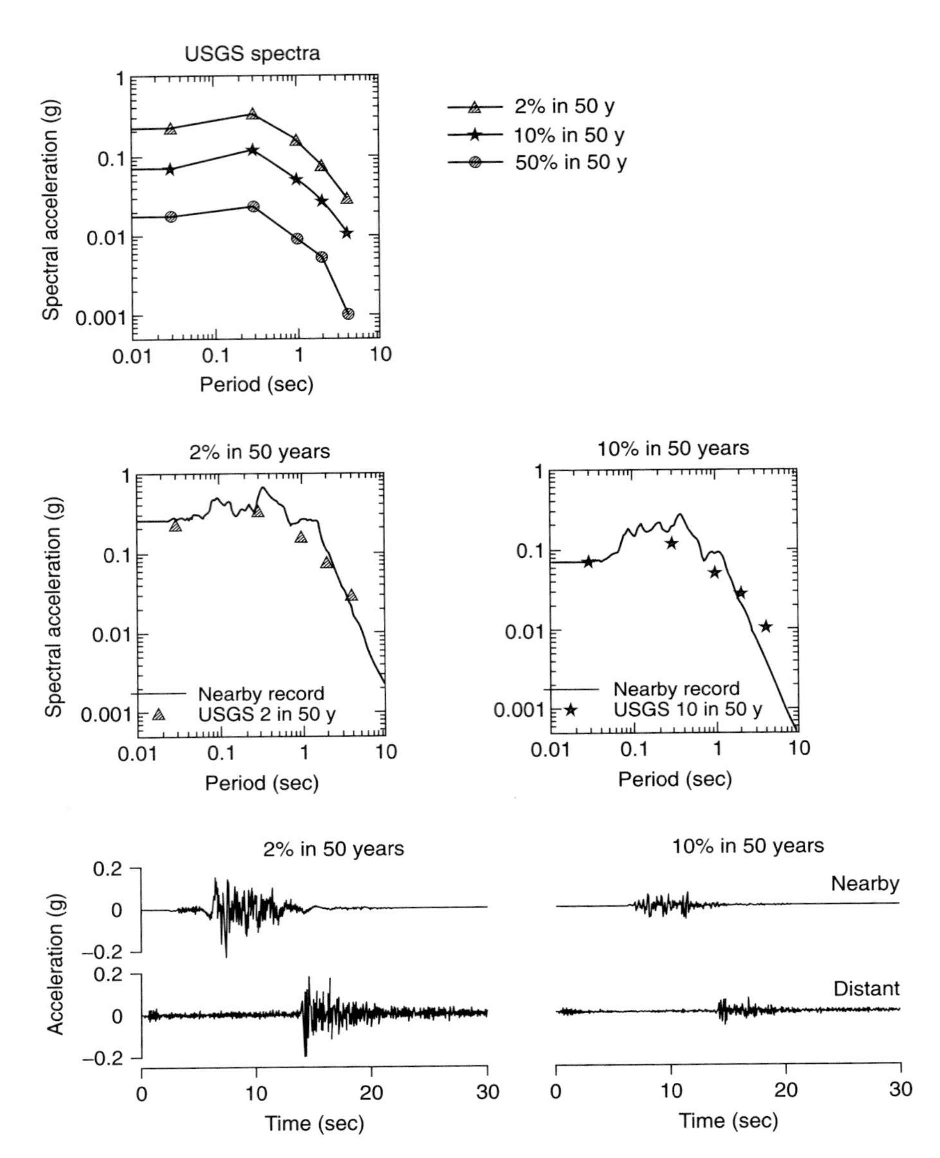

FIGURE 3 *Top:* Probabilistic response spectra for three frequencies of exceedance from the USGS National Seismic Hazard Maps, modified for soil site conditions. *Middle:* Response spectra of scaled simulations of small nearby earthquakes selected to represent the probabilistic spectra for 2% in 50 years and 10% in 50 years. *Bottom:* Scaled simulations of small nearby earthquakes and larger more distant earthquakes used to represent the probabilistic spectra for 2% in 50 years and 10% in 50 years (from Somerville *et al.*, 1997).

in 1996 (Frankel *et al.*, 1996; 2000), which form the basis for the NEHRP Provisions (FEMA, 1998), which are used in building codes in the United States. The horizontal axis of Figure 2 shows the peak acceleration, the vertical axis on the left shows the annual frequency of exceedance of the peak acceleration, and the axis on the right shows its inverse, the corresponding average return period.

The hazard curve in Figure 2 is for peak acceleration (response spectral acceleration at zero period) on soil, but the PSHA can produce analogous hazard curves for response spectral acceleration for a suite of periods. From these hazard curves, a suite of response spectra corresponding to several return periods can be constructed, as shown for Boston at the top of Figure 3 (from Somerville *et al.*, 1997). The products of a PSHA are ideally suited for performance-based design because they specify the largest ground motions that are expected to occur for a range of different annual probabilities (or return periods).

For performance-based design, the ground motions may increasingly need to be specified not only by response spectra but also by suites of strong-motion time histories for input into time-domain nonlinear analyses of structures. This is because pushover analyses based on spectral displacement or time-domain nonlinear analyses based on strong-motion time histories are preferred in evaluating seismically induced building damage and failure. Time histories whose response spectra are scaled to the target response spectra (derived from the USGS maps) at the top of Figure 3 are shown at the bottom of Figure 3. The match between the target spectra and the spectra of the time histories is shown in the middle of Figure 3.

The probabilistic response spectra (top of Fig. 3) represent the aggregated contribution of a range of earthquake magnitudes occurring at various rates on each of several discrete faults or seismic source zones located at various distances from the site, and they include the effect of random variability in the ground motions for a given magnitude and distance. However, in order to provide ground-motion time histories that represent the response spectrum, we must choose one or more discrete combinations of magnitude, distance, and eta to represent the probabilistic ground motion. The parameter eta is defined as the number of standard deviations above or below the median ground-motion level for that magnitude and distance that is required to match the probabilistic spectrum. The magnitude, distance, and eta values are estimated through deaggregation of the probabilistic seismic hazard (McGuire, 1995). The most convenient way to display the deaggregation of the hazard at a particular site is a map showing for each location the combinations of magnitude and distance that contribute to the hazard at the site (Bazzurro and Cornell, 1999). This kind of map facilitates the selection of a suite of magnitudes and distances that are most representative of the hazard at the site.

The nonlinear response of structures is strongly dependent not only on the level of the ground motion but also on the phasing of the input ground motion, and on the detailed nature of its spectrum. Therefore, a comprehensive measure of nonlinear response requires the use of multiple realistic time histories having phasing and response spectral peaks and troughs that represent the full range of magnitude–distance combinations that affect the seismic hazard. Thus, the shapes of the response spectra of individual time histories ideally should not be modified in the scaling procedure. Spectrally modified time histories may be appropriate when only a few nonlinear structural response analyses are possible. In this case, it may be preferable for the response spectrum of the time history to be modified to match the target spectrum because the response spectrum of just one time history or those of a few time histories may not provide a very good match to the target probabilistic spectrum.

6. Seismic Risk Assessment

Seismic risk typically is defined as the probability of loss due to seismic events. Thus, taking seismic risk typically means taking a chance on an estimated probability of incurring some loss from seismic events. The expected level of loss is sometimes used rather than the probability of loss. The term *loss* is defined in many ways with some common ones being human life loss, monetary loss, and loss of function. A collection of papers on current approaches to loss estimation can be found in *Earthquake Spectra*, volume 13, number 4 (1997).

A seismic risk evaluation for a given structure at a given site is initiated by first defining the loss to be considered for the given conditions, which should induce critical thinking about the problem to be addressed. Next, seismic hazard at the site is probabilistically evaluated in terms of selected hazard parameters such as the modified Mercalli intensity (MMI), peak ground acceleration, response spectrum, and fault displacement for the site. Given a level of seismic hazard, the probability of damage caused by that level of seismic hazard is evaluated for the structure. Finally, the probability of loss is evaluated for each level of damage. If desired, the expected loss can then be calculated.

The assessment of seismic hazard for ground motions typically uses PSHA, as discussed earlier. In the past, the ground-motion hazard has most commonly been represented by MMI. However, more quantitative parameters such as peak ground acceleration and response spectral acceleration are increasingly being used in seismic risk assessment. The PSHA provides the probability for each level of seismic hazard. Sometimes this probability is normalized with respect to the duration or magnitude of the earthquake if the structure of interest is sensitive not only to the specified peak ground-motion parameters but also to the duration or magnitude of earthquake.

The probability of damage suffered by the structure for each given level of seismic hazard is now typically evaluated using a fragility function. Thus, a fragility function describes the conditional probability of exceeding various damage states, ranging from none to complete, given a postulated seismic hazard level. These damage probabilities are then used to compute the probabilities of losses or the expected values of losses.

Until recently, most damage estimation was done using modified Mercalli intensity. Although MMI is nominally a measure of ground-motion severity, it is actually described largely in terms of damage states at the higher intensity levels based on often-unreliable human observations. Further, its relationship to instrumental ground-motion parameters (such as peak acceleration and peak velocity) is based on empirical correlations. While MMI provides a direct link between ground-motion severity and damage, it has limited value for performance-based seismic engineering where structural behavior for a specified quantitative level of ground motion needs to be evaluated.

New methods of seismic risk assessment that embody the conceptual framework of performance-based design were incorporated in the NEHRP Guidelines for the Rehabilitation of Buildings (FEMA 273). These guidelines use the capacity spectrum method to estimate the expected damage state of a structure. This method is based on the relationship between spectral acceleration and spectral displacement, which is used to represent the earthquake demand on the structure. The demand is compared with the capacity of the structure, which is expressed as a relationship between the lateral earthquake force acting on the structure and the resulting roof displacement, for example. To facilitate the comparison, the force is converted to spectral acceleration and the roof displacement is converted to spectral displacement. The intersection of the demand and capacity curves is an indicator of the expected damage state of the structure.

Recent earthquakes have demonstrated the complex interactions that exist between the various components of the infrastructure of modern urban environments. The loss of a single bridge, dam, power substation, or telephone exchange can have far-reaching secondary effects. The ability to anticipate these effects would provide a basis for earthquake hazard mitigation. Communities could use estimated losses to prioritize mitigation efforts in the most needed areas (such as the reinforcing of a particular structure) and develop contingency plans for emergency measures (such as identifying alternate transportation routes). Government agencies and financial institutions could use the information on estimated losses to minimize the social and financial impact of the earthquake disaster. A seismic risk assessment that includes life cycle costs can be used to make decisions on seismic design or the cost-benefit impact of seismic codes or regulations.

In order to realize these objectives, the Federal Emergency Management Agency (FEMA) in the United States has recently developed a loss estimation methodology called HAZUS (Whitman *et al.*, 1997). This methodology uses an inventory of buildings and infrastructure broken down into different structural categories, and a map of the ground motion of the scenario earthquake specified as response spectral acceleration. Fragility functions that relate the ground-motion level to damage states for different categories of structures are used to estimate damage states, based on the capacity spectrum method of FEMA 273 described above. This information is used to estimate direct losses, including direct costs for repair and replacement of buildings and infrastructure, and direct costs associated with loss of function, casualties, displacement of people from buildings, quantity of debris generated, and regional economic impacts. Functionality losses of buildings, critical facilities, lifelines, and infrastructure are also estimated, together with the extent of induced hazards such as fire, flooding, and release of hazardous materials. These results, while not probabilistically framed, provide a general sense of what could be expected under the postulated scenario earthquake.

7. Earthquake Engineering Practice

Earthquake engineering practice in general is strongly influenced by seismic codes that are currently in force. Therefore, as codes become more theoretically sound, as evidenced for example by the performance-based approach adopted in FEMA 273/302, practitioners in earthquake engineering will need to become more familiar with the theoretical background of related seismic disciplines. In addition to some aspects of engineering seismology and seismic geology, earthquake engineers need to become more familiar with the principles of geotechnical earthquake engineering, structural dynamics, or other related fields, depending on their specialty. This trend toward performance-based seismic evaluations has already been manifested in engineering practice, and it is expected to continue for the foreseeable future. Balancing and complementing this increasingly theoretically sound practice will be the continuing lessons learned from actual earthquake experience. The healthy growth of earthquake engineering practice depends on the tempering of sound theory by actual experience in earthquakes.

Consistent with this general trend, we expect an increasing trend in earthquake engineering practice toward the use of probabilistically estimated response spectral values in seismic design and retrofit. In the past, building codes tended to approximate ground-motion hazard by simple regional zoning, e.g., the seismic zone factor Z in the Uniform Building Code in the United States. Until recently, this caused the use of site-specific response spectral values based on a probabilistic approach to be relatively rare, and to usually be associated with cases that had special features. However, in recent years, probabilistically based response spectral values in some form have been increasingly used in the United States and other developed countries. With the future building codes in the United States using probabilistically estimated site-specific response spectral maps, earthquake engineering practitioners are expected to increasingly rely on probabilistically estimated site-specific response spectra to evaluate seismic hazards.

Thus, the increasing trend should be that, for nearly all categories of structures, seismic hazards are described in probabilistic terms, for example, as the annual probability of exceeding a specified level of ground motion. The probabilistic approach recognizes that there is a finite chance of exceeding any hazard level, but it recognizes that there must be some balance between costs and risks. The probabilistic approach provides a rational way of establishing an acceptable level of the risk, which ultimately must be made by the society at large, the regulator, or the owner of the facility.

As performance-based seismic design and retrofitting practice become more prevalent, a significant paradigm shift is expected in earthquake engineering practice. The shift will be from the overall load-reduction approach to the damage-versus-deformation approach reflected in current approaches to performance-based seismic design. The current professional experience of many structural earthquake engineers is based on some form of the load-reduction approach. Therefore, some readjustment in thinking will be required in order to take full advantage of performance-based design.

The current, commonly used load-reduction approach in the seismic design of buildings usually requires capacity-based seismic foundation design. However, the shift toward a performance-based approach will require more deformation-based seismic foundation design. Geotechnical earthquake engineers need to refocus their efforts on seismically induced deformation, both transient and permanent, rather than capacity alone. Thus, the expected shift in approach will affect not only structural earthquake engineers but also geotechnical earthquake engineers.

Performance-based seismic evaluations will increase the need for the various component disciplines of earthquake engineering practice to become more rationally linked. This will increase the importance of communication among practitioners not only within each discipline, but also between practitioners of related disciplines at the interface between their disciplines. One such example is soil-structure-interaction (SSI) analysis. Currently, SSI effects are often ignored in earthquake engineering practice because they are not significant in many cases. However, they also are often ignored because such refinements are considered unnecessary even in those cases where the effects of SSI may be significant. Even for modest-sized commercial buildings, whenever the foundation-ground system is "soft" in relation to the overlying building, the effects of SSI may be quite significant. Potential conservatism resulting from ignoring such effects may not be consistent with the intent of the performance-based approach, or with potential economic impacts.

The trend toward performance-based design will have the desirable effect of forcing the use of more theoretically sound seismic evaluations in designing or retrofitting structures. However, the final judge of seismic design and retrofitting efforts is the performance of the resulting structures under actual seismic events. Because the performance-based approach is relatively new, future earthquakes will test the consequences of this approach, and studies of actual seismic performance will be needed in order to make appropriate adjustments to the approach. Implementation of such studies and dissemination among practitioners of the results of such studies will be very important.

8. Seismic Building Codes

A detailed description of seismic design provisions and guidelines that are in effect in the United States is provided in *Earthquake Spectra*, volume 16, number 1 (Hamburger and Kircher, 2000; see also Chapter 68 by Borcherdt *et al.*). In the following, we outline the evolution of the seismic provisions of building codes into their current form, and anticipate their future evolution within the framework of performance-based seismic engineering.

Seismic building codes have traditionally had the goal of protecting life safety by preventing major damage under a postulated earthquake shaking condition. This goal was accomplished by using an equivalent static lateral force (lateral force) applied to the building and designing each component based on the resulting shears and moments. Representing the effects of earthquake shaking by lateral force was probably the most intuitive choice at the time when data on seismic ground shaking and seismic performance of buildings were scarce. Initially, the lateral force was dependent on the weight of the building and a region-dependent empirical coefficient. As the data on seismic motion and building performance started to accumulate, response spectral analysis of these data made the lateral force dependent on more factors. These factors included the fundamental period of the building, foundation soil type, and the structural type. This last factor, usually called the R factor, reflected the ductility associated with the overall structural system. The actual value of R depended on the structural system and the formulation of the lateral force equation, but, typically, it was larger than 1 and lower than about 12. The R factor was a load-reduction factor to reduce the lateral load to reflect in an approximate way the ductile behavior of the structural system. The lateral load was usually applied to the building as a base shear, which was appropriately distributed to various story levels of the building. In addition to the lateral load, the importance of certain aspects of structural detailing was increasingly emphasized.

As the quality and quantity of strong-motion recordings of earthquakes has grown over the years, the representation of the nature and level of ground motions in building codes has steadily improved. Strong ground motions recorded during the 1989 Loma Prieta earthquake provided the impetus for a major revision of the soil factor in the equation for the base shear. This led to two sets of amplitude-dependent site amplification factors derived from six classifications of site subsurface conditions (Borcherdt, 1994). These factors represent the nonlinear behavior of soils during strong ground motions in both the short- and long-period ranges. These factors were incorporated into the 1994 NEHRP seismic provisions and then into the 1997 UBC. Strong ground motions recorded during the 1994 Northridge and 1995 Kobe earthquakes emphasized that near-fault ground motions have distinct pulselike characteristics that are not represented in the code lateral force equation. The effect of these pulse motions was incorporated in the 1997 UBC in an approximate way by introducing near-fault factors.

In early building codes, the seismic hazard was treated as being uniform over large regions defined as seismic zones, although it was recognized that there are wide variations in seismic hazard within these regions. For example, within a given seismic zone, the seismic hazard is higher close to active faults than at large distances from them. This has led to the use of probabilistic methods for characterizing the seismic hazard on a national scale. For example, the United States National Seismic Hazard Maps (Frankel *et al.*, 1996) provide probabilistic estimates of ground motions on a grid of sites for a reference site condition. These ground motions are provided for several annual probabilities of exceedance, and are thus directly useful in the developing trend toward performance-based design that is discussed earlier and further below. The ground-motion maps show regional variations that reflect variations in seismic potential and seismic wave attenuation throughout the country. The ground-motion estimates are provided for several periods, from which a response spectrum can be constructed. The MCE ground-motion maps of the 1997 NEHRP provisions are based in part on these probabilistic ground-motion maps.

Although the lateral force equation in building codes continues to be refined, it cannot fully account for all aspects of ground motion and the dynamic response of structures. For complex or important structures, codes require that ground-motion time histories, derived either from recorded accelerograms or from strong-motion simulations, be used in dynamic calculations of building response.

Current trends in the development of new building codes have all embraced the concept of performance-based design, and conceptual frameworks of that approach have been developed in SEAOC Vision 2000, FEMA 273, and EERI (1998). Unlike the traditional approach that has been described above, performance-based design requires an explicit evaluation of the structure's performance at each of several ground-motion levels corresponding to a set of performance objectives. The performance objectives may range from continued function of the building during relatively small, but frequent ground motions; to limiting damage below the life safety threshold in severe, but less frequent ground motions; to prevention of collapse for very severe, but infrequent ground motions. Each performance objective is associated with an annual probability of occurrence, with increasingly undesirable performance characteristics caused by increasing levels of strong ground motion having decreasing annual probability of occurrence, as represented by the diagonal lines shown in Figure 1. Probabilistic estimates of seismic hazard are ideally suited to performance-based design, because they specify the ground motions that are expected to occur for a range of different annual probabilities of occurrence (return periods).

With the development of performance-based building codes, the trend is toward evaluating the "displacement" of the structure as a direct indicator of seismic performance, and away from the traditional load-reduction method. This new trend is desirable because it is more rational and allows buildings to be designed to various damage levels. It also moves the codes away from the global, poorly defined R-factor concept to component ductility, which is easier to address.

The first code to reflect performance-based design in the United States is FEMA 273, which addresses the retrofit of existing buildings. FEMA 273 contains a series of four alternative analysis procedures: a linear static procedure using a distribution of forces over the height of the building; a linear dynamic procedure using a response spectrum or time history input; a nonlinear static "pushover" procedure using a target displacement; and a nonlinear dynamic procedure using a time history input. Acceptance criteria are used to decide which procedures are acceptable in a given situation.

The performance of buildings during earthquakes is often very sensitive to the detailing that is used in the structure. In order to model the performance of an existing structure, it may be necessary to realistically model the behavior of these details, in addition to the global behavior of the structure. This indicates the increasingly detailed and complex requirements that are being placed on structural analysis in order to satisfy the objectives of performance-based design.

9. Policy Decisions

Policy decisions regarding seismic issues are made by a variety of governmental agencies at various levels ranging from federal to state and local. Some agencies focus on certain types of facilities, such as water-retaining structures, hospitals, and schools, while others have a much broader jurisdiction. These agencies are assisted by various professional organizations such as the Earthquake Engineering Research Institute (EERI) and the Applied Technology Council (ATC).

The enforcement and continual upgrading of seismic building codes by many municipalities and other agencies throughout the seismically active portions of the world reflects the single most important impact of policy decisions regarding seismic issues. As the performance-based approach takes hold, the impacts of these codes on additional aspects of seismic code development and implementation are likely to continue to increase. At the same time, this shift to a performance-based approach will challenge the awareness and the past inertia of all the parties involved in the process. The performance-based approach not only is new but also involves more decisions, such as the selection of performance levels.

Potential policy decisions involve such major areas as funding for seismic research, distributing guidelines on seismic issues, preearthquake planning and preparedness, postearthquake emergency response, and seismic hazard mitigation programs, which include building codes. For example, seismic policy decisions by the California Division of Mines and Geology (CDMG) have profound impacts on schools, hospitals, and other facilities in California. Similarly, at the federal level, seismic policy decisions made by the Federal Emergency Management Agency (FEMA) have profound impacts on strategies involved in the post-earthquake recovery following major seismic disasters. Governmental and other agencies at various levels are continually making seismic policy decisions in these areas.

To facilitate such processes, there exist some umbrella organizations such as the Western States Seismic Policy Council (WSSPC). The WSSPC includes representatives of the earthquake programs of 13 western states. The essential mission of WSSPC is "to provide a forum to advance earthquake hazard reduction programs throughout the Western Region and to develop, recommend, and present seismic policies and programs through information exchange, research, and education." A member of WSSPC can make a motion on a policy issue, with supporting data, to the Board of Directors of WSSPC. Members of WSSPC can discuss and study the motion, which may become a policy recommendation by vote. This process provides for checks and balances as well as for eventual consensus support for each proposed seismic policy issue.

An umbrella organization such as WSSPC serves an important function in optimizing the process of policy decisions regarding seismic issues, because those decisions require a long-term perspective to achieve effective funding and eventual payoff. Public spending on seismic issues voted by public officials usually will not see noticeable payoff during the tenure of the officials. The situation is similar with seismic research. Furthermore, because earthquakes for any given region are generally infrequent, public policies regarding seismic issues are not conducive to sustained public awareness, interest, and support. Public interest sharply rises after a major seismic event, but dissipates rather quickly with time. For these and other reasons, policy decisions regarding seismic issues need to be championed by strong nonpolitical entities.

In general, policy decisions always have economic implications. Governmental and regulatory agencies are aware of the importance of addressing the potential impact of seismic policy decisions on their constituencies. An increasing number of them are performing risk (cost-benefit) assessments to evaluate the impact of their policy decisions on society. This trend is desirable under the current economic environment.

References

Abrahamson, N. A., and K. Shedlock (1997). Overview, ground motion attenuation relations. *Seism. Res. Lett.* **68**, 9–23.

Abrahamson, N. A., and P. G. Somerville (1996). Effects of the hanging wall and foot wall on ground motions recorded during the Northridge Earthquake. *Bull. Seism. Soc. Am.* **86**, S93–S99.

Abrahamson, N. A., and R. R. Youngs (1992). A stable algorithm for regression analysis using the random effects model. *Bull. Seism. Soc. Am.* **82**, 505–510.

Anderson, J. G., and S. E. Hough (1984). A model for the shape of the Fourier amplitude of acceleration at high frequencies. *Bull. Seism. Soc. Am.* **84**, 1969–1993.

Atkinson, G. (1993). Source spectra for earthquakes in eastern North America. *Bull. Seism. Soc. Am.* **83**, 1778–1798.

Atkinson, G., and D. Boore (1997). Some comparisons between recent ground motion relations. *Seism. Res. Lett.* **68**, 24–40.

Atkinson, G. M., and W. J. Silva (1997). An empirical study of earthquake source spectra for California earthquakes. *Bull. Seism. Soc. Am.* **87**, 97–113.

Bazzuro, P., and A. C. Cornell (1999). Disaggregation of seismic hazard. *Bull. Seism. Soc. Am.* **9**, 501–520.

Bertero, V. V., S. A. Mahin, and R. A. Herrera (1978). Aseismic design implications of near-fault San Fernando earthquake records. *Int. J. Earthquake Eng. Struct. Dyn.* **6**, 31–42.

Boatwright, J., and G. Choy (1992). Acceleration source spectra anticipated for large earthquakes in eastern North America. *Bull. Seism. Soc. Am.* **82**, 660–682.

Boore, D. M. (1983). Stochastic simulation of high-frequency ground motions based on seismological models of the radiated spectra. *Bull. Seism. Soc. Am.* **73**, 1865–1894.

Boore, D. M., W. B. Joyner, and T. E. Fumal (1997). Equations for estimating horizontal response spectra and peak acceleration from western North American earthquakes: a summary of recent work. *Seism. Res. Lett.* **68**, 128–153.

Borcherdt, R. D. (1994). Estimates of site-dependent response spectra for design (methodology and justification). *Earthquake Spectra* **10**, 617–653.

Brune, J. (1970). Tectonic stress and the spectra of seismic shear waves from earthquakes. *J. Geophys. Res.* **75**, 4997–5009.

Campbell, K. W. (1997). Empirical near-source attenuation relationships for horizontal and vertical components of peak ground acceleration, peak ground velocity, and pseudo-absolute acceleration response spectra. *Seism. Res. Lett.* **68**, 4570.

Cohee, B. P., P. G. Somerville, and N. A. Abrahamson (1991). Ground motions from hypothesized $M_w = 8$ subduction earthquakes in the Pacific Northwest. *Bull. Seism. Soc. Am.* **81**, 28–56.

Cornell, C. A. (1968). Engineering seismic risk analysis. *Bull. Seism. Soc. Am.* **58**, 1583–1606.

Crouse, C. B., and J. W. McGuire (1996). Site response studies for purpose of revising NEHRP seismic provisions. *Earthquake Spectra* **12**, 407–439.

Earthquake Engineering Research Institute (1998). "Action Plan for Performance Based Design, Report to FEMA."

Electric Power Research Institute (1993). "Guidelines for determining design basis ground motions." EPRI TR-102293.

Federal Emergency Management Agency (1998). "1997 Recommended Provisions for Seismic Regulations for New Buildings and Other Structures." FEMA 302, 337 pp.

Federal Emergency Management Agency (1997). "NEHRP Guidelines for the Seismic Rehabilitation of Buildings." FEMA 273.

Frankel, A., C. Mueller, T. Barnhard, D. Perkins, E. Leyendecker, N. Dickman, S. Hanson, and M. Hopper (1996). National Seismic Hazard Maps, June 1996. *Open-File Report 96-532*, US Geological Survey, Golden, Colorado.

Frankel, A. D., C. S. Mueller, T. P. Barnhard, E. V. Leyendecker, R. L. Wesson, S. C. Harmson, F. W. Klein, D. M. Perkins, N. C. Dickman, S. L. Hanson, and M. G. Hopper (2000). USGS National Seismic Hazard Maps, June 1996. *Earthquake Spectra* **16**, 1–19.

Graves, R. W. (1996). Simulating seismic wave propagation in 3D elastic media using staggered-grid finite-differences. *Bull. Seism. Soc. Am.* **86**, 1091–1106.

Graves, R. W., A. Pitarka, and P. G. Somerville (1998). Ground motion amplification in the Santa Monica area: effects of shallow basin edge structure. *Bull. Seism. Soc. Am.* **88**, 1224–1242.

Gutenberg, B., and C. F. Richter (1964). *Seismicity of the Earth and Associated Phenomena* (2nd Edition). Princeton University Press, Princeton, N.J.

Hall, J. F., T. H. Heaton, M. W. Halling, and D. J. Wald (1995). Near-source ground motion and its effect on flexible buildings. *Earthquake Spectra* **11**, 569–605.

Hamburger, R. O., and C. A. Kircher, Editors (2000). "Seismic Design Provisions and Guidelines," *Earthquake Spectra* **16**(1), 307 pp. [This Theme Issue is reproduced on the attached Handbook CD.]

Helmberger, D. V. (1983). Theory and application of synthetic seismograms. In: *Earthquakes: Observation, Theory and Interpretation* (H. Kanamori, and E. Boschi, Eds.), North-Holland, Amsterdam, 174–221.

International Council of Building Officials (ICBO) (1997). "Uniform Building Code," Whittier, CA.

Kawase, H. (1996). The cause of the damage belt in Kobe: "The Basin-Edge Effect," constructive interference of the direct S-wave with the basin-induced diffracted/Rayleigh waves. *Seism. Res. Lett.* **67**, 25–34.

Kulkarni, R. B., R. R. Youngs, and K. J. Coppersmith (1984). Assessment of confidence intervals for results of seismic hazard analysis. *Proc. 8th World Conf. on Earthquake Engineering*, Vol. I, pp. 263–270.

Kulkarni, R. B., K. Sadigh, and I. M. Idriss (1979). Probabilistic evaluation of seismic exposure. *Proc. Second U.S. National Conf. on Earthquake Engineering*, pp. 90–99.

Martin, G. R. (Editor) (1994). "Workshop on Site Response during Earthquakes and Seismic Code Provisions." Proceedings NCEER, SEAOC, BSSC: University of Southern California, Los Angeles, California, November 1992.

Martin, G. R., and R. Dobry (1994). Earthquake site response and seismic code provisions. *NCEER Bulletin* **8**(4), 1–6.

McGuire, R. K. (1995). Probabilistic seismic hazard analysis and design earthquakes: closing the loop. *Bull. Seism. Soc. Am.* **86**, 1275–1284.

Pitarka, A., K. Irikura, T. Iwata, and H. Sekiguchi (1998). Three-dimensional simulation of the near-fault ground motion for the 1995 Hyogo-ken Nanbu (Kobe), Japan, earthquake. *Bull. Seism. Soc. Am.* **88**, 428–440.

Sato, T., R. W. Graves, and P. G. Somerville (1999). Three-dimensional finite difference simulations of long-period strong ground motion in the Tokyo metropolitan area during the 1990 Odawara earthquake (Mj 5.1) and the great Kanto earthquake (Ms 8.2) in Japan. *Bull. Seism. Soc. Am.* **89**, 579–607.

Schwartz, D. P., and K. J. Coppersmith (1984). Fault behavior and characteristic earthquakes from the Wasatch and San Andreas faults. *J. Geophys. Res.* **89**, 5681–5698.

Seed, R. B. (1994). "Workshop on Site Response during Earthquakes and Seismic Code Provisions," Proceedings NCEER, SEAOC, BSSC: University of Southern California, Los Angeles, California, November 18–20 (G. M. Martin, Ed.).

Seed, H. B., and I. M. Idriss (1969). Influence of soil conditions on ground motions during earthquakes. *J. Soil Mech. Foundation Div., Am. Soc. Civil Eng.* **95**(SM1), 99–137.

Senior Seismic Hazard Analysis Committee (SSHAC, 1996). "Probabilistic seismic hazard analysis: a consensus." Report to USDOE, USNRC, EPRI.

Silva, W. J., N. Abrahamson, G. Toro, and C. Costantino (1997). "Description and validation of the stochastic ground motion model." Report to Brookhaven National Laboratory, Contract No. 770573.

Somerville, P. G., R. W. Graves, and C. K. Saikia (1995). Characterization of Ground Motions during the Northridge Earthquake of January 17, 1994, Program to Reduce the Earthquake Hazards of Steel Moment Frame Buildings, SAC Report 95-03.

Somerville, P., C. K. Saikia, D. Wald, and R. Graves (1996). Implications of the Northridge earthquake for strong ground motions from thrust faults. *Bull. Seism. Soc. Am.* **86**, S115–S125.

Somerville, P. G., N. F. Smith, R. W. Graves, and N. A. Abrahamson (1997). Modification of empirical strong ground motion attenuation relations to include the amplitude and duration effects of rupture directivity. *Seism. Res. Lett.* **68**, 199–222.

Somerville, P., N. Smith, S. Punyamurthula, and J. Sun (1997). Development of Ground Motion Time Histories for Phase 2 of the FEMA/SAC Steel Project. Report No. SAC/BD-97/04.

Somerville, P. G., K. Irikura, R. Graves, S. Sawada, D. Wald, N. Abrahamson, Y. Iwasaki, T. Kagawa, N. Smith, and A. Kowada (1999). Characterizing earthquake slip models for the prediction of strong ground motion. *Seism. Res. Lett.* **70**, 59–80.

Structural Engineers' Association of California (1996). "Recommended Lateral Force Requirements and Commentary," 1996, (6th Edition).

Toro, G. R., N. A. Abrahamson, and J. F. Schneider (1997). Modeling of strong ground motions from earthquakes in Central and Eastern North America: best estimates and uncertainties. *Seism. Res. Lett.* **68**, 41–57.

Trifunac, M. D. (1990). How to model amplification of strong earthquake motions by local soil and geological site conditions. *Int. J. Earthquake Eng. Struct. Dyn.* **19**, 833–846.

Wald, D. J., and T. H. Heaton (1994). Spatial and temporal distribution of slip of the 1992 Landers, California earthquake. *Bull. Seism. Soc. Am.* **84**, 668–691.

Ward, S. N. (1994). A multidisciplinary approach to seismic hazard in Southern California. *Bull. Seism. Soc. Am.* **84**, 1293–1309.

Wells, D. L., and K. J. Coppersmith (1994). Analysis of empirical relationships among magnitude, rupture length, rupture area and surface displacement. *Bull. Seism. Soc. Am.* **84**, 974–1002.

Whitman, R. V., T. Anagnos, C. A. Kircher, H. J. Lagorio, R. S. Lawson, and P. Schneider (1997). Development of a national earthquake loss estimation methodology. *Earthquake Spectra* **13**, 643–661.

Youngs, R. R., and K. J. Coppersmith (1985). Implications of fault slip rates and earthquake recurrence models to probabilistic seismic hazard estimates. *Bull. Seism. Soc. Am.* **75**, 939–964.

Youngs, R. R., N. A. Abrahamson, F. I. Makdisi, and K. Sadigh (1995). Magnitude-dependent variance of peak ground acceleration. *Bull. Seism. Soc. Am.* **85**, 1161–1176.

66

Advances in Seismology with Impact on Earthquake Engineering

S. K. Singh
Instituto de Geofísica, UNAM, México City, México
M. Ordaz
Instituto de Ingeniería, UNAM, México City, México
J. F. Pacheco
Instituto de Geofísica, UNAM, México City, México

1. Introduction

The purpose of earthquake engineering is to design structures in such a way that they perform satisfactorily when subjected to earthquake loads that take place with certain return periods. Since providing lateral strength to a structure is expensive, there is always a trade-off between safety and cost: the higher the required safety, the higher the cost. Moreover, in modern seismic design the quality of a structural solution frequently depends on the detailed knowledge the designer has on the characteristics of the ground motions that the structure will suffer.

Hence, in order to do a rational earthquake design, in the sense of providing good structural solutions as economically as possible, engineers need to know three basic things: (1) where earthquakes occur, (2) how often earthquakes of given sizes (usually magnitudes) occur in their source regions, and (3) the characteristics of the ground motions at a site if an earthquake of a given magnitude were to occur at a given location. These are the main questions whose answers earthquake engineering has historically sought from seismology. The first two are related to seismotectonics, seismic source studies, earthquake catalogs, maximum magnitudes, etc. The third deals with attenuation studies and estimation of ground motion.

This chapter will emphasize those recent advances in seismology (including geodesy, which may be called the zero-frequency seismology) that have had or will have direct impact on earthquake engineering. Topics include geodynamics, seismotectonics, seismicity, seismic source studies, propagation and attenuation of seismic waves, site effects, and estimation of ground motions during future earthquakes. These topics are covered in greater detail in other chapters of the book. Here we will emphasize their connection and impact on earthquake engineering.

It will be impossible to do justice to all works in seismology with impact on earthquake engineering. What follows reflects, to a great extent, our own experience.

2. Seismicity, Plate Tectonics, and Seismic Gaps

In order to perform hazard analyses, earthquake engineers need to know where earthquakes occur. Most of the answers have been provided by plate tectonics studies, which show that earthquakes take place mostly at plate boundaries (interplate earthquakes). Evaluation of seismic risk became more accurate after recognizing that places close to plate boundaries presented larger hazard from earthquakes than other areas farther from plate boundaries (Kelleher *et al.*, 1973).

However, large damaging earthquakes do happen inside the so-called rigid plates (intraplate events). Some of these earthquakes occur in the subducted plates, while others take place in the interior of the upper plate. Earthquakes of the latter type have caused extensive damage in central India, eastern Australia, and eastern United States. Identification of these intraplate earthquake sources has also been possible due to contributions from seismology.

Advances in the quality and quantity of seismological instrumentation have given rise to better earthquake location and accurate estimation of seismic source parameters, including magnitudes. Worldwide broadband digital networks offer good-quality

INTERNATIONAL HANDBOOK OF EARTHQUAKE AND ENGINEERING SEISMOLOGY, VOLUME 81B *ISBN: 0-12-440658-0*

teleseismic records for events with magnitudes as low as 5.0 for routine processing. Recently, regional broadband networks have also implemented routine determination of earthquake source parameters (Ritsema and Lay, 1993; Romanowicz *et al.*, 1993; Thio and Kanamori, 1995). Detailed knowledge of the location of active faults, and the mode of slippage along them, is of importance for earthquake hazard estimation. Blind thrust faults are an example of how these studies help to assess the seismic hazard from faults that cannot be mapped at the surface (Stein and Yeats, 1989).

If we assume that the entire relative plate slip on an interface occurs during earthquakes and the size of the largest earthquake on a given plate segment is roughly constant and is known, then it is straightforward to compute the recurrence period of such events. It is a consequence of the fact that total slip during smaller earthquakes is negligible compared to the slip during the largest earthquake. Davies and Brune (1971) calculated seismic slip rates from catalogs and compared them with predicted slip rates from plate-motion models. Their findings suggested that most of the accumulated interseismic slip from relative plate motion is accounted for by the slip released from earthquakes along the plate boundary. Later studies by Peterson and Seno (1984), Jackson and McKenzie (1988), and Pacheco *et al.* (1993) also compared seismic slip rates with plate motion rates, finding a discrepancy between the two values for many areas. These studies show that the ratio of seismic slip to plate slip, on many plate boundaries, is less than one-half. There are segments along the San Andreas Fault in California where the fault slips almost continuously without producing any large or significant earthquake. Thus, the recurrence times estimated from plate slip rates might give a lower bound for the return period of large earthquakes. The fraction of plate slip released aseismically varies from one region to another. Given the limited extent of global earthquake catalogs and errors in the measurements of seismic moments for large and old earthquakes, estimations are very crude and not very accurate (McCaffrey, 1997).

Plate tectonics theory and the possibility that the recurrence period of large earthquakes along a plate boundary may be roughly constant gave rise to the seismic-gap hypothesis (e.g., McCann *et al.*, 1979). According to this hypothesis, large to great earthquakes have larger probabilities of occurrence in those places where no large earthquake has occurred in recent times. The longer the time since the last big event, the larger the probability of occurrence of a large earthquake. It is assumed that gap-filling earthquakes are characteristic of a given active area or fault system. A characteristic earthquake is the largest earthquake that ruptures the same fault segment with similar magnitude during repeated seismic cycles. Seismic gaps have been identified along almost all the active plate boundaries. The seismic-gap hypothesis has had an impact on the hazard analysis in many regions (e.g., Rosenblueth *et al.*, 1989; Working Group on California Earthquake Probabilities, 1995). It should be pointed out, however, that there is still controversy over the validity of the hypothesis (Kagan and Jackson, 1991; Nishenko and Sykes, 1993; Kagan and Jackson, 1993).

3. Geodesy and Earthquake Hazard

As mentioned above, various studies have shown that the ratio of seismic slip to plate slip on many plate boundaries is less than one-half. This observation suggests that aseismic slip occurs during interseismic periods. Furthermore, there are segments of plate boundaries with no history of large earthquakes. Either these segments are slipping aseismically or the strain is being accumulated and will be released in the future during a large or great earthquake. Clearly, this knowledge is essential in seismic hazard estimation. For this reason and for understanding the physical process involved in the earthquake cycle, it is of critical importance to know the spatiotemporal variation of strain during interseismic, coseismic, and postseismic phases, which puts into perspective the importance of geodetic measurements. Geodesy has traditionally played a significant role in geodynamics. However, recent advances in space geodesy promise to revolutionize our understanding of geodynamics, in general, and earthquake cycle, in particular. The Global Positioning System (GPS) uses Earth-orbiting satellites to give the relative location of a point with an accuracy of a few millimeters. GPS receivers are being used in permanent as well as portable modes to solve specific problems. Coseismic displacements, which have been measured for many earthquakes, provide important constraints on the spatial distribution of the slip on the fault. Two examples are the Landers earthquake of 1992 (Hudnut *et al.*, 1994), M_w 7.3, where M_w stands for moment magnitude (Kanamori, 1977; see Chapter 44 by Utsu), and the Colima-Jalisco earthquake of 1995, M_w 8.0 (Melbourne *et al.*, 1997). A novel and powerful geodetic technique, called the Synthetic Aperture Radar (SAR) interferometry (see Chapter 37 by Feigl), produced a spectacular image of the coseismic displacement field following the Landers earthquake (Massonet *et al.*, 1993). The technique uses repeat-pass radar backscatter images. It holds great promise to provide dense spatial coverage of coseismic displacement fields. In some cases, the GPS data has shown very slow (over a timescale of ~1 year) postseismic slip whose cumulative seismic moment roughly equals the seismic moment released during the rupture of the mainshock (e.g., Heki *et al.*, 1997). Such slow slip would account for the deficit in the seismic slip as compared to the plate slip.

The geodetic data are bound to provide spectacular and surprising results in the near future. In the last few years, over 1000 permanent GPS stations have become operational in Japan. This is the most dense regional network currently in operation. Just 1 year of data from this dense network (10- to 25-km spacing between stations) is sufficient to obtain relative motions of the sites (see, e.g., Report of the Coordinating Committee for Earthquake Prediction, 1999). The results obtained from this and other networks will have direct impact on seismic hazard estimation.

One can visualize dense networks of GPS receivers monitoring crustal movements in seismically active areas. Such networks would provide needed information regarding the accumulation and release of strain energy, leading to time-dependent seismic hazard estimation. They may also help to map hidden faults and active faults in stable continental areas. Ward (1994) provides an interesting application of geodetic data, along with geology, paleoseismology, observational seismology, and synthetic seismicity, to assess seismic hazard in areas far from known faults in southern California.

4. Size of Earthquakes, Size Distribution, and Characteristic Earthquakes

Ishimoto and Iida (1939) and Gutenberg and Richter (1944) observed that the size distribution of earthquakes followed a power law. This distribution is usually called the Gutenberg and Richter (G-R) relation (see Chapter 43 by Utsu), which can be expressed as

$$\log N(M) = a - bM \tag{1}$$

where N is the number of earthquakes with magnitude $\geq M$ for a given region. For worldwide catalogs, the value of b is close to 1. As most regional catalogs cover a relatively short observation period, it is difficult, in most areas, to empirically determine the recurrence time of large earthquakes or the maximum magnitude expected for a given region. If earthquakes obeyed the G-R relation, then an indication of recurrence times for large earthquakes could be obtained from a catalog of smaller events. This practice has become common in earthquake hazard analysis.

Different magnitude scales and incompleteness of earthquake catalogs have been a major hindrance in the reliable estimation of a and b values. With advances in the earthquake source theory, it became clear that the usual magnitude scales saturate (see, e.g., Brune and Engen, 1969; Kanamori, 1977, 1983), in the sense that, beyond some value, the magnitude does not grow as the size of the earthquake increases. Introduction of the concept of seismic moment, M_0, (Aki, 1966) and its associated moment-magnitude scale, M_w (Kanamori, 1977) has been of great importance in hazard analysis. This magnitude, which is defined in terms of amplitudes of seismic waves at very long period (theoretically at infinite period), does not saturate with increasing size of the earthquake, unlike other magnitudes measured at shorter periods. With this magnitude scale, the historical problems that inconsistent magnitude scales and their saturation for large earthquakes posed are slowly being solved.

The modified worldwide catalog, based on the M_w scale, suggests a difference in the b value for small and large earthquakes (Pacheco *et al.*, 1992). This observation implies a breakdown of self-similarity. Due to the finite width of the seismogenic zone, large earthquakes are confined within the surface of the Earth and the transition zone where the rock behavior changes from brittle to plastic (Scholz, 1990). Hence, large earthquakes can grow only along strike, while small earthquakes can grow in any direction within the fault plane. The difference in b values between small and large earthquakes causes an overestimation of the recurrence time for large earthquakes as compared with the recurrence time predicted from a G-R distribution with the b value obtained from catalogs of small events.

Usually, earthquake catalogs are complete above a threshold magnitude, which is larger the farther the catalog extends in the past. For instance, the worldwide catalog is possibly complete for magnitude ≥ 7 since 1906 (when many Wiechert seismographs became operational), but it may be complete for magnitude ≥ 5.4 since 1963 (when the World Wide Standard Seismograph Network was installed). Statistical procedures have been developed to merge subcatalogs that are complete beyond different magnitude thresholds for different periods of time (e.g., Rosenblueth and Ordaz, 1987).

Historical and instrumental records of seismicity are too short to define clearly the magnitude-frequency distribution for large earthquakes. Some estimates of recurrence times are now obtained from the geologic record. Studies like trenching along faults, measurement of displaced drainage channels, and terrace uplift patterns are part of a new discipline called paleoseismology (see Chapter 30 by Grant). A classical example is the work by Sieh (1978), who studied a section of late Holocene sediments cut by the San Andreas Fault, 50 km northeast of Los Angeles, and reported an average recurrence period of 160 years of large/great earthquakes along this segment of the fault.

In some very active fault segments, or where geological data have extended the seismic record to one or more seismic cycles, it has been reported that earthquakes of magnitudes close to the maximum magnitude, M_{max}, occur more frequently than expected from G-R relations (Singh *et al.*, 1983; Wesnousky *et al.*, 1983; Schwartz and Coppersmith, 1984; Davison and Scholz, 1985). While seismicity of an entire fault zone, composed of a number of separate fault segments, may be described by the G-R distribution, some individual fault segments display a characteristic earthquake distribution when viewed independently. The characteristic earthquake, in this sense, is similar to the one used in the seismic-gap hypothesis. The frequency-size distribution is no longer represented by the G-R relationship but is well described by the gamma distribution (Main, 1995). It is possible, however, that the characteristic earthquake behavior is not valid for all very active faults, or, at least, it cannot be satisfactorily demonstrated. Some of the studies have used geologically based slip-rate information to infer the frequency of M_{max}, under the assumption that all the slip was seismic, which may not be true, since the only data that can be used to test this hypothesis are properly treated historical and instrumental seismicity and a chronology of individual paleoseismic events. Although there is much debate about this model—whether sparse data at the high magnitudes cause the distribution to deviate from the G-R distribution or not—the implications for earthquake hazard are very important (e.g., Youngs and Coppersmith, 1985). If large earthquakes are more frequent than predicted from short-term seismicity, then by using G-R relationships alone we are underestimating the hazard.

Historical and geologic observations rule out a strictly periodic model for earthquake recurrence. A given segment of a fault might rupture with variable slip and/or variable rupture length. Different models have been proposed to explain the observations (Sieh, 1981; Schwartz and Coppersmith, 1984). The variable slip model proposes that at a given point on the fault, the displacement varies from one event to the next, giving rise to variable earthquake sizes from one cycle to the next one, while the slip rate is the same for the whole fault. The uniform slip model postulates that the displacement at a given point is always the same from one cycle to the next, and both the size of the large earthquakes and the slip rates are constant. For the characteristic earthquake model, the displacement at a given point is constant from one event to the next, but the slip rates vary along the length of the fault, and the size of the large event is constant. The available data do not permit a clear preference between the models. In the last two models, the maximum earthquake size can be inferred from studying past ruptures. In the first model, however, the maximum size of a future earthquake cannot be determined from past earthquakes. In this case, the rupture of the whole segment has to be taken as the upper bound for the maximum earthquake size.

5. Source Studies

The general acceptance of spectral scaling of earthquake sources (Aki, 1967; Brune, 1970) and a grasp of theoretical bases of some empirical relations in seismology (Kanamori and Anderson, 1975) has had great impact in seismic hazard estimation. These works have contributed, among other things, to an understanding of the connection between different magnitude scales (Kanamori, 1983) and the role of seismic moment as a measure of earthquake source strength, to the successful use of the ω^2 source spectrum (along with M_0 and a stress parameter) to estimate ground motions, and in the synthesis of ground motion using empirical or theoretical Green's functions. Analysis of a great number of earthquakes has shown a variation of stress drop of only about two orders of magnitude ($\sim$1 to 100 bars) over a range of M_0 of 10^{16} to 10^{30} dyne-cm (see, e.g., Kanamori, 1994).

Scaling relations show that M_0 of small earthquakes, which do not rupture the entire width of the seismogenic zone, will scale as the cube of the rupture length, L. There is, however, still some doubt about the scaling for very large earthquakes. One model, the L-model, will predict $M_0 \propto L^2$ (Scholz, 1982). Another model, the W-model, which states that the slip is scale-invariant, will predict $M_0 \propto L$ for large earthquakes. In both cases, the estimated ground motions of earthquakes that rupture the whole width of the seismogenic layer will differ from those of earthquakes that do not. There is still a lively debate on this issue (see, e.g., Romanowicz and Rundle, 1993; Sornette and Sornette, 1993; Romanowicz and Rundle, 1994). It is clear, however, that these models have implications for earthquake hazard analysis, since predicted strong ground motions from future large earthquakes would depend on the model used.

Initially, the determination of source parameters (location, seismic moment, and focal mechanism) was a tedious process. However, advances in seismic source theory, seismic instrumentation, communication technology, and computers now permit fast and reliable estimation of source parameters of moderate and large earthquakes (see Chapter 50 by Sipkin). For example, the Harvard Centroid Moment Tensor (CMT) solution of all significant earthquakes, which provides centroid location and moment tensor, is available within a few hours of the earthquake. The National Earthquake Information Center (NEIC) of the USA and the Earthquake Research Institute in Japan also routinely report moment tensor solutions. The University of Michigan determines source time functions of large earthquakes in near real time. Thus, gross and robust source parameters of important earthquakes, such as seismic moment, focal mechanism, and source duration, along with location and depth, now become available very quickly. These estimates provide the framework for earthquake engineers to interpret damage data.

Detailed studies of source process are also becoming more common, thanks to high-quality local and regional seismic and geodetic data. These data, along with teleseismic records, are inverted to map the spatial and temporal distribution of slip on the fault. Two examples are the Landers, California, earthquake of 1992 (M_w 7.3) and the Kobe, Japan, earthquake of 1995 (M_w 6.9). As seen from Figure 1, the joint inversion of seismic and geodetic data provides a detailed image of the rupture (Wald and Heaton, 1994). It is important to point out that the inversion using an individual data set is fairly consistent with the inversion based on the complete data set (see, e.g., Yoshida *et al.*, 1996). This figure shows a complex pattern with areas of large slip (broken asperities) with adjacent patches of no slip. There are two possible explanations for the lack of slip: (a) these are areas where the deformation occurs by creep, and (b) these are barriers, which would rupture in the next sequence. The implication of the first explanation is that the rupture would repeat and lead to a characteristic earthquake sequence. In the second explanation, the rupture pattern would change from cycle to cycle, leading to a noncharacteristic earthquake sequence. As mentioned earlier, it would be very useful in the estimation of seismic hazard to know which of these two explanations is valid in a region. This knowledge can only be gained from detailed mapping of slip during earthquakes and geodetic measurements during the interseismic phase of an earthquake cycle. Detailed knowledge of the slip on the fault also provides elements to postulate future rupture scenarios that, in turn, are needed to simulate ground motions from earthquakes.

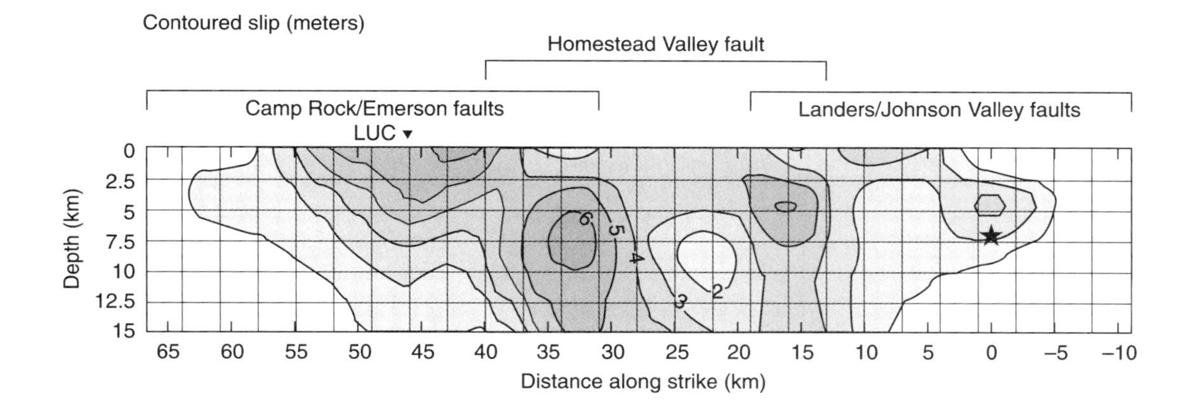

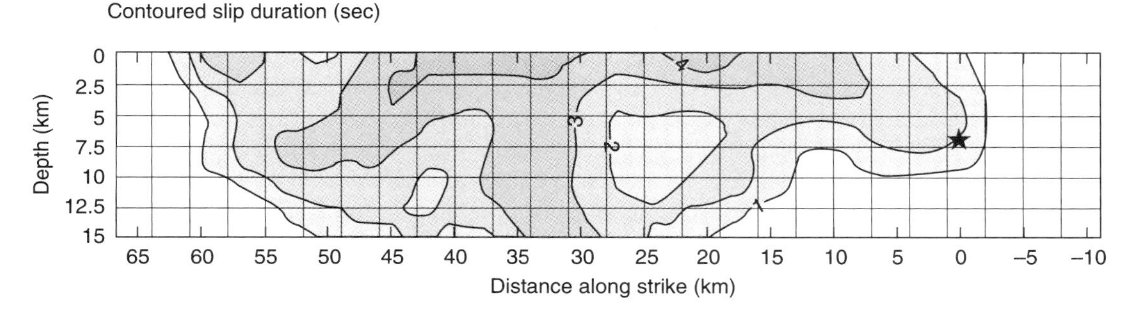

FIGURE 1 Slip distribution (top, in meters) and duration (bottom, in sec) of the 1992 Landers earthquake (M_w 7.3) determined from joint inversion of strong-motion, teleseismic, and geodetic data. Star denotes the hypocenter (after Wald and Heaton, 1994).

6. Estimation of Strong Ground Motion

Estimation of strong ground motion on firm ground—that is, without inclusion of site effects—is still mainly carried out with the use of attenuation relations. These are usually relations between magnitude, distance, and some measures of earthquake intensity with engineering significance, such as peak ground acceleration, A_{max}, peak ground velocity, V_{max}, or response spectral ordinates.

In the early 1970s, attenuation relations were mainly derived on empirical grounds. Observed values of, say, A_{max}, were fitted to simple functional forms, with free parameters whose values were chosen to minimize the misfit between observed and computed values (see, for instance, Esteva and Villaverde, 1973). In recent years, other methods have been adopted. We will discuss some of the more interesting ones.

Many of the attenuation studies are essentially curve fitting (see, for example, the January–February 1997 issue of *Seismological Research Letters*; Chapter 60 by Campbell). This produces in many cases parameters that are optimal in the statistical sense, but that might have unrealistic or inadmissible values in the light of seismological theory. Also, the functional forms usually chosen for data fitting often produce numerically unstable solutions, especially for a sparse data set (Ordaz *et al.*, 1994). Several things have been done to overcome these difficulties.

Joyner and Boore (1981) proposed a two-step regression procedure. In the first step, geometrical spreading was fixed, and the parameter associated with anelastic attenuation was derived. In the second step, scaling with magnitude was obtained. This procedure made regressions more stable and helped to solve the problem of correlation between magnitude and distance, but more interestingly, it used seismological knowledge to restrict the universe of possible solutions to the regression problem by fixing the geometrical spreading on a theoretically reasonable ground.

On different lines of thought, Hanks and McGuire (1981) related root mean square (rms) acceleration to ω^2-source spectra modified by anelastic and geometric attenuation, through Parseval's theorem, and, using results from random vibration theory, related rms acceleration to A_{max}. Boore (1983) extended these results to predict V_{max} and response spectra by generating time series of filtered and windowed Gaussian noise whose amplitude spectrum approximated the modified ω^2-source spectrum.

Although Boore's results are attenuation relations, for them to be useful to reproduce a given strong-motion database, some parameters, like the quality factor, Q, and stress-drop, need to be calibrated. This approach proved especially powerful for regions with scarce strong-motion data, such as the eastern United States (e.g., Boore and Atkinson, 1987; Toro and McGuire, 1987) and the Indian peninsular shield region (Singh *et al.*, 1999). Even with very limited databases (where early regressions could not be done), since the implied functional forms are theoretically based, the attenuation relations are more realistic.

These ideas are part of what could be called the semiempirical approach: Underlying functional forms are chosen that comply with accepted seismological theories, and free parameters are then selected to minimize the difference between the predicted and the observed values of ground motion. Additionally, since the free parameters have physical meaning, the universe of possible solutions is further constrained.

The statistical use of information from both data and seismological theory can be formally expressed with the use of Bayes' theorem (see Chapter 82 by Vere-Jones and Ogata). Attenuation relations have been derived with this technique (Veneziano and Heidari, 1985; Ordaz *et al.*, 1994). Before examining the data, prior probability distributions are fixed on the free parameters, relying upon seismological knowledge. These distributions are then updated with observed data, ending with estimates that combine both theory and data.

With the increase of strong-motion instrumentation in the world, new and more precise techniques to study attenuation from the engineering viewpoint have come out. For instance, it has been possible to obtain nonparametric attenuation equations (Anderson and Quaas, 1988), and attenuation equations have been derived for general classifications of soil types. Furthermore, site-specific attenuation relations are not uncommon these days (e.g., Singh *et al.*, 1987; Ordaz *et al.*, 1994; Reyes, 1998) as illustrated in Figure 2, where the predicted 5% damped response spectra are shown for a firm-ground site in Mexico City.

Before the 1980s, interest was concentrated in estimation of A_{max} and V_{max}. Although the engineering importance of frequency content of the motion was recognized many years ago, techniques to draft the full spectrum with peak values alone (Newmark and Hall, 1969) are now considered too crude, especially in view of the wide variety of site effects that exist in the real world. In recent years, the need of more precise descriptions of frequency content has given rise to frequency-dependent attenuation relations, either purely empirical (e.g., Sadigh *et al.*, 1997) or semiempirical (e.g., Boore, 1983; Toro and McGuire, 1987). Most useful for engineering purposes are those relating magnitude and distance to response spectral values (typically, pseudoaccelerations or pseudovelocities for 5% of critical damping). So far, spectral attenuation relations have been constructed processing individually the observed values corresponding to a single period. A more general statistical technique, in which correlation among periods is recognized, is yet to be constructed. In this technique, attenuation relations for several frequencies would be simultaneously derived, to take advantage of the fact that knowledge of spectral ordinates at a given period contains some information about the values at neighboring periods.

The increasing complexities of engineering structures and the advances in inelastic structural models have called, in the recent years, for better methods of generation of synthetic accelerograms. The early simulation techniques used in engineering applications were based on filtering band-limited white noise (e.g., Jennings *et al.*, 1969), the filter characteristics being empirical. However, the most important advances come from applications of seismological theory. For instance, since the general acceptance of the ω^2-source model with constant stress-drop (Aki, 1967; Brune, 1970), the methods in which a synthetic accelerogram is produced by scaling the observed one by a constant have almost been abandoned. Boore (1983) proposed a simple method in which artificial accelerograms are obtained with band-limited white noise, filtered in frequency and modulated in time. This technique produces accelerograms that satisfy some seismological restrictions, like appropriate duration and frequency-dependent scaling with magnitude.

Hartzell (1978) proposed the use of recordings of small earthquakes as empirical Green's functions (EGF) to synthesize recordings of large events. In this approach, the accelerogram of the target event is the sum of signals coming from subevents, which are basically a delayed version of the EGF. The rupture of the subevents needs to be specified in size, in time, and

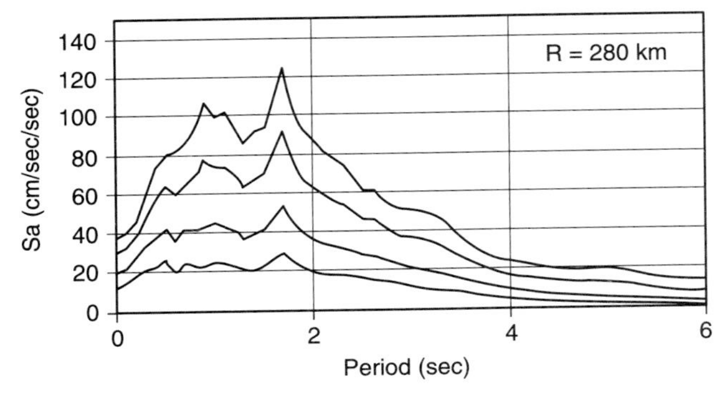

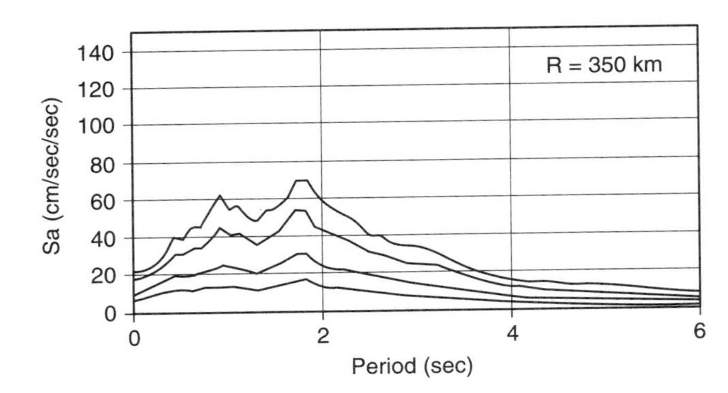

FIGURE 2 Site-specific response spectra (pseudoaccelerations, 5% damping) for station CU in Mexico City, for moment magnitudes 7, 7.5, 8, and 8.3. *Left:* hypocentral distance 280 km; *right:* hypocentral distance 350 km.

sometimes also in space. Purely deterministic and homogeneous rupture scenarios lead, in many cases, to unrealistic high-frequency content, requiring the introduction of some randomness. Irikura (1983), using source scaling relations and a finite source model has extensively applied this method. Several simpler procedures have been devised to give the required randomness to the source. In some of them, this randomness is achieved by assigning various probability distributions to the delay times between the subevents of the source (e.g., Joyner and Boore, 1986; Wennerberg, 1990; Ordaz *et al.*, 1995). Some others make the size of the subevent a random variable as well (Zeng *et al.*, 1994). Another possibility is the use of random ruptures with empirical source functions along with theoretical Green's functions (Somerville, 1993). However, in all of them, restrictions from seismological theory are present, which makes artificial accelerograms more reliable than they would be without the restrictions. An example of synthesis of ground motion using EGF is shown in Figure 3. In general, methods based on empirical Green's functions that require fixing many parameters are limited to academic use. For engineering purposes, simpler techniques—even if less precise—are generally preferred.

Until recently, due to the scarcity of near-source recordings, attenuation relations were poorly constrained for small distances. Although there is still a paucity of strong-motion data close to the fault, available information suggests a saturation of A_{max} with magnitude. For example, Singh *et al.* (1989) report that horizontal A_{max} at hard sites above the Mexican subduction zone, at a distance of 16 km from the fault surface, reaches a value of about 0.5 g for $M \sim 5.5$ and does not increase with increasing magnitude. Similarly, the A_{max} data for $6.5 \leq M \leq 7.4$, compiled by Heaton *et al.* (1995) at distances of 5 km or less from the surface projection of the rupture area, also show little increase with magnitude. The median peak acceleration for this data set is 0.81 g. More recent attenuation relations have started including this feature (e.g., Youngs *et al.*, 1997). However, damage to some buildings is more related with V_{max} or U_{max}, the peak ground displacement, which occur at longer period and are expected to increase with magnitude. Reliable data on long-period ($T > 5$ sec) near-source strong motion are very scarce, so usual attenuation relations become highly unreliable. An attractive alternative is the deterministic modeling, since at these periods U_{max} is relatively insensitive to crustal heterogeneities and source complexities.

Near-source ground motions at periods greater than 1 sec have been successfully modeled for many earthquakes. These studies show that the duration of the slip at any given point of the fault is much less than the total rupture duration (Heaton, 1990). Even during large earthquakes, the slip duration at a given point on the rupture is only a few seconds (Fig. 1, bottom), much smaller than the expected duration from simple earthquake models, which require all points to move at least until the rupture stops. Since the slip on parts of the fault may exceed a few meters over a few seconds, the particle velocity on the fault may exceed 100 cm/sec (for example, during the Chi-Chi, Taiwan, earthquake of 1999; Shin *et al.*, 2000; Teng *et al.*, 2001). The short duration of slip is an important constraint in the computation of velocity and

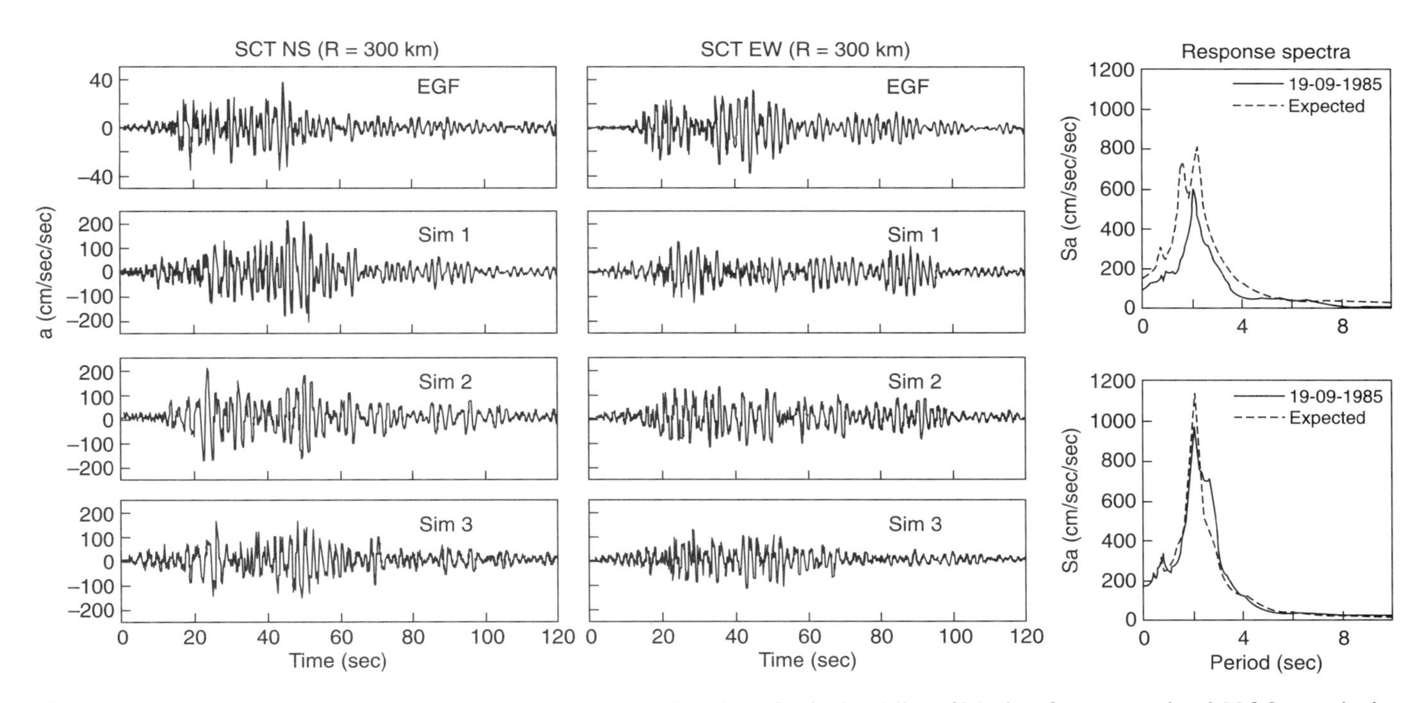

FIGURE 3 Prediction of expected ground motions at the lakebed station SCT, in the Valley of Mexico, from a postulated M 8.2 event in the Guerrero gap, Mexico. Empirical Green's function: earthquake of 25 April 1989, M 6.9. Expected pseudoacceleration response spectrum (5% damping) from simulations is compared with that observed during the 19 September, 1985, Michoacán earthquake. The summation of EGF follows the method given by Ordaz *et al.* (1995).

displacement seismograms in the near-source region of large earthquakes. Computation of synthetic seismograms in the near-source region, under plausible rupture scenarios, may become a routine practice in the design of earthquake-resistant structures situated close to a fault, and compensate for the lack of long-period data in the near-source region. An example of such computations can be found in Heaton *et al.* (1995), where the authors compute expected ground motions from a postulated $M = 7$ earthquake on a hidden fault below the Los Angeles basin.

The strong-motion data near the toe of shallow thrust earthquakes are very sparse since the toe lies offshore, near the trench, in the subduction zones. However, estimation of strong motion near the toe may be critical in such regions as northern India. Foam rubber experiments (Brune, 1996a) and the 2D lattice numerical model (Shi *et al.* 1998) suggest that anomalously high accelerations near the toe of the hanging wall, resulting from trapping of waves in the upper wedge, may be expected, which is supported at least by recordings during some reverse-faulting earthquakes (Brune, 1996a). These results may have significant impact on seismic hazard estimation above thrust and reverse faults.

Recently, it has been pointed out that precariously balanced rocks act as low-resolution seismoscopes and have the potential to constrain ground motions and, hence, useful information about seismic risk (Brune and Whitney, 1992; Brune, 1996b). This field is new and exciting. The methodology to use this information in the seismic hazard estimation is currently in development.

7. Site Effects

It has been known for over one hundred years that damage during earthquakes is related to the local geological site conditions. Some dramatic, recent examples come from the Valley of Mexico during the 1985 Michoacán earthquake, the town of Leninakan during the 1988 Armenian earthquake, and the damage patterns during the Loma Prieta earthquake of 1989, the Northridge earthquake of 1994, and the Kobe earthquake of 1995. These earthquakes emphasize the necessity of estimating the effect of local site conditions on the expected ground motion (see e.g., Chapter 61 by Kawase).

Figure 4 shows an example of site effects in the Valley of Mexico during the coastal earthquake of September 14, 1995 ($M_w = 7.3$), located about 300 km from the valley. For geotechnical purposes, the valley is divided into three zones: (1) the lakebed zone, which consists of a 10- to 100-m deposit of highly compressible clay with high water content, underlain by dense sands; (2) the hill zone, basically formed by surface layer of lava flows or volcanic tuffs; and (3) the transition zone, composed of alluvial sandy and silty layers with occasional intervals of clay layers. As seen in Figure 4, the waves are greatly amplified in the lakebed zone as compared to the hill zone, the relative spectral amplification reaching values of up to about 50 at some frequencies between 0.2 and 0.7 Hz (Singh *et al.*, 1988a,b). Furthermore, seismic waves in the lakebed zone have very long duration and exhibit a harmonic character.

Site response has been estimated from (a) recordings of earthquakes, (b) microtremor measurements, and (c) theoretical computations.

Perhaps the most reliable estimation of site response is obtained when earthquake recordings are available at the surface and at downhole sensors, or at the surface and a nearby rock site, which is taken as the reference site. In the latter case, the epicentral distance should be much greater than the station separation so that the input motion may be assumed identical at both sites. In this case, the spectral ratio of the S-wave group defines the site response. Figure 5 shows site response of some lakebed sites of the Valley of Mexico with respect to the hill-zone site of CU. Note that the site response is relatively large and very stable. The question arises whether the reference site is free of site effect. If the ground motion can be reliably predicted at the reference site, then the response of this site is of no consequence. An example is the Valley of Mexico where the reference site is CU. Because of the relatively large number of recordings available at this site, attenuation relations permit reliable estimation of ground motion at Ciudad Universtaria (CU) from future earthquakes (Singh *et al.*, 1987; Castro *et al.*, 1988; Ordaz *et al.*, 1994). The ground motion in the valley is estimated from the Fourier acceleration spectrum at CU, the known site, response of instrumented lakebed sites (about 100 in the valley, see Fig. 5) with respect to CU, along with interpolation for other sites (Pérez-Rocha, 1998), and application of random vibration theory (Singh *et al.*, 1988a,b; Ordaz *et al.*, 1988; Reinoso and Ordaz, 1999). This procedure is the basis of a computer code, called Program Z, which is currently being used among building designers in Mexico City to estimate site-specific response spectra, and is also the basis of near real time shake maps for Mexico City. Note that CU, the reference site, is amplified by a factor of 10 at some frequencies (0.2–1.0 Hz), as shown by Ordaz and Singh (1992). Indeed, there appears to be no hard rock site in the Valley of Mexico that is free from large site response (Singh *et al.*, 1995; Shapiro *et al.*, 1997).

The S-wave site amplifications may also be estimated from the observed spectra of one or more earthquakes, assuming a geometrical spreading model and, in some cases, the shape of the source spectra (e.g., Andrews, 1986; Boatwright *et al.*, 1991; Castro *et al.*, 1990; Humphrey and Anderson, 1992; Ordaz and Singh, 1992). In this method, the reference site is a selected rock site, the average of a few rock sites, or the average of all sites where the data are recorded.

Coda waves have been used to estimate the site amplification (Tsujiura, 1978; Phillips and Aki, 1986; Margheriti *et al.*, 1994; Su and Aki, 1995). This amplification corresponds to average site effect of shear waves and is relative to an unknown average reference station.

The methods outlined above give site amplification relative to a reference rock site or to an average rock site. As mentioned

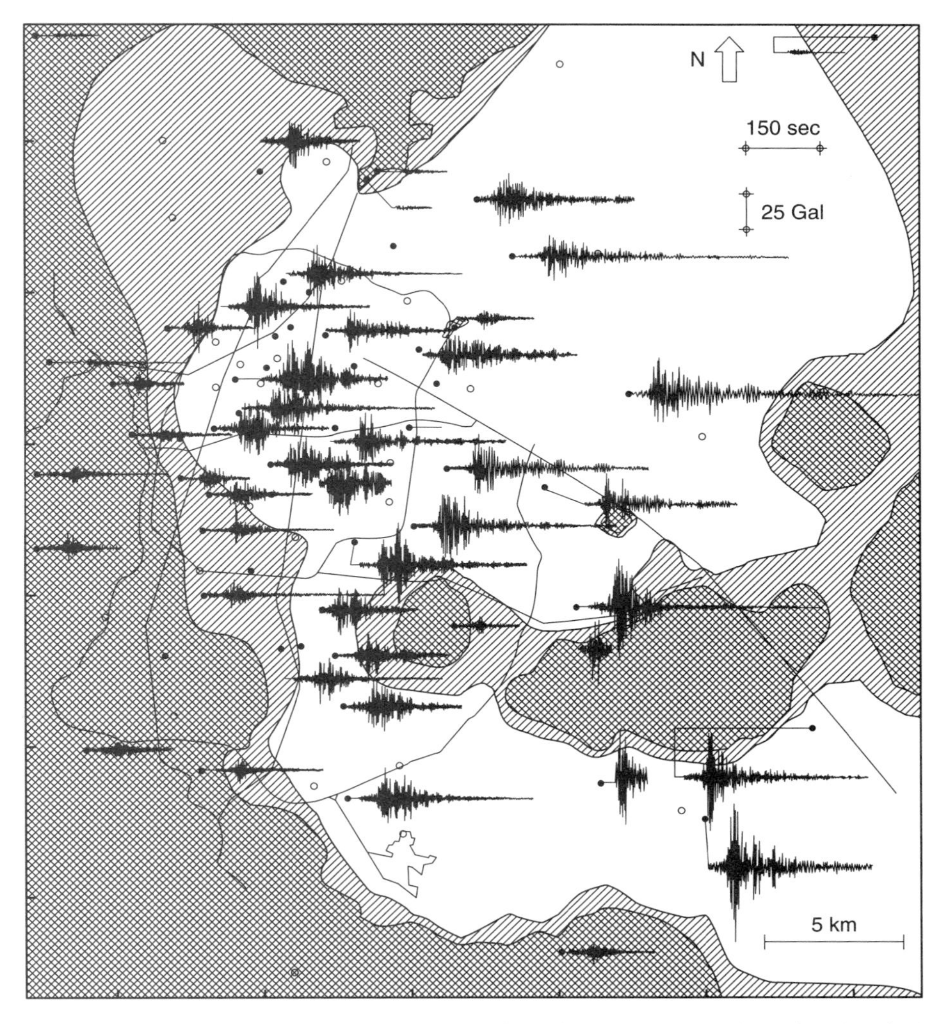

FIGURE 4 N–S accelerograms in the Valley of Mexico from the Copala earthquake of 14 September, 1995 (M_w 7.3), located a distance of about 300 km. Darkly shaded area: hill zone; lightly shaded area: transition zone; unshaded area: lakebed zone. Note that in the hill zone the accelerations are much lower and shorter in duration than in the lakebed zone.

above, if the ground motion can be reliably predicted at the reference site, then, for practical purposes, the question of site response of the reference site is not important. If, however, synthetic seismograms are to be used to predict the ground motion, then it becomes important to know the response of the reference site. Su *et al.* (1998) proposed a synthetic Green's function generated in an appropriate regional, layered, elastic, crustal model as the reference. The spectral ratio of recorded data from small earthquakes to synthetic seismograms defines the site response. This site response can then be used as correction to the synthesized ground motion for the target event. Some observations suggest that the spectral ratios change both in amplitude and shape with increasing motion levels, due to nonlinear soil response. This topic is discussed later.

Three-component recordings at the same site have been used to infer the site response, thus eliminating the need for a reference station. Lermo and Chávez-García (1993) explored the possibility of Fourier spectral ratio of horizontal to vertical component of S waves of a local or a regional earthquake as a measure of the site response. The method seems to give satisfactory results (Lermo and Chávez-García, 1993; Field and Jacob, 1995). This approach is analogous to the receiver function technique for determining the Earth's structure using teleseismic P waves (Langston, 1979). The receiver function method assumes that the deconvolution of the vertical component of P wave from the corresponding radial component removes the source and the path effect and gives the impulse response of the structure below the station.

A fast and inexpensive method, which may be especially useful in areas of low seismicity, is based on the recordings of microtremors (Kanai, 1957). The basic assumption is that the input signal is white noise and the response of a sedimentary site

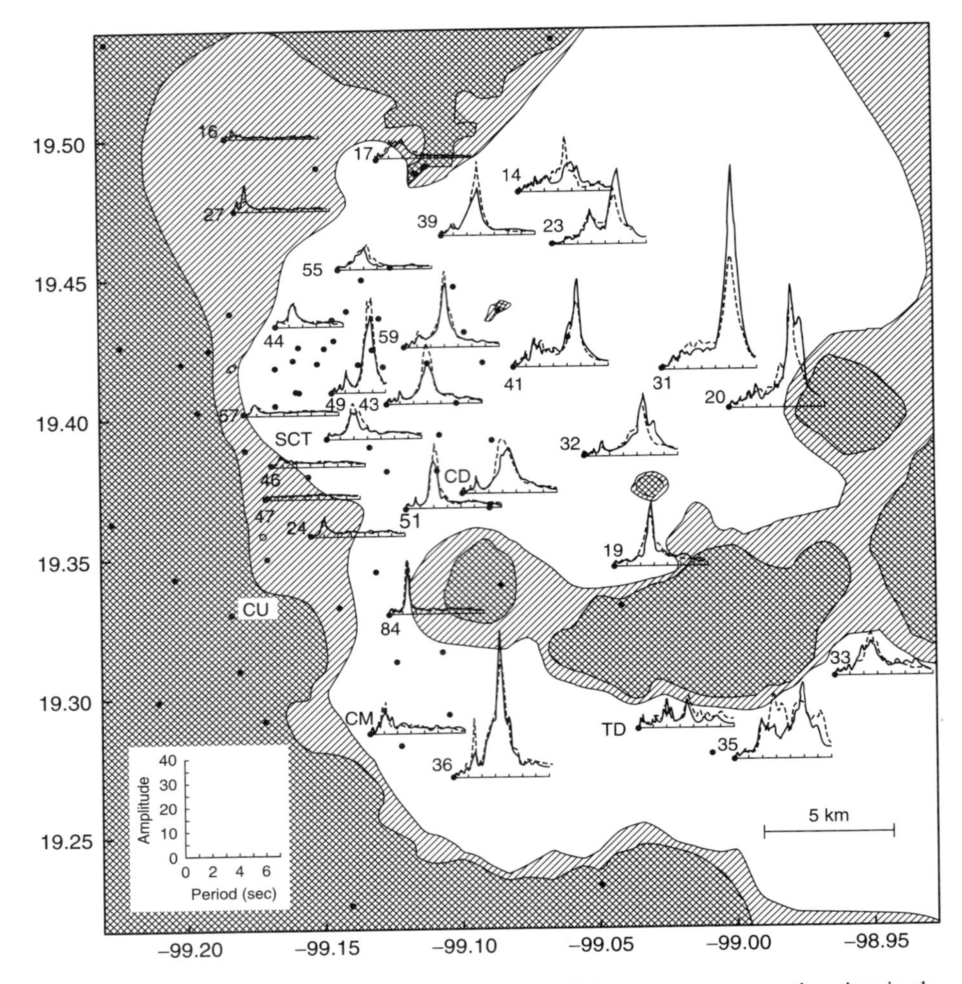

FIGURE 5 Average spectral ratios with respect to CU at some representative sites in the lakebed zone of the Valley of Mexico (after Reinoso and Ordaz, 1999).

to ambient noise is similar to its response from incident seismic waves. Some studies report success in estimating the dominant period of the site from the peak in the spectra of the horizontal components of the noise (e.g., Ohta *et al.*, 1978; Lermo *et al.*, 1988). Attempts have been made to infer site response from the spectral ratio of microtremors at the site of interest with respect to a close rock site, with mixed results (e.g., Lermo *et al.*, 1988; Field *et al.*, 1990; Rovelli *et al.*, 1991; Morales *et al.*, 1991; Gutiérrez and Singh, 1992). Finally, horizontal- to vertical-component spectral ratio of microtremors has been used to infer site response. This technique, proposed by Nakamura (1989), seems useful in identifying the fundamental frequency of the site (e.g., Lermo and Chávez-García, 1994; Field and Jacob, 1995).

Figure 6 compares four techniques to estimate site response at the lakebed site of SCT1 in the Valley of Mexico (modified from Lermo and Chávez-García, 1994). The spectral ratio with respect to the hill-zone site of CU is well reproduced by the horizontal- to vertical-component spectral ratio of earthquake recordings at SCT1 (the "receiver" function technique). The site response estimated from microtremor data appears to be less precise but still gives some useful information about its response.

The theoretical response of a layered crust (often a layer over half-space) has been extensively used in engineering applications. With recent advances in numerical methods and speed of computers, it is now possible to compute response of complex valleys and topographies (Kawase and Aki, 1989; Sánchez-Sesma and Campillo, 1991). Such studies are very useful in understanding the physics of the site response. Also, the results from generic models can be quite illuminating. However, detailed 3D structure and geotechnical information will be needed to predict the response of a given site. This may be possible only in some areas. An example is the Los Angeles region. Wald and Graves (1998) compared long-period ($\geq$2 sec), observed strong-motion data in this region from the 1992 Landers (M_w 7.3) with synthetic seismograms computed in three different 3D models, recently derived for southern California. The comparison only partially validates the use of a geologically constrained

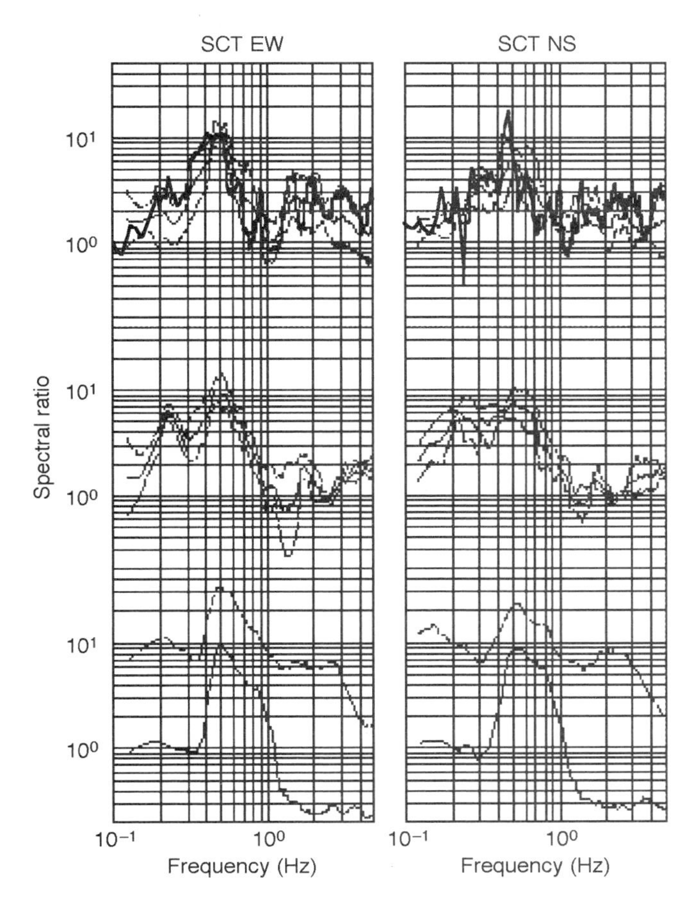

FIGURE 6 Comparison of different techniques to infer response of SCT, a site located in the lakebed zone of the Valley of Mexico. *Top frame:* spectral ratio with respect of the reference hill-zone site of CU from earthquake recordings. Thin continuous line: average; dashed lines: ± one standard deviation; thick line: 19 September 1985 (M_w 8.0), Michoacán earthquake. *Middle frame:* spectral ratio of horizontal to vertical component of earthquake recordings at SCT. Continuous line: average; dashed lines: ± one standard deviation. *Bottom frame:* Dashed line: average spectral ratio of microtremor at SCT with respect to CU; continuous line: average spectral ratio of horizontal to vertical component of microtremor at SCT.

3D structure in predicting long-period ground motions in the Los Angeles region. Olsen and Archuleta (1996) used a 3D finite-difference method to simulate ground motion in the metropolitan area of Los Angeles and compared their simulations with recordings of the Northridge earthquake, finding a reasonable agreement.

Based on laboratory experiments on soil samples, geotechnical engineers have long been aware of nonlinear behavior of the subsoil during strong ground motion. It is only recently, however, that much of the seismological community has come to accept it. Several seismological studies have analyzed weak and strong ground motion at the same surface site and, in some cases, motions in borehole arrays. From the analysis of the Loma Prieta earthquake data, Chin and Aki (1991) reported a pervasive nonlinear effect. Nonlinear behavior of stiff-soil site has been reported from the analysis of Northrigde earthquake sequence (Field *et al.*, 1997; Su *et al.*, 1998). Su *et al.* (1998) found that nonlinearity during the Northridge mainshock was significant at sites where A_{max} was above 0.3 g, V_{max} was above 20 cm/sec, or the peak strain exceeded 0.06%. There is still some doubt about the pervasiveness of the nonlinear behavior. For example, the spectral ratios of earthquake recordings at sites in the lakebed zone of Mexico City with respect to a reference hill-zone site show little nonlinear behavior (Fig. 6) even at $A_{max} \sim 0.2$ g, $V_{max} \sim 50$ cm/sec, and maximum shear strain of about 1% recorded during the 1985 Michoacán earthquake (Singh *et al.*, 1988a,b; Ordaz and Faccioli, 1994; Singh *et al.* 1997). Similarly, Gutiérrez and Singh (1992) found no clear evidence of nonlinear behavior at soft sites (sand and clays) in Acapulco for $A_{max} \sim 0.3$ g. To complicate matters, O'Connell (1999) reports that linear-wave propagation models, which incorporate weakly heterogeneous, random 3D crustal velocity variations, may explain the reduction in relative amplification at soft sites. As can be noted, the nonlinear behavior of subsoil during strong ground motion is presently a field of vigorous seismological research. More recordings of weak and strong motions at surface sites and in borehole arrays are needed to resolve the outstanding issues of *in situ* nonlinear response.

8. Conclusions

Solutions to earthquake engineering problems, either design of structures, or simulation of accelerograms, or estimation of seismic hazard, are today very different than they were before a strong interaction between engineers and seismologists started in the 1970s. This interaction, which gave birth to the new discipline of engineering seismology, gathered force due to the strict design requirements in building nuclear power plants, and is flourishing at present. Today, the difference between strong motion—previously a domain of engineers—and weak motion—seismologists' realm—has disappeared, due to progress in instrumentation and to the awareness that both weak- and strong-motion data are useful to understand the nature of earthquakes. Indeed, today it is inconceivable to study detailed rupture processes of large earthquakes without near-source strong-motion data. Similarly, it is unimaginable to synthesize an accelerogram without imposing restrictions obtained from seismological theory.

This chapter discusses some of the advances in seismology that bear on earthquake engineering, and it has also highlighted outstanding problems in seismology that are not resolved. Intense multidisciplinary research is foreseen using such diverse tools as space-based geodetic measurements (see, e.g., Chapter 37 by Feigl) and paleoseismology (see, e.g., Chapter 30 by Grant) to solve these issues. In doing so, seismology will keep contributing to the advancement of earthquake engineering.

References

Aki, K. (1966). Generation and propagation of G waves from the Niigata earthquake of June 16, 1964. *Bull. Earthquake Res. Inst., Tokyo Univ.* **44**, 23–88.

Aki, K. (1967). Scaling law of seismic spectrum. *J. Geophys. Res.* **72**, 1217–1231.

Anderson, J. G., and R. Quaas (1988). The Mexico earthquake of September 19, 1985—Effect of magnitude on the character of strong ground motion: an example from the Guerrero, Mexico network. *Earthquake Spectra* **4**, 635–646.

Andrews, D. J. (1986). Objective determination of source parameters and similarity of earthquakes of different sizes. In: *Earthquake Source Mechanics*, Maurice Ewing Series, S. Das, J. Boatwright, and C. H. Scholz (Editors), American Geophysical Union, Washington, D.C., 259–268.

Boatwright, J., J. B. Fletcher, and T. E. Fumal (1991). A general inversion scheme for source, site, and propagation characteristics using multiply recorded sets of moderate-sized earthquakes. *Bull. Seismol. Soc. Am.* **81**, 1754–1782.

Boore, D. M. (1983). Stochastic simulation of high-frequency ground motions based on seismological models of radiated spectra. *Bull. Seismol. Soc. Am.* **73**, 1865–1884.

Boore, D. M., and G. M. Atkinson (1987). Stochastic prediction of ground motion and spectral response parameters at hard-rock sites in Eastern North America. *Bull. Seismol. Soc. Am.* **77**, 440–467.

Brune, J. N. (1970). Tectonic stress and spectra of seismic shear waves from earthquakes. *J. Geophys. Res.* **75**, 4997–5009.

Brune, J. N. (1996a). Particle motion in a physical model of shallow angle thrust faulting. *Proc. Indian Acad. Sci.* **105**, 197–206.

Brune, J. N. (1996b). Precariously balanced rocks and ground-motion maps for southern California. *Bull. Seismol. Soc. Am.* **86**, 43–54.

Brune, J. N., and G. R. Engen (1969). Excitation of mantle Love waves and definition of mantle wave magnitude. *Bull. Seismol. Soc. Am.* **59**, 923–933.

Brune, J. N., and J. W. Whitney (1992). Precariously balanced rocks with rock varnish: paleoindicators of maximum ground acceleration? *Seism. Res. Lett.* **63**, 21.

Castro, R., S. K. Singh, and E. Mena (1988). An empirical model to predict Fourier amplitude spectra of horizontal ground motion in Ciudad Universitaria, Mexico City from coastal earthquakes. *Earthquake Spectra* **4**, 675–686.

Castro, R. R., J. G., Anderson, and S. K. Singh (1990). Site response, attenuation, and source spectra of S waves along the Guerrero, Mexico subduction zone. *Bull. Seismol. Soc. Am.* **80**, 1481–1503.

Chin, B. K., and K. Aki (1991). Simultaneous study of the source, path and site effects on strong ground motion during the 1989 Loma Prieta earthquake: a preliminary result on pervasive nonlinear site effects. *Bull. Seismol. Soc. Am.* **81**, 1859–1884.

Davies, G., and J. N. Brune (1971). Regional and Global Fault Slip Rates from Seismicity. *Nature* **229**, 101–107.

Davison, F. C., and C. H. Scholz (1985). Frequency-moment distribution of earthquakes in the Aleutian Arc: a test of the characteristic earthquake model. *Bull. Seismol. Soc. Am.* **75**, 1349–1361.

Esteva, L., and R. Villaverde (1973). Seismic risk, design spectra, and structural reliability. *Proc. 5th World Conf. Earthq. Eng., Rome*, 2586–2596.

Field, E. H., and K. H. Jacob (1995). A comparison and test of various site-response estimation techniques, including three that are not reference-site dependent. *Bull. Seismol. Soc. Am.* **85**, 1127–1143.

Field, E. H., S. H. Hough, and K. H. Jacob (1990). Using microtremor to assess potential earthquake site response. *Bull. Seismol. Soc. Am.* **80**, 1456–1480.

Field, E. H., P. A. Johnson, I. A. Beresnev, and Y. Zeng (1997). Nonlinear ground-motion amplification by sediments during the 1994 Northridge earthquake. *Nature* **390**, 599–602.

Gutenberg, B., and C. Richter (1944). Frequency of earthquakes in California. *Bull. Seismol. Soc. Am.* **34**, 185–188.

Gutiérrez, C., and S. K. Singh (1992). A site effect study in Acapulco, Guerrero, Mexico: comparison of results from strong motion and microtremor data. *Bull. Seismol. Soc. Am.* **82**, 642–659.

Hanks, T. C., and R. K. McGuire (1981). The character of high-frequency strong ground motion. *Bull. Seismol. Soc. Am.* **71**, 2071–2095.

Hartzell, S. H. (1978). Earthquake aftershocks as Green's functions. *Geophys. Res. Lett.* **5**, 1–4.

Heaton, T. H. (1990). Evidence for and implications of self-healing pulses of slip in earthquake rupture. *Phys. Earth Planet. Interiors* **64**, 1–20.

Heaton, T. H., J. F. Hall, D. J. Wald, and M. W. Halling (1995). Response of high-rise and base-isolated buildings to a hypothetical M_w 7.0 blind thrust earthquakes. *Science* **206**, 206–211.

Heki, K., S. Miyazaki, and H. Tsuji (1997). Silent fault slip following an interplate thrust earthquake at the Japan trench. *Nature* **386**, 595–598, 1997.

Hudnut, K. W., Y. Bock, M. Cline, P. Fang, J. Freymueller, K. Gross, D. Jackson, S. Larsen, M. Lisowski, Z. Shen, and J. Svarc (1994). Cosesimic displacements in the 1992 Landers earthquake sequence. *Bull. Seismol. Soc. Am.* **84**, 625–645.

Humphrey, J. R., and J. G. Anderson (1992). Shear-wave attenuation and site response in Guerrero, Mexico. *Bull. Seismol. Soc. Am.* **82**, 1622–1645.

Ishimoto, M., and K. Iida (1939). Observations sur les sisms enregistre par le microseismograph contruite dernierment (I). *Bull. Earthquake Res. Inst. Tokyo Univ.* **17**, 443–478.

Irikura, K. (1983). Semi-empirical estimation of strong ground motions during large earthquakes, *Bull. Disaster Prevention Res. Inst., Kyoto Univ.* **32**, 63–104.

Jackson, J., and D. McKenzie (1988). The relationship between plate motions and seismic moment tensors, and the rates of active deformation in the Mediterranean and Middle East. *Geophys. J. R. Astr. Soc.* **93**, 45–73.

Jennings, P. C., G. W. Housner, and N. C. Tsai (1969). Simulated earthquake motions for design purposes, *Proc. 4th World Conf. Earthq. Eng.* Chilean Association on Seismology and Earthquake Engineering, Santiago, Chile, Vol. I, 1969, pp. A1-145 to A1-160.

Joyner, W. B., and D. M. Boore (1981). Peak horizontal acceleration and velocity from strong-motion records including records from the Imperial Valley, California, earthquakes. *Bull. Seismol. Soc. Am.* **71**, 2011–2038.

Joyner, W. B., and D. M. Boore (1986). On simulation of large earthquakes by Green's functions addition of smaller earthquakes. In: *Earthquake Source Mechanics*, S. Das, J. Boatwright, and C. H. Scholz (Editors), Maurice Ewing Series 6, American Geophysical Monograph **37**, 269–274.

Kagan, Y. (1993). Statistics of characteristic earthquakes. *Bull. Seismol. Soc. Am.* **83**, 7–24.

Kagan, Y., and D. Jackson (1991). Long-term earthquake clustering. *Geophys. J. Int.* **104**, 117–133.

Kagan, Y., and D. Jackson (1993). Reply. *J. Geophys. Res.* **98**, 9917–9920.

Kanai, K. (1957). The requisite conditions for predominant vibration of ground. *Bull. Earthquake Res. Inst. Tokyo Univ.* **31**, 457.

Kanamori, H. (1977). The energy release in great earthquakes. *J. Geophys. Res.* **82**, 2981–2987.

Kanamori, H. (1983). Magnitude scale and quantification of earthquakes. *Tectonophysics* **93**, 185–199.

Kanamori, H. (1994). Mechanism of earthquakes. *Ann. Rev. Earth Planet. Sci.* **22**, 207–237.

Kanamori, H., and D. L. Anderson (1975). Theoretical basis of some empirical relations in seismology. *Bull. Seismol. Soc. Am.* **65**, 1073–1095.

Kawase, H., and K. Aki (1989). A study of the response of a soft basin for incident S, P, and Rayleigh waves with special reference to the long duration in Mexico City. *Bull. Seismol. Soc. Am.* **79**, 1361–1382.

Kelleher, J., L. Sykes, and J. Oliver (1973). Possible criteria for predicting earthquake locations and their application to major plate boundaries of the Pacific and Caribbean. *J. Geophys. Res.* **78**, 2547–2585.

Langston, C. (1979). Structure under Mount Rainier, Washington, inferred from teleseismic body waves. *J. Geophys. Res.* **84**, 4749–4762.

Lermo, J., M. Rodríguez, and S. K. Singh (1988). Natural period of sites in the Valley of Mexico from microtremor measurements and strong motion data. *Earthquake Spectra,* **4**, 805–814.

Lermo, J., and F. J. Chávez-Garcia (1993). Site effect evaluation using spectral ratios with only one station. *Bull. Seismol. Soc. Am.* **83**, 1574–1594.

Lermo, J., and F. J. Chávez-Garcia (1994). Are microtremors useful in site response evaluation. *Bull. Seismol. Soc. Am.* **84**, 1350–1364.

Main, I. G. (1995). Earthquakes as critical phenomena: implications for probabilistic seismic hazard analysis. *Bull. Seismol. Soc. Am.* **85**, 1299–1308.

Margheriti, L., L. Wennerberg, and J. Boatwright (1994). A comparison of coda and S-wave spectral ratios as estimates of site response in the southern bay area. *Bull. Seismol. Soc. Am.* **84**, 1815–1830.

Massonnet, D., M. Rossi, C. Carmona, F. Adragna, G. Peltzer, K. Feigl, and T. Rabaute (1993). The displacement field of the Landers earthquake mapped by radar interferometry. *Nature* **369**, 227–230.

McCaffrey, R. (1997). Statistical significance of the seismic coupling coefficient. *Bull. Seismol. Soc. Am.* **87**, 1069–1073.

McCann, W. R., S. P. Nishenko, L. R. Sykes, and J. Krause (1979). Seismic gaps and plate tectonics: Seismic potential for major boundaries. *Pure Appl. Geophys.* **177**, 1082–1147.

McGuire, R. K. (1978). A simple model for estimating amplitude spectra of horizontal ground acceleration. *Bull. Seism. Soc. Am.* **68**, 803–822.

Melbourne, T., I. Carmichael, C. DeMets, K. Hudnut, O. Sanchez, J. Stock, G. Suárez, and F. Webb (1997). The geodetic signature of the M 8.0 Oct. 9, 1995, Jalisco subduction earthquake. *Geophys. Res. Lett.* **24**, 715–718.

Morales, J., F. Vidal, J. A. Peña, G. Alguacil, and J. M. Ibañez (1991). Microtremor study in the sediment-filled basin of Zafarraya, Granada (Southern Spain). *Bull. Seismol. Soc. Am.* **81**, 687–693.

Nakamura, Y. (1989). A method for dynamic characteristics estimation of subsurface using microtremor on ground surface. *QR Railway Tech. Res. Inst.* **30**, 1.

Newmark, N. H., and W. H. Hall (1969). Seismic design criteria for nuclear reactor facilities, *Proc. 4th World Conf. Earthq. Eng.*, Santiago, Chile.

Nishenko, S, and L. R. Sykes (1993). Comment on "Seismic Gap Hypothesis: Ten years after" by Y. Y. Kagan and D. D. Jackson. *J. Geophys. Res.* **98**, 9909–9916.

O'Connell, D. H. R. (1999). Replication of apparent nonlinear wave propagation models. *Science* **283**, 2045–2050.

Ohta, Y., H. Kagami, N. Goto, and K. Kudo (1978). Observation of 1- to 5-second microtremors and their application to earthquake engineering, Part I: Comparison of long-period accelerations at the Tokachi-Oki earthquake of 1968. *Bull. Seismol. Soc. Am.* **68**, 767–779.

Olsen, K. B., and Archuleta, R. J. (1996). Three-dimensional simulation of earthquakes on the Los Angeles fault system. *Bull. Seismol. Soc. Am.* **86**, 575–586.

Ordaz, M., and S. K. Singh (1992). Source spectra and spectral attenuation of seismic waves from Mexican earthquakes, and evidence of amplification in the hill zone of Mexico City. *Bull. Seismol. Soc. Am.* **82**, 24–43.

Ordaz, M., S. K. Singh, E. Reinoso, J. Lermo, J. M. Espinosa, and T. Domínguez (1988). Estimation of response spectra in the lake bed zone of the valley of Mexico during the Michoacán earthquake. *Earthquake Spectra,* **4**, 815–834.

Ordaz , M., S. K. Singh, and A. Arciniega (1994). Bayesian attenuation regressions: an application to Mexico City. *Geophys. J. Int.* **117**, 335–344.

Ordaz, M., J. Arboleda, and S. K. Singh (1995). A scheme of random summation of an empirical Green's function to estimate ground motion from future large earthquakes. *Bull. Seismol. Soc. Am.* **85**, 1635–1647.

Ordaz, M., and E. Faccioli (1994). Site response analysis in the Valley of Mexico: selection of input motion and extent non-linear soil behavior. *Int. J. Earthquake Eng. Struct. Dyn.* **23**, 895–908.

Pacheco, J. F., C. H. Scholz, and L. R. Sykes (1992). Changes in frequency-size relationship from small to large earthquakes. *Nature* **355**, 71–73.

Pacheco, J. F., L. R. Sykes, and C. H. Scholz (1993). Nature of seismic coupling along simple plate boundaries of the subduction type. *J. Geophys. Res.* **98**, 14133–14159.

Peterson, E. T., and T. Seno (1984). Factors affecting seismic moment release rates in subduction zones. *J. Geophys. Res.* **89**, 10233–10248.

Pérez-Rocha, L. E. (1998). Respuesta sísmica estructural: efectos de sitio e interacción sueloestructura, Ph.D. thesis, UNAM, Mexico.

Phillips, W. S., and K. Aki (1986). Site amplification of coda waves from local earthquakes in central California. *Bull. Seismol. Soc. Am.* **76**, 627–648.

Reinoso, E., and M. Ordaz (1999). Spectral ratios for Mexico City from free-field recordings. *Earthquake Spectra* **15**, 273–296.

Report of the Coordinating Committee for Earthquake Prediction (1999). Velocity of crustal deformation of Japan. Edited by Geographical Survey Institute, Ministry of Construction, vol. 61, 555–573.

Reyes, C. (1998). El estado límite de servicio en el diseño sísmico de edificios, Ph.D Thesis, School of Engineering, National Autonomous University of México.

Ritsema, J., and T. Lay (1993). Rapid source mechanism determination of large ($M_w \geq 5$) earthquakes in the western United States. *Geophys. Res. Lett.* **20**, 1611–1614.

Romanowicz, B., D. Dreger, M. Pasyanos, and R. Uhrhammer (1993). Monitoring of strain release in central and northern California using broadband data. *Geophys. Res. Lett.* **20**, 1643–1646.

Romanowicz, B., and J. Rundle (1993). On scaling relations for large earthquakes. *Bull. Seismol. Soc. Am.* **83**, 1294–1297.

Romanowicz, B., and J. Rundle (1994). Reply to Comment on "On scaling relations for large earthquakes," by B. Romanowicz and J. Rundle, from the perspective of a recent nonlinear diffusion equation linking short term deformation to long term tectonics. *Bull. Seismol. Soc. Am.* **84**, 1684–1685.

Rosenblueth E., and M. Ordaz (1987). Use of seismic data from similar regions. *Int. J. Earthquake Eng. Struct. Dyn.* **15**, 619–634.

Rosenblueth, E., M. Ordaz, F. J. Sánchez-Sesma, and S. K. Singh (1989). Design spectra for Mexico's Federal District. *Earthquake Spectra* **5**, 258–272.

Rovelli, A., S. K. Singh, L. Malagnini, A. Amato, and M. Cocco (1991). Feasibility of the use of microtremors in estimating site response during earthquakes: some test cases in Italy. *Earthquake Spectra* **7**, 551–561.

Sadigh, K., Chang, C.-Y., Egan, J. A., Makdishi, F., and Youngs, R. R. (1997). Attenuation relationships for shallow crustal earthquakes based on California strong-motion data. *Seism. Res. Lett.* **68**, 180–189.

Sánchez-Sesma, F. J., and M. Campillo (1991). Diffraction of P, SV, and Rayleigh waves by topographic features: a boundary integral formulation. *Bull. Seismol. Soc. Am.* **81**, 2234–2253.

Scholz, C. H. (1982). Scaling laws for large earthquakes: consequences for physical models. *Bull. Seismol. Soc. Am.* **72**, 1–14.

Scholz, C. H. (1990). *The Mechanics of Earthquakes and Faulting*. Cambridge University Press, Cambridge, 433 pp.

Schwartz, D., and K. Coppersmith (1984). Fault behavior and characteristic earthquakes: Examples from Wasach and San Andreas faults. *J. Geophys. Res.* **89**, 5681–5698.

Shapiro, M. N., M. Campillo, A. Paul, S. K. Singh, D. Jongmans, and F. J. Sánchez-Sesma (1997). Surface wave propagation across the Mexican Volcanic Belt and the origin of the long-period seismic-wave amplification in the Valley of Mexico. *Geophys. J. Int.* **128**, 151–166.

Shi, B., A. Anooshehpoor, J. N. Brune, and Y. Zeng (1998). Dynamics of thrust faulting: 2D lattice model. *Bull Seismol. Soc. Am.* **88**, 1484–1494.

Shin, T. C., K. W. Kuo, W. H. K. Lee, T. L. Teng, and Y. B. Tsai (2000). A preliminary report on the 1999 Chi-Chi (Taiwan) earthquake. *Seism. Res. Lett.* **71**, 24–30.

Sieh, K. (1978). Prehistoric large earthquakes produced by slip on the San Andreas Fault at Pallet Creek, California. *J. Geophys. Res.* **83**, 3907–3939.

Sieh, K. (1981). A review of geological evidence for recurrence times of large earthquakes. In: *Earthquake Prediction, an International Review* (D. Simpson and P. Richards, Eds.), M. Ewing Series 4, American Geophysical Union, Washington, D.C., pp. 209–216.

Singh, S. K., M. Rodriguez, and L. Esteva (1983). Statistics of small earthquakes and frequency of occurrence of large earthquakes along the Mexican subduction zone. *Bull. Seismol. Soc. Am.* **73**, 1779–1796.

Singh, S. K., E. Mena, R. Castro, and C. Carmona (1987). Empirical prediction of ground motion in Mexico City from coastal earthquakes. *Bull. Seismol. Soc. Am.* **77**, 1862–1867.

Singh, S. K., J. Lermo, T. Domínguez, M. Ordaz, J. M. Espinosa, E. Mena, and R. Quaas (1988a). A study of relative amplification of seismic waves in the Valley of Mexico with respect to a hill zone site (CU). *Earthquake Spectra,* **4**, 653–674.

Singh, S. K., E. Mena, and R. Castro (1988b). Some aspects of source characteristics of 19 September 1985 Michoacán earthquake and ground motion amplification in and near Mexico city from the strong motion data. *Bull. Seismol. Soc. Am.* **78**, 451–477.

Singh, S. K., M. Ordaz, M. Rodríguez, R. Quaas, E. Mena, M. Ottaviani, J. G. Anderson, and D. Almora (1989). Analysis of near-source strong motion recordings along the Mexican subduction zone. *Bull. Seismol. Soc. Am.* **79**, 1697–1717.

Singh, S. K., R. Quaas, M. Ordaz, F. Mooser, D. Almora, M. Torres, and R. Vasquez (1995). Is there truly a "hard" rock site in the Valley of Mexico? *Geophys. Res. Lett.* **22**, 481–484.

Singh, S. K., M. Santoyo, P. Bodin, and J. Gomberg (1997). Dynamic deformations of shallow sediments in the Valley of Mexico, Part II: Single station estimates. *Bull. Seismol. Soc. Am.* **87**, 540–550.

Singh, S. K., M. Ordaz, R. S. Dattatrayam, and H. K. Gupta (1999). A spectral analysis of the May 21, 1997, Jabalpur, India, earthquake ($M_w = 5.8$) and estimation of ground motion from future earthquakes in the Indian shield region. *Bull. Seismol. Soc. Am.* **89**, 1620–1630.

Somerville, P. (1993). Engineering applications of strong ground motion simulation. *Tectonophysics* **218**, 195–219.

Sornette, D., and A. Sornette (1994). Comment on "On scaling relations for large earthquakes," by B. Romanowicz and J. Rundle, from the perspective of a recent nonlinear diffusion equation linking short term deformation to long term tectonics. *Bull. Seismol. Soc. Am.* **84**, 1679–1683.

Stein, R., and R. Yeats (1989). Hidden earthquakes. *Science* **260**, 48–57.

Su, F., and K. Aki (1995). Site amplification factor in central and southern California determined from coda waves. *Bull. Seismol. Soc. Am.* **85**, 452–466.

Su, F., J. G. Anderson, and Y. Zeng (1998). Study of weak and strong ground motion including nonlinearity from the Northridge, California, earthquake sequence. *Bull. Seismol. Soc. Am.* **88**, 1411–1425.

Teng, T. L., Y. B. Tsai, and W. H. K. Lee (Editors) (2001). Dedicated Issue on the Chi-Chi (Taiwan) Earthquake of September 20, 1999. *Bull. Seismol. Soc. Am.* **91**, 893–1395.

Thio, H. K., and H. Kanamori (1995). Moment-tensor inversion for local earthquakes using surface waves recorded at TERRAscope. *Bull. Seismol. Soc. Am.* **85**, 1021–1038.

Toro, G. R., and R. K. McGuire (1987). An investigation into earthquake ground motion characteristics in eastern North America. *Bull. Seismol. Soc. Am.* **77**, 468–489.

Tsujiura, M. (1978). Spectral analysis of coda waves from local earthquakes. *Bull. Earthquake Res. Inst. Tokyo Univ.* **53**, 1–48.

Veneziano, D., and M. Heidari (1985). Statistical analysis of attenuation in the Eastern United States, in methods of ground motion estimation for the Eastern United States. EPRI Research Project RP-2556-16, Palo Alto, CA, USA.

Wald, D. J., and T. H. Heaton (1994). Spatial and temporal distribution of slip for the 1992 Landers, California, earthquake. *Bull. Seismol. Soc. Am.* **84**, 668–691.

Wald, D. J. and R. W. Graves (1998). The seismic response of the Los Angeles basin, California, *Bull. Seismol. Soc. Am.* **88**, 337– 356.

Ward, S. N. (1994). A multidisciplinary approach to seismic hazard in southern California. *Bull. Seismol. Soc. Am.* **84**, 1293–1309.

Wennerberg, L. (1990). Stochastic summation of empirical Green's functions. *Bull. Seismol. Soc. Am.* **80**, 1418–1432.

Wesnousky, S. G., C. H. Scholz, K. Shimazaki, and T. Matsuda (1983). Earthquake frequency distribution and the mechanics of faulting. *J. Geophys. Res.* **88**, 9331–9340.

Working Group on California Earthquake Probabilities (1995). Seismic hazard in Southern California: probable earthquakes, 1994–2024. *Bull. Seismol. Soc. Am.* **85**, 379–439.

Yoshida, S., K. Koketsu, B. Shibazaki, T. Sagiya, T. Kato, and Y. Yoshida (1996). Joint inversion of near- and far-field waveforms and geodetic data for the rupture process of the 1995 Kobe earthquake. *J. Phys. Earth* **44**, 437–455.

Youngs, R. R., and K. J. Coppersmith (1985). Implications of fault slip rates and earthquake recurrence models to probabilistic seismic hazard estimates. *Bull. Seismol. Soc. Am.* **75**, 939–964.

Youngs, R. R., S.-J. Chiou, W. J. Silva, and J. R. Humphrey (1997). Strong ground motion attenuation relationships for subduction zone earthquakes. *Seism. Res. Lett.* **97**, 58–73.

Zeng, Y., J. G. Anderson, and G. Yu (1994). A composite source model for computing realistic synthetic strong ground motions. *Geophys. Res. Lett.* **21**, 725–728.

67

An Introduction to the Earthquake Response of Structures

Paul C. Jennings
California Institute of Technology, Pasadena, California, USA

1. Introduction

The intent of this study is to provide an introduction to the dynamics of earthquake response of buildings and other structures. The emphasis is on linear response, not because all significant earthquake response is within the linear range—certainly major structural damage and collapse are highly important nonlinear phenomena—but because an understanding of linear response provides an important background for formulating and understanding the complex and approximate methods of analysis used to describe nonlinear behavior. In addition, because the typical intent of earthquake-resistant design is to restrict damage to levels that are not significant safety hazards, linear analysis is often applicable to earthquake response because many structures that suffer only slight to moderate damage are capable of being described using equivalent linear properties.

When an earthquake moves the base of a structure, its masses experience forces of inertia as the structure attempts to follow the motion of the ground. The resultant motions of the structure depend in a complex way on the amplitude and other features of the ground motion, the dynamic properties of the structure, and the characteristics of the materials of the structure and its foundation. By Newton's laws of motion the force on each mass is equal to the amount of mass times the total acceleration, which is conveniently partitioned into the acceleration with respect to the ground, plus the acceleration of the ground. If written in terms of the displacement of the structure relative to its base, the resulting equations of motion typically reduce to the familiar form of those for a structure with an immovable base subject to external forces. The equivalent external forces are applied at each mass point and are equal to the mass times the acceleration of the ground. Hence, earthquake forces for most structures are characterized by the time history of the ground acceleration, rather than the ground velocity or displacement. (This is not the case for extended structures such as large bridges and dams, nor for cases in which soil–structure interaction is important.) The magnitude of the induced force is proportional to the structural mass, which explains why heavy but weak construction, such as unreinforced masonry, fares so poorly during strong shaking.

1.1 Important Characteristics of Earthquake Excitation

Because the time history of the acceleration of the ground characterizes the forces strong earthquakes exert on structures, it is clearly central to the advancement of earthquake-resistant design to know what forms potentially damaging accelerations can take. The first measurement of strong-motion acceleration was made during the destructive Long Beach, California, earthquake of March 10, 1933 (Neumann, 1935), by the earthquake engineering program of the U.S. Coast and Geodetic Survey. Special instruments had to be developed to record potentially damaging motions as the seismometers used at that time for scientific study of earthquakes would be forced off-scale by such strong shaking and typically would only record the onset of the motion. Since the recordings of the Long Beach earthquake, recording programs have steadily but slowly grown to the point where thousands of significant records of ground motion and structural response have been obtained, including records from nearly all the highly seismic areas of the world. These measurement programs, the records they have obtained, and the techniques used to process the records are described in other chapters of this Handbook. In spite of the success of the measurement programs, significant gaps persist in the knowledge of strong ground motion. The infrequency of truly strong ground shaking at a given location, the difficulties of providing adequate instrumental coverage in the areas of strongest shaking, and the varied effects of source mechanisms and site conditions

INTERNATIONAL HANDBOOK OF EARTHQUAKE AND ENGINEERING SEISMOLOGY, VOLUME 81B *ISBN: 0-12-440658-0*

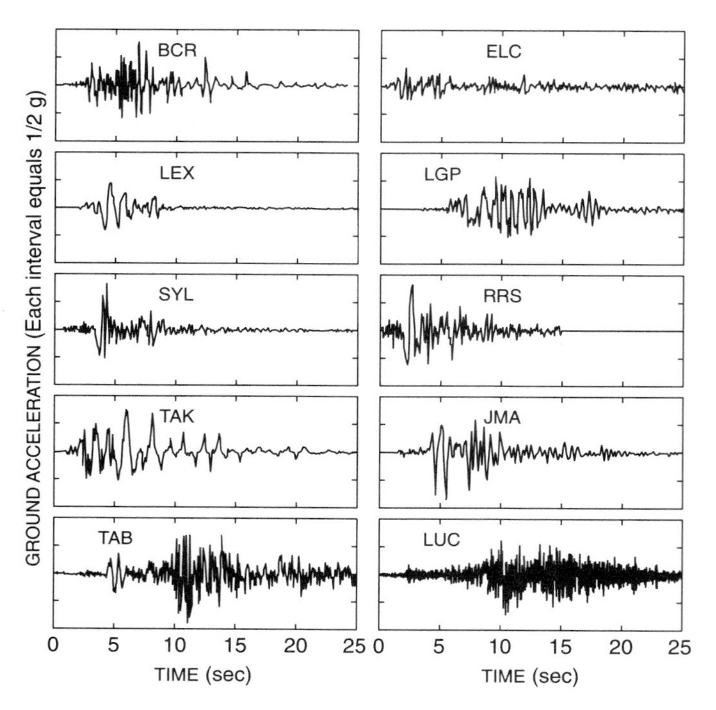

FIGURE 1 Strong ground accelerations recorded close to the causative fault. BCR is the 230° component of the Bonds Corner record from the 1979 M_W 6.4 Imperial Valley earthquake; ELC is the well-known N–S component of the 1940 M_W 6.9 event; LEX (Lexington Dam) and LGP (Los Gatos Presentation Center) are from the 1989 M_W 6.9 Loma Prieta earthquake; SYL (Sylmar County Hospital free-field) and RRS (Rinaldi Receiving Station) are from the 1994 M_W 6.7 Northridge earthquake; TAK (Takatori Station of Japan Railways) and JMA (Japan Meteorological Association) are from the 1995 M_W 6.9 Kobe event; TAB is from the 1978 M_W 7.4 Tabas, Iran, earthquake; and LUC is from the Lucerne Valley site in the 1992 M_W 7.2 Landers earthquake. Except for the first two records, the accelerations shown are horizontal components resolved to the direction of maximum peak-to-peak velocity (Hall, 1999).

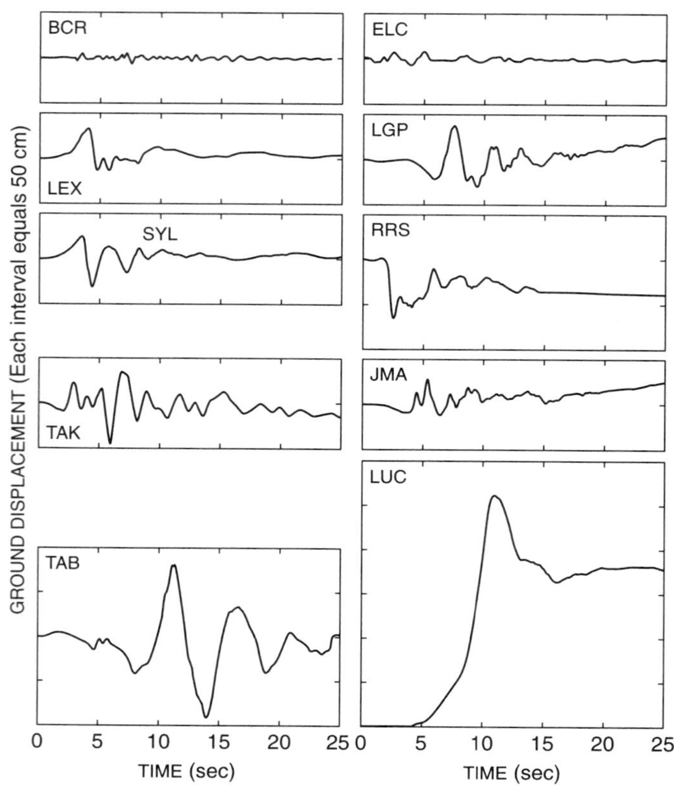

FIGURE 2 Ground displacements calculated from accelerations recorded near the causative fault. The sites are the same as those shown in Figure 1 (Hall, 1999).

on the resultant motion combine to make our present state of knowledge of strong earthquake forces limited and incomplete. In particular, only a few records exist to illustrate the nature of damaging ground motions during great earthquakes, and the few records obtained in the near field of major (magnitude 6 and 7) earthquakes indicate that large, potentially very destructive, pulses in the ground motion can occur (Hall *et al.*, 1995; Campbell, 1997; Somerville *et al.*, 1997).

In broad terms, strong-motion acceleration is characterized by amplitude, duration, and frequency content. Figure 1 shows some important accelerograms that illustrate these concepts. Figures 2 and 3 show ground displacements and velocities, respectively, determined from integration of recorded accelerations. As can be seen in these figures, integration smoothes the records of ground shaking. Acceleration highlights the high-frequency content of the ground motion, while the velocity emphasizes mid-period motions in the range around 1- to 2-second periods. The displacement is typically dominated by periods of a few to several seconds. Modern accelerographs are quite accurate; doubly integrated records obtained close to faults, such as the LUC trace in Figure 2, even show the permanent displacements associated with fault rupture.

The amplitude of the ground motion is clearly an important parameter; it is appealing, but difficult, to measure the overall strength of the motion well with a single number. A common measure of amplitude is the peak acceleration, which is easily and accurately determined directly from the accelerogram, without data processing. The peak ground acceleration also gives the maximum response of structures that are essentially rigid. This simple measure has proved to be useful, but it can be quite imprecise in that narrow spikes of acceleration can occur in acceleration records, spikes which determine the peak acceleration but do not have significance for the vast majority of structural responses.

The peak ground velocity has also been used as a general measure of strong-motion amplitude (Campbell, 1997; Hall *et al.*, 1995). Velocity is an easier concept to visualize than acceleration, particularly for the general public, and it has the additional advantage of being associated with a period range of more general importance to earthquake response than does the peak acceleration. However, unlike peak acceleration, peak velocity does not directly give the response of a particular class of structures and it requires data processing to determine its value.

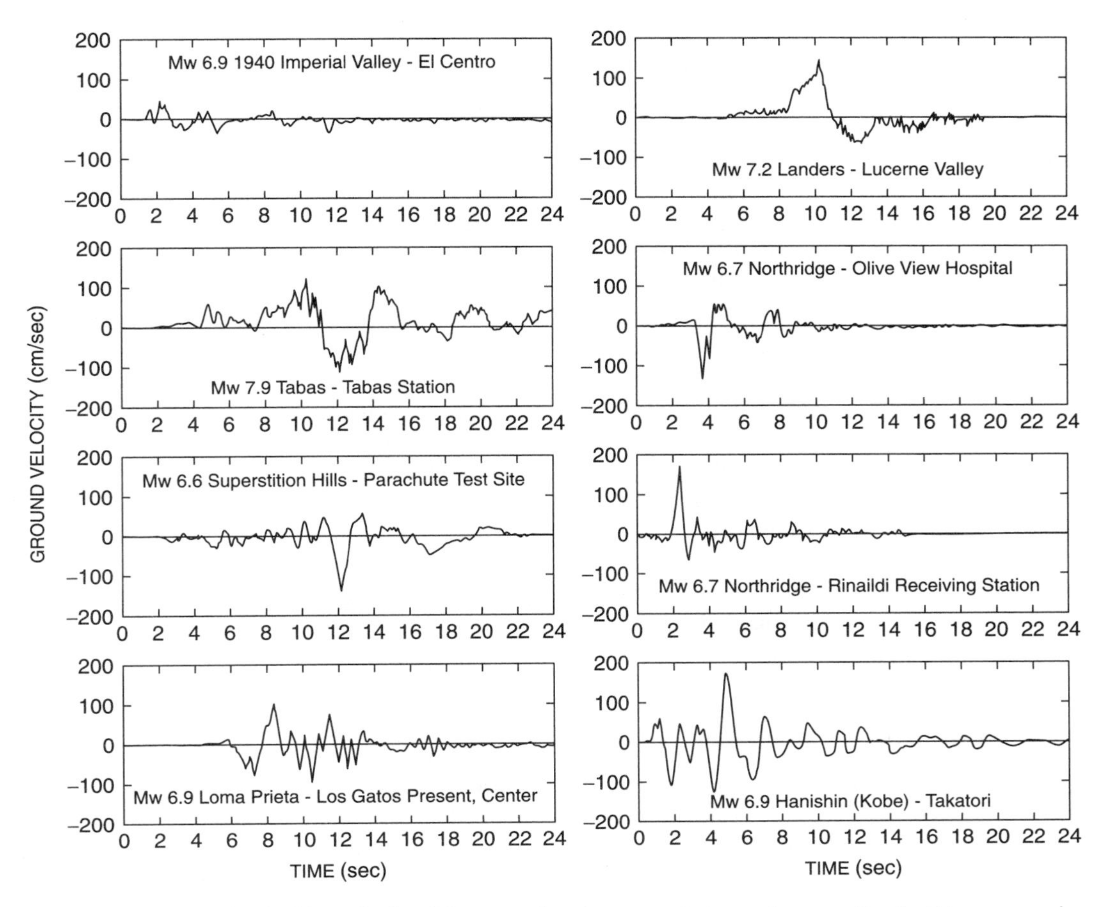

FIGURE 3 Ground velocities calculated from accelerations recorded near the causative fault in some major earthquakes. Six of the velocities shown are associated with the same sites as the records in Figures 1 and 2 (Hall *et al.*, 1995).

Peak ground displacement is the easiest parameter to understand, but it is typically associated with motions of such long period that it does not give an informative measure of the potential effects of strong shaking. For example, most structures can simply ride along with ground displacements with periods of many seconds, independent of the amplitude. Exceptions to this general case would include suspension bridges, large fluid storage tanks, and buried pipelines.

The limitations of simply using peak values have led to the development of other measures of amplitude. The most common of these for engineering use are spectral techniques, discussed in a subsequent section of this chapter.

The duration of the acceleration is another obviously important property of the ground motion. From the engineering viewpoint, the duration of motion amplifies damage or its potential. Brittle unreinforced masonry structures that can withstand one or two large oscillations may fail under prolonged motions, repairable cracks in nonductile concrete columns can progress to major failures, and while a second or so of shaking of wet, sandy soil can be withstood, repeated oscillations can induce liquefaction. Also, the movement of rocking and sliding equipment and other items continues to grow under prolonged shaking. If the level of response of a structure is not large enough to be damaging to its structural and nonstructural elements, then the duration of shaking is not so important in a strictly technical sense. However, the duration of shaking is always important from the viewpoint of those subjected to the motion, as anyone who has experienced the motions of a major earthquake can testify.

Frequency content is a term used to characterize the way the acceleration contains energy at different frequencies and is thereby more hazardous to some structures than to others. For example, accelerations recorded at some sites in Mexico City during the earthquake of 1985 (Singh *et al.*, 1988) were nearly sinusoidal, with a period of several seconds, indicating a very concentrated frequency content in a band that unfortunately coincided with the natural frequencies of many tall buildings. However, these same motions were not particularly hazardous to stiff, short buildings, which were able to follow the long-period movements of the ground without significant internal deformation. The frequency content of strong-motion acceleration is commonly measured by the response spectrum, which gives a spectrum of the maximum responses of simple structures with a range of natural frequencies and a specific damping, and by the well-known Fourier spectrum. These methods are discussed later in this chapter.

Another important characteristic of strong ground acceleration that has become more apparent in recent years as records have accumulated is the existence of large pulses within near-field motions. Such pulses can be seen in some of the accelerograms in Figure 1 and in the velocity plots in Figure 3. On the other hand, as also seen in Figure 1, some records do not show this feature prominently, but are better characterized as broadband random motions. Pulse-like motions have been observed in the Parkfield earthquake of 1966, the Imperial Valley earthquake of 1979, as well as the more recent Landers (1992), Northridge (1994), Kobe (1995), and Chi-Chi, Taiwan (1999), earthquakes, among others. Some of the important historical accelerograms, including El Centro (1934 and 1940), Taft 1952, and Tokachi-oki (1968) showed some pulses, but were more suggestive of broadband motion. Most of the records in this later group were obtained at moderate epicentral distances, rather than in the near field. Because the structural response to a pulse of acceleration is different in important ways from the response to random motion, understanding the source and effects of large pulses has become an important effort in earthquake engineering and engineering seismology (Hall *et al.*, 1995).

1.2 Earthquake Response

From the viewpoint of function and safety, the amplitude of earthquake response of a structure is perhaps best measured against the capacity of the structure to resist motions of different amplitudes, first without damage, then with damage, but without serious threat to the integrity of the structure, and finally against the level of response at which collapse is imminent. The amplitudes and frequency contents of earthquake ground motions corresponding to these levels of structural response obviously differ for different structures.

The most direct way to increase earthquake safety is to raise the strength of the structure, i.e., to increase the level of response associated with the onset of yielding or brittle failure. However, it is often more economical to reduce the risk by providing sufficient ductility that the structure can withstand many cycles of response in the yielding range without significant loss of strength. This is the basic approach embodied in building codes for earthquake-resistant buildings, including those of wood-frame, steel-frame, concrete, and masonry construction. Another approach is to lower the level of response by modifying the energy input into the structure by the ground motion. This can be accomplished by the addition of active or passive dampers to the structure, or by the placement of isolators between the foundation of the building and the ground (Housner *et al.*, 1997). Some important applications of this relatively new approach include hospitals, historic structures not amenable to structural retrofit, and highway bridges. Base isolation effectively reduces the fundamental frequency of the structure involved and is most effective for bridges and for stiff buildings. For example, a four-story hospital with a first natural frequency of two Hertz can be modified to have a first natural frequency of one-half Hertz. The accelerations within the structure are reduced by this change, thereby increasing the probability that the hospital can resume function quickly after the earthquake.

Structures of different types can be sensitive to very different features of earthquake motion. Short buildings are sensitive to the high-frequency excitation generally concentrated near the causative fault, while tall buildings are sensitive to long-period motions, which can exist over large areas. Long bridges and large dams are sensitive to the spatial variations in the ground acceleration, while suspension bridges can be affected, in addition, by motions with ultra-long periods, ten seconds or more. Pipelines can be affected by traveling waves of motion, while buildings on isolators are very sensitive to the relative displacement that can develop at the perimeter of the structure.

Many of these problems and topics go beyond linear structural response; however, linear analysis provides the first step in addressing nearly all of them.

2. The Response Spectrum

The response spectrum (Benioff, 1934; Housner, 1952) is perhaps the most basic tool of earthquake engineering. It directly gives the maximum response of single-degree-of-freedom structures to particular ground motions and, through modal analysis, also is applicable to the earthquake response of multi-degree-of-freedom structures. In addition, it describes the frequency content of ground motions in a way that is more fundamental to earthquake response than traditional Fourier analysis, because of the inclusion of the effects of damping. The response spectrum also forms the basis for the construction of design spectra, such as those in building codes and in design criteria for major projects.

As illustrated in Figure 4, the ordinates of a response spectrum are given by the maximum absolute value of the displacement,

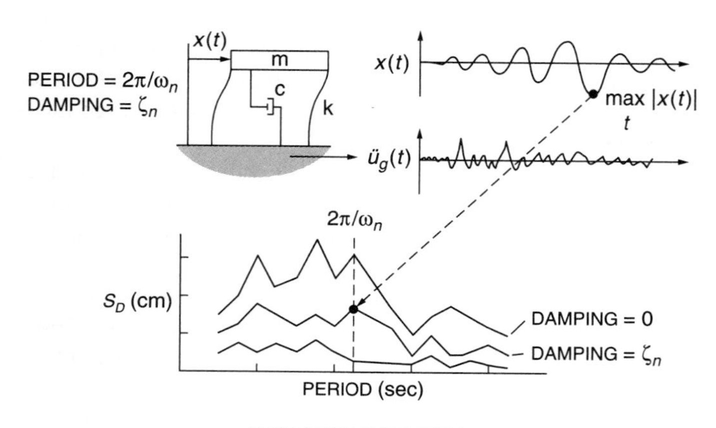

FIGURE 4 Response spectra are determined from the earthquake motion of a single-degree-of-freedom oscillator. The ordinates of the response spectrum are determined by the maximum absolute value of the response quantity of interest as a function of the natural period and damping of the oscillator.

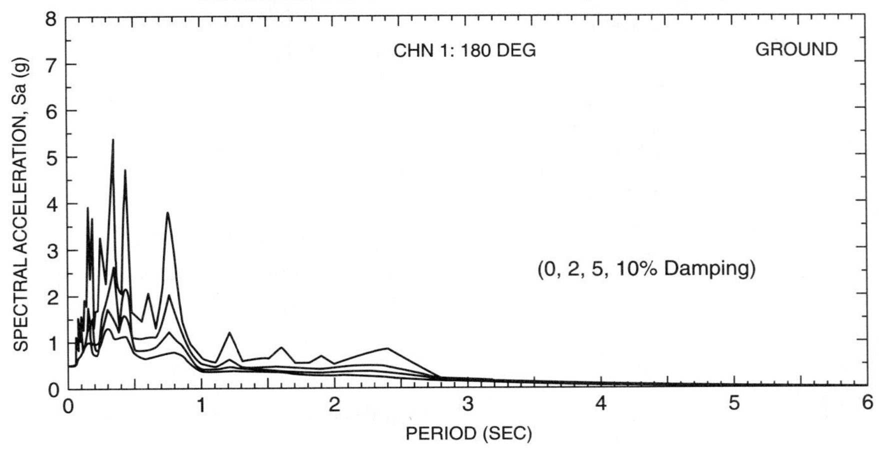

FIGURE 5 Acceleration spectra from the E–W component of motion recorded at the ground level of a 10-story building during the Northridge earthquake of January 17, 1994 (Darragh *et al.*, 1995).

velocity or acceleration of a single-degree-of-freedom oscillator with specified natural period and damping. Common response spectra include those for the relative displacement and velocity, and for the total acceleration of the oscillator; these are denoted by S_D, S_V, and S_A, respectively. The most commonly used response spectra are the acceleration spectra, S_A, and the pseudovelocity spectrum, $S_{PV} = \omega_n S_D$. An example of an acceleration spectrum is given in Figure 5.

In terms of the relative displacement, $x(t)$, the oscillator in Figure 4 has the equation of motion

$$\ddot{x} + 2\omega_n\zeta\dot{x} + \omega_n^2 x = -\ddot{u}_g(t) \tag{1}$$

in which

$$\omega_n = \sqrt{\frac{k}{m}}, \qquad \zeta = \frac{c}{2\sqrt{km}}. \tag{2}$$

Assuming the oscillator is initially at rest, the solution to Eq. (1) is

$$x(t) = \frac{-1}{\omega_n\sqrt{1-\zeta^2}} \int_0^t \ddot{u}_g(\tau)\, e^{-\omega_n\zeta(t-\tau)} \sin\omega_n\sqrt{1-\zeta^2}\,(t-\tau)\, d\tau. \tag{3}$$

In this notation the ordinates of response spectra are defined by

$$\begin{aligned} S_D(\omega_n, \zeta) &= \max_t |x(t)|, \\ S_V(\omega_n, \zeta) &= \max_t |\dot{x}(t)|, \\ S_A(\omega_n, \zeta) &= \max_t |\ddot{x}(t) + \ddot{u}_g(t)|. \end{aligned} \tag{4}$$

In practice, the time interval is effectively extended beyond the end of the accelerogram used as excitation, to include the possibility that the maximum value might occur in the subsequent free vibrations.

2.1 Properties of the Response Spectra

Consideration of the equation of motion, Eq. (1), and the properties of the oscillator in Figure 4 reveals the following limits as the oscillator stiffness becomes very large, i.e., as $\omega_n \to \infty$:

$$S_D \to 0, \qquad S_V \to 0, \qquad S_A \to \max_t |\ddot{u}_g|. \tag{5}$$

Thus, in this case the oscillator simply moves with the ground. On the other hand, when the stiffness of the oscillator becomes very slight, the mass tends to become isolated from the ground motion. Thus as $\omega_n \to 0$:

$$S_D \to \max_t |u_g(t)|, \quad S_V \to \max_t |\dot{u}_g(t)|, \quad S_A \to 0. \tag{6}$$

In most applications the amount of damping, ζ, is less than 10% of critical and consequently the earthquake response near its maximum is nearly sinusoidal at the natural frequency, ω_n. This condition, plus the fact that $\dot{x}(t)$ is zero when $|x(t)|$ is at its maximum, allows the following approximations to apply:

$$\begin{aligned} &S_V \approx \omega_n S_D, \\ &S_A \approx \omega_n S_V, \\ &S_A \approx \omega_n^2 S_D = \omega_n S_{PV}, \quad \text{and} \\ &\frac{1}{2} m S_V^2 \approx \frac{1}{2} k S_D^2. \end{aligned} \tag{7}$$

These approximations are not necessarily accurate for very high or very low frequencies, or for cases in which the response is not approximately sinusoidal. The relations do permit, however, a convenient graphical portrayal of response spectra, as shown in Figure 6. In this figure, the logarithm of the pseudovelocity, S_{PV},

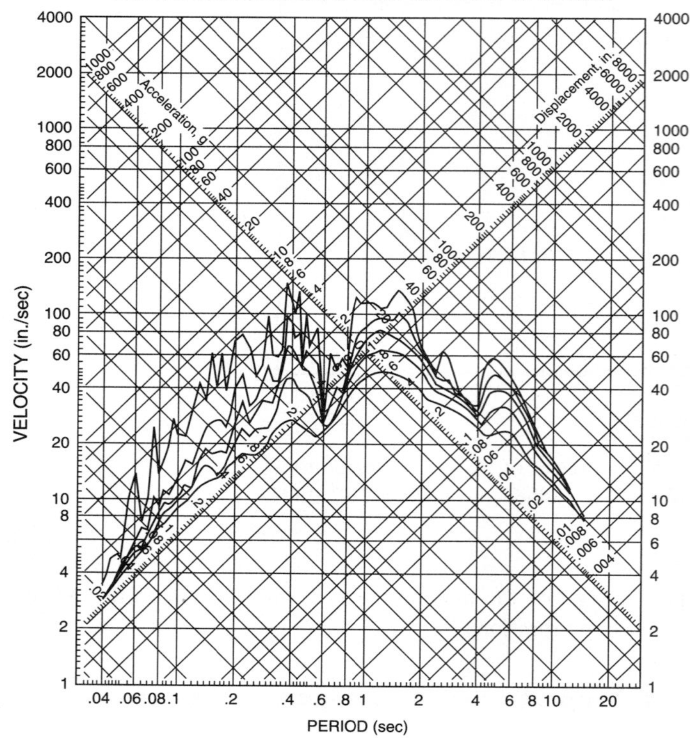

FIGURE 6 A tripartite plot of the pseudovelocity spectrum of the S14°W component of ground motion recorded on the abutment of Pacoima Dam during the San Fernando earthquake of February 9, 1971 (Earthquake Engineering Research Laboratory, 1973b). The different curves are for damping values of 0, 2, 5, 10, and 20 percent of critical damping.

is plotted against the logarithm of the oscillator period, $T_n = 2\pi/\omega_n$. On such a plot, invented by Edward Fisher (Housner 1997), the maximum displacement, pseudovelocity, and total acceleration of the oscillator can be read conveniently from the same graph. In this tripartite logarithmic representation, lines with a positive slope of unity have a constant acceleration spectra, S_A, whereas lines with slopes of negative unity correspond to a constant displacement spectra, S_D. Because the pseudovelocity is plotted, the ordinates of the plot approach the maximum absolute value of the ground acceleration for very small periods, whereas for very large periods the values approach a line with slope of negative unity, which represents the maximum ground displacement.

2.2 Relation to Fourier Spectra

The Fourier spectrum of an earthquake ground motion, $\ddot{u}_g(t)$, and its undamped response spectra are closely related (Benioff, 1934; Kawasumi, 1956; Hudson, 1962; Jennings, 1974, 1983). For an accelerogram of duration, T_d, the amplitude and phase of its Fourier transform can be written as

$$|F(\omega)| = \left[\left\{\int_0^{T_d} \ddot{u}_g(t)\cos\omega t\,dt\right\}^2 + \left\{\int_0^{T_d} \ddot{u}_g(t)\sin\omega t\,dt\right\}^2\right]^{\frac{1}{2}},$$

$$\Phi(\omega) = \tan^{-1}\frac{-\int_0^{T_d}\ddot{u}_g(t)\sin\omega t\,dt}{\int_0^{T_d}\ddot{u}_g(t)\cos\omega t\,dt}. \tag{8}$$

For comparison, if one first introduces the energy associated with relative motion of the undamped oscillator used to define response spectra,

$$E(t) = \frac{1}{2}kx^2 + \frac{1}{2}m\dot{x}^2, \tag{9}$$

then the energy per unit mass can be written as

$$\left[\frac{2E(t)}{m}\right]^{\frac{1}{2}} = \left[\left\{\int_0^{t} \ddot{u}_g(t)\cos\omega t\,dt\right\}^2 + \left\{\int_0^{t} \ddot{u}_g(t)\sin\omega t\,dt\right\}^2\right]^{\frac{1}{2}}, \tag{10}$$

in which, for convenience, the subscript has been dropped from the natural frequency, ω_n. A phase angle between the two components of energy can also be defined by

$$\Phi(\omega) = \tan^{-1}\frac{-\int_0^{t}\ddot{u}_g(t)\sin\omega t\,dt}{\int_0^{t}\ddot{u}_g(t)\cos\omega t\,dt}. \tag{11}$$

The only difference between Eq. (8) and Eqs. (10) and (11) is the time at which the integrals are evaluated. Thus, the Fourier spectrum of the acceleration is equal to twice the energy per unit mass of the undamped oscillator, Eq. (10), evaluated at the end of the motion. The properties of the response spectra for the undamped oscillator imply that

$$\omega S_D \geq |F(\omega)| \quad \text{and} \quad S_V \geq |F(\omega)|. \tag{12}$$

The maximum response of the undamped oscillator often occurs near the end of the motion, so equality is frequently approached in Eq. (12).

This result shows the close relation between Fourier spectra and the undamped response spectra, but it also shows that the Fourier spectra do not include the effect of damping, a quantity fundamental to earthquake response.

2.3 The Role of Damping

The damping of the oscillator in Figure 4 generally, but not always, reduces the response to earthquake excitation. Because of the change in frequency that the damping introduces, it is possible for very small values of damping to increase the response over the undamped case and for spectral curves for different dampings to cross. In practice, this happens rarely and the amounts are insignificant. If earthquake acceleration were a stationary random process, the average square of steady-state response would be inversely proportional to the damping (Crandall and Mark, 1963; Lutes and Sarkani, 1997). This dependence appears to approximate the effects of damping in most cases of response to earthquakes of significant duration (several natural periods, T_n); that is, response spectra in such cases tend to vary as the inverse square root of the damping value. Damping is less effective in reducing the response to pulse-like excitations. For example, if the oscillator in Figure 4 is subjected to a half sine pulse at its natural frequency, 5% damping only reduces the peak response by about 7.5% from the undamped case. In the limiting case of a very short pulse (of unit area), Eq. (3) produces a maximum response of

$$|x|_{\max} = \frac{1}{\omega_n\sqrt{1-\zeta^2}} e^{-\zeta\pi/(2\sqrt{1-\zeta^2})}. \tag{13}$$

This result shows that small values of damping are relatively ineffective in reducing the maximum response to this type of excitation; for all natural frequencies the maximum response for $\zeta = 0.05$ is only 7.6% less than that for $\zeta = 0$. Doubling the damping from 0.05 to 0.10 decreases the response only by another 7.3%, whereas for steady-state response to stationary random excitation, the reduction would be 29%.

Truly viscous damping can appear in the mounting systems for mechanical equipment and in some special structures such as passively damped buildings and bridges, but it is not present in typical buildings. In the case of typical buildings, viscous damping is usually used in analysis and design to approximate the combined effects of such mechanisms as material damping, nonstructural damage, and low levels of yielding and structural damage.

2.4 Calculation of Response Spectra

There are many ways to calculate response spectra inasmuch as the solution to Eq. (1) or the evaluation of Eq. (3) can be developed by any applicable numerical procedure. One method that is fast and convenient is based on the exact value of the state vector of the oscillator in response to piecewise continuous excitation (Iwan, 1960; Nigam and Jennings, 1969). In this approach, the acceleration record is assumed to be linear between the digitization points, typically $\Delta t = 0.01$ sec, and the state vector is evaluated incrementally by

$$\underline{x}_{i+1} = \mathbf{A}(\zeta, \omega_n, \Delta t)\underline{x}_i + \mathbf{B}(\zeta, \omega_n, \Delta t)\underline{a}_i, \tag{14}$$

in which

$$\underline{x}_i = \begin{Bmatrix} x_i \\ \dot{x}_i \end{Bmatrix}, \quad \underline{a}_i = \begin{Bmatrix} a_i \\ a_{i+1} \end{Bmatrix}, \quad i = 0, 1, 2, 3, \cdots, s. \tag{15}$$

In Eqs. (14) and (15) the a_i are the $s+1$ ordinates of the acceleration record, ω_n and ζ are the natural frequency and damping from Eq. (1), and **A** and **B** are two-by-two matrices whose elements are rather lengthy, but which only require evaluation once for each spectral calculation (Nigam and Jennings, 1969). Not only is this approach relatively fast, the calculation is exact within the easily understood assumption of piecewise linearity.

2.5 Spectrum Intensity

A response spectrum such as shown in Figure 5 portrays the maximum response of oscillators with a range of periods and a number of damping values. The area under one of the curves is an indication of the strength of the earthquake acceleration measured, in effect, by the average response of a family of oscillators with that value of damping. Housner (1952) originally defined the spectrum intensity, *SI*, by

$$SI(\zeta) = \int_{0.1}^{2.5} S_V(\zeta, T_n)\, dT_n, \tag{16}$$

and calculated values for $\zeta = 0$ and $\zeta = 0.20$. Housner used the spectrum intensities to compare the strengths of different acceleration records and to scale records as a means for constructing an average spectral shape for use in earthquake-resistant design (Housner, 1959).

A related approach is to measure the strength of earthquake excitation by the energy of response of the oscillator used to define the response spectrum (Arias, 1970). Beginning with the energy imparted to the oscillator,

$$W_n(\omega_n, \zeta) = -\int_0^{\infty} \ddot{u}_g(t)\dot{x}(t)\, dt, \tag{17}$$

Arias then integrated over the entire range of oscillators to form a measure of the intensity of the earthquake motion, as follows:

$$W(\zeta) = \int_0^{\infty}\int_0^{\infty} \ddot{u}_g(t)\,\dot{x}(t)\, dt\, d\omega_n. \tag{18}$$

Equation (18) was evaluated by substituting for $\dot{x}(t)$ from Eq. (3) to yield

$$W(\zeta) = \frac{\cos^{-1}\zeta}{\sqrt{1-\zeta^2}} \int_0^{\infty} [\ddot{u}_g(t)]^2\, dt. \tag{19}$$

The effect of damping is contained in the coefficient of the integral in Eq. (19). It is a weak function of the damping, only varying from $\pi/2 = 1.57$ for $\zeta = 0$ to 1.40 for $\zeta = 0.20$. The integral in Eq. (19) is the total energy in the excitation, related to the Fourier transform of the excitation by Parseval's equation.

$$\int_0^{\infty} [\ddot{u}_g(t)]^2\, dt = \frac{1}{2\pi}\int_0^{\infty} |F(\omega)|^2\, d\omega \tag{20}$$

Because of the close relation between response and Fourier spectra, it is perhaps not surprising that the intensity developed by Arias, Eq. (19), is closely correlated with the spectrum intensity, Eq. (16).

The total energy, the spectrum intensity, and the Arias intensity are broadband measures of the strength of earthquake acceleration and are used to compare and scale accelerograms. Other commonly used measures for this purpose are the peak acceleration, peak velocity, and the response spectrum ordinate for a representative value of period and damping.

2.6 Limitations of Response Spectra

Some information is lost when the response spectrum is calculated from an accelerogram because the accelerogram cannot be reconstructed from the spectrum as can be done, for comparison, from the Fourier spectrum. In addition, the response spectrum gives only the maximum response; it does not indicate whether this maximum is an isolated single peak or a value that was approached many times during the course of the response. The difference in these two cases can be very important in assessing the accumulation of damage and in understanding other aspects of the response of yielding structures, or in estimating the potential of soils to liquefy.

This feature suggests the indirect way the duration of the excitation affects the response spectrum. If the acceleration resembles a stationary random process, the average square of the response of an undamped oscillator will increase linearly with time, whereas the square of the average damped response asymptotically approaches a constant value (Caughey and Stumpf, 1961). As would be expected, heavily damped oscillators reach their asymptotes faster than lightly damped ones. This behavior is seen in the response spectra of long accelerograms, wherein the response spectra for different values of damping tend to be separated. In particular, the undamped response spectrum is significantly greater than the spectra for even small values of damping. On the other hand, for short accelerograms, especially pulse-like excitations, the response spectra for different values of damping, including the undamped case, are not well separated for periods greater than that of the excitation. An example of these effects is shown in Figure 7. This partial and indirect reflection of the duration of the acceleration is one of the limitations of the response spectrum.

2.7 Application to Single-Degree-of-Freedom Structures

Because the response spectrum is based on the response of a single-degree-of-freedom oscillator, it is directly applicable to simple structures and systems that can be modeled by one degree of freedom. Such applications include in-plane vibrations of one-story buildings, the planar response of small or stiff bridges, and the planar response of machinery. With simple modifications, the response spectrum can also be applied to the fundamental mode response of more complex systems, including buildings, bridges, and fluid storage tanks. For example, Housner (1957, 1970) has shown how to determine the mass, stiffness, and height of a single-degree-of-freedom structure that has the same period, and gives the same force and moment at the base of the structure, as the fundamental mode of a multi-degree-of-freedom system.

When more general techniques are needed, the tools of matrix algebra or continuum mechanics are usually employed.

3. Dynamics of Multi-Degree-of-Freedom Structures

3.1 Fundamental Concepts

The earthquake response of linear multi-degree-of-freedom structures can be introduced by considering the in-plane response of the simplified three-story building shown in Figure 8. It is assumed that the story masses are concentrated at the floor levels. The frame, deforming as indicated in the figure, provides the restoring forces. The beams and columns can bend, but are rigid axially. Under these assumptions, the equations of motion of the structure are

$$\begin{bmatrix} m_1 & 0 & 0 \\ 0 & m_2 & 0 \\ 0 & 0 & m_3 \end{bmatrix} \begin{Bmatrix} \ddot{u}_1 \\ \ddot{u}_2 \\ \ddot{u}_3 \end{Bmatrix} + \begin{bmatrix} k_{11} & k_{12} & k_{13} \\ k_{21} & k_{22} & k_{23} \\ k_{31} & k_{32} & k_{33} \end{bmatrix} \begin{Bmatrix} u_1 \\ u_2 \\ u_3 \end{Bmatrix} = - \begin{Bmatrix} \ddot{u}_g m_1 \\ \ddot{u}_g m_2 \\ \ddot{u}_g m_3 \end{Bmatrix} \quad (21)$$

in which $\ddot{u}_g$ is the acceleration of the ground and u_j is the lateral deflection of the jth mass with respect to the base. An element, k_{ij}, of the stiffness matrix, **K**, is defined as the static force at the ith level resulting from a unit displacement in the jth level, with all other displacements held at zero. The elements of **K** are found by analysis of the deformation of the structural frame (see, for example, Sack, 1984). By the reciprocal theorem (Timoshenko, 1955), $k_{ij} = k_{ji}$, $i, j = 1, 2, 3$. If the floor beams are effectively rigid in bending, the frame deforms only in shear and $k_{13} = k_{31} = 0$.

Equation (21) can be written in matrix notation, a form that lends itself to generalization to more degrees of freedom and more components of response, as follows:

$$\mathbf{M}\ddot{\underline{u}} + \mathbf{K}\underline{u} = -\mathbf{M}\underline{1}\ddot{u}_g. \quad (22)$$

In keeping with Eq. (21), the mass matrix, **M**, is diagonal and the stiffness matrix, **K**, is symmetric, $k_{ij} = k_{ji}$ (and for practical cases, both matrices are positive definite, as well). The column vector $\underline{1}$ has unity for each of its elements.

The earthquake response of the structure is addressed by first determining properties of free vibrations. If the righthand side

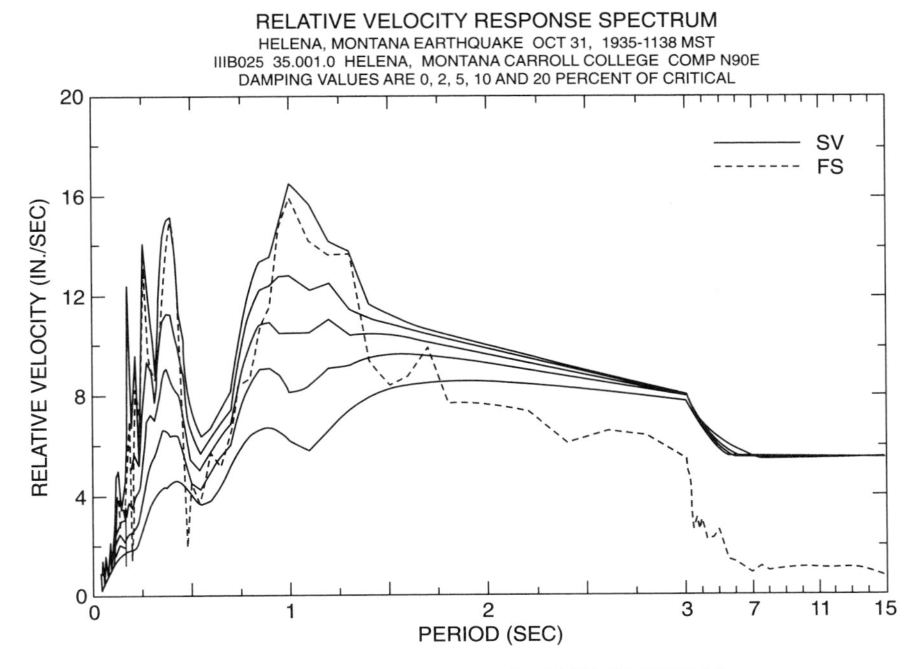

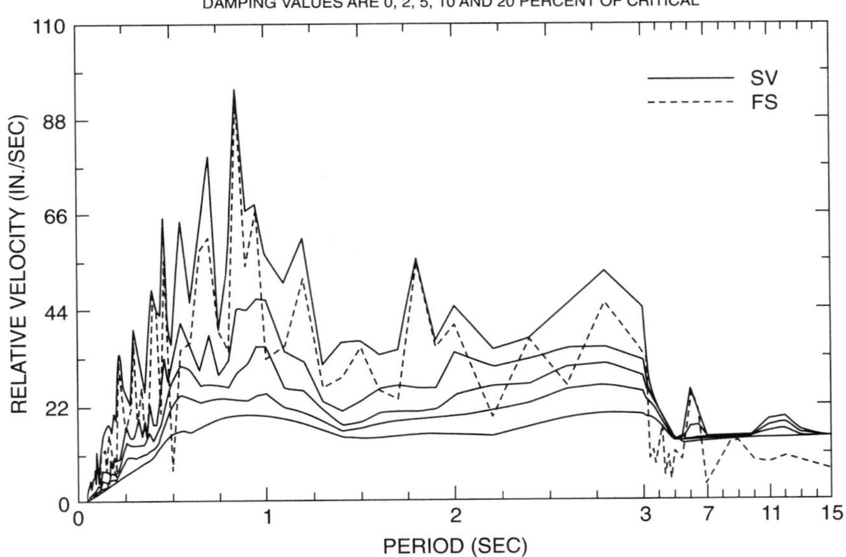

FIGURE 7 Effect of duration on response spectra. The acceleration that produced the top response spectra was only a few seconds long, and included a pulse with a period a little over one second, whereas the lower response spectra come from a record with over 20 seconds of strong shaking. Note the relatively greater effects of damping in the lower spectra for periods greater than $1\frac{1}{2}$ seconds. Also, but less obvious, the lower spectra show relatively greater separation between the undamped and 2% damped spectra than is the case in the upper spectra. The dashed curve labeled FS is the absolute value of the Fourier spectrum of the accelerogram and the different solid curves are for damping values of 0, 2, 5, 10, and 20% of critical damping.

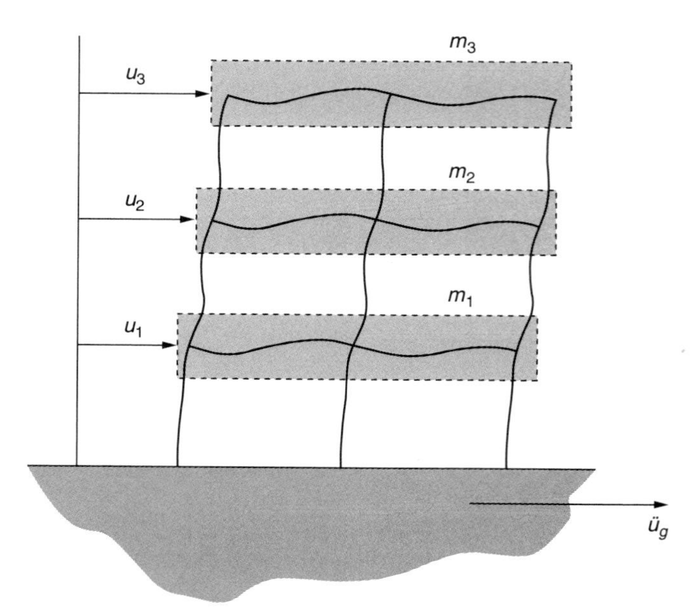

FIGURE 8 Planar earthquake response of an undamped multi-degree-of-freedom structure. The masses are taken to be concentrated at the floor levels, and the girders are assumed to be inextensible.

of Eq. (22) is set to zero and free vibrations are assumed, i.e.,

$$\underline{u} = \underline{\phi} \sin \omega t, \tag{23}$$

then one is quickly led to the classical eigenvalue problem,

$$[\mathbf{K} - \omega^2 \mathbf{M}]\underline{\phi} = 0. \tag{24}$$

In order for a nontrivial solution, $\underline{\phi}$, to exist, the determinant of the coefficient matrix must vanish. For the structure in Figure 8,

$$\begin{vmatrix} k_{11} - \omega^2 m_1 & k_{12} & k_{13} \\ k_{12} & k_{22} - \omega^2 m_2 & k_{23} \\ k_{13} & k_{23} & k_{33} - \omega^2 m_3 \end{vmatrix} = 0. \tag{25}$$

When evaluated, the determinant in Eq. (25) is a cubic in ω^2, whose three solutions, ω_1, ω_2, and ω_3, are the three frequencies at which free vibrations can exist. For each of these frequencies, a unique solution, $\underline{\phi}_j$, the mode shape, can be found, which describes the displacement of the three masses within a multiplicative constant (i.e., $\alpha\underline{\phi}_j$ is also a solution for α any scalar constant). Note that using $-\omega_j$ generates no new solutions and that the uniqueness of the mode shape only to within a multiplicative constant implies that only the relative, not absolute, values of the displacements of the masses are specified by the mode shapes. If the structure in Figure 8 were to have n degrees of freedom rather than 3, the determinant of Eq. (25) would be of order n in ω^2 and would produce, in general, n different natural frequencies and mode shapes.

Mode shapes and associated natural frequencies and damping values are the principal quantities obtained from experimental studies of buildings and other structures. Figures 9 and 10 show

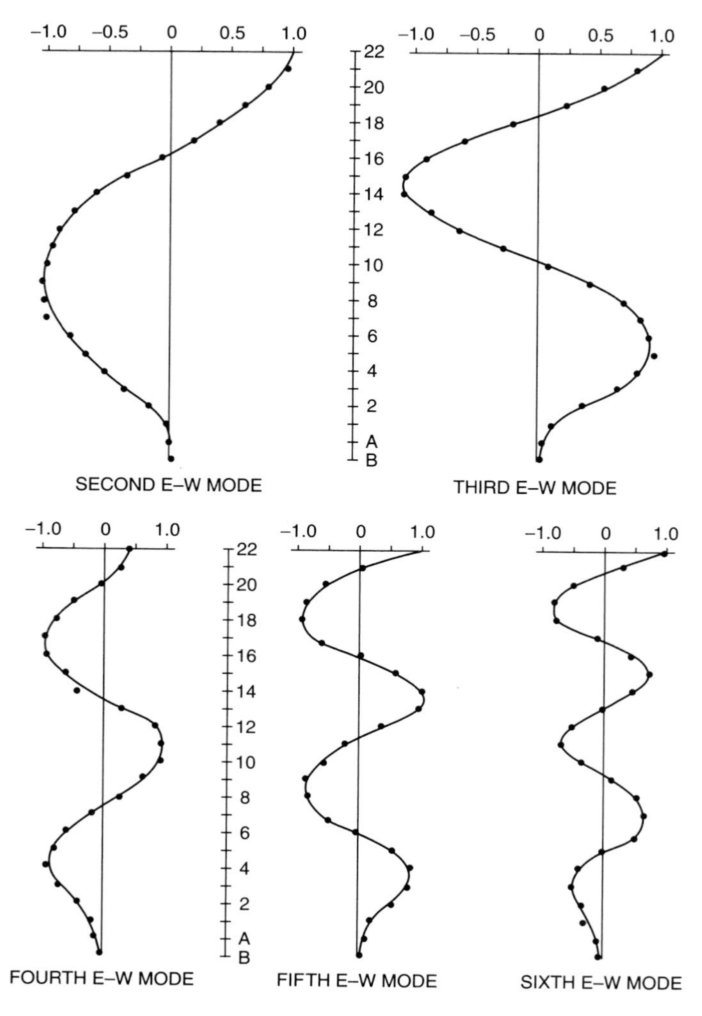

FIGURE 9 Measured mode shapes of a 22-story steel-frame building. Shown are the primary components of the second through sixth translational modes in the E–W direction. These modes also exhibit minor components in N–S translation and in torsion. The fundamental mode, not shown, resembled a straight line, with significant E–W and N–S components and a small amount of torsion (Jennings *et al.*, 1971).

Source: *Earthquake Engineering and Structural Dynamics*, Jennings *et al*, 1971. © John Wiley & Sons Limited. Reproduced with permission.

results determined during two such experiments (Jennings *et al.*, 1971; Foutch, 1976).

For convenience, the mode shapes are usually normalized; that is, the multiplicative constant is fixed. In experimental work it is customary to do this by setting the displacement in the mode shape equal to unity at the top of the structure or by setting the largest value of the mode shape equal to unity, while in analytical work it is more common to scale the mode shapes such that

$$\underline{\phi}_j^T \mathbf{M}\, \underline{\phi}_j = 1. \tag{26}$$

A key property possessed by the mode shapes is their orthogonality with respect to both the mass and stiffness matrices

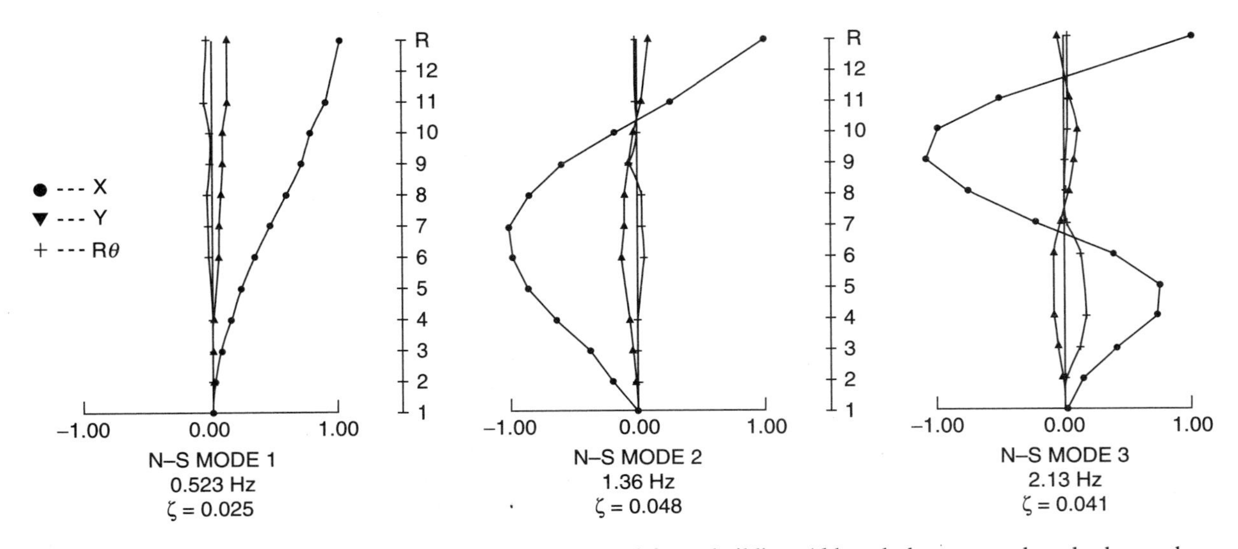

FIGURE 10 First three N–S mode shapes of a 12-story steel-frame building. Although the measured mode shapes show primarily N–S translation, each mode has some E–W and torsional components. R is the radius of gyration of the floor area (Foutch, 1976).

(see, for example, Chopra, 1995).

$$\underline{\phi_i^T}\, \mathbf{M}\, \underline{\phi_j} = 0 \quad \text{and} \quad \underline{\phi_i^T}\, \mathbf{K}\, \underline{\phi_j} = 0, \qquad i \neq j. \tag{27}$$

For Eq. (27) to hold, it is sufficient, but not necessary, for the natural frequencies to be distinct. The same analysis shows that when $i = j$,

$$\omega_j^2 = \frac{\underline{\phi_j^T}\, \mathbf{K}\, \underline{\phi_j}}{\underline{\phi_j^T}\, \mathbf{M}\, \underline{\phi_j}}. \tag{28}$$

Physically, the orthogonality of the mode shapes can be viewed as a consequence of the fact that the free vibrations at the different natural frequencies, Eq. (23), are independent. Each such vibration is determined by its initial conditions and, once set into motion, continues without interacting with the vibrations of other modes. Thus, the structure can vibrate freely in a single mode or in any linear combination of all the modes, each acting independently. Another important consequence of this modal independence is that the n mode shapes of an n-degree-of-freedom structure form a mathematical basis for the space of possible configurations of the structure (Franklin, 1968). That is, any static or dynamic deflection of the structure can be expressed as a linear combination of the mode shapes.

Returning to earthquake excitation, Eq. (22), with the addition of a damping matrix, **C**,

$$\mathbf{M}\, \underline{\ddot{u}} + \mathbf{C}\, \underline{\dot{u}} + \mathbf{K}\, \underline{u} = -\mathbf{M}\, \underline{1}\, \ddot{u}_g. \tag{29}$$

The solution is developed by assuming, as was done for free vibrations, that the earthquake response is a sum, each term being a mode shape multiplied by a function of time, $\xi_j(t)$, called the normal coordinate, so that

$$\begin{aligned} \underline{u} &= \underline{\phi_1}\, \xi_1(t) + \underline{\phi_2}\, \xi_2(t) + \cdots + \underline{\phi_n}\, \xi_n(t), \quad \text{or} \\ \underline{u} &= \Phi\, \underline{\xi}, \end{aligned} \tag{30}$$

in which Φ is the matrix composed of the mode shapes $\underline{\phi_1}$ through $\underline{\phi_n}$. Equation (30) is substituted into Eq. (29) and the result premultiplied by Φ^T to produce

$$\Phi^T \mathbf{M} \Phi\, \underline{\ddot{\xi}} + \Phi^T \mathbf{C} \Phi\, \underline{\dot{\xi}} + \Phi^T \mathbf{K} \Phi\, \underline{\xi} = -\Phi^T \mathbf{M}\, \underline{1}\, \ddot{u}_g. \tag{31}$$

Because of the orthogonality of the mode shapes [Eq. (27)], the first and third matrices in Eq. (31) are diagonal and to preserve the uncoupling of the individual equations in Eq. (31) it is assumed that the damping matrix is such that the second matrix is also diagonal. Using Eq. (28), defining a modal damping, ζ_j, by

$$2\omega_j \zeta_j = \frac{\underline{\phi_j^T}\, \mathbf{C}\, \underline{\phi_j}}{\underline{\phi_j^T}\, \mathbf{M}\, \underline{\phi_j}}, \tag{32}$$

and a modal participation factor, α_j, by

$$\alpha_j = \frac{\underline{\phi_j^T}\, \mathbf{M}\, \underline{1}}{\underline{\phi_j^T}\, \mathbf{M}\, \underline{\phi_j}}, \tag{33}$$

Eq. (31) becomes

$$\ddot{\xi}_j + 2\omega_j \zeta_j \dot{\xi}_j + \omega_j^2 \xi_j = -\alpha_j \ddot{u}_g, \quad j = 1, 2, \cdots, n. \tag{34}$$

Equation (34) states that the normal coordinate of each mode responds to the earthquake like a single-degree-of-freedom oscillator. Each mode experiences a scaled version of the excitation, determined by the participation factor, α_j. The solutions to the n equations in Eq. (34) can be calculated and at each point in time the modal responses can be added to produce the total response of the structure according to Eq. (30). From $\underline{u}(t)$ any details of the response, such as inter-level deflections and member forces, can be calculated in turn. Alternatively, one can exploit the fact that Eq. (34) is the same as Eq. (1), the equation that defines the response spectrum, except for the factor α_j. This similarity means that spectrum techniques can be applied directly to each of the independent modes of a complex structure and the results combined for application to the entire structure. Before this procedure is discussed, the parameters in Eqs. (29) and (34) will be examined.

3.2 Participation Factors

Examination of Eq. (33) shows that the participation factors are not unique, but depend on how the mode shapes are normalized. For example, doubling the amplitude of the mode shape halves the participation factor. Recombination by Eq. (30), however, removes the effect of this nonuniqueness, producing a physical response, $\underline{u}$, that is independent of the way the modes are normalized.

The physical meaning of the participation factor is clarified by summing $\alpha_j \underline{\phi_j}$ over the n modes,

$$\sum_{j=1}^{n} \alpha_j \underline{\phi_j} = \underline{1}, \tag{35}$$

which can be validated by premultiplying by $\underline{\phi_j^T}\mathbf{M}$. This result, which is independent of the way the mode shapes are normalized, states that the participation factors prescribe the combination of mode shapes that adds up to a unit displacement of each mass of the structure. If the mode shapes are normalized so the displacements at the nth level are unity, then the participation factors themselves add to unity and each one gives the fraction of the earthquake excitation that goes to that particular mode.

3.3 Mass and Stiffness

The elements of the mass and stiffness matrices of Eq. (29) are found from the geometry of the elements of the structure and the densities of the materials from which they are composed. For a building, for example, knowledge of the beams, columns, floor slabs, and how they are joined together allows the assembly of the stiffness matrix (Sack, 1984; Wilson, 1998; Venkatraman and Patel, 1970). This information, plus knowledge of the mass of architectural and nonstructural elements of the building and the mass of its contents is used to construct the mass matrix. Constructing the mass and stiffness matrices is a very significant task for a major civil engineering structure, one that takes both effort and judgment. Note, too, that a fairly accurate knowledge of the structure is required before these matrices can be calculated, even approximately.

3.4 Damping

For low to moderate levels of excitation, recorded responses show structures do respond to earthquakes like the complex, damped linear systems that are governed by Eqs. (29) and (34). However, the damping matrix, **C**, in Eq. (29), unlike the mass and stiffness matrices, cannot be determined effectively by calculation. The mechanism of damping in structures appears to be a complicated combination of material damping, friction and rubbing in structural and nonstructural joints, and microcracking in structural and nonstructural elements. For larger amplitudes of shaking, the energy absorbed by first, nonstructural, and then, structural damage contributes to the effective damping shown by the response. Although few, if any, of the mechanisms involved are truly viscous, the combined effect is modeled fairly well by the choice of appropriate values of modal dampings in Eq. (34). The values are a general function of the amplitude of response and the complexity of the structure. A bare steel frame or tower might show damping near 0.5% of critical for small motions, a value not much larger than that of a tuning fork. The modal dampings found from vibration tests of buildings tend to be in the 1.5 to 3% range, while the response to non-damaging earthquakes motions tends to be in the 3 to 5% range. System identification studies of motions at the level of moderate nonstructural and some structural damage have indicated effective modal damping values of 10% or more of critical (McVerry, 1979; Beck, 1978). Five percent is a commonly used value in structural design, wherein a linear analysis of the structure in response to reduced earthquake forces is often done.

Continuing this approach to damping through the modal values, it is natural to ask if the modal dampings can be set arbitrarily and, if so, how **C** can be found. Beginning with n arbitrary modal dampings, ζ_j, a diagonal matrix, **Z**, can be formed as follows:

$$\mathbf{Z} = \text{diag} : \left[\frac{2\omega_j \zeta_j}{\underline{\phi_j^T} \mathbf{M} \underline{\phi_j}} \right]. \tag{36}$$

If the damping matrix, **C**, is then taken as

$$\mathbf{C} = \mathbf{M}\Phi\mathbf{Z}\Phi^T\mathbf{M}, \tag{37}$$

and substituted into Eq. (29), it is quickly seen that **C** will be diagonalized when Eq. (30) is employed and premultiplication of Eq. (29) by Φ^T is performed. Although this development shows that arbitrary modal damping is possible without losing the orthogonality of the modes, some of the elements of Eq. (37) can imply positive, rather than dissipative, "dampers" between different points of the structure.

For structures in which only two or so modes are important to the earthquake response, a related, simpler approach can be used. This technique, attributed to Lord Rayleigh (Meirovitch, 1967), takes the damping matrix to be a linear combination of the mass and stiffness matrices, as follows:

$$\mathbf{C} = \alpha\mathbf{M} + \beta\mathbf{K}. \tag{38}$$

If this **C** is substituted into Eq. (29), the values of ζ_j in Eq. (34) become

$$\zeta_j = \frac{\alpha}{2\omega_j} + \frac{\beta\omega_j}{2}. \tag{39}$$

The two free parameters, α and β, allow the damping in two modes of response to be selected arbitrarily; the others then are fixed. Typically, for buildings the lowest two modes are used to determine the constants because of the importance of these modes in earthquake response. This implies, unless β is zero or extremely small, that higher modes of the structure will be very heavily damped, inasmuch as Eq. (39) states that for large frequencies, the damping increases linearly with frequency. For other structures, the constants could be determined by criteria that try to give reasonable values of damping over a range of frequencies most important to seismic response. As a function of ω, Eq. (39) defines a curve that is concave upward, with a minimum value of $\sqrt{\alpha\beta}$ at $\omega = \sqrt{\alpha/\beta}$. The damping rises to twice the minimum value when the frequency reduces to $0.27\sqrt{\alpha/\beta}$ or rises to $3.7\sqrt{\alpha/\beta}$. Outside of this range, the form of Eq. (39) indicates that the damping may grow rapidly with further reductions or increases in frequency, depending on the values of α and β.

Caughey and O'Kelley (1965) have generalized Rayleigh's result and have established that the necessary and sufficient condition for the existence of classical normal modes ($\Phi^T \mathbf{C} \Phi$ diagonal) is that the matrices $\mathbf{M}^{-1}\mathbf{K}$ and $\mathbf{M}^{-1}\mathbf{C}$ commute.

3.5 Determining Parameters from Response

Although damping cannot be reliably calculated, it can be determined from observed experimental response rather directly, at least for the lower modes of the structure (e.g., Jennings *et al.*, 1971; Trifunac, 1970). On the other hand, the mass and stiffness, which can be calculated relatively reliably, are only observed indirectly. Measured natural frequencies and mode shapes are insufficient to determine **M** and **K**, as can be seen from Eq. (24), which shows that the same mode shapes and frequencies will result if the mass and stiffness matrices are both multiplied by an arbitrary constant. A simpler problem results if **M**, positive definite, is assumed known and it is desired to determine **K** from the measured dynamic properties of the structure. Making the additional assumption that all n natural frequencies and mode shapes have been measured, the orthogonality relations can be used to determine **K**. Taking, for convenience, mode shapes normalized according to Eq. (26), and forming the diagonal matrix

$$\Omega = \text{diag} : \left[\omega_j^2\right], \tag{40}$$

the existence of the required inverses permits one to write

$$\mathbf{K} = \Phi^{T^{-1}} \Omega \Phi^{-1}. \tag{41}$$

The implications of Eq. (41) can be illustrated by considering the special case in which **M** is proportional to the identity matrix. In this case $\Phi^{-1} = \Phi^T$ and Eq. (41) simplifies to

$$\mathbf{K} = [\mathbf{k}_{ij}] = \left[\sum_{l=1}^{n} \omega_l^2 \phi_{il} \phi_{jl}\right]. \tag{42}$$

Equation (42) indicates that the elements of the stiffness matrix, which physically are identified with structural components such as walls, beams, or columns, individually or in combination, are in this representation composed of sums of products. Each product contains a square of a natural frequency and two modal ordinates. The modal ordinates will all be less than or equal to unity, but the square of the natural frequency may be a large number. The important diagonal elements of **K** are sums of positive terms, which become progressively larger, in general, as n increases, implying that the most significant contributions will come from the highest modes. The off-diagonal terms will be composed of terms with both positive and negative algebraic signs. From structural mechanics (Sack, 1970; Venkatraman and Patel, 1970), each of the j columns of the stiffness matrix can be envisioned as the force profile required to produce a unit deflection at the jth level or mass point, with zero deflections at the other mass points. Thus, in a typical case, most of the off-diagonal terms of **K** will be smaller than the diagonal terms, quite small if they are far from the diagonal. This means in the representation of Eq. (42) that a typical off-diagonal term will be a small number, found from an ill-conditioned sum of numbers, some relatively large, with opposite algebraic signs.

The general features of the problem will not change if **M** is not proportional to the identity matrix. These considerations indicate why the use of measured earthquake responses or data from vibration tests to determine the stiffness of an individual structural member, or changes in that stiffness, are not straightforward matters, although the results would obviously be quite useful. Research in this area of structural health monitoring or damage detection is one of the more active research topics in earthquake engineering. Instrumental arrays have been installed or are in planning for just this purpose, and significant analytical research has been done (Housner *et al.*, 1997; Kobori *et al.*, 1999).

4. Spectrum Techniques for Multi-Degree-of-Freedom Structures

4.1 An Equivalent Oscillator

Because the equation for the earthquake response of the individual modes of a multi-degree-of-freedom structure, Eq. (34), is essentially the same as the response of a single-degree-of-freedom oscillator to the ground motion, Eq. (1), one way to introduce spectrum techniques to the response of multi-degree-of-freedom structures is to extend this analogy by constructing a simple oscillator that represents a mode of the more complicated system to the degree possible (Housner, 1970). This is illustrated in Figure 11, which shows an idealized building vibrating in a single mode along with the equivalent single-degree-of-freedom structure.

The simple oscillator is described by five parameters: the mass, m; the stiffness, k; the damper, c; the height, h; and the deflection, y. The conditions of equivalence are that the oscillator have the same natural frequency and fraction of critical damping as the mode, that its motion produce the same maximum

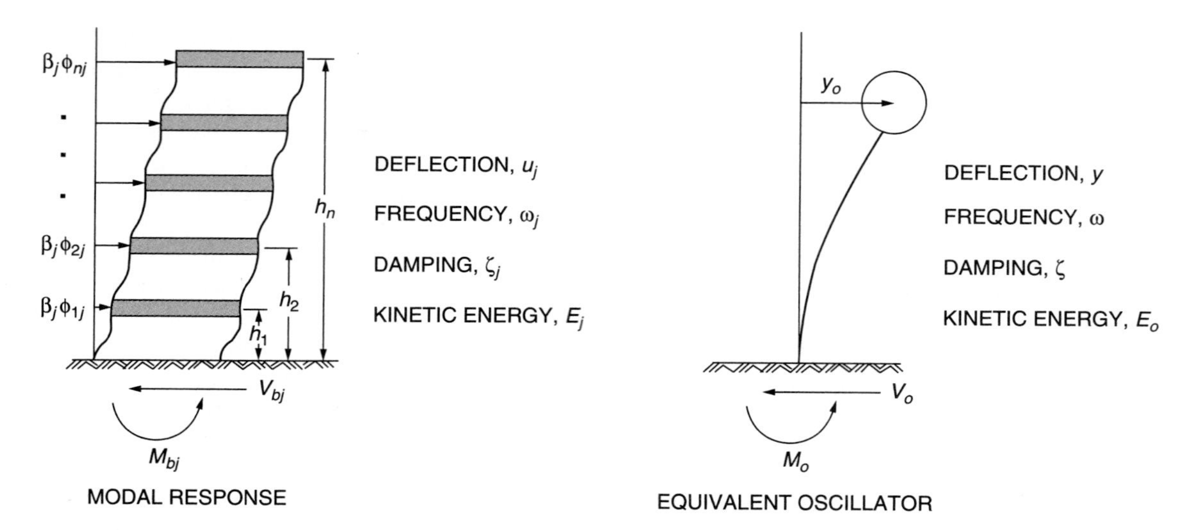

FIGURE 11 Vibration of a multi-degree-of-freedom structure in one of its modes and the equivalent single-degree-of-freedom oscillator modeling the response of the mode.

base shear and base moment as the mode, and that its kinetic energy be the same as that of the modal response. As illustrated in Figure 11, the base shear and base moment are the force and moment exerted on the foundation of the structure by its motion; the maximum values of these resultants play a significant role in earthquake-resistant design. For the oscillator under the convenient condition of free vibration,

$$\begin{array}{ll}
\text{Deflection} & y = y_0 \sin \omega t, \\
\text{Natural frequency} & \omega^2 = k/m, \\
\text{Damping factor} & \zeta = c/2\sqrt{km}, \\
\text{Base shear} & V_0 = m\omega^2 y_0 \sin \omega t, \\
\text{Base moment} & M_0 = hm\omega^2 y_0 \sin \omega t, \text{ and} \\
\text{Kinetic energy} & E_0 = \dfrac{1}{2} m\omega^2 y_0^2 \cos^2 \omega t.
\end{array} \tag{43}$$

For the multi-degree-of-freedom structure responding in a single, normalized j^{th} mode, the corresponding quantities are

$$\begin{array}{ll}
\text{Deflection} & \underline{u_j} = \beta_j \, \underline{\phi_j} \sin \omega_j t, \\
\text{Natural frequency} & \omega_j^2 = \dfrac{\underline{\phi_j^T} \mathbf{K} \, \underline{\phi_j}}{\underline{\phi_j^T} \mathbf{M} \, \underline{\phi_j}}, \\
\text{Damping factor} & \zeta_j = \dfrac{1}{2\omega_j} \dfrac{\underline{\phi_j^T} \mathbf{C} \, \underline{\phi_j}}{\underline{\phi_j^T} \mathbf{M} \, \underline{\phi_j}}, \\
\text{Base shear} & V_{bj} = \sum_{i=1}^{n} m_i \ddot{u}_i = \beta_j \omega_j^2 \sin \omega_j t \underline{\phi_j^T} \mathbf{M} \underline{1}, \\
\text{Base moment} & M_{bj} = \sum_{i=1}^{n} h_i m_i \ddot{u}_i = \beta_j \omega_j^2 \sin \omega_j t \underline{\phi_j^T} \mathbf{M} \underline{h}, \text{ and} \\
\text{Kinetic energy} & E_j = \dfrac{1}{2} \sum_{i=1}^{n} m_i \dot{u}_i^2 = \dfrac{1}{2} \beta_j^2 \omega_j^2 \cos^2 \omega_j t \underline{\phi_j^T} \mathbf{M} \, \underline{\phi_j}.
\end{array} \tag{44}$$

In Eq. (44), β_j is the amplitude of the modal response, $\underline{h}$ is a vector whose elements are the heights, h_i, of the masses, and n is both the number of modes and the number of stories, as indicated in Figure 11. Applying the conditions of equivalence produces the following parameters for the simple oscillator, expressed in terms of the properties of the mode under consideration:

$$\begin{array}{ll}
y_0 = \dfrac{\beta_j}{\alpha_j}, & c = \alpha_j \dfrac{\underline{\phi_j^T} \mathbf{C} \, \underline{\phi_j}}{(\underline{\phi_j^T} \mathbf{M} \, \underline{\phi_j})^{1/2}}, \\
m = \alpha_j \underline{\phi_j^T} \mathbf{M} \, \underline{1}, & \\
k = \omega_j^2 m, \quad \text{and} & h = \dfrac{\underline{\phi_j^T} \mathbf{M} \, \underline{h}}{\underline{\phi_j^T} \mathbf{M} \, \underline{1}}.
\end{array} \tag{45}$$

in which α_j is the participation factor, Eq. (33). With these properties, the response of the oscillator to earthquake forces is equivalent to that of the selected mode of the structure in the sense that the shear and moment reactions at the bases are identical and the vector of deflections of the structure is given by

$$\underline{u_j} = \alpha_j y_0 \, \underline{\phi_j}. \tag{46}$$

The equivalence implies that spectrum techniques applied to the oscillator can be carried over to the modal responses of the structure. In a typical application the process might go as follows. The masses of the structure would be determined and the frequency and mode shape of interest, often those of the fundamental mode, would be calculated or approximated. The fraction of critical damping at the expected level of response would be estimated from experience. This information would be sufficient to determine the participation factor of the mode and the natural period, mass, stiffness, height, and damping factor for the simple oscillator (the damper coefficient, c, would generally not be required). The response of the oscillator, y_0, then would be determined from a response spectrum of the excitation, and the base shear and base moment would be found from Eq. (43); these are also the base shear and moment of the modal response. The participation factor provides the relation between the amplitudes of the oscillator and the structure, via Eq. (46). With the deflection profile of the structure determined, interstory deflections, accelerations at floor levels, and other modal response quantities could be found.

4.2 Maximum Modal Responses

The following analysis is presented within the context of the planar vibrations of structures such as shown in Figures 8 and 11, with one mass per floor level. In many cases, the results apply directly to more complex structures, but modifications or extensions are required in other cases.

The similarity between Eqs. (1) and (34) implies from Eq. (4) that

$$\begin{aligned}\max_t |\xi_j(t)| &= \alpha_j S_D(\omega_j, \zeta_j),\\ \max_t |\dot{\xi}(t)| &= \alpha_j S_V(\omega_j, \zeta_j), \quad \text{and}\\ \max_t |\ddot{\xi}(t) + \alpha_j \ddot{u}_g(t)| &= \alpha_j S_A(\omega_j, \zeta_j).\end{aligned} \tag{47}$$

Similarly, a formal solution for each normal coordinate can be adapted from Eq. (3), as follows:

$$\xi_j(t) = \frac{-\alpha_j}{\omega_j\sqrt{1-\zeta_j^2}} \int_0^t \ddot{u}_g(\tau)\, e^{-\omega_j\zeta_j(t-\tau)} \sin \omega_j\sqrt{1-\zeta_j^2}(t-\tau)\, d\tau. \tag{48}$$

Equation (30) is used to compose the physical displacement $\underline{u}$ by adding the independent modal responses. If the normal coordinates are calculated at each point in time, then $\underline{u}(t)$ results. However, if Eqs. (47) are applied, only maximum values of the normal coordinates are determined and because these maxima do not, in general, occur at the same time, approximate techniques must be used to estimate the maximum values of physical quantities.

Letting $\underline{u_j}$ be the vector of physical displacements associated with the response of the jth normal coordinate, and writing $\underline{u_j} = \underline{\varphi_j}\, \xi_j(t)$ according to Eq. (30), application of Eqs. (30) and (47) and the definition of the pseudovelocity spectrum produces

$$\max_t \underline{u_j}(t) = \frac{\alpha_j S_{PV}(\omega_j, \zeta_j)}{\omega_j} \underline{\phi_j}, \quad j = 1, 2, 3, \cdots, n. \tag{49}$$

The maximum of $\underline{u_j}(t)$ is understood to occur at the time when the ordinates of the vector reach their maximum absolute values, i.e., the time at which $|\xi_j(t)|$ reaches its maximum value. Because response spectra are based on absolute values, the algebraic signs of S_{PV} and α_j in Eq. (49) are irrelevant. Similarly, the vector $\underline{\phi_j}$ can be multiplied by -1 without changing the overall result, but the algebraic signs of the elements of $\underline{\phi_j}$ and $\underline{u_j}$ with respect to each other are important and should be retained.

Equation (49) is a basic result. The maximum displacement contributed by the jth mode is itself important and it also permits the calculation of interstory deflections. Furthermore, if multiplied by the stiffness matrix, $\mathbf{K}$, it produces the maximum force profile associated with the jth mode. This, in turn, allows the calculation of interstory shear forces and forces in individual structural members.

Response spectra can also be used to find the maximum acceleration in the structure contributed by an individual mode. If $\underline{a}$ denotes the vector of total accelerations, i.e., that of the base plus that relative to the base, then

$$\underline{a}(t) = \underline{\ddot{u}}(t) + \ddot{u}_g(t)\underline{1}. \tag{50}$$

Substituting for $\underline{\ddot{u}}$ from Eq. (30) and applying Eq. (35) gives

$$\underline{a}(t) = \sum_{j=1}^{n} \ddot{\xi}_j(t)\underline{\phi_j} + \ddot{u}_g(t)\sum_{j=1}^{n} \alpha_j \underline{\phi_j}. \tag{51}$$

Combining the terms in Eq. (51),

$$\underline{a}(t) = \sum_{j=1}^{n} [\ddot{\xi}_j(t) + \alpha_j \ddot{u}_g(t)]\underline{\phi_j}, \tag{52}$$

separates the total acceleration into its modal components. Letting $\underline{a_j}(t)$ be the component of the jth mode and applying Eqs. (47) and (7) produces

$$\max_t \underline{a_j} = \alpha_j \omega_j S_{PV}(\omega_j, \zeta_j)\underline{\phi_j}, \quad j = 1, 2, 3, \cdots, n. \tag{53}$$

The algebraic signs of the terms in Eq. (53) play the same role as they do in Eq. (49). Because it can describe the modal acceleration at the point of attachment to the structure, Eq. (53) is important for the design of mechanical equipment, architectural components, and other contents of the structure, as well as for assessing the motion experienced by the occupants.

Equation (53) also provides an alternate way to calculate the maximum modal shear forces and moments at any given floor of the structure; the masses must be known, but the stiffness matrix is not required. Multiplying the acceleration of each floor by the associated mass produces a vertical force profile, with the maximum inertia force at the ith level from the jth mode being $m_i \alpha_j \omega_j S_{PV}(\omega_j, \zeta_j)\phi_{ij}$. Summing the forces in the profile above the ith level produces the maximum modal shear force at that level, V_{ij},

$$V_{ij} = \alpha_j \omega_j S_{PV}(\omega_j, \zeta_j) \sum_{l=i+1}^{n} m_l \phi_{lj}. \tag{54}$$

The base shear force, V_{bj}, is particularly important because of its central role in earthquake-resistant design and its use as a basic index of the demands placed by the earthquake excitation on a structure,

$$V_{bj} = \alpha_j \omega_j S_{PV}(\omega_j, \zeta_j) \sum_{l=1}^{n} m_l \phi_{lj}. \tag{55}$$

For the structures under consideration, in which $\mathbf{M}$ is diagonal, the sum in Eq. (55) is $\underline{\phi_j^T}\mathbf{M}\underline{1}$, and by use of Eq. (33), the maximum modal base shear simplifies to

$$V_{bj} = \alpha_j^2 \omega_j S_{PV}(\omega_j, \zeta_j) \underline{\phi}_j^T \mathbf{M} \underline{\phi}_j. \tag{56}$$

Using the same profile of inertia forces and the elements of the vector $\underline{h}$ introduced in Eq. (44) and Figure 11, the maximum moment at the ith level from the jth mode of response, M_{ij}, is

$$M_{ij} = \alpha_j \omega_j S_{PV}(\omega_j, \zeta_j) \sum_{l=i+1}^{n} (h_l - h_i) m_l \phi_{lj}. \tag{57}$$

The base moment becomes

$$M_{bj} = \alpha_j \omega_j S_{PV}(\omega_j, \zeta_j) \sum_{l=1}^{n} h_l m_l \phi_{lj}, \tag{58}$$

which for the structures under consideration is

$$M_{bj} = \alpha_j \omega_j S_{PV}(\omega_j, \zeta_j) \underline{\phi}_j^T \mathbf{M} \underline{\mathbf{h}}. \tag{59}$$

The base and lower-level moments are important because of the tensile and compressive strains they impose on the columns and walls of the structure, particularly those at the edges. Column failures in tension and tensile failures of anchor bolts have been observed in very strong shaking, including Caracas in 1967 (Sozen *et al.*, 1968) and in both the recent Northridge (Hall, 1994) and Kobe (Chung, 1996) earthquakes.

4.3 Combination of Modal Responses

For the planar buildings under consideration, the number of modes is equal to the number of stories; hence for multi-story structures it is usually necessary to approximate the maximum total response from the maximum responses of two or more significantly excited modes. As noted previously, the maximum responses of different modes do not, in general, occur at the same time, but it is possible for the responses of different modes to be correlated to various degrees. The modal responses tend to be uncorrelated if the natural frequencies are well separated and the excitation is chaotic and long in comparison to the natural periods. On the other hand, if the natural frequencies are close together, or if the excitation is a short pulse of ground motion, or is quasi-harmonic, some modal responses can be highly correlated.

All of the commonly applied methods of combining the modal maxima treat the possible correlation of the responding modes, with varying degrees of sophistication. The simplest, most conservative, approach is simply to add the contributions of the modes. This forms an upper bound to the combination by assuming, in effect, that the maxima all occur simultaneously with the most unfavorable algebraic signs. Letting R_j be the maximum response of the variable of interest from the response of the jth mode, n be the number of modes considered (possibly less than the total number), and letting R_D be the combined, or design value, this approach gives

$$R_D = |R_1| + |R_2| + \cdots + |R_n|, \tag{60}$$

wherein the absolute value signs are included to emphasize the additive nature of each modal contribution. Equation (60) gives a worst-case combination, which is often useful in judging other, more likely, ways of combining the modal responses, but it is rarely used in design because it is so conservative.

A way of combining the modal responses that is often used in design and is embodied in some building codes is the familiar SRSS (Square Root of the Sum of the Squares) approach. Rosenblueth (1951) first put forward this method, which is based on concepts from random vibrations. The equation corresponding to Eq. (60) is

$$R_D = \left[R_1^2 + R_2^2 + \cdots + R_n^2\right]^{1/2}. \tag{61}$$

Depending on the possible correlation of modal responses, Eq. (61) is not necessarily conservative, but it does ensure that R_D will always be larger than the largest contribution from any single mode.

A related approach, credited to the Office of Naval Research and reportedly used in the design of submarines, is

$$R_D = R_{j_{\max}} + \left[\sum_{\substack{j=1 \\ j \neq j_{\max}}}^{n} R_j^2\right]^{1/2}. \tag{62}$$

This approach approximates the combined response by adding the largest modal response, $R_{j_{\max}}$, which will certainly occur, to an SRSS estimate of the other modal contributions at that time. Equation (62) has the feature that R_D will always exceed the sum of the absolute values of the two largest modal responses. Hence, it is more conservative than the SRSS combination.

The degree of correlation among the modes is addressed more fully by a method called the Complete Quadratic Combination, CQC (Der Kiureghian, 1981; Chopra, 1995). In this approach the correlation between the modes is addressed explicitly by introducing correlation coefficients, ρ_{ij}, which vary between zero and unity. By definition, when the subscripts are equal, $\rho_{jj} = 1$. The modes are combined by

$$R_D = \left[\sum_{j=1}^{n} \sum_{i=1}^{n} \rho_{ij} R_j R_i\right]^{1/2}. \tag{63}$$

Combining the terms of Eq. (63) with equal indices gives

$$R_D = \left[\sum_{j=1}^{n} R_j^2 + \sum_{j=1}^{n} \sum_{\substack{i=1 \\ i \neq j}}^{n} \rho_{ij} R_j R_i\right]^{1/2}. \tag{64}$$

This form shows that the CQC method gives a result that extends the SRSS approach by adding or subtracting measures of the correlated modal responses. Chopra (1995) presents formulas for the correlation coefficients developed by Der Kiureghian (1981) and by Rosenblueth and Elorduy (1969) and discusses their behavior. With the notation $\beta_{ij} = \omega_i/\omega_j$, Der Kiureghian's result is

$$\rho_{ij} = \frac{8(\zeta_i\zeta_j)^{1/2}(\zeta_i + \beta_{ij}\zeta_j)\beta_{ij}^{3/2}}{\left(1-\beta_{ij}^2\right)^2 + 4\zeta_i\zeta_j\beta_{ij}\left(1+\beta_{ij}^2\right) + 4\left(\zeta_i^2+\zeta_j^2\right)\beta_{ij}^2}. \quad (65)$$

Equation (65) shows the correlation coefficients to be functions of the frequency ratios and the modal dampings. Frequency ratios close to unity and larger values of modal damping increase the correlation coefficients, while large and small frequency ratios and low dampings rapidly decrease the coefficients toward zero.

5. Response in the Frequency Domain

The Fourier transforms of $\xi_j(t)$ and $\ddot{u}_g(t)$ in Eq. (34) can be denoted by

$$X_j(\omega) = \int_0^{\infty} \xi_j(t)e^{-i\omega t}\,dt \quad \text{and} \quad \ddot{U}_g(\omega) = \int_0^{\infty} \ddot{u}_g(t)e^{-i\omega t}\,dt, \quad (66)$$

wherein it is assumed that $\xi_j(t)$ and $\ddot{u}_g(t)$ tend to zero for large time so the transforms will exist. If it is assumed further that the structure is at rest at $t = 0$, then the Fourier transform of Eq. (34) can be solved for $X_j(\omega)$,

$$X_j(\omega) = \frac{-\alpha_j\ddot{U}_g(\omega)}{\left(\omega_j^2-\omega^2\right) + i2\omega\omega_j\zeta_j}. \quad (67)$$

From Eq. (52),

$$\underline{a}_j(t) = \underline{\phi}_j[\ddot{\xi}_j(t) + \alpha_j\ddot{u}_g(t)]. \quad (68)$$

Letting the Fourier transform of $\underline{a}(t)$ be $\underline{A}(\omega)$, it can be found from Eqs. (52), (67), and (68) that

$$\underline{A}(\omega) = \ddot{U}_g(\omega)\sum_{j=1}^{n}\underline{A_j}(\omega), \quad (69)$$

with

$$\underline{A_j}(\omega) = \alpha_j\underline{\phi_j}\left[\frac{1+i2\frac{\omega}{\omega_j}\zeta_j}{\left(1-\frac{\omega^2}{\omega_j^2}\right)+i2\frac{\omega}{\omega_j}\zeta_j}\right]. \quad (70)$$

The summation in Eq. (69) is the transfer function between the ground acceleration and the acceleration at the different levels of the structure. Equation (70) is the component of this transfer function contributed by the jth mode. These components are complex functions and both the amplitude and phase of the modal contribution are functions of the frequency ratio, ω/ω_j, and the damping factor, ζ_j. From Eq. (70), the ith level of the vector $\underline{A_j}(\omega)$ has the following expression for amplitude and phase:

$$|A_j(\omega, x_i)| = |\alpha_j\phi_{ij}|\frac{\left[1+\left(2\frac{\omega}{\omega_j}\zeta_j\right)^2\right]^{1/2}}{\left[\left(1-\frac{\omega^2}{\omega_j^2}\right)^2+\left(2\frac{\omega}{\omega_j}\zeta_j\right)^2\right]^{1/2}}, \quad \text{and} \quad (71)$$

$$\Phi_j(\omega) = \tan^{-1}\frac{-2\frac{\omega^3}{\omega_j^3}\zeta_j}{\left(1-\frac{\omega^2}{\omega_j^2}\right)+\left(2\frac{\omega}{\omega_j}\zeta_j\right)^2}. \quad (72)$$

For each mode, Eq. (71) defines the ith component of a vector, while the modal phase in Eq. (72) is independent of the position, x_i. Equations (71) and (72) are the same as the amplitude ratio and phase shift describing base excitation of a simple oscillator (Thompson, 1965, Section 3.4), within a multiplicative factor. The amplitude, $|A_j(\omega, x_i)|$ at the ith level is illustrated in Figure 12(a), along with a sketch of $|\underline{A}(\omega)|/\ddot{U}_g(\omega)|$ for the ith level of the structure in Figure 12(b). The height of the individual peaks in Figure 12(b) depends on the modal damping and the product of the participation factor and the modal ordinate, ϕ_{ij}. Note that the complex nature of Eq. (69) prevents $|\underline{A}(\omega)|$ at a particular level from being simply the sum of the absolute values of the individual modal components.

The natural frequencies of the structure determine the locations of the peaks in Figure 12(b), while the sharpness of the peaks is determined by the modal dampings. The relative amplitudes of peaks at different levels in the structure at a particular natural frequency are determined primarily by the mode shape associated with that frequency. These features of the response in the frequency domain have been used to determine the dynamic properties of the lower modes of structures from their earthquake response (McVerry, 1979; McVerry *et al.*, 1979). The complex quantities $\underline{A}(\omega)$ and $\ddot{u}_g(\omega)$ can be calculated from measured accelerograms allowing, at least conceptually, the determination of the transfer function and its parameters, the individual natural frequencies, dampings, and mode shapes. Because the absolute values of the modal components of the transfer function, $|A_j(\omega, x_i)|$, are, in general, nonzero at all frequencies, simple filtering in the frequency domain will not necessarily isolate the modal components well. In the references mentioned it proved preferable to begin by determining the best estimate of the properties of the dominant first mode, then to subtract the estimate of its response from the total before proceeding to find the properties of the second mode, etc.

This type of analysis in the frequency domain can also be applied to the response of structures to ambient excitation (Trifunac, 1970, 1972; Ward and Crawford, 1966). Under favorable circumstances, the properties of several modes can be determined by this approach.

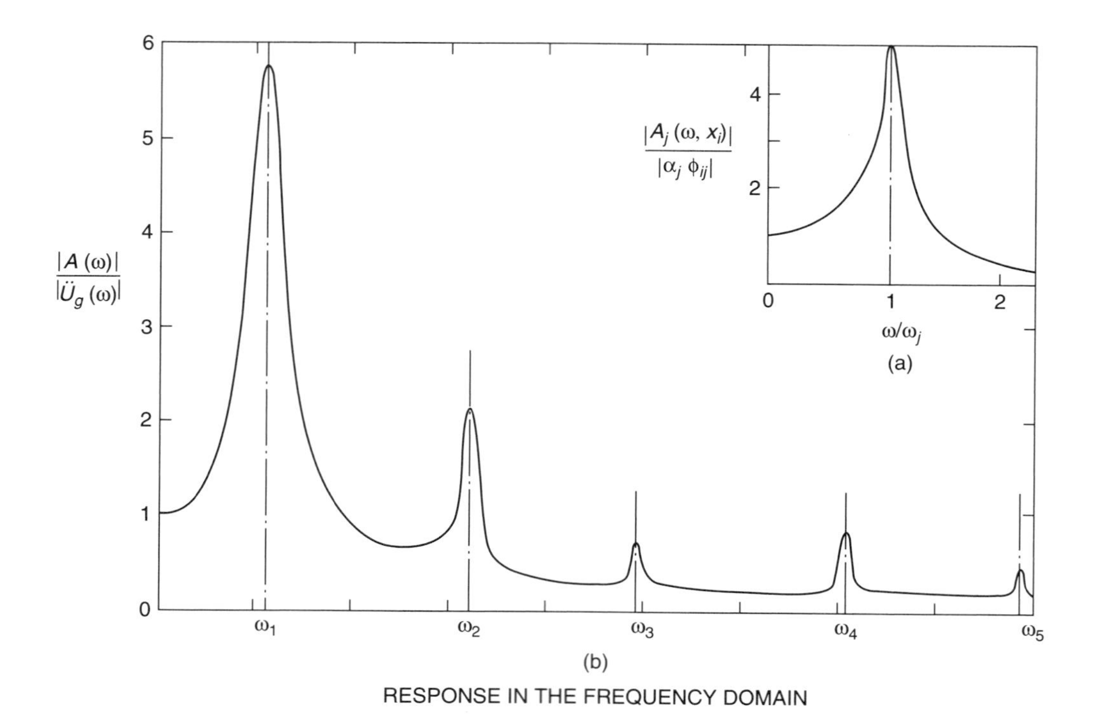

FIGURE 12 Transfer functions for determining response viewed in the frequency domain. (a) The modulus of a term representing a single mode. (b) The modulus for a multi-degree-of-freedom structure.

6. Earthquake Response of a Uniform Shear Beam

6.1 Introduction

It is useful to study the earthquake response of the cantilevered, uniform shear beam shown in Figure 13 for three reasons. First, the dynamic properties are sufficiently simple that it serves as a convenient vehicle for illustrating many of the fundamental features of earthquake response of linear structures. Second, it introduces the analysis of continuous systems, systems that have some features different from the discrete systems considered so far. For example, the uniform shear beam can be analyzed both by modal analysis, as done herein, and by wave propagation methods (Hall *et al.*, 1995; Jacobsen and Ayre, 1958). Finally, the uniform shear beam serves as an approximate model for the earthquake response of a very important class of structures: tall, regular buildings. For these reasons, the uniform shear beam has been employed often in earthquake engineering studies (e.g., Biot, 1932; Rosenblueth, 1951; Jennings, 1969, 1997; Newmark and Rosenblueth 1971; Iwan, 1996).

For application to tall buildings, it should be noted that the uniform stiffness distribution is not a very accurate representation of that of the buildings, whose stiffnesses tend to decrease with height, but the observed natural frequencies of tall buildings are well modeled by those of the shear beam. The mode shapes of the shear beam are quite similar to the observed mode shapes of some tall buildings, particularly those in the 5- to 15-story range. On the other hand, because of the effects of axial deformations in the columns, very tall or very slender buildings often show a fundamental mode shape that is closer to a straight line than to the concave inward fundamental mode shape of the shear beam. However, even when the mode shapes of a tall building differ significantly from those of the uniform shear beam, its properties can still be useful in studying the earthquake response of the real structure (Jennings, 1969, 1997).

With the notation in Figure 13, the equation of motion for the displacement, $u(x, t)$, as a function of the vertical location, x, is

$$\frac{\partial^2 u}{\partial t^2} - \frac{G}{\rho}\frac{\partial^2 u}{\partial x^2} = -\ddot{u}_g(t). \tag{73}$$

The problem is approached by assuming a product solution, $u(x, t) = \phi(x)\xi(t)$, to the free vibration version of Eq. (73) ($\ddot{u}_g = 0$). This leads to

$$\frac{\ddot{\xi}(t)}{\xi(t)} = \frac{G}{\rho}\frac{\phi''(x)}{\phi(x)}, \tag{74}$$

which can only be satisfied if each fraction is equal to a constant, taken as $-\omega^2$. Two equations result, as follows:

$$\ddot{\xi}(t) + \omega^2\xi(t) = 0, \quad \text{and} \quad \phi''(x) + \frac{\rho\omega^2}{G}\phi(x) = 0. \tag{75}$$

The solutions are

$$\begin{aligned} \xi(t) &= C_1 \sin\omega t + C_2 \cos\omega t \\ \phi(x) &= C_3 \sin\omega\sqrt{\rho/G}x + C_4 \cos\omega\sqrt{\rho/G}x. \end{aligned} \tag{76}$$

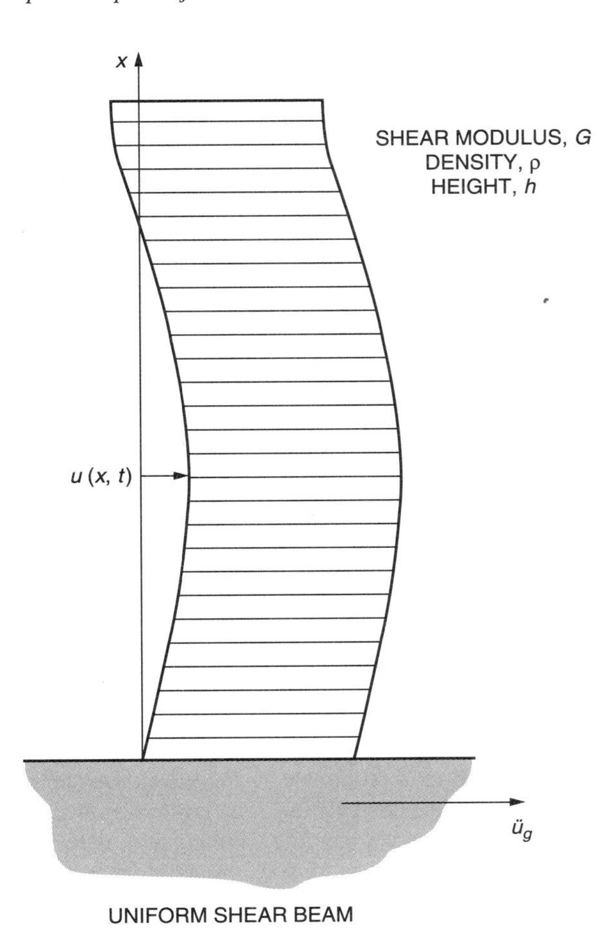

FIGURE 13 A cantilevered shear beam responding to earthquake excitation.

The constants C_1 and C_2 are determined by the initial velocity and displacement of the free vibrations of the beam, while the constants C_3 and C_4 are determined from the boundary conditions

$$\phi(0) = 0 \quad \text{and} \quad \phi'(h) = 0. \tag{77}$$

The boundary conditions can be satisfied by taking C_3 and C_4 equal to zero, but this implies the trivial solution $u(x, t) = 0$. However, a non-trivial solution results if C_4 is taken as zero and the other boundary condition is satisfied by taking ω such that

$$\cos \omega\sqrt{\rho/G}h = 0. \tag{78}$$

Equation (78) is satisfied by taking the argument of the cosine function to be $\pi/2$, plus any integer multiple of π [adding negative multiples of π satisfies Eq. (78), but does not produce more solutions of Eq. (76)]. The ω's that meet this condition are the natural frequencies of the shear beam and the corresponding functions $\phi(x)$ from Eq. (76) are the following mode shapes:

$$\omega_j = \sqrt{\frac{G}{\rho}}\frac{(2j-1)\pi}{2h}, \quad j = 1, 2, 3, \cdots$$
$$\phi_j(x) = \sin\frac{(2j-1)\pi x}{2h}. \tag{79}$$

In this normalization the sinusoidal mode shapes are such that $\phi_j(h) = (-1)^{j+1}$. It is seen, too, that the ratios of the higher natural frequencies to the fundamental go as the odd integers.

6.2 Earthquake Response

In analogy to Eq. (30), the earthquake response is addressed by taking

$$u(x, t) = \sum_{j=1}^{\infty} \phi_j(x)\,\xi_j(t). \tag{80}$$

Substituting into Eq. (73) and applying Eqs. (79) yields

$$\sum_{j=1}^{\infty} \left[\ddot{\xi}_j(t) + \omega_j^2\xi_j(t)\right]\phi_j(x) = -\ddot{u}_g(t). \tag{81}$$

The mode shapes of Eq. (79) are orthogonal, that is,

$$\int_0^h \phi_j(x)\phi_k(x)\,dx = 0, \quad k \neq j,$$
$$= \frac{h}{2}, \quad k = j. \tag{82}$$

Equation (81) can be multiplied by $\phi_k(x)$ and the resulting equation integrated from zero to h. Assuming interchange of the operations of summation and integration is permissible, use of Eq. (82) produces

$$\ddot{\xi}_j(t) + \omega_j^2\xi_j(t) = -\alpha_j\ddot{u}_g, \quad j = 1, 2, 3, \cdots, \tag{83}$$

in which

$$\alpha_j = \frac{4}{(2j-1)\pi}. \tag{84}$$

Rayleigh damping, in the form of Eq. (39), can be added to Eq. (83) by the addition to Eq. (73) of terms proportional to $\frac{\partial u}{\partial t}$ and $\frac{\partial^3 u}{\partial x^2 \partial t}$, or arbitrary modal damping can be postulated, without reference to the implications to Eq. (73), as follows:

$$\ddot{\xi}_j(t) + 2\omega_j\zeta_j\dot{\xi}_j(t) + \omega_j^2\xi_j(t) = -\alpha_j\ddot{u}_g, \quad j = 1, 2, 3, \cdots. \tag{85}$$

These results show that the earthquake response of the shear beam is essentially the same as that of the discrete structures considered earlier, except for the number of modes that can possibly be involved.

6.3 Maximum Modal Responses

Equations (47) apply to Eq. (85), and one can write in analogy to Eqs. (49) and (53),

$$\max_t u_j(x,t) = \frac{\alpha_j S_{PV}(\omega_j, \zeta_j)}{\omega_j}\phi_j(x), \tag{86}$$

and

$$\max_t a_j(x,t) = \alpha_j \omega_j S_{PV}(\omega_j, \zeta_j)\phi_j(x), \tag{87}$$

in which $a_j(x,t)$ is the acceleration contributed by the jth mode. Using the participation factors, mode shapes, and natural frequencies of the shear beam from Eqs. (79) and (84), the maximum displacement and acceleration at the top of the beam reduce to

$$\begin{aligned} |u_{hj}|_{\max} &= \max_t u_j(h,t) = \frac{4S_{PV}(\omega_j, \zeta_j)}{\omega_1 \pi (2j-1)^2}, \quad \text{and} \\ |a_{hj}|_{\max} &= \max_t a_j(h,t) = \frac{4\omega_1 S_{PV}(\omega_j, \zeta_j)}{\pi}. \end{aligned} \tag{88}$$

The maximum modal shear force as a function of x is found by integrating a differential force consisting of the mass, $\rho A\, d\eta$, times the acceleration of Eq. (87), from x to the top of the beam.

$$V_j(x) = \rho A \alpha_j \omega_j S_{PV}(\omega_j, \zeta_j) \int_x^h \phi_j(\eta)\, d\eta \tag{89}$$

With the specific properties of the shear beam this becomes

$$V_j(x) = \frac{8\omega_1 W S_{PV}(\omega_j, \zeta_j)}{\pi^2 g(2j-1)} \cos\frac{(2j-1)\pi x}{2h}, \tag{90}$$

in which W is the total weight of the beam and g is the acceleration of gravity. Equation (90) gives the maximum base shear produced by the jth mode,

$$|V_{bj}|_{\max} = \frac{8\omega_1 W S_{PV}(\omega_j, \zeta_j)}{\pi^2 g(2j-1)}. \tag{91}$$

The maximum modal moment as a function of x is found by integrating the moment of the differential inertia force used to find the maximum modal shear force, as follows:

$$M_j(x) = \rho A\, \alpha_j\, \omega_j S_{PV}(\omega_j, \zeta_j) \int_x^h (\eta - x)\phi_j(\eta)\, d\eta. \tag{92}$$

Evaluating Eq. (92) with the properties of the shear beam,

$$M_j(x) = \frac{16\omega_1 W h S_{PV}(\omega_j, \zeta_j)}{\pi^3 g(2j-1)^2}\left[(-1)^{j+1} - \sin\frac{(2j-1)\pi x}{2h}\right]. \tag{93}$$

The maximum base moment from the jth mode then becomes

$$|M_{bj}| = \frac{16\omega_1 W h S_{PV}(\omega_j, \zeta_j)}{\pi^3 g(2j-1)^2}. \tag{94}$$

The drift ratio is another important parameter of response that can be evaluated for the shear beam. In buildings the drift ratio at a particular level is the relative interfloor deflection divided by the interstory height. For the continuous shear beam this tangent of a small angle is given by $d(x,t) = \frac{\partial u}{\partial x}$. The modal maximum of the drift ratio can be found from Eq. (86) and the properties of the beam, with

$$\max_t |d(x,t)| = \frac{2S_{PV}(\omega_j, \zeta_j)}{\omega_1 h(2j-1)} \cos\frac{(2j-1)\pi x}{2h}. \tag{95}$$

In practice, the most important drift ratio is often that at the base (first floor) of the structure. For the shear beam the cosine is unity at the base, giving

$$|d_{bj}|_{\max} = \max_t |d(x,t)| = \frac{2S_{PV}(\omega_j, \zeta_j)}{\omega_1 h(2j-1)}. \tag{96}$$

The simple form of the maximum modal contributions to important parameters of earthquake response allows their explicit combination by some of the methods given earlier.

6.4 Combination of Modal Responses

The sum of the absolute values and the SRSS methods are next used to illustrate how the modal responses of the shear beam may be combined to approximate its total response (Jennings, 1969). To simplify the presentation, it is further assumed that all the modes have the same damping value, ζ, and, unless stated otherwise, that $S_{PV}(\omega_j, \zeta)$ is a constant. Many average pseudovelocity spectra, particularly for large earthquakes, rise quickly as a function of period to a constant value, and continue at this level as the period increases to values of several seconds. This behavior is also reflected in building codes, many of which embody a similar property by specifying an acceleration spectrum for design forces that is inversely proportional to period for large periods. Because it is reasonably accurate at long periods, the principal effect of the assumption that the response spectrum is a constant is to overstate the contributions of the higher modes to the total response. However, with some exceptions, this conservative assumption is not significant, as the higher modes tend to contribute progressively less to parameters of the total response because of their decreasing participation factors.

Beginning with the displacement at the top of the beam, the maximum modal contributions, Eq. (88), combine according to the sum of the absolute value and SRSS methods to give, respectively,

$$|u_h|_{\max} \le \frac{4S_{PV}}{\pi\omega_1}\left[1 + \frac{1}{9} + \frac{1}{25} + \frac{1}{49} + \cdots\right], \tag{97}$$

and

$$|u_h|_{\max} \approx \frac{4S_{PV}}{\pi\omega_1}\left[1 + \frac{1}{81} + \frac{1}{625} + \frac{1}{2401} + \cdots\right]^{1/2}. \tag{98}$$

The results show the strong dominance of the fundamental mode in the estimated displacement at the top of the beam and also the way that the SRSS method heavily discounts the smaller modal contributions in comparison to the worst case of Eq. (97). The bracketed sums in these two equations sum to $\pi^2/8$ and $\pi^4/96$, respectively, so

$$|u_h|_{\max} \leq 0.50\frac{\pi S_{PV}}{\omega_1},$$

and (99)

$$|u_h|_{\max} \approx 0.41\frac{\pi S_{PV}}{\omega_1}.$$

Equation (99) indicates that the maximum possible response is 22% greater than the estimate given by the SRSS approach.

Applying the combination methods to the base shear gives

$$|V_b|_{\max} \leq \frac{8\omega_1 W S_{PV}}{\pi^2 g}\left[1+\frac{1}{3}+\frac{1}{5}+\frac{1}{7}+\cdots\right] \tag{100}$$

and

$$|V_b|_{\max} \approx \frac{8\omega_1 W S_{PV}}{\pi^2 g}\left[1+\frac{1}{9}+\frac{1}{25}+\frac{1}{49}+\cdots\right]^{1/2}. \tag{101}$$

The bracketed series in Eq. (100) is mathematically divergent, but is made finite by limiting the number of contributing modes. If, for example, five modes are included in Eq. (100) and $\pi^2/8$ is introduced into Eq. (101), then

$$|V_b|_{\max} \leq 1.45\frac{\omega_1 W S_{PV}}{g},$$

and (102)

$$|V_b|_{\max} \approx 0.90\frac{\omega_1 W S_{PV}}{g}.$$

In the case of base shear, the highest possible value for five contributing modes is over 60% more than that given by the SRSS approximation. This suggests that the variance of the base shear can be rather large and indicates that the SRSS method may be somewhat unconservative when the modal maxima are of the same order of magnitude.

The corresponding combinations for the base moment are

$$|M_b|_{\max} \leq \frac{16\omega_1 W h S_{PV}}{\pi^3 g}\left[1+\frac{1}{9}+\frac{1}{25}+\frac{1}{49}+\cdots\right] = 2.0\frac{\omega_1 W h S_{PV}}{\pi g}, \tag{103}$$

and

$$|M_b|_{\max} \approx \frac{16\omega_1 W h S_{PV}}{\pi^3 g}\left[1+\frac{1}{81}+\frac{1}{625}+\frac{1}{2401}+\cdots\right]^{1/2} = 1.63\frac{\omega_1 W h S_{PV}}{\pi g}. \tag{104}$$

The two expressions for the base moment show the dominance of the fundamental mode, following a pattern similar to that shown by the deflection at the top of the beam.

The various modes combine in a quite different way to form the maximum acceleration at the top of the beam. Combining the modal accelerations of Eq. (88) by the two methods yields

$$|a_h|_{\max} \leq \frac{4\omega_1 S_{PV}}{\pi}[1+1+1+1+\cdots], \tag{105}$$

and

$$|a_h|_{\max} \approx \frac{4\omega_1 S_{PV}}{\pi}[1+1+1+1+\cdots]^{1/2}. \tag{106}$$

These results indicate that all modes can contribute equally to the acceleration at the top of the beam if the pseudovelocity spectrum is the same for all modes. To achieve more realistic approximations, the assumption that S_{PV} is a constant has to be revised in the case of the acceleration on the top of the beam. One such revision is to make the assumption that at some specific frequency, for example, 2 Hz, S_{PV} changes from a constant value to one that is inversely proportional to frequency, i.e., to assume a constant acceleration spectrum for high frequencies as is used in many building codes. This assumption, plus a limit to the number of modes considered, allows the series in Eqs. (105) and (106) to be summed (Jennings, 1997). Alternatively, the number of modes considered can simply be reduced to practical numbers. If, for example, five modes are considered, Eqs. (105) and (106) become

$$|a_h|_{\max} \leq 6.4\omega_1 S_{PV},$$

and (107)

$$|a_h|_{\max} \approx 2.8\omega_1 S_{PV}.$$

The marked difference in the two results suggests that the actual values may vary widely and that the SRSS method may be rather unconservative, particularly in cases where modes may be strongly correlated.

The modal drift ratios at the base of the beam can be combined in the same ways, yielding a form similar to that for base shear, as follows:

$$|d_b|_{\max} \leq \frac{2S_{PV}}{\omega_1 h}\left[1+\frac{1}{3}+\frac{1}{5}+\frac{1}{7}+\cdots\right] = 3.6\frac{S_{PV}}{\omega_1 h}, \tag{108}$$

and

$$|d_b|_{\max} \approx \frac{2S_{PV}}{\omega_1 h}\left[1+\frac{1}{9}+\frac{1}{25}+\frac{1}{49}+\cdots\right]^{1/2} = 2.2\frac{S_{PV}}{\omega_1 h}. \tag{109}$$

As was done for base shear, five modes were used to evaluate the series in Eq. (108), giving the coefficient 3.6.

6.5 Application to the Response of Tall Buildings

The earthquake response of a uniform shear beam can be used to gain first approximations for the corresponding results for tall, regular buildings. The frequency ratios of the shear beam closely model those observed for tall buildings and the mode shapes are fair representations of the average modes of tall buildings, with some exceptions (Jennings, 1997). The most obvious differences between the uniform shear beam and regular tall buildings are the variation with height of the lateral stiffness present in buildings and, for many buildings, a fundamental mode shape that differs from a half-sine wave. This does not negate the usefulness of the shear beam, because differences in the fundamental mode shape can be accounted for by modifying the first term in the preceding equations (Jennings, 1969).

For the results of the previous section to apply, the building should be tall enough that its fundamental period is a few seconds. This ensures that at least two or three modes are in the range where the assumption of constant S_{PV} is applicable. The parameters needed to evaluate the expressions are S_{PV}, which characterizes the earthquake excitation, the fundamental frequency or period of the building and, for the maximum drift ratio, the height. The weight, W, also appears in the equations, but is not often needed, as normalized versions of base shear, $|V_b|_{\max}/W$, and the base moment, $|M_b|_{\max}/Wh$, usually suffice. With the required parameters found or estimated, the formulas of the preceding section can be evaluated numerically for application to the tall building under examination.

The results can also be used to investigate the general trends of the earthquake response of tall, regular buildings. As the height of a building with constant cross-section increases, the weight of the building tends to increase linearly with height, as does the fundamental period. For example, one of the earliest approximate formulas for the fundamental period of steel-frame buildings was $T_n = 0.1N$, with N being the number of stories. Applying these two proportional relations to the combination of modal responses, Eqs. (97) through (108), shows that for large height, the base shear and the first-floor drift ratio tend to be independent of height, whereas the acceleration at the top of the building tends to decrease inversely with height. However, the top acceleration still can be stronger than the motion at ground level and the increasing number of participating modes with increasing height modulates this last trend. The deflection at the top of the structure and the base moment tend to increase linearly with height. For comparison, in the case of wind excitation, the base moment increases faster than the square of the height because the affected area and the wind velocity both increase with height.

7. Some Special Topics in Linear Response

7.1 Torsional Response

The planar response discussed previously applies to structures with at least one axis of symmetry. More generally, structures can possess mode shapes with components in the two horizontal directions, and in torsion. Each of the two horizontal components of ground shaking will excite these complex modes. In some instances, vertical components of the mode shapes will exist as well, creating the possibility of modes excitable by each of the three components of ground shaking.

This problem can be introduced by considering the earthquake response of the two-story structure shown in Figure 14.

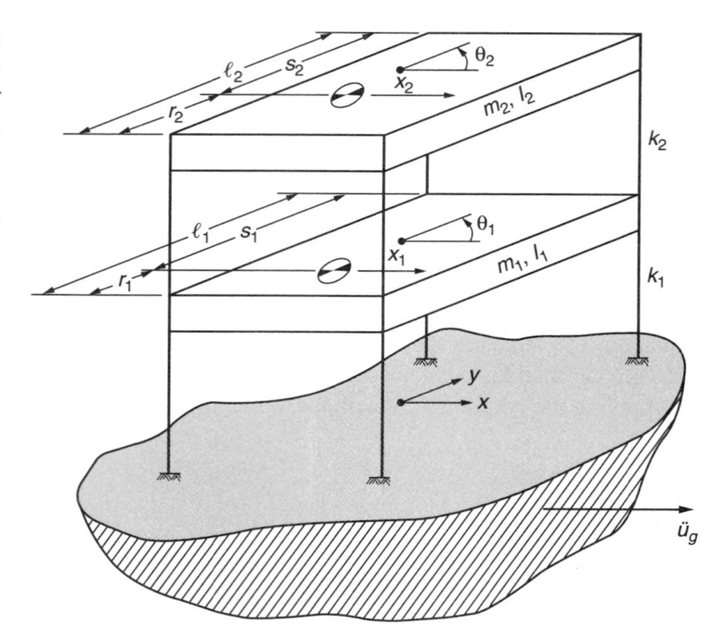

FIGURE 14 An idealized two-story structure exhibiting torsional response to ground motion. Because the centers of mass of the floors are not coincident with the centers of rigidity, shaking in the x-direction will cause the floor masses to rotate.

As seen in the figure, l_1 and l_2 are the widths of the floors in the y-direction and r_1 and r_2 locate the centers of mass of the floors, m_1 and m_2, in that direction. For convenience, distances $s_1 = l_1 - r_1$ and $s_2 = l_2 - r_2$ are also introduced. The structure in Figure 14 is symmetric about the y-axis, but the possible offsets of the centers of mass of the two floors in the y-direction mean that the x-component of the ground motion, $\ddot{u}_g(t)$, can excite both torsional and translational response. The ground motion in the y-direction will excite only an independent translational response. In Figure 14, k_1 and k_2 are the stiffnesses in the x-direction of the columns in the first and second floors, respectively. All columns at a given level are assumed to have the same properties; if they were to differ, the structure would exhibit a torsional component of response even if the two masses were symmetrically aligned. The deflection of the structure is described by the relative displacements of the centers of mass of the floors with respect to the base, x_1 and x_2, and by the floor rotations θ_1 and θ_2. The centroidal moments of inertia of the masses are denoted by I_1 and I_2.

Lagrange's equations for the generalized coordinates, q_i (Housner and Hudson, 1959), in the form

$$\frac{d}{dt}\left(\frac{\partial T}{\partial \dot{q}_i}\right) + \frac{\partial V}{\partial q_i} = 0, \quad q_i = x_1, x_2, \theta_1 \text{ and } \theta_2, \tag{110}$$

are used to derive the equations of motion of the structure in response to the ground motion $\ddot{u}_g(t)$. In Eq. (110), T is the kinetic energy of absolute motion (i.e., measured with respect to an inertial frame of reference) and V is the potential energy stored in the deformation of the structure. These expressions are found to be

$$
\begin{aligned}
T &= \frac{1}{2}m_1(\dot{x}_1+\dot{u}_g)^2+\frac{1}{2}m_2(\dot{x}_2+\dot{u}_g)^2+\frac{1}{2}I_1\dot{\theta}_1^2+\frac{1}{2}I_2\dot{\theta}_2^2,\\
V &= k_1(x_1+r_1\theta_1)^2+k_1(x_1-s_1\theta_1)^2\\
&\quad + k_2[(x_2+r_2\theta_2)-(x_1+r_1\theta_1)]^2\\
&\quad + k_2[(x_2-s_2\theta_2)-(x_1-s_1\theta_1)]^2.
\end{aligned}
\tag{111}
$$

Applying Eq. (110) produces four equations of motion, which can be placed in the matrix form

$$
\begin{bmatrix} m_1 & 0 & 0 & 0\\ 0 & m_2 & 0 & 0\\ 0 & 0 & I_1 & 0\\ 0 & 0 & 0 & I_2\end{bmatrix}
\begin{Bmatrix}\ddot{x}_1\\ \ddot{x}_2\\ \ddot{\theta}_1\\ \ddot{\theta}_2\end{Bmatrix}
+[\mathbf{K}]\begin{Bmatrix}x_1\\ x_2\\ \theta_1\\ \theta_2\end{Bmatrix}
=-\ddot{u}_g(t)\begin{Bmatrix}m_1\\ m_2\\ 0\\ 0\end{Bmatrix},
\tag{112}
$$

in which the stiffness matrix, **K**, is given by

$$
[\mathbf{K}]=
\begin{bmatrix}
4(k_1+k_2) & -4k_2 & -2(k_1+k_2)(s_1-r_1) & 2k_2(s_2-r_2)\\
-4k_2 & 4k_2 & 2k_2(s_1-r_1) & -2k_2(s_2-r_2)\\
-2(k_1+k_2)(s_1-r_1) & 2k_2(s_1-r_1) & 2(k_1+k_2)\left(r_1^2+s_1^2\right) & -2k_2(r_1r_2+s_1s_2)\\
2k_2(s_2-r_2) & -2k_2(s_2-r_2) & -2k_2(r_1r_2+s_1s_2) & 2k_2\left(r_2^2+s_2^2\right)
\end{bmatrix}.
\tag{113}
$$

The equations describing translational and rotational motions are seen to become uncoupled when $r_1=s_1$ and $r_2=s_2$, but in the general case the four equations are coupled, implying that each of the four mode shapes will contain both translational and rotational components. Mathematically, however, Eq. (112) is associated with a classical eigenvalue problem meeting the conditions of Eq. (22), so natural frequencies, mode shapes, and participation factors can be found by the same methods used for planar response. Moreover, spectrum methods can be used to find the maximum response of the structure in the various modes and these values can be combined to estimate the maximum response of the structure.

These remarks also hold for more complicated cases occurring when both the mass and stiffness distributions are asymmetric, and when the number of floors is some general number, n. In such circumstances, it will generally be the case that each of the $3n$ modes will include a rotational component and two translational components. Consequently, the modes can be excited by both translational components of the ground motion and by rotational motions as well, if they are significant. Because of possible correlation among the components of the ground motion, this feature adds an extra complicating factor to the estimation of maximum values of modal responses.

7.2 Response of Extended Structures

Some structures, including long bridges, large earth dams, and major manufacturing facilities, are so long in at least one of their dimensions that different points on the base can be subjected to significantly different earthquake excitations. A simple example

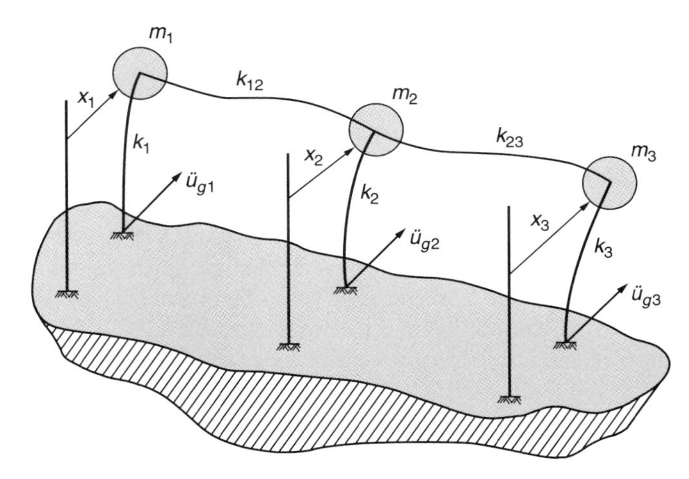

FIGURE 15 A simple structure extended in plan. The supports of this three-degree-of-freedom idealization of the earthquake response of a bridge are sufficiently far apart that the three piers are subject to different ground motions.

to introduce this type of problem is shown in Figure 15, which can be viewed as an idealized version of a bridge supported on three piers subjected to transverse earthquake excitation. The relative deflections of the masses, m_1, m_2, and m_3, at the tops of the piers are represented by x_1, x_2, and x_3, respectively, and the stiffness constants of the piers are denoted by k_1, k_2, and k_3. The resistance of the bridge deck to transverse deformation is represented by the stiffness coefficients k_{12} and k_{23}. The piers are far enough apart that the ground motions at their bases, $\ddot{u}_{g1}(t)$, $\ddot{u}_{g2}(t)$, and $\ddot{u}_{g3}(t)$, may be significantly different. The equations of motion for the generalized coordinates $q_i=x_1, x_2$, and x_3 are again found by applying Eq. (110). In this case the kinetic and potential energies are given by

$$
\begin{aligned}
T &= \frac{1}{2}m_1(\dot{u}_{g1}+\dot{x}_1)^2+\frac{1}{2}m_2(\dot{u}_{g2}+\dot{x}_2)^2+\frac{1}{2}m_3(\dot{u}_{g3}+\dot{x}_3)^2,\\
V &= \frac{1}{2}k_1x_1^2+\frac{1}{2}k_2x_2^2+\frac{1}{2}k_3x_3^2+\frac{1}{2}k_{12}[(u_{g1}+x_1)\\
&\quad -(u_{g2}+x_2)]^2+\frac{1}{2}k_{23}[(u_{g2}+x_2)-(u_{g3}+x_3)]^2.
\end{aligned}
\tag{114}
$$

Applying Lagrange's equations and placing the resulting equations of motion in matrix form yields

$$
\begin{aligned}
&\begin{bmatrix} m_1 & 0 & 0\\ 0 & m_2 & 0\\ 0 & 0 & m_3\end{bmatrix}
\begin{Bmatrix}\ddot{x}_1\\ \ddot{x}_2\\ \ddot{x}_3\end{Bmatrix}\\
&\quad+\begin{bmatrix} k_1+k_{12} & -k_{12} & 0\\ -k_{12} & k_2+k_{12}+k_{23} & -k_{23}\\ 0 & -k_{23} & k_3+k_{23}\end{bmatrix}
\begin{Bmatrix}x_1\\ x_2\\ x_3\end{Bmatrix}\\
&=-\begin{Bmatrix}m_1\ddot{u}_{g1}\\ m_2\ddot{u}_{g2}\\ m_3\ddot{u}_{g3}\end{Bmatrix}
+\begin{Bmatrix}-k_{12}(u_{g1}-u_{g2})\\ k_{12}(u_{g1}-u_{g2})-k_{23}(u_{g2}-u_{g3})\\ k_{23}(u_{g2}-u_{g3})\end{Bmatrix}.
\end{aligned}
\tag{115}
$$

Equation (115) states that the structure is excited by two features of the ground motion. The first term of excitation is the inertial loading associated with the acceleration of the bases of the piers, while the second term is the internal forces induced by differential displacements of the supports. If the ground motion at the three supports is the same, Eq. (115) quickly reduces to the expected form. Because free vibrations of the structure are defined by Eq. (115) with the right-hand side set to zero, the natural frequencies and mode shapes of the structure in Figure 15 are the same as the corresponding case of uniform base motion and, further, the solution of the earthquake response, Eq. (115), can be expressed as the sum of these mode shapes times normal coordinates, i.e., as in Eq. (30). Using Eq. (30), normalizing the mode shapes by Eq. (26), and adding modal damping allows Eq. (115) to become

$$\begin{aligned}&\mathbf{I}\,\ddot{\underline{\xi}} + diag : [2\omega_j\zeta_j]\dot{\underline{\xi}} + \Omega\,\underline{\xi}\\ &= -\Phi^T\mathbf{M}\begin{Bmatrix}\ddot{u}_{g1}\\ \ddot{u}_{g2}\\ \ddot{u}_{g2}\end{Bmatrix} + \Phi^T\begin{Bmatrix}-k_{12}(u_{g1}-u_{g2})\\ k_{12}(u_{g1}-u_{g2})-k_{23}(u_{g2}-u_{g3})\\ k_{23}(u_{g2}-u_{g3})\end{Bmatrix}. \quad (116)\end{aligned}$$

Letting the components of the last vector in Eq. (116) be given by f_1, f_2 and f_3, and expanding the matrix multiplication yields a convenient form for the equations of motion,

$$\begin{aligned}&\ddot{\xi}_j(t) + 2\omega_j\zeta_j\dot{\xi}(t) + \omega_j^2\xi_j(t)\\ &= -\sum_{k=1}^{3}\phi_{jk}m_k\ddot{u}_{gk} + \sum_{k=1}^{3}\phi_{kj}f_k, \quad j=1,2,3. \quad (117)\end{aligned}$$

Although developed for a simple example with three degrees of freedom, Eqs. (116) and (117) suggest the form that the equations of motion of more complicated extended structures might take. It is seen that, in general, the ground displacement as well as the ground acceleration must be known to evaluate the equations and therefore, that standard response spectrum methods are not directly applicable.

If the ground motions at the bases of the piers are generally similar, with small differences, it is convenient to define the motions by their differences, $\delta_j(t)$, from the average base motion, $u_g(t)$; that is, $u_{g1}(t) = u_g(t) + \delta_1(t)$, etc. With this notation, Eq. (116) becomes

$$\begin{aligned}&\mathbf{I}\,\ddot{\underline{\xi}} + diag : [2\omega_j\zeta_j]\dot{\underline{\xi}} + \Omega\,\underline{\xi}\\ &= -\ddot{u}_g(t)\begin{Bmatrix}\alpha_1\\ \alpha_2\\ \alpha_2\end{Bmatrix} - \Phi^T\mathbf{M}\begin{Bmatrix}\ddot{\delta}_1(t)\\ \ddot{\delta}_2(t)\\ \ddot{\delta}_2(t)\end{Bmatrix} + \Phi^T\begin{Bmatrix}f_1\\ f_2\\ f_3\end{Bmatrix}. \quad (118)\end{aligned}$$

In Eq. (118), the components f_j are determined by the $\delta_j(t)$, and the α_j are the modal participation factors from Eq. (33). Equation (118) poses the problem in the form of the response to a uniform, average base motion, plus two correction terms associated with the differences, often small, among the base motions of the extended structure. The first correction term will tend to be acceleration-like, whereas the second correction term will usually have a much lower frequency content, because it is based on ground displacements.

The interested reader is referred to Chopra (1995) for further information on this topic, including an alternate way of partitioning the response.

7.3 Soil–Structure Interaction

The foundation conditions of a structure can affect its response to earthquake motion, depending on the mass and stiffness of the structure and the flexibility of the supporting soil (Newmark and Rosenblueth, 1971; Bielak and Jennings, 1973; Luco, 1980). Generally speaking, the largest effects are associated with stiff, massive structures founded on soft soils, and the least effects occur for light, flexible structures on hard rock foundations. When significant, soil–structure interaction modifies the response in several important ways. For planar motions of building-like structures, for example, a structure with classical normal modes for fixed-based conditions will lose that property and two new coordinates and two new natural frequencies will be introduced. The new coordinates are the translation and rotation of the structure's base. For relatively rigid structures on soft soils, one of the new frequencies will mainly be associated with a rocking motion of the structure, with some in-phase translation of the base and some first mode-like deformation within the structure. At the other new frequency the structure and foundation exhibit more complicated motion. The first new frequency is typically the lowest of the system, while the second is typically among the highest. For less rigid structures on firmer soils, the fundamental frequency of the soil–structure system is typically associated with a motion that is principally first mode-like motions of the structure with respect to the base, plus some in-phase rocking and translation of the base. More complex forms of soil–structure interaction are also highly important for other types of structures, for example, bridges, dams, and harbor facilities (Okamoto, 1984).

The foundation soil or rock exhibits many of the properties of an elastic medium. In particular, the forces exerted by the building on the soil produce waves of motion that radiate from the foundation and can be detected at large distances under favorable conditions. During forced vibration tests of the nine-story Millikan Library on the Caltech campus in Pasadena, the induced motions were measured by a seismograph on top of Mount Wilson, 6.7 miles away (Jennings, 1970). The waves emanating from the foundation carry energy away from the system, energy that in most cases is lost in the earth. This loss of energy from the structure–soil system is termed radiation damping. It is present even if the foundation is perfectly elastic (if effectively infinite in extent); any material damping in the foundation is an additional effect. Because the contact stresses on the foundation during rocking are of opposite directions, while those of translation are unidirectional, radiation damping from rocking tends to be much less than that associated with translation.

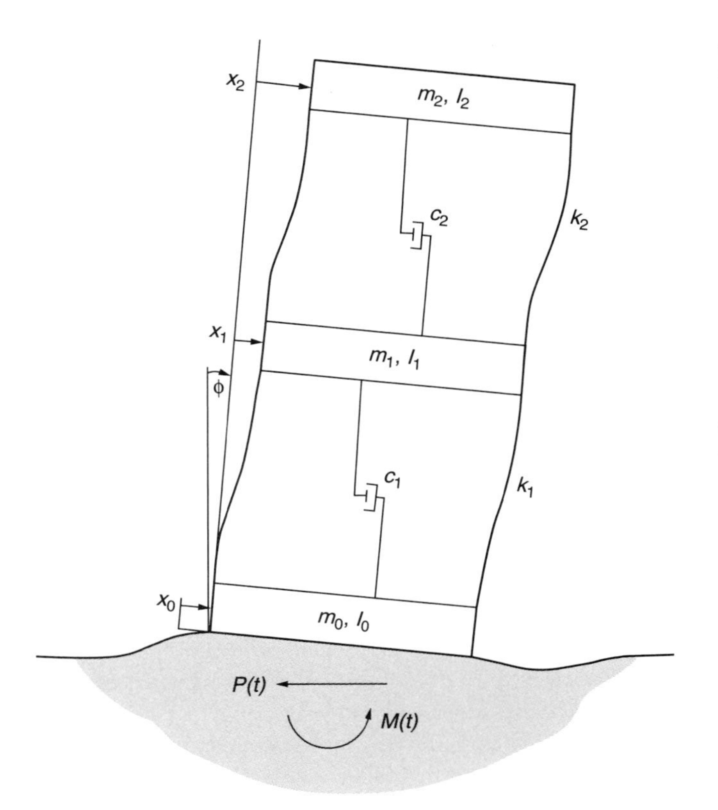

FIGURE 16 Model of a two-story building including the effects of soil–structure interaction. The resistance of the soil to deformation is described by the force, $P(t)$, and the moment, $M(t)$.

The equations of motion governing soil–structure interaction can be introduced by examination of the two-story structure shown in Figure 16. This figure resembles Figure 8, with the addition of a base mass, m_0, coordinates x_0 and ϕ, which describe the motion of the base with respect to an inertial frame, and the reactive force and moment of the soil, $P(t)$ and $Q(t)$, respectively. The coordinates x_1 and x_2 continue to measure the deflection of the structure with respect to its base. Additional structural parameters that enter are the centroidal moments of inertia of the three masses, I_0, I_1, and I_2, and the interstory heights, h_1 and h_2. The ground motion is characterized by the "free-field" motion, $\ddot{u}_g(t)$. This could be, for example, in the form of an incoming, vertically propagating *SH* wave. More generally, the free-field motion is conceived as the motion that would exist at the base in the absence of the structure.

Lagrange's equations are again used to derive the equations of motion, this time in the form

$$\frac{d}{dt}\left(\frac{\partial T}{\partial \dot{q}_i}\right) + \frac{\partial V}{\partial q_i} = F_i(t), \quad q_i = x_1, x_2, x_0, \phi. \tag{119}$$

The $F_i(t)$ are the generalized forces associated with the four generalized coordinates, $q_i(t)$, found by determining the virtual work, δW, performed during a virtual displacement of the coordinates, δq_i. The kinetic and potential energies and the virtual work are found to be

$$\begin{aligned} T &= \frac{1}{2}m_0(\dot{u}_g + \dot{x}_0)^2 + \frac{1}{2}m_1(\dot{u}_g + \dot{x}_0 + \dot{x}_1 + h_1\dot{\phi})^2 \\ &\quad + \frac{1}{2}m_2(\dot{u}_g + \dot{x}_0 + \dot{x}_2 + h_2\dot{\phi})^2 + \frac{1}{2}(I_0 + I_1 + I_2)\dot{\phi}^2, \\ V &= \frac{1}{2}k_1x_1^2 + \frac{1}{2}k_2(x_2 - x_1)^2, \quad \text{and} \\ \delta W &= -P(t)\delta x_0 - Q(t)\delta\phi - c_1\dot{x}_1\delta x_1 \\ &\quad - c_2(\dot{x}_2 - \dot{x}_1)(\delta x_2 - \delta x_1). \end{aligned} \tag{120}$$

Applying Lagrange's equations yields the following equations of motion:

$$\begin{aligned} [\mathbf{M}]\begin{Bmatrix} \ddot{x}_1 \\ \ddot{x}_2 \\ \ddot{x}_0 \\ \ddot{\phi} \end{Bmatrix} + [\mathbf{C}]\begin{Bmatrix} \dot{x}_1 \\ \dot{x}_2 \\ \dot{x}_0 \\ \dot{\phi} \end{Bmatrix} + [\mathbf{K}]\begin{Bmatrix} x_1 \\ x_2 \\ x_0 \\ \phi \end{Bmatrix} + \begin{Bmatrix} 0 \\ 0 \\ P(t) \\ Q(t) \end{Bmatrix} \\ = -\ddot{u}_g(t)\begin{Bmatrix} m_1 \\ m_2 \\ m_0 + m_1 + m_2 \\ m_1h_1 + m_2h_2 \end{Bmatrix}, \end{aligned} \tag{121}$$

wherein

$$[\mathbf{M}] = \begin{bmatrix} m_1 & 0 & m_1 & m_1h_1 \\ 0 & m_2 & m_2 & m_2h_2 \\ m_1 & m_2 & m_0 + m_1 + m_2 & m_1h_1 + m_2h_2 \\ m_1h_1 & m_2h_2 & m_1h_1 + m_2h_2 & m_1h_1^2 + m_2h_2^2 + I_0 + I_1 + I_2 \end{bmatrix}$$

$$[\mathbf{C}] = \begin{bmatrix} c_1 + c_2 & -c_2 & 0 & 0 \\ -c_2 & c_2 & 0 & 0 \\ 0 & 0 & 0 & 0 \\ 0 & 0 & 0 & 0 \end{bmatrix}, \quad [\mathbf{K}] = \begin{bmatrix} k_1 + k_2 & -k_2 & 0 & 0 \\ -k_2 & k_2 & 0 & 0 \\ 0 & 0 & 0 & 0 \\ 0 & 0 & 0 & 0 \end{bmatrix}. \tag{122}$$

The equations of motion can be placed in a more familiar form if the soil reactions, $P(t)$ and $Q(t)$ are assumed to be composed of spring and damper-like elements, i.e.,

$$\begin{Bmatrix} P(t) \\ Q(t) \end{Bmatrix} = \begin{bmatrix} k_{hh} & k_{hm} \\ k_{mh} & k_{mm} \end{bmatrix}\begin{Bmatrix} x_0 \\ \phi \end{Bmatrix} + \begin{bmatrix} c_{hh} & c_{hm} \\ c_{mh} & c_{mm} \end{bmatrix}\begin{Bmatrix} \dot{x}_0 \\ \dot{\phi} \end{Bmatrix}. \tag{123}$$

With this substitution, the damping and stiffness matrices become

$$\begin{aligned} [\mathbf{C}_S] &= \begin{bmatrix} c_1 + c_2 & -c_2 & 0 & 0 \\ -c_2 & c_2 & 0 & 0 \\ 0 & 0 & c_{hh} & c_{hm} \\ 0 & 0 & c_{hm} & c_{mm} \end{bmatrix}, \\ [\mathbf{K}_S] &= \begin{bmatrix} k_1 + k_2 & -k_2 & & \\ -k_2 & k_2 & & \\ & & k_{hh} & k_{hm} \\ & & k_{mh} & k_{mm} \end{bmatrix}, \end{aligned} \tag{124}$$

and the equations of motion reduce to

$$[\mathbf{M}]\begin{Bmatrix}\ddot{x}_1\\ \ddot{x}_2\\ \ddot{x}_0\\ \ddot{\phi}\end{Bmatrix}+[\mathbf{C}_S]\begin{Bmatrix}\dot{x}_1\\ \dot{x}_2\\ \dot{x}_0\\ \dot{\phi}\end{Bmatrix}+[\mathbf{K}_S]\begin{Bmatrix}x_1\\ x_2\\ x_0\\ \phi\end{Bmatrix} = -\ddot{u}_g(t)\begin{Bmatrix}m_1\\ m_2\\ m_0+m_1+m_2\\ m_1h_1+m_2h_2\end{Bmatrix}. \tag{125}$$

The matrices $\mathbf{C}_S$ and $\mathbf{K}_S$ in Eq. (125) can be taken as symmetric without loss of generality, but the resulting equations are still such that the system does not possess classical normal modes; a further assumption about the damping matrix is required to achieve that property. For example, the system for $\mathbf{C}_S = 0$ can be solved, which does have classical normal modes. These modes can then be applied to Eq. (125), as was done in Eq. (31), and any nonzero off-diagonal terms in the transformed damping matrix can be set to zero. Making a simplification such as this reduces the two-story building on a flexible foundation as shown in Figure 16 to a system with four natural frequencies and four orthogonal modes of vibration. In general, each mode will exhibit interstory deflections plus translation and rotation of the base. Equation (125) shows that the earthquake affects the interstory deflections, x_1 and x_2, just as it does for the fixed-base case; these equations are unaffected if x_0 and ϕ both approach zero. The base coordinate, x_0, is excited by the motion as though the structure were rigid, and the rocking motion is excited by the moments of the inertial forces of the two elevated masses, again as if the structure were rigid. The induced motions, of course, are coupled by the off-diagonal terms of the mass, damping, and stiffness matrices, a coupling that makes the response different from the fixed-base case.

The simplified version of the system just discussed illustrates many of the features of structures founded on flexible foundations, but the assumption that the soil reactions can be characterized by spring and damper elements with constant properties is quite restrictive and must be applied with care, typically over a narrow frequency range within the response. This limitation arises because some of the constant values of the equivalent springs and dampers best suited to model the flexibility of the foundation depend significantly on the frequency of the motion. Values might be found, for example, that effectively model the effects of interaction for the fundamental mode of the system, but the same values may not closely approximate the effects of interaction for the higher natural frequencies. Consequently, more accurate solutions of soil–structure interaction problems are usually attempted by applying the Fourier or Laplace transforms to the equations of motion (Jennings and Bielak, 1973; Luco, 1980). In their transformed form, the soil reactions, $P(t)$ and $Q(t)$, can be represented by spring and damper-like elements in a form similar to Eq. (123), with the elements of the matrices having frequency-dependent values. Elasto-dynamic problems such as the vibration of a rigid disc on an elastic half-space provide analytical or numerical values of these frequency-dependent impedances (e.g., Veletsos and Wei, 1971; Luco and Westmann, 1971).

8. Introduction to Advanced Methods of Analyses

8.1 Nonlinear Analysis

As mentioned earlier, analysis of the response of structures that yield significantly or experience major damage or failure is beyond the capability of linear analysis. Nonlinear methods must be used to study these problems. For single-degree-of-freedom structures, such as shown in Figure 4, the nonlinear version of the equation of motion can be conveniently written as

$$\ddot{x} + \omega_n^2 g(x, \dot{x}, t) = -\ddot{u}_g(t), \tag{126}$$

which reduces to Eq. (1) when $g(x, \dot{x}, t) = x + 2\frac{\zeta}{\omega_n}\dot{x}$. The resistance function, $g(x, \dot{x}, t)$, includes both nonlinear stiffness properties, such as hysteresis, as well as any linear or nonlinear effects of velocity-dependent damping.

Because of its complexity, very few practically important cases of Eq. (126) can be solved, and solutions are typically found by numerical methods or by approximate methods such as equivalent linearization (Caughey, 1960; Jennings, 1968; Iwan, 1970). Exact solutions can be developed to Eq. (126) if the form of the resistance is simple, for example, elasto-plastic or bilinear hysteretic, and if the earthquake motion is also highly idealized (e.g., Iwan, 1961).

Not many structures have the simple geometry shown in Figure 4, and direct numerical integration of the equations of motion is virtually the only method available for determining the nonlinear earthquake response of multi-degree-of-freedom structures. Although such analyses are done in the design of some special structures such as offshore drilling platforms or monumental structures, most such analyses are done as research studies (see, for example, Murphy, 1973; Hall *et al.*, 1995) and rarely comprise a part of the design calculations for the design of buildings.

8.2 Finite Element Analysis

As the geometrical complexity of structures increases, finite element methods increasingly become the methods of choice for either linear or nonlinear analyses. Standard matrix methods can treat the linear, three-dimensional response of multi-story buildings with rigid floors and multiple columns per level, but the use of beam, column, and joint elements becomes more attractive if the flexibility of the girders or floor system must be incorporated, or if yielding of the structural frame must be considered. Similarly, a structural shear wall can be modeled as a shear and bending beam and its stiffness at each level can be coupled to the matrix describing the stiffness of the structural frame, but if yielding of the wall is of interest, finite element methods would commonly be employed from the outset. Corresponding situations occur in the study of other structures, including bridges, dams, storage tanks, etc. As the geometrical complexity of the structure increases or the need to consider yielding or other nonlinear behavior increases, simple linear models become increasingly less relevant and recourse to finite element methods must be made. The reader is referred to Chapter 69 by Hall for examples of finite element studies of earthquake response.

The finite element approach is fundamentally very powerful. In the case of a building, for example, the detailed linear or nonlinear behavior of every beam, column, joint, and floor slab could be determined by using thousands of elements each, in principle. In theory, the same detail could also be modeled for the foundation and surrounding soil. At the present time, application of finite element methods is limited by the ability to characterize accurately nonlinear material behavior such as cracking and yielding; by the ability to connect elements of different types, such as those for reinforcing steel and concrete; by the ability to consolidate the results of calculations and to portray them usefully; by the limitations of available software; and eventually, by the sheer size of the numerical operations that must be performed. As future advances reduce these limitations, it is expected that finite element methods will become more and more accurate and their present wide use will become even more prevalent.

9. Conclusion

The results and approaches presented in this chapter differ from those typically used in the design and analysis of earthquake-resistant structures. However, in many cases they illustrate the general influences of parameters of the structure and the ground motion in ways that are not possible in calculations for design or in individual numerical calculations of the response of realistic, but more complicated structures. It is the author's hope that this feature will be especially useful to students as they develop their understanding of earthquake response.

References

Arias, A. (1970). A measure of earthquake intensity. In: "Seismic Design for Nuclear Power Plants," (R. J. Hansen, Ed.), pp. 438–483. Massachusetts Institute of Technology, Cambridge, Massachusetts.

Beck, J. L. (1978). Determining models of structures from earthquake records. *Earthquake Engineering Research Laboratory Report 78-01*, California Institute of Technology, Pasadena, California.

Benioff, H. (1934). The physical evaluation of seismic destructiveness. *Bull. Seismol. Soc. Am.* **24**, 398–403.

Bielak, J., and P. C. Jennings (1973). Dynamics of building-soil interaction. *Bull. Seismol. Soc. Am.* **63**, 9–48.

Biot, M. A. (1932). "Transient Oscillations in Elastic Systems," doctoral thesis, California Institute of Technology, Pasadena, California.

Campbell, K. W. (1997). Empirical near-source attenuation relationships for horizontal and vertical components of peak ground acceleration, peak ground velocity, and pseudo-absolute acceleration response spectra. *Seismol. Res. Lett.* **68**, 154–179.

Caughey, T. K. (1960). Random excitation of a system with bilinear hysteresis. *J. Appl. Mech., Am. Soc. Mech. Eng.* **27**, 649–652.

Caughey, T. K., and H. S. Stumpf (1961). Transient response of a dynamic system under random excitation. *J. Appl. Mech., Am. Soc. Mech. Eng.* **28**, 563–566.

Caughey, T. K., and M. E. J. O'Kelley (1965). Classical normal modes in damped linear systems. *J. Appl. Mech., Am. Soc. Mech. Eng.* **32**, 583–588.

Chopra, A. K. (1995). *Dynamics of Structures.* Prentice-Hall, Englewood Cliffs, New Jersey.

Chung, R., Ed. (1996). *The January 17, 1995 Hyogoken-Nanbu (Kobe) Earthquake, NIST Special Publication 901*, National Institute of Standards and Technology, U.S. Department of Commerce, Washington, D.C.

Crandall, S. H., and W. D. Mark (1963). *Random Vibration in Mechanical Systems.* Academic Press, New York and London.

Darragh, R., M. Huang, T. Cao, V. Graizer, and A. Shakal (1995). Los Angeles Code-Instrumented Building Records from the Northridge, California Earthquake of January 17, 1994: Processed Release No. 2. California Strong Motion Instrumentation Program, California Department of Conservation, Division of Mines and Geology, Office of Strong Motion Studies, Sacramento, California.

Der Kiureghian, A. (1981). A response spectrum method for random vibration analysis of MDF systems. *Earthq. Eng. Struct. Dyn.* **9**, 419–435.

Earthquake Engineering Research Laboratory (1972). Analysis of Strong Motion Earthquake Accelerograms, Vol. III-Response Spectra, Part A Accelerograms IIA001 through IIA020. *Earthquake Engineering Research Laboratory Report 72-80.* California Institute of Technology, Pasadena, California.

Earthquake Engineering Research Laboratory (1973a). Analysis of Strong Motion Earthquake Accelerograms, Vol. III-Response Spectra, Part B Accelerograms IIB021 through IIB040. *Earthquake Engineering Research Laboratory Report 73-80.* California Institute of Technology, Pasadena, California.

Earthquake Engineering Research Laboratory (1973b). Analysis of Strong Motion Earthquake Accelerograms, Vol. III-Response Spectra, Part C Accelerograms IIC041 through IIC055. *Earthquake Engineering Research Laboratory Report 73-81.* California Institute of Technology, Pasadena, California.

Franklin, J. N. (1968). *Matrix Theory.* Prentice-Hall, Englewood Cliffs, New Jersey.

Foutch, D. A. (1976). A Study of the Vibrational Characteristics of Two Multistory Buildings, *Earthquake Engineering Research Laboratory Report 76-03.* California Institute of Technology, Pasadena, California.

Hall, J. F., Technical Ed. (1994). Northridge Earthquake January 17, 1994, Preliminary Reconnaissance Report. Earthquake Engineering Research Institute, Oakland, California.

Hall, J. F. (1999). Personal communication.

Hall, J. F., T. H. Heaton, M. W. Halling, and D. J. Wald (1995). Near-source ground motions and its effects on flexible buildings. *Earthq. Spectra* **11**, 569–605.

Housner, G. W. (1952). Spectrum intensities of strong-motion earthquakes. In: "Proceedings of the Symposium on Earthquake and Blast

Effects on Structures." Earthquake Engineering Research Institute, Berkeley, California.

Housner, G. W. (1957). Dynamic pressures on accelerated fluid containers. *Bull. Seismol. Soc. Am.* **47**, 15–35.

Housner, G. W. (1959). Behavior of structures during earthquakes. *J. Eng. Mech. Div., Am. Soc. Civil Eng.* **85**(EM4), 109–129.

Housner, G. W., and D. E. Hudson (1959). *Applied Mechanics-Dynamics*. 2nd ed. D. Van Nostrand Company, Princeton, New Jersey.

Housner, G. W. (1970). Design spectrum. In: "Earthquake Engineering," (R. L. Weigel, Ed.), Prentice-Hall, Englewood Cliffs, New Jersey.

Housner, G. W. (1997). "Connections—The EERI Oral History Series—George W. Housner" (Stanley Scott, interviewer). Earthquake Engineering Research Institute, Berkeley, California.

Housner, G. W., and P. C. Jennings (1982). *Earthquake Design Criteria*. Earthquake Engineering Research Institute, Berkeley, California.

Housner, G. W., L. A. Bergman, T. K. Caughey, A. G. Chassiakos, R. O. Claus, S. F. Masri, R. E. Skelton, T. T. Soong, B. F. Spencer, and J. T. P. Yao (1997). Structural control: past, present and future. *J. Eng. Mech. Div., Am. Soc. Civil Eng.* **123**, 897–971.

Hudson, D. E. (1962). Some problems in the application of spectrum techniques to strong-motion earthquake analysis. *Bull. Seismol. Soc. Am.* **52**, 417–430.

Iwan, W. D. (1960). "Digital Calculation of Response Spectra and Fourier Spectra." Unpublished note, California Institute of Technology, Pasadena, California.

Iwan, W. D. (1961). "The Dynamic Response of Bilinear Hysteretic Systems," doctoral thesis, California Institute of Technology, Pasadena, California.

Iwan, W. D. (1970). Steady-state response of yielding shear structure. *J. Eng. Mech. Div., Am. Soc. Civil Eng.* **96**(EM6), 1209–1228.

Iwan, W. D. (1996). The drift demand spectrum and its application to structural design and analysis. *Paper No. 1116, Proc. 11th World Conf. Earthq. Eng.*, Acapulco, Mexico.

Jacobsen, L. S., and R. S. Ayre (1958). *Engineering Vibrations*. McGraw-Hill, New York, Toronto, London.

Jennings, P. C. (1968). Equivalent viscous damping for yielding structures. *J. Eng. Mech. Div., Am. Soc. Civil Eng.* **94**(EM1), 103–116.

Jennings, P. C. (1969). Spectrum techniques for tall buildings. *Proc. Fourth World Conf. Earthq. Eng.*, Vol. II, Section A-3, pp. 61–74. Santiago, Chile.

Jennings, P. C. (1970). Distant motions from a building vibration test. *Bull. Seismol. Soc. Am.* **60**, 2037–2043.

Jennings, P. C. (1974). Calculation of selected ordinates of Fourier spectra. *Earthq. Eng. Struct. Dyn.* **2**(3), 281–293.

Jennings, P. C. (1983). Engineering seismology. In: "Earthquakes: Observation, Theory and Interpretation" (H. Kanamori and E. Boschi, Eds.), pp. 138–173. North Holland, Amsterdam-New York-Holland.

Jennings, P. C. (1997). Earthquake Response of Tall Regular Buildings, *Earthquake Engineering Research Laboratory Report 97-01*. California Institute of Technology, Pasadena, California.

Jennings, P. C., and J. Bielak (1973). Dynamics of building-soil interaction. *Bull. Seismol. Soc. Am.* **63**, 9–48.

Jennings, P. C., R. B. Matthiesen, and J. B. Hoerner (1971). Forced Vibration of a 22-Story Steel Frame Building. Earthquake Engineering Research Laboratory, California Institute of Technology, and Earthquake Engineering and Structures Laboratory, University of California, Los Angeles, Pasadena, California.

Kawasumi, H. (1956). Notes on the theory of vibration analyzer. *Bull. Earthq. Res. Inst.*, Tokyo Univ. **34**(1), 1–8.

Kobori, T., Y. Inoue, K. Sato, H. Iemura, and A. Nishitani, Eds. (1999). "Proceedings of the Second World Conference on Structural Control," Vol. I–III. John Wiley & Sons, Chichester, New York.

Luco, J. E., and R. A. Westmann (1971). Dynamic response of circular footings. *J. Eng. Mech. Div., Am. Soc. Civil Eng.* **97**(EM5), 1381–1395.

Luco, J. E. (1980). Seismic safety margins research program (Phase I) linear soil-structure interaction. In: "Soil–Structure Interaction: The Status of Current Analysis Methods and Research" (prepared by J. J. Johnson). NUREG/CR-1780 UCRL 53011, January 1981.

Lutes, L. D., and S. Sarkani (1997). *Stochastic Analysis of Structural and Mechanical Vibrations*. Prentice-Hall, Upper Saddle River, New Jersey.

McVerry, G. H. (1979). Frequency Domain Identification of Structural Models from Earthquake Records. *Earthquake Engineering Research Laboratory Report 79-02*. California Institute of Technology, Pasadena, California.

McVerry, G. H., J. L. Beck, and P. C. Jennings (1979). Identification of linear structural models from earthquake records. *Proc. 2nd U.S. Nat. Conf. Earthq. Eng.* Stanford, California.

Meirovitch, L. (1967). *Analytical Methods in Vibration*. The Macmillan Company, New York, London.

Murphy, L. M., Scientific Coordinator (1973). "San Fernando, California, Earthquake of February 9, 1971," Vol. I. U. S. Department of Commerce, National Oceanic and Atmospheric Administration, Environmental Research Laboratories, Washington, D.C.

Neumann, F. (1935). "United States Earthquakes, 1933," Coast and Geodetic Survey, U.S. Department of Commerce, United States Government Printing Office, Washington, D.C.

Newmark, N. M., and E. Rosenblueth (1971). *Fundamentals of Earthquake Engineering*. Prentice-Hall, Englewood Cliffs, New Jersey.

Nigam, N. C., and P. C. Jennings (1969). Calculation of response spectra from strong-motion earthquake records. *Bull. Seismol. Soc. Am.* **59**, 909–922.

Okamoto, S. (1984). *Introduction to Earthquake Engineering*, 2nd ed. University of Tokyo Press, Tokyo.

Rosenblueth, E. (1951). "A Basis for Aseismic Design of Structures," doctoral thesis, University of Illinois, Urbana, Illinois.

Rosenblueth, E., and J. Elorduy (1969). Responses of linear systems to certain transient disturbances. *Proc. Fourth World Conf. Earthq. Eng.*, Vol. I, session A1, pp. 185–196. Santiago, Chile.

Sack, R. L. (1984). *Structural Analysis*. McGraw-Hill, New York.

Singh, S. K., E. Mena, and R. Castro (1998). Some aspects of source characteristics of the 19 September 1985 Michoacan earthquake and ground motion amplification in and near Mexico City from strong motion data. *Bull. Seismol. Soc. Am.* **78**, 451–477.

Somerville, P. G., N. F. Smith, R. W. Graves, and N. A. Abrahamson (1997). Modification of empirical strong motion attenuation relations to include the amplitude and duration effects of rupture directivity. *Seismol. Res. Lett.* 68, 199–222.

Sozen, M. A., P. C. Jennings, R. B. Matthiesen, G. W. Housner, and N. M. Newmark (1968). Engineering Report on the Caracas Earthquake of 29 July, 1967. National Academy of Sciences, Washington, D.C.

Thomson, W. T. (1965). *Vibration Theory and Applications*. Prentice-Hall, Englewood Cliffs, New Jersey.

Timoshenko, S. (1955). *Strength of Materials, Part I, Elementary Theory and Problems*, 3rd ed. D. Van Nostrand Company, Toronto.

Trifunac, M. D. (1970). Ambient Vibration Test of a Thirty-Nine Story Steel Frame Building, *Earthquake Engineering Research Laboratory Report 70-02*. California Institute of Technology, Pasadena, California.

Trifunac, M. D. (1972). Comparison between ambient and forced vibration experiments. *Earthq. Eng. Struct. Dyn.* **1**(2), 133–150.

Veletsos, A. S., and T. Y. Wei (1971). Lateral and rocking vibrations of footings. *J. Soil Mech. Foundation Div., Am. Soc. Civil Eng.* **97**(SM9), 1227–1248.

Venkatraman, B., and S. A. Patel (1970). *Structural Mechanics with Introduction to Elasticity and Plasticity*. McGraw-Hill, New York.

Ward, H. S., and R. Crawford (1966). Wind induced vibrations and building modes. *Bull. Seismol. Soc. Am.* **56**, 793–813.

Wilson, E. L. (1998). *Three Dimensional Static and Dynamic Analysis of Structures*. Computers and Structures, Inc., Berkeley, California.

68

Seismic Design Provisions and Guidelines in the United States: A Prologue

Roger D. Borcherdt
US Geological Survey, Menlo Park, CA, USA
Ronald O. Hamburger
EQE International, Oakland, CA, USA
Charles A. Kircher
Kircher & Associate, Palo Alto, CA, USA

1. Introduction

Seismic design provisions and guidelines are the basis for reduction of potentially devastating losses of life and property from earthquakes. Six tragic earthquakes since 1985, affecting Mexico, Armenia, the United States, Japan, Turkey, and Taiwan, caused combined property losses exceeding $320 billion and loss of lives exceeding 143,900. These losses emphasize the need to improve the earthquake resistance of the built environment in zones of high seismic risk. With present population exceeding 290 million in major earthquake zones and world population expected to increase by 2 to 4 billion people in the next 50 years, losses from future earthquakes can be expected to reach even greater levels if worldwide improvements in earthquake resistance are not made.

The last decade of the twentieth century saw an unprecedented series of improvements in seismic design provisions and guidelines developed and implemented in the United States. These improvements, stimulated by important lessons learned from recent earthquakes, are based on recent evaluations of seismic hazard, advances in technology, and new concepts involving performance-based design. They provide a new set of standards for earthquake-resistant design, construction, and retrofit for application in regions with seismic hazard levels ranging from high to very low.

Improvements in the provisions and guidelines for new buildings in the United States are manifest in the most recent versions of the National Earthquake Hazard Reduction Program provisions (1997 NEHRP) and the Uniform Building Code provisions (1997 UBC). Consensus concerning the improvements has indicated that these documents serve as the basis for the new 2000 International Building Code (IBC) provisions. The consolidation of these standards represents a significant milestone toward development of a uniform set of provisions for earthquake-resistant design and construction of new buildings.

Improving the earthquake resistance of existing buildings is a major obstacle to the reduction of future earthquake losses worldwide. Recent efforts in the United States have also led to the development and implementation of new standards and guidelines for existing buildings based on performance-based design concepts applicable to most building types in regions with different seismic hazard levels. Ongoing implementation of these provisions together with those for concrete frame and wall construction is expected to have a significant impact on losses to existing buildings from future earthquakes in the United States.

Vulnerabilities of bridges and overhead transportation structures demonstrated by recent earthquakes have stimulated the development of an extensive new set of design and retrofit criteria and guidelines. These provisions, as developed by one of the largest state transportation departments in the United States, represent a major step forward in improving earthquake safety for transportation facilities.

The basis for these new provisions as recommended in the United States is the theme of a special issue published by the Earthquake Engineering Research Institute (Earthquake Spectra, Vol. 16, No. 1, February 2000), entitled "Seismic Design

INTERNATIONAL HANDBOOK OF EARTHQUAKE AND ENGINEERING SEISMOLOGY, VOLUME 81B *ISBN: 0-12-440658-0*

Provisions and Guidelines." The Theme Issue is reproduced in its entirety on the attached Handbook CD #2. Abstracts of each manuscript are included at the end of this Prologue to facilitate access.

The Preface of the Theme Issue (Hamburger and Kircher) provides a historical review of efforts leading to the present status of provisions for new and existing buildings and overhead transportation structures in the United States. This insightful review provides an overview and identifies references critical to the development of the most recent provisions.

2. Ground Motion Estimation

The Theme Issue provides the basis for the estimation of earthquake ground shaking as presently adopted in the 1997 NEHRP and the 2000 IBC provisions. It describes the scientific basis for the new national seismic hazard maps developed for the United States (Frankel *et al.*, 1996). It provides the procedures used to incorporate the seismic hazard maps into the new national "Design Maps" or "*Maximum Considered Earthquake Maps (MCE)*," as the maps adopted in the 1997 NEHRP and the 2000 IBC provisions depicting ground motion parameters for sites on "firm to hard rock" (Leyendecker *et al.*). The basis for the new site factors (F_a and F_v) and the site classes (A–E) as presently adopted in each set of provisions is described in detail as the justification for extending MCE ground motion estimates to sites underlain by soil (Dobry *et al.*). In addition, the Theme Issue provides the background for the new near-source factors (N_a and N_v) developed and adopted as an interim solution for the 1997 UBC estimates of ground shaking derived from the earlier seismic zone maps (Peterson *et al.*).

3. Provisions for New Buildings and Other Structures

The Theme Issue provides the background for the 1997 NEHRP and the 1997 UBC provisions for new buildings and other new structures. Features of the 1997 NEHRP provisions as they evolved from tentative provisions published in 1978 are described (Holmes). The basis for the many changes to model code force provisions for elements of structures and nonstructural components as adopted in both sets of provisions are provided in detail (Drake and Bragagnolo). Background for the present provisions for other types of structures, including earth-retaining structures, tanks, and water-treatment structures is discussed (Sprague and Legatos). The basis for major changes in concrete-related provisions as adopted in both sets of provisions and further modified for incorporation into the 2000 IBC are presented (Ghosh). The background for new provisions as being adopted in the 2000 IBC are presented for the seismic design of composite and hybrid concrete–steel structures (Deierlein). The guidelines and provisions for steel-framed buildings as developed following the Northridge earthquake—interim guidelines (FEMA 267), updated (FEMA 267a), incorporated into the 1997 American Steel Institute of Steel Construction, guidelines, and being adopted in part into the 2000 IBC provisions—are discussed (Malley *et al.*). Special design requirements for regions of moderate seismicity are discussed in detail (Nordenson and Bell).

4. Provisions for Existing Structures

The Theme Issue provides the background for recent up-to-date provisions for seismic rehabilitation of existing buildings. It provides the basis for the *NEHRP Guidelines and Commentary for the Seismic Rehabilitation of Buildings* developed on the basis of new performance-based engineering design concepts for all building types (Shapiro *et al.*). As an additional supplement the Theme Issue includes an overview of a thorough set of recommendations concerning the "*Seismic Evaluation and Retrofit of Concrete Buildings*" developed by the state of California (Comartin *et al.*). It includes a summary of a performance-based methodology to evaluate, quantify, and repair earthquake damage to concrete and masonry wall buildings (Maffei *et al.*).

5. Seismic Design Criteria for Bridges

The Theme Issue provides the background for design criteria for bridges as it has been developed in the last two decades by the California Department of Transportation. These criteria are based on the earthquake performance of bridges as observed during several damaging California earthquakes of the last two decades. The overview provides a summary of the most recent demand/capacity methodology developed to improve the earthquake performance of bridges.

6. Abstracts

Abstracts for the Theme Issue, "Seismic Design Provisions and Guidelines" (Hamburger and Kircher, Eds., 2000, *Earthquake Spectra*, vol. 16, no. 1, 316 pp.) are provided here to assist the reader. The Theme Issue is reproduced in its entirety on the attached Handbook CD #2, under the directory of \68Borcherdt.

6.1 Ground Motion Estimation

6.1.1 USGS National Seismic Hazard Maps

by A. D. Frankel, C. S. Mueller, T. P. Barnhard, E. V. Leyendecker, R. L. Wesson, S. C. Harmsen, F. W. Klein, D. M. Perkins, N. C. Dickman, S. L. Hanson, and M. G. Hopper

The U.S. Geological Survey (USGS) recently completed new probabilistic seismic hazard maps for the United States, including Alaska and Hawaii. These hazard maps form the basis of the probabilistic component of the design maps used in the 1997 edition of the NEHRP Recommended Provisions for Seismic Regulations for New Buildings and Other Structures, prepared by the Building Seismic Safety Council and published by FEMA. The hazard maps depict peak horizontal ground acceleration and spectral response at 0.2, 0.3, and 1.0 sec periods, with 10%, 5%, and 2% probabilities of exceedance in 50 years, corresponding to return times of about 500, 1000, and 2500 years, respectively. In this paper we outline the methodology used to construct the hazard maps. There are three basic components to the maps. First, we use spatially smoothed historic seismicity as one portion of the hazard calculation. In this model, we apply the general observation that moderate and large earthquakes tend to occur near areas of previous small or moderate events, with some notable exceptions. Second, we consider large background source zones based on broad geologic criteria to quantify hazard in areas with little or no historic seismicity, but with the potential for generating large events. Third, we include the hazard from specific fault sources. We use about 450 faults in the western United States (WUS) and derive recurrence times from either geologic slip rates or the dating of prehistoric earthquakes from trenching of faults or other paleoseismic methods. Recurrence estimates for large earthquakes in New Madrid and Charleston, South Carolina, were taken from recent paleoliquefaction studies. We used logic trees to incorporate different seismicity models, fault recurrence models, Cascadia great earthquake scenarios, and ground-motion attenuation relations. We present disaggregation plots showing the contribution to hazard at four cities from potential earthquakes with various magnitudes and distances.

6.1.2 Development of Maximum Considered Earthquake Ground Motion Maps

by Edgar V Leyendecker, R. Joe Hunt, Arthur D. Frankel, and Kenneth S. Rukstales

The 1997 NEHRP Recommended Provisions for Seismic Regulations for New Buildings use a design procedure that is based on spectral response acceleration rather than the traditional peak ground acceleration, peak ground velocity, or zone factors. The spectral response accelerations are obtained from maps prepared following the recommendations of the Building Seismic Safety Council's (BSSC) Seismic Design Procedures Group (SDPG). The SDPG-recommended maps, the Maximum Considered Earthquake (MCE) Ground Motion Maps, are based on the US Geological Survey (USGS) probabilistic hazard maps with additional modifications incorporating deterministic ground motions in selected areas and the application of engineering judgment. The MCE ground motion maps included with the 1997 NEHRP Provisions also serve as the basis for the ground motion maps used in the seismic design portions of the 2000 International Building Code and the 2000 International Residential Code. Additionally, the design maps prepared for the 1997 NEHRP Provisions, combined with selected USGS probabilistic maps, are used with the 1997 NEHRP Guidelines for the Seismic Rehabilitation of Buildings.

6.1.3 New Site Coefficients and Site Classification System Used in Recent Building Seismic Code Provisions

by R. Dobry, R. D. Borcherdt, C. B. Crouse, I. M. Idriss, W. B. Joyner, G. R. Martin, A S. Power, E. E. Rinne, and R. B. Seed

Recent code provisions for buildings and other structures (1994 and 1997 NEHRP Provisions, 1997 UBC) have adopted new site amplification factors and a new procedure for site classification. Two amplitude-dependent site amplification factors are specified: F_a for short periods and F_v for longer periods. Previous codes included only a long-period factor S and did not provide for a short-period amplification factor. The new site classification system is based on definitions of five site classes in terms of a representative average shear wave velocity to a depth of 30 m ($\overline{V}_s$). This definition permits sites to be classified unambiguously. When the shear wave velocity is not available, other soil properties such as standard penetration resistance or undrained shear strength can be used. The new site classes, denoted by letters A–E, replace site classes in previous codes denoted by S1–S4. Site classes A and B correspond to hard rock and rock, Site Class C corresponds to soft rock and very stiff/very dense soil, and Site Classes D and E correspond to stiff soil and soft soil. A sixth site class, F, is defined for soils requiring site-specific evaluations. Both F_a and F_v are functions of the site class, and also of the level of seismic hazard on rock, defined by parameters such as A_a and A_v (1994 NEHRP Provisions), S_s and S_1 (1997 NEHRP Provisions), or Z (1997 UBC). The values of F_a and F_v decrease as the seismic hazard on rock increases due to soil nonlinearity. The greatest impact of the new factors F_a and F_v as compared with the old S factors occurs in areas of low-to-medium seismic hazard. This paper summarizes the new site provisions, explains the basis for them, and discusses ongoing studies of site amplification in recent earthquakes that may influence future code developments.

6.1.4 Active Fault Near-Source Zones Within and Bordering the State of California for the 1997 Uniform Building Code

by Mark D. Petersen, Tousson R. Toppozada, Tianqing Cao, Chris H. Cramer, Michael S. Reichle, and William A. Bryant

The fault sources in the Project 97 probabilistic seismic hazard maps for the state of California were used to construct maps for defining near-source seismic coefficients, N_a and N_v, incorporated in the 1997 Uniform Building Code (ICBO, 1997). The near-source factors are based on the distance from a known active fault that is classified as either Type A or Type B. To determine the near-source factor, four pieces of geologic information are required: (1) recognizing a fault and determining whether or not the fault has been active during the Holocene, (2) identifying the location of the fault at or beneath the ground surface, (3) estimating the slip rate of the fault, and (4) estimating the maximum earthquake magnitude for each fault segment. This paper describes the information used to produce the fault classifications and distances.

6.2 Provisions for New Buildings and Other Structures

6.2.1 The Seismic Provisions of the 1997 Uniform Building Code

by Robert E. Bachman and David R. Bonneville

Currently the most widely accepted code regulations in the United States for seismic design of structures and nonstructural components are those found in the Uniform Building Code (UBC). The UBC seismic requirements were significantly revised in the 1997 edition. Among the issues addressed in the UBC revisions are near-source effects and ground-acceleration-dependent soil site amplification factors for both short- and long-period structures. Also, the design force levels in the 1997 UBC are based on strength design rather than the previously used allowable stress design. Other significant changes include introduction of a redundancy/reliability factor, a more realistic consideration of story drift and deformation compatibility, and new equations for equivalent static forces for both structural and nonstructural components. This paper traces the recent history of the code development and describes the major elements of the 1997 UBC seismic provisions.

6.2.2 The 1997 NEHRP Recommended Provisions for Seismic Regulations for New Buildings and Other Structures

by William T. Holmes

This paper follows the evolution of the NEHRP Recommended Provisions for Seismic Regulations for New Buildings and Other Structures from the development of ATC 3-06 to the 1997 edition of the document. The features of the 1997 NEHRP Provisions are described in detail. Complementary information about the NEHRP Provisions is found in several other papers in this volume of Earthquake Spectra. Subject areas covered elsewhere are specifically referenced in this paper.

6.2.3 Model Code Design Force Provisions for Elements of Structures and Nonstructural Components

by Richard M. Drake and Leo J. Bragagnolo

With the publication of the 1997 Uniform Building Code (UBC) and the 1997 NEHRP Recommended Provisions for the Seismic Regulations for New Buildings and Other Structures, there has been a significant change in the earthquake design force provisions for buildings, structures, elements of structures, and nonstructural components. Engineers and architects need to become informed regarding a variety of earthquake design force provisions, primarily those published in the UBC and those developed as part of the NEHRP Provisions. Both sources provide design force provisions for the building structural system and separate design force provisions for elements of structures and nonstructural components. This paper describes the development, evolution, and application of the earthquake design force provisions for elements of structures and nonstructural components.

6.2.4 Nonbuilding Structures Seismic Design Code Developments

by Harold O. Sprague and Nicholas A. Legatos

The building code development process has traditionally given little effort to developing the seismic design process of nonbuilding structures. This has created some unique problems and challenges for the structural engineers who design these types of structures. The intended seismic performance requirements for "building" design are based on life safety and collapse prevention. Structural elements in buildings are allowed to yield as a method of seismic energy dissipation. The seismic performance of nonbuilding structures varies depending on the specific type of nonbuilding structure. Nonlinear behavior in some nonbuilding structures is unacceptable, while other nonbuilding structures may be allowed to yield during an earthquake. Nonbuilding structures comprise a vast myriad of structures constructed of all types of materials, with markedly different dynamic characteristics, and with a wide range of performance requirements. This paper discusses the development of codes, design practices, and future of the seismic design criteria for nonbuilding structures.

6.2.5 Major Changes in Concrete-Related Provisions—1997 UBC and Beyond

by S. K. Ghosh

US seismic codes are undergoing profound changes as of this writing. Changes from the 1994 to the 1997 edition of the Uniform Building Code (UBC) (ICBO, 1994, 1997) are many and far-reaching in their impact. The 1997 edition of the National Earthquake Hazards Reduction Program (NEHRP) Recommended Provisions for Seismic Regulations for New Buildings (BSSC, 1998) contains further evolutionary changes in seismic design requirements beyond those of the 1997 UBC. The latter document forms the basis of the seismic design provisions of the first edition of the International Building Code (IBC), published in the spring of 2000. This paper first discusses the major changes that have been made in the concrete-related provisions from the 1994 to the 1997 edition of the UBC. The paper gives background to these changes, provides essential details on them, and indicates how they have been incorporated (at times with significant modifications) into the 1997 NEHRP Provisions and the 2000 IBC. The ACI 318-99, Building Code Requirements for Structural Concrete (ACI, 1999), is adopted by reference into the 2000 IBC This entails further changes in concrete-related provisions beyond the 1997 UBC. Some of the more important of these changes are discussed here. A small number of amendments and additions to the ACI 318-99 provisions are included in the 2000 IBC. The more important of these are also outlined in this paper.

6.2.6 New Provisions for the Seismic Design of Composite and Hybrid Structures

by Gregory G. Deierlein

While there have been significant advances in the design and construction of composite steel–concrete building structures, their use in regions of high seismicity has been hindered by the lack of design criteria in building codes and specifications. This has prompted initiatives in the Building Seismic Safety Council and the American Institute of Steel Construction to develop seismic design provisions for composite structures. The 1997 edition of the AISC Seismic Provisions includes a new section with requirements for composite steel–concrete structures that is cross-referenced by the general seismic loading and design criteria in the 1997 NEHRP Provisions and the 2000 International Building Code. Intended to complement existing provisions for steel, reinforced concrete, and composite structures in the AISC-LRFD Specification and the ACI 318 Building Code, these new provisions provide an important resource for seismic design of composite structural systems, members, and connections.

6.2.7 Seismic Design Guidelines and Provisions for Steel-Framed Buildings: FEMA 267/267A and 1997 AISC Seismic Provisions

by James O. Malley, Charles J. Carter, and C. Mark Saunders

One of the important surprises of the Northridge earthquake of January 17, 1994, was the widespread and unanticipated brittle fracture of welded steel beam-to-column connections. Although no casualties or collapses occurred during the Northridge earthquake as a result of these connection failures, and many WSMF buildings were not damaged at all, a wide spectrum of brittle connection damage did occur, ranging from minor cracking to completely severed columns. This paper summarizes two of the most important documents that have been developed in response to the damage suffered to steel moment frame buildings in the Northridge earthquake. The first, FEMA 267, Interim Guidelines: Evaluation, Repair, Modification and Design of Welded Steel Moment Frame Structures, was generated from studies undertaken as part of a project initiated by the US Federal Emergency Management Agency (FEMA) to reduce the earthquake hazards posed by steel moment-resisting frame buildings. The second document addressed in this paper is the 1997 edition of the American Institute of Steel Construction (AISC) Seismic Provisions for Structural Steel Buildings (commonly referred to as the AISC Seismic Provisions), which incorporates the new information generated by the FEMA-sponsored project and other investigations on the seismic performance of steel structures, and has been adopted by reference into the 2000 International Building Code (IBC).

6.2.8 Seismic Design Requirements for Regions of Moderate Seismicity

by Guy J. P. Nordenson and Glenn R. Bell

The need for earthquake-resistant construction in areas of low-to-moderate seismicity has been recognized through the adoption of code requirements in the United States and other countries only in the past quarter century. This is largely a result of improved assessment of seismic hazard and examples of recent moderate earthquakes in regions of both moderate and high seismicity, including the San Fernando (1971), Mexico City (1985), Loma Prieta (1989), and Northridge (1994) earthquakes. In addition, improved understanding and estimates of older earthquakes in the eastern United States such as Cape Ann (1755), La Malbaie, Quebec (1925), and Ossippe, New Hampshire (1940), as well as monitoring of micro-activity in source areas such as La Malbaie, have increased awareness of the earthquake potential in areas of low-to-moderate seismicity. Both the hazard and the risk in moderate seismic zones (MSZs) differ in scale and kind from those of the zones of high seismicity. Earthquake hazards mitigation measures for new and existing construction need to be adapted from those prevailing in regions of high seismicity in recognition of these differences. Site effects are likely to dominate the damage patterns from earthquakes, with some sites suffering no damage not far from others, on soft soil, suffering near collapse. A number of new seismic codes have been developed in the past quarter century in response to these differences, including the New York City (1995) and the Massachusetts State (1975) seismic codes. Over the same period, the national model building codes that apply to most areas of low-to-moderate seismicity in the United States, the Building Officials and Code Administrators (BOCA) Code and the Southern Standard Building Code (SSBC), have incorporated up-to-date seismic provisions. The seismic provisions of these codes have been largely inspired by the National Earthquake Hazard Reduction Program (NEHRP) recommendations. Through adoption of these national codes, many state and local authorities in areas of low-to-moderate seismicity now have reasonably comprehensive seismic design provisions. This paper will review the background and history leading up to the MSZ codes, discuss their content, and propose directions for future development.

6.2.9 NEHRP Guidelines and Commentary for the Seismic Rehabilitation of Buildings

by Daniel Shapiro, Christopher Rojahn, Lawrence D. Reaveley, James R. Smith, and Ugo Morelli

Based on the conclusion that the primary barrier to widespread seismic rehabilitation of buildings in the United States was the lack of a consensus-backed, nationally applicable, professionally accepted rehabilitation standard, the Federal Emergency Management Agency supported the development of the NEHRP Guidelines and Commentary for the Seismic Rehabilitation of Buildings (FEMA 273 and 274). A six-year effort by a team of experienced professional practitioners and university researchers who were motivated to produce a standard that specifically addressed the differences in designing for seismic resistance in new buildings, as opposed to existing buildings, resulted in the NEHRP Guidelines and Commentary for the Seismic Rehabilitation of Buildings. These NEHRP Guidelines will provide the tools for design professionals of varying expertise in seismic design to design economical and appropriate seismic rehabilitation for buildings of essentially any size, commonly used building material, and configuration.

6.3 Provisions for Existing Structures

6.3.1 Seismic Evaluation and Retrofit of Concrete Buildings: A Practical Overview of the ATC 40 Document

by Craig D. Comartin, Richard W. Niewiarowski, Sigmund A. Freeman, and Fred M. Turner

The Applied Technology Council (ATC), with funding from the California Seismic Safety Commission, developed the document, Seismic Evaluation and Retrofit of Concrete Buildings, commonly referred to as ATC 40. This two-volume, 612-page report provides a recommended procedure for the seismic evaluation and retrofit of concrete buildings. Although the focus is specifically on concrete buildings, the document provides information on emerging techniques applicable to most building types. This paper provides an introduction and overview of that document. The conceptual basis of the procedures is performance-based design using nonlinear static structural analysis. The ATC 40 document comprises a practical guide to the entire evaluation and retrofit process. Topics include performance objectives, seismic hazard, determination of deficiencies, retrofit strategies, quality assurance procedures, nonlinear static analysis using the capacity spectrum method, modeling recommendations, foundation effects, and response limits.

6.3.2 Evaluation of Earthquake-Damaged Concrete and Masonry Wall Buildings

by Joe Maffei, Craig D. Comartin, Brian Kehoe, Gregory R. Kingsley, and Bret Lizundia

Efforts at improving earthquake recovery policies have been hampered by a lack of criteria and standards for evaluating and repairing damaged buildings. The Applied Technology Council has developed a performance-based methodology for the evaluation of earthquake-damaged concrete wall buildings and masonry wall buildings, recently published as FEMA 306/307/308. The methodology provides a way to quantify damage in terms of loss of seismic performance capability. It also provides guidelines for remedial measures to restore or improve seismic performance capability. In this methodology, the expected future seismic performance of a building is evaluated in its pre-event, damaged, and repaired conditions. Following the nonlinear static analysis procedure, displacement demands and capacities of the structure are used as indices of seismic performance. Identifying the governing mechanism of nonlinear deformation and the behavior mode of a structure and its components is shown to be a necessary first step toward evaluating expected seismic performance, interpreting indications of damage, and assessing their significance. The methodology provides a technical resource for understanding how buildings respond seismically on both global and component levels, and gives a basis for formulating post-earthquake policies.

6.4 Seismic Design Criteria for Bridges

6.4.1 Caltrans' New Seismic Design Criteria for Bridges

by Mark Yashinsky and Thomas Ostrom

Caltrans' Seismic Design Criteria (SDC) has been adopted as the minimum seismic standard for ordinary bridges on California's highways. The SDC is a compilation of new and existing seismic criteria that had been previously documented in a variety of Caltrans documents. The SDC extends the capacity design philosophy introduced in the 1980 Caltrans Bridge Design Specifications. The most significant departure from the previous procedure is that ductile members are now designed by comparing the displacement demand to the displacement capacity. The demands are generated by a linear elastic analysis, and the capacities are determined from a curvature analysis that incorporates the nonlinear behavior of the structural elements. The demand/capacity methodology supplants the previous method based on reducing the elastic dynamic forces by a force reduction factor. In this paper, the significant features of Caltrans' SDC are described.

Acknowledgments

A complete copy of the Theme Issue is provided on the attached Handbook CD #2 in PDF format with the permission of the Earthquake Engineering Research Institute. The permission for reproduction of the Theme Issue is gratefully acknowledged.

References

Hamburger, R. O., and C. A. Kircher, eds., (2000). Seismic Design Provisions and Guidelines, *Earthquake Spectra* **16**(1), 316 pp.

69

Finite Element Analysis in Earthquake Engineering

John F. Hall
California Institute of Technology, Pasadena, California, USA

1. Introduction

The finite element method is a powerful numerical analysis technique that has been widely applied in earthquake engineering for modeling the response of structures. The method derives its power from the variety of elements, such as beams, shells, and springs, that can be combined together to represent complex systems. Example applications include buildings, bridges, earth and concrete dams, off-shore towers, pipelines, and tanks. Often, interaction between the structure and its supporting foundation and soil or even a fluid must be considered, and these complete systems can also be modeled with finite elements.

Engineers routinely use in-house or commercially available software for computing natural frequencies and mode shapes, needed in linear response spectrum analysis, and for performing time history integration of the full matrix equations of motion of linear structures. Some recent design procedures call for nonlinear static "push over" analysis (Applied Technology Council, 1996), and finite element programs for such computations are now finding significant use. On the other hand, nonlinear time history analysis is only slowly being accepted in practice. Reasons are numerous and include longer run times, difficulty of defining valid constitutive relations for materials under cyclic loading, increased number of material parameters to be selected, greater need to investigate sensitivity of results to choice of material parameters, possible problems encountered in running available software, such as lack of convergence, and poor understanding by some potential users of the theories involved. Although nonlinear time history analysis of important base-isolated structures is not uncommon, in these cases the nonlinearity is often confined to the isolators and is relatively well known and well behaved. A notable exception to the limited use of nonlinear time history analysis has been the work carried out under the program to retrofit seismically vulnerable bridges in California. This effort by the California Department of Transportation in association with private engineers and university researchers has been quite sophisticated and surely has advanced the state of the art (see Section 9).

This chapter treats finite element analysis at the level of nonlinear time history computations. Linear time history analysis and nonlinear static pushover can be viewed as special cases. Response spectrum techniques are not covered. Nor are derivations of the matrices and vectors of individual element types included, with the exception of the planar beam, which serves as an example. For these other topics, the reader is referred to finite element texts (Bathe, 1982; Cook *et al.*, 1989).

The following three sections of this chapter present general theory: equations of motion and their solution in Section 2, earthquake loading and superstructure/substructure partitioning in Section 3, and fluid interaction in Section 4. The next three sections contain background material for an example building analysis in Section 8. In Section 5, the planar beam element is derived. Planar frame modeling is the subject of Section 6, and the use of planar frames to model complete buildings is discussed in Section 7. Finally, to demonstrate a state-of-the-art nonlinear time history analysis, some details of the retrofit design effort for the San Diego–Coronado Bay Bridge are presented in Section 9.

2. Equations of Motion

The finite element method is the most widely used procedure in earthquake engineering for analysis of a structure. It is a discretization technique whereby the structure is divided into pieces, or elements, within which the solution is interpolated from nodal degrees of freedom. Solution of a matrix equation yields values for the degrees of freedom as functions of time. Types of elements include beam, plate, shell, plane stress and

INTERNATIONAL HANDBOOK OF EARTHQUAKE AND ENGINEERING SEISMOLOGY, VOLUME 81B ISBN: 0-12-440658-0

strain, three-dimensional bricks and other shapes, as well as axial and rotational springs. A beam element is discussed in a subsequent section of this chapter.

Application of the finite element method to the linear equations of elasticity results in the matrix equation of motion (Bathe, 1982), as follows:

$$[M]\{\ddot{a}(t)\} + [C]\{\dot{a}(t)\} + [K]\{a(t)\} = \{F(t)\}, \tag{1}$$

where t denotes time; the dot indicates time differentiation; $[M]$, $[C]$, and $[K]$ are symmetric mass, damping, and stiffness matrices, respectively; and $\{F(t)\}$ is a vector of forces applied at the nodes. Equation (1) contains individual equations of motion corresponding to each degree of freedom $a_i(t)$. These are statements that the externally applied force $F_i(t)$ equals the internal force applied by the elements to degree of freedom $a_i(t)$ resulting from inertia, damping, and stiffness of the structure. $[M]$ and $[K]$ are based on actual mass and stiffness properties, whereas $[C]$ is usually constructed on an *ad hoc* basis to produce desired amounts of damping in important modes of the system. For an earthquake analysis, $\{F(t)\}$ consists of static gravity loads $\{F^G\}$ and time-dependent loads $\{F^E(t)\}$ derivable from the earthquake ground motions.

Construction of $[M]$ and $[K]$ involves assembly from element mass and stiffness matrices. This means that the connectivity among the degrees of freedom is localized, and as a result, nodes can often be numbered so that all nonzero terms in $[M]$ and $[K]$ fall into a relatively narrow profile about the diagonal. A narrow profile is important for an efficient solution process. Furthermore, $[M]$ can usually be well approximated as a diagonal matrix, which corresponds physically to the mass being lumped at the nodes.

A common choice for $[C]$ is Rayleigh damping.

$$[C] = \frac{2\xi}{\omega_a + \omega_b}[\omega_a\omega_b M + K], \tag{2}$$

where the frequencies ω_a and ω_b (rad/sec) are chosen to cover the frequency range of the response as well as possible so that the important modes of the structure have fractions of critical damping close to the desired value ξ (Fig. 1). The $[K]$-proportional part of $[C]$ generates damping to resist velocities of the nodes relative to each other, while the $[M]$-proportional part resists total velocities. Physically, the latter term is unrealistic, but its inclusion allows wider control over the modal damping values. Other formulas for $[C]$ offer even more control, but they result in $[C]$ being a full matrix (Clough and Penzien, 1993), which is a computational disadvantage. With Rayleigh damping, the envelope of the profiles of the nonzero terms in $[M]$ and $[K]$ is imparted to $[C]$.

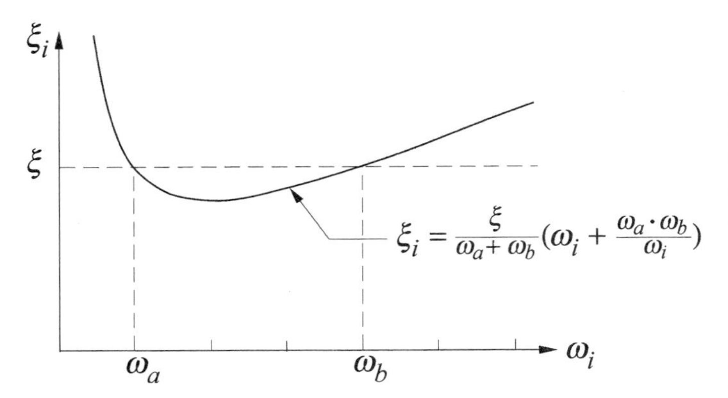

FIGURE 1 Actual damping fraction ξ_i for mode i as a function of frequency ω_i of mode i when $[C]$ is given by Eq. (2).

If Eq. (1) is being solved through a transformation to modal coordinates, then damping values can be assigned directly to each modal equation, and a damping matrix $[C]$ is not needed. However, the focus of this chapter is on nonlinear behavior for which modal analysis cannot be applied.

In earthquake response, nonlinearity in the stiffness forces can occur for even moderate ground motions, and it can be pronounced under strong shaking. Nonlinear stiffness forces are not expressible by the product $[K]\{a(t)\}$, so the equations of motion are written instead as (Bathe, 1982)

$$[M]\{\ddot{a}(t)\} + [C]\{\dot{a}(t)\} + \{R(t)\} = \{F(t)\}, \tag{3}$$

where $\{R(t)\}$ is the vector of stiffness forces. Computation of $\{R(t)\}$ may have to incorporate material nonlinearities such as yielding and cracking as well as geometric nonlinearities, which include large rotational motions and effect of stress on stiffness.

Solution of Eq. (1) or Eq. (3) involves time integration to generate the response at discrete times at interval Δt, i.e., at Δt, $2\Delta t$, etc. Initial conditions are required at time 0, usually the structure carrying gravity loads in the at-rest state. A procedure for this static solution is presented next, followed by a discussion of two popular time integration schemes: central difference and constant average acceleration.

In the linear case, the static displacements from gravity loads are computed from a solution of the equilibrium equations, as follows:

$$[K]\{a(0)\} = \{F^G\}. \tag{4}$$

In the presence of stiffness nonlinearity, equilibrium is expressed as

$$\{R(0)\} = \{F^G\}, \tag{5}$$

which must be solved by an iterative process involving linearizations about successive approximations to the solution. For iteration k, based on current approximate displacements $\{a^k(0)\}$ and corresponding stiffness forces $\{R^k(0)\}$, the linearization statement is

$$\{R(0)\} = \{R^k(0)\} + \left[K_T^k(0)\right]\{\Delta a^k\}, \tag{6}$$

where $[K_T^k(0)]$ is a tangent stiffness matrix that relates infinitesimal increments of $\{R^k(0)\}$ and $\{a^k(0)\}$. Substitution of Eq. (6) into Eq. (5) results in

$$\left[K_T^k(0)\right]\{\Delta a^k\} = \{F^G\} - \{R^k(0)\} \tag{7}$$

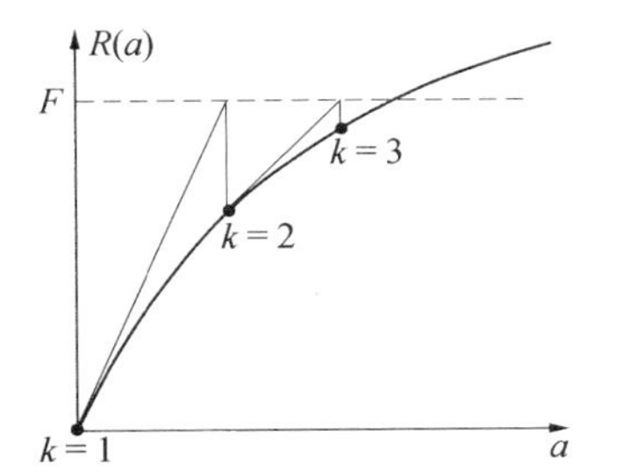

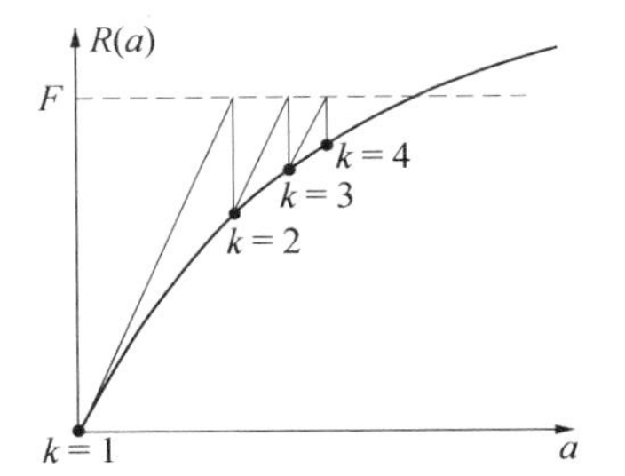

FIGURE 2 Iterative solution to $R(a) = F$ for a softening system. Comparison between using the current tangent (left) and the same initial tangent for every iteration (right).

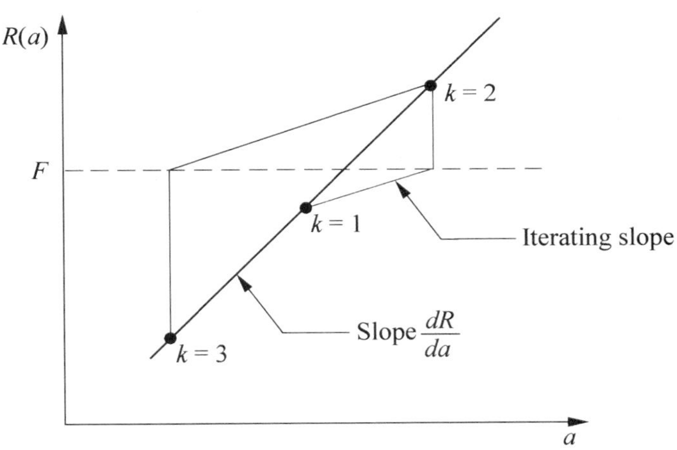

FIGURE 3 Iterative solution to $R(a) = F$ using an iterating slope less than current tangent dR/da. Convergence fails if the iterating slope is less than $\frac{1}{2}\, dR/da$.

from which the displacement increments $\{\Delta a^k\}$ are computed. The updated displacements are

$$\{a^{k+1}(0)\} = \{a^k(0)\} + \{\Delta a^k\}, \tag{8}$$

and the corresponding stiffness forces $\{R^{k+1}(0)\}$ are found from $\{R^k(0)\}$ by following the subsequent actual nonlinear behavior through the increment. The tangent matrix $[K_T^k(0)]$, having been assembled from element tangent matrices, has the same profile as the linear stiffness matrix $[K]$. Often, the tangent matrix is symmetric; exceptions include cases of frictional sliding and nonassociated plasticity.

Equations (7) and (8) define the iteration process, and iterations continue until convergence. The first iteration produces $\{a^2(0)\}$ and $\{R^2(0)\}$, and $k = 1$ in Eqs. (7) and (8). $\{a^1(0)\}$ and $\{R^1(0)\}$ are zero vectors representing the initial unloaded state; $[K_T^1(0)]$ is the initial tangent (elastic) stiffness matrix. Convergence occurs when norms of $\{\Delta a^j\}$ and $\{F^G - R^{j+1}(0)\}$ after some iteration j fall below specified tolerances.

Several variations of the preceding iterative process can improve computational performance and convergence. One of these is load stepping, whereby the gravity loads are applied in steps with iteration to convergence within each step before the next one is applied. Another is the use of the same iterating matrix, say $[K_T^\ell(0)]$ from iteration ℓ, over several iterations or even load steps. This reduces the number of matrix factorizations required in the solution of Eq. (5), but often at the expense of more iterations (Fig. 2). Because the static solution to gravity loads seldom involves significant nonlinearity, the use of the initial elastic stiffness matrix for all iterations may be effective. However, if this matrix ends up representing a system softer than the current state, which can happen for cable structures that stiffen under gravity load, the process may fail to converge (Fig. 3).

Time integration by the central difference method is based on the following time-stepping formulas:

$$\{\ddot{a}(t)\} = \frac{1}{(\Delta t)^2}\{a(t+\Delta t) - 2a(t) + a(t-\Delta t)\} \tag{9a}$$

$$\{\dot{a}(t)\} = \frac{1}{2\cdot\Delta t}\{a(t+\Delta t) - a(t-\Delta t)\}. \tag{9b}$$

Substitution into Eq. (3) yields

$$\left[\frac{1}{(\Delta t)^2}M + \frac{1}{2\cdot\Delta t}C\right]\{a(t+\Delta t)\} = \{F(t)\} - \{R(t)\} + \frac{2}{(\Delta t)^2}[M]\{a(t)\} + \left[-\frac{1}{(\Delta t)^2}M + \frac{1}{2\cdot\Delta t}C\right]\{a(t-\Delta t)\}, \tag{10}$$

which is solved once per time step. Thus, the solution can be advanced a time step while only having to know the stiffness forces at the current time. This is explicit time integration and also the reason iteration is not required for nonlinear stiffness.

For stable integration with central difference, the time step Δt must be less than a critical value Δt_{cr}, which is proportional to the shortest modal period T_{min} in the discretization (Bathe, 1982). Because T_{min} is usually very small, a practical, but not sufficient, requirement for employing the central difference method is that only a diagonal matrix be present on the left side of Eq. (10). Although $[M]$ can usually be made diagonal, this condition for the damping matrix is more severe. One remedy is to use the less accurate backward difference formula

$$\{\dot{a}(t)\} = \frac{1}{\Delta t}\{a(t) - a(t-\Delta t)\} \tag{11}$$

for the velocity, which removes $[C]$ from the left side altogether. T_{min} can even be zero, for example, if rotational degrees of freedom are present with no assigned mass, as is common in analysis of frames comprised of beam elements. If T_{min} is zero, the central difference method cannot be applied. And even if T_{min} is not zero, it can often be so small as to render central difference ineffective, even with a diagonal left-side matrix.

Time integration by the constant average acceleration method is unconditionally stable, which means that Δt can be selected solely for accuracy considerations (Bathe, 1982). Generally, then, Δt can be larger than for the central difference method; in which case, fewer time steps are required. However, stiffness-based terms appear on the left side with no possibility of diagonalization, necessitating solution of simultaneous equations. Additionally, iterations are needed in nonlinear cases because the equations of motion are being satisfied implicitly at the future time $t + \Delta t$ when the stiffness forces are unknown, rather than at the current time t as for central difference.

Constant average acceleration, an implicit time integration scheme, uses the following time-stepping relations:

$$\{\ddot{a}(t+\Delta t)\} = \frac{4}{(\Delta t)^2}\{a(t+\Delta t) - a(t)\} - \frac{4}{\Delta t}\{\dot{a}(t)\} - \{\ddot{a}(t)\} \tag{12a}$$

$$\{\dot{a}(t+\Delta t\} = \{\dot{a}(t)\} + \frac{1}{2}\{\ddot{a}(t) + \ddot{a}(t+\Delta t)\}\Delta t. \tag{12b}$$

To define the kth iteration in the step from t to $t+\Delta t$, replace $a(t+\Delta t)$ in Eq. (12a) by $\{a^k(t+\Delta t)\} + \{\Delta a^k\}$, linearize the stiffness forces as

$$\{R(t+\Delta t)\} = \{R^k(t+\Delta t)\} + \left[K_T^k(t+\Delta t)\right]\{\Delta a^k\}, \tag{13}$$

and substitute Eqs. (12) and (13) into Eq. (3) written at time $(t+\Delta t)$ to obtain

$$\begin{aligned}\left[\frac{4}{(\Delta t)^2}M + \frac{2}{\Delta t}C + K_T^k(t+\Delta t)\right]\{\Delta a^k\} \\ = \{F(t+\Delta t)\} - \{R^k(t+\Delta t)\} \\ - \left[\frac{4}{(\Delta t)^2}M + \frac{2}{\Delta t}C\right]\{a^k(t+\Delta t)\} \\ + \left[\frac{4}{(\Delta t)^2}M + \frac{2}{\Delta t}C\right]\{a(t)\} \\ + \left[\frac{4}{\Delta t}M + C\right]\{\dot{a}(t)\} + [M]\{\ddot{a}(t)\}.\end{aligned} \tag{14}$$

The notation $^k(t+\Delta t)$ refers to the state reached in iteration $k-1$, except $^1(t+\Delta t)$ means (t). To carry out iteration k, Eq. (14) is solved for $\{\Delta a^k\}$; the new displacement approximation is found as

$$\{a^{k+1}(t+\Delta t)\} = \{a^k(t+\Delta t)\} + \{\Delta a^k\} \tag{15}$$

and the updated stiffness forces $\{R^{k+1}(t+\Delta t)\}$ are computed from $\{R^k(t+\Delta t)\}$ by following the actual nonlinear behavior through the increment. After convergence, $\{\ddot{a}(t+\Delta t)\}$ and $\{\dot{a}(t+\Delta t)\}$ are obtained from Eq. (12) using the last approximation to $\{a(t+\Delta t)\}$, and the next time step commences.

Because earthquake response involves load reversals, the tangent stiffness matrix $[K_T(t)]$ will not necessarily be known for path-dependent material behavior because it depends on the loading/unloading nature of the subsequent increment. Use of an incorrect tangent can lead to deviation from the correct path and even divergence. As in the static solution, a tangent matrix from a previous iteration, or even the initial elastic matrix $[K_T^1(0)]$, can sometimes be used to improve computational performance.

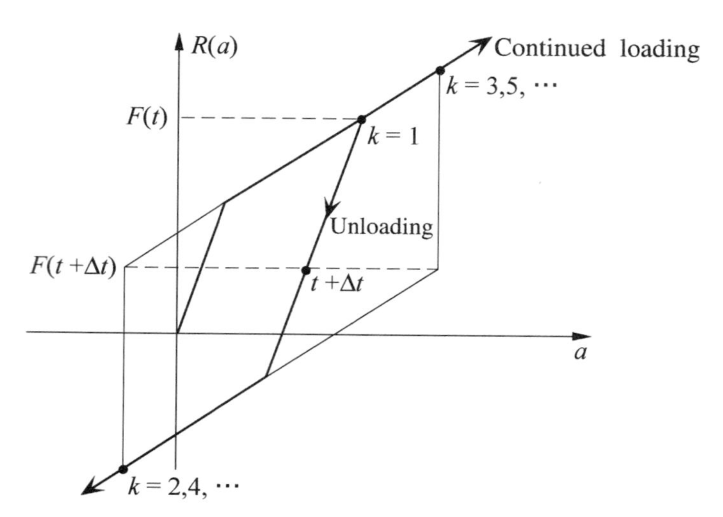

FIGURE 4 Iterative solution to $R(a) = F$ for an unloading step from t to $t+\Delta t$. Use of the loading tangent, rather than the unloading tangent, can lead to a nonconverging loop.

Because difficulties can occur when the iterating matrix is softer than the current tangent, and because unloading can be a stiffening response, it is often beneficial to start each time step with several iterations using $[K_T^1(0)]$ and then switch to a current tangent (Fig. 4). Stiffening behavior also occurs when previously buckled braces straighten out and when members develop significant tensile forces. Convergence is particularly hard to achieve when frictional sliding is present. For difficult conditions, a feature that automatically detects convergence problems and then subdivides the current time step is recommended. A smaller Δt, in addition to reducing the errors from using an incorrect tangent over the time step, "stiffens" the left-side matrix by increasing the contributions from inertia and damping, as is evident from Eq. (14).

Inclusion of common types of nonlinearity for the stiffness forces $\{R(t)\}$, such as yielding, places limits on the magnitude of these forces. However, the use of linear damping through $[C]$ means there are no upper bounds on the damping forces. In cases where a structure undergoes high-velocity response, the ratio of damping to other forces can become unrealistically high. One solution is to incorporate nonlinearity into the damping, but there is presently no accepted procedure for doing this. Linear Rayleigh damping will be used throughout this chapter.

The ability to solve a nonlinear structural dynamics problem is still hampered today by limited knowledge of complex material behavior. For example, proper analysis of a reinforced concrete structure should involve material models for bond degradation between the reinforcing steel and concrete, flexural and shear cracking, shear transfer across cracks, and softening of concrete in compression as a function of the amount of confining steel present. These features all need further study, especially with regard to cyclic loading. Many of the material models used by modern computer programs are highly simplified compared to actual structural behavior.

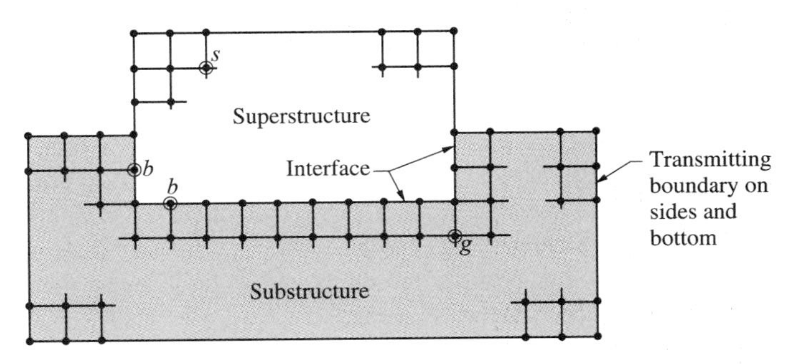

FIGURE 5 Division of structure, foundation, and ground into degrees of freedom of the superstructure off the interface (s), the interface itself (b), and the substructure off the interface (g). Substructure is shown shaded.

3. Earthquake Loading

The equations of motion from the previous section are now partitioned between superstructure and substructure as shown in Figure 5. Superstructure degrees of freedom off the interface with the substructure are denoted by s; substructure degrees of freedom off the interface are denoted by g; and interface degrees of freedom are denoted by b. Furthermore, matrices and vectors arising from the substructure discretization are given an over bar. Substructure matrices can contain transmitting boundary terms designed to absorb body and surface waves at the sides and bottom of the substructure.

As formulated here, the substructure must remain linear, but any portion of the foundation and surrounding ground can be made part of the superstructure, which is allowed to be nonlinear. Earthquake loading is defined at the interface in the form of free-field motions, i.e., those that would occur during the earthquake if the superstructure portion of the domain were absent. The intent here is to allow a nonlinear superstructure, spatially nonuniform free-field motions, and interaction of the superstructure with a linear substructure.

The free-field ground motions can be considered to be caused by forces $\{\bar{F}_b^E\}$ and $\{\bar{F}_g^E\}$ applied to the substructure. These forces are used to generate the earthquake in the model and are given by

$$\begin{Bmatrix} \bar{F}_b^E(t) \\ \bar{F}_g^E(t) \end{Bmatrix} = \begin{bmatrix} \bar{M}_{bb} & \bar{M}_{bg} \\ \bar{M}_{gb} & \bar{M}_{gg} \end{bmatrix} \begin{Bmatrix} \ddot{a}_b^{ff}(t) \\ \ddot{a}_g^{ff}(t) \end{Bmatrix} + \begin{bmatrix} \bar{C}_{bb} & \bar{C}_{bg} \\ \bar{C}_{gb} & \bar{C}_{gg} \end{bmatrix} \begin{Bmatrix} \dot{a}_b^{ff}(t) \\ \dot{a}_g^{ff}(t) \end{Bmatrix} + \begin{bmatrix} \bar{K}_{bb} & \bar{K}_{bg} \\ \bar{K}_{gb} & \bar{K}_{gg} \end{bmatrix} \begin{Bmatrix} a_b^{ff}(t) \\ a_g^{ff}(t) \end{Bmatrix}, \tag{16}$$

where the superscript *ff* denotes free-field ground motions. The free-field motions $\{a_g^{ff}(t)\}$ off the interface are arbitrary and do not affect the solution because linearity is assumed in the substructure. Putting the superstructure and substructure discretizations together and applying the static gravity loads acting on the superstructure and the earthquake generating loads from Eq. (16) result in

$$\begin{bmatrix} M_{ss} & M_{sb} & 0 \\ M_{bs} & M_{bb} + \bar{M}_{bb} & \bar{M}_{bg} \\ 0 & \bar{M}_{gb} & \bar{M}_{gg} \end{bmatrix} \begin{Bmatrix} \ddot{a}_s(t) \\ \ddot{a}_b(t) \\ \ddot{a}_g(t) \end{Bmatrix} + \begin{bmatrix} C_{ss} & C_{sb} & 0 \\ C_{bs} & C_{bb} + \bar{C}_{bb} & \bar{C}_{bg} \\ 0 & \bar{C}_{gb} & \bar{C}_{gg} \end{bmatrix} \begin{Bmatrix} \dot{a}_s(t) \\ \dot{a}_b(t) \\ \dot{a}_g(t) \end{Bmatrix} + \begin{Bmatrix} R_s(t) \\ R_b(t) \\ 0 \end{Bmatrix} + \begin{bmatrix} 0 & 0 & 0 \\ 0 & \bar{K}_{bb} & \bar{K}_{bg} \\ 0 & \bar{K}_{gb} & \bar{K}_{gg} \end{bmatrix} \begin{Bmatrix} a_s(t) \\ a_b(t) \\ a_g(t) \end{Bmatrix} = \begin{Bmatrix} F_s^G \\ F_b^G \\ 0 \end{Bmatrix} + \begin{Bmatrix} 0 \\ \bar{F}_b^E(t) \\ \bar{F}_g^E(t) \end{Bmatrix}, \tag{17}$$

where $\{R_s(t)\}$ and $\{R_b(t)\}$ are the nonlinear stiffness forces for the superstructure.

Equation (17) can be integrated in time as discussed in the previous section. For implicit time integration such as constant average acceleration, which factors the left-side matrix, the degrees of freedom should be numbered to localize their coupling as much as possible. This minimizes the profile of the matrix about the diagonal and, thereby, minimizes the factorization effort. The grouping of the degrees of freedom as implied by the partitioning of Eq. (17) would not be optimum; the partitioning is used just to clarify the formulation.

Caution is advised in two respects when using the formulation of Eq. (17). First, because the right-side earthquake forces are based not only on the free-field accelerations, but on the velocities and displacements as well [refer to Eq. (16)], spatial variations in the free-field motions along the interface should be checked to make sure they are realistic. For example, long-period errors can arise if the free-field velocities and displacements are determined by integrating recorded accelerograms. Second, a mass-proportional damping term, as present in Rayleigh damping, will generate damping forces proportional to the total velocities, which include the ground velocity component. Cases are possible where increasing the amount of mass-proportional damping actually increases the stiffness forces generated in some parts of the superstructure (Wilson *et al.*, 1999). A general way to handle this problem is not available at present.

One special case that can be put in a more convenient form is of free-field motions uniform along the interface as defined by x, y, and z components $u^{ff}(t)$, $v^{ff}(t)$, and $w^{ff}(t)$, respectively. The response is decomposed as follows:

$$\begin{Bmatrix} a_s(t) \\ a_b(t) \\ a_g(t) \end{Bmatrix} = \begin{Bmatrix} a_{su}^{ff}(t) \\ a_{bu}^{ff}(t) \\ a_g^{ff}(t) \end{Bmatrix} + \begin{Bmatrix} a_s^{rel}(t) \\ a_b^{rel}(t) \\ a_g^{rel}(t) \end{Bmatrix}, \tag{18}$$

where the subscript u denotes uniform and where $\{a_{su}^{ff}(t)\}$ and $\{a_{bu}^{ff}(t)\}$ contain $u^{ff}(t)$, $v^{ff}(t)$, and $w^{ff}(t)$ in positions corresponding to x, y, and z translational degrees of freedom, respectively, and zeroes for rotational degrees of freedom. On the right side of Eq. (18), the first term is the free-field motion for the b and g degrees of freedom and a compatible rigid motion for the s degrees of freedom. The second term is the unknown response relative to the first term. Equations (16) and (18) are substituted into Eq. (17) to give

$$\begin{bmatrix} M_{ss} & M_{sb} & 0 \\ M_{bs} & M_{bb}+\bar{M}_{bb} & \bar{M}_{bg} \\ 0 & \bar{M}_{gb} & \bar{M}_{gg} \end{bmatrix} \begin{Bmatrix} \ddot{a}_s^{rel}(t) \\ \ddot{a}_b^{rel}(t) \\ \ddot{a}_g^{rel}(t) \end{Bmatrix} + \begin{bmatrix} C_{ss} & C_{sb} & 0 \\ C_{bs} & C_{bb}+\bar{C}_{bb} & \bar{C}_{bg} \\ 0 & \bar{C}_{gb} & \bar{C}_{gg} \end{bmatrix} \begin{Bmatrix} \dot{a}_s^{rel}(t) \\ \dot{a}_b^{rel}(t) \\ \dot{a}_g^{rel}(t) \end{Bmatrix} + \begin{Bmatrix} R_s(t) \\ R_b(t) \\ 0 \end{Bmatrix} + \begin{bmatrix} 0 & 0 & 0 \\ 0 & \bar{K}_{bb} & \bar{K}_{bg} \\ 0 & \bar{K}_{gb} & \bar{K}_{gg} \end{bmatrix} \begin{Bmatrix} a_s^{rel}(t) \\ a_b^{rel}(t) \\ a_g^{rel}(t) \end{Bmatrix} = \begin{Bmatrix} F_s^G \\ F_b^G \\ 0 \end{Bmatrix} - \begin{bmatrix} M_{ss} & M_{sb} & 0 \\ M_{bs} & M_{bb} & 0 \\ 0 & 0 & 0 \end{bmatrix} \begin{Bmatrix} \ddot{a}_{su}^{ff}(t) \\ \ddot{a}_{bu}^{ff}(t) \\ 0 \end{Bmatrix} \quad (19)$$

from which the relative response can be found by time integration. In computing the stiffness forces $\{R_s(t)\}$ and $\{R_b(t)\}$ for the superstructure, only the relative response needs to be considered because rigid motions do not generate deformations and stresses. In Eq. (19), this condition has also been applied to the damping forces by omitting the term

$$\begin{bmatrix} C_{ss} & C_{sb} & 0 \\ C_{bs} & C_{bb} & 0 \\ 0 & 0 & 0 \end{bmatrix} \begin{Bmatrix} \dot{a}_{su}^{ff}(t) \\ \dot{a}_{bu}^{ff}(t) \\ 0 \end{Bmatrix}$$

from the right side. This is an improvement over Eq. (17) because mass proportional damping in Eq. (19) now will only generate damping forces proportional to the velocities relative to the uniform free-field interface velocities. Another advantage of Eq. (19) is that only the free-field accelerations at the interface, not velocities and displacements, are needed to define the earthquake loading.

As thus formulated, the earthquake loading is defined by free-field ground motions, i.e., those occurring with the superstructure absent. Because linearity of the substructure is assumed, only the free-field motions at the interface between the superstructure and substructure are relevant. Considerations in determining the free-field motions include the following:

- Earthquake magnitude, fault type, location of the site relative to the fault, geologic structure between site and fault, surficial soil conditions.
- Length of the interface as compared to the shortest important wavelength in the ground motion. Uniform ground motion can be assumed if the interface is relatively short; such is often the case for buildings. For long bridges, significant differences in free-field ground motions can occur at the piers.
- Influence of the structural foundation, if it is included in the substructure domain. Piles and caissons stiffen the soil and can alter the free-field ground motions.
- The shape of the interface. Exclusion of the structural foundation from the substructure means that the interface will follow an excavation, and this irregular shape of the interface complicates the free-field motions. If the structure is located in a canyon, as are dams and some bridges, then the interface is naturally irregular.

Computation of free-field motions can be a complex process, beyond the scope of the present discussion. However, a few examples will demonstrate some different situations.

Figure 6a shows a dam in a canyon, and the interface between the superstructure and substructure has been placed at the actual interface between the dam and the rock. This assumes the rock remains linear but allows nonlinear behavior anywhere in the dam. The earthquake source could be represented by an incident wave of specified type, direction, and time history, the latter possibly based on an actual earthquake record from a rock site. Then, the free-field problem becomes one of determining the response of the substructure to this incident wave (Figure 6b). Typically, considerable variation in the computed free-field motions occur around the canyon, necessitating that the dam response be calculated from Eq. (17), which allows the free-field motions to be nonuniform. Interaction with water in the reservoir is discussed in Section 4.

The system in Figure 7a consists of a bridge supported by rock abutments and short piers at each end and by a caisson under the center tower. The caisson passes through soil to the rock below. Included in the substructure are the rock and a portion of the soil away from the caisson. Although the substructure is assumed to be linear, some moderate nonlinearity in this part of the soil can be tolerated if appropriate equivalent linear material properties are employed. The superstructure includes not only the bridge but the caisson and adjacent soil in order to be able to model possibly stronger soil nonlinearity caused by lateral motion of the caisson. The free-field problem considers the substructure only and could again make use of an incident wave traveling through the rock (Figure 7b). The bridge response would be computed from Eq. (17).

The final example is a building on level ground supported by slender piles. As shown in Figure 8a, the interface between substructure and superstructure is placed at the base of the building,

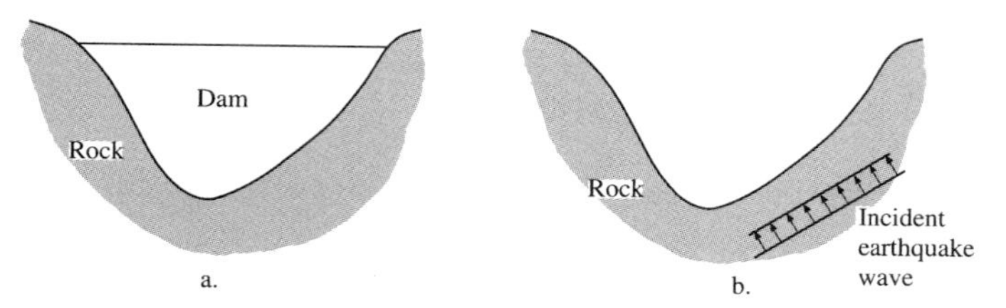

FIGURE 6 Dam in a rock canyon (a). Nonuniform free-field ground motion is computed as solution to an incident wave (b). Region designated as substructure is shown shaded.

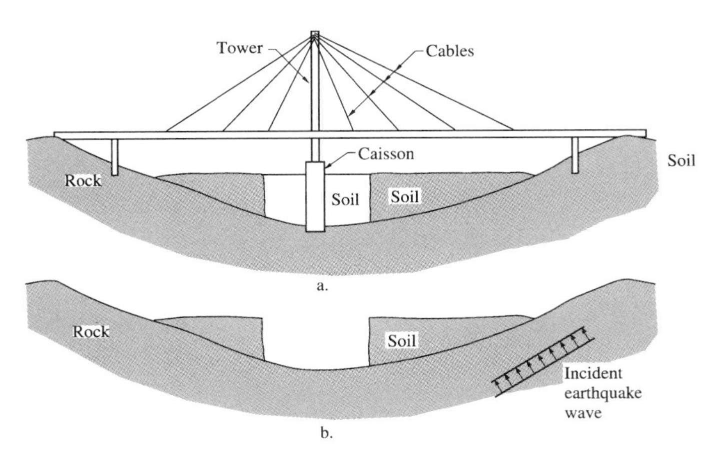

FIGURE 7 Bridge on caisson foundation in soil and rock (a). Nonuniform free-field ground motion is computed as solution to an incident wave (b). Region designated as substructure is shown shaded.

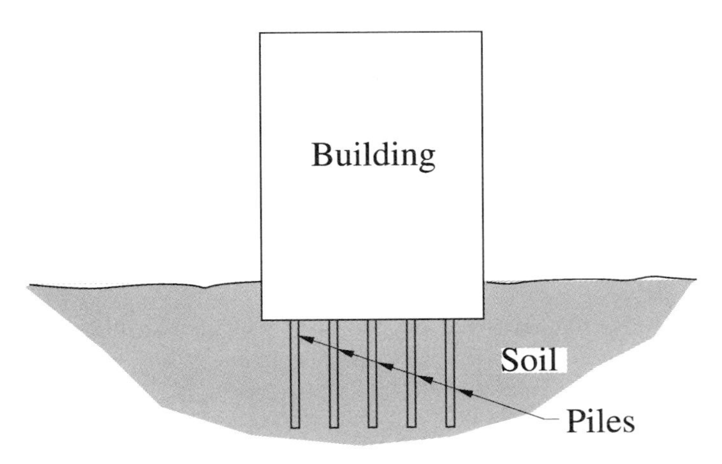

FIGURE 8 Building on slender pile foundation in soil. Free-field ground motions are assumed uniform and selected as an existing record. Region designated as substructure is shown shaded.

assuming linearity in the soil and foundation. An existing earthquake record would typically be selected for the free-field motion and taken as uniform over the interface, with the assumption that the free-field motion is unaffected by the slender piles. Building response would be computed from Eq. (19). In this case, because no substructure domain terms appear on the right side of Eq. (19), it is tempting to allow the substructure stiffness forces on the left side to be nonlinear. For example, a simple model of the foundation and soil would use nonlinear springs supporting the interface degrees of freedom. However, unlike the case of the preceding bridge example, this procedure does not account for the earthquake loading of a structure being affected by nonlinear behavior of its foundation and adjacent soil. Errors associated with such a procedure are not well quantified, so caution is advised.

Computational requirements for nonlinear earthquake analysis of structures using Eqs. (17) or (19), especially for three-dimensional systems, can be excessive. Therefore, approximations are often used to reduce the solution effort. For example, because the substructure is defined to be in the range of linear behavior, it can often be well represented by matrices $\bar{M}^*_{bb}$, $\bar{C}^*_{bb}$, and $\bar{K}^*_{bb}$, which involve only the interface degrees of freedom. The concept is analogous to static condensation of the g degrees of freedom but with approximate consideration of dynamic effects. Eqs. (17) and (16) are modified to

$$\begin{bmatrix} M_{ss} & M_{sb} \\ M_{bs} & M_{bb} + \bar{M}^*_{bb} \end{bmatrix} \begin{Bmatrix} \ddot{a}_s(t) \\ \ddot{a}_b(t) \end{Bmatrix} + \begin{bmatrix} C_{ss} & C_{sb} \\ C_{bs} & C_{bb} + \bar{C}^*_{bb} \end{bmatrix} \begin{Bmatrix} \dot{a}_s(t) \\ \dot{a}_b(t) \end{Bmatrix} + \begin{Bmatrix} R_s(t) \\ R_b(t) \end{Bmatrix} + \begin{bmatrix} 0 & 0 \\ 0 & \bar{K}^*_{bb} \end{bmatrix} \begin{Bmatrix} a_s(t) \\ a_b(t) \end{Bmatrix} = \begin{Bmatrix} F^G_s \\ F^G_b \end{Bmatrix} + \begin{Bmatrix} 0 \\ \bar{F}^E_b(t) \end{Bmatrix} \quad (20)$$

and

$$\{\bar{F}^E_b(t)\} = [\bar{M}^*_{bb}]\{\ddot{a}^{ff}_b(t)\} + [\bar{C}^*_{bb}]\{\dot{a}^{ff}_b(t)\} + [\bar{K}^*_{bb}]\{a^{ff}_b(t)\}. \quad (21)$$

Equation (19) is modified similarly. Should the substructure model include only stiffness and stiffness-proportional damping, then $[\bar{M}^*_{bb}] = [0]$, and $[\bar{C}^*_{bb}]$ and $[\bar{K}^*_{bb}]$ can be obtained through a true static condensation.

Elimination of the g degrees of freedom in the substructure matrices can produce full coupling among the b degrees of freedom through the * matrices. This increases the effort for matrix operations, especially factorization. To deal with this when using an implicit time integration scheme that requires factoring the left-side matrix, one can omit those terms of the * matrices that lie outside the profile determined by the superstructure alone when assembling the left-side matrix, but use full assembly on the right side. Errors from the omitted terms get eliminated in the iterations. Thus, the effort for an iteration is reduced significantly, but a few more iterations may be needed.

Additional information on formulations for earthquake loading of systems containing structure, foundation, soil, and rock components can be found in Clough and Penzien (1993), Wolf (1988), Ad Hoc Committee on Soil-Foundation-Structure Interaction (1998), and Tan and Chopra (1995).

4. Presence of Fluid

The earthquake response of structures such as dams, off-shore platforms, reservoir intake towers, and tanks is affected by dynamic pressures generated in the fluid (liquid) at the contact surface with the solid component (structure, foundation, and ground). To consider this effect, a fluid domain must be added to the formulations of the previous section.

A fluid domain can be discretized into finite elements using displacement degrees of freedom in a way similar to plane or three-dimensional solid elements (Wilson and Khalvati, 1983). The main difference is that only volumetric strain energy is included. Because this leaves the fluid domain with some zero energy modes of deformation, the addition of constraints based on the irrotational condition may be necessary. Assembly of fluid elements is the same as for solid elements because the degrees of freedom are the same, except that a tangential slip condition using double nodes needs to be maintained along the solid–fluid interface. Large displacements of the fluid relative to the solid along the interface can be treated by nodal coordinate updating and a special algorithm to update the nodal connectivity across the interface. However, except for large sloshing in tanks, the relative motions are usually small enough so that the original nodal connectivity can be maintained.

For fluid elements with displacement degrees of freedom, the formulations of the previous section apply by considering the fluid domain to be part of the superstructure. A transmitting boundary may be needed to replace a far extent of the fluid domain that is truncated from the model. Nonlinear material behavior in the fluid in the form of cavitation, for which the tangential volumetric stiffness reduces to zero when the pressure drops to the vapor pressure, can be included.

If the fluid motions are not large, then a different and more efficient procedure that uses a single degree of freedom per node is possible. The fluid response is described by the wave equation written in terms of the dynamic pressure $p^d(t)$, as follows:

$$\frac{\partial}{\partial x}\left(\frac{1}{\rho}\frac{\partial p^d(t)}{\partial x}\right)+\frac{\partial}{\partial y}\left(\frac{1}{\rho}\frac{\partial p^d(t)}{\partial y}\right)+\frac{\partial}{\partial z}\left(\frac{1}{\rho}\frac{\partial p^d(t)}{\partial z}\right)=\frac{1}{\kappa}\ddot{p}^d(t), \tag{22}$$

where κ is the fluid compressibility and ρ is the density. Boundary conditions include

$$\frac{\partial p^d(t)}{\partial z}=-\frac{1}{g}\ddot{p}^d(t) \tag{23}$$

at the free surface, which is a linearized condition for surface waves, and

$$\frac{1}{\rho}\frac{\partial p^d(t)}{\partial n}=-(\ddot{u}\bar{i}+\ddot{v}\bar{j}+\ddot{w}\bar{k})\cdot\bar{n} \tag{24}$$

along accelerating boundaries. In the above, pressure is positive in compression; z is in the vertical direction (upward positive); g is the acceleration due to gravity; $\bar{n}$ is a unit vector outward and normal to the boundary of the fluid domain; and $\ddot{u}$, $\ddot{v}$, and $\ddot{w}$ are x, y, and z components of acceleration in the solid at the fluid boundary. Equation (23) can represent moderate but not large surface waves. The dynamic pressure $p^d(t)$ does not include the static component p^s, which must be computed separately from

$$\frac{\partial p^s}{\partial z}=-\rho g \tag{25}$$

through the depth and

$$p^s=0 \tag{26}$$

at the free surface.

To develop the finite element equations of motion when a fluid domain with pressure degrees of freedom is present, these degrees of freedom are partitioned into those on the free surface, denoted by o, and those below the free surface, denoted by q. Degrees of freedom of the solid part are partitioned s, b, and g as

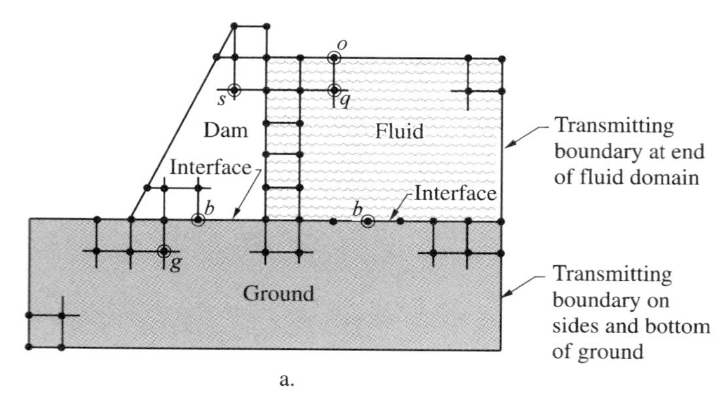

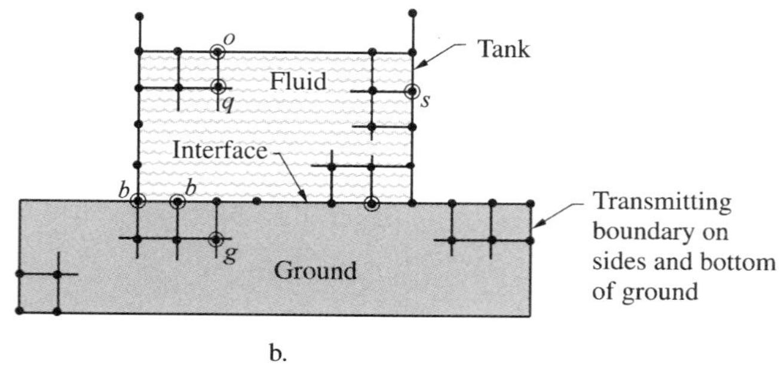

FIGURE 9 Discretizations for a dam (a) and a tank (b) showing partitioning of the degrees of freedom. Substructures are shown shaded. Fluid domains are rippled.

before, except the interface degrees of freedom b are extended along the ground where the fluid is in contact with the ground. Figure 9 illustrates the partitioning for a dam and a tank. The solid and fluid discretizations must match along their interface.

The matrix equation of motion for the combined structure, foundation, ground, and fluid domains will be written without the s, b, g partitioning for the solid component in order to save space. The result is

$$\begin{aligned}&\begin{bmatrix}M+\bar{M} & 0 & 0\\ -B_o & G_{oo}+\hat{G}_{oo} & G_{oq}\\ -B_q & G_{qo} & G_{qq}\end{bmatrix}\begin{Bmatrix}\ddot{a}(t)\\ \ddot{p}^d_o(t)\\ \ddot{p}^d_q(t)\end{Bmatrix}\\ &+\begin{bmatrix}C+\bar{C} & 0 & 0\\ 0 & 0 & 0\\ 0 & 0 & 0\end{bmatrix}\begin{Bmatrix}\dot{a}(t)\\ \dot{p}^d_o(t)\\ \dot{p}^d_q(t)\end{Bmatrix}+\begin{Bmatrix}R(t)\\ 0\\ 0\end{Bmatrix}\\ &+\begin{bmatrix}\bar{K} & -A_o & -A_q\\ 0 & H_{oo} & H_{oq}\\ 0 & H_{qo} & H_{qq}\end{bmatrix}\begin{Bmatrix}a(t)\\ p^d_o(t)\\ p^d_q(t)\end{Bmatrix}\\ &=\begin{Bmatrix}F^G\\ 0\\ 0\end{Bmatrix}+\begin{bmatrix}A_q\\ 0\\ 0\end{bmatrix}\{p^s_q\}+\begin{Bmatrix}\bar{F}^E\\ 0\\ 0\end{Bmatrix}\end{aligned} \tag{27}$$

where the H, G, and $\hat{G}$ terms originate from the left side of Eq. (22), from the right side of Eq. (22), and from Eq. (23), respectively. The A and B terms implement the boundary conditions that connect the fluid and solid domains across their interface. Fluid pressures are converted to loads acting on the solid by the A terms. Accelerations of the solid are converted to "loads" on the fluid domain [according to Eq. (24)] by the B terms. The coupling between the solid and fluid degrees of freedom exists only along the interface and is localized, so there are many zeroes in the A and B terms. Except for the nonzero A and B terms, the three partitioned matrices on the left side of Eq. (27) are symmetric. Details of the derivation of the fluid domain terms and those connecting the fluid and solid domains are given by Hall and Chopra (1980).

Time integration of Eq. (27) requires a few modifications from what was described in Section 2 of this chapter. For an explicit method such as central difference, the left-side matrix cannot be diagonalized because of the B terms. Nevertheless, because the equations for the solid component in Eq. (27) do not involve time derivatives of the pressures, $\{a(t+\Delta t)\}$ can be computed independently from

$$[M+\bar{M}]\{\ddot{a}(t)\}+[C+\bar{C}]\{\dot{a}(t)\}+\{R(t)\}+[\bar{K}]\{a(t)\}$$
$$-[A_o]\{p_o^d(t)\}-[A_q]\{p_q^d(t)\}=\{F^G\}+[A_q]\{p_q^s\}+\{\bar{F}^E\} \quad (28)$$

after substitution of Eq. (9), and then $\{\ddot{a}(t)\}$ can be computed from the time-stepping formula. Finally, the pressures at time $t+\Delta t$ are found from

$$\begin{bmatrix} G_{oo}+\hat{G}_{oo} & G_{oq} \\ G_{qo} & G_{qq} \end{bmatrix}\begin{Bmatrix} \ddot{p}_o^d(t) \\ \ddot{p}_q^d(t) \end{Bmatrix}+\begin{bmatrix} H_{oo} & H_{oq} \\ H_{qo} & H_{qq} \end{bmatrix}\begin{Bmatrix} p_o^d(t) \\ p_q^d(t) \end{Bmatrix}$$
$$=\begin{bmatrix} B_o \\ B_q \end{bmatrix}\{\ddot{a}(t)\} \quad (29)$$

after applying the time-stepping formula to the second-derivative pressures and diagonalizing the $\hat{G}$ and G terms in the resulting left-side matrix. For an implicit time integration scheme such as constant average acceleration, the A and B contributions to the left-side matrix can be made symmetric by appropriately scaling the equations because it can be shown that

$$[B_o]=-[A_o]^T \quad \text{and} \quad [B_q]=-[A_q]^T. \quad (30)$$

The displacement and pressure degrees of freedom should be ordered to localize the coupling and narrow the profile of the left-side matrix as much as possible. For either explicit or implicit time integration, nonlinearity in the form of cavitation in the fluid can be incorporated.

The solution effort for Eq. (27) can be reduced by eliminating some of the degrees of freedom. This has been discussed in the previous section with regard to the substructure, leading to $[\bar{M}_{bb}^*]$, $[\bar{C}_{bb}^*]$, and $[\bar{K}_{bb}^*]$. For the fluid domain, if Eq. (23) is replaced by

$$p^d(t)=0 \quad (31)$$

at the free surface to neglect the effect of surface waves, then the pressure degrees of freedom $\{p_o^d(t)\}$ are zero and can be omitted. If the water can be assumed to be incompressible, then the G terms are zero, and any of the q degrees of freedom can be eliminated by static condensation. Both of these assumptions are usually valid for offshore platforms and reservoir intake towers under seismic excitation. For dams, effects of surface waves are small as well, but fluid compressibility can be important. The reverse is true for tanks; fluid compressibility can be neglected but sloshing needs to be modeled, either by Eq. (23) or by a more exact procedure if the amplitudes are large. Elimination of pressure degrees of freedom assumes linear material behavior in the fluid, so cavitation cannot then be included.

If both surface waves and fluid compressibility are neglected, then the pressure degrees of freedom can be eliminated entirely. In Eq. (27), $\{p_q^d(t)\}$ is solved in terms of the solid accelerations and substituted back in to give

$$[M+\bar{M}+M^A]\{\ddot{a}(t)\}+[C+\bar{C}]\{\dot{a}(t)\}+\{R(t)\}+[\bar{K}]\{a(t)\}$$
$$=\{F^G\}+[A_q]\{p_q^s\}+\{\bar{F}^E\}, \quad (32)$$

where the fluid is represented by a symmetric matrix of added mass terms given by

$$[M^A]=[A_q][H_{qq}]^{-1}[A_q]^T. \quad (33)$$

Terms of $[M^A]$ are associated only with degrees of freedom of the solid along the interface with the fluid. An alternate procedure is to eliminate the o degrees of freedom and use only the q degrees of freedom on the solid–fluid interface. With these pressure degrees of freedom denoted by q', the modified version of Eq. (27) is

$$\begin{bmatrix} M+\bar{M} & 0 \\ -B_{q'} & 0 \end{bmatrix}\begin{Bmatrix} \ddot{a}(t) \\ \ddot{p}_{q'}^d(t) \end{Bmatrix}+\begin{bmatrix} C+\bar{C} & 0 \\ 0 & 0 \end{bmatrix}\begin{Bmatrix} \dot{a}(t) \\ \dot{p}_{q'}^d(t) \end{Bmatrix}$$
$$+\begin{Bmatrix} R(t) \\ 0 \end{Bmatrix}+\begin{bmatrix} \bar{K} & -A_{q'} \\ 0 & H_{q'q'}^* \end{bmatrix}\begin{Bmatrix} a(t) \\ p_{q'}^d(t) \end{Bmatrix} \quad (34)$$
$$=\begin{Bmatrix} F^G \\ 0 \end{Bmatrix}+\begin{bmatrix} A_{q'} \\ 0 \end{bmatrix}\{p_{q'}^s\}+\begin{Bmatrix} \bar{F}^E \\ 0 \end{Bmatrix},$$

where $[H_{q'q'}^*]$ is $[H_{qq}]$ with the noninterface pressure degrees of freedom condensed out, and the A and B terms are as before except the zeroes corresponding to the noninterface fluid degrees of freedom have been removed. Each pressure degree of freedom in Eq. (34) can conveniently be assigned to the adjacent solid node for the purpose of ordering the degrees of freedom.

The added mass matrix $[M^A]$ of Eq. (32) couples the solid degrees of freedom along the entire solid–fluid interface, and the H^* term of Eq. (34) couples all pressure degrees of freedom along this interface. During implicit time integration, some efficiency can be gained by using approximations with reduced coupling for the left-side matrix, the associated error being removed in the iteration process. For the added mass matrix $[M^A]$ in Eq. (32), a diagonal approximation is possible

(Fenves *et al.*, 1989). For $[H^*_{q'q'}]$ in Eq. (34), the partial assembly method discussed in the previous section is effective. The goal in both cases is to gain efficiency in factoring the left-side matrix without requiring significantly more iterations to convergence.

For a tank, the q degrees of freedom can be condensed to the q' ones along the interface similar to Eq. (34), but retaining the $\hat{G}$ terms and the o degrees of freedom. Additionally, the uniform ground motion assumption can be introduced. The result, also without the s, b, g partitioning in the solid, is

$$\begin{bmatrix} M+\bar{M} & 0 & 0 \\ -B_o & \hat{G}_{oo} & 0 \\ -B_{q'} & 0 & 0 \end{bmatrix} \begin{Bmatrix} \ddot{a}^{rel}(t) \\ \ddot{p}^d_o(t) \\ \ddot{p}^d_{q'}(t) \end{Bmatrix} + \begin{bmatrix} C+\bar{C} & 0 & 0 \\ 0 & 0 & 0 \\ 0 & 0 & 0 \end{bmatrix} \begin{Bmatrix} \dot{a}^{rel}(t) \\ \dot{p}^d_o(t) \\ \dot{p}^d_{q'}(t) \end{Bmatrix}$$
$$+ \begin{Bmatrix} R(t) \\ 0 \\ 0 \end{Bmatrix} + \begin{bmatrix} \bar{K} & -A_o & -A_{q'} \\ 0 & H^*_{oo} & H^*_{oq'} \\ 0 & H^*_{q'o} & H^*_{q'q'} \end{bmatrix} \begin{Bmatrix} a^{rel}(t) \\ p^d_o(t) \\ p^d_{q'}(t) \end{Bmatrix} \qquad (35)$$
$$= \begin{Bmatrix} F^G \\ 0 \\ 0 \end{Bmatrix} + \begin{bmatrix} A_{q'} \\ 0 \\ 0 \end{bmatrix} \{p^s_{q'}\} - \begin{bmatrix} M \\ -B_o \\ -B_{q'} \end{bmatrix} \{\ddot{a}^{ff}_u(t)\},$$

where the H^* terms result from the condensation process. For implicit time integration, partial assembly of the H^* terms into the left-side matrix is again effective.

A sampling of structure/fluid formulations in earthquake engineering is contained in Fok and Chopra (1985), Camara (2000), Goyal and Chopra (1989), and Haroun (1980).

5. A Beam Element for Planar Bending

A planar beam element is developed here and used in later sections of this chapter. The formulation demonstrates how to include both material nonlinearity in the form of plastic hinges and axial yielding and geometry nonlinearity by updating nodal coordinates and incorporating the effect of axial force. Assumptions are appropriate for a structural steel beam; some would need modification to represent a reinforced concrete member.

Figure 10 shows several basic concepts. The relation σ–ε between axial stress and strain is assumed to be elastic-plastic

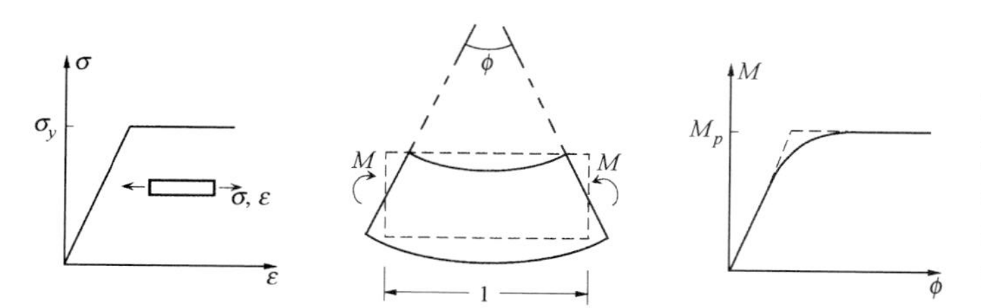

FIGURE 10 Basic concepts for yielding and bending of a beam under uniform moment, M.

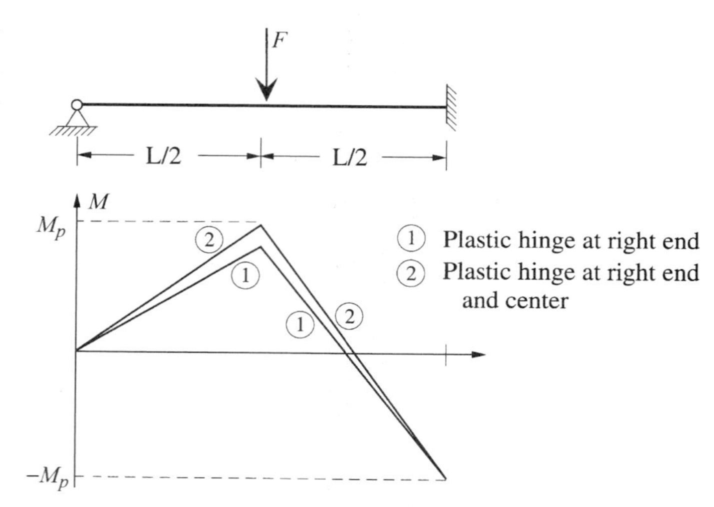

FIGURE 11 Moment diagram sequence as load F increases. Beam is pinned at left end and fixed at right end.

with yield stress σ_y. When a length of beam of such material is subjected to a uniform bending moment M, axial stresses and strains occur, linear with depth, which cause the beam to bend in constant curvature φ. The M–φ diagram has a smooth transition from elastic to plastic behavior because the yielding spreads gradually into the cross section. When the section is fully yielded, the moment equals M_p, the plastic moment capacity. For a beam of constant cross section with varying moment along its length, yielding begins at the location of maximum moment and gradually spreads into the cross section as well as along the length. As the loads increase, eventually the cross section where the moment is maximum becomes fully yielded, and then a concentrated rotation (kink) develops there.

If the M–φ relation for a beam is idealized as elastic-plastic (dashed line in Figure 10), then, under loading that produces varying moment along the length, a kink begins when the moment at a location reaches M_p without prior yielding there. These kinks are called plastic hinges even though they act as hinges only incrementally when $|M|$ equals M_p. Figure 11 shows the moment diagram sequence for a beam that forms two plastic hinges during loading. The hinge at the right end forms first, and the later formation of the center hinge creates a mechanism. The maximum value the load F can attain in the example is $6M_p/L$, as can be computed from statics.

Because a plastic hinge is an idealization that occurs over a zero length of a beam, the relationship between moment and kink angle at the hinge location is actually rigid-plastic. All parts of the beam between plastic hinges remain elastic. During an earthquake, unloading can occur, in which case the moment at a plastic hinge reduces. However, the kink angle does not change until the moment again reaches the hinge strength, either M_p or $-M_p$. Figure 12 shows the M_p vs. kink angle relation for a plastic hinge under cyclic loading exhibiting rigid-plastic behavior. The kink angle is denoted by κ.

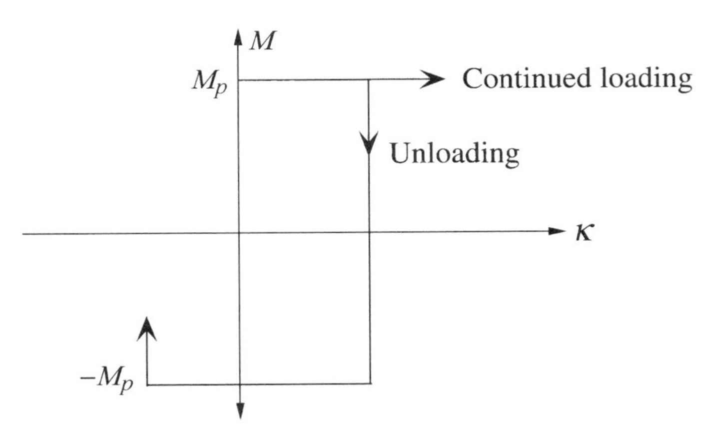

FIGURE 12 Rigid-plastic relation for moment M and kink angle κ at a plastic hinge.

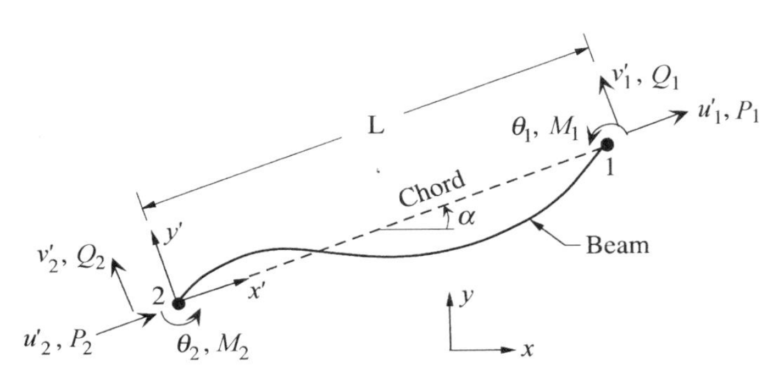

FIGURE 13 Beam element with degrees of freedom and end forces and moments in local coordinate system.

The beam element to be developed has two nodes, one at each end. These nodes are the connection points for the beam elements that make up a frame structure, and it is at these locations where forces are applied and where plastic hinges are allowed to form. If significant loads are applied to a beam between its connection points, then the beam should be discretized with multiple elements. For example, the beam in Figure 11 should be represented by two beam elements with nodes at each support and at the load application point. The limitation to nodal loading means that the axial and shear forces, P and Q, in a beam element are constant along its length and that the bending moment M varies linearly except for a part produced by the axial force P acting through the lateral deflections. If this additional moment is enough to cause the maximum moment and, hence, a possible plastic hinge to occur between nodes, then multiple elements should be used to discretize this length. An example is a brace that buckles laterally under axial compression. If the connections of the brace to the structure are pinned, then a plastic hinge could occur approximately at midspan. Therefore, a two-element discretization with a center node would suffice.

For a planar structural model, each beam node has two translational and one rotational degrees of freedom (Figure 13). For the local coordinate system x'–y', u'_i and v'_i are translations at node i in the x' and y' directions, and θ_i is the rotation about the perpendicular direction.* As the nodal coordinates are updated, the directions of x' and y' change; x' always follows the chord connecting nodes 1 and 2. Corresponding to the u'_i, v'_i, and θ_i degrees of freedom are nodal forces P_i and Q_i and moment M_i, respectively. The nodal quantities are collected in vector form for an element as

$$\{R'\}^e = \langle P_1 \quad Q_1 \quad M_1 \quad P_2 \quad Q_2 \quad M_2\rangle^T \tag{36}$$

and

$$\{a'\}^e = \langle u'_1 \quad v'_1 \quad \theta_1 \quad u'_2 \quad v'_2 \quad \theta_2\rangle^T. \tag{37}$$

Increments in these two vectors are related by

$$\{\Delta R'\}^e = [K'_T]^e \{\Delta a'\}^e, \tag{38}$$

where $[K'_T]^e$ is the element tangent stiffness matrix.

Before $\{R'\}^e$ and $[K'_T]^e$ can be assembled into $\{R\}$ and $[K_T]$ of Eqs. (7), (10), and (14), they must be transformed to the global x–y system for which the degrees of freedom are u_i and v_i in the x and y directions and θ_i about the perpendicular direction. Note that θ_i is the same in local and global coordinates. The transformations make use of a matrix $[T]$, which relates the local and global degrees of freedom, i.e.,

$$\{a'\}^e = [T]\{a\}^e, \tag{39}$$

where

$$\{a^e\} = \langle u_1 \quad v_1 \quad \theta_1 \quad u_2 \quad v_2 \quad \theta_2\rangle^T. \tag{40}$$

The matrix $[T]$ is defined as

$$[T] = \begin{bmatrix} \cos\alpha & \sin\alpha & 0 & 0 & 0 & 0 \\ -\sin\alpha & \cos\alpha & 0 & 0 & 0 & 0 \\ 0 & 0 & 1 & 0 & 0 & 0 \\ 0 & 0 & 0 & \cos\alpha & \sin\alpha & 0 \\ 0 & 0 & 0 & -\sin\alpha & \cos\alpha & 0 \\ 0 & 0 & 0 & 0 & 0 & 1 \end{bmatrix}, \tag{41}$$

where α is the angle from x to the updated x' based on current coordinates. The quantities $\{R'\}^e$ and $[K'_T]^e$ transform to global coordinates according to

$$\{R\}^e = [T]^T\{R'\}^e \tag{42}$$

and

$$[K_T]^e = [T]^T[K'_T]^e[T], \tag{43}$$

where Eq. (42) is based on equilibrium, and Eq. (43) results from applying the incremental versions of Eqs. (39) and (42) to Eq. (38).

The procedure that determines $\{R'\}^e$ and $[K']^e$ in an iteration of a static load step or a time step is now described. This is an updating process based on $\{\Delta a'\}^e$, which is found from the computed increments $\{\Delta a\}^e$ using the incremental version of Eq. (39). For purposes of the discussion, it is convenient to have the plastic hinges form just inside the nodes. As shown in

* For convenience, the notation $^k(t+\Delta t)$, denoting the time step and iteration number, is dropped in this section.

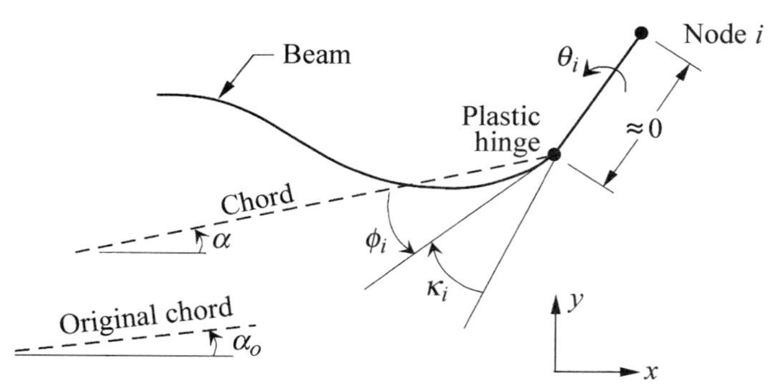

FIGURE 14 Definition of kink angle κ_i in terms of node rotation θ_i, chord rotation $\alpha - \alpha_o$, and beam rotation ϕ_i relative to chord just inside plastic hinge.

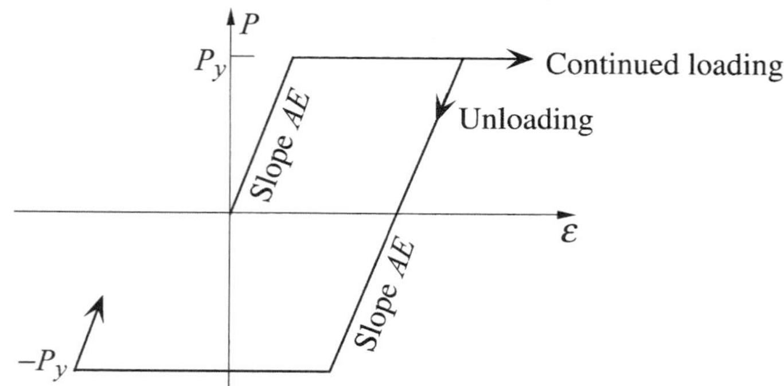

FIGURE 15 Elastic-plastic relation for axial load P and axial strain ε.

Figure 14, the rotation of the end of the beam at node i is θ_i; the rotation just inside the hinge relative to the chord x' is ϕ_i; α_o is the original chord angle; and the kink angle is κ_i given by

$$\kappa_i = \theta_i - \phi_i - (\alpha - \alpha_o). \tag{44}$$

The relation between M_1, M_2, and ϕ_1, ϕ_2 can be obtained by solving the beam differential equation (Salmon and Johnson, 1980). The result is

$$\begin{bmatrix} a & b \\ b & c \end{bmatrix} \begin{Bmatrix} \phi_1 \\ \phi_2 \end{Bmatrix} = \begin{Bmatrix} M_1 \\ M_2 \end{Bmatrix}, \tag{45}$$

where

$$a = c = \frac{EI}{L} \frac{\omega \sin\omega - \omega^2 \cos\omega}{2 - 2\cos\omega - \omega \sin\omega} \quad \text{for } P < 0 \tag{46a}$$

$$= \frac{4EI}{L} \quad \text{for } P = 0 \tag{46b}$$

$$= \frac{EI}{L} \frac{-\omega \sinh\omega + \omega^2 \cosh\omega}{2 - 2\cosh\omega + \omega \sinh\omega} \quad \text{for } P > 0 \tag{46c}$$

$$b = \frac{EI}{L} \frac{-\omega \sin\omega + \omega^2}{2 - 2\cos\omega - \omega \sin\omega} \quad \text{for } P < 0 \tag{46d}$$

$$= \frac{2EI}{L} \quad \text{for } P = 0 \tag{46e}$$

$$= \frac{EI}{L} \frac{\omega \sinh\omega - \omega^2}{2 - 2\cosh\omega + \omega \sinh\omega} \quad \text{for } P > 0 \tag{46f}$$

and E = Young's modulus, I = moment of inertia of the beam cross section, L is the original length of the beam element, and $\omega = L\sqrt{|P|/EI}$. The preceding relation is valid for elastic behavior between the hinges, small strains, and small deflections of the beam relative to the current chord x'. Shear deformations are neglected, but the formulas for a, b, and c include the nonlinear effect of the axial force P on bending stiffness. P is positive in tension and for $P = P_{cr} = -\pi^2 \frac{EI}{L^2}$, the Euler buckling load, the determinant of the matrix $[\begin{smallmatrix} a & b \\ b & c \end{smallmatrix}]$ is zero. Note that out-of-plane buckling, which is sometimes critical, cannot be modeled with the planar formulation.

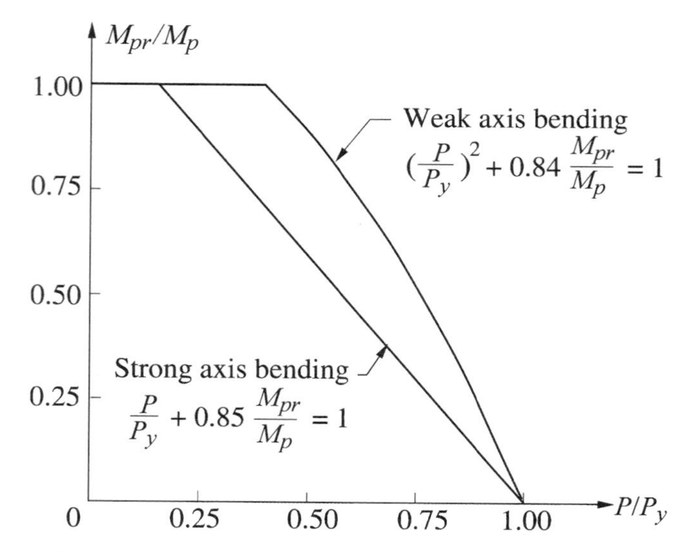

FIGURE 16 Reduced plastic moment strength M_{pr} as a function of axial force P for a steel I section (Dowrick, 1987).

The axial force P depends on the axial strain history as shown in Figure 15. The behavior is assumed to be elastic-plastic with the yield capacity P_y equal to the cross-sectional area A of the beam times the material yield strength σ_y. The element length should be such that $|P_{cr}| > P_y$, using a multiple-element discretization of a member if necessary. Dependence of the bending moment capacity M_p on the presence of axial load can be accounted for as shown in Figure 16, for example, which is appropriate for a steel I section. The reduced value of M_p is denoted by M_{pr}.

For use in the tangent stiffness matrix, the incremental relations

$$\begin{bmatrix} a_T & b_T & 0 \\ b_T & c_T & 0 \\ 0 & 0 & d_T \end{bmatrix} \begin{Bmatrix} \Delta\theta_1 - \Delta\alpha \\ \Delta\theta_2 - \Delta\alpha \\ \Delta u_1' - \Delta u_2' \end{Bmatrix} = \begin{Bmatrix} \Delta M_1 \\ \Delta M_2 \\ \Delta P \end{Bmatrix} \tag{47}$$

need to be defined. The terms a_T, b_T, and c_T depend on the conditions for the plastic hinges. A hinge can be active under positive moment ($M = M_{pr}$, condition denoted by A+), active

TABLE 1 Nine Possible Hinge Cases for a Beam Element

Hinge Cases		Values For		
Node 1	Node 2	a_T	b_T	c_T
I	I	a	b	c
I	A+ or A−	$a - b^2/c$	0	0
A+ or A−	I	0	0	$c - b^2/a$
A+ or A−	A+ or A−	0	0	0

under negative moment ($M = -M_{pr}$, condition denoted by A−), or inactive ($|M| < M_{pr}$, condition denoted by I). A summary appears in Table 1 for nine possible hinge cases at the two nodes. If axial yielding is taking place, then $|P| = P_y$ and $d_T = 0$; otherwise $d_T = \frac{AE}{L}$. The state at the beginning of an iteration is the basis for selecting a specific tangent relation. This choice may be incorrect, for example, if an active hinge unloads; this is one reason why iteration is required.

To form $[K'_T]^e$, Eq. (47) is combined with

$$\begin{Bmatrix} \Delta\theta_1 - \Delta\alpha \\ \Delta\theta_2 - \Delta\alpha \\ \Delta u'_1 - \Delta u'_2 \end{Bmatrix} = [S]\{\Delta a'\}^e \tag{48}$$

and

$$\{\Delta R'\}^e = [S]^T \begin{Bmatrix} \Delta M_1 \\ \Delta M_2 \\ \Delta P \end{Bmatrix} + \frac{P}{L}\langle 0 \quad \Delta v_1 - \Delta v_2 \quad 0 \quad 0 \quad \Delta v_2 - \Delta v_1 \quad 0\rangle^T \tag{49}$$

to construct a 6×6 matrix, where

$$[S] = \begin{bmatrix} 0 & -\frac{1}{L} & 1 & 0 & \frac{1}{L} & 0 \\ 0 & -\frac{1}{L} & 0 & 0 & \frac{1}{L} & 1 \\ 1 & 0 & 0 & -1 & 0 & 0 \end{bmatrix}. \tag{50}$$

The second term in Eq. (49) is an additional effect of axial load on equilibrium, which involves the v_1 and v_2 degrees of freedom. An approximate form of $[K'_T]^e$, linearized in P and corresponding to the no hinge case, is

$$[K'_T]^e = \begin{bmatrix} \frac{AE}{L} & 0 & 0 & -\frac{AE}{L} & 0 & 0 \\ 0 & \frac{12EI}{L^3} + \frac{6}{5}\frac{P}{L} & -\frac{6EI}{L^2} - \frac{P}{10} & 0 & -\frac{12EI}{L^3} - \frac{6}{5}\frac{P}{L} & -\frac{6EI}{L^2} - \frac{P}{10} \\ 0 & -\frac{6EI}{L^2} - \frac{P}{10} & \frac{4EI}{L} + \frac{2PL}{15} & 0 & \frac{6EI}{L^2} + \frac{P}{10} & \frac{2EI}{L} - \frac{PL}{30} \\ -\frac{AE}{L} & 0 & 0 & \frac{AE}{L} & 0 & 0 \\ 0 & -\frac{12EI}{L^3} - \frac{6}{5}\frac{P}{L} & \frac{6EI}{L^2} + \frac{P}{10} & 0 & \frac{12EI}{L^3} + \frac{6}{5}\frac{P}{L} & \frac{6EI}{L^2} + \frac{P}{10} \\ 0 & -\frac{6EI}{L^2} - \frac{P}{10} & \frac{2EI}{L} - \frac{PL}{30} & 0 & \frac{6EI}{L^2} + \frac{P}{10} & \frac{4EI}{L} + \frac{2PL}{15} \end{bmatrix}, \tag{51}$$

which is valid for positive or negative P.

Following an iteration in which displacement increments $\{\Delta a\}^e$ and then $\{\Delta a'\}^e$ are computed, the correct plastic hinge case out of the nine possible must be identified before the corresponding tangent matrix $[K'_T]^e$ can be constructed. This identification takes place during the updating of $\{R'\}^e$, the process for which is now described.

From the updated coordinates of nodes 1 and 2, the current length of the beam element is determined, leading to the increment in axial strain $\Delta\varepsilon$. The updated axial force P is found by following the P–ε relation in Figure 15; then $P_1 = -P_2 = P$. The next step deals with the moments and tests each of the nine plastic hinge cases to determine the correct one. Consider as an example a test for A+ at node 1 and I at node 2. If this case is the correct one, then $M_1 = M_{pr}$, and ϕ_2 can be updated with $\Delta\phi_2 = \Delta\theta_2 - \Delta\alpha$. These values of M_1 and ϕ_2 allow ϕ_1 and M_2 to be computed from Eq. (45), and $\Delta\kappa_1$ can then be computed from the incremental version of Eq. (44). If $\Delta\kappa_1 \geq 0$ and $|M_2| < M_{pr}$, the case is validated. However, if $\Delta\kappa_1 < 0$ or $|M_2| \geq M_{pr}$, then the case fails and the next one is tested; only one case will pass. After M_1 and M_2 are determined, updated shear forces are found from

$$Q_1 = -Q_2 = -\frac{M_1 + M_2}{L} \tag{52}$$

where L is the updated length.

For use in an iterative scheme within a load step or time step, the displacement increments employed to update $\{R'\}^e$ and find $[K'_T]^e$ are total increments accumulated since the beginning of the step. This avoids artificial unloading during the iterations.

Improvements to the planar beam element can be made by adding shear deformations (Przemieniecki, 1968) and by making the M–κ relation at a plastic hinge more realistic by including strain hardening and degradation effects on M_p such as weld fracture and buckling of flanges for steel sections. Strain hardening can be incorporated simply by placing a rotational spring in parallel with a plastic hinge to make the M–κ relation as shown in Figure 17 (Hall and Challa, 1995). The stiffness of the rotational spring can be conveniently expressed as $h \cdot \frac{6EI}{L}$, where

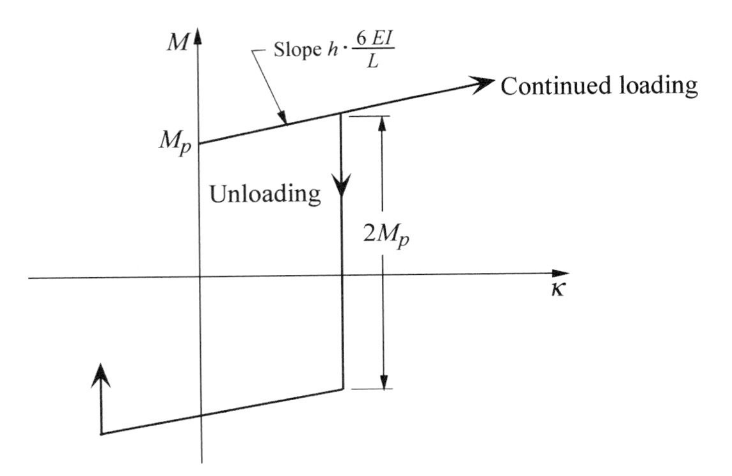

FIGURE 17 Modified relation for moment M and kink angle κ at a plastic hinge to consider strain hardening.

the term $\frac{6EI}{L}$ is the elastic rotational stiffness of the beam when $M_1 = M_2$. A value $h = 0.025$ agrees reasonably well with test results for steel I sections (Hall and Challa, 1995). In addition, different plastic hinge strengths M_{p1} and M_{p2} can be assigned to the nodes. This is useful when a pin connection exists at node i of the beam element, in which case M_{pi} is set to zero.

For three-dimensional applications, the planar beam element can be generalized to accommodate twisting and out-of-plane bending. Such an element has three translational and three rotational degrees of freedom at each node. Some situations that require three-dimensional modeling are discussed in Section 7 of this chapter.

A more complex formulation for a beam, known as a fiber element, either planar or three-dimensional, divides the beam longitudinally into segments and the segments into fibers (Hall and Challa, 1995; Kaba and Mahin, 1984). Each fiber follows a uniaxial hysteretic stress–strain law, which can be as realistic as desired. This element can naturally capture the M–P strength interaction, gradual spread of yielding within the cross section and along the length of the beam, residual stresses, strain hardening, and cracking of brittle materials such as concrete. A reinforced concrete beam would have some fibers representing the longitudinal reinforcing steel and others representing the concrete. Such fibers could also be added to the top of a steel beam to model composite action with a concrete slab. Shear behavior is usually assumed to be elastic in a fiber element, a deficiency for reinforced concrete, which can have important nonlinear deformation modes involving shear. Including nonlinear shear behavior for reinforced concrete as well as other features such as slip of reinforcing bars are current research topics.

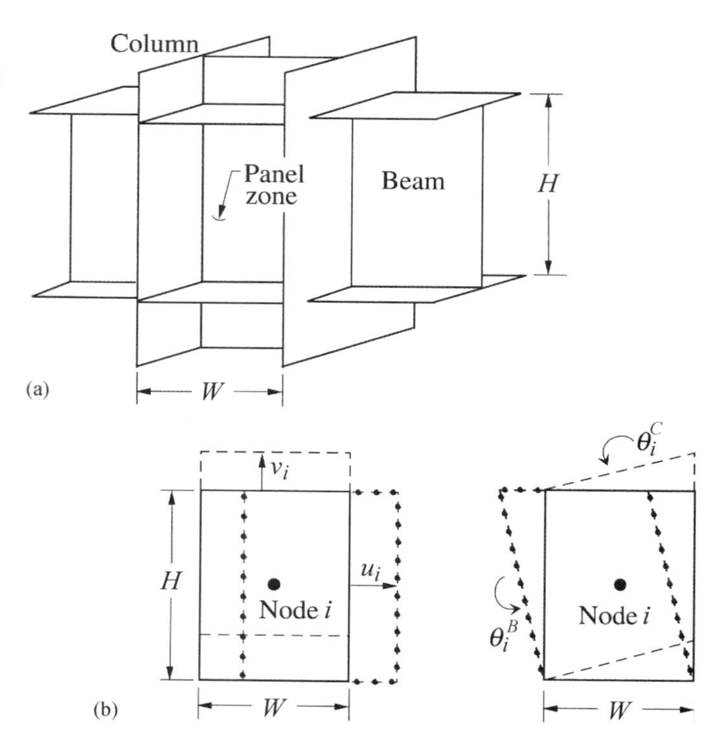

FIGURE 18 Perspective view of a joint in a steel moment frame (a). The four nodal degrees of freedom of a flexible joint region (b).

6. Planar Frame Modeling

A planar frame is an assemblage of horizontal beams and vertical columns, with or without inclined braces, where all members lie in a single plane. Such frames are commonly used in buildings to resist lateral loads from wind and earthquake. A frame building typically has several planar frames oriented in perpendicular directions, at least two in each direction. Planar frames with beams and columns but no braces resist lateral loads through flexure and, hence, are termed moment frames. Connections of beams to columns, referred to as moment connections, are designed to fully transmit forces and moments and are usually quite rigid. Braced frames employ inclined braces to form triangular grids with beams and columns; such grids have inherent rigidity. Lateral loads applied to a braced frame are resisted by axial forces that develop in the members. The beam element developed in the previous section can be used in horizontal, vertical, or inclined positions and, thus, can represent not only beams but columns and braces as well.

The simplest frame computer model merely connects beam elements at their nodes, but this fails to account for the finite dimensions and flexibility of actual joint regions. Under lateral loading, the joint regions in a moment frame where beams and columns are connected can be highly stressed in shear and can undergo significant deformation and even yield. To capture these effects, special shear elements are employed. A single node can still be used at each intersection of a beam and column; however, two rotational degrees of freedom are needed in addition to the two translational degrees of freedom. At node i, the degrees of freedom are x translation u_i, y translation v_i, beam-ends rotation θ_i^B, and column-ends rotation θ_i^C.*

These concepts are illustrated in Figure 18 for steel I sections. The joint region, termed the panel zone, is the short length of column within the depth H of the beam. Width of the panel zone is the column width W. Strength and stiffness of the panel zone owe mainly to the column web; the contribution from the column flanges is smaller. The derivation that follows is appropriate for steel construction as depicted in Figure 18 and omits the flange contribution of the column. A formulation for reinforced concrete joints has similar features but with the complication associated with slipping of reinforcing bars anchored in a joint.

Under loading, a steel panel zone deforms approximately as a parallelogram with an average shear strain γ given by $\theta_i^C - \theta_i^B$. Corresponding to γ is a shear stress τ, and the panel zone behavior is thus defined by the τ–γ relation. As shown in Figure 19,

* The notation ${}^k(t+\Delta t)$ is again omitted in this section.

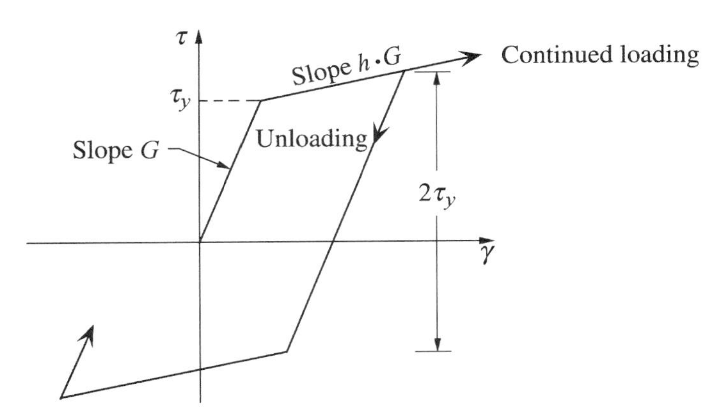

FIGURE 19 Bilinear relation for panel zone shear stress τ and shear strain γ.

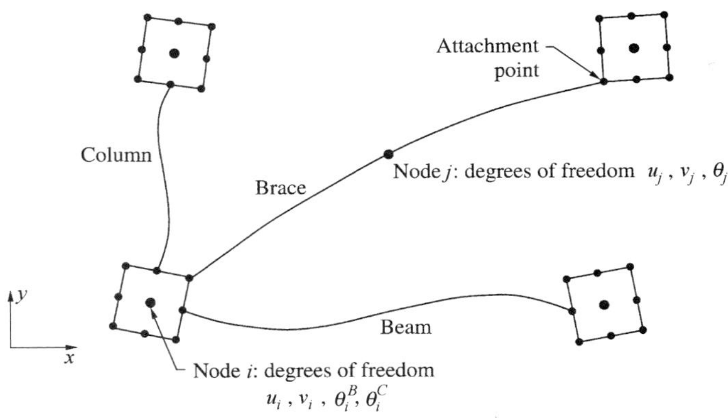

FIGURE 21 Connectivity of beam elements and joint elements for the cases of beam, column, and brace.

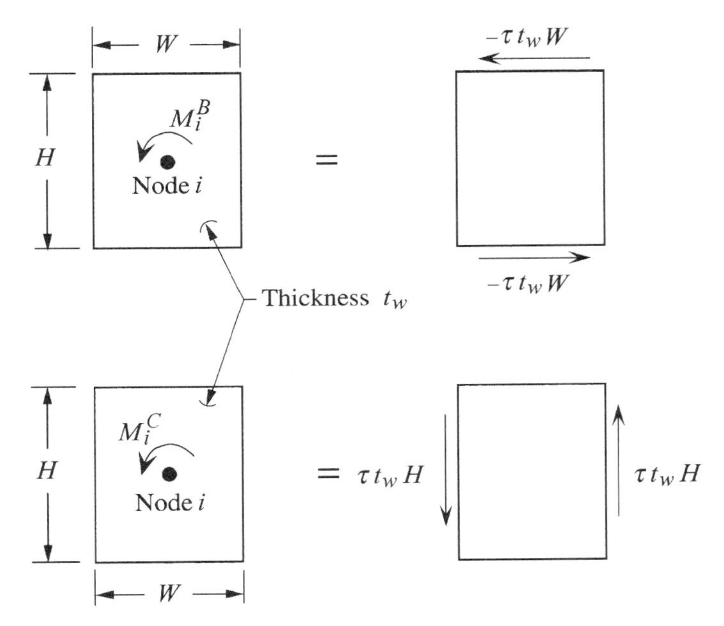

FIGURE 20 Definition of panel zone moments in terms of double couples.

a bilinear hysteretic behavior is assumed, which is characterized by elastic shear modulus G, second slope $h \cdot G$, and shear strength τ_y. Tests show that $h = 0.035$ and $\tau_y = 0.55\sigma_y$ are reasonable (Challa and Hall, 1994). In incremental form,

$$\Delta\tau = G_T \Delta\gamma, \tag{53}$$

where the tangent shear modulus is the current slope from Figure 19.

Associated with the angles θ_i^B and θ_i^C are panel zone moments M_i^B and M_i^C, which are related to the shear stress τ through double couples as shown in Figure 20. The moment M_i^{pz} of the panel zone is given by

$$M_i^{pz} = M_i^C = -M_i^B = \tau\, t_w HW, \tag{54}$$

where t_w is the column web thickness. This thickness should include any doubler plates welded to the column web to increase the panel zone strength and stiffness.

Combining all of the above leads to

$$\{\Delta R\}^{PZ} = [K_T]^{PZ}\{\Delta a\}^{PZ}, \tag{55}$$

where

$$[K_T]^{PZ} = t_w HW \begin{bmatrix} G_T & -G_T \\ -G_T & G_T \end{bmatrix} \tag{56}$$

$$\{R\}^{PZ} = \langle M_i^B \; M_i^C \rangle^T \tag{57}$$

$$\{a\}^{PZ} = \langle \theta_i^B \; \theta_i^C \rangle^T. \tag{58}$$

G_T is either the elastic shear modulus G, or, during yielding, G_T equals hG. Updating of $\{R\}^{PZ}$ from increments $\{\Delta a\}^{PZ}$ begins by converting $\Delta\theta_i^B$ and $\Delta\theta_i^C$ into the shear strain increment $\Delta\gamma$, then follows the τ–γ relation in Figure 19 to update τ, then converts τ to the two moments M_i^B and M_i^C from Eq. (54). The quantities $\{R\}^{PZ}$ and $[K_T]^{PZ}$ are assembled directly into $\{R\}$ and $[K_T]$ of Eqs. (7), (10), and (14) without any transformation.

Beam elements can connect to joints and to other beam elements (Figure 21). As shown in the figure, the frame degrees of freedom are expressed in the global x–y system and are associated with joints (u_i, v_i, θ_i^B, θ_i^C at node i) as well as nodes along the length of a member when more than one element is used to discretize the member (u_j, v_j, θ_j at node j). Consideration of the actual dimensions of the joints on the geometry of the connectivity is desirable, and this makes necessary a two-step transformation between the local beam degrees of freedom and the global ones of the joints. The transformations are applied to $\{R'\}^e$ and $[K'_T]^e$ of a beam element, computed in local x'–y' coordinates as described in the previous section but using the attachment points to the joints as if these points were actual nodes. The first step uses the matrix $[T]$ from Eq. (41) as before to transform the beam degrees of freedom from local to global coordinates. For the second step, as an example, consider the beam of Figure 22 modeled by a single element; each end is

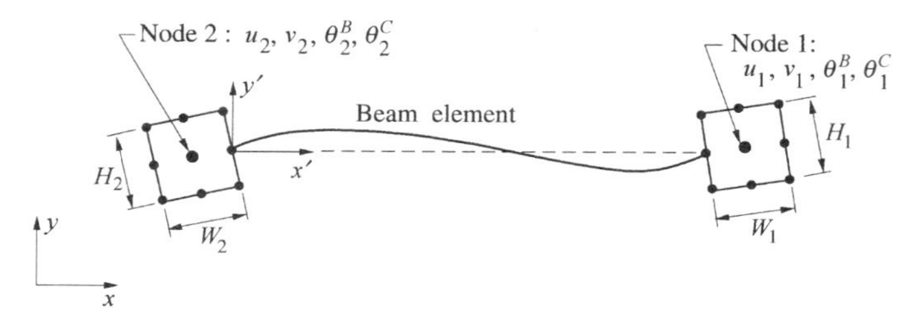

FIGURE 22 Beam element connected to sides of joint elements.

attached to a side of a joint. In this case, the second transformation is accomplished with matrix $[B]$ given by

$$[B] = \begin{bmatrix} 1 & 0 & 0 & \frac{W_1}{2}\sin\theta_1^C & 0 & 0 & 0 & 0 \\ 0 & 1 & 0 & -\frac{W_1}{2}\cos\theta_1^C & 0 & 0 & 0 & 0 \\ 0 & 0 & 1 & 0 & 0 & 0 & 0 & 0 \\ 0 & 0 & 0 & 0 & 1 & 0 & 0 & -\frac{W_2}{2}\sin\theta_2^C \\ 0 & 0 & 0 & 0 & 0 & 1 & 0 & \frac{W_2}{2}\cos\theta_2^C \\ 0 & 0 & 0 & 0 & 0 & 0 & 1 & 0 \end{bmatrix}. \tag{59}$$

This matrix, together with $[T]$, relates the local beam degrees of freedom to the global frame ones as

$$\{\Delta a'\}^e = [T][B]\{\Delta a\}^e, \tag{60}$$

where

$$\{a\}^e = \langle u_1 \quad v_1 \quad \theta_1^B \quad \theta_1^C \quad u_2 \quad v_2 \quad \theta_2^B \quad \theta_2^C \rangle^T, \tag{61}$$

and where $\{a'\}^e$ is given by Eq. (37). The matrix $[B]$ also considers the geometry of the deformed joint shape and, as given in Eq. (59), assumes that initially each joint has horizontal top and bottom edges and vertical sides. Matrix $[B]$ would be constructed differently for a beam, column, or brace, and it also depends on whether one or both ends of the element are connected to a joint.

Transformation of $\{R'\}^e$ and $[K'_T]^e$ results in

$$\{R\}^e = [B]^T[T]^T\{R'\}^e \tag{62}$$

and

$$[K_T]^e = [B]^T[T]^T[K'_T]^e[T][B], \tag{63}$$

which can be assembled into Eqs. (7), (10), and (14). Thus, when the actual dimensions of joints are accounted for, Eqs. (62) and (63) replace Eqs. (42) and (43). Equation (60) is also used to find $\{\Delta a'\}^e$ from the computed increments $\{\Delta a\}^e$ to begin the updating process for $\{R'\}^e$ and $[K'_T]^e$ in an iteration of a static load step or a time step.

The arrangement just described, where the nodal degrees of freedom of beam elements connecting to a joint are compatible with the joint degrees of freedom, is appropriate for moment frames where columns are continuous through a joint and beams are attached by moment connections. For a braced frame, full moment connections are usually not employed, and the attachments of the beams and braces to a joint are often considered to be pinned. As stated in the previous section, setting M_p to zero at a node where a pin connection exists produces the desired condition of zero moment without introducing additional degrees of freedom.

To summarize, the planar frame model combines beam elements and joint shear elements. Beam elements, as described in the previous section, represent the actual beams, columns, and braces in the frame. Plastic hinging, axial yielding, and effect of axial load on bending stiffness have been included. Shear elements are employed for the joints in the frame to model their flexibility and yielding as well as to represent their actual dimensions. As the solution progresses, the nodal coordinates and joint angles are updated, and the current geometry is used in the formulation. This accounts for the so-called P–Δ effect, i.e., the destabilizing effect of gravity acting through the lateral displacements of a building. Modeling of member buckling requires a multi-element discretization within a member; only in-plane buckling has been considered.

7. Modeling of Complete Frame Buildings

The framing in a building supports vertical gravity loads as well as lateral loads from wind and earthquake. Generally, only a part of the framing in the form of moment or braced frames comprises the lateral load resisting system. The building plan shown in Figure 23a has four moment frames, one each along lines 1, 5, A, and C, and each spanning two bays. These frames have moment connections between the beams and columns. In other locations, beams are pinned to the columns so as to transfer shear forces but not bending moment. (In steel construction, a pin connection between a beam and a column is made by attaching only the beam web to a column.) The interior columns in Figure 23a primarily carry gravity loads because all of their beam connections are pinned. Other important elements of the structural system are the floors and roof, which act as diaphragms to tie the framing together at each level.

The planar frame concept defined in the previous section, when combined with various enhancements, is able to model a significant variety of complete building systems. A simple case is when separate planar analyses can be used for orthogonal directions of loading. This requires a symmetric system so that twisting does not occur. For the building in Figure 23a, with symmetric structure and mass distribution, a half-building model appropriate for north-south ground motions plus gravity loads and the vertical component of ground motion appears in Figure 23b. The moment frame along line 5 as well as the pin-connected frame along line 4 and a half slice of the pin-connected frame along line 3 are included. Each frame is present separately except as linked by springs, which connect horizontal degrees of freedom between corresponding pairs of nodes at each level. One set of springs connects nodes on line 5 to nodes on line 4, and another set connects nodes on line 4 to nodes on line 3. These springs represent the in-plane shear stiffness of the floors and roof, which act as diaphragms. Often, these diaphragms

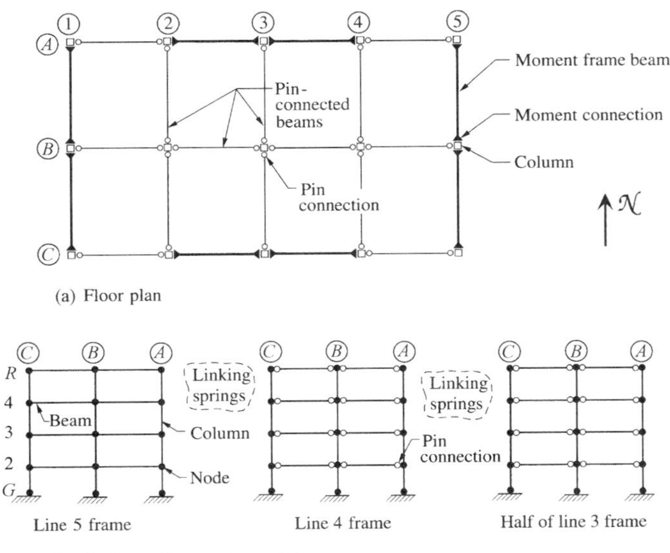

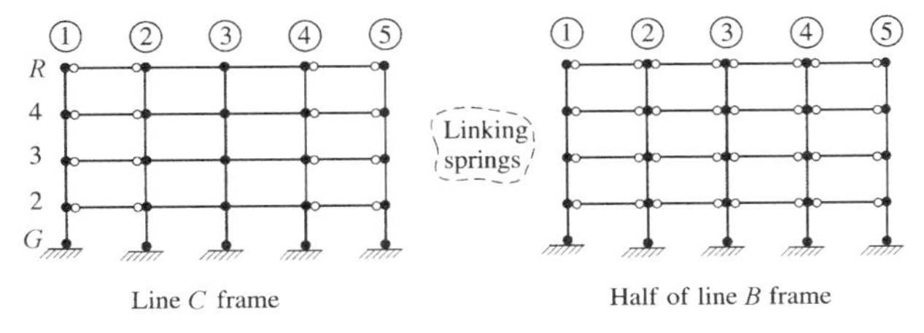

FIGURE 23 Symmetric building, which can be analyzed using separate planar frame models for north-south ground motion and east-west ground motion. Gravity loads and vertical ground motion are also applied to each model. Details of linking springs in parts b and c are not shown.

are stiff enough in plane that they can be assumed rigid; then, the linking springs can be assigned high stiffness values. Out-of-plane stiffness of the floors and roof is neglected. The pin connections in frame lines 3 and 4 are modeled by setting the beam M_p to zero for any node where a pin exists. However, not much is gained by including the pin-connected columns along lines 3 and 4 individually, and a more efficient procedure is to lump the line 4 columns into a single full-height one, together with their associated masses and gravity loads, similar for the line 3 columns. This maintains the column stiffness contribution, the inertia forces, and the P–Δ effect from the tributary gravity loads. With a rigid diaphragm assumption, a single equivalent column could be used for both lines 3 and 4 since their lateral displacements would be equal at each level.

In the perpendicular direction of the example building, a model consisting of the moment frame along line C plus a half slice of the pin-connected frame along line B would be appropriate for east-west and vertical ground motions plus gravity loads (Figure 23c). Linking springs are again used to connect the two frames. Column lumping could be applied to the pin-connected columns at the ends of line C and along line B.

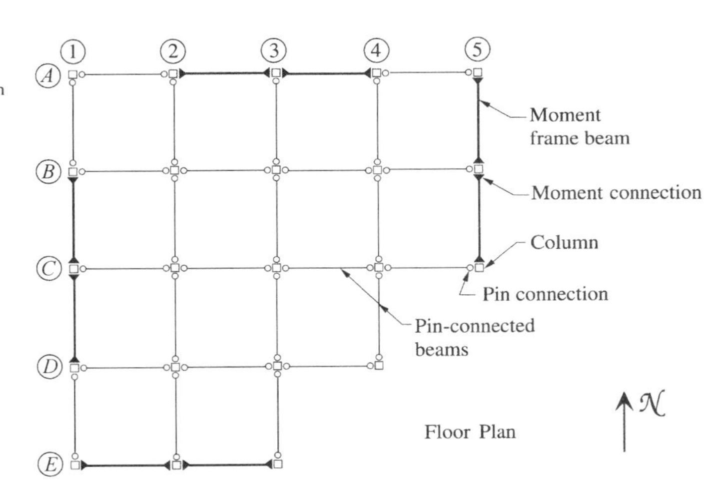

FIGURE 24 Unsymmetrical building that can be analyzed using a single model of planar frames placed in a three-dimensional setting.

Another building design containing two double-bay moment frames in each direction is shown in plan in Figure 24. However, the lack of symmetry in this building makes twisting likely during earthquake response. For example, north-south ground motion would directly load the moment frames along lines 1 and 5, and the perpendicular moment frames along lines A and E would receive load indirectly through twisting of the building. Thus, a three-dimensional treatment is in order. However, the planar frame concept can still be used by placing the moment frames in a three-dimensional setting with correct location and orientation. Each of these frames would contribute stiffness to the system only in its own plane and use only its in-plane degrees of freedom. Constraints are imposed on how the frames move based on the floors and roof deflecting rigidly in their own planes (rigid diaphragm assumption). Lumping of the pin-connected columns is not possible when twisting is present because the lateral displacements of each column will be different. To eliminate the associated degrees of freedom, these columns can be omitted and their stiffnesses approximately added back in the form of story shear springs. The P–Δ effects from the gravity loads carried by the omitted columns can be accounted for without using any extra degrees of freedom through forces applied to the model at the original column locations. Thus, the unsymmetric example building in Figure 24 can be represented by a single model with pseudo three-dimensional characteristics. All three components of ground motion are applied simultaneously. Details are given by Carlson (1999).

In the previous examples of Figures 23 and 24, the moment frames are separated from one another and linked only by springs or constraints representing the floor and roof diaphragms. Intersecting frames create shared columns and require a higher level of three-dimensional treatment. In the example building in Figure 25, column $A1$ will undergo bi-axial bending from the moments applied by beams connecting from the two directions, and the shear forces in these beams will both contribute to the axial force in the column. Such cases having orthogonal frame

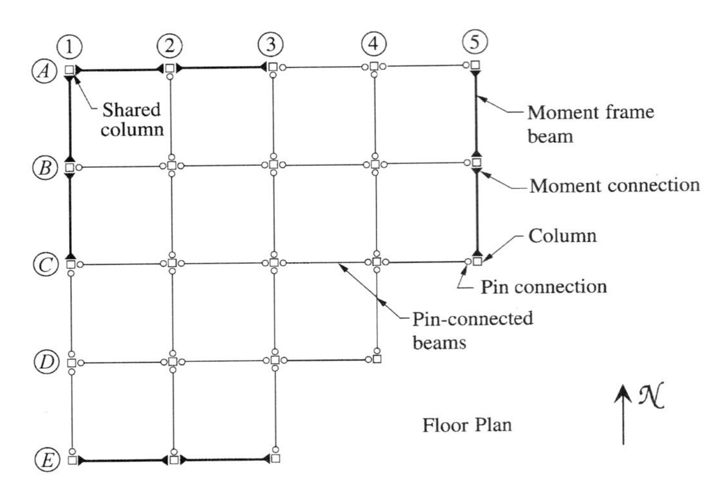

FIGURE 25 Building containing a shared column between two intersecting frames. Column $A1$ requires full three-dimensional treatment.

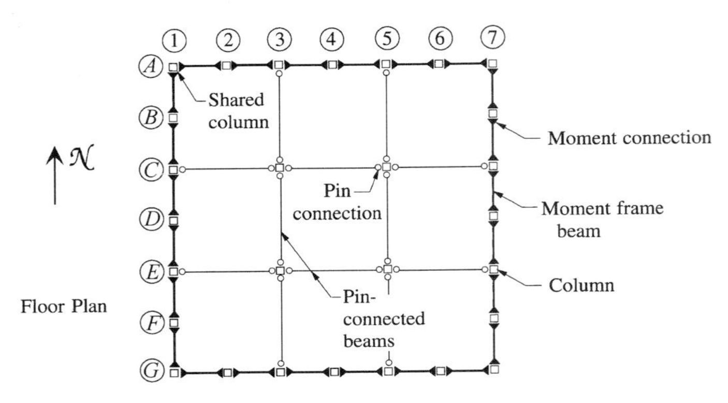

FIGURE 26 Building containing intersecting perimeter frames for which the global flange effect may be important.

intersections can be treated using planar frames except for the shared columns, which are modeled as three-dimensional elements (Carlson, 1999). Full three-dimensional treatment is possible, but economy often requires that the number of degrees of freedom present in the model be reduced by employing planar frames.

Another three-dimensional feature of intersecting frames is the global flange effect. Although present in the previous example, this effect is pronounced for a tall building with closely spaced moment-frame or braced-frame columns. A plan of such a building with perimeter moment frames intersecting at the corners is shown in Figure 26. Under north-south ground motion, for example, the moment frames along lines A and G behave somewhat like the flanges of a box beam, while the moment frames along lines 1 and 7 behave like the webs. Much of the overturning moment is carried by axial forces in the columns, which are distributed in plan like bending stresses in a box beam. In fact, tall buildings are often designed with this scheme in mind because considerable stiffness can be gained by making the structure act as a vertical cantilever beam with flange action. Such behavior is efficiently modeled using intersecting planar frames with three-dimensional shared columns, as discussed for the previous example.

Moment frames that intersect at angles other than right angles will in most cases have to be modeled with full three-dimensionality. This is also true even for orthogonal frames when moment connections are used in the interior, creating many columns that are shared between intersecting frames. Nevertheless, the planar frame concept as modified to accommodate a limited number of three-dimensional columns shared between orthogonal frames has the capability to model a majority of moment-frame and braced-frame buildings.

8. Example Building Analysis

A steel moment-frame building (Figure 27) with six stories aboveground and a one-story basement is analyzed in this section for combined gravity and earthquake loads. The building design uses A36 steel and meets the requirements of the 1994 Uniform Building Code (International Conference of Building Officials, 1994) for typical office dead and live loads, Zone 4 (highest seismicity), soil S_D (stiff soil), importance factor $I = 1$, R factor of 12, and design period of 1.16 seconds. As shown in the figure, the building is rectangular in plan with three 7.32-m-wide (24 feet) bays running north-south and four 9.15-m-wide (30 feet) bays

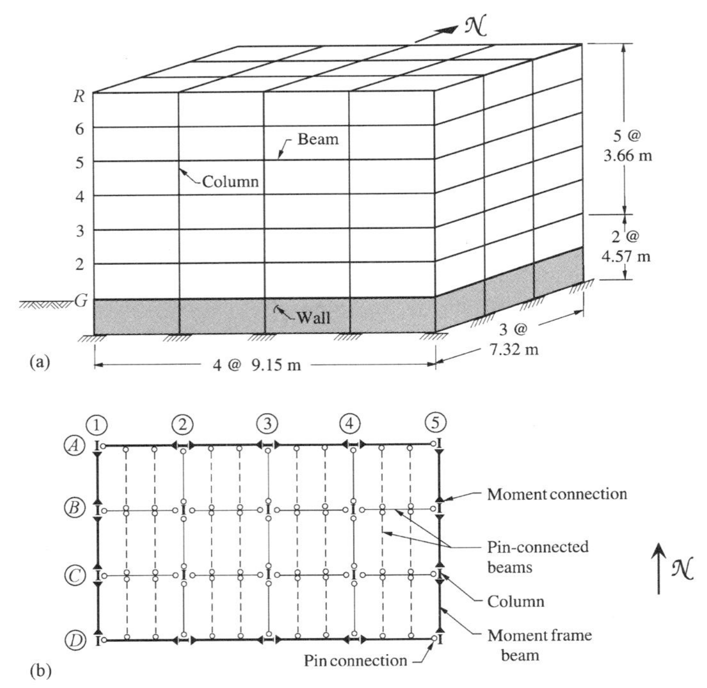

FIGURE 27 Example building to be analyzed: perspective view (a) and floor plan (b). Orientations of I-section columns are as shown in plan.

TABLE 2 Member Designations for the Example Building. Column Orientations Are as Shown in Figure 27

Beam Schedule

Level	Lines *A*, *D*	Lines 1, 5
G, 2, 3	W27 × 114	W24 × 103
4, 5	W24 × 103	W24 × 84
6, *R*	W24 × 68	W24 × 55

Column Schedule

Story	Corner	*B*1, *C*1, *B*5, *C*5	*A*2, *A*3, *A*4 *D*2, *D*3, *D*4	Interior
Bsmt., 1, 2	W14 × 159	W27 × 161	W27 × 194	W14 × 132
3, 4	W14 × 132	W27 × 129	W27 × 146	W14 × 74
5, 6	W14 × 99	W24 × 84	W24 × 103	W14 × 74

running east-west. Lateral forces are resisted by moment frames on the perimeter and by concrete walls in the basement story. Interior framing is designed for gravity loads only with pin connections between the beams and columns. To avoid bi-axial bending in the corner columns, pins are also used to connect the east-west running beams to these columns. Member designations for the beams and columns according to the AISC (American Institute of Steel Construction, 1989) are listed in Table 2. Panel zones are reinforced with doubler plates of thickness 0.95 cm (3/8 inch) for the corner columns and 1.27 cm (1/2 inch) for the other moment-frame columns.

The analysis presented here is based on planar frame modeling using the beam element developed previously. Separate planar models are employed for north-south ground motion and for east-west ground motion. Because of symmetry, only halves of the building are included. Depictions of the computer models are shown in Figure 28. The north-south model consists of the 3-bay frame on line 5 (east frame) connected by pinned beams to an equivalent column representing the rest of the east half of the building. These pinned beams serve as linking springs. Properties of the equivalent column are computed by summing full contributions from all columns on line 4 and half contributions from columns on line 3. These contributions include structural properties and the tributary gravity loads and masses. Basement walls are not modeled explicitly, but by lateral supports at ground level because these walls are stiff in plane. The east-west model consists of the 4-bay frame on line *D* (south frame) and another equivalent column representing the rest of the south half of the building. The floor and roof diaphragms are assumed to be rigid.

As discussed in the previous section, the use of separate models for north-south and east-west ground motions is appropriate when the two orthogonal sets of planar frames are separated from one another. For the present example, each corner column actually participates in two orthogonal moment frames and is thus a shared column. However, because of their pin connections to the east-west running beams, the corner columns receive most of their load under north-south ground motion as participants in the line 1 or line 5 moment frames. Thus, separate models can be used as an approximation, subject to verification.

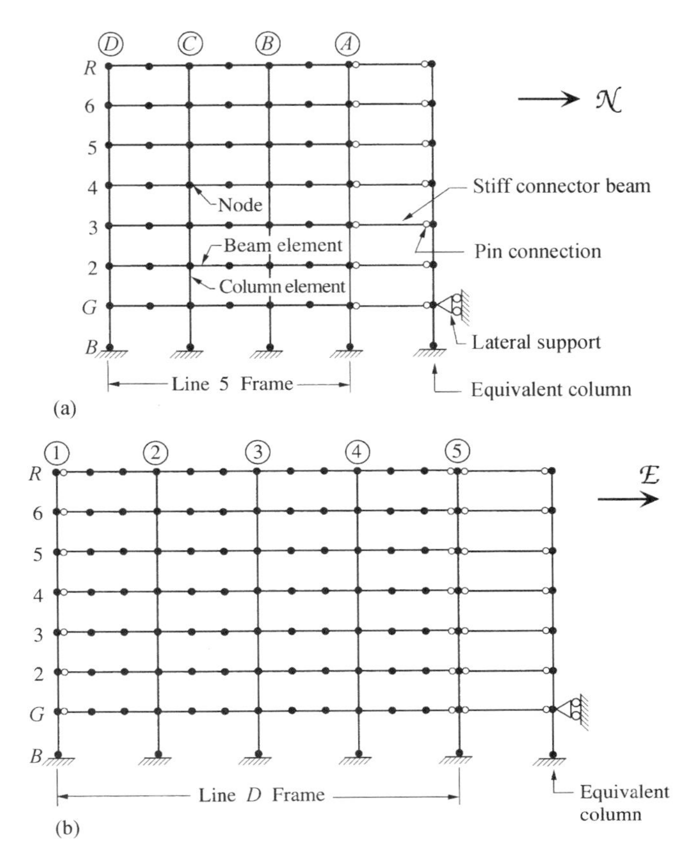

FIGURE 28 Computer models of example building for north-south ground motion (a) and east-west ground motion (b). Gravity loads and vertical ground motion are also applied to each model.

For the analyses, gravity loads are based on 3.83 kPa (80 psf) roof dead load, 4.55 kPa (95 psf) floor dead load, 0.24 kPa (5 psf) roof live load, 0.96 kPa (20 psf) floor live load, and 1.68 kPa (35 psf) cladding load on the exterior face of the building. In the interior framing, these loads are carried by secondary floor and roof beams running north-south and spaced at 3.05-meter (10 feet) intervals, as shown in Figure 27. The line *D* frame model contains nodes at third points of its beams to pick up reactions from the north-south running beams. The line 5 frame model contains nodes at the midpoints of its beams to pick up a smaller amount of gravity load. Tributary floor or roof areas for determining the amounts of gravity load to be applied to the nodes of the two computer models are shown in Figure 29.

Because the columns are fairly stiff, buckling is not expected, and formation of plastic hinges should be confined to the frame joints. Therefore, each column segment spanning one-story height is modeled with a single element. Bases of all columns are taken to be fixed, and the foundation is assumed to be rigid.

The equation of motion for uniform ground motion [Eq. (19), specialized for rigid foundation] is integrated by the constant average acceleration method with a time step of 0.02 seconds. Nodal masses are based on the same tributary areas as used for the gravity loads, but the roof live load contribution is not

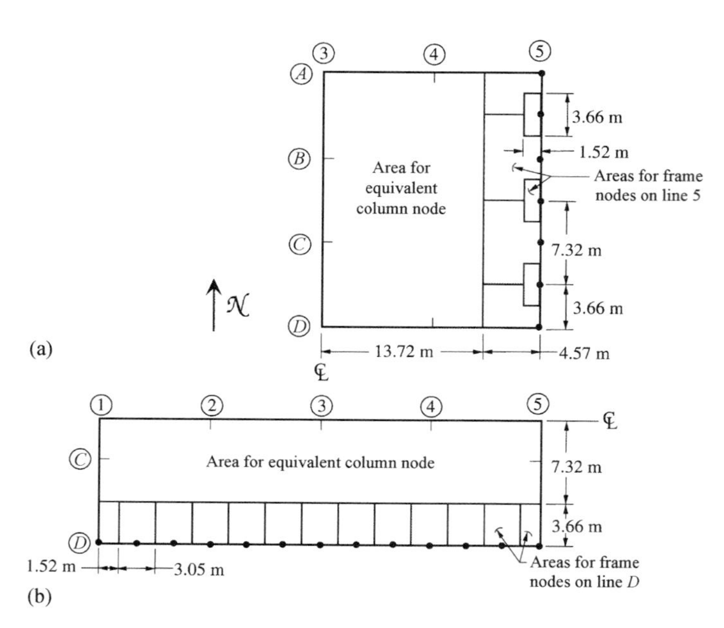

FIGURE 29 Tributary areas for gravity load computation for nodes of the north-south model (a) and east-west model (b).

considered and the floor live load contribution is reduced to 0.48 kPa (10 psf). The mass matrix is diagonal, with the mass at each node applied to both the horizontal and vertical degrees of freedom. Rayleigh damping is employed to give 5% of critical at periods of 0.4 seconds and 1.6 seconds. Computed periods for the fundamental modes of the linear system (P–Δ effect included) are 1.58 seconds for the north-south model (line 5 frame plus equivalent column) and 1.55 seconds for the east-west model (line D frame plus equivalent column).

Other features of the analysis include those discussed in previous sections: plastic hinge strength M_{pr} reduced under presence of axial load P as given in Figure 16, plastic hinge strain hardening at $h = 0.025$, effect of axial load on bending stiffness, axial strength P_y, panel zone flexibility and yielding with second slope at $h = 0.035$, actual dimensions of joints, and updating of nodal coordinates and joint angles. Further, capability for elastic shear deformations is added to the beam elements (Hall and Challa, 1995). A realistic strength of 290 MPa (42 ksi) is used for the A36 steel. Composite action between beams and slabs is not included.

The ground motion recorded at the Olive View Hospital free-field site during the 1994 Northridge earthquake is employed in the analyses. Horizontal and vertical components of ground displacement, velocity, and acceleration are shown in Figure 30, along with the pseudoacceleration response spectra and displacement response spectra in Figure 31. The particular horizontal component selected is the one that maximizes the peak-to-peak ground velocity. In the analysis of each model, this horizontal component is applied together with gravity loads and the vertical component of ground motion.

Maximum plastic hinge rotations and panel zone plastic shear strains (in excess of yield) from the analyses are shown in Figure 32. The units are percent of a radian (radians times 100).

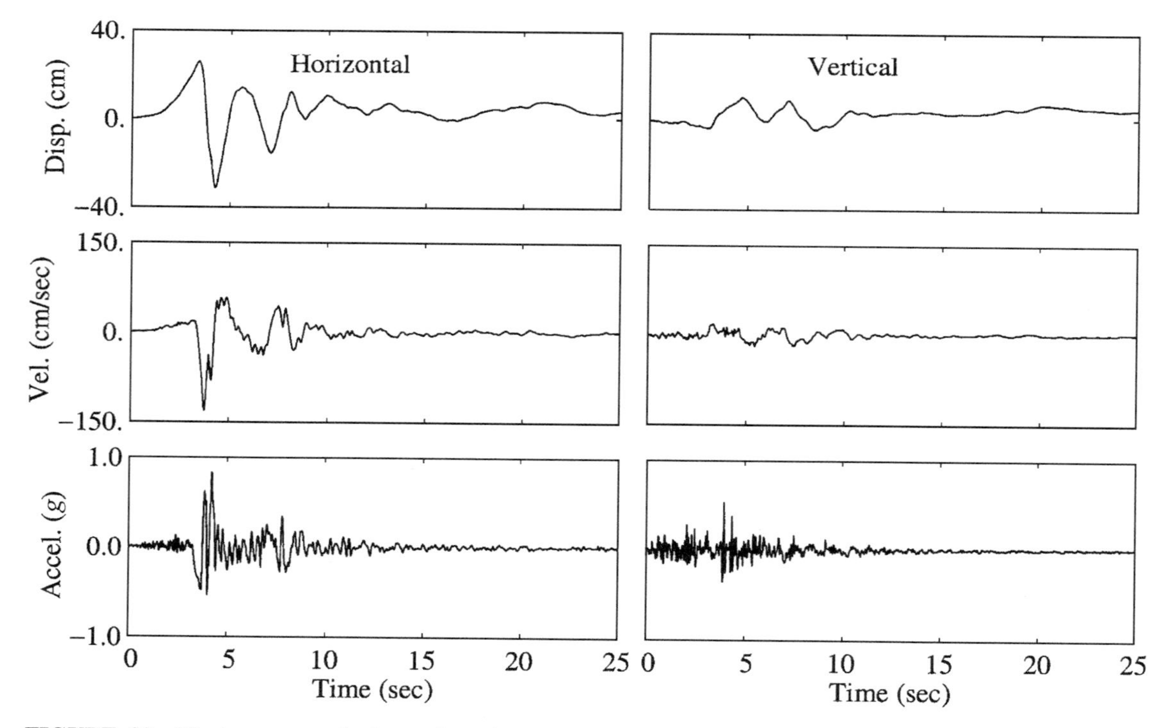

FIGURE 30 Displacement, velocity, and acceleration time histories for the ground motions recorded at the Olive View Hospital free-field site during the Northridge earthquake.

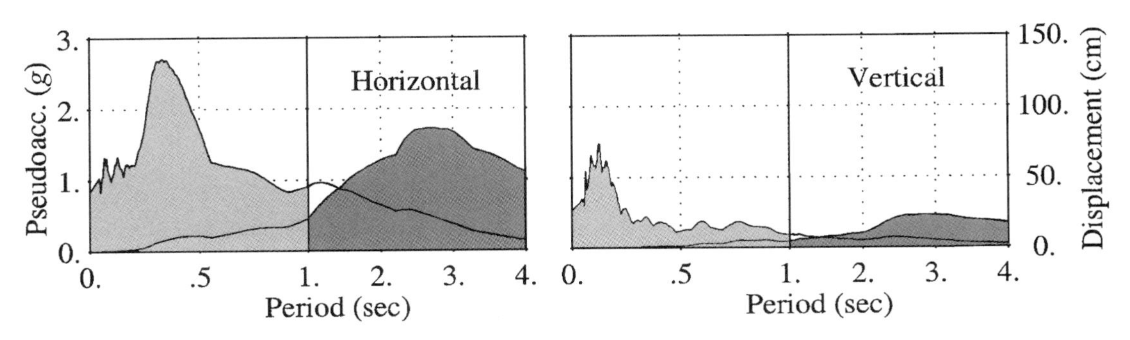

FIGURE 31 Pseudoacceleration (light gray) and displacement (dark gray) response spectra for the Olive View Hospital free-field ground motions (5% damping).

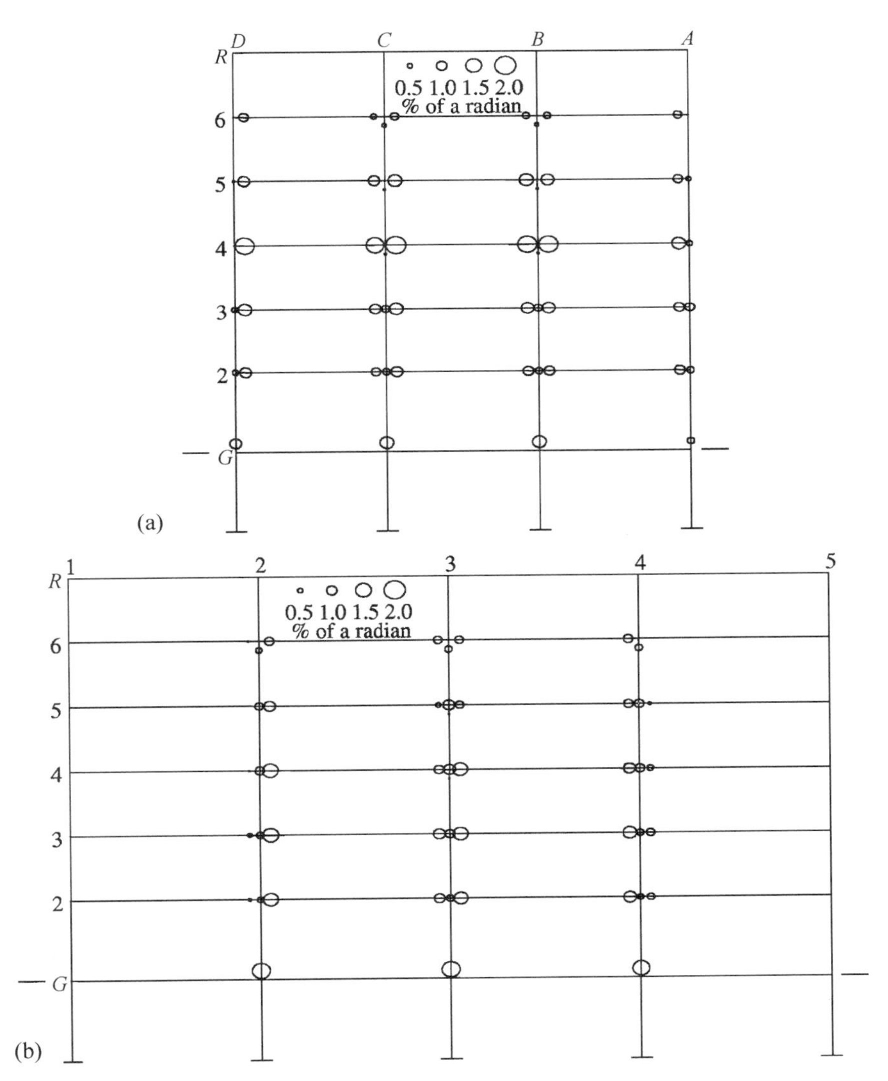

FIGURE 32 Maximum inelastic rotations of plastic hinges and panel zones for the frames on lines 5 (a) and D (b).

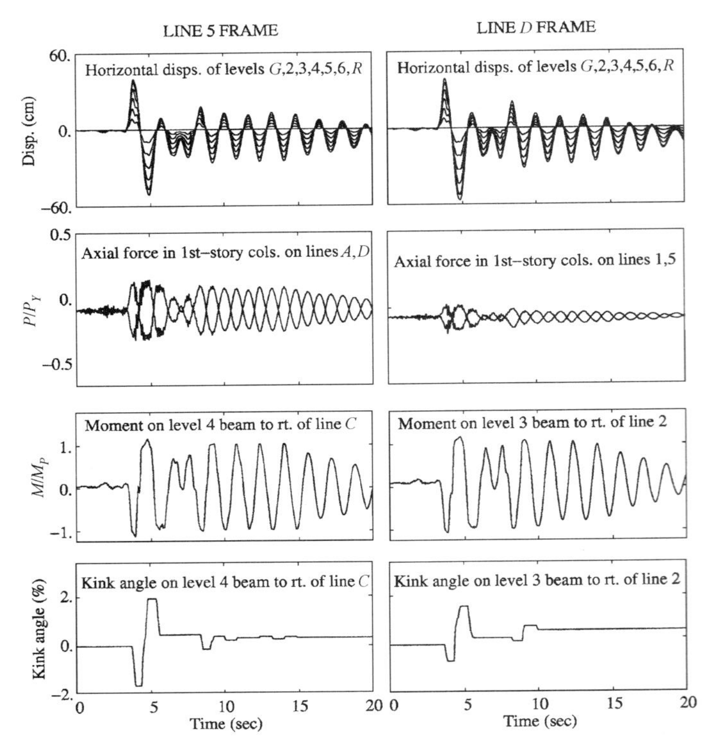

FIGURE 33 Selected time histories for the frames on lines 5 (left) and D (right).

The largest plastic hinge rotations occur in the first-story columns at ground level and in the beams from the 2nd to 5th floors. All values are below 2%, which would be considered a successful performance given the severity of the ground motion. Panel zone shear strains are small due to the strengthening provided by the doubler plates. Almost no column yielding occurs above ground level because the design employs the strong column-weak beam concept, which forces most of the yielding into the beams or panel zones.

Selected time history responses are plotted in Figure 33. Horizontal displacements of each level show a peak roof displacement of 57 cm for east-west response along with a peak drift of 3.2% in the third story. Values for north-south response are slightly smaller. Drift is story displacement divided by the story height. Axial forces in the first-story corner columns show larger response for north-south ground motion, but the values are well below the axial yield strength and do not cause any appreciable reduction in the column plastic moment capacity. Time histories of moment and kink angle are plotted for beam locations where the maximum kink angle is largest: 4th floor just to the right of the line *C* column for the north-south model and 3rd floor just to the right of the line 2 column for the east-west model. Maximum kink angles at these two locations are 1.9% and 1.5%, respectively. The moments exceed M_p due to the strain hardening feature. Predominant periods of response evident in Figure 33 are 1.60 seconds for north-south response and 1.57 seconds for east-west response. These slightly exceed the periods of the elastic structure due to the nonlinearity present.

9. Bridge Case History

As a result of the 1989 Loma Prieta earthquake, the California Department of Transportation has undertaken a major program to retrofit seismically vulnerable bridges in the state. Finite element analyses have been used extensively to develop the retrofit designs. The most challenging analyses are being carried out for the long crossings over bay waters. These major bridges require many degrees of freedom due to their great extents, and are also complex structurally, are typically located close to major faults where the design ground motions are strong, and often are founded at least partly on soft soil. Because of the high cost to retrofit these bridges, the performance criteria for the safety evaluation earthquake accept significant damage, which means that even the finite element models for the retrofitted structures must be nonlinear. Further, long bridges have numerous expansion joints, and these are inherently nonlinear due to impacts in the joints as they close during earthquake shaking. The soft soil contributes still another source of nonlinearity.

In this section, some details of the retrofit design of the San Diego–Coronado Bay Bridge are presented. The information was supplied by ANATECH Corporation of San Diego, California, who also performed the analyses (Dameron *et al.*, 1997; ANATECH Corp. and McDaniel Engineering/J. Muller Intl. Joint Venture, 1999). Mr. Robert Dameron of ANATECH assisted in the preparation of this material.

The San Diego–Coronado Bay Bridge (Figs. 34 and 35) is 2.6 km long and consists of girders on concrete piers founded on piles over water and spread footings over land. The central portion of the bridge is a three-span continuous steel box girder extending for a length of 570 m at a height 65 m above the water surface. Other spans are either side-by-side plate girders in an alternating anchor/suspended span arrangement or simply supported precast concrete girders. The safety evaluation earthquake was postulated to be a magnitude 7 to 7.5 event within 1.5 km of the bridge but with some possibility of fault rupture under the bridge. Seismological consultants supplied three-component sets of free-field ground motions, which were different from one pier to the next because of spatial variations.

Computational considerations make a detailed finite element model of a structure as large and complex as the subject bridge impractical. Therefore, a two-level modeling strategy is necessary. For analysis of the entire structure, a coarse finite element discretization was used to predict overall dynamic response. This was supplemented by much more detailed discretizations of various components to define local nonlinear behavior beyond the capability of the coarse global model. In general, the global

FIGURE 34 The San Diego–Coronado Bay Bridge.

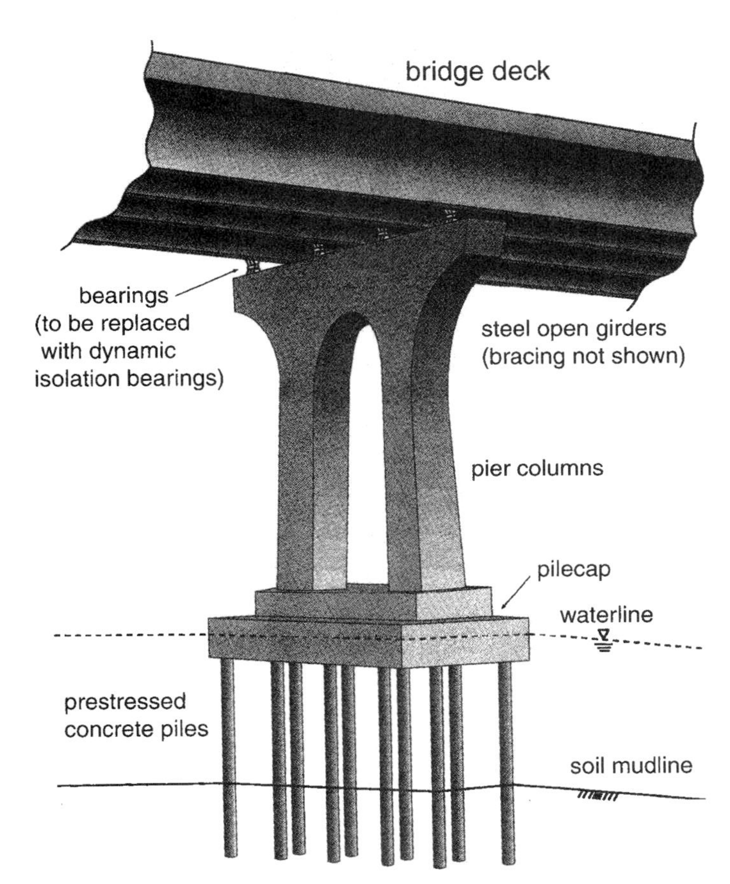

FIGURE 35 Sketch of pier, pile cap, and piles supporting a segment of the steel plate girder deck.

model was used to predict deformation demands, which were compared to deformation capacities determined from the detailed component models.

Detailed analyses made of representative piers (Figure 36) under statically applied lateral displacements using ANATECH's in-house software revealed important nonlinear behavioral modes such as locations of potential shear failures and plastic hinges. With this information, pier retrofits were designed to limit nonlinearity to plastic hinging at the tops and bottoms of the columns, which greatly facilitated modeling at the global level. A detailed analysis of a deck segment (Figure 37) identified a lateral buckling mechanism owing to inadequate cross bracing between the steel plate girders. With retrofitted bracing and replacement of some bearings with rubber isolators, the deck could be treated elastically in the global model where it was represented by a single line of beam elements whose properties were derived from the detailed deck models. Other detailed models were constructed of the piles, pile caps, existing bearings, and hanger mechanisms for the suspended plate girder spans and used to guide the treatment of these components in the global model.

A portion of the global model at a pier supporting a base-isolated segment of the plate girder deck is shown in Figure 38. Because such a global model consists of beam elements, springs (restrainers, bumpers, and isolators), rigid links, and dampers, it is often referred to as a stick model. Nonlinear features in the portion of the model above the pile cap include hysteretic shear behavior in the isolators, compression-only contact in the bumpers, tension-only contact in the longitudinal restrainers, the damper force–velocity relation, and plastic hinging in

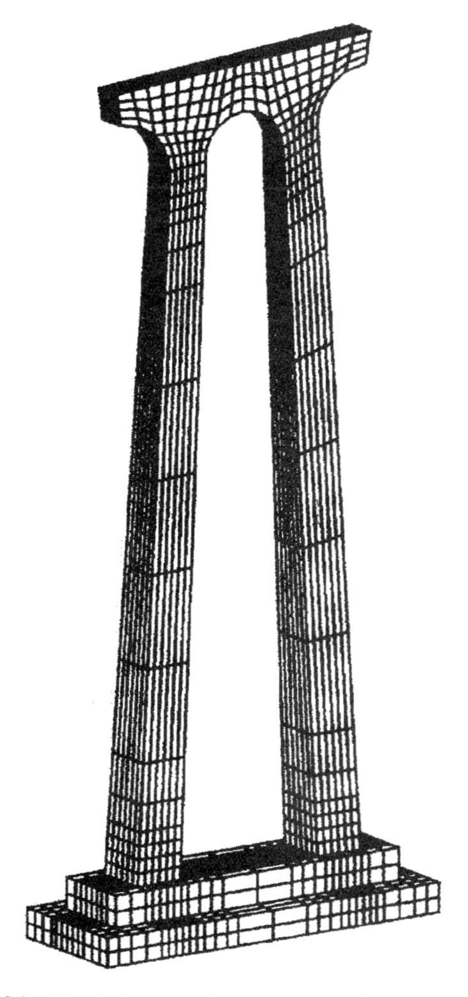

FIGURE 36 Detailed finite element model of a pier and pile cap. The elements are three-dimensional solids with nonlinear constitutive behavior for concrete and capability for including reinforcing steel.

the columns. To capture the finite length of the plastic hinge region, as determined from the detailed pier models, beam elements with distributed flexural yielding were employed. Geometric nonlinearity was also included, and this feature is especially important for the hangers because they can undergo large rotations. Below the pile cap, the piles and soil were originally modeled using beam elements for the piles and nonlinear horizontal and vertical springs for the soil. Each spring had one end connected to a pile and the other end constrained to move at the free-field motion for the respective pier. To reduce the number of degrees of freedom associated with the piles and soil, some stiffnesses were linearized using secant values, which permitted many degrees of freedom to be condensed out. Important nonlinear mechanisms were retained, including pile uplift, which can occur when the foundation is subjected to large overturning moment.

A sketch of the full global model containing 35 spans is shown in Figure 39. Besides the features already discussed, this model includes an abutment, steel box and precast concrete girder deck segments, other types of expansion joints and bearings, and spread footings under the land-based piers. There are approximately 12,000 degrees of freedom. Analyses using the global model were run with ADINA (ADINA R&D Inc., 1994), a commercially available structural analysis program.

Separate analyses of the San Diego–Coronado Bay Bridge were carried out for vulnerability assessment, retrofit design, and checking of the final design by an independent engineering company. Numerous analyses were required in the design phase because many retrofit variations had to be evaluated for performance and cost effectiveness. There were also several sets of ground-motion time histories that had to be considered, and uncertainty in the location of possible fault displacement under the bridge meant different locations had to be examined. It was therefore a necessity that the global model be efficient so that each computer run could be completed in a reasonable amount of time.

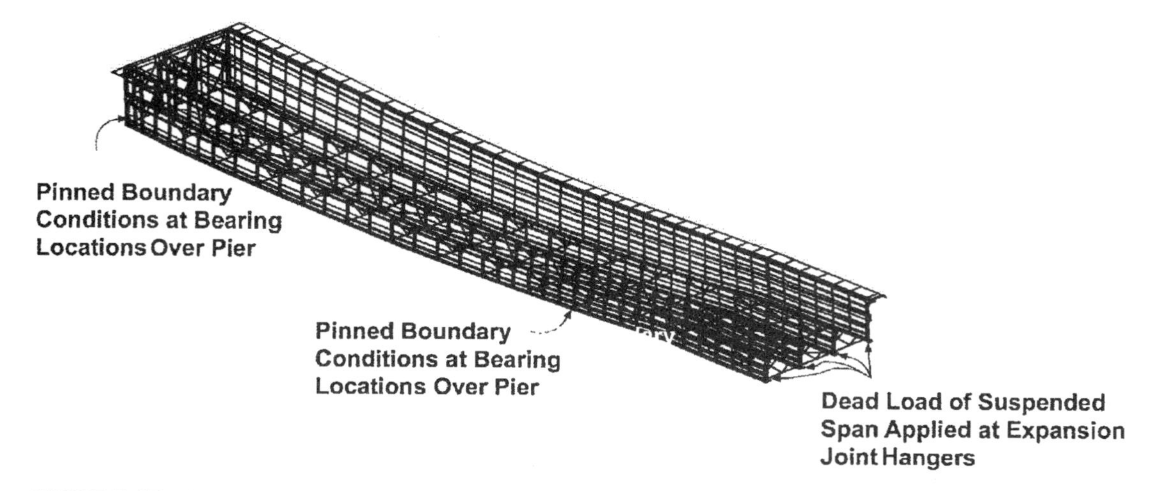

FIGURE 37 Detailed finite element model of steel plate girder deck with cross bracing between girders and concrete roadway on top. Elements are shells for the plate girders and roadway and beams for the bracing.

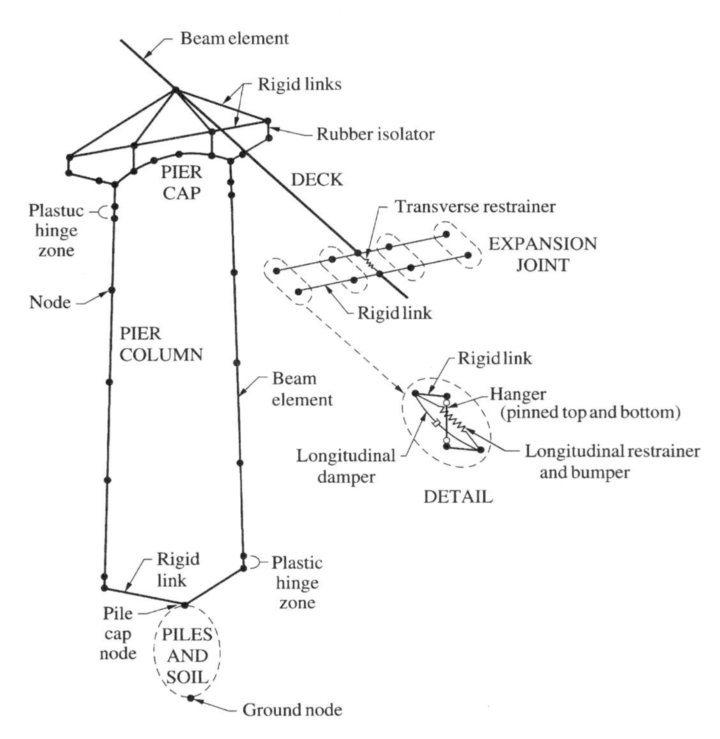

FIGURE 38 Portion of global finite element model at a pier supporting base-isolated plate girder deck (four girders wide) showing anchored and suspended spans. Details of piles and soil are not shown.

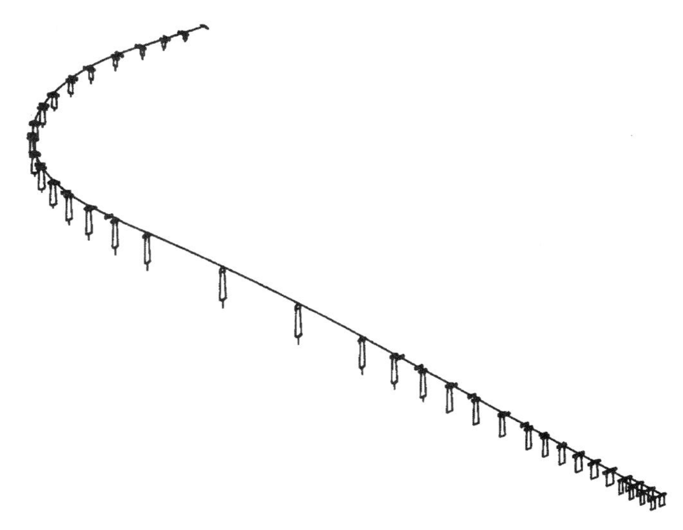

FIGURE 39 Global model of San Diego–Coronado Bay Bridge. Individual nodes are not shown.

10. Final Remarks

This chapter has presented an overview of finite element analysis as applied to earthquake engineering. The most general formulations include material and geometric nonlinearities, dynamic response computation, and interactions between structure, foundation, soil, and fluid. All of these aspects have been discussed.

Advances in finite element analysis continue to be made and there is every reason to believe that improvements will continue to be made in the future. One motivation is the need to predict the response of structures to strong ground shaking with greater confidence than is now possible. Such improvements will be necessary, for example, before performance-based design procedures can become a reality. Achieving such a result requires

- Improved definition of material and component behavior under cyclic loading so that better models can be incorporated into finite element computer codes. This is especially true for degrading behavior found, for example, in nonductile concrete buildings or liquefiable soil.
- More information on the response characteristics of subsystems such as bridge expansion joints or pile groups in soft soil so that finite element treatments can be verified and calibrated using real data.
- Better computational algorithms that exhibit robust convergence under highly nonlinear conditions.

In order for the usage of sophisticated finite element programs in engineering practice to expand, other improvements are also needed. These include more efficient pre- and post-processors to aid in preparation of input data and visualization of results. Computer speeds must also increase significantly, and much potential is available through parallel processing. So, there are certainly many areas where productive research can be undertaken.

References

Ad Hoc Committee on Soil-Foundation-Structure Interaction (1998). Seismic soil-foundation-structure interaction, Report to California Department of Transportation, Sacramento, CA.

ADINA R&D Inc. (1994). *ADINA User's and Theory Manuals*. Watertown, MA.

American Institute of Steel Construction (1989). *Manual of Steel Construction, Allowable Stress Design*. Chicago, IL.

ANATECH Corporation and the McDaniel Engineering/J. Muller International Joint Venture (1999). San Diego–Coronado Bay Toll Bridge Final Peer Review Report, February.

Applied Technology Council (1996). *ATC 40, Seismic Evaluation and Retrofit of Concrete Buildings*. Redwood City, CA.

Bathe, K.-J. (1982). Finite Element Procedures in Engineering Analysis. Prentice-Hall, Inc., Englewood Cliffs, NJ.

Camara, R.J. (2000). A method for coupled arch-dam-foundation-reservoir seismic behaviour analysis, *Earthq. Eng. Struct. Dynam.* 29(4), 441–460.

Carlson, A. E. (1999). Three-Dimensional Nonlinear Inelastic Analysis of Steel Moment-Frame Buildings Damaged by Earthquake Excitations. Report No. EERL 99-02, Earthquake Engineering Research Laboratory, California Institute of Technology, Pasadena, CA.

Challa, V. R. M., and J. F. Hall (1994). Earthquake collapse analysis of steel frames. *Earthq. Eng. Struct. Dynam.* **23**(11), 1199–1218.

Clough, R. W., and J. Penzien (1993). *Dynamics of Structures*. McGraw-Hill, Inc., New York.

Cook, R. D., D. S. Malkus, and M. E. Plesha (1989). *Concepts and Applications of Finite Element Analysis*. John Wiley & Sons, Inc., New York.

Dameron, R. A., V. P. Sobash, and I. P. Lam (1997). Nonlinear seismic analysis of bridge structures. Foundation-soil representation and ground motion input. *Computers and Structures* **64**(5), 1251–1269.

Dowrick, D. J. (1987). *Earthquake Resistant Design*. John Wiley & Sons, New York.

Fenves, G. L., S. Mojtahedi, and R. B. Reimer (1989). ADAP-88 A Computer Program for Nonlinear Earthquake Analysis of Concrete Arch Dams. Report No. UCB/EERC-89/12, Earthquake Engineering Research Center, University of California, Berkeley, CA.

Fok, K.-L., and A. K. Chopra (1985). Earthquake Analysis and Response of Concrete Arch Dams. Report No. UCB/EERC-85/07, Earthquake Engineering Research Center, University of California, Berkeley, CA.

Goyal, A., and A. K. Chopra (1989). Earthquake Analysis and Response of Intake-Outlet Towers. Report No. UCB/EERC 89/04, Earthquake Engineering Research Center, University of California, Berkeley, CA.

Hall, J. F., and A. K. Chopra (1980). Dynamic Response of Embankment, Concrete-Gravity and Arch Dams Including Hydrodynamic Interaction. Report No. UCB/EERC-80/39, Earthquake Engineering Research Center, University of California, Berkeley, CA.

Hall, J. F., and V. R. Murty Challa (1995). Beam-column modelling. *J. Eng. Mechanics* **121**(12), 1284–1291.

Haroun, M. A. (1980). Dynamic Analysis of Liquid Storage Tanks. Report No. EERL 80-04, Earthquake Engineering Research Laboratory, California Institute of Technology, Pasadena, CA.

International Conference of Building Officials (1994). *Uniform Building Code*. Whittier, CA.

Kaba, S. A., and S. A. Mahin (1984). Refined Modelling of Reinforced Concrete Columns for Seismic Analysis. Report No. UCB/EERC-84/03, Earthquake Engineering Research Center, University of California, Berkeley, CA.

Przemieniecki, J. S. (1968). *Theory of Matrix Structural Analysis*. McGraw-Hill Book Co., New York.

Salmon, C. G., and J. E. Johnson (1980). *Steel Structures, Design and Behavior*. Harper & Row, New York.

Tan, H., and A. K. Chopra (1995). Earthquake Analysis and Response of Concrete Arch Dams. Report No. UCB/EERC 95/07, Earthquake Engineering Research Center, University of California, Berkeley, CA.

Wilson, E. L., and M. Khalvati (1983). Finite elements for the dynamic analysis of fluid solid systems. *Int. J. Numer. Meth. Eng.* **19**(11), 1657–1668.

Wilson, E. L., G. R. Morris, and M. I. Suharwardy (1999). Numerical Errors Inherent in Seismic Structural Analysis Using Absolute Displacement Loading. October 1999 draft manuscript.

Wolf, J. P. (1988). *Soil-Structure Interaction Analysis in Time Domain*. Prentice Hall, Inc., Englewood Cliffs, NJ.

70

Liquefaction Mechanisms and Induced Ground Failure

T. Leslie Youd
Brigham Young University, Provo, Utah, USA

1. Introduction

The destructive power of liquefaction was abruptly brought to the world's attention by two large earthquakes in 1964—the March 27 Great Alaskan ($M_w = 9.2$) and the June 16 Niigata, Japan ($M_w = 7.5$), events. Liquefaction-induced lateral spread during the Alaskan earthquake distorted more than 250 highway and railway bridges, damaging most beyond repair and causing several to collapse (Fig. 1). Flow failures in shoreline areas during this same event carried port facilities out to sea, while returning waves overran coastlines causing additional deaths and destruction. In total, more than 50% of the Alaskan earthquake damage was caused by liquefaction-induced ground failure (Youd, 1978). During the Niigata earthquake, liquefaction reduced bearing strength beneath many buildings, causing settlement and tipping (Fig. 2). At other localities, lateral spreads collapsed bridges, severed pipelines, and wreaked havoc on pile foundations and other underground structures (Fig. 3). Research following those two events has clarified mechanisms controlling liquefaction, produced procedures for predicting its occurrence, and generated methods for mitigating damaging effects.

Field and laboratory tests indicate that liquefaction and subsequent ground deformation comprise complex phenomena that are difficult to model either physically or analytically. Thus, empirical procedures have become the standard of practice for evaluation of liquefaction resistance, prediction of ground deformation, and design of remedial measures. This chapter reviews mechanisms controlling liquefaction, subsequent soil deformation, and methods of analysis commonly applied in engineering practice.

2. Mechanisms of Liquefaction and Ground Deformation

The formal definition of soil liquefaction is "the transformation of a granular soil from a solid state to liquefied state as a consequence of increased pore water pressure and reduced effective stress" (Comm. on Soil Dynamics, 1978). Liquefaction occurs as seismic waves propagate through saturated granular sediment layers, which induces cyclic shear deformation and collapse of loose particulate structures (Fig. 4). As collapse occurs, contacts between grains are disrupted and loads previously carried through particle-to-particle contacts are transferred to the interstitial pore water. This load transfer generates increased pore water pressure and concomitant decrease of intergranular or effective stress. As pore water pressures increase, the sediment layer softens, allowing greater deformation and an accelerated rate of collapse of particulate structures. When the pore pressure reaches a certain critical level, the effective stress approaches zero and the granular sediment begins to behave as a viscous liquid rather than a solid, and liquefaction has occurred.

With the soil in a liquefied and softened condition, ground deformations occur readily in response to static or dynamic loading. The amount of deformation is a function of loading conditions, amplitudes and frequencies of seismic waves, the thickness and extent of the liquefied layer, the relative density and permeability of the liquefied sediment, and the permeability of surrounding sediment layers. Post-liquefaction shear behavior of liquefied sediment has been studied in statically and cyclically loaded laboratory tests. For example, stress–strain and pore pressure curves from three triaxial compression tests are plotted on Figure 5. The test specimens, composed of Ottawa

INTERNATIONAL HANDBOOK OF EARTHQUAKE AND ENGINEERING SEISMOLOGY, VOLUME 81B ISBN: 0-12-440658-0

FIGURE 1 Bridge buckled by 0.7 m of liquefaction-induced compression during the 1964 Alaska earthquake (USGS Photo).

sand, were prepared at different relative densities, saturated, and then loaded to failure by slowly increasing the axial load.

The response curves from Test 4-4 illustrate the behavior of loose granular materials. As that specimen was loaded, pore pressures increased to a level approaching the lateral confining pressure. At that point, the granular specimen liquefied and large flow deformation occurred, as indicated by the darkened segment of the curve. In the liquefied condition, about 20% shear strain occurred in approximately one second. Granular materials that liquefy and undergo large or unlimited flow deformation are termed "contractive" because the granular structure is sufficiently loose that shear at constant volume can occur with pore pressures remaining steady or rising slightly. Cyclic loading of contractive materials may also trigger liquefaction and flow failure. For flow failure to occur, ground slope or other loading conditions must induce static shear stresses that are larger than the shear resistance of the liquefied soil. Otherwise, flow would not occur. During flow, liquefied soil generates a small, but finite shear resistance, termed the "residual strength" or the "steady-state strength."

Many contractive soil layers have lain in a stable condition beneath sloping ground for years, only to suddenly fail as a consequence of seismically triggered liquefaction. For example, Figure 6 is a photograph of a flow failure that carried an embankment and roadway into Lake Merced during the 1957 Daly City, California, earthquake ($M = 5.2$). That embankment had been in use for many years and was stable until shaken by the 1957 earthquake.

Moderately dense granular soils are also capable of liquefying, but the density of the particulate packing is too compact to allow unlimited flow deformation. The behavior of a moderately dense soil is illustrated by the curves from Test 4-7 (Fig. 5). As with Test 4-4, pore pressures rose during initial shear deformation due to local collapses within the particulate structure. At a shear strain of about one percent, pore pressures reached a critical level, triggering liquefaction and flow deformation. The specimen deformed though an axial strain of about 6% in a fraction of a second. At that point, however, dilation developed within the specimen, which reduced the pore pressures, increased the shear resistance, and arrested the flow. This arrest caused the sample to revert back to a solidified condition. Continued loading of the specimen produced additional dilation-induced decrease of pore pressure and concomitant increase in shear resistance.

The phenomenon of dilation is illustrated diagrammatically in Figure 4C. Shear of granular soils requires particles to either roll over or slide past one another. For contractive soils, deformation causes ever-decreasing volume of the granular packing

FIGURE 2 Apartment buildings tilted by liquefaction-induced loss of bearing strength during the 1964 Niigata, Japan, earthquake.

FIGURE 3 Piles sheared by 1.2 m of lateral spread displacement during the 1964 Niigata, Japan, earthquake.

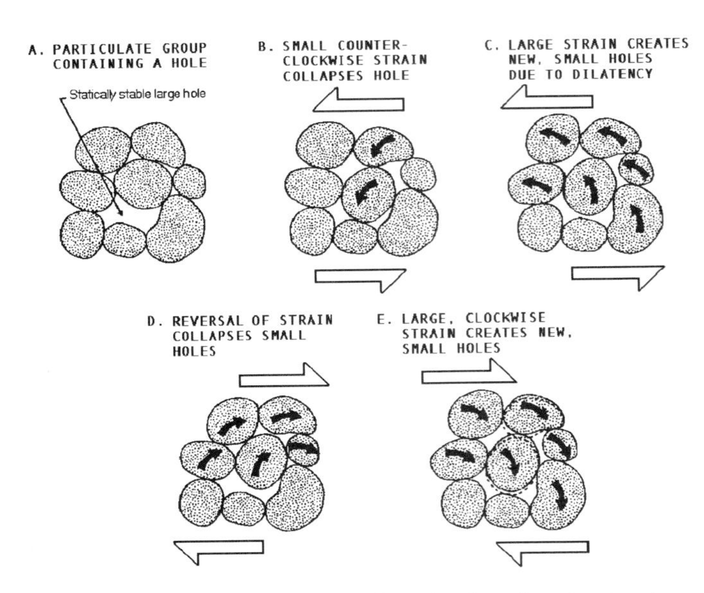

FIGURE 4 Schematic representation of packet of sand grains showing packing changes that occur during cyclic shear deformation. Modified from Youd, T. L., 1977, Packing changes and liquefaction susceptibility: American Society of Civil Engineers, *Journal of the Geotechnical Engineering Division*, V. 103, no. GT8, p. 918–922. Reproduced by permission of the American Society of Civil Engineers.

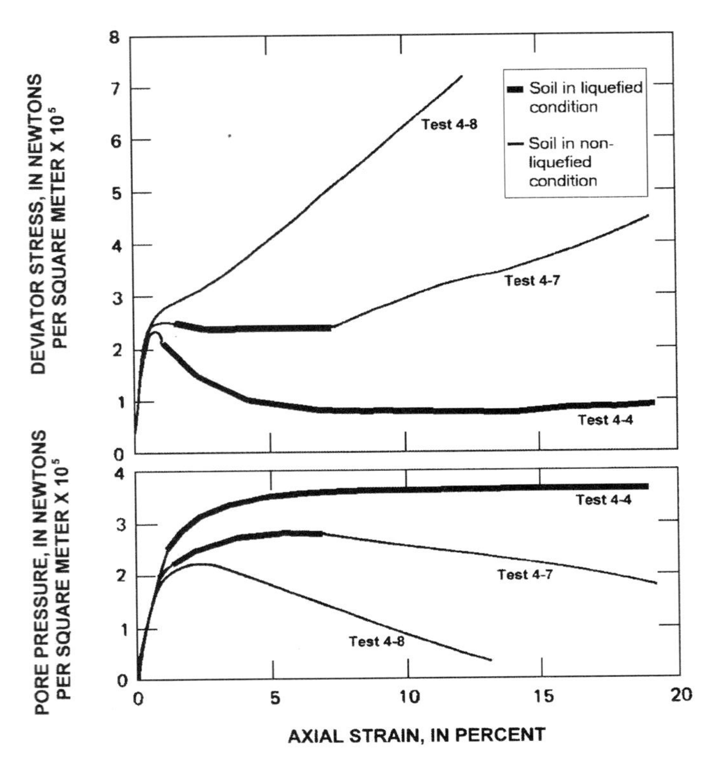

FIGURE 5 Stress–strain and pore pressure curves from undrained triaxial compression tests on saturated Ottawa sand (data from Castro, 1969; curves modified from Youd, 1975).

(where drainage can occur), until a steady-state volume condition is reached, termed the "critical void ratio." For moderately dense to dense soils, shear deformation causes particles to climb over one another, increasing void space in the particulate packing. Alternatively, the granular particles could fracture or crush during shear, preventing dilation. Most minerals in granular sediments, such as quartz, do not break easily, however, which forces grains to roll over one another and dilation to occur. For saturated materials sheared under undrained conditions, the specimen volume remains constant, with the tendency to dilate generating greatly reduced pore pressure and increased shear resistance. Such strength gain is illustrated by the stress–strain curves for Test 4-7 (Fig. 5). In this instance, the shear strength of the specimen more than doubled during shear in the post-liquefaction state.

Moderately dense soils can deform a finite amount in a liquefied state before dilatency reduces pore pressures and reverts the material back into a solidified condition. Figure 4C depicts a packet of granular particles in a dilatent state with particles forced to roll or slide over one another, tending to create small voids. Additional loading forces the particles more tightly against each other, increasing the dilative action. This condition is termed dilative arrest, which prevents further flow deformation in the direction of loading. A reversal of loading, however, reverses the direction of movement and releases the dilative action (Fig. 4D). This release repressurizes the water in the voids and reliquefies the granular soil. Flow deformation of the reliquefied material then occurs in the reverse direction, until movement is again halted by dilative arrest at the opposite extremity of the deformation excursion (Fig. 4E). During an earthquake, many reversals of stress may occur, generating repeated episodes of liquefaction, flow deformation, and dilative arrest. This repeated activation and halting of flow is termed cyclic mobility or cyclic liquefaction. Cumulative displacements from cyclic action can be substantial and damaging over the duration of a large earthquake. Such materials are classed as dilative, however, because unlimited flow deformations are prevented by dilative actions.

Test specimen 4-8 (Fig. 5) was too dense to liquefy, although pore pressures did rise during initial loading of the specimen. At larger strains, the specimen dilated strongly, causing greatly reduced pore pressures and high shear resistance. In this instance, dilation occurred before pore pressure could rise to the level required to trigger liquefaction. Cyclic shear loading may generate increased pore pressure and a transient liquefied condition in dense soils, but large flow deformation is prevented by the strong dilative tendency. Thus, liquefaction of dense soils does not create a significant earthquake hazard.

3. Liquefaction-induced Ground Failure

As noted previously, only when liquefaction leads to large ground deformation or ground failure is it hazardous to constructed works. Liquefaction may lead to any one of several forms of

FIGURE 6 Liquefaction-induced flow slide of highway embankment into Lake Merced during the 1957 Daly City, California, earthquake (photo by M.G. Bonilla, US Geological Survey).

ground failure, depending on surface loads, site geometry, and the depth, thickness, and extent of the liquefied layer. Ground failures are divided into two general categories depending on whether induced ground movements are primarily lateral or vertical.

3.1 Ground Failures Characterized by Lateral Ground Displacement

Ground failures associated with lateral ground displacements are of three types: flow failure, lateral spread, and ground oscillation (Youd, 1975). The bounds between these failure types are transitional, with the type of failure and amount of displacement dependent on local site conditions.

3.1.1 Flow Failure

Flow failure is the most catastrophic type of ground failure caused by liquefaction. Flow failure occurs on steep slopes (greater than 6% or 3.5°) underlain by contractive soils. These failures are characterized by large lateral displacements (several meters or more) and severe internal disruption of the failure mass. Structures founded on or within, or overridden by, the mobilized soil are usually destroyed. Figure 6 illustrates a flow failure that developed in a highway fill at the western edge of Lake Merced in San Francisco during the 1957 Daly City earthquake. Figure 7 shows a near-breach of the Lower San Fernando Dam due to a flow failure that carried the crest and much of the embankment into the reservoir during the 1971 San Fernando earthquake. As a consequence of this near-breach, 80,000 people were evacuated from their homes in the heavily populated San Fernando Valley downstream from the dam.

3.1.2 Lateral Spread

Lateral spread is characterized by horizontal displacement of soil layers down gentle slopes or toward a free face, such as an incised river channel (Fig. 8). Displacement occurs in response to a combination of static gravitational and earthquake-generated inertial forces acting on sediments within and above the liquefied zone. Displacement may be due to flow or cyclic mobility, depending on the relative density of the sediment. During failure, the surface layers commonly break into large blocks, which transiently jostle back and forth and up and down in the form of ground waves (ground oscillation) as they progressively migrate horizontally. Displacements usually do not exceed a few meters, but at localities where ground conditions are particularly vulnerable and shaking is intense, larger displacements have occurred.

Lateral spread displacement commonly creates open fissures and other extensional features at the head, shear deformation

FIGURE 7 Flow failure in the Lower San Fernando Dam during the 1971 San Fernando, California, earthquake that carried crest and much of embankment into the impounded reservoir (photo by T.L. Youd, US Geological Survey).

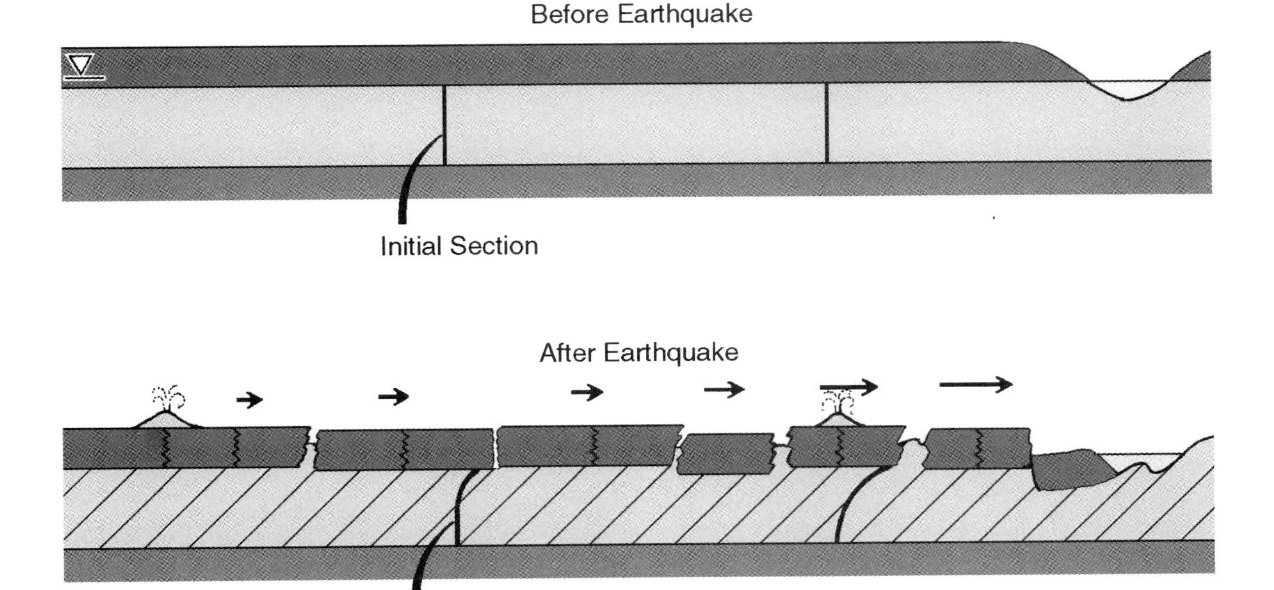

FIGURE 8 Diagram of a lateral spread showing shear deformation in the liquefied layer and migration of surface deposits down gentle slope or toward a free face (after Youd, 1984).

along the margins, and compressional features at the toe. These movements have pulled apart, sheared, or compressed foundations, pipelines, bridges, pavements, and other man-made structures. Figure 9 shows the Marine Sciences Laboratory at Moss Landing, California, that was pulled apart by a lateral spread movement of about 1.5 m during the 1989 Loma Prieta earthquake. Figure 1 shows an Alaska railway bridge that was compressed and buckled by a lateral spread that carried floodplain deposits toward a shallowly incised river channel. Figure 3 shows piles beneath a building that were fractured and offset by about 1.2 m due to lateral spread during the 1964 Niigata, Japan, earthquake.

3.1.3 Ground Oscillation

Ground oscillation occurs on flat terrain in response to inertial forces acting on decoupled soil materials above or within the liquefied zone (Fig. 10). This decoupling allows large transient ground motions or ground waves to develop, but permanent displacements are usually small and chaotic. Observers of ground oscillation commonly note slow-moving ground waves, up to a meter high, accompanied by opening and closing of fissures. During the 1989 Loma Prieta earthquake, ground oscillation broke water, gas, and sewer lines, and fractured pavements, curbs, and foundations in the Marina District of San Francisco (Fig. 11) (Pease and O'Rourke, 1997). However, the ground oscillations did not generate a consistent pattern of permanent lateral ground displacements that would be indicative of a lateral spread.

3.2 Ground Failures Characterized by Vertical Displacements

Liquefaction may produce large vertical ground displacements from three sources as described in the following paragraphs.

3.2.1 Loss of Bearing Strength

When the soil supporting a building or other structure liquefies and loses strength, soil deformations may occur, allowing the structure to settle and tip (Fig. 12). For example, during the Niigata, Japan, earthquake of 1964, spectacular bearing failures occurred at the Kawagishicho apartment complex, with several four-story buildings settling and tipping by as much as 60 degrees (Fig. 2). In this instance, liquefaction developed in a sand layer several meters below ground and then propagated upward through the overlying layers. The rising wave of liquefied soil created a quick condition, a form of liquefaction, weakening the soil supporting the buildings and allowing the structures to slowly settle and tip.

3.2.2 Buoyant Rise of Buried Vessels

Tanks, pipelines, unloaded timber piles, and other lightweight buried structures commonly buoyantly rise when the surrounding soil liquefies and behaves as a dense liquid. For example, the fuel tank shown in Figure 13, which was about one-third full at the time of the 1993 Hokkaido Nansei Oki, Japan, earthquake, floated to the ground surface as a consequence of liquefaction.

FIGURE 9 Marine Science Laboratory, Moss Landing, California, pulled apart by 1.5 m of lateral spread displacement during the 1989 Loma Prieta, California, earthquake (photo by T.L. Youd, Brigham Young University).

Before Earthquake

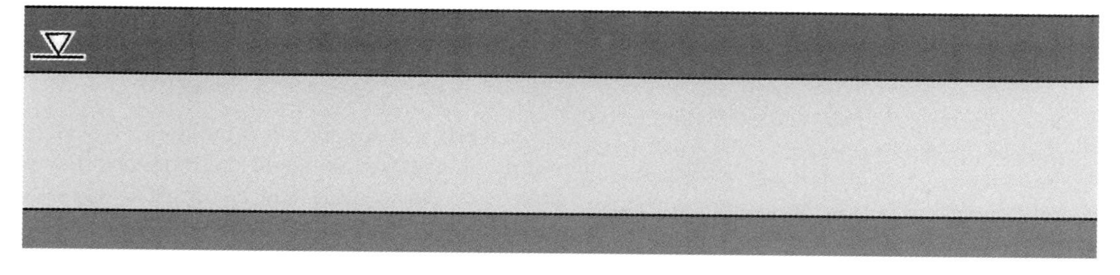

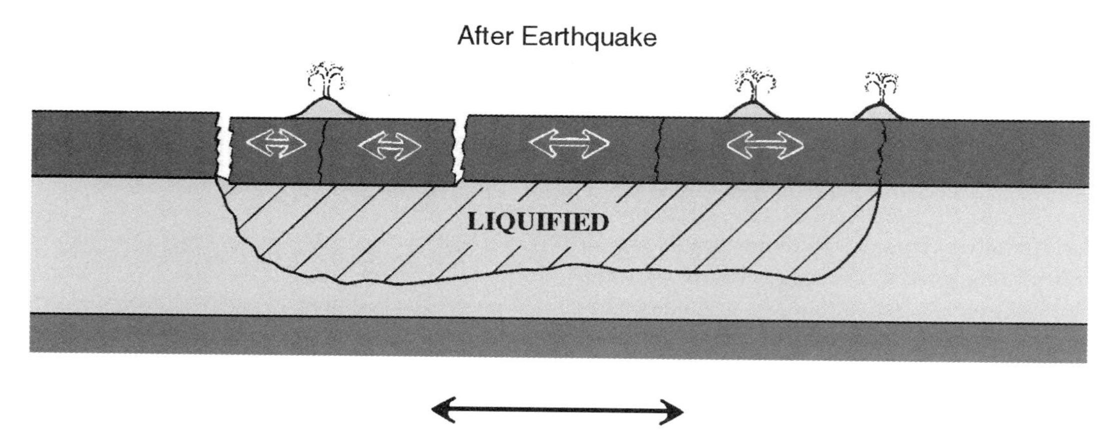

FIGURE 10 Diagram of ground oscillation showing decoupled blocks of intact soil that oscillate back and forth due to inertial forces and softening of liquefied soil (after Youd, 1984).

FIGURE 11 Buckled pavement, over-thrust curbs, and tilted utility poles in Marina district of San Francisco caused by liquefaction and ground oscillation during the 1989 Loma Prieta, California, earthquake (photo Courtesy of T.D. O'Rourke, Cornell University).

3.2.3 Ground Settlement

As noted previously, shear deformations generated by earthquake shaking tend to compact granular soils (Fig. 4B), leading to ground settlement. In dry sediments, compaction occurs quickly as air is driven from the voids. In saturated sediments with restricted drainage, pore pressures rise, causing softening of the soil, larger shear deformations, and enhanced compaction. Thus, settlements in saturated soils usually occur more slowly but are also usually larger than in unsaturated soils. Settlements commonly range from a few percent of the thickness of loose liquefiable layers to a fraction of a percent for dense sediments. Uniformly thick layers of homogeneous sediment usually compact and settle rather evenly with little consequent damage. Where granular layers vary in thickness, contain discontinuities, or where soil properties vary locally,

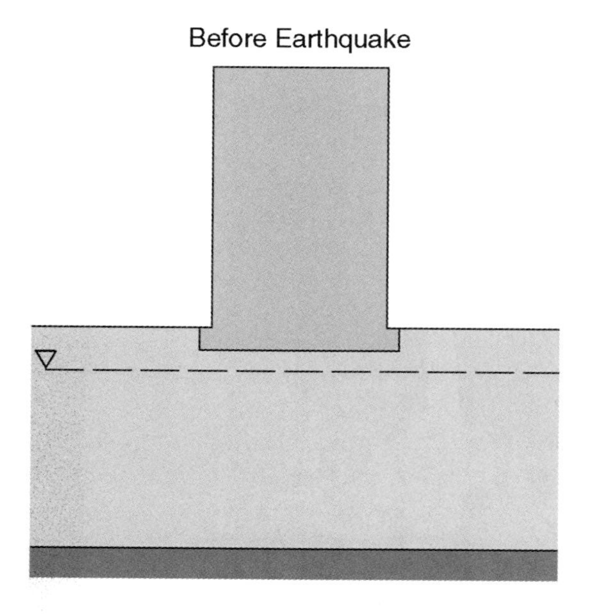

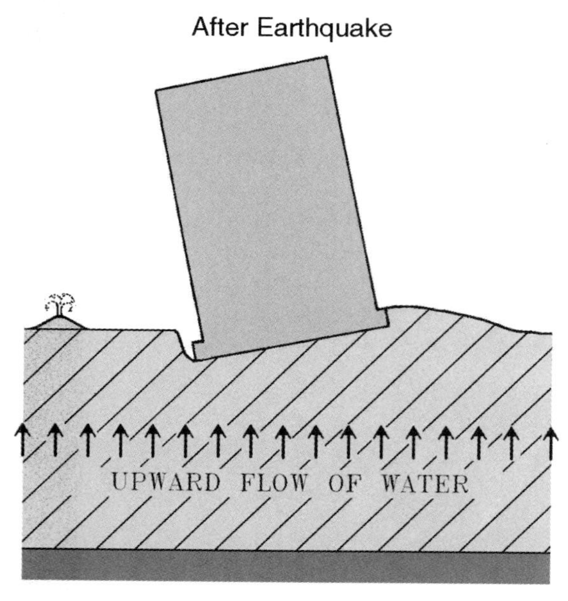

FIGURE 12 Diagram of bearing failure showing upward migration of pore water creating a liquefied or quick condition beneath structure (after Youd, 1984).

differential settlements are induced that typically are very damaging.

4. Evaluation of Liquefaction Hazard

4.1 Regional Procedures

Liquefaction does not occur randomly in nature but is limited to a rather narrow range of sedimentary, seismic, and hydrologic environments. Mapping or preliminary screening of liquefaction hazard is often based on mapping of susceptible environments. The general procedures for regional evaluation of liquefaction susceptibility listed here are summarized from the author's treatment of this topic in a monograph entitled "Screening Guide for Rapid Assessment of Liquefaction Hazard at Highway Bridge Sites" (Youd, 1998).

4.1.1 Geologic Evaluation

Sediments most susceptible to liquefaction are recently deposited, uncemented, granular materials that lie beneath the free groundwater table. Because density and cementation generally increase with age, liquefaction resistance also generally increases with age of the deposit. The geologic criteria listed in Table 1 are commonly used by engineers and geologists for preliminary assessment of liquefaction hazard and for mapping of liquefaction susceptibility (Youd, 1991). These criteria indicate that saturated late Holocene deposits are usually highly susceptible to liquefaction and that resistance generally increases with age through the Holocene and into the Pleistocene eras. Pre-Pleistocene sediments are generally immune to liquefaction and are rated at very low susceptibility.

The mode of deposition also influences liquefaction susceptibility. Sediments sorted into fine- and coarse-grained layers by wind, stream, or wave actions are generally more susceptible to liquefaction than unsorted sediments, such as residual soils and glacial tills. Processes that compact or consolidate sediments, such as dessication and glacial overriding, also reduce liquefaction susceptibility.

4.1.2 Seismic Evaluation

A certain threshold of seismic energy is required to generate liquefaction. That threshold is primarily a function of the density and stiffness of the granular material and the duration of earthquake shaking. Even the most susceptible natural materials require some seismic energy to generate a liquefied condition. The seismic factors most commonly used by engineers to characterize seismic energy propagating through a site are peak horizontal acceleration at ground surface (a_{max}) and earthquake magnitude (M). As applied by engineers, a_{max} is a measure of the intensity of the earthquake shaking and M is a measure of the duration. Minimal values of earthquake magnitude and peak acceleration required to generate liquefaction and damaging ground deformations are listed in Table 2 (Youd, 1998). These threshold values are generally conservative and are only applicable for screening evaluations. Site-specific analyses are required to evaluate liquefaction hazard for individual localities.

4.1.3 Water Table Evaluation

Only saturated sediments or sediments likely to become saturated need to be considered as susceptible to liquefaction. (A few flow landslides have developed in loose dry sediment, such as catastrophic flow slides that occurred in deep loess

FIGURE 13 Fuel tank that buoyantly rose to the surface due to liquefaction of surrounding soil deposits during the 1993 Hokkaido Nansei Oki, Japan, earthquake (photo by T.L. Youd, Brigham Young University).

deposits during the 1920 Kansu, China, earthquake (Close and McCormick, 1922), but these instances are very rare and can be neglected in routine evaluations.) Sediments beneath a free groundwater table, permanent or perched, are generally saturated. Because liquefaction resistance increases with overburden pressure and age of sediment, both of which generally increase with depth, liquefaction resistance usually increases markedly with depth. Thus the deeper the water table, the greater the natural resistance to liquefaction. Geotechnical investigations at sites of past liquefaction indicate that about 90% of episodes of liquefaction in natural sediments have developed in areas where the water table is shallower than 10 m and that few episodes have occurred in areas with a water table deeper than 15 m. (These relationships are not valid for man-made fills, such as embankment dams, dikes, or causeways.) Based on these observations, the water table depth relationships listed in Table 3 are suggested by Youd (1998).

4.2 Site-specific Evaluation of Liquefaction Resistance

4.2.1 Soil Type

Liquefaction is generally restricted to coarse-grained sediments (silts, sands, and gravels) that are sufficiently loose and uncemented so that they easily compact during seismic shaking. Clay particles in the granular matrix bond the coarser particles together and inhibit seismic compaction. Thus clay-rich sediments are generally resistant to liquefaction. Based on observed behavior of cohesive soils in areas of strong ground shaking, primarily reported from China, Seed and Idriss (1982) developed the criteria listed in Table 4. These criteria, commonly referred to as "the Chinese criteria," are widely used in the geotechnical profession and provide generally conservative predictions of liquefaction behavior. Youd and Gilstrap (1999) tested the applicability of the Chinese criteria at 19 sites where liquefaction did or did not occur during recent (1964–1994) earthquakes and where detailed site data have been compiled. Out of the 186 clay-content values reported from the 19 sites, all but 2 were from soil layers that did or did not liquefy in agreement with the 15% criterion in Table 4. Similarly, all reported liquid limits greater than 35% were from layers that did not liquefy. However, these layers were also characterized by clay contents greater than 15%. Nevertheless, the data compiled by Youd and Gilstrap generally verify the validity of the criteria listed in Table 4.

4.2.2 Penetration Resistance

Over the past 25 years, a procedure termed the "simplified procedure" has evolved for evaluating liquefaction resistance of soils. This procedure has become the standard of practice in North America and throughout much of the world. A parallel and generally similar procedure has been developed in Japan and is

TABLE 1 Estimated Susceptibility of Sedimentary Deposits to Liquefaction during Strong Seismic Shaking (after Youd and Perkins, 1978)

Type of deposit (1)	Distribution of cohesionless sediments in deposit (2)	Likelihood that cohesionless sediments, when saturated, would be susceptible to liquefaction (by age of deposit)			
		<500 yr (3)	Holocene (4)	Pleistocene (5)	Pre-Pleistocene (6)
		(a) Continental Deposits			
River channel	Locally variable	Very high	High	Low	Very low
Flood plain	Locally variable	High	Moderate	Low	Very low
Alluvial fan and plains	Widespread	Moderate	Low	Low	Very low
Marine terraces and plains	Widespread	—	Low	Very low	Very low
Delta and fan-delta	Widespread	High	Moderate	Low	Very low
Lacustrine and playa	Variable	High	Moderate	Low	Very low
Colluvium	Variable	High	Moderate	Low	Very low
Talus	Widespread	Low	Low	Very low	Very low
Dunes	Widespread	High	Moderate	Low	Very low
Loess	Variable	High	High	High	Unknown
Glacial till	Variable	Low	Low	Very low	Very low
Tuff	Rare	Low	Low	Very low	Very low
Tephra	Widespread	High	High	?	?
Residual soils	Rare	Low	Low	Very low	Very low
Sebka	Locally variable	High	Moderate	Low	Very low
		(b) Coastal Zone			
Delta	Widespread	Very high	High	Low	Very low
Estuarine Beach/	Locally variable	High	Moderate	Low	Very low
High wave energy	Widespread	Moderate	Low	Very low	Very low
Low wave energy	Widespread	High	Moderate	Low	Very low
Lagoonal	Locally variable	High	Moderate	Low	Very low
Fore shore	Locally variable	High	Moderate	Low	Very low
		(c) Artificial Fill			
Uncompacted fill	Variable	Very high	—	—	—
Compacted fill	Variable	Low	—	—	—

Modified from Youd, T. L., and Perkins, D. M., 1978, Mapping of liquefaction induced ground failure potential: American Society of Civil Engineers, *Journal of the Geotechnical Engineering Division*, V. 104, no. GT4, p. 433–446. Reproduced by permission of the American Society of Civil Engineers.

TABLE 2 Minimum Earthquake Magnitudes and Peak Horizontal Ground Accelerations, with Local Site Amplification, that are Capable of Generating Liquefaction in Very Susceptible Natural Deposits (after Youd, 1998)

	Liquefaction Hazard for Bridge Sites	
Earthquake Magnitude[1] M_w	Soil Profiles Types[2] I and II (stiff sites)	Soil Profile Types[2] III and IV (soft sites)
$M < 5.2$	Very low hazard for $a_{max} < 0.4$ g	Very low hazard for $a_{max} < 0.1$ g
$5.2 < M < 6.4$	Very low hazard for $a_{max} < 0.1$ g	Very low hazard for $a_{max} < 0.05$ g
$6.4 < M < 7.6$	Very low hazard for $a_{max} < 0.05$ g	Very low hazard for $a_{max} < 0.025$ g
$M > 7.6$	Very low hazard for $a_{max} < 0.025$ g	Very low hazard for $a_{max} < 0.025$ g

[1]Magnitudes are at the upper end of the magnitude intervals given by Hanson and Perkins (1995).
[2]Soil Profile Types, as defined in Standard Specifications for Highway Bridges (AASHTO, 1992).

TABLE 3 Relative Liquefaction Susceptibility of Natural Sediments as a Function of Groundwater Table Depth (after Youd, 1998)

Groundwater table depth	Relative liquefaction susceptibility
<3 m	Very High
3 m to 6 m	High
6 m to 10 m	Moderate
10 m to 15 m	Low
>15m	Very Low

TABLE 4 Criteria for assessing liquefiability of fine-grained soils (modified from Seed and Idriss, 1982)

Criteria required for liquefaction of fine-grained soils (All three criteria must be met for soil to be liquefiable)
• Clay Fraction (Percent Finer than 0.005 mm) < 15%
• Liquid Limit (LL) < 35%
• Moisture Content (MC) > 0.9 LL

incorporated in the Japanese Bridge Code (Iwasaki, 1986). That procedure is widely used in Japan and some other areas.

Following the disastrous earthquakes of 1964, Seed and Idriss (1971) developed the "simplified procedure" for evaluation of liquefaction resistance. This procedure has been updated periodically since that time with landmark papers by Seed (1979), Seed and Idriss (1982), Seed *et al.* (1985), NRC (1985), Iwasaki (1986), and Youd and Idriss (1997). Application of the simplified procedure requires calculation of two primary variables—the seismic demand placed on a soil layer, expressed in terms of cyclic stress ratio (CSR), and the capacity of the soil to resist liquefaction, expressed in terms of cyclic resistance ratio (CRR). The factor of safety against liquefaction in terms of these two variables is written as

$$FS = CRR/CSR \qquad (1)$$

Seed and Idriss (1971) formulated the following equation for calculation of CSR:

$$CSR = (\tau_{av}/\sigma'_{vo}) = 0.65(a_{max}/g)(\sigma_{vo}/\sigma'_{vo})r_d \qquad (2)$$

where a_{max} is the peak horizontal acceleration generated by the earthquake, g is the acceleration of gravity, σ_{vo} and σ'_{vo} are total and effective overburden stresses, respectively, and r_d is a stress reduction coefficient. Recommended procedures for evaluation of these variables are given by Youd *et al.* (2001).

The Japanese Bridge Code prescribes a very similar formula for the factor of safety against liquefaction, as follows:

$$FS = R/L, \qquad (3)$$

where L is the dynamic stress placed on the soil layer by the earthquake shaking

$$L = (a_{max}/g)(\sigma_{vo}/\sigma'_{vo})r_d \qquad (4)$$

and R is the soil liquefaction capacity factor.

Several procedures have been applied to determine CRR or R. One plausible method is to extract undisturbed soil specimens from field sites and then cyclically load those specimens in the laboratory to model seismic loading conditions in the field. Unfortunately, specimens of granular soils retrieved with standard drilling and sampling procedures are too disturbed to yield meaningful laboratory test results. Only through specialized sampling techniques, such as ground freezing, can sufficiently undisturbed specimens be retrieved. The cost of such procedures is prohibitive for all but the most critical projects. To avoid the difficulties associated with sampling and laboratory testing, field tests have become the state-of-practice for routine investigations.

Several field tests have gained common usage for evaluation of liquefaction resistance, including the standard penetration test (SPT), the cone penetration test (CPT), shear-wave velocity measurements (V_s), and the Becker penetration test (BPT). The standard penetration test has been the most commonly used test in foundation investigations for buildings, bridges, and other structures. Recently the cone penetration test has gained widespread use and has become the preferred initial test by many investigators (Robertson and Wride, 1997). Procedures for calculation of CRR or L from SPT or CPT data are beyond the scope of this paper, but can be found in papers by Youd *et al.* (2001) for North America and many other areas and by Iwasaki (1986) for Japanese practice.

5. Assessment of Ground Failure Hazard

Damage occurs when liquefaction leads to intolerable ground displacements or ground failure. Failure types include flow failure, inertially induced embankment deformation, lateral spread, loss of bearing strength, and ground settlement. Analyses of displacements generated by these failure types are generally complex and require the expertise of geotechnical specialists. Details of these procedures are beyond the scope of this chapter; thus only generalized descriptions are given here.

5.1 Slope or Embankment Instability Leading to Flow Failure

Liquefaction reduces the shear strength of contractive soil layers, which may lead to instability and catastrophic flow failure in embankments or steep slopes. Flow failure occurs when the shear resistance of the liquefied soil is reduced to a level that is less than the gravitational shear stress acting on the slope or embankment. Evaluation of flow failure potential begins with a standard limit–equilibrium analysis, but with the steady state or residual strength assigned to all liquefiable layers. Commercial computational programs are available to aid this analysis. If the static factor of safety is less than 1, catastrophic flow failure is possible. If the factor of safety is greater than 1, the embankment may be classed as statically stable against catastrophic flow failure, but may still undergo intolerable deformation as noted in the following paragraph.

5.2 Embankment or Slope Deformation

Liquefaction-induced slope or embankment deformations occur as a consequence of soil softening and yielding in response to inertial forces generated by an earthquake. Large deformation does not often occur during a single cycle of loading, but cumulative deformation over several cycles of loading may lead to intolerable displacements. Embankment and slope deformations are not easily modeled because of the complexity of stress–strain relationships that are influenced by pore pressure fluctuations generated by compression and dilation of the particulate structure during shear. The inhomogeneous composition of most natural soils adds to the difficulty of modeling the *in situ* stress–strain behavior.

5.3 Lateral Spread

If a site is stable against flow failure and excess deformation, the next mode of ground displacement that should be evaluated is lateral spread. Several analytical and mechanistic models are proposed for estimating lateral spread displacement. For example, the mechanistic model of Byrne (1991) has been used to estimate ground displacements for several important projects. For routine analyses, however, the empirical procedure of Bartlett and Youd (1995) has been widely applied. Bartlett and Youd derived their empirical model from multiple linear regression (MLR) of a large data set of measured ground displacements, topographic data, and subsurface soil profiles and cross sections. The final model predicts lateral displacement as a function of earthquake magnitude, distance from the seismic energy source, the thickness of loose granular layers with an $(N_1)_{60}$ less than 15 (approximately the cumulative thickness of liquefiable layers), and the mean grain sizes and fines contents of materials in the loose granular layers. The Bartlett and Youd procedure has been updated by Youd *et al.* (2002) to correct some errors and improve the predictive performance of the MLR model.

5.4 Analysis of Ground Settlement

Earthquake shaking is an effective compactor of granular soils—dry, moist, or saturated—leading to vertical ground displacement or ground settlement. Laboratory studies show that compaction is enhanced by liquefaction of granular materials (Lee and Albaisa, 1974). Tokimatsu and Seed (1987) and Ishihara and Yoshimine (1992) compiled data from laboratory tests and case histories of earthquake-induced ground settlements to formulate a simplified empirical criteria for estimation of ground settlement. These procedures are used for estimating liquefaction-induced settlement for routine engineering investigations. Predicted settlements from the two procedures are comparable for most sites.

The premise of these procedures is that earthquake shaking generates cyclic shear strains that compact granular soils. Where drainage cannot occur rapidly, the tendency to compact also generates transient pore water pressures that prevent an immediate decrease in volume. However, as pore pressures later dissipate, the layer consolidates, producing volumetric strain and ground settlement. Tokimatsu and Seed correlate volumetric strains with cyclic stress ratio (CSR) as a measure of the intensity of earthquake shaking and $(N_1)_{60}$, which correlates with relative density of the sand. Ishiara and Yoshimine correlate volumetric strain with a factor of safety against liquefaction. The final steps are to divide the sediment profile into layers, estimate the volumetric strain for each layer, multiply the thickness of each layer by the estimated volumetric strain, and then sum the incremental settlements.

5.5 Loss of Bearing Strength

The remaining possible liquefaction-induced hazard is loss of foundation bearing strength. Loss of bearing strength leads to penetration of foundations (deep or shallow) into the liquefied sediment. A standard bearing capacity analysis may be used to assess foundation stability for shallow foundations, but with residual soil strengths assigned to all liquefiable layers. The axial load capacity of piles is commonly calculated using standard procedures, but with zero strength assigned to all liquefiable layers. The lateral load resistance of piles is usually estimated by multiplying the calculated lateral resistance without liquefaction by a factor ranging from 0.1 to 0.3, depending on the thickness and penetration resistance of the liquefiable layers.

5.6 Lack of Ground Failure Potential

If all of the analyses just noted indicate an adequate factor of safety against failure, the site may be classed as non-hazardous and immune to detrimental effects of liquefaction, even though liquefaction of some subsurface layers may occur. Conversely, if the analyses indicate an inadequate factor of safety for any mode of failure, unacceptable soil deformation may occur leading to structural damage and possible casualties. In such instances, mitigative measures are required to reduce the risk.

6. Mitigative Measures

Where a liquefaction hazard has been identified, mitigative measures are required to eliminate or reduce the hazard to an acceptable level. These measures may include any one or a combination of the following actions: (1) avoidance of the hazard through zoning restrictions or relocation of facilities to safer sites; (2) strengthening of the structure to withstand the effects of liquefaction; (3) strengthening of the ground to prevent liquefaction and damaging ground deformations; and (4) evaluation and acceptance of the risk where hazard to life and limb is minimal. All of these measures have been used effectively to reduce damage. For example, well-reinforced shallow foundations force differential lateral displacements into shear of soil layers beneath the foundation rather than fracture of the foundation and superstructure. Similarly, pile or other deep foundations that transfer structural loads to competent layers beneath the liquefiable sediments have proven effective in preventing damage to structures at sites where little or no lateral ground movement was generated. Insurance has been used as a protection against large financial loss by many individuals and industries at localities where the threat of injury or loss of life is small.

7. Conclusion

The extensive damage suffered at many localities during past earthquakes has brought the importance of liquefaction as an earthquake hazard to the world's attention. Research over the past 40 years has identified mechanisms controlling the liquefaction process, developed procedures for predicting liquefaction and ground failure occurrence, and provided measures to mitigate potential damage. These mechanisms, procedures, and measures are reviewed in this chapter.

References

AASHTO (1992). *Standard Specifications for Highway Bridges*, 15th ed. American Association of State Highway and Transportation Officials, Washington, D.C.

Bartlett, S. F., and T. L. Youd (1995). Empirical prediction of liquefaction-induced lateral spread. *J. Geotech. Eng. Div., ASCE* **121**, 316–329.

Byrne, P. M. (1991). A model for predicting liquefaction induced displacement. In "Proc. 2nd International Conference on Recent Advances in Geotechnical Earthquake Engineering and Soil Dynamics," pp. 1027–1035. St. Louis, Missouri.

Castro, G. (1969). Liquefaction of Sands. In: "Harvard Soil Mechanics," Series **81**, p. 112, Harvard University, MA.

Close, U., and E. McCormic (1922). Where the mountains walked. *National Geographic* **41**, 445–464.

Committee on Soil Dynamics (1978). Definition of terms related to liquefaction. *J. Geotech. Eng. Div., ASCE* **104**, 1197–1200.

Hanson, S. L., and D. M. Perkins (1995). Seismic sources and recurrence rates as adopted by USGS staff for the production of the 1982 and 1990 probabilistic ground motion maps for Alaska and the conterminous United States. *Open-File Report* **95-257**, *US Geol. Surv.*, Denver, Colorado.

Ishihara, K., and M. Yoshimine (1992). Evaluation of settlements in sand deposits following liquefaction during earthquakes. *Soils and Foundations* **32**, 173–188.

Iwasaki, T. (1986). Soil liquefaction studies in Japan; state-of-the-art. *Soils and Foundations* **26**, 1–60.

Lee, K. L., and A. Albaisa (1974). Earthquake-induced settlements in saturated sands. *J. Soil Mechan. Foundations Div., ASCE* **100**, 387–400.

National Research Council (NRC) (1985). *Liquefaction of Soils During Earthquakes*. National Academy Press, Washington, D.C.

Pease, J. W., and T. D. O'Rourke (1997). Mapping of liquefiable layer thickness for seismic hazard assessment. *J. Geotech. Geoenviron. Eng.* **123**, 46–56.

Robertson, P. K., and C. E. Wride (1997). Cyclic liquefaction and its evaluation based on the SPT and CPT. In: "NCEER Workshop on Evaluation of Liquefaction Resistance of Soils, National Center for Earthquake Engineering Research Technical Report," *NCEER 97-0022*, pp. 41–87.

Seed, H. B. (1979). Soil liquefaction and cyclic mobility evaluation for level ground during earthquakes. *J. Geotech. Eng. Div., ASCE* **105**, 201–255.

Seed, H. B., and I. M. Idriss (1971). Simplified procedure for evaluating soil liquefaction potential. *J. Soil Mechan. Foundations Div., ASCE* **97**, 1249–1273.

Seed, H. B., and I. M. Idriss (1982). "Ground Motions and Soil Liquefaction during Earthquakes." Monograph of the Earthquake Engineering Research Institute, Oakland, California, 134 pp.

Seed, H. B., K. Tokimatsu, L. F. Harder, and R. F. Chung (1985). Influence of SPT procedures in soil liquefaction resistance evaluations. *J. Geotech. Eng. Div., ASCE* **111**, 1425–1445.

Tokimatsu, K., and H. B. Seed (1987). Evaluation of settlements in sands due to earthquake shaking. *J. Geotech. Eng. Div., ASCE* **113**, 861–878.

Youd, T. L. (1975). Liquefaction, flow, and associated ground failure. In: "Proceedings of the U.S. National Conference on Earthquake Engineering," pp. 146–155. Ann Arbor, Michigan.

Youd, T. L. (1977). Packing changes and liquefaction susceptibility. *J. Geotech. Eng. Div., ASCE* **103**, 918–922.

Youd, T. L. (1978). Major cause of earthquake damage is ground failure. *Civil Engineering* **48**, 47–51.

Youd, T. L. (1984). Geologic effects—liquefaction and associated ground failure. In: "Proceedings of the Geologic and Hydrologic Hazards Training Program," pp. 210–232. *Open-File Report 84-760, U.S. Geol. Surv.*, Menlo Park, California.

Youd, T. L. (1991). Mapping of earthquake-induced liquefaction for seismic zonation. In: "Proceedings of the 4th International Conference on Seismic Zonation," pp. 111–147. Stanford, California.

Youd, T. L. (1998). Screening guide for rapid assessment of liquefaction hazard at highway bridge sites. In: "Multidisciplinary Center for Earthquake Engineering Research Technical Report" *MCEER-98-0005*, p. 58.

Youd, T. L., and S. G. Gilstrap (1999). Liquefaction and deformation of silty and fine-grained soils. In: "Proceedings of the 2nd International Conference on Earthquake Geotechnical Engineering," pp. 1013–1020. Lisbon, Portugal.

Youd, T. L., and D. M. Perkins (1978). Mapping of liquefaction induced ground failure potential. *J. Geotech. Eng. Div., ASCE* **104**, 433–446.

Youd, T. L., C. M. Hansen, and S. F. Bartlett (2002). Revised MLR equations for prediction of lateral spread displacement. *J. Geotech. Eng. Div., ASCE* **128**, no 12, in press.

Youd, T. L., I. M. Idriss, R. D. Andrus, I. Arango, G. Castro, J. T. Christian, R. Dobry, W. D. L. Liam Finn, L. F. Harder, Jr., M. E. Hynes, K. Ishihara, J. P. Koester, S. S. C. Liao, W. F. Marcuson, III, G. R. Martin, J. K. Mitchell, Y. Moriwaki, M. S. Power, P. K. Robertson, R. B. Seed, and K. H. Stokoe, II (2001). Liquefaction resistance of soils: Summary report from the 1996 NCEER and 1998 NCEER/NSF workshops on evaluation of liquefaction resistance of soils. *J. Geotech. Eng. Div., ASCE* **127**, pp. 817–833.

71

Advances in Analysis of Soil Liquefaction during Earthquakes

Jean-Pierre Bardet
University of Southern California, Los Angeles, California, USA

1. Introduction

Soil liquefaction has caused major damage during past earthquakes. During the 1964 Niigata, Japan, earthquake, soil liquefaction forced buildings to settle and tilt, underground storage tanks to float to the ground surface, and large ground areas to spread laterally and damage underground lifelines, bridges, and port quaywalls. In Chapter 70 on soil liquefaction, Youd reviewed the basic mechanisms of liquefaction and the ground failures induced by liquefaction. Based on laboratory experiments on saturated soil specimens, soil liquefaction was understood to be the result of an increase in pore water pressure associated with a decrease in soil frictional resistance during earthquake shakings. Youd gave examples of liquefaction damage to buildings, pile foundations, and buried structures during past earthquakes; summarized some engineering approaches used in North America to evaluate liquefaction hazards at regional and site levels based on geological, SPT, and CPT data; and provided an empirical equation for assessing the amplitude of liquefaction-induced lateral spreads.

This chapter complements Youd's chapter on liquefaction by reviewing the recent advances in understanding and analyzing soil liquefaction. Following the introduction, the second section reviews the case histories of soil liquefaction on which engineering practice is largely based. The third section reviews the physical modeling of liquefaction using soil specimens in the laboratory and reduced scale models in shaking tables and centrifuges. The fourth section examines the numerical modeling of liquefaction. The last section covers the latest research advances on soil liquefaction.

2. Liquefaction Case Histories

Our engineering understanding of soil liquefaction during earthquakes is based largely on case histories of liquefaction occurrence and related ground deformation, documented from past earthquakes.

2.1 Case Histories of Liquefaction Occurrence

As listed in Table 1, a large number of liquefaction case histories have been documented after earthquakes in China, Japan, South America, and North America. Various types of data have been collected from these case histories and organized into databases (e.g., Seed and Idriss, 1971; Harder, 1991) in an effort to define the evidences for liquefaction (e.g., ground deformations and sand boils) and the circumstances for which liquefaction did occur or did not occur. These databases are constantly revised in order to remove possible uncertainties on data collected and add new entries after recent earthquakes. Such databases are the foundation of the engineering liquefaction analyses as described by Youd and Idriss (1998) and in Chapter 70 by Youd. In these databases, the geometry and properties of soil deposits are mostly characterized by Standard Penetration Tests (SPT), a common soil investigation technique in geotechnical engineering. However, new databases have been developed based on other types of soil deposit characteristics including shear-wave velocity (e.g., Andrus *et al.*, 1999) and cone penetration test (CPT) data (e.g., Robertson and Wride, 1998; Stark and Olson, 1995).

INTERNATIONAL HANDBOOK OF EARTHQUAKE AND ENGINEERING SEISMOLOGY, VOLUME 81B ISBN: 0-12-440658-0

TABLE 1 List of Earthquakes with Documented Case Histories of Liquefaction

Earthquake	Reference
1891 Mino-Owari, Japan	Kishida (1969)
1906 San Francisco, USA	Hamada and O'Rourke (1992)
1923 Kanto, Japan	Kodera (1964)
1944 Tohnankai, Japan	Kishida (1969)
1948 Fukui, Japan	Kishida (1969)
1960 Tokachi-Oki, Japan	Ohsaki (1970)
1964 Niigata, Japan	Kishida (1969); Koizumi (1966); Ishihara *et al.* (1979); Youd and Kiehl, 1996
1971 San Fernando, USA	Hamada and O'Rourke (1992)
1975 Haicheng, China	Xie (1979); Shengcong *et al.* (1983)
1976 Guatemala	Seed *et al.* (1979)
1976 Tangshan, China	Xie (1979); Shengcong *et al.* (1983)
1977 Argentina	Idriss *et al.* (1979)
1978 Miyagiken-Oki, Japan	Tohno *et al.* (1981); Ishihara *et al.* (1980)
1979 Imperial Valley, USA	Youd and Bennett (1983)
1981 Westmorland, USA	Bennett *et al.* (1984)
1983 Borah Peak, Idaho	Youd *et al.* (1985)
1987 Superstition Hills, California	Youd and Holzer (1994); Scott and Hushmand (1995)
1989 Loma Prieta, USA	Holzer (1998); Bennett *et al.* (1999); Holzer *et al.* (1994)
1993 Hokkaido Nansei-oki, Japan	Isoyama (1994)
1994 Northridge earthquake, USA	Holzer et al.(1999); Bardet and Davis (1996); Davis and Bardet (1996)
1995 Hyogoken-Nanbu, Japan	Hamada *et al.* (1996); Ishihara *et al.* (1996)

Almost all the case histories on soil liquefaction have been documented after the events took place. There are very few case histories for which the two main factors controlling liquefaction (i.e., pore pressure and ground acceleration) were actually both recorded at the liquefaction sites during earthquakes. To our knowledge, only one case history of liquefaction was fully documented with simultaneous acceleration and pore pressure records during an actual earthquake, i.e., at the Wildlife site during the 1987 Imperial Valley earthquake (Holzer *et al.*, 1989). A few controversies, however, clouded the measured time history of pore pressure at the Wildlife site (e.g., Scott and Hushmand, 1995; Youd and Holzer, 1994, 1995), and deserve consideration in the analysis of the recordings. There have been liquefaction case histories in which the time history of acceleration, but not of pore pressure, was recorded by several downhole strong-motion instruments located at the ground surface and at various depths (i.e., Iwasaki and Tai, 1996), which allowed researchers (e.g., Elgamal *et al.*, 1996) to calculate the average shear-strain behavior of the liquefying soils using inverse methods. Several sites throughout the world have now been instrumented with pore pressure sensors, and are likely to yield valuable information on soil liquefaction (e.g., Zeghal and Elgamal, 1994) during future earthquakes.

2.2 Case Histories of Liquefaction-Induced Ground Deformation

Soil liquefaction has especially devastating effects on structures when it triggers lateral spreads. During past earthquakes, large areas of ground were observed to shift laterally due to soil liquefaction (Figs. 1 and 2). These liquefaction-induced lateral ground deformations have amplitudes ranging from a few centimeters to tens of meters in the case of flow slides. They can take place under gently sloping ground conditions (0.1 to 6%). Liquefaction-induced ground deformations are generally observed close to open faces or in gently sloping ground. These deformations are usually driven by a combination of transient and static shear stresses and attributed to the loss of shear strength of underlying saturated soils. During the 1995 Hyogoken-Nanbu, Japan, earthquake, liquefaction-induced lateral ground deformation was one of the major causes of damage to lifelines and pile foundations of buildings and bridge piers along the Kobe shoreline (Hamada *et al.*, 1996; Karube and Kimura, 1996; Matsui and Oda, 1996; Tokimatsu *et al.*, 1996). As shown in Figure 2, lateral ground deformations can cause bending and axial compression in buried pipes, which may damage pipes during earthquakes.

Some of the liquefaction-induced lateral spreads were observed to take place after the end of earthquake shaking. During the 1971 San Fernando earthquake, the liquefaction-induced slide in the lower San Fernando Dam took place about twenty minutes after the earthquake shaking ended (Seed *et al.*, 1975). Similar delayed deformations were also observed as liquefaction-induced subsidence during the 1964 Niigata, Japan, earthquake (Kobayashi *et al.*, 1999). These delayed deformations are most likely the result of the redistribution of excess pore pressure, which was generated during the earthquake shaking.

A large number of case histories of liquefaction-induced lateral spreads have been compiled in Hamada and O'Rourke (1992). Some examples of case histories include those from the 1964 Niigata earthquake (Youd and Kiehl, 1996); 1971

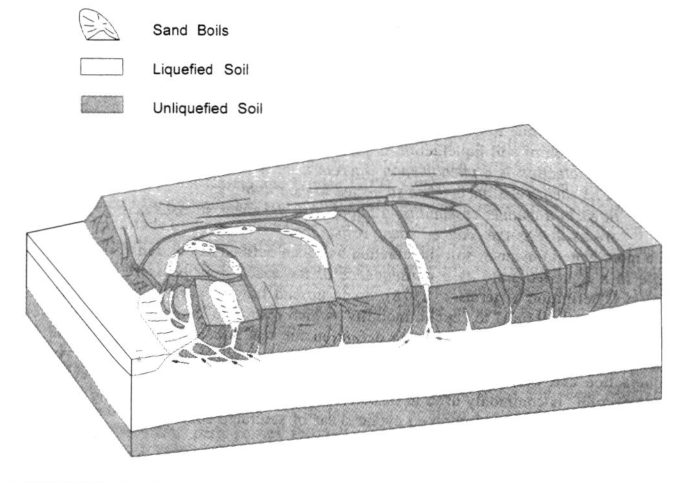

FIGURE 1 Schematic description of a lateral spread resulting from soil liquefaction during an earthquake (Rauch, 1997).

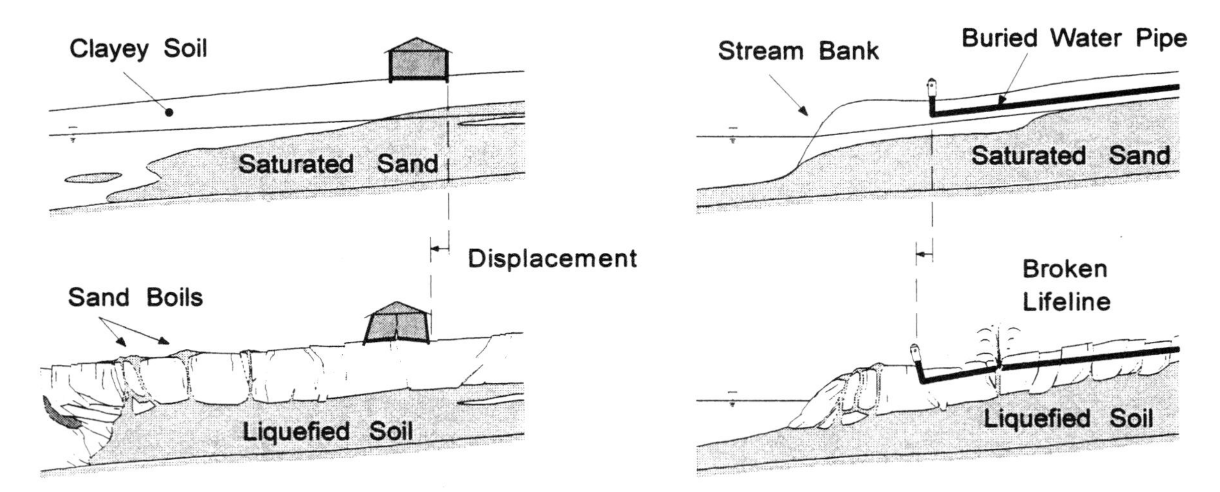

FIGURE 2 Illustration of liquefaction-induced ground deformation and associated damage to buried pipelines for ground-slope and free-face cases (Rauch, 1997).

San Fernando earthquake; 1983 Nihonkai-Chubu earthquake; 1983 Borah Peak, Idaho, earthquake (Youd *et al.*, 1985); 1987 Superstition Hills, California, earthquake (Youd and Holzer, 1994); 1989 Loma Prieta earthquake (Holzer *et al.*, 1994); 1993 Hokkaido Nansei-oki earthquake (Isoyama, 1994); 1994 Northridge earthquake (Holzer *et al.*, 1999; Bardet and Davis, 1996; Davis and Bardet, 1996); and the 1995 Hyogoken-Nanbu earthquake (Hamada *et al.*, 1996; Ishihara *et al.*, 1996).

Permanent lateral ground deformations after earthquakes are generally measured using ground surveying and aerial photographs. Ground surveys are commonly used in assessing the damage to constructed facilities. They are based on well-established optical techniques, and are sometimes integrated with global positioning system (GPS) techniques. Ground surveying is accurate (<5 mm) but unfortunately limited to areas that had survey monuments prior to earthquakes. Examples of ground surveying can be found in Bardet and Davis (1996). The main drawback of ground surveys is that they are confined to areas of limited extent, and may be missing the global modes of deformation of larger areas.

Aerial photogrammetry has been used to display comprehensive fields of permanent displacement after earthquakes (e.g., Hamada *et al.*, 1996). Figure 3 shows an example of permanent ground displacement obtained using aerial photographs in Niigata, Japan, after the 1964 Niigata, Japan, earthquake (Hamada and O'Rourke, 1992). The determination of ground displacement requires aerial stereo photographs taken before and after the earthquake. The accuracy of the measured amplitude of ground displacements depends on the photograph scale (i.e., flight altitude), timing of photographs, and presence of cultural objects (e.g., street corners and fire hydrants) that can be tracked on photographs taken before and after the earthquake (e.g., Sano, 1998). The best measurements correspond to low-altitude photographs taken of urban and industrial areas just before and after the earthquake. Unpopulated and rural areas usually yield less reliable tracking points especially when they are covered with vegetation. In general, low-altitude photographs from before the earthquake are more difficult to find.

Various types of data collected from case histories of liquefaction-induced lateral ground deformation have been compiled in databases: Bartlett and Youd (1992), Rauch and Martin (2000), and Bardet *et al.* (1999). As illustrated in Color Plate 27, the information from these case histories (e.g., ground displacement, soil borehole, and topographical and seismic data) can be efficiently stored, displayed, and analyzed using geographic information systems (GIS).

3. Empirical Modeling of Liquefaction

In geotechnical earthquake engineering practice, liquefaction-induced ground deformations are usually estimated using three different types of empirical models that predict separately (1) the occurrence of liquefaction, (2) ground settlement, and (3) lateral ground deformation.

3.1 Empirical Modeling of Liquefaction Occurrence

Figure 4 shows an example of empirical modeling of liquefaction occurrence based on the database of Harder (1991), which has entries originating from 17 earthquakes. The database entries are plotted in terms of earthquake loading and soil resistance to liquefaction. The soil layer properties are characterized by SPT normalized blow counts N_{160}, which is calculated from the blow count N after accounting for the diversity of SPT equipment and testing depths (e.g., Youd and Idriss, 1998). By definition, a SPT blow count N corresponds to the number of blows required for a standard sampler to penetrate one foot into the ground

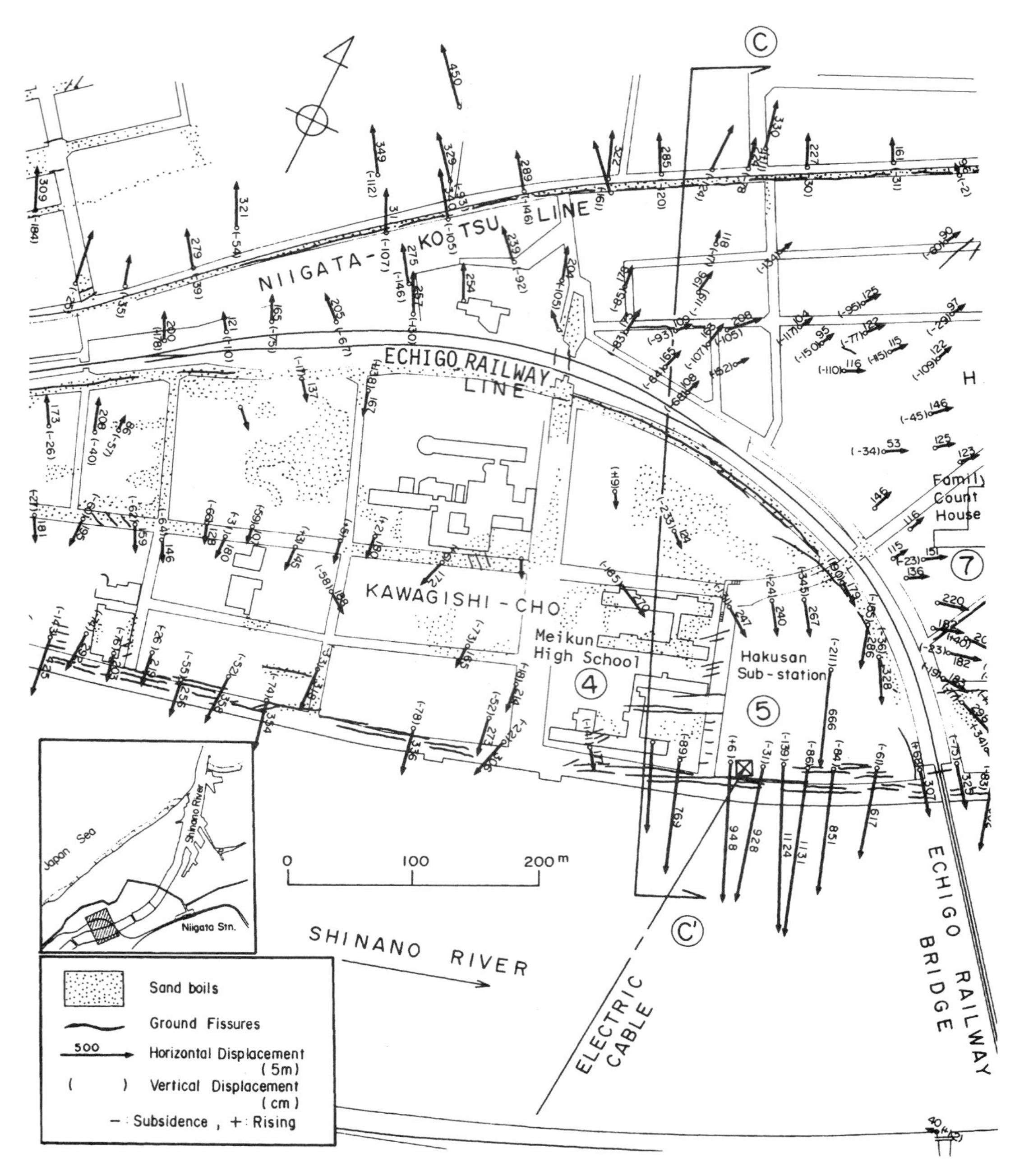

FIGURE 3 Examples of ground displacement vectors measured in Kawagachi-cho, Niigata, Japan, after the 1964 Niigata, Japan, earthquake using aerial photographs (Hamada and O'Rourke, 1992).

(e.g., Seed *et al.*, 1985). N_{160} is an indication of the soil density in the field, and is usually larger for denser soils. The loading applied by the earthquake to soils is characterized by the cyclic stress ratio τ/σ_0' where τ is the equivalent cyclic stress amplitude and σ_0' the initial vertical effective stress. As detailed in Seed and Idriss (1971) and Youd and Idriss (1998), τ is calculated from the peak ground acceleration, earthquake magnitude, and depth. τ/σ_0' characterizes the earthquake loading that may induce liquefaction, whereas N_{160} characterizes the resistance of soils to liquefaction. As shown in Figure 4, the solid data points corresponding to liquefaction are differentiable from the hollow ones corresponding to no liquefaction. A borderline can be drawn to separate the cases of liquefaction from those of no liquefaction. Youd and Idriss (1998) give up-to-date specific details of the procedures and equations for these borderlines, which depend on the amount of fine contents in soil deposits.

Similar empirical methods have been proposed to determine the occurrence of liquefaction based on different characterization of earthquake loading and soil layer properties. These methods rely on different selection of earthquake loading

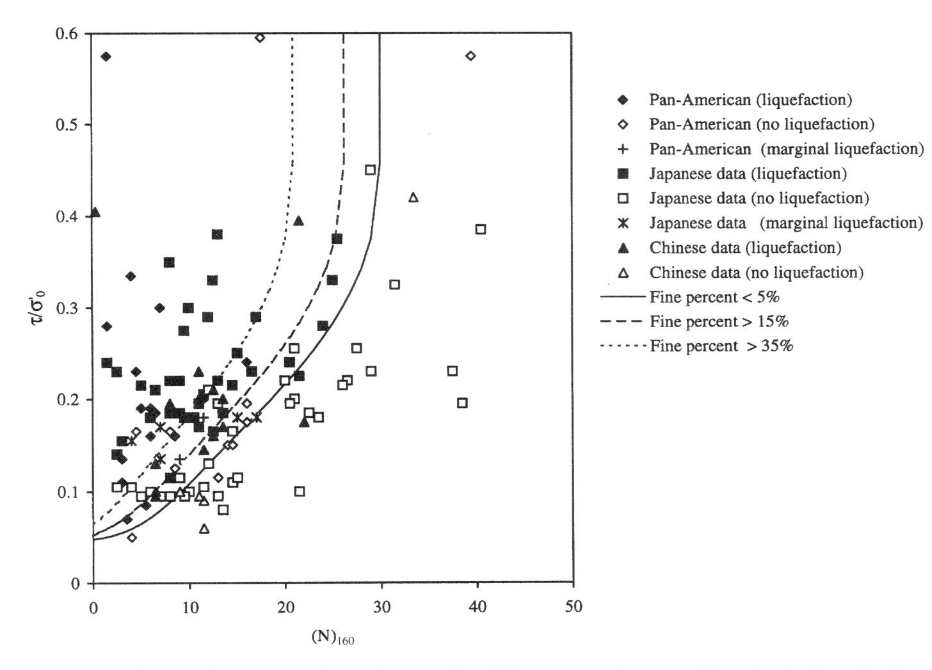

FIGURE 4 Cyclic stress ratio and normalized blow count for case histories in liquefaction occurrence database (Harder, 1991).

and/or soil resistance to liquefaction. Soil resistance was characterized in terms of Cone Penetration Test (CPT) resistance (e.g., Robertson and Wride, 1998) and shear-wave velocity (e.g., Andrus *et al.*, 1999). Earthquake loading was defined using energy (e.g., Berrill and Davis, 1985; Trifunac, 1995). These recent methods seem to better differentiate the cases of liquefaction and no-liquefaction occurrences. Based on revised databases for the case histories of liquefaction documented in terms of either SPT or CPT boreholes, neural-network models (e.g., Goh, 1994, 1996) and probabilistic models (e.g., Liao, 1980; Liao *et al.*, 1988; Liao and Lum, 1998; Toprak *et al.*, 1999) have also been proposed for characterizing the occurrence of liquefaction; the latter models are designed for assessing the probability of liquefaction occurrence at selected sites.

3.2 Empirical Modeling of Liquefaction-Induced Lateral Spread

In addition to models for liquefaction occurrence, empirical models were proposed for assessing the amplitude of liquefaction-induced lateral spreads.

3.2.1 Youd and Perkins (1987)

Youd and Perkins (1987) presented a method for mapping liquefaction severity index (LSI), which is related to the extent and severity of liquefaction phenomena within liquefaction-susceptible soils. The LSI model, which presents some similarities to attenuation curves for peak ground acceleration, proposes an index for liquefaction associated with the "worst" geologic conditions of recent loose Holocene deposits and shallow water table. The LSI coefficient is related to the amplitude of horizontal ground deformation to distance from seismic energy source and moment magnitude as follows:

$$\log LSI = -3.49 - 1.86 \log R + 0.98 M_w \quad (1)$$

where R is the horizontal distance (km) to seismic energy source, and M_w is the earthquake moment magnitude. The LSI model is derived from experiences with Japanese and western United States earthquakes, and cannot be extrapolated to other parts of the world (e.g., the eastern US).

3.2.2 Bartlett and Youd (1992) and Youd *et al.* (1999)

Bartlett and Youd (1992) devised two separate Multiple Linear Regression (MLR) models: one for ground slope of infinite extent, and the other for free face. The principles of MLR modeling are extensively described in statistics textbooks (e.g., Draper and Smith, 1981). The Bartlett and Youd (1992) model was recently revised and its revised MLR equations are fully described in Youd *et al.* (1999) and Chapter 70 by Youd. As shown in Figure 5, the model predictions are scattered within the lines with 1:0.5 and 1:2 slope, while they should fall close to the line with a 1:1 slope for a perfect prediction.

3.2.3 Hamada *et al.* (1986)

Hamada *et al.* (1986) predicted the amplitude of horizontal ground deformation only in terms of slope and thickness of liquefied layer, as follows:

$$D = 0.75 H^{0.5} S^{0.33} \quad (2)$$

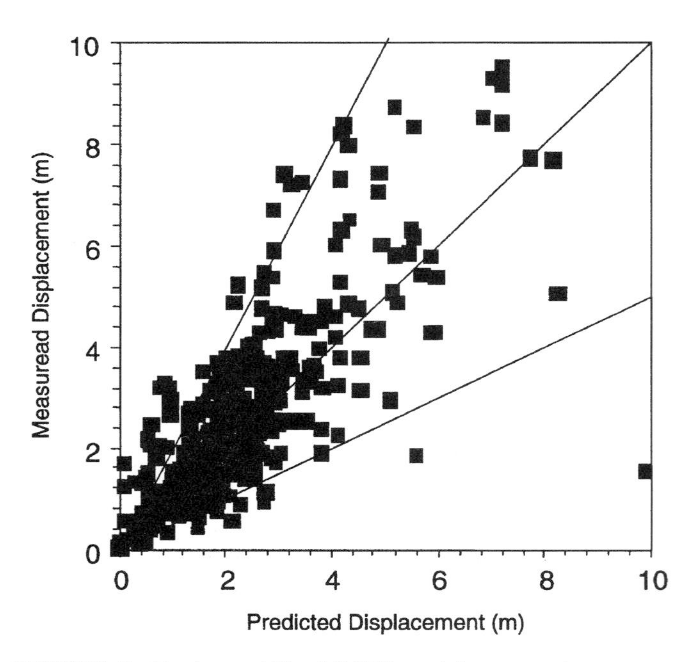

FIGURE 5 Bartlett and Youd (1992) model: measured versus predicted liquefaction-induced lateral displacement (data points from the database of Bartlett and Youd, 1992).

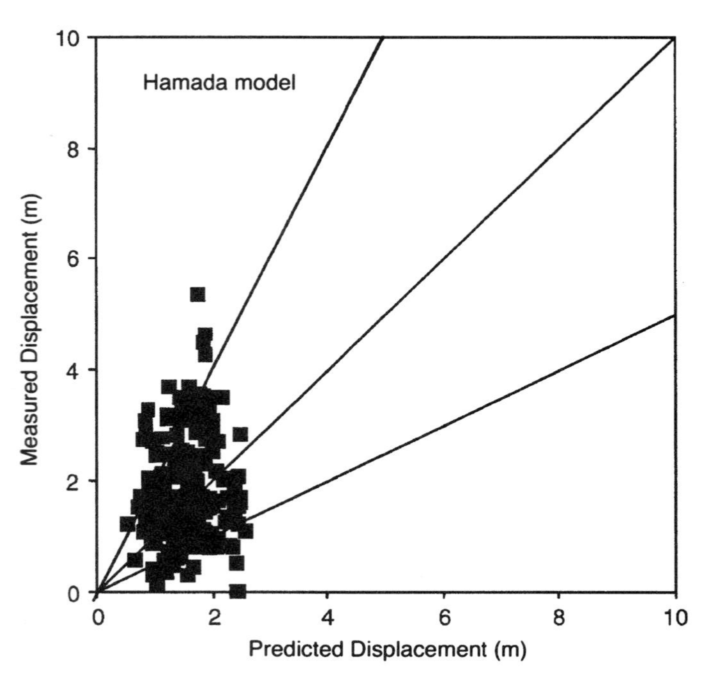

FIGURE 6 Hamada *et al.* (1986) model for ground slope: measured versus predicted liquefaction-induced lateral spreads (data points from the database of Bartlett and Youd, 1992).

where D is the horizontal displacement (m), S is the slope (%) of ground surface or base of liquefied soil, and H is the thickness (m) of liquefied soil. Equation (2) is only based on topographic and geotechnical parameters (i.e., S and H). Figure 6 compares measured displacements and those calculated using Eq. (2) for the ground-slope entries in the Bartlett and Youd (1992) database. In view of the model simplicity, the predicted values are in reasonable agreement with measured values.

3.2.4 Rauch and Martin (2000)

In contrast to other modelers, Rauch and Martin (2000) considered that the displacement vectors associated with liquefaction-induced lateral spreads could not be treated as individual data points, but had to be regrouped in small sets indicative of common liquefaction-induced slides. Rauch and Martin fitted the average and standard deviation of lateral ground displacement for the separate slides they were able to delineate. They proposed three models, referred to as regional, site, and geotechnical, which apply to large regions down to smaller construction sites and depend on a different number of input parameters. The regional average model, which has four input parameters all related to earthquake ground motion, is

$$D = (D_R - 2.21)^2 + 0.149, \text{ and}$$
$$D_R = (613M_w - 13.9R_f - 2420A_{\max} - 11.4T_d)/1000 \qquad (3)$$

where D is the average horizontal displacement (m); R_f is the shortest horizontal distance (km) to fault rupture; M_w is the moment magnitude; $A_{\max}$ is the peak horizontal acceleration (g) at ground surface; and T_d is the duration (sec) of strong earthquake motions ($>0.05g$). The site average model, which accounts for the site geometry, is

$$D = (D_R + D_S - 2.44)^2 + 0.111, \text{ and}$$
$$D_S = (0.523L_{\text{slide}} + 42.3S_{\text{top}} + 31.3H_{\text{face}})/1000 \qquad (4)$$

where L_{slide} is the length (m) of slide area from head to toe; S_{top} is the average slope (%) across the surface of lateral spread; and H_{face} is the height of free face (m) measured vertically from toe to crest. Finally, the geotechnical average model, which requires a liquefaction analysis (e.g., Youd and Idriss, 1998), is

$$D = (D_R + D_S + D_G - 2.49)^2 + 0.124, \text{ and}$$
$$D_S = (50.6\, Z_{\text{FSmin}} - 86.1Z_{\text{liq}})/1000 \qquad (5)$$

where Z_{FSmin} is the average depth (m) corresponding to minimum factor of safety; and Z_{liq} is the average depth (m) to the top of the liquefied layer. Figure 7 shows the comparison of measured versus calculated average displacements, using entries in the Rauch and Martin (2000) database. The scatter between predictions and observations is larger than in Figures 5 and 6.

3.2.5 Bardet *et al.* (1999) Models

Bardet *et al.* (1999) developed a four-parameter MLR model to provide a first-order approximation of liquefaction-induced displacement over large areas, and to avoid the uncertainties arising from the determination of soil properties such as mean grain size

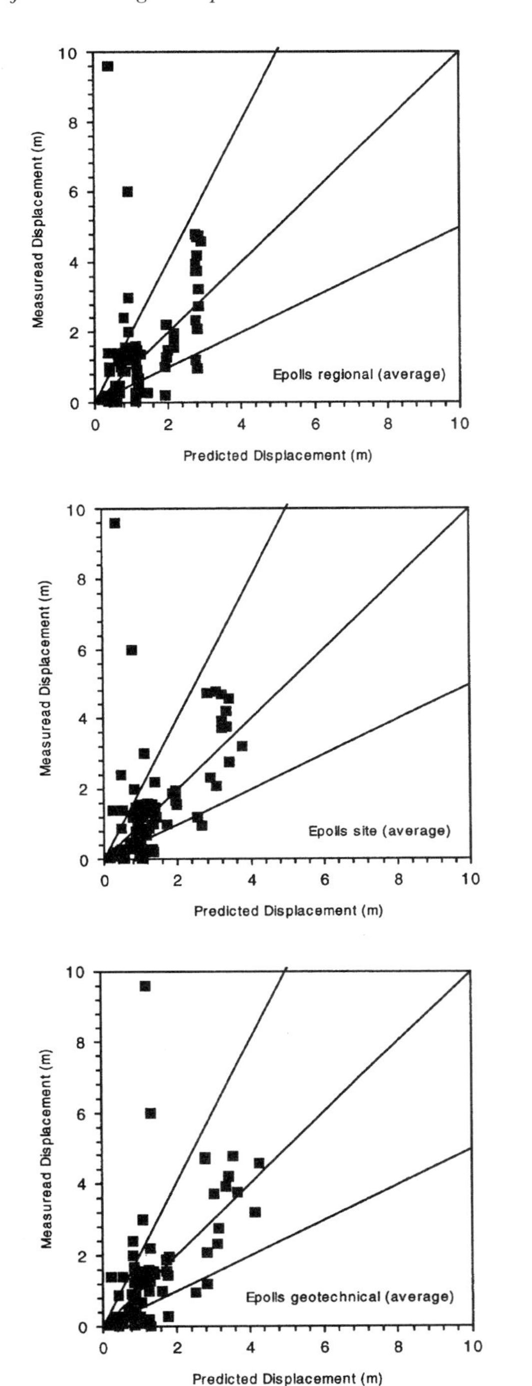

FIGURE 7 Regional, site, and geotechnical Rauch (1997) models: measured versus calculated average displacements [data points from Rauch (1997) database].

and fine contents. Their model has the following equation:

$$\begin{aligned}\text{Log}(D + 0.01) = &-6.815 - 0.465^F + 1.017M \\ &+ -0.278\,\text{Log}(R) + -0.026R \\ &+ 0.497^F\,\text{Log}(W) + 0.464^G\,\text{Log}(S) \\ &+ 0.558\,\text{Log}(T_{15})\end{aligned} \tag{6}$$

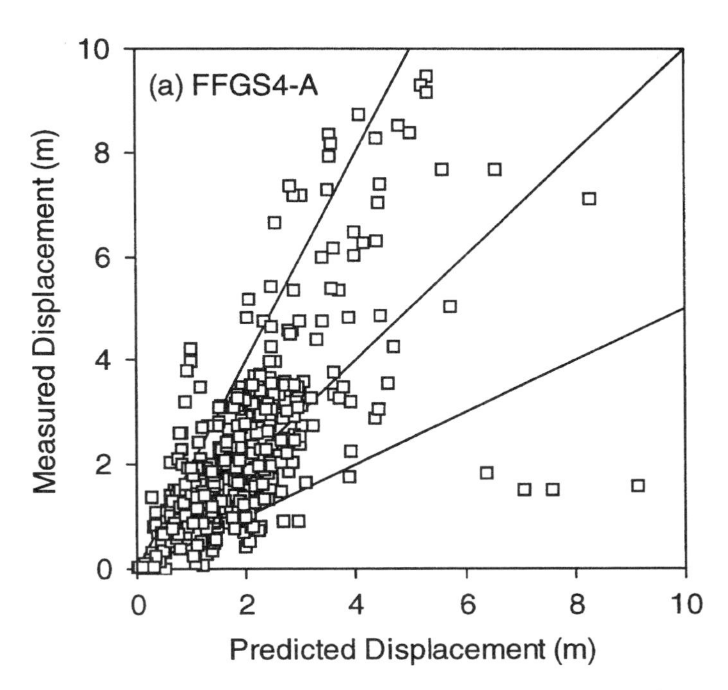

FIGURE 8 Measured versus predicted displacements for four-parameter model (Bardet *et al.*, 1999).

where D is the horizontal displacement (m); M is the moment magnitude; R the nearest horizontal distance (km) to seismic energy source or fault rupture; S the slope (%) of ground surface; W the free-face ratio (%); and T_{15} the thickness (m) of saturated cohesionless soils with $N_{160} < 15$. Equation (6) applies to both free-face and ground-slope cases. In the free-face cases, the coefficient with the superscript G is set equal to zero, while in the ground-slope cases, the two coefficients with the superscript F are set equal to zero. Figure 8 shows the measured displacements plotted against those predicted by the model. The points are more scattered about the 1:1 line than in Figure 5. As expected, the four-parameter model predicts measured ground displacement less accurately than a six-parameter model.

The four-parameter model, however, has advantages for mapping of ground deformation over large areas, mainly because the other two parameters (i.e., fine contents and mean grain size) used by Youd *et al.* (1999) are difficult to determine over large areas made of heterogeneous soils. Figure 9 and Color Plate 28 show the observed and predicted displacements, surface map, borehole location, and spatial distribution of the MLR parameters, thickness T_{15} and free-face ratio for a site in Niigata, Japan, during the 1964 Niigata earthquake. The values of T_{15} and free-face ratio W were calculated over 50 by 34 grid points evenly spaced at a 10.5-m grid interval. As shown in Color Plate 28(b), the contours indicate that the amplitude of predicted displacements decrease uniformly with distance from the free face. However, the contours in Color Plate 28(a) display a different and more concentrated spatial distribution of observed ground deformation, which was caused by the local failure of the embankment along the Shinano river.

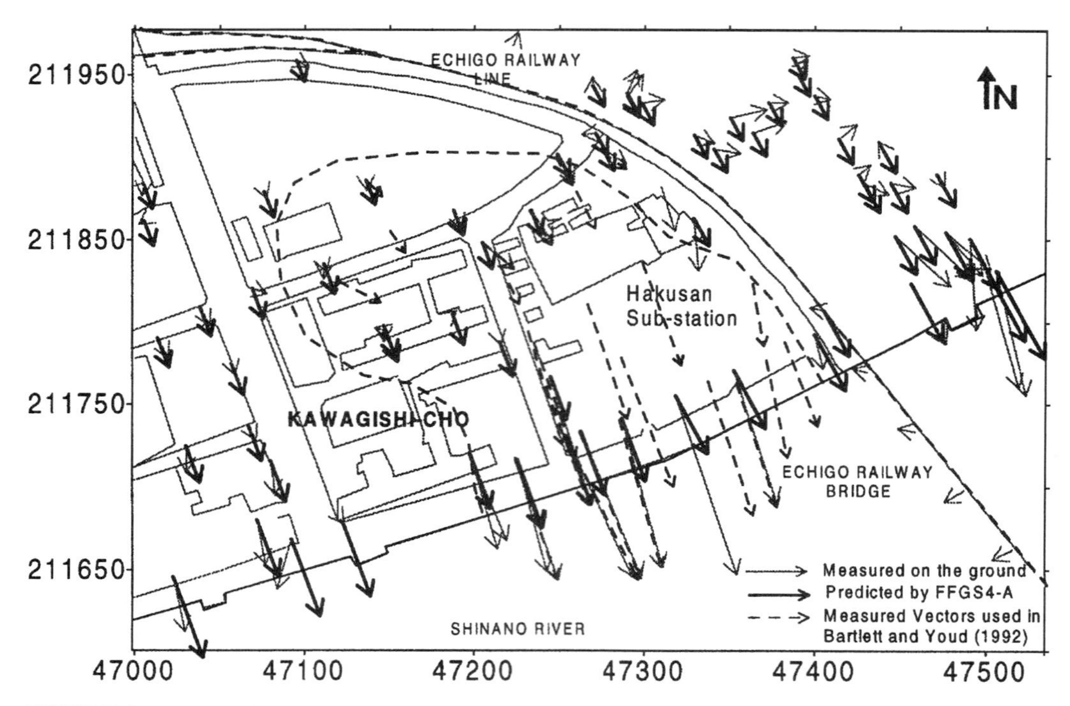

FIGURE 9 Measured (Hamada, 1999) and predicted (Bardet *et al.*, 1999) liquefaction-induced displacements for slide G-10 FF′ (Kawagishi-cho) at Niigata, Japan, during the 1964 Niigata earthquake. (Coordinates are in meters.)

4. Physical Modeling of Liquefaction

Soil liquefaction has been investigated successfully in great depth by means of physical models using specimens in the laboratory and reduced-scale models in shaking tables and centrifuges.

4.1 Laboratory Investigations of Soil Liquefaction

Laboratory experiments are useful for investigating the aspects of soil behavior related to liquefaction under controlled conditions. Laboratory experiments attempt to simulate the field conditions undergone by saturated soils during earthquakes. These experiments usually involve cyclic shear loading of saturated soil specimens in an undrained condition, with measurement of cyclic stresses and strains as well as of buildup of excess pore pressure and accumulated permanent shear strains. Ishihara (1996) reviewed the different types of equipment used in the laboratory, including triaxial test apparatus, simple shear test apparatus, and hollow-cylindrical torsional test apparatus. Ishihara (1996) pointed out the significant effects of sample preparation (e.g., wet tamping, dry deposition, and water sedimentation) on sample behavior. Soil behavior relevant to liquefaction was investigated in the laboratory mainly for remolded specimens, because those can be prepared identically and tested under various loading conditions. Table 2 lists a few laboratory data sets on remolded samples relevant to soil liquefaction. To our knowledge, complete data sets on undisturbed cohesionless samples taken from the field are not available, mainly due to sampling considerations (e.g., freezing techniques, Ishihara, 1996).

TABLE 2 Laboraty Data Sets on Laboratory Soil Specimens Relevant to Liquefaction

Name	Type of Test	Reference
Nevada Sand	T, S	Arulmoli *et al.* (1992)
Bonnie Silt	T, S	Arulmoli *et al.* (1992)
Monterey Sand	S	Wolfe *et al.* (1977)
Fuji River Sand	T, Ts	Tatsuoka and Ishihara (1974)
Toyoura Sand	T, Ts	Ishihara *et al.* (1996)
Fontainebleau Sand	T	Luong (1980)
Banding Sand	T	Castro (1969)

T = Triaxialtest; Ts = Torsionaltest; S = Simple shear test.

4.1.1 Preliquefaction Behavior

The behavior of saturated sands leading to liquefaction has been extensively studied in the laboratory under undrained conditions (e.g., Ishihara, 1993, 1996). Figure 10 illustrates the typical response of saturated sand observed in cyclic undrained simple shear tests. During the applied shear cycles of Figure 10(b), the pore pressure gradually increases [Fig. 10(c)] and the effective vertical stress decreases until it reaches a zero value [Fig. 10(f)]. As the effective vertical stress approaches zero, the soil sample undergoes large cyclic shear strains [Fig. 10(a)]. The magnitudes of these shear strains are controlled by stress-dilatancy, which

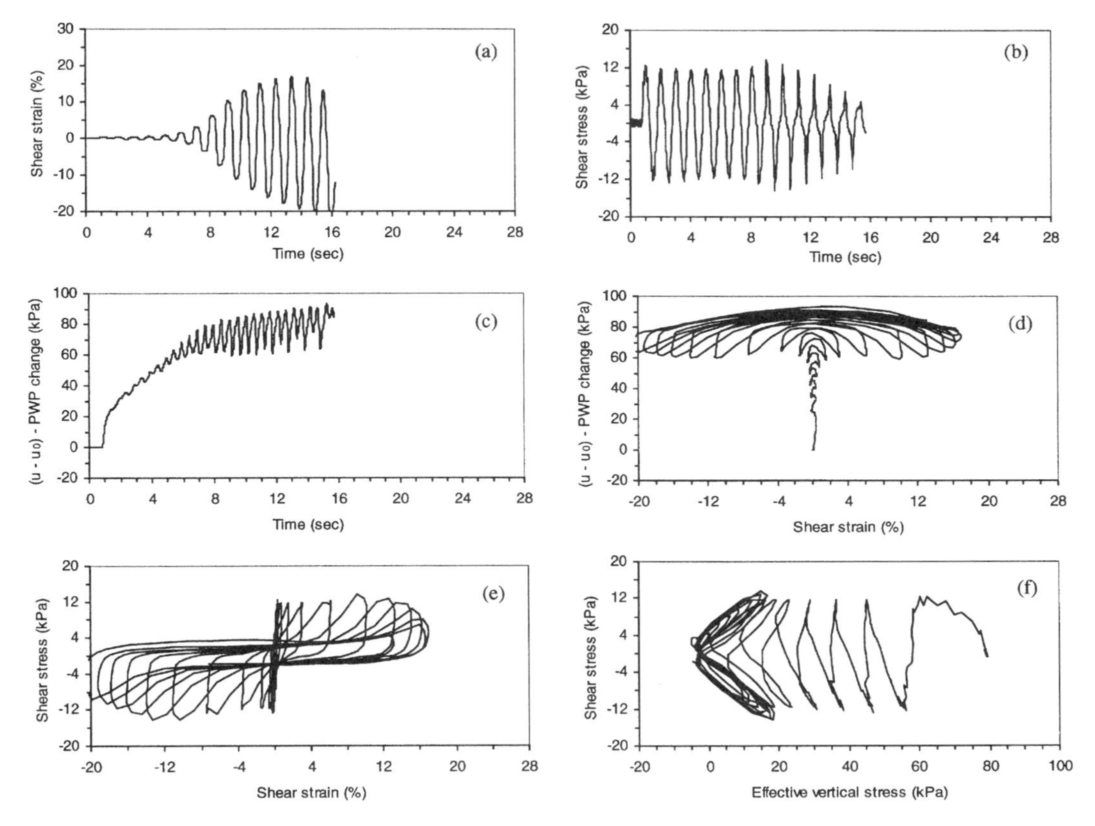

FIGURE 10 Cyclic simple shear test at 80 kPa confining pressure on 60% relative density Nevada Sand. Time histories of (a) shear strain; (b) shear stress; (c) pore pressure change; (d) pore pressure-strain response; (e) stress-strain response; and (f) effective stress path (Arulmoli *et al.*, 1992).

increases the effective vertical stress and therefore the shear strength of sands. The large transient strains as observed in the laboratory in Figure 10(a) were identified as possible sources of damage to the buried lifelines in the Marina District during the 1989 Loma Prieta earthquake (Pease and O'Rourke, 1997). As shown in Figure 10(b), the shear stress cycles are symmetric without a static offset, and the resulting cyclic shear strain remains approximately symmetric without a significant permanent bias. However as shown in Figure 11, in the case of unsymmetric stress cycles during an undrained triaxial test, the resulting axial strain with the laboratory results of Figure 11 imply that shear strains may accumulate in the field due to shear stress cycles during earthquake shaking. This cyclic accumulation of permanent shear strain (i.e., ratcheting) is certainly a plausible interpretation for some observed liquefaction-induced lateral spreads.

4.1.2 Postliquefaction Volumetric Deformation

Besides the typical response of Figures 10 and 11, there is another aspect of soil behavior in the laboratory that is relevant to liquefaction-induced ground deformation. This other behavior takes place after the initial liquefaction during cyclic loadings, and is referred to as post-liquefaction behavior. The volumetric and shear components of post-liquefaction deformations have been modeled separately. Tokimatsu and Seed (1987) and Ishihara (1996) investigated experimentally the post-liquefaction volumetric strains of saturated sands and proposed methods for estimating settlements of liquefied soil layers after earthquakes. As shown in Figure 12, the amount of post-liquefaction volumetric deformation depends on not only the initial relative density D_r, but also the factor of safety against liquefaction F_l:

$$F_l = (\tau_{av,l}/\sigma'_v)/(\tau_{av}/\sigma'_v) \quad (7)$$

where $\tau_{av,l}/\sigma'_v$ is the ratio of shear stress amplitude $\tau_{av,l}$ to effective vertical stress σ'_v required for liquefaction, and the applied ratio of shear stress amplitude τ_{av} to effective vertical stress. As shown in Figure 12, Ishihara (1996) proposed a correlation between relative density, normalized standard penetration test (SPT) blow count, and normalized cone penetration test (CPT) resistance, which is useful to compute the amount of post-liquefaction settlement.

4.1.3 Postliquefaction Shear Deformation

Post-liquefaction shear deformations in saturated sands were experimentally investigated using torsional tests (e.g., Yasuda *et al.*, 1994; Shamoto *et al.*, 1997) and triaxial tests (e.g., Nakase *et al.*, 1997). Both torsional and triaxial tests indicate that

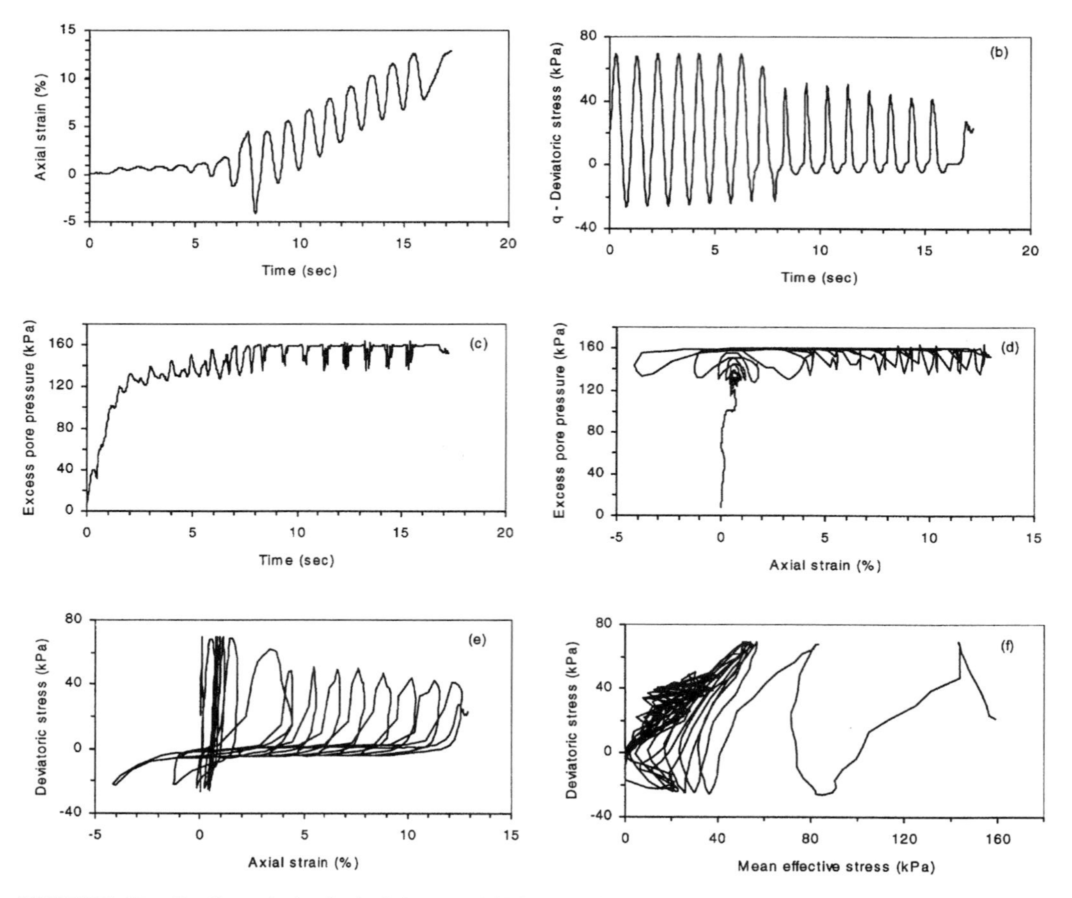

FIGURE 11 Cyclic undrained triaxial test at 160 kPa confining pressure on 40% relative density Nevada Sand. Time histories of (a) axial strain; (b) deviator stress; (c) excess pore pressure; (d) excess pore pressure–strain response; (e) stress–strain response; and (f) effective stress path (Arulmoli *et al.*, 1992).

liquefied soils regain shear strength beyond some shear-strain threshold γ_L. Figure 13 shows an example of such a rehardening during a monotonic undrained torsional shear test following cyclic liquefaction (Peiris and Yoshida, 1996). The liquefied sand gradually regains shear strength when it is sheared beyond some threshold shear strain γ_L. The value of γ_L depends on the relative density D_r and the amount of cyclic shear strain prior to liquefaction (or the factor of safety against liquefaction). The strain threshold γ_L is related to the maximum displacement of lateral spreads, and is useful for predicting the upper bound of liquefaction-induced lateral displacements.

4.2 Centrifuge Modeling of Liquefaction

The main rationale for centrifuge testing in geotechnical engineering is to preserve the nonlinear characteristics of pressure-dependent soil behavior. For this purpose, the same gravitational normal stresses are applied to corresponding points of model and prototype, hence preserving the same stiffness and strength properties in model and prototype (e.g., Ko, 1988). In centrifuge modeling, reduced-scale models are constructed to represent larger prototypes and subjected to a centrifugal acceleration n generated by the rotation of the centrifuge arm. In the case of centrifuge earthquake modeling, the model is shaken in flight by a hydraulic shaker through its base (Figs. 14, 15, and 16).

As shown in Table 3, all physical quantities in the centrifuge models are scaled to determine the prototype responses. In these scaling relations, the stresses and strains have identical scaling in the model and prototype, which preserves the nonlinear stress–strain relationships of soils. Lengths are scaled with n. Time in dynamic processes is scaled with $1/n$, whereas time in diffusion processes is scaled with $1/n^2$. Consequently, the prototypes involving simultaneously dynamic and diffusion processes require special considerations, as in the case of liquefaction-induced lateral displacements. The diffusion time can be properly scaled by introducing viscous additives into the interstitial water (e.g., Ko, 1988).

4.2.1 Liquefaction of Soil Deposits

Hushmand *et al.* (1988) simulated in the centrifuge the liquefaction of sand deposits during earthquakes. They modeled a 10-m-thick uniform soil deposit of infinite lateral extent using a laminar box, which is a stack of aluminum rings with

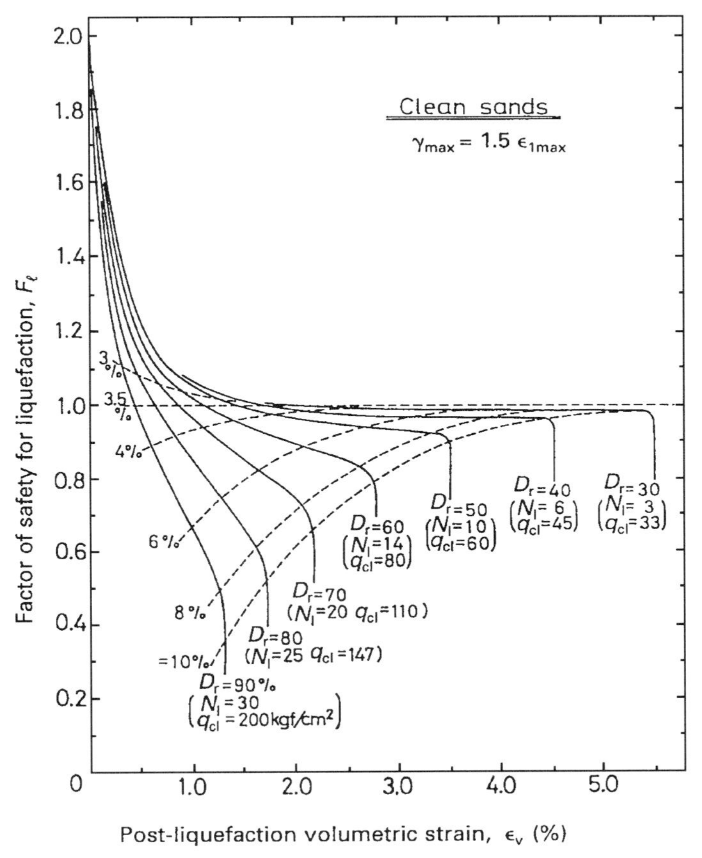

FIGURE 12 Chart for determination of post-liquefaction volumetric strain of clean sands as a function of factor of safety (Ishihara, 1996).

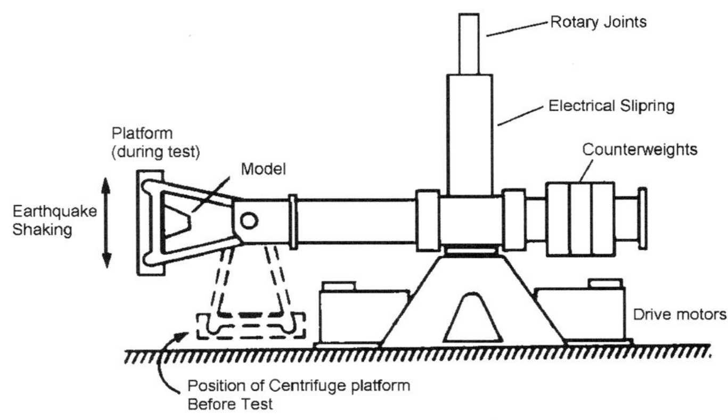

FIGURE 14 Sketch of RPI geotechnical centrifuge (Dobry *et al.*, 1995).

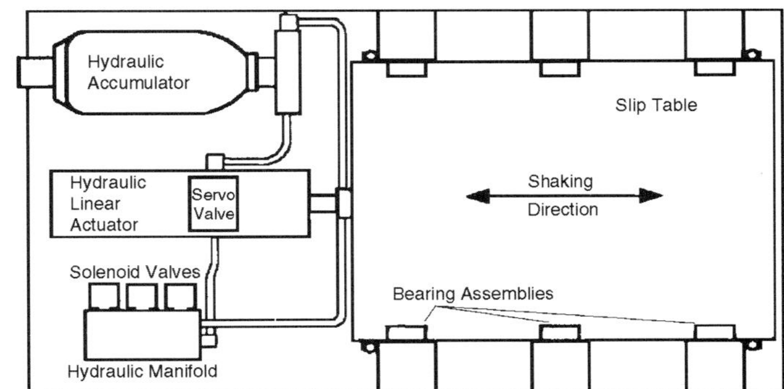

FIGURE 15 Schematic view of in-flight shaker at RPI (Dobry *et al.*, 1995).

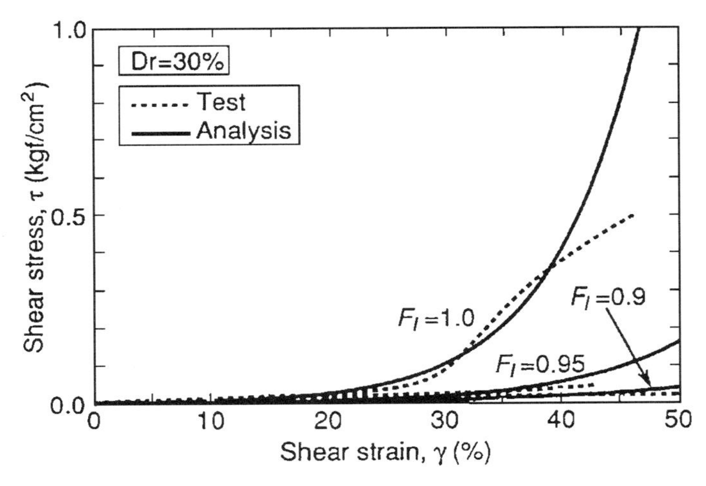

FIGURE 13 Stress–strain response of Toyoura sand at 30% relative density in the post-liquefaction range during undrained torsional tests (data after Yasuda *et al.*, 1994; model after Yoshida, 1996).

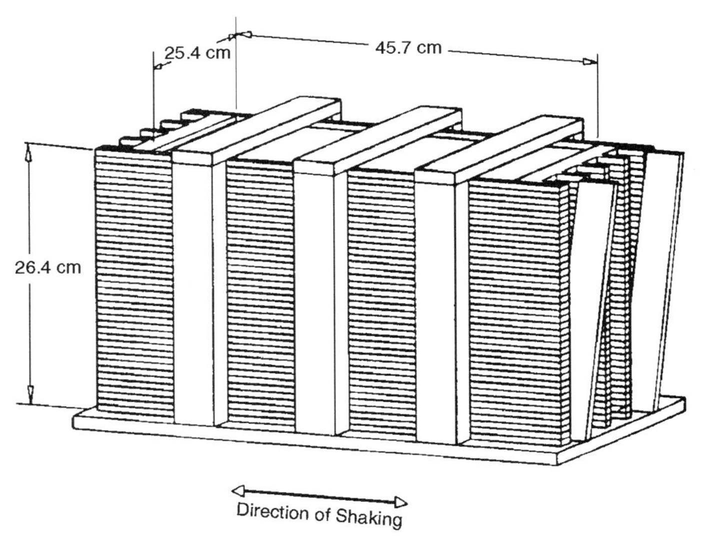

FIGURE 16 Schematic view of a laminar box (Dobry *et al.*, 1995).

TABLE 3 Scaling Relations Used for Centrifuge Modeling (Dobry *et al.*, 1995)

Parameter	Model Units	Prototype Units
Length	$1/n$	1
Velocity	1	1
Acceleration	n	1
Stress	1	1
Strain	1	1
Time		
Dynamic	$1/n$	1
Diffusion	$1/n^2$	1
Frequency	n	1

n = centrifuge acceleration in g.

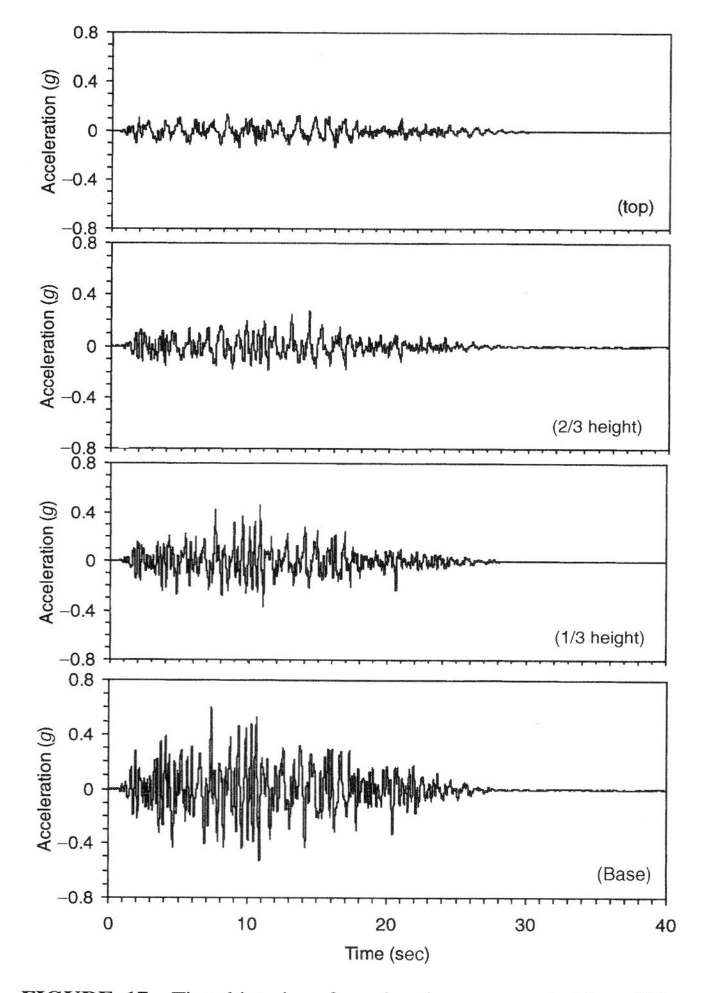

FIGURE 17 Time histories of acceleration measured at four different depths in the centrifuge model of saturated sand deposit (Hushmand *et al.*, 1988).

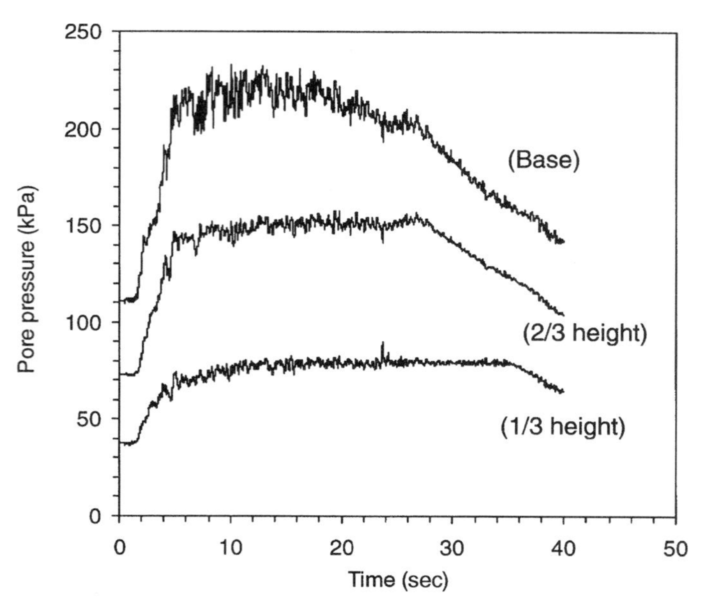

FIGURE 18 Time histories of prepressure measured at three different depths in the centrifuge model of saturated sand deposit (Hushmand *et al.*, 1988).

very small friction between them. The soil deposit was made of Nevada sand having a relative density of 53%. The model was instrumented with accelerometers, displacement transducers, and pore pressure sensors. The input ground-motion acceleration was applied to the base of the container. Figures 17 and 18 show the time histories of acceleration and pore pressure recorded during a centrifuge experiment. There are three distinct phases in the pore pressure records. In the first phase as the shaking starts, the excess pore pressure increases rapidly until it reaches its maximum value (i.e., initial effective vertical stress) and the soil liquefies. In the second phase during the shaking duration, the complete soil column remains liquefied and the pore pressure remains equal to its maximum value. In the third phase as the shaking intensity decreases, the excess pore pressure starts to diffuse from the top and consolidation takes place. The acceleration recordings indicate that the liquefied soil column lost some of its shear strength, transmitted a fraction of the input acceleration, and underwent large lateral displacement.

4.2.2 Liquefaction-Induced Lateral Spread

Dobry *et al.* (1995) simulated lateral spreads in the centrifuge. As shown in the example of Figure 19, lateral spreads were modeled in inclined laminar boxes using a 10-m-thick Nevada sand deposit, which has a gentle slope inclination of 2%. The sand had a relative density of 40% and was saturated with water. As shown in Figure 19, the laminar box was shaken by a sine-like input ground motion at its base. The pore pressure, acceleration, and lateral displacement were recorded at various depths. As shown in Figure 19, (1) the lateral ground deformation is associated with unsymmetric spikes of ground acceleration in the uphill direction, (2) there are negative pore pressure spikes that increase the effective stress and stop the downhill ground deformation, and (3) the lateral ground deformation does not

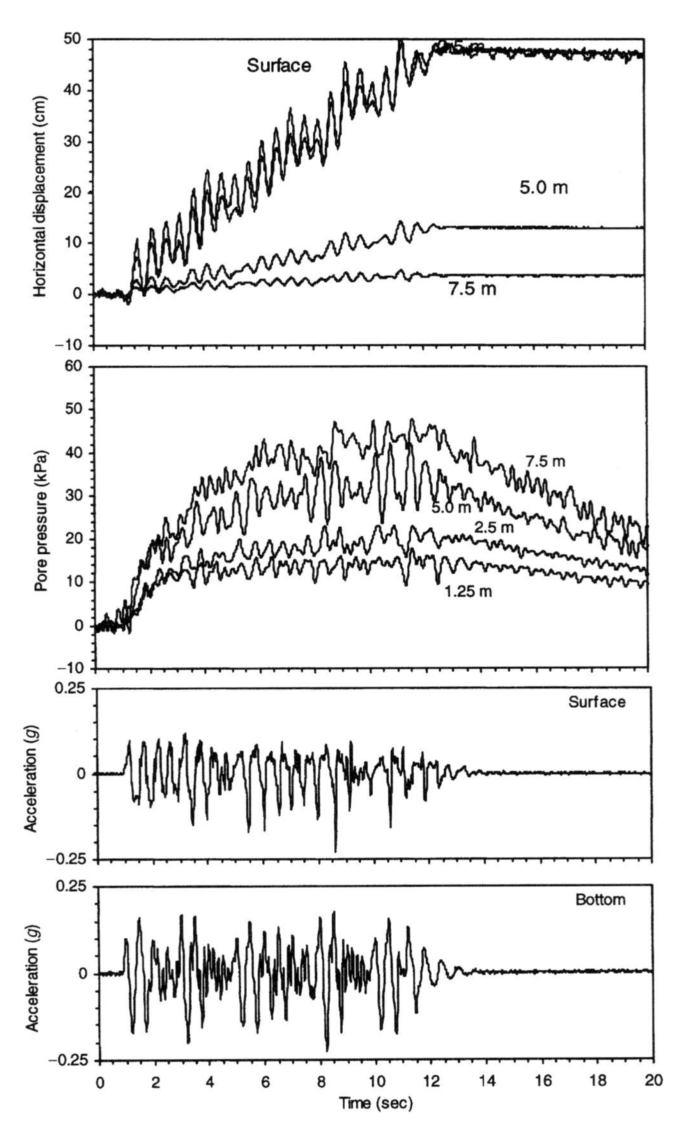

FIGURE 19 Time histories of lateral displacement, excess pore pressure, and accelerations at various depths in lateral spread centrifuge model (Dobry *et al.*, 1995).

continue after the shaking. In another set of centrifuge experiments, Dobry and Abdoun (1999) simulated delayed ground deformations using a series of small-amplitude shakings following the main shaking, which is reminiscent of aftershocks after main earthquake events.

At present, centrifuge tests model liquefaction-induced ground deformation as a ratcheting phenomenon, which stops with the end of ground shaking. This approach does not encompass all types of delayed ground deformations and failures, such as those observed after the 1964 Niigata earthquake. The modeling of these delayed deformations may require proper scale of diffusion and use of viscous additives. In spite of these present limitations, the centrifuges are useful tools for modeling liquefaction-induced lateral spreads.

4.3 Shaking Table Investigations

Liquefaction-induced lateral spreads have been extensively investigated using shaking tables (e.g., Towhata *et al.*, 1996). In these types of experiments, a reduced-scale model of soil deposits is subjected to short pulses or a continuous time history of acceleration simulating the earthquake ground motion. Figure 20 shows typical results of ground deformation obtained in shaking table tests. The acceleration, pore pressure, and displacement are measured at various locations in the model, which is 2 m long and 50 cm high. In the impact test [Figure 20(a)], the model is subjected to a very short acceleration pulse. The pore pressure rises very quickly, and the deformation takes place over an extended period of time after the impulse, in disagreement with centrifuge observations. In the shaking test [Figure 20(b)], the model is subjected to a sine-like base acceleration. The pore pressure gradually builds up until the soil liquefies. The ground deformations are progressive and stop with the base acceleration, as observed in centrifuge experiments.

In shaking table tests, Kokusho *et al.* (1998) reported the formation of a water film beneath a thin layer of silt sandwiched between sand layers. When such a film appeared, the soil mass above the silt layer was observed to glide in the downward direction not only during but also after the shaking. When the film did not form, the lateral flow took place mainly during the shaking. This shaking table experiment indicates the influence of drainage conditions on liquefaction-induced deformations.

4.4 Summary from Physical Modeling

Towhata *et al.* (1996) proposed an explanation for the simultaneous or delayed timing of ground deformation and ground shaking, which reconciles the observations in laboratory, shaking table, and centrifuge experiments. As illustrated in Figure 21, due to seismic shaking, the pore pressure rises until the soil liquefies. The soil remains liquefied over some time interval, then gradually solidifies and consolidates as pore pressure decreases, depending on drainage conditions. The timing of the liquefaction-induced deformation depends on the time history of driving stress and shear strength. The deformation starts when the shear strength is smaller than the driving shear stress, and stops when the shear strength is larger than the driving shear stress. In the case of rapid drainage, the shear strength is likely to be regained after the shaking ends, and the deformation will stop with the shaking. In the case of slower drainage, the shear strength may be regained much more slowly, and the deformation may extend after the shaking.

Based on centrifuge observations (e.g., Fig. 19), the lower amplitude range of liquefaction-induced deformation is obtained for relatively dense sands, corresponds to cyclic ratcheting during transient earthquake loading, and is controlled by transient shear stress, number of loading cycles, relative density, and stress-dilatancy. Based on laboratory experiments (e.g., 15), the upper amplitude range of liquefaction-induced deformation takes place for looser sands, and is controlled by a

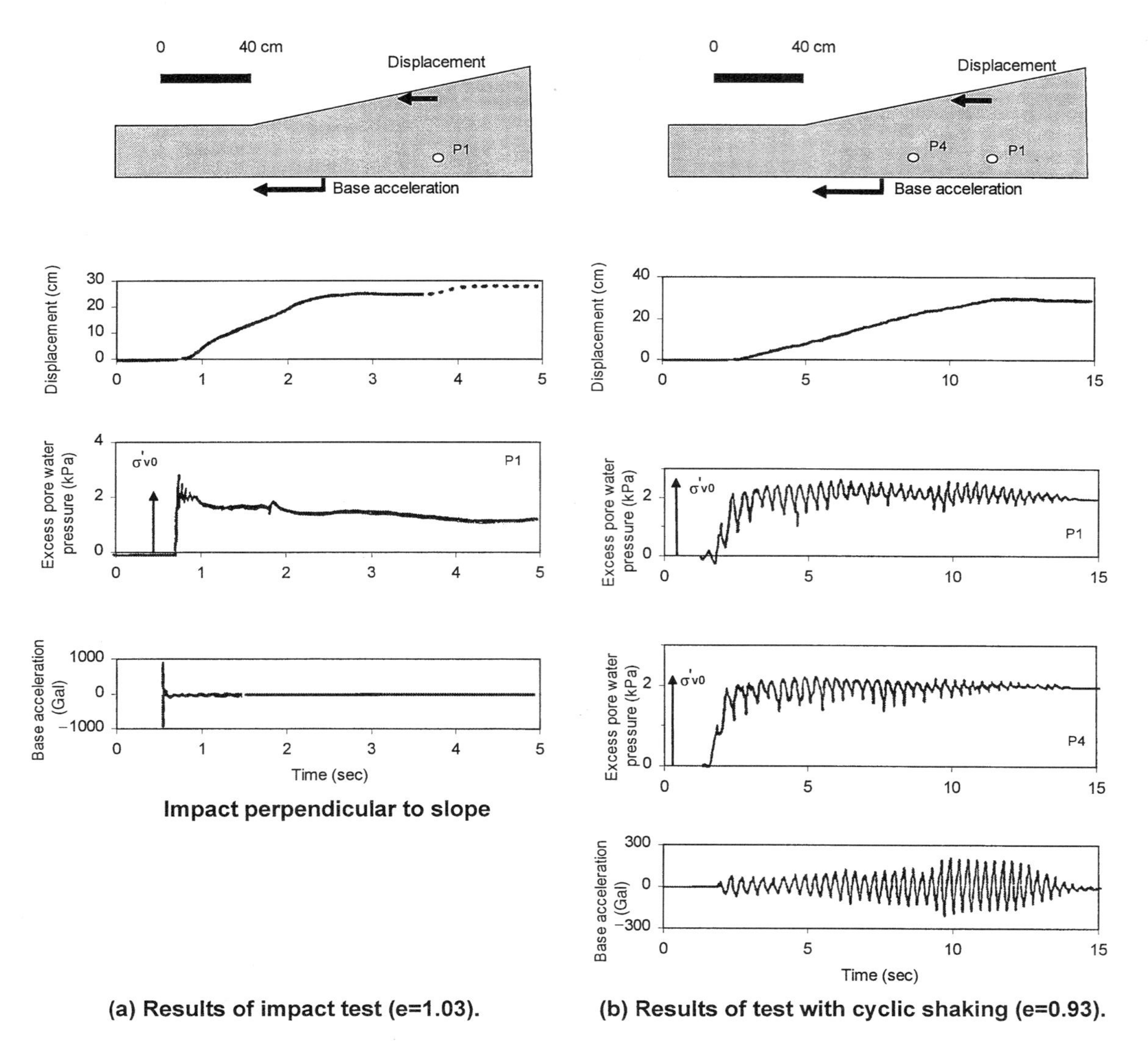

FIGURE 20 Time history of acceleration, pore pressure, and lateral displacement observed in shaking table test: (a) impact test, and (b) cyclic shaking (Towhata *et al.*, 1996).

post-liquefaction regain of shear strength induced by stress-dilatancy. This regain of shear strength is observed for a threshold shear strain γ_L, which depends on the initial relative density and the maximum amplitude of cyclic shear strain.

5. Numerical Modeling of Liquefaction

Since the early work of Housner (1958), many numerical models have been proposed for simulating soil liquefaction. They fall into two main categories: (1) simplified when they simulate only a few particular aspects of soil liquefaction for practical engineering applications, and (2) generalized when they attempt to account for a large number of soil liquefaction features.

5.1 Simplified Modeling

Simplified models have been proposed to simulate specific aspects of soil liquefaction, including (1) pore pressure buildup during earthquakes and (2) liquefaction-induced ground deformation.

5.1.1 Strain-Based Pore Pressure Models

Simplified models were proposed for determining the pore pressure buildup leading to liquefaction in saturated sands (e.g., Peacock and Seed, 1968; Martin *et al.*, 1975). These one-dimensional models relate the pore pressure increase to the amplitude of shear strain during simple shear loadings of constant amplitude. They describe the experimental results on which they are based rather well, and have been extensively used in soil

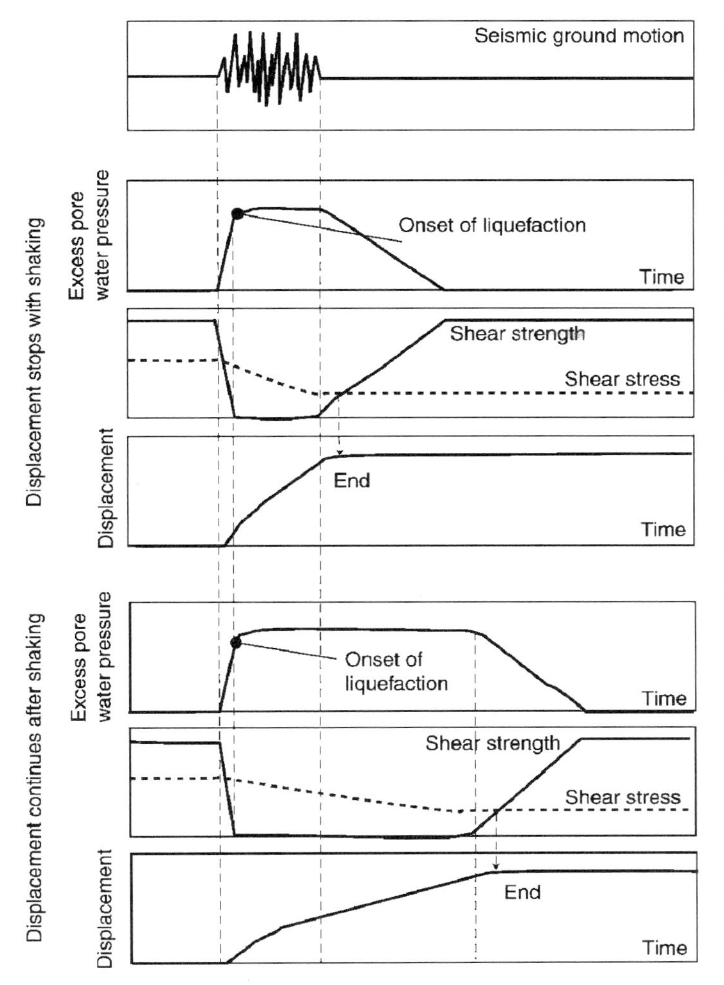

FIGURE 21 Possible explanations for the continuation and cessation of liquefaction-induced ground deformation after the earthquake shaking (after Towhata *et al.*, 1997).

liquefaction analysis based on peak ground acceleration and one-dimensional shear beam models (Seed and Idriss, 1971). These models are difficult to apply for generalized multidimensional and random cyclic loading. However, these simplified models clearly recognized the significant effects of pore pressure on the stress–strain response of saturated soils and prompted the development of more rational effective stress methods as described hereafter.

5.1.2 Energy-Based Pore Pressure Model

Nemat-Nasser and Sokooh (1979) introduced an energy concept for predicting soil liquefaction. They related the pore pressure buildup during laboratory cyclic loading to the total energy dissipated by the hysteretic stress–strain loop. Berrill and Davis (1985) verified the applicability of this energy concept in predicting the pore pressure generated in the case of undrained triaxial cyclic tests, and Figueroa *et al.* (1994) in the case of random cyclic torsional tests. The excess pore pressure Δu can be related to dissipated energy through

$$\Delta u/\sigma_0' = \alpha W_N^{\beta} \tag{8}$$

where σ_0' is the initial effective stress; W_N is a dimensionless energy term, and α and β are material constants. Liquefaction is predicted (i.e., $\Delta u/\sigma_0' = 1$) when the input earthquake energy exceeds some threshold energy. The energy-based pore pressure models were used to develop methods for predicting soil liquefaction during earthquakes (e.g., Berrill and Davis, 1985; Law and Cao, 1990; Trifunac, 1995), which departs from the traditional liquefaction analysis (e.g., Youd and Idriss, 1998).

5.1.3 Newmark Sliding Block Model for Lateral Spread

Lateral spreads have been modeled based on the concept that Newmark (1965) originally intended for calculating the deformations of earth dams during earthquakes.

Yegian *et al.* (1991) proposed that the amplitude D of permanent displacement is

$$D = N_{\text{eq}} T^2 a_{\text{p}} f(a_{\text{y}}/a_{\text{p}}) \tag{9}$$

where N_{eq} is the number of cycles of equivalent uniform base motion, T is the period (sec), a_{y} is the yield acceleration, a_{p} is the peak acceleration (g), and f is the dimensionless function depending on base motion. Baziar *et al.* (1992) proposed that D depends on peak velocity $v_{\max}$

$$D = N_{\text{eq}} v_{\max}^2 / a_{\max} / f(a_{\text{y}}/a_{\max}) \tag{10}$$

where $a_{\max}$ is the peak acceleration. Jibson (1994) proposed that D (cm) depends on the Arias intensity I_{a} (m/s):

$$\text{Log}\, D = 1.46 \log I_{\text{a}} - 6.642\, a_{\text{y}} + 1.546 \tag{11}$$

where a_{y} is the yield acceleration (g).

The Newmark-based models assume that the deformation takes place on a well-defined failure surface, the yield acceleration remains constant during shaking, and the soil is perfectly plastic. However, these assumptions do not hold in the case of liquefied soils and lateral spreads, because (1) the shear strain in liquefied soil does not concentrate within a well-defined surface, (2) the shear strength (and yield acceleration) of saturated soils varies during cyclic loading as pore pressure varies, and (3) soils are generally not perfectly plastic materials, but commonly harden and/or soften.

Byrne (1991) extended the concepts of Newmark's sliding block by introducing a resisting force varying with displacement. The liquefied mass stops when the initial kinetic energy is dissipated by the resisting force. For resisting forces that vary linearly with displacement, Byrne (1991) found final displacement 2 or 3 times greater than standard Newmark estimates.

5.1.4 Minimum Potential Energy Model

Towhata *et al.* (1997) developed a minimum potential energy model, which determines the final position of soil layers that liquefy. The model was successfully applied to model various shaking table test results, was extended to three dimensions (Orense and Towhata, 1996), and was applied to simulate case histories of liquefaction-induced lateral ground deformation during the Niigata, 1964, earthquake (Figure 22).

5.1.5 Viscous Models

Hamada *et al.* (1994, 1999) proposed similitude laws based on viscous models for simulating the liquefaction-induced deformation of soils and the forces they apply to buried structures (e.g., piles). The viscous properties of liquefied soils are difficult to measure experimentally as they vary with time and depend on drainage conditions. Yashima *et al.* (1997) attempted to simulate the flow failure of a liquefied embankment using a viscous model and a computational fluid mechanics code.

5.2 Generalized Modeling: Effective Stress Approach

One of the most promising and versatile methods for modeling soil liquefaction is the effective stress approach. It relies on the principles of continuum mechanics, constitutive equations, and effective stress, and treats engineering problems as boundary value problems. The nonlinear and irreversible stress–strain responses of soils are modeled using effective stress constitutive models. The nonlinear coupled equations of stress equilibrium and water diffusion are solved using numerical techniques (e.g., finite elements).

5.2.1 Effective Stress Constitutive Model

The stresses in saturated soils are characterized using effective stress (e.g., Terzaghi, 1943; Schrefler *et al.*, 1990) as follows:

$$\sigma'_{ij} = \sigma_{ij} - p\delta_{ij} \tag{12}$$

where σ_{ij} is the total stress; σ'_{ij} is the effective stress; and p is the interstitial fluid pressure. The nonlinear effective stress models are usually formulated using incremental constitutive equations as follows:

$$d\sigma'_{ij} = C_{ijkl}d\varepsilon_{kl} \tag{13}$$

where C_{ijkl} is a fourth-order tensor that characterizes the tangential material moduli and varies with stress and strain; $d\sigma'_{ij}$ is the increment of effective stress; and $d\varepsilon_{ij}$ is the strain increment. Many constitutive models have been proposed to simulate soil liquefaction. Table 4 lists only a few of these models, and

TABLE 4 Partial List of Constitutive Models for Effective Stress Liquefaction Analysis

General Framework	Model
Bounding surface plasticity	Bardet (1993) Pastor *et al.* (1985, 1990) Yogachandran (1991) Wang *et al.* (1990)
Multiple-yield surface plasticity	Prevost (1985) Para-Colmenares (1996)
Kinematic plasticity	Oka *et al.* (1992) Aubry *et al.* (1982)
Strain plasticity	Iai (1992)
Spatial mobilized plane	Matsuoka and Sakakibara (1987)

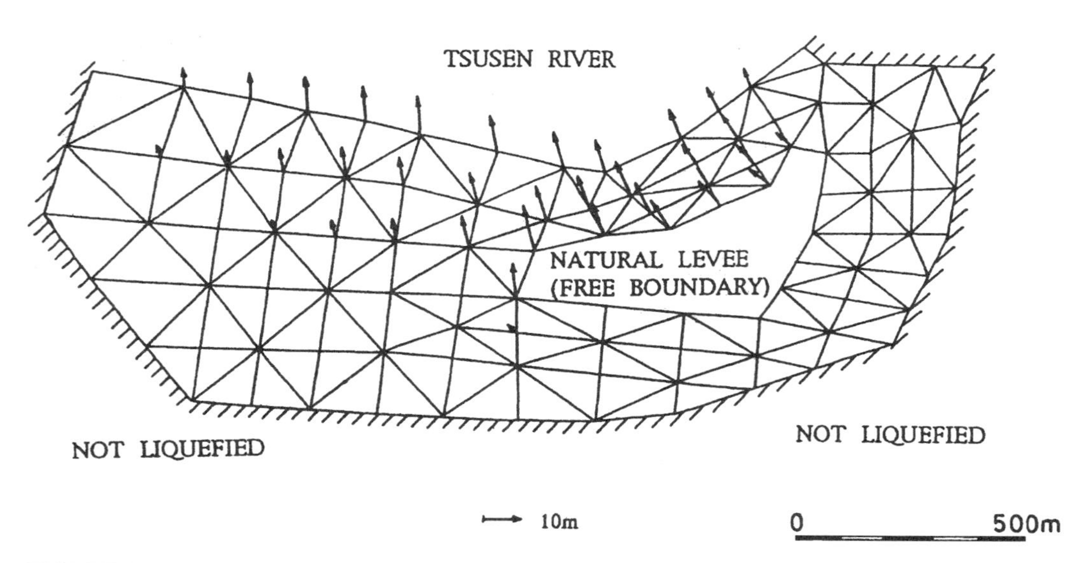

FIGURE 22 Three-dimensional simulation of liquefaction-induced lateral displacement of gentle slope in Oogata area of Niigata after the 1964 Niigata earthquake using the minimum potential energy model of Towhata *et al.* (1997).

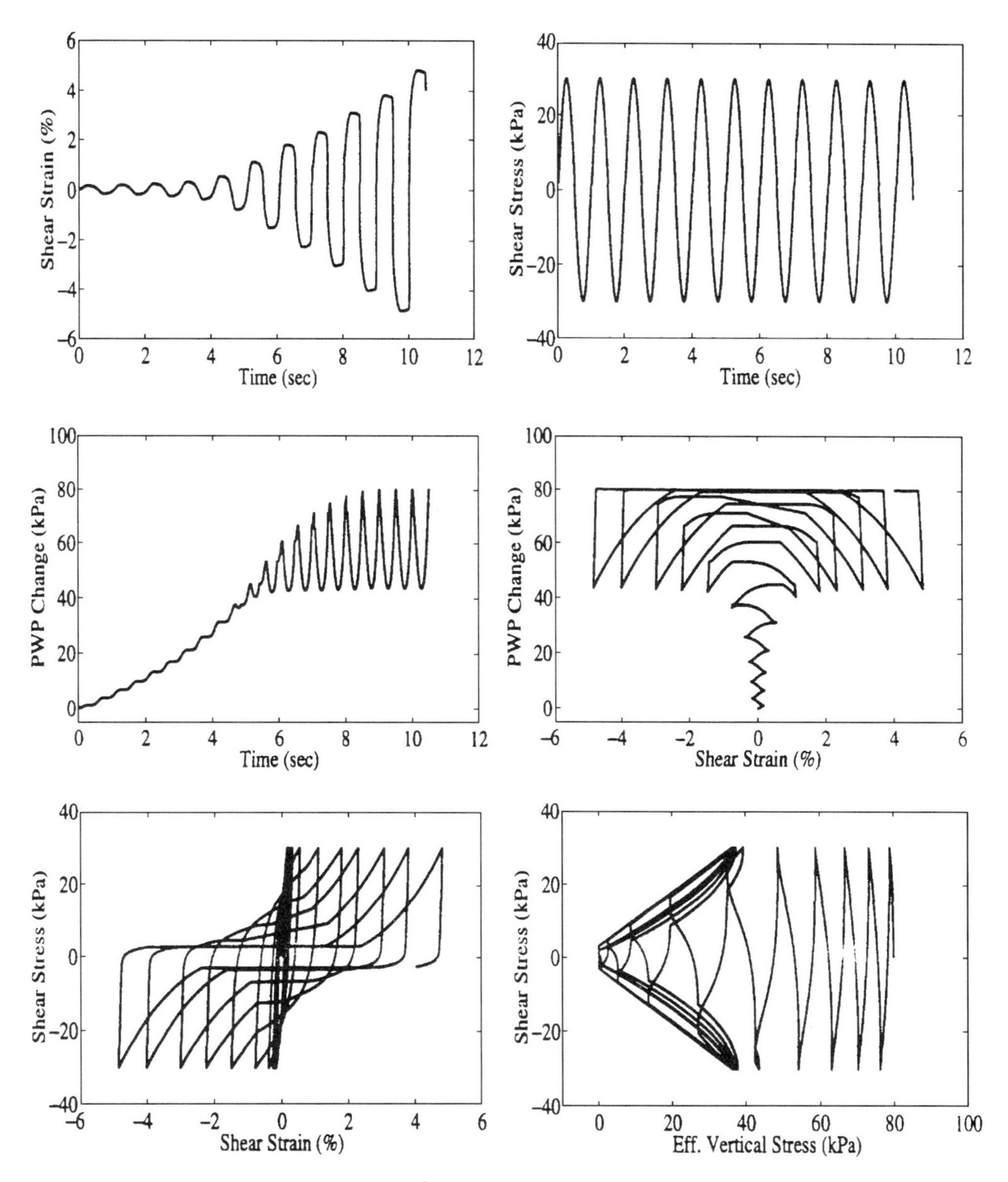

FIGURE 23 Example of simulated stress–strain response during cyclic simple shear test (Parra-Colmenares, 1996).

groups them according to their main theoretical framework. These models simulate in different ways the irreversible volumetric strain responsible for pore pressure buildup and liquefaction during cyclic loadings. They rely on either kinematic hardening for single- or multiple-yield surface plasticity, or isotropic hardening for bounding surface plasticity. Figure 23 is an example of stress–strain response simulated with a constitutive model, which can be compared to the laboratory-measured responses of Figures 11 and 12. As shown in Figures 11, 12, and 13, effective stress constitutive models can represent many nonlinear and irreversible features of soil behavior observed in laboratory experiments.

Not all models are capable of simulating all aspects of soil liquefaction equally well—post-liquefaction behavior and low-mean effective pressure behavior of soils are especially problematic. Most models are only applicable to model the cyclic generation of pore pressure and strain (Fig. 11). Very few models (e.g., Peiris and Yoshida, 1996) are capable of describing the large-strain post-liquefied stress–strain response of saturated sands as shown in Figure 13.

5.2.2 Effective Stress Method for Liquefaction Analysis

The incremental stress–strain relations [Eq. (13)] describe the nonlinear behavior of soil specimens in the particular cases of the uniform stress and strain states of laboratory experiments. These nonlinearities are generalized for the nonuniform stress and strain states encountered in engineering problems using governing equations and boundary conditions, i.e., formulating the engineering problems as boundary value problems. Biot (1941, 1962) first formulated the governing equations for describing the mechanical coupling between the soil porous matrix and its interstitial water. As summarized in Bardet and Sayed (1993), the original Biot formulation was modified in many ways. As described in Zienkiewicz and Shiomi (1984), soil liquefaction analyses can be derived from several different approximations of these governing equations for two-phase materials and different selections of nodal variables and spatial interpolations. Details of these rather lengthy formulations are omitted hereafter. One of the most commonly used is referred to as u-p formulation. The governing equations are (Zienkiewicz and Shiomi, 1984)

$$\sigma'_{ji,j} - \rho \frac{d^2 u_i}{dt^2} + \rho b_i + \frac{\partial p}{\partial x_i} = 0 \quad i = 1, 2, 3 \tag{14}$$

where b_i is the earth gravity; ρ is the unit mass of the saturated soil; t is time; and x_i $(i = 1, 2, 3)$ are the spatial coordinates. The fluid pressure p obeys the flow conservation equation (e.g., Schrefler *et al.*, 1990)

$$\frac{\partial}{\partial x_i}\left(-k\frac{\partial p}{\partial x_i} + k\rho_w b_i\right) + \frac{\partial}{\partial x_i}\left(\frac{du_i}{dt}\right) - \frac{1}{\Theta}\frac{dp}{dt} = 0 \tag{15}$$

where ρ_w is the fluid unit mass, and k is the hydraulic permeability. Θ is the bulk compressibility,

$$\frac{1}{\Theta} = \frac{n}{K_f} + \frac{1-n}{K_s} \tag{16}$$

where n is the porosity, K_s is the solid grain bulk modulus, and K_f is the fluid bulk modulus. Equations 14 and 15 apply to nearly incompressible solid grains and interstitial fluid, but not to three-phase materials with liquid–gas capillary tension (Schrefler *et al.*, 1990).

As described in Zienkiewicz and Shiomi (1984), the continuous fields of solid displacement $\mathbf{u}$ and water pressure $\mathbf{p}$ are approximated by discretized fields using shape functions $\mathbf{N_u}$ and $\mathbf{N_p}$, as follows:

$$\mathbf{u} = \mathbf{N_u}\bar{\mathbf{u}} \quad \text{and} \quad \mathbf{p} = \mathbf{N_p}\bar{\mathbf{p}} \tag{17}$$

where $\bar{\mathbf{u}}$ and $\bar{\mathbf{p}}$ are the nodal solid displacement and water pressure, respectively. Many investigators (e.g., Zienkiewicz and Taylor, 1991) recommend particular interpolations for $\mathbf{u}$ and $\mathbf{p}$ for avoiding numerical problems associated with incompressible fluid. After application of the principle of virtual work, Eqs. 14 and 15 become the following matrix problem:

$$\begin{pmatrix}\mathbf{M} & \mathbf{0}\\ \mathbf{0} & \mathbf{0}\end{pmatrix}\begin{pmatrix}\ddot{\bar{\mathbf{u}}}\\ \ddot{\bar{\mathbf{p}}}\end{pmatrix} + \begin{pmatrix}\mathbf{0} & \mathbf{0}\\ \mathbf{Q}^T & \mathbf{S}\end{pmatrix}\begin{pmatrix}\dot{\bar{\mathbf{u}}}\\ \dot{\bar{\mathbf{p}}}\end{pmatrix} + \begin{pmatrix}\mathbf{K} & -\mathbf{Q}\\ \mathbf{0} & \mathbf{H}\end{pmatrix}\begin{pmatrix}\bar{\mathbf{u}}\\ \bar{\mathbf{p}}\end{pmatrix} = \begin{pmatrix}\mathbf{f}_u\\ \mathbf{f}_p\end{pmatrix} \tag{18}$$

where $\mathbf{f}_u$ and $\mathbf{f}_p$ are the generalized forces arising from boundary conditions. $\mathbf{M}$, $\mathbf{C}$, $\mathbf{S}$, $\mathbf{K}$, $\mathbf{Q}$ and $\mathbf{H}$ are the following matrices:

$$\mathbf{M} = \int_\Omega \rho \mathbf{N_u^T N_u}\, d\Omega; \quad \mathbf{Q} = \int_\Omega \mathbf{B^T m N_p}\, d\Omega; \quad \mathbf{S} = \int_\Omega \mathbf{N_p^T}\frac{1}{\Theta}\mathbf{N_p}\, d\Omega;$$
$$\mathbf{K} = \int_\Omega \mathbf{B^T D B}\, d\Omega; \quad \text{and} \quad \mathbf{H} = \int_\Omega \nabla \mathbf{N_p^T k} \nabla \mathbf{N_p}\, d\Omega \tag{19}$$

where $\mathbf{m}$ is equal to [1,1,1,0,0,0]; $\mathbf{B}$ is the solid strain matrix; ∇ represents the gradient operator; and the superscript ^{T} denotes a matrix transpose. Equation (18) is a nonlinear hyperbolic matrix problem because matrix $\mathbf{K}$ varies nonlinearly with soil stresses. The transient problem may be integrated with respect to time using various types of time discretization algorithms (Hughes, 1987). At each time step, the nonlinear problem is solved using a nonlinear matrix solver such as Newton-Raphson iterations (e.g., Zienkiewicz and Taylor, 1991).

Figures 24 and 25 illustrate an example of liquefaction analysis using the effective stress approach. The quaywall structure of Figure 24 was shaken by the 1995 Hyogoken-Nanbu earthquake. As a result of soil liquefaction, the quaywall tilted and moved toward the sea. As shown in Figure 25, the engineering problems of Figure 24 were discretized using a finite element mesh and the material stress–strain responses were simulated using Iai model (Iai *et al.*, 1992). The earthquake loading was specified at the bottom boundary using the time history of a nearby earthquake recording. The computer program used in the analysis is FLIP (Iai *et al.*, 1992). As shown in Figures 24 and 25, the results of the numerical calculation are consistent with the observed mode of deformation of the quaywall.

5.2.3 VELACS

The feasibility of effective stress methods for modeling soil liquefaction was investigated in a National Science Foundation-sponsored project called VELACS, an acronym that stands for VErification of Liquefaction Analysis using Centrifuge Studies. Numerical analysts were asked to predict the response of some predefined boundary value problems, before those were actually modeled and executed in several centrifuges. The analysts were provided with the soil stress–strain responses measured in the laboratory (Arulmoli *et al.*, 1992), and the geometry, boundary conditions, and applied loads and displacements (Arulanandan and Scott, 1993). The comparison of numerical simulation and centrifuge results can be found in Arulanandan and Scott (1993–1994). These exercises can be repeated by those interested, as all VELACS input and output data are available from the Internet at *http://geoinfo.usc.edu/gees*. VELACS revealed a scatter in numerical predictions, which was attributed to many factors including differences in constitutive modeling, numerical codes, treatment of boundary value problems, and calibration of model constants from laboratory data. VELACS was an instructive research exercise, which pointed out that liquefaction analyses

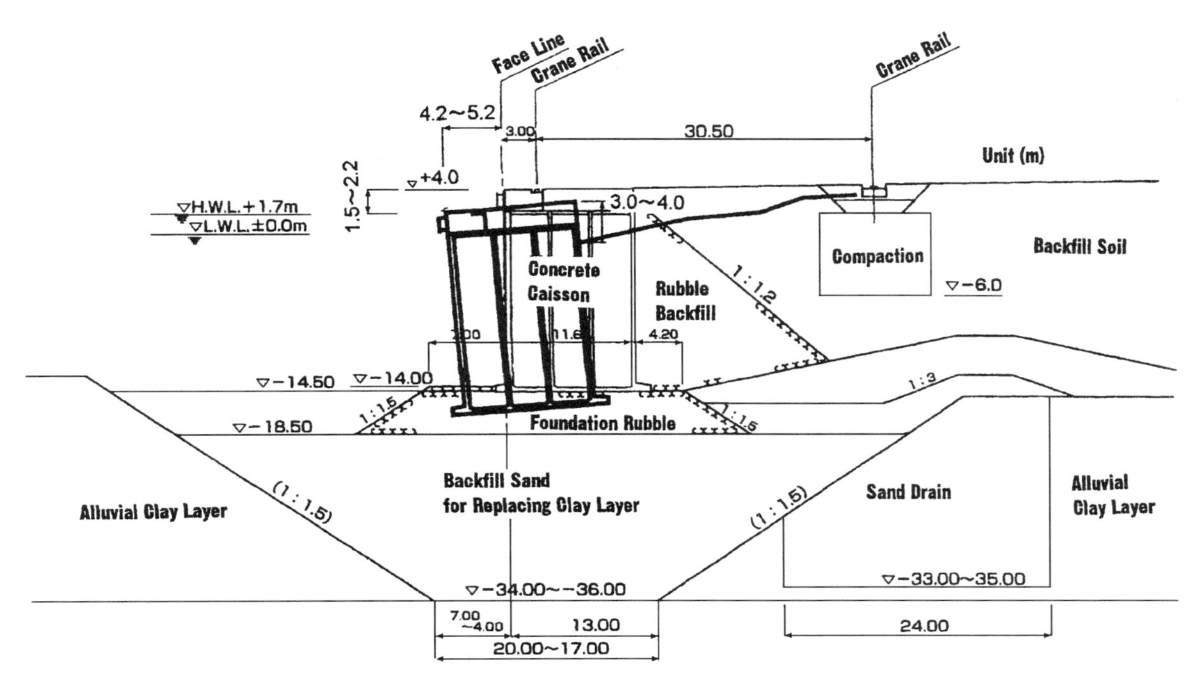

FIGURE 24 Cross section and deformation of a quaywall at Kobe Port (RC-5, Rokko Island) after the 1995 Hyogoken-Nanbu earthquake (Iai *et al.*, 1999).

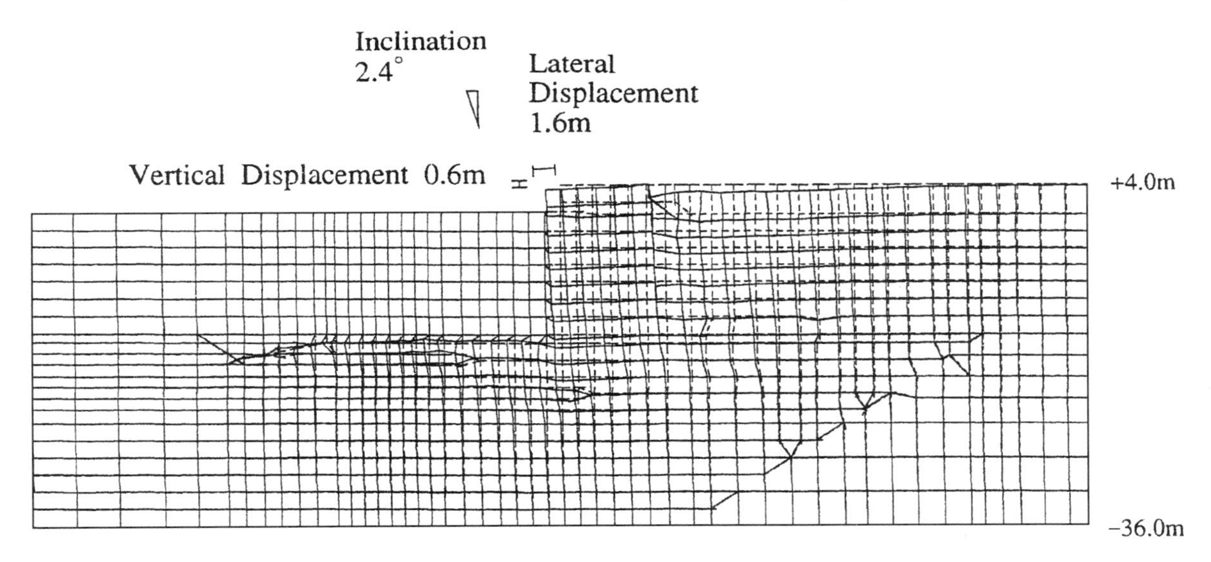

FIGURE 25 Computer deformation of quaywall of Figure 24 obtained from effective stress method (Iai *et al.*, 1999).

based on the effective stress approach are powerful engineering tools when properly used.

6. Recent Advances on Liquefaction

In addition to the numerical modeling previously described, there are two recent developments in soil liquefaction: soil liquefaction as a material instability, and computer simulations with discrete element methods.

6.1 Liquefaction as a Material Instability

Nonlinear pervious solids that have connected pores filled with an interstitial fluid (i.e., two-phase materials) can become mechanically unstable as shown by Rice (1975) and Vardoulakis (1985, 1986). The instability analyses that neglected interstitial fluid flow (e.g., Darve, 1994; Di Prisco and Nova, 1994; Nova, 1994; Lade, 1989) found only instabilities resulting from solid nonlinearities (e.g., Bardet, 1991; Hill and Hutchinson, 1975; Vardoulakis and Sulem, 1995). The analyses with interstitial

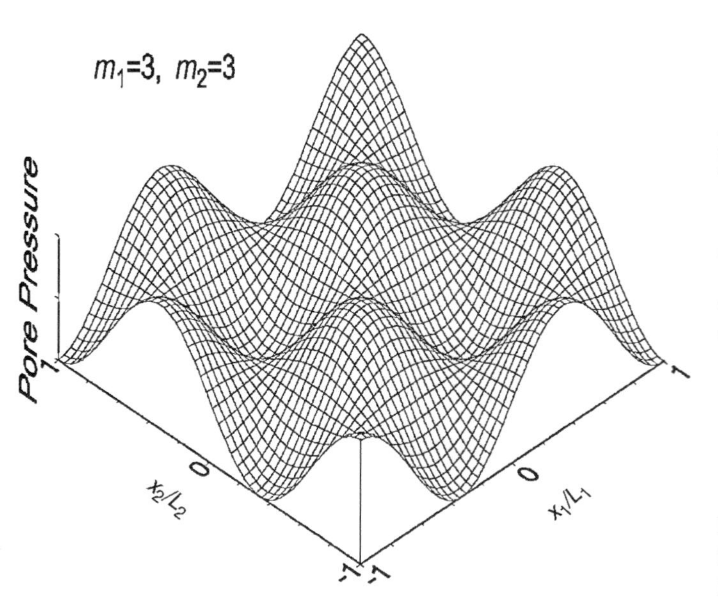

FIGURE 26 Distribution of pore pressure at emergence of two-phase instability in simulated plane strain undrained compression on an elastic-plastic model (Iai and Bardet, 2000).

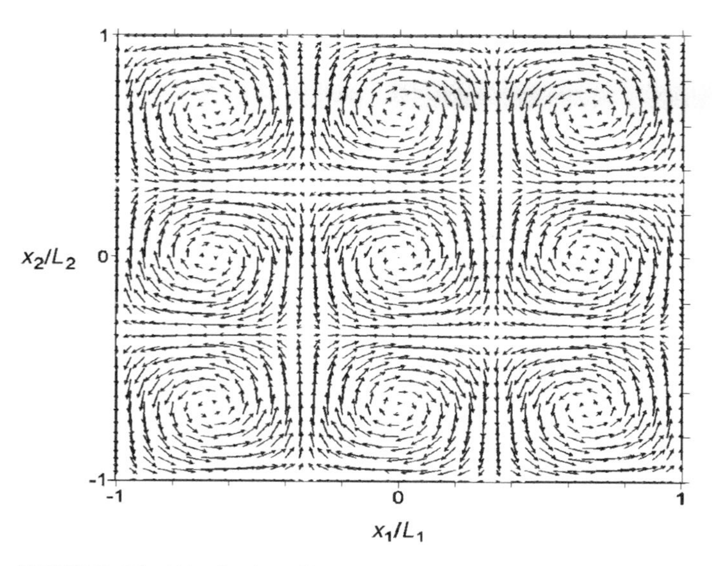

FIGURE 27 Distribution of incremental displacements in solid at the emergence of two-phase instability in simulated plane strain undrained compression on an elastic-plastic model (Iai and Bardet, 2000).

fluid flow (e.g., Bardet and Shiv, 1995; Iai and Bardet, 2000) discovered new types of instability (i.e., two-phase instability) resulting from the coupling between nonlinear soils and interstitial fluid flow. As illustrated in Figures 26 and 27, two-phase instabilities are characterized by (1) a nonhomogenous field of solid deformation with vortex-like patterns growing exponentially with time and (2) a corresponding nonuniform flow of interstitial water exhibiting source–sink flow patterns. As shown in Figure 28, two-phase instability may occur slightly before the peak deviator stress during the undrained compression of elastic-

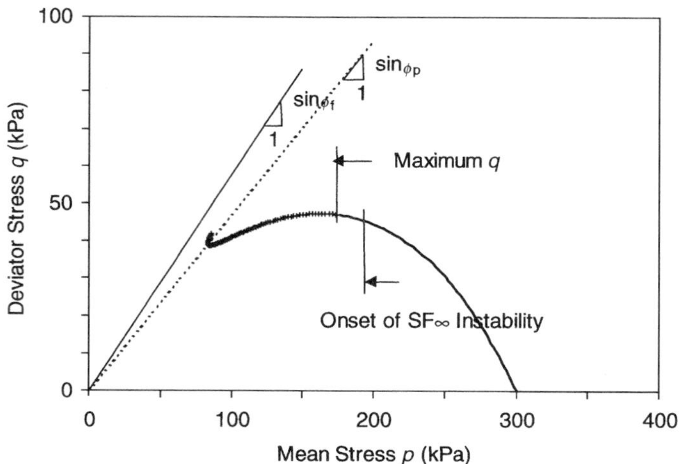

FIGURE 28 Emergence of two-phase instability represented on effective stress path during a plane strain undrained simulation on elastic-plastic (Iai and Bardet, 2000).

plastic soils. Two-phase instability is different from strain localization, strain softening, and constitutive singularities, and has practical implications on the numerical modeling of soil liquefaction. It may disrupt the numerical liquefaction analyses using elastic-plastic soil models (e.g., Bardet, 1996), and thus prompts the need for further analytical and experimental investigations.

6.2 Discrete Element Simulations of Soil Liquefaction

Soil liquefaction has been simulated using assemblies of discrete particles. The first discrete modeling of soils can be traced to Hertz (1882) and Reynolds (1885). Dantu (1957) and Schneebli (1955) modeled soils as assemblies of rigid rods, and noticed striking similarities between the mechanical responses of these mechanical analogs and real soils. This early work was first followed by theoretical and experimental investigations on the stress–strain response of assemblies of discrete particles (e.g., Duffy and Mindlin, 1957; Deresiewicz, 1958; Dobry and Ng, 1992), and then later by computer simulations for dry media (e.g., Cundall and Strack, 1979) and saturated media (e.g., Nakase, 1999). These last simulations account for the generation and diffusion of pore pressure between the particles.

As shown in the example of Figure 29 (Nakase, 1999), the soil was modeled using an assembly of 3000 circular disks with diameter ranging from 2.4 to 6.6 mm and a mean grain diameter of 5 mm. The disks were numerically packed in a space 14 cm high and 41 cm wide under the action of gravity. Two different initial packings were constructed to simulate the response of dense and loose materials, respectively. The vertical boundaries are periodic (Cundall and Strack, 1979), and the cyclic sinusoidal shear force is applied through the top rigid platen, which is also subjected to constant confining pressure (i.e., σ_c = 98 kPa). As shown in Figures 30 and 31, the simulated material responses of assemblies of discrete particles are similar to the

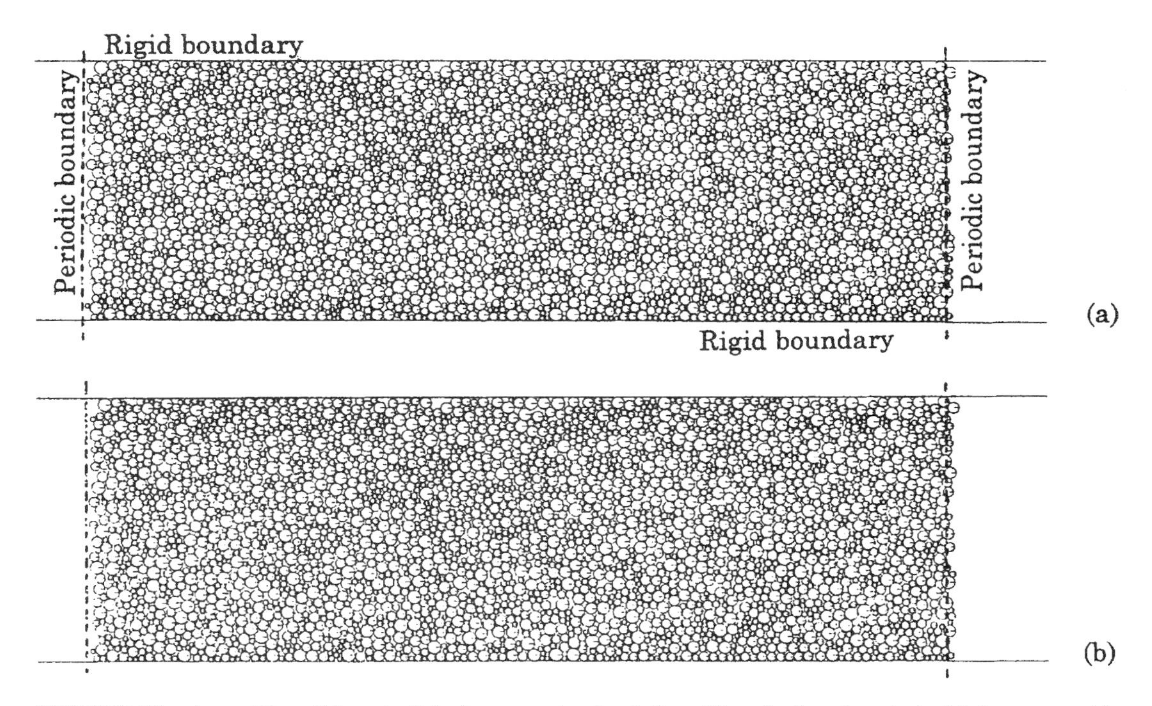

FIGURE 29 Assemblies of discrete disks for computer simulation of liquefaction shear tests: (a) dense assembly, and (b) loose assembly (Nakase, 1999).

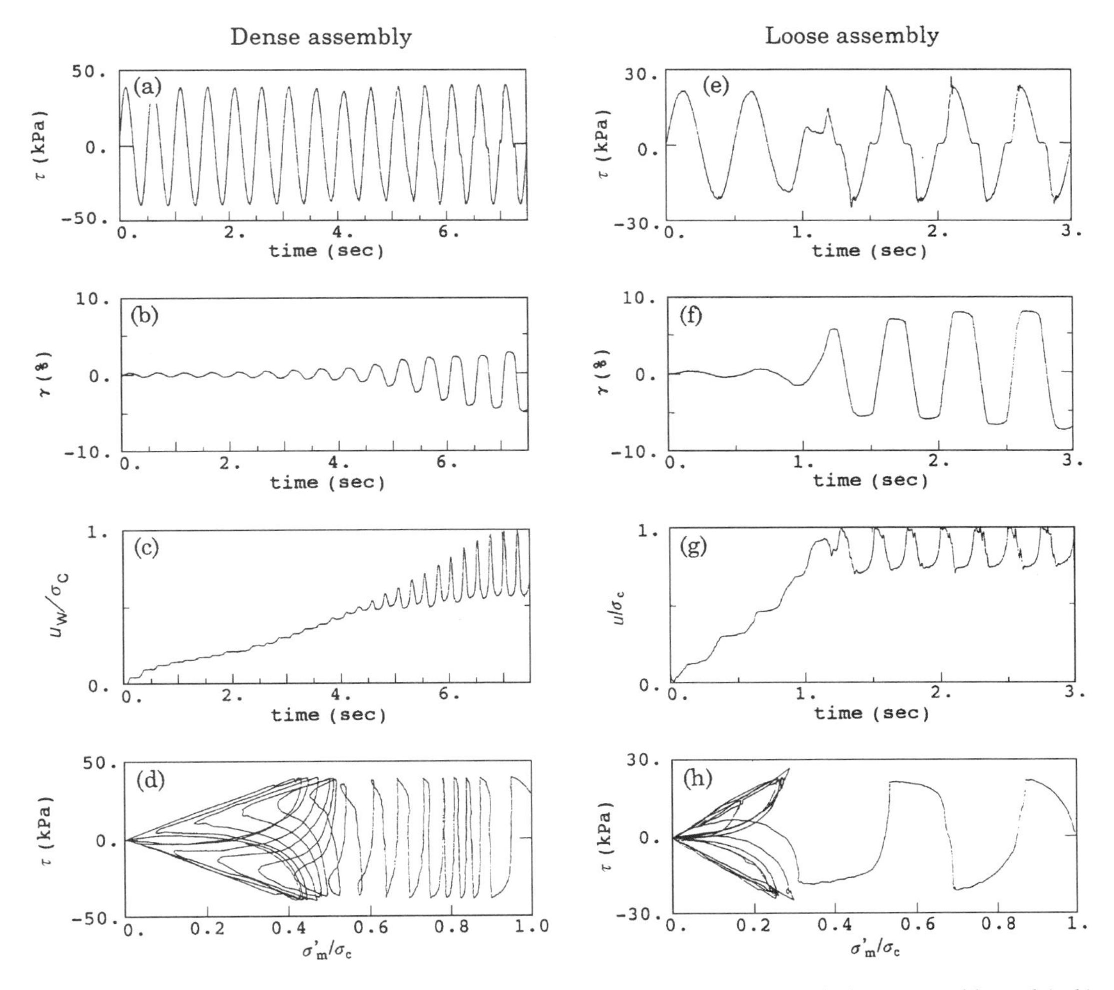

FIGURE 30 Simulated results of liquefaction shear tests on discrete disks: (a–d) dense assembly, and (e–h) loose assembly (Nakase, 1999).

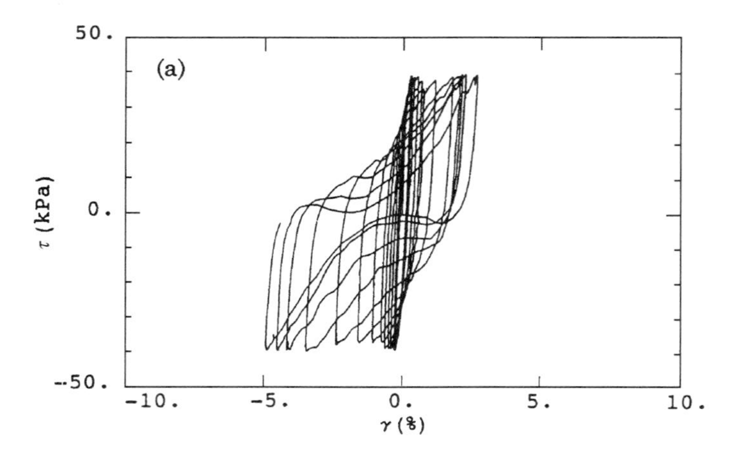

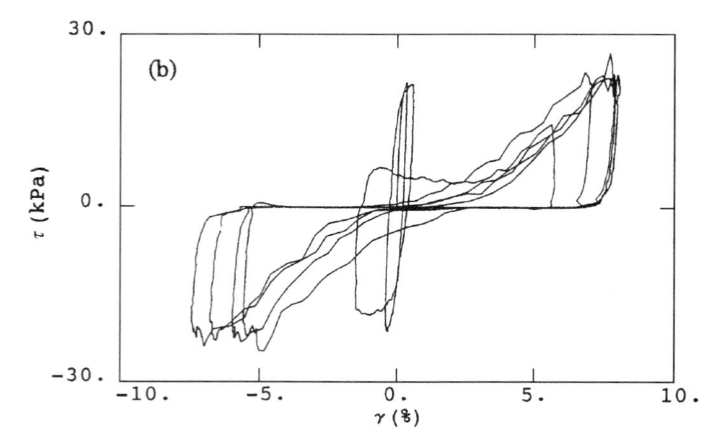

FIGURE 31 Simulated stress–strain curves of a liquefaction shear test on discrete disks: (a) dense assembly, and (b) loose assembly (Nakase, 1999).

observed responses of actual soil specimens in the laboratory (Figure 11). As shown in Figure 30(a) and 30(e), the shear stress τ is calculated at the bottom platen. The shear strain γ gradually increases with the excess pore pressure u_w, while the resulting mean effective stress σ'_m decreases. Finally, the specimens display cyclic mobility at different extents as observed in Figure 11. These computer simulations are based on a limited number of physical processes, and are therefore useful to understand the fundamentals of soil liquefaction. These numerical approaches are at the forefront of liquefaction research and promise to yield new understanding about the mechanisms of lateral spreads.

References

Andrus, R. D., K. H. Stokoe, II, and R. M. Chung (1999). Draft guidelines for evaluating liquefaction resistance using shear wave velocity measurement and simplified procedures. National Technical Information Service (NTIS), Technology Administration, US Department of Commerce. Springfield, NISTIR 6277, 134 pp.

Arulanandan, K., and R. F. Scott, Eds. (1993 and 1994). Verification of numerical procedures for the analysis of soil liquefaction problems. In: "Proc. Int. Conf. on the Verification of Numerical Procedures for the Analysis of Soil Liquefaction Problems," Volumes 1 and 2. A. A. Balkema, Rotterdam, the Netherlands.

Arulmoli, K., K. K. Muraleetharan, M. M. Hossain, and L. S. Fruth (1992). VELACS: Verification of liquefaction analysis by centrifuge studies, Laboratory Testing Program, Soil Data Report. The Earth Technology Corporation, Long Beach, CA, Report for the National Science Foundation.

Aubry, D., J. C. Hujeux, F. Lassoudiere, and Y. Meimon (1982). A double memory model with multiple mechanisms for cyclic soil behavior. In: "Proc. Int. Symposium on Numerical Models in Geomechanics," pp. 3–13. A.A. Balkema, Rotterdam, the Netherlands.

Bardet, J. P. (1986). A bounding surface plasticity model for sands. *J. Eng. Mech., ASCE* **112**(11), 1198–1217.

Bardet, J. P. (1991). Analytical solutions for the plane strain bifurcation of compressible solids. *J. Appl. Mech., ASME* **58**, 651–657.

Bardet, J. P. (1996). Finite element analysis of two-phase instability for saturated porous hypoelastic solids under plane strain loadings. *Eng. Computations* **13**(7), 29–48.

Bardet, J. P., and C. A. Davis (1996). Performance of San Fernando dams during the 1994 Northridge earthquake. *J. Geotech. Eng. Div., ASCE* **122**(7), 554–564.

Bardet, J. P., and H. Sayed (1993). Velocity and attenuation of compression waves in nearly saturated soils. *Soil Dynam. Earthq. Eng.* **12**(7), 391–402.

Bardet, J. P., and A. Shiv (1995). Plane-strain instability of saturated porous media. *J. Engrg Mech., ASCE* **121**(6), 717–724.

Bardet, J. P., Q. Huang, and S. W. Chi (1993). Numerical prediction for Model No. 1; Numerical prediction for Model No. 3; and Numerical prediction for Model No. 7. In: "Proceedings of the International Conference on the Verification of Numerical Procedures for the Analysis of Soil Liquefaction Problems" (K. Arulanandan and R. F. Scott, Eds.), pp. 67–86, 483–488, and 829–834. A. A. Balkema., Rotterdam, the Netherlands.

Bardet, J. P., J. Hu, T. Tobita, and N. Mace (1999). Database of case histories on liquefaction-induced ground deformation. Report, Civil Engineering Department, University of Southern California, Los Angeles.

Bardet, J. P., N. Mace, and T. Tobita (1999). Liquefaction-induced ground deformation and failure. Report, Civil Engineering Department, University of Southern California, Los Angeles.

Bardet, J. P., N. Mace, T. Tobita, and J. Hu (2000). Large-scale modeling of liquefaction-induced ground deformation. *Earthq. Spectra, EERI*, in press.

Bartlett, S. F., and T. L. Youd (1992). Empirical analysis of horizontal ground displacement generated by liquefaction-induced lateral spreads. Technical report NCEER-92-0021, National Center for Earthquake Engineering Research, State University of New York, Buffalo, 5-14-15.

Bartlett, S. F., and T. L. Youd (1995). Empirical prediction of lateral spread displacement. *J. Geotech. Eng. Div., ASCE* **121**(4), 316–329.

Baziar, M. H., R. Dobry, and A. W. M. Elgamal (1992). Engineering evaluation of permanent ground deformation due to seismically-induced liquefaction. Technical report NCEER-92-0007, National Center for Earthquake Engineering Research, State University of New York, Buffalo, 2 Vols.

Bennett, M. J., P. V. McLaughlin, J. S. Sarmiento, and T. L. Youd (1984). Geotechnical investigation of liquefaction sites, Imperial Valley, California. *Open File Report 84-252, US Geol. Surv.*, Menlo Park, California.

Bennett, M. J., and J. C. Tinsley, III (1995). Geotechnical data from surface and subsurface samples outside of and within liquefaction-related ground failures caused by the October 17, 1989, Loma Prieta earthquake, Santa Cruz and Monterey Counties, California. *Open File Report 95-663, US Geol. Surv.*, Menlo Park, California.

Berrill, J. B., and R. O. Davis (1985). Energy dissipation and seismic liquefaction of sands: a revised model. *Soils and Foundations* **25**(2), 106–118.

Biot, M. A. (1941). General theory of three-dimensional consolidation. *J. Appl. Phys.* **12**, 155–164.

Biot, M. A. (1962). Mechanics of deformation and acoustic propagation in porous media. *J. Acoust. Soc. Am.* **34**, 1254–1264.

Byrne, P. M. (1991). A model for predicting liquefaction induced displacement. In: "Proc. Second Int. Conf. on Recent Advances in Geotechnical Earthquake Engineering and Soil Dynamics," Vol. 2, pp. 1027–1035. St. Louis, Missouri.

Cao, Y. L., and K. T. Law (1992). Energy dissipation and dynamic behaviour of clay under cyclic loading. *Canadian Geotech. J.* **29**, 103–111.

Castro, G. (1969). "Liquefaction of Sands." Ph.D. thesis, Harvard University, Harvard Soil Mechanics Series, No. 81.

Cundall, P. A., and O. D. L. Strack (1979). A discrete numerical model for granular assemblies. *Géotechnique* **29**, 47–65.

Dafalias, Y. F., and L. R. Herrmann (1982). Bounding surface formulation of soil plasticity. In: "Soil Mechanics—Transient and cyclic loads" (G. Pande and O.C. Zienkiewicz, Eds.), pp. 253–282. John Wiley and Sons, London.

Dantu, P. (1957). Contribution à l'étude mécanique et géometrique des milieux pulvérulents. In: "Proc. 4th Int. Conf. of Soil Mechanics and Foundation Eng." London, UK.

Darve, F. (1994). Stability and uniqueness in geomaterials constitutive modelling. In: "Localization and Bifurcation Theory for Soils and Rocks" (R. Chambon, J. Desrues, and I. Vardoulakis, Eds.), pp. 73–88. A. A. Balkema, Rotterdam, the Netherlands.

Davis, C., and J. P. Bardet (1996). Performance of two reservoirs during the 1994 Northridge earthquake. *J. Geotech. Eng. Div., ASCE* **122**, (8), 613–622.

Davis, R. O., and J. B. Berrill (1982). Energy dissipation and seismic liquefaction of sands. *Earthq. Eng. Struct. Dynam.* **10**(2), 59–68.

Deresiewicz, H. (1958). Stress-strain relations for a simple model of a granular medium. *J. Appl. Mech., ASME* 403–406.

Di Prisco, C., and R. Nova (1994). Stability problems related to static liquefaction of loose sand. In: "Localization and Bifurcation Theory for Soils and Rocks" (R. Chambon, J. Desrues, and I. Vardoulakis, Eds.), pp. 59–70. A.A. Balkema, Rotterdam, the Netherlands.

Dobry, R., and T.-T. Ng (1992). Discrete modeling of stress-strain behavior of granular media at small and large strain. *Eng. Computations* **9**, 129–143.

Dobry, R., V. Taboada, and L. Liu (1995). Centrifuge modeling of liquefaction effects during earthquakes. In: "Proc. 1st Int. Conf. on Earthq. Geotech. Eng., IS-Tokyo." Keynote and Theme lectures, pp. 129–162.

Dobry, R., and T. Abdoun (1999). Post-triggering response of liquefied sand in the free field and near foundations. In: "Proceedings of ASCE Specialty Conference on Geotechnical Earthquake Engineering and Soil Dynamics" (P. Dakoulas, M. Yegian, and R. D. Holtz, Eds.), Vol. 1, pp. 270–300. ASCE Special Publication No. 75.

Draper, N. R., and H. Smith (1981). *Applied Regression Analysis*, 2nd ed. John Wiley & Sons, New York, 709 pp.

Duffy, J., and R. D. Mindlin (1957). Stress-strain relations and vibrations of a granular medium. *J. of Appl. Mech., ASME* **79**, 585–593.

Elgamal, A. W., M. Zeghal, and E. Parra (1996). Liquefaction of reclaimed island in Kobe, Japan. *J. Geotech. Eng. Div., ASCE* **122**(1), 39–49.

Figueroa, J. L., A. S. Saada, L. Lian, and M. N. Dahisaria (1994). Evaluation of soil liquefaction by energy principles. *J. Geotech. Eng. Div., ASCE* **120**(9), 1554–1569.

Goh, A. T. C. (1994). Seismic liquefaction potential assessed by neural networks. *J. Geotech. Eng. Div., ASCE* **120**(9), 1467–1480.

Goh, A. T. C. (1996). Neural-network modeling of CPT seismic liquefaction data. *J. Geotech. Eng. Div., ASCE* **122**(1), 70–73.

Hamada, M. (1999). Similitude law for liquefied ground flow. In: "Proceedings of the 7th U.S. Japan Workshop on Earthquake Resistant Design of Lifeline Facilities and Countermeasures for Soil Liquefaction" (T. D. O'Rourke, J. P. Bardet, and M. Hamada, Eds.), pp. 191–205. NCEER.

Hamada, M., and T. D. O'Rourke (1992). Case histories of liquefaction and lifeline performance during past earthquakes. Technical report NCEER-92-0001, National Center for Earthquake Engineering Research, State University of New York, Buffalo, 2 Vols.

Hamada, M., and K. Wakamatsu (1998). Liquefaction-induced ground displacement triggered by quaywall movement. *Soils and Foundations*, Special Issue Vol. 2, Japanese Geotech. Soc., pp. 85–96.

Hamada, M., S. Yasuda, R. Isoyama, and K. Emoto (1986). Study on liquefaction induced permanent ground displacement. Report for the Association for the Development of Earthquake Prediction.

Hamada, M., H. Sato, and T. Kawakami (1994). A consideration for the mechanism for liquefaction-related large ground displacement. In: "Proceedings of the 5th U.S. Japan Workshop on Earthquake Resistant Design of Lifeline Facilities and Countermeasures for Soil Liquefaction" (T. D. O'Rourke and M. Hamada, Eds.), pp. 217–232. NCEER Report 94-0026.

Hamada, M., R. Isoyama, and K. Wakamatsu (1996). Liquefaction-induced ground displacement and its related damage to lifelines facilities. *Soils and Foundations*, Special issue, Japanese Geotech. Soc., pp. 81–97.

Hamada, M., R. Isoyama, and K. Wakamatsu (1996). The 1995 Hyogoken-Nanbu (Kobe) Earthquake, Liquefaction, Ground Displacement and Soil Condition in Hanshin Area. Association for Development of Earthquake Prediction, The School of Science and Engineering, Waseda University, Japan Engineering Consultants.

Harder, L. (1991). Evaluation of Liquefaction Potential Using the SPT. Seismic Short Course, Evaluation and Mitigation of Earthquake Induced Liquefaction Hazards, University of Southern California, Los Angeles.

Hertz, H. (1882). "Über die Berührung fester elastische Körper and über die Harte (On the contact of rigid elastic solids and on hardness)." *Verhandlungen des Vereins zur Beförderung des Gewerbefleisses*, Leipzig.

Hill, R., and J. W. Hutchinson (1975). Bifurcation phenomena in the plane tension test. *J. Mech. Phys. Solids* **23**, 239–264.

Holzer, T. L. (1998). The Loma Prieta, California, earthquake of October 17, 1989—Liquefaction. In: "Strong Ground Motion and Ground failure" (T. L. Holzer, Ed.), *US Geol. Surv. Professional Paper 1551-B*, pp. 314.

Holzer, T. L., T. L. Youd, and T. C. Hanks (1989). Dynamics of liquefaction during the 1987 Superstition Hills, California, earthquake. *Science* **244**, 56–59.

Holzer, T. L., J. C. Tinsley, M. J. Bennett, and C. S. Mueller (1994). Observed and predicted ground deformation—Miller Farm lateral spread, Watsonville, California. In: "Proceedings of the 6th U.S. Japan Workshop on Earthquake Resistant Design of Lifeline Facilities and Countermeasures for Soil Liquefaction" (M. Hamada and T. D. O'Rourke, Eds.), pp. 79–98. NCEER.

Holzer, T. L., M. J. Bennett, J. C. Tinsley, D. J. Ponti, and R. V. Sharp (1996). Causes of ground failure in alluvium during the Northridge, California, earthquake of January 17,1994. In: "Proceedings of the 6th U.S. Japan Workshop on Earthquake Resistant Design of Lifeline Facilities and Countermeasures for Soil Liquefaction" (M. Hamada and T. D. O'Rourke, Eds.), pp. 345–360. NCEER.

Holzer, T. L., M. J. Bennett, D. J. Ponti, and J. C. Tinsley (1998). Permanent ground deformation in alluvium during the 1994 Northridge, California, earthquake. *Open-file Report 98-118, US Geol. Surv.* Menlo Park, California.

Holzer, T. L., M. J. Bennett, D. J. Ponti, and J. C. Tinsley (1999). Liquefaction and soil failure during the 1994 Northridge earthquake. *J. Geotech. Geoenviron. Eng., ASCE* **125**(6), 438–452.

Housner, G. (1958). The mechanism of sand blows. *Bull. Seismol. Soc. Am.* **48**, 155–161.

Hughes, T. J. R. (1987). *The Finite Element Method*. Prentice-Hall, Inc., Englewoods Cliffs, NJ.

Hushmand, B., R. F. Scott, and C. B. Crouse (1988). Centrifuge liquefaction tests in a laminar box. *Géotechnique* **38**(2), 253–262.

Iai, S., and J. P. Bardet (2000). Plane strain instability of saturated elastoplastic soils. *Geotechnique,* in press.

Iai, S., Y. Matsunaga, and T. Kameoka (1992). Strain space plasticity model for cyclic mobility. *Soils and Foundations* **32**(2), 1–15.

Iai, S., K. Ichii, H. Liu, and T. Morita (1998). Effective stress analyses of port structures. *Soils and Foundations*, Japanese Geotech. Soc., Special Issue, pp. 97–114.

Iai, S., K. Ichii, H. Sato, and H. Liu (1999). Residual displacement of gravity quaywalls, parameter study through effective stress analysis. In: "Proceedings of the 7th U.S. Japan Workshop on Earthquake Resistant Design of Lifeline Facilities and Countermeasures for Soil Liquefaction" (T. D. O'Rourke, J. P. Bardet, and M. Hamada, Eds.), pp. 549–563. NCEER.

Idriss, I. M., I. Arango, and G. Brogan (1979). Study of liquefaction in November 23, 1977 earthquake, San Juan province, Argentina. Woodward-Clyde Consultants.

Inagaki, H., S. Iai, T. Sugano, H. Yamazaki, and T. Inatomi (1996). Performance of caisson type quay walls at Kobe Port. *Soils and Foundations*, Japanese Geotech. Soc., Special issue, pp. 119–136.

Ishihara, K. (1993). Liquefaction and flow failure during earthquakes. The 33rd Rankine lecture, *Geotechnique* **43**(3), 351–415.

Ishihara, K. (1996). *Soil Behavior in Earthquake Geotechnics*. Oxford Science Publications, Clarendon Press, Oxford, UK.

Ishihara, K., M. L. Silver, and H. Kitagawa (1979). Cyclic strength of undisturbed sands obtained by a piston sampler. *Soils and Foundations* **19**(3).

Ishihara, K., Y. Kawase, and M. Nakajima (1980). Liquefaction characteristics of sands deposits at an oil tank site during the 1978 Miyagike-Oki earthquake. *Soils and Foundations* **20**(2).

Ishihara, K., K. Shimizu, and Y. Yamada (1981). Pore water pressures measured in sand deposit during an earthquake. *Soils and Foundations* **20**(4).

Ishihara, K., S. Yasuda, and H. Nagase (1996). Soil characteristics and ground damage. *Soils and Foundations*, Japanese Geotech. Soc., Special Issue, pp. 109–118.

Ishihara, K., K. Yoshida, and M. Kato (1997). Characteristics of lateral spreading in liquefied deposits during the 1995 Hanshin-Awaji earthquake. *J. Earthq. Eng.* **1**(1), 23–55.

Isoyama, R. (1994). Liquefaction-induced ground failures and displacements along the Shiribeshi-toshibetsu River caused by the 1993 Hokkaido Nansei-oki earthquake. In: "Proceedings of the 5th U.S. Japan Workshop on Earthquake Resistant Design of Lifeline Facilities and Countermeasures for Soil Liquefaction" (T. D. O'Rourke and M. Hamada, Eds.), pp. 1–26. NCEER.

Iwasaki, T., K. Kawashima, and K. Tokida (1978). Report of the Miyagiken-Oki earthquake of June, 1978. Public Works Research Institute, Ministry of Construction, Report No. 1422 (in Japanese).

Iwasaki, Y., and M. Tai (1996). Strong motion records at Kobe Port Island. *Soils and Foundations*, Japanese Geotech. Soc., Special Issue, pp. 29–40.

Jibson, R. (1994). Predicting earthquake-induced landslide displacement using Newmark's sliding block analysis. Transportation Research Record 1411, Transportation Research Board, Washington, D.C., pp. 9–17.

Karube, D., and M. Kimura (1996). Damage to foundation of railways structures. *Soils and Foundations*, Japanese Geotech. Soc., Special Issue, pp. 201–210.

Kishida, H. (1969). Characteristics of liquefied sands during Mino-Owari, Tohnankai, and Fukui earthquakes. *Soils and Foundations* **9**(1), 75–92.

Ko, H. Y. (1988). Summary of the state-of-the-art in centrifuge model testing. In: "Centrifuges in Soil Mechanics" (W. H. Craig, R. G. James, and A. N. Schofield, Eds.), pp. 11–18. Balkema, Rotterdam, the Netherlands.

Kobayashi, Y., I. Towahata, and A. A. Acacio (1999). Analysis on liquefaction-induced subsidences of shallow foundations. In: "Proceedings of the 7th U.S. Japan Workshop on Earthquake Resistant Design of Lifeline Facilities and Countermeasures for Soil Liquefaction" (T. D. O'Rourke, J. P. Bardet, and M. Hamada, Eds.), pp. 565–577. NCEER.

Kodera, J. (1964). Earthquake damage and the ground of pier foundations, Part 1. *Tsuchi-to-Kiso* **12**(3) (in Japanese).

Koizumi, Y. (1966). Changes in density of sand subsoil caused by the Niigata earthquake, *Soils and Foundations* **6**(2).

Kokusho, T., K. Watanabe, and T. Sawano (1998). Effect of water film on lateral flow failure of liquefied sand. In: "Proceedings of the 11th European Conference on Earthquake Engineering," pp. 1–8. A.A. Balkema, Rotterdam, the Netherlands.

Lade, P. V. (1989). Experimental observations of stability, instability and shear planes in granular materials. *Ingenieur Archiv* **59**, 114–123.

Law, K. T., and Y. L. Cao (1990). An energy approach for assessing seismic liquefaction potential. *Canadian Geotech. J.* **27**, 213–233.

Liang, L., J. L. Figueroa, and A. S. Saada (1995). Liquefaction under random loading: Unit energy approach. *J. Geotech, Eng., ASCE* **121** (11), 776–781.

Liao, S. S. C. (1986). Statistical modeling of earthquake-induced liquefaction. Ph.D. Thesis, Department of Civil Engineering, Massachusetts Institute of Technology, 470 pp.

Liao, S. S. C., and K. Y. Lum (1998). Statistical analysis and application of the magnitude scaling factor in liquefaction analysis. In: "Geotechnical earthquake engineering and soil dynamics III," pp. 410–421. Seattle, WA.

Liao, S. S. C., D. Veneziano, and R. V. Whitman (1988). Regression models for evaluating liquefaction probability. *J. Geotech. Eng. Div., ASCE* **114**(4), 389–411.

Luong, M. P. (1980). Stress-strain aspects of cohesionless soils under cyclic and transient loadings. In: "Proceedings of International Symposium on Soils under Transient and Cyclic Loadings, Swansea" (G. N. Pande and O. C. Zienkiewicz, Eds.), pp. 315–324. A.A. Balkema, Rotterdam, the Netherlands.

Martin, G. R., W. D. L. Finn, and H. B. Seed (1975). Fundamentals of liquefaction under cyclic loading. *J. Geotech. Eng. Div., ASCE* **101**(5), 423–438.

Matsui, T., and K. Oda (1996). Foundation damage of structures. *Soils and Foundations*, Japanese Geotech. Soc., Special Issue, pp. 189–200.

Matsuoka, H., and K. Sakakibara (1987). A constitutive model for sands and clays evaluating principal stress rotation. *Soils and Foundations*, Japanese Geotech. Soc. **27**(4), 73–88.

Nakase, H. (1999). A simulation study on liquefaction using DEM. In: "Mechanics of granular materials" (M. Oda and K. Iwashita, Eds.), pp. 178–182. A.A. Balkema, Rotterdam, the Netherlands.

Nakase, H., A. Hiro-oka, and T. Yanagihata (1997). Deformation characteristics of liquefied loose sand by triaxial compression tests. In: "Proceedings of IS-Nagoya 97, Deformation and Progressive Failure in Geomechanics" (A. Asaoka, T. Adachi, and F. Oka, Eds.), pp. 559–564. Pergamon, Elsevier Science.

National Research Council (1985). *Liquefaction of Soils during Earthquakes*. Committee on Earthquake Engineering, National Academy Press, Washington, D.C.

Nemat-Nasser, S., and A. Shokooh (1979). A unified approach to densification and liquefaction of cohesionless sand in cyclic shearing. *Canadian Geotech. J.* **16**, 659–678.

Newmark, N. M. (1965). Effects of earthquakes on embankments and dams. *Géotechnique* **15**(2), 139–160.

Nova, R. (1991). A note on sand liquefaction and soil stability. In: "Constitutive Laws for Engineering Materials" (C. S. Desai *et al.*, Eds.), pp. 153–156. American Society of Mechanical Engineers.

Oh-Oka (1976). Drained and undrained stress-strain behavior of sands subjected to cyclic shear stress under nearly plane strain condition. *Soils and Foundations*, Japanese Soc. Geotech. Eng. **16**(3), 19–31.

Ohsaki, Y. (1970). Effects of sand compaction on liquefaction during Tokachi-Oki earthquake. *Soils and Foundations*, Japanese Soc. Geotech. Eng. **10**(2).

Oka, F., A. Yashima, M. Kato, and K. Sekigushi (1992). A constitutive model for sand based on the nonlinear kinematic hardening rule and its applications. In: "Proc. 10th World Conf. on Earthq. Eng.," pp. 2529–2534.

Orense, R. P., and I. Towhata (1996). Three-dimensional analysis on lateral displacement of liquefied subsoil. *Soils and Foundations.*

O'Rourke, T. D., and P. A. Lane (1989). Liquefaction hazards and their effects on buried pipelines. Technical report NCEER-89-0007, National Center for Earthquake Engineering Research, State University of New York, Buffalo.

Parra-Colmenares, E. J. (1996). Numerical modeling of liquefaction and lateral ground deformation including cyclic mobility and dilation response in soil systems. Ph.D. Thesis, Rensselaer Polytechnic Institute, Troy, NY.

Pastor, M., O. C. Zienkiewicz, and K. H. Leung (1985). Simple model for transient soil loading in earthquake analysis. *Int. J. Anal. Numer. Methods Geomech.* **9**, 453–498.

Pastor, M., O. C. Zienkiewicz, and A. H. C. Chan (1990). Generalized plasticity and the modeling of soil behaviour. *Int. J. Analytical Numerical Methods in Geomech.* **14**, 151–190.

Peacock, W. H., and H. B. Seed (1968). Sand liquefaction under cyclic loading simple shear conditions. *J. Geotech. Eng. Div., ASCE* **94**(3), 689–708.

Pease, I. W., and T. D. O'Rourke (1997). Seismic response of liquefaction sites. *J. Geotech. Geoenviron. Eng., ASCE* **123**(1), 37–45.

Pease, J. W., and T. D. O'Rourke (1993). Liquefaction hazards in the Mission District and south of Market Stress areas, San Francisco, California. *External Contract No. 14-08-001-G2128, US Geol. Surv.*

Peiris, T. A., and N. Yoshida (1996). Modeling of volume change characteristics of sand under cyclic loading. In: "Proc. Eleventh World Conf. Earthq. Eng.," Acapulco, Mexico, Paper No. 1087.

Prevost, J. H. (1985). A simple plasticity theory for frictional cohesionless soils. *J. Soil Dynam. Earthq. Eng.* **4**(1), 9–17.

Pyke, R. M., L. A. Knuppel, and K. L. Lee (1978). Liquefaction of hydraulic fills. *J. Geotech. Eng. Div., ASCE* **104**(11).

Rauch, A. F. (1997). EPOLLS: An empirical method for predicting surface displacement due to liquefaction-induced lateral spreading in earthquakes. Ph.D. thesis, Virginia Polytechnic Institute, Virginia.

Rauch, A. F., and J. R. Martin (2000). EPOLLS model for predicting average displacement on lateral spreads, *J. Geotech. Geoenviron. Eng., ASCE* **126**(4), 360–371.

Reynolds, O. (1885). On the dilatancy of media composed of rigid particles in contact, with experimental illustration. *Philosophical Magazine, Series 5*, **20**, 469–481.

Rice, J. R. (1975). On the stability of dilatant hardening of saturated rock masses. *J. Geophys. Res.* **80**, 1531–1536.

Robertson, P. K., and C. E. Wride (1998). Evaluating cyclic liquefaction potential using the CPT. *Canadian Geotech. J.* **35**(3), 442–459.

Sano, Y. (1998). GIS evaluation of Northridge earthquake ground deformation and water supply damage. Master thesis, Cornell University, Ithaca, New York, January.

Sasaki, Y., I. Towahata, K. Tokida, K.Yamada, H. Matsumoto, Y. Tamari, and S. Saya (1992). Mechanism of permanent displacement of ground caused by seismic liquefaction. *Soils and Foundations* **32**(3), 76–96.

Schneebli, G. (1955). Une analogie mécanique pour les terres sans cohésion. *Proc. Acad. Sci., Paris, France*, **243**, 256.

Schrefler, B. A., L. Simoni, L. Xikui, and O. C. Zienkiewicz (1990). Mechanics of partially saturated porous media. In: "Numerical Methods and Constitutive Modeling in Geomechanics" (C. S. Desai and G. Gioda, Eds.), pp. 169–209. Springer-Verlag.

Scott, R. F., and B. Hushmand (1995). Piezometer performance at Wildlife liquefaction site, California. *J. Geotech. Eng. Div., ASCE*, Discussion, **121**(12), 912–919.

Seed, H. B., and P. De Alba (1986). Use of SPT and CPT Tests for evaluating the liquefaction resistance of sands. In: "Use of In-Situ Tests in Geotechnical Engineering," ASCE Geotechnical Special Publication No. 6, 281–302.

Seed, H. B., and I. M. Idriss (1971). Simplified procedure for evaluating soil liquefaction potential. *J. Soil Mech. Foundation Div., ASCE* **97**(9), 1249–1274.

Seed, H. B., K. L. Lee, I. M. Idriss, and F. I. Makdisi (1975). The slides in the San Fernando Dams during the earthquake of February 9, 1971. *J. Geotech. Eng. Div., ASCE* **101**(7), 651–688.

Seed, H. B., I. Arango, C. K. Chan, A. Gomez-Masso, and R. G. Ascoli (1979). Earthquake-induced liquefaction near Lake Amatitlan, Guatemala. EERC Report No. UCB/EERC-79/27, September, University of California, Berkeley.

Seed, H. B., F. K. Tokimatsu, L. F. Harder, and R. M. Chung (1985). Influence of SPT Procedures in Soil Liquefaction Resistance Evaluation. *J. Geotech. Eng. Div., ASCE* **111**, 1425–1445.

Shamoto, Y., J.-M. Zhang, and S. Goto (1997). Mechanism of large post-liquefaction deformation in saturated sand. *Soils and Foundations,* Japanese Geotech. Soc., **37**(2), 71–80.

Shamoto, Y., J.-M. Zhang, and K. Tokimatsu (1997). New charts for predicting large residual post-liquefaction ground deformation. Extended abstract for *the 8th Int. Conf. on Soil Dynam. and Earthq. Eng.* (SDEE-97), 2 pp.

Shamoto, Y., J.-M. Zhang, and K. Tokimatsu (1998). Methods for evaluating residual post-liquefaction ground settlement and horizontal displacement. *Soils and Foundations,* Special Issue Vol. 2, Japanese Geotech. Soc., pp. 69–84.

Shengcong, F., and F. Tatsuoka (1983). Soil liquefaction during Haicheng and Tangshan earthquake in China. In: "Report of Japan-China Cooperative Research on Engineering Lessons from Recent Chinese Earthquakes" (C. Tamura, T. Katayama, and F. Tatsuoka, Eds.). Institute of Industrial Science, University of Tokyo.

Shibata, T., F. Oka, and Y. Ozawa (1996). Characteristics of ground deformation due to liquefaction. *Soils and Foundations*, Japanese Geotech. Soc., Special issue, pp. 65–79.

Simcock, K. J., R. O. Davis, J. B. Berrill, and D. Mullenger (1983). Cyclic triaxial tests with continuous measurement of dissipated energy. *Geotech. Testing J., ASTM* **6**(1), 35–39.

Stark, T. D., and S. C. Olson (1995). Liquefaction resistance using CPT and filed case histories. *J. Geotech. Eng. Div., ASCE* **121**(12), 856–869.

Stokoe, K. H., J. M. Roesset, J. G. Biecherwale, and M. Aoouad (1988). Liquefaction potential of sands from shear wave velocity. In: "Proc. 9th World Conf. Earthq. Eng.," Tokyo, Japan, Vol. III, pp. 213–218.

Taguchi, Y., A. Tateishi, F. Oka, and A. Yashima (1995). A cyclic elastoplastic model for sand based on the generalized flow rule and its application. In: "Proc. 5th Int. Sym. Numer. Methods in Geomech.," Davos, pp. 57–62.

Tatsuoka, F., and K. Ishihara (1974). Yielding of sands in triaxial compression. *Soils and Foundations* **14**(2), 63–76.

Terzaghi, K. V. (1943). *Theoretical Soil Mechanics*. John Wiley and Sons, New York.

Tohno, I., and S. Yasuda (1979). Liquefaction of the ground during the 1978 Miyagiken-oki earthquake. *Soils and Foundations* **21**(3), 18–34.

Tokimatsu, K., and H. B. Seed (1987). Evaluation of settlements in sand due to earthquake shaking. *J. Geotech. Eng. Div., ASCE* **113**(8), 861–878.

Tokimatsu, K., and Y. Yoshimi (1983). Empirical correlation of soil liquefaction based on SPT N-Value and Fines Content. *Soils and Foundations* **23**(4).

Tokimatsu, K., H. Mizumo, and M. Kakurai (1996), Building damage associated with geotechnical problems. *Soils and Foundations,* Japanese Geotech. Soc., Special Issue, pp. 219–234.

Toprak, S., T. L. Holzer, M. J. Bennett, and J. C. Tinsley (1999). CPT- and SPT-based probabilistic assessment of liquefaction potential. In: "Proceedings of the 6th U.S. Japan Workshop on Earthquake Resistant Design of Lifeline Facilities and Countermeasures for Soil Liquefaction" (T. D. O'Rourke, J. P. Bardet, and M. Hamada, Eds.), pp. 69–86. MCEER.

Towhata, I., H. Toyota, and W. M. Vargas (1995). Dynamics in lateral flow of liquefied ground. In: "Proc. 10th Asian Reg. Conf. on Soil Mechan. and Foundation Eng., Beijing," Vol. 2, pp. 245–247.

Towhata, I., H. Toyota, and, W. M. Vargas (1996). A transient study on lateral flow on liquefied ground. In: "Proc. 7th Int. Sym. on Landslides, Trondheim, Norway," Vol. 2, pp. 1047–1054. A.A. Balkema, the Netherlands.

Towhata, I., R. P. Orense, and H. Toyota (1997). Mathematical principles in prediction of lateral ground displacement induced by seismic liquefaction. *Soils and Foundations.*

Trifunac, M. D. (1995). Empirical criteria for liquefaction in sands via standard penetration tests and seismic wave energy. *Soil Dynam. Earthq. Eng.* **14**(6), 419–426.

Vardoulakis, I. (1985). Stability and bifurcation of undrained, plane rectilinear deformation on water-saturated granular soils. *Int. J. Numer. Anal. Methods in Geomech.* **9**, 399–414.

Vardoulakis, I. (1986). Dynamic stability analysis of undrained simple shear on water-saturated granular soils. *Int. J. Numer. Anal. Methods in Geomech.* **10**, 177–190.

Vardoulakis, I., and J. Sulem (1995). *Bifurcation Analysis in Geomechanics*. Chapman & Hall.

Wang, Z. L., Y. F. Dafalias, and C. K. Chen (1990). Bounding surface hypoplasticity model for sand. *J. Eng. Mech., ASCE* **116**, 983–1001.

Wolfe, W. E., M. Annaki, and K. L. Lee (1977). Soil liquefaction in cyclic cubic test apparatus. In: "Proc. 6th World Conf. Earthq. Eng.," New Dehli, pp. 2151–2156.

Xie, J. (1979). Empirical criteria for sand liquefaction. In: "Proc. Second U.S. Nat. Conf. Earthq. Eng.," Stanford University, Stanford.

Yashima, A., F. Oka, J. M. Konrad, R. Uzuoka, and Y. Taguchi (1997). Analysis of a progressive flow failure in an embankment of compacted fill. *Proc. IS-Nagoya* (A. Asaoka, T. Adachi, and F. Oka, Eds.), pp. 599–604. Pergamon Press.

Yasuda, S., T. Masuda, N. Yoshida, H. Kiku, S. Itafuji, K. Mine, and K. Sato (1994). Torsional shear and triaxial compression tests on deformation characteristics of sands before and after liquefaction. In: "Proc., 5th U.S.-Japan Workshop on Earthquake Resistant Design of Lifeline Facilities and Countermeasures Against Soil Liquefaction," Salt Lake City. Technical Report NCEER-94-0026, National Center for Earthquake Engineering Research, pp. 249–266.

Yasuda, S., K. Ishihara, and S. Iai (1997). A simple procedure to predict ground flow due to liquefaction behind quaywalls. Extended abstract for the 8th Int. Conf. on Soil Dynam. and Earthq. Eng. (SDEE-97), 2 pp.

Yegian, M. K., E. A. Marciano, and V. G. Gharaman (1991). Earthquake-induced permanent deformation: probabilistic approach. *J. Geotech. Eng. Div., ASCE* **117**(1), 35–50.

Yogachandran, C. (1991). Numerical and centrifugal modeling of seismically induced flow failures. Ph.D. thesis, University of California, Davis.

Yoshida, N., S. Tsujino, and K. Inadomaru (1995). Preliminary study on the settlement of ground after liquefaction. In: "Proc. 29th Japan Nat. Conf. of Soil Mech. and Foundation Eng.," pp. 859–860 (in Japanese).

Youd, T. L., and J. L. Bennett (1983). Liquefaction sites, Imperial Valley, California. *J. Geotech. Eng. Div., ASCE* **109**(3), 440–457.

Youd, T. L., and T. L. Holzer (1994). Piezometer performance at Wildlife liquefaction site, California. *J. Geotech. Eng. Div., ASCE* **120**(6), 975–995.

Youd, T. L., and T. L. Holzer (1995). Reply to Discussion by R. F. Scott and B. Hushmand of "Piezometer performance at Wildlife liquefaction site, California," *J. Geotech. Eng. Div., ASCE* **121**(12), 919.

Youd, T. L., and I. M. Idriss (1998). Proceedings of NCEER workshop on evaluation of liquefaction resistance of soils. National Center for Earthquake Engineering Research, State University of New York, Buffalo.

Youd, T. L., and S. J. Kiehl (1996). Ground deformation characteristics caused by lateral spreading during the 1995 Hanshin-Awaji earthquake. In: "Proceedings of the 6th U.S. Japan Workshop on Earthquake Resistant Design of Lifeline Facilities and Countermeasures for Soil Liquefaction" (M. Hamada and T. D. O'Rourke, Eds.), pp. 221–242. NCEER.

Youd, T. L., and D. M. Perkins (1987). Mapping of Liquefaction Severity Index. *J. Geotech. Eng. Div., ASCE* **113**, 1374–1392.

Youd, T. L., E. L. Harp, D. K. Keefer, and R. C. Wilson (1985). The Borah Peak, Idaho earthquake of October 28, 1983—liquefaction. *Earthq. Spectra, EERI* **2**(6), 71–90.

Youd, T. L., C. M. Hansen, and S. F. Barlett (1999). Revised MLR equations for predicting lateral spread displacement. In: "Proceedings of the 6th U.S. Japan Workshop on Earthquake Resistant Design of Lifeline Facilities and Countermeasures for Soil Liquefaction" (T. D. O'Rourke, J. P. Bardet, and M. Hamada, Eds.), pp. 99–114. NCEER.

Zeghal, M., and A.W. Elgamal (1994). Analysis of site liquefaction using earthquake records. *J. Geotech. Eng. Div., ASCE* **120**(6), 996–1017.

Zienkiewicz, O. C., and T. Shiomi (1984). Dynamic behavior of saturated porous media: the generalized Biot formulation and its numerical solution. *Int. J. Numer. Analy. Methods in Geomech.* **8**, 71–96.

Zienkiewicz, O. C., and R. L. Taylor (1991). *The Finite Element Method.* McGraw-Hill, New York.

Earthquake Prediction and Hazards Mitigation

72

Earthquake Prediction: An Overview

Hiroo Kanamori
California Institute of Technology, Pasadena, California, USA

1. Introduction

Because earthquakes occur suddenly, often with devastating consequences, earthquake prediction is a matter of great interest among the public and emergency service officials. However, the term "earthquake prediction" is often used to mean two different things. In the common usage, especially among the public, "earthquake prediction" means a highly reliable, publicly announced, short-term (within hours to weeks) prediction that will prompt some emergency measures (e.g., alert, evacuation, etc.). Exactly how reliable this type of prediction should be depends on the social and economic situations of the region involved. The issue is whether the quality of prediction is good enough to benefit the society in question. Allen (1976) lists six attributes required for this type of prediction: (1) It must specify a time window. (2) It must specify a space window. (3) It must specify a magnitude window. (4) It must give some sort of indication of the author's confidence in the reliability of the prediction. (5) It must give some sort of indication of the chances of the earthquake occurring anyway, as a random event. (6) It must be written and presented in some accessible form so that data on failures are as easily obtained as data on success.

In the second usage, "earthquake prediction" means a statement regarding the future seismic activity in a region, and the requirement for high reliability is somewhat relaxed in this usage. In a way, this is a more general scientific prediction of a physical system, and as such it is nothing but a study of "physics of earthquakes." The reliability of a specific prediction depends on the level of our understanding of the process, and the amount and quality of data we have. Because the basic physical process of earthquakes is now reasonably well understood, and high-quality geophysical data are being collected, it should be possible to make some predictions regarding the future seismic activity in a region on the basis of whatever geophysical parameters are observed and their interpretations. This type of prediction or forecast also has important social implications on time scales of months and years. However, it is best to distinguish it from the short-term prediction previously described. In either case, a good scientific understanding of the process is a prerequisite to useful practical prediction.

The subject of earthquake prediction has been a matter of intense debate (e.g., Nature Debate, *http://helix.nature.com/debates/earthquake/equake-frameset.html*), and there does not seem to be a general consensus. Recent papers by Geller *et al.* (1997a; 1997b), Scholz (1997), Wyss (1997b), Aceves and Park (1997), and Geller (1997) demonstrate the diversity of opinion on this subject.

In this article, we review the current thinking on the scientific aspect of earthquake prediction research (i.e., with the emphasis on earthquake prediction in the second usage), and discuss its social implications. In this article, no distinction is made in the use of terms "prediction" and "forecast." We note, however, that "prediction" is often used for a statement on a specific earthquake, and "forecast" is more commonly used for a statement on the general seismic behavior of a region in the future.

Many books and review papers are available on this subject (to mention a few, Rikitake, 1976, 1982; Wyss, 1979, 1991; Vogel, 1979; Isikara and Vogel, 1982; Unesco, 1984; Mogi, 1985; Gupta and Patwardhan, 1988; Olson *et al.*, 1989; Lomnitz, 1994; Gokhberg *et al.*, 1995; Sobolev, 1995; Knopoff, 1996; Sykes *et al.*, 1999; and Rikitake and Hamada, (2001), so we do not discuss the details of the individual cases of earthquake prediction, or the specifics of methodology. Also, some national reports (e.g., Sobolev and Zavyalov, 1999; Kuznetsov, 1999; Zhang and Liu, 1999; and Zhu and Wu, 1999) have recently summarized activities in their respective countries.

2. The Earthquake Process

In a narrow sense, an earthquake is a sudden fracture in the Earth's interior, together with the resulting ground shaking; in a broad sense, it is a long-term complex stress accumulation and release process occurring in a highly heterogeneous medium.

INTERNATIONAL HANDBOOK OF EARTHQUAKE AND ENGINEERING SEISMOLOGY, VOLUME 81B ISBN: 0-12-440658-0

Advances have been made in understanding crustal deformation and stress accumulation processes, rupture dynamics, rupture patterns, friction and constitutive relations, interaction between faults, fault-zone structures, and nonlinear dynamics. Thus, it should be possible to predict to some extent the seismic behavior of the crust in the future from various measurements taken in the past and at present. However, the incompleteness of our understanding of the physics of earthquakes in conjunction with the obvious difficulty in making detailed measurements of various field variables (structure, strain, etc.) in the Earth makes accurate deterministic short-term predictions difficult.

Earthquakes occur, moreover, in a complex crust–mantle system. This system includes some distinct structures such as the seismogenic zone and faults, as well as highly heterogeneous structures with all length scales. The distinct structures are responsible for the deterministic behavior of earthquakes, but the interactions between different parts of the complex system result in the chaotic behavior of earthquake sequences. Many studies have demonstrated that even a simple mechanical model of earthquakes exhibits a very complex behavior, suggesting that earthquakes have the characteristics of a chaotic process (Burridge and Knopoff, 1967; Otsuka, 1972; Turcotte, 1992). Because of this chaotic component of the process, it would be difficult to predict earthquakes in a deterministic way; predictions can be made only in a statistical sense and only with considerable uncertainty (Turcotte, 1992).

Several processes are especially responsible for the uncertainties. If we assume that plate motion is statistically stationary, then the stress changes due to plate motion can be estimated with relatively small uncertainties. However, the stress in the crust also changes with time on a local scale. For example, the stress on a fault can be affected by nearby earthquakes. Because earthquakes occur on a complex array of faults, the crustal stress field is irregular on a local scale and determination of future earthquake locations would inevitably be uncertain.

The strength of the crust may change as a function of time, too. For example, migration of fluids in the crust could change the local strength of the crust and affect the occurrence of earthquakes (e.g., Raleigh *et al.*, 1976; Ingebritsen and Sanford, 1998). Our knowledge of hydrological processes in the crust is limited, and the temporal variation of the strength of the crust is difficult to predict, so again we have large uncertainties in the timing of the occurrence of earthquakes.

Prediction of the size (magnitude) of an earthquake is also uncertain, because a small earthquake may trigger another event in the adjacent area, cascading to a much larger event (e.g., Brune, 1979). Although the extent of a "stressed" area (e.g., a seismic gap) may ultimately determine the maximum size of the earthquake, the growth of rupture is likely to have some stochastic elements. Any small earthquake may grow into the maximum earthquake determined by the size of the gap, or may stop halfway, depending on small variations in the mechanical properties of rocks in the fault zone. For example, along the Nankai trough, southwest Japan, two adjacent segments broke in two separate $M \approx 8$ earthquakes in 1944 and 1946. However, in 1854 these segments broke in two $M = 8+$ earthquakes 32 hours apart, and in 1707, they broke simultaneously (Ando, 1975). Physically or geologically, each one of these sequences can be considered a single earthquake, but whether it occurs in two distinct events in a single sequence, 32 hours apart, or 2 years apart would have very different social consequences, and it's difficult to determine why these segments broke in these three different sequences.

Another important process is triggering by external effects. Hill *et al.* (1993) observed significant seismic activities in many geothermal areas soon after the 1992 Landers, California, earthquake. Although the detailed mechanism is still unknown, it appears that the interaction between fluid in the crust and strain changes caused by seismic waves from the Landers earthquake was responsible for sudden weakening of the crust. If sudden weakening of the crust resulting from dynamic loading plays an important role in triggering earthquakes, deterministic predictions of the initiation time of an earthquake will be difficult.

In the following, we discuss some issues of long-term forecast and short-term prediction separately, because they have very different social implications. Sometimes forecasts on intermediate time scales are treated separately from long-term forecasts, but here we treat them together as long-term forecasts. There is no generally used definition of "short-term," "intermediate-term," and "long-term" predictions, and we generally follow the definition given by Sykes *et al.* (1999) as a useful guideline: Immediate alert (0 to 20 sec), Short-term prediction (hours to weeks), Intermediate-term prediction (1 month to 10 years), Long-term prediction (10 to 30 years), Long-term potential (>30 years). The actual usage, however, may vary depending on the specific circumstances.

3. Long-Term Forecast

The basis of long-term forecast is the elastic rebound theory (Reid, 1910). If the stress accumulates at a constant rate, and the strength of the crust is constant, one would expect a relatively regular recurrence of earthquakes on a given segment of fault. However, due to previously mentioned fault interactions, weakening of crust due to increase in pore pressures, or some nonlinear processes, the actual occurrence can be more irregular than would be expected from the simple elastic rebound theory. Even with this difficulty, long-term forecasts are useful because considerably large uncertainties can be tolerated for long-term applications. In general, such forecasts are easier for the places where stress accumulation rate is faster (e.g., plate boundaries with fast plate motion) than for the places with slower stress accumulation rates.

3.1 Seismic Gap Method

The seismic gap method is most frequently used for long-term forecasts. The basic premise is that large earthquakes occur more or less regularly in space and time as a result of gradual stress buildup and sudden stress release by failure. Imamura (1928) documented historical earthquakes in the Nankai trough, southwest Japan, and on the basis of regularity of the occurrence of large earthquakes, he forecast large earthquakes in this area. In fact, two large earthquakes with $M \approx 8$ occurred in this area in 1944 and 1946. A similar idea was used by Fedotov (1965) for the Kamchatka and Kurile Is. regions, and by Mogi (1968) for all of Japan.

Kelleher (1970), Kelleher *et al.* (1973), and others used this method more formally in the framework of plate tectonics. A portion of a plate boundary that has historically experienced large earthquakes, but not recently (e.g., 30 years), is more likely to produce a large earthquake in the next few decades than those places that have recently experienced large events—this portion of a plate boundary is called a seismic gap. This method has been used to forecast earthquakes on subduction zones and some strike-slip plate boundaries such as the San Andreas fault. Long-term forecasts made with the gap method are generally considered to have been successful for several large ($M > 7.5$) earthquakes (e.g., the 1972 Sitka, Alaska, earthquake (Kelleher, 1970), the 1973 Nemuro-Oki, Japan, earthquake (Utsu, 1970), the 1978 Oaxaca, Mexico, earthquake (Kelleher *et al.*, 1973), the 1985 Valparaiso, Chile, earthquake (Kelleher, 1972; Nishenko, 1985)), but the method is subject to all the uncertainties previously mentioned, and is not meant to be used for definitive forecasts. In fact, there were several cases in which the forecast with this method caused confusing results. Wyss and Wiemer (1999) describe the case for the 1986 Andreanof Is. earthquake, and emphasize that oversimplified models can lead to a confusing result, and slip history and tectonic difference of fault segments need to be considered in assessing the seismic potential of gaps.

The application of the gap method to smaller earthquakes is subject to even larger uncertainties and its usefulness is somewhat questionable. For smaller events, the uncertainty in the earthquake locations is often comparable to the size of the gap so that the location of the gap becomes ambiguous. Also implicit in this method is the assumption that approximately the same segment along subduction zones fails repeatedly in approximately the same fashion, but many examples have demonstrated that this is not always the case. The rupture patterns vary significantly from sequence to sequence (e.g., the 1906 Colombia earthquake (Kelleher, 1972; Kanamori and McNally, 1982)). This spatial variability is probably a result of complex interactions between different parts of plate boundaries.

Also, large earthquakes occur not only on the main plate boundary but also in the areas adjacent to it, thereby adding complexity to the spatial and temporal pattern of seismicity. Examples are the 1933 Sanriku earthquake ($M_w = 8.4$) and the 1994 Shikotan earthquake ($M_w = 8.3$), both of which occurred within the subducting oceanic plate, but not directly related to the stress accumulation process on the subduction boundary. Because of these difficulties, there have been controversies on the usefulness of the gap method, especially for events with $M < 7.5$ (Jackson and Kagan, 1991, 1993; Nishenko and Sykes, 1993).

Despite these difficulties, the long-term forecast of large earthquakes using the gap method is useful for understanding the long-term behavior of seismic zones, as long as the forecast is interpreted with caution. Given the complex earthquake rupture process, use of oversimplified models—whether in favor of or against the gap hypothesis—should be avoided.

Following are three examples of long-term forecasts that illustrate the use of this method.

3.1.1 Parkfield Earthquake Prediction Experiment

Moderate earthquakes with $M \approx 6$ have occurred on the San Andreas fault near Parkfield, California, in 1922, 1934, and 1966. Further studies revealed that similar-size earthquakes occurred earlier in 1857, 1881, and 1901 in approximately the same area. Several seismologists noted that these events seem to have occurred relatively regularly, with an average interval of about 22 years (Bakun and McEvilly, 1984). Also, the pattern of foreshocks of the 1966 event is strikingly similar to that of the 1934 event. This regularity and similarity led some seismologists to believe that these Parkfield earthquakes are "characteristic" earthquakes occurring repeatedly at approximately the same location on the San Andreas fault. If that regularity had continued into the future, the next event would have been expected to occur sometime around 1988. It is to be noted, however, that (1) the locations of these events are not accurately known, (2) the record before 1900 is uncertain, (3) the 1857 event is an immediate foreshock of the $M \approx 8$ Fort Tejon earthquake and is not an isolated event like the other events, and (4) the range of inter-event intervals is actually fairly large, 12 to 32 years.

Despite these complexities, according to one estimate, the probability of the next characteristic Parkfield earthquake occurring before 1993 was 0.95. A focused earthquake prediction experiment (Parkfield Earthquake Prediction Experiment) began in 1985 (Bakun and McEvilly, 1984; Bakun and Lindh, 1985), and many instruments (seismometers, creep-meters, strainmeters, laser ranging devices, etc.) were installed to monitor various seismological and geophysical parameters with the hope of capturing precursory phenomena before the next Parkfield earthquake. Although many interesting results on seismicity, velocity structures, wave propagation characteristics, and fault slip patterns have been obtained from the data recorded with these instruments, the predicted earthquake has not occurred yet. This result demonstrates that the earthquake process in the Earth's crust is complex, involving many parameters, and predictions on the basis of a simple model with a small number of parameters are inevitably uncertain. In retrospect, many reasons can be given to explain why the predicted earthquake did not occur. For example, (1) the past Parkfield earthquakes may not have occurred on exactly the same segment of the San Andreas fault (Segall and Harris, 1987) and the characteristic earthquake model cannot be used, (2) earthquake activities in the adjacent areas (e.g., Coalinga) may have significantly decreased the stress loading rate on the San Andreas fault near Parkfield, thereby delaying the occurrence of the predicted event (e.g., Miller,

1996), (3) the earthquake process is more random than is usually assumed in the characteristic earthquake model.

Aside from the scientific issues, this type of experiment has merit in allowing the local governments to develop a useful protocol for informing state and local officials about time-dependent hazard levels (NEPECWG, 1994). However, this should not be confused with the scientific issues associated with the Parkfield experiment.

Other issues on the Parkfield prediction are discussed in Savage (1993) and Michael and Langbein (1993).

3.1.2 The 1989 Loma Prieta Earthquake

The 1989 Loma Prieta earthquake ($M_w = 6.9$) occurred in an area where long-term or intermediate-term forecasts of a large earthquake had been made by several seismologists. Most of these forecasts were based on the elastic rebound theory: The next earthquake is likely to occur when the strain released in the previous one has been restored.

The Loma Prieta earthquake occurred near the southeastern end of the rupture zone of the 1906 San Francisco earthquake. The amount of surface break associated with the 1906 San Francisco earthquake suggested that the slip during the 1906 San Francisco earthquake was not large enough to completely release the accumulated strain (i.e., slip deficit) along the Santa Cruz Mountain segment of the San Andreas fault. Also, this segment exhibited a distinct absence of small earthquakes over a distance of some 40 km, a pattern often thought to appear before large earthquakes. On the basis of these observations, as well as the fault geometry, several forecasts had been made for the rupture length, magnitude, and approximate timing, expressed in probabilistic terms. What actually occurred came very close to these forecasts. For example, Sykes and Nishenko (1984) forecast an $M = 7.0$ earthquake with a probability of 0.19 to 0.95 in 20 years. Likewise, Lindh (1983) made a forecast for an $M = 6.5$ earthquake with a probability of 0.30 in 20 years. Scholz (1985) estimated that a rupture of the 75-km-long slip-deficient segment could occur in 60 to 110 years, resulting in an $M = 6.9$ earthquake. By the usual standard for intermediate-term predictions, these forecasts are considered very accurate.

However, one puzzling aspect of these forecasts is that the Loma Prieta earthquake did not seem to have occurred on the San Andreas fault in a strict sense. The fault plane of the Loma Prieta earthquake inferred from the earthquake mechanism and the aftershock distribution is dipping about 70°SW, and does not coincide with that of the San Andreas fault. The fault slip motion had a large vertical component, and is different from what is expected of the San Andreas fault. Because the very basis of the forecast was the slip deficit on the San Andreas fault, the forecast would lose its logical basis if the Loma Prieta earthquake did not occur on the San Andreas fault. Furthermore, the strain data suggest significant amounts of slip at depth on the San Andreas fault during the 1906 earthquake (Thatcher and Losowski, 1987). If this is the case, there was no slip deficit on the San Andreas fault to begin with. Another argument on the basis of the crustal deformation pattern suggests that the repeat time of the Loma Prieta type earthquakes could be several thousand years, rather than several hundred years (Valensise and Ward, 1991). If these arguments are valid, the 1989 Loma Prieta earthquake was not the predicted earthquake, but was a relatively rare event on a structure different from the San Andreas fault that happened to have occurred at about the predicted time.

It would be fair to say that the situation is more complex than either of these simple arguments indicates. The overall regional strain deficit could have existed and the preearthquake seismicity in the area may have justified some forecast, but considering the complexity in the fault structure in the area, one would not expect a very simple scenario as presented to work all the time. For more details, see Harris (1998a).

3.1.3 Tokai, Japan, Earthquake Prediction

Large earthquakes have repeatedly occurred along the Nankai Trough along the southwestern coast of Japan. The sequence during the past 500 years includes large ($M \approx 8$) earthquakes in 1498, 1605, 1707, 1854, and 1944–1946, with an average interval of about 120 years (Ando, 1975). In the early 1970s, several Japanese seismologists noticed that the 1944–1946 sequence was somewhat smaller than the two previous events, and suggested that the rupture during the 1944–1946 sequence did not reach the northeastern part of the Nankai trough (called the Suruga trough), thereby leaving this portion as a mature seismic gap (this argument is similar to that for the forecast of the Loma Prieta earthquake) (Ishibashi, 1977). There is some evidence that the rupture of both the 1854 and 1707 earthquakes extended all the way to the Suruga trough.

With this argument, this portion of the Nankai trough became known as the Tokai gap, with a potential of causing an $M \approx 8$ earthquake in the near future. In 1978, the Japanese government introduced the Large-Scale Earthquake Countermeasures Act, and embarked on an extensive project to monitor the Tokai gap. Many institutions deployed all kinds of instruments for monitoring geophysical activities, and detailed plans for emergency relief efforts were made.

It is more than 20 years since the project began, but the predicted Tokai earthquake has not occurred yet. It is possible that the predicted earthquake is yet to occur, or the deformation pattern near the corner of the plate boundary is so complex that a simple recurrence model cannot be used. In the latter case, the predicted event may not occur in the near future. Because the Tokai earthquake prediction did not have any specific prediction time window, it is hard to assess the significance of this prediction effort at this time. However, many seismologists now seem to agree that accurate forecasts are difficult even for this plate boundary with a seemingly regular historical earthquake sequence.

The complex geometry of the plate boundary (e.g., segmentation) and the effects of large earthquakes in the adjacent areas may have affected the state of stress on the plate boundary.

3.2 Stress Transfer

A question is often raised regarding whether the stress on a particular fault is changing in the direction to promote failure or not. In addition to secular loading by plate motion, the stress on a fault is affected by past earthquakes in adjacent areas. If the size and mechanism of earthquakes in the adjacent areas are known, we can compute the stress changes on the fault on a time scale of a few decades. This concept has been tested by Smith and Van de Lindt (1969), Rybicki (1973), Yamashina (1979), and Das and Scholz (1981). For the 1968 Borrego Mountain, California, earthquake, a significant aftershock cluster occurred in the area where shear stress was increased by the mainshock. This concept was more rigorously applied to several recent earthquakes (the 1992 Landers earthquake (Stein *et al.*, 1992; Harris and Simpson, 1992; Jaume and Sykes, 1992), the 1994 Northridge earthquake (Stein *et al.*, 1994), and the 1995 Kobe earthquake (Toda *et al.*, 1998)). In some cases (e.g., the 1992 Big Bear earthquake ($M = 6.4$) which occurred soon after the Landers earthquake; some aftershocks of the Landers, Northridge, and Kobe earthquakes), the hypothesis of triggering by stress transfer is well demonstrated. In other cases, the situation is not that obvious (Hardebeck *et al.*, 1998). The method has also been used to understand long-term seismicity in California as a result of loading and unloading of faults caused by large earthquakes in the area (Deng and Sykes, 1997; Harris and Simpson, 1996).

In general, if the geometry of the fault system, the loading mechanism, and the structure and properties of the crust are known in an area, it should be possible to compute the regional stress changes and infer the seismic behavior of the entire area (Rybicki *et al.*, 1985). Stress transfer between different faults is an important mechanism controlling regional seismicity on decadal time scales (see Harris, 1998b; Stein, 1999; Chapter 73 by Harris), but the lack of detailed knowledge of the initial stress condition and the model parameters makes it difficult to make definitive forecasts of future seismicity.

3.3 Seismicity Patterns

The change in the stress or strength of the crust may manifest itself as spatial and temporal changes in seismicity patterns such as quiescence, increase, and doughnut patterns (Mogi, 1969; Utsu, 1970; Wyss and Habermann, 1979; Habermann, 1981). For some earthquakes, seismic quiescence had been identified before the occurrence (e.g., 1973 Nemuro-Oki earthquake (Utsu, 1970); the 1978 Oaxaca, Mexico, earthquake, (Ohtake *et al.*, 1977); Bear Valley, California (Wyss and Burford, 1987), 1986 Andreanof Islands earthquake (Kisslinger, 1988)). In some retrospective studies, seismicity patterns were related to the occurrence of several large earthquakes (e.g., the 1906 San Francisco earthquake (Ellsworth *et al.*, 1981), the 1868, 1906, and 1989 earthquakes in the San Francisco Bay Area (Sykes and Jaume, 1990)), but the type of pattern may depend on the regional tectonic structure, fault geometries, and the loading system; it is unclear at present how to quantitatively relate seismicity patterns to an impending earthquake. In some cases, the completeness of seismicity catalogs used for identifying seismicity patterns was questioned (e.g., Whiteside and Habermann, 1989).

Another general approach along this line is a formal assessment of earthquake potential primarily using earthquake catalogs (e.g., Gelfand *et al.*, 1976; Keilis-Borok *et al.*, 1988; Keilis-Borok, 1996). This approach is based on systematic examinations of earthquake catalogs to identify relations between some seismicity patterns (such as clustering, quiescence, and sudden increase in activity) and past large earthquakes, and using these relations to forecast future seismic activities on intermediate time scales. The method is being tested (e.g., Kossobokov *et al.*, 1999; Rotwain and Novikova, 1999), but its usefulness for practical purposes is yet to be determined.

Another use of seismicity data is to relate the temporal variation of cumulative seismic moment of earthquakes in a region to a behavior of a system that evolves toward a critical point. Summaries of the method are found in Jaume and Sykes (1999), and a review of accelerating seismic energy release prior to large earthquakes, and its relation to cellular automata models, is in Sammis and Smith (1999).

4. Short-Term Prediction

For the average citizen and the public, "Earthquake Prediction" means a short-term prediction of a specific earthquake on a relatively short time scale, e.g., a few weeks. Such prediction must specify the time, place, and magnitude of the earthquake in question with sufficiently high reliability and probability (Allen, 1976). However, for the reasons mentioned previously, any such short-term prediction, if made, is bound to be very uncertain. Even uncertain predictions may be useful for those places where the social and economical environments are relatively simple and false alarms can be socially tolerated. However, in modern highly industrialized cities with complex lifelines, communication systems, and financial networks, such uncertain predictions could inadvertently damage local and global economies, so they are generally not useful unless the society involved is willing to accept the potential loss that could be inflicted by false alarms.

Despite this difficulty, many attempts to observe precursory phenomena for the purpose of short-term earthquake prediction have been made.

4.1 Precursors and Anomalous Phenomena

The term "precursor" means two different things. In a restricted usage, "precursor" implies some anomalous phenomenon that always occurs before an earthquake in a consistent manner. This is the type of precursor one would wish to find for short-term earthquake prediction. As far as we know, universally accepted precursors that occur consistently before every major earthquake have not yet been found.

In contrast, "precursor" is often used in a second sense to mean some anomalous phenomena that may occur before large earthquakes. Because an earthquake may involve nonlinear preparatory processes before failure, it is reasonable to expect a precursor of this type. However, it may not always occur before every earthquake, or even if it occurs, it may not always be followed by a large earthquake. Thus, in this case, the precursor cannot be used for a definitive earthquake prediction. Nevertheless, it is an interesting physical phenomenon worthy of scientific study. Foreshocks are a good example of a precursor of this type. Some large earthquakes were preceded by distinct foreshock activity, but many earthquakes do not have foreshocks. Also, a group of small earthquakes can occur without any major earthquake following it.

These precursors may be identified in retrospective studies, but it would be very difficult to identify some anomalous observations as a precursor of a large earthquake before its occurrence. Even if an anomaly were detected, it would be difficult to use it for accurate predictions of the size and timing of the impending earthquake, considering the stochastic nature of earthquakes.

Many anecdotal or qualitative reports on earthquake precursors can be found in the literature (Rikitake, 1986). Systematic efforts to detect precursors began in the 1960s. These efforts included measurements of seismicity, strain, seismic velocities, electric resistivity and potential, radio-frequency emission, ground water level, and ground water chemistry.

Encouraging reports of large (about 10%) precursory changes in the ratio of seismic P velocity to S velocity were made for several earthquakes (Aggarwal *et al.*, 1973; Whitcomb *et al.*, 1973). Similar changes had been reported earlier in the former Soviet Union (Semenov, 1969; Nersesov, 1970) and China. These changes were interpreted as manifestations of rock-dilatancy and fluid diffusion in micro-cracks just before failure (Scholz *et al.*, 1973). However, many precise measurements using not only earthquakes but also controlled sources, performed following the initial reports, failed to verify the large changes in the velocity reported by earlier studies (e.g., McEvilly and Johnson, 1973, 1974). In most cases, the velocity changes, if detected at all, were less than 1% or below the experimental noise level.

Similarly, large changes in ground-water chemistry, especially the concentration of radon, were reported before several large earthquakes in the former Soviet Union and China. Some results in Japan, especially the change before the 1978 Izu-Oshima earthquake, are considered significant by some (Wakita *et al.*, 1980). Tsunogai and Wakita (1995) and Igarashi *et al.* (1995) reported intriguing changes in the chloride ion and radon concentrations in ground water before the 1995 Kobe, Japan, earthquake. However, the results in the United States were generally not encouraging, and most geochemical monitoring efforts have been discontinued. It is probably fair to say that the negative results from seismic velocity ratio and radon monitoring in the United States may not be entirely definitive because of the lack of instruments very close to the epicenters of large earthquakes, but most seismologists would agree that these precursors, if they exist, are not easily detectable.

Several intriguing hydrological precursors have been reported (Roeloffs, 1988), but more complete documentation of the data needs to be made before they can be used for a definitive interpretation of crustal processes leading to seismic failure.

An intriguing observation of very low-frequency (0.1 to 10 Hz) radio (RF) emission was reported for the 1989 Loma Prieta, California, earthquake (Fraser-Smith *et al.*, 1990). The level of RF emission detected by an antenna located at about 7 km from the epicenter increased far above the background level about 3 hours before the earthquake. The emission also increased 12 days and 1 day before the earthquake. Although the exact cause of this emission is not established, this observation is probably one of the clearest anomalous signals detected before a large earthquake.

Crampin *et al.* (1999) recently made an interesting observation of shear-wave splitting. They used temporal variations of shear-wave splitting to correctly forecast the time and magnitude of an $M = 5$ earthquake in Iceland.

Efforts to detect slow strain precursors have been extensive in California, but no obvious strain precursors have been detected (e.g., Johnston *et al.*, 1990, 1994). It is important to note that a slow strain change was observed in 1993 near San Juan Bautista, California (Linde *et al.*, 1996), but no large earthquake followed it. For some subduction-zone earthquakes (e.g., the 1960 Chilean earthquake, Kanamori and Cipar (1974); Cifuentes and Silver (1989), the 1983 Akita-Oki, Japan, earthquake (Linde *et al.*, 1988), and the 1944 Tonankai, Japan, earthquake (Mogi, 1984)), slow deformations prior to the mainshock have been reported, but the instrumental data are not complete enough to make definitive cases.

A prediction method using changes in electric potential has been extensively used for prediction of earthquakes in Greece (e.g., Varotsos and Lazaridou, 1991), but its validity is presently vigorously debated (Lighthill, 1996; Geller, 1996).

Although many precursors have been reported, the study made by a committee under the International Association of Seismology and the Earth's Interior (IASPEI) (Wyss, 1991) concluded that only 3 out of 31 precursors subjected to review qualified as such. Although this type of evaluation depends on the criteria used, it is reasonable to say that reliable predictions using this type of precursor seem to be difficult at present (see also Wyss, 1997a).

Despite the limited value of "precursors" for short-term earthquake prediction, studies of such preparatory processes are important for a better understanding of the physics leading up to seismic failure in the Earth's crust, and careful, systematic, and quantitative investigations may be warranted.

One intriguing example of a short-term prediction is that of the 1975 Haicheng, China, earthquake. A destructive earthquake ($M = 7.3$) occurred near Haicheng, China, on February 4, 1975. More than 1 million people lived near the epicenter. It has been widely reported that this earthquake was successfully predicted. Unfortunately, the Cultural Revolution was still taking place in 1975, and detailed information did not emerge in peer-reviewed scientific literature. Thus, it is not possible to assess this prediction with complete objectivity.

Judging from the various reports on the Haicheng earthquake, it appears that very extensive foreshock activity, including a few hundred instrumentally recorded events, played the most important role in motivating mass evacuation, which saved many thousands of lives. However, it is unclear (1) how many false alarms had been issued before the final evacuation, (2) whether the evacuation was done under the direction of the local government or by more spontaneous decision by the local units or residents, and (3) what the total number of casualties was (the estimate ranges from 0 to 1300). Without knowing these details, it is unclear whether the methodology used in this prediction would work consistently for earthquake prediction purposes in other places, especially other countries with different economic and social environments.

Another example in which somewhat uncertain short-term predictions were actually used for practical purposes is the 1997 earthquakes in Jiashi, Xinjiang, China (Zhang *et al.*, 1999). Anomalous changes in several geophysical parameters were used to predict some events in an active swarm sequence. Again, it is unclear whether similar methods are practical in other countries with different economic and social environments.

5. Strategy for Seismic Hazard Mitigation

Given the uncertainty and indeterminacy described thus far, the important question becomes how can we effectively utilize the physics-based forecasts to reduce the threat of earthquakes? The usefulness of such forecasts depends on the time scale involved for such assessments. To effectively mitigate the impact of future earthquakes, seismic hazards need to be addressed on several different time scales. On the time scale of decades, land use regulations and building codes need to be improved. On the time scale of a few years, earthquake preparedness measures should be encouraged at personal and community levels. On shorter time scales, months to days, accurate earthquake predictions of size, location, and time would be required.

In the following we discuss the mitigation strategies with different time scales.

5.1 Regional Seismic Hazard Assessment

The seismic gap method can be extended to a general methodology for assessing regional seismic hazards. The strain rate in a region can be computed from the rate of plate motion. If we assume that a certain fraction, η, of the strain is relieved by earthquakes, one can estimate the overall rate of earthquake occurrence in the area. Then, if we assume that the size distribution is governed by the conventional magnitude–frequency relation (this relation is often called the Gutenberg-Richter relation, or Ishimoto-Iida relation), or one of its variants, we can estimate the average return period of earthquakes with a given magnitude. If we assume that $\eta = 1$, then the method gives the upper bound of the regional seismic hazard. Although it is in general difficult to determine the value of η, the regional seismicity and geological data can be used to place some constraints on the model and improve the estimate. Because seismological, geodetic, and geological data contain information on different time scales, we can make a more comprehensive estimate of regional seismic hazard by combining all of these data. However, it is not easy to determine the real uncertainty of the estimate because the database is limited and many assumptions are implicitly or explicitly made.

The long-term seismic hazard is usually expressed in maps portraying the likelihood of earthquake occurrence or of specific parameters such as the probability of exceedance of given levels of ground shaking over a certain period (e.g., 30 years). In some cases the hazard value is time-dependent in the sense that it depends on the time since the last large earthquake in the region (e.g., Working Group on California Earthquake Probabilities, 1988). In other cases, the hazard value is estimated on the basis of integrated geological and seismological data for a region, and is time-independent in the sense that it does not change with time from some specific earthquakes in the region (e.g., Working Group on California Earthquake Probabilities, 1995, 1999; Frankel *et al.*, 1996). This type of long-term hazard estimate is important for various seismic hazard reduction measures such as the development of realistic building codes, retrofitting existing structures, and land-use planning. The US building codes are based on time-independent hazard maps.

As already mentioned, earthquake hazard assessment is fundamentally a predictive effort, and is subject to all the uncertainties caused by the limited amount of data and the models used. Also, when it is used for practical purposes, its limitations and uncertainties need to be carefully communicated to the practitioners and public so that the scientific information is properly interpreted and used. The distinction between hazard (the physical phenomenon such as ground shaking, magnitude of expected earthquake, etc) and risk (the likelihood of human and property loss that can result from the hazard) needs to be clearly made, and the hazard information should be judiciously used in conjunction with the concept of acceptable risk (i.e., how safe is safe enough?) appropriate for the question being addressed. More details on this topic are discussed in Chapter 65 by Somerville and Moriwaki.

5.2 Intermediate-Term Strategy

Although short-term prediction with high probability of success is not possible at present, nor in the foreseeable future, there is a possibility that improved physical measurements of various parameters of the crust can be used to identify the areas where the state of the crust is close to failure. For example, continuous monitoring of strain changes in the Earth's crust with GPS will provide us with critical information on where strain is accumulating rapidly, and where aseismic deformation is taking place (Heki, 1997; Kato *et al.*, 1998; Ozawa *et al.*, 1999).

Although the physics is still not well understood, some seismicity patterns, changes in electromagnetic properties, changes in discharge rate and chemistry of ground water, and episodic changes in strain could provide important information on future earthquake activity in the area. A probabilistic use of these precursory changes could provide a useful means for forecasting the likelihood of earthquakes in a region (Aki, 1981). However, it is important to note that the investigations of these phenomena should be considered exploratory at present, and overly optimistic statements on the usefulness of these studies for operational earthquake prediction should be avoided.

5.3 Short-Term Strategy

A short-term prediction, if any is made, is bound to be very uncertain. For certain purposes (e.g., deployment of strong-motion instruments, Yamaoka *et al.*, (1999)), short-term predictions with large uncertainties can be useful. However, even if such short-term predictions should become possible, large earthquakes in densely populated urban areas are likely to cause extensive damage and disruption to society.

5.4 Real-Time Strategy

To minimize the immediate impact of large earthquakes, a mitigation strategy using real-time technology has been implemented. Despite the fact that engineering designs of individual structures have improved significantly, modern urban and suburban regions as a whole are more vulnerable to earthquakes than ever. Effective seismic hazard reduction depends on taking full advantage of the recent technical advances in seismological methodology and instrumentation, computer, and telemetry technology. In highly industrialized communities, rapid earthquake information is critically important for emergency services agencies, utilities, communications, financial companies, and media to make quick reports and damage estimates, and to determine where emergency response is most needed. The recent earthquakes in Northridge, California, and Kobe, Japan, clearly demonstrated the need for such information. Several systems equipped to deal with these needs have been already implemented. With the improvement of seismic sensors and communication systems, a significant increase in the speed and reliability of such a system is possible, so that it will eventually have the capability of estimating the spatial distribution of strong ground motion within seconds after an earthquake. Some facilities could receive this information before ground shaking begins. This would allow for clean emergency shutdown or other protection of systems susceptible to damage, such as power stations, computer systems, and telecommunication networks.

The idea of using rapid earthquake information for emergency operations is not new, and several systems have been developed in Japan, Mexico, Taiwan, and the United States. The following are some examples.

5.4.1 Japan

In the late 1950s, simple seismometers were installed for a railway alarm system. Since the operation of the Bullet Train system started in 1964, an automatic system to stop or slow trains during strong earthquakes has been developed (Nakamura and Tucker, 1988). The most recent system, UrEDAS, utilizes a sophisticated seismic detection/location algorithm, and is currently used by the Japanese railway system. Also, some utility companies have developed real-time ground-motion detection systems for emergency services for their own facilities. Recently, the City of Yokohama embarked on a project to deploy a real-time 150-station strong-motion network.

5.4.2 Mexico

In 1985, a $M = 8.1$ earthquake in the Michoacan seismic gap, about 320 km west of Mexico City, caused very heavy damage in Mexico City. Because a similar large earthquake is expected within the next few decades in the Guerrero seismic gap, about 300 km southwest of Mexico City, a seismic alert system, SAS (Seismic Alert System), was developed as a public early warning system in 1991. This system has a specific objective: to detect $M > 6$ earthquakes in the Guerrero gap with a seismic network deployed in the gap area, and issue an early warning of strong ground motion to the residents and authorities in Mexico City. Because it takes about 100 sec for seismic waves to travel from the Guerrero area to Mexico City, this system could provide an early warning with up to 60 sec lead time. A $M = 7.3$ earthquake occurred on September 14, 1995, in the Guerrero gap, and this system successfully broadcast an alarm on commercial radio stations in Mexico City about 72 sec prior to the arrival of strong ground motion (Espinosa-Aranda *et al.*, 1995; Chapter 76 by Espinosa-Aranda).

5.4.3 Taiwan

Two prototype earthquake early warning systems have been implemented in Taiwan, one for a local area near Hualien, and another for the entire island (Lee *et al.*, 1996; Teng *et al.*, 1997; Chapter 64 by Shin *et al.*). These systems use a state-of-the-art seismic network technology, and are designed to provide critical information on earthquakes and resulting ground motions for various emergency and recovery operations (Wu *et al.*, 1999). During the recent Chi-Chi, Taiwan, earthquake ($M_w = 7.6$), these systems rapidly distributed critical information on the earthquake and ground motions to various emergency services groups in Taiwan to help minimize the impact of the earthquake (Tsai, 2000; Wu *et al.*, 2000).

5.4.4 United States

In 1990, the California Institute of Technology (Caltech) and the US Geological Survey (USGS) Pasadena Office initiated the CUBE (Caltech/USGS Broadcast of Earthquakes) project in southern California (Kanamori *et al.*, 1991). It was realized that closely coordinated efforts between academia, governments, and private companies are essential for effective earthquake mitigation in modern metropolitan areas such as Los Angeles. One of the important objectives of the CUBE project is to promote closely coordinated efforts between various organizations with a rapid and reliable earthquake information system during major earthquake sequences. In 1993, the University of California at Berkeley and the US Geological Survey Office in Menlo Park developed the REDI (Rapid Earthquake Data Integration Project) system to broadcast earthquake data in central California (Gee *et al.*, 1996; Chapter 77 by Gee *et al.*). In 1997, the US Geological Survey, the California Institute of Technology, and the California Division of Mines and Geology cooperated to create TriNet, the next-generation seismic information system for southern California (Chapter 78 by Hauksson *et al.*).

6. Conclusion

Despite the progress made in understanding the physics of earthquakes, the predictions of earthquake activity we can make today are inevitably very uncertain, mainly because of the highly complex nature of earthquake process. The question becomes whether such uncertain predictions are useful for society—Are such uncertain predictions societally acceptable and beneficial? Long-term forecast is important for various seismic hazard reduction measures, such as the development of realistic building codes, retrofitting existing structures, and land-use planning. Short-term predictions with the current level of uncertainty may not be beneficial for a highly industrialized and economically sophisticated society. However, such predictions could be useful for those places where the social and economical environments are such that false alarms can be socially tolerated. Whether short-term predictions with uncertainty are useful or not for a given society should be investigated not only by seismologists, but also by engineers, social scientists, government officials, and industry representatives of the society in question.

The next question, given this uncertainty, is, Can we use our present knowledge of earthquakes effectively for mitigation of seismic risk? There are many ways in which we can improve our ability to minimize the impact of earthquakes. One such approach is "real-time seismology," which includes not only technical development, but also promotion of coordinated efforts between scientists, engineers, government officials, and the general public through the use of real-time systems.

In order to make such efforts more effective, it is important to understand the basic physics of earthquakes, and solid basic research should be promoted. However, it is also important to be aware that more knowledge may not necessarily lead to better prediction capability. We may only understand better why it is so difficult to accurately predict short-term earthquake behavior. Because earthquake prediction is a matter of serious concern among the public and emergency services officials, it is the important responsibility of scientists to communicate to them what is possible and what is not possible at present.

Acknowledgments

I thank Paul Spudich, Evelyn Roeloffs, Max Wyss, and an anonymous reviewer for helpful comments.

References

Aceves, R. L., and S. K. Park (1997). Cannot earthquakes be predicted? *Science* **278**, 488.

Aggarwal, Y. P., L. R. Sykes, J. Armbruster, and M. L. Sbar (1973). Premonitory changes in seismic velocities and prediction of earthquakes. *Nature* **241**, 101–104.

Aki, K. (1981). A probabilistic synthesis of precursory phenomena. In: "Earthquake Prediction, An International Review" (D. W. Simpson and P. G. Richards, Eds.) pp. 566–574. Am. Geophys. Un., Washington, D.C.

Allen, C. R. (1976). Responsibilities in earthquake prediction. *Bull. Seismol. Soc. Am.* **66**, 2069–2074.

Ando, M. (1975). Source mechanisms and tectonic significance of historical earthquakes along the Nankai trough, Japan. *Tectonophysics* **27**, 119–140.

Bakun, W. H., and A. G. Lindh (1985). The Parkfield, California, earthquake prediction experiment. *Science* **229**, 619–624.

Bakun, W. H., and T. V. McEvilly (1984). Recurrence models and Parkfield, California, earthquakes. *J. Geophys. Res.* **89**, 3051–3058.

Brune, J. N. (1979). Implications of earthquake triggering and rupture propagation for earthquake prediction based on premonitory phenomena. *J. Geophys. Res.* **84**, 2195–2198.

Burridge, R., and L. Knopoff (1967). Model and theoretical seismicity. *Bull. Seismol. Soc. Am.* **57**, 341–371.

Cifuentes, I. L., and P. G. Silver (1989). Low-frequency source characteristics of the great 1960 Chilean earthquake. *J. Geophys. Res.* **94**, 643–663.

Crampin, S., T. Volti, and R. Stefánsson (1999). A successfully stress-forecast earthquake. *Geophys. J. Int.* **138**, F1–F5.

Das, S., and C. Scholz (1981). Off-fault aftershock clusters caused by shear stress increase. *Bull. Seismol. Soc. Am.* **71**, 1669–1675.

Deng, J., and L. R. Sykes (1997). Evolution of the stress field in southern California and triggering of moderate-size earthquakes: A 200-year perspective. *J. Geophys. Res.* **102**, 9859–9886.

Ellsworth, W. L., A. G. Lindh, W. H. Prescott, and D. G. Herd (1981). The 1906 San Francisco earthquake and the seismic cycle. In: "Earthquake Prediction, An International Review" (D. W. Simpson and P. G. Richards, Eds.), pp. 126–140. Am. Geophys. Un., Washington, D.C.

Espinosa-Aranda, J. M., A. Jimenez, G. Ibarrola, F. Alcantar, A. Aguilar, M. Inostroza, and S. Maldonado (1995). Mexico City Seismic Alert System. *Seismol. Res. Letters* **66**, 42–53.

Fedotov, S. A. (1965). Regularities of the distribution of strong earthquakes in Kamchatka, the Kuril Islands, and northeast Japan. *Trudy Inst. Fiz. Zemli., Acad. Nauk, SSSR* **36**, 66–94.

Frankel, A., C. Mueller, T. Barnhard, D. Perkins, E. Leyendecker, N. Dickman, S. Hanson, and M. Hopper (1996). National seismic hazard maps: Documentation June 1996. *Open File Report 96-532, US Geol. Surv.* Denver, Colorado, 110 pp.

Fraser-Smith, A. C., A. Bernardi, P. R. McGill, M. E. Ladd, R. A. Heliwell, and O. G. Villard (1990). Low-frequency magnetic field measurements near the epicenter of the Ms 7.1 Loma Prieta earthquake. *Geophys. Res. Lett.* **17**, 1465–1468.

Gee, L. S., D. S. Neuhauser, D. S. Dreger, M. E. Pasyanos, R. A. Uhrhammer, and B. Romanowicz (1996). Real-time seismology at UC Berkeley: The rapid earthquake data integration project. *Bull. Seismol. Soc. Am.* **86**, 936–945.

Gelfand, I. M., S. A. Guberman, V. I. Keilis-Borok, L. Knopoff, F. Press, E. Y. Ranzman, I. M. Rotwain, and A. M. Sadovsky (1976). Pattern recognition applied to earthquake epicenters in California. *Phys. Earth Planet. Inter.* **11**, 227–283.

Geller, R. J. (1996). Debate on evaluation of the VAN method: Editor's introduction. *Geophys. Res. Lett.* **23**, 1291–1293.

Geller, R. J. (1997). Earthquake prediction: a critical review. *Geophys. J. Int.* **131**, 425–450.

Geller, R. J., D. D. Jackson, Y. Y. Kagan, and F. Mulargia (1997a). Earthquakes cannot be predicted. *Science* **275**, 1616–1617.

Geller, R. J., D. D. Jackson, Y. Y. Kagan, and F. Malaria (1997b). Cannot earthquakes be predicted? *Science* **278**, 488–490.

Gokhberg, M. B., V. A. Morgounov, and O. A. Pokhotelov (1995). *Earthquake Prediction, Seismo-Electromagnetic Phenomena.* Gordon and Breach Publishers, Singapore, 193 pp.

Gupta, S. K., and A. M. Patwardhan (Eds.) (1988). *Earthquake Prediction: Present Status.* University of Poona, Pune, 280 pp.

Habermann, R. E. (1981). Precursory seismicity patterns: Stalking the mature seismic gap. In: "Earthquake Prediction, An International Review" (D. W. Simpson and P. G. Richards, Eds.), pp. 29–42. Am. Geophys. Un., Washington, D.C.

Hardebeck, J. L., J. J. Nazareth, and E. Hauksson (1998). The static stress change triggering model: Constraints from two southern California aftershock sequences. *J. Geophys. Res.* **103**, 24427–24437.

Harris, R. A. (1998a). Forecasts of the 1989 Loma Prieta, California, earthquake. *Bull. Seismol. Soc. Am.* **88**, 898–916.

Harris, R. A. (1998b). Introduction to special section: Stress triggers, shadows, and implications for seismic hazard. *J. Geophys. Res.* **103**, 24347–24358.

Harris, R. A., and R. W. Simpson (1992). Changes in static stress on southern California faults after the 1992 Landers earthquake. *Nature* **360**, 251–254.

Harris, R. A., and R. W. Simpson (1996). In the shadow of 1857—The effect of the great Ft. Tejon earthquake on subsequent earthquakes in southern California. *Geophys. Res. Lett.* **23**, 229–232.

Heki, K., and Y. Tamura (1997). Short-term afterslip in the 1994 Sanriku-Haruka-Oki earthquake. *Geophys. Res. Lett.* **24**, 3285–3288.

Hill, D. P., P. A. Reasenberg, A. Michael, W. J. Arabaz, G. Beroza, *et al.* (1993). Seismicity remotely triggered by the magnitude 7.3 Landers, California, earthquake. *Science* **260**, 1617–1623.

Igarashi, G., S. Saeki, N. Takahata, K. Sumikawa, S. Tasaka, Y. Sasaki, M. Takahashi, and Y. Sano (1995). Ground-water radon anomaly before the Kobe earthquake in Japan. *Science* **269**, 60–61.

Imamura, A. (1928). On the seismic activity of central Japan. *Japan J. Astron. Geophys.* **6**, 119–137.

Ingebritsen, S. E., and W. E. Sanford (1998). *Groundwater in Geologic Processes.* Cambridge University Press, Cambridge, 341 pp.

Ishibashi, K. (1977). Re-examination of a great earthquake expected in the Tokai district, central Japan—possibility of the "Suruga Bay earthquake." *Rept. Coord. Comm. Earthq. Pred.* **17**, 126–132 (in Japanese).

Isikara, A. M., and A. Vogel (Eds.) (1982). *Multidisciplinary Approach to Earthquake Prediction.* Friodr. Vieweg & Sohn, Braunschweig, 578 pp.

Jackson, D. D., and K. K. Kagan (1991). Seismic gap hypothesis: Ten years after. *J. Geophys. Res.* **96**, 21419–21431.

Jackson, D. D., and K. K. Kagan (1993). Reply: to Comment on "Seismic gap hypothesis: Ten years after." *J. Geophys. Res.* **99**, 9917–9920.

Jaume, S. C., and L. R. Sykes (1992). Change in the state of stress on the southern San Andreas fault resulting from the California earthquake sequence of April to June 1992. *Science* **258**, 1325–1328.

Jaume, S., and L. R. Sykes (1999). Evolving towards a critical point: A review of accelerating seismic moment/energy release prior to large and great earthquakes. *Pure Appl. Geophys.* **155**, 279–306.

Johnston, M. J. S., A. T. Linde, and M. T. Gladwin (1990). Near-field high resolution strain measurements prior to the October 18, 1989, Loma Prieta Ms 7.1 earthquake. *Geophys. Res. Lett.* **17**, 1777–1780.

Johnston, M. J. S., A. T. Linde, and D. C. Agnew (1994). Continuous borehole strain in the San Andreas fault zone before, during, and after the 28 June 1992, Mw 7.3 Landers, California, earthquake. *Bull. Seismol. Soc. Am.* **84**, 799–805.

Kanamori, H., and J. J. Cipar (1974). Focal processes of the Great Chilean earthquake May 22, 1960. *Phys. Earth Planet. Inter.* **9**, 128–136.

Kanamori, H., and K. C. McNally (1982). Variable rupture mode of the subduction zone along the Ecuador-Colombia coast. *Bull. Seismol. Soc. Am.* **72**, 1241–1253.

Kanamori, H., E. Hauksson, and T. Heaton (1991). TERRAscope and CUBE Project at Caltech. *EOS, Trans. Am. Geophys. Un* **72**, 564.

Kato, T., G. S. El-Fiky, E. W. Oware, and S. Miyazaki (1998). Crustal strains in the Japanese islands as deduced from dense GPS array. *Geophys. Res. Lett.* **25**, 3445–3448.

Keilis-Borok, V. I. (1996). Intermediate-term earthquake prediction. *Proc. Natl. Acad. Sci. USA* **93**, 3748–3755.

Keilis-Borok, V. I., L. Knopoff, I. M. Rotwain, and C. R. Allen (1988). Intermediate-term prediction of occurrence times of strong earthquakes. *Nature* **335**, 690–694.

Kelleher, H. A. (1970). Space-time seismicity of the Alaska-Aleutian seismic zone. *J. Geophys. Res.* **75**, 5745–5756.

Kelleher, J. A. (1972). Rupture zones of large South American earthquakes and some predictions. *J. Geophys. Res.* **77**, 2087–2103.

Kelleher, J., L. Sykes, and J. Oliver (1973). Possible criteria for predicting earthquake locations and their application to major plate boundaries of the Pacific and the Caribbean. *J. Geophys. Res.* **78**, 2547–2585.

Kisslinger, C. (1988). An experiment in earthquake prediction and the 7 May 1986 Andreanof Islands earthquake. *Bull. Seismol. Soc. Am.* **78**, 218–229.

Knopoff, L. (1996). Earthquake Prediction: The Scientific Challenge. *Proc. Natl. Acad. Sci. USA*, **93** 3719–3720.

Kossobokov, V. G., L. L. Romashkova, V. I. Keilis-Borok, and J. H. Healy (1999). Testing earthquake prediction algorithms: statistically significant advance prediction of the largest earthquakes in the Circum-Pacific, 1992-1997. *Phys. Earth Planet. Inter.* **111**, 187–196.

Kuznetsov, I. V. (1999). Earthquake mechanisms and dynamics of seismicity. In: "1995–1998 National Report," pp. 41–48. Russian Academy of Sciences, Moscow.

Lee, W. H. K., T. C. Shin, and T. L. Teng (1996). Design and implementation of earthquake early warning systems in Taiwan. In: "Eleventh World Conference on Earthquake Engineering," Paper No. 2133, Elsevier Sci. Ltd., Amsterdam.

Lighthill, J. (Ed.) (1996). *A Critical Review of VAN. World Scientific*, Singapore.

Linde, A. T., K. Suyehiro, S. Miura, I. S. Sacks, and A. Takagi (1988). Episodic aseismic slip, stress distribution and seismicity. *Nature* **334**, 513–515.

Linde, A. T., M. T. Gladwin, M. J. S. Johnston, R. L. Gwyther, and R. G. Bilham (1996). A slow earthquake sequence on the San Andreas fault. *Nature* **383**, 65–68.

Lindh, A. G. (1983). Preliminary assessment of long-term probabilities for large earthquakes along selected fault segments of the San Andreas Fault system in California. *Open-File Report 83-63, US Geol. Surv.*, Menlo Park, California.

Lomnitz, C. (1994). *Fundamentals of Earthquake Prediction*. John Wiley & Sons, New York, 326 pp.

McEvilly, T. V., and L. R. Johnson (1973). Earthquakes of strike-slip in central California: Evidence on the question of dilatancy. *Science* **182**, 581–584.

McEvilly, T. V., and L. R. Johnson (1974). Stability of P and S velocities from central California quarry blasts. *Bull. Seismol. Soc. Am.* **64**, 343–353.

Michael, A., and J. Langbein (1993). Earthquake prediction lessons for Parkfield Experiment. *EOS, Trans. Am. Geophys. Un* **74**, 145, 153–155.

Miller, S. A. (1996). Fluid-mediated influence of adjacent thrusting on the seismic cycle at Parkfield. *Nature* **382**, 799–802.

Mogi, K. (1968). Sequential occurrences of recent great earthquakes. *J. Phys. Earth* **16**, 30–36.

Mogi, K. (1969). Some features of recent seismic activity in and near Japan (2) Activity before and after great earthquakes. *Bull. Earthq. Res. Inst.*, Tokyo Univ., **47**, 395–417.

Mogi, K. (1984). Temporal variation of crustal deformation during the days preceding a thrust-type great earthquake—The 1944 Tonankai earthquake of Magnitude 8.1, Japan. *Pure Appl. Geophys.* **122**, 765–780.

Mogi, K. (1985). *Earthquake Prediction*. Academic Press, Tokyo, 355 pp.

Nakamura, Y., and B. E. Tucker (1988). Japan's earthquake warning system, Should it be imported to California? *California Geology* **41**, 33–41.

National Earthquake Prediction Evaluation Council Working Group (1994). Earthquake Research at Parkfield, California, 1993 and beyond. *US Geol. Surv. Circular 1116*. US Government Printing Office, Washington D.C., 14 pp.

Nersesov, I. (1970). Earthquake prognostication in the Soviet Union. *Bull. New Zealand Soc. Earthq. Eng.* **3**, 108–119.

Nishenko, S. P. (1985). Seismic potential for large and great interplate earthquakes along the Chilean and southern Peruvian margins of South America: A quantitative reappraisal. *J. Geophys. Res.* **90**, 3589–3615.

Nishenko, S. P., and L. R. Sykes (1993). Comment on "Seismic gap hypothesis: Ten years after." *J. Geophys. Res.* **98**, 9909–9916.

Ohtake, M., T. Matumoto, and G. V. Latham (1977). Seismicity gap near Oaxaca, Southern Mexico as a probable precursor to a large earthquake. *Pure Appl. Geophys.* **115**, 375–385.

Olson, R. S., B. Podesta, and J. M. Nigg (1989). *The Politics of Earthquake Prediction*. Princeton Univ. Press, Princeton, 187 pp.

Otsuka, M. (1972). A chain-reaction-type source model as a tool to interpret the magnitude–frequency relation of earthquakes. *J. Phys. Earth* **20**, 35–45.

Ozawa, T., T. Tabei, and S. Miyazaki (1999). Interplate coupling along the Nankai Trough off southwest Japan derived from GPS measurements. *Geophys. Res. Lett.* **26**, 927–930.

Raleigh, C. B., J. H. Healy, and J. D. Bredehoeft (1976). An experiment in earthquake control at Rangely, *Colorado. Science* **191**, 1230–1237.

Reid, H. F. (1910). The mechanism of the earthquake. In: "The California Earthquake of April 18, 1906." *Report of the State Earthquake Investigation Commission, 2*, pp. 1–192, Carnegie Institution, Washington, D.C.

Rikitake, T. (1976). *Earthquake Prediction*. Elsevier, Amsterdam, 357 pp.

Rikitake, T. (1982). *Earthquake Forecasting and Warning*. D. Reidel Publishing, Dordrecht, 402 pp.

Rikitake, T. (1986). *Earthquake Premonitory Phenomena: Database for Earthquake Prediction*. Tokyo University Press, Tokyo, 232 pp.

Rikitake, T., and K. Hamada (2001). Earthquake Prediction. In: "Encyclopedia of Physical Science and Technology," 3rd ed., Vol. 4, pp. 743–760. Academic Press, San Diego.

Roeloffs, E. A. (1988). Hydrologic precursors to earthquakes: A review. *Pure Appl. Geophys.* **126**, 177–209.

Rotwain, I., and O. Novikova (1999). Performance of the earthquake prediction algorithm CN in 22 regions of the world. *Phys. Earth Planet. Inter.* **111**, 207–213.

Rybicki, K. (1973). Analysis of aftershocks on the basis of dislocation theory. *Phys. Earth Planet Inter.* **7**, 409–422.

Rybicki, K., T. Kato, and K. Kasahara (1985). Mechanical interaction between neighboring active faults—Static and dynamic stress field induced by faulting. *Bull. Earthq. Res. Inst.* Tokyo Univ., **60**, 1–21.

Sammis, C. G., and S. W. Smith (1999). Seismic cycles and the evolution of the stress correlation in cellular automaton models of finite fault networks. *Pure Appl. Geophys.* **155**, 307–334.

Savage, J. C. (1993). The Parkfield prediction fallacy. *Bull. Seismol. Soc. Am.* **83**, 1–6.

Scholz, C. H. (1985). The Black Mountain asperity: Seismic hazard of the southern San Francisco Peninsula, California, *Geophys. Res. Lett.* **12**, 717–719.

Scholz, C. (1997). Whatever happened to earthquake prediction? *Geotimes* **42**(5), 16–19.

Scholz, C. H., L. R. Sykes, and Y. P. Aggarwal (1973). Earthquake prediction: a physical basis. *Science* **181**, 803–810.

Segall, P., and R. Harris (1987). Earthquake deformation cycle on the San Andreas fault near Parkfield, California. *J. Geophys. Res.*, **92**, 10511–10525.

Semenov, A. M. (1969). Variation in the travel-time of transverse and longitudinal waves before violent earthquakes. *Izvestia, Acad. Sci. USSR Phys. Solid Earth* (English Translation), *No. 4*, 245–248.

Smith, S. W., and W. Van de Lindt (1969). Strain adjustments associated with earthquakes in southern California. *Bull. Seismol. Soc. Am.* **59**, 1569–1589.

Sobolev, G. A. (1995). *Fundamental of Earthquake Prediction*. Electromagnetic Research Center, Moscow, 161 pp.

Sobolev, G. A., and A. D. Zavyalov (1999). Earthquake prediction. In: "1995-1998 National Report," pp. 22–28. Russian Academy of Sciences, Moscow.

Stein, R. S., G. C. P. King, and J. Lin (1992). Change in failure stress on the southern San Andreas fault system caused by the 1992 magnitude=7.4 Landers earthquake. *Science* **258**, 1328–1332.

Stein, R. S., G. C. P. King, and J. Lin (1994). Stress triggering of the 1994 M=6.7 Northridge, California, earthquake by its predecessors. *Science* **265**, 1432–1435.

Stein, R. S. (1999). The role of stress transfer in earthquake occurrence. *Nature* **402**, 605–609.

Sykes, L. R., and S. C. Jaume (1990). Seismic activity on neighbouring faults as a long-term precursor to large earthquakes in the San Francisco Bay area. *Nature* **348**, 595–599.

Sykes, L. R., and S. P. Nishenko (1984). Probabilities of occurrence of large plate rupturing earthquakes for the San Andreas, San Jacinto, and Imperial Faults, California, 1983–2003. *J. Geophys. Res.* **89**, 5905–5927.

Sykes, L. R., B. E. Shaw, and C. H. Scholz (1999). Rethinking earthquake prediction. *Pure Appl. Geophys.* **155**, 207–232.

Teng, T. L., L. Wu, T. C. Shin, Y. B. Tsai, and W. H. K. Lee (1997). One minute after: Intensity map, epicenter, and magnitude. *Bull. Seismol. Soc. Am.* **87**, 1209–1219.

Thatcher, W., and M. Lisowski (1987). Long-term seismic potential of the San Andreas fault southeast of San Francisco, California. *J. Geophys. Res.* **92**, 4771–4784.

Toda, S., R. S. Stein, P. A. Reasenberg, J. H. Dieterich, and A. Yoshida (1998). Stress transferred by the 1995 Mw=6.9 Kobe, Japan, shock: Effect on aftershocks and future earthquake probabilities. *J. Geophys. Res.* **103**, 24543–24565.

Tsai, Y. B., and M. Y. Huang (2000). Strong ground motion characteristics of the Chi-Chi, Taiwan earthquake of September 21, 1999. *Earthq. Eng. & Eng. Seismol.* **2**, 1–21.

Tsunogai, U., and H. Wakita (1995). Precursory chemical changes in ground water: Kobe earthquake, Japan. *Science* **269**, 61–63.

Turcotte, D. L. (1992). *Fractals and Chaos in Geology and Geophysics*. Cambridge University Press, Cambridge, 221 pp.

Unesco (1984). *Earthquake Prediction*. Terra Scientific Publishing Co., Tokyo, 995 pp.

Utsu, T. (1970). Large earthquakes near Hokkaido and the expectancy of the occurrence of a large earthquake off Nemuro. *Rept. Coord. Comm. Earthq. Predict.* **7**, 7–13.

Valensise, G., and S. N. Ward (1991). Long-term uplift of the Santa-Cruz coastline in response to repeated earthquakes along the San Andreas fault. *Bull. Seismol. Soc. Am.* **81**, 1694–1704.

Varotsos, P., and M. Lazaridou (1991). Latest aspects of earthquake prediction in Greece based on seismic electric signals. *Tectonophysics* **188**, 321–347.

Vogel, A. (Ed.) (1979). *Terrestrial and Space Techniques in Earthquake Prediction Research.* Friodr. Vieweg & Sohn, Braunschweig, 712 pp.

Wakita, H., Y. Nakamura, K. Notsu, M. Noguchi, and T. Asada (1980). Radon anomaly: a possible precursor of the 1978 Izu-Oshima-Kinkai earthquake. *Science* **207**, 882–883.

Whitcomb, J. H., J. D. Garmany, and D. L. Anderson (1973). Earthquake prediction: Variation of seismic velocities before the San Fernando earthquake. *Science* **180**, 632–635.

Whiteside, L., and R. E. Habermann (1989). The seismic quiescence prior to the 1978 Oaxaca earthquake IS NOT a precursor to that earthquake (abstract). *Eos, Trans. Am. Geophys. Un* **70**, 1232.

Working Group on California Earthquake Probabilities (1988). *Probabilities of large earthquakes occurring in California on the San Andreas fault*. Government Printing Office, Washington, D.C.

Working Group on California Earthquake Probabilities (1995). Seismic Hazards in Southern California: Probable Earthquakes, 1994 to 2024. *Bull. Seismol. Soc. Am.* **85**, 379–439.

Working Group on California Earthquake Probabilities (1999). Earthquake Probabilities in the San Francisco Bay Region: 2000 to 2030—A Summary of Findings, *Open File Report 99-517, US Geol. Surv.*, Menlo Park, California. [Online version 1, *http://geopubs.wr.usgs.gov/open-file/of99-517/*].

Wu, Y. M., J. K. Chung, T. C. Shin, N. C. Hsiao, Y. B. Tsai, W. H. K. Lee, and T. L. Teng (1999). Development of an integral earthquake early warning system in Taiwan—Case for the Hualien area earthquakes. *Terrestrial, Atmospheric and Ocean Sciences* **10**, 719–736.

Wu, Y. M., W. H. K. Lee, C. C. Chen, T. C. Shin, T. L. Teng, and Y. B. Tsai (2000). Performance of the Taiwan Rapid Earthquake Information Release System (RTD) during the 1999 Chi-Chi (Taiwan) earthquake, *Seismol. Res. Lett.* **71**, 338–343.

Wyss, M. (Ed.) (1979). *Earthquake Prediction and Seismicity Patterns* (Reprinted from *Pure Appl. Geophys., Vol. 117*, No. 6, 1979). Birkhauser Verlag, Basel, 237 pp.

Wyss, M. (Ed.) (1991). *Evaluation of Proposed Earthquake Precursors*. Am. Geophys. Un., Washington, D.C., 94 pp.

Wyss, M. (1997a). Second round of evaluations of proposed earthquake precursors. *Pure Appl. Geophys.* **149**, 3–16.

Wyss, M. (1997b). Cannot earthquakes be predicted. *Science* **278**, 487–488.

Wyss, M., and R. O. Burford (1987). A predicted earthquake on the San Andreas fault, California. *Nature* **329**, 323–325.

Wyss, M., and R. E. Habermann (1979). Seismic quiescence precursory to a past and a future Kurile islands earthquake. *Pure Appl. Geophys.* **117**, 1195–1211.

Wyss, M., and S. Wiemer (1999). How can one test the seismic gap hypothesis. *Pure Appl. Geophys.* **155**, 259–278.

Yamaoka, K., T. Ooida, *et al.* (1999). Detailed distribution of accelerating foreshocks before a M5.1 earthquake in Japan. *Pure Appl. Geophys.* **155**, 335–354.

Yamashina, K. (1979). A possible factor which triggers shallow intraplate earthquakes. *Phys. Earth Planet. Inter.* **18**, 153–164.

Zhang, G., and J. Liu (1999). Earthquake Prediction in China. In: "1995-1998 China National Report on Seismology and Physics of the Earth's Interior," pp. 75–88, China Meteorological Press, Beijing.

Zhang, G., L. Zhu, X. Song, Z. Li, M. Yang, N. Su, and X. Chen (1999). Predictions of the 1997 strong earthquakes in Jiashi, Xinjiang, China. *Bull. Seismol. Soc. Am.* **89**, 1171–1183.

Zhu, C., and Z. Wu (1999). The physics of earthquake prediction. In: "1995–1998 China National Report on Seismology and Physics of the Earth's Interior," pp. 89–98. China Meteorological Press, Beijing.

73

Stress Triggers, Stress Shadows, and Seismic Hazard

Ruth A. Harris
US Geological Survey, Menlo Park, California, USA

1. Introduction

Large earthquakes nucleate, propagate, and terminate. That much we do know. But what about the physics of why, when, and where these actions occur? The complete picture of earthquake mechanics remains unresolved. To help solve at least part of the complex puzzle sooner rather than later, I have focused here on the role of stress changes in earthquake mechanics. In this chapter I tackle the topics of earthquake promotion, or triggering, and earthquake delay, or deferment, due to earthquake-generated changes in stress. Among the questions that I attempt to answer: Can we understand and adequately model the mechanics of local inter-earthquake and intra-earthquake triggering and prevention? Do earthquake-induced static or dynamic stress changes trigger subsequent earthquakes? Is there a triggering threshold? Can stress shadow (regions where faults are relaxed) calculations be used to estimate where and when future earthquakes will not occur? With this chapter I delve into some earthquake interaction topics that have drawn special attention. I primarily concentrate on research that has performed quantitative estimates of earthquake-generated stress changes and applied these calculations to the study of earthquake interactions. For a review of earthquake triggering by man-made and natural mechanisms, the reader is referred to Chapter 40 by McGarr *et al.*

2. Background

Although human-induced earthquakes, through activities such as fluid injection and withdrawal, mining, and hydrocarbon recovery, have been recognized for decades (e.g., Carder, 1945; Healy *et al.*, 1968; Kovach, 1974; Raleigh *et al.*, 1972, 1976; Kisslinger, 1976; Segall, 1992; McGarr and Simpson, 1997; Chapter 40 by McGarr *et al.*), the impact of interactions between natural earthquakes has not been as well understood. Pioneering papers by Smith and Van de Lindt (1969), Rybicki (1973), Yamashina (1978, 1979), Das and Scholz (1981), and Stein and Lisowski (1983) presented calculations of mainshock static stress changes affecting subsequent earthquake locations, but these preliminary determinations were not formally adopted by the scientific community for use in earthquake hazard assessment. Part of the reason may have been that the static stress changes discussed were thought to be small, of the order of 0.1 MPa (1 bar), a value that is just a fraction of the stress-drop during an earthquake. Also, the original results were predominantly qualitative in nature and hard to judge quantitatively. In the subsequent 15–20 years, a large body of additional research has been added to the original work.

There is now an internationally distributed data set of static stress changes, generated by large earthquakes, influencing the timing and locations of subsequent "natural" earthquakes. Stress-change calculations have been performed for earthquakes in the Asal Rift of northeastern Africa (Jacques *et al.*, 1996), in Chile (Delouis *et al.*, 1998), in Greece (Hubert *et al.*, 1996), in Italy (Nostro *et al.*, 1997, 1998; Troise *et al.*, 1998), in Japan (Yamashina, 1978, 1979; Rybicki *et al.*, 1985; Kato *et al.*, 1987; Yoshioka and Hashimoto, 1989a,b; Okada *et al.*, 1990; Pollitz and Sacks, 1995, 1997; Toda *et al.*, 1998), along the Macquarie Ridge (Das, 1992), in Mexico (Mikumo *et al.*, 1998; Singh *et al.*, 1998; Mikumo *et al.*, 1999), in New Zealand (Robinson, 1994), in Turkey (Roth, 1988; Nalbant *et al.*, 1996; Stein *et al.*, 1997; Nalbant *et al.*, 1998; Hubert-Ferrari *et al.*, 2000), and in the United States in Alaska (Taylor *et al.*, 1998), California (Smith and Van de Lindt, 1969; Rybicki, 1971, 1973; Das and Scholz, 1981; Mavko, 1982; Stein and Lisowski, 1983; Mavko *et al.*, 1985; Li *et al.*, 1987; Simpson *et al.*, 1988; Oppenheimer *et al.*, 1988; Hudnut *et al.*, 1989; Michael, 1991; Reasenberg and Simpson, 1992; Harris and Simpson, 1992; Jaumé and Sykes,

ISBN: 0-12-440658-0

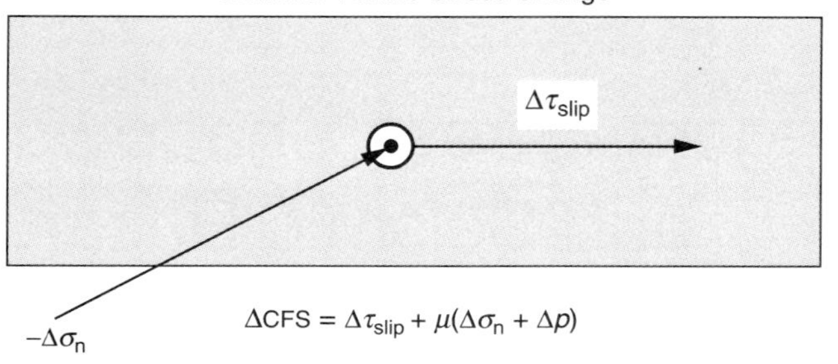

FIGURE 1 ΔCFS, the change in Coulomb failure stress, or Coulomb stress increment, is used to evaluate if one earthquake brought another earthquake's fault closer to, or farther from, failure. If $\Delta CFS > 0$, the first earthquake brought the second earthquake closer to failure, if $\Delta CFS < 0$, the first event sent the second event farther away from failure, and into a stress shadow. The stress shadow lasts for the length of time that it takes the second fault plane to recover from the stress increment. One manner of recovery is through long-term tectonic loading (e.g., Simpson and Reasenberg, 1994; Deng and Sykes, 1997a,b; Lienkaemper *et al.*, 1997). ΔCFS is resolved onto the fault plane and in the slip direction of the second earthquake, at the hypocenter of the second earthquake; $\Delta\tau_{slip}$ is the change in shear stress due to the first earthquake resolved in the slip direction of the second earthquake. $\Delta\sigma_n$ is the change in normal stress due to the first earthquake, resolved in the direction orthogonal to the second fault plane. $\Delta\sigma_n > 0$ implies increased tension; μ is the coefficient of friction and Δp is the change in pore pressure. $\Delta\tau_{slip}$ and $\Delta\sigma_n$ are often calculated using Okada's (1992) formulations. Many recent authors have simplified this equation by dropping the explicit calculation of Δp and using an apparent coefficient of friction, μ'. The simplified equation [Eq. (4)] and assumptions behind μ' are discussed in the text.

1992; Stein *et al.*, 1992; Du and Aydin, 1993; Oppenheimer *et al.*, 1993; Stein *et al.*, 1994; Simpson and Reasenberg, 1994; King *et al.*, 1994; Bennett *et al.*, 1995; Harris *et al.*, 1995; Harris and Simpson, 1996; Deng and Sykes, 1996; Jaumé and Sykes, 1996; Deng and Sykes, 1997a,b; Bürgmann *et al.*, 1997; Reasenberg and Simpson, 1997; Hardebeck *et al.*, 1998; Harris and Simpson, 1998; Fox and Dziak, 1999; Parsons *et al.*, 1999; Perfettini *et al.*, 1999; Stein, 1999), and Nevada (Hodgkinson *et al.*, 1996; Caskey and Wesnousky, 1997).

Many of the authors have used calculations of a Coulomb stress increment (Fig. 1) calculated from an elastic-dislocation model of the mainshock (e.g., Chinnery, 1963; Weertman and Weertman, 1964; Okada, 1992) and have examined the geographical pattern of subsequent earthquakes relative to the pattern of change in Coulomb failure stress. Almost all of these studies profess to find a positive correlation between the number (or rate) of aftershocks, or the occurrence of subsequent mainshocks, and regions of calculated stress increase. Many of the studies also show a deficiency (or rate decrease) of aftershocks, or subsequent mainshocks, in regions of calculated stress decrease.

3. The Landers Earthquake

Perhaps the most publicized calculations of static stress changes were those inferred for the 1992 M7.3 Landers earthquake in southern California (Harris and Simpson, 1992; Henyey, 1992; Jaumé and Sykes, 1992; Stein *et al.*, 1992). This large earthquake occurred near the populous San Bernardino region of the San Andreas fault (Sieh *et al.*, 1993). Soon after the earthquake's location and mechanism were determined, it became apparent that the estimated static-stress effect of the Landers earthquake was to reduce the normal stress on part of the San Andreas fault. Calculations made shortly after the Landers earthquake estimated that Landers had hastened the next large earthquake on the San Andreas fault by about 1–2 decades (Harris and Simpson, 1992; Jaumé and Sykes, 1992; Stein *et al.*, 1992), if its nucleation point were to fall in the area of reduced normal stress. Unfortunately, the timing of the next large San Andreas fault earthquake could not be assigned an absolute value because no one knew precisely when the next San Andreas earthquake was due to occur, or where it might nucleate, even without the normal stress change. In the meantime, afterslip along the Banning fault, which lies in close proximity to the San Andreas, has been inferred (Shen *et al.*, 1994), and the San Andreas fault has, as of August 2002, not yet slipped in the next large or great earthquake.

I now address a range of failure criteria that has been used to model earthquake interactions, both in the vicinity of the Landers earthquake and elsewhere. I start with the simplest model, Coulomb failure stress, then explore more complex hypotheses. Table 1 provides a snapshot of some of the theories by very briefly describing the required parameters, mentioning a few of the successes and failures, and listing some related papers.

4. Coulomb Failure Stress

Spatial patterns of static stress changes calculated using Coulomb failure assumptions (e.g., Jaeger and Cook, 1969) (Fig. 1) seem to correlate well with spatial patterns of aftershocks; more aftershocks commonly occur where the change in Coulomb stress is positive than where the change is negative. A similar pattern has been observed for aftershock rates, which increased in regions of elevated Coulomb stress and decreased in regions of diminished stress (e.g., Reasenberg and Simpson, 1992, 1997; Toda *et al.*, 1998). With Coulomb failure, earthquakes that occur in regions of increased static stress are said to have been advanced toward failure by the positive stress increment. The amount of advance, Δt, equals $\Delta CFS/\dot{\tau}$ where ΔCFS is the change in Coulomb failure stress (Fig. 1), and $\dot{\tau}$ is the long-term stressing rate. Examples of inferred advancement (where the subsequent earthquake(s) have already occurred) are the cascade of earthquakes this century on the North Anatolian fault in Turkey (Stein *et al.*, 1997; Nalbant *et al.*, 1998; Hubert-Ferrari *et al.*, 2000) and the 1954 Rainbow Mountain-Fairview Peak-Dixie Valley, Nevada, earthquakes (Hodgkinson *et al.*, 1996).

TABLE 1 Summary of Stress-Change Approaches

Method	Parameters Required	Successes	Problems	Authors
Static Coulomb Failure Stress (Elastic) ΔCFS	Mainshock static slip model, μ', and $\Delta\sigma$, $\Delta\tau$, $\dot{\tau}$, on known fault planes and known slip directions[1]	ΔCFS > 0 explains locations of aftershocks that do occur. ΔCFS < 0 predicts shadows (timing and locations). May give rupture extent.	Many ΔCFS > 0 faults don't experience subsequent large earthquakes, so it's hard to use ΔCFS > 0 as a predictive tool. Does not predict size of impending aftershock. Does not explain delayed failure.	Smith & Van de Lindt [1969]; Rybicki [1973]; Yamashina [1978; 1979]; Stein & Lisowski [1983]; Simpson *et al.* [1988]; Yoshioka & Hashimoto [1989a,b]; Reasenberg & Simpson [1992]; Hardebeck *et al.* [1998]; Harris & Simpson [1998]; Nalbant *et al.* [1998]; Nostro *et al.* [1998]; Taylor *et al.* [1998]; Toda *et al.* [1998]; Anderson & Johnson [1999]; Fox and Dziak [1999]; Parsons *et al.*, 1999; Perfettini *et al.*, 1999; Astiz *et al.* [2000], Hubert-Ferrari *et al.* [2000], etc. (See text for more authors)
Dynamic Coulomb Failure Stress (Elastic) ΔCFS(t)	Mainshock dynamic fault slip model, μ', and $\Delta\sigma(t)$, $\Delta\tau(t)$ on known fault planes and known slip directions[1]	May predict rupture lengths, given fault geometry	Doesn't explain long delays (>tens of seconds) between subevents. Needs more testing.	Harris *et al.* [1991]; Harris & Day [1993]; Hill *et al.* [1993]; Gomberg & Bodin [1994]; Spudich *et al.* [1994; 1995]; Kase & Kuge [1998]; Harris & Day [1999]; Magistrale & Day [1999]
Static Rate-and-State	Mainshock static slip model, $\Delta\sigma$, $\Delta\tau$, σ, τ, $\dot{\tau}$, A, B, D_c, H, time of last event, recurrence interval (to determine slip speed)	Seems to predict aftershock duration	Needs more testing. Rate-and-state parameters defined in the lab, but not known for the Earth.	Dieterich [1994]; Dieterich & Kilgore [1996]; Roy & Marone [1996]; Stein *et al.* [1997]; Gross & Bürgmann [1998]; Gomberg *et al.* [1998]; Harris & Simpson [1998]; Toda *et al.* [1998]
Dynamic Rate-and-State	Mainshock dynamic fault slip model, $\Delta\sigma(t)$, $\Delta\tau(t)$, σ, τ, $\dot{\tau}$, A, H, time of last event, slip speed	May explain remote triggering	Needs more testing. Still need to define rate-and-state parameters in the Earth. Inertial terms not yet included in models.	Dieterich [1987]; Gomberg *et al.* [1997; 1998]; Belardinelli *et al.* [1999]
Static Coulomb Failure Stress (Viscoelastic)	Mainshock slip model, Maxwell relaxation time, relaxing layer thickness	May explain time delays between mainshock and subsequent events, also irregular recurrence intervals	Needs more testing, also needs more geodetic data to confirm viscoelastic parameters	Dmowska *et al.* [1988]; Roth [1988]; Ghosh *et al.* [1992]; Ben-Zion *et al.* [1993]; Taylor *et al.* [1996]; Pollitz & Sacks [1997]; Deng *et al.* [1998]; Freed & Lin [1998]; Kenner & Segall [1999]; Deng *et al.* [1999], etc.
Fluid Flow	Mainshock slip model, permeability tensor	May explain time delays between mainshock and subsequent events	May not be successful at predicting both the spatial and temporal aftershock pattern	Li *et al.* [1987]; Hudnut *et al.* [1989]; Noir *et al.* [1997], etc.

[1] If the aftershock fault planes are not known, then some authors assume optimally oriented faults; this requires knowledge of the background stress directions.

4.1 Stress Shadows

Similarly, Coulomb stress-change theory has been successfully applied to situations where faults were relaxed, the result of a negative change in Coulomb failure stress, $\Delta CFS < 0$. For cases where a fault is relaxed or put into a stress shadow (Harris and Simpson, 1993, 1996), one can perform simple determinations of the time that it should take for long-term tectonic loading to recover the static-stress change. Once again, the time change, now a delay, is simply expressed as $\Delta CFS/\dot{\tau}$ and is the time required to bring the fault back to its state of stress before it was relaxed.

ΔCFS time-delay calculations were performed by Simpson *et al.* (1988), after it was noted that creepmeters along the right-lateral San Andreas fault moved left-laterally following the nearby 1983 magnitude 6.7 Coalinga earthquake in central California. Simpson *et al.* (1988) estimated that the 1983 Coalinga earthquake delayed the next moderate Parkfield earthquake on the San Andreas fault in central California by at least one year. This technique of using Coulomb stress changes to estimate time delays has also been applied to larger earthquakes. Simpson and Reasenberg (1994) and Jaumé and Sykes (1996) calculated the effects of the great 1906 San Francisco earthquake on nearby active faults. The 1906 earthquake, which ruptured the San Andreas fault in central and northern California, relaxed many of the San Francisco Bay area's faults and delayed subsequent large earthquakes for decades. Simpson and Reasenberg (1994) determined that after 1906, large earthquakes on nearby faults disappeared, then later reappeared at a time consistent with models of long-term tectonic reloading. Simpson and Reasenberg (1994) also made estimates of the effects of the large 1989 Loma Prieta earthquake on nearby San Francisco Bay area faults. Lienkaemper *et al.* (1997) validated these estimates by calculating that a recent resumption of creep on the Hayward fault is consistent with tectonic erosion of the 1989 stress shadow. We now examine the intricacies of the ΔCFS calculations by going back to "first principles," then explore the range of parameters used in such calculations.

4.2 Coulomb Stress Change Defined

Using Coulomb failure assumptions (e.g., Jaeger and Cook, 1969), one can define a Coulomb failure stress, CFS, such that

$$\text{CFS} = |\vec{\tau}| + \mu(\sigma + p) - S \qquad (1)$$

where $|\vec{\tau}|$ is the magnitude of the shear traction on a plane, σ is the normal traction (positive for tension) on the plane, p is the fluid pressure, S is the cohesion, and μ is the coefficient of friction. If one assumes that μ and S are constant over time, then a change in CFS is

$$\Delta\text{CFS} = \Delta|\vec{\tau}| + \mu(\Delta\sigma + \Delta p) \qquad (2)$$

Note that the first term on the right side of this equation implies an isotropic failure plane—a more realistic assumption might require the plane to always fail by slipping in a constant rake direction, in which case the first term becomes $\Delta\tau_{\text{rake}}$ or $\Delta\tau_{\text{slip}}$ (Fig. 1), which is the change in shear stress in the rake direction. The advantage of using changes in stress is that, oftentimes, absolute values of stress are not known, but stress-change values can be calculated fairly readily from information about the geometry and slip direction of an earthquake rupture. (The exact details of geometry and slip also become less important the farther one goes from the rupture.) If a preferred rake direction and a fault-plane orientation are known for a fault that has experienced a stress change, then $\Delta\tau_{\text{rake}}$ can be calculated directly from the stress-change tensor and Eq. (2), and using $\Delta\tau_{\text{rake}}$ or $\Delta\tau_{\text{slip}}$ becomes a preferred method of calculating Coulomb stress changes. If the fault plane is assumed to be isotropic, then calculation of $\Delta|\vec{\tau}|$ requires some knowledge of the preexisting "regional" stress field, although simplifying assumptions can be made (e.g., Oppenheimer *et al.*, 1988; King *et al.*, 1994).

Following Simpson and Reasenberg (1994), approximating the effects of pore-pressure changes on the change in Coulomb failure stress, requires one to make some assumptions. One possible end-member assumption is that the medium is homogeneous and isotropic. Then, for the undrained situation immediately after the static stress changes have occurred, but before fluids have had a chance to flow freely, Rice and Cleary (1976) and Roeloffs (1988, 1996) have shown that

$$\Delta p = -\beta' \Delta\sigma_{\text{kk}}/3 \qquad (3)$$

where β' for rock is similar to Skempton's (1954) coefficient β that was determined for soils and depends on the bulk modulii of the material and the fraction of volume that the fluid occupies (Rice and Cleary, 1976); σ_{kk} is the sum of the diagonal elements of the stress tensor. From Skempton (1954), β theoretically ranges from 0 (dry soil) to 1 (fully saturated soil). β' has been measured to range from 0.47 [Indiana limestone at 30 MPa external stress (Hart, 1994)], to approximately 1.0, with values of 0.7–1.0 often reported (e.g., Berge *et al.*, 1993; Green and Wang, 1995). Wang (1997) has reported laboratory observations suggesting that terms in addition to those appearing in Eq. (3) may be needed.

A more complex assumption (as summarized by Simpson and Reasenberg, 1994) is that the fault-zone materials are more ductile than the surrounding materials, as in the model of Rice (1992), so that $\sigma_{xx} = \sigma_{yy} = \sigma_{zz}$ in the fault zone, and therefore $\Delta\sigma_{\text{kk}}/3 = \Delta\sigma$. In this case, with the assumptions that the medium is homogeneous and isotropic both outside of the fault zone and inside the more ductile fault zone, one can obtain

$$\Delta\text{CFS} = \Delta|\vec{\tau}| + \mu'\Delta\sigma \qquad (4)$$

where $\mu' = \mu(1 - \beta')$.

It has become common in the literature to state the Coulomb failure stress-change equation, i.e., Eq. (4), without detailing the assumptions regarding pore-fluid behavior. The parameter μ' is often called the apparent coefficient of friction and is intended to include the effects of pore fluids as well as the material properties of the fault zone. This strategy is mostly an attempt to cover up our lack of knowledge about the role of pore fluids. Strictly speaking, although a constant μ' can account for instantaneous pore-fluid behavior, in some cases, like the Rice (1992) model, this may not be true in general. Comparing Eq. (4) with Eqs. (2) and (3), for example, it is noted that for this case, μ' is a function of $\Delta\sigma_{\text{kk}}$ and $\Delta\sigma$:

$$\mu' = \mu\left(1 - \frac{\beta'}{3}\frac{\Delta\sigma_{\text{kk}}}{\Delta\sigma}\right) \qquad (5)$$

Although it is convenient to lump our ignorance of pore-fluid behavior into a redefined "apparent coefficient of friction," μ', we run the risk of missing some important clues in interpreting our data. Beeler *et al.* (2001) have deprecated the use of μ' in favor of explicit use of μ and Δp for this reason and the interested reader is referred to that paper for a thorough discussion of the friction coefficient. Nonetheless, in the discussion that follows, we will use μ' in the nonrigorous way that has become commonplace, but caution the reader that it is probably best to stick with Eqs. (2) and (3) when performing stress-change calculations.

A study of the stress changes produced by earthquakes in the Harvard catalog (Kagan, 1994) and earthquakes from the combined Harvard, Preliminary Determination of Epicenters (PDE), and California Institute of Technology/US Geological Survey (CIT/USGS) southern California catalogs (Kagan and Jackson, 1998) suggest $\mu' \approx 0$. Reasenberg and Simpson (1992) showed that $\mu' = 0.2$ best fit the Loma Prieta aftershock data, and Gross and Bürgmann (1998), who used a different technique to estimate μ', also found low values most appropriate. A range of μ', between 0 and 0.6, is preferred by Deng and

Sykes (1997b). These numbers are for different time periods, and for different tectonic settings. The Reasenberg and Simpson (1992) study covered a few years of Loma Prieta aftershocks in the San Francisco Bay area of California, whereas Deng and Sykes (1997b) covered a decade of small earthquakes in southern California. Parsons *et al.* (1999) propose that different faults may be described with different values of μ'. They analyzed the pattern of seismicity following the 1989 Loma Prieta, California, earthquake and inferred a higher value of μ' for the San Francisco Bay area's minor oblique (right-lateral thrust) faults, and a lower value for the major strike-slip faults such as the San Andreas fault. Some have proposed that μ' in the CFS equations may actually change with time, due to migrating pore fluids (e.g., Reasenberg and Simpson, 1992; Harris and Simpson, 1992; Jaumé and Sykes, 1992).

4.3 Fluid Flow

The migrating pore fluids and changing pore-pressure hypothesis explains the physics behind some human-induced seismicity (e.g., Raleigh *et al.*, 1972, 1976; Seeber *et al.*, 1998; Chapter 40 by McGarr *et al.*). This hypothesis has also been used to explain inter-earthquake relations. Nur and Booker (1972) suggested that mainshock-induced pore-pressure changes may control the timing of aftershocks. Similarly, Li *et al.* (1987) attempted to observe a spatial and temporal shift in aftershock locations due to pore-fluid flow, and Hudnut *et al.* (1989) used pore-pressure changes to explain an 11-hour delay between two adjacent earthquakes in 1987 that occurred on conjugate fault planes in southern California. Jaumé and Sykes (1992) discussed the potential effects of pore-fluid flow in the aftermath of the 1992 Landers earthquake, but the implications have not yet been tested. Noir *et al.*'s (1997) study of an earthquake sequence in Central Afar also supports the fluid-flow hypothesis. On the opposite side, some workers have proposed that fluid flow cannot be the sole explanation for aftershock triggering, because aftershocks appear to occur predominantly at the edges of mainshock high-slip regions (Mendoza and Hartzell, 1988). Scholz (1990) also argued that aftershocks would not be occurring simultaneously over the entire mainshock rupture plane if fluid flow alone were responsible.

4.4 Coulomb Stress-Change Magnitude Threshold

It appears that static stress changes as low as 0.01 MPa (0.1 bar) can affect the locations of aftershocks (Reasenberg and Simpson, 1992; King *et al.*, 1994; Hardebeck *et al.*, 1998; Anderson and Johnson, 1999). This value is just a fraction of the stress drop during earthquakes, which is one reason that Coulomb stress changes are said to "enhance" or "encourage" the occurrence of an earthquake, as opposed to generating the earthquake (but also see Gomberg *et al.*, 1997, for another viewpoint). Can static stress changes smaller than 0.01 MPa (0.1 bar) also trigger, or delay, earthquakes? This remains an unresolved issue. In some instances smaller stress changes do appear correlated with patterns of seismicity occurrence (e.g., Harris and Simpson, 1996; Deng and Sykes, 1997b; Nalbant *et al.*, 1998), but there are often not enough events to allow a rigorous statistical test. One exception is the thorough quantitative study by Anderson and Johnson (1999), who examined the 1987 Superstition Hills, California, earthquake sequence. They found no evidence for stress triggering of the aftershocks except when the mainshocks-induced stress changes exceeded 0.1–0.3 bars or the aftershocks occurred from 1.4 to 2.8 years after the mainshocks.

4.5 Coulomb Stress Changes: Static Near-Field Studies

Although the majority of Coulomb failure models assume that one is far enough away from the first earthquake so that one can ignore fine details of its slip distribution [this distance is quantified by Hardebeck *et al.* (1998) for the Landers and Northridge earthquakes], a few papers have ambitiously tackled the near-field topic. These include studies by Beroza and Zoback (1993), Dodge *et al.* (1995, 1996), Nostro *et al.* (1997), Caskey and Wesnousky (1997), Kilb *et al.* (1997), and Perfettini *et al.* (1999). Nostro *et al.* (1997) examined how the static stress changes generated during the first parts of the 1980 Irpinia (Italy) earthquake triggered the subsequent coseismic fault rupture. Caskey and Wesnousky (1997) modeled the detailed surface slip distribution during the 1954 Fairview Peak and Dixie Valley, Nevada, earthquakes. They suggest that the static stress changes induced by the faults that ruptured first may have influenced the amounts of slip on the subsequently ruptured faults, at least at the Earth's surface. Perfettini *et al.* (1999) propose a similar observation for Lake Elsman, California, earthquakes, which preceded the 1989 Loma Prieta rupture.

Crider and Pollard (1998) examined fault interaction in the near field, concentrating on the generic case of interacting normal faults. They used geologic field observations and numerical modeling to determine how, where, and if individual normal faults may grow and link up to become more complex structures. Their findings may assist in the development of "cascade" models for normal-faulting earthquakes, i.e., determining how large a normal fault earthquake can become. The Crider and Pollard results may also prescribe when and where to expect sizable normal-faulting aftershocks.

In a study of the specific case of low-angle normal faults, Axen (1999) found that although the record of mainshocks on these shallowly dipping faults may be sparse, earthquakes do occur on these faults and damaging subevents (and subsequent main events) on the low-angle faults can also be triggered by slip on steeper-dipping normal faults. He cites examples from earthquakes in Nevada, Italy, and Turkey. Axen's (1999) examples include the effects of pore-pressure changes, following the stress change models of Sibson (1986) and Harris and Day (1993).

Taylor *et al.* (1998) modeled ΔCFS generated by three great subduction earthquakes. They examined whether ΔCFS was consistent with the position, timing, and, in contrast to other studies, the mechanisms of aftershocks in the upper plate for each of the great earthquakes. For the most part, ΔCFS appears to have been a plausible explanation, for the simplified case of mainshock slip isolated on a single asperity.

4.6 Coulomb Failure Stress: Volcanic Eruptions and Earthquake Triggering

Just as earthquakes change the stress at nearby points, so does magma transport (Stuart and Johnston, 1975). Therefore, the Coulomb failure hypothesis for earthquake triggering can also be tested by looking at the interaction between volcanic and earthquake activity. In the early 1980s, Thatcher and Savage (1982) conducted such a study. Thatcher and Savage calculated Coulomb stress changes and showed how magma chamber inflation helped trigger large earthquakes on the Izu Peninsula, Japan. In a recent investigation, Nostro *et al.* (1998) examined the historical record of Vesuvius volcanic eruptions and southern Apennine earthquakes in Italy to test if Coulomb failure stress changes could explain the interevent temporal and spatial patterns. They found a suggestive relationship but could not definitively show that Coulomb failure stress explained all of the observations.

4.7 Aseismic Slip Models

In opposition to many of the Coulomb failure stress studies discussed in this paper, some near-field studies of earthquakes have come to different conclusions than their far-field counterparts. Both Beroza and Zoback (1993) and Kilb *et al.* (1997) examined the near-field static stress effect of the 1989 M_w6.9 Loma Prieta mainshock on its nearby aftershocks. Beroza and Zoback (1993) concluded that the calculated (near-field) mainshock static stress changes did not do a better than random job at determining the aftershock mechanisms. One possibility they proposed was that dynamic strains or pore-pressure changes influenced the locations of aftershocks. Kilb *et al.* (1997) examined a similar data set, with the addition of an assumed homogeneous-background stress field to the calculated stress changes. Kilb *et al.* (1997) concluded that it was possible to use a static-stress-change model and simulate the observed aftershock mechanism diversity, but only if the initial stress level in the crust was quite low, lower than would be required to have generated the Loma Prieta mainshock. Therefore, Kilb *et al.* (1997) also concluded that some other factors must explain the aftershock pattern. Their considered options included dynamic stresses, pore-pressure changes, an inhomogeneous background stress field, the driving stresses of aseismic deep slip, and also the possibility that the initial assumptions of fault geometry were not accurate.

In a study of the Upland earthquake sequence (1988–1990) in southern California, Astiz *et al.* (2000) show that calculated Coulomb stress changes generated by the moderate earthquakes do not match the aftershock pattern. Whereas the aftershocks primarily occur on one side of the fault plane, the Coulomb stress pattern for optimally oriented faults predicts symmetry. Similarly, Hardebeck *et al.* (1998) had difficulty matching the pattern of aftershocks following the 1994 Northridge, California, earthquakes with the Coulomb stress pattern inferred for the mainshock.

Dodge *et al.* (1995, 1996) have also suggested that static stress-triggering increments in the near-field do not explain all earthquake interactions. In a departure from the usual studies of large earthquakes triggering smaller ones, Dodge *et al.* (1995, 1996) looked at the relationship between initial smaller events and the subsequent larger earthquake in a sequence. Dodge *et al.* (1995) examined foreshocks of the 1992 Landers, California, earthquake and found that the largest foreshocks probably did not trigger the beginning (an M_W4.4 subevent) of the Landers mainshock. Their calculations showed predominantly zero or negative cumulative Coulomb failure stress at the site of the M_W 4.4 subevent as a result of the occurrence of the largest foreshocks. They used this result to conclude that fault-zone geometry (e.g., Segall and Pollard, 1980; Reasenberg and Ellsworth, 1982), rather than static stress-triggering increments, controlled where the Landers mainshock nucleated, and that the entire Landers foreshock sequence may have been driven by aseismic creep (also see Deng and Sykes, 1997a, for a different viewpoint). That aseismic creep controls the rupture process is the key idea of earthquake-instability models (e.g., Stuart and Tullis, 1995; Wesson and Nicholson, 1988). Dodge *et al.* (1996) extended their previous work (Dodge *et al.*, 1995) by looking at five more foreshock–mainshock sequences in California, in addition to Landers, and found that four out of six (including Landers) of the foreshock–mainshock sequences did not follow a simple Coulomb failure (static) stress-change model of earthquake triggering.

Perfettini *et al.* (1999) arrived at the same result as Dodge *et al.* (1996) for the relationship between the moderate-magnitude foreshocks of 1988 and 1989 and the 1989 Loma Prieta, California, mainshock. The foreshocks do not appear to have pushed the mainshock hypocenter toward failure. To step out of this conundrum, Perfettini *et al.* observed that the maximum mainshock slip did occur in the region of increased foreshock stress change. Therefore Perfettini *et al.* proposed that the foreshocks did influence the mainshock rupture process, although the connection between the foreshocks and mainshock hypocenter did not follow traditional expectations.

4.8 Rate-and-State Friction

Laboratory-derived constitutive equations have also been applied to the study of earthquake interactions. One such set of equations is those termed "rate-and-state," because in the laboratory observations that the equations describe, fault strength is dependent on slip rate and on slip and time history of the fault.

Rate-and-state equations allow for a time delay before the onset of failure. They also describe how some faults may slide stably, whereas others experience "stick-slip" motion. A number of authors have derived or considered rate-and-state friction models (see Scholz,1998, for a review), but for this chapter I summarize the results of only two papers: Dieterich (1994) and Dieterich and Kilgore (1996) have made simple testable predictions of the effects of stress changes on aftershock timing and locations using rate-and-state equations.

Whereas Coulomb friction invokes the simple Eq. (1) to describe the failure limit, rate-and-state formulations use a more complex relation that also describes the evolution of quantities as they approach failure. For example, a simplified, one-state-variable form of rate-and-state friction is (Dieterich and Kilgore, 1996)

$$\tau = \sigma\left[\mu_o + A\ln\left(\frac{\dot{\delta}}{\dot{\delta}*}\right) + B\ln\left(\frac{\theta}{\theta*}\right)\right] \quad (6)$$

where $\dot{\delta}$ is the sliding speed, θ is a state variable that can be interpreted to represent the effects of contact time between the two surfaces of a fault, and μ_0, A, and B are empirically determined coefficients. $A - B > 0$ leads to stable sliding; $B - A > 0$ leads to instability (stick-slip behavior). The asterisked terms are normalizing constants.

In the rate-and-state formalism, an earthquake nucleates on a fault when the sliding speed, $\dot{\delta}$, "runs away" and increases dramatically above interseismic values (to speeds on the order of centimeters per second). Dieterich [1994, equation (A13)] gives the time to failure in terms of the sliding speed and other variables as follows:

$$\mathrm{t} = \frac{A\,\sigma}{\dot{\tau}}\ln\left(1 + \frac{\dot{\tau}}{H\sigma\dot{\delta}}\right) \quad (7)$$

where

$$H = \frac{-k}{\sigma} + \frac{B}{D_c} \quad (8)$$

and t is the time to failure, $\dot{\tau}$ is the long-term stressing rate, k is the effective stiffness for source nucleation, B is a fault constitutive parameter, and D_c is the characteristic sliding distance (see Table 2 for more definitions).

If an external factor, such as a static stress change generated by a nearby earthquake, perturbs the sliding speed, it also changes the eventual nucleation time or time to failure. Following Dieterich (1994), during the long interval of self-driven accelerating slip and before an externally applied stress change, a fault's sliding speed depends on the initial stress and state. After a stress step occurs, the slip speed depends on the new stress and state. Dieterich [1994; equation (A17)] explicitly describes how a stress change perturbs the sliding speed (and thereby, the time to failure) as follows:

$$\frac{\dot{\delta}}{\dot{\delta}_o} = \left(\frac{\sigma}{\sigma_o}\right)^{\alpha/A}\exp\left(\frac{\tau}{A\,\sigma} - \frac{\tau_o}{A\,\sigma_o}\right) \quad (9)$$

TABLE 2 Definitions

τ	Shear stress
τ_o	Initial shear stress before an external stress step
$\Delta\tau$	Shear stress change
$\Delta\tau_{\text{slip}}$	Shear stress change resolved in the slip direction
σ, σ_n	Normal stress*
σ_o	Initial normal stress before an external stress step*
$\Delta\sigma$, $\Delta\sigma_n$	Normal stress change*
S	Cohesion
$\dot{\tau}$	Tectonic loading rate
CFS	Coulomb Failure Stress
ΔCFS	Change in Coulomb Failure Stress
Δt	Change in time to failure due to a change in stress
β	Skempton's coefficient for soils
β'	Skempton-like coefficient for rock
p	Pore pressure
Δp	Change in pore pressure
μ	Coefficient of friction
μ'	Apparent coefficient of friction
μ_o	Empirically determined rate-and-state coefficient
t	Time to failure
D_c	Critical slip distance (meters)
k	Spring stiffness
H	A combination of rate-and-state parameters [Eq. (7)]
B	An empirically determined rate-and-state coefficient
A	An empirically determined rate-and-state coefficient
α	An empirically determined rate-and-state parameter
θ	State variable
$\theta*$	A normalizing constant
$\dot{\delta}$	Slip speed
$\dot{\delta}_o$	Initial slip speed before an external stress step
$\dot{\delta}*$	A normalizing constant

*Note: The Coulomb failure and rate-and-state equations use different sign conventions for the normal stress. In the CFS equations, negative σ indicates compression; whereas in the rate-and-state equations of Dieterich [1994], for example, positive σ indicates compression.

where $\dot{\delta}_o$, τ_o and σ_o are the sliding speed, shear stress, and effective normal stress (which includes pore-pressure effects) on the nearby fault before the stress change, and $\dot{\delta}$, τ, and σ are the sliding speed, shear stress, and effective normal stress (positive for compression) after the stress change. α is a sum of parameters governing normal-stress dependence of the state variables. We note that Eqs. (7) and (9) are approximations to more complicated forms and assume that the fault is at the stage where slip is accelerating so that the slip speed, $\dot{\delta}$, greatly exceeds a steady-state speed, D_c/θ (Dieterich, 1994).

Harris and Simpson (1998) used the preceding equations and examined how a rate-and-state stress shadow should appear, and compared it with a Coulomb failure stress shadow. They applied the shadow calculation to sites in the San Francisco Bay area of California and found that the Dieterich (1994) rate-and-state formulations are consistent with Bay area seismic history, for a very wide range (several orders of magnitude) of the rate-and-state parameters.

Gomberg *et al.* (1997, 1998) have used rate-and-state theory to examine the opposite effect, stress triggering. Gomberg *et al.* (1997) suggest that earthquakes may be triggered by transient, oscillatory loads (also see Scholz, 1998, for an opposing viewpoint) or that these loads may trigger earthquakes that would not have occurred without the stress changes generated by another earthquake. That is, some earthquakes were not only clock-advanced by a previous event, but were actually created. Gomberg *et al.* (1998) study the effect of loading history on triggering, on time scales of both the earthquake cycle (tens or hundreds of years) and of seismic waves (seconds to minutes). They also explicitly compare predictions of the Dieterich rate-and-state frictional model with those of Coulomb shear stress change calculations (assuming constant normal stress).

Dieterich's (1994) equations can also predict aftershock rates. Gross and Kisslinger (1997) solved for some variables in the rate-and-state equations (e.g., for A and σ) by examining the aftershocks of the 1992 Landers earthquake. Gross and Bürgmann (1998) followed a similar tactic and investigated the aftershocks of the 1989 M6.9 Loma Prieta earthquake. Toda *et al.* (1998) used Dieterich's (1994) formulations and investigated the aftershock rates of the 1995 Kobe earthquake, using the data to determine values for some of the rate-and-state parameters, and also to determine earthquake probabilities based on the stress-change calculations.

5. Dynamic Stresses

5.1 Far-Field (Distant) Triggering

In addition to static stress changes, dynamic or transient stress changes may be capable of triggering earthquakes. We have evidence from a few large and great earthquakes that distant triggering has occurred (Lawson, 1908; Hill *et al.*, 1993; Anderson *et al.*, 1994; Gomberg and Bodin, 1994; Gomberg, 1996; Gomberg and Davis, 1996; Steeples and Steeples, 1996; Wen *et al.*, 1996; Gomberg *et al.*, 1998; Papadopoulos, 1998; Singh *et al.*, 1998; etc.). Lomnitz (1996) suggested, from the distances where the triggering is observed, that the triggering is due to deep-seated flow in the earth, in a response to the triggering earthquake. Alternatively, dynamic strains may be the cause of some of the distant triggering, but large regions of the Earth's crust are not responding to these strain changes by producing triggered earthquakes. Parkfield, California, may be the most famous case of this situation. With Parkfield we have one of the world's more expected (or overdue) earthquakes, and yet it has not been triggered by dynamic strains from recent nearby or distant earthquakes (Fletcher and Spudich, 1998; Spudich *et al.*, 1994, 1995). Spudich *et al.* (1995) found that peak dynamic stress of 0.1 MPa (1 bar) after the Landers earthquake did not trigger an expected Parkfield, California, earthquake, although distant triggering due to Landers appears to have occurred in other locations (e.g., Hill *et al.*, 1993; Anderson *et al.*, 1994; Bodin and Gomberg, 1994; Gomberg and Bodin, 1994; Gomberg, 1996; Gomberg and Davis, 1996).

A recent earthquake in the Mojave Desert, which occurred 20–30 km east of the Landers earthquake, may provide more clues about dynamic triggering processes. The October 16, 1999, magnitude 7.1 Hector Mine earthquake produced higher amplitude seismic waves at seismometers to the south of the earthquake than at seismometers to the north of the earthquake, the opposite of the 1992 Landers earthquake, which generated higher amplitude waves to the north. There was also significantly more distant triggered seismicity to the south following Hector Mine and to the north following Landers. Therefore, the dynamic waves generated by each of the earthquakes played a role in determining the subsequent patterns of triggered seismicity (Scientists from US Geological Survey, Southern California Earthquake Center, and California Division of Mines and Geology, 2000).

In addition to earthquakes, nature provides a longer period dynamic-stressing process with which to study earthquake triggering, the Earth's tides. Vidale *et al.* (1998) examined the effects of tidal stresses in earthquake triggering. Through inspection of tidal stresses at the times of $>$13,000 earthquakes, Vidale *et al.* found that earthquakes are randomly distributed throughout the tidal cycle. Therefore, dynamic stress rates as large as 10^{-3} MPa/hour (10^{-2} bar/hour) are not preferentially triggering earthquakes. This is quite significant because tidal stressing rates are much higher than the tectonic loading rates we usually associate with earthquake occurrence. Vidale *et al.*'s findings were corroborated by Lockner and Beeler's (1999) laboratory experiments, which showed that only about 1% of earthquakes should correlate with tidal triggering, and that triggering is controlled by both amplitude and frequency of the stresses.

5.2 Near-Field Triggering

The aforementioned studies were for the far-field, where dynamic stress changes greatly exceed static stress changes, but in the near-field, both may be important (e.g., Rybicki *et al.*, 1985). In the near-field, inter-earthquake dynamic effects (stress waves) may be determining the size of the mainshock itself (Harris *et al.*, 1991; Harris and Day, 1993, 1999; Kase and Kuge, 1998; Belardinelli *et al.*, 1999; Magistrale *et al.*, 1999). This is observed in situations where one earthquake triggers another event within seconds. When this intra-event triggering occurs along the fault strike or even close by, the result is a much larger earthquake or a multiple event. The resulting implications for seismic hazard cannot be overstated.

Cotton and Coutant (1997) have pointed out that dynamic Coulomb failure calculations, which are similar to the commonplace static calculations, except that the normal and shear stresses can vary as a function of time [e.g., Harris and Day (1993)], and static Coulomb failure calculations do not produce the same spatial patterns for unilateral ruptures. Unilateral ruptures are ruptures that preferentially propagate in one direction along the strike or dip, as opposed to bilateral ruptures, which propagate more symmetrically. That is, $\Delta \text{CFS}(t) = \Delta\tau_s(t) + \mu[\Delta\sigma_n(t) + \Delta p(t)]$ for a unilateral rupture does not produce the same pattern as the static stress change calculation $\Delta \text{CFS} = \Delta\tau_s + \mu(\Delta\sigma_n + \Delta p)$. It is not clear from the Cotton and

Coutant (1997) study if the near-field static or dynamic stress changes are more important in triggering aftershocks.

Spontaneous-rupture modeling, where both inertia and the physics of earthquake rupture are included and not just assumed, appears to be the appropriate next step toward understanding near-field dynamic triggering. To date, only a few authors have ventured into the complex arena of multi-fault, spontaneous-rupture, dynamic triggering. Harris *et al.* (1991), Harris and Day (1993), Bouchon and Streiff (1997), Kame and Yamashita (1997), and Kase and Kuge (1998) used two dimensions, and Harris and Day (1999) and Magistrale and Day (1999) used three dimensions, to attempt to model how an earthquake spontaneously propagating on one fault can trigger earthquakes on another, noncoplanar fault. As seen with the Landers earthquake, which nucleated on one fault then successively triggered rupture on four more faults (Wald and Heaton, 1994), near-field dynamic stress changes appear to be a key to modeling multi-fault events.

5.3 Viscoelastic Models

At the other end of the spectrum from dynamic triggering in the near-field are models that include the effects of long-term loading and relaxation of the lithosphere and asthenosphere. Most of the static-stress-change models discussed previously are elastic approximations of the Earth's crust and upper mantle and do not explicitly include long-term viscoelastic behavior. A few authors have added a viscoelastic effect, to better model the complete earthquake cycle, or to account for a long time between the initial event(s) and a "triggered" event (e.g., Freed and Lin, 1998). Dmowska *et al.* (1988) used a 1-D model to look at stress fluctuations during the earthquake cycle in coupled subduction zones, to show when and where large aftershocks and subsequent mainshocks could occur. Taylor *et al.* (1996) used 2-D models of earthquake cycles in a generic subduction zone to illustrate the effect of viscoelastic relaxation in both the mantle and shallow portion of the thrust zone on the timing of seismicity in the outer rise and at intermediate depth. In another tectonic setting, Pollitz and Sacks (1997) used viscoelastic triggering and proposed that the 1995 Kobe, Japan, earthquake was "triggered" by two earthquakes that occurred 50 years before Kobe.

Viscoelastic analyses have also been performed for earthquakes in transform faulting regions. Roth (1988) examined the seismicity in the western part of the North Anatolian fault zone, using both elastic and viscoelastic models. He found good agreement between the spatial and temporal pattern of $M \geq 6$ earthquakes for both types of models. Ben-Zion *et al.* (1993) included viscoelastic loading to show how or if two great earthquakes on the San Andreas fault, 1857 Ft. Tejon and 1906 San Francisco, California, may have modulated the timing of moderate Parkfield, California, earthquakes. Parkfield is located on a section of the San Andreas fault between the two great ruptures. Ben-Zion *et al.* (1993) concluded that the closer great earthquake, 1857, most likely had influenced the timing of moderate Parkfield events, but that 1906 was too distant to greatly impact the timing. Ghosh *et al.* (1992) examined the generic 2-D case of two parallel strike-slip faults in an elastic layer overlying a viscoelastic halfspace. They showed how creep on one or both of the faults would change the potential of earthquakes on neighboring faults. The Ghosh *et al.* study showed an additional time-dependence of the shear stresses that is not present in an elastic analysis. Similarly, Kenner and Segall (1999) propose that the simple elastic models commonly in use do not do justice to the Earth's time-dependent material properties. They examine the case of the 1906 San Francisco earthquake and show how a 2-D model that includes a viscoelastic lower crust might perturb the duration and magnitude of a stress shadow on the Hayward fault from elastic estimates.

There is still a debate about the role of viscoelastic behavior in the earthquake cycle. Some authors (e.g., Bock *et al.*, 1997; Donnellan and Lyzenga 1998) propose that afterslip continues on the fault plane of the earthquake, whereas others (e.g., Deng *et al.*, 1998, 1999) propose that a viscoelastic response in the lower crust better explains postseismic behavior. Which of these mechanisms is actually in effect has implications for the durations and patterns of the calculated stress changes.

6. Stress Changes over Multiple Earthquake Cycles

Many models of static or dynamic stress changes influencing subsequent earthquake occurrence apply to a particular tectonic setting and for one earthquake cycle or less. Some research has been conducted to evaluate longer-term effects, including multiple earthquake cycles. Goes (1996) examined the recurrence of $M \geq 7$ earthquakes in the Middle America Trench, Japan, Alaska, Chile, and along the San Andreas fault. She found a marked aperiodicity in the recurrence times of these earthquakes, which she attributed to a significant amount of stress interaction between fault segments.

Using synthetic seismicity models, and running for hundreds or hundreds of thousands of earthquake cycles, fault interaction has been shown to have a large impact on the timing and location of simulated earthquakes (Yamashita, 1995). Ward (1992) did such a calculation to simulate earthquake statistics in the Middle America Trench, Ward and Goes (1993) looked at a similar problem for the San Andreas fault, and Ward (2000) has examined simulated earthquake histories for the San Francisco Bay area of central California. In an exercise for the Wellington area of New Zealand, Robinson and Benites (1996) found that the simulations could generate clustering of large earthquakes, and sometimes long periods of quiescence, depending on which faults produced the large earthquake first in the sequence. All of the authors found that fault (earthquake) interactions significantly changed the timing (and locations) of the simulated events from those that would have occurred without the effect of stress interactions.

6.1 Critical Earthquake Concept

There are also hypotheses that present ideas about earthquake patterns vastly different from simple elastic or viscoelastic fault interaction models. Bowman *et al.* (1998) present and evaluate a model of precursory seismicity that is based on the ideas of statistical physics and a critical point hypothesis (Sornette and Sammis, 1995). They determine that the cumulative strain released by foreshocks in a given region increases as a power-law time to failure before the ultimate "mainshock" event. Bowman *et al.* (1998) propose that the affected length scales are greater than those due to simple elastic interactions between earthquakes. Instead, the mainshock magnitude scales with the size of the regional fault network, which could be hundreds of kilometers or more, for great earthquakes. Therefore, in the critical earthquake concept, seemingly unrelated, far-separated earthquakes are actually making the eventual mainshock possible by smoothing out the stress field at long wavelengths.

7. Implications for Seismic Hazard

7.1 Probability Estimates

Although many of the calculations performed to date enjoy at least moderate credibility within the scientific community, potential benefits to the public may be lost if the results are not converted into societally useful numbers. In an attempt to remedy this, Cornell *et al.* (1993), Stein *et al.* (1997), and Toda *et al.* (1998) have taken their static-stress-change models for nearby faults and converted the results into earthquake probability estimates. Toda *et al.* (1998) performed the calculations for Japan in the wake of the 1995 Kobe earthquake.

The August 17, 1999, magnitude 7.4 earthquake in Izmit, Turkey (Toksöz, 2002), which killed more than 15,000 people (Barka, 1999; Toksöz *et al.*, 1999), may be the first widely publicized case where preearthquake probability estimates had included stress-change calculations. Stein *et al.* (1997) incorporated the stressing history of previous known earthquakes on the North Anatolian fault and an assumed deep slip (fault loading) rate into rate-and-state friction formulations. They then estimated the probability of future earthquake occurrence on a number of stretches of the North Anatolian fault, including a section near Izmit. Shortly after the August 1999 earthquake, a number of scientists claimed success in forecasting the Izmit event using stress-change calculations, including Stein *et al.* (1997) and Nalbant *et al.* (1998).

Following the August Izmit earthquake, a few scientists updated their stress-change calculations and presented their preferred locations for a subsequent large earthquake. Although a magnitude 7.1 earthquake did occur November 12, 1999, at the eastern end of the Izmit rupture, a site anticipated by Barka (1996; 1999), most other forecasts have concentrated on a rupture in the Marmara Sea region, to the west. This future quake is envisioned to put Turkey's population center of Istanbul at high risk (e.g., Barka *et al.*, 1999; Toksoz *et al.*, 1999; Hubert-Ferrari, 2000).

7.2 What We Have Learned

From our extensive, internationally distributed set of inter- and intra-earthquake stress-change calculations, we have been able to glean some understanding of earthquake behavior. We have learned that small, medium, and large aftershocks generally do occur in regions that were "stressed up" by a mainshock, and that damaging earthquakes generally do not occur in stress-shadowed regions that remain relaxed following the stress-change effects of nearby great earthquakes. This latter effect is most obvious in the wake of earthquakes such as the 1906 San Francisco earthquake (Ellsworth *et al.*, 1981; Simpson and Reasenberg, 1994; Jaumé and Sykes, 1996; Harris and Simpson, 1998), the 1857 Ft. Tejon earthquake (Harris and Simpson, 1996; Deng and Sykes, 1997a), and the 1891 Nobi earthquake (Pollitz and Sacks, 1995). All of these great earthquakes shut off subsequent large earthquakes for decades on faults that were relaxed by the great events.

Smith and Van de Lindt (1969) and King *et al.* (1994) have extended the stress-shadow idea to a role in rupture termination. They argue that the extent of earthquake rupture may be controlled by the static-stress effects of previous events. At least in a few cases, it appears that a large earthquake starting in a region of static stress increase may have terminated when it encountered a region of static-stress decrease. This may have been what happened for the M6.7 Big Bear, California, earthquake (King *et al.*, 1994).

7.3 The Magnitude Problem

If we study great earthquakes, we can use simple models such as Coulomb failure to explain where and when no (or few) large aftershocks occurred. After a great or large earthquake, we can also study the large aftershocks that did occur, and use simple models, such as Coulomb failure, or complex models, such as rate-and-state friction, to explain their occurrence. Ideally, we would have a hint beforehand of where and when to expect damaging aftershocks. Instead, what both of these criteria provide is just a range of nucleation sites, but no magnitude information. This is similar to many of the models currently in use—they determine where nucleation could occur, but do not a priori place a limit on the extent of rupture (magnitude), except as a statistical phenomenon (e.g., Dieterich *et al.*, 1994). One type of quasi-static model that does look at the potential maximum magnitude is that of Miller *et al.* (1996). In the Miller *et al.* (1996) hypothesis, the effect of a nearby earthquake reducing the normal stress on a fault is to delay a large earthquake from occurring, whereas small earthquakes could still occur. The effect of a nearby earthquake increasing the normal stress is to advance the time to the next large earthquake. This hypothesis does, however, need more testing to determine whether or not it is applicable to most crustal faults (e.g., see Harris and Simpson, 1998).

The maximum-magnitude question is most likely also a dynamics problem. Once a large aftershock has been nucleated, is its magnitude already predetermined (e.g., Ellsworth and Beroza, 1995), or do geometrical fault features control the rupture extent (e.g., Sibson, 1985; Wesnousky, 1988; Harris and Day, 1993), or is the answer in the along-strike and along-dip strength heterogeneity (e.g., Beroza and Mikumo, 1996; Day *et al.*, 1998)? It is clear that although we have learned much about earthquake patterns, we are just beginning to scratch the surface when it comes to incorporating earthquake physics into our models (Rice and Ben-Zion, 1996).

8. Future Research

The research performed to date still leaves many questions unanswered. Some believe that static-stress calculations have predictive power in estimating where future large earthquakes will occur (e.g., Nalbant *et al.*, 1998). If that is true, then what can we do with the information that an earthquake-producing fault has received an increase in stress? And, even if the stress increase does generate an immediate crustal response, can we determine if it will be aseismic or seismic? These were some of the questions after the 1992 Landers earthquake and they remain as questions today.

An answer may be uncovered using the method of Robinson and Benites (1995, 1996). They ran 200,000 computer simulations to test how the major faults in the Wellington region of New Zealand may interact. Although the first earthquake in the sequence may not be predicted, their study shows how this first event would affect the timing and locations of subsequent large damaging earthquakes. Their 3-D modeling effort could be a benchmark for other studies of tectonically active regions, and a model to test with geologic trenching and geodetic observations.

Acknowledgments

Versions of this manuscript benefited from the helpful reviews of Nick Beeler, Massimo Cocco, Renata Dmowska, Joan Gomberg, Hiroo Kanamori, Art McGarr, Bob Simpson, David Simpson, Bill Stuart, Lynn Sykes, and an anonymous reviewer.

References

Anderson, G., and H. Johnson (1999). A new statistical test for static stress triggering: application to the 1987 Superstition Hills earthquake sequence. *J. Geophys. Res.* **104**, 20153–20168.

Anderson, J. G., J. N. Brune, J. N. Louie, Y. Zeng, M. Savage, G. Yu, Q. Chen, and D. dePolo (1994). Seismicity in the western Great Basin apparently triggered by the Landers, California, earthquake, 28 June 1992. *Bull. Seismol. Soc. Am.* **84**, 863–891.

Astiz, L., P. M. Shearer, and D. C. Agnew (2000). Precise relocations and stress-change calculations for the Upland earthquake sequences in Southern California. *J. Geophys. Res.* **105**, 2937–2853.

Axen, G. J. (1999). Low-angle normal fault earthquakes and triggering. *Geophys. Res. Lett.* **26**, 3693–3696.

Barka, A. (1996). Slip distribution along the North Anatolian fault associated with large earthquakes of the period 1939–1967. *Bull. Seismol. Soc. Am.* **86**, 1238–1254.

Barka, A. (1999). The 17 August 1999 Izmit earthquake. *Science* **285**, 1858–1859.

Beeler, N. M., R. W. Simpson, D. A. Lockner, and S. H. Hickman (2001). Pore-fluid pressure, apparent friction and Coulomb failure. *J. Geophys. Res.* **106**, 30701–30713.

Belardinelli, M. E., M. Cocco, O. Coutant, and F. Cotton (1999). The redistribution of dynamic stress during coseismic ruptures: evidence for fault interaction and earthquake triggering. *J. Geophys. Res.* **104**, 14925–14946.

Bennett, R. A., R. E. Reilinger, W. Rodi, Y. Li, and M. N. Toksöz (1995). Co-seismic fault slip associated with the 1992 Mw6.1 Joshua Tree, California, earthquake: implications for the Joshua Tree-Landers earthquake sequence. *J. Geophys. Res.* **100**, 6443–6461.

Ben-Zion, Y., J. R. Rice, and R. Dmowska (1993). Interaction of the San Andreas fault creeping segment with adjacent great rupture zones and earthquake recurrence at Parkfield. *J. Geophys. Res.* **98**, 2135–2144.

Berge, P. A., H. F. Wang, and B. P. Bonner (1993). Pore pressure buildup coefficient in synthetic and natural sandstones, *Int. J. Rock Mech. Mining Sci. Geomech. Abs.* **30**, 1135–1141.

Beroza, G. C., and T. Mikumo (1996). Short slip duration in the presence of heterogeneous fault properties. *J. Geophys. Res.* **101**, 22449–22460.

Beroza, G. C., and M. D. Zoback (1993). Mechanism diversity of the Loma Prieta aftershocks and the mechanics of mainshock-aftershock interaction. *Science* **259**, 210–213.

Bock, Y., S. Wdowinski, P. Fang, J. Zhang, S. Williams, H. Johnson, J. Behr, J. Genrich, and J. Dean (1997). Southern California permanent GPS geodetic array; continuous measurements of regional crustal deformation between the 1992 Landers and 1994 Northridge earthquakes. *J. Geophys. Res.* **102**, 18013–18033.

Bodin, P., and J. Gomberg (1994). Triggered seismicity and deformation between the Landers, California, and Little Skull mountain, Nevada, earthquakes. *Bull. Seismol. Soc. Am.* **84**, 835–843.

Bouchon, M., and D. Streiff (1997). Propagation of a shear crack on a nonplanar fault; a method of calculation. *Bull. Seismol. Soc. Am.* **87**, 61–66.

Bowman, D. D., G. Ouillon, C. G. Sammis, A. Sornette, and D. Sornette (1998). An observational test of the critical earthquake concept. *J. Geophys. Res.* **103**, 24359–24372.

Bürgmann, R., P. Segall, M. Lisowski, and J. Svarc (1997). Postseismic strain following the 1989 Loma Prieta earthquake from GPS and leveling measurements. *J. Geophys. Res.* **102**, 4933–4955.

Carder, D. S. (1945). Seismic investigations in the Boulder Dam area, 1940–1944, and the influence of reservoir loading on local earthquake activity. *Bull. Seismol. Soc. Am.* **35**, 175–192.

Caskey, S. J., and S. G. Wesnousky (1997). Static stress changes and earthquake triggering during the 1954 Fairview Peak and Dixie Valley earthquakes, central Nevada. *Bull. Seismol. Soc. Am.* **87**, 521–527.

Chester, F. M. (1995). A rheological model for wet crust applied to strike-slip faults. *J. Geophys. Res.* **100**, 13033–13044.

Chinnery, M. A. (1963). The stress changes that accompany strike-slip faulting. *Bull. Seismol. Soc. Am.* **53**, 921–932.

Choy, G. L., and J. R. Bowman (1990). Rupture process of a multiple main shock sequence: analysis of teleseismic, local, and field observations of the Tennant Creek, Australia, earthquakes of January 22, 1988. *J. Geophys. Res.* **95**, 6867–6882.

Cornell, C. A., S.-C. Wu, S. R. Winterstein, J. H. Dieterich, and R. W. Simpson (1993). Seismic hazard induced by mechanically interactive fault segments. *Bull. Seismol. Soc. Am.* **83**, 436–449.

Cotton, F., and O. Coutant (1997). Dynamic stress variations due to shear faulting in a plane-layered medium. *Geophys. J. Int.* **128**, 676–688.

Crider, J. G., and D. D. Pollard (1998). Fault Linkage: 3D mechanical interaction between echelon normal faults. *J. Geophys. Res.* **103**, 24373–24391.

Das, S. (1992). Reactivation of an oceanic fracture by the Macquarie Ridge earthquake of 1989. *Nature* **357**, 150–153.

Das, S., and C. Scholz (1981). Off-fault aftershock clusters caused by shear stress increase? *Bull. Seismol. Soc. Am.* **71**, 1669–1675.

Day, S. M., G. Yu, and D. J. Wald (1998). Dynamic stress changes during earthquake rupture. *Bull. Seismol. Soc. Am.* **88**, 512–522.

Delouis, B., H. Philip, L. Dorbath, and A. Cisternas (1998). Recent crustal deformation in the Antofagasta region (northern Chile) and the subduction process. *Geophys. J. Int.* **132**, 302–338.

Deng, J., and L. R. Sykes (1996). Triggering of 1812 Santa Barbara earthquake by a great San Andreas shock: implications for future seismic hazards in southern California. *Geophys. Res. Lett.* **23**, 1155–1158.

Deng, J., and L. R. Sykes (1997a). Evolution of the stress field in southern California and triggering of moderate-size earthquakes: a 200-year perspective. *J. Geophys. Res.* **102**, 9859–9886.

Deng, J., and L. R. Sykes (1997b). Stress evolution in southern California and triggering of moderate-, small-, and micro-size earthquakes. *J. Geophys. Res.* **102**, 24411–24435.

Deng, J., M. Gurnis, H. Kanamori, and E. Hauksson (1998). Viscoelastic flow in the lower crust after the 1992 Landers, California, earthquake. *Science* **282**, 1689–1692.

Deng, J., K. Hudnut, M. Gurnis, and E. Hauksson (1999). Stress loading from viscous flow in the lower crust and triggering of aftershocks following the 1994 Northridge, California, earthquake. *Geophys. Res. Lett.* **26**, 3209–3212.

Dieterich, J. (1994). A constitutive law for rate of earthquake production and its application to earthquake clustering. *J. Geophys. Res.* **99**, 2601–2618.

Dieterich, J. H. (1987). Nucleation and triggering of earthquake slip: effect of periodic stresses. *Tectonophysics* **144**, 127–139.

Dieterich, J. H., and B. Kilgore (1996). Implications of fault constitutive properties for earthquake prediction, *Proc. National Acad. Sci.* **93**, 3787–3794.

Dmowska, R., J. R. Rice, L. C. Lovison, and D. Josell (1988). Stress transfer and seismic phenomena in coupled subduction zones during the earthquake cycle. *J. Geophys. Res.* **93**, 7869–7884.

Dodge, D. A., G. C. Beroza, and W. L. Ellsworth (1995). Foreshock sequence of the 1992 Landers, California, earthquake and its implications for earthquake nucleation. *J. Geophys. Res.* **100**, 9865–9880.

Dodge, D. A., G. C. Beroza, and W. L. Ellsworth (1996). Detailed observations of California foreshock sequences: implications for the earthquake initiation process. *J. Geophys. Res.* **101**, 22371–22392.

Donnellan, A., and G. A. Lyzenga (1998). GPS observations of fault afterslip and upper crustal deformation following the 1994 Northridge, earthquake. *J. Geophys. Res.* **103**, 21285–21298.

Du, Y., and A. Aydin (1993). Stress transfer during three sequential moderate earthquakes along the central Calaveras fault, California. *J. Geophys. Res.* **98**, 9947–9962.

Ellsworth, W. L., and G. C. Beroza (1995). Seismic evidence for a seismic nucleation phase. *Science* **268**, 851–855.

Ellsworth, W. L., A. G. Lindh, W. H. Prescott, and D. G. Herd (1981). The 1906 San Francisco earthquake and the seismic cycle. In: "Earthquake Prediction—An International Review," *Maurice Ewing Series 4*, pp. 126–140. American Geophysical Union, Washington, D.C.

Fletcher, J. B., and P. Spudich (1998). Rupture characteristics of the three M 4.7 (1992–1994) Parkfield earthquakes. *J. Geophys. Res.* **103**, 835–854.

Fox, C. G., and R. P. Dziak (1999). Internal deformation of the Gorda Plate observed by hydroacoustic monitoring. *J. Geophys. Res.* **104**, 17603–17615.

Freed, A. M., and J. Lin (1998). Time-dependent changes in failure stress following thrust earthquakes. *J. Geophys. Res.* **103**, 24393–24409.

Ghosh, U., A. Mukhopadhyay, and S. Sen (1992). On two interacting creeping vertical surface-breaking strike-slip faults in a two-layer model of the lithosphere. *Phys. Earth Planet. Inter.* **70**, 119–129.

Goes, S. D. B. (1996). Irregular recurrence of large earthquakes: an analysis of historic and paleoseismic catalogs. *J. Geophys. Res.* **101**, 5739–5749.

Gomberg, J. (1996). Stress/strain changes and triggered seismicity following the M_W7.3 Landers, California, earthquake. *J. Geophys. Res.* **101**, 751–764.

Gomberg, J., and P. Bodin (1994). Triggering of the $M_S = 5.4$ Little Skull mountain, Nevada, earthquake with dynamic strain. *Bull. Seismol. Soc. Am.* **84**, 844–853.

Gomberg, J., and S. Davis (1996). Stress/strain changes and triggered seismicity at The Geysers, California. *J. Geophys. Res.* **101**, 733–749.

Gomberg, J., M. L. Blanpied, and N. M. Beeler (1997). Transient triggering of near and distant earthquakes. *Bull. Seismol. Soc. Am.* **87**, 294–309.

Gomberg, J., N. Beeler, M. Blanpied, and P. Bodin (1998). Earthquake triggering by transient and static deformations. *J. Geophys. Res.* **103**, 24411–24426.

Green, D. H., and H. F. Wang (1986). Fluid pressure response to undrained compression in saturated sedimentary rock. *Geophysics* **51**, 948–956.

Gross, S., and R. Bürgmann (1998). The rate and state of background stress estimated from the aftershocks of the 1989 Loma Prieta, California, earthquake. *J. Geophys. Res.* **103**, 4915–4927.

Gross, S., and C. Kisslinger (1997). Estimating tectonic stress rate and state with Landers aftershocks. *J. Geophys. Res.* **102**, 7603–7612.

Hardebeck, J. L., J. J. Nazareth, and E. Hauksson (1998). The static stress change triggering model: constraints from two southern California aftershock sequences. *J. Geophys. Res.* **103**, 24427–24437.

Harris, R. A. (1998). Introduction to special section: stress triggers, stress shadows, and implications for seismic hazard. *J. Geophys. Res.* **103**, 24347–24358.

Harris, R. A., and S. M. Day (1993). Dynamics of fault interaction: parallel strike-slip faults. *J. Geophys. Res.* **98**, 4461–4472.

Harris, R. A., and S. M. Day (1999). Dynamic 3D simulations of earthquakes on en echelon faults. *Geophys. Res. Lett.* **26**, 2089–2092.

Harris, R. A., and R. W. Simpson (1992). Changes in static stress on southern California faults after the 1992 Landers earthquake. *Nature* **360**, 251–254.

Harris, R. A., and R. W. Simpson (1993). In the shadow of 1857; an evaluation of the static stress changes generated by the M8 Ft. Tejon, California, earthquake, *EOS Trans. Am. Geophys. Un.* **74**, Fall Meeting Supplement, 427.

Harris, R. A., and R. W. Simpson (1996). In the shadow of 1857-effect of the great Ft. Tejon earthquake on subsequent earthquakes in southern California. *Geophys. Res. Lett.* **23**, 229–232.

Harris, R. A., and R. W. Simpson (1998). Suppression of large earthquakes by stress shadows—Implications of rate and state friction laws—two examples. *J. Geophys. Res.* **103**, 24439–24451.

Harris, R. A., R. J. Archuleta, and S. M. Day (1991). Fault steps and the dynamic rupture process: 2-d numerical simulations of a spontaneously propagating shear fracture. *Geophys. Res. Lett.* **18**, 893–896.

Harris, R. A., R. W. Simpson, and P. A. Reasenberg (1995). Influence of static stress changes on earthquake locations in southern California. *Nature* **375**, 221–224.

Hart, D. J. (1994). Laboratory measurements of a complete set of poroelastic moduli for Berea sandstone and Indiana limestone. M.S. thesis, University of Wisconsin, Madison, 44 pp.

Healy, J. H., W. W. Rubey, D. T. Griggs, and C. B. Raleigh (1968). The Denver earthquakes. *Science* **161**, 1301–1310.

Henyey, T. (1992). Earthquakes: Californians feel the tension. *Nature* **360**, 206–207.

Hill, D. P. *et al.* (1993). Seismicity remotely triggered by the magnitude 7.3 Landers, California, earthquake. *Science* **260**, 1617–1623.

Hodgkinson, K. M., R. S. Stein, and G. C. P. King (1996). The 1954 Rainbow Mountain-Fairview Peak-Dixie Valley earthquakes: a triggered normal faulting sequence. *J. Geophys. Res.* **101**, 25459–25471.

Hubert-Ferrari, A., A. Barka, E. Jacques, S. S. Nalbant, B. Meyer, R. Armijo, P. Tapponier, and G. C. P. King, Seismic hazard in the Marmara Sea following the 17 August 1999 Izmit earthquake. *Nature* **404**, 269–273.

Hubert, A., G. King, R. Armijo, B. Meyer, and D. Papanastasiou (1996). Fault re-activation, stress interaction, and rupture propagation of the 1981 Corinth earthquake sequence. *Earth and Planet. Sci. Lett.* **142**, 573–585.

Hudnut, K. W., L. Seeber, and J. Pacheco (1989). Cross-fault triggering in the November 1987 Superstition Hills earthquake sequence, southern California. *Geophys. Res. Lett.* **16**, 199–202.

Jacques, E., G. C. P. King, P. Tapponnier, J. C. Ruegg, and I. Manighetti (1996). Seismic activity triggered by stress changes after the 1978 events in the Asal Rift, Djibouti. *Geophys. Res. Lett.* **23**, 2481–2484.

Jaeger, J. C., and N. G. W. Cook (1969). *Fundamentals of Rock Mechanics.* Methuen and Co., London, 513 pp.

Jaumé, S. C., and L. R. Sykes (1992). Change in the state of stress on the southern San Andreas fault resulting from the California earthquake sequence of April to June 1992. *Science* **258**, 1325–1328.

Jaumé, S. C., and L. R. Sykes (1996). Evolution of moderate seismicity in the San Francisco Bay region, 1850 to 1993; seismicity changes related to the occurrence of large and great earthquakes. *J. Geophys. Res.* **101**, 765–789.

Kagan, Y. Y. (1994). Incremental stress and earthquakes. *Geophys. J. Int.* **117**, 345–364.

Kagan, Y. Y., and D. D. Jackson (1998). Spatial aftershock distribution. *J. Geophys. Res.* **103**, 24453–24467.

Kame, N., and T. Yamashita (1997). Dynamic nucleation process of shallow earthquake faulting in a fault zone. *Geophys. J. Int.* **128**, 204–216.

Kase, Y., and K. Kuge (1998). Numerical simulation of spontaneous rupture processes on two non-coplanar faults: the effect of geometry on fault interaction. *Geophys. J. Int.* **135**, 911–922.

Kato, T., K. Rybicki, and K. Kasahara (1987). Mechanical interaction between neighboring active faults—An application to the Altera fault, central Japan. *Tectonophysics* **144**, 181–188.

Kenner, S., and P. Segall (1999). Time-dependence of the stress shadowing effect and its relation to the structure of the lower crust. *Geology* **27**, 119–122.

Kilb, D., M. Ellis, J. Gomberg, and S. Davis (1997). On the origin of diverse aftershock mechanisms following the 1989 Loma Prieta earthquake. *Geophys. J. Int.* **128**, 557–570.

King, G. C. P., R. S. Stein, and J. Lin (1994). Static stress changes and the triggering of earthquakes. *Bull. Seismol. Soc. Am.* **84**, 935–953.

Kisslinger, C. (1976). A review of theories of mechanisms of induced seismicity. *Engin. Geol.* **10**, 85–98.

Kovach, R. L. (1974). Source mechanisms for Wilmington oil field, California, subsidence earthquakes. *Bull. Seismol. Soc. Am.* **64**, 699–711.

Lawson, A. C. (1908). *The California Earthquake of April 18, 1906.* Report of the State Earthquake Investigation Commission, Volume I, Carnegie Inst. of Washington, 451 pp.

Li, V. C., S. H. Seale, and T. Cao (1987). Postseismic stress and pore pressure readjustment and aftershock distributions. *Tectonophysics* **144**, 37–54.

Lienkaemper, J. J., J. S. Galehouse, and R. W. Simpson (1997). Creep response of the Hayward fault to stress changes caused by the Loma Prieta earthquake. *Science* **276**, 2014–2016.

Lockner, D. A., and N. M. Beeler (1999). Premonitory slip and tidal triggering of earthquakes. *J. Geophys. Res.* **104**, 20133–20151.

Lomnitz, C. (1996). Search of a worldwide catalog for earthquakes triggered at intermediate distances. *Bull. Seismol. Soc. Am.* **86**, 293–298.

Magistrale, H., and S. Day (1999). 3D simulations of multi-segment thrust fault rupture. *Geophys. Res. Lett.* **26**, 2093–2096.

Mavko, G. M. (1982). Fault interaction near Hollister, California. *J. Geophys. Res.* **87**, 7807–7816.

Mavko, G. M., S. Schulz, and B. D. Brown (1985). Effects of the 1983 Coalinga, California, earthquake on creep along the San Andreas fault. *Bull. Seismol. Soc. Am.* **75**, 475–489.

McGarr, A., and D. Simpson (1997). A broad look at induced and triggered seismicity. In: "Rockbursts and Seismicity in Mines" (S. Lasocki and S. Gibowicz, Eds.), pp. 385–396. Balkema, Rotterdam.

Mendoza, C., and S. H. Hartzell (1988). Aftershock patterns and main shock faulting. *Bull. Seismol. Soc. Am.* **78**, 1438–1449.

Michael, A. J. (1991). Spatial variations of stress within the 1987 Whittier Narrows, California, aftershock sequence: new techniques and results. *J. Geophys. Res.* **96**, 6303–6319.

Mikumo, T., T. Miyatake, and M. A. Santoyo (1998). Dynamic rupture of asperities and stress change during a sequence of large interplate earthquakes in the Mexican subduction zone. *Bull. Seismol. Soc. Am.* **88**, 686–702.

Mikumo, T., S. K. Singh, and M. A. Santoyo (1999). A possible stress interaction between large thrust and normal faulting earthquakes in the Mexican subduction zone. *Bull. Seismol. Soc. Am.* **89**.

Miller, S. A., A. Nur, and D. L. Olgaard (1996). Earthquakes as a coupled shear stress—high pore pressure dynamical system. *Geophys. Res. Lett.* **23**, 197–200.

Nalbant, S. S., A. A. Barka, and Ö. Alptekin (1996). Failure stress change caused by the 1992 Erzincan earthquake ($M_S = 6.8$). *Geophys. Res. Lett.* **23**, 1561–1564.

Nalbant, S. S., A. Hubert, and G. C. P. King (1998). Stress coupling between earthquakes in northwest Turkey and the North Aegean sea. *J. Geophys. Res.* **103**, 24469–24486.

Noir, J., E. Jacques, S. Békri, P. M. Adler, P. Tapponnier, and G. C. P. King (1997). Fluid flow triggered migration of events in the 1989 Dobi earthquake sequence of central Afar. *Geophys. Res. Lett.* **24**, 2335–2338.

Nostro, C., M. Cocco, and M. E. Belardinelli (1997). Static stress changes in extensional regimes: an application to southern Apennines (Italy). *Bull. Seismol. Soc. Am.* **87**, 234–248.

Nostro, C., R. S. Stein, M. Cocco, M. E. Belardinelli, and W. Marzocchi (1998). Two-way coupling between Vesuvius eruptions and southern Apennine earthquakes (Italy) by elastic stress transfer. *J. Geophys. Res.* **103**, 24487–24504.

Nur, A., and J. R. Booker (1972). Aftershocks caused by pore fluid flow? *Science* **175**, 885–887.

Okada, Y. (1992). Internal deformation due to shear and tensile faults in a half-space. *Bull. Seismol. Soc. Am.* **82**, 1018–1040.

Okada, Y., and K. Kasahara (1990). Earthquake of 1987, off Chiba, central Japan and possible triggering of eastern Tokyo earthquake of 1988. *Tectonophysics* **172**, 351–364.

Oppenheimer, D. H., P. A. Reasenberg, and R. W. Simpson (1988). Fault plane solutions for the 1984 Morgan Hill, California, earthquake sequence: evidence for the state of stress on the Calaveras fault. *J. Geophys. Res.* **93**, 9007–9026.

Oppenheimer, D. H. *et al.* (1993). The Cape Mendocino, California, earthquakes of April 1992: subduction at the triple junction. *Science* **261**, 433–438.

Papadopoulos, G. A. (1998). An unusual earthquake time cluster in the Greek mainland during May–June 1995. *J. Geodynamics* **26**, 261–269.

Parsons, T., R. S. Stein, R. W. Simpson, and P. A. Reasenberg (1999). Stress sensitivity of fault seismicity: a comparison between limited-offset oblique and major strike-slip faults. *J. Geophys. Res.* **104**, 20183–20202.

Perfettini, H., R. S. Stein, R. W. Simpson, and M. Cocco (1999). Stress transfer by the 1988–1989 $M = 5.3$, 5.4 Lake Elsman foreshocks to the Loma Prieta fault: unclamping at the site of peak mainshock slip. *J. Geophys. Res.* **104**, 20169–20182.

Pollitz, F. F., and I. S. Sacks (1995). Consequences of stress changes following the 1891 Nobi earthquake, Japan. *Bull. Seismol. Soc. Am.* **85**, 796–807.

Pollitz, F. F., and I. S. Sacks (1997). The 1995 Kobe, Japan, earthquake: a long-delayed aftershock of the offshore 1944 Tonankai and 1946 Nankaido earthquakes. *Bull. Seismol. Soc. Am.* **87**, 1–10.

Raleigh, C. B., J. H. Healy, and J. D. Bredehoeft (1972). Faulting and crustal stress at Rangely, Colorado. In: "Flow and Fracture of Rocks" (Griggs volume), *Am. Geophys. Un. Geophys. Monograph 16*, pp. 275–284. Washington, D.C.

Raleigh, C. B., J. H. Healy, and J. D. Bredehoeft (1976). An experiment in earthquake control at Rangely, Colorado. *Science* **191**, 1230–1237.

Reasenberg, P., and W. L. Ellsworth (1982). Aftershocks of the Coyote Lake, California, earthquake of August 6, 1979: a detailed study. *J. Geophys. Res.* **87**, 10637–10655.

Reasenberg, P. A., and R. W. Simpson (1992). Response of regional seismicity to the static stress change produced by the Loma Prieta earthquake. *Science* **255**, 1687–1690.

Reasenberg, P. A., and R. W. Simpson (1997). Response of regional seismicity to the static stress change produced by the Loma Prieta earthquake. In: "The Loma Prieta, California, Earthquake of October 17, 1989—Aftershocks and Postseismic Effects" (P. A. Reasenberg, Ed.), pp. D49–D71. *US Geol. Surv. Prof. Paper 1550-D.*

Rice, J. R. (1992). Fault stress states, pore pressure distributions and the weakness of the San Andreas fault. In: "Fault Mechanics and Transport Properties of Rock: A Festschrift in Honor of W.F. Brace" (B. Evans and T.-F. Wong, Eds.), pp. 475–503. Academic Press, San Diego.

Rice, J. R., and Y. Ben-Zion (1996). Slip complexity in earthquake fault models, *Proc. Nat. Acad. Sciences* **93**, 3811–3818.

Rice, J. R., and M. P. Cleary (1976). Some basic stress diffusion solutions for fluid-saturated elastic porous media with compressible constituents. *Rev. Geophys. Space Phys.* **14**, 227–241.

Robinson, R. (1994). Shallow subduction tectonics and fault interaction: the Weber, New Zealand, earthquake sequence of 1990–1992. *J. Geophys. Res.* **99**, 9663–9679.

Robinson, R., and R. Benites (1995). Synthetic seismicity models of multiple interacting faults. *J. Geophys. Res.* **100**, 18229–18238.

Robinson, R., and R. Benites (1996). Synthetic seismicity models for the Wellington region, New Zealand: implications for the temporal distribution of large events. *J. Geophys. Res.* **101**, 27833–27844.

Roeloffs, E. A. (1988). Fault stability changes induced beneath a reservoir with cyclic variations in water level. *J. Geophys. Res.* **93**, 2107–2124.

Roeloffs, E. A. (1996). Poroelastic techniques in the study of earthquakes-related hydrologic phenomena. *Adv. Geophys.* **37**, 135–195.

Roth, F. (1988). Modelling of stress patterns along the western part of the North Anatolian fault zone. *Tectonophysics* **152**, 215–226.

Roy, M., and C. Marone (1996). Earthquake nucleation on model faults with rate-and state-dependent friction: effects of inertia. *J. Geophys. Res.* **101**, 13919–13932.

Rybicki, K. (1971). The elastic residual field of a very long strike-slip fault in the presence of a discontinuity. *Bull. Seismol. Soc. Am.* **61**, 79–82.

Rybicki, K. (1973). Analysis of aftershocks on the basis of dislocation theory. *Phys. Earth Planet. Inter.* **7**, 409–422.

Rybicki, K., T. Kato, and K. Kasahara (1985). Mechanical interaction between neighboring active faults—static and dynamic stress field induced by faulting. *Bull. Earthq. Res. Inst. Univ. Tokyo* **60**, 1–21.

Scholz, C. H. (1990). *The Mechanics of Earthquakes and Faulting.* Cambridge University Press, Cambridge, 439 pp.

Scholz, C. H. (1998). Earthquakes and friction laws. *Nature* **391**, 37–42.

Scientists from the U.S. Geological Survey, Southern California Earthquake Center, and California Division of Mines and Geology (2000).

Preliminary Report on the 16 October 1999 M 7.1 Hector Mine, California, earthquake. *Seismol. Res. Lett.* **71**, 11–23.

Seeber, L., J. G. Armbruster, W.-Y. Kim, C. Scharnberger, and N. Barstow (1998). The 1994 earthquakes in the Cacoosing Valley near Reading, PA: a shallow rupture controlled by preexisting structure and by quarry unloading. *J. Geophys. Res.* **103**, 24505–24521.

Segall, P. (1992). Induced stresses due to fluid extraction from axisymmetric reservoirs. *Pure Appl. Geophys.* **139**, 535–560.

Segall, P., and D. D. Pollard (1980). Mechanics of discontinuous faults. *J. Geophys. Res.* **85**, 4337–4350.

Shen, Z.-K., D. D. Jackson, Y. Feng, M. Cline, M. Kim, P. Fang, and Y. Bock (1994). Postseismic deformation following the 1992 Landers earthquake, California, 28 June 1992. *Bull. Seismol. Soc. Am.* **84**, 780–791.

Sibson, R. H. (1985). Stopping of earthquake ruptures at dilational fault jogs. *Nature* **316**, 248–251.

Sibson, R. H. (1986). Rupture interactions with fault jogs. In: "Earthquake Source Mechanics," Maurice Ewing Series 6, pp. 157–167. *American Geophysical Union*, Washington, D.C.

Sieh, K. E. (1978). Slip along the San Andreas fault associated with the great 1857 earthquake. *Bull. Seismol. Soc. Am.* **68**, 1421–1448.

Sieh, K. E. *et al.* (1993). Near-field investigations of the Landers earthquake sequence, April to July 1992. *Science* **260**, 171–176.

Simpson, R. W., and P. A. Reasenberg (1994). Earthquake-induced static stress changes on central California faults. In: "The Loma Prieta, California, Earthquake of October 17, 1989—Tectonic Processes and Models" (R. W. Simpson, Ed.), pp. F55–F89. *US Geol. Surv. Prof. Paper 1550-F*.

Simpson, R. W., S. S. Schulz, L. D. Dietz, and R. O. Burford (1988). The response of creeping parts of the San Andreas fault to earthquakes on nearby faults: two examples. *Pure Appl. Geophys.* **126**, 665–685.

Singh, S. K., J. G. Anderson, and M. Rodriguez (1998). Triggered seismicity in the Valley of Mexico from major Mexican earthquakes. *Geofis. Int.* **37**, 3–15.

Skempton, A. W. (1954). The pore-pressure coefficients A and B. *Géotechnique* **4**, 143–147.

Smith, S. W., and W. Van de Lindt (1969). Strain adjustments associated with earthquakes in southern California. *Bull. Seismol. Soc. Am.* **59**, 1569–1589.

Spudich, P., L. K. Steck, M. Hellweg, J. B. Fletcher, and L. Baker (1994). Transient stresses at Parkfield, California, produced by the M7.4 Landers earthquake of June 28, 1992: implications for the time-dependence of fault friction. *Ann. Geofis.* **37**, 1807–1822.

Spudich, P., L. K. Steck, M. Hellweg, J. B. Fletcher, and L. Baker (1995). Transient stresses at Parkfield, California, produced by the M7.4 Landers earthquake of June 28, 1992: observations from the UPSAR dense seismograph array. *J. Geophys. Res.* **100**, 675–690.

Steeples, D. W., and D. D. Steeples (1996). Far-field aftershocks of the 1906 earthquake. *Bull. Seismol. Soc. Am.* **86**, 921–924.

Stein, R. S. (1999). The role of stress transfer in earthquake occurrence. *Nature* **402**, 605–609.

Stein, R. S., and M. Lisowski (1983). The 1979 Homestead Valley earthquake sequence, California: control of aftershocks and postseismic deformation. *J. Geophys. Res.* **88**, 6477–6490.

Stein, R. S., G. C. P. King, and J. Lin (1992). Change in failure stress on the southern San Andreas fault system caused by the 1992 magnitude = 7.4 Landers earthquake. *Science* **258**, 1328–1332.

Stein, R. S., G. C. P. King, and J. Lin (1994). Stress triggering of the 1994 M = 6.7 Northridge, California, earthquake by its predecessors. *Science* **265**, 1432–1435.

Stein, R. S., A. A. Barka, and J. H. Dieterich (1997). Progressive failure on the North Anatolian fault since 1939 by earthquake stress triggering. *Geophys. J. Int.* **128**, 594–604.

Stuart, W. D., and M. J. S. Johnston (1975). Intrusive origin of the Matsushiro earthquake swarm. *Geology* **3**, 63–67.

Stuart, W. D., and T. E. Tullis (1995). Fault model for preseismic deformation at Parkfield, California. *J. Geophys. Res.* **100**, 24079–24099.

Taylor, M. A. J., G. Zheng, J. R. Rice, W. D. Stuart, and R. Dmowska (1996). Cyclic stressing and seismicity at strongly coupled subduction zones. *J. Geophys. Res.* **101**, 8363–8381.

Taylor, M. A. J., R. Dmowska, and J. R. Rice (1998). Upper plate stressing and back arc seismicity in the subduction earthquake cycle. *J. Geophys. Res.* **103**, 24523–24542.

Thatcher, W., and J. C. Savage (1982). Triggering of large earthquakes by magma-chamber inflation, Izu Peninsula, Japan. *Geology* **10**, 637–640.

Thatcher, W., G. Marshall, and M. Lisowski (1997). Resolution of fault slip along the 470-km long rupture of the great 1906 earthquake and its implications. *J. Geophys. Res.* **102**, 5353–5367.

Toda, S., R. S. Stein, P. A. Reasenberg, J. H. Dieterich, and A. Yoshida (1998). Stress transferred by the 1995 $M_W = 6.9$ Kobe, Japan, shock: effect on aftershocks and future earthquake probabilities. *J. Geophys. Res.* **103**, 24543–24565.

Toksöz, M. N. (Ed.) (2002). The Izmit, Turkey, earthquake of 17 August 1999. *Bull. Seismol. Soc. Am.* **92**, 1–526.

Toksöz, M. N., R. E. Reilinger, C. G. Doll, A. A. Barka, and N. Yalcin (1999). Izmit (Turkey) earthquake of 17 August 1999: first report. *Seismol. Res. Lett.* **70**, 669–679.

Troise, C., G. De Natale, F. Pingue, S. M. Petrazzuoli (1998). Evidence for static stress interaction among earthquakes in South-Central Apennines (Italy). *Geophys. J. Int.* **134**, 809–817.

Vidale, J., D. Agnew, M. Johnston, and D. Oppenheimer (1998). Absence of earthquake correlation with earth tides; an indication of high preseismic fault stress rate. *J. Geophys. Res.* **103**, 24567–24572.

Wald, D. J., and T. H. Heaton (1994). Spatial and temporal distribution of slip for the 1992 Landers, California, earthquake. *Bull. Seismol. Soc. Am.* **84**, 668–691.

Wang, H. F. (1997). Effects of deviatoric stress on undrained pore pressure response to fault slip. *J. Geophys. Res.* **102**, 17943–17950.

Ward, S. N. (1992). An application of synthetic seismicity calculations in earthquake statistics: the Middle America trench. *J. Geophys. Res.* **97**, 6675–6682.

Ward, S. N. (2000). San Francisco bay area earthquake simulations: a step toward a standard physical earthquake model. *Bull. Seismol. Soc. Am.* **90**, 370–386.

Ward, S. N., and S. D. B. Goes (1993). How regularly do earthquakes recur? A synthetic seismicity model for the San Andreas fault. *Geophys. Res. Lett.* **20**, 2131–2134.

Weertman, J., and J. R. Weertman (1964). *Elastic Dislocation Theory*. MacMillan, New York, 213 pp.

Wen, K.-L., I. A. Beresnev, and S.-N. Cheng (1996). Moderate-magnitude seismicity remotely triggered in the Taiwan region by large earthquakes around the Philippine Sea plate. *Bull. Seismol. Soc. Am.* **86**, 843–847.

Yamashina, K. (1978). Induced earthquakes in the Izu Peninsula by the Izu-Hanto-Oki earthquake of 1974, Japan. *Tectonophysics* **51**, 139–154.

Yamashina, K. (1979). A possible factor which triggers shallow intraplate earthquakes. *Phys. Earth Planet. Inter.* **18**, 153–164.

Yamashita, T. (1995). Simulation of seismicity due to ruptures on noncoplanar interactive faults. *J. Geophys. Res.* **100**, 8339–8350.

Yoshioka, S., and M. Hashimoto (1989a). A quantitative interpretation on possible correlations between intraplate seismic activity and interplate great earthquakes along the Nankai trough. *Phys. Earth Planet. Inter.* **58**, 173–191.

Yoshioka, S., and M. Hashimoto (1989b). The stress field induced from the occurrence of the 1944 Tonankai and the 1946 Nankaido earthquakes, and their relation to impending earthquakes. *Phys. Earth Planet. Inter.* **56**, 349–370.

74

The GSHAP Global Seismic Hazard Map

Domenico Giardini
Swiss Federal Institute of Technology, Zurich, Switzerland
Gottfried Gruenthal
Geo-Forschungs Zentrum, Potsdam, Potsdam, Germany
Kaye Shedlock
US Geological Survey, Golden, Colorado, USA
Peizhen Zhang
China Seismological Bureau, Beijing, China

1. Introduction

One of the most frightening and destructive phenomena of nature is a severe earthquake and its aftereffects. Catastrophic earthquakes account for 60% of worldwide casualties associated with natural disasters. Economic damage from earthquakes is increasing, even in technologically advanced countries with some level of seismic zonation, as shown by the 1989 magnitude 6.9 Loma Prieta, California (more than $ 6 billion), 1994 magnitude 6.7 Northridge, California (more than $ 25 billion), 1995 magnitude 6.8 Kobe, Japan (more than $ 100 billion), and 1999 magnitude 7.4 Turkey (more than 1.5% of the GNP of Turkey) earthquakes.

Earthquakes are the expression of the continuing evolution of the Earth and of the deformation of its crust. Earthquakes occur worldwide, and while the largest events concentrate on plate boundary areas and active plate interiors, large and moderate earthquakes may take place, if rarely, in all continental areas. Vulnerability to disaster is increasing as urbanization and developments occupy more areas that are prone to the effects of significant earthquakes. The growth of megacities in seismically active regions around the world often includes the construction of seismically unsafe buildings and infrastructures, due to an insufficient knowledge of existing seismic hazard. Mitigation of the effects of earthquakes, including the loss of life, property damage, and social and economic disruption, depends on reliable estimates of seismic hazard. National, state, and local governments, decision makers, engineers, planners, emergency response organizations, builders, universities, and the general public require seismic hazard estimates for land use planning, improved building design and construction (including adoption of building construction codes), emergency response preparedness plans, economic forecasts, housing and employment decisions, and many more types of risk mitigation.

The Global Seismic Hazard Assessment Program (GSHAP) was designed to provide a useful global seismic hazard framework and to serve as a resource for national and regional agencies, by coordinating national efforts in multi-national regional-scale projects, by reaching a consensus on the scientific methodologies for seismic hazard evaluation and by ensuring that the most advanced methodologies are available worldwide through technology transfer and educational programs. A list of acronyms used in this chapter is given in Table 1.

The GSHAP was implemented in the 1992–1999 period. All regional activities were completed by 1998 and all regional results and the GSHAP Map of Global Seismic Hazard are published (Giardini and Basham, 1993; Giardini, 1999) and are available freely on the GSHAP homepage at *http://seismo.ethz.ch/GSHAP/*. Earthquake catalogs compiled by the GSHAP are archived on the attached Handbook CD-ROM #2 under the directory of \74Giardini.

This report summarizes the development and achievements of the GSHAP and the compilation of the Global Seismic Hazard Map.

INTERNATIONAL HANDBOOK OF EARTHQUAKE AND ENGINEERING SEISMOLOGY, VOLUME 81B ISBN: 0-12-440658-0

TABLE 1 A List of Acronyms

ADRIA	Adria plate GSHAP test area
AGSO	Australian Geological Survey Organization
CAUCAS	INTAS Ct. 94-1644: Test area for seismic hazard assessment in the Caucasus
CERESIS	Centro Sismologico Regional para la America del Sur
CSB	China Seismological Bureau, Beijing, formerly the State Seismological Bureau
DACH	GSHAP test area covering Germany, Austria, and Switzerland
ETHZ	Swiss Federal Institute of Technology, Zurich
GFZ	Geo-Forschungs Zentrum, Potsdam
IAEE	International Association of Earthquake Engineering
IASPEI	International Association of Seismology and Physics of the Earth Interior
ICSU	International Council for Science
IDNDR	UN International Decade for Natural Disaster Reduction
IGCP	International Geological Correlation Program
ILP	International Lithosphere Program
ING	Istituto Nazionale di Geofisica, Rome
INTAS	International Association for the cooperation with countries of the former Soviet Union
ISC	International Seismological Centre
ISS	International Seismological Service
IUGG	International Union of Geodesy and Geophysics
IUGS	International Union of Geological Sciences
PAIGH	Pan-American Institute of Geography and History
pga	peak ground acceleration
pgv	peak ground velocity
PILOTO	European Union Ct. 94-0103: Pilot project for regional earthquake monitoring and seismic hazard assessment (Euro-Mediterranean and Andean regions)
PSHA	Probabilistic Seismic Hazard Assessment
QSEZ-CIRPAN	European Union CIPA Ct. 94-0238: Quantitative Seismic Zoning of the Circum Pannonian Basin
RELEMR	Reduction of Earthquake Losses in the Eastern Mediterranean Region project
SA	Spectral Acceleration
SHA	Seismic Hazard Assessment
SESAME	UNESCO/IUGS International Geological Correlation Program n. 382: Seismotectonics and Seismic Hazard Assessment of the Mediterranean
UNESCO	UN Educational, Scientific and Cultural Organization
USGS	United States Geological Survey
WMO	World Meteorological Organization
WSSI	World Seismic Safety Initiative

2. The Assessment of Seismic Hazard

Seismic hazard is defined as a probabilistic measure of earthquake ground shaking at a given location. The assessment of seismic hazard is the first step in the evaluation of the seismic risk, obtained by combining the seismic hazard with local site effects (anomalous amplifications tied to soil conditions, local geology, and topography) and with the vulnerability factors (type, value, and age of buildings and infrastructures, population density, land use, date and time of the day). Frequent large earthquakes in remote areas result in high seismic hazard but pose low risk; moderate earthquakes in densely populated areas entail low hazard but high risk.

Early attempts at constructing seismic hazard maps provided estimates of the severity of ground shaking or damage from known or likely earthquakes. These maps were soon improved by including the frequency of occurrence of the shaking levels depicted. Modern seismic hazard assessment (SHA) began in the late 1960s, with the publication of a series of papers describing and applying the probabilistic seismic hazard assessment (PSHA) method. By the mid-1970s, the United States and many other countries established national PSHA programs and began producing national probabilistic seismic hazard (PSH) maps. Not predictors of the occurrence, or recurrence, of specific earthquakes, PSH maps are predictors of likely levels of ground shaking from earthquakes during specific time windows. By the 1990s, half of the countries of the world had produced at least one national seismic hazard map, and the early pioneers had well-developed national programs to update their seismic hazard maps routinely (see Chapter 65 on seismic hazards and risk assessment in engineering practice by Somerville and Moriwaki).

In its simplest form, a probabilistic seismic hazard assessment is a specific solution of the "Total Probability Theorem," as follows:

$$P[G] = \iint P[G|m \text{ and } r]\, f_M(m)\, f_R(r)\, dm\, dr \qquad (1)$$

where P is probability, G is the event of interest, and m and r are independent random variables that influence G. Simply put, the probability that event G occurs is calculated by multiplying the conditional probability of event G given the occurrence of events m and r, by the (independent) probabilities of events m and r, integrated over all values of m and r. For hazard mapping, G represents the exceedance of a specific level of ground motion at a site of interest during an earthquake, m is magnitude, r is distance. So, the probability of strong shaking at a site is dependent on the magnitude and distance of all possible earthquakes in the surrounding area. Because uncertainties in the parameters and modeling techniques may be explicitly incorporated into the analysis, PSH analysis is applicable anywhere, including areas where only rudimentary geological, geophysical, and geotechnical data are available. PSH assessments improve as the quality of the data and methods improve.

Deterministic, or scenario, seismic hazard (DSH) assessments provide relatively detailed maps of the distribution of shaking from the largest possible earthquake, or series of earthquakes, believed likely to occur in a specific region. DSH assessments require that the regional seismicity, geology, and geophysical and geotechnical data be well quantified. The probability of occurrence of the largest possible earthquake, or series of earthquakes, determines the usefulness of DSH assessments (see Chapter 62 on using engineering seismology tools in ground shaking scenarios by Faccioli and Pessina).

Modern seismic hazard assessment programs commonly map ground motion parameters such as maximum intensity, peak ground acceleration (PGA), peak ground velocity (PGV), and several spectral accelerations (SA). Each ground motion mapped corresponds to a portion of the bandwidth of energy radiated from an earthquake. Peak ground acceleration and 0.2–0.5 sec SA correspond to short-period energy that will have the greatest effect on short-period structures (buildings up to about 7 stories tall, the most common building size in the world). Longer-period SA maps (1.0 sec, 2.0 sec, etc.) depict the level of shaking that will have the greatest effect on longer-period structures (10+ story buildings, bridges, etc.). Fifty years is the most commonly chosen exposure window. There are three commonly mapped probability levels of exceedance: 2%, 5%, or 10% (98%, 95%, or 90% chance of non-exceedance, respectively). These probability levels of exceedance are useful concepts in engineering, but not readily understood by non-engineers. In general, the larger the probability of exceedance, the more likely the ground motions will occur. For example, a map of ground motions with a 10% chance of exceedance in 50 years will depict the ground motions from those earthquakes most likely to occur. Because small earthquakes are more likely than large earthquakes, a 10% chance of exceedance in 50 years map will depict the more frequent, smaller ground motions likely during the exposure time of interest. Alternatively, a map of ground motions with a 2% chance of exceedance in 50 years will depict the ground motions from the likely events and from the less likely, which are usually larger, earthquakes.

From their inception, seismic hazard maps have served as critical input to building codes. Historically, maps of PGA values have formed the basis of seismic zone maps that were included in US building codes, including the US Uniform Building Code, which included seismic provisions specifying the horizontal force a building should be able to withstand during an earthquake. The newly adopted International Building Code includes maps of short- and long-period spectral accelerations (0.2 sec SA and 1.0 sec SA).

There are three major elements of modern probabilistic seismic hazard assessment as implemented by GSHAP (Cornell, 1968; McGuire, 1993; Giardini and Basham, 1993): (1) the characterization of seismic sources; (2) the characterization of ground motion attenuation; and (3) the actual calculation of hazard values. Variations in application of each element of seismic hazard assessment lead to differences in the estimated hazard.

2.1 Characterization of Seismic Sources

The first element of seismic hazard assessment, the characterization of seismic sources, involves answering three questions: Where do earthquakes occur? How often do earthquakes occur? How big can we expect these earthquakes to be?

The global evaluation of seismic hazard requires the characterization of the earthquake cycle over recurrence times spanning from 10 to 10^3 years in active tectonic areas and up to 10^3–10^5 years in areas of slow crustal deformation. Approximately 90% of all earthquakes occur along the plate boundaries. Figure 1 shows the plate boundaries and the locations of earthquakes with magnitudes $\geq$5.3 recorded from 1964 to 1999 (see also Color Plate 15 in Chapter 41 by Engdahl and Villasenor). There are three types of plate boundaries: transform faults, subduction

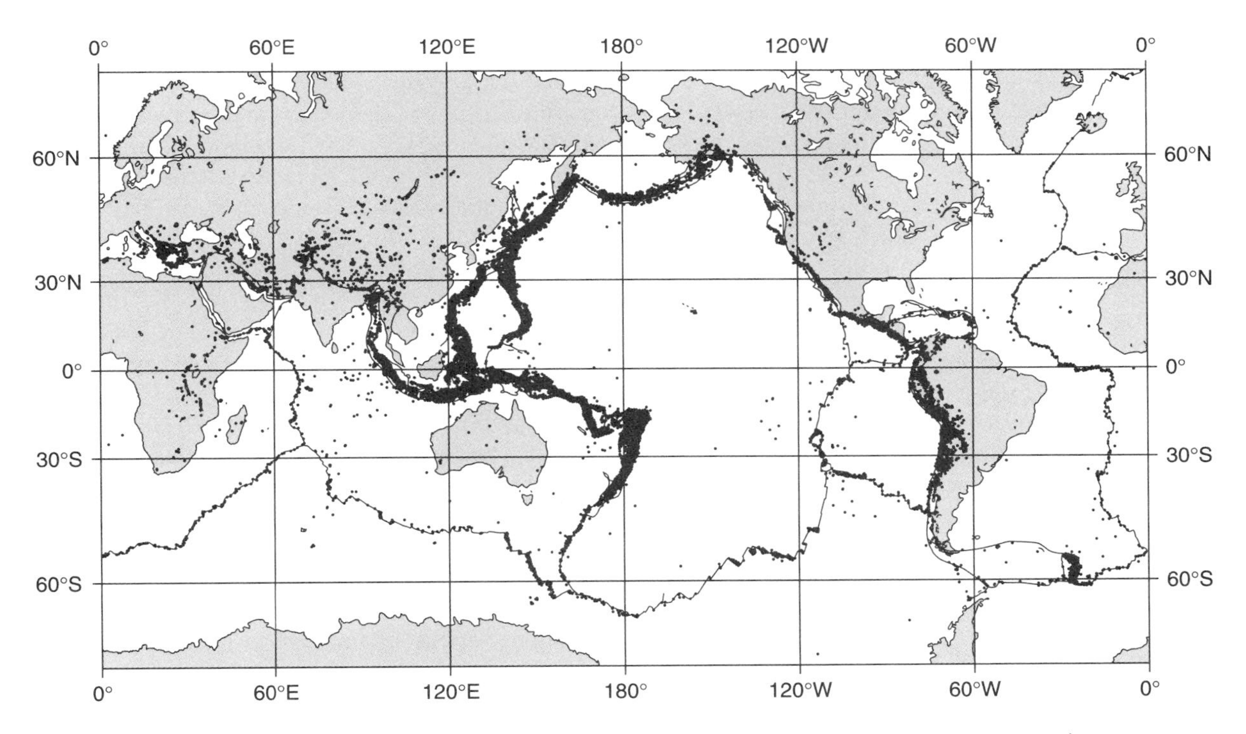

FIGURE 1 Plate boundaries and global seismicity. The plate boundaries are drawn in black. The continents are gray. Earthquakes with magnitudes $\geq$5.3 recorded from 1964 to 1999 are plotted in circles.

zones, and spreading zones. Transform fault boundaries are where plates slide past one another. Transform fault earthquakes tend to be shallow (within the mid-to-upper crust, or less than about 20 km deep) and occur along fairly linear patterns. The San Andreas fault, along the coast of California and northwestern Mexico, is a transform fault plate boundary.

Subduction zones are where one plate overrides, or subducts, another. The overriding plate pushes the subducting plate down into the Earth, where it melts. Once they melt, the lighter rocks in the subducted plate move up through and heat up rocks in the overriding plate, forming active volcanoes. There are multiple seismic sources in subduction zones: intraplate earthquakes within both the subducting and overriding plates, and interplate earthquakes, along the fault surface between the two plates. Most subduction zones occur along the edges of continents, where oceanic crust is being subducted beneath continental crust. Large and great subduction zone earthquakes tend to be deep (tens to hundreds of km) and the subduction zones along coasts are tens to hundreds of km offshore. Thus, energy released in large subduction zone earthquakes has begun to attenuate (weaken) before it reaches onshore population centers. Well-known oceanic-continental subduction zones include those along the west coast of South America and along the east coast of Japan. A notable exception is the India-Eurasia plate boundary. Continental India is colliding with, and subducting beneath, continental Eurasia.

Spreading zones are plate boundaries where two plates are moving apart from each other. Magma (molten rock) rises, adding new material to both plates. Earthquakes in spreading zones are shallow. Most spreading zones are oceanic; for example, Eurasia and North America are moving apart along the mid-Atlantic Ridge. A rare example of an on-shore spreading zone is Iceland, which sits astride the mid-Atlantic Ridge.

The Earth averages about 18 large (magnitude 7.0–7.9) and one great (magnitude 8.0 or above) earthquake among the tens of thousands of earthquakes every year. While large and great earthquakes are a major source of seismic hazard, they are not always the most important source. Shallow, moderate earthquakes (magnitude 5.0–7.0) that occur near population centers can cause considerable shaking and damage. For example, the 1994 Northridge, California, and 1995 Kobe, Japan, earthquakes were shallow, moderate earthquakes.

As a first step in SHA, the locations of all instrumentally recorded earthquakes are collected in seismicity catalogs. These catalogs are the fundamental tool used to determine where, how often, and how big earthquakes are likely to be. However, the instrumental recording of earthquakes is a twentieth-century phenomenon, while the physical processes that drive earthquakes occur on much longer time scales. Seismicity catalogs may be extended hundreds to thousands of years backward in time by including historical and paleoseismic data. Shaking, casualty, and damage reports from historical earthquakes (documented earthquakes that occurred prior to instrumental recording) are analyzed in a variety of ways in order to estimate their locations and magnitudes. Buried ground surfaces, submerged forests, exhumed fault traces and other paleoseismic (ancient) data are mapped, dated, and analyzed in order to estimate the ages and spatial extent of the earthquakes or other tectonic activity that created them. All of these data are combined in extended seismicity catalogs.

Even when the catalogs are extended backward, seismicity statistics are based on geologically short catalogs, so other deformation data and geological evidence are examined (geomorphology, rates of crustal deformation from land and space geodesy, neotectonic and geodynamic modeling) to supplement the historical record of seismicity in building a seismic source model covering earthquake cycles up to a few thousand years. Geodetic monitoring can reveal regional strain accumulation in currently aseismic regions, as well as better constrain deformation rates in seismically active areas. Regional strain accumulation may be spatially interpolated or partitioned to estimate (or better quantify) earthquake magnitudes and recurrence intervals on known faults. Measurements of strain accumulation in aseismic regions may be used to establish upper- and/or lower-bound estimates of possible earthquake magnitudes and recurrence intervals.

The results from seismic monitoring, the historic record, geodetic monitoring, and the geologic record are combined to characterize seismic sources. Although many interpretations of the wide range of input data are possible, only two different earthquake source characterization methods are used for PSH assessments: the delineation of seismic source zones (fault or area) and the historic parametric method.

The delineation of seismic source zones involves specifying the geographical coordinates of an area (polygonal) or fault (linear/planar) source. The hazard is assumed to be uniform within each polygon or along each fault segment and may be described using the parameters of minimum (damage threshold) and maximum magnitude earthquakes and the rate of seismicity, derived from the Gutenberg-Richter (GR) relationship (see Chapter 43 on statistical features of seismicity by Utsu) as follows:

$$\log N = a - bM \tag{2}$$

where M is magnitude, N is the number of earthquakes of magnitude M or greater, and a and b are constants.

The historic parametric method determines seismicity rates (again based on the GR equation) for each point of a grid through the spatial smoothing of historical seismicity. Historic parametric applications commonly supplement these seismicity rates with specific scenario earthquakes and background seismicity source zones.

The multiple methods and wide range of interpretations of the data result in uncertainties associated with source characterization. Various schemes are invoked to either explicitly or implicitly include these uncertainties in seismic hazard calculations. For example, multiple source zone models may be defined. Source zone boundaries may be drawn around historical seismicity clusters (a "historical" model), around geologic and/or tectonic structures, or a combination of both. Hazard calculations from each model may then be combined using various schemes (e.g., logic trees, weights, etc.) that produce a mean (or median) hazard value.

2.2 Characterization of Strong Ground Motion

The second element of an earthquake hazard assessment is estimates of expected ground motion at a given distance from an earthquake of a given magnitude. These estimates are usually equations, called attenuation relationships (see Chapter 60 by Campbell), that express ground motion as a function of magnitude and distance (and occasionally other variables, such as type of faulting). Ground motion attenuation relationships may be determined in two different ways: empirically, using previously recorded ground motions, or theoretically, using seismological models to generate synthetic ground motions that account for the source, site, and path effects. There is overlap in these approaches, however, because empirical approaches fit the data to a functional form suggested by theory and theoretical approaches often use empirical data to determine some parameters.

Earthquake magnitude, style of faulting, source-to-site distance, and local site conditions (site classification) must be clearly defined in order to estimate ground motions. Moment magnitude (M_w) is the preferred magnitude measure, because it is directly related to the total amount of energy released during the earthquake (see Chapter 41 by Engdahl and Villasenor). Style of faulting needs to be specified because, within 100 km of a site, strike-slip earthquakes generate smaller ground motions than reverse and thrust earthquakes, except for $M_w \geq 8.0$. The geometry of the source-to-site distance measures must be clearly specified, because different attenuation relationships have been derived using different geometries. There are also several site classification schemes, ranging from a description of the physical properties of near-surface material to very quantitative characterizations. Seismic hazard maps are calculated for a specific site classification (hard rock, soft rock, stiff soil, soil, soft soil, etc.). Hazard values calculated for rock/stiff soil sites (the most common site classifications) are lower than hazard values calculated for soil sites. Often, hazard values for soil sites may be estimated from the rock/stiff soil site values commonly depicted on hazard maps through multiplication by a specified factor, but these are no more than rough estimates.

2.3 Calculation of Expected Ground Motion Values

The third element of hazard assessment is the actual calculation of expected ground motion values. Once sources are characterized and attenuation functions are selected, the likely ground motion from each possible source (earthquake) is calculated for every point on a grid. Each of these single-source ground motion values has the same probability of occurrence as the earthquake that produces it. This calculation of site-specific ground motion values is performed for every possible source that can affect that site. All of these calculations are turned into an annual frequency of occurrence, and exceedance, of various levels of the ground motion parameter of interest. The final hazard values are determined by summing over the time period of interest.

The most commonly mapped seismic ground motions are accelerations, which are measures of the rate of change of velocity. Prior to an earthquake, objects, the ground, and people are "at rest": our velocities are at or near zero and thanks to the rotational and orbital effects of the Earth, the acceleration of gravity holds us in place. During an earthquake, objects, the ground, and people are suddenly in motion: our velocity increases rapidly from zero. That increase, or acceleration, acts against the acceleration of gravity. Non-anchored objects and people may slide, shake, or, when the acceleration exceeds that of gravity, become airborne (briefly become weightless). Buildings and anchored objects will be shaken, and the larger the acceleration, the more violent the shaking. The acceleration values depicted in seismic hazard maps are directly related to the lateral forces specified in seismic building code provisions (see Chapter 68 by Borcherdt *et al.*). The acceleration of gravity and the accelerations depicted in seismic hazard maps are measured in meters-per-second-per-second, or m/sec^2. Hazard map values are presented in either m/sec^2 or $\%g$, where g is the acceleration of gravity ($\sim$9.78 m/sec^2). A mapped value of $50\%g$ means the acceleration from the earthquake is half that of gravity. A shaking level of about $10\%g$ is the damage threshold for old or non-earthquake-resistant structures close to the location of an earthquake. But the relationship between shaking level and damage is variable, depending on many factors, including the distance from the earthquake, type of building, site classification, and more. In general, however, the higher the shaking level, the greater the damage potential.

3. The GSHAP Implementation

Following the ICSU request to provide scientific input for UN/IDNDR demonstration activities, on August 1991, the ILP initiated the planning and preparation for the GSHAP. The UN/IDNDR Scientific and Technical Committee endorsed the GSHAP as a Decade demonstration project on March 1992. The GSHAP was launched with a Technical Planning Meeting in Rome (June 1992) to focus the consensus of the scientific community on the development of a multi-national and multidisciplinary approach to seismic hazard assessment, and to define the schedule and structure of the program. The first year was devoted to the definition of the regional structure, to the establishment of the program in the international scientific and engineering communities, to the coordination with other UN/IDNDR activities, and to the establishment of a funding strategy. In July 1993 the GSHAP Planning Volume was published (Giardini and Basham, 1993), containing a global compilation of the existing status quo in seismic hazard assessment and the technical guidelines for the GSHAP implementation.

The first implementation phase (1993–95) established multinational test areas for seismic hazard assessment in regions of high seismotectonic significance. After a program evaluation at the IUGG XXI General Assembly in Boulder, Colorado (August 1995), the second implementation phase extended the GSHAP coverage to more test areas and regions, covering most of the world. All regional results were presented and evaluated in August 1997 at the IASPEI General Assembly in Thessaloniki. The final phase (1997–99) focused on completion of the regional hazard assessment, on the compilation of all regional databases and results, on the compilation of the GSHAP map of global seismic hazard and on the dissemination of GSHAP products and materials. The GSHAP map of global seismic hazard and the GSHAP Summary Volume (Giardini, 1999) were published in December 1999.

The assessment of seismic hazard in different areas of the globe is hindered by political boundaries and technical limitations related to the quality and availability of the basic data needed for seismic hazard assessment, which varies greatly around the world (McGuire, 1993). The GSHAP aimed at reducing these limitations, by assessing hazard in a coordinated, multidisciplinary and multi-national approach, and implementing a common seismotectonic probabilistic method on a regional and global scale. To achieve a global dimension, GSHAP established a mosaic of regions under the coordination of regional centers. The goal was to start first to compute the seismic hazard in selected test areas, and then to expand to cover whole continents and finally the globe. This strategy has been maintained in many of the originally established ten regions, while elsewhere the activities focused directly on key test areas under the coordination of large working groups. Some areas, specifically the Mediterranean and the Middle East, have been covered by a mosaic of overlapping projects, while elsewhere (i.e., parts of Africa and of the Western Pacific rim) the hazard mapping was obtained only at the end of the program by using published materials. In specific cases GSHAP allied with existing hazard projects with similar purpose and methodologies, to avoid duplications and to strengthen the across-boundary cooperation (i.e. in the Balkans and Near East). For each region or test area a working group of national experts was established, covering the different fields required for seismic hazard assessment, to produce common regional catalogues and databases and to assess regional hazard.

The following list and the map in Color Plate 29 illustrate the global coverage of GSHAP, separating GSHAP Regions (outlined in black), Test Areas (blue), and Cooperating Projects (green):

GSHAP Regions: *1. Central-North America, 2. South America (CERESIS), 3. Central-Northern Europe, 6. Middle East (Iran), 7. Northern Eurasia, 8. Eastern Asia, 10. South-West Pacific;*

GSHAP Test Areas: Northern Andes (PILOTO), Caucasus (CAUCAS), Adriatic Sea (ADRIA), East African Rift, India-China-Tibet-Myanmar-Bangla Dash, Ibero-Maghreb;

Cooperating Projects: Mexico-C. America-Caribbean-S. America (PAIGH-IDRC), Circum Pannonian Basin (QSEZ-CIRPAN), Eastern Mediterranean (RELEMR, USGS/UNESCO), Mediterranean (SESAME).

The GSHAP implementation and activities were supervised by a Steering Committee listing renowned experts in seismic hazard assessment and earthquake engineering from around the world: H. Gupta (Chair, India), P. Basham (Secretary, Canada), N. Ambraseys (UK), D. BenSari (Morocco), M. Berry (ILP), E. Engdahl (IASPEI), M. Ghafori-Ashtiany (Iran), A. Giesecke (CERESIS), P. Grandori (Italy), A. Green (ILP), D. Mayer-Rosa (Switzerland), R. McGuire (USA), R. Punong-Bayan (Philippines), B. Rouban (UNESCO), G. Sobolev (Russia), G. Suarez (Mexico), and P. Zhang (China).

GSHAP was globally coordinated in the 1992–97 period by ING, Roma, and in the final 1997–99 period by ETH, Zurich.

A key factor in the GSHAP implementation has been the technical workshops (58) organized in 1992–98, to bring together national experts from all the disciplines involved in the assessment of seismic hazard, with up to a hundred and more participants.

Another key element of the GSHAP implementation was the activities and tasks devoted to the improvement of the global practice of seismic hazard assessment. Among these are the digital scanning and processing of the ISS and BCIS Bulletins to produce a uniform global catalog of large earthquakes for this century (see Chapter 41 by Engdahl and Villasenor), the adoption of a common seismotectonic probabilistic approach for global application, and the adoption of common software for the assessment of seismic hazard. In addition, GSHAP maintained close cooperation and ties with international programs and agencies, among these the ILP projects "World Map of Major Active Faults," " Earthquakes of the Late Holocene," and "Earthquake and Megacities Initiative," IASPEI and its commissions, UN organizations (UNESCO, WMO, IUGG, ICSU, IUGS), and programs in earthquake engineering (IAEE, WSSI).

4. The GSHAP Map of Global Seismic Hazard

The GSHAP Global Seismic Hazard Map (Color Plate 30) integrates the results obtained in the regional areas. Four GSHAP regional centers acted as focal points to collect and merge the existing results in large continental areas: USGS, Colorado, for the Americas; GFZ, Potsdam, for Europe-Mediterranean-Africa-Middle East; CSB, Beijing, for Central-Eastern Asia; AGSO, Canberra, for Australia-Western Pacific margin. An editorial committee prepared common technical specifications for the final compilation of the regional reports, the databases, and the hazard maps. Global coordination was provided by ETH, Zurich. The map has been assembled at USGS by D. Giardini, G. Gruenthal, K. Shedlock, and P. Zhang, and published at USGS, Golden, Colorado.

The global map closely follows the maps produced by the GSHAP regions and test areas, and in particular the regional compilations. Some adjustments were required in order to render the map more homogeneous; as a result, significant differences with the regional maps are seen in the Philippines and the Himalayan provinces of India and Iran. The global compilation entailed only the integration of the regional hazard maps and no attempt was made to compile the regional earthquake catalogues and seismic source models in homogeneous global datasets. The GSHAP global map depicts hazard in terms of peak ground acceleration with a 90% chance of non-exceedance in 50 years. Peak ground acceleration (PGA) is a short-period ground motion parameter that is proportional to force and is the most commonly mapped ground motion parameter, because current building codes that include seismic provisions specify the horizontal force a building should be able to withstand during an earthquake; short-period ground motions affect short-period structures (e.g. one-to-two story buildings), which are the largest class of structures in the world. The site classification is rock everywhere except Canada and the United States, which assume rock/firm soil reference ground conditions.

The map colors chosen to delineate the hazard roughly correspond to the actual level of the ground motion associated with the 475-year average return period. The cooler colors therefore represent lower hazard while the warmer colors represent higher hazard. Specifically, white and green correspond to weak ground motion (0–8%g, where g equals the acceleration of gravity); yellow and orange correspond to moderate ground motion (8–24%g); pink and red correspond to strong motion (24–40%g); and dark red and brown correspond to very strong motion (>40%g). Approximately 70% of the Earth's continental landmasses have low hazard (PGA) values, 22% have moderate hazard (PGA) values, 6% have high hazard (PGA) values, and 2% have very high hazard (PGA) values. In general, the largest seismic hazard is found in areas that have been, or are likely to be, the sites of the largest plate boundary earthquakes.

5. Conclusions

The GSHAP Global Seismic Hazard Map is the first reference map for seismic hazard on a global scale, expressing the probability of ground shaking in a parameter of engineering interest (PGA), and the first obtained by the close collaboration of the scientists responsible for national seismic hazard zonations. As a result of their participation in GSHAP, more than half of the countries of the world now have an upgraded and improved hazard map. Also because of GSHAP, the global standards in seismic hazard assessment have markedly improved in the 1990–2000 period, with specific regard to the implementation of multidisciplinary information, the refinement of the databases, and standardization of the assessment of earthquake hazards. National hazard maps have improved in developed countries involved in across-border cooperation (i.e., in Europe) as well as in Third-World countries with no previous experience in SHA (i.e., the African Rift).

GSHAP promoted multi-national cooperation in all continents, with particular emphasis in critical border areas. Some examples are S. Africa worked together with the African Rift framework in a regional scientific program; Russia, Turkey, and Iran cooperated in the Caucasus; China and India cooperated over many years in a sensitive border area; the Andean countries worked together under a unified framework program.

All regional results and the GSHAP Map of Global Seismic Hazard are published (Giardini and Basham, 1993; Giardini, 1999) and are available freely on the GSHAP homepage at *http://seismo.ethz.ch/GSHAP/*. Earthquake catalogs compiled by the GSHAP are archived on the attached Handbook CD-ROM #2 under the directory of \74Giardini.

Acknowledgments

The GSHAP was implemented with the support of many international projects and organizations, national scientific agencies, and research institutions. The primary effort involved more than 500 scientists willing to devote their time, knowledge, and strength to this international endeavour.

Financial support for international activities was provided by ING Roma, ETH Zurich, NATO, ILP, ICSU, UNESCO, IASPEI, the European Council, the European Union, the International Geological Correlation Program, the UN/IDNDR Secretariat, and INTAS. Many national agencies actively supported the participation of their scientists in the GSHAP and the organization of regional activities and meetings. The economic support of ING, Roma, ETH, Zurich, and USGS, Golden Colorado for the publication of the GSHAP Volumes and of the GSHAP Global Map of Seismic Hazard is gratefully acknowledged. Some proprietary software for hazard computation (FRISK88M) was supplied freely to GSHAP by R. McGuire of RiskEngineering Inc.

References

Cornell, C. A. (1968). Engineering seismic risk analysis. *Bull. Seismol. Soc. Am.* **18**, 1583–1606.

Giardini, D. (Ed.) (1999). The Global Seismic Hazard Assessment Program (GSHAP): 1992–1999. *Annali Geofis.* **42**(3-4), 272 pp.

Giardini, D., and P. Basham (Eds.) (1993). The Global Seismic Hazard Assessment Program (GSHAP). *Annali Geofis.* **36**(3-4), 288 pp.

McGuire, R. K. (1993). The practice of earthquake hazard assessment. Special Volume, IASPEI, 284 pp.

75

The Sociological Dimensions of Earthquake Mitigation, Preparedness, Response, and Recovery

Paul W. O'Brien
California State University, Stanislaus, California, USA
Dennis S. Mileti
University of Colorado, Boulder, Colorado, USA

1. History and Development of Societal Earthquake Research

Although disasters have plagued humans almost since the dawn of their emergence on Earth (Genesis 6–8 and Exodus 8–10 treat such disasters as acts of God) and although many disasters (the London fire [1666], the Lisbon earthquake [1755], the Chicago fire [1871], the San Francisco earthquake [1906], to name but four) have been chronicled, no investigation of the social implications of such events occurred until 1920 when Samuel H. Prince submitted a dissertation to Columbia University on the Halifax disaster of 1917 during which more than 2000 people perished as the result of an explosion on a munitions ship (Prince, 1920).

It was not until more than 20 years later, during World War II, that any further research was undertaken on how people respond to disasters of magnitude. At that time the United States Army produced a series of studies (Strategic Bombing Surveys) that looked at the effectiveness of bombing on populations during World War II. During the 1950s, fear of nuclear war lead the US government to fund the Civil Defense Program to construct an early-warning system should the United States be attacked. Much of that program was based on social science research; these early studies on warning systems (Moore *et al.*, 1963; Fritz and Marks, 1954; Fritz and Mathewson, 1956) form the foundation for the modern subfield of disaster research in the social sciences.

During the late 1950s, the focus changed from man-made disasters (e.g., bombing of cities and related civil defense questions) to natural disasters such as earthquakes, fires, floods, hurricanes and tornadoes, mud slides, and so on. Early studies (Fritz and Williams, 1957; Moore *et al.*, 1963; Dynes and Quarantelli, 1968; Quarantelli, 1970) set out to discover the social factors that could describe what happens in a natural disaster.

The disciplines of sociology and geography were at the forefront of these studies, with the stated goals of understanding the societal dynamics before, during, and immediately after the disaster experience, and how rebuilding was accomplished (Fritz and Marks, 1954).

The first assessment of the field was completed by Gilbert White and Eugene Haas at the University of Colorado in the early 1970s (White and Haas, 1975). In their landmark study, they evaluated all disaster-related research up to that point and gave their assessment on research that was still missing in the field. The second assessment, detailed in a new book, *Disasters by Design* (Mileti, 1999), gives an assessment of the field of disaster research in the 25 years after the first assessment. The impact of this second assessment is too early to predict, but it has the possibility of altering the course of future research for decades to come, as with the first assessment. Mileti argues that a holistic approach to disasters—especially mitigation—should be the main focus of research.

Second, under the leadership of its current director, James Lee Witt, the Federal Emergency Management Agency (FEMA), the lead federal agency involved with domestic disasters, is now

INTERNATIONAL HANDBOOK OF EARTHQUAKE AND ENGINEERING SEISMOLOGY, VOLUME 81B ISBN: 0-12-440658-0

also focusing on mitigation and a more proactive involvement in dealing with disasters than was evident in the past. Witt is the first director of FEMA to have direct hands-on, applied experience.

Sociologists and other social scientists view natural disasters as a series of interrelated stages, as the title of this chapter suggests: mitigation, preparedness, response, and recovery and reconstruction. Each stage will be taken up in Section 3.

2. Societal Units of Analysis

Social scientists have identified several "natural" societal groupings (or "units of analysis") by which they can organize their research in the field: the individual, family, organizations, and communities.

Individuals and families are the two main groupings for research investigations on disasters. These two populations are usually at the forefront when one investigates the impacts of such events. Considerable research has been done with populations concerning earthquakes, tornadoes, hurricanes, and other disasters.

Organizations also form a vital link in the data collection on disasters. Organizations such as schools, churches, banks, and retail stores participate in two main ways during and after disasters: (1) they are directly impacted by events, and (2) they play an important role in the postevent process of reconstruction.

Finally communities, as societal groupings, form the third unit of analysis. It is at the community level that one can investigate more macro-level societal forces than with individuals. Some of the forces are culture, political dynamics, and economic dynamics.

3. Current Status of Research

3.1 Mitigation

3.1.1 Background

Mitigation is one of the major avenues available in preparing for possible disasters. Mitigation includes policies and actions taken before an event that are intended to minimize the extent of damage and injury when an event occurs. As a general rule, mitigation does not include emergency response planning or postevent actions. The following (taken from Mileti and Sorensen, 1990: 1–3) are some of the possible mitigation activities available: (1) land-use management, (2) insurance, (3) engineered protection works, (4) construction standards, (5) disaster response plans, and (6) emergency warning systems.

Land-use controls involve restricting the use of land, for example, based on estimates of likely shaking intensities in future earthquakes. In many earthquake-prone areas in California, property must be evaluated by a geologist before construction is permitted in order to determine whether the land conforms to certain standards depending on proposed use.

The major land-use law in California, the Alquist-Priolo Special Studies Zones Act, states in part:

> No structure for human occupancy, identified as a project under Section 2621.6 of the Act, shall be permitted to be placed across the trace of an active fault. Furthermore, as the area within fifty (50) feet of such active faults shall be presumed to be underlain by active branches of that fault unless proven otherwise by an appropriate geologic investigation and report as specified in Section 3603 (d) of this subchapter, no such structures shall be permitted in this area. (Barclay's California Code of Regulations 1990, p. 442)

Earthquake insurance also plays an important role in dealing with the risk of damage. A national earthquake insurance program has been proposed. The program envisioned is similar to the National Flood Insurance Program and would create a federally backed insurance fund that could be drawn upon in the event of an earthquake with catastrophic losses. Earthquake insurance enables earth-quake-induced losses to be distributed across all persons who have purchased insurance rather than born only by those persons involved.

Engineering designs that are resistant to earthquakes have been mandated for new structures in California that perform special functions. Examples of these special structures include the following:

1. Hospitals or other medical facilities having surgery or emergency treatment areas
2. Fire and police stations
3. Municipal government disaster operation and communication centers deemed to be vital in emergencies
4. Buildings where the primary occupancy is for assembly use for more than 300 people

(Office of the State Architect 1986, p. 139)

Others areas of the United States have used a variety of techniques in their mitigation efforts, be it wind damage, coastal erosion, or flooding. Construction standards are another strategy for dealing with hazards. Building codes, for example, are attempts to reduce the amount of injuries, loss of life, and structural damage experienced in earthquakes. Older, precode buildings generally tend to suffer the most damage in an earthquake. Structures built in compliance with newer, improved designs are better able to withstand earthquake-induced shaking.

3.1.2 Synthesis of What Is Known

For decades people, organizations, and government agencies both in the United States and in other countries have attempted to prepare for the effects of hazards. The building of dams and levees for flood protection, irrigation, and drinking water began in the last century. The building of breakwaters as safe harbors also has a long history. In "tornado alley" of the United States, we see the building of storm cellars and other devices to protect human life and livestock. Finally, building codes have been implemented and revised continually as reactions in large part to losses suffered across the country as a result of disastrous events.

Many societal trends are forcing a reevaluation of current disaster response models both in the United States and abroad. One is the increasing growth of the United States population. More people are in danger every year by the sheer growth in real population numbers. Not only is the population growing, but people tend to migrate to the coasts of the country. In the United States, for example, large numbers of people continue to move to hazard-prone areas such as the coastal regions of California, Texas, Florida, and the eastern seaboard. In the United States today, fully 70% of Americans live in these areas, and it is these areas that are most susceptible to earthquakes, tornadoes, hurricanes, and flooding. Thus, losses will continue to escalate across the county without some type of intervention.

3.2 Preparedness

3.2.1 Background

Preparedness is another critical component of behavior under investigation. Designing sustainable hazards mitigation will not eliminate the need for emergency preparedness and response to the destruction and loss of human life related to disasters.

One of the core problems in looking at disaster preparedness is attempting to define exactly what constitutes a disaster. This discussion in the field has grown louder as increasing numbers of researchers from various disciplines investigate a whole host of phenomena. Opinions range from those of Dr. Gilbert White at the University of Colorado, who views disasters as social vulnerability, to those of Dr. Wolf Dombrowsky at the Katastrophenforshungsstelle (Disaster Research Institute) at Christian-Albrechts University in Kiel, Germany, who states that they are the result of human activities.

3.2.2 Synthesis of What Is Known

Socioeconomic factors have a profound impact on household preparedness. For example, households with higher socioeconomic status tend to be better prepared than others. Purchasing insurance, assembling first aid kits, storing food and water, rearranging furniture, and making a household disaster plan are examples of household preparedness activities. But many people take no preparedness action at all, even in hazard-prone areas. Although some of the factors that affect preparedness are known, there is still need for further research into understanding social psychological processes involved in making critical decisions. In other words, researchers know who prepares, but not yet exactly why (Mileti and Fitzpatrick, 1993).

A good deal is known about how public risk information and communication can overcome obstacles to foster significant amounts of household preparedness. Public awareness campaigns designed to improve household preparedness have generally been undertaken in the context of growing awareness about a near-term threat such as the probability of an earthquake on the San Andreas Fault in Parkfield, California. Less is known about what types of incentives will motivate people to increase and sustain preparedness efforts during periods of relative normalcy.

In the Parkfield, California, example, prediction and warning systems have been researched extensively as ways to alert those at risk to an impending disaster so that they can engage in some sort of preevent protective action. Much thought has resulted in exactly what to tell the public in both a pre- and postevent warning situation (Mileti, Fitzpatrick, and Farhar 1990).

As with households, knowledge about organizational preparedness and the factors that encourage it is still quite uneven. Considerably more is known about preparedness among public sector organizations than others, but even that is far from comprehensive.

There is agreement that preparedness among local emergency management agencies in the nation has improved significantly. Over time, more community organizations appear to have become interested, and planning is becoming more integrated. As in the 1960s and 1970s, local emergency management agencies remain diverse in their organization and operations, including different jurisdictions and responsibilities, relationships with other emergency-relevant organizations, and resources available. They are, however, well adapted to their local situations. There is also considerable variation and lack of standardization in preparedness across local emergency management agencies (Drabek, 1994).

Four general factors have been suggested as contributing to successful local emergency management agencies: the existence of persistent hazards; integration of the emergency management office into the day-to-day activities and structure of local government; extensive relationships with other community organizations; and concrete outputs to the community, such as the maintenance of an emergency operations center (Wenger *et al.*, 1986).

In spite of the improvement, problems remain. There is still a tendency to base plans on disaster myths rather than on accurate knowledge, plan in isolation, emphasize command and control (or a top-down organizational pattern), focus on the written products (like plans) rather than the process, succumb to overconfidence based upon successful response to routine emergencies, and accept the low priority of disaster planning as an excuse for inaction. Further, preparedness has been found to be fragmented, rather than integrated across different local organizations and sectors; as a result, planning for disasters tends to be done in isolation.

Police departments, especially smaller ones, tend not to devote much internal energy to disaster planning due to limited resources. When they do plan, police agencies tend to do so in isolation from other community organizations; few have adopted an interorganizational approach to disasters. The police appear to believe that disasters can be handled through the expansion of everyday emergency procedures (Wenger *et al.*, 1989).

Fire departments have improved their preparedness levels and expanded their disaster- and crisis-related tasks beyond firefighting. In particular, they tend to be involved in planning for the provision of emergency medical services. Nevertheless, like police departments, fire departments show a tendency to plan internally (Wenger *et al.*, 1989). Most of what is known about preparedness for the provision of emergency medical service (EMS) providers comes from studies that were conducted during the mid- to late-1970s. Those studies show that, like the fire and police units, the EMS providers tend to plan in isolation and believe that expansion of everyday activities can cover a disaster situation. There is evidence that hospitals and health-care organizations are not prepared to advise people about or treat victims of earthquake-related chemical hazards and disasters.

Three factors have been consistently identified in the research as having positive influence on disaster preparedness among nonemergency organizations. First, larger organizations are more likely to have both greater resources and a more urgent need for strategic planning, and thus a greater concern for disaster preparedness. Second, the level of perceived risk of organizational or department managers is positively correlated with preparedness. Third, the extent to which managers report seeking information about environmental hazards is positively correlated with organizational preparedness (Drabek, 1994).

Until relatively recently, business preparedness was virtually never investigated by researchers. The research that does exist indicates that private firms are less than enthusiastic about disaster preparedness, even in hazard-prone areas. The strongest predictor of preparedness levels among businesses is size, followed by previous disaster experience, and owning rather than leasing the business property (Tierney, 1996).

States possess broad authorities and play a key role in disaster preparedness and response, both supporting local jurisdictions and coordinating with the federal government on a wide range of disaster-related tasks. Federal resources usually cannot be mobilized in a disaster situation without a formal request from the governor, and states have a number of their own resources at their disposal for use in emergencies, including the National Guard. States are required to develop their own disaster plans, and they typically also play a role in training local emergency responders. States have responsibilities for environmental protection and the delivery of emergency medical services, and state emergency management duties have broadened in recent years as a result of legislation like the Superfund Amendments and Reauthorization Act (SARA) Title III, which requires states to coordinate the chemical emergency preparedness activities of Local Emergency Planning Committees (LEPCs).

In the mid-1970s the National Governors' Association (NGA) compiled detailed information from 57 states, commonwealths, and territories on their emergency preparedness and response activities. In the nearly 20 years since the NGA report, there have been only a few scattered studies on preparedness measures undertaken by states, and their scope has been limited. Few comprehensive studies have been done. And what states do undoubtedly makes a difference at the local level. However, without research that takes an in-depth look at what states and localities are actually doing, researchers can conclude little about their role in the preparedness process.

The picture is scarcely better at the national level. Much of the knowledge in hand about federal government preparedness comes from detailed case studies that either focus on the federal government at a particular point in time or assess changes in federal policies and programs that have taken place over time.

In the United States, federal disaster preparedness evolved out of an earlier concern for civil defense. At times, this emphasis made implementing preparedness measures difficult. National-level preparedness initiatives tend also to be shaped by dramatic events. For example, the Three Mile Island nuclear accident stimulated federal action to encourage extensive evacuation planning for areas around nuclear power plants, and the Bhopal disaster was a major factor influencing the content of SARA Title III, which mandated local emergency planning. Similarly, many provisions in the Oil Pollution Act of 1990 were a direct response to the 1989 Exxon Valdez oil spill. The federal response plan had already been developed before Hurricane Andrew in 1992, but the delayed and uncoordinated response to that disaster prompted calls for improved response planning.

One key message in the research literature is that federal preparedness is influenced and constrained not only by institutional power differentials but also by the nature of the intergovernmental system itself—the nature of federalism; the complexity of agencies, responsibilities, and legislation; and the difficulty of effective interagency coordination.

Disaster response activities are taken when a disaster strikes and include emergency sheltering, search and rescue, care of the injured, firefighting, damage assessment, and other emergency measures. Disaster responders must also cope with response-generated demands such as the need for coordination, communication, ongoing situation assessment, and resource mobilization during the emergency period.

3.3 Response

3.3.1 Background

Responses are actions taken following a disaster and can include a variety of actions such as search and rescue, care of the injured, emergency sheltering, damage assessment, firefighting, and crowd control. This response period is a time when emergency management is in control of getting the necessary materials together to handle the crisis. This period of response investigates how communities, citizens, and emergency professionals interact to bring about a successful beginning of reconstruction, once all emergency needs have been met.

3.3.2 Synthesis of What Is Known

The response period has been the most-studied phase of disaster. Conceptual frameworks, research designs, and the items studied range widely across studies, making generalization difficult. Some topics have been studied extensively, while others have received little emphasis. Studies that focus on small groups like households are much more common than others.

Most knowledge about postdisaster sheltering and housing comes from research done in the last 15 years. Little is known about housing patterns across social classes, racial/ethnic groups, and family types. Postdisaster sheltering and housing encompass both physical and social processes, and different phases as the disaster progresses into the recovery period (Quarantelli, 1982a).

Housing patterns after disasters are influenced by predisaster conditions, such as interorganizational mobilization and communication; preimpact community conflict; resource and power differentials; and both victim and community sociodemographic characteristics. Predisaster social ties influence where people go when they flee their homes; those who have small or weak social networks are more likely to use public shelters, while others go to family members and friends. Preexisting social inequities, including differences in income and household resources, home ownership, insurance, and access to affordable housing, also have a significant impact on housing options after disasters (Quarantelli, 1982b).

The literature on US disasters consistently shows that social solidarity remains strong in even the most trying of circumstances. There is no doubt that disasters engender prosocial, altruistic, and adaptive responses during the emergency period immediately after the disaster's impact.

Behavior in disaster situations is adaptive, and considerable continuity exists between pre- and postdisaster behavior patterns. Rather than being dazed and in shock, residents of disaster-stricken areas are active and willing to assist one another. Important response work typically is performed by community residents themselves.

Volunteer activity increases at the time of disaster impact and remains widespread during the emergency period. It can emerge spontaneously, or be institutionalized as part of an organization that takes on volunteers like the Red Cross. In addition, there are so-called "permanent" emergency volunteers, who regularly get involved in response activities.

New groups invariably form (or emerge) during and after disasters, usually in situations characterized by a lack of planning; ambiguity over legitimate authority; exceptionally large disaster tasks (like search and rescue); a legitimizing social setting; a perceived threat; a supportive social climate; and the availability of certain nonmaterial resources. Political and social inequality may also drive emergence.

Emergent citizen groups are typically composed of a small active core, a larger supporting circle, and a still larger number of nominal supporters. They usually have few monetary resources, but volunteer time and commitment are more important than financing. Sometimes emergent groups will transform themselves after the disaster to address more general community needs and persist as an organization (Barton, 1969; Kreps, 1991).

Organizations responding in disaster situations face a number of challenges. They must mobilize; assess the nature of the emergency; prioritize goals, tactics, and resources; coordinate with other organizations and the public; and overcome operational impediments. All of these activities must be accomplished under conditions of uncertainty, urgency, limited control, and limited access to information. In the absence of prior interorganizational and community planning, each affected agency will tend to perform its disaster-related tasks in an autonomous, uncoordinated fashion. One of the challenges of disaster planning and management is to overcome the natural tendency organizations have to maintain their independence and autonomy and encourage them to have a broader interorganizational and communitywide focus such as mitigation (Kreps, 1991).

Early emergency response research pointed to difficulties local agencies had in actually managing the response in disaster situations. That situation evidently has improved over time, although those improvements have been uneven.

Local emergency management agencies are widely varied in their assigned responsibilities; in their relationships with other community organizations; in how they carry out their emergency-related tasks; and in the amount and kinds of crisis-relevant resources under their control. They have many different patterns of organization, ranging from those in which emergency management agencies are weak, isolated, or bypassed during the response period to those that are well-institutionalized and embedded in a communitywide emergency management system.

The use of emergency operations centers (EOCs) in the management of disaster response operations has now become common. In general, they work best when they are adequately staffed and supplied, have management and communications systems, survive the disaster impact, and have clearly delineated functions.

Studies on response seldom focus specifically on these organizations. It is known that changes within police and fire departments are more likely when a disaster is extensive, where resource levels are low, and where there has been little prior planning; that decision making becomes more diffuse during disasters than in nondisaster times; and that problems with communication and convergence are common. Fire departments were found to undergo fewer organizational changes during disasters and in general to have fewer problems in disaster situations than police departments. Both fire and police departments prefer a high degree of autonomy in their everyday operations, and these patterns carry over into disaster situations (Wenger *et al.*, 1989).

As is the case with the other crisis-relevant organizations, there is not a large body of work on how providers of emergency medical services (EMS) perform in disasters. As noted above, postdisaster search and rescue activities are typically performed by non-EMS people. The transportation of disaster victims to hospitals is almost invariably uncoordinated; there is usually an oversupply of EMS resources (especially transportation resources like ambulances) after disaster impact; triage tends to be "informal, sporadic, and partial," and central coordination of emergency care activities is rare (Quarantelli, 1983, p. 76).

Until recently, studies on the response of private-sector organizations in disaster situations were virtually nonexistent, and to date very few systematic studies have been done. Thus, little is known about how private organizations actually respond when faced with disaster-related demands. Existing studies tend to focus on particular types of organizations and rather narrow topics; small and nonrepresentative samples also limit their generalizability or the ability to refer to all private-sector entities.

Media interest in disasters is recent but increasing, driven both by large-scale disasters that became major media events and by a growing interest, particularly in the communications field, in studying disaster reporting.

On the one hand, disasters are framed by news organizations in ways that can be misleading and especially oversimplified. The media can convey erroneous impressions about the magnitude and even the location of disaster damage, as occurred, for example, when San Francisco was characterized as virtually in ruins after the Loma Prieta earthquake, when in fact the city was only selectively damaged. To the extent they perpetuate myths about disaster behavior, the media convey unrealistic impressions about disaster-related needs and problems, potentially leading both the public and decision makers to worry about the wrong things.

That notwithstanding, the media also make a strong positive contribution in disaster situations. By reporting extensively on disasters and the damage they create, the media can help speed up assistance to disaster-stricken areas, and postdisaster reporting can also provide reassurance to people who are concerned about the well-being of their loved ones. Good science reporting can educate the public about hazards, and in-depth stories can help provide the basis for informed hazard-reduction decisions.

Networks of organizations operating during disaster response differ in terms of which organizations are central and how they are structured; the mix of organizations involved; lack of cohesiveness, particularly interorganizational communications; ambiguity of authority; and poor utilization of special resources (Drabek, 1986).

Interorganizational and community response to disasters has been found to be marked by poor coordination among responding organizations and between the public and private sectors. Problems with obtaining and disseminating accurate information are common, and the involvement of extracommunity organizations further complicates response. Situational factors (time of day, day of the week, whether an emergency occurred during daylight or nighttime) are important in either facilitating or hindering response.

The disaster-stricken community has been described as altruistic, therapeutic, consensus-oriented, and adaptive. Early studies documented enhanced community solidarity and morale after disaster impact, suspension of predisaster conflicts, a leveling of status, increased community participation, and shifting in community priorities to emphasize central tasks, such as the protection of human life.

Little of the more recent research conducted on this topic contradicts this, though some exceptions have been found. Disasters can become occasions for organized resistance against established institutional structures and bureaucratic procedures. After the Loma Prieta earthquake, for example, preexisting community conflicts rapidly reemerged. Ethnic solidarity was a major factor in that resistance (Simile, 1995).

Currently, very little is known about state-level disaster response. One study of intergovernmental coordination found great variability in response among different disasters. Response activities analyzed in that study tended to be judged more positively when they were initiated and managed by lower levels of government. Also, both actual and perceived government effectiveness were related to disaster magnitude, the extent to which governmental agencies were prepared, and the public's capacity to cope with disaster impacts (Schneider, 1995).

At the personal and household level, ethnic and minority status, gender, language, socioeconomic status, social attachments and relationships, economic resources, age, and physical capacity all have an impact on the propensity of people to take preparedness actions, evacuate, and take other mitigation measures. In addition, people use a wide variety of decision-making processes, not all rational (Bolin, 1986).

Household preparedness activities are more likely to be undertaken by those who are routinely most attentive to the media (that is, those who are educated, female, and white); are more concerned about other types of social and environmental threats; have personally experienced disaster damage; are responsible for the safety of school-age children; are linked with the community through long-term residence, home ownership, or high levels of social involvement; have received some sort of disaster education; and can afford to take the steps necessary to get prepared.

For organizations, governments, and also people in general, mandates and legal incentives can in some instances induce preparedness, proper response, and other actions. For example, it is unlikely that formal disaster plans would have become almost universal at the local level if they had not been required.

The same factors that influence preparedness at the household level also exist for organizations: Hazards have low salience for most organizations except when there is an imminent threat; disaster-related problems must compete with more pressing concerns; organizations in financial difficulty will downplay preparedness unless it is essential; the resources necessary to deal with it may not be adequate; there is a lack of clear and measurable performance objectives; public and official support is inadequate; and the local emergency management organization lacks expertise.

Three additional factors help influence preparedness and response: the governmental context, disaster experience, and the progress in professionalization of emergency management, discussed below.

The organization, effectiveness, and in particular the diversity of preparedness and response efforts in the United States are in large measure a consequence of the structure of hazard management policy, which is in turn embedded in the intergovernmental system. A number of US studies have pointed out the difficulties inherent in conducting hazard reduction policy in a social context in which responsibility for different aspects of the problem is diffused among governmental levels and agencies and in which both accountability and the ability to implement policy are weak. In addition, the legislative authority is diffuse, and often adopted in the wake of a particularly compelling disaster. There is debate (and often confusion) over the respective roles of governments, and they tend to change over time. Emergency management's own goal has expanded in the last 20 years, beyond the civil preparedness mode that had shaped it for decades.

In the United States, disaster preparedness and response are primarily the responsibility of local governments. However, emergency management is typically not a priority at that level, and local capacity and financial resources are easily overwhelmed. Very little research exists on the ways in which this and other aspects of governmental structure and policies influence preparedness and response activities.

In general, research suggests that prior disaster experience is a major predictor of higher levels of preparedness and more effective response, both for households and for organizations, largely because it leads to greater awareness of the consequences of disasters and the demands that disasters generate.

Recently, there has been a trend toward professionalization of emergency management. A generation ago, the position of local civil defense director was not considered a full-time job, and the skills needed to perform it were not well defined. Since the 1970s, full-time local emergency managers have become more common, and their roles have broadened beyond civil defense and immediate postimpact emergency activities. With the notion of comprehensive emergency management, the role has expanded to include mitigation and recovery.

Today there is wide acceptance of the idea that managing disasters requires specialized knowledge, skills, and training. The process of professionalization has been accompanied by the formation of organizations and associations concerned with the training of and awarding credentials to emergency management specialists, the development of specialized publications, and the spread of professional meetings and training.

For example, FEMA's Emergency Management Institute provides instruction in emergency management for state and local officials, emergency managers, volunteer organization personnel, and practitioners in related fields. Each state emergency management office has a FEMA-funded training officer who coordinates federally funded training programs throughout the state.

More colleges and universities are offering emergency management courses and undergraduate degrees in emergency management. The National Coordinating Council on Emergency Management (NCCEM, now the International Association of Emergency Managers), awarded certification in emergency management to over 600 people between 1993 and 1998.

3.4 Recovery and Reconstruction

3.4.1 Background

Recovery and reconstruction form the last phase in response to a disaster. It is during this period that people, organizations, and government agencies attempt to reconstruct what has been lost by repairing physical aspects of the built environment.

Although recovery can include portions of reconstruction, at the same time it connotes a more human aspect that includes caring for the injured and mourning the loss of friends and family. This process typically begins several days following the event and can take years to complete.

3.4.2 Synthesis of What Is Known

Since the late 1970s, new models of recovery have been developed; long-term disaster impacts and the effects of government policy on community recovery have been examined; and the relationship between recovery and mitigation has been examined. It has been observed that sustainability may hold the key to incorporating mitigation into recovery and reconstruction.

The term *recovery* has been used interchangeably with reconstruction, restoration, rehabilitation, and postdisaster redevelopment. Regardless of the term used, the meaning has historically implied putting a disaster-stricken community back together. Early views of recovery almost exclusively saw recovery as the reconstruction of the damaged physical environment. Its phases went all the way from temporary measures to restore community functions, through replacement of capital stocks to predisaster levels and returning the appearance of the community to normal, to the final phase, which involved promoting future economic growth and development. Researchers discovered that communities try to reestablish themselves in forms similar to predisaster patterns, and that the resulting continuity and familiarity in postdisaster reconstruction enhance psychological recovery.

But other researchers viewed disasters as opportunities to address long-term material problems in local housing and infrastructure. They recast reconstruction into a developmental process of reducing vulnerability and enhancing economic capability (Anderson and Woodrow 1989).

The contemporary perspective is that recovery is not just a physical outcome but a social process that encompasses decision making about restoration and reconstruction activities. This perspective highlights how decisions are made, who is involved in making them, what consequences those decisions have on the community, and who benefits and who does not. The process approach also stresses the nature, components, and activities of related and interacting groups in a systemic process and the fact that different people experience recovery differently. Rather than viewing recovery as linear with "value-added" components, this approach views the recovery process as probabilistic and recursive. This perspective has shifted the emphasis to examining differential group involvement and away from cataloging reconstruction constraints (Berke *et al.*, 1993).

Researchers have demonstrated that there are patterns in the recovery process. Recovery is a set of actions that can be learned about and implemented deliberately (Rubin, 1995). It is possible to anticipate some of the main concerns, problems, and issues during recovery and to plan accordingly. Local participation and initiative must be achieved.

Some of the key deterrents to speedy recovery have been identified through research, namely outside donor programs that exclude local involvement; poorly coordinated and conflicting demands from federal and state agency-assisted programs; staff who are poorly prepared to deal with aid recipients; top-down, inflexible, and standardized approaches; and aid that does not meet the needs of the needy (Berke *et al.*, 1993).

Research on recovery and reconstruction has yielded information about households and families, organizations and businesses, and communities as a whole. Most research has focused on family recovery and has sought to answer questions like: What type of families are most disrupted by disaster? What family types recover most quickly? What things account for different rates of family recovery? Different models of family recovery have been used, and this has led researchers to investigate different things. Most, however, have examined how recovery is affected by the family's socioeconomic status and other demographic characteristics, its position in the life cycle, race or ethnicity, real property losses, employment loss, loss of wage earner(s), the family's capacity (economic reserves, extended family support and assistance), and the use of extrafamilial assistance programs. Past studies also defined recovery differently: how successful victims were in regaining predisaster income levels; the relationship between magnitude of losses and available economic resources; whether families eventually returned to their original damaged home or a comparable one; and whether the victim family believed that it had recovered.

Despite the differences in the methods they used, researchers have agreed that the following factors affect family recovery. Note that many of these parallel the influences on preparedness and response, described later in this chapter.

Linkages to extended family are strengthened immediately after disasters, and this lasts well into recovery. Extended kin groups provide assistance to relatives. Socioeconomic status, race, ethnicity, and even gender are interrelated in complex and different ways. Ethnic and racial minority groups are typically disproportionately poorer and thus disproportionately more vulnerable to disaster and to the negative impacts of long-term recovery. Poorer families have more difficulty recovering from disasters and also have the most trouble acquiring extrafamilial aid.

Rural victims were more likely to use their kin group as a source of emergency shelter than urban families. In rural areas, high-income victims have fewer losses than low-income families; but in urban areas, income seems to make no difference. Rural families are less likely to receive extrafamilial assistance than urban victims.

Businesses have many of the same characteristics as households; for example, they vary in size, financial resources, and age; they are physically housed in structures that are more or less vulnerable; and they differ in the resources they control. Some businesses are obviously less vulnerable to disaster and more capable of recovering than others. Businesses play vital community roles, but research to date has not documented the effect of business closure on family and community recovery. Obviously, the longer businesses are closed, the greater the economic strain on employees' families, and the greater the impact on local government revenues. The character of the community may be altered as people leave to market, shop, bank, and use recreational facilities elsewhere, or if, for example, their children go to schools at a greater distance from their homes.

There are many components of community recovery—residential, commercial, industrial, social, and lifelines—and there are various degrees of recovery. Some aspects of community life, such as tax revenues and property values, may take years to return to normal.

Successful recovery efforts typically include strong local community participation and the integration of the community into regional and national networks. This requires not only strong local governmental capacity but also a cohesive system of public, private, and volunteer groups that are integrated within the community. State and national governments can provide relief and recovery assistance to communities, but with assistance comes increased interaction among officials across government levels and increased state and federal intervention into local decision making. The intergovernmental relations can be facilitated through strong local leaders, local government capacity to take action on its own behalf, and local officials who understand state and federal disaster relief programs (Rubin, 1995).

With each disaster, new knowledge is gained about how to plan more effectively for recovery and reconstruction. But this information and experience has not been systematically collected, nor has it been synthesized into a coherent body of knowledge.

Regardless of the severity of a disaster or the level of assistance from higher levels of government, the primary responsibility for recovery ultimately falls on local governments. Yet little information exists to guide local decisions.

It is possible to educate and train public officials to cope effectively with recovery in their jurisdictions, but planning for recovery has been minimal in the United States. This is changing. For example, FEMA's Emergency Management Institute now offers courses on recovery, and some states now offer recovery training to local officials. Domestic and international case studies do exist, largely due to the efforts of groups like the United Nations Disaster Relief Organization, and the Office of Foreign Disaster Assistance of the US Agency for International Development, both of which have assisted many nations in the task of rebuilding.

After disasters, critical policy choices emerge, forcing unwelcome decisions upon local government about whether to rebuild quickly or safely. Postdisaster recovery and reconstruction planning and management commonly reflect an effort to balance certain ideal objectives with reality. Recovery is characterized by wanting to return rapidly to normal, to increase safety, and to improve the community.

Time is by far the most compelling factor influencing local government recovery decisions, actions, and outcomes. Viewed theoretically, policy decisions might range from major acquisition of hazardous lands at one end to minor modifications in the construction code at the other. Viewed practically, real decisions are likely to be severely limited by economic pressure and pressure to decide quickly. The pressures to restore normalcy in response to victims' needs and desires are so strong that safety and community improvement goals—modifying land use, retrofitting damaged buildings, creating new parks, or widening existing streets—are often compromised or abandoned.

The notion of predisaster planning for postevent recovery is a relatively new and powerful concept. When further developed, tested, and evaluated, such knowledge may help many communities mitigate current hazards before a disaster and recover more quickly and safely afterward. Preevent planning organizes community processes for more timely and efficient postevent action, clarifies key recovery roles and responsibilities, identifies potential financing, minimizes duplicative or conflicting efforts, avoids repetition of other communities' mistakes, and achieves greater public safety and community improvement. Most important, it can help communities "think on their feet" and thus be flexible enough to adapt their postdisaster actions to the actual conditions.

Interest in preevent planning for postdisaster recovery and reconstruction has grown within the past decade. For example, local, state, and private entities in California have developed several training exercises, manuals, and guidelines for earthquake recovery. The City of Los Angeles Emergency Operations Organization's plan was generally helpful in recovering from the Northridge earthquake, although there were some problems in implementation (Spangle Associates, 1997).

To be effective, local recovery plans should have the following components.

- *Community involvement.* In the absence of consensus, recovery can become politicized and foster conflict. The plan must have been fully discussed and agreed to, and accepted by the community before the disaster.
- *Information.* Plans are only as good as the information they contain on (1) the characteristics of the hazards and the areas likely to be affected; (2) the population's size, composition, and distribution; (3) the local economy; (4) the resources likely to be available after disaster; (5) the powers, programs, and responsibilities of local, state, and federal governments; (6) existing land-use patterns and building stock location and characteristics; and (7) local infrastructure.
- *Organization.* Plans should anticipate the need for topic-specific groups or organizations. Having an official rebuilding/restoration organization, for example, in place immediately after a disaster helps ensure that wiser decisions will be made.
- *Procedures.* Following standard political processes for making recovery decisions consumes time, which delays action. Construction procedures, permitting, and code review should be streamlined (but not relaxed) after disasters. If possible, recovery plans should incorporate a vision of hazard reduction after disasters through altered land-use patterns.
- *Damage evaluation.* Plans should be made for the mobilization, deployment, and coordination of building damage inspection; standards for repair and reconstruction should be set in advance; and relocation should be an explicit option.
- *Finances.* Information should be on hand about basic government and private programs for obtaining recovery, relocation, acquisition, and other funding. For example, it can be extremely expensive to alter land-use patterns after disaster, and in major recent cases, relocation occurred only when the funds were provided by the federal government.

Many normal planning procedures are suspended when disaster-stricken communities start to recover and reconstruct themselves. Multiple and potentially conflicting goals are being sought simultaneously but at a faster rate than normal. Extraordinary teamwork is required among various local government departments. Planners must shift from an otherwise slow, deliberative, rule-oriented procedure to one that is more flexible, free-wheeling, and team-oriented. For example, after the 1991 Oakland Hills fire, planners emerged from the early recovery with a different perception of their roles, heightened awareness of the complexities of implementing hazard mitigation within a turbulent postevent milieu, and greater acceptance of team-oriented responsibilities. Planners interviewed after that event revealed a feeling that Oakland had been essentially unprepared for the recovery issues it faced. They thought that a recovery plan formulated before the disaster not only might have reduced the time needed to sort through the various policy decisions that instead had to be dealt with from scratch but also would have helped them anticipate the pressures they encountered (Topping, 1998).

Disaster recovery and reconstruction have been quite aptly described as a set of processes in search of a policy. There are various federal loan and assistance programs to assist long-term recovery, and the provisions of the federal tax code that allow for exempting disaster losses provide a hidden form of longer-term financial assistance. Like many aspects of policymaking for disasters, these programs have been built up over the years in response to specific events and problems; they are not the result of an intentional policy for ensuring speedy and sensible recovery (May, 1985).

The absence of a federal policy for recovery assistance is explained by several factors. In withholding appropriations for such purposes as part of the Disaster Relief Act Amendments of 1974, federal policymakers cited concern over the potential costs of funding recovery programs and about their effectiveness. Some subsequent research failed to find any long-term economic impacts from most disasters, further undermining the case for federal funding for long-term recovery. Perhaps most important, as disaster costs have mounted, federal policymakers have been reluctant to open up new avenues for relief assistance.

Special federal grants or loans after disasters have been documented in most instances as important in hastening recovery. Federal funds for recovery and reconstruction tend to be a small but nonetheless important part of recovery financing. They can cause problems, however, if recovery decisions or actions are delayed in order to tailor them to the requirements of federal programs. It is also important to note that the economic conditions of a city have a lot to do with the pace of recovery.

4. How Knowledge Has Been Applied

The postdisaster period is an opportunity to upgrade the quality of construction to better resist subsequent events and to begin to think through ways of mitigating future damage. Recovery processes can also be used as opportunities to advance programs already in place, such as those addressing urban renewal, traffic bottlenecks, architectural incompatibilities, and nonconforming uses. Since the early 1980s FEMA has required postdisaster mitigation planning. With amendments to the Disaster Act in 1988, the rules were modified to allow a greater percentage of disaster relief monies for funding mitigation programs. This provides an important link between recovery processes and disaster mitigation efforts.

This seemingly sensible approach to recovery and mitigation is not always easy to achieve, however. The findings from studies of the implementation of planning and reconstruction recommendations after the 1964 Alaska earthquake, of land-use planning after other earthquakes, and of local recovery in the early 1980s after several presidentially declared disasters point to a strong bias on the part of decision makers toward maintaining the status quo. These sagas are largely ones of missed opportunities. Time after time, local leaders failed to take advantage of the recovery period to reshape their devastated communities in a way that would improve their ability to withstand future disasters. Part of the explanation lies with the complexities of recovery and the fact that compressed time periods make it difficult to evaluate options systematically. But the lack of clear goals at federal, state, or local levels; the complexity of acting in concert with multiple entities; and the absence of institutional capacity brought about by advance planning all help undermine recovery targeted toward mitigation or sustainability.

In some cases, of course, dramatic changes in the postdisaster character of the area are nearly impossible to implement—as in the case of a damaging earthquake in a heavily urbanized area. The patterns of urban land use and the extensive investment in infrastructure make it impractical to do much besides reconstruct within much the same framework, although perhaps with improved building-by-building construction.

At the same time, there are many instances in which mitigation measures were implemented during recovery. Some land-use changes were put in place after the 1964 Alaska earthquake, for example, leaving an "earthquake park" open space. Several small communities, with significant federal financial assistance, were relocated away from flood-prone areas after the 1993 Midwest floods.

Two important factors bring about postdisaster improvements. One is the existence of a preexisting plan or ongoing process for reshaping the community. This provides a commitment to follow through and the necessary preexisting knowledge of potential options. The second factor is the availability of outside funding to help bring about the desired changes.

Achieving patterns of rebuilding that generally keep people and property out of harm's way is increasingly viewed as an essential element of any disaster recovery program. Rebuilding that fails to acknowledge the location of high hazard areas is not sustainable, nor is housing that is not built to withstand predictable physical forces. Indeed, disasters should be viewed as providing unique opportunities for change—not only for building local capability for recovery, but for long-term sustainable development as well.

Integrating sustainable development with disaster recovery requires some shifts in current thinking, land use, and policy. Some broad guidelines for developing recovery strategies that promote sustainable hazards mitigation are listed below, based partly on principles suggested by Mitchell (1992).

- Sustainable hazards mitigation calls for the adoption of a much longer time frame in recovery decision making. Particularly foolhardy are short-term actions that destroy or undermine natural ecosystems, and that encourage or facilitate long-term growth and development patterns that expose more people and property to hazards.
- Substantial attention must be paid during rebuilding to future potential losses from hazards.
- The principal resources available for recovery linked to sustainability are the people themselves and their local knowledge and expertise. By mobilizing this resource, positive results can be achieved with modest outside assistance.
- Technical assistance and training from outside organizations serve to reinforce and strengthen local organizational capacity, which in turn permits the implementation of recovery, reconstruction, mitigation, and various developmental activities.
- Postdisaster construction and land-use policies must recognize that the extreme natural events that may cause disasters are recurrent events, and thus impose limits on redevelopment.

- Features of the natural environment that serve important mitigation functions, such as wetlands and sand dunes, should taken into account during rebuilding.
- Scientific uncertainty about occurrence, prediction, or vulnerability should not postpone structural strengthening or hazard avoidance during rebuilding.
- Postdisaster reconstruction that does not account for future natural disasters is an inefficient investment of recovery resources.
- Individuals and groups across all sectors should participate in recovery planning.

Four developments stand out from the research and experience of the last two decades. First, effective preparedness and response activities help save lives, reduce injuries, limit property damage, and minimize all sorts of disruptions that disasters cause. As such, preparedness and response are vital to society's ability to survive extreme natural and technological events over the long term. This contribution to disaster resiliency is by far the strongest link between preparedness and response and sustainable hazards mitigation.

Second, the theoretical approach to disaster preparedness and response (and research on it) has changed dramatically in the last 20 years. It has moved from a "functional" view of disasters to one that recognizes the tremendous influence social norms and public perceptions and expectations have on the occurrence, effects of, and recovery from disasters. This broader view, which acknowledges the variability and subjectivity of people's outlooks, will be invaluable as the US population becomes more diverse. There are many details to flesh out to complete the understanding of the varied ways in which different people or groups prepare and respond to disasters. In addition, there is much to learn from research done in other countries and from cross-cultural work.

Third, a great deal has been learned about who prepares for disasters (households with higher socioeconomic status and those of nonminorities), but why they do so is still a mystery. Socioeconomic differences are now recognized to play a large role in determining whether and how people get ready for disasters and react once they have occurred. There is considerable understanding of the way information about hazards and preparedness recommendations can foster appropriate behavior. But knowledge about all of these issues and factors is incomplete.

Finally, it is hoped that over the past 20 years the myth of human dysfunction in the immediate emergency period after disasters has been thoroughly dispelled. In case after case, people have shown that their response to disasters is characterized by altruism, an enhanced sense of community, helpfulness, resourcefulness, and extraordinary resiliency.

There has been a shift in conceptualizing disaster recovery over the last two decades. Recovery was viewed then as a linear phenomenon with stages and end products. These approaches have been augmented by viewing recovery as a process of interaction and decision making among a variety of groups and institutions, including households, organizations, businesses, the broader community, and society. A further shift may be needed—that of giving less attention to restoring damaged structures and more attention to decision-making processes at all levels as they relate to the recovery process.

Research has documented that most local disaster plans need to be extended to address recovery and reconstruction more directly and explicitly. There are many postdisaster opportunities for rebuilding in safer ways and in safer places. Disaster plans could start to identify such opportunities in advance.

A growing number of studies have found that locally based recovery approaches are most effective. The definitive characteristic of the bottom-up approaches is that the principal responsibility and authority for carrying out the program rests with a community-based organization, supplemented as needed by technical and financial assistance from outside the community.

5. Needed Next Steps

A new emphasis is being focused on sustainable hazards management: Planners must go beyond the traditional reaction following an event. New emphasis must be placed on modeling possible problems and failures, and responding to such events proactively. Throughout human history, natural disasters have been viewed as acts of God against which humans were helpless. It has only been in the last half-century that agencies have begun efforts to mitigate the extent of damage suffered during natural disasters. By discouraging construction altogether in areas prone to disasters or by ensuring improved construction and requiring retrofitting of existing structures in such areas, federal, state, and local agencies have learned that massive destruction and societal upheaval need not always occur.

Critical in the area of disaster studies is the need for multidisciplinary research to be conducted. Disasters impact many areas of society, for example, the social fabric, the physical infrastructure, and the underlying causes for the disaster (e.g., an earthquake or flood). What is needed now and into the foreseeable future is the ability of teams of disaster researchers across disciplines to investigate problems from their various discipline-specific perspectives. Only in this way can a rational wide-ranging problem-solving methodology be successful. One of the primary tasks today in the disaster community is to design more disaster-resilient communities. The keyword from many sectors of the field is mitigation. In looking at past events throughout this century in the United States and beyond, one realizes that the costs of disasters are on a steady increase. This comes at a time when government seems to have fewer resources to protect endangered citizens. One promising avenue is the hope placed on a variety of mitigation techniques specific to the event. In addition, continued support for research by both the public and private sectors across disciplines will ensure that decisions are based on a rational framework.

References

Anderson, M., and P. Woodrow (1989). *Rising from the Ashes: Development Strategies in Times of Disaster.* Westview Press, Boulder.

Barclay's California Code of Regulations (1990). Government Printing Office, Sacramento.

Barton, A. H. (1969). *Communities in Disaster.* Anchor Books, New York.

Berke, P. R., J. Kartez, and D. Wenger (1993). Recovery after disaster: Achieving sustainable development, mitigation and equity. *Disasters* **17**(2), 93–109.

Bolin, Robert C. (1986). Disaster impact and recover: A comparison of black and white victims. *Int'l. J. of Mass Emer. and Dis.* **4**(1): 35–50.

Drabek, T. E. (1986). *Human Systems Responses to Disaster.* Springer-Verlag, New York.

Drabek, T. E. (1994). *Disaster Evacuation Behavior: Tourists and Other Transients.* Springer-Verlag, New York.

Dynes, R. R., and E. L. Quarantelli (1968). Group behavior under stress: A required convergency of organizational and collective behavior. *Sociol. Social Res.* **52**, 416–429.

Fritz, C., and E. S. Marks (1954). The NORC studies of human behavior in disaster. *J. Social Issues* **10**(3), 26–41.

Fritz, C. E., and J. H. Mathewson (1956). "Convergence Behavior: A Disaster Control Problem." Special report for Committee on Disaster Studies, National Research Council, Washington, D.C.

Fritz, C., and H. B. Williams (1957). The human being in disasters: A research perspective. *Ann. Am. Acad. Political Social Sci.* **309**, 42–51.

Kreps, G. A. (1991). Organizing for emergency management. In: *Emergency Management: Principles and Practice for Local Government* (T. E. Drabek and G. J. Hoetmer, Eds.), pp. 30–54. International City Management Association, Washington, D.C.

May, P. J. (1985). *Recovery from Catastrophes: Federal Disaster Relief Policy and Politics.* Greenwood Press, Westport.

Mileti, D. S. (1999). *Disasters by Design: A Reassessment of Natural Hazards in the United States.* Joseph Henry Press, Washington, D.C.

Mileti, D. S., C. Fitzpatrick, and B. C. Farhar. (1990). "Risk Communication and Public Response to the Parkfield Earthquake Prediction Experiment, Final Report." Hazards Assessment Laboratory, Fort Collins.

Mileti, D. S., and J. Sorensen (1990). *Communication of Emergency Public Warnings: A Social Science Perspective and State-of-the-Art Assessment.* Oak Ridge National Laboratory, Oak Ridge, TN.

Mileti, D. S., and C. Fitzpatrick (1993). *The Great Earthquake Experiment: Risk Communication and Public Action.* Westview Press, Boulder.

Mitchell, K. (1992). Natural hazards and sustainable development. Paper presented at the Natural Hazards Research and Applications Workshop, Boulder, Colorado.

Moore, H. E., F. L. Bates, M. V. Layman, and V. J. Parenton (1963). "Before the Wind: A Study of Response to Hurricane Carla." National Research Council Disaster Study No. 19, National Academy of Sciences, Washington, D.C.

Office of the State Architect (1986). "California State Accessibility Standard, Title 24 CAC, Interpretive Manual. July 19." Government Printing Office, Sacramento, CA.

Prince, S. H. (1920). Catastrophe and social change: Based upon a sociological study of the Halifax disaster. *Studies History Econ. Public Law* **94**, 1–152.

Quarantelli, E. L. (1983). *Delivery of Emergency Medical Services in Disasters: Assumptions and Realities.* Irvington, New York.

Quarantelli, E. L. (1982a). General and particular observations on sheltering and housing in American disasters. *Disasters* **6**, 277–281.

Quarantelli, E. L. (1982b). Inventory of disaster field studies in the social and behavioral sciences: 1919–1979. Disaster Research Center, Columbus.

Quarantelli, E. L. (1970). The community general hospital. *Am. Behav. Scientists* **13**(3), 380–391.

Rubin, C. B. (1995). Physical reconstruction: Timescale for reconstruction. In: "Wellington after the Quake: The Challenge of Rebuilding Cities." Centre for Advanced Engineering and the New Zealand Earthquake Commission, Christchurch, New Zealand.

Schneider, S. K. (1995). *Flirting with Disaster: Public Management in Crisis Situations.* Armonk, N.Y.: M. E. Sharpe.

Simile, C. (1995). "Disaster Settings and Mobilization for Contentious Collective Action: Case Studies of Hurricane Hugo and the Loma Prieta Earthquake." Doctoral dissertation, Department of Sociology and Criminal Justice, University of Delaware.

Spangle Associates with Robert Olson Associates, Inc. (1997). "The Recovery and Reconstruction Plan of the City of Los Angeles: Evaluation of Its Use after the Northridge Earthquake." Spangle Associates, Portola Valley, CA.

Tierney, K. J. (1996). "Business Impacts of the Northridge Earthquake." Disaster Research Center, Delaware.

Topping, K. C. (1998). Model recovery and reconstruction ordinance. In: "Pre-Event Planning for Post-Disaster Recovery. Planners Advisory Service Report Prepared for the Federal Emergency Management Agency," J. Schwab *et al.*, Eds, American Planning Association, Chicago, IL.

Wenger, D. E., E. L. Quarantelli, and R. R. Dynes. (1989). "Disaster Analysis: Police and Fire Departments." DRC Final Project Report No. 37, Disaster Research Center, University of Delaware.

Wenger, D., E. L. Quarantelli, and R. R. Dynes. (1986). "Disaster Analysis: Emergency Management Offices and Arrangements." DRC Final Project Report No. 34, Disaster Research Center, University of Delaware.

White, G., and E. Haas (1975). *Assessment of Research on Natural Hazards.* MIT Press, Cambridge.

76

The Seismic Alert System of Mexico City

J. M. Espinosa-Aranda
Centro de Instrumentación y Registro Sísmico, México, D.F., México
F. H. Rodríquez
Universidad Nacional Autónoma de México, México, D. F., México

1. Introduction

In Mexico we have learned from experience the disastrous consequences of earthquakes. The *M* 8.1 Michoacán earthquake of September 19, 1985, at 7:19 A.M., killed about 10,000 people, injured 30,000, and left 50,000 families homeless in Mexico City. It collapsed 371 high buildings including 37 public schools and large modern hospitals like the Social Security Medical Center and the State General Hospital. The collapse of the main central telephone building cut off all the international and national long-distance communications for three days.

Not only Mexico City was affected; the states of Michoacán, Colima, and Jalisco also had losses. In Ciudad Guzman, a town of Jalisco State, 580 adobe houses collapsed. The high cost in lives was due in part to the soil conditions and structural characteristics of buildings, but also to the lack of preparedness for rapid response in case of big earthquakes (Esteva, 1988).

Seismological research shows a high probability that a strong earthquake similar in size to the one in 1985 could occur in the Guerrero Gap, between the cities of Acapulco and Zihuatanejo (Singh *et al.*, 1981, 1982; Anderson *et al.*, 1994). The consequences of this earthquake for Mexico City would be great because of soil conditions (Ordaz *et al.*, 1995) and the structural characteristics of buildings.

In January 1986, after the big earthquakes that struck Mexico City in 1985, the Mexican National Council of Science and Technology (CONACyT), together with the National Research Council (NRC) from the United States, issued research recommendations to learn about the damages suffered in this seismic event (CONACyT, 1986). The societal and technical studies were a main reason to promote the development of an earthquake early warning system, in response to seismologists' concern in the early 1980s about the seismic hazard growing in the Guerrero Gap.

The Mexico City Government Authorities have supported the Civil Association, Centro de Instrumentación y Registro Sísmico AC (**CIRES**), the development and implementation of the Mexican Seismic Alert System (SAS) since 1989. Using near-the-source earthquake sensors to warn in advanced remote sites by electric signals, propagated much faster than seismic waves (300,000 km/s compared with 8 km/s), as suggested by J.D Cooper in the *San Francisco Daily Evening Bulletin*, Nov, 3, 1868 (Nakamura, 1996). This project started operation in August 1991 for evaluation with a few users. By the end of 1992 it was an experimental project broadcasting the early warning to some public elementary schools, applying it in 25 buildings where the earthquake warning signal sounds when the SAS magnitude forecast is greater than 5.

The planning for the dissemination and education program of the early warning signal was conducted with six public deliberations carried out in 1992 (Fundación Javier Barros Sierra, 1992). The opinions of public and private organizations for emergency response, government officials, lifeline administrators, disaster researchers, individuals, groups of citizens, private organizations, and the general population were summarized in the conclusions of this study. The result was the design of experimental evaluation programs for the education of SAS users.

The programs include ways for disseminating the early warning, type of expected response from the people such as evacuation and actions to be carried out, and the evaluation of possible attitudes toward the usefulness of the early warning. Also included are plans to deploy the early warning signal in the subway system, government offices, gas stations, and oil-pumping plants.

After the successful seismic detection that generated early warning signals, between 65 to 73 sec in advance, for two Guerrero earthquakes, *M* 5.8 and *M* 6, on May 14, 1993 (Espinosa-Aranda *et al.*, 1995), as well as after two years of initial operation

INTERNATIONAL HANDBOOK OF EARTHQUAKE AND ENGINEERING SEISMOLOGY, VOLUME 81B ISBN: 0-12-440658-0

with nearly 100 seismic events registered, the Mexico City Government Authorities announced the start of SAS as a public service on August 1, 1993. Since this date, the early warning signal has been issued with the support of many of the local AM/FM commercial radio stations in the Mexico City area. The SAS general warning signal is issued only if the magnitude forecast is greater than 6 in accordance with the government recommendations, to protect the population still not trained in evacuation drills.

The fact that most major earthquakes that are likely to cause damage in Mexico City will come from the Guerrero coast 280 km away, and the low requirement of SAS development, maintenance, and operation, give this project a high-return benefit and social value for the Mexico City population. Between January 1990 and December 1997, the system cost $3 million for development, installation, and operation. The 1999 maintenance budget was $600,000, which included the cost of some spare parts and technological enhancement. This remarkable low cost has, in a sizeable part, an explanation. The personnel in charge of starting the SAS project in 1986, had at that time, more than 15 year accumulated experience in similar tasks, drawn from their duties in SISMEX (acronym of Seismotelemetric Information System of Mexico), the first successful strong-motion telemetry network that was uninterruptedly operating since 1970 at the Instituto de Ingeniería (Prince, *et al.*, 1973). In fact, they were part of the team responsible both of the circuitry design, hardware construction and network development, as well as the installation and operation of SISMEX.

The 1993 administrative decision to start the SAS as a public service opened the challenge of how to educate and prepare a population of 20 million people to use the early warning signal. The effectiveness of a public earthquake early warning system focusses on the ability to provide alert as well as to have an adequate population response.

The issuing of the seismic alert is only one element in the process where the preparedness of city residents is fundamental. Drills and education can determine the populations proper response using the SAS alert signals.

2. The Mexico City Early Warning System

The Seismic Alert System consists of four parts: the Seismic Detection System, a Dual Telecommunications System, a Central Control System, and a Radio Warning System for users. The seismic detection system consists of 12 digital strong-motion field stations located along a 300-km stretch of the Guerrero coast, arranged 25 km apart. Each field station includes a microcomputer that continually processes local seismic activity that occurs within a 100-km radial coverage area around each station. An algorithm locally detects and estimates magnitude of an incoming earthquake within 10 seconds of its initiation (Espinosa-Aranda, 1989). If it is estimated as $M > 6$ or $M > 5$, a general or restricted warning message is sent. At least two stations must confirm the occurrence of the earthquake before the warning signal is automatically broadcast from the SAS.

The Dual Telecommunications System consists of a VHF central radio relay station, located near Acapulco, and three UHF radio relay stations located between the Guerrero coast and Mexico City. Two seconds are required for information sent by one of the field stations on the Guerrero coast to reach Mexico City; this data is sent digitally coded.

The Central Control System continually receives information on the operational status of the field stations and communication relay stations, as well as the actual detection of an earthquake in progress. Information received from the stations is processed automatically to determine magnitude and is used in the decision to issue a public alert. In order to generate an early warning signal, two thresholds are defined: general alert ($M > 6$) and restricted alert ($M > 5$). When an earthquake category is determined, a warning message is automatically broadcast by the UHF radio transmitter located in the central control station.

The Radio Warning System for users disseminates the seismic early audio warnings via commercial radio stations and audio-alerting mechanisms to residents of Mexico City, public schools, government agencies with emergency response functions, key utilities, public transit agencies, and some industries. Public and private entities are equipped with specially designed radio receivers to obtain the SAS alert. In each place there is a person in charge of the SAS receivers. His duties are to check the status of the receiver and coordinate all the activities of disaster prevention such as exercises and drills of evacuation.

2.1 Coverage of the Radio Warning System

The radio receivers installed in the commercial AM/FM radio stations, schools of the National Ministry of Public Education (*Secretaría de Educación Pública*), the public housing complex of El Rosario, and the Mexico City commuter rail organization (METRO) are very important due to their social impact.

There have been radio receivers installed in the commercial AM/FM radio stations since September 1995. These receivers were equipped with special audio control systems designed by CIRES to switch over the standard audio program from the radio stations to a 60-sec prerecorded message of early warning. This message consists of a clearly identifiable special tone and the statement "seismic alert, seismic alert" in Spanish (*alerta sísmica, alerta sísmica*). This statement is automatically broadcast without the intervention of human operators. Earlier, in some stations a cassette had to be inserted into the broadcast equipment in order to play the alert message, resulting in the loss of valuable time. The warning message does not contain technical information, specific guidance of protective actions, or a description of the potential dangers or severity of the earthquake.

Since 1992, the *Secretaría de Educación Pública* has been an active participant in the development of the system. Although only 27 schools have been equipped with SAS radio receivers, the education authorities have designated personnel who, among other duties, permanently monitor the radio stations during classroom hours and manually turn on a siren in case of an early warning. The estimated scope of people covered with this method in the schools is 1.9 million students and teaching personnel.

El Rosario is a densely populated public housing project inhabited by 200,000 people. The area is characterized by low-rise multiunit apartment buildings constructed between 1960 and 1970 and surrounded by open areas and recreational facilities. El Rosario has a public audio-warning system connected to the SAS. At this location, a system of powerful loudspeakers is installed in a tower. The early warning is broadcast as a clearly audible signal. It is estimated that the number of inhabitants receiving the warning signal is 10,000.

The Mexico City commuter rail organization, METRO, uses the early warning system to command trains to travel at reduced speed and stop at the next station upon receipt of a warning notification. It is estimated that 600,000 persons are traveling during rush hours.

There are radio receivers in the Civil Protection Agency, the Public Works Department (*Secretaría General de Obras*), the Central Command of the Mexican Army (*Estado Mayor del Ejército*), some public buildings in Mexico City, the public electric utility (CFE), universities, the offices of the XXXVI District of the Mexican Army, the Central Agency for Disaster Prevention (CENAPRED), the police department of the state of Mexico, and the law enforcement task force (*Procuraduria General de Justicia*).

2.2 Internet

The official report of SAS alert is sent by fax, and e-mail and is published on the Internet. The e-mails are received 2 minutes after the SAS issued the early warning signal, the information in the home page is updated in 2 minutes, and the fax distribution to 250 users takes 20 minutes.

3. Societal Response to the Earthquake Early Warning Signal

After the September 14, 1995, *M* 7.3 Copala earthquake (Anderson *et al.*, 1995), when the SAS issued an alert signal to the public (see Figure 1) with a 72-sec advantage, it was estimated that the number of people warned during this earthquake was 4,389,000 (Espinosa-Aranda *et al.*, 1996). The response of the people to the early warning signal is summarized in Table 1, where there are two groups of people: those who received training in a long-term plan and those who had no training.

The frequency of evacuation drills (see Table 1) shows that children at public schools receive intensive training in a systematic manner, with evacuation drills at least each month. Other groups that received training with at least two evacuation drills per year were the residents of the El Rosario neighborhood.

TABLE 1 People in Mexico City Reached by the SAS Earthquake Early Warning Signal on September 14, 1995

Users	Type of user	%	People warned	Drills per year
Children at public schools	Trained	44	1,970,000	10
Listeners of radio stations	Not trained	46	2,000,000	?
Users at public places with alert METRO	Not trained	9	400,000	0
Users in El Rosario complex	Trained	1	10,000	2
Total	—	100	4,389,000	—

3.1 Disaster Mitigation School Program

The program of rapid response for public and private schools for children in Mexico City started after the September 1985 earthquake. Since 1986, the National Ministry of Public Education (*Secretaría de Educación Pública*; SEP) has sponsored the development and implementation of a systemic earthquake hazard reduction plan in Mexico City at all school levels, through their Emergency and Personal Security Program for Schools (*Programa de Seguridad y Emergencia Escolar*). The goal of this program is to improve the response of children to a variety of disasters, including earthquake disasters. It is focused primarily on school-age youths, and it applies to all local schools in Mexico City (National Ministry of Public Education, 1997).

As a result of this program, the ensuing evacuations after the Copala earthquake, according to education officials, were orderly and well coordinated. Also, comparative research about the children's response to the early warning was conducted on two private schools in Mexico City (Arjonilla, 1998).

The major goals of the SEP hazard reduction plan are as follows:

- Development of an Emergency and Personal Security School Program for children (*Programa de Seguridad y Emergencia Escolar*)
- Implementation of security committees in all schools
- Evaluation of disaster impact
- Development of a School Disaster Action Plan

The activities promoted under the Emergency and Personal Security School Program include developing instructional materials such as booklets, posters, guides, and videos; programs for the evaluation of student performance; character education programs; and community service projects with the Mexico City Civil Protection authorities. The evaluation plans include developing computer monitoring programs for controlling the evolution of the program. A budget allocated to this project provides financial support to carry out all activities.

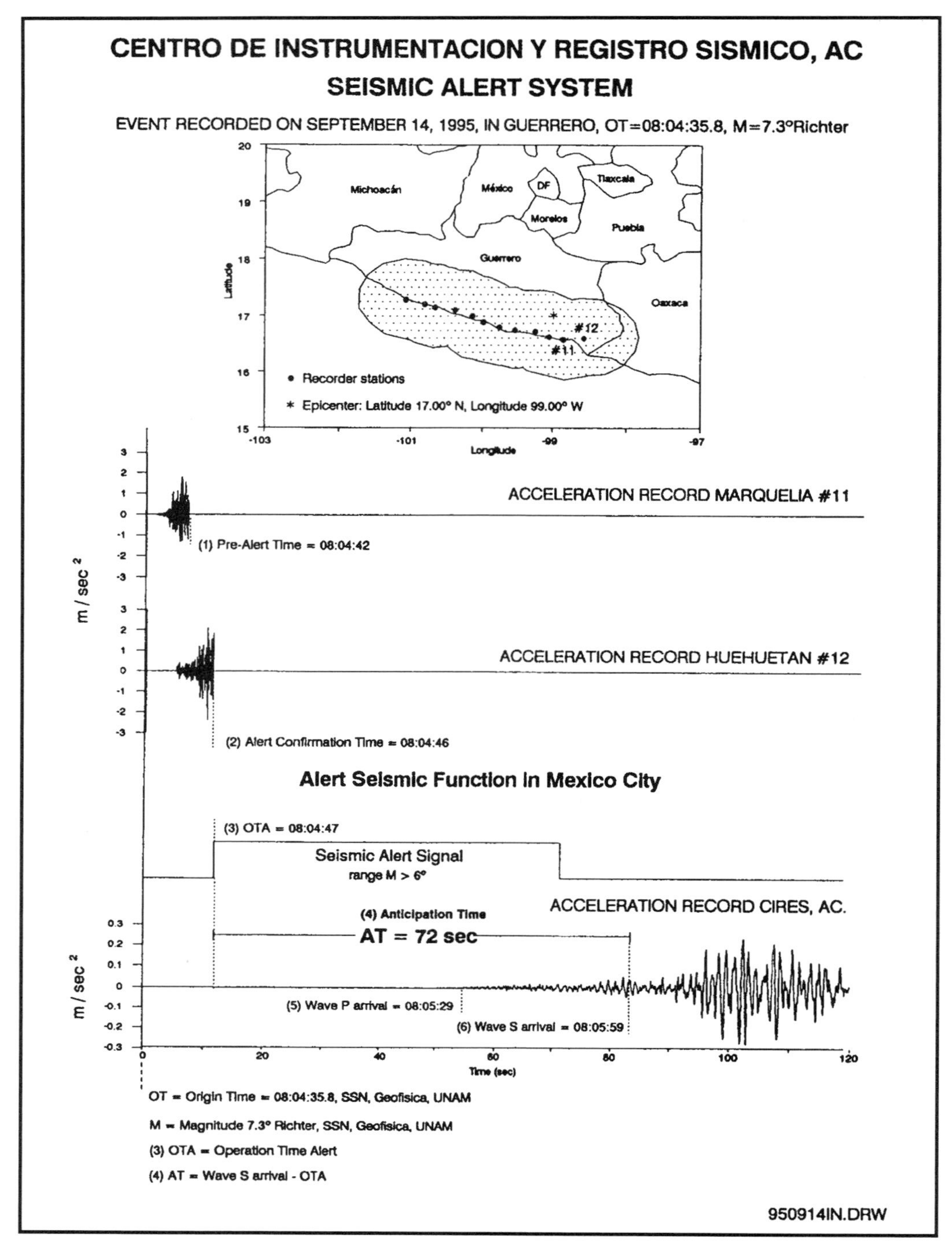

FIGURE 1 Time diagram of early warning signal advantage in Mexico City, SAS issued in September 14, 1995. (from W. H. K. Lee, and J. M. Espinosa-Aranda, 1998).

The security committees are formed by teachers, children, parents, and school authorities. There are working teams for evacuation, first aid, rescue, and search of buried people. The team members cooperate and share responsibility as group leaders, assistants, and volunteers.

Evaluation of disaster impact comprises the description of an earthquake disaster risk scenario at every school; the list of all buildings that could be destroyed, damaged, or affected by a strong earthquake; the number and distribution of children and the type of evacuation that they can make; plans for evacuation

TABLE 2 Results of the Emergency and Personal Security School Program

		School security		Committees disaster impact		Disaster action plan	
School term	Total schools	TOTAL	%	TOTAL	%	TOTAL	%
1989–1990	5585	5505	98.6	831	14.9	659	11.8
1990–1991	5730	5618	98.0	4215	73.6	4224	73.7
1991–1992	5641	5475	97.1	4622	81.9	4481	79.4
1992–1993	5596	5407	96.6	5079	90.8	4919	87.9
1993–1994	5876	5775	98.3	5330	90.7	5330	90.7
1994–1995	6207	6058	97.6	5839	94.1	4724	76.1
1995–1996	6332	6246	99.0	6223	98.0	6223	98.0

for that situation; and the establishment of safety zones for students.

The School Disaster Action Plan is carried out by practicing response to earthquakes with evacuation drills. In 1992, this plan started to use the restricted-range early warning signals of the SAS in one set of 25 schools (National Ministry of Public Education, 1995). Before installing the SAS receivers, teachers and parents were informed about the purpose of SAS. Today these drills are carried out in almost 6223 schools, most of them with no SAS radio receivers. In those schools with no SAS receivers, in accordance with the SEP official recommendations, at least two of the service personnel are in charge of listening to AM/FM radio receivers during work hours and activating local sirens if the SAS general warning is issued.

Table 2 shows the results obtained from the application of this program between 1990 and 1996. Until 1996 the total number of schools in the program was 6332, with 99% of them applying the Emergency and Personal Security School Program. The average number of drills per school was between 0.8 and 1.5 per month. Typical evacuation times are 60 sec for children at preschool, where all the buildings have one level; 80 sec for primary school, and between 45 and 90 sec for secondary school. During the evacuation, children have slogans like "I don't run, I don't push, and I don't scream."

The restricted warning signals are sent to the 25 original selected schools when the SAS seismic magnitude forecast is greater than 5, one every two months on average, and the general warning, when the SAS seismic magnitude forecast is greater than 6, once a year on average.

The school population includes students from all educational levels: children ages 41 days to 5 years from day nurseries, ages 4 to 6 from kindergarten or preschool, ages 6 to 11 from elementary school, ages 12 to 14 from secondary schools, ages 12 or above from technical secondary schools and TV secondary, ages 15 or above from postsecondary, and all ages in special education for children with disabilities.

3.2 Universities

With the same system, between 1995 and 1997, others universities like the Universidad Autónoma Metropolitana, Tecnologico de Monterrey, and the Faculty of Engineering of the National University of Mexico started using the SAS warning signals on their campus.

At the Universidad Autónoma Metropolitana campus Iztapalapa, a SAS radio receiver was installed in June 1997. The population reached with loudspeakers is 15,000 students with a nearby population of 25,000. After installation, an internal program for hazard mitigation was applied at the campus. To date, seven alerts have been issued during working hours and evacuation has been carried out in an average time of 30 to 60 sec. Surveys show that since the SAS installation the neighbor population to the university has been responding to the early warning signal. Now 90% of the population responds to the early warning signal.

3.3 Community Trained Groups

Community-based organizations at the big housing complex El Rosario and the local city government carry out disaster prevention activities such as earthquake response simulations. Since 1995, El Rosario has had a public audio warning system with high-powered loudspeakers installed in towers, controlled by the SAS. Residents of El Rosario created a community organization for training residents in evacuations and preventive actions.

3.4 Groups with No Training

The average Mexico City adult resident who listens to the alert signals on the commercial radio stations has no training in evacuation or earthquake mitigation. The government of Mexico City in 1995 disseminated a brochure to 2 million households, and for some time a spot was transmitted repeatedly during the day on the radio stations to promote earthquake education and rapid response.

These procedures have been recommended for communicating risk (Mileti, 1990; see also Chapter 75 by O'Brien and Mileti). However, earthquake drills for the average city resident have not been carried out frequently.

Another important user group is the people traveling at the Mexico City METRO who do not hear the SAS warning signal, because the signal is issued to operators who travel and stop the trains at the next station, where they open the doors and wait for the seismic effects and further instructions.

3.5 Mexico City SAS Technical Incidents

Since the SAS public operation service began in August 1993, the early warning system has had three problems. The first incident was a missed alarm during the October 24, 1993, magnitude 6.7 earthquake. The second was a false alarm broadcast to the public on November 16, 1993, at 19:20 local time. The third incident occurred on May 31, 1995, when a magnitude 4.6 earthquake struck in the Guerrero and Oaxaca coast at 6:49:47 local time and a restricted early warning to schools was issued by SAS, but by human error a school-type SAS receiver was installed in one AM/FM broadcasting station, generating confusion. Although no warning was intended to be issued to the public, at that moment, one of the most popular morning news programs in the city was on the air, and the chief reporter announced that a big earthquake was about to strike Mexico City.

All these failures were due to human errors. These historical incidents caused some panic and anger, but no person was hurt or injured because of the false or missed alarm. Situations like uncontrolled emotional response causing deaths due to a false warning did not happen. During all the operation of the SAS since 1991, there have been many minor failures, but only the three mentioned above were of importance.

An initial attempt to calculate the reliability of the SAS was made using the data from the operation results of the system from September 1991 to July 1993; this estimation was $R = 0.9764$ (Jimenez *et al.*, 1993). This estimation did not include the radio receivers of users that receive the early warning signal. When extended until year 2000, the reliability is above 99%. The reliability enhancement was due mainly to the completion of the redundant communication path between Guerrero and Mexico City.

4. SAS Performance

Since August 1991, after 11 years of continuous operation, the SAS has recorded 1373 seismic events in the Guerrero coast: 12 of them strong enough to trigger general alerts in Mexico City, 46 restricted ones, and one false general alert. The performance and reliability of SAS is continuously monitored (Jimenez *et al.*, 1993). There are two scenarios in the issue of the early warning signal; one for earthquakes striking at night and the other for earthquakes striking in the daytime, during working hours.

One of the factors for the success of SAS during the Copala earthquake was probably that it occurred at 8:04 in the morning, the daytime scenario, with the majority of people awake and the children at school, with an estimated 4,389,000 people covered by the warning signal. In a nighttime scenario, this situation could be less favorable: Only the people at the El Rosario housing complex or the people who live near the schools or universities with loudspeakers installed could be reached by the earthquake early warning sound.

The SAS does not get the same results, day or night, during those months when the children are on vacation, but we are starting the installation of commercial telecontrolled warning receivers, capable of being turned on when the SAS issues its alert signals.

5. Conclusion

The SAS lesson in Mexico City during the Copala earthquake helps us to identify the societal response strengths and weaknesses of the earthquake early warning signal.

The main goal of the SAS is life safety. The long-term plan of hazard mitigation of the SEP has created earthquake awareness in the children who have assisted during these years at the various levels of school. Even though they did not suffer the disastrous consequences of the 1985 earthquake, they are more aware than most people who lived through that event, who still are not trained. The SEP program ensures that all students, teachers, parents, and school officials will have the training and support they need to carry out an adequate drill response.

There is a need to identify how the untrained general public responds to the early warning signal. The ability of the general public that listens to the radio to respond to early warning signals should be improved. Improvised response and decisions of untrained people should be avoided.

More involvement is needed from federal and local government decision makers to carry out a general program to improve awareness in the average person and promote mechanisms that could be used when an earthquake strikes at night. There is a need to generate more awareness about the seismic risk in Mexico City.

Fortunately, the Guerrero Gap big earthquake has not occurred, and we can propose and discuss new ideas and programs to prepare the untrained people.

Acknowledgments

We are greatly in debt with Daniel Ruiz, former Head of the Public Works in the Mexico City, Federal District Department, who encouraged and sponsored the initial development, and implementation of the Mexican Seismic Alert System, from January 1990 to November 1997, also to Cesar Buenrostro, new Minister of Public Works in the Federal District Government who, since 1998, accepted the challenge to continue promoting and sponsoring this seismic early warning, as one of the public services in Mexico City. With the collaboration of the National Ministry of Public Education, Secretaría de Educación Pública (SEP), since 1992, we started the experimental use of the SAS radio receivers in many schools of Mexico City, and since 1993, with the support of the Asociación de Radiodifusores del Valle de México the SAS warning signal is issued as public service. We express special gratitute to Professor W.H.K. Lee, who encouraged us to present this particular earthquake early warning system approach. But this system would not exist without the discussion support and the invaluable ideas of Emilio Rosenblueth (in memoriam), leader of the Fundación Javier Barros Sierra A.C.

References

Anderson, J., R. Quaas, S. K. Singh, J. M. Espinosa-Aranda, A. Jimenez, J. Lermo, J. Cuenca, S. F. Sanchez, R. Meli, M. Ordaz, S. Alcocer, B. Lopez, L. Alcantara, E. Mena, and C. Javier (1995). The Copala Guerrero, Mexico earthquake of September 14, 1995 (MW = 7.4): A preliminary report. *Seism. Res. Lett.* **66**(6), 11–19.

Anderson, J., J. Brune, J. Prince, R. Quaas, S. K. Singh, D. Almora, P. Bodin, M. Onate, J. R. Vazquez, and J. M. Velasco (1994). Guerrero Mexico, Accelerograph Array: Summary of data: 1988. *Geophys. Int.* **33**, 341–371.

Arjonilla, E. (1998). Estudio comparativo sobre el sismo del 14 de septiembre de 1995, Consejo Mexicano de Ciencias Sociales, A.C.

CONACyT, NRC (1986). Investigacion para aprender de los sismos de septiembre 1985 en Mexico: Informe Tecnico preparado por comites conjuntos del Consejo Nacional de Ciencia y Tecnologia (Mexico) y el National Research Council (EUA).

Espinosa-Aranda, J. M. (1989). Evaluación de un algoritmo para detectar sismos de subducción. Memorias de los VIII y VII Congresos Nacionales de Ingeniería Sísmica e Ingeniería Estructural, Acapulco, Gro., México.

Espinosa-Aranda, J. M., A. Jimenez, G. Ibarrola, F. Alcantar, A. Aguilar, M. Inostroza, and S. Maldonado (1995). Mexico City Seismic Alert System, *Seism. Res. Lett.* **66**(6), 42–53.

Espinosa-Aranda, J. M., A. Jimenez, G. Ibarrola, F. Alcantar, A. Aguilar, M. Inostroza, and S. Maldonado (1996). Results of the Mexico City Early Warning System. *11th World Conference on Earthquake Engineering*, Paper No. 2132.

Esteva, L. (1988). The Mexico earthquake of September 19, 1985—Consequences, lessons, and impact on research and practice. *Earthquake Spectra* **4**, 413–426.

Fundacion Javier Barros Sierra (1992). Aprovechamiento de la Alerta Sismica, File report on Public deliberations carried out in fall 1992.

Jimenez, A., J. M. Espinosa, F. Alcantar, and J. Garcia (1993). Analisis de confiabilidad del Sistema de Alerta Sismica, X Congreso Nacional de Ingenieria Sismica, Puerto Vallarta, Jal, Mexico, 629–634.

Lee, W. H. K., and J. M. Espinosa-Aranda (1998). Earthquake early warning systems: current status and perspectives. In "Proceedings of International IDNDR-Conference on Early Warning Systems for the Reduction of Natural Disasters," Potsdam, Germany.

Mileti, D. S. (1990). Communicating public earthquake risk information, prediction and perception of natural hazards. *Proc. Symposium*, p. 143–152, Perugia, Italy.

Nakamura, Y. (1996). Real-time information system for hazards mitigation. Proceeding 11th World Conference Earthquake Engineering, Paper No. 2134.

National Ministry of Public Education (*Secretaria de Educacion Publica*), Mexico (1995). Evaluacion de la Operacion del Programa Piloto Sistema de Alerta Sismica en Planteles de Educacion Basica. Direccion General de Operacion de Servicios Educativos en el Distrito Federal.

National Ministry of Public Education (*Secretaria de Educacion Publica*), Mexico (1997). Memorias de Gestion Educativa 1994–1997.

Ordaz, M., F. Sanchez-Sesma, S. K. Singh (1995). La respuesta sismica en el Valle de Mexico (observaciones y modelos), Ingenieria Civil, 317, Mexico, septiembre 1995.

Prince, J., F. H. Rodríguez, E. Z. Jasorski, and G. A. Kilander (1973). A Strong Motion Radio Telemetry Netework, Proc. V World Conference on Earthquake Engineering, Rome.

Singh, S. K., L. Astiz, and J. Havskov (1981). Seismic gaps and recurrence periods of large earthquakes along the Mexican subduction zone: a reexamination. *Bull. Seism. Soc. Am.* **71**, 827–843.

Singh, S. K., J. M. Espindola, J. Yamamoto, and J. Havskov (1982). Seismic potential of the Acapulco-San Marcos region along the Mexican subduction zone. *Geophys. Res. Lett.* **9**, 633–636.

77

The Rapid Earthquake Data Integration Project

Lind Gee, Douglas Neuhauser, Douglas Dreger, Robert Uhrhammer, and Barbara Romanowicz
Berkeley Seismological Laboratory, University of California, Berkeley, CA, USA
Michael Pasyanos
Lawrence Livermore National Laboratory, Livermore, CA, USA

1. Introduction

Interest in rapid access to earthquake information has grown enormously in the last few years. In addition to satisfying inquiries from the public and the media, rapid notification programs provide valuable information for earthquake disaster response. Recognizing the importance of this information for seismic hazard mitigation, efforts to design and implement systems to provide earthquake parameters in a timely manner have expanded over the last 10 years at both the regional and the national level (Heaton, 1985; National Research Council, 1991; Kanamori *et al.*, 1991; Buland and Person, 1992; Romanowicz *et al.*, 1992; Ekstrom, 1992; Ammon *et al.*, 1993; Bakun *et al.*, 1994; Malone, 1994) as well as abroad (Nakamura and Tucker, 1988; Espinosa-Aranda *et al.*, 1995; Wu *et al.*, 1997; USGS, 1999). See also Chapter 64 by Shin *et al.*, Chapter 76 by Espinosa-Aranda, and Chapter 78 by Hauksson *et al.*

In 1991, the Berkeley Seismological Laboratory (BSL) initiated a program to upgrade and expand the Berkeley Digital Seismic Network (BDSN) in northern and central California. Today, the BDSN consists of 20 broadband, high-dynamic range stations (Fig. 1). Each of the BDSN sites is equipped with a broadband seismometer, a strong-motion accelerometer, and a 24-bit digital datalogger, and the data are transmitted to UC Berkeley continuously using dedicated digital telemetry. In order to avoid data loss during utility disruptions, each site has a 3-day supply of battery power and is also accessible via a dial-up phone line. The combination of high-dynamic range sensors and digital dataloggers ensures that the BDSN has the capability to record the full range of earthquake motion for both rapid response and research studies. Additional information on the BDSN is provided in Chapter 87.

In parallel with the upgrade of the BDSN, the BSL has developed a number of analysis methods to expedite the determination of earthquake parameters based on the broadband data. New procedures for estimating earthquake location (Dreger *et al.*, 1998), the seismic moment tensor (Romanowicz *et al.*, 1993; Dreger and Romanowicz, 1994), and finite-fault parameters (Dreger and Kaverina, 1999; 2000) have been developed. These developments form the core of the REDI system.

In northern California, earthquake monitoring is the shared responsibility of the USGS Menlo Park and the BSL. In this chapter, we describe the developments at the BSL that contribute to the overall operation of the joint earthquake notification system.

2. System Design

Similar to most automated earthquake processing systems, REDI operations can be divided into two major elements: event identification and event processing. The event identification element includes operations such as phase picking, event association, and event selection. The event processing element is separated into several stages, with each earthquake assessed for a particular type of processing based on its location and size.

2.1 Event Identification

The REDI processing system is designed to operate in two modes (Fig. 2). In one configuration, the system uses phase picks from the BDSN to detect and locate earthquakes (Gee *et al.*, 1996a). In the second configuration, the REDI system operates as part of the northern California "Rapid Earthquake Location Service"

INTERNATIONAL HANDBOOK OF EARTHQUAKE AND ENGINEERING SEISMOLOGY, VOLUME 81B ISBN: 0-12-440658-0

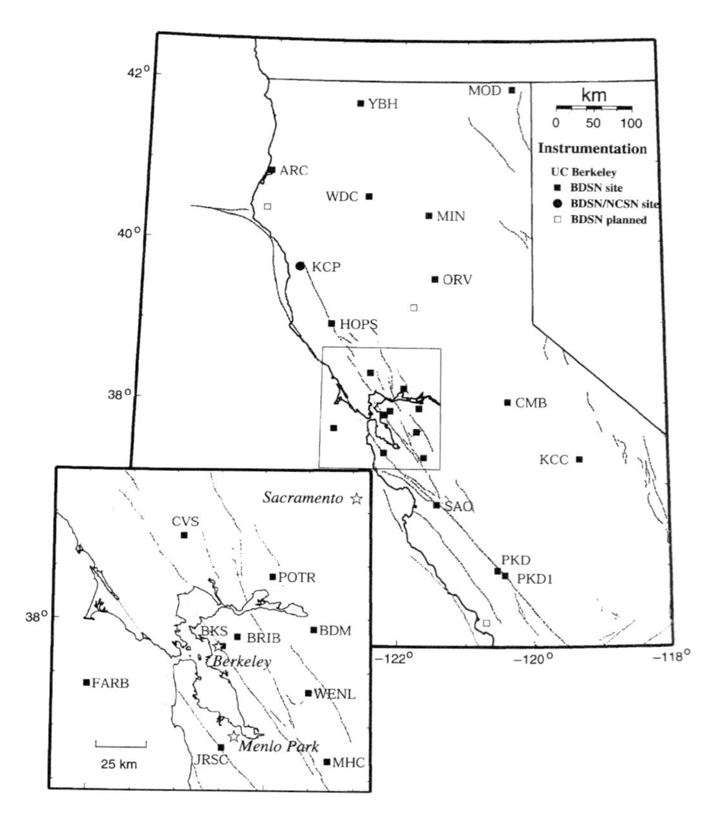

FIGURE 1 Map illustrating the distribution of BDSN stations used in REDI processing. Filled squares indicate the location of current BDSN sites, while open squares indicate planned sites. Stars in the inset map indicate the locations of Sacramento, Berkeley, and Menlo Park.

(Gee *et al.*, 1996b) and is coordinated with the Earthworm/Earlybird system operated by the USGS Menlo Park. In this mode, the Earthworm/Earlybird system performs the event identification component and drives the REDI processing with the resulting event information.

Although rarely exercised during normal operations, the stand-alone capability of the REDI system is critical for situations when communications between Berkeley and Menlo Park are disrupted. The importance of rapid notification for emergency response requires that each component have the capability to operate independently.

2.1.1 Stand-Alone Operation

2.1.1.1 Phase Picking

In the initial development of the REDI system (Gee *et al.*, 1996a), event detections from the Murdock–Hutt–Halbert (MHH) algorithm (Murdock and Hutt, 1983) were used as phase picks. These detectors are run on the vertical-component data from the broadband and strong-motion sensors to identify events and trigger the HH and HL channels (Table 1). The time of the detection is generally a reliable P-arrival time. As an event detector, however, the MHH algorithm turns off for some time period following a trigger before returning to detection mode. As implemented on the BDSN dataloggers, the system will retrigger 55 sec after the preliminary detection on the HHZ or HLZ channels if the event is ongoing, generating additional detections that are not necessarily associated with arriving phases.

The REDI Project has used the MHH detections for several years with good results. However, since the MHH detector is

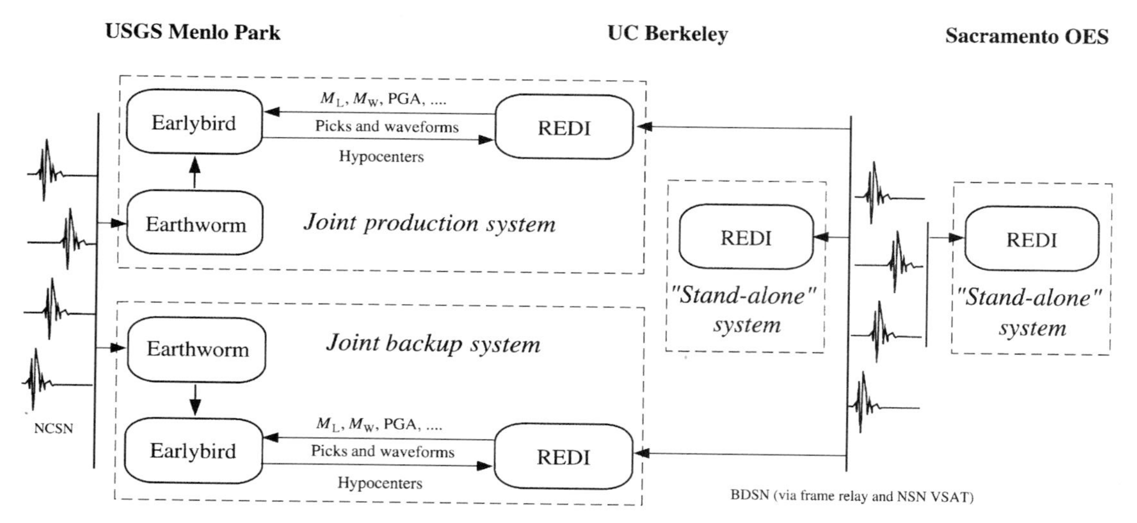

FIGURE 2 Schematic diagram illustrating the connectivity between the real-time processing systems at the USGS Menlo Park and UC Berkeley. The two Earthworm/Earlybird and REDI system modules form the primary and secondary joint notification units for northern California, respectively. The third REDI system in UC Berkeley is a stand-alone system in case communication with the USGS is disrupted. The REDI system in Sacramento OES is a stand-alone system that performs independent data acquisition and processing and is designed to provide earthquake information during a major Bay Area earthquake.

TABLE 1 Typical data streams used in REDI processing from the broadband and strong-motion sensors of the BDSN, with channel name, sampling rate, and processing application. In the case of the HH and HL channels, sampling rates are either 80 or 100 sps, depending on the datalogger

Sensor	Channel	Rate (sps)	Application
Broadband	HH	80.0/100.0	Picking
Broadband	BH	20.0	Picking; M_L, M_E
Broadband	LH	1.0	Moment tensor; Finite fault
Strong-motion	HL	80.0/100.0	Peak ground motions

based on a single component (generally the vertical), it does not utilize the full power of the three-component broadband systems to detect later arrivals, identify phase type, and estimate arrival azimuth. This is particularly an issue for a sparse network such as the BDSN, where the use of information in addition to the P-wave is necessary to obtain a reliable location.

In order to enhance automatic phase picking on the BDSN waveforms, we have implemented an algorithm using a characteristic function for each three-component station of the form

$$f(t) = av(t)^2 + b(\delta v(t)/\delta t)^2 + c(\delta\theta(t)/\delta t)^2 + d(\lambda(t))^2 \quad (1)$$

where $v(t)$ is the summed velocity of the three components, $\delta v(t)/\delta t$ is the time derivative of the velocity, $\theta(t)$ is the change of the solid angle of the velocity vector, $\lambda(t)$ is maximum eigenvalue from the polarization decomposition, and a, b, c, and d are weighting factors (Gee *et al.*, 2000). A high-pass filter at 100 sec is applied to the data in order to remove long-period noise from the time series before forming the characteristic function. No other filter is applied to the waveform data, which limits the detection threshold in areas of significant noise.

A trigger or pick is declared when the ratio of the short-term average to the long-term average (STA/LTA) of the characteristic function exceeds a treshold value. This threshold is set on a station-by-station basis to account for variation in the background noise level. The picking algorithm performs well on local and regional events of magnitude 3.5 and higher, which is an appropriate threshold for the sparse broadband network. We are currently running it on the REDI test platform, using both BH and HH data. In addition to providing arrival time estimates, the picker computes the azimuth of the incoming arrival based on the polarization decomposition. Although the picker produces computationally stable estimates of azimuth, comparison with the observed azimuth shows an uncertainty of $\pm 20°$. While most of the scatter is in the picker estimates, the analyst estimates show considerable variation. We believe that some of the scatter is due to 3-D structure, although a significant component is introduced by the FIR (Finite Impulse Response) filters in the Quanterra dataloggers (Scherbaum and Bouin, 1997; Uhrhammer *et al.*, 1999). The FIR filters also influence the resolution of the pick time, and we are reviewing the possibility of removing the FIR filter in near real time.

2.1.1.2 Association

Phase detections may be generated by a variety of mechanisms besides seismic sources. Sifting through picks generated by cultural, meteorological, and telemetry noise to find earthquakes is a difficult task, which may be further complicated by energetic aftershock sequences and swarms. The REDI system uses an association program, which was developed as a component of the Earthworm system at the USGS Menlo Park (Johnson *et al.*, 1995).

The Earthworm association algorithm, known as "binder," uses a phase stacking approach to identify events. Since it was originally designed for use with a dense, short-period network, we have had to tune the configuration parameters in order to optimize it for the BDSN. This has required adjustments in the parameters controlling the size of the association grid, the assignment of weights based on the quality and phase type of the pick, and residual taper as a function of distance. This last parameter has been particularly important, as the criteria for associating picks will differ if the average interstation spacing is 100, rather than 10 km.

2.1.1.3 Selection

The main task of the associator is to analyze phase picks as they are received and combine them into probable events. Since phase information is acquired continuously, the associator's "view" of an event is never static and event information is constantly updated as additional arrivals are analyzed. Therefore, a separate program identifies events ready to be processed.

The selection program, known as "bruno" or "the watchdog," is based on the Earthworm "report" module. It uses criteria based on temporal evolution of the event and the number of associated arrivals in order to decide when an event is ready to be processed. The temporal criteria attempt to maintain a balance between the reliability of the solution and the rapidity of notification. Although the associator may locate an event within seconds based on only five picks, the quality of the location may be improved by waiting for additional phase information. The selection program provides two measures of delay. The first assesses the total time from the initial association, while the second marks the time since the last change in the event. These temporal criteria can be overridden by two additional parameters based on the number of associated phases. The "min picks" parameter is designed to reduce the number of small or "garbage" events clogging the system by preventing selection unless a minimum number of phases are associated. The "max picks" parameter allows bruno a shortcut around the temporal controls by allowing an event to be selected when the required number of associated arrivals is met.

2.1.2 Joint Operation

The "stand alone" event identification described above operates continuously on the REDI processing computers. However, on a routine basis, the REDI system uses event information

TABLE 2 Current REDI processing stages. The location procedures in stage 1 are skipped in normal operation of the joint system, but they are utilized if problems prevent the flow of hypocenter information from the USGS. The weight is the importance assigned to the stage to prioritize event processing. "Slots" identifies the number of reserved processes for a stage, while "Max" indicates the maximum number of invocations of a particular stage that may run simultaneously. P1–P3 refer to the different processing polygons. P1 is the normal northern California region, P2 is an expanded area including southern Oregon and western Nevada, and P3 is a southern California polygon

Stage	Weight	Slots	Max	P1	P2	P3	Processing
1	1.00	1	99	All events	$M \geq 4.0$	$M \geq 5.0$	Revised location
2	0.91	1	99	$M \geq 3.0$	"	"	Local and energy magnitude
3	0.80	0	99	$M \geq 4.5$	$M \geq 5.0$	$M \geq 5.5$	Peak ground motions
4	0.75	0	1	$M \geq 3.5$	"	"	Seismic moment tensor
5	0.75	0	1	$M \geq 6.0$	$M \geq 6.0$	$M \geq 6.0$	1-D finite fault
6	0.75	0	1	"	"	"	2-D finite fault

derived from the Earthworm/Earlybird system in Menlo Park. The joint notification system was implemented on April 16, 1996, merging the programs in Menlo Park and Berkeley into a single earthquake notification system.

On the USGS side, incoming analog data from the NCSN are digitized, picked, and associated as part of the Earthworm system (Johnson *et al.*, 1995). Preliminary locations, based on phase picks from the NCSN, are available within seconds, based on the association of a few arrivals, while final locations and preliminary coda magnitudes are available within 2–4 min. Earthworm reports events—both the "quick-look" 25-station hypocenters (without magnitudes) and the more final solutions (with coda magnitudes)–to the Earlybird alarm module in Menlo Park. This module sends the Hypoinverse archive file to the REDI system for additional processing, generates pages to USGS and BSL personnel, and updates the northern California web pages.

The data transfer between the USGS and the BSL utilizes a dedicated T1 (1.5 Mbit/s) frame-relay circuit. This "intranet" or private network between the BSL and USGS provides a high-bandwidth connection for exchanging waveform and parametric data. The systems are configured to use the Internet if the frame-relay circuit fails.

2.2 Event Processing

The event processing component of REDI may be driven either by the "stand alone" system or by the Earthworm/Earlybird system in Menlo Park. In day-to-day operation, each REDI unit receives events from the Earthworm/Earlybird system while simultaneously maintaining its own list of earthquakes based on the BDSN data. While operating in one state, the events generated by the other are stored and can be retrieved for later processing.

Both sources of events are structured to allow "age" control of the earthquakes. In general, the system is configured to ignore "old" earthquakes (i.e., earthquakes whose origin time differs from current time by some amount) in order to expedite processing. However, the system may be configured to process all events, providing the capability to rerun earthquake sequences for testing. In general, the REDI system excludes events more than 4 hours old and they are placed in an "ignore" directory. These ignored earthquakes may be scheduled for processing manually.

Once an event is selected, it is queued for analysis by the event scheduler. The scheduler (known as "shepherd") oversees event processing from initiation to notification and decides which stages of analysis are required. REDI currently supports six levels of processing (Table 2): (1) computation of revised location; (2) computation of local and energy magnitude; (3) estimation of peak ground motions; (4) determination of seismic moment, centroid depth, and faulting mechanism; (5) and (6) estimation of fault length, width, and distribution of slip. Stages 5 and 6 are new additions to REDI and are currently operating on a test system.

The scheduler utilizes information about an earthquake's size and location to decide which stages are appropriate. The processing criteria are defined using geographical polygons. Within a polygon, each stage has a magnitude threshold. This structure allows the REDI system to process smaller events within northern California and larger events at regional distances. Three polygons are currently configured, and the structure can be expanded to additional regions. Events that do not meet the regional criteria are saved in an "ignore" directory and may be scheduled manually for processing.

Nearly every stage relies on waveform data from the BDSN. In order to facilitate the processing, 30 min of the most recent waveform data are stored in a shared memory region (Kanamori *et al.*, 1999) on the REDI processing computer. This includes the HL, BH, and LH data streams, since different modules require different data (Table 1).

Access through the shared memory region greatly expedites access to the waveform data and accelerates REDI processing. If the requested data are not available in the shared memory region (either because an event is "late" or because of telemetry problems), the system trys to fulfill the request from files written by the data acquisition system or, for older data, the Northern California Earthquake Data Center archive. This capability is part of the design for testing the system with older events.

Each processing stage also has its own timing and priority. The data needs vary from stage to stage and the REDI system uses a simple scheme to schedule each stage, based on the elapsed time (T_w) since the event occurred:

$$T_w = T_0 + (X - X_0)/V_0 \tag{2}$$

where T_0 is a fixed time delay in seconds, X_0 is a distance offset in km, X is the distance between the epicenter and the network centroid in kilometers, and V_0 is the velocity of the slowest waves of interest in kilometers per second. This formulation allows for complete flexibility in the scheduling of each stage. So, for example, the local magnitude calculation is scheduled based on the distance/velocity criteria (in order to make sure that the S waves have arrived) while the moment tensor computations are based on an absolute time delay (5 min of waveform data are required for all stations).

REDI processing is hierarchical—for an event, only a single stage of processing is active at any given time. However, the system is configured to handle multiple events simultaneously. In order to avoid overloading the real-time computers during a major earthquake or an active swarm, the REDI system prioritizes the events for processing and provides control over the number and type of processing activities running at any time. Each stage is assigned a weight, and the absolute priority for an event at a particular stage is the stage weight times the current event magnitude (e.g., $P_i = M_i W_j$ where P_i is priority of the ith event and M_i is its current estimated magnitude, and W_j is the weight of stage j). As illustrated in Table 2, the system places the highest weight on stage 1, decreasing to 0.75 for stages 4, 5, and 6. The prioritization is designed to expedite the processing of larger earthquakes, while providing a balance between the various stages of processing. For example, this scheme prioritizes computing a moment tensor for a M 5.0 earthquake over the relocation of a M 3.75 event.

Shepherd uses the time delay and the priority to schedule events for processing. In order to control the demand on the processing computers, the REDI system defines a maximum number of processes that may run simultaneously, reserves slots for certain types of processing, and limits the number of any particular stage (Table 2). This additional control is important considering the cpu-intensive nature of certain processing stages, such as the moment tensor and finite-fault estimation. We have configured the REDI system to permit a maximum of four processes and to reserve one of the four slots for relocation and one for magnitude estimation. No limits on the number of processes in stages 1–3 are set (i.e., the system can run four relocation processes simultaneously), but the seismic moment tensor (and the future finite-fault procedures) is limited to one running process to avoid swamping the system.

Although this level of structure may seem excessive, our experience with swarms indicates that it is important to provide this control. An alternative is to distribute the processing stages among a number of computers.

2.2.1 Stage 1—Revised Location

The processing in this stage depends on the source of the earthquake information. For events generated by the Earthworm/Earlybird system, no relocation is necessary and this stage is used only for notification purposes. For events selected from the stand-alone system, this stage generates a revised location using the standard regional earthquake location program of the BSL (Uhrhammer, 1982). Once relocated, the magnitude of the event is recomputed based on the first few cycles of the P-wave amplitude (Hirshorn *et al.*, 1993).

2.2.2 Stage 2—Local and Energy Magnitude

In this stage, waveform data from the BDSN are processed to compute local and energy magnitude. The data request is formulated based on the current estimated magnitude of the event. Expected Wood–Anderson amplitudes are calculated for each station, based on the event's location and magnitude. If the predicted Wood–Anderson amplitudes exceed 0.5 mm, the waveform data for that station are extracted. This condition was imposed to speed processing by analyzing only relevant stations and to avoid contamination by noise. It reflects the sparse distribution of the BDSN (an average spacing of 50–100 km), which limits the lowest magnitude level to approximately 3.

The processing algorithm checks whether the data maximum exceeds 85% of the maximum of the voltage limits of the Quanterra datalogger and discards those channels. However, this test does not detect clipping from exceeding the limits of the phase-locked loop's feedback capabilities. This condition occurs when the force on the inertial seismometer mass generated by ground acceleration exceeds the force that the feedback loop can generate. However, this condition is not easily detectable in the raw time domain signal since it can occur at any time, and it is not simply related to the velocity-proportional output of the seismometer.

In the initial development of the REDI project, only broadband data were used in this computation, since the strong-motion data were transmitted only when triggered. However, the increased bandwidth of digital telemetry has made it possible to transmit the HL data continuously. This change has allowed us to modify the extraction software to supplement the broadband waveforms with strong-motion data, avoiding the problem of clipping in the broadband data from either of the cases described above.

Following the traditional definition of local magnitude, only the horizontal components are used to compute Wood–Anderson synthetic amplitudes (Uhrhammer *et al.*, 1996). Amplitudes less than 0.001 mm are rejected, and readings from different sensors at the same site are averaged in order to avoid overweighting a particular station. The event magnitude is determined from the mean of the station magnitudes, after the application of station corrections.

The local magnitude computation is quite rapid. Even when all BDSN stations are used, the M_L computation is completed in less than 1 min. However, the problem of the saturation of the local magnitude scale is well known, and we have relied on moment tensor algorithms in stage 4 to provide a more robust estimate of earthquake size. As described below, the moment tensor procedures only return an estimate of M_W when there is substantial agreement between two methodologies. In order to ensure a rapid and robust magnitude estimate, we have recently expanded the capabilities of the REDI system to include energy magnitude.

To implement energy magnitude, we calibrated the algorithm of Kanamori *et al.* (1993) for northern California. Using a suite of 152 earthquakes, we determined estimates of seismic energy release using

$$e_{ij} = 23.6 \times 10^5 r^2 [r_0 q(r_0)/rq(r)]^2 \int \sum v_{ij}(t)^2 dt \quad (3)$$

where e_{ij} is the energy of earthquake j at station i, r is the hypocentral distance, r_0 is 8 km, $q(r)$ is the attenuation curve for distance r, v_{ij} is the observed velocity, and the expression includes correction terms for the radiation pattern. In the REDI implementation, the integration is performed over the P and S waveforms. The revised attenuation relation of Kanamori *et al.* (1993) provides a good fit to the northern California data and no additional distance correction is required. On the other hand, significant station variations were observed. In order to reduce the influence of site effects, we chose the station BKS as a reference station (in keeping with the practice for local magnitude) and set its station correction to 1. We then determined site-specific scale factors for the network: $e_{1j} = e_{ij}s_i$ where e_{1j} is the estimate of the seismic energy at BKS for earthquake j and s_i is the correction factor for station i.

Averaging over the suite of 152 earthquakes, the correction factors range from 0.14 to 3.7 and correlate, in general, with the local magnitude station corrections (Table 3). This is a larger variation in these corrections than observed by southern California, which may reflect the greater variability of geologic conditions at BDSN sites. The coastal stations tend to have similar corrections to BKS, but the stations in the Sierra stand out with their large corrections (CMB, KCC, ORV, WDC, YBH).

Using southern California events, Kanamori *et al.* (1993) obtained a linear relationship between the $\log E$ and local magnitude [$M_E = 0.51(\log E - 9.05)$]. We have found that this describes the northern California events well, and we have added the energy magnitude estimation to the current stage for local magnitude. Under this revised processing, both M_L and M_E are calculated for every qualified event (earthquakes with coda magnitude greater than 3.0).

TABLE 3 Station corrections for determination of energy release. For comparison, the M_L station adjustments (δ_{M_L}) are included in the table (Uhrhammer *et al.*, 1996). Asterisk indicates a preliminary estimate of the adjustment

Station	s_i	No. of obs	δ_{M_L}
ARC	0.2808	126	+0.209
BDM	0.5177	49	−0.023*
BKS	1.0000	127	−0.035
BRIB	0.7515	89	+0.176
BRK	1.6369	97	+0.198
CMB	2.8105	127	+0.240
CVS	1.0835	78	−0.189*
FARB	1.9465	92	−0.150
HOPS	1.1581	114	+0.324
JRSC	0.8514	124	+0.139
KCC	3.6114	96	+0.369
MHC	0.8681	126	+0.128
MIN	0.5680	121	−0.107
ORV	3.7033	127	+0.428
PKD	0.6153	96	−0.033*
PKD1	0.1420	27	−0.198
POTR	0.2266	41	+0.159*
SAO	1.3975	125	+0.314
WDC	2.5639	126	+0.484
WENL	0.4048	82	+0.089*
YBH	2.1010	126	+0.499

2.2.3 Stage 3—Peak Ground Motions

Once a revised magnitude is determined, the next stage in the REDI processing is the determination of peak ground motions. This stage is activated for events of M 3.5 and higher. At this stage, waveform data from all sites with strong-motion instruments are extracted. If the returned data fit specified group velocity windows, then the instrument response is deconvolved and the data are high-pass filtered above 10 sec. The peak ground acceleration, velocity, and displacement are determined, along with the time of each peak. The processing software also determines the duration of shaking at four levels: 5%, 10%, 20%, and 50% g.

2.2.4 Stage 4—Seismic Moment Tensor

The REDI system was expanded to include the automatic estimation of the seismic moment tensor in 1997. In addition to providing valuable information on the source (relevant for estimating tsunamigenic potential or for identifying magma movement), the determination of the moment tensor is the first step in procedures for estimating the distribution of slip and prediction of ground shaking.

The REDI system uses two separate methods of determining moment tensor solutions. One is based on the time-domain inversion of complete waveform data (Dreger and Romanowicz, 1994), while the second is based on the frequency-domain inversion of surface waves (Romanowicz *et al.*, 1993). Initially automated in 1994 (Pasyanos *et al.*, 1996), these methods utilize LH data from the BDSN and are now fully incorporated in REDI processing.

As implemented in the REDI processing, both inversion methodologies are run for every qualifying event (earthquakes with M_L greater than 3.5). Each algorithm produces an estimate of the seismic moment, the moment tensor, the centroid depth, and solution quality. The REDI system uses the individual solution qualities to compute a weighted average of moment magnitude, to compare the mechanisms using normalized root-mean-square of the moment tensor elements (Pasyanos *et al.*, 1996), and to determine a "total" mechanism quality. The comparison or correlation between the two solutions provides an objective measure of the difference in the estimated source mechanisms:

$$C = 1 - \sqrt{\sum_{i=1}^{3}\sum_{j=1}^{3}(\delta M'_{ij})^2/8} \qquad (4)$$

where M'_{ij} is M_{ij}/M_0. The $\sqrt{8}$ normalization factor causes C to range from a value of 1 (perfect correlation) to 0 (double-couple mechanisms of exactly the opposite sense of motion). In the current implementation, the resulting estimate of moment magnitude is used in preference to local or energy magnitude when the solution quality exceeds 50%. The moment tensor is distributed automatically only when the correlation between the two methodologies exceeds 0.5 and the solution quality is greater than 50% (where 100% is the highest quality).

A review of the REDI moment tensor solutions in the last two years indicates that the automatic determination compares favorably to the final solutions obtained by BSL scientists. Out of the total 220 events processed during 1997–1999, 119 or slightly more than half had a correlation of greater than 0.5 (Fig. 3 and Table 4). The ratio of well-correlated mechanisms increases with event size, and reaches 74% for events of $M \geq 4.5$. Our experience indicates that most of the discrepancy between the methods for smaller earthquakes may be attributed to noisy waveforms or other problems related to the signal-to-noise ratio. A comparison of the reviewed or final solutions indicates that approximately 80% are well correlated, independent of magnitude.

Of the events of magnitude 4.5 and higher, only 9 events show discrepancies between the two solutions. In general, the complete waveform method obtains a better variance reduction than the surface wave method for these events. Yet, the strike of one of the nodal planes obtained by the surface wave methodology is similar to one of the nodal planes by the complete waveform approach. The poorer surface-wave solution may reflect the inherent problem of determining the slip angle in surface-wave inversion for shallow events.

The 9 discrepant events were located in the Cape Mendocino and Mammoth Lakes region of northern California. In the current implementation, the Green's function for the complete waveform method is chosen from a group calculated every 5 km in distance between 40 and 400 km, and every 3 km in the depth range from 5 to 39 km using one of two possible 1-D models. These models do not currently account for the 2- or 3-D structure along specific source–receiver paths. For the Cape Mendocino region, which is located in a complex tectonic region, we have tested several existing 1-D models and are working to develop an improved model. This is an ongoing project, as the calibration of path effects requires significant effort. For the Mammoth Lake region there are indications that the discrepancy of the solutions may be caused by the source characteristics rather than the path effects unmodeled in the present (wave-length) scale. To obtain more stable solutions for these regions, 2-D or 3-D structural models may be necessary.

TABLE 4 Correlation Between the Two Automated Moment Tensor Solutions for Events from 1997 to 1999 As a Function of Magnitude

M	No. of events	$C \geq 0.5$
≥3.5	220	119 (54%)
≥4.0	92	57 (62%)
≥4.5	32	23 (74%)

2.2.5 Stages 5 and 6—Finite-Fault Parametrization

We have recently expanded the REDI processing environment to include the estimation of finite-fault parameters. This approach is based on the use of theoretical Green's functions and is an extension of the finite-source inversions that are commonly performed on local strong-motion data (e.g., Wald *et al.*, 1996; Cohee and Beroza, 1994). Using a precomputed set of Green's functions for an appropriate 1-D velocity model, it is possible to consider an arbitrarily oriented source using parameters obtained from the automated moment tensor analysis and a directivity model of an expanding circular rupture with a constant rupture velocity and dislocation rise time.

Based on the development of Kaverina *et al.* (1997) and Dreger and Kaverina (1999), we have implemented the finite-fault estimation procedure in two REDI stages. In stage 5, BDSN broadband waveform data are prepared for inversion, and rough estimates of the fault dimensions are derived using the empirical scaling relationships of Wells and Coppersmith (1994). Using these parameters to constrain the overall dimensions of the extended source, the process tests the two possible fault planes (determined in stage 4) over a range of rupture velocities by performing a series of inversions using a line-source representation.

Each line-source computation is quite rapid. Calculations for the 1992 Landers and 1994 Northridge earthquakes required approximately 3.5 min to test seven rupture velocities for the two different planes (Dreger and Kaverina, 1999). In tests based on the Landers and Northridge earthquakes, the fault plane was clearly defined by the variance reduction from the line-source inversions. In addition to the identification of the fault plane and apparent rupture velocity, this stage yields preliminary estimates of the rupture length, dislocation rise time, and the distribution of slip in one dimension.

The second component of the finite-fault parameterization uses the best-fitting fault plane and rupture velocity from stage 5 to obtain a more refined image of the fault slip through a full two-dimensional inversion. If stage 5 fails to identify the probable

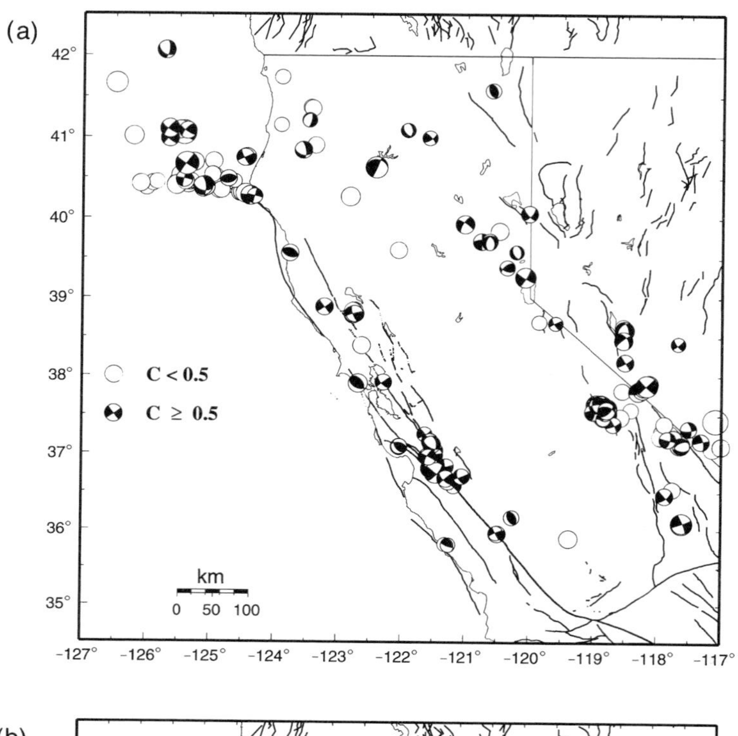

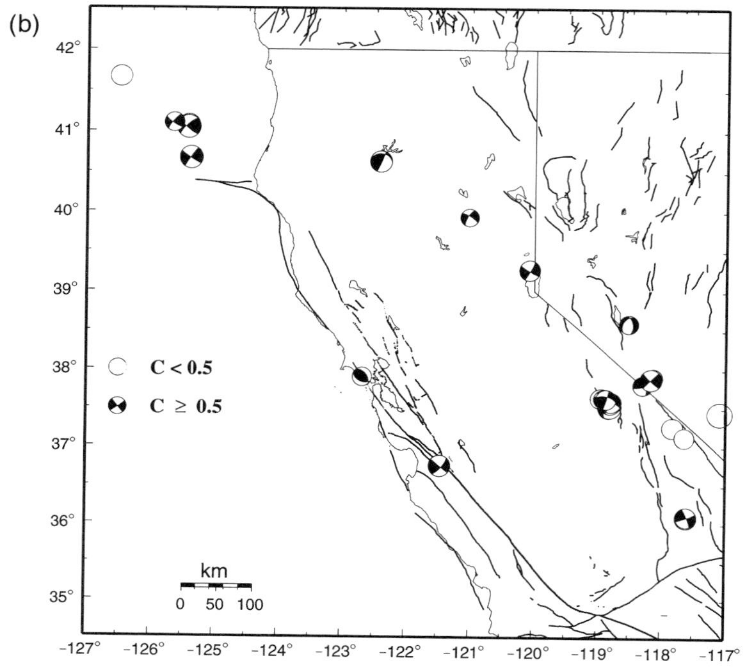

FIGURE 3 Maps showing the correlation between the two automatic moment tensor solutions. (a) displays all events from 1997 to 1999, while (b) shows events with $M \geq 4.5$. Larger events tend to have better-correlated moment tensor solutions.

fault (due to insufficient separation in variance reduction), stage 6 will compute the full inversion for both fault planes. In the present implementation, the full inversion requires an additional 20–30 min per plane, depending on the resolution, on a Sun UltraSPARC1/200e.

This extension of REDI processing is running on a development system at the BSL, where it is being tested. We anticipate migrating these modules to the production systems in the fall of 2000. At the time of the 1999 Hector Mine earthquake, the REDI implementation was not complete. However, the approach was tested using several regional-distance stations from TriNet. Dreger and Kaverina (2000) first used the line-source inversion to identify the causative fault plane and then obtained the distribution of slip on the NW-trending plane using the best-fitting rupture velocity. The image of the finite rupture obtained by this procedure agreed well with aftershock locations and observed surface rupture.

2.2.6 New Developments

There are several new developments of the REDI system under way. To complement the finite-fault parametrization in stages 5 and 6, we are working toward the estimation of fault rupture using GPS (Murray *et al.*, 1996). Geodetic measurements of coseismic displacements provide important constraints on the extent and magnitude of fault slip that are poorly resolved by seismic data. Even greater information can be gained by combining the often complementary geodetic and seismic observations (e.g., Wald *et al.*, 1996). We are developing and implementing methods for determining coseismic displacements from the GPS data in near real time and for rapidly combining this information with seismic observations to provide better finite-fault and strong-motion estimates. As planned, this system would parallel the REDI processing and provide an independent estimate of the source parameters.

The quantification of a finite rupture is important for emergency response as it may be used to identify communities directly affected by the faulting process. In addition, these parameters may be used to predict areas of strong shaking following a major earthquake. A slip map can be used to generate synthetic seismograms on a fixed grid in the hypocentral region and processed to yield values of PGA, PGV, and spectral acceleration. The predicted values may be combined with observed data and contoured to generate maps of strong ground shaking and fully account for an extended rupture.

These developments complement the "ShakeMap" efforts in southern California (Wald *et al.*, 1999), where the dense station distribution of TriNet allows for "data-driven" maps of strong ground shaking to be produced within a few minutes after an event. In this approach, observed data values are interpolated assuming a point source centered at the strong-motion centroid (Kanamori, 1993) and the attenuation relationship of Joyner and Boore (1981). Values are then corrected for site conditions based on a Quaternary/Tertiary/Mesozoic geological map. This implementation is fast, and maps can be produced within a few minutes after an event. This methodology is extremely successful in areas with a dense concentration of strong-motion stations with real-time or near real time telemetry and has been implemented in the San Francisco Bay Area (Boatwright *et al.*, 1999). In areas with limited strong-motion instrumentation, the point-source approximation will limit the accuracy of these maps, due to the greater reliance on predicted values.

We are working toward the development of a system in which a series of ground-motion maps will be generated following major earthquakes, based on the available information. For example, the TriNet ShakeMaps would represent the first-generation map produced immediately following the event. This map could be updated within a few minutes based on the mechanism and an attenuation relationship with a directivity term (e.g., Somerville *et al.*, 1997) following the line-source inversion. A third map could be produced after the full 2-D distribution of slip is obtained. Essentially, this spectrum of maps accounts for improved prediction of ground motions as more information about the source becomes available. This is particularly important for areas with limited strong-motion instrumentation, such as northern California outside of the San Francisco Bay Area.

A final aspect of current REDI development is focused on the flow of information within the system. At the present, REDI uses simple flat files to store results. Whenever a stage is initiated, it reads a summary file containing the results from previous stages. We are in the process of replacing the flat file system with a database. A database will provide more flexibility in terms of information flow and will allow us to implement more distributed computing. This is particularly important from the perspective of generating "ShakeMaps" as they are cpu-intensive products.

2.3 Event Notification

A critical component to any automated processing system is the distribution of information. As part of the USGS-BSL earthquake notification system, the REDI software is designed to perform notification at each event processing stage and currently uses radio paging and the Internet to distribute information. As in the Caltech-USGS Broadcast of Earthquakes (CUBE) system, the REDI system distributes messages by pager for use with computer displays as well as for notification of key personnel. The REDI system also distributes messages by e-mail and uses the WWW to make information available to the general public.

In order to facilitate the distribution of pages, we have established a direct line with the commercial paging service. This avoids the problem of dialing the provider during periods of high demand. The dedicated line is supplemented by dial-out phone lines, and the paging system is verified daily through test notifications. In order to avoid calling individual pagers, we have established "capcodes" for mass notification. Multiple pagers can be programmed with the same capcode, which makes it possible to reach a large number of pagers with a single notification.

Over the last few years, the WWW has become an increasingly important medium for distributing earthquake information. Data from the REDI project are incorporated into the "California Recent Earthquakes" Web page at *http://quake.usgs.gov/recenteqs/* and in the northern California "ShakeMaps" Web page at *http://quake.wr.usgs.gov/study/effects/*.

3. System Implementation

At present, two Earthworm/Earlybird systems in Menlo Park feed two REDI processing systems at UC Berkeley (Fig. 2). One of these systems is the production or paging system; the

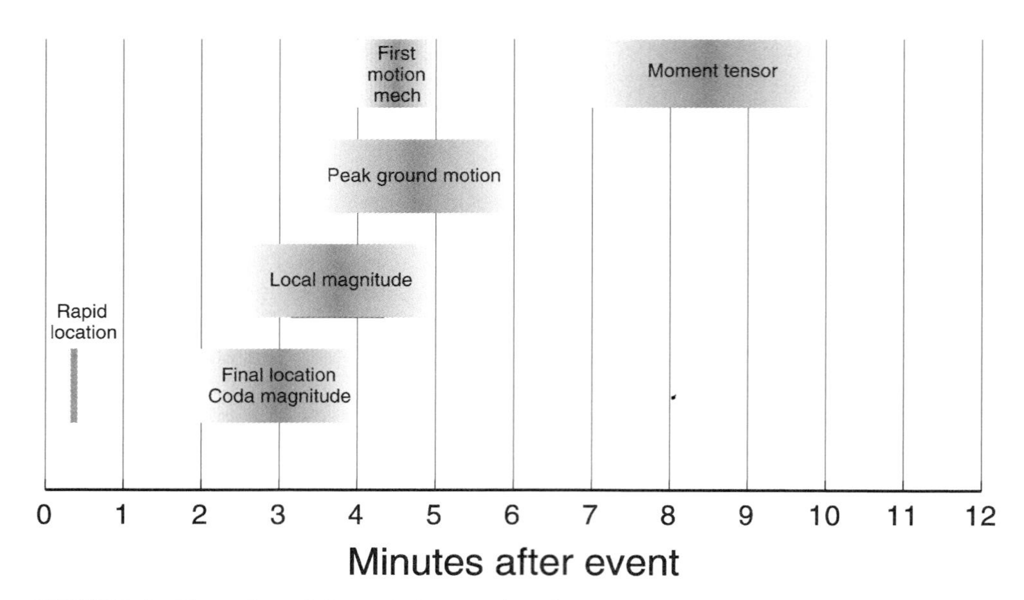

FIGURE 4 Illustration of the typical processing times associated with the current USGS-BSL earthquake notification system in northern California.

other is set up as a hot backup. The second system is frequently used to test new software developments before migrating them to the production environment.

In the current configuration, these two REDI systems (a Sun Ultra 2170 and a Sun Ultra 2200) perform both data acquisition for the BDSN as well as REDI processing. Each machine acquires data from approximately half the network, divided in order to minimize the impact if one of the acquisition systems fails. Waveform data is shared between the two computers, allowing each REDI processing system access to the entire network. In addition, the BSL operates a third REDI system, which uses BDSN picks to form an independent list of associated events. This third system provides redundancy in case the communication links with the USGS Menlo Park are disrupted. A fourth system, discussed below, is installed at the OES Warning Center in Sacramento in order to provide a backup processing facility outside of the Bay Area.

This structure has facilitated automatic earthquake processing in northern California (Fig. 4). The dense network and Earthworm/Earlybird processing environment of the NCSN provides rapid and accurate earthquake locations, low magnitude detection thresholds, and first-motion mechanisms for smaller earthquakes. The high-dynamic-range dataloggers, digital telemetry, and broadband and strong-motion sensors of the BDSN and REDI analysis software provide reliable magnitude determination, moment tensor estimation, peak ground motions, and source rupture characteristics. Robust preliminary hypocenters are available about 25 seconds after the origin time, while preliminary code magnitudes follow within 2–4 min. Estimates of local magnitude are generally available 30–120 sec later, and other parameters, such as the peak ground acceleration and moment magnitude, follow within 1–4 min.

As noted earlier, earthquake information from the joint notification system is distributed by pager, e-mail, and the WWW. The first two mechanisms "push" the information to recipients, while the current Web interface requires interested parties to actively seek the information. Consequently, paging and, to a lesser extent, e-mail are the preferred methods for emergency response notification. The Northern California Web site has enjoyed enormous popularity since its introduction and provides a valuable resource for information whose bandwidth exceeds the limits of wireless systems and for access to information that is useful not only in the seconds immediately after an earthquake but in the following hours and days as well.

3.1 Monitoring System State of Health

An important component of the REDI system is the monitoring of heartbeats and data flow from critical systems and processes. Because this is inherently a distributed system, it is critical to monitor the "health" of every component. In the REDI system, the monitor program is a master scheduler that can perform several types of monitoring. Its configuration file tells it how often to perform each instance of each type of monitoring, to whom (and how) alarm conditions should be reported, whether error notification should be repeated on a regular basis for persistent problems, and whether a notification message should be generated when an alarm condition is cleared.

The monitor program can perform the following types of monitoring operations.

1. *Computer and network:* The monitor program can "ping" other computers to ensure that they are alive and raise an alarm if no response is received within a specified amount of time.

2. *Disk resources:* The monitor program can directly monitor available disk resources and raise an alarm if available disk space drops below a specified level.
3. *Process table:* The monitor program can check to see that specified processes are running and raise an alarm if that process dies.
4. *Subsystem monitoring:* The monitor program can run any specified program that examines the state of health of a particular subsystem of the REDI processing or BDSN data acquisition. If the program runs successfully without generating any output, it indicates that the subsystem is OK. Otherwise, the output text from the program identifies the error condition of the subsystem.

Each specific monitoring function is described by a single line in the monitor configuration file that contains the monitoring type, monitoring command and arguments, a notification list consisting of a list of users and notification methods, how often the monitoring command should be run, and alarm notification flags that indicate how often renotification of outstanding alarm conditions should occur, and whether to generate an "alarm cleared" message when the alarm condition has been cleared.

The monitor program "remembers" the previous state of each monitoring command that it runs. When an alarm condition is raised by a specific command, it performs the specified alarm notification and marks that system to be in an alarm state. It will continue to run that monitor command, but it will not generate a new alarm message each time. The configuration file specifies how often (if at all) a new alarm notification message should be generated for each alarm condition, and whether a notification message should be generated when an alarm condition is cleared.

The REDI system uses both e-mail and alphanumeric pages to distribute alarm notification messages. New distribution mechanisms can be simply added by specifying a new distribution keyword and the corresponding distribution program to the monitor configuration file. New subsystem monitoring can be easily added by creating a subsystem monitor program that conforms to the monitor specification, and installing a scheduling line for it in the monitor configuration file.

The REDI system currently monitors two subsystems. The first is the availability of BDSN waveform data. The BDSN waveform data used by the REDI processing system is placed in a shared memory region for rapid access by the various processing programs. A monitoring program computes the latency of the most recent sample of data for each data channel and generates an alarm if either a single channel of data or a combination of channels exceeds specified latency values. This subsystem monitor alerts us to data telemetry or network problems that affect our data acquisition system.

The second monitored subsystem is BSL-USGS communication. All of the programs that exchange data between the BSL computers and the USGS Earthworm and Earlybird computers exchange heartbeat messages at regular intervals. A heartbeat message serves two functions: It informs the remote program that the computer is still alive *and* that data can be delivered over the active IP socket in a timely fashion. The time of each local and remote heartbeat message and the time when each distinct type of data was last received from the remote system is logged in a shared memory system or time-stamp file. A subsystem monitoring program is run every minute to detect whether there is a problem with any aspect of data exchange between the REDI system and the USGS systems. This subsystem monitoring will notify us of a network outage between the REDI and USGS computers, a loss of synchronization between any pair of REDI and USGS processes, and any other condition that causes the REDI system to not receive expected data from the USGS computers.

3.2 Reliability and Redundancy

The implementation of the two REDI systems as part of the joint notification system in northern California (Fig. 2) provides some form of redundancy. However, there are a number of potential issues that have an impact on the reliability and redundacy of the system.

The communication between the USGS and BSL systems is based on a T1 frame-relay connection between Menlo Park and Berkeley. When this dedicated link fails, the system "falls back" to using the Internet. Both of these landlines are vulnerable to disruption during an earthquake. Similarly, most of the incoming BDSN data flows through a second T1 frame-relay connection to the BSL. Loss of this T1 would interrupt data flow to the REDI processing system. While a few stations use alternate means such as radio and NSN satellite, the incoming T1 is a significant point of failure.

Recognizing this—and conscious of the BSL's location within a kilometer of the Hayward fault—we have established a backup processing facility in the Sacramento headquarters of the California Office of Emergency Services. This site receives data from six BDSN stations and operates independently of the UC Berkeley operations. Although utilizing a small subset of the BDSN, the system has the capability to detect and locate moderate-size earthquakes.

To set up this prototype redundant processing facility, we installed a Sun Ultra 140 with a Vanguard 320 FRAD in Sacramento. A 56-Kbps frame-relay connection was established and divided into seven 8-Kbps circuits. Six of these circuits are used for data acquisition from the stations YBH, ORV, CMB, SAO, BKS, and HOPS, while the seventh circuit is used for a connection between the OES site and UC Berkeley. Although 8 Kbps is a fairly "narrow" telemetry pipe (and we initially experienced some problems in establishing these circuits because of the limited bandwidth), it is sufficient to transmit continuous data at 100 sps from the strong-motion sensors, and 20 and 1 sps from the broadband seismometers installed at these sites. In effect, these six stations are supporting two data acquisition processes, one in Berkeley and one in Sacramento. The data acquisition in Sacramento does not depend on any operations in Berkeley.

The OES system has been operational since late September 1998, using the same REDI software installed at UC Berkeley.

Preliminary analysis indicates that the facility is operating well. In the 4 months from September to December 1998, a number of small events and eight earthquakes of magnitude 4 and higher were processed. In nearly all cases, the location based on the subset of BDSN stations was within a few kilometers of the joint notification determination. Problems have been experienced with events significantly outside the network, and part of this difficulty is due to the lack of *S*-wave picks to constrain the location outside of the network. Our work with the three-component picker holds promise for rectifying this situation as well as providing estimates of arrival azimuth that can be used to further constrain the location.

3.3 System Performance

Northern California has not experienced the same seismic challenges as southern California in recent years. However, the August 12, 1998, earthquake near San Juan Bautista provides a simple illustration of the joint notification system (Uhrhammer *et al.*, 1999). Although this is a relatively minor event (M_D 5.3, M_L 5.3, M_W 5.2; the only reported injuries were a teenager falling from a bunk bed and a painter falling from a ladder [San Francisco Chronicle, 8/13/98]), it is notable as the first event for which the REDI system paged strong ground motion data.

The first alert of the event at the BSL came when the threshold signal detector at the station BKS triggered (except, of course, for those personnel who felt the earthquake) at 22 sec after the origin. The BSL monitors the threshold detector at BKS as a preliminary indicator of an event of interest. Triggering at a velocity of 10 microns/sec, the "seismic alarm" notifies BSL personnel by pager and serves as a "wake-up call" (literally, in this case) that an event is in progress. This seismic alarm page is one example of continuous data processing at the BSL.

A "quicklook" location from the Earthworm system, utilizing data from the Northern California Seismic Network at the USGS Menlo Park, was produced within 30 sec, based on *P*-wave detections at the nearest 25 stations. A final automatic solution with a coda magnitude of 5.3 was transmitted to the REDI processing system at UC Berkeley 217 sec after the event occurred and posted on the Web page.

The REDI processing system utilized BDSN data for the determination of local magnitude, ground motions, and the seismic moment tensor. Using the preliminary magnitude as a guide, the REDI system processed BDSN waveforms from 14 stations to determine a local magnitude of 5.4. The Earthworm location and the REDI local magnitude were paged 272 seconds after the event, and the magnitude information was transmitted to the USGS. Data from BDSN strong-motion instruments were processed, and peak ground accelerations that exceeded 0.5% *g* were distributed by pager (12% *g* was observed at the BDSN station SAO, which is located about 1 km from the epicenter). In the final step of automated processing for this event, the seismic moment tensor and moment magnitude were determined. The automatic moment magnitude of 5.1, the strike-slip mechanism (strike 130/dip 89/rake −164), and the centroid depth of 8 km compare extremely well with the reviewed solution of magnitude 5.1, mechanism (strike 129/dip 85/rake −171), and depth of 8 km. All REDI processing was complete a total of 11 minutes after the origin of the event.

Although a small event, this earthquake was widely felt. The automatic earthquake information produced by the UCB/USGS system provided valuable information to the emergency response managers. For example, the Peninsula Commute Service, which operates train service on the San Francisco Peninsula, was able to decide that only 1 train of the 14 running from Gilroy to San Francisco needed to be stopped while the track was inspected.

4. Conclusions

The Rapid Earthquake Data Integration System has been developed for the automated estimation of earthquake parameters using data from a sparse, broadband network. The system is designed on the basis of a staged hierarchy of processing, with the goal of providing control of the type and number of processes running at any time. The current processing capabilities include the determination of local and energy magnitude, peak ground motions, and the seismic moment tensor. Current development efforts are focused on extending REDI processing to include the determination of finite-fault parameters and the generation of maps of strong ground shaking.

Acknowledgments

The REDI system was developed in cooperation with related projects at the USGS and Caltech. We have benefited from the software development of the Earthworm group in Menlo Park and Golden and from the TriNet group in Pasadena. We are grateful to the efforts of David Oppenheimer, Bruce Julian, and Lynn Dietz toward the establishment and operation of the joint earthquake notification system in northern California. We thank Hiroo Kanamori, Egill Hauksson, Tom Heaton, and Phil Maechling for fruitful exchanges of experience and ideas. Steve Fulton at UC Berkeley has put considerable effort into the development of the picking algorithm and computed the station corrections for the energy magnitude.

References

Ammon, C., A. Velasco, and T. Lay (1993). Rapid Estimation of rupture directivity: application to the 1992 Landers (Ms=7.4) and Cape Mendocino (Ms=7.2) California Earthquakes. *Geophys. Res. Lett.* **20**, 97–100.

Bakun, W. H., F. G. Fischer, E. G. Jensen, and J. VanSchaack (1994). Early Warning System for aftershocks. *Bull. Seism. Soc. Am.* **84**, 359–365.

Boatwright, J., H. Bundock, J. Luetgert, D. Oppenheimer, L. Baker, J. Fletcher, J. Evans, C. Stephens, K. Fogelman, L. Gee, D. Carver, V. Graizer, C. Scrivner, and M. Mclaren (1999). Implementing ShakeMap in the San Francisco Bay Area. *EOS, Trans. Am. Geophys. Un.* **80**, F701.

Buland R., and W. Person (1992). Earthquake early alerting service. *EOS, Trans. Am. Geophys. Un.* **73**(43), 69.

Cohee, B., and G. Beroza (1994). A comparison of two methods for earthquake source inversion using strong motion seismograms. *Ann. Geophys.* **37**, 1515–1538.

Dreger, D., and A. Kaverina (1999). Development of Procedures for the Rapid Estimation of Ground Shaking Task 7: Ground Motion Estimates for Emergency Response. PEER Final Report. University of California, Berkeley, Seismological Laboratory, 32 pp.

Dreger, D., and A. Kaverina (2000). Seismic remote sensing for the earthquake source process and near-source shaking: a case study of the October 16, 1999 Hector Mine Earthquake. *Geophys. Res. Lett.* **27**, 1941–1944.

Dreger, D., and B. Romanowicz (1994). Source characteristics of events in the San Francisco Bay region. *Open-File-Report 94-176*, pp. 301–309, US Geological Survey, Menlo Park, California.

Dreger, D., R. Uhrhammer, M. Pasyanos, J. Franck, and B. Romanowicz (1998). Regional and far-regional earthquake locations and source parameters using sparse broadband networks: a test on the Ridgecrest sequence. *Bull. Seism. Soc. Am.* **88**, 1353–1362.

Ekstrom, G. (1992). A system for automatic earthquake analysis. *EOS, Trans. Am. Geophys. Un.* **73**(43), 70.

Espinosa-Aranda, J., A. Jimenez, G. Ibarrola, F. Alcantar, A. Aguilar, M. Inostroza, and S. Maldonado (1995). Mexico City Seismic Alert System. *Seism. Res. Lett.* **66**, 42–53.

Gee, L., S. Fulton, D. Dreger, A. Kaverina, D. Neuhauser, M. Murray, and B. Romanowicz (2000). Recent Enhancements to the Rapid Earthquake Data Intgeration System (abstract). *Seism. Res. Lett.* **71**, 232.

Gee, L., D. Neuhauser, D. Dreger, M. Pasyanos, B. Romanowicz, and R. Uhrhammer (1996a). The Rapid Earthquake Data Integration System. *Bull. Seism. Soc. Am.* **86**, 936–945.

Gee, L., A. Bittenbinder, B. Bogaert, L. Dietz, D. Dreger, B. Julian, W. Kohler, A. Michael, D. Neuhauser, D. Oppenheimer, M. Pasyanos, B. Romanowicz, and R. Uhrhammer (1996b). Collaborative Earthquake Notification in Northern California (abstract). *EOS, Trans. Am. Geophys. Un.* **77**, F451.

Heaton, T. H. (1985). A model for a seismic computerized alert network. *Science* **228**, 987–990.

Hirshorn, B., A. G. Lindh, R. V. Allen, and C. Johnson (1993). Real time magnitude estimation for a prototype early warning system (EWS) from the P-wave, and for earthquake hazards monitoring from the coda envelope. *Seism. Res. Lett.* **64**, 48.

Johnson, C., A. Bittenbinder, B. Bogaert, L. Dietz, and W. Kohler (1995). Earthworm: a flexible approach to seismic network processing. *IRIS Newsletter* **XIV**(2), 1–4.

Joyner, W., and D. Boore (1981). Peak horizontal accelerations and velocity from strong-motion records including records from the 1979 Imperial Valley, California, earthquake. *Bull. Seism. Soc. Am.* **71**, 2011–2038.

Kanamori, H. (1993). Locating earthquakes with amplitude: application to real-time seismology. *Bull. Seism. Soc. Am.* **83**, 264–268.

Kanamori, H., E. Hauksson, and T. Heaton (1991). TERRAscope and CUBE project at Caltech. *EOS, Trans. Am. Geophys. Un.* **72**, 564.

Kanamori, H., P. Maechling, and E. Hauksson (1999). Continuous monitoring of ground-motion parameters. *Bull. Seism. Soc. Am.* **89**, 311–316.

Kanamori, H., J. Mori, E. Hauksson, T. Heaton, L. Hutton, and L. Jones (1993). Determination of earthquake energy release and M_L using TERRASCOPE. *Bull. Seism. Soc. Am.* **83**, 330–346.

Kaverina, A., D. Dreger, and M. Antolik (1997). Toward automating finite fault slip inversions for regional events. *EOS, Trans. Am. Geophys. Un.* **78**, F45.

Malone, S. (1994). A review of seismic data access techniques over the Internet, *EOS*, **75**, 429.

Murdock, J., and C. Hutt (1983). A new event detector designed for the Seismic Research Observatories. *Open-File Report 83-0785*, US Geological Survey, 39 pp.

Nakamura, Y., and B. E. Tucker (1988). Japan's earthquake warning system: Should it be imported to California? *California Geology* **41**, 33–40.

National Research Council (1991). *Real-time Earthquake Monitoring*. National Academy Press, Washington, D.C.

Romanowicz, B., D. Dreger, M. Pasyanos, and R. Uhrhammer (1993). Monitoring of strain release in central and northern California using broadband data. *Geophys. Res. Lett.* **20**, 1643–1646.

Romanowicz, B., G. Anderson, L. Gee, R. McKenzie, D. Neuhauser, M. Pasyanos, and R. Uhrhammer (1992). Real-time seismology at UC Berkeley. *EOS, Trans. Am. Geophys. Un.* **73**(43), 69.

Pasyanos, M., D. Dreger, and B. Romanowicz (1996). Toward real-time estimation of regional moment tensors. *Bull. Seism. Soc. Am.* **86**, 1255–1269.

Scherbaum, F., and M.-P. Bouin (1997). FIR filter effects and nucleation phases. *Geophys. J. Int.* **130**, 661–668.

Somerville, P. G., N. F. Smith, R. W. Graves, and N. A. Abrahamson (1997). Modification of empirical strong ground motion attenuation relations to include the amplitude and duration effects of rupture directivity. *Seism. Res. Lett.* **68**, 199–222.

Uhrhammer, R. (1982). The optimal estimation of earthquake parameters. *Phys. Earth Planet. Interiors* **30**(2/3), 105–118.

Uhrhammer, R., S. Loper, and B. Romanowicz (1996). Determination of local magnitude using BDSN broadband records. *Bull. Seism. Soc. Am.* **86**, 1314–1330.

Uhrhammer, R., L. S. Gee, M. Murray, D. Dreger, and B. Romanowicz (1999). The M_w 5.1 San Juan Bautista, California earthquake of 12 August 1998. *Seism. Res. Lett.* **70**, 10–18.

US Geological Survey (1999). An Assessment of Seismic Monitoring in the United States—Requirement for an Advanced National Seismic System. *US Geol. Surv. Circular 1188*, 55 pp.

Wald, D., T. Heaton, and K. Hudnut (1996). The slip history of the 1994 Northridge, CA, earthquake determined from strong motion, teleseismic, GPS and leveling data. *Bull. Seism. Soc. Am.* **86**, s49–s70.

Wald, D., V. Quitoriano, T. Heaton, H. Kanamori, C. Scrivner, and C. Worden (1999). TriNet "ShakeMaps": Rapid generation of peak ground motion and intensity maps for earthquakes in southern California. *Earthquake Spectra* **15**, 537–556.

Wells, D. L., and K. J. Coppersmith (1994). New empirical relationships among magnitude, rupture length, rupture width, rupture area, and surface displacement. *Bull. Seism. Soc. Am.* **84**, 974–1002.

Wu, Y.-M., T.-C. Shin, C.-C. Chen, Y.-B. Tsai, W. H. K. Lee, and T. L. Teng (1997). Taiwan Rapid Earthquake Information Release System. *Seism. Res. Lett.* **68**, 931–943.

78

TriNet: A Modern Ground-Motion Seismic Network

Egill Hauksson
Seismological Laboratory, California Institute of Technology, Pasadena, California, USA
Lucile M. Jones
US Geological Survey, Pasadena, California, USA
Anthony F. Shakal
Office of Strong Motion Studies, Division of Mines and Geology, Sacramento, California, USA

1. Introduction

In the early 1990s, southern California experienced the 1992 M_W 7.3 Landers and the 1994 M_W 6.7 Northridge earthquakes. Both events vividly demonstrated the need for improved instrumentation to capture data from such large and damaging earthquakes. In particular, the Southern California Seismic Network (SCSN) did not perform as well as the US Geological Survey (USGS) and California Institute of Technology (Caltech) had hoped. Signals from the short-period sensors were clipped for the mainshocks and large aftershocks. The large volumes of data saturated the computers at the central site, and providing data for research and information for emergency response was difficult at best (Kanamori *et al.*, 1997). Similarly, in the 1994 Northridge earthquake, the lack of strong-motion instruments in the vicinity of damaged structures made it difficult to evaluate the cause of damage to those structures (e.g., Krawinkler *et al.*, 1995).

To improve the instrumentation for recording ground motions in southern California, the USGS, Caltech, and California Division of Mines and Geology (CDMG) began planning the TriNet project in early 1995 (Heaton *et al.*, 1996; Mori *et al.*, 1998a; and Mori *et al.*, 1998b). The Governor's Office of Emergency Services (OES) encouraged the TriNet agencies to cooperate and form the TriNet partnership. Thus, OES fostered the development of new technologies and information products, needed for improving emergency response and measuring the effectiveness of long-term mitigation.

New developments in technology also made cooperation between weak- and strong-motion networks feasible and desirable. The TERRAscope, a network of 24 digital stations added to the SCSN beginning in 1988, was equipped with both broadband and strong-motion sensors and provided data on-scale for both the 1992 Landers and the 1994 Northridge earthquakes (Kanamori *et al.*, 1991). Further, the analog film instruments used by CDMG for strong-motion recording were being surpassed in data quality by new digital strong-motion recorders. New CDMG digital strong-motion instruments could communicate via dial-up telephone line to a central facility and thus provide information near real time (Shakal *et al.*, 1995). The TriNet agencies realized that a coordinated broadband and strong-motion network was technically possible and that combining resources could lead to new high-quality products for scientific engineering and emergency management applications.

TriNet also benefited from the development of other seismic networks around the world. The Mexico City early warning system that has been in operation since 1991 provides helpful insights about how to provide seismic alerts (Espinosa-Aranda *et al.*, 1995; Chapter 76 by Espinosa-Aranda). Extensive development of strong-motion networks in Taiwan and associated near real time processing of data show how TriNet-like systems can aid in emergency response (Teng *et al.*, 1997; Chapter 64 by Shin *et al.*).

As new capabilities are added to TriNet, they are being brought online as quickly as possible. This makes it possible to test these capabilities in a realistic environment. The purpose of this chapter is to document (1) the objectives of TriNet; (2) the TriNet model for cooperation between regional seismic and strong-motion networks; and (3) the present status of the implementation of TriNet.

INTERNATIONAL HANDBOOK OF EARTHQUAKE AND ENGINEERING SEISMOLOGY, VOLUME 81B ISBN: 0-12-440658-0

2. Objectives

The broad mission of TriNet is to mitigate the impact of future large earthquakes in southern California. To accomplish this mission, a wide spectrum of activities is required, including data collection and processing, information distribution, and seismological and earthquake engineering research and data analysis. The following are the five major TriNet activities or objectives directed at accomplishing this mission.

2.1 Seismic Network and Strong-Motion Program

TriNet agencies will operate a resilient seismic network and a strong-motion program to record all earthquake ground motions of interest in southern California. The instrument spacing must be dense enough to document the true distribution of ground motions, at scale of a few kilometers, and robust enough to operate successfully in even the largest possible earthquakes.

2.2 Rapid Distribution of Information

TriNet will distribute information about an earthquake rapidly after occurrence to save lives and property, by facilitating rapid, more effective decision making and mitigating actions such as focussed search and rescue, fire prevention, and deployment of engineers and inspectors for building inspection. TriNet will determine and broadcast accurate estimates of earthquake parameters, such as magnitude and location within 1 minute of the ending of an earthquake rupture and will distribute preliminary estimates of the ground shaking (ShakeMap) within 3 minutes of rupture initiation for moderate and large events. These estimates will be updated as further information is available.

2.3 Database of Earthquake Information

TriNet will create an easily accessible database of earthquake information in southern California for seismological and engineering research. This will include a catalog of parameters for all earthquakes of magnitude 1.8 and greater onshore, and 2.5 offshore within the SCSN reporting area. A waveform database of high-fidelity on-scale data from broadband sensors and accelerometers will be maintained by Caltech, and a strong-motion database with data from moderate to major earthquakes will be maintained by CDMG. All data will be easily accessible to researchers and practitioners.

2.4 Prototype Alert System

TriNet will develop a prototype alert system that will inform that an earthquake has begun before damaging shaking arrives at more distant sites. TriNet will conduct social science research on the use of such data to be ready to implement when funds for future TriNet enhancements, including sufficient stations, are obtained.

2.5 Interagency Cooperation

The TriNet agencies will cooperate with other agencies working to mitigate the earthquake hazard in southern California with the recording, analysis, and distribution of information. This includes the California Governor's Office of Emergency Services, the Federal Emergency Management Agency, the Southern California Earthquake Center, and the US Council of the National Seismic System. TriNet will ensure that software systems created under this project will be available to other regional seismic networks working toward similar goals.

3. Organization

The organization of the TriNet project is structured to give equal representation to the TriNet members and to facilitate accomplishment of the objectives of the project. The TriNet project is headed by a steering committee with representatives from CDMG, Caltech, USGS, OES, and the TriNet Advisory Committee (TAC). The TAC advises the steering committee, with three representatives from outside organizations, appointed by each of the three TriNet agencies, and one representative of OES and one of SCEC. In addition, one member of the advisory committee is a member at large. Being an external, user-oriented committee, TAC has provided important guidance to the implementation of the TriNet project.

The TriNet administrative working group (TAWG) has one representative from each of the three TriNet institutions. The TAWG manages ongoing administrative issues related to the implementation of TriNet. For instance, TAWG coordinates site selection for stations, data exchange, and development of joint products.

Two subcommittees report to TAWG. The ShakeMap working group implements ShakeMap, a rapidly distributed map of ground shaking intensities, through cooperative efforts from all three TriNet institutions. The TriNet outreach committee advises the TriNet outreach manager about relevant issues and makes recommendations to TAWG.

4. New Approaches and Technologies

To accomplish its mission, TriNet is applying new technologies enabling a much greater dynamic range of sensors and bandwidth of digital communication links to provide accurate and vital information within minutes of the occurrence of a large earthquake.

Caltech's Seismological Laboratory and the USGS are adding 150 new stations to the existing seismic network, the Southern

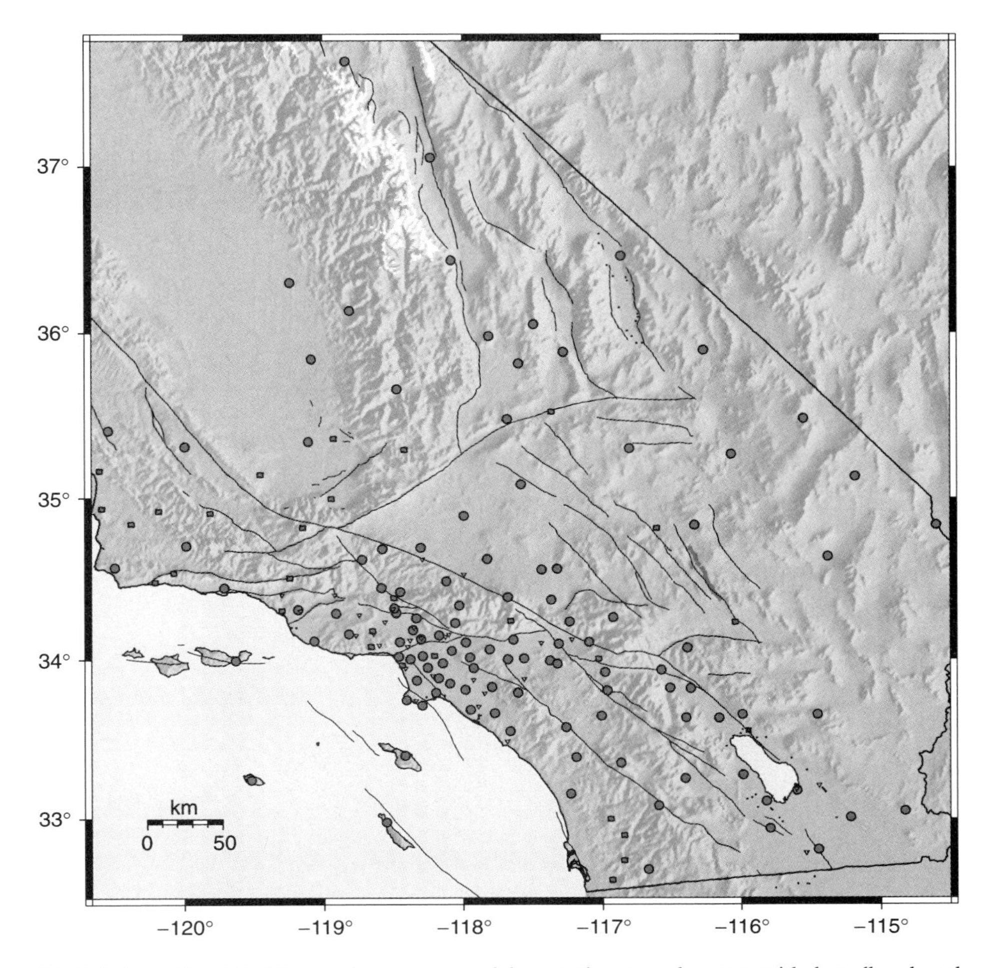

FIGURE 1 SCSN/TriNet stations connected by continuous telemetry with broadband and strong-motion sensors. Included are 120 existing stations (circles), 40 planned stations (squares), and 30 stations with strong-motion sensors only (inverted triangles).

California Seismic Network, with real-time data transmission and processing capability and new digital dataloggers, broadband sensors, and strong-motion sensors (Fig. 1).

CDMG is upgrading and expanding strong-motion instrumentation at 400 sites in southern California as part of their existing California Strong Motion Instrumentation Program (CSMIP). The CDMG TriNet stations are equipped with phone lines and configured to dial out to Sacramento when triggered by earthquake shaking (Fig. 2). The USGS National Strong Motion Program (USGS/NSMP) has upgraded 75 of their current southern California sites to retrieve near real time ground-motion parameters. Both the CSMIP and the USGS/NSMP projects collect data for improving building codes as well as for ShakeMap.

4.1 Broadband and Strong Motions

TriNet is designed to fill the roles of both traditional seismic networks and strong-motion programs and builds on the experience of TERRAscope (Kanamori *et al.*, 1991). Stations equipped with both broadband and strong-motion sensors record ground motions on-scale from ground noise to 200% *g*. In addition, by co-locating strong and weak motion and telemetering strong-motions, we achieve: (1) accurate real-time magnitudes; (2) near real time mapping of strong shaking for emergency response; and (3) expanded studies of site effects and source mechanisms using weak motions from strong-motion sites. Some of the advantages of a coordinated broadband and strong-motion network are the capability of generating ShakeMap, improved geographical distribution of sites, and efficiencies gained through sharing of software and operations strategies.

The broadband sensors are intended to record site response over a wide frequency band. In particular, unusual site response may occur at stations in the basins related to resonance at long wavelength (corresponding to periods of 3 to 10 sec) because of the basin shape or unusually soft ground conditions. The broadband sensors are well suited to collect such a database quickly because they record on-scale data for both small- and moderate-sized regional earthquakes.

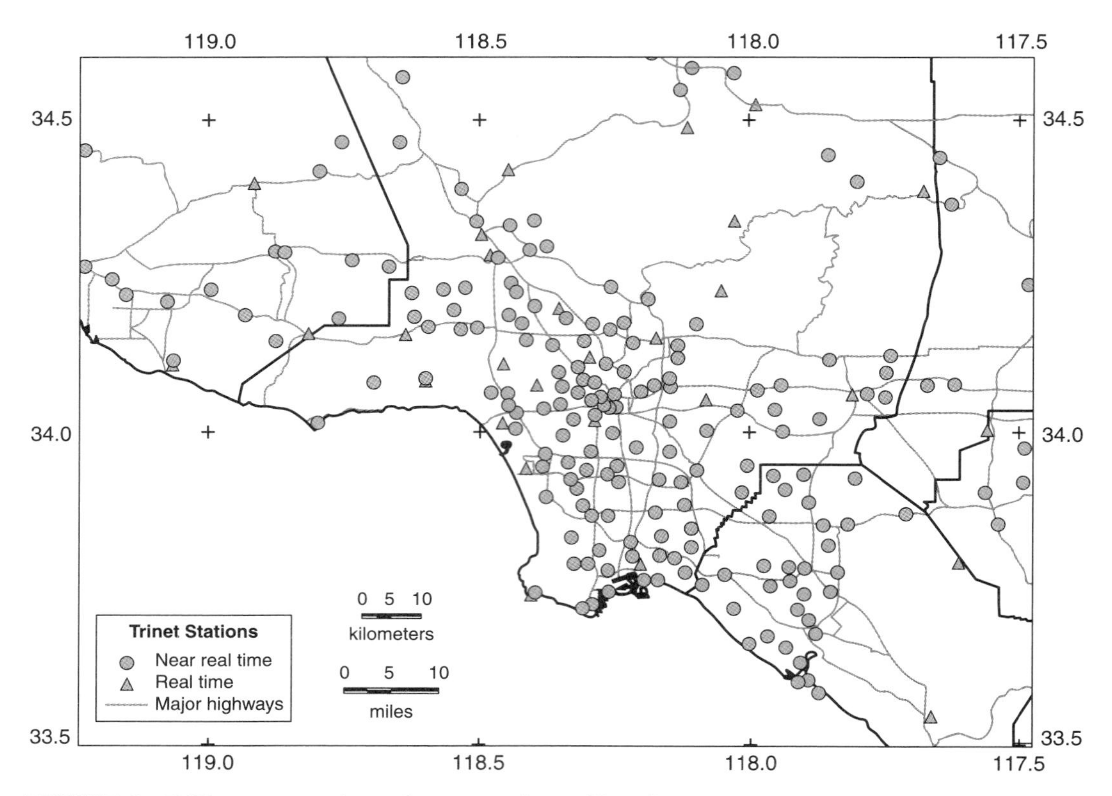

FIGURE 2 TriNet strong-motion and strong-motion and broadband stations in the greater Los Angeles area operated by the TriNet agencies. CDMG/TriNet and Caltech/USGS TriNet stations located outside the greater Los Angeles area are not shown. Both planned and installed stations are shown.

Stations equipped with only strong-motion sensors and digital recorders are capable of recording down to moderate-sized earthquakes, thus creating a database of ground motion for each site much more quickly than was possible in the past. This capability also makes it unnecessary to deploy portable recorders at these sites following a large mainshock.

4.2 Data Processing and Databases

Two data processing and archiving centers handle the TriNet data. In Sacramento, the CDMG processes event-triggered near real time strong-motion data from the 400 strong-motion TriNet stations. Caltech and USGS in Pasadena acquire real-time data streams from 150 broadband and strong-motion stations and 50 strong-motion stations.

The CDMG acquisition software uses commercial communication software to communicate with the field stations. A series of computers are used and provide redundancy, robustness, and capacity. Processing software parallels the software used in the traditional processing performed at CDMG for years, but it is upgraded to run in an automated, robust mode. Data products for the earthquake engineering community include acceleration, velocity and displacement time series, and response spectra, and they are made available from the CSMIP strong-motion database. The data are also available for seismological applications, through the Caltech-maintained seismological database in formats and conventions most often used in seismological research.

The Caltech/USGS software design of the real-time data acquisition and processing system is built around a commercial database system from Oracle Corporation. The software systems are being developed in cooperation with UC Berkeley, USGS Menlo Park, and other regional networks. The commercial database as the center of the real-time acquisition system provides: (1) replication technology for system robustness and rapid redundancy; (2) commercial standards for data archiving and retrieval; (3) automatic archiving; and (4) data accessible on the World Wide Web as soon as they are recorded.

Caltech/USGS perform real-time processing on more than 1200 data channels. The data are processed in RAM before they are archived to disk to facilitate rapid dissemination of information such as hypocenter and magnitude.

4.3 Digital Telemetry

In general, digital telemetry facilitates both station maintenance and rapid data retrieval. The state of health can be checked instantaneously while data are being sent either continuously or on demand.

The CDMG stations use dial-out telemetry with regular phone lines. This allows routine assessment of state of health as well as near real time retrieval of strong-motion data following significant earthquakes. This approach reduces maintenance costs by limiting field visits to those instruments and stations actually requiring repair or service, rather than periodically visiting all field sites at a fixed schedule. CDMG has stations throughout southern California, in various phone company jurisdictions, and high-baud-rate, dependable connections are generally available. Cellular telephone connections are used at sites where conventional lines are unavailable or too expensive. In addition, Integrated Services Digital Network (ISDN) lines are used for data connection to certain arrays or well-instrumented structures in urban areas. In a major earthquake, phone communication may become unavailable at certain stations in the area of strongest shaking, but the large number of stations in the nearby area allows the shaking pattern to be defined for the ShakeMap. In any case, the recorded data is stored on-site and can be recovered later.

All of the Caltech/USGS TriNet stations are connected with Internet Protocol (IP) communications links to the central site in Pasadena. This makes it possible to transmit User Datagram Protocol (UDP) packets of continuous data streams to the central computer. As data are being transmitted we can also access the station for state-of-health information or for software upgrades.

Several different telemetry technologies are employed to implement the IP links, including frame-relay digital telephone lines, spread spectrum radios, and digital microwave. The use of different telemetry technologies allows for improved robustness by eliminating some single points of failure.

4.4 Automation

TriNet acquires and processes all data streams and distributes earthquake parameters automatically. To maximize the use of these data for emergency response in the Caltech/USGS TriNet element, the real-time data are used to generate parametric information automatically. This information is sent without human intervention to users. A duty seismologist checks the validity before confirming or canceling the information after the fact. Automatically distributed parameters include: (1) hypocenter; (2) origin date and time; (3) magnitude; (4) various hypocentral quality parameters; (5) peak accelerations at the 10 closest stations; and (6) ShakeMap.

The CDMG TriNet element records, processes, and transmits to the ShakeMap computers strong-motion data in a fully automated way. A dedicated frame-relay circuit between Sacramento and Pasadena ensures a reliable, high-speed connection. The data from the CDMG stations are combined with the parametric data produced by the real-time network system at Pasadena and transferred to a dedicated computer that generates the ShakeMap products automatically.

In the future we plan to distribute alerts that an earthquake is under way. These alerts may arrive at some sites just before the strong ground shaking. We are developing the capability to estimate the distribution of shaking for an earthquake under way as the beginning of an alert system. We plan to work with a small group of potential users to understand how this type of shaking information can be most useful.

4.5 Reliability

One of the objectives of TriNet is to operate successfully during and after a major earthquake. To accomplish this, many redundancies are built into the system. All TriNet stations have on-site recording capability for strong motion, and in some cases for broadband data, as well as data transmission or telemetry to a central site. To increase the reliability of TriNet, several different methods of telemetry are used to transmit the data to two different central sites. The CDMG/TriNet stations use commercial telephone, cellular, and ISDN service to transmit data to Sacramento. The Caltech/USGS stations use a variety of independent digital communications such as frame relay, spread spectrum radios, and wide area computer networks made available by the local utilities, microwave, and the Internet to transmit data to Pasadena. At the central site in Pasadena two data acquisition computers located in two different buildings receive the data. Stations with high-bandwidth communications links send data to both acquisition computers. Digital strong-motion stations send data to a third, dedicated computer that redistributes the data to the two acquisition computers. Backup power is provided for all data acquisition equipment at the central site and at the field recording sites.

At the CDMG site in Sacramento, a bank of independent personal computers handles incoming calls from stations triggered by earthquake shaking. The computers take incoming calls in sequence. Field stations that cannot reach a computer in Sacramento at a given moment are programmed to try again after randomized delay intervals, starting from a few seconds. Any one of the computers may fail without affecting the rest. Backup power is provided for the computers, and a dedicated high-speed communication line connects Sacramento and Pasadena.

The Caltech/USGS TriNet software is stress tested using synthetic data generated from recordings of the M_W 7.3 Landers earthquake. In particular, this makes it possible to detect potential problems in the parametric software that is broadcast as part of the rapid notification. The early versions of ShakeMap are generated in Pasadena using the real-time data from the Caltech/USGS stations (Wald *et al.*, 1999b). Later updates are provided within 30 minutes by incorporating the CDMG amplitudes that are forwarded to Caltech. To increase reliability, CDMG also generates independent ShakeMaps based on all available data in Sacramento. These maps are made available on the Internet and via RIMS, an emergency information system operated by OES, to emergency responders.

Although many redundancies have been built into TriNet, it is possible that a major earthquake may make the TriNet central site in Pasadena inoperable. To cover such a catastrophe, we will have a remote backup site, probably located in northern California. Such a backup site would receive data directly from the remote stations, and thus would not be dependent on the presence of the Pasadena central site.

5. Products

TriNet products consist of raw data, processed data, and parametric information derived from the data. The data and the derived parametric information that are made available to users become available in different time frames.

5.1 Prototype Alert System

A prototype seismic alert system to determine whether the concept is truly feasible for deployment in southern California is still in the design stage. We plan to develop the capability to experiment with seismic alerting by distribution of information that an earthquake is under way before the shear waves have arrived at sites where damage may occur (Heaton, 1985; Monastersky, 1998).

The alerts may provide an alert time of seconds to several tens of seconds before the most damaging shaking arrives. In general, this is useful mostly for large or major distant earthquakes and will probably not provide a valuable alert for nearby moderate-sized earthquakes. In southern California, the greatest benefit would be gained from an alert in Los Angeles of an M 7.5–8.0 earthquake occurring on the San Andreas fault. Some regions of very poor soil conditions could benefit from seismic alerts because ground motions from both nearby and distant earthquakes can be amplified and may constitute a significant hazard.

Some potential uses for alerts are stopping trains, interrupting transporting of toxic chemicals, securing databases, alerting schoolchildren, and bringing elevators to the closest floor. The present level of funding for TriNet, however, will not allow the level of robustness and redundancy needed for a regionwide deployment of a seismic computerized alert system.

5.2 Rapid Notification

TriNet is working to ensure that emergency response professionals, the media, seismologists, and earthquake engineers in southern California will have the information they need immediately after an earthquake. The real-time data transmission from the 200 TriNet stations makes it possible for TriNet to provide rapid notification and improve the Caltech-USGS Broadcast of Earthquakes (CUBE) (Kanamori *et al*., 1991). This notification is provided via pagers, posted on the Internet in text format and on maps, and via e-mail.

The most basic products for an earthquake are the date, time, location, depth, magnitude, and some quality factors associated with these parameters. The goal is to distribute this information within 1 min of the origin time of a moderate- to large-sized earthquake. The determination of the magnitude of a major earthquake on the San Andreas fault may take more than a minute, if the rupture is not completed by that time.

ShakeMap, a map of the distribution of ground shaking following an $M > 3.5$ earthquake, is a new product developed with TriNet funding. The ShakeMap provides several contour maps of instrumental intensity, peak acceleration, peak velocity, and selected response spectral parameters. The objective of TriNet is to post on the TriNet Web site (www.trinet.org) the first version of ShakeMap within 3 min of a significant earthquake. Within 30 min, an updated version is provided with significant enhancement of data from the 400 CDMG TriNet stations.

TriNet uses recursive data filtering developed by Kanamori *et al.* (1999) to generate real-time amplitude values for TriNet. The actual production and implementation of ShakeMap is discussed by Wald *et al.* (1999b).

5.3 Frequency-Dependent Ground-Motion Maps

One of the objectives of TriNet is to develop frequency-dependent ground-motion maps to facilitate land-use planning. This task uses ground motions from many earthquakes to determine shaking levels in several different frequency bands. These maps will be used to complement other hazards maps based on geological and other types of information.

5.4 Data Archives

The TriNet data are available to all users through WWW interfaces at *http://www.trinet.org*. The Caltech/USGS TriNet data are archived and distributed by the Southern California Earthquake Data Center (*www.scecdc.scec.org*), which is operated by Caltech. Datasets of all broadband velocity waveforms and strong-motion waveforms for all earthquakes of interest are maintained and are available in standard seismological formats (i.e., mini-SEED).

The CDMG TriNet data are routinely archived and distributed through the CSMIP Web interface as part of TriNet. These data are stored in standard CSMIP and COSMOS formats for engineering use.

Following a major earthquake in southern California, two datasets in CSMIP and SEED format are assembled for the mainshock. A strong-motion dataset that includes data from all TriNet stations is quality checked by CDMG and distributed via the TriNet Web site. These data are also mirrored at Web sites that traditionally distribute strong-motion data such as CDMG and USGS/NSMP. Caltech and USGS in Pasadena make a seismological dataset available via the SCEC Data Center. The seismological dataset also includes broadband velocity waveforms.

5.5 Research

TriNet data are being used in both applied and basic research in seismology and earthquake engineering. The applied research is aimed at earthquake hazards issues such as understanding site effects and other parameters that may affect the development of potentially damaging ground motions. The fundamental research that is not directly supported by TriNet is focused on improving understanding of earthquake physics and earth structure. The high quality and high density of digital data enable more accurate and detailed studies of earthquake hazards, seismotectonics, crustal velocities, source processes, and characteristics of the interior of the Earth, among many other important research topics. Two sample record sections are shown in

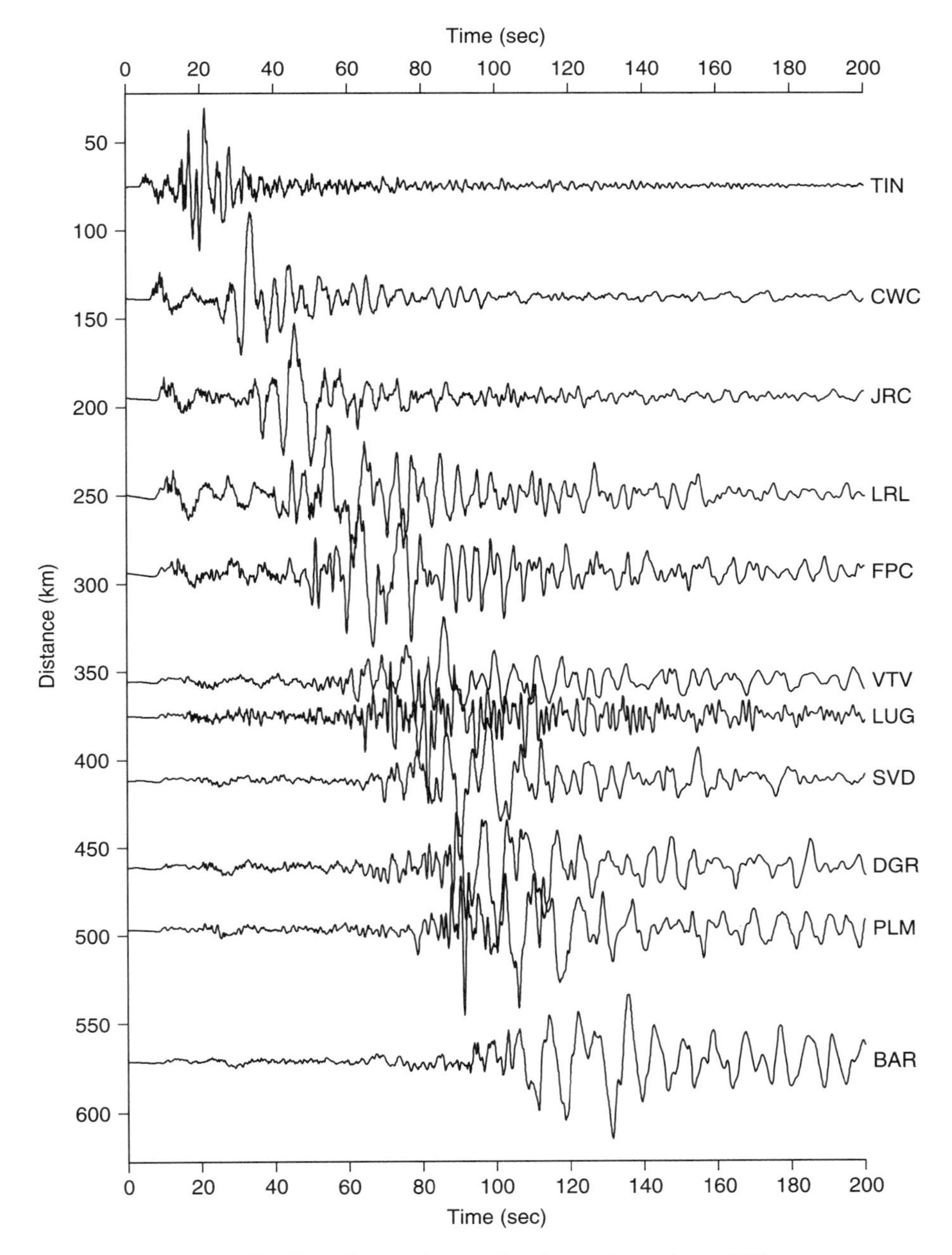

FIGURE 3 The availability of on-scale broadband waveforms from TriNet opens up new areas for research. Selected stations recording an *M* 5.7 earthquake in Mammoth Lakes (north of the TriNet region). All the regional waveforms from 60 km to 600 km are recorded on-scale.

Figures 3 and 4. The regional event that was located near Mammoth Lakes shows crustal phases as these pass through the crust of southern California. The teleseism shows both crustal and upper mantle phases across northern Mexico and southern California.

5.6 Communications and Outreach

TriNet has an active communications and outreach program that works with users on the refinement of TriNet products and education on the uses of theproducts. The outreach program includes data utilization studies in both earthquake engineering and social science, and public policy aspects of the prototype alert system. As an example, a workshop on ShakeMap development was held in November 1998 to get user feedback about future development of ShakeMap (Goltz and Wald, 1998). At this workshop, the users stressed that they preferred a timely product over a highly refined product that may take a long time to prepare.

To facilitate the development and deployment of the prototype computerized alert system, TriNet is funding studies by

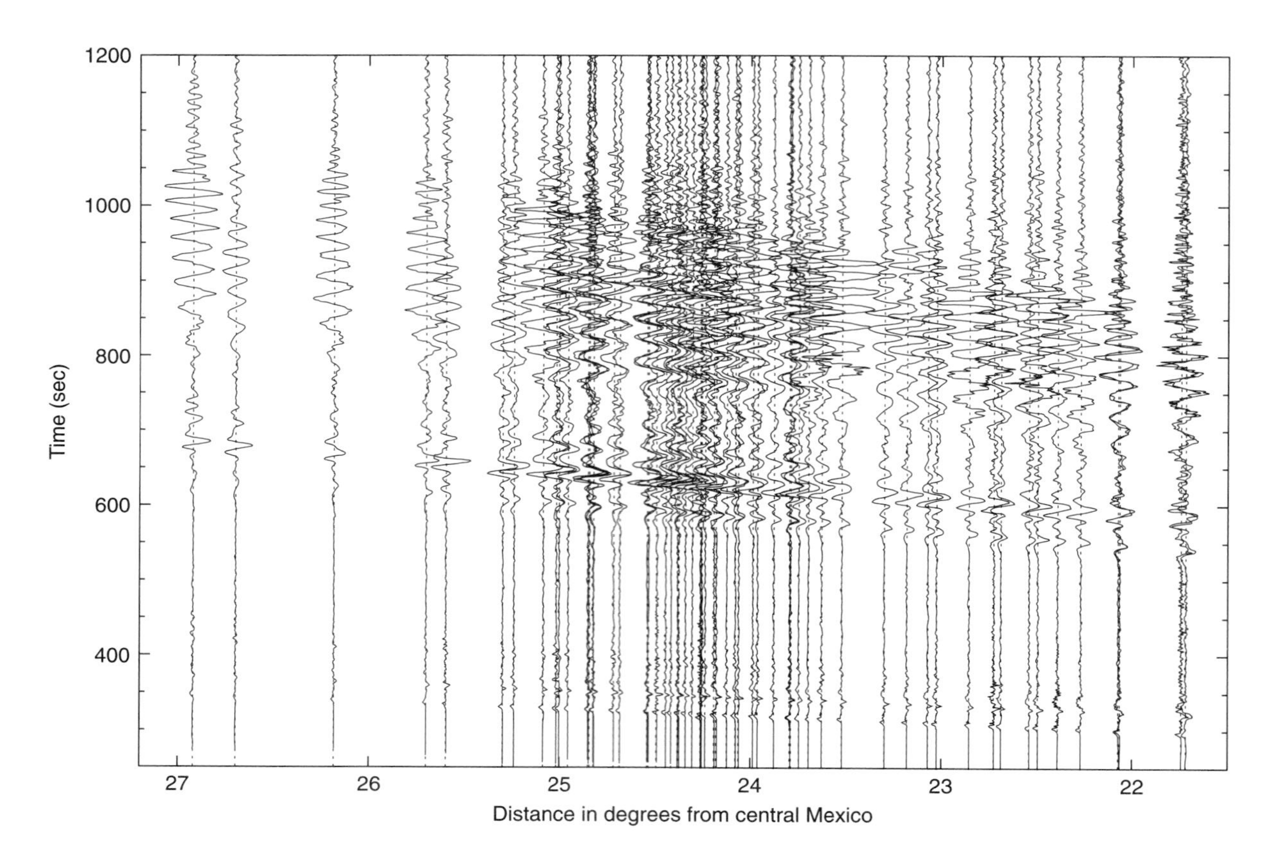

FIGURE 4 Teleseismic waveforms recorded from the M_W 6.7 earthquake in central Mexico on June 15, 1999. Eighty stations recorded the event on-scale. The development of the wave field across southern California is clearly recorded, allowing for studies of detailed Earth structure.

experts in social science and public policy. These studies address how local governments, utilities, and others could use information about earthquake shaking either immediately after or just before the shaking arrives at their sites. The subcontractor uses focus groups to perform studies that include identification of potential users and actions. The contractor also developed recommendations for the format of possible alert messages and approach to training and education based on how these issues are handled in other equivalent emergency situations. An important part of these studies is identification and analysis of public policy issues raised by an earthquake alert and identification of what agency should issue the alerts. The contractor also assisted with developing recommendations and plans for deployment of a prototype computerized seismic alert system.

6. Engineering Data Utilization

As part of the TriNet effort, the Data Utilization project of CDMG was expanded to accelerate the process of improving seismic design codes and provisions based on the analysis of recorded strong-motion data. TriNet records a wealth of data to allow improved understanding of ground motion for engineering design. In addition, an important existing set of data for code improvement is that recorded by code-required instruments in buildings in the Los Angeles area during the Northridge earthquake. For many of the buildings, Northridge was the strongest motion experienced since they were built; in a significant number, damage occurred to the steel moment frame. The utilization of these data to uncover the lessons to be learned from the response of steel and other buildings during strong shaking was made possible through TriNet by a digitization project in which many of these analog records were digitized and processed. Strong shaking records from over 100 buildings have been digitized, and several data utilization projects are under way. An important example is the work by Naeim *et al.* (1999) on detailed analyses of several steel-framed buildings damaged in the Northridge shaking, which will result in proposed code revisions. The ground motion itself is also important for engineering design, especially the ability to infer strong shaking from studies of weak motion, as studied by Su *et al.* (1999). Focused data interpretation studies of this type are a key part of bringing closure to the efforts to record ground motion under TriNet, bringing the data to bear on the problem of improving seismic design and zoning practices.

7. TriNet Performance during the 1999 M_W 7.1 Hector Mine Earthquake

The Hector Mine earthquake provided a good test and evaluation of the performance of TriNet in general and the TriNet ShakeMap. As of October 1999, there were about 120 USGS-Caltech TriNet real-time stations online. In addition, about 200 CDMG TriNet and other strong-motion stations whose data is accessed via a dial-out protocol were operational. For the Hector Mine earthquake, a location and preliminary M_L estimate of 6.6 were obtained approximately 90 sec after the event, and an energy magnitude of M 7.0 was obtained approximately 30 sec later.

Earthquake location and magnitude estimates for significant events are broadcast via e-mail, the World Wide Web, and pager messages via the CUBE system. The first CUBE page following Hector Mine, including the M_L 6.6 value, broadcast 90 sec after the event and was received approximately 3.5 min after the event; an updated page with the estimate M 7.0 was received approximately 30 sec later.

TriNet produces "ShakeMap," which involves rapid (within 3–5 min) generation of maps of instrumental data acquisition combined with newly developed relationships between recorded ground-motion parameters and expected shaking intensity values (Wald *et al.*, 1999b). Production of maps is automatic, triggered by any significant earthquake in southern California (Wald *et al.*, 1999a). The first ShakeMap for the Hector Mine earthquake was produced within 4 min of the event. Initially, ShakeMap ground motions in the near-source region were estimated using ground-motion regression from a point location at the epicenter. During the hours following the earthquake, information about fault dimensions became available (in the form of aftershock distribution, source rupture models, and observed surface rupture) and were incorporated into ShakeMap. Thus, the fault location and dimensions served as the basis for ground-motion estimation in the near-source region in part because the source region is remote and only one TriNet station was located within 25 km of the surface rupture.

TriNet has provided funds to improve loss estimation tools used by the California Office of Emergency Services (OES) to include ShakeMap as input data. Although ShakeMap data files were used for rapid loss estimation calculations for the Hector earthquake, the results were not of significant societal impact because the event occurred in an uninhabited region and only caused moderate shaking in sparsely populated regions of the Mojave Desert.

8. Conclusions

TriNet is a new approach to gathering and disseminating information about earthquakes. It is a multifunctional system that allows a diverse group of users and researchers to accomplish differing goals within the framework of one network. TriNet provides data for research in seismology, for research in earthquake engineering, and for improving decision making in emergency response.

TriNet is not only about capturing data from a major earthquake. It is also about how to get diverse groups of stakeholders and networks to work together toward a common goal. The ShakeMap is an example of a product that has emerged from TriNet, not only because TriNet has new technical capabilities, but because several diverse user groups have agreed that they can all use a common product. The recommendations and review by the TAC advisory committee have significantly improved the quality of these products. The emergency responders want the instrumental intensity ShakeMap for immediate potential impact assessment. The earthquake engineers want spectral response ShakeMaps for inferring potential damage to structures. The seismologists want ground velocity maps to infer rupture direction and site effects. TriNet has formed the necessary umbrella to allow all of these groups to accomplish their goals. TriNet and its products are larger than the sum of its parts as demonstrated in the 1999 M_W 7.1 Hector Mine earthquake.

Acknowledgments

Funding for TriNet has been provided by the Federal Emergency Management Agency (FEMA) and the California Governor's Office of Emergency Services (OES) through the Hazards Mitigation Grant Program established following the 1994 Northridge earthquake. The required 25% cost-sharing is provided by California Division of Mines and Geology, California Institute of Technology, California Trade and Commerce Agency, Caltrans, Ida H. L. Crotty, Federal Emergency Management Agency, GTE California, Donna and Greg Jenkins, Pacific Bell/CalREN, Southern California Edison, Sun Microsystems Inc., Times Mirror Foundation, and others. Funding has also been provided by the United States Geological Survey from their special Northridge funds. We thank the TriNet staff in both Pasadena and Sacramento for making TriNet become a reality. J. Polet and J. Chen provided record sections and L. Gee provided a detailed review. Division of Geological and Planetary Sciences, contribution number 8675.

References

Espinosa-Aranda., J. M., A. Jimenez, G. Ibarrola, F. Alcantar, A. Aguilar, M. Inostroza, and S. Maldonado (1995). Mexico City Seismic Alert System. *Seism. Res. Lett.* **66**, 42–53.

Goltz, J., and D. Wald (1998). TriNet ShakeMap Workshop, California Institute of Technology, November 2, 1998: Report of proceedings and recommended actions; unpublished report; available from *http://www.trinet.org*.

Heaton, T. H. (1985). A model for seismic computerized alert network. *Science* **28**, 987–990.

Heaton, T. H., R. Clayton, J. Davis, E. Hauksson, L. Jones, H. Kanamori, J. Mori, R. Porcella, and T. Shakal (1996). The TriNet Project. *Proc. 11th World Conf. Earthq. Eng., Paper No. 2136, Acapulco, Mexico.*

Kanamori, H., E. Hauksson, and T. Heaton (1991). TERRAscope and CUBE Project at Caltech. *EOS, Trans. Am. Geophys. Un.* **72**, 564.

Kanamori, H., E. Hauksson, and T. Heaton (1997). Real-time seismology and earthquake hazard mitigation. *Nature* **390**, 461–464.

Kanamori, H., P. Maechling, and E. Hauksson (1999). Continuous monitoring of ground-motion parameters. *Bull. Seism. Soc. Am.* **89**, 311–316.

Krawinkler, H., J. C. Anderson, and V. V. Bertero (1995). Steel buildings. In: "Northridge Earthquake of January 17, 1994 Reconnaissance Report," W. T. Holmes and P. Somers (Editors), Volume 2: Supplement C to volume 11 of *Earthquake Spectra*, pp. 25–47, Earthquake Engineering Research Institute, Oakland, CA.

Monastersky, R. (1998). Racing the waves: Seismologists try to catch quake tremors quickly enough to save lives. *Science News* **153**, 169–153.

Mori. J., H. Kanamori, J. Davis, E. Hauksson, R. Clayton, T. Heaton, L. Jones, T. Shakal, and R. Porcella (1998a). Major improvements in progress for southern California earthquake monitoring. *EOS, Trans. Am. Geophys. Un.* **79**, 217, 221.

Mori. J., H. Kanamori, J. Davis, E. Hauksson, R. Clayton, T. Heaton, L. Jones, T. Shakal, and R. Porcella (1998b). Major improvements in progress for southern California earthquake monitoring (Reprint of Eos Article, includes color figures), *Calif. Geology* **51**(6), 14–17.

Naeim, F., R. Lobo, K. Skliros, and M. Sgambelluri (1999). Seismic performance of four instrumented steel moment resisting buildings during the January 17, 1999 Northridge earthquake. *SMIP99 Proceedings*, pp. 51–81, Calif. Div. Mines and Geology, Sacramento, CA.

Shakal, A., C. Peterson, *et al.* (1995). CSMIP near real time strong motion monitoring system: Rapid data recovery and processing for event response. *SMIP95 Proceedings*, pp. 1–10, Calif. Div. Mines and Geology, Sacramento, CA.

Su, F., Y. Zeng, J. Anderson, and V. Graizer (1999). Site response of weak and strong ground motion including non linearity. *SMIP99 Proceedings*, pp. 1–21, Calif. Div. Mines and Geology, Sacramento, CA.

Teng, T.-L., L. Wu, T.-C. Shin, Y.-B. Tsai, and W. H. K. Lee (1997). One minute after: Strong-motion map, effective epicenter, and effective magnitude. *Bull. Seism. Soc. Am.* **87**, 1209–1219.

Wald, D. J., V. Quitoriano, T. H. Heaton, H. Kanamori, C. W. Scrivner, and C. B. Worden (1999a). TriNet "ShakeMaps": Rapid generation of peak ground motion and intensity maps for earthquakes in southern California. *Earthquake Spectra* **15**(3), 537–556.

Wald, D. J., V. Quitoriano, T. H. Heaton, and H. Kanamori (1999b). Relationships between peak ground acceleration, peak ground velocity and modified Mercalli intensity in California. *Earthquake Spectra* **15**(3), 557–564.

National and International Reports to IASPEI

79

Centennial National and Institutional Reports: Seismology and Physics of the Earth's Interior

79.1 General Introduction

Carl Kisslinger
CIRES, University of Colorado, Boulder, CO, USA

The history of IASPEI and its contributions to world science is, essentially, the story of contributions by IASPEI's Member Countries to the development of seismology and the physics of the Earth's interior during the past 100 years. In many nations in highly seismic settings, documentation of the study of earthquakes extends back to their early recorded histories. For the century of its existence, IASPEI, often in collaboration with other IUGG Associations, has played a vital role in promoting and fostering cooperative efforts among research groups from diverse nations, encouraging scientific advances in less-developed nations, and providing a forum for international exchange of ideas and dissemination of research findings. The actual planning, initiation, and successful completion of the research programs under the aegis of IASPEI have been the products of the efforts of thousands of individual scientists and research teams, working through their research institutions, universities, and national laboratories.

This chapter comprises summaries extracted and excerpted from the National and Institutional Reports submitted by those nations and research entities that chose to prepare one. A few recently independent nations that are not now formally affiliated with IUGG and IASPEI, but were formerly part of a Member Country, did contribute reports, for which the editors are grateful.

These summaries were edited and prepared in final form in order to achieve uniformity of style and format. The revised versions were sent to the original authors for approval. Some editing was done to provide consistency in transliterating into the Latin alphabet from other alphabets or writing systems. We did retain British or American spellings of English words, as preferred by the authors.

These summaries are the products of much thoughtful effort by the authors, often teams of cooperating scientists within the nation. Practical limitations on the size of a printed volume require that we offer only capsule versions as guides to important stories of science in progress. The complete national and institutional reports, with figures, maps, tables, bibliographies, and biographies of key scientists, as well as some attached materials, such as earthquake catalogs and historically important documents, constitute a volume much greater than the printed Handbook. All of this material is reproduced on the accompanying Handbook CD#2 so that this storehouse of valuable historical and contemporary information is preserved and easily accessible.

These reports convey the development of our science and the contributions made by each nation and institution, often in the context of the international exchanges and cooperation that are characteristic of the earth sciences. An examination of the past 100 years of worldwide IASPEI-related activities also provides some insight into the impacts of political and economic history on the efforts of dedicated individuals and their organizations during the often troubled and turbulent 20th century. Many of the accounts begin in the first decade of the 20th century, in the years before World War I. These reveal the effects of the disruptions caused by that war and World War II, as well as the changes wrought by subsequent political upheavals and the restructuring of nations in many parts of the world. The step-by-step transition of seismology during the 1920s and 1930s to a quantitative,

INTERNATIONAL HANDBOOK OF EARTHQUAKE AND ENGINEERING SEISMOLOGY, VOLUME 81B ISBN: 0-12-440658-0

physics-based science, and the associated advances in understanding the evolution, structure, and physics of the Earth's interior, are documented in the combined chronicles of the national programs and research organizations that produced these advances.

Many of the institutions and some of the nations reviewed here came into existence after World War II and more recently, as a result of reorganization of older research entities and of nations themselves. These individual histories may cover only a relatively short period of time, but are often rooted in a distinguished past that holds keys to the development of contemporary geophysics. For example, some of the present Member Countries arose with the breakup of the Austro-Hungarian monarchy after World War I, others from the creation of independent nations in post-colonial Africa and southeast Asia, and, recently, from the political developments in the former Union of Soviet Socialist Republics and Yugoslavia. The summaries and reports are arranged alphabetically by the English names of the nations, but integrating the reviews of nations now independent but parts of regional groupings or former parts of larger national entities can provide a more complete and historically informative picture of the contributions to science of a particular geographic region.

The narratives in these national and institutional reports demonstrate, in the context of scientific progress, that the histories of nations and institutions are the stories of the combined efforts of thousands of dedicated individuals and of the national structures that support their work. The achievement of the goals and dreams of innovators often depends on their determination to succeed in the face of difficult circumstances. Finally, taken together, these reports confirm that the science of the Earth would not be at its present high level without the international cooperation fostered by IASPEI, IUGG, and other multi-national organizations.

The following sub-chapters contain national and institutional reports submitted to IASPEI from 56 countries: Albania, Argentina, Armenia, Australia, Austria, Azerbaijan, Belgium, Bolivia, Brazil, Bulgaria, Canada, Chile, China (Beijing), China (Taipei), Croatia, Czech Republic, Denmark, El Salvador, Estonia, Ethiopia, France, Georgia, Germany, Greece, Hungary, Iceland, India, Indonesia, Iran, Israel, Italy, Japan, Kazakhstan, Kyrgyzstan, Luxembourg, Macedonia, Mexico, Moldova, New Zealand, Norway, Philippines, Poland, Portugal, Russia, Slovakia, Slovenia, South Africa, Spain, Sweden, Switzerland, Turkey, Ukraine, United Kingdom, United States of America, Uzbekistan, and Vietnam.

79.2 Albania

The editors received a report, "The Development of Seismology in Albania," from Dr. Betim Muço, the IASPEI National Correspondent, Seismological Institute of Academy of Sciences of Albania. We extracted and excerpted this summary. The full report is archived on the attached Handbook CD#2, under the directory of \7902Albania. It describes the most important milestones of development of seismology in Albania, its progress and contributions to the world community of scientists in their challenging war against earthquakes. It also contains a brief description of research projects and publications. An inventory of the Albanian Seismic Network is included in Chapter 87, and biographic sketches of some Albanian seismologists may be found in Appendix 3 of this Handbook.

1. History and Introduction

The history of seismology in Albania began, for the most part, in the second half of the 20th century. The first seismic map of the country is dated 1942, prepared for the needs of Italian troops that occupied Albania during World War II.

The first catalog of Albanian earthquakes was prepared in 1951 and a seismic map of Albania was compiled for construction purposes in 1952. The construction and installation of the seismological station of Tirana in 1968 was an important milestone in the development of seismology in Albania. Some works carried out from 1950 through 1975 were: The Seismic Map of Albania on the scale of 1:1.250.000, the Catalogue of Earthquakes of Albania, and the Seismotectonic Map of Albania on the scale of 1: 1.000.000.

The year 1976 might be considered as the beginning of the Albanian Seismological Network (ASN). The practice of organization of work in these stations and homogenization of magnitude formulas and other routine procedures were objects of a long and continuous work from the responsible staff of the Center. Some new directions were opened for the studies of the Seismological Center of Albania. The most important works produced during this period are: Seismic Zonation of Albania in the scale 1:500.000, The Earthquake of April 15, 1979, Microzonations Studies of different Cities of Albania , and The Series of Earthquakes of Nikaj-Merturi.

Great changes since 1991 followed on the opening and democratization of Albania. New opportunities have been created for the Seismological Center of Albania to communicate and cooperate with similar institutions worldwide. Since January 1994, the Seismological Center has been renamed the Seismological Institute.

2. Research Activities

The Seismological Institute of Albania remains the only institution of the country in charge of seismic monitoring, studies of seismology, seismotectonics, engineering seismology, and earthquake engineering. The Albanian Seismic Network (ASN) has ten seismological stations under operation, mostly with digital dial-up instruments. The Seismological Institute continues to compile the Monthly Seismological Bulletins distributed to interested institutions and organizations abroad. Since 1968 the data of these bulletins are included in the monthly publication of the International Seismological Center, Newbury, Great Britain.

During the years, the Institute has collaborated with many neighboring similar institutions, such as the Institute of Earthquake Engineering and Engineering Seismology, Skopje, FYROM; Department of Geophysics of Aristotle University of Thessaloniki, Thessaloniki, Greece; National Geophysical Institute, Rome, Italy; Department of Earth Sciences of University

INTERNATIONAL HANDBOOK OF EARTHQUAKE AND ENGINEERING SEISMOLOGY, VOLUME 81B ISBN: 0-12-440658-0

of Trieste, Trieste, Italy; Seismological Institute of Montenegro, Podgorica, Montenegro. In Albania, the Seismological Institute has strong ties with the Faculty of Geology and Mining of Polytechnic University of Tirana; Albanian Geological Service; Faculty of Civil Engineering of Polytechnic University of Tirana; Geological Institute of Oil, Fier; Council of Regulations of Territory of Albania, and others.

3. Heads/Directors of Center/Institute

1968–1972: Head, Seismological Station of Tirana: Eduard Sulstarova
1973–1993: Director, Seismological Center of Tirana: Eduard Sulstarova
1993 (March to August): Director, Seismological Center of Tirana: Siasi Koçiu
1993 (September)–1997 (August): Director, Seismological Institute: Betim Muço
1997 (September)–1999 (July): Director, Seismological Institute: Shyqyri Aliaj
1999 (July)–2001 (September): Director of Seismological Institute: Siasi Koçiu
2001 (September)–present: Director of Seismological Institute: Shyqyri Aliaj

4. Some Important Publications of the Seismological Institute of Albania

1. Sulstarova, E. and Koçiaj, S. (1975). The Catalogue of Earthquakes of Albania.
2. Sulstarova, E., Koçiaj, S. and Aliaj, Sh. (1972). The Seismic Map of Albania on the scale 1:1.250,000.
3. Aliaj, Sh., Koçiaj, S and Sulstarova, E. (1973). The Seismotectonic Map of Albania on the scale 1:1.000,000.
4. Sulstarova, E., Koçiaj, S. and Aliaj, Sh. (1980). The Seismic Zonation of Albania on the scale 1:500,000.
5. The Earthquake of April 15, 1979 (papers collected from the specialists from Albania, Montenegro, and other countries on the Symposium organized for the abovementioned earthquake, on April 15, 1980, at Shkodra) (1983). 8 Nentori Publishing House (Skender Dede, Ed.).
6. Muço, B. (1983). The Series of Earthquakes of Nikaj-Merturi.
7. Group of authors (1989). The Technical Design and Construction Code.
8. The periodical publications of the Institute (different authors, 1984–1990).
9. AJNTS, (2001). Deterministic Approach of Seismic Zonation of Some Balkan Countries, No. 10. (B. Muço and P. Fuga, Eds.).

79.3 Argentina

The editors received the Argentine National Centennial Report from Dr. Claudia L. Ravazzoli, Secretary of the Argentine Subcommittee of Seismology. The report is written in Spanish by Dr. Nora Sabbione, the National Correspondent to IASPEI, and Dr. Ravazzoli. The full report is archived as a computer readable file on the attached Handbook CD#2, under the directory of \7903Argentina. A summary that was extracted and excerpted from the full report is shown here. An inventory of seismic networks in Argentina is included in Chapter 87 of this Handbook.

1. Introduction

Investigation of seismological activities in Argentina goes back to 1897, due to the efforts of several institutions. At present, the following centers are working on this subject: Instituto de Prevención Sísmica (INPRES) of the Argentine Geological Mines Survey; the Instituto de Investigaciones Antisísmicas "Ing. Algo Bruschi" (IDIA) at the Faculty of Engineering of the National University of San Juan; the Instituto Sismologico "Fernando Volponi" at the Faculty of Exact, Physics and Natural Sciences of the National University of San Juan; the Instituto Antártico Argentino; and the Department of Seismology and Meteorological Information and the Department of Applied Geophysics at the Faculty of Astronomical and Geophysical Sciences of the National University of La Plata. These groups cover different topics in seismological activity, such as network of seismological stations, data acquisition and interpretation, hypocenter localization, prevention, teaching activities, basic and applied research, earthquake engineering and many problems related to the numerical simulation of seismic wave propagation in real media.

2. Institutions in Seismology and Related Sciences

2.1 San Juan

(1) INPRES (Instituto de Prevención Sísmica)-SEGEMAR

Address: Roger Balet 47 norte-5400 San Juan–Argentina
e-mail: <giuliano@inpres.gov.ar, bufaliza@inpres.gov.ar>

(2) IDIA (Instituto de Investigaciones Antisísmicas "Ing. Algo Bruschi"), Facultad de Ingeniería - Universidad Nacional de San Juan

Address: Avda. del libertador Gral. San Martín 1290 (o), 5400 San Juan–Argentina
e-mail: <idia@unsj.edu.ar>

(3) Instituto Sismológico Fernando Volponi, Facultad de Ciencias Exactas, Físicas y Naturales Universidad Nacional de San Juan

Address: Meglioli 1160 sur-5400, Rivadavia-San Juan–Argentina
e-mail: <triep@andes.unsj.edu.ar>

2.2 La Plata

(1) Depto. de Sismología Facultad de Ciencias Astronómicas y Geofísicas, Universidad Nacional de La Plata

INTERNATIONAL HANDBOOK OF EARTHQUAKE AND ENGINEERING SEISMOLOGY, VOLUME 81B ISBN: 0-12-440658-0

Address: Paseo del bosque s/n - 1900 - La Plata, Pcia. de Buenos Aires. Argentina
e-mail: <nora@fcaglp.fcaglp.unlp.edu.ar>

(2) Depto. de Geofísica Aplicada, Facultad de Ciencias Astronómicas y Geofísicas, Universidad Nacional de La Plata

Address: Paseo del bosque s/n - 1900 - La Plata, Pcia. de Buenos Aires. Argentina
e-mail: <claudia@fcaglp.fcaglp.unlp.edu.ar>

2.3 Buenos Aires

(1) IAA (Instituto Antártico Argentino), Dirección Nacional del Antártico

Address: Cerrito 1248 - 1010 - Buenos Aires - Argentina
e-mail: <febrer@antar.org.ar>

A more detailed report in Spanish, which includes a list of publications and of researchers, is included on the attached Handbook CD#2, under the directory of \7903Argentina.

79.4 Armenia

The editors received three institutional reviews from the Republic of Armenia: (1) Institute of Geophysics and Engineering Seismology, the National Academy of Sciences of the Republic of Armenia, from its director, Prof. S. M. Hovhannesyan; (2) National Survey for Seismic Protection, from its Chief Seismologist, Dr. Avetis Arakelyan; and (3) Georisk, from its Executive Director, Dr. Arkady Karakhanian.

These full reports were edited to conform to our style and are archived as computer-readable files on the attached Handbook CD#2, under the directory of \7904Armenia. A brief summary that was extracted and excerpted from each of these reports is shown here. Biographies and biographical sketches of some Armenian scientists and engineers may be found in Chapter 89, and in Appendix 3 of this Handbook, respectively.

1. Institute of Geophysics and Engineering Seismology (IGES)

The IGES was founded in 1961 in Leninakan. The principal disciplinary divisions of the Institute are geophysics, engineering seismology, and earthquake-resistant building. Scientific activities focus on seismic hazard assessment, investigation of earthquake focal mechanism and physics; search for effective earthquake precursors; working out of new methods and improvement of well-known ones in mining geophysics; investigation of seismic properties of the ground; improvement of General Seismic Zonation (GSZ) and Microseismic Zonation (MSZ) methods; working out of methods for assessment of quantitative parameters of seismic influence; working out of new methods for assessment of earthquake resistance of buildings; and improvement of modeling of the seismic process concerning similarity theory and practice. The most productive investigations in the sphere of geophysics are revealing the deep geological structure within the territory of Armenia on the basis of magnetometric, gravimetric, and aeromagnetometric data and precise determination of certain regions.

One of the significant achievements from the early period of the Institute's activities is the working out of a method for engineering analysis of seismic forces. The theoretical postulates of this methodology fundamentally influenced creation, development, and improvement of methods for seismic force determination, creation of earthquake-registering equipment (multi-pendulum seismometers), and seismic and microseismic zonation methods. After the Spitak earthquake, enormous work has been done in the sphere of investigation of earthquake consequences, with precise determination and engineering analysis of the destruction of and damage to buildings. Investigations have been carried out in cities and settlements for earthquake intensity evaluation, seismic hazard evaluation in the territory of Armenia, and some other scientific and experimental problems. The Institute also took part in compiling the Armenian Constructive Norms (ACN).

The full report, including biographical sketches of key personnel, may be found on the attached Handbook CD#2, under the directory of \7904Armenia.

2. National Survey for Seismic Protection (NSSP)

The NSSP was founded under the Government of the Republic of Armenia in July 1991. NSSP was given special governmental status and ministerial powers. The basic goal of NSSP is seismic risk reduction in Armenia. The NSSP has developed two Strategic National Programs on "Seismic Risk Reduction in Armenia" and "Seismic Risk Reduction in Yerevan City." The programs, adopted by the government of the Republic of Armenia in July 1999, are designed for 30 years. All the ministries and other governmental, non-governmental, and private

ISBN: 0-12-440658-0

organizations will implement these National Programs under the general coordination of NSSP, as assigned by the government.

The Seismic Risk Reduction Strategy includes seismic hazard and risk assessment; vulnerability reduction in urban areas, including reinforcement and upgrading of existing buildings, design of new codes and standards; public awareness, people education and training; early warning and notification; partnership establishment, involving public and private organizations; risk management, including emergency response and rescue operations; disaster relief and people rehabilitation; insurance; and state disaster law and regulations.

The NSSP has a staff of nearly 900, of which 300 are scientists and 300 are engineers and technicians. The NSSP combines all seismic risk reduction related scientific studies, including geophysics, geology, geochemistry, geodesy and earthquake engineering, sociology, and psychology. The NSSP has a unique multi-parameter observation network for monitoring of Earth lithosphere and atmosphere, consisting of 150 stations. The NSSP is divided into three regional departments with appropriate centers organized according to research directions. Main investigations in Earth sciences and earthquake engineering are implemented by the Seismic Hazard Assessment United Center and Earthquake Engineering Center respectively. The Seismic Hazard Assessment United Center implements all activities, concerning seismological investigations and data processing and analysis. The Earthquake Engineering Center is responsible for development and introduction of new technologies for building reinforcement and vulnerability assessment.

In 1998 in Geneva, the Armenian NSSP was awarded the "Certificate of Distinction" of the United Nations Sasakawa Disaster Prevention Award in appreciation for its distinguished contribution to disaster prevention, mitigation, and preparedness, and for furthering the goals of the International Decade for Natural Disaster Reduction.

Some details of the history of seismological studies in Armenia and significant achievements of the Armenian NSSP in the field of seismic risk reduction, as well as biographical sketches of key personnel, are given in the full report on the attached Handbook CD#2, under the directory of \7904Armenia.

3. Georisk

Georisk is a scientific research company that emerged and acts today as a research center under the Institute of Geological Sciences, Armenian National Academy of Sciences. Recently, Georisk has established its private status as a closed joint-stock company undertaking independent professional research and providing consultation on the assessment of seismic and other types of natural hazards, and estimation of their environmental impacts. Georisk has accumulated extensive experience in the field and is heavily involved in international cooperation (NSF, NATO, INTAS, and PICS grants, amid others).

The main research areas include active tectonics and seismotectonics; paleoseismicity; earthquake geology; tectonic stress; rock failure and earthquakes; coseismic and aseismic slips, and slip rates according to GPS monitoring and geological data; historical seismicity and macroseismicity; seismogeodynamics, geodynamics, and earthquake prediction; seismic hazard/risk assessment for large industrial facilities (Armenian NPP and others) and urban areas (Yerevan, Stepanakert, Tebriz, etc); seismic zoning, seismotectonic and geodynamic modeling; geochemistry and seismicity of active fault zones and attributable environmental effects; triggering of volcanic activity by strong seismicity; local seismic culture, influence of strong seismicity on the development of archaeo-cultures, national traditions, social and psychological aspects of society development.

A table of scientific projects undertaken by Georisk, a comprehensive publication list, a catalog of strong earthquakes in the territory of Armenia, and biographical sketches of Georisk personnel are included in the full report on the attached Handbook CD#2, under the directory of \7904Armenia.

79.5 Australia

The editors received the Australian Centennial Report to IASPEI from Kevin McCue, Coordinator of the Australian National Committee to IASPEI. The full report was compiled by McCue with contributions from G. Gibson, B.L.N. Kennett, C.T.J. Bubb, D. Love, P. Mora, I.D. Ripper, C. Sinadinovski, and V. Wesson. It is archived as a computer-readable file on the attached Handbook CD#2, under the directory of \7905Australia. An inventory of two seismic networks in Australia is included in Chapter 87. Biographies and biographical sketches of some Australian scientists and engineers may be found in Chapter 89, and in Appendix 3 of this Handbook.

1. Introduction

The Australian continent is an entirely intraplate observing platform in a region of the globe surrounded by oceans and intense seismicity. Australia is, therefore, ideally placed for monitoring earthquakes worldwide. The first seismograph in Australia, a Gray-Milne instrument, was installed at Melbourne in 1888. Today, Geoscience Australia (formerly the Australian Geological Survey Organisation, AGSO) in Canberra is the hub of a modern telemetered broadband national network, supplemented by local area networks operated by state governments, universities, and private infrastructure owners. More than 200 seismographs are currently installed in Australia and the Australian Antarctic Territory.

Recent strong earthquakes, including five of the world's ten fault-scarp forming events of the 20th century, have provided clear evidence of significant earthquake hazard in Australia. In 1961 Australian engineers developed the first earthquake regulations for what was then the Territory of Papua New Guinea, which straddles the Australian/Pacific plate boundary. The Ms 6.8 Meckering, Western Australia, earthquake in 1968 was the impetus for development of the first earthquake code for Australia, Australian Standard AS121-1979.

2. Earthquake Monitoring

Milne seismographs of the first worldwide network were installed in Perth in 1901, Melbourne in 1902, Sydney in 1906, and Adelaide in 1909. Jesuits set up the widely renowned Riverview Observatory (RIV) in Sydney in 1909 as part of their worldwide network (see Chapter 3 of this Handbook). The undamped Milne recorders were upgraded in the 1920s but the IGY, 1957/58, triggered the real advance in seismology in Australia.

The Australian Government Bureau of Mineral Resources (later Geoscience Australia) acquired some of the Carnegie Institute seismographs at the end of the IGY, which grew into the current network of 2 arrays, 9 broad-band, and 15 short-period stations telemetered into GA Canberra by telephone line and satellite. Three of these stations are in the Australian Antarctic Territory (MAW, MCQ, and CSY), the data telemetered via ANARESAT to Canberra. A further 15 stations will be added to the telemetered network in the next 2 years.

A prototype earthquake alarm and tsunami warning system is operating at GA—rostered duty seismologists are paged and after checking waveforms via the web or dial-up lines alert Emergency Management Australia and government and media contacts. The Seismology Research Centre also operates a local earthquake alert system in eastern Australia via an independent network of seismographs, computers, and pagers. Networking of these two complementary systems is in progress.

In addition to the telemetered network, GA has another 30 digital dial-up stations equipped with seismometers and/or accelerometers in the cities (a cooperative network with the states known as JUMP for Joint Urban Monitoring Network). Other instruments were strategically placed at gaps in the telemetered network following the 1989 Newcastle, New South Wales, earthquake, which killed 13 persons and caused damage of more than $1000M.

The full report presents the evolution of the Australian National Network since 1901, descriptions of major earthquakes

INTERNATIONAL HANDBOOK OF EARTHQUAKE AND ENGINEERING SEISMOLOGY, VOLUME 81B ISBN: 0-12-440658-0

in the country by region or state, development of state networks, work on monitoring nuclear explosions, the development of the data base, including historical earthquakes, and paleoseismological investigations. Sketches of the contributions of major figures, such as K. Bullen, D. Denham, F. Stacey, and A. Hales, as well as of the current scientific leadership in seismology, are included. A comprehensive bibliography of published Australian research results supplies the documentation for these contributions.

3. Earthquake Engineering

Earthquake engineering practice actually commenced in Australia after the second World War as engineers from the Commonwealth Department of Works were faced with designing buildings, dams, roads, airports, and bridges in the (League of Nations) mandated Trust Territory of Papua New Guinea. Large earthquakes were known to be relatively frequent in Papua New Guinea. Earthquake design regulations were not required for structures in Australia until much later.

The first Australian document to deal with earthquakes was Comworks Technical Instruction 5-A-21, issued in December 1961. It was amended in 1964 and again in 1965. The new series S3 was issued in July 1969. It was called *The Design of Buildings in Areas Subject to Earthquakes*. The "areas subject" did not include Australia. Then in 1968, a magnitude 6.8 earthquake demolished the small wheatbelt town of Meckering, some 130 km east of the capital of Western Australia, Perth, a city of 600,000 people. It caused minor damage in Perth and a 35-km-long surface thrust fault with up to 2m vertical offset, which buckled railway lines, pipelines, and roads. Engineers understood then that earthquakes did occur in Australia, too. An Australian National Committee for Earthquake Engineering (ANCEE) was formed and began to write a set of guidelines for earthquake engineering design. These guidelines became the first Australian Earthquake Code, which was published by the Australian Standards Association as AS2121 in 1979.

This Standard was used voluntarily by the Commonwealth Government, South Australian and West Australian State Governments, and several local Government Councils, but somehow was not called up by the Building Code of Australia (BCA), which would have made it mandatory everywhere.

Standards Australia, as it became known, reconstituted a committee to rewrite AS2121 and develop a new hazard map in 1989 and this was put on fast track after the Newcastle earthquake in December of that year. The Australian Loading Code, AS1170.4, was published in 1993 and was called up by the BCA that year. This was always meant to be an interim code until a new joint Standard between Australia and New Zealand is developed.

The Australian Earthquake Engineering Society (AEES) was established as a professional society of the Institution of Engineers, Australia, in 1990 following the destructive Newcastle earthquake of December 28, 1989. The Society affiliated with IAEE in the same year, although first Professor Stan Shaw and then Charles Bubb were Australian delegates to IAEE from 1968 when the ANCEE formed. AEES has some 200 members, holds an annual conference, and publishes the conference proceedings and a quarterly newsletter.

4. Seismological and Related Organizations and Institutions

The full report contains brief descriptions of the following organizations and institutions, with the Internet addresses at which details may be found:

- The Australian Earthquake Engineering Society (www.aees.org.au)
- Geoscience Australia (formerly The Australian Geological Survey Organisation) (www.ga.gov.au)
- Australian National University (rses.anu.edu.au/~seismology)
- The New South Wales Department of Works and Services
- The Department of Primary Industries and Energy, South Australia (www.mines.sa.gov.au/earthquakes.htm)
- The Seismology Research Centre, a division of Environmental Systems and Services Pty Ltd. (www.esands.com./)
- The University of Queensland (quakes.earth.uq.edu.au)
- The University of Tasmania (www.geol.utas.edu.au/geol/)
- Melbourne University (www.civag.unimelb.edu.au/)
- The University of Adelaide (www.civeng.adelaide.edu.au/)

79.6 Austria

The editors received an institution review, "Geophysical Services at the Central Institute for Meteorology and Geodynamics," from Dr. Wolfgang A. Lenhardt, the IASPEI National Correspondent, Seismological Service, Department of Geophysics, Central Institute for Meteorology and Geodynamics, Vienna, Austria. The full report is archived as a computer-readable file on the attached Handbook CD#2, under the directory of \7906Austria. An inventory of the Austrian seismic network is included in Chapter 87. Earthquake engineering in Austria is reviewed in Chapter 80.2, and biographic sketches of some Austrian scientists and engineers may be found in Appendix 3 of this Handbook.

1. Introduction

The Austrian Central Institute for Meteorology and Geodynamics (ZAMG), founded in 1851, is based in Vienna. In the beginning, the main area of activity concerned meteorological and geomagnetic observations. After a strong earthquake hit Ljubljana, Slovenia, in 1895, the Austrian Academy of Sciences founded the Seismological Service of Austria. The Service became part of ZAMG in 1904 and since then the Department of Geophysics at ZAMG has been responsible for the continuous observation of earthquakes.

Presently, the Department of Geophysics consists of four sub-divisions: Seismology, Geomagnetics, Applied Geophysics, and Seismic Network and Computer Technology. In addition, a super-conducting GWR gravimeter has been in operation since 1995. The new Conrad Observatory, south of Vienna, began operation in 2001. This observatory will serve as the seismological and gravimetric base in Austria. Within the observatory, four boreholes have been prepared, to be used for testing and comparing geophysical instruments under very well-controlled conditions.

In 2001, ZAMG celebrated its 150th anniversary with an excellent publication (Hammerl *et al.*, 2001).

2. Historical Overview

1842–1857: First geomagnetic survey of the Austrian Empire by Karl Kreil
1851: "K.K. Central-Anstalt für Meteorologie und Erdmagnetismus" founded
1852: First geomagnetic observatory in Vienna
1890: Second geomagnetic survey
1895: Earthquake Commission founded by the Austrian Academy of Sciences
1904: First seismograph in operation in Vienna. Publication of first seismological bulletin
1904: Institute renamed to "Zentralanstalt für Meteorologie und Geodynamik"
1930: Third geomagnetic survey
1935: The Departments of Seismology and Geomagnetics are merged into the Department of Geophysics
1960: Fourth geomagnetic survey
1977: Participation in the first airborne geomagnetic survey of Austria
1982: Start of Archeo Prospections®, a special method for detecting buried ancient objects
1989: First digital seismic network in Tyrol
1992: First strong-motion station in Austria installed
1993: Participation in EXERCISE 93—army training for earthquake rescue teams
1993: Start of VibroScan®, a special method for simulating ground vibrations
1994: Installation of two GPS stations near Innsbruck in Tyrol
1995: First seismic broadband station in Austria in operation. Opening of ACORN (Austrian-Carpathian Online Research Network)

INTERNATIONAL HANDBOOK OF EARTHQUAKE AND ENGINEERING SEISMOLOGY, VOLUME 81B ISBN: 0-12-440658-0

1995: Gravimetric Earth Tide equipment in operation
1998: Completion of 5th geomagnetic survey
2001: 150th anniversary of the Central Institute for Meteorology and Geodynamics, and publication of Hammerl *et al.* (2001)
2002: Official opening of the CONRAD Observatory

3. Modern Earthquake Monitoring

The first digital short-period seismic stations were installed around Innsbruck in Tyrol in 1989. Additional broadband stations are now in operation in Molln, Upper Austria; Kölnbrein, Carinthia; Arzberg, Styria; Hochobir, Carinthia; and Damüls, Voralberg. All stations are directly linked to the data processing centre in Vienna. Data are, therefore, immediately available for further evaluation. Preparations are underway for further extension of the seismic network.

The first strong-motion sensor was installed in Vienna in 1992. Up to January 2003, 24 sensors were in operation in the country, with a plan to install additional strong-motion stations to improve the coverage of the most seismically active regions.

Seismic data are not only exchanged with Austria's neighboring countries, but also submitted to the World Data Center A in Colorado, US, the International Seismological Center in the United Kingdom, the European Mediterranean Seismological Centre in Paris, France, and the Comprehensive Test Ban Treaty Organization in Vienna. Access to the data is available with the automatic data request manager via the email address <autodrm@zamg.ac.at>.

4. Current Research Activities

The Geophysical Service of Austria conducts historical earthquake research, evaluates recent earthquakes on a routine basis, updates the earthquake catalog and determines the earthquake hazard for construction of large dams for the country, provides hazard maps for the Austrian building code, carries out geomagnetic base surveys, determines soil conditions with classical geophysical methods, and surveys archaeological sites, among other activities. These activities, along with senior research staff, are briefly described in the full report on the attached Handbook CD#2, under the directory of \7906Austria.

References

Hammerl, C., W. Lenhardt, R. Steinacker, and P. Steinhauser (Eds.), (2001). "Die Zentralanstalt für Meteorologie und Geodynamik 1851–2001: 150 Jahre Meteorologie und Geophysik in Österreich," Graz: Leykam Buchverlagsgesellschaft m.b.H., pp. 891 + 2 CDs (in German, some articles have English abstracts).

79.7 Azerbaijan

The editors received two institutional reviews from the Azerbaijan Republic: (1) the Scientific Center of Seismology of the Presidium of Azerbaijan Academy of Sciences, from its Director, Prof. Ikram Kerimov, and (2) the Center of Seismological Survey of the Academy of Science of Azerbaijan, from its General Director, Prof. Arif Gassanov. These two reports are archived as computer-readable files on the attached Handbook CD#2, under the directory \7907Azerbaijan. The second report is written in Russian and the editors are grateful to Dr. Tatyana Rautian for translating the critical part of it into English. Biographical sketches of some Azerbaijan scientists may be found in Appendix 3 of this Handbook.

The following summary was extracted and excerpted from these two reports.

1. Scientific Center of Seismology of the Presidium of Azerbaijan Academy of Sciences

The report presents a short description of the work carried out in the Center during the past 20–25 years. Long-standing research is dedicated to study of the weak high-frequency Earth microseisms. This research has led to new physical notions about the processes taking place in the lithosphere, has shown the nonlinear character of the propagation and interaction with the medium of the weak signals of wave origin, and has allowed the development of criteria for diagnosis of the state of the medium.

The full report claimed that "Previously unknown regularity of changes in microseisms before an earthquake has been established, which stipulated that at distances that exceed the size of the epicentral zone a multi-stage increasing in the intensity of microseisms is registered, with simultaneous decreasing of their main frequency, and there arise recurring impulses (zugs) of micro-oscillations that are increasing in intensity and decreasing in time between their appearances, which are polarized in the direction of the epicenter of the future earthquake."

This discovery was claimed "to make possible the explanation of many natural processes and serve as a stimulus for beginning the working out of a nonlinear theory of relation of weak signals with the medium. . . . The results achieved may be used for solving on a new methodological basis a wide spectrum of problems: seismic events forecasting; diagnostics of medium stress state; revelation of the effects of induced seismicity; ecological control of large industry and other activity for minimizing and preventing the negative influence of induced effects on ecology and the environment; seismic zoning and microzonation, taking into account the local dynamics of the medium and the changing mosaic of breakage with time; mapping of large inhomogeneities; and creation of unified seismological and geophysical systems."

The professional staff of the Scientific Center has had a good opportunity to work in 35–40 different seismic and aseismic regions. During the 10 years of research and experiments aimed at revealing the Earth–Space geophysical system in the 1980s, we have been able to compare seismic behavior with many other geophysical fields. This has allowed us to create new methods of analysis, control, and managing of the medium state. We have looked into questions of the influence of human activity on the medium and problems of related ecology and environment. Unlike approaches considering mechanical and chemical pollution, we consider this problem from the point of view of energetic pollution, accumulation of additional stress in the environment caused by non-controlled industrial activity. Such processes unfold with ever-increasing speed and intensity. Analysing the data accumulated during tens of years, we can state that many increased numbers of events of different nature have an induced characteristic.

Another fundamental scientific result was claimed to "lead to the construction of a new model of earth structure, with a newly discovered layer in the mantle at the depth of 795–1505 km."

INTERNATIONAL HANDBOOK OF EARTHQUAKE AND ENGINEERING SEISMOLOGY, VOLUME 81B ISBN: 0-12-440658-0

The following scientists participated in the research reported in the full report: I. G. Kerimov (Director of the Center and participant in all phases of the work), N.A. Ahmedov (Deputy Director), A.N. Ali-zadeh, D.M. Nagiev, V.D. Abbasov, G.E. Muzaffarov, V.A. Shlyakhovskiy, M.A. Mustafaev, M.A. Ragimov, S.I. Kerimov, E.R. Alieva, I. Zalov, N.A. Belova, T.D. Ismailova, T.H. Velieva, V.I. Timoshenko. Biographical sketches of these scientists can be found in the full report on the Handbook CD.

2. Center of Seismological Survey of the Academy of Science of Azerbaijan

The Center was organized in 1980 for the development of seismological monitoring and earthquake prediction. The departments of the Center were created for registration of local and teleseismic earthquakes, study of the seismic regime, determination of parameters of earthquake sources for moderate and felt earthquakes, and the observation of variations of geophysical fields: the variations of geomagnetic fields, electromagnetic fields in the range of radio frequencies, the non-tidal variation of gravity. The observations were also organized for the hydro-, gas- and radio-geochemical fields. The mathematical methods and programs were created for processing and interpretation of the data obtained.

The determinations of main earthquake parameters (origin time, coordinates, depth, magnitude) as well as fault plane solutions are made routinely. The small earthquakes are used for the monitoring the temporal and space dynamics of seismicity. The pattern of small earthquakes is supposed to contain information about the locations of future large earthquakes. The temporal variations of geophysical and geochemical fields are considered as probable precursors of large earthquakes. These variations are the main subject of the investigations for the earthquake prediction method.

The catalog of earthquakes in the Azerbaijan and surrounding areas was completed for the 1980–1999 time period. It includes the main parameters of earthquakes and the fault plane solutions. The analysis was made of the main peculiarities of the seismic regime. Typical changing of geophysical and geochemical fields before the earthquakes with magnitude $\geq$4 were found, and the full report claimed: "The concentration of hydro-, gas- and radiometric fields changes before such earthquakes reach 200–1000 percent at distances of 800 km."

The most active scientists are A.G. Gassanov (Director), G.D. Etirmishli (Deputy Dirctor), A.R. Aliev (Deputy Director), A.G. Rsaev (Deputy Director), R.A. Keramova, T.Ya. Mamedov, and K.R. Kolesnichenko. Biographical sketches of these scientists can be found in the full report on the Handbook CD.

79.8 Belgium

The editors received a report, "Seismology and Physics of the Earth's Interior in Belgium," from Dr. R. Verbeiren, IASPEI National Correspondent, Royal Observatory of Belgium, Ringlaan 3, B_1180 Brussels, Belgium. This report is archived as a computer file on the attached Handbook CD#2, under the directory \7908Belgium. An inventory of the Belgium Seismic Network is included in Chapter 87. Biographies and biographical sketches of some Belgium seismologists and geophysicists may be found in Chapter 89, and in Appendix 3 of this Handbook, respectively. A summary that is extracted and excerpted from the full report is shown here.

1. Introduction

The full report summarizes the developments and achievements in the fields of earthquake seismology and physics of the Earth's interior in Belgium during the twentieth century, with a list of key references.

1.1 Seismology

Earthquake seismology started in Belgium as a private initiative in 1899, located at the Royal Observatory of Belgium (ROB) in Uccle (Brussels). In 1904 the observation facilities and instruments were donated to the ROB and in 1913 monitoring of seismicity became an official mission of the ROB. Since then the division of seismology took it as its duty to execute all relevant studies in order to understand more about local seismicity and seismotectonics of the region. Between 1958 and 1970 three new official stations were installed. Since 1985 a network of 25 digital stations has been deployed, and in 2000 a network of 12 accelerometer stations was created. Also beginning in 1985, a number of research projects were initiated on historical seismicity, paleoseismology, and seismic hazard assessment. These research areas are described briefly in the full report, with maps and lists of key references.

1.2 Physics of the Earth's Interior

From the foundation of the ROB, the study of the Earth's rotation vector (time and polar motion) was one of its basic tasks. Due to the increasing accuracy in the determination of its value, Prof. P. Melchior introduced, between 1950 and 1960, Earth tides research at the ROB as an alternative way to understand the rotational behaviour of the Earth with respect to its internal structure. Since then, he and his successors have studied the influence of the liquid core, the core mantle boundary, and the eigenmodes of the Earth on the observed values.

2. Current Research in Seismology

The basic activity of the seismology division of ROB still consists of the monitoring of present-day regional and worldwide seismicity. To accomplish this task, the division manages 5 broadband stations and 25 short-period stations, numbers that vary from time to time. The division is responsible for the maintenance of the sites, acquisition systems, and seismometers. Routine work consists of detecting and locating local events, measuring phases of teleseismic events, and sending them to the relevant international agencies. Current seismological research is primarily centered around the analysis and interpretation of local seismicity and its relation with the seismotectonic regime of the region. The main topics are described next.

2.1 Present-day Seismicity

Local earthquakes down to magnitude 1 are analyzed carefully to detect active faults, determine their geometry, and the dimensions and regime of the seismogenic zone.

2.2 Historical Seismicity

After the first studies of P. Alexandre in 1985, it became obvious that the existing historical catalogs needed revision.

INTERNATIONAL HANDBOOK OF EARTHQUAKE AND ENGINEERING SEISMOLOGY, VOLUME 81B ISBN: 0-12-440658-0

He detected many mistakes and mislocations due to an inadequate interpretation of historical archives, and discovered unused materials that needed investigation. More earthquakes from the 19th and 20th century need reinvestigation with the recently developed European macroseismic scale EMS98. In 2000 a study of the important 1692 Verviers earthquake ($M_S \sim 6.0$ to 6.5) was started, and the collection of reports and articles of more recent events for which the division did not have any material has also begun.

2.3 Paleoseismology

After several reconnaissance trips in 1995 in the east of the country, superficial evidence was found for the occurrence of medium-large earthquakes ($M_S \sim 6.0$) along the western fault scarp of the Roer Graben. This was the starting point of extensive paleoseismological research projects of the region: The project PALEOSIS (http://www.astro.oma.be/PALEOSIS/index.htm) from 1997 till 1999, and the project SAFE (http://www.astro.oma.be/SEISMO/SAFE/index.htm) from 2001 till 2002, both sponsored by the European Commission. The aims are to identify paleoearthquakes along selected active faults in the studied regions, to estimate the magnitude and recurrence time of these events, and to establish a formulation of the results directly usable for seismic hazard analysis.

2.4 Seismotectonics

As the rates of deformation along the Lower Rhine and Roer Graben differ considerably between estimations based on geological and geodetic evidence, long-term observational programs have been started in 1998. Permanent GPS stations have been installed on both sides of the Western fault scarp of the Roer Graben to measure the rate of displacement and an absolute gravity profile crossing the Ardennes Massif and the Roer Graben to measure an eventual uplift or subsidence of the region. This profile is and will be reoccupied every six months for at least a period of five to ten years.

2.5 Seismic Hazard

Because until 1998 all seismic hazard information was based on the 1975 map of maximum macroseismic intensity, a plan to compute an up-to-date seismic hazard map based on almost one century of instrumental observations and the improved catalog of historical seismicity was projected in cooperation with the Institute of Earthquake Engineering and Engineering Seismology in Skopje, Macedonia. Realizing that several parameters needed for the seismic hazard assessment are insufficiently known and in the framework of the EUROCODE 8 regulations, a study was launched to determine the site effects in the country on different kinds of soils and regions. The registrations of the newly installed accelerograph network will also be used in the near future for this effect.

3. Current Research in Physics of the Earth's Interior

Research is devoted to the computation of the response of the Earth to external and internal forces with special attention to

- Free and forced oscillations of the inner core.
- The topography of the core–mantle boundary.
- Normal modes and eigenfunctions of the Earth.

4. Biographical Sketches

The full report on the Handbook CD includes biographies of Oscar Somville (1880–1980), Charles Charlier (1897–1953), Jean-Marie Van Gils (1918–1989), and Paul Melchior (1925–).

79.9 Bolivia

The editors received an institutional review of the Observatorio San Calixto of La Paz, Bolivia, from Rodolfo Ayala Sánchez. The full report is archived as a computer-readable file on the attached Handbook CD#2, under the directory of \7909Bolivia. The following summary is extracted and excerpted from the full report. An inventory of this seismic observatory is included in Chapter 87. Biographies and biographical sketches of some Bolivian seismologists may be found in Chapter 89, and in Appendix 3 of this Handbook, respectively.

1. Introduction

Observatorio San Calixto (OSC) of La Paz, Bolivia, was founded May 1, 1913, by Pierre Marie Descotes, S.J., and became the first seismological station of first order in South America. OSC is organized in five sections: seismology, infrasound, meteorology, astronomy, and applied geophysics. OSC has operated in the course of its history several seismic and infrasonic networks; these nets have contributed to the advance of seismology in this part of the hemisphere. The directors and the research staff members of OSC have contributed to the development of science and education in Bolivia, mainly in geophysics.

2. History

The Society of Jesus consented to the resolution of the Second General Assembly of the International Seismological Association in 1911, recommending the installation of a seismic station of first order in La Paz, Bolivia, for the study and the investigation of the seismic activity of South America. Father P.M. Descotes, S.J. (1877–1964), who attended this assembly, was chosen for this task by the Society. In 1892, a meteorological station was installed in the school, and this was the platform for the future OSC. For the seismological assignment, at the end of 1911, Brother Esteban Tortosa was sent to the city of La Paz to carry out the preliminary studies. Brother Tortosa, with the help of Brother José Lizarralde, chose the best location inside the block occupied by the "Colegio San Calixto," the crypt of the church, destined originally to be the cemetery of the Jesuits. Here they installed the first test seismograph. Built by themselves, it consisted of a bifilar horizontal pendulum of 450-kg mass, of modified Mainka type. It worked for nearly two years, with time measured by a simple electric pendulum clock, which controlled the time marks on smoked paper, attached to a drum moved by the mechanism of an alarm clock.

On November 13, 1912, Father Descotes arrived at La Paz, with the objective of accomplishing his task. To manage this task he was prepared by appropriate studies carried out with Father Sánchez Navarro-Neumann. Descotes found a test seismograph in operation on his arrival at La Paz. Immediately, having limited personal and economic resources, Descotes began to design and to supervise the construction of the necessary instruments, consisting of more powerful seismographs, that were put into operation on May 1, 1913, official date of the foundation of OSC. This day marks the beginning of the bulletins of seismic data that have been issued without interruption until the present. At the end of 1913, the time service also began, owing to the necessity of having exact time for the seismic records, an essential base of all seismological observatories. For the time service, the instruments included Leroy electric chronometers, a Claude & Driencourt prism astrolabe of the firm of Jobin of Paris, and a meridian telescope of T. Cooke & Sons. Other chronographs and chronometers were later acquired in 1918.

In 1932, OSC had four different sections: meteorology, seismology, astronomy, and the time service of the Republic of Bolivia. The section of seismology had high-gain mechanical seismographs, and the Galitzin-Wilip seismometers, placing the OSC on a par with similar observatories in the Americas and among the best in the world. In seismology, the great merit of

INTERNATIONAL HANDBOOK OF EARTHQUAKE AND ENGINEERING SEISMOLOGY, VOLUME 81B ISBN: 0-12-440658-0

Father Descotes was his fidelity and ability to interpret the seismograms.

Another important activity of Descotes was his teaching in "Colegio San Calixto," where he taught physics and chemistry to several generations of students that included future presidents and ministers of Bolivia. On July 29, 1958, owing to his precarious state of health, Father Descotes was transferred to the House of Probation of the Society of Jesus in Santa Vera Cruz, 7 km southeast of the city of Cochabamba, Bolivia, continuing as Director of OSC and conducting from there the preparation of the seismic bulletins until his death in 1964.

In 1959, to help in the work of the administration of OSC, Father Ramón Cabré, S.J., doctor in physical sciences, was named Vice-Director. In 1964, after the death of P.M. Descotes, Father Cabré was named the Director of OSC. Cabré was the founder and the first Director of the "Centro Regional de Sismología para América del Sur" (CERESIS, see institutional report, Chapter 81), with headquarters in Lima, Peru, from 1965 to 1968. In November 1993, Father Cabré left the observatory, after having served for 34 years, as Director for 29 years.

On November 4, 1993, an Australian, Father Lawrence A. Drake, S.J., doctorate in geophysics, assumed the direction of OSC. He brought with him 40 years of work experience in Riverview Observatory of Sydney, Australia, acting as director for 27 years, and 20 years of investigation and teaching in geophysics and geology in Macquarie University, Sydney.

3. Functions and Organization of OSC

The main functions of OSC are

- To monitor the seismic activity of Bolivia and bring the seismic catalog of the country up to date;
- To operate and to maintain the National Seismological Net, including the International Monitoring System (IMS) stations of the Comprehensive Test Ban Treaty Organization (CTBTO);
- To inform and provide technical advice to the authorities, media, and public on aspects related to seismology;
- To transfer knowledge and new technologies in the field of seismology to the country. In November and December 1999, new systems of acquisition, processing, and transmission of data via satellite by means of antennas in several stations and in OSC were installed, as well as the IMS infrasonic and auxiliary seismic stations, in cooperation with DASE and within the framework of the CTBTO.
- A continuous publication of related scientific works in geophysics, mainly seismology, on two fundamental topics: seismicity and seismic hazard of Bolivia and the internal structure of the Andean region.

OSC is organized by a Directory integrated by six members—a team composed by one director, investigators, technicians, and administrative personal.

The OSC has three main areas of investigation:

(1) Seismology: Studies of the internal structure of the crust and mantle below Bolivia, seismotectonics, seismic hazard and risk
(2) Infrasound: The study of the infrasonic signals produced by natural events (shallow earthquakes, volcanic explosions) and artificial events (nuclear tests)
(3) Applied Geophysics: Studies focused on electric-resistivity sounding and seismic prospecting.

A more detailed history of OSC is given in the full report on the Handbook CD. It also includes biographies for P.M. Descotes and R. Cabre Roige, biographical sketches for L.A. Drake, R. Ayala Sanchez, E. Minaya R., and A.J. Vega Benavidez, an extensive list of publications, and a description of the seismic and infrasonic networks. Current information about OSC is available at its Web site: http://observatoriosancalixto.org.bo/.

79.10 Brazil

The editors received a report, "Seismology at the Institute of Astronomy, Geophysics and Atmospheric Sciences of the São Paulo University (IAG/USP), São Paulo, Brazil," from Jesus Berrocal, a member of the Editorial Advisory Board of this Handbook. The full report was originally written in Portuguese with an English abstract. We are grateful to Jorge Martins for helping us in the editing process. The author subsequently submitted an English version of the full report, and it is archived as a computer-readable file on the attached Handbook CD#2, under the directory of \7910Brazil. The following summary is extracted and excerpted from the full report, including several selected references to some Brazilian seismological research. Biographical sketches of some Brazil scientists may be found in Appendix 3 of this Handbook.

1. Introduction

Because the Brazilian territory is located in the intraplate region of the South American plate, it is stable from the seismological point of view. For this reason and because of its short historical record (since the beginning of the 16th century), its tradition in seismology was very limited until the 1970s. In the full report, the attempts to implant seismology in Brazil since Don Pedro I, emperor of Brazil in the 19th century, are reviewed, followed by the installation of the RDJ seismographic station in 1906 at the National Observatory of Rio de Janeiro. In 1964, NAT, a station of WWSSN, was installed in Natal, Rio Grande do Norte, and then was operated by a research group of the Brazilian Navy. Routine analysis of the seismograms was not carried out at this station nor at Rio de Janeiro at that time (Berrocal and Assumpção, 1982).

2. South American Array System (SAAS)

Seismology was not effective in Brazil until the installation of the South American Array System (SAAS) at the end of the 1960s by the South American Seismological Expedition organized by Edinburgh University, Scotland-UK. SAAS was installed around Brasilia, the newest Brazilian capital, with the support of the British Institute of Geological Sciences, the Brazilian National Research Council (CNPq), the United States Coast and Geodetic Survey (USCGS), and the UNESCO Department of Earth Sciences, with the collaboration of the South American Seismological Research Center (CERESIS) and the University of Brasilia (UnB). The permanent stage of SAAS started to operate in 1972.

3. Research in Geophysics

While SAAS was being implanted, interest of some Brazilian universities rose. Advancement of applied geophysics at the federal universities of Bahia and Para and physics of the Earths' interior at the University of São Paulo allowed the creation of the Department of Geophysics in 1973.

Two important facts helped the development of seismology in Brazil: one was the implementation of nuclear power plants in Brazil beginning in the 1970s, and the other was the need to monitor instrumentally cases of seismicity induced by the impoundment of hydroelectric reservoirs.

4. Brazilian Institutions in Seismological Research

A brief description of the Brazilian institutions that do seismology as part of their programs is given in the full report: The Seismological Observatory at the Brasilia University (SIS/UnB), the Department of Theoretical and Experimental Physics of the Universidade Federal do Rio Grande do Norte (UFRN), the National Observatory (ON/CNPq), and the Institute of Technological Research (IPT/SP).

INTERNATIONAL HANDBOOK OF EARTHQUAKE AND ENGINEERING SEISMOLOGY, VOLUME 81B ISBN: 0-12-440658-0

5. Seismological Activities at IAG/USP

A description of the seismological activities carried on by the Seismology Group of the IAG/USP is given in the full report. Those activities include the participation of that group since its beginning in 1973, when the Department of Geophysics of IAG/USP was also established. The principal research lines executed by the group are mentioned, including some references that are listed at the end of the report. A seismicity of Brazil map it is also shown.

References

Assumpção, M. (1983). A regional magnitude scale for Brazil. *Bull. Seismol. Soc. Am.*, **73**, 237–246.

Assumpção, M. (1992). The regional intraplate stress field in South America. *J. Geophys. Res.*, **97**, 11889–11903.

Assumpção, M. (1998). Seismicity and stresses in the Brazilian Passive Margin. *Bull. Seismol. Soc. Am.*, **88**, 160–169.

Assumpção, M. (1998). Focal mechanism of small earthquakes in SE Brazilian shield: a test of stress models of the South American plate. *Geophys. J. Int.*, **133**, 490–498.

Berrocal, J., and M. Assumpção (1982). Seismology in Brazil. *Earthquake Inf. Bull.*, **14**, 19–21.

Berrocal, J., and C. Fernandes (1993). Seismic hazard in Brazil. In: "The Practice of Earthquake Hazard Assessment" (McGuire, Ed.), IASPEI/ESC, pp. 38–41.

Berrocal, J., and R. Masse (1989). A proposal to establish a seismological data base and a digital seismic network for South America. *Revista Geofisica*, **31**, 221–260.

Berrocal, J., M. Assumpção, R. Antezana, C.M. Dias Neto, R. Ortega, H. França, and J. Veloso (1984). "Sismicidade do Brasil." IAG-USP/CNEN, São Paulo, 320 pp.

Berrocal, J., M. Assumpção, R. Antezana, C.M. Dias Neto, H. França, and R. Ortega (1983). Seismic activity in Brazil in the period 1560–1980. *Earthquake Prediction Research, Tokyo-Japan*, **2**, 191–208.

Berrocal, J., C. Fernandes, A. Bueno, N. Seixas, and A. Bassini (1993). Seismic activity in Monsuaba-RJ, Brasil, between December 1988 and February 1989. *Geophys. J. Int.*, **113**, 73–82.

Berrocal, J., C. Fernandes, A. Bassini, and J.R. Barbosa (1996). Earthquake hazard assessment in Southeastern Brazil. *Rev. Geof. Inter.*, **35**, 257–272.

Ferreira, J.M., R.T. Oliveira, M. Assumpcão, J.A. Moreira, R.G. Pearce, and M. Takeya (1995). Correlation of seismicity and water level in the Açu Reservoir—An example from Northeast Brazil. *Bull. Seismol. Soc. America*. **85**, 1483–1489.

Takeya, M., J.M. Ferreira, R.G. Pearce, M. Assumpção, J.M. Costa, and C.M. Sophia (1989). The 1986–1988 intraplate earthquake sequence near João Câmara, Northeast Brazil—evolution of seismicity. *Tectonophys.*, **167**, 117–131.

79.11 Bulgaria

The editors received this brief report, "Seismology in Bulgaria," from Ms. Rumiana Glavcheva, Head of the Seismological Department and Institute's Scientific Council Secretary, Geophysical Institute of the Bulgarian Academy of Science. It had been edited and returned to the author for revisions. The revised report is presented here. An inventory of the Bulgarian seismic network is included in Chapter 87. Earthquake engineering in Bulgaria is reviewed in Chapter 80.3, and biographic sketches of some Bulgarian seismologists and earthquake engineers may be found in Appendix 3 of this Handbook. Please also visit the Web site of the Geophysical Institute of the Bulgarian Academy of Science at: http://www.geophys.bas.bg/ for more information.

1. Introduction

Bulgarian seismology began in 1891, when a network of correspondents for observation of felt earthquakes was organized and Spas Watzof, the director of the Central Meteorological Station, Sofia, initiated the collection of macroseismic reports. In 1903, this young Bulgarian effort in seismology became a part of the International Seismological Association. The scientific world received published annual reports on earthquake impacts in Bulgaria over the course of 75 years.

2. Macroseismic Data Collection

Going through the 17 annual volumes edited by Watzof, one can find that even at his time Watzof was concerned with some important topics, now considered quite common: "seismic centre" location (now, epicentre determination and source zones), density of observation points and territorial coverage (topics of great importance for both non-instrumental and instrumental data acquisition), accumulation of as many witness reports as possible from a given locality (statistical nature of modern intensity scales), and so on.

The seismological survey employees coming after academician Watzof's death (1928) carried on with the macroseismic data collection. Thus, annual reports on felt earthquakes were issued until 1965, though changeable in type. These were of two parts, descriptions and macroseismic bulletins, or bulletins only, or presenting isoseismal maps in addition to the bulletins. Archives of initial descriptions were formed as well. After 1965, many studies of moderate-size earthquakes and atlases of isoseismal maps were published. Field observations were carried out for a careful investigation of the consequences of the 1977 Vrancea earthquake, 1986 North Bulgarian earthquake series, as well as some weaker seismic events. The contribution of Bulgaria to long-term seismicity investigation of the Balkan region made in the framework of some recently completed European projects is worthy of mention.

3. Instrumental Seismology

Instrumental seismology started its development in Bulgaria in 1905, when a Bosch-Omori seismograph was installed in Sofia. This action was motivated by the occurrence of several violent earthquakes, some of them followed by long-lasting aftershock sequences: the 1892 earthquake in northeast Bulgaria (maximum intensity 8 according to Medvedev-Sponheuer-Karnik (MSK) scale, magnitude around 7); the Black Sea earthquake in 1901 (intensity 9 MSK in northeast Bulgaria, magnitude 7 or more); and two transfrontier, Bulgaria-Macedonia, earthquakes (maximum intensity in Bulgaria up to 10 MSK, magnitudes about 7–7.5). Two facts concerning the earliest stage of Bulgarian instrumental seismology are noteworthy: (1) the first bulletin, which systematized the registrations during 1905, appeared immediately after the instrument installation; and (2) selected Bulgarian registrations were used by A. Mohorovičić and H.F. Reid

INTERNATIONAL HANDBOOK OF EARTHQUAKE AND ENGINEERING SEISMOLOGY, VOLUME 81B ISBN: 0-12-440658-0

in their world-recognized works. The Sofia station remained the only instrumented site in Bulgaria up to 1961, although its equipment was modernized. The seismic network appeared and was developed to cover the seismogenic areas of Bulgaria in the 1960s and 1970s.

Since August 1, 1960, seismology in Bulgaria has been developing within the Seismological Department of the Geophysical Institute, Bulgarian Academy of Sciences. The National Seismological System, established according to a Ministerial Council decision in 1980, belongs to the same department (more information is in Chapter 87 of this Handbook). The seismological department is in charge of all the earthquake-connected problems in the country, including express information in the case of felt events. Bulgarian seismology contributes to more detailed documentation of seismic phenomena in the Balkan Region. It also contributes to solving practical tasks addressed to important national initiatives.

4. Current Research Activities

Scientific and practical activities of contemporary Bulgarian seismology include the following tasks: monitoring of natural, artificial, and induced seismic events in Bulgaria and its surroundings; earthquake statistics (general, aftershocks, etc); seismic zonation of the Central Balkans; long-term seismicity; earthquake source mechanism; kinematic and dynamic specification of large sources, seismotectonics; seismic energy attenuation; tsunami modeling (Black Sea); seismic hazard evaluation; seismic waves propagation (kinematic and dynamic aspects); seismic monitoring and site-effects studies in areas of state importance (nuclear power plants, dams, lifelines, etc); delineating earth structures and dynamic processes in tectonically active regions by seismological data; searching for earthquake precursors of seismic, electromagnetic, or other geophysical nature.

79.12 Canada

Dr. Colin Thomson, the Canadian National Correspondent to IASPEI, provided the editors a list of addresses of 18 Canadian institutions that participated in the Quadrennial Report to IUGG. Invitations were sent to all these institutions, and we received six institutional reviews as replies: (1) "A Century of Seismology at the Dominion Observatory, Ottawa" by Allison L. Bent and John Adams, (2) "The Pacific Geoscience Centre and One Hundred Years of Seismological Studies on Canada's West Coast" by John F. Cassidy, Garry C. Rogers, and Roy D. Hyndman, (3) a report from F. Walter Jones, Institute for Geophysical Research, University of Alberta, Edmonton, Alberta, (4) a report from Alan E. Beck, Department of Earth Sciences, University of Western Ontario, London, Ontario, (5) a report from Karen Cunningham, Department of Earth and Atmospheric Science, York University, Toronto, Ontario, and (6) a report from Henry V. Lyatsky, Lyatsky Geoscience Research & Consulting Ltd., Calgary, Alberta.

These submitted reports, slightly edited and re-formatted, are archived as computer-readable files on the attached Handbook CD #2, under the directory of \7912Canada. The following summary is extracted and excerpted from these reports. Earthquake engineering in Canada is reviewed in Chapter 80.4. An inventory of the Canadian National Seismic Network is included in Chapter 87, and biographical sketches of some Canadian scientists and engineers may be found in Appendix 3 of this Handbook.

1. A Century of Seismology at the Dominion Observatory, Ottawa

Website: http:/www.seismo.nrcan.gc.ca/

Since 1906, when the first seismograph was installed in Ottawa, the Dominion Observatory, now part of the Geological Survey of Canada (GSC), has played a lead role in earthquake monitoring and research in Canada. The seismograph network has grown to over 120 stations, the data from most of which are transmitted in near real time to Ottawa for processing and archiving. Seismologists at the GSC in Ottawa are responsible for locating earthquakes in eastern and northern Canada, maintaining a database of all located Canadian earthquakes, and responding to felt and significant earthquakes.

Since 1976, the responsibility for western Canadian seismicity has moved to the Pacific Geosciences Centre, near Victoria (see Section 2). A principal objective of the seismology group is to ensure the safety of the Canadian public in the event of a major earthquake. To this end, seismic hazard assessment has been a focal point of research since the 1950s, when the first Canadian seismic hazard maps were produced. Seismic hazard assessment has progressed from a purely seismicity-based process to include geological and paleoseismic data, spectral (rather than peak) ground motions, and improved estimates of uncertainty. Much of the basic research is used ultimately to improve the hazard assessment.

Fields of research include the re-analysis of historical earthquakes, detailed source studies of recent earthquakes, focused studies of individual seismic source zones using integrated geophysical data sets, and the evaluation of ground motion relations and attenuation. Canadian seismologists have also undertaken research in other areas, such as the structure of the Earth's interior and nuclear test ban seismology. Test ban research has been focused on using magnitude as a discriminant. The Ottawa seismology group is responsible for the operation of the seismology, infrasound, and hydroacoustic facilities that comprise Canada's contribution to the International Monitoring System set up under the current Comprehensive Test Ban Treaty. For current information on the National Earthquake Hazards Program and the Nuclear Test Ban Treaty Verification Program, please visit the Web site, http:/www.seismo.nrcan.gc.ca/.

INTERNATIONAL HANDBOOK OF EARTHQUAKE AND ENGINEERING SEISMOLOGY, VOLUME 81B ISBN: 0-12-440658-0

2. The Pacific Geoscience Centre and One Hundred Years of Seismological Studies on Canada's West Coast

Web site: URL: www.pgc.nrcan.gc.ca/

The west coast of Canada includes several plate boundaries and thus is a seismically active area. The Government of Canada deployed the first seismograph in the region in 1898, in Victoria, British Columbia. The first high-gain short-period seismograph, sensitive to local seismicity, was installed in 1948. As of December 1999, more than 50 stations, including 18 three-component broadband instruments, 100 strong motion instruments, and 9 continuous GPS tracking sites are operating in Western Canada to monitor earthquake activity and crustal deformation. Most of the instrumentation is located in the densely populated and seismically active southwest corner of British Columbia.

The modern earthquake research program on the west coast began in 1951, when W. G. Milne was transferred to the Victoria Geophysical Observatory from the Dominion Astrophysical Observatory in Ottawa. His work over the next 20 years played a key role in the development of the earthquake hazard maps of Canada for use in the National Building Code.

In 1976, the Pacific Geoscience Centre (PGC) was formed near Victoria as a subdivision of the Pacific Division of the Geological Survey of Canada, bringing together seismologists and marine geologists. During the following 24 years the PGC has assembled earth scientists covering a broad range of disciplines, many providing input to earthquake hazards, including earthquake seismology, seismic reflection and refraction, geodynamics, potential field studies, deep-sea, nearshore, and coastal marine studies, heat flow studies, computer modeling, paleomagnetism, absolute gravity, sedimentology, and geological studies. The PGC is thus enabled to tackle earth science questions that can only be solved using multidisciplinary techniques. Some examples are estimating the megathrust earthquake potential, mapping the structure and tectonics of the Pacific margin and the Canadian Cordillera, applying modern high-resolution marine techniques to mapping the sea floor, including active faults, using satellite technology to study crustal deformation, and using modern seismic data available in near real time to provide information on earthquakes and related hazards.

In addition to details on these research activities, the full report presents a complete listing of earthquake-related publications (covering the time period 1976–2000) by the PGC staff and those involved in earthquake studies on Canada's west coast.

3. The Institute for Geophysical Research, University of Alberta, Edmonton, Alberta

The Institute for Geophysical Research is composed of members who are engaged in geophysical research in three University departments: the Department of Earth and Atmospheric Sciences, the Department of Mathematical and Statistical Sciences, and the Department of Physics. In the 2002–2003 academic year, there were 26 faculty members from these 3 departments, 7 Professors Emeriti (3 active), 10 Research Associates, 21 Postdoctoral Fellows, 10 technical staff, and 77 PhD and MSc students.

The purpose of the Institute is to foster interaction and collaboration among individuals from the three departments, to advance geophysical studies within the University, to assist in the dissemination of knowledge, to help train leaders in areas of importance for the future well-being of our environment, and to promote interaction between the University and the broader community.

The Institute publishes an annual report and, three times a year, publishes a newsletter, "inuksuk." A twice-yearly competitive Institute Student Conference Travel Awards Program, supported by industry, provides travel funds for students to attend and participate in conferences and workshops. The Institute holds an Annual Symposium that features talks by Institute graduate students together with two or three invited speakers. The Institute also sponsors seminars by both local speakers and speakers from other institutions, including universities, industry, and government.

4. Department of Earth Sciences, University of Western Ontario, London, Ontario

The Department of Earth Sciences of the University of Western Ontario was formed in June 1993 by the merger of the Departments of Geology and Geophysics. The evolution of the present structure began in 1958, when the first Department of Geology and Geophysics in Canada was created by the appointment of Robert J. Uffen, then of the Department of Physics, to the Department of Geology as head of the sub-department of Geophysics, with Harold Reavely as head of the sub-department of Geology. In 1958, Alan E. Beck, of Physics, was added to Geophysics and the department was split into two full departments, achieving another first for Canada and the University. The Department of Geophysics grew in the following decade, with appointments of faculty specializing in paleomagnetism, shock-wave physics, heat flow, age determination, seismology, geochemistry, and rock and mineral physics. The full-time faculty was quite stable for years, but supplemented by many arrivals and departures of visiting professors. The first faculty retirements in 1988/89 foreshadowed a decrease in the level of activity.

The initial development of the research program was concentrated on paleomagnetism and geothermics, areas in which the University already had some reputation. Seismology was added as an area of concentration because of interest in the subject and its significance for the nation. Age determination capability was added because it was needed in connection with the paleomagnetic research. A program in rock and mineral physics at temperatures and pressures representative of the lower crust and upper mantle was developed to fill a gap in Canadian, as well as worldwide, research. Exploration geophysics was covered by distributing responsibilities among the seismologists, potential field specialists, etc.

Some achievements of the research through the years include experimental verification of a thermally induced magnetic reversal (Carmichael); demonstration that major earthquakes could produce sudden changes in polar wobble (Smylie and Mansinha); application of fractal theory to the interpretation of crustal seismic transects (Mereu); and the realization that perturbations to the subsurface temperature profiles could be used to infer past temperature variations (Beck).

IASPEI held its 21st General Assembly at the University of Western Ontario in 1981. Information about the present faculty and activities may be found at: http://www.uwo.ca/earth/.

5. Department of Earth and Atmospheric Science, York University, Toronto, Ontario

Web site: http://www.eas.yorku.ca/

The research staff of five in this department share a common focus in their research in that they are all concerned with global geophysical dynamics. Research activities include mantle convection, core dynamics and Earth rotation, geomagnetism, experimental and theoretical geophysical fluid dynamics, and remote sensing. The teaching program is enhanced by the fact that development of instrumentation and interpretation in many areas of geophysics has its roots in the Toronto area.

Accomplishments of the research personnel include numerical models of mantle convection that have examined a number of issues related to thermal and dynamic history of the Earth (G. T. Jarvis and co-workers). D. E. Smylie, with graduate students and international collaborators, has used global observations from superconducting gravimetry to model dynamics of the inner core. The solution of the Ekman boundary layer problem of a sphere oscillating in a contained rotating fluid has recently led to a very precise measurement of the viscosity in the layer just outside the inner core boundary. K. D. Aldredge and co-workers have studied fluctuations of the paleomagnetic field and tidally forced instability of the Earth's fluid outer core.

Current activities are directed to the geophysical fluid dynamics of the Earth's fluid core, the application of geographical information systems to analysis of earth and environmental systems, geodynamics and mantle convection, global earth dynamics, and remote sensing and geodynamics. Specific topics under investigation are given in the full report.

6. Lyatsky Geoscience Research & Consulting Ltd.

Web site: http:// www.telusplanet.net/public/lyatskyh

Lyatsky Geoscience Research & Consulting Ltd. is a freelance geoscience and consultancy firm founded in 1986 and based in Calgary, Alberta, Canada. The firm employs potential-field, seismic, remote-sensing, and geological information to delineate fault networks in the crust. This information is then utilized to assess natural hazards, and to tie the faults to the genesis and distribution of mineral and hydrocarbon deposits in a predictive fashion that assists practical exploration.

Trained in both geology and geophysics at the University of British Columbia, H. Lyatsky has worked in many parts of the Canadian Shield, Alberta and Williston basins, Mackenzie Basin, the east and west coasts of Canada, Canadian and U.S. Cordillera, the Beaufort Sea, the North Sea, Ukraine, and Azerbaijan. Experience has demonstrated that the firm's horizontal gradient vector method of processing potential-field data is especially useful for delineating fault networks.

The ability to correlate geophysical anomalies with known geological information has proved very fruitful for interpreting the structure and evolution of sedimentary basins. The reexamination of modern tectonics along North America's western continental margin suggests the conventional earthquake risk estimates in the Vancouver–Seattle region are significantly exaggerated.

79.13 Chile

The editors received the "Chilean National Centennial Report to IASPEI" from Prof. Edgar G. Kausel, IASPEI National Correspondent, Department of Geophysics, University of Chile, Santiago, Chile. This report is archived as a computer-readable file on the attached Handbook CD, under the directory of \7913Chile. The following summary is extracted and excerpted from the full report. Earthquake engineering in Chile is reviewed in Chapter 80.5, and biographical sketches of scientists and engineers from Chile may be found in Appendix 3 of this Handbook.

1. Introduction

The west coast of South America is one of the most seismically active regions of the world. Subduction of the Nazca Plate under the South American Plate takes place at a rate of about 10 cm/year. More than 50 percent of this subduction process occurs along the Chilean coast. As a result at least one great earthquake of magnitude 8 or larger takes place in Chile every decade, and any given coastal portion is struck by events of this size once per century.

This high seismicity is known since Diego de Almagro and Pedro de Valdivia conquered this territory in the first half of the 16th century. Letters and reports to the Spanish authorities in Europe describing great earthquakes are abundant since then. In addition, famous naturalists such as Charles Darwin, María Graham, and many others became interested in the description and study of this phenomenon.

However, it is only during the first decade of the 20th century that a formal institution was founded, the National Seismological Service, as a consequence of the disastrous 1906 Valparaiso earthquake. Ferdinand de Montessus de Ballore became the first Director of the Service. A national seismic network was installed, and systematic recording, analysis, and cataloging of earthquakes was initiated, tasks that have continued without interruption until today. One of the best catalogs of historical earthquakes is now available, describing great and large events since 1540. A list of the largest earthquakes is shown in Table 1 of the full report on the Handbook CD. A few decades after its foundation (1929), the Seismological Service became part of the University of Chile, as it is at present (Institute of Seismology and later Department of Geophysics).

2. Earthquake Monitoring

The first record of an earthquake in Chile was obtained by I. M. Gillis, chief of the U.S. Astronomical Expedition to the Southern Hemisphere, 1849–1852. Gillis recorded in Santiago the earthquake of April 2, 1851, by means of a seismoscope, and he gave time of initiation and maximum amplitude, which indicates that the instrument had some device for time determination. A similar seismoscope was installed in La Serena, some 400 km north of Santiago. The instruments apparently operated from November 1849 to September 1852. With these two instruments Gillis concluded the existence of important differences in seismicity between Santiago and La Serena.

Later, J. I. Vergara, Director of the Astronomical Observatory of the University of Chile (1873–1881) and Rector (President) of the University of Chile (1888–1889), described the earthquake of July 7, 1873, felt in various cities. Using arrival times observed by telegraph officers (meridian times), he attempted to determine the seismic wave velocity. Some years later, in 1881, in a report to the Ministry of Education he asked for funds to acquire appropriate instruments to "record earthquakes so frequently felt in our country."

At the end of the 19th century, the Clark brothers, engineers in charge of the construction of the transandean railway from Chile to Argentina, acquired a seismoscope that was installed in the Santiago Astronomical Observatory located in Quinta Normal park. It was made in England, and consisted of an inverted pendulum without time control. The instrument operated until 1908, and recorded the great 1906 Valparaiso earthquake.

INTERNATIONAL HANDBOOK OF EARTHQUAKE AND ENGINEERING SEISMOLOGY, VOLUME 81B ISBN: 0-12-440658-0

The first national seismic network installed by Montessus in 1908 consisted of stations in Santiago (2 horizontal and 1 vertical Wiechert, 2 Bosch-Omori, 2 Stiatessi, and 1 seismoscope), Tacna, Copiapó, Osorno, and Punta Arenas (2 horizontal Wiechert in each city), and 30 seismoscopes in other towns throughout Chile.

The seismometers of Osorno and Punta Arenas operated temporarily in Juan Fernandez Islands, Easter Island, and the Antarctic continent. One of the Wiechert's seismographs was installed by the Charcot expedition in Petermann Island for nine months and Deception Island for fifteen days. One of the Bosch-Omori's (100 kg) was installed in Easter Island, Port Mataveri, for the period April 25, 1911, to May 5, 1912. Sixty-five earthquakes were recorded at epicentral distance up to 3664 km. None of the sixty-five events was felt by residents of the island.

The National Seismological network evolved in a very irregular way to the network that is at present in operation.

3. Research Activities

After the death of Montessus in 1923, Carlos Bobillier (1923–1935), Enrique Donoso (1935–1941), and Federico Greve (1941–1958) occupied the directorship of the Institute of Seismology. In 1955 Cinna Lomnitz (1958–1962) radically modified the objectives of the Institute by incorporating research groups in seismology, prospecting geophysics, and meteorology. Enrique Gajardo (1962–1963), Edgar Kausel (1964–1972, 1973–1982), and in various periods of two years Peter Welkner, José Rutllant, Humberto Fuenzalida, Edgar Kausel, and Patricio Aceituno succeeded to the position. The present director is Diana Comte (1999–).

In Chile, seismology and physics of the Earth's interior are mainly taught and studied by the Department of Geophysics of the University of Chile. Only recently, some other groups are being established at Universidad de la Frontera (Temuco) and Universidad de Tarapaca (Arica). A large number of studies have been made during the history of the Seismological Service—Department of Geophysics, related principally with basic theoretical research, seismicity, source process, strong motion, seismic hazard, velocity structure, and geometry of the downgoing slab. In addition, the Department is responsible for the Master's degree in Geophysics since 1966. Due to the good level of this degree, the Department was considered by the Organization of American States (OAS) as a "center of excellence" in geophysics. Many seismologists with Ph.D. degrees obtained in different universities of Europe, America, and Japan have worked or are working in the Department of Geophysics: Cinna Lomnitz, Armando Cisternas, Edgar Kausel, Lautaro Ponce, Alfredo Eisenberg, Raul Madariaga, Raul Husid, John Bannister, Juan Enrique Luco, Carlos Leiva, Sergio Barrientos, Mario Pardo, Diana Comte, Emilio Vera, Klaus Bataille, and Jaime Campos.

The earthquake engineering group of the Department of Civil Engineering, University of Chile, has been actively involved in strong motion seismology and structural engineering. This group is also responsible for teaching the earthquake engineering courses for the Civil Engineering degree. Well-known civil engineers are or have been part of this group: Rodrigo Flores, Arturo Arias, Joaquín Monge, Mauricio Sarrazín, Rodolfo Saragoni, María O. Moroni, Maximiliano Astroza, and Rubén Boroschek. Present research of this group is related to structural engineering, earthquake-resistant design, strong motion, soil amplification, base isolation, seismic hazard, attenuation, seismic risk, building code, and soil mechanics. The group operates a strong motion network with accelerometers distributed along Chile from Arica to Puerto Montt. Activities related with seismic codes and base isolation studies are conducted by the Department of Structural Engineering of the Pontificia Universidad Católica de Chile.

A national association of seismology and earthquake engineering ACHISINA (Asociación Chilena de Sismología e Ingeniería Antisísmica) with more than 100 members has been active since 1964. Among other activities, ACHISINA is responsible for the organization of national and international workshops and meetings, development of seismic codes, and earthquake structural engineering practice.

79.14 China (Beijing)

The editors received the report, "Seismology in China in the 20th Century," from Dr. Z.L. Wu, the IASPEI National Correspondent, Institute of Geophysics, China Seismological Bureau, 5 Minzu Xueyuan Nan Road, Beijing 100081, China. The authors of this report are Y.T. Chen, Z.L. Wu, and L.L. Xie. This report (with appendices) is archived as computer-readable files on the attached Handbook CD#2, under the directory of \7914China(Beijing). Earthquake engineering in China is reviewed in Chapter 80.6, and an inventory of seismic networks in China is included in Chapter 87. Biographies and biographical sketches of some Chinese scientists and engineers may be found in Chapter 89, and in Appendix 3 of this Handbook, respectively. Please note that the Chinese order of name is used here, i.e., family name first and then the given names.

1. Introduction

Because it has suffered from intensive seismic disasters, China has a long history of studying earthquakes. The earliest written record of a Chinese earthquake can be traced back to 1831 BC. In AD 132, Zhang Heng invented the first seismoscope in the world.

Modern seismology was introduced to China early in the 20th century. Wong Wenhao's articles in the 1920s were the pioneer works in modern Chinese seismology. The first seismological observatory installed and managed by Chinese seismologists was the Chiufeng Seismic Station in Peiping (now Beijing) in 1930. The first Chinese-made seismograph was the "Yong-Ni (Wong Wenhao) seismograph" designed by Li Shanbang and Qin Xinling in 1943. In 1946, Fu Chengyi (C.Y. Fu) made significant contribution to the theory of seismic waves. But generally, before the 1950s progress in Chinese seismology was slow partly due to the continuous warfare.

2. Seismology after 1949

Since the establishment of the People's Republic of China in 1949, there has been a great development in Chinese seismology. In 1956, *Study of Seismicity in China and Preparedness against Seismic Disasters* was included in the *National Twelve-year Long-term Project of the Development of Science and Technology*. In the proposal four topics were proposed: development of seismograph networks and seismological instrumentation, intensity zonation and study of the regional features of seismicity, study of the effect of seismic strong ground motion on buildings and development of earthquake engineering, and study of earthquake prediction. The last topic was one of the earliest scientific proposals in the world on the study of earthquake prediction.

In 1957 the *Chronological Table of Chinese Historical Earthquakes* was published. In 1960 the first edition of the *Catalogue of Chinese Earthquakes* was published. This comprehensive study is regarded as one of the most important contributions of Chinese seismology to the world. Historical records in eastern China are shown to be complete down to magnitude 5 for over 500 years since the Ming Dynasty (AD 1368–1644), being one of the longest complete earthquake records before the era of instrumental seismology.

The period of 1966 to 1976 was a seismically active period on the Chinese mainland, during which nine strong earthquakes occurred with magnitude larger than 7. Chinese seismologists carried out extensive studies and experiments in earthquake prediction in this decade, with both success and failure. The successful prediction of the February 4, 1975, Haicheng, Liaoning, M_S 7.3 earthquake was regarded as the first and up to now one of the few successful predictions in human history. The State Seismological Bureau, which was founded in 1971 (and renamed as the China Seismological Bureau in 1998), played an important role in coordinating seismological work nationwide.

INTERNATIONAL HANDBOOK OF EARTHQUAKE AND ENGINEERING SEISMOLOGY, VOLUME 81B *ISBN: 0-12-440658-0*

3. Major Accomplishments

Seismology in China from the 1950s to the 1990s accomplished the following. First, nationwide and regional seismograph networks were installed and operated; seismological instrumentation was developed to a considerable extent. Second, the Beijing-Tianjin-Tangshan-Zhangjiakou earthquake prediction experiment site and the western Yunnan earthquake prediction experiment site were developed; the system for earthquake prediction study and mitigation of seismic disasters was formed, which gave full consideration to the seismicity in China and the characteristics of modern Chinese society. Third, through a large amount of geophysical explorations and geological investigations, the lithospheric structure, stress field, and tectonics of the Chinese mainland and its surrounding regions were studied, providing the study of earthquakes with a fundamental geodynamic background. Fourth, field investigation was carried out for almost every strong earthquake whose mezoseismal region was accessible, with emphases on the geology of earthquakes, engineering seismology, aftershock observation, and earthquake prediction study, accumulating valuable observational data and experiences. Last, the study of active faults and historical earthquakes provided useful information for the long-term prediction of earthquakes and the estimation of seismic hazards.

Since the 1980s, the booming economy in China, the rapid development of urbanization, and the construction of large engineering projects have provided seismologists with both new opportunities and new challenges. In the meantime, Chinese seismologists have become more active in the international seismological communities.

4. Recent Advances

From 1996 to 2000, the China Seismological Bureau (CSB) carried out a nationwide project in upgrading the national and regional seismograph networks into broadband digital recordings. From 1996 to 2000, the CSB led a nationwide project in GPS measurement, aiming to monitor crustal movements and to predict earthquakes. Starting from 1999, the CSB has led a national project of fundamental research studying the mechanism and prediction of continental earthquakes. As an endeavor to involve the public into the reduction of seismic disasters, the *Law of the People's Republic of China on Protecting Against and Mitigating Earthquake Disasters* was enacted in 1997. In 1999, a comprehensive project for geophysical monitoring and earthquake emergency response was launched in the "capital circle" region in and around Beijing.

5. China National Report on CD

The complete *China National Report* is archived on the attached Handbook CD#2, under the directory of \7914China(Beijing), and includes five appendices: (1) Biographies of six Chinese pioneers in studying earthquakes—Zhang Heng (or Chang Heng), Weng Wenhao (W.H. Wong), Li Shanbang (S.P. Lee), Gu Gongxu (K.H. Ku), Fu Chengyi (C.Y. Fu), and Liu Huixian (H.X. Liu); (2) *The China National Report on Seismology and Physics of the Earth's Interior, 1995–1998* for the XXII General Assembly of IUGG, Birmingham, UK, July 1999; (3) Table of contents of the *Acta Seismologica Sinca*, volume 1 through 22 (1979–2000). This journal is the major scientific journal on earthquake research published in China and its English edition has been available since 1990; (4) *The Concise Catalogue of Earthquakes in and around China (2300 BC to AD 2000; magnitude ≥ 6)*; and (5) *The Catalog of Earthquakes in China and its Vicinity (2300 BC to AD 2000, magnitude ≥ 5).*

6. The China Seismological Bureau

Address: 63 Fuxing Ave., Beijing 100036, China.

6.1 Management Responsibilities

The China Seismological Bureau (formerly State Seismological Bureau, established in 1971) is under the leadership of the State Council, coordinating and managing the affairs related to earthquake disaster prevention and reduction in China. The major responsibilities include

- Working out, supervising, and implementing working principles, policies, laws and regulations, and technical standards related to earthquake disaster reduction and prevention.
- Formulating development strategies and intermediate- and long-term plans and annual programs related to earthquake disaster reduction and prevention.
- Managing the affairs related to earthquake monitoring and prediction.
- Managing the affairs related to nationwide and regional seismic intensity zonation and seismic hazard assessment.
- Working out and promulgating nationwide and regional earthquake resistance design standards, checking, supervising, and monitoring earthquake resistance design standards for national critical engineering projects.
- Organizing and managing studies of earthquake sciences to promote the development of earthquake science and technology.
- Educating the citizens and publicizing seismic hazard prevention and reduction, jointly with other related organizations.
- Managing and coordinating affairs related to international cooperation and exchange in earthquake science and technology.

6.2 Infrastructure

The headquarters of the China Seismological Bureau is organized into six departments: Administration office, Department of Personnel and Education, Department of Planning and Finance, Department of Monitoring and Prediction, Department of Science and Technology Development and International Cooperation, and Department of Disaster Reduction and Legislations.

There are provincial seismological bureaus in 29 provinces, autonomous regions, and 4 municipal cities (Beijing, Shanghai, Tianjin, and Chongqing), each of which is under CSB-dominated dual leadership. In earthquake-prone areas, there are also seismological bureaus and/or offices at lower levels, which are under the local government-dominated dual leadership.

There are 14 major CSB-affiliated institutions: Center for Analysis and Prediction, Institute of Geophysics, Institute of Geology, Institute of Crustal Deformation and Dynamics, Institute of Engineering Mechanics in Harbin, Wuhan Institute of Seismology, Lanzhou Institute of Seismology, Geophysical Exploration Center (Zhengzhou), First Deformation Monitoring Center (Tianjin), Second Deformation Monitoring Center (Xi'an), Comprehensive Observation Center, Earthquake Information and Data Center, Seismological Press, College of Disaster Prevention Techniques and Service Center.

Upon the occurrence of a large earthquake, the China Seismological Bureau will be subject to the Office of the Commanding Headquarters for Disaster Relief and Emergency Management under the State Council.

In the Bureau, there are about 10,000 scientists and technicians working in different areas related to earthquake disaster reduction, such as seismological network management and earthquake monitoring and prediction, engineering seismology and earthquake engineering, seismic hazard analysis and assessment, data transmission, processing and management and so forth.

6.3 Institute of Geophysics, China Seismological Bureau

Address: 5 Minzudaxue Nanlu, Haidian District, Beijing 100081, China.

The predecessor of the Institute of Geophysics, China Seismological Bureau, is the Institute of Geophysics, Chinese Academy of Sciences (Academia Sinica), which was founded on April 6, 1950. In 1978 the Institute became part of the State Seismological Bureau (SSB), which changed to the China Seismological Bureau (CSB) in 1998.

The Institute is a national research institution responsible for comprehensive research and applications in seismology, geomagnetism, and physics of the Earth's interior, observational seismology, engineering seismology, and socio-seismology. It has a staff of 947, including 4 members of the Chinese Academy of Sciences, 1 member of the Chinese Academy of Engineering, 159 full research professors, associate professors, and/or senior engineers, 267 assistant professors and/or engineers, 91 assistants and technicians. Forty-two experts in the Institute are awarded special subsidies by the government for their outstanding contributions in science and technology. The Institute is authorized by the Academic Degree Committee of the State Council to confer M.Sc. and Ph.D. degrees in solid earth geophysics. The Institute is also authorized to provide postdoctoral positions.

For nearly half a century the Institute has been carrying out research using the methods of physics, high technology, and social sciences. Various achievements have been obtained in the fields of seismotectonics, seismogeneisis, observational seismology, earthquake prediction study, structure and dynamics of the Earth's interior, theoretical and experimental studies of the physics of seismic source, theoretical and experimental geomagnetism, seismic zonation, engineering seismology and earthquake engineering, evaluation of seismic vulnerability, prediction of urban seismic disasters, earthquake countermeasures, instrumentation and seismological network, legislation of seismic disaster mitigation, standardization in seismology, earthquake insurance, and geoelectricity, contributing to the economic and societal significance of seismological researches in China. During the most recent twenty years the Institute has won the National Award of Science and Technology Progress (the Guojia Keji Jinbu Jiang) second place for 2 items and third place for 4 items, the National Science Award (the Guojia Ziran Kexue Jiang) third place for 1 item, the National Bibilio-Award (the Guojia Youxiu Tushu Jiang) for 2 items, the SSTC Invention Authentication (the SSTC Faming Zhengshu) for 1 item, the 1978 National Science Congress Award (the Quanguo Kexue Dahui Jiang) for 18 items, the National Mechanical Industry Congress Award (the Quanguo Jixie Gongye Kexue Dahui Jiang) for 1 item, and the SSB Award of Science and Technology Progress (the SSB Keji Jinbu Jiang), first place for 14 items, second place for 34 items, and third place for 52 items.

In the development of seismology in China, the Institute is a pioneer in several aspects: the first generation of Chinese-made seismographs; the first agency for the management, quality control, standardization, and international data exchange of the national seismological and geomagnetic networks; the first Chinese regional telemetered seismograph network (the Beijing Telemetered Seismograph Network); the first joint telemetered networks in China (the North China Joint Telemetered Seismograph Network); pioneering intensity zoning for national construction projects; the first generation of intensity zoning mapping of China; the first institution in China studying nuclear explosion seismology; leading in the study of seismotectonics and physics of seismic source in China; the first laboratory of physics of seismic source in China; initiating the study of earthquake prediction in China; the earliest study of induced seismicity in China; the first nationwide geomagnetic survey and the first generation of the 1:7,000,000 national geomagnetic map; the first magnetic-field-free laboratory in China; the first laboratory of seismoelectricity in China; the beginnings of the study of socio-seismology, seismological standardization, and earthquake insurance in China; organizing domestic and/or international projects of seismotectonics, strong motion seismology, and geodynamics.

The Institute has 15 research divisions, a geophysical observatory (the Beijing National Earth Observatory, former Beijing Geophysical Observatory), a seismological instrument laboratory, a center for computer and networking, a center for facilities and supplies, and a R&D center, as well as administration and faculty. The Institute also accommodates the Secretariat of the Asian Seismological Commission (ASC) of IASPEI, the Seismological Society of China (SSC), and the Chinese Geophysical Society (CGS). The Institute Library has over 200,000 books and journals in solid-Earth geophysics published since 1869. The Institute Archive manages the scientific archives and seismological and geomagnetic data. Extensive international cooperation is being carried out in the Institute. In a recent ten-year span, over 100 person-years of scholars from other countries and/or regions have come to the Institute for scientific cooperation. The Institute in turn has sent hundreds of person-years of scholars abroad to participate in international and/or regional symposia and to make professional visits. At present there are about 20 international cooperative projects underway in the Institute.

The Institute is engaged in routine observation, fundamental research, scientific development, and professional education. It also hosts the editorial offices of academic journals and/or technique reports such as *Acta Seismologica Sinica* (Bulletin of the Seismological Society of China, Dizhen Xuebao, Chinese Edition), *Acta Seismologica Sinica* (Bulletin of the Seismological Society of China, English Edition), *Recent Developments in World Seismology (Guoji Dizhen Dongtai), Translated World Seismology (Shijie Dizhen Yicong), Seismological and Geomagnetic Observation and Research (Dizhen Dici Guance yu Yanjiu), Journal of the Theory and Application of Computer Tomography (CT Lilun yu Yingyong Yanjiu), Annual Report of the National Seismograph Network of China, Annual Report of Earthquakes in China, Annual Report of Geomagnetic Observation in China*, and *Observational Report of Geomagnetic Storms.*

6.4 The Institute of Engineering Mechanics, China Seismological Bureau

Address: 29 Xuefu Road, Harbin 150080, China.

The Institute of Engineering Mechanics (IEM) was planned in 1952 and formally established in 1954 under the Chinese Academy of Sciences and was renamed The Institute of Engineering Mechanics, China Seismological Bureau in 1984.

IEM is the biggest and earliest research institute in China conducting research on earthquake engineering and engineering seismology. Since the beginning of the 1970s, the Institute has further emphasized its main efforts at earthquake engineering and safety engineering in response to the rapid economic development of the country and the urgent need for natural disaster reduction at home and abroad.

IEM is the pioneer in China in almost all aspects of the field of earthquake engineering. It initiated earthquake engineering research in 1955 and has been making significant contributions to national earthquake engineering research as well as to practice. For many years, IEM has endeavored to communicate and cooperate with the international earthquake engineering communities. Delegations and scientists from many countries have visited IEM or attended the various scientific and technical conferences organized by IEM; IEM has also sent abroad quite a number of delegations and visiting scholars for good-will visits, conferences, lecturing, and joint research. Through such activities, IEM has gained an increasing international reputation.

At present, IEM has 12 technical divisions covering almost all disciplines of earthquake engineering. The personnel of the Institute number 515, of which about two-thirds are scientific and technical staff, including 93 senior members. There is also a sizable group of graduate students enrolled in the Institute for Master and Doctor degree.

6.5 Institute of Geology, China Seismological Bureau

Address: De Wai Qijiahuozi, Beijing 100029, China.

The Institute of Geology, China Seismological Bureau, is a comprehensive research institution in solid earth sciences. Major scientific research fields include active tectonics, current crustal movement, structure of Earth's interior, geodynamics, tectonophysics, origin and process of earthquakes, geochronology, seismic zonation and engineering safety assessment, seismic hazard prevention and reduction.

The Institute was established in 1978 from several divisions of the Institute of Geology, Academia Sinica, which was founded in 1950, and subordinated to the State Seismological Bureau in 1971. Professors Hou Defeng, Zhang Wenyou, Gao Wenxue, Ma Xingyuan, Liu Ruoxin, and Ma Zonjin were the directors of the Institute. In 1998 the State Seismological Bureau (SSB) was renamed the China Seismological Bureau (CSB) and hence, as an affiliation to CSB, the Institute is named the Institute of Geology, CSB, and is headed by Professor Liu Qiyuan.

There are 393 employees in the Institute, 227 of them scientists, including 2 members of the Chinese Academy of Sciences, 1 member of the Chinese Academy of Engineering, and 132 senior scientists. The Institute consists of 9 research divisions and 2 service centers: (1) Division of Underground Fluid and Remote Sensing, (2) Division of Intermediate and Long-Term Earthquake Prediction and Seismic Zonation, (3) Division of Earthquake Dynamics, (4) Division of Active Tectonics, (5) Division of Earthquake Engineering and Advanced Technology, (6) Division of Geophysics of Solid-Earth, (7) Division of Tectonophysics, (8) Division of Comprehensive Studies on Natural Disasters, (9) Division of Neochronology, (10) Center for Science and Technology Information, and (11) Center for Computing and Network.

Besides the main task of earthquake hazards prevention and reduction, the Institute is also actively engaged in national key scientific projects. Significant achievements have been made, as illustrated by 10 national award-winning projects and 114 ministry award-winning projects, including the important outcomes of scientific research in compiling *Lithospheric Dynamics Atlas of China* and the *Geoscience Transect of Eastern China*. The Institute has also made significant contribution to the national economy and construction by undertaking the seismic safety assessment for more than 150 engineering projects, including the Dayawan and the Qinshan nuclear power plants, and the Three-gorge hydropower project.

The Institute has advanced prospecting equipment, and 14 laboratories including the Laboratory of Tectonophysics and Laboratory of Neogeochronology. Authorized by the State Council, the Institute is one of the institutions that confer Doctoral and Master Degrees in the fields of structural geology, solid earth geophysics, and geochemistry.

The Institute publishes a scientific journal, *Seismology & Geology*, and has established close relationships and international cooperation with many countries around the world, especially with the scientists and research institutions in the United States, Russia, Japan, Canada, United Kingdom, France, Germany, and Australia. More than 400 people have been abroad for international cooperation during the past years.

6.6 Lanzhou Institute of Seismology, China Seismological Bureau

Address: 410 Donggang West Road, Lanzhou 730000, China.

Lanzhou Institute of Seismology, China Seismological Bureau (CSB), is situated in Lanzhou City of Gansu Province, which is located at the geographic center of China, on the northeast margin of the Qinghai-Xizang Plateau and the north segment of the North–South seismic belt in China. It is one of the earliest earthquake research institutions of China, with a history of nearly forty years.

Lanzhou Institute of Seismology, directly under the China Seismological Bureau, is a comprehensive research center in the field of earthquake science in northwestern China. There are 409 scientists and technicians now, among them 105 senior researchers. It consists of 9 research divisions, possessing a large number of advanced instruments, computers, and laboratories, and managing a series of digital seismic monitoring networks and digital strong motion observation networks, such as digital seismographs and digital observation networks of the earthquake precursors in the Gansu Province.

The Institute has made fruitful contributions to seismological research and the national economy through construction in northwestern China. It has undertaken many key projects assigned by the government and CSB. In the last decade, it has won about 60 national or provincial scientific achievement prizes.

The Institute conducts research in geophysics, seismogeology, engineering seismology, geochemistry, and earthquake prediction. Various branches interact with each other so that observation, prediction, and applied and theoretical studies of seismological science are fully combined. The main studies involve seismogenic environments and tectonic backgrounds, physics of seismic source, deep seismic sounding, focal mechanism, digital seismology, deep geophysical sounding, seismogeology, loess dynamics, natural disasters, and so on. In addition, the Institute engages in methodological study of comprehensive earthquake prediction based on precursors of electromagnetism, underground fluid, seismicity, etc., and applied study of engineering seismology, prediction and evaluation of seismic hazards such as ground motion parameters zonation, seismic risk analysis, and so on.

Since 1980, the Institute has carried out efficient cooperation and exchanges with more than ten countries such as Belgium, US, Germany, France, Russia, etc. Meanwhile, tens of scientists of the Institute visited more than ten foreign countries for international scientific exchange and cooperative research.

The Institute is one of the professional education and training centers in the seismological field in China, one of the authorized units of the Master degree approved by the Academic Degree Committee, the State Council of China. In the last 2 decades, more than 130 postgraduates have been conferred a Master degree by the Institute.

79.15 China (Taipei)

The editors received from Dr. Y. T. Yeh, the IASPEI National Correspondent, a report prepared by the National IUGG Committee of China (Taipei). It consists of six institutional reviews and is archived as computer-readable files on the attached Handbook CD#2, under the directory of \7915China(Taipei). We extracted and excerpted the following summary. An inventory of seismic networks in the Taiwan area is included in Chapter 87, a review of the strong motion programs is given in Chapter 64, and some biographical sketches may be found in Appendix 3 of this Handbook.

1. Institute of Earth Sciences, Academia Sinica

Web site: http://www.earth.sinica.edu.tw/

The Institute of Earth Sciences (IES) is located in Nankang, Taipei. The history of the Institute begins in 1970, with a request from the National Science Council to T. L. Teng and F. T. Wu for an earthquake research plan. The Seismology Task Force was founded in 1971 and the Seismology Group, attached to the Institute of Physics, Academia Sinica, with Y. B. Tsai as head, was founded in 1973. The Institute of Earth Sciences was formally inaugurated in 1982, with Y. B. Tsai as the Director. At present, the Institute's research efforts are concentrated in seismology, tectonophysics, geochemistry, and mineral and rock physics. It also contributes significantly to graduate education in Taiwan and offers expertise to earthquake disaster prevention and risk analyses. The Institute has 16 research fellows, 9 associate research fellows, 5 assistant research fellows, 11 postdoctoral fellows, and 15 research assistants. The names and research specialties of these scientists, as well as details of the research topics and a list of references, are given in the full report on the Handbook CD.

2. Seismological Observation Center, Central Weather Bureau

Web site: http://www.cwb.gov.tw/

This Center superseded the former Geophysical Department of the Central Weather Bureau (CWB) in 1989, with the mission of observing earthquake activity in the Taiwan area and preventing possible earthquake destruction.

A project, "Enhancement in Earthquake Observation: Taiwan Earthquake Observational Network Establishing Program," was executed promptly. The Seismology Center completed a real-time, 3-component, short-period seismic network of 75 stations in the early 1990s. In addition, the Center initiated the extensive Taiwan Strong Motion Instrumentation Program during 1992–1997, which consisted of 637 digital free-field accelerographs, 56 building arrays in 9 major metropolises, and a real-time system of telemetered digital accelerographs. A Phase II program, beginning in 1998, succeeded in upgrading the real-time system of digital accelerographs to be fully automatic in disseminating information of location, magnitude, and shake map of strong earthquakes occurring in the Taiwan region to the officials within 1 to 2 minutes. See Chapter 64 for more details.

This real-time system operated successfully during the disastrous 1999 Chi-Chi earthquake ($M_W = 7.6$), and CWB collected the most extensive set of seismic data of a major earthquake in the 20th century. The number of digital free-field accelerographs has since increased to over 750, and the number of building arrays to 60 (with a total of about 2,000 accelerometers in buildings and bridges). A permanent Global Positioning System network was initiated in 1993 and now consists of 16 stations.

Further details of the CWB programs and a list of references are given in the full report on the Handbook CD (which also includes a short article, "Seismological Observation and Service in Taiwan up to 1970," by Ming-Tung Hsu). Please also see the "Dedicated Issue, Chi-Chi, Taiwan Earthquake of 20 September

INTERNATIONAL HANDBOOK OF EARTHQUAKE AND ENGINEERING SEISMOLOGY, VOLUME 81B ISBN: 0-12-440658-0

1999" (*Bull. Seismol. Soc. Am.*, **91**, 893–1395, 2001, and the attached CD-ROM Supplement of Seismic Data).

3. Institute of Geophysics, National Central University

Web site: http://www.ncu.edu.tw/

This Institute was founded in 1962, the first academic institution in Taiwan devoted to education in geophysics. It now has 17 full-time faculty members and approximately 40 graduate students in Master and Ph.D. programs. The names and research specialties of the faculty are given in the full report on the Handbook CD.

From 1962 to 1986, scientific activities were focused on oil exploration and seismology. In recent years, there has been a growing interest in engineering applications of geophysics and pedagogy in earth sciences. Experimental and theoretical research consists of four main topics: seismology, the seismic method of exploration, gravity and magnetics, and geoelectric studies. On average, about ten Master and Ph.D. dissertations are finished each year. The details of the research efforts in each of the four topics are given in the full report, as well as a list of publications by Institute personnel, on the Handbook CD.

4. Institute of Seismology and Applied Geophysics, National Chung Cheng University

This Institute was founded in 1990 as the Institute of Seismology in response to the frequent earthquakes affecting the southwestern area of Taiwan. Applied geophysics was added in 1994. The faculty of nine full-time and one part-time members has the ultimate objective of facilitating research and instruction in the areas of earthquake mechanism and the interior structure of the Earth, the solution of environmental problems, development of techniques for seismic hazard mitigation and earthquake prediction, and the training of specialists in these fields. The equipment and facilities of the Institute are given in the full report on the Handbook CD.

5. Institute of Oceanography, National Taiwan University

Web site: http://www.oc.ntu.edu.tw/

This Institute was established in 1968 as the first oceanographic institution in Taiwan. The Institute offers M.S. and Ph.D. programs in physical, chemical, biological, and geological oceanography, marine geophysics, and fisheries. Education and research activities in seismology are carried out by faculty in the Division of Marine Geology and Geophysics.

The first experiment of an ocean bottom seismometer was carried out in offshore Taiwan in 1985, in collaboration with the Institute of Earth Sciences, Academia Sinica, and the Hawaii Institute of Geophysics. Marine multi-channel seismic reflection studies have been conducted routinely by Institute faculty since 1989. Other recent projects and biographical sketches of some Institute personnel are given in the full report on the Handbook CD.

6. National Center for Research on Earthquake Engineering

Web site: http:/www.ncree.gov.tw/

The National Center for Research on Earthquake Engineering (NCREE) was established in 1990. The major mission of the Center is to promote earthquake engineering research and to upgrade seismic-resistant design standards.

The research in seismic engineering takes four directions: (1) innovative technologies for earthquake engineering, (2) advanced building systems, (3) seismic hazard mitigation program, and (4) seismic engineering renewal-retrofit technology. The long-term goals of the Center are (1) to undertake high-profile progressive research in the selected areas, (2) to transform the Center from a national one into a major international hub of earthquake engineering research, (3) to supervise orientation of domestic earthquake engineering research, and (4) to play a leading role in modification and upgrading of earthquake-resistant design regulations.

More details are given in the full report on the Handbook CD, and in the NCREE website.

79.16 Croatia

The editors received the report, "Seismology in Croatia—National Centennial Report to IASPEI," from Dr. Marijan Herak, IASPEI National Correspondent. The report is written by Marijan Herak, Dragutin Skoko, and Davorka Herak (Geophysical Institute, University of Zagreb), and is archived as a computer-readable file in the attached Handbook CD#2, under the directory of \7916Croatia. The following summary is extracted and excerpted from the report.

An inventory of the Croatian Seismic Network is included in Chapter 87, and a review of earthquake engineering in Croatia is given in Chapter 80.7. The biography of Andrija Mohorovičić (1857–1936) is given in Chapter 89, and biographical sketches of some Croatian scientists and engineers may be found in Appendix 3 of this Handbook.

1. Introduction

Seismology in Croatia has deep roots that extend back into the 19th century. It developed at first almost exclusively as the result of the efforts of Andrija Mohorovičić. In later years, the number of active seismologists varied, but never exceeded twelve. All were associated with the Geophysical Institute in Zagreb. This small community bore the responsibility for maintaining high standards in seismological research and education. Seismology is taught at the Department of Geophysics of the Faculty of Science and Mathematics, University of Zagreb, at both the undergraduate and postgraduate levels.

2. Seismology in Croatia until 1960

Immediately after the great Zagreb earthquake of November 9, 1880, the Yugoslav Academy of Science and Arts in Zagreb established the "Committee for observation of earthquake-related phenomena" with the task to study Croatian earthquakes and methodically collect all related data. In the first volume of its Papers, the Academy published the extensive report on the Zagreb earthquake. The Academy's Committee later also published all available information on Croatian earthquakes from AD 361 to 1906.

In January 1892, Andrija Mohorovičić became the director of the Meteorological Observatory in Zagreb (later the Geophysical Institute). In 1893 he initiated systematic collection of earthquake data, been regularly published since 1906 in the Seismic Reports. Following the strong earthquake of December 17, 1901, near Zagreb, Kišpatić and Mohorovičić made every effort to have a seismic station set up in Zagreb. After the first seismoscopes and seismographs were obtained in the beginning of the 20th century, Mohorovičić realized the importance of accurate time keeping and initiated regular observation of times of passage of stars through the local meridian.

From the seismograms of the strong earthquake in the Kupa Valley (Pokupsko) on October 8, 1909, Mohorovičić showed the existence of the boundary between the Earth's crust and mantle—now called the "Mohorovičić discontinuity." His studies of the Pokupsko earthquake also yielded the procedure of unique location of earthquake focus and the analytical expression for the increase of elastic wave velocity with depth.

After Mohorovičić retired, the seismological research lost most of its momentum, and for 20 years no seismological papers were published due to political turmoil. The seismographic observations in Zagreb, however, continued with practically no interruptions, and all seismograms are still available in the archive of the Geophysical Institute. After World War II, the Geophysical Institute was incorporated into the Faculty of Science and Mathematics of Zagreb University. Most of the efforts are dedicated to construction of local travel-time curves based on empirical data, and to the problems related to locations of the earthquakes.

 ISBN: 0-12-440658-0

3. Seismology in Croatia (1961–1999)

Following the damaging Makarska-Hvar earthquake of 1962, seismological research was intensified again. Croatian seismologists published papers in cooperation with foreign colleagues, dealing with location of earthquakes and seismicity of Central Adriatic and other Croatian regions, and analyzed the problem of optimum distribution of seismic stations. The first seismic maps of the Croatian territory were published, as well as studies dealing with microzonation of cities and larger urban areas. Inclusion of Yugoslavia (and thus also Croatia) into the UNESCO project "Survey of the seismicity of the Balkan region" in the 1970s provided Croatian seismologists the opportunity to participate in international teams studying the seismicity of southeastern Europe. One of the main outcomes is the catalog of earthquakes and the collection of macroseismic maps. In this period papers were published on the seismotectonics of the Croatian region, and the relation between magnitude, intensity, and maximal acceleration was investigated. The increased seismic activity of the Friuli region (Italy) in 1976, and the Montenegro earthquake of 1979 prompted macroseismic and neotectonic studies, analyses of the aftershock sequences, investigations of temporal and spatial variation of seismicity, estimating parameters of seismic forces, strong motion analyses, and more.

In 1985 the Croatian Seismological Survey was founded and became responsible for collection, analysis, and archiving of seismological data. In the second half of the century, especially in the last three decades, the scientific interest of Croatian seismologists broadened and international cooperation intensified. New techniques and methods have been introduced, and instrumentation has been substantially improved. Studies include improvement of earthquake location techniques, revision and updating of the Croatian Earthquake Catalog, quantification of earthquakes (especially M_S), seismotectonical considerations, historical seismology, determination of velocity distribution including studies of anisotropy, attenuation of seismic waves (coda Q), earthquake statistics and prediction (CN algorithm), earthquake hazard and risk, and seismic zonation of Croatia.

4. Seismological Instrumentation

The first seismological station on Croatian territory was opened in Pula in 1900, equipped at first with a Vicentini, and afterward also with Conrad and Wiechert seismographs. The station was closed in 1918, after the collapse of the Austro-Hungarian Empire, and it is not known what happened to the instruments and the seismogram archive. Besides Pula, the following stations also operated in the first half of the 20th century: Rijeka (1901–1918, seismograph Vicentini), Sinj (1914–1924, seismograph Conrad), Šibenik (1926–1940, seismograph Conrad, temporary operation), Dubrovnik (1927–1929, seismograph Conrad), Zagreb (1906–).

The Zagreb (ZAG) seismological station was founded when the first seismological instrument, the Agamemnone electrical seismoscope, was acquired by the Academy in 1901. A Vicentini seismograph was installed in the basement of the Geophysical Institute in Zagreb on April 6, 1906. The instrument was operational for 18 years. Dissatisfied with the instrument's performance, Mohorovičić purchased the Wiechert seismograph with a mass of 80 kg that recorded horizontal ground motion and installed it in Zagreb in 1908. Soon afterward, in 1909, a new horizontal instrument was obtained with a mass of 1000 kg. The vertical Wiechert instrument (1200 kg) was acquired in 1932. These seismographs operated with almost no interruption until 1983, when they were moved to the Institute's new location on Horvatovac, Zagreb. These seismographs were restored and are now in operating condition.

Electromagnetic seismographs (Sprengnether, SKM-3, Vegik) were obtained in early 1970s, in the framework of the UNESCO project "Survey of the Seismicity of the Balkan region." At that time, a strong motion network was also installed. The new instruments were used to open new permanent stations Puntijarka (PTJ, 1972) and Hvar (HVAR, 1973), and later also Dubrovnik (DBK, 1986) and Rijeka (RIY, 1988). The first 16-bit digital instrument was installed in Zagreb in 1989. The modern Croatian seismic network is based on 7 broadband seismometers with 24-bit digitizers, which were obtained in 1999. (See the "Croatia" entry in Chapter 87.)

79.17 Czech Republic

The editors received the "National Centennial Report of the Czech Republic" from Dr. Vladimír Čermák, President of the National IUGG Committee. This report is archived as a computer-readable file in the attached Handbook CD#2, under the directory of \7917Czech. We extracted and excerpted the following summary. An inventory of seismic networks in the Czech Republic is included in Chapter 87. Biographies and biographical sketches of some Czech scientists may be found in Chapter 89, and in Appendix 3 of this Handbook, respectively.

1. Introduction

The earth sciences, including geophysics and seismology, have a long and rich history within the territory of the present-day Czech Republic. The natural sciences have been taught since 1622 at Charles University, Prague, founded in 1348 as the first educational center in central Europe. Geophysical measurements began in 1839. In 1908 the permanent seismological station in Cheb began operation under the leadership of G. Irgang, in order to record earthquake swarms occurring in nearby Vogtland.

Modern geophysics in the country is inseparable from the almost 80-year-long history of the Geophysical Institute of the Academy of Sciences of the Czech Republic. The first director of the Institute, provisionally associated with the Faculty of Science of Charles University, was Václav Láska (1862–1943). The Institute became one of the state research institutions of Czechoslovakia, newly formed as an independent country after World War I. As a center for seismological, geomagnetic, and general geophysical research, the Institute also represented Czechoslovakian geophysics internationally with contributions, especially from Alois Zátopek (1907–1985), and Vít Kárník (1926–1994).

As a result of the separation of Czechoslovakia in 1993 into the Czech and Slovak Republics, the Czechoslovak Academy of Sciences ceased to exist, and the Academy of Science of the Czech Republic was created. Presently, seismology and earthquake engineering research are carried out in the following four institutions.

2. The Geophysical Institute, Academy of Sciences, Prague

Web site: http://www.ig.cas.cz/

The basic activities of the Geophysical Institute include observatory and field measurements for continuous and episode monitoring of various geophysical fields in the Czech territory as well as in adjacent parts of Central Europe. There is good cooperation with the worldwide data network services and data centres. The Institute is formally organised in five scientific departments, which cover the major geophysical disciplines. The seismological unit, by far the largest, is further sub-divided into four laboratories: seismic observations and interpretations, theoretical seismology, tectonics and geodynamics, and seismic modelling. The other units cover gravimetry, heat flow and radioactivity, geomagnetism, and geoelectricity. More detailed information about the Institute, including a list of publications, is given in the full report on the Handbook CD.

3. The Institute of Rock Physics, Academy of Sciences, Prague

Web site: http://www.irsm.cas.cz/

This institute evolved from the Institute for Scientific Research of Coal, founded in 1927. The original orientation toward mining problems has been replaced with induced seismic phenomena in seismology. Current research is mainly on slope stability and rock movements, including studies of geological factors in

INTERNATIONAL HANDBOOK OF EARTHQUAKE AND ENGINEERING SEISMOLOGY, VOLUME 81B ISBN: 0-12-440658-0

regions selected for nuclear waste depositories and underground gas storage.

In 1979 the Mining Institute merged with the Geological Institute of the Academy of Sciences, under the new name of Institute of Geology and Geotechnics of the Czechoslovak Academy of Sciences. In March 1990 this Institute split into the Institute of Geotechnics of the Czechoslovak Academy of Sciences, and the Geological Institute of the Czechoslovak Academy of Sciences. More details about the Institute, including a list of publications, are given in the full report on the Handbook CD.

4. The Department of Geophysics, Charles University, Prague

Web site: http://seis.karlov.mff.cuni.cz/

The Department is engaged in the following research topics and more details (including references) are given in the full report on the Handbook CD.

4.1. Theoretical and Numerical Research of Seismic Waves. Theoretical and numerical research of seismic waves propagating in complex, 2-D and 3-D, isotropic and anisotropic structures has a long tradition. The research has been performed in close cooperation with the Geophysical Institute, Czech Acad. Sci., and several other institutions.

4.2. Earthquakes, Ground Motions. The earthquake research has concentrated on numerical modeling of seismic ground motions, including the source, path, and site effects. Seismic station Prague (code PRA), in existence since 1924, was upgraded in the 1960s by long-period seismographs, and recently by a broadband CMG-3T instrument. Mainly due to the work of A. Zatopek, station PRA contributed substantially to the development of European seismology. Since 1997, Charles University has been running three broadband CMG-3T stations in the Corinth Gulf, Greece, as a joint project with the Seismological Laboratory, University of Patras. The data are provided to ORFEUS data center.

4.3. Earth Structure. The structural seismic research has dealt with (1) properties of the channel waves Lg and Rg, (2) travel times and amplitudes of PKP waves, (3) dispersion of surface waves along several long profiles in Central Europe, and (4) the influence of the Earth's structure on the amplitudes of body waves (to improve magnitude calibration curves).

5. The Institute of Physics of the Earth, Masaryk University Brno, Brno

Web site: http://www.ipe.muni.cz/

The Institute of Physics of the Earth (IPE) was originally formed in 1984, and since 1992 has been part of the Masaryk University (MU) in Brno. Regional seismic monitoring has become the most important task for the IPE, and international cooperation started with the German Regional Seismic Network (GRSN) and with the GFZ, Potsdam, in particular within the GEOFON project. Extensive cooperation was also established with Austria (Central Institute of Meteorology and Geodynamics—ZAMG) in the field of regional seismic monitoring. The IPE currently operates two seismic networks (Network Kraslice in Western Bohemia and Network Temelín in Southern Bohemia) and four modern broadband stations (VRAC, MORC, JAVC, and KRUC). The National Data Centre for the CTBTO was established in the IPE. The Institute operates an auxiliary seismic station that is a part of the International Monitoring System in Vranov near Brno (AS026). More details, including a list of publications, are given in the full report on the Handbook CD.

79.18 Denmark

The editors received a brief Danish report from Dr. Søren Gregersen, IASPEI National Correspondent, Seismology Section, National Survey and Cadastre, Copenhagen, Denmark. The report was edited and is shown here. An inventory of the Danish seismographic stations is included in Chapter 87. Biographies and biographical sketches of some Danish seismologists and geophysicists may be found in Chapter 89, and in Appendix 3 of this Handbook, respectively.

1. Introduction

Earthquakes in Denmark are so rare that the seismologists know the few important ones by heart, (e.g., 1759 in Norway/Sweden felt in nearly all of Denmark, 1841 in northern Jylland, 1869 in the middle of Sjaelland, and 1930 near Copenhagen). The felt reports on these old earthquakes were collected and published by various geology professors, and the files are presently in the Geological Museum of University of Copenhagen.

In the beginning of the 20th century the army lieutenant colonel E.G. Harboe (1845–1919) became interested in seismology, and he had seismograph stations installed in Greenland (Disko Island, 1907–1912) and Iceland (Reykjavik from 1909). He also was able to get Denmark admitted into IASPEI in 1914. During the first two decades of the 1900s he collected and described the Danish earthquakes. But it was not until 1926 that a systematic seismic service was established in Denmark. The director of the Danish Geodetic Institute convinced the Carlsberg foundation to support the purchase and operational cost of, first, a seismograph in Copenhagen from 1926, and then two in Greenland.

2. Inge Lehmann (1888–1993)

In 1926, the newly employed mathematician Inge Lehmann became involved with seismology. She was immediately sent to seismological key points in Europe: Gutenberg in Göttingen and Jeffreys in Cambridge. For a quarter of a century she was the seismology institution in Denmark, with technical help from army non-commissioned officer H. Rasmussen and, since 1946, Erik Moeller, and only occasional academic help. At her disposal, she had the Copenhagen (COP) seismograph station with Wiechert and Galitzin-Wilip seismographs, and from 1936 a Benioff seismograph. In Greenland she had Wiechert seimographs in Ivigtut (IVI) from 1927 to 1953, and various replacements, which never came to function properly, from 1955 to 1960. The other seismograph station in Greenland was placed in Scoresbysund on the east coast from 1928.

Inge Lehmann was much concerned with travel-time studies, where she advocated the personal collection and reading of seismograms from various stations rather than the collection of many readings done by various analysts. Inge Lehmann is best known for her discovery that the earth has an inner core, but she also studied the structure of the upper mantle. At a depth of 220 km she proposed a discontinuity, which is often referred to as the Lehmann discontinuity. Inge Lehmann and her work are presented on the home page of the National Survey and Cadastre, Denmark (www.kms.dk) under research and development/seismology (Danish version). At this home page is a link to a home page at UCLA, where Inge Lehmann's work is presented by Leon Knopoff. She also investigated the Danish earthquakes occurring during her time and she made a comprehensive overview of Danish earthquakes.

3. Henry Jensen (1915–1974)

In 1953 Henry Jensen took over the seismology department. His special field was seismic noise, microseisms. He was much involved in the International Geophysical Year (IGY) in 1957–1958, and the establishment of a new seismograph station in northern Greenland, Station Nord (NOR) for IGY. Together with the new seismologist Joergen Hjelme, he established

INTERNATIONAL HANDBOOK OF EARTHQUAKE AND ENGINEERING SEISMOLOGY, VOLUME 81B ISBN: 0-12-440658-0

(1962–1964) and continued operation of four stations of the World Wide Standardized Seismograph Network (WWSSN): Copenhagen (COP) in Denmark, Nord (NOR) in northern Greenland, Kap Tobin (KTG), which replaced the Scoresbysund seismograph, and Godhavn (GDH) on Disko Island in western Greenland, where Harboe had his station in the beginning of the century. In this time of build-up, Erik Hjortenberg joined Danish seismology, still housed in the state institution, Danish Geodetic Institute. The fields of investigation of the three seismologists were mainly explosionstudies of crustal structure and microseisms.

4. Recent Advances

In the beginning of the 1970s new topics were taken up by Søren Gregersen, who came into seismology when Henry Jensen started teaching in Copenhagen University and left his position in the Geodetic Institute. Several portable short-period seismographs were used for studies of the Danish earthquakes and the responsible stresses. Studies of surface waves in laterally inhomogeneous media and of earthquakes in Greenland were carried out. Technical assistance has been by Hans Peter Rasmussen and John Christensen.

A new generation of seismologists has come into Danish seismology with the introduction of digital seismographs (1998–1999). Tine B. Larsen has brought in the field of mantle dynamics, including plumes. Large-scale lithospheric structure is being investigated in Greenland as well as in Denmark and Peter Voss has become involved in travel-time tomography. At the disposal of the seismology group are now eight permanent stations, four each in Denmark and Greenland. In 1990 the Danish Geodetic Institute was merged with other institutions to create the National Survey and Cadastre–Denmark, in which there is a Seismology Section.

The old Copenhagen (COP) location is maintained. It is situated on limestone in old fortifications. In an abandoned limestone mine in central Jylland is Moensted (MUD). 50 km south of Copenhagen, in an abandoned chalk pit, is Lille Linde (LLD). Only in one place in Denmark is it possible to place the seismograph on bedrock, in the small island of Bornholm (BRD). Scoresbysund (SCO) in eastern Greenland has in 1982 come into operation again, replacing KTG. Danmarkshavn (DAG) replaced NOR in 1972. It is now a digital GEOFON station as well as a WWSSN station. In western Greenland the old Godhavn location has been abandoned for the more accessible Sondre Stromfjord (SFJ), in cooperation with IRIS/GEOFON. A new seismograph is in operation at Narsarsuaq (NRS), southern Greenland.

The data from the stations in Denmark are accessible through AUTODRM, while those in Greenland are less accessible, through tape via mail. These seismograph stations are waiting for better communication in the future.

Selected References

Gregersen, S. (1989). The seismicity of Greenland. In: "Earthquakes at North-Atlantic passive margins: Neotectonics and postglacial rebound" (S. Gregersen and P.W. Basham, Eds.), pp. 345–353. Kluwer Academic Press.

Gregersen, S., J. Hjelme, and E. Hjortenberg (1998). Earthquakes in Denmark. *Bull. Geol. Soc. Denmark*, **44**, 115–127.

Hjelme, J. (1996). History of seismological stations in Denmark with Greenland. In: "Seismograph recordings in Sweden, Norway—with arctic regions, Denmark—with Greenland, and Finland," (R. Wahlstorm, Ed.), Seismological Department, Uppsala University, Sweden.

Hjortenberg, E. (1987). Seismic noise sampled in a world wide experiment. *Phys. Earth Planet. Inter.*, **47**, 1–3.

Jensen, H. (1961). Statistical studies on the IGY microseisms from Koebenhavn and Nord. Danish Geodetic Institute Bulletin no. 39, 67 pp.

Lehmann, I. (1936). P′ (pronounced P prime). *Publ. Bur. Centr. Seism. Int. A 14*, 87–115.

Lehmann, I. (1956). Danish earthquakes (in Danish). *Bull. Geol. Soc. Denmark*, **13**, 88–103.

Lehmann, I. (1960). Structure of the upper mantle as derived from the travel times of seismic P and S waves. *Nature,* **186**, 956.

Vaccari, F., and S. Gregersen (1998). Physical description of Lg-waves in the inhomogeneous crust of the continents. *Geophys. J. Int.*, **135**, 711–720.

79.19 El Salvador

Although El Salvador is not an IASPEI member, the editors invited Walter Salazar (Tokyo Institute of Technology and Universidad Centroamericana José Simeón Cañas) to prepare a review of seismology and earthquake engineering in El Salvador based on Salazar (1999). The full report is archived as a computer-readable file on the attached Handbook CD#2, under the directory of \7919ElSalvador. We extracted and excerpted the following summary. An inventory of seismic networks in El Salvador is included in Chapter 87, and biographical sketches of some El Salvadoran seismologists may be found in Appendix 3 of this Handbook.

1. Introduction

El Salvador is a small (area $\approx$ 21,000 km^2) country in Central America. The population is about 6 million and is highly concentrated in the capital city, San Salvador, where 1.8 million persons are living now. The earthquakes on January 13 and February 13, 2001, showed the vulnerability of El Salvador (e.g., EERI, 2001; Lomnitz and Rodríguez-Elizarrarás, 2001). Many people were buried in landslide areas, others lost their houses and their relatives, and sad stories about the earthquakes are numerous. The full report on the Handbook CD attempts to explain the fundamental research that national and international researchers have done for El Salvador. Some suggestions are made for more effective seismic risk mitigation.

2. Tectonics and Seismicity

The tectonic regime in Central America and the Caribbean area is the result of the interaction of the Pacific, North America, Cocos, Nazca, and South America plates. The main source of seismicity that affects El Salvador is in the subduction zone. The Middle American Trench is located 125 km away from the coast, where the Cocos plate begins to submerge underneath the Caribbean plate (Dewey and Suárez, 1991), generating earthquakes up to magnitude of about 8. The most recent destructive earthquakes, June 19, 1982, and January 13, 2001, were reported as intra-plate shocks with normal fault mechanism in the Cocos plate.

Another important seismogenic source is the volcanic chain (Stoiber and Carr, 1977; White, 1991). Seismicity is concentrated in the upper 25 km of the crust and within a nearly continuous belt of 20 km along the axis of the principal Quaternary volcanoes, parallel to the subduction trench. Earthquakes in the volcanic chain are moderate, reaching a magnitude of 6.7, and are interpreted as the result of a right-lateral shear zone, driven by an oblique component of convergence between the Caribbean and Cocos plates. The most recent destructive upper-crustal events on October 10, 1986 (Harlow *et al.*, 1993), and February 13, 2001, are reported as having a strike-slip mechanism.

3. Strong Ground Motions

The first strong motion network in El Salvador was under the Centro de Investigaciones Geotécnicas (CIG), which belongs today to the Ministry of Environment and Natural Resources. Another strong motion network is operated by Geotérmica Salvadoreña (GESAL) in a geothermal plant at Berlin, Department of Usulután (using digital instruments). A new strong motion digital network deployed by the Universidad Centroamerica José Simeón Cañas (UCA) consists of ten digital instruments with time-code generators (Bommer *et al.*, 1997). The main objective of the networks is to cover the seismicity from both the volcanic chain and the subduction zone. A very good set of strong motion data resulted from the recent destructive earthquakes of January 13 and February 13, 2001, in the near and far field (Cepeda and Salazar, 2001; López-Casado *et al.*, 2001).

Recently, the Japan International Cooperation Agency (JICA) is supporting a new strong motion network in the whole country, with the collaboration of the Servicio Nacional de Estudios Territoriales (SNET), within the Ministry of Environment and

INTERNATIONAL HANDBOOK OF EARTHQUAKE AND ENGINEERING SEISMOLOGY, VOLUME 81B ISBN: 0-12-440658-0

Natural Resources. The Spanish Agency for International Cooperation (AEIC) has also donated new strong motion accelerographs for El Salvador. The current strong motion network operators are SNET (taking over from CIG, which no longer exists), GESAL, and UCA. They have established a cooperative agreement to share their data.

4. Seismic Risk Assessment

A seismic catalog was compiled by Salazar *et al.* (1997) as part of the project Seismic Risk Assessment in El Salvador, supported by the European Economic Community. This catalog includes earthquakes from all of the recognized seismogenic zones from 1898 to July 1994. During the preparation of this catalog, some inconsistent events that might have a strong influence on seismic hazard assessment were identified. Two examples are described in the full report.

Four studies have developed seismic hazard assessment for El Salvador. The results are presented in a map of iso-acceleration curves for the 10 percent probability of exceedance and 50 years of lifetime. The main elements contributing to seismic risk in the metropolitan region of San Salvador have been identified and reviewed in detail. The most important seismic risk countermeasures have been the three seismic codes developed since 1966. The details of the projects (with figures and a list of references) are presented on the Handbook CD. Additional details may be found on the Web site: http://www.geo.mtu.edu/~raman/GSASalvador.html.

Selected References

Bommer, J.J., A. Udías, J.M. Cepeda, J.C. Hasbun, W.M. Salazar, A. Suárez, N.N. Ambraseys, E. Buforn, J. Cortina, R. Madariaga, P. Méndez, J. Mezcua, and D. Papastamatiou (1997). A new digital accelerograph network for El Salvador. *Seismol. Res. Letters,* **68**, 426–437.

Cepeda, J., and W. Salazar (2001). Los terremotos de enero y febrero: resultados y análisis en base a la red de acelerógrafos de la UCA. *Estudios Centroamericanos* 627/628, 11–27.

Dewey, J., and G. Suárez (1991). Seismotectonics of the Middle America. In: "Neotectonics of North America" (Slemmons *et al.*, Eds.), pp. 309–321. Geol. Soc. Am. Decade Map, **1**.

EERI (2001). Preliminary observations on the El Salvador earthquakes of January 13th and February 13th, 2001. *EERI Newsletter* 35(7), July, 12 pp. Also distributed on CD by the Earthquake Engineering Research Institute (http:///www.eeri.org).

Harlow, D., R. White, M. Rymer, and S. Alvarado (1993). The San Salvador earthquake of 10 October 1986 and its historical context. *Bull. Seismol. Soc. Am.,* **83**, 1143–1154.

Lomnitz, C., and S. Rodríguez-Elizarrarás (2001). El Salvador 2001: earthquake disaster and disaster preparedness in a tropical volcanic environment. *Seismol. Res. Lett.,* **72**, 346–351.

López-Casado, C., B. Benito, J.J. Bommer, M. Ciudad-Real, and J.A. Peláez (2001). Análisis de los acelerogramas registrados en los terremotos de El Salvador de 2001. *Memorias del Segundo Congreso Ibero-americano de Ingeniería Sísmica*, Madrid.

Salazar, W.M., N.N. Ambraseys, and J.J. Bommer (1997). Compilación de un catálogo sísmico para El Salvador y zonas aledeñas. *Proceedings, Seminar on the Evaluation and Mitigation of the Seismic Risk in Central America Area, September 22–27,* pp. 67–76. Universidad Centroamericana José Simeón Cañas, San Salvador, El Salvador.

Salazar, W. (1999). Research Activities on Seismology and Earthquake Engineering in El Salvador, Central America. Research reports on earthquake engineering, Tokyo Inst. Tech., Japan. No. 70, 1–17.

Stoiber, R., and M. Carr (1977). Geological setting of destructive earthquakes in Central America. *Geol. Soc. Am. Bull.,* **88,** 151–156.

White, R.A. (1991). Tectonic implications of upper-crustal seismicity in Central America. In: "Neotectonics of North America" (Slemmons *et al.*, Eds.), pp. 323–338. Geol. Soc. Am. Decade Map, **1**.

79.20 Estonia

The editors received the report, "Seismic Monitoring Program in Estonia," from Tarmo All, the IASPEI National Correspondent. Olga Heinloo and Tarmo All from Geological Survey of Estonia, Kadaka tee 82, EE12618, Tallinn, Estonia, wrote this report. It is archived as a computer-readable file on the attached Handbook CD#2, under the directory of \7920Estonia. We extracted and excerpted the following summary. Biographical sketches of some Estonian scientists may be found in Appendix 3 of this Handbook.

1. Early History

At the end of 1896 G. V. Levitski installed apparatus with two Rebeur-Paschwitz horizontal pendulums into the cellar of the Yuryev University Astronomy Observatory and started to register earthquakes. In November 1897 he installed another Zöllner's horizontal pendulum, made by Repsold. In 1897, 80 earthquakes were registered in the Yuryev Seismological Station (25 powerful and very powerful; 43 days the station did not work). Thus, regular registration of earthquakes was started in Estonia.

A more detailed account (by Olga Heinloo) of G. V. Levitski (1852–1918) and his successor, A. Orlov, and the Yuryev (Tartu) Seismological Station in 1896–1912 is given in Appendix 1 of the full report. A historical overview of instrumental-seismic observations in Estonia was published by Heinloo *et al.* (1996) and is given in Appendix 2 of the full report on the Handbook CD.

2. The Galitzin-Wilip Seismograph

The scientific activity of the historic persons Boris B. Galitzin (1862–1916), Hugo Masing (1873–1939), and Johann Wilip (1870–1942) is analyzed in detail by Olga Heinloo in Appendix 3 of the full report on the Handbook CD.

The year of birth for the Galitzin-Wilip (G-W) seismograph was 1925 at the Hugo Masing Werkstatt für Wissenschaftliche Instrumente in Tartu, Estonia. Twenty-five sets of seismographs (one vertical and two horizontal components), were manufactured during 1926–1939, and sent to twenty-two seismic stations of the world on their orders. The G-W vertical seismograph is based on J. Wilip's doctoral thesis "A Galvanometrically Registering Vertical Seismograph with Temperature Compensation," defended in Tartu in 1930. In this thesis Wilip gave the description, underlying theory, basis of temperature compensation, regulation and establishment of constants of the newest seismographs, already sent to five European and four American seismic stations.

The second period in the life of the Tartu seismic station began on January 1, 1931. The head of this station was the first Estonian professor of physics, Johann Wilip, who installed in the station a full set of G-W seismographs. J. Wilip's seismic station was in operation until the beginning of World War II in 1939. The seismograms of the Wilip's seismic station from the years 1931–1939 are preserved in the archives of Tartu University Library.

3. The Present Seismic Monitoring Program

On October 25, 1987, the Institute of Geochemistry and Geophysics of the Byelorussian Academy of Sciences installed a CM-3 seismograph in the cellar of the former Tartu Observatory. The previous measurements using seismographs carried out in Tartu (1896–1912 and 1931–1939) are discussed more thoroughly in the paper by A. Heinloo *et al.* (1996). At the same time the seismological station in Tallinn started working. The seismographs were installed to observe the seismic activity of Estonian territory. This information is of importance for designing and building nuclear power plants, high-rise buildings, and other large construction projects. At Tartu Seismic Station the

INTERNATIONAL HANDBOOK OF EARTHQUAKE AND ENGINEERING SEISMOLOGY, VOLUME 81B ISBN: 0-12-440658-0

tapes were changed twice a day and were sent to Minsk. The measurements were financed by the Byelorussian Academy of Sciences. Since 1991, the studies are financed from the state budget of the Estonian Republic. Since 1994 both stations are in the possession of the Geological Survey of Estonia (EGK).

The first computer was installed in September 1995, after which the elaboration of the seismogram database and data processing began. The database comprises all seismic events recorded on Estonian territory since 1991. In 1995 the catalogue of distant earthquakes recorded at Tallinn and Tartu Seismic Stations in 1991–1994 was compiled (EGK Depository of Manuscript Reports).

On June 28, 1996, GeoForschungsZentrum Potsdam (GFZ) installed modern seismological equipment in the cellar of Tartu Observatory: a very-broadband digital seismograph. Since then, Tartu Seismic Station has participated in the GEOFON (GeoForschungsNetz) Project led by GFZ. GEOFON comprises 32 seismic stations; the seismograms recorded at these stations can be observed via the Internet at http://www.gfz-potsdam.de/geofon/gfn_liveseis.html. Also in 1996, the Tallinn seismological station was reinstalled to the Suurupi beacon (15 km from town) where there is no high-frequency background conditioned by town noise. Since 1999, the data processing is performed digitally. As a result of this, the accuracy of measurements has improved and the number of annually recorded earthquakes has more than doubled.

Tartu Seismic Station continuously encounters difficulties when recording weak seismic events. The problems are due to considerable microseismic background at the station, caused by traffic and big trees. Transferring the station to a location with microseismic background of lesser intensity is now in preparation.

References

Heinloo, A., O. Heinloo, and H. Sildvee, 1996. Historical overview of instrumental-seismical observations, *Estonia. Bull. Geol. Surv. Estonia,* **6**(1), 34–38.

EGK Depository of Manuscript Reports. The catalogues of distant earthquakes recorded in Estonia in annual reports of instrumental seismological monitoring.

79.21 Ethiopia

The editors received a brief report of Ethiopia from Dr. Laike M. Asfaw, Geophysical Observatory, Addis Ababa University, Addis Ababa, Ethiopia. This report, with a list of references, is archived as a computer-readable file on the attached Handbook CD#2, under the directory \7921Ethiopia. We extracted and excerpted the following summary.

1. Introduction

Seismological observation started in Ethiopia in 1959 at the Geophysical Observatory, on the campus of University College of Addis Ababa (now Addis Ababa University). The first director of the Observatory was a Jesuit, Father Pierre Gouin. The Observatory was initially established for geomagnetic observation in 1957.

The only major earthquake sequence ever to shake Addis Ababa in living memory occurred from June to September 1961 and was recorded by the seismographs at the Observatory. At the time the panic caused by the earthquakes was moderated by the information that was being issued continually by the Observatory in Addis Ababa. As a result of the central role played by the Geophysical Observatory in the earthquake disaster of 1961, seismological observation has been given considerable importance. The extent of damage caused by the earthquake sequence of 1961 was assessed during a field trip to the region and the macroseismic information that accumulated since then formed an important data base on which estimates of regional seismic attenuation relation was based.

Initially, Wood-Anderson and Willmore seismometers were used with photographic recording. In 1963, with the establishment of a WWSSN station, AAE, more systematic study of local earthquakes was made while fulfilling the routine operation of the AAE. Routine interpretation of seismograms and determination of earthquake parameters for local and regional events led to the publication of several research papers and a book, "Earthquake History of Ethiopia and the Horn of Africa," by Gouin (1979). This book contains a substantial number of analyses of seismograms from the AAE and macroseismic information from field assessment of damage. The first seismic zoning map of Ethiopia was published in 1976 in relation to the national effort to draft a code for earthquake-resistant structures.

In 1982 the building of an indigenous research capacity in the field of seismology, with a view to reducing earthquake disaster through infrastructure building and training, was initiated with the support of SAREC (Swedish Agency for Research Cooperation with Developing Countries). Between 1982 and 2000 several technicians and a number of Ph.D.s have been trained under this program.

2. Seismic Instrumentation

Through SAREC support, several analog and digital seismographs have been acquired for mobile and permanent deployments. Seismometers ranging from broadband (Guralp CMG-3T) to short-period (S-13) and including semi-broadband (Lennartz LE-3S/5D) sensors are now operating at the Geophysical Observatory. The recorders range from dial-up to stand-alone, with equipment byTeledyne Geotech (1984), Lennartz (1984), Nanometric (1994), and Reftek (1999). Strong motion accelerographs have also been purchased from Kinemetrics (1996).

In 1994 the Guralp broadband seismometer was used with the IRIS/USGS Quanterra recorder. Broadband recording went on until 1997, when the IRIS-MEDNET equipment was extended to include a very broadband Streckeisen seismometer. The new equipment system is classified as an auxiliary station by the CTBTO monitoring system.

The former AAE is still active with short-period three-component recording. The new seismometer site operated jointly with the IRIS/MEDNET is in a tunnel 20 kilometers southwest of the AAE and is linked to the central station in the Geophysical Observatory via telemetery.

INTERNATIONAL HANDBOOK OF EARTHQUAKE AND ENGINEERING SEISMOLOGY, VOLUME 81B ISBN: 0-12-440658-0

3. Recent Advances

Several research publications have been possible as a result of the seismological support given by SAREC. This made it possible to produce several papers on source mechanism (e.g., Kebede *et al.*, 1989), seismicity (e.g., Kebede, and Kulhanek, 1991) and seismic hazard (Asfaw, 1992; 1993; Kebede and Asfaw, 1996).

Currently, the Observatory in Addis Ababa exchanges data with several African countries and also sends data to the Euro-Mediterranean seismological center and the International Seismological Centre at Newbury, UK.

Apart from the seismological activities at the Geophysical Observatory of Addis Ababa University, there is a government plan to establish a national earthquake infrastructure. The National Meteorological Services Agency of Ethiopia is the body entrusted with the responsibility of establishing and operating seismological networks and issuing earthquake information.

References

Asfaw, L.M. (1992). Seismic risk at a site in the East African rift system. *Tectonophys.*, **209**, 301–309.

Asfaw, L.M. (1993). Seismic hazard in Ethiopia. In: "The Practice of Earthquake Hazard Assessment" (R. Mcguire, Ed.), pp. 102–104. IASPEI/ESC Publication.

Gouin, P. (1979). "The Earthquake History of Ethiopia and the Horn of Africa." International Development Research Centre, Ottawa, Canada.

Kebede, F., W. Kim, and O. Kulhanek (1989). Dynamic source parameters of the March–May, 1969, Serdo Earthquake Sequence in central Afar, Ethiopia, deduced from Teleseismic Body Waves. *J. Geophys. Res.*, **94**, 5603–5614.

Kebede, F., and O. Kulhanek (1991). Recent seismicity of the East African Rift system and its implications, *Phys. Earth Planet. Inter.*, **68**, 259–273.

Kebede, F., and L.M. Asfaw (1996). Seismic hazard assessment for Ethiopia and the neighboring countries, *Sinet (Ethiop. J. Sci.)*, **19**(1), 15–50.

79.22 France

The editors received a collection of nine reports from Dr. Anne Deschamps, the French National Correspondent to IASPEI, Géosciences Azur, CNRS/UNSA, Nice. These reports reviewed the development and research activities of French institutions working in the sciences related to seismology and the physics of the Earth's interior. We extracted and excerpted a summary as shown here. These nine reports are archived as computer-readable files on the attached Handbook CD#2, under the directory of \7922France. An inventory of two French seismic networks is included in Chapter 87. Biographies and biographical sketches of some French scientists and engineers may be found in Chapter 89, and in Appendix 3 of this Handbook, respectively.

1. One Century of Seismology in Strasbourg

This section was submitted by Michel Cara, Ecole de Physique du Globe, Université Louis Pasteur, Strasbourg. The University of Strasbourg was closely associated with the birth of seismology in Europe at the end of the 19th century. The original Strasbourg station was founded by G. Gerland in 1900 as the "central seismographic station of the German Empire." In 1992, this historical station was transformed into a museum, where 17 seismometers made between 1895 and 1960 are displayed.

G. Gerland, Professor of Geography at the University of Strasbourg, played a key role in promoting seismology as a new science. He founded the journal "Beiträge zur Geophysik," which was published from 1887 to 1990. As director of the station from 1900 to 1910, he organized the first and second international meetings in seismology there (April 11–13, 1901; July 24–28, 1903). These meetings were the founding meetings for the International Association of Seismology and the "Central International Bureau of Seismology."

After a few years of interruption during World War I, the activities of the "International Association of Seismology" were restarted at the first meeting of IUGG in 1922. Strasbourg remained the location of the Central Bureau, with E. Rothé as director. He was also the director of the newly formed "Institut de Physique du Globe," and director of the French "Bureau Central Sismologique pour la France et les Colonies," which was founded by an official governmental act in 1921. One of the earliest decisions taken by E. Rothé was to equip the French oversea territories with seismological observatories.

After World War II, Jean Pierre Rothé succeeded his father at the head of both the Institut de Physique du Globe de Strasbourg and the Central International Bureau. Then, after 71 years of existence under German and French leadership, the central bureau was closed in 1975. In 1976, the Euro-Mediterranean Seismological Center (EMSC) was founded under the responsibility of Stephan Mueller of the ETH in Zürich and Elie Petterschmitt in Strasbourg. EMSC continues in some way the activity of the International Central Bureau, but for Europe and the Mediterranean area only. In 1993, EMSC was moved to Bruyères le Chatel, south of Paris, thanks to support from the Laboratoire de Detection et de Géophysique (LDG) of the Commissariat de l'Energie Atomique (CEA).

Scientific activity at the International Bureau has changed with time. Important scientific materials have been collected and published during the past century. During the German period, catalogs of earthquakes (Série B. with macroseismic information), and the collection of seismograms were the most visible part of Bureau activity. After 1924, the scientific publications were divided as "Série A" (scientific papers) and "Série B" (earthquake monographs and earthquake catalogues).

Studying seismicity has been a long tradition at Strasbourg University. The first regional telemetered seismic network in France was, as a result, developed there. It was installed in the late 1960s, within the upper Rhine Graben. The technology used by the Strasbourg group was adapted from that developed at the

INTERNATIONAL HANDBOOK OF EARTHQUAKE AND ENGINEERING SEISMOLOGY, VOLUME 81B ISBN: 0-12-440658-0

French Atomic Commission of Energy, where a seismic network covering all France was created in the early 1960s for detecting nuclear tests.

A robust telemetered array of seismometers was developed soon after by Strasbourg seismologists for the purpose of monitoring the seismicity of a gas field in Lacq, in the southwest of France. After the Friuli, 1976, earthquake, a new portable telemetered network was designed in the Strasbourg Institute. The team in charge of source studies has engaged in field operations to monitor the aftershock activity and to help install permanent telemetered networks in several parts of the world.

Seismic tomography is another field of seismology that has been developing rapidly in Strasbourg during the past 15 years. Using surface-wave tomography, P-wave tomography, or deep crustal tomography based on seismic methods, the Institut de Physique du Globe de Strasbourg has become a place where many efforts are made for imaging the structure of the Earth's interior.

The century-old tradition of seismological observations led the Institut de Physique du Globe de Strasbourg to develop new observational means, such as the "Réseau National de Surveillance Sismique," RéNaSS, to better monitor seismicity in France after 1982. Also, in close cooperation with the Institut de Physique du Globe de Paris, GEOSCOPE, a global broadband seismological network was created after 1983. The Strasbourg group took charge of several GEOSCOPE stations in the southern hemisphere (Antarctica, Indian Ocean, New Caledonia).

The Strasbourg University remains the headquarters of the French Central Bureau of Seismology (BCSF), which was created in 1921. On a regular basis, BCSF publishes the final information on earthquakes occurring in France. BCSF relies in large part on the facilities of the operational center of the "Réseau National de Surveillance Sismique" linked to nearly 80 short-period seismometers in France.

2. The French National Network of Seismic Survey

This section was submitted by Michel Granet, EOST, Institut de Physique du Globe de Strasbourg. Despite the small number and the relatively moderate size of earthquakes, seismic risk must be considered in France and destructive events could happen in the future. This implies a particular effort of our country to carry out seismic monitoring of the territory, to identify active faults with a quantitative estimation of their potential of rupture.

Real-time monitoring of French seismicity is the main task of two complementary networks: the National Network of Seismic Survey (RéNaSS, INSU-CNRS) and the network of the Commissariat de l'Energie Atomique (LDG-CEA, see Section 8), whose initial goal was to detect nuclear tests. This report presents RéNaSS.

Since the mid-1970s, the National Institute of the Science of the Universe (INSU-CNRS) supports a national program to monitor the seismicity of France. The different regional seismic networks were progressively joined to build the federated network called RéNaSS. Now, RéNaSS gathers all the regional seismic networks and isolated seismological stations installed and maintained by the Observatories of the Science of Universe (OSU), the Institute of Physics of the Earth of Paris, and the seismological laboratories of different universities. The objectives of RéNaSS include (1) real-time monitoring of French seismicity, including the distribution of earthquake parameters (latitude, longitude, depth, origin time and, when available, the focal mechanism) as a "routine service" or as a "rapid determination service," and (2) recording, collecting, and archiving the seismic data (seismograms of local and teleseismic events and earthquakes parameters) for scientific purposes.

The main expected outputs of RéNaSS are (1) to inform the Home Office, the Civil Protection Services, and the concerned Prefectures in case of a major and destructive earthquake in France (this activity is labeled as a "public service" mission by the Home Office and imposed a 24-hour duty); (2) to contribute to a better knowledge of seismic hazard; (3) to further enhance the knowledge of seismotectonic movements; (4) to obtain a seismic tomography beneath France; and (5) to contribute to the study of the deep structures of the Earth.

RéNaSS is a federation of regional seismic networks, isolated seismological stations, and broadband observatories whose main components are connected via a central site, located in Strasbourg, near the office of the French Seismological Central Bureau (BCSF). The regional networks are the responsibility of different seismic departments from either the CNRS or the universities. The current organization is as follows: (1) The national headquarter of RéNaSS is in charge of the rapid determination of the earthquake parameters, the collection, the archiving and the distribution of original and processed data (arrival times, source parameters); and (2) The regional headquarters have the technical and scientific responsibility of the regional seismic networks; they are also the main partners of local and regional administrative councils. At present, RéNaSS consists of the following:

A short-period university network consisting of 113 seismological stations gathered in 7 regional networks. Some isolated stations complement the network. These regional dense seismic networks are geographically located in areas where the historical and instrumental seismicity is the most important, thus allowing a very precise location of earthquakes: Upper Rhine Graben, Alps, region of Provence, Nice hinterland area, Pyrenees, French Massif Central, region of Charentes. There are 4 real-time SP networks. The transmission of seismological data is achieved using radio-telemetry from the remote station in the field to a central station equipped with a continuous data recording system. There are 3 regional networks off-line. The data are transmitted to the regional center or to the central site at Strasbourg using either switched phone links or the METEOSAT satellite.

A broadband network equipped with a high-dynamic range acquisition system. The national broadband network is currently composed of 13 stations equipped with Streckeisen STS2 sensors and 2 "observatories" equipped with Streckeisen STS1 sensors.

A central site that is in charge of real-time monitoring of seismicity and routine processing of the short-period and broadband networks. The connection with central or regional headquarters is made via the switched phone lines.

The national headquarters at Strasbourg (the so-called central site) is responsible for the daily utilization of the 69 stations linked to Strasbourg. Approximately 2500 local earthquakes are processed during each year. The transmission of the parametric data is done daily, weekly and monthly, as well as by a classical paper bulletin and as a computer report via Internet. At Strasbourg, raw data (seismograms) and parameter data (phase pickings, earthquake location and magnitude) are immediately accessible on the Web site (http://renass.u-strasbg.fr).

The national headquarter of RéNaSS is also responsible for the fast determination of large events, and has to inform the civil authorities when an earthquake has been felt. These messages are transmitted in French or English, depending on the recipient. Also, in the case of a major teleseismic event, they send refined earthquake parameters (location and magnitude [M_b and/or M_{sz}]) and dispatch all available data.

3. Geodynamics: an Overview of Activity of Some French Laboratories

This section was submitted by Claude Froidevaux, Départment T.A.O., Ecole Normale Supérieure, Paris. It reviews various French contributions to the key geodynamics problems of lithospheric stresses and deformation, the dynamics of the Earth's mantle, and thermal convection in the mantle.

One of the first mechanical problems that arose in the framework of plate tectonics is how thick is a lithospheric plate? There are two distinct answers to this: One refers to the thermal state of the lithosphere, the second one is purely mechanical and derives a thickness from the observation of the elastic response of the lithosphere under a load. In this report the author selects some achievements in the field of mantle dynamics, based on the work of research teams originally located at the University of South-Paris, and now at Grenoble University, and at Ecoles Normales Supérieures in both Lyon and Paris. Some contributions from the teams in Toulouse are also referenced.

The first part of the report deals essentially with the dynamical deformation of the lithosphere. Some fundamental problems for which significant progress has been made are models of lithospheric deformation, lithospheric stretching instabilities for a highly nonlinear rheology, and the causes of the step-like structure of divergent plate boundaries.

The segment on mantle dynamics reviews concepts related to the driving mechanism of plate tectonics, such as the idea of continental drift derived from geological observations. For many years the proposed surface mobility could not be thought of as linked with a plausible physical mechanism. Plate tectonics, by comparison, derived from converging geophysical evidences. The driving mechanism of plate tectonics and the corresponding deep mantle circulation became clearer during the 1980s on the basis of the first global 3-D mantle structures derived from seismic tomography. Basic themes discussed are mantle circulation induced by internal loads, predicted plate velocities, geoid and radial viscosity profiles, and the question of one- or two-layered mantle convection.

The third part, on thermal convection in the mantle, treats the thermal regime and subsidence of oceanic plates, the CHABLIS model for lithospheric cooling, factors contributing to continental breakup, and the long-term stability of the continental lithosphere.

4. Seismology and Seismotectonics at the Observatoire Midi-Pyrénées

This section was submitted by Annie Souriau, Observatoire Midi-Pyrénées, Toulouse. The Observatoire Midi-Pyrénées is located in Toulouse, in southwest France, not far from the Pyrenees mountain range. It is part of the University Paul-Sabatier, and it also has close relationships with the Centre National d'Etudes Spatiales (CNES), the most important national space center in Europe. These two affiliations have motivated the development, in 1980, of a small group in seismology, whose activities proceed in close relationship with the other scientific fields, in particular geodesy, geodynamics, geology, and tectonics. Research activities focus mostly on three topics: Global Earth and planetary structure and dynamics, satellite techniques applied to seismology and seismotectonics, and, because of the geographic location of Toulouse, seismotectonics of the Pyrenees range. The group has also an educational activity, teaching at various levels at the University and a dozen other schools and supervising student research projects from undergraduate to postdoctoral levels. In addition, it has an observatory activity for the seismic monitoring of the French Pyrenees. Details of the achievements in each of the three main topics, with references, are presented in the full report on the Handbook CD.

Global Earth studies have first been motivated by a desire for better understanding of the relationships between Earth structure and data provided by satellite geodesy, in particular geoid and Earth rotation parameters. Several studies have concerned the properties of Earth models derived from seismological data and their relationships with the geoid. Some studies have also focused on the excitation of the Chandler wobble.

Satellite geodesy, satellite imaging, and radar interferometry have appeared as very powerful tools for investigating continental deformations associated with seismic cycles or with aseismic creep. Space techniques, seismic measurements, and field observations are combined to come to a better understanding of tectonics in seismically active regions.

Studies of the seismotectonics of the Pyrenees range have been motivated by historical seismicity reports of nine earthquakes of intensity VIII since the end of the Middle Ages. The analysis of propagation times of both regional events and teleseisms has made it possible to obtain a 3D tomographic image of the Pyrenees down to 200 km. This image, which is the first one obtained for a whole mountain range, reveals interesting features related to the buildup of the range. Specific studies have concerned the distribution of seismicity in space and time, in particular in the Arette region where long series are available, and the relationships between structure and seismicity.

Members of the observatory, with their specialties, are given in the full report on the Handbook CD.

5. The Seismological Department of IPG of Paris

This section was submitted by Jean-Paul Montagner, Department of Seismology, IPG Paris. The Department of Seismology (hereafter referred to as the SeismoLab), is one of the scientific departments of IPG (Institute of Physics of the Globe of Paris), associated with CNRS (National Scientific Research Center) and the University of Paris VII—Denis Diderot. Its primary activity is related to the investigation of natural systems of the Earth, by using seismic phenomena. SeismoLab is sweeping the whole spectrum of seismological studies, from observational seismology to seismology applied to oil exploration, as well as theoretical geophysics. The two fields of excellence of the SeismoLab regard the investigation of the structures of the Earth at all scales and all depths (tomographic studies from the center of the Earth up to the crust) and the understanding of natural processes such as earthquakes and volcanic eruptions. Due to the complexity of the natural phenomena under investigation, our department is closely interacting with other laboratories of IPG, particularly in tectonics, geodesy, and geodynamics as well as in geochemistry, geomagnetism, fluid mechanics, acoustics, and mineral physics.

IPG in Paris has a long-standing tradition in instrumentation, field experiments, observatory maintenance, and Earth processes theoretical and numerical modeling. Following the example of the astronomical observatories, the Institutes of Physics of the Earth were created in 1921. At that time, the IPGs were in charge of atmospheric, geomagnetic, seismic, and volcanic monitoring of the French territories. IPG-Paris gradually abandoned its activity in external geophysics and most of its scientific activity is now devoted to the inner Earth. New laboratories (geochemistry, marine geophysics, then tectonics, geomaterials) joined IPG in order to reinforce its scientific activity and to follow the profound evolution of Earth Sciences, where the diversity and complexity of natural phenomena were recognized.

In 1991, the IPG of Paris became an independent Institute with three fundamental duties: fundamental research in Earth Sciences, education and diffusion of knowledge, and monitoring of seismic, magnetic, and volcanic activity through global or regional observatories. It is presently divided into five scientific departments (Seismology, Geochemistry, Geomagnetism, Mineral Physics, and Tectonics) and three cross-cutting departments, responsible for Research and Monitoring facilities, in order to make the interactions between research laboratories easier: (1) The SeismoLab is actively involved in the cross-cutting departments of IPG, (2) the department of observatories (Global Seismological Network GEOSCOPE; Seismic and deformation monitoring observatories of Antilles, La Réunion Island and Djibouti), and (3) the Department of Spatial Studies (Martian seismic exploration, VBB seismometer, Netlander), the Department of numerical modeling and simulation of natural phenomena, and the seismotectonic group (GPS measurement network, SAR interferograms).

The SeismoLab is involved as well in more specific scientific programs, related to earthquake and volcanic processes, the understanding and modeling of the propagation of seismic waves in complex media and their application to tomographic imaging and structural studies. Among the different fields of geophysics, seismology is probably the most efficient field for providing quantitative information on the inner structure of the Earth and on the most spectacular manifestations of its dynamics: earthquakes and volcanic eruptions.

The primary activity of the SeismoLab is related to the investigation of natural systems of the Earth by using seismic phenomena. The Seismological Laboratory has a matrix structure in which the scientific activity is organized around five research laboratories where researchers are gathered according to their thematic orientation: (1) Experimental Seismology (Head: Alfred Hirn), (2) Global Seismology and Planetology (Head: Jean-Paul Montagner), (3) Seismogenesis (Head: Pascal Bernard), (4) Rock Mechanics (Head: François Cornet), and (5) Tomography and Geophysical Modeling (Head: Jean-Pierre Vilotte).

The SeismoLab is as well in charge of the development and maintenance of the global seismic observatory GEOSCOPE, and the geophysical monitoring of several European volcanoes (seismicity, ground deformation measurements by InSAR and GPS). The SeismoLab is in charge of several French, European, and international scientific programs and is collaborating with many research and academic institutions around the world.

Only the main recent and ongoing projects in each of these research laboratories are listed in the full report on the Handbook CD and the reader is referred to the list of publications. The scientific activity of the SeismoLab is presently focused on instrumental developments (broadband seismometers, stress sensors, multiparameter ocean bottom stations and observatories, satellite monitoring and spatial missions), the understanding of Earth processes, by carrying out field experiments, theoretical and numerical simulation and modeling, in a multiscale and multidisciplinary approach.

Current information about the Institute of Physics of the Globe of Paris may be found at its Web site at: http://www.ipgp.jussieu.fr/.

6. The Laboratoire de Geophysique Interne et Tectonophysique

This section was submitted by Denis Hatzfeld, Observatoire des Sciences de l'Univers de Grenoble, Université Joseph Fourier, Grenoble. The full report on the Handbook CD offers more details of reseach projects developed in the laboratory.

The Laboratoire de Géophysique Interne et Tectonophysique (LGIT) is one of the departments of the Observatoire des Sciences de l'Univers de Grenoble, which also includes departments in astrophysics, planetology, glaciology, hydrology, geology, oceanography, and signal processing. The LGIT is a joint laboratory between the University of Grenoble and the National Scientific Research Center (CNRS). It was founded in 1975 by four seismologists but very quickly increased in number and broadened its scope in geophysics. In 2003, LGIT is a laboratory of about 50 scientists, 20 engineers, and 30 Ph.D. students, divided into 5 different teams: Seismology and Earth Structure, Seismic Risk, Fluids and Dynamics, Environmental Geochemistry, and Geodynamo.

The main research themes in seismology at LGIT are as follows:

6.1 Rupture Mechanism During Major Earthquakes

We have been studying the source of large earthquakes by inversion of geodetic and seismological data. We also developed theoretical models of the rupture process. The observations seem to indicate a scaling law of frictional parameters controlling the initiation. We are currently working with mathematicians to develop a rigorous theoretical renormalization framework of the earthquake initiation.

6.2 Imaging of the Lithosphere and the Deep Earth

Investigating the structure of cratonic lithosphere or mountain belts is one of the keys to a better understanding of the origin and the dynamics of continents. We have focused our effort on these two specific classes of problems, and our team organized or participated in a number of temporary seismological experiments in Tibet (*e.g.*, Tien Shan, Qinghai, Western Kunlun), in the Central Andes, in Iran (Zagros), as well as in Northern Europe on the Tornquist Zone and the cratonic lithosphere of Finland.

6.3 Quantification of Small-scale Heterogeneity at Depth

Techniques enable us to map and quantify the different scale lengths of heterogeneities inside the Earth. As such, mesoscopic physics offers an attractive alternative characterization of heterogeneous structures. Our present interest is in the development of methods based on interference effects in random media, such as coherent backscattering or persistent correlations in diffuse fields. These investigations may be applied to the monitoring of the dynamics of active earth systems such as volcanoes.

6.4 Shallow Seismic Exploration for Natural Hazard and Environmental Applications

We closely associate theoretical developments of electro-kinetic coupling, numerical simulations, and field experiments. Through this approach it is possible to characterize fluid contents of the soil, which is an important issue in numerous environmental problems.

6.5 Numerical Simulation of Wave Propagation in Complex Media and Rupture Process

Our group has put a constant effort into developing new techniques of modeling. It now has a strong reputation for its numerical developments, in particular for semi-analytic methods of wave-field computation such as discrete wavenumber and boundary integral equations methods.

6.6 Instabilities: Faulting, Volcanoes, and Landslides

The implications of the different friction laws on rupture dynamics and the use of global analyses allow a quantification of the reaction of the upper crust to perturbations, specifically in terms of seismic activity.

6.7 Continental Deformation and Seismotectonics

We have been studying crustal deformation and the related mantle structure mainly in two different regions: Greece and Iran. One approach to complement GPS surveys has been to install dense networks of portable stations for a period of several weeks. We contributed in such work to precise mapping of the active faulting and the strain pattern. This was combined with studies of crustal properties and mantle anisotropy to look for a possible relation between crustal and mantle deformation. One of the obvious outputs of these studies is to better evaluate seismic hazard.

6.8 Engineering seismology and seismic hazard

We contribute to development of the future generation of seismic hazard and risk assessment methods, combining crustal strain-rate mapping from seismic and geodetic data (GPS), the knowledge of local site effects, and the modeling of spectral ground shaking. We contribute to the development of new ground motion models (empirical strong motion attenuation relations and finite faults numerical simulation). Specific research work is also performed to analyze, with civil engineers, building vulnerability.

LGIT is also strongly involved in observations and instrumentation. Since its start, the seismological groups of LGIT have been very much involved in instrumentation development, experimental studies, and national networks. It created the regional seismological network (SISMALP, 44 telemetred short-period stations) that surveys and locates the seismicity of the Alps. SISMALP is complemented by 4 broadband stations (ROSALP). LGIT was a leader for the constitution of the Lithoscope mobile seismological instrument pool, of about 70 portable stations, which is the national instrument for crustal and lithospheric studies. It was also in charge of the constitution of the national Strong Motion Mobile Network (25 mobile stations). LGIT is in charge of developing the new instrument for High Resolution Imagery (IHR) that is going to be deployed on volcanoes and active faults. Finally, it was charged by the Ministry of Environment to create the National Strong Motion Network (RAP) and constitute a working group on Seismic Hazard with Civil Engineers, the Atomic Commission, and the Geological Survey. The RAP is a modern, telemetred network of about 100 stations equipped with sensitive instruments. Data are archived and distributed free in LGIT.

7. IRD's Fifty Years of Research in Geosciences: A Review

This section was submitted by Bernard Pontoise, Institut de Recherche pour le Développement (IRD), Paris. Apart from the initiatives of isolated individuals in the 19th century, French research in tropical environments developed only gradually until just after the end of World War I, when the government decided that one way to revive the French economy was to fully exploit the colonies in Africa, Madagascar, the Far East, and the Pacific. Scientific research was to play a leading role in this. As the minister for the colonies declared in Parliament in 1919, "It will be by relying on science and creating experimental fields and laboratories that the colonies will achieve a better understanding of their resources and be able to exploit them rationally."

At scientific conferences held in 1931 and 1937, the need to structure this budding colonial research emerged very clearly. That structure was set by act of parliament on October 11, 1943. It was named the Colonial Scientific Research Bureau—Orsc. The ministry for the navy and colonies was made responsible for Orsc, and the Director of the CNRS was appointed president of the organization. Its mission was to pilot, co-ordinate, and monitor research in the territories for which the navy and colonies ministry was responsible. After the liberation of France and the re-establishment of Republican law, the creation of Orsc was first invalidated along with all the legislation passed by the Vichy government, then reconfirmed in a statute signed by General Charles de Gaulle on November 22, 1944.

During the first years of its existence, Orsc set out to form a body of researchers prepared to work overseas; to develop top-level scientific training specializing in tropical issues; and to set up a network of multi-purpose research centers capable of having an impact in the colonies of French West and Equatorial Africa, Madagascar, and the Pacific.

By 1946, the first scientists recruited by Orsc, ranging in discipline from entomology, botany, and phytopathology to soil science, hydrology, sociology, and geography, went out to fulfil their assignments in the field—to explore and to promote economic use of these tropical lands, in many parts of which no scientific investigation had ever been carried out.

Between 1949 and 1953, the name of the Office changed twice. Orsc became Orsom (*Office de la recherche scientifique outre-mer*) and then, under a decree issued in November 1953, Orstom (*Office de la recherche scientifique et technique outre-mer*).

In 1998, Orstom was renamed IRD, with a new organization strengthening the different research units. Through research and study of the warmest, and often poorest, areas of the earth, IRD has been contributing for fifty years to better understanding of our planet and to making it habitable for the greatest number. This has been and still is IRD's vocation.

The IRD Geoscience approach is presented in the full report on the Handbook CD according to thematic topics: continental lithosphere in Africa and Latin America; the oceanic lithosphere; subduction zones, including seismic and volcanic hazards; geophysical observatories, and applied geophysics.

8. Activities of CEA/DASE in the Domain of Seismology

This section was submitted by Bernard Massinon, Département Analyse, Surveillance, Environnement, Bruyères le Châtel. DASE has been working for a long time in the domain of geophysics and especially in seismology. One of its divisions, LDG, is in charge of the geophysical studies and activities associated with monitoring and environment. This division was created and managed by Professor Yves Rocard, the physicist who started to develop detection seismology in France in the 1960s.

More recently, DASE played an important role in the Group of Scientific Experts of the Conference of Disarmament, who designed the International Monitoring System of the CTBTO and defined its specifications. DASE is presently contributing to the building of this International Monitoring System of the Provisory Technical Secretariat by installing stations on the French territories and also on some foreign countries within the frame of collaborations.

DASE is running a large-aperture digital seismic network in France of more than 40 stations, with data sent in real time by VSAT facilities to Bruyeres le Chatel, where its data processing center is located. This information is provided to the Rénass for French seismicity monitoring and to the EMSC for European seismicity monitoring.

Some other studies have been developed, for example, in the field of seismic hazard and earthquake precursors, and in seismic data processing. These various activities, of humanitarian and economical interest, are developed by several research teams of DASE owning adapted instrumental sites, a rock mechanic laboratory, modeling tools, teledetection and interferograms processing facilities working together in collaboration with the Institut de Physique du Globe, CNRS, and University.

Details of specific DASE projects are given in the full report on the Handbook CD.

9. Seismic Prospecting: French Industrial Research and Activities

This section was submitted by G. Grau, CGG, Paris and D. Michon, IFP, Paris. The French groups most active in research and development of the seismic methods of prospecting are Compagnie Générale de Géophysique (CGG), a service company, and Institut Français du Pétrole (IFP), a public research institution. They have been working in a concerted effort with the research departments of the French oil companies, Elf and Total, or their mother companies. Concerted and carefully planned research and development efforts have helped put CGG among the top service companies and the oil company TotalFina-Elf at the fourth rank for reserves and production.

The most famous achievement of the French school, as far as wave theory is concerned, is the publication by L. Cagniard, then director of the mother company of CGG, of his *Réflexion et réfraction des ondes séismiques progressives* (1939). Cagniard's method was the source of numerous further studies in wave propagation. A thorough review of French research and achievements in the theory underlying seismic prospecting, data acquisition in marine and land operations, and data processing and interpretation is given in the full report on the Handbook CD.

79.23 Georgia

The editors received a national centennial report of Georgia from Dr. Tamaz L. Chelidze, Director of the Institute of Geophysics, Georgian Academy of Sciences, 1 Alexidze str, 380093 Tbilisi, Georgia. This report is archived as a computer-readable file on the attached Handbook CD#2, under the directory of \7923Georgia. Earthquake engineering in Georgia is reviewed in Chapter 80.8. Biographies and biographical sketches of some Georgian scientists and engineers may be found in Chapter 89, and in Appendix 3 of this Handbook, respectively.

1. Introduction

Macroseismic data on the strong Caucasian earthquakes beginning from the 11th century are recorded in old Georgian manuscripts. The instrumental period began January 6, 1899, when the first seismograms were recorded by the Reuber-Ehlert seismograph with horizontal pendulum. In the period from 1902 to 1909 in the Caucasus, besides Tiflis, nine seismic stations of II class were established. The data obtained in this period were used for seismological research. For example, Wiechert and Zoeppritz extensively used the Caucasian data for plotting master curves of seismic wave arrivals.

2. The Tiflis Seismic Station

By 1903 the Tiflis seismic station was equipped with most seismographs existing at that time: Reuber-Ehlerts system, consisting of three horizontal pendulums with photographic registration, Cancani seismograph with vertical pendulum, Mein's horizontal pendulum, and Bosch's heavy horizontal pendulum. Later on Zöllner's and Galitzin's seismographs were deployed in Tbilisi. All seismographs in Tbilisi worked without interruption until 1916, and the station was functioning again after the civil war in Russia. Beginning in 1900, materials from Tiflis seismic station were published as monthly bulletins.

3. Early Accomplishments

E. Bjuss compiled the first macroseismic maps of the Southern Caucasus (1931) and the first detailed description of seismicity of the region in three volumes (1948, 1952, 1955). The first theoretical and experimental hodographs (travel-time tables) for the Caucasian region were suggested by G. Tvaltvadze, A. Tskhakaja, T. Lebedava, *et al.* Beginning from 1933 the seismological research was carried out by the newly organized Institute of Geophysics.

In 1930–1945 Niko Muskhelishvili developed a new approach to the solution of problems of elasticity, based on the theory of functions of complex variable. The outstanding Georgian mathematician, V. Kupradze, in 1953 gave the integral formulation of the Huygen's principle for steady-state vector elastic waves. He and, independently, P.M. Morse and H. Feshbach extracted from the Stokes-Love solution the fundamental singular solution of elastodynamics, otherwise known as the elastic Green's tensor G_{ij}.

4. Recent Advances

For a long time Tbilisi has been the regional center for seismological research in the Caucasus. Outstanding contributions to the progress of seismology in Georgia have been made in the past four decades. A computer database of Caucasian Earthquakes from ancient times to the present was created in the 1970s (O. Gotsadze). Extensive studies of conditions of surface wave generation and propagation led to discovery of new fundamental phenomena. In particular, the existence of surface waves reflected and refracted from the lateral heterogeneities (faults) was predicted theoretically and discovered experimentally (D. Sikharulidze). Some fundamental forward and inverse seismological problems, such as true motion reconstruction and delineation of seismogenetic lineaments, were analyzed by M. Alexidze. Seismicity, seismic regime, and seismotectonic

INTERNATIONAL HANDBOOK OF EARTHQUAKE AND ENGINEERING SEISMOLOGY, VOLUME 81B ISBN: 0-12-440658-0

deformation of the Caucasus were thoroughly investigated by E. Jibladze. Percolation (fractal) model of fracture was first suggested by Georgian scientists in 1979 (T. Chelidze). The percolation model explains many experimental seismological and geophysical data, related to pre-, co- and post-seismic processes and the earthquake prognosis problem. New probabilistic GIS-based maps of seismic hazards in Georgia have been developed in recent years on the basis of a modern refined scheme of active faults and a new seismic catalog of the Caucasian region compiled by internationally accepted methods (Z. Javakhishvili).

5. Research Cooperation

Georgian seismology traditionally has been connected with Russian seismologists: P. Nikiforov, G. Gamburtsev, E. Savarensky, Yu. Riznichenko, M. Sadovsky, V. Keilis-Borok, A. Nikolaev, G. Sobolev, N. Shebalin, A. Vvedenskaja, and many others. There has been active collaboration between Russia and Georgia, and many Georgian seismologists have advance training in Russia. There are also very close ties with Azerbaijani and Armenian colleagues.

6. Biographies and Biographical Sketches

The full report on the Handbook CD includes biographical sketches of Tamaz Lucka Chelidze, Eleonora Alexander Jibladze, David Ilia Sikharulidze, Zurab Javakhishvili, and Otar Varazanashvili, and memorials of outstanding seismologists of Georgia: Merab Alexidze (1930–1993), Otar Gotsadze (1929–1993), Peter Mandjgaladze (1944–1993), Alexander Tskhakaia (1902–1970), Evgeni Bjuss (1885–1965), Guri Tvaltvadze (1907–1970), and Georgi Murusidze (1915–1988).

79.24 Germany

The editors received the "German National Report" from Dr. Rainer Kind, IASPEI National Correspondent, GeoForschungsZentrum, Potsdam, Germany. This report is archived as computer-readable files on the attached Handbook CD#2, under the directory of \7924Germany. It consists of four parts: (1) "The Early German Contributions to Modern Seismology" by J. Schweitzer (NORSAR, Norway); (2) Six short reviews of contributions by German scientists since about 1960 by R. Kind (GeoForschungsZentrum, Potsdam), P. G. Malischewsky (Friedrich-Schiller Universität, Jena), J. Mechie (GeoForschungsZentrum, Potsdam), and G. Müller (Johann Wolfgang Goethe-Universität Frankfurt); (3) Biographies and biographical notes by J. Schweitzer and associates; and (4) "Prof. Dr. Beno Gutenberg—The Bibliography" compiled by J. Schweitzer. We extracted and excerpted the following summary.

An inventory of seismic networks in Germany is included in Chapter 87. Biographies and biographical sketches of some German scientists and engineers may be found in Chapter 89, and in Appendix 3 of this Handbook, respectively.

1. The Early German Contributions to Modern Seismology (J. Schweitzer)

Scientists working in Germany played central roles in the development of modern geophysics. As seismology and the study of the physics of the Earth's interior emerged as quantitative disciplines during the late 19th and early 20th centuries, essential contributions were made to both the observational and theoretical aspects at a number of research centers within Germany. This section reviews the development of these centers and the research results from the 1880s to the 1930s. The history of the earliest period of German seismology is given in Davison, *The Founders of Seismology* (Cambridge University Press, 1927) and Tams, Materialien zur Geschicte der deutschen Erdbebenforschung bis zur Wende des 19. zum 20. Jahrhundert (*Neues Jahrbuch für Geologie und Paläontologie*, 1952).

1.1 Early Pioneers

The earliest years are marked by the work of some of the best-known names in seismology, related geophysics, and mathematics applied to geophysical problems: Rebeur-Paschwitz, Gerland, Sieberg, A. Schmidt, Wiechert, Geiger, Gutenberg, Herglotz, Zoeppritz, Angenheister, Mintrop, and Radon. Biographical information about these scientists and details of their contributions to building the foundations of modern seismology, as well as to IASPEI itself, are given in the full report on the Handbook CD. Other researchers whose names are well known to seismologists are included, as well as a comprehensive bibliography. Only a few milestones on this long and productive road will be summarized here.

The first key event was the recognition by Ernst von Rebeur-Paschwitz of the recorded signal from a teleseismic event; waves from a Japanese earthquake were seen by his horizontal pendulums in Germany. He quickly realized the significance of this observation for investigation of the interior of the Earth and proposed the creation of a global network of seismographic stations for this purpose.

Emil Wiechert, who had transferred his interests from conventional physics to the constitution of the Earth, was a major figure in the development of German and world seismology. Of Wiechert's many contributions, two that stand out in this context are (1) the creation of the world's first Institute of Geophysics in Göttingen, and (2) the development and deployment of his inverted pendulum seismograph, especially the 80 kg version, that made implementation of Rebeur-Paschwitz's vision of a global network feasible. The Göttingen institute became a nucleus for the development of modern seismology. The long

INTERNATIONAL HANDBOOK OF EARTHQUAKE AND ENGINEERING SEISMOLOGY, VOLUME 81B ISBN: 0-12-440658-0

series of papers, *Über Erdbebenwellen*, by Wiechert and his colleagues, and the many other publications by this group (listed in the bibliography on the Handbook CD), formed the basis for much of seismological theory and data interpretation.

1.2 Gerland at Strasbourg

After the early death of Rebeur-Paschwitz in 1895, Gerland promoted the idea of an international bureau to collect seismic observations worldwide, and the idea of an international society as a discussion forum and to coordinate all seismological research. Gerland was also interested in establishing Strasbourg as the center for seismology in Germany, and the "Rebeur-Ehlert pendulum" for installation worldwide. These seismographs were built by the factory "J. und A. Bosch" in Strasbourg (later in Hechingen, near Tübingen). Later, this factory also became known as the designer and manufacturer of the "Bosch-Omori seismographs," and of the "Mainka pendulum."

During the last years of the 19th century, numerous seismic stations were installed worldwide, and in 1899, Gerland succeeded in setting up the first central seismological institute for Germany in Strasbourg. He managed to get the institute jointly financed by the Reichsland Elsass-Lothringen and the German Empire and became the first director of this institute, officially called "Kaiserliche Hauptstation für Erdbebenforschung" (Imperial Central Station for Earthquake Research). The institute became responsible for the organization of the planned homogenization and extension of the German seismological network. In addition, it became the central address in Germany for collecting all earthquake observations and for recording the seismicity of the world with all important available seismic instruments.

During the 7th International Geographic Conference in Berlin 1899, Gerland proposed the formation of an international seismological society with the following objectives: The society should (1) as far as possible promote a systematic macroseismic investigation of all countries, especially of those without seismic stations and therefore with relatively unknown seismicity, (2) as far as possible organize a homogeneous microseismic observation system, and (3) concentrate its publications, which should be published as independent supplements to *Beiträge zur Geophysik*. The 7th International Geographic Conference supported this proposal and an "International Permanent Commission for Earthquake Research" was spontaneously founded in Berlin with 54 members.

1.3 The First International Conference on Seismology, Strasbourg 1901

With support by the German foreign office and the organization of the International Geographic Congresses, Gerland issued invitations to the First International Conference on Seismology in Strasbourg (April 11–13, 1901), with the objective of a vote for establishing an international seismological society. Official delegates came from Belgium, Germany, Japan, Italy, Austria, Hungary, Russia, and Switzerland, in addition to numerous individuals, mostly members of the International Permanent Commission formed earlier in Berlin. Instead of a society of individuals, as planned by Gerland, the conference proposed an International Association of States with a Permanent Commission, a General Assembly, and a Central Bureau; a structure similar to the International Geodetic Association.

However, it became clear that establishing such an international association would take some time. The non-German delegates asked the conference for a vote to install a preliminary Central Bureau at Gerland's "Hauptstation für Erdbebenforschung" in Strasbourg, which was accepted. To support the work of the Central Bureau, all participants obliged themselves to send their earthquake observations to Strasbourg. Emil Rudolph, a fellow of Gerland's institute, was nominated as secretary of the conference, and he received the duty of publishing the proceedings of the conference. In 1902, the proceedings were published as *I. Ergänzungsband der Beiträge zur Geophysik* in French and German.

1.4 The Second International Conference on Seismology, Strasbourg 1903

The Second International Seismological Conference also met in Strasbourg (July 24–28, 1903) and the proceedings were again published in German and French. This time, 19 states were officially represented with delegates in Strasbourg: Argentina, Austria, Belgium, Bulgaria, Chile, Congo, Germany, Great Britain, Hungary, Italy, Japan, Mexico, Netherlands, Portugal, Romania, Russia, Spain, Sweden, Switzerland, and the United States of America. During the conference, the statute of an International Seismological Association of States with its structure was defined: The association should have a Permanent Commission, a General Assembly, which should meet every four years, and a Central Bureau. The statute became part of the draft for an international convention to be signed by the states. This convention was intended to take effect on April 1, 1904, and to be valid for the next 12 years (i.e., to March 31, 1916). It was planned to locate the main office of the Central Bureau in Strasbourg with Gerland as its director. The main duty of the Central Bureau was to be collecting and editing of all microseismic and macroseismic observations.

1.5 The International Seismological Association

The International Seismological Association (ISA) was founded by the signing of the convention by April 1, 1904, by the following 18 states: Belgium, Bulgaria, Chile, Congo, Germany, Greece, Hungary, Italy, Japan, Mexico, Netherlands, Norway, Portugal, Romania, Russia, Spain, Switzerland, and the United States of America. However, Argentina, Austria, Denmark, France, Great Britain, Serbia, and Sweden supported the Association in principle, but could not agree with four articles of the convention. These articles were changed in their favor at the Third International Conference on Seismology in Berlin on August 15, 1905. At the same conference, the first Permanent Commission was formed and L. Palazzo from Rome, Italy, was elected as its first president (see Chapter 4 by Adams).

The name of the Association is either the "International Seismological Association" as translated from "Internationalen Seismologischen Assoziation," or the "International Association of Seismology" as translated from "Association Internationale de Sismologie," because the official languages of the Association were originally German and French.

During the following years, many different meetings of the Permanent Commission took place, and the first General Assembly of the ISA was organized in The Hague, Netherlands (September 21–25, 1907). During this assembly, the Permanent Commission also had a meeting and could add as new members the delegates of the states that had signed the convention after 1904: Canada, France, Great Britain, and Serbia. The ISA then had 22 national members. The Central Bureau in Strasbourg published the results of its bulletin work and single investigations about specific scientific questions, as requested by the Permanent Commission. For instance, the Central Bureau organized and evaluated a concourse on seismographs for the meeting in The Hague. For such work, international guests came to Strasbourg and supported the permanent employees of the Central Bureau and the "Hauptstation für Erdbebenforschung." If one reads the yearly reports about the work in Strasbourg as published in *Beiträge zur Geophysik*, one gets the impression that the Central Bureau and the "Hauptstation für Erdbebenforschung" were a common institution using the same resources, but officially, they were separate organizations.

Gerland was the Director of the Central Bureau until his retirement as Professor of the University in 1910. Then Oskar Hecker became Director of the "Hauptstation für Erdbebenforschung," and in parallel, Director of the Central Bureau of the ISA. The most important duty of the Central Bureau in Strasbourg was to publish compilations of seismic observations: Emil Rudolph was able to compile the first bulletin with associated observations from several stations for the years 1895–1897. The next worldwide catalog was compiled for the year 1903. Then it was Elmar Rosenthal, who published the first international bulletin in the name of the International Seismological Association for the year 1904. During the following years, Siegmund Szirtes published phase readings from internationally distributed seismic stations in catalogs and bulletins for the years 1905–1908. Macroseismic observations were collected and published in bulletin form by Adolf Christensen, Robert Lais, Emilio Oddone, Emil Rudolph, Erwin Scheu, August Sieberg, and Georg Ziemendorff (see also Chapter 88 by Schweitzer and Lee). The production of these bulletins stopped due to World War I and the dissolution of the International Seismological Association.

In 1913, Argentina became the 23rd member of IAS. The yearly reports of Hecker, published in *Gerlands Beiträge zur Geophysik*, indicate a normal, routine development of the ISA, its Central Bureau, and a fruitful international cooperation in seismology. Nevertheless, World War I changed everything.

As far as it is visible in the yearly reports, the Central Bureau tried to keep business as usual during the war. After World War I, Strasbourg again became part of France and the responsibility for the Central Bureau of the ISA became part of the postwar problems between France and Germany. The Central Bureau and the "Hauptstation für Erdbebenforschung," most of the instruments, the scientific material and libraries of both institutes, and their employees became inaccessible for German seismology. So, German seismology after World War I was without a central institute until 1923, when Hecker became the director of the "Reichsanstalt für Erdbebenforschung" in Jena.

After World War I, it became clear that international scientific cooperation had to be reorganized. The International Research Council founded the International Union for Geodesy and Geophysics (IUGG) in 1919, and seismology was a section of this Union. In April 1922, the Third General Assembly and the 5th Conference of the Permanent Commission of the ISA had a meeting in Brussels and the International Seismological Association dissolved itself on April 24, 1922. One month later, at the first General Assembly of the IUGG in Rome, the new Seismology Section of the IUGG was established. This time, German seismologists could not participate or become members of this Section.

1.6 German Seismology after World War I

Without the possibility of participating in international seismological conferences, the German seismologists needed a forum for discussion. Emil Wiechert invited German seismologists to a special meeting in September 1922 at the yearly conference of the "Deutsche Naturforscher und Ärzte" (German natural scientists and physicians) in Leipzig, and Hecker proposed at this meeting the foundation of the "Deutsche Seismologische Gesellschaft." The 24 seismologists present spontaneously followed the idea and founded this society, which became the nucleus of the "Deutsche Geophysikalische Gesellschaft" founded two years later at the second meeting of the "Deutsche Seismologische Gesellschaft" in Innsbruck, Austria. At the end of the 1920s, the common scientific interests between individual seismologists became more important than "old political" problems, and after solving the financial problem of paying the yearly fees to the IUGG, Germany became an official member of the IUGG in 1937.

1.7 Beno Gutenberg

One of Wiechert's students was Beno Gutenberg. From his early days in Göttingen, through his distinguished career in Frankfurt and at the California Institute of Technology, Gutenberg stands as one of the foremost leaders in geophysics and especially in seismology. He had remarkable ability to extract correct interpretations from the mostly imperfect data available to him in the early days (such as his value for the radius of the Earth's core). His work through the decades touched on every aspect of seismology and the physics of the Earth's interior. The Fourth section of the German National Report on the Handbook CD contains a comprehensive listing of his publications and is in itself a summary of progress in many aspects of solid-earth geophysics in the first half of the 20th century.

1.8 Biographies

The full report on the CD includes biographies and biographical notes of the following notable scientists (with connections to Germany), who have contributed significantly to seismology and/or physics of the Earth's interior: (1) Angenheister, Gustav Heinrich (1878–1945); (2) Benndorf, Hans (1870–1953);) (3) Bock, Günter (1944–2002); (4) Borne, Georg von dem (1867–1918); (5) Ehlert, Reinhold (1871–1899); (6) Gauss, Carl Friedrich (1777–1855); (7) Geiger, Ludwig Carl (1882–1966); (8) Gerland, Georg Cornelius Karl (1833–1919); (9) Gutenberg, Beno (1889–1960); (10) Hecker, Oskar August Ernst (1864–1938); (11) Herglotz, Gustav (1881–1953); (12) Mainka, Karl (1873–1943); (13) Mintrop, Ludger (1880–1956); (14) Müller, Gerhard (1940–2002); (15) Rebeur-Paschwitz, Ernst von (1861–1895); (16) Rosenthal, Elmar (1873–1919); (17) Rudolph, Emil (1854–1915); (18) Sieberg, August (1875–1945); (19) Tams, Ernst (1882–1963); (20) Wegener, Alfred Lothar (1880–1930); (21) Wiechert, Emil (1861–1928); and (22) Zoeppritz, Karl (1881–1908).

2. Recent German Contributions

This section includes six topical reviews of contributions by German scientists since about 1960.

2.1 Modern Broadband Seismology (R. Kind)

Germany has a long tradition in broadband seismology. After early work in both East and West Germany in the 1970s, the Gräfenberg (GRF) array, near Nuremberg, became the world's first major digital broadband array, fully operational in 1980. The first commercially produced Wielandt-Streckeisen seismometers were deployed in the GRF array in 1982. The array is now operated as an activity of the Federal Institute for Geosciences and Natural Resources (BGR, Hannover). The German Regional Seismic Network (GRSN), 16 stations forming a modern national network, was installed between 1991 and 1996. The 45-station global GEOFON network, founded in 1992 by the GeoForschungsZentrum Potsdam, is a contribution to the international seismic network.

2.1.1 The Gräfenberg (GRF) Array

(http://www.szgrf.bgr.de/)

The 13 stations of the GRF array are located within an area of about 50 by 100 km east of the city of Nuremberg. It became fully operational in April 1980, although continuous recordings of the first subarray are available since 1976. A summary of the GRF activities (including examples of research using broadband data) after 10 years of its operation is given by B. Buttkus (ed.), *Ten Years of the Gräfenberg Array* (Geologisches Jahrbuch Reihe E, Heft 35, 135 pp., 1986). The array is operated by the Seismologisches Zentralobservatorium (SZGRF), which is part of the Bundesanstalt für Geowissenschaften und Rohstoffe (BGR). It is supported by the Deutsche Forschungsgemeinschaft (DFG).

2.1.2 The German Regional Seismic Network (GRSN)

(http://www.szgrf.bgr.de/)

The German Regional Seismic Network was installed between 1991 and 1996 by the German universities supported by the DFG and BGR. It forms a modern national seismograph network with 16 open stations distributed over the whole country and central data archiving at the SZGRF in Erlangen. Here the first Streckeisen STS-2 broadband sensors with extended high-frequency range were installed and the open station concept was realized, where all users could directly log into the station for data retrieval.

2.1.3 The Geoforschungsnetz of the GFZ (GEOFON)

(http://www.gfz-potsdam.de/geofon/)

The global GEOFON project was founded by the GFZ in 1992. Presently (summer 2001) about 45 broadband stations contribute to the network. Most stations are operated in close cooperation with seismological institutions in host countries. The data are transmitted to GFZ and archived at Potsdam. In addition, the GFZ operates a pool of mobile seismic stations including at present about 70 broadband sensors.

2.2 Free Oscillations (G. Müller)

Since 1972, the Black Forest Observatory (BFO) has recorded ground deformations over a band of frequencies extending from seismic waves through free oscillations to tidal deformations. At a depth of 150 m, it is one of the quietest observatories in the world for long-period monitoring by broadband seismometers, a tidal gravimeter, borehole pendulums, and strainmeters. A few highlights from the free-oscillation research at BFO are mentioned (e.g. observations of oSo, oSn, oTn, and volcanic eruptions); more details are given in the full report on the Handbook CD.

2.3 Surface Waves (second part of the 20th century) (P.G. Malischewsky)

Because of political reasons, the seismological research in Germany in the second part of the 20th century is characterized by independent developments in East and West Germany, respectively. Nevertheless, in this period interesting results concerning seismic surface waves were achieved in both theoretical and practical aspects.

Surface wave research in East Germany during the 1960s through the 1980s yielded new geological results. In addition, the effects of anisotropy and attenuation on surface wave propagation, as well as their reflection, refraction, and scattering, were studied in detail. Research in West Germany led to new knowledge of crustal structure, including the bedrock of the Alps. Theoretical results on the polar phase shift of Rayleigh waves and the structural interpretation of non-plane waves were achieved. Applications of surface wave observations to practical problems have also been furthered. More details with references can be found in the full report on the Handbook CD.

2.4 New Seismic Phases (G. Müller)

Since about 1960, the advent of networks and arrays of identical analog and digital seismic stations facilitated the detection of weak seismic phases that were either unknown before or were expected, but not observed. Familiar examples are the steep-angle reflection *PKiKP* from the inner core or underside reflections from mantle discontinuities. In a few cases, German seismologists contributed to such detection.

An example is the discovery of *P* waves associated with the phase *SKS* that are diffracted along the core–mantle boundary, either before entrance into the core or after exiting (SKPdiffS). Other observations are *P* reflections some 300 km above the core–mantle boundary, SKS anisotropy and *P*-to-*S* conversions (receiver functions) from the Moho and upper mantle discontinuities (see the full report on the Handbook CD for details).

2.5 The Reflectivity Method of Seismogram Synthesis (R. Kind)

Since the complete analytical solution of the problem of propagation of elastic waves in a half-space by Lamb in 1904, progress toward layered models was slow until the age of computers. The matrix formulation of N.A. Haskell and the decomposition of the wave field into generalized rays were combined by K. Fuchs and G. Müller to create the reflectivity method of computing synthetic seismograms. R. Kind extended the method for the case of a buried source, which permitted application to earthquake seismograms, including depth phases and surface waves (see full report on the Handbook CD for more details and references).

2.6 Controlled Source Seismology (J. Mechie)

The earliest German work in controlled source seismology is summarized in the account of Mintrop's contributions in the first part of the full report on the Handbook CD. The post-World War II studies began with the recording of the large explosions at Helgoland (1947) and Haslach (1948). From the 1960s on, seismic refraction profiles, some up to 2000 km long, have been carried out, often as cooperative projects involving more than one western European country. Important results were also obtained from profiles completed in the former East Germany.

Some deep reflection studies were also completed as an early cooperative program between the prospecting industry and the universities. For at least 10 years after 1958 the recording time of routine seismic prospecting surveys was extended from 5 to 12 seconds. A statistical interpretation of the number of reflections as a function of two-way travel-time revealed, for the first time using deep reflection studies, the Moho reflection and mid-crustal reflections. At the same time as data were being collected, methods were also being developed in Germany to interpret these data. One significant advance, which occurred at the end of the 1960s, was the development of the reflectivity method by K. Fuchs and G. Müller, so that amplitudes (i.e. the full wavefield) and not only travel times could be interpreted.

During the 1980s, much German effort went into the completion of the European Geotraverse, which produced a picture of the crust and lithosphere along a north–south line from the North Cape in Norway, through Sweden, Denmark, Germany, Switzerland, Italy, Corsica, and Sardinia, to Tunisia. Within the framework of the KTB (Kontinentales Tiefbohrprogramm) in Germany in the 1980s, several coincident near-vertical incidence and wide-angle reflection profiles were carried out and one 3-D near-vertical incidence reflection survey centered at the deep drill hole was completed. As a result of the coincident near-vertical incidence and wide-angle reflection measurements, considerable effort was put into modeling the reflective properties of the lower crust.

In the 1990s much of the work under the DEKORP banner has been undertaken outside Germany, including the URSEIS95 project in the Urals and the ANCORP96 project in the Andes. In the Urals some of the deepest reflections (45 seconds of two-way travel-time, equivalent to about 170 km depth) ever imaged by the near-vertical incidence reflection method were observed, while in the Andes the deepest-ever image of a subduction boundary (80 km depth) was achieved by the same method. In 1999 the TRANSALP project from Germany through Austria into Italy was completed, while in 1996 a reflection profile across the North German basin was carried out.

Mainly during the 1980s and 1990s, the German groups participated often as leading partners in many international seismic projects in many parts of the world. Notable examples include the Afro-Arabian rift system, the Andes, Urals, Tibet, the north Atlantic/Iceland region, the continental margin of Namibia, the

Baltic Sea/Gulf of Bothnia region, the northeastern Pacific, and along the Pacific margin of central and South America.

Within the framework of the URSEIS95 project, complementary wide-angle data demonstrated the presence of a 15–18-km-thick crustal root beneath the central part of the southern Urals orogen. Wide-angle seismic data collected in the central Andes (northern Chile, northern Argentina, and southern Bolivia) by German groups since the 1980s have delineated the crustal structure of this part of the orogen, including a 65–70-km-thick crust under the Altiplano and the Western and Eastern Cordilleras. The wide-angle seismic data collected in the Kenya rift in 1985, 1990, and 1994 revealed significant lateral variations in structure both along and across the rift. The crust thins along the rift axis from 35 km in the south to 20 km in the north, corresponding to an increase in extension from south to north. Across the rift margins there are abrupt changes in Moho depth (5–10 km) and uppermost mantle velocity (7.5–7.7 km/sec beneath the rift itself and 8.0–8.1 km/sec beneath the rift flanks). At present, German groups are still very actively involved in various seismic projects throughout the world, including Tibet, the Andes, Rumania, Indonesia, and the Middle East. More details with references can be found in the full report on the Handbook CD.

79.25 Greece

The editors received two reports from Greece: (1) "Seismology in Greece," from Drs. G. F. Karakaisis and C. B. Papazachos, Department of Geophysics, School of Geology, Aristotle University of Thessaloniki, GR-54006 Thessaloniki, and (2) the institutional review from Prof. P. Varotsos of The Solid Earth Physics Institute, University of Athens, Athens. These two reports are archived as computer-readable files on the attached Handbook CD#2, under the directory of \7925Greece. We extracted and excerpted the following combined summary from these reports. An inventory of seismic networks in Greece is included in Chapter 87. Biographies and biographical sketches of some Greek scientists and engineers may be found in Chapter 89, and in Appendix 3 of this Handbook, respectively.

1. History of Greek Seismology

Greece exhibits the highest seismic activity in western Eurasia. On the average, an earthquake of magnitude 6.3 occurs every year in the broader Aegean area (34–43°N and 18–30° E). This high seismicity is attributed to the convergence of the Aegean lithosphere and the eastern Mediterranean lithosphere in a north–south direction. The rate of this convergence is ~4 cm/yr, and is mainly due to the overriding of the Aegean lithosphere on the Mediterranean lithosphere (~3–3.5 cm/yr). It is also partly due to the northward motion of the African plate (0.5–1cm/yr) with respect to the Eurasian plate and subduction of the eastern Mediterranean lithosphere under the southern Aegean lithosphere.

The first historically documented information in Greece on an earthquake dates back to the 6th century BC. In 1895, a new law was passed by which the National Observatory of Athens was organized in three departments: the Astronomical, the Meteorological, and the Geodynamic. The last was responsible for the study of earthquakes. In 1898 the first seismograph, Agamemnone type, was installed in Athens, while the first reliable seismometer, Mainka-type, was installed in 1911 in Athens and is still in operation. In 1962, the seismological station of Athens became one of the stations of the World-Wide Standardized Seismograph Network (WWSSN).

During the last four decades several permanent networks of seismological stations were established in Greece. In addition to the two seismological centers in Athens (the Geodynamic Institute of the Observatory of Athens, and the Seismological Laboratory of the University of Athens), two new seismological centers were established in Thessaloniki: the Geophysical Laboratory of the University of Thessaloniki (1977) and the Institute of Engineering Seismology and Earthquake Engineering (1979). The Seismological Laboratory of the University of Patras has been established recently, as well as the Solid Earth Physics Institute, University of Athens.

2. Seismicity

Early seismicity studies of the 20th century for the broader Aegean area had been of qualitative character. Earthquake catalogs, which included all the basic focal parameters of earthquakes in this area, were compiled during the last four decades. On the basis of these catalogs, numerous semi-quantitative and quantitative studies have been carried out.

In one of the recent studies a new method has been proposed for the reliable estimation of the two most commonly used parameters of seismicity, b and a. On the basis of these parameters, the most probable maximum magnitude of shallow earthquakes in the broader Aegean area for a period of 1 year is 6.3, whereas a magnitude 7 event occurs about every 6 years.

INTERNATIONAL HANDBOOK OF EARTHQUAKE AND ENGINEERING SEISMOLOGY, VOLUME 81B ISBN: 0-12-440658-0

3. Crustal and Upper Mantle Structure

In the first systematic study of the crustal structure of the Aegean by the use of body and head waves, an average crustal thickness of 42 km was proposed, with a relatively low P_n wave velocity of 7.87 km/sec. From surface wave dispersion, an average crustal thickness for southeastern Europe (35–45 km) was determined. Results from seismic refraction experiments suggested an average crustal thickness for the Aegean and the Hellenides mountain range that varied between 30 and 44 km, respectively, whereas for the southern Aegean (Cretan sea) it ranged between 20 and 25 km and less than 20 km for the Ionian Sea.

Since the mid-1990s important results have been determined using travel-time and surface-wave tomography. The subduction beneath the Aegean area is visible and extends up to ~800–900 km. The presence of slab detachment in the subducted lithosphere on the western segment of the subduction can also be seen. Both these results have important geotectonic implications, because the large length of the subduction imposes constraints on the age of the subduction.

4. Earthquake Prediction

Research work in Greece related to earthquake prediction mainly concerns the investigation of seismic sequences, seismicity regularities, and earthquake precursors.

Research on seismic regularities includes observations on seismic gaps, on seismic quiescence, on migration of the seismic activity, on the behavior of the seismic activity during the seismic cycle, and on other types of seismicity regularities. During the last decade several studies have been carried out on time-dependent seismicity, which led to the development of the *regional time and magnitude predictable model*, which has been used to determine probabilities of occurrence of strong earthquakes during the next years in Greece and almost all areas of the world where earthquakes occur. Intensive work is focused lately on the critical earthquake concept and specifically on the *accelerating moment release model*. Research on earthquake precursors is mainly focused on preseismic variations of the electric field (see Section 11). However, the results are subject to vigorous debate.

5. Seismic Hazard

The basic work on seismic hazard in Greece concerns the geographical distribution of the maximum observed intensity, the maximum expected macroseismic intensity, peak ground acceleration or velocity, and the duration of the strong ground motion.

In one of the most recent seismic hazard studies, an improved procedure was followed in which all new information concerning seismicity, seismotectonics, attenuation laws of macroseismic intensities, and local site conditions were taken into account. The results of the time-independent approach were given in terms of the expected macroseismic intensity at 144 sites in the studied area for a mean return period of 475 years (probability of exceedance 10 percent in 50 years) while, for the first time, the time-depended hazard at these sites was estimated and expressed in terms of occurrence probability of strong seismic motion (I ≥ VII) for the period 1996–2010.

6. Institute of Geodynamics, National Observatory of Athens

Address: P.O. Box 20048, 118 10 Athens, Greece.
Web site: http://www.gein.noa.gr/English/

J. Schmidt, the first Director of the Observatory, was the first to collect data about the earthquakes occurring in Greece. In 1893, the Greek Seismological Survey was formed. Later, the Survey became the core of the Geodynamics Institute. Observations were done in Athens using two Brassart seismoscopes and around the various Greek provinces using personnel employed in the local meteorological stations. The first instruments were put in operation in 1897. They were five seismographs of "Agamemnon" type that were installed in selected locations between 1899 and 1902. In 1910 the Observatory installed a two-component Mainka seismograph. Since then many seismographs have been installed. In 1962 a WWSSN station was installed in Athens. Various seismographs were installed by the Observatory in many Greek locations during the 1960s and 1970s.

In March 1983 a telemetry network started operating between Athens, Valsamata, Kozani, Hagia Paraskevi, and Neapolis stations. The new network consists of vertical-component seismographs of thermal recording type, using Geotech S-13 seismometers. At the Athens station two horizontal-component seismometers were also installed. In February 1988 the Athens WWSSN station was converted to thermographic recording. In the spring of the same year twelve new telemetry stations were installed in Kerkyra (Corfu), Karpenissi, Pelio, Rhodopi, Samos, Laconia, Karpathos, Rhodes, Ithomi, Vamos, Poligiros, and Kastellorizo.

In 1993 the installation of the digital network started. The first stations are Athens, Ioannina, Valsamata, Ithomi, Samos, Poligiros, Karpenissi, Pelio, and Thera (Santorini). In 1994 in collaboration with ING (Rome), a new station was installed in Anogia (Crete) that became part of the Mediterranean Network. This station started operating in 1995 and in 1999 became part of the global system of monitoring nuclear explosions that is operated by the United Nations. Since 1998 new seismological stations have been progressively installed in a number of locations. In addition, the rest of the stations were converted to digital, real-time transmission with advanced broadband instruments.

The research activities of the Geodynamics Institute cover a broad spectrum of seismology, physics of the earth's interior, and geophysics. It has particular interest in research activities in the fields of seismicity, focal mechanisms of earthquakes, properties of the seismic source, propagation of seismic waves, strong ground motion, seismic hazard, microzonation studies, crust and mantle structure, earthquake prediction, seismotectonics, paleoseismology, tsunamis, geological remote sensing, and applied geophysics. The Observatory staff is listed in the full report on the Handbook CD.

7. Seismological Laboratory, National & Kapodistrian University of Athens

Address: Ilissia, Panepistimioupoli, 157 84 Athens, Greece.

Web site: http://dggsl.geol.uoa.gr/images/e_setn.html

The Seismological Laboratory belongs to the Department of Geophysics and Geothermy, Faculty of Geology, National and Kapodistrian University of Athens, and was established in 1929. Its major functions have been the education of students, and the monitoring of seismicity of the area of Greece, in cooperation with the Geodynamic Institute of the National Observatory of Athens.

The Laboratory operates 3 modern digital telemetric station networks: CORNET and VOLNET in the areas of the Gulf of Coring and central Greece. The new, also telemetric seismic network, ATHENET, covers an area of approximately 100 km radius around Athens, and consists of 8 satellite stations equipped with broadband seismometers and on-site 24-bit digitizers. Data are transmitted to the central processing station, situated at the Laboratory building. In addition to the permanent networks, the Laboratory has many portable instruments.

During the last 60 years the Seismological Laboratory has greatly contributed to scientific advances in the field of Seismology in Greece. In the course of its activities over the years, a better understanding of the seismicity, seismic hazard, and seismic risk of the area of Greece was accomplished. Several local seismic networks were deployed, monitoring the seismicity and determining the seismotectonic regime of certain seismogenic areas. Moreover, applications of theoretical seismology to Greek data resulted in revised and/or local seismic intensity and seismic energy attenuation laws and in the determination of seismic source parameters for a large number of earthquakes At the same time, a large number of microzonation studies aiming at the mitigation of seismic risk were performed. In recent years, major advances were also made in the field of engineering seismology. The Laboratory welcomes a large number of physics and geology undergraduates. The personnel are listed in the full report.

8. Department of Geophysics, Aristotle University of Thessaloniki

Address: GR-54006 Thessaloniki, Greece.

The Laboratory of Geophysics of the Aristotle University of Thessaloniki was established in 1976 by presidential degree, together with the Chair of Geophysics. Since 1983 the Laboratory belongs to Department of Geophysics, which is one out of five departments of the School of Geology of the university.

The Department of Geophysics was the first educational and research geophysics institution in northern Greece as well as the first such institution in Greece outside Athens. Since its establishment, the Department has rapidly developed research, educational, and broader social activity in problems related to seismology, physics of the Earth's interior, and applied geophysics. The educational activity is for both undergraduate and post-graduate students.

Since 1981 a telemetric network of 1 central and 16 permanent peripheral seismological stations has been installed and is in operation continuously. The staff members perform many useful functions including participating in the development of the New Seismic Code of Greece, advising the authorities for the antiseismic construction of several important technical structures, giving lectures on seismic risk and other problems, and giving reliable earthquake information. The members of the staff are listed in the full report on the Handbook CD.

9. Institute of Engineering Seismology and Earthquake Engineering

Address: 46 Georgikis Sholis Avenue, P.O. Box 53 Foinikas, GR-55102 Thessaloniki, Greece.

Web site: http://www.itsak.gr/

The Institute of Engineering Seismology and Earthquake Engineering (ITSAK) was established in Thessaloniki (Greece) in 1979, after the disastrous earthquake ($M = 6.5$) that struck the city on June 20, 1978. Today, ten researchers (all with doctoral degrees) work at ITSAK. The main objective of the Institute is applied research in the fields of engineering seismology, soil dynamics, and earthquake engineering, aiming at upgrading the Greek Seismic Code and mitigating damage from earthquakes.

The Institute's principal results include (1) deployment and operation of a free-field strong-motion network, which today is nationwide, consisting of 78 accelerographs and covering most of the country's major towns; (2) compilation of an extensive databank of strong motion recordings, comprising Greek, European, and other accelerograms around the world; (3) organization of a specialized library in the fields of engineering seismology, earthquake engineering, and soil dynamics, (4) construction of design spectra that were incorporated in the Greek Seismic Code; (5) instrumentation and monitoring of the dynamic behavior of buildings and large-scale engineering structures; (6) deployment of a dense 3D accelerograph array close to Thessaloniki; and (7) study of aftershock sequences of strong earthquakes occurring in Greece. The research staff and their functions are listed in the full report on the Handbook CD.

10. The University of Patras Seismological Laboratory

Address: Rio-Patra, 26500, Greece.

Web site: http://santorini.geology.upatras.gr/

The University of Patras Seismological Laboratory (UPSL) was established in 1990. A small five-station permanent seismological network around the Patras gulf was installed in 1992. It has evolved into one of the largest seismological networks in Greece. The Patras Seismological Lab Seismic Network (PATNET) covers the western Greece area, the Peloponnese, and the Southern Aegean. It consists of the base station at the University of Patras, four broadband stations, and thirty-four short-period stations. UPSL operates an earthquake prediction center, which consists of the simultaneous monitoring of various parameters and their comparison with the recorded seismicity. Parameters that are currently measured are electro-telluric variations, variations of groundwater temperature and level, electromagnetic variations at various spectra bands, and seismic parameters such as attenuation and seismicity.

In 1997 UPSL started developing a seismic reflection unit, which today has evolved into a fully independent seismic section. During 2000, UPSL acquired two big Vibroseis trucks of 50,000 lb impact force each, and has recently developed close links with the oil industry for the implementation of new geophysical technologies in hydrocarbon exploration. The personnel are listed in the full report on the Handbook CD.

11. The Solid Earth Physics Institute, University of Athens

The Solid Earth Physics Institute (S.E.P.I.) was founded at the end of 1994. The objectives are research activities, development of new technologies, and technology dissemination concerning the continuous recording and study of the physical properties of the earth's solid crust. Emphasis is given to the quantitative estimation of the physical properties that precede abrupt tectonic movements. Furthermore, S.E.P.I. is aiming at the education of researchers, students, governmental officials, etc., as well as at cooperation with the Greek and European industry on new methods related with the detection, acquisition, transmission, and storage of scientific data.

The current activities of S.E.P.I. include field measurements of low frequency ($\leq$1Hz) electrical precursors, seismic electric signals (SES), RF disturbances, as well as laboratory measurements and theoretical studies. Since 1982, continuous measurements of the Earth's electric field have been carried out in Greece. Data were transmitted to the central station located at an Athens suburb, and hence the SES could be recognized on a real-time basis. Since 1995, several stations were reactivated and the number of dipoles at each of them was significantly increased. Since 1990, dataloggers were installed at several stations collecting data from short dipole (50–400 m) and long dipole (2–20 km) arrays.

The real-time data collection allows recognition of the electrical precursor(s) well before an earthquake occurrence so that predictions can be issued in advance. Since May 15, 1988, all predictions have been sent to the Greek authorities and several scientific institutes in other countries. Each prediction usually contains (1) the date and the station(s) at which the SES were recorded; (2) estimation of the expected epicentral area; and (3) estimation of the magnitude of the impending earthquake. See the full report on the Handbook CD for further technical details and a list of references.

79.26 Hungary

The editors received the "National Report of Hungary" from Dr. P. Márton, National Correspondent to IASPEI, Geophysics Department, Eötvös Loránd University, H-1117 Budapest, Pázmány Péter stny. 1/C, Hungary. This report, which consists of eight separate reviews and biographies of the deceased and biographical sketches of the living, is archived as a computer-readable file on the attached Handbook CD#2, under the directory of \7926Hungary. We extracted and excerpted a summary as shown here. Biographies and biographical sketches of some Hungarian seismologists and geophysicists may be found in Chapter 89, and in Appendix 3 of this Handbook, respectively.

1. Introduction

The report starts with a concise summary of the history of the development of earthquake research in Hungary. Special emphasis is placed on the "heroic" period, marked by F. Schafarzik and R. Kövesligethy. The latter was General Secretary of the International Association of Seismology (1905–1922). The role of L. Egyed in rebuilding and modernizing the seismologic service in the mid-1950s and 1960s is also highlighted, as well as the more recent developments toward an up-to-date data acquisition system and monitoring network for seismic safety. Hungarian seismological stations since 1900 are shown in Table 1.

2. Seismic Safety (Attila Meskó)

This contribution is concerned with the seismic safety of two large constructions, the Paks Nuclear Power Plant (PNPP) and the Gabcikovo (Nagymaros) Barrage System (G(N)BS). Although seismic activity in Hungary is low to moderate, earthquakes with magnitude between 6.0 and 6.5 have occurred during the last 1500 years. As far as the PNPP is concerned, the presence of a capable fault can be ruled out with high probability, the power plant meets stability standards against the operation basis earthquake, and the vulnerable part of the power plant can be reinforced to meet stability standards. The earthquake hazard of the G(N)BS originates from three seismogenic zones nearby, and the maximum magnitudes for the 1000-year-recurrence period were estimated between 5.8 and 6.8.

3. Seismic Hazard Analysis (Zoltán Bus)

This contribution summarizes the results of a deterministic hazard computation for Hungary using synthetic seismograms to an upper frequency limit of 1 Hz for the determination of the expected maximum displacement, velocity, and design ground acceleration (DGA). The conclusions are (1) a considerable part of the seismic hazard originates from the seismogenic zones of the neighboring countries, (2) the highest DGA reaches 0.14 g (corresponding approximately to degree VIII on the MSK-64 intensity scale), and (3) three of the largest towns are particularly subject to high seismic risk.

4. Hypocenter-Velocity Tomography (Zoltán Bus)

This review describes a numerical experiment for the determination of the velocity structure beneath the Hungarian part of the Pannonian Basin by hypocenter-velocity tomography. Both the method applied and the results are presented. The main

INTERNATIONAL HANDBOOK OF EARTHQUAKE AND ENGINEERING SEISMOLOGY, VOLUME 81B ISBN: 0-12-440658-0

TABLE 1 Hungarian Seismological Stations Since 1900

Code	Name	Operating period	Instrument	
BUD	Budapest	1902–	1902–1905	Strasbourg (horiz.), Bosch
			1906–1912	Bosch, Vicentiny-Konkoly, Wiechert 1000 kg
			1913–1945	Wiechert 1000 kg
			1946–1949	Wiechert 1000 kg, Krumbach 100 kg
			1950–1962	Wiechert 1000 kg
			1962–1972	VEGIK (Z), Kirnos (3 comp.)
			1973–1980	Kirnos (3 comp.), Modified Kirnos, Ullmann-Teupser
			1981–1990	Kirnos (3 comp.), Modified Kirnos
CJR	Kolozsvár (now Cluj Napoca, Romania)	1911–1912	1911–1912	Mainka 210 kg
DEB	Debrecen	1963–1964		?
GYL	Gyula	1995–	1995–	Kinemetrics SS-1 + SSR-1
HRB	Ógyalla (now Hurbanovo, Slovakia)	1903-1912	1903	Strasbourg
		1938–1945	1904	Bosch (horiz.)
			1905–1908	Bosch, Vicentiny-Konkoly
			1909–1910	Bosch
			1911–1912	Mainka 210 kg
			1938–1945	Mainka 210 kg
JOS	Jósvafõ	1970–1985	1970–1985	Modified Kirnos (Z)
KAL	Kalocsa	1900(?)–1912	1900(?)–	Brassart (Z)
		1940–1963	1903–1906	Rossi avisatore
			1907–1908	?
			1910–1963	Wiechert 200 kg (horiz.)
KEC	Kecskemét	1937–1942	1937–1942	Wiechert 80 kg (Z)
		1951–1972	1951–1966	Krumbach 100 kg (horiz.)
			1967–1972	Kirnos (3 comp.), Quervain-Piccard 25 kg
PSZ	Piszkéstetõ	1964–	1964	Modified Kirnos (Z), Kirnos (Z)
			1965–1966	Modified Kirnos (Z), Kirnos (3 comp)
			1967–1976	Modified Kirnos (Z)
			1977–1990	Short period seismograph (Z)
			1990–	Streckeisen STS-2 VBB
RIY	Fiume (now Riyeka, Croatia)	1903–1912	1903-1904	Vicentiny (horiz.)
			1905	Vicentiny (Z.)
			1906–1912	Vicentiny (horiz.)
SOP	Sopron	1963–	1995–	Kinemetrics SS-1 + SSR-1
SZE	Szeged	1909–1911	1909–1911	Mainka
		1939–1963	1939	Mainka 135 kg
			1940–1963	Mainka 210 kg (horiz.)
TIH	Tihany	1986		
TIM	Temesvár (now Timisoara, Romania)	1901–1912	1901-1902	Cacciatore
			1903–1904	Agamennone avisatore, Rossi-Forel avisatore
			1905	Vicentiny-Konkoly, Rossi-Forel avisatore
			1906–1912	Vicentiny-Konkoly
UZH	Ungvár (now Uzhgorod, Ukraina)	1909–1912	1909–1912	Bosch 10 kg
		1940–1942	1940-1942	Wiechert 80 kg
ZAG	Zágráb (now Zagreb, Croatia)	1906–1908	1906–1908	Vicentiny-Konkoly

conclusion is that the velocity distribution obtained for the upper 29 km is acceptable for the whole area and is in agreement with the *a priori* geophysical and geological knowledge.

5. Seismic Tomography (Zoltán Wéber)

This contribution is on seismic travel-time tomography, describing particularly the author's proposals to improve the efficiency of this popular method of subsurface velocity determination. These concern the shortest path ray tracing, application of the simulated annealing method for finding the global minimum of a suitably chosen cost function, and model discretization for finding an optimum, irregular triangular cell parameterization that best suits the ray path geometry leading to better-conditioned tomographic equations. The paper ends with the description of a novel inverse technique by Wéber to estimate interval velocities from zero-offset and finite-offset vertical seismic profiles.

6. Re-evaluation of the 1763 Komárom Earthquake (Győző Szeidovitz)

The author has made an attempt to re-evaluate the Komárom earthquake of June 28, 1763. Using essentially contemporaneous data about the number of farm houses and the average cost of their reconstruction, an isoseismal map could be constructed and the maximum intensity (9 on MSK-64) and focal depth (4–10 km) estimated.

7. Mantle Convection (László Cserepes)

The author reviews the progress made in numerical modeling of mantle convection during the last 30 years with special emphasis of his (and colleagues') contributions to better understanding of the dynamics of the Earth's lithosphere and mantle.

8. Seismic Lithosphere and Asthenosphere Investigations in Hungary (Károly Posgay, Tamás Bodoky, and Endre Hegedűs)

This contribution is a full-length paper and is basically a review of the seismic investigations carried out in this country for the region of the lithosphere–asthenosphere boundary. As a result of these investigations, fundamental inferences could be made about the structure and tectonic evolution of the studied zones.

79.27 Iceland

The editors received a summary report from Dr. Ragnar Stefánsson, IASPEI National Correspondent, Department of Geophysics, Icelandic Meteorological Office, Reykjavik, Iceland. The report was edited and is shown here. Biographical sketches of some Icelandic scientists and engineers may be found in Appendix 3 of this Handbook.

1. Introduction

High seismic and volcanic activity has for a long time fostered much interest in seismology in Iceland. Iceland is located on an extremely active part of the mid-Atlantic ridge. Earthquakes reaching magnitude 7 occur mainly on two east–west transform zones, one in southwest Iceland, and the other along the north coast of the country. Active volcanism is confined to north–south elongated spreading zone segments, most densely near central Iceland, above the inferred location of the Iceland mantle plume. The east–west spreading rate across the mid-Atlantic ridge, near Iceland, is about 1.8 cm/year. The velocity is modified locally in time and space by mantle-plume activity and rifting episodes.

Historic earthquake evidence since about 1700 constitutes a major resource for contemporary understanding of seismic hazards. The destruction caused by a sequence of large earthquakes in southwest Iceland in 1896 was studied in detail, and documents describing these events contain unique information for understanding the neotectonic processes.

2. Earthquake Monitoring

The International Seismological Association took the initiative to install a Mainka seismograph in 1909, with a second component added in 1914. The records were sent to Strasbourg for analysis. Only a small fraction of records from this period is available. Recordings were resumed by the Icelandic Meteorological Office (IMO) in 1925, operating the same Mainka instruments. Since then and until 1970 IMO was the only institution monitoring seismic signals in Iceland. During the 1950s and 1960s, seismic equipment was modernized, most notably by the WWSSN seismic station installed in Akureyri in 1964.

In the 1970s the Science Institute, University of Iceland, led the build-up of a seismic network, optimized to record local seismicity in the seismically most active parts of Iceland. The stations were one-component writing records onto paper. The older networks have since 1991 gradually been replaced by the SIL system, built up and operated by the IMO. The SIL system currenly consists of 42, three-component digital seismic stations, located throughout Iceland (Stefánsson *et al.*, 1993; Böðvarsson *et al.*, 1996). The SIL network is focused on detecting microearthquakes, down to magnitude zero, and provides information on hypocenters and source characteristics in near real time. An alert system has been operated in Iceland since 1992, based mostly on the automatic evaluation of seismicity by the SIL network. The system issues alerts if certain parameters pass predefined limits for specified regions.

Additional measurements related to strain buildup and earthquake release have gradually been introduced during the last decades by various institutions. These measurements include observations of radon changes, volumetric strain, GPS, InSAR, and hydrological observations.

During the last decade the Engineering Research Institute, University of Iceland, has been installing a network of strong motion accelerometers within some of the most earthquake-prone areas.

3. Crust and Upper Mantle Structure

Studies of crust and upper mantle structure started in Iceland in the 1950s, based on active seismic experiments, as well as on local and teleseismic signals of the permanent networks. Since then intensive research work has been ongoing in this field, culminating in several international projects (e.g., Allen *et al.*, 2002).

INTERNATIONAL HANDBOOK OF EARTHQUAKE AND ENGINEERING SEISMOLOGY, VOLUME 81B ISBN: 0-12-440658-0

4. Mitigating Earthquake and Volcanic Risks

Based on seismological evidence and historical data, the first earthquake hazard assessment map of Iceland was published in 1958. With the combined development of seismometry and models for strain buildup and earthquake release, methods of hazard assessments have evolved significantly. Such advances have enabled the recent (2002) hazard maps and building standards within Eurocode 8.

Since 1988 several earthquake prediction research projects have been carried out in Iceland, with considerable international participation, revealing significant new understanding of earthquake processes.

Based on automatic monitoring and key research findings, a service is in operation in Iceland that provides information and, when possible, warnings to the public and civil protection agencies. The Department of Geophysics of the IMO is at the center of this service with respect to earthquakes and volcanic eruptions. It also utilises observations from other institutions and consultations with other geoscientists. On some occasions, useful warnings have been made before volcanic eruptions and earthquakes, helping to facilitate the development of an "Early warning and information system (EWIS)," which is a new tool for fast application of geophysical observations and scientific knowledge for the purpose of mitigating earthquake and volcanic risks (Halldorsson, 1996; Stefánsson *et al.*, 1999).

References

Allen, R.M., *et al.* (2002). Plume-driven plumbing and crustal formation in Iceland. *J. Geophys. Res.*, **107**(B8), ESE-4, 1–19.

Böðvarsson, R., S.Th. Rögnvaldsson, S. Jakobsdóttir, R. Slunga, and R. Stefánsson (1996). The SIL data acquisition and monitoring system. *Seismol. Res. Lett.*, **67**(5), 35–46.

Halldorsson, P. (1996). Seismic hazard assessment. In: "Seismic and Volcanic Risk" (B. Thorkelsson and M. Yeroyanni, Eds.), pp. 25–32. Proceedings of the workshop on monitoring and research for mitigating seismic and volcanic risk, Reykjavik, Iceland, October 20–22, 1994. European Commission, Environment and Climate Programme.

Stefánsson, R., R. Bödvarsson, R. Slunga, P. Einarsson, S. Jakobsdóttir, H. Bungum, S. Gregersen, J. Havskov, J. Hjelme, and H. Korhonen (1993). The SIL project, background and perspectives for earthquake prediction in the South Iceland seismic zone. *Bull. Seismol. Soc. Am.*, **83**, 696–716.

Stefánsson, R., F. Bergerat, M. Bonafede, R. Bödvarsson, S. Crampin, P. Einarsson, K. Feigl, Á Gudmundsson, F. Roth, and F. Sigmundsson (1999). Earthquake-prediction research in a natural laboratory—PRENLAB. In: "Seismic risk in the European Union II" (M. Yeroyanni Ed.), pp. 1–39. Proceedings of the review meeting, Brussels, Belgium, November 27–28, 1997. European Commission.

79.28 India

The editors received the "Centennial National Report of India: Contributions in Seismology (1900–2000)" from Dr. Harsh Gupta, Chairman of the National IUGG Committee. The report is written by Harsh Gupta (Department of Ocean Development, New Delhi), R .K. Chadha, S.C. Bhatia, D. Srinagesh, D.S. Ramesh, M. Ravi Kumar, P. Mandal, N.P. Rao, and Vineet Gahalaut (National Geophysical Research Institute, Hyderabad), and S. Basu (Department of Earthquake Engineering, Indian Institute of Technology, Roorkee). It is archived as a computer-readable file in the attached Handbook CD#2, under the directory of \7928India. We extracted and excerpted the following summary.

Earthquake engineering in India is reviewed in Chapter 80.9, and an inventory of the Indian national seismic network is included in Chapter 87. Biographies and biographical sketches of some Indian scientists and engineers may be found in Chapter 89, and in Appendix 3 of this Handbook, respectively.

1. Introduction

The Great Assam earthquake of June 12, 1897, triggered the onset of earthquake studies in India during British times. The first seismological observatory was installed at Calcutta in December 1898, followed by Bombay and Madras in the next few months. While the seismological investigations were initiated in the first half of the 20th century, the second half saw the renaissance of the seismological research in the country. In the early 1990s, the occurrence of the Latur earthquake of September 30, 1993 provided the real impetus for entering into the digital era of seismology.

Because basic research in seismology enhances our understanding of the earthquake phenomenon and accordingly has significant long-term implications in terms of human suffering, Indian seismologists have concentrated their efforts since the 1950s in comprehending complex seismic phenomena by applying simple analytical tools to obtain closed form solutions. They addressed problems in the field of propagation of elastic waves in complex and inelastic media, nature of seismic sources, mysteries of Earth's free oscillations, duality of normal modes and rays, to name a few.

During the last 100 years, several research institutes, government departments, and universities in the country began pursuing studies on earthquakes. Most efforts have gone into improving the detection and location capabilities of earthquakes with more sophisticated digital short-period and broadband instruments in the Peninsular shield of India as well as in the seismically active Himalayan region. This has provided data to improve our understanding of inter- and intraplate earthquakes with reference to the Indian plate. Under the GSHAP program, the seismic hazard map prepared for the Indian plate region, comprising the Himalaya, Northeast India, the Indian shield, South China, Nepal, Burma and Andaman regions, delineated 86 potential seismic source zones, the highest risk being in the Burmese arc, the Northeastern India, and the Hindukush regions with peak ground acceleration (PGA) values of the order of 0.35–0.4 g.

2. Significant Contributions

The significant contributions in seismology from India during the past four decades include (1) understanding the mechanism of reservoir-triggered earthquakes through extensive seismic surveillance in the Koyna region, (2) structure of peninsular and extra-peninsular India from controlled source and passive seismological experiments such as body-wave and surface-wave modeling and tomographic inversion studies, (3) studies of the source mechanism and moment tensor solutions along with numerical modeling of the stress fields to comprehend the active tectonics of the Indian plate margins and the shield region, and (4) multi-disciplinary geophysical studies in the source region of the Latur earthquake of September 30, 1993, together with the results from deep drilling to know the seismogenesis of the Stable Continental Region earthquakes.

INTERNATIONAL HANDBOOK OF EARTHQUAKE AND ENGINEERING SEISMOLOGY, VOLUME 81B ISBN: 0-12-440658-0

3. Recent Advances

With the upgrading from analog to digital broadband seismometry in the country during the past decade, understanding about the structure and dynamics of the Earth has changed dramatically. Indian scientists are now in a position to understand the seismic structure and the source in a better way. The techniques used for oil exploration are now being extended to probe the lithospheric structure in greater detail. The lithospheric architecture deduced from velocity images is integrated with complex surface geology to understand the evolution of the Indian continental masses. The improved characterization of earthquake sources for large and moderate earthquakes in the Koyna region in western India, in terms of fault geometry and rupture dynamics, has opened up new vistas for tectonic studies. Although long-term forecast of earthquake potential now appears to be a possibility, medium- and short-term predictions are still a distant reality. In this scenario, major thrust will now be shifted from earthquake forecasting to earthquake hazard assessment and mitigation. From this point of view, strong motion records and ground amplification information are now available that can contribute to designing of safer earthquake-resistant critical structures such as dams and nuclear power plants in the country.

These achievements have been made possible through intensive research under various time-targeted projects at different organizations. The government of India's current emphasis on research and development to create wealth and improve the quality of life has led to the formation of projects relevant to the present-day needs of the country. The full report, on the Handbook CD#2, provides, under eight subject headings, information on the contributions made by the Indian scientific community in the field of seismology during the last 100 years. In addition the specific missions and past achievements are summarized, along with a comprehensive bibliography of publications by Indian scientists and some biographical sketches.

4. Institutional Reviews

4.1 India Meteorological Department, New Delhi

The India Meteorological Department (IMD), New Delhi, is the oldest organization to be associated with seismology research in India since the establishment of a seismological observatory at Alipur (Calcutta) on December 1, 1898. Before 1947, national seismological observatories were located only at five places, viz., Calcutta, Bombay, Kodaikanal, Hyderabad, and Delhi. By 1970, this network had increased to 18 permanent national observatories. Among these, the observatories at New Delhi, Shillong, Pune, and Kodaikanal had 3-component short-period and 3-component long-period seismographs and were a part of the World Wide Standard Seismological Network (WWSSN) of the USGS since 1963–1964. In the mid-1960s, IMD also started manufacturing Wood Anderson seismographs, electromagnetic seismographs, and photographic recorders. Presently, with 45 national observatories, IMD is primarily responsible for providing immediate information on earthquakes occurring in the country to the government of India and to the media. In addition to accumulation of seismological data for the last 100 years, IMD provides monthly bulletins containing the phase data to national and international organizations after complete analysis. IMD has also provided seismological inputs to the Bureau of Indian Standard Institution for preparation of seismic zoning of India.

During early 1997, ten national observatories have been upgraded to the standard of the Global Seismograph Network. These are at Pune, Karad, Thiruvanthapuram, Chennai, Visakhapatnam, Bilaspur, Bokoro, Bhopal, Bhuj, and Ajmer. In these upgraded observatories high-quality broadband digital data of ground velocity and acceleration are being obtained. These observatories are connected to a Central Receiving Station (CRS) at New Delhi through telephone. At CRS the data is downloaded from these stations in a near-real-time mode and analyzed. Important functions of IMD are (1) monitoring of earthquakes and determination of parameters, (2) research work in theoretical and applied seismology, (3) cooperation with other national and international organizations in seismology, (4) imparting training in seismology to departmental and non-departmental persons, (5) updating the epicentral map and providing seismic inputs for preparation of seismic zoning map, (6) giving earthquake certificates, seismicity reports for construction, (7) seismological inputs for disaster and mitigation, (8) replying to all queries regarding seismology on behalf of the government, including the questions raised in Parliament and State Assemblies.

4.2 National Geophysical Research Institute, Hyderabad

The National Geophysical Research Institute (NGRI), Hyderabad, is a premier geo-scientific organization in India. The Institute, with about 750 scientific and highly skilled technical support and administrative staff, has built up a record of scientific excellence and technical competence in most disciplines of solid earth geophysics.

NGRI houses a permanent seismological observatory with 3-component short-period and long-period seismographs, similar to that of World Wide Standard Seismic Network (WWSSN) type. The data from this observatory are contributed daily to the USGS for global monitoring of earthquakes. NGRI observatory was upgraded to include GEOSCOPE in 1988 and GARNET in 1998. NGRI also operates two telemetered networks, one in Northeast India and one in south India, in addition to several seismic stations near river valley projects in India and a few broadband stations in the Indian Peninsular Shield region.

NGRI has been actively engaged in seismological research with over 50 scientists engaged in study on various aspects of seismology. Pioneering work in the field of "Reservoir Induced Seismicity" (RIS) was done by the NGRI scientists during the early 1970s, and since then much progress has been made toward understanding this phenomenon. During the past 4 decades, significant contributions were made to understanding the mechanism of reservoir-triggered earthquakes through extensive seismic surveillance in the Koyna region, with digital instruments. Other areas in which NGRI has made contributions are understanding the structure of peninsular and extra-peninsular India from controlled-source and passive seismological experiments such as body wave and surface wave modeling studies, tomographic inversion, etc. Studies of the source mechanism and moment tensor solutions along with numerical modeling of the stress fields have enabled comprehension of the active tectonics of the Indian plate margins and the shield region. Multi-disciplinary geophysical studies in the source region of the Latur earthquake, together with the results from deep drilling, have thrown light on the seismogenesis of the SCR earthquakes. A state-of-the-art seismic hazard map of India and adjoining regions prepared under the Global Seismic Hazard Assessment program provides inputs for planning of safer engineering structures and mitigation of seismic hazard. These achievements have been made possible through intensive research under various time-targeted projects. The thrust areas on which the Institute is currently concentrating are (1) data management and dissemination, (2) earthquake diagnostics and travel time tables for the Indian region, (3) crust and upper mantle structure of the Indian shield, (4) dynamics of earthquake source processes in the Indian region, (5) earthquake hazard assessment and earthquake prognostics, (6) theoretical seismology, and (7) high-resolution controlled-source seismological experiments.

4.3 Bhabha Atomic Research Centre, Trombay, Mumbai

The major activities of the Seismology Section of Bhabha Atomic Research Centre (BARC) center on the two arrays at Gauribidanur, Karnataka and another at Delhi. In addition, BARC has installed a seismic array with a telemetered link near Bhatsa Dam in western Maharashtra to monitor reservoir-related and regional seismic activities. Seismic activity around nuclear power plants in the country is routinely monitored by the organization.

A field-processor-based digital seismic telemetry system has been developed for acquisition of high-quality digital seismic data from a seismic array system. Using special broadband seismic sensors along with industrial-grade processors, digital UHF wireless link employing advanced communication protocols, and a matching multi-channel processor-based receiving unit, this system is capable of providing continuous data in a wide dynamic range. The system is to be established at Gauribidanur array, which would substantially augment the array's capability.

Development of signal processing methods such as artificial neural network (ANN) based algorithms for seismic source discrimination, spectral methods for seismic source analysis, and synthetic seismogram simulations are being carried out for various applications. The ANNs developed in the section have shown excellent capability to detect and to identify weak explosion signals from various global test sites.

A satellite-based communication system for rapid on-line transfer of seismic waveforms and extracted parametric data in near real time has been established at Gauribidanur and BARC, Trombay. This system that allows remote login and round-the-clock access greatly facilitates exchange of vital seismic data. In addition, scientists are also involved in understanding the seismotectonics of the Andaman Nicobar–Burmese arc region, Peninsular shield, and around Delhi regions.

4.4 Geological Survey of India

The Geological Survey of India (GSI) is involved in carrying out seismological investigations in areas of regional hazard assessment, seismotectonic evaluation of critical areas, paleoseismology, micro and macroseismic activities, and neotectonics, and is currently involved in the preparation of a seismotectonic atlas on 1:1 million scale in both the Himalayan region and the Peninsular shield of India. At present, GSI is in the process of establishing a digital telemetered seismic network at Khandwa in central India where earthquake swarms have been reported in the recent past.

The GSI also operates seismic stations in mobile mode to monitor the aftershocks of damaging earthquakes in the country in collaboration with other institutions like IMD, NGRI, and the Wadia Institute of Himalayan Geology. The Geological Survey of India has been extensively involved in the study of various earthquakes during 1990s, including Uttar Kashi earthquake of 1991, Latur 1993, Jabalpur 1997, and Chamoli 1999. Earlier, GSI studied all great earthquakes in India beginning from the 1897 Great Assam earthquake of June 12.

The scientists of GSI are also associated with seismic tomographic studies in northeast India. Their work has brought out interesting results of the velocity structure beneath the Shillong Plateau and Assam Valley areas. The low-velocity zones are inferred to be fault zones that are seismically active in the region. The GSI (Northern region), in a collaborative pilot project with BRGM, France, took up seismic hazard assessment of Kangra-Dharamshala area in Himachal Pradesh in north India. Utilizing a probabilistic approach, a detailed seismotectonic appraisal of north India has been made and ground acceleration maps of the region for different return periods generated.

4.5 Indian Institute of Technology, Roorkee

The Department of Earthquake Engineering, University of Roorkee (now, Indian Institute of Technology, Roorkee), has been associated with all the works related to the development and deployment of strong motion instruments in India since the early 1960s. Initially, the Structural Response Recorder (SRR) was developed, followed by the Roorkee Earthquake School

Accelerograph (RESA) in the late 1960s. The primary emphasis of the strong motion studies in India was to develop a methodology for determination of earthquake design parameters for important projects as dictated by the need of the country.

A.R. Chandrasekaran in 1987 presented the status of deployment of strong motion recorders in India. A description of development of analog accelerographs and method of data processing used to correct the recorded accelerogram was also presented. Status of the strong motion program, performance of instruments (SRR and accelerograph), and achievements of the strong motion instrument network installed and maintained by the Department of Earthquake Engineering, University of Roorkee was presented by Chandra *et al.* in 1996.

4.6 Regional Research Laboratory, Jorhat, Assam

In seismically active Northeast India, comprising the continent–continent collision boundary along Himalaya in the north and the Burmese subduction zone in the east, the Regional Research Laboratory (RRL) operates a six-element short-period seismic station network with its central station at Jorhat in addition to several stand-alone systems. The data provides information to study the seismotectonics of the region. In Manipur, a short-period analog seismograph and triaxial digital strong motion accelerograph are being operated by Department of Geology, Manipur University. In 1996 three more observatories are added to the existing station.

4.7 Universities and Other Departments

In the wake of recent damaging earthquakes in India, the Department of Science and Technology (DST) has provided funds to various universities to participate in the seismicity programs of the country. These universities are Benarus Hindu University (BHU) and Kumaun in Uttar Pradesh, Kurushetra in Harayana, Manipur in Northeast India, Delhi University and Indian School of Mines in Bihar. In addition to universities, DST also provides funding for seismology research in India to various institutions such as IMD, NGRI, and Wadia Institute of Himalayan Geology (WIHG), Dehradun.

Several networks of seismic stations are run by different river valley project authorities to monitor seismicity around water reservoirs and critical structure such as nuclear power plants.

79.29 Indonesia

The editors received a report, "Indonesia's Country Report on Earthquakes and Engineering Seismology," from Prof. M.T. Zen, Laboratory of Geodynamics, Geophysical Engineering Department, Bandung Institute of Technology, Bandung 40132, Indonesia. It is compiled by M.T. Zen, S. Sukmono, H. Mochtar, and A. Soehaimi. The edited version of the report is archived as a computer-readable file on the attached Handbook CD#2, under the directory of \7929Indonesia. We extracted and excerpted a brief summary here.

1. Introduction

The Republic of Indonesia consists of 17,508 islands, which together with the surrounding sea make up the Maritime Continent of Indonesia. Of the 8 million km^2 that constitute the national territory, only one-quarter is land. The Maritime Continent came into being through the convergence of the Pacific, Eurasian, and Indian Ocean-Australian plates. One consequence is that, geodynamically, it is one of the most active regions on Earth.

2. Seismicity and Seismotectonics

Studies of seismicity and seismotectonics have led to the definition of seismic source zones, a map of which was published by Kertapati *et al.* (1998). Only a small fraction of the territory has been labeled "stable" or "zone of infrequent small earthquakes." A comprehensive bibliography on the seismicity and seismotectonics of Indonesia is given in the full report on the Handbook CD.

3. Disaster Mitigation

Because most of Indonesia is subject to strong earthquakes, it has been imperative that seismic hazards be studied and mitigated, and there is a National Natural Disaster Mitigation Program in Indonesia. The National Coordinating Board for the Mitigation of Natural Hazards is headed by the Coordination Minister of People's Welfare. Working groups under this board include seismology, volcanology, landslides, flood control, forest fire control, and other industrial hazards. Ten related institutions are directly connected to the Board. The complete text of a paper, *Development of Seismic Hazard Map for Indonesia*, by the Indonesian Seismic Hazard Committee, 1999, is included in the full report on the Handbook CD.

4. Research Activities

The institutions involved in or interested in earthquake studies and engineering seismology include the National Coordinating Board for the Mitigation of Natural Hazards; the National Agency of Meteorology and Geophysics; the National Space Agency; the National Mapping Agency; the Agency for the Assessment and Application of Technology; the Geological Research and Development Center; the Human Settlement Development Center (Department of Public Works); the Bandung Institute of Technology (three departments); the Laboratory of Geotechnology (Indonesia Institute of Sciences); and Wiratman Associates (a private enginnering company). The personnel leading the activities in each of these organizations, as well as the main responsibilities, are given in the full report on the Handbook CD.

Earthquake and engineering seismology research is carried out in the Civil Engineering Department and the Geophysical Engineering Department of the Bandung Institute of Technology. The main topics are seismic risk analysis and earthquake zonation, both nationwide and microzonation for the city of Jakarta; liquefaction potential for the proposed north Jakarta reclamation; and seismic hazard and countermeasures for Bandung Municipality. Research on building safety from the earthquake disaster point of view is a program of the Institute for Human Settlement. The National Agency for the Assessment

INTERNATIONAL HANDBOOK OF EARTHQUAKE AND ENGINEERING SEISMOLOGY, VOLUME 81B *ISBN: 0-12-440658-0*

and Application of Technology carries out a seismic hazard assessment program, including the detection of movements on active faults, preparing a hazard map and contingency planning for the Sunda Strait region, and optimizing an early warning system (especially for volcanoes).

The engineering seismology programs are mainly concerned with data based on field observations. Emphasis is on the delineation of earthquake zones and microzonation in large cities.

5. List of Academic and Government Institutions

The following institutions, scientists, and engineers are involved in, interested in, or concerned with earthquakes and engineering seismology.

Bakornas Bencana Alam (National Coordinating Board for the Mitigation of Natural Hazard), under the Coordinating Minister for People's Welfare. Mr. H. B. Burhanudin. Coordination and supervision of all activities related to the mitigation of natural hazards.

BMG (National Agency of Meteorology and Geophysics). Dr. Gunawan Ibrahim. Operation of seismic stations distributed over the whole of Indonesia.

LAPAN (National Space Agency). Prof. Dr. Ir. Harjono Djojodihardjo. Research and development in space sciences and technology, including remote sensing technology.

Bakosurtanal (National Mapping Agency). Prof. Dr. Ir. Joenil Kahar. Coordination of all activities related to mapping.

BPP-Teknologi (National Agency for the Assessment and Application of Technology). Dr. Ir. Indroyono Soesilo, Dr. Ridwan Djamaludin. Assessment and application of technology for economic development.

GRDC (Geological Research and Development Center). Engkon Kertapati, Asdani Soehaimi, Herman Mochtar, Gurning. Geological mapping and research and development in the earth sciences, including the mitigation of geological hazards.

Human Settlement Development Center, Department of Public Works. Mr. Soepardiyono. Research and development in human settlement, building codes, and building safety.

Bandung Institute of Technology. Research and education.

- Civil Engineering Department: Prof. Dr. Ir. Azis Djajaputra.
- Geophysical Engineering: Prof. M.T. Zen, Dr. Agus Laesanpura, Dr. Sigit Sukmono, Dr. Darharta, Dr. Iwan T. Taib.
- Meteorology and Geophysics: Dr. Nanang T. Puspito, Dr. Sri Widiyantoro.

Laboratory of Geotechnology, Indonesia Institute of Sciences. Dr. Hery Haryono, Dr. Djedi Widarto.

References

Kertapati, E.K., A. Soehaimi, A. Djuhanda, and I. Effendi. "Seismotectonic Map of Indonesia," 2nd Ed. Geological Research and Development Centre, Bandung, 1998.

79.30 Iran

The editors received a brief summary of seismology in Iran, from Prof. Mohammad Reza Gheitanchi, IASPEI National Correspondent, Institute of Geophysics, Tehran University, Tehran, Iran. We present the edited version here (Section 1). We also received a brief report on "IIEES Mission and Major Achievements in Recent Years" from Prof. Mohsen Ghafory-Ashtiany, President of the International Institute of Earthquake Engineering and Seismology (IIEES), Tehran, Iran. It is archived as a computer-readable file in the attached Handbook CD#2, under the directory of \7930Iran. We extracted and excerpted a summary as shown in Section 2.

An inventory of seismic networks in Iran is included in Chapter 87, and biographical sketches of some Iranian scientists and engineers may be found in Appendix 3 of this Handbook.

1. Seismology in Iran

The principal institute, dealing with seismic activities as well as monitoring and reporting earthquake parameters, has been the Institute of Geophysics in Tehran University. The Institute is officially responsible for issuing prompt and instantaneous reports of strong earthquakes for the government. Monitoring of earthquakes in this Institute goes back to 1959, by means of two seismic stations in Tehran and Shiraz cities. In 1965 three WWSSN stations were installed in Tabriz, Shiraz, and Mashad. The last one was upgraded to SRO in 1978.

The Iranian long period array (ILPA) was another cooperative work, which started its operation in 1976. It consists of seven boreholes, broadband seismic telemetered stations, all located in southwest of Tehran, six on a circle, and one in the center, about fifty kilometers apart. During the same period, more stations are set up in Kermanshah, Mahabad, Kashan, Minoodasht, and Broojen, which are major cities in the active areas of the nation.

After the June 20, 1990, Rudbar disastrous earthquake (M_S 7.3), the government approved, on the request of Tehran University, developing local seismic networks in active provinces of Iran. In 1994 the installations started with 2 networks in Tehran and northwest of Iran. Tehran network at the first stage has 14 short period telemetric stations which are products of Nanometrics from Canada. A similar network has been installed in Tabriz, which is in northwest Iran. It has 8 seismic stations at the first stage. Those two networks are expected to be upgraded to several more stations as well as a few broadband seismometers. During the past 5 years seismic networks have been installed in Yazd (central Iran), Guchan (north of Khorasan Province), Sarry (Mazandaran Province in southern Caspian Sea), and Esphehan (central Iran). Each of them has 4 telemetered short period seismic stations and is planned to have more stations in future. Each network is designed to be able to keep in touch, for data exchange, with the other networks, based on the telephone lines. Several other seismic networks in Hamadan, southern Khorasan, and Fars Province along the Zagros are under construction. By operation of those networks it will be possible to locate local earthquakes in all active parts of Iran.

In addition to the Geophysics Institute of Tehran University, the Ferdowsi University (http://seismo.um.ac.ir/) in Khorasan Province and the International Institute of Earthquake Engineering and Seismology (http://www.iiees.ac.ir/) in Tehran are trying to install some broadband seismic stations in Khorasan and the other parts of the nation.

Strong motion accelerographs have been deployed in Iran since the 1970s by the Building and Housing Research Center (BHRC) of the Ministry of Construction (http://www.bhrc.gov.ir/). Recently the number of related instruments has grown to about 1000.

2. International Institute of Earthquake Engineering and Seismology

The International Institute of Earthquake Engineering and Seismology (IIEES) is a comprehensive international research

INTERNATIONAL HANDBOOK OF EARTHQUAKE AND ENGINEERING SEISMOLOGY, VOLUME 81B ISBN: 0-12-440658-0

institute in the field of earthquake studies. Based on the 24th UNESCO General Conference Resolution DR/250 and Iranian government approval in 1989, it was established as an independent institute under Iran's Ministry of Science, Research and Technology. The main goal of IIEES is seismic risk reduction and mitigation both in Iran and the region by promoting research and education in science and technology related to seismotectonics, seismology, and earthquake engineering. IIEES activity in research covers all aspects of the earthquake problem, from tectonic study to retrofitting complex structures, and education ranging from public education to a Ph.D. program in earthquake engineering. IIEES is composed of the following divisions: Seismology, Geotechnical Earthquake Engineering, Structural Earthquake Engineering, Risk Management, Graduate School, Public Education and Information.

IIEES action plans are seismotectonic and seismological research, developing the Iranian National Seismic Network as well as mobile seismic networks, providing online post-event information to the disaster management authorities, conducting research in geotechnical microzonation and seismic safety of structures, developing aseismic design methods, guidelines, and codes, vulnerability and risk assessment of cities, promoting earthquake safety, prevention, and preparedness, education and training by offering M.S. and Ph.D. graduate program and specialized courses, providing consultancy to the government and industries, and cooperation and exchange with international and regional organizations and institutions. The IIEES recent contributions and accomplishments are as follows:

Seismotectonic Studies: major active faults map of Iran (scale 1:1,000,000) and of Tehran (scale 1:100,000), paleoseismological study of active faults such as Tabriz, North Tehran, and Mosha faults, investigation of active tectonics of Alborz, Zagros, and Makran, detailed study of the special and important sites, such as the petrochemical city in Asaluyeh, power plants, etc.

Seismology Investigations: seismic monitoring by a broadband national seismic network, seismicity catalog of Iran (historical catalog and post 1900), global and local GPS measurements, and micro-seismic monitoring in various seismotectonic regions of Iran, such as Zagros Belt (Busheher to Yazd) and Alborz, for the purpose of crustal and upper mantle studies.

Seismic Hazard Analysis and Geotechnical Studies: processing and analysis of the strong motion data recorded in Iran since 1975, development of attenuation relationship, seismic hazard zoning map of Tehran, geotechnical microzonation of Tehran, microzonation of various provinces and industrial cities, topographical effects on free-field strong motions, behavior of compacted composite clays, ambient and forced vibration studies on two embankment dams, and guideline for seismic design of soil structures.

Investigations of Building, Lifeline and Special Structures: seismic vulnerability functions for typical Iranian buildings, guidelines for vulnerability assessment of common building type, strengthening schemes for the typical Iranian steel structures, seismic vulnerability of Tehran, economical study on earthquake-resistant structures, upgrading the technical knowledge of engineers, studies concerning all lifeline systems, vulnerability assessment of the Tehran Water System, guidelines for seismic design of a gas network, forced vibration analysis of typical bridges, offshore structures, concrete dams, etc., vulnerability assessment of oil and chemical facilities, and a risk awareness program for oil and chemical industry managers.

More details of the preceding activities, as well as activities for public education and awareness, graduate studies and international cooperation are given in the full report on the Handbook CD. More current information is available at the IIEES Web site: http://www.iiees.ac.ir/.

79.31 Israel

The editors received two reports from Israel. The first is "Earthquake and Engineering Seismology in Israel" from Dr. Avi Shapira, IASPEI National Correspondent, Geophysical Institute of Israel, Lod 71100, Israel. The second is "Seismology and Physics of the Earth's Interior at The Weizmann Institute of Science, Rehovot, Israel, 1954-2000" from Prof. Ari Ben-Menahem, Department of Computer Science and Applied Mathematics of the Weizmann Institute of Science, Rehovot 76100, Israel.

Both reports are archived as computer-readable files on the attached Handbook CD#2, under the directory \7931Israel. An inventory of the Israel Seismic Network is included in Chapter 87. A biography of C.L. Pekeris is included in Chapter 89, and biographical sketches of some scientists from Israel may be found in Appendix 3 of this Handbook.

1. Earthquake and Engineering Seismology in Israel (A. Shapira)

Earthquakes are phenomena long known in the Holy Land. The most recent catalogs are those of Ben-Menahem (1979, 1981) and Amiran, *et al.* (1994). The earthquake catalog that covers the seismic activity along the Dead Sea fault system was recently compiled and unified through a joint effort of the Seismology Divisions of the Geophysical Institute of Israel and the Natural Resources Authority of Jordan.

C. Lomnitz, in collaboration with the Geological Survey of Israel (GSI), installed the first seismometers in Israel in 1954, located in Jerusalem, Safed (in the Galilee) and in Haifa. Milestones in seismicity monitoring of the region are the installation in 1964 of the WWSSN station in Jerusalem (JER), operated by the GSI, and the establishment of the Adolpho Bloch Geophysical Observatory of the Weizmann Institute of Science near Eilat (EIL) in 1969. A. Ben-Menachem of the Weizmann Institute of Science and E. Arieh of the Geological Survey of Israel performed a pioneering work in studying the seismic activity of the Middle East in general and that of the Dead Sea transform in particular, and set the foundations for modern seismological research in Israel. A revised and updated earthquake catalog of earthquakes in Israel and adjacent areas, 1900–1980, by Arieh, *et al.* (1985) is the first comprehensive and unified compilation of instrumental data.

The Seismology Division of GII operates the seismic monitoring systems of Israel, provides earthquake hazard assessments to the private and the public sectors, and is involved in a wide spectrum of disaster reduction projects. From a small network of 6 stations in 1979, the Israel Seismic Network (ISN) has expanded to a network of 32 seismic stations. Y. Bartov and his colleagues from the Geological Survey of Israel have compiled a map that presents the faults that are assumed active. This information, together with the seismological data, was used to identify the seismogenic zones in and around Israel. These studies have lead to the updating of the Israeli building code.

The first mandatory anti-seismic building code was introduced in 1971. It was modified in 1995 and awaits another modification in 2003. In the last decade (1992–2002), the Seismology Division conducted a series of site response investigations across Israel. These studies revealed the great effects that the soft soils may have on the amplitudes and frequency content of the seismic ground motions. An organized effort to enhance preparedness and improve emergency management in a case of an earthquake catastrophe is now being implemented.

Israeli seismologists are now actively involved in research work associated with improving seismic monitoring, enhancing detection capability of seismic networks, and location accuracy and discrimination between man-made seismic events and earthquakes. The Seismology Division of GII has conducted the Dead Sea Calibration Shots Experiment that was used to calibrate travel-time models for the East Mediterranean and the Middle East. The full report on the Handbook CD includes a list of Israeli institutions related to earthquake research and preparedness.

INTERNATIONAL HANDBOOK OF EARTHQUAKE AND ENGINEERING SEISMOLOGY, VOLUME 81B ISBN: 0-12-440658-0

2. Seismology and Physics of the Earth's Interior at The Weizmann Institute of Science, Rehovot, Israel, 1954–2000 (A. Ben-Menahem)

Instrumental seismology began in Israel in May 1954: At the initiative of Profs. Beno Gutenberg and Hugo Benioff from Caltech Pasadena, US, two short-period seismograph systems with photographic recording were installed in Jerusalem and Safed under the auspices of the National Physical Laboratory (NPL) of the Research Council at the Prime Minister's Office.

Cinna Lomnitz, then Benioff's graduate student at Caltech (now professor at the National University of Mexico), was sent to Israel to install the equipment and train Ari Ben-Menahem (then research assistant at NPL) in the operation of the stations and their maintenance. When Lomnitz returned to Caltech in May 1954, regular seismic recording commenced under the directorship of Ben-Menahem, who became the first instrumental seismologist in Israel. In 1958 Ben-Menahem left for graduate studies at Caltech, and returned in 1965 as a professor of geophysics at the Weizmann Institute of Science.

In 1965, A. Ben-Menahem and C. L. Pekeris decided to establish a modern geophysical laboratory in Israel, capable of recording a wide variety of signals. A donation by Adolpho Bloch was essential to covering the initial budget. A suitable site was found in a granite massif 12 km north of Eilat. The equipment included strainmeters, tiltmeters, short- and long-period seismographs, and a microbarograph. From 1976 to 1982, the Bloch Observatory was supplemented with a network of 22 stations to record microearthquakes along the Dead Sea fault.

During 1954–2000, research in seismology and physics of the Earth's interior was conducted at the Weizmann Institute of Science by two groups.

A group headed by Prof. C. L. Pekeris conducted research centered on the mathematical geophysics topics: (1) theoretical seismograms—Lamb's problem, (2) free oscillations of the Earth—eigenperiods, (3) tides of the ocean, and (4) internal constitution of the Earth.

A group headed by Prof. A. Ben-Menahem did research centered on analyses of data acquired from the Adolpho Bloch Geophysical Observatory in Eilat and worldwide networks. Their main topics of research were (1) fault dynamics and mechanism of earthquakes, (2) seismotectonics and seismicity along the Dead Sea transform province, (3) explosion seismology, (4) wave propagation in complex media, (5) earthquakes and the Earth's rotation—the Chandler wobble, (6) inversion of seismic data, (7) theoretical seismograms for realistic source and Earth models, (8) amplitudes of free oscillations from seismic sources, and (9) history of singular geophysical events.

References to publications documenting the results of this research and biographical sketches of Pekeris and Ben-Menahem are included with the full report on the Handbook CD.

References

Amiran, D., E. Arieh, T. Turcotte (1994). Earthquakes in Israel and adjacent areas: Macroseismic observations since 100 B.C.E. *Israel Exploration Journal*, **44**, 261–305.

Arieh, E., D. Arzi, N. Bendik, A. Eckstein, R. Issakov, B. Reich, A. Shapira (1985). Revised and updated catalog of earthquakes in Israel and adjacent areas (1900–1980). *GII report Z6/1216/83(3).*

Ben-Menahem, A., (1976). Dating of Historical Earthquakes by mud-profiles of lake-bottom sediments. *Nature*, **262**, 201–202.

Ben-Menahem, A., (1979). Earthquake Catalogue for the Middle East. 92 BC–1980 AD. *Boll. Geofisica,* **21**, 245–310.

Ben-Menahem, A., (1991). 4000 years of seismicity along the Dead-Sea Rift. *J. Geophys. Res.,* **96**, 17325–17351.

Ben-Menahem, A., E. Aboodi, and R.L. Kovach, (1977). Rate of seismicity of the Dead-Sea Region over the past 4000 years. *Phys. Earth Planet. Inter.,* **14**, 17–27.

Ben-Menahem, A., and D. Brooke, (1982). Earthquake Risk in the Holy Land. *Boll. Geofisica*, **24**, 175–203.

79.32 Italy

The editors received a summary of "Italian Seismology in the 20th Century" from Dr. Gianluca Valensise, Istituto Nazionale di Geofisica e Vulcanologia, Rome. It is written by Enzo Boschi (IASPEI National Correspondent) and Gianluca Valensise, and an edited version is shown here. We also received two reports from Prof. Giuliano F. Panza, Department of Earth Sciences, University of Trieste, Via Weiss, 4, I-34127 Trieste, Italy. They are titled: (1) "Seismology at the Department of Earth Sciences, University of Trieste" by G. F. Panza and P. Suhadolc, and (2) "Structure and Non-Linear Dynamics of the Earth (SAND Group), The Abdus Salam International Centre for Theoretical Physics, Trieste, Italy" by G. F. Panza, A. Soloviev, and A. Aoudia. These two reports are archived as computer-readable files on the attached Handbook CD#2, under the directory of \7932Italy. We extracted and excerpted a summary here.

An inventory of seismic networks in Italy is included in Chapter 87. Biographies and biographical sketches of some Italian scientists and engineers may be found in Chapter 89, and in Appendix 3 of this Handbook, respectively.

1. Italian Seismology in the 20th Century (E. Boschi and G. Valensise)

It has been over a century since Italy established what was to become a leading tradition for seismology and seismological observation worldwide. At the time of the earthquake catastrophe of the Messina Straits in 1908, Italy was already one of the most densely monitored countries in the world, many of the instruments having been designed and built by Italian seismologists. The 1908 tragedy also prompted new efforts for the rating and classification of earthquake effects, ultimately leading to the Mercalli-Cancani-Sieberg scale (1912). Important and exciting theoretical and observational advancements were regularly disseminated through the Bollettino della Società Sismologica Italiana, one of the world's oldest geophysical journals (from 1895).

Subsequent political and economic circumstances did not favor the development of Italian seismology, which slowed its pace despite a series of further catastrophic earthquakes in the years 1915–1920 and 1930–1933. Things were to take a sudden twist in 1936, when Sicilian physicist Antonino Lo Surdo founded the Istituto Nazionale di Geofisica (I.N.G.) as a division of the National Research Council (C.N.R.), which was then under the guidance of Guglielmo Marconi. Lo Surdo, a renowned experimental physicist, became interested in seismology and geophysics after the 1908 Messina earthquake had killed most of his family members.

I.N.G.'s main task was "Promoting, executing and coordinating studies and research on geophysical phenomena and their applications," which it has done for many years. The new institution, however, gained momentum only after WW II, when it became a national leader, not only for seismology but also for research on the Earth's geomagnetism and on the ionosphere. After leaving C.N.R. in 1945, I.N.G. gradually expanded its observational network by installing seismometers in various Italian universities and other geophysical instrumentation in specifically built observatories. At the end of the 1960s I.N.G. started creating a National Seismological Network, partly based on its own facilities and partly on local networks run by the universities and other institutions. This process culminated in 1982, when modern telephone-wire telemetry allowed about 20 seismic traces to be recorded and analyzed in real time in the Rome headquarters for civil defense purposes.

In 1958 the Italian Ministry of Public Education established the Osservatorio Geofisico Sperimentale di Trieste, which carries on a long tradition of geophysical and oceanographic observations and was put in charge of promoting seismic exploration both inland and at sea. Research on the structure of the lithosphere and on the seismicity of active volcanoes was carried out

INTERNATIONAL HANDBOOK OF EARTHQUAKE AND ENGINEERING SEISMOLOGY, VOLUME 81B ISBN: 0-12-440658-0

by C.N.R. divisions created at the end of the 1960s. Following the large Friuli (northern Italy) earthquake of 1976, the Servizio Sismico Nazionale of the Ministry of Public Affairs took on the important task of revising and enforcing updated and more detailed seismic safety regulations.

The catastrophic 1980 Irpinia earthquake also prompted significant innovations, concerning both basic research and its immediate applications. Starting in 1982 I.N.G. was profoundly reorganized and its research potential substantially extended. Significant research achievements came as a result of a massive project funded by Italy's electricity company for siting nuclear facilities country-wide. In 1986 C.N.R.'s Gruppo Nazionale Difesa dai Terremoti took on the task of reassessing the seismic hazard and promoting new policies for risk mitigation on a national scale. In 1988 I.N.G. launched MedNet, a state-of-the-art broadband network straddling the opposite shores of the Mediterranean. The seismological community substantially increased its visibility in the main geophysical journals and its participation in the main international forums.

The new millennium brought about another major step in the organization of Italy's seismological research. With a decree signed on January 10, 2001, the Italian government gave birth to the Istituto Nazionale di Geofisica e Vulcanologia (I.N.G.V.). A young but dynamic institution formerly under the C.N.R., along with older distinguished research centers such as former I.N.G. and Osservatorio Vesuviano, I.N.G.V. merged into one of the largest institutions worldwide for the investigation of natural phenomena and mitigation of their adverse effects.

Further information may be found on the I.N.G.V. Web site, http://www.ingv.it/.

2. Seismology at the Department of Earth Sciences, University of Trieste (G. F. Panza and P. Suhadolc)

Seismology in Trieste dates back to Austro-Hungarian times, when the first seismograph was installed in the city, but the station was abandoned soon after World War I. A seismological observatory with an autonomous administration, Istituto Geofisico di Trieste, started operating on March 8, 1931. It was equipped with Wiechert seismographs, which were disconnected after the WWSSN TRI-117 station started to operate on July 29, 1963. TRI-117 operated continuously at the Osservatorio Geofisico Sperimentale, now Istituto Nazionale di Oceanografia e Geofisica Applicata (OGS), until December 1996, when the WWSSN was closed.

Research in seismology started at the Istituto di Geodesia e Geofisica (Institute of Geodesy and Geophysics), University on Trieste, during the 1960s and early 1970s, with a series of publications by Antonio Marussi, Director of the Institute, and Michele Caputo and Maria Zadro, professors, related to free oscillations of the Earth. Research in seismology was fostered by the occurrence of the Friuli 1976 earthquake. Claudio Ebblin, a researcher at the Institute, and Maria Zadro, published some papers on still-unexplained, possible earthquake precursory signals. Claudio Chiaruttini, a researcher at the Institute, first investigated strong motion data in collaboration with L. Siro. In particular they proposed the first PGA attenuation with distance relation for the Friuli area. Chiaruttini also studied general attenuation and tectonic properties of the Friuli area.

In the late 1970s Peter Suhadolc, who started his research in seismology at the Osservatorio Geofisico Sperimentale after the 1976 earthquake, was invited to collaborate with the Institute as researcher in seismology. He held seismology lectures at the "Scuola di perfezionamento in fisica," a graduate school in physics, and investigated the September 16, 1977, Friuli earthquake both in terms of the spatial distribution and time evolution of its aftershocks.

In 1980 Giuliano Francesco Panza was invited to Trieste as the Chair of geophysical prospecting, subsequently moving to the Chair of seismology. He started the Seismology and Physics of the Earth's Interior group and introduced broadband seismology in Italy (VBB station TTE). Michele Cuscito and Claudio Chiaruttini, researchers, and a few students formed the initial core of the group. In collaboration with Fred Schwab, Panza started to develop the theory and related computer codes, based on Leon Knopoff's method, to automatically compute, given a stratified anelastic medium, the frequency-domain information for all modes in a phase velocity–frequency range needed to calculate synthetic seismograms. Peter Suhadolc soon joined him in this endeavour, followed later by Giovanni Costa and Franco Vaccari, two Ph.D. students (now researchers). In 1984–1986 the first synthetic seismograms computed via modal summation were used to analyse deep seismic sounding data and waveforms of the Borrego Mountain event. In 1984 the work on programming was essentially finished. Suhadolc and Chiaruttini, under the guidance of Panza, applied the synthetic seismograms in engineering seismology to supply a theoretical explanation for the variability of strong ground motion; in particular, the increase of the PGA with distance, due to the supercritical S-wave Moho reflection.

A few years later, Donat Faeh, now at ETH (Zurich), joined the group as a Ph.D. student, and under Panza's supervision, and a hybrid method was developed that gives the possibility to compute seismograms in laterally heterogeneous anelastic structures, combining modal summation and finite-difference schemes. Theory and related codes for the *analytical* description of wave propagation in laterally varying media have been developed by Panza in collaboration with Soren Gregersen, Franco Vaccari, Fred Schwab, Tatiana Yanovskaya, and Fabio Romanelli, a Ph.D. student (now researcher). This kind of research, recently extended to tsunami waves, is still in progress.

The collaboration between G. F. Panza, L. Knopoff, and A. Levshin has led to work on the determination of structural models of the Earth from dispersion curves of surface waves.

This research, done in collaboration with the late Stephan Mueller from ETH (Zurich) and with Gildo Calcagnile from the University of Bari, permitted delineation of the lithosphere-asthenosphere system in the Mediterranean and discovery of deep roots (vertical subduction) under the Alps. The follow-up of this research, work on the geodynamic models for the Mediterranean and in the Italian area in particular, is still in progress.

An impetus in earthquake studies related to earthquake prediction problems gained momentum with the fruitful collaboration with the research group led by V.I. Keilis-Borok from the International Institute of Earthquake Prediction Theory and Mathematical Geophysics (Russian Academy of Sciences, Moscow, Russia). Keystones in this process were workshops regularly held at the ICTP since December 1983. The collaboration with the International Institute of Earthquake Prediction Theory and Mathematical Geophysics has been subsequently extended to the study of seismic wave generation, propagation, and modeling, involving Tatiana Yanovskaya from St. Petersbourg and Alexander Gusev from Petro-Kamchatskii. The seismology group closely collaborates with the "Structure and Non-Linear Dynamics of the Earth" (SAND Group) established at ICTP in 1991 (see next section). In particular, monitoring of seismicity is made with intermediate-term medium-range prediction algorithms CN and M8 with the active participation of Vladimir Kossobokov, Ira Rotwain, Lina Romashkova, and Antonella Peresan. Collaboration with ICTP also materialized in the development of the International IUGS-UNESCO-IGCP Project 414 "Realistic Modeling of Seismic Input for Megacities and Large Urban Areas."

The collaboration of Giuliano F. Panza with Jan Sileny, from Prague, and of Peter Suhadolc with Shamita Das, from Oxford, has led, with the active engagement of Angela Sarao, to the development of waveform inversion methods suitable to retrieve the full moment tensor of point sources and the rupture parameters of extended sources, respectively. Full moment inversions have been intensively performed in volcanic areas (Etna, Campi Flegrei, and Vesuvio) with the purpose of identifying possible seismic precursors of eruptions. Surface-wave tomography has been performed to define crustal and mantle magma reservoirs in volcanic areas (Campi Flegrei and Vesuvio).

In the beginning of the1990s, in the framework of two European Commission projects, the Friuli Accelerometric Network (RAF) was installed in the Friuli (NE Italy) seismic area. This was the first digital strong motion network with absolute timing at a regional scale established in Italy.

In recent years work has been performed on Italian seismicity and important Italian earthquakes, in particular, at the Alps-Dinarides contact. Active tectonics, GIS-based analysis, and GPS measurements are the latest fields of study carried on in the framework of relevant national projects funded by MIUR. In the same framework a large project, led by G. F. Panza, of surface-wave tomography of the Mediterranean area is still in progress.

A quite active project is under development in the framework of the Programma Nazionale Ricerche in Antartide (PNRA) for the comparative analysis of the lithosphere–asthenosphere geophysical properties in the Scotia sea, Caribbean, and Tyrrhenian sea and for the study of the stress regime in the Scotia sea region.

International training has always been an important task of the group, especially in connection with ICTP activities. Several trainees from developing countries have been hosted by the group, supported by the ICTP Training and Research in Italian Laboratories programme. The group grew and is flourishing with the help of several graduate students who got their Ph.D. and stayed as post-docs or collaborators. More details about the Department, including a chronological list of publications in seismology (1960–2001), are given in the full report on the Handbook CD. The Department Web site is: http://www.dst.units.it/.

3. Structure and Non-Linear Dynamics of the Earth (SAND Group) (G. F. Panza, A. Soloviev and A. Abdelkrim)

The group "Structure and Non-Linear Dynamics of the Earth" (SAND Group), is based at the Abdus Salam International Centre for Theoretical Physics (the Abdus Salam ICTP) in Trieste, Italy. The activities of the Centre have been described in the September 2001 issue of *Physics Today*. The SAND Group activities were initiated in 1991 to carry out research and educational activities in physics of the solid Earth. The objectives of the group are to (1) develop a new theoretical base and computational framework for the study of critical phenomena in the Earth's lithosphere, with special attention to their predictability, (2) develop a new approach for seismic risk mitigation on the basis of 3-D modeling of Earth structure and earthquake sources, through the study of wave propagation in three-dimensionally heterogeneous, inelastic, and anisotropic media, and (3) transfer the developed methodology to scientists of the Third World, which is achieved through joint research, with special attention to training the potential leaders, and by combining the workshops with subsequent individual projects.

The activities under SAND Group are divided into two main lines: (1) Nonlinear dynamics of the Earth's lithosphere, and (2) Structure of the Earth with application to seismic risk mitigation.

The first research line is led by V.I. Keilis-Borok (Russian Academy of Sciences, Moscow, Russia) and includes the following practical goals: (1) improving our understanding of earthquake mechanisms and dynamics of seismicity, (2) advancing earthquake prediction techniques, and (3) advancing probabilistic seismic hazard assessment. The earthquake-generating lithosphere is regarded as a hierarchical, nonlinear, dissipative system and earthquakes are considered as critical transitions in it. Models of dynamics of seismicity are developed and their integration with the phenomenology of earthquake occurrence should help to overcome the difficulties connected with the absence of fundamental constitutive equations and the impossibility of direct measurements at the depth in the lithosphere, where the earthquakes originate.

The second research line is led by G.F. Panza (University of Trieste, Italy). As a vehicle for carrying out the Abdus Salam ICTP mandate to serve science and scientists of developing countries, the Task Group II-4 "Three-Dimensional Modeling of the Earth's Tectosphere" (3DMET) of the IUGG-IUGS Commission of the Lithosphere/International Lithosphere Program (ILP) has been organized. 3DMET has originated the UNESCO-IUGS-IGCP project "Realistic Modeling of Seismic Input for Megacities and Large Urban Areas," presently in progress, centered at the Abdus Salam ICTP. This project addresses the problem of pre-disaster orientation: hazard prediction, risk assessment, and hazard mapping, in connection with seismic activity and man-induced vibrations. Activities are in progress in the following key urban areas and megacities: Algiers, Al Hoceima, Antananarivo, Bangalore, Beijing, Bucharest, Budapest, Cairo, Catania, Damascus, Delhi, Kathmandu, Ljubljana, Meknès, Mexicali, Mexico City, Naples, Rome, Santiago de Chile, Santiago de Cuba, Silistra, Sofia, Thessaloniki, Tijuana, and Zagreb. This choice is representative of a broad spectrum of seismic hazard levels.

The full report on the Handbook CD includes an extensive reference list and a chronological list of SAND Group publications, 1991–2001. The Group's Web site is: http:\\www.ictp.trieste.it/sand.

79.33 Japan

We received the Centennial Report of Japan from the Japanese National Committee for IASPEI (edited by M. Ishida). This report has five sections: (1) Introduction by K. Shimazaki, (2) Historical development of seismology by T. Utsu, (3) Biographies of 62 deceased Japanese seismologists and earthquake engineers by S. Miyamura, (4) Biographical sketches of 384 Japanese seismologists, earthquake engineers, and earth scientists, and (5) 27 institutional reports including the Seismological Society of Japan. The complete report is archived as computer-readable files on the attached Handbook CD#2, under the directory of \7933Japan.

A summary was extracted and excerpted from the complete report and is presented in the following sub-sections. Please also see Chapter 80.10 for the "Centennial Report of Japan in Earthquake Engineering" submitted by the Japanese National Committee for IAEE. An inventory of Japanese seismic networks is included in Chapter 87. Biographies and biographical sketches of some Japanese scientists and engineers may be found in Chapter 89, and in Appendix 3 of this Handbook, respectively.

1. Introduction (K. Shimazaki)

The Japan National Committee for IASPEI has prepared this national report encompassing the programs and activities in seismology, the physics of the Earth's interior, and earthquake engineering during the past 100 years. The report includes a review of the historical developments in seismology and reports from 20 institutions or research organizations, some of which are universities with more than one relevant department or research institute, for a total of 27 institutional reports.

The oldest Japanese earthquake described in an historical document occurred in 416 and classical literature contains many references to earthquakes. However, the history of modern seismology in Japan started in 1872 when G. F. Verbeck, a Dutch professor at Kaiseigakko (former University of Tokyo), and E. Knipping, a German meteorologist, independently began seismological observations in Tokyo (Usami and Hamamatsu, 1967). The Seismological Society of Japan, the world's oldest, was established in 1880.

The formation of new organizations devoted to earthquake studies has, unfortunately, been associated with the occurrence of seismic disasters in the country. In 1892 the Imperial Earthquake Investigation Committee was established for investigation of the prevention of earthquake disasters, in the wake of the devastating M8.0 Nobi earthquake of 1891 (commonly known as the Mino-Owari earthquake outside of Japan.), The Committee was created especially for investigating whether there are any means of predicting earthquakes and what structures and materials are best fitted to resist such shocks, themes that remain the subjects of research. They proceeded through (1) collection of facts concerning earthquakes, seismic waves, eruptions, etc.; (2) geologic investigations of the cause of earthquakes and relation of earthquakes to geological formations, etc.; (3) investigations connected with the nature of earthquakes, especially the law of propagation of waves; (4) magnetic observations; (5) underground temperature observations; (6) gravity observations; (7) latitude observations; (8) testing the strength of various building materials; and (9) testing various structures and joints (Kikuchi, 1897).

The great Kanto earthquake of 1923 spurred the development of the Earthquake Research Institute at the University of Tokyo in 1925. After the Hyogoken-Nanbu (Kobe) earthquake in 1995, the Headquarters of Earthquake Research Promotion was established in the office of the Prime Minister. Since then observation systems have been upgraded and now Japan is probably the most densely instrumented earthquake country in the world.

Historical developments of seismology in Japan were not necessarily straightforward and sometimes took a roundabout path. One good example is a study of the earthquake source. A good start was made by Shida's pioneering work in 1917 (Shida, 1929), which was followed by a theoretical study by Nakano (1923). Combining these works with some observational results indicating that one of the nodal lines runs parallel to the fault strike, we might conclude that our present view on the

INTERNATIONAL HANDBOOK OF EARTHQUAKE AND ENGINEERING SEISMOLOGY, VOLUME 81B ISBN: 0-12-440658-0

earthquake source was conceived more than 70 years ago. But somehow the actual history shows a more complicated path of development. An important item we consider at present might be seen as trivial in the future. Actually, many accomplishments described in an earlier review (e.g., Terada and Matuzawa, 1926) do not necessarily impress us at present. Most of the works represent one of many steps in the progress of research, which may not be distinctive retrospectively.

Many excellent reviews of seismology in Japan have been published, but most of them are in Japanese and may not be easily accessible from abroad. They include special issues of the Journal of the Seismological Society of Japan, i.e., the special issues "Seismology in Japan" and "A Hundred Years of Japanese Seismology" edited by Y. Sato (1967; 1981) and "Seismology in the 1980s" edited by Y. Fukao (1991). Several books in Japanese also provide good guides for an early history of seismology in Japan (Hagiwara, 1982; Ikegami, 1987; Fujii, 1967; Hashimoto, 1983; Yamashita, 1989). Exceptions are "Seismology in Japan" (Geller *et al.*, 1995), which is a 500-page summary of research and observational programs carried out by the seismological community in Japan in recent years, and earlier reviews by Terada and Matuzawa (1926) and Kawasumi (1937). Wadati (1989) briefly described his research career in "Born in a country of earthquakes" and a life of John Milne is depicted in "Father of Modern Seismology" (Herbert-Gustar and Nott, 1980; Usami, 1982). Also "Seismic Activity in Japan—Regional perspectives on the characteristics of destructive earthquakes—(excerpt)" (Earthquake Research Committee, 1998) shows seismicity of Japan with color maps prepared for lay readers. "Seismology and Earthquake Engineering in Japan" (Committee for Study on Seismicity and Seismic Hazard, 1998) contains descriptions of Japanese regulations related to building codes, earthquake warning, and earthquake disaster countermeasures.

Responding to the call for contributions from the IASPEI Committee on Education, the Japan National Committee for IASPEI (chairman: K. Shimazaki; members: M. Ando, N. Fujii, A. Hasegawa, K. Hirahara, M. Ishida, K. Oike, Y. Okada, and H. Shimamura) decided to compile this national centennial report in 1999. Dr. Mizuho Ishida was in charge of editorial works and all members of the committee participated in reviewing the manuscript. Without Dr. Ishida's endeavor this national report would not have been completed. The following institutional reports are listed in alphabetical order of the English names of the parent organizations.

The full Japanese National Report, including the details of the following institutional report summaries, with histories, organizational structures, bibliographies and biographies, is to be found on the attached Handbook CD#2, under the directory of \7933Japan.

References

Committee for Study on Seismicity and Seismic Hazard (1998). "Seismology and Earthquake Engineering in Japan," Property and Casualty Insurance Rating Organization of Japan, 237 pp.

Earthquake Research Committee (1967). Seismic activity in Japan—Regional perspectives on the characteristics of destructive earthquakes—(excerpt), Science and Technology Agency.

Fujii, Y. (1967). "Seismology in Japan" (translated from Japanese title), Kinokuniyashoten, 239 pp.

Fukao, Y. (editor) (1991). Seismology in the 1980s, Special Issue, *J. Seismol. Soc. Japan*, **44**, 1–405.

Geller, R. J., K. Hirahara, M. Ishida, I. Nakanishi, and E. Ohtani (Eds.) (1995). Seismology in Japan, *J. Phys. Earth*, **43**, No.3, 4, and 5.

Hagiwara, T. (1982). "Hundred Years of Seismology" (translated from Japanese title), Univ. Tokyo Press, 233 pp.

Hashimoto, M. (1983). "Opening of Seismology in Japan—Life of Seikei Sekiya, a pioneer" (translated from Japanese title), Asahishinbunsha, 261 pp.

Herbert-Gustar, A. L., and P. A. Nott (1980). "John Mile, Father of Modern Seismology," Paul Norbury Publications Ltd., Kent, England, 199 pp.

Ikegami, R. (1987). "Searching for a Hypocenter: Development of modern seismology" (translated from Japanese title), Heibonsha, Tokyo, 258 pp.

Kawasumi, H. (1937). An historical sketch of the development of knowledge concerning the initial motion of an earthquake, *Publications du Bureau central seismologique international, series A, Travaux Scientifique*, **15**, 1–76.

Kikuchi, D. (1897). Preface, *Pub. Earthq. Invest. Comm.*, **1**, 1–3.

Nakano, H. (1923). Notes on the nature of the forces which give rise to the earthquake motions, *Seismol. Bull. Central Meteorological Observatory Japan*, **1**, 92–122.

Sato, Y. (Ed.) (1967). Seismology in Japan (in Japanese), Special Issue, *J. Seismol. Soc. Japan*, **20**, No. 4, 1–326.

Sato, Y. (Ed.) (1981). A Hundred Years of Japanese Seismology (in Japanese), *J. Seismol. Soc. Japan*, **34**, 1–207.

Shida, T. (1929). Recollection of researches on the rigidities of the earth and of its crust and earthquake motions (translated from Japanese title), *Toyogakugeizasshi*, **45**, 275–289.

Terada, T., and T. Matuzawa (1926). A historical sketch of the development of seismology in Japan, Scientific Japan past and present, Third Pan-Pacific Science Congress, Tokyo.

Usami, T., and O. Hamamatsu (1967). History of earthquakes and seismology (in Japanese), In Sato Y. (Ed.), Seismology in Japan, Special Issue, *J. Seismol. Soc. Japan*, **20**, No.4, 1–34.

Wadati, K. (1989). Born in a country of earthquakes, Ann. Rev. Earth Planet. Sci., 17, 1–12.

Yamashita, F. (1989). "A Life of Akitsune Imamura; a pioneer of earthquake prediction" (translated from Japanese title), Seijisha, Tokyo, 316 pp.

2. Historical Development of Seismology in Japan (T. Utsu)

The development of seismographs and the installation of a nationwide seismic network during the late 19th and early 20th centuries in Japan contributed greatly to the investigation of spatial and temporal distributions of earthquakes, including the first confirmation of deep-focus earthquakes. The dense network also facilitated earthquake mechanism studies from radiation patterns of seismic waves, which led to the double-couple source model in the 1930s. Theoretical studies of seismic wave generation and propagation have been intensively carried out in Japan since the 1920s. Other important studies include studies of crustal deformation related to major earthquakes, statistical studies of earthquake occurrences (e.g., the power-law distribution in space, time, and energy), studies of Earth structure (e.g., lateral heterogeneity beneath island arcs). Advances in seismology of Japan after the establishment of plate tectonics are not described, because the contributions in this period are too numerous and diverse to review individually, but most of them have received international recognition.

This review is presented with the following main sections: the late 19th century, the early 20th century before the 1923 Kanto earthquake, between the 1923 Kanto earthquake and World War II, two decades preceding the establishment of plate tectonics, a brief note on seismology in Japan in the past three decades. The full version is to be found on the Handbook CD with the list of references documenting this history.

3. Biographies of Deceased Japanese Seismologists and Earthquake Engineers (S. Miyamura)

Biographies of the following 62 deceased Japanese seismologists and earthquake engineers are included in the full report on the Handbook CD: (1) Ban, Shizuo (1896–1989), (2) Haeno, Seizô (1906–1942), Hagiwara, Takahiro (1908–1999), (4) Hattori, Ichizô (1851–1929), (5) Hisada, Toshihiko (1914–1988), (6) Homma, Shôsaku (1913–1953), (7) Honda, Hirokichi (1906–1982), (8) Iida, Kumiji (1909–2000), (9) Imamura, Akitsune (1870–1948), (10) Inouye, Win (1905–2000), (11) Ishimoto, Mishio (1893–1940), (12) Kawasumi, Hiroshi (1904–1972), (13) Keimatsu, Mitsuo (1907–1976), (14) Kikuchi, Dairoku (1855–1917), (15) Koto, Bunjiro (1856–1935), (16) Kubo, Keizaburo (1922–1995), (17) Kusakabe, Shirôta (1875–1924), (18) Matuzawa, Takeo (1902–1989), (19) Miki, Haruo (1923–2000), (20) Minai, Ryô'ichirô (1930–1991), (21) Minakami, Takeshi (1909–1985), (22) Minami, Kazuo (1907–1984), (23) Minami, Tadao (1940–1999), (24) Mononobe, Nagaho (1888–1941), (25) Musha, Kinkichi (1891–1962), (26) Muto, Kiyoshi (1903–1989), (27) Naito, Tachû (1886–1970), (28) Nakagawa, Kyôji (1912–1992), (29) Nakamura, Kazuaki (1932–1987), (30) Nakamura, Saemon-Taro (1891–1974), (31) Nakano, Hiroshi (1894–1929), (32) Nasu, Nobuji (1899–1983), (33) Nishimura, Eiichi (1907–1964), (34) Ohsaki, Yorihiko (1921–1999), (35) Omori, Fusakichi (1868–1923), (36) Omote, Syun'ichirô (1912–2002), (37) Osawa, Yutaka (1927–1991), (38) Otsuka, Yanosuke (1903–1950), (39) Sano, Toshikata (1880–1956), (40) Santo (Akima until 1952), Tetsuo (1919–1997), (41) Sassa, Kenzo (1900–1981), (42) Satô, Yasuo (1918–1996), (43) Sekiya, Seikei (1854–1896), (44) Sezawa, Katsutada (1895–1944), (45) Shida, Toshi (1876–1936), (46) Suyehiro, Kyôji (1877–1932), (47) Suzuki, Ziro (1923–1997), (48) Takahasi, Ryutarô (1904–1993), (49) Takeyama, Kenzaburô (1908–1986), (50) Tanabashi, Ryô (1907–1974), (51) Taniguchi, Tadashi (1900–1995), (52) Tatsuno, Kingo (1854–1919), (53) Terada, Torahiko (1878–1935), (54) Terazawa, Kwan'ichi (1882–1969), (55) Tsuboi, Chuji (1902–1982), (56) Tsuboi, Yoshikatsu (1907–1990), (57) Tsubokawa, Ietsune (1918–1994), (58) Tsuya, Hiromichi (1902–1988), (59) Uchida, Yoshikazu (1885–1972), (60) Umemura, Hajime (1918–1995), (61) Wadati, Kiyoo (1902–1995), and (62) Watanabe, Hikaru (1934–2000).

4. Biographical Sketches of Japanese Seismologists, Earthquake Engineers, and Earth Scientists

Biographical sketches of the following 384 Japanese scientists and engineers in seismology, earthquake engineering, and the physics of the Earth's interior are included in the full report on the Handbook CD: Abe, K.; Adachi, Y.; Aida, T.; Akaogi, M.; Aoi, S.; Aoki, H.; Aoyama, H.; Arima, F.; Asada, T.; Asano, S.; Aso, T.; Dan, K.; Dohi, H.; Eguchi, T.; Finn, W.D.L.; Fujii, Shigeru; Fujii, Shunji; Fujii, T.; Fujii, Y.; Fujimoto, H.; Fujisawa, K.; Fujita, E.; Fujita, Y.; Fujiwara, H.; Fujiwara, T.; Fukao, Y.; Fukazawa, K.; Fukuyama, E.; Funahara, H.; Furukawa, Y.; Furumoto, M.; Furumoto, Y.; Furumura, T.; Furuya, M.; Geller, R.J.; Gu, J.; Guan, B.; Hakuno, M.; Hamada, K.; Hamada, M.; Hamada, N.; Hamano, Y.; Hara, Tadashi; Hara, Tatsuhiko; Harada, T.; Hasegawa, A.; Hashimoto, M.; Hayashi, H.; Heki, K.; Hibino, H.; Higashihara, H.; Hirahara, K.; Hirai, T.; Hiraishi, H.; Hirao, K.; Hirasawa, T.; Hirata, N.; Hirono, T.; Hisada, Y.; Hisano, M.; Honda, S.; Honjo, Y.; Honkura, Y.; Hori, A.; Hori, S.; Horikoshi, K.; Horiuchi, S.; Hoshiba, M.; Hoshiya, M.; Hotta, H.; Hurukawa, N.; Hyodo, M.; Iai, S.; Ichii, K.; Ichikawa, M.; Ichinose, T.; Ida, Y.; Ide, S.; Iemura, H.; Igarashi, A.; Igarshi, G.; Ihzuka, T.; Iidaka, T.; Iio, Y.; Ikeda, R.; Ikeda, T.; Ikarashi, K.; Ikenaga, M.; Ikeuchi, T.; Imoto, M.; Inaba, Y.; Inaoka, S.; Inoue, T.; Irie, Y.; Irikura, K.; Isezaki, N.; Ishibashi, K.; Ishibashi, Y.; Ishida, M.; Ishihara, K.; Ishii, H.; Ishii, T.; Ito, H.; Ito, Y.; Iwasaki, R.; Iwasaki, T.; Iwata, T.; Izumi, H.; Kabeyasawa, T.; Kameda, H.; Kameoka, H.; Kanai, K.; Kanda, J.; Kaneoka, I.; Kaneta, K.; Kasahara, J.; Kasahara, K.; Kasai, K.; Katayama, T.; Kato, D.; Kato, N.; Kato, T.; Kawakatsu. H.; Kawamura, S.; Kawano, M.; Kawasaki, I.; Kawase, H.; Kawashima, K.; Kikuchi, M.; Kimura, K.; Kinoshita, H.; Kinoshita, M.; Kinoshita, S.; Kitagawa, Y.; Kiyono, J.; Kobayashi, K.; Kobayashi, Y.; Kobori, T.; Kodera, J.; Koike, T.; Kojima, K.; Koketsu, K.; Kokusho, T.; Kono, S.; Kosuga, M.; Kosugi, M.; Koyama, J.; Koyama, S.; Kubo, T.; Kudo, K.; Kuge, K.; Kumagai, H.; Kunii, K.; Kushiro, I.; Kuwamura, H.; Kuwano, J.; Maeda, T.; Maehara, Y.; Maekawa, K.; Maruyama, K.; Maruyama, T.; Maseki, R.; Masuda, H.; Matsuda, T.; Matsumoto, S.; Matsu'ura, M.; Matsu'ura, R.S.; Matsumori, T.; Matsuura, T.; Matsuzawa, T.; Meguro, K.; Midorikawa, M.; Midorikawa, S.; Mikami, A.; Mikami, H.; Misono, S.; Mita, A.; Miura, F.; Miwa, S.;

Miyamoto, A.; Miyata, M.; Miyatake, T.; Miyamura, S.; Mizoue, M.; Mizutani, H.; Mochizuki K.; Mogi, K.; Mori, J.; Morikawa, H.; Morimoto, R.; Morita, S.; Morita, T.; Morita, Y.; Moriya, T.; Motosaka, M.; Murakami, H.; Murakami, S.; Murakami, Y.; Nagai, M.; Nagao, T.; Nagashima, I.; Nagumo, S.; Nakada, M.; Nakamura, H.; Nakamura, Takaaki; Nakamura, Toshiharu; Nakamura, Tsuneyoshi; Nakamura, Yukiko; Nakamura, Yutaka; Nakanishi, I.; Nakashima, M.; Narahashi, H.; Nariyuki, Y.; Negishi, H.; Nishida, R.; Nishimura, T.; Nishitani, A.; Nishiyama, M.; Noguchi, H.; Noguchi, S.; Nojima, N.; Noritomi, K.; Nozu, A.; Numata, A.; Obara, K.; Ogata, Y.; Ogawa, T.; Ohashi, Y.; Ohmachi, T.; Ohminato, T.; Ohnaka, M.; Ohno, K.; Ohno, M.; Ohtake, M.; Ohtani, E.; Ohtsuka, S.; Oike, K.; Okada, H.; Okada, Y.; Okamoto, Shin, Okamoto, Susumu, Okubo, S.; Okumura, K.; Omura, K.; Otani, S.; Ozima, M.; Pulido, N.E.; Rikitake, T.; Sadohara, S.; Sagiya, T.; Saito, M.; Sakai, S.; Sakamoto, I.; Sakamoto, S.; Sakashita, K.; Sasai, Y.; Sasaki, K.; Sasatani, T.; Satake, K.; Sato, Haruo; Sato, Hiroshi; Sato, R.; Sato, Tamao; Sato, Toshiaki; Sato, Toshinori; Sato, Y.; Satoh, T.; Sawada, T.; Sekiguchi, H.; Seno, T.; Shiba, Y.; Shibazaki, B.; Shima, E.; Shimada, S.; Shimamoto, E.; Shimamoto, T.; Shimamura, H.; Shimazaki, K.; Shimoda, I.; Shimozuru, D.; Shinozuka, M.; Shiobara, H.; Shiohara, H.; Shiotani, T.; Shirai, N.; Shoji, G.; Soda, S.; Sotomura, K.; Suetsugu, D.; Sugano, T.; Sugito, M.; Suita, K.; Sun, W.; Suyehiro, K.; Suyehiro, S.; Suzuki, K.; Suzuki, S.; Suzuki, Y.; Tada, M.; Tagawa, H.; Tagawa, K.; Takada, T.; Takagi, A.; Takahashi, E.; Takahashi, Y.; Takaki, M.; Takano, K.; Takei, Y.; Takenaka, H.; Takeo, M.; Takeuchi, Y.; Takewaki, I.; Takiguchi, K.; Takimoto, K.; Takizawa, H.; Tanaka, A.; Tanaka, H.; Tanaka, Sachito; Tanaka, Satoru; Tanaka, Yasuhiko; Tanaka, Yasuo; Tanaka, Yoshihiro; Taniguchi, H.; Tanioka, Y.; Tateishi, A.; Tazime, K.; Tazoh, T.; Terada, K.; Toki, K.; Tokimatsu, K.; Tokusu, M.; Tomoda, Y.; Toramaru, A.; Towhata, I.; Tsuboi, S.; Tsuji, B.; Tsukuda, T.; Tsumura, K.; Uchiyama, Y.; Uebayashi, H.; Uemura, K.; Uenishi, K.; Uetani, K.; Ukawa, M.; Umeda, Y.; Umehara, Y.; Unjoh, S.; Urano, K.; Urayama, M.; Usami, T.; Utada, H.; Utsu, T.; Uwabe, T.; Uyeda, S.; Uyeshima, M.; Uzawa, T.; Wakita, H.; Watanabe, F.; Watanabe, H.; Watanabe, J.; Watanabe, K.; Yagi, T.; Yamada, M.; Yamamoto, K.; Yamamoto, T.; Yamanaka, H.; Yamano, M.; Yamashina, K.; Yamashita, T.; Yamazaki, F.; Yashima, A.; Yasuda, S.; Yokoi, T.; Yokoo, Y.; Yomogida, K.; Yoshida, A.; Yoshida, S.; Yoshimura, C.; Yoshimura, M.; Yoshii, T.; Yoshioka, N.; Yukutake, T.; Zhang, F.; and Zhao, D.

5. Institutional Reports

5.1 Seismological Society of Japan (M. Ishida)

Web site: http://www.soc.nii.ac.jp/ssj/

The Seismological Society of Japan (SSJ), the first seismological society in the world, was first established in April 1880 and continued until 1892. The number of members was 117 on December of 1881, 62 of them foreigners. The second SSJ was reactivated in 1929 and started to publish "Zisin" in January 1929. The activities of the Society, however, were suspended during the period from 1943 to 1947 and were resumed on January 1948. The number of members was 321 in October 1948. The Japanese language name had been just "The Seismological Society" until April 1993. The current name of the Society is "The Seismological Society of Japan." The number of members increased year by year and was more than 2400 at the end of 1999.

The main purpose of the first SSJ was the furtherance of studies on earthquakes and volcanoes and collection of facts relating to seismic and volcanic phenomena. The Society also worked to make seismometers and recording systems. These activities have mostly continued to the present. The main purpose of the present SSJ is to promote research on earthquakes, the structure of the Earth's interior, and other related phenomena, to exchange and disseminate information on seismology, and to contribute to reduction and prevention of earthquake hazards. After the 1995 Hyogo-ken Nanbu earthquake (M7.2), the SSJ was partially reorganized in order to intensify the contribution to reduction and prevention of earthquake hazards.

The detailed review on the Handbook CD includes (1) history of SSJ, (2) activities of SSJ, (3) memberships of SSJ, and (4) list of SSJ presidents.

5.2 Association for the Development of Earthquake Prediction (K. Hamada)

Web site: http://www.adep.or.jp/

The Association for the Development of Earthquake Prediction (ADEP) was established on January 22, 1981, as a juridical foundation. Since then, many surveys and much research on seismotectonics, earthquake prediction, seismic ground motions, earthquake disasters, and countermeasures against earthquake disasters have been consigned to ADEP from the national government, prefecture governments, and commercial companies. These works are carried out based on the latest expertise in cooperation with leading scientists and the products are highly evaluated. ADEP has been making every endeavor to return the advanced consequences in the academic field to general society through issues practically needed.

The great Hyogoken-Nanbu (Kobe) earthquake disaster in 1995 motivated the establishment of the Headquarters for Earthquake Research Promotion (the Headquarters) in the Prime Minister's office in order to promote earthquake-related observation, surveys, and research comprehensively on a nationwide scale. The Earthquake Research Center was established in ADEP in order to assist the Headquarters, and started its activity in the later 1995.

The Science and Technology Agency enacted a grant for Promotion of Research Facilities in Deep Vault in the Crust in 1996. Following this system, ADEP started the Tono Research Institute of Earthquake Science in 1997 in Gifu Prefecture. Research on inland earthquakes and countermeasures against earthquake disasters is carried out in this institute. ADEP is also preparing a supportive organization for maintenance and data analysis of the seismic observation networks of government research institutes.

The main projects include (1) research on earthquake prediction and disaster prevention, (2) financial support for research on earthquake prediction and disaster prevention, (3) research on responses of structures and ground during earthquakes and on earthquake resistance, (4) popularization of knowledge on earthquake prediction and disaster prevention, (5) support to the activities of the Headquarters.

5.3 Ehime University, Department of Earth Sciences (T. Irifune)

Web site: http://www.ehime-u.ac.jp/~cutie/english/intro.html

The Ehime University was established some 20 years ago. Since then many excellent young earth scientists have been hired in the fields of geophysical and mineralogical sciences in many parts of Japan. Some of them were recruited by other major universities, and are now among the leading earth scientists in our country. These include Profs. M. Toriumi, T. Murakami, T. Koyaguchi, M. Ogawa (University of Tokyo), K. Fujino (Hokkaido Univ.), E. Ohtani (Tohoku Univ.), I. Kawabe (Nagoya Univ.), T. Takeshita (Hiroshima Univ.), and S. Yoshioka (Kyushu Univ.).

Research in the geophysical sciences group focuses on the experimental and computer simulation studies of the Earth's deep interior. This group consists of Prof. I. Ohno (the inventor of the RPR method for determination of elastic properties of tiny crystals), T. Irifune (working on mantle mineralogy with multi-anvil high-pressure devices), D. Zhao (expert in precise 3-D imaging with seismological tomography), H. Mori (TEM and X-ray mineralogist on meteorite and Earth materials), and T. Inoue (high-pressure mineral physicist focusing on the effect of volatiles).

5.4 Geographical Survey Institute (T. Tada)

Web site: http://www.gsi.go.jp/

The Geographical Survey Institute (GSI), Ministry of Construction, is a governmental organization that is responsible for surveying and mapping in Japan. The Army Land Survey of the General Staff Office is a predecessor of GSI before 1945.

GSI and its predecessor Army Land Survey have conducted nationwide horizontal and vertical control surveys and local geodetic re-surveys after destructive earthquakes. The survey results have revealed secular deformation of the Japanese Islands such as distribution of strain accumulation and conspicuous crustal movements associated with large earthquakes such as the 1923 Kanto earthquake, the 1946 Nankaido earthquake, and so on. All these results have made contributions to understanding processes associated with earthquake occurrences and volcanic eruptions in Japan.

After the first plan for earthquake prediction research by the Geodesy Council started in 1964, the geodetic surveys have been conducted in accordance with the earthquake prediction research plan. Other than the nationwide surveys and local re-surveys after large earthquakes, GSI has conducted intensive geodetic surveys in selected areas such as the Tokai area to monitor crustal movements.

Recently, GSI has been positively utilizing space technologies for the geodetic surveys. Up to present GSI has deployed nearly 1,000 permanent automatic GPS (Global Positioning System) observation stations all over Japan. Therefore, it has become possible to monitor crustal movements in the Japanese Islands continuously. GSI is also actively participating in internationally cooperative observations by Very Long Baseline Interferometry (VLBI), which contributes to understanding of global crustal dynamics. Moreover, GSI has been carrying out a study on applications of differential interferometric SAR (Synthesized Aperture Radar) for detection of crustal deformation since 1994. Using interferometric SAR techniques, detailed crustal deformations associated with earthquakes, volcanic activities, and other geophysical phenomena have been obtained.

In addition to the survey and geodetic observations, GSI has been conducting an extensive modeling study of geodetic data to deduce physical process related to various tectonic activities. Crustal deformations associated with major earthquakes have been studied and seismic deformations of the Japanese Islands and volcanic activities have been extensively modeled. Currently GSI is carrying out a simulation research concerning crustal deformation, which is associated with various crustal activities of the Japanese Islands to contribute to mitigation of seismic as well as volcanic hazard.

GSI is also in charge of managing the Coordinating Committee for Earthquake Prediction (CCEP), which was established in 1969 to exchange information on observations and research for earthquake prediction among universities and governmental organizations and to investigate the results.

The detailed review on the Handbook CD includes reviews of (1) activities before the earthquake prediction research program (triangulation, leveling and tide observation, and re-surveys after destructive earthquakes), (2) activities after the start of the earthquake prediction research program (horizontal control survey, leveling, and other activities), and (3) recent activities (continuous GPS observations—GEONET, VLBI, and SAR interferometry, simulation and modeling), and (4) Coordinating Committee for Earthquake Prediction (CCEP).

5.5 Geological Survey of Japan (T. Noda)

Web site: http://www.gsj.go.jp/

The Geological Survey of Japan (GSJ) is the only national institute for integrated earth science research in Japan. It was established in 1882 with the aims of making geological maps of the country and undertaking research related to the exploration of mineral resources. GSJ has contributed to the rapid progress in geophysical and geochemical exploration methods for minerals since the 1950s. In response to changing social needs, since the 1970s it has expanded scientific activities into the additional fields of marine geology, geothermal research, local and global environmental issues, and prediction and prevention of natural hazards. GSJ encourages international collaboration with many countries and communities, and dissemination of geoscientific information to the public.

There are many scientists who are engaged in seismology and earthquake-related research in GSJ. They work mainly in the following departments: 1) Earthquake Research Department containing a department-attached senior researcher and the three research sections Active Fault Research Section/ Tectonophysics Section/ Seismotectonics Section, 2) Geophysics Department, and 3) Geothermal Research Department.

The Earthquake Research Department aims to predict earthquakes and characterize their features for the prevention and mitigation of disasters. The tragic 1995 Hyogo-ken Nanbu earthquake has greatly increased social demands for earthquake research. On July 1, 1997, the department was established in order to study active tectonics based on surveys of active faults, fracture processes using rock-fracturing experiments in the laboratory, and short-term earthquake prediction by monitoring ground water.

The Geophysics Department has developed various techniques for precise imaging of the subsurface structure of the Earth. Research crustal dynamics is also carried out using up-to-date simulation techniques. Understanding the structure, physical characteristics, and processes of the Earth's crust is essential for the exploration and development of energy and mineral resources, and mitigation and prediction of natural hazards.

The Geothermal Research Department carries out basic research on geology, geochemistry, and geophysics in geothermal fields in order to understand thermal phenomena within the Earth's crust. It also conducts research projects concerning deep-seated geothermal resources. Research for the development of exploration technology in geothermal reservoirs is carried out in collaboration with the New Energy and Industrial Technology Development Organization (NEDO) under the auspices of the New Sunshine Project promoted by MITI. The department is responsible for evaluation and analysis of survey data from NEDO and for supplementary tasks to NEDO's projects.

The report of GSJ on the Handbook CD contains a review of the Earthquake Research Department and the Geophysics Department, with major developments and accomplishments, current activities, research staff members, and recent major bibliographic references.

5.6 Hokkaido University, Department of Earth and Planetary Physics (H. Shimamura)

Web site: http://www.ep.sci.hokudai.ac.jp/

At Hokkaido University, solid geophysics study started in 1952 when the Department of Geophysics (DG) in the Faculty of Science was founded. In DG, two laboratories, Volcanology and Seismology, and Applied Geophysics, engaged in seismological research and education. Scientific personnel who supported DG in seismology are Professors T. Matuzawa (1960–1963), K. Tazime (1956–1983), Hiroshi Okada (1964–1998), J. Koyama (1995–), and K. Yomogida (1998–). Associate Professors include S. Sakuma (1954–1958), T. Utsu (1964–1971), N. Den (1959–1972), K. Abe (1973–1983), H. Shimamura (1972–1979 (who moved to ISV)), I. Nakanishi (1983–1994), T. Moriya (1970–), and T. Sasatani (1973–). DG has had a close relation with ISV (and its predecessors) and personnel were exchanged between DG and ISV (and its predecessors). In 1994, the Department of Geophysics was reorganized into a part of the Department of Earth and Planetary Sciences, by unifying with the Department of Geology.

Intensive investigations were made in many fields. They were about seismic wave theory, attenuation of seismic waves beneath the Japanese Island Arcs, seismicity in and around Hokkaido and the Southern Kurile islands, and crust and upper mantle structure in the ocean basin using the ocean bottom seismometers (OBS) and explosive source, mechanisms of the large earthquake in the subduction and deep seismic zones, precise upper crustal structure beneath an earthquake swarm region in Matsushiro, Nagano Prefecture using explosive sources, microearthquake survey in Hokkaido island by portable high-gain seismometers with long-term recorders, and crustal structure on the Hidaka mountains.

Some scientific highlights of these investigations are the partial success of forecasting the 1973 Nemuro-oki Earthquake, M 7.6, development of the original ocean bottom seismometer (which was inherited by ISV's predecessor, LOBS), and discovery of the three-dimensional collision structure of the Hidaka mountains, in southern Hokkaido. The portable high-gain seismometers with long-term tape recorders have been developed by Dr. T. Moriya, which helped seismicity studies and controlled source studies very much.

5.7 Hokkaido University, Institute of Seismology and Volcanology (H. Shimamura)

Web site: http://www.eos.hokudai.ac.jp/

The Institute of Seismology and Volcanology (ISV) was established in April 1998. The Institute carries out geophysical research, mainly on seismology and volcanology. Research on geomagnetism and geodesy are also undertaken. This institute is responsible to the Japanese national projects of earthquake prediction and volcanic eruption prediction.

ISV consists of four laboratories: seismological observation research, ocean bottom seismology, volcanological research, and subsurface structure. The new institute was a reorganization of six laboratories belonging to the Faculty of Science, described following, and an expansion to new fields. The original laboratories were Usu Volcano Observatory, Urakawa Seismological Observatory, Sapporo Seismological Observatory, Erimo Geophysical Observatory, Research Center for Earthquake Prediction, and Laboratory for Ocean Bottom Seismology.

The report of DG and ISV on the Handbook CD includes a review of (1) research activities before the establishment of ISV, the Research Center for Earthquake Prediction (RCEP, 1976–1998), and the Laboratory for Ocean Bottom Seismology (LOBS, 1979–1998); (2) current activities of the Laboratory for Seismological Observation Research, the Laboratory for Ocean Bottom Seismology, the Laboratory for Subsurface Structure, and the Laboratory for Volcanology; and (3) a list of current research members.

5.8 Hot Springs Research Institute of Kanagawa Prefecture (T. Tanada)

Web site: http://www.pref.kanagawa.jp/osirase/05/0325/

The Hot Springs Research Institute of Kanagawa Prefecture (HSRI), located in Odawara, was established in 1969. Operated by the prefecture government, HSRI has extended its activities into four main fields of geoscience: hydrology, geology, volcanology, and seismology. It provides research findings and services related to earthquakes to agencies and the people of Kanagawa prefecture.

Notable studies on seismological topics conducted at HSRI are (1) investigation of the relation between age of fossil trees submerged in Ashinoko and landslides induced by large earthquakes; (2) ground-water monitoring for earthquake prediction by an amateur network (the Catfish Club); (3) seismicity of Hakone volcano; (4) study of the anticipated M7 class earthquake, named the Western Kanagawa Prefecture Earthquake, using reconstructed seismic and crustal deformation network; and (5) study of radon concentration around active faults in Kanagawa Prefecture.

The research staff includes Director General K. Hirano (chemistry), Research Director R. Kuraishi (chemistry), and Researchers M. Oyama (hydrology and volcanology), H. Ito (seismology), T. Tanada (seismology), K. Itadera (hydrology), Y. Daita (chemistry and seismology), K. Murase (seismology), and T. Tanbo (seismology).

5.9 International Institute of Seismology and Earthquake Engineering, Building Research Institute, Ministry of Construction (D. Suetsugu)

Web site: http://iisee.kenken.go.jp/

The main task of the International Institute of Seismology and Earthquake Engineering (IISEE) is to conduct training programs for researchers of developing countries to mitigate earthquake disaster. Since 1962 we have provided more than 1000 researchers from 79 countries with training. Every year we perform the Seismology and Earthquake Engineering Regular Course, which is 11 months long, and the Global Seismological Observation Course, which is about 2 months long and specialized for detection and discrimination of underground nuclear tests. We organize the Seminar Course every 2 years, focused on some up-to-date topics mainly for following up ex-participants of the Regular Course. We perform the Individual Course, which is occasionally held for participants to study some specific subjects for less than one year. The IISEE members are involved also in advanced research projects of seismology and earthquake engineering, which helps us update curriculum of the training courses. Details of history, training programs, facilities, and research projects are also presented in http://iisee.kenken.go.jp/.

The IISEE members are involved in the following research projects at present: (1) development of integrated analysis technique of crustal motion data; (2) Superplume Project (an international research project on whole Earth dynamics); (3) modeling the generation process of inland crustal earthquakes; (4) developing codes for a vector-parallel supercomputer; (5) nationwide strong-motion observation; (6) dense strong motion instrument array in Sendai; (7) strong motion instrument network in the metropolitan area; (8) new engineering framework for performance-based design of building structures; (9) revision of the Japanese building standard law; (10) performance statement system of residential buildings; (11) new technologies for seismic retrofit; (12) hybrid/composite structural system; (13) smart materials and structural systems; (14) development of earthquake and tsunami disaster mitigation technologies and their integration for the Asia-Pacific Region; and (15) planning of urban residential area and construction management techniques in developing countries in the circum-Pacific region.

The IISEE issues the following publications on a yearly basis: (1) Bulletin of the International Institute of Seismology and Earthquake Engineering (in English), (2) Reports of Individual Studies by Participants at IISEE (in English), (3) IISEE Year Book (in English), (4) IISEE Annual Report (in Japanese), and (5) IISEE Brochure (in English).

The report of IISEE on the Handbook CD includes a history of IISEE, outline of training courses, curriculum of training courses, information about application for training courses and facilities, a list of staff members (with a list of UNESCO expert lecturers for the Seismology and Earthquake Engineering Regular Course), and a list of selected publications (in English) of IISEE staff members (1994–1999).

5.10 Japan Meteorological Agency (N. Hamada)

Web site: http://www.jma.go.jp/JMA_HP/jma/indexe.html

The efforts for prevention and mitigation of disasters caused by earthquakes and tsunamis have been continued as one of Japan's most important national programs. The Japan Meteorological Agency (JMA) is a governmental organization responsible for tsunami forecasting (warnings/advisories), the short-term prediction of a large-scale earthquake, and information services for earthquakes, tsunamis, and volcanic activities. The information services are used as a trigger for prompt disaster preventive and relief actions of related agencies. To take these responsibilities,

JMA has maintained nationwide seismological and volcanological observation networks and cooperated with related national and international organs and universities. The development of instruments and observation systems, investigations of earthquakes, and publication of the national earthquake catalogs of Japan have also been important tasks of JMA. The JMA national earthquake catalogs have been available since 1923 and may be the longest and most nearly complete regional earthquake catalogs in the world. The catalogs have been used in many seismological studies. Seismograms obtained by the networks play an important role in the study of seismology and earthquake engineering. Macroseismic observation of JMA has a century-long history and the observed data have been used as a key to interpret historical documents of earthquakes. Contemporary state-of-the-art technologies and experience in seismology were applied to the development of various instruments such as short- and long-period seismographs, strong motion seismographs of a mechanical and electromagnetic type, ocean bottom seismographs, and Seismic Intensity meters.

Current seismological activity of the Japan Meteorological Agency includes (1) an earthquake monitoring network of about 180 seismological stations at around 60-km intervals throughout Japan, (2) seismic intensity observations with instrumental seismic intensity meters, (3) tsunami forecast and tsunami observation, (4) earthquake prediction program, (5) investigation of seismicity in and around Japan.

The headquarters of the Japan Meteorological Agency is divided into five departments: Administration, Weather Forecast, Observations, Seismological and Volcanological, and Climate and Marine Weather. The Seismological and Volcanological department has four divisions: Administration Management, Earthquake and Tsunami Observation, Earthquake Prediction and Information, and Volcanological. Matsushiro Seismological Observatory belongs to the Earthquake and Tsunami Observation division. There are five district meteorological observatories and a meteorological observatory: Sapporo, Sendai, Tokyo, Osaka, Fukuoka, and Okinawa. There are Senior Seismological Information Officers in Sapporo, Sendai, Osaka, and Fukuoka district observatories. Except Tokyo, each district meteorological observatory and Okinawa Meteorological Observatory has a Seismological and Volcanological Section (SVS). SVS of each district meteorological observatory has a staff of about 20 members, including a head, and SVS of Okinawa observatory has 16 staff including a head. The Seismological and Volcanological Department in the head office of JMA and each Seismological and Volcanological Section operate a Regional Tsunami Warning Center. The Meteorological Research Institute has 9 research departments including a Seismology and Volcanology Research Department. The Seismology and Volcanology Research Department has a director and 19 researchers. The department performs research and development useful for improving the seismological and volcanological services of JMA.

The report of JMA on the Handbook CD includes (1) history of the Japan Meteorological Agency in seismology, and (2) major contributions in seismology and physics of the Earth's interior made by JMA personnel.

5.11 Kagoshima University, Nansei-Toko Observatory for Earthquakes and Volcanoes (T. Kakuta)

Web site: http://leopard.sci.kagoshima-u.ac.jp/noev/home.htm

Through 1998, NOEV has deployed 20 telemeter stations, including temporary ones, from South Kyushu to the northern part of the Ryukyu Islands. Though seismic stations are not yet sufficient along the islands, detectability of events and accuracy of the locations for earthquakes in and around Kyushu are greatly increased. In the seas surrounding the region, moreover, several temporary observations have been carried out by using ocean-bottom seismometers in collaboration with a research group of Japanese universities. By locating many events accurately, several discontinuities perpendicular to the arc system have been detected in the subducting seismic plane. We have also clearly identified crustal earthquakes distributed in a narrow zone nearly perpendicular to the arc system. They give an important clue to investigate the seismotectonics in the region.

In recent years, several large earthquakes occurred along the Kyushu-Ryukyu arc. Some of them are interplate earthquakes, as those east off Tanegashima Island on October 18, 1996 (M6.2), southern Hyuga-nada on October 19 and on December 3, 1996 (M6.6), etc. The 1995 Amami-Oshima-Kinkai Earthquakes as well as the event east off Tanegashima Island on January 24, 1999 (M6.2), are intraplate earthquakes. Crustal earthquakes are the 1994 Northern Kagoshima Earthquake (M5.7), the 1996 Southern Tanegashima Earthquake (M5.7), and the 1997 Northwestern Kagoshima Earthquakes (M6.5, M6.3). We observed them not only by the regional stations but also by distributing temporarily seismic stations around the focal region, and succeeded in clarifying their characteristics in detail, especially aftershock distributions and focal mechanisms.

The report on the Handbook CD includes a summary of major developments and accomplishments in seismic network, OBS observations, survey of crustal structures, GPS observations, split subduction models of the Philippine Sea plate, and studies of crustal earthquakes, a list of research staff members, and a list of publications.

5.12 Kyoto University, Laboratory of Solid Earth Physics (K. Oike, A. Okada, and S. Takemoto)

Web site: http://www.kugi.kyoto-u.ac.jp/index-E.html

Seismological studies in Kyoto University were started in 1907, under Toshi Shida. He founded the Kamigamo Geophysical Observatory in 1909. The first laboratory of geophysics was established in the Department of Physics in 1918 and became the Department of Geophysics in 1921. After the Disaster Prevention Research Institute was established in 1951, research and education in seismology in the university were developed through cooperation of the Institute and the faculty of Science.

The Laboratory of Solid Earth Physics was established in 1995 as the successor to three existing laboratories, as part of the reform of Kyoto University. It now has three research sections: the seismological research section, chaired by Kazuo Oike; the research section of the physics of the Earth's crust chaired by Atsumasa Okada; and the research section of geodesy, chaired by Shuzo Takemoto. The Laboratory has responsibility for the education of students of the graduate school, which has the master's course and the doctor's programs. The capacity of the master's course of solid-earth physics is about twenty students. The education is done in cooperation with the professors of the Earthquake Prediction Research Center and the research group of the earthquake strong motion and others of the Disaster Prevention Research Institute, Kyoto University.

5.13 Kyoto University, Disaster Prevention Research Institute (M. Ando and K. Irikura)

Web site: http://www.rcep.dpri.kyoto-u.ac.jp/

The Disaster Prevention Research Institute, Kyoto University, was established in 1951. It carries out research on a variety of problems related to the prevention or reduction of natural disasters. The Institute commenced its work at laboratories in Kyoto and in the Abuyama Seismological Observatory in Takatsuki with three research sections and sixteen research staff. It has since established more and more research sections and attached corresponding facilities to address the increased research needs that have arisen as a consequence of changes in social conditions and the diversification of natural disaster potential.

By the end of 1999, the Institute reorganized itself into five research divisions and five research centers. Integrated Management of Disaster Risk; Earthquake Disaster Prevention; Geo-Disasters; Fluvial and Marine Disasters; Atmospheric Disasters; Research Center for Disaster Environment; Research Center for Earthquake Prediction; Sakurajima Volcano Research Center; Water Resources Research Center; and Research Center for Disaster Reduction Systems. Seismology and physics of the Earth's interior have been studied at Kyoto University in both the Graduate School of Science and DPRI. Related research on earthquake seismology has also been conducted at both organizations. The Regional Center for Earthquake Prediction, Abuyama Seismological Observatory, and other two observatories were part of the Faculty of Science, presently Graduate School of Science. In 1990, these sections were combined and reorganized into the Research Center for Earthquake Prediction (RCEP) as an attached institution of DPRI to promote extensive and effective research on earthquake mechanisms and earthquake prediction. A course in Seismology in the Department of Geophysics, Faculty of Science was also established to fulfill education and basic research.

Engineering seismology and earthquake engineering are studied at the Research Division of Earthquake Disaster Prevention. Volcanology is studied at the Sakurajima Volcano Research Center.

The research staff members of the Institute are also affiliated with the Graduate Schools of Science and Engineering. Many graduate students come to the Institute to carry out their studies under supervision of its staff members.

The report of the Institute on the Handbook CD includes a review of (1) Research Center for Earthquake Predication with a list of research staff members, and (2) Division of Earthquake Disaster Prevention with a list of research staff members.

5.14 Kyushu University, Department of Earth and Planetary Sciences (S. Suzuki and H. Takenaka)

Web site: http://www.gaea.kyushu-u.ac.jp/

Kyushu University is located in Fukuoka City, northern part of Kyushu Island, Japan. The Faculty of Sciences was established in April 1939, starting with the Departments of Physics, Chemistry, and Geology, to which were added those of Mathematics (in 1942) and Biology (in 1949). The Department of Geology was reorganized as the Department of Earth and Planetary Sciences in 1990, including four geophysical laboratories in the Department of Physics. For educational and research purposes, the Faculty has the Institute of Seismology and Volcanology (formerly the Shimabara Earthquake and Volcano Observatory in Nagasaki prefecture), the Amakusa Marine Biological Laboratory in Kumamoto prefecture, and the Cryogenic Laboratory.

The Department of Earth and Planetary Sciences has two major missions, which are research and education. The educational mission includes both undergraduate and graduate levels leading to B.Sc., M.Sc., and D.Sc. degrees. A strong emphasis is focused on graduate education by recent reforming of the university. Research seeks advances in Earth and planetary sciences. There are 14 professors, 12 associate professors, and 15 research associates. The research fields are seismology and physical volcanology, dynamics of the Earth's interior, evolution of the solar system, geomagnetosphere physics, atmospheric science, Earth and planetary fluid dynamics, mineralogy, chemical geodynamics, metallogenic geochemistry, marine geoscience, organic geoscience, and paleobiology.

The Laboratory of Seismology and Physical Volcanology in the Department of Earth and Planetary Sciences has staff members Professor Sadaomi Suzuki, Associate Professor Hiroshi Takenaka, and Research Associate Nobuki Kame. In June 2000 there were four doctoral graduate students, six masters students, and six undergraduate students. The report of the laboratory on the Handbook CD lists all of the English-language publications of the Laboratory during 1998–2000.

5.15 Kyushu University, Institute of Seismology and Volcanology (S. Suzuki and H. Shimizu)

Web site: http://www.sevo.kyushu-u.ac.jp/

The Institute of Seismology and Volcanology, which was reorganized from Shimabara Earthquake and Volcano Observatory (SEVO), is located at the eastern foot of Unzen Volcano in the Shimabara peninsula, western Kyushu. The former institution of SEVO is Shimabara Volcano Observatory, which was established in April 1971 for the purpose of research on the prediction of volcanic eruptions by geochemical and seismological approaches. The observatory was reorganized to SEVO in April 1984, in order to expand a sphere of research into fundamental studies on earthquake prediction. In 1985, SEVO started to install the seismic network that could detect microearthquake activity in the Kyushu district.

In 1990, Mt. Fugen of Unzen Volcano began to erupt after 198 years of dormancy. A dacite lava dome appeared in 1991, and had been growing until February 1995, with frequent occurrence pyroclastic flows. The observation system of SEVO was reinforced soon after the 1990 eruption. Observatory staff carried out intensive geophysical, geochemical, and geological investigations at Unzen Volcano in cooperation with the Joint University Research Group and national institutes.

Two large inter-plate earthquakes of M6.6 occurred in the Hyuganada region in October and December 1996, and disastrous inland earthquakes (M6.5, M6.3) took place in the Kagoshima prefecture in 1997. In April 2000, SEVO was reorganized to the Institute of Seismology and Volcanology in order to promote earthquake prediction research in Kyushu as the regional observation center. The institute also became a center for studying the tectonics of the Kyushu-Ryukyu Arc. In particular, back-arc tectonics is one of the most important subjects of the Institute.

5.16 Nagoya University, Research Center for Seismology and Volcanology (N. Fujii)

Web site: http://www.seis.nagoya-u.ac.jp/

This research center, RCSV, was reestablished in April 1999, with Prof. Naoyuki Fujii as Director. The major purposes of this center are to understand the nature of earthquake occurrence and volcanic eruption processes mainly from geophysical and geochemical observations together with experimental approaches, and to make predictions of such processes. This center is responsible for the Japanese national projects of earthquake prediction and volcanic eruption prediction.

RCSV consists of three sub-divisions: research in structural changes, detection of subsurface material movements by seismological and geodetic observations, and development of observational techniques. The new research center was a reorganization of five original laboratories belonging to Faculty of Science, Nagoya University, and an expansion to new fields. The original laboratories, described in detail in the full report, were Inuyama Seismological and Crustal Movement observatories, Mikawa Crustal Movement observatory, Research Center for Earthquake Prediction, and Takayama Seismological observatory.

Recent major topics include: (a) detailed descriptions of changes in earthquake swarm activities around aftershock area of the 1984 western Nagano prefecture earthquake (M6.8), and findings of the accelerating foreshocks before a M5.1 earthquake, (b) after-slip of a moderate-size earthquake detected by several strain meters surrounding the epicenter, (c) temporal changes in the ground deformations in the Tokai region, (d) combined observations of a dense GPS network, SAR Interferometry, and gravity changes in earthquake swarm areas, (e) development of new instruments, such as a controlled seismic source.

5.17 Nagoya University, Department of Earth and Planetary Sciences (N. Fujii)

Web site: http://www.seis.nagoya-u.ac.jp/

At Nagoya University, solid geophysics study started in 1952 as a division of geophysics, in the Department of Earth Sciences, Faculty of Science, which had been founded in 1949. In 1969, the division of seismology was established in a cooperative relationship with the Research Center for Seismology and Volcanology (RCSV and its predecessors). Both divisions of geophysics and seismology engaged in research on planetary physics, physics of Earth's interior, tectonophysics and seismology and in education of undergraduate and graduate students.

The two divisions of solid geophysics (physics of Earth's interior) and seismology had a close relation with RCSV in both research and education for graduate and undergraduate students. In 1996, the two divisions were unified into the division of Earth and Planetary Physics, when the Department of Earth and Planetary Sciences was reorganized to belong to the Graduate School of Science.

The fields of investigations covered by the division of Earth and Planetary Physics include physico-chemical states (i.e., past and future, interior and surroundings), system analysis of environmental problems, origin and evolution of Earth and planets by theoretical and experimental approaches. Geodynamics, tectonophysics, and the structure of the Earth and the moon are also intensively investigated, together with seismic waves, earthquake source mechanisms, and computer simulations of earthquake-generating processes including experimental rock physics. Applied seismology and earthquake predictions are cooperative programs with RCSV.

Some of the scientific highlights are new developments of ultra-high pressure and temperature apparatus, an accurately controlled seismic source (ACROSS) and new findings of the incessant excitation of the Earth's free oscillations, post-seismic deformations of inland destructive earthquakes, a slow thrust slip event of about one year duration beneath the Bungo Channel, southwest Japan, and so on.

5.18 National Research Institute for Earth Science and Disaster Prevention, Science and Technology Agency (M. Ishida)

Web site: http://www.bosai.go.jp/

The National Research Institute for Earth Science and Disaster Prevention (NIED) was originally established on April 1, 1963 in Tokyo as the "National Research Center for Disaster Prevention (NRCDP)," affiliated to the Science and Technology Agency (STA), Prime Minister's Office after the 1959 Isewan-Typhoon, by which about 5000 lives were lost in western Japan. The NRCDP moved from Tokyo to Tsukuba Science City in April 1978 and reorganized as the NIED in 1990.

NIED has been conducting research on earthquakes, volcanic eruptions, landslides, meteorological disasters, snow avalanches, and other natural disasters. The goal of our research is to mitigate natural disasters. Great efforts have been made to understand the physical mechanism of earthquakes and crustal activities throughout Japan as well as to promote engineering technology in these general areas. A large-scale earthquake simulator was constructed in Tsukuba Science City in 1970. The basis of the present nationwide observational network for studying earthquakes and crustal activities was initiated in 1973 when the borehole station, at a depth of about 3000 m and equipped with a three-component short-period seismograph and two-component tiltmeter, was completed at Iwatsuki, in the metropolitan area. The short-period borehole seismic network, broadband seismic network, and strong motion seismic network have been enforced with social demand during this thirty-year period.

More than 600 short-period borehole seismic stations, including four 3000 m deep and fifteen 2000 m deep borehole seismic stations, about 40 borehole tiltmeter and 10 borehole strainmeter stations, 6 ocean bottom seismometers, 30 broadband seismic stations, and approximately 1500 strong motion seismic stations have been deployed so far.

NIED also is facilitating research efforts for natural disaster prevention in Earth sciences with the application of advanced technologies to clarify the mechanism of global climatic changes and the relationships between the changes and their effects on natural disaster occurrence and the consequences.

5.19 Okayama University, The Institute for Study of the Earth's Interior (E. Ito)

Web site: http://ultra3.misasa.okayama-u.ac.jp/home_j.shtml

The Institute for Study of the Earth's Interior (ISEI) of Okayama University was established on April 1, 1985, at Misasa, Japan, by reorganizing the Institute for Thermal Spring Research. The primary objectives of the ISEI are to investigate the origin, evolution, and dynamics of the Earth, and also to collaborate with external researchers in relevant fields by making available to them the Institute's facilities. Research in high-pressure mineral physics at Misasa began in the early 1970s when E. Ito installed a split-sphere type of high-pressure apparatus. He made pioneering contributions to understanding the mineralogical constitution of the deep mantle by studying the phase relations of geophysically important silicates at high pressures and temperatures.

Upon the reorganization of the ISEI in 1995, the mineral physics group was substantially strengthened by the addition of new staff members with diverse interests. Accordingly, new methodologies were induced and new facilities were installed. For example, new methods have been developed using the diamond anvil cell (DAC), elastic spectroscopy, electrical conductivity, and NMR. Subsolidus and supersolidus experiments on the Earth's materials have been carried out in both the multi-anvil press and the DAC, in order to clarify the state of the deep mantle and to constrain material fractionation and core segregation in an early magma-ocean. Important constituents of the deep mantle are synthesized and characterized in terms of crystal chemistry and physical properties such as elasticity and electrical conductivity. Recently, in situ X-ray experiments using synchrotron radiation have occupied an important part of our activity: i.e., precise determination of the stability relations and equations of state for mantle phases, as well as viscosity measurement of molten silicate by means of X-ray radiography. This variety of information will be combined with seismological and geochemical constraints on the Earth's interior to solve problems directly related to the origin, evolution, and dynamics of the Earth.

5.20 Science and Technology Agency, Headquarters for Earthquake Research Promotion (T. Fukui)

Web site: http://www.jishin.go.jp/main/index.html

In light of the great Hyogo-ken Nanbu earthquake disaster, which occurred on January 17, 1995, the Special Measure Law on Earthquake Disaster Prevention was passed on June 9 and implemented on July 18, in the same year. This law provides maintenance of the system to promote earthquake research. On the basis of this law, the Headquarters for Earthquake Research Promotion was set up in the Prime Minister's Office. The Headquarters is composed of the Minister of State for Science and Technology as the director and the vice-ministers of relevant ministries and agencies. The Policy Committee and the Earthquake Research Committee were established in the Headquarters. They are composed of staffs of relevant ministries and agencies and people of experience or academic standing. Since the establishment of the Headquarters, earthquake research in Japan has been promoted with close cooperation of relevant ministries and agencies.

Headquarters operates under the following mandate: (1) planning comprehensive and basic policies on promotion of earthquake research, (2) coordinating administrative works related to earthquake research such as budgets for relevant ministries and agencies, (3) planning comprehensive survey and observation related to earthquakes, (4) collecting, arranging, analyzing, and comprehensively evaluating the results of earthquake research and relevant surveys by relevant ministries and agencies universities, and other relevant organizations, and (5) public relations based on the comprehensive evaluations.

The Policy Committee is to carry out missions 1, 2, 3, and 5 and the Earthquake Research Committee is to handle mission 4. Both committees have some subcommittees which deliberate detailed issues. Headquarters has planned a fundamental observation plan and dense earthquake observation network all over Japan. Moreover, Headquarters decided comprehensive and basic policy for earthquake research promotion, and according to the policy, maps of estimated strong ground motion are in progress. The system for the Earthquake research committee to evaluate earthquakes and provide official announcement to the public was established. Headquarters is promoting earthquake research in Japan more strongly; earthquake disasters should be mitigated by results of the current earthquake research.

5.21 Tohoku University, Solid Earth Physics Laboratory, Department of Geophysics (M. Ohtake)

Web site: http://zisin.geophys.tohoku.ac.jp/

Geophysical studies started at Tohoku University with the creation of the Chair of Geophysics in 1922, which was occupied by Shirota Kusakabe (1922–1924), and by Saemontaro Nakamura (1924–1945). In 1945, the Chair was expanded to the Division of Geophysics, composed of the Chairs of Seismology, Geo-Electro Magnetism, and Meteorology. In 1949, regular publication of the "Tohoku Geophysical Journal" (Fifth Series of the "Science Report of Tohoku University," which has published important scientific achievements at the Department) started. As of 1999, five laboratories covering various fields of geophysics are under the umbrella of the Department of Geophysics, and the Laboratory of Solid Earth Physics continues the half-century endeavor to promote research and education on seismology and the physics of Earth's interior at Tohoku University. The professors of seismology have been Saemontaro Nakamura (1945–1951), Hirokichi Honda (1951–1960), Ziro Suzuki (1961–1986) (President of IASPEI, 1983–1987), Hiroyuki Hamaguchi (1986–1988), Masakazu Ohtake (1988–), and Haruo Sato (1997–).

In the past five decades, the laboratory promoted intensive research to deepen the scientific understanding of an earthquake by theoretical and observational approaches. One of the most important contributions was theoretical research to establish the double couple mechanism of an earthquake. The elaborate works are summarized in Honda (1962). This was followed by a proposal of circular crack model of Sato and Hirasawa (1973). In the early 1950s, this laboratory conducted a number of pioneering observations of microearthquakes and artificial explosions by using newly developed high-sensitivity seismometers of moving-coil type. The results of observations are fully reflected in a series of articles on earthquake statistics by Suzuki (1953, 1955, 1958, 1959). Current research includes studies of seismotectonics, generation of high-frequency seismic waves, triggering mechanisms of earthquakes, and seismic events associated with volcanic activity.

5.22 Tohoku University, Research Center for Prediction of Earthquakes and Volcanic Eruptions (T. Hirasawa)

Web site: http://aob-new.aob.geophys.tohoku.ac.jp/index_e.html

The Research Center for Prediction of Earthquakes and Volcanic Eruptions was re-organized in 1998. It is descended from Mukaiyama Observatory, which was established in 1912 under Shirota Kusakabe for seismological, meteorological, and astronomical observations. In 1931 it moved from Mukaiyama to Yagiyama, both in the city of Sendai. In 1952 it was renamed Seismological Observatory. In 1967, the observatory further moved to the present Aobayama Campus of Tohoku University and changed its name to Aobayama Seismological Observatory. The observatory was supervised by Saemontaro Nakamura, Hirokichi Honda, Yoshio Kato, and Akio Takagi.

The direct predecessor of the present center is Observation Center for Earthquake Prediction, which was established in 1974 mainly for observational studies of earthquake phenomena in response to the national project of earthquake prediction. In 1987, it was re-organized and named Observation Center for Prediction of Earthquakes and Volcanic Eruptions to make itself responsible for the national project of volcanic eruption prediction as well as for that of earthquake prediction. The center was directed by Akio Takagi (1974–1989) and Tomowo Hirasawa (1989–present). The center developed networks for seismic and crustal deformation observations in northeastern Honshu, Japan, and all the data observed have been telemetered to the center in Sendai. The observed seismic data of high quality revealed the double-planed structure of the deep seismic zone in the subducting Pacific plate in the northeastern Japan arc. It is now well-known that earthquakes occurring on the upper plane are characterized by down-dip compression and earthquakes on the lower plane are by down-dip tension (Hasegawa *et al.*, 1978). Further, the analysis of strain data obtained by the crustal deformation network disclosed the migration of maximum shear strain at a speed of 40km/year approximately toward the motion direction of the Pacific plate relative to the overriding continental plate (Ishii *et al.*, 1980).

Current research includes seismic structure and seismotectonics in the northeastern Japan convergent margin, deep structure of arc volcanoes, interplate coupling, and laboratory and numerical simulations of earthquake processes.

5.23 Tokai University, Earthquake Prediction Research Center (S. Uyeda and T. Nagao)

Web site: http://yochi.iord.u-tokai.ac.jp/

The Earthquake Prediction Research Center (EPRC) was established under the Institute of Ocean Research and Development, Tokai University, in April 1995, the year of the great Hyogo-ken Nanbu (Kobe) earthquake. That was a mere coincidence because the establishment of EPRC was decided in 1994, well before the Kobe event. EPRC is located on the campus of the School of Marine Science and Technology, Tokai University, in Shimizu City, which is inside the expected epicentral area of the predicted Tokai Earthquake. There are four researchers and technicians as of the end of 1999.

The main objective of the EPRC is to investigate the possibility of short-term earthquake prediction by using mainly electromagnetic methods, recognizing the possible usefulness of the recently reported electromagnetic precursors to earthquakes. EPRC aims at systematic research in various frequency ranges and clarification of physical mechanisms through cooperation with experts around the world. EPRC has no seismic network of its own because our main interest is not earthquakes themselves but what happens before earthquakes.

Just after the establishment of EPRC, the Science and Technology Agency (STA) of Japan decided to start "Earthquake Frontier Research Program." The initiation of this program was an after-effect of the Kobe Earthquake. The "Earthquake Frontier Research Program" consists of five sub-programs. One of the sub-programs, i.e., the RIKEN International Frontier Research Program on Earthquakes (IFREQ), decided to set up its headquarters at EPRC, which became the main contractor of the program. Up to the present, almost all of the research activity of EPRC has been based on IFREQ.

5.24 The University of Tokyo, Department of Earth and Planetary Sciences (M. Matsu'ura and Y. Hamano)

Web site: http://www.eps.s.u-tokyo.ac.jp/

The Department of Earth and Planetary Science, Graduate School of Science, was formed on April 1, 2000, by the merger of four departments related to the earth and planetary science: the Department of Earth and Planetary Physics, the Department of Geology, the Department of Mineralogy, and the Department of Geography. Three subgroups (earthquake physics, structure of the Earth's interior, and dynamics of the Earth's interior) in the Solid Earth Science group and one subgroup (analysis of the Earth and planetary system) in the Earth and Planetary System Science group cover the research fields related to IASPEI. Although the department is quite new, the previously existing four departments have histories going back to the 19th century.

Modern seismology was started in 1880 at a small seismological research laboratory affiliated to the Faculty of Science, the University of Tokyo, by an invited foreign professor, James A. Ewing. This is the root of the Department of Earth and Planetary Science. Another invited foreign professor, John Milne, in the Faculty of Engineering, and a young Japanese physicist, Seikei Sekiya, joined the seismological research laboratory. Professors Milne and Ewing greatly contributed to the establishment of the Seismological Society of Japan in 1880. The seismological research laboratory was authorized as a unit of education and research in 1885 and the Seismological Laboratory, chaired by Prof. Sekiya, was established.

In 1893 the Chair of seismology was established and held by Prof. Sekiya. He soon became ill and died in 1896. A young Lecturer, Fusakichi Ohmori, succeeded S. Sekiya as a professor in seismology. Professor Ohmori contributed to the development of seismology in Japan through activity in the Imperial Earthquake Investigation Committee. In November 1923, shortly after the great Kanto earthquake, Prof. Ohmori died and Akitsune Imamura became a professor and held the Chair of seismology. In October, the Department of Seismology was established in the Faculty of Science. Two years later, the Earthquake Research Institute was also established in the University of Tokyo, and the Imperial Earthquake Investigation Committee ceased to exist. In 1929 Professor Imamura organized the Seismological Society, and started to publish its bulletin "Zisin." Professor Imamura retired in 1931, and Associate Professor Takeo Matuzawa (promoted to a professor in 1936) held the Chair of seismology through 1960.

In 1941 the Department of Seismology ceased to exist, and the Department of Geophysics (the Geophysical Institute) was newly established. The Department of Geophysics was started with the four Laboratories of seismology, meteorology, oceanography, and geodynamics. In 1953 the fifth Laboratory, geomagnetism and geoelectricity, was added. In this department Professor Matuzawa held the Chair of seismology and Professor Chyuji Tsuboi held the Chair of geodynamics. After the retirement of Prof. Matuzawa, Professor Hirokichi Honda (1961–1966), Professor Toshi Asada (1966–1980), and Professor Ryosuke Sato (1981–1989) held the Chair of seismology. After the retirement of Professor Tsuboi in 1963, Professor Hitoshi Takeuchi held the Chair of geodynamics and changed the name of the Laboratory to physics of the earth and planetary interiors. Professor Takeuchi retired in 1982, and Professor Mineo Kumazawa took over.

The aim of the Department of Earth and Planetary Science is to construct a center of excellence for education and research in Earth and planetary sciences in Japan. The Department has about 110 faculty members responsible for education and research in various fields of Earth and planetary science. About half of the faculty members hold appointments in the Graduate School of Science of the Hongo campus, and the remaining faculty members belong to the Laboratory for Earthquake Chemistry of the School of Science, the Earthquake Research Institute, the Ocean Research Institute, the Center for Climate System Research, the Institute for Solid State Physics, the Graduate School of Arts and

Sciences at the Komaba campus of the University, the Center for Spatial Information Science, the Graduate School of Frontier Sciences, and the National Institute of Space and Aeronautical Sciences. About 110 students are admitted to the Master course, and about 50 to the Doctoral course every year.

The Department of Earth and Planetary Science is divided into five major groups: Oceanic and Atmospheric Science group (OAS), Space and Planetary Science group (SPS), Earth and Planetary System Science group (EPSS), Solid Earth Science group (SES), and Geosphere and Biosphere Science group (GBS). A brief summary of the research field of each group is given in the Department's report on the Handbook CD.

Research and education in solid Earth physics are mainly covered by the Solid Earth Science group and the Earth and Planetary System Science group. Earthquake physics subgroup (chaired by Professor M. Matsu'ura) in SES conducts research and education in the physics of earthquake generation, crust–mantle dynamics, theory and application of geophysical data inversion, and computer simulation of earthquake generation cycles. Structure of the Earth's interior subgroup (chaired by Professor R. Geller) in SES conducts theoretical and computational study of seismic wave propagation, and aims to determine the 3-D structure of the Earth's interior by inversion analysis of seismic waveform data. The main research subject of dynamics of the Earth's interior subgroup (chaired by Professor Y. Ida) in SES is the modeling of mantle dynamics and physical processes involved in various magmatic and volcanic phenomena. Analysis of the Earth and planetary system subgroup (chaired by Professor Y. Hamano) in EPSS aims to specify operational processes of the interactions between multiple domains of the Earth and planetary system by means of theory, observation, and experiments, and to explore the mechanism to stabilize and maintain present condition of the system. The subgroup is presently focused on the investigation of dynamics and evolution of the Earth and planetary interiors such as the fluid motion of the core.

5.25 The University of Tokyo, Earthquake Research Institute (T. Yoshii)

Web site: http://www.eri.u-tokyo.ac.jp/

The Earthquake Research Institute (ERI) was established in November 1925 as a part of Tokyo Imperial University. The great disasters caused by the Kanto Earthquake of 1923 provided the trigger for establishment. The decade and a half after the formation of ERI was the time of the rise of modern seismology in Japan. During this period, the Institute continuously expanded its organization and played a major role with its remarkable research in wide fields of seismology, volcanology, geology, and earthquake engineering.

After World War II, the Institute was re-established as one of the research institutes of the University of Tokyo. Following the nationwide cooperative Earthquake Prediction Program started in 1965 and Volcanic Eruption Prediction Program started in 1974, the Institute played a core role in execution of the programs, as well as serving as the central institute in Japan for fundamental geophysical and engineering research concerning earthquakes and volcanoes.

In the last few decades, various cooperative studies, such as seismic observations in several inland areas, seismic and geophysical observations in the ocean, application of Global Positioning System (GPS), seismic and geophysical observations by a network covering the whole of the western Pacific, and experiments on volcanic structure and magma supply system, have been planned and conducted as joint research of many universities and institutions in Japan.

To achieve the further promotion of these projects, the Earthquake Research Institute was reorganized in 1994 as a shared institute of the University of Tokyo. The reorganization of the Institute formed four research divisions and four research centers, provided positions for visiting professors, and formulated a system of various kinds of cooperative studies. In April 1997, the Ocean Hemisphere Research Center was established to develop and operate global multi-disciplinary observations in the Pacific hemisphere.

After the disastrous Hyogo-ken Nanbu earthquake of 1995, earthquake prediction research in Japan changed its direction on a large scale. In this situation, the Earthquake Research Institute, the biggest institute for studies of solid Earth sciences in Japan, is quite important in new research.

A full report of the Institute is given on the Handbook CD. It includes (1) major developments and accomplishments of the Earthquake Research Institute with references to relevant publications, (2) current activities of the Earthquake Research Institute in the Division of Earth Mechanics, Division of Global Dynamics, Division of Monitoring and Computational Geoscience, Division of Disaster Mitigation Science. Earthquake Prediction Research Center, Earthquake Observation Center, Earthquake Information Center, Volcano Research Center, Ocean Hemisphere Research Center, and Selected Topics of Cooperative Researches, (3) research staff of the Earthquake Research Institute, including a list of professors and associate professors, and (4) a complete index of the Bulletin of the Earthquake Research Institute (1926–1999).

5.26 The University of Tokyo, Institute for Solid State Physics (T. Yagi)

Web site: http://www.issp.u-tokyo.ac.jp/index.html

The Institute for Solid State Physics (ISSP) of The University of Tokyo was established on April 1, 1957, upon the recommendation of the Science Council of Japan and with the concurrence of the Ministry of Education and the Science and Technology Agency. S. Kaya was the director during this period of establishment, and most research at the institute focused on topics of solid state physics and materials science. Research in mineral physics began in 1961 when S. Akimoto joined the High Pressure Research Division. He constructed a tetrahedral-anvil type high-pressure apparatus, and started a study of the olivine-spinel transformation in silicates. He made many pioneering contributions on this topic, and extended the research into various topics regarding the physics of the Earth's interior. After his retirement in 1986, T. Yagi succeeded him as director of the laboratory. He has extended the research into high pressure *in situ* observations, rather than quench experiments, using both multi-anvil type and diamond-anvil type high-pressure apparatus combined with synchrotron radiation. Very precise studies of the equations of state and phase relations of various silicates have been carried out up to conditions of the Earth's lower mantle. Yagi's group now belongs to the Division of New Materials Science, and further information and a list of publications are available at the Institute Website.

5.27 Yokohama City University, Seismological Laboratory (S. Tsuboi)

Web site: http://www.seis.yokohama-cu.ac.jp/

Seismology in Yokohama City University officially started in 1997, when the Seismological Laboratory was established. This laboratory maintains the high-density strong motion seismograph network in Yokohama City. However, active works have been conducted in earthquake source studies and experimental rock mechanics since the 1970s. Scientific personnel who have supported this work are M. Kikuchi (1982–1996), N. Yoshioka (1986–), M. Saito (1997–), S. Tsuboi (1999–), and Y. Ishihara (1991–).

The activities of the Seismological Laboratory mainly consist of observational and experimental seismology. Yokohama City University has operated four broadband seismograph stations in the southern Kanto district since 1991. Earthquake source studies using these broadband seismograms have been conducted for both regional and teleseismic earthquakes. In addition to the broadband seismograph network, the high-density strong motion seismograph network (at 1–2 km spacing) has been established in Yokohama city as a part of Real-time Assessment of Earthquake Disaster in Yokohama (READY) since 1997. READY is the first system ever built in Japan for earthquake hazard mitigation. Strong motion seismograms recorded by this network are used to estimate crustal structure in Yokohama City. Experimental studies have been done in rock mechanics. The main target of these experiments is to investigate friction property along the earthquake fault surface.

79.34 Kazakhstan

The editors received two reports from Kazakhstan: (1) a brief "Institutional Centennial Report" from Dr. Nadezhda N. Belyashova, Director of the Institute of Geophysical Research, National Nuclear Center, Republic of Kazakhstan, and (2) "About Some Methods of Earthquake Short-Term Prediction Around Almaty Megapolis" from Dr. M. Khaidarov, Director, Scientific Forecasting Center of the Emergency Agency of the State Agency, Republic of Kazakhstan. These two reports were edited and archived as computer-readable files on the attached Handbook CD#2, under the directory \7934Kazakhstan. We extracted and excerpted a summary here. A review of earthquake engineering in Kazakhstan is given in Chapter 80.11, and biographical sketches of some scientists and engineers of Kazakhstan may be found in Appendix 3 of this Handbook.

1. Institute of Geophysical Research, National Nuclear Center

Address: Meridian Site 490021 Kurchatov East Kazakhstan Republic of Kazakhstan

The Institute of Geophysical Research was established in 1993 soon after the formation of the independent Republic of Kazakhstan. The Institute was created from the Kazakhstani subdivisions of the Institute for Physics of Earth, Russian Academy of Sciences (observatory "Borovoye"), and Ministry of Geology (South Affiliate of the Special Regional Geophysical Expedition) of the former USSR. After the removal of military organizations from Kazakhstan in 1994, the Institute inherited the seismic stations that monitored nuclear testing from 1960 to 1992.

At present, the Institute consists of three geophysical observatories—"Kurchatov" (East Kazakhstan), "Borovoye" (Central Kazakhstan), "Kaskelen" Southern Kazakhstan)—two seismic stations—"Aktyubinsk" (Western Kazakhstan), "Makanchi" (South-eastern Kazakhstan)—Data Center (Almaty), and two research departments—research on geo-environmental processes, and engineering investigations. The main activities of the Institute are related to observational seismology, Earth's structure study, and earthquake engineering.

1.1 Observational Seismology

The Kazakhstani seismic network consists of permanent and mobile stations (arrays), and the Center for Acquisition and Processing of Seismic Information established in 1999. The permanent observational network consists of three seismic arrays (a large-aperture array—"Borovoye," medium-aperture array—"Kurchatov," small-aperture array—"Makanchi") and one three-component station ("Aktyubinsk"), equipped with broadband digital and short-period seismometers. The permanent network and the data center are part of the International Monitoring System (IMS) created to verify compliance with the Comprehensive Test Ban Treaty (CTBT). In addition, this network is part of the Global Seismological Network (GSN). The mobile part of the observational network consists of 40 seismic stations "KARS." The network performance enables continuous round-the-clock monitoring of seismic events ($m_b \geq 4.0$ at distances over 3000 km, $m_b < 4.0$ at distances less than 3000 km).

New technologies for detection, location, and discrimination of seismic and acoustic events are being developed. The instrumentation for seismic monitoring is being updated from analogous to digital. Telecommunication networks using satellites are being implemented between the seismic stations and the data center. Calibration of the Kazakhstani seismic stations using chemical explosions is aimed at elimination of the Semipalatinsk Test Site (STS), and other ground truth data.

1.2 Earth's Structure Study

The velocity model of Central Kazakhstan is being developed using the records of 25 100-ton calibration explosions conducted

INTERNATIONAL HANDBOOK OF EARTHQUAKE AND ENGINEERING SEISMOLOGY, VOLUME 81B ISBN: 0-12-440658-0

at the STS in 1998–1999, the data of deep seismic sounding, acquired in past years, and the records of industrial shots and local earthquakes. The upper mantle and lithosphere structure is being studied using converted waves of earthquakes.

1.3 Earthquake Engineering

The data of the observational network are being used for seismic zoning of Kazakhstan, and to develop a modern building code. Seismic investigations for earthquake-resistant design and construction of large industrial structures are being conducted (e.g., a new atomic station is being designed). The dynamic processes and a prognosis of geo-environmental consequences at the sites of underground nuclear explosions are being studied. Seismic observations are carried out to provide safety after elimination of underground nuclear infrastructure (tunnels and boreholes at the STS).

1.4 International Collaborations

The Institute collaborates with Kazakhstani and foreign organizations working in the field of seismological research: Institute of Seismology, Kazakhstan; Institute for Dynamics of the Geospheres, Institute of Physics of Earth, Russia; Lamont-Doherty Earth Observatory, Columbia University, USA; IRIS Consortium, US Geological Survey, USA; NORSAR, Norway; and others.

2. Scientific Forecasting Center, Emergency Agency of the State Agency

The Scientific Forecasting Center "PROGNOZ" (SFC) is one of the subdivisions of the Emergency Agency of the Republic of Kazakhstan. The network of nine seismic sites of SFC "PROGNOZ" was created in 1989–1990, intended for monitoring the seismic situation in the southeast region of Kazakhstan. This network is incorporated into the unified seismological network of the Republic of Kazakhstan. The staff of SFC consists of 86 persons, of whom 9 are scientific employees with advanced degrees.

The observing region is North Tien-Shan. It is situated in the intra-continental area of the Eurasian mainland. North Tien-Shan is a highly seismic dangerous area. During the last 150 years, there were 4 great earthquakes (magnitude > 8) in Central Asia, 2 of which (Chilikskoe 1889, $M = 8.3$, depth $= 40$ km and Keminskoe 1911, $M = 8.2$, depth $= 25$ km) occurred in the territory of south Kazakhstan near Almaty. One other strong earthquake (Vernenskoe 1887, $M = 7.3$, depth $= 20$ km) occurred in the same region.

This territory has a high density of population at present and a short-term prediction of earthquakes is considered to be very important. According to the majority opinion the territory of Tien-Shan has entered a new period of seismic activation. The government of the Republic of Kazakhstan and the Emergency Committee have undertaken a number of measures to amplify the scientific and practical works in the field of earthquakes and earthquake forecasting.

The primary task of SFC is a well-timed notification of the government, the Emergency Agency, and other interested organizations of dangerous changes in the seismic situation. This task is realized by using new nontraditional methods of seismic monitoring. Information on the seismic situations is sent four times a day under dangerous situations, i.e., at the time of formation of a precursor. The method of short-term forecasting is based on the registration of the anisotropy of the microseismic background in separate frequency ranges through a system of horizontal torsion pendulums.

Some results on the analysis of temporal sequences of correlation of earthquakes and torsion oscillations were presented recently at two international forums: XXVI General Assembly of European Seismological Commission (ESC) in Tel-Aviv and The World Forum of Big Cities Protection in Istanbul in 1998. More details are given in the full report on the Handbook CD.

79.35 Kyrgyzstan

The editors received a report from Prof. A.T. Turdukulov, Director of the Institute of Seismology of the National Academy of Science, Asanbai 52/1, Bishkek, 720060 Kyrgyz Republic. The edited report is archived as a computer-readable file on the attached Handbook CD#2, under the directory of \7935Kyrgyzstan. We extracted and excerpted a summary as shown here. Earthquake engineering in Kyrgyzstan is reviewed in Chapter 80.12 of this Handbook. An inventory of the seismic network in the Kyrgyz Republic is included in Chapter 87. Biographies and biographical sketches of some scientists and engineers of Kyrgyzstan may be found in Chapter 89, and in Appendix 3 of this Handbook, respectively.

1. Introduction

The Institute of Seismology (KIS) was founded in 1975. KIS activity involves joint cooperation with the world scientific community in the fields of seismological observations, seismic tomography, seismotectonics, and seismic hazard assessment. The Director is Prof. Asker Turdukulov.

2. Organization

The Institute consists of seven laboratories, organized in three divisions according to the scientific main themes of KIS, and one Seismological Expedition. The Division of Seismozoning, headed by Prof. K. Abdrakhmatov, is made up of the laboratories of Seismotectonics, Regional Seismology, and Deep Structure of Source Zones. The Division of Earthquake Prediction, headed by Prof. E. Mamyrov, includes the laboratories of Hydrogeochemical and Tectonophysical Methods of Earthquake Prediction, Geophysical Methods of Earthquake Prediction, and Seismological Methods of Earthquake Prediction. The third division is the Division of Engineering Seismology and Microzoning Studies, consisting of a single laboratory.

3. Research Activities

The main themes of research are (1) seismic risk assessment and seismic engineering, (2) Earth deep structure and seismic potential, (3) seismic hazard analysis, and (4) earthquake prediction.

Currently, 29 regional analog seismic stations, strong ground motion recorders, and 10 telemetric stations of a joint American–Kyrgyz earthquake monitoring network (KNET) and 1 IRIS global network station, as well as geomagnetic stations and hydrogeochemical, hydrogeodynamic stations, operate within the Kyrgyz republic. This joint network (KNET) has been operating since 1991. A temporary American broadband network of 14 digital stations operated in Kyrgyzstan during the time period 1997–2000.

More details, including biographies and biographical sketches, are given in the full report on the Handbook CD.

INTERNATIONAL HANDBOOK OF EARTHQUAKE AND ENGINEERING SEISMOLOGY, VOLUME 81B ISBN: 0-12-440658-0

79.36 Luxembourg

The editors received the "National Report of the Grand Duchy of Luxembourg" from Ing. J.A. Flick. This report is archived as a computer-readable file on the attached Handbook CD#2, under the directory of \7936 Luxembourg. We extracted a summary as shown here. An edited version of the biographical sketches of P. Melchior and J. Flick may be found in Appendix 3 of this Handbook.

The program of research in seismology and related earth science began in Luxembourg in 1963. The guidance and assistance of Prof. Paul Melchior of the Royal Observatory of Belgium (ROB) was essential to the initial success of these developments. The first installation was that of an Askania gravimeter in the Casemates du St. Esprit, by P. Melchior. A three-component Galitzin-Wilip seismograph system was added in 1967 by J. M. Van Gils, also of the ROB. J. A. Flick of Luxembourg was appointed as the local coordinator in charge of daily maintenance.

Because the Casemates site was unacceptably noisy and subject to thermal disturbances, a new site was selected in an old gypsum mine near the village of Walferdange, leading to the establishment of the Walferdange Underground Laboratory for Geodynamics (WULG). Following the 1970 General Assembly of the European Seismological Commission, P. Melchior, J. Flick, F. Barlier, and M. Lefèbvre created the Journées Luxembourgeoises de Géodynamique, the JLG. This forum, in which European scientists can informally discuss geodynamics, now meets twice each year for three days. Research results are presented at a level that is accessible to a wide variety of participants.

In 1988 the European Center for Geodynamics and Seismology was created by the Council of Europe in Luxembourg and in 1994 a Convention was signed between the ECGS and the Grand Duchy of Luxembourg. The ECGS promotes programs in geodynamics research related to the study of tectonic deformations in connection with earthquakes. The ECGS also supports regular meetings, colloquia, and workshops, and has assisted in the further development of the WULG. About 19 "Cahiers bleus" (Proceedings of the workshops) have been published.

Some major projects that have been part of the evolution of the solid Earth research efforts in Luxembourg include the Earthquake Zoning Map of Northwest Europe (1976), the east–west seismic profile (1979), the Seismic Control Network (begun in 1988), the development of the transfrontier group (begun in 1994), and the Eifel Plume Project (1997). Further information about these and other advances, and biographies of P. Melchior and J. Flick, are given in the full report on the Handbook CD.

INTERNATIONAL HANDBOOK OF EARTHQUAKE AND ENGINEERING SEISMOLOGY, VOLUME 81B ISBN: 0-12-440658-0

79.37 Macedonia

The editors received the "National Report of the Republic of Macedonia" from Vera Cejkovska, Lazo Pekevski, and Ljupco Jordanovski, Seismological Observatory, St. Cyril and Methodius University, P.O. Box 422, 1000 Skopje, Republic of Macedonia. This report is archived as a computer-readable file on the attached Handbook CD#2, under the directory of \7937Macedonia. We extracted and excerpted a summary as shown here. Earthquake engineering in Macedonia is reviewed in Chapter 80.13 of this Handbook. An inventory of the Macedonian seismic stations is included in Chapter 87. Biographical sketches of some Macedonian scientists and engineers may be found in Appendix 3 of this Handbook.

1. Introduction

The organization and development of activities in the field of earthquake seismology in the Republic of Macedonia during the 20th century have been basically related to the development of the historical and political conditions in the Balkans. The larger geographic region of Macedonia was the last of the Balkan territories that was liberated from the Turks in 1912–1913. Until then, seismological observation in Macedonia had been reduced to compilation of macroseismic data on felt strong earthquakes by the Turkish administration or individual foreign investigators. After the Balkan Wars (1912–1913), part of Macedonia, called by its geographical name of Vardar Macedonia, became part of Serbia, and after World War I, part of the Kingdom of Yugoslavia. It was in 1913 that the Seismological Service of Serbia began to officially extend its activities in Vardar Macedonia. In 1944, the Republic of Macedonia was established in the territory of Vardar Macedonia as one of the six constituent republics of the Federal Republic of Yugoslavia. In 1991, after the disintegration of this federal state, complete independence of the Republic of Macedonia was proclaimed.

2. Seismological Monitoring

The first independent seismological monitoring in the Republic of Macedonia started on July 1, 1957, with foundation of the seismological station in Skopje (SKO). Two Mainka mechanical seismographs (EW and NS components, pendulum mass of 450 kg) and a Wiechert contact timing device represented the first equipment of the station. In February 1963, the Conrad mechanical seismograph was installed. After the devastating Skopje earthquake of July 26, 1963, a Vegik analog short-period electromagnetic seismograph was installed. Lehner-Griffith, Willmore (short period), Press-Ewing (long period) electromagnetic seismographs and an AR-240 strong motion recorder were put into operation in March 1966. In the 1960s, the Seismological Station in Skopje founded two new seismological stations, Valandovo (VAY, 1966) and Ohrid (OHR, 1967). These stations were equipped with Lehner-Griffith short-period electromagnetic seismographs.

In 1966, the seismological station in Skopje became the Seismological Observatory within the University of Skopje. In 1976, the Seismological Observatory in Skopje became a section of the Faculty of Physics in Skopje, while in 1984 it became an institute within the Faculty of Natural Sciences and Mathematics in Skopje. Following the new worldwide trends of development of instrumental seismology, the Seismological Observatory in Skopje started in 1990 to build a telemetric network of SSR-1 digital seismological stations, with SS-1 (short period) and WR-1 (wide-range period) seismometers.

3. Recent Advances

Nowadays, the Seismological Observatory at the Faculty of Natural Sciences and Mathematics, St. Cyril and Methodius University is the only institution in the Republic of Macedonia that is authorized and obliged to perform seismological

INTERNATIONAL HANDBOOK OF EARTHQUAKE AND ENGINEERING SEISMOLOGY, VOLUME 81B ISBN: 0-12-440658-0

service in the Republic. It is organized into four sections: microseismics, macroseismics, seismological instrumentation, and computer science laboratory. The Observatory systematically monitors instrumentally seismic activity in the territory of the Republic of Macedonia and the bordering areas and also records the regional and teleseismic earthquakes. In cases of felt earthquakes in the Republic of Macedonia, the Observatory compiles and processes the data on the macroseismic effects of earthquakes. The instrumental and macroseismic data are compiled, stored, analyzed, and published in seismological bulletins and catalogs for the international exchange of seismological data and for scientific, teaching, and civil engineering purposes. The Observatory also performs scientific research, education, and applications in the field of seismology and geophysics.

More detailed information, including a list of research staff members and publications, is given in the full report on the Handbook CD.

79.38 Mexico

The editors received the "National Report of Mexico" from Prof. Cinna Lomnitz, IASPEI National Correspondent, Instituto de Geofisica, Universidad Nacional Autónoma de Mexico, Mexico City, Mexico. This report is archived as a computer-readable file on the attached Handbook CD#2, under the directory of \7938Mexico. An edited version of the summary submitted by the author is shown here. An inventory of seismic network in Mexico is included in Chapter 87 of this Handbook. A biography of Emilio Rosenblueth is included in Chapter 89, and biographical sketches of some Mexican scientists and engineers may be found in Appendix 3 of this Handbook.

1. Introduction

Earthquake history in Mexico starts around 1422, the date of the earliest recorded Mexican earthquake and three years before the foundation of Mexico City. Earthquakes were not feared as much as they are today because the construction methods (lightweight homes and pyramids) were less vulnerable to seismic effects. Earthquake science in Aztec times was relatively advanced compared to the state of knowledge in Asia or Europe. The fall of Mexico City to the Spanish siege occurred in 1521, but the Spanish had no better scientific explanation for earthquakes than the Aztec priests. Eventually, however, some distinguished Mexican scientists, such as J.A. Alzate (18th century), argued that earthquake hazard might have something to do with the soft black mud on which Mexico City was built. They were not far from the truth.

2. Earthquake Monitoring

The earliest Mexican seismic records stem from the 1880s. The administration of General Porfirio Diaz strongly supported the sciences, especially astronomy and the earth sciences. Mexico was one of the founding members of the International Association of Seismology, the present-day IASPEI. In 1904 the government purchased, at great expense, several sets of modern seismographs including Wiechert and Bosch-Omori instruments. As a result, we were fortunate in obtaining an outstanding record of the 1906 San Francisco earthquake written on smoked paper at the Tacubaya station.

The National Seismological Service was inaugurated by President Diaz on September 5, 1910, as a part of the festivities of the 100th anniversary of Mexican Independence. From 1910 through 1968 the Seismological Service used the travel-time tables published by Omori in 1909. In 1920 President Obregón decided to recall José Vasconcelos from exile to head the National University. In 1929, as a result of the autonomy movement, the National Seismological Service was formally transferred to the university together with the Geological Service and the Astronomical Observatory. Since 1949 the National Seismological Service is a part of the Institute of Geophysics of the National University (UNAM). No formal research in seismology was done in the early years of the Seismological Service, except for the publication of isoseismal maps and some seismicity surveys of selected states.

3. Seismology at UNAM

In 1968 the director of the Institute of Geophysics, Professor Ismael Herrera, hired a research professor with formal academic training in seismology. Plate tectonics had been born the year before, and Mexico was an exciting new research area as it is located at the intersection of five tectonic plates with entirely different styles of interaction.

The funding of science became a priority with the creation of the National Council of Scientific Research in 1971. One of the first grants of the Council was awarded to a digital seismic network to be built at the university. The first digital interfaces were constructed from scratch by university technicians and engineers

INTERNATIONAL HANDBOOK OF EARTHQUAKE AND ENGINEERING SEISMOLOGY, VOLUME 81B ISBN: 0-12-440658-0

around 1975. These were our RESMAC stations. A microwave link was established between the long-distance telephone exchange in Mexico City and the UNAM campus, and the signals were unscrambled at our end by large, clumsy computers. A half-dozen RESMAC stations are still running as part of the national seismic network.

In the early 1970s Professor Shri Krishna Singh joined the staff. He has been the head of the Seismology Department at UNAM, and its most distinguished and prolific scientist, until the present. The Department now includes 15 research professors, 5 of whom originally joined the department from abroad. One of our professors is on leave as a leading official of the Nuclear Test-Ban Treaty Organization in Vienna. A group of young research professors from the department moved to a new campus in Juriquilla, in the state of Querétaro. They have now joined the geologists and geophysicists to form a new and lively UNAM Center of Geosciences there.

Rapid and reliable information on Mexican earthquakes can be found at the Web site of the National Seismological Service, URL: http://www.ssn.unam.mx/.

4. Earthquake Engineering at UNAM

The vulnerability of Mexico to earthquake disasters was rapidly evolving as a result of the introduction of multistory apartment buildings in the 1940s. The late Emilio Rosenblueth was asked to develop the Mexico City Building Code along the general lines of the California code. He played a crucial role in the successive modifications of the Mexican codes up to and including the present 1987 version.

Earthquake engineering at UNAM is and remains a full-fledged division of the Institute of Engineering. At present there are 10 research professors in the division including the current institute director, Prof. Francisco Sánchez Sesma. In addition there are 15 engineers in the department of seismic instrumentation. They are in charge of strong motion instruments operating in Mexico; more than 120 such stations operate in Mexico City alone, in cooperation with City Hall and the Barros Sierra Foundation.

5. Other Institutions

The Center of Scientific Research and Higher Education (CICESE) at Ensenada, Baja California, was created in 1971 as a graduate school under the National Science Council. It has a Seismology Department with 15 research professors. CICESE and UNAM currently account for most of the scientific production in the earth sciences in Mexico but emerging groups of earthquake researchers are also found in the academic communities of the Federal District, Jalisco, Puebla, Nuevo León, and other states.

Shortly after the disastrous 1985 Michoacán earthquake the National Center of Disaster Prevention (CENAPRED) was created by the federal government with the assistance of Japan. In addition to public information, CENAPRED is mainly engaged in monitoring and technological development; much of its leading scientific personnel is on loan from UNAM. Concurrently, a number of federal and state agencies have created special emergency outfits able to assist the population in the event of an earthquake.

In 1984 the Mexican scientific community created an original leadership institution, the National Research System (SNI). SNI is now a part of the National Science Council (CONACYT); it currently has 7,000 members, including more than 100 seismologists. Most of them are also members of the Mexican Geophysical Union, a 40-year-old organization that organizes a well-attended annual meeting in Puerto Vallarta, Jalisco, in November. The Mexican Geophysical Union is closely associated with the American Geophysical Union and Mexican seismologists also tend to be members of the Seismological Society of America. Their scientific contributions may often be found in the Bulletin of the SSA or in the journal of the Mexican Geophysical Union, Geofísica Internacional, published quarterly since 1960.

79.39 Moldova

The editors received a report from Prof. Anatol Drumea, Director of the Institute of Geophysics and Geology, Moldavian Academy of Sciences, Academie str., 3, Kishinev, MD-2028, Moldova. It is written by A. Drumea, V. Ginsari, and A. Zaicenco. The edited report is archived as a computer-readable file on the attached Handbook CD#2, under the directory of \7939Moldova. We extracted and excerpted the following summary.

1. Introduction

Seismic observations in Moldova on a regular basis started in 1949, when, on December 20, the first seismogram was recorded at the seismic station Kishinev. The year 1963 could be considered the starting point of the scientific investigations into earthquake engineering, when the first volume of scientific publications was issued dedicated to problems of tectonics and seismology of Moldova, prepared by the group of young scientists of the Institute of Geology and Mineral Resources of the Academy of Sciences of Moldavian Soviet Socialist Republic (MSSR).

The Institute of Geophysics and Geology (IGG) was founded in 1967 on the basis of the Institute of Geology and Minerals and the regional seismic station "Kishinev." The research priorities of the Institute are monitoring of seismicity of the Vrancea zone, seismic hazard and risk assessment, microzonation, GIS technologies, and mathematical models in earthquake engineering. The present director is Dr. Vasilii Alkaz. The staff has numbered from 100 to120 in the 1970s and 1980s to 50 in the 1990s. Currently the staff consists of 22 seismologists (including staff of seismological stations), 8 of them with Ph.D. degrees. The seismological section consists of (1) Laboratory for Seismology, (2) Laboratory of Survey of Seismic Effects, and (3) the Center of Experimental Seismology.

2. Research Activities

The territory of the Republic of Moldova is influenced by earthquakes of intermediate depth from the Vrancea seismic zone, situated in Romania. The strongest of these earthquakes are distributed in the depth interval of 80–150 km, with maximum magnitude of 7.5–7.8. The most significant seismic effect, maximum intensity VIII–IX on the scale of XII, is observed in Romania and Moldova. Statistical information about seismic activity of the Vrancea zone is available since the year 1000. On average, strong earthquakes of magnitude $M \geq 6$ occur five times or more per century. Some of them (November 10, 1940, March 4, 1977, August 31, 1986) caused casualties and considerable damage.

The main mission of the seismological section is monitoring seismicity for the territory of Moldova, and conducting seismotectonic investigation, seismic hazard assessment, long-term earthquake prediction research, and engineering seismology. These investigations have resulted in maps of macro- and microzonation for seismic-resistant construction and are the basis for taking measures in reducing the consequences of strong earthquakes.

3. Seismic Network

The seismic network of Moldova consists of five seismic stations, situated in Kishinev, Cahul, Leovo, Soroky, and Djurjuleshti. Kishinev is the base station for the network; the other four provide regional data. Station Kishinev was established in 1949 by the Institute for Earth Physics, USSR Academy of Sciences, to provide supplementary data on parameters of Carpathian earthquakes. Station Cahul started its observations in 1978 and provides additional information for studying of characteristics of earthquakes from the Vrancea zone. Stations Leovo (1982) and Soroki (1983) were established in connection with structural changes in the Soviet network in 1979 for work on earthquake forecasts. Djurjuleshti was installed in 1988. Basic data of the seismic stations and equipment installed are shown in Table 1 and Table 2.

INTERNATIONAL HANDBOOK OF EARTHQUAKE AND ENGINEERING SEISMOLOGY, VOLUME 81B ISBN: 0-12-440658-0

TABLE 1 Data on Seismic Motion Recording Instruments in Moldova (Main Channels)

Name of station	Type	Component	Frequency-amplitude Characteristics		Speed of film mm/min
			V_{max}	T_{max}	
Kishinev	SKM-3	NS, EW, Z	3000	0.6-1.9	60
	SKD	NS, EW, Z	1000	0.2-19	30
	SKD csr	NS, EW, Z	100	0.2-15	30
	SD-2	Z	1200	17-52	15
	SD-2 csr	Z	33	26-67	15
Cahul	S5S	NS, EW, Z	1000	0.1-3.5	30
	S5S csr	NS, EW, Z	50	0.1-4.0	30
Leova	S5S	NS, EW, Z	2000	0.2-1.6	30
	S5S csr	NS, EW, Z	100	0.4-2.0	30
Soroca	SKM-3	NS, EW, Z	25000	0.1-1.3	60
			30000	0.2-0.7	60
	S5S csr	NS, EW, Z	400	0.1-4.0	60
Giurgiulesti	S5S	NS, EW, Z	2000		

TABLE 2 Data on Strong Motion Recording Instruments

Seismic station	Channels of record	Coef. of amplification
Kishinev	Seismograph S5S+M012/60Hz	Vo1 = 200, Vo2 = 5
	Velocitymeter S5S+M010/60Hz	Vov = 2 sec
	Accelerometer OSP+M010/40Hz	Voa = 0.01 sec^2
	Accelerometer OSP+M010A/40Hz	Voa = 0.15 sec^2
	Accelerometer SSRZ	Voa = 1.6×10^{-3} sec^2
	Accelerometer UAR	Voa = 1.6×10^{-3} sec^2
	Seismoscope SBM	
	Seismoscope AIS-3M	
Cahul	Accelerometer SSRZ	Voa = 1.6×10^{-3} sec^2
	Seismoscope SBM	
Leova	Seismograph VBP+M001/10Hz	Vo = 20
	Accelerometer OSP+GB-IV	Voa = 0.03 sec^2
Soroca	Seismograph S5S+M012	V_0 = 50
	Accelerometer OSP+M010/40Hz	Voa = 0.05 sec^2
	Accelerometer SSRZ	Voa = 1.6×10^{-3} sec^2
	Seismoscope SBM	
Giurgiulesti	Accelerometer SSRZ	Voa=1.6×10^{-3} sec^2

4. Recent Advances

The first personal computers appeared in the Institute in the late 1980s and allow processing vast amounts of data, and constructing graphical presentations of the results, especially maps. A database of the seismological information has been created in the Institute, including the catalog of the earthquakes and focal mechanisms of the studied region, macroseismic information. The statistical algorithms for interpretation of seismic intensity and seismic impact and alternative models of its assessment are considered in probabilistic representation of seismic hazard.

In late 1980s and early 1990s the group of seismologists from the Institute performed a series of works for determining the velocities of seismotectonic deformations. The basis of that method, which provided quantitative description of the contribution of seismicity into tectonic deformation, is the model of "seismic flow" suggested by Yuri V. Riznichenko in 1977. The velocity of the seismotectonic deformation was calculated and graphically represented in maps (with isolines) for the Vrancea zone and the Carpathian-Balkan region, as well as for the Caribbean region and Southern Sandwich islands.

In 1990–1995 the Laboratory of Seismology performed the investigation of the horizontal discontinuities of the upper mantle for Moldova and neighboring Romania by analysis of teleseismic P-wave propagation. The Laboratory of Survey of Seismic Effects has launched a project aimed at utilizing GIS technology for storing and processing of the available information. This project allows constructing of seismic macrozonation maps in digital format, and certain advances in seismic risk and seismic microzonation studies.

More detailed information on the activities and recent advances, with a list of past and current staff members and a representative bibliography, are given in the full report on the Handbook CD.

79.40 New Zealand

The editors received the New Zealand National Report from Warwick Smith and Martin Reyners, Institute of Geological and Nuclear Sciences, P.O. Box 30368, Lower Hutt, New Zealand. This report, including a list of references, is archived as a computer-readable file on the attached Handbook CD#2, under the directory of \7940NewZealand. A summary extracted and excerpted from the full report is shown here. For earthquake engineering in New Zealand, please see Chapter 80.14 submitted by the New Zealand Society for Earthquake Engineering to the International Association for Earthquake Engineering. Appendix 3 of this Handbook includes biographical sketches of some seismologists and earthquake engineers of New Zealand.

1. Introduction

New Zealand owes its seismicity to the fact that it straddles the boundary between the Pacific and Australian plates. Historical shallow earthquakes have reached magnitude 8.2. The largest events to occur in the 20th century were the Buller earthquake of 1929 ($M_S = 7.8$) and the Hawke's Bay earthquake of 1931 ($M_S = 7.8$). The latter was New Zealand's most damaging earthquake. By far the most active period was 1929 to 1934, with five shallow earthquakes reaching magnitude 7 or more. This extraordinarily active period stimulated research into the nature and causes of earthquakes, and also saw the beginnings of anti-seismic design provisions for domestic and commercial buildings.

2. Research Organizations

The Seismological Observatory has conducted almost all of the seismological recording and most of the related research in New Zealand. Seismographs were already operating in New Zealand at the beginning of the 20th century: Wellington had an instrument since about 1884, and two Milne instruments were in use from 1888. From 1900, systematic recording was maintained using Milne, and later Milne-Shaw, seismographs. Recording of strong ground motion and research in engineering seismology were initiated in the early 1960s by the Physics and Engineering Laboratory. The Institute of Geophysics at Victoria University of Wellington has had an active programme in seismological research since its establishment in 1971. The New Zealand Society for Earthquake Engineering has been active in promoting research since it was formed in 1968, as has the New Zealand Geophysical Society since its formation in 1980.

Directors of the Seismological Observatory have had various titles during the 20th century. These titles are not important, but the incumbents were as follows: G. Hogben (1900–1912), C.E. Adams (1912–1936), R.C. Hayes (1936–1960), F.F. Evison (1960–1964), R.D. Adams (1964–1978), G.A. Eiby (1978–1979), W.D. Smith (1979–1995), and T.H. Webb (1995–2002).

The Institute of Geophysics of the Victoria University of Wellington has been led by F.F. Evison (1971–1988), J.H. Ansell (1988–1993), and E.G.C. Smith (1994–present).

3. Seismograph Network and Earthquake Research

The seismograph network in New Zealand has steadily expanded during the 20th century, with the addition of Wood-Andersons during the 1930s and 1940s, a substantial upgrading in the mid-1960s using Willmore short-period seismographs with photographic recording, installation of radio telemetered networks from the 1970s, a conversion to digital recording in the 1980s,

INTERNATIONAL HANDBOOK OF EARTHQUAKE AND ENGINEERING SEISMOLOGY, VOLUME 81B ISBN: 0-12-440658-0

and a start made on satellite telemetry from broadband stations to the Observatory in the late 1990s. Similarly, the strong motion network has expanded since 1960 to over 230 instruments, including 100 digital accelerographs. Please see the "New Zealand" entry in Chapter 87 and its archived files on the attached Handbook CD#2 for information about the seismograph network and seismic data in New Zealand.

Most seismological studies done in New Zealand have had a local focus, given the high seismicity rate. Research on local earthquakes has provided a wealth of information on the structure of the country. Studies of intermediate depth earthquakes by Eiby in the late 1950s and early 1960s established that the uppermost mantle beneath the North Island and northern South Island is laterally inhomogeneous, well before the advent of plate tectonics theory. More recent studies have used dense deployments of portable seismographs to study the structure of the plate boundary in detail. In particular, studies of the Hikurangi subduction zone have revealed marked changes in the structure of both the subducted and overlying plates along the strike. These have been combined with the results of earthquake mechanism studies to infer changes in plate coupling along the subduction zone.

Many of New Zealand's largest earthquakes have been studied in detail, using both field and instrumental data. The first quantitative estimates of seismic hazard were made in the mid-1930s. These were based solely on seismicity, but also addressed the issues of attenuation of intensity with epicentral distance, and microzoning effects. Estimates of seismic hazard have evolved over the years, and probabilistic seismic hazard maps for the country are now based on the distribution and long-term recurrence behaviour of active faults as well as the spatial distribution of earthquakes observed in historical time.

4. Other Seismological Contributions

New Zealand seismologists have also made significant contributions in the fields of earthquake prediction, whole Earth studies, wave propagation, microzoning, reservoir-induced seismicity, and the detection of nuclear explosions. New Zealand's proximity to Antarctica has resulted in studies there. The active Taupo Volcanic Zone in the central North Island has stimulated a large amount of seismological research, including studies of volcanic tremor at Mt Ruapehu. The expansion of the strong motion database has resulted in a number of studies of the attenuation of strong ground motion throughout the country. Also, comprehensive studies of original reports of damage in historical earthquakes have allowed the estimation of damage ratios as a function of MM intensity, appropriate to New Zealand construction.

5. Further Information

Further information on the organizations involved in seismology in New Zealand, and their current activities, can be found at the following Web sites:

- Institute of Geological & Nuclear Sciences: http://www.gns.cri.nz/.
- Institute of Geophysics, School of Earth Sciences, Victoria University of Wellington: http://www.geo.vuw.ac.nz/.
- New Zealand Geophysical Society: http://www.geo.vuw.ac.nz/science.orgs/nzgs.html/.
- New Zealand Society for Earthquake Engineering: http://www.nzsee.org.nz

79.41 Norway

The editors received the "National Centennial Report of Norway" from Dr. Hilmar Bungum, IASPEI National correspondent of the Norwegian National Committee for IUGG, NORSAR, 2027 Kjeller, Norway. This report is archived as a computer-readable file on the attached Handbook CD#2, under the directory of \7941Norway. It is organized as follows: (1) a detailed account of the history of Norwegian seismic stations and a description of the present seismic network, by M. Sellevoll, H. Bungum, and J. Havskov, and (2) short reviews of some of the key Norwegian institutes involved in seismology, physics of the Earth's interior, and earthquake engineering, including obituaries and biographical sketches.

We extracted and excerpted the following summary. An inventory of seismic networks in Norway is included in Chapter 87 of this Handbook. Biographies and biographical sketches of some Norwegian scientists and engineers may be found in Chapter 89, and in Appendix 3 of this Handbook, respectively.

1. Introduction

Seismology in Norway started in the 1880s with macroseismic observations (questionnaires), a practice which has continued without interruption at the University of Bergen ever since. The history of earthquake recording dates back to 1905, when the first seismograph was installed in Bergen. The instrumentation has now grown to a network of many single stations and different types of seismic arrays, the latter operated by NORSAR. In relation to the fairly modest seismicity of the country, Norway has for a long time maintained comparatively strong efforts in this field.

Another important reason for the strong position of solid earth physics in Norway has been the fact that the NORSAR array, intended for research within seismological verification, was installed in the country in the late 1960s. Activity in this field has been strong ever since, tied to the establishment of NORSAR as a permanent institution, which later expanded into other fields of solid earth geophysics. One of the main tasks now is to monitor the comprehensive test ban treaty signed in 1996.

A third reason for the build-up of solid earth physics in Norway and a greatly increased demand for geophysicists has been the discovery of large petroleum resources off shore. The geophysicists are largely educated at Norwegian universities. This development has also vitalized the research environments within many fields of science only indirectly related to exploration geophysics.

2. Institute of Solid Earth Physics, University of Bergen

The Institute of Solid Earth Physics, University of Bergen, originated in 1899 with the national responsibility of Bergen Museum to collect macroseismic data in Norway. The University of Bergen was founded in 1946 out of the activities of the Bergen Museum, and a separate Seismological Observatory was created in 1960. In 1990 geomagnetism and paleomagnetism were also included in the activities and the name was changed to Institute of Solid Earth Physics. Today a scientific staff of 15 faculty members carries out research and offers a curriculum in seismology (3), petroleum geophysics (9), and paleomagnetism (3).

A Bosch-Omori seismograph began operation in the basement of Bergen Museum on May 25, 1905, and operated until 1959. This was the first instrument in Norway until a Wiechert horizontal seismograph was installed at the same site in 1929, supplemented two years later by a corresponding vertical seismograph. The Wiechert seismographs were operated until January 1968.

Installation of modern instrumentation started with three WWSSN stations (Kongsberg, Kirkenes, and Ny-Ålesund), and with three short-period stations (Jan Mayen, Tromsø, and Bergen) after 1960. At the same time the institute operated a pilot array station (Lillehammer) in preparation for the NORSAR array, which became operational in 1971.

INTERNATIONAL HANDBOOK OF EARTHQUAKE AND ENGINEERING SEISMOLOGY, VOLUME 81B ISBN: 0-12-440658-0

During the 1980s, a national network of short-period and broadband stations was deployed in Norway, including upgrading of the Jan Mayen and Ny-Ålesund (Svalbard) stations. Today, the institute operates a total of 22 stations, most of which have two-way data transfer via Internet, thus permitting near-real-time data collection and analysis. Network management has led to significant software development (Seisan) now used in a number of countries. Also, a small portable digital field station has been developed (sold by the Swiss company Geosig).

Research in seismology is directed toward seismicity, seismic hazard studies, network operation, classification of seismic events, and synthetic seismogram analysis in 2D and 3D. The institute has integrated international M.Sc. and Ph.D. programs in all 3 disciplines, and currently 17 students from 13 countries are enrolled in seismology alone.

3. Department of Geophysics, University of Oslo

The Solid Earth Physics group at the Department of Geophysics of the University of Oslo has a tradition of working with problems related to seismic wave propagation. A major member of the group was Professor Durk Doornbos, who headed the group between 1985 and 1993. An important part of the research performed in the group is still focused on developing new methodologies for modeling elastic wave propagation in complex structures and for inverting elastic wave data, with applications in seismology and in seismic prospecting.

Research activity on surface waves was started by the group at the beginning of the 1970s with the pioneering work by Bjørn Gjevik on surface wave propagation in laterally heterogeneous structures using ray theory, and with the analysis of the interaction of surface waves with lithospheric discontinuities. Surface wave propagation is still an active research theme, with the development of the concept of radiation modes to complement the normal modes, and the modeling of surface wave propagation in complex structures.

Mapping the structure of the core-mantle boundary (CMB) region is one of the most important contributions of the group to seismological research, with analysis of the topography of the CMB using short-period P waves, and analysis of diffracted S waves to study the radial structure, heterogeneity, and anisotropy in the D" layer.

A series of works on multiple scattering by irregular topographies that include methodological developments and applications to study the CMB and analyses of the effect of the Earth's topography on seismological recordings have been carried out. Different works on inversion have been performed in the group. The most noticeable and still-appicable theme concerns diffraction tomography, with the development of new methods based on a combination of wave and ray theory, in order to overcome the limitations of traditional travel-time tomography.

4. Norwegian Seismic Array (NORSAR)

NORSAR was established in 1968. It has been affiliated with the Research Council of Norway since 1970, and it was established as an independent research foundation as of July 1, 1999. NORSAR conducts research, development, and consulting within various fields of seismology and applied geophysics, including seismological verification. NORSAR is today one of the world's largest seismological observatories, with more than 30 years of experience with advanced seismological processing and data analysis techniques.

NORSAR has been a main contributor to the technology presently being implemented at the International Data Centre (IDC) in Vienna for verification of the Comprehensive Nuclear Test Ban Treaty (CTBT) signed in 1996. NORSAR scientists were instrumental in developing the concept of regional seismic arrays in the 1980s, a technology that today constitutes a backbone in the International Monitoring System (IMS), consisting of 170 seismic stations being prepared for data transmission to the IDC. NORSAR is now operating the CTBT-related National Data Center (NDC) for Norway.

The basis for these efforts has been the establishment and operation by NORSAR of a number of seismic arrays in Norway (NORSAR, NORES, ARCES) and at Svalbard (SPITS). In addition NORSAR has participated in establishing and operating several arrays in other countries, including Russia (APATITY), Finland (FINESS), Germany (GERESS), and Sweden (HAGFORS), and is still receiving data in real time from all of these installations. NORSAR also established the Northern Norway Seismic Network together with University of Bergen's Institute of Solid Earth Physics, and has cooperated closely with that institute both with exchange of data and on research projects.

The backbone of the research activities at NORSAR over 30 years has been verification research, which in turn has led its scientists into a number of research fields tied to the Earth's interior structure and composition, from core to crust, to seismic sources and wave propagation characteristics, at teleseismic, regional, and local distances, as well as to a variety of problems tied to applied and observational seismology.

Starting in the mid-1970s, NORSAR became gradually more engaged in engineering seismology, notably earthquake hazard analysis and the various components included therein, such as strong motion data analysis and inversions, seismic source studies, seismo-geological assessments, and microzonation studies. While the basis for this work was the need for seismic hazard assessment for the oil industry in the North Sea, NORSAR has later extended this to include most continents and a variety of industrial projects, including power plants and lifelines. This activity, including about 110 projects, has in turn often been tied to significant research efforts within geophysics, including regional and local seismotectonics.

Since the beginning of the 1980s NORSAR has also engaged itself increasingly in seismic prospecting and related seismic modeling, with activities ranging from theoretical studies to the development of commercial software. Based on ray tracing and other wavefield modeling techniques, both 2D and 3D solutions have been developed, applicable for modeling of seismic propagation through complex and anisotropic geological structures at various scales, including petroleum reservoirs.

The NORSAR staff consists of scientists (~15) and software engineers (~15) who are actively participating in research projects within array seismology, seismological verification, seismotectonics and earthquake hazard, and seismic exploration techniques. Over the 30 years since 1970, NORSAR scientists have produced close to 700 scientific contributions, a large number of which have been published in international scientific journals.

5. Norwegian Geotechnical Institute (NGI)

NGI is a private foundation for research and advisory services, located in Oslo, Norway. NGI was established in 1953 and currently employs a staff of 140 persons of whom 98 have university degrees.

Design, instrumentation, and analysis of foundations for buildings, bridges, offshore and harbor structures, evaluation of geotechnical aspects related to tunnels, rock caverns, and reservoirs, dams and environmental engineering, and landslide hazards are our major fields of expertise. NGI became an early international centre for geotechnical research. Trademarks of NGI research, now covering all aspects of geotechnical engineering, are to meet the industry's needs for practical and reliable design tools and foundation solutions.

In the field of earthquake engineering and earthquake-triggered landslide hazard, core competency within NGI is related to the following areas: (1) seismic hazard evaluation (in cooperation with NORSAR), (2) evaluation of local site response, (3) field and laboratory techniques for evaluation of dynamic soil and rock properties, (4) microzonation studies, (5) stability of slopes and embankments under earthquake loading, (6) seismic response analysis and earthquake-resistant design of dams, and (7) liquefaction potential evaluation.

Examples of major assignments include (1) development of National Application Document for Eurocode-8, (2) participation in the 8-year project "Earthquake hazard mitigation in Central America," (3) participation in microzonation projects for Managua (Nicaragua), Central Valley (Costa Rica), and David (Panama), (4) evaluation of seismic stability of submarine slopes in the North Sea and Caspian Sea, (5) participation in the EU research project "Long-period earthquake risk in Europe" within the 4th Frame Programme, and (6) numerous on-shore and off-shore site response and liquefaction potential studies.

6. Norwegian Association for Earthquake Engineering

The seismic hazard in Norway and adjacent areas is considered to be moderate. The largest earthquake of the century occurred in 1904 and had its epicenter in the outer part of the Oslofjord. Its magnitude on the Richter scale is estimated as 5.4. This earthquake and several earthquakes along the western coastline, having magnitudes of 4 and more, have led to a large interest in earthquake-related phenomena and public support for seismic data collection. When the petroleum industry started to explore for oil in the 1960s, the information about seismicity and the experience of the seismologists were particularly useful.

As the huge oil-producing installation on the Norwegian continental shelf required investments in the billion of dollars range, engineering seismology and design of earthquake-resistant platforms became new disciplines in Norway. Furthermore, onshore petroleum terminals and plants were designed to resist earthquake loads.

Sufficient interests were present in 1988 to establish the Norwegian Association for Earthquake Engineering (NAEE). The society is at present involved, jointly with the national Council for Building Standardization, in preparing design codes taking earthquake effects realistically into account for building of different classes of importance. This work is based on CEN's Eurocode 8. Furthermore, the society initiated a study to develop a seismic zonation map of Norway and its continental shelf. This study was financed with the support of oil companies and public works organizations and was finalized in 1998. At present all onshore and offshore oil-related installations and major public works are checked as to their earthquake resistance.

Furthermore, over a period of more than 10 years the oil industry has partly financed a seismic data acquisition program, realizing the importance of proper data to establish an observational basis for seismic hazard assessment. This program includes collection of seismic data from a number of seismic stations located along the Norwegian Coast.

Finally, NAEE realizes the importance of engineering seismology for a small nation that intends to be an international actor offering its technology on the worldwide market. The present board members of NAEE are professor Ove T. Gudmestad, Statoil (Chairman), Dr. Rune Dahlberg, Det Norske Veritas, (Oslo), Prof. Jens Havskov, University of Bergen, and Prof. Svein Remseth, NTNU, Trondheim.

79.42 Philippines

The editors received an institutional review of the Philippine Institute of Volcanology and Seismology from its director, Dr. Raymundo S. Punongbayan, PHIVOLCS Bldg. Carlos P. Garcia Avenue, University of the Philippines Campus, Diliman, Quezon City, Philippines. This review, including biographical sketches of several key researchers, is archived as a computer-readable file on the attached Handbook CD#2, under the directory of \7942Philippines. A summary that was extracted and excerpted from the full report is shown here. The Web site for the Philippine Institute of Volcanology and Seismology is: http://www.phivolcs.dost.gov.ph/

1. Introduction

The Philippines are situated in a zone where at least 3 lithospheric plates coalesce. More than 200 volcanoes are located within the country and it experiences at least 5 felt earthquakes a day. Written descriptions of effects of earthquakes were done as early as 1589. In 1868 Jesuit priests at the Manila Observatory began intermittent recording of earthquakes (see Chapter 3 of this Handbook). Regular, continuous-recording instruments were installed in 1877 and were upgraded after 1880 (see Chapter 88 of this Handbook). The Weather Bureau was established later to perform official duties similar to the Manila Observatory. After World War II, new stations were established in Manila, Baguio, and Davao. The last two sites became WWSSN stations in 1962.

The violent eruption of Hibok-Hibok Volcano in 1951 made the Philippine nation realize the necessity to seriously monitor and conduct studies on active volcanoes in the country. The Commission on Volcanology (COMVOL) was created in 1952. While gathering knowledge and expertise in volcanology, the Commission shifted part of its attention to the energy needs of the country in the 1960s, pioneering in geothermal energy exploration.

In 1976, a UNDP-funded project designed to establish a seismological network in Southeast Asia was implemented in the Philippines through the Philippine Atmospheric, Geophysical Astronomical Services Administration (PAGASA). Under the project, a twelve-station network was established in the Philippines.

2. History and Functions of PHIVOLCS

In 1982, the National Science Development Board (NSDB) and its agencies were reorganized into the National Science and Technology Administration (NSTA). The Commission on Volcanology was restructured and renamed the Philippine Institute of Volcanology (PHIVOLC), with a redefined set of goals and objectives. In 1984, seismology, formerly a concern of PAGASA, was transferred to the Institute and PHIVOLC was renamed the Philippine Institute of Volcanology and Seismology or PHIVOLCS. The NSTA of 1982 was structurally and functionally transformed into a Department of Science and Technology (DOST).

In 1987, PHIVOLCS was mandated to perform the following functions: (1) predict the occurrence of volcanic eruptions and earthquakes and their related geotectonic phenomena; (2) determine how eruptions and earthquakes will occur and also areas likely to be affected; (3) exploit the positive aspects of volcanoes and volcanic terranes in the furtherance of the socioeconomic development efforts of the government; (4) generate sufficient data for forecasting volcanic eruptions and earthquakes; (5) formulate appropriate disaster-preparedness plans; and (6) mitigate hazards of volcanic activities through an appropriate detection, forecast, and warning system.

PHIVOLCS is headed by a Director whose office coordinates and supervises the overall operation of the Institute. PHIVOLCS has four Scientific and Technological (S&T) Research and Development (R&D) Divisions, namely (1) Volcano Monitoring

INTERNATIONAL HANDBOOK OF EARTHQUAKE AND ENGINEERING SEISMOLOGY, VOLUME 81B ISBN: 0-12-440658-0

Station Code*	Station Name	Address
AAP	Antique Seismic Station	Old Capitol Bldg., DOST 6 San Jose, Anini-y, Antique
BBP	Basco Seismic Station	PHIVOLCS Bldg., Basco, Batanes
BCP	Baguio Seismic Station	Dairy Farm Compound, Sto. Tomas Road, Baguio City
BAG	Baguio – Manila Observatory	Manila Observatory, Baguio City
BIP	Bislig Seismic Station	Post I, Tabon Hill Top Forest Drive Terminal, Bislig, Agusan del Norte
CTB	Cotabato Seismic Station	Old Hospital Site, PC Hill, Cotabato City
CGP	Cagayan Seismic Station	Malasag, Cagayan de Oro City
CVP	Cal-lao Seismic Station	Aggugadan, Peñablanca, Cagayan
DMP	Davao Seismic Station	Brgy. Sto Niño, Tugbok District, Davao City
DCP	Dipolog Seismic Station	Sicayab, Dipolog City
GSP	General Santos Seismic Station	MSU-Tambler Campus, General Santos City (near Parker Volcano)
GQP	Guinayangan Seismic Station	Brgy., Calimpac, Guinayangan, Quezon
KAP	Kalibo Seismic Station	Provincial Capitol Site, Kalibo, Aklan.
KCP	Kidapawan Seismic Station	De Manezod St., Kidapawan, North Cotabato
LLP	Lapu-Lapu Seismic Station	City Hall Compound, Lapu-Lapu City
LQP	Lucban Seismic Station	Brgy., Ayuti, Lucban, Quezon
MMP	Masbate Seismic Station	Municipal Compound, Masbate, Masbate
PCP	Palayan Seismic Station	Brgy. Atate, Palayan City
PLP	Palo Seismic Station	Arado, Palo, Leyte
PIP	Pasuquin Seismic Station	INAC, Pasuquin, Ilocos Norte
PGP	Puerto Galera Seismic Station	Puerto Galera, Oriental Mindoro
PPR	Puerto Princesa Seismic Station	PSC Compound, Tiniguiban, Puerto Princesa, Palawan
RCP	Roxas Seismic Station	Brgy. Milibili, Roxas City
SCP	Surigao Seismic Station	Capitol Site, Surigao City
SNP	Sibulan Seismic Station	Brgy. San Antonio, Sibulan, Negros Oriental
SZP	Santa Seismic Station	Namalangan, Santa, Ilocos Sur
TGY	Tagaytay Seismic Station	Akle St., Tagaytay City
TBP	Tagbilaran Seismic Station	Uptown Housing Obojan District, Tagbilaran City
ZMP	Zamboanga Seismic Station	150-C San Jose Road, Zamboanga City

*Station codes are PHIVOLCS codes and may not be international codes.

and Eruption Prediction Division; (2) Seismological Observation and Earthquake Prediction; (3) Geology and Geophysics Research and Development; and (4) Geologic Disaster Awareness and Preparedness.

3. Volcano Observatories and Seismological Stations

PHIVOLCS currently monitors 6 of the 22 active volcanoes in the Philippines: (1) Mayon Volcano, (2) Taal Volcano, (3) Pinatubo Volcano, (4) Bulusan Volcano, (5) Kanlaon Volcano, and (6) Hibok-Hibok Volcano. It also operates the seismological stations in the table above.

4. Some Recent Advances

In April 2000, Phase I of the JICA Grant-Aid Project on the Improvement of the Earthquake and Volcano Monitoring System in the Philippines was implemented, wherein the instrumentation of the 34 seismic stations of PHIVOLCS was upgraded into digital recording systems. This upgrading led to a significant improvement on the quality of data being collected by PHIVOLCS for a more accurate determination of earthquake hypocenters and magnitudes. It is now possible to record both weak and strong motions in each station.

Monitoring capabilities for short-term volcanic eruption prediction are being developed by setting up appropriate networks of instruments within an active volcano to monitor and detect precursors of volcanic unrest. This includes seismic monitoring networks, periodic geodetic measurements, and SO_2 monitoring using Correlation Spectrometer (COSPEC). In the event of immediate volcanic unrest and/or large earthquakes, PHIVOLCS has developed a Quick Response Team whose aim is to respond promptly to such crises.

More details about recent activities in (1) volcano and earthquake monitoring, (2) volcano and seismic hazards mapping, and (3) volcano and disaster preparedness plans and ecotourism, as well as staffing are given in the full report on the Handbook CD.

79.43 Poland

The editors received four institutional reports of Poland from Dr. S. J. Gibowicz, IASPEI National Correspondent, Institute of Geophysics, Polish Academy of Sciences, Warsaw, Poland. These reports are archived as a computer-readable file on the attached Handbook CD#2, under the directory of \7943Poland. We extracted and excerpted the following summary. An inventory of the Polish Seismological Network is included in Chapter 87. Biographies and biographical sketches of some Polish scientists and engineers may be found in Chapter 89, and in Appendix 3 of this Handbook, respectively.

1. Institute of Geophysics, Polish Academy of Sciences (S.J. Gibowicz)

The Institute of Geophysics of the Polish Academy of Sciences, established in 1952, is heir to a long tradition of geophysical research in Poland and continues basic research in solid earth physics. The world's first chair of geophysics was established in 1895 by Prof. M. P. Rudzki at the Jagiellonian University in Krakow. The first seismic station on Polish territory was installed in 1903 in Krakow and the first seismic network to monitor mining-induced seismicity was established in Upper Silesia at the end of the 1920s.

The Institute is strongly engaged in both theoretical and observational seismological research, working through three departments: Seismology, Earth's Interior Dynamics, and Polar and Marine Research. The seismic network in Poland is composed of seven teleseismic and regional stations, with very broadband instruments at four stations. The Department of Polar and Marine Research has operated the Hornsund seismic station on Spitsbergen since 1978 and operated the H. Arctowski Antarctic station from 1979 to 1997.

Two research topics of special interest in the Department of Seismology are deep structure of the crust and upper mantle in Poland, neighboring countries, and polar areas, and seismicity induced by mining. The determination of deep crustal structure of the transeuropean suture zone between the Precambrian and Paleozoic platforms in central Europe has been one of the main aims of studies carried out during the past 30 years. The data from some 6500 km of deep seismic profiles have been collected and interpreted. Between 1976 and 1991 the Institute organized 3 expeditions to the Spitsbergen region and 4 to the West Antarctic. Seismic measurements were performed along profiles with a total length of 2000 km in the Spitsbergen area and 4000 km in the Antarctic.

Seismicity induced by mining is a well-known phenomenon that has been long studied in Poland. Studies of seismic events in mines are based on records from underground local networks run by the mining industry. The investigation of source mechanisms and source parameters is based on moment tensor inversions and spectral analysis of seismic waves. Non-shearing components of source mechanisms are of special interest.

Theoretical research related to earthquakes is performed mostly in the Department of the Earth's Interior Dynamics. Models of premonitory processes leading to rock damage are of special interest. A continuum description of complex deformations in inhomogeneous media, including the analysis of defect distribution and related stress evolution, enabled the formulation of a theory of earthquake premonitory and fracture rebound processes. Much attention is paid to the interaction of different physical fields and objects that could manifest themselves in precursory phenomena. Consideration of the thermodynamics of fracture processes and phase transitions has opened new prospects in the research.

A listing of the research staff of the three departments, a comprehensive list of references to published research, and biographies of Z. H. Droste and M. P. Rudzki are given in the full report on the Handbook CD.

INTERNATIONAL HANDBOOK OF EARTHQUAKE AND ENGINEERING SEISMOLOGY, VOLUME 81B ISBN: 0-12-440658-0

2. Department of Geophysics, University of Mining and Metallurgy, Krakow (J. Jarzyna)

The Department of Geophysics of the Faculty of Geology, Geophysics and Environmental Protection of the University of Mining and Metallurgy has been engaged in research and teaching of mining geophysics since the early 1970s. The research work is focused on problems related to induced seismicity and rock bursts observed in underground mines of the Upper Silesian coal basin and Lubin-Glogow copper region. The main goals are to achieve better understanding of seismic source mechanisms in mines and to join the common efforts of geophysicists and engineers to predict rock bursts and mitigate their effects on production and surface structures. To reach these goals, the research groups of the department apply a wide range of geophysical methods, including seismology, seismoacoustic techniques, a microgravity method, and geoelectric tecniques.

The department's research on mining-induced seismicity encompasses statistical and fractal analysis of seismic data for the assessment of seismic hazard and rock burst proneness. The activity in engineering seismology comprises works on constructing monitoring equipment and developing algorithms and software for interpretation of vibrations generated by mining seismic sources. Methods for predicting mining seismic events using acoustic emission have also been developed. Continuous recording of gravity changes due to longwall advances are correlated with seismoacoustic emission and seismic activity. Electrical resistivity measurements have been made in mine galleries and at the surface of several coal mines to determine relationships between stress changes in the rock mass and changes in electrical conductivity.

A listing of the research staff, a comprehensive list of references to published research, and a biography of E. W. Janczewski are given in the full report on the Handbook CD.

3. Department of Applied Geology, University of Silesia (W. M. Zuberek)

The department, headed by Prof. Waclaw M. Zuberek, consists of three sections: Physics of the Earth (Head; Prof. W. M. Zuberek); Applied Geophysics (Head: Prof. A. F. Idziak), and Geology of Mineral Deposits (Head: Prof. E. Konstantynowicz). The primary objective of the department is to educate students at all levels, undergraduate, graduate, and post graduate, with courses offered in M.Sc. programs in geology, geography, and geophysics. The department members also teach advanced courses in a Ph.D. program in Earth Sciences.

The main research program of the department includes mining seismology, acoustic emission in rocks under mechanical and thermal loadings, memory effects in rocks during deformation, dynamical processes of rock fracture (rock bursts and coal bumps, mine tremors), tectonophysics, environmental geophysics for the Section of the Physics of the Earth. The Section of Applied Geophysics has specialized in shallow geophysical prospecting, engineering geophysics, geophysical investigation of fractured rock massifs, mining seismology, and tectonophysics. The Section of Geology of Mineral Deposits carries out activities in geology and economics of deposits, mining geology, geological prospecting, protection of resources, and environmental protection in mining areas. The department cooperates with many universities and research institutes in Poland, the Czech Republic, and the US. The major achievement of the department, which is staffed mostly with young researchers in the process of scientific development, is the realization of several government projects, getting several Ph.D. degrees, and successful organization of several courses on mining geophysics.

4. Earthquake Engineering (Edward Maciag)

Although Poland is considered to be an aseismic country, seismic waves from strong earthquakes, e.g., in Romania, induce weak vibrations, especially in tall buildings. In addition, tremors accompanying excavations in coal and copper mining and quarry operations may cause substantial damage to structures, especially old masonry buildings. Maximum accelerations of surface vibrations from rock bursts exceed 0.1 g. The need to design buildings in the regions of mining-induced vibrations has required research and analysis of the influence of these tremors. The research has been conducted for a few years, primarily at the Institute of Structural Mechanics, Krakow University of Technology and, to a lesser extent, at the Main Institute of Mining in Katowice. A list of publications on these topics since 1986 is included in the full report on the Handbook CD.

79.44 Portugal

The editors received a report, "Centennial Seismology Development in Portugal," from Prof. Luis A. Mendes-Victor, IASPEI National Correspondent, Centro de Geofisica da Universidade de Lisboa, Lisbon, Portugal. This report is archived as a computer-readable file on the attached Handbook CD#2, under the directory of \7944Portugal. We extracted and excerpted the following summary. An inventory of seismic networks in Portugal is included in Chapter 87. Biographical sketches of some scientists and engineers of Portugal may be found in Appendix 3 of this Handbook.

1. Introduction

In Portugal, the observed development in seismology has been, since the beginning of the 20th century and for almost seven decades, triggered by the occurrence of large earthquakes. A brief analysis of the development of seismometry and seismology in Portugal through this century is presented here, with reference to the present and former Portuguese territories.

2. Portuguese Mainland

Large earthquakes have affected Portugal in the past. The deployment in 1902 of a "seismic device" consisting of two Milne pendulums was the first initiative to study seismic activity in the Portuguese territory. Since then the development of seismology has been motivated by concern about the effects produced by strong earthquakes. Several initiatives have been taken in the aftermath of such earthquakes, especially those of April 23, 1909, February 28, 1969, and January 1, 1980. After 1910, the seismographic network was increased nationwide. After 1946, the Serviço Meteorológico Nacional was responsible for this network, a responsibility given in 1976 to the Instituto Nacional de Meteorologia e Geofisica (INMG).

Courses in geophysical sciences were created in 1946 at the universities of Lisboa, Porto, and Coimbra. The observatories of those three universities, renamed as Instituto Geofisico, supported geophysical studies and research in Portugal. At present, the University of Lisbon is the only one continuing to offer graduate and post-graduate studies in the geophysical sciences. Since 1977, several projects directed to better seismic coverage of the Portuguese economic zone and the improvement of the seismic network have been undertaken by the Centro de Geofisica da Universidade de Lisboa (CGUL) and by the Instituto de Meteorologia (formerly Instituto Nacional de Meteorologia e Geofisica).

With the sponsorship of the European Commission (EC), the CGUL installed the first broadband seismic station at Évora in 1992. The deployment of two ocean-bottom seismometers (Improvement of Seismic Observation at the Azores Network (RIOSA) is underway and a warning system for civil protection is foreseen. A representative list of Portuguese publications on seismological subjects and some historical photographs are included in the full report on the Handbook CD.

3. Azores and Madeira

The beginning of a "seismic network" occurs in the Azores, in 1902, with the installation of two horizontal Milne seismographs in Ponta Delgada (São Miguel) and Horta (Faial). To provide better coverage of the archipelago a third station was installed at Angra do Heroismo (Terceira) in 1932.

In 1963, a seismographic station from the WWSSN network was installed at Ponta Delgada Observatory (SMN Azores) identical to the one installed in Porto. After the January 1, 1980, Azores earthquake, the INMG and the local university have run 19 seismographic stations at almost all the islands. They are mainly devoted to seismic monitoring of the Azores islands, with the exception of 8 seismic stations that are also monitoring the site of a geothermal area of São Miguel island.

INTERNATIONAL HANDBOOK OF EARTHQUAKE AND ENGINEERING SEISMOLOGY, VOLUME 81B ISBN: 0-12-440658-0

In July 1992, an OBS experiment was done at the Azores by the CGUL, in cooperation with Hokkaido University, for the study of the Azores Triple Junction. A French network from the Institut de Physique du Globe de Paris has been deployed at the Azores islands in order to provide better coverage of the area under study.

In 1997 the IRIS Consortium, through the IRIS Global Seismograph Network (GSN), has installed at Chã de Macela, São Miguel, a very broadband seismic station. The parent organization is, in Portugal, the Instituto de Meteorologia.

CGUL is also involved in a project to install two permanent seismic stations on the ocean bottom near the Azores Archipelago (Project RIOSA). These stations improve the geometry of the Azores seismic network, contributing to the investigation of active processes on the plate's boundary, for different space and time scales. These stations will also contribute to the seismic alert system (SIVISA) maintained by the Institute of Meteorology and the University of Azores. CGUL will support a new international project proposal for the installation of permanent seismic stations on the ocean floor, as well as broadband stations in some islands of the Azores. This project will be developed in collaboration with French, Italian, and American institutions.

Madeira has been equipped with an Agamennon seismoscope since 1910. After the February 28, 1969, earthquake the SMN installed a short-period seismic station in Funchal (1973). Unfortunately this situation was maintained for only a few years.

4. The Former Colonies

Seismology studies started in Mozambique in November 1951, with the deployment of a Wiechert horizontal seismograph in Campos Rodrigues Observatory in Lourenço Marques (nowadays Maputo). The station was improved with the acquisition of new seismographs, vertical Weichert, two horizontal Wood-Anderson and one vertical Benioff, which began to work on a regular basis in January 1958.

Although Lourenço Marques was the main seismograph center, the equipment was moved to Changalane (60 km away from Lourenço Marques) for technical reasons in September 1961. More equipment was then acquired, such as two horizontal Benioff seismographs and one vertical Benioff. Station Tete began operation in August 1970. A geophysical center in Nanpula was established, equipped with a seismic station. Three Benioff seismographs were deployed, similar to those installed in Changalane and Tete.

To study Angola's seismicity, a seismographic network was planned with stations located in Luanda, Sá da Bandeira, Oncócua, and Dundo. From January 1959 until 1966, the seismographic station of Luanda was provisionally established in the basement of the radio-sounding building, near the airport. In June 1967 the new station of Luanda-Belas began to operate. The seismographic station of Sá da Bandeira was installed in March 1962, near the strongest seismicity area of Angola. From March 7 until December 6, 1962, 299 earthquakes were recorded, a huge number compared with the 75 that were registered in the Luanda-Belas Station.

In August 1962, the US Coast and Geodetic Survey proposed the installation of one WWSSN station in Angola. Although the first choice was Luanda, the Sá da Bandeira seismic station was finally chosen. It became one of the best seismic stations in Africa on October 26, 1964. The Oncócua seismographic station was installed in 1969 with Benioff seismometers.

Diamonds Mining Company, in cooperation with Direcção Geral do Serviço Meteorológico Nacional, built the seismographic station in Dundo on February 28, 1968. This station was equipped with three Benioff seismometers. Seismogram analysis results were published in "Bulletin Seismique d'Angola." The "Anuario Sismologico de Portugal" contains macroseismic data about earthquakes that occurred in Angola.

79.45 Russia

The editors received the "Russian National Centennial Report" from Dr. M.V. Nevsky, IASPEI National Correspondent, Schmidt United Institute of Physics of the Earth, Russian Academy of Sciences (RAS), Bol. Gruzinskaya str., 10, Moscow 123995, Russia. This report is edited by M.V. Nevsky and A.D. Zavyalov, and consists of (1) introduction (M.V. Nevsky), (2) early history of seismological observations in Russia (A.V. Nikolaev, E.N. Sedova), (3) centennial reviews of 13 research institutions (A.D. Zavyalov, Coordinator), and (4) National Geophysical Committee of Russia (A.D. Povzner). The full report is archived as 3 computer-readable files on the attached Handbook CD#2, under the directory of \7945Russia. In addition, the editors also received five institutional reviews, which are also archived as computer-readable files in the same directory.

We extracted and excerpted the following summary from the Russian National Report and the five additional institutional reviews. Earthquake engineering in Russia is reviewed in Chapter 80.15 of this Handbook. An inventory of seismic networks in Russia is included in Chapter 87. Biographies and biographical sketches of some Russian scientists and engineers may be found in Chapter 89, and in Appendix 3 of this Handbook, respectively.

1. General Information

The Russian National Centennial Report is devoted to the Centennial of the International Association of Seismology and Physics of the Earth Interior, and includes the review of the most important results in Russian seismology of the 20th century. The report contains a brief account of the history of seismology in Russia, beginning with fundamental works of one of the founders of modern seismology, B.B. Galitzin, and his colleagues. A brief summary is given for the main achievements of Russian seismology in the design of seismological instruments, development of seismological observation systems, seismic wave propagation theory, earthquake source physics, and structural seismology, including the study of the crust, the mantle, and the Earth's core. Progress in complex seismological and seismotectonic methods of seismic hazard assessment and seismic zoning, as well as in long-term and middle-term strong earthquake prediction, is also presented.

Extensive reviews of research institutes working in various fields of modern seismology and of the National Geophysical Committee of Russia form the "General Information" part of the Russian National Report. It is archived as Ch7945CDpartA.pdf on the Handbook CD.

Thirteen institutions contributed to the National Report, and they cover about 90 percent of all present theoretical and experimental seismological studies in Russia. These institutional reviews contain numerous bibliography and biographical sketches of Russian seismologists and geophysicists, who made valuable contributions to the development of seismology in the 20th century. The first institutional review by the Schmidt United Institute of Physics of the Earth is archived as Ch7945CDpartB.pdf on the Handbook CD. The remaining twelve institutional reviews are archived as Ch7945CDpartC.pdf on the Handbook CD.

2. National Geophysical Committee of Russia (A. D. Povzner)

The National Geophysical Committee of Russia (NGCR) was formed by the reorganization of the former Soviet Geophysical Committee (SGC) after the dissolution of the USSR. The NGCR includes sections corresponding to the IUGG structure and is primarily engaged in the providing of relations to the IUGG and its associations. After the death of V.V. Belousov in 1990, G. A. Sobolev became the Chairman of the NGCR, and its scientific secretary is Yu. S. Tyupkin.

The history of the role of the NGCR in the promotion and development of geophysics is, therefore, first the story of the Soviet Geophysical Committee. This committee played a major role in the realization of the ideas and activities of IASPEI in the former Soviet Union and continues to do so for Russia as the

INTERNATIONAL HANDBOOK OF EARTHQUAKE AND ENGINEERING SEISMOLOGY, VOLUME 81B ISBN: 0-12-440658-0

NGCR. The details of the many activities of the former Soviet Geophysical Committee are provided in the full report on the Handbook CD.

Unlike many other academic committees and councils, the SGC did restrict its activity to the development of scientific recommendations. The SGC provided the execution of particular works, with the involvement of research institutes and higher schools of the country. Some examples are the multidisciplinary geophysical expeditions to the East African rift zone and Iceland and the publication of the All-Union series on "Study Results According to International Geophysical Projects" and of collections of papers on seismology, geothermics, and other IASPEI disciplines. Also, the SGC hosted the General Assemblies of IUGG and IASPEI Assemblies. Another important activity was the acquisition and distribution of observation and study results through the World Data Centers.

The account of the contributions of the SGC, now the NGCR, on the CD is organized according to the following topics: (1) studies in the framework of international geophysical projects; (2) representation of the interests of the Academy of Sciences, formerly of the USSR, now of Russia, in the IUGG and IASPEI; and (3) centralized international data exchange and publication of study results.

3. Schmidt United Institute of Physics of the Earth, RAS (M.V. Nevsky)

The Schmidt United Institute of Physics of the Earth (UIPE) of the Russian Academy of Sciences (RAS) was created in 1928, when the Seismological Institute of the Soviet Union Academy of Sciences was organized in Leningrad. P.M. Nikiforov was appointed the director. The Seismological Institute and the Institute of Theoretical Geophysics were united in 1947, to form the Geophysical Institute, with O.Yu. Schmidt and G.A. Gamburstev as successive directors. A division of the Geophysical Institute became the Institute of Physics of the Earth (IPE) in 1956, with M.A. Sadovsky as the director from 1960 to 1989. IPE was reorganized as UIPE in 1993, with V.N. Strakhov as the director.

UIPE is the largest center of Russian geophysics. Presently, 2 academicians, 6 RAS corresponding members, about 100 doctors of science, and over 230 candidates of science work at the institute. About 60–70 percent of these research workers are engaged in problems of seismology, structure and physics of the Earth's interior, and geodynamics.

Several generations of the institute scientists made fundamental contributions to seismology in the 20th century. Investigations of seismological instrumentation (D.P. Kirnos, E.S. Borisevich, D.A. Kharin, A.V. Rykov *et al.*) have been and still are of major importance. E.F. Savarensky, I.L. Nersesov, N.V. Shebalin, N.V. Kondorskaya, O.E. Starovoit *et al.* developed methods and organized the seismic network in the former USSR. G.A. Gamburtsev, Yu.V. Riznichenko, I.P. Kosminskaya, N.I. Pavlenkova, S.M. Zverev *et al.* developed the method of deep seismic sounding (DSS). Effective methods to study the Earth's crust, mantle, and core have been developed by Yu.V. Riznichenko, E.F. Savarensky, L.P. Vinnik *et al.* Recently, extensive investigations into physical-mathematical foundations of seismic tomography are being investigated by L.P. Vinnik, A.V. Nikolaev *et al.*

Substantial contributions to theoretical seismology have been made by the institute's scientists in the physics of wave processes (N.V. Zvolinsky *et al.*), earthquake source theory (B.V. Kostrov, A.V. Vvedenskaya, L.V. Nikitin *et al.*), analysis of seismotectonic deformations (Yu.V. Riznichenko, S.L. Yunga *et al.*), and geomechanics (V.N. Nikolaevsky, L.V. Nikitin *et al.*). The experimental study of earthquake focal mechanisms provided a significant contribution to the physics of crustal deformation and seismic processes (A.V. Vvedenskaya, S.L. Yunga, L.M. Balakina *et al.*).

Theoretical investigations in the physics of earthquakes conducted in the institute were accompanied by experiments on the simulation of source processes (G.A. Sobolev, S.D. Vinogradov *et al.*) and by extensive geological observations in epicentral zones of large earthquakes. Long-term investigations of the UIPE scientists in seismotectonics (B.A. Petrushevsky, I.E. Gubin, V.N. Krestnikov, G.I. Reisner, V.N. Sholpo *et al.*) have been vital to the development of historical-geological methods employed in the study of seismic processes and seismic hazard prediction.

Investigations in seismic hazard assessment and seismic zoning have led to the development of a complex of methods for the construction of seismic zoning maps: New zoning maps of the former USSR territory were constructed and the older ones were updated (Yu.V. Riznichenko, I.L. Nersesov, N.V. Shebalin, V.I. Ulomov, V.I. Bune *et al.*). Recently, a new series of general seismic zoning maps of the Russia territory (OSR-97) were completed under the leadership of V.I. Ulomov, using a modified methodological approach.

Investigations for earthquake prediction were initiated by G.A. Gamburtsev in the early 1950s. The UIPE scientists proposed and implemented the creation of multidisciplinary geophysical research areas for developing the earthquake prediction database (M.A. Sadovsky, I. L. Nersesov, V.I. Myachkin, G.A. Sobolev *et al.*). The experimental data gathered from these areas have been used for developing physical foundations of intermediate- and short-term earthquake prediction. Progress has been achieved in the development of intermediate-term prediction algorithms, which were successfully implemented on a real-time scale (G.A. Sobolev *et al.*).

Investigations in engineering seismology were carried out at the institute for many years (S.V. Medvedev, I.L. Nersesov, N.V. Shebalin, V.V. Shteinberg *et al.*). Recently this topic was transferred to other institutes of RAS and other departments.

During the last 20 years, a comprehensive study of physical properties of the geological medium has been carried out. A new, hierarchical block model of the medium was proposed (M.A. Sadovsky, V.F. Pisarenko *et al.*), the effects of time variation in seismic properties of the medium were studied with reference to its stress–strain state (A.V. Nikolaev, M.V. Nevsky *et al.*), and nonlinear effects of the medium were experimentally estimated (A.V. Nikolaev *et al.*). Methods for use in the seismic monitoring of the crust, particularly with regard to the induced seismicity effects have been developed (A.V. Nikolaev, A.G. Gamburtsev *et al.*).

For many years the leading scientists of UIPE have been engaged in the training of specialists for many research and production institutions in Russia, CIS countries, and other foreign states. Over the past 30 years, hundreds of candidate and doctor dissertations were prepared under the guidance of UIPE seismologists.

The full UIPE report contains detailed information about the investigations in traditional fields of seismology such as seismometry (A.V. Rykov), structural seismology (N.I. Pavlenkova, L.P. Vinnik, and G.L. Kosarev), theory of earthquake source (G.S. Kushnir), seismotectonics (G.I. Reisner), seismic zoning (V.I. Ulomov), and earthquake prediction (G.A. Sobolev). The report also presents short biographic notes of the UIPE eminent seismologists. It is archived as Ch49-45CDpartB.pdf on the Handbook CD.

4. International Institute of Earthquake Prediction Theory and Mathematical Geophysics, RAS (I.V. Kuznetsov)

The International Institute of Earthquake Prediction Theory and Mathematical Geophysics was founded in 1989 on the basis of the Department of Computational Geophysics of the Institute of Physics of the Earth, USSR Academy of Sciences. The Institute has 125 employees, among them 12 D.Sc.s and Professors, 40 Ph.D.s, and 52 University graduates. The faculty of the Institute has maintained cooperation with the leading research institutions from more than 20 countries. The Institute participates in many international projects and conducts annual international workshops on mathematical geophysics and earthquake prediction. The founding director is V. I. Keilis-Borok and the current director, since 1998, is A. A. Soloviev.

The expertise of the Institute covers the application of modern mathematics to seismology and related earth sciences. The Institute is known for pioneering results in the following fields: (1) determination of earthquake mechanism; (2) wave propagation; (3) signal processing; (4) verification of compliance with nuclear test ban treaty; (5) structure of the Earth; (6) seismic risk; (7) geophysical fluid dynamics; (8) magnetic dynamo; (9) seismicity and earthquake prediction.

The application of chaos theory to the dynamics of the lithosphere is the most recent innovation. This approach has led to development of earthquake prediction algorithms (A.M. Gabrielov, V.I. Keilis-Borok, and M.G. Shnirman; C.R. Allen, V.I. Keilis-Borok, V. Kossobokov, S.C. Bhatia, *et al.*), and to a new understanding of instability of large systems in general. Potential applications range from geological disasters to megacities and socio-economic systems (V.I. Keilis-Borok, A.J. Lichtman *et al.*) to a new understanding of instability of large systems in general.

The Institute has the following laboratories and research groups: (1) nonlinear dynamics and earthquake prediction (V.I. Keilis-Borok), (2) mathematical problems of nonlinear dynamics (Ya. Sinai), (3) statistical methods in geophysics (V. Pisarenko), (4) interpretation of wave fields (B. Bukchin), (5) digital seismological registration (Yu. Kolesnikov), and (6) system software (N. Galkina).

The Institute's current fields of research include (1) scenarios and symptoms of instability in hierarchical nonlinear systems, (2) mathematical modeling of the dynamics of tectonic blocks-and-faults systems, (3) new generation of earthquake prediction algorithms, (4) geological instability of platforms, (5) monitoring of stress/strain fields, (6) interaction with civil protection authorities in estimation of seismic risk and in earthquake prediction, (7) geophysical fluid dynamics: mantle plumes and formation of sedimentary basins, (8) elastodynamics: inversion and 3D free oscillations, (9) high-resolution spectral analysis, and (10) new broadband seismograph.

5. Institute of Volcanology, Far East Branch, RAS (S.A. Fedotov)

The Institute of Volcanology (IV) of the Far East Branch (FEB), RAS, has carried out the main part of seismological research in the area of Kamchatka and Commander Islands during the 20th century. A number of major structures are part of this globally important tectonic region: Kamchatka, the northern element of the Kurile-Kamchatka island arc; the adjacent Commander Islands, the western end of the Aleutian Island arc; the conjunction zone between Kurile-Kamchatka and the Aleutian Island arc; and the eastern end of the earthquake belt stretching from the Arctic Ocean through Yakutiya to the Northern Kamchatka, indicating the presumed boundary between the Asian and American lithospheric plates.

The seismic activity in Kamchatka reaches the highest level on Earth. More than 40 percent of earthquakes on the Russian territory and the most powerful tsunami occur in Kamchatka. Most of the population of the peninsula lives in the areas with predicted intensity of earthquakes over 9. The problems of seismic zoning, engineering seismology, and earthquake prediction are of major importance in the region. Scientific investigations of earthquakes, volcanoes, and geodynamics have been carried out in this area in the 18th,19th, and 20th centuries.

Seismological studies in the 20th century can be divided into four main stages: instrumental observations at remote stations

(1897–1946); the beginning of seismological observations of Kamchatka volcanoes (1947–1960); the period of detailed seismological studies of the Institute of Volcanology, FEB RAS, and the Institute of Physics of the Earth, Academy of Sciences USSR (1961–1993); the period when the seismic network was transferred from the Institute of Volcanology, FEB RAS to the Geophysical Survey (GS), RAS (1994–1999). The main efforts in creating the Kamchatka seismic network, working out the observation methods and seismicity studies, were undertaken by the Institute of Volcanology, FEB RAS, during the third stage, which lasted about one-third of a century. During 1961–1993 the research scientists of the Institute worked on the problems of interrelations between seismicity, deep structure and volcanism of Kamchatka, earthquake prediction, and other key problems of seismology.

At the end of the 20th century, the main goals of the future studies of the Institute of Volcanology and other research institutions in Kamchatka are (1) development of seismological and volcanic observation systems; (2) study of the nature of mechanism and regularities of the seismic, volcanic, and geodynamic processes; and (3) continuation of research in earthquake and volcanic eruption prediction. The importance of this task is obvious from the seismic situation in the region. Comparison of accumulated and yielded seismic energy in the Kamchatka seismogenic zone shows that at the end of the 20th century the accumulated energy is sufficient to produce several earthquakes with magnitude about 8.

The seismological research in the Institute of Volcanology was supported by the Academy of Sciences of the USSR (later, Russia) and the local authorities, and personally by the first director of the Institute of Volcanology, B. I. Piip, and director of the Institute of Physics of the Earth, M. A. Sadovsky. Hundreds of research scientists and technical personnel participated in the seismological studies of the Institute during the 20th century. Among them are G.S. Gorshkov, P.I. Tokarev, I.P. Kuzin, A.M. Bargasarova, L.S. Shumilina, I.G. Simbireva, A.A. Gusev, V.M. Zobin, T.S. Lepskaya, V.I. Gorel'chik, V.D. Feofilaktov, E.I. Gordeev, and V.A. Gavrilov. More details can be found in the full report on the Handbook CD.

6. Institute of Computational Mathematics and Mathematical Geophysics Siberian Branch, RAS (A.S. Alekseev and B.G. Mikhailenko)

The Institute was founded in 1964 as an academic institute for the development of computational mathematics and methods of mathematical simulation in the following fields: (1) physics of atmosphere and ocean, (2) geophysics of the solid Earth, (3) physics, (4) chemistry, and (5) information systems. Until 1997, the institute was named "The Computing Center" of the Siberian Branch of the Russian Academy of Sciences. G.I. Marchuk was the first director from 1964 to 1980. A.S. Alekseev was the director of the institute from 1980 to 1999, and B.G. Mikhailenko is the director since 1999.

The Institute has 8 departments. On average, they consist of 3–4 laboratories. The staff of each laboratory consists of 25–30 scientists. The total number of researchers is approximately 200 people. Among them are 32 full Professors and 130 Doctors of Philosophy in Mathematics and Physics.

Two departments deal with problems of geophysics, specifically, seismology and seismic prospecting. These are the Department of Mathematical Problems of Geophysics and the Department of Geophysical Informatics. The former has 5 laboratories, 48 researchers, 5 Professors, and 24 Doctors of Sciences. A.S. Alekseev is head of the Department. The latter has 3 laboratories, 22 researchers, 2 Professors, and 7 Doctors of Sciences. B.M. Glinsky is head of the Department.

The main directions of work of these two departments are (1) the theory, mathematical models, and numerical methods for solving forward dynamic problems of seismology and seismic prospecting for inhomogeneous, anisotropic, and fractured media; (2) the theory, mathematical models, and numerical methods for solving inverse kinematic problems of seismics for 2D and 3D media of general types; (3) the theory, mathematical models, and numerical methods for solving inverse dynamic problems of seismics for 1.5D, 2D, and 3D elastic media of general types; (4) the theory, mathematical models, and numerical methods for solving direct and inverse combined dynamic problems for multidisciplinary geophysical observations; development of a concept of multidisciplinary earthquake prediction, construction of a model of integral geophysical precursor; (5) mathematical simulation of the processes of generation, propagation in the ocean, and run-up of tsunami waves to the shore; geoinformation systems for investigation and prediction of tsunami waves; and (6) theoretical and experimental development of methods of active (vibrational) seismology.

The full report of the Institute on seismology contains the main results of the Institute in theoretical and applied seismology, a list of basic references, and biographical sketches.

7. Institute for Dynamics of Geospheres, RAS (V. V. Adushkin and A. A. Kalmykov)

The Institute for Dynamics of Geospheres (IDG), RAS, was created in 1991, on the basis of the Special Sector of Institute of Physics of the Earth (IPE), RAS. Beginning in 1947, the Special Sector has investigated geophysical aspects of powerful impacts on the environment. The Institute has gained a wealth of experience in durable observations, model studies, and analysis and theoretical calculations of explosive processes in various media, including nuclear explosions, the exploitation of explosion energy in the national economy, the provision of explosive work safety, and the development of geophysical equipment.

The Institute currently carries out basic and applied studies of physical processes in the inner and outer shells of the Earth, occurring under the influence of high-energy sources, as well as the study of effects of these processes on the human environment. The Institute has highly skilled specialists with wide experience in geophysics, geomechanics, geodynamics, seismology, explosion physics, development of measuring systems, and the creation of computer programs for calculating global geophysical phenomena.

The Institute consists of 15 laboratories whose staff includes 1 academician of RAS, 1 corresponding member of RAS, 22 doctors of sciences, and 68 candidates of sciences. Among the specialists of the IPE Special Sector and the IDG were such famous seismologists as M.A. Sadovsky, I.P. Pasechnik, and S.D. Kogan, and well-known seismologists O.K. Kedrov and D.D. Sultanov continue their work in the Institute. The Institute has developed experimental techniques and equipment (testing units) for studying explosions, a high-pressure press system, laboratory testing units for studying seismic waves from explosions in various conditions, and testing units of intense electromagnetic (laser, ultraviolet, and microwave) and plasma action. This allowed much room for a wide range of research and model laboratory experiments in geophysics.

In 1991, a new scientific direction "Geophysics of Strong Disturbances" (GSD) was formed in the Institute, in order to examine dynamic processes in the upper and lower geospheres, which are associated with the impact of intense natural and anthropogenic sources on these geospheres. The purpose of the GSD studies is to develop an integrated model for global operation of interacting geospheres and their behavior under natural and anthropogenic disturbances.

Seismological investigations have traditionally played a significant role in basic developments of the Institute. These investigations are currently conducted in the Laboratories of Experimental Geophysics (headed by the V.V. Adushkin), Explosion Seismology (headed by V.M. Ovchinnikov), and Deformational Processes in the Crust (headed by G.G. Kocharyan), as well as in the Group of Experimental Seismology (headed by D.D. Sultanov). The investigations are performed in three main directions: explosion seismology, the internal structure of the Earth, and the geomechanics of block structures of the crust.

The report of the Institute contains more details about the research programs, a list of basic references, and biographical sketches.

8. Institute of the Earth's Crust, Siberian Branch, RAS (N.A. Logachev, K.G. Levi, and V.A. Potapov)

The Institute of the Earth's Crust of the Siberian Branch of the Russian Academy of Sciences evolved from the Irkutsk seismic station. The founding and development of seismology in Siberia and its principal achievements are associated with A.A. Treskov, who may be considered the first Siberian seismologist. Prof. Treskov directed the Irkutsk station for 37 years, from 1926, and from 1932 combined its work with the Institute of the Earth's Crust, Siberian Branch, USSR Academy of Sciences. He served as head of the Seismological Laboratory from 1963 until his death. He also taught at Irkutsk institutions of higher education. Prof. Treskov's early work concentrated on the interpretation of deep-focus earthquake data and on problems of crust and mantle structure. In 1947, he developed a teleseismic method for determining crustal thickness on the basis of waves reflected from the base of the crust.

A significant change in Siberian seismology occurred in 1959, when responsibility for the Irkutsk, Kabansk, and Kyahta stations was transferred from the Geophysical Institute, USSR AS, to the Institute of the Earth's Crust (then called the East Siberian Geological Institute). A department of geophysics was organized, under Prof. Treskov, leading to the expansion of the network of seismographic stations and the enhancement of the group of researchers engaged in seismological studies. The expanded program included problems of earthquake energy determination and focal mechanisms. Among the many projects undertaken in the late 1950s was the comprehensive seismogeological examination of the regions of the largest Transbaikalia and Mongolian earthquakes, a joint project of the East Siberian Branch, USSR AS, the Irkutsk State University, and the Institute of Physics of the Earth, USSR AS. In this period, V.P. Solonenko laid the basis for the paleoseismic earthquake prediction method, since further developed.

In 1970, a number of new laboratories were created: Engineering Seismology (V. Pavlov), Seismology (A.A. Treskov), Regional Seismicity (S.I. Golenetsky), and Seismogeology (V.P. Solonenko). The research plan of the Laboratory of Engineering Seismology included the development of seismic microzoning and seismic impact prediction in permafrost soil conditions. The theoretical results were further developed and applied to assessing engineering-seismological conditions of structures at station settlements along the Baikal-amur Main Road (BAM). A seismic zoning map for the BAM construction territory was prepared (V.P. Solonenko).

In the 1980s and 1990s, seismologists of the Institute carried out the joint Russian-Mongolian examination for seismic zoning of Mongolian regional centers and industrial structures. A major contribution was the 1997 map of seismic zoning of East Siberia.

The Section of Geophysics and Recent Geodynamics of the Institute is presently composed of six laboratories engaged in seismology, seismogeology, seismogeodynamics, and applied seismology. Studies of the stress field dynamics and destruction in the lithosphere and earthquake prediction for the Baikal rift and adjacent areas are the main research directions of the Laboratories of Recent Geodynamics, Tectonophysics, and Seismogeology. The Laboratory of Multidisciplinary Geophysics works mostly on problems of structural seismology. The Laboratory of General and Engineering Seismology is directed toward source seismology, seismic regime of the Baikal seismic zone, engineering seismology, and seismic impact prediction.

9. Institute of Geophysics, Siberian Branch, RAS (S.V. Gol'din, N.D. Zhalkovsky, and P.G. Dyad'kov)

The Institute of Geophysics (IG), Siberian Branch, RAS, was organized on the basis of the Department of Geophysics, Institute of Geophysics and Geology (IGG), Siberian Branch, USSR AS. Seismological investigations have been carried out since 1959, initially under the leadership of E.E. Fotiadi, then the head of the Department of Geophysics, IGG. He formed the Laboratory of Physics of the Crust and Regional Geophysics to conduct instrumental observations in the Altai Sayan fold area. These observations gave the first view of the distribution of the main epicentral zones in this region. In 1965, a regional seismic network was established, capable of monitoring all earthquakes down to magnitude 3 in the territory. With the flow of new information, the IGG management organized a Laboratory of Seismology, headed by V.N. Gaisky.

From 1965 to 1975, the work of the laboratory focused primarily on (1) macroseismic and instrumental data for the seismicity of the Altai Sayan fold area and adjacent territories; (2) the relations between the seismicity and geological structure and evolution of the region; and (3) elucidation of the degree of similarity of seismic processes at different energy levels, including microearthquakes.

In the latter half of the 1970s, the IGG seismological investigations were substantially expanded. The execution of the program during 1975–1985 produced important data for the deep structure, tectonics, recent movements, seismicity, and related topics in the Baikal-Amur Rail Line region and the Baikal rift zone as a whole. Important results about the conditions of earthquake occurrence and the scientific basis for earthquake prediction were obtained through the cooperative activities of the Laboratory of Deep Seismic Studies, Laboratory of Electromagnetic Fields, Laboratory of Natural Geophysical Fields, and Laboratory of Seismology. Coordination and overall leadership was provided by N.N. Puzyrev.

A major development in experimental investigations of earthquake preparation and recent geodynamics was the establishment in 1966–1968 of the test area known, since the 1980s, as the Baikal prognostic test area. Tiltmeter observations in the southern part of this test area since 1985 and strainmeter observations since 1990 provide monitoring of deformations in the junction zone of the Main Sayan fault and the southern end of the Baikal basin. In the early 1990s, a 100-ton vibrator was installed at the test site for systematic sounding of the crust with both longitudinal and transverse waves. Rapid variations in the crustal stress were detected during the seismic activity in the Baikal test area during 1989–1995.

Much attention has been given to theoretical seismology throughout the history of the IGG and the IG, especially problems of seismic wave propagation and the solution of geophysical inverse problems. The Institute of Geophysics presently continues theoretical and experimental studies on all of the aforementioned subjects, with focus on the stress–strain state of the crust in seismically active areas and on physical models of earthquake sources and processes.

10. Laboratory of Seismology and Geodynamics, Institute of Oceanology, RAS (L.I. Lobkovsky and I.P. Kuzin)

The Laboratory of Sea Seismology was created in 1987 at the Institute of Oceanology by S.L. Soloviev. L.I. Lobkovsky was placed in charge of the laboratory in 1994. The name was changed to the Laboratory of Seismology and Geodynamics in 1997, but retained the studies of sea seismology as its mission.

Soloviev developed the laboratory on the basis of his well-known expertise in the scientific study of tsunami. In 1961–1968 and 1971–1977 he led the data-gathering survey for tsunami in the USSR Far East. During 1971–1979 he also chaired the Inter-Association Committee for Tsunami of IUGG. His principal motive for forming the laboratory was to initiate a system of seismological observations on the bottom of oceans and seas. Because most existing seismological stations are located within continents, far from the most seismically active regions of the Earth, the existing system of observations is obviously inadequate with respect to the objects being studied.

A technical basis for solving the problem of ocean-floor seismological investigations was provided by the development of ocean-bottom stations in the United States and Japan in the 1960s. From the mid-1960s, buoy stations designed by Rykunov and Sedov at Moscow State University were dominant in the USSR. A self-floating sea-bottom instrument designed at the Sakhalin Multidisciplinary Research Institute did not gain wide acceptance. The key initial task of the laboratory was to develop a technical basis for seafloor investigations.

The second stage of activity of the laboratory was the testing of newly developed stations in the Indian Ocean (1983–1984) and the Sea of Japan (1985). From 1987, systematic studies of seafloor seismicity began in the Aegean, Tyrrhenian, Ionian, and Laptev Seas (1987–1989), and in the Atlantic and Pacific Oceans (1988, 1991, and 1994). All of this work was supervised by Soloviev. A number of remarkable results on seismicity distribution and crust and upper mantle structure in the eastern Mediterranean resulted from this work.

The basic direction of the future development of sea seismology is the improvement of seafloor station design, observation systems, and the processing of seismological information. The enhancement of frequency and dynamic range, the combining of analog and digital recording, and the increase in the duration of autonomous operation are some of the main points in the further development of the observing system. Other plans for improvement include the use of borehole installations and the use of cable connections to shore or to drilling-platform-type structures. The improvement of data processing is essentially related to the implementation of computer systems. A list of Soloviev's key publications is included with other details on the Handbook CD.

11. Geophysical Survey, RAS (O.E. Starovoit)

Several major periods can be identified in the history of instrumental seismic observations in Russia. The most recent period, 1991 to the present time, began with the breakup of the Soviet Union. At that time, the unity of the seismic network was disturbed, and the former republics (CIS countries) entered into a time of difficult economic circumstances. Some seismic stations began to cease activity, and the seismic monitoring, even in seismic regions, was under threat of suspension. Some CIS countries have not yet fully recovered from this critical situation.

In Russia, by government decisions of 1993–1994, a federal system of seismological observations and earthquake prediction was established, a federal mission-oriented program developed, and a government client was defined for the development, modernization, and perfection of seismic observations in the country. The client is the State Committee (now Ministry) for Civil Defense, Emergency, and Removal of Natural Disaster Consequences (Russian Emergency Ministry, (EMERCOM of Russia).

To solve the tasks assigned by government decisions, on the initiative of N.P. Laverov, the RAS Presidium made a decision to organize a Geophysical Survey of the Russian Academy of Science, based on the following principles: (1) the Survey must unite and coordinate the activity of all the RAS experimental expeditions and parties; i.e., a unified system of seismic observations must be built up in the country; (2) for solving current tasks (monitoring, data processing, publication of catalogs and bulletins, operation of the express information survey, and others), the Survey provides basic research on Earth sciences with experimental data and functions following the plans agreed with the research institutes of RAS; (3) the Survey interacts with international and national seismological centers with the purpose of data exchange and integration into the world system of seismic observations; and (4) the Survey has an independent legal status.

The Survey was charged with the following basic tasks: (1) investigations of seismic monitoring of the solid Earth, including the development of methods and facilities of production, acquisition, and treatment of seismological data; (2) continuous seismic monitoring on the Russian territory and in its individual regions for seismic zoning and earthquake prediction, with express notification of central and local executive authority and departments concerned about earthquakes and possible consequences of predicted earthquakes; (3) execution of seismic observations for submarine earthquakes in the Pacific regions and the timely determination of the probability of tsunami generated by these earthquakes; (4) providing for studies in the academic research institutes with seismological and geophysical data for solving basic problems in Earth sciences; (5) providing of the participation of the Russian Academy of Sciences in international seismological projects and global system of seismological observations; (6) preparation of a seismological data bank and providing of interregional and international data exchange; and (7) effective interaction with the REM institutions entering in the Federal Survey of Seismological Observations and Earthquake Prediction.

An up-to-date system of seismic observations has been established and functions under the Russian Academy of Sciences. The system has connections to international and national seismic centers for global seismic monitoring and more effective monitoring on the Russian territory. The development of the Russian teleseismic networks is directed toward the creation of additional stations and their instrumentation with modern digital facilities for data recording and exchange, toward the perfection of information-processing centers, and toward the wide use of telecommunication facilities for transmission of and access to data, with the purpose of bettering express data processing and publishing seismological catalogs and bulletins.

The full report of the Geophysical Survey on the Handbook CD contains a detailed description of the history of instrumental seismic observations in Russia and the seismic network of the RAS, as well as a list of basic references.

12. Department of Physics of the Earth, Lomonosov Moscow State University (G.I. Petrunin, V.I. Trukhin, and V.B. Smirnov)

The Department of Physics of the Earth was organized in 1945 as the Department of Seismology and Physics of the Crust within the Faculty of Physics, Moscow State University. Prof. V.F. Bonchkovsky, Director of the Seismological Institute, USSR AS, became the first head of the department. He was joined by two young staff members of the Seismological Institute, E.F. Savarensky and D.P. Kirnos, both of whom became well known internationally.

During its early years, the department had no laboratory facilities within the Faculty of Science. Students conducted their research at the Geophysical Institute, USSR AS, mainly in two areas: research on crustal tilts and deformations, under V.F. Bonchkovsky; and elastic wave propagation in the Earth, supervised by E.F. Savarensky. The first lectures on terrestrial magnetism were given in 1949 and this subject developed as one of the principal research themes of the department.

Students and staff participated in a number of important field expeditions in 1952 and 1953. It became imperative to increase the number of students specializing in seismology in order to meet the needs of the newly formed Seismic Service throughout the vast territory of the USSR and the urgent need for specialists for the task of detecting nuclear explosions. The staff members eagerly awaited the move to the new building of the Faculty of Physics on Lenin Hills, where special rooms for laboratories for seismic and geomagnetic investigations had been designed. This move occurred in the spring of 1953 and four modern laboratories were set up: Physics of Seismic Waves, Terrestrial Magnetism, Tilts and Deformations, and Origin and Evolution of the Earth. The department was staffed by prominent scientists, including O. Yu. Schmidt.

G.A. Gamburtsev initiated experimental seismological research within the Geophysical Division by the organization of the Laboratory of Experimental Seismology in 1953–1955. The investigation of microseisms and the modeling of seismic wave propagation were also begun during this same period under the leadership of E.F. Savarensky and L.N. Rykunov. The modeling research was aimed at establishing dynamic criteria for assessing the mechanical properties of materials in the Earth's core. The investigations of microseisms evolved through several stages. Tripartite seismic stations developed in the Laboratory of Seismic Waves were used to locate the sources of microseisms and to assess the direction of their propagation. At the same time, research directed by Savarensky led to the first determinations of the energy of elastic waves generated by earthquakes.

In the late 1950s, V.A. Magnitsky became head of the department, which then became even more active. Gravimetric investigations of the Caucasus and the Crimea provided new data on the deep tectonics of these regions. Pioneering seismic studies of the sea floor were carried out. Further contributions to theoretical seismology were made. Geomagnetic studies related to global geophysics were pursued into the 1960s and 1970s. During this same period, staff members of the department, in collaboration with the Institute of Physics of the Earth, performed a series of theoretical studies in solid earth geophysics. The department also intensified its teaching activities. By the 1980s there were 12 specialized courses in the department.

In recent years the range of research topics has been supplemented by new tasks, including new studies in the Laboratories of Geomagnetism and Geothermy. Research in the department is conducted within a number of state programs, as well as with grant support received by the staff members. The department celebrated its 50th anniversary in 1995. It has educated more than 450 geophysicists and specialists in solid earth studies, including a number of prominent scientists.

13. Department of Physics of the Earth, St. Petersburg State University (T.B. Yanovskaya)

Seismological investigations in the Department of Physics of the Earth, St. Petersburg State University (SPSU), began in the late 1950s, on the initiative and under the support of E.F. Savarensky. At that time, he gave lectures on seismology and supervised diploma-oriented research of the students of the department. In those years, E.M. Lin'kov (1928–1997) initiated works on experimental seismology and studies of long-period oscillations of the Earth. Unique instruments were designed and studies were carried out on the properties of a magnetically controlled magnetron converter-vacuum valve, which was first applied in seismological practice as a parametric converter. The development of domestic seismometry in this direction led to the advent of one of the first locally produced long-period seismometers. These convincingly recorded the free oscillations of the Earth, with a period of 55 min, from the Alaska earthquake in March 1964.

Subsequent work on the improvement of the effective sensitivity of seismic channels in the low-frequency range allowed detection and analysis of previously unknown oscillations of the Earth, with periods of the order of one hour and longer. In the mid-1980s, a new direction was taken: analysis of very-long-period (seismogravitational) oscillations of the Earth. The studies of such oscillations under the supervision of L.N. Petrova revealed their relation to large-scale geodynamic processes and their involvement in the interaction between different geospheres. The monitoring of oscillations with periods from 0.5 to 5 hr is being continued with the use of a seismogravimetric complex having very-long-period detecting channels. These use photoelectirc converters and an automated system of digital data acquisition.

In the late 1960s to early 1970s, theoretical seismology investigations were initiated under the supervision of T.B. Yanovskaya. Studies were conducted on solving inverse problems of seismology and on the theory of surface wave propagation in laterally inhomogeneous media. E.N. Its and T.B. Yanovskaya developed a new approach to estimating the fields of surface waves reflected and refracted at vertical contacts simulating the block boundaries in the crust. The method of ray seismic tomography was developed (T.B. Yanovskaya and P.G. Ditmar) and applied in studying lateral inhomogeneities of the crust and upper mantle in different regions of the world. The computer programs for surface wave tomography, developed in the Department, are widely used in a number of seismological institutions of the world, in the United States, Italy, China, Greece, Bulgaria, Spain, and other countries.

From the mid-1990s, the Department has taken part in the NARS-DEEP project, an initiative of Utrecht University and under the support of INTAS. In 1996, two broadband stations with autonomous detection were installed at Saint Petersburg and Pskov, and recently, another station was established at Petrozavodsk. In addition, the GEOFON equipment was installed in 1998, based on the Pulkovo seismic station. These stations made it possible to examine the little-studied transition region between the Russian Platform and Alpine fold zone in western Europe, as well as to study the weak seismicity in the southern Baltic Shield and the northern part of Lake Ladoga.

The full report of the Department on the Handbook CD includes a review of the most important scientific results:

(1) observations of long-period oscillations, (2) theory of seismic wave propagation, (3) solution of inverse problems and seismic tomography, and (4) spectral quantification of earthquakes; a list of basic references, and a biographical sketch.

14. Mining Institute of the Kola Scientific Center, RAS (A.A. Kozyrev)

The Mining Institute of the Kola Scientific Center, RAS was founded in 1961, on the basis of the Khibiny, Tietta Station, which was built by A.E. Fersman in 1930. The first directors of the Institute were N.A. Voronkov and M.D. Fugzan. From 1962 to 1980, the Institute was headed by I.A. Turchaninov. From 1980 to present, the director is N.N. Mel'nikov. The Kola Institute is a unique academic mining organization providing scientific support for the improvement of the technical level in development of mineral resources and the underground expanse of the Russian Northwest, with the purpose of development of regional productive forces and environmental protection.

The Institute carries out multidisciplinary investigations of the following scientific topics: study of the properties and stress state of rock masses; study of physicotechnical and engineering geology problems of underground construction; underground building and the efficient use of the national underground regions, including soundness of the design of economical and safe atomic power stations in rocky masses; and others.

The Institute is organized into 9 research laboratories and 2 departments. At present, 1 academician of RAS, 20 doctors of sciences, and 40 candidates of sciences work at the Institute. The Institute holds leading positions in geomechanics, underground building, integrated use of mineral resources, and other subjects. The Institute has enduring scientific contacts with many countries, including Norway, Sweden, Finland, Germany, France, Belgium, the US, Canada, Japan, China, and Mongolia.

The most important achievements of the Mining Institute in recent years include (1) validation of an integrated technique for determination of the stress state of a fresh rock mass, and development of new methods for estimating mass state from measurements in deep geological prospecting boreholes; (2) development of a scientific initiative on the variation in the geodynamic regime of a tectonically stressed rocky mass under the influence of large-scale mining works; and (3) development and implementation of a system of instrumental observations as part of the program of the Kola geodynamic test area to search for precursors of strong seismic phenomena in the upper crust. Some precursors of rock-tectonic bursts and anthropogenic earthquakes in the Khibiny and Lovozerskii masses were recently found, manifesting themselves in anomalous alternating strain values and in the activity of seismoacoustic emission in the area of large seismic event preparation.

15. Mining Institute, Ural Branch, RAS (A.A. Malovichko)

The Perm region (West Urals) occupies a unique place on the seismological map of Russia. In 1873, specifically for the Perm province, A.P. Orlov first gathered and analyzed data for earthquakes and their effects. His book "On the Earthquakes in Pre-Ural Countries" was the first Russian monograph devoted to earthquake problems. The organized study of regional seismicity began only about 100 years later. In 1991, on the initiative of the Mining Institute, Ural Branch, RAS, the regional scientific and technical program "Seismicity of the West Urals," supported by the Perm region administration, was begun.

Since 1993, all the works on the natural and anthropogenic seismicity of the region have been carried out and coordinated by a specialized laboratory organized in the Mining Institute, under the guidance of A.A. Malovichko. Its staff consists of 15 persons, including 1 doctor of sciences and 2 candidates of sciences. Furthermore, 12 persons have part-time jobs at the laboratory. The laboratory conducts studies in the following main directions: (1) a multidisciplinary study of unstable geodynamic zones on various scales, with the purpose of detailed seismic zoning; (2) development of theoretical and methodical basics of monitoring systems for observations of natural and anthropogenic seismicity; and (3) methods of spatial-temporal prediction of large seismic events to provide a high degree of safety in operation of mining enterprises and important industrial, engineering, and civil objects on the surface and underground.

The most important results obtained by the laboratory are (1) the quantitative features of the seismic regime in the Upper Kama potassium salt deposit (UKPSD) were revealed allowing for its natural and anthropogenic nature; (2) temporal features of the formation of sources of large natural and anthropogenic earthquakes were revealed at the developed potassium mines and the period of earthquake recurrence was estimated; and (3) the fine structure of anthropogenic seismicity has been revealed, which is associated with the trigger effect of global (lunar–solar tides and non-uniform rotation of the Earth) and technological causes.

16. Geodynamic Research Center, Institute Hydroproject (A. Strom)

The Geodynamic Research Center (GRC), a branch of the organization "Institute Hydroproject" (before 1995: the Geophysical Department of Hydroproject Institute), has worked in the field of engineering seismology since the end of the 1960s. Seismic hazard assessment and evaluation of seismic design parameters were performed for most of the high dams designed in seismically active regions of the former Soviet Union and for a large number of similar projects abroad. The design of seismically stable structures and verifying of seismic stability of the existing structures is conducted as well. During recent years similar complex investigations have been performed also for thermal and nuclear power plants and other types of structures.

The GRC is the leading agency of the Russian Ministry of Fuel and Energy in the fields of geodynamic monitoring and safety of the hydraulic and thermal power plants. Complex monitoring of the potentially hazardous geodynamic processes has been carried out in the areas of several high dam sites such as Ingouri, Rogun, Toktogul, Nurek, Chirkey, Sayano-Shushenskaya, and Zeya HPP.

Along with the preceding applied investigations, several studies have been performed in the fields of seismotectonics, seismology, engineering seismology, and earthquake engineering, in particular, the analysis of earthquake source dimensions as a function of earthquake magnitude; compilation of a database of earthquake surface rupture parameters that includes data on more then 300 rupturing events all over the world, and its analysis; paleoseismological studies in different regions of the former USSR and Mongolia; compilation of the seismic model of the Earth; compilation of the catalog of Caucasus region earthquakes, with $M \geq 4.0$ ($K \geq 11$) from the ancient times to 1997 [available on the Internett: http//zeus.wdbc.rssi.ru/wdcb/sep/caucasus/welcomru.html].

The Center also conducted research on discrimination of quarry and industrial explosions from catalogs of natural earthquakes; study of the topographic effects on seismic vibrations; elaboration of design ground motions assessment methodology; development of the mathematical models of the behavior of dams and thermal power plants under static and seismic loads; development of the appropriate software for the control systems for seismic-proof structures; design of special measures to guarantee seismic stability of dams and power plants; assessment of the stability of potentially unstable slopes and landslides during earthquakes and elaboration of measures for their stabilization; geodynamic monitoring of natural and technogenic processes in the rock massifs endangering safety of large power plants, especially in the regions of high geodynamic risk; design of seismological and geophysical instruments, in particular the Automatic Digital Recorder of Seismic Events (ADRSS) and Multichannel Portable Digital Seismic Station (SP-002).

In addition to current activities, a list of research staff members, and a comprehensive bibliography, the full report of the center on the Handbook CD includes details of many projects, a database of surface rupture parameters, discrimination of the quarry and industrial explosions from natural earthquakes, a catalog of Caucasus region earthquakes, with ($K \geq 11$) from ancient times to 1997, design ground motions determination, numerical modeling of "Dam-Foundation-Reservoir" systems' behavior during earthquakes, and seismological instrumentation.

17. Laboratory of Seismology, Institute of Volcanic Geology and Geochemistry, Far East Branch, RAS (A.A. Gusev)

This laboratory is the successor of the Laboratory of Seismic Prognosis (LSP) of the Institute of Volcanology, which was reorganized in 1991. LSP was created in 1978, with Alexander Gusev as chief and the following research staff. V.M. Pavlov (1978–1997), I.R. Abubakirov (1984–1997), V.K. Lemzikov (1978–1986), A.G. Petukhin (1991–1997), and E.M. Guseva (1991–1997).

The main lines of research during 1978–1999 were (1) earthquake source: tectonophysical nature of the earthquake source process, inverse problems for the source; incoherent (high-frequency, noise-like) radiation from an earthquake source and its properties; (2) random scattering of seismic waves: theory and numerical (Monte Carlo) modeling of scattered wave fields; sounding of the lithosphere by means of scattered waves; developing techniques for identification of temporal variations of parameters of scattered waves, including those that have the potential for earthquake forecasting; (3) engineering seismology and seismic hazard assessment, especially for Kamchatka: (a) data processing techniques and analysis of strong ground motion, its amplitudes, spectra, duration; analysis of macroseismic intensity; (b) estimating recurrence periods for amplitudes; (c) seismic zonation, general and detailed; (d) issuing proposals on spectral design loads for aseismic design of structures on Kamchatka; and (4) practical real-time earthquake forecasting: (a) monitoring the seismic situation on Kamchatka by means of original techniques based on analysis of scattered seismic waves (coda); (b) real-time identification of precursory anomalies and issuing of forecasts.

The full report of the Laboratory on the Handbook CD includes the most significant results of studies for 1978–1999, a list of publications, and biographical sketches.

18. Institute of Lithosphere of Marginal Seas, RAS (G.L. Koff)

The Institute of the Lithosphere of Marginal Seas (outer and inner seas) of the Russian Academy of Sciences (ILRAS) was created in 1978. It has since been directed in turn by A.V. Sidorenko, A.L. Yanshin, and, until the present, N.A. Bogdanov. A laboratory for the analysis of engineering-geological problems of Moscow and other large urban agglomerations, now called the Laboratory of Seismogeological Risk Problems, was formed in the institute in 1979, under the direction of G.L. Koff. Under his direction, idealized problems of methods for the estimation of seismic vulnerability and seismic risk have been studied in this laboratory. Methods of seismic risk mapping, including creation of databases and geographical information systems for the estimation of risk of injury, were designed. Problems of estimation, differentiation, and mapping of fault zones are studied in a laboratory for the estimation of seismogeological hazard.

The laboratory employs 15 persons, including 3 professors (doctors of science) and 4 candidates of science. Employees of this laboratory have defended 2 doctoral and 4 candidate theses over the last 20 years. Ten monographs and more than 60 articles (15 in English) have been published.

19. Kamchatkian Seismological Department, Geophysical Service, RAS (I.R. Abubakirov, E.I. Gordeev, A.A. Gusev, V.M. Pavlov, V.A. Saltykov)

The Kamchatkian Seismological Department (KSD) of the Russian Academy of Sciences was organized in 1979 as a division of the Institute of Volcanology of the Russian Academy of Sciences. The main purpose for creating KSD was to combine observational, applied, and theoretical seismology. The high level of seismicity gave the unique possibility to study near-field radiation of earthquake source, temporal and spatial distribution of microseismicity, structure and earthquake prediction problems. There have been seven earthquakes with magnitude greater than 7 since 1992. The analysis of strong motion data enabled the estimation of the specific Kamchatka inter-relationship among peak acceleration, distance, and magnitude.

Research in engineering seismology on Kamchatka was begun by installation of the first strong motion instruments in 1964–1969 by the Pacific Seismological Expedition. The two largest earthquakes, of Nov. 24, 1971 ($M_W = 7.7$, $H = 100$ km), and Dec. 15, 1971 ($M_W = 7.7$), both were recorded near the epicenter, each by a single instrument. From 1980, a strong motion network has operated under the management of the Kamchatkian Seismological Department. More $M7+$ events were recorded (Aug 17, 1983, $M_W = 7$; Dec 28, 1984, $M_W = 6.9$, etc.), and also a number of smaller ones.

In 1984–1988, the dedicated software for strong motion data processing and analysis was developed by A.A. Gusev *et al.* and applied to data. Based on these efforts, the first version of recommended design loads for the eastern coast of Kamchatka was proposed. Later, the systematic analysis of local data enabled estimating for the first time the specific Kamchatka inter-relationship among peak acceleration, distance, and magnitude by A.A. Gusev *et al.*

Seismic hazard studies were begun in 1993, aimed at compilation of a new seismic zoning map for northern Kamchatka. A new methodology was proposed based on the combination of Monte Carlo simulation of earthquake catalog and calculation of earthquake effects based on the finite earthquake source model. Dedicated software was designed and used to attain this goal. Later, a modified version of this software was used for seismic zoning of all of Russia, both in phases of UNESCO's GSHAP project and also for designing the new official seismic zoning map of Russia.

20. Yakutsk Seismological Department, Geophysical Service, Siberian Branch, RAS (S.V. Shibaev)

The Experimental-Methodical Seismological Department (EMSD), Geophysical Service, of the Siberian Branch of the Russian Academy of Sciences, was organized in 1979 as a division of the Institute of Geological Science, SBRAS. Scientific staff members of EMSD include B.M. Kosmin, B.S. Imaev, I.P. Imaeva, D.M. Peresypkin, K.V.Timirshin, A.F. Petrov.

The EMSD conducted seismic observation in a wide region from the Arctic (Laptev Sea) to Tynda, Amursky region, and from the Olenek River to Okhotsk Sea. There were 22 seismic stations in the late 1970s, but there are only 14 at present. All seismic stations have digital records. Seismographs of Russian manufacture (SDAS, Obninsk, Baikal-22, Novosibirsk), and also the digital seismic station of Michigan State University (USA) are available. The IRIS digital instruments were installed in 1993 at the base station Yakutsk and at Tiksi. Current seismic station information is contained on a FTP server of the department.

The earthquake catalog is representative of the whole region for K $\geq$ 11–12 ($M_S \geq$ 4–4.5). At the eastern side (near Okhotsk Sea) for K $\geq$10–11 ($M_S \geq$ 3.5–4). In the Aldan region, where the stations Ust-Nyukzha, Chulman, Aldan, Tynda, Nerungri, Chagda are located close to each other, earthquakes with K $\geq$ 8 ($M_S \geq 2$) are almost never missed. In the southeast area (there are the stations Artyk, Moma, Ust-Nera) the lower limit of K for earthquakes detected is K $\geq$ 9 ($M_S \geq$ 2.5–3).

At the moment the EMSD is carrying out research on specifying seismic maps (OCP-97) in the Yakutia territory, regional seismology, and seismotectonics (active faults, paleoseismology).

79.46 Slovakia

The editors received the "Slovak National Centennial Report to IASPEI" from Prof. Peter Moczo, the IASPEI Correspondent of the Slovak National Committee for IUGG. This report is written by Peter Moczo, Miroslav Bielik, Peter Labák, and Martin Bednárik, Geophysical Institute, Slovak Academy of Sciences, Dúbravská cesta 9, 84528 Bratislava, Slovak Republic. The full report is archived as a computer-readable file on the attached Handbook CD#2, under the directory of \7946Slovakia. We extracted and excerpted the following summary. An inventory of the Slovak national seismic network is included in Chapter 87.

1. Introduction

The territory of the Slovak Republic, overlain mostly by the western Carpathian Mountains, is characterized by moderate seismic activity. Since 1034, more than 590 felt earthquakes with epicenters within this territory are documented in catalogs. Because more than 550 of them occurred after the strong 1763 Komárno earthquake ($I_O \approx 8$–9° on the MSK-64 intensity scale), the actual number of felt earthquakes during the documented period must be considerably higher. The Komárno earthquake and the 1858 Žilina earthquake ($I_O \approx 7$–8° on the EMS-98 intensity scale) strongly stimulated interest in the earthquake activity in the western Carpathians in the Austro-Hungarian monarchy (to which the present Slovak territory then belonged).

2. Earthquake Monitoring

Recording of earthquakes began in January 1, 1902, at the seismograph station in Ógyalla (now Hurbanovo). That station is still operating and with its mechanical Mainka seismograph, operating since 1909, is one of the oldest stations in Europe. Earthquake activity, including the January 9, 1906, Dobrá Voda earthquake ($I_O \approx 8$–9° on the MSK-64 intensity scale), was studied mainly by Hungarian seismologists until the Czechoslovak Republic was proclaimed in 1918. The Czech seismologists then took over. The Cabinet of Geophysics of the Slovak Academy of Sciences, established in 1953 and in 1965 renamed the Geophysical Institute of the Slovak Academy of Sciences, became a national institution for geophysical and seismological research in Slovakia. Its importance increased after the separation of Czechoslovakia into the Czech and Slovak Republics.

The present Slovak national seismic network includes 5 stations. Except for the historic Hurbanovo (HRB) station, all of them are equipped with 3-component, short-period velocimeters and digital registration. The Bratislava (ZST) station is equipped with broadband instruments and on-line telemetry. The local EBO network (six stations) and the EMO network (seven stations) are deployed around the Jaslovské Bohunice and Mochovce Nuclear Power Plants, respectively. All stations are equipped with 3-component, short-period velocimeters and digital registration. The seismic network is undergoing significant reconstruction and 6 new stations are being built.

3. Current Research Activities

The present Geophysical Institute of the Slovak Academy of Sciences has four scientific departments: (1) Seismology, (2) Geomagnetism, (3) Gravimetry and Geodynamics, (4) Physics of the Atmosphere. Professor Peter Moczo, is the head, and Dr. Peter Labák, Dr. Jozef Kristek, Mgr. Erik Bystrický, Mgr. Miriam Kristeková, Dr. Ladislav Halada, Dr. Monika Kováčová, Mgr. Peter Franek, Mgr. Martin Gális, Mgr. Andrej Cipciar, and Lucia Fojtíková are researchers and Ph.D. students in the Department of Seismology.

The Department operates the national network of seismic stations, participates in monitoring in two local seismic networks, analyses historical earthquakes and the recent earthquake activity on the Slovak territory, performs seismic hazard studies for the whole territory as well as for important localities, develops computational methods for numerical modeling of

INTERNATIONAL HANDBOOK OF EARTHQUAKE AND ENGINEERING SEISMOLOGY, VOLUME 81B ISBN: 0-12-440658-0

seismic ground motion and site effects of earthquakes, investigates anomalous seismic motion in local surface geologic structures, and develops methods of seismic signal analysis. Peter Moczo and his colleagues have contributed to the application of the finite-difference method of modeling seismic wave propagation and earthquake ground motion, including ground motion for realistic viscoelastic models of local geologic structures.

The Geophysical Institute serves as a national CTBTO (Comprehensive Test Ban Treaty Organization) data center. Researchers of the department participate in several international research grant projects and closely cooperate with seismologists in the Czech Republic, Austria, France, US, Switzerland, Japan, and other countries.

In 2001 the Seismology and Geodynamics Lab in the Department of Geophysics, Faculty of Mathematics, Physics and Informatics, Comenius University in Bratislava was established. Prof. Peter Moczo, head of the department, Dr. Eva Kačmariková, and students focus on seismic source dynamics. Dr. Peter Labák and Dr. Jozef Kristek also teach seismological courses in the department.

With regard to research on the deep structure of the lithosphere, Miroslav Bielik and his colleagues have contributed to the application of 2D and 3D quantitative interpretation of gravity anomalies in the western Carpathians and adjacent regions, development of density modeling methods, calculation of the stripped gravity maps, study of local and regional isostasy in various areas of continental lithosphere, and integrated geophysical modeling and rheological predictions of the lithosphere with implications for tectonic scenarios. The results have been used to define seismogenic zones in the region, following the division of the western Carpathians into the principal neotectonic blocks. This pioneer step in the definition of the neotectonic evolution of the western Carpathians was based on the contemporary knowledge of deep-seated structures defined by the deep seismic refraction sounding (DSS) and gravimetry.

Recently, a large effort is devoted to the investigation of the structure of the continental lithosphere, involving integrated application of geophysical, geological, and petrological studies. V. Bezák, M. Bielik, P. Konečný, P. Kubeš, J. Šefara *et al.* developed new models of the structure and geodynamics of the western Carpathian lithosphere along profiles crossing this orogenic belt.

A. Lankreijer, M. Bielik, S. Cloetingh, and D. Majcin used extrapolation of failure criteria, lithology, and temperature models to predict the rheology of the lithosphere for two sections crossing the Carpathians and surrounding regions. Calculations suggest significant lateral variation in rheology of different tectonic units, with important implications for tectonic evolution.

The full report on the Handbook CD provides the details of these studies and a comprehensive bibliography.

79.47 Slovenia

The editors received a brief National Centennial Report for Slovenia from Dr. Andrej Gosar, the Slovenian National Correspondent to IASPEI, Geophysical Survey of Slovenia, Dunajska 47, SI-1000, Ljubljana, Slovenia. The full report is archived as a computer-readable file on the attached Handbook CD, under the directory of \7949Slovenia. A summary that was extracted and excerpted from it is shown here. An inventory of the Slovenian National Seismic Network is included in Chapter 87. Biography of Albin Belar, a notable Slovenian seismologist in the early 20th century, is included in Chapter 89. Biographical sketches of some Slovenian seismologists may be found in Appendix 3 of this Handbook.

1. First Seismological Station

In 1897, two years after the great Ljubljana earthquake, the first seismological station began operating in Ljubljana, the capital of Slovenia, which at that time was a part of the Austro-Hungarian monarchy. Most of the credit for it goes to a pioneer of Slovene seismology, Albin Belar, a professor of chemistry and natural sciences, who not only founded and ran the observation station, but also designed and built the instruments. Through cooperation with recognised experts in Europe and elsewhere, he managed to provide the station with an enviable level of equipment. Unfortunately, after WW I, Belar was dismissed from service and most of the equipment confiscated. He continued to record earthquakes with some instruments until 1930 at his private observatory at Podhom near Bled.

The official earthquake observation station in Ljubljana began operation in 1924. Its modest equipment was augmented in 1925 by a Wiechert seismograph, which remained in constant use until the start of WW II, but the seismograms from this period have not survived.

2. The Astronomical-Geophysical Observatory

The first attempts to revive seismology in Slovenia after WW II took place in 1949, when the Wiechert seismograph was once again set up. In 1952 the University of Ljubljana began building the Astronomical-Geophysical Observatory (AGO) on Golovec hill in Ljubljana. Building was completed in 1958 and the renovated Wiechert seismograph was set up in the basement. Since then there has been continuous recording of earthquakes in Slovenia.

In 1966 the AGO managed to acquire instruments that conformed to the standards of the World Wide Standard Seismograph Network. For the recording of earthquakes in the near vicinity they selected the Lehner & Griffith three-component short-period system made by Teledyne. For registration of more distant earthquakes, a system of a long-period vertical seismometer and two horizontal seismometers, all made by Sprengnether, was purchased. These instruments are still in use today. In 1971 some Willmore Mk-II short-period instruments were installed that are still in function today, and a Vegik seismometer fitted with a Sprengnether recorder, which is no longer in use. This period was rounded off with some Soviet-made three-component seismographs given to the AGO by UNESCO.

3. The Seismological/Geophysical Survey of Slovenia

In 1980 the seismological section of the observatory broke away from the University and the Seismological Survey of the Republic of Slovenia was founded by the government. Its first task was to establish a network of seismological stations, since at the time there were only two: LJU in Ljubljana and CEY near Cerknica. In 1985 seismological station VOY at Vojsko commenced operation. This was followed by station VBY in Bojanci (1986).

INTERNATIONAL HANDBOOK OF EARTHQUAKE AND ENGINEERING SEISMOLOGY, VOLUME 81B ISBN: 0-12-440658-0

From 1954 to 1994 Slovene seismology was developed and headed by Vladimir Ribaric, who also published the first earthquake catalog of Slovenia, covering the time period 792–1993.

In 1994 the Seismological Survey was renamed the Geophysical Survey of Slovenia. At present the national earthquake monitoring network consists of seven seismological stations. In addition to those already mentioned there are also stations BISS in Bistriski Jarek, CESS at Cesta above Krsko, and DOBS in Dobrina, all since 1996. The network also includes a number of accelerographs. In 1999 the project of modernization of the Slovenian seismological network started with a final goal of establishing a dense network of broadband stations with real-time data analysis.

4. Historical Earthquakes

Historical earthquakes are a constant topic of the research and re-evaluation process. The macroseismic practice in Slovenia began after the 1895 Ljubljana earthquake, when the Earthquake Commission of the Imperial Academy of Sciences in Vienna was established. Since that time, macroseismic data for the earthquakes felt on the Slovenian territory are collected more or less continuously. In 1986 a network of the permanent voluntary observers of macroseismic effects was established, which today counts more than 5000 volunteers.

5. Biography of Albin Belar (1864–1939)

Albin Belar was born on February 21, 1864, in Ljubljana, Slovenia (Austro-Hungarian monarchy). He received a Ph.D. degree in Physics and Natural Science from the University of Graz. He was Professor of Chemistry and Natural Science, State High School of Ljubljana (Slovenia) from 1896 to 1908, and the Statal School Supervisor for German Primary and High Schools in Carniola—Slovenia from 1908 to 1918. He was also a Member of the Seismological Commission of Imperial Academy of Science in Vienna for Dalmatia (Croatia).

Selected publications of Albin Belar are included in the full report on the attached Handbook CD. His major seismological contributions are (1) establishment of the first seismological station in Austro-Hungarian monarchy (1897) in the State High School of Ljubljana (Slovenia); (2) construction of different seismographs, including the Belar-Luckmann horizontal and vertical seismographs; (3) construction (together with baron Codelli) of a wireless receiver for the reception of accurate time signals applied in seimological research (1910); (4) serving as editor and publisher of the monthly journal "Die Erdbebenwarte" with appendix "Neueste Erdbebennachrichten" (1901–1910); and (5) operation of his private seismological station at Podhom near Bled (Slovenia) after dismissal from his post (1921–1930).

79.48 South Africa

The editors received a report, "Earthquakes, Seismic Hazard and Earth Structure in South Africa," from Professor C. Wright, National Correspondent to IASPEI. The authors are Prof. Wright (Bernard Price Institute of Geophysical Research, The University of the Witwatersrand, Johannesburg) and Dr. L.M. Fernandez (Council for Geoscience, Geological Survey of South Africa, Pretoria). The full report is archived as a computer-readable file on the attached Handbook CD#2, under the directory of \7948SouthAfrica. An inventory of two seismic networks in South Africa is included in Chapter 87. Biographical sketches of some South African scientists may be found in Appendix 3 of this Handbook. The following summary was extracted and excerpted from the full report.

1. Introduction

In the years before modern digital seismology, many of the contributions to earthquake seismology came from both the Geological Survey of South Africa (now the Council for Geoscience) and the Bernard Price Institute of Geophysical Research, University of the Witwatersrand. Most of the recorded seismic activity is associated with deep gold mining on the periphery of the Witwatersrand Basin. An unusual aspect of South African seismology has consequently been the use of earthquake activity in the main gold mining areas to determine the structure of the crust and uppermost mantle in much the same way as large explosions have been used in other parts of the world. The regular occurrence in time and place, and the comparatively large magnitudes of these mine tremors, has made such studies possible. Information on the deep seismological structure of the South African lithosphere remained scanty until the deployment by R.W.E. Green of eight broadband seismic stations within the Kaapvaal craton over a period of eighteen months. The international Kaapvaal Craton programme with its massive broadband seismic experiment conducted across Botswana, South Africa, and Zimbabwe is leading to new insights into the evoluton of the Kaapvaal and Zimbabwe cratons, with American and southern African groups involved in the analysis and interpretation of the data. The Lesotho Highlands Water Project involving construction of the Katse dam and its impoundment in 1995, has also led to induced seismicity that is currently being investigated.

The development of seismology in South Africa has relied on workers who have spent much of their careers in South Africa, and also on many who spent their early careers in South Africa, and who left the country for distinguished careers elsewhere. However, because of the small size of the seismological community in South Africa, significant contributions to the study of South African earthquakes and applying earthquake records to the study of Earth structure have been made by scientists who have been employed in major overseas research establishments.

The full report provides a comprehensive overview of work on the development of permanent and temporary networks of seismic stations for monitoring of South African seismicity, the use of the recorded data to better understand earthquake processes and Earth structure beneath southern Africa, and the development of methodologies and results for seismic hazard assessment within South Africa. However, many of the developments within the country, particularly in understanding the seismic source, have come from within the mine seismology community, whose contributions are as applicable to the study of tectonic earthquakes as to the understanding of mining-induced seismicity. These contributions have been covered, but in less detail than studies of natural seismic activity.

It is difficult to separate South African seismological contributions from those of its immediate neighbours (Botswana, Lesotho, Mozambique, Namibia, Swaziland, and Zimbabwe), none of whom are providing separate reports for this Handbook. A few comments on seismic monitoring and earthquake activity in these countries have therefore been included. Neotectonic and paleoseismic studies are also of great importance in seismic

INTERNATIONAL HANDBOOK OF EARTHQUAKE AND ENGINEERING SEISMOLOGY, VOLUME 81B ISBN: 0-12-440658-0

hazard evaluation, but are covered only briefly to keep the size of the report manageable.

2. Seismicity of South Africa

Historical records of the seismicity of South Africa go back to 1620, and the first seismograph station was installed in Johannesburg in 1910, though few permanent stations existed until after 1950. Early seismic studies were concentrated in the Geological Survey in Pretoria (now the Council for Geoscience), the University of Cape Town, the University of the Witwatersrand, and the Hermanus Magnetic Observatory. A unique aspect of South African seismicity is that most of the located events are associated with deep gold mining on the margin of the Witwatersrand Basin. This led to extensive research on mine seismicity within the Bernard Price Institute of Geophysical Research, University of the Witwatersrand, starting in 1939, and subsequent use of the high level of seismic activity to synthesize long-range seismic refraction profiles for determination of the structure of the crust and uppermost mantle. The high concentrations of seismic activity in and around the deep gold mines also led to the development of seismic networks for the mining industry, starting in the 1960s, which are presently deployed in many of the deep mines or mining areas to monitor activity around stopes. These networks now focus on assessing risk to mining operations and miners, and on evaluating methods of mine design.

Concentrations of natural seismic activity occur in an arc extending from the Koffiefontein area of the Free State through Lesotho and northeast through KwaZulu Natal to the Mozambique Channel, in Namaqualand, and also in the Ceres area of the Western Cape, where an earthquake of magnitude $m_b = 6.3$ occurred on September 29, 1969. Earthquakes of magnitude greater than 4.5 have also occurred in the Klerksdorp and Welkom gold mining areas, where more extensive faulting occurs than in other areas of gold mining. Seismic risk studies have been undertaken in the Council for Geoscience, which also runs the present South African network, and contributes to international efforts to monitor the nuclear test ban treaty. Related studies of neotectonics have been undertaken by university scientists, the Atomic Energy Corporation of South Africa, and the Council for Geoscience. Several deployments of temporary networks of seismic stations have also been undertaken over the last fifty years to determine earth structure, to better understand seismicity associated with gold mining, and to monitor seismicity induced by impoundment of dams.

3. The South African National Digital System

The South African Network of Seismological Stations consists of mostly digital instrumentation. In the early version of the digital network developed by the South African Geological Survey, the data collected by an event-triggered PC program were first stored on the hard disk of the PC, and later transmitted by telephone line to the central station. Synchronization of the network was done automatically by the central station each time that telephonic connection was established. The present version of the digital network uses GPS as a timing device, and telephone lines have been replaced by cellular units and by e-mail facilities. The central station has remote control of the field station recording parameters, such as the number of samples per second, triggering parameters, and sensitivity.

Three of the present stations, Boshof, Sutherland, and SANAE are provided with broadband instruments. The first has the instruments located in a deep borehole, and, as a Seismic Primary three-component station of the Comprehensive Nuclear Test Ban Treaty Organisation (CTBTO) network, the data are transmitted in real time to the central station in Pretoria and then to the International Data Centre in Vienna. The second station is part of the IRIS network. The third is located at the South African Antarctic Base, SANAE, and is an Auxiliary Station of the CTBTO. The information collected by the national network is analysed by the Seismological Section of the Council for Geoscience, and the results published in monthly bulletins (quarterly bulletins since 1998) and in annual catalogs.

79.49 Spain

The editors received the "National Report from Spain" from Dr. Julio Mezcua, the IASPEI National Correspondent, Instituto Geográfico Nacional, Madrid, Spain. This report (including three useful appendices and seven institutional reviews) is archived as a set of computer-readable files on the attached Handbook CD#2, under the directory \7949Spain. An inventory of seismic networks in Spain is included in Chapter 87, and biographical sketches of some Spanish scientists and engineers may be found in Appendix 3 of this Handbook. A summary that is extracted and excerpted from the full report is shown following.

1. Introduction

The National Committee for Geodesy and Geophysics took responsibility under the direction of the National Correspondent of the IASPEI section for collecting the information provided by the different national agencies and universities about their history, activities, and accomplishments in seismology in the 20th century. To have a general overview of principal milestones in both theoretical and instrumental seismology in Spain, Mezcua and Batlló wrote a comprehensive review of main achievements in the different areas of Spanish seismology from the early days up to the present. A selected bibliography is included with this paper, where original writings or very well-documented references made possible the reading of the original sources, which can be considered to contribute to highlighting the important developments in seismology.

The full report given on the attached Handbook CD#2 includes three publications of historical significance: (1) "Introduccion al Analisis de Sismogramas" by Gonzalo Payo, for its recommended use for beginners in the interpretation of seismic phases in analog recordings in seismology. (2) "Catalogo—Inventario de Sismografos Antiguos (A Catalog of Old Spanish Seismographs)" by Josep Batlló, because it is an exhaustive description of the instruments and its characteristics that were in operation in Spain, and (3) "Catalogue of Digital Historical Seismograms (1912–1962)" by Elena Samardjieve, Gonzalo Payo, and José Badal.

The full report also includes reviews of the following institutions in Spain with lists of publications. Only a brief summary is given here.

2. Instituto Andaluz de Geofísica y Prevención de Desastres Sísmicos

Web site: http:// www.ugr.es/iag/iag.html

The Andalusian Institute of Geophysics and Seismic Disasters Prevention is a research institute of the University of Granada, created in 1989 from a previous group around the former Cartuja Observatory. Its members are teachers of the University of Granada and of the University of Almeria, postgraduate students, and technical staff.

The institute has deployed at present a microseismicity telemetered network of thirteen stations, ten accelerographs, and four broadband (BB) instruments in the area. Additionally, the institute operates three other BB installed in the South Shetland islands (Antarctica). As portable equipment, there are seven small arrays, which may be deployed and configured in a very versatile way for detailed local field surveys. Most of these instruments (except the sensors) have been designed and built at the Institute.

The institute members are teachers of the University of Granada and of the University of Almeria, postgraduate students, and technical staff. The main research lines are centred on the seismological knowledge of the Betic Cordilleras and surrounding areas (Alboran Sea), an active continental boundary of the Euroasian plate in contact with the African plate, and where a number of destructive and damaging earthquakes have occurred in the past centuries. Within this objective, studies on spatial and temporal seismicity, seismotectonics, focal mechanisms and source parameters, seismic attenuation, wave velocity

INTERNATIONAL HANDBOOK OF EARTHQUAKE AND ENGINEERING SEISMOLOGY, VOLUME 81B ISBN: 0-12-440658-0

tomography, surface wave dispersion, local site response, and seismic zonation and microzonation have been carried out. Other active lines of work are volcanic seismicity (Antarctic and Italian volcanoes), shallow geophysical exploration (for example, applied to archaeology), and seismic instruments development.

3. Geophysical Research at the Institute of Earth Sciences "J. Almera"—CSIC

This report is written by J.J. Dañobeitia and J. Gallart, Department of Geophysics, Institute of Earth Sciences "J. Almera"—CSIC, c/Lluis Solé i Sabarís, s/n, 08028 Barcelona, Spain. Research in geophysics at the Institute of Earth Sciences "J. Almera" (IJA), Barcelona, Spain, started in 1987, following a mandate from the Spanish Research Council (CSIC) to Prof. E. Banda to promote and develop there an advanced research group. Since then, the Department of Geophysics (DG) of the IJA has been devoted to the study of the Earth's lithosphere. Its main aims are to derive constraints on the structure, physical properties, and geodynamic evolution of the continental and oceanic lithosphere. To achieve these goals, integrated geophysical methods are considered and developed. Special emphasis is given to seismic methods such as reflection and refraction profiling, tomography, anisotropy, and seismotectonics, complemented by potential fields (gravity, magnetics, geoide), heat flow, structural geology, and numerical modeling of geodynamic processes. A huge variety of thematic studies have been undertaken, from the deep architecture of mountain belts and continental margins, across intraplate volcanic activity, to active processes generated at mid-oceanic ridges. From the beginning, the DG also was intended to deliver academic formation at a Ph.D. level, and nowadays the DG incorporates young Spanish scientists that were working all over Europe and the United States.

The research at the DG is basically accomplished through projects funded by Spanish and European agencies. So far, major geophysical studies of lithospheric structure have been focussed on (a) collision processes at orogenic belts; (b) onshore–offshore mapping of continental margins and oceanic basins, and (c) intraplate volcanism at oceanic islands. Main projects and results in such different environments are summarized hereafter. All the projects to be mentioned have been led by or with a relevant participation of members of the DG. More details can be found in the full report, with an extensive bibliography.

4. The Ebre Observatory

Web site: http://www.readysoft.es/home/observebre/

This report is written by Arantza Ugalde, Observatori de l'Ebre, Horta Alta, 38, 43520-Roquetes (Tarragona), Spain. The Society of Jesus founded the Ebre Observatory in 1904 at the town of Roquetes (northeastern Iberian Peninsula). Its geographical coordinates are 40° 49.23' N and 0° 29.60' E. Although since the establishment of the Observatory it has been devoted to the study of solar-terrestrial physics, a seismological section was created, because at that time it was believed that there could exist a relation between solar activity and seismicity. It is, therefore, the oldest seismic station in Catalonia and the third in Spain after San Fernando (1898) and Cartuja (1902), both located in the southern part of the Iberian Peninsula. The Ebre Observatory seismic station (EBR) has been operating different seismographs almost without interruption since its foundation, and its records, almost completely preserved, are of key importance for the instrumental study of western Mediterranean seismicity. More than 100,000 original seismograms of all the seismographs operating at EBR have been kept up to the present, as well as some old seismographs and a large amount of written materials concerning the seismic station (calibrations, seismic bulletins, etc.). The Instituto Geográfico Nacional (IGN) microfilmed the whole collection of seismograms and related documents in 1995.

The main tasks developed at the seismic section of the Ebre Observatory concern observation and research. Observation includes control of the correct operation of the seismic instruments and their day-to-day maintenance. The seismic events recorded are also analysed and seismic bulletins are elaborated weekly. These bulletins are sent by electronic mail to the IGN, the ICC, and the Institut d'Estudis Catalans (IEC). The Ebre readings are also sent to the International Seismological Centre (ISC).

The main research lines developed at the Observatory are related to seismicity, seismic wave attenuation, and soil amplification phenomena. Some studies have been developed by digitising old seismic records in order to study new parameters related to the source (magnitude, seismic moment, etc.). A local magnitude formula has also been elaborated for moderate magnitude earthquakes for regional distances from the Observatory by J. Vila *et al.* in 1996. Seismic wave attenuation studies mainly concern coda attenuation and tomography, which has allowed us to characterise the lithosphere characteristics in several regions of the world (e.g., L.G. Pujades, J.A. Canas, A. Ugalde *et al.*, in 1997 and 1998). Finally, microtremor analysis in order to estimate soil amplification factors is another subject of study developed at the Observatory (A. Ugalde, A. Alfaro *et al.* in 1998 and 1999).

The academic tasks carried out are also important, because the Ebre Observatory, as a centre of the Ramon Llull University, gives a doctorate program in Geophysics.

5. Royal Naval Institute and Observatory in San Fernando

The Royal Naval Institute and Observatory in San Fernando (ROA), which is the oldest astronomical observatory in Spain, had its beginnings in the 18th century. The study of seismology has been one of the traditional fields of the Geophysics Department since 1898, when the first seismograph was installed at the request of John Milne—the pendulum number 6, the oldest seismograph in Spain, with a period of 18 seconds. Since then, several instruments have been installed at the ROA and at the present time, there are (1) a short-period seismic net, (2) a long-period station, and (3) a VBB net.

The short-period seismic net, called "San Fernando Observatory-Gibraltar Strait Net," is composed of nine telemetric stations deployed in the southwest of Andalucía since 1987. Nowadays, a new digital station is being developed in order to replace the present analog stations in the near future.

Since the Milne Pendulum, several instruments have been installed at San Fernando Observatory—horizontal pendulums of the Mainka type, developed and built in this Observatory (1912), vertical pendulum with a mass of 100 kg, and a 2-second period (1921), Alfani Seismographs (5 and 8 seconds of period, 1933) and finally, a 3-component Sprengnether station (1966), replaced by a new Sprengnether 5500 Station (1978).

From 1996, a VBB digital network has been deployed, the so-called ROA/UCM VBB Network, as a cooperative effort of the Royal Naval Institute and Observatory in San Fernando, and the Geophysical Department Complutense University of Madrid (UCM), and with the collaboration of the Geoforschungszentrum (GFZ) of Potsdam (Germany). The network is actually formed by four stations: SFUC (June 1996) located SW Spain, CART (December 1997) located southeast Spain, MAHO (June 1999) located at Baleares Islands, and MELI (December 1999) located in North Africa. These four stations provide good coverage of the seismicity in the Alborán Sea and Gibraltar Strait area, and they are integrated, as cooperative stations, in the GEOFON Network.

From 1973 to 1985, ROA, together with other Spanish institutions, has participated in a Deep Seismic Profiles program in the Iberian Peninsula, under the European Geotransverse Project. In 1989, ROA took part in the deep seismic ILIHA (Iberian Lithosphere soundings Project). In 1994, ROA, together the UCM Geophysical Department, and with the collaboration of the IGN, carried out a seismic profile in southwestern Spain in order to reach a better understanding of the structures of this region. Also, magnetic profiles were carried out on the Cádiz Gulf.

6. Servei Geològic de Catalunya, Institut Cartogàfic de Catalunya

Web site: http://www.icc.es/

This report is written by Antoni Roca, Head of the Geological Survey of Catalonia, Parc de Montjuïc, 08038 Barcelona, Spain. The Geological Survey of Catalonia, a unit of the regional government of Catalonia, has carried out activities in geophysics since 1982. In 1995 it became a Unit of the Institut Cartogràfic de Catalunya, which is the institution from the Autonomous Government of Catalonia in charge of cartography and related geosciences. The main working areas of the Geological Survey are geology and geological cartography, engineering geology, seismology, applied geophysics, snow avalanche forecasting, and cartography and study of other natural hazards.

The activities in the field of seismology and earthquake engineering are focused on two main subjects: (1) study and survey of regional seismicity, and (2) seismic risk assessment and earthquake engineering. One of the main activities is the operation of a dense seismic network for the surveillance of seismicity in the territory of Catalonia (northeastern part of Spain). Periodic bulletins are published and distributed and information is also made available through the Internet.

7. Department of Geotechnical Engineering and Geosciences, UPC

Web site: http://www-etc.upc.es/sismologia/indice.html

This report is written by Euardo Alonso, Head, Department of Geotechnical Engineering and Geosciences, Jordi Girona, 1-3, módulo D2, UPC-Campus Nord. 08034 Barcelona, Spain. The main research subjects of the Department are those related to geotechnical engineering, seismology and earthquake engineering, engineering geology, groundwater hydrology, mechanical behavior of engineering materials, and terrestrial photogrammetry. Among other graduate and undergraduate courses, the Department organizes master and doctoral programs in earthquake engineering and structural dynamics. The Department also includes an Applied Geophysics Service, which works mainly in geophysics applied to civil and geology engineering.

Scientific programs include seismic risk assessment, seismic tomography, coda Q, objective regionalization methods, and high-resolution geophysical prospecting.

8. Escuela Universitaria de Ingeniería Técnica Topográfica, UPM

The Surveying School at the Universidad Politécnica de Madrid (UPM) has been actively engaged in research related to earthquake engineering for the last four years. The main areas of research undertaken by UPM cover assessment of seismic hazard and seismic risk, with particular reference to the Iberian Peninsula. Some of the specific topics developed are historical seismicity, characterization of strong ground motion, site effect, contribution of local and distant sources in seismic hazard, and the correlation of strong motion parameters and earthquake damage to structures.

Previous topics have been developed throughout several research projects. The main ones are: SEISMODOC and INTAS/GEORGIA 97-0870, both supported by the European

Commission; DAÑOS, financed by the Spanish Nuclear Safety Council and the National Enterprise for Radioactive Waste Disposal; and ESPECTRO, supported by the National Geographic Institute of Spain. In the frame of these projects, some results are the creation of the strong motion Data Bank MFS-Daños, compiling records from all over the world; compilation of an electronic library of historical documents related to European seismicity prior to 1945; the development of a Geographical Information System for seismic hazard studies in Spain, as well as different results related to ground motion models and potential damage parameters. Research activities include seismic hazard assessment, MFS-strong motion databank and database, and observed vulnerability and damage. Major developments and accomplishments in seismic engineering are described, together with the research programs, composition of the working group, and the bibliographic references in the full report.

79.50 Sweden

The editors received the report "Seismology in Sweden" from Prof. Ota Kulhánek, IASPEI National Correspondent, Department of Earth Sciences, Uppsala University, 75236 Uppsala, Sweden. This report is archived as a computer-readable file on the attached Handbook CD#2, under the directory of \7950Sweden. We extracted and excerpted a summary as shown here. An inventory of the Swedish national seismic network is included in Chapter 87. A biography of Markus Båth and biographical sketches of some Swedish seismologists and geophysicists may be found in Chapter 89, and in Appendix 3 of this Handbook, respectively.

1. Introduction

Seismological observations and research in Sweden have been carried out essentially at two institutions, the University of Uppsala and the Defence Research Institute in Stockholm. Activities and research orientation at these two institutions differ from each other to a large extent. This report describes only contributions made by the University of Uppsala.

In the past, the Uppsala group has worked on the observational side, operating the Swedish Seismographic Station Network, including the analysis of records. Much of the research started as an empirical approach on various tasks, based on Swedish and other seismograms. Theory and the implications of the work usually entered at a later stage. During the past decade or so, worldwide large data sets have become easily available and the earlier approach is no longer typical of current work.

2. Earthquake Monitoring

The birth of seismological instrumental observations in Sweden dates to October 4, 1904, when a 1000-kg Wiechert-type inverted pendulum mechanical instrument began operating in Observatory Park, central Uppsala. The installation of the seismograph was a result of a fruitful cooperation between the universities in Göttingen and Uppsala. Uppsala meteorology professor H.H. Hildebrandsson (1838–1925) cooperated with Göttingen professor E. Wiechert (1861–1928) through works on atmospheric electricity. Due to this relationship and also to results of the International Seismological Conference in Strasbourg in 1901, Hildebrandsson succeeded in establishing seismological observations in Uppsala. Wiechert, who in 1898 became director of the Geophysical Institute in Göttingen, had started seismological observations by the end of the 19th century. F. Åkerblom (1869–1942), the assistant of Hildebrandsson, visited Göttingen and after his return to Uppsala, prepared the installation of a seismograph within the premises of the Meteorological Institute. The instrument was deployed in October 1904 under the supervision of G. Bartels, an engineer from Göttingen. Åkerblom later became professor of meteorology and the first to carry out seismological research in Sweden.

More systematic seismological research was resumed in the Meteorological Institute in the 1940s. From 1949 to 1961, the work was done in the Seismological Laboratory, led by Markus Båth (1916–2000). This laboratory became an independent university institute in 1961, still led by Båth until his retirement in 1976, when O. Kulhánek became director. Further reorganizations within the university culminated in 1998 when all 13 geo-groups were integrated into the Department of Earth Sciences, with seismology as one of the programmes. There have usually been between 10 and 15 persons, including visiting scientists and graduate students, associated with the Seismology Programme. In 1961, seismology became a university subject at Uppsala, on the graduate level.

In 1983, the central station in Uppsala was equipped with a three-component, short-period Teledyne-Geotech system with S-13 sensors and hot-stylus recording. In 1989, a three-component broadband system with leaf-spring Wielandt-Streckeisen STS-1 seismometers was deployed, together with a data acquisition system with 16-bit resolution and 20 Hz

INTERNATIONAL HANDBOOK OF EARTHQUAKE AND ENGINEERING SEISMOLOGY, VOLUME 81B ISBN: 0-12-440658-0

sampling. Considerable efforts have been made to convert the analog system into a modem digital system. After many years of negotiations with various authorities, in 1997, the University of Uppsala finally provided financial means for a new network. It was put into operation in autumn 1998 and consists of six stations at approximately the same locations as the existing stations. All stations except Uppsala are now equipped with Guralp CMG-3ESPD broadband seismometers with digital output. Responses of the seismometers are flat to velocity in the period range from 0.02 to 30 sec. The sampling frequency is 100 Hz. The Uppsala station is operating an STS-2 instrument since November 1999. The data acquisition system used is the so-called SIL system by R. Bodvarsson *et al.* The new system has been named the Swedish National Seismic Network, SNSN, to distinguish it from the analog network, which was discontinued in December 1998. More details about the history of seismographic observations can be found in the full report on the Handbook CD.

3. Research and International Activities

The full report provides details of the research activities, with a brief list of references. The Uppsala group has produced many hundreds of scientific publications during the past half-century. The research was on both earthquake source properties and propagation path properties. Of special interest are several paleo-earthquakes in northern Sweden with fault lengths of up to 160 km and vertical displacements of up to 15 m, caused by the elastic rebound following the last deglaciation about 9000 years ago. The faults were probably formed by single events that ruptured through most of the crust. The largest event, $M_w = 8.2$, was larger than other known stable-continent earthquakes.

The Uppsala group has produced or cooperated in compilations of a number of earthquake catalogs and performed many seismicity studies (Fennoscandia, Tanzania, Kenya, Turkey, Thailand, Costa Rica, Panama, Zimbabwe, Ethiopia, Egypt, Korea, among others). The catalogs have formed the basis for further studies (tectonic relations, energy release) but also served for engineering applications (hazard evaluation).

During the last decade or so, interests have been focused more on studies of dynamic source parameters, physics of the earthquake sources, spatio-temporal patterns of seismic energy release, etc. The Uppsala group investigated separately events in Canada, Costa Rica, Cyprus, Egypt, Ethiopia, Greece, Honduras, Korea, Panama, Tanzania, southern Africa, Sweden, Thailand, and Zimbabwe.

During 1992–2000, the Uppsala group coordinated the project "Seismotectonic Regionalization of Central America," sponsored by the Swedish Agency for Research Cooperation with Developing Countries. A number of studies concerning East Africa have been conducted since the mid-1980s. A seismological field experiment designed to determine the anisotropic structure of the lithosphere was carried out in south-central Sweden in 1991, jointly with colleagues from France and the former Czechoslovakia. The seismology programme at Uppsala has always operated on an international basis. It has not been unusual for up to eight nationalities and five continents to be represented simultaneously.

Staff members have been active in international seismology, including European and other international scientific bodies. Markus Bårth was vice president of IASPEI, 1956–1957, and its Associate General Secretary, 1957–1963. O. Kulhánek served as president of ORFEUS, "Observatories and Research Facilities for European Seismology" 1989–1999, and chaired the IASPEI working group on the Manual of Seismogram Interpretation, 1985–1993. The University also offers M.Sc. and Ph.D. degrees in seismology. The faculty members have also published advanced textbooks for the benefit of both instruction and research.

79.51 Switzerland

The editors received the "Swiss National Centennial Report to IASPEI," written by J. Ansorge, T. Kohl (IASPEI National Correspondent), and D. Mayer-Rosa. All authors are from the Institute of Geophysics, Swiss Federal Institute of Technology (ETH), Zurich, Switzerland. This report is archived as a computer-readable file on the attached Handbook CD#2, under the directory of \7951Switzerland. We extracted and excerpted a summary as shown here. An inventory of the Swiss seismic network is included in Chapter 87 of this Handbook. Biographies and biographical sketches of some Swiss scientists and engineers may be found in Chapter 89, and in Appendix 3 of this Handbook, respectively.

1. Introduction

The prime objectives of the Swiss Report are to summarize the developments and achievements of earthquake seismology, earthquake engineering, lithospheric structure, and geothermal research in Switzerland. The report includes historical reviews, major results, main present activities, and lists of references to provide the reader with basic information and a starter for further detailed information. Many of the past and present activities are also part of the International Lithosphere Program (ILP), currently presided over by A.G. Green at Zurich. The report was compiled by J. Ansorge (Structure of the Lithosphere), Th. Kohl (Geothermal Research), and D. Mayer-Rosa (Earthquake Seismology), all at the Institute of Geophysics, ETH Hoenggerberg, CH-8093 Zurich, Switzerland.

2. Earthquake Monitoring

Earthquakes have been systematically reported in Switzerland since the 16th century, though single major earthquakes are mentioned much earlier. One of the first collections of observations on earthquakes and other natural phenomena may be found in Conrad Lycosthenes, "Prodigiorum ac ostentorum chronicon," of 1557. Johann Jakob Scheuchzer included a "historical description of all earthquakes which have been felt in Switzerland from time to time" in his "Beschreibung der Naturgeschichte des Schweizerlandes" of 1706. Scheuchzer recognized certain areas of significant earthquake activity in Switzerland, such as the St. Gall Rhine Valley and the Cantons of Basel and Glarus. An extensive list of earthquakes in Switzerland is contained in Elie Bertrand's "Memoires historiques et physiques sur le tremblement de terre" of 1757. Certainly the most systematic and complete record on historical earthquakes was compiled by G.H. Otto Volger, in 1857, entitled "Untersuchungen ueber das Phaenomen der Erdbeben in der Schweiz." The description of effects is remarkably objective and exhaustive for that time. He also developed a seven-degree macroseismic scale to quantify the effects of one of the strongest earthquakes in this millennium, the 1855 Wallis earthquake. For the first time in Switzerland, he related earthquakes to geologic structures and discussed their geographic and statistic occurrence. In addition, A. Candreia, in 1905, wrote a chronicle on earthquakes in the eastern part of Switzerland for the time span 829–1879, which is recognized as another very reliable source of information.

In 1878 the Swiss Earthquake Commission was founded following suggestions by F. A. Forel, A. Forster, and A. Heim. Hence, Switzerland was the first country (followed by Italy in 1879 and Japan in 1880) to establish a national institution for the systematic observation of earthquakes. The first action of this commission was to organize a macroseismic service for the whole country. During this time, Rossi and Forel created a macroseismic scale with 10 degrees, initially called "Swiss-Italian Scale." It was widely adopted and in use until 1964, when the European Seismological Commission recommended the MSK scale and later the new EMS scale with 12 degrees.

The first seismological station was established in Zurich in 1911 and equipped with two 450-kg Mainka horizontal

INTERNATIONAL HANDBOOK OF EARTHQUAKE AND ENGINEERING SEISMOLOGY, VOLUME 81B ISBN: 0-12-440658-0

pendulums (gain 200) and an 80-kg Wiechert vertical instrument (gain 130). It was mainly due to the Earthquake Commission and A. de Quervain that the early stations in Switzerland could be financed and installed. In the same year a second station (Mainka horizontal) was established in the Observatory of Neuchâtel and a third station in 1915 in Chur (Bosch-Omori horizontal), following an initiative by A. Kreis.

The Swiss Earthquake Commission was succeeded in 1913 by the Swiss Seismological Service (SSS), which has been part of the Institute of Geophysics, ETH Zürich, since 1956. The SSS is in charge of all national earthquake-related aspects, including seismic networks, and compiles historical and instrumental earthquake catalogs. Instrumental recording began in 1911 with two stations and has been extended to the present status of 36 short-period and broadband stations. To meet engineering requirements, 60 evenly distributed accelerometer stations and an additional 30 installed in large hydroelectric dams are in operation.

3. Research Activities

Earthquake-related research in Switzerland includes seismicity, seismotectonic investigations (which are closely related to the research on lithospheric structure), seismic hazard and microzonation studies, and contributions to the Comprehensive Nuclear Test Ban Treaty Organization. In addition, the SSS has participated in and/or carried out a number of international missions on demand of either international agencies or foreign national authorities. These missions included mainly the design and installation of seismic networks, follow-up training, and cooperative research activities. A major role was taken in the recently completed Global Seismic Hazard Assessment Program (GSHAP), as part of the International Decade of Natural Disaster Reduction on European and global scales.

Activities to unravel the lithospheric structure in the Alps have been based almost entirely on cooperative research among various national or international institutions, with emphasis on geophysical and geological aspects. Geophysical research toward the deep structure has been done mostly by the Institute of Geophysics, ETH. All other university geoscience institutes contributed indispensable geological, petrological, and geotectonic data. National cooperation culminated in two major research programs, the earlier of which was concentrated on the Swiss Geotraverse Basel-Chiasso. An extensive cooperative and multidisciplinary project, "The Deep Structure of the Swiss Alps," closely related to the European Geotraverse (EGT), followed in 1985. Substantial structural differences among three traverses through the western and central Alps appear, especially in the lower crust. Research toward a 3D lithospheric image has led to a quantitatively assessed model of the Alpine crust–mantle boundary that shows two offsets that divide the interface into a European, an Adriatic, and a Ligurian Moho, with the European Moho subducting below the Adriatic, and the Adriatic underthrusting the Ligurian Moho.

Geothermal research started early in the 19th century and was coupled with rail tunneling projects. The first estimates of heat flow for continental Europe and of the influence of topography and rock radioactivity as a heat source were achieved. Modern heat flow determinations began in the mid-20th century. Among the influences that have been assessed, a widely used petrophysical relation between heat production and P-wave velocity has been established. Major achievements have been realized with numerical modeling, which enables the calculation of the full 3D subsurface temperature field by accounting for complex thermal mechanisms such as caused by nonlinear behavior of thermal conductivity, fluid advection, and transient effects or topography.

Biographies of Alfred de Quervain (1879–1927), Ernst Wanner (1900–1955), Fritz Gassmann (1899–1990), and Stephan Mueller (1930–1997), and selected publications that are related to various research activities presented here are provided in the full report on the Handbook CD.

79.52 Turkey

The IASPEI National Committee of Turkey chose not to prepare a national report, and supplied us with a list of institutions that conduct seismological and geophysical research in Turkey. Invitations were sent to all these institutions, but due to the disastrous Izmit earthquake of August 17, 1999 (see, e.g., *Bull. Seismol. Soc. Am.*, vol. 92, no. 1, 2002), we received only one reply, from Prof. Haluk Eyidogan, Department of Geophysics, Istanbul Technical University, Istanbul, Turkey. The available information is archived as a computer-readable file on the attached Handbook CD#2, under the directory of \7952Turkey. We extracted and excerpted the following summary. An inventory of seismic networks in Turkey is included in Chapter 87, and biographical sketches of some scientists and engineers of Turkey may be found in Appendix 3 of this Handbook.

1. Institutions that Conduct Seismological and Geophysical Research

Atatürk University, Earthquake Research Center

Address: Atatürk Universitesi Deprem Arastirma Merkezi, Erzurum, Turkey.

Contact: Dr. Salih Bayraktutan; E-mail: <eyidogan@itu.edu.tr>.

Bogazici University, Kandilli Observatory

Address: Bogazici Universitesi Kandilli Rasathanesi, Cengelköy Isatanbul, Turkey.

Contact: Prof. A. Mete Isikara; E-mail: <isikara@boun.edu.tr >.

Dokuz Eylül University, Izmir

Address: Dokuz Eylül Universitesi Mühendislik Mimarlik Fak. Jeofizik Müh. Böl., Bornova, Izmir, Turkey.

Contact: Prof. Güngör Taktak.

General Directorate of Disasters Affairs

Address: Afet Isleri Gn. Md. Deprem Aras. Dai. Eskisehir yolu 10.km, 06530 Ankara, Turkey.

Contact: Mr. Oktay Ergünay; E-mail: <iravul@sismo.deprem.gov.tr>.

General Directorate of Electrical Power Resources Survey and Development Administration

Address: Elektrik Isleri Etüd Idaresi Gn. Md. Eskisehir yolu 7. Km., Ankara, Turkey.

Contact: Mr. Osman Ilhan; E-mail: <elektriketut@eie.gov.tr>.

General Directorate of Mineral Research and Exploration

Address: MTA Jeolojik Etüd ve Temel Arastirmalar Dai. Bsk., Ankara, Turkey.

Contact: Mr. Erkan Muftuoglu

General Directorate of State Hydrolic Works

Address: Devlet Du Isleri Gn. Md. Jeoteknik Hizmetler ve Y.A.S. Dai. Bsk., Yücetepe, Ankara, Turkey.

Contact: Dr. Erdal Sekercioglu; E-mail: <yas@dsi.gov.tr>.

Hacettepe University, Geological Engineering Department

Address: H.U. Jeoloji Müh. Böl. Beytepe Kampüsü, Ankara, Turkey.

Istanbul Technical University, Sismoloji ve Sismotektonik Alt BirimiAddress: I.T.U. Maden Fakultesi. YBYK UY-Gar Merkezi Sismoloji ve Sismotektonik Alt Birimi, 34469 Maslak, Istanbul, Turkey.

Contact: Assoc. Prof. Argun Kocaoglu; E-mail: <kocaoglu@itu.edu.tr>.

Istanbul Technical University, Department of Geophysical Engineering

Address: Istanbul Teknik Universitesi, Maden Fakultesi, Jeofizik Muhendisligi Bolumu, 34469 Maslak, Istanbul, Turkey.

Contact: Prof. Haluk Eyidogan; E mail: <eyidogan@itu.edu.tr>.

Istanbul University, Department of Geophysical Engineering,

Address: Istanbul Universitesi Müh. Fak. Jeofizik Müh. Böl. Avcilar Kampüsü 34840, Istanbul, Turkey.

INTERNATIONAL HANDBOOK OF EARTHQUAKE AND ENGINEERING SEISMOLOGY, VOLUME 81B ISBN: 0-12-440658-0

Contact: Prof. Ömer Alptekin, or Prof. Demir Kolcak; E-mail: <alptekin@istanbul.edu.tr>, <kolcak@istanbul.edu.tr>.

Karadeniz Technical University, The Faculty of Engineering and Architecture

Address: Karadeniz Teknik Universitesi, Mühendislik Mimarlik Fakültesi, Jeofizik, Mühendisligi, Bölümü 61080 Trabzon, Turkey.

Contact: Dr. Kenan Gelisli.

Kocaeli University, Department of Geophysical Engineering

Address: K.U. Mühendislik Mimarlik Fak. Jeofizik Müh. Böl., Kocaeli, Turkey.

Contact: Prof. Özer Kenar.

Middle East Technical University

Address: ODTU Jeoloji Müh. Böl., Ankara, Turkey.

Contact: Prof. Nurkan Karahanoglu; E-mail: <nurkan@rorqual.cc.metu.edu.tr>.

Süleyman Demirel University, Earthquake and Geotechnique Research Center

Address: Süleyman Demirel Universitesi, Müh. Fak. Jeofizik Müh. Böl., Isparta, Turkey.

Contact: Dr. Ergün Türker; E-mail: <deprem@mmf.sdu.edu.tr>.

The Scientific and Technical Research Council of Turkey

Address: TUBITAK Yapi Arastirma Grubu, Istanbul cad. No.88, Ankara, Turkey.

Contact: Prof. Ergin Atimtay.

The Scientific and Technical Research Council of Turkey, Marmara Research Center

Address: TUBITAK Marmara Arastirma Merkezi Yerbilimleri Bölümü, Gebze, Kocaeli, Turkey.

Contact: Prof. Mustafa Aktar; E-mail: <aktar@boun.edu.tr>.

Turkish Atomic Energy Agency

Address: Atom Enerjisi Kurumu Baskanligi, Eskisehir yolu 13. Km., Ankara, Turkey.

Contact: Prof. Cengiz Yalcin.

2. Department of Geophysics, Istanbul Technical University (H. Eyidogan)

Web site: http://www.geop.itu.edu.tr/

The Department of Geophysics, Mining Faculty, Istanbul Technical University, was established in 1974 and offers education and research programs for the B.Sc., M.Sc., and Ph.D. degrees. The Department has four full-time professors, one part-time professor, six associate professors, two research associates, and nine research assistants. The academic activities are shared by these academic staff members, who are mainly involved in projects on land and marine seismic prospecting for shallow and deep geological structures, investigation of Earth resources, earthquakes and crustal deformations in Anatolia and the surroundings.

The Department of Geophysics has recently initiated a program to reconstruct the ongoing educational system and aims at developing good quality English-speaking students equipped with state-of-the-art knowledge and experience. The main academic and engineering activities include (1) shallow and deep sea seismic studies; (2) seismic investigations of the structure of coasts and bays; (3) investigation of crustal structure; (4) geophysical investigation of environmental problems; (5) source mechanism of earthquakes and seismotectonics; (6) seismic hazard and seismic zonation, attenuation of seismic waves and site response investigation; (7) monitoring of microearthquakes, induced seismicity, and aftershock studies; (8) installation and maintenance of strong motion accelerographs; (9) paleomagnetic and archeomagnetic studies; (8) mineral and coal field research using geophysical methods; (10) local and regional gravity and magnetic mapping; and (11) magneto-telluric investigations.

79.53 Ukraine

The editors received an institutional review from Drs. Juriy M. Wolfman and Bella Pustovitenko, Seismology Department, Geophysics Institute, National Academy of Science of Ukraine, Studencheskaya St. 3, Simferopol, 95001 Ukraine. This review is archived as a computer-readable file on the attached Handbook CD#2, under the directory of \7953Ukraine. We extracted and excerpted the following summary. Biographical sketches of some scientists of Ukraine may be found in Appendix 3 of this Handbook.

1. Introduction

Ukraine is situated at the boundary of two major geostructures, the East European platform and the Mediterranean fold belt. Seismic hazards for Ukraine are seismic source zones within its territory, as well as those in immediate proximity to its borders. The four distinct Ukrainian seismotectonic provinces that have been defined on the basis of characteristics of geological structure and the degree of most recent tectonic activity and seismicity are described in the full report.

2. Seismological Observations

Instrumental seismological observations are made by specialized subdivisions of the S. I. Subbotin Geophysics Institute of the National Academy of Science of Ukraine, located in Kiev. The Seismology Department, based in Simferpol, carries out monitoring of the Crimea-Black Sea region; The Seismicity Department of the Carpathian region, in Lvov, is responsible for seismicity studies to the west of Ukraine. The main aims of these subdivisions are to investigate the fundamental properties and characteristics of the seismic process in the Crimea-Black Sea and Carpathian regions and to estimate the seismic hazard in Ukraine, with emphasis on inhabited localities and sites of especially important and potentially hazardous ecological features.

At present, seismic monitoring is accomplished with 22 stations, located in the Crimea and the Carpathian region. 11 of which are equipped with digital seismographs. The Crimean regional system has existed since 1928, since the destructive magnitude 6.8 earthquake of 1927. Information about analog and digital seismic stations of Ukraine is given in Table 1 and Table 2, respectively. The evolution of the monitoring system and the instrumentation are described in the full review on the Handbook CD.

3. Research Activities

A number of fundamental problems of seismology are under investigation. In the sphere of instrumental observations, updating of observation methods and the search for new methods of interpretation, parameterization, and unification of data are actively pursued. Investigation of seismicity and the seismic regime, with the elaboration of new ways of studying the space–time structure of seismicity, is another topic. Processes in seismic sources and source zones are also studied. The properties of the environment in the Black Sea-Crimea region have been investigated with the help of seismological methods. The most important applied task of the Seismology Department is the estimation of the seismic hazard of those territories subject to the effects of earthquakes. Both general and detailed seismic zoning are needed for general economic planning purposes in the first case and especially for evaluating seismic hazard for areas of important facilities, such as nuclear power stations, in the second. Some details of these five areas of research, with an extensive list of references, are given in the full review on the Handbook CD.

INTERNATIONAL HANDBOOK OF EARTHQUAKE AND ENGINEERING SEISMOLOGY, VOLUME 81B ISBN: 0-12-440658-0

TABLE 1 Information About Analog Seismic Stations of the Ukraine

			Coordinates			Equipment		
Stations	Code	Start	°N	°E	m	Type	Vmax	Tmax
Lvov	LVV	1899	49.82	24.03	320	MP	1050	0.2–20
						LP	850	17–55
Uzgorod	UZH	1934	48.63	22.29	160	MP	940	0.2–20
Uzgorod-pavilion	UZH2	1963	48.66	22.33	170	SP	38000	0.5–0.8
Kosov	KOV	1961	48.31	25.06	450	MP	1050	0.2–19
						SP	25000	0.3–0.8
Rahov	RAK	1956	48.05	24.19	500	SP	32800	0.5–0.8
Mezgorie	MEZ	1961	48.51	23.51	440	SP	31100	0.5–0.8
Morshin	MORS	1978	49.13	23.89	260	SP	14200	0.5–1.0
Simferopol	SIM	1928	44.95	34.12	275.0	SP	16000	0.1–0.8
						MP	1000	0.2–20
						LP	1000	18–55
Sevastopol	SEV	1928	44.54	33.68	42	SP	20000	0.1–0.7
						SP*	300000	0.1–0.4
Yalta	YAL	1928	44.49	34.15	25	SP	20000	0.2–0.6
Alushta	ALU	1951	44.68	34.40	61	SP	20000	0.2–0.5
Feodosiya	FEO	1927	45.02	35.39	40	SP	10000	0.1–0.7
Sudak	SDK	1988	44.89	35.00	110	SP	20000	0.2–0.6
Kerch	KRCH	1997	45.31	36.46	50	SP	10000	0.1–0.5
Donuzlav	DNZ	1998	45.46	33.10	40	SP	10000	0.1–0.8

* Equipment with electronic amplifier.

TABLE 2 Information About Digital Seismic Stations of the Ukraine

			Coordinates			Equipment	
Stations	Code	Start	°N	°E	m	Dynamic range, db	Frequency ange, Hz
Lvov	LVV	1999	49.82	24.03	320	100	0.02–5
Nignee Selishie	NSL	1998	48.20	23.46	250	100	0.2–15
Trosnik	TRS	1998	48.10	22.96	126	100	0.2–15
Gorodok	GRD	2000	49.18	26.50	200	100	0.2–15
Mukacheve	MKC	1999	48.45	22.69	152	100	0.2–15
Koroliove	KOR	1998	48.16	23.14	150	100	0.2–15
Kiev	KIGF	2000	50.47	30.36	312	100	0.2–15
Kiev-Iris	Kiev	1994	50.69	29.2	83.6	100	0.2–15
Yalta	Ylt	2000	44.49	34.15	23.6	100	0.01–20
Simferopol	Sym	2000	44.95	34.12	275.0	100	0.01–20

79.54 United Kingdom

The editors received the "UK National Report to IASPEI" from Dr. R.M.W. Musson, British Geological Survey, West Main Road, Edinburgh, EH9 3LA, UK. The report was compiled by R.M.W. Musson from material by R. Severn (University of Bristol), G. Helffrich (University of Bristol, and IASPEI National Correspondent), and A. Douglas (AWE Blacknest). It is archived as a computer-readable file on the attached Handbook CD#2, under the directory of \7954UnitedKingdom. We extracted and excerpted the following summary. An inventory of the UK Seismograph Network is included in Chapter 87 of this Handbook. Biographies and biographical sketches of some UK scientists and engineers may be found in Chapter 89, and in Appendix 3 of this Handbook, respectively.

1. Introduction

The United Kingdom was in the forefront of the early development of seismology. Much early work on the theoretical underpinnings of seismology, the development and worldwide distribution of seismic instrumentation, and, probably most importantly, the gathering and organising of the information in a way that would improve understanding of the Earth, was carried out in the British Isles. In the earlier part of the 20th century global earthquake monitoring was effectively the responsibility of one man, John Milne.

1.1 John Milne and the Shide Circulars

Milne's early work in the development of instrumental seismology was accomplished while in Japan in the late 19th century. He returned to Britain in 1895 and set up at the observatory in the Isle of Wight at Shide. From there, he arranged for the distribution of Milne-design seismographs to many observatories around the world, thus forming the basis of the first global recording network. These observatories sent back readings to Milne, who collated them and issued the first world earthquake bulletins, known as the Shide Circulars (see Chapter 88 of this Handbook). These were published by the British Association for the Advancement of Science (BAAS). The first circular, containing data from April 1899 to February 1900, was issued in 1900.

Milne continued to work at Shide and supervise this network until his death in 1913. In his last years he collaborated with J.J. Shaw of West Bromwich in the improvement of instrument design, the result being the Milne-Shaw seismograph, which incorporated electromagnetic damping. This quickly replaced the original Milne instrument and was used throughout the world for many years.

1.2 H. H. Turner and the International Seismological Summary

For a few years after Milne's death his work at Shide was continued by J. H. Burgess and S. Pring, under the supervision of H. H. Turner at Oxford University. Between 1918 and 1919 the work was transferred to Oxford. Turner and Burgess had re-titled the Shide Circulars as the Shide Bulletins, and reformatted them so that the data were grouped by earthquake rather than by station. With the transfer of the work to Oxford, this style of organisation of reporting continued, and the result was then published (starting with data for 1918, published in 1923) as the International Seismological Summary (ISS). Turner died in 1930 but the work of preparing the ISS continued at Oxford in the hands of F. A. and E. F. B. Bellamy and J. S. Hughes. In 1946 the ISS bureau moved to Kew Observatory, and continued its work under the administrative direction of Harold Jeffreys, the calculations being done by Hughes and his colleagues. After Jeffreys, R. Stoneley took over as director, and finally P. L. Willmore took over in 1963.

INTERNATIONAL HANDBOOK OF EARTHQUAKE AND ENGINEERING SEISMOLOGY, VOLUME 81B ISBN: 0-12-440658-0

1.3 ISS and the Bulletin of International Seismological Centre

In January 1964, the ISS became the International Seismological Research Centre (ISRC), which was very soon changed to the International Seismological Centre (ISC). The publications of the ISC were still issued under the ISS title up to and including the 1963 volume (published in 1969); the data from 1964 onward were published as the Bulletin of the ISC. Willmore was succeeded by Edward Arnold as ISC director in 1970, and in 1975 the ISC moved from Edinburgh to Newbury, Berkshire. Anthony Hughes, who had been Assistant Director of the ISS at Oxford, took over as director in 1977. In 1986 ISC moved to Thatcham, also in Berkshire, where it remains to this day. Hughes retired in 1998 and was succeeded by the present director, Ray Willemann.

1.4 Travel-time Tables and Other Advances

A corollary of the work of the ISS was the compiling of standard travel-time tables, necessary for earthquake location, but also a key to answering questions about the Earth's interior. The first usable travel-time tables had been produced by R.D. Oldham, who discovered that the Earth had a central core. However, the compilation of travel-time tables is most associated with the names of Harold Jeffreys and Keith Bullen, who started work on this subject at Cambridge in 1931. At that time the standard tables were those of Zöppritz, modified by Jeffreys, which were recognised as being in error by up to twenty seconds in places. The Jeffreys-Bullen "Revised Travel-Time Tables" were first published in 1935. The definitive edition appeared in 1948, which became the international standard for many years. Jeffreys was also the first person to recognise (in 1926) that the Earth's core is in fact liquid. This work involved the first use of absolute wave amplitudes in the estimation of earthquake energy, prefiguring the later development of earthquake magnitude by Richter.

Another important advance related to the structure of the Earth was the discovery of deep-focus earthquakes, first identified by Turner in 1922 from abnormal delays of P waves in some antipodal records. This was confirmed with better data by Stoneley and F.J. Scrase in 1931. Stoneley is also known for his discovery in 1924 of a type of interface wave that now bears his name, and he made an important contribution to the theory of anisotropic effects on surface waves, in 1949.

Some important landmark discoveries with implications for seismology are made in related disciplines. One of the most important advances in the second half of the 20th century relating to understanding earthquakes was the proposal of plate tectonics in 1967 by two geophysicists at Cambridge, Dan Mckenzie and Bob Parker. This new paradigm of the crustal processes in the earth provided a way of thinking about seismogenic processes that has become fundamental to modern seismology, especially in areas such as seismic hazard assessment. Other work in geophysics was also important in bringing about this improved understanding of crustal processes, including the work of Fred Vine and Drum Matthews and work by geophysicists like Edward Bullard, Patrick Blackett, and Keith Runcorn that contributed to the establishing of continental drift as a reality.

Mathematical advances have also been important. It was Augustus Love, at Oxford, who established the existence of tangentially polarised surface waves, which are now known as Love waves. It was also Love who, in 1904, first solved the Navier equation for a seismic source in an infinite solid. Horace Lamb, another British mathematician, first solved the problem (also in 1904) of how waves emanate from a vertical impulse applied to the surface of a solid. In 1923 James Jeans developed the asymptotic normal mode theory, another mathematical advance with seismological implications.

1.5 Earthquake Engineering

The beginnings of UK interest in earthquake engineering occurred in the 1958–1968 period, as part of the work of the Arch Dams Committee of the Institution of Civil Engineers. Its early years coincided with the development of the finite element method, which itself was dependent on the development of the digital computer, and it was these two influences that allowed arch dams to be analysed for earthquake loads, particularly at University College Swansea and the University of Bristol. In the 1960s the Institution of Civil Engineers became a focus for researchers in earthquake engineering, who, in 1968, inspired by G. Wood of Ove Arup and Partners, created the Society for Earthquake and Civil Engineering Dynamics (SECED), the British National Section of the International Association for Earthquake Engineering. By this time, all the large UK firms of consulting engineers had created small centres of expertise, which were to develop in the following 40 years into bodies having international eminence in earthquake engineering design. Theoretical developments at the basic level continued in several universities, principally at Imperial College and Bristol, and simple shaking tables were constructed at both. In 1980 the Science and Engineering Research Council (SRC) were persuaded by the UK Construction Industry that earthquake engineering studies should be energetically promoted in UK universities. As a result of this, the Universities of Nottingham and Cambridge became involved, and a new 6-axis, 3 by 3 m shaking table was constructed at Bristol during 1983–1985.

From 1990 onward, the European Union's research programmes played a major role in developing UK competence in earthquake engineering in two respects. First, in its Research Networks initiatives, Imperial College has played a significant part in research topics such as reinforced concrete frames, infilled frames, composite (steel/concrete) structures, seismic actions, and innovative design concepts. For the new 2000–2003

Research Network, involving thirteen European research groups, Imperial College will be the Co-ordinating Partner.

The second major influence of the EU on earthquake engineering in the UK arises from its Large Facilities Programme. Shaking tables were designated as one such Large Facility, and the Bristol table was included in this description. Since 1992 it has been the coordinator of a European group of facilities and has used modern developments in control engineering to significantly improve the fidelity with which dynamic test facilities can be controlled. This has opened up many new research areas in experimental aspects of earthquake engineering, most notably the ability to study the behaviour of test-pieces in the nonlinear range of material behaviour.

Since 1997, research at the University of Oxford has been concerned with the development of sub-structuring techniques, in which the major part of the structure is modelled computationally, and a smaller, more interesting, part is studied experimentally. The interface between the two parts is the major topic of research.

The detailed report is given on the Handbook CD, including reviews of the following institutions in greater depth.

2. The International Seismological Centre (ISC)

Address: Pipers Lane, Thatcham, Berkshire RG19 4NS, England, UK (R. J. Willemann, Director)

The International Seismological Centre is a non-governmental organisation charged with the final collection, analysis, and publication of standard earthquake information. Earthquake readings are received from almost 3,000 seismograph stations representing every part of the globe. The Centre's main task is to re-determine earthquake locations making use of all available information, and to search for new earthquakes, previously unidentified by individual agencies. It is the successor to the International Seismological Summary as the final arbiter of global earthquake locations. It was created out of the ISS in 1963, and was based at first in Edinburgh, but is now located at Thatcham, near Newbury, Berkshire.

The initial group of 7 supporting member countries is now increased to more than 50. Member institutions include national academies, government departments, and universities. Each member contributing at least a minimum unit of subscriptions appoints a representative to the Centre's Governing Council, which meets every two years to decide the Centre's policy and operational programme. Representatives from UNESCO and the International Association of Seismology and Physics of the Earth's Interior also attend these meetings. The Governing Council appoints the Director, Deputy Director, and a small Executive Committee to oversee the Centre's operations.

The analysis of the earthquake data is undertaken in monthly batches, with a delay of 22 months to allow the information used to be as complete as possible. The results are published as monthly bulletins and on CD-ROM.

3. The British Geological Survey

Address: West Mains Road, Edinburg, EH9 3LA, UK (A. Walker, Programme Leader)

The British Geological Survey (BGS) is the national geological survey of the UK, founded in 1835. It is administered as an independent component body of the Natural Environment Research Council (NERC), which is the UK's leading body for basic, strategic, and applied research in the environmental sciences. The BGS is the UK's national centre for earth science information and expertise. The Global Seismology and Geomagnetism Group (now, Earthquake and Forensic Seismology and Geomagnetism Programme) within the BGS is the national agency for seismological monitoring, and maintains a network across the country of over 140 stations, supported by a Customer Group led by the Department of the Environment, Transport and the Regions. The programme is led by Alice Walker. Results from the monitoring network are published on the Web and as annual bulletins. Instrumental earthquake monitoring of British earthquakes by BGS started in 1970; since 1974 macroseismic monitoring has also been undertaken of all significant British earthquakes.

The BGS maintains, under Roger Musson, the National Seismological Archive as part of the same programme. This serves as a repository for all surviving records and material from previous seismological observatories in the UK. Included are all known historical seismograms recorded in the UK, and collections of papers from some British seismologists, as well as collections of copies of material relating to historical British earthquakes. A catalogue of historical British earthquakes has been published.

BGS also hosts the National Data Centre for the Comprehensive Test Ban Treaty monitoring project, under David Booth, and is a focus for seismological research in several areas, such as seismic hazard and risk, and seismic anisotropy, especially in its application to the hydrocarbon industry.

4. Forensic Seismology Group, AWE Blacknest

Address: Blacknest, Brimpton, Nr Reading, Berkshire RG7 4RS (Peter Marshall, Group Leader)

The AWE Blacknest Group is responsible for advising the UK Government on forensic seismology. To enable the advice to be soundly based, the group carries out research on seismological methods of detecting, locating, and identifying earthquakes and explosions and on methods of estimating the yield of explosions from the seismic waves they generate. The group was set up in the late 1950s as part of the UK Atomic Weapons Research Establishment and moved to Blacknest (a large country house near Reading, Berkshire) in March 1961, from where it was led by Alan Douglas until his retirement at the end of 2000. It is now led by Peter Marshall.

The early work of the group concentrated on the use of seismometer arrays for detecting the short-period *P* waves from earthquakes and explosions at long range (>3000 km). It was found that (provided the signals could be detected) *P* signals recorded at such distances are much easier to interpret and thus the source of the signals more likely to be identified than with recordings made at shorter distances. The work resulted in the establishment by the group of medium-aperture arrays in Scotland (Eskdalemuir, EKA) and, as cooperative projects between the UK and the host countries, at Yellowknife, Canada (YKA), Gauribidanur, India (GBA), and Warramunga, Australia (WRA). These arrays have proved to be some of the best seismological stations in the world and all, it is hoped, will eventually become part of the International Monitoring System being set up to verify the Comprehensive Test Ban.

The recordings from the arrays have been much used for seismological research in general as well as for forensic seismology. Other areas where the group has made significant contributions are in improvements to epicentre and seismic (body and surface wave) magnitude estimation, broadband seismology, the analysis and synthesis of short-period *P* seismograms, and seismological methods for estimating the yield of underground explosions.

5. Other Institutions

A number of universities in the UK are active in seismological research, either earthquake or applied. These include Birmingham, Bristol, Cambridge, Durham, East Anglia, Edinburgh, Keele, Leeds, Leicester, Liverpool, London, Oxford, and Southampton. The study of geophysics, including seismology, is also promoted by the British Geophysical Association, a joint association of two learned societies, the Geological Society and the Royal Astronomical Society. An award from the Natural Environment Research Council has recently been made to a consortium of universities (Leicester, Leeds, Cambridge, and Royal Holloway, London) for the acquisition of a new generation of land seismic equipment. The ability to produce increasingly high-resolution seismic images of the interior of the Earth is experiencing a new revolution via present digital technology coupled with positional and timing information from satellites, and this initiative will enable the wider UK seismological community to be part of this revolution.

79.55 United States of America

The editors assembled a centennial report for the United States of America (US) by inviting about 50 US institutions and organizations engaging in seismological research or its promotion and publication to submit reviews of their histories and activities. About one-third of the invited institutions, including many of the major ones in the US, responded. Their reviews are archived as computer-readable files in the attached Handbook CD#2, under the directory of \7955USA. We extracted and excerpted the following summary from these reports.

Earthquake engineering in the US is reviewed in Chapter 80.16 of this Handbook. An inventory of seismic networks in the US is included in Chapter 87. Biographies and biographical sketches of some American scientists and engineers may be found in Chapter 89, and in Appendix 3 of this Handbook, respectively.

1. Introduction (C. Kisslinger)

The centennial national report of the United States of America begins with a review of the origins and evolution of seismology and related geophysics during the first part of the 20th century, followed by the series of institutional reports that describe their early history, as well as their structures, projects, and achievements since 1960. The rationale behind this plan stems from the fact that most of the major developments in the science have taken place in the latter part of the century, a period of rapid growth and progress coming from the efforts of individuals and teams working in governmental, industrial, and academic settings. An attempt to summarize this entire era in a single comprehensive review by one or a few authors would almost certainly have resulted in the omission of important contributions or the misplacing of emphasis. Therefore, each institution or organization was invited to tell its own story. The 15 sections of this summary following the early history were extracted from the reports of those institutions and organizations that chose to reply.

Because not all US institutions and organizations responded to the invitation to submit a contribution to this report, the following list of entities engaging in research on seismology and physics of the Earth's Interior, or promoting such activities, was compiled from an Internet search. Although this list is extensive, it was not possible to be all-inclusive by this technique. The Web site address at which more information may be found for each institution is given in parentheses. Please note that Web site addresses may change in the future.

American Geophysical Union (http://www.agu.org/)
Arizona State University, Dept. of Geosciences (http://geology.asu.edu/)
Boston College, Dept. of Geology and Geophysics (http://www.bc.edu/geology/)
Boston University, Dept. of Earth Sciences (http://www.bu.edu/es/)
Brown University, Dept. of Geological Sciences (http://www.geo.brown.edu/)
California Geological Survey (http://www.consrv.ca.gov/cgs/)
California Institute of Technology, Div. of Geological and Planetary Sciences (http://www.gps.caltech.edu/)
Carnegie Institution of Washington, Dept. of Terrestrial Magnetism (http://www.ciw.edu/DTM.html)
Colorado School of Mines, Dept. of Geophysics (http://www.geophysics.mines.edu/)
Columbia University, Lamont-Doherty Earth Observatory (http://www.ldeo.columbia.edu/)
Cornell University, Dept. of Earth and Atmospheric Sciences (http://www.geo.cornell.edu/)
Duke University, Div. of Earth and Ocean Sciences (http://www.env.duke.edu/eos/)
Florida International University, Dept. of Earth Sciences (http://www.fiu.edu/orgs/geology/)
Georgia Institute of Technology, School of Earth and Atmospheric Sciences (http://www.eas.gatech.edu/)

INTERNATIONAL HANDBOOK OF EARTHQUAKE AND ENGINEERING SEISMOLOGY, VOLUME 81B ISBN: 0-12-440658-0

Harvard University, Dept. of Earth and Planetary Sciences (http://www-eps.harvard.edu/)
Idaho State University, Dept. of Geosciences (http://www.isu.edu/departments/geology/)
IRIS, The Incorporated Research Institutions for Seismology (http://www.iris.washington.edu/)
Kansas State University, Dept. of Geology (http://www.ksu.edu/geology/)
Lawrence Livermore National Laboratory, Inst. of Geophysics and Planetary Physics (http://www.llnl.gov/urp/IGPP/)
Los Alamos National Laboratory, Geophysics and Seismology (http://www.ees.lanl.gov/Capabilities/geophys/)
Massachusetts Institute of Technology, Dept. of Earth, Atmospheric and Planetary Sciences (http://www-eaps.mit.edu/)
Michigan State University, Dept. of Geological Sciences (http://www.glg.msu.edu/)
Michigan Technological University, Dept. of Geological and Mining Engineering and Sciences (http://www.geo.mtu.edu/)
National Earthquake Information Center (http://neic.usgs.gov/)
National Geophysical Data Center (http://www.ngdc.noaa.gov/)
New Mexico Institute of Mining & Technology, Geophysics (http://www.ees.nmt.edu/Geop/)
Northwestern University, Dept. of Geological Sciences (http://www.earth.nwu.edu/)
Oregon State University, College of Oceanic and Atmospheric Sciences (http://www.oce.orst.edu/)
Pennsylvania State University, Dept. of Geological Sciences (http://www.geosc.psu.edu/)
Princeton University, Dept. of Geosciences (http://www.geophysics.princeton.edu/)
Rensselaer Polytechnic Institute, Dept. of Earth and Environmental Sciences (http://www.rpi.edu/dept/geo/)
Rice University, Earth Science Dept. (http://terra.rice.edu/department/)
Saint Louis University, Dept. of Earth and Atmospheric Sciences (http://www.eas.slu.edu/)
San Diego State University, Dept. of Geological Sciences (http://www.geology.sdsu.edu/)
Seismological Society of American (http://www.seismosoc.org/)
Southern California Earthquake Center (http://www.scec.org/)
Southern Methodist University, Dept. of Geological Sciences (http://www.geology.smu.edu/)
Stanford University, School of Earth Sciences (http://pangea.stanford.edu/)
State University of New York at Binghamton, Dept. of Geological Sciences (http://www.geol.binghamton.edu/)
State University of New York at Stony Brook, Dept. of Geosciences (http://pbisotopes.ess.sunysb.edu/geo/)
Texas A&M University, Dept. of Geology and Geophysics (http://geoweb.tamu.edu/)
U.S. Geological Survey, Earthquake Hazards Program (http://earthquake.usgs.gov/)
University of Alabama, Department of Geological Sciences (http://www.geo.ua.edu/)
University of Alaska Fairbanks, Dept. of Geology and Geophysics (http://www.uaf.edu/geology/)
University of Arizona, Dept. of Geosciences (http://www.geo.arizona.edu/)
University of California, Berkeley, Seismological Laboratory (http://www.seismo.berkeley.edu/seismo/)
University of California, Los Angeles, Dept. of Earth and Space Sciences (http://www.ess.ucla.edu/)
University of California, Riverside, Inst. of Geophysics and Planetary Physics (http://www.igpp.ucr.edu/)
University of California, San Diego, Inst. of Geophysics and Planetary Physics (http://www.igpp.ucsd.edu/)
University of California, Santa Barbara, Inst. for Crustal Studies (http://www.crustal.ucsb.edu/ics/)
University of California, Santa Cruz, Earth Sciences Dept. (http://emerald.ucsc.edu/)
University of Colorado, Boulder, CIRES (http://cires.colorado.edu/)
University of Georgia, Dept. of Geology (http://www.gly.uga.edu/)
University of Hawaii at Manoa, Dept. of Geology and Geophysics (http://www.soest.hawaii.edu/GG/)
University of Illinois at Urbana Champaign, Dept. of Geology (http://hercules.geology.uiuc.edu/)
University of Kansas, Dept. of Geology (http://www.geo.ukans.edu/)
University of Memphis, Center for Earthquake Research and Information (http://www.ceri.memphis.edu/)
University of Miami, Div. of Marine Geology and Geophysics (http://www.rsmas.miami.edu/divs/mgg/)
University of Michigan, Dept. of Geological Sciences (http://www.geo.lsa.umich.edu/)
University of Minnesota, Dept. of Geology and Geophysics (http://www.geo.umn.edu/)
University of Missouri—Columbia, Dept. of Geological Sciences (http://www.missouri.edu/~geolwww/)
University of Montana, Montana Bureau of Mines and Geology (http://www.mbmg.mtech.edu/)
University of Nevada, Reno, Seismological Laboratory (http://www.seismo.unr.edu/)
University of Oregon, Dept. of Geological Sciences (http://darkwing.uoregon.edu/~dogsci/)
University of South Carolina, Seismology (http://www.seis.sc.edu/)
University of Southern California, Dept. of Earth Sciences (http://www.usc.edu/dept/earth/)
University of Texas at Austin, Institute of Geophysics (http://www.ig.utexas.edu/)
University of Texas at Dallas, Dept. of Geosciences (http://www.utdallas.edu/dept/geoscience/)
University of Texas at El Paso, Dept. of Geological Sciences (http://www.geo.utep.edu/)
University of Utah, Dept. of Geology and Geophysics (http://www.mines.utah.edu/~wmgg/)

University of Washington, Earth and Space Sciences Dept. (http://www.geophys.washington.edu/)
University of Wisconsin, Madison, Dept. of Geology and Geophysics (http://www.geology.wisc.edu/home.html)
Virginia Polytechnic Institute and State University, Dept. of Geological Sciences (http://www.geol.vt.edu/)
Washington University in St. Louis, Dept. of Earth and Planetary Sciences (http://epsc.wustl.edu/)
Woods Hole Oceanographic Institution, Geology and Geophysics (http://www.whoi.edu/G&G/)
Yale University, Dept. of Geology and Geophysics (http://love.geology.yale.edu/)

2. Seismology and Physics of the Earth's Interior in the US (1900–1960) (C. Kisslinger and B.F. Howell, Jr.)

This review of the history of IASPEI-related science in the United States during the first sixty years of the 20th century was written by Carl Kisslinger, University of Colorado at Boulder, and Benjamin F. Howell, Jr., Pennsylvania State University. The following summary was excerpted by Kisslinger in his role as editor for the centennial national reports.

Seismology and solid earth geophysics were in their infancy in the United States before the California earthquake of 1906. Only a few individuals, especially at the University of California, Berkeley, and in the US Geological Survey, recognized the scientific significance of these disciplines. The 1906 earthquake led to major advances, mostly at California institutions. The Seismological Society of America was founded shortly after the earthquake and became a major avenue for dissemination of the results of earthquake research and for communication between seismologists and earthquake engineers. The Carnegie Institution of Washington and the California Institute of Technology played important roles in the early development of the science. Seismological observatories were established by universities and the federal government, with leadership provided by the US Coast and Geodetic Survey (USCGS) and the Jesuit Seismological Association. The last major observatory development within the era covered by this review was the creation of the Worldwide Standardized Seismograph Network, part of a program of fundamental research in support of the quest for methods to monitor underground nuclear explosions.

American inventors contributed new types of seismographs, including the Wood-Anderson torsion instrument and the Benioff seismograph and strainmeter, as well as significant improvements to Galitzin-type instruments. The USCGS had the responsibility for assembling data on US seismicity for most of this period. In addition, the USCGS developed a program of strong ground motion measurements in California, thus providing essential input to earthquake engineers. The major research topics investigated by American geophysicists were physics of seismic sources (earthquakes and explosions), theory and interpretation of seismic waves, tsunamis and microseisms, internal structure and composition of the Earth, physical conditions and material properties in the interior. Observations that became central to the development of the plate-tectonics model of geological processes after 1960 were accumulated during this time. Programs of higher education in geophysics evolved in American universities. The era closed with three developments that enhanced national interest and support for these disciplines: the creation of the Vela-Uniform program of research on monitoring nuclear explosions, the Alaskan earthquake of 1964, which led to a major expansion of research on earthquake hazards, and the adoption of computer technology for geophysical data acquisition and numerical analysis.

Although American scientists and engineers were active in every aspect of seismology during the sixty-year interval, among the specific achievements the following, some already mentioned, are noteworthy. Many of these achievements were based on the foundations laid by earlier research in Europe and Japan, and all have been followed by decades of further development.

- The formulation of the fault origin and the elastic rebound model of earthquake occurrence
- The determination of earthquake focal mechanisms by analysis of body wave polarities
- Advances in the theory and interpretation of dispersed surface waves
- Advances in the theory, observation, and interpretation of body waves, leading to more detailed models of the Earth's crust and deep interior
- Observations and interpretation of the Earth's normal modes
- Studies of the physical properties of rocks at elevated pressure and temperature
- The creation of the Worldwide Standardized Seismograph Network
- The integration of earthquake data into earthquake-resistant design of structures
- Early recording of strong ground motion that produced records that served as a world standard for many years.

The evolution of US programs in all of these areas, accompanied by a list of references documenting this history, is presented in the full review on the Handbook CD.

3. California Institute of Technology, Seismological Laboratory (H. Kanamori)

Web site: http://www.gps.caltech.edu/seismo/

The Seismological Laboratory of the California Institute of Technology was established in 1921 under the auspices of the Carnegie Institution of Washington, and is one of the world's leading centers for geophysical and seismological research.

In 1926, the laboratory began cooperative operations with the new geology department at Caltech and embarked on a project of installing a network of Wood-Anderson seismographs around southern California. By the early 1970s, the Seismological Laboratory's seismic network included 30 stations distributed in this region. At the same time, the US Geological Survey (USGS) was directed by Congress to study local earthquakes in California. Thus began the cooperative program between the Seismological Laboratory and USGS that has developed into the present Caltech-USGS Southern California Seismic Network, which operates more than 200 short-period seismic stations, and is providing earthquake information to the public.

Although this network is valuable in locating earthquakes, it has not until recently been capable of recording the types of motions that dominate strong ground shaking, primarily because the intensity of shaking from nearby earthquakes is so great that most seismometers go "off-scale." Moreover, the post-1970 cooperative network does not record horizontal motions, so key aspects of the information that seismologists could use to study fault zones and the nature of earthquake sources has simply been unavailable.

In 1987, the Laboratory deployed a modern broadband seismograph system at Pasadena. This system was expanded to a broadband seismographic network TERRAscope (1988) and later to TriNet (1997), which now has more than 150 broadband, high-dynamic-range seismographs in southern California. As part of this upgrade, digital, broadband data from the seismic stations are now telemetered continuously to the Seismological Laboratory. These instruments allow data to be recorded from the largest to smallest of earthquakes, from high to low frequency, at hundreds of stations around California, thereby increasing by orders of magnitude the amount of information that can be obtained on the physics of shaking. Significantly, these new instruments stay "on-scale" even close in to the largest of earthquakes, thereby dramatically increasing the opportunity to characterize and utilize "strong motion" data to understand earthquakes, their effects, and their relation to crustal structure and dynamics. In addition, by recording horizontal motions over a wide frequency range, even small local earthquakes can now be used to address fundamental questions about earthquake-generated motions.

The present TriNet system is a joint network between Caltech, USGS, and the State of California, and provides ShakeMap, real-time ground-motion data for use by emergency service agencies (see Chapter 78 of this Handbook). The high-quality waveform data from TriNet are extensively used for earthquake studies as well as earth structure research. The Seismological Laboratory now has about 12 professors and Ph.D. researchers, and 20 Ph.D. graduate students involved in many aspects of geophysical and seismological research including the structure, physics, chemistry of Earth's interior, and physics of earthquakes.

The full report on the Handbook CD includes a list of publications by staff members and visiting scholars. It was compiled by Viola Carter and Michele May from the reprints file maintained at the Lab, and contains over 2,000 entries from 1911 to 2000 (includes some papers from the decade before the Lab was founded).

4. Carnegie Institution of Washington, Department of Terrestrial Magnetism, Seismology (1945–2000) (Louis Brown)

Web site: http://www.ciw.edu/DTM-seismology.html

The Carnegie Institution of Washington's Department of Terrestrial Magnetism (DTM) entered seismology in 1948 and has remained active in that field to the present. The department's scientific work can be grouped into four chronologically overlapping divisions: (1) explosion seismology for studying the structure of the continental crust, (2) earthquake seismology conducted with arrays of narrow-band instruments deployed for specific studies, (3) the invention, development, and application of the borehole strainmeter, and (4) earthquake seismology carried out with arrays of both permanent and portable broadband instruments for regional and global analyses.

Merle Tuve initiated the department's work in explosion seismology by obtaining the cooperation of the Navy, with which Tuve had formed important wartime bonds, to obtain substantial quantities of surplus explosives for ocean detonation. He was also able to interest mining companies in coordinating large shots so that seismometers could be advantageously placed. These studies quickly involved other research groups, and many instruments were fielded during summer expeditions. The most important result of the work, which extended over more than a decade, was the unavoidable conclusion that the structure of the Earth's crust is more complex than had been originally thought.

Beginning with an expedition for the International Geophysical Year (1957), which involved ocean and mine shots, the Andes began to figure strongly in the department's research. The collaborations instituted with South American scientists at that time resulted in the establishment of a number of permanent stations suitable for observation of earthquakes. DTM seismologists expanded their studies early in the next decade with data from the World-Wide Standardized Seismograph Network.

In 1968 Selwyn Sacks and Dale Evertson invented the borehole strainmeter, a device using hydraulic amplification to provide very wide bandwidth and low noise for determining the buildup of strain in rock. The instrument has gone through several stages of development and has returned data on a variety of phenomena, including the discovery of slow earthquakes and other strain transients in seismic zones. Quite unexpectedly, Alan Linde discovered that suitably sited strainmeters could detect subsurface magma movement prior to a volcanic eruption.

The enormous capabilities of modern computing and the widespread use of three-component broadband seismometers have altered substantially the way the DTM group approaches its science. The global network allowed Paul Silver to discover patterns of seismically anisotropic regions of the mantle reflecting the strains imparted by tectonic deformation, past and present. David James and Paul Silver deployed portable arrays of such instruments to determine regional-scale structures in the continental mantle. The first high-resolution tomographic upper-mantle image of a mantle plume was carried out in Iceland with portable broadband seismometers from DTM.

5. Columbia University, Lamont-Doherty Earth Observatory

Web site: http://www.ldeo.columbia.edu/

The following two contributions, by J. Oliver and L. Sykes, are excerpts from talks given in connection with the 50th anniversary of Lamont (LDEO) in 1999. With the permission of the authors, the full texts of these personal histories that convey much of the history of the development of seismology at Lamont-Doherty, and a complete publication list (from 1949 to 2001), prepared under the supervision of P. Richards, are included on the Handbook CD. Additional information about LDEO, including a historical summary, may be found on the Web site http://www.ldeo.columbia.edu/.

5.1 BEEYOU: A Personal History of the Early days of Seismology at Lamont-Doherty Earth Observatory (Jack Oliver)

Beeyou—-! Beeyou!—-! That is an attempt to imitate the sound of a crack in the ice of a frozen lake heard at some distance from the crack. The study of such noises by Maurice Ewing during the 1930s led to the early stages in the development of the program in earthquake seismology at the Lamont Geological Observatory, which would eventually make some important contributions to the plate tectonics revolution of the 1960s. Ewing recognized that the character of that sound was the result of dispersion. Then, during WWII, Ewing studied the seismic waves from small explosive sources in shallow water, which also formed dispersive wave trains, in this case as a consequence of propagation.

Based on his wartime experience, Ewing, who joined Columbia University at the end of World War II, began to study the seismic waves that were generated by large earthquakes, and that traveled across oceanic paths. He was joined by Frank Press, then a graduate student. They were able to explain, as a consequence of dispersion, the entire train of earthquake-generated surface waves. They also developed the Press-Ewing seismograph for recording all three components of seismic waves of long periods. One reason for the move of Ewing's research group in geophysics from the Columbia campus to the Lamont Geological Observatory when it was initiated in 1949 was the need for a quiet site at which to operate earthquake seismographs. Soon one of the finest seismograph stations in the world was operating at Lamont, and the instruments began to produce observations that led to identification of seismic waves that had previously not been understood, and portions of the Earth were explored in ways previously not possible.

Ewing and Press proposed that a global network of ten such seismographs be installed for the IGY, 1957–1958. When Press left Lamont in 1955 to take a position at Caltech, I was made head of the earthquake seismology program and inherited the task of installing that global network. Soon a unique set of abundant data of quality and quantity came to Lamont .The observation program expanded even further. Instruments were installed in a deep mine, on the sea floor, on the Moon, and in temporary arrays for observing microearthquakes in Nevada and Alaska and deep earthquakes in Tonga-Fiji. Many unexpected discoveries were made based on the wonderful new observational data on the Earth and on earthquakes from those monitoring systems.

In the 1950s, we discovered waves generated by underground nuclear explosions that had not been detected by seismographs elsewhere. As a result, I became involved in political discussions in Geneva on a nuclear test ban, and that was the start of a long Lamont effort on this topic (see the following article by Lynn Sykes). The government research program that arose from the test ban negotiations funded, among other things, a grand 100-station seismograph network, partly patterned after Lamont's IGY network. Lamont archived copies of all the seismic data from that new network, following again Ewing's inductive style with its emphasis on learning from observations, and that data storehouse would be a factor in the contributions soon to come by Lamont seismologists to the plate tectonics revolution.

Participation by Lamont seismologists in the early stages of the plate tectonics revolution went something like this: Earthquake surface wave studies, as well as seismic refraction studies based on explosive sources, had shown that the crust of the deep sea basins differed from that of the continents and hence was not subsided continental crust. Marie Tharp and Bruce Heezen, while working on their now-famous physiographic map of the oceans, found the median rift of the mid-ocean ridge system, and also found that most oceanic earthquakes occurred beneath that rift. Using seismicity as a key piece of information, with Ewing they described the globe-encircling extent and continuity of that mid-ocean rift system. Harry Hess's hypothesis on seafloor spreading was widely known at Lamont shortly after he proposed it, but it languished for a time. Then the geomagnetic group brought Lamont's huge supply of data on sea-floor magnetics to bear on the problem. In so doing, they strengthened the case for the hypothesis. The news of their successes spread to the earthquake seismology group, particularly Lynn Sykes and me, through the efforts of Jim Heirtzler and John Foster. (See the accompanying paper by Lynn Sykes for his contribution to this history of the confirmation of the seafloor spreading hypothesis.)

The critical question then arose: If new material was being added to the surface of the Earth at the ridges, how was the Earth accommodating this new added area? We had, at Lamont, the data to resolve the problem. Bryan Isacks and I had initiated a project to study deep earthquakes in the Tonga-Fiji area. We found that the zone of deep earthquakes that dipped beneath the arc marked a somewhat thicker zone of very efficient seismic wave propagation. We were able to draw the cross section that shows the lithosphere dipping beneath Tonga and hence being underthrust there to depths of at least 700 km, the basis for the modern subduction model. We were thus more or less onto the concept of the plates at that time. We called our model the "mobile lithosphere" model. The term "plate tectonics" was not yet in use. It was Jason Morgan at Princeton who developed the global pattern of the plates and the geometrical basis for their motions. His paper and our paper were both presented at the famous AGU meeting of 1967. The subsequent contributions to the development of the plate-tectonics model by many Lamont scientists and others of that era are described in the full paper on the accompanying Handbook CD.

Ewing's emphasis on observation and on archiving of data, and on the inductive style of science, was a key factor in the earthquake seismology program. Furthermore, Lamont, with its diversity of activities and its wide variety of earth scientists, provided a stimulating and fertile environment for all. Early Lamont also had an especially fine esprit de corps.

The details of the research efforts summarized here and, especially, the contributions of Lamont graduate students and post-doctoral researchers, may be found on the Handbook CD.

5.2 "Seismology, Plate Tectonics and the Quest for a Comprehensive Nuclear Test Ban Treaty: A Personal History of 40 years at LDEO" (Lynn Sykes)

In 1958, a group of experts from eastern and western countries gathered in Geneva to discuss ways to verify a treaty to halt all nuclear testing. At their first meeting they had much information on nuclear explosions in the atmosphere but data on only one very small underground nuclear explosion code-named RAINIER, which was conducted by the United States in 1957. Concurrently, 1957–1958 also marked the International Geophysical Year (IGY), the ambitious coordinated multinational scientific effort to study the Earth. For the IGY, Lamont had deployed about a dozen seismograph stations around the world with new instruments developed by Frank Press and Maurice Ewing. At one of those stations at Lamont, Jack Oliver observed seismic waves generated by the largest underground nuclear explosion conducted by the US in 1958, the 19-kiloton BLANCA test. The US government had quickly recognized seismology's potential for detecting and identifying underground nuclear tests, and a panel of technical experts recommended in 1957 that the US should greatly expand funding of seismology to increase fundamental understanding and to develop better instrumentation. That funding program, called Vela Uniform, almost instantaneously transformed seismology from a sleepy, poorly supported scientific backwater to a field flooded with new funds, professionals, students, and excitement. The modest seismograph network established by Lamont during IGY became the prototype for the World-Wide Standardized Seismograph Network (WWSSN), installed in 1963 with funds from the Vela program.

The early 1960s also marked an era in which the frontiers of seismology expanded rapidly. A wealth of new data poured in from the IGY network. When I arrived at Lamont, records were just coming in of the great Chilean earthquake of May 1960. Those data, along with records from a new strainmeter installed in a deep abandoned mine at Ogdensburg, N.J., and from instruments operated by Caltech and UCLA, provided unequivocal evidence that the quake had caused the entire planet to vibrate as a unit. Another of my officemates, Lee Alsop, a new post-doc, was on the forefront of studying this new discovery of the existence of Earth's free oscillations, one of the greatest breakthroughs in geophysics.

For my Ph.D. thesis I decided to use the data from the IGY stations to analyze how seismic surface waves propagated across oceanic areas. But to accomplish this task, it was essential to have much more accurate data than were available at the time on the locations and origin times of earthquakes in oceanic areas. Using an IBM 7090 computer at the NASA Goddard Space Center near Columbia, which was then the world's largest and fastest computer, I began amassing a data set of more precisely located earthquakes that had occurred over several recent decades, largely along the mid-Atlantic ocean ridge system and in the southern hemisphere. I discovered that many of the calculated locations of earthquakes in the southern oceans were in error by as much as 500 kilometers. My revised, more precise locations were far less scattered around the ocean floor. Instead they aligned in an intriguing zig-zag pattern. The earthquakes coalesced along the centerlines of mid-oceanic ridges but then suddenly changed direction by nearly 90°, zagging for a certain distance and then just as abruptly zigging perpendicularly again to resume their original direction. In 1963 I speculated that one of these direction changes marked a great fracture zone that intersected the East Pacific Rise. A bathymetric survey confirmed the existence of what is now known as the Eltanin Fracture Zone.

I turned my efforts to detailed mapping of earthquakes along mid-oceanic ridges and island arcs. J. Tuzo Wilson proposed the concept of transform faults as an essential feature of seafloor spreading. My new data on earthquake mechanisms on the active sections of these transform faults, combined with magnetic data by Jim Heirtzler of Lamont and Walter Pitman, strongly supported the proposed processes of seafloor spreading, transform faulting, and continental drift. As new data were gathered, Bryan Isacks, Jack Oliver, and I decided to write a comprehensive paper, *Seismology and the New Global Tectonics*, which in 1968 pulled together information from many seismological studies that related to the plate tectonics hypothesis.

The new understanding of the locations and properties of earthquakes following from the plate tectonics theory allowed for better discrimination between earthquakes and explosions. In 1974 I was invited to join a US team going to Moscow for negotiations of what became the Threshold Test Ban Treaty (TTBT). I thought then that the TTBT was another step toward a Comprehensive Test Ban Treaty that would outlaw all nuclear testing. I had no idea that 22 years would pass before a CTBT would finally be approved by the United Nations and signed by the leaders of 149 nations.

In the late 1980s, Paul Richards of Lamont and I joined others on an advisory panel that provided information to the Office of Technology and Assessment of the US Congress. In 1988, the OTA conducted the first major independent review on seismic verification of nuclear testing. In 1992, Congress passed legislation limiting nuclear weapons testing. In 1996 the CTBT was signed, with an extensive international monitoring system set up to verify it.

During the four decades since I came to Lamont as a graduate student, seismology has finally fulfilled its promise to verify a nuclear test ban. Plate tectonics was a nice bonus.

6. National Geophysical Data Center, NOAA (A.M. Hittelman and J.A. Ikelman)

Web site: http://www.ngdc.noaa.gov/

The National Geophysical Data Center (NGDC) was created by the US Department of Commerce in 1965, as part of the Environmental Science Services Administration, Washington, D.C. It grew out of scattered activities, some of which had their origins in the International Geophysical Year, 1957–1958. In 1970, the National Oceanic and Atmospheric Administration (NOAA) was formed within the US Department of Commerce, and NGDC was placed under NOAA. At that time several additional geophysical data activities within the Department were absorbed by NGDC. In 1972, the Data Center moved to Boulder, Colorado. Currently, NGDC and the other NOAA data centers are part of NOAA's National Environmental Satellite and Data Information Services, reflecting the new emphasis on science augmented by satellite data.

NGDC took on some of the activities originally initiated by the former US Coast and Geodetic Survey, including management of seismological databases and programs. The years since 1965 produced a number of contributions to seismology and the physics of the Earth's interior. World Data Center-A activities in seismology (including tsunamis) have introduced a wider net of international cooperation and data exchange.

NGDC programs in historical seismogram data rescue, and compilations of earthquake, tsunami, geothermal, and volcano data have enriched the decades of the late 20th century. NGDC has contributed databases, publications, maps, and technological research. NGDC has also contributed to awakening public awareness of natural hazards' effects and mitigation through educational products and information on the World Wide Web. Throughout the years, NGDC has accessed emerging technology for the benefit of the scientific discipline.

The overall mission of the National Geophysical Data Center is to provide data and information in the fields of solid earth geophysics, marine geology and geophysics, solar–terrestrial physics, and paleoclimatology. Part of this task is to rescue and preserve historical data, and to create databases that are well documented, reliable, and easy to access.

For current activities of the National Geophysical Data Center, plus information on products and services relating to natural hazards, go to http://www.ngdc.noaa.gov/seg. Address: NOAA/National Geophysical Data Center, 325 Broadway, Boulder, Colorado, 80305, USA.

7. New Mexico Institute of Mining and Technology (A.R. Sanford, R.C. Aster, J.W. Schlue, H.J. Tobin, and K.W. Lin)

Web site: http://www.ees.nmt.edu/Geop/

A program of seismological research was initiated at New Mexico Tech in the late 1950s. Our review presents a chronological history from 1957 through 1999 that documents the milestones and accomplishments of the program. Two major long-term and ongoing projects during the over 40 years of research were (1) the physical characteristics of a highly unusual midcrustal magma body in the central Rio Grande rift beneath Socorro, New Mexico, and (2) the seismicity of New Mexico. Our review includes a section on the Socorro Magma Body that presents observations which best constrain its depth, extent, internal structure, relation to seismicity, and other geophysical properties. A section on the regional seismicity studies (1) describes development of catalogs based on instrumental data (1962–1998) and pre-instrumental data (1869–1961), (2) presents and discusses the geographic distribution of earthquakes in New Mexico, and (3) presents and discusses a seismic hazard map for the region.

The full report is given on the Handbook CD and includes milestones and accomplishments of the seismological research program, seismicity of New Mexico from 1869 through 1998, and references.

8. Saint Louis University, Earthquake Seismology (B.J. Mitchell and R.B. Herrmann)

Web site: http://www.eas.slu.edu/

Seismological studies at Saint Louis University began in 1909 and the Department of Geophysics, the first in the western hemisphere, formed in 1925. Soon after founding the department, Father James B. Macelwane, S.J., for whom the Macelwane medal of the AGU is named, revitalized the previously extant

Jesuit Seismological Service network. He assumed leadership of a new national organization, the Jesuit Seismological Association (JSA), and began directing research over a wide variety of seismological topics at Saint Louis University. SLU's long history of seismograph operation had begun with the installation of a Wiechert seismograph in 1909 and most recently includes deployments of modem digital broadband instruments over a broad portion of the central United States and over a more concentrated area closer to the New Madrid seismic zone. The department has an archive of seismograms from SLU instruments beginning in 1909, as well as a collection of worldwide seismic station bulletins consisting of over 400,000 pages on microfilms.

Early departmental researchers, using records from SLU and other instrumentation, made several important discoveries. Graduate students, working with Fr. Macelwane, discovered a second-order velocity discontinuity at about 950 km deep in the mantle and a zone of very low velocities in the lowermost mantle just above the core–mantle boundary. Macelwane and J.E. Ramirez, S.J., a post-graduate researcher, found that microseisms, previously unexplained, were traveling waves that were generated by storms at sea. Because of their research accomplishments many SLU graduate students have become important researchers and leaders at various institutions throughout the world. Research by current departmental seismology faculty and students addresses all facets of surface-wave dispersion and attenuation, source processes, crustal structure using receiver functions, CTBT monitoring, regional variations of seismic Q and its relation to crustal evolution, crust/upper mantle structure (including velocity, anisotropy, and Q), engineering seismology, anisotropy, and computational seismology.

The full report is given on the Handbook CD and includes a short history of SLU seismology, outlook for the future, and references. It also includes a chronological list of Saint Louis University publications in seismology (1911–1999).

9. Seismological Society of America (S. Newman)

Web site: http://www.seismosoc.org/

The Seismological Society of America (SSA) is a scientific society devoted to the advancement of earthquake science. Founded in 1906 in San Francisco, the Society now has over 2,000 members throughout the world representing a variety of technical interests: seismologists and other geophysicists, geologists, engineers, insurers, and policymakers in preparedness and safety.

The objects of the Seismological Society of America are

- Promote research in seismology, the scientific investigation of earthquakes and related phenomena.
- Promote public safety by all practical means.
- Enlist the interest of engineers, architects, contractors, insurers, and property owners in the obligation to protect the community against disasters due to earthquakes and earthquake fires by showing that it is reasonably practicable and economical to build for security.
- Inform the public by appropriate publications, lectures, and other means to an understanding of the fact that earthquakes are dangerous chiefly because we do not take adequate precautions against their effects, whereas it is possible to insure ourselves against damage by proper studies of their geographic distribution, historical sequence, activities, and effects on buildings.

SSA has been publishing its *Bulletin of Seismological Society of America* (BSSA) since 1911, and more recently, the *Seismological Research Letters* (SRL). In addition, it organizes an annual meeting. A full report of the Society is given on the Handbook CD, and includes the general history of SSA by P. Byerly, and a list of the contents of the Bulletins of Seismological Society of America (1911–2002). For more information, please visit: http://www.seismosoc.org/.

10. Southwest Research Institute, Center for Nuclear Waste Regulatory Analyses

The Center for Nuclear Waste Regulatory Analyses (CNWRA) was established in 1987 at Southwest Research Institute (SwRI) in San Antonio, Texas. Its principal purpose is to assist the Nuclear Regulatory Commission (NRC) in licensing the nation's first geological repository for high-level radioactive waste, which is proposed to be developed at Yucca Mountain (YM), Nevada. The role of the CNWRA has since expanded significantly to include solving complex technical problems for industry and providing comprehensive technical support to the NRC in commercial and federal site decommissioning, spent nuclear fuel (SNF) storage and transportation, and uranium recovery programs. The CNWRA has a staff of approximately 80 located at the SwRI facilities and a technical support office in Rockville, Maryland.

In the area of seismology and earthquake engineering, the CNWRA staff has extensive expertise in seismic source characterization; evaluation and mitigation of seismic hazards; numerical modeling for earthquake mechanics, seismic design of surface and subsurface facilities, and quantitative hazard assessment; and analyses to support development of regulations and regulatory guidance. As part of its original mission, the CNWRA has been investigating various aspects of earthquake and seismic engineering for the proposed geological repository at YM over the last 12 years. The CNWRA applied seismic tomographic methods to delineate thermal anomalies. A series of analog "sand-box" models was developed to investigate the kinematic and geometric aspects of crustal deformation related to growth of pull-apart basins along strike-slip fault systems. Numerical modeling was conducted to better understand earthquake mechanics, the initiation and propagation of slip on listric normal faults, and the associated energy dissipation. The CNWRA is conducting a variety of geological and geophysical investigations aimed at seismic source characterization and developing specialized computer software and hardware for such investigations. The CNWRA used a multiphased approach to understand the stability of and improve design techniques for underground facilities under combined thermal and dynamic load and used field investigations to guide laboratory studies and numerical modeling of the stability of emplacement drifts in jointed rock masses.

The CNWRA has also conducted extensive technical evaluations of seismic hazard assessment and seismic design to support NRC licensing activities related to the SNF dry storage facilities at the Idaho National Engineering and Environmental Laboratory and the Skull Valley Indian Reservation in Utah and NRC oversight of the gaseous diffusion plant at Paducah, Kentucky.

The CNWRA conducts internal research and development projects to strengthen its capabilities. For example, supported by the SwRl Internal Research and Development program, the CNWRA developed the implications of global-positioning-system-derived strain rates in assessing seismic and volcanic hazards, and microseismic activity as an indicator of macro-failure of rock. Currently, SwRl is investigating the seismic behavior of storage canisters and casks for deployment in high-seismicity sites around the world.

The full report is given on the Handbook CD and includes a description of contributions in physics of the Earth's interior, seismic source characterization and hazard analyses and engineering seismology, biographical sketches of the key research staff, and bibliography.

11. Stanford University, History of Geophysics (R.L. Kovach and G.C. Beroza)

Web site: http://pangea.stanford.edu/

Stanford University has a long tradition in earthquakes, seismology, and studies of the physics of the Earth's interior. John Casper Branner came to Stanford in 1891 as the first professor hired for the university. Besides serving as university president, he was in charge of the geology program from 1892 until 1915. He was witness to the 1906 San Francisco earthquake, producing many of the photographs for the famous report on The California Earthquake of April 18, 1906. He served as president of the Seismological Society of America from 1910 to 1912 and was a founder of its bulletin, published from 1911 to 1934 by the Stanford University Press. Bailey Willis, known as the Earthquake Professor, came to Stanford as Chairman of the Department of Geology in 1915, arguing strongly for the development of seismology as a separate field of study. He served as President of the Seismological Society of America for 6 years from 1921 to 1926. He made several significant contributions dealing with earthquakes in California and was vociferous about earthquake hazards in the state until his death in 1949. A formal Department of Geophysics was not created until 1957, although geophysics was a part of the formal earth sciences curriculum.

The full report is given on the Handbook CD and includes seismic stations operated by Stanford, historical timeline of faculty, and a brief description of current research activities.

12. United States Geological Survey, Earthquake Hazards Program (R. A. Page)

Web site: http://earthquake.usgs.gov/

The United States Geological Survey (USGS) was established in 1879 with responsibility for "classification of the public lands, and examination of the geological structure, mineral resources, and products of the national domain." Earthquakes were an early subject of study, both as an agent sculpting the landscape of the western United States and as a violent geologic phenomenon capable of great destruction. Through the first several decades of the 20th century, USGS scientists investigated many damaging earthquakes, including the 1886 Charleston, South Carolina,[1] and 1906 San Francisco, California,[2] earthquakes. Not until the 1960s, when the second largest earthquake of the 20th century struck southern Alaska in 1964,[3] did the USGS initiate a sustained program to improve scientific understanding of earthquakes in order to reduce future earthquake losses. In 1965, an earthquake research center was established in California, and the seeds were planted for a formal earthquake research program bringing together geologic, seismological, and geophysical disciplines.

The 1970s were a decade of great growth for earthquake research in the United States. In 1973, the earthquake monitoring, reporting, and research activities of the National Oceanic and Atmospheric Administration were transferred into the USGS and integrated into the Earthquake Hazard Reduction Program.[4] The major elements of the program were earthquake hazard mapping and risk evaluation, earthquake prediction, earthquake modification and control, seismic engineering, earthquake information services, post-earthquake studies, and application of earth science knowledge to reducing earthquake losses. The talents and capabilities of universities, state agencies, and the private sector were enlisted in pursuit of program goals. Effort under the USGS program expanded sharply when the National Earthquake Hazards Reduction Program was established in 1977. The components of the expanded USGS program largely remained the same, but the balance of emphasis shifted among the program elements.[5] Research on methods to predict the occurrence of earthquakes became the largest element and remained so through most of the 1980s. Hazard assessment and mapping continued as a major component; fundamental studies into the causes and mechanisms of earthquakes were nurtured; research on earthquake modification and control focused on induced seismicity.

Major progress was achieved on several fronts through the 1980s, but earthquake prediction remained an elusive goal.

The great damage in central California caused by the Loma Prieta earthquake in 1989 led to a refocusing of the USGS program with emphasis shifted from predicting the occurrence of earthquakes to predicting their effects.[6] Earthquake potential studies and hazard assessment were expanded in four regions in particular—Pacific Northwest, San Francisco Bay, southern California, and Central Mississippi Valley. The 1994 Northridge, California, and 1995 Kobe, Japan, earthquakes further demonstrated the vulnerability of modern urban centers to even moderate-sized earthquakes. To help reduce future losses, the USGS Earthquake Hazards Program at the end of the 20th century is focused on three activities:[7] (1) producing maps, reports, and planning scenarios and demonstrating their application to risk assessment and loss reduction; (2) collecting, interpreting, and disseminating information on US and significant foreign earthquakes in support of disaster response, scientific research, global security, earthquake preparedness, and public education; and (3) pursuing fundamental research to understand earthquake occurrence and effects in order to develop and improve methods of hazard assessment and strategies for loss reduction.

The staff of the USGS Earthquake Hazards Program currently numbers about 300. The Program Coordinator resides at the USGS National Center in Reston, Virginia. Most of the personnel are located in Menlo Park, California, at the USGS Western Regional Center, or in Golden, Colorado, near the USGS Central Regional Center. Personnel are also based at four field offices—Memphis, Tennessee; Albuquerque, New Mexico; Pasadena, California; and Seattle, Washington. All except Albuquerque are located at universities. The full report on the Handbook CD contains biographical sketches of many current and former staff of the Earthquake Hazards Program and some personnel of the closely associated Volcano Hazards Program, as well as a few difficult-to-obtain earthquake-related USGS reports. Biographies are included for the following deceased personnel: W.B. Joyner, H.P. Liu, R.B. Matthiesen, F.A. McKeown, L.C. Pakiser, and S.W. Stewart.

Notes

[1] Dutton, C.E. (1887). The Charleston earthquake of August 31, 1886: U.S. Geological Survey Ninth Annual Report, Washington, D.C.

[2] Lawson, A.C. (Chairman) (1908). The California earthquake of April 18, 1906: Report of the State Earthquake Investigation Commission, Carnegie Institution of Washington, Washington, D.C.

[3] Multiple authors (1965–1970). The Alaska earthquake, March 27, 1964: U.S. Geological Survey Professional Papers 541, 542, 543, 544, 545, 546.

[4] Wallace, R.E. (1974). Goals, strategy, and tasks of the Earthquake Hazard Reduction Program: U.S. Geological Survey Circular 701, 27 pp.

[5] Hamilton, R.M. (1978). Earthquake Hazards Reduction Program—Fiscal Year 1978 Studies Supported by the U.S. Geological Survey: U.S. Geological Survey Circular 780, 36 pp.

[6] Page, R.A., D.M. Boore, R.C. Bucknam, and W.R. Thatcher (1992). Goals, opportunities and priorities for the USGS Earthquake Hazards Reduction Program: U.S. Geological Survey Circular 1079, 60 pp.

[7] Page, R.A., J. Mori, E.A. Roeloffs, and E.S. Schweig, (1997). Earthquake Hazards Program Five-Year Plan 1998–2002: U.S. Geological Survey Open-File Report 98-143, 37 pp.

13. University of California, Berkeley Seismological Laboratory (B. Romanowicz)

Web site: http://www.seismo.berkeley.edu/seismo/

The Berkeley Seismological Laboratory (BSL), formerly Berkeley Seismographic Station, is the oldest Organized Research Unit on the campus of the University of California, Berkeley (UCB). Its mission is unique in that, in addition to research and education in seismology, it is responsible for providing timely information on earthquakes (particularly in northern California) to the UCB constituency, the public, and various state and private organizations. The BSL is both a research center and a facility/data resource. One unique activity of the BSL has been the operation, maintenance, and progressive modernization of facilities comprising a regional seismic network and an associated data archive, serving a broad community of users, both inside and outside the UCB campus.

U. C. Berkeley installed the first seismograph in the western hemisphere at Mount Hamilton in 1887. From that beginning and continuous evolution of instruments and methods of data analysis, the BSL currently operates a regional network of 21 sites, the Berkeley Digital Seismic Network, 2 borehole local seismic networks at Parkfield, Calif. (10 sites), and along the Hayward fault (8 land sites plus 10 sites under San Francisco Bay bridges, in cooperation with Lawrence Livermore National Laboratory), and 18 stations of a cooperative permanent GPS network. Data from the digital network, the GPS network, and part of the Hayward fault network are telemetered continuously and in real time over telephone "frame-relay" connections and serve as input for the joint US Geological Survey/UCB earthquake notification program and the development of enhanced procedures (e.g., finite fault modeling, "shake maps") within the Rapid Earthquake Data Integration program (see Chapter 77 of this Handbook). These data, together with other data pertaining to earthquakes and tectonics in central and northern California, are archived at the Northern California Earthquake Data Center, an on-line archive located at the BSL and the USGS, Menlo Park, accessible to outside users over the Internet. Some details of these facilities and related projects are given in the full report on the Handbook CD.

Research at the BSL focuses in two areas: regional tectonics and earthquake seismology, and "global seismology," the use of seismic data to probe the deep structure of the Earth. In recent years, prominent contributions have been made, among others, to the study of space–time distribution of microseismicity at Parkfield, with implications for earthquake scaling relations and recurrences; real-time, automatic methods for the retrieval of source parameters using sparse broadband seismic networks, including moment tensors, and, most recently, space–time distribution of slip on the fault plane of major earthquakes; volcanic seismic sources at the Long Valley Caldera; global mantle tomography; fine-scale structure of the deep mantle and core.

Highlights of the current research program, in addition to the continuation of work in the areas just mentioned, include collaboration with engineers for the estimation of ground shaking following a strong earthquake in the San Francisco Bay Area and its consequences on the built environment, in collaboration with several federal and state agencies; combining seismic, geodetic, and InSAR data for the study of the Hayward fault; and a pilot experiment, involving multiparameter observations, to study tectonic strain fluctuations in space and time at ten sites in the San Francisco Bay Area. The BSL is a founding member of the recently established California Integrated Seismic Network.

The contributions of the BSL and its staff to undergraduate and graduate earth science courses, as well as to K–12 outreach in the San Francisco Unified School District, are described in the full report, as are other public relations activities. More complete descriptions of the research efforts mentioned, as well as a bibliography of recent publications and biographical sketches of current personnel are also on the Handbook CD. It also includes a published article by Bruce A. Bolt on "One hundred years of contribution of the University of California Seismographic Stations". Further information is found on the Internet at http://www.seismo.berkeley.edu/seismo/.

14. University of Colorado at Boulder: Seismology and Physics of the Earth's Interior (C. Kisslinger)

Web site: http://cires.colorado.edu/

Solid earth geophysics was introduced at the University of Colorado at Boulder (CU) in 1940 and a seismograph station was set up in the Geology Building soon after. A geophysics program was established in the Department of Geological Sciences in 1960, with the first research in paleomagnetism and crustal deformation. In 1967, the Cooperative Institute for Research in Environmental Sciences (CIRES) was established through the efforts of J. C. Harrison as a joint enterprise of the University and the US Department of Commerce. The solid earth sciences research program in CIRES, which grew out of several pre-existing activities at the University, is only one part of the CIRES agenda,. The seismology program started with the appointment of C. Kisslinger as Director of CIRES in 1972. E.R. Engdahl, M. Wyss, and H. Spetzler were early appointees as Fellows of CIRES in solid earth science. The group has grown over the years with the appointment of additional Fellows and the collaboration with other CU faculty personnel. Significant related research is carried out in several academic departments, outside of CIRES, but often with productive collaboration.

Research on IASPEI-related topics has fallen into three general categories: source physics and seismotectonics, earthquake prediction research, and a broad range of studies of the Earth's internal structure, composition, and dynamic processes. A fundamental philosophy is that the research progresses best through close interaction of theorists, observational geophysicists, and laboratory experimentalists.

From 1974 to the mid-1990s, much research on seismotectonics, related plate tectonics, and prediction was based on data provided by the Central Aleutians Seismic Network, based on Adak Island, Alaska, supported by the US Geological Survey, with great assistance from the US Navy facilities on Adak. Other areas for major work on these subjects have been east and central Asia, the eastern Mediterranean, and Hawaii, mid-continent, and California, US.

The observational studies were accompanied by laboratory investigations of rock failure in which laser holography was the tool for tracking sample deformation leading to rupture. In support of the source physics studies, theoretical analyses and computer simulations of earthquake processes and wave generation by explosions have been carried out (C.B. Archambeau, J. Rundle). Field observations of crustal deformation by point strain, tilt, and creep measurements, and recent GPS campaigns to monitor regional scale crustal deformation (R. Bilham) have contributed to the seismotectonics efforts.

Studies of the Earth's interior have included both crust and upper mantle investigations and a variety of studies of structure and processes in the deeper interior. Crust and upper mantle work (A. Sheehan, C. Jones) has concentrated on the Rocky Mountain and Sierra Nevada regions, but has also included seafloor studies. Work on the deep interior (J. Wahr, M. Ritzwoller) has included studies of the core–mantle boundary and the distribution of inelastic attenuation and wave velocities in the mantle and core. Here also, the global research has been supported by laboratory studies of attenuation and wave velocities.

The full report on the Handbook CD provides details of these research efforts, as well as a list of CIRES Visiting Fellows and Research Associates in solid earth studies and a bibliography of representative publications.

15. University of Texas at Austin, Institute of Geophysics (K.K. Ellins and C. Frohlich)

Web site: http://www.ig.utexas.edu/

Texas, with its historical association with the energy industry, its large and economically important continental shelf and slope, and its direct access to the world's oceans through the Gulf of

Mexico, is a natural location for an Earth sciences research program of global scope that includes a strong effort in marine geology and geophysics. The University of Texas Institute for Geophysics (UTIG) fills this role, and is known internationally as a leading academic research group in geology and geophysics. Originally founded in 1972, UTIG is an Organized Research Unit within The University of Texas System under the auspices of the College of Natural Sciences of the University of Texas at Austin. Organized Research Units enhance the university's ability to fulfill its mission of education and research by providing opportunities for faculty and staff to undertake research programs, often involving graduate and undergraduate students, that do not fit easily into the framework of an academic department. UTIG research activities are carried out all over the world and include large-scale, multi-investigator, multi-institutional field programs requiring extensive coordination and logistical support. The importance of geophysical measurements and their mathematical interpretation in the exploration for petroleum and economically useful minerals has also led to valuable partnerships between UTIG and industry.

The Institute for Geophysics presently hosts 21 research scientists, 1 research fellow, 3 post-doctoral fellows, and 2 emeritus professors. The support staff of 31 people includes engineers, systems analysts, technicians, and administrative personnel. There are between 20 and 30 graduate students involved in UTIG research as partial fulfillment of their Masters and Ph.D. degree requirements—most through the Department of Geological Sciences at The University of Texas at Austin. Each year about 10 undergraduate students engage in research at the Institute. UTIG also plays a role in K–12 education through formal education programs and informal outreach efforts.

The full report is given on the Handbook CD and includes an institutional history, scientific contributions in (i) lunar and planetary seismology, (ii) seismic reflection profiling, (iii) ocean bottom seismograph program, (iv) computation and inverse theory, (v) earthquake seismology, (vi) tectonics, and (vii) aerogeophysics, institute publications, and research staff profiles.

16. University of Utah (W.J. Arabasz)

Web sites: http://www.seis.utah.edu/
http://www.mines.utah.edu/geo/
http://www.civil.utah.edu/

Seismology at the University of Utah dates from the installation of two Bosch-Omori seismographs on campus in 1907, motivated by the recognition of earthquake dangers in Utah and neighboring parts of the Intermountain West following modern settlement of the region in the mid-1800s. Scientific interest in earthquakes and the seismotectonic framework and geodynamics of this region arose naturally at the University, because of proximity to active Basin-Range normal faulting (notably Utah's 340-km-long Wasatch fault zone), historical and instrumental seismicity ($M \leq 7.5$) within the 1500-km-long Intermountain Seismic Belt, and the nearby Yellowstone hotspot.

During the first half of the 1900s, seismology (chiefly observational) grew slowly at the University. A Department of Geophysics was formed in 1952 and was merged in 1968 into the current Department of Geology and Geophysics, thus beginning the modern era of multifaceted seismology and earthquake-related studies. During recent decades, significant contributions both in earthquake and active-source seismology have been made by faculty scientists (currently five), together with associated students and staff, in the Department of Geology and Geophysics.

The diverse contributions primarily relate to seismicity, seismotectonics, and earthquake hazard studies of the Intermountain Seismic Belt—particularly the Wasatch Front region and the Yellowstone volcanic area—tectonophysics; seismic refraction, reflection, and tomographic imaging of crustal and lithospheric structure; geodynamics of the Yellowstone hotspot; geodetic studies using Global Positioning System (GPS) technology; mining-induced seismicity; 2D and 3D forward modeling and seismic imaging methods; near-surface environmental and engineering geophysics; and earthquake ground-motion modeling in 3D basins.

Formal teaching and research in earthquake engineering at the University of Utah began about 1980 and has gained significant emphasis in the University's Department of Civil and Environmental Engineering since 1993. Notable areas of current faculty specialization, involving six structural and geotechnical engineers, include soil and structural dynamics, nonlinear modeling of structural response, technologies for seismic strengthening and rehabilitation of existing buildings and bridges, and seismic improvement and stabilization of soil foundations and embankments.

Further information is available at these Web sites: (http://www.seis.utah.edu/), (http://www.mines.utah.edu/geo/), (http://www.civil.utah.edu/).

17. Weston Observatory of Boston College (R.W. Ott, S.J. and J.E. Ebel)

Web site: http://www.bc.edu/westonobservatory/

Weston Observatory is the oldest continuously operating geophysical research institute under private sponsorship in northeastern North America. Research at the Observatory has focused chiefly on geophysics and geology, with principle emphasis being on seismology, geomagnetics, and regional tectonics.

In 1928, Fr. Francis A. Tondorf, S.J. at Georgetown University donated a pair of Bosch-Omori seismometers to Weston College, the seminary for the New England Jesuit Province. Fr. Henry M. Brock, S.J was appointed to be the first Director of the Station. The first earthquake was recorded in January of 1931. Fr. G.A. O'Donnall, S.J. served as Director of the Station from 1935 to

1940 and was followed by Fr. M.J. Ahern, S.J. who held the position from 1940 to 1949. In 1935, Fr. Ahern bought a set of 100-kilogram Benioff instruments with long- and short-period recording assemblies.

From the late 1930s to the 1960s seismic field crews from the Observatory were used to map out the best routes for highways; in siting power plants, stone quarries, damsites, and airports; to assist in archeological studies; and to find water sources for many of the small towns around Boston.

Construction was begun on a new building for the Observatory in 1946, and the seismographs were moved to the new site in 1949 along with the offices, laboratories, and shops. Other buildings were added to the campus in the 1960s and 1970s to provide more space for classrooms, laboratories, and offices. The Observatory has housed the Catherine B. O'Connor Geoscience Library since 1960. Fr. Ahern was succeeded as Director in 1949 by Fr. Daniel Linehan, S.J. In 1947 the Observatory became affiliated with Boston College. At that time the University formed a Department of Geophysics and offered a masters degree in this field. The Department was centered at the Observatory and Fr. Linehan was appointed the first Chairman of the Department.

In the mid 1950s geomagnetic research was begun at the Observatory. The research was directed toward the collection, reduction, and evaluation of geomagnetic field data and electrical field phenomena. A magnetic network, which consisted of between four and six mobile stations, was operated in widely scattered locations around the United States.

In 1958 Boston College founded an undergraduate Department of Geology with Fr. James W. Skehan, S.J., the Assistant Director of Weston Observatory, as the first Chairman. The chairmanship of the Department of Geophysics was assigned to Fr. John F. Devane, S.J. in 1963, and in 1968 the two formerly separate but cooperating academic units were combined into the Department of Geology and Geophysics with Fr. Skehan as Chairman. Weston Observatory remained a separate facility until a further consolidation in 1977, when the Observatory became an integral part of the Department.

On November 22, 1961, the US Coast and Geodetic Survey established at Weston one of the first of the 123 stations of the World Wide Standardized Seismic Network (WWSSN). The 6 seismometers installed then continue in operation today.

The New England Seismic Network was established in 1962 and became an integral part of the Northeastern Seismic Network. It began with the installation of four three-component seismic stations in northern New England. It consisted of more than thirty stations, throughout New England in the early 1980s, under Dr. John Ebel, the present Director of the Observatory. Today the New England Seismic Network is operated by Weston Observatory in cooperation with the Earth Resources Laboratory at MIT. All of the Weston Observatory stations are PC-based with on-site recording, three-component broadband sensors, and dial-up telephone telemetry or direct Internet links to the central station.

Also, in this last decade, new initiatives were begun in the field of Geoscience Information Systems. A geographic information center is housed at the Observatory and is used as a research tool to assist investigations in seismic hazard, geotechnical engineering, geology, and environmental research. A research program is being developed in geotechnical engineering, particularly in problems related to nonlinear soil behavior in earthquake shaking. The Observatory also houses a Paleobotany and Palynology Laboratory, which is engaged in research on the origin and early evolution of land plants based on fossil spores from lower Paleozoic rocks from around the world. Geologists at Weston Observatory, in their study of the assembly and breakup of supercontinents through time, map and analyze the regional geology of selected localities in terranes surrounding the Atlantic Ocean.

79.56 Uzbekistan

The editors received two reports from Uzbekistan: (1) "Seismology in Uzbekistan in the XX-th Century: Report to IASPEI" from Dr. K. Abdullabekov, Director of the Institute of Seismology, Academy of Science, Tashkent, Uzbekistan; and (2) an institutional review of the Institute of Geology and Geophysics, Academy of Science, Tashkent, Uzbekistan from its Deputy Director, Dr. Bakhtiar Nurtaev. The first report was compiled by Abdullabekov, M. Bakiev, M. Usmanova, and S. Tyagunov. The second report was written in Russian with an abstract and five biographical sketches in English. Both reports are archived as computer-readable files on the attached Handbook CD#2, under the directory of \7956Uzbekistan. We extracted and excerpted the following summary.

Earthquake engineering in Uzbekistan is reviewed in Chapter 80.17. Biographies and biographical of some scientists and engineers of Uzbekistan may be found in Chapter 89, and in Appendix 3 of this Handbook, respectively.

1. Introduction

More than half of the territory of Uzbekistan falls into the zone of seismic intensity VII according to the MSK seismic intensity scale. Its eastern and southeastern parts, where large cities and the capital city, Tashkent, as well as numerous industrial units, are located, are in intensity zones VIII and IX. From historical manuscripts there are macroseimic data available on more than 500 earthquakes with magnitude greater than 5.

Development of instrumental seismology in Uzbekistan started in 1892 with the installation of 14 seismoscopes of the Russian Geographical Society on the territory of Tashkent Astronomical Observatory. In 1900 the construction of underground facilities for seismic equipment began. In May 1901 Repsold-Zelner instruments were installed and regular observations at fixed seismographic stations were started on July 13, 1901. The results of seismic observations of the station "Tashkent" were published in international seismological bulletins till 1913, filling a noticeable place in the world-system of seismological observations.

The failure of all Central Asian seismic stations at the times of the Chatkal earthquake of November 3, 1946, and the Ashkhabad earthquake of October 5, 1948, significantly stimulated the development of seismological science in Uzbekistan. Currently the seismic network of the Institute of Seismology consists of 23 seismic stations. These stations are provided with analog instruments (SKM-3, SM-3, and VEGIK). In addition, eight stations are equipped with digital instruments ACSS-1 developed by scientists of the Complex Experimental Methodical Expedition of the Institute of Seismology.

2. Institute of Seismology

The Institute of Seismology of the Academy of Sciences of Uzbekistan was established soon after the destructive Tashkent earthquake of April 26, 1966. The first head of the institute was G.A. Mavlyanov (1966–1985). From 1985 to 1988 the institute was led by O.M. Borisov. Since 1989 it is under the leadership of K.N. Abdullabekov.

The major theme of the research program is development of scientific fundamentals and methods of seismic zonation and earthquake prediction. A great volume of work on producing new maps of seismic zoning and microzoning is being carried out, as well as searching for earthquake precursors and forecasting. The investigations are mostly concentrated in the Tashkent, Fergana, and Kyzyl-Kum test geodynamic fields. The discovery of the phenomenon of changes in chemical and gas composition of ground waters (elements and isotopes) in the course of an earthquake was made in the Institute of Seismology. Scientists of the Institute were awarded the Beruni State Prize of the Republic of Uzbekistan for development maps of general seismic zoning of Uzbekistan and seismic microzoning of urban areas. Seismic microzoning maps for 26 large cities of

INTERNATIONAL HANDBOOK OF EARTHQUAKE AND ENGINEERING SEISMOLOGY, VOLUME 81B ISBN: 0-12-440658-0

Uzbekistan have been compiled and put into practice for anti-seismic design and construction.

The Institute consists of the Complex Experimental and Methodological Expedition, the Magnetic-Ionospheric Station, and 11 research laboratories. In addition to the 23 permanent seismic stations, there are 20 complex prognostic stations where permanent observations for seismometric, magnetometric, pulse electromagnetic, deformational, and hydro-geochemical parameters are being conducted.

The staff of the Institute includes 400 persons, 72 of them scientists. Among them there are 13 Doctors of Sciences (D.Sc.) and 38 Candidates of Sciences (Ph.D.). At present there are 4 post-doctorate and 15 post-graduate students taking part in thematic studies. Some of the current goals of the Institute are to extend the seismological network, increase the number of complex prognostic stations, and strengthen the fundamental and applied investigations on the problems of seismic zoning, earthquake risk management, physics of earthquake source, and earthquake prediction.

Details of the history of seismological investigations in Uzbekistan and of the programs within the research subdivisions, with biographical sketches and publication lists for the personnel, are given in the full report on the Handbook CD.

3. Institute of Geology and Geophysics

The Institute of Geology and Geophysics, Academy of Sciences of Uzbekistan, is a leading research organization in Uzbekistan dealing with the problems of geology, mineralogy, and geochemistry of mineral deposits, petrology of igneous, metamorphic, and metasomatic rocks, glaciology, as well as theory and methods of geophysical investigation of the Earth interior. The Institute, founded in 1939, consists of 11 laboratories, with 230 employees. The Director is F.A. Usmanov.

The research of the Department of Geophysics includes complex geophysical investigations of the deep structure of the Earth in different geotectonic structural zones of Central Asia, as well as analysis of deep tectonic structure of ore fields, deposits, and seismogenic zones. Studies include the definition of the physical state of substances in different horizons of the crust and nature of intracrustal boundaries. The estimation of the presence of mineral resources is attempted through definition of the properties of geophysical fields (gravitation, magnetic, temperature, and seismicity), the nature of deep geophysical processes of recent orogenesis and its links with seismicity and questions of seismic hazard assessment of territories at sites of critical plants (hydro and power stations).

Induced seismicity in areas of water reservoirs, oil and gas fields, and the development and installation of automatic systems of seismic monitoring on dams, as well as analysis of dynamic parameters of seismic impacts are investigated.

Other subjects of the basic research effort are the distribution of seismic heterogeneities in the upper part of the Earth's crust in ore areas, definition of parameters of the upper levels (up to depth of 10 km) of the crust, and also research of near-surface (up to 1 km) structures by the method of exchange waves, construction of detailed seismic sections under the profiles and their geologic interpretation. Parameters of the seismic regime for defining the seismogenic zones in Uzbekistan with an estimation of their hazard level are also investigated.

More details are given in the full report on the Handbook CD.

79.57 Vietnam

The editors received the report, "Developments and Main Research Results in Seismology and Physics of the Earth's Interior," from Prof. Dinh Xuyen Nguyen, President, National Committee for IASPEI, National Centre for Natural Science and Technology of Vietnam. This report is archived as a computer-readable file on the attached Handbook CD#2, under the directory of \7957Vietnam. We extracted and excerpted the following summary. An inventory of the Vietnam National Seismic Network is included in Chapter 87. Biographical sketches of some Vietnamese scientists may be found in Appendix 3 of this Handbook.

1. Introduction

Seismological observations and research began in Vietnam with the International Geophysical Year (IGY), 1957–1958, although the first seismic station had been established in 1924 in the Phulien Geophysical Observatory. In 1957 the Vietnam National Committee for the International Geophysical Year was founded and made responsible for implementing the national IGY Program. Two seismic stations were built at that time in Sapa and Phulien. At the same time the first research works on seismicity in Vietnam were initiated. The responsible organization for these observation and studies was the Department of Geophysics in the Vietnam Meteorological Survey. In 1986, this Department of Geophysics became the Institute of Geophysics of the Vietnam National Center for Scientific Research (now, The Vietnam National Center for Natural Science and Technology). The Department of Geophysics initially and the Institute of Geophysics since 1986 have been the main institutions in Vietnam responsible for observation and study on Seismology and Physics of the Earth's Interior.

At present, there is one telemetry array of nine stations and nine 3-component stations in Vietnam. All of them are short-period, PC-based digital recording stations. Based on the distribution of earthquakes, the telemetry array and six of the 3-component stations are in the north; three of the 3-component stations are in the south. The fundamental parameters of the network are given in the full report on the Handbook CD. The seismological network in Vietnam is monitoring all the seismic events of magnitude larger than 3 in the north, and all the events of magnitude larger than 4 in the entire territory of Vietnam.

2. Accomplishments

The main National Projects completed in recent years are as follows:

- Seismicity and seismic zonation of North Vietnam (1968).
- Seismicity and seismic zonation of Vietnam (1985).
- Study of Earth's crust deep structure using geophysical data (1980).
- The UNDP Projects VIE/ 84/ 011 "Reinforcement and Modernization of Seismological Service of Vietnam" (1986) and VIE/ 93/ 002 "Reinforcement and Modernization of Seismological Network of Vietnam" (1993).
- Study of Induced seismicity in the big reservoir regions (1992).
- Study of measures for seismic disaster mitigation in Vietnam (1996).

Besides these projects many studies on Engineering Seismology, Earth's crust structure, and velocity structure, and related topics were also completed. The main results of the above-mentioned projects are

- The seismicity manifestation laws in Vietnam were clarified and on this basis a series of seismic zoning maps (Seismic Hazard maps) of the territory of Vietnam were compiled.
- The seismic station network of Vietnam was reinforced and modernized. Now it consists of 16 seismic stations, among them 8 stations centralized into a seismic telemetry system.

INTERNATIONAL HANDBOOK OF EARTHQUAKE AND ENGINEERING SEISMOLOGY, VOLUME 81B ISBN: 0-12-440658-0

- A series of Earth's crustal structure maps (e.g., map of discontinuities in the Earth's crust, tectonic fault map, Earth's crustal velocity structure models) were compiled and used for study of lithosphere geodynamic laws.
- The maps of seismic microzonation of many cities and important construction sites were compiled and used for earthquake-resistant design and planning.

3. International Cooperation

In the process of development of observation and study the international cooperation in the field of Seismology and Physics of Earth's interior was also successfully developed. Now, Vietnam is an official member of IUGG, IASPEI, IAGA, ASC, and has good cooperation with Japan, France, China, the United States, Canada, Germany, ASEAN countries, and many international organizations such as ISC, AGU, NEIS, etc.

During August 19–31, 2001, Vietnam successfully hosted the Joint Scientific Assembly of the International Association of Geomagnetism and Aeronomy (IAGA) and the International Association of Seismology and Physics of the Earth's Interior (IASPEI) in Hanoi, Vietnam. It was well attended by hundreds of scientists from around the world.

At present, Vietnam concentrates efforts to the studies for seismic prevention and mitigation (Seismic Hazard and Risk Assessment), Earth structure and geodynamics, as well as earthquake prediction.

The full report on the Handbook CD includes a list of completed and ongoing projects, biographical sketches of the key personnel, and a list of publications. In addition, the full report includes five overviews of the main results of seismological and geophysical research in Vietnam: (1) Vietnam Seismological Network (by Le Tu Son), (2) seismicity and seismic hazard assessment in Vietnam (by Nguyen Dinh Xuyen and Tran Thi My Thanh), (3) seismic activity in the Hoa Binh Region after water filling of the reservoir (by Nguyen Ngoc Thuy and Nguyen Dinh Xuyen), (4) study of engineering seismology in Vietnam (by Nguyen Dinh Xuyen), and (5) gravity field and structure of the Earth's interior (by Cao Dinh Trieu).

80

Centennial National and Institutional Reports: Earthquake Engineering

80.1 General Introduction

Sheldon Cherry
Department of Civil Engineering, University of British Columbia, Vancouver, BC, Canada

1. Background

In 1999, the International Association of Earthquake Engineering (IAEE) warmly accepted an invitation from the International Association of Seismology and Physics of the Earth's Interior (IASPEI) to collaborate on a project that was to celebrate and mark the occasion of the hundredth anniversary of IASPEI—the production of a centennial publication entitled the "International Handbook of Earthquake and Engineering Seismology." IAEE was asked to obtain a directory of earthquake engineering research institutions (government, university, industrial, and appropriate professional societies, etc.) in each member country of IAEE, which at that time involved approximately fifty countries. These institutions were to be the centers in which earthquake engineering research was being performed and/or in which earthquake engineering societies existed. The overall goal here was to prepare a brief summary of the history, mission, activities, and accomplishments of earthquake engineering as a discipline throughout the world. It was agreed that the directory would include the names of the earthquake engineering research staff members in each institution, together with their areas of expertise, and individual contact sources (postal and email addresses; and fax and telephone numbers), as well as those of their institutions and professional societies, including Web sites (if available) and administrative addresses.

2. Procedure Used

The following procedure was used to meet the stated goal:

1. IAEE, through its Central Office, sought the assistance of the national delegates representing their member countries on IAEE's General Assembly of Delegates, to secure the desired material. This was normally accomplished by arranging for the directors of the active institutions/societies in each member country to prepare a review of its history, mission, activities, and accomplishments over the period of its existence.
2. These institutional reviews were not restricted in length. They were to be placed, in full, as computer files on a CD attached to the Handbook. The reviews were required to include an abstract that would be edited for the printed volume of the Handbook. As well, each Institution's research staff members were encouraged to include their personal biographical sketch as a part of their Institutional Directory. Such sketches were to contain (i) basic personal/professional information; (ii) major earthquake engineering contributions; and (iii) a selected list of publications. Biographical sketches of an institution's retired researchers were similarly eligible for inclusion in the Directory. Biographical sketches on behalf of the world's

INTERNATIONAL HANDBOOK OF EARTHQUAKE AND ENGINEERING SEISMOLOGY, VOLUME 81B ISBN: 0-12-440658-0

notable deceased earthquake engineers were also invited; it was anticipated that the combined submissions from all member countries would be limited to about 100 individuals.

3. The submission of other useful institutional information on earthquake engineering, such as available software, laboratory resources, newsletters etc. was also encouraged.

It is felt that this type of requested material will complement and supplement the extensive reviews of selected topics in earthquake engineering contained in the Handbook, and will serve as a valuable source of historical resource material.

3. Reports Received

We received a report on earthquake engineering from each of the following countries: Austria, Bulgaria, Canada, Chile, China, Croatia, Georgia, India, Japan, Kazakhstan, Kyrgyzstan, Macedonia, New Zealand, Russia, United States of America, and Uzbekistan. A summary of each of these reports is given in the following sections.

Some aspects of earthquake engineering activities are included in the reports on seismology in Chapter 79 for the following IAEE members: Argentina, Armenia, Australia, Chile, China, El Salvador, Greece, Hungary, India, Indonesia, Iran, Israel, Kazakhstan, Japan, Mexico, Norway, Poland, Russia, Slovakia, Switzerland, Taiwan, United Kingdom, and Vietnam.

The remarkable advances in engineering seismology documented in this Handbook are primarily the result of the skill and efforts of engineers and scientists working in national organizations and institutions throughout the world. Their stories are told in the technical chapters on specific topics. This present chapter is intended to summarize the activities of the earthquake engineering organizations and institutions that responded to the call for submissions. In the spirit of this Centennial Handbook, the emphasis is on past history and on activities current at the time of writing. Information on the most recent activities of many of these organizations and institutions may be found on their Web sites.

80.2 Austria

A Report of the Austrian Association for Earthquake Engineering to IAEE was submitted by Prof. Rainer Flesch and Dr. Wolfgang Lenhardt. This report contains the reviews of four Austrian institutions, which are engaged in teaching and research in earthquake engineering. It also includes biographical sketches of key researchers. We extracted and excerpted a summary of these institutions, which is provided here. Readers are encouraged to read the full report, which is archived as a computer-readable file on the attached Handbook CD#2, under the directory of \8002Austria. Biographical sketches of some Austrian scientists and engineers may be found in Appendix 3 of this Handbook.

1. Research Unit Construction, Arsenal Research

Address: Arsenal Research, Faradaygasse 3, Vienna, A-1030, Austria.
Web site: http://www.arsenal.ac.at/
Head: Prof. Rainer Flesch
<flesch.r@arsenal.ac.at >

Arsenal Research was founded in 1950 as a research institution of the government. In the early years it consisted of institutes for Machinery Engineering, Electrotechnical Engineering, and Geotechnical Engineering. The institution was always oriented toward practical applications involving strong measurement and testing facilities. A working group on vibrations and acoustics was also initiated at a very early stage of the institution's history. Over the years, Arsenal Research became the premier non-university research institution of the Republic. In 1997, Arsenal was transformed into a private limited company. In 1999, it became a subsidiary company of Austrian Research Centers Seibersdorf, which is also a non-university research organization in which the Republic of Austria is a shareholder. Following an internal reorganization in 1998, Arsenal was changed to a "flat hierarchy" and the new name, "Arsenal Research," was born. Arsenal Research now has four operative units: traffic, energy, construction, and environment.

The Construction Unit covers the fields of road engineering, acoustics, vibration and shock testing, vibration and structure-borne noise protection, soil mechanics and geotechnical engineering, structural dynamics and earthquake engineering, and structural monitoring. There are several working fields related to structural dynamics and earthquake engineering, as follows:

1. Earthquake-resistant design of structures (bridges, buildings, dams, etc).
2. Dynamic in-situ testing of structures in order to evaluate the dynamic properties of existing structures; FE modeling of structures; model updating; assessment of earthquake capacity of existing structures; and techniques for retrofit and seismic upgrading.
3. Structural monitoring; safety inspection via measurement of vibrations; and quality assessment.
4. Vibration and shock tests in the laboratory.
5. Vibration and structure-borne noise protection, especially in railway engineering; prognosis of vibration and structure-borne noise; and techniques for noise and vibration reduction.

2. Seismological Service, Central Institute for Meteorology and Geodynamics

Address: Hohe Warte 38, Vienna, A-1190, Austria.
Web site: http://www.zamg.ac.at/
Contact: Dr. Wolfgang Lenhardt
<lenhardt@zamg.ac.at>, and Dr. Gerald Duma
<duma@zamg.ac.at>

The Central Institute for Meteorology and Geodynamics (ZMAG) was founded in 1851. In the beginning, the main area

INTERNATIONAL HANDBOOK OF EARTHQUAKE AND ENGINEERING SEISMOLOGY, VOLUME 81B ISBN: 0-12-440658-0

of the Institute's activity concerned meteorological and geomagnetic observations. After a strong earthquake hit Ljubljana in Slovenia in 1895, the Austrian Academy of Sciences founded the Seismological Service of Austria, which became part of ZAMG in 1904. At this time, the Department of Geophysics at ZAMG also became responsible for the continuous observation of earthquakes.

Today, the Department of Geophysics at ZAMG consists of three sub-divisions: Seismology, Geomagnetics, and Environmental and Engineering Geophysics. The Seismological Service of Austria conducts historical earthquake research, locates recent earthquakes, updates the earthquake catalog of Austria, and determines the earthquake hazard for large dam construction in Austria; it also provides hazard maps for the Austrian Building Code. The first strong motion sensor was installed in Vienna in 1992. By 1999, eighteen sensors of this type were in operation in Austria. It is planned to install additional strong motion stations to improve the coverage of the most seismic active regions of the country. Please see Chapter 79.6 for a more detailed account of seismological research.

3. Institute of Rational Mechanics, Technical University of Vienna

Address: Wiedner-Hauptstr. 8-10/E201, Vienna, A-1040, Austria.
Head: Prof. Franz Ziegler
<fz@hp720.allmech.tuwien.ac.at>

A one-semester course (intermediate graduate level) on "Structural Dynamics" with emphasis on earthquake engineering is offered regularly (alternatively in German and English) by F. Ziegler. This course is accompanied by training lessons (hands-on computer modeling, by C. Adam) and a laboratory course (involving a fully computerized shaking table for models up to 100 kg, by R. Heuer). The latter is taught in small groups of four students, who obtain hands-on experience with various test specimens (including vibration absorbers).

Continuous and skeletal structures (also under condition of fluid–structure interaction) are analyzed under deterministic and random seismic excitation. The inelastic performance of ductile structures is considered by the method of "plastic sources acting in the elastic background structure," a locally developed routine. The response of computer models is compared to experimental data obtained from tests on small-scale structural models. Plastic drift, the P–D effects and structure–soil interaction are modeled analytically and measured experimentally under single- and double-point excitation. Additional research topics include studies on

1. Vibration absorbers developed for inelastic structures (computer and experimental modeling).
2. Earthquake safety of large dams and weirs. This topic has received special attention. In-situ testing has been successfully performed. The actual dynamic properties of arch dams with reservoirs filled to various levels have been identified.
3. Seismic waves. The ray integral method has been developed into a working tool to investigate parallel and dipping single- or double-surface layers. Synthetic seismograms have been produced for actual and assumed dipping angles (up to quarter spaces) for line and point sources. Time random sources have been considered within correlation theory. Emphasis is on application of symbolic manipulation programs.
4. Graduate students have completed master theses and doctoral dissertations using computer modeling and/or laboratory equipment.

4. Institute of Structural Concrete, Technical University of Graz

Address: Lessingstraße 25, Graz, A-8010, Austria.
Head: Prof. Lutz Sparowitz
Specialist for Earthquake Engineering: Prof. Rainer Flesch

Courses offered consist of (1) structural dynamics and earthquake engineering, lecture, 2 hours per week, one semester (instruction in German); and (2) structural dynamics and earthquake engineering, practical calculations, 1 hour per week, one semester (instruction in German).

Course contents include theory of vibrations, dynamic loads, methods of analysis, dynamic material behavior, soil dynamics, fundamental rules of earthquake engineering, isolation techniques, dynamics of buildings, bridges, and dams, and railway-induced vibrations.

Research in the field of earthquake engineering and structural dynamics is also carried out in close cooperation with Arsenal Research, a subsidiary of Austrian Research Centers Seibersdorf, a non-university research organization.

80.3 Bulgaria

A report of the Bulgarian Association for Earthquake Engineering was submitted by the Bulgarian National Committee on Earthquake Engineering. It consists of a directory of earthquake engineering research institutions, and two institutional reviews. The complete report is archived as a computer-readable file on the attached Handbook CD#2, under the directory \8003Bulgaria. We extracted and excerpted the following summary. Seismology in Bulgaria is reviewed in Chapter 79.11, and biographical sketches of some Bulgarian scientists and engineers may be found in Appendix 3 of this Handbook.

1. Directory of Earthquake Engineering Institutions

1.1 Central Laboratory for Seismic Mechanics and Earthquake Engineering (CLSMEE) at the Bulgarian Academy of Sciences

Ludmil Tzenov, Dr.Sc., Professor, Eng., Head of CLSMEE: Dynamics of soil–structure interaction, dynamic behavior of structures, design seismic loading, seismic-resistant structures, seismic codes, seismic protection of historical monuments.

Svetoslav Simeonov, Ph.D., Assoc. Prof., Eng., Vice Director of CLSMEE: Soil liquefaction, structural dynamics, wave propagation, soil–structure interaction, full-scale dynamic. experiments.

Ivanka Paskaleva, Ph.D., Assoc. Prof., Eng., Scientific Secretary of CLSMEE: Wave propagation, soil–structure interaction, seismic risk, seismic macrozonation, rock mechanics.

Elena Vasseva, Ph.D., Assoc. Prof., Eng.: Structural mechanics; reinforced concrete structures, earthquake engineering and seismic-resistant structures, dynamics and stability of structures, limit state analysis, computer-aided design, fracture mechanics.

Marin Kostov, Ph.D., Assoc. Prof., Eng.: Wave propagation, soil–structure interaction, probabilistic structural mechanics; seismic risk, seismic macrozonation, seismic fragility.

Silvia Dimova, Ph.D., Assoc. Prof., Eng.: Passive seismic protection, friction devices, seismic fragility, soil–structure interaction, design seismic loading, irregular structures, structural mechanics.

1.2 National Building Center (NBC), Ministry of Regional Development and Public Works

Mihail Kmetov, Ph.D., Dr.Sc., Professor, Head of NBC: Nonlinear analysis of RC structures.

Evgueni Milchev, Ph.D., Dr.Sc., Assoc. Professor: Geometrical nonlinearity and stability of structures; earthquake engineering.

Georgi Blagoev, Ph.D., Professor: Reliability of structures, experimental mechanics, and testing of structures.

1.3 University of Architecture, Civil Engineering and Geodesy

Nikola Ignatiev, Ph.D., Prof., Eng.: Reinforced concrete structures, earthquake engineering and seismic-resistant design of structures, dynamics of structures.

Petar Sotirov, Ph.D., Prof., Eng.: Experimental tests of seismic behaviour of structures and structural members, soil–structure interaction, seismic safety of steel structures.

Zdravko Petkov Bonev, Ph.D., Assoc. Prof., Eng.: Statistics, stability, and dynamics of structures, fracture mechanics, numerical methods and modeling, endochronic theory of concrete plasticity, and evaluation of inelastic seismic response of frames.

Rossitza Gancheva, Ph.D., Assoc. Prof., Eng.: Reinforced concrete frame members and walls; analytical hysteresis modeling; nonlinear investigation of reinforced concrete frame-type structures under seismic excitations.

Todor Karamanski, Ph.D., Prof., Eng.: Structural dynamics, soil–structure interaction.

INTERNATIONAL HANDBOOK OF EARTHQUAKE AND ENGINEERING SEISMOLOGY, VOLUME 81B ISBN: 0-12-440658-0

Tanko Ganev, Ph.D., Prof., Eng.: Structural dynamics, soil–structure interaction.

1.4 Research Institute of Building

Mincho Dimitrov, Ph.D., Prof., Eng.: Seismic reliability of structures, experimental testing of structures and structural members.

Georgi Diankov, Ph.D., Assoc. Prof., Eng.: Seismic reliability of structures, experimental testing of structures and structural members.

2. Central Laboratory for Seismic Mechanics and Earthquake Engineering at the Bulgarian Academy of Sciences

Address: Acad. G. Bonchev St., Block 3, Sofia 1113, Bulgaria.
Director: Prof. Ludmil Tzenov.

Established in 1982 as the outcome of a working group that had been in operation since 1968, the Central Laboratory for Seismic Mechanics and Earthquake Engineering (CLSMEE) is the only scientific institute in Bulgaria whose main goal is seismic risk reduction for the purpose of (1) minimizing casualties, (2) ensuring safety of material and cultural valuables, and (3) ensuring functioning of lifeline systems and activities.

The activities of CLSMEE include (1) seismic risk assessment of urban areas, buildings, and structures, (2) analyses of the response of the soil–structure–equipment system to dynamic excitations, (3) monitoring of strong ground motions, (4) seismic safety of all types of buildings and structures, (5) elaboration of standard documents for design and construction in seismic regions, and (6) training of scientific and engineer specialists and providing earthquake knowledge for the population.

The most significant case studies carried out by CLMSEE include (1) Kozloduy Nuclear Power Plant and the site of the Belene Nuclear Power Plant; Maritza Iztok thermal power plant; (2) dams including the Chaira, Kardjaly, Draginovo, Tzankov Kamak, and Sofia metropolitan dams; (3) the salt body of Provadia; and (4) monuments of cultural heritage: the Rila Monastery, the church St. Sofia, the buildings of the Bulgarian national bank in Varna and Plovdiv, the Antique Theater in Plovdiv, the Tombul Mosque, renaissance houses, and more.

Projects involving international collaboration include (1) chairmanship and participation in international projects of "Seismic risk reduction in the Balkan region," "Industrial structural systems in seismic regions," "Seismic safety of cultural monuments," "Ultimate capacity of the reinforced concrete structures of NPP to seismic excitations," and (2) Collaboration with foreign and international organizations. The national representative of Bulgaria in EAEE and IAEE is Ludmil Tzenov. More details about equipment and research staff members and biographical sketches are given in the full report on the Handbook CD.

3. National Building Center, Ministry of Regional Development and Public Works

Address: 47, "Chr. Botev" Blvd. 1606 Sofia, Bulgaria.
Head: Prof. Mihail Kmetov.

The National Building Center (NBC) was established by decree of the Council of Ministers as a budget-supported unit to the Ministry of regional development and public works, and started its activities in March 1992. Its approved "main activities" include scientific, applied, and development functions.

The NBC's work falls into three main categories: (1) a normative basis for building design is being developed. NBC houses the Bulgarian Technical Committee of Standardization–TCS 56. The Director of NBC is also the chairman of TCS 56; (2) an information computer system for the registration and analysis of software developed in Bulgaria and abroad; and (3) representing the Ministry in the work of some international organizations. More details about NBC's activities and research staff members are given in the full report on the Handbook CD.

80.4 Canada

A brief report on Canadian earthquake engineering research was submitted by Prof. Donald L. Anderson, President of the Canadian Association for Earthquake Engineering/L'association canadienne du génie parasismique (CAEE/ACGP). Seismology in Canada is reviewed in Chapter 79.12. Biographical sketches of some Canadian earthquake engineers and seismologists may be found in Appendix 3 of this Handbook.

1. Introduction

The Canadian Association for Earthquake Engineering/L' association canadienne du génie parasismique (CAEE/ACGP) (http://cee.carleton.ca/CAEE) is the national organization for earthquake engineering in Canada. It publishes a quarterly newsletter, represents Canada on the IAEE, sponsors the Canadian Conference on Earthquake Engineering, which is held every four years, and will host the 13WCEE to be held in Vancouver in 2004. It was incorporated federally in 1993 and has about 150 members.

2. Brief History

Earthquake engineering research in Canada got its start in the 1960s, and Canada was a founding member of IAEE. The Canadian National Committee for Earthquake Engineering (CANCEE) was formed in 1963 as an Associate Committee of the National Research Council and is responsible for the earthquake provisions in the National Building Code of Canada. It was also originally charged with the task of encouraging research in earthquake engineering and the education of practitioners in earthquake engineering matters.

The first Symposium on Earthquake Engineering was held in Canada at the University of British Columbia in 1965. It was chaired by Prof. Sheldon Cherry and sponsored by the host University, the Engineering Institute of Canada, and the Association of Professional Engineers of British Columbia. In 1971, under the auspices of CANCEE, the First Canadian Conference on Earthquake Engineering was held in Vancouver, again at the University of British Columbia. Since that time Conferences have been held every four years in different locations in Canada, with the most recent being the Eighth Conference held in Vancouver in 1999 under the auspices of CAEE/ACGP.

The Associate Committees of the National Research Council were reorganized in 1991. CANCEE was retained as a committee for the purpose of providing input on seismic matters to the National Building Code of Canada; it could no longer sponsor conferences or represent Canada on the IAEE. CAEE/ACGP was formed to take over these roles, as well as to foster research and communication among practicing engineers and researchers.

3. Research Institutions

Earthquake engineering research in Canada is mainly carried out in the universities, with some material-specific research in public/private laboratories. Seismological research is performed in universities, federal government laboratories, and large public/private corporations. Funding mainly comes from the Natural Sciences and Engineering Research Council (NSERC), government contracts, and from commercial testing.

Listed here are universities and other centres currently involved in earthquake engineering research. For up-to-date information and details on the facility and personnel, the reader is encouraged to consult the indicated Web sites.

Four regions in Canada presently have universities with earthquake engineering as a program or research specialty. They are

1. Vancouver, British Columbia—University of British Columbia, Civil Engineering (http://www.civil.ubc.ca/earthquake.htm).

INTERNATIONAL HANDBOOK OF EARTHQUAKE AND ENGINEERING SEISMOLOGY, VOLUME 81B ISBN: 0-12-440658-0

2. Hamilton, Ontario—McMaster University, Civil Engineering (http://www.eng.mcmaster.ca/civil/vibs).
3. Ottawa, Ontario—(i) University of Ottawa, Civil Engineering (http://by.genie.uottawa/cvg/), (ii) Carleton University, Civil Engineering (http://www.civeng.carleton.ca). The University of Ottawa and Carleton University have formed the Ottawa Carleton Earthquake Engineering Research Centre (OCEERC) (http://by.genie.uottawa.ca/oceerc).
4. Montreal and area, Quebec—(i) Ecole Polytechnique, Civil Engineering (http://www.struc.polymtl.ca), (ii) McGill University, Civil Engineering (http://www.mcgill.ca/civil), (iii) Concordia University, Civil Engineering (http://www.encs.concordia.ca/bce), and (iv) University of Sherbrooke, Civil Engineering (http://www.gci.usherb.ca/groupes.crgp).

The following universities offer some graduate courses in earthquake engineering and have been involved in earthquake engineering research:

1. University of Alberta, Civil Engineering (http://www.civil.ualberta.ca).
2. University of Toronto, Civil Engineering (http://www.civ.toronto.ca).

Other institutions or laboratories involved in earthquake research in Canada are

1. Forintek Canada Ltd. (http://www.forintek.ca)—Forintek specializes in research on timber structures.
2. Geological Survey of Canada (GSC) (http://www.gsc.nrcan.gov.ca)—The GSC performs seismological studies for CANCEE and monitors the strong ground motion network in Canada.

80.5 Chile

A report of the institutions involved with earthquake engineering in Chile was submitted by Rafael Riddell, National delegate of Chile to the IAEE. We extracted/excerpted the summarized version of this report shown here. Readers are encouraged to read the full report, which is archived as a computer-readable file on the attached Handbook CD#2, under the directory of \8005Chile. Seismology in Chile is reviewed in Chapter 79.13, and biographical sketches of some Chilean earthquake engineers and seismology may be found in Appendix 3 of this Handbook.

1. Chilean Earthquake Engineering and Seismology Association

Address: Avda. Blanco Encalada 2120 Piso 4 Santiago, Chile.
Web site: http://www.dgf.uchile.cl/achisina/

The Chilean Earthquake Engineering and Seismology Association was created in 1963. It has more than 100 members, and the support of more than 50 institutions from the private and public sector. Its objectives are (1) to support and spread knowledge, development, and research on seismology and earthquake engineering; (2) to coordinate work and studies and to spread the results to the national and international communities; (3) to develop national and international meetings and to promote the national and international exchange of knowledge; and (4) to represent the Chilean national organization on the International Association for Earthquake Engineering and other appropriate national and international agencies.

The Association's activities include (1) permanent contact and exchange of work and research with groups and individuals at the national and international level; (2) promotion and development of national meetings (eight national meetings have been held to date at four-year intervals and the 4th World Conference on Earthquake Engineering was held in Chile in 1969); (3) development of manuals, codes, and standards for seismic design and analysis (Codes NCh433 and NCh2369); and (4) publication of technical material. The Association's present officers are Tomas Guendelman B., President; Edgar Kausel V., Vice President; Paulina Gonzalez S., Secretary; Alberto Maccioni Q., Treasurer; Elias Arze L., Treasurer; and Ruben Boroschek K., Executive Secretary. It also has the following Regional Presidents: Mario Duran L., Patricio Bonelli C., and Giuliano M.

The full report on the Handbook CD also contains biographical sketches of Elias Arze Loyer, Silvana Cominetti, Diana Patricia Comte Selman, Rodrigo Flores Alvarez, Paulina Gonzalez, Tomas Guendelman Bedrack, Mario David Guendelman Bebrak, Edgar G. Kausel, and Jorge Eduardo Lindenberg Bustos.

2. Universidad Catolica de Chile, College of Engineering, Department of Structural and Geotechnical Engineering

Address: Casilla 306 - Correo 22, Santiago, Chile.
Web site: http://www.ing.puc.cl/

The University was founded in 1888. Initial courses in Mathematics and Construction were the seed for the School of Civil Engineering founded in 1897. Today, the College of Engineering has more than 2700 students and 90 full-time professors, most of whom hold Ph.D. degrees obtained at prestigious universities abroad. Research and instruction in the Department of Structural and Geotechnical Engineering is centered on the fundaments of earthquake engineering. The Department has four Laboratories: Static Tests, Dynamic Tests and Vibration Control, Soil Mechanics, and Strong Motion Seismology.

Members of the Department include Rafael Riddell (Head), Ernesto Cruz, Juan Carlos De la Llera, Pedro Hidalgo, Rodrigo Jordan, Carl Luders, Fernando Rodriguez, Hernan Santa Maria, Jorge Troncoso, Michel Van Sint Jan, Jorge Vasquez, Marco Ceroni, Ricardo Marcoleta, and Daniel Lowener.

INTERNATIONAL HANDBOOK OF EARTHQUAKE AND ENGINEERING SEISMOLOGY, VOLUME 81B ISBN: 0-12-440658-0

The Department offers basic courses for Civil Engineers, and advanced courses for the Structural and Geotechnical Engineering Diplomas, as well as for the Master of Engineering and Doctoral degrees in Structural and Geotechnical Engineering (the only accredited Doctoral program in these fields in the country). Main research subjects are Structural Dynamics, Ground Motion Characteristics, Seismic Response, Earthquake Analysis and Design, Seismic Risk, Seismic Isolation and Vibration Control, Seismic Behavior and Repair of Reinforced Concrete and Masonry Structures.

The full report on the Handbook CD also includes biographical sketches of Ernesto F. Cruz, Juan Carlos de la Llera, Pedro Hidalgo, Rafael Riddell, Fernando Rodriguez-Roa, Jorge H. Troncoso, Michel Van Sint Jan, and Jorge Vasquez.

3. University of Chile, Structural and Construction Division, Department of Civil Engineering

Address: Avda. Blanco Encalada 2120 Piso 4
Santiago, Chile.
Web site: http://www.fcfmuchile.cl/

The Structural and Construction Division, Department of Civil Engineering, University of Chile, was created in 1953. Its objectives are (1) teaching of earthquake engineering, structures and construction for civil engineering students; and (2) research on earthquake and structural engineering, including accelerograph networks, seismic risk, seismic zonation, characterization of earthquake motions, engineering seismology, site effect, masonry design, analysis and design of shear and frame concrete buildings, analysis and design of steel structures, seismic simulation, and passive energy dissipation.

Its activities include (1) publication of research reports and papers in conferences and journals; (2) participation in the development of national codes on earthquake design; (3) Organization of national and international conferences; (4) serving as home of the Chilean Earthquake Engineering and Seismology Association (ACHISINA); (5) serving as home of the Iberoamerican Association of Earthquake Engineering (AIBIS); and (6) responsibility for the Chilean Accelerograph Network.

Members include G. Rodolfo Saragoni H. (Head), Arturo Arias S., Maximiliano Astroza I., Gregorio Azocar G., Felipe Beltran, Ruben Boroschek K., Sergio Cordero, Ricardo Herrera M., Maria O. Moroni Y., Mauricio Sarrazin A., and Pedro Soto M.

The full report on the Handbook CD contains biographical sketches of Maximiliano Astroza I., Ruben Luis Boroschek, Maria Ofelia Moroni, and G. Rodolfo Saragoni Huerta.

80.6 China

A report to IAEE on earthquake engineering in China was submitted by Prof. Li-Li Xie, Secretary-General of the China Association of Earthquake Engineering. We extracted/excerpted a summary of the report as shown here. Readers are encouraged to read the full report, which is archived as a computer-readable file on the attached Handbook CD#2, under the directory of \8006China. Seismology in China is reviewed in Chapter 79.14. Biographies and biographical sketches of some Chinese earthquake engineers and seismologists may be found in Chapter 89, and in Appendix 3 of this Handbook, respectively.

1. Chinese Research Institutions in Earthquake Engineering

At present, there are 24 major institutions in China engaging in earthquake engineering research. They are listed here with their specialties and contact information.

Beijing Institute of Architectural Design and Research

Address: 62 Nanlishi Road, Beijing, 100045, China.

Specialties: Performance of buildings.

Beijing General Municipal Engineering Design and Research Institute

Address: B. 2 Yuetan Nanjie, Beijing, 100045, China.

Specialties: Performance of bridges.

China Northwest Building Design Institute

Address: 173 West 7th Road, Xi'an, 710003, China.

Specialties: Performance of buildings.

Dalian University of Technology

Address: Dalian University of Technology, DaLian, 116024, China.

Specialties: Performance of buildings; performance of dams; performance of bridges; geo-technical earthquake engineering; computational facilities available and experimental facilities available (3×3M shaking table).

Department of Civil Engineering, Taiyuan University of Technology

Address: 79 Yingze Street, Taiyuan, 030024, China.

Specialties: Performance of buildings.

Department of Civil Engineering, Zhengzhou Institute of Technology

Address: 97 Wenhua Road, Zhengzhou, China.

Specialties: Structural response control; performance of dams.

Department of Vibration and Aseismic Engineering, China Academy of Railway Sciences

Address: 2 Daliushu Road, Beijing, 100081, China.

Specialties: Seismic design code for railway systems, seismic hazard analysis, performance of bridges; geo-technical earthquake engineering; computational facilities available and experimental facilities available.

Guangdong Seismological Bureau

Address: 81 Xianliezhong Road, Guangzhou, 510070, China.

Specialties: Seismic hazard analysis; Engineering seismology; strong motion instrumentation; performance of buildings.

Harbin Institute of Technology

Address: 92, West Dazhi Street, Harbin, 150001, China
Web site: http://www.hit.edu.cn/

Specialties: Seismic design code and guideline for buildings; performance of buildings and offshore platform; structural control and isolated structures; structural health monitoring;

INTERNATIONAL HANDBOOK OF EARTHQUAKE AND ENGINEERING SEISMOLOGY, VOLUME 81B ISBN: 0-12-440658-0

computational facilities and experimental facilities available (3×3M shaking table and pseudo-dynamic testing installation).

Hehai University

Address: 1 Xinkang Road, Nanjing, 210098, China.

Specialties: Performance of dams.

Huazhong University of Science and Technology

Address: 1037 Luoyu Road, Wuhan, China.

Specialties: Isolated buildings.

Institute of Aseismic Engineering Structures, Wuhan University of Technology

Address: 122 Luoshi Road, Wuhan, 430070, China.

Specialties: Computational facilities available; structural response control; computational facilities and experimental facilities (3×3M shaking table) available.

Institute of Earthquake Engineering, China Academy of Building Research

Address: Xiaohuangzhuang, Anwai, Beijing, 100013, China.
Web site: http://www.cabr.ac.cn/

Specialties: Seismic design code for buildings; Engineering Seismology; strong motion instrumentation; performance of buildings; geo-technical earthquake engineering; computational facilities available and experimental facilities available (3×3M shaking table); isolated structure and structural response control.

Institute of Earthquake Engineering, China Institute of Water Resources and Hydropower Research

Address: 20 Chegongzhuang West Road, Beijing, 100044, China.

Specialties: Performance of dams.

Institute of Earthquake Engineering, Jiangsu Seismological Bureau

Address: 3 Weigang, Nanjing, 210014, China.

Specialties: Seismic hazard analysis, strong motion instrumentation, seismic performance of buildings, dams and bridges, geotechnical earthquake engineering.

Institute of Engineering Mechanics, China Seismological Bureau

Address: 29 Xuefu Road, Harbin, 150080, China.
Web site: http://iem.net.cn/

Specialties: Seismic zonation and seismic hazard analysis; seismic design code and guidelines for buildings, bridges, transportation, communication, and nuclear power plants; strong motion instrumentation, data process and data archiving; seismic performance of various structures (such as buildings, dams, bridges, nuclear power plants, industrial facilities, and underground structures); lifeline earthquake engineering, geotechnical earthquake engineering; base isolation and structural response control; earthquake disaster mitigation and emergency response; social and economic aspect of seismic disaster reduction; computational facilities and experimental facilities (5×5m shaking table) available.

Institute of Geophysics, China Seismological Bureau

Address: PO Box 8116, Beijing 100081, China
Web site: http:// www.eq-igp.ac.cn/

Specialties: Engineering seismology; geo-technical engineering; computational facilities available.

Institute of Seismology, Gansu Seismological Bureau

Address: 410 Donggang West Road, Lanzhou, 730000, China.

Specialties: Geo-technical earthquake engineering.

Shandong Building Scientific Research Institute

Address: 13 Wuyingshan Road, Jinan, 250031, China.

Specialties: Performance of buildings.

Shanghai Nuclear Engineering Research and Design Institute

Address: 29 Hongcao Road, Shanghai, 200233, China.

Specialties: Performance of nuclear power plants.

Southeast University

Address: 2 Sipailou, Nanjing, 210096, China.

Specialties: Performance of buildings; structural control; computational facilities available.

Tianjin University

Address: 92 Weijin Road, Tianjin, 300072, China.

Specialties: Performance of buildings; geo-technical engineering; computational facilities available.

Tongji University

Address: 1239 Siping Road, Shanghai, China.

Specialties: Seismic hazard analysis; strong motion instrumentation; seismic performance of buildings and bridges; geotechnical earthquake engineering; computational facilities and experimental facilities (4×4M shaking table) available; seismic performance of nuclear power plants; structural response control; seismic design code for buildings, bridges, and nuclear power plants.

Tsinghua University

Address: Tsinghua University, Beijing, 100084, China.

Specialties: Engineering seismology; performance of buildings; performance of bridges; computational facilities available;

experimental facilities available; performance of nuclear power plants; structural response control.

2. Institute of Engineering Mechanics, China Seismological Bureau

2.1 History and Introduction

The Institute of Engineering Mechanics (IEM) was formally established in 1954 under the Chinese Academy of Sciences. In 1984 it was renamed as the Institute of Engineering Mechanics of China Seismological Bureau.

IEM is the biggest and oldest research institute in China conducting research on earthquake engineering and engineering seismology. Since the beginning of the 1970s, the Institute has further focused its main efforts on earthquake engineering and safety engineering, in response to the rapid economic development of the country and the urgent needs of natural disaster reduction at home and abroad.

The total personnel in IEM amounts to 460, of which about two-thirds are scientific and technical staff, including 108 professors and senior engineers, 3 members of the Chinese Academy of Engineering, and 2 members of the Chinese Academy of Sciences. As the leading research institute in China on earthquake engineering, IEM has actively sponsored national and international scientific and technical activities. IEM has also established effective communication and collaboration links with international earthquake engineering communities through cooperation projects, visits by scholars, and technical seminars. During the last decade, IEM has successfully developed and executed a number of joint research projects with its counterparts in the United States, Japan, Russia, New Zealand, Australia, Southern American countries, and the European Community. Dozens of international workshops and symposia were organized and hundreds of foreign scholars and experts were invited to visit the Institute. There are more than 30 countries all over the world with which IEM has established formal academic exchange relations.

IEM is authorized by the Academic Degree Committee of the State Council to confer M.Sc. and Ph.D. degrees in the areas of Earthquake and Disaster Engineering, and Geo-Engineering. The Institute is also authorized to provide post-doctoral positions. At present, IEM offers the following academic degrees: (1) Masters degrees in Earthquake and Disaster engineering; Geo-Engineering; Structural mechanics; Solid mechanics; and (2) Doctoral degrees in Earthquake and Disaster engineering, and Geo-Engineering.

Since the recovery of the Chinese academic education system in 1978, over 146 students have been awarded Master's degrees by IEM, while another 79 students have been awarded Doctoral degrees. About 50 postgraduate students are enrolled at IEM each year.

Since 1985, staff members of IEM have been awarded 263 prizes in recognition of their research achievements, among which 17 are prizes of "the Development of Science and Technology Award" at the national level, and 74 are prizes at the ministerial and provincial levels. Three influential quarterlies are edited and published in Chinese by IEM, namely Earthquake Engineering and Engineering Vibration, Journal of Natural Disasters, and World Information on Earthquake Engineering.

The China Association of Earthquake Engineering and the Professional Commission of Earthquake Engineering of the Chinese Society of Seismology have established their secretariats at IEM. IEM is also a collective member of the China Civil Engineering Society and the Chinese Society of Vibration Engineering. Many staff members are invited regularly to contribute and present their research findings in technical workshops organized by various professional societies and associations in the country.

2.2 Accomplishments

In the field of earthquake engineering, IEM is the pioneer in China in almost all aspects. It initiated earthquake-engineering research in 1955 and has made great achievements and significant contributions to China's earthquake engineering research and practice, as follows:

- Developed and manufactured its own digital strong motion accelerographs and established the national strong motion instrument network on free field and in structures.
- Designed, installed, and operates the biggest earthquake simulator (5m×5m in size and six degrees of freedom) in China.
- Involved in preparing the *National Seismic Zoning Map* and in developing seismic hazard analysis approaches.
- Drafted the first seismic code in China and has been involved in preparing several seismic codes for buildings, bridges, dams, nuclear power plants, etc.
- Developed the first national map of earthquake loss assessment.
- Has the largest professional library (about two hundred thousand books and reports) in the country.
- Has taught and trained several hundreds of talented scientific professionals who started their careers at IEM and who continue to make achievements in their profession after moving to other institutions.

For many years, IEM has endeavored to communicate and cooperate with the international earthquake engineering communities. Delegations and scientists from many countries have visited IEM or attended the various scientific and technical conference organized by IEM; it has also sent quite a number of delegations and visiting scholars abroad for good-will visits, conferences, and to present lectures and undertake cooperative research. Through such activities, IEM has gained an increasing international reputation.

2.3 Divisions and Laboratories

At present, IEM has 12 technical divisions covering almost all disciplines of earthquake engineering, including strong earthquake motion observation, engineering seismology, building earthquake engineering, lifeline earthquake engineering, disaster prevention, geo-technical earthquake engineering, instrument and equipment for earthquake engineering, and information services.

The Key Laboratory of Earthquake Engineering and Engineering Vibration, China Seismological Bureau, is established and maintained at IEM, encompassing three sub-laboratories: earthquake motion simulation; soil dynamics; and ultra-low frequency vibration calibration. A National Center of Strong Seismic Motion Data Processing and Analysis was established at IEM in 1998.

The full report on the Handbook CD also contains a biography of Liu Huixian (1912–1992), and biographical sketches of Jinren Jiang, Xing Jin, Zhenpeng Liao, Xiaxin Tao, Qianxin Wang, Lili Xie, Zhiqian Yin, and Jingqing Zhu. Other leading professors include: Jingshan Bo, Xun Guo, Feng Hong, Fulu Men, Jingjiang Sun, Xiaoxin Wang, Junfei, Xie, Jianguo Xiong, Xiaoming Yuan, Yifan Yuan, Zhendong Zhao, and Minzheng Zhang.

3. Institute of Geophysics, China Seismological Bureau

The predecessor of the Institute of Geophysics, China Seismological Bureau (IGCSB), is the Institute of Geophysics, Chinese Academy of Sciences (Academia Sinica), which was founded April 6, 1950. In 1978 the Institute became part of the State Seismological Bureau (SSB), which became known as the China Seismological Bureau (CSB) in 1998.

One of the divisions in this institute is involved in seismic hazard analysis and in the national seismic zoning program. Recently this institute also developed some methodologies for urban seismic disaster mitigation. Its main recognized accomplishments in this area are the publication of the Chinese historical earthquake catalogue; the preparation of the first seismic zoning map in terms of seismic intensity and peak ground motion in China. The full report on the Handbook CD also contains a biographical sketch of Yuxian Hu.

4. Institute of Earthquake Engineering, China Institute of Water Resources and Hydropower Research

The Institute of Water Resources and Hydropower Research (IWHR) was established in 1958 by merging three organizations, namely, the Water Resources Research Institute under the Ministry of Water Resources, the Hydropower Research Institute under the Ministry of Electric Power Industry, and the Institute of Hydraulic Research under the Chinese Academy of Sciences. Disbanded in 1969 and reorganized in February 1978, the Institute is now affiliated with the Chinese Academy of Sciences, the Ministry of Water Resources, and the State Electric Power Corporation. It was renamed as the China Institute of Water Resources and Hydropower Research in 1994.

The Institute has a long history, which can be traced back to 1933. At that time, the Tianjin Hydraulic Laboratory was the pioneer research institute in the field of hydraulic engineering in China. It became a component of the Institute after it was merged with the Water Resources Institute in 1949.

The Institute is composed of 14 research departments and centers. It boasts a staff of 1,145, 54 percent of which are scientific and technical personnel, including 268 senior engineers and 222 engineers. Among these, 2 are academicians of the Chinese Academy of Sciences and 2 are academicians of the Chinese Academy of Engineering. The Institute is accredited and authorized by the State Council to offer programs of graduate studies for Doctor's and Master's degrees. From the 1950s to 1997, 329 Masters and 35 Doctors degrees have been awarded. The Institute's leading Professors include Houqun Chen.

5. School of Civil Engineering, Harbin Institute of Technology

The School of Civil Engineering consists of a provincial key lab and 4 ministerial key disciplines of which the Civil Engineering Department was given a positive evaluation for its excellence in the state specialty assessment activities carried out in 1995 and 2000. The School is now assuming and cooperatively sponsoring one of the "ninth Five-Year Plan" key projects of the National Natural Science funds, an additional 12 projects of National Natural Science funds, and 34 other projects. It has made great achievements in the fields of large span space structures, high-level steel structures and light steel, concrete-filled steel tube structures, prestressed concrete structures, seismic engineering, structural vibration control, diagnostic technology of structures, groundwork and foundation engineering, structural and engineering optimization, and analysis and design of structure reliability. Leading Professors include Hui Li, Ji Liu, Jinping Ou, Guangyuan Wang, Huanding Wang, Lili Xie, and Kexu Zhang.

6. School of Civil Engineering, Tongji University

The School of Civil Engineering is the strongest of its kind in the country, with the largest number of teaching and research staff. It consists of 7 departments and research institutes offering

5 undergraduate programs, 13 master degree programs, 7 doctoral programs, and 1 post-doctoral mobile research station. There are 93 full professors, 134 associate professors, 2,800 undergraduate students, 981 master degree and doctoral candidates, and 11 international students in the School. This School is the leading one in China in terms of comprehensive strength in the civil engineering discipline. The equipment in the Civil Engineering Disaster Reduction State Lab has reached international standards, and the research conducted in the Lab has been internationally recognized. The full report on the Handbook CD also contains biographical sketches of Lichu Fan, Guo-hao Li, Jie Li, Menglin Lou, Xilin Lu, and Zhixin Xu. Other leading Professors include Qifeng Luo, Zaiyong Zhang, and Bolong Zhu.

80.7 Croatia

A report on earthquake engineering was submitted by the Croatian Society for Earthquake Engineering. It is excerpted in the following sections. The report also includes biographical sketches of Drazen Anicic and Riko Rosman. The complete report is archived as a computer-readable file in the attached Handbook CD#2, under the directory \8007Croatia. Seismology in Croatia is reviewed in Chapter 79.16, and biographical sketches of Croatian engineers and scientists may be found in Appendix 3 of this Handbook.

1. Croatian Society for Earthquake Engineering (CSEE)

Address: c/o Civil Engineering Institute of Croatia, 10000 Zagreb, P.O. Box 283, 1, Rakusina, Croatia.
E-mail: <danicic@zg.igh.hr>

The Croatian Society for Earthquake Engineering (CSEE), Zagreb, was founded in 1965 and was part of the the Yugoslav Association for Earthquake Engineering until the dissolution of Yugoslavia. CSEE has been a member of the European and International Association for Earthquake Engineering since 1992. CSEE is a nonprofit organization, which works to promote earthquake engineering as a science and profession, and helps government bodies in implementing new European earthquake codes.

CSEE has no permanent staff. The executive body of 7 members is elected at society conferences every 4 years. CSEE has about 80 members, 15 of them actively involved in earthquake engineering research at universities and institutes. Their main fields of interest are seismology, engineering geology, soil mechanics, structural dynamics, earthquake-resistant design of buildings, repair and strengthening of existing buildings and monuments. Research results are published in national journals and in the proceedings of national and international conferences. Officers of the CSEE are (1) President: Mihaela Zamolo, (2) Vice-President: Riko Rosman, and (3) Secretary: Drazen Anicic.

2. Directory of Research Institutions

2.1 Civil Engineering Institute of Croatia

Address: 1000 Zagreb, P.O. Box 283, 1, Rakusina, Croatia.
Web site: http://www.igh.hr/

Drazen Anicic, <danicic@zg.igh.hr>, structural dynamics, earthquake-resistant building design, seismic strengthening and repair.
Mihaela Zamolo, <mzamolo@zg.igh.hr>, structural dynamics, earthquake-resistant building design, seismic strengthening and repair.
Dragan Moric, <dmoric@zg.igh.hr>, structural dynamics, earthquake-resistant building design, seismic strengthening and repair.

2.2 Faculty of Civil Engineering, University of Zagreb

Address: 10000 Zagreb, 26 Kaciceva, Croatia.

Zorislav Soric, structural dynamics, earthquake-resistant building design, seismic strengthening and repair.
Josip Dvornik, structural dynamics, earthquake-resistant building design.
A. Szavits-Nossan, earthquake-resistant building design, earthquake-resistant engineered structure design, soil mechanics.
Sonja Zlatović, soil mechanics.

2.3 Faculty of Civil Engineering, University of Split

Address: 21000 Split, 15 Matice Hrvatske, Croatia.

Antun Mihanovic, structural dynamics, earthquake-resistant building design.

ISBN: 0-12-440658-0

Pavo Marovic, structural dynamics, earthquake-resistant building design.

2.4 Faculty of Civil Engineering, University of Rijeka

Address: 51000 Rijeka, 5, Viktora Cara Emina, Croatia.

Mehmed Causevic, <Mehmed@master.gradri.hr>, structural dynamics, earthquake-resistant building design.

2.5 Faculty of Civil Engineering, University of Osijek

Address: 31000 Osijek, 16 Drinska, Croatia.

Vladimir Sigmund, <vladimir.sigmund@public.srce.hr>, structural dynamics, earthquake-resistant building design.

2.6 Geophysical Institute, PMF Faculty, University of Zagreb

Address: 10000 Zagreb, Bb. Horvatovac, Croatia.

Vladimir Kuk, seismic monitoring and databases, seismic hazard analysis, geophysics.

2.7 Institute for Geology

Address: 10000 Zagreb, Sachsova 2, Croatia.

Bozidar Biondic, seismic hazard analysis, geophysics.

80.8 Georgia

A report of the Institute of Structural Mechanics and Earthquake Engineering, Georgian Academy of Sciences, was submitted by Prof. Guram Gabrichidze, Director, Georgian Engineering Academy. The report also includes biographies of Shio German Napetvaridze (1914–1987), Emil Aleksey Sekhniashvili (1924–1992), and Kiriak Samsonovich Zavriev (1891–1978), and biographical sketches of Michael Aleksander Marjanishvili and Vladislav Boris Zaalishvili. The complete report is archived on the attached Handbook CD#2, under the directory \8008Georgia. We extracted and excerpted the following summary.

Seismology in Georgia is reviewed in Chapter 79.23. Biographies and biographical sketches of some Georgian engineers and scientists may be found in Chapter 89, and in Appendix 3 of this Handbook.

1. Institute of Structural Mechanics and Earthquake Engineering, Georgian Academy of Sciences

Address: M. Alexidze str. bld. 8, Tbilisi, 380093, Georgia.
Web site: http://www/acnet.ge/seismo.htm

The Institute was founded in 1947 as an outgrowth of the Bureau of Antiseismic Construction of the Georgian Academy of Sciences. It was called the Institute of Construction until 1962, when it was given its present name, the Institute of Structural Mechanics and Earthquake Engineering. It has also been known as the K. Zavriev Institute of Construction Mechanics and Seismostability.

The first Director of the Institute was Kiriak Zavriev, academician of Georgian Academy of Sciences. He served in this position from the founding of the Institute until his death in 1978. Academician Zavriev advanced the dynamic theory of seismic resistance in 1928–1929, and initiated the use of lightweight concrete, containing natural light aggregates, for the seismic-resistant design of structures in all regions of the former USSR.

The Institute has special departments in engineering seismology and seismicity theory, earthquake engineering, structural mechanics, applied theory of elasticity, theory of ultimate equilibrium, prestressed structures, physico-chemical mechanics of concrete, mechanics of polymers, testing design, technical-economic research, and technical information. The Institute has carried out research on seismic microzoning in regions containing densely populated settlements and building sites; this led to the development of microzoning maps for these regions.

Considerable work has been done on the application of modern techniques of structural mechanics (finite element and finite difference methods) for the purpose of finding solutions to seismic problems. These techniques were applied in the earthquake-resistant design of hydrotechnical structures and led to appropriate clauses in five existing earthquake-engineering codes, and to a "Manual for Design of Hydrotechnical Structures" that was used in the design of large hydro-power stations.

The Institute is the main establishment involved in the development of the solution to seismic problems arising in transport and other network structures. It was responsible for developing "Recommendations on design and project of transport structures, erected on the BAM route." The Institute also carried out experimental research on the seismic resistance of bridge abutments and supporting walls along the BAM route. Models of these structures were tested on the most powerful seismic platform in the former USSR. The results of this research were given to the institutions involved in this project, for their use in designing the BAM route structures.

The Institute pays great attention to the survey and analysis of the consequences of most significant earthquakes, both in Georgia and abroad. The analysis of earthquake consequences verified the seismic specifications that appear in the Soviet codes for buildings and structures. The Institute made considerable contributions toward the development of these codes.

INTERNATIONAL HANDBOOK OF EARTHQUAKE AND ENGINEERING SEISMOLOGY, VOLUME 81B ISBN: 0-12-440658-0

In the 1960s, the Institute was involved in the development of the design of arch dams. A complex technique was prepared for the design of these structures under all kinds of loads and for various rocky foundations and geometric configurations, and special attention was devoted to their dynamic response.

The Institute conducts experimental research to check the assumptions that are the basis of developed design methods, and to examine the validity of rationally proposed spatial structures. Tests are carried out both on full-sized structures and their models. The Institute also renders assistance to design and building organizations, supports efficient research, and maintains contacts with different scientific research institutions.

In collaboration with the Institute of Physics of the Earth, the Institute carries out research on the assessment of seismic hazard. It also undertakes collaborative scientific research with the Latvian Research Institute for construction, and has established close cooperation with relevant institutes in Azerbaijan and Armenia.

The Institute actively collaborates with such international research societies as the International Association for Earthquake Engineering (IAEE), the International Association on Shell Structures (IASS), and the International Federation of Prestressed Reinforced Concrete (FIP). A symposium organized by this federation that centered on the use of prestressed reinforced concrete in earthquake engineering was held in Tbilisi in 1972.

The main research trends of the Institute include

- Design of the special complex structures exhibiting physical and geometric nonlinearity, and elastic and plastic material properties experiencing creep under complex static and dynamic loading;
- Elaboration of numerical techniques for seismic microzoning;
- Development of the dynamic theory for the seismic resistance of buildings;
- Study of the strength, deformability, and physical nature of concrete failure—structural properties of cement stone, investigation of the environmental effects, and the character of the loading;
- Study of reinforced concrete and polymer materials.

80.9 India

The editors received a "List of Universities and Institutions in India engaged in Earthquake Engineering Research and Application" from Prof. H.R.Wason, Secretary of the Indian Society of Earthquake Technology and Professor, Department of Earthquake Engineering, Indian Institute of Technology, Roorkee, India. Additional information was also provided by Vipin Chandra Joshi, on behalf of Prof. Wason.

These materials are archived as a computer-readable file on the attached Handbook CD#2, under the directory of \8009India. We extracted and excerpted the following summary from these sources. Seismology in India is reviewed in Chapter 79.28. Biographies and biographical sketches of some Indian engineers and scientists may be found in Chapter 89, and in Appendix 3 of this Handbook, respectively.

1. Indian Society of Earthquake Technology

The Indian Society of Earthquake Technology (ISET) came into being in November 1962 with the late Prof. Jai Krishna as its founding President. ISET was registered in December 1964 under Society's Registration Act 1961. The aims and objectives of the Society are as follows:

- To promote research, development, and awareness work in the field of Earthquake Technology.
- To provide a necessary forum for scientists and engineers of various specializations to come together and exchange ideas on the problems of Earthquake Technology.
- To disseminate scientific and engineering knowledge in the field of Earthquake Technology.
- To honour pioneering and meritorious contributions in the field of Earthquake Technology.

The Society began publishing a periodical entitled "Bulletin of the Indian Society of Earthquake Technology" in 1964. The name of the Bulletin was changed to "ISET Journal of Earthquake Technology" in March 1998. In addition to the Journal, the Society also publishes a quarterly "News Letter." It has also published a number of books that further its aims and objectives. In particular, the Society has published a "Catalogue of Earthquakes in India and its Neighbourhood," and has reprinted the International Association for Earthquake Engineering's 1986 publication "A Manual of Earthquake Resistant Non-Engineered Construction." The Society is responsible for organizing the ISET Annual Lecture, which is delivered by an eminent scientist, distinguished engineer, or specialty expert.

The Society has evoked wide interest since its inception. Its current membership, which includes Individual Fellows and Members, as well as Institution Members, stands at 1200. Since its inception, the Society has been involved in the organization of Symposia in Earthquake Engineering, which are held quadrennially. In addition, the Society has been active in the organization of several conferences, including the Sixth World Conference on Earthquake Engineering, held in New Delhi in 1977. ISET is also engaged in various activities intended to create an awareness of earthquakes and earthquake engineering by the general public, as well as by the scientific/engineering communities. From time to time, the Society organizes short-term training courses dealing with earthquake technology.

The recent (magnitude 7.6) Bhuj earthquake of January 26, 2001, was the largest earthquake to occur in India since its independence in 1947. Following this event, ISET organized a special, three-day Workshop on "Recent Earthquakes of Chamoli and Bhuj," which was held in Roorkee May 24–26, 2001. Ninety-eight delegates attended this Workshop, including representatives from 47 organizations. A total of 60 technical papers covering different themes were presented during the Workshop.

ISET is a founding member of the International Association for Earthquake Engineering (IAEE) and represents India on this world forum. The Society's President serves as India's National Delegate on the General Assembly of Delegates of the IAEE, and its Vice-President serves as the Deputy National Delegate,

INTERNATIONAL HANDBOOK OF EARTHQUAKE AND ENGINEERING SEISMOLOGY, VOLUME 81B ISBN: 0-12-440658-0

in the absence of the President. In view of his distinguished service to Earthquake Engineering, the late Prof. Jai Krishna was elected as an Honorary Member of the IAEE. Professor Krishna served the IAEE as its President from 1973 to 1980. A number of ISET Members have served on the Executive Committee of the IAEE.

ISET plays a role in the formulation of various Indian Standard Codes of Practice related to earthquake engineering; the President of ISET is a member of the relevant committee responsible for drafting the earthquake design section of the code.

2. Department of Earthquake Engineering, Indian Institute of Technology, Roorkee

The main objectives of the Department of Earthquake Engineering are (1) teaching and training, (2) research, (3) providing technical knowledge (consultancy), (4) developmental work, and (5) dissemination of knowledge through the holding of symposia, special lectures, and courses. The Department has served as a national resource for dealing with problems related to earthquake engineering.

The Department of Earthquake Engineering, at what was originally the University of Roorkee and is now named the Indian Institute of Technology, Roorkee, is the only one of its kind in India, and amongst a few in the world. In 1969, this Department became the successor to the School of Research and Training in Earthquake Engineering (SRTEE), which was established in 1960 as an activity within the Structural Engineering Section of the Department of Civil Engineering at the then University of Roorkee. The Department has been actively engaged in teaching, research, development, and consultancy work in four major fields: Structural Dynamics, Soil Dynamics, Seismology and Seismotectonics, and Instrumentation. It has contributed very significantly to the field of earthquake disaster mitigation over the last 40 years.

The Department has been instrumental in the development of expertise in earthquake engineering and engineering seismology, in formulating Indian Standard Codes for earthquake-resistant design, and in the establishment of the Indian Society of Earthquake Technology. The academic activity of the department includes the offering of teaching programmes leading to master of engineering degrees and to post-graduate diplomas in Earthquake Engineering and Earthquake Technology, as well as research programmes leading to a Ph.D. degree in Earthquake Engineering.

Areas in which active research is being pursued include seismic analysis and design of structures; vibration isolation; nonlinear finite element analysis, machine foundations; dynamic characteristics of soils; soil liquefaction; rock slope stability; microzonation; seismic hazard analysis; seismic risk; engineering seismology; site-dependent earthquake parameters; microearthquake investigations employing local seismological networks; and strong ground motion studies. Some of the equipment and computer software used for earthquake engineering studies have been developed in-house. Major research entrusted to the Department includes the study of prototype well-foundations, and the deployment of strong motion and telemetered arrays in the Garhwal Himalaya, and around the KAPP nuclear power plant site, to study the various characteristics of seismicity. The Department of Earthquake Engineering has played a key role in providing consultancy services for the seismic design of almost all major engineering projects in the country, with the aim of mitigating disasters caused by earthquakes. These projects include nuclear power plants, major dams, bridges, petrochemical and other industrial complexes, and multistory buildings. In addition to the holding of various workshops and symposia for disseminating knowledge related to earthquake engineering, the Department has regularly organized Symposia on Earthquake Engineering on a quadrennial basis. The faculty of the Department has made outstanding contributions in developing the technical know-how for carrying out earthquake engineering studies. The faculty is highly qualified and well versed in the methods and techniques needed for earthquake-resistant design. In addition to conducting pure pedagogical research, these methods and techniques are being regularly applied to solve intricate problems, which have direct relevance to industry. Faculty members have served as President and Directors of the IAEE, and have acted as UNESCO experts for rendering technical advice in the field of earthquake engineering.

More details about the Department, including a biography of the late Jai Krishna, can be found in the full report on the Handbook CD.

3. List of Universities and Institutions

The following is a list of organizations in India engaged in earthquake engineering research and application. Contact information is included in the full report on the Handbook CD.

1. Department of Earthquake Engineering, Indian Institute of Technology, Roorkee - 247 667.
2. Department of Civil Engineering, Indian Institute of Technology, Roorkee - 247 667.
3. Department of Earth Sciences, Indian Institute of Technology, Roorkee - 247 667.
4. Water Resources Development Training Centre, Indian Institute of Technology, Roorkee - 247 667.
5. Indian Society of Earthquake Technology, Dept. of Earthquake Eng., Indian Institute of Technology, Roorkee - 247 667.
6. Central Building Research Institute, Roorkee - 247 667 (U.P.).
7. U.P. Irrigation Research Institute, Roorkee - 247 667.

8. Structural Engineering Research Centre, C.S.I.R. Campus, Taramani, Chennai - 600 113.
9. Structural Engineering Research Centre, Central Govt. Encl., Sector 19, Kamla Nehru Nagar, Ghaziabad - 201001.
10. Central Water Commission, Sewa Bhawan, R.K. Puram, New Delhi - 110 066.
11. Central Water & Power Research Station, P.O. Khadakwasala, Research Station, Pune - 411024.
12. India Meteorological Department, Mausam Bhavan, Lody Road, New Delhi - 110 003.
13. Regional Research Laboratory, Jorhat - 785 006 (Assam).
14. Indian Institute of Technology, Hauzkhas, New Delhi - 110 016.
15. Indian Institute of Technology, Powai, Mumbai - 400 076.
16. Indian Institute of Technology, Kanpur - 208 016.
17. Indian Institute of Technology, Panbazar, Guwahati - 781001.
18. Bhabha Atomic Research Centre, Trombay, Mumbai - 400 085.
19. Kumaon Engineering College, Dwarahat, Distt. Almora - 263 653 (U.P.).
20. Disaster Management Institute, (Housing & Environment Department, Govt. of M.P.), Paryavaran Parisar, E-5, Arera Colony, PB No. 563, Bhopal - 462 016 (M.P.).
21. National Institute of Rock Mechanics (Ministry of Mines, Govt. of India), Champion Reefs P.O., Kolar Gold Fields, Karnataka - 563 117.
22. Bharat Heavy Electricals Limited, Advanced Research Projects Division, 7th Floor, H.T. House, K.G. Marg, New Delhi - 110 001.
23. Department of Science and Technology, Technology Bhavan, New Mehraulli Road, New Delhi - 110 016.
24. Wadia Institute of Himalayan Geology, 33 General Mahadeo Singh Road, Dehradun - 248 001 (U.P.).
25. Nuclear Power Corporation, Vikram Sarabhai Bhavan, 6th Floor, Central Avenue, Anushaktinagar, Mumbai - 400 094.
26. Maharashtra Engineering Research Institute, Meri Campus, Dindori Road, Nashik - 422 004.
27. Visvesvaraya Regional Engineering College, Nagpur - 440 011.
28. Housing and Urban Development Corporation Ltd., HUDCO Bhawan, Lodhi Road, New Delhi - 110 003.

80.10 Japan

The editors received the "Japan National Report on Earthquake Engineering" compiled by Prof. Kenzo Toki, National Delegate of Japan of the International Association for Earthquake Engineering. The report contains a directory of 80 Japanese institutions engaging in earthquake engineering and related research, and is divided into 3 parts: (1) universities, (2) companies, and (3) public organizations. Over 440 researchers are listed with their specialties and contact information. Some institutions also provided their brief histories, activities, and accomplishments. In addition, biographies of 4 deceased engineers (Hajime Umemura, 1918–1995; Kiyoshi Muto, 1903–1989; Ryo Tanabashi, 1907–1974; and Toshikata Sano, 1880–1956), and biographical sketches (including lists of selected publications) of over 170 researchers are included in the report. The full report is archived as three computer-readable files on the attached Handbook CD#2, under the directory of \8010Japan.

We extracted and excerpted the following summary of this material in the form of a directory, and details are given in the full report on the Handbook CD. A review of research in seismology and physics of the Earth's interior in Japan is given in Chapter 79.33 of this Handbook. Biographies and biographical sketches of some Japanese engineers and scientists may be found in Chapter 89, and in Appendix 3 of this Handbook, respectively.

1. Universities

1.1 Fukui University, Graduate School of Architecture and Civil Engineering

Address: 3-9-1, Bunkyo, Fukui 910-8507, Japan.

Researchers include (1) Kengo Tagawa: Dynamics of structures, structural control, seismic design methodology; and (2) Keisuke Kojima: Earthquake response analysis of geotechnical structures. Their biographical sketches are given in the full report on the Handbook CD.

1.2 Fukuyama University, Department of Architecture and Graduate School of Engineering

Address: Sanzo, 1 Gakuen-Cho, Fukuyama 729-0292, Japan.
Web site: http://www.fukuyama-u.ac.jp/

Researchers include (1) Koichi Minami: Concrete structures, steel concrete structures; (2) Teruo Kamada: Wooden structures, earthquake activity; and (3) Akio Nakayama: Steel structures, brittle failure.

1.3 Gifu University, Department of Civil Engineering

Address: 1-1 Yanagido, Gifu 501-1193, Japan.
Web site: http://www.cive.gifu-u.ac.jp/

The Department of Civil Engineering was established on May 1, 1947, in the Gifu Technical School, which was founded as the Gifu Prefectural Higher Technical School in 1942. In April 1952, Gifu Technical School became a national university and the engineering departments were reestablished as the Faculty of Engineering, Gifu University. The Graduate School of Engineering, which is made up of six divisions, including the Division of Civil Engineering, was founded in 1967. A "Geological and Structural Diagnostics" research section was founded in 1997 as one of the administrative units within the Department of Civil Engineering. The major topics of this research section are (1) comprehensive seismic hazard analysis for regional earthquake disaster prevention, (2) ground motion prediction for given fault parameters, (3) effective stress-based three-dimensional soil liquefaction analysis, (4) systems approach for the prevention and mitigation of urban earthquake disaster, (5) assessment of seismic performance of network facilities and their function, (6) inverse analysis of geotechnical and groundwater problems, and (7) reliability design.

INTERNATIONAL HANDBOOK OF EARTHQUAKE AND ENGINEERING SEISMOLOGY, VOLUME 81B ISBN: 0-12-440658-0

Researchers include (1) Yuusuke Honjyo: Geotechnical engineering, structural analysis; (2) Masata Sugito: Strong ground motion, response analysis of ground; (3) Atsushi Yashima: Liquefaction analysis, soil–structure interaction; (4) Feng Zhang: Liquefaction analysis, soil–structure interaction; (5) Nobuoto Nojima: Urban earthquake disaster prevention, seismic risk analysis; and (6) Yoshinori Furumoto: Response analysis of ground, strong ground motion. Their biographical sketches are given in the full report on the Handbook CD.

1.4 Hachinohe Institute of Technology

Address: 88 Ohbiraki, Myo, Hachinohe, Aomori 031-8501, Japan.
Web site: http://www.hi-tech.ac.jp/

Researchers include (1) Naomi Sakajiri: Effects of surface geology on strong ground motion; and (2) Yukitake Shioi: Seismic design of foundations, soil dynamics.

1.5 Hokkaido University, Graduate School of Engineering

Address: Nishi 8 Kita 13 Kita-Ku, Sapporo 060-8629, Japan
Web site: http://www.civil.hokudai.ac.jp/bri/jbri.html

Researchers include Toshiro Hayashikawa: Dynamics of structures, structural control, nonlinear dynamic analysis.

1.6 Kagoshima University, Faculty of Engineering

Address: 1-21-40 Kohrimoto, Kagoshima 890-0065, Japan

Researchers include (1) Kazuo Matsumura: Dynamics of structures, seismic risk analysis; (2) Yasuhiro Uchida: Steel structures, composite structures; (3) Shinichi Shioya: Reinforced concrete structures; (4) Kenji Kawano: Dynamics of offshore structures, dynamic soil–structure interaction; (5) Susumu Yoshihara: Dynamics of offshore structures; (6) Venkataramana Katta: Dynamics of offshore structure.

1.7 Kanto-Gakuin University, Department of Architecture

Address: 4834 Mutsuura-cho, Kanazawa-ku, Yokohama 236-8501, Japan.
Web site: http://www.kanto-gakuin.ac.jp/

Researchers include (1) Eiji Makitani: Seismic retrofit of RC structures, shear transfer of precast concrete connections; (2) Norio Abeki: Seismic microzonation by microtremor observation; (3) Alan Burden: Earthquake-resistant design of shell and suspension structures; and (4) Hideyuki Takashima: Seismic response of a single-layer lattice shell.

1.8 Kinki University, Kyushu School of Engineering, Department of Architecture

Address: 11-6 Kayanomori, Iizuka, 820-8555, Japan.
Web site: http://mulukhiya.cc.fuk.kindai.ac.jp/~arch/00stff98.html

Researchers include (1) Koichi Abe: Concrete engineering, structural design, steel structures; (2) Masayuku Ono: Aseismic walls, R/C structures, earthquake engineering; and (3) Akira Hiramats: Earthquake response, R/C structures, structural analysis, restoring force characteristics.

1.9 Kinki University, School of Science & Engineering, Department of Architecture

Address: 3-4-1 Kowakae, Higashi-Osaka, Osaka 577-8502, Japan.
Web site: http://www.kindai.ac.jp/

Researchers include (1) Toshiyuki Kubota: Reinforced concrete structures; (2) Eiji Tateyama: Steel structures; and (3) Masahide Murakami: Timber structures.

1.10 Kyoto University, Disaster Prevention Research Institute

Address: Gokasho, Uji, Kyoto 611-0011, Japan.
Web site: http://www.dpri.kyoto-u.ac.jp/

The Disaster Prevention Research Institute (DPRI), which is affiliated with Kyoto University, was established in 1951. It carries out research on a variety of problems related to the prevention or reduction of natural disasters. The Institute commenced its work with 3 research sections. It has increased the number of research sections and attached facilities to meet the diversified research needs arising from the advance in disaster science, and the change in the situation of disasters resulting from changes in social conditions. By the end of 1995 it had set up 16 research sections, 4 research centers, 5 observatories, and 2 laboratories. In 1996, the Institute was reorganized into 5 research divisions and 5 research centers in order to achieve greater flexibility when responding to changing research needs, and to widen the research fields by linking physical and engineering research with human, social and management science.

Simultaneously with this reorganization, the Institute was opened to all researchers from other universities who are concerned with the investigations of disaster. This collaboration will be fulfilled through the establishment of joint research projects and research meetings proposed by researchers from both within and outside the Institute. The Institute cooperates with the Graduate School of Science and Engineering of Kyoto University. Students come to the Institute to accomplish their studies under supervision of the Institute's staff members.

The DPRI is organized as follows: (1) Director; (2) Advisory Committee; (3) Committee for Joint Research; (4) Faculty Meetings; (5) Research Divisions: Integrated Management of Disaster Risk, Earthquake Disaster Prevention, Geo-Disaster, Fluvial and Marine Disaster, and Atmospheric Disaster; (6) Research Centers: Disaster Environment, Earthquake Predictions, Sakurajima Volcano, Water Resources, and Disaster Reduction Systems; (7) Division of Technical Affairs; (8) Administrative Office; (9) General Affairs Division; and (10) Accounting Division.

Researchers of the DPRI include (1) Hiroyuki Kameda: Professor, Earthquake engineering, disaster management, GIS; (2) Yoshiyuki Suzuki: Professor, Earthquake engineering, stochastic systems, structural control; (3) Satoshi Tanaka: Research Associate, Lifeline earthquake engineering; (4) Kojiro Irikura: Professor, Earthquake ground motion, seismic risk; (5) Tadanobu Sato: Professor, Dynamics of foundation structures, structural health monitoring, structural control; (6) Haruo Kunieda: Professor, Dynamics of shells and spatial structures; (7) Taijiro Nonaka: Professor, Plastic analysis, structural mechanics; (8) Koji Matsunami: Associate Professor, Applied seismology; (9) Masayoshi Nakashima: Associate Professor, Analysis and design of steel structures; (10) Sumio Sawada: Associate Professor, Input earthquake motion; (11) Tomotaka Iwata: Research Associate, Strong motion seismology; (12) Shigehiro Morooka: Research Associate, Dynamics of building structures; (13) Riki Honda: Research Associate, Dynamics of ground; (14) Masataka Ando: Professor, Earth structure, Quaternary seismology (now at Nagoya University); (15) Norihiko Sumitomo: Professor, Geomagnetism and tectonomagnetism; (16) Tamotsu Furusawa: Professor, Crustal movement, data processing; (17) Mitsuhiko Shimada: Professor, Experimental solid earth physics; (18) James Jiro Mori: Professor, Seismic waveform analysis, crust and upper mantle structure; (19) Yasuhiro Umeda: Professor, Earthquake prediction, source process; (20) Kazuo Matsumura: Associate Professor, Geophysical data processing; (21) Fumiaki Takeuchi: Associate Professor, Seismic activity; (22) Naoto Oshiman: Associate Professor, Tectonomagnetism and Earth interior; (23) Kiyoshi Ito: Associate Professor, Seismicity and crustal structure; (24) Kunihiko Watanabe: Active faults, earthquake prediction; (25) Takashi Yanagidani: Associate Professor, Experimental rock mechanics; (26) Manabu Hashimoto: Associate Professor, Crustal movements, tectonics; (27) Takuo Shibutani: Waveform analysis, earthquake statistics; (28) Hikaru Doi: Research Associate, Crustal movements; (29) Kunihiro Shigetomi: Research Associate, Crustal movements; (30) Fumio Ohya: Research Associate, Crustal movements, geodetic surveys; (31) Hiroshi Katao: Research Associate, Microseismicity, crustal structure; (32) Wataru Morii: Research Associate, Microseismicity; (33) Tadashi Konomi: Research Associate, Microseismicity; (34) Kensuke Onoue: Research Associate, Crustal movements, ground deformations; (35) Masahiro Teraishi: Research Associate, Crustal movements; (36) Shiro Ohmi: Research Associate, Seismic data processing; (37) Kajuro Nakamura: Research Associate, Gravity changes, crustal movements, GPS; and (38) Peiliang Xu: Research Associate, Inversion problems, geodesy.

1.11 Kyoto University, Graduate School of Civil Engineering

Address: Yoshida Honmachi, Sakyo, Kyoto 606-5801, Japan.
Web site: http://www.kuciv.kyoto-u.ac.jp/welcome-e.html

The Department of Civil Engineering was established in 1897, and is one of the oldest departments in the Faculty of Engineering. The Department of Transportation Engineering was founded in 1963. In 1996, in order to cope with the growth in and complication of civil engineering related fields, these two departments were reorganized as the departments of Civil Engineering and of Civil Engineering Systems. The former deals mainly with basic problems in civil engineering, and the latter emphasizes a systems approach to infrastructure systems.

Programs of study leading to the master's and doctoral degrees are available in the major civil engineering fields. The requirement for the Master's degree of engineering is 30 credits in the major field. A Master's thesis is also required under an adviser's research guidance. In all cases, the Graduate School of Civil Engineering seeks to provide students with the broadest and most complete education consistent with the demands of their prospected careers.

Researchers include (1) Kenzo Toki: Soil–structure interaction, strong ground motion; (2) Hirokazu Iemura: Dynamics of structures, hybrid structures, structural control; (3) Junji Kiyono: Simulation of earthquake ground motion, evacuation behavior; (4) Akira Igarashi: Structural control; and (5) Yoshikazu Takahashi: Concrete structures. Their biographical sketches are given in the full report of the Handbook CD.

1.12 Kyoto University, School of Architecture

Address: Yoshida Honmachi, Sakyo-ku, Kyoto 606-8501, Japan.
Web site: http://www.archi.kyoto-u.ac.jp/

The School of Architecture was founded in 1920 and by 1921 had four professorial chairs. With the recognition of the importance of research and education in this field, the School was enlarged and had eight chairs by 1965. Furthermore, the rapid specialization of architecture and the social needs of architects brought about the foundation of the School of Architectural Engineering with six chairs in 1964. In 1991, three chairs of the two Schools, one as a core chair and the other two as affiliated chairs, participated in the Department of Global Environment Engineering, which was founded as a new type of department, independent of the undergraduate schools. The two Schools, their corresponding departments, and the Department of Global Environment Engineering worked together for the sake of education and research up until 1995.

In 1996, the two Schools merged to meet the new national education policy, which prescribed systematic and comprehensive education through four undergraduate years of university, and to fulfill the expectations concerning the role of graduate schools in higher education and top-level research. The new School took the traditional name of the School of Architecture, but it is a completely new organization with new policies.

Researchers include (1) Bunzo Tsuji: Constitutive models for steel and concrete, stability of steel structures, ultimate strength and collapse mechanism of composite structures; (2) Hiroaki Nagaoka: Foundation engineering, construction of foundations; (3) Fumio Watanabe: Reinforced and prestressed concrete structures, composite structures, seismic design; (4) Koji Uetani: Structural mechanics, stability theory of structures, limit state design of building structures; (5) Masahiro Kawano: Earthquake engineering, building geoenvironment engineering, building foundations and geotechnical engineering; (6) Ichiro Inoue: Steel structures; (7) Shigeru Fujii: Mechanical properties of structural materials, numerical analysis of reinforced concrete members, bond and anchorage of reinforced concrete; (8) Izuru Takewaki: Seismic-resistant design of structures, soil–structure interaction, structural control; (9) Minehiro Nishiyama: Prestressed concrete structures, dynamic analysis, seismic design; (10) Makoto Ohsaki: Optimum design of architectural systems, structural mechanics, and computational design methods; (11) Hidekazu Nishizawa: Steel structures, structural planning for buildings, conservation engineering; (12) Keiichiro Suita: Steel structures; (13) Masahiro Yamazaki: Design of foundations, construction of foundations; (14) Susumu Kono: Reinforced concrete structures; (15) Hiroshi Tagawa: Stability analysis and limit state design of building structures, optimum design of architectural systems; and (16) Yuichi Sato: Application of continuous fiber to reinforced concrete.

Biographical sketches of Bunzo Tsuji, Masahiro Kawano, Fumio Watanabe, Koji Uetani, Shigeru Fujii, Izuru Takewaki, Minehiro Nishiyama, Keiichiro Suita, Hiroshi Tagawa, Yuichi Sato, Susumu Kono, Yoshitsura Yokoo, Takuji Kobori, Kiyoshi Kaneta, Tsuneyoshi Nakamura, and Shiro Morita are given in the full report on the Handbook CD, as well as the biography of Ryo Tanabashi (1907–1974).

1.13 Kyushu Institute of Design, Department of Environmental Design

Address: 4-9-1 Shiobaru, Minami-ku, Fukuoka 815-8540, Japan.
Web site: http://www.kyushu-id.ac.jp/

Researchers include (1) Masamichi Ohkubo: Seismic structural design, reinforced concrete buildings; and (2) Tomokazu Yoshioka: Structural engineering.

1.14 Kyushu Sangyo University, Faculty of Engineering

Address: 2-3-1 Matsukadai, Higashi-ku, Fukuoka 813-8503, Japan.
Web site: http://www.ip.kyusan-u.ac.jp/

Kyushu Sangyo University (KSU) was founded in 1960. The Faculty of Engineering, KSU, was established in 1963 with three departments: Mechanical Engineering, Electrical Engineering, and Industrial Chemistry. One year later, two more departments, Civil Engineering and Architecture, were added. The Master's Degree Programs started in 1973 and 1975. In 1996 the two Doctoral Degree Programs were added. The campus of KSU is set in an area of 500,000 square meters. There are more than 30 buildings on the campus, including all the necessary facilities to accommodate comfortably more than 15,000 students. Fukuoka, with approximately 1.2 million inhabitants, is the largest metropolitan area and most rapidly growing city in western Japan.

The KSU Realtime Earthquake Information System and the Two-Directional Dynamic Testing Table were introduced in 1999 to help teach subjects concerning earthquake engineering in mechanical engineering, civil engineering, and architecture. Researchers include (1) Hidemori Narahashi: Strong ground motion, real-time earthquake information; and (2) Syun'itiro Omote: Earthquake hazard mitigation. Their biographical sketches are given in the full report on the Handbook CD.

1.15 Maebashi Institute of Technology, Department of Civil Engineering

Address: 460-1, Kamisadori-cho, Maebashi-shi, Gunma 371-0816, Japan.
Web site: http://www.maebashi-it.ac.jp/

Researchers include Makoto Nasu: Ground–structure interaction, mechanism of earthquake damage.

1.16 Miyazaki University, Department of Civil and Environmental Engineering

Address: 1-1 Gakuen Kibanadai Nishi, Miyazaki 889-2192, Japan.

Researchers include Takanori Harada: Dynamic soil–structure interaction, strong ground motion. His biographical sketch is given in the full report on the Handbook CD.

1.17 Musashi Institute of Technology, Graduate School of Civil Engineering

Address: 1-28-1 Tamatsutsumi, Setagaya, Tokyo 158-8557, Japan.
Web site: http://slpc14.sl.civil.musashi-tech.ac.jp/main-a.html

Researchers include Toshiyuki Katada: Nonlinear surface ground motion, soil dynamics.

1.18 Nagaoka University of Technology, Civil and Environmental Engineering

Address: 1603-1 Kamitomioka, Nagaoka, Niigata 940-2188, Japan.
Web site: http://www.nagaokaut.ac.jp/index-e.html

Researchers include (1) Kyuichi Maruyama: Seismic design of reinforced concrete structures; (2) Masatsugu Nagai: Seismic design of steel structures; and (3) Satoru Ohtsuka: Soil dynamics. Their biographical sketches are given in the full report on the Handbook CD.

1.19 Nagoya University, Department of Architecture

Address: Furo-cho, Chikusa-ku, Nagoya 464-8603, Japan.
Web site: http://www.sharaku.nuac.nagoya-u.ac.jp/

Researchers include (1) Nobuo Fukuwa: Urban disaster mitigation, soil–structure interaction, structural control; (2) Jun Tobita: Dynamics of structures and ground, microtremor studies, system identification; and (3) Masaru Nakano: Source mechanisms of earthquakes.

1.20 Nagoya University, Department of Civil Engineering

Address: Furo-cho, Chikusa-ku, Nagoya 464-8603, Japan.
Web site: http://www.civil.nagoya-u.ac.jp/

Researchers include (1) Yoshihiro Sawada: Earthquake observation, geophysical exploration; and (2) Hideki Nagumo: Effect of surface geology on seismic motion.

1.21 National Defense Academy, Department of Civil Engineering

Address: 1-10-20 Hashirimizu, Yokosuka 239-8686, Japan.
Web site: http://www.nda.ac.jp/

Researchers include Hiroshi Sato: Natural disaster prevention engineering, and his biographical sketch is given in the full report on the Handbook CD.

1.22 Nihon University, College of Industrial Technology

Address: 1-2-1, Izumi-cho, Narashino-city, Chiba 275-8575, Japan.

Researchers include (1) Choshiro Tamura: Strong earthquake motion, earthquake resistance of underground structures and fill-type dams; and (2) Katsuyoshi Yamabe: Strong ground motion, vibration characteristics of buildings.

1.23 Nihon University, College of Science and Technology

Address: 1-8 Kanda-Surugadai, Chiyoda, Tokyo 101-8308, Japan.
Web site: http://www.cst.nihon-u.ac.jp/

Researchers include (1) Hiroshi Akiyama: Earthquake-resistant limit state design, dynamic buckling, seismic isolation; (2) Shinji Ishimaru: Structural control, earthquake-resistant design; (3) Hiromi Adachi: RC structures, retrofitting, pseudo-dynamic testing; (4) Kazufumi Hanada: Soil–structure interaction, forced vibration tests; and (5) Hiroo Shiojiri: Dynamics of structures, seismic isolation.

1.24 Osaka Institute of Technology

Address: 5-16-1 Ohmiya, Asahi, Osaka 535-8585, Japan.
Web site: http://www.oit.ac.jp/english/index-e.html

Researchers include Yoshihiro Takeuchi and Hirotoshi Uebayashi. Their biographical sketches are given in the full report on the Handbook CD.

1.25 Ritsumeikan University, Department of Civil Engineering

Address: 1-1-1 Noji Higashi, Kusatsu, Shiga 525-8577, Japan.
Web site: http://www.ritsumei.ac.jp/se/rv/index-e.html

Researchers include Kazuyuki Izuno: Seismic design of bridges.

1.26 Science University of Tokyo, Department of Civil Engineering

Address: Noda City 278-8510, Japan.

Researchers include (1) Sigeaki Morichi: Strong ground motion; and (2) Terumi Touhei: Strong ground motion.

1.27 Senshu University Hokkaido College, Graduate School of Civil Engineering

Address: 1610 Bibai, Hokkaido 079-0197, Japan.
Web site: http://www.senshu-hc.ac.jp/Scripts/File

Researchers include Takakichi Kaneko: Strong ground motion, seismic waveform analysis, earthquake prediction, monitoring procedures.

1.28 Tohoku-gakuin University, Department of Civil Engineering

Address: 1-13-1 Chuou, Tagajo, Miyagi 985-0873, Japan.
Web site: http://www.civil.tohoku-gakuin.ac.jp/

Researchers include Shigeru Hiwatashi: Steel structures, steel bridges, ultimate strength.

1.29 Tohoku Institute of Technology, Dept. of Civil Engineering and Dept. of Architecture

Address: 35-1 Yagiyama Kasumicho, Sendai 982-0381, Japan.
Web site: http://www.tohtech.ac.jp/

Researchers include (1) Makoto Kamiyama: Strong ground motion, microtremor studies; (2) Shigeya Kawamata: Dynamics of structures, structural control; (3) Jiro Suzuya: Steel structures; (4) Reiji Tanaka: Reinforced concrete structures; and (5) Yoshihiro Abe: Dynamics of structures, micro-tremor measurement.

1.30 Tohoku University, Disaster Control Research Center

Address: Aoba-ku, Aoba-yama 06, Sendai 980-8579, Japan.
Web site: http://www.disaster.archi.tohoku.ac.jp/

Researchers in the Center include (1) Masato Motosaka: Soil–structure interaction, ground motion studies; and (2) Michio Hosoi: Reinforced concrete structures. A biographical sketch of Motosaka in given in the full report on the Handbook CD.

1.31 Tohoku University, Department of Civil Engineering

Address: Aoba 6, Aoba-ku, Sendai 980-8579, Japan.
Web site: http://www.civil.tohoku.ac.jp/

Researchers of the Department include (1) Eiji Yanagisawa: Liquefaction, earthquake-resistant design of dams; and (2) Motoki Kazama: Liquefaction, amplification of seismic motion in soft ground.

1.32 Tokyo Denki University, College of Science and Engineering

Address: Hatoyama, Hiki-gun, Saitama 350-0394, Japan.

Researchers include Susumu Yasuda: Soil dynamics.

1.33 Tokyo Institute of Technology

Address: 2-12-1 O-Okayama-Meguro-ku, Tokyo 152-8551, Japan.

Researchers include (1) Hisato Hotta: Associate Professor, Dept. of Architecture and Building Engineering; (2) Kikuo Ikarashi: Associate Professor, Dept. of Architecture and Building Engineering; (3) Kazuhiko Kawashima: Professor, Dept. of Civil Engineering; (4) Koji Kimura: Professor, Dept. of Mechanical and Environmental Informatics; (5) Jiro Kuwano: Associate Professor, Dept. of Civil Engineering; (6) Toshiyuki Ogawa: Professor, Dept. of Architecture and Building Engineering; (7) Gaku Shoji: Research Associate, Dept. of Civil Engineering; (8) Katsuki Takiguchi: Professor, Dept. of Mechanical and Environmental Informatics; and (9) Kohji Tokimatsu: Professor, Dept. of Architecture and Building Engineering. Biographical Sketches of Hisato Hotta, Kikuo Ikarashi, Kazuhiko Kawashima, Koji Kimura, Jiro Kuwano, Toshiyuki Ogawa, Gaku Shoji, Katsuki Takiguchi, and Kohji Tokimatsu are given in the full report on the Handbook CD.

1.34 Tokyo Metropolitan University, Graduate School of Architectural Engineering

Address: Minamiosawa 1-1, Hachioji, Tokyo 192-0397, Japan.

Researchers include (1) Takao Nishikawa: Structural engineering, seismic engineering; (2) Shinji Yamazaki: Structural engineering, steel structures; (3) Manabu Yoshimura: Earthquake engineering, reinforced concrete structures; and (4) Kazuhiro Kitayama: Reinforced concrete structures, seismic design.

1.35 Tottori University, Department of Civil Engineering

Address: 4-101, Koyamacho-Minami, Tottori 680-8552, Japan.
Web site: http://www.cv.tottori-u.ac.jp/

The Department of Civil Engineering was founded in 1967 within the Faculty of Engineering, Tottori University. In 1980 it was divided into the Department of Civil Engineering and Department of Ocean Civil Engineering. The new Department of Civil Engineering was established in 1989, through the restructuring of these two, old departments. The education and research carried out in this new department aims at improvement of the quality of life through research ranging from the design and construction of highways at present to cities and resorts in the future. In recent years, a great deal of importance has been attached to education and research on earthquake engineering.

Researchers include (1) Tomoyuki Ikeuchi: Steel structures, structural dynamics; (2) Hitoshi Morikawa: Earthquake engineering, stochastic mechanics; and (3) Ryohei Nishida: Engineering seismology. Their biographical sketches are given in the full report on the Handbook CD.

1.36 University of Tokushima, Faculty of Engineering

Address: 2-1 Minami-Josanjima, Tokushima, Tokushima 770-8506, Japan.
Web site: http://www.ce.tokushima-u.ac.jp/index-eg.html

Researchers include (1) Kiyoshi Hirao: Seismic design methods, inelastic demand spectra; (2) Tsutomu Sawada: Strong ground motion, earthquake damage prediction; (3) Yoshifumi Nariyuki: Seismic disaster risk assessment, seismic design and retrofit of bridges; (4) Atsushi Mikami: Soil–structure interaction, seismic isolation. Their biographical sketches are given in the full report on the Handbook CD.

1.37 University of Tokyo, Department of Architecture

Address: 7-3-1 Hongo, Bunkyo-ku, Tokyo 113-8656, Japan.
Web site: http://www.arch.t.u-tokyo.ac.jp/

The Department of Architecture was founded in 1877 within the Faculty of Engineering, University of Tokyo. A young architect from England, Josiah Conder (1852–1920), was invited to serve as an instructor. From the experience of the 1891 Nohbi earthquake, safety was realized to be important for the design of buildings in this country, and structural engineering became an important part of education and research in the department of architecture. Mr. Tamisuke Yokogawa started to lecture on Steel Construction in the Department of Architecture in January 1904. Mr. Toshikata Sano (1880–1956), also known as Riki Sano, started to lecture on "Calculations for Building Construction" in January 1905 and on "Iron Construction and Reinforced Concrete" in January 1906.

Professor Sano proposed the use of a seismic coefficient in the design of earthquake-resistant buildings in his doctoral thesis entitled "Earthquake Resisting Structure of Houses," in 1914. Professor Kiyoshi Muto (1903–1989) proposed a simple and practical procedure for the frame analysis of structures under lateral loading in "Architectural Institute of Japan Standard for Structural Calculations of Reinforced Concrete Structures" in 1933. Professor Muto was the key person that made the high-rise Kasumigaseki Building possible in Japan in 1967. After the 1968 Tokachi-oki earthquake, Professor Hajime Umemura (1918–1995) directed research to prevent brittle shear failure of reinforced concrete columns and the revision of design requirements of tie spacing in the Building Standard Law. He initiated research to develop practical procedures for evaluating the earthquake-resisting capacity of existing buildings.

Following the 1956 World Conference on Earthquake Engineering in Berkeley, Professor Muto organized the second World Conference on Earthquake Engineering in Tokyo and Kyoto in 1960 and proposed the formation of the International Association for Earthquake Engineering. He became the founding President of the Association in 1963. Professor Umemura was elected to serve as the seventh President of the Association in 1984 and subsequently organized the ninth World Conference in Japan in 1988. Professor Hiroyuki Aoyama (1932–) became a Director of the Association in1996, and its Executive Vice-President in 2000.

Researchers of the Department include (1) Yuichiro Inaba: Steel structures; (2) Ryoji Iwasaki: Structural dynamics; (3) Hitoshi Kuwamura: Steel structures; (4) Taizo Matsumori: Earthquake engineering, reinforced concrete structures; (5) Yoshimitsu Ohashi: Timber structures; (6) Shunsuke Otani: Earthquake engineering, reinforced concrete structures, response analysis, earthquake-resistant building design; (7) Isao Sakamoto: Timber structures, nonstructural elements; (8) Hitoshi Shiohara: Earthquake engineering, reinforced concrete structures, earthquake-resistant building design; and (9) Tsuyoshi Takada: Structural dynamics, probabilistic risk assessment, structural safety and reliability, stochastic mechanics, earthquake engineering, reliability-based design, performance-based design.

1.38 University of Tokyo, Graduate School of Frontier Sciences, Institute of Environmental Studies

Address: 7-3-1 Hongo, Bunkyo-ku, Tokyo 113-8656, Japan.

The Institute of Environmental Studies was founded in 1999 as a department in the Graduate School of Frontier Sciences. Major buildings are to be constructed in a few years on its campus in Kashiwa, 40 km north of Tokyo. Five courses are offered at the postgraduate education level, and approximately 40 of the Institute's 65 professors have been transferred to it from the Graduate School of Engineering. Some of the professors share the undergraduate education course on Socio-Cultural and Socio-Physical Environmental Studies with the Department of Architecture, Faculty of Engineering. The field of Information and Built Environment, chaired by Professor Jun Kanda, deals with structural safety and disaster mitigation problems in terms of environmental issues. Kanda's specialties are structural reliability, earthquake engineering, wind engineering, socio-economics. His biographical sketch is included in the full report on the Handbook CD.

1.39 University of Tokyo, Institute of Industrial Science, International Center for Disaster-Mitigation Engineering

Address: 4-6-1 Komaba, Meguro-ku, Tokyo 153-8505, Japan.
Web site: http://incede.iis.u-tokyo.ac.jp/default.html

The International Center for Disaster-Mitigation Engineering (INCEDE) was established on April 12, 1991, in the Institute of

Industrial Science (IIS) of the University of Tokyo, as Japan's contribution toward the United Nations International Decade for Natural Disaster Reduction. At present INCEDE emphasizes the disciplines of urban earthquake disaster-mitigation engineering, hydrology/water resources, and remote sensing/Geographic Information Systems. The Institute of Industrial Science, to which INCEDE belongs, is an autonomous Institute of the University of Tokyo. It has about 1000 staff, graduate students, and visiting researchers in various engineering disciplines working in 110 different laboratories. In addition to INCEDE's full-time staff, 8 laboratories of the IIS, specializing in the aforementioned disciplines, act as cooperative members of INCEDE. INCIDE's role is catalytic in nature, in that it synthesizes the activities of these groups in a concerted effort toward disaster mitigation.

Researchers include (1) Kimiro Meguro: Urban earthquake disaster mitigation engineering; and (2) A. S. Herath: Water resources engineering. A biographical sketch of Kimiro Meguro is given in the full report on the Handbook CD.

1.40 Utsunomiya University, Architectural and Civil Engineering Department

Address: 7-1-2, Yoto, Utsunomiya 321-8585, Japan.
Web site: http://www.utsunomiya-u.ac.jp/index.html

The Department of Architecture and Civil Engineering, Faculty of Engineering of Utsunomiya University, was established in 1977. This Department educates and trains students as engineers or research workers who can contribute to society by utilizing new construction technology from housing to national land development. The Department offers two courses: Architecture and Civil Engineering. The Architecture Course deals with planning, design, production, and construction of dwellings, buildings, regions, and cities.

In the early years, activity in earthquake engineering was limited to analytical and observational studies of the interaction between soils and buildings. Experimental and theoretical studies on the structural behavior of steel and composite building structures, such as shear walls using corrugated deck plates, and various types of beam-to-column connections of steel structures, were initiated in 1987. In addition, from 1995 studies were started on the earthquake resistance of timber structures.

Researchers include (1) Atsuo Tanaka: Structural design of steel structures, connections in steel structures; (2) Yasutaka Irie: Soil–building interaction, dynamic characteristics of structures; (3) Akinori Nakajima: Seismic design of steel bridge structures, dynamic problem of composite structures; and (4) Hiroshi Masuda: Structural design of steel structure, connections in steel structures. Their biographical sketches are given in the full report on the Handbook CD.

1.41 Waseda University, Department of Architecture

Address: 3-4-1 Okubo, Shinjuku, Tokyo 169-8555 Japan.
Web site: http://www.arch.waseda.ac.jp/

Researchers include (1) Yasuo Tanaka: Reinforced concrete columns, dynamic behavior of large floating structures, cable structures, shell structures; (2) Satsuya Soda: Performance-based seismic design, structural control with passive and semi-active dampers; and (3) Akira Nishitani: Active and semi-active structural response control, structural system identification, structural safety and reliability. Biographical sketches for Yasuo Tanaka and Satsuya Soda are given in the full report on the Handbook CD.

1.42 Yamaguchi University, Faculty of Engineering

Address: 2-16-1 Tokiwadai Ube, Yamaguchi 755-8611, Japan.

Yamaguchi University was founded in 1949 as a national university by combining various colleges and schools, some of which had histories going back to the 19th century. During the second half of the 20th century, it grew to a sizeable university. Today, the University has a staff of about 1990, including about 270 full professors, and more then 9,000 students.

The origin of the Faculty of Engineering goes back to the Ube Higher Technical School, which was established in 1939. In 1949, this school was incorporated as the Faculty of Engineering of Yamaguchi University, and was initially composed of 4 departments: Mechanical Engineering, Mining, Industrial Chemistry, and Civil Engineering. By April 1990, the Faculty was made up of 10 departments with a total of 47 full professors and an admission capacity of 420 students. In October 1990, the Faculty and its associated technical college were united to form 7 new departments. With this reorganization, the Faculty has grown to an admission capacity of 640 and has an academic staff including 69 full professors.

Researchers of the Faculty include (1) Tadayoshi Aida: Vibration engineering; (2) Toshihiko Aso: Earthquake engineering; (3) Masayuki Hyodo: Geotechnical engineering, soil dynamics; (4) Fusanori Miura: Earthquake engineering, disaster prevention systems; (5) Hitomi Murakami: Disaster mitigation planning; (6) Ayaho Miyamoto: Bridge engineering, structural concrete engineering, advanced systems engineering; (7) Hideaki Nakamura: Structural concrete engineering; (8) Kouichi Takimoto: Earthquake engineering (Research Associate); and (9) Tetsuro Yamamoto: Earthquake-resisting ground engineering. Their biographical sketches are given in the full report on the Handbook CD.

1.43 Yokohama National University, Artificial and Built Environment Systems

Address: 79-5, Tokiwadai, Hodogaya-ku, Yokohama 240-8501, Japan.

Researchers include (1) Suminao Murakami: Urban safety planning, building safety planning; and (2) Satoru Sadohara: Urban safety planning, building safety planning, urban infrastructure planning. Their biographical sketches are given in the full report on the Handbook CD.

2. Companies

2.1 Central Research Institute of Electric Power Industry

Address: 1646 Abiko, Abiko-shi, Chiba 270-1194, Japan.

The Central Research Institute of Electric Power Industry (CRIEPI), a nonprofit organization, was established in 1951 at Komae City in Japan to serve as the comprehensive central research institute for the electric power industry. CRIEPI is developing technologies for solutions to the many problems associated with the global energy and resources industry, the environment, and sustainable economic development, from a long-term perspective. The mission of CRIEPI involves the following three goals: (1) harmony between energy and the environment, (2) cost reduction and greater reliability, and (3) creation of a society that uses energy efficiently.

The 1999 research budget for the Abiko Research Laboratory, CRIEPI, totaled 4.95 billion yen, which funded the following activities: exploratory and basic research (26.3%); applied and development research (23.2%); common research (10.2%); and commissioned research projects (from government) (40.4%). In that same year CRIEPI carried out 344 projects, which can be characterizsd as follows: exploratory and basic research (19.5%); applied and development research (17.2%); contract research projects (from utilities) (57.8%); and commissioned research projects (from government) (5.5%).

The CRIEPI staff numbered 226 in 1999 and included the following specialists: electrical engineering (0.4%), civil engineering (31.4%), geology (10.6%), mechanical engineering (5.3%), chemistry (1.8%), biology (21.7%), environmental science (13.7%), research assistants (5.8%), and administrative staff (6.2%).

Research activities of the Geotechnical and Earthquake Engineering Department in FY 1999 consisted of (1) Exploratory and Basic Research: seismic reliability analysis of structures, estimating methods of design earthquake motions, prediction of earthquake-induced deformations for ground–structure systems, evaluation methods for geotechnical properties of the ground, and technology for geoenviromental evaluation of the ground; and (2) Applied and Development Research: development of seismic risk management for electric power facilities, studies of three-dimensional base isolation systems, development of seismic performance evaluation of asphalt-type facing for rock-fill dams, and demonstration study on CAES in unlined rock caverns.

Researchers include (1) Koichi Nishi: Geotechnical engineering; (2) Katsuhiko Ishida: Engineering seismology; (3) Teruyuki Ueshima: Earthquake engineering; (4) Hiroshi Ito: Rock engineering; (5) Shunji Sasaki: Engineering seismology; (6) Okamoto Toshiro: Geotechnical engineering; (7) Jun'ichi Tohma: Earthquake engineering; (8) Kazuta Hirata: Earthquake engineering, probabilistic structural reliability; (9) Yukihisa Tanaka: Geotechnical engineering; (10) Takashi Nozaki: Rock engineering; (11) Tetsuyuki Kataoka: Geotechnical engineering; (12) Keizo Ohtomo: Earthquake engineering; (13) Koichi Shin: Rock mechanics and rock engineering; (14) Mamoru Kanatani: Geotechnical engineering; (15) Kiyotaka Sato: Earthquake engineering, local site effects; (16) Hiroshi Yajima: Earthquake engineering; (17) Hitoshi Tochigi: Earthquake engineering; (18) Shuichi Yabana: Structural dynamics, seismic isolation; (19) Hideo Komine: Environmental geotechnics; (20) Sadanori Higashi: Engineering seismology, site effects; (21) Yojiro Ikegawa: Rock engineering; (22) Yoshiaki Shiba: Engineering seismology, strong ground motion; (23) Yasuki Ootori: Earthquake engineering; (24) Yoshiharu Shumuta: Earthquake engineering; (25) Tadashi Kawai: Geotechnical engineering; (26) Kenji Kanazawa: Earthquake engineering; (27) Tetsuji Okada: Rock engineering; (28) Akihiro Matsuda: Earthquake engineering; (29) Hiroaki Kobayakawa: Rock engineering; (30) Hiroaki Sato: Earthquake engineering, wave propagation; (31) Tsutomu Kanazu: Concrete engineering, physical properties of concrete; (32) Shinichi Matsuura: Structural engineering, FEM analysis, buckling; (33) Yutaka Hagiwara: Structural engineering, buckling, computational science; (34) Tatsumi Endoh: Concrete engineering, design for reinforced concrete structures; (35) Kosuke Yamamoto: Structural engineering, computational science, maintenance; (36) Takuro Matsumura: Concrete engineering, durability of concrete, steel corrosion; (37) Yukihiro Toyoda: Structural engineering, computational mechanics; (38) Tatsuo Nishiuchi: Concrete engineering, dam safety analysis, concrete dam engineering; (39) Michiya Sakai: Structural engineering, computational science, buckling analysis; (40) Mikio Shimizu: Structural engineering, cable dynamics, galloping; (41) Jun Matsui: Concrete engineering, micro-structure analysis of concrete; (42) Takeshi Yamamoto: Concrete engineering, concrete material science; (43) Kiyoshi Saito: Structural engineering, computational science, maintenance; (44) Tomomi Ishikawa: Structural engineering, wind engineering; (45) Toyofumi Matsuo: Concrete engineering, fracture mechanics for concrete; (46) Masato Nakajima: Structural engineering, reliability and probabilistic safety analysis; (47) Yoshinori Miyagawa: Concrete engineering, FEM analysis of reinforced concrete structures; (48) Yuzo Shiogama: Structural engineering, computational science, maintenance.

2.2 CRC Research Institute, Inc.

Address: 2-7-5, Minamisuna, Koto-Ku, Tokyo 136-8581, Japan.
Web site: http://www.crc.co.jp/

Activities of the Seismic Design Analysis Department at the CRC Research Institute include (1) dynamic analysis of bridges, buildings, mechanical components, atomic power stations; (2) natural analysis; (3) seismic response analysis; (4) mechanical vibration analysis; (5) soil–structure interaction response analysis; (6) liquefaction analysis of soil; and (7) wind response analysis.

Researchers include Hiroyuki Kameoka: Soil-structure interaction, dynamics of structures, liquefaction analysis. His biographical sketch is given in the full report on the Handbook CD.

2.3 Electric Power Development Co., Ltd., Chigasaki Research Center

Address: 9-88 Chigasaki 1-Chome, Chigasaki, Kanagawa, 253-0041, Japan.
Web site: http://www.epdc.co.jp/

Researchers include Yoshiaki Ariga: Earthquake engineering, dam engineering, earthquake observations, shaking table tests.

2.4 Hazama Corporation, Technical Research Institute

Address: Nishimukai, Karima, Tsukuba, Ibaraki, 305-0822, Japan.
Web site: http://www.hazama.co.jp/

Hazama Corporation has been one of Japan's leading construction companies for more than 100 years. Established in 1889 with a mandate to contribute to society through its construction-related business activities, Hazama has evolved into a global-scale, total engineering constructor.

The Technology Department of Hazama was set up in 1945, and was renamed the Technical Research Institute in 1975. The Institute was moved to Yono City, Saitama Prefecture, in 1977. The Institute moved toTsukuba Science City in Ibaraki Prefecture, to mark Hazama's centenary celebration in 1991. The Corporation is conducting a vast array of research and development projects related to environmental protection, increased project safety, and improved cost ratios and overall performance.

Hazama Technical Research Institute occupies more than 70,000 square meters of space, which contains a main administration building, 7 laboratories, and areas for field experiments. Its Structural Engineering Laboratory includes a reaction floor and wall testing system and a shake table testing system. The length of the reaction floor is 32.5 meters, and its 2 reaction walls enable 2-direction loading. The reaction wall can resist 10,000tfm of bending moment and 4,000tf of shear force. The laboratory carries out advanced seismic experiments and real-time, high-speed loading tests utilizing an online computer-controlled actuator system.

The facility's high-performance shaking table can accommodate specimens up to a weight of 80t and simulate seismic behavior of large-scale specimens with a high degree of accuracy. Its horizontal and vertical axes can be controlled simultaneously. The facility can carry out vibration tests on any type of structure.

Researchers include (1) Takashi Inoue: Seismic hazard analysis, strong ground motions, soil–structure interaction; (2) Yoshio Ito: Structural seismic design, structural control; (3) Akio Hori: Structural analysis, three-dimensional structures, large deformation analysis, steel structures, concrete-filled tubes; (4) Shigeki Sakai: Strong ground motion, seismic hazard analysis, soil–structure interaction; (5) Tsunehisa Matsuura: Reinforced concrete structures, high-strength concrete, seismic capacity evaluation; (6) Yasuhiko Tanaka: Environmental vibrations, health monitoring of structures, structure-borne sound; (7) Kazuhiko Urano: Soil–structure interaction, dynamic nonlinear analysis, liquefaction; and (8) Yuji Adachi: Liquefaction, seismic response analysis of structures, soil–structure interaction. Their biographical sketches are given in the full report on the Handbook CD.

2.5 Kajima Corporation, Kobori Research Complex Inc.

Address: 6-5-30, Akasaka, Minato-ku, Tokyo 107-8502, Japan.

The Kobori Research Complex was established on November 1, 1986, in Tokyo. Its organizational structure is as follows: (1) President: Takuji Kobori, (2) Managing Director: Toshihiko Kubota, (3) Planning and Administration Department, (4) Earthquake and Soil Dynamics Department, and (5) Structural Dynamics and Control Department.

The Kobori Research Complex carries out research in the following fields: (1) Soil Dynamics and Earthquake Engineering: assess the impact of ground motion history on global and local seismic response; obtain dynamic interaction effect between soil and building; estimate the influence of liquefaction of soils; and analyze three-dimensional nonlinear earthquake response of structures; (2) Structural Response Control Technology: reduce the earthquake response of buildings using active dynamic dampers; evaluate the efficiency of hybrid control systems for structures; strengthen the seismic safety of existing buildings by active or/and passive dampers; and analyze and evaluate earthquake response of structures; (3) Earthquake Risk Reduction Technology: assess the risk of urban cities before and after a large earthquake; support the decision for earthquake recovery of business performance; and estimate the fragility function of machinery, assemblies, and structures.

Kobori Research Complex has carried out many successful projects of the following types: (1) structural design of tall buildings: office buildings, reinforced concrete high-rise housings, and hotels; (2) earthquake response analysis of mega-structures: power plants, athletic stadiums, and a broadcasting station; (3) research and development on reactor technology for nuclear power stations: reactor buildings, turbine buildings, and miscellaneous buildings; (4) structural response control of buildings: office buildings, hotels, and a ski-dome; and (5) structural strengthening of existing buildings: city halls, dormitories, and apartment houses.

Researchers include (1) Takuji Kobori: Earthquake engineering, structural control; (2) Hiroo Kanayama: Dynamics of structures, earthquake response of buildings; (3) Mitsuo Sakamoto: Structural control, earthquake response of buildings; (4) Kenji Miura: Soil–structure interaction, soil dynamics; (5) Toshikazu Yamada: Structural control, earthquake response of buildings; (6) Masamitsu Miyamura: Earthquake hazard mitigation, strong ground motion; (7) Tomohiko Arita: Structural control, strengthening of buildings; (8) Masayuki Takemura: Strong ground motion, source mechanism; and (9) Akira Endoh: Solid mechanics, structural engineering. A biographical sketch of Takuji Kobori is given in the full report on the Handbook CD.

2.6 Kajima Corporation, Kajima Technical Research Institute

Address: 2-19-1, Tobitakyu, Chofu-shi, Tokyo 182-0036, Japan.
Web site: //www.kajima.co.jp/tech/katri/index.html

Since its establishment in 1840, Kajima Corporation has played a leading role in the construction industry and has influenced the growth of buildings and infrastructures in Japan. This long history as a pioneering company delivering the construction innovations demanded by the changing eras has been made possible by strong technological capabilities fostered by continuous research and development. Kajima Technical Research Institute (KaTRI) was established in 1949 as the first R&D institute in Japan's construction industry and constitutes the core of all R&D activities in Kajima. As Japan is an earthquake-prone country, the major mission and resources of KaTRI are directed toward earthquake engineering as well as to engineering seismology.

KaTRI has made remarkable accomplishments in earthquake engineering R&D since the early 1960s, thanks especially to the leadership of the late Dr. Kiyoshi Muto. In addition to KaTRI, Dr. Muto integrated all the talents in the corporation, such as the Information Processing Center and the Civil/Building Design Departments, when developing techniques for the digitalization of analog strong ground motion records. His software for dynamic response analysis of structures became an important practical tool for the earthquake-resistant design of structures. The Kasumigaseki Building in 1968, Japan's first skyscraper, and the Shiinamachi Building in 1973, the first high-rise reinforced concrete building in Japan, plus special structures such as nuclear power plants, bridges, etc., all made use of this development in the following decades.

Related testing facilities, such as KaTRI's large-scale structural laboratory, shaking table laboratory, and geo-technical laboratory, were completed during the 1980s. These facilities enabled static, pseudo-dynamic, cyclic/dynamic, and centrifuge tests to be conducted to verify and confirm the seismic performance of structures, including newly developed elements and systems.

In the early 1970s, KaTRI became the first non-public institution, to begin to collect data on strong earthquake motions and their effect on both structures and the ground, for the purpose of studying and clarifying the dynamic behavior of structures, and the influence of soil–structure interaction on that behaviour. Earthquake waves were measured not only at the ground surface, but also in the deep rock stratum. The network of such observations was expanded in the Tokyo Bay area and the Tokai area in the1980s. The results of these studies were used to assess the input design ground motions for the earthquake design of structures, for the consideration of fault models, and for early warning systems.

In 1992, KaTRI also started to set up network observations in the Hollister and Borrego areas in California. The data obtained from these horizontal and vertical array observations led to research studies by various researchers on engineering seismology. KaTRI's achievements have been presented to governmental authorities through contracted clients, such as electric power companies, as well as to international institutions and conferences, and have become the basis of upgraded guidelines and design regulations for Japan's high-rise buildings and nuclear power plants. As of 1999, KaTRI's network observatory involves a total of 600 components situated at 71 points in the aforementioned areas.

The number of researchers in KaTRI exceeds 300, and its fiscal budget is approximately US$60 million as of 1999. More than half of its researchers are, to some extent, engaged in earthquake engineering. Researchers in the geotechnical, structural, seismic, and soil-foundation groups of the Civil and Building Engineering Departments are fully engaged in this capacity. About 50 researchers in the earthquake motion, vibration control, and risk assessment groups of the Advanced Technology Department are intensively engaged in the field of earthquake engineering and engineering seismology with Dr. Etsuzo Shima, a supervisor of Kajima. Names of major researchers are listed here. Detailed information of the resources, annual reports, and testing facilities open to the public are available at http://www.kajima.co.jp/.

Researchers include (1) Etsuzo Shima: Seismic zoning methodology, strong ground motion wave propagation of deep underground structures; (2) Tsunehisa Tsugawa: Dynamics of structures, concrete and steel structures, earthquake disaster reduction; (3) Masanori Niwa: Strong ground motion, soil–structure interaction; (4) Tetsuo Takeda: Dynamics of structures, pre-stressed concrete structures, hybrid structures; (5) Yasuo Murayama: Reinforced concrete (RC) structures; (6) Yoshihiro Hishiki: Dynamics of structures, pre-stressed concrete structures, hybrid structures; (7) Kaoru Mizukoshi: Earthquake risk assessment; (8) Shoji Uchiyama: Soil–structure interaction; (9) Katsuya Igarashi: Dynamics of structures, RC structures; (10) Shin-ichi Yamanobe: Dynamics of structures, RC structures; (11) Kumiko Suda: RC structures; (12) Hiroshi Hayashi: Soil structures; (13) Katsuya Takahashi: Strong ground motion; (14) Naoto Ohbo: Strong ground motion, soil–structure interaction; (15) Yukio Naito: Structural dynamics, soil–structure interaction; (16) Kaeko Yahata: Soil–structure interaction; (17) Tomonori Ikeura: Strong ground motion, fault models; (18) Susumu Ohno: Strong ground motion, wave propagation; (19) Toshio Kobayashi: Dynamics of structures; (20) Yoshiyuki Kasai: Dynamics of structures, impact loading; (21) Naoki Tanaka: Steel structures and structural analysis; (22) Norio Suzuki: RC structures and structural analysis; and (23) Toshiyuki Fukumoto: Hybrid structures. A biographical sketch of Etsuzo Shima is given in the full report on the Handbook CD.

2.7 Kajima Corporation, Building/Civil Engineering Design Departments and Information Processing Center

Address: 5-30, Akasaka 6-chome, Minato-ku, Tokyo 107-8502, Japan.
Web site: http://www.kajima.co.jp/welcome.html

Researchers include (1) Tadashi Sugano: Dynamic analysis, concrete and steel structures; (2) Eiji Fukuzawa: Dynamic analysis, concrete and steel structures; (3) Yutaka Isozaki: Dynamic analysis, concrete and steel structures; (4) Masaaki Yamamoto: Reliability-based design; (5) Yasuaki Shimizu: Reliability-based design, structural dynamics; (6) Tomoyasu Hamada: Soil dynamics, liquefaction; (7) Yoshio Sunasaka: Structural dynamics, strong ground motion; (8) Takashi Miyashita: Analysis of structures; (9) Kiyoshi Masuda: Soil–structure interaction; (10) Hiroshi Morikawa: Analysis of structures; (11) Takashi Yoshikiyo: Soil dynamics; (12) Hachiro Ukon: Stochastic analysis; (13) Fumio Sasaki: Soil–structure interaction; and (14) Motomi Takahashi: Analysis of structures.

2.8 Kansai Electric Power Co. Inc., Technical Research Center

Address: 11-20 Nakoji 3-chome, Amagasaki City, Hyogo 661-0974, Japan.
Web site: http://www.kepco.co.jp/

Researchers include Yoshio Soeda: Strong ground motion.

2.9 Kawasaki Steel Corporation, Structure Research Laboratories

Address: 351 Naganuma-cho, Inage-ku, Chiba, 263-0005, Japan.

The Structural Research Laboratories of Kawasaki Steel Corporation were established in 1970, with the mission of research and development of new steel structures and their members.

Our research laboratories have contributed in various engineering areas including civil engineering (new steel piling system, new steel foundation, segments), architecture, offshore engineering (mega-float structures), pipeline engineering (new seismic design), and structural experiments (tensile, compression, buckling, fatigue, and vibratory tests).

Researchers include (1) Takeshi Koike: Lifeline, seismic design of underground structures; (2) Sachito Tanaka: Seismic design and analysis on foundations, bridges, and quay wall; (3) Kenji Uemura: Steel dampers of buildings; (4) Takumi Ishii: Welding structures, beam-to-column connections; (5) Kazuyoshi Fujisawa: Steel building materials, steel dampers of buildings; (6) Shinya Inaoka: Steel building materials, seismic design of steel buildings; and (7) Yukio Murakami: Steel building materials, hybrid columns of buildings. Their biographical sketches are given in the full report on the Handbook CD.

2.10 Mitsui Construction Co., Ltd., Technical Research Institute

Address: 518-1 Komaki, Nagareyamam-shi, Chiba 270-0132, Japan.
Web site: http://www.mcc.co.jp/

Researchers include (1) Toshiyuki Noji: Structural engineering, earthquake engineering; (2) Tomio Arii: Soil mechanics, foundation engineering; (3) Hisayuki Yamanaka: Structural engineering, aseismic structures; (4) Masaharu Tanigaki: Dynamics of structures, aseismic structures; (5) Tetsuya Yamada: Structural engineering; (6) Yuichi Hirata: Structural engineering, dynamics of structures, vibration control; (7) Hideyuki Kosaka: Structural engineering, dynamics of structures, vibration control; (8) Kuniaki Yamagishi: Earthquake engineering, dynamics of structures; and (9) Kenji Tano: Structural engineering, aseismic structures.

2.11 NEWJEC Inc., Earthquake Engineering Div., Technology Development Dept.

20-19 Shimanouchi 1-chome, Chuo-ku, Osaka 542-0082, Japan.
Web site: http://www.newjec.co.jp/

Researchers include (1) Kimihiko Kunii: Dynamic analysis, seismicity, dynamic soil properties; (2) Masaru Urayama: Dynamic analysis, seismicity, dynamic soil properties; (3) Mitsuhiro Tokusu: Activity of faults, geological structure; (4) Masayuki Yamada: Strong ground motion, surface-wave propagation, rock–structure interaction; (5) Toshiyuki Hirai: Strong ground motion; and (6) Tadashi Hara: Liquefaction. Their biographical sketches are given in the full report on the Handbook CD.

2.12 NKK Corporation, R&D Division

Address: Minami-watarida, Kawasaki 210-0855, Japan.
Web site: http://www.lab.keihin.nkk.co.jp/

Modern steels have been developed to satisfy the requirements of consumers in terms of function, shape, and appearance, and this development continues unceasingly. Steel is ideally suited to recycling, and thus is environment-friendly. NKK's mission is to reveal to the world the unlimited possibilities of steel, an old yet new material.

A precursory research center of NKK Corporation began in 1948. It had a department to study the earthquake-resistant design of steel structures in order to develop new materials and steel members and to respond to customers' requirements. The current R&D division was organized in 1978 and since 1988 it consists of three research centers: Material and Processing, Applied Technology, and Engineering.

NKK's technologies are utilized in many situations: buildings, bridges, pipelines, pressure vessels, storage tanks, ships, and marine structures. Technological experience accumulated over many years is an essential prerequisite for the construction of durable structures. Demands for improved seismic resistance in structures have grown considerably, particularly since the 1995 Great Kobe earthquake. NKK was one of the first organizations to become involved in this problem, and its research and development have produced significant results in terms of structural materials, design, and construction methodology.

Research on seismic design and disaster prevention measures is carried out in the Applied Technology Research Center and Engineering Research Center. The Applied Technology Research Center has a civil and building research department, and the Engineering Research Center has an energy research department. Researchers in these Centers are engaged in projects concerned with structural steel members for buildings, bridges, bridge piers, foundation piles, structures in harbor and riverside areas, such as caissons, offshore structures, as well as pipeline systems such as city gas, natural gas, petroleum and water pipelines, pressure vessels, storage systems, cryogenic station systems, and so on.

Researchers at the Applied Technology Research Center (Civil and Building Research Department) include (1) Takashi Okamoto: Piles, bridge piers, tanks, soil–fluid–structure interaction, seismic design; (2) Shigeki Ito: Steel buildings, elasto-plastic behavior, seismic design; (3) Hideaki Nagayama: Caisson, shield tunnel, hybrid structure, structural mechanics; (4) Hitoshi Ito: Steel building, elasto-plastic behavior, optimum design; (5) Koji Sekiguchi: Foundations, soil–structure interaction, soil mechanics; (6) Masashi Kato: Bridges, elasto-plastic behavior, large deformations; and (7) Hisaya Kamura: Steel buildings, structural control, seismic analysis.

Researchers at the Engineering Research Center (Energy Research Department) include (1) Nobuhisa Suzuki: Pipelines, pipe–soil interaction, large deformations, seismic design; (2) Akihiko Kato: Pipelines, pressure vessels, fatigue, fracture mechanics; (3) Yoshimi Ono: Pipelines, pressure vessels, fatigue, fracture mechanics; and (4) Eiji Matsuyama: Pipelines, pipe–soil interaction, large deformations, seismic design.

2.13 NTT, Power and Building Facilities Inc., Research and Development Dept.

Address: 3-9-11, Midori-cho, Musashino-shi, Tokyo 180-8585, Japan.
Web site: http://www.ntt-f.co.jp/

Researchers include Hiroshi Dohi: Strong ground motions, site effects, soil–structure interaction. His biographical sketch is given in the full report on the Handbook CD.

2.14 NTT, Access Network Service Systems Laboratories

Address: 1-7-1 Hanabatake, Tsukuba, Ibaraki, 305-0805, Japan.
Web site: http://www.at-net.ne.jp/RandD/

Researchers include (1) Kazuhiko Fujihashi: Seismic countermeasures for underground telecommunication facilities; (2) Masaru Okutsu: Seismic countermeasures for underground telecommunication facilities; and (3) Koji Komatsu: Seismic countermeasures for underground telecommunication facilities.

2.15 Obayashi Corporation, Technical Research Institute

Address: 640, Shimokiyoto 4-chome, Kiyose-shi, Tokyo 204-0011, Japan.
Web site: http://www.obayashi.co.jp/Giken/

Researchers include (1) Toshikazu Takeda: Seismic design of structures, reinforced concrete, base isolation structures; (2) Yutarou Omote: Seismic design of structures, reinforced concrete; (3) Katsuyoshi Imoto: Reinforced concrete structures, finite element analysis; (4) Tetsuo Suzuki: Seismic design of structures, base isolation structures, structural control; (5) Yozo Goto: Earthquake engineering on lifeline and industrial structures, earthquake geotechnical engineering; (6) Toshio Kikuchi: Seismic hazard and risk assessment, lifeline structures, seismic protective systems for bridges; (7) Takashi Matsuda: Soil–structure interaction, centrifuge model tests, earthquake geotechnical engineering; (8) Joji Ejili: Engineering seismology, strong ground motions, local site effects; (9) Hajime Ohuchi: Seismic design of structures, cable stayed bridges, retrofit techniques of structures, hybrid structures; (10) Motoyuki Okano:

Hybrid structures, carbon fiber retrofit techniques; (11) Yuzuru Yasui: Soil–structure interaction, earthquake engineering, structural control; (12) Akira Teramura: Base isolation, structural control; (13) Kunio Wakamatsu: Earthquake engineering, engineering seismology; (14) Mitsuru Kageyama: Structural control, earthquake engineering; (15) Yasuhiko Takahashi: Steel structures, seismic design, structural control; (16) Hiroaki Eto: Three-dimensional nonlinear seismic response frame analysis, reinforced concrete buildings; (17) Matsutaro Seki: Base isolation structures, seismic strengthening, structural control; (18) Kohzo Kimura: Advanced composite materials, seismic strengthening; (19) Tatsuo Nakayama: Shear and torsion of reinforced concrete, steel plate reinforced concrete; (20) Hideo Katsumata: Seismic strengthening, structural control, shaking table tests; (21) Arihide Nobata: Engineering seismology, seismic hazard assessment, strong ground motions; (22) Takao Seki: Soil–structure interaction, engineering seismology, base isolation structures; (23) Hirokazu Sugimoto: Steel structures, seismic strengthening, structural control; and (24) Kazuhiro Naganuma: Finite element analysis, earthquake-resistant design, reinforced concrete structures.

2.16 Oiles Corporation, Technical Development Department

Address: 1000 Hakari-Chyo, Ashikaga-City, Tochigi, Japan.
Web site: http://www.netspace.or.jp/~oiles

Researchers include (1) Ikuo Shimoda: Base isolation, vibration control, damping; and (2) Masayoshi Ikenaga: Base isolation, damping. Their biographical sketches are given in the full report on the Handbook CD.

2.17 Osaka Gas Co., Ltd.

Address: 1-2 Hiranomachi 4-chome, Chuo-ku, Osaka 541-0046 Japan.

Researchers include (1) Daihachi Okai: Civil & Architectual Technology Team Engineering Department: Manager; (2) Yasuo Ogawa: Earthquake resistance of pipelines; and (3) Takeyoshi Nishizaki: Earthquake resistance of LNG tanks.

2.18 Oyo Corporation, Earthquake Engineering Department, Oyo Technical Center

Address: 2-61-5 Toro-Cho, Omiya, Saitama 330-8632, Japan.
Web site: http://www.oyo.co.jp/

Researchers include (1) Keiji Tonouchi: Manager; (2) Akio Yamamoto: Strong ground motion, ground liquefaction; (3) Shukyo Segawa: Seismic microzonation; and (4) Hideaki Shinohara: Seismic microzonation, strong ground motions.

2.19 Pacific Consultants Co. Ltd., Technical Research Institute

Address: 1-7-5, Sekido, Tama, Tokyo 206-8550, Japan.
Web site: http://www.pacific.co.jp/

Researchers include (1) Akio Hayashi: Dynamics of structures, seismic isolation; and (2) Kazutoshi Yamamoto: Seismic response of structure, dynamic characteristics of soil and ground stability.

2.20 Shimizu Corporation, Institute of Technology

Address: 4-17, Etchujima, 3-Chome, Koto-Ku, Tokyo 135-8530, Japan.

Researchers include Takashi Tazoh: Soil–structure interaction, seismic response of soft soil deposits, dynamic behavior of pile foundations. His biographical sketch is given in the full report on the Handbook CD.

2.21 Shinozuka Research Institute

Address: 3F Koushin Bldg. 4-5-1, Nishi-Shinjuku, Shinjuku-ku, Tokyo 160-0023, Japan.
Web site: http://www.shinozukaken.co.jp/

Researchers include (1) Masanobu Shinozuka: Structural safety and reliability; (2) Masaru Hoshiya: Structural reliability and design, random vibrations; (3) Masanori Hamada: Life-line earthquake engineering, soil liquefaction; (4) Kentaro Sotomuri: Aseismic design; (5) Takaaki Nakamura: Fluid–structure interaction; and (6) Toshiharu Nakamura: Hybrid structures. Their biographical sketches are given in the full report on the Handbook CD.

2.22 Sumitomo Construction Co. Ltd., Institute of Technology & Development

Address: 1726 Niragawa, Minamikawachi-machi, Kawachigun, Tochigi 329-0432, Japan
Web site: http://www.sumiken.co.jp/

Researchers include (1) Fumiaki Arima: Earthquake response control of buildings, vibration isolation technology, damping technology; and (2) Hiroshi Mikami: Dynamics of underground structures, countermeasures against soil liquefaction. Their biographical sketches are given in the full report on the Handbook CD.

2.23 Taisei Corporation, Taisei Research Institute

Address: 344-1 Nase-cho, Totsuka-ku, Yokohama 245-0051, Japan.

The Technical Research Center was established in June 1958. Its first experimental facilities were established in Tokyo in June

1960. In August 1979 the center moved to Yokohama, where a new shaking table was installed. The Centrifuge Testing Laboratory was opened in December 1990.

The mission of Taisei is threefold: (1) To understand the mechanism of earthquakes, the transmission of vibrations, and the response of structures; (2) To develop numerical tools to simulate the earthquake ground motion and the response of structures; and (3) To develop rational seismic design procedures and methods to improve seismic performance.

Activities and accomplishments of Taisei include (1) Seismology: Monitoring and analysis of strong earthquake ground motions, numerical simulation of earthquake ground motions, and studies of design earthquake motions; (2) Geotechnical earthquake engineering: Studies of nonlinear ground response and soil–structure interaction, studies of soil liquefaction and the development of countermeasures, development of design methodologies and strengthening methods for earth structures, and experiments using the centrifuge testing laboratory; (3) Seismic design practice: 2D, 3D response analysis of structures during very large earthquakes, development of design methodologies for underground structures, foundations, and superstructures; development of base isolation systems for buildings, bridges, and computer floors, development of structural control system for buildings, and experiments using the shaking table; (4) Seismic risk management: Analysis of earthquake hazard, development of computer programs for evaluation of seismic risk, and survey of earthquake damage.

Taisei uses several software packages in its research as listed following.

(1) FDAP: Frequency domain three-dimensional dynamic analysis program: We can evaluate the two- or three-dimensional behavior of fluid, structure, and ground systems in frequency domain during earthquakes. FDAP also has an equivalent linear technique to confirm nonlinear behavior under large earthquakes.

(2) TDAP: Time domain three-dimensional dynamic analysis program: We can evaluate the two- or three-dimensional behavior of fluid, structure, and ground systems in time domain during earthquakes. TDAP can analyze the nonlinear response of the objective system considering various kind of nonlinear models for the structure, soil, and interface element between the soil and structure.

(3) LIQCA: Coupled analysis program of liquefaction: LIQCA was developed by Kyoto University and Gifu University in order to analyze the dynamic behavior of liquefied soil and the dynamic interaction between liquefied soil and structures. LIQCA has the constitutive law for sand, which is derived with the elasto-plastic theory using the concept of the nonlinear kinematic hardening rule, and is assembled with a coupled FEM-FDM numerical method using the two-phase mixture theory by Biot. We have 2-dimensional and 3-dimensional versions of LIQCA and we can evaluate 2-dimensional and 3-dimensional problems for liquefaction.

(4) DIANA-SD: Dynamic Interaction Approaches and Nonlinear Analysis using Stress Density model: This is a FEM code for a two-phase medium such as a liquefiable soil. The Stress–Density model is incorporated as a constitutive model for soil. Complex behaviors of liquefied soil can be simulated and the response of piles and superstructures can be evaluated with 2-D and 3-D FEM versions.

(5) SYSTEM-E: Convenient evaluation system of design earthquake motion: A computer-aided earthquake wave generation system runs on an engineering workstation. The system includes the following three functions: (i) a function based on seismic hazard analysis, (ii) a function based on design spectra, and (iii) a semi-empirical Green's function. The system also has a database about historical earthquakes, active fault data, and design earthquake motions.

(6) SRM: Seismic Risk Management system: The SRM system proposes an appropriate strategy for seismic risk by considering the balance between the investment in seismic safety and the seismic-induced risk. The system has the following functions: (1) seismic loss estimation for structures, (2) reliability-based maintenance planning, and (3) GIS-based seismic hazard estimation.

(7) DIMPLE: Three-dimensional finite difference program for wave propagation analysis in large scale, modeling complex soil structures at depths to tens of km under plains and basins containing heterogeneous fault sources. This staggered-type FDM program is run on a vector supercomputer to compute 300 sec. time traces over the 127 km by 150 km Kanto plain model in three days.

Taisei Corporation uses several advanced experimental apparatuses as follows:

(1) Triaxial Earthquake Simulator: This simulator reproduces three-dimensional strong ground earthquake motions with high accuracy for testing structures, instruments, and furniture. Its characteristics are: (i) Size of shaking table: 4 m $\times$ 4 m; (ii) Capacity of shaking table: 196 kN; (iii) Direction of shaking: horizontal 2 direction and vertical direction; (iv) Maximum acceleration: horizontal 1.0 G, vertical 1.0 G; (v) Maximum velocity: horizontal 100 cm/sec, vertical 50 cm/sec; and (vi) Maximum displacement: horizontal 20 cm, vertical 10 cm.

(2) Centrifuge Machine: This equipment simulates the possible behavior of soil structures up to 100 meters underground by using scaled models. An electrohydraulic shaker can be mounted to simulate earthquake motion. Its characteristics are radius of arm: 6.90 m, maximum centrifuge acceleration: 200 G, and maximum capacity: 3.92 kN.

(3) Structural Testing Machine: This machine is used to confirm the strength of the full-scale members of a structure. Its capacity is 1000 tons for compression, and 500 tons for tension.

Researchers include (1) Soichi Kawamura: Earthquake ground motions, soil–structure interaction, seismic diagnosis and strengthening, base isolation and response control, seismic risk management; (2) Shunji Fujii: Earthquake ground motions, soil liquefaction, soil–structure interaction, base isolation and structural control, seismic risk management; (3) Masayoshi Hisano: Base isolation and structural control, dynamic response of structures, seismic risk management; (4) Masayoshi Takaki: Soil–structure interaction, base isolation, vibration isolation and control; (5) Toshiro Maeda: Earthquake ground motion, design input ground motion, earthquake observation and analysis, soil–structure interaction; (6) Ichiro Nagashima: Structural control and base isolation, nonlinear ground response and soil liquefaction, soil–structure interaction; (7) Hiroshi Hibino: Earthquake ground motion, base isolation and response control; (8) Shigehiro Sakamoto: Stochastic FEM, reliability, seismic risk management, numerical simulation of flow; (9) Takafumi Iizuka: Seismic risk management, nonlinear FEM; (10) Chiaki Yoshimura: Earthquake ground motion; (11) Hideki Funahara: Nonlinear ground response and soil liquefaction, soil–structure interaction; (12) Yasuo Uchiyama: Earthquake ground motion, disaster prevention; (13) Ryota Maseki: Structural control and base isolation, system identification; (14) Hiromitsu Izumi: Soil–pile–structure interaction, dynamics of underground structures; (15) Kazuaki Watanabe: Earthquake motions, dynamic soil–structure interaction; (16) Yukio Shiba: Earthquake motions, dynamic soil–structure interaction; (17) Akira Tateishi: Effective stress analysis, dynamic soil–structure interaction; (18) Tetsushi Uzawa: Seismic risk management, reliability assessment for structures; (19) Tadafumi Fujiwara: Behavior of gravity type structures, soil improvement against liquefaction; (20) Kenichi Horikoshi: Soil improvement against liquefaction, piled raft foundations; (21) Susumu Okamoto: Seismic isolation systems for bridges, ground motions; (22) Katsuyuki Sakashita: Earthquake ground motions, soil–structure interaction; and (23) Yoshihiro Tanaka: Nonlinear seismic response analyses, constitutive models of concrete and soil damage evaluation and risk management, development of innovative new type bridge structures. Their biographical sketches are given in the full report on the Handbook CD.

2.24 Takenaka Corporation, Building Design Department, Osaka Main Office

Address: 2-3-10, Nishi Hon-machi, Nishi-ku, Osaka, Japan

Researchers include Kunio Terada, and his biographical sketch is given in the full report on the Handbook CD.

2.25 Tobishima Corporation, Research Institute of Technology

Address: 5472 Kimagase, Sekiyado-Machi, Chiba 270-0222, Japan.
Web site: http://www.tobishima.co.jp/

Researchers include (1) Shigeru Miwa: Strong ground motion, soil–structure interaction, liquefaction; (2) Atsunori Numata: Geotechnical engineering, liquefaction; (3) Takaaki Ikeda: Strong ground motion, geotechnical engineering, liquefaction; (4) Eiji Shimamoto: Geotechnical engineering, liquefaction; (5) Tomoki Shiotani: Geotechnical engineering, acoustic emission; (6) Baoqi Guan: Strong ground motion; and (7) Kouji Ohno: Geotechnical engineering, liquefaction. Their biographical sketches are given in the full report on the Handbook CD.

2.26 Toda Corporation, Civil Engineering

Address: 7-1, Kyobashi 1-Chome, Chuo-ku, Tokyo 104-8388, Japan.
Web site: http://www.toda.co.jp/

Researchers include Masayoshi Ishida, Civil Engineering & Technical Dept.

2.27 Tokyo Electric Power Co., Power Engineering R & D Center, Seismic Design Group

Address: 4-1, Egasaki-cho, Tsurumi-ku, Yokohama 230-8510, Japan.
Web site: http://www.rd.tepco.co.jp/syagai/english/common/mokuji.html

Researchers include (1) Hideaki Saito: Seismic design of NPP, RC structures; (2) Hideo Tanaka: Earthquake design spectra, seismic hazard; (3) Masayoshi Shimada: Performance design, seismic damage assessment; (4) Masayuki Takeuchi: Structural system on buildings of NPP, structural mechanism; (5) Tomiichi Uetake: Strong ground motion, site effects, array observations; (6) Satoshi Shimura: Performance design, evaluation of soil properties; (7) Tomohiro Fujita: SC structure, seismic isolation, structural control; (8) Miwa Sasanuma: Soil–structure interaction, structural health monitoring; (9) Tomohiko Hiroshige: Shaking table tests, seismic damage assessment; (10) Satoru Takahashi: GIS, Seismic damage assessment, earthquake observations; and (11) Masatomo Kikuchi: Dynamic analysis of piles, seismic damage assessment.

2.28 Yachiyo Engineering Co. Ltd., Seismic Engineering Department

Address: Nakameguro 1-10-21, Meguro-award, Tokyo 153-8639, Japan.
Web site: http://www.yachiyo-eng.co.jp/e/home.htm (English), http://www.yachiyo-eng.co.jp/ (Japanese)

The Seismic Engineering Department of Yachiyo Engineering Co, Ltd. investigated the damage to bridges caused by past earthquakes in Japan during the period 1923–1964 under the direction of Dr. Kodera. Some of the results were applied to the seismic design of new bridges to avoid the failure of girders. For example, it is recommended that large displacements of the girders be restricted, even at the movable support, and to avoid large transverse displacements.

The Seismic Engineering Department of Yachiyo Engineering Co, Ltd. experimentally investigated a new type of shock absorber using viscous material installed at the movable support of continuous girder bridges. Two continuous girder bridges fitted with shock absorbers suffered damage during the 1978 Miyagioki earthquake. Seismic response analysis of those bridges was carried out using the earthquake acceleration record obtained near the bridge site. The results of the analysis were in good agreement with the actual damage sustained by the bridge. Severe damage concentrated at the fixed support structure was avoided.

The dynamic analysis of reinforced concrete piers in the elasto-plastic range were studied and the results were used to propose a practical design method in 1980. Many series of tests of RC piers under cyclic loading were performed as a result of a commission from the Public Works Research Institute N.O.C. The test data were used as the basis for revising the seismic design of bridges.

The shear failure of the RC columns of railway rigid frame viaducts during the 1995 Kobe earthquake was investigated. The shear capacity S_U and the maximum shear force S_M, which corresponds to the flexural strength of the structure, were calculated neglecting the axial forces due to the earthquake. Both the columns of rigid frame railway viaducts and the piers of highway bridges, which suffered catastrophic shear failure, had an index $S_U/S_M < 1$. Those with $S_U/S_M > 1$ suffered no significant shear failures. On the basis of these tests results, we proposed that the ratio S_U/S_M be used as an index of vulnerability to shear failure. This proposal was adopted by the Ministry of Transportation and used to assess the priority for retrofitting the existing viaducts.

The Seismic Engineering Department of Yachiyo Engineering Co., Ltd. has participated in Joint Research Investigating the Seismic Isolation of Long Span Bridges Using Damping Devices (1994–1996), and has carried out studies on the effect of the damping devices. It has also participated in the Seismic Engineering Software Workshop (1997–1998), and has performed studies on modeling, setting conditions of input data, and computational methods for nonlinear dynamic analysis. Our areas of expertise are seismic engineering, seismic design consulting, seismic retrofitting design, seismic upgrade design, seismic capacity evaluation for structures and earth structures, nonlinear dynamic analysis and seismic diagnosis, environmental vibration measurement and investigation, etc.

Researchers include (1) Juro Kodera; (2) Yasufumi Umehara; (3) Yasuo Maehara; (4) Kyoichi Sasaki; (5) Yutaka Ishibashi; (6) Jin Watanabe; (7) Shizue Misono; (8) Yoshito Fujita; (9) Hideyuki Suzuki; (10) Yusuke Ogura; (11) Yoshiki Ishikawa; (12) Rie Ozawa; (13) Yoshiaki Kobayashi; and (14) Kazushi Nagoya. Biographical sketches of Juro Kodera, Yasufumi Umehara, Yasuo Maehara, Kyouichi Sasaki, Yutaka Ishibashi, Jin Watanabe, Shizue Misono, Yoshito Fujita, are given in the full report on the Handbook CD.

3. Public Organizations

3.1 Hanshin Expressway Public Corporation

Address: 4-1-3 Kyutarocho, Chuo, Osaka 541-0056, Japan.
Web site: http://west.park.or.jp/hanshin-expressway/

Hanshin Expressway Public Corporation (HEPC) has been involved in urban expressway development around the Kansai urban area since its founding in 1962. Currently, the total length of the Hanshin Expressway that is open to traffic is 221.2 km, and the average daily traffic volume is 0.95 million vehicles. HEPC is now constructing about 30 km of expressway.

HEPC has conducted research on the seismic design of highway viaducts and underground structures, and on retrofitting methods for piers. It has also established seismic design standards based on the Japanese Design Specifications of Highway Bridges. Contact persons are Masahiro Koike, Manager of the Design Division of the Engineering Department, and Kiyoshi Miyawak, Manager of the Maintenance Engineering Division of the Maintenance and Facility Department.

3.2 Honsyu Shikoku Bridge Authority, Long-Span Bridge Engineering Center

Address: 4-1-22, Onoedori, Chuo-ku, Kobe 651-0088, Japan.
Web site: http://www.hsba.go.jp/

Researchers include Nobuyuki Kashima: Aseismic design for long-span bridges.

3.3 Japan Association of Representative General Contractors, Central Research Institute for Construction Technology

Address: 1-6-34 Konan Minato-ku, Tokyo, 108-0075, Japan.
Web site: http://www.chugiken.or.jp/

The Japan Association of Representative General Contractors (JARGC) was established on January 26, 1976, in Tokyo. Its member corporations are the backbone general contractors in Japan, with capital of more than hundred million yen.

The Central Research Institute for Construction Technology (CRICT) was founded on October 1, 1987, as an affiliated organization of JARGC, together with its general contractor members. CRICT devotes itself to research on seismic design of buildings and construction technologies. Many cooperative research projects have been carried out between CRICT and universities and other research institutes. The CRICT Joint Research Center was completed in June 1992, and is used as the laboratory of CRICT and its members.

The work of the CRICT falls into five main categories.

1. Development of High-Rise Reinforced Concrete Buildings. This project research was carried out from 1990 to 1996. CRICT and its member corporations studied the seismic-resisting design methodology and construction technologies for 25-story to 30-story high-rise reinforced concrete buildings.
2. Earthquake Response Characteristics of Base-Isolated Buildings. The first phase of this project took place from 1988 to 1992. The second phase began in 1995. CRICT and its member corporations studied the seismic-resisting design methodology and response characteristics of base-isolated buildings. Tests of the base isolators subjected to tension were performed in the Joint Research Center from 1996 to 1997.
3. Development of Open Building System. This is a current project. Unlike general buildings, the open building system consists of the support structure and the infill system (dwelling unit). The support should possess a sufficient long service life and be flexibly adaptable for renewal of the infill system, which may occur several times during the service life of the support structure. In our research, a precast, prestressed concrete structure is used as the support structure. A base-isolation system was inserted in the system.
4. Seismic Behavior of Prestressed Concrete High-Rise Buildings. This is a cooperative research project between CRICT and Building Research Institute, Ministry of Construction of Japan. The specifications for prestressed concrete buildings with heights greater than 31m have not been specified in the seismic-resisting design code of Japan. The object of this research is to investigate the earthquake response characteristics of prestressed concrete high-rise buildings. The earthquake response characteristics of the buildings have been investigated by numerical analysis. The tests on beam–column connections were performed recently.
5. Collapse Mechanism of Reinforced Concrete Buildings with Soft First Story. This cooperative research project between CRICT and Building Research Institute, Ministry of Construction of Japan, started in October 1998. One of the objectives of this study is to evaluate the earthquake resisting capacity of existing reinforced concrete buildings with soft first stories in Japan, since many structures of this type collapsed in the 1995 Kobe earthquake. It is expected that the research results will contribute to a new design code for such structures in Japan. Numerical earthquake response analyses of 6-, 10-, and 14-story buildings with soft first stories were carried out. A substructure pseudo-dynamic test of the 6-story building specimen was performed in August 1999.

Researchers include (1) Shin Okamoto: Earthquake engineering; (2) Kyozo Fukazawa: RC structures; and (3) Jianhua Gu: Structural engineering. Their biographical sketches, and the organization chart of CRICT, are given in the full report on the Handbook CD.

3.4 Agency of Industrial Science and Technology, MITI, National Institute for Resources and Environment

Address: 16-3 Onogawa, Tsukuba, Ibaraki 305-8569, Japan.
Web site: http://www.aist.go.jp/

Researchers include Masayuki Kosugi: Rock discontinuity behavior, rock mechanics, fracture mechanics. His biographical sketch is given in the full report on the Handbook CD.

3.5 Ministry of Transport, Port and Harbour Research Institute

Address: 3-1-1 Nagase, Yokosuka 239-0826, Japan.
Web site: http:/www.phri.go.jp/

Researchers include (1) Susumu Iai: Performance of port structures, effective stress analysis; (2) Koji Ichii: Liquefaction-induced deformations, performance-based design; (3) Masahumi Miyata: Mechanics of granular materials; (4) Toshikazu Morita: Deformation of port structures during earthquakes; (5) Tsuyoshi Nagao: Reliability-based design and limit state design method; (6) Atsushi Nozu: Engineering seismology, response of port structures; (7) Yukihiro Sato: Strong motion observations; (8) Takahiro Sugano: Performance of port structures, shaking table tests; and (9) Tatsuo Uwabe: Coupled hydrodynamic response of port structures. Biographical sketches of Koji Ichii, Masafumi Miyata, Toshikazu Morita, Atsushi Nozu, Takahiro Sugano, and Tatsuo Uwabe are given in the full report on the Handbook CD.

3.6 Ministry of Construction, Public Works Research Institute

Address: 1 Asahi, Tsukuba-shi, Ibaraki-ken 305-0804 Japan.
Web site: http://www.pwri.go.jp/

Researchers include (1) Masahiko Yasuda: Earthquake engineering, structural dynamics, long-span bridges; (2) Hideki Sugita: Earthquake engineering, earthquake hazard mitigation planning; (3) Keiichi Tamura: Strong ground motion, soil liquefaction; (4) Shigeki Unjoh: Seismic design and retrofit of infrastructure systems, structural control and isolation; (5) Osamu Matsuo: Soil dynamics; (6) Yasuhiko Wakizaka: Active faults, deep underground structures, seismic hazard maps; and (7) Yoshikazu Yamaguchi: Embankment dam.

3.7 RIKEN, Earthquake Disaster Mitigation Research Center

Address: 2465-1 Mikiyama, Miki, Hyogo 673-0433, Japan.
Web site: http://www.miki.riken.go.jp/

The Earthquake Disaster Mitigation Research Center (EDM) was established in January 1998 under the framework of the RIKEN (The Institute of Physical and Chemical Research), and is located in a building in the Mikiyama Forest Park of Hyogo Prefecture. This site was selected with the firm support of the Hyogo Prefectural Government. Based on lessons learned from the 1995 Kobe earthquake, the EDM aims to carry out multidisciplinary research, encompassing the physical, engineering, and social sciences. The main purpose of the EDM is to produce "frontier research on earthquake disaster mitigation for urban regions." The major research activities are performed by three research teams: the disaster process simulation team, the disaster information system team, and the structural performance team.

Researchers include (1) Hiroyuki Kameda: Earthquake engineering, disaster management GIS; (2) Haruo Hayashi: Earthquake disaster psychology, info-communication systems; (3) Fumio Yamazaki: Earthquake engineering, engineering seismology, urban disaster management, stochastic structural mechanics; (4) Tetsuo Kubo: Earthquake engineering, structural engineering, structural dynamics; and (5) Hitoshi Taniguchi: Earthquake engineering, engineering seismology. Their biographical sketches are given in the full report on the Handbook CD.

3.8 Shizuoka Prefectural Government, Observation and Research Office, Disaster Prevention Bureau

Address: 9-6 Ohtecho, Shizuoka 420-0853, Japan

Director: Kunio Ozawa.

3.9 Technology Center of Metropolitan Expressway

Address: 3-10-11 Toranomon, Minato-ku, Tokyo 105-0001, Japan.
Web site: http://www.tecmex.or.jp/ (Japanese only)

80.11 Kazakhstan

The editors received a report of the Kazakh Research and Experimental Design Institute of Aseismic Engineering and Architecture (KazNIISSA), from its Director, Dr. Marat Umarbaevich Ashimbaev, Bld.53, Mynbaev st., 480057, Kazakhstan. This report is archived as a computer-readable file on the attached Handbook CD#2, under the directory of \8011Kazakhstan. We extracted and excerpted the following summary. A review of seismology in Kazakhstan is given in Chapter 79.34, and biographical sketches of some Kazakh engineers and scientists may be found in Appendix 3 of this Handbook.

1. Introduction

The Kazakh Research and Experimental Design Institute of Aseismic Engineering and Architecture (KazNIISSA) was developed from the former State Head Territorial Design Institute "Kazakhsky Promstroiniiproekt," which was founded in 1956. In 1964 the latter Institute was reorganized into the State Design and Research Institute "Kazakhsky Promstroiniiproekt." In 1990, following a decision by the Government of the Republic of Kazakhstan, the Research Sector of this Institute became the independent Institute KazNIISSA.

KazNIISSA is a research center that deals with problems of aseismic building, construction, and architecture development in the Republic of Kazakhstan. Its mission is to develop unified scientific and technical strategies for the government in the field of building/construction development. Its main function is to deal with complex problems of aseismic building/construction, including

- Research on the seismic stability of buildings and construction; analysis of current housing systems in terms of their seismic resistance; developing designs for the strengthening of new buildings and construction and for the reconstruction of buildings/construction damaged as a result of earthquakes or other destructive forces; assessment of the expected extent of damages and expenses incurred in connection with potential earthquakes
- Tests of buildings and construction.
- Elaboration of governmental normative documents on aseismic building/construction development.
- Inspection and certification of buildings and construction for their seismic resistance.
- Planning and undertaking experimental studies of new types of designs for aseismic buildings and construction.
- Inspection and engineering analysis of building/construction damage as a result of earthquakes; preparation of scientifically proven recommendations and designs to be applied during the aftermath of earthquake disasters.

KazNIISSA's personnel consists of 50 employees, including 4 doctors of technical sciences, 9 candidates of technical sciences, and 30 scientific workers and engineers.

2. Main Results and Achievements

The Institute has made considerable progress in both theoretical and experimental studies. Unique experiments were carried out in the Medeu Area in 1966–1967. The vibrations of 20 buildings and other construction due to applied artificial seismic loading were recorded on a seismogram.

A network of engineering and seismometric support stations (ESS) for securing instrumental data on the behaviour of buildings during earthquakes was put into service in 1968 and is still in operation. This network consists of seventeen stations mounted on buildings of various construction types, number of stories and site ground conditions. Only one station in the CIS territory is furnished with strain-measuring equipment to register the strain imposed on buildings during earthquakes.

The Institute personnel have been studying ground liquefaction problems due to destructive earthquakes since 1966. This phenomenon has been investigated in Tashkent (1966), Gazly

INTERNATIONAL HANDBOOK OF EARTHQUAKE AND ENGINEERING SEISMOLOGY, VOLUME 81B ISBN: 0-12-440658-0

(1976 and 1984), Kayrakkum (1983), Zhalanash-Tup (1978), the Carpathians (Romania, Moldavia) (1986), Spitak (1988), and South-Kazakhstan (Zaisan) (1990). In addition, specialists of the Institute have been involved in the study of earthquake aftereffects in Romania, Northern Iran, USA (Loma-Pieta), and Turkey (1999).

The Institute has prepared standards documents that regulate engineering and building development in the Republic of Kazakhstan. Among the most important documents are

- Manual for the design of industrial buildings using reinforced concrete framing in seismic zones
- Recommendations for the calculation and design of construction in seismic zones that is to be built from volumetrical blocks
- Recommendations for the strengthening of frame-building columns in seismic zones
- Recommendations for the design of buildings with seismo-isolated kinematic bases
- Recommendations for the calculation of the reliability of multistoried buildings subject to earthquake load
- Provisions for the arrangement of examination, design, and restoration of buildings in the aftermath of earthquake disasters
- Construction Norms and Rules of the Republic of Kazakhstan (RK CN&R) B.1.2-4-98 "Building Development in Seismic Zones"
- Construction Norms in the Republic of Kazakhstan (RK CN) 5.2.2-7-95 "Housing System Development in Almaty and Adjacent Territories with Regard to Seismic Micro-Zoning"
- Catalogue on Building Damage as a Result of Earthquakes. Earthquake in Zaysan, 1990. (Russian and English)
- Theoretical proposals on upgrading of constructive designs of multistoried frame industrial buildings using reinforced concrete framing in seismic zones (constructive and calculation measures to improve standard models).

3. Current Research

At present, investigations are being carried out to improve construction standards in seismic zones and to introduce refinements in the details of microseismic zoning maps (as a Supplement); methods of assessing buildings subject to seismic loading, having regard to the factual description of building behavior during earthquakes and the variation of seismic load parameters; construction norms and rules for foundations and substructures in special ground conditions.

Biographical sketches of key research investigators, with a selected list of publications, are included in the full report on the Handbook CD.

80.12 Kyrgyzstan

The editors received a report of the Kyrgyz Scientific-Research and Design Institute Construction "KyrgyzNIIPstroitelstva" from its Director, Dr. Seitbek T. Imanbekov, Cholpon-Atinskaya Str. 2, 720571 Bishkek, VPZ, Kyrgyz Republic. This report is archived as a computer-readable file on the attached Handbook CD#2, under the directory of \8012Kyrgyzstan. We extracted and excepted the following summary from the material received. A review of seismology in Kyrgyzstan is given in Chapter 79.35, and biographical sketches of some Kyrgyz engineers and scientists may be found in Appendix 3 of this Handbook.

1. Introduction

The Institute KyrgyzNIIPstroitelstva Gosarhsroiinspekcii, under the Government of the Kyrgyztan Republic, is the main organization of the Republic that deals with seismic-resistant construction.

The Institute was founded in 1981 in relation to the USSR's major Institutes in Moscow in the areas of seismic-resistant construction, reinforced concrete, and bases and foundations (CNIISK by name Kucherenko, NIIGB, and NIOSP by name Gersevanova). It is now located in the territory of the Kyrgyztan Republic.

The Institute includes eight main scientific divisions: (1) Department of Theoretical and Experimental Research in Seismic-Resistant Construction, (2) Department of Seismic-Resistant Construction, (3) Department of Building Inspections, (4) Department of Seismic Protection of Buildings, (5) Department of New Constructions, (6) Department of New Building Materials, (7) Department of Engineering Construction and Environmental Protection, and (8) the Osh Branch of the Institute. There are 60 scientific employees in the Institute, 11 of them Doctors of Technical Sciences. A total of 6 people, working in the Institute, are now studying at the postgraduate and thesis levels.

More than 600 scientific works have been published by the Institute since its initiation (18 years). In addition, 24 standards, codes, and other documents have been developed during this same period. The research and studies of the Institute are applied in the building industry. The Institute constantly cooperates in the areas of seismic-resistant construction and seismic risk evaluation with research institutes of the CIS countries—Russia (CNIISK by name Kucherenko, Scientific and Technical Center on Seismic Resistance Construction and Engineering Protection from Natural Hazards), Kazakhstan (KazNIISA), Uzbekistan (TashZNIIEP), and others.

The Institute participates in development of projects financed by international organizations—ISTC, UN, and NATO—in the field of seismic risk reduction.

The Institute consists of an administrative building and an experimental workshop with a traveling crane and a laboratory space equipped with force actuators for applying loads and instruments for detecting deformations and motions. The Institute has a special test facility for the dynamic testing of real buildings and structures under seismic design loads.

2. Major Developments

National codes on earthquake engineering have been developed by the Institute. These include (1) SNiP 2.01.02-94 KR (Building Code) for construction in the territory under an expected seismic intensity greater than 9; (2) SNiP 31.01-95 KR (Building Code) for changing the functions of existing residential buildings; (3) SNiP 2.01.01.93 KR (Building Code) for construction on the soil–geological territory conditions of Bischkek; (4) SNiP 13-01-98 KR (Building Code) for instructions on engineering

INTERNATIONAL HANDBOOK OF EARTHQUAKE AND ENGINEERING SEISMOLOGY, VOLUME 81B ISBN: 0-12-440658-0

inspections and the definition of deterioration of engineering buildings; (5) SNiP 22-01-98 KR (Building Code) for estimating earthquake-resistant capability of existent buildings; (6) SNiP KR 31.01-99 (Building Code) for functional changes to areas of existing apartment buildings; (7) SNiP 10-01-99 (Building Code) for standard principles of normal construction practice; (8) RDC 31-01-99 (Building Code) for engineering inspection of buildings and construction or reconstruction in the territory of Kyrgyz Republic; (8) Temporal recommendations for building in the territory contiguous to the Issyk-Ata fault in Bishkek city; (9) Research of seismic loads, taking into account regional seismological conditions.

The collection of papers of the Institute include (1) Collection of papers KyrgyzNIIPS. Bishkek, 1993; (2) Collection of papers KyrgyzNIIPS 1994–1995, Bishkek, Ilim, 1995; (3) Collection of papers KyrgyzNIIPS 1996–1997, Bishkek, Ilim, 1997; and (4) Collection of papers KyrgyzNIIPS 1998–1999, Bishkek, Ilim, 1999.

3. Current Activities

The Institute is currently conducting research in the following areas: (1) development of the theoretical bases of seismic-resistant construction and evaluation of seismic risk, (2) experimental research on building construction and products, (3) development and perfecting of standard norms of rules, recommendations, and manuals of seismic-resistant building construction, (4) introduction of innovative construction and seismic isolation in the practice of construction, (5) development of new building materials and constructions, (6) engineering inspection and evaluation of seismic resistance of existing buildings, (7) mitigation of and other natural hazards earthquake consequences, (8) reliability of engineering systems.

Biographical sketches of Seitbek T. Imanbekov, Svetlana K. Uranova, and Ulugbek T. Begaliev are given in the full report on the Handbook CD.

80.13 Macedonia

We received a report of the Institute of Earthquake Engineering and Engineering Seismology (IZIIS), University St. Cyril and Methodius, Skopje, Macedonia, from its director, Prof. Kosta Talaganov. This report is archived as a computer-readable file on the attached Handbook CD#2, under the directory of \8013Macedonia. A review of seismology in Macedonia is given in Chapter 79.37, and biographical sketches of some Macedonian engineers and scientists may be found in Appendix 3 of this Handbook.

1. Introduction

The address of IZIIS is Salvador Aliende Str. 73, P.O. Box 101, 91000 Skopje, Macedonia. Prof. Kosta Talaganov is the Director, and Prof. Mihail Garevski is the Deputy Director.

The Institute of Earthquake Engineering and Engineering Seismology (IZIIS) was established in 1965 as an institution within the University St. Cyril and Methodius to organize research and training in the area of earthquake engineering and engineering seismology. In following these tasks, the Institute received support from the United Nations and its specialized agencies UNDP and UNESCO for many years following the disastrous Skopje earthquake of July 26, 1963. While meeting the immediate needs of reconstruction of the city, the Institute created conditions favorable for permanent progress in research and training of scientific staff and engineers.

The Institute is organized in specialized research sections, and covers the following research activities: regional studies (seismology, seismotectonics, and geophysics); local soil studies and soil dynamics; vulnerability analysis; seismic stability of engineering structures; seismic stability of building structures and engineering materials; nuclear engineering and reliability of structures; control engineering; dynamic testing in the laboratory; and informatics and computer science.

Out of the 56 scientific staff members, 15 have Ph.D. degrees and 23 have Master's degrees in technical sciences, while the remaining 18 staff members are graduate civil engineers, geologists, electrical engineers, physicists, economists, and philologists. Fifteen professors, associate professors, and assistant professors from the Institute are lecturing and supervising the post-graduate studies. Most of the scientific staff of the Institute have, at one time or the other, worked at the universities in the United States, Japan, and other countries, as visiting professors or research fellows, or as UN experts in earthquake engineering or engineering seismology in developing countries.

2. Training and Research Activities

Economic development and the threat of catastrophic earthquakes have resulted in an increasing demand for specialists in earthquake engineering and engineering seismology at the master's and doctoral degree levels. IZIIS, as an International Training Center, organizes post-graduate studies in various branches and at various levels. During its 30-year experience of training earthquake engineers and specialists in engineering seismology, courses have been organized for 250 candidates for the Master's degree in technical sciences. In addition, the Institute organizes special courses for engineers from Macedonia and developing countries.

The scientific research activities of the Institute are aimed basically at defining the technical basis for earthquake mitigation reduction. A large number of projects have been carried out on the following subjects: (1) study of strong earthquake occurrence, definition and improvement of the methods of earthquake risk assessment, study of the seismicity and evaluation of the risk of future earthquake damage; (2) determination of the dynamic properties of materials and structures, in order to establish consistent scientific criteria for the stability of civil engineering structures under various dynamic effects; (3) establishment of economic and technical criteria for evaluating the consequences of strong earthquakes and optimizing the economic value of

INTERNATIONAL HANDBOOK OF EARTHQUAKE AND ENGINEERING SEISMOLOGY, VOLUME 81B ISBN: 0-12-440658-0

structures with respect to earthquake intensity, frequency of occurrence, expected lifetime and purpose; (4) development of technology for economic and rational construction using industrialized construction methods and systems, adequate materials, pre-cast elements and structural systems; and (5) establishment of the scientific and technical basis for updating and improving the Construction Code, compatible with the level of economic development of the country.

3. Research Sections

3.1 Natural and Technological Hazards and Geotechnics

Research focuses on geophysical, geotechnical, and geomechanical investigations; earthquake geotechnical engineering; geotechnical hazards, risk, preventive structures and measures; geotechnical and seismic monitoring; and foundations for structures. Research staff members include Prof. Dr. Vladimir Mihailov, Head of section, and Research Assistants Dusan Aleksovski, Gavril Mirakovski, and Vlatko Sesov.

3.2 Building Engineering and Materials

Research focuses on structural safety of buildings, building rehabilitation, building materials, historic buildings, new technologies in building engineering, and vulnerability of building structures. Research staff members include Prof. Predrag Gavrilovic, Head of section; Assoc. Prof.: Golubka Necevska Cvetanovska; Scientific Advisor: Dr. Nikola Nocevski; Assist. Prof.: Zivko Bozinovski and Veronika Sendova; and Research Assistants: Blagojce Stojanoski, and Roberta Petrusevska.

3.3 Engineering Structures and Systems Engineering

Research focuses on transportation facilities, dams and special engineering structures, new technologies in the field of engineering structures, seismic vibration control, seismic vulnerability control and systems engineering, structural diagnosis, and rehabilitation of structures. Research staff members include Prof. Danilo Ristic, Head of section; Assist. Prof.: Vlado Micov and Viktor Hristovski; and Research Assistants: Nikola Sesov, and Nikola Zisi.

3.4 Risk and Disaster Management

Research focuses on natural hazards and risk studies, measures for risk reduction, insurance policies, preparedness planning, emergency management, socio-economic impacts, search and rescue techniques, and vulnerability and risk of engineering and social lifeline systems. Research staff members include Prof. Zoran Milutinovic, Head of section, and Research Assistants Goran Trendafilovski and Tatjana Olumceva.

3.5 Informatics

Research focuses on engineering software development and application, database development, computer networking, and development of expert systems. Research staff members include Eng. Slobodan Micajkov, Head of section, and Research Assistants Nikola Hadzitosev and Andrej Sendov.

3.6 Dynamic Testing Laboratory

Research focuses on experimental mechanics, development of experimental methods and techniques, dynamic behavior under earthquake, wind, and blast, new technologies for construction, construction materials, and transmission lines and special bridge structures. Research staff members include Prof. Ljubomir Taskov, Head of section; Prof. Dimitar Jurukovski; and Research Assistants: Lidija Krstevska and Zoran Rakicevic.

3.7 Geology

Research focuses on geological, neotectonic, and seismotectonic investigations. Research staff members include Scientific Advisor: Dr. Radojko Petkovski, Head of section, and Research Assistants Miodrag Manic, Lenka Timioska, and Biserka Dimiskovska.

3.8 Industrial and Energetic Systems

Research focuses on analysis of industrial systems, nuclear and power plants analysis, reliability studies, new technologies in construction, engineering materials, and dynamic reliability of mechanical and electrical components. Research staff members include Assoc. Prof. Dimitar Petrovski, Head of section, Assist. Prof. Snezana Stamatovska, and Research Assistants Violeta Mircevska and Aleksandar Paskalov.

4. Coordinated and Other Activities

Coordinated activities include multidisciplinary projects, quality assurance systems, and standards and codes evaluation and recommendations. Details, as well as laboratory equipment, postgraduate studies, and biographical sketches of researchers, are given in the full report on the Handbook CD.

80.14 New Zealand

The editors received a summary of the New Zealand Society for Earthquake Engineering from its President, David Brunsdon. It is presented here. A review of seismology in New Zealand is given in Chapter 79.40, and biographical sketches of some New Zealand engineers and scientists may be found in Appendix 3 of this Handbook.

1. Background

The New Zealand Society for Earthquake Engineering (NZSEE) was formed in April 1968. The Society's formation and objectives originated from the 3rd World Conference on Earthquake Engineering, held in New Zealand in 1965. In its first year, the Society launched its Bulletin and three research and study groups, plus a reconnaissance team—a level of activity that set the scene for the years ahead! The mission of the society is *to gather, shape, and apply knowledge to reduce the impact of earthquakes on our communities.*

The Society's formal objectives are

- To foster the advancement of the science and practice of earthquake engineering across all disciplines, and
- To promote cooperation among scientists, engineers, and other professionals in the broad field of earthquake engineering through interchange of knowledge, ideas, results of research, and practical experience.

Membership of the Society includes professional engineers, scientists, insurers, emergency managers, and others having an interest in earthquake phenomena or in the effects of earthquakes. The Society currently has approximately 640 members. This is a slight decrease from the peak level of 660 in 1997. About 150 of these members are based overseas, and a further 45 are student members.

2. Principal Activity Areas

Following a comprehensive strategic planning process, the work of NZSEE has been structured under nine principal activity areas, as follows:

- Membership
- Communications
- Publications
- Promotion of research & development
- Conferences
- Learning from earthquakes
- International involvement
- Promoting earthquake mitigation
- Promoting earthquake preparedness.

Current and future activity highlights include communications, conferences, and promoting earthquake mitigation.

Communications—the quarterly production of the Society's Bulletin. Papers in the Bulletin are peer reviewed and cover the full spectrum from seismology and geology to applied earthquake engineering. The Society has just commenced production of an electronic newsletter to members that provides topical news between the Bulletin issues.

Conferences—In addition to running its annual conferences every March, the Society hosted the 7th Pacific Conference on Earthquake Engineering in Christchurch February 13–15, 2003.

Promoting earthquake mitigation—The Society is currently running Technical Project Groups (or Committees) on the following topics: (1) assessment and improvement of the structural performance of earthquake risk buildings, (2) seismic design of storage tanks, and (3) building in close proximity to active faults.

The output from each of these groups will be a best practice guideline document for practitioners. Information on previous project reports can be obtained from the Secretary.

Please visit the NZSEE Web site for more information at: http://www.nzsee.org.nz/.

INTERNATIONAL HANDBOOK OF EARTHQUAKE AND ENGINEERING SEISMOLOGY, VOLUME 81B ISBN: 0-12-440658-0

80.15 Russia

The editors received the "Report of the Russian National Committee for Earthquake Engineering to IAEE" from its Chairman, Prof. J. Eisenberg, TsNIISK, 6, 2nd Institutsky Street, Moscow, 109428, Russia. This report is archived as a computer-readable file on the attached Handbook CD#2, under the directory of \8015Russia. We extracted and excerpted a summary of this material, as shown here.

A review of seismology in Russia is given in Chapter 79.45 of this Handbook. Biographies and biographical sketches of some Russian engineers and scientists may be found in Chapter 89, and in Appendix 3 of this Handbook, respectively.

1. Research Institutions in Russia and in Some Former USSR Countries—1999

1.1 Earthquake Engineering Research Center (EERC), TsNIISK, State Construction Committee, Russian Federation

Prof. J.M. Eisenberg, Director
Specialties: Seismic risk and optimal design, seismoisolation, code development

Dr. V.I. Smirnov, Vice-Director
Specialties: Seismoisolation, dynamic tests, nonlinear analysis

Dr. A.M. Melentyev, Head of Department
Specialties: Seismic risk assessment, field earthquake studies, repair and retrofit of structures

1.2 Transportation University

Prof. Dr. T.A. Belash, Department Head
Specialties: Seismoisolation

Prof. Dr. A.M. Uzdin
Specialties: Transportation structures earthquake reliability

1.3 Soil and Foundation Research Institute, State Construction Committee

Prof. Dr. V.A. Ilyichev, Director
Specialties: Soil–structure interaction

Prof. Dr. L.R. Stavnitser, Head, Laboratory
Specialties: Soil and foundation earthquake design

1.4 Moscow State Construction University

Prof. Dr. Yu.T. Chernov
Specialties: Dynamic analysis

Prof. Dr. A.V. Zabegaev
Specialties: RC elements earthquake resistance

1.5 Reinforced Concrete Research Institute

Prof. Dr. A.S. Zalesov, Head, Laboratory
Specialties: RC elements inelastic behavior

1.6 Center on Earthquake Engineering and Natural Disaster Reduction (CENDR), State Construction Committee, Russian Federation

Dr. M.A. Klyachko, Director
Specialties: Seismic risk; analysis earthquake scenario
[Editor's note: A report from CENDR was submitted to IASPEI, and it is archived as a computer-readable file on the attached Handbook CD, under the directory of \8015Russia.]

INTERNATIONAL HANDBOOK OF EARTHQUAKE AND ENGINEERING SEISMOLOGY, VOLUME 81B ISBN: 0-12-440658-0

1.7 Research Center, Ministry of Defense, Russian Federation

Prof. Dr. V.S. Belyaev, Head of Center
Specialties: Seismoisolation of nuclear facilities, shake table tests

1.8 Russian Association for Earthquake Engineering, Dynamic, Natural and Man-Made Loads Protection

Prof. Dr. J.M. Eisenberg, President
Prof. Dr. V.S. Belyaev, Vice-President
Dr. A. M. Melentyev, Executive Director
Dr. V.I. Smirnov, Member of Executive Committee

1.9 Research Institute of Lithosphere, Russian Academy of Sciences

Prof. Dr. G.L. Koff, Head of Laboratory
Specialties: Seismic microzonation, seismic risk analysis

1.10 Moscow Elastic Energy University

Prof. Dr. V.V. Bolotin, Head, Department
Specialties: Probabilistic methods in earthquake engineering safety of structures

1.11 Armenian National Academy of Sciences

B.K. Karapetyan, Executive Secretary
E.E. Khachiyan, Head of Department
M.G. Melkumyan, Vice Director, National Seismic Survey

1.12 Research Institute for Structural Mechanics and Seismic Stability, Georgian Academy of Sciences

G.K. Gabrichidze, Director

1.13 Kirgiz Laboratory of Earthquake Engineering, Ministry of Construction

Dr. S.K. Uranova, Head of Laboratory

1.14 Uzbekistan Academy of Sciences

T.R. Rashidov, Director of Research Institute Tashkent, Uzbekistan

2. Earthquake Engineering Research Center (EERC), Central Research Institute for Building Structures (TsNIISK)

TsNIISK has existed since 1929. The Earthquake Engineering Research Center dates from the same time, under different names (Department, Laboratory, or Center). EERC TsNIISK is officially recognized as the leading institution in the area of earthquake engineering research and Seismic Building Codes development. The first Soviet Seismic Building Code was adopted as a state law in 1941. The most active scientists in the field of earthquake engineering in Russia and in other former USSR republics at the earlier time were A. Zshoher, V. Pildish, V.A. Bykhovsky, and I.I. Goldenblat (Russia); K.S. Zavriev (Georgia); and A.G. Nazarov (Armenia).

In 1957, the dynamic standard response spectra design method was incorporated in the USSR code (even earlier than in the California Code) in the form suggested by I.L. Korchinsky of the EERC. Powerful exciters and shaking tables were constructed in EERC in the 1960s and 1970s. S.V. Polyakov was among the experts who initiated the RC prefabricated large panel wall building investigations and implementation in the USSR.

J.M. Eisenberg pioneered the research and implementation of seismoisolation—in particular, adaptive seismoisolation—in USSR and Russia, beginning in the early 1970s. Hundreds of seismoisolated buildings of different types have been constructed in Russia and the former USSR. Eisenberg also developed new probabilistic mathematical models of seismic inputs, which take into account the uncertainties of the seismological information.

Prominent contributions have been made by other experts of the EERC: I.I. Goldenblat, N.A. Nicolaenko, M.P. Barshein, A.M. Zharov, B.A. Kirikov, and many others.

Prominent achievements of EERC include (1) development and implementation of the dynamic standard response spectra method by I.L. Korchinsky (1905–1993) into the Russian Seismic Building Code, (2) development and implementation in Russia and the USSR of RC prefabricated large panel highly earthquake-proof buildings by S.V. Polyakov (1918–1992), (3) philosophy of aseismic design, probability approaches, and nonlinear analysis by I.I. Goldenblat (1907–1990), (4) spatial structural dynamic analysis by N.A. Nikolaenko (1929–1988), and (5) adaptive seismoisolation, simple and low-cost, investigations, development and implementation, fuse-elements seismoisolation, new mathematical models of seismic inputs, shaking table and exciter tests by J.M. Eisenberg.

More details, including collaborations and a biographical sketch of Jacob M. Eisenberg, are given in the full report on the Handbook CD.

3. Russian National Committee for Earthquake Engineering

The Russian National Committee for Earthquake Engineering (RUNCEE) was established in 1972 as the Soviet National Committee in Earthquake Engineering. The main goal of the Committee is the participation in the International Association for Earthquake Engineering (IAEE) and in the European Association for Earthquake Engineering (EAEE) activity in the organization of international and regional conferences, seminars, workshops, exhibitions, as well as other forms of international cooperation in earthquake engineering.

The Chairman and Soviet National Delegate for 15 years, from 1972 to 1987, was S.V. Polyakov. The second Chairman was N.N. Skladnev, from 1987 to 1994. J.M. Eisenberg has been the Chairman of RUNCEE from 1994 to the present. Soviet and Russian directors of the IAEE from the period 1972 to the present were K.S. Zavriev, S.V. Polyakov, A.I. Martemyanov, I.N. Burgman, N.N. Skladnev, and J.M. Eisenberg.

N.N. Skladnev was the President of EAEE from 1990 to 1994, and V.S. Polyakov and J.M. Eisenberg were Vice-Presidents of that organization between 1994 to 1998. The IX European Conference in Earthquake Engineering was held in Moscow in 1990 and was organized by RUNCEE and EAEE. RUNCEE issues a Journal of Earthquake Engineering six times a year.

The Executive Bureau of RUNCEE consists of J.M. Eisenberg, Chairman and National Delegate IAEE and EAEE; V.S. Beliaev, Vice-Chairman and Deputy National Delegate for EAEE; V.I. Ilyichev, Vice-Chairman; V.I. Smirnov, Vice-Chairman; A.M. Melentyev, Secretary General; V.P. Abarykov, Member; M.A. Klyachko, Member and Deputy National Delegate for IAEE; and G.S. Shestoperov, Member.

More details about RUNCEE, including biographical sketches of V.V. Bolotin and O. A. Savinov, are given in the full report on the Handbook CD.

4. Georgian Academy of Sciences

In 1920, a destructive earthquake occurred in Kartli, Georgia. A team of Georgian engineers and geologists participated in the studies of the consequences of that earthquake. The results of that investigation were summarized in L. Koniushevsky's monograph "Kartli earthquake in February 20, 1920." Systematic research in the field of structures in Georgia started at this time, and primarily focused on the seismic resistance of structures. This work was carried out under the guidance of Kiriak Zavriev, a young engineer trained in Russia.

Zavriev developed the basic principles of earthquake-resistant dynamic theory in the following monographs: "Earthquake resistance dynamic theory" (1936), "Seismic resistance theory" (1937), "Dynamics of structures" (1946), and "Stability and dynamics of structures" (1959, with G. Kartsivadze). These were written in collaboration with co-authors, among them A.G. Nazarov. In the 1930s, under the guidance of Zavriev, who supervised the studies in Transcaucasia, the first Soviet Building Codes for design in seismic regions, as well as many other regulatory documents, including specifications and codes for design of reinforced concrete and steel structures, were developed.

In 1941 the Bureau of Antiseismic Structures was created by the Academy of Sciences of Georgia under Zavriev's guidance. It was reorganized as the Institute of Construction Works in the 1947. Since 1962, it has been called the "Institute of Structural Mechanics and Earthquake Engineering." The Institute has united prominent experts in the sphere of earthquake engineering. It soon become the center of research in the field of seismic-resistant structures throughout the territory of the former USSR.

During their years at the Institute, Sh. Djabua and A. Churaian authored many monographs: "An album of details on seismic structures of dwelling houses" (1950), "Some problems of precast reinforced concrete application in earthquake engineering" (1956), "Panel connection in vertical joints of seismic-resistant large-panel buildings" (1968) and more. Sh. G. Napetvaridze studied seismic resistance of structures and published "The problems of seismic resistance theory" (1956). G. Kartsivadze published the book "Seismic resistance of man-made transportation structures" (1974).

The researchers of the Institute are active in international projects. More details, including biographical sketches of Kiriak S. Zavriev (1891–1978), and Armen G. Nazarov (1908–1983), are given in the full report on the Handbook CD. See also Chapter 80.8.

5. Uzbekistan Academy of Sciences

Studies of engineering seismology and earthquake engineering began in Uzbekistan in the 1940s. In 1947 the Uzbekistan Academy of Sciences was founded, headed by M.T. Urazbaev.

The basic principles of earthquake engineering were developed for important project designs, and a number of recommendations, regulations, and codes were introduced. This direction of activity was intensified after an earthquake in 1946. A number of "aseismic" scientists have been working in Uzbekistan at the Uzbek School of Seismic Construction since 1966 (M.T. Urazbaev, V.T. Rasskazovsky, Yu.R. Leiderman, V.K. Kabulov, T.R. Rashidov, and others).

The Tashkent earthquake of 1966 served as a new challenge for the development of the Tashkent School. At present a great number of researchers are working in the Institute of Mechanics and Seismic Stability of Structures, Institute of Seismology, Tashkent Architecture Institute, and other institutions.

The principal results in the field of earthquake engineering in Uzbekistan are (1) analysis of the aftermath of strong ground motions, including the Tashkent earthquakes of 1946 and 1966.

(On the basis of this analysis, a monograph was published containing recommendations for the new city-buildings in Tashkent); (2) development of physical methods of design of structures on seismic stability, as well as recommendations for building regulations in regions having high seismicity; (3) formulation of a new seismodynamic theory of underground structures, serving as the basis of the earthquake-resistant design of underground structures. Tashkent Metro, and Metros in other cities were built on the basis of this method, as well as various communication systems in seismic regions. Experiments and tests involving spatial structures and communication systems are unique and important. Results of these works are included in regulation codes; (4) implementation of a network of seismometric stations on unique buildings and structures, serving as a constant laboratory; (5) building of new types of buildings on rubber-metal supports; and (6) inclusion of results of the research for Tashkent city by the Tashkent School in the International project RADIUS, Secretariat IDNDR of UN on seismic risk.

More details are given in the full report on the Handbook CD. See also Chapter 80.17.

80.16 United States of America

The editors assembled a centennial report on earthquake engineering for the United States of America with the help of Susan Tubbesing, Executive Director of the Earthquake Engineering Research Institute (EERI). About a dozen major institutions and organizations engaging in earthquake engineering research or its promotion and publication in the United States were invited to submit reviews of their histories and activities. About half of the invited institutions responded. Their reviews are archived as computer-readable files in the attached Handbook CD#2, under the directory of \8016USA. We extracted and excerpted the following summary from these reports.

Seismology in the United States is reviewed in Chapter 79.55. Biographies and biographical sketches of some American engineers and scientists may be found in Chapter 89, and in Appendix 3 of this Handbook, respectively.

1. Introduction

Early earthquake engineering research in the United States has been reviewed by George Housner in Chapter 2 of this Handbook. The Earthquake Engineering Research Institute (EERI) is the US national member of IAEE. It has over 2500 individual members and maintains an excellent website (http://www.eeri.org/) with current information of the profession. There are many institutions in the United States that are engaged in earthquake engineering research, or that provide information on this subject. Most of them belong to regional centers, and/or the Consortium of Universities for Research in Earthquake Engineering (CUREE). We compiled the following list with Web site address from an Internet search.

Applied Technology Council (http://www.atcouncil.org/)
California Geological Survey (http://www.consrv.ca.gov/cgs/)
California Seismic Safety Commission (http://www.seismic.ca.gov/)
Center for Earthquake Research and Information (CERI), University of Memphis (http://www.ceri.memphis.edu/)
Consortium of Organizations for Strong-Motion Observation Systems (http://www.cosmos-eq.org/)
Consortium of Universities for Research in Earthquake Engineering (CUREE) (http://www.curee.org/)
Earthquake Engineering Center for the Southeastern United States (ECSUS), Virginia Tech (http://ecsus.ce.vt.edu/)
Earthquake Engineering Research Center (EERC), University of California, Berkeley (http://eerc.berkeley.edu/)
Earthquake Engineering Research Institute (EERI) (http://www.eeri.org/)
Earthquake Engineering Research Laboratory, California Institute of Technology (http://www.eerl.caltech.edu/)
Federal Emergency Management Agency (FEMA) (http://www.fema.gov/)
John A. Blume Earthquake Engineering Center, Stanford University (http://blume.stanford.edu/index.html)
Mid-America Earthquake Center (MAE), University of Illinois (http://mae.ce.uiuc.edu/)
Multidisciplinary Center for Earthquake Engineering Research (MCEER), SUNY Buffalo (http://mceer.buffalo.edu/)
National Information Service for Earthquake Engineering (http://www.nisee.org/)
National Institute of Standards and Technology (http://www.nist.gov/)
Pacific Earthquake Engineering Research Center (PEER) (http://peer.berkeley.edu/)
Southern California Earthquake Center (http://www.scec.org/)
UC Berkeley NEES Site (http://nees.berkeley.edu/)
U.S. Geological Survey, Earthquake Hazards Program (http://earthquake.usgs.gov/)

In addition, we extracted and excerpted a summary of the reports received from these organizations in the following sections.

INTERNATIONAL HANDBOOK OF EARTHQUAKE AND ENGINEERING SEISMOLOGY, VOLUME 81B ISBN: 0-12-440658-0

2. Applied Technology Council (A.G. Brady)

Web site: http://www.atcouncil.org/

The Applied Technology Council (ATC) is a nonprofit corporation founded to protect life and property through the advancement of science and engineering technology. With a focus on seismic engineering, and a growing involvement in wind and coastal engineering, ATC's mission is to develop state-of-the-art, user-friendly resources and engineering applications to mitigate the effects of natural and other hazards on the built environment. ATC fulfills a unique role in funded information transfer by developing nonproprietary consensus opinions on structural engineering issues. ATC also identifies and encourages needed research and disseminates its technological developments through guidelines and manuals, seminars, workshops, forums, and electronic media, including an interactive Web site and other emerging technologies.

Since its inception in the early 1970s, the Applied Technology Council has developed numerous highly respected, award-winning, technical reports that have dramatically influenced structural engineering practice. Please refer to the Applied Technology Council on the accompanying Handbook CD for a complete list of ATC reports and other products.

With offices in California and Washington, DC, ATC's corporate personnel include an executive director, senior-level project managers and administrators, and technical and administrative support staff. The organization is guided by a distinguished Board of Directors comprised of representatives appointed by the American Society of Civil Engineers, the National Council of Structural Engineers Associations, the Structural Engineers Association of California, the Western Council of Structural Engineers Associations, and four at-large representatives. Projects are performed by a wide range of highly qualified consulting specialists from professional practice, academia, and research—a unique approach that enables ATC to assemble the nation's leading specialists to solve technical problems in structural engineering. Funding for ATC projects is obtained through government agencies and from the private sector in the form of tax-deductible contributions.

3. California Institute of Technology: Earthquake Engineering (W.D. Iwan)

Web site: http://www.eerl.caltech.edu/

Earthquake engineering at Caltech has a long and rich tradition. From the initial deployment of strong motion instruments to the introduction of seismic design codes and the drafting of landmark legislation creating the Federal Earthquake Hazard Mitigation Program, the Caltech earthquake engineering faculty has played a leadership role. The current faculty and students are building on this heritage.

Current research by faculty and students in the earthquake engineering group is focused on such areas as understanding the nature and distribution of strong ground motion resulting from nearby earthquakes, understanding the seismic performance of a wide spectrum of structural systems ranging from dams to wood-frame buildings, developing techniques to improve seismic decision making in the presence of large uncertainties, developing simplified methods of performance-based engineering analysis and design, and studying new approaches to active control of civil structures. In addition to its involvement in basic earthquake engineering research, the faculty continues to be active in the development of seismic safety policy at both the local and federal levels.

The Earthquake Engineering Research Laboratory and associated Albert Niu Lin Laboratory of Structural Engineering provide specialized equipment and instrumentation for a wide range of earthquake studies. Available laboratory equipment includes a geotechnical centrifuge with a built-in shake table and both electro-magnetic and hydraulic shake tables for structural model testing. In addition, the laboratory maintains a set of rotating eccentric mass building shakers that have been used extensively in field-testing of full-scale buildings, dams, and bridges.

The nine-story Millikan Library building on campus provides a unique test bed for structural, geotechnical, and seismological research. One of the most extensively instrumented and studied buildings in the US, this structure is currently being outfitted with a state-of-the-art real-time structural health and damage monitoring system. This system is being designed as a prototype system that can be installed in structures throughout the world. The motion of the building can be continuously monitored in real time from any location of the world over the Internet.

The Caltech earthquake engineering group also maintains a branch of the National Information Service for earthquake engineering (NISEE). This resource center contains a wealth of information on Earthquake Engineering and related subjects. Included are an extensive collection of books and reports, as well as a large number of pictures and slides of earthquake damage. The Service also supports the Caltech Strong Motion Center, which archives important strong motion data and distributes a simple accelerograph data analysis package for use in earthquake engineering education and by professionals.

A complete list of published reports of earthquake engineering and related laboratories is given in the full report on the Handbook CD. For more current and detailed information, please visit the Web site at: http://www.eerl.caltech.edu/.

4. Consortium of Organizations for Strong-Motion Observation Systems (J.C. Stepp)

Web site: http://www.cosmos-eq.org/

Recognizing that the development of public policies and specific actions to protect public safety in earthquakes depend heavily on the timely availability of accurate measurements of the strong ground shaking on the ground as well as measurements of the response of structures to strong shaking, the Consortium of Organizations for Strong Motion Observation Systems (COSMOS) was organized in 1999 for the purpose of coordinating strong motion observation networks and advancing the use of strong motion measurements in research and practice. A fundamental principal on which COSMOS was founded is that member strong motion programs may, through COSMOS, benefit from other members' programs to address mutual strong motion data acquisition and dissemination issues, advance recording and instrument technologies, improve maintenance techniques, develop data archiving and dissemination technologies, and maintain ongoing communication with strong motion data users. COSMOS is international in scope, membership being open to all organizations that operate strong motion observation networks or have strong motion databases, to academic institutions and researchers, to government agencies that use strong motion data, and to organizations and individuals engaged in earthquake safety practice.

Members of COSMOS provide financial support for the Organization's programs, and work groups constituted of members implement the programs. Current programs fall into three groupings: coordination of member strong motion programs for the purpose of advancing strong motion measurements, development of guidelines and standards, and development and management of a strong motion data archiving and dissemination infrastructure. Coordination is facilitated through ongoing communication among programs through standing committees, active work groups, and an annual meeting. The development of guidelines and standards is achieved by work groups through the process of assimilating the appropriate technical basis, preparing a draft guideline or standard, and obtaining broad review by COSMOS member organizations and individual members, and resolution of review comments.

The current flagship activity of COSMOS is the development and operation of a virtual data dissemination system for strong motion data—the COSMOS Strong-Motion Virtual Data System <http://db.cosmos-eq.org>. The goal of the VDC is to make strong motion data and derivative products available at one access location with sufficient efficiency that the System can serve as an effective resource for earthquake safety practitioners as well as for emergency response and recovery following damaging earthquakes. The VDC links database archives of member strong motion programs and, as made available, holds strong motion data from observation networks that do not maintain database archives, directly in the VDC server. This architecture permits efficient dissemination of strong motion data through the Internet from multiple data archives by accessing a single URL. All strong motion data users throughout the world can freely access the VDC through the Internet.

More information about COSMOS and a list of publications are given in the full report on the Handbook CD. For more current information about COSMOS, please visit its Web site at http://www.cosmos-eq.org/

5. Consortium of Universities for Research in Earthquake Engineering (R.K. Reitherman)

Web site: http://www.curee.org/

California Universities for Research in Earthquake Engineering (CUREe) is a nonprofit organization, incorporated in 1988, made up of professors and researchers at eight California universities, as well as other individual members from academia, industry, and the professions that relate to earthquakes. The institutional members of CUREe are California Institute of Technology, Stanford University, University of California (Berkeley, Davis, Irvine, Los Angeles, and San Diego), and University of Southern California.

CUREe's purposes are (1) identifying new ways research can solve earthquake problems by collecting and synthesizing information and making it easily accessible, (2) establishing national and international hazard research relationships, (3) performing earthquake engineering and related research, (4) managing research consortia and cooperative programs, and (5) educating experts, practitioners, students, and the public.

Typical projects involve several universities, researchers, and students, emphasizing engineering along with contributions from other areas of expertise that are relevant to particular project topics, such as earth science, planning, and architecture. Involvement of practicing engineers in producing and implementing research are key CUREe goals. Representative current projects include

- The SAC Joint Venture (SEAOC-ATC-CUREe), a FEMA-funded effort to solve the problem of fractures in welded steel moment frames, which surfaced in the 1994 Northridge earthquake. This large and multi-year (1994–1998) cooperative research project is producing revised design guidelines and supporting information for steel structure design.
- The CUREe-Caltech Woodframe Project, a three-year project devoted to improving the performance of woodframe buildings in earthquakes, including a program of focused testing and analysis, investigations of buildings that have undergone strong earthquakes (especially the 1994 Northridge Earthquake), translation of the project's findings into codes and standards, consideration of economic aspects such as insurance, and an education and outreach component.
- Joint US–Japan research with Kajima Corporation on a variety of advanced research topics, currently in its fourth phase and tenth year.
- Activities of the US Panel on Structural Control, including the First and Second World Conferences on Structural Control, funded by NSF.
- Carrying out for the National Earthquake Hazards Reduction Program a national conference and comprehensive compendium of research results relating to the 1994 Northridge Earthquake.

- China–US Bilateral Workshop on Seismic Codes, funded by NSF.
- A series of symposia co-sponsored with universities honoring lifelong contributions in earthquake engineering (Caltech-CUREe Housner Symposium, 1995; UC Berkeley EERC-CUREe Bertero Symposium, 1997, Stanford-CUREe Shah Symposium 1997, and in cooperation with Japanese sponsors, the 2000 Kobori Symposium).

6. Earthquake Engineering Research Institute (S.K. Tubbesing)

Web site: http://www.eeri.org/

The Earthquake Engineering Research Institute (EERI), founded in 1949, is an international, nonprofit, professional association comprised of more than 2,500 engineers, geoscientists, building officials, architects, planners, social scientists, and others actively working in the earthquake hazard reduction field. EERI's main objective is to reduce earthquake risk by advancing the science and practice of earthquake engineering, improving understanding of the impact of earthquakes on the physical, social, economic, political, and cultural environment, and by advocating comprehensive and realistic measures for reducing the harmful effects of earthquakes. Today, as the field of earthquake hazard mitigation expands, EERI takes on tasks that reflect its unique interdisciplinary membership: fostering communication between different disciplines and bridging the gap between new knowledge, design practice, and risk reduction policies.

EERI is probably best known for its field investigations and reports of the effects of destructive earthquakes. With NSF support, EERI has coordinated hundreds of post-earthquake reconnaissance investigations to maximize learning from destructive earthquakes. Under this project, EERI has published reconnaissance reports identifying lessons learned in earthquakes throughout the world.

Much of the work of EERI is conducted under the auspices of two dozen technical committees. The committees plan seminars, annual meetings, national and specialty conferences. They direct the projects of the Endowment Fund, review student papers, select fellowship recipients, oversee the Institute's electronic and traditional publications, raise funds, and prepare statements of policy.

EERI sponsors major technical conferences to provide a forum for the exchange of information between researchers in diverse but related disciplines. In addition, every four years EERI organizes the US National Conference on Earthquake Engineering. EERI Annual Meetings provide structured programs on current issues and informal exchanges between researchers, practitioners, and government policy makers. Invited speakers present research results, share their practical ideas and experiences, discuss hazards reduction policy, and assess the effects of recent earthquakes. EERI sponsors technical seminars throughout the country. These seminars inform practitioners of the most recent advances in research and practice.

In the fall of 1993, EERI created an Endowment Fund. The Fund stimulates new and unique ventures, identifies gaps in research, improves application and practice, and facilitates public policy to reduce earthquake risks. A series of White Papers publishes the results of numerous Endowment Fund projects.

In 1984, the Institute introduced *Earthquake Spectra*, a quarterly journal devoted to current research pertaining to earthquake risk reduction, which is provided to members and subscribers. *Spectra* is intended to serve the informational needs of many active professions: engineers, code officials, geologists and seismologists, planners and public officials. Members also receive a monthly *Newsletter* containing information about the Institute's activities, a calendar of meetings, publications, and relevant information from around the world. EERI's other publications include seminar, workshop, and conference proceedings, Monographs, the Design and Oral Histories series, Endowment Fund White Papers, other special publications, and a range of audiovisual materials, including slide sets, CD-ROMS, and videos.

Student Chapters have been established at universities throughout the country to encourage participation of students in earthquake-related areas of research and professional practice. Since 1991, EERI has offered two scholarship programs, funded by FEMA, to encourage the transfer of research to practice and advance the goals of the National Earthquake Hazard Reduction Program. One supports graduate study and the other is a mid-career fellowship. EERI has also established Regional chapters to focus attention on regional issues of seismic risk, design, and public policy. A Distinguished Lecturer is named annually and invited to present his/her lecture to EERI Student and Regional chapters.

The full report of EERI, including a list of EERI publications and 7 monographs as computer-readable files, are archived on the attached Handbook CD#2, under the directory of \8016USA\EERI.

7. Multidisciplinary Center for Earthquake Engineering Research (J. Stoyle)

Web site: http://mceer.buffalo.edu/

The Multidisciplinary Center for Earthquake Engineering Research (MCEER) is a national center of excellence that develops and applies knowledge and advanced technologies to reduce earthquake losses. Headquartered at the University at Buffalo, State University of New York, MCEER was originally established in 1986 by the National Science Foundation (NSF) as the country's first National Center for Earthquake Engineering Research (NCEER).

Comprising a consortium of researchers from numerous disciplines and institutions throughout the United States, the Center's mission is to reduce earthquake losses through research and the application of advanced technologies that improve engineering, pre-earthquake planning, and post-earthquake recovery strategies. Toward this end, the center coordinates a nationwide program of multidisciplinary team research, education, and outreach activities. Funded principally by NSF, New York State, and the Federal Highway Administration (FHWA), the Center derives additional support from the Federal Emergency Management Agency (FEMA), other state governments, academic institutions, foreign governments, and private industry.

MCEER research focuses on improving seismic assessment and performance of buildings, highways, and other infrastructure, and response and recovery systems. Projects foster team collaboration among academic researchers; professionals in engineering, design and other related disciplines; government officials; manufacturers; and additional stakeholders in both the public and private sectors.

The Center also sponsors an education program that provides learning opportunities for students at the K–12 and university undergraduate and graduate levels, and for practitioners seeking specialized training through continuing education courses. Outreach activities include broad-based dissemination of information and technology through research reports, national and international conferences, industry partnerships, and an Information Service that provides convenient access to published, recorded, and online materials on engineering, geology, and social, political, and economic aspects of earthquakes.

MCEER has also established numerous cooperative research programs with institutions outside the US, including those in Japan, the People's Republic of China, Mexico, and Taiwan, to exchange findings and advance earthquake hazards mitigation and loss reduction principles.

MCEER is comprised of a consortium of researchers from numerous disciplines and member institutions throughout the United States. The member institutions include (1) Cornell University, School of Civil and Environmental Engineering, (2) EQE International, Center for Advanced Planning and Research, (3) NJIT New Jersey Institute of Technology, (4) Rensselaer Polytechnic Institute, Departments of Civil and Environmental Engineering, (5) State University of New York at Buffalo, Department of Civil, Structural and Environmental Engineering, (6) The Pennsylvania State University, Department of Energy, Environmental and Mineral Economics, (7) University of Delaware, Disaster Research Center, (8) University of Houston, Department of Civil and Environmental Engineering, (9) University of Nevada/Reno, Civil Engineering Department, (10) University of Notre Dame, Department of Civil Engineering and Geological Sciences, (11) University of Pennsylvania, Wharton School of Business Risk Management & Decision Processes Center, (12) University of Southern California, Department of Civil and Environmental Engineering, (13) University of Washington, Department of Geography, and (14) Virginia Polytechnic Institute and State University, College of Engineering.

The MCEER Web site offers a wealth of information about its various programs, and links to other sites of interest regarding earthquake hazard mitigation. The Web site address is *http://mceer.buffalo.edu/*.

8. Pacific Earthquake Engineering Research Center (C.D. James)

Web site: http://peer.berkeley.edu/

The Pacific Earthquake Engineering Research Center (PEER) is an Earthquake Engineering Research Center administered under the US National Science Foundation Engineering Research Center program.

The mission of PEER is to develop and disseminate technology for design and construction of buildings and infrastructure to meet the diverse seismic performance needs of owners and society. Current approaches to seismic design are indirect in their use of information on earthquakes, system response to earthquakes, and owner and societal needs. These current approaches produce buildings and infrastructure whose performance is highly variable and may not meet the needs of the owners and society. The PEER program aims to develop a performance-based earthquake engineering approach that can be used to produce systems of predictable and appropriate seismic performance.

To accomplish its mission, PEER has organized a program built around research, education, and technology transfer. The research program merges engineering seismology, engineering, and socioeconomic considerations in coordinated studies to develop fundamental information and enabling technologies that are tested and refined using field laboratories and applied in demonstration projects. Primary emphases of the research program at this time are on older existing concrete buildings, bridges, and highways, electrical utilities, and ports and harbors. The education program promotes engineering awareness in the general public and trains undergraduate and graduate students to conduct research and to implement research findings developed in the PEER program. The technology transfer program involves practicing earthquake professionals, government agencies, and specific industry sectors in PEER programs to promote implementation of appropriate new technologies.

PEER has since 1997 been established as a consortium of lead, core, and affiliated institutions with more than 100 faculty participants at 18 universities. The Lead Institution is the University of California, Berkeley. The Core Institutions are the California Institute of Technology, Stanford University, the University of California (Davis, Irvine, Los Angeles, and San Diego), the University of Southern California, and the University of Washington. Affiliated Institutions exist in the states of Alaska, California, Hawaii, Nevada, Oregon, Utah, and Washington, the most highly seismic regions of the United States.

More information about PEER staff personnel, the research organization, and its work are given in the full report on the Handbook CD. For more current information about PEER, please visit its Web site at http://peer.berkeley.edu/

80.17 Uzbekistan

The editors received a brief summary of the "Uzbek Scientific-Research Institute for Typical and Experimental Design of Residential and Public Construction of the State Committee of Architecture and Construction, Joint Stock Venture AO UzLITTI," from its Director, Dr. S.A. Khodjaev, 17 Niyazova Street, Tashkent 700095, Uzbekistan. The edited summary is shown here. A review of seismology in Uzbekistan is given in Chapter 79.56, and biographical sketches of some Uzbek engineers and scientists may be found in Appendix 3 of this Handbook.

1. Introduction

AO UzLITTI is the largest institution in Central Asia carrying out a variety of complex studies in the field of seismic-resistant design and construction of civil buildings. These investigations include scientific research, laboratory and field experiments, and all types of design documentation, taking into consideration the specific conditions of Uzbekistan. The present Director of AO UzLITTI is S.A. Khodjaev, the Deputy Director is A.M. Kamilov, and the Scientific Secretary is A. Azhidinov.

2. Current Activities

The main scientific focus and current activities of AO UzLITTI are

- Development of design codes for earthquake loads, taking into account regional peculiarities of seismic hazard.
- Development of approaches for the assessment and mitigation of seismic risk.
- Investigation of structural response by analytical/experimental methods on models and through in-situ tests during the earthquakes, engineering analysis of the consequences of earthquakes.
- Development of new construction systems exhibiting increased safety, methods of aseismic srengthening of existing dwellings (including architectural monuments), seismic-resistant design of buildings.
- Development of codes and standards.

3. Accomplishments

Reports on the engineering analysis of the consequences of strong earthquakes in the former USSR are stored at the Institute. These include Tashkent (1966), Gazli (1976 and 1984), Nazarbek (1980), Spitak (1988), Kamashi (2000) and others. (Work by Rasskazovsky V.T., Uzlov S.T., Rzhevsky V.A., Asamov H.A., Khakimov Sh.A.Gamburg Yu.A.Ibragimov R.S., Konobeeva L.V., Shirin V.V., Avanesov G.A., and Plahty K.A.)

Many typical technical solutions for the reinforcement (retrofit) of buildings damaged during earthquakes have been developed at the Institute. Some of these strengthened buildings, for example, large panel buildings, were later tested by earthquakes. (Work by Hamburg Yu.A., Uzlov S.T, Khakimov Sh.A., Asamov H.A., Shirin V.V., Rzhevsky V.A., Konobeeva L.V., Mukhamedshin L.A., Yanbulatov R.K., and Plahty K.A.)

AO UzLITTI is a main developer of recommendations for codes and standards of design in earthquake-prone regions of Uzbekistan. New topics that appear in the new code, KMK 2.01.03-96 "Construction in earthquake prone areas," include (1) validation of the code to account for the design of buildings and facilities in zones having seismic intensities of 7, 8, 9, and higher (MSK-98 units); (2) requirements to reconstruct/retrofit buildings; limit state design of buildings; (3) consideration of the regional peculiarities of the Republic, including territories having seismic intensities greater than 9 degrees; (4) requirements for the provision of the seismic safety of residential houses that contain weak materials, volume blocks, and non-bearing elements;

INTERNATIONAL HANDBOOK OF EARTHQUAKE AND ENGINEERING SEISMOLOGY, VOLUME 81B ISBN: 0-12-440658-0

and (5) rehabilitation and retrofitting of existing buildings. (Work by Rzhevsky V.A., Tsypenyuk I.F., Khakimov Sh.A., Asamov H.A., Gamburg Yu.A.,Uzlov S.T, Shirin V.V., Rzhevsky. V.A., Kamilov A.M., Mukhamedshin L.A., Filyavich V.N., Ibragimov R.S., and Plahty K.A.)

Seismic-resistant construction systems have been developed at the Institute, including modular frame reinforced concrete buildings having planar and cross shapes, large panel systems with enlarged spans, volume block buildings, and non-welded frame systems exhibiting increased safety. (Work by Yanbulatov R.K, Mukhamedshin L.A., Gorbatsky M.E., Burdman V.I., Farsyan B.V., Khakimov Sh.A., Tursunbaeva C.N., Tsypenyuk I.F., Rzhevsky V.A., Gamburg Yu.A., Uzlov S.T, and Plahty K.A.)

The Institute is presently conducting investigations for the development of methods of assessing and mitigating the seismic risk and vulnerability of buildings in Uzbekistan cities, and for the development of non-welded RC frame systems and seismic-resistant construction for rural areas of individual houses made from weak materials.

AO UzLITTI has been nominated by the Government of Uzbekistan as the leading organization in the Republic in the field of seismic-resistant design/construction, and the rehabilitation and strengthening of buildings and facilities.

81

Centennial Reports: International Organizations

81.1 General Introduction

Carl Kisslinger
CIRES, University of Colorado, Boulder, CO, USA

The remarkable advances of the past century in earthquake and engineering seismology documented in this Handbook are primarily the result of the skill and efforts of scientists and engineers working in the national organizations and institutions. Their stories are told in Chapter 79 and in the technical chapters on specific topics. However, the activities of these individuals, research teams, and their organizations have been strongly encouraged and enhanced by a number of international scientific entities created in order to build bridges among the national efforts and to facilitate multi-disciplinary research projects. Reports from those especially linked to IASPEI are presented in this chapter. In the spirit of this centennial handbook, the emphasis is on past history and on activities current as of the time of writing. For many of these organizations, information on the most recent aspects may be found on their Web sites.

Several other organizations with strong ties to the objectives and interests of IASPEI, although truly international, have recognized connections to or are identified with particular countries. Their stories are told in the national reports of those countries. One example is the International Seismological Centre. The history, essential facts, and accomplishments of the ISC, the world authority for publication of earthquake locations, born and nurtured in Great Britain, are given in Chapter 4, International Seismology, and in the national report of the United Kingdom in Chapter 79.54. Another multinational contributor to the works included in this Handbook is the International Institute of Seismology and Earthquake Engineering (IISEE). Now located in Tsukuba, Japan, IISEE was originally funded jointly by UNESCO and the Japanese government, now by the Japanese government. An institutional report on IISEE is included within the Japanese national report in Chapter 79.33. Similarly, an institutional report on the International Institute of Earthquake Engineering and Seismology (IIEES) located in Tehran, Iran, is included within the Iranian national report in Chapter 79.30.

A fourth organization included elsewhere in the Handbook is the International Centre for Theoretical Physics, Trieste, Italy. The activities of ICTP in seismology are described in a section of the Italian national report, Chapter 79.32.

The nine reports presented here are only brief accounts of the history, organization, and achievements of these bodies in their furtherance of science and technology. Most reports include the Web site addresses and lead to other sources of information from which more details can be learned. The IASPEI report also includes a listing of all the IASPEI committees and commissions, including those that did not submit a report for publication here.

In addition to the non-governmental organizations included in this chapter, one inter-governmental organization stands out as a major contributor to the work of IASPEI: the United Nations Educational, Scientific, and Cultural Organization (UNESCO). Although we do not include any report on UNESCO activities in earthquake and engineering seismology in this chapter, numerous examples of financial and organizational support to seismological projects are to be found in the preceding chapters of the Handbook. UNESCO has not only cooperated with IASPEI in planning and partial financing of international projects, it has also contributed directly to the work of a number of the international bodies reviewed in this chapter.

Summaries are presented in this chapter for the following international organizations:

- International Association for Earthquake Engineering (IAEE)

INTERNATIONAL HANDBOOK OF EARTHQUAKE AND ENGINEERING SEISMOLOGY, VOLUME 81B ISBN: 0-12-440658-0

- International Association of Seismology and Physics of the Earth's Interior (IASPEI)
- International Heat Flow Commission (IHFC)
- International Union of Geodesy and Geophysics (IUGG)
- IASPEI Committee on Education
- IASPEI-IAEE Joint Working Group on Effects of Surface Geology on Seismic Motion
- Centro Regional de Sismologia para America del Sur (CERESIS)
- GeoHazards International
- Global Alliance for Disaster Reduction

81.2 International Association for Earthquake Engineering

The editors received this summary report of the International Association for Earthquake Engineering (IAEE) from its immediate past president, Prof. Sheldon Cherry, Department of Civil Engineering, University of British Columbia, Vancouver, BC, Canada. There have been productive joint working groups of IAEE and the International Association of Seismology and Physics of the Earth's Interior (IASPEI) (see for example, Chapter 81.7). The present IASPEI Handbook project is in collaboration with IAEE. Photographs of all IAEE presidents and secretaries-general are shown in Figures 1 and 2. In addition, we selected photographs of six eminent deceased earthquake engineers (J.A. Blume, J. Freeman, H.X. Liu, T. Naito, N. Newmark, and S.V. Polyakov) in Figures 2 and 3. A group photo (taken while Kyoji Suyehiro was giving lectures at California Institute of Technology in 1932) is also shown in Figure 3.

1. Introduction

The 1st World Conference on Earthquake Engineering was held in Berkeley, California, in 1956 to commemorate the 50th anniversary of the 1906 San Francisco earthquake. A computer-readable file of the proceedings is archived on the Handbook CD#2, under the directory of \8102IAEE. As a result of an initiative taken during the 2nd World Conference on Earthquake Engineering in Tokyo, Japan, in 1960, the International Association for Earthquake Engineering (IAEE) was officially established in February 1963, with Professor Kiyoshi Muto (Japan) serving as its founding President. The Association's Central Office was located in Tokyo and Professor Kazuo Minami (Japan) served as IAEE's first Secretary-General.

The objective of the Association is to promote international cooperation among scientists, engineers, and other professionals in the broad field of earthquake engineering through the interchange of knowledge, ideas, and the results of research and practical experience. The Association accomplishes its objective by (a) holding World Conferences; (b) promoting and facilitating the interchange of information; and (c) promoting and facilitating the extension of technical cooperation.

2. Membership of IAEE

Membership in the Association is through National Organizations of Earthquake Engineering following formal application to the IAEE Secretary-General. National Organizations must agree to further the objectives of the IAEE and to abide by its Statutes. At present, there are 54 Member Organizations of IAEE. The authority of the Association is exercised through (a) the General Assembly of Delegates, composed of one National Delegate appointed by each National Organization; (b) the Executive Committee, composed of the IAEE elected officers, Directors, and Consultative Members; and (c) the Officers, composed of the President, Executive Vice President, Vice President, and Secretary-General. The General Assembly of Delegates and the Executive Committee of the Association meet during each World Conference. The President presides over these meetings and, with the assistance of the Secretary-General, acts on behalf of the Association in developing policies and procedures initiated by the Executive Committee for consideration and approval by the General Assembly of Delegates, and in accordance with IAEE Statutes.

3. World Conferences on Earthquake Engineering

IAEE generally sponsors the World Conferences on Earthquake Engineering (WCEE) on a quadrennial basis. Following the first

INTERNATIONAL HANDBOOK OF EARTHQUAKE AND ENGINEERING SEISMOLOGY, VOLUME 81B ISBN: 0-12-440658-0

K. Muto (Japan)
IAEE President, 1963-1965

J. E. Rinnie (USA)
IAEE President, 1965-1969

G. W. Housner (USA)
IAEE President, 1969-1973

E. Rosenblueth (Mexico)
IAEE President, 1973-1977

J. Krishna (India)
IAEE President, 1977-1980

D. E. Hudson (USA)
IAEE President, 1980-1984

H. Umemura (Japan)
IAEE President, 1984-1988

G. Grandori (Italy)
IAEE President, 1988-1992

T. Paulay (New Zealand)
IAEE President, 1992-1996

FIGURE 1 Photographs of IAEE Presidents: Muto, Rinnie, Housner, Rosenblueth, Krishna, Hudson, Umemura, Grandori, and Paulay.

S. Cherry (Canada)
IAEE President, 1996-2002

L. Esteva (Mexico)
IAEE President, 2002-2006

K. Minami (Japan)
IAEE Secretary-General
1965-1977

Y. Osawa (Japan)
IAEE Secretary-General
1977-1988)

T. Katayama (Japan)
IAEE Secretary-General
1988-2002

H. Iemura (Japan)
IAEE Secretary-General
2002-

John A. Blume (1909-2002)

John Freeman (1855-1932)

Huixian Liu (1912-1992)

FIGURE 2 Photographs of IAEE Presidents: Cherry, and Esteva; photographs of IAEE Secretaries-General: Minami, Osawa, Katayama, and Iemura; and photographs of Blume, Freeman, and Liu.

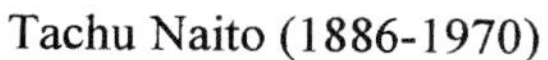

Tachu Naito (1886-1970) Nathan Newmark (1910-1981) Svyatoslav V. Polyakov (1918-1992)

Kyoji Suyehiro (1877-1932) giving lecture at California Institute of Technology, 1932.

Left to Right: J.P. Buwalda, R.R. Martel, K. Suyehiro, B. Gutenberg, and John Anderson.

FIGURE 3 Photographs of Naito, Newmark, and Polyakov; and a group photo of Kyoji Suyehiro *et al.* at California Institute of Technology, 1932.

two world conferences, subsequent conferences were held as follows: 3WCEE, Auckland/Wellington, New Zealand 1965; 4WCEE, Santiago, Chile 1969; 5WEE, Rome, Italy 1973; 6WCEE, New Delhi, India 1977; 7WCEE, Istanbul, Turkey 1980; 8 WCEE, San Francisco, USA 1984; 9WCEE, Tokyo/Kyoto, Japan 1988; 10 WCEE, Madrid, Spain 1992; 11WCEE, Acapulco, Mexico 1996; and 12WCEE, Auckland, New Zealand 2000. The 13WCEE will be held in Vancouver, Canada, in 2004. Extensive Proceedings of all Conferences held to-date have been published and provide a valuable reference source and history of the modern development of earthquake engineering.

4. Officers and Honorary Members of IAEE

The present President of IAEE is Professor Luis Esteva (2002–2006), Mexico, and the Secretary-General is Professor Hirokazu Iemura (2002–), Japan. The immediate Past-President is Professor Shel Cherry (2002–2004), Canada, and the immediate past Secretary-General is Dr. Tsuneo Katayama (1988–2002), Japan. A complete list of the former Presidents and Secretaries-General of IAEE is given here.

Former Presidents:

Prof. Kiyoshi Muto (1963–1965) Japan
Mr. John E. Rinnie (1965–1969) USA
Prof. George W. Housner (1969–1973) USA
Prof. Emilio Rosenblueth (1973–1977) Mexico
Prof. Jai Krishna (1977–1980) India
Prof. Donald E. Hudson (1980–1984) USA
Prof. Hajime Umemura (1984–1988) Japan
Prof. Giuseppe Grandori (1988–1992) Italy
Prof. Thomas Paulay (1992–1996) New Zealand
Prof. Sheldon Cherry (1996–2002) Canada

Former Secretaries-General:

Prof. Kazuo Minami (1965–1977) Japan
Prof. Yutaka Osawa (1977–1988) Japan
Dr. Tsuneo Katayama (1988–2002) Japan

IAEE has elected the following honorary members: Nicholas N. Ambraseys (UK), John A. Blume* (USA), Ray W. Clough (USA), Luis Esteva (Mexico), Rodrigo Flores A. (Chile), Giuseppe Grandori (Italy), George W. Housner (USA), Donald E. Hudson* (USA), Kiyoshi Kanai (Japan), Takuji Kobori (Japan), Jai Krishna* (India), Kazuo Minami* (Japan), Kiyoshi Muto* (Japan), Tachu Naito* (Japan), Nathan Newmark* (USA), Shunzo Okamoto (Japan), Robert Park (New Zealand), Thomas Paulay (New Zealand), S.V. Polyakov* (USSR), Joseph Penzien (USA), Jakim Petrovski (Macedonia), John E. Rinne* (USA), Emilio Rosenblueth* (Mexico), Hajime Umemura* (Japan), Geoffrey B. Warburton (UK), and Rifat Yarar (Turkey). (The * indicates deceased honorary members.)

5. Biographies

A biography for John A. Blume (1909–2002), Donald E. Hudson (1916–1999), Jai Krishna (1912–1999), Kazuo Minami (1907–1984), Kiyoshi Muto (1903–1989), Tachu Naito (1886–1970), Nathan Newmark (1910–1981), Yutaka Osawa (1927–1991), Emilio Rosenblueth (1926–1994), and Hajime Umemura (1918–1995) may be found in Chapter 89 of this Handbook.

A biographical sketch for Nicholas N. Ambraseys, Sheldon Cherry, Luis Esteva, Rodrigo Flores A., George W. Housner, Hirokazu Iemura, Kiyoshi Kanai, Tsuneo Katayama, Takuji Kobori, Shunzo Okamoto, Robert Park, Thomas Paulay, and Joseph Penzien can be found in Appendix 3 of this Handbook.

6. IAEE Central Office

The IAEE Central Office is located at 5-26-20, Shiba, Minato-ku, Tokyo 108, Japan; Fax: +81-3-3453-0428; Email: <secretary@iaee.or.jp>. For more detailed information, including the Associations Statutes, a list of Member National Organizations and their designated National Delegates, present and past IAEE Executive Committee members, including Consultative Members, and IAEE publications, etc., please visit the IAEE Web site at http://www.iaee.or.jp/.

82

Statistical Principles for Seismologists

David Vere-Jones
Victoria University, Wellington, New Zealand
Yosihiko Ogata
Institute of Statistical Mathematics, Tokyo, Japan

From the founding father, Edmund Halley, to Harold Jeffreys in the last century, geophysicists have been associated with key developments in statistics. This strong tradition, born of the necessity of extracting useful information from large quantities of unpromising data, is still an important part of geophysics, and of seismology in particular. A good grasp of statistical principles, including modeling as well as computational and data-handling aspects, is important in most areas of modern geophysical research.

This Chapter in the printed volume is an abridged version of Part I of the full manuscript in the attached Handbook CD #2. It contains a condensed exposition of basic principles of probability and statistics, up to the questions of parameter estimation, model selection, and model testing.

The full manuscript in the attached Handbook CD #2 consists of two separate but companion parts. Part I is a slightly extended version of the printed Chapter, with an additional section on stochastic process and other models, and a list of useful distributions. Part II is wholly contained in the Handbook CD #2. It contains examples of the application of statistical methods to a number of key problems in seismology and related earth sciences, built mainly around the work of the second author. They include examples of the use of the ETAS model to study quiescence, aftershocks, and background seismicity. A special feature is the systematic use of the information-based selection procedures AIC and ABIC, which provide, in particular, a powerful approach to smoothing and inverse problems.

This Chapter in the printed volume has the relatively modest, quite deliberate, but still substantial aim of presenting a condensed summary of basic statistical principles. The range of potential applications of statistics in seismology is so huge, and the space available so tight, that any other aim seems unrealistic. This limitation does mean, however, that not all statistical problems in seismology can be tackled through the techniques outlined in this Chapter. Now, as often in the past, problems in seismology are extending the boundaries of new statistical thinking. In particular, the reader will find in this Chapter relatively little discussion of statistical techniques for dealing with data from processes that exhibit self-similar or fractal type behaviour. Such problems are typical of those currently attracting attention within statistics itself. There are brief mentions within this printed Chapter, and more detailed treatments, at least for selected problems, and from one statistical viewpoint, in Part 2 of the full manuscript in the attached Handbook CD #2. The new techniques do not discredit or replace the old ones; rather they extend the arsenal of weapons available within statistics, and the range of situations that can be handled. Most of the basic principles remain unaltered.

1. Design Considerations and the Presentation of Data

1.1 Design Considerations in Seismology

Seismology is primarily an observational science; its scientists rarely have the opportunity to control the levels of variables (such as temperature, stress, or depth) that might influence their results. Consequently, classical design considerations, which dominate the subjects of agricultural trials or sample surveys, play a relatively minor role in most seismological experiments. This does not mean, however, that principles of design have no place in seismology. Some aspects of the experiments can be controlled: the number of observations or length of observation period; the

INTERNATIONAL HANDBOOK OF EARTHQUAKE AND ENGINEERING SEISMOLOGY, VOLUME 81B *ISBN: 0-12-440658-0*

quality and consistency of data collection; the division of time or space regions into parts for comparison; the need for replication and careful verification of results; avoidance of (or at the very least, making allowance for) the use of the same data to both set up the experiment (in terms of selection of subregions, etc.) and to vindicate its conclusions; and so on. All of these are vitally important aspects. While only common sense principles are required, attention to them is likely to be more effective in avoiding unsound conclusions than the use of the latest and most sophisticated techniques. As any statistical consultant will confirm, the most valuable time to consult a statistician is before the experiment commences, not after the data has been collected.

1.2 Methods of Data Display

The groundwork for any statistical analysis is laid by the careful summary and display of data. If the data consist of a set of n real-valued observations $(x_1, x_2, \ldots, x_n)$, some of the more important *summary statistics* are the sample mean

$$\bar{x} = (1/n)\sum_i x_i, \tag{1}$$

the sample variance

$$s^2 = [1/(n-1)]\sum_i (x_i - \bar{x})^2, \tag{2}$$

the higher sample moments

$$m_k = (1/n)\sum_i x_i^k, \tag{3}$$

the sample median (midpoint), quartiles, and range. For long-tailed distributions, moments become unreliable, and either alternative measures, such as the quartiles, should be preferred, or a preliminary transformation of the data (e.g., taking logs, as in converting seismic moment to magnitude) should be undertaken to bring the data closer to normality. However, the moments of the transformed variables may have a very different physical connotation to the moments of the original variables, so that transformations do not solve all problems.

More important than these summary figures are various methods of graphic display. These are important, first, because the eye is more powerful than the intellect in discerning patterns, and the analyst needs all the help he or she can get, and, second, because one good diagram rates more than many words in conveying the results of an analysis to potential users. If the data admit of a clear interpretation, then a display will make that interpretation visually transparent. The art of data presentation lies in finding that display.

Some idea of the wealth of display options now available can be obtained from recent texts on *exploratory data analysis* (Tukey, 1977; Atkinson, 1982) and *graphical statistics* (Gnanadesikan, 1981; Cleveland, 1993). The development of modern statistical packages such as *SPlus* or *Statgraph* is closely linked to the need to rapidly produce a wide range of graphs and other possible displays (Chambers and Hastie, 1992). Here we shall have space to list only some of the basic forms.

We first recall that data are generally classified as *discrete* if in principle they take on only a finite set of values (commonly integers), *continuous* if in principle they vary over a continuous range of values, and *multivariate* if each observation contains several components. They can also be classified according to their role, into *response* variables and *explanatory* variables.

The *histogram* is perhaps the most venerable of display methods. It is used to depict continuous data or grouped discrete data. The data range is divided into cells or bins, and the height of each cell adjusted so that its *area* (rather than its *height*, as in a *bar-plot*—the distinction is important if cells are of unequal width) is proportional to the number of observations. If the cell entries are normalized to total unity (i.e., they represent proportions rather than numbers), the histogram can be regarded as a rough estimate of an underlying probability density. Default options, which automatically select the number of cells in a histogram and their boundaries, are convenient, but can produce misleading results. The freedom to specify cell boundaries is frequently useful, and sometimes essential.

The *empirical cumulative distribution function* (c.d.f.) $\hat{F}(x)$, specifies the cumulative proportion of observations not exceeding x. Although less informative at first glance, the empirical c.d.f. is more stable than the histogram, does not depend on arbitrarily chosen cell boundaries, and provides a *consistent* estimate of $F(x)$, in the sense that for given x the sample values converge to the true value as the sample size increases. By contrast, even a smoothed density estimate can be made consistent only by a careful balancing of bandwidth against sample size. Read in reverse (y-axis to x-axis), $\hat{F}(x)$ provides a visual plot of the *quantiles* of the distribution, i.e., the values of x having a given percentage of the data to their left.

The c.d.f. complement, the *empirical survivor function*,

$$\hat{S}(x) = 1 - \hat{F}(x), \tag{4}$$

defines the *tail* of the empirical distribution: the proportion of observations exceeding a particular value x. Plots of its logarithm against x (or against some transform such as $\log x$) can help to discriminate between different types of tail behaviour. The Gutenberg-Richter plot of log-frequency against magnitude is a well-known example of this device in seismology. Its linear form shows that the distribution of magnitudes has (to a first approximation) an exponential tail. It also corresponds to a linear plot of log-frequency against log-energy or log-seismic moment, and hence indicates that energies and moments have distributions with a power-law tail.

Another useful plot is the *Q-Q plot*, obtained by plotting (for different values of the proportion p) the pairs q_p, q_p^*, where q_p is the quantile of one distribution (say the empirical distribution) and q_p^* the corresponding quantile of a second distribution. If the two distributions are identical, the points so obtained will line up exactly along the line $y = x$. Departures from this line provide information about how and where the two distributions differ.

A *box and whisker plot* is a neat device for displaying summary measures of location and spread. In the standard version, a box is drawn extending from the lower to the upper quartile of the data set, with a line across the middle to mark the median. The whiskers extend beyond the ends of the box to mark the *range* of the data set (i.e., its minimum and maximum values).

The *scatter plot* is the basic tool for displaying bivariate data (x_i, y_i). It consists simply of the cloud of points on the x–y plane. For multivariate data, a quick appraisal of dependence relations can be obtained by plotting on the same page (for easy comparison) the scatter plots of all possible pairs of coordinates. If both variables are random, the scatter plot may be superposed on a smoothed contour plot for the estimated bivariate density (see kernel estimates following). If the x variable has an explanatory character, one may be interested in the *mean* or *median regressions* of y-on-x plots of the mean or median values of the y variable for selected ranges of the x variable. Also, a box-plot may be produced for each set of y values corresponding to a particular range of x values illustrating roughly how the distribution as well as its summary properties vary over the selected range of x values.

The main weakness of a histogram is its visual dependence on the choice of cells. Many smoothing methods have been suggested to overcome this defect, most stemming from the idea of a *moving average* such as the *kernel estimate*,

$$\hat{f}(y) = \frac{1}{n}\int k_h(y - x)\,dN(x) = \frac{1}{n}\sum_{i=1}^{n} k_h(y - x_i). \quad (5)$$

Here the kernel k_h is a fixed, usually symmetric, probability density function, h is its *bandwidth*, controlling the degree of smoothing (e.g., the standard deviation in a normal kernel), and $dN(x)$ represents *counting measure*, which adds one term to the sum on the right every time a data value x_i is passed. The left side is to be regarded as an estimate of an underlying true or population density. Typically, the density is estimated on a dense grid of points and the values passed to a contour-plotting or other graphical routine.

Similar ideas can be used for plotting smoothed versions of regression lines, where a typical kernel estimate might take the form

$$\hat{M}(Y \mid x) = \frac{\sum_i y_i\, k_h(x - x_i)}{\sum_i k_h(x - x_i)}. \quad (6)$$

Many other approaches to smoothing data are possible, including the use of smoothing and thin-plate splines, fitting orthogonal polynomials, etc. Some general references are Silverman (1986) and Härdle (1990). Part II of the full manuscript on the attached Handbook CD #2 contains examples of the use of ABIC, the Bayesian version of Akaike's Information Criterion, to determine optimal smoothing regimes for contour plots of seismicity or b values.

2. Fundamental Probability Concepts

2.1 Probability, Uncertainty, and Information

Probability has been interpreted as the strength of belief associated with members of a family of propositions, the asymptotic relative frequency of different outcomes from a repeated experiment, and an abstract measure associated with members of a family of admissible subsets. Although the differences in interpretation are far from trivial, they all lead to essentially the same mathematical structure, the canonical example of which is a non-negative, additive function defined on subsets of a finite set. Some insight into the fundamental role of this model comes from a theorem of R. Cox (1946), roughly to the effect that, under mild conditions, any logically consistent structure for numerically evaluating uncertainty, or "degrees of belief," is necessarily equivalent to the structure of classical probability. Cox's result provides rather strong justification for using the probability approach in preference to other postulated systems for quantifying and manipulating uncertainties (see Lindley, 1965, 1971, and Jaynes, 1983, for further commentaries).

Probability is not the only candidate for the primary concept in the theory. From Huygens (1657) to Whittle (1992), there has been a strong minority view that holds that the notion of expectation, rather than probability, should play this role. This approach has the advantage, not shared by probability itself, that it can be used to develop both classical and quantum probability theories (see Whittle, 1992).

Another different (but ultimately equivalent) approach to describing uncertainty stems from Boltzmann's fundamental work on *entropy* in statistical mechanics. Boltzmann related the degree of disorder of the state of a physical system to the logarithm of its probability. If, for example, the system has n non-interacting and identical particles, each capable of existing in each of K equally likely states, the leading term in the logarithm of the probability of finding the system in a configuration with n_1 particles in state 1, n_2 in state 2, etc., is given by the *Boltzmann entropy*,

$$\mathcal{H}_\pi = -\sum_{i=1}^{K} \pi_i \log \pi_i, \quad (7)$$

where $\pi_i = n_i/n$.

In information theory, minus the logarithm of the probability of a symbol (essentially the number of characters required to represent the probability efficiently in a binary code) is defined to be the *information* conveyed by transmitting that symbol. In this context, the entropy can be interpreted as the expected information conveyed by transmitting a single symbol from an alphabet in which the symbols occur with the probabilities π_k. Although information theory remains more restricted in scope than probability theory, information ideas surface in many parts of both probability and statistics.

In this chapter we shall interpret probability as a *measure* applied to subsets of a probability space. Of the many texts devoted to this theory, Feller (1968), Gnedenko (1962), and Whittle (1992) combine insight and mathematical depth with a minimum of abstract mathematics. Loève (1963) and Billingsley (1995) provide more advanced treatments.

2.2 Basic Notions

The first stage in developing a probabilistic treatment of some phenomenon is to select a universal set Ω (*probability space, outcome space*) to determine the scope of the discussion. The elements ω of this set, the *elementary events*, are interpreted as the distinct possible outcomes of the conceptual experiment under consideration. Though elementary in one sense, they may be quite complex in another; for example, each one may be a possible time history for an evolving physical system.

Selected sets of outcomes are then interpreted as *events*. If the probability space is uncountable (as are the real numbers) some subsets may be so irregular as to defy any attempt to assign a probability to them, and cannot be counted as events. At least, the class of events is assumed to be rich enough to form an *algebra* (i.e., unions, intersections, and complements of events remain events).

A *probability measure*, denoted by Pr throughout this chapter, is then chosen to assign probabilities to events. The following requirements, which form the starting point for an axiomatic treatment, should be satisfied:

A1 (non-negativity): $Pr(A) \geq 0$;
A2 (finite additivity): if A, B are disjoint events, $Pr(A \cup B) = Pr(A) + Pr(B)$;
A3 (normalization): $Pr(\Omega) = 1$;
A4 (countability requirements): If $\{A_i; 1 = 1, 2 \ldots\}$ is a sequence of disjoint events, then their union is an event, and $Pr(\cup_{i=1}^{\infty} A_i) = \sum_{i=1}^{\infty} Pr(A_i)$.

Two events A, B are said to be *independent* if

$$Pr(A \cap B) = Pr(A)\,Pr(B). \tag{8}$$

The essential idea is that the occurrence of B has no effect on the probability of occurrence of A. This is made explicit by the introduction of the *conditional probability* of A given the occurrence of B, written $Pr(A \mid B)$: provided $Pr(B) \neq 0$,

$$Pr(A \mid B) = Pr(A \cap B)/Pr(B). \tag{9}$$

Then A and B are independent if and only if either $Pr(B) = 0$ or $Pr(A \mid B) = Pr(A)$.

The defining relation Eq. (9) can be extended to form a *chain rule* for conditional probabilities. The next stage is

$$Pr(A \cap B \cap C) = Pr(A)\,Pr(B \mid A)\,Pr(C \mid A \cap B). \tag{10}$$

Situations involving chains of conditional probabilities arise in considering possible precursors to large earthquakes (see Part II, §8 in the attached Handbook CD #2 for one example) and more generally in any model for a system evolving randomly in time (e.g., the point-process models discussed in Part II, §3 in the attached Handbook CD #2).

A simple extension of assumption *A2* yields the *rule of total probability*: if $B_1, B_2, \ldots, B_K$ is a class of *disjoint events* covering the whole of Ω (i.e., if the B_i form a *partition* of Ω), then

$$Pr(A) = \sum_{k=1}^{K} Pr(A \cap B_k) = \sum_{k=1}^{K} Pr(A \mid B_k)\,Pr(B_k). \tag{11}$$

This in turn yields Bayes' rule for interchanging the arguments in a conditional probability,

$$Pr(B_i \mid A) = \frac{Pr(A \cap B_i)}{Pr(A)} = \frac{Pr(A \mid B_i)\,Pr(B_i)}{\sum_{k=1}^{K} Pr(A \mid B_k)\,Pr(B_k)}. \tag{12}$$

A *random variable* is a numerical quantity determined by the outcome of the random experiment. It is therefore a mapping, say $X(\omega)$, from the outcome space Ω into the real line. The mapping must be *measurable*—i.e., sufficiently well-behaved for the set $A_x = \{\omega : X(\omega) \leq x\}$ to be an event for every x. This implies, among other properties, that if X is a random variable, so also are most functions of X, such as e^X, X^2, $|X|$.

The *cumulative distribution function* (c.d.f.) of a random variable X,

$$F_X(x) \equiv Pr(A_x) = Pr[X(\omega) \leq x], \tag{13}$$

can then be validly defined. It is right continuous, and monotonic increasing from 0 to 1.

A random variable and its c.d.f. can be classified as *discrete* or *continuous*, according as the c.d.f. is a step function, or increases continuously from 0 to 1. A continuous c.d.f. is said to be *absolutely continuous* when a *probability density function* $f_X(x)$ exists, such that

$$F_X(x) = \int_{-\infty}^{x} f_X(u)\,du. \tag{14}$$

Under mild conditions it can be obtained by differentiating the c.d.f.

Not all random variables are either discrete, or absolutely continuous, or even a mixture of these. There also exist *singular continuous* distributions, which are highly irregular, but important in the discussion of fractal structures. An example is the distribution of the "random cascade," defined by a series of the form

$$Y = U_1 + U_2/d + U_3/d^2 + \cdots, \tag{15}$$

where d is an integer and the U_i are independently and identically distributed according to a discrete distribution on the integers $0, 1, 2, \ldots, d-1$. The Cantor distribution ("Devil's staircase") arises when $d = 3$ and each U_i has probability $1/2$ of taking the values 0, 2.

In the multivariate case, a *random vector* $\vec{X}$, of dimension k say, is a mapping from Ω into $\mathcal{R}^k$ with the property that, for all

$\vec{x}$, $\{\omega : \vec{X}(\omega) \leq \vec{x}\}$ is an event. The resulting function

$$F_{\vec{X}}(\vec{x}) = Pr(\vec{X} \leq \vec{x}) \tag{16}$$

is called the *multivariate distribution function* of the vector $\vec{X}$, or (equivalently) the *joint distribution function* of its components. Its *marginal distributions* are the distributions of the individual components, taken singly or in smaller subsets. A special case of considerable importance arises when the joint distribution function can be represented as the product of its (1-dimensional) marginal distributions, as follows:

$$F_{\vec{X}}(\vec{x}) = F_{X_1}(x_1)\, F_{X_2}(x_2) \cdots F_{X_k}(x_k). \tag{17}$$

This is the necessary and sufficient condition for the component variables to be *mutually independent*.

When densities of a k-variate distribution exist, they are with respect to Lebesgue measure (hypervolume) in k–D space; deriving them from the distribution function requires one differentiation for each component. Special techniques, generally involving the introduction of specially adapted coordinate systems, are needed for dealing with distributions that live on lower dimensional subsets, or manifolds, such as the surface of a sphere in 3-D space.

More generally again, a *stochastic process* is a family of random variables indexed by time, which we shall write either as $X(t)$, or $X(n)$, depending on whether time is continuous (t) or discrete (n). The process is a *random field* if the indexing variable belongs to 2- or 3-dimensional space. The stochastic properties of the process are defined by its *finite-dimensional (fi-di) distributions*: the joint distributions of the process values $\{X(t_i);\ i = 1, 2, \ldots, m\}$, for all possible integers m and configurations of time points $t_1, t_2, \ldots, t_m$. The distributions are not written down exhaustively, but a rule is given describing how the more complex joint distributions can be built up. The mean and covariance functions

$$m(t) = E[X(t)]; \quad C(s,t) = Cov[X(s), X(t)] \tag{18}$$

are important attributes of the process, but in general are not sufficient to determine it completely. In an abstract treatment, each possible *realization* of the process is thought of as a single point in a space of functions or sequences. It is this space of functions that serves as the probability space, and on which is defined the underlying probability measure.

2.3 Expectations

An *expected value* is a weighted average of the general form

$$\text{Expected value} = \text{sum (value} \times \text{probability)}. \tag{19}$$

The sum can be taken either over the different possible outcomes in probability space, yielding

$$E(X) = \int_{\Omega} X(\omega)\, Pr(dw), \tag{20}$$

or as a sum over the different values of the random variable, leading to

$$E(X) = \int_{-\infty}^{\infty} x\, dF_X(x). \tag{21}$$

The first representation, which corresponds to the *ensemble average* in statistical mechanics, shows that the expectation acts *linearly* on random variables; the second (which reduces to a sum in the discrete case and an integral against the density function when the latter exists) provides the usual way of computing the expectation. It is true, even if not quite obvious, that the two representations necessarily yield the same result.

The X in equations (20) and (21) can be replaced by any continuous (or even measurable) function $g(X)$; note however that the expectation is said to exist only when the integral of $|g(X)|$ converges. The higher moments $E(X^k)$ of X, its *central moments* $E[\{X - E(X)\}^k]$ [the *variance* $Var(X)$ is the case $k = 2$] and cross-moments such as $E(XY)$ or $Cov(X, Y) = E[\{X - E(X)\}\{Y - E(Y)\}]$, are all examples of expectations. They are the model versions of the sample moments introduced in the first section.

The idea of conditioning extends to the notions of *conditional distribution* and *conditional expectation* of one random variable, given the value of another. For example, if X and Y have a discrete joint distribution p_{ij}, the conditional expectation of X, given the value j of Y, is defined to be

$$E[X \mid Y = j] = \frac{\sum_i i p_{ij}}{\sum_i p_{ij}}. \tag{22}$$

Limiting arguments justify the analogous form for a bivariate distribution with density $f(x, y)$. Defining the conditional expectation in general requires *Radon-Nikodym derivatives*; see, for example, Billingsley (1995).

The rule of total probability extends to become the *iterated expectation rule*

$$E(X) = E[E(X \mid Y)]. \tag{23}$$

Here the inner expectation is taken over X, for a given value of Y, and the outer expectation is taken over the values of Y. This is one of the most important formulae in the subject, with a wealth of applications. From it derive the *convolution formula* for the density of the sum of two independent random variables,

$$f_{X+Y}(x) = \int_{-\infty}^{\infty} f_X(x-u)\, f_Y(u)\,du; \tag{24}$$

the formulae for the mean and variance of the sum of a random number N of i.i.d. random variables X_i,

$$\begin{aligned} E\left[\sum_{i=1}^{N} X_i\right] &= E(X)\, E(N); \\ Var\left[\sum_{i=1}^{N} X_i\right] &= E(X)^2\, Var(N) + Var(X)\, E(N), \end{aligned} \tag{25}$$

and many others.

An important group of characteristics defined by expectations are the integral transforms of a distribution, namely the *characteristic function*,

$$\phi_X(s) = E(e^{isX}) = \int_{-\infty}^{\infty} e^{isx} dF_X(x), \quad (26)$$

the *moment generating function*, $E(e^{sX})$, and the *Laplace transform*, $E(e^{-sX})$. The *probability generating function* or *z-transform*

$$P_X(z) = E(z^X) = \sum_{k=0}^{\infty} p_k z^k \quad (27)$$

plays a similar role for nonnegative discrete random variables.

The first three transforms all (under suitable assumptions) generate the moments of the distribution, as, for example,

$$\phi(s) = 1 + isE(X) - \frac{s^2}{2!}E(X^2) - i\frac{s^3}{3!}E(X^3) + \cdots. \quad (28)$$

The p.g.f. generates the probabilities if expanded about $z = 0$, while expanded about $z = 1$ it generates the *factorial moments*

$$E(X^{[k]}) \equiv E[X(X-1)\ldots(X-k+1)]. \quad (29)$$

Another key property is that when independent random variables are added, their transforms multiply. For example, if X and Y are independent,

$$\phi_{X+Y}(s) = \phi_X(s)\phi_Y(s). \quad (30)$$

2.4 Limit Theorems

Much work in the early history of probability was devoted to two key results, the *Law of Large Numbers* (*LLN*) and the *Central Limit Theorem* (*CLT*). The LLN started life as Bernoulli's theorem—in a large number of independent, identical trials, the sample proportion will, with high probability, be very close to the probability of success in a single trial. A more general version of this theorem asserts the convergence of sample averages to the expected value: if $X_1, X_2 \ldots$ form an i.i.d. sequence with $E(X)$ finite, then for any $\epsilon > 0$,

$$Pr\left(\left|\frac{1}{n}\sum_{i=1}^{n} X_i - E(X)\right| > \epsilon\right) \to 0. \quad (31)$$

A stronger version (the strong LLN) asserts that in fact convergence occurs with probability 1. If $E(X)$ does not exist, the sample means will not stabilize as n increases, but will either diverge to $\pm\infty$, or show increasing fluctuations; this lack of stability is an important point to bear in mind when dealing with heavy-tailed data such as seismic moments or insurance claims (see, for example, Adler *et al.*, 1998).

Independence is a further important assumption for the LLN, but similar results hold more generally. In fact the LLN is nothing other than a special case of the equality of time average (over n) and ensemble average (over ω) asserted by the *ergodic hypothesis* in statistical mechanics. Any general statement relating the limit of time averages to expectations is an *ergodic theorem*. Typical requirements for equality are that the X_i form a stationary sequence with finite mean, and satisfy some kind of weak independence (or *mixing*) condition. (See, for example, Billingsley, 1995.)

The CLT started as an attempt by de Moivre to refine Bernoulli's estimates for the LLN. He used a version of Stirling's formula to show that the remainder in Eq. (31) can be approximated by what we now recognize as the integral of a normal density. In due course this became the assertion that for an i.i.d. sequence the distribution of the normalized sums

$$\sum_{i=1}^{n} \frac{X_i}{\sigma\sqrt{n}}, \quad \sigma^2 = Var(X_i), \quad (32)$$

converge to the standard normal $N(0, 1)$ distribution.

Again the theorem has been much generalized, both by relaxing the independence assumptions, and by asking whether some other normalization might work if the variance was infinite. In the latter case, new limit distributions arise, namely the class of *stable distributions*, characterized by the requirement that the sum of k such variables have, apart from a scale factor $d(n)$ and a possible shift in location, the same distribution as one of the original variables. The scale factor is necessarily of the form n^{α}, with $0 < \alpha \leq 2$. The definition is reminiscent of self-similarity, and of the definition of a fractal set, and in fact the stable distributions play an important role in both these contexts (see Mandelbrot, 1977).

This generalization in turn extends to *triangular arrays*, where the limiting random variables Y are *infinitely divisible*: For all integers k, Y can be written as the sum of k i.i.d. components (see, e.g., Gnedenko, 1962; Feller, 1968).

3. Principles of Modeling and Inference

3.1 What Is a Probability Model?

An opposition is sometimes made between a physical model and a probability model. The opposition is, or should be, a false one. What is in question is the need to extend the physical model so that it describes the full variability of the observed data, including that part which, to quote Jeffreys (1939) "is called 'error,' and usually quickly forgotten or altogether disregarded in physical writings." As Jeffreys goes on to point out, however, properly handling this part of the variation is an important element of the scientific method. Without it, there is no adequate way to gauge the improvements brought about by proposed new laws, or the inadequacy of old ones. The essence of probability modeling is finding the right way of linking the concepts of the physical theory to random variables with appropriate distributions and dependency relations. In seismology, in particular, many details of the physical processes generating earthquakes are not accessible to direct observation; such features have to be modeled by approximating stochastic processes if the results are to be fitted to data such as earthquake catalogs or strain measurements.

In practice, much probability modeling has the limited scope of describing or predicting some feature of immediate interest. In such contexts it may be helpful to keep two general rules in mind: The model should embody sensible physical ideas about the processes it is supposed to represent; and it should be as simple as is consistent with providing adequate predictions for the purpose to hand. Observance of these simple rules would go a long way toward avoiding some of the more serious abuses of statistical modeling procedures.

In trying to develop probability models for new situations, it is sometimes helpful to have a handy catalog of useful components from which such models can be made. Such components include in particular the commonly occurring distributions and their properties. Full dictionaries of different types of distributions have been compiled by Evans *et al.* (1993), Johnson *et al.* (1994). A much smaller and less detailed list is given in the Appendix (Handbook CD #2). Beyond this, useful ingredients include the simplest stochastic processes, such as Poisson and renewal processes, Markov chains, Brownian motion and related diffusion processes, etc. These lie beyond the scope of this introduction, but a number of models relevant to modeling catalog data are described in Part II, particularly §3-§7 in the attached Handbook CD #2.

3.2 Model Fitting

Model fitting involves finding that member of a chosen class of models which in some sense is "closest" to a given data set (sample). Usually the class of models is a *parametric family*, the members of which are defined up to a small number of numerical parameters. Estimating these is the task of *parametric estimation*. *Non–parametric estimation* corresponds to selecting a model from a class defined by a much broader characteristic, such as continuity. We shall illustrate the issues that arise in parametric estimation by considering the problem of selecting a distribution, from within a parametric family, to fit a set of assumed independent and identically distributed (*i.i.d.*) observations. Similar considerations apply to more complex problems, such as fitting a stochastic process model.

The most important technique is the *method of maximum likelihood*. The likelihood is the joint probability (or probability density in the continuous case) of the full set of sample values. The parameter values that maximize this joint probability are the required estimates. In the i.i.d case the likelihood is the product of the density functions, evaluated at each of the sample values.

Although it is not obvious at first sight, this method can be cast in the form of minimizing a distance, namely the *Kullback-Leibler distance*, between the true density f_0 and the candidate density $\hat{f}$ from the parametric family under consideration. This distance takes the form

$$\begin{aligned}\mathcal{D}(f_0, \hat{f}) &= \int f_0(x) \log \left[\frac{f_0(x)}{\hat{f}(x)}\right] dx \\ &= \int f_0(x) \log f_0(x)\, dx - \int f_0(x) \log \hat{f}(x)\, dx.\end{aligned} \tag{33}$$

Although the true f_0 is unknown, the first term in the right-hand expression acts as an unknown constant, while the second can be written as $E[\log \hat{f}(X)]$, which can be approximated by the average of the sample values

$$n^{-1} \sum \log \hat{f}(x_i) = n^{-1} \log L. \tag{34}$$

At least in an average sense, therefore, the distance will be minimized when the likelihood is maximized. The Kullback-Leibler distance is also useful in understanding what happens when the true model lies outside the family being fitted. As the sample size increases, the model selected by maximum likelihood will then converge toward the member of the parametric family closest, in the sense of this distance, to the true distribution.

To consider a specific example, suppose that $x_1, \ldots, x_n$ is believed to be an i.i.d. sample from the Pareto(a, α) distribution with density

$$f(x) = (\alpha/a)(1 + x/a)^{-(\alpha+1)}. \tag{35}$$

The model is then fully specified up to the values of a, α. Because of the assumed independence, the likelihood is the product of the density functions, evaluated at each of the sample values. Its logarithm (the usual quantity passed to the maximization routine) then takes the form

$$\begin{aligned}&\log L(a, \alpha; x_1, \ldots, x_n) \\ &\quad = -(\alpha + 1) \sum_1^n \log \left(1 + \frac{x_i}{a}\right) - n \log \left(\frac{\alpha}{a}\right).\end{aligned} \tag{36}$$

In this example the maximum can be found by differentiation with respect to a, α, and setting the derivatives to zero. This yields two equations, of which one can be solved analytically, while the other requires numerical solution.

A general-purpose optimization routine, for use in more complex situations, is an essential adjunct to practical model fitting in statistics, and is now included in most statistical packages. Some skill is needed in using such routines, as they can be sensitive to the choice of parameterization and to the initial conditions. Situations can also arise in which no local maximum exists, or many local maxima exist. In nonstandard examples, plots of the likelihood surface, or of 2- or 3-dimensional projections of the surface, may help to explain otherwise unexpected results.

Other estimation methods include the *method of minimum chi-squared*, in which those parameter values are selected that make the χ^2 distance [Eq. (43)] a minimum, and more *ad hoc* methods, which involve matching selected characteristics from the model and the sample. The best known of these latter is the *method of moments*, whereby the model moments are matched to the sample moments until just enough equations are obtained to estimate the unknown parameters uniquely. Once again there are problems if the distribution is heavy-tailed. In the Pareto example considered, for example, the method of moments leads to the two equations

$$\bar{X} = \frac{a}{(\alpha - 1)}, \quad \hat{m}_2 = \frac{a^2}{(\alpha - 1)(\alpha - 2)}. \tag{37}$$

Solving these always leads to estimates in which $\alpha > 2$, even if the data is generated by simulation from a distribution with $\alpha < 2$. The true second moment is then infinite, while the sample second moment, and hence the estimate of α, will depend unstably on the few largest sample values. In such a situation it is preferable to equate the model and sample averages of functions other than the powers of the random variable (in effect resorting to a transformation of the variable), or to use the *method of quantiles*, whereby selected quantiles are matched rather than selected expectations.

The *method of least squares* occupies a rather special role in this context. In essence, it is a method of curve-fitting that is particularly appropriate when the data is known to have the form

$$\text{observation} = \text{parametric curve} + \text{error}, \tag{38}$$

and the errors are normally distributed. In this case, the least squares estimates turn out to be the same as the maximum likelihood estimates, and enjoy many special advantages. In other situations, where the observations have a different type of distribution, least squares methods may not be optimal, and can lead to quite poor estimates. This is particularly the case for heavy-tailed data sets, where the occasional large values have a disproportionate influence on the sum of squares. Simple least squares should also be avoided in situations where one is fitting a cumulative distribution, as in estimating the b-value from a cumulative Gutenberg-Richter plot. Here simple least squares not only leads to unstable estimates, but quite wrongly estimates the standard error of the parameter values.

Maximum likelihood methods should be preferred when the model is reasonably good. If there are suspected discrepancies from the model (e.g., if the data seems to contain *outliers*), a more robust method (less sensitive to deviations from the model assumptions) may be preferred. For example, a median or truncated mean may be preferred to the mean of the full data set as an estimate of location (see, e.g., Huber, 1981). Bayesian methods are outlined in the final section.

3.3 Precision of Estimates

Estimated parameter values will show some variability from one sample to another, even when the underlying model is fixed. In general, the deviation of the estimate $\hat{\theta}$ from the true value θ_0 will have a systematic component, or *bias*, $E(\hat{\theta}) - \theta_0$, and a random component that may be measured by its *standard error*, $\sqrt{Var}(\hat{\theta})$. The estimate is *mean-square consistent* if $E[(\hat{\theta} - \theta_0)^2] \to 0$, in which case both the bias and the standard error tend to zero. In nonstandard situations where the means and variances of the estimates may fail to exist, more general definitions will be needed (see, for example, Lehmann, 1983). The famous *Cramer-Rao lower bound* asserts that for a given sample size, and under some smoothness conditions (see, for example, Rao, 1973 or Serfling, 1980 for more precise statements), there is a universal lower bound to the precision that can be achieved by any unbiased estimate: If $\hat{\theta}$ is unbiased, and based only on the sample values, then

$$Var(\hat{\theta}) \geq \left[E\left(\frac{\partial^2 \log L(\theta, \vec{x})}{\partial \theta^2} \right) \right]^{-1}, \tag{39}$$

where the derivative is evaluated at θ_0.

In the i.i.d. case, the log-likelihood is the sum of n terms each with the same expectation. In this case, the standard error $\sqrt{Var}(\hat{\theta})$ decreases as $n^{-1/2}$, which is the best possible rate of convergence under standard conditions.

The expectation that appears in the right side of Eq. (39) is called the *Fisher information*, usually denoted by $\mathcal{F}$ and not to be confused with the *Shannon information*, or entropy, referred to earlier. It measures the sharpness of the peak of the expected log-likelihood; the sharper the peak, the more precise the estimate. In the multivariate case, $\mathcal{F}$ is a matrix, namely, the expected value of the *Hessian matrix* (matrix of second derivatives) of the sample log-likelihood function.

Under some regularity conditions, the inverse of the Fisher information, $\mathcal{F}^{-1}$, provides both a lower bound and an asymptotic form for the variance of the maximum likelihood estimates. This implies that a maximum likelihood estimate is asymptotically efficient, in the sense that the ratio of its variance to the smallest achievable variance approaches unity. At the same time such estimates are asymptotical unbiased, and asymptotically normally distributed. These are the main theoretical reasons for preferring maximum likelihood estimates when the model is appropriate. Conversely, the biggest drawback to likelihood methods is their sensitivity to departures from the assumed model. That is, likelihood estimates can be quite misleading if the data used in the estimation have characteristics that differ substantially from data produced by the assumed model.

If the Fisher information is hard to obtain directly, a convenient alternative is to approximate it by the numerical value of the Hessian matrix, which is often available as a by-product of the optimization routine. This is usually acceptable, because an error in the standard error is of second-order importance. If the convergence is poor, however, the Hessian matrix obtained in this way may be unreliable.

In a limited number of situations, an exact *confidence interval* for the estimate can be found, namely an interval (L, U), estimated from the sample, with the property that

$$Pr[(L, U) \text{ covers } \theta_0] = 1 - \alpha \tag{40}$$

where α is a specified *confidence level*. Because this equality has to hold simultaneously for all values of the unknown θ_0, it is only in rather exceptional situations that such an interval can exist. The best known of these is the confidence interval for the mean μ of a normal distribution with known variance σ^2. For an i.i.d. sample, this takes the form

$$\bar{X} \pm \frac{Z_{1-\alpha/2}}{\sigma/\sqrt{n}}. \tag{41}$$

Here Z_p refers to the $100p$-th percentile of the standard normal distribution. If the variance is unknown, σ^2 is replaced by the sample variance s^2 and Z_p is replaced by the corresponding percentile t_p from the t-distribution with $(n - 1)$ degrees of freedom. Equation (41) is also used approximately in the many situations where the parameter estimate is approximately normal with known standard error; this standard error then replaces $\sigma/\sqrt{n}$ in Eq. (41). One should beware, however, of assuming too readily that standard conditions will prevail; at the very least, large samples and smooth dependence on the parameters are required.

In multivariate examples, the role of a confidence interval is replaced by that of a *confidence region*. These are useful in 2-parameter problems, where a general-purpose approximate confidence region can be based on the likelihood function $L(\theta)$. It takes the form of the contour line

$$2 \log L(\theta) = 2 \log L(\hat{\theta}) - \chi^2_{1-\alpha/2}. \tag{42}$$

In higher dimensions, such regions can be formulated in principle, but become impractical, and one is reduced to examining the parameters individually or in pairs. High correlation between parameter estimates is a cause of instability, and an obstacle to clear interpretation, both suggesting the desirability of reparameterization.

3.4 Model Testing

The *model testing* stage involves checking the agreement between selected features of the model and the corresponding features of the sample. A test is specified by computing an appropriate "distance" (the *test statistic*) between the sample and model features selected. Because the data are subject to random sampling fluctuations, some discrepancy between observed and model values is to be expected. The problem then is to find a guide as to what might constitute a "tolerable" discrepancy. Such a guide can be obtained from the distribution of the test statistic under the assumption that the selected model is the true one. A particular quantile of this *sampling distribution*—conventionally the 95% percentile—is then chosen to represent a "noise level," and the observed value of the test statistic is considered tolerable ("not significant") until it exceeds that level. The complement of the percentage selected—i.e., 5% in the case of a 95%ile—is called the *significance level* of the test. The complement of the percentile reached by the observed value of the test statistic is called its *p-value* or *observed significance level* (*OSL*). The test is *1-sided* if departures in only one direction are tested, otherwise it is *2-sided*.

For example, in testing whether the value μ_0 is the mean of a normal distribution with known variance σ^2, the normalized distance $|Z| = |\bar{x} - \mu_0|/(\sigma/\sqrt{n})$ is often used as the (2-sided) test statistic. It is considered acceptable at the 5% level if its value does not exceed 1.96, which is the 95% percentile of the distribution of $|Z|$ under the given assumptions, and the 97.5% percentile of the standard normal distribution itself. If the observations happened to give $|Z| = 2$, the p-value would be $Prob(|Z| \geq 2) = 0.0456$. If σ^2 is not known, it is replaced by the sample estimate s^2, and the t-distribution is used in place of the normal distribution.

In the formal description of such a test, the assumed model is called the *null hypothesis*, and contrasted to an *alternative hypothesis* (such as $\mu \neq \mu_0$ in the preceding example). The role of the alternative hypothesis is really to guide the investigator in the choice of a suitable test statistic (e.g., 1-sided or 2-sided). Because almost any sample is likely to yield at least some feature—measured by an appropriate test statistic—that differs significantly from what might be expected from the model, problems arise if the test is suggested by the data themselves. The moral is that the alternative hypothesis, which suggests the test statistic to be used, should be selected *in advance of seeing the data*. It should reflect the purpose of the statistical analysis.

A common trap arises if the data are allowed to determine the test used; standard significance levels then no longer apply, and special techniques must be developed (such as those used in so-called change-point problems) to make allowance for this feature. This may prove to be a difficult task, particularly when the test concerned involves complex spatial boundaries that are partially dictated by the characteristics of the data being studied. In such situations difficult and extensive simulations provide one of the few manageable techniques for obtaining significance levels.

The probability of the selected level being exceeded if a particular alternative is true is called the *power* of the test against that alternative. Ideally, tests would have high power against all alternatives, but in reality one has to accept a compromise—high power against a narrow class of alternatives, or lower power against a wider class of alternatives. The t-test is used to discriminate against alternatives with different values of the mean, but would be of no use in testing against alternatives with the same mean but higher variances. By contrast, the chi-squared test, based on Eq. (43), will ultimately (with enough data) discriminate against differences in both mean and variance, but will be less powerful than the t-test in discriminating against alternatives with different means.

The value of this hypothesis-testing framework in scientific work is somewhat debatable. In particular, there is no need to insist that a particular alternative must be *accepted* if the null hypothesis is *rejected*. This terminology, as well as the use of a fixed significance level, derives from the context of "acceptance sampling," i.e., repeated sampling from batches in a mass production process, hardly the best of models for scientific research. The role of a statistical test in scientific work should be in helping the researcher to assess the fit of the model to the data, not in forcing the acceptance or rejection of the model outright.

The best known tests, such as the t-test, are usually against specific alternatives such as changes in the sample mean or variance when the underlying distributional form is kept constant (e.g., normal). They form the core of modern statistical procedures; for reference one needs a familiar and well-tried text, such as Hald (1952) or Hogg and Tanis (1989). We mention in addition two overall tests for how well the data from an i.i.d. sample fits a given type of distribution. The first is the *chi-squared test*, which uses the χ^2 distance

$$D_{\chi^2} = \sum \frac{(observed - expected)^2}{expected} = \sum \frac{observed^2}{expected} - n, \tag{43}$$

to measure the discrepancy between model and sample. The terms in the sum relate to the observed and expected numbers of observations in each of a finite set of cells, the number "expected" being obtained by multiplying the total number of observations, n, by the probability allotted to that cell by the distribution currently in view. Supposing the cells are fixed as the sample size increases, the asymptotic distribution of this quantity is χ^2 with $K - d - 1$ degrees of freedom, where K is the number of cells, and d is the number of fitted parameters (preferably fitted by minimizing the χ^2-value, but any asymptotically equivalent method, including maximum likelihood, will do). For optimal results the cells should be chosen to have equal expected numbers, not equal lengths.

The *Kolmogorov-Smirnov test* is based on the *Kolmogorov-Smirnov distance*, D_{KS}. This is determined by first plotting values of the model c.d.f. $F(x)$ (abcissa) against those of the empirical c.d.f. $\hat{F}(x)$ (ordinate) for a series of values of x. The resulting curve passes through the points (0, 0) and (1, 1) and the maximum vertical departure from the line $y = x$ provides the distance in question. Its distribution is independent of F, so for any F the observed value of D_{KS} can be compared directly to tabulated percentiles of the Kolmogorov-Smirnov statistic (e.g., Hogg and Tanis, 1989).

A very general model-testing framework is provided by the *likelihood ratio tests*, which apply to *nested hypotheses*. Here one family of parametric models is embedded within a larger family, and it is desired to test whether a model from the smaller family might be considered adequate—as when a linear regression is tested against quadratic or higher order regressions. The test statistic used is the *log-likelihood ratio*. This may be written in the form

$$\Delta = \log L(\hat{\theta}_2) - \log L(\hat{\theta}_1), \tag{44}$$

where $(\hat{\theta}_1, \hat{\theta}_2)$ are the (vector) parameter estimates from the smaller and larger families, respectively. The asymptotic theory asserts that, for large samples, and under some additional conditions (basically smooth dependence on the parameters), 2Δ has a χ^2 distribution on $d_2 - d_1$ degrees of freedom, where d_1, d_2 are the dimensions of $\hat{\theta}_1, \hat{\theta}_2$ respectively.

In the example considered earlier, the test could be used to discriminate between the Pareto distribution and the simpler exponential distribution, which arises as the special case where $\alpha \to \infty$, $a/\alpha = constant = \lambda^{-1}$. First the Pareto model is fitted, and the numerical value of the likelihood at the maximum is obtained. Then the best fitting exponential model is obtained, and its likelihood is calculated. Finally, the two values are substituted into the formula for Δ, and compared with the percentiles of the χ^2 distribution on 1 degree of freedom.

3.5 Model Selection

In principle, model-fitting ideas can be extended to the problem of selecting not only within but also between model families. In particular, the principle that the model preferred should be that giving the highest likelihood is not restricted to applications within a parametric family.

There are, however, some major precautions that must be observed before the extension is taken too literally. For example, with n observations, it should be possible to obtain a perfect fit by any model with n parameters. Hence some account must be taken of the different numbers of parameters in the competing model classes. Moreover, if likelihood ideas are to be used, the asymptotic theory should apply equally to all model classes considered.

Two standard approaches to model selection are the use of *model selection criteria*, such as the Akaike Information Criterion (AIC), and *stepwise regression* techniques.

The AIC is based on an extension of the asymptotic theory for maximum likelihood estimates, and is subject to similar limitations to other applications of that theory: essentially distributions with smooth dependence on their parameter values. Analysis suggests that when the true model is *overfitted*—more parameters are fitted than exist in the true model—both the maximum log-likelihood and the Kullback-Leibler distance to the true model will be inflated by random quantities. On average, it turns out that the difference

$$AIC = -2 \max(\log \text{ likelihood}) + 2k \tag{45}$$

should show a rapid decrease down to the model with the correct number of parameters, then a slower increase. On average, therefore, the model for which AIC is a minimum should be closest to the true model. Further details, with many examples, are given in the texts by Sakamoto *et al.* (1986), and in Part II (in the attached Handbook CD #2) where AIC methods are systematically applied to model selection problems in seismology and geophysics.

An apparent disadvantage of AIC methods is that for a fixed model they do not consistently estimate the correct number of parameters as the sample size increases; rather there is always some probability of selecting a model with too many parameters. Its proponents would argue that this is not a fault, but an indication that AIC offers a practical compromise between overfitting on the one hand, and being responsive to the data on the other. A possibly more serious restriction is that AIC assumes standard asymptotic likelihood theory, and may need modification where that does not apply.

An alternative for handling nested families is the use of a stepwise procedure. Typically, such a procedure would use the likelihood ratio test, at a specified level, to establish the value of adding in each extra parameter. The selection can be carried out bottom up (adding parameters until no significant increase is achieved), or top down (dropping parameters until a significant worsening is observed), and both approaches have their proponents. As always, the choice of an appropriate significance level adds an arbitrary element to the model selection process. None of the procedures work well in situations where a large set of different combinations of parameters have to be compared. In such a case a Bayesian approach, in which the analysis is guided by some prior expectations of the most likely model structures, may prove more effective.

4. Simulation, Prediction, and Bayesian Methods

4.1 Simulation and Prediction

The availability of easy-to-use simulation procedures has brought about a near-revolution in the use of probability models. For many purposes (although certainly not all), the question "Is this model well understood?" might as well be phrased, "Can the model be simulated?" Predictions, significance levels, confidence intervals, scenarios to be anticipated under variation of the model parameters—all are readily constructed from a simulation model. Moreover, formulating a simulation model requires essentially the same analysis as formulating the model in mathematical terms; the contrast is in whether its features will be derived logically or by numerical experimentation.

Virtually all probability simulations start from *pseudo-random numbers*—sequences of real numbers that match as closely as possible the properties of an i.i.d. sequence of uniform (0, 1) random variables. Most methods currently in use are based on refinements and elaborations of the *linear congruential method*, which computes successive values deterministically from an initial *seed* S. The ability to select and record the initial seed is an important means of ensuring repeatability: Starting the simulation from the same seed produces *exactly* the same output; no other details need to be recorded.

The major statistical packages now have excellent random number generators, but this was not always the case, and there are many cautionary tales about large-scale simulation studies vitiated by poor-quality random number generators. From uniform random numbers, random numbers following other distributions are produced by two basic methods. The *inversion method* requires generating U uniform and solving (numerically or analytically) $F(x) = U$. In simple cases the solutions can be written down: For example, $X = -\mu \log U$ generates an exponential with mean μ, and $Y = c(U^{-1/\alpha} - 1)$ generates a Pareto(c, α) distribution. For discrete distributions the usual procedure is to first build up a table of the partial sums $F(k) = \sum_{j=0}^{k} p_j$, then generate a uniform random variable U, and set $X = k$ if U lies in the interval $F(k) < U \leq F(k+1)$.

The *thinning method* depends on first finding an auxiliary density function $g(x)$, from which random numbers are easy to compute, and which dominates the given density in the sense $Cg(x) \geq f(x)$ for all x and some $C > 0$. Then random variables Y, U are generated according to g and the uniform (0, 1) distribution respectively, and Y is *accepted* if $U \leq f(Y)/Cg(Y)$, otherwise rejected. The output consists of the accepted values.

The resulting random numbers can be considered a sample from the initial distribution, and either studied as such, or fed into more complex models. By generating many such samples and averaging, *Monte Carlo* estimates may be produced for char-acteristics of interest. Ripley (1988) provides a systematic treatment.

The techniques we have described for simulating sequences of independent random variables do not require great changes to carry over to the general problem of simulating dependent sequences (Markov and renewal processes, point processes with history-dependent intensities, etc.). The essential difference is that the distribution of the next random element will in general depend on the past history of the process. Provided this dependence can be made explicit, simulation can proceed in a sequential manner, following the techniques already described. The growing importance of this viewpoint is shifting emphasis onto models or model descriptions with explicit dependence on the past. Examples of such models for simulating seismicity patterns, in both time and space-time contexts, are outlined in Part II, §6-8 and Appendix A.2 in the attached Handbook CD #2.

4.2 Prediction

Prediction, at least if that term is understood sufficiently broadly, is often the underlying goal of a statistical exercise. A deterministic model predicts a single value, but a statistical model incorporates rules not only for forecasting systematic effects, but also for forecasting the variation about those effects. Its end-point is a probability distribution for the quantity being predicted. Anything that stops short of that, such as a single best prediction value, is short-changing the user (though the user may not know how to profit from the extra information).

Testing predictions, if the predictions are based on a statistical model, comes back in the end to testing the model. One has confidence in the predictions insofar as one has confidence in the model on which the predictions are based. The most limited situation is that in which the predictions are purely empirical and based only on past occurrences of the same phenomenon, with no physical or structural justification. The model is then one of repeated trials with unknown dependence structure. This is not a good paradigm for earthquake prediction, because dependence between the trials can greatly alter the significance levels, and takes much data to determine. Moreover, earthquakes are infinitely varied, so a repeated trials framework is at best a poor approximation to the real situation. It seems necessary to look toward statistical models that embody substantial insights into the underlying physical processes before prediction schemes for earthquakes become practically effective.

The statistical prediction techniques most commonly used tend to fall into two categories. The best-known and longest established methods are those involving linear prediction techniques in time series. Time series play such a major role in geophysics, and in seismology in particular, that they would need a special chapter to review adequately. Some among many general references are Brillinger (1981), Bath (1974), and Hannan (1970). Here we look just at one special issue, namely the development of linear predictors based on *time-domain models* (the contrast is with the spectral methods developed for *frequency-domain models*).

The principal time-domain models are the *moving average* (*MA*) model

$$X(t) = \sum_{k=0}^{r} a_k \epsilon(t-k), \tag{46}$$

the *autoregression* (*AR*) model

$$X(t) = \sum_{k=1}^{s} b_k X(t-k) + \epsilon(t), \tag{47}$$

and the combined *ARMA* model

$$\sum_{k=0}^{s} b_k X(t-k) = \sum_{k=0}^{r} a_k \epsilon(t-k) \tag{48}$$

with $a_0 = b_0 = 1$, where in all cases the $\epsilon(t)$ are assumed to form an orthogonal sequence of zero-mean, constant variance random variables, which must be $N(0, \sigma^2)$ for the strict-sense theory (requiring specified distributions as well as specific parameters such as means and variances) to apply. The models can be fitted to data using either method-of-moments (wide-sense) or maximum likelihood (strict-sense) methods. A model selection method such as AIC may be used to locate the most effective representation for a given data series. The reason that these models are particularly important for prediction is that the AR model has a natural 1-step predictor of the form

$$\hat{X}(t+1) = \sum_{k=1}^{s} b_k X(t-k+1). \tag{49}$$

Under standard assumptions, the actual value X_{t+1} will be normally distributed about the 1-step predictor, with zero mean and known variance. Moreover, because there are general theorems to the effect that under appropriate conditions the other ARMA forms have an equivalent AR representation, it will be clear that the ARMA models provide a flexible and practical basis for fitting, simulating, and predicting many types of time series. (For technical details, see the references quoted, or Box and Jenkins, 1976).

The assumptions underpinning the use of these prediction methods are rather crucial, however. The fact that it is possible to use linear representations so widely is based on a very special (indeed, a characterizing) property of the multivariate normal distribution: The mean of the conditional distribution of one of the components, given values of the others, is a linear combination of these given values. This is the famous linear regression property of the multivariate normal distribution. If the distributions happen not to be multivariate normal, then the conditional distributions will not in general have this linear regression property, and the best predictors, even in a least-squares sense, will not in general be linear predictors.

Because many types of data, in seismology as much as elsewhere, are non-normal, a more general method of finding predictive distributions is needed. One such general method is based on Monte Carlo simulations. It depends on the model structure being specified in such a way that the distribution of the next value of the process, given the past, is an explicit function of the past history. Then simulation methods can be used to simulate the next value, this in turn can be fed into the formulae, which determine the distribution of the next random element, and so on; from the last observation on, the simulation can keep going, using a combination of past data and the additional values it generates itself. The simulations then provide empirical distributions for all quantities that might be required for prediction purposes: probabilities, expected values, variances, etc., and not just for the future process values, but for any functionals of those values that may be of interest.

This approach is both general and powerful. Extensive numerical computations are involved, but this is ceasing to be a limitation except in the most complex or extensive models. Even under the normality assumptions referred to earlier, where there are analytic formulae for predicting forward the expected values of the random variables themselves, simulation methods may still be needed to compute the distributions of nonlinear functionals, such as a maximum or the time to the next value exceeding a given level.

4.3 Bayesian Methods

Bayesian methods provided the first and still the most logically attractive approach to the "inverse problem" of probability: Given the data, establish probability statements concerning the parameters. Bayes, Laplace, and those who followed them, recognized that such a program is not logically feasible unless there is some starting point for treating the parameters as random variables. This starting point, the *prior distribution*, is like an axiom: It has no real justification other than the willingness of the user to take it up. Hence the price one pays for adopting this approach, for all its logical appeal, is a degree of subjectivity. Proponents would claim that some element of this kind is unavoidable in statistical inference; that it rarely causes a problem in operational terms; and that it is more than offset by the power and flexibility of the resultant theory. Wherever one stands in this debate, there is no doubt that Bayesian methods form an important component of modern statistics, and an indispensable one in problems involving costs and decisions.

The Bayesian approach to estimation is summarized in the formula

$$Posterior = \frac{prior \times likelihood}{marginal}. \tag{50}$$

The prior is a distribution for the unknown parameters θ (which may be a vector), with density say $\pi(\theta)$. It summarizes the knowledge concerning θ that is available before the data is seen. The likelihood $L(\vec{x} \mid \theta)$ is here best regarded as the conditional distribution of the data, given the parameters. The denominator is the marginal distribution for the data, which has the form of a mixture distribution

$$f(\vec{x}) = \int L(\vec{x} \mid \theta)\pi(\theta)d\theta. \tag{51}$$

The posterior distribution combines the information about θ from both prior knowledge and the data, and forms the basis for any further inference or decisions concerning θ. The equation itself is a continuous version of Bayes' formula [Eq. (12)].

The chief controversy concerning Bayesian methods has always related to the interpretation and choice of the prior distribution. Here many different viewpoints—*subjective Bayesian, objective Bayesian, empirical Bayesian*, among others—are still possible. The physicists, from Laplace onward, have led the field in the search for objective priors, particularly priors that would most appropriately reflect a state of "pure ignorance." The *maximum entropy* approach of Jaynes and co-workers (Jaynes, 1983) is a powerful modern attack on this problem. The ABIC method developed by Akaike and colleagues in Tokyo is another variant on the same theme. Designed for situations where there are many parameters to be estimated, it involves a two-stage approach. The prior, which is used to steer the estimation procedures away from implausible solutions, contains its own unknown parameters (hyperparameters). These are selected by applying the ordinary AIC procedures to the marginal distribution of the data [the denominator in Eq. (50)]. Further details and several examples are given in Part II; see in particular, §2, §10, and Appendix 2.1 in the attached Handbook CD #2.

In more conventional applications, a convenient choice, when it is available, is a *conjugate prior*, with the property that the prior and the posterior have the same form up to change in the parameter values. For example, if the data is a count N with a Poisson (μ) distribution, and μ has a gamma distribution $\Gamma(\alpha, \lambda)$, the posterior distribution is $\Gamma(\alpha + N - 1, \lambda + 1)$. Its mean can be written in the form of a *credibility estimate*—a weighted average of prior and current values:

$$\mu_{post} = p\mu_{prior} + (1 - p)N, \quad p = \frac{\lambda}{\lambda + 1}, \tag{52}$$

Other useful conjugate pairs are normal-normal (for the mean μ), beta-binomial, gamma-exponential, inverse gamma-normal (for the variance σ^2).

If, in classical inference, parametric inference is based on the likelihood, in Bayesian inference it is based on the posterior density. The mode (maximum) of the posterior density defines the *maximum a posteriori* (*MAP*) estimate, and provides the closest analogy to the maximum likelihood estimate. Many results for *MAP* estimates, including numerical methods, asymptotic normality, and the form of the asymptotic variances, carry over directly from that context. Another important class comprises the *Bayes estimates*, which arise when a cost is put on the error in estimating the parameter values. For a given cost function, and a given prior, the associated Bayes estimate minimizes the expected posterior loss. For example, the posterior mean is the Bayes estimate associated with the quadratic cost function $(\hat{\theta} - \theta_0)^2$, while the posterior median is the Bayes estimate for a cost function proportional to the absolute difference $|\hat{\theta} - \theta_0|$. In general, whatever estimate is used, the influence of the prior decreases with increasing sample size, until Bayesian and classical estimates coincide.

Bayesian confidence intervals can be constructed by drawing a horizontal line through the mode of the posterior density, and dropping its level until the area (probability) under the density function, between the points of intersection of the line and the density, is equal to the desired confidence level. This construction leads to a confidence interval having minimum width for the prescribed confidence level. Because it incorporates prior information, the Bayesian confidence interval is generally shorter than its classical counterpart.

Bayesian concepts are equally relevant to the problem of prediction. With an i.i.d. sequence, for example, one might think of predicting the next observation by its expected value, using the best-fitting member of the parametric family under consideration. But this takes no account of the possible errors in the parameter estimates. The Bayesian approach is to form a *predictive distribution* as a mixture of the members of the parametric family, weighting the members according to the posterior density. The resulting distribution can be used to determine point estimates and confidence intervals for the predicted quantity, or probability statements about its value. The approach can be extended readily to the prediction methods based on simulation described in the previous section. The main difference is that for each new run of the simulation, new parameter values should be selected from the posterior distribution.

Bayesian estimation and prediction can both be subsumed under the more general heading of Bayesian methods in decision theory; see texts such as Raiffa and Schaeffer (1961), Ferguson (1967), or Lehmann (1983), for introductions to these ideas.

References

Adler, R. J., R. E. Feldman, and M. S. Taqqu (Eds.) (1998). *A Practical Guide to Heavy Tails*. Birkhauser, Boston.

Atkinson, A. C. (1982). *Plots, Transformations and Regression: An Introduction to Graphical Methods of Diagnostic Regression Analysis*. Oxford University Press, Oxford.

Båth, M. (1974). *Spectral Analysis in Geophysics*. Elsevier, New York.
Billingsley, P. (1995). *Probability and Measure*, 3rd ed. Wiley, New York.
Bishop, Y. M., S. E. Fienberg, and P. W. Holland (1975). *Discrete Multivariate Analysis: Theory and Practice*. M.I.T. Press, Cambridge.
Box, G. E. P., and G. M. Jenkins (1976). *Time Series Analysis, Forecasting and Control*, 2nd ed. Holden-Day, San Francisco.
Brillinger, D. R. (1981). *Time Series: Data Analysis and Theory*, 2nd ed. Holt, Rinehart and Winston, New York.
Chambers, J. M., and T. J. Hastie (1992). *Statistical Models in S*. Wadsworth & Brooks/Cole, Pacific Grove.
Cleveland, W. S. (1993). *Visualizing Data*. Hobart Press, Summit, New Jersey.
Cook, R. D., and S. Weisberg (1982). *Residuals and Influence in Regression*. Chapman and Hall, New York.
Cox, D. R., and D. V. Hinkley (1974). *Theoretical Statistics*. Chapman and Hall, London.
Cox, R. T. (1946). Probability, frequency, and reasonable expectation. *Am. J. Phys.* **14**, 1–13.
Cramer, H. (1946). *Mathematical Statistics*. Princeton University Press, Princeton (many subsequent editions).
Daley, D. J. and D. Vere-Jones (1988). *An Introduction to the Theory of Point Processes*. Springer, New York.
Evans, M. N., N. Hastings, and B. Peacock (1933). *Statistical Distributions*, 2nd ed. Wiley, New York.
Feller, W. (1968). *An Introduction to Probability Theory and Its Applications*, Vol. I and II, 3rd ed. Wiley, New York.
Ferguson, T. S. (1967). *Mathematical Statistics; a Decision Theoretic Approach*. Academic Press, New York.
Fisher, R. A. (1925). *Statistical Methods for Research Workers*. Oliver and Boyd, London (many subsequent editions).
Gnanadesikan, R. (1981). *Statistical Data Analysis of Multivariate Observations*. Wiley, New York.
Gnedenko, B. V. (1962). *Theory of Probability*. Chelsea, New York.
Hald, A. (1952). *Statistical Theory with Engineering Applications*, 2nd ed. Wiley, New York.
Hannan, E. J. (1970). *Multiple Time Series*. Wiley, New York.
Härdle, W. (1990). *Applied Non-Parametric Regression*. Cambridge University Press, Cambridge.
Hogg, R. V., and E. A. Tanis (1989). *Probability and Statistical Inference*. MacMillan, New York.
Huber, P. (1981). *Robust Statistics*. Wiley, New York.
Huygens, C. (1657). *De Ratiociniis in Aleae Ludo*. In: Part 5 of "Exertitationes Mathematica" (F. van Schooten, Ed.). Leiden.
Jaynes, E. T. (1983). *Papers on Entropy Methods*. (R. D, Rosenkrantz, Ed.). Reidel, Dordrecht.
Jeffreys, H. (1939). *Theory of Probability*. University Press, Oxford (many subsequent editions).
Johnson, N. L., S. Kotz, and N. Balakrishnan (1994). *Distributions in Statistics; Discrete Distributions; Continuous Univariate Distributions 1; Continuous Univariate Distributions 2; Continuous Multivariate Distributions*. Wiley, New York (in various editions with coauthors).
Khinchin, A. (1957). *Mathematical Foundations of Information Theory*. Dover, New York.
Lehmann, E. (1983). *Theory of Point Estimation*. Wiley, New York.
Lindley, D. V. (1965). *Introduction to Probability and Statistics from a Bayesian Viewpoint. Vol 1: Probability; Vol 2: Inference*. Cambridge University Press, Cambridge.
Lindley, D. V. (1971). *Making Decisions*. Wiley, London.
Loève, M. (1977). *Probability Theory*, 4th ed. Springer, New York.
Mandelbrot, B. (1977). *Fractals: Form, Chance and Dimension*. Freeman, San Francisco.
McCullagh, P., and J. A. Nelder (1983). *Generalised Linear Models*. Chapman and Hall, London.
Raiffa, H., and R. Schlaifer (1961). *Applied Statistical Decision Theory*. Harvard University Press, Boston.
Rao, C. R. (1973). *Linear Statistical Inference and Its Applications*. Wiley, New York.
Ripley, B. (1988). *Stochastic Simulation*. Wiley, New York.
Sakamoto, Y., M. Ishiguro, and G. Kitagawa (1986). *Akaike-Information Criterion Statistics*. Reidel, Dordrecht.
Serfling, R. J. (1980). *Approximation Theorems of Mathematical Statistics*. Wiley, New York.
Silverman, B. (1986). *Density Estimation for Statistics and Data Analysis*. Chapman and Hall, London.
Silvey, D. (1970). *Statistical Inference*. Penguin, Harmondsworth.
Tukey, J. W. (1977). *Exploratory Data Analysis*. Addison-Wesley, New York.
Whittle, P. (1992). *Probability via Expectation*, 3rd ed. Springer, Berlin.

Editor's Note

Due to space limitations, the full manuscript of this Chapter (Part I by Vere Jones and Ogata, and Part II by Ogata) is placed as computer-readable files in the attached Handbook CD #2, under the directory of \82VereJones. Please see also Chapter 16, Probabilistic approach to inverse problems, by Mosegaard and Tarantola.

83

Seismic Velocities and Densities of Rocks

Nikolas I. Christensen and Darrell Stanley
University of Wisconsin, Madison, Wisconsin, USA

1. Introduction

Much of our knowledge of crustal and upper mantle structure and composition comes from seismic refraction and reflection investigations. Crustal and upper mantle velocities have been determined on a worldwide basis (e.g., Holbrook *et al.*, 1992; Christensen and Mooney, 1995; Mooney *et al.*, 1998). These studies have often provided earth scientists with information on compressional and shear-wave velocities at various crustal depths, as well as velocity gradients and anisotropies in the form of azimuthal variations and shear-wave splitting from various localities. The velocities are used by seismologists to infer mineralogy, porosity, pore fluid types, temperature, and present or paleolithospheric stress originating from mineral and crack orientations. In addition, seismic reflections originate from contrasts of acoustic impedances, defined as products of velocity and density. To interpret this seismic data requires a detailed knowledge of velocities and densities of rocks, which are provided by laboratory investigations.

This chapter summarizes velocities in many lithologies believed to be important constituents of the lithosphere, experimental aspects of velocity measurements, and velocity–density relationships. The velocities presented in this paper are from a database obtained over a time span of three decades from the rock physics laboratory currently in operation at the University of Wisconsin–Madison. For additional summaries of rock seismic properties, the reader is referred to the articles of Birch (1960), Christensen (1982), Gerbande (1982), Rudnick and Fountain (1995), and Mavko *et al.* (1998).

2. Experimental Technique

Early measurements of elastic wave velocities in rocks were based on the resonant frequencies of vibration of cylindrical bars. These studies were particularly successful in measuring shear velocities (e.g., Birch and Bancroft, 1938). By the early 1950s, advances in electronics made it practical to determine velocities by observing the travel times of pulses in cylindrical rock specimens, usually at the ultrasonic frequency (e.g., Hughes and Jones, 1950). This method of measurement is well suited for studies at high pressures and is presently the standard technique used to obtain both compressional and shear-wave velocities in rocks.

In the pulse transmission technique, transducers are placed on both ends of a rock cylinder usually 2.54 cm in diameter and 4 to 6 cm in length. For velocity measurements as a function of confining pressure, the sample is jacketed with copper foil to prevent high-pressure oil from entering microcracks and pores. Transducers and brass backing pieces are assembled to the ends of the jacketed cylinder, and rubber tubing is slipped over the assembly to exclude the pressure fluid from the transducers and rock specimen. The brass backing pieces, which are attached to the transducers, are ungrounded and the copper jacket surrounding the sample is grounded to the pressure vessel.

The sending transducer converts the input electrical pulse of 50 to 500 V and 0.1 to 10 μsec width, to a mechanical signal, which is transmitted through the rock. The receiving transducer changes the wave back to an electrical pulse, which is amplified and displayed on an oscilloscope screen (Fig. 1). Once the system is calibrated for time delays, the travel time of the elastic wave through the specimen is measured directly on the oscilloscope or with the use of a mercury delay line (Birch, 1960, Christensen, 1985). A major advantage of the delay line is that it increases precision, because the gradual onset of the first arrival from the sample is approximated by the delay line. The rock velocity V_r is given by:

$$V_r = \ell_s V_{hg} / \ell_{hg} \quad (1)$$

INTERNATIONAL HANDBOOK OF EARTHQUAKE AND ENGINEERING SEISMOLOGY, VOLUME 81B ISBN: 0-12-440658-0

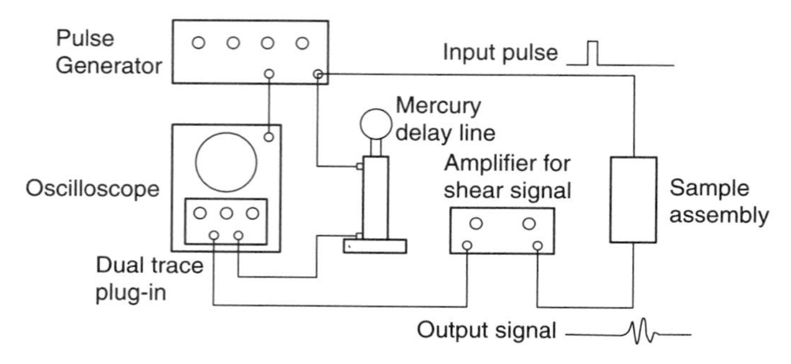

FIGURE 1 Schematic diagram of electronics for velocity measurements.

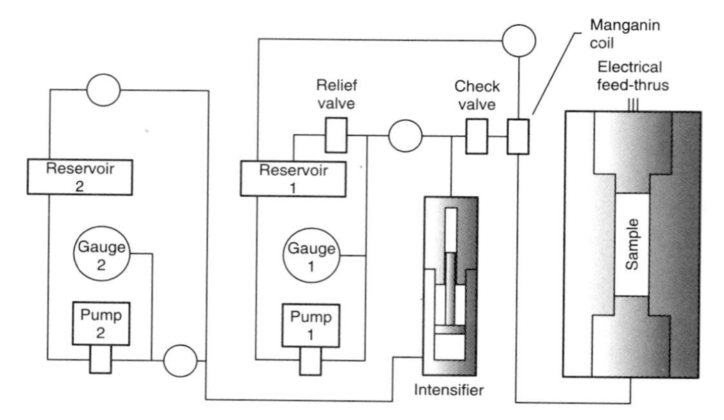

FIGURE 2 Pressure system for velocity measurements to 1000 MPa.

where ℓ_s is the sample length, ℓ_{hg} is the length of the mercury column, and V_{hg} is the velocity of mercury (1.446 km/s at 30°C).

The accuracy of the velocity measurements depends primarily on the signal quality. The transducer response is a function of its natural frequency and damping, which along with propagation through the rock cylinder results in a gradual onset of the first arrival. Signals are much better at high pressures, where cracks in the rocks have been closed, and the transducers are firmly coupled to the rock core face. The accuracy of velocity measurements at pressures above 100 MPa is usually within 0.5% for low-porosity igneous and metamorphic rocks and approximately 1% for most well-indurated sedimentary rocks (Christensen, 1985). The precision of the measurements at elevated pressures is 0.1%.

The pressure system used for the velocity measurements presented in this chapter is illustrated in Figure 2. The interior dimensions of the pressure vessel are 4.5 cm in diameter and 28 cm in length. Two air pumps operate the intensifier, which is capable of generating hydrostatic pressures in excess of 1 GPa. The check valve allows recharging of the high-pressure side of the intensifier. Confining pressure is measured by monitoring the change in resistivity of a manganin coil in contact with the pressure media, a synthetic di-ester base fluid, on the high-pressure side of the intensifier.

3. Rock Velocities

Compressional and shear-wave velocities are summarized as a function of pressure in Tables 1 and 2 for many common rock types. For the igneous and metamorphic lithologies, velocities were measured at intervals of 20 MPa from 0 to 100 MPa and at 100 MPa intervals from 100 MPa to 1000 MPa for both increasing and decreasing pressure. A typical data set from a mafic granulite facies rock is shown in Figure 3. The large rise in velocity below 200 MPa has been shown to be the result of crack closure produced by the hydrostatic pressure (e.g., Birch, 1960). At elevated pressures, the velocity–pressure curves are linear and related to the intrinsic effect of pressure on the mineral components.

Because of their high porosities, sedimentary rocks velocities in Table 2 are reported only to 200 MPa. The porosities of the sedimentary rocks are usually significantly higher than the igneous and metamorphic lithologies. The pores of many sedimentary rocks will not close with the application of pressure (e.g., Hughes and Jones, 1951; Wyllie *et al.*, 1956), and often application of high pressures produces crushing and significant hysteresis between increasing and decreasing velocity measurements.

Several of the igneous and metamorphic rock categories of Table 1 are similar to those summarized by Christensen and Mooney (1995) and Christensen (1996). Notable additions include velocities of rhyolite, obsidian, peridotite, and blueschist, as well as ultra high pressure coesite eclogite from Dabieshan, China. The rhyolite samples are from Bishop, California, and the blueschist samples were collected from western Washington and Alaska.

The sedimentary rocks in Table 2 are all well-consolidated, typical mid-continent platform lithologies. Sandstone and dolostone samples were collected from the UPH 2 and UPH 3 drill

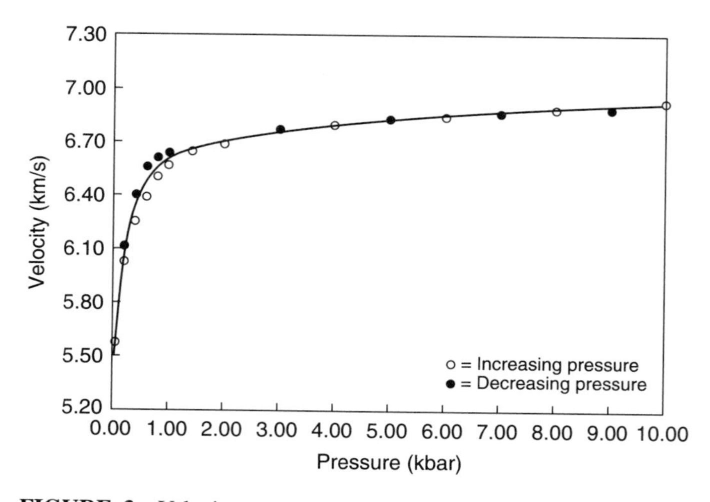

FIGURE 3 Velocity measurements as a function of confining pressure for a mafic granulite.

TABLE 1 Average Compressional (V_P) and Shear (V_S) Wave Velocities as a Function of Pressure and Densities (ρ) for Igneous and Metamorphic Rock Types*

					200 MPa		400 MPa		600 MPa		800 MPa		1000 MPa	
				ρ	V_P	V_S	V_P	V_S	V_P	V_S	V_P	V_S	V_P	V_S
Igneous rocks														
Plutonic	Granite (GRN)	$S = 108$	Avg.	2652	6.246	3.669	6.296	3.692	6.327	3.706	6.352	3.717	6.372	3.726
		$R = 38$	S.D.	23	0.128	0.116	0.121	0.104	0.121	0.100	0.123	0.096	0.124	0.095
	Diorite (DIO)	$S = 24$	Avg.	2810	6.497	3.693	6.566	3.717	6.611	3.733	6.646	3.745	6.675	3.756
		$R = 8$	S.D.	85	0.161	0.120	0.144	0.110	0.134	0.106	0.120	0.101	0.120	0.099
	Gabbro (GAB)	$S = 174$	Avg.	2968	7.138	3.862	7.200	3.888	7.241	3.905	7.273	3.918	7.299	3.929
		$R = 58$	S.D.	69	0.252	0.129	0.255	0.125	0.258	0.124	0.261	0.124	0.263	0.124
	Diabase (DIA)	$S = 45$	Avg.	2936	6.712	3.729	6.756	3.748	6.782	3.757	6.800	3.762	6.814	3.766
		$R = 15$	S.D.	91	0.266	0.170	0.254	0.156	0.249	0.151	0.244	0.144	0.243	0.141
Volcanic	Rhyolite (RHY)	$S = 9$	Avg.	2050	3.950	2.164	4.162	2.346	4.299	2.462	4.402	2.524	4.479	2.557
		$R = 3$	S.D.	29	0.113	0.180	0.098	0.175	0.095	0.173	0.095	0.171	0.095	0.170
	Andesite (AND)	$S = 30$	Avg.	2627	5.533	3.034	5.712	3.097	5.814	3.130	5.885	3.155	5.940	3.177
		$R = 10$	S.D.	71	0.260	0.208	0.227	0.207	0.224	0.204	0.226	0.207	0.229	0.248
	Basalt (BST)	$S = 252$	Avg.	2882	5.914	3.217	5.992	3.246	6.044	3.264	6.084	3.279	6.118	3.291
		$R = 145$	S.D.	139	0.546	0.302	0.544	0.293	0.543	0.291	0.542	0.288	0.542	0.288
	Obsidian (OBS)	$S = 2$	Avg.	2383	5.810	3.540	5.795	3.515	5.780	3.490	5.766	3.475	5.750	3.456
		$R = 1$	S.D.	13	0.000	0.014	0.007	0.007	0.000	0.014	0.005	0.010	0.010	0.010
Metamorphic rocks														
Low grade	Metagraywacke (MGW)	$S = 27$	Avg.	2682	5.829	3.406	5.950	3.448	6.028	3.474	6.089	3.495	6.139	3.512
		$R = 9$	S.D.	52	0.348	0.291	0.309	0.273	0.283	0.265	0.234	0.253	0.231	0.249
	Slate (SLT)	$S = 27$	Avg.	2807	6.156	3.301	6.240	3.351	6.297	3.384	6.342	3.410	6.379	3.432
		$R = 9$	S.D.	21	0.103	0.106	0.099	0.096	0.094	0.090	0.081	0.081	0.075	0.077
	Phyllite (PHY)	$S = 57$	Avg.	2738	6.243	3.543	6.305	3.569	6.343	3.585	6.373	3.597	6.398	3.608
		$R = 19$	S.D.	46	0.116	0.140	0.093	0.139	0.090	0.140	0.087	0.140	0.086	0.140
	Zeolite facies basalt	$S = 54$	Avg.	2915	6.319	3.413	6.400	3.444	6.454	3.464	6.495	3.480	6.530	3.493
	(ZFB)	$R = 18$	S.D.	83	0.270	0.152	0.262	0.152	0.259	0.153	0.256	0.155	0.255	0.156
	Prehnite-pumpellyte	$S = 36$	Avg.	2835	6.353	3.545	6.436	3.575	6.492	3.595	6.535	3.610	6.571	3.623
	facies basalt (PFB)	$R = 12$	S.D.	108	0.414	0.237	0.370	0.219	0.354	0.214	0.334	0.207	0.326	0.204
	Greenschist facies	$S = 36$	Avg.	2978	6.820	3.883	6.884	3.911	6.925	3.929	6.957	3.943	6.983	3.955
	basalt (GFB)	$R = 12$	S.D.	86	0.227	0.131	0.220	0.134	0.221	0.137	0.223	0.141	0.225	0.143
Medium grade	Granitic gneiss (GGN)	$S = 72$	Avg.	2643	6.010	3.501	6.145	3.553	6.208	3.583	6.245	3.607	6.271	3.627
		$R = 24$	S.D.	46	0.184	0.167	0.135	0.143	0.122	0.137	0.107	0.130	0.101	0.128
	Tonalite gneiss (TGN)	$S = 156$	Avg.	2742	6.180	3.552	6.256	3.585	6.302	3.606	6.337	3.622	6.366	3.636
		$R = 52$	S.D.	68	0.189	0.166	0.171	0.150	0.166	0.148	0.161	0.141	0.160	0.139
	Mica quartz schist	$S = 87$	Avg.	2824	6.267	3.526	6.370	3.579	6.433	3.610	6.482	3.634	6.523	3.654
	(MQS)	$R = 29$	S.D.	122	0.307	0.232	0.310	0.227	0.314	0.228	0.321	0.219	0.324	0.217
	Amphibolite (AMP)	$S = 78$	Avg.	2996	6.866	3.909	6.939	3.941	6.983	3.959	7.018	3.974	7.046	3.987
		$R = 26$	S.D.	85	0.224	0.151	0.199	0.136	0.197	0.133	0.197	0.131	0.197	0.130
	Quartzite (QRZ)	$S = 24$	Avg.	2652	5.963	4.035	6.012	4.048	6.045	4.052	6.070	4.053	6.091	4.054
		$R = 8$	S.D.	8	0.074	0.048	0.076	0.042	0.077	0.041	0.078	0.040	0.079	0.040
	Marble (MBL)	$S = 21$	Avg.	2721	6.916	3.653	6.944	3.707	6.961	3.743	6.974	3.770	6.985	3.794
		$R = 7$	S.D.	12	0.085	0.188	0.080	0.181	0.080	0.180	0.080	0.179	0.081	0.179
High grade	Felsic granulite (FGR)	$S = 87$	Avg.	2758	6.411	3.608	6.474	3.631	6.514	3.646	6.545	3.657	6.571	3.667
		$R = 29$	S.D.	79	0.132	0.134	0.127	0.125	0.127	0.125	0.129	0.126	0.131	0.127
	Mafic granulite (MGR)	$S = 102$	Avg.	2971	6.839	3.767	6.902	3.799	6.942	3.820	6.973	3.836	7.000	3.849
		$R = 34$	S.D.	82	0.182	0.121	0.181	0.115	0.184	0.113	0.190	0.111	0.193	0.111
	Mafic garnet granulite	$S = 81$	Avg.	3111	7.110	3.974	7.197	4.007	7.249	4.026	7.290	4.040	7.324	4.052
	(MGG)	$R = 27$	S.D.	104	0.184	0.122	0.164	0.108	0.154	0.105	0.154	0.104	0.154	0.104
	Mafic eclogite (MEC)	$S = 51$	Avg.	3485	8.001	4.481	8.078	4.524	8.127	4.553	8.166	4.575	8.198	4.5–94
		$R = 17$	S.D.	67	0.156	0.015	0.160	0.141	0.156	0.143	0.150	0.150	0.149	0.147
	Coesite eclogite (CEC)	$S = 15$	Avg.	3544	8.048	4.567	8.146	4.610	8.187	4.629	8.214	4.642	8.236	4.652
		$R = 5$	S.D.	52	0.169	0.108	0.134	0.100	0.120	0.100	0.113	0.101	0.107	0.102
	Anorthositic granulite	$S = 30$	Avg.	2763	6.931	3.736	7.003	3.766	7.049	3.784	7.085	3.798	7.114	3.810
	(AGR)	$R = 10$	S.D.	63	0.134	0.105	0.138	0.105	0.140	0.107	0.145	0.110	0.147	0.112
	Dunite (DUN)	$S = 36$	Avg.	3310	8.299	4.731	8.352	4.759	8.376	4.771	8.390	4.778	8.399	4.783
		$R = 12$	S.D.	14	0.091	0.118	0.083	0.116	0.083	0.116	0.084	0.116	0.085	0.116
	Blueschist (BSC)	$S = 9$	Avg.	3021	7.021	3.876	7.128	3.930	7.188	3.960	7.230	3.981	7.264	3.998
		$R = 3$	S.D.	16	0.040	0.033	0.031	0.032	0.030	0.029	0.031	0.029	0.032	0.030
	Restite (RST)	$S = 9$	Avg.	2874	6.586	3.352	6.757	3.414	6.844	3.441	6.906	3.461	6.954	3.477
		$R = 3$	S.D.	124	0.435	0.638	0.448	0.667	0.457	0.680	0.464	0.690	0.471	0.698

* Densities in kg/m^3 and velocities in km/s; Avg., average; S.D., standard deviation; S, number of specimens (cores); R, number of rocks.

TABLE 2 Average Compressional (V_P) and Shear (V_S) Wave Velocities as a Function of Pressure and Densities (ρ) for Sedimentary Rock Types*

				ρ	10 MPa V_P	10 MPa V_S	50 MPa V_P	50 MPa V_S	100 MPa V_P	100 MPa V_S	200 MPa V_P	200 MPa V_S
Sedimentary rocks												
Carbonate	Dolostone (DOL)	$S = 49$	Avg.	2661	6.017	3.306	6.264	3.438	6.345	3.486	6.423	3.522
		$R = 28$	S.D.	127	0.382	0.246	0.406	0.242	0.428	0.229	0.430	0.227
	Limestone (LIM)	$S = 31$	Avg.	2659	5.659	3.064	5.891	3.172	5.995	3.200	6.092	3.241
		$R = 29$	S.D.	151	0.665	0.311	0.641	0.274	0.650	0.298	0.654	0.288
Clastic	Sandstone (SSN)	$S = 110$	Avg.	2301	4.551	2.821	4.792	3.003	4.863	3.052	4.948	3.103
		$R = 59$	S.D.	168	0.519	0.313	0.505	0.317	0.506	0.336	0.521	0.344
	Shale (SHL)	$S = 14$	Avg.	2333	3.584	2.273	3.778	2.349	3.918	2.405	3.987	2.474
		$R = 7$	S.D.	68	0.215	0.124	0.208	0.127	0.197	0.128	0.197	0.128
	Siltstone (SLS)	$S = 12$	Avg.	2659	4.621	2.849	5.024	3.020	5.227	3.108	5.430	3.207
		$R = 6$	S.D.	41	0.168	0.077	0.080	0.100	0.067	0.125	0.075	0.140

* Densities in kg/m^3 and velocities in km/s; Avg., average; S.D., standard deviation; S, number of specimens (cores); R, number of rocks.

holes, located in northern Illinois (Coates *et al.*, 1983). Limestone, siltstone, shale, and additional sandstone samples were collected from the Thorn Hill sedimentary section located in Tennessee (Christensen and Szymanski, 1991). Additional shale samples were collected from a quarry in southern Indiana (Johnston and Christensen, 1995).

Compared with velocity measurements as a function of pressure, the determination of velocity temperature derivatives for rocks has not received as much attention by experimentalists. It has been well known since the early studies of Ide (1937) that the application of temperature to a rock at atmospheric pressure results in the formation of cracks, which often produce permanent damage to the rock. Because of this, reliable measurements of the temperature derivatives of velocity require confining pressures high enough to prevent crack formation. In general, pressures of 200 MPa are sufficient for temperature measurements to 300°C.

Several techniques have been used to measure the effects of temperature on rock velocities (e.g., Birch, 1943; Fielitz, 1971; Stewart and Peselnick, 1977; Christensen, 1979; Kern and Richter, 1981; Ramananantoandro and Manghnani, 1978). These studies have used either resonance techniques or more frequently the pulse transmission method. In general, $(\delta V_P/\delta T)_P$ for common rocks ranges from -0.3×10^{-3} to -0.6×10^{-3} km/s/°C, and $(\delta V_S/\delta T)_P$ varies between -0.2×10^{-3} and -0.4×10^{-3} km/s/°C. Temperature derivatives of compressional and shear-wave velocities for some common rocks are given in Tables 3 and 4.

Since increasing temperature decreases velocities and increasing pressure increases velocities, velocity gradients in homogeneous crustal regions depend on the geothermal gradient. The change of velocity with depth is given by

$$dV/dZ = (\delta V/\delta P)_T dP/dZ + (\delta V/\delta T)_P dT/dZ \quad (2)$$

where V is velocity, Z is depth, T is temperature, and P is pressure. In regions with normal geothermal gradients (25°–40°C/km), dV/dZ is approximately zero (Christensen, 1979).

TABLE 3 Compressional Velocity Temperature Derivatives for Some Common Rocks

Rock	dV_P/dT ($\times 10^{-3}$ km/s/°C)	Reference
Granite, Massachusetts	−0.39	Christensen, 1979
Gabbro, Mid Atlantic Rise	−0.57	"
Basalt, East Pacific Rise	−0.39	"
Anorthosite, Quebec	−0.41	"
Dunite, Washington	−0.56	"
Granulite, New Jersey	−0.49	"
Peridotite, Italy	−0.49	Kern and Richter, 1981
Sillimanite Gneiss, Norway	−0.36	"

TABLE 4 Shear Velocity Temperature Derivatives for Some Common Rocks

Rock	dV_S/dT ($\times 10^{-3}$ km/s/°C)	Reference
Norite, Germany	−0.15	Kern and Richter, 1981
Dunite, Norway	−0.35	"
Peridotite, Italy	−0.39	"
Granite, Germany	−0.21	"
Biotite Gneiss, Norway	−0.17	"
Amphibolite, Norway	−0.21	"

High heat flow regions, however, where geothermal gradients are high, will have crustal velocity reversals as long as compositional changes with depth are minimal.

3.1 Igneous Rocks

Classification schemes for most rock types allow for considerable variations in mineralogy for a given rock type. Because of this, rocks correctly classified as granite, gabbro, and so on show significant variations in velocities and densities; thus, the standard deviations given in Table 1 often reflect this variability.

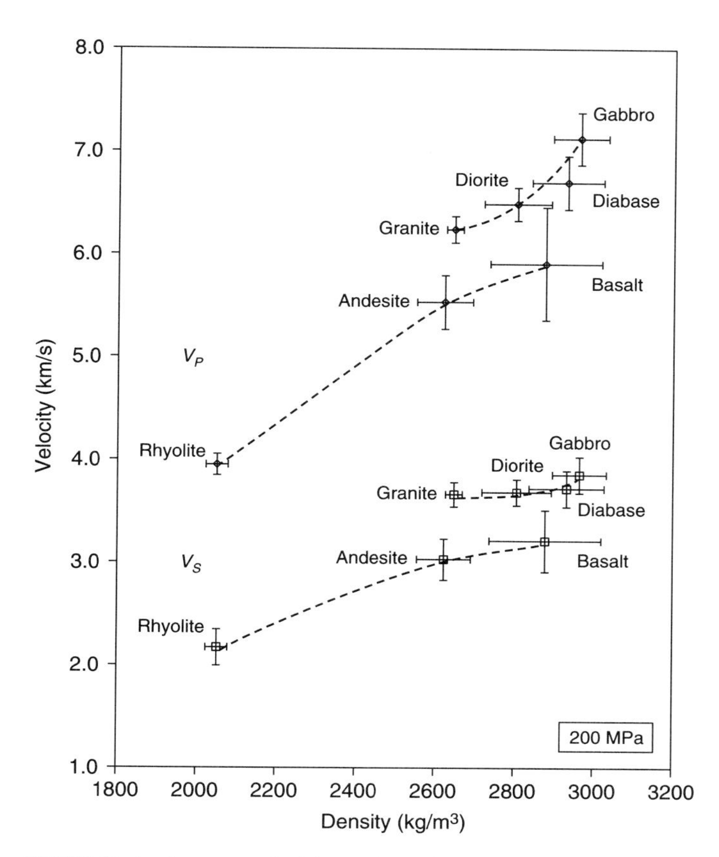

FIGURE 4 Average velocities and standard deviations versus average densities for common igneous rocks.

In addition, the classification of the more common igneous rocks allows for a gradual change in mineralogy as one goes from granite to gabbro, where quartz content decreases, feldspar becomes more calcic, and relatively fast minerals such as amphibole and pyroxene become increasingly abundant. The result is an increase in velocities and densities.

In Figure 4, velocities of the common igneous rocks are plotted against their densities at 200 MPa. For the plutonic rocks, V_P increases gradually from about 6.3 km/s in granite to 6.5 km/s in dorite and 7.2 km/s in gabbro. Shear-wave velocities for these rocks show a less dramatic increase from approximately 3.7 km/s in granite to 3.85 km/s in gabbro. Quartz has a relatively high shear-wave velocity (low Poisson's ratio), thus V_S of granite is almost as high as that of diorite.

The volcanic rocks have lower velocities and densities of chemically equivalent plutonic rocks (Figure 4). Also note the relatively large range in velocities and densities between rhyolite and basalt. These differences are for the most part related to abundant alteration common to the volcanic rocks, which lowers velocities and densities. In addition, stiff vesicles will not close under the range of hydrostatic pressures for which velocities are reported in this chapter and thus contribute significantly to the lowering of velocities and densities.

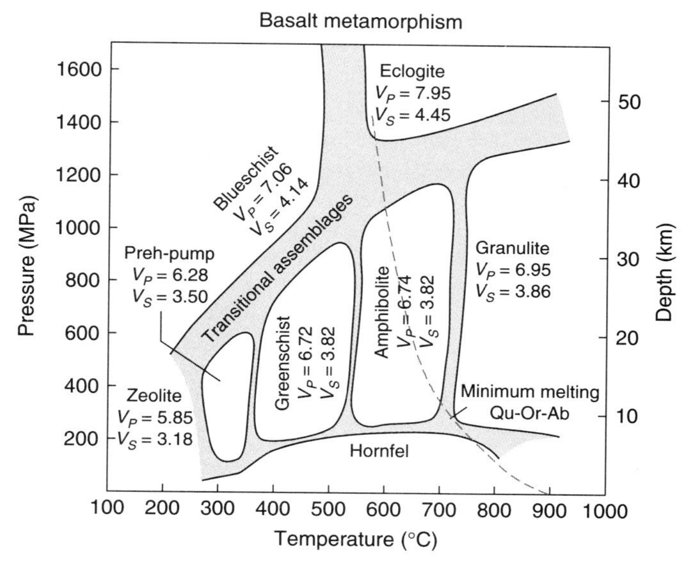

FIGURE 5 Average velocities at temperature and pressure for metamorphosed basalt.

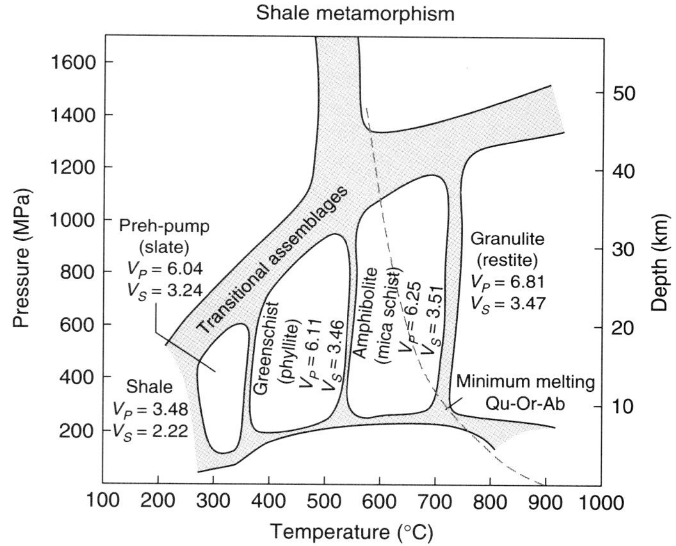

FIGURE 6 Average velocities at temperature and pressure for metamorphosed shale.

3.2 Metamorphic Rocks

For deep crustal studies, velocities in metamorphic lithologies are likely to be of more significance than velocities in igneous rocks. The mineralogies of metamorphic rocks and hence their velocities and densities are functions of many variables, the most important being bulk chemistry, pressure and temperature history, and the availability of water during metamorphism. Earlier studies have shown that velocities generally increase systematically with progressive metamorphism (e.g., Christensen, 1996).

Two important parent rock types for many of the metamorphic rocks of Table 1 are basalt and shale. Figures 5 and 6 show a simplified metamorphic facies classification after Yardley (1989) for metamorphic lithologies exclusive of the low-pressure, high-temperature hornfels. Average velocities at average pressures and temperatures are given for each metamorphic facies for basalt (Fig. 5) and shale (Fig. 6). The velocities are from Table 1 and have been corrected for temperature (Tables 3 and 4). Note that basalt compressional and shear-wave velocities along a "normal" geotherm will increase with depth from 5.85 and 3.18 km/s at near-surface pressure and temperature conditions to 6.95 and 3.86 km/s at lower crustal granulite facies depths. The high-pressure blueschist and eclogite facies assemblages have average velocities higher than the mafic granulite rocks.

Velocities resulting from progressive metamorphism of shale also increase with increasing metamorphic grade. The change in velocities going from shale to slate is quite significant (V_P and V_S increase 74% and 47%, respectively). Although velocities in the pelitic assemblages are usually much lower than the metamorphic basaltic equivalent, at granulite conditions the pelitic "restites," which have lost significant felsic components, have lower crustal velocities.

3.3 Sedimentary Rocks

This study includes five major sedimentary rock types divided into clastic and carbonate categories. The carbonate rocks include dolostone and limestone, and the clastic rocks include sandstone, shale, and siltstone. Figure 7 shows the average and range of compressional and shear-wave velocities for the five rock types at 200 MPa confining pressure.

The carbonate rocks have the highest average compressional and shear-wave velocities at 6.4 and 3.5 km/s for dolostone and 6.1 and 3.2 km/s for limestone. The wide range in compressional wave velocity (~1.0 and 1.5 km/s) and shear-wave velocity (~0.5 and 0.7 km/s) for the dolostone and limestone samples, respectively, is mainly due to variations in porosity and quartz content.

The clastic rocks have lower average velocities than the carbonates. The sandstones have an average compressional wave velocity of 4.9 and shear-wave velocity of 3.1 km/s, shale has velocities of 4.0 and 2.5 km/s, and siltstone has velocities of 5.4 and 3.2 km/s. The sandstone samples also show wide ranges of compressional wave velocity (~1.8 km/s) and shear-wave velocity (~0.8 km/s). This is due primarily to variations in carbonate and clay content as well as variations in porosity. The shale and siltstone samples have much smaller ranges in velocity and in general have smaller variations in mineralogy and porosity. It should be noted, however, that there are significantly fewer siltstone and shale samples in our compilation.

Figure 8 shows velocity versus density for all of the sedimentary samples included in this study. Here the dolostone and

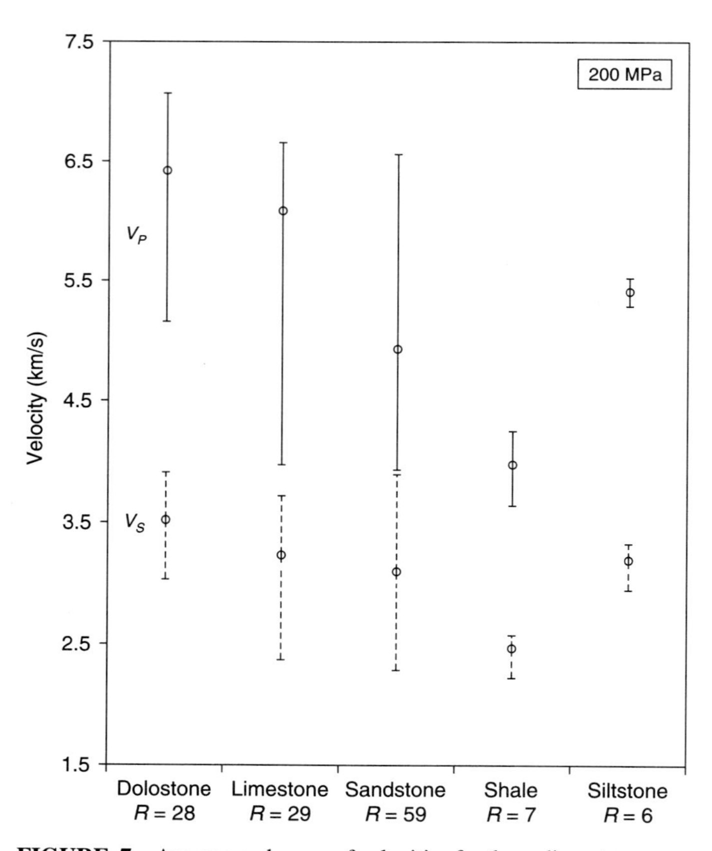

FIGURE 7 Average and range of velocities for the sedimentary samples included in this study. R is the number of rocks for each lithology.

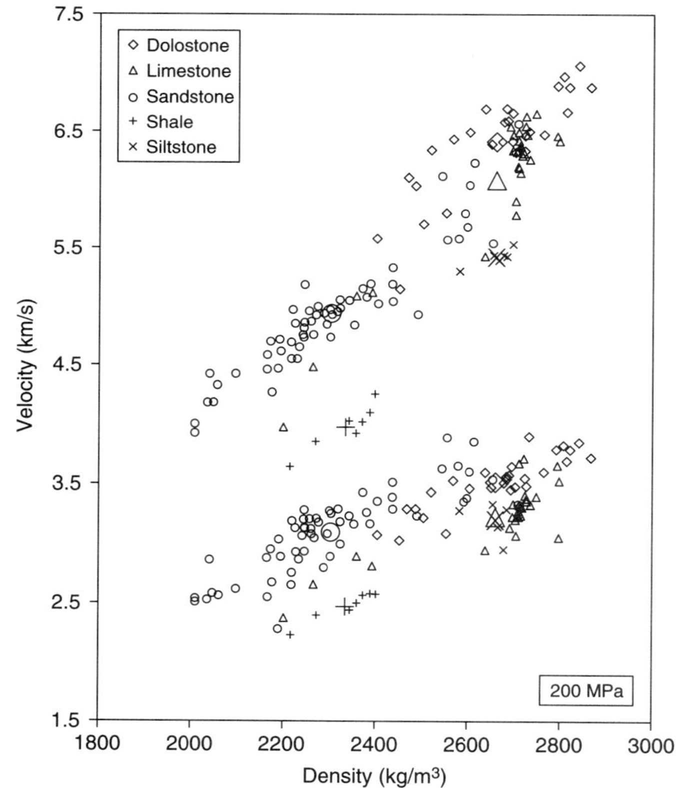

FIGURE 8 Velocity versus density for the sedimentary samples included in this study. Large symbols are the average values.

limestone samples occupy the region of high velocity and density, while the sandstone and shale samples are in the low-velocity, low-density region. It should be noted that the shale samples have densities similar to the sandstones, but significantly lower velocities. Also, the siltstone samples have similar densities and shear-wave velocities to the carbonate samples but significantly lower compressional wave velocities. The wide ranges of velocity and density values for the carbonates and the sandstone samples are also illustrated in Figure 7.

The shale samples have significant anisotropy, with the fast direction for compressional waves being parallel to bedding. The fast direction for shear waves is with propagation parallel to bedding and vibration direction also parallel to bedding. Anisotropy in the shale samples can be as high as 30% for both compressional and shear-wave velocity. For further discussion on shale anisotropy, see Johnston and Christensen (1995).

4. Velocity–Density Relationships

The relationship between seismic velocity and density has important implications for multidisciplinary geophysical studies involving seismic and gravity methods. A detailed examination of the density and seismic velocity (at elevated pressures) of rocks of different lithologies can help in the determination of velocity from gravity data or vice versa. This relationship can also be used to determine lithology from either gravity or seismic data. Finally, velocity–density relationships provide information about the acoustic impedances of different rock types, which are important in reflection seismology exploration.

Previous examples of velocity–density relationships include the Nafe-Drake curve (Nafe and Drake, 1957), which is based on experimental velocity data on unconsolidated sediments, sedimentary, metamorphic, and igneous rocks from various sources as well as a theoretical solution for the upper mantle. Birch's (1961) law relates velocity, density, and mean atomic weight for igneous and metamorphic rocks, and Christensen and Salisbury (1975) provide regression line solutions for various oceanic rocks. Gardner *et al.* (1974) define an empirical relationship between velocity and density for a variety of sedimentary rock types. Finally, Christensen and Mooney (1995) present average velocities and average densities for several igneous and metamorphic rocks as well as linear and nonlinear regression line parameters for several depths.

In this chapter we present an empirical relationship between velocity (at elevated pressures) and density for sedimentary, igneous, and metamorphic rocks. Velocity and density data used are averages of 2096 cores from 772 rock samples (Tables 1 and 2). Figure 9 shows a plot of compressional and shear-wave velocity at 200 MPa versus density for the lithologies discussed earlier in this chapter. This figure also includes linear trend lines and their equations for both the compressional and the shear-wave data. The data used for this plot are from Table 1.

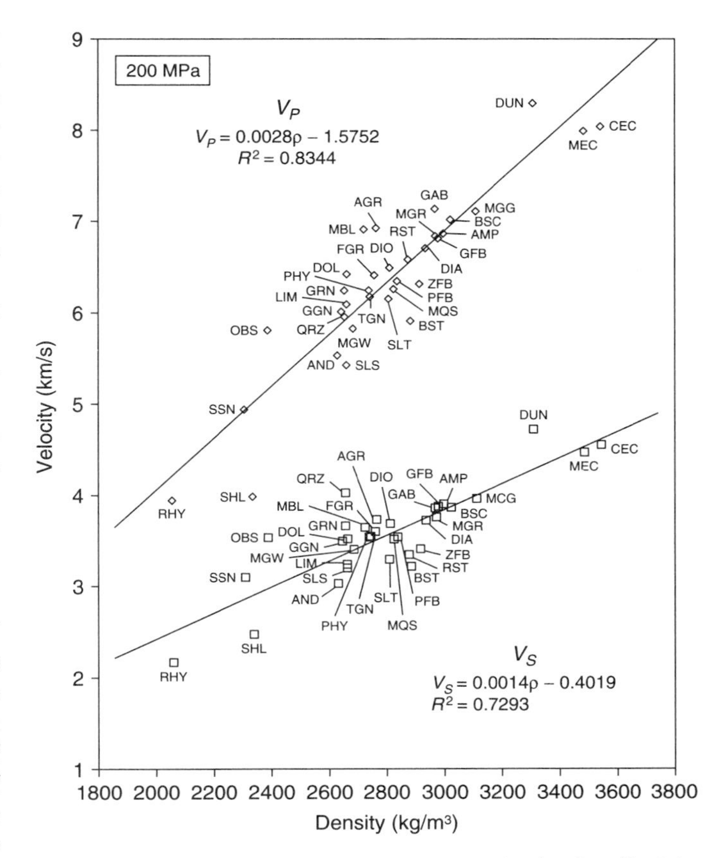

FIGURE 9 Average velocity versus average density for all of the lithologies included in this study.

5. Conclusions

This chapter has been intended to give a summary of laboratory-derived velocities of several different rock lithologies. We have also attempted to correlate important rock properties with velocity, such as density, metamorphic grade, and mineralogy.

All of the velocities and densities reported here were measured in the same rock physics lab currently located at the University of Wisconsin–Madison. We realize that there are many rock types not included in this summary. We feel, however, that those included are a good representation of common lithologies encountered in most crustal and upper mantle geophysical investigations.

References

Birch, F. (1943). Elasticity of igneous rocks at high temperatures and pressures. *Geol. Soc. Am. Bull.* **54**, 263–285.

Birch, F. (1960). The velocity of compressional waves in rocks to 10 kilobars, Part 1. *J. Geophys. Res.* **65**, 1083–1102.

Birch, F. (1961). The velocity of compressional waves in rocks to 10 kilobars, Part 2. *J. Geophys. Res.* **66**, 2199–2224.

Birch, F., and D. Bancroft (1938). The effect of pressure on the rigidity of rocks. *J. Geol.* **46**, 59–87.

Christensen, N. I. (1979). Compressional wave velocities in rocks at high temperatures and pressures, critical thermal gradients, and crustal low-velocity zones. *J. Geophys. Res.* **84**, 6849–6857.

Christensen, N. I. (1982). Seismic velocities. In: *Handbook of Physical Properties of Rocks* (R. S. Carmichael, Ed.), Vol. 2, pp. 1–228, CRC Press, Boca Raton, FL.

Christensen, N. I. (1985). Measurements of dynamic properties of rock at elevated temperatures and pressures. In *Measurement for Rock Properties at Elevated Pressures and Temperatures* (H. J. Pincus and E. R. Hoskins, Eds.), pp. 97–107, Am. Soc. for Test. and Mater., Philadelphia, PA.

Christensen, N. I. (1996). Poisson's ratio and crustal seismology. *J. Geophys. Res.* **101**, 3139–3156.

Christensen, N. I. and W. D. Mooney (1995). Seismic velocity structure and composition of the continental crust: A global view. *J. Geophys. Res.* **100**, 9761–9788.

Christensen, N. I. and M. H. Salisbury (1975). Structure and composition of the lower oceanic crust. *Rev. Geophys.* **13**, 57–86.

Christensen, N. I., and D. L. Szymanski (1991). Seismic properties and the origin of reflectivity from a classic Paleozoic sedimentary sequence, Valley and Ridge province, southern Appalachians. *Geol. Soc. Am. Bull.* **103**, 277–289.

Coates, M. S., B. C. Haimson, W. J. Hinze, and W. R. Van Schmus (1983). Introduction to the Illinois Deep Hole project. *J. Geophys. Res.* **88**, 7267–7275.

Fielitz, K. (1971). Elastic wave velocities in different rocks at high pressure and temperature up to 750°C. *Z. Geophys.* **37**, 943–956.

Gardner, G. H. F., L. W. Gardner, and A. R. Gregory (1974). Formation velocity and density—The diagnostic basics for stratigraphic traps. *Geophysics*, **39**, 770–780.

Gerbande, H. (1982). Elastic wave velocities and constants of elasticity of rocks and rock forming minerals. In: *Physical Properties of Rocks* (G. Angenheister, Ed.), pp. 1–140, Landolt-Börnstein, Springer.

Holbrook, W. S., W. D. Mooney, and N. I. Christensen (1992). The seismic velocity structure of the deep continental crust. In: *Lower Continental Crust* (D. M. Fountain, R. Arculus, and R. Kay, Eds.), pp. 1–43, Elsevier, New York.

Hughes, D. S., and H. J. Jones (1950). Variation of elastic moduli of igneous rocks with pressure and temperature. *Geol. Soc. Amer. Bull.* **61**, 843–856.

Hughes, D. S., and H. J. Jones (1951). Elastic wave velocities in sedimentary rocks. EOS, *Trans. Am. Geophys. Un.* **32**, 173–178.

Ide, J. M. (1937). The velocity of sound in rocks and glasses as a function of temperature. *J. Geol.* **45**, 689–716.

Johnston, J. E., and N. I. Christensen (1995). Seismic anisotropy of shales. *J. Geophys. Res.* **100**, 5991–6003.

Kern, H., and A. Richter (1981). Temperature derivatives of compressional and shear-wave velocities in crustal and mantle rocks at 6 kbar confining pressure. *J. Geophys. Res.* **49**, 47–56.

Mavko, G., T. Mukerji, and J. Dvorkin (1998). *The Rock Physics Handbook*, pp. 289–303, Cambridge University Press, MA.

Mooney, W. D., G. Laske, and T. G. Masters (1998). CRUST 5.1: A global crustal model at 5° × 5°. *J. Geophys. Res.* **103**, 727–747.

Nafe, J. E., and C. L. Drake (1957). Variation with depth in shallow and deep water marine sediments of porosity, density and the velocity of compressional and shear waves. *Geophysics* **22**, 523–552.

Ramananantoandro, R., and M. H. Manghnani (1978). Temperature dependence of the compressional wave velocity in an anisotropic dunite; measurements to 500 degrees C at 10 kbar. *Tectonophysics* **47**, 73–84.

Rudnick, R. L., and D. M. Fountain (1995). Nature and composition of the continental crust; a lower crustal perspective. *Rev. Geophys.* **33**, 267–309.

Stewart, R., and L. Peselnick (1977). Velocity of compressional waves in dry Franciscan rocks to 8 kbar and 300 degrees C. *J. Geophys. Res.* **82**, 2027–2039.

Wyllie, M. J. R., A. R. Gregory, and L. Caruso (1956). Elastic wave velocities in heterogeneous and porous media. *Geophysics* **21**, 41–70.

Yardley, B. W. (1989). *An Introduction to Metamorphic Petrology*, John Wiley, New York, 248 pp.

84

Worldwide Nuclear Explosions

Xiaoping Yang, Robert North, and Carl Romney
Science Applications International Corporation, Arlington, Virginia, USA
Paul G. Richards
Columbia University, Palisades, New York, USA

1. Introduction

The first nuclear test, Trinity, exploded near Alamogordo, New Mexico, USA, on 16 July 1945, marked the beginning of the nuclear weapons era. There is evidence that 2039 additional explosions were conducted by seven countries (China, France, India, Pakistan, the Soviet Union, the United Kingdom, and the United States) between 1945 and 1998, according to information in the Nuclear Explosion Database at the Center for Monitoring Research (CMR) (Yang *et al.*, 2000a; *http://www.cmr.gov*). For completeness, the CMR database also includes the possible but disputed occurrence of an atmospheric nuclear explosion on 22 September 1979, with the "responsible country" listed as "Unknown." This chapter summarizes information on worldwide nuclear explosions extracted from the CMR database.

The CMR Nuclear Explosion Database contains comprehensive data relevant to nuclear monitoring research (e.g., origin time, location, and yield) on nuclear explosions worldwide. To ensure the completeness of information, data have been collected from a variety of sources, ranging from government announcements to media reports. The database have been maintained and updated on a regular basis as new data has become available and as errors have been corrected. In general, an event in the database has several "origins" (data on location, time of occurrence, and confidence bounds). A preferred origin for any given event has been selected based on the most complete and accurate information on the coordinates and time of each explosion. When possible, ground truth (GT) categories have been assigned to the data source to reflect the location accuracy. GTX refers to events with location accuracy better than X km (Yang *et al.*, 2000b). While the accuracy of the origin time is also crucial information, at this time the focus is limited to location accuracy when defining GT events.

Table 1 gives the number of events by country in alphabetical order, and event locations are shown in Figure 1. The two nuclear weapons that the United States exploded over Japan (in August 1945) are also included. Official listings are available for all explosions conducted by the United States (DOE/NV-209, 1997; NV209), and for all nuclear explosions conducted by the Soviet Union (Mikhailov *et al.*, 1996; RFAE). Partial listings of nuclear explosions are available from official publications of China (China Today, 1988; CT), France (CEA/DAM), and the United Kingdom (in NV209). These listings confirm the nuclear nature of each listed event, and frequently the yield has also been given. However, they may not include origin times or event coordinates. Such data have been obtained for most events from other sources, official and unofficial and of varying quality, as described in the following sections. In the absence of coordinates or time for a given event, nominal coordinates or an origin time of 00:00:00 were assigned (there are exceptions for USSR events on 6 June 1964, 8 October 1978, and 7 July 1987 whose announced or estimated origin times were 00:00:00).

2. US and UK Nuclear Explosions

The United States Department of Energy (NV209) listed all nuclear tests (e.g., date, location, yield range, and explosion type) conducted by the United States; these took place from July 1945 to September 1992. Among the 1056 events are 24 joint underground tests with the United Kingdom (US-UK) conducted in Nevada. The United States also provided official information on many tests for inclusion in Preliminary Determinations of Epicenters (PDE) or bulletins of the International Seismological Centre (ISC), including origin time and coordinates. This information is identified by authors AEC, ERDA, or DOE.

Griggs and Press (1961; G+P), published authoritative data on times and coordinates of most US tests through 1958. Springer

INTERNATIONAL HANDBOOK OF EARTHQUAKE AND ENGINEERING SEISMOLOGY, VOLUME 81B ISBN: 0-12-440658-0

and Kinnaman (1971; 1975; SPRINGER) published such information on time and coordinates for subsequent US or joint US-UK underground nuclear explosions through 1973. The accuracy of the origin information (latitude, longitude, time) for these later tests is expected to be well within ±0.1 km in location and ±0.005 sec in time. Other preferred data sources include Gutenberg and Richter (1946; G+R) for the Bikini Baker event, Gutenberg (1946) for the time of the Trinity event, and Johnson *et al.* (1981) for the Swordfish event.

Authoritative information on the 21 UK atmospheric tests has been published by the US Department of Defense and Atomic Energy Commission (Glasstone, 1964), giving date, yield range, and location name. Peter Marshall of the UK Atomic Weapons Establishment has kindly provided confirmation of 18 of these, and also added name, height of burst, and yield (BLACKNEST). Along with the 24 joint US-UK underground tests given in NV209, the event list is believed to be complete for the UK nuclear explosions.

TABLE 1 Worldwide Nuclear Explosions by Country

Country	Number of events	Official listing
China	45	CT (partial)
France	198	CEA/DAM (partial)
India	3	—
Pakistan	2	—
Soviet Union	715	RFAE
United Kingdom	45 (24 joint USA-UK)	BLACKNEST/NV209 (partial)
United States	1032	NV209
Unknown/disputed	1	—
Total	2041	

3. USSR Nuclear Explosions

The Ministries for Atomic Energy and for Defense of the Russian Federation (RFAE) released official information (including date, location name, yield range, and explosion type) on all USSR nuclear weapons tests and peaceful nuclear explosions (PNEs) that took place from 1949 through 1990. There are 715 tests in total. Bocharov *et al.* (1989; BOCHAROV) published authoritative information on time and coordinates for 96 underground USSR nuclear explosions at the Semipalatinsk Test Site. The accuracy of the origin information (latitude, longitude, time) for these tests is thought to be well within ±0.1 km in location and ±0.05 sec in time. Sultanov *et al.* (1999; SULTANOV) provide a more diverse spectrum of data on Soviet PNEs (most are accurate to ±1 km in location and ±1 sec in time, but some are accurate only on the order of 10–25 km). Other preferred data sources include Khristoforov (1996; KHRISTOFOROV) for seven underwater/near-surface explosions in Novaya Zemlya, AWE-JED (U.K. Atomic Weapons Establishment, 1994) and Richards (2000; RICHARDS) for other Novaya Zemlya events, Khalturin *et al.* (2001) for some small events in Semipalatinsk, and Murphy and Jenab (1992) for the USSR Joint Verification Experiment (JVE) event.

4. French Nuclear Explosions

There is no authoritative source of location information for all of the French tests. The first four tests, conducted in the Sahara, were surface or tower shots reported only to be near Reggan (Glasstone, 1964). However, data on the 13 underground tests in the Sahara, as reported by Bolt (1976; BOLT), originally published by Duclaux and Michaud (1970) through the Academy

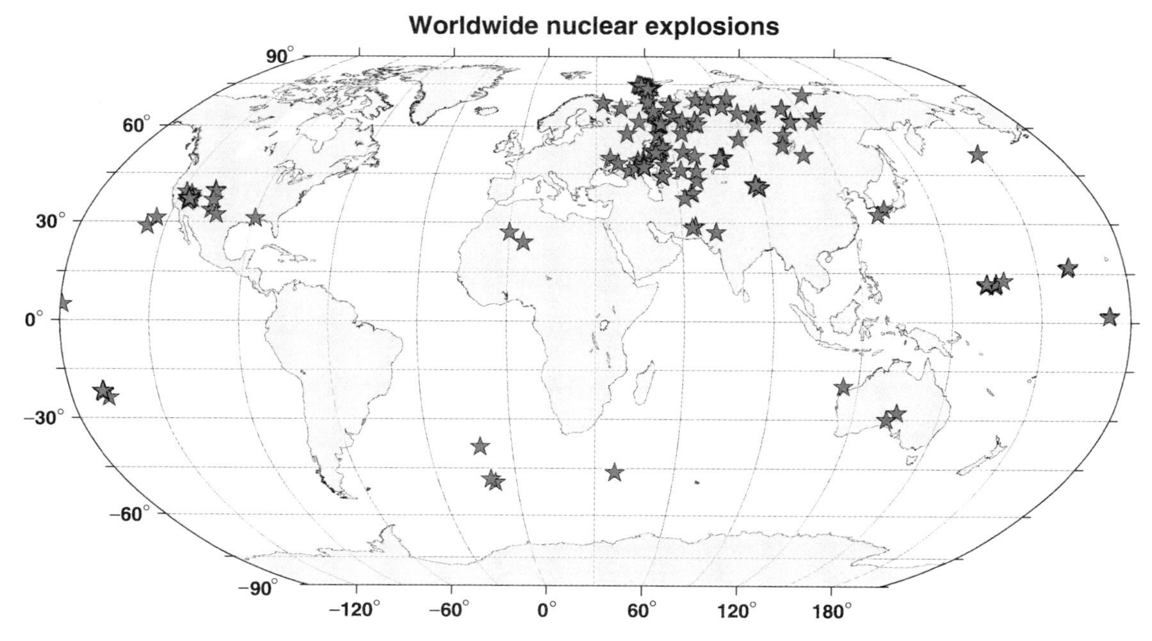

FIGURE 1 Locations for worldwide nuclear explosions in the CMR Nuclear Explosion Database.

of Science, Paris, almost certainly must be derived from authoritative sources.

An official list of French nuclear tests at the Pacific Test Center (CEA/DAM) gives information for 175 events in French Polynesia from 1966 to 1991, including date and time (accurate to the minute and in most cases to the second), event name, location name, test mode, and released nuclear energy range. Since that time, six more probable tests have been conducted at that site. For the better-recorded events at Mururoa, the origins obtained by author AWE-JED (U.K. Atomic Weapons Establishment, 1993) probably possess location accuracies of the order of ±5 km.

5. Chinese, Indian, and Pakistani Nuclear Explosions

For the Chinese events there is no authoritative information on origin time and location. Thirty-two of the 45 Chinese tests have been confirmed by an official publication of the Chinese government (China Today, 1988). For the Chinese tests in the CMR database, preferred data sources include AWE-JED (U.K. Atomic Weapons Establishment, 1993) and GUPTA-JED (Gupta, 1995). They are probably fairly accurate (at least on the order of ±5 km in location), and the latter were referenced to locations based on satellite data. Two AWE-JED events (6 October 1983; 3 October 1984) may possess GT2 quality.

For the Indian and Pakistani explosions, satellite data analyses yielded locations accurate to ±1 km or better (Gupta and Pabian, 1996; GUPTA+PABIAN; Barker *et al.*, 1998; BARKER), except that the preferred origin for the 13 May 1998 event is based on the news media.

6. Event List of Preferred Origins of Selected Events

The preferred origins of selected events in the CMR Nuclear Explosion Database are listed in Table 2 in chronological order. We have selected some highlights from each country's program, but have had to make a best judgment in some cases when there was incomplete information for a country. The complete list of explosions is on the attached Handbook CD, under the directory of \84Yang.

For each country, when applicable, we chose to include the first test, first underground, first underwater, highest yield, highest underground yield, largest cratering, first salvo, first high-altitude/near-space, last atmospheric, and the last test. The data format is time (GMT; month/day/year hour:minute:second), latitude (degree; north positive), longitude (degree; east positive), depth (km; downward positive), body-wave magnitude (mb), yield (kt), maximum yield (kt), explosion code, event name, and preferred author.

The four-character explosion code used in this table indicates the type and setting of the explosion: character 1 for type of explosion (N-nuclear); character 2 for air (A), water (W), or underground (U); character 3 for confirmed (C) or presumed (P); and character 4 for US (U), USSR (S), France (F), China (C), India (I), United Kingdom (G), or Pakistan (P). The five-character code is the same as the four-character code except that characters 4 and 5 indicate the joined countries: USA and UK. The 10-character code indicates multiple shots: e.g., NUCS_SALVO. The character "_" is used for unknown.

In conformity with treaties between the United States and the Soviet Union, a salvo is defined, for multiple explosions for peaceful purposes, as two or more separate explosions where a period of time between successive individual explosions does not exceed 5 sec and where the burial points of all explosive devices can be connected by segments of straight lines, each of them connecting two burial points, and the total length does not exceed 40 kilometers. For nuclear weapons tests, a salvo is defined as two or more underground nuclear explosions conducted at a test site within an area delineated by a circle having a diameter of 2 km and conducted within a total period of time of 0.1 sec.

The event list combines information on magnitude and yield from several origins of each event. If there is no mb value for the preferred origin, the value listed is the maximum of all the origins for that event, whose mb's are obtained from networks with good azimuthal distributions (e.g., ISC, PDE). The yield information is taken from the official sources, NV209 for the US events, RFAE for the USSR events, CEA/DAM for the French events, and NV209 and BLACKNEST for the UK events. When there is no official source above for an event, the yield and maximum yield are values from the preferred origin.

When both a yield and a maximum yield are listed, these values should be taken to mean a reported range in yield. For salvos, these values are sums of the values reported for each of the individual explosions.

7. Summary and Conclusions

We have listed information on all 2041 worldwide nuclear explosions (including one unknown/disputed event) conducted by seven countries, extracted from the CMR database. Until recently, none of the nuclear weapons states had published a complete list of accurate locations and origin times for all of the nuclear explosions it has carried out. But revision 15 of the DOE NV209 (2000) and Springer *et al.* (2002) have recently published complete hypocentral information, including origin time, latitude, longitude, and depth, for all US and joint US-UK tests. This information will replace the current NV209 in the CMR database at a later date.

All the 1032 events by the United States are confirmed (NV209) nuclear explosions. For the 45 events by the United Kingdom, 18 events are confirmed atmospheric nuclear explosions (BLACKNEST), 24 events are confirmed underground nuclear explosions jointly with the United States, and three others are given by Glasstone (1964). All the 715 events by the USSR are confirmed (RFAE) nuclear explosions. For the 198 events by France, 175 are confirmed nuclear explosions at the Pacific Test Center between 1966 and 1991 (CEA/DAM), six additional recent events at that location have been recorded by the PIDC (Prototype International Data Center; http://www.cmr.gov)

TABLE 2 Selected Event Lists of Preferred Origins

Date	Time	Latitude	Longitude	Depth	mb	Yield	Ymax	Type	Name	Source	Comments
07/16/45	11:29:21.0	33.6753	−106.4747	−0.03		21		NACU	Trinity	G+P	US first test
07/24/46	21:34:59.8	11.5833	165.5000	0.00		21		NWCU	Baker	G+R	US first underwater test
08/29/49	00:00:00.0	50.0000	78.0000	0.00		22		NACS	Joe-1	RFAE	USSR first test
10/03/52	00:00:00.0	−20.0000	115.0000	0.00		25		NACG	Hurricane	BOLT	UK first test
02/28/54	18:45:00.0	11.6908	165.2736			15000		NACU	Bravo	G+P	US highest yield test
09/21/55	00:00:00.0	70.7000	54.5600	0.01		3.5		NWCS	—	KHRISTOFOROV	USSR first underwater test
09/19/57	16:59:59.5	37.1958	−116.2031	0.24		1.7		NUCU	Rainier	G+P	US first underground test
04/28/58	19:05:00.0	1.6700	−157.2500	0.00		3000		NACG	grapple	BOLT	UK highest yield test
08/01/58	10:50:05.6	16.7439	−169.5333	−76.81		3800		NACU	Teak	G+P	US first high altitude test
09/23/58	17:58:00.0	1.6700	−157.2500	0.00		25		NACG	grapple	BOLT	UK last atmospheric test
02/13/60	07:04:00.0	27.0000	0.0000	0.00				NAPF	gerb._bleue	BOLT	French first test
10/11/61	07:39:59.9	49.7727	77.9950	0.00		1		NUCS	—	BOCHAROV	USSR first underground test
10/30/61	08:33:27.8	73.8000	53.5000	0.00		50000		NACS	—	PDE	USSR highest yield
11/07/61	11:29:59.9	24.0571	5.0521	0.00				NUPF	agate	BOLT	French first underground test
12/10/61	19:00:00.0	32.2636	−103.8658	0.36		3		NUCU	Gnome	SPRINGER	US first PNE
03/01/62	19:10:00.1	37.0413	−116.0287	0.36		9.5		NUCUG	Pampas	SPRINGER	UK first underground test
07/06/62	17:00:00.1	37.1770	−116.0454	0.19	4.40	104		NUCU	Sedan	SPRINGER	US largest cratering test
11/01/62	00:00:00.0	49.0000	46.0000	0.00		300		NACS	—	RFAE	USSR first high-altitude test
11/04/62	00:00:00.0	17.0000	−169.0000				20	NACU	Tightrope	NV209	US last atmospheric test
12/25/62	00:00:00.0	73.0000	55.0000	0.00		8.5		NACS	—	RFAE	USSR last atmospheric test
08/23/63	13:20:00.1	37.1250	−116.0354	0.25			40	NUCU_SALVO	Kohocton	SPRINGER	US first Salvo
10/16/64	07:00:00.0	41.5000	88.5000	0.00		0	20	NAPC	—	BOLT	Chinese first test
01/15/65	06:00:00.8	49.9350	79.0090	0.18	5.80	140		NUCS	Chagan 1004	SULTANOV	USSR largest cratering; USSR first PNE
03/30/65	08:00:00.0	52.9000	56.5000	1.38		4.6		NUCS_SALVO	Butanel-1	SULTANOV	USSR first salvo
09/10/65	17:12:00.0	37.0780	−116.0167	0.46	5.16	20	200	NUCUG	Charcoal	SPRINGER	UK highest underground yield
08/24/68	18:30:00.5	−22.2280	−138.6440	0.00	4.95	1000		NACF	Canopus	AWE-JED	French highest yield
09/08/68	19:00:01.0	−21.8210	−138.9750	0.00	4.91	1000		NACF	Procyon	AWE-JED	French highest yield
09/22/69	16:14:59.2	41.3760	88.3180		5.20	19.2		NUPC	—	GUPTA-JED	Chinese first underground
11/06/71	22:00:00.1	51.4719	179.1069	1.79	6.80		5000	NUCU	Cannikin	SPRINGER	US highest underground yield
09/12/73	06:59:54.8	73.3140	55.0560	0.00	6.97	1950	14500	NUCS_SALVO	—	RICHARDS	USSR highest underground yield
05/18/74	02:34:55.0	27.0950	71.7520	0.11	5.00	12		NUCI	—	GUPTA+PAB IAN	India first test
05/23/74	13:38:30.2	37.1245	−116.0789	0.47	4.80	20	200	NUCUG	Fallon	ERDA	UK highest underground yield
09/14/74	23:30:00.0	−22.0000	−139.0000	0.00			1000	NACF	Verseau	CEA/DAM	French last atmospheric test
11/17/76	06:00:12.7	40.6960	89.6270		4.70	4000		NAPC	—	GUPTA-JED	Chinese highest yield
07/25/79	17:57:00.0	−21.8800	−138.9400	0.00	6.11		150	NUCF	Tydee	AWE-JED	French highest underground yield (based on mb)
09/22/79	03:00:00.0	−46.3600	37.5700	0.00				NAP-	—	NEWSPAPER	Unknown/disputed event
10/16/80	04:30:29.7	40.7190	89.6510		4.44	1000		NAPC	—	GUPTA-JED	Chinese last atmospheric test
10/24/90	14:57:58.5	73.3310	54.7570	0.00	5.70	20	230	NUCS_SALVO	—	RICHARDS	USSR last test
11/26/91	18:35:00.1	37.0965	−116.0696	0.00	4.60		20	NUCUG	Bristol	DOE	UK last test
09/23/92	15:04:00.0	37.0210	−115.9890	0.00	4.40		20	NUCU	Divider	NV209	US last test
01/27/96	21:29:57.8	−22.2400	−138.8200	0.00	5.30			NUPF	—	PDE	French last test
07/29/96	01:48:57.8	41.8200	88.4200	0.00	4.90			NUPC	—	PDE	Chinese last test
05/11/98	10:13:44.2	27.0780	71.7190	0.00	5.20			NUCI_SALVO	Shakti-1	BARKER	India first salvo
05/13/98	00:00:00.0			0.00		.6		NUCI_SALVO	Shakti-4	NEWSPAPER	India
05/28/98	10:16:17.6	28.8300	64.9500	0.00	4.87			NUCP_SALVO	—	BARKER	Pakistan first test
05/30/98	06:54:54.9	28.4900	63.7300	0.00	4.60			NUCP	—	PDE	Pakistan

between 1995 and 1996, and 17 events before July 1966 are probable nuclear explosions at test sites in the Sahara. For the 45 Chinese events, all 32 events between 1964 and 1984 are confirmed nuclear tests (China Today, 1988), and 13 probable nuclear tests have been detected seismically at the Chinese test site since 1984. Two Indian and Pakistani tests, respectively, have also been recorded by teleseismic stations.

We cannot guarantee that all events are actually nuclear explosions, or that our compilation of the nuclear explosion list is complete. Neither can we guarantee the accuracy of all of the descriptive information on these events, even though we believe the compilers of this information have taken great care. Given the volume of the data, the chance of some errors is high. Accordingly, it is ultimately the user's responsibility to assess the accuracy and completeness of the data, and to determine if they are sufficient for the user's intended purpose. Work to eliminate discrepancies and duplications, and to add new information as it becomes available, will continue.

Acknowledgments

The following individuals contributed information/help to the CMR Nuclear Explosion Database: Spiro Spiliopoulos and Lesley Hodgson at the Australian Geological Survey Organization, Vladimir Belov and Yury Kraev at the Special Monitoring Service of the Ministry of Defense of the Russian Federation, Jannine Ford at the US Department of Energy, Peter Marshall at the UK Atomic Weapons Establishment, B. Massinon at the Departement Analyse et Surveillance de l'Environnement in France, Jack Murphy at Maxwell Technology, Vitaly Khalturin at Lamont-Doherty Earth Observatory of Columbia University, NORSAR, and Hans Israelsson, Keith McLaughlin, Richard Stead, and Beth Turner at CMR. Victor Kirichenko at SAIC Moscow office and Peter Marshall at Blacknest have provided especially helpful reviews of this paper.

References

Barker, B., M. Clark, P. Davis, M. Fisk, M. Hedlin, H. Israelsson, V. Khalturin, W.-Y. Kim, K. McLaughin, C. Meade, J. Murphy, R. North, J. Orcutt, C. Powell, P. Richards, R. Stead, J. Stevens, F. Vernon, and T. Wallace (1998). Monitoring nuclear tests. *Science* **281**, 1967–1968.

Bocharov, V. S., S. A. Zelentsov, and V. N. Mikhailov (1989). Characteristics of 92 underground nuclear explosions at the Semipalatinsk Test Site. *Atomaya Energia* **87** (3) (in Russian).

Bolt, Bruce A. (1976). *Nuclear Explosions and Earthquakes: The Parted Veil.* W. H. Freeman, San Francisco.

CEA/DAM. The atolls of Mururoa and Fangataufa (French Polynesia), Part II: Nuclear testing—mechanical, lumino-thermal and electromagnetic effects. Commissariat a l'Energie Atomique—Direction des Applications Millitaires.

China Today (1988). Nuclear Industry, JPRS-CST-88-008.

DOE/NV-209 (1997). United States nuclear tests, July 1945 through September 1992, Rev. 14, U.S. Department of Energy.

DOE/NV-209 (2000). United States nuclear tests, July 1945 through September 1992, Rev. 15, U.S. Department of Energy.

Duclaux, F., and L. Michaud (1970). Conditions experimentales des tirs nucleaires souterrains Francais au Sahara, 1961–1966. *C.R. Acad. Sci., Paris* **270**, Serie B, 189–192.

Glasstone, S. (1964). The Effects of Nuclear Weapons. US Government Printing Office, Washington, D.C. (revised).

Griggs, D. T., and F. Press (1961). Probing the earth with nuclear explosions. *J. Geophys. Res.* **61**, 237–258.

Gupta, V. (1995). Locating nuclear explosions at the Chinese test site near Lop Nor. *Science and Global Security* **5**, 205–244.

Gupta, V., and F. Pabian (1996). Investigating the allegations of Indian nuclear test preparations in the Rajasthan desert. *Science and Global Security* **6** (2).

Gutenberg, B. (1946). Interpretation of records obtained from the New Mexico atomic bomb test, July 16, 1945. *Bull. Seism. Soc. Am.* **36**, 327–330.

Gutenberg, B., and C. F. Richter (1946). Seismic waves from atomic bomb tests. *Trans. Am. Geophys. Union* **27**, 776–778.

Johnson, C. T., W. P. de la Houssaye, and T. McMillian (1981). Hydroacoustic signals at long ranges from Shot SWORDFISH, U.S. Navy Electronics Laboratory, A Bureau of Ships Laboratory, San Diego, California, NEL Report 1212 (EX), Extracted Version.

Khalturin, V., T. Rautian, and P. G. Richards (2001). A study of small magnitude seismic events during 1961–1989 on and near the Semipalatinsk Test Site, Kazakhstan. *Pure Appl. Geophys.* **158**, 143–171.

Khristoforov, B. (1996). About the control of the underwater and above water nuclear explosions by hydroacoustic methods, Institute for Dynamics of Geosphere, Russian Academy of Sciences, Final report for the project SPC-95-4049, Moscow, October 11.

Mikhailov, V. N., *et al.* (1996). USSR nuclear weapons tests and peaceful nuclear explosions, 1949 through 1990. RFNC-VNIEF, Sarov.

Murphy, J., and Jenab (1992). Development of a comprehensive seismic yield estimation system for underground nuclear explosions. Maxwell Technology, PL-TR-92-2076, SSS-TR-92-13129.

Richards, P. G. (2000). Accurate estimates of the absolute location of underground nuclear tests at the northern Novaya Zemlya Test Site. Proc. Second Workshop on IMS Location Calibration, Oslo, Norway.

Springer, D. L., and R. L. Kinnaman (1971). Seismic source summary for U.S. underground nuclear explosions, 1961–1970. *Bull. Seism. Soc. Am.* **61**, 1073–1098.

Springer, D. L., and R. L. Kinnaman (1975). Seismic source summary for U.S. underground nuclear explosions, 1971–1973, *Bull. Seism. Soc. Am.* **65**, 343–349.

Springer, D., G. Pawloski, J. Ricca, R. Rohrer, and D. Smith (2002). Seismic source summary for all U.S. below-surface nuclear explosions, *Bull. Seism. Soc. Am.* **92**, 1806–1840.

Sultanov, D. D., J. R. Murphy, and Kh. D. Rubinstein (1999). A seismic source summary for Soviet Peaceful Nuclear Explosions. *Bull. Seism. Soc. Am.* **89**, 640–647.

U.K. Atomic Weapons Establishment (1993–1994). AWE Reports No. O 12/93 (China), No. O 11/93 (France), No. O 2/94 (Novaya Zemlya), H.M. Stationery Office.

Yang, X., R. North, and C. Romney (2000a). CMR nuclear explosion database (Revision 3). CMR Technical Report, CMR-00/16, Arlington, Virginia.

Yang, X., I. Bondár, and C. Romney (2000b). PIDC Ground Truth (GT) database (Revision 1). CMR Technical Report, CMR-00/15, Arlington, Virginia.

85

Earthquake Seismology Software

85.1 Overview

J. Arthur Snoke
Department of Geological Sciences, Virginia Tech, Blacksburg, Virginia, USA
Mariano García-Fernández
Institute of Earth Sciences "Jaume Almera" / C.S.I.C., Barcelona, Spain

1. The Evolution in the Use of Computers in Earthquake Seismology

As is documented throughout this Handbook, our knowledge of Earth structure and the earthquake process has expanded exponentially during the last half of the 20th century. This progress has been reflected in a still-increasing abundance of computer-based programs and packages for recording ground motion, locating and analyzing earthquakes, developing data and models for earthquake hazard mitigation, determining the internal structure of the Earth, sharing and displaying data, and organizing large databases. We begin this overview of earthquake seismology software by reflecting briefly on some of the initiatives and developments in computers and computing.

1.1 Computer Hardware Development

Computers in 1960 that filled a room and drew enough power to light a house were slower and had less capacity than ones that fit in a pocket in 2000. In the early 1970s, a rule of thumb for minicomputers such as the Digital Equipment PDP 11/34 was "$1 K for 1 KB of computer memory." (That computer also had a limit of 64 KB of memory accessible for a single task, which limited the size and type of projects that could be handled—which had a lasting effect on the programming style for many researchers who started out on such machines.) From the 1960s through the early 1990s, much of the computing at universities and scientific research centers was done on mainframe computers that could be accessed first via cards and then with terminals and dial-up modems. The conversion from tubes to solid-state components and the development of microprocessors led first to minicomputers and then to the desktops and finally to the laptops and handheld computers of today. Another advance is the use of computers in the field for data acquisition, processing, and telemetry. For further details on the history of computers in the sciences, see, for example, Kaufmann and Smarr (1993) and J. A. N. Lee's continually updated history on the Web at *http://ei.cs.vt.edu/~history/*.

1.2 Evolution in Use of Computers

Seismology, through DARPA (see Chapter 1 by Agnew), deserves some credit in helping to bring about a paradigm shift in the 1960s from thinking of computers as simply arithmetic engines to thinking of them as a communication medium between people (Hauben, 1994). Prior to the development of the Internet, the only way researchers could exchange data and programs was using the same kind of hard media that were used to communicate with computers: This evolved from punch cards, to 9-track magnetic tapes and cartridges, to floppy disks, to 8 mm or 4 mm tapes and CDs. In the past 25 years, direct data transfer has evolved from KERMIT over dial-up modems, to FTP, to the World Wide Web. The current state-of-the-art programs and procedures are described in Chapter 85.4 by van Eck *et al.* on ORFEUS, and Chapter 86 by Ahern on the FDSN and the IRIS Data Management System.

1.3 Station Coverage with Reliable Timing

The WWSSN, in the early 1960s, provided for the first time worldwide coverage by a single network for earthquake locations, which resulted in good azimuthal and fair-to-good depth

INTERNATIONAL HANDBOOK OF EARTHQUAKE AND ENGINEERING SEISMOLOGY, VOLUME 81B ISBN: 0-12-440658-0

control for major earthquakes. The availability and accessibility of such data provided an impetus for the development of seismology software for handling the data and for using that data to get information about both earthquakes and Earth structure. The history of seismic networks and arrays along with the evolution of access to their data can be found in several chapters in the Handbook.

2. Early Computer-Based Studies

Pioneering works using computers in seismology include those on surface-wave data inversion (e.g., Dorman and Ewing, 1962) and those related to earthquake location (e.g., Bolt, 1960; Flinn, 1960; Nordquist, 1962; Eaton, 1969)—many of which used the pre-computer-developed Jeffreys-Bullen travel-time tables (Jeffreys and Bullen, 1958). These programs were developed mainly at universities and government laboratories in the United States.

In the earliest days of computers in seismology, many seismologists worked in isolation and may never have published their code, so any overview of the history of seismology programs is inevitably incomplete. W. H. K. Lee (personal communication, 2000) says that Jerry Eaton should be credited with opening up seismological software to others. Eaton was among the first researchers to include the source code, and he is arguably the first to work out the math for computing travel times and derivatives for a source inside multiple flat layers over a half-space. Eaton's derivation is reproduced by Lee and Stewart (1981).

During the 1970s and the 1980s, there were several major projects that dealt with compilation of seismological algorithms. These include volumes 11–13 of the Academic Press series Methods in Computational Physics, devoted to seismology (Bolt, 1972a; 1972b; 1973), the two volumes of *Computer Programs in Earthquake Seismology* by R. B. Herrmann (1978), and the activities of the IASPEI Subcommission on Algorithms. The IASPEI subcommission convened a Symposium on Earthquake Algorithms at the 21st IASPEI General Assembly, London, Canada, in July 1981 (Engdahl, 1982), which was summarized in a report of the World Data Center A (Engdahl, 1984). At the 23rd IASPEI General Assembly in Tokyo, in August 1985, the Subcommission on Algorithms was commissioned with the task of sponsoring publication of a book compiling a selection of basic and advanced methods and algorithms, including available computer codes on tape or disk (Doornbos, 1988). Other computer code algorithms were also published in the engineering and geophysics literature, and in books on numerical analysis (e.g., Claerbout, 1985; Press *et al.*, 1986) and inverse theory (e.g., Menke, 1984; Tarantola, 1987; Parker, 1994).

The seismological community has played an important role in expanding the use of personal computers (PCs) since 1981. A Committee on PCs was formed within the American Geophysical Union (AGU) in 1986 (Lee *et al.*, 1986). In 1988, IASPEI established a Working Group on Personal Computers, under their Commission on Practice, to promote sharing of seismological software among scientists worldwide. A Working Group on Personal Computers in Seismicity Studies was created in 1994 under Subcommission A of the European Seismological Commission. The activities of those and similar working groups included the organization of symposia and workshops at national and international meetings with demonstrations of existing hardware/software applications (e.g., García-Fernández *et al.*, 1992). The publication and distribution of seismological software were among the major objectives of the working groups, and their endeavors led to the creation of the IASPEI Seismological Software Library (see Chapter 85.2 by Lee) and the IASPEI PC Shareware Library (see Chapter 85.3 by García-Fernández).

Some of the earlier computer programs developed for seismological applications have been so widely used that they are still standards. Among these are earthquake location programs HYPO71 (Lee and Lahr, 1975; Chapter 85.17 by Lee *et al.*), HYPOELLIPSE (Lahr, 1979; Chapter 85.7 by Lahr and Snoke), and HYPOINVERSE (Klein, 1978; Chapter 85.8 by Klein); codes for determining focal mechanisms that formed the FPFIT/FPPLOT/FPPAGE package by Reasenberg and Oppenheimer (1985) and FOCMEC by Snoke *et al.*, 1984; Chapter 85.12 by Snoke); and Kennett's (1988; Chapter 85.10 by Kennett) reflectivity code for computing synthetic seismograms. In earthquake hazard analysis, EQRISK by McGuire (1976) and SEISRISK III by Bender and Perkins (1987) remain standard codes.

In engineering seismology, early programs that became standards include Volumes I and II for digitization and processing of strong-motion accelerograms (Trifunac and Lee, 1973); BAP to process and plot digitized strong-motion earthquake records (Converse and Brady, 1992); and SHAKE (Schnabel *et al.*, 1972) or QUAD4 (Idriss *et al.*, 1973) for soil-response analysis. An early and still popular strong-motion simulation program is SIMQKE (Gasparini and Vanmarcke, 1976).

3. Contents of Subchapters for Chapter 85

The subchapters of Chapter 85 contain summaries of selected earthquake seismology software packages or collections of software packages. The collections described in Chapter 85.2 by Lee include references, and, in Chapter 85.4 by van Eck *et al.*, there are links to many software packages, both freeware and commercial.

Examples of complete seismic data analysis packages are SAC/SAC2000 (Tull, 1987; Chapter 85.5 by Goldstein *et al.*) and SEISAN/SEISNET (Chapter 85.6 by Havskov and Ottemöller, and references therein). Recent simulation programs for incorporating the seismic source and propagation path include SMSIM (Chapter 85.13 by Boore), which uses stochastic methods, and COMPSYN and ISOSYN (Chapter 85.14 by Spudich and Xu), which use complete response and far-field ray-theory Green's functions, respectively, to model the full-wave field from a finite-fault rupture in a vertically layered Earth model.

The subfolder for Chapter 85.17 by Lee *et al.* on the attached Handbook CD contains, for archival purposes, the source and manual for the earthquake location program HYPO71, which ran on mainframe computers with input from punch cards and output (usually) directly to a printer. Also included is HYPO71PC (Lee and Valdes, 1985), which is the HYPO71 program adapted to run on a PC. Reflecting the evolution in the technology, input and output in HYPO71PC are both done via files.

Subchapters not already referenced in this overview are Pujol's software for joint hypocenter determination using local events (Chapter 85.9), Sambridge's nonlinear inversion by direct search using the Neighbourhood Algorithm (Chapter 85.15), Dreger's time-domain seismic moment tensor inversion (Chapter 85.11), Clévédé and Lognonné's higher-order perturbation theory: 3D synthetic seismogram package (Chapter 85.16), and Wielandt's software for seismometer calibration and signal analysis (Chapter 85.18).

Choices for the computer language, platform, and operating system have evolved with time. One of us (jas) worked with Lahr's HYPOELLIPSE program (Chapter 85.7 by Lahr and Snoke) on four platforms: IBM/CMS, Honeywell-Multics, Digital-Vax, Sun-Sunos and Sun-Solaris, and there is also a PC version. Currently (June 2002), PC and UNIX are the preferred platforms, but Linux is gaining favor and the newest operating system on the Mac (OS X) is essentially a flavor of UNIX. Many older programmers still program in FORTRAN, but the current trend is to write code in C++ or Java (see Chapter 85.4 by van Eck *et al.*, and Chapter 85.5 by Goldstein *et al.*). One reason, therefore, for not publishing executable code on the attached CD is that it may be obsolete within a few years.

A few of the packages use subroutines from the *Numerical Recipes* collections—a published set of subroutines available in C, Fortran 77, or Fortran 90 (Press *et al.*, 1986). The publisher of *Numerical Recipes* has not given permission for machine-readable source code to be included on the CD, so the code must be acquired separately.

This chapter in the hard-copy volume of the Handbook includes only summaries about each package or collection. For the 14 packages, the attached Handbook CD contains documentation along with (in most cases) source code and sample runs. Executable code will *not* be included on the CD, but the current version can be obtained from provided Internet sites or by contacting the author.

Navigation among the subfolders for Chapter 85 on the attached Handbook CD is done via a Web browser—which we have found to be the most cross-platform compatible method. The top file in the subfolder on the attached CD for Chapter 85 contains instructions and suggestions about accessing compressed files that are stored on the CD.

4. Additional Software Packages

The collection published here is far from comprehensive. Many of the areas not covered here are available from Web sites. These include synthetic seismogram computation and surface wave inversion (Herrmann, 1996), receiver function analysis (Ammon, 1997), calculation and display of displacement, coseismic strain, and Coulomb failure stress changes (Toda *et al.*, 1998), and GMT—Generic Mapping Tools (Wessel and Smith, 1991 and 1998). Several very useful programs for educational purposes are available from the PEPP (Princeton Earth Physics Project) Software Archives at *http://lasker.princeton.edu/software.html*. Specific compilations of earthquake engineering software are available from NISEE (National Information Service for Earthquake Engineering) at *http://nisee.berkeley.edu/*, and from GGSD (Geotechnical and Geoenvironmental Software Directory) at *http://www.ggsd.com/*.

References

Ammon, C. J. (1997). Receiver-Function Analysis, *http://www.essc.psu.edu/~ammon/HTML/RftnDocs/rftn01.html*. Penn State University.

Bender, B., and D. M. Perkins (1987). SEISRISK III: A Computer Program for Seismic Hazard Estimation. *US Geol. Surv. Bulletin* **1772**, 48 pp.

Bolt, B. A. (1960). The revision of earthquake epicenters, focal depths and origin-times using a high-speed computer. *Geophys. J.* **3**, 433–440.

Bolt, B. A. (Ed.) (1972a). *Seismology: Surface Waves and Earth Oscillations. Methods in Computational Physics, Volume 11*, Academic Press, San Diego.

Bolt, B. A. (Ed.) (1972b). *Seismology: Body Waves and Sources. Methods in Computational Physics, Volume 12*, Academic Press, San Diego.

Bolt, B. A. (Ed.) (1973). *Geophysics. Methods in Computational Physics, Volume 13*, Academic Press, San Diego.

Claerbout, J. F. (1985). *Imaging the Earth's Interior.* Blackwell, Oxford.

Converse, A. M., and A. G. Brady (1992). BAP: Basic Strong-Motion Accelerogram Processing Software; Version 1.0. *Open-File Report, 92-296A*, US Geological Survey, Menlo Park, California.

Doornbos, D. J. (1988). *Seismological Algorithms: Computational Methods and Computer Programs.* Academic Press, London.

Dorman, J., and M. Ewing (1962). Numerical inversion of seismic surface wave dispersion data. *J. Geophys. Res.* **67**, 5227–5241.

Eaton, J. P. (1969). HYPOLAYR, a computer program for determination of hypocenters of local earthquakes in an earth consisting of uniform flat layers over a half space. *Open-File Report*, US Geological Survey, Menlo Park, California.

Engdahl, E. R. (Ed.) (1982). "Earthquake Algorithms." Special Issue *Phys. Earth Planet. Interior* **30**, 85–271.

Engdahl, E. R. (1984). "Documentation of earthquake algorithms." Report SE-35, World Data Center A, Boulder, Colorado.

Flinn, E. A. (1960). Local earthquake location with an electronic computer. *Bull. Seism. Soc. Am.* **50**, 467–470.

Flinn, E. A. (1960). Local earthquake location with an electronic computer. *Bull. Seism. Soc. Am.* **50**, 467–470.

García-Fernández, M., A. Roca, and G. Poupinet (Eds.) (1992). "Applications of Personal Computers in Seismology." Proceedings of the Workshop W58 of the 22nd General Assembly of the European Seismological Commission. Servei Geologic de Catalunya, Barcelona.

Gasparini, D. A., and E. H. Vanmarcke (1976). Simulated Earthquake Motions Compatible with Prescribed Response Spectra. *Dept. of Civil Engineering, Research Report R76-4*, Massachusetts Institute of Technology, Cambridge, Massachusetts.

Hauben, M. (1994). History of ARPANET, *http://www.dei.isep.ipp.pt/docs/arpa.html.*

Healy, J. H., V. I. Keilis-Borok, and W. H. K. Lee (Eds.) (1997). "Algorithms for Earthquake Statistics and Prediction," IASPEI Seismological Software Library Volume 6, Seismological Society of America, El Cerrito, California.

Herrmann, R. B. (1978). Computer Programs in Earthquake Seismology, Volume 1: General Programs (NTIS PB 292 462), and Volume 2: Surface Wave Programs (NTIS PB 292 463). Edited by R. B. Herrmann, Department of Earth and Atmospheric Sciences, Saint Louis University.

Herrmann, R. B. (1996). Computer Programs in Seismology, at *http://www.eas.slu.edu/People/RBHerrmann/ComputerPrograms.html.* Saint Louis University.

Idriss, I. M., J. Lysmer, R. Hwang, and H. B. Seed (1973). QUAD-4—A Computer Program for Evaluating the Seismic Response of Soil Structures by Variable Damping Finite Element Procedures. *Earthquake Eng. Res. Center, Report No. UCB/EERC-73/16.* University of California, Berkeley.

Jeffreys, H., and K. E. Bullen (1958). "Seismological Tables." British Association for the Advancement of Science, Gray Milne Trust, Office of the British Association, Burlington House, London, England.

Kaufmann, W. J., and L. L. Smarr (1993). "Supercomputing and the Transformation of Science." Scientific American Library, Volume 43, A division of HPHLP, New York.

Kennett, B. L. N. (1988). Systematic approximations to the seismic wave field. In: *Seismological Algorithms: Computational Methods and Computer Programs* (D. J. Doornbos, Ed.) pp. 89–168, Academic Press, London. The current version is discussed in Chapter 85.10 by Kennett.

Klein, F. W. (1978). Hypocenter location program HYPOINVERSE. *Open-File Report 78-694*, US Geological Survey, Menlo Park, California, 113 pp. The current version is discussed in Chapter 85.8 by Klein.

Lahr, J. C. (1979). HYPOELLIPSE: A computer program for determining local earthquake hypocentral parameters, magnitude and first motion pattern. *Open-File Report 79-431*, US Geological Survey, Menlo Park, California, 57 pp. The current version (*Open-File Report 99-23*) is discussed in Chapter 85.7.

Lee, W. H. K., and J. C. Lahr (1975). HYPO71 (Revised): A computer program for determining hypocenter, magnitude, and first motion pattern of local earthquakes. *Open-File Report 75-311*, US Geological Survey, Menlo Park, California, 113 pp. The program is described in Chapter 85.17 by Lee *et al.*, and the complete source code is included on the attached Handbook CD.

Lee, W. H. K., and S. W. Stewart (1981). "Principles and Applications of Microearthquake Networks." *Advances in Geophysics, Supplement 2*, Academic Press, New York. This out-of-print book is reproduced in the subfolder for Chapter 17 by Lee in the Handbook CD.

Lee, W. H. K., and C. M. Valdes (1985). HYP071PC: A personal computer version of the HYPO71 earthquake location program. *Open-File Report, 85-749*, US Geological Survey, Menlo Park, California, 43 pp. The program is described in Chapter 85.17, and the complete source code is included in the attached Handbook CD.

Lee, W. H. K., J. C. Lahr, and R. E. Habermann (1986). Applications of personal computers in geophysics. *EOS, Trans. Am. Geophys. Un.* **67**, 1321–1323.

McGuire, R. (1976). EQRISK. Evaluation of Earthquake Risk to Site. Fortran Computer Program for Seismic Risk Analysis. *Open-File Report, 76-67*, US Geological Survey, Denver, Colorado, 92 pp.

Menke, W. (1984). *Geophysical Data Analysis: Discrete Inverse Theory.* Academic Press, Orlando (revised edition: 1989).

Nordquist, J. M. (1962). A special-purpose program for earthquake location with an electronic computer. *Bull. Seism. Soc. Am.* **52**, 431–437.

Parker, R. L. (1994). *Geophysical Inverse Theory.* Princeton University Press, Princeton.

Press, W. H., B. P. Flannery, S. A. Teukolsky, and W. T. Vetterling (1986). *Numerical Recipes.* Cambridge University Press, Cambridge. Subsequent editions were published in 1989, 1992, and 1999 and include example books as well as CDs and diskettes with subroutines in Fortran 77, Fortran 90, Pascal, and C. Online versions of these can be obtained through http://www.nr.com/.

Reasenberg, P. A., and D. H. Oppenheimer (1985). FPFIT, FPPLOT and FPPAGE: Fortran computer programs for calculating and displaying earthquake fault-plane solutions. *Open-File Report, 85-739*, US Geological Survey, Menlo Park, California, 109 pp.

Schnabel, P. B., J. Lysmer, and H. B. Seed (1972). SHAKE: A computer program for earthquake response analysis of horizontally layered sites. Earthquake Eng. Res. Center, Report No. UCB/EERC72/12, Univ. of California, Berkeley, 102 pp.

Snoke, J. A., J. W. Munsey, A. C. Teague, and G. A. Bollinger (1984). A program for focal mechanism determination by combined use of polarity and *SV-P* amplitude ratio data. *Earthquake Notes* **55**(3), 15. The current version is discussed in Chapter 85.12.

Tarantola, A. (1987). *Inverse Problem Theory.* Elsevier, New York.

Toda, S., R. S. Stein, P. A. Reasenberg, J. H. Dieterich, and A. Yoshida (1998). Stress transferred by the 1995 Mw=6.9 Kobe, Japan shock: Effect on aftershocks and future earthquake probabilities. *J. Geophys. Res.* **103**, 24543–24565. See also *http://quake.wr.usgs.gov/research/deformation/modeling/coulomb/coulomb20/software.html.*

Trifunac, M. D., and V. W. Lee (1973). Routine Computer Processing of Strong Motion Accelerograms. Earthquake Eng. Res. Lab. Report 73-03, California Institute of Technology, Pasadena.

Tull, J. E. (1987). Sac—Seismic Analysis Code. Tutorial Guide for new users. Lawrence Livermore National Laboratory, Livermore, CA, UCRL-MA-112835. The current version is discussed in Chapter 85.5.

Wessel, P., and W. H. F. Smith (1991). Free Software helps Map and Display Data. *EOS, Trans. Am. Geophys. Un.* **72**, 441 and 445–446.

Wessel, P., and W. H. F. Smith (1998). New, improved version of the Generic Mapping Tools Released. *EOS, Trans. Am. Geophys. Un.* **79**, 579. http://gmt.soest.hawaii.edu/.

85.2 The IASPEI Seismological Software Library

W. H. K. Lee
US Geological Survey, Menlo Park, California, USA (retired)

1. Introduction

Since computers became widely available in the early 1960s, seismologists have been using them for data acquisition, processing, and analysis, as well as theoretical computation and modeling. For example, the book by Doornbos (1988) contains a collection of seismological algorithms with the corresponding computer programs available on tape or disk from the World Data Center A for Solid Earth Geophysics. The introduction of personal computers in the early 1980s further revolutionized the use of computers for scientific research. Instead of expensive mainframe computers that required a large staff to operate, inexpensive personal computers allowed creative applications to be implemented by individuals with a shoestring budget.

The author was conducting experiments in a quarry for the Defense Advanced Research Projects Agency in the mid-1980s and implemented a simple real-time seismic monitoring system for fieldwork using the personal computer technology. At about the same time, many geophysicists started using personal computers and the author was appointed to chair a committee on personal computers by the American Geophysical Union (AGU). However, the author was not able to persuade AGU to publish software to be prepared by the committee because AGU correctly identified that it would be a money-losing proposition as the market for geophysical software is too small. The committee was soon disbanded.

In 1988, IASPEI established a Working Group on Personal Computers to promote the sharing of seismological software and hardware information among seismologists, and the author was appointed as its chair. An editorial advisory board was created, consisting of Hiroo Kanamori (chair), R. D. Adams, V. Cerveny, E. R. Engdahl, Y. Fukao, R. B. Herrmann, E. Kausel, V. Keilis-Borok, B. L. N. Kennett, and S. K. Singh. In addition to the author, the Working Group consisted of John Lahr and Frank Scherbaum as associate editors, and Mariano García-Fernández as the Shareware Library editor (See Chapter 85.3 by García-Fernández).

To achieve its stated goal, the author realized that seismological software must somehow be published. Unfortunately, funding agencies, such as the US National Science Foundation and the US Geological Survey, were not persuaded. As a last resort, the author approached the Seismological Society of America (SSA). SSA agreed to collaborate with IASPEI to publish the IASPEI software on the conditions that (1) the software project be self-supporting, (2) an SSA-designated examiner must give prepublication approval, and (3) SSA members can buy the software volume at half price.

Since IASPEI itself does not have any money to support the project, the project depended on volunteers, and the author (as the editor) did everything possible to minimize production costs. Robin Adams, then secretary-general of the IASPEI, persuaded UNESCO to order copies for free distribution in the developing countries, and S. K. Singh, a member of the editorial advisory board, persuaded the Third World Academy to do the same. Also, many SSA members ordered the software. Within about a year, the IASPEI software publication project became self-supporting.

During the 12 years after the Working Group was established, six volumes were published. Each IASPEI software volume includes the executable code and examples on floppy diskettes, as well as printed documentation for IBM-compatible or Intel-based personal computers running the Microsoft DOS operating system. These volumes are available for sale (until the existing stocks are exhausted) from the Seismological Society of America, 201 Plaza Professional Building, El Cerrito, CA 94530, USA (Phone: 1-510-525-5474; Fax 1-510-525-7204; e-mail: *krowe@seismosoc.org*; Web site: *http://www.seismosoc.org*).

INTERNATIONAL HANDBOOK OF EARTHQUAKE AND ENGINEERING SEISMOLOGY, VOLUME 81B *ISBN: 0-12-440658-0*

Source-code packages are available from the SSA for the first four volumes of the IASPEI Software Library. URL *http://lbutler.geo.uni-potsdam.de/service_p.htm* has the source code of Volume 5. To remedy the fact that the source code for Volume 6 was not published, we archive it in the subfolder for Chapter 85.2 on the attached Handbook CD.

2. A Brief Summary of the IASPEI Software Library Volumes

2.1 Volume 1: "Realtime Seismic Data Acquisition and Processing" (Lee, 2000a)

Volume 1 (1st Edition, 1989; 2nd Edition, 1994; Y2K Version, 2000) contains programs for real-time seismic data acquisition, processing, and analysis. It includes a description of the hardware implementation for a PC-based real-time seismic system and programs to perform seismic data acquisition, interactive picking of seismic phases, filtering, spectral and coda Q analysis, and earthquake location. The second edition of this volume contains software revisions and several new programs. Included are supports for the USGS digital telemetry standards and PC-SUDS format. The Y2K version includes updates and additional programs for Y2K compliance.

Worldwide, over 200 seismic networks and arrays have used the Volume 1 software. This software contributed to the success of predicting the Mount Pinatubo volcanic eruption (Kerr, 1991) and to the rapid reporting (within 102 sec) of accurately determined location and magnitude with a shake map of the disastrous Chi-Chi (Taiwan) earthquake in 1999 (Shin *et al.*, 2000; Wu *et al.*, 2000).

2.2 Volume 2: "Plotting and Displaying Seismic Data" (Lee, 2000b)

Volume 2 (1st Edition, 1990; 2nd Edition, 1994; Y2K Version, 2000) is a companion to the first volume. It contains 12 computer programs for plotting seismic data on the monitor screen and generating hard-copy plots. It includes a three-dimensional data-viewing program for rotating, enlarging, or shrinking objects and data points. The second edition of this volume contains several new programs for plotting seismic waveform data in PC-SUDS format. The Y2K version includes updates and additional programs for Y2K compliance.

2.3 Volume 3: "Digital Seismogram Analysis and Waveform Inversion" (Lee, 1994a)

Volume 3 contains two major programs: SeisGram by Anthony Lomax for interactive analysis of digital seismograms, and SYN4 by Robert McCaffrey, Geoffrey Abers, and Peter Zwick for inversion of teleseismic body waves. This volume (published in 1991; updated in 1994 to support PC-SUDS format) is a toolbox for seismological research, especially on broadband digital data, and has been used in several training courses worldwide.

2.4 Volume 4: "Bibliographic References and BSSA Database" (Lee, 1994b)

Volume 4 (published in 1994) is a toolbox for managing bibliographic information and includes programs to automate the reference preparation in manuscripts and to manage a user's own references. It also includes a bibliographic database of all papers published in the *Bulletin of the Seismological Society of America* (1911–1993) and some frequently cited articles for seismologists. Recently, the Seismological Society of America established the BSSA Web Index, which is up-to-date and searchable online, by using the BSSA data prepared in Volume 4 and extending the coverage to the current issue. It is open to everyone on the SSA Web site: *http://www.seismosoc.org/*.

2.5 Volume 5: "Programmable Interactive Toolbox for Seismological Analysis" (Scherbaum and Johnson, 1992)

Volume 5 is a Programmable Interactive Toolbox for Seismological Analysis (PITSA) by Frank Scherbaum and James Johnson, and includes a course on "First Principles of Digital Signal Processing for Seismologists." The manual and PC-version code were published in 1992. A version of PITSA for Sun workstations is available from the IRIS Data Management Center at *ftp://ftp.iris.washing.edu/pub/software/PITSA/*. This volume has been used in several training courses worldwide.

2.6 Volume 6: "Algorithms for Earthquake Statistics and Prediction" (Healy *et al.*, 1997)

Volume 6 contains three software packages: SASeis by Tokuji Utsu and Yosihiko Ogata, UpDate by P. N. Shebalin, and M8 by V. G. Kossobokov. SASeis contains ten programs for statistical analysis of seismicity. M8 is an algorithm using pattern-recognition techniques for intermediate-term earthquake prediction (based on an the analysis of smaller-magnitude earthquakes from an earthquake catalog), and Update is an algorithm to aid the preparation of input data for the M8 program.

3. Discussion

The rapid changes in computer hardware and software have rendered many of the published IASPEI software library volumes obsolete. Although the speed of personal computers now is hundreds of times faster than before, the complexity of operating systems makes software development far more difficult. As financial support for seismological research becomes more competitive, most seismologists can no longer afford to spend as much time on such volunteer work as would be needed to maintain and update the software library. The IASPEI Software Library achieved some modest success, but it is being phased out—in fact, the IASPEI Working Group on Personal Computers was disbanded in 1999.

The traditional way of publishing seismological software on paper and disks is just too expensive to be supported by a limited sale. If authors are willing to share their software, it can now be easily made available via the Internet.

References

Doornbos, D. J. (1988). *Seismological Algorithms: Computational Methods and Computer Programs*. Academic Press, London, 469 pp.

Healy, J. H., V. I. Keilis Borok, and W. H. K. Lee (Editors) (1997). "Algorithms for Earthquake Statistics and Prediction." IASPEI Software Library, Vol. 6, Seismological Society of America, El Cerrito, CA, 221 pp. and 4 diskettes.

Kerr, R. A. (1991). A job well done at Pinatubo Volcano. *Science* **263**, 514.

Lee, W. H. K. (Editor) (2000a). "Realtime Seismic Data Acquisition and Processing." IASPEI Software Library, Vol. 1 (2nd Edition, Y2K Version), Seismological Society of America, El Cerrito, CA, 286 pp. and 3 diskettes.

Lee, W. H. K. (Editor) (2000b). "Plotting and Displaying Seismic Data." IASPEI Software Library, Vol. 2 (2nd Edition, Y2K Version), Seismological Society of America, El Cerrito, CA, 254 pp. and 2 diskettes.

Lee, W. H. K. (Editor) (1994a). "Digital Seismogram Analysis and Waveform Inversion." IASPEI Software Library, Vol. 3 (Updated Version), Seismological Society of America, El Cerrito, CA, 183 pp. and 3 diskettes.

Lee, W. H. K. (Editor) (1994b). "Bibliographic References and BSSA Database." IASPEI Software Library, Vol. 4, Seismological Society of America, El Cerrito, CA, 263 pp. and 1 diskette.

Scherbaum, F., and J. Johnson (1992). "Programmable Interactive Toolbox for Seismological Analysis." IASPEI Software Library, Vol. 5, Seismological Society of America, El Cerrito, CA, 269 pp. and 2 diskettes.

Shin, T. C., K. W. Kuo, W. H. K. Lee, T. L. Teng, and Y. B. Tsai (2000). A preliminary report on the 1999 Chi-Chi (Taiwan) earthquake. *Seism. Res. Lett.* **71**, 24–30.

Wu, Y. M., W. H. K. Lee, C. C. Chen, T. C. Shin, T. L. Teng, and Y. B. Tsai (2000). Performance of the Taiwan Rapid Earthquake Information Release System (RTD) during the 1999 Chi-Chi (Taiwan) earthquake. *Seism. Res. Lett.* **71**, 338–343.

85.3 IASPEI PC Shareware Library

Mariano García-Fernández
Institute of Earth Sciences, "Jaume Almera" / C.S.I.C., Barcelona, Spain

1. Introduction

In 1988, IASPEI established a Working Group on Personal Computers under the Commission on Practice, chaired by W. H. K. Lee, to promote sharing of seismological software among scientists worldwide. A Working Group on Personal Computers in Seismicity Studies was created in 1994 under Subcommission A of the European Seismological Commission (ESC), chaired by M. García-Fernández and N. Voulgaris. The activities of those and similar working groups included the organization of symposia and workshops at national and international meetings at which there were demonstrations of existing hardware/software applications. The publication and distribution of seismological software were among the main activities of the IASPEI Working Group on PCs. Toward achieving these goals, the IASPEI Seismological Software Library (SSL; see Chapter 85.2 by Lee) was created and started its publication in 1989.

2. The IASPEI PC Shareware Library (PCSL)

The IASPEI Working Group on PCs also established a PC Shareware Library (PCSL), edited by M. García-Fernández, to provide fast and wide distribution of software that, unlike the IASPEI Software Library, was not peer reviewed. The PCSL was first introduced in 1991 during the workshop "Applications of Personal Computers in Geophysics," held at the XX General Assembly of the International Union of Geodesy and Geophysics (IUGG) in Vienna, Austria. The PCSL was published on diskettes that were freely copied at scientific meetings, or mailed upon request to the editor or the regional corresponding members (C. D. N. Collins in Australia, New Zealand, and the South Pacific; D. Zhang in China; M. G. Al-Ibiary in Egypt; N. Voulgaris in Greece; G. Zonno in Italy; and S. Baris in Turkey).

A PCSL Special Volume was presented at the 27th General Assembly of IASPEI, Wellington, New Zealand, January 1994, including six diskettes with a total of 21 programs amounting to 8 megabytes in compressed form, covering different seismological applications and utilities from earthquake location to expert systems for earthquake prediction. A PCSL 2nd Edition, substantially larger, was published in 1997 including 31 programs amounting to 30 megabytes in compressed form. At that time, the large size of the library made it difficult to keep it published on diskettes, so its distribution was revamped taking advantage of the Internet facilities.

ORFEUS, the nonprofit organization to coordinate and promote digital, broadband seismology in the European-Mediterranean area, offered its Web site for this purpose, and since early 1998 the PCSL is available for download at *http://orfeus.knmi.nl/* (see Chapter 85.4 by van Eck *et al.*).

The IASPEI PC Shareware Library includes the following software:

- 2D&3D (D. Crossley). Contouring and mesh plotting Fortran routines.
- BOLAS (F. Núñez-Cornú). Interactive focal mechanism calculation.
- CONVSEIS (M. C. Oncescu and M. Rizescu). Digital seismic-data format conversion.
- ESEP-PC (Zhuang Kunyuan). Expert system for earthquake prediction.
- EWS (P. Uniyal and A. Manglik). Early warning system for viruses.
- F-E Code (J. B. Young). FORTRAN and C routines for Flinn-Engdahl regions code.
- FPFIT/FPPLOT/FPPAGE (P. A. Reasenberg and D. Oppenheimer). Calculation of earthquake fault-plane solutions.
- FREEFORM (T. Habermann). Data file format conversion.

INTERNATIONAL HANDBOOK OF EARTHQUAKE AND ENGINEERING SEISMOLOGY, VOLUME 81B *ISBN: 0-12-440658-0*

- GEOLABS (D. Crossley). Interactive geophysical programs for students.
- GUNVI (R. Roberts *et al.*) Analysis of single-station three-component digital seismograms.
- HYPOCENTER (B. R. Lienert). Regional and teleseismic earthquake location.
- HYPOELLIPSE (J. C. Lahr). Local and near-regional earthquake location.
- HYPO-GM (I. Amanatashvili *et al.*). Regional earthquake location.
- HYPOINVERSE (F. W. Klein). Earthquake location.
- IASP91-PC (B. R. Lienert). IASPEI 91 Seismological Tables for PC.
- LOCA/LOCAL (A. Nava). Interactive local earthquake location.
- MON1 (V. Zhuravlev). Interactive analysis of three-component digital seismograms.
- PAISA (J. A. Jaramillo). Off-line utility for XDETECT (SSL Vol. 1).
- PASTA (Y. V. Roslov). Spectral analysis tools for PITSA (SSL Vol. 5).
- PLOT4PC (D. Crossley). CALCOMP-compatible Fortran plotter library.
- QNONLIN (J. M. Ibañez *et al.*). Coda-Q calculation by nonlinear fit.
- RAYAMP (D. Crossley). Interactive 2-D seismic ray tracing.
- REDSIS (R. Ortiz *et al.*). Portable seismic-data acquisition system.
- Seismic/Eruption (A. Jones). Space-time display of seismicity and volcanoes.
- Seismic Lineaments (J. Gogiashvili). Reconstruction of the true spatial distribution of seismicity from data distorted by random errors.
- Seismic Waves (A. Jones). Animation of seismic waves synchronised with waveforms.
- SELCATAL (M. García Fernández). Spatial, time and, size earthquake catalog selection.
- SIZ2 (E. Sulstarova). Local earthquake location.
- SparseNet (M. Joswig). Signal detection and analysis for seismic networks by AI techniques.
- T-GRAF (G. Lombardo *et al.*). 2-D, 3-D, and cross-section plots of earthquake hypocenters.
- ZMAP (S. Wiemer). Interactive analysis of space-time seismicity changes.

The complete PCSL 2nd Edition manual is in subfolder 85.3 on the attached Handbook CD.

85.4 ORFEUS Seismological Software Library

Torild van Eck and Bernard Dost
Royal Netherlands Meteorological Institute, De Bilt, The Netherlands
Manfred Baer
Eidgenössische Technische Hochschule, Zürich, Switzerland

1. Introduction

The Observatories and Research Facilities for EUropean Seismology (ORFEUS; *http://orfeus.knmi.nl/*), funded by European countries, is the nonprofit foundation that aims at coordinating and promoting digital, broadband seismology in the European-Mediterranean area. The working group on software is one of the initiatives within this organization, and the Library was started in 1997 (van Eck, 1997).

The ORFEUS Seismological Software Library is an electronic searchable catalog of freely available software for seismological analysis and research. The library is accessible through the Internet and is maintained by the ORFEUS working group on software.

The main goal of the ORFEUS Seismological Software Library is to provide an organized overview of the software that is made available in the academic community either through the Internet (Web sites or FTP) or on personal request. We further hope that by facilitating software access we will stimulate faster improvements and further developments in seismological software. Presently, this is happening in both format conversion software and relevant Java developments.

2. Background and Motivation

Software exchange among seismologists at academic institutes and observatories has generally been wide-ranging and generous. Early efforts to coordinate and gather algorithms and software within the seismological community, i.e., the Seismological Algorithms (Doornbos, 1988) and the IASPEI PC Shareware Library (see Chapter 85.3 by García-Fernández), proved therefore very successful.

Programs like HYPO71 (Lee and Lahr, 1975; Chapter 85.17 by Lee *et al.*), HYPOINVERSE (Klein, 1978; Chapter 85.8 by Klein), HYPOCENTER (Lienert *et al.*, 1986), SAC/SAC2000 (Chapter 85.5 by Goldstein *et al.*) and GMT (Wessel and Smith, 1991) have shown a general software evolution pattern consisting of three steps: First, newly developed software remains available only among a selected group of people. Second, if the software has been proven to be useful in a number of research projects, it is made freely available for noncommercial use. Third, once widely distributed, others make improvements and modifications. The corresponding implementations and distribution may vary from full anarchy resulting in, for example, the many versions and flavors of HYPO71, to strictly controlled, as with SAC and GMT.

The rapid new developments within standard packages and fast developments of new packages for seismological, numerical, statistical, and graphical applications are difficult to follow through books and/or hearsay from colleagues. Since the middle of the 1990s the widely implemented File Transfer Protocol (FTP) and the World Wide Web (WWW) offer excellent tools to maintain a dynamic library with up-to-date information on relevant software.

3. Library Structure

For practical reasons, the Library is presently divided into four sections:

1. "Software links," mainly referring to UNIX/Linux compatible software
2. "PC-shareware links," mainly referring to DOS/Windows compatible software

INTERNATIONAL HANDBOOK OF EARTHQUAKE AND ENGINEERING SEISMOLOGY, VOLUME 81B ISBN: 0-12-440658-0

3. "Conversion software," format conversion software for seismological waveform data
4. "Java/CORBA links," mainly referring to Java applications and related OO software

In those cases where the software author has no Internet or FTP site available, ORFEUS offers our FTP site as a depository. The conversion software, for example, is largely available directly from our FTP site. Otherwise, we simply provide links to the author's site. When possible, the second method is preferable, as it enables the author to make easy and quick updates and provide additional information/restrictions for the potential user. The links are checked on a regular basis to ensure that they remain valid.

Our simply providing the links has the disadvantage that we have very little influence regarding the quality standards of the sites. We can but assume that a self-regulating mechanism will keep the standards of the programs and documentation at a high level. Well-programmed and well-documented software will receive a wider distribution and will encourage improvements, resulting in more citations for the program author and fewer requests for assistance. We also encourage people to cite the program authors when using the programs.

The PC-shareware links is a further development of the IASPEI PC Shareware Library initiated and, until recently, maintained by Mariano García-Fernández. The IASPEI PC Shareware Library is presently being updated and integrated within the PC-shareware links. This work is supervised by the PC-software working group within the FDSN working group on software and is discussed in more detail in Chapter 85.3 by García-Fernández.

Commercial software is referenced only when it becomes relevant for freely available software. Examples are MATLAB and the various flavors of Linux.

Presently, our library contains close to 200 entries from about 20 countries with a major contribution from the United States.

4. Public Licensing

As our catalog is freely accessible through the Internet, the software may be prone to commercial misuse. In order to keep this to a minimum, we recommend that authors include the GNU General Public License (*http://www.gnu.org/copyleft/*). This is a relatively simple procedure and provides a fairly good protection against misuse and conversion to proprietary software.

5. Object Oriented Software

Object oriented (OO) software development has only recently been introduced into seismological applications. In spite of many obvious advantages, shown in the few available programs, progress within earthquake-related seismological OO software developments have so far been slow. This is mainly because older (working) programs written in FORTRAN or C often need to be completely restructured and rewritten to take advantage of OO, which may require significant individual efforts—both in learning the new languages (for programmers who still think in FORTRAN or C) and in rewriting the code. An advantage of OO languages is their modular form, which enables many programmers to share the same modules and consequently enable faster software developments. OO software developments may accelerate once standard modules become available (Lomax, 2000). In other scientific fields, this process has already started. Within seismology, some modest first efforts in CORBA are becoming available (Sanders and Crotwell, 2000). We anticipate that the Seismological Software Library will play a significant role in developing future OO software in seismology, specifically in Java and CORBA.

6. Contributions

Authors of seismological software are encouraged to submit their software (links) to ORFEUS. Every suggested link is checked and quickly reviewed for relevancy, but otherwise no restrictions apply. ORFEUS is also willing to provide assistance in making the software available if this should be necessary. Corrections and improvements are always welcome and will be quickly implemented. Users can be kept abreast of new development at ORFEUS through our newsletters found at *http://orfeus.knmi.nl/*

References

Doornbos, D. J. (Editor) (1988). *Seismological Algorithms: Computational Methods and Computer Programs*. Academic Press, London, 469 pp.

Klein, F. W. (1978). Hypocenter location program—HYPOINVERSE: Users guide to versions 1, 2, 3 and 4. *Open-File Report 78-694*, US Geological Survey, Menlo Park, California, 113 pp. The current version is discussed in Handbook Chapter 85.8.

Lee, W. H. K., and J. C. Lahr (1975). HYPO71 (revised): A computer program for determining hypocenter, magnitude, and first motion pattern of local earthquakes. *Open File Report 75-311*, US Geological Survey, Menlo Park, California, 113 pp. [The program is described in Chapter 85.17, and this OFR is included in the subfolder for Chapter 85.17 on the attached Handbook CD.]

Lienert, B. R., E. Berg, and L. N. Frazer (1986). HYPOCENTER: An earthquake location method using centered, scaled, and adaptive least squares. *Bull. Seism. Soc. Am.* **76**, 771–783.

Lomax, A. (2000). The ORFEUS Java workshop: Distributed computing in earthquake seismology. *Seism. Res. Lett.* **71**, 589–592.

Sanders, M., and P. Crotwell (2000). Fissures CORBA Framework for the IRIS Consortium. 2AB Inc. Report, 124 pp., *http://www.seis.sc.edu/software/Fissures/Fissures2000.pdf*.

van Eck, T. (1997). The ORFEUS Seismological Software Library. *Seism. Res. Lett.* **66**, 952–953.

Wessel, P., and W. H. F. Smith (1991). Free software helps map and display data. *EOS, Trans. Am. Geophys. Un.* **72**, 441.

85.5 SAC2000: Signal Processing and Analysis Tools for Seismologists and Engineers

Peter Goldstein, Doug Dodge, Mike Firpo, and Lee Minner
Lawrence Livermore National Laboratory, Livermore, California, USA

1. Introduction

SAC2000, or SAC, as many users refer to it, is a primary signal processing and analysis tool for a large portion of the international seismological research and engineering communities including academic, government, and business institutions. SAC has extensive, well-documented, well-tested, and well-maintained data processing and analysis capabilities, a macro programming language that allows users to develop new analysis techniques and customized processing programs, and the ability to do both batch and interactive processing. SAC's strengths also include the ability to process a diverse range of data types. Its extensive usage (more than 400 institutions worldwide) has made it much easier for researchers to develop collaborative research projects. SAC is relatively easy to use and is available on a variety of hardware platforms. Part of its popularity is due to its user-oriented development philosophy, which has led to consistent, backward-compatible development, guided by user input and needs.

Development of SAC2000 began in 1993 as a follow-on to the development of the original Seismic Analysis Code (SAC) (Tull, 1987; Tapley and Tull, 1992). Motivation for this development included the recognition that existing tools could not handle the anticipated amount of readily available seismic data, the desire to improve processing and analysis capabilities by taking advantage of state-of-the-art seismological and computational techniques, and the need to improve upon the efficiency, memory utilization, and portability of the existing tools (Goldstein and Minner, 1995; Goldstein *et al.*, 1998). To accomplish these goals the original 100,000+ line SAC code was translated from FORTRAN to C, and significant additions, modification, and improvements have been made to address development needs. Primary support for this development has been through the Department of Energy and Lawrence Livermore National Laboratory's treaty monitoring program with significant collaboration with the seismic research community. A primary development goal for SAC2000 has been to meet the seismic signal processing and analysis needs of the DOE treaty monitoring R&D teams and the rest of the treaty monitoring R&D community.

SAC2000's extensive signal processing capabilities include data inspection, signal correction, and quality control; unary and binary data operations; travel-time analysis; spectral analysis including high-resolution spectral estimation, spectrograms, and binary sonograms; and array and three-component analysis. These capabilities have proved useful for solving a number of geophysical problems, including estimation and analysis of strong ground motions; earthquake, explosion, and volcanic source studies; seismic discrimination and identification studies; magnitude estimation; travel-time analysis; studies of wave propagation phenomena such as path and site effects; and investigations of Earth structure. It has also proved useful for analysis of other geophysical data such as measurements of electromagnetic or hydro-acoustic phenomena.

More detailed descriptions of SAC's development history and capabilities are described in the subfolder for Chapter 85.5 on the attached Handbook CD. Further information can be found in the introduction and tutorial sections of the help package on the SAC2000 Web pages at *http://www.llnl.gov/sac*. SAC2000 is available on a variety of hardware platforms using the UNIX or Linux operating systems. These include Sun Solaris, PC and Mac Linux, Compaq/Dec Alpha, IBM RS6000, and SGI.

INTERNATIONAL HANDBOOK OF EARTHQUAKE AND ENGINEERING SEISMOLOGY, VOLUME 81B ISBN: 0-12-440658-0

Access to SAC2000 can be obtained by contacting Peter Goldstein via e-mail at *peterg@llnl.gov*.

2. Future Work

Future plans for SAC include continued maintenance and selected upgrades while simultaneous collaborations with the seismic community are undertaken to reengineer SAC as an object oriented program with an interface that will give programmers direct access to SAC's methods. The interface will allow SAC to be used as a "seismological compute engine" from within other programs. Users will be able to develop analysis systems within SAC using its macro processing capabilities, and then implement "production" versions of the systems through the programming interface. The goal of such a reengineering is to provide more flexible and efficient tools for the analysis of large databases or distributed data sets.

Examples of selected upgrades include incorporation of a Coda magnitude command, incorporation of the high-resolution spectral estimation subprocess commands with the main body of the code, enhancements to our interface to the GMT mapping tools (Wessel and Smith, 1991), and new array analysis capabilities. When feasible, we will also continue work to make SAC more compatible with other data formats (currently supported formats include CSS3.0, GSE2.0, PC suds, SEGY, and ASCII text) and tools such as MATLAB *www.mathworks.com* and the TauP toolkit (Crotwell *et al.*, 1999).

References

Crotwell, H. P., T. J. Owens, and J. Ritsema (1999). The TauP Toolkit: Flexible Seismic Travel-time and Ray-path Utilities. *Seism. Res. Lett.* **70**(2), 154–160.

Goldstein, P., and L. Minner (1995). A status report on the development of SAC2000: a new seismic analysis code. UCRL–ID–121523.

Goldstein, P., D. Dodge, M. Firpo, and S. Ruppert (1998). What's new in SAC2000? Enhanced processing and database access. *Seism. Res. Lett.* **69**(3), 202–204.

Tapley, W. C., and J. E. Tull (1992). SAC command reference manual version 10.6*e*. Lawrence Livermore National Laboratory, Livermore, California, M282REV–4.

Tull, J. E. (1987). SAC—Seismic Analysis Code. Tutorial guide for new users. Lawrence Livermore National Laboratory, Livermore, California, UCRL–MA–12835.

Wessel, P., and W. H. F. Smith (1991). Free software helps map and display data. *EOS, Trans. Am. Geophys. Un.* **72**, 441.

85.6 SEISAN Earthquake Analysis and SEISNET Network Automation Software

Jens Havskov
Department of Geoscience, University of Bergen, Bergen, Norway
L. Ottemöller
British Geological Survey, Edinburgh, United Kingdom

1. Introduction

In seismology, a wealth of data acquisition and processing systems are available, and a seismic observatory typically uses several systems for both data acquisition and processing and perhaps yet another system for research-related tasks. A common problem is the lack of a proper database structure, which prevents effective use of the data. The goal of SEISAN and SEISNET is to automate data retrieval from different data acquisition systems, whether local or remote, through SEISNET and provide a common platform for data processing and storage through SEISAN. The two systems are integrated so that SEISNET collects data directly into the SEISAN database and uses SEISAN programs for preliminary processing.

Both SEISAN and SEISNET rely heavily on public domain software, and both can be described as a system to integrate known programs and data acquisition systems into a common system. The software packages (source code only) are in the subfolder for Chapter 85.6 on the attached Handbook CD. The full distributions (including executables) are available at *http://www.geouib.no/seismo/software/software.html*.

2. SEISAN

The main goal of SEISAN (Havskov and Ottemöller, 1999) is to organize data from all kinds of seismic stations into a simple database and to provide all the tools needed for routine processing. A secondary goal is to facilitate research tasks by taking advantage of having parametric and waveform data in a unified database. Several research-type programs that work directly on the database are included.

A third goal is that SEISAN must work in an identical manner under Sun, Linux, and MS Windows and that data can be moved among the three platforms without any modification.

The main capabilities can be summarized as follows:

- Routine processing: phase picking, hypocenter location, and magnitudes
- Determination of source parameters: fault-plane solution, stress drop, etc.
- Crustal structure: velocities, layer thickness, and attenuation
- Seismic catalogs: ISC data, database management, completeness, statistics, etc.
- Seismic hazard: attenuation, catalogs, and soil response

SEISAN works with five-letter station codes and is fully year 2000 compatible. In order to facilitate the use of SEISAN with any kind of data and to use programs outside SEISAN, several format-conversion programs are included.

SEISAN does not have advanced signal processing capabilities, and therefore the UNIX versions have been interfaced with PITSA (Scherbaum *et al.*, 1999; see also Chapter 85.2 by Lee) and SAC (Goldstein *et al.*, 1998; Chapter 85.5 by Goldstein *et al.*) so that, when using the SEISAN database program, both programs can be invoked directly.

2.1 System Requirements for SEISAN

SEISAN runs on Sun Solaris, Linux, and PC (Windows 95/NT or higher). SEISAN does not have any special hardware

INTERNATIONAL HANDBOOK OF EARTHQUAKE AND ENGINEERING SEISMOLOGY, VOLUME 81B ISBN: 0-12-440658-0

requirements and no commercial software is needed to run a compiled version of SEISAN. For graphical hard copies, a PostScript laser printer is required. The programs are mostly written in FORTRAN, a few in C, and source codes for all the main SEISAN programs are included. The programs have been compiled and linked with system compilers and linkers on Sun, GNU compilers on Linux, and Digital Visual Fortran 5.0 on the PC. The compilers on PC and Sun are commercial, while the GNU compilers are public domain. SEISAN can be set to use color or black and white on both screen and laser printer.

3. SEISNET

The SEISNET (Ottemöller and Havskov, 1999) software is made to combine seismic stations of various types with communication capabilities into a network. The main routines carried out by SEISNET are transfer of parametric data, network event detection, transfer of waveform data, and automatic determination of epicenter location and magnitude. The data are stored in a central SEISAN database. Alternatively, the automatic waveform data transfer can be based on given hypocenter and origin time. An important design goal has been to build software that can support various types of acquisition systems and communication methods. SEISNET was designed for networks with data acquisition on remote systems and therefore does not work in real time.

The following list gives an overview of the main functions that are automated in SEISNET:

- Retrieval of detection information from seismic stations
- Retrieval of epicentral information provided by seismic centers
- Retrieval of waveform data from seismic stations
- Retrieval of waveform data using AutoDRM (Kradolfer, 1996)
- Network event detection
- Automatic phase identification, hypocenter location, and magnitude determination
- Transfer of waveform data from selected field stations based on a given hypocenter location and origin time

3.1 System Requirements for SEISNET

SEISNET is available for the platforms Sun Solaris and Linux. The software is based on freely available software only. SEISAN has to be installed, since the two packages are closely integrated. SEISNET requires the Expect, Tcl, and Tk software, which are available for the three UNIX platforms. In addition, the C-Kermit software is required. Information on location and installation of all the packages is given in the manual.

The data acquisition systems supported at present (1999) are Quanterra (e.g., as used on the IRIS stations) and SEISLOG (Utheim *et al.*, 2001) for QNX and Windows. SEISNET also supports any kind of data acquisition system that creates event files and is accessible through FTP. The only requirement is that date and start time are given in the event file names. In addition, the software supports automatic data retrieval through AutoDRM. More systems will be added as the need arises. Methods of communication supported are TCP/IP and modem dial-up.

References

Havskov, J., and L. Ottemöller (1997). Electronic Seismologist—SeisAn Earthquake Analysis Software. *Seism. Res. Lett.* **70**, 532–534.

Goldstein, P., D. Dodge, M. Firpo, and S. Ruppert (1998). What's new in SAC2000? Enhanced processing and data base access. *Seism. Res. Lett.* **69**, 202–205.

Kradolfer, U. (1996). AutoDRM—The first five years. *Seism. Res. Lett.* **67**, 30–33.

Natvik, Ø., T. Utheim, and J. Havskov (1999). SEISLOG: A seismic data acquisition system for Windows95/NT, user manual. University of Bergen.

Ottemöller, L., and J. Havskov (1999). SeisNet: A general purpose virtual seismic network. *Seism. Res. Lett.* **70**, 522–528.

Scherbaum, F., J. Jones, and A. Rietbrock (1999). Programmable Interactive Toolbox for Seismological Analysis, PITSA user manual.

Utheim, T., and J. Havskov (1999). The SEISLOG Data Acquisition System, user manual. University of Bergen.

Utheim, T., and J. Havskov, and Oe. Natvik (2001). SEISLOG Data Acquisition Systems. *Seism. Res. Lett.* **72**, 77–79.

85.7 The HYPOELLIPSE Earthquake Location Program

John C. Lahr
US Geological Survey, Golden, Colorado, USA
J. Arthur Snoke
Department of Geological Sciences, Virginia Tech, Blacksburg, Virginia, USA

1. Summary of Background and Purpose

The earthquake location program HYPOELLIPSE, first published in 1979 (Lahr, 1979) and subsequently updated with added features (Lahr, 1999), was initially developed to locate crustal and subcrustal earthquakes of southern Alaska using arrival times recorded by a sparse regional seismograph network. This is a more difficult problem than locating shallow earthquakes within a dense network, which is the problem generally faced within California, so HYPOELLIPSE includes many more user-adjustable parameters than does HYPO71 (Lee and Lahr, 1975), the earthquake location program developed for California that is the ancestor of HYPOELLIPSE.

Travel times may be computed from velocity models or travel-time tables. The program HYPOTABLE, which is included with this distribution, can create spherical-earth travel-time tables for use with HYPOELLIPSE. This allows the program to be used with stations beyond the distance at which significant travel-time errors would be introduced by a flat-layer travel-time model. As is demonstrated in the documentation, for shallow-focus events, systematic errors of at least 0.1 sec are found for distances beyond the P_g–P_n crossover distance, while for deep-focus events similar errors occur at smaller epicentral distances (Snoke and Lahr, 2001).

A relatively new feature of HYPOELLIPSE is the ability to work in areas with large topographic relief. Previous versions assumed that all of the stations were located at the same elevation and that small elevation differences could be accounted for by station elevation corrections. The basic assumptions of this approach break down when there are significant topographic variations, especially for very shallow earthquakes. We faced this problem working with a dense seismic network on Redoubt volcano in Alaska. HYPOELLIPSE currently allows stations to be "embedded" within the velocity model, and will correctly compute travel time and take-off angles even for stations that are at a lower elevation than the earthquake.

2. Hardware and Software Requirements

Fortran source code is included in the subfolder for Chapter 85.7 on the attached Handbook CD for both UNIX and PC-DOS platforms. The UNIX source code has been successfully compiled and linked on Sun Sunos and Sun Solaris computers, and will probably also compile easily on other UNIX or Linux platforms. On the PC-DOS platform the Microsoft Fortran Optimizing Compiler Version 5.10 has been used. (This compiler is the predecessor of Digital Virtual Fortran, which in turn became Compaq Virtual Fortran.) Executable code for Sun Solaris and for DOS platforms is available from *http://geohazards.cr.usgs.gov/iaspei_pgms/hypoellipse/*, the HYPOELLIPSE Web site.

A complete program manual, as well as sample data sets and detailed instructions for running them with HYPOELLIPSE, are included on the attached Handbook CD. The HYPOELLIPSE Web site will include program updates made subsequent to the production of the CD.

Acknowledgments

We thank J. Dewey for his helpful review of this chapter.

INTERNATIONAL HANDBOOK OF EARTHQUAKE AND ENGINEERING SEISMOLOGY, VOLUME 81B *ISBN: 0-12-440658-0*

References

Lee, W. H. K., and J. C. Lahr (1975). HYPO71 (Revised): A computer program for determining hypocenter, magnitude, and first motion pattern of local earthquakes. *Open-File Report 75-311*, US Geological Survey, Menlo Park, California, 113 pp.

Lahr, J. C. (1979). HYPOELLIPSE: A computer program for determining local earthquake hypocentral parameters, magnitude, and first motion pattern. *Open-file Report 79-431*, US Geological Survey, Menlo Park, California, 310 pp.

Lahr, J. C. (1999). HYPOELLIPSE: A computer program for determining local earthquake hypocentral parameters, magnitude, and first-motion pattern (Y2K Compliant Version). *Open-file Report 99-023*, US Geological Survey, Golden, Colorado, paper and online editions, 112 pp.

Snoke, J. A., and J. C. Lahr (2001). Locating earthquakes: At what distance can the earth no longer be treated as flat? *Seism. Res. Lett.* **72**, 538–541.

85.8 The HYPOINVERSE2000 Earthquake Location Program

Fred W. Klein
US Geological Survey, Menlo Park, California, USA

Following in the tradition of USGS earthquake location programs like HYPOLAYR (Eaton, 1969), HYPO71 (Chapter 85.2 by Lee), and HYPOELLIPSE (Chapter 85.7 by Lahr and Snoke), the HYPOINVERSE2000 program determines earthquake locations and magnitudes from seismic network data like first-arrival *P* and *S* arrival times, amplitudes, and coda durations. The present version of HYPOINVERSE2000 is in routine use by the Hawaiian Volcano Observatory, The Northern California Seismic Network, the Nevada network, and many other networks. It is the standard location program supplied with the Earthworm seismic acquisition and processing system and has thus gotten wide use. It is Y2000 capable.

1. General Features

HYPOINVERSE has the following general features:

- *Command environment*. HYPOINVERSE2000 is command driven. Commands can be entered from the keyboard at the prompt or from a file. Parameters can be supplied on the command line, or they can be omitted and the program will give prompts showing what the current values are. HYPOINVERSE2000 is normally a stand-alone program that reads and writes files, but it has been implemented in Earthworm as a callable subroutine. There are a variety of convenience features including rapid binary station files and initialization, and interactive earthquake location and editing.
- *File formats*. All data files are ASCII, column-oriented files in a variety of formats from HYPO71 to HYPOINVERSE2000 archive and summary. Many users prefer the archive format for both input and output because it has all of the event (summary), station (phase) data, and controls such as trial or fixed hypocenter in it. All fields of all file types are described in the manual. Both 2- and 4-digit year formats are supported. Some input formats are probably compatible with other programs, and if not, translation programs should not be difficult to write. Files have several 1-letter comment fields for phases such as arrival impulsiveness, polarity, station remark, and data source code; and 1-letter comment fields for magnitude types (M_D, M_L, etc.); and an event code for quarry blast, felt earthquake, long period, etc. A 10-digit identification number can also be defined. Station codes can be 10 letters long and are divided into three parts.
- *Crustal models*. Models are in a flat Earth. Stations are at the surface but can have delays. Earthquake depths are thus below the local surface, but they are only an approximation if there is steep terrain like a conical volcano. Crustal models may either have homogeneous layers (with a model specified only by a table of velocities and depths), or have linear gradients within layers. Linear gradient models are generated ahead of time by the program TTGEN and are stored in fast travel-time tables. Up to 35 models may be assigned to different geographic regions with smoothly varying travel times and station delays interpolated between them.
- *Magnitudes*. Two different amplitude magnitudes based on a local magnitude formula from a variety of analog and digital, displacement or velocity stations may be computed for each event. Two different coda duration magnitudes based on a variety of formulas may be computed for each event. Available coda formulas can include terms for duration (log and linear), depth, and distance; a bilinear equation with different terms above and below a certain duration; extra terms derived by Jerry Eaton for Northern California; and a lapse-time relation. An external magnitude may be passed through without modification.

INTERNATIONAL HANDBOOK OF EARTHQUAKE AND ENGINEERING SEISMOLOGY, VOLUME 81B ISBN: 0-12-440658-0

- *Station gain history*. Files with time-varying station gains and magnitude corrections are supported.
- *Station weighting*. Phases and stations may be individually weighted. Residual and distance weighting can suppress high-residual or distant stations.
- *Inversion*. The program uses singular-value decomposition for the least-squares inversion that permits a variety of controls such as eigenvalue cutoff and damping.
- *Output results*. In addition to the usual outputs are an error ellipse specified by its principal axes (calculated from the covariance matrix of the inversion), horizontal and vertical errors derived from the maximum projections of these error axes, a preferred event magnitude chosen from the five magnitudes available, the station importance from the inversion (that considers all factors such as final weights and station geometry), and the predominant source of data if the event is merged from different networks. A 3-letter region code is derived from the location and depth and is placed on the summary line. This code can come from a user-supplied subroutine, but subroutines for Hawaii and northern California are built in.

2. Hardware and Software Requirements

HYPOINVERSE2000 is written in Fortran 77 with (minor) extensions that compiles on UNIX (Sunos or Solaris), VMS (VAX or ALPHA), and PC OS2. The UNIX executable does not require that Fortran libraries be on the computer because it is compiled with the -Bstatic option. The UNIX executable is about 2 megabytes in size, which includes a station table of 4000 stations, earthquakes of 1000 phases, and 36 crustal models. A simple change of a single parameter statement and recompiling will reduce these arrays if necessary. Operating system differences are isolated into a few small subroutines that are substituted at make/link time. Similar platforms could probably be supported with only a few code modifications.

3. Documentation and Source Code

The (FORTRAN) source files, auxiliary files including documentation, and input plus output files for test runs are in the subfolder for Chapter 85.8 on the attached Handbook CD. The program gradually evolves as new needs arise, so the user is advised to check the anonymous FTP site, *ftp://swave.wr.usgs.gov/pub/outgoing/klein/hyp2000*, for updates.

The Hypoinverse documentation is now published as an open-file report (Klein 2002). See the site *http://geopubs.wr.usgs.gov/open-file/of02-171/* for downloadable versions of the documentation.

References

Eaton, J. P. (1969). HYPOLAYR, a computer program for determining hypocenters of local earthquakes in an earth consisting of uniform flat layers over a half space. *Open-File Report*, US Geological Survey, Menlo Park, California, 155 pp.

Klein, F. W. (2002), User's guide to Hypoinverse, a Fortran program to solve for earthquake locations and magnitudes. *Open-File Report 02-171*, US Geological Survey, Menlo Park, California, 123 pp.

85.9 Software for Joint Hypocentral Determination Using Local Events

Jose Pujol
CERI, University of Memphis, Memphis, Tennessee, USA

1. Background

The introduction of the joint epicentral determination (JED) technique (Douglas, 1967) was a major step toward the improvement of earthquake locations when using velocity models that did not account for the presence of lateral velocity variations. When the variations are significant, the locations may be in error by a few to tens of kilometers, depending on the type of data used (local, regional, teleseismic) and the number and distribution of stations. The JED technique was extended to include the computation of event depths by Dewey (1972) and is known as the joint hypocentral determination (JHD) technique. The basic feature of both techniques is the joint location of a number of events and the simultaneous determination of a correction term for each of the stations. For a given group of events, each "station correction" will be the same for all the events. To improve the efficiency of the computations, several approaches have been used (see Pujol, 2000, for a review). The JHD software described here is based on the approach of Pavlis and Booker (1983), as modified by Pujol (1988). The basic features of the algorithm are outlined below.

In matrix form, the JHD problem has the following expression:

$$\mathbf{W}_j\mathbf{r}_j = \mathbf{W}_j\mathbf{A}_j d\mathbf{X}_j + \mathbf{W}_j d\mathbf{s}; \qquad j = 1, M \tag{1}$$

(Pujol, 2000), where j identifies an event, $\mathbf{r}_j$ is the vector of travel-time residuals, $\mathbf{A}_j$ is the matrix of derivatives corresponding to the standard single-event location problem, $d\mathbf{X}_j$ is the vector of origin time and hypocenter adjustments, $d\mathbf{s}$ is the vector of station corrections adjustments, and $\mathbf{W}_j$ is a diagonal matrix of weights (≤ 1). This matrix generalizes the matrix $\mathbf{S}_j$ of Pujol (1988), which involves ones and zeros only. For theoretical and computational convenience, the number of rows of $\mathbf{A}_j$, $\mathbf{r}_j$, $d\mathbf{s}$, and $\mathbf{W}_j$ are all equal to the number of stations (N). If a station did not record event j, that station is assigned a weight of zero in $\mathbf{W}_j$. The condition that $\mathbf{W}_j$ be diagonal is discussed below.

Equation (1) shows the coupling between the earthquake locations and the station corrections. The decoupling of these two contributions is accomplished in two steps. Following Pavlis and Booker (1983), the first step is to apply the singular-value decomposition to $\mathbf{W}_j\mathbf{A}_j$:

$$\mathbf{W}_j\mathbf{A}_j = \mathbf{U}_j\Lambda_j\mathbf{V}_j^T = \mathbf{U}_{jp}\Lambda_{jp}\mathbf{V}_{jp}^T \tag{2}$$

The matrix Λ_j is diagonal with elements equal to the singular values of $\mathbf{W}_j\mathbf{A}_j$. The right side of Eq. (2) comes from the relation $\mathbf{U}_j = (\mathbf{U}_{jp}|\mathbf{U}_{jo})$, where p is the number of nonzero singular values and the vertical bar indicates matrix partition. A similar relation applies to $\mathbf{V}_j$. For the earthquake location problem $p \leq 4$, although to avoid nonunique solutions only matrices $\mathbf{W}_j\mathbf{A}_j$ with $p = 4$ should be used.

The second step is to introduce Eq. (2) in Eq. (1) and to use the fact that $\mathbf{U}_{jo}^T\mathbf{U}_{jp} = \mathbf{O}$, where $\mathbf{O}$ is the zero matrix. This gives

$$\mathbf{U}_{jo}^T\mathbf{W}_j d\mathbf{s} = \mathbf{U}_{jo}^T\mathbf{W}_j\mathbf{r}_j; \qquad j = 1, M \tag{3}$$

The M equations (3), one for each event, can be written as a single matrix equation:

$$\mathbf{S}\, d\mathbf{s} = \mathbf{R} \tag{4}$$

Matrix $\mathbf{S}$ has N columns and $M \times (N-4)$ rows, and can be very large, but equation (4) can be solved very efficiently using the least-squares method, which requires solving

$$\mathbf{S}^T\mathbf{S}\, d\mathbf{s} = \mathbf{S}^T\mathbf{R} \tag{5}$$

or, equivalently,

$$\left(\sum_{j=1}^{M}\mathbf{W}_j\mathbf{U}_{jo}\mathbf{U}_{jo}^T\mathbf{W}_j\right) d\mathbf{s} = \sum_{j=1}^{M}\mathbf{W}_j\mathbf{U}_{jo}\mathbf{U}_{jo}^T\mathbf{W}_j\mathbf{r}_j \tag{6}$$

INTERNATIONAL HANDBOOK OF EARTHQUAKE AND ENGINEERING SEISMOLOGY, VOLUME 81B   ISBN: 0-12-440658-0

Equation (5) is solved by computing the generalized inverse $\mathbf{Z}^{\dagger}$ of $\mathbf{S}^T\mathbf{S}$:

$$d\mathbf{s} = \mathbf{Z}^{\dagger}\mathbf{S}^T\mathbf{R} \quad (7)$$

Once $d\mathbf{s}$ has been computed, the $d\mathbf{X}_j$'s are obtained from Eq. (1):

$$\mathbf{W}_j\mathbf{A}_j d\mathbf{X}_j = \mathbf{W}_j\mathbf{r}_j - \mathbf{W}_j d\mathbf{s}; \qquad j = 1, M \quad (8)$$

Equation (8) is solved by damped least squares. The $d\mathbf{s}$ and $d\mathbf{X}_j$'s thus obtained are used to update the station corrections and initial estimates of origin times and hypocentral coordinates, and a new iteration is started.

2. Main Features of the Software

The software is intended for use with data recorded by networks having apertures of a few hundred kilometers at most. This restriction is due to the use of a raytracing subroutine that assumes flat horizontal layering. The software can compute both P- and S-wave corrections. This is done by generating a new station list with two different fictitious stations (one for P and one for S arrivals) with the same actual location. However, although the P and S wave corrections are computed separately, they are not independent because they are coupled via the V_p/V_s ratio.

The software is very efficient for two reasons. One is the use of the least-squares solution [Eq. (6)] and the other is the fact that the matrix $\mathbf{W}_j$ is diagonal. The latter condition requires that for each event all the stations be listed in the same order. As this is not usually the case, this situation is taken care of as follows. For each event, a temporary matrix of derivatives is generated from the actual matrix of derivatives in such a way that each row of the temporary matrix always corresponds to the same station (this is done by a reordering of indexes). When a station did not record an event, the corresponding row is made of zeros.

Two parameters are used to assess the numerical stability of the computations. One of the parameters is the sum of the station corrections, which for each iteration should be equal to the sum of the initial values of the initial corrections. The second parameter is the magnitude of the largest singular value of $\mathbf{S}^T\mathbf{S}$, which cannot exceed the number of events. If these conditions are not satisfied, it will be necessary to inspect the data carefully to determine the cause of the problem. In the author's experience, this situation may arise when the events are too clustered. Another important feature of the software is that it requires neither a master event nor specifying in advance any of the station corrections. The trade-off between origin times and station corrections that characterizes the JHD problem manifests itself in the condition that the average of the station corrections remain constant through all the iterations (Pujol, 1988).

Good-quality data are required for successful results. If the data are of poor quality, it may be impossible to find a consistent set of station corrections. Although this is fairly obvious, some people have tried to use the software with catalog data with disappointing results. On the other hand, the JHD technique can be used to detect inconsistent arrivals in a data set (Viret *et al.*, 1984). For example, it has been used to identify converted waves incorrectly reported as S waves (Pujol *et al.*, 1998, Ratchkovsky *et al.*, 1997).

3. Brief Overview of Results Obtained with the Software

The main conclusion of careful analyses of several data sets is that for relatively clustered events the JHD technique generally produces a considerable improvement over the locations obtained using single-event location techniques. In addition, the JHD station corrections give a clear indication of the presence of lateral velocity variations. This result was not unexpected, but what was really surprising was the magnitude of the corrections, which in some cases spanned ranges of 1 sec or more for P waves and 2 sec or more for S waves. These observations correspond to data recorded by networks with apertures of about 200 km or less. These wide ranges are rarely seen in the station residuals obtained using single-event location, and they have a significant effect on earthquake location. This fact is particularly important when one is interested in the accurate identification of active faults. A summary of results is included on the attached Handbook CD.

4. Hardware and Software Requirements

The source code in the subfolder for Chapter 85.9 on the attached Handbook CD is written in Fortran 77 with a few VAX/VMS extensions. The software has been compiled, linked, and run on Sun platforms under Sunos and Solaris. The subfolder also includes input and output files, all extensively documented.

Acknowledgments

The development of the software was supported by the State of Tennessee through the Centers of Excellence program. The raytracing function *ttinvr* included in the software was provided by Dr. R. Crosson. Dr. J. A. Snoke provided the EISPACK matrix routines *tred2* and *imtql2* and greatly improved the presentation of the documentation. CERI contribution No. 392.

References

Dewey, J. (1972). Seismicity and tectonics of western Venezuela. *Bull. Seism. Soc. Am.* **62**, 1711–1751.

Douglas, A. (1967). Joint epicentre determination. *Nature* **215**, 47–48.

Pavlis, G., and J. Booker (1983). Progressive multiple event location (PMEL). *Bull. Seism. Soc. Am.* **73**, 1753–1777.

Pujol, J. (1988). Comments on the joint determination of hypocenters and station corrections, *Bull. Seism. Soc. Am.* **78**, 1179–1189.

Pujol, J. (2000). Joint event location—The JHD technique and applications to data from local seismic networks. In: *Advances in Seismic Event Location* (C. Thurber and N. Rabinowitz, Eds.), pp. 163–204. Kluwer Academic Publishers.

Pujol, J., R. Herrmann, S. C. Chiu, and J. M. Chiu (1998). Constrained relocation of New Madrid seismic zone earthquakes. *Seism. Res. Lett.* **69**, 56–68.

Ratchkovsky, N., J. Pujol, and N. Biswas (1997). Relocation of earthquakes in the Cook Inlet area, south-central Alaska, using the joint hypocenter determination method. *Bull. Seism. Soc. Am.* **87**, 620–636.

Viret, M., G. Bollinger, J. A. Snoke, and J. Dewey (1984). Joint hypocenter relocation studies with sparse data sets—A case history: Virginia earthquakes. *Bull. Seism. Soc. Am.* **74**, 2297–2311.

85.10 Synthetic Seismogram Calculation Using the Reflectivity Method

Brian L. N. Kennett
Research School of Earth Sciences, Australian National University, Canberra, Australia

1. Background

The response of a stratified medium to excitation by a source can be extracted by using the technique of separation of variables. For a horizontally stratified medium this leads to a representation in terms of the frequency-wavenumber domain, and seismograms can be calculated by performing a numerical double integral (see, e.g., Kennett, 1983).

For weakly attenuative medium, there are some advantages to working in terms of slowness rather than wavenumber since the frequency-independence of the reflection and transmission properties at interfaces can then be exploited.

2. The Algorithm

The algorithm implemented in the package is described in some detail in Kennett (1988) and was originally released as part of the *Seismological Algorithms* tape.

The synthesis of seismograms is achieved by building the response in the frequency-slowness domain and then undertaking a slowness integral for each specified range from the source, followed by an inversion to the time-domain using a Fast Fourier transform. The integration over slowness is carried out using the trapezium rule because this can cope with the complexities of the response function as well as the oscillatory Bessel functions that include the horizontal phase information.

A point source is taken to be embedded in a set of uniform isotropic layers with slight attenuation—$0.0001 < Q^{-1} < 0.05$—underlain by a uniform half-space. A cylindrical coordinate system is employed with the origin passing through the source. The response is built up from the reflection and transmission properties of different portions of the stratification. Within in each portion of the layering the reflection and transmission matrices are calculated using a simple recursive scheme, which allows control of the level of internal multiples within each layer.

The structure is divided into two parts by the introduction of a separation level, and different forms of the response are used depending on whether the source lies above or below this level. This allows different treatments of the propagation properties in the upper and lower zones. In each case, a wide variety of approximations to the response can be generated, so that, e.g., it is possible to include all free-surface reflections or to suppress them entirely. The radiation at the source can be confined to just P or S waves, and a similar choice may also be made at the receiver.

The flexibility of the representation means that it is possible to produce complete synthetic seismograms as well as concentrate on particular groups of phases.

The structure of the numerical integrals for the displacement as a function of surface position and time can be illustrated with the case of the vertical component:

$$u_z(r, \phi, t) = \sum_m \int_{\omega_1}^{\omega_2} \mathrm{d}\omega\, \omega^2 \times \int_{p_1}^{p_2} \mathrm{d}p\, U(p, \omega) S(z_s, m) J_m(\omega p r) \mathrm{e}^{\mathrm{i}m\phi}$$

where r is range to station, ϕ is azimuth to station, and z_s is source depth. The medium response $U(p, \omega)$ calculated by composition of reflection and transmission matrices does not depend on angular order m. This dependence is contained entirely in the source term $S(z_s, m)$.

INTERNATIONAL HANDBOOK OF EARTHQUAKE AND ENGINEERING SEISMOLOGY, VOLUME 81B ISBN: 0-12-440658-0

In order to minimize effort in the numerical integrals, the asymptotic forms of the Bessel functions are used so that there is a common integral for all the azimuthal orders at the source. The range of integration in both frequency ω and slowness p is under the control of the user. It is possible to break up a large slowness range into pieces and to use differing slowness increments in each and then add the resulting seismograms.

3. The Reflectivity Code

The code contributed to this volume is the ECR package, which is the author's implementation of the reflectivity approach for horizontally stratified media. It is in the subfolder for Chapter 85.10 on the attached Handbook CD.

The complete source code is provided. Coding is "standard" Fortran 77 with no compiler enhancements. The package has been successfully run on a wide variety of platforms since its release in 1988 (under the name `snert`). The code was written when memory was expensive and so the limits on the size of the slowness and frequency arrays are conservative. The size of the arrays can easily be increased by editing the common blocks.

Run times and computer memory depend on whether or not adjustments are made to the basic version as well as on the computer platform being used. For an upper-mantle example with 70 layers using 1200 slowness and 250 active frequencies with seismogram output for 32 sites, the computation time is about 6 min on Sun Sparc Ultra 5 with approximately 16 MB of memory used.

Information on constructing suitable input files is provided in the documentation on the CD volume. Care needs to be taken to ensure adequate sampling in both slowness and frequency so that the numerical integrals converge. Oscillatory seismogram traces indicate undersampling in slowness.

Seismograms are written out in a simple format (`zst`) using unformatted records. It should be a simple task to adapt it to alternative formats.

References

Kennett, B. L. N. (1983). *Seismic Wave Propagation in Stratified Media.* Cambridge University Press, Cambridge.

Kennett B. L. N. (1988). Systematic approximations to the seismic wave field. In: *Seismological Algorithms: Computational Methods and Computer Programs* (D. J. Doornbos, Ed.), pp. 237–259, Academic Press, London.

85.11 TDMT_INV: Time Domain Seismic Moment Tensor INVersion

Douglas S. Dreger
University of California, Berkeley, California, USA

1. Background and Purpose

This distribution is a set of programs and shell scripts for determining the seismic moment tensor by inversion of complete, three-component seismic waveforms. Included are source code for programs to compute the required Green's function catalog, various utilities, and the moment tensor inverse routine. The frequency-wavenumber integration program, FKRPROG, written by Chandan Saikia of URS, is included with permission and is used to compute the Green's functions for appropriate one-dimensional seismic velocity models. A users' manual and tutorial exercises are provided to demonstrate usage. In addition, an example "wrapper" program is included to illustrate how this package may be automated. The distribution has been built on PCs running Linux and Sun workstations running Solaris. It compiles using the GNU C (gcc) and GNU Fortran (g77) compilers.

This software package has been in use at the University of California, Berkeley Seismological Laboratory (BSL) since 1993 and is employed to automatically investigate all $M_L > 3.5$ events in northern California (e.g., *www.seismo.berkeley.edu/~dreger/mtindex.html*). The package has also been successfully implemented at the Japan National Research Institute for Earth Science and Disaster Prevention (NIED; *http://argent.geo.bosai.go.jp/*). In addition, it has been used by a number of individual researchers in the United States, Europe, and Asia.

2. Hardware and Software Requirements

In the subfolder for Chapter 85.11 on the attached Handbook CD, the distribution is provided as a gzip-compressed tar file. Untarring and uncompressing this file produces a directory structure with a root directory named MTPACKAGE. The source code is located in the MTCODE subdirectory and the command "make all" builds the software package on either PC Linux or Sun Solaris systems. The software uses subroutines from *Numerical Recipes in C* (Press *et al.*, 1988), which must be obtained separately by the user.

References

Dreger, D. S., H. Tkalcic, and M. Johnston (2000). Dilational processes accompanying earthquakes in the Long Valley Caldera. *Science* **288**, 122–125.

Dreger, D., and B. Savage (1998). Aftershocks of the 1952 Kern County, California, Earthquake Sequence. *Bull. Seism. Soc. Am.* **89**, 1094–1108.

Fukuyama, E., M. Ishida, D. Dreger, and H. Kawai (1998). Automated seismic moment tensor determination by using on-line broadband seismic waveforms. *Zishin* **51**, 149–156 (in Japanese with English abstract).

Fukuyama, E., and D. Dreger (2000). Performance test of an automated moment tensor determination system for the future "Tokai" earthquake. *Earth Planets Space* **52**, 383–392.

Dreger, D., R. Uhrhammer, M. Pasyanos, J. Franck, and B. Romanowicz (1998). Regional and far-regional earthquake locations and source parameters using sparse broadband networks: A test on the ridgecrest sequence. *Bull. Seism. Soc. Am.* **88**, 1353–1362.

Pasyanos, M. E., D. S. Dreger, and B. Romanowicz (1996). Towards Real-Time Determination of Regional Moment Tensors. *Bull. Seism. Soc. Am.* **86**, 1255–1269.

Press, W. H., S. A. Teukolsky, W. T. Vetterling, and B. P. Flannery (1988). *Numerical Recipes in C.* Cambridge University Press, Cambridge.

Romanowicz, B., D. Dreger, M. Pasyanos, and R. Urhammer (1993). Monitoring of strain release in central and northern California using broadband data. *Geophys. Res. Lett.* **20**, 1643–1646.

Uhrhammer, R., L. Gee, M. Murray, D. Dreger, and B. Romanowicz (1999). The Mw5.1 San Juan Bautista Earthquake of 12 August, 1998. *Seism. Res. Lett.* **70**, 10–18.

INTERNATIONAL HANDBOOK OF EARTHQUAKE AND ENGINEERING SEISMOLOGY, VOLUME 81B *ISBN: 0-12-440658-0*

85.12 FOCMEC: FOCal MEChanism Determinations

J. Arthur Snoke
Department of Geological Sciences, Virginia Tech, Blacksburg, Virginia, USA

1. Summary of Background and Purpose

The FOCMEC package is in the subfolder for Chapter 85.12 on the attached Handbook CD. The package contains programs for determining and displaying double-couple earthquake focal mechanisms. Input data are polarities (P, SV, SH) and/or amplitude ratios (SV/P, SH/P, SV/SH). The main program, Focmec, performs an efficient, systematic search of the focal sphere and reports acceptable solutions based on selection criteria for the number of polarity errors and errors in amplitude ratios. The search of the focal sphere is uniform in angle, with selectable step size and bounds. The selection criteria for both polarities and amplitudes allow correction or weightings for near-nodal solutions. Applications have been made to finding best-constrained fault-plane solutions for suites of earthquakes recorded at local to regional distances, analyzing large earthquakes observed at teleseismic distances, and using recorded polarities and relative amplitudes to produce waveform synthetics.

Input data can include up to a total of 500 polarities and amplitude ratios—the maximum number of input data is a compile-time option. The program Focmec produces two output files: a complete summary of information about all acceptable solutions, and a shorter summary file with one line for each acceptable solution that can be used as an input to other programs for display or further analysis. Another program in the package, Focplt, produces focal-sphere plots based on the Focmec input data (polarities, ratios) alone or superimposed on solutions (fault planes, compression and tension axes, SV and SH nodal surfaces). Other auxiliary programs include one to create input files for Focmec, a program that converts from among various ways of presenting a double-couple solution, a program that calculates the radiation factors for an input mechanism, and programs for displaying and printing plot files.

Instructions are included for compiling and running the programs, and there are two data sets with scripts and documentation for running the programs.

2. Hardware and Software Requirements

Executables (and built libraries) are provided for two Sun Microsystems operating systems: Sunos 4.1.4 and Solaris 2.x. Programs and subroutines are coded in Fortran 77. The FORTRAN coding is "standard" with few Sun Fortran enhancements—the coding has been changed very little from that used in earlier versions that ran on PDP 11/34, VAX/VMS, and IBM/CMS platforms. The code has been compiled and linked with only a few modifications (mostly in the I/O) on PC and other UNIX platforms. Scripts for running the program are in UNIX csh format and are heavily commented. The plotting package, as written, requires the SAC Fortran library, but higher-level calls follow the CALCOMP convention so should be easily adapted to other plotting packages.

3. The FOCMEC Home Page

The FOCMEC home page is at *http://www.geol.vt.edu/outreach/vtso/focmec/*. Included are instructions on obtaining the latest versions of the package, including executables. Executables for current versions of all programs are included for Sun Solaris 2.8, for current versions of all executables except the plotting program Focplt on a PC (Windows 98 or later), and for all programs

INTERNATIONAL HANDBOOK OF EARTHQUAKE AND ENGINEERING SEISMOLOGY, VOLUME 81B ISBN: 0-12-440658-0

as of June 2000 on Sunos 4.1.x. An Update History page on the Web site itemizes the chronology of major changes in the package.

Relevant References

Aki, K., and P. G. Richards (1980). *Quantitative Seismology: Theory and Methods*. W. H. Freeman, San Francisco. Pages 105–109 contain a comprehensive discussion of the theory behind the determination of a seismic moment tensor and the double-couple solution.

Dreger, D. S., H. Tkalĉić, and M. Johnston (2000). Dilational processes accompanying earthquakes in the Long Valley Caldera. *Science* **288**, 122–125. This study includes a review of the methodology for inverting full waveforms to get all six components of the moment tensor, and it includes a description and application of the statistical methods used to estimate the probability of non-double-couple solutions from a full-waveform dataset for an event. See also Chapter 85.11 in this Handbook.

Dziewonski, A. M., T. A. Chou, and J. H. Woodhouse (1981). Determination of earthquake source parameters from waveform data for studies of regional and global seismicity. *J. Geophys. Res.* **86**, 2825–2852. This paper describes the method used in full-waveform inversion for large ($M \geq 5.5$) earthquakes to produce the Centroid Moment Tensor (CMT) solution published for all large earthquakes.

Herrmann, R. B. (1975). A student's guide for the use of *P* and *S* wave data for focal mechanism determination. *Earthquake Notes* **46**, 29–39. Covers much the same territory as Aki and Richards, but at a less advanced level. Included are equations for the conversions among different representations of a double-couple focal mechanism.

Snoke, J. A. (1989). Earthquake mechanisms. In: *The Encyclopedia of Solid Earth Geophysics* (D. E. James, Ed.), pp. 239–245. Van Nostrand Reinhold, New York. A primer on double-couple focal mechanisms.

Stein, S., and E. Klosko (2002). Earthquake Mechanisms and Plate Tectonics. Chapter 7 of this Handbook. This chapter shows how earthquake mechansims can be used to constrain earthquake slip vectors and the stresses in various tectonic settings.

85.13 SMSIM: Stochastic Method SIMulation of Ground Motion from Earthquakes

David M. Boore
US Geological Survey, Menlo Park, California, USA

1. Summary of Background and Purpose

A simple and powerful method for simulating ground motions is based on the assumption that the amplitude of ground motion at a site can be specified in a deterministic way, with a random phase spectrum modified such that the motion is distributed over a duration related to the earthquake magnitude and to distance from the source (Boore, 2003). This method of simulating ground motions often goes by the name "the stochastic method." It is particularly useful for simulating the higher-frequency ground motions of most interest to engineers, and it is widely used to predict ground motions for regions of the world in which recordings of motion from potentially damaging earthquakes are not available. This simple method has been successful in matching a variety of ground-motion measures for earthquakes with seismic moments spanning more than 12 powers of ten. One of the essential characteristics of the method is that it distills what is known about the various factors affecting ground motions (source, path, and site) into simple functional forms that can be used to predict ground motions. This provides a means by which the results of more rigorous studies on specific factors can be incorporated into practical predictions of ground motion.

SMSIM is a set of programs for simulating ground motions based on the stochastic method. Separate programs are included for random-vibration and time-domain simulations, but an effort has been made to make the input and output parameter files the same for both applications. The programs include a number of drivers that call subroutines for computing various measures of ground motion using random-vibration calculations (GM_RV.FOR) and time-domain calculations (ACC_TS.FOR and GM_TD.FOR); the time-domain subroutines call subroutines contained in the module TD_SUBS.FOR. In addition, RVTDSUBS.FOR and RECIPES.FOR contain routines that are common to both applications. RECIPES.FOR contains programs from *Numerical Recipes* (Press *et al.*, 1992), with minor modifications for a few of the subroutines. (The *Numerical Recipes* programs must be acquired separately; RECIPES.FOR indicates which routines are needed and what modifications must be made to a few of the routines.) A revised version of the SMSIM manual (Boore, 1996, 2000) is included as a PDF file in the software distribution. The latest version of the programs and manual can be found at *http://quake.usgs.gov/~boore/*.

The drivers provided in this report produce peak acceleration, peak velocity, peak displacement, Arias intensity, and response spectra for a range of oscillator periods. The time-domain drivers include the option of saving to a file a specified subset of the suite of time series used to obtain peak motions. Some of the drivers obtain a single magnitude and distance interactively from the user; others read the magnitudes and distances from a control file and compute motions for a table of magnitudes and distances.

The modules were designed so that the drivers can be easily modified to produce the ground-motion parameters for other combinations of magnitude, distance, or input parameters.

Programs are also given to compute Fourier spectral amplitudes corresponding to the model specified by the input-parameter file, either directly in the frequency domain or from Fourier transformation of the simulated time series (to provide an independent check of the calculations).

Also included in the package are a set of programs for computing site response for SH waves. These include site amplifications

INTERNATIONAL HANDBOOK OF EARTHQUAKE AND ENGINEERING SEISMOLOGY, VOLUME 81B ISBN: 0-12-440658-0

using the square root of the effective seismic impedance (sometimes known as the "quarter-wavelength approximation"), as well as amplifications based on plane-wave propagation in a stack of constant-velocity layers. The square-root impedance amplifications are a useful first-order approximation to the complete amplification, and the program (SITE_AMP) provides a useful way to digitize a velocity model made up of a series of linear velocity gradients into a consistent set of constant velocity layers, which can then be used in a plane-wave-propagation program such as NRATTLE.FOR (written by C. Mueller with modifications by R. Herrmann, and included in SMSIM with their permission).

The programs do *not* include a graphics user interface; they are intended to be run within a command prompt window (such as a DOS window in Windows 9x). The input is from ASCII files or from the screen. The outputs of the programs are ASCII files, some of which are arranged in columnar format for easy importation into spreadsheets or graphics programs.

An annotated list of programs and subroutines, as well as instructions for compiling and using the programs, is included in the manual, contained in the subfolder for Chapter 85.13 on the attached Handbook CD. The manual contains extensive descriptions of the input parameters as well as several sets of input parameters, including those used to obtain the sample output in the manual and those used by Atkinson and Boore (1995) and by Frankel *et al.* (1996) to simulate ground motions in eastern North America.

2. Hardware and Software Requirements

Executables can be built or obtained from the author to run in a DOS window on a PC. The programs and subroutines are coded in Fortran 77, with a few widely used extensions, including the use of dynamically allocatable arrays in the time series applications (instructions are given in the latest version of the manual as well as in the comments in relevant programs [search for "ALLOCATABLE:"] for easily disabling this Fortran 90 extension).

Relevant References

Atkinson, G. M., and D. M. Boore (1995). Ground motion relations for eastern North America. *Bull. Seism. Soc. Am.* **85**, 17–30. Used the stochastic method to simulate ground motions for a grid of distances and magnitudes and fit simple functional form to this set of simulated motions to obtain equations for ground motion as a function of magnitude and distance.

Boore, D. M. (1996). SMSIM—Fortran programs for simulating ground motions from earthquakes: version 1.0. *Open-File Report 96-80-A*, US Geological Survey, Menlo Park, California, 73 pp. The first version of SMSIM.

Boore, D. M. (2000). SMSIM—Fortran programs for simulating ground motions from earthquakes: version 2.0—A revision of OFR 96-80-A. *Open-File Report 00-509*, US Geological Survey, Menlo Park, California, 53 pp. Available in PDF format at *http://geopubs.wr.usgs.gov/open-file/of00-509/.*

Boore, D. M. (2003). Prediction of ground motion using the stochastic method. *Pure Appl. Geophys.* (in press). A review of the stochastic method as I have developed it, including extensive references to applications of the method.

Frankel, A., C. Mueller, T. Barnhard, D. Perkins, E. Leyendecker, N. Dickman, S. Hanson, and M. Hopper (1996). National seismic hazard maps: Documentation June 1996. *Open-File Report 96-532*, US Geological Survey, Menlo Park, California, 69 pp. Used SMSIM to produce ground motion used in constructing seismic hazard maps for the central and eastern United States. The input parameter file used for these calculations is included in SMSIM.

Press, W. H., S. A. Teukolsky, W. T. Vetterling, and B. P. Flannery (1992). *Numerical Recipes in Fortran 77* (2nd Edition). Cambridge University Press, Cambridge.

85.14 Software for Calculating Earthquake Ground Motions from Finite Faults in Vertically Varying Media

Paul Spudich
US Geological Survey, Menlo Park, California, USA
Lisheng Xu
Institute of Geophysics, China Seismological Bureau, Beijing, China

1. Introduction

We have prepared two Fortran 77 software packages, COMPSYN and ISOSYN, for calculating earthquake ground motions caused by ruptures occurring on planar faults of finite spatial extent embedded in vertically varying media. These are in the subfolder for Chapter 85.14 on the attached Handbook CD. Both packages contain all necessary Fortran source files, as well as source files for visualizing the results of the calculations in device-independent PostScript graphic output files. Also included in the packages are documentation and sample input and output files that duplicate the example of Spudich and Archuleta (1987, fig. 20). All software must be run in a command-line-oriented operating system and has been successfully run under DOS and VMS.

2. Complete Synthetic Software

The COMPSYN (*Comp*lete *Syn*thetic) software uses the numerical techniques of Spudich and Archuleta (1987, pp. 247–263) to evaluate the representation theorem integrals on a fault surface. The Green's functions for wave propagation are calculated using a modified version of the Discrete Wavenumber/Finite Element (DWFE) method of Olson *et al.* (1984). The COMPSYN package has two primary strengths. First, the Green's functions include the complete response of the Earth structure, so that all P and S waves, surface waves, leaky modes, near-field terms, and static displacements are included in the calculated seismograms. Second, the codes are computationally fairly fast (typical computation times are minutes) compared to fully three-dimensional codes, so that the user can simulate ground motions from many hypothetical rupture models in minimal time. COMPSYN can be used to simulate ground motions for frequencies up to a few hertz from faults as long as about 100 km at distances to about 200 km. However, the effect of anelastic attenuation is impossible to include in the COMPSYN package.

3. Isochrone-integration Synthetic Applications

The ISOSYN (*Iso*chrone-integration *Syn*thetic) applications use the isochrone-integration technique of Bernard and Madariaga (1984) and Spudich and Frazer (1984) as their basic calculational algorithm. These applications are appropriate for calculating high-frequency ground motions near large earthquake ruptures. Specifically, these codes calculate the P and S wave motions that result from use of far-field ray-theory Green's functions in the representation theorem. The ISOSYN package has two primary strengths. First, it is computationally very fast (typical computation times of seconds), so that in principle the user can simulate ground motions from thousands of hypothetical rupture models in minimal time. Second, the graphical applications display many physical quantities on the fault that help the users understand and interpret the source effects, such as directivity and radiation patterns, that affect the calculated seismograms. An early example of such interpretive output is shown in Spudich and Oppenheimer (1986). ISOSYN is typically used to simulate

INTERNATIONAL HANDBOOK OF EARTHQUAKE AND ENGINEERING SEISMOLOGY, VOLUME 81B ISBN: 0-12-440658-0

ground motions up to frequencies of 10–20 Hz from faults up to about 100 km long at distances up to about 100 km. Because the ISOSYN package does not include the effects of anelastic attenuation, reflected waves, surface waves, or near-field terms in the Green's functions, the ISOSYN applications are most appropriately applied to situations in which the ground motions are expected to be dominated by direct *P* and *S* waves from the source. While the direct far-field *P* and *S* waves are always calculated correctly by this application, these waves do not dominate the expected real Earth response in some situations, such as (1) large distances from shallow sources, where surface waves and scattered body waves are strong; (2) near faults at low frequency, where near-field terms are strong; and (3) at very high frequency, where the effect of anelastic attenuation is strong.

Published examples of applications of the COMPSYN package are Archuleta (1984), Spudich and Archuleta (1987), and Guatteri and Spudich (2000). Additional published examples of applications of the ISOSYN package are found in Beroza and Spudich (1988), Beroza (1991), Cocco and Pacor (1993), and Guatteri and Cocco (1996).

References

Archuleta, R. J. (1984). A faulting model for the 1979 Imperial Valley earthquake. *J. Geophys. Res.* **89**, 4559–4585.

Bernard, P., and R. Madariaga (1984). A new asymptotic method for the modeling of near field accelerograms. *Bull. Seism. Soc. Am.* **74**, 539–558.

Beroza, G. C. (1991). Near source modeling of the Loma Prieta earthquake: Evidence for heterogeneous slip and implications for earthquake hazard. *Bull. Seism. Soc. Am.* **81**, 1603–1621.

Beroza, G. C., and P. Spudich (1988). Linearized inversion for fault rupture behavior: Application to the 1984 Morgan Hill, California, earthquake. *J. Geophys. Res.* **93**, 6275–6296.

Cocco, M., and F. Pacor (1993). The rupture process of the 1980 Irpinia, Italy, earthquake from the inversion of strong motion waveforms. *Tectonophysics* **218**, 157–177.

Guatteri, M., and M. Cocco (1996). On the variation of slip direction during earthquake rupture: Supporting and conflicting evidence from the 1989 Loma Prieta earthquake. *Bull. Seism. Soc. Am.* **86**, 1935–1951.

Guatteri, M., and P. Spudich (2000). What can strong motion data tell us about slip-weakening fault friction laws? *Bull. Seism. Soc. Am.* **90**, 98–116.

Olson, A., J. Orcutt, and G. Frazier (1984). The discrete wavenumber/finite element method for synthetic seismograms. *Geophys. J. R. Astr. Soc.* **77**, 421–460.

Spudich, P., and R. Archuleta (1987). Techniques for earthquake ground motion calculation with applications to source parameterization of finite faults. In: *Seismic Strong Motion Synthetics* (B. A. Bolt, Ed.), pp. 205–265. Academic Press, Orlando.

Spudich, P., and L. N. Frazer (1984). Use of ray theory to calculate high frequency radiation from earthquake sources having spatially variable rupture velocity and stress drop. *Bull. Seism. Soc. Am.* **74**, 2061–2082.

Spudich, P., and D. Oppenheimer (1986). Dense seismograph array observations of earthquake rupture dynamics (S. Das, J. Boatwright, and C. Scholz, Eds.), *Am. Geophys. Un. Geophys. Monograph 37*, 285–296.

85.15 Nonlinear Inversion by Direct Search Using the Neighbourhood Algorithm

Malcolm Sambridge
Research School of Earth Sciences, Australian National University, Canberra, Australia.

1. Background

The Neighbourhood Algorithm (NA) was introduced by Sambridge (1999a,b) as a direct search method for nonlinear inversion. This approach is applicable to a wide range of inversion problems, particularly those where the relationship between the observables (data) and the unknowns (a finite set of model parameters) is rather complex, e.g., fitting of seismic waveforms for Earth structure or source parameters.

The approach is divided into two stages. In the first, known as the search stage, one samples a multidimensional parameter space for combinations of parameters (models) that provide satisfactory fit to observed data. In the second, known as the appraisal stage, one tries to extract information from the complete ensemble of models collected, e.g., on resolution and trade-offs. The search algorithm is in the same class of techniques as Genetic Algorithms (GA) and Simulated Annealing (SA), in that it uses randomized decisions to drive the search and avoids the need for calculation of derivatives of the data misfit function. These techniques are often associated with global optimization problems. The NA differs from previous techniques in that it requires just two control parameters to be tuned, and the search process is driven by only the rank of models with respect to the data misfit criterion, and not the misfit itself. This allows considerable flexibility because any combination of data-fit criteria, or other information, can be used to rank models. Recently the NA has been applied to hypocenter location (Sambridge and Kennett, 2001) and seismic source characterization (Marson-Pidgeon *et al.*, 2000).

2. The Algorithm

The NA makes use of simple geometrical concepts to search a parameter space. The basic idea is illustrated with a simple example. Figure 1 shows results from a two-parameter problem in which the NA has been used to maximize a multipeaked fitness function in a plane. The top left panel shows the initial set of 10 points distributed quasi-randomly. At each stage, the entire parameter space is partitioned into a set of Voronoi cells (nearest neighbor regions), one about each previously sampled model. These cells are used to guide subsequent sampling in a randomized fashion. As iterations proceed, the algorithm concentrates sampling in promising regions. In this example, the NA distributes 20 new points in the best 20 Voronoi cells at each iteration. The top right panel shows the Voronoi cells after 100 points, and the bottom left after 500 points have been added. Clearly, all four prominent maxima are well sampled. In fact, the global maximum was located after 442 samples. Sambridge (1999a) has demonstrated that even though the NA is based on simple geometrical principles, it results in a highly self-adaptive search, and it remains computationally practical even in much higher dimensional spaces (e.g., 10–100).

3. The NA-Sampler Package

The code contributed to this volume is the NA-sampler package, which is the author's implementation of the Neighbourhood Algorithm for a multidimensional parameter space search.

INTERNATIONAL HANDBOOK OF EARTHQUAKE AND ENGINEERING SEISMOLOGY, VOLUME 81B ISBN: 0-12-440658-0

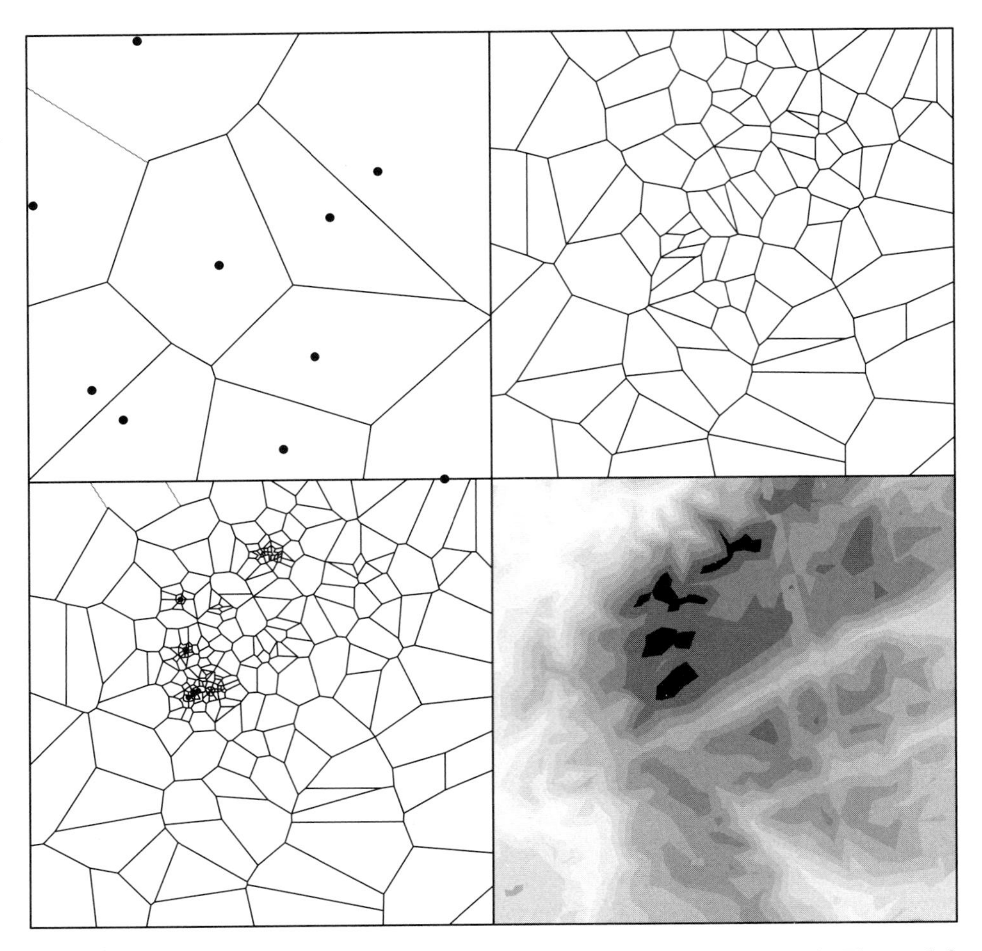

FIGURE 1 The figure shows three stages of Neighbourhood Algorithm search. The top left panel shows the initial 10 uniformly randomly distributed points and their corresponding Voronoi cells. The top right panel shows the Voronoi cells of the first 100 points generated by the NA, and the bottom left panel shows the Voronoi cells for 500 points. The bottom right panel shows the true fitness landscape; darker shades are higher fitness.

It consists of a set of user-callable subroutines and an example driver program. The latter applies the NA algorithm to the estimation of crustal seismic velocity profiles from inversion of seismic receiver functions (RFI). Several independent "utility" programs are also provided that manipulate the output files of the main program and plot some results using either X or PostScript. Some of these programs are specific to the RFI problem, and others may be more generally useful.

To apply the NA to a new problem, one needs to supply three subroutines and a main driver program that may be trivially simple, as in the case of the supplied example. These routines are compiled together with the NA code. The program is controlled by a single ASCII input file in which the two tunable parameters and other options are set. The output is in the form of a simple summary file and a direct-access binary file containing the models generated by the search algorithm together with their supplied data fit. This file can be analyzed with a plot program provided or converted to an ASCII file and assessed separately by the user. In the receiver function inversion example, one of the three user routines is used to write out all velocity models, together with the observed and predicted receiver functions from the best-fit velocity profile. Several programs are provided that display models and visually compare the data fit.

4. Hardware and Software Requirements

The NA-sampler package is in the subfolder for Chapter 85.15 on the attached Handbook CD. The complete source code is provided in a combination of Fortran 77 and C. Fortran coding is "standard" with no Sun Fortran or other compiler enhancements. Detailed HTML instructions are included for installing and running the programs. To date, the package has been compiled and tested on a Sun workstation running Solaris 2.4 with both native and GNU gcc/g77 compilers, on a Compaq Alpha running Tru64unix, and on an SGI platform running IRIX 6.5. In each case, the test examples produced exactly the same results.

5. The NA Home Page

The NA home page is at *http://rses.anu.edu.au/~malcolm/na/na.html*. Included are instructions on obtaining the latest versions of NA-sampler as well as the author's implementation of the appraisal-stage algorithm NA-Bayes.

References

Marson-Pidgeon, K., B. L. N. Kennett, and M. Sambridge (2000). Source depth and mechanism inversion at teleseismic distances, using a neighbourhood algorithm. *Bull. Seism. Soc. Am.* **90**, 1369–1383.

Sambridge, M. (1999a). Geophysical Inversion with a Neighbourhood Algorithm I: Searching a parameter space. *Geophys. J. Int.* **138**, 479–494.

Sambridge, M. (1999b). Geophysical Inversion with a Neighbourhood Algorithm II: Appraising the ensemble. *Geophys. J. Int.* **138**, 727–746.

Sambridge, M., and B. Kennett (2001). Seismic event location: Nonlinear inversion using a Neighbourhood Algorithm. *Pure Appl. Geophys.* **158**, 241–257.

85.16 Higher Order Perturbation Theory: 3D Synthetic Seismogram Package

Éric Clévédé and Philippe Lognonné
Institut de Physique du Globe de Paris, France

The Higher Order Perturbation Theory (unofficially referred to as HOPT) has been developed by Philippe Lognonné starting at the end of the 1980s. The package presented here is the result of 10 years of theoretical work and implementation. This first version as a package includes the totality of the up-to-date codes written during these years. The present set of libraries and programs allows the computation of normal modes and seismograms for a wide variety of 3D Earth models, with only a few required parameter settings. Optimization, new options, and new interfaces are in current developments to improve the efficiency and usability of the codes.

The codes run successfully on Sun Solaris 2.7. Disk space of at least 1 GB is recommended for practical use of these codes. The main memory-consuming program (*pertuanel*) needs around 600 MB, which may be achieved with swap space. The *minos* program provided in the distribution has been written by F. Gilbert, G. Masters, and J. Woodhouse.

Please consult our Web page at *http://www.ipgp.jussieu.fr/~clevede/HOPT/* for bug reports and updates.

We strongly recommend the reading of the reference articles (Lognonné and Romanowicz, 1990; Lognonné, 1991; Clévédé and Lognonné, 1996; and Chapter 10 of this Handbook) before attempting to use this package. We strongly advise that this version of the package not be used as a black box!

This package is presented in full (except for executables) in the subfolder for Chapter 85.16 on the attached Handbook CD. Included are a README file, documentation (in PostScript and PDF formats), souce code, installation scripts, input file examples, and working directories with input files for the 1D PREM (Dziewonski and Anderson, 1981) and the 3D SAW12D (Li and Romanowicz, 1996) models.

References

Clévédé, É., and P. Lognonné (1996). Fréchet derivatives of coupled seismograms with respect to an anelastic rotating Earth. *Geophys. J. Int.* **124**, 456–482.

Dziewonski, A., and D. L. Anderson (1981). Preliminary Reference Earth Model. *Phys. Earth Planet. Interiors* **25**, 297–356.

Li, X. D., and B. Romanowicz (1996). Global mantle shear velocity model developed using non-linear asymptotic coupling theory. *J. Geophys. Res.* **101**, 22245–22272.

Lognonné, P., and B. Romanowicz (1990). Fully coupled Earth's vibrations: the spectral method. *Geophys. J. Int.* **102**, 365–395.

Lognonné, P. (1991). Normal modes and seismograms in an anelastic rotating Earth. *J. Geophys. Res.* **96**, 20309–20319.

INTERNATIONAL HANDBOOK OF EARTHQUAKE AND ENGINEERING SEISMOLOGY, VOLUME 81B ISBN: 0-12-440658-0

85.17 The HYPO71 Earthquake Location Program

W. H. K. Lee
US Geological Survey, Menlo Park, California, USA (retired)
John C. Lahr
US Geological Survey, Golden, Colorado, USA
C. M. Valdes
Universidad Nacional Autónoma de México, Mexico City, Mexico

1. Introduction

HYPO71, a computer program for determining hypocenter, magnitude, and first-motion pattern of local earthquakes, was first released in 1971. It is perhaps the first earthquake location program that achieved worldwide usage, as evidenced by the fact that about 1000 copies of the HYPO71 manual were requested and distributed (Lee, 1990). It is included here for historical documentation.

2. History

Although Geiger (1912) introduced an earthquake location procedure based on the least squares in 1910, it was not a practical procedure until digital computers became common in the 1960s. In the early 1960s, many seismologists around the world wrote earthquake location programs based on Geiger's method, such as HYPOLAYR (Eaton, 1969), which included a listing of the source code. For ease of routine data processing of a large regional seismic network, the HYPO71 program was written with an emphasis on a simple user interface for batch processing. The original HYPO71 program was dated December 21, 1971, and a user's manual was released (Lee and Lahr, 1972).

In order to generalize HYP071 for worldwide usage and to correct a few programming "bugs," the program HYP071 was revised on November 25, 1973, and a note on "Revisions of HYP071" was released on January 30, 1974. In 1975, all reprints of the original HYP071 manual were exhausted, and the authors took this opportunity to release a revised manual, called HYP071 (Revised) (Lee and Lahr, 1975). When the personal computers were introduced in the early 1980s, a version of the HYP071 program called HYPO71PC was released by Lee and Valdes (1985). This version was also published in the IASPEI Software Library (see Chapter 85.2 by Lee). HYPO71PC differs from HYPO71 in that input and output are both done through files, whereas input for HYPO71 was from punch cards and output was (generally) directly to the printer. Since the HYPO71PC program was written in standard FORTRAN, it can be compiled on most computers with no modifications.

3. Files on the Attached Handbook CD

The subfolder for Chapter 85.17 on attached Handbook CD contains the following:

1. Discussion.html—an HTML file including a brief discussion on earthquake location and some additional references.
2. HYPO71manual.pdf—a PDF file of the HYPO71 (Revised) manual (Lee and Lahr, 1975; listings of the computer files are omitted because readers can list these files themselves).
3. HYPO71.for—the FORTRAN source code of the HYPO71 program.

INTERNATIONAL HANDBOOK OF EARTHQUAKE AND ENGINEERING SEISMOLOGY, VOLUME 81B ISBN: 0-12-440658-0

4. HYPO71PC.for—the FORTRAN source code of the HYPO71 program for IBM-compatible personal computers.
5. HYPO71.inp—an input test file for the HYPO71PC program.
6. HYPO71.out—output results from running HYPO71PC using HYPO71.inp as the input file.
7. Lee1972.pdf—a PDF file of Lee *et al.* (1972) on estimating local magnitude using signal duration (commonly called the coda magnitude).

4. Discussion

Our experience indicates that locating local earthquakes accurately requires considerable efforts (see the discussion given in the Discussion.html file on the CD-ROM). As discussed in Lee and Stewart (1981, pp. 130–139), earthquake location is a nonlinear problem, and there is no foolproof method to locate an earthquake because the input data may *not* be sufficient to constrain the problem. The HYPO71 program does not solve the equations in the Geiger's method by the traditional matrix inversion techniques as almost all other earthquake programs do. It uses a multiple stepwise regression method (Draper and Smith, 1966) to adjust hypocenter parameters only if it is statistically significant above a prescribed critical F-value. Obviously, this method reduces to the traditional technique if one sets the critical F-value to zero. We believe that by studying the output results of a HYPO71 solution, one can gain some insight into what are the difficulties in reducing the residuals in solving a nonlinear problem by iterations.

References

Draper, N. R., and H. Smith (1966). *Applied Regression Analysis*. John Wiley & Sons, New York, 407 pp.

Eaton, J. P. (1969). HYPOLAYR: a computer program for determining hypocenters of local earthquakes in an earth consisting of uniform flat layers over a half space. *Open-File Report*, US Geological Survey, Menlo Park, California, 155 pp.

Geiger, L. (1912). Probability method for the determination of earthquake epicenters from the arrival time only (translated from Geiger's 1910 German article). *Bull. St. Louis Univ.* **8**(1), 56–71.

Lee, W. H. K. (1990). Replace the HYPO71 format? *Bull. Seism. Soc. Am.* **80**, 1046–1047.

Lee, W. H. K., and J. C. Lahr (1972). HYP071: A computer program for determining hypocenter, magnitude, and first motion pattern of local earthquakes. *Open-File Report*, US Geological Survey, Menlo Park, California, 100 pp.

Lee, W. H. K., and J. C. Lahr (1975). HYP071 (Revised): A computer program for determining hypocenter, magnitude, and first motion pattern of local earthquakes. *Open-File Report 75-311*, US Geological Survey, Menlo Park, California, 113 pp. [This OFR is included in the subfolder for Chapter 85.17 on the attached Handbook CD.]

Lee, W. H. K., and S. W. Stewart (1981). *Principles and Applications of Microearthquake Networks*. Academic Press, New York, 293 pp. [This is included in the subfolder for Chapter 17 on the attached Handbook CD.]

Lee, W. H. K., and C. M. Valdes (1985). HYP071PC: A personal computer version of the HYPO71 earthquake location program. *Open-File Report 85-749*, US Geological Survey, Menlo Park, California, 43 pp.

Lee, W. H. K., R. E. Bennett, and K. L. Meagher (1972). A method of estimating magnitude of local earthquakes from signal duration. *Open-File Report*, US Geological Survey, Menlo Park, California, 28 pp. [This is included in the subfolder for Chapter 85.17 on the attached Handbook CD.]

85.18 Software for Seismometer Calibration and Signal Analysis

Erhard Wielandt
University of Stuttgart, Stuttgart, Germany

1. Introduction

Fortran source code of six computer programs mentioned in "Seismometry" by E. Wielandt (Chapter 18 in Part A of this Handbook) is included in the subfolder for Chapter 85.18 on the attached Handbook CD. Current versions, including some executables, can be downloaded from the Web site: *http://www.geophys.uni-stuttgart.de/downloads/*. These are stand-alone programs for calibrating and testing seismometers and for standardizing noise data; they do not form a package for general seismic processing. A README file with explanations, a set of test data, and output files are provided with each program. No graphics are included.

2. Programs

- CALEX: Determines parameters of the transfer function of a seismometer from the response to an arbitrary input signal (which must be recorded together with the output signal). The transfer function is implemented in the time domain as an impulse-invariant recursive filter. The inversion uses the method of conjugate gradients (only moderately efficient but quite failsafe).
- DISPCAL: Determines the generator constant of a horizontal or vertical seismometer from an experiment where the seismometer is moved stepwise on the table of a machine tool or a mechanical balance.
- NOISECON: Converts noise specifications into several sets of standard and nonstandard units and compares them to the USGS New Low Noise Model (Peterson, 1993). This interactive program is available in BASIC, FORTRAN, and C. A 32-bit MS Windows executable *noisecon.exe* can be downloaded from the FTP site.
- SINFIT: Fits a sinewave to two signals (normally input and output signals of an amplifier, filter, or seismometer). The first signal is supposed to be an undisturbed reference signal, and the frequency of the sinewave is determined from the first signal alone. The relative amplitude and phase between the input sinewave and the second signal are then determined with frequency fixed.
- TILTCAL: Determines the generator constant of a horizontal seismometer from an experiment in which the seismometer is tilted stepwise.
- UNICROSP: Estimates seismic and instrumental noise separately from the coherency of the output signals of two seismometers.

Reference

Peterson, J. R. (1993). Observations and modelling of background seismic noise. *Open-File Report 93-322*, US Geological Survey, Albuquerque, New Mexico, 94 pp.

INTERNATIONAL HANDBOOK OF EARTHQUAKE AND ENGINEERING SEISMOLOGY, VOLUME 81B ISBN: 0-12-440658-0

86

The FDSN and IRIS Data Management System: Providing Easy Access to Terabytes of Information

Timothy K. Ahern
IRIS Data Management Center, Seattle, Washington, USA

1. Introduction

In 1985, the Commission on Practice of the International Association for Seismology and Physics of the Earth's Interior (IASPEI) formed a Working Group on Digital Data Exchange. Soon after that the Federation of Digital Broad-Band Seismograph Networks (FDSN) formed, and as its first mission it assumed the responsibility for the development of the standard format for the exchange of digital earthquake data.

During this time period the spirit of international cooperation in seismology was reaching an all-time high. Broadband seismometers had brought together the long-period and short-period communities, and seismologists were very open to the concept of free and open exchange of data. Traditionally, seismological cooperation had been frustrated by a plethora of seismic data recording instruments, seismic data formats, and variable request mechanisms. The FDSN introduced an international organization that allowed cooperation and coordination between those installing and operating seismic networks on a scale that had not previously occurred.

The FDSN maintains a World Wide Web presence at *http://www.fdsn.org*, and additional information can be found by referencing that location. At the beginning of the year 2000, the FDSN had 18 institutional members[1] from 16 countries. Countries represented include Australia, Canada, Chile, China, Czech Republic, France, Germany, Israel, Italy, Japan, Mexico, Netherlands, Poland, Switzerland, Taiwan, and the United States. Additionally, several countries participated in the FDSN without being official members.[2]

Membership in the FDSN is open to all organizations that operate seismic networks that record broadband data. FDSN members agree to coordinate station siting and provide free and open access to the seismological data their networks generate. This chapter attempts to summarize the various activities of the FDSN as they pertain to the activities of data generation and distribution. Due to my close affiliation with IRIS, it will have greater information related to IRIS, but this in no way is intended to diminish the level of activity at all the other FDSN centers.

2. The FDSN Terms of Reference

The FDSN is an international association affiliated with IASPEI. It conducts its business and was established under what are called the FDSN Terms of Reference. The FDSN Terms of Reference follow.

The International Seismological Community recognizes the new opportunities within its field for improved understanding of the internal structure and dynamical properties of the Earth provided by recent developments in seismograph network technology.

It also recognizes that rapid access to seismic data networks of modern broad-band digital instruments wherever they might be is now possible.

The developments include greatly improved broad-band seismographic systems that capture the entire seismic wave field with high fidelity, efficient and economical data communications and storage and widely available, powerful computing facilities.

The federation is open to all national and international programs committed to the deployment of broad-band seismographs and willing to contribute to the establishment of an optimum global system with timely data exchange.

INTERNATIONAL HANDBOOK OF EARTHQUAKE AND ENGINEERING SEISMOLOGY, VOLUME 81B ISBN: 0-12-440658-0

I. *Goals*

In view of the above and to take advantage of existing developing global and regional networks the "Federation of Digital Broad-Band Seismograph Networks (FDSN)" is formed to provide a forum for:

- *developing common minimum standards in seismographs (e.g. bandwidth) and recording characteristics (e.g. resolution and dynamic range);*
- *developing standards for quality control and procedures for archiving and exchange of data among component networks;*
- *coordinating the siting of additional stations in locations that will provide optimum global coverage.*

II. *Institutional Frame*

The Federation is an independent international association and is affiliated with the International Association for Seismology and Physics of the Earth's Interior (IASPEI).

III. *Membership and Organization*

- *Membership in the FDSN is open to national and international programs committed to both the development and operation of the broad-band digital networks and achievement of the goals of the Federation.*
- *The structure of the FDSN includes a steering committee and an executive committee.*
- *The members of the FDSN steering committee will consist of one representative per network who will be appointed or selected, by the network, from within the organization they represent.*
- *Between meetings of the steering committee, the activities of the federation will be coordinated by an elected executive committee whose membership may not exceed one representative from any national or international program.*
- *The FDSN executive committee, elected for a four year term, will be headed by a chairman, assisted by not more than three members, and a Secretary. The chairman of the executive committee will preside over the meetings of the steering committee. Members of the executive committee shall be elected from within the steering committee or the member organizations.*
- *The FDSN steering committee will form all necessary working groups or special technical committees as required to reach the objectives of the FDSN.*
- *The FDSN steering committee will meet at least once a year. Special meetings may be called by the chairman as appears necessary for the progress of the FDSN.*
- *Concerning all recommendations made and actions to be taken, each member of the FDSN will have one vote, and an affirmative vote by 2/3 members present at an FDSN steering committee meeting will be required.*
- *Five members of the FDSN will constitute a quorum for FDSN steering committee meetings.*
- *No fees are imposed but voluntary contributions may be requested to cover costs for communications.*
- *Any member may resign at any time by giving written notice to the chairman.*

3. Activities of the FDSN

The FDSN meets once per year usually during the time of another international meeting. Much of its work is conducted in its four working groups: (1) station-siting working group, (2) data formats and data exchange working group, (3) software coordination working group, and (4) a working group that focuses on CTBT issues.

3.1 The FDSN Seismic Network

The station-siting working group allows the member networks to coordinate the location of new seismic stations. It selects a subset of stations in the member networks to compose the FDSN Seismic Network. These stations are selected to provide a network that will provide optimal global coverage of the Earth. This working group is also charged with coordinating instrumentation standardization. Figure 1 shows the current set of FDSN stations that have been identified. Most of the FDSN stations are currently installed, while others are only planned.

3.2 The SEED Format

The working group on data formats and data exchange has been responsible for the development of the Standard for Exchange of Earthquake Data (SEED)[3] format, the most comprehensive format for data exchange in existence. Members agree to distribute their data in this format and in some cases assist other countries to get their data into SEED, with ORFEUS taking a lead in helping FDSN members in this area. This working group also coordinates activity in providing standard data access tools with which requests for seismic data can be sent to any FDSN Data Center. This group has coordinated SPYDER® autoDRM and is currently working on a Networked Data Center concept called NetDC.

3.3 FDSN Software Coordination

The software coordination group was established at the FDSN meeting held in Thessaloniki, Greece, in 1997. Its goal is to coordinate software activities both in the area of existing software and coordinating new approaches to software development. This group will be considering new object oriented approaches to software including Java and Common Object Request Broker Architecture (CORBA).

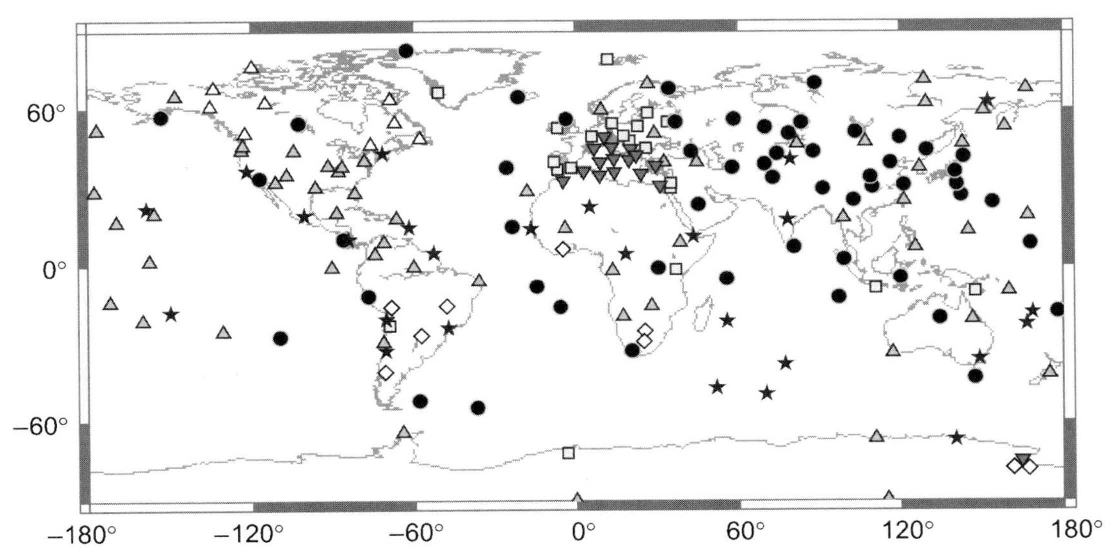

FIGURE 1 The map shows the stations of the FDSN network. These stations are a subset of stations belonging to the various members of the FDSN that were selected from a larger set of stations. These stations were selected primarily to give a good geographical distribution of seismic stations with broadband recording characteristics. Data from these stations are included on the FDSN CD-ROM. It should be noted that some of these stations are not yet in existence, but FDSN networks are in the process of installing them.

3.4 FDSN Products

The FDSN has produced two primary products at this time. The first is the FDSN CD-ROM. All member networks send continuous data to the FDSN Continuous Data Archive at the IRIS DMC. Currently there are 172 stations that are designated FDSN stations. The USGS National Earthquake Information Center (NEIC) sends requests for time windows of data for all events larger than magnitude 5.7, indicating the hypocenter information, the stations, channels, and specific time windows for each station. The IRIS DMC then extracts the designated windows from the continuous archives and builds a single SEED event volume, containing the hypocenter information, one volume for each event. These SEED volumes are then forwarded to the NEIC, which masters and produces the CD-ROMs for distribution. At the present time there have been only four CD-ROMs produced, covering the period January–December 1990. The SEED volumes for all of the 1991 data have been forwarded to the NEIC and should be available soon. Current production of the 1992 CD-ROMs is under way.

The second major product of the FDSN has been the FDSN station book. This is a very extensive book containing information for more than 250 seismic stations around the world. The book contains information about station contacts, network operators, locations, geology, instrumentation type, types of data recorded, pictures of the area, response curves, and estimates of the seismic noise at each site. The FDSN station book is available via the World Wide Web at *http://www.fdsn.org*, on a CD-ROM available from IRIS at no charge, or as a 600-page, full-color book at a nominal charge from IRIS. Contributions to the FDSN station book were made by the various networks, and the book itself was produced by IRIS.

4. Characteristics of FDSN Data

The FDSN emphasizes collection of broadband data whose signal content goes from a few hertz to several thousands of seconds. The dominant sensor used is the Streckeisen STS-1 three-component seismometer for surface installations, and the Teledyne Geotech KS36000 or 54000 for borehole installations. These sensors are often augmented with strong-motion seismometers or sensors that are sensitive at higher frequencies for special purposes such as monitoring regional signals as in the Comprehensive Test Ban Treaty monitoring regime. Figure 2 shows the range of periods and accelerations that typical FDSN instrumentation can now record on-scale. The low-gain seismometers can record signals of up to $2g$ acceleration. Frequencies from 0.03 seconds to several thousand seconds can be recorded. The very broadband seismometers can record signals from about 0.15 seconds up to several thousand seconds, and the high-frequency sensors can record small-magnitude events at 0.03 seconds.

Figure 2 clearly shows the concept at the heart of the FDSN: The traditional World Wide Seismic System Network (WWSSN) short-period and long-period responses both lie well within the recording capabilities of the Very Broad Band seismic channel of current FDSN instrumentation.

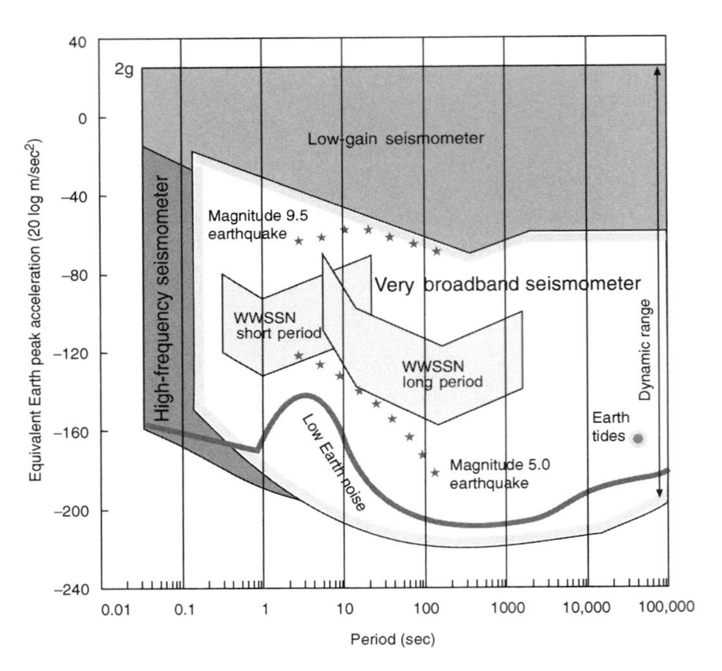

FIGURE 2 This figure shows the recording characteristics of IRIS and similar FDSN recording systems. The very broadband seismometers can record frequencies from about 5 Hz to periods of several thousand seconds. The WWSSN long-period and short-period data channels can be derived from the VBB seismic channel, and in essence this defines the term VBB in the FDSN context. Other sensors such as low-gain seismometers and high-frequency seismometers are added to some FDSN stations to extend the recording frequencies of the data (modified from R. Butler, IRIS).

In general, FDSN seismic stations record three components of ground motion at several different sampling rates, all of which are derived from the single seismometer. Traditional channels are recorded at 20 samples per second, 1 sample per second, 1 sample per 10 seconds, and 1 sample per 100 seconds. The FDSN design goal is to record the data channels with 24 bits of resolution, although some recording systems have implemented dual-gain 16-bit recording. The data are usually compressed using a first- or second-difference scheme that provides from 4 to 7 times compression. Most FDSN stations record between 10 megabytes and 30 megabytes of data per day. These characteristics are state of the art in seismometry today.

5. FDSN Data Centers

In general, each FDSN network operates a data center that provides quality control of the data and data distribution services for data from their network. Activities within the FDSN coordinate data center activities as well in an effort to make the task of requesting data as simple as possible. In general, the FDSN data centers cooperate in several types of data products and request mechanisms. The following sections itemize some of the more common products and request mechanisms that can be found at FDSN data centers.

At the present time, those organizations that operate major FDSN data centers that distribute their data in SEED format include IES in Taipei; CSB in Beijing; GSC in Ottawa; GEOFON in Pottsdam; GEOSCOPE in Paris; Graefenberg in Erlangen, Germany; MEDNET in Rome; IRIS in Seattle; USGS in Golden, USA; ORFEUS in deBilt, the Netherlands; and ERI in Tokyo. Many of these centers have very sophisticated data management systems from which worldwide seismologists can receive data directly. The IRIS DMC, as the first FDSN Data Center for Continuous Data, offers data from all of the above networks and is the only data center with such a complete data collection; however, not all stations from all networks are available through IRIS. Data are often available in a more timely manner from data centers operated by the specific network. For instance, GEOSCOPE data can be received directly from Paris much sooner than it can from the IRIS DMC. Additionally, some data centers such as ORFEUS have data available from a very broad set of stations in Europe that are not available from any other data center. Interested seismologists should consult the FDSN Web page to determine the specific data available from each FDSN data center.

5.1 SPYDER®—*S*ystem to *P*rovide *Y*ou *D*ata from *E*arthquakes *R*apidly

The SPYDER® (registered trademark of The IRIS Consortium) system was developed by Dr. Steve Malone and his colleagues at the University of Washington for the IRIS Data Management System. The NEIC in Golden, Colorado, sends alert messages by e-mail to the SPYDER system whenever an event is detected. A control system called Badger, at one of three primary FDSN data centers (GEOFON, IRIS DMC, ORFEUS), receives these messages and determines what stations should be contacted and time windows of data recovered. This determination is made based upon the size of the event and the event–station distance. Installed around the world are 13 different SPYDER nodes. These nodes respond to messages from Badger and are charged with accessing the various stations, generally those in close geographic proximity. Currently, SPYDER nodes are running in Australia, New Zealand, Taiwan, Japan, Netherlands, Germany, France, and the United States. Some countries operate more than one node. The SPYDER nodes recover data from the designated stations and then return the data to the originating Badger system using the Internet. The connections to the individual stations vary widely: Some are dial-up connections over telephone lines, some are made through Internet connections, and some make use of satellite communication systems such as Inmarsat. ORFEUS, GEOFONE, and IRIS work together to ensure that each of these locations has a complete set of SPYDER data.

SPYDER data are not quality controlled, and users must therefore be careful. All of the data for each SPYDER event are assembled into a single SEED volume and placed in anonymous FTP areas of various data centers. Older SPYDER systems also have an interface that is supported via telnet connections. Newer SPYDER interfaces are generally made through Web interfaces, one of which, WILBER, is discussed later.

5.2 FARM—Fast Archive Recovery Method

As mentioned previously, SPYDER data volumes are available in SEED format but contain data that have not been quality controlled. SPYDER volumes obviously only contain data from FDSN stations that have communication connections installed. FARM volumes are produced by some FDSN data centers, such as the IRIS DMC, and contain data that have come through the entire quality control path within the various FDSN data centers. FARM volumes contain all the data from all stations that generated data, forwarded the data to the FDSN data center, had quality control procedures applied to them, and were archived at that center. FARM volumes generally contain much more data than equivalent SPYDER volumes, both in terms of the number of included stations and in terms of the length of the time windows contained in the volumes. In general, FARM volumes contain 1 hour of Very Broadband data (20 samples per second) for three components, several hours of Long Period Data (1 sample per second), and any triggered high-frequency channels. FARM products come bundled with a variety of graphics including pseudo record sections, waveform GIF images, table-of-contents files, and of course the SEED volume for the various events. Events of moment magnitude (M_w) 5.8 and greater are included in the FARM data, and events of magnitude 5.5 or larger are included if the event is deeper than 100 km.

In general, there are about 250 FARM events per year, and at the present time there are about 150 gigabytes of FARM data in the IRIS DMC anonymous FTP area. FARM products are made about 2 to 4 months behind real time and when possible replace the SPYDER products that preceded them.

Each FDSN data center is responsible for the production of the various FARM products for their respective networks. The IRIS DMC has produced FARM products for the GDSN and IRIS GSN networks beginning with data in 1972 and continuing to the present time. Eventually, all FDSN networks will produce their own set of FARM volumes and make them available through any FDSN data center that wishes to have copies online.

When FARM products were originally conceived, the volumes tended to be small. As more and more stations, some with channels sampled at higher rates, became available, the size of the volumes increased, exceeding 100 megabytes in some cases. This makes FTPing them impractical. For this reason, FARM products can be distributed on tape to those requesting them. Additionally, new interfaces such as WILBER and WEED, which will be discussed later, can access the FARM data products.

5.3 Archived Data at the IRIS DMC

The FDSN data centers all provide access to any data contained in their archives. The various data centers have a wide variety of storage management solutions varying from relatively small optical systems, to RAID disk systems with significant capacity, to very large tape-based mass storage systems. The largest storage system that exists within the FDSN is installed at the IRIS Data Management Center, with a capacity of 50 terabytes. The IRIS DMC has a special place in the structure of the FDSN; it is the FDSN Archive for Continuous Data. This means that all FDSN member networks send data from at least the FDSN network stations to the IRIS DMC where the data are stored in the same manner that IRIS's own data are stored.

Figure 3 shows the growth in the archive at the IRIS DMC, clearly indicating the contribution that the FDSN networks have made to the IRIS DMC holdings. The diagram shows the data holdings at the IRIS DMC as of July 31, 2000. Although there are 43 different seismic networks whose data are archived, the networks have been grouped into broad categories. The IRIS GSN is the lowermost category (5.6 terabytes), the next band is data from other FDSN networks (1.0 terabytes), the next band is data from IRIS-operated networks in the former Soviet Union (JSP) (0.9 terabytes), the next band is various regional networks (1.7 terabytes), and the uppermost category is the IRIS PASSCAL program (4.9 terabytes). The total archive size at the IRIS DMC is 14.1 terabytes.

The IRIS DMC has developed a fairly sophisticated method of archiving data. Initially, data are stored on a large 1.5 terabyte disk-based RAID system. The data are stored on that RAID for about 4 months during which time the data are edited and repaired in any necessary way. After this holding period, the data are then moved into the tape-based mass storage system in a very well sorted order. Since most data requests are made for specific events, the IRIS DMC stores data in a time-sorted order in which all data for a given network for a given day are generally placed as sequential files on the tape media. Then when a request for earthquake data is made, normally only one or two tapes need to be mounted and so data are recovered quite quickly. Since some requests are all data from a single station for many years, the IRIS DMC also stores the data in a station-sort order. If we did not do that, then station-oriented requests could cause every tape to be mounted. This dual-sort order also allows us to protect the online holdings since each seismogram is stored twice in the storage system. For disaster recovery purposes, a copy of the time-sorted data is made to DLT tapes that are shipped off-site to a location in Colorado for safekeeping.

Although this chapter has emphasized the FDSN data, it should be clear that many data centers have large holdings of seismic data that are not from permanent stations recording data from FDSN-like installations. For the IRIS DMC, the Program for Array Seismic Studies of the Continental Lithosphere (PASSCAL) data are the most notable example of this. These data are from temporary deployments of recording systems equipped to record signals from regional or teleseismic events.

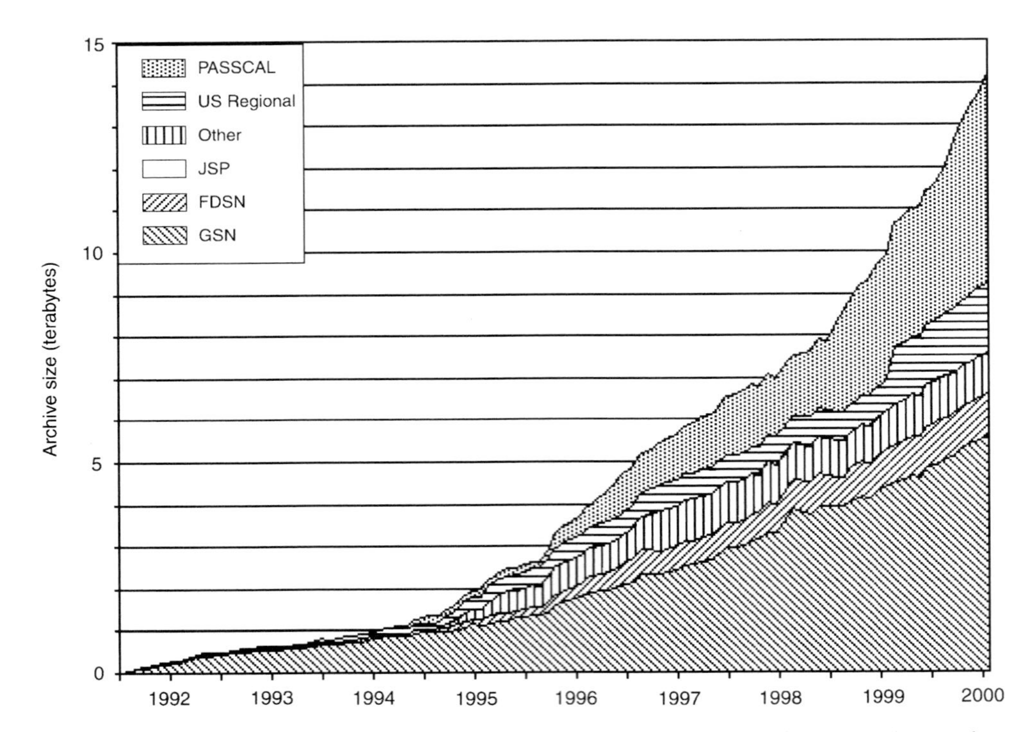

FIGURE 3 The IRIS DMC acts as both the IRIS archive and the archive for the continuous data from stations belonging to the FDSN network. The diagram shows the growth in the archive at the IRIS DMC since the DMC was established in Seattle, Washington, in 1992. The lowest portion is data from the IRIS GSN and USGS GDSN networks. The next higher band represents data from the FDSN network stations. The middle band is data from the IRIS JSP networks installed in the former Soviet Union. The next band represents data from regional networks and some historical networks such as ILPA. The upper band represents data from the PASSCAL program. There were about 14.1 terabytes of compressed seismic data in the IRIS DMC mass storage systems as of July 31, 2000.

These data are acquired by individual scientists who convert the data to SEED format and transfer it to the IRIS DMC. To a large extent, the data from the IRIS PASSCAL program help to fill in the seismic data between GSN stations, and therefore the two data sources complement each other.

A variety of request tools have been developed to allow users to recover data from online data sets such as SPYDER and FARM as well as making customized requests for data directly from archive holdings at FDSN data centers.

6. Generic Data Request Tools Used by the FDSN

The FDSN has been instrumental in standardizing the format in which seismic waveform data are distributed. It has also attempted to coordinate the data access methods available to the research community. A brief summary of three methods will be presented next. The first method is the autoDRM system developed by Urs Kradolfer of the Swiss Seismological Service. The second is the BREQ_FAST mechanism pioneered by IRIS. The final method is NetDC, a new concept of networking data centers that is currently being developed within the FDSN.

6.1 The AutoDRM Method

The United Nations (UN) formed the Group of Scientific Experts (GSE) to study how seismology could be used to monitor a Comprehensive Test Ban Treaty. The GSE conducted several Technical Tests (TT), the third of which was the GSETT-3 experiment. During this experiment the AutoDRM mechanism of making requests for waveforms was developed. The AutoDRM is an e-mail–based method of requesting data as well as returning the data to the requestor via e-mail. AutoDRMs officially support the transfer of SEED data as well, but most implementations rely on data transfer via FTP. Complete instructions on how to use the AutoDRM can be obtained by sending e-mail to *autodrm@seismo.ifg.ethz.ch.*

In the body of the e-mail message, put the following:

```
BEGIN 2.0
GUIDE
E_MAIL        your_email.address
STOP
```

A copy of the AutoDRM guide will be returned to you. AutoDRMs vary in their capabilities but always allow access

```
BEGIN                         Start of a new request
DATE1 199307190800            Start of time interval: 1993Jul19 08:00 GMT
DATE2 199307192030            End of time interval  : 1993Jul19 20:30 GMT
WAVEF OSS                     \ Send all waveforms of the stations OSS and
WAVEF SLE                     / SLE recorded within the time interval
CALIB OSS                     Send calibration data of station OSS
EMAIL was@gsehub.css.gov      E-mail address where output should be sent to
STOP                          End of the request
```

FIGURE 4 This is an example AutoDRM message requesting data from stations OSS and SLE. Data will be returned by e-mail in a compressed ASCII format.

```
.NAME Joe Seismologist
.INST Small University
.MAIL 101 Fast Lane, Middletown, USA  89432
.EMAIL joe@small.edu
.PHONE 555 555-1212
.FAX   555 555-1213
.MEDIA Electronic
.ALTERNATE MEDIA 1/2" tape - 6250
.ALTERNATE MEDIA EXABYTE
.LABEL Joe's FIRST Request
.SOURCE ~NEIC PDE~Jan 1990 PDE~National Earthquake Information Center - USGS DOI~
.HYPO ~1990 01 02 20 21 32.62~ 13.408~ 144.439~135.0~18~216~Mariana Islands~
.MAGNITUDE ~5.7~mb~
.END
GRFO IU 1995 01 02 00 18 10.4 1995 01 02 00 20 10.4 1 SHZ
ANTO IU 1995 01 02 02 10 36.6 1995 01 02 02 12 36.6 1 SH?
GRFO IU 1995 01 02 02 10 37.1 1995 01 02 02 12 37.1 1 SH?
SEE  CD 1995 01 02 14 45 08.9 1995 01 02 14 47 08.9 1 SHZ
BDF  IU 1995 01 04 02 42 13.4 1995 01 04 02 44 13.4 1 SHZ
NNA  II 1995 01 04 02 41 57.5 1995 01 04 02 43 57.5 1 BHZ
PFO  TS 1995 01 04 02 41 57.5 1995 01 04 02 43 57.5 1 BHZ
PFO  II 1995 01 04 02 41 57.5 1995 01 04 02 43 57.5 1 BHZ
KMI  CD 1995 01 04 02 41 57.5 1995 01 04 02 43 57.5 1 BHZ
SSE  CD 1995 01 04 02 18 25.4 1995 01 04 02 20 25.4 2 B?? SHZ
PAS  TS 1995  1  4  2 10 49    1995  1  4  2 12 49    3 BH? SHZ L??
```

FIGURE 5 An example BREQ_FAST file is shown in this figure. The various lines that identify the requestor are preceded with a **.**TAG such as **.**NAME, **.**MAIL, etc. The data lines begin with the name of the seismic station, followed by the network code of that station. The e-mail message is sent to an account such as *BREQ_FAST@iris.washington.edu*, and the data are returned according to what media was specified. Most often requests are now sent by FTP.

to waveform data in a fairly convenient and timely manner. AutoDRMs also support station inventory information, catalog and bulletin information, and channel response information at some installations.

An example of an AutoDRM data request is presented in Figure 4.

6.2 BREQ_FAST Request Mechanisms

The IRIS DMC developed the first e-mail–based seismic data request mechanism in 1988. This mechanism was termed BREQ_FAST, indicative of a way to make *B*atch *REQ*uests for data and get the data *FAST*. Figure 5 provides an example of what a BREQ_FAST request looks like.

The BREQ_FAST format is simply a file format that allows users to make requests for data in SEED volumes. The initial few lines of a BREQ_FAST file contain information about the user, their name and address, type of media they wish to receive, and an identifying label for the SEED volume. If they wish to, they may include information about the hypocenter related to the data and the hypocenter information will also be included in the resulting SEED data volume. After the .END statement, lines that specify station channel time windows can be specified.

The IRIS DMC services roughly 50,000 requests for seismic data each year. BREQ_FAST is the most common data request mechanism used by the international seismic community for data from the IRIS DMC.

The BREQ_FAST mechanism has been adopted by several FDSN data centers as a request mechanism. Those data centers supporting BREQ_FAST include

- IRIS DMC, Seattle
- GEOFONE Data Center, Potsdam
- Central Weather Bureau, Taipei
- Ocean Hemispheres Data Center, Tokyo
- Northern California Earthquake Data Center, Berkeley

6.3 Networked Data Centers

A new concept is evolving within the FDSN, a concept of truly distributed but well-coordinated access among data centers called NetDC. A user sends a NetDC-formatted request message to any of the participating data centers. The NetDC software extracts portions of the request that reference other data centers and forwards that portion of the request to the authoritative data center. Some portions of the request would normally also be processed at the data center originally receiving the request. The various data centers extract the requested data or information and return it to the originating data center. That data center then merges the information or data as needed and returns it to the requester.

The NetDC system is totally peer based. There is no central data center. Submitting a request to any of the NetDC centers should result in the same information being returned to the requester.

The NetDC system is software that interfaces to the existing software within a data center. It works with the system that is already in place, rather than trying to totally replace the existing system at a data center. IRIS produces a software system called the Portable Data Collection Center (PDCC) system, which is a fairly complete system to implement standard data quality control tools and data conversion to the SEED format at any location at which it is installed. IRIS distributes a complete system that includes PDCC and NetDC to any data center that desires it.

The NetDC format is a superset of the BREQ_FAST format. Certain control commands were added to BREQ_FAST, but the two formats are very similar. Additionally, IRIS has committed to supporting the IMS1.0 AutoDRM format as well. Since AutoDRM formats do not have mechanisms to control the routing between data centers, AutoDRM requests to a NetDC-enabled system can only access information at the receiving data center.

The NetDC system has been installed at four data centers:

- IRIS DMC, Seattle, USA
- NCEDC, Berkeley, USA
- ORFEUS Data Center, deBilt, the Netherlands
- GEOSCOPE Data Center, Paris, France

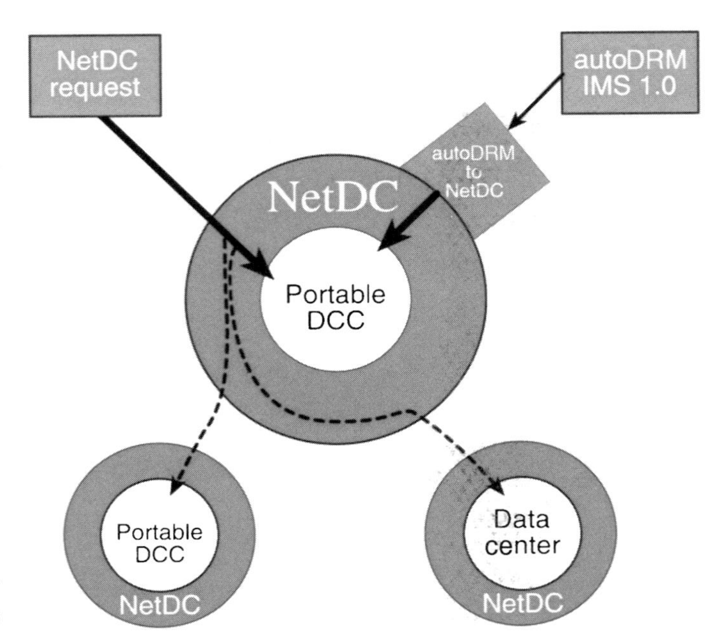

FIGURE 6 The networked data center concept. NetDC is a system that accepts requests for data from researchers via e-mail. The requested information may or may not be at the receiving data center. The NetDC system is a message and information routing system that coordinates the retrieval of information across geographically separated data centers. This gives the impression to the user that there is only one data center, when in fact there can be an unlimited number.

The MEDNET Data Center in Rome is presently evaluating the system.

Figure 6 shows how the NetDC system can be operated in conjunction with an existing data center system or how it can be packed with the IRIS PDCC package.

7. Request Tools at the IRIS DMC

Although IRIS supports the tools mentioned above, a variety of other, IRIS DMC–specific request tools have also been developed for accessing data in the IRIS DMC. Early portions of this chapter have shown that there are three major data types at the IRIS DMC:

- The non-quality-controlled SPYDER dataset
- The quality-controlled FARM data volumes
- Data in the continuous archive

The SPYDER and FARM datasets are stored online in disk-based RAID systems that ensure high availability and reliability, and do not require DMC staff involvement to provide access to data. The continuous archive is stored in a robotic system (near line) but is a resource with controlled access. The tools are necessarily quite different for access to these different sources. The online datasets are data for individual events, whereas the archive is continuous, allowing event windows to be extracted. Tools that allow users to specify station-channel-time window specification are necessarily directed toward the continuous

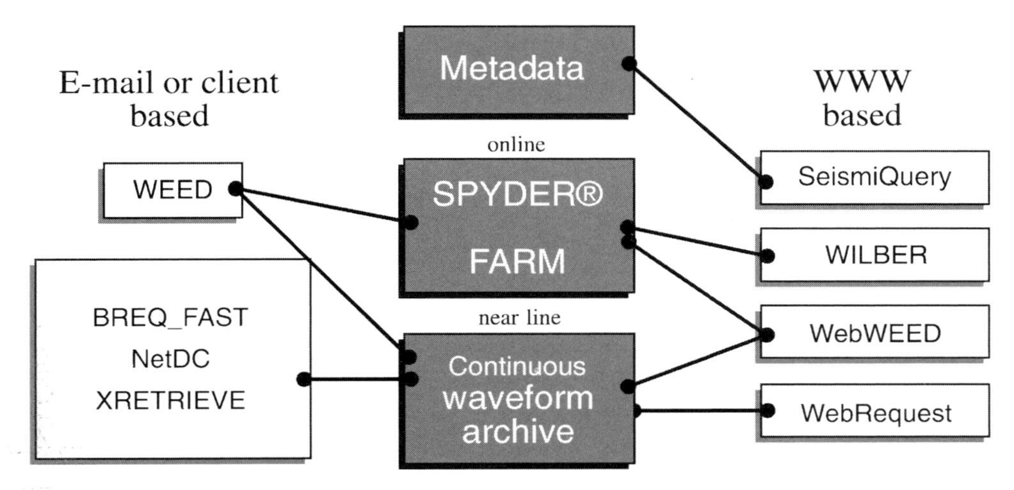

FIGURE 7 IRIS DMC access tools. The DMC has developed eight different data access methods or tools. Four of these tools (BREQ_FAST, NetDC, XRETRIEVE and WebRequest) only allow access to the data in the continuous archive and are based upon the user specifying station-channel-time windows for the data. One tool, WILBER, only allows access to the online datasets. Two tools (WEED and WebWEED) allow access to the continuous archive and the online datasets. WILBER, WEED, and WebWEED are all based upon events. The SeismiQuery tool allows access to the metadata information in the Oracle DBMS.

archive. Tools that are event based can usually be directed toward the FARM and SPYDER datasets but allow extraction of data from the archive as well. Figure 7 shows the various tools that can be used to generate data requests for IRIS data and which tool is most appropriate for each kind of request.

BREQ_FAST and NetDC are e-mail–based tools; specifically formatted files are sent to an e-mail address at IRIS to make a data request. WEED is a program that runs on a requestor's own computer and can make complex requests for data based upon event characteristics, station–event relationships, or station characteristics. WEED does submit the request to the DMC using e-mail, but this is done automatically. XRETRIEVE is a client program that contacts the DMC server for current information or to make a request.

WILBER, WebWEED, and WebRequest are all tools that are accessed using a browser connected to the IRIS DMC Web server. SeismiQuery is a Web-based tool that provides a graphical user interface (GUI) method to make queries into the Oracle DBMS maintained by the IRIS DMC.

Depending on the type of data request being considered, there is usually a request tool that makes the task of requesting the data very simple. An IRIS DMC Data Access Tutorial is available from IRIS or can be found at *http://www.iris.washington.edu/manuals/tutorial.html*. This tutorial will help you determine which tool to use and how to use the selected tool.

8. The Use of the IRIS DMC

Although IRIS is largely funded by the National Science Foundation of the United States, it is a resource that is of benefit to the entire global community of seismologists. Between 15 and 20% of all customized data requested goes to researchers outside the United States. It should be emphasized that US researchers benefit directly by having free and open access to data from FDSN member networks, whose data centers provide reciprocal support for data distribution. When the IRIS DMC was first being contemplated, it was estimated that roughly 200 data requests would be serviced per year. The most significant measure of the success of the IRIS DMC is the large number of data shipments made in the past 5 years as shown in Figure 8.

One of the primary motivations for having data from FDSN network stations at a single data center is that integrated data volumes can be produced. By having the data at the IRIS DMC, IRIS can systematically produce data volumes from the FDSN network for all of the larger events in a given year. The generation of the FDSN CD-ROM volumes is a direct result of this data integration effort.

We believe that the number of customized requests will stay at a manageable level and that much of the future growth will be accommodated using systems like WILBER and WEED that quickly access online datasets in a reliable and rapid manner.

Figure 9 shows the total amount of data that has been shipped from the IRIS DMC by year. In 2000, the IRIS DMC shipped 1.2 terabytes of data to scientists and data centers in response to data requests. More than 20,000,000 individual seismograms were shipped in 2000.

9. Future Cooperation within the Seismological Community

A conceptual revolution in broadband seismology began with the formation of the FDSN. Not only did the installation of broadband seismometers merge the acquisition activities of

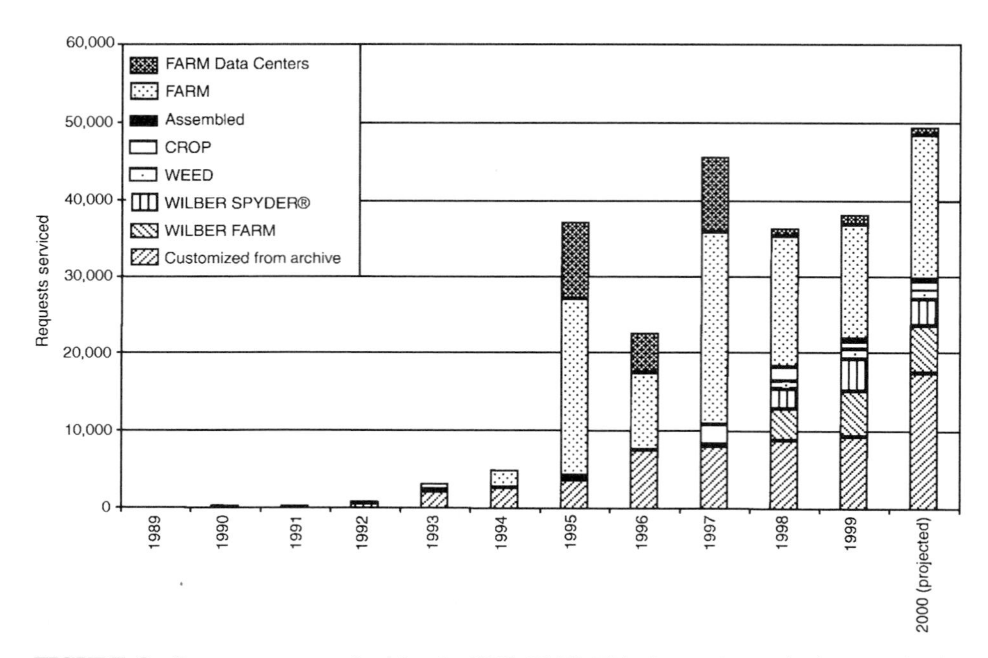

FIGURE 8 Data requests serviced by the IRIS DMC. This figure shows the increase in the number of data requests serviced by the IRIS DMC between 1989 and 2000 (projected). Customized shipments are those that access data from the large near-line storage systems, WILBER is a Web-based tool that accesses the online datasets of SPYDER® and FARM, and the WEED category shows all requests generated by the WEED access tool and may access the primary archive, or the SPYDER® or FARM data repositories. CROP is another tool that accesses FARM/SPYDER® datasets. FARM shipments represent the total number of FARM products that are shipped by FTP or by tape to either scientists or other data centers.

short-period and long-period seismologists, it provided a forum for coordinating activities across international boundaries. The decision to distribute data in a standard exchange format removed significant barriers that had impeded data exchange earlier. The willingness to openly share data, and where possible provide access directly to the seismic station, revolutionized acquisition of seismic data. Today hundreds of seismic stations deliver their data over the Internet in a near real time and totally open manner.

At the heart of the FDSN lies standardization. The characteristics of the seismometers, the standardization in sampling rates, channel names, exchange formats, and standardization of data request mechanisms have all contributed to the advances in seismology during the past two decades. By taking the lead in installing state-of-the-art instrumentation and developing data exchange standards and request mechanisms, the FDSN has changed the manner in which seismology is done.

We anticipate the FDSN will be involved in future developments in the area of deployment of ocean-bottom seismometers through the International Ocean Network (ION). We also anticipate that the stations operated by members of the FDSN will begin recording multisensor data from a wide variety of geophysical, atmospheric, and environmental sensors. We anticipate that the FDSN and its members will play an important role in coordinating the activities of networks of multiple scales including global, national, and regional networks as well as making contributions in the deployment and data management of temporary stations such as IRIS PASSCAL.

The present state of the FDSN represents an unprecedented and successful advance in seismic data management. However, it is only one step toward a more ambitious data management system that remains to be developed for the 21st century.

10. Summary

The amount of data now available from FDSN data centers is vast. The FDSN serves as the only international body that coordinates the activities of networks that are deploying seismic stations on both national and international scales. Its work in the development of a standard data exchange format, SEED, and in developing standardized data request tools is very important if the task of data gathering is to remain simple for the average seismological researcher. Although we know that the tools and methods of accessing data will continue to evolve, it is most probable that the amount of data will continue to increase. IRIS and members of the FDSN will continue to make improvements in their systems to ensure that the high-quality data of the FDSN member networks continues to be made available in a timely and easy-to-use manner.

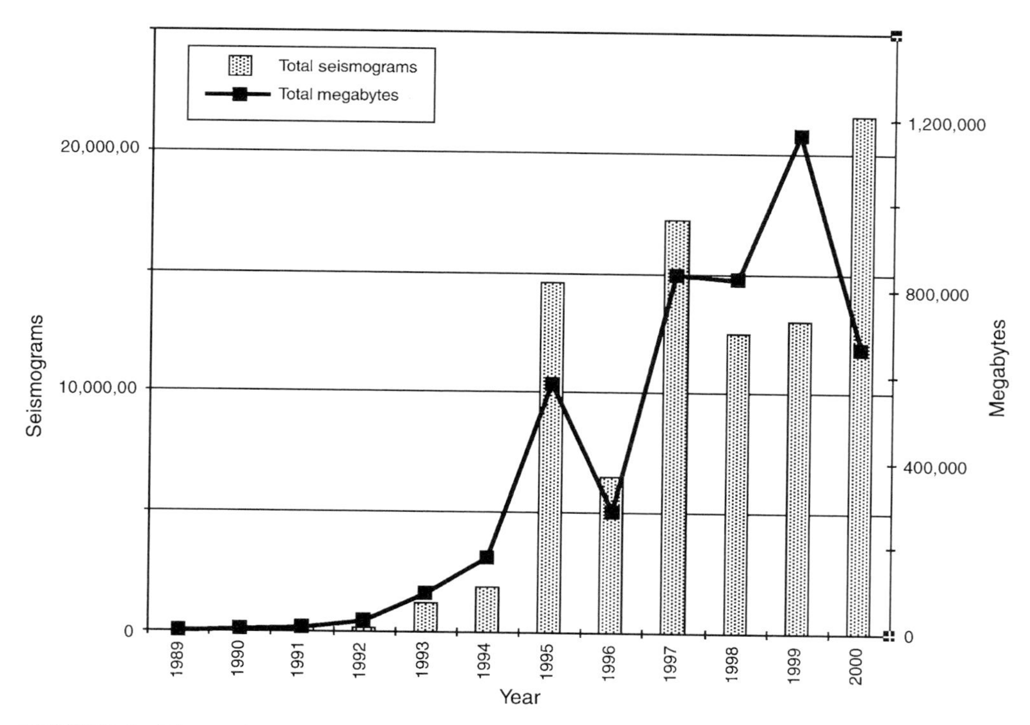

FIGURE 9 The number of megabytes and seismograms shipped in customized requests and in FARM shipments. The columns represent the total number of seismograms shipped and range between 14 million and 22 million per year. The line represents the total number of bytes sent. Between 0.6 and 1.2 terabytes of data are now being sent from the DMC to seismologists on an annual basis. These numbers do not include the amount of data shipped using any of the online WILBER or WEED request mechanisms that request data from the online datasets of FARM and SPYDER®. This projection was made in July 2000.

Acknowledgments

The Incorporated Research Institutions for Seismology (IRIS) is a consortium of over 95 research institutions with major commitments to research in seismology and related fields. IRIS operates a facilities program in observational seismology and data management sponsored by the National Science Foundation through its Division of Earth Sciences. The activities of IRIS including the work of the IRIS Data Management System are funded through NSF grant number EAR-95-29992.

Notes

1. Current institutional members are Australian Geological Survey/AGSO (Australia), Geological Survey of Canada/GSC (Canada), University of Chile (Chile), Institute of Earth Sciences/IES (Taiwan China), China Seismological Bureau/CSB (China), Geophysical Institute, Institute of Physics of the Earth/IPE (Czech Republic), Project Geoscope (France), GEOFON/GFZ(Germany), Seismologisches Zentralobservatory, Graefenberg (Germany), Geological Survey of Israel (Israel), MEDNET/Istituto Nazionale di Geofisica (Italy), Earthquake Research Institute/ERI University of Tokyo (Japan), Science and Technology Agency/STA-NIED (Japan), UNAM (Mexico), ORFEUS (The Netherlands), Polish Academy of Sciences, Institute of Geophysics (Poland), Eidgenossische Technische Hochschule (Switzerland), Incorporated Research Institutions for Seismology/IRIS (United States), United States Geological Survey/USGS (United States).
2. Other countries that have been involved in the FDSN, but are not presently active, include Russia, Brazil, South Africa, Iran, Saudi Arabia, and India. Certainly, other countries have also attended meetings but on an irregular basis.
3. A significant amount of the capitalization costs for the IRIS GSN, PASSCAL, and Data Management System have also been contributed by the US Department of Defense and the US Department of Energy. Smaller amounts of support have been received from various foundations and private corporations.

87

Global Inventory of Seismographic Networks

John C. Lahr
US Geological Survey, Denver, Colorado, USA
Torild Van Eck
Royal Netherlands Meteorological Institute, De Bilt, The Netherlands

1. Introduction

An untold amount of time, effort, and resources has been devoted to the operation of regional and global seismographic (or simply, seismic) networks and the generation of earthquake catalogs during the 20th century. This chapter documents and archives basic information about most of the world's seismic networks and provides a valuable starting point for future seismotectonic and hazard studies.

Each network was asked to respond to the following survey questions in order to provide important basic data, such as the duration of recording, the region monitored, the equipment used, and the seismologists and organizations involved. They were also asked to contribute a short abstract summarizing their network. Both the survey responses and the abstracts are archived as computer-readable files on the attached Handbook CD. The survey questions are listed in the next section.

2. Survey Questions

The survey questions were as follows:

1. NAME:
   (Give the official name of your network and the dates for which it has operated, such as "Southern Alaska Seismic Network, 1971–1989.")
2. STAFF:
   (Include one principal author for this report and their contact information [mail and e-mail addresses, phone number]. Also list staff members who have contributed to the success of the network.)
3. GEOGRAPHICAL LOCATION:
   (Describe the area covered by your network and the approximate geographical boundaries.)
4. NETWORK CODE:
   (Give the FDSN, or NEIC, or similar 2–4 letter network code used in reports, if there is one.)
5. INSTITUTION(S):
   (List the name and address of the institution(s) operating this network. Include WWW homepage address, if any.)
6. STATIONS:
   (Give summary of number and type of stations in network, such as # analog short-period, # digital broad-band, # of 3-comp sites, # strong-motion. If you record stations operated and recorded by another network, mention these separately.)
7. RECORDING:
   (Indicate type of recording system(s). Include sample rate, dynamic range, continuous or triggered, software, and/or other characteristics.)
8. DATA ANALYSIS:
   (Indicate the nature of your data processing including whether your network operates an automatic location and reporting system, how quickly the data are processed and quality checked, and to whom the results are distributed.)
9. DATA AVAILABILITY:
   (Indicate where data and analysis results are published and/or made available to others, including catalogs, phase data, and waveform data.)
10. OTHER PRODUCTS:
    (Indicate any other products provided or produced by your network such as hazard maps or plans, educational

INTERNATIONAL HANDBOOK OF EARTHQUAKE AND ENGINEERING SEISMOLOGY, VOLUME 81B *ISBN: 0-12-440658-0*

TABLE 1 Seismographic Networks Listed by Country (See text for abbreviations. This table is included on the attached Handbook CD. Many items in this table have links to files on the CD, and e-mail addresses are included for each contact person.)

Country	Network name	Years of operation	No. of stations	Approx. No. of earthquakes	Contact
Albania	Albanian Seis. Net.	1968–	sp:11	500/yr	Betim Muco
Antarctica	Mount Erebus Volcano Obs.	1980–	sp:10; bb:1	600 Strombolian expl./yr	Rick Aster
Argentina	Strong Ground Motion Nat. Net.	1970–	See CD		Juan Carlos Suarez
	Seis. Stations Nat. Net.	1969–	See CD	15/mo	Norberto Pueblo
Australia	The Seis. Research Centre Seis. Net. (SRC)	1976–	100	300/yr	Adam Pascale
	South Australian (ADE)	1909–	sp:15; 3c:5	7,000 since 1940	David Love
Austria	Vienna (AGS)	1902–	lp:2; sp:5; bb:5; sm:17	Available by autoDRM	Wolfgang Lenhardt
Belgium	Belgian Seis. Net.		sp:28		Thierry Camelbeeck
Bolivia	Obs. San Calixto of La Paz (OSC)	1986–	sp:9; lp:3; bb:1	180/yr ($M_L > 3.5$)	R. Rodolfo Ayala Sánchez
Bulgaria	Nat. Telemetered System for Seis. Info., Sofia (SOF)	1980–	sp:14 & 2 local sp networks & MedNet:1	1,500/yr	Rumiana Glavcheva
Canada	Nat. Seis. Net. (CN)	1897–	sp:87; bb:40; sm:56		Jim Lyons
Cape Verde Islands	Fogo Volcano Net. (VIGIL)	1999–	sp:5; bb:2	50/mo	Joao Fonseca
China (Beijing)	Nat. Net. (NSSN)	1978–	sp:63; lp:33; bb:66	3,200/yr	Liu Ruifeng
	China Digital Seis. Net. (CDSN)	1986–	bb:11; sm:9		Huang Jing
China (Taipei)	Broadband Array for Seismology (BATS)	1996–	bb:14; mobile:15	3,500/yr	Honn Kao/ Bor-Shouh Hwang
	Taiwan Rapid Earthquake System	1995–	dig. sm:79		Yih-Min Wu
	Taiwan Seis. Net. (TSN)	1897–1991– (dig.)	sp:75; bb:20; dig. sm:700	Approx. 15,000/yr	Chien-Hsin Chang
Colombia	Nat. Seis. Net. (RSNC)	1993–	sp:20; sm:130; bb:1	2,500/yr	Anibal Ojeda/Nelson E. Pulido H.
	Seis. Net. at Galeras Volcano Obs.	1989–	sp:14; bb:4; sm:1		Diego M. Gómez
	Nevado Del Ruiz (See Machin-Cerro)	1985–1988			Carlos Arturo Garzon J.
	Machin-Cerro Bravo Vol. Seis. Net.	1988–	sp:15; bb:1		Carlos Arturo Garzon J.
Costa Rica	Nat. Seis. Net. (RSN)	1974–	sp:18	>20,000	Wilfredo Rojas
	Arenal-Miravalles Seis. Net. (OSIVAM)	1994–	sp:12; sm:4		Rafael Barquero
Croatia	Croatian Seis. Net. (ZAG)	1900–	sp:4; lp:2; sm:14	1,541 in catalog (373 BC–1999)	Marijan Herak
Czech Republic	Czech Nat. Seis. Net. (CNSN)	1908–	bb:10	15,000 (catalog & bulletin)	Jan Zednik
	Western Bohemian Net. (WEBNET)	1992–	sp:10	See survey	Josef Horalek
Denmark	Danish Nat. Data Center	1927–	vbb:7; sp:2	2,000/yr	Peter Voss
Republic of Djibouti	Observatoire Géophysique d'Arta (ARTA)	1976–	sp:16	3,254 located in 1999	Alain Hauser
Dominican Republic	Red Sismica (INDRHI)	1978–	sp:15		Ing. Luis Odonel Gomez
Ecuador	Seis. Net. of Austro (Southern Ecuador)	1995–	sp:7; sm:8		Ricardo Peñaherrera Leon
	Seis. Net. of N. Ecuador (EPN)	1988–	See CD	2,000–5,000/yr	Alexandra Alvarado
Egypt	Nat. Seis. Net. (ENSN)	1975–	sp:51; bb:5		Tarek Rashad Kebeasy
El Salvador	Berlin Geothermal Net. (GESAL)	1996–	sp:9; sm:2	800/yr	Lic. José Antonio Rivas
	Central American Univ. Accelerogr. Net. (UCA)	1984–	sm:10	800/yr	José Mauricio Cepeda
Finland	Nat. Seis. Net.	1924–	sp:13	595 since 1610	Pasi Lindblom

(continued)

TABLE 1 (*continued*)

Country	Network name	Years of operation	No. of stations	Approx. No. of earthquakes	Contact
France	Nat. Net. of Seis. Survey (ReNaSS)		sp:106; bb:6	Local: 1,600/yr Tele.: 600/yr	Michel Granet
	Piton de la Fournaise Vol. Obs. (OVPF)	1980–	sp:24	700/yr	Jean-Louis Cheminee
French West Indies	Soufrière de Guadeloupe (GUAD)	1950–	sp:22		Jean-Christophe Komorowski
Germany	Net. of Seis. Station Bensberg (BNS)	1954–	sp:10; lp:1; sm:8		Klaus-G. Hinzen
	Seis. Obs.: Ruhr-Univ. Bochum (BUG)	1981–	sp:5; bb:1		Michael L. Jost
	Baden-Wuerttemberg Seis. Net. (LEDBW)	1893–	sp:30; sm:15	Local 100/yr Reg. & Tele: 1,000/yr	Wolfgang Bruestle
	Grafenberg Array (GRF)	1976–	bb:13		Klaus Stammler
	Northrhine-Westphalia Net. (GLA NRW)	1980–	sp:8; sm:3	500 eqs./yr 8,000 expls./yr	Rolf Pelzing
	Muenchen Fuerstenfeldbruck Obs.	1974–	sp:5; bb:2		Eberhard Schmedes
	Regional Seis. Net. (GRSN)	1991–	bb:16		Klaus Stammler
	Jena Seis. Net. (MOX)	1900–	sp:3; bb:4		Diethelm Kaiser
Greece	Nat. Obs. of Athens (ATH)	1964–	bb:32; sp:8; sm:50	3,000/yr 55,000 $M_L \geq 2.5$	George Stavrakakis
	University of Thessaloniki (THE)	1981–	sp:16	20,580	Basil C. Papazachos
	University of Patras	1991–	sp:48; bb:3; sm:3		Akis Tselentis
India	India Meteorological Department	1898–	sp & bb: 212; sm: 140		Kusala Rajendran
Iran	Khorasan Seis. Net. (KHSN)	2000–	bb:6		Jafar Shoja-Taheri
	Nat. Seis. Net. (INSN)	1989–	bb:18	350/yr	Mohammad Mokhtari
	Mazandaran Digital Seis. Net. (MDSN)		sp:4		Behzad Pourmohammad
Israel	Israel Seis. Net. (ISN)	1981–	sp:13; bb:8; sm:63		Rami Hofstetter
Italy	Friuli Accelerograph Net., Univ. of Trieste (RAF)	1995–	bb:10; vbb:2	140/yr for mag. 1.5 to 5.6	Giovanni Costa
	Marchesan Seis. Net. (RSM)	1978–	sp:16	500/yr	Giancarlo Monachesi
	Reg. Seis. Net. of NW Italy (RSNI)	1967–	sp:11; bb:6	25,000	Stefano Solarino
	Mediterranean Net. Rome (MEDNET)	1989–	15		Salvatore Mazza
	Nat. Strong Motion Net.	1972–	sm:235; dig. sm:79		Paolo Marsan
Jamaica	Jamaica Seis. Net. (JSN)	1961–	sp:11; bb:1; sm:3	1,600	Margaret Grandison
Japan	Aso Volcanological Lab., Kyoto University	1928–	sp:31; bb:8; sm:1		Yasuaki Sudo
	Japan Meteorological Agency (JMA)	1926–	114–625	500–75,000/yr	Nobuo Hamada
	Nansei-Toko Obs. for Earthquakes & Volcanoes (NOEV)	1993–	sp:11	18,000/yr	Toshiki Kakuta
Kuwait	KNSN	1997–	sp:7; bb:1; 3c:8		Jasem Al-Awadhi
Kyrgyz Republic	Kyrgystan Inst. of Seis. (KNET)	1927–	sp:33; lp:23; bb:7	See catalog	Ilyasov Bektash
Lithuania	Seis. Net. of Lithuania	1970–	sp:4; sm:21		Saulius Sliaupa
Macedonia	Macedonia Republic Seis. Obs. (SORM)	1957–	See CD		Lazo Pekevski
Malaysia	Malaysian Seis. Net.	1976–	sp:19; sm:3		Jalan Sultan
Mexico	Nat. Seis. Service, Mexico (SSN)	1910–	sp:30; bb:20	700/yr	Javier Pacheco
	Mexico, Baja Califonia (RESNOM)	1980–	sp:12; lp:1		Luis Munguía-Orozco
Montenegro	Montenegro Seis. Obs. (PDG)	1982–	sp:9; bb:1	1,500/yr	Branislav Glavatovic
Morocco	Nat. Net. (CNRM)	1937–	sp:38	400/yr	Ben Aissa Tadili
Nepal	Nat. Net. of Napal	1979–	sp:21		Gautam Umesh
Netherlands	Seis. Net. of the Netherlands (SNN)	1904–	sp:12; bb:4; sm:13	500/yr	Hein Haak
New Zealand	The Seismological Obs.	1884–	sp:60; bb:4; sm:270	188,000	Terry Webb

(*continued*)

TABLE 1 *(continued)*

Country	Network name	Years of operation	No. of stations	Approx. No. of earthquakes	Contact
Nicaragua	Nicaraguan Seis. Net.	1975–1985 1992–	sp:29; bb:1; sm:19		Wilfried Strauch
Norway	Norwegian Nat. Seis. Net. (NNSN) NORSAR (NOR)	 1971–	bb:5; sp:19 See CD	 4,500/yr	Jens Havskov Jan Fyen
Peru	Seis. Net. of Peru (RSNP)	1931–	sp:20; bb:11; sm:5		Hernando Tavera
Poland	Polish Seismological Net.	1986–	vbb:4; bb:1; sp:2; lp:1	Mine blasts: 500/yr; few eqs.	Pawel Wiejacz
Portugal	Lisboa Instituto de Meteorologia (LIS)	1961–	sp:14; mp:14; bb:1	4,200 since 1961	Maria Luisa Senos
	Lisboa, U. of the Azores (PDA, CVUA)	1997–	sp:28; lp:1; vlp:1; mp:12; bb:1	6,200 since 1997	J. Luis Gaspar
	Inst. Geofísico, Univ. de Lisboa		sp:2; bb:2; sm:1		Luis Mendes-Victor
Puerto Rico	Puerto Rico Seis. Net. (PRSN) Puerto Rico and Virgin Island Strong Motion Program (PRVISMP)	1987–	sp:16; bb:11; sm:36	600/yr	Christa G. von Hillebrandt-Andrade
Romania	Romanian Seis. Net. (RO)	1980–	sp:17; sm:21	2,800 1984–2000	Constantin Ionescu
Russia	Teleseismic Net. of Russia (RTSN)	1912–	bb:19; sp,mp,lp:24	2,500–3,000/yr	Oleg Starovoit
	N Caucasus Regional Seis. Net. (NCRSN)	1932–	sp:7; mp:7; bb:1	150–500/yr	Oleg Starovoit
	Caucasus Mineral Waters Local Seis. Net. (CMWLSN)	1982–	sp:8; bb:1	120–140/yr	Oleg Starovoit
	Dagestan Local Seis. Net. (DLSN)	1980–	sp:15; sm:7	1,000–2,000/yr	Marat G. Daniyalov
	Kamchatka Regional Seis. Net. (KRSN)	1961–	sp:34; sm:25	70,000	Evgenii Gordeev
	Sakhalin Seis. Net. (SAKH)	1947–	sp:14; lp:9; sm:3; dig.:8	1,050/yr	Dmitry Kuznetsov
	Magadan Regional Seis. Net.	1978–	bb:2; sp:7	See CD	Gounbina Larissa Vladilenovna
	Baikal Regional Seis. Net. (BEMSE)	1901–	sp:21	3,500/yr	Vladimir Naiditch
	Altai-Saiyan Regional Seis. Net. (ASEMSE)	1962–	sp:20; bb:3		Alexandr F. Emanov
	Yakutiya Seis. Net., (EMSP)	1956–	sp:7; bb:4	800–5,000/yr located ($M > 2$)	A. G. Larionov
Slovak Republic	Nat. Seis. Net. (BRA)	1902–	sp:5; mp:1; bb:1	2,000/yr	Martin Bednarik
Republic of Slovenia	Nat. Seis. Net. (LJU)	1897–	7	2,500/yr, 400 local	Mladen Zivcic
South Africa	AngloGOLD, west WITS OPERATIONS	early 1990s– (dig.)	5	2,000/day	Dragan Amidzic
	Nat. Seis. Net. (PRE)	1949	sp:25; lp:2	1,000/yr	Gerhard Graham
Spain	Andalusina Net. (RSA)	1903–1983– (bb)	sp:13; bb:4; sm:12	2,500/yr	Jesus M. Ibanez
	Catalan Seis. Net.	1985–	sp:11; bb:3; sm:7	200/yr	Xavier Goula
	Royal Naval Institute, San Fernando (ROA)	1986– (sp) 1996– (bb) 1898– (lp)	sp:9; lp:1; bb:4	500/yr	J. Martin Davila
	Nat. Geographic Inst. (IGN), Spain and Canary Islands Net.	1996–	sp:37; bb:5; sm:63	1,000/year	Francisco Vidal Sanchez
	Lab. of Geophysical Studies "Eduard Fontserè," Institute of Catalan Studies (CA)	1984– (sp) 1995– (bb)	sp:3; bb:1		Josep Vila
Sweden	Swedish Nat. Seis. Net.	1994–1998– (dig.)	bb:6	5–10/day	Ronald Arvidsson
Switzerland	Swiss Seis. Net.	1911–	sp:27; bb:27; sm:62		Manfred Baer

(continued)

TABLE 1 *(continued)*

Country	Network name	Years of operation	No. of stations	Approx. No. of earthquakes	Contact
Turkey	IZINET	1993–	sp:13		Serif Baris
	Tubitak Net.	1993–	See CD	>5,000	Mehmet Ergin
United Kingdom	UK Seismograph Net.	1969–	sp:162; bb:1; sm:17	50/yr	Brian J. Baptie
United States	See below				
Venezuela	Nat. Seis. Net.	1982–	sp:14; bb:1; sm:55	120/yr	Herbert Rendon
Vietnam	Nat. Net. (VNSN)	1975–	sp:15		Le Tu Son
West Indies	Eastern Caribbean Seis. Net.	1950–	sp:27; bb:12; sm:5		John B. Shepherd
	Montserrat Vol. Obs.	1995– (sp) 1996– (bb)	sp:10; bb:4		Glenn Thompson
Zambia	Zambia Nat. Net.		sp:5; lp:1; bb:1		Daniel Lombe
United States	Pacific Tsunami Warning Center Hawaii Net., HI (PTWC)		sp:11; bb:1		Barry Hirshorn
	PG&E Central Coast Seis. Net. (CA)	1986–	sp:20		Marcia McLaren
	Princeton Earth Physics Project Net. (PEPP)	1997–	bb:40	140/yr	Robert Phinney
	US Nat. Seis. Net. (US)		bb:34; sm:28		Ray Buland
	US Nat. Strong-Motion Program Net.	1932–	sm:675 (70% dig.)		Ron Porcella/ Chris Stephens
	West Coast and Alaska Tsunami Warning Center (AT)	1967–	sp:21; lp:5; bb:2		Tom Sokolowski
	The following ***United States*** *networks are ordered by state.*				
United States	Alaska Earthquake Info. Center, AK (AK)		sp:191; bb:9		Roger Hansen
	Southern Alaska, AK	1971–1989	sp:50	55,000	Kent Fogleman
	Arizona Earthquake Info. Center, AR (AR)		sp:7; bb:1;	1,000	Doug Bausch
	Anza Array, CA (AZ)	1982–	sp:1; bb:13;	42,000	Frank Vernon
	Berkeley Digital Seis. Net., CA (BK)	1910–1990– (dig.)	bb:25; sm:25	25,000	Lind Gee
	California Dept. of Water Resources, CA (CDWR)		sp:20; sm:54		Dave Kessler
	California Strong Motion Instrument Program (CSMIP)	1972–	sm:600		Anthony F. Shakal
	Northern California Seis. Net., CA (NC)	1967–	sp:376; sm:26	428,000	David Oppenheimer
	Southern California Seis. Net., CA (SC)	1932–	sp:163; bb:79; sm:10	411,000	Kate Hutton
	Delaware Geological Survey Seis. Net., DE (DGS)	1971–	sp:3		Stefanie Baxter
	Georgia Tech Seis. Net., GA (GTSN)		sp:6; bb:1		Tim Long
	Hawaiian Vol. Obs. Seis. Net., HI (early) (HVO)	1912–1985	See CD	17,738 (1912–1959); 107,974 (1960–1998)	Fred Klein
	Hawaiian Vol. Obs. Seis. Net., HI (current) (HVO)	1912–	sp:53; bb:13; sm:4	200,000	Paul Okubo
	Boise State Univ. Net., ID (BU)		sp:19	250 for $M \geq 2.8$	James Zollweg
	Idaho Nat. Eng. and Envi. Lab., ID (INL)		sp:24	8,500	Suzette Jackson Payne
	Ricks College—Teton Seis. Net., ID (RC)		sp:4		Edmund J. Williams
	New England Seis. Net., Boston College, MA (NE/BC)	1975–	bb:11	5,000 500 (1980–1998)	John E. Ebel
	New England Seis. Net., MIT, MA (NE/MIT)		sp:4; bb:1	5,000	Charles Doll, Jr.

(continued)

TABLE 1 *(continued)*

Country	Network name	Years of operation	No. of stations	Approx. No. of earthquakes	Contact
United States	Cooperative New Madrid Seis. Net., MO (NM/SLU)	1975–	sp:10; bb:6		Robert B. Herrmann
	New Madrid, MO (BILL)		sp:1; bb:5		Brian J. Mitchell
	Montana Regional Seis. Net., MT (MB)	1982–	sp:31; bb:1	15,000	Michael Stickney
	Central N Carolina Seis. Net., NC (SE/UNC)		sp:12; bb:1		Chris Powell
	Los Alamos Seis. Net., NM (LANL)		sp:7; sm:6		Leigh House
	New Mexico Tech Seis. Net., NM (SC)	1997–	sp:17	1,900	Rick Aster
	W. Great Basin & S. Great Basin Seis. Nets., NV (NN/SN)	1900–	sp:102; bb:9	50,000	Ken Smith
	Lamont-Doherty Cooperative Seis. Net., NY (LD)	1949–	sp:22; bb:6; sm:6	50/yr	Won-Young Kim
	Puerto Rico Seis. Net., PR (PRSN)		sp:16; bb:11; sm:30	6,800	Christa G. von Hillebrandt-Andrade
	Coop. New Madrid Seis. Net. & S. Appalachian Coop. Seis. Net., TN (NM/CERI)	1974–	sp:75	4,000	Mitchell M. Withers
	East Tennessee Seis. Net., TN (TN)	1978–	sp:17; sm:4	1,100	Jeffrey W. Munsey
	S Carolina Seis. Net. (SE/USE)	1974–	sp:22		Pradeep Talwani
	West Texas Seis. Net., TX (WTSN)		sp:6; bb:1		Diane I. Doser
	Univ. of Utah Reg. Seis. Net., UT (UU/WY)	1962–	sp:82; bb:8; sm:2	Utah: 21,675 (1962–1998) Yellowstone: 18,512 (1984–1998)	Susan J. Nava
	Virginia Tech Reg. Seis. Net., VA (SE/BLA)		sp:8; bb:1		Martin Chapman
	Pacific Norwest Seis. Net., WA (UW)	1969–	sp:134; bb:10; sm:16	88,100 (sample)	Steve Malone

TABLE 2 Global Seismographic Networks (See text for abbreviations. This table is also placed as a computer-readable file on the attached Handbook CD. Many items in this table have links to files on the CD, and e-mail addresses are included for each contact person.)

Network name	Years of operation	No. of stations	Contact
GEOSCOPE	1982–	bb:28	Genevieve Roult
Global Seis. Net. of IRIS (GSN)	1986–	sp:100; bb:119; sm:94	Rhett Butler
CTBTO	1997–	See CD	Sergio Barrientos
Public Seis. Net. (PSN)		sp:43; bb:51; sm:2	Steve Hammond
(GEOFON)	1993–	bb:39; sm:5	Winfried Hanka

outreach material or training, special displays, or support functions for other organizations.)

11. NEW OR PROPOSED DEVELOPMENTS:
    (Indicate here planned upgrades or additions to your network, including station upgrades, new telemetry plans, new recording equipment, etc.)
12. APPROXIMATE NUMBER OF EARTHQUAKES:
    Give the approximate number of earthquakes you have located.
13. ACKNOWLEDGMENTS:
    (List agencies providing support for network.)

3. Tables of Network Summaries

Unless otherwise noted, each network listed in Tables 1 and 2 has answered the survey questions and submitted a short abstract describing their network. The abstracts and survey answers, along with additional information from many networks, are included in full on the attached Handbook CD. Table 1 is a list of seismographic networks listed by country. Table 2 lists the global networks. The following general abbreviations are used:

Approx.	Approximate
eqs.	earthquakes
expl.	explosion
Info.	Information
Mag (M)	Magnitude
mo	month
Nat.	National
Net.	Network
No.	Number
Obs.	Observatory
Reg.	Regional
Seis.	Seismographic
Tele.	Teleseismic
Vol.	Volcano
yr	year

Abbreviations for types of seismic instruments are as follows:

3c	three-component
bb	broad band
dig.	digital
lp	long period
mp	medium period
sm	strong motion
sp	short period
vbb	very broad band
vlp	very long period

Acknowledgments

We wish to thank all of the network operators that contributed to this chapter. In addition, we thank Steve Malone for compiling the information on US networks and Willie Lee for developing the concept of this summary chapter.

Editor's Note

We found it very difficult to "synchronize" the information collected over a few years for this chapter with the national and institutional reports collected over the same period. Readers should also consult Chapter 79, "Centennial National and Institutional Reports: Seismology," edited by Kisslinger; Chapter 80, "Centennial National and Institutional Reports: Earthquake Engineering," edited by Cherry; and Chapter 81, "International Organization Reports to IASPEI," edited by Kisslinger. For old seismographic stations and bulletins (from the 1890s up to 1920), please see Chapter 88 by Schweitzer and Lee.

For readers interested in earthquake catalogs, please see also Chapter 41, "Global Seismicity: 1900–1999," by Engdahl and Villasenor, and Chapter 42, "A List of Deadly Earthquakes in the World (1500–2000)," by Utsu.

A large amount of information has been submitted by seismographic networks for this chapter. It is archived on the attached Handbook CD under the directory \87Lahr.

88

Old Seismic Bulletins to 1920: A Collective Heritage from Early Seismologists

Johannes Schweitzer
NORSAR, Kjeller, Norway
W. H. K. Lee
US Geological Survey, Menlo Park, California, USA (retired)

1. Introduction

Scientists began systematic instrumental observation of earthquakes in the latter part of the 19th century. Several authors (e.g., Ehlert, 1898; Berlage, 1930; Dewey and Byerly, 1969; and Ferrari, 1990, 1992) describe the history of the development of an adequate instrumentation for seismology. In the 1880s, scientists in Italy, Japan, and Germany began to record more or less continuously the ground motion with their newly developed seismographs (Figure 1). The value of these old seismograms for modern seismology has been emphasized by Kanamori (1988), and many examples can be found in Lee *et al.* (1988). Because of the limited ability to reproduce the original (analog) paper seismograms, seismologists had to describe their observations in words and numbers. It took about 20 years beyond the 1880s to develop an adequate procedure. By the early 1900s, first-order features of the seismic records were deciphered. The recording instruments were able to produce seismograms in which one could distinguish between the onsets of all three wave types (i.e., *P*, *S*, and surface waves), and a common vocabulary for the description of these records was developed (Borne, 1904; Chapter 79.24.1 on Handbook CD #2 by Schweitzer). Figures 2 and 3 show early bulletins from Göttingen where Borne developed this notation.

Today, we can follow the developments in early seismology by comparing seismic bulletins; at the beginning in the 1870s, we find phenomenological descriptions of macroseismic effects of earthquakes, and by the late 1890s, we find lists with physically measured parameters (i.e., measured onset times, dominant periods, and amplitudes of seismic waves). Unfortunately, not many of the early seismograms survived, so few are accessible today for reanalysis with modern knowledge about wave propagation in the Earth. The bulletin publications about these early seismograms are therefore often the only source of information we have now. They were "published" in many forms, including handwritten logs (Figure 2), mimeographed sheets (Figure 3), and printed reports (Figure 4), and they were collected in special series of seismic data compilations. In addition, seismic readings and seismograms were often published in various scientific reports after some disastrous earthquakes. For example, Figure 5 consists of three pages taken from Omori (1905) published in Japanese. We translated a few critical words into English and added them in Figure 5. The upper two frames in Figure 5 are seismic readings (for vertical, east–west, and north–south components) for eight earthquakes at the Tainan Observatory (an auxiliary station of the Taihoku Meteorological Observatory, located in the southwestern part of Taiwan). The seventh earthquake (magnitude 6.3, given by Utsu in Chapter 42) occurred on November 6, 1904. Seismograms (east–west component) from Gray-Milne seismographs at five Taiwan stations for this earthquake are shown in the lower frame.

2. The First Earthquake Bulletins

During the last half of the 19th century, scientists in many countries began to systematically collect data of macroseismically observed earthquakes and the locations of these events, known only on the basis of such data. In some countries, scientists and/or their governments established special committees or commissions to do this work.

INTERNATIONAL HANDBOOK OF EARTHQUAKE AND ENGINEERING SEISMOLOGY, VOLUME 81B ISBN: 0-12-440658-0

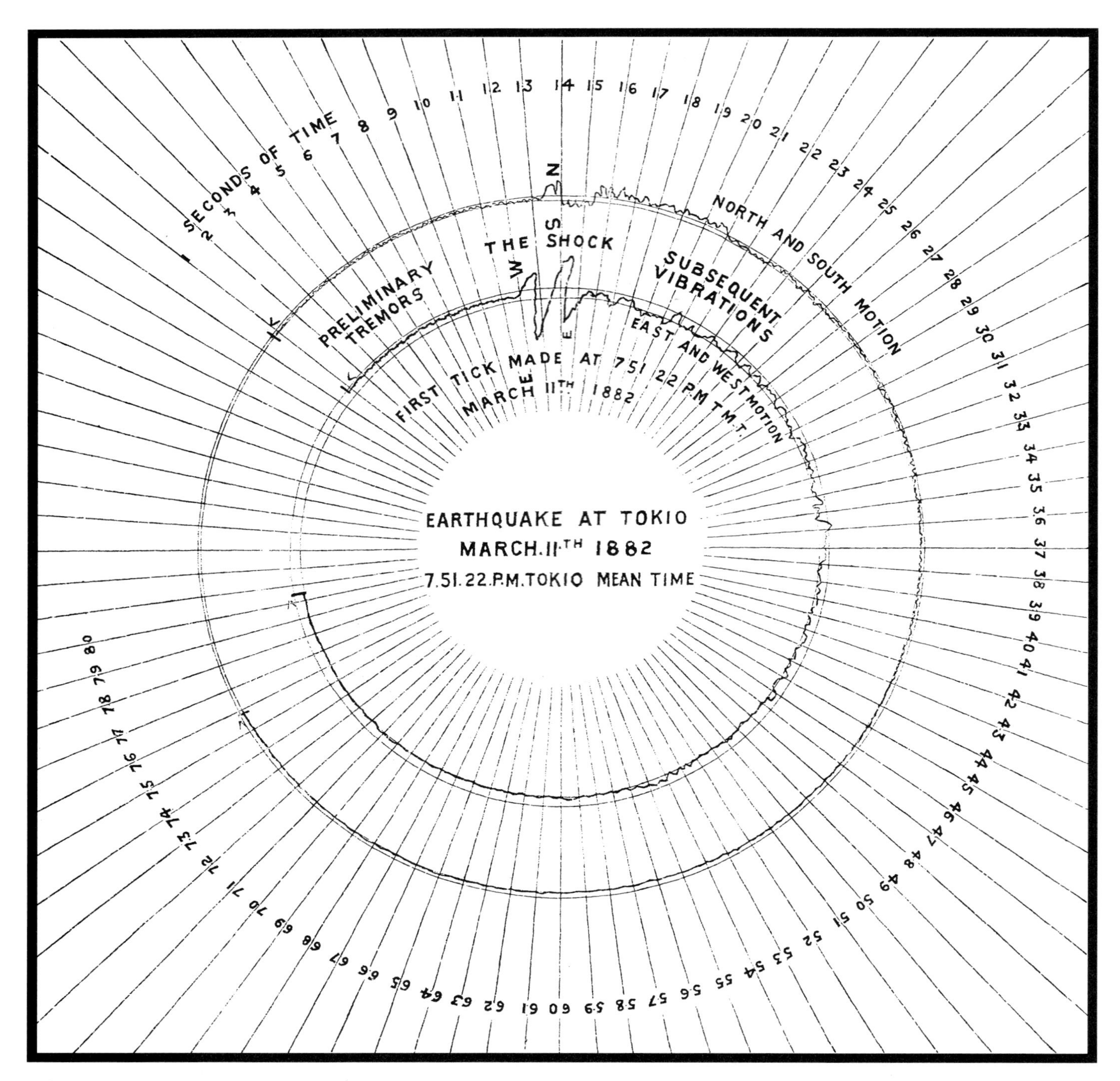

FIGURE 1 One of the oldest seismograms as recorded by Milne in Tokio (Tokyo) in 1882 (an inverted image after Milne, 1883).

2.1 Early Earthquake Observations

As far as we know, the "Erdbebenkommission der Schweizerisch Naturforschenden Gesellschaft" (Earthquake Commission of the Swiss Society for Natural Scientists) was the first such commission, founded in Switzerland in 1878 (Sieberg, 1904; see Chapter 79.51 on Handbook CD #2 by Ansorge *et al.*). These commissions or committees published earthquake lists, often updated every year. After the beginning of instrumental observations of earthquakes with early seismoscopes or seismographs, these lists were improved by adding better source time information (e.g., usually the onset times of the surface wave train at the instrument) and observed maximum amplitudes for locally or regionally recorded earthquakes. These observations were often published together with data from other natural phenomena (e.g., from meteorology or volcanology). Design details of

Geophysikalisches Institut - Göttingen 18.

No 28. 1911 Juli 10 9h – Juli 17 9h (Greenwich.)

Datum	Ch	Ph	Zeiten (Greenwich) h m s	T s	A_E μ	A_N μ	A_Z μ	Bemerkungen
Juli 11	I	e	21 41,1	?	–	–	–	
		i	21 44 50	8	–	3,0	2,0	
		L	22					
		F	23¼					
12	II n	e	4 21 16	6?	–	–	–	Herd ca 10000 Km
		iP	21 29	6	2½	2½	10	entfernt.
		PR1	4 25 19	(6)	6,0	6,0	5	
		PRII	4 27 40	6	6,0	6,0	4	
		iS	4 31 27	9	18	5	8	
		PS	4 32 19	8-9	9	14	3	
		SRI?	4 37 51	10-11	5,0	4,0	5,0	Die SR I,II u.s.w sind nicht
		eL	4 49	50				nicht mit Sicherheit zu erkennen.
		M_1	4 51-56 m	60	450	450	350	bemerkenswert die ausser-
		M_2	4 58	42	280	280	230	ordentlich grosse Periode
		M_3	5 2	30	210	380	200	in M_1 u. M_2.
		M_4	5 8-9 m	27	360	440	300	
		F	10					
13	I n	e	?					Beginn fällt in den
		eL	9 27,5	20	–	–	–	Bogenwechsel.
		M	9 31,7	18	5	5	7	
		F	10					
14	I	eL	2 38,4	24	7	4	–	andauernde seismi-
		F	3 20					sche Unruhe.

Ansel.

FIGURE 2 An example of a handwritten bulletin from station Göttingen (see number 69 in Table 1) that shows the weekly bulletin number 28 of the year 1911. A. Ansel, who also signed the bulletin, made the analysis. Special results (besides the onset parameter readings) were written in the column "Bemerkungen" (Remarks): on July 12, was an event at approximately 10,000 km distance, for which he observed relative large periods for the surface waves and had problems in identifying the onsets for SS (SRI) and SSS (SRII); on July 13, the first onsets were not analyzed because they arrived at the station during the change of the registration paper; and on July 14, continuing microseisms were observed. Note the delay of about 2 months between the last reported time and the time received by the library of the US Weather Bureau in Washington, D.C.

Geophysikalisches Jnstitut-Göttingen

24

N67/39 191 Septemberll 9 h -Oktober 2 9 h (Greenwich-Zeit)

Datum	Ch	Ph	Zeiten h m sec	T sec	A_E μ	A_N μ	A_Z μ	Bemerkungen.
Sept13	I	e	22 34,0	-	-	-	-	
		M?	34,6	14	6	3	-	
		F	42					
15	Iu		13 23 42	4	-	-	1	
		PR1	27 45	4	-	-	1	
		iS	34 43	18	20	5	-	
		eL	50					
		M	14 11,4	20	40	15	20	
		F	16 00					
17	I	e	3 36,4	-	-	-	-	3 Beben mit scharfem
		iP1	37 27	3-4	-	-	2	Einsatz in der ZKom-
		iP2	39 45	4	-	-	2	ponente.Horizentalpen
		iP3	51 24	4	-	-	2	del ausser Betrieb.
22	Ir	iP	5 12 12	4	-	-	2	Starke mikroseismi-
		eL	34,9					sche Bewegung ver
		M	40,7	24	10	12	10	deckt die Einsätzed.
		F	6 50					H.Komp. von P u. S.
26		e	14 46,5					
		M	49,4	20	6	7	-	
		F	15 20					
29		i	12 17,6	6	-	-	3	
		F	30					

gez. Ansel.

FIGURE 3 An example of a hand-typed bulletin from station Göttingen (see number 69 in Table 1) that shows the weekly bulletin numbers 37 to 39 (of the year 1911) on one page! During these 3 weeks, only six reportable seismic events were observed at Göttingen. Again, special analysis results are written in the column "Bemerkungen" (Remarks): on September 17, the onsets were described as "three sharp onsets on the vertical component," when the horizontal components were out of operations, and on September 22, the microseisms were so strong that they masked the *P* and *S* onsets on the horizontal components. In this example, the delivery time from Göttingen, Germany, to Washington, D.C., was much faster (about 3 weeks).

RECORDS OF MILNE SEISMOGRAPH NO. 16, AT CHRISTCHURCH, NEW ZEALAND.

Latitude, 43° 31′ 50″ S.; longitude, 172° 37′ 18″ E. Time employed: Greenwich mean civil time.

P.T., preliminary tremors less than 2 mm. complete amplitude; A.T., after-tremors less than 2 mm. complete amplitude; B., E., beginning and end of vibrations not less than 2 mm.; Amp., half-range in millimetres; 1 mm. boom motion = 0″.43. B. and E. signify beginning and end of amplitudes exceeding 1 mm., the range being 2 mm.

C. COLERIDGE FARR, Observer.

Date.		P.T. (from)	B.	Maxima.		Amp.	E.	A.T. (till)	B.P.	Remarks.
				From	To					
1902.						Mm.			Secs.	
Jan.	1	05.44·5	06.09·7	06.16·0	..	5·1	06.23·0	08.40·0	20	
"	2	..	..	15.04·5	15.07·3	1·0	..	..	..	Very sharp tremor.
"	8	23.59·7	..	00.07·7	..	1·1	..	06.47·0	18	Magnetographs much affected from 11 h. 00 m. to 11 h. 39 m.
"	12	22.39·5	23.05·47	23.08·0	..	2·6	23.10·0	00.07·0	18	
"	15	06.30·2	..	06.36·5	..	0·4	..	06.42·0	18	
"	17	06.18·5	..	06.46·8	..	0·4	..	07.18·0	18	
"	18	23.49·3	..	00.09·7	00.28·3	0·3	..	12.39·0	18	
"	21	10.13·0	..	10.22·3	..	0·6	..	10.35·5	18	
"	21	22.32·3	..	22.41·5	..	0·4	..	23.02·0	18	
"	22	06.17·0	..	06.23·0	..	0·3	..	06.28·2	18	
"	24	..	..	10.06·5	..	0·5	..	..	18	Very sharp tremor.
"	24	23.34·7	23.39·7	23.48·6	..	16·7	..	..	18	The largest "shock" recorded to this date, but not felt personally.
				23.49·5	23.51·5	17·0+				
					23.55·6	7·5				
					23.57·8	5·2	23.49·1	27.30·0		
"	26	04.30·7	..	04.36·6	..	0·7	..	04.43·0		
"	26	05.34·0	..	05.36·0	05.41·0	0·3	..	05.44·0	18	
"	28	07.26·0	..	07.28·0	..	0·3	..	..	18	

FIGURE 4 An example of an early seismic station bulletin published in a scientific journal, which shows the bulletin of station Christchurch in New Zealand (see number 143 in Table 1) of January 1902. Note the descriptive character of this early listing without any phase names.

the instruments were also described together with the observations. Thus, we can obtain amplification characteristics for such old instruments, although the instruments themselves may no longer exist (Ferrari, 1990, 1992).

One of the earliest such periodic publications regarding earthquake observations was the "Bolletino del Vulcanismo Italiano," issued in Italy 1874–1911. Then in 1879, the "Uffizio Centrale di Meteorologia e Geodinamica" (Official States Bureau for Meteorology and Geodynamics) was founded in Rome to centralize the earthquake observations for the whole country (Sieberg, 1904).

In Manila, Philippines, the first systematic investigations of earthquakes started in 1865 with the founding of the "Observatorio de Manila" where a "Servicio Seismológico" was established. The monograph "La Seismología en Filipinas" is a good example of the stepwise change of earthquake catalogs from earthquake listings to an early type of station bulletin (Masó, 1895; see Chapter 3 by Udias and Stauder).

The first nationwide working service for earthquake observations in Japan was established after the founding of the Seismological Society of Japan in 1880. Figure 1 shows the seismogram of an earthquake on March 11, 1882, as recorded by John Milne in Tokio (Tokyo), Japan. As far as we know, this is one of the earliest examples of a seismogram, which by chance had been preserved only because it was published in Milne's report to the British Association for the Advancement of Science (Milne, 1883). From 1886 on, the Seismological Society of Japan published their "Seismological Bulletin of Japan."

In 1892, the "Shinsai-Yobô-Chôsa-Kwai" (Imperial Earthquake Investigation Committee) was founded, and it published many detailed reports in Japanese and in foreign languages (mostly in English) (Sieberg, 1904; see Chapter 79.33.2 on Handbook CD #2 by Utsu).

In Russia, the Permanent Central Seismic Commission was organized in 1900 under the then Imperial Academy of Sciences of Russia (see Chapter 79.45.1.2 on Handbook CD #2 by Nikolaev and Sedova). This Commission published many important volumes in seismology (including both instrumental and noninstrumental seismic observations) from 1902 to 1919. In particular, many important contributions of Boris Galitzin to modern instrumental seismology were first published in this periodical. Figure 6 shows an example of four seismograms recording the P onsets of two different events. Each event is recorded both with his horizontal pendulum and with a new vertical seismometer presented in his paper (Galitzin, 1910). Galitzin commented in particular on the fact that P onsets are much more observable on vertical than on horizontal components.

Earthquake commissions, committees, or equivalent institutions were also founded in Austria (1895), in Bulgaria (1892), in some of the German states (Baden, 1880; Bayern, 1879; Württemberg, 1886; and Sachsen, 1875), in Greece (1859), in Hungary (1882), and in Norway (1887) (Sieberg, 1904; Tams, 1950). Before these organizations built their own seismic stations, they collected macroseismic data mostly from the public and regularly published reports about all observed earthquakes.

Tainan Observatory 臺南測候所

年月日(明治)	發震時 {(前)ハ午前 (後)ハ午後}	水平 繼續時間 初期微動	水平 繼續時間 主要部	水平 繼續時間 全地震	水平 震動ノ平均振動期 初期微動 (東西)	水平 震動ノ平均振動期 初期微動 (南北)	水平 震動ノ平均振動期 主要部 (東西)	水平 震動ノ平均振動期 主要部 (南北)	水平 震動ノ平均振動期 終期 (東西)	水平 震動ノ平均振動期 終期 (南北)
年 月 日	時 分 秒	分 秒	分 秒	分 秒	秒	秒	秒	秒	秒	秒
34. 9. 13	8. 29. 37 (前)	(東西)15.6 / (南北)14.0	—	4. 30	0.78	0.72	1.14	1.25	1.08	1.19
35. 3. 20	9. 59. 34 (前)	0. 24	(東西) 52 / (南北) 55	5. 15	0.97	0.91	0.75 / 1.11 / **1.75**	0.74 / 1.19	1.03	0.84 / 1.35
36. 6. 7*	5. 7. 17 (後)	0. 19	(最强) 7	8. 00	—		—		—	1.32
37. 4. 24*	2. 38. 44 (後)	0. 12	(最强) 8.6	6. 00	0.61 / —	0.60 / **1.37**	**1.16** / 1.54	**0.95** / 1.46 / 1.86	0.94	1.33
37. 5. 2	5. 12. 26	(東西) 9.8 / (南北) 8.6	(東西) 25 / (南北) 29	3. 45	**0.33** / **1.15**	**0.25** / 0.52	0.43 / **1.41**	0.36 / **0.83** / 1.24	1.00	1.21
37. 5. 2	—	(東西) 6.0 / (南北) 5.5	(東西)19.8 / (南北)17.6	1. 23	—		0.51 / **0.96**	1.18	0.93	0.62 / 1.09
37. 11. 6*	4. 26. 30 (前)	7.2	13	7. 00	0.35 / **1.45**	**1.38**	0.91 / 1.47 / **1.86**	0.52 / 0.95 / **1.47**	1.74 / 1.21	1.49 / 1.20
37. 11. 15	5. 27. 00 (後)	8.6	39	2. 00	—		1.01	1.24	0.99	1.23

地震觀測摘要(續キ) Earthquake Observations

年月日(明治)	水平動 全振幅 初期微動 東西	水平動 全振幅 初期微動 南北	水平動 全振幅 主要部 (東西)	水平動 全振幅 主要部 (南北)	顯著振動 全振幅	顯著振動 方向	上下動 繼續時間 初期微動	上下動 繼續時間 主要部	上下動 繼續時間 繼續	上下動 平均振動期 初期微動	上下動 平均振動期 主要部	上下動 平均振動期 終期	上下動 全振幅 初期微動	上下動 全振幅 主要部
	ミリメートル		ミリメートル		ミリメートル		秒	秒	分秒	秒	秒	秒	ミリメートル	ミリメートル
34. 9. 13	0.4	0.52	0.86	1.1	—	—	—	—	—	—	—	—	—	—
35. 3. 20	0.68	0.30	3.9	3.7	第一動 4.8 / 第二動 5.3	S51°E / N47°W	24	45	2. 30	0.62	**0.93** / 0.39	0.71	0.15	1.04
36. 6. 7*	—		—		—	—	—	—	—	—	—	—	—	—
37. 4. 24*	3.9 / —	**3.1** / **6.2**	23.8(?) / 11.4	13.7	14.5	S52°E	10.	—	—	0.092 / **0.55** / 0.82	1.17	—	1.32	2.62 / 3.7
37. 5. 2	0.8	0.6	2.3	1.8	—	—	11	15	1. 23	0.37	**0.40** / 0.88	0.48	0.24	0.44
37. 5. 2	0.1	—	0.28	0.31	—	—	5.6	14.5	0. 37	—	0.39	0.45	—	0.03
37. 11. 6*	1.2	2.1	15.4	13.2	第一動 7.6 / 第二動16.6 / 第三動16.2	N35°W / S71°E / N66°W	7.4	23.8 / (最强) 10.6	2. 44	0.093 / 0.40 / **0.92**	0.48 / 0.66 / **1.03**	0.47 / 0.90 / 1.24	1.05	1.3
37. 11. 15	—		0.52	0.4	—	—	—	—	—	—	—	—	—	—

第十七圖

明治三十七年十一月六日激震東西地動計ノ記象

Earthquake on November 6, 1904 (EW Seismograph Records)

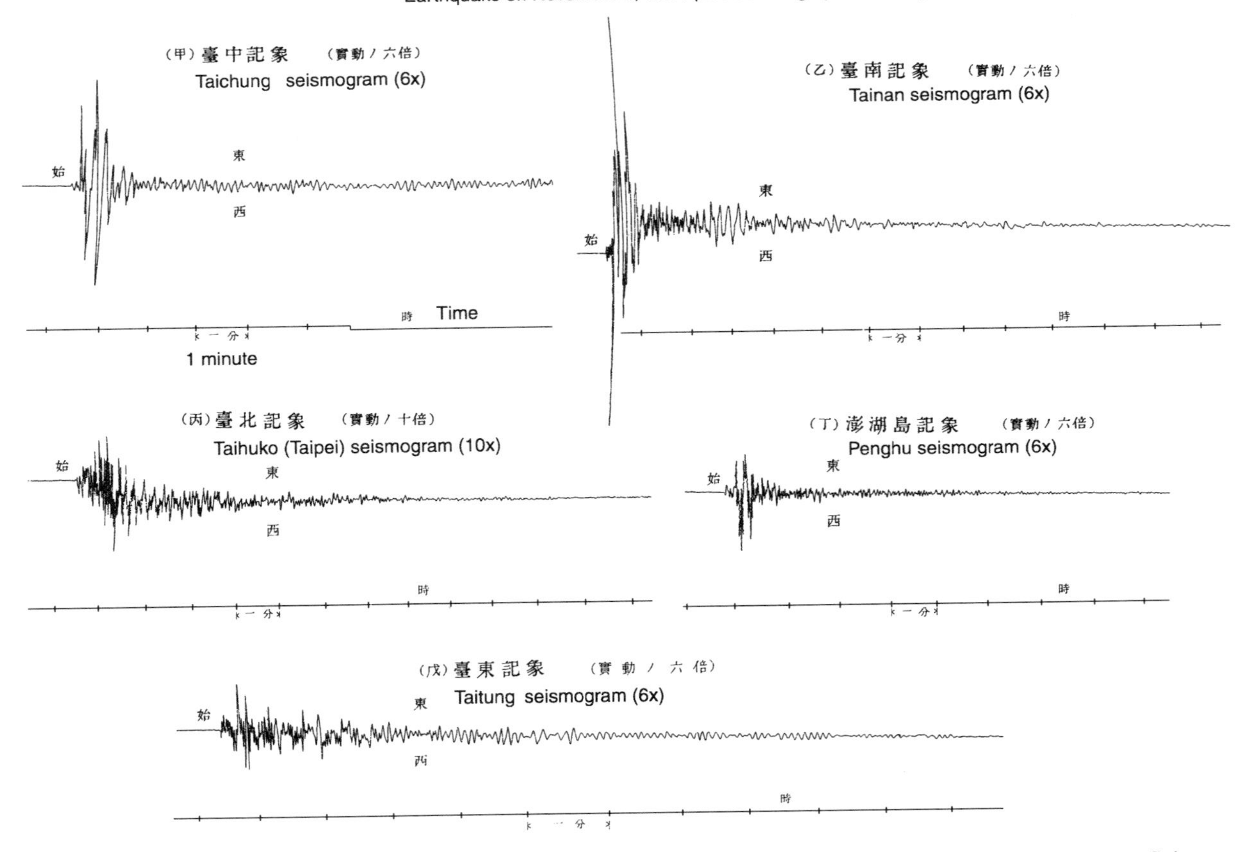

FIGURE 5 The upper two frames showing seismic readings for eight earthquakes recorded at the Tainan Observatory. Seismograms from Gray-Milne seismographs at five Taiwan stations for the November 6, 1904, earthquake are shown in the lower frame. See text for more explanations.

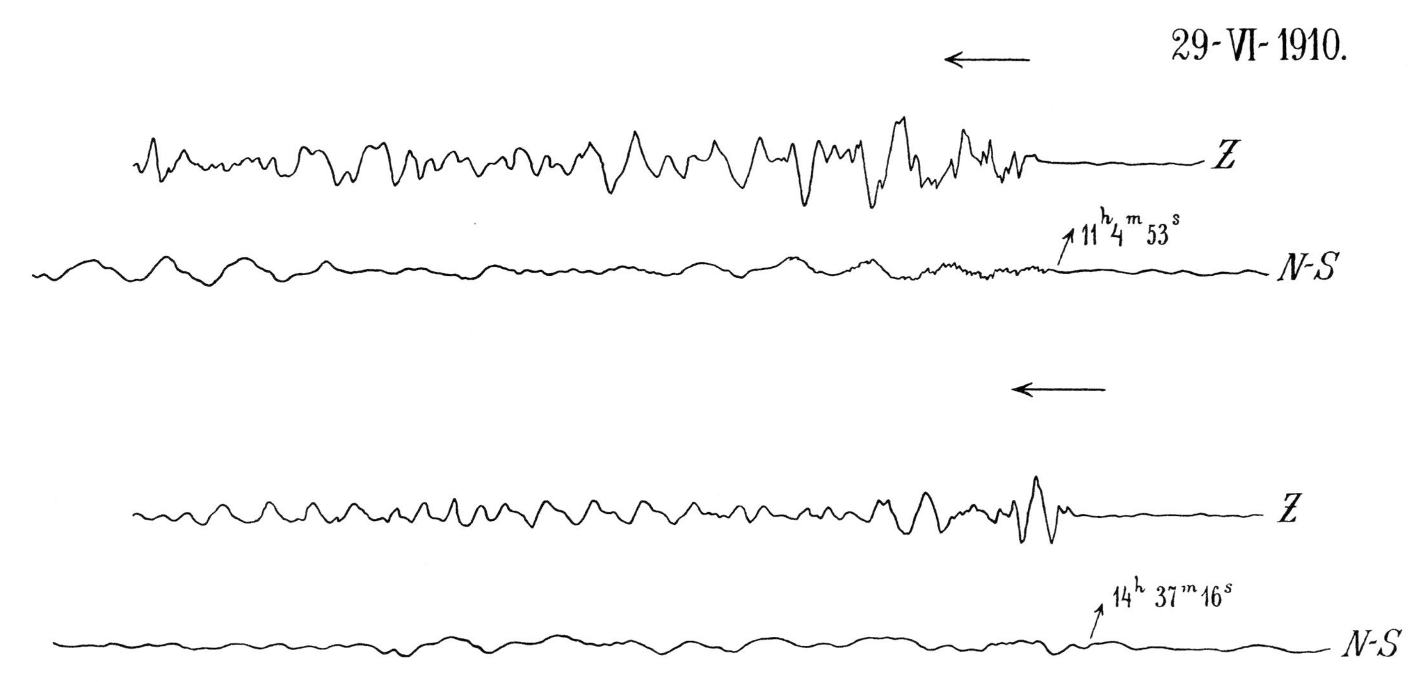

FIGURE 6 Four seismograms showing *P* onsets of two earthquakes recorded with a vertical and a horizontal component of Galitzin seismographs (from Galitzin, 1910).

2.2 The First Teleseismic Observation

During the last decade of the 19th century and the first decade of the 20th century, seismographs became more sensitive, and many seismic stations or observatories were established all over the world. A key discovery was the first observation of a teleseismic signal from an earthquake at about 80° epicentral distance in 1889 by Ernst von Rebeur-Paschwitz (Rebeur-Paschwitz, 1889; Chapter 1 by Agnew; Chapter 79.24.1 by Schweitzer on Handbook CD #2; Chapter 79.33.2 on Handbook CD #2 by Utsu). Rebeur-Paschwitz continued to search for earthquakes and in particular for teleseismic signals in his recordings of ground movement, and in 1892, he published a monograph about his instrument (Rebeur-Paschwitz, 1892). Figure 7 shows a plate with seismograms from this publication.

One important step in the direction of more readable seismograms was the introduction of damping. One of the first seismologists experimenting with damping his horizontal pendulum was Rebeur-Paschwitz (1892). However, the common application of damping came later. As an early example, we show in Figure 8 a plate showing seismograms recorded in 1899 by Emil Wiechert (1899) from regional (his Figures 9–10) and teleseismic (his Figures 11–14) distances. Afterwards, seismologists started to investigate the problems of elastic wave propagation through the Earth and to decipher step by step the principal structure of our planet. For these investigations, all kinds of instrumental earthquake observations were essential, and we can observe an intense international exchange of parameter data of recorded seismograms among many seismic observatories and/or more central institutes for seismological research. In parallel, the network of seismic stations became denser and denser. When the First International Seismological Conference was held in Strasbourg, the participants collected a worldwide list of known seismic stations. A colored map showing the distribution of these stations, with additional information on the type of equipment installed, was published in the proceedings of the conference (Weigand, 1902); we reproduce this map as additional material on the attached Handbook CD #3 (under the directory of \88Schweitzer). According to a listing in Sieberg (1904), at least 108 seismic stations were in operation at the beginning of 1904.

2.3 The BAAS and the "Shide Circulars"

The British Association for the Advancement of Science (BAAS) must be credited for fostering (and providing some financial support for) all kinds of scientific investigations and, in particular, for investigating earthquakes and related phenomena. Committees were appointed, and their reports were systematically published in the annual volumes of the "Report of the British Association for the Advancement of Science" for over 100 years, from 1832 until the early 1950s. Early reports include "Instruments to Record Earthquakes in Scotland and Ireland" (BAAS, 1841, 1842, 1843, 1844), "Facts of Earthquake Phenomena" (BAAS, 1850, 1851, 1852, 1854, 1858), and "Earthquakes and Seismometers" (BAAS, 1854).

Although Milne was teaching in Japan in 1880 (see biography of John Milne in Chapter 89), he persuaded the BAAS to appoint a committee "for the purpose of investigating the earthquake phenomena of Japan." This Committee initially consisted of only two persons: A. C. Ramsay and John Milne (as Secretary), and their reports began to appear in the BAAS Reports (1882).

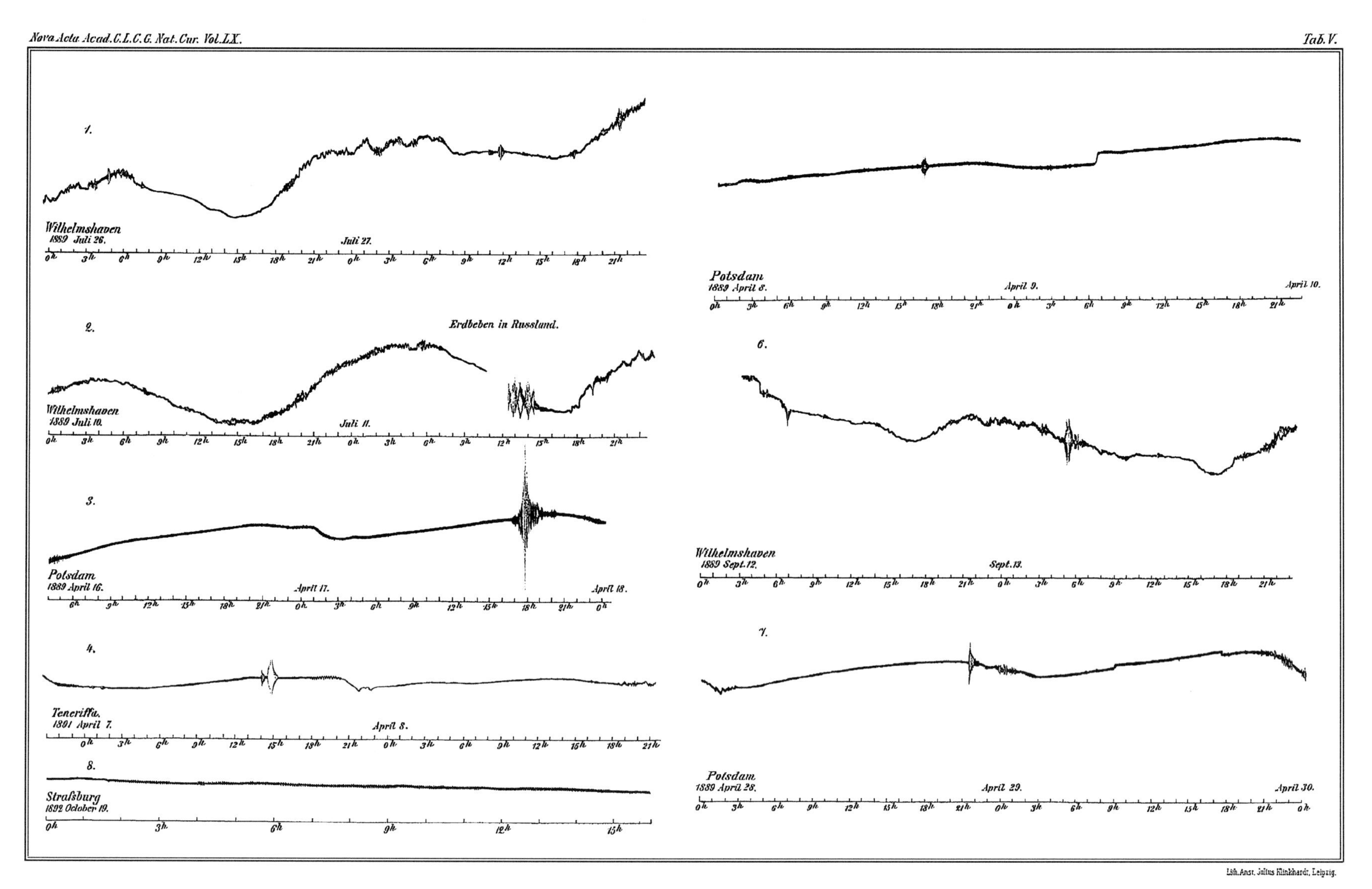

FIGURE 7 Plate No. V from Rebeur-Paschwitz (1892) showing early teleseismic observations, including the first known teleseismic seismogram from April 17, 1889 (# 3) as recorded in Potsdam, Germany, and an earthquake in Russia on July 10, 1889 (# 2), observed in Wilhelmshaven, Germany.

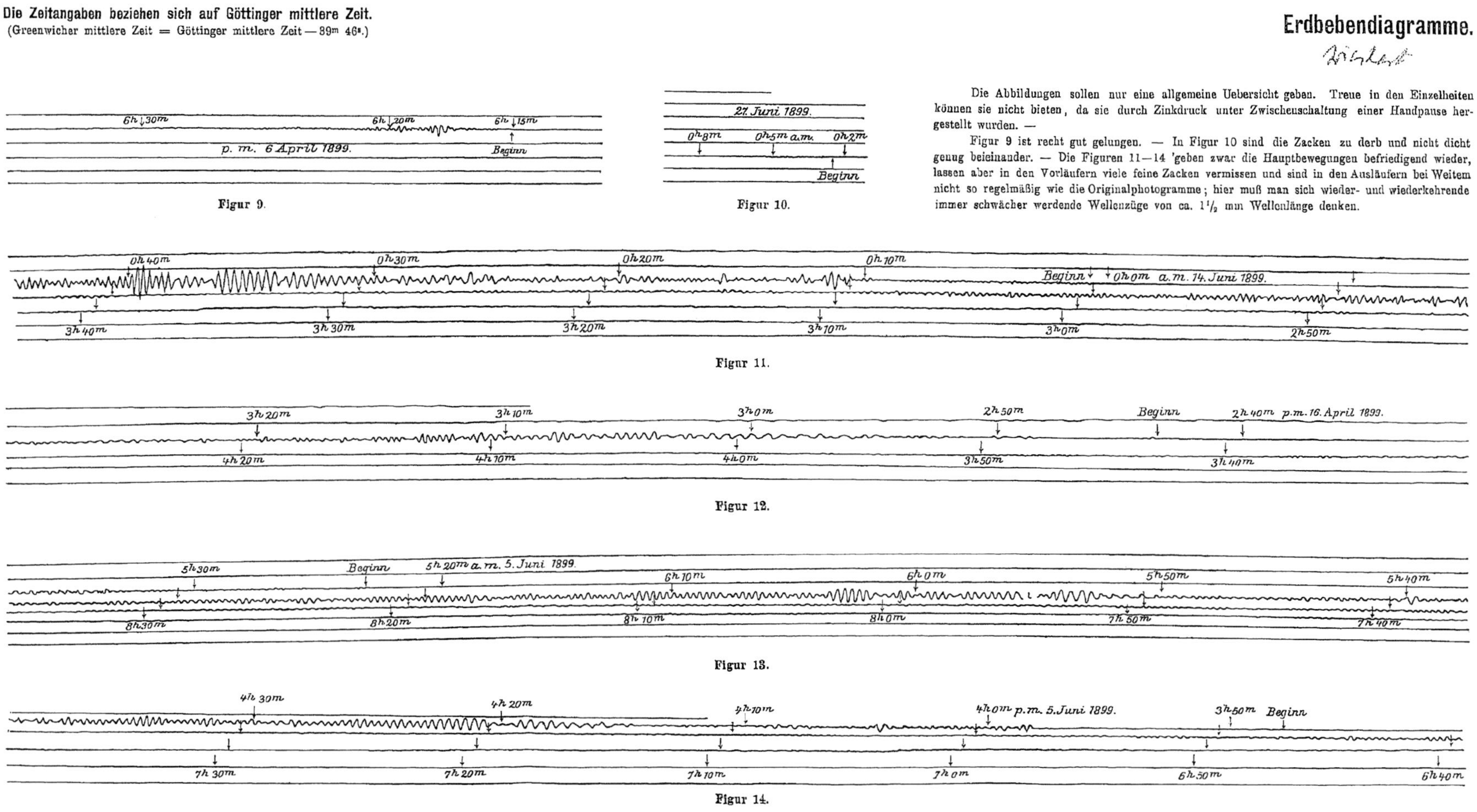

Die Zeitangaben beziehen sich auf Göttinger mittlere Zeit.
(Greenwicher mittlere Zeit = Göttinger mittlere Zeit — 39m 46s.)

Erdbebendiagramme.

Die Abbildungen sollen nur eine allgemeine Uebersicht geben. Treue in den Einzelheiten können sie nicht bieten, da sie durch Zinkdruck unter Zwischenschaltung einer Handpause hergestellt wurden. —

Figur 9 ist recht gut gelungen. — In Figur 10 sind die Zacken zu derb und nicht dicht genug beieinander. — Die Figuren 11—14 geben zwar die Hauptbewegungen befriedigend wieder, lassen aber in den Vorläufern viele feine Zacken vermissen und sind in den Ausläufern bei Weitem nicht so regelmäßig wie die Originalphotogramme; hier muß man sich wieder- und wiederkehrende immer schwächer werdende Wellenzüge von ca. 1½ mm Wellenlänge denken.

6h 30m 6h 20m 6h 15m
p. m. 6 April 1899. Beginn

Figur 9.

27. Juni 1899.
0h 8m 0h 5m a.m. 0h 2m
Beginn

Figur 10.

0h 40m 0h 30m 0h 20m 0h 10m Beginn 0h 0m a.m. 14. Juni 1899.
3h 40m 3h 30m 3h 20m 3h 10m 3h 0m 2h 50m

Figur 11.

3h 20m 3h 10m 3h 0m 2h 50m Beginn 2h 40m p.m. 16. April 1899.
4h 20m 4h 10m 4h 0m 3h 50m 3h 40m

Figur 12.

5h 30m Beginn 5h 20m a.m. 5. Juni 1899. 6h 10m 6h 0m 5h 50m 5h 40m
8h 30m 8h 20m 8h 10m 8h 0m 7h 50m 7h 40m

Figur 13.

4h 30m 4h 20m 4h 10m 4h 0m p.m. 5. Juni 1899. 3h 50m Beginn
7h 30m 7h 20m 7h 10m 7h 0m 6h 50m 6h 40m

Figur 14.

FIGURE 8 A collection of seismograms published by Wiechert (1899). The seismograms were recorded in Göttingen with a damped horizontal pendulum. Figures 9–10 show regional events, and Figures 11–14 show teleseismic events of unknown source. In the added text (in German), Wiechert wrote that the original seismograms were transferred by hand to the printing plate and that some features of the original records were lost. However, he mentioned that Figure 9 looks very much like the original and that in Figures 11–14 the main character of the records was mostly preserved.

In later years, it became the Committee for the Investigation of Earthquake and Volcanic Phenomena of Japan, and a total of 14 reports were published in the annual volumes of the "Report of the British Association for the Advancement of Science" (BAAS, 1881–1895).

In 1891, the BAAS appointed a committee (with Charles Davison as the secretary) to investigate the earthquake tremors occurring in the British isles and published 5 reports in the annual volumes of "Report of the British Association for the Advancement of Science" (BAAS, 1891–1895). After John Milne returned to England in 1895, these two BAAS committees were combined to form the Committee on Seismological Investigation, with Davison and Milne as secretaries.

John Milne (and Rebeur-Paschwitz in 1895) recognized not only the need to deploy seismographs worldwide to monitor earthquakes but also the need to collect seismic readings for any cooperating stations so that a database is available for earthquake location and seismological research. The first 18 reports of the joint BAAS Committee (BAAS, 1896–1913) were mostly written by John Milne and chronicled the Committee's efforts in establishing the first worldwide seismograph network, collecting the seismic readings and locating the earthquakes. In 1899, the numbers of seismic readings became too great to be published in the BAAS annual volumes; from then on, they were issued as supplements to the Reports and were called the "Shide Circulars." These Circulars (see Figure 9) were collections of bulletins from stations distributed worldwide and mostly equipped with Milne's type of seismographs (see Chapter 4 by Adams). The list of contributing seismic stations changed, but during that time the total number of stations continually increased. Until the end of 1912, the biannual "Shide Circulars" contained seismic readings from about 30 regularly reporting stations, almost all in the then British Empire. However, many of these stations also published their own station bulletins in addition to the "Shide Circulars." As an example, we show in Figure 4 the bulletin of Christchurch, New Zealand, for January 1902.

Despite Milne's urging, fewer than a third of all operating seismic stations worldwide cooperated, as it was a voluntary effort. Furthermore, the "Shide Circulars" contained only onset time readings (and maximum amplitude and duration) from large earthquakes that were observed at several stations. The locations of these events were published in the "Reports of the British Association for the Advancement of Science" as earthquake lists until 1910 (18th Report). These publications stopped in 1913 when John Milne died. Despite their obvious shortcomings, the "Shide Circulars" remain the only continuous compilation of early station bulletins worldwide from 1899 through 1912, and they total over 1,000 pages. Many of these early seismic stations are no longer in existence, and their station bulletins are only accessible through the "Shide Circulars." Even the "Shide Circulars" themselves are difficult to find today in a complete set.

After Milne's death in 1913, the efforts were continued under Turner's leadership (see biography of H. H. Turner in Chapter 89), eventually leading to the publication of the "International Seismological Summary."

2.4 The Central Bureau of ISA

In 1899, the "Kaiserlische Hauptstation für Erdbebenforschung" (Imperial Central Station for Earthquake Research) in Germany was founded in Strasbourg. At this institution and at the later co-located "Zentral-Bureaus der Internationalen Seismologischen Assoziation" (Central Bureau of the International Seismological Association, or ISA), Emil Rudolph and colleagues worked on international earthquake catalogs and bulletins (see Chapter 79.24.1 by Schweitzer on Handbook CD #2). Rudolph was able to compile an initial bulletin with associated observations from several stations for the years 1895–1897 (Rudolph, 1903a, 1903b) and a special catalog for observations at European stations of Japanese earthquakes during the years 1893–1897 (Rudolph, 1904). The next worldwide catalog was compiled for the year 1903 (Rudolph, 1905). In 1907, Elmar Rosenthal published (in the name of the International Seismological Association) the first international bulletin for the year 1904, which was based on seismic data from 139 seismic stations, 54 of them in Japan. The Japanese readings, extended by observations from stations on Formosa (Taiwan), the Philippines, and Indonesia, were used by Rudolph (1907) to compile a special regional catalog of East Asian earthquakes in 1904.

During the following years, Siegmund Szirtes published bulletins for the years 1905–1908. The 1908 list of known seismic stations, published by the ISA, contained 199 entries (Szirtes, 1908), and the catalog (updated in 1912) listed 265 already named stations (Szirtes, 1912c). However, not all of the named stations were regularly reporting seismic readings to the data-collecting organizations. Many stations did not publish their results regularly, and quite a few stations "disappeared" without leaving any useful bulletin materials.

The ISA also systematically collected worldwide macroseismic data. These macroseismic observations were published in bulletin form by Rudolph (1905), Oddone (1907), Christensen and Ziemendorff (1909), Scheu (1911a, 1911b), Scheu and Lais (1912), and Sieberg (1917). However, the production of these bulletins stopped due to World War I and the dissolution of the International Seismological Association (see Chapter 4 by Adams).

2.5 The "International Seismological Summary" (ISS)

After Milne's death in 1913, Herbert Hall Turner continued the work of Milne and published the "Bulletin of the British Association of the Advancement of Science, Seismology Committee" for the years 1913–1917. After World War I, the international cooperation in seismology was reorganized at the International Union of Geodesy and Geophysics (IUGG) conference in Rome (May 1922), and Turner's work was recognized. The newly established Seismology Section of IUGG, which was named the International Association of Seismology (IAS) in 1930 at its General Assembly in Stockholm, asked the British Association to publish an international bulletin in the name of the IAS (see Figure 10 for the first two pages of

British Association for the Advancement of Science.

Circular No. 1, *issued by the Seismological Committee*, Professor J. W. JUDD, *F.R.S.* (*Chairman*), Mr. JOHN MILNE, *F.R.S.*, *Shide, Isle of Wight* (*Secretary*).

CONTENTS.

Registers from similar Horizontal Pendulums (*Milne type*) 1899–1900.

These registers are for the most part continuous with those published by the Seismological Investigation Committee in the Reports of the British Association, 1896 to 1899.

The time employed is Greenwich mean civil time, expressed in hours, minutes, and decimals of minutes.

D. first P.T.'s.—This should refer to the duration of the first preliminary tremors or the first uniform thickening of the trace.

Amplitude indicates half the complete range of the maximum motion. Where this exceeds one millimeter it is expressed in millimeters and in seconds of arc. Values less than one millimeter refer to a thickening of the line, and indicate half its width.

These registers refer to Shide, Kew, San Fernando (Spain), Cape of Good Hope, Mauritius, Cairo, Calcutta, Madras, Bombay, Tokio, and Batavia.

To the Tokio register there is appended a list of local earthquakes, a few of which are common to the records of horizontal pendulums in distant countries.

It is hoped that corresponding registers will be received from Toronto, Victoria (B.C.), Arequipa, Swarthmore College (Philadelphia), Mexico, Cordova (Argentina), New Zealand, Beyrut, Honolulu, Trinidad, and Paisley.

2

The Register from Shide, Isle of Wight, England.
Observer, JOHN MILNE; *Assistant*, SHINOBU HIROTA.

Shide No.	Date		Commencement	D. of first P.T.'s	Maximum	Amplitude		Total duration	Remarks
					1899.				
			H. M.	M.	H. M.	MM.	SEC.	H. M.	
273	April	4	2 8·5	—	—	0·5	—	0 10	—
274	,,	5	24 0·0	—	—	—	—	—	—
275	,,	6	About 8 34·3	—	—	0·75	—	0 5	—
276	,,	7	17 40·5	—	—	0·5	—	0 13	—
277	,,	12	16 35·6	—	—	—	—	—	—
278	,,	12	18 8·7	9·3	18 25·2	2·0	0·94	2 40	Reinforcement at 20h. 28·4m.
279	,,	13	4 28·5	—	{ 4 33·5 4 37·6 }	1·5	0·70	0 29	Tremors make commencement uncertain.
280	,,	13	11 9·8	—	—	—	—	—	—
281	,,	15	5 0·0	—	—	—	—	—	—
282	,,	16	About 14 2·7	15·5	{ 14 18·3 14 21·4 }	4·5	2·11	1 48	—
283	,,	17	2 47·3	14·5	—	0·75	—	0 48	Tremors make commencement uncertain.
284	,,	28	20 46·9	—	—	0·75	—	0 12	—
285	May	2	15 32·1	—	—	0·5	—	0 10	—
286	,,	8	4 39·1	—	—	0·5	—	1 0	A second max. at 5h. 10·6m.
287	,,	14	14 47·3	—	—	0·3	—	0 20	—
288	,,	15	13 51·4	—	—	0·5	—	0 28	—
289	,,	17	19 54·3	—	20 49·2	0·5	—	1 15	A second max. at 20h. 49·2m.
290	June	4	19 36·5	—	—	0·25	—	0 5	—
291	,,	5	4 46·9	6·2	—	2·25	1·05	2 46	—
292	,,	5	15 17·7	—	15 39·5	1·25	0·58	0 50	—
293	,,	9	12 8·8	—	—	0·25	—	0 50	—
294	,,	14	11 21·1	8·0	11 55·2	5·5	2·58	3 0	—
295	,,	17	1 22·5	—	—	0·25	—	1 0	—
296	,,	19	9 10·8	—	—	0·25	—	0 45	—
297	,,	19	12 56·3	—	—	0·25	—	0 15	—
298	,,	24	17 56·2	—	—	0·5	—	0 25	—
299	,,	29	23 14·3	—	23 59·3	0·5	—	1 5	—
300	July	2	12 59·0	—	—	0·5	—	0 5	—
301	,,	2	20 14·5	—	—	0·25	—	0 4	—
302	,,	3	16 37·2	—	—	0·25	—	0 4	—
303	,,	7	9 25·6	—	—	0·25	—	0 6	—
304	,,	8	14 2·4	—	—	0·25	—	0 5	—
305	,,	9	19 27·5	—	20 3·7	0·75	—	1 10	—
306	,,	10	22 57·5	—	—	0·25	—	0 25	—
307	,,	11	7 39·8	15·0	7 54·8	1·0	0·41	1 20	—
308	,,	12	1 42·9	—	2 1·9	1·25	0·51	1 20	—
309	,,	14	13 36·6	15·6	14 3·6	3·0	1·23	3 0	—
310	,,	17	11 17·7	—	—	0·5	—	1 5	—
311	,,	17	17 16·7	—	—	0·5	—	0 2	—
312	,,	19	12 48·8?	—	—	0·25	—	—	—
313	,,	20	10 18·8	—	—	0·5	—	0 45	Time uncertain.
314	,,	21	11 54·7	—	—	0·25	—	0 3	—
315	,,	26	12 52·3	—	—	0·25	—	0 3	—
316	,,	26	15 10·2	—	—	0·25	—	0 3	—

FIGURE 9 The first two pages from No. 1 of the "Shide Circulars."

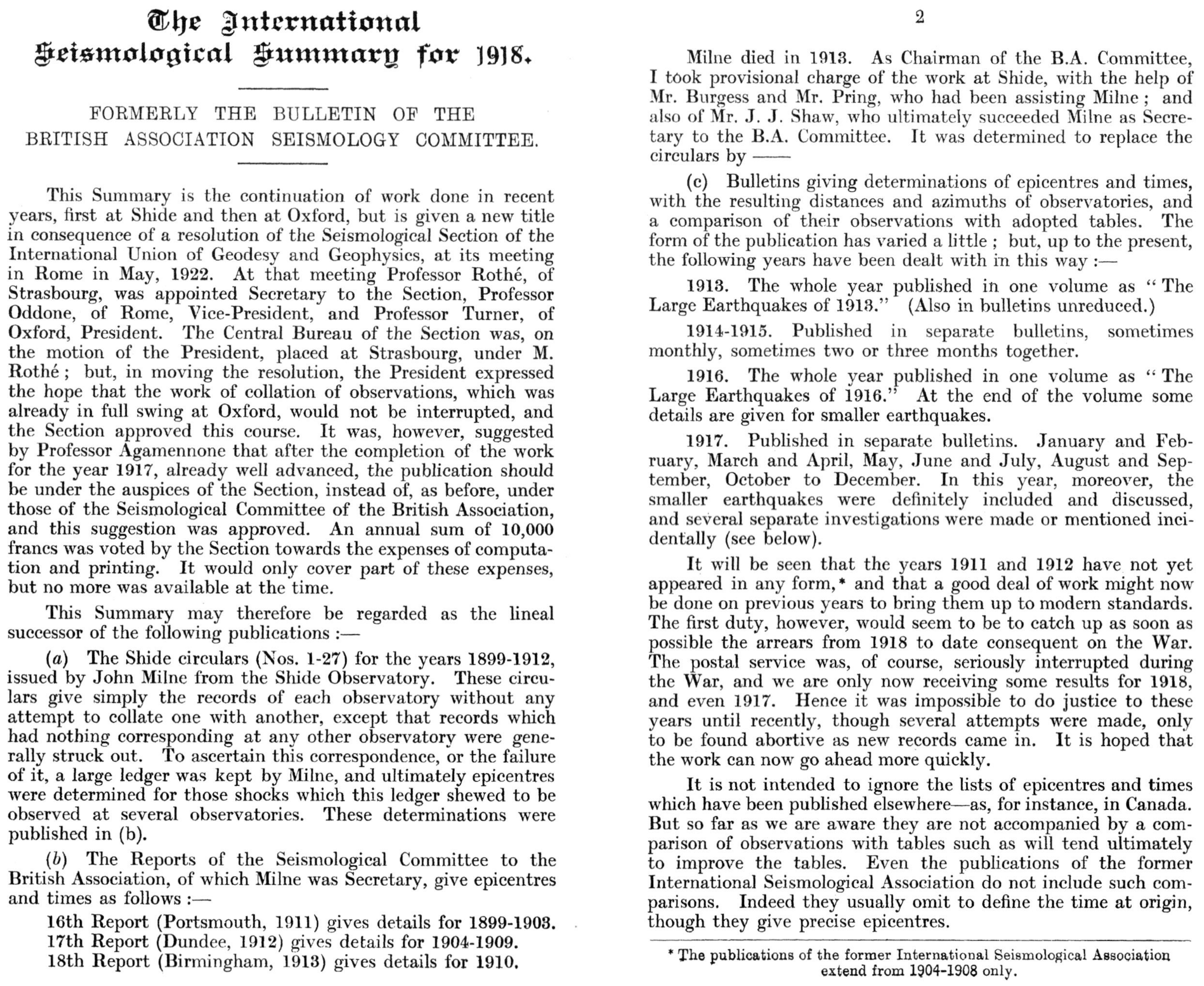

The International Seismological Summary for 1918.

FORMERLY THE BULLETIN OF THE BRITISH ASSOCIATION SEISMOLOGY COMMITTEE.

This Summary is the continuation of work done in recent years, first at Shide and then at Oxford, but is given a new title in consequence of a resolution of the Seismological Section of the International Union of Geodesy and Geophysics, at its meeting in Rome in May, 1922. At that meeting Professor Rothé, of Strasbourg, was appointed Secretary to the Section, Professor Oddone, of Rome, Vice-President, and Professor Turner, of Oxford, President. The Central Bureau of the Section was, on the motion of the President, placed at Strasbourg, under M. Rothé; but, in moving the resolution, the President expressed the hope that the work of collation of observations, which was already in full swing at Oxford, would not be interrupted, and the Section approved this course. It was, however, suggested by Professor Agamennone that after the completion of the work for the year 1917, already well advanced, the publication should be under the auspices of the Section, instead of, as before, under those of the Seismological Committee of the British Association, and this suggestion was approved. An annual sum of 10,000 francs was voted by the Section towards the expenses of computation and printing. It would only cover part of these expenses, but no more was available at the time.

This Summary may therefore be regarded as the lineal successor of the following publications :—

(*a*) The Shide circulars (Nos. 1-27) for the years 1899-1912, issued by John Milne from the Shide Observatory. These circulars give simply the records of each observatory without any attempt to collate one with another, except that records which had nothing corresponding at any other observatory were generally struck out. To ascertain this correspondence, or the failure of it, a large ledger was kept by Milne, and ultimately epicentres were determined for those shocks which this ledger shewed to be observed at several observatories. These determinations were published in (b).

(*b*) The Reports of the Seismological Committee to the British Association, of which Milne was Secretary, give epicentres and times as follows :—

16th Report (Portsmouth, 1911) gives details for 1899-1903.
17th Report (Dundee, 1912) gives details for 1904-1909.
18th Report (Birmingham, 1913) gives details for 1910.

2

Milne died in 1913. As Chairman of the B.A. Committee, I took provisional charge of the work at Shide, with the help of Mr. Burgess and Mr. Pring, who had been assisting Milne; and also of Mr. J. J. Shaw, who ultimately succeeded Milne as Secretary to the B.A. Committee. It was determined to replace the circulars by ——

(c) Bulletins giving determinations of epicentres and times, with the resulting distances and azimuths of observatories, and a comparison of their observations with adopted tables. The form of the publication has varied a little; but, up to the present, the following years have been dealt with in this way :—

1913. The whole year published in one volume as "The Large Earthquakes of 1913." (Also in bulletins unreduced.)

1914-1915. Published in separate bulletins, sometimes monthly, sometimes two or three months together.

1916. The whole year published in one volume as "The Large Earthquakes of 1916." At the end of the volume some details are given for smaller earthquakes.

1917. Published in separate bulletins. January and February, March and April, May, June and July, August and September, October to December. In this year, moreover, the smaller earthquakes were definitely included and discussed, and several separate investigations were made or mentioned incidentally (see below).

It will be seen that the years 1911 and 1912 have not yet appeared in any form,* and that a good deal of work might now be done on previous years to bring them up to modern standards. The first duty, however, would seem to be to catch up as soon as possible the arrears from 1918 to date consequent on the War. The postal service was, of course, seriously interrupted during the War, and we are only now receiving some results for 1918, and even 1917. Hence it was impossible to do justice to these years until recently, though several attempts were made, only to be found abortive as new records came in. It is hoped that the work can now go ahead more quickly.

It is not intended to ignore the lists of epicentres and times which have been published elsewhere—as, for instance, in Canada. But so far as we are aware they are not accompanied by a comparison of observations with tables such as will tend ultimately to improve the tables. Even the publications of the former International Seismological Association do not include such comparisons. Indeed they usually omit to define the time at origin, though they give precise epicentres.

* The publications of the former International Seismological Association extend from 1904-1908 only.

FIGURE 10 The first two pages from the introduction of the first issue of the "International Seismological Summary for 1918" by H. H. Turner, summarizing the efforts by the British Association for the Advancement of Science in collecting and analyzing seismic readings around the world.

the introduction by H. H. Turner, and Figure 11 for two sample pages). From the issue for the year 1918, the bulletin was named the "International Seismological Summary," or ISS (Rothé, 1981; Chapter 1 by Agnew; Chapter 4 by Adams; Chapter 79.24.1 on Handbook CD #2 by Schweitzer). The ISS was relatively worldwide in scope, covering the largest events from the beginning of the 1920s, and also contains some smaller and regional events. Then in 1963, the International Seismological Centre (ISC) was organized, and in 1964, the ISS became the "Bulletin of the International Seismological Centre." A summary of the current activities of the ISC is given in Willemann and Storchak (2001). See also Chapter 81.5.

3. The Value of Old Bulletins for Modern Seismology

The earthquake lists and bulletin data from the early seismic stations not only document the history of seismology but also have intrinsic scientific value. When Beno Gutenberg and Charles Francis Richter (see biographies of B. Gutenberg and C. F. Richter in Chapter 89) developed the surface-wave magnitude scale in the late 1930s, they analyzed some original seismograms; in addition, they retrieved amplitude data from numerous station bulletins (because the ISS does not contain the amplitude information). However, since the last edition of their famous book *Seismicity of the Earth* (Gutenberg and Richter, 1954), their published event magnitudes have been recalculated in studies for many regions and for many single events by numerous authors. Even today, these old bulletins are needed for event relocation as our technique for earthquake location improves. They are also useful for magnitude analysis in order to establish a consistent magnitude scale. For example, the very early bulletin data from the mostly undamped seismographs before 1900 can be used to determine event magnitudes (Abe, 1994). For all seismic risk studies, a catalog of earthquakes must be as complete as possible. Here the old bulletins are essential to extend the instrumental observation period as far back in time as possible.

In Chapter 41 of this Handbook, Engdahl and Villasenor present a global earthquake catalog for the period from 1900 through to 1999. However, they could not relocate any earthquake before 1918 because the ISS began in 1918 and the seismic bulletins before 1918 were too difficult to assemble; Gutenberg and Richter (1954) already recognized this problem. We hope that this chapter and the computer files of early seismic bulletin materials up to 1920 will help future research on the pre-1918 earthquakes.

4. The Status of the Collections of Old Seismic Bulletins

As mentioned above, not many of the original seismograms are available today. Since seismic bulletins were often published in small numbers, old seismic bulletins are also difficult to find due to organizational changes at the seismic stations, natural catastrophes, accidents, and wars. Many old seismic stations no longer exist, and the existing stations often do not have a complete set of their own bulletins. Fortunately, the international exchange of bulletins during the early days of seismology led to a worldwide spread of bulletins, and they were collected and are preserved at a few places today.

However, many of these old bulletins are in very bad physical condition due to paper decay and/or damaged due to usage and poor storage conditions. So, an idea came up to preserve these old seismic bulletins, especially from the early days up until 1920. We have tried to obtain the originals or good-quality copies so that they can be scanned into computer image files. One huge collection of old seismic bulletins was accumulated over 30 years by W. H. K. Lee at the US Geological Survey (USGS) in Menlo Park by combining materials from four major collections: (1) Seismology Center of the National Oceanic and Atmospheric Administration (NOAA), (2) New Zealand, (3) St. Louis University, and (4) Seismology Branch of the USGS. Unfortunately, space was no longer available at the USGS in 1996, and only a fraction of this collection (mostly seismic bulletins from before World War II) was saved by storing them at Lee's home. This salvaged collection is now the backbone of a planned scanning project.

Other important sources are the bulletins collected at the University of Hamburg, Germany, in the British National Seismological Archive held at the British Geological Survey (Henni *et al.*, 1999; Lovell and Henni, 1999; Henni *et al.*, 2000); the microfilms of the bulletins held at St. Louis University (Herrmann *et al.*, 1983) before shipping to the USGS; and the bulletins collected in Strasbourg at the former ISA. However, all these collections have many gaps: Some known station bulletins are missing, and the contents of the collection in Strasbourg are still unknown due to lack of manpower and time to make a complete inventory of this archive. During the summer and autumn of 2000, we tried to fill in existing gaps as much as possible and to collect bulletin material from missing stations. All this newly collected material is now stored at NORSAR.

The task to publish a (mostly) complete collection of the bulletin material on the attached CDs in this Handbook became very difficult due to the lack of funding. Some missing material could not be located easily and the old bulletins could not be scanned fast enough, as this project depends on voluntary work. In addition, after the systemic search during the last few years, the available material became far too voluminous to be scanned by one volunteer (Shirley L. Lee) to meet the deadline for this Handbook. Therefore, we decided to document the state of the collection as of December 2002.

Table 1 contains the already collected and the newly located bulletin materials. Many stations published their seismogram readings several times in early weekly or monthly listings, and sometimes also in yearly collections. In this table, only one entry was added per station and time, to indicate for which time period bulletin materials are available. Detailed references of station bulletins are given in the notes at the end of this chapter.

5

1918 JANUARY, FEBRUARY, & MARCH.

Jan. 1d. 15h. 2m. 10s. Epicentre 38°·0N. 23°·5E. (as on 1914 Oct. 17d. 6h.)

	△	P.	O – C.	L.	M.
	°	m. s.	s.	m.	m.
Athens	0·2	e 0 4	+1	0·6	0·8
Zagreb	9·6	e 2 32	+8	—	6·0
Helwan	10·4	8 50	?L	(8·8)	—
De Bilt	19·0	—	—	e 10·8	—

Jan. 1d. Records also at 0h. (San Fernando and Eskdalemuir), 7h. (Mizusawa), 12h. (La Paz).

Jan. 2d. Records at 3h. (Algiers), 4h. (De Bilt, Bidston, Rio Tinto, La Paz, and Helwan), 7h. (Helwan), 10h. (La Paz), 18h. and 19h. (Batavia), 20h. and 21h. (Monte Cassino), 23h. (Manila).

Jan. 3d. Records at 0h. (Lick and Eskdalemuir), 4h. (Port au Prince), 6h. and 8h. (Helwan), 13h. (Zi-ka-wei, Manila (3), Bombay, Edinburgh, and Colombo), 14h. (Eskdalemuir, Zagreb, Bidston, and Manila (2)), 15h. (Manila (2)), 16h. (Manila and Harvard), 17h. (Manila (2)), 18h. (Manila (2)), 19h. and 20h. (Manila), 22h. (Manila (3)).

1918. Jan. 4d. 4h. (I) 30m. 5s. (II) 32m. 25s. Epicentre 10°·5N. 91°·0W.

A = –·017, B = –·983, C = +·183 ; D = –1·000, E = +·017 :
G = –·003, H = –·183, K = –·983.

		△	Az.	P.	O – C.	S.	O – C.	L.	M.
		°	°	m. s.	s.	m. s.	s.	m.	m.
II Balboa Heights	E.	11·4	97	3 11	+21	—	—	6·6	7·9
II	N.	11·4	97	3 19	+29	—	—	6·7	—
II Tacubaya		11·9	319	e 2 33	–25	—	—	—	—
I Vieques		25·9	70	e 6 48	?PR₁	—	—	—	—
I Tucson		28·5	323	e 6 49	+36	12 15	+67	e 15·9	17·3
I Cheltenham		30·9	22	—	—	12 53	+63	e 16·5	17·6
I Georgetown		31·0	21	e 6 57	+19	12 26	+35	e 16·3	—
I Washington		31·0	21	8 20?	?	13 0	+69	15·7	—
I Ann Arbor		32·4	10	6 49	–45	13 7	+11	16·7	18·9
I Ithaca		34·3	19	—	—	—	—	e 16·8	—
I Toronto		34·7	15	—	—	—	—	e 17·5	—
II		34·7	15	—	—	—	—	e 17·0	18·3
II La Paz		35·2	140	i 7 38	+23	i 13 34	+36	18·7	19·3
I Harvard		36·3	25	7 9	–15	12 51	–23	e 16·9	19·6
I Northfield		37·2	22	—	—	e 12 55	–32	e 18·9	—
I Ottawa		37·3	18	—	—	e 13 39?	+11	18·9	—
II Lick		38·3	319	e 7 55	+15	—	—	—	—
II Berkeley		39·1	319	e 8 9	+22	—	—	—	21·3
I Victoria		46·5	331	—	—	15 56?	+21	20·0	33·0
II Honolulu		64·9	289	e 19 35	?S	(19 35)	+11	e 31·9	34·9
II Coimbra		77·7	50	7 35?	?	21 35?	–22	41·3	43·5
II Rio Tinto		79·2	53	24 35	?S	(24 35)	+141	—	53·6
II San Fernando		79·6	55	24 5	?S	(24 5)	+106	43·6	50·1
II Eskdalemuir		80·1	35	12 12	– 8	22 13	–11	36·6	46·6
I Edinburgh		80·1	35	21 35	?S	(21 35)	–49	(33·3?)	51·9
II Bidston		80·3	37	11 35	–46	22 47	+20	—	47·3
II Granada		81·6	53	12 13	–15	22 5	–37	—	—
I Kew		82·2	39	—	—	—	—	—	49·9
II Paris		84·4	41	—	—	e 22 39	–33	39·6	47·6
II De Bilt		84·9	38	—	—	e 22 47	–31	34·6	49·3
II Uccle		85·2	39	e 12 29	–20	—	—	e 40·6	—
I Barcelona		85·3	49	—	—	—	—	41·0	49·9
II Rocca di Papa		93·0	47	—	—	(e23 53)	–52	23·9	56·2
II Vienna		93·4	40	e 13 12	–22	—	—	—	—
II Zagreb	N.E.	94·0	42	e 13 14	–24	e 24 15	–41	45·6	56·6
II	N.W.	94·0	42	i 13 20	–18	e 24 11	–45	—	50·6
II Helwan		111·4	52	28 35	?S	(28 35)	+54	—	—
II Melbourne		123·2	231	—	—	—	—	e 62·6	68·5
I Mauritius		148·6	112	—	—	—	—	70·5	78·7

For Notes see next page.

6

NOTES TO JAN. 4d. 4h. (I) 30m. 5s. (II) 32m. 25s.

Additional records: Tucson MN = +17·1m., LN = +10·4m. Vieques eLN = +12·2m., MN = +18·0m. Cheltenham LN = +15·9m., MN = +17·0m. Ann Arbor SN? = +13m.37s., MN = +19·9m., PE = +6m.7s., SE = +12m.25s., LE = +14·6m., M = +18·9m. Toronto (II) Li = +17·6m. La Paz RP = +9m.10s., M = +20·8m. Harvard T_0 = 4h.29m.55s. Ottawa eLN = +16·4m., LN = +23·9m. Victoria L = +26·3m. Coimbra L = +38·6m. San Fernando S = +35m.5s., MN = +49·1m. De Bilt MN = +35·7m. Rocca di Papa MN = +90·4m. Zagreb T_0 = 4h.32m.50s.

1918. Jan. 4d. 15h. 48m. 45s. Epicentre 6°·5S. 153°·5E.

(as on 1913 Sept. 3d. 20h.).

A = –·889, B = +·443, C = –·113 ; D = +·446, E = +·895 :
G = +·101, H = –·051, K = –·994.

	△	Az.	P.	O – C.	S.	O – C.	L.	M.
	°	°	m. s.	s.	m. s.	s.	m.	m.
Sydney	27·4	184	10 51	?S	(10 51)	+ 3	14·5	16·4
Riverview	27·4	184	i 6 8	+ 6	i 10 51	– 3	e 14·6	16·2
Melbourne	32·3	193	—	—	12 3	–10	17·2	20·2
Manila	38·5	303	e 7 37	– 5	—	—	—	—
Batavia	46·4	268	e 8 15	–28	—	—	—	—
Zi-ka-wei	48·7	323	e 8 59	+ 1	—	—	—	—
Honolulu	55·1	58	e 15 3	?	e 20 9	?SR₁	e 23·0	32·4
Berkeley	89·4	52	—	—	e 30 19	?SR₁	—	—
Victoria	90·5	41	37 28	?	—	—	46·3	50·2
Mauritius	93·8	250	40 39	?L	—	—	(40·6)	48·8
Toronto	120·8	42	—	—	—	—	e 66·6	75·2
Helwan	121·0	301	31 15	?	—	—	—	—
Edinburgh	127·3	344	56 15	?L	—	—	(56·2)	87·2
De Bilt	127·5	336	—	—	—	—	59·2	62·3
Bidston	129·4	342	—	—	—	—	—	51·2
La Paz	132·8	120	e 19 37	[+12]	—	—	67·4	72·2

Additional records: Riverview PS = +11m.16s., MN = +16·1m., MZ = +24·0m., T_0 = 15h.48m.56s. Perth (△ = 43°·3) records merely 16h.3m.50·5s. to 17h.8m.58·2s. De Bilt MN = +65·3m. Eskdalemuir (△ = 127°·8) records 16h.30m. to 17h.30m.

Jan. 4d. Records also at 0h. (Zagreb), 1h. (Manila (2)), 3h. (Manila), 4h. (Tacubaya), 5h. (La Paz), 6h. (Colombo and Manila (2)), 7h. (Manila (2)), 8h. (Helwan), 11h. (Manila), 13h. (Algiers), 14h. (Manila), 17h. (Manila and Paris), 19h. (Manila and La Paz).

Jan. 5d. Records at 1h. (La Paz), 5h. (Helwan), 7h. (La Paz), 8h. (Helwan), 13h. (Monte Cassino, Helwan, and Manila), 14h. (Manila), 19h. (Helwan, Zagreb, Pola, and La Paz), 21h. (Monte Cassino), 23h. (Manila).

Jan. 6d. Records at 1h. (Manila), 7h. and 14h. (Helwan), 16h. (La Paz and Simla), 21h. (Taihoku), 22h. (Zurich).

Jan. 7d. Records at 4h. (Colombo), 5h. (Edinburgh), 13h. (Algiers), 18h. (Manila and Melbourne), 22h. (Taihoku and Zi-ka-wei), 23h. (Helwan).

Jan. 8d. Records at 0h. (Taihoku), 9h. (Balboa Heights), 10h. (Monte Cassino), 13h. (Helwan), 14h. (Edinburgh), 18h. (La Paz).

Jan. 9d. Records at 3h. (Taihoku and Zi-ka-wei), 4h. (San Fernando and Helwan), 5h. (La Paz), 6h. (Athens), 7h. (Zi-ka-wei), 11h. (Batavia), 19h. (San Fernando).

Jan. 10d. Records at 6h. (Mizusawa), 7h. (Zagreb), 8h. (Algiers), 9h. (Athens, Manila, and Rocca di Papa), 13h. (Manila), 16h. (Colombo), 19h. (La Paz).

Jan. 11d. Records at 1h. (La Paz), 3h. (Colombo), 4h. (Helwan and Colombo), 6h. (Colombo), 7h. (La Paz), 12h. (Riverview and Marseilles), 15h. (Colombo), 17h. (La Paz).

FIGURE 11 Two sample pages from the first issue of the "International Seismological Summary for 1918."

TABLE 1 Early Seismic Stations (Up to 1920) in Our Collections of Seismic Bulletins and Earthquake Lists

	Seismic Station	Coordinates		Country	Time Period
		Lat	Lon		
1	Alger, Bousareah	36.80	3.04	Algeria	1910–1920
2	Discovery	−77.84	166.75	Antarctica	1902–1903
3	Chacaritos (Chacarita, Buenos Aires, Buenos Ayres) #	−34.59	−58.48	Argentina	09.1906–10.1906 07.1908–1908
4	Córdoba #	−31.42	−64.20	Argentina	03.1899–01.1905
5	La Plata	−34.91	−57.93	Argentina	11.1907–1920
6	Pilar-Córdoba #, *	−31.67	−63.88	Argentina	03.1905–06.1907 1908–06.1909 1913–1914
7	Southern Andes (Mendoza)	−32.88	−68.85	Argentina	03.1908–1912
8	Adelaide #	−34.97	138.58	Australia	1909–06.1909 06.1910–1915
9	Cocos, Keeling Islands #	−12.20	96.90	Australia	1911, 1913–1914
10	Perth #, *	−31.95	115.84	Australia	10.1901–1911 1913–1915
11	Riverview	−33.83	151.16	Australia	03.1909–1920
12	Sydney #	−33.87	151.20	Australia	07.1906–1920
13	Austria			Austria	1896–1920
14	Graz	47.08	15.45	Austria	1907–1919
15	Innsbruck	47.26	11.38	Austria	1913–1915
16	Judenburg	47.17	14.67	Austria	1916
17	Kremsmünster *	48.06	14.13	Austria	1899–1908
18	Wien (Vienna) *	48.25	16.36	Austria	1905–1920
19	Baku (Bakou, Baky)	40.38	49.87	Azerbaijan	10.1906–1908 1910, 1912–1916
20	Balachany (Balakhan, Balakhany)	40.45	49.92	Azerbaijan	03.1907–1910
21	Šemakha (Chemakha, Shemakha)	40.63	48.63	Azerbaijan	1902–1916
22	Zurnabad (Zournabath, Zuraband)	40.52	46.27	Azerbaijan	12.1908–10.1909 1912
23	D'Uccle *	50.80	4.36	Belgium	1901–1920
24	La Paz	−16.27	−68.12	Bolivia	05.1913–1920
25	Bosnia and Herzegovina			Bosnia and Herzegovina	1896–1912
26	Sarajevo *	43.87	18.43	Bosnia and Herzegovina	1904–1913
27	Fernando de Noronha #	−3.83	−32.42	Brazil	03.1911–1915
28	Rio de Janeiro *	−22.91	−43.17	Brazil	1906–1920
29	Bulgaria			Bulgaria	1901–1912
30	Sofia (Sofija) *	42.69	23.33	Bulgaria	04.1905–1911 04.1916–1920
31	Manitoba (St. Boniface), Winnipeg	49.54	−97.07	Canada	02.1910–1912
32	Ottawa	45.39	−75.72	Canada	1906–?
33	Toronto #, *	43.66	−79.39	Canada	09.1897–1920
34	Victoria, British Columbia #, *	48.52	−123.42	Canada	10.1898–1915
35	St. Vincent (São Vicente) #	16.50	−24.00	Cape Verde	11.1910–06.1912 1913–1914
36	Copiapó	−27.35	−70.35	Chile	Before 1917–?
37	Chile			Chile	1906–1908
38	Santiago de Chile	−33.45	−70.66	Chile	1910–1917 1919–1920
39	Taihoku (Formosa/Taiwan) *	25.07	121.47	China (Taipei)	1897–1920
40	Tsingtau (Qingdao), Kiautschou (Jia zhou)	36.07	120.32	China (Beijing)	1909–03.1910 1911–1913
41	Zi-Ka-Wei (Hsu-chia-hui) *	31.18	121.43	China (Beijing)	1900–1920
42	San José	9.94	−84.08	Costa Rica	?–06.1903
43	Croatia and Slavonia			Croatia	1903, 1906–1920
44	Pula (Pola) *	44.87	13.85	Croatia	1896–1918
45	Rijeka (Fiume) *	45.34	14.39	Croatia	1906–1912
46	Zagreb (Agram) *	45.82	15.98	Croatia	1903, 1906–1920
47	Cheb (Eger)	50.08	12.38	Czech Republic	11.1908–06.1919

(continued)

TABLE 1 *(continued)*

	Seismic Station	Coordinates		Country	Time Period
		Lat	Lon		
48	Disko (Godhavn)	69.25	−53.53	Denmark (Greenland)	10.1907–05.1912
49	Quito	−0.20	−78.50	Ecuador	1904, 1906
50	Helwan, Abbasia (Abbassieh), Cairo #, *	29.86	31.34	Egypt	10.1899–1920
51	Tartu (Dorpat, Jurjew, Iouriev, Iouriev) *	58.38	26.72	Estonia	1897, 1902–09.1908
52	Fiji	−20.00	178.00	Fiji	1914
53	Besançon	47.25	5.99	France	Before 1917–?
54	Clermont-Ferrand	45.43	2.97	France	11.1913–?
55	France			France	1901–1919
56	Grenoble	45.19	5.74	France	1893–1906
57	Marseille	43.31	5.39	France	10.1910–?
58	Paris (Parc St. Maur)	48.81	2.49	France	06.1909–1920
59	Puy-de-Dôme	45.77	2.97	France	1910–10.1913
60	Strasbourg (Strassburg, Straßburg) #, *	48.58	7.76	France	02.1892–08.1893 04.1895–1897 07.1900–03.1904 05.1905–1916 1919–1920
61	Akhalkalaki	41.42	43.49	Georgia	04.1903–09.1909
62	Batumi (Batoum, Batum)	41.67	41.64	Georgia	04.1903–09.1909
63	Borjomi (Boržom, Borjom, Borshom)	41.85	43.39	Georgia	1903–09.1909
64	Tbilisi (Tblissi, Tiflis) #, *	41.72	44.79	Georgia	1900–1909 03.1910–1916
65	Aachen	50.78	6.08	Germany	09.1906–1914
66	Biberach	48.09	9.79	Germany	1914–1917
67	Bochum	51.49	7.23	Germany	12.1909–1921
68	Darmstadt, Jugenheim	49.76	8.65	Germany	1911–1914
69	Göttingen *	51.55	9.96	Germany	01., 07.1903–10.1914
70	Hamburg *	53.56	9.98	Germany	10.1900–1920
71	Heidelberg (Königstuhl)	49.40	8.73	Germany	06.1909–09.1916
72	Hohenheim	48.72	9.22	Germany	04.1905–1920
73	Jena *	50.95	11.58	Germany	04.1905–05.1913
74	Leipzig	51.34	12.39	Germany	03.1902–1910
75	München *	48.15	11.61	Germany	08.1905–1908 1911–1920
76	Nördlingen	48.85	10.49	Germany	1912–1920
77	Plauen	50.48	12.15	Germany	1906–1910
78	Potsdam #, *	52.38	13.07	Germany	04.1889–09.1889 1897–01.1898 01.–03.1899 04.1902–1920
79	Ravensburg	47.78	9.61	Germany	02.1919–1920
80	Taunus	50.22	8.45	Germany	07.1913–06.1914
81	Wilhelmshaven	53.53	8.15	Germany	04.1889–10.1889
82	Accra (Akkra)	5.52	−0.18	Ghana	1915
83	Athénai (Athens)	37.97	23.72	Greece	1900–10.1915 07.1918–1920
84	Port-au-Prince	18.56	−72.36	Haiti	1901–1920
85	Budapest *	47.48	19.02	Hungary	1902–11.1920
86	Hungary			Hungary	1885, 1886 1894–1913, 1920
87	Kalocsa	46.53	18.98	Hungary	1910–1920
88	Reykjavik	64.14	−21.91	Iceland	1910, 1913
89	Bombay (Colaba) #, *	18.90	72.82	India	09.1898–1920
90	Calcutta (Alipore) #, *	22.54	88.37	India	1899–06.1912 1913–1914
91	Kodaikánal #, *	10.23	77.47	India	02.1900–1920
92	Madras #	13.07	80.25	India	05.1898–11.1899
93	Simla	31.10	77.18	India	06.1905–11.1908 1914–1916

(continued)

TABLE 1 (*continued*)

	Seismic Station	Coordinates		Country	Time Period
		Lat	Lon		
94	Indonesia			Indonesia	1881–1920
95	Jakarta (Batavia) #, *	−6.18	106.84	Indonesia	06.1898–1920
96	Koeta Radja (Banda Atjeh, Kutaradja), Sumatra	5.57	95.25	Indonesia	1904
97	Cork #	51.88	−8.47	Ireland	1912–1918
98	Limerick	52.67	−8.63	Ireland	1913–1914
99	Benevento	41.12	14.8	Italy	02.1895–?
100	Casamicciola (Isola d'Ischia) #	40.75	13.90	Italy	02.1895–1897 04.1898–03.1899
101	Catania #, *	37.51	15.10	Italy	04.1895–03.1899 1908–06.1909 1911–1915
102	Chiavari	44.32	9.32	Italy	1913–1914
103	Firenze (Quarto-Castello) *	43.82	11.22	Italy	11.1898–1908
104	Firenze (Querce)	43.79	11.28	Italy	03.1895–1906 1911–1914, 1920
105	Firence (Ximeniano) *	43.78	11.26	Italy	02.1985–1920
106	Genova	44.42	8.92	Italy	02.1895–?
107	Italy			Italy	~ 1894–1914 1917–1920
108	Messina *	38.20	15.56	Italy	1904–1907
109	Mileto (Morabito)	38.60	16.05	Italy	1908–1912
110	Mineo (Guzzanti)	37.25	14.73	Italy	1895–1910
111	Moncalieri, close to Torino (Turin)	45.00	7.70	Italy	12.1906–1920
112	Monte Cassino (Montecassino)	41.49	13.82	Italy	1911, 1913–1917 1919
113	Padova (Padua)	45.40	11.87	Italy	02.1895–09.1897 1899, 01.–06.1901 01.–06.1902 1903–03.1904 1907–1915
114	Pavia *	45.18	9.17	Italy	02.1895–1897
115	Pisa (Capannoli, Baldini)	43.72	10.38	Italy	1910–1912
116	Portici (Napoli)	40.80	14.33	Italy	02.1895–?
117	Reggio di Calabria	38.10	15.65	Italy	1895–?
118	Rocca di Papa #, *	41.76	12.72	Italy	1895–03.1899 1915
119	Roma	41.90	12.48	Italy	1895–1897 1912–1920
120	Siena	43.32	11.31	Italy	1894–05.1896 01.–06.1897
121	Trieste #, *	45.65	13.76	Italy	1898–1903 1907–1918
122	Valle di Pompei (Pompei)	40.75	14.50	Italy	1908–1920
123	Venezia	45.43	12.34	Italy	1907–1908 1919–1920
124	Japan #			Japan	1899–04.1903
125	Kobe *	34.69	135.18	Japan	1919–1920
126	Mizusawa *	39.13	141.13	Japan	1902–1920
127	Nagasaki	32.73	129.87	Japan	04.1913–1920
128	Osaka *	34.68	135.52	Japan	1882–1920
129	Tokyo, Central Meteorological Office (Hitotsubashi) #	35.69	139.76	Japan	05.1895–12.1896 1898–1902, 1904 1920
130	Tokyo, University (Hongo) #, *	35.71	139.76	Japan	07.1899–1920
131	Wjernoje (Vernyï), Alma Ata	43.28	76.94	Kazakhstan	1906–1907
132	Inch'on (Tyosen, Zinsen, Chemulpo, Jinsen)	37.48	126.63	Korea	1918–1920
133	Beirut #	33.90	35.47	Lebanon	1904–06.1910 03.1911–1912 1914

(*continued*)

TABLE 1 (*continued*)

	Seismic Station	Coordinates		Country	Time Period
		Lat	Lon		
134	Ksara	33.82	35.89	Lebanon	1|910–1911 1913–1914
135	Valetta #	35.90	14.51	Malta	07.1906–1912 05.1914–04.1916
136	Mauritius (Pamplemousses) #, *	−20.09	57.55	Mauritius	09.1898–1920
137	Mazatlán	23.19	−106.39	Mexico	08.–12.1910
138	México	19.33	−99.19	Mexico	1888–1920
139	Oaxaca	17.02	−96.71	Mexico	08.–12.1910
140	Tacubaya *	19.40	−99.19	Mexico	1906–1912 1917, 1920
141	Yucatán	21.00	−89.00	Mexico	1905–1913
142	De Bilt	52.10	5.18	Netherlands	06.1904–10.1904 04.1908–1920
143	Christchurch #, *	−43.53	172.63	New Zealand	11.1901–1911, 1915
144	Wellington #, *	−41.29	174.77	New Zealand	10.1900–10.1912 1915–12.1920
145	Bergen *	60.38	5.33	Norway	06.1905–1920
146	Norway			Norway	1834–1920
147	Balboa Heights	8.96	−79.56	Panama	03.1915–1920
148	Lima #	−11.97	−77.03	Peru	06.1907–1908 01.–06.1911 1913–1914
149	Manila *	14.58	120.98	Philippines	1868–1895 1898–1920
150	Kraków (Krakau) *	50.06	19.94	Poland	12.1903–1917
151	Krietern-Breslau (Wroclaw)	51.12	17.04	Poland	02.1908–06.1908 09.1909–1910 1912–06.1914
152	Lisboa (Lisbon)	38.72	−9.13	Portugal	1920
153	Ponta Delgada, St. Miguel, Açores #, *	37.74	−25.69	Portugal	07.1903–1915
154	Portugal			Portugal	1903
155	Vieques #, *	18.15	−65.45	Puerto Rico	07.1904–1920
156	Bucharest (Bucarest, Bucureşti, Bucaresci)	44.41	26.10	Romania	05.1903–1911
157	Romania			Romania	1839–1909
158	Timisoara (Temesvar)	45.74	21.22	Romania	1906–1912
159	Derbent	42.00	48.40	Russia	04.1905–09.1909
160	Ekaterinburg (Ekaterinbourg, Jeckaterinburg, Sverdlovsk)	56.81	60.64	Russia	1907–09.1908 1913–1914
161	Irkutsk (Irkoutsk) #, *	52.27	104.31	Russia	12.1901–03.1909 1912–1915
162	Kabansk (Kamensk)	52.05	106.62	Russia	1904–09.1908
163	Königsberg (Kaliningrad)	54.83	20.50	Russia	04.1912–05.1914
164	Krasnojarsk (Krasnoïarsk)	56.02	92.87	Russia	1903–1906
165	Maritouy (Marituj)	51.76	104.11	Russia	11.–12.1908?
166	Pavlovsk (Pawlowsk, near St. Petersburg)	59.68	30.45	Russia	1903–1904 1913–07.1914
167	Pulkovo (Pulkovwa, near St. Petersburg)	59.77	30.32	Russia	12.1906–02.1907 11.1907–05.1908 1911–06.1914
168	Russia			Russia	1902–09.1908 1911–1912
169	Tchita (Tšita)	52.02	113.50	Russia	1904–09.1908
170	Apia *	−13.81	−171.78	Samoa	1905–1920
171	Mahé #	−4.61	55.49	Seychelles	06.1911–1914
172	Hurbanovo (Ógyalla, Stará Ďala) *	47.87	18.19	Slovakia	1902–1912
173	Ljubljana (Laibach) *	46.04	14.53	Slovenia	1897–06.1914
174	Cape of Good Hope (Capetown) #, *	−33.93	18.48	South Africa	07.1899–1915

(*continued*)

TABLE 1 *(continued)*

	Seismic Station	Coordinates		Country	Time Period
		Lat	Lon		
175	Cartuja (Granada) *	37.19	−3.60	Spain	1903–09.1916
176	Ebro (Tortosa) *	40.82	0.49	Spain	04.1905–06.1909 1910–1920
177	Fabra (Barcelona)	41.42	2.12	Spain	1907–03.1913 04.1914–1920
178	Puerto Orotava (Puerto de la Cruz), Tenerife	28.42	−16.52	Spain	12.1890–04.1891
179	Río Tinto #	37.77	−6.63	Spain	1911–1915
180	San Fernando #, *	36.47	−6.21	Spain	02.1898–1920
181	Spain			Spain	1899–1928
182	Urania	41.40	2.15	Spain	06.1913–07.1914
183	Colombo #	6.90	79.87	Sri Lanka	1906, 07.1909–1915
184	Uppsala (Upsala) *	59.86	17.63	Sweden	10.1904–05.1905 07.1906–1920
185	Switzerland			Switzerland	1879–1920
186	Zürich	47.37	8.58	Switzerland	11.1911–1920
187	Port of Spain (St. Clair) #, *	10.67	−61.50	Trinidad and Tobago	1901–09.1908 06.1909, 06.1912
188	Harpoot (Harput, Kharput, Kharpert, Elâziğ)	38.72	39.27	Turkey	1907–1909
189	Istanbul (Constantinople)	41.00	29.00	Turkey	1894–04.1896 01.–04.1897
190	Turkey			Turkey	1894–04.1896 01.–04.1897
191	Charkow (Kharkov)	50.00	36.25	Ukraine	04.1894–1897 1909
192	Czernovitz (Czernowitz, Tschernowzy)	48.30	25.93	Ukraine	04.1913–07.1914
193	Lwiw (Lemberg, Lwów)	49.82	24.03	Ukraine	1900–1903 08.1910–1920
194	Makejewka (Makejevka, Makeyevke)	48.03	37.98	Ukraine	03.1912–11.1913
195	Nicolajew (Nikolajew, Nikolaev) #	46.97	31.97	Ukraine	02.1892–1900 1902, 1904–09.1908
196	Uschhorod (Uschgorod, Ungvar, Uzhorod)	48.62	22.30	Ukraine	1911–1912
197	United Kingdom			United Kingdom	1913–1920
198	Ascension Island #	−7.95	−14.35	United Kingdom	11.1910–1914
199	Bidston (Liverpool) #, *	53.40	−3.07	United Kingdom	09.1897–03.1899 1901–1919
200	Duce (Aberdeen)	57.22	−2.17	United Kingdom	1914–1915
201	Blackford Hill (Edinburgh) #, *	55.93	−3.18	United Kingdom	08.1896–04.1898 12.1900–1915
202	Eskdalemuir #	55.31	−3.21	United Kingdom	07.1909–1916 1920
203	Haslemere (Hazlemere, Frensham Hall Observatory) #	51.08	−0.72	United Kingdom	12.1906–1915
204	Kew (London) #, *	51.47	−0.31	United Kingdom	04.1898–1915
205	Newport, Gwent	51.59	3.00	United Kingdom	1914–1915
206	Paisley (Coats Observatory) #, *	55.85	−4.43	United Kingdom	1902–1913
207	Shide, Isle of Wight #, *	50.69	−1.29	United Kingdom	08.1895–1915
208	St. Helena #	−15.92	−5.73	United Kingdom	02.1911–06.1911 1914
209	Stonyhurst College (Blackburn) #	53.84	−2.47	United Kingdom	07.1909–1918 1920
210	United Kingdom			United Kingdom	1889–1907
211	Warley (Birmingham)	52.47	1.88	United Kingdom	1914
212	West Bromwich #	52.52	−1.98	United Kingdom	08.1908–06.1912 1913–1915
213	Woodbridge Hill, Guildford #	51.25	−0.59	United Kingdom	1910–1915
214	Albany, New York	42.65	−73.76	USA	1906–07.1912
215	Ann Arbor, Michigan (Detroit Observatory)	42.28	−83.73	USA	08.1909–1915

(continued)

TABLE 1 *(continued)*

	Seismic Station	Coordinates Lat	Lon	Country	Time Period
216	Baltimore #, *	39.30	−76.62	USA	04.1901–06.1911 1913
217	Berkeley *	37.87	−122.26	USA	1906, 1910–1920
218	Buffalo (Canisius), New York	42.93	−78.85	USA	10.1910–1911 1915–1920
219	Cambridge, Massachusetts (Harvard University)	42.38	−71.12	USA	04.1908–09.1920
220	Cheltenham, Maryland #, *	38.73	−76.85	USA	12.1904–1920
221	Cleveland, Ohio	41.49	−81.71	USA	1911–03.1916
222	Denver (Sacred Heart), Colorado	39.68	−105.00	USA	10.1909–01.1910 1911–1920
223	Fordham, New York	40.96	−73.89	USA	1915–02.1918
224	Georgetown, Washington, D.C.	38.91	−77.07	USA	1915–1920
225	Honolulu, Hawaiian Islands #, *	21.32	−158.06	USA	04.1903–1920
226	Honolulu, Oahu College, Hawaii #	21.3	−157.9	USA	12.1901–01.1901
227	Ithaca, New York	42.45	−76.48	USA	1912–08.1917
228	Kilauea, Hawaii	19.42	−155.29	USA	From 1914
229	Lawrence, Kansas	38.96	−95.25	USA	03.1915–1920
230	Mobile, Alabama (Spring Hill)	30.69	−88.14	USA	1910–1912 10.1918–1920
231	Mt. Hamilton, California (Lick Observatory)	37.34	−121.64	USA	1901, 06.1903, 1904 05.1911–1915
232	New Orleans (Loyola), Louisiana	29.95	−90.12	USA	1911
233	Northfield, Vermont	44.17	−72.68	USA	1915–1920
234	Santa Clara, California	37.44	−121.95	USA	?.1910–1919
235	Seattle, Washington	47.65	−122.31	USA	From 03.1910
236	Sitka, Alaska #, *	57.06	−135.50	USA	05.1904–1920
237	Spokane (Gonzaga), Washington	47.64	−117.46	USA	?.1910–09.1911
238	St. Louis, Missouri	38.63	−90.23	USA	12.1910–1920
239	Tucson, Arizona	32.31	−110.78	USA	09.1910–1920
240	USA			USA	1897–1906
241	Washington, D.C. *	38.96	−77.06	USA	06.1903, 1904 1906–1907 10.1914–1920
242	Tashkent (Taschkent, Taškent) *	41.33	69.30	Uzbekistan	1902–1909 09.1912–05.1914
243	Belgrad (Beograd, Belgrade) *	44.82	20.46	Yugoslavia	1901–10.1911
244	Serbia			Yugoslavia	1901–1907
245	Binza (Leopoldville)	−4.37	15.25	Zaire	1909–1920

This table contains our knowledge (as of December 2002) about seismic bulletins and earthquake lists. The time period is approximately from the beginning of instrumental readings until the end of 1920, as far as we have them in originals or in copies or as we found them somewhere cited in literature or listed in library catalogs. The names of the seismic stations are listed under their respective countries where they are located today, with the countries alphabetical. However, the names of stations and towns/villages also changed over time. In these cases, we tried to give all name versions used in the bulletins and the current names. For some entries, a country name is used as the station name, which means that during this time period, centralized earthquake lists and/or bulletins were published. The references found and further remarks for each entry are given in the Appendix, Notes for Table 1, at the end of this chapter.

All stations marked with a pound sign (#) contributed their readings at least once to the early bulletins published by the BAAS (BAASRP, 1896–1899, and Shide Circulars, 1899–1912). All stations marked with a star (*) observed the 1906 San Francisco earthquake and contributed a seismogram copy to Reid (1908). A very valuable resource to get an overview of bulletin production and research activities at North American seismic observatories was the 25th-year anniversary commemorative volume of the Jesuit Seismological Association, edited by James Bernard Macelwane, which contains a complete bibliography related to their 18 seismic stations in the United States and Canada (Macelwane, 1950).

The 1906 San Francisco earthquake was observed worldwide at most of the seismic stations. The State Earthquake Investigation Commission compiled observations of this earthquake from 96 stations and published them in a special report (1908) together with copies of the seismograms and station descriptions. We indicate in Table 1 all stations used in this report with an asterisk. However, as mentioned before, not all stations were regularly publishing bulletin data; therefore the list of Reid (1908, Volume II, page 61) contains 18 stations for which we so far have not been able to find bulletins or other

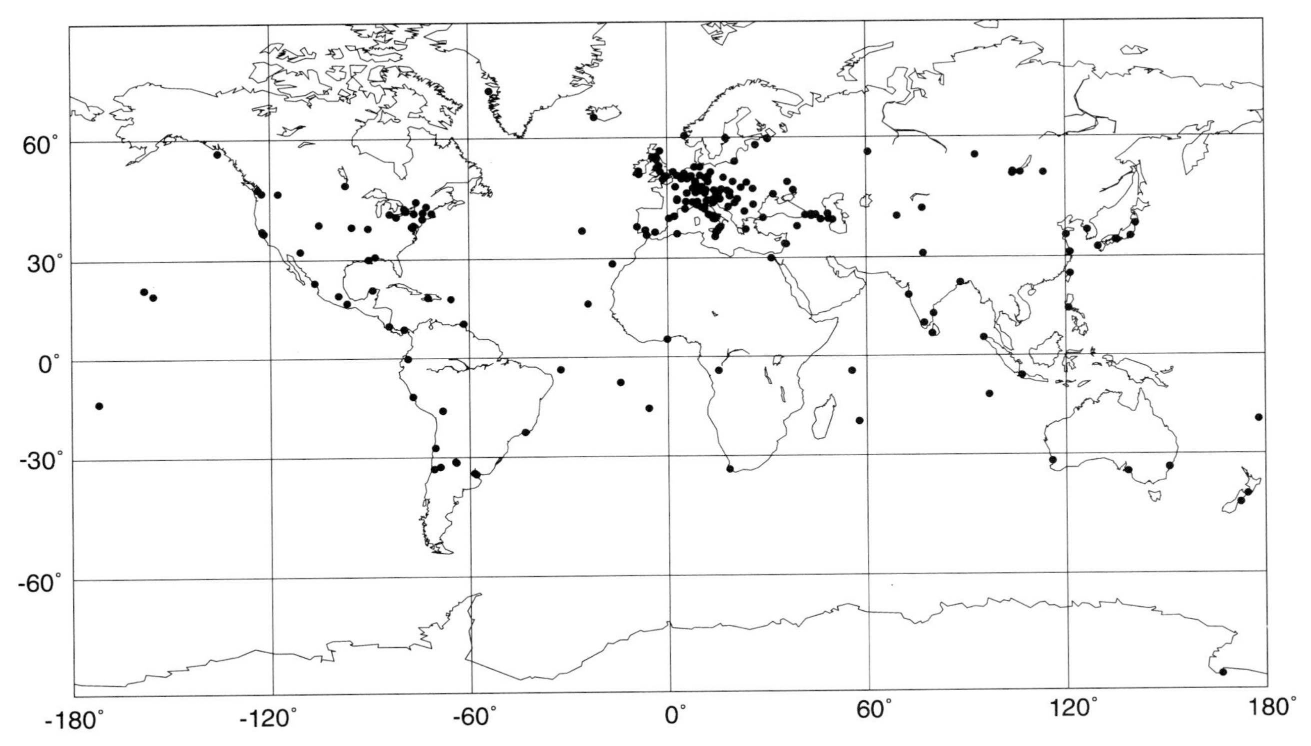

FIGURE 12 A map showing all stations for which we could locate seismic bulletin material with earthquake observations from before 1921.

related material. Figure 12 shows a map with all stations for which we have located bulletin materials (see Table 1 for further details).

5. Samples of Old Seismic Bulletins on the Attached Handbook CD #3

On the attached Handbook CD #3 (under the directory of \88Schweitzer), we included most of the critical materials from the old seismic bulletins up to 1920. Although they constitute only a fraction of the old bulletin materials, these critical materials provide a significant data set of instrumental observations from the 1880s to 1920. Our selected materials are mostly published in English, but some important materials in French, German, Japanese, and Russian are also included.

Our selection criteria are rather subjective, mostly from the information density consideration (i.e., we tried to archive as many seismic readings as possible given a finite amount of space on the CD-ROM). We prefer (1) compilations of multiple station bulletins to single station bulletins and (2) published bulletins to handwritten logs, mimeographed sheets, or preliminary bulletins. However, some seismic bulletins from a few stations of important geographical locations are also included.

5.1 Selections from the BAAS Publications

From the BAAS publications, we selected the following items for scanning and archived the images as computer-readable files on the Handbook CD #3 under the directory of \88Schweitzer:

- "Catalogues of Earthquakes recorded at the Central Meteorological Observatory in Tokio" in the annual BAAS volumes. In 1883, the first of the Gray-Milne seismographs was constructed and began operation in Tokyo, Japan. These catalogs of seismic readings were included in the reports of the BAAS Committee for Investigation of Earthquake Phenomena of Japan, and then in the reports of the BAAS Committee for Seismological Investigation. Only the first catalog was scanned and archived as an example.
- "Report of the British Association for the Advancement of Sciences, Committee on Earth Tremors" (BAAS, 1893). This report contains an appendix, "Account of Observations Made with the Horizontal Pendulum," by Ernst von Rebeur-Paschwitz. This account in English is of great historical interest concerning the first identification of a teleseismic event.
- "Reports of the British Association for the Advancement of Sciences, Committee on Seismological Investigation," 1st Report (1896) to 18th Report (1913).

- "Shide Circulars" for 1899–1912, published by John Milne.
- "Bulletins of the British Association for the Advancement of Sciences—Seismological Committee" for 1913–1917.

5.2 Selections from the ISA Publications

These samples include the bulletin publications of the International Seismological Association before World War I:

- Emil Rudolph (Rudolph, 1903a, 1903b), with seismogram readings for 1895–1898.
- Bulletins published by the ISA for 1904–1908.

5.3 Selection from the ISS

We include on the Handbook CD #3 the first 3 years of the "International Seismological Summary" (1918–1920).

5.4 Selection from the National Bulletin of Italy

From the year 1895 on, the "Regio Ufficio Centrale di Meteorologia e Geodinamica" in Rome started to publish all earthquake observations (microseismical and macroseismical) in Italy as a supplement to the "Bollettino della Società Sismologica Italiana." To document this early national effort, we include on the Handbook CD #3 the whole first volume of this bulletin, containing local, regional, and teleseismic observations in Italy during 1895 (Baratta, 1895).

5.5 Selections from the Publications of the Russian Permanent Central Seismic Commission

In Russia, the Permanent Central Seismic Commission issued a series of publications from 1903 to 1919. We selected all the seismic bulletins for the years 1902, 1911, and 1912. Their data for 1903–1907 were included in the ISA publications. For the year 1908, we have so far found only an incomplete copy, missing the data from the last quarter of the year. We could not find any nationwide Russian seismic bulletins for 1909–1910 or for 1913–1920.

5.6 Selections from the US Weather Bureau Publications

The US Weather Bureau had the official responsibility for earthquake monitoring in the United States from the end of 1914 until 1925 (when the task was transferred to the US Coast and Geodetic Survey). Seismic readings of US stations, as well as of stations in Canada, Puerto Rico, and the Panama Canal Zone, were published monthly in the "Monthly Weather Review." We have selected the Seismology Section of the "Monthly Weather Review" by the US Weather Bureau (1914–1920).

5.7 Selections from Single Station Bulletins

We also included, as examples of single station bulletins, the listings of Göttingen (Germany), Honolulu (Hawaii), La Paz (Bolivia), Osaka (Japan), Sitka (Alaska), and Zi-Ka-Wei (China). We hope to publish all our collected seismic bulletins on CDs in the future after the collection is more complete and the scanning work is finished.

Acknowledgments

We thank all the following friends, colleagues, and station operators who helped us complete the collection of old bulletins by searching their libraries or archives and copying and shipping requested issues: Kayihan Aric, Manfred Baer, Josep Batlló, Wolfgang Brüstle, Thorsten Büßelberg, José Martín Davila, Bruno De Simoni, Bernard Dost, Lawrence Drake, Torild van Eck, Misha Elashvili, Graziano Ferrari, Edmund Fiegweil, Silvia Filosa, Paul Henni, Rolf Herber, Diethelm Kaiser, Rainer Kind, Michael Korn, Ota Kulhanek, Wolfgang Lenhardt, Thomas Meier, Luis Alberto Mendes-Victor, Jose Morales-Sato, Claudia Piromallo, Ines Rio, Joachim Ritter, Eberhard Schmedes, Götz Schneider, Dmitry Storchak, Bernd Tittel, Liane Tröger, Arantza Ugalde, Roland Verbeiren, Erhard Wielandt, and Mladen Živčić. We also wish to thank the staff of the SISMOS Project ("Istituto Nazionale di Geofisica e Vulcanologia"), for allowing us to publish their scanned version of the Italian bulletin of 1895 on the attached Handbook CD#3; Ritsuko S. Matsu'ura, for supplying us with an image file of Omori (1905); and Carl Kisslinger, for making comments on an earlier draft. This is NORSAR contribution No. 717.

References

Abe, K. (1994). Instrumental magnitudes of historical earthquakes, 1892 to 1898. *Bull. Seismol. Soc. Am.*, **84**, 415-425.

BAASSC: See British Association for the Advancement of Science, Seismological Committee (1913-1917).

Baratta, M. (1895). *Notizie sui terremoti avvenuti in Italia durante l'anno 1895.* Regio Ufficio Centrale di Meteorologia e Geodinamica, 230 pp., appendix to *Bollettino della Società Sismologica Italiana*, **1**.

Berlage, H.P. Jr. (1930). Seismometer. In: *Handbuch der Geophysik, Band 4: Erdbeben.* (Gutenberg, B., Ed. and Au.). Verlag Bornträger, Berlin, 299-526.

Borne, G. von dem (1904). Seismische Registrierungen in Göttingen, Juli bis Dezember 1903. *Nachrichten von der Königlichen Gesellschaft der Wissenschaften zu Göttingen, Mathematisch-physikalische Klasse*, 440-464.

British Association for the Advancement of Science, Committee on Seismological Investigations (1896). First Report. In: Report of the 1896 Meeting of the British Association for the Advancement of Science, **76**, 180-230. [BAASRP 1896]

British Association for the Advancement of Science, Committee on Seismological Investigations (1897). Second Report. In: Report of the 1897 Meeting of the British Association for the Advancement of Science, **77**, 129-206. [BAASRP 1897]

British Association for the Advancement of Science, Committee on Seismological Investigations (1898). Third Report. In: Report of the 1898 Meeting of the British Association for the Advancement of Science, **78**, 179-276. [BAASRP 1898]

British Association for the Advancement of Science, Committee on Seismological Investigations (1899). Fourth Report. In: Report of the 1899 Meeting of the British Association for the Advancement of Science, **79**, 161-238. [BAASRP 1899]

British Association for the Advancement of Science, Seismological Committee (BAASSC) (1900-1912). Circular Nos. 1-27. These circulars are generally known as the "Shide Circulars."

British Association for the Advancement of Science, Seismological Committee (BAASSC) (1913-1917). [Monthly] Bulletin, 3-month issues between January 1913 and December 1917.

Christensen, A., and G. Ziemendorff (1909). *Les tremblements de terre ressentis pendant l'année 1905*. Publications du Bureau Central de l'Association Internationale de Sismologie, Série B, Catalogues, Strasbourg, 543 pp.

Dewey, J., and P. Byerly (1969). The early history of seismometry (to 1900). *Bull. Seismol. Soc. Am.*, **59**, 183-227.

Ehlert, R. (1898). Zusammenstellung, Erläuterung und kritische Beurtheilung der wichtigsten Seismometer mit besonderer Berücksichtigung ihrer praktischen Verwendbarkeit. *Beiträge zur Geophysik*, **3**, 350-475. This paper is reproduced on the attached Handbook CD #3 (under the directory of \88 Schweitzer).

Ferrari, G., editor (1990). *Gli strumenti sismici storici Italia e contesto europeo—historical seismic instruments, Italy and the European framework*. Istituto Nazionale di Geofisica, Bologna, 198 pp.

Ferrari, G., editor (1992). *Two hundred years of seismic instruments in Italy 1731-1940*. Istituto Nazionale di Geofisica, Bologna, 156 pp.

Galitzin (Golicyn), B. (1910). Ueber einen neuen Seismographen für die Vertikalkomponente der Bodenbewegung. Separatdruck aus den *Nachrichten der Seismischen Kommission*, **IV**, part 2 (*Comptes Rendus des Séances de La Commission Sismique Permanente*, **Tome 4**, Livraison II), 34 pp.

Gutenberg, B., and C.F. Richter (1954). *Seismicity of the Earth and Associated Phenomena*, 2nd edition, revised. Princeton University Press, Princeton, New Jersey, ix + 310 pp.

Henni, P.H.O., J.H. Lovell, and K.I. Lawrie (1999). Seismological bulletins held in the National Seismological Archive (NSA). Version 2. British Geological Survey, Technical Report WL/99/16, Edinburgh.

Henni, P.H.O., J.H. Lovell, and K.I.G. Lawrie (2000). UK historical seismograms and bulletins held in the NSA. British Geological Survey, Technical Report WL/99/21, Edinburgh.

Herrmann, R.B., M. Whittington, and H. Meyers (1983). Historical station bulletin microfilming project, preliminary inventory. National Geophysical Data Center, NOAA, Boulder, Colorado, 59 pp.

International Seismological Summary for ..., yearly volumes 1918-1962.

ISA: See Christensen and Ziemendorff (1908), Lais (1913), Oddone (1907), Rosenthal (1907), Scheu (1911a,b, 1917), Scheu and Lais (1912), Sieberg (1917), Szirtes (1908, 1909a,b, 1910a,b, 1912a,b, 1913a,b).

ISS: See International Seismological Summary.

Kanamori, H. (1988). Importance of historical seismograms for geophysical research. In: *Historical Seismograms and Earthquakes of the World*, (W.H.K. Lee, H. Meyers, and K. Shimazaki, Eds.). Academic Press, San Diego, pp. 16-23.

Lais, R. (1913). *Catalogue général des tremblements de terre, ressentis par l'homme et registres par des instruments pendant l'année 1907*. Publications du Bureau Central de l'Association Internationale de Sismologie, Série B, Catalogues, Strasbourg.

Lee, W.H.K., H. Meyers, and K. Shimazaki, editors (1988). *Historical Seismograms and Earthquakes of the World*. Academic Press, San Diego. This out-of-print book is reproduced as computer-readable file on the attached Handbook CD #1 (under the directory of \17Lee1).

Lovell, J.H., and P.H.O. Henni (1999). *Historical seismological observatories in the British Isles (pre-1970)*. Version 3. British Geological Survey, Technical Report WL/99/13, Edinburgh.

Macelwane, J.B., editor (1950). *Jesuit Seismological Association 1925-1950, Twenty-Fifth Anniversary Commemorative Volume*. Central Station, Saint Louis University, St. Louis, MO, xi + 347 pp.

Masó, P. Miguel Saderra (1895). *La Seismología en Filipinas—Datos para el estudio de terremotos del Archipiélago Filipino*. Observatorio de Manila, Dirigido por los Padres de la Compañía de Jesús. Tipo-Litográfico de Ramírez y Compañía, Manila, 124 pp. This book is reproduced as computer-readable file on the attached Handbook CD #3 (under the directory of \88Schweitzer).

Milne, J. (1883). Report of the Committee [on the earthquake phenomena of Japan]. In: Report of the Fifty-Third Meeting of British Association for the Advancement of Science Held at Southport in September 1883, pp. 211-215.

Oddone, E. (1907). *Les tremblements de terre ressentis pendant l'année 1904*. Publications du Bureau Central de l'Association Internationale de Sismologie, Série B, Catalogues, Strasbourg, xii + 361 pp.

Omori, F. (1905). On the Formosa earthquakes. Report of the Imperial Earthquake Investigation Committee, No. 54, 223 pp., in Japanese.

Rebeur-Paschwitz, E. von (1889). The earthquake of Tokio, April 18, 1889. *Nature*, **40**, 294-295. This paper is reproduced in Chapter 79.24.1 on Handbook CD #2 by Schweitzer.

Rebeur-Paschwitz, E. von (1892). Das Horizontalpendel und seine Anwendung zur Beobachtung der absoluten und relativen Richtungs-Aenderungen der Lothlinie. *Nova Acta der Ksl. Leop.-Carol. Deutschen Akademie der Naturforscher,* **Band LX**, Nr. 1, Halle, 216 pp.

Rebeur-Paschwitz, E. von (1895). Horizontalpendel-Beobachtungen auf der Kaiserlichen Universitäts-Sternwarte zu Strassburg 1892-1894. *Beiträge zur Geophysik,* **2**, 211-536.

Reid, H.F. (1908). *Mechanics of Earthquakes*. Volume II of: *State Earthquake Investigation Commission (Andrew C. Lawson, Chairman). The California earthquake of April 18, 1906*. Report in two volumes and atlas, Carnegie Institution of Washington, Washington. (This volume was photographically reprinted in 1969.)

Rosenthal, E. (1907). *Katalog der im Jahre 1904 registrierten seismischen Störungen*. Publications du Bureau Central de l'Association Internationale de Sismologie, Série B, Catalogues. Veröffentlichungen des Zentralbureaus der Internationalen Seismologischen Assoziation. Serie B, Kataloge. Strassburg, xii + 145 pp.

Rothé, J.-P. (1981). Fifty years of history of the International Association of Seismology (1901-1951). *Bull. Seismol. Soc. Am.*, **71**, 905-923.

Rudolph, E. (1903a). Die Fernbeben des Jahres 1897. *Beiträge zur Geophysik*, **5**, 1-93.

Rudolph, E. (1903b). Seismometrische Beobachtungen. *Beiträge zur Geophysik*, **5**, 94-169.

Rudolph, E. (1904). Seismometrische Beobachtungen über japanische Fernbeben in den Jahren 1893–1897. *Beiträge zur Geophysik*, **6**, 377-434.

Rudolph, E. (1905). *Katalog der im Jahre 1903 bekannt gewordenen Erdbeben. Beiträge zur Geophysik*, ***Ergänzungsband III***. Verlag Wilhelm Engelmann, Leipzig, xvii + 674 pp.

Rudolph, E. (1907). Ostasiatischer Erdbebenkatalog. *Beiträge zur Geophysik*, **8**, 114-218.

Scheu, E. (1911a). *Catalogue général des tremblements de terre, ressentis par l'homme et registres par des instruments pendant l'année 1906*. Publications du Bureau Central de l'Association Internationale de Sismologie, Série B, Catalogues, Strasbourg, 45 pp.

Scheu, E. (1911b). *Catalogue régional des tremblements de terre ressentis pendant l'année 1906*. Publications du Bureau Central de l'Association Internationale de Sismologie, Série B, Catalogues, Strasbourg, 112 pp.

Scheu, E. (1917). *Catalogue régional des tremblements de terre ressentis pendant l'année 1908*. Publications du Bureau Central de l'Association Internationale de Sismologie, Série B, Catalogues, Strasbourg, 194 pp.

Scheu, E., and R. Lais (1912). *Catalogue régional des tremblements de terre ressentis pendant l'année 1907*. Publications du Bureau Central de l'Association Internationale de Sismologie, Série B, Catalogues, Strasbourg, 123 pp.

Shide Circular: See British Association for the Advancement of Science.

Sieberg, A. (1904). *Handbuch der Erdbebenkunde*. Vieweg Verlag, Braunschweig, xvii + 360 pp.

Sieberg, A. (1917). *Catalogue général des tremblements 1908*. Publications du Bureau Central de l'Association Internationale de Sismologie, Série B, Catalogues, Strasbourg, 69 pp.

Szirtes, S. (1908). *Coordonnées des stations sismiques du globe et tableaux auxiliaires pour les calculs sismiques*. Publications du Bureau Central de l'Association Internationale de Sismologie, Série A, Mémoires. Veröffentlichungen des Zentralbureaus der Internationalen Seismologischen Assoziation. Serie A, Abhandlungen. Strassburg, 23 pp.

Szirtes, S. (1909a). *Katalog der im Jahre 1905 registrierten seismischen Störungen I. Teil*. Publications du Bureau Central de l'Association Internationale de Sismologie, Série B, Catalogues. Veröffentlichungen des Zentralbureaus der Internationalen Seismologischen Assoziation. Serie B, Kataloge. Strassburg, iv + 193 pp.

Szirtes, S. (1909b). *Katalog der im Jahre 1905 registrierten seismischen Störungen II. Teil*. Publications du Bureau Central de l'Association Internationale de Sismologie, Série B, Catalogues. Veröffentlichungen des Zentralbureaus der Internationalen Seismologischen Assoziation. Serie B, Kataloge. Strassburg, vii + 68 pp.

Szirtes, S. (1910a). *Katalog der im Jahre 1906 registrierten seismischen Störungen, I. Teil, Die schwächeren und weniger ausgeprägten Störungen (III B)*. Publications du Bureau Central de l'Association Internationale de Sismologie, Série B, Catalogues. Veröffentlichungen des Zentralbureaus der Internationalen Seismologischen Assoziation. Serie B, Kataloge. Strassburg, iv + 110 pp.

Szirtes, S. (1910b). *Katalog der im Jahre 1906 registrierten seismischen Störungen, II. Teil, Die grossen und gut ausgeprägten Störungen (III B)*. Publications du Bureau Central de l'Association Internationale de Sismologie, Série B, Catalogues. Veröffentlichungen des Zentralbureaus der Internationalen Seismologischen Assoziation. Serie B, Kataloge. Strassburg, viii + 86 pp.

Szirtes, S. (1912a). *Katalog der im Jahre 1907 registrierten seismischen Störungen*. Publications du Bureau Central de l'Association Internationale de Sismologie, Série B, Catalogues. Veröffentlichungen des Zentralbureaus der Internationalen Seismologischen Assoziation. Serie B, Kataloge. Strassburg, x + 120 pp.

Szirtes, S. (1912b). *Registrierungen der besser ausgeprägten seismischen Störungen des Jahres 1907, Ergänzung zum seismischen Katalog*. Publications du Bureau Central de l'Association Internationale de Sismologie, Série B, Catalogues. Veröffentlichungen des Zentralbureaus der Internationalen Seismologischen Assoziation. Serie B, Kataloge. Strassburg, v + 111 pp.

Szirtes, S. (1912c). Geographische Koordinaten der seismischen Stationen nebst Hilfstabellen. *Gerlands Beiträge zur Geophysik*, **11**, Kleine Mitteilungen, 177-199.

Szirtes, S. (1913a). *Katalog der im Jahre 1908 registrierten seismischen Störungen*. Publications du Bureau Central de l'Association Internationale de Sismologie, Série B, Catalogues. Veröffentlichungen des Zentralbureaus der Internationalen Seismologischen Assoziation. Serie B, Kataloge. Strassburg, viii + 133 pp.

Szirtes, S. (1913b). *Registrierungen der besser ausgeprägten seismischen Störungen des Jahres 1908, Ergänzung zum seismischen Katalog*. Publications du Bureau Central de l'Association Internationale de Sismologie, Série B, Catalogues. Veröffentlichungen des Zentralbureaus der Internationalen Seismologischen Assoziation. Serie B, Kataloge. Strassburg.

Tams, E. (1950). Materialien zur Geschichte der deutschen Erdbebenforschung bis zur Wende des 19. zum 20. Jahrhundert Teil II und Teil III. *Mitteilungen des Deutschen Erdbebendienstes*, **Sonderheft 2**, Akademie Verlag, Berlin 1950, 169 pp.

U.S. Weather Bureau (US-WB). Monthly Weather Review. [Note: A monthly issue of mostly meteorological data. A "Seismology Section" began in the December 1914 issue, and from 1915 through 1924 contained the non-instrumental and instrumental observations by stations in the United States (including Hawaii, Puerto Rico, and Panama Canal Zone) and Canada for the years 1914–1920.]

Weigand, B. (1902). Ausbreitung der mikroseismischen Beobachtungen. In: *Verhandlungen der vom 11. bis 13. April 1901 zu Straßburg abgehaltenen Ersten Internationalen Seismologischen Konferenz*. (Rudolph, E. Ed.). *Beiträge zur Geophysik*, **Ergänzungsband I**, Wilhelm Engelmann, Leipzig, 1902, viii + 439 pp., 183-188.

Wiechert, E. (1899). Seismometrische Beobachtungen im Göttinger Geophysikalischen Institut. *Nachrichten von der Königl. Gesellschaft der Wissenschaften zu Göttingen, Mathematisch-physikalische Klasse*, 195-208.

Willemann, R.J., and D.A. Storchak (2001). Data collection at the International Seismological Centre. *Seismol. Res. Lett.*, **72**, 440-461.

Appendix Notes for Table 1

The following Notes contain references for the listed seismic stations in Table 1, as far as the bulletins as originals or as copies could be checked or additional bibliographic references could be found. The Notes are numbered, corresponding to the numbers shown in Table 1. References in *italics* mean that this bulletin or publication is located somewhere but that we have no copy in our collection. One may deduce from volume numbers and bulletin titles that additional material may also have existed. However, if we have not seen originals or any type of citation, we have not made such conclusions and have omitted such entries.

The phrase *BGS* means that these bulletins can be found in the National Seismological Archive of the British Geological Service (Henni *et al.*, 1999; Lovell and Henni, 1999; Henni *et al.*, 2000); the phrase *in Hamburg* means that these bulletins were located in the collection of the seismological station of the University in Hamburg, Germany; the phrase *at ISC* means that these bulletins were located in the collection at the International Seismological Centre in Thatcham, England; the phrase *at KNMI* means that these bulletins were located in the library of the KNMI (Royal Netherlands Meteorological Institute) in De Bilt, The Netherlands; and the phrase *NOAA* means that these bulletins were listed in the report of the microfilm project of Herrmann *et al.* (1983). [Note: After the microfilming, the NOAA collection was sent to the US Geological Survey, Menlo Park. However, only a small fraction of this collection was saved by W.H.K. Lee at his home at present.]

The bulletins of the ISA (1904–1908) contain reference lists, and the BAASSC contains listings of contributing stations for the time period 1913–1915. Such references were used as information that a station bulletin for specific time periods existed. However, these references were only given here when no other (complete) bulletin information was available.

We cannot exclude the possibility that, for some stations, the different references given for the same time period will later show up again because the quality and style of found citations changed with time. This list of references reflects the international nature of seismology; we found bulletins published in numerous different languages. Although we tried to cite as accurately as possible, we are sure that we copied errors from old citation lists and may also have added new errors. This will be true in particular for page numbers, spelling errors, and missing accents in languages unknown to the authors. However, we also have examples in which today's spelling is different from that of 100 years ago.

1. Alger (Algiers, Bousareah)

Since 1910 included in "Annales du Bureau ..." of France, see 55.
1911–1919 included in "Bulletin mensuel du ..." of France, see 55.
03.1919–12.1920 included in Rothé (1920, 1922), see 60.

2. Discovery

Bernacchi, L.C., and J. Milne (1908). Earthquakes and other earth movements recorded in the Antarctic regions, 1902–1903. In "National Antarctic Expedition 1901–1904. Physical Observations." Royal Society, London, 1908, 37-96.

3. Chacaritos (Chacarita, Buenos Aires, Buenos Ayres)

At ISC: 08.–09.1906.
Shide Circular **20**, 325-326 (1908-07-06–1908-12-28).

4. Córdoba

Shide Circular **2**, 48-57 (1899-03-06–1900-06-21); **6**, 172-181 (1900-06-26–1901-12-21); **8**, 259-266 (1902-01-07–1903-06-30); **10**, 315-318 (1903-07-01–1903-12-28); **14**, 85-86 (1904-07-01–1905-01-19); **17**, 209-210 (1904-01-03–1904-06-30).

5. La Plata

Negri, G. (1909). Organización del servicio sísmico y sus primeros resultados (Noviembre de 1907 á Diciembre de 1908). Obs. Astr. de la Universidad Nacional de La Plata. N. S. ***2****, 54 pp.*
NOAA: 1907–1920.

6. Pilar-Córdoba

Shide Circular **14**, 86-88 (1905-03-02–1905-12-29); **15**, 128-134 (1906-01-02–1906-12-25); **17**, 210-211 (1907-01-02–1907-06-26); **20**, 327-331 (1908-01-06–1909-06-29). Used in the BAASSC (1913–1917) 1913–1914.

7. Southern Andes (Mendoza)

Loos, P.A. (1907). Estudios de seismología. Los movimientos seísmicos en Mendoza. Anales del Ministerio de Agricultura ***3****, 38 pp.*
Loos, P.A. (1908). Informe sobre los movimientos seísmicos observados en las provincias de Mendoza y San Juan, durante el mes de marzo de 1908. B. del Ministerio de Agricultura ***9****, 237-243.*
Loos, P.A. (1910). Meteorología y Seismologia. La Viticultura Argentina ***1****, 43-50. Boletín de la Sociedad Sismológica Sud-Andina: publicado bajo la Protección de los Gobiernos de San Juan y de Mendoza 1912.*

8. Adelaide

Shide Circular **23**, 41-44 (1909-01-01–1909-06-29); **25**, 157-158 (1911-07-03–1911-12-31); **26**, 222-225 (1910-07-03–1911-06-28); **26**, 225-226 (1912-01-01–1912-06-14); **27**, 276-277 (1912-07-07–1912-12-28).
Used in the BAASSC (1913–1917) 1913–1915.

9. Cocos, Keeling Islands

Shide Circular **24**, 91 (1911-01-01–1911-06-15); **25**, 155-156 (1911-07-12–1911-12-31).
Used in the BAASSC (1913–1917) 1913–1914.

10. Perth

Shide Circular **5**, 140 (1901-10-08–1901-12-31); **6**, 167 (1902-01-01–1902-06-30); **7**, 216-217 (1902-07-02–1902-12-25);

8, 247-248 (1903-01-02–1903-06-09); **9**, 290 (1903-08-12–1903-12-31); **10**, 311 (1904-01-17–1904-06-27); **11**, 343-344 (1904-07-04–1904-12-23); **12**, 21-23 (1905-01-08–1905-06-30); **13**, 56-58 (1905-07-01–1905-12-21); **14**, 95-97 (1906-01-03–1906-06-28); **15**, 147-151 (1906-07-02–1906-12-26); **16**, 177 (1907-01-01–1907-06-27); **17**, 218-219 (1907-07-09–1907-12-25); **18**, 248-249 (1908-01-04–1908-05-20); **19**, 283-284 (1908-07-03–1908-12-30); **20**, 333-334 (1909-01-03–1909-06-28); **22**, 412-413 (1909-07-04–1909-12-23); **22**, 413 (1910-01-30–1910-06-29); **23**, 44 (1910-07-07–1910-12-30); **24**, 95 (1910-03-11–1910-06-29); **24**, 95 (1911-01-02–1911-06-28); **25**, 156-157 (1911-07-04–1911-12-31).

06. + 12.1903 used in Strasbourg's "Von den Instrumenten . . ." and "Verzeichnis der im . . . ," see 60.

Used in the BAASSC (1913–1917) 1913–1915.

*The Perth Observatory Bulletin **1**, 1913, Perth Observatory.*

11. Riverview

Riverview College Observatory (Sydney), New South Wales: Seismological Bulletin (monthly): 03.1909–12.1910, 01.1912–06.1912, 01.1913–03.1916, *04.1916–06.1916, 01.1917, 05.1917–06.1917*, 07.1917–03.1918, *04.1918–06.1918*, 07.1918–11.1918, 12.1918, 01.1919–04.1919, *01.1920–03.1920*, 04.1920–12.1920 (in italics: at ISC).

Used in the BAASSC (1913–1917) 1913–1915.

BGS: 1909–1920.

NOAA: 01.–12.1920.

12. Sydney

Shide Circular **16**, 178-182 (1906-07-23–1907-06-27); **17**, 219-220 (1907-07-01–1907-12-30); **18**, 249 (1908-01-11–1908-06-19); **19**, 284 (1906-08-16–1906-08-22); **19**, 285 (1908-06-30–1908-12-31); **20**, 334-335 (1909-01-03–1909-06-27); **21**, 372-373 (1909-07-05–1909-12-28); **22**, 414-415 (1910-01-01–1910-06-30); **24**, 96-98 (1910-07-03–1911-06-28); **25**, 158-160 (1911-07-03–1911-12-03); **26**, 226-227 (1912-02-16–1912-06-18); **27**, 277-278 (1912-07-07–1912-12-16).

In Hamburg: Seismic Bulletins 1910–1920.

Used in the BAASSC (1913–1917) 1913–1915.

13. Austria

*Mojsisovics, Edmund v. (1897). Bericht über die Organisation der Erdbeben-Beobachtung nebst Mittheilungen über während des Jahres 1896 erfolgte Erdbeben. Mittheilungen der Erdbeben-Commission der kaiserlichen Akademie der Wissenschaften in Wien **I**. In: Sitz. Ber. der kaiserlichen Akademie der Wissenschaften in Wien **106**, mathematisch-naturwissenschaftliche Classe, Abth. 1, Heft II.*

*Mojsisovics, Edmund v. (1898–1900). Allgemeiner Bericht und Chronik der im Jahre 1897 innerhalb des Beobachtungsgebietes erfolgten Erdbeben, Mittheilungen der Erdbeben-Commission der kaiserlichen Akademie der Wissenschaften in Wien **V**. Ibid., **107**, Heft V; Ibid. Jahre 1898, **X**., **108**, Heft IV; Ibid. Jahre 1899, **XVIII**., **109**, Heft III.*

*Mojsisovics, Edmund v. (1901–1904). Allgemeiner Bericht und Chronik der im Jahre 1900 im Beobachtungsgebiete eingetretenen Erdbeben. Mitteilungen der Erdbeben-Kommission der kaiserlichen Akademie der Wissenschaften in Wien, Neue Folge **II**; Ibid., Jahre 1901, Neue Folge **X**;. Ibid. Jahre 1902, Neue Folge **XIX**; Ibid. Jahre 1903, Neue Folge **XXV**.*

*In Hamburg: Allgemeiner Bericht und Chronik der im Jahre 1904 in Österreich beobachteten Erdbeben. K. K. Zentralanstalt für Meteorologie Geodynamik in Wien, Offizielle Publikation No. **1**, Wien 1906, 7 + 155 pp; Ibid. Jahre 1905, No. **2**, Wien 1907, 6 + 219 pp; Ibid. Jahre 1906, No. **3**, Wien 1908, 6 + 199 pp; Ibid. Jahre 1907, No. **4**, Wien 1909, 10 + 209 pp; Ibid. Jahre 1908, No. **5**, Wien 1910, 6 + 281 pp; Ibid. Jahre 1909, No. **6**, Wien 1911, 7 + 188 pp; Ibid. Jahre 1910, No. **7**, Wien 1912, 6 + 162 pp; Ibid. Jahre 1911, No. **8**, Wien 1914, 6 + 153 pp; Ibid. Jahre 1912, No. **9** + **10**, Wien 1915, 5 + 179 pp; Ibid. Jahre 1914, No. **11**, Wien 1917, 6 + 130 pp; Ibid. Jahre 1915, No. **12**, Wien 1919, 9 + 135 pp; Ibid. Jahren 1916–1921, No. **13**, Wien 1922, 40 pp.*

*In Hamburg: Conrad, Viktor. Die zeitliche Verteilung der in den Österreichischen Alpen- und Karstländern gefühlten Erdbeben in den Jahren 1897 bis 1907. Mitteilungen der Erdbeben-Kommission der kaiserlichen Akademie der Wissenschaften in Wien, Neue Folge **XXXVI**.*

*In Hamburg: Conrad, Viktor. Die zeitliche Verteilung der in den Jahren, 1897 bis 1907 in den Österreichischen Alpen- und Karstländern gefühlten Erdbeben (ein Beitrag zum Studium der sekundär auslösenden Ursachen der Erdbeben, II. Mitteilung). Ibid., Neue Folge **XLIV**.*

14. Graz

*Benndorf, H. (1908). Die Erdbebenstation am physikalischen Institut der Universität Graz. Mitteilungen des Naturwissenschaftlichen Vereines für Steiermark **44**, 234–*

Rožič, J., and N. Stücker (1909). Erster Bericht über seismische Registrierungen in Graz im Jahre 1907. Ibid., **45**, 237-256.

Rožič, J. (1910). Zweiter Bericht über seismische Registrierungen in Graz im Jahre 1908. *Ibid.*, **46**, 362-381.

Stücker, N., and A. Fritsch (1911). Dritter Bericht über seismische Registrierungen in Graz im Jahre 1909. *Ibid.*, **47**, 219-241.

Stücker, N. (1911). Vierter Bericht über seismische Registrierungen in Graz im Jahre 1910. *Ibid.*, **47**, 242-267.

Stücker, N. (1912). Fünfter Bericht über seismische Registrierungen in Graz im Jahre 1911. *Ibid.*, **48**, 248-273.

Stücker, N. (1912). Die mikroseismische Bewegung in Graz in den Jahren 1907–1911. *Ibid.*, **48**, 274-281.

Stücker, N. (1913). Sechster Bericht über seismische Registrierungen in Graz im Jahre 1912. *Ibid.*, **49**, 237-263.

Stücker, N. (1913). Die mikroseismische Bewegung in Graz im Jahre 1912. *Ibid.*, **49**, 264-266.

Stücker, N. (1914). Siebenter Bericht über seismische Registrierungen und die mikroseismische Bewegung in Graz im Jahre 1913. *Ibid.*, **50**, 1-33.

Stücker, N. (1915). Achter Bericht über seismische Registrierungen und die mikroseismische Bewegung in Graz im Jahre 1914. *Ibid.*, **51**, 1-23.

Stücker, N. (1916). Neunter Bericht über seismische Registrierungen und die mikroseismische Bewegung in Graz im Jahre 1915. *Ibid.*, **52**, 35-91.

Stücker, N. (1917). Zehnter Bericht über seismische Registrierungen und die mikroseismische Bewegung in Graz im Jahre 1916. *Ibid.*, **53**, 263-309.

Stücker, N. (1918). Elfter Bericht über seismische Registrierungen und die mikroseismische Bewegung in Graz im Jahre 1917. *Ibid.*, **54**, 301-342.

Zwölfter Bericht über seismische Registrierungen in Graz [1918].

Dreizehnter Bericht über seismische Registrierungen in Graz [1919].

Wöchentlicher Erdbebenbericht, 1911–07.1914.

15. Innsbruck

Seismische Aufzeichnungen Innsbruck, Institut für Kosmische Physik, 01.–09., 11.–52.1913, 01.–26.1914.

At ISC: 01.1913–06.1914, 08.1914–05.1915, 08.–12.1915.

Used in the BAASSC (1913–1917) 1913–1914.

16. Judenburg

BGS: 1916.

17. Kremsmünster

Schwab, P. Franz (1900). Bericht über Erdbebenbeobachtungen in Kremsmünster. Mittheilungen der Erdbeben-Commission der kaiserlichen Akademie der Wissenschaften in Wien **XV**. In: Sitz. Ber. der kaiserlichen Akademie der Wissenschaften in Wien **109**, mathematisch-naturwissenschaftliche Klasse, Abth. 1, Heft II, 51 pp.

In Hamburg: Schwab, P. Franz. Bericht über die Erdbebenbeobachtungen in Kremsmünster im Jahre 1900. Mitteilungen der Erdbeben-Kommission, Neue Folge ***IV****.*

In Hamburg: Schwab, P. Franz. Ibid., Jahre 1901, Neue Folge ***XII****.*

Schwab, P. Franz (1903). Bericht über die Erdbebenbeobachtungen in Kremsmünster im Jahre 1902. *Ibid.*, Neue Folge **XXI**, 23 pp.

In Hamburg: Schwab, P. Franz. Ibid., Jahre 1903, Neue Folge ***XXVI****.*

At KNMI: Schwab, Fr. (1908). Erdbeben-Beobachtungen in Kremsmünster. Jahresbericht des Vereinsmuseums Fransico-Carolinum. Linz 1908, 33 pp. [1904–1907].

At KNMI: Schwab, Fr. (1909). Beilage zu den Erdbeben-Beobachtungen in Kremsmünster 1904–1907. Ibid., Linz 1909, 6 pp. [1908].

Used in the BAASSC (1913–1917) 1913–1914.

BGS: 1906–1907.

18. Wien (Vienna)

In Hamburg: Conrad, Viktor. Beschreibung des seismischen Observatoriums der k. k. Zentralanstalt für Meteorologie und Geodynamik in Wien. Mitteilungen der Erdbeben-Kommission der kaiserlichen Akademie der Wissenschaften in Wien, Neue Folge ***XXXIII****, 28 pp.*

In Hamburg: Monatliche Mitteilungen der k. k. Zentralanstalt für Meteorologie und Geodynamik. Wien 1905.

Wöchentliche Erdbebenberichte, 01.1906–52.1906; 01.1906–52.1907; 01.1906–52.1908.

In Hamburg: Conrad, Viktor. Seismische Registrierungen in Wien, k. k. Zentralastalt für Meteorologie und Geodynamik, im Jahre 1909 (mit einigen Hilfstabellen zur Analyse von Bebendiagrammen). Mitteilungen der Erdbeben-Kommission der kaiserlichen Akademie der Wissenschaften in Wien, Neue Folge ***XXXIX****.*

Schneider, Rudolf. Seismische Registrierungen in Wien, k. k. Zentralanstalt für Meteorologie und Geodynamik, im Jahre 1910. *Ibid.*, Neue Folge **XLI**, 49 pp.; *Ibid.*, Jahre 1911, Neue Folge **XLV**, 55 pp.; *Ibid.*, 1912, Neue Folge **XLVII**.

Wien, Zentralanstalt für Meteorologie und Geodynamik, Seismische Aufzeichnungen, 1913–1920 (handwritten or typed lists).

NOAA: yearly reports 1912, 1913.

19. Baku (Bakou, Baky)

Included in: Seismische Monatsberichte des Physikalischen Observatoriums zu Tiflis, 10.1906–12.1907, see 64.

Bulletin of Imperial Russia 1907–1908, see 168.

Renholm, Edwin (1913). Seismometrische Beobachtungen in Baku und Balachany in der Zeit vom 1. Januar bis 31. Dezember 1910.

Wöchentliches Bulletin der Nobel'schen Seismischen Station Baku: 1–3, 6–17, 19–52, 1913; 1–8, 10–22, 1914.

In Hamburg: Seismometrische Beobachtungen in Baku und Balachany, 1908, 1910.

In Hamburg: Erdbebenberichte Baku, 1912–1914.

At ISC: 04.1912–11.1914.

Monthly seismological bulletin of Baku seismological station 1912–1916.

Used in the BAASSC (1913–1917) 1913–1915.

20. Balachany (Balakhan, Balakhany)

Included in: Seismische Monatsberichte des Physikalischen Observatoriums zu Tiflis, 1902–12.1909, see 64.

Monthly bulletin of Tiflis seismological station 1910–1916, see 64.

Bulletin of Imperial Russia 1907–1908, see 168.

Renholm, Edwin (1913). Seismometrische Beobachtungen in Baku und Balachany in der Zeit vom 1. Januar bis 31. Dezember 1910.

In Hamburg: Seismometrische Beobachtungen in Baku und Balachany, 11.1908–1910.

21. Šemakha (Chemakha, Shemakha)

Included in: Seismische Monatsberichte des Physikalischen Observatoriums zu Tiflis, 01.1903–09.1909, see 64.

Bulletin of Imperial Russia 1904–1908, see 168.

22. Zurnabad (Zournabath, Zuraband)

10.1908–10.1909, see 64.

Bulletin of Imperial Russia 1912, see 168.

23. D'Uccle

Bulletin Mensuel de la Station Géophysique d'Uccle (Station Ernest Solvay) (1901) Mai 1901, 3 pp.; (1901) Juin 1901, 3 pp.; (1901) Juillet 1901, 2 pp.; (1901) Août 1901, 3 pp.;

(1901) Septembre–Octobre 1901, 3 pp.; (1902) Novembre 1901–Juin 1902, 14 pp.; (1903) Avril a Décembre 1902, 19 pp.; (1904) Janvier a Mars 1903, 15 pp.; (1904) Avril a Décembre 1903, 19 pp.

Observations Séismologique faites a Uccle (1906–1910). Annales de l'Observatoire Royal de Belgique en 1904 et 1905, Nouvelle Sérié, Physique du Globe **3**, 393-398; *Ibid.*, en 1906, **3**, 399-421; *Ibid.*, en 1907, **4**, 122-138; *Ibid.*, en 1908, **4**, 258-264; *Ibid.*, en 1909, **5**, 98-108.

Observations Séismologique faites a Uccle (1911–1914). Annales de l'Observatoire Royal de Belgique en 1910, 181-194; *Ibid.*, en 1911, 195-216; en 1912, 217-266; en 1913, 267-321.

Bulletin Sismique de l'Observatoire Royal Belgique, 1914–1920.

In Hamburg: Annales de l'Observatoire Royal de Belgique. Nouvelle Sérié, Physique du Globe, 1914–1918.

01.–12.1916 published in BAASSC (1916) as supplement to "Large Earthquakes of 1916."

24. La Paz

Boletín de la Estación Sismológica del Colegio de San Calixto (PP. Jesuitas), La Paz (Bolivia). Single leaves, 05.–11.1913.

In Hamburg: Bulletin 1913–1920.

Boletín mensual del Observatorio Meteorológico y Sísmico del Colegio de San Calixto, dirigido por PP. de la Compañía de Jesús, 1918. In: Boletín de la Dirección Nacional de Estadística y Estudios Geográficos, República de Bolivia, ***2. Ép. 1 1918****.*

Quadros mensuales del Observatorio Meteorológico y Sísmico del Colegio de San Calixto, dirigido por PP. de la Compañía de Jesús, 1920. Ibid., ***2. Ép. 3 1920****.*

At ISC: 05.1914–04.1917, 06.1917–12.1920.

Used in the BAASSC (1913–1917) 1913–1915.

BGS: 1916–1920.

NOAA: 05.1913–03.1914, 05.1914–12.1920.

25. Bosnia and Herzegovina

In Hamburg: Erdbebenbeobachtungen in Bosnien und der Herzegovina 1896–1912.

Harisch, O. (1907). Zusammenstellung der Ergebnisse der in den Jahren 1904 und 1905 in Bosnien und Herzegovina stattgefundenen Erdbebenbeobachtungen. Ergebnisse der meteorologischen Beobachtungen an der Landesstation in Bosnien und der Hercegovina in den Jahren 1904 und 1905. Sarajevo 1907, pp 81.

Zusammenstellung der Ergebnisse der in Bosnien und Herzegovina stattgefundenen Erdbebenbeobachtungen im Jahre 1906. Sarajevo 1907; Jahre 1907. Sarajevo 1908; Jahre 1908. Sarajevo 1909, pp. 17.

Harisch, O. (1910). Zusammenstellung der im Jahre 1909 in Bosnien und der Herzegovina stattgefundenen Erdbebenbeobachtungen. Ergebnisse der meteorologischen Beobachtungen an der Landesstation in Bosnien und der Hercegovina im Jahre 1909. Sarajevo 1910, pp. 27.

Zusammenstellung der Ergebnisse der im Jahre 1910 in Bosnien und der Herzegovina stattgefundenen Erdbebenbeobachtungen. Ibid., Jahre 1910. Sarajevo 1911, pp. 19.

Harisch, O. (1912). Zusammenstellung der Ergebnisse der im Jahre 1911 in Bosnien und Herzegovina stattgefundenen Erdbebenbeobachtungen. Sarajevo 1912, pp. 23.

26. Sarajevo

At ISC: 01.1906–04.1910, 06.1910–06.1912, 10.1912–12.1913.

Used in the BAASSC (1913–1917) 1913.

NOAA: 1904, 1905, 09.1910–12.1913.

27. Fernando de Noronha

Shide Circular **24**, 83 (1911-03-24–1911-06-25); **25**, 146 (1911-07-03–1911-12-16); **26**, 207-209 (1911-12-21–1912-06-18); **27**, 273-274 (1912-07-16–1912-12-24).

Used in the BAASSC (1913–1917) 1913–1915.

28. Rio de Janeiro

Bolletin Mensal do Observatorio do Rio de Janeiro. Rio de Janeiro 1906, 1907, 1908.

In Hamburg: Boletín Sismológico 1906–1920.

NOAA: 1906–1920.

29. Bulgaria

In Hamburg: Zemletresenjia o Bulgarija: otcet za zabelezenite zemletresenija prez . . . godina (Tremblements de terre en Bulgarie) ***1****(1902)–****21*** *(1920).*

Watzof, Spas: Tremblements de terre en Bulgarie. Nr. ***4*** *Liste des tremblements de terre observés pendant l'année 1903.*

Watzof, Spas (1907–1909). Ibid., Nr. ***7****, 1906, 56 pp.; Ibid., Nr.* ***8****, 1907, 78 pp.; Ibid., Nr.* ***9****, 1908, 95 pp.*

NOAA: 1917–1920.

30. Sofia (Sofija)

Watzof, Spas (1907–1912). Bulletin Sismographique de l'Institut Météorologique Central de Bulgarie Nr. **1**. Enregistrements à Sofia de 16 avril à 31 décembre 1905, 56 pp.; *Ibid.*, Nr. **2**, 1 janvier à 31 décembre 1906, 34 pp.; *Ibid.*, Nr. **3**, 1 janvier à 31 décembre 1907, 43 pp.; *Ibid.*, Nr. **4**, 1 janvier à 30 juin 1908, 24 pp.; *Ibid.*, Nr. **5**, 1 juillet à 31 décembre 1908, 21 pp.; *Ibid.*, Nr. **6**, 1 janvier à 31 décembre 1909, 43 pp.; *Ibid.*, Nr. **7**, 1 janvier 1910 jusquáu 31 décembre 1911, 23 pp.

NOAA: 1917–1920.

31. Manitoba (St. Boniface), Winnipeg

Rousseau, Ferdinand A. (1910, 1911). Jesuit seismological service record of the St. Boniface earthquake station, Canada, November–December 1910; January–June 1911. Monthly Weather Review, 1910, 1911.

Blain, Joseph (1910, 1911, 1912). Jesuit seismological service record of the St. Boniface earthquake station, Manitoba, Canada for February–October, 1910; July–December, 1911; January–April, 1912. Monthly Weather Review 1910, 1911, and 1912.

NOAA: 01.1911–03.1912.

32. Ottawa

Annual Report in Terrestrial Magnetism and Gravity. Report of the Chief Astronomer, Dominion astronomical observatory, Supplement from 1906.
Klotz, O. (1910). Seismology, Terrestrial Magnetism and Gravity. Report of the Chief Astronomer 1909, Supplement No. 1, Ottawa 1910.
Mimeographed monthly bulletins from 01.1910.

33. Toronto

BAASRP **1899**, 170–172 (1897-09-20–1899-06-05).
Shide Circular **2**, 35-37 (1899-06-05–1899-12-31); **3**, 66-69 (1900-01-02–1900-12-25); **4**, 102-103 (1901-01-04–1901-06-24); **5**, 127-129 (1901-07-11–1901-12-31); **6**, 152-153 (1902-01-01–1902-06-11); **7**, 203 (1902-07-05–1902-12-28); **8**, 236-237 (1903-01-04–1903-06-16); **9**, 280 (1903-07-21–1903-12-23); **10**, 305 (1904-01-03–1904-06-27); **11**, 347 (1904-07-10–1904-12-21); **12**, 23 (1905-01-20–1905-06-30); **13**, 40 (1905-07-06–1905-12-18); **14**, 76-77 (1906-01-06–1906-06-30); **16**, 174-175 (1906-07-08–1907-06-22); **17**, 214-215 (1907-07-01–1907-12-30); **18**, 247 (1908-01-11–1908-06-18); **19**, 278 (1908-07-08–1908-12-28); **20**, 323 (1909-01-12–1909-06-27); **21**, 359-360 (1909-07-07–1909-12-09); **22**, 401-402 (1910-01-01–1910-06-29); **23**, 30 (1910-07-03–1910-12-30); **24**, 85 (1911-01-01–1911-06-30); **25**, 147-149 (1911-07-01–1911-12-31); **26**, 210-212 (1912-01-04–1912-06-29); **27**, 274-275 (1912-07-07–1912-12-28).
06. + 12.1903 used in Strasbourg's "Von den Instrumenten . . ." and "Verzeichnis der im . . . ," see 60.
1904 In: *Reid, Harry Fielding. Records of Seismographs. Terrestrial Magnetism and Atmospheric Electricity, 1905.*
Klotz, O. (1909). Seismograph Records. J. R. Astr. Soc. Canada, 152–156.
Results of meteorological, seismological, and magnetical observations, Toronto Observatory, 1911, 1921.
Used in the BAASSC (1913–1917) 1913–1915.
See US-WB, 1914–1920.
BGS: 1907–1920.
At ISC: 01.1910–02.1911, 04.1911–12.1920.
NOAA: 08.1910–01.1916, 03.1916–12.1920.

34. Victoria, British Columbia

BAASRP **1899**, 172-173 (1898-10-11–1899-05-25).
Shide Circular **2**, 38-39 (1899-06-05–1899-12-31); **3**, 70-74 (1900-01-02–1900-12-25); **4**, 103-105 (1901-01-04–1901-06-24); **5**, 129-131 (1901-07-11–1901-12-31); **6**, 153-154 (1902-01-01–1902-06-16); **7**, 204 (1902-07-05–1902-12-31); **8**, 237-238 (1903-01-04–1903-06-25); **9**, 281 (1903-07-12–1903-12-30); **10**, 306 (1904-01-03–1904-06-27); **11**, 348 (1904-07-10–1904-12-21); **12**, 24 (1905-01-10–1905-06-30); **13**, 41 (1905-07-06–1905-12-18); **14**, 77-78 (1906-01-06–1906-06-24); **16**, 176 (1906-07-08–1907-06-13); **17**, 215-216 (1907-07-01–1907-12-30); **18**, 247-248 (1908-01-11–1908-06-18); **19**, 279 (1908-07-08–1908-12-28); **20**, 324 (1909-01-11–1909-06-27); **21**, 360 (1909-07-07–1909-12-09); **22**, 402-403 (1910-01-01–1910-06-29); **23**, 31 (1910-07-03–1910-12-30); **24**, 85-86 (1911-01-01–1911-06-19); **25**, 149-150 (1911-07-01–1911-12-29); **26**, 212-213 (1912-01-04–1912-06-29); **27**, 275-276 (1912-07-07–1912-12-28).
06. + 12.1903 used in Strasbourg's "Von den Instrumenten . . ." and "Verzeichnis der im . . . ," see 60.
1904 In: *Reid, Harry Fielding. Records of Seismographs. Terrestrial Magnetism and Atmospheric Electricity, 1905.*
ISA: 1906.
Used in the BAASSC (1913–1917) 1913–1915.
See US-WB, 1914–1920.

35. St. Vincent (São Vicente)

Shide Circular **23**, 32 (1910-11-15–1910-12-25); **24**, 70 (1911-01-03–1911-06-15); **26**, 206 (1911-07-05–1912-06-10).
Used in the BAASSC (1913–1917) 1913–1914.

36. Copiapó

Included in "Boletín del servicio sismológico de Chile . . . ," see 37.

37. Chile

*In Hamburg: Montessus de Ballore, F. de (1909). Boletín del servicio sismológico de Chile **1**. Años de 1906, 1907, 1908. Santiago di Chile 1909, 200 pp.*
*Montessus de Ballore, F. de (1910). Boletín del servicio sismológico de Chile **2**. Primer Semestre de 1909. Anales de la Universidad **125**, Santiago di Chile 1910, 819-918.*
In Hamburg: 1912–1917.

38. Santiago de Chile

Included in "Boletín del servicio sismológico de Chile . . . ," see 37.
BGS: 1910–1913.
NOAA: 1919–1920.

39. Taihoku (Formosa/Taiwan)

Taihoku Meteorological Observatory was established in 1896. A Gray-Milne seismograph was installed in 1897, and an Omori seismograph in 1903. Auxiliary stations in Taiwan equipped with Gray-Milne seismographs include Tainan (1898), Taitung (1903), Taichung (1903), and Penghu (1898).
Omori, F. (1905). [Note: This report contains many seismic readings from stations in Taiwan (Formosa) from about 1897 to the end of 1904.]
1904 used in Rudolph (1907).
02.–03.1904 used in Strasbourg's "Verzeichnis der im . . . ," see 60.
Cheng, S.N., C.C. Chang, C.F. Wu, Y.T. Yeh, and T.C. Shin (1997). Compilation of Earthquake Data in Taiwan during the Period of Japanese Occupation (I) and (II). Report IESCR97-003 amd IESCR97-004, Institute of Earth Science of the Academia Sinica and Central Weather Bureau, Taipei, Taiwan, 1352 pp. [Note: These two reports contain many seismic readings from stations in Taiwan from 1904 to 1943.]

40. Tsingtau (Qingdao), Kiautschou (Jia zhou)

Seismische Registrierungen des Kaiserlichen Observatoriums in Tsingtau, Heft I. Januar 1909 bis 1. April 1910.

In Hamburg: Seismische Registrierungen in Tsingtau 1911–1913.
NOAA: 1911–1913.
Used in the BAASSC (1913–1917) 1913.

41. Zi-Ka-Wei (Hsu-chia-hui)

Bulletin sismologique de Zi-ka-wei pour l'année 1900–1903.
*Observatoire Magnétique Météorologique et Sismologique de Zi-Ka-Wei. Bulletin des observ. sismologique **31**, Année 1905, Fascicule C Sismologie. Chang-Hai 1908, 29 pp.*
Bulletin sismologique de Zi-ka-wei pour l'année 1906 et tableaux résumés des années 1904, 1905 et 1906. Bulletin des Observations **33**, Année 1907, Fascicule C Sismologie. Observatoire Magnétique Météorologique et Sismologique de Zi-Ka-Wei (Chine), Chang-Hai 1912.
Bulletin des Observations **34**, Année 1908, Fascicule C Sismologie. Observatoire Magnétique Météorologique et Sismologique de Zi-Ka-Wei (Chine), Chang-Hai 1912.
*Bulletin des Observations **35**, Année 1909, ibid.*
Bulletin des Observations **36**, Année 1910, *ibid.*,1914.
Zi-Ka-Wei (Chine) Bulletin Sismique de l'Observatoire de Zi-Ka-Wei, prés Chang-hai, Chine. Single leafs 01.1906–12.1920.
1904 used in Rudolph (1907).
03.1904 used in Strasbourg's "Verzeichnis der im . . . ," see 60.
At ISC: 07.1909–11.1919, 01.–06.1920.
Used in the BAASSC (1913–1917) 1913–1915.

42. San José

*Boletín del Instituto Físico-Geográfico de Costa Rica (San José) **1**(1901)–**3**(1903).*

43. Croatia and Slavonia

Bulletin Hebdomadaire des Observatoires sismiques de la Hongrie et la Croatie, Avril 1909, No. 16-18.
See also 46.

44. Pula (Pola)

Veröffentlichung des Hydrographischen Amtes der Kaiserlichen und Königlichen Kriegsmarine in Pola, Gruppe 2, Jahrbuch der meteorologischen, erdmagnetischen und seismischen Beobachtungen, N.F. 1896–1899.
In Hamburg: Jahrbuch der meteorologischen, erdmagnetischen und seismometrischen Beobachtungen in Pola, 1900.
Jahrbuch der meteorologischen, erdmagnetischen und seismometrischen Beobachtungen in Pola, 1901. *Ibid.*, 1902. *Ibid.*, 1903.
Kesslitz, W. (1905). Jahrbuch der meteorologischen, erdmagnetischen und seismometrischen Beobachtungen in Pola. Neue Folge **9**.
Kesslitz, W. (1907). *Ibid.*, 1906.
Kesslitz, W. (1908). *Ibid.*, Neue Folge **12,** Beobachtunge des Jahres 1907. Veröffentlichung des Hydrographischen Amtes der Kaiserlichen und Königlichen Kriegsmarine in Pola.
Kesslitz, W. (1909). *Ibid.*, Neue Folge **13**, Beobachtunge des Jahres 1908.
Jahrbuch der meteorologischen, erdmagnetischen und seismometrischen Beobachtungen in Pola, 1909. *Ibid.*, 1910. *Ibid.*, 1911. *Ibid.*, 1912.*Ibid.*, 1913. *Ibid.*, 1914. *Ibid.*, 1915.
In Hamburg: Jahrbuch der meteorologischen, erdmagnetischen und seismometrischen Beobachtungen in Pola, 1916. Ibid., 1917.
Seismische Aufzeichnungen (handwritten), 1910–1914.
At ISC: 01.1907–12.1917.
NOAA: 1901, 1904–1915, 1918.

45. Rijeka (Fiume)

1906–1912 in Kövesligethy (1907, 1909, 1913), see 86.

46. Zagreb (Agram)

*Kišpatić, M. (1904). Dvadeset i prvo Potresno Izvješće za godinu 1903 (21. Erdbebenbericht für das Jahr 1903). Südslawische Akademie der Wissenschaften und Künste **158**, Agram 1904.*
Godišnje Izvješće Zagrbačkog Meteorološkog Opservatorija za Godinu 1906 (Jahrbuch des Meteorologischen Observatoriums in Zagreb (Agram) für das Jahr 1906), Dio IV, Potresi u Hrvatskoj i Slavonij god. 1906, IV. Teil, Erdbeben in Kroatien und Slavonien im Jahre 1906. Ibid., 1907.
Mohorovičić, A. (1910). Godišnje Izvješće Zagrbačkog Meteorološkog Opservatorija za Godinu 1908 (Jahrbuch des Meteorologischen Observatoriums in Zagreb (Agram) für das Jahr 1908), Dio IV, Potresi u Hrvatskoj i Slavonij god. 1908, IV. Teil Erdbeben in Kroatien und Slavonien im Jahre 1908, 55 pp.
Godišnje Izvješće Zagrbačkog Meteorološkog Opservatorija za Godinu 1909. Ibid., im Jahre 1909.
1906–1909 in Kövesligethy (1907, 1909), see 86.
In Hamburg: Zagreb, 1906–1920.
Seismische Aufzeichnungen: weekly lists 01.1913–No. 21, 1914.
01.–12.1916 published in BAASSC (1916) as supplement to "Large Earthquakes of 1916."
NOAA: 1913–07.1914, 08.1918–12.1918.
At ISC: 01.1913–08.1914, 08.–12.1918, 1502–1938 (?).
Used in the BAASSC (1913–1917) 1914.

47. Cheb (Eger)

*Uhlig, V. Bericht über die seismischen Ereignisse des Jahres 1900 in den deutschen Gebieten Böhmens. Mitteilungen der Erdbeben-Kommission, Neue Folge **III**.*
Irgang, Georg (1908/09). Jahresbericht der k. k. Staats-Oberrealschule in Eger 1908/09.
Irgang, Georg (1911/12). Seismische Registrierungen der Erdbebenwarte in Eger vom 20. November 1908 bis 31. Dezember 1911. Jahresbericht der k. k. Staats-Oberrealschule in Eger 1911/12.
Irgang, Georg (1912). Seismische Registrierungen in Eger vom 20. Nov. 1908 bis 31. Dez. 1911. Wiener Anzeiger 1912, 195-198.

Irgang, Georg (1912/13). Seismische Registrierungen der Erdbebenwarte in Eger im Jahre 1912. Jahresbericht der k. k. Staats-Oberrealschule in Eger 1912/13, 1-12.

Irgang, Georg (1913/14). Seismische Registrierungen der Erdbebenwarte in Eger vom 1. Januar 1913 bis 30. April 1914. *Ibid.*, 1913/14, 29-50.

Irgang, Georg (1927). Bericht über die seismischen Registrierungen der Erdbebenwarte in Eger in der Zeit vom 1. Mai 1914 bis Juli 1919. Státní Ústav pro Geofysiku 1927, 1-6.

Bulletin of the Seismological Station Cheb May 1914–June 1919, handwritten station book, 36 pp.

48. Disko (Godhavn)

Harboe, E.G. (1912). Das Erdbebenobservatorium auf der Disko-Insel. G. Gerlands Beiträge zur Geophysik **11**, Kleine Mitteilungen, 9-28.

Harboe, E.G. (1915–1918). Gerlands Beiträge zur Geophysik **14**, Kleine Mitteilungen, 24-31.

49. Quito

ISA: 1904, 1906.

50. Helwan (Hulwān, Abbasia, Abbassieh, Cairo)

Shide Circular **1**, 15 (1899-10-06–1899-11-21); **2**, 40-42 (1900-01-05–1900-06-22); **3**, 75-76 (1900-07-20–1900-12-27); **4**, 106 (1901-01-04–1901-06-15); **5**, 132 (1901-09-08–1901-12-24); **6**, 156 (1902-01-07–1902-06-20); **7**, 205-206 (1902-07-05–1902-12-16); **8**, 239 (1903-01-04–1903-06-04); **10**, 313-315 (1903-07-01–1904-07-27); **11**, 335 (1904-08-11–1904-12-20); **12**, 17 (1905-01-13–1905-06-30); **13**, 47-48 (1905-07-09–1905-12-26); **14**, 84 (1906-01-21–1906-06-24); **15**, 127 (1906-07-14–1906-12-27); **16**, 173 (1907-01-01–1907-02-24); **17**, 205-208 (1907-05-07–1907-12-30); **18**, 234-236 (1908-01-02–1908-06-28); **19**, 267-269 (1908-06-30–1908-12-28); **20**, 313-315 (1909-01-02–1909-06-29); **21**, 363-365 (1909-07-01–1909-12-28); **22**, 404-407 (1910-01-01–1910-06-30); **24**, 75-80 (1910-07-07–1911-06-30); **25**, 138-144 (1911-07-01–1911-12-31); **26**, 198-204 (1912-01-03–1912-06-28); **27**, 260-265 (1912-07-01–1912-12-31).

05., –10.1903 used in Strasbourg's "Von den Instrumenten . . ." and "Verzeichnis der im . . . ," see 60.

At ISC: 07.1914–11.1917, 01.1918–12.1918 (Helwan).

Helwan Observatory Bulletins, Volume 1, 1911–1920 (?).

Ismail, A. (1960). Near and local earthquakes of Helwan (1903–1950), Bulletin of Helwan Observatory ***49***.

Used in the BAASSC (1913–1917) 1913–1915.

51. Tartu (Dorpat, Jurjew, Iouriev, Iouriev)

Rudolph (1903a, b).

Bulletin of Imperial Russia 1902–1908, see 168.

Orloff, A. (1906). Über die Seismogramme des Zöllnerschen Horizontalpendels. Sitzungsberichte der Dorpatischen Naturforschenden Gesellschaft 1906 ***15***, *3.*

52. Fiji

Used in the BAASSC (1913–1917) 1914.

53. Besançon

Included in "Annales du Bureau . . ." of France, see 55.

54. Clermont-Ferrand

Since 11.1913 included in "Annales du Bureau . . ." of France, see 55.

55. France

Annales du Bureau Central Météorologique de France, Bulletin Sismologique, 1910–1919.

In Hamburg: Bulletin Mensuel du Bureau Central Météorologique de France, 1901–1913.

At ISC: 1913.

56. Grenoble

Reboul, P. (1907). Notes sur la sismologie et les séismes régistrés en Dauphiné (1893–1906). Travaux du Lab. Géol. Univ. de Grenoble ***8***, *1905–1907, 97-110.*

Observations used in 1895 as reference in the national bulletin of Italy, see 107.

57. Marseille

Since 10.1910 included in "Annales du Bureau . . ." of France, see 55.

58. Paris (Parc St. Maur)

Included in "Annales du Bureau . . ." of France, see 55.

Observations sismologique de la Station du Parc Saint-Maur, 1915–1920.

Bulletin Sismique (handwritten), 1920.

Used in the BAASSC (1913–1917) 1913–1915.

BGS: 1911–1920.

59. Puy-de-Dôme

Included in "Annales du Bureau . . ." of France, see 55.

60. Strasbourg (Strassburg, Straßburg)

Rebeur-Paschwitz (1895).

Ehlert, Reinhold (1898). Horizontalpendelbeobachtungen im Meridian zu Straßburg i. E. von April bis Winter 1895. Beiträge zu Geophysik **3**, 131-215.

Rudolph (1903a, b).

Shide Circular **4**, 101 (1901-01-07–1901-05-28); **5**, 127 (1901-06-13–1901-09-28).

Bericht No. 1, Erdbeben im Juli 1900–Bericht No. 6, Erdbeben im Dezember 1900.

Bericht No. 1, Erdbeben im Januar 1901–Bericht No. 12, Erdbeben im Dezember 1901.

Beobachtungen an Erbebenmessern Januar 1902–Dezember 1902. *Ibid.,* Januar 1903–April 1903.
Von den Instrumenten der Hauptstation aufgezeichnete Seismogramme nebst Angaben anderer Stationen über dieselben Beben. Mai 1903. *Ibid.,* Juni 1903.
Verzeichnis der im Juli 1903 registrierten Beben, soweit solche bis 01. II. 1904 zur Kenntnis der Hauptstation gelangt sind.
Verzeichnis der im August 1903 registrierten Beben, soweit solche bis 14. III. 1904, *ibid.*
Verzeichnis der im September 1903 registrierten Beben, soweit solche bis 24. III. 1904, *ibid.*
Verzeichnis der im Oktober 1903 registrierten Beben, soweit solche bis 28. IV. 1904, *ibid.*
Verzeichnis der im November 1903 registrierten Beben, soweit solche bis 28. V. 1904, *ibid.*
Verzeichnis der im Dezember 1903 registrierten Beben, soweit solche bis 28. VI. 1904, *ibid.*
Verzeichnis der im Januar 1904 registrierten Beben, soweit solche bis 22. XI. 1904, *ibid.*
Verzeichnis der im Februar 1904 registrierten Beben, soweit solche bis 15. II. 1905, *ibid.*
Verzeichnis der im März 1904 registrierten Beben, soweit solche bis 15. III. 1905, *ibid.*
Mainka Carl (1910). Seismometrische Beobachtungen in Strassburg i. E. in der Zeit vom 1. Januar bis 31. Dezember 1905. Beträge zur Geophysik **10**, 387-467.
Wöchentlicher Erdbebenbericht der Kaiserlichen Hauptstation für Erdbebenforschung zu Straßburg. No. 1 (1905, Mai 29)–No. 31 (Dezember, 31 1905) (handwritten).
Wöchentlicher Erdbebenbericht der Kaiserlichen Hauptstation für Erdbebenforschung zu Straßburg i. Els. für das Jahr 1906. No. 1–52 (handwritten), with “Anleitung zum Beobachten von Erdbeben,” Beilage zu Nr. 108 der Straßburger Korrespondenz, Dienstag den 11. September 1906, “Makroseismischer Monats-Bericht, April 1906,” and “Makroseismischer Monats-Bericht, Mai 1906.” *Ibid.,* Jahr 1907. No. 1–52 (handwritten), with “Beilage zum Wochenbericht No. 1 des Jahres 1907” and “Makroseismische Nachrichten.” *Ibid.,* No. 1–52, 1908 (handwritten), with “Zur Bestimmung der Epizentren.”
Kaiserlichen Hauptstation für Erdbebenforschung in Straßburg i. E., Makroseismische Nachrichten No. 5–31, 1908 (handwritten).
Wöchentlicher Erdbebenbericht der Kaiserlichen Hauptstation für Erdbebenforschung zu Straßburg i. E. 1909. No. 1–51, 1909 (handwritten).
Kaiserlichen Hauptstation für Erdbebenforschung in Straßburg i. E., Makroseismische Nachrichten No. 1–13, 1909 (handwritten).
Monatliche Übersicht der an der Kaiserlichen Hauptstation für Erdbebenforschung in Straßburg i. E. bekannt gewordenen Erdbeben. No. 1, Juli–No. 6, Dezember, 1909.
Wöchentlicher Erdbebenbericht der Kaiserlichen Hauptstation für Erdbebenforschung zu Straßburg i. E. 1910. No. 1–6, 9–30 (handwritten).
Seismometrische Aufzeichnungen der Kaiserl. Hauptstation für Erdbebenforschung i. Straßburg i. Els. No. 31–52, 1910 (handwritten).
Monatliche Uebersicht der an der Kaiserl. Hauptstation für Erdbebenforschung in Straßburg i. E. bekannt gewordenen Erdbeben. No. 1, Januar–No. 3, März, 1910.
Monatliche Uebersicht über die seismische Tätigkeit der Erdrinde nach den der Kaiserl. Hauptstation für Erdbebenforschung in Straßburg i. E. zugegangenen Nachrichten. No. 4, April–No. 12, Dezember, 1910.
Seismometrische Aufzeichnungen der Kaiserl. Hauptstation für Erdbebenforschung i. Straßburg i. Els. No. 1–52, 1911.
Monatliche Uebersicht über die seismische Tätigkeit der Erdrinde nach den der Kaiserl. Hauptstation für Erdbebenforschung in Straßburg i. E. zugegangenen Nachrichten. No. 1, Januar–No. 12, Dezember, 1911.
Strassburg i. E., Seismische Aufzeichnungen der Kaiserlichen Hauptstation für Erdbebenforschung, No. 1–52, 1912.
Mitteilungen über Erdbeben im Jahre 1912. Hauptstation für Erdbebenforschung früher in Straßburg, zurzeit in Jena. Jena 1920, 26 pp., handwritten.
Strassburg i. E., Seismische Aufzeichnungen der Kaiserlichen Hauptstation für Erdbebenforschung, No. 1–52, 1913.
Mitteilungen über Erdbeben im Jahre 1913. Hauptstation für Erdbebenforschung früher in Straßburg, zurzeit in Jena. Jena 1920, 26 pp., handwritten, partly typed [January–June, 1913].
Strassburg i. E., Seismische Aufzeichnungen der Kaiserlichen Hauptstation für Erdbebenforschung, No. 1–52, 1914.
Strassburg i. E., Seismische Aufzeichnungen der Kaiserlichen Hauptstation für Erdbebenforschung, Aperiodische Pendel mit galvanometrischer Registrierung nach Galitzin (analyzed by Beno Gutenberg). 02.–07., 12.1914–05.1915.
Strassburg i. E., Seismische Aufzeichnungen der Kaiserlichen Hauptstation für Erdbebenforschung, No.1–23 (December, 31) 1915. *Ibid.,* No. 1–24 (December, 27), 1916.
In Hamburg: Rothé, E. (1920). Annuaire de l'Institut de Physique du Globe 1919, Deuxième Partie: Sismologie, 16 pp.
In Hamburg: Rothé, E. (1922). Ibid., 1920, 59 pp.

61. Akhalkalaki

Included in: Seismische Monatsberichte des Physikalischen Observatoriums zu Tiflis, 04.1903–09.1909, see 64.
Bulletin of Imperial Russia 1904–1908, see 168.

62. Batumi (Batoum, Batum)

Included in: Seismische Monatsberichte des Physikalischen Observatoriums zu Tiflis, 04.1903–09.1909, see 64.
Bulletin of Imperial Russia 1904–1908, see 168.

63. Borjomi (Boržom, Borjom, Borshom)

Included in: Seismische Monatsberichte des Physikalischen Observatoriums zu Tiflis, 01.1903–09.1909, see 64.
Bulletin of Imperial Russia 1904–1908, see 168.

64. Tbilisi (Tblissi, Tiflis)

Ezemesjacnyja svedenija o zemletrjasenijach, otmecennych trojnym gorizontal'nym majatnikom Reber-Elert v Tiflisskoj Fiziceskoj Observatorii, 1900–1902. Ezemesjacnyj

sejsmiceskij bjulleten' Tiflisskoj Fiziceskoj Observatorija, 04.1903–10.1909.
In Hamburg: Seismische Monatsberichte des Physikalischen Observatoriums zu Tiflis, 1900–1909.
Shide Circular **8**, 254–256 (1903-01-02–1903-07-27); **9**, 292 (1903-08-02–1903-11-29).
Bulletin of Imperial Russia 1902–1908, see 168.
06., 09.1903, 01.–03.1904 used in Strasbourg's "Von den Instrumenten . . ." and "Verzeichnis der im . . . ," see 60.
Wöchentliche Erdbebenberichte des Physikalischen Observatoriums zu Tiflis (1910–1911).
Monthly bulletin of Tiflis seismological station 1910–1916.
In Hamburg: Erdbebenberichte Tiflis, 1912–1914.
At ISC: 01.–06.1906, 03., 05.–08.1910, 01.1911–06.1914.
Used in the BAASSC (1913–1917) 1913–1914.
NOAA: 07.1910–12.1911, 02.1913.
BGS: 1913.

65. Aachen

Erdbebenstation der Technischen Hochschule in Aachen. Handwritten monthly reports: 09.1906–07.1914.

66. Biberach

Published together with Hohenheim (see 72) in Mack, Karl (1915, 1916, 1917).

67. Bochum

Mintrop, L. (1909): Die Beobachtungen der Bochumer Erdbeben station in der Zeit vom 1. Dezember 1908 bis 1 juli 1909. Glückauf, **45**, 1006-1009.
Beobachtungen der Erdbebenstation der Westfälischen Berggewerschaftskasse in der Zeit vom 20. Dez. 1909 bis zum 19. Dez. 1910. Glückauf ***46***, 1910.
Weekly reports published in Glückauf until 1945.

68. Darmstadt, Jugenheim

Beobachtungen der seismischen Station 01.–06.1912 (printed).
At ISC: 01., 05., 06., 08.–10.1911, 04.–06.1912, 03.– 04.1913, 01.–04.1914.

69. Göttingen

Linke, Frank, and Emil Wiechert (1903). Monatsberichte über seismische Registrierungen in Göttingen, 1903 Januar. Königlich Geophysikalisches Institut zu Göttingen, 4 pp.
Borne, Georg von dem (1904). Seismische Registrierungen in Göttingen, Juli bis Dezember 1903. Nachrichten von der Königlichen Gesellschaft der Wissenschaften zu Göttingen, Mathematisch-physikalische Klasse, 440-464.
Schering, H. (1905). Seismische Registrierungen in Göttingen im Jahre 1904. *Ibid.,* 181-200.
Angenheister, Gustav Heinrich (1906). Seismische Registrierungen in Göttingen im Jahre 1905 (including: Ernst Wiechert: Uebersicht über die registrierenden Seismometer der Station (pp. 377-379). *Ibid.,* 357-416.
Zoeppritz, Karl (1908). Seismische Registrierungen in Göttingen im Jahre 1906. *Ibid.,* 129-190.
Geiger, Ludwig (1909a). Seismische Registrierungen in Göttingen im Jahre 1907 mit einem Vorwort über Die Bearbeitung der Erdbebendiagramme. *Ibid.,* 107-123, and 124-151.
Geiger, Ludwig (1909b). Seismische Registrierungen in Göttingen im Jahre 1908 mit einem Vorwort über Hilfsmittel zur Berechnung der wahren Bodenschwankung. *Ibid.,* 152-165, and 166-203.
Geiger, Ludwig (1913). Seismische Registrierungen in Göttingen im Jahre 1909. *Ibid.,* 365-391.
Geiger, Ludwig (1914). Seismische Registrierungen in Göttingen im Jahre 1910. *Ibid.,* 245-271.
Ansel, A. (1913). Seismische Registrierungen in Göttingen im Jahre 1911. *Ibid.,* 289-325.
Geophysikalisches Institut-Göttingen, weekly reports: No. 1–52, 1912; No. 1–52, 1913; No. 1–41 (October 11), 1914 (all in single leaves).

70. Hamburg

Mittheilungen der Horizontalpendel-Station Hamburg, monthly reports: 10.1900–06.1903.
Mitteilungen der Hauptstation für Erdbebenforschung am Physikalischen Staatslaboratorium zu Hamburg 01.1903–09.1905, 11.1905–03.1906–12.1907.
05.1903–03.1904 used in Strasbourg's "Von den Instrumenten . . ." and "Verzeichnis der im . . . ," see 60.
Die seismischen Registrierungen in Hamburg. Mitteilungen der Hauptstation für Erdbebenforschung am Physikalischen Staatslaboratorium zu Hamburg, 1. April–31. Dezember 1908, weekly reports.
Tams, E. (1909). Die seismischen Registrierungen in Hamburg vom 1. April 1908 bis zum 31. Dezember 1908. Mitteilungen aus dem physikalischen Staatslaboratorium in Hamburg. 6. Beiheft zum Jahrbuch der Hamburgischen Wissenschaftlichen Anstalten **26 1908**.
Tams, E. (1910). Die seismischen Registrierungen in Hamburg vom 1. Januar 1909 bis zum 31. Dezember 1909 nach den Beobachtungen der Hauptstation für Erdbebenforschung am physikalischen Staatslaboratorium in Hamburg. Mitteilungen aus dem physikalischen Staatslaboratorium in Hamburg. 5. *Ibid.,* **27 1909**.
Mitteilungen der Hauptstation für Erdbebenforschung am Physikalischen Staatslaboratorium zu Hamburg 1. Januar bis 31. Dezember 1909, weekly reports. *Ibid.,* 1910. *Ibid.,* 1911.
Monatliche Mitteilungen der Hauptstation für Erdbebenforschung am Physikalischen Staatslaboratorium zu Hamburg, 01.1912–12.1913, 08.1914–03.1915, 01.1918–12.1920.
At ISC: 04.–12.1915.

71. Heidelberg (Königstuhl)

Grossherzogliche Sternwarte Heidelberg, or Astrophysikalisches Institut Koenigstuhl-Heidelberg, or Koenigstuhl-Sternwarte Heidelberg, or Grossherzoglich Badische Sternwarte Heidelberg: monthly reports: 06.1909–12.1913, 02.1914–09.1916.

72. Hohenheim

Mack, Karl (1907). Die Erdbebenwarte in Hohenheim und ihre Einrichtung und Erderschütterungen in Hohenheim während des Zeitraums vom 1. April 1905 bis 31. Dezember 1906. Deutsches Meteorologisches Jahrbuch 1906, Württembergisches Teilheft, 8 pp.

Mack, Karl (1908–1917). Nachrichten von der Hohenheimer Erdbebenwarte aus dem Jahr 1907 und Erderschütterungen in Hohenheim während des Jahres 1907. Deutsches Meteorologisches Jahrbuch 1907, Württembergisches Teilheft, 10 pp.; *Ibid.*, Jahrbuch 1908, 12 pp.; *Ibid.*, Jahrbuch 1909, 11 pp.; *Ibid.*, Jahres 1910., 16 pp.; *Ibid.*, Jahres 1911, 23 pp.; *Ibid.*, Jahres 1912, 12 pp.; *Ibid.*, Jahres 1913, 13 pp.; *Ibid.*, Jahres 1914, 20 pp.; *Ibid.*, Jahres 1915, 19 pp.; *Ibid.*, Jahres 1916, 1917, 1918, 45 pp.

Mack, Karl (1923?). Nachrichten von der Hohenheimer Erdbebenwarte und Erderschütterungen in Württemberg während des Jahres 1919, Manuscript 28 pp.

Mack, Karl (1923?). *Ibid.*, Jahres 1920, Manuscript 13 pp.

73. Jena

Monatliche Erdbebenberichte der Seismischen Station zu Jena, 04.1905–12.1911.

Monatliche Erdbebenberichte der Hauptstation für Erdbebenforschung zu Jena, 1912.

Pechau, W. (1914/15). 1913 Erdbebenaufzeichnungen der Hauptstation für Erdbebenforschung Jena.

74. Leipzig

Credner, H. (1898). Die sächsischen Erdbeben während d. Jahre 1889 bis 1897 insbesondere das sächsisch-böhmische Erdbeben vom 24. October bis 29. November 1897. Sächsische Gesellschaft der Wissenschaften zu Leipzig, 83 pp.

Credner, H. (1907). Die sächsischen Erdbeben während der Jahre 1904 bis 1906. Ber. math.-phys. Kl. Kgl. sächs. Gesellschaft der Wissenschaften zu Leipzig 59, 333-355.

Etzold, Franz (1902). Das Wiechertsche astatische Pendelseismometer der Erdbebenstation Leipzig und die von ihm gelieferten Seismogramme von Fernbeben. *Ibid.*, **54**, 283-326.

Etzold, Franz (1903a). Die von Wiecherts astatischem Pendelseismometer in der Zeit vom 15. Juli bis 31. Dezember 1902 in Leipzig gelieferten Seismogramme von Fernbeben. *Ibid.*, **55**, 22-38.

Etzold, Franz (1903b). Bericht über die von Wiecherts astatischem Pendelseismometer in Leipzig vom 1. Januar bis 30. Juni 1903 registrierten Fernbeben und Pulsationen. *Ibid.*, **55**, 296-321.

Etzold, Franz (1904a). Vierter Bericht der Erdbebenstation Leipzig. Die in Leipzig vom 1. Juli 1903 bis 30. April 1904 von Wiecherts astatischem Pendelseismometer registrierten Erdbeben und Pulsationen. *Ibid.*, **56**, 289-295.

Etzold, Franz (1904b). Fünfter Bericht der Erdbebenstation Leipzig. I Die in Leipzig vom 1. Mai bis 31. Oktober 1904 registrierten Erdbeben und Pulsationen. II Über die Aufzeichnung der infolge des Läutens der Kirchenglocken zu Leipzig erzeugten Bodenschwingungen. *Ibid.*, **56**, 302-310.

Etzold, Franz (1906). Sechster Bericht der Erdbebenstation Leipzig. I. Die in Leipzig vom 1. November 1904 bis 31. Dezember 1905 und die in Plauen vom 17. August bis 31. Dezember 1905 aufgezeichneten Seismogramme. II. Die in Leipzig und Plauen vom 1. November 1904 bis 31. Dezember 1905 aufgezeichneten, nicht von Erdbeben herrührenden Bewegungen. *Ibid.*, **58**, 81-105.

Etzold, Franz (1907a). Siebenter Bericht der Erdbebenstation Leipzig. I. Die in Leipzig und Plauen vom 1. Januar bis 31. Dezember 1906 aufgezeichneten Seismogramme. II. Die in Leipzig und Plauen vom 1. Januar bis 31. Dezember 1906 aufgezeichneten pulsatorischen Bewegungen. *Ibid.*, **59**, 2-34.

Etzold, Franz (1907b). Achter Bericht der Erdbebenstation Leipzig. I. Die in Leipzig und Plauen vom 1. Januar bis 30. Juni 1907 aufgezeichneten Seismogramme. II. Die in Leipzig und Plauen vom 1. Januar bis 30. Juni 1907 aufgezeichneten pulsatorischen Bewegungen. *Ibid.*, **59**, 356-370.

Etzold, Franz (1908a). Neunter Bericht der Erdbebenstation Leipzig. I Die in Leipzig vom 1. Juli bis 31. Dezember 1907 aufgezeichneten Seismogramme. II Die in Leipzig und Plauen vom 1. Januar bis 30. Juni 1907 aufgezeichneten pulsatorischen und sonstige nicht seismischen Bewegungen. *Ibid.*, **60**, 57-78.

Etzold, Franz (1908b). Zehnter Bericht der Erdbebenstation Leipzig. I Die in Leipzig und Plauen vom 1. Januar bis 30. Juni 1908 aufgezeichneten Seismogramme. II Die in Leipzig und Plauen vom 1. Januar bis 30. Juni 1907 aufgezeichneten pulsatorischen und sonstige nicht seismischen Bewegungen. *Ibid.*, **60**, 223-239.

Etzold, Franz (1909). Elfter Bericht der Erdbebenstation Leipzig. Die in Leipzig und Plauen vom 1. Juni bis 31. Dezember 1908 aufgezeichneten Seismogramme. *Ibid.*, **61**, 62-91.

Etzold, Franz (1910). Zwölfter Bericht der Erdbebenwarte zu Leipzig. Die in Leipzig und Plauen während des Jahres 1909 aufgezeichneten Seismogramme. *Ibid.*, **62**, 3-31.

Etzold, Franz (1911). Dreizehnter Bericht der Erdbebenstation Leipzig. Die in Leipzig und Plauen während des Jahres 1910 aufgezeichneten Seismogramme. *Ibid.*, **63**, 291-315.

75. München

Messerschmitt, J.B. (1907). Die Registrierungen der letzten großen Erdbebenkatastrophen auf der Erdbebenstation in München. Mitteilungen der geographischen Gesellschaft in München **2**, 197-235.

Messerschmitt, J.B. (1909a). *Ibid.*, **4**, 127-131.

Messerschmitt, J.B. (1909b). Magnetische Beobachtungen in München aus den Jahren 1901 bis 1905 und Erdbebenregistrierungen vom Jahre 1905. Veröffentlichungen des Erdmagnetischen Observatoriums und der Erdbebenhauptstation bei der Königlichen Sternwarte in München **2**, 38-43.

Messerschmitt, J.B. (1909c). Registrierungen einiger südeuropäischer Erdbeben auf der Münchener Erdbebenstation. Sitzungsberichte der Königlich Bayerischen Akademie der Wissenschaften, Mathematisch-Physikalische Klasse 1909, 16. Abhandlung, 13 pp.

ISA: Messerschmitt, J.B. Seismische Beobachtungen in München 1908.

Brunhuber, A., and J.B. Messerschmitt (1910). Die Beobachtungen der beiden sächsisch-böhmischen Erdbebenschwärme vom Oktober und November 1908 im nordöstlichen Bayern und die Registrierungen auf der Münchener Erdbebenstation. Berichte des naturwissenschaftlichen Vereins zu Regensburg **12** (1907–1908).

München. Seismische Aufzeichnungen der Königlich Bayerischen Erdbeben-Hauptstation: Handwritten reports in single leaves: No. 1–No. 29 1911, No. 1–No. 30 1912, No. 1–No. 26 1913 and typescript copy No. 1–No. 16 1913, handwritten No. 1–No. 28 1914, No. 1–No. 28 1915.

München. Seismische Aufzeichnungen der Königlichen Erdbebenwarte München (Sternwarte): typescript No. 1–No. 21 1914, handwritten No. 1–No. 17 1916, No. 1–No. 18 1917, No. 1–No. 15 1918, No. 1–No. 14 1919, No. 1–No. 13 1920.

76. Nördlingen

Nördlingen. Seismische Aufzeichnungen der Königlich Bayerischen Erdbebenzweigstation I. Handwritten reports in single leaves: No. 1–No. 9 1912, No. 1–No. 9 1913, 10 pp. without numbering for 1914 and typescript copy No. 1–No. 7 1914, No. 1–No. 14 1915, No. 1–No. 11 1916, No. 1–No. 8 1917, No. 1–No. 8 1918, No. 1–No. 7 1919, No. 1–No. 6 1920.

77. Plauen

Published together with the bulletins of Leipzig, see 74.

78. Potsdam

Rebeur-Paschwitz (1889, 1892, 1895).

Rudolph (1903a, b).

BAASRP **1897**, 171 (1897-01-03–1897-03-06); **1899**, 184 (1897-10-02–1898-01-06).

Seismometrische Beobachtungen in Potsdam in der Zeit vom 1. April bis 31. Dezember 1902. Veröffentlichung des Königlich Preuszischen Geodätischen Institutes, Neue Folge.

Seismometrische Beobachtungen in Potsdam in der Zeit vom 1. Januar bis 31. Dezember 1903. *Ibid.*

Hecker, Oskar (1905). Seismometrische Beobachtungen in Potsdam in der Zeit vom 1. Januar bis 31. Dezember 1904. *Ibid.,* Neue Folge **21**, 6 + 119 pp.

Seismometrische Beobachtungen in Potsdam in der Zeit vom 1. Januar bis 31. Dezember 1905. *Ibid.*

Hecker, Oskar (1907–1910). Seismometrische Beobachtungen in Potsdam in der Zeit vom 1. Januar bis 31. Dezember 1906. *Ibid.,* Neue Folge **30**, 59 pp.; *Ibid.,* 1907, Neue Folge **35**, 64 pp.; *Ibid.,* 1908. *Ibid.,* Neue Folge **42**, 37 pp.

Meissner, O. (1910–1912). Seismometrische Beobachtungen in Potsdam in der Zeit vom 1. Januar bis 31. Dezember 1909. *Ibid.,* Neue Folge **47**, 26 pp.; 1910, Neue Folge **50**, 27 pp.; 1911, Neue Folge **55**, 46 pp.

Seismometrische Beobachtungen in Potsdam in der Zeit vom 1. Januar bis 31. Dezember 1912; *Ibid.*, 1913.; *Ibid.*, 1914, Neue Folge **64**, 25 pp; *Ibid.,*1915, Neue Folge **67**, 21 pp.; *Ibid.,* 1916, Neue Folge **73**, 19 pp.; *Ibid.,* 1917 and 1918., Neue Folge **76**, 25 pp.

Meissner, Otto, J. Picht, and R. Berger (1926). Seismometrische Beobachtungen in Potsdam in der Zeit vom 1. Januar 1919 bis 31. Dezember 1924. *Ibid.,* Neue Folge **96**, 18 pp.

79. Ravensburg

Mack, Karl (1923?). Erderschütterungen in Ravensburg während des Jahres 1919, Manuscript together with Hohenheim (see 72), 3 pp.

Mack, Karl (1923?). Erderschütterungen in Ravensburg während des Jahres 1920, Manuscript together with Hohenheim (see 72), 9 pp.

80. Taunus

Nachrichten des Taunus-Observatoriums des Physikalischen Vereins zu Frankfurt am Main–Seismische Aufzeichnungen der Von Reinach'schen Erdbebenwarte. Monthly reports July 1913–June 1914.

81. Wilhelmshaven

Rebeur-Paschwitz (1889, 1892, 1895).

82. Accra (Akkra)

Used in the BAASSC (1913–1917) 1914.

83. Athénai (Athens)

*In Hamburg: Annales de l'Observatoire National d'Athènes **1** (1896)–**3*** (1912).

*In Hamburg: Eginitis, D. (1905). Etude des séismes survenus en Grèce pédant les années 1900–1903. Annales de l'Observatoire National d'Athènes **IV**, 135, 488.*

*In Hamburg: Eginitis, D. (1910). Etude des séismes survenus en Grèce pédant les années 1904–1908. Ibid., **V**, 62-67, 586-587.*

*In Hamburg: Eginitis, D. (1912). Etude des séismes survenus en Grèce pédant les années 1909–1912. Ibid., **VI**, 36, 318-320.*

At ISC: 01.1912–09.1914, 11.1914–10.1915, 07.– 12.1918, 04.1919–12.1920.

84. Port-au-Prince

Bulletin annuel de la Société Astronomique et Météorologique de Port-au-Prince, Haïti, 1901–1910.

Used in the BAASSC (1913–1917) 1914.

Mouvements sismique. Microsismes enregistrés á Port-au-Prince par le pendule vertic. Macrosismes. Bull. Semestrial de l'Observatoire Mét. du Séminaire-Collége St. Martial, Port-au-Prince. Juillet–Décembre 1910 (1911), 149-150. Ibid., Janvier–Juin 1911 (1912), 59-61. Ibid., Juillet–Décembre 1911 (1912), 147–162.

In Hamburg: Port-au-Prince Haiti Bulletin Semestrial 1910–1920.

85. Budapest

In Hamburg: Berichte der Erdbebenwarte Budapest 1902–1905.

05., 06., 08., 09., 11., 12.1903, 02., 03.1904 used in Strasbourg's "Von den Instrumenten . . . " and "Verzeichnis der im . . . ," see 60.
1906–1912 in Kövesligethy (1907, 1909, 1913), see 86.
Kövesligethy, Radó (1920). Rapport sur les observations sismologiques faites pendant les années 1913–1919 á l'observatoire de Budapest.
L'Observatoire de Budapest, 1913–1920.
Used in the BAASSC (1913–1917) 1913–1914.
BGS: 1894–1914.
NOAA: Rapport sur . . . 01.1906–12.1920, monthly 01.1912–04.1913, 10.1913–06.1914.

86. Hungary

In Hamburg: Shaferzik: Ungarische Erdbeben 1885–1886.
In Hamburg: Az 1894/95 evekben Magyarorszagon eszlelt foeldrengesek, Die in den Jahren 1894/95 in Ungarn beobachteten Erdbeben.
In Hamburg: Az 1896/99 evekben Magyarorszagon eszlelt foeldrengesek, Die in den Jahren 1896/99 in Ungarn beobachteten Erdbeben.
In Hamburg: Réthly, A. (1909). Az 1900, 1901 1902 évi Magyarországi földrengések: a M. Kir. Földmivelésügyi Ministerium fenhatósága alatt álló M. Kir. Orsz. Meteorológiai és Földmágnességi Inézet hivatalos kiadványa, Die Erdbeben in Ungarn im Jahre 1900, 1901 und 1902. Offizielle Publikation der dem Kgl. Ung. Ackerbauministerium unterstehenden Kgl. Ung. Reichsanstalt für Meteorologie und Erdmagnetismus, pp. 130.
In Hamburg: Réthly, A. (1906). Az 1903. Ibid., Az 1904. Ibid. Az 1905.
In Hamburg: Réthly, A. (1907). Az 1906. Ibid., 143 pp.
Kövesligethy, Radó (1906). Rapport annuel sur les observations sismiques des pays de la sainte couronne de Hongrie (1906 Évi jelentés a Magyar szent korona országainak földrengési állomasairól).
In Hamburg: Réthly, A. (1908). Az 1907 évi Magyarországi földrengések: a M. Kir. Földmivelésügyi Ministerium fenhatósága alatt álló M. Kir. Orsz. Meteorológiai és Földmágnességi Inézet hivatalos kiadványa, Die Erdbeben in Ungarn im Jahre 1907. Offizielle Publikation der dem Kgl. Ung. Ackerbauministerium unterstehenden Kgl. Ung. Reichsanstalt für Meteorologie und Erdmagnetismus.
Bulletin Hebdomadaire des Observotois sismiques de la Hongrie et la Croatie, Avril 1909, No. 16-18.
*Réthly, A. (1912). Die in Ungarn im Jahre 1911 beobachteten Erdbeben. Földtani Közlöny **42**, 82-92.*
In Hamburg: Avis et bulletin macroseismique de Hongrie 1906–1913.
Kövesligethy, Radó (1909). Rapport sur les observations faites pendant les années 1907 et 1908 aux observatoires sismiques des pays de la sainte couronne de Hongrie (Levö földrengési observatoriumokban 1907 es 1908 országani területén. Években vegzett megfigyelésekröl).
Kövesligethy, Radó (1913). Rapport sur les observations faites pendant les années 1909–1912 aux observatoires sismologique de Hongrie.
At ISC: 1896–1907, 1912, 1913, 1920.
NOAA: 1903–1913.

87. Kalocsa

1906–1912 in Kövesligethy (1907, 1909, 1913), see 86.
BGS: 1912–1920.

88. Reykjavik

At ISC: 01.–12.1910.
NOAA: 1910.
BGS: 1910.
Bulletin sismique de la Station Internationale de Reykjavik, 1910.
Übersicht über die an der Internationalen Erdbebenstation in Reykjavik registrierten Erdbebenstörungen, 1913 Januar–April. Mitteilungen des Zentralbureaus der Internationalen Seismologischen Assoziation **1**, **Nr. 2** (1913), 40. Supplement to: Gerlands Beträge zur Geophysik **13** (1914). *Ibid.*, 1913 Mai–Juli, **1**, **Nr. 4** (1914), 97–100. Supplement to: Gerlands Beträge zur Geophysik **13** (1914). *Ibid.*, 1913 August–Dezember, **2**, **Nr. 1** (1915), 19-21. Supplement to: Gerlands Beträge zur Geophysik **14** (1915–1918).

89. Bombay (Colaba)

BAASRP **1899**, 176-177 (1898-09-13–1899-03-23).
Shide Circular **1**, 13 (1898-09-22–1899-12-31); **2**, 44-45 (1900-01-04–1900-06-04); **3**, 85-86 (1900-07-29–1900-01-25); **4**, 108 (1901-01-07–1901-06-24); **5**, 135 (1901-08-06–1901-12-31); **6**, 159-160 (1902-01-01–1902-06-16); **7**, 208 (1902-07-09–1902-12-30); **8**, 242 (1903-01-04–1903-06-08); **9**, 287 (1903-08-11–1903-12-28); **10**, 309 (1904-01-20–1904-06-27); **11**, 333 (1904-07-18–1904-12-20); **12**, 15 (1905-01-22–1905-06-14); **13**, 43 (1905-07-02–1905-12-10); **14**, 80 (1906-01-06–1906-06-24); **15**, 122 (1906-07-14–1906-12-26); **16**, 171 (1907-01-01–1907-06-25); **17**, 202 (1907-07-09–1907-12-30); **18**, 239 (1908-01-11–1908-06-28); **19**, 274 (1908-07-13–1908-12-28); **20**, 315-316 (1909-01-04–1909-06-28); **21**, 366 (1909-07-07–1909-12-22); **22**, 409 (1910-01-01–1910-06-29); **23**, 36 (1910-07-07–1910-12-30); **24**, 88 (1911-01-01–1911-06-28); **25**, 152 (1911-06-30–1911-12-31); **26**, 216-217 (1912-01-04–1912-06-29); **27**, 266 (1912- 07-07–1912-12-28).
Magnetical, Meteorological, Atmospheric Electric and Seismographic Observations made at the Government Observatories, Bombay and Alibag in the years 1898–1899, 1900–1905, 1906–1910, 1911–1915, 1916– 1920.

90. Calcutta (Alipore)

BAASRP **1899**, 177-178 (1899-01-18–1899-03-26).
Shide Circular **1**, 11 (1899-09-29–1899-12-06); **3**, 84-85 (1900-07-07–1900-12-31), **4**, 108 (1901-01-08–1901-06-26); **5**, 134 (1901-07-04–1901-12-26); **6**, 158-159 (1901-12-31–1902-06-11); **7**, 207 (1902-07-09–1902-12-16); **8**, 240-242 (1903-01-04–1903-06-23); **9**, 286 (1903-07-01–1903-12-23);

10, 309 (1904-01-16–1904-06-27); **11**, 332 (1904-07-23–1904-12-20); **12**, 14 (1905-01-08–1905-06-30); **13**, 42-43 (1905-07-06–1905-12-18); **14**, 79-80 (1906-01-06–1906-06-29); **15**, 121-122 (1906-07-13–1906-12-23); **16**, 169 (1907-01-02–1907-06-30); **17**, 201 (1907-07-04–1907-12-25); **19**, 272-273 (1908-01-11–1908-12-18); **20**, 317-318 (1908-01-22–1908-06-22); **21**, 367-368 (1909-07-07–1909-12-23); **23**, 35-36 (1910-01-01–1910-12-30); **24**, 86-87 (1910-07-05–1911-06-17); **25**, 151-152 (1911-07-04–1911-12-31); **26**, 215-216 (1912-01-04–1912-06-29).
06., 12.1903 used in Strasbourg's "Von den Instrumenten ..." and "Verzeichnis der im ...," see 60.
Used in the BAASSC (1913–1917) 1913–1914.

91. Kodaikánal

Shide Circular **2**, 44 (1900-02-08–1900-06-21); **3**, 86-88 (1900-07-12–1900-12-30); **4**, 109 (1901-01-07–1901-06-24); **5**, 135-137 (1901-07-16–1902-01-01); **6**, 160-161 (1902-01-01–1902-06-28); **7**, 208-209 (1902-07-05–1902-12-28); **8**, 243-244 (1903-01-06–1903-06-22); **9**, 287-288 (1903-07-16–1903-12-28); **10**, 310 (1904-03-02–1904-06-29); **11**, 333-334 (1904-07-23–1904-12-20); **12**, 15-16 (1905-01-22–1905-06-30); **13**, 44 (1905-07-06–1905-12-10); **14**, 81 (1906-01-06–1906-06-24); **15**, 123 (1906-07-10–1906-12-26); **16**, 171-172 (1907-01-02–1907-06-25); **17**, 202 (1907-09-02–1907-12-30); **18**, 240 (1908-01-11–1908-06-30); **19**, 274-275 (1908-07-13–1908-12-28); **20**, 316-317 (1909-01-22–1909-06-27); **21**, 367 (1909-07-07–1909-12-29); **22**, 410 (1910-01-01–1910-06-29); **23**, 37 (1910-07-07–1910-12-30); **24**, 89 (1911-01-01– 1911-06-17); **25**, 153-154 (1911-07-04–1911-12-31); **26**, 218 (1912-01-04–1912-06-26); **27**, 266-267 (1912-07-07–1912-12-28).
06., 12.1903 used in Strasbourg's "Von den Instrumenten ..." and "Verzeichnis der im ...," see 60.
At ISC: 01.1907–06.1907.
Bulletin Kodaikanal Observatory, 1908–1920.
Used in the BAASSC (1913–1917) 1913–1915.

92. Madras

BAASRP **1899**, 175 (1898-05-21–1899-02-10).
Shide Circular **1**, 12 (1899-03-17–1899-11-24).

93. Simla

*In Hamburg: Patterson, J. (1909). The Simla seismograms obtained between June 1905 and November 1908. Memoirs of the Indian Meteorological Department **20**, part 3, 33-.*
01.–12.1916 published in BAASSC (1916) as supplement to "Large Earthquakes of 1916."
Used in the BAASSC (1913–1917) 1914–1915.

94. Indonesia

At KNMI: Vulkanische Verschijnslen en Aardbevingen in den Oost-Indischen Archipel, waargenomen gedurende het jaar: 1898–1920. Natuurk. Tijdschr. voor Ned.-Indiï.
In Hamburg: Indonesia, 1900–1913.
BGS: 1881–1920.

95. Jakarta (Batavia)

BAASRP **1899**, 178-179 (1898-06-04–1899-03-12).
Shide Circular **1**, 18–21 (1899-01-03–1900-01-21); **2**, 45-47 (1900-01-24–1900-06-26); **3**, 88-90 (1900-07-08–1900-12-28); **4**, 110 (1901-01-04–1901-06-25); **5**, 137-138 (1901-07-04–1901-12-30); **6**, 161-162 (1902-01-03–1902-06-29); **7**, 209-210 (1902-07-06–1902-12-28); **8**, 244-246 (1903-01-01–1903-06-25); **9**, 288 (1903-07-03–1903-12-31); **10**, 310-311 (1904-01-01–1904-06-22); **11**, 334-335 (1904-07-01–1904-12-23); **12**, 16 (1905-01-08–1905-06-30); **13**, 45-47 (1905-07-03–1905-12-28); **14**, 82-83 (1906-01-15–1906-06-30); **15**, 124-126 (1906-07-01–1906-12-27); **16**, 172-173 (1907-01-01–1907-06-27); **17**, 204-205 (1907-07-01–1907-12-30); **18**, 242-243 (1908-01-05–1908-06-30); **20**, 322-323 (1908-07-01–1908-12-28); **21**, 369-371 (1909-01-01–1909-06-30).
Observations made at the Royal Magnetical and Meteorological Observatory at Batavia, 1900–1906.
1904 used in Rudolph (1907).
05.–07., 09.–12.1903, 01., 03.1904 used in Strasbourg's "Von den Instrumenten ..." and "Verzeichnis der im ...," see 60.
*Bemmelen, W. van. Observations made at the Royal Magnetical and Meteorological Observatory at Batavia 1906. Volume **29**.*
Bemmelen, W. van. Seismological Bulletin, Batavia Observatory, Java, 1909–1911.
Erdbeben Bericht, Koninklijk Magnetisch en Meteorologisch Observatorium, Batavia, 1909.
Seismological Bulletin, Royal Magnetical and Meteorological Observatory, Koninklijk Magnetisch en Meteorologisch Observatorium Batavia, 1910–1920.
Used in the BAASSC (1913–1917) 1913–1915.
BGS: 1881–1920.
NOAA: 1900–1920.

96. Koeta Radja, Sumatra

1904 used in Rudolph (1907).
01.–03.1904 used in Strasbourg's "Verzeichnis der im ...," see 60.

97. Cork

Shide Circular **26**, 192 (1912-01-04–1912-05-23); **27**, 256 (1912-08-09–1912-12-09).
Annual Bulletins of the University 1912–1918.

98. Limerick

Used in the BAASSC (1913–1917) 1913–1914.

99. Benevento

From 02.1895 included in the national bulletin of Italy, see 107.

100. Casamicciola (Isola d'Ischia)

From 02.1895 included in the national bulletin of Italy, see 107.
BAASRP **1897**, 170 (1896-06-14–1897-02-19); **1899**, 187 (1898-04-15–1899-03-07).
Rudolph (1903a, b).

101. Catania

From 04.1895 included in the national bulletin of Italy, see 107.
BAASRP **1898**, 198 (1897-03-23–1898-01-29); **1899**, 188-189 (1898-03-29–1899-03-23).
Rudolph (1903a, b).
Arcidiacono, S. (1903). Sui recenti terremoti etnei, Bullettino dell'Accademia Gioenia di scienze naturali in Catania, Fascicolo LXXIX–Dicembre 1903.
At ISC: 01.–07.1908, 10.1907–06.1909.
Bulletino Sismologico from 10.1913.
NOAA: 1911–1915.

102. Chiavari

At ISC: 01.1913–12.1914.

103. Firenze (Quarto-Castello)

In Hamburg: Stiattesi, D. Raffaello (1900–1903). Spoglio delle osservazioni sismiche dal 1° Novembre 1898 al 31° Ottobre 1899 (anno meteorico 1899). Bollettino sismografico dell' Osservatorio di Quarto (Firenze), 79 pp.; Ibid., 1° Novembre 1899 al 31° Ottobre 1900 (anno meteorico 1900), 62 pp.; Ibid., 1° Novembre 1900 al 31° Luglio 1901, 71 pp.; Ibid., 1° Agosto 1901 al 31° Luglio 1902, 73 pp.; Ibid., 1° Agosto 1902 al 30° Novembre 1903, 84 pp.
At KNMI: Stiattesi, D. Raffaello (1909). Spoglio delle osservazioni sismiche dal 1° Dicembre 1903 al 30° Novembre 1906. Ibid., 115 pp.
11.1903 Used in Strasbourg's "Verzeichnis der im . . . ," see 60.
ISA: 1904–1908.
BGS: 1903–1906.

104. Firenze (Querce)

At KNMI: 1898–1906.
At ISC: 01.1905–04.1906, 01.1911–05.1914, 09.–10. 1914, 01.–12.1920.
From 03.1895 included in the national bulletin of Italy, see 107.
Used in the BAASSC (1913–1917) 1913–1914.

105. Firence (Ximeniano)

From 02.1895 included in the national bulletin of Italy, see 107.
In Hamburg: Osservazioni dell'anno 1901. Bollettino sismologico dell'Osservatorio Ximeniano dei PP. Delle Scuole Pie di Firenze, Anno 1, Firenze 1902, 103 pp.; dell'anno 1902, 46 pp.; dell'anno 1903; dell'anno 1904.
In Hamburg: Osservatorio Ximeniano, Registrazioni sismiche 1901–1905.
05.1903–03.1904 used in Strasbourg's "Von den Instrumenten . . . " and "Verzeichnis der im . . . ," see 60.
ISA: 1904–1908.
Registrazioni sismografiche all'Osservatorio Ximeniano Firenze, 1901–1903.
Osservazioni dell'anno 1909. Bollettino sismologico dell'Osservatorio Ximeniano dei PP. Delle Scuole Pie di Firenze; Ibid., dell'anno 1914.
Registrazioni sismiche, Osservatorio Ximeniano dei Padri delle Scuole Pie, No 13, 1913.
Pubblicazioni dell'Osservatorio Ximeniano dei Padri Scolopi, 1913–1920.
BGS: 1901–1920.
NOAA: 01.–04.1908, 11.1908–04.1909, 12.1909, 02.– 06.1910, 07., 08., 12.1913, 01., 03., 05., 06., 1914, 1919, 1920.

106. Genova

From 02.1895 included in the national bulletin of Italy, see 107.

107. Italy

1894 and earlier: the editorial notice to Baratta (1895) contains the following sentence: "La presente pubblicazione tiene luogo del ***Supplemento*** *che, con le notizie sismiche a tutto il 1894, fu unito* ***Bollettino quotidiano del R. Ufficio Centrale di Meteorologia e di Geodinamica****." (This publication replaces the* ***Supplemento*** *which, together with the seismic news until 1894, was joint to the daily* ***Bollettino quotidiano del R. Ufficio Centrale di Meteorologia e di Geodinamica****.)*
Many Italian station readings for the years 1894–1897 were included in Rudolph (1904).
Baratta, Mario (1895). Notizie sui terremoti avvenuti in Italia durante l'anno 1895. Regio Ufficio Centrale di Meteorologia e Geodinamica, 230 pp. Appendice al Bollettino della Società Sismologica Italiana **1**.
Palazzo, Luigi (1896). Notizie sui terremoti avvenuti in Italia durante l'anno 1896, 171 pp. *Ibid.*, **2**.
Agamennone, Giovanni (1897–1900). Notizie sui terremoti osservati in Italia durante l'anno 1897, 318 pp. Appendice al Bollettino della Società Sismologica Italiana **3**; l'anno 1897 (2.° semestre), 255 pp., **4**; *Ibid.*, l'anno 1898, 301 pp., **5**.
Cancani, Adolfo (1900–1904). Notizie sui terremoti osservati in Italia durante l'anno 1899, 293 pp. Appendice al Bollettino della Società Sismologica Italiana, **6**; *Ibid.*, l'anno 1900, 259 pp., **7**; l'anno 1901, 540 pp., **8**; *Ibid.*, l'anno 1902, 559 pp., **9**.
Agamennone, Giovanni (1904–1905). Notizie sui terremoti osservati in Italia durante l'anno 1903, 585 pp. *Ibid.*, **10**.
Monti, Virgilio (1906–1907). Notizie sui terremoti osservati in Italia durante l'anno 1904, 583 pp., Appendice al Bollettino della Società Sismologica Italiana, **11**; *Ibid.*, l'anno 1905, 658 pp., **12**.
Martinelli, Giuseppe (1908–1914 [1923]). Notizie sui terremoti osservati in Italia durante l'anno 1906, 521 pp. Appendice al Bollettino della Società Sismologica Italiana, **13**; *Ibid.*, l'anno 1907, 567 pp., **14**; *Ibid.*, l'anno 1908, 645 pp., **15**; *Ibid.*, l'anno 1909, 629 pp., **16**; *Ibid.*, l'anno 1910, 647 pp., **17**; *Ibid.*, l'anno 1911, 587 pp., **18**.
Cavasino, Alfonso (1934–1935). Notizie sui terremoti osservati in Italia durante l'anno 1912. Regio Ufficio Centrale di

Meteorologia e Geodinamica, 431 pp.; *Ibid.*, l'anno 1913, 438 pp.

Cavasino, Alfonso (1927). Bollettino sismico anno 1917, Fasc. 1, Microsismi, Regio Ufficio Centrale di Meteorologia e Geodinamica, 127 pp.; *Ibid.*, anno 1918, 102 pp.; *Ibid.*, anno 1919, 127 pp.; *Ibid.*, anno 1920, 136 pp.

Ingrao, G. (1927a). Bollettino sismico anno 1917, Fasc. 2, Macrosismi, Regio Ufficio Centrale di Meteorologia e Geodinamica, 46 pp.; *Ibid.*, anno 1918, 22 pp.; *Ibid.*, anno 1919, 26 pp.; *Ibid.*, anno 1920, 27 pp.

108. Messina

In Hamburg: Annuario dell'anno 1906. Osservatorio di Messina. Istituto di Fisica Terrestre e Meteorologia delle R. Università. Messina 1904–1907.

109. Mileto (Morabito)

Labozzetta, R. Bolletino sismologico dell'Osservatorio "Morabito" nel seminario di Mileto (Calabria) ***1****, 1908.*
At ISC: 01.–06.1908, 10.1908–12.1909, 07.1910– 12.1912.
BGS: 1910–1912.

110. Mineo (Guzzanti)

From 01.1895 included in the national bulletin of Italy, see 107.
In Hamburg: Osservatorio Meteorico-Geodinamico "Guzzanti" in Mineo, Bollettino Mensile 1900–1910.

111. Moncalieri, close to Torino (Turin)

At ISC: 12.1906–12.1920.
In Hamburg: Moncalieri Bollettino Osservazioni sismiche 1907–1920.
BGS: 1908–1920.
NOAA: 12.1908–12.1920.
Used in the BAASSC (1913–1917) 1914–1915.

112. Monte Cassino (Montecassino)

Used in the BAASSC (1913–1917) 1913–1914.
NOAA: 1911, 1914–07.1916, 11.1916–12.1917, 01.–03. 1919.

113. Padova (Padua)

From 02.1895 included in the national bulletin of Italy, see 107.
Rudolph (1903a, b).
NOAA: 09.1895–06.1901, 01.–06.1902, 1903, 1904, 08.1910–08.1912, 10.1912–09.1913, 03.1914– 09.1915.
At ISC: 01.–12.1903.
05.1903–03.1904 used in Strasbourg's "Von den Instrumenten . . ." and "Verzeichnis der im . . . ," see 60.
Bolletino mensile delle registrazioni dei microsismografi dell'Istituto di Fisica della R. Universita Padova. A. R. I. Veneto di Sc., lett. ed Arti ***66****, 1906/07; Ibid.,* ***67****, 1907/08.*
In Hamburg: Bolletino Mensile di Padova 1907–1913.
Used in the BAASSC (1913–1917) 1913–1915.

114. Pavia

From 02.1895 included in the national bulletin of Italy, see 107.
Rudolph (1903a, b).

115. Pisa (Capannoli, Baldini)

BGS: Bolletino sismologico dell'Osservatorio Baldini. Capannoli, Pisa, 1910–1912.
At ISC: 03.–09., 11.1910–02.1911, 07.–12.1911.
NOAA: Osservatorio Baldini 07.–12. 1911.

116. Portici (Napoli)

From 02.1895 included in the national bulletin of Italy, see 107.

117. Reggio di Calabria

From 01.1895 included in the national bulletin of Italy, see 107.

118. Rocca di Papa

From 01.1895 included in the national bulletin of Italy, see 107.
Rudolph (1903a, b).
BAASRP **1897**, 172 (1896-06-14–1897-02-19); **1898**, 194 (1897-05-23–1898-01-29); **1899**, 186 (1898-04-15–1899-03-07).
Used in the BAASSC (1913–1917) 1915.
BGS: 1912–1920.
NOAA: 1917–1920.

119. Roma

From 01.1895 included in the national bulletin of Italy, see 107.
Rudolph (1903a, b).
At ISC: 07.–12.1920.

120. Siena

Vicentini, G. (1894). Osservazioni sismiche. Atti della R. Academia dei Fisiocritici, Siena 1894, (4) ***V****.*
Vicentini, G. (1894). Movimenti sismici. Ibid.
Cinelli, Modesto (1894). Sulle registrazione del microsismografo Vicentini avute a Siena dal 15 luglio al 31 ottobre 1894. Ibid.
Lussana, Silvio (1895). Osservazioni sismiche fatte col microsismografo Vicentini. Ibid., Siena 1895, ***VI****[with observations 11.1894–04.1895].*
Vicentini, G. (1896–1897). Sugli apparecchi impiegati nello studio delle ondulazioni del suolo. Ibid., Siena 1896–1897, ***VIII*** *[with observations 03.1894–09. 1896].*
From 02.1895 included in the national bulletin of Italy, see 107.
Rudolph (1903a, b).

121. Trieste

1899, 185 (1899-01-22–1899-03-25).
Mazelle, Eduard (1899). Die Einrichtung der seismischen Station in Triest und die vom Horizontalpendel aufgezeichneten

*Erdbebenstörungen von Ende August 1898 bis Ende Februar 1899. Mitteilungen der Erdbeben-Kommission der kaiserlichen Akademie der Wissenschaften in Wien **XI**. In: Sitz. Ber. der kaiserlichen Akademie der Wissenschaften in Wien **108**, mathematisch-naturwissenschaftliche Classe, Abth. 1, Heft V.*

*Mazelle, Eduard (1900). Erdbebenstörungen zu Triest, beobachtet am Rebeur-Ehlert'schen Horizontalpendel vom 1. März bis Ende Dezember 1899. Mitteilungen der Erdbeben-Kommission der kaiserlichen Akademie der Wissenschaften in Wien **XVII**. Ibid., **109** , Abth. 1, Heft II.*

*In Hamburg: Mazelle, Eduard: Erdbebenstörungen zu Triest, beobachtet am Rebeur-Ehlert'schen Horizontalpendel im Jahre 1900. Mitteilungen der Erdbeben-Kommission der kaiserlichen Akademie der Wissenschaften in Wien, Neue Folge **V**; Ibid., im Jahre 1901, Neue Folge **XI**.*

Mazelle, Eduard (1903–1906). Erdbebenstörungen zu Triest, beobachtet am Rebeur-Ehlert'schen Horizontalpendel im Jahre 1902. Mitteilungen der Erdbeben-Kommission der kaiserlichen Akademie der Wissenschaften in Wien, Neue Folge **XX**, 87 pp.; *Ibid.*, Jahre 1903, Neue Folge **XXX**, 37 pp.

At ISC: 01.1907–03.1912, 07., 08.1912, 12.1912–12.1918.

ISA: 1907, 1908 weekly listings.

Used in the BAASSC (1913–1917) 1913–1914.

NOAA: 10.1910–07.1914.

122. Valle di Pompei (Pompei)

In Hamburg: Bollettino Valle di Pompei 1908–1920.

At ISC: 01.–12.1909.

Used in the BAASSC (1913–1917) 1913–1915.

BGS: 1908–1908.

NOAA: 08.1910–08.1914, 01.–04.1915, 09.–12.1916, 01.1918–12.1920.

123. Venezia

Risultato delle Osservazioni sismografiche eseguite nell'osservatorio del Seminario Patriarcale di Venezia nell'anno 1907. R. Istituto Veneto di Scienze, Lettere et Arti, Venezia 1907–08 (1909).

Risultato delle Osservazioni sismografiche eseguite nell'osservatorio del Seminario Patriarcale di Venezia nell'anno 1908. Ibid., 1908–09 (1910).

In Hamburg: Osservatorio di Venezia, Bollettino mensile 1919–1925.

124. Japan

Central Meteorological Observatory stations in Shide Circular **1**, 27-28 (1900-02-17); **8**, 266-270 (1900-03-12–1902-11-21); **9**, 294-296 (1902-12-31–1903-04-19).

Central Meteorological Observatory: Annual Report of the Central Meteorological Observatory of Japan. Part II on the Earthquakes in the year 1900, Tokio, 1909.

Bulletin of the Imperial Earthquake Investigation Committee **1**, **No. 3**, Tokyo 1907.

Omori, F. (1908). List of stronger Japan earthquakes 1902–1907. Bulletin of the Imperial Earthquake Investigation Committee **2**, 58–88.

125. Kobe

Seismological Bulletin of the Kobe Observatory, 1919, 1920.

126. Mizusawa

Seismological Observations at Mizusawa for the period between 1902–1967. International Latitude Observatory of Mizusawa 1984, 380 pp.

127. Nagasaki

Seismic Bulletin in Nagasaki, 04.1913–12.1920.

128. Osaka

The Seismological Bulletin in Osaka from 1882 to 1929, the Osaka Meteorological Observatory 1931, 132 pp.

Used in the BAASSC (1913–1917) 1913–1915.

129. Tokyo, Central Meteorological Office (Hitotsubashi)

Chaplin, W.S. An examination of the earthquakes recorded at the Meteorological Observatory, Tokio. Trans. Asiatic Soc. of Japan **6**, part 4.

BAASRP **1897**, 133-137 (1895-05-06–1896-12-17); **1899**, 188–191 (1898-01-27–1899-01-29).

Shide Circular **1**, 29–30 (1899-01-29–1899-12-31); **3**, 90–92 (1900-01-15–1900-12-25); **5**, 141–144 (1901-01-01–1901-12-26); **7**, 223–225 (1902-01-01–1902-12-31).

1904 used in Rudolph (1907).

Tokyo, Japan, Seismic Bulletin of the Central Meteorological Observatory of Japan, 01.–48, 50.–52.1920.

Omori, F. (1903). Observations of earthquakes at Hitotsubashi (Tokyo), Earthquake Inv. Comm. **13**, 1, 143.

130. Tokyo, University (Hongo)

Shide Circular **1**, 24-26 (1899-07-24–1900-02-17); **7**, 218-223 (1900-02-25–1902-12-28); **9**, 293-294 (1902-12-31–1903-06-02); **11**, 344-345 (1903-07-09–1904-12-30); **15**, 144-147 (1905-01-13–1906-12-12); **20**, 319-321 (1907-01-02–1908-12-28); **24**, 91-93 (1909-01-28–1910-12-26).

Bulletin data from station Tokyo-Hongo in: Publications of the earthquake investigation committee in foreign languages **21**, Tokyo 1905, 102 pp.

NOAA: 1901–1920.

131. Wjernoje (Vernyï), Alma Ata

Bulletin of Imperial Russia 1906, 1907, see 168.

132. Inch'on (Tyosen, Zinsen, Chemulpo, Jinsen)

NOAA: 1918–1920.

133. Beirut

Shide Circular **11**, 339-340 (1904-01-20–1904-12-23); **12**, 19 (1904-12-31–1905-06-30); **13**, 48-49 (1905-07-06–1905-12-17); **14**, 89 (1906-01-06–1906-06-29); **15**, 134-135 (1906-07-06–1906-12-26); **17**, 208-209 (1907-01-01–1907-12-23); **18**, 237 (1908-01-08–1908-06-27); **19**, 269 (1908-07-13–1908-12-28); **20**, 308 (1909-01-05–1909-06-22); **21**, 362 (1909-06-27–1909-12-09); **22**, 403-404 (1910-01-16–1910-06-17); **23**, 39 (1910-01-16–1910-06-17); **26**, 204-205 (1911-03-02–1912-06-27); **27**, 265 (1912-07-07–1912-12-09).
Used in the BAASSC (1913–1917) 1914.

134. Ksara

Berloty, B. Bulletin sismique. Observatoire de Ksara 1910–1911.
At ISC: 01.1913–06.1914.
Annales de l'Observatoire de Ksara (Liban), Observations, Section séismologique [?–1921].
Used in the BAASSC (1913–1917) 1913–1914.
NOAA: 04.1911–12.1911, 01.1913–05.1914.

135. Valetta

Shide Circular **15**, 117-118 (1906-07-07–1906-12-22); **16**, 165-166 (1907-01-01–1907-06-25); **17**, 198 (1907-07-01–1907-12-30); **18**, 234 (1908-01-11–1908-06-27); **19**, 265-266 (1908-07-03–1908-12-30); **20**, 307-308 (1909-01-01–1909-06-27); **21**, 361-362 (1909-07-01–1909-12-29); **22**, 407-408 (1910-01-01–1910-06-28); **24**, 73-74 (1910-07-07–1911-06-28); **25**, 137-138 (1911-07-01–1911-12-31); **26**, 197 (1912-01-20– 1912-06-28); **27**, 259-260 (1912-07-07–1912-12-28).
Used in the BAASSC (1913–1917) 1914–1915.
NOAA: 07.–09.1910, 11.1910–12.1911, 03.–12.1912, 03.–05.1914, 07.1914–04.1916.

136. Mauritius (Pamplemousses)

BAASRP **1899**, 179–181 (1898-09-19–1899-03-12).
Shide Circular **1**, 16-18 (1898-09-19–1899-09-29); **3**, 77-83 (1899-10-19–1900-12-25); **4**, 112-113 (1901-01-07–1901-04-27); **6**, 164-166 (1901-05-07–1901- 09-30); **7**, 212 (1901-10-03–1901-12-31); **8**, 250 (1899-02-28–1899-12-25); **8**, 251-253 (1903-01-02–1903-06-25); **9**, 290 (1903-07-02–1903-09-08); **12**, 25-28 (1902-01-01–1902-12-16); **14**, 99-102 (1903-10-01–1905-12-19); **16**, 185-186 (1906-01-02–1906-12-19); **18**, 251-252 (1907-01-02–1907-12-15); **20**, 311-312 (1908-01-11–1908-12-23); **22**, 417-418 (1909-01-03–1909-12-23); **23**, 45 (1910-01-01–1910-06-30); **24**, 80-81 (1910-07-03–1910-12-30); **26**, 220-222 (1911-01-01–1911-12-31); **27**, 269-271 (1912-01-01–1912-12-24).
In Hamburg: Claxton, T. Results of the Observations at the Royal Alfred Observatory Mauritius, 1899–1909.
06.,07.1903 used in Strasbourg's "Von den Instrumenten ..." and "Verzeichnis der im ...," see 60.
At ISC: 1896, 1920.
BGS: 1896–1920.
Used in the BAASSC (1913–1917) 1913–1915.

137. Mazatlán

See 140.

138. México

In Hamburg: México 1902–1912.
*Aguilera, J.G. (1909). Catálogo de los temblores macroseísmos sentidos en la Republica Mexicana durante los años de 1904 á 1908. Parérgones del I. Geol. de México. **II**, 10, 387–467.*
Boletín mensual del Observatorio Meteorológico Central de México, 01.1888–07.1916. Boletín mensual del Observatorio Meteorológico y Sysmológico Central de México, 08.1916–03.1918.
Boletín mensual del Servicio Meteorológico Mexicano, 04.1918–12.1920.

139. Oaxaca

See 140.

140. Tacubaya

ISA: 1906–11.1908.
*Catálogo de los temblores (macroseismos) sentidos en la República Mexicana y microseismos registrados en la Estación Seismológica Central, Tacubaya, D.F., durante il segundo semestre de 1909. Paregones del Instit. Geológ. de Mexico. **3 (1911)**, Nr. 8, 437–496.*
*Catálogo de los temblores (macroseismos) sentidos en la República Mexicana y microseismos registrados en la Estación Seismológica Central, Tacubaya, D.F., durante el año de 1910. –Mircoseismos registrados en las Estaciones Seismológica de Mazatlán y Oaxaca, de Agosto á Diciembre de 1910. Ibid., **3 (1911)**, Nr. 10, 527–587.*
NOAA: 1907–1912, 01., 07.–12.1917, 1920.

141. Yucatán

In Hamburg: Yucatán 1905–1913.

142. De Bilt

Seismische Registrierungen in De Bilt, **1,** 26 Juni–5 Oktober 1904, 16 April 1908–1913. Koninklijk Nederlandsch Meteorologisch Instituut No. 108, Utrecht, 24 + 94 pp., 1915.

Seismische Registrierungen in De Bilt, **2**, 1914. *Ibid*, 16 + 24 pp., 1916.

Seismische Registrierungen in De Bilt, **3**, mit einem Anhang: Die mikroseismische Bewegung April 1908–1915. *Ibid.*, 16 + 112 pp., 1917.

Seismische Registrierungen in De Bilt, **4** (1916), 6 + 102 pp., 1918; *Ibid.*, **5** (1917). 16 + 92 pp., 1920; *Ibid.*, **6** (1918), 8 + 84 pp., 1921; *Ibid.*, **7** (1919), 11 + 68 pp., 1922; *Ibid.*, **8** (1920), 11 + 62 pp., 1923.

143. Christchurch

Farr, C. Coleridge (1902). Records of Milne Seismograph No. 16, at Christchurch, from November 1901. Transactions of the New Zealand Institute **34 (1901)**, 604.

Farr, C. Coleridge (1903). Records of Milne Seismograph No. 16, at Christchurch, New Zealand. *Ibid.*, **35 (1902)**, 593–597.

Skey, H. F., and George Hogben (1905–1913). Records of the Milne Seismographs Nos. 16 and 20, taken at Christchurch and Wellington [data for 1903–1904]. Transactions of the New Zealand Institute, **37 (1904)**, 582-589; *Ibid.* [data for 1905], **38 (1905)**, 568-574; *Ibid.* [data for 1906–1911], **44 (1912)**, 441-457.

Shide Circular **6**, 169-171 (1901-11-23–1902-06-26); **7**, 218 (1902-07-06–1902-12-25); **9**, 292-293 (1903-01-04–1903-06-21); **10**, 312-313 (1903-07-02–1904-03-28); **11**, 341 (1904-04-01–1904-06-27); **12**, 20-21 (1904-07-12–1904-12-31); **13**, 58-59 (1905-01-07–1905-06-30); **14**, 98 (1905-12-28); **15**, 151-152 (1906-06-30); **16**, 182-184 (1907-06-27); **17**, 221 (1907-12-30); **18**, 250 (1908-06-18); **19**, 286 (1908-12-28); **20**, 335-336 (1909-01-01–1909-06-28); **21**, 373-374 (1909-07-01–1909-12-28); **22**, 415-416 (1910-01-10–1910-06-29); **27**, 278-279 (1911-04-06–1911-12-23).

BGS: New Zealand Journal of Science and Technology: 1915.

144. Wellington

Hogben, George (1902). Records of Milne Seismograph No. 20, at Wellington, from October 1900, to December 1901 (inclusive). Transactions of the New Zealand Institute **34 (1901)**, 598-606.

Hogben, George (1903). Records of Milne Seismograph No. 20, at Wellington, from January 1902, to December 1902 (inclusive). *Ibid.*, **35 (1902)**, 598-606.

1903–1911 see Christchurch, New Zealand (143).

Shide Circular **6**, 168-169 (1900-10-07–1902-06-26); **8**, 256-258 (1902-07-01–1902-12-31); **11**, 341 (1903-01-14–1903-06-10); **11**, 342 (1904-03-26–1904-12-11); **22**, 415 (1910-05-29–1910-06-29); **24**, 98 (1910-07-29–1911-05-05); **26**, 228 (1912-06-07–1912-10-17).

*Hector Observatory Bulletin, Wellington **36**, 1915.*

NOAA: 12.1920.

145. Bergen

See Norway (146) 1905–1920.

Bulletin Sismique de l'Institut Geologique de Bergen Museum (1915-1922), several issues per year.

146. Norway

Thomassen, T. C. (1889). Berichte über die, wesentlich seit 1834, in Norwegen eingetroffenen Erdbeben. Bergens Museums Aarsberetning for 1888, No. 4, 52 pp.

Reusch, Hans (1888). Jordskjælv i Norge 1887. Forhandlinger i videnskabsselskabet i Christiania **8**, 10 pp.

Thomassen, T. C. (1891). Jordskjælv i Norge 1888–1890. Bergens Museums Aars beretning for 1890 No. 3, 56 pp.

Thomassen, T. C. (1894). Jordskjælv i Norge 1891–1893. Bergens Museums Aarbog for 1893, Afhandlinger og Aarsberetning No. 3, 57 pp.

Reusch, Hans (1895). Jordskjælv i 1894. In: Jordskjælv i Norge, Tre Afhandlinger, Forhandlinger i videnskabsselskabet i Christiania 1895 No. 10, 3-11 pp.

Rekstad, J. (1899). Jordskjælv i Norge aarene 1895–1898. Bergens Museums Aarbog 1899 No. 4, 40 pp.

Kolderup, Carl Frederik (1899–1913). Jordskjælv i Norge i [1899–1912]. Bergens Museums Aarbog 1899, No. 9, 46 pp.; *Ibid.*, 1900, No. 8, 12 pp.; *Ibid.*, 1901, No. 14, 21 pp.; *Ibid.*, 1902, No. 11, 35 pp.; *Ibid.*, 1903, No. 15, 25 pp.; *Ibid.*, 1905, No. 4, 35 pp.; *Ibid.*, 1906, No. 3, 37 pp.; *Ibid.*, 1907, No. 12, 49 pp.; *Ibid.*, 1908, No. 129, 49 pp.; *Ibid.*, 1909, No. 10, 33 pp.; *Ibid.*, 1910, No. 8, 22 pp.; *Ibid.*, 1911, No. 16, 21 pp.; *Ibid.*, 1912, No. 11, 38 pp.; *Ibid.*, 1913, No. 12, 19 pp.

Kolderup, Carl Frederik (1914–1922). Jordskjælv i Norge i [1913–1920], Fra Bergens Museums jordskjælvsstation 1914–1915, No. 16, 18 pp.; *Ibid.*, 1914–1915, No. 17, 11 pp.; *Ibid.*, 1917–1918, No. 10, 11 pp.; *Ibid.*, 1921–1922, No. 2, 26 pp.

147. Balboa Heights

03.1915–12.1920 included in US-WB.

148. Lima

Shide Circular **17**, 212-213 (1907-06-01–1907-12-30); **18**, 244-246 (1908-01-04–1908-06-28); **19**, 277 (1908-07-16–1908-12-13); **24**, 84 (1911-01-03–1911-06-18).

Used in the BAASSC (1913–1917) 1913–1914.

149. Manila

Masó, P. Miguel Saderra (1895). La Seismología en Filipinas–Datos para el estudio de terremotos del Archipiélago Filipino. Observatorio de Manila, Dirigido por los Padres de la Compañía Jesús. Tipo-Litográfico de Ramírez y Compañía, Manila 1895, 124 pp.

Manila Meteorological Observatory verified observations (1894), January to December.

Manila Observatory Monthly Bulletins (1897), January to December.

In Hamburg: Observatoria de Manila Boletín 1898–1901,

1904 used in Rudolph (1907).

07., 08., 10.1903, 01.–03.1904 used in Strasbourg's "Verzeichnis der im . . . ," see 60.

Algué, J. (1907–1909). Bulletins for the year [1906–1908]. Department of the Interior, Weather Bureau, Manila Central Observatory, Manila, 1907; *Ibid.*, 1908; *Ibid.*, 1909.

Manila Observatory–Philippines Island, Seismological Bulletin, 07.1910.

Manila Central Observatory, Bulletin, . . . , The Government of the Philippines Islands, Weather Review, 01.1919–12.1920 [monthly bulletins].

In Hamburg: Philippines Weather Bureau Bulletin 1901–1920.

At ISC: 08.–10.1909, 12.1909–06.1914, 06.–08.1915, 10.–11.1915, 01.–08.1916, 09.–12.1920.

Used in the BAASSC (1913–1917) 1913–1915.

150. Kraków (Krakau)

Rudzki, M.P. (1904–1909). Seismische Beobachtungen in [1903–1908]. Resultate der meteorologischen und seismologischen Beobachtungen an der k. k. Sternwarte in Krakau, 1903; *Ibid.*, in 1904; *Ibid.*, in 1905; *Ibid.*, in 1906.; *Ibid.*, in 1907; *Ibid.*, in 1908; *Ibid.*, in 1909.

Seismologische Beobachtungen in [1910–1913], an der k. k. Sternwarte Krakau.

Seismische Aufzeichnungen: 01.1913–06.1914.

At ISC: 01.1910–12.1917.

151. Krietern-Breslau (Wroclaw)

Seismische Berichte der Erdwarte zu Krietern, Kreis Breslau: handwritten weekly reports February 1–June 3, 1908.

Frölisch, O. (1910). Seismische Beobachtungen der Kgl. Erdbebenwarte in Krietern bei Breslau. August bis November 1910. Kohle und Flötz, Kattowitz, 1910.

Seismischer Bericht der Erdwarte Krietern Kreis Breslau, monthly reports: 09.1909–12.1910, 01.–02.1912.

Krietern-Breslau, Seismische Aufzeichnungen der Königlichen Erdwarte, monthly reports: 03.1912–08.1913, 11.1913–06.1914.

152. Lisboa (Lisbon)

Anais do Observatório Central Meteorológico do Infante D.Luis **58**, Part III, Observações Sismológicas 1920, 7 pp.

153. Ponta Delgada, St. Miguel, Açores

Shide Circular **9**, 284 (1903-07-27–1903-12-23); **10**, 307 (1904-01-20–1904-06-25); **11**, 330-331 (1904-07-04–1904-12-28); **12**, 12-13 (1905-01-19–1905-06-27); **13**, 39 (1905-07-06–1905-12-17); **14**, 75 (1906-01-21–1906-06-29); **15**, 120 (1906-07-13–1906-12-26); **16**, 168 (1907-01-02–1907-06-26); **17**, 200 (1907-08-05–1907-12-30); **18**, 238 (1908-02-14–1908-06-29); **19**, 270-271 (1908-07-08–1908-12-28); **20**, 309 (1909-01-15–1909-06-08); **21**, 359 (1909-07-06–1909-12-10); **22**, 400-401 (1910-01-01–1910-06-29); **23**, 33 (1910-07-20–1910-12-29); **24**, 69 (1911-01-01–1911-06-25); **25**, 132-133 (1911-07-01–1911-12-16); **26**, 193 (1912-01-24–1912-06-16); **27**, 256 (1912-07-01–1912- 12-09).

12.1903 used in Strasbourg's "Verzeichnis der im . . . ," see 60.

Used in the BAASSC (1913–1917) 1913–1915.

154. Portugal

*Choffat, P. (1904). Les tremblements de terre de 1903 en Portugal. "Communicações" du Service Géologique Du Portugal **5**. 1904.*

155. Vieques, Porto Rico (Puerto Rico)

In Hamburg: Results of Observations Vieques, Puerto Rico, 1903–1910, 1911–1918.

1904 in: *Reid, Harry Fielding. Records of Seismographs. Terrestrial Magnetism and Atmospheric Electricity, 1905.*

Shide Circular **13**, 50-52 (1904-07-10–1905-06-30).

ISA: 1906, 1907, 1908.

Hazard, Daniel L. (1912–1922). Results of Observations made at the United States Coast and Geodetic Survey Magnetic Observatory at Vieques, Porto Rico, 1909 and 1910, Department of Commerce and Labor, Coast and Geodetic Survey, pp. 92-93; *ibid.*, 1911 and 1912, pp. 100-102; *ibid.*, 1913 and 1914, pp. 100-102; *ibid.*, 1915 and 1916, pp. 98-99; *ibid.*, 1917 and 1918, pp. 100-103.

See US-WB, 1914–1920.

156. Bucharest (Bucarest, Bucuresti, Bucaresci)

03.1904 used in Strasbourg's "Verzeichnis der im . . . ," see 60.

Bulletin of Imperial Russia 1904, 1905, see 168.

Included in 157.

Bulletin Sismologique, Observatoire Astron. et Mét. de Bucarest 1910–1911.

At ISC: 05., 08., 11., 12.1903, 02., 04., 07., 08.1904, 01.–06.1906, 02.–08.1907, 01.1910–12.1911.

NOAA: 1910–1911.

157. Romania

*Hepites, Ştefan C. (1893a). Registrul cutremurelor de pămâdi Românie (1839–1892). Analele Institutului Meteorologic al României, **6**, B55-B68.*

*Hepites, Ştefan C. (1893-1900). Registrul cutremurelor de pămâdi Românie [1893–1900]. Analele Institutului Meteorologic al României, **8**, B13-B31; Ibid., **10**, B58-B85; Ibid., **11**, B205-B208; **12**, B224-B283; Ibid., **13**, B203-B207; **14**, B233-B235; Ibid., **15**, B110-B114; Ibid., **16**, B123-B127.*

*Hepites, Ştefan C. (1901). Arhiva seismică a României 1901. Ibid., **17**, B317-B342.*

05., 06., 08., 09., 11., 12.1903, 03.1904 used in Strasbourg's "Von den Instrumenten . . . " and "Verzeichnis der im . . . ," see 60.

*At KNMI: Hepites, Ştefan C. (1905). Materiale pentru sismografia României. XI seismele din Anul 1904. Anale le Academiei Române, Serie II, Tom. **27**, 175-185.*

*At KNMI: Hepites, Ştefan C. (1906). Materiale pentru sismografia României. XII seismele din Anul 1905. Ibid. Tom. **28**, 333-337.*

*At KNMI: Hepites, Ştefan C. (1907). Materiale pentru sismografia României. XIII seismele din Anul 1906. Ibid., Tom. **29**, 172-216.*

In Hamburg: Hepites, Ştefan C. (1907). Arhiva seismică a României 1902–1906 (Archive sismique de Roumanie, Années

*1902–1906). Analele Institutului Meteorologic al României **18**, partea 2, Anul 1902, B189-B303.*
In Hamburg: Mouvements sismiques en Roumanie pendant la période 1907–1909. Bul. Mensuel. Observatoire Astron. et Mét. de Bucarest 1910, 21 pp.

158. Timisoara (Temesvar)

1906–1912 in Kövesligethy (1907, 1909, 1913), see 86.
BGS: 1907–1909.

159. Derbent

Included in: Seismische Monatsberichte des Physikalischen Observatoriums zu Tiflis, 03.1905–09.1909, see 64.
Bulletin of Imperial Russia 1906–1908, see 168.

160. Ekaterinburg (Ekaterinbourg, Jeckaterinburg, Sverdlovsk)

Bulletin of Imperial Russia 1907–1908, see 168.
Ezenedel'nyj bjulletin Sejsmiceskoj Stancii Pervogo Razrjada pri Observatorii v Ekaterinburge, 01.1913– 04.1914.
In Hamburg: Erdbebenberichte Ekaterinburg 1913–1914.
At ISC: 10.1913–06.1914.
Used in the BAASSC (1913–1917) 1914.
NOAA: 1913–1914.
BGS: 1913.

161. Irkutsk (Irkoutsk)

Shide Circular **5**, 141 (1901-12-05–1901-12-31); **7**, 214-216 (1902-01-01–1902-12-28); **8**, 246-247 (1903-01-02–1903-06-26); **9**, 289 (1903-07-08–1903-12-28); **10**, 318-319 (1904-01-07–1904-06-30); **11**, 346 (1904-07-01–1904-12-30); **13**, 44-45 (1905-04-03–1905-06-30); **15**, 136-139 (1905-07-06–1906-09-29); **17**, 203 (1906-10-02–1906-12-27); **18**, 240-242 (1907-01-01–1907-12-30); **19**, 275-276 (1908-01-11–1908-09-23); **20**, 318-319 (1908-10-01–1909-03-22).
Bulletin of Imperial Russia 1902–1908, see 168.
06., 12.1903 used in Strasbourg's "Von den Instrumenten . . ." and "Verzeichnis der im . . . ," see 60.
In Hamburg: Ikoutsk Bulletin Sismique 1902–1905, 1912–1914.
Station seismique de I-re classe d'Irkoutsk, Bulletin hebdomadaire, 12–27, 29–31, 34–41, 1912. *Ibid.,* 1–3, 5–16, 19, 22, 23, 1913.
At ISC: 01.1912–06.1914.
Used in the BAASSC (1913–1917) 1913–1915.
NOAA: 06.1912–01.1914, 05., 06., 1914.
BGS: 1913.

162. Kabansk (Kamensk)

Bulletin of Imperial Russia 1904–1908, see 168.

163. Königsberg (Kaliningrad)

Mitteilungen der Hauptstation für Erbebenforschung im Gr. Raum des Geologischen Instituts zu Königsberg Preußen. Monthly reports: 04.1912–05.1914.

164. Krasnojarsk (Krasnoïarsk)

Bulletin of Imperial Russia 1903–1906, see 168.

165. Maritouy (Marituj)

Bulletin of Imperial Russia, 11.–12.1908 (?), see 168.

166. Pavlovsk (Pawlowsk, near St. Petersburg)

Bulletin of Imperial Russia 1903, 1904, see 168.
NOAA: 01.1913–07.1914?

167. Pulkovo (Pulkovwa, near St. Petersburg)

Galitzin, Boris (1908). Seismometrische Beobachtungen in Pulkowa. Comptes Rendus des Séances de La Commission Sismique Permanente, **Tome 3**, Livraison I, 117-172.
Galitzin, Boris (1909). Seismometrische Beobachtungen in Pulkowa. Zweite Mitteilung. *Ibid.,* **Tome 3**, Livraison II, 5-119.
Wilip, J. Die zentrale seismische Station in Pulkova. *Ibid.,* **Tome 5**, Livraison II, 133-170 [data for 1911].
Wöchentliche Erdbebenberichte in Pulkovwa, 1–52, 1912. *Ibid.,* 1–11, 13–52, 1913. *Ibid.,* 1–7, 9–14, 17–25, 27–28, 1914.
In Hamburg: Wöchentliche Erdbebenberichte Pulko, 1913–1914.
At ISC: 01.1912–07.1914.
NOAA: 1912.
Used in the BAASSC (1913–1917) 1913–1914.

168. Russia

The Russian bulletins 1902–1907 were published as supplement to the "Comptes Rendus des Séances de La Commission Sismique Permanente" published by "Académie Impériale des Séances de Russie" and in addition as a quarterly journal with own page numbering.
Levitski, G. (1903a). Bulletin de la Commission Centrale Sismique Permanente Année 1902, Janvier–mars, 61 pp.; Avril–juin, 42 pp.; Juillet–septembre, 51 pp.; Octobre–décembre, 50 pp.
Levitski, G. (1903b). Bulletin de la Commission Centrale Sismique Permanente Année 1902, Janvier–juin. Comptes Rendus des Séances de La Commission Sismique Permanente, **Tome 1**, Livraison II, pp. 1-104; *ibid.,* Juillet–décembre, **Tome 1**, Livraison III, pp. 105-206.
Levitski, G. (1903–1904). Bulletin de la Commission Centrale Sismique Permanente Année 1903, Janvier–mars, 73 pp.; Avril–juin, 45 pp.; Juillet–septembre, 40 pp.; Octobre–décembre, 75 pp.
Levitski, G. (1904). Bulletin de la Commission Centrale Sismique Permanente Année 1903, Janvier–décembre. Comptes

Rendus des Séances de La Commission Sismique Permanente, Tome 2, Livraison I, pp. 1–235.

Levitski, G. (1905). Bulletin de la Commission Centrale Sismique Permanente Année 1904, Janvier–mars, 63 pp.; Avril–juin, 45 pp.; Juillet–septembre, 48 pp.; Octobre–décembre, 51 pp.

Levitski, G. (1906). Bulletin de la Commission Centrale Sismique Permanente Année 1904, Janvier–décembre. Comptes Rendus des Séances de La Commission Sismique Permanente, **Tome 2**, Livraison II, pp. 1-207.

Levitski, G. (1906–1907). Bulletin de la Commission Centrale Sismique Permanente Année 1905, Janvier–mars, 55 pp.; Avril–juin, 48 pp.; Juillet–septembre, 139 pp.; Octobre–décembre, 63 pp.

Levitski, G. (1907). Bulletin de la Commission Centrale Sismique Permanente Année 1905, Janvier–décembre. Comptes Rendus des Séances de La Commission Sismique Permanente, **Tome 2**, Livraison III, pp. 1-307.

Levitski, G. (1907–1908). Bulletin de la Commission Centrale Sismique Permanente Année 1906, Janvier–mars, 64 pp.; Avril–juin, 44 pp.; Juillet–septembre, 52 pp.; Octobre–décembre, 69 pp.

Levitski, G. (1908). Bulletin de la Commission Centrale Sismique Permanente Année 1906, Janvier–décembre. Comptes Rendus des Séances de La Commission Sismique Permanente, **Tome 3**, Livraison I, pp. 1-229.

Levitski, G. (1908–1909). Bulletin de la Commission Centrale Sismique Permanente Année 1907, Janvier–mars, 55 pp.; Avril–juin, 79 pp.; Juillet–septembre, 50 pp.; Octobre–décembre, 88 pp.

Levitski, G. (1909). Bulletin de la Commission Centrale Sismique Permanente Année 1907, Janvier–juin. Comptes Rendus des Séances de La Commission Sismique Permanente, **Tome 3**, Livraison II, pp. 1-135.

Levitski, G. (1910a). Bulletin de la Commission Centrale Sismique Permanente Année 1907, Juillet–décembre. Comptes Rendus des Séances de La Commission Sismique Permanente, **Tome 3**, Livraison II, pp. 105-206.

Levitski, G. (1910b). Bulletin de la Commission Centrale Sismique Permanente Année 1908, 112 pp. [but missing October–December].

Nikiforov, P. (1912). Bulletin de la Commission Centrale Sismique Permanente 1911, 26 + 26 pp.

Nikiforov, P. (1914). Bulletin de la Commission Centrale Sismique Permanente 1912, 26 + 52 pp.

BGS: 1907–1912.

169. Tchita (Tšita)

Bulletin of Imperial Russia 1904–1908, see 168.

170. Apia

At ISC: 01.–04.1905, 06.1905–12.1907, 04., 05., 07., 08.1907, 03.–09.1908, 11.1908–10.1909.

ISA: 1906, 1907, 1908 monthly.

Wegener, Kurt (1912). Die seismischen Registrierungen am Samoa-Observatorium der Kgl. Gesellschaft der Wissenschaften zu Göttingen in den Jahren 1909 und 1910. Nachrichten von der Königlichen Gesellschaft der Wissenschaften zu Göttingen, Mathematisch-physikalische Klasse, 267-384.

Weekly listings 01.–25., 31.–52.1910, 01.1911–September 1913.

At ISC: 03.1910–10.1911, 10., 11., 1912, 01., 05.–09. 1913.

NOAA: 01.1909–12.1913.

Angenheister, Gustav (1923). Liste der wichtigsten am Samoa-Observatorium 1913/20 registrierten Erdbeben. Nachrichten von der Königlichen Gesellschaft der Wissenschaften zu Göttingen, Mathematisch-physikalische Klasse aus dem Jahre 1922, 53-55.

171. Mahé

Shide Circular **24**, 84 (1911-06-01–1911-06-15); **25**, 147 (1911-07-04–1911-12-31); **26**, 219 (1912-01-04– 1912-02-21); **27**, 268 (1912-09-11–1912-12-05).

Used in the BAASSC (1913–1917) 1913–1914.

172. Hurbanovo (Ógyalla, Stará Dala)

1902–1912, all observations from Ógyalla were published together with the Hungarian observations, see 86.

05., 06., 08., 09., 11., 12.1903, 02., 03.1904 used in Strasbourg's "Von den Instrumenten ..." and "Verzeichnis der im ...," see 60.

173. Ljubljana (Laibach)

*Seidl, Ferdinand (1898). Die Erderschütterungen Laibachs in den Jahren 1851 bis 1886, vorwiegend nach den handschriftlichen Aufzeichnungen K. Deschmanns. Mitteilungen der Erdbeben-Kommission **VI**. In: Sitz. Ber. Akademie der Wissenschaften Wien **107**, mathematisch-naturwissenschaftliche Klasse, Abt. 1, Heft VI.*

Belar, Albin (1897/1898). Ueber Erdbebenbeobachtung in alter und gegenwärtiger Zeit und die Erdbebenwarte in Laibach. Jahresbericht der k. k. Staats-Oberrealschule in Laibach 1897/1898, 43 pp.

ISA: 1906, 2907 + 1908 weekly listings.

Neuste Erdbebennachrichten 09.1901–12.1908 as published in "Die Erdbebenwarte" **1** (1901/1902)–**8** (1909), ***9** (1909/1910).* These monthly reports contain readings from the station in Laibach and from other stations as far as they reported to the institute in Laibach. The reported onsets were as far as possible associated to common events.

08., 09., 11.1903 used in Strasbourg's "Verzeichnis der im ...," see 60.

*In Hamburg: Achitsch, A. Seismische Aufzeichnungen in Laibach, gewonnen an der Erdbebenwarte im Jahre 1913. Mitteilungen der Erdbeben-Kommission der Kaiserlichen Akademie der Wissenschaften in Wien. Neue Folge **48**.*

Wöchentliche Erdbebenberichte, No. 41, 1910–No. 24, 1914.

At ISC: 01.1907–01.1912.

174. Cape of Good Hope (Capetown)

BAASRP **1899**, 182 (1899-07-14–1899-07-31).

Shide Circular **1**, 22-24 (1899-08-04–1900-02-24); **2**, 43 (1900-03-06–1900-06-21); **3**, 76-77 (1900-07-29–1900-12-25); **4**, 107 (1901-01-08–1901-06-17); **5**, 133-134 (1901-08-09–1902-01-01); **6**, 157-158 (1902-01-12–1902-06-11); **7**, 206 (1902-06-11–1902-12-13); **8**, 239-240 (1903-01-03–1903-06-29); **9**, 284-285 (1903-07-01–1903-12-23); **10**, 308 (1904-01-02–1904-06- 29); **11**, 331-332 (1904-07-01–1904-12-22); **12**, 13 (1905-01-01–1905-06-30); **13**, 42 (1905-07-06–1905-11-08); **14**, 74 (1906-01-21–1906-06-24); **15**, 119 (1906-07-06–1906-12-26); **16**, 167 (1907-01-01–1907-06-27); **17**, 198-199 (1907-07-01–1907-12-30); **18**, 238-239 (1908-01-11–1908-06-03); **19**, 271-272 (1908-07-28–1908-12-30); **20**, 309-311 (1909-01-02–1909-06-30); **21**, 365-366 (1909-07-06–1909-12-20); **22**, 408–409 (1910-01-01–1910-06-25); **23**, 33-34 (1910-07-07–1910-12-30); **24**, 82 (1911-01-01–1911-06-28); **25**, 144-145 (1911-07-01–1911-12-31); **26**, 206-207 (1912-01-01–1912-06-28); **27**, 271-272 (1912-07-07–1912-12-09).

06., 12.1903 used in Strasbourg's "Von den Instrumenten . . . " and "Verzeichnis der im . . . ," see 60.

Used in the BAASSC (1913–1917) 1913–1915.

175. Cartuja (Granada)

Sección seísmica, Enero 1903, Observatorio astronómico, geodinámico y meteorológico de Granada, Enero 1903, 29.

Sección geodinámica, Febrero 1903, *Ibid.,* Febrero 1903, 3-5.

Sección sísmica, Marzo 1903–Diciembre 1903, *Ibid.,* monthly reports.

Sección sísmica, Enero 1904–Diciembre 1904, *Ibid.,* monthly reports.

03.1904 used in Strasbourg's "Verzeichnis der im . . . ," see 60.

Sismogramas obtenidos en el año 1905. Sección sísmica, Año III Enero 1905, Observatorio astronómico, geodinámico y meteorológico de Granada, Enero 1905, 7-10.

Sección sísmica, Enero 1906–Diciembre 1906, *Ibid.,* monthly reports, missing November.

Sección sísmica, Enero 1907–Diciembre 1907, *Ibid.,* monthly reports, missing March.

Boletín mensual de la estación sismológica del observatorio de Cartuja (Granada). 07.1908–09.1916.

Navarro-Neumann, M.S. (1907). Os terremotos observados san o auxilio de instrumentos. Brotéria, Rev. Bimensal ***6****, 217-250.*

Navarro-Neumann, M.S. (1908–1914). Bulletin Sismique de Cartuja. Bulletin de la Société belge d'Astronomie, [Météorologie, Géodésie et Physique de Globe] **13**, 81-84, 121-124, 160-163, 206-208, 240-242, 330-333, 378-381, 416-418; *Ibid.,* **14**, 37-38, 134-138, 175-176, 212-214, 255-257, 335-337, 413-416, 483-486, 529-531; *Ibid.,* **15**, 37-38, 88-90, 138-140, 181-182, 221-223, 258-260, 302-305, 350-352, 376-378, 435-437, 470-472, 512-514; *Ibid.,* **16**, 33-35, 77-80, 127-128, 174-176, 199-201, 234-235, 271-274, 271-274, 313-317, 358-359, 401-403, 432-435; *Ibid.,* **17**, 32-34, 76-78, 118-119, 162-163, 245-250, 282-286, 310-314, 356-357, 399-400; *Ibid.,* **18**, 30-31, 105-109, 141-142, 174-175, 196-197, 252-253, 276-277, 314-317, 346-348, 377-379; *Ibid.,* **19**, readings until April or May.

Used in the BAASSC (1913–1917) 1913–1915.

NOAA: 11.1908–12.1915, 01.–03., 06.–09.1916.

176. Ebro (Tortosa)

Observatorio del Ebro, Observaciones seismograficas: monthly reports 04.–11.1905, 01.–12.1906, 02.–12.1907.

Observatorio del Ebro, handwritten bulletin book: 01.–10.1906, 04., 05.1907, 01.–10.1908, 02., 03., 05., 06.1909.

Boletin Mensual del Observatorio del Ebro, Observatorio de Física Cósmica del Ebro (Bulletin de L'Observatoire de L'Ebre), Vol. I., No 1–12, 1910. *Ibid.,* Vol. II., No 1–12, 1911. *Ibid.,* Vol. III., No 1–12, 1912. *Ibid.,* , Vol. IV., No 1–9, 1913.

In Hamburg: Boletín mensual de Observatorio del Ebro, 1913–1920.

177. Fabra (Barcelona)

Comas Sola, José (1908). Estadistica sismologica de 1907, en Barcelona, Memorias de la Real Academia de Ciencias y Artes de Barcelona, Vol. 5, 505-509.

Comas Sola, José (1908). Observaciones sísmicas durante el Año 1907, *Ibid.,* Tercera época1**156** Vol. 6 Núm. 31, 510-518.

Comas Sola, José (1909). Observaciones sísmicas efectuadas durante el Año 1908, *Ibid.,* Tercera época1**171** Vol. 7 Núm. 13, 532-543.

Comas Sola, José (1910–1912). Observaciones sísmicas durante el Año [1909–1911], *Ibid.,* Tercera época1**190** Vol. 8 Núm. 15, 286-298; *Ibid.,* Tercera época1 **205** Vol. 8 Núm. 30, 545-557; *Ibid.,* Vol. 10 Núm. 12, 235-249.

Comas Sola, José (1913). Resumen Sísmico de 1912 y de 1913 (basta el 17 de abril de este último año), *Ibid.,* Tercera época **1234** Vol. 10 Núm. 27, 556-569.

Comas Sola, José (1915–1921). Observaciones sísmicas durante el Año [1914–1920], Memorias de la Real Academia de Ciencias y Artes de Barcelona.

178. Puerto Orotava (Puerto de la Cruz), Tenerife

Rebeur-Paschwitz (1892, 1895).

179. Río Tinto

Shide Circular **24**, 72-73 (1911-01-01–1911-06-17); **25**, 136-137 (1911-07-04–1911-12-31); **26**, 193-194 (1912-01-04–1912-06-28); **27**, 259 (1912-07-07– 1912-12-24).

Used in the BAASSC (1913–1917) 1913–1915

180. San Fernando

Observaciones séismicas, Años 1898 y 1899. Anales del Instituto y Observatorio de Marina en San Fernando, Sección 2.ª, Observaciones Meteorológicas, Magnéticas y Séismicas Año 1899, San Fernando 1900, 1-2.

Observaciones séismicas, Año 1900, *Ibid.,* San Fernando 1901, 1-2.

Observaciones séismicas, Año 1901, *Ibid.,* San Fernando 1902, 151-152.

Observaciones séismicas, Año 1902, *Ibid.,* San Fernando 1903, 151-153.

Observaciones séismicas, Año 1903, *Ibid.,* San Fernando 1904, 151-153.

Observaciones séismicas, Año 1904, *Ibid.*, San Fernando 1905, 151-155.
05.–11.1903, 01.–03.1904 used in Strasbourg's "Von den Instrumenten ..." and "Verzeichnis der im ...," see 60.
Observaciones séismicas, Año [1905–1920]. Anales del Instituto y Observatorio de Marina en San Fernando, Sección 2.ª, Observaciones Meteorológicas, Magnéticas y Séismicas Año [1905–1920], San Fernando [1906–1921].
BAASRP **1899**, 174 (1898-02-18–1899-07-17).
Shide Circular **1**, 14 (1899-07-17–1899-12-31); **2**, 40 (1900-01-05–1900-06-21); **3**, 74-75 (1900-07-29–1900-12-25); **4**, 106 (1901-01-07–1901-06-24); **5**, 131-132 (1901-08-06–1901-12-31); **6**, 155 (1902-01-01–1902-06-22); **7**, 205 (1902-07-05–1902-12-30); **8**, 238 (1903-01-05–1903-06-10); **9**, 283 (1903-07-02–1903-12-29); **10**, 306-307 (1904-01-20–1904-06-27); **11**, 329-330 (1904-07-01–1904-12-31); **12**, 10-12 (1905-01-01–1905-06-30); **13**, 38 (1905-07-06–1905-12-29); **14**, 73-74 (1906-01-06–1906-06-30); **15**, 115-117 (1906-07-20–1906-12-29); **16**, 163-165 (1907-01-01–1907-06-30); **17**, 197 (1907-07-01–1907-12-30); **18**, 233 (1908-01-11–1908-06-27); **19**, 265 (1908-07-08–1908-12-29); **20**, 306 (1909-01-05–1909-06-11); **21**, 358 (1909-07-07–1909-12-29); **22**, 398-400 (1910-01-01–1910-06-30); **23**, 28-29 (1910-07-03–1910-12-29); **24**, 70-72 (1911-01-01–1911-06-25); **25**, 133-136 (1911-07-01–1911-12-31); **26**, 194-197 (1912-01-04–1912-06-29); **27**, 257-258 (1912-07-07–1912-12-29).

181. Spain

In Hamburg: Rodríguez, José Galbis (1932). Catálogo sísmico de la zona comprendida entre los meridianos 5° E. Y 20° W. De Greenwich y los paralelos 45° y 25° N. Velasco, Madrid 1932, 10 + 807 pp.

182. Urania

*Bulletin from 06.–12.1913, 02.–07.1914 published in: Servicio Sismologico de la Sociedad Astronomica de España y America. Boletin de la Sociedad Astronomica de España y America **III**, 101-102, 117-118, 135-136, 150, 160. Ibid., **IV**, 30-31, 79-80, 95-96, 110.*

183. Colombo

Shide Circular, **16**, 170 (1906-01-31–1906-12-29); **21**, 368-369 (1909-07-07–1909-12-16); **22**, 410-411 (1910-01-01–1910-06-29); **23**, 38 (1910-06-30–1910-12-30); **24**, 89-90 (1911-01-01–1911-06-17); **25**, 155 (1911-07-04–1911-12-31); **26**, 219 (1912-01-04– 1912-06-27); **27**, 267-268 (1912-07-07–1912-12-28).
Used in the BAASSC (1913–1917) 1913–1915.

184. Uppsala (Upsala)

Åkerblom, Filip (1906). Seismische Registrierungen in Upsala Oktober 1904–Mai 1905. Nachrichten von der Königlichen Gesellschaft der Wissenschaften zu Göttingen, Mathematisch-physikalische Klasse, 125-140.
Åkerblom, Filip (1913). Observations séismographiques faites à l'observatoire météorologique d'Upsala de juillet à décembre 1906, 18 pp.
Koraen, Tage (1914). Observations séismographiques faites à l'observatoire météorologique d'Upsala de janvier 1907 à août 1912, 122 pp.
Landin, Sven (1917). Observations séismographiques faites à l'observatoire météorologique d'Upsala de septembre 1912 à avril 1917, 72 pp.
Upsala 1917–1920.

185. Switzerland

Heim, Albert (1881). Die Schweizerischen Erdbeben vom November 1879 bis Ende 1880, Jahrbuch des Tellurischen Observatoriums zu Bern **1881**, 26 pp.
*Die Erdbeben der Schweiz in den Jahren 1888–1891. Annalen der Schweizerischen Meteorologischen Zentralanstalt **1891**.*
Früh, J. (1893). Die Erdbeben der Schweiz im Jahre 1892. Annalen der Schweizerischen Meteorologischen Zentralanstalt **1892**, 16 pp.
Früh, J. (1894). Die Erdbeben der Schweiz im Jahre 1893. *Ibid.*, **1893**, 6 pp.
*Früh, J. (1895). Die Erdbeben der Schweiz im Jahre 1894. Ibid., **1894**.*
Früh, J. (1896–1905). Die Erdbeben der Schweiz im Jahre [1895–1904]. *Ibid.*, **1895**, 14 pp.; *Ibid.*, **1866**, 18 pp.; *Ibid.*, **1897**, 9 pp.; *Ibid.*, **1898**, 13 pp.; *Ibid.*, **1899**, 3 pp.; *Ibid.*, **1900**, 3 pp.; *Ibid.*, **1901**, 8 pp.; *Ibid.*, **1902**, 3 pp.; *Ibid.*, **1903**, 4 pp.; *Ibid.*, **1904**, 4 pp.
Quervain, A. de (1906–1909). Die Erdbeben der Schweiz im Jahre [1905–1907]. *Ibid.*, **1905**, 13 pp.; *Ibid.*, **1906**, 5 pp.; *Ibid.*, **1907**, 7 pp.
Quervain, A. de (1909). Die Erdbeben der Schweiz im Jahre 1908 und die Schallverbreitung der Dynamitexplosion an der Jungfraubahn, 15. November. *Ibid.*, **1908**, 11 pp.
Quervain, A. de (1910–1913). Die Erdbeben der Schweiz im Jahre [1909–1913]. *Ibid.*, **1909**, 7 pp.; *Ibid.*, **1910**, 13 pp.; *Ibid.*, **1911**, 8 pp.
Quervain, A. de (1913). Die Erdbeben der Schweiz im Jahre 1912, Die im Jahre 1912 auf der Erdbebenwarte bei Zürich registrierten Nahebeben, Ueber Herdtiefenbestimmung aus herdnahen Stationen und die dabei erforderliche und erreichbare Zeitgenauigkeit. *Ibid.*, **1912**, 12 pp.
Jahresbericht des Schweizerischen Erdbebendienstes, 1913–1920. *Ibid.*

186. Zürich

See Switzerland 185.

187. Port of Spain (St. Clair)

Shide Circular **4**, 111 (1901-01-07–1901-06-25); **5**, 140-141 (1901-07-01–1901-12-31); **6**, 166 (1902-01-01–1902-06-28); **7**, 212-213 (1902-07-01–1902-12-18); **8**, 253-254 (1903-01-03–1903-06-20); **9**, 291 (1903-07-01–1903-12-23); **10**, 312 (1904-01-12–1904-06-26); **11**, 343 (1904-07-24–1904-12-21); **12**, 21 (1905-03-17–1905-06-25); **13**, 50 (1905-07-09–

1905-12-28); **14**, 90-91 (1906-01-25–1906-06-30); **15**, 139-140 (1906-08-06–1906-12-26); **16**, 174 (1907-02-12–1907-06-13); **17**, 209 (1907-10-16–1907-12-30); **18**, 244 (1908-01-20–1908-05-31); **19**, 276-277 (1908-07-08–1908-09-12); **21**, 361 (1909-06-07–1909-06-08); **26**, 210 (1909-06-08–1909-06-18).

Seismograph records, Bull. Dept. Agriculture VIII, 64-67, Governm. Print. Office, Trinidad, 1908.

06., 12.1903 used in Strasbourg's "Von den Instrumenten . . ." and "Verzeichnis der im . . . ," see 60.

188. Harpoot (Harput, Kharput, Kharpert, Elâzig)

Riggs, H.H.: Euphrat College, monthly reports 1907, 1908, 1909.

189. Istanbul (Constantinopel)

See Turkey 190.

190. Turkey

Agamennone, G. (1895). Liste des tremblements de terre qui ont eu lieu dans l'Empire Ottoman pendant l'année 1894. Bulletin météorologique et séismique de l'Ob- servatoire Impérial de Constantinople année: 1894, 1-6.

Agamennone, G. (1896). Liste des tremblements de terre qui ont eu lieu en Orient et en particulier dans l'Empire Ottoman. L'activité sismique en Orient et en particulier dans l'Empire Ottoman pendant l'année 1895. *Ibid.*, 1-68.

Agamennone, G. (1896). Liste des tremblements de terre qui ont eu lieu en Orient et en particulier dans l'Empire Ottoman. L'activité sismique en Orient et en particulier dans l'Empire Ottoman pendant l'année 1896. *Ibid.*, 1-26 (January–April 1896).

Salih Zéky (1897). Liste des tremblements de terre observés en Orient et en particulier dans l'Empire Ottoman pendant le mois des Janvier et Février 1897. *Ibid.*, Janvier et Février 1897, 1-3.

Salih Zéky (1897). Liste des tremblements de terre observés en Orient et en particulier dans l'Empire Ottoman pendant le mois des Mars et Avril 1897. *Ibid.*, Mars et Avril 1897, 1-2.

191. Charkow (Kharkov)

*Levitski, G.V. (1894). Quelques résultats d'observations effectuées á l'Observatoire Astronomique de l'Université de Kharkov avec les pendules Rebeur-Paschwitz. Comm. Soc. Math. Kharkov **4**, 5-6, 206-208.*

Lewitzky, G. (1896). Ergebnisse der auf der Charkower Universitätssternwarte mit dem v. Rebeur'schen Horizontalpendel angestellten Beobachtungen. Charkow, 1896, 63 pp.

Struve, L.: Ergebnisse der auf der Charkower Universitätssternwarte mit dem v. Rebeur'schen Horizontalpendel angestellten Beobachtungen. Charkow, 28 pp.

Observations used in 1895 as reference in the national bulletin of Italy, see 107.

1893–1897 used in Rudolph (1904).

Rebeur-Paschwitz (1895).

Rudolph (1903a, b).

Kudrewitsch, B. (1911). Resultate aus den im Jahre 1909 angestellten Beobachtungen an den v. Rebeur- Paschwitzchen Horizontalpendeln der Charkower Sternwarte. Publikation der Charkower Universtitätssternwarte 1911, Nr. 6, 1-35.

192. Czernovitz (Czernowitz, Tschernowzy)

At ISC: 01.1913–07.1914.

Used in the BAASSC (1913–1917) 1913–1914.

193. Lwiw (Lemberg, Lwów)

*In Hamburg: Láska, W. Bericht über die Erdbebenbeobachtungen in Lemberg. Mitteilungen der Erdbeben-Kommission der kaiserlichen Akademie der Wissenschaften in Wien, Neue Folge **I**.*

*In Hamburg: Láska, W. Die Erdbeben Polens. Des historischen Teiles 1. Abteilung. Ibid., Neue Folge **VIII**.*

*In Hamburg: Láska, W. Bericht über die Erdbeben- Beobachtungen in Lemberg während des Jahres 1901. Ibid., Neue Folge **IX**.*

Láska, W. (1903). Bericht über die seismologischen Aufzeichnungen des Jahres 1902 in Lemberg. *Ibid.*, Neue Folge **XXII**, 37 pp.

*In Hamburg: Láska, W. Jahresbericht des Geodynamischen Observatoriums zu Lemberg für das Jahr 1903, nebst Nachträgen zum Katalog der polnischen Erdbeben. Ibid., Neue Folge **XXVIII**.*

Wöchentliche Erdbebenberichte, Seismisches Observatorium Lemberg, k.k. Technische Hochschule, 34.1910–31.1918.

Wöchentliche Erdbebenberichte, Seismisches Observatorium Lemberg (Lwów), Technische Hochschule, 32.1918–52.1920.

194. Makejewka (Makejevka, Makeyevka)

At ISC: 03.1912–11.1913.

Used in the BAASSC (1913–1917) 1913.

195. Nicolajew (Nikolajew, Nikolaev)

Rebeur-Paschwitz (1895).

1893–1897 used in Rudolph (1904).

Rudolph (1903a, b).

Kortazzi, J. (1894). Account of observations made with the horizontal pendulum at Nicolajew, Report of 64th Meeting, British Association of the Advancement of Science, 155-158.

Observations used in 1895 as reference in the national bulletin of Italy, see 107.

Kortatsii I.E. (1895). Observations à l'aide du pendule horizontal Rebeur-Paschwitz à l'Observatoire de Nicolaev. Soc. Astron. Russe Publ. **4**, 24-25.

BAASRP **1897**, 169 (1896-06-14–1897-03-15); **1898**, 196 (1897-05-05–1898-01-29); **1899**, 183-184 (1898-03-06–1899-03-12).

Kortazzi, J. (1900). Les perturbations du pendule horizontal à Nicolajew en 1897, 1898 et 1899. Beiträge zur Geophysik **4**, 383-405.

Kortazzi, J. (1903). Les perturbations du pendule horizontal à Nicolajew en 1900. Beiträge zur Geophysik **5**, 663-666.
Bulletin of Imperial Russia 1902, 1904–1908, see 168.

196. Uschhorod (Uschgorod, Ungvar, Uzhorod)

1911–1912 in Kövesligethy (1913), see 86.

197. United Kingdom

See 210.

198. Ascension Island

Shide Circular **23**, 34-35 (1910-11-09–1911-03-10); **24**, 83 (1911-03-15–1911-06-25); **25**, 145 (1911-07-06–1911-12-16); **26**, 205 (1912-01-31–1912-06-08); **27**, 272 (1912-07-07–1912-12-09).
Used in the BAASSC (1913–1917) 1913–1914.

199. Bidston (Liverpool)

BAASRP **1898**, 195 (1897-09-06–1898-03-28); **1899**, 185-186 (1898-03-28–1899-03-02).
Shide Circular **4**, 98-99 (1901-01-08–1901-06-24); **5**, 123-124 (1901-06-30–1902-01-01); **6**, 150-151 (1902-01-01–1902-06-26); **7**, 199-200 (1902-07-01–1902- 12-28); **8**, 234-235 (1903-01-01–1903-06-25); **9**, 276-277 (1903-07-02–1903-12-31); **10**, 303 (1904-01-02–1904-06-27), **11**, 327 (1904-07-01–1904-12-28); **12**, 8 (1905-01-08–1905-06-30); **13**, 35 (1905-07-01–1905-12-31); **14**, 69-70 (1906-01-02–1906-06-27); **15**, 112-113 (1906-07-04–1906-12-26); **16**, 159-160 (1907-01-01–1907-06-30); **17**, 194 (1907-07-01–1907-12-30); **18**, 230 (1908-01-04–1908-06-28); **19**, 260-261 (1908-06-30–1908-12-30); **20**, 298-300 (1909-01-03–1909-06-30); **21**, 346-348 (1909-07-01–1909-12-30); **22**, 383-384 (1910-01-01–1910-07-10); **23**, 12-13 (1910-07-12–1910-12-30); **24**, 56-58 (1911-01-01–1911-06-28); **25**, 113-114 (1911-07-03–1911-12-31); **26**, 174-175 (1912-01-04–1912-06-29); **27**, 245-247 (1912-07-01–1912-12-29).
06., 12.1903 used in Strasbourg's "Von den Instrumenten ..." and "Verzeichnis der im ...," see 60.
Used in the BAASSC (1913–1917) 1913–1915.
BGS: 1901–1919.

200. Duce (Aberdeen)

Used in the BAASSC (1913–1917) 1914–1915.

201. Blackford Hill (Edinburgh)

BAASRP **1897**, 168 (1896-08-25–1897-03-18); **1898**, 194 (1897-06-03–1898-01-29); **1899**, 186 (1898-03-05–1898-04-15).
Shide Circular **4**, 99-100 (1900-12-20–1901-06-24); **5**, 124-126 (1901-07-18–1901-12-31); **6**, 151-152 (1902-01-01–1902-06-22); **7**, 200-201 (1902-07-05–1902-09-28); **8**, 235-236 (1903-01-04–1903-06-25); **9**, 277-278 (1903-07-04–1903-12-28); **10**, 304 (1904-01-07–1904-06-27); **11**, 328 (1904-07-10–1904-12-21); **12**, 9 (1905-01-06–1905-06-30); **13**, 36 (1905-07-06–1905-12-31); **14**, 70-71 (1906-01-01–1906-06-26); **15**, 113-114 (1906-07-10–1906-12-26); **16**, 161 (1907-01-01–1907-06-26); **17**, 195 (1907-07-01–1907-12-30); **18**, 231 (1908-01-08–1908-06-27); **19**, 261-262 (1908-07-01–1908-12-28); **20**, 300-302 (1909-01-03–1909-06-30); **21**, 348-350 (1909-07-01–1909-12-31); **22**, 365-387 (1910-01-01–1910-06-30); **23**, 21-24 (1910-07-02–1910-12-30); **24**, 64-66 (1911-01-01–1911-06-28); **25**, 117-120 (1911-07-01–1911-12-31); **26**, 189-192 (1912-01-03–1912-06-29); **27**, 252-254 (1912-07-01–1912-12-29).
06., 12.1903 used in Strasbourg's "Von den Instrumenten ..." and "Verzeichnis der im ...," see 60.
Used in the BAASSC (1913–1917) 1913–1915.

202. Eskdalemuir

Shide Circular **21**, 353 (1909-07-07–1909-12-10); **22**, 389-390 (1910-01-01–1910-06-29); **23**, 26-27 (1910-07-07–1910-12-30); **24**, 68-69 (1911-01-01–1911-06-21); **25**, 124-127 (1911-07-01–1911-12-31); **26**, 183-187 (1912-01-03–1912-06-29); **27**, 247-251 (1912-07-01–1912-12-28).
Used in the BAASSC (1913–1917) 1913–1915.
BGS: 10.1913–08.1916, 01.–10.1920.
Annual report in Meterorological Year Book, pt. 3, no. 2.

203. Haslemere (Hazlemere, Frensham Hall Observatory)

Shide Circular **16**, 162-163 (1906-12-17–1907-06-25); **17**, 196 (1907-07-01–1907-12-15); **18**, 232 (1908-01-03–1908-06-23); **19**, 264 (1908-07-03–1908-12-29); **20**, 305 (1909-01-03–1909-06-28); **21**, 356-357 (1909-07-05–1909-12-23); **22**, 397-398 (1910-01-01–1910-06-29); **23**, 19-21 (1910-07-03–1910-12-30); **24**, 63-64 (1911-01-01–1911-06-28); **25**, 122-124 (1911-07-01–1911-12-31); **26**, 181-183 (1912-01-04–1912-06-29); **27**, 241-243 (1912-07-01–1912-12-31).
Used in the BAASSC (1913–1917) 1913–1915.

204. Kew (London)

BAASRP **1899**, 166-170 (1898-02-27–1899-04-03).
Shide Circular **1**, 4-11 (1899-04-04–1899-12-31); **2**, 33-35 (1900-01-05–1900-06-30); **3**, 64-65 (1900-07-01–1900-12-25); **4**, 96-97 (1901-01-07–1901-06-24); **5**, 121-122 (1901-07-07–1901-12-31); **6**, 149 (1902-01-01–1902-06-11); **7**, 197-198 (1902-06-16–1902-12-28); **8**, 232-233 (1903-01-04–1903-06-29); **9**, 275-276 (1903-07-04–1903-12-28); **10**, 302 (1904-01-07–1904-06-27); **11**, 326 (1904-07-01–1904-12-28); **12**, 6-7 (1905-01-03–1905-06-30); **13**, 33-34 (1905-07-03–1905-12-17); **14**, 67-68 (1906-01-21–1906-06-27); **15**, 110-111 (1906-07-08–1906-12-26); **16**, 158-159 (1907-01-02–1907-06-25); **17**, 193 (1907-07-04–1907-12-30); **18**, 229 (1908-01-13–1908-06-27); **19**, 259 (1908-07-01–1908-12-28); **20**, 297-298 (1909-01-02–1909-06-27); **21**, 345-346 (1909-07-03–1909-12-22); **22**, 382-383 (1910-01-01–1910-06-30); **23**, 11-12 (1910-07-03–1910-12-30); **24**, 55-56 (1911-01-01–1911-06-28); **25**, 111-112 (1911-07-01–1911-12-31); **26**, 172-173 (1912-01-04–1912-06-29); **27**, 237-239 (1912-07-07–1912-12-29).

06., 12.1903 used in Strasbourg's "Von den Instrumenten . . ." and "Verzeichnis der im . . . ," see 60.
Used in the BAASSC (1913–1917) 1913–1915.
Original observatory handwritten bulletins held at BGS: 04.1899–1912, 1914.

205. Newport, Gwent

Used in the BAASSC (1913–1917) 1914–1915.

206. Paisley (Coats Observatory)

Shide Circular **7**, 201-202 (1902-01-01–1902-12-30); **9**, 278-280 (1903-01-04–1903-12-31); **10**, 304-305 (1904-01-01–1904-06-27); **11**, 329 (1904-07-01–1904-12-20); **12**, 10 (1905-01-20–1905-06-30); **13**, 37-38 (1905-07-02–1905-12-17); **14**, 71-72 (1906-01-01–1906-06-24); **15**, 114-115 (1906-07-08–1906-12-23); **16**, 162 (1907-01-01–1907-06-13); **17**, 195 (1907-07-04–1907-10-23); **18**, 231 (1908-01-01–1908-06-03); **19**, 262-263 (1908-07-08–1908-12-29); **20**, 302-305 (1909-01-03–1909-06-30); **21**, 350-352 (1909-07-01–1909-12-31); **22**, 387-389 (1910-01-01–1910-06-30); **23**, 24-26 (1910-07-03–1910-12-31); **24**, 66-67 (1911-01-01–1911-06-28); **25**, 127-129 (1911-07-04–1911-12-31); **26**, 187-189 (1912-01-04–1912-06-29); **27**, 254-255 (1912-07-03–1912-12-28).
12.1903 used in Strasbourg's "Verzeichnis der im . . . ," see 60.
Used in the BAASSC (1913–1917) 1913.
BGS: 1902–02.1909.

207. Shide, Isle of Wight

BAASRP **1896**, 184-196 (1895-08-19–1896-03-22); **1897**, 149-151 (1896-06-14–1897-03-18); **1898**, 191-193 (1897-03-23–1898-02-16); **1899**, 162-166 (1898-02-27–1899-04-03).
Shide Circular **1**, 2-4 (1899-04-04–1899-12-31); **2**, 32-33 (1900-01-04–1900-06-28); **3**, 62-63 (1900-07-02–1900-12-25); **4**, 94-95 (1901-01-07–1901-06-24); **5**, 116-120 (1901-07-02–1901-12-31); **6**, 146-148 (1902-01-01–1902-06-22); **7**, 194-197 (1902-07-01–1902-12-20); **8**, 228-232 (1902-12-28–1903-06-25); **9**, 272-274 (1903-07-01–1903-12-28); **10**, 298-301 (1904-01-03–1904-06-26); **11**, 322-325 (1904-07-01–1904-12-30); **12**, 2-6 (1905-01-03–1905-06-30); **13**, 30-33 (1905-07-01–1905-12-29); **14**, 62-67 (1906-01-02– 1906-06-27); **15**, 106-110 (1906-07-04–1906-12-26); **16**, 154-158 (1907-01-01–1907-06-30); **17**, 188-192 (1907-07-01–1907-12-30); **18**, 224-228 (1908-01-01–1908-06-27); **19**, 254-258 (1908-07-01–1908-12-30); **20**, 288-296 (1909-01-02–1909-06-30); **21**, 338-344 (1909-07-01–1909-12-31); **22**, 376-381 (1910-01-01–1910-06-30); **23**, 2-10 (1910-07-02–1910-12-31); **24**, 49-54 (1911-01-01–1911-06-30); **25**, 100-110 (1911-07-01–1911-12-31); **26**, 164-172 (1912-01-01–1912-06-30); **27**, 230-237 (1912-07-01–1912-12-29).
06., 12.1903 used in Strasbourg's "Von den Instrumenten . . ." and "Verzeichnis der im . . . ," see 60.
Monthly Bulletin from 01.1913–
Used in the BAASSC (1913–1917) 1913–1915.

208. St. Helena

Shide Circular **24**, 82 (1911-02-18–1911-06-25).
Used in the BAASSC (1913–1917) 1914.

209. Stonyhurst College (Blackburn)

Shide Circular **21**, 353-355 (1909-07-06–1909-12-31); **22**, 393-395 (1910-01-01–1910-06-30); **23**, 15-17 (1910-07-02–1910-12-31); **24**, 60-61 (1911-01-01–1911-06-28); **25**, 115-116 (1911-07-01–1911-12-31); **26**, 175-177 (1912-01-04–1912-06-29); **27**, 243-245 (1912-07-01–1912-12-29).
Earthquake records by Milne seismograph meridian boom, Stonyhurst College Observatory 1913.
At ISC: 07.1914–08.1916, 01.1917–02.1918, 1920.
BGS: 07.1909–02.1917, 1920.
NOAA: 07.1909–03.1916, 09.1916–03.1917, 1920.
Used in the BAASSC (1913–1917) 1913–1915.

210. United Kingdom

*Davison Charles (1891–1893). On the British earthquakes of [1889–1892], Geolog. Mag. 8, 57-67, 306-316, 364-372; Ibid., Geolog. Mag. 8, 450-455; Ibid., Geolog. Mag. **9**, 299-305; Ibid., Geolog. Mag. **10**, 291-302.*
*Davison Charles (1900). On some minor British earthquakes of the years 1893–1899. Geolog. Mag. **17**, 164-177.*
Davison, Charles (1903). On the British earthquakes of the years 1889–1900. Beiträge zur Geophysik **5**, 242-313.
Davison, Charles (1908). On the British earthquakes of the years 1901–1907. Beiträge zur Geophysik **9**, 441-504.
BGS: Meteorological office, 1913-1920.

211. Warley (Birmingham)

Used in the BAASSC (1913–1917) 1914.

212. West Bromwich

Shide Circular **21**, 355-356 (1909-07-03–1909-12-10); **22**, 395-397 (1910-01-01–1910-06-30); **23**, 17-19 (1910-07-02–1910-12-30); **24**, 61-62 (1911-01-02– 1911-06-17); **25**, 129-132 (1911-07-01–1911-12-31); **26**, 177-178 (1912-01-04–1912-06-29).
Used in the BAASSC (1913–1917) 1913–1915.
09.1908–06.1911 station bulletins in Lapworth Museum.

213. Woodbridge Hill, Guildford

Shide Circular **22**, 391-392 (1910-01-01–1910-06-30); **23**, 14-15 (1910-07-05–1910-12-29); **24**, 58-59 (1911-01-01–1911-06-28); **25**, 120-122 (1911-07-01–1911-12-31); **26**, 178-180 (1912-01-04–1912-06-29); **27**, 239-241 (1912-07-01–1912-12-24).
Used in the BAASSC (1913–1917) 1913–1915.
BGS: 1910–1915.
NOAA: 1912, 1915.

214. Albany, New York

Annual reports in "Report of the Director of the State Museum," 1906–1912, from: Woodworth, J.B. (1917), details see 219.

215. Ann Arbor (Detroit Observatory), Michigan

Publications of the Astronomical Observatory of the University of Michigan ***1****, 1912, 54-72.*
NOAA: 01.1912–12.1915.

216. Baltimore

Shide Circular **4**, 112 (1901-04-05–1901-06-13); **5**, 138-139 (1901-04-05–1902-01-01); **6**, 163-164 (1902-01-01–1902-05-25); **7**, 211 (1902-06-08–1902-12-16); **8**, 249 (1903-01-04–1903-06-02); **9**, 281-282 (1903-07-09–1903-12-30); **11**, 336-338 (1904-01-01–1904-12-30); **12**, 18 (1905-01-04–1905-04-04); **13**, 49 (1905-06-14–1905-12-27); **14**, 90 (1906-01-03–1906-02-19); **15**, 135-136 (1906-05-17–1906-12-26); **17**, 213-214 (1907-01-01–1907-07-01); **18**, 246 (1907-07-20–1907-11-21 and 1908-02-01–1908-05-15); **20**, 324- 325 (1908-06-30–1908-12-28 and 1909-01-23–1909- 06-03); **22**, 416-417 (1909-06-27–1909-12-09); **23**, 32 (1910-01-22–910-12-23); **24**, 84 (1911-01-03–1911-06-15).
06.–03.1904 used in Strasbourg's "Von den Instrumenten . . . " and "Verzeichnis der im . . . ," see 60.
Reid, Harry Fielding. Records of Seismographs. Terrestrial Magazine ***10****, No. 2, 4.*
1904 in: Reid, Harry Fielding. Records of Seismographs. Terrestrial Magnetism and Atmospheric Electricity, 1905.
Used in the BAASSC (1913–1917) 1913.

217. Berkeley

Preliminary report of the State Earthquake Investigation Commission, 1906.
H.O. Wood, Records from Oct. 30, 1911, in Univ. Calif. Publications, Bull. of the Seismographic Stations.
The registration of earthquakes at the Berkeley Station and at the Lick Observatory Station, 1910–1920.
Used in the BAASSC (1913–1917) 1915.
BGS: 1911–1920.
NOAA: 11.1910–12.1920.

218. Buffalo (Canisius College), New York

Wessling, Henry J. (1910). Jesuit seismological service record of the earthquake station, Canisius College, Buffalo, June–September 1910. Monthly Weather Review 1910.
Repetti, William C. (1910, 1911). Jesuit seismological service record of the Canisius College, Buffalo. Monthly Weather Review October–December 1910; January–August 1911.
Ahern, Michael J. (1911). Jesuit seismological service record of the Canisius College earthquake station, Buffalo, September–December 1911. Monthly Weather Review 1911.
NOAA: 05.–09.1911.
Curtin, John A. (1915, 1916). Seismological Report, Canisius College, Buffalo, New York, Monthly Weather Report, January–December 1915; January–May, November 1916.

219. Cambridge (Harvard University), Massachusetts

Woodworth, J.B. (1909). Report of the Harvard seismographic station. Annual Report. Museum of Comparative Zoölogy. Cambridge, Mass., 1908–1909, 28-32.
Woodworth, J.B. (1910). Second annual report of the Harvard seismographic station. Ibid., 1910, 27-34.
See US-WB, 1914–1920.
Woodworth, J.B. (1917). Seventh annual report of the Harvard seismographic station. Annual Report. Museum of Comparative Zoölogy. Cambridge, Mass., 1917, 111-161; including "Seismographic Stations," 151-161.
At ISC: 01.–05.1913, 07.1913–05.1916, 03.–06.1917, 01.1918–11.1919, 01.–05., 10.–12.1919, 06.–09.1920.
Used in the BAASSC (1913–1917) 1913–1915.
NOAA: annual bulletins 04.1908–12.1915.
NOAA: preliminary 01.1911–05.1916, 03., 04., 06.1917, 01.1918–05.1919, 01.–06., 08., 09.1920.

220. Cheltenham, Maryland

Shide Circular **13**, 52-53 (1904-12-02–1905-06-30).
1904 in: Reid, Harry Fielding. Records of Seismographs. Terrestrial Magnetism and Atmospheric Electricity, 1905.
Hazard, Daniel L. (1910–1922). Results of Observations made at the [United States] Coast and Geodetic Survey Magnetic Observatory at Cheltenham, Maryland, 1905 and 1906, Department of Commerce and Labor, Coast and Geodetic Survey, pp. 92–94 [earthquake data for December 1904–1906]; *ibid.*, 1907 and 1908, pp. 90-92; *ibid.*, 1909 and 1910, pp. 90-93; *ibid.*, 1911 and 1912, pp. 96-98; *ibid.*, 1913 and 1914, pp. 96-97; *ibid.*, 1915 and 1916, pp. 110-111; *ibid.*, 1917 and 1918, pp. 114-116; *ibid.*, 1919 and 1920, pp. 94-96.
See US-WB, 1914–1920.

221. Cleveland, Ohio

Printed bulletins since 03.1911.
NOAA: 01.1911–03.1916.

222. Denver (Sacred Heart College), Colorado

Forstall, Armand W. (1909, 1910, 1911, 1915, 1916, 1917, 1918*). Jesuit seismological service record of the Sacred Heart College earthquake station of Denver, Colorado. Monthly Weather Review October 1909–January 1910; 1–12 and Special 5 (1911),* January–December 1915; 1916; January–July, September–December 1917; February–June, August–November 1918.
Forstall, Armand W. (1919, 1920). Seismological report, Sacred Heart College earthquake station. Monthly Weather Review April–December 1919; January–April, June. September–December, 1920.
NOAA: 08.1911–12.1912, 03.1913–12.1915, 03.–06., 08., 09.1916, 11.1916–02.1917, 04.1919.

223. Fordham, New York

Repetti, William C. (1915, 1916, 1918). Seismological record, Fordham University, New York. Monthly Weather Review

February–June, September–November 1915; April, June, September–October 1916; January–February 1918.

Sullivan, Daniel H. (1917, 1918). Seismological Report, Fordham University, New York. Monthly Weather Review January, December 1917; April, May 1918.

See US-WB, 1914–1920.

224. Washington, D.C. (Georgetown University)

Tondorf, Francis Anthony (1915, 1916, 1917, 1918, 1919, 1920). Seismological Report of Georgetown University. Monthly Weather Report January, April–December 1915; 1916; 1917; January–November 1918; January–June, August–October, December 1919; 1920.

Tondorf, Francis Anthony (1915, 1916, 1917). Georgetown University seismological bulletin, No. 1–9, 1915; No. 10–21, 1916; No. 22–25, 1917.

Tondorf, Francis Anthony (1916, 1917, 1918, 1919, 1920). The registration of earthquakes and press dispatches on earthquakes, Georgetown University, 13th series, 1916; 14th series, No. 4, 1917; 15th series, No. 2, 1918; 16th series, No. 2, 1919; 17th series, 1920.

At ISC: 1919, 1920.

NOAA: 12.1915–12.1920.

BGS: 1916–1920.

225. Honolulu, Hawaiian Islands

1904 in: Reid, Harry Fielding. Records of Seismographs. Terrestrial Magnetism and Atmospheric Electricity, 1905.

Hazard, Daniel L. (1910–1922). Results of observations made at the United States Coast and Geodetic Survey Magnetic Observatory near Honolulu, Hawaii, 1905 and 1906. Department of Commerce and Labor, Coast and Geodetic Survey, pp. 90-95 [earthquake data for October 1903–1906]; *ibid.*, 1907 and 1908, pp. 90-95; *ibid.*, 1909 and 1910, pp. 90-94; *ibid.*, 1911 and 1912, pp. 94-99; *ibid.*, 1913 and 1914, pp. 98-105; *ibid.*, 1915 and 1916, pp. 96-101; *ibid.*, 1917 and 1918, pp. 98-104; *ibid.*, 1919 and 1920, pp. 92-107.

Shide Circular **13**, 55-56 (1904-07-23–1905-07-01); **14**, 91-94 (1903-04-03–1904-06-27); **15**, 140-143 (1905-07-03–1906-12-26); **17**, 216-218 (1907-01-01–1907-12-30); **19**, 280-282 (1908-01-01–1908-12-28); **20**, 332-333 (1909-01-03–1909-06-28); **21**, 371-372 (1909-07-07–1909-12-30); **22**, 411-412 (1910-01-01–1910-06-30); **23**, 39-41 (1910-07-03–1910-12-30); **24**, 93-94 (1911-01-01–1911-06-28); **25**, 160-161 (1911-07-01–1911-12-31); **26**, 214-215 (1912-01-04–1912-06-29); **27**, 279-280 (1912-07-07–1912-12-29).

See US-WB, 1914–1920.

226. Honolulu (Oahu College), Hawaii

Shide Circular **6**, 181 (1901-12-30–1902-01-24); station moved to 225.

227. Ithaca, New York

Mimeographed bulletins from 01.1912.

NOAA: 03.1912–02.1915, 06.1905–08.1917.

228. Kilauea, Hawaii

Weekly bulletin of Hawaiian volcano observatory ***2****, 1914 from: Woodworth, J.B. (1917), details see 219.*

229. Lawrence, Kansas

See US-WB, 1914–1920.

230. Mobile (Spring Hill), Alabama

Ruhlmann, Cyril (1910, 1911, 1912). Jesuit seismological service record of the earthquake station at Mobile, Alabama, No. 1–6, 1910; No. 1–12, 1911; special 3 and No. 13–146, 1912. Monthly Weather Review, 1910–1912.

Ruhlmann, Cyril (1918, 1919, 1920). Seismological report of Spring Hill College, Mobile, Alabama. Monthly Weather Review October 1918; April, December 1919; January 1920.

At ISC: 05.1920–12.1920.

NOAA: 1911, 01.–08., 12.1912, 04.1919–12.1920.

231. Mt. Hamilton (Lick Observatory), California

Lick Observatory Bulletin, ***1****, 1901.*

06.1903 used in Strasbourg's "Von den Instrumenten . . . ," see 60.

1904 in: *Reid, Harry Fielding. Records of Seismographs. Terrestrial Magnetism and Atmospheric Electricity, 1905.*

1910–1920, see Berkeley 217.

Used in the BAASSC (1913–1917) 1914–1915.

232. New Orleans (Loyola University), Louisiana

Frankhauser, Joseph B. (1911). Jesuit seismological society record of the Loyola University earthquake station, New Orleans, Louisiana. Monthly Weather Review, No. 6–12, 1912.

233. Northfield, Vermont

See US-WB, 1914–1920.

234. Santa Clara, California

Ricard, Jerome S. (1910, 1911, 1912, 1913). Jesuit seismological service record of the earthquake station, Santa Clara, California., No. 1–29, 1910; No. 48–85, 1911; special 36 and No. 86–109, 1912; No. 110–124, 1913. Monthly Weather Review 1910–1913.

Printed bulletin from 06.1911.

Ricard, Jerome S. Record of the seismograph station, University of Santa Clara, California, 1914–1919.

At ISC: J.S.A. Sta. Clara, California 09.1911–05.1914, 09.–11.1914, 01.1915–06.1916.

Used in the BAASSC (1913–1917) 1913–1914.

NOAA: 06.1911–06.1915, 09.1915–06.1916.

235. Seattle, Washington

From 03.1910 occasional reports on observed seismic onsets (cited after Woodworth, J.B., 1917, details see 219).

236. Sitka, Alaska

Shide Circular **13**, 53-54 (1904-05-02–1904-12-20).
1904 in: *Reid, Harry Fielding. Records of Seismographs. Terrestrial Magnetism and Atmospheric Electricity, 1905.*
Registrierungen im Jahre 1905, Sitka, Alaska. In: Szirtes (1909b), p. 68.
Hazard, Daniel L. (1910–1922). Results of observations made at the United States Coast and Geodetic Survey Magnetic Observatory at Sitka, Alaska, 1905 and 1906, Department of Commerce and Labor, Coast and Geodetic Survey, pp. 92-94 [earthquake data for May 1904–1906]; *ibid.*, 1907 and 1908, pp. 92-94; *ibid.*, 1909 and 1910, pp. 92-95; *ibid.*, 1911 and 1912, pp. 96-100; *ibid.*, 1913 and 1914, pp. 98-99; *ibid.*, 1915 and 1916, pp. 94-95; *ibid.*, 1917 and 1918, pp. 100-101; *ibid.*, 1919 and 1920, p. 98.
Typewritten bulletins 1913, 1914.
See US-WB, 1914–1920.

237. Spokane (Gonzaga College), Washington

Bacigalupi, Eugene M. (1910, 1911). Jesuit seismological service record of the Gonzaga College earthquake station, Spokane, Washington, No. 1–15, 1910; No. 1–6, 1911. Monthly Weather Review 1910, 1911.
NOAA: 01.–03., 07.–09.1911.

238. St. Louis, Missouri

Joliat, Joseph S. (1911). Seismology in St. Louis University–earthquakes registered in St. Louis University observatory during 1910. Bull. St. Louis University **7**, 32-33.
Goesse, John B. (1912). Record of the earthquake station, St. Louis University.
Corey, Anthony H., and J.B. Goesse (1912). Jesuit seismological service record of the earthquake station, St. Louis University. Bull. St. Louis University **8**, 77-86.
Goesse, John B. (1914). Record of the earthquake station, St. Louis University.
Goesse, John B., G.E. Rueppel, and J. Roubik (1914). Geophysical observatory, seismological and meteorological departments. Earthquake records for 1913. Bull. St. Louis University ***10***, *54-60, 61-89.*
Goesse, John B. (1915–1917). Seismological report, geophysical observatory, St. Louis University. Monthly Weather Review January–July, September–December 1915; January–June, August, October–November 1916; January–August, November, December 1917.
Goesse, John B. (1915, 1916, 1917). Record of the earthquake station, St. Louis University.
Goesse, John B. (1918, 1919). Seismological report, geophysical observatory, St. Louis University. Monthly Weather Review January–November 1918; April–May 1919.
Rueppel, George E. (1919, 1920). Seismological Report, St. Louis University, 1919–1920.
At ISC: 11.1910–08.1913, 10.–12.1913.
NOAA: 01.1910–12.1920.
Used in the BAASSC (1913–1917) 1913–1915.

239. Tucson, Arizona

Hazard, Daniel L. (1912–1923). Results of observations made at the United States Coast and Geodetic Survey Magnetic Observatory near Tucson, Arizona, 1909 and 1910, p. 58. Department of Commerce, United States Coast and Geodetic Survey; *ibid.*, 1911 and 1912, pp. 100-103; *ibid.*, 1913 and 1914, pp. 100-102; *ibid.*, 1915 and 1916, pp. 98-100; *ibid.*, 1917 and 1918, pp. 98-100; *ibid.*, 1919 and 1920, pp. 95-97.
See US-WB, 1914–1920.
NOAA: 09.1910–12.1920.

240. USA

McAdie, A.G. (1907). Catalogue of earthquakes on the Pacific Coast, 1897–1906. Smithonian Miscellaneous Collections ***49***, *No. 1721, 24 pp.*
Report of the State Earthquake Investigation Commission 1.1906, 2.1906, Carnegie Institution Publication ***87.***

241. Washington, D.C.

ISA: U.S. Weather Bureau, Monthly Weather Review 1904.
1904 in: *Reid, Harry Fielding. Records of Seismographs. Terrestrial Magnetism and Atmospheric Electricity, 1905.*
06.1903 used in Strasbourg's "Von den Instrumenten . . . ," see 60.
ISA: Marvin, C.F. (1907). Distant earthquakes recorded at the Weather Bureau during the year 1906. Monthly Weather Review ***36***, No. 13, *Washington 1907.*
See US-WB, 1914–1920.

242. Tashkent (Taschkent, Taškent)

05.–08., 12.1903, 02., 03.1904 used in Strasbourg's "Von den Instrumenten . . . " and "Verzeichnis der im . . . ," see 60.
Notice mensuelles sur le tremblements de terre, Observatoire de Tachkent 1908.
Bulletin of Imperial Russia 1902–1908, see 168.
At ISC: 03.–08.1907, 10.1907–03.1908, 09.–12.1909, 01.–12.1913, 02.–05.1914.
Seismological Bulletin (weekly): 1.1912 (September 14.1912)–66.1913, 2.–18.1914.
Used in the BAASSC (1913–1917) 1913–1915.
NOAA: 09.1912–05.1914.
BGS: 1912–1913.

243. Belgrad (Beograd, Belgrade)

ISA: 1906–1908 weekly listings.
At ISC: 01.–10.1907, 01.1908–10.191.
NOAA: 1901–1906, 01., 04.–07.1909, 01.–06.1910.

244. Serbia

In Hamburg: Mikailović, J. Die Erdbeben in Serbien 1901–1907.

245. Binza (Leopoldville)

NOAA: 1909–1920.

Appendix: Seismographs 1856–1910

The most important part in studying seismic waves of the Earth is of course the instrumentation used to make these measurements. Therefore, we wish to provide information about the most important instruments in seismology, i.e., seismoscopes and seismographs. A comprehensive overview of early seismoscopes and seismographs that were in use before 1900 can be found in Ehlert (1898). This article contains numerous sketches of these instruments, illustrating their principles and describing their advantages and disadvantages. The classic 1914 monograph, *Vorlesungen über Seismologie* by B. Galitzin, contains many figures illustrating most of the important seismographs invented in the first decade of the 20th century and of which many were in use for during the following decades. Charles R. Hutt (USGS) contributed a set of pictures of modern seismometers and recorders used at seismic stations world-wide. The above cited materials and additional information are given as computer readable files on the attached Handbook CD #3, under the directory of \88Schweitzer.

The following ten figures illustrate some of most popular seismographs that were introduced from about 1856 to 1910. For references of these figures, please see the archived files on the CD.

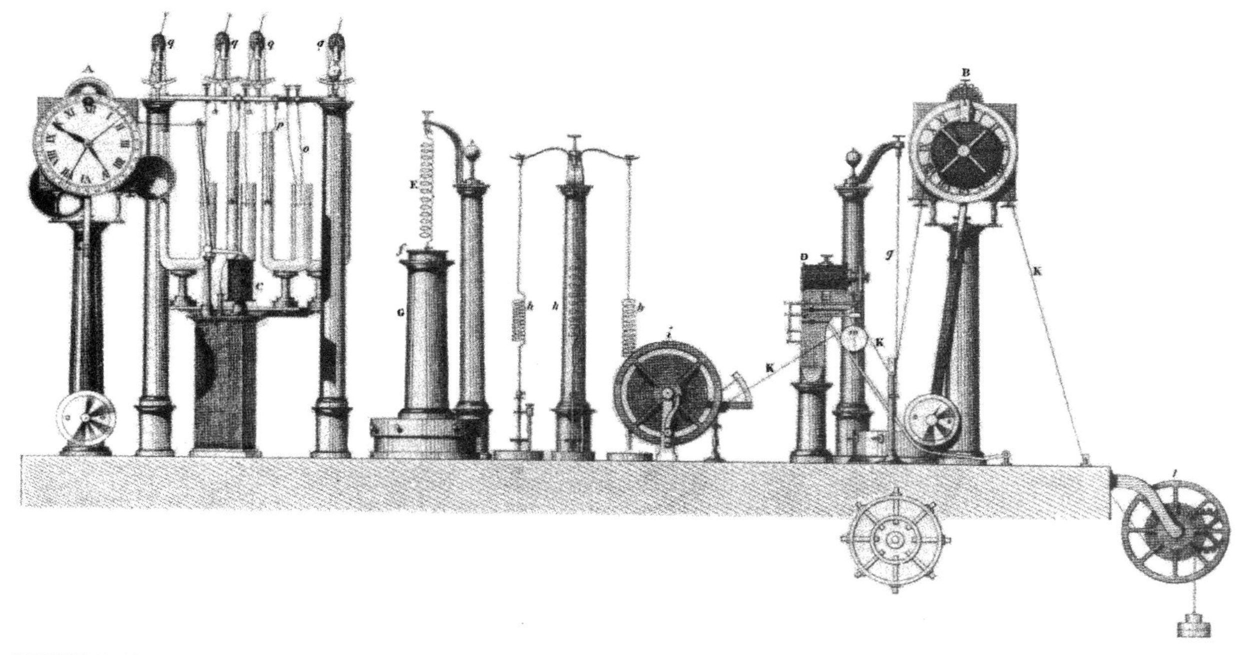

FIGURE 13 Palmieri's "Seismgrafo elettro-magnetico" (after Ferrari, 1990).

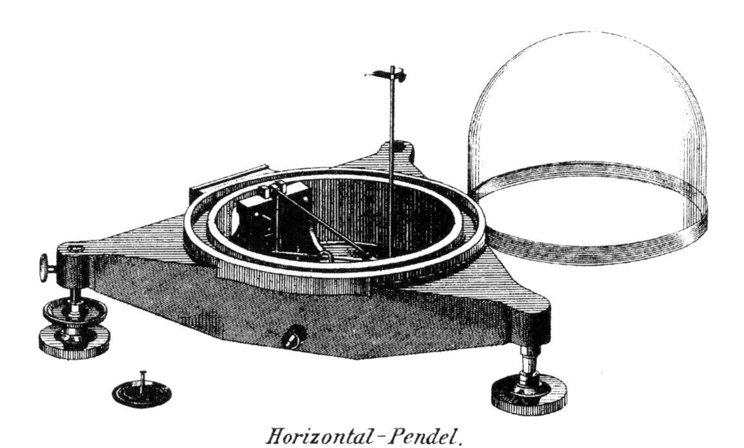

FIGURE 14 The drawing shows a Rebeur-Paschwitz Horizontal Pendulum as it was installed in Potsdam, Germany and in Wilhelmshaven, Germany in 1889 (after Rebeur-Paschwitz, 1892).

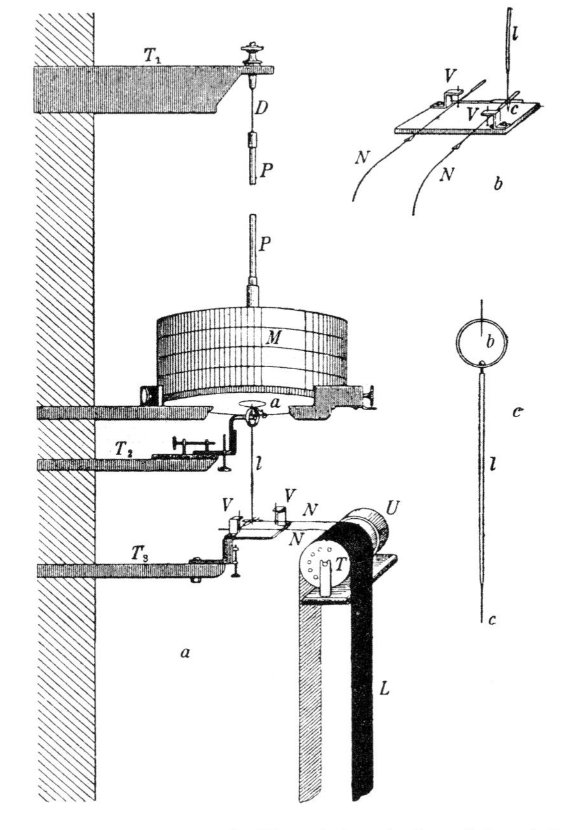

FIGURE 16 A drawing of a Vicentini vertical pendulum (after Galitzin, 1914).

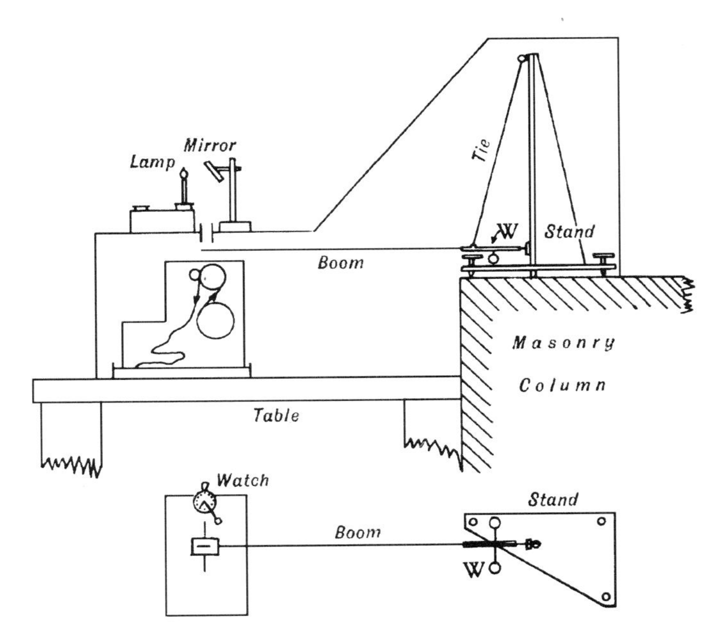

FIGURE 15 The Milne horizontal seismograph (after Dewey and Byerly, 1969).

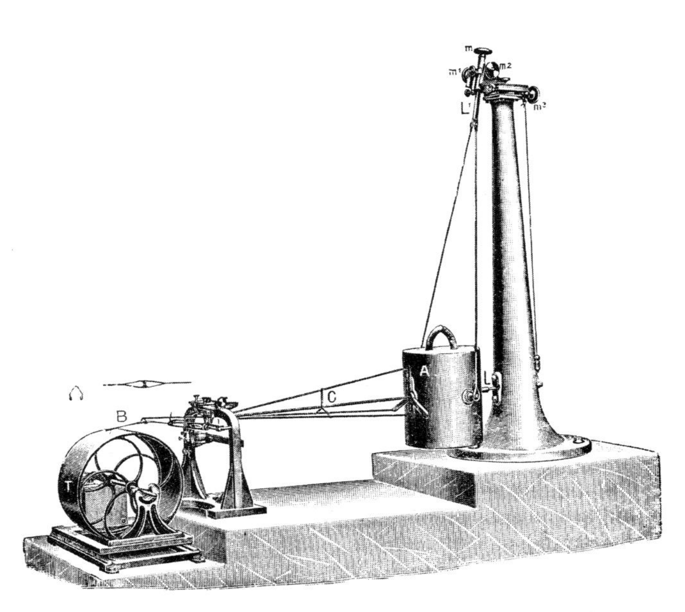

FIGURE 17 A drawing of a Bosch-Omori *Strassburger Schwerpendel* (after Galitzin, 1914). This type of Omori's seismograph was built by the factory *J. und A. Bosch* in Strasbourg.

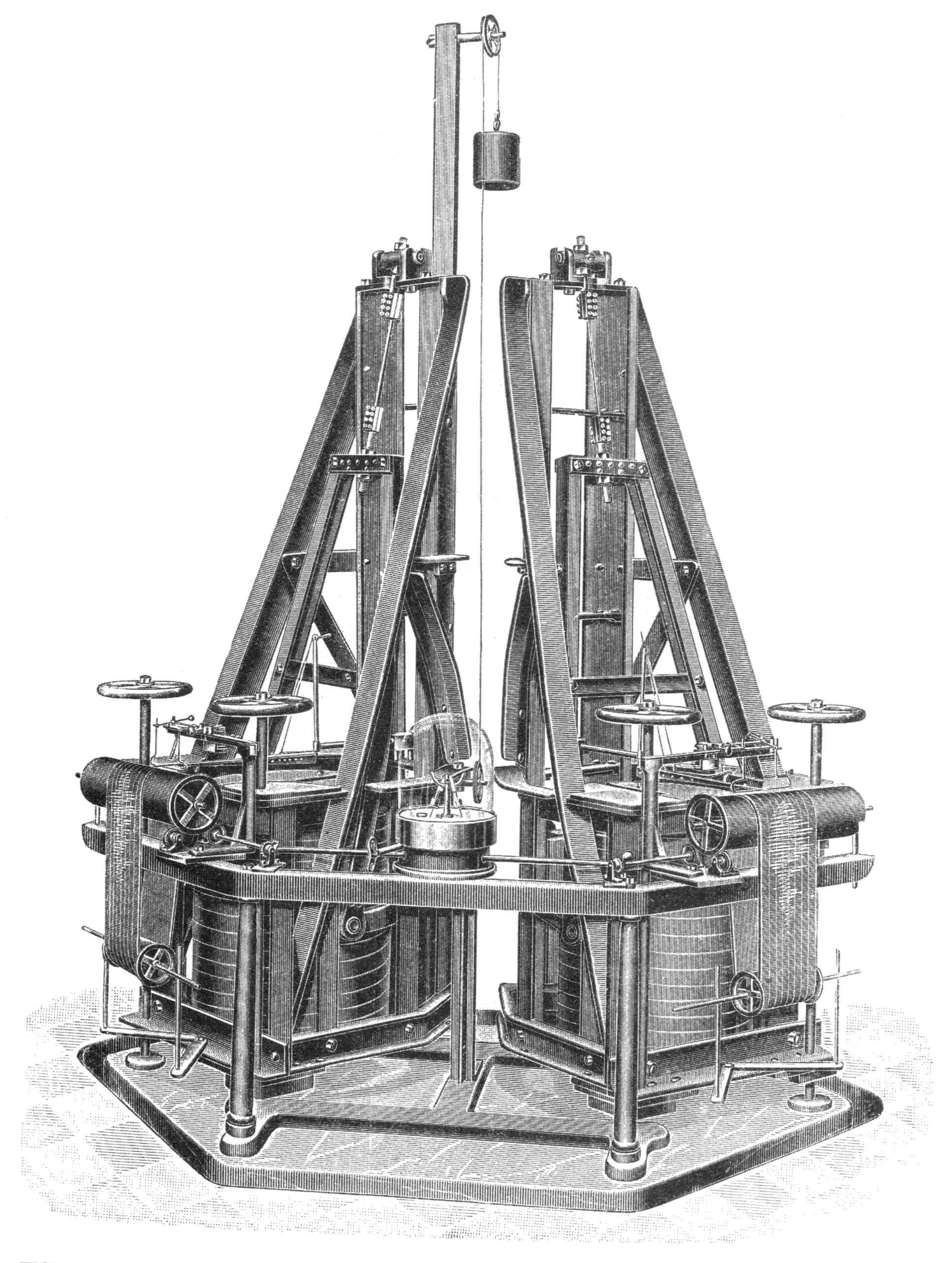

FIGURE 18 A drawing of a complete set of Mainka Horizontal Pendula (after Galitzin, 1914). These Mainka seismographs were also built by the factory *J. und A. Bosch* in Strasbourg.

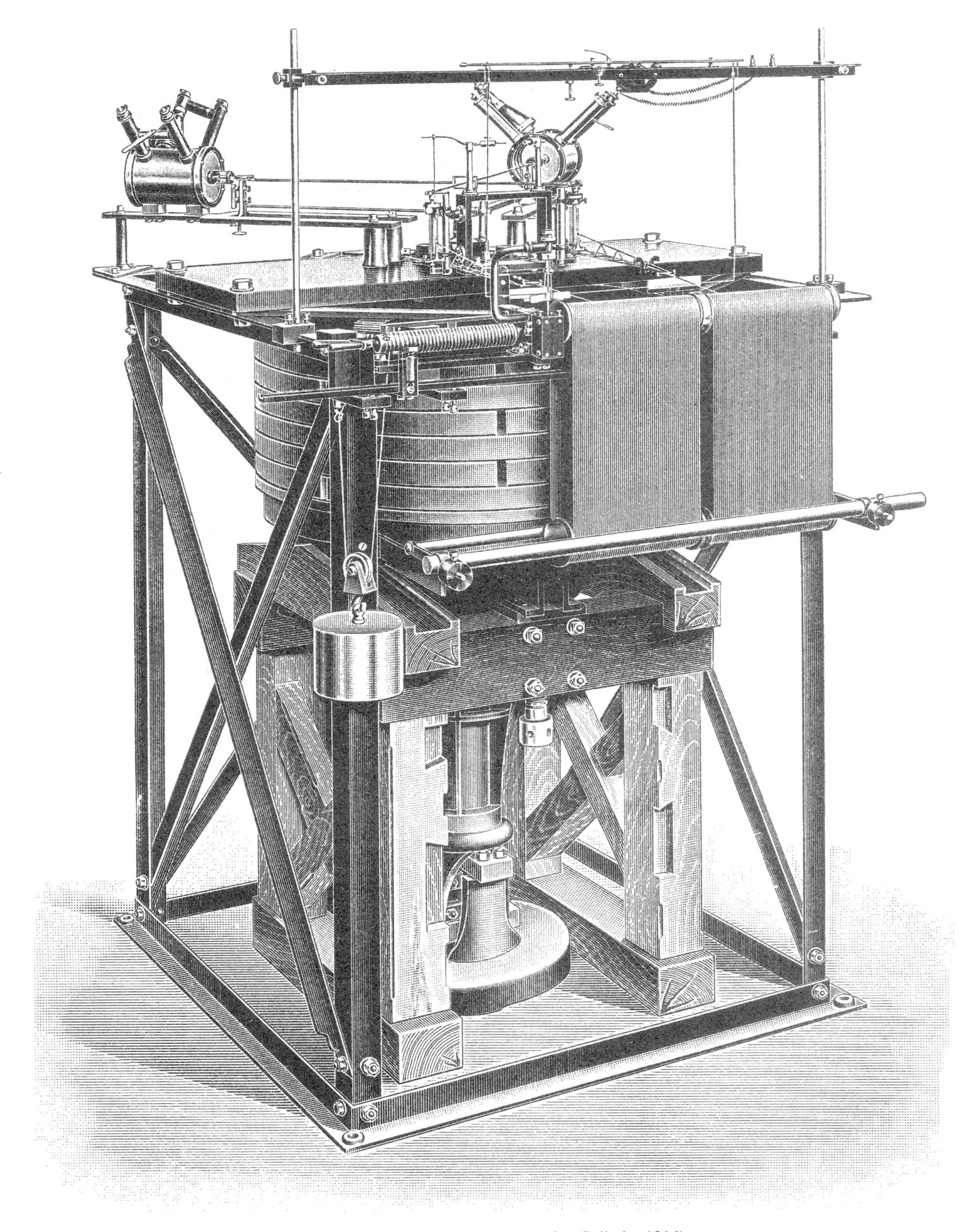

FIGURE 19 A drawing of a Wiechert Astatic Horizontal Pendulum (after Galitzin, 1914).

FIGURE 20 A drawing of a Wiechert Vertical Pendulum (after Galitzin, 1914).

FIGURE 21 A drawing of Mintrop/Wiechert Mobile Seismometers built to observe higher frequencies in all three components (after Galitzin, 1914).

FIGURE 22 A picture of a Galitzin Vertical Pendulum (after Galitzin, 1914).

89

Biographies of Interest to Earthquake and Engineering Seismologists

B. F. Howell, Jr.
Pennsylvania State University, University Park, Pennsylvania, USA

This chapter contains abbreviated and edited biographies of many individuals, nearly all deceased, who may be of interest to earthquake and engineering seismologists. Each biography emphasizes contributions to seismology and studies of the Earth's interior, or earthquake engineering, with only glancing attention to other fields. The sources for these biographies were the individuals' colleagues, usually prepared as memorials. A very few living individuals of advanced age have been included, as indicated by no closing date.

Biographical sketches submitted by living individuals themselves are indexed in Appendix 3 of this Handbook, and are archived on the attached Handbook CD#3, under the directory \BioSketches. Regrettably, during the preparation of this Handbook, some sketches have had to be adapted and moved to this chapter upon the death of their authors.

A brief summary of the history of seismology is given in Chapter 1 of Part A of this Handbook. Ben-Menahem (1995) presented "A concise history of mainstream seismology," and discussed its "origins, legacy, and perspectives." Many scientists and engineers have contributed to the advance of seismology, and it is believed appropropriate to include biographies of some of these contributors in this Handbook. Except for Davison (1927), there seems to be no published volume containing biographies of seismologists worldwide. This chapter represents a modest attempt to summarize over 300 biographies pertaining to seismology. Due to space limitations, only brief accounts can be given. Readers are encouraged to read the cited published sources, and the cited files archived on the attached Handbook CDs, to gain a better sense of the achievements and qualities of their professional forebears.

Most of the biographies included here were submitted in the national and institutional reports (see Chapters 79 and 80). Full versions of these biographies are archived in the national and institutional reports on the attached Handbook CDs (see Appendix 4 of this Handbook for a table of contents of the CDs). A condensed version of a submitted biography was first prepared. It was then edited to follow a more uniform style. However, several countries did not include biographies. Some important deceased individuals from these countries were selected and their biographies were prepared either from published sources or by our invited colleagues. Included are very brief biographies of some well-known scientists (such as Gauss and Newton) who are frequently cited in this Handbook.

References to more complete biographies are given at the end of each biography entry. Consulted sources used in preparing this chapter are given in two forms: One includes the complete source reference, and the other the author(s) and year of publication. The latter form is used for either a specific file archived on the attached Handbook CDs, or a published source that is applicable to more than one biography. Full citations for these cases are given in the list of references at the end of this chapter.

In order to make each biography as self-contained as possible, publications by that individual are given within the entry.

The following abbreviations are used for sources that publish biographical notices:

1. BSSA: Bulletin of the Seismological Society of America
2. SRL: Seismological Research Letters, published by the Seismological Society of America
3. NCAB: National Cyclopedia of American Biography
4. GSA: Geological Society of America
5. NAS: U.S. National Academy of Sciences
6. Chapter: refers to the numbered chapter of this Handbook.

INTERNATIONAL HANDBOOK OF EARTHQUAKE AND ENGINEERING SEISMOLOGY, VOLUME 81B *ISBN: 0-12-440658-0*

A

Adams, Leason Heberling (1887–1969)

American physical chemist and geochemist. He studied phase equilibria of rocks and thermodynamic properties of glass and minerals, and is noted for co-originating the method of estimating density in the earth (Williamson, E.D., and L.H. Adams, Density distribution in the Earth, *J. Wash. Acad. Sci.*, **18**, 413–428, 1923). Adams was born in Cherryvale, Kansas, and received the B.S. degree from University of Illinois in 1906 and an honorary D.Sc. from Tufts College in 1941. He was a chemist with the US Geological Survey in 1909–1910, then employed at Carnegie Institution of Washington from 1910 to 1957, being director of its Geophysical Laboratory from 1936 to 1952. Adams was president of the American Geophysical Union 1944–1947, and was awarded the Union's Bowie Medal in 1950. [*NAS Biogr. Mem.* **57**:2–33, 1987]

Agalarova, Alyaviya Bekhbud (1933–1992)

Azerbaijani seismologist. Agalarova graduated in physics from Azerbaijan State University in 1961 and worked in its Geology Institute until 1980 on Azerbaijan seismicity. From 1980 to 1992 she headed the institute's Division of Calculation, working on collecting, analyzing, and classifying Azerbaijan earthquakes and estimating future seismic energy flux for this region. In 1975 she defended a candidacy thesis on Azerbiajan seismicity at the Institute of Physics of the Earth of the USSR Academy of Sciences, and thereafter worked on crustal dynamics, focal mechanisms, and seismic zoning of the Caucasus region. [Chapter 79.7]

Agamennone, Giovanni (1858–1949)

Italian seismologist. He was most noted for the design of early seismographs, but also worked on field surveys of earthquakes, analysis of seismograms, wave-propagation studies, magnetic effects accompanying earthquakes, and early attempts at earthquake prediction. Agamennone studied physics at University of Rome. He started his career in seismology in 1886 as an assistant at the Ischia Geodynamic Observatory, later moving to the Central Office of Meteorology and Geodynamics in Rome. He was one of the founders, in 1895, of the Italian Seismological Society. From 1895 to 1897 he organized the Turkish Government Seismic Service at Istanbul, installing seismographs of his own design. In 1899 he returned to Italy as the director of the Rocca di Papa Geodynamic Observatory. In 1929 he became director of the Rome Central Office of Meteorology and Geodynamics. [Ferrari (2003)]

Alexidze, Merab (1930–1993)

Georgian geophysicist who worked on direct and inverse problems of gravity and seismology, creating models of Earth structure of the Caucasus and adjoining regions. He designed a program for hypocenter depth determination and worked on seismic tomography. He was a professor and member of the Georgian Academy of Sciences and director of its Institute of Geophysics starting in 1985. [Chapter 79.23]

Alfani, Guido (1876–1940)

Italian geophysicist who studied the effects of Italian earthquakes and dealt with anti-seismic construction. He designed numerous meteorological and seismic instruments for the Ximenian Observatory, including the use of radio for time signals. He was also a popular lecturer on seismology all over Italy, and wrote many newspaper articles on seismology, meteorology, and physics. Alfani was born in Florence and received a religious education. He became an assistant at the Ximenian Observatory in Florence in 1900, and took over its management in 1905. He was appointed a visiting professor of seismology at the University of Florence in 1909. [Ferrari (2003)]

Angenheister, Gustav Heinrich (1878–1945)

German seismologist. He started investigating surface waves in 1906, discovering that their velocities are greater under the oceans than the continents, but that their attenuation is also greater there, depending on the period. He studied the relation of volcanic to seismic events in the Tonga region. At Göttingen he focused on crustal structures and engineering seismology, developing portable seismometers and a shaking table to calibrate them. He used quarry blasts to measure local crustal structures. He investigated proposed building sites, the eigenfrequencies of pillars, and the dispersion of surface waves.

Born in Kleve, Germany, Augenheister studied mathematics and natural sciences at Heidelberg, Munster, Munich, and Berlin, where he received the Ph.D. in 1902 for a thesis on elasticity of metals. He was an assistant at the Institute of Physics at Heidelberg University from 1903 to 1904, then moved to the Institute of Geophysics at Göttingen in 1905. The following year he relocated to Apia, Samoa, where he worked in the Samoan Observatory, first as an observer and from 1914 to 1921 as director, carrying out meteorological and seismological observations. This work was interrupted in 1910 by a trip to Iceland to study geomagnetic perturbations and polar lights. He returned to Germany in 1921, where he held the position of observer at the Geophysical Institute at Potsdam and honorary professor at the Technical University of Berlin. In 1929 he moved to Göttingen as director of the Institute of Geophysics. He was editor of the Zeitschrift fur Geophysik and a geophysics editor for the Handbuch fur Experimental Physik. [Ritter *et al.* (2003)]

Auden, John Bicknel (1903–)

British geologist who worked on a variety of geological projects in India, including economic geology surveys and geology and glaciology of the Himalaya Mountains. He was the author of the memoir on the 1934 Bihar-Nepal earthquake (*Mem. Geol. Surv. India*, **73**, 1939) and studied the seismicity of the Koyna Reservoir, Maharashtra, Damodar Valley, and Shahkot areas. Auden was born at York, England, and graduated from Cambridge University; he received a D.Sc. from Cambridge in 1947. In 1926 he was appointed assistant superintendent of the Geological Survey of India, and became superintending geologist in 1945. Beginning in 1934, his attention shifted to water supply and engineering problems. He worked for the Land and Water Resources Division of the United Nations Food and Agriculture Organization from 1960 to 1962. He served for two years as foreign secretary of the Geological Society of India and was elected to the Indian National Science Academy. [Narula (1998)]

B

Baclund, Oscar Andreevich (1846–1916)

Swedish-born Russian astronomer. He was very supportive of seismological research in Russia in the early days, although his research was largely on celestial mechanics. Baclund graduated from Uppsala University in 1872. He worked at Stockholm Observatory from 1874 to 1878 and at Pulkovo Observatory from 1879 to 1916, serving as the director after 1895. He was a member of the Permanent Central Seismic Commission of the St. Petersburg Imperial Academy from 1899 to 1916. [Chapter 79.45]

Ban, Shizuo (1896–1989)

Japanese pioneer in research on prestressed concrete and its anti-seismic properties. He emphasized the importance of the ultimate-strength-design method for reinforced concrete structures. He helped to establish and develop the Japan General Building Research Center. Ban was born in Tokyo and graduated in architecture from Tokyo Imperial University in 1922. He was appointed associate professor of Kyoto Imperial University in 1922, promoted to professor in 1933, and retired in 1959. He became a member of the Japanese Academy in 1971. [Miyamura (2003)]

Banerji, Sudhangshu Kumar (1893–1966)

Indian geophysicist who studied microseisms as related to weather, Indian earthquakes, focal depth of earthquakes, natural earth currents, and meteorological problems. Banerji was born in Bengal and received the D.Sc. from the University of Calcutta in 1918. He was professor of mathematics there from 1919 to 1922, director of the India Meteorological Department's (IMD's) Colaba and Alibag Observatories from 1922 to 1932, and director of the IMD from 1944 to 1950. He was secretary of the Calcutta Mathematical Society in 1918, physical science secretary of the Asiatic Society from 1918 to 1922, and a founding fellow of the Indian Academy of Sciences in 1935. [Chapter 79.28]

Baratta, Mario (1868–1935)

Italian seismologist and geographer. In 1901 he published *The Earthquakes of Italy* describing the 1364 earthquakes known to have occurred in Italy for the years 1 to 1898. He updated this in 1935. He published a detailed monograph on the 1908 Messina, Italy, earthquake. Baratta was born in Voghera and graduated in natural sciences from the University of Pavia in 1890. He was an assistant at the Royal Central Office of Meteorology and Geodynamics at Rome from 1891 to 1896. Starting in 1903 he taught geography at the University of Pavia. [Ferrari (2003)]

Båth, Markus (1916–2000)

Swedish seismologist who established the Swedish seismic network and played a leading role in the seismology of nuclear testing. Båth was born in Katrineholm, and his career was exclusively with the Science Faculty of Uppsala University, starting with the B.S. (1939), M.S. (1940), Fil. Dr. (1943), and D.Sc. (1949). He served as an assistant at the Meteorological Institute (1939–1949), and was an assistant professor of seismology and meteorology (1949–1961), associate professor (1961–1967), and professor (1967–1981) in seismology, and professor emeritus (1981–2000). He established the first Seismological Institute in Sweden, which he directed from 1967 to 1976. Starting in the 1940s, he installed a seismographic network consisting of six permanent stations covering the whole of Sweden. He inspected their seismograms daily. He provided rapid announcements of underground nuclear explosions in the 1960s and 1970s.

During visits to the California Institute of Technology Båth worked with Gutenberg and Richter. He cooperated with UNESCO, and was vice-president (1956–1957), then associate general secretary (1957–1963) of the International Association of Seismology. He was awarded the Gutenberg medal of the European Geophysical Society for outstanding seismological research in 1998, and received an honorary doctorate from Uppsala University in 1999. As a teacher, he was an excellent lecturer and trained a number of outstanding seismologists. Båth's exceptional theoretical skills are manifested in his three books. A prolific scientific writer, he wrote a total of 336 scientific contributions plus book reviews, technical reports on seismological aspects of nuclear testing and earthquake hazards, and many popular Earth sciences contributions. [Husebye (2002); Kulhanek (2002)]

Bayes, Thomas (1702–1761)

English theologian and mathematician who, in a posthumous publication, was the first to use probability induction and establish a mathematical basis of probability. The "Bayesian method" accepts not only the interpretation of a probability as the limit of an experimental histogram, but also as the representation of subjective information. Bayes was born in London, educated privately, and ordained as a Nonconformist minister. In the late 1720s he became minister of the Presbyterian Chapel in Tunbridge Wells near London. See Chapter 16 (pp. 244–245) of Part A of this Handbook for a brief discussion of Bayes theorem. [James and James (1976); Britannica (2003)]

Belar, Albin (1864–1939)

Slovenian seismologist noted for designing seismometers, as a pioneer in using time signals transmitted by radio, for studies of individual earthquakes, and for publishing a monthly journal "Die Erdbebenwarte." Belar earned a B.S. in chemistry and natural sciences in 1890 from the University of Vienna and a Ph.D. in physics and natural sciences from the University of Graz. He was an assistant professor at the Austro-Hungarian Naval Academy at Rijeka from 1890 to 1896, professor of chemistry and natural sciences at Ljubljana State High School from 1896 to 1908, a state school supervisor from 1908 to 1918, and a member of the Seismological Commission of the Vienna Academy of Sciences. In 1897 he established the first seismological station in Austro-Hungary, in Ljubljana. He operated his own private seismological station from 1921 to 1930. [Chapter 79.47]

Beloussov, Vladimir Vladimirovich (1907–1990)

Russian geologist best known for his work in structural geology and tectonics. Beloussov started study at Moscow State University in 1927, moved to Leningrad in 1930, and received the doctorate there in 1938 for research on the geology of the Caucasus region. His first job was with the Geological Prospecting Office and later the State Radium Institute, working on the geology of helium. He started teaching at Leningrad University in 1939, being promoted to professor in 1940. During World War II he moved to Kazan, where he worked on radioactivity as a source of the Earth's internal heat. In 1943 he started work at the Geological Prospecting Institute at Moscow, soon thereafter entering the Theoretical Geophysical Institute of the USSR Academy of Sciences, where he organized the Geophysical Laboratory.

In 1948 Beloussov led the Soviet delegation to the International Geological Congress in London. He taught tectonophysics at Moscow State University from 1953 until his death. He was head of the Soviet delegation to the 1954 IUGG General Assembly. He was elected first vice-president of IUGG in 1957 and president in 1960. He was active in organizing the work of the International Geophysical Year. His ideas are well presented in his book, *Basic Problems in Geotectonics*, the English edition published in 1962. [Chapter 79.45]

Benioff, Victor Hugo (1899–1968)

American seismologist noted for his studies of the distribution of deep earthquakes in Wadati-Benioff zones along tectonic-plate boundaries, and for plotting patterns of strain rebound as estimated from magnitudes. He designed the variable reluctance seismometer that was widely used, and he studied problems of stress relaxation and the Earth's free oscillations using the Benioff strain seismometer. Benioff was born in Los Angeles and received the B.A. degree from Pomona College in 1921. He was an assistant at Lick Observatory from 1923 to 1924, then in 1925 became one of the first employees of what is now the Caltech Seismological Laboratory. He was appointed assistant professor in 1937, associate professor in 1938, and professor in 1950. He was president of the Seismological Society of America in 1958–1959. [*BSSA* **58**:1701–1703, 1968; Goodstein (1991)]

Benndorf, Hans (1870–1953)

Austrian geophysicist who first calculated travel times for a spherical, horizontally layered body and introduced the ray-parameter (Benndorf's law). He was a professor for physics in Vienna and from 1904 in Graz; his main work was on atmospheric electricity. [Schweidler (1941); Kertz (2002)]

Beránek, Břetislav (1922–1981)

Czech seismologist. Born in Vinary, Beránek graduated from Charles University in Prague, where he was a student of Zátopek. Beránek started his career in oil prospecting for the Czechoslovak Oil Wells Company, and after its incorporation into Geofyzika Brno he served as its deputy director for research. He worked intensively in the field of deep seismic sounding in the territory of Czechoslovakia, organized international cooperation within the Upper Mantle Project and in the Geodynamics Project, and was an active member of the European Seismological Commission of IASPEI. [Chapter 79.17]

Berg, Joseph Wilbur, Jr. (1920–1997)

American seismologist. Berg's research activities included earthquake and explosion seismology and crustal and upper-mantle structure. He was a professor at University of Tulsa (1954–1955), University of Utah (1955–1960), and Oregon State University (1961–1966). During 1960–1964, he worked at the US Institute of Defense Analysis, where he helped establish the World-Wide Standardized Seismograph Network. He was Executive Secretary of the Office of Earth Sciences (1966–1987), US National Academy of Sciences–National Research Council. Berg was elected President of the Seismological Society of America in 1968–1969. [SRL 69:297, 1998]

Berlage, Hendrik Petrus, Jr. (1896–1968)

Dutch geophysicist and meteorologist. His interests included seismometers, deep focus earthquakes in Indonesia, climatology (in particular monsoon prediction in Indonesia), and the origin and development of Earth and Moon. He worked in De Bilt, at the Institut de Physique du Globe in Strasbourg, in Indonesia (Batavia (Jakarta) and Bandan) from 1925, and from 1950 again in De Bilt. In 1930, he contributed with a classical paper on seismometers to Gutenberg's *Handbuch der Geophysik,* **4**. [Snelders (1989)]

Berson, Inna Solomonovya (1914–1974)

Russian geophysicist. She was a pioneer in developing both refraction and reflection methods of seismic surveying and interpretation of such surveys, including borehole surveys. She also led seismic prospecting expeditions in different areas of the Soviet Union for many years. Berson graduated in geophysics from Dnepropentrovsk Mining Institute in 1937 and received a doctorate in 1952. Starting in 1952 she was head of the Seismic Methods of Prospecting Department of the USSR Academy of Sciences. [Chapter 79.45]

Bertelli, Timoteo (1826–1905)

Italian seismologist. His research was primarily on microseisms, which he studied by directly watching the motions of a pendulum, using a magnifying lens. He correlated the variations in the level of the microseismic movements with the seasons, meteorological activity, and distant earthquakes. Bertelli was born in Bologna and was ordained a priest in Naples in 1850. He taught mathematics and physics at a variety of schools in Italy. He was teacher at a boarding school in Florence from 1857 until his death, except for serving as director of the Vatican Specola in Rome from 1895 to 1898. [Ferrari (2003)]

Bertrand, Elie (1712–1790)

Swiss naturalist and geologist who described effect of the 1755 Lisbon earthquake and compiled the first catalog of Swiss earthquakes. Bertrand was a pastor at Berne. A man of wide-ranging interests, he was a member of the academies of Berlin, Florence, and Lyon. His major publications include *Mémoires Historiques et Physiques sur des Tremblements de Terre* (La Haye, 1757, 328 pp.). [Davison, 1927]

Bevis, John (1693–1771)

British physician, astronomer, and naturalist. Educated at Christ Church, Oxford, Bevis practiced medicine while pursuing optics and astronomy for pleasure. In 1757 he anonymously published a book, *The History and Philosophy of Earthquakes* (London, 351 pp.). [Davison, 1927]

Birch, Albert Francis (1903–1992)

American geophysicist whose many accomplishments all stemmed from studies of the physical properties of rocks and minerals. He was a pioneer in determining the composition, structure, and temperature of the Earth's interior (see, e.g., "Elasticity and constitution of the Earth's interior," *J. Geophys. Res.*, **57**, 227–286, 1952). He measured the compressibility and seismic velocities of rocks as a function of temperature and pressure, and extrapolating his data to high pressures, showed that they explained the observed variations of wave velocities in the Earth. He developed the Birch-Murnaghan equation of state to predict and explain variations in physical properties with depth in the Earth. He predicted that the liquid outer core could not be pure iron but must contain lighter elements, but that the inner core could be nearly pure iron. He was principal author of the first edition of the *Handbook of Physical Constants* (*GSA Special Paper 36*, Geological Society of America, 1942), a compilation of a wide variety of physical constants needed for geological and geophysical calculations.

Birch was also a pioneer in heat-flow measurements of the Earth, and served as the first chairman of the International Heat Flow Committee (see Chapter 81.4). He was president of the Geological Society of America in 1963–1964; he received its Day Medal in 1950 and Penrose Medal in 1969. From the American Geophysical Union he received the Bowie Medal in 1960. Birch shared the Vetlesen Prize with Bullard in 1968.

Birch was born in Washington, D.C., and received the B.S. (1924), M.A. (1929), and Ph.D. (1931) degrees in physics at Harvard University. From 1932 to 1974 he was a research assistant, assistant professor, associate professor, and professor at Harvard University. [*NAS Biogr. Mem.* **74**:1–25, 1998; Birch (1979)]

Bjuss, Evgeni (1885–1965)

Georgian geophysicist who worked on seismicity, seismic zoning, and earthquake recurrence. Bjuss started as an astronomer but in 1914 became head of the seismic station at Baku, moving in 1921 to the Seismic Department of the Tbilisi Geophysical Observatory, which later became the Department of Seismology of the Institute of Geophysics of the Georgian Academy of Sciences. A lecturer on seismology and astronomy at Tbilisi State University, his principal publication was the three-volume *Seismic Conditions of the Caucasus*. [Chapter 79.23]

Blume, John A. (1909–2002)

American earthquake engineer. Blume is noted for his pioneering work in the development and application of new design concepts and analysis of buildings and structures in response to earthquakes. Born in Gonzales, California, he received his A.B. degree (1933) and his Degree of Engineer (1935) from Stanford University. His thesis was on the dynamic response of buildings.

In 1935 and 1936 he worked as a construction engineer on the San Francisco–Oakland Bay Bridge. Later he worked for the Standard Oil Company of California and the structural engineering design firm of H.J. Brunnier.

In 1945, Blume established John A. Blume and Associates (JAB), which soon became a preeminent consulting firm in structural and earthquake engineering. JAB designed or analyzed scores of special earthquake projects, among them the two-mile-long Stanford Linear Accelerator, the Embarcadero Center in San Francisco, and the Diablo Canyon Nuclear Power Plant. Blume helped found the Earthquake Engineering Research Institute. He published over 150 paper, articles, and books, a remarkable number for a person in private practice, and was elected to the US National Academy of Engineering in 1969. [Blume, 1994]

Bock, Günter (1944–2002)

German geophysicist who died prematurely in a plane crash on Nov. 6, 2002. He listed his seismological interests as (1) development of models on reservoir-induced seismicity, (2) investigation of the effect of subducting lithosphere on travel times of body waves, (3) study of the Earth's mantle using body waves and shear wave splitting analysis, (4) development of methods for the rapid determination of earthquake source parameters, and (5) the study of source characteristics of intermediate-depth and very deep earthquakes. Bock was born in Paderborn, earning a diploma in geophysics at Munich University in 1971 and a Ph.D. in geophysics at Karlsruhe University in 1978. He was a research associate at Karlsruhe University (1971–1978) and a research scientist at the Gesellschaft für Strahlen- und Umweltforschung (1978–1979). He then spent time in Australia, first as a research fellow at the Australian National University (1979–1983) and as lecturer and senior lecturer at the University of New England (1984–1992). He was a senior research scientist at the GeoForschungsZentrum Potsdam starting in 1992. [Bock (2002)]

Born, Max (1882–1970)

German physicist and a founder of quantum mechanics. Born is best remembered by theoretical seismologists as the originator of the Born approximation method, in which perturbation theory is applied to problems concerning the scattering of atomic particles. Born was born in Breslau, Germany (now Wroclaw, Poland), and was educated at various schools before graduating from the University of Göttingen in 1907. He became professor of theoretical physics at Göttingen in 1921. Born fled Germany in 1933 and settled first in England. He was elected to the Tait chair of natural philosophy at the University of Edinburgh in 1936 and retired to Germany in 1953. See Chapter 52 (pp. 863–864). [Britannica (2003)]

Borne, Georg von dem (1867–1918)

German seismologist who developed, together with Wiechert, the first nomenclature for seismic phases (P, S, and L waves). He developed first ideas about prospection seismology, worked on sound propagation in the atmosphere, and founded the seismic station in Breslau-Krietern (today Wroclaw, Poland). [Schweitzer (2003)]

Branner, John Casper (1850–1922)

American geologist. He wrote on earthquakes in California and Brazil and on the resources of Brazil, but is also remembered for his role in leading and building institutions. Branner was born at New Market, Kentucky, and received the B.S. degree at Cornell in 1882 and the Ph.D. from Indiana University in 1885. He worked for the Geological Survey of Pennsylvania from 1883 to 1885, was professor of geology at Indiana University for a year, state geologist of Arkansas from 1887 to 1892, then professor of geology at Stanford University from 1885 to 1915, serving as vice-president from 1899 to 1913 and president from 1913 to 1915. He was a member of Lawson's commission studying the 1906 San Francisco earthquake, a charter member of the Seismological Society of America and chairman of the society's organizing committee, its president from 1910 to 1912, the first chairman of the BSSA Publication Committee (1911–1922), and president of the Geological Society of America in 1904. He was a member of the US National Academy of Sciences and received the Hayden Medal of the Philadelphia Academy of Sciences in 1911. [*BSSA* **12**:1–11, 1922]

Brillouin, Léon Nicolas (1889–1969)

French physicist who contributed to many areas of physics, including quantum mechanics and information theory. He is best known by theoretical seismologists as the "B" in the WKBJ method (applied in ray theory; see Chapter 8 (p. 90) of Part A of this Handbook). Brillouin was born at Sevres and was educated at the Ecole Normale Supérieure and the University of Munich. He taught physics as a Sorbonne professor and was appointed the chair of theoretical physics at the Collège de France in 1932. He emigrated to the United States in 1947. He taught at Harvard and Columbia and served as a research director of IBM. [Holmes (1990)]

Brown, Robert (1773-1858)

Scottish botanist. In 1827 Brown observed erratic motion of pollen grains in water under a microscope, now known as Brownian motion. Brown was born in Montrose, studied medicine at Edinburgh, and served as a medical officer in Ireland. In 1801 he accompanied the Flinders expedition to Australia, making his reputation as a great collector. From 1806 to 1822 he was librarian of the Linnean Society, then a fellow, then president from 1849 to 1853. [*Proc. Roy. Soc. Lond.* **9**:527–532, 1859; Daintith and Gjertsen (1999)]

Bullard, Edward Crisp (1907–1980)

English geophysicist and one of the best-known geophysicists of his generation. Bullard's experimental and theoretical work contributed to advances in many areas of geophysics. Born in Norwich, he became a research student at the Cavendish laboratory in Cambridge in 1929 under Blackett. Later he worked with Massey on electron scattering (until 1931). Then he became demonstrator in the Department of Geodesy and Geophysics in Cambridge and worked with Conyngham and Jeffreys with a pendulum apparatus over the East African rift. In 1936 he got the Smithson Fellowship of the Royal Society and developed a short-period seismometer, with which he examined the basement in southeast England. In 1938 he began measuring heat flow in South Africa. In November 1939 he became a research officer for the admiralty and developed preventive measures against magnetic mines. In 1941 he was elected a member of the Royal Society. Bullard became head of the physics department at the University of Toronto in 1948. Here he encouraged the development of radioactive age determination of rocks, and became acquainted with the use of a large computer for his research on the generation of the Earth's magnetic field.

In the late 1940s, Bullard developed an instrument for measuring heat flow at sea. He was the director of the UK Physical Laboratory from 1949 to 1954, but still found time to publish several papers (e.g., "The flow of heat through the floor of the Atlantic Ocean," *Proc. Roy. Soc. Lond. A*, **222**, 408–429, 1954). He returned to the Department of Geodesy and Geophysics at Cambridge in 1955, and made it one of the best places of the world to study geophysics, and a major center for the development of plate tectonics. Bullard was instrumental in establishing the International Heat Flow Committee in 1963 (see Chapter 81.4). [Holmes (1990); Bullard (1974)]

Bullen, Keith Edward (1906–1976)

Australian applied mathematician and geophysicist who made significant contributions to seismology and physics of the Earth's interior. He published, with Jeffreys, the famous J-B travel-time tables in 1940. Born in Auckland, New Zealand, Bullen was educated at the universities of Auckland, Melbourne, and Cambridge. In 1946 he became professor of applied mathematics at the University of Sydney. Bullen wrote prolifically. There are 290 papers in his list of publications. The topics are diverse. Apart from the many research papers there are scientific biographies, articles in encyclopedias and dictionaries of science, and articles on education, especially mathematical education. His first book, *Introduction to the Theory of Seismology*, was published by the Cambridge University Press in 1947 and has been a standard text for seismology ever since. His short monograph, *Seismology*, was published by Methuen in 1954. His last book, *The Earth's Density*, was published in 1975; it covers a wider range of geophysics than its title would suggest because the problem of the Earth's density distribution is so intimately related to seismological information on the Earth's interior. Bullen served as president of IASPEI from 1954 to 1957. A photo of Bullen is included in Figure 2 of Chapter 81.3 of this Handbook. [*Biogr. Mem. Australian Acad. Sci., Records of AAS*, **4**(2), 1979]

Bychovsky, Victor Arnoldovich (1907–1982)

Russian engineering seismologist whose research interests included soil problems as well as dynamic and structural engineering problems. His early work (1936) was on seismicity scales. He created and headed the Earthquake Engineering Laboratory of the Central Research Institute of Structures in Moscow. He developed the Seismic Building Code for active seismic areas of the USSR in 1957. He was co-author in 1961 of *Fundamentals of Seismic Design* and wrote over 150 papers on seismology and earthquake engineering. [Rzhevsky (2002)]

Byerly, Perry (1897–1978)

American seismologist who developed the method of determining fault parameters from the polarities of first motions on the seismogram. He authored an early textbook: *Seismology* (Prentice-Hall, New York, 1942). He was the first to point out the significance of what came to be called the "20-degree discontinuity." He used both earthquakes and blasts to study crustal structure. He was the first to demonstrate by seismic studies that there is a root under the Sierra Nevada Mountains. He was the first to demonstrate that water-well pressure can be used to record earthquakes, and he studied the energy of earthquakes, dispersion of seismic waves, discontinuities in the Earth, earthquake swarms, and the T-phase of Hawaiian earthquakes as well as studies of many individual earthquakes and quarry blasts.

Byerly was born in Clarinda, Iowa, and received the A.B. (1921) and Ph.D. (1924) degrees in physics at the University of California, Berkeley, plus an honorary LL.D. in 1966. He was an instructor of physics at the University of Nevada from 1924 to 1925, then became the first director of Berkeley's Seismograph Station, which he expanded into a sixteen-station network including the first use of telemetering of data to the central station of the network in the early 1960s. He was appointed assistant professor of seismology in 1927, associate professor in 1931, and professor in 1941. He was head of Berkeley's Department of Geological Sciences from 1949 to 1954. He was secretary of the Seismological Society of America from 1930 to 1956 and president in 1957–1958. He was president of IASPEI from 1960 to 1963. A photo of Byerly is included in Figure 2 of Chapter 81.3 of this Handbook. He was elected to the US National Academy of Sciences in 1946 and was chairman of the Academy's Section of Geophysics and chairman of the National Research Council panel on seismology and gravity for the International Geophysical Year (1957–1958). [*BSSA* **69**:928–945, 1979]

C

Cabré-Roige, Ramón, S.J. (1922–1997)

Spanish-born Bolivian seismologist whose research pertained to the seismicity and seismotectonics of Bolivia and South America. Cabré was born in Barcelona and studied physical sciences at the University of Barcelona, where he received his doctoral degree in 1959. Shortly thereafter he moved to La Paz, Bolivia, where he became sub-director, then, on the death of the director Pierre Descotes, director of the Jesuit-run Observatorio de San Calixto. He held this position until 1993. During his thirty years as director, the observatory prospered: The seismographic station was selected as one of the WWSSN stations and was subsequently upgraded to HGLP-ASRO status in 1972. Cabré was founder of CERESIS (Regional Center of Seismology of South America) and was its director until 1968 (see Chapter 81.8). He was named to the Bolivian National Academy of Science in 1971. Between 1971 and 1974 he was vice president of the Geophysics Commission of the Panamerican Institute of Geography and History. From 1971 to 1975 he was a member of the Executive Committee of IASPEI. [Stauder (1999a)]

Cagniard, Louis (1900–1971)

French geophysicist who developed the theory of seismic waves in 1937. His classic memoir, *Reflection et Refraction de Ondes Seismiques Progressives*, was translated into English by E.A. Flinn and C.H. Dix and published by McGraw-Hill, New York, in 1962. He also introduced the magnetotelluric method in geophysical prospecting in 1953. His name is most frequently associated with the "Cagniard-De Hoop" methods in seismology (see Aki and Richards, 2002). Cagniard founded the applied geophysics department of the Paris Science Faculty in 1946, and was a director of the Compagnie Générale de Géophysique. [Cara (2003)]

Caloi, Pietro (1907–1978)

Italian geophysicist who worked in a wide variety of fields including hydrodynamics of large bodies of water, slow movements of the Earth's crust, microseisms produced by ice and atmospheric pressure, Rayleigh and Somigliana waves, channeling of seismic waves in low-velocity layers, wave attenuation, core–mantle boundary problems, fault dynamics, studies of individual earthquakes, and the philosophy of science. Caloi graduated in mathematics from the University of Padua in 1929, studied astronomy for a year there, and then was employed as an assistant at the Geophysical Institute of Trieste where he helped install a new seismic station. In 1935 he became chair in seismology at the University of Rome. In 1937, he also joined the National Institute of Geophysics, first as a principal geophysicist, then in 1948 as chief geophysicist, and in 1951 as director of its observatory. He helped found the European Seismological Commission in 1951, serving as its secretary. From 1951 to 1954 he was Vice-President of IASPEI. He was a member of the Accademia Nazionale dei Lincei. [Ferrari (2003)]

Cancani, Adolfo (1856–1904)

Italian geophysicist. His major contribution was the modification of the Mercalli intensity scale based on ground accelerations that doubled with each step increase in the scale. He was one of the first to recognize that seismic waves were elastic waves. Cancani was born in Rome and graduated in physics in Rome in 1884. He joined the Central Office for Meteorology and Geophysics (COMG) in 1888. He moved to the Rocca di Papa Geodynamic Observatory in 1892, then rejoined COMG in 1899 as an assistant in the geodynamics department. He was a founding member of the Italian Seismological Society (1895). He designed and built numerous seismic instruments, and at the 1903 International Seismological Conference in Strasbourg was appointed to the Commission on Instrumentation. He was also a noted teacher, holding a visiting professorship at the Royal University of Modena at the time of his death. [Ferrari (2003)]

Carder, Dean Samuel (1897–1973)

American seismologist. His early research was on building-vibration measurements and reservoir-induced earthquakes. He also worked on seismic waves from nuclear explosions, microseisms, and crustal structure. Carder was born in Medford, Oregon, and received a B.S. degree in mining engineering from Oregon State College in 1921, an M.S. in geology from Idaho in 1925, and a Ph.D. in seismology from the University of California, Berkeley, in 1933. He joined the US Coast and Geodetic Survey in 1933, rising to chief seismologist in 1957. He received the Colbert medal from the American Society of Military Engineers in 1956. He was president of the Seismological Society of America in 1961–1962. [*BSSA* **64**:246–248, 1974]

Cauchy, Augustin Louis (1789–1857)

French mathematician who, after Euler, was the most prolific mathematician in history. In seismology Cauchy is best known for the "Cauchy problem": finding a solution of the wave equation for given initial conditions on the wave function and its time derivative defined for the whole space (see Chapter 9 of this Handbook). Cauchy was born in Paris and educated at the Ecole Polytechnique, where he became professor of mechanics in 1816 after several years as an engineer. He lost this position in 1831 and held positions in Turin and Prague. He regained his position after the revolution of 1848. His research included pioneering work in waves, elasticity, and group theory. [Bell (1937); James and James (1976); Britannica (2003)]

Cavalleri, Giovanni Maria (1807–1874)

Italian physicist who designed early seismoscopes based on pendulums of different lengths and natural periods of oscillation. These were used to record the times of earthquakes, their strength, direction, and duration of shaking. Cavalleri was born at Crema, became a Barnabite father, and taught literature and physics at the College of Monza. [Ferrari (2003)]

Cecchi, Filippo (1822–1887)

Italian physicist who designed five seismic instruments, including the first to record on smoked paper. Cecchi was born in Ponte Buggianese and studied physics and mathematics at Florence. At the age of 17 he entered the Order of Pius Schools. As a physics teacher at San Giovannino he became interested in instrument design, working on telegraphic and electromagnetic apparatus, a thermometer, a barometer, and a lightning grounder. In 1872 he became director of the Ximenian Observatory, where he developed a network of meteorological stations. In 1884 he received a gold medal at the National Exhibition at Turin. [Ferrari (2003)]

Chakrabarthy, Subodh Kumar (1909–1987)

Indian mathematician who worked on calibration of seismometers, individual earthquake reports, the earthquake source mechanism, and geomagnetic storms. Born in Bengal, Chakrabarthy received the D.Sc. from the University of Calcutta in 1943. He was a lecturer at City College of Calcutta from 1935 to 1945, director of the Colaba and Alibag Observatories of the Indian Meteorological Department from 1945 to 1948, professor and head of the mathematics department at Bengal Engineering College from 1949 to 1963, and professor and head of the Department of Applied Mathematics at the University of Calcutta from 1963. He was head of the Calcutta Mathematical Society from 1970 to 1972, received the Mouat Medal of the University of Calcutta in 1943, and won the Elliot Prize of the Asiatic Society in 1944. He became a fellow of the Indian Academy of Sciences in 1949 and president of the Mathematical Section of the Indian Science Congress in 1954. [Chapter 79.28]

Chang, Heng (78–139)

See Zhang, Heng.

Charlier, Charles (1897–1953)

Belgian seismologist who published the first systematic study of Belgian earthquakes plus a report of the Heligoland explosion. He also rebuilt the Uccle seismic station. Charlier received the Ph.D. in physical and mathematical sciences in 1927 at the Catholic University of Louvain. At the Royal Observatory of Belgium he was appointed an assistant in 1923, adjunct astronomer in 1936, and astronomer in 1945. He was head of its Seismological Division from 1945 to 1953, a founding member of the European Seismological Commission in 1949, a vice-president of IASPEI from 1948 to 1951, and a co-organizer of the 1951 IUGG General Assembly meeting in Brussels. [Chapter 79.8]

Christoffel, Elwin Bruno (1829–1900)

German mathematician who invented the process of covariant differentiation, and introduced the "Christoffel symbols." Christoffel was born in Montjoie (now Monschau) and earned his doctorate at the University of Berlin in 1856. Between 1859 and 1892 he taught at the University of Berlin, the Polytechnicum in Zurich, the Gewerbsakademie in Berlin (now the University of Technology of Berlin), and the University of Strasbourg. The Christoffel matrix in seismology is a symmetric real-valued matrix, of which eigenvalues give the seismic velocities and eigenvectors give the polarization directions of the waves. See Chapters 9 and 53 of this Handbook. [James and James (1976)]

Cloud, William K. (1910–1984)

American seismologist who developed the strong motion recording systems of the US Coast and Geodetic Survey in the United States and in South America and studied the accelerograms recorded by these systems. Cloud was born in Tucson, Arizona, and received the B.S. degree in mechanical engineering from the University of Arizona in 1934. He worked for the US Department of Agriculture from 1934 to 1942, served in the navy during World War II, and worked for the Seismological Field Service of the US Coast and Geodetic Survey from 1946 to 1971, serving as its chief from 1952 to 1971. He was associate research scientist at the University of California, Berkeley, from 1971 to 1976. He was treasurer of the Seismological Society of America from 1953 to 1970, and secretary from 1971 to 1981. [*BSSA* **75**:897–899, 1985]

Comninakis, Panagiotis (1931–2001)

Greek seismologist. He was born on Lesbos island, graduated in physics from the University of Athens, and in 1957 became an assistant at the Geodynamic Institute of the National Observatory of Athens, rising to senior researcher in 1975 after earning a Ph.D. for a thesis on the seismology of the Aegean area. He worked on the crustal structure of the Aegean area and was an early proponent of the subduction of the African lithospheric plate under the Aegean. He compiled earthquake catalogs of the Aegean area, including arrival times of seismic waves and damage reports. [Chapter 79.25]

Conrad, Viktor (1876–1962)

Austrian-American seismologist and meteorologist who discovered that the Earth crust is often divided into an upper and a lower crust (Conrad discontinuity) and invented the

Conrad seismometer. Conrad was professor for cosmic physics in Chernovits, Bucovina, and Vienna. After World War 1, he became editor of *Gerlands Beiträge zur Geophysk* until he had to emigrate to the United States after the occupation of Austria by Germany in 1938. In the United States, he worked mostly on climatological and bioclimatological problems. [Hader (1962); Kertz (2002)]

Coulomb, Charles Augustin de (1736–1806)

French engineer and physicist best known for formulating "Coulomb's Law" in electricity. The unit of electric charge is named in his honor. Coulomb was born in Angoulême and trained as a military engineer, graduating in 1761. During the next 20 years he practiced applied mechanics with a rigorous mathematical technique, also making important progress in instrumentation and friction theory (see a discussion of Coulomb failure criterion in Chapter 32 of this Handbook). In 1781 he was elected to the Académie des Sciences and gave up engineering to work in physics. He made fundamental contributions to electromagnetics and other fields. [Britannica (2003)]

Coulomb, Jean (1904–1999)

French mathematician, geophysicist, and seismologist. In 1935, Coulomb and Grenet published an important paper in seismometry, "Nouveaux principes de construction des séismographes électromagnétiques," *Ann. phys., Sér.* **11**(3), 321–369. Coulomb wrote a number of books, including *la Constitution physique de la Terre* in 1952 (*The Physical Constitution of the Earth* by Jean Coulomb and Georges Jobert, Hafner, New York, 1963), and *l'Expansion des fonds océaniques et la dérive des continents* in 1969. As part of the famous "Nicolas Bourbaki" book project by a group of French mathematicians at the Ecole Normale Supérieure in 1935, he was in charge of the chapter on spherical functions. In 1937 he was appointed director of the Institute of Meteorology and Physics of the Earth Institut in Algiers. In 1941 Coulomb was appointed professor at the Science Faculty in Paris and director of the IPG in Paris. From 1957 to 1962 he was director of the large national research organization CNRS, and from 1962 to 1967 he was at the head of the national space center (CNES), now a very large and successful space-industry-oriented organization. Coulomb was elected to the French Academy of Science in 1960 and served as its president from 1976 to 1978. He was also the president of IUGG from 1967 to 1971. [Cara (2003)]

Cox, Allan Verne (1926–1987)

American geophysicist best known for establishing the time scale for magnetic reversals with Richard Doell and Brent Dalrymple, and a major contributor to the plate tectonics revolution. Cox was born in Santa Ana, California, and was educated at the University of California, Berkeley. From 1959 to 1967 he was a research geophysicist at the US Geological Survey in Menlo Park, California. He joined the faculty of Stanford University in 1968 and served as the dean of its School of Earth Sciences from 1979 until his unexpected death in 1987. See Chapter 6 of this Handbook. [McGraw-Hill (1980)]

D

Daly, Reginald Aldworth (1871–1957)

Canadian-born American geologist. He was a pioneer student of the elastic moduli of the outer layers of the Earth and an influential writer of books. He worked on areal geology, origin of coral reefs, and classification of igneous rocks. He originated the idea that submarine canyons are produced by turbidity currents. Daly was born at Napanee, Ontario, Canada, and received the A.B. degree from Victoria College, Ottawa in 1891 and the A.M. (1891) and Ph.D. (1893) from Harvard University. He was professor of geology at Harvard from 1912 to 1942 and head of the geology department from 1912 to 1925. He modernized the Harvard seismograph station (replacing Woodworth's old station, which was in the museum on the campus). He authored *Our Mobile Earth* (1921), *Igneous Rocks and the Depths of the Earth* (1914 and 1923), *The Changing World of the Ice Age* (1934), *The Architecture of the Earth* (1942), and *Strength and Structure of the Earth* (1940). Daly was a member of the US National Academy of Sciences. [*NAS Biog. Mem.* **34**:31–64, 1960]

Darwin, George Howard (1845–1912)

English astronomer who contributed to the studies of Earth's interior by his research on Earth's rotation and tidal friction. Darwin was born at Down, the second son of Charles Darwin. He was educated at Trinity College, Cambridge, and became the Plumian professor of astronomy in 1883. He published *The Tides and Kindred Phenomena of the Solar System* in 1898. [*Obit. Proc. Roy. Soc. Lond. A*, **89**:i–xiii, 1913]

Davidson, George (1825–1911)

American astronomer and geodesist. Davidson was born in Nottingham, England. He worked for the US Coast and Goedetic Survey from 1845 to 1895 and was professor of geography at the University of California, Berkeley, from 1898 to 1905. He was the first president of the Seismological Society of America. He conducted geodetic surveys and studied Pacific Ocean currents. He wrote the Coast Pilot of the west coast of the United States and 200 memoirs. [*BSSA* **2**:1, 1912; *NAS Bull.* **7**:227]

Day, Arthur Lewis (1869–1960)

American physicist and geophysicist. He worked on mineral equilibria at high temperatures, and extended the measurable temperature scale to above 1150 degrees C. He studied volcanos and hot springs, and was the first to find water in volcanic emissions. Born in Brookfield, Massachusetts, Day received the A.B. (1892) and Ph.D. (1894) degrees from Yale University. He was a physical geologist with the US Geological Survey from 1890 to 1905, and became the first Director of the Carnegie Institution of Washington Geophysical Laboratory from 1906 to 1936. He was Chairman of the Carnegie Advisory Committee on Seismology (1921–1936) that recommended the establishment of what is now the Seismological Laboratory of the California Institute of Technology. He started Carnegie research in seismology and the radioactive content of ocean samples. He was Secretary of the US National Academy of Sciences from 1913 to 1919 and Vice-President in 1933–1934. He was President of Geological Society of America in 1936 and Bowie Medalist of American Geophysical Union in 1940. [*NAS Biog. Mem.* **47**:27–48, 1975; Goodstein (1991)]

Dezső Csomor (1919–1984)

Hungarian seismologist. Csomor studied the seismicity and seismic activity cycles of Hungary and adjoining areas, published a map of seismic activity of these areas, and studied the structure of the crust beneath Hungary. He also studied the seismic waves from blasts. Born in Perbeléd, Csomor received the M.S. in mathematics and physics from Pázmány Péter University in 1947, and the Ph.D. in technical sciences from the Hungarian Academy of Sciences in 1974. He was a Research Fellow at Eötvös Loránd University from 1942 to 1971 and a Research Fellow at the Budapest Seismological Observatory from 1971 to 1979. [Chapter 79.26]

Degenkolb, Henry John (1913–1989)

American structural engineer. The president of Henry J. Degenkolb Engineers, he was an authority on earthquake-resistant design of structures, hazards to buildings, and seismic provisions of building codes. Degenkolb was born in Peoria, Illinois, and received the BSCE degree from the University of California, Berkeley, in 1936. He was a founding member of the California Seismic Safety Commission, a member of the President's Task Force on Earthquake Hazards Reduction in 1970–1971, and a member of the California Building Standards Commission from 1971 to 1985. [*BSSA* **81**:1044–1047, 1991; Degenkolb (1994)]

de Quervain, Alfred (1879–1927)

Swiss geophysicist. Born at Muri, de Quervain earned a Ph.D. in meteorology at the University of Bern in 1903. He began his career with a meteorological mission to Russia in 1901, followed by meteorological research at Strasbourg, France. From 1903 until his death he was an adjunct at the Swiss Central Meteorological Agency in Zürich. He participated in meteorological expeditions to Greenland in 1909 and 1912. He was also head of the Swiss Seismological Service beginning in 1913, and constructed and installed permanent seismic stations in Zürich (1911) and at Chur and Neuchâtel (1927). He also constructed a pioneer portable seismograph for recording aftershocks of earthquakes and used it to determine focal depths and crustal velocities. [Chapter 79.41]

De Rossi, Michele Stefano (1834–1898)

Italian seismologist. He is best known for his role in developing the Rossi-Forel intensity scale. De Rossi was born in Rome and graduated from the University of Rome in 1855. He began organizing a network of seismic observatories in 1873, and started publication of the Bulletin of Italian Volcanism in 1874, in which reports of Italian earthquake studies were an important part. He proposed a scale for evaluating intensity in 1874, eventually combining it with the scale developed by Forel in Switzerland to become the widely used Rossi-Forel intensity scale (see Chapter 49). He advocated simple and inexpensive instruments and designed several seismoscopes. In 1882 the Italian government established the Rome Geodynamics Observatory with De Rossi as its director. In 1890 he designed the Rocca di Papa Seismic Observatory, of which he became director. [Ferrari (2003)]

Descotes, Pierre M., S.J. (1877–1964)

French-born Bolivian seismologist and a pioneer of South American seismology. Descotes was born in Savoy in 1877 and after completing his studies spent some time working with Manual Sánchez Navarro-Neumann in the Jesuit-run seismological observatory of Cartuja in Granada. In 1911 the 2nd General Assembly of the International Seismological Association recommended that the Jesuits install a seismological station in the central part of South America. In response, Descotes moved to La Paz in 1912, where he founded the Observatorio de San Calixto. His exacting accuracy and rigor in the observations he reported made the station one of the most respected in the southern hemisphere. In 1949 Gutenberg and Richter acknowledged that "La Paz at once became and still remains the most important single seismological station of the world." [Stauder (1999a)]

Dietz, Robert Sinclair (1914–1995)

American oceanographer, originator of modern ideas on sea-floor spreading and a pioneer student of impact craters on the Earth. He discovered the Emperor Seamounts and with Picard helped design the bathyscaphe for deep-ocean studies. He worked on marine geology, plate tectonics, underwater sound, sedimentation and structure of the sea floor, and submarine processes. Dietz was born in Westfield, New Jersey, and received a B.S. degree in 1937 and M.S. in 1941 in geology from the University of Idaho. He was at the Illinois Geological Survey from 1935 to 1937 and Scripps Institute of Oceanography from 1937 to 1939. He joined the US Army Air Force from 1941 to 1946, moving to the Naval Electronics Laboratory from 1946 to 1952. After Fulbright study at Tokyo University in 1952–1953, he was at the US Office of Naval Research from 1954 to 1958, then the US Geological Survey during the important years from 1958 to 1965. Dietz was at NASA from 1965 to 1967, the University of Florida from 1967 to 1970, and the Atlantic Oceanographic and Meteorological Laboratory from 1970 to 1977. He finished his career as professor of geology at Arizona State University from 1977 to 1995. [*GSA Memorial* **29**:25–27, 1999; Dietz (1994)]

Dirac, Paul Adrien Maurice (1902–1984)

English theoretical physicist best known for his seminal role in quantum mechanics, and whose Dirac delta function is indispensable in theoretical seismology. Dirac was born in Bristol, England, educated at Cambridge University, and became the Lucasian professor of mathematics in 1932 (a post he held until his retirement in 1969). [Holmes (1990); Britannica (2003)]

Doornbos, Durk Jakob (1943–1993)

Norwegian geophysicist whose research focused on the deep interior of the Earth, particularly the D" layer and the core–mantle boundary. Doornbos graduated from the University of Groningen, Netherlands, in 1955 and earned the doctorate at University of Utrecht, Netherlands, in 1974, serving as senior lecturer there until 1980, when he moved to NORSAR. In 1985 he was appointed a professor in the geophysics department of the University of Oslo. He was elected to the National Academy of Sciences of Norway in 1988. He studied the scattering of PKP waves as related to lateral inhomogenieties of the lower mantle, diffracted *P* and *S* waves as related to the velocity gradient in the D" layer, scattering of waves at layer boundaries, topography of the core–mantle boundary, and core phases. He was Norwegian delegate to the IASPEI and in 1991 was chairman of the Norwegian National Committee for the IUGG. From 1983 to 1987 he chaired the IASPEI Commission for Earthquake Algorithms. He helped to write the Science Plan for the IUGG Committee on Study of the Earth's Deep Interior and was its chairman from 1991 to 1993. [Chapter 79.41]

Doppler, Christian Johann (1803–1853)

Austrian physicist. He is best known as the discoverer of the Doppler effect, which is observed for waves radiated from a moving harmonic oscillator with a given frequency. The observed frequency is shifted from the source frequency depending on the direction with respect to the moving direction. See Chapter 37 of this Handbook. Doppler was born in Salzburg and studied mathematics at the Vienna Polytechnic and the University of Vienna. In 1835 he took a post at the Technical Academy in Prague, becoming professor in 1841. In 1850 he was named the founding director of the Insitute of Physics at Vienna University. [Daintith and Gjertsen (1999)]

Drakopoulos, John (1937–1999)

Greek seismologist whose research was on seismicity, earthquake sequences (foreshocks and aftershocks), microzonation, seismic hazard, earthquake prediction, and electric field phenomena preceding earthquakes. Drakopoulos was born in Arcadia and graduated in 1960 in physics from the National University of Athens (NUA), receiving an M.Sc. in electronics in 1965 and a Ph.D. in seismology in 1968. After a year of study in Japan, he was appointed senior lecturer at NUA in 1974 and became professor in 1979. He served as director of the geophysics department from 1982 to 1994 and was director of the Institute of Geodynamics of the National Observatory of Athens and president of its Earthquake Protection and Planning Organization. He served as president of the School of Geology for three terms starting in 1983 and vice-president of the NUA for two terms starting in 1994. [Chapter 79.25]

Droste, Zofia Halina (1930–1994)

Polish seismologist. Born in Warsaw, Droste earned her M.Sc. degree in theoretical physics from the University of Warsaw in 1954. Her Ph.D. thesis was on "An analysis of dynamic properties of the far field displacement generated by mining tremors in Upper Silesia." She was an associate professor in the Institute of Geophysics of the Polish Academy of Sciences. Her research interests were seismic events involved in mining operations, the dislocation mechanism in earthquake generation, and the energy distribution in seismic waves. [Chapter 79.43]

Duke, C. Martin (1917–1988)

American earthquake engineer. C. Martin Duke was Professor of Engineering at the University of California at Los Angeles (UCLA) between 1947 and 1980. He received a B.S. in civil engineering, with honors, in 1939 from the University of California, Berkeley, and an M.S. degree there in 1941. After World War II, Duke served a two-year stint in engineering practice. He joined the UCLA faculty as an assistant professor in 1947. He rose through the academic ranks to professor and also served in various administrative posts, notably as chairman of the Department of Engineering.

The 1952 Kern County Earthquake began his interest in earthquake engineering, and in 1956 he took a sabbatical leave to work with colleagues in Japan. His early research included enumeration of lessons learned from earthquake damages in the 1957 Mexico City and 1960 Chilean Earthquakes. In the late 1960s and early 1970s, Professor Duke worked on research projects concerning the effects of earthquakes on water resources and on investigations of source, path, and site effects on the nature of ground shaking. He served as president of the Earthquake Engineering Research Institute (EERI) from 1970 to 1973 and he managed EERI's "Learning from Earthquakes" project immediately after the San Fernando Earthquake of February 9, 1971. He led the preparation of a comprehensive report on this notable event. This contribution was acknowledged by the American Society of Civil Engineers (ASCE) in 1973 when Professor Duke received the Ernest E. Howard Award.

During the San Fernando earthquake investigation Duke realized that much more information was needed about the earthquake response of energy, water, transportation, and communication systems. The ASCE Technical Council on Lifeline Earthquake Engineering was established, primarily due to his leadership.

Martin Duke was an active participant in post-earthquake investigations. In 1979, he led an EERI team to investigate the December 1972, Managua, Nicaragua, earthquake and he also led a similar expedition to Mindanao in the Philippines after the earthquake in 1976. These investigations helped him formulate recommendations for EERI studies to maximize learning from destructive earthquakes. [Jennings, 2003].

Dutton, Clarence Edward (1841–1912)

American geologist who participated in early surveys of the western United States. He originated the term "isostasy." He authored the monographs *The Charleston Earthquake of August 31, 1886* (in 1889) and *Earthquakes in the Light of the New Seismology* (in 1904). He was born at Wallingford, Connecticut, and received the A.B. degree at Yale University in 1860. [*NAS Biog. Mem.* **32**:132–145, 1958]

E

Egyed, László (1914–1970)

Hungarian geophysicist. Egyed worked on focal depth and velocity determinations and earth structure, and was known as one of the exponents of the expansion hypothesis of the Earth. He wrote *Fundamentals of Geophysics* in 1955 (Tankönyvkiadó, Budapest, 535 pp.) and *Physics of the Earth* in 1956 (Akadémiai Kiadó, Budapest, 365 pp.), both in Hungarian. He was born in Fogaras, and received the M.S. in 1936 and the Ph.D. in 1938 in mathematics and physics from Pázmány Péter University. He was a Senior Lecturer there from 1937 to 1941, and worked for the Hungarian-American Oil Company from 1941 to 1948. He joined the faculty of Eötvös Loránd University in 1947 and served as Senior Lecturer (until 1950), Reader, Professor and Head of the Geophysics Department (until 1970), and Dean of the Faculty of Natural Sciences (1966–1970). He received the Kossuth Prize in 1957 and was elected to the Hungarian Academy of Sciences in 1960. [Chapter 79.26]

Ehlert, Reinhold (1871–1899)

German seismologist who studied and worked in Strasbourg under supervision of Gerland. Ehlert improved the Reuber-Paschwitz pendulum (Reuber-Ehlert pendulum) and investigated in detail all the seismometers of his era. He died very young due to a skiing accident in the Swiss Alps. [Gerland (1899); Schweitzer (2003)]

Eiby, George Allison (1918–1992)

New Zealand's foremost authority on earthquakes. He devoted much effort to fostering the development of seismology by encouraging the enhancement of recording networks, the development of understanding about the Earth's structure and the occurrence of earthquakes, and the social issues related to perception of risk and preparedness. His work on the upper mantle was seminal, showing that the occurrence of deep earthquakes under New Zealand is related to lateral heterogeneities in structure. This work predated developments in Plate Tectonics, and was based on detailed analysis of epicentres located and plotted manually, long before such work was facilitated by dense networks and modern computer analysis techniques. Eiby was born in Wellington, educated at Victoria University of Wellington, and became a cadet in the Department of Scientific and Industrial Research in 1939, training in seismology and positional astronomy. Apart from service with the Royal Air Force during World War II he remained with the DSIR until 1979, when he retired as superintendent of the Seismological Observatory. [Reyners (1992); Smith (1992)]

Epinatyeva, Antonina Michaylovna (1914–1998)

Russian geophysicist. She was a pioneer in developing seismic prospecting methodology, noted for her research on the nature of seismic refracted and reflected waves in the Earth's crust. Epinatyeva graduated in geophysics from Moscow State University in 1939 and spent her career in seismic surveying for the Institute of Physics of the Earth of the USSR Academy of Sciences. [Chapter 79.45]

Euler, Leonhard (1707–1783)

Swiss mathematician and the most prolific mathematician in history. He introduced Euler's equation, Euler angles, and many others. Born in Basel, Euler entered the University of Basel in 1720, at the age of 14, completed his Master's degree in philosophy in 1723, and completed his studies in 1726. In 1727 he joined the St. Petersburg Academy of Science and became professor of physics in 1730. In 1741 he joined the Berlin Academy as director of mathematics, and in 1766 he returned to St. Petersburg. After his death the St. Petersburg Academy continued to publish Euler's unpublished work for nearly 50 more years. [Bell (1937); James and James (1976); Britannica (2003)]

Ewing, William Maurice (1906–1974)

American geophysicist whose contributions involved the oceans, the crust, seismometry, and the Moon. He is also remembered for his leadership at a crucial time in geoscience. Ewing was born in Lockney, Texas, and earned his B.A. (1926), M.A. (1927), and Ph.D. (1931) in physics from Rice University. His earliest seismic work was on propagation of seismic waves in various materials such as rocks and ice, carried out while an instructor at Lehigh University. This was followed by refraction measurements along the US Atlantic coast. In 1937 he began a series of gravity studies at sea (mostly with Worzel). He moved to Woods Hole Oceanographic Institution in 1940, where he developed new apparatus for underwater photography. In his studies of sound propagation he discovered the SOFAR sound channel in the oceans, and showed that the T-phase on seismograms was a pulse propagated through the SOFAR channel. In 1946 he moved to Columbia University and in 1949 became director of the Lamont Geological Observatory, later renamed Lamont-Doherty and today the Lamont-Doherty Earth Observatory. With Doherty's grant he bought the ships Vema and Robert D. Conrad, and he then became a leading participant in all phases of oceanic research. In 1949 he began refraction studies of the ocean floor, finding the Moho discontinuity at 5 km depth without any layer of continental (sialic) rocks above it. His work on dispersion of elastic waves in water-covered areas led to the book (with Press and Jardetsky), *Elastic Waves in Layered Media* (McGraw-Hill, New York, 1957), and to better understanding of the layers of the crust and upper mantle that differed between the oceans and the continents. He developed new seismometers (with Frank Press) capable of recording in the previously neglected period range 10 to 60 seconds. Beginning in 1960, he developed methods of obtaining reflection profiles at sea using small explosive charges. He improved the design of sonic depth recorders and used them to profile thousands of miles of ocean floor. He explained the series of transatlantic cable breaks following the 1929 Grand Banks earthquake as due to turbidity currents stirred up by the earthquake. This led to the realization that turbidity currents are a main source of continental slope and deep-sea deposits. His ocean-depth profiles outlined the structure of the mid-Atlantic ridge, and (with Heezen and Tharp) he traced the extent of the deep depressions along its crest. He pointed out that mid-ocean earthquakes followed these belts, and suggested that earthquakes most commonly occurred where his surveys showed that there were fault scarps along them. This work is summed up in GSA Special Paper 65, *The Floor of the Oceans*. Ewing was a great collector of cores and magnetic profiles along the routes that his ships traveled. He studied microseisms (with Press and Donn), and he was involved with early studies of lunar seismology. He held the vice-presidency and presidency of both the American Geophysical Union and Seismological Society of America. He was a member of the National Academy of Sciences Mohole and Deep Sea Drilling Project committees and (with Worzel) was co-chief scientist on leg 1 of the surveys. He mentored many students who went on to distinguished careers as geophysicists. [*NAS Biog. Mem.* **51**:119–192, 1980]

F

Fermat, Pierre de (1601–1665)

French mathematician whose best-known contribution to seismology is Fermat's principle of least time. Fermat was born at Beaumont-de-Lomagne and was educated at home and then at Toulouse in preparation for the magistracy. In 1631, he became commissioner of requests and was promoted to a royal councillorship in the local parliament of Toulouse in 1648. His recreation was mathematics, but he made significant contributions in many areas of mathematics as an amateur. He invented analytic geometry independently of Descartes and shared with Pascal the creation of the mathematical theory of probability. He was the leading number theorist in his time. His famous "last theorem" was proved only recently after more than 300 years of efforts by countless mathematicians. See Chapter 52 (pp. 864–865) of this Handbook. [Bell (1937); Britannica (2003)]

Fisher, Ronald Aylmer (1890–1962)

English mathematician whose work underlies much of modern scientific analysis. Fisher was born at East Finchley and was educated at Cambridge University. In 1919 he joined the Rothamsted Experimental Station at Herpenden as a statistician. He became the Galton Professor of Eugenics at the University College, London, in 1933 and then the Arthur Balfour Professor of Genetics at Cambridge. He made extensive contribution to genetics, the design of experiments (by introducing the concept of randomization), and statistical procedure (analysis of variance). See Chapters 16 and 82 of this Handbook. {McGraw-Hill (1980)]

Flinn, Edward Ambrose (1931–1989)

American geophysicist who worked on underground nuclear test detection, application of digital-signal processing techniques and statistical methods to seismology, characterization of seismic geographic regions into Flinn-Engdahl zones, and waves through the inner core. Flinn was born in Oklahoma City, Oklahoma, and received a B.S. from Massachusetts Institute of Technology in 1953 and a Ph.D. from Caltech in 1960. He worked for United Electrodynamics from 1960 to 1968 and Teledyne Geotech from 1968 to 1974. He became Chief of NASA's Geodynamics Program and Director of NASA's Division of Space Sciences Lunar Program. He was secretary general of the IUGG Inter-Union Committee on the Lithosphere from 1980 to 1986. [Engdahl (2002)]

Fock (or Fok), Vladimir Aleksandrovich (1898–1974)

Russian physicist and mathematician, some of whose contributions are useful tools in seismology. Fock was born in

St. Petersburg and was educated at Petrograd University in physics and mathematics. He worked simultaneously in various research and educational institutions, including Leningrad University, and the Lebedev Physics Insitute and the Institute of Physics Problems of the Soviet Academy of Sciences. He contributed to many areas of physics, including quantum mechanics, the Fock functional, the Fock space, and the Hatree-Fock method. [Holmes (1990)]

Forbes, James David (1809–1868)

Scottish physicist who designed an early seismic instrument. Forbes was born in Edinburgh and was educated at the University of Edinburgh preparing for law practice, but at the same time published important scientific papers anonymously using the name "Δ". He became professor of natural philosophy from 1833 to 1859 and was named principal of St. Andrews University in 1859. He was noted for his research on heat conduction and glaciers. As a member of the British Association's Committee for "obtaining instruments and registers to record shocks of earthquakes in Scotland and Ireland," he devised an inverted pendulum design to study the small earthquakes occurring near Comrie, Scotland (Forbes, 1844). [*Obit. Proc. Roy. Soc. Lond.***19**:i–ix, 1870; Debus (1968)]

Forel, François-Alphonse (1841–1912)

Swiss geologist best known for his contribution to the ten-degree Rossi-Forel scale of 1883, the first internationally used scale for earthquake intensity. Forel was born in Morges Vaud, educated as a physician, and was professor of physiology and general anatomy at the University of Lausanne from 1869 to 1895. He was a founder of limnology, and studied glaciers, seiches, and earthquakes extensively. See Chapter 49 (p. 810) of this Handbook. [Debus (1968)]

Fourier, Joseph (1768–1830)

French mathematician and physicist. He is best known as the originator of Fourier series and the Fourier transform, and laid the foundation for the study of heat conduction. Born in Auxerre, Fourier entered the Benedictine abbey of St Benoit-sur-Loire for training as a priest in 1787, but left in 1789. After a few years of involvement in the French revolution, he entered the Ecole Normale in Paris in 1795. Shortly thereafter he began teaching at the Ecole Polytechnique, and in 1797 succeeded Lagrange to the chair of analysis and mechanics. The next year he joined Napoleon's expedition to Egypt and served in the government for the next two decades. He was elected to the Académie des Sciences in 1817. [Bell (1937); James and James (1976); Britannica (2003)]

Freeman, John Ripley (1855–1932)

American civil and hydraulic engineer, and a pioneer in earthquake engineering and insurance. Freeman became interested in earthquakes at the age of 70, and persuaded Kyôji Suyehiro (the first director of the Earthquake Research Instute of the Tokyo Imperial University) to lecture on engineering seismology in America. He also persuaded the US Coast and Geodetic Survey (responsible for earthquake monitoring in the United States at that time) to develop a strong motion accelerograph and to implement a strong-motion monitoring program (see Chapter 2 in Part A of this Handbook). He authored the book *Earthquake Damage and Earthquake Insurance* (McGraw-Hill, New York, 1932).

Freeman was born in West Bridgeton, Maine. He studied engineering at the Massachusetts Institute of Techology, and received his degree in civil engineering in 1876. He joined the Inspection Department of Factory Mutal Fire Insurance Companies and studied water sprinkler systems for fire protection. He became the chief engineer of the Charles River Dam in 1903, and served as the president of the American Society of Civil Engineering in 1922. [*NCAB* **36**, 1950.]

Fu, Chengyi (1909–2000)

Chinese geophysicist. Fu was born in Beijing, earned a Ph.D. from California Institute of Technology in 1944, and was an assistant professor there from 1944 to 1946, working on the theory of seismic waves at a boundary. From 1947 to 1949 he was a professor of Solid Earth Geophysics at the Institute of Meteorology of the Academia Sinica. He became professor at its Institute of Geophysics in 1949 and a member of the Academy in 1956. From 1953 to 1956 he was professor of geophysics at the Beijing College of Geology. In 1956 he sponsored a national project for seismology as a part of the National Twelve-Year Plan for the Development of Science and Technology in China, with emphasis on earthquake prediction. He was head of the Division of Solid Earth Geophysics at Peking University from 1953 to 1956, head of the Division of Physics of the Earth's Crust at the University of Science and Technology of China from 1964 to 1966, and dean of the Department of Geophysics and Space Physics there from 1967 to 1983. He founded the Research Division of Seismic Source Physics in the Institute of Geophysics of the Chinese Academy of Sciences. [Chapter 79.14]

G

Gaiskii, Valentin (1923–1975)

Russian seismologist. He worked on problems in seismicity, microearthquakes, energy of earthquakes, and focal dimensions. Gaiskii graduated from the State University of Irkutsk in 1949 and earned the Ph.D. in physics and mathematics at the Institute of Physics of the Earth, USSR Academy of Sciences, in Moscow in 1953 and achieved the rank of professor there in 1967. He was employed at Irkutsk Seismic Station from 1949 to 1954, was manager of the Tadzhik Institute of Seismology and Aseismic Construction of the Tadzhik Academy of Sciences from 1955 to 1965, laboratory manager of the Institute of Geology, Geophysics and Mineralogy of the Siberian Branch of the USSR Academy of Sciences in Novosibirisk from 1965 to 1975, and its deputy director from 1972 to 1975. [Chapter 79.45]

Galanopoulos, Aggelos (1910–2001)

Greek seismologist. Galanopoulos wrote numerous papers on seismology, seismotectonics, and seismic hazard of Greece. Born in Achaia (Peloponnese), he graduated from the School of Sciences of the National University of Athens, and received a Ph.D. degree in seismology in 1937. In 1949 he became Director of the Institute of Geodynamics of the National Observatory of Athens, where he served until 1978. Under his guidance, the Institute of Geodynamics was equipped with modern seismographs installed in several high seismicity regions of Greece, while the collection of macroseismic data was organized. He was a member of the National Academy of Athens. [Chapter 79.25]

Galitzin (Golitzin), Boris B. (1862–1916)

Russian physicist and the founding father of Russian seismology. Galitzin, a prince of the Imperial Russia Empire who traced his ancestry to the princes of Lithuania, graduated as a cadet from the St. Petersburg Naval College in 1880. After naval service for eight years, he went to Strasbourg University and graduated in philosophy in 1890. He became a privatdocent in Moscow and then professor of physics in Jurjef, before his promotion to Petrograd in 1893. His earlier scientific papers were mainly on the properties of gases and liquids, and the critical state, but his work also covered other branches of general physics, including molecular forces and radiation energy. After some controversy regarding his work, he shifted his interests first to meteorology and then seismology. He made meteorological observations and observed the 1906 solar eclipse in Nova Zemlya. In 1902 he developed the first electromagnetic seismograph, making possible a much greater magnification of ground motions. He developed the first seismic recording network in Russia, consisting of seven principal stations using electrodynamic seismometers and fourteen regional stations with mechanical seismographs. He proposed locating epicenters using only P–S times to get distance and the ratio of north–south to east–west recorded amplitudes and first motions to get direction, as well as the more currently used method based on the arrival times at several stations. He studied microseisms, seismic-wave absorption, earthquake energy, and the theory of Rayleigh waves. He studied earthquake precursors in the form of ground deformations, geoelectric and hydrogeologic phenomena, including developing apparatus to record them. He was elected to the Russian Academy of Science in 1909, received an honorary doctorate from Manchester University in 1910, and was elected president of the International Seismological Association in 1911. He became a corresponding member of the Goettingen Academy of Sciences in 1913 and a foreign member of the London Royal Society in 1916. During World War I he headed the Russian Meteorological Department until his death of pneumonia. Among his extensive publications is a book, *Vorlesungen uber Seismometrie* (Lessons in Seismology), published in Russian in 1911, and a German version translated by O. Hecker was published in 1914. This classic book is archived on the Handbook CD#3, under the directory of \88Schweitzer. A photo of Galitzin is included in Figure 1 of Chapter 81.3 of this Handbook. [Klotz (1917); Chapter 79.45]

Galli, Ignazio (1841–1920)

Italian geophysicist who studied atmospheric electricity and reported 148 cases of luminous phenomena accompanying earthquakes. He also studied seismic vibrations using seismoscopes. Galli was born at Velletri and taught in the middle school there. In 1867 he founded an observatory there. He received a gold medal at the National Exhibition of Turin in 1884. [Ferrari (2003)]

Galperin, Evsey Iosifovich (1920–1990)

Russian geophysicist noted for developing the method of vertical seismic profiling, polarization method, and borehole seismology. Galperin graduated from Moscow Geological Prospecting Institute in 1949. He started work for the Institute of Physics of the Earth of the USSR Academy of Sciences in 1946. From 1956 to 1958 he headed the Integrated Pacific Ocean Expedition of the International Year. He also developed improved seismic methods of surveying. [Chapter 79.45]

Gamburtsev, Grigory Aleksandrovich (1903–1955)

Russian geophysicist who made contributions to seismic survey methods, seismicity studies, and earthquake prediction. From 1921 to 1932 Gamburtsev studied optics and gravimetry and the Kursk region magnetic anomalies. From 1932 to 1948 he worked in seismic prospecting and wrote *Seismic Methods of Prospecting*. He was director of the Geophysical Institute of the USSR Academy of Science. From 1945 to 1951 he worked on exploration for uranium ores. He developed methods of high-frequency seismic surveying and, with Riznichenko, Berzon and Epinatyeva, developed the correlation method of refracted waves, methods of low-frequency seismic surveying, shear and converted waves, deep seismic sounding, and seismoacoustic methods. After 1948 he began studies of seismicity and earthquake prediction. He organized studies of the Garm seismic complex in Tajikistan. At the USSR Academy of Science, he was chairman of the Council of Seismology, scientific head of the Northern Tyan Shan Geophysical Expedition, deputy Academician-Secretary, a member of the Bureau of the Department of Physical and Mathematical Sciences, head of the department of earthquake physics, and professor at the Moscow State University. In 1941 he was awarded the State Prize of the USSR for the development of a method and instruments for seismic prospecting, an Order of the Red Banner of Labor in 1945, and an Order of Lenin in 1953. In 1946 he became the Correspondent Member of the USSR Academy of Science and in 1953 an Academician. [Chapter 79.45]

Gassmann, Fritz (1899–1990)

Swiss geophysicist who worked on the theory of transmission of seismic waves, particularly in porous and anisotropic media, on intensity of Swiss earthquakes, and on seismic prospecting. Gassmann was born in Zürich and studied at the Swiss Federal Institute of Technology (SFIT) in Zürich, receiving a diploma in mathematics and physics in 1923 and a Ph.D. in physics in 1926. He worked as an assistant at the Swiss Seismological Service from 1920 to 1928, qualified as a lecturer and began teaching at SFIT in 1928, was promoted to associate professor in 1942 and professor in 1952. He was head of the Institute of Geophysics at SFIT from 1942 to 1962 and head of the Swiss Seismological Service from 1956 to 1959. He was a visiting professor at Purdue University in 1952 and at the University of Illinois in 1962. [Chapter 79.51]

Gauss, Carl Friedrich (1777–1855)

German mathematician who has been called "the prince of mathematicians" and ranks with Archimedes and Newton among the greatest mathematians. He had a long and brilliant career in pure and applied mathematics (including probability and statistics). In particular, he applied mathematical methods to solve problems in astronomy, geodesy, geomagnetism, and physics. Gauss introduced the method of least squares (also independently derived by Adrien Legendre), which has been widely used in science, including earthquake location. Gauss was born in Brunswick and obtained his doctorate from the University at Helmstedt at the age of 22. He became professor of astronomy and director of the observatory at the University of Göttingen in 1807, and remained there the rest of his life. [Bell (1937); James and James (1976); Bühler (1981); Kertz (2002)]

Geiger, Ludwig Carl (1882–1960)

Swiss physicist. He worked on travel-time charts with Zoeppritz, identifying angles of incidence of *P, PP, PPP, S, SS*, and *SSS* and worked with Gutenberg calculating ray paths in the Earth, finding their depths of penetration (to 2550 km) and identifying discontinuities in the upper mantle. He calculated values of Poisson's ratio for the upper mantle and developed the "Geiger method" of locating earthquakes using only first arrivals of *P* waves. Geiger studied natural science at Basel, Berlin, Heidelberg, and Göttingen, receiving the Ph.D. in 1906 for a thesis on spectroscopy. From 1907 to 1912 he was an assistant at Wiechert's Institute of Geophysics at Göttingen, where he worked on spectroscopy and seismology. He was in charge of the Göttingen seismic bulletins and worked on the interpretation of seismic observations. In 1913 he moved to Samoa as an observer at the Samoan Observatory, where he was director until 1914 when Angenheister took charge. While in Samoa he conducted geoelectric, geomagnetic, meteorological, and astronomical observations including studies of the zodiacal light. When the British captured Samoa during World War I, Geiger returned to Switzerland and became a manager of his family's pharmaceutical company. [Ritter and Schweitzer (2003a); Schweitzer (2003)]

Gerland, Georg Cornelius Karl (1833–1919)

German geographer, anthropologist, geophysicist, musician. In 1889, he founded the first journal explicitly naming geophysics in its title, the *Beiträge zur Geophysik*, later called *Gerlands Beiträge zur Geophysik*. He organized the first two international conferences on seismology in 1901 and 1903 in Strasbourg and, following the fundamental ideas of Rebeur-Paschwitz, he was the founder of the first international seismological organization, the International Seismological Association (ISA), forerunner of IASPEI. Until 1910 Gerland was director of the ISA Central Bureau in Strasbourg. A photo of Gerland is included in Figure 1 of Chapter 81.3. [Schweitzer (2003)]

Gherzi, Ernesto, S.J. (1886–1976)

Italian seismologist. Gherzi was born in San Remo and entered the Society of Jesus in 1903. In 1910 he arrived in China, where he was a Professor of Physics at the University of Aurora. He returned to Europe for further studies, then came back to China in 1920, where he joined the staff of the Jesuit-run Zikawei Observatory. There he worked in the seismology and meteorology departments. In meteorology his principal work was the study of typhoons. In 1924 he was one of the first to assign the origin of certain microseisms to the fluctuations of the atmospheric pressure in the vicinity of a tropical cyclone. He continued this line of investigation of microseisms and their atmospheric origins for many years. He also studied the layers of the ionosphere and their relation to the formation of tropical storms. After the Jesuits were expelled from China in 1949, Gherzi was invited by the Portugese government to organize the Macao Observatory. Subsequently, after a few short stays at other Jesuit universities, he came to Montreal in 1954, where he helped organize the Geophysical Observatory of the College Jean de Brebeuf. There, until his death in 1976, he worked on atmospheric electricity, solar radiation, and seismology. He was a member of the Pontifical Academy of Science, the Academy of Science of

Lisbon, and the Academy of Science of New York. [Stauder (1999a)]

Gilbert, Grove Karl (1843–1918)

American geologist. Best known for his studies of Basin and Range geology, he examined the effects of the 1872 Owens Valley earthquake and recognized the role of earthquakes in raising the Basin and Range mountains. He warned the citizens of Utah of the dangers of earthquakes on the Wasatch fault. He was a member of Lawson's committee studying the effects of the 1906 San Francisco earthquake and made extensive studies along the San Andreas fault. Gilbert was born in Rochester, New York, and graduated from the University of Rochester in 1862. He served on the Wheeler, Powell, and Harriman expeditions to explore the western United States starting in 1871. He was with the US Geological Survey from 1879 to 1918, serving as Chief Geologist in 1889–1892. [Pyne (1980); *NCAB* **13**:46; *Bull. GSA* **31**:26-64]

Girlanda, Antonino (1913–1988)

Italian seismologist. As long-time director of the Messina Sesimological Observatory, he studied deep earthquakes in the Tyrrhenian Sea area, the 20-degree discontinuity, and induced microseisms. Girlanda graduated in mathematics and physics from the University of Messina in 1941 and spent his career there, rising to visiting professor of seismology. From 1947 to 1983 he was director of the Messina Seismological Observatory. He designed seismographs for the observatory, and served on the European Seismological Commission subcommittees on calibration of seismographs and alpine explosions. [Ferrari (2003)]

Goiran, Agostino (1835–1909)

Italian natural scientist. He studied the seismic activity at Monte Baldo between 1870 and 1876 and in 1880 published *The Seismic History of the Province of Verona*. He also constructed a seismic observatory in Maffei. Goiran was born in Nice, France, and received a teaching diploma in physics and mathematical sciences at the University of Turin. Starting in 1869 he taught physics and natural history at the Royal High School Maffei and at the Royal Female College at Verona. He became a member of the Verona Academy of Agriculture, Commerce and Arts in 1874. His main scientific interest was the flora of Verona. He set up a seismic observatory at the Royal High School Maffei, equipping it with a seismoscope and a tromometer. [Ferrari (2003)]

Goldenblat, Ioseph Victorovich (1907–1990)

Russian engineer whose work emphasized the importance of the nonlinear behavior of buildings under seismic stress. Goldenblat received the D.Sc. degree and was a professor at the Central Research Institute of Structures in Moscow. His interests spanned dynamics, thermodynamics, structural mechanics, and theoretical physics, but he is best known for his work in earthquake engineering. He helped to develop the USSR Seismic Building Code. [Rzhevsky (2002)]

Golitzin, Boris B. (1862–1916)

See Galitzin (Golitzin), Boris B.

Gorshkov, Georgy Petrovich (1909–1984)

Russian engineering seismologist who studied problems of construction of earthquake-resistant buildings, paleoseismology, recent tectonic movements, and the geology of earthquake occurrence. Gorshkov graduated from Leningrad Mining Institute in 1931 and started work at the Institute of Seismology of the USSR Academy of Sciences where he studied the Zanzegur, Armenia, earthquake. He compiled a series of zoning maps of all of the USSR. From 1949 to 1984 he worked both at Moscow State University and on the Committee of Seismology of the Academy. [Chapter 79.45]

Gotsadze, Otar (1929–1993)

Georgian seismologist who studied the seismicity of the Caucasus region, identified seismic lineaments as related to important structures, estimated earthquake recurrence rates, improved the Tbilisi and Garedji seismic observatories, developed a computerized database of Caucasus earthquakes, and prepared a seismic zoning map of Georgia. Gotsadze held a doctorate in physical-mathematical sciences and was head of the Department of Regional Seismology of the Institute of Geophysics of the Georgian Academy of Sciences. [Chapter 79.23]

Grablovitz, Giulio (1846–1928)

Italian geophysicist who founded and directed the Geodynamics Observatory at Casamicciola. He designed seismographs, studied seismic waves, and researched the relation of earthquakes to the tides. Grablovitz was born in Trieste and set out to be an astronomer. In 1895 he published a tidal schedule and proposed new kinds of tide gauges. He was appointed to the Royal Geodynamics Commission in 1885 and given the task of founding and directing the Geodynamics Observatory at Casamicciola. He was a founding member of the Italian Seismological Society and received a gold medal at the Milan Exhibition of 1906

Green, George (1793–1841)

English baker/miller and mathematician. George Green was born in July 1793 in the village of Sneinton, near Nottingham, and his father was a baker. He went to Robert Goodacre's school in

1801 when he was eight years old, and left the school in midsummer 1802 in order to work at his father's bakery (and mill later). His only schooling, therefore, consisted of four terms, and he somehow learned mathematics and physics by himself.

On 14 December 1827 he published an advertisement in the Nottingham review: "In the Press, and shortly will be published, by subscription, An Essay on the Application of Mathematical Analysis to the Theories of Electricity and Magnetism. By George Green...." The *Essay* was published in March 1828, and there were 51 subscribers who could hardly have understood a word in it. In the preface Green indicated that his "limited sources of information" prevented his giving a proper historical sketch of the mathematical theory of electricity, and he cited few references. The *Essay* begins with introductory observations emphasizing the central role of the potential function. The general properties of the potential function are subsequently developed and applied to electricity and magnetism. The formula connecting surface and volume integrals, now known as Green's theorem, was introduced in the *Essay,* and also "Green's function," now widely used in mathematical physics and also in seismology. Unfortunately, the importance of Green's *Essay* was not recognized at that time because nobody with sufficient mathematical skills to appreciate its importance had seen the work. However, Sir Edward Bromhead, a local landowner, read the *Essay* and encouraged Green to write three more papers: two on electricity (published by the Cambridge Philosophical Society in 1833 and 1834) and one on hydrodynamics (published by the Royal Society of Edinburgh in 1836). Green became solely responsible for the family business in 1829 when his father died. He spent his days working at the mill and continued his research late at night in the uppermost room of the mill. In 1833, Green let out his mill and moved to Cambridge to study as an undergraduate at Caius College, Cambridge. He was by now forty years old. Green graduated in Mathematics in 1837, and published six more papers before being elected as a fellow of the college in 1839. Unfortunately his health began to fail and he died in Sneinton on 31st May 1841. Lord Kelvin (William Thomson) recognized the importance of Green's 1828 *Essay* in 1845, and was responsible for republishing the work, with an introduction, in 1850. [Cannell (2001); Holmes (1990); James and James (1976); Britannica (2003)]

Griffith, Alan Arnold (1893–1963)

British aeronautical engineer whose theoretical studies of brittle fracture are important in the earthquake generation process. Griffith earned a B.Eng. in mechanical engineering (1914), M.Eng. (1917), and D.Eng. (1921) at the University of Liverpool. In 1920 he published a seminal article on the theory of the brittle fracture ("The phenomena of rupture and flow in solids," *Phil. Trans. Roy. Soc. Lond. A,* **221**, 163–198, 1920). His work on stress concentration led him to realize that cracks can theoretically produce infinite stress at their ends. He pointed out that cracks could cause rupture. Subsequently, seismologists recognized the importance of Griffith's theory in the earthquake-generating process. During the late 1920s, Griffith and Frank Whittle independently made the first practical proposals for the use of gas turbine engines in aircraft. Griffith concentrated on developing an axial flow compressor, and in 1929 he proposed a gas turbine engine driving a propeller, the so-called turbo-prop engine. At Rolls Royce (1939–1960) he designed turbojet engines, and in the 1950s, vertical take-off aircraft. He developed the remarkable "flying bedstead," which first flew in 1954. [*Biogr. Mem. Fell. Roy. Soc. Lond.*, **10**:117–136, 1964]

Griggs, David Tressel (1911–1974)

American geophysicist. Born in Columbus, Ohio, Griggs earned his first degree at Ohio State University in 1932, then was a junior fellow at Harvard University (1934–1941) and professor at Massachusetts Institute of Technology (1941–1942). During the war years he worked at Los Alamos Laboratory (1942–1946), then moved to the University of California, Los Angeles (1948–1974). His earliest work was on mechanical properties of rocks, especially creep. His scale models of mountain building showed that contraction was not a possible cause, but convection was reasonable. He helped set up the RAND Corporation and was the first head of its physics department (1946–1948). He worked on earthquake mechanisms, and developed equipment for high-pressure high-temperature experiments (50 kilobars). He demonstrated hydrolytic weakening of quartz and silicate systems. [*NAS Biog. Mem.* **64**:113–133, 1994]

Gu, Gongxu (1908–1992)

Chinese geophysicist whose work was primarily in exploration geophysics. Gu was born in Jaishan, graduated from Datong University in Shanghai in 1929, and received the M.Sc. from Colorado School of Mines in 1936. He was appointed a research fellow at the Chinese National Academy in Peiping in 1936, then professor at the Academy's Institute of Geophysics and Meteorology in 1950. He was elected to the Chinese Academy of Sciences in 1955, president of the Chinese Geophysical Society, and editor-in-chief of Acta Geophysica Sinica from 1957 to 1964, chairman of the Chinese delegation to IUGG in 1977, president of the Seismological Society of China, and editor-in-chief of Acta Seismologica Sinica in 1979. His book, *Introduction to Geophysical Exploration*, was published in Beijing in 1990. [Chapter 79.14]

Gubin, Igor Evgenyevich (1906–2001)

Russian geophysicist. His major scientific field has been the relation between seismic and geologic processes and the mitigation of the effects of earthquakes. He compiled geologic maps of the Tadjik and adjacent regions and studied their seismicity. He conducted geologic investigations of the Tien-Shan and Pamir regions, and developed seismicity maps of eighteen regions showing the relation of geologic structures to seismic zones. Gubin graduated from Leningrad Language Institute in 1931, then Leningrad Mining Institute in 1935. Starting in 1931, he worked at the All-Union Geological Institute of the USSR Ministry of Geology. After 1938 he headed a geologic sector of the USSR Academy of Sciences Tadjik base. He completed the candidate's exam in 1943 and received the doctorate in 1961. Since 1945 he worked at the Academy's Institute of Physics of the Earth. He became a Member-Correspondent of the Academy in 1976. [Chapter 79.45]

Gutenberg, Beno (1889–1960)

German-born American geophysicist. Privat-Dozent (Instructor) 1924–1930 and Ausserordentlicher (Adjunct) Professor 1926–1930, University of Frankfurt. Professor of Geophysics and Meteorology from 1930, and Director, Seismological Laboratory from 1947, California Institute of Technology. The foremost observational seismologist of the twentieth century. Contributed many important and lasting discoveries of the structure of the solid Earth and its atmosphere. Pioneered the use of travel-time curves to determine earth structure. From 1914 onward, he constructed increasingly accurate travel-time curves for *P* and *S* waves and many later phases. Gutenberg and Richter's four monumental papers, "On Seismic Waves" (*Gerl. Beit. zur Geophys.* **43**, 56–133, 1934; **45**, 280–360, 1935; **47**, 73–131, 1936; **54**, 94–136, 1939) represent the foundation of modern observational seismology, being an exhaustive study of the properties of body waves, surface waves, amplitudes, etc.

He also pioneered the use of amplitudes (of *P* and *PP*) to determine Earth structure. He made the first direct identification of the existence of the core of the Earth; his determination of the depth to the core–mantle boundary in 1914 remains one of the most accurately known dimensions in the Earth's structural catalog. He searched for core *S* phases and showed that if they existed, they must have an extremely low wave velocity. One year after Lehmann's discovery of the inner core, he made precise the radius of the inner core and determined its *P*-wave velocity, with an accuracy that is the same as the estimates today. He used the dispersion of Love and Rayleigh waves to confirm that surface wave velocities across continents were slower than across the oceans; from surface wave dispersion he derived an oceanic crustal thickness of 5 km (1924). He made the first identification of the mantle low-velocity zone and placed it at a depth of about 100 km. He discovered the low-velocity zone in the lower crust. He was the first to identify long-period G-waves as Love waves and observed the first globe-circling G-waves.

From his work on the differences of the structure of the Earth under continents and oceans, he was convinced of the likelihood of continental drift and convection in the mantle, and in 1927 developed a theory of flow in the mantle. Gutenberg proposed that the continents formed a single supercontinent in Cretaceous times and that climates depended on the geography of the continents. He constructed a temperature profile for the mantle and estimated the temperature at the core surface. He constructed the first density model for the mantle (1923), and a model for the structure of the atmosphere. From 1910 until his death he was concerned with the nature of microseisms, and proposed that microseisms are caused by differential loading of the ocean bottom by distant storms.

Gutenberg and Richter determined the magnitude-scale for surface waves, and were the first to show that the strongest earthquakes of the world had magnitudes around $M = 8.5$. He determined the energy-magnitude formula; his coefficient of 1.5 remains the standard today. Gutenberg and Richter showed that the magnitude-frequency distribution was log-linear and had a b value of 1.

He was author of the *Lehrbuch der Geophysik* (1929) and editor of the compendious *Handbuch der Geophysik*, which spanned all of the then-important geophysical subjects from seismology and Earth structure to the structure of the atmosphere, space physics and geochemistry; he contributed major reviews to the Handbuch on seismology, Earth structure, and the structure of the atmosphere. He authored *Physics of the Earth's Interior* (Academic Press, New York, 1959) and (with Richter) *Seismicity of the Earth* (Princeton Univ. Press, Princeton, 1949; 2nd edition, 1954), and was editor of and significant author in *The Internal Constitution of the Earth*.

Gutenberg received many honors. He was a member of the US National Academy of Sciences, the American Academy of Arts and Sciences, and many foreign academies. He was President of the Seismological Society of America (1945–1947) and President of IASPEI (1951–1954). He was the recipient of the Bowie Medal of the American Geophysical Union, and a number of other medals and prizes. [Knopoff (2000); Goodstein (1991); A complete bibliography of Gutenberg may be found in Chapter 79.24.4 on Handbook CD, under the directory of \7924Germany.]

Guzzanti, Corrado (1852–1934)

Italian geophysicist. He was born in Mineo, Sicily, and became director of the Post and Telegraph Office there. In 1882 he set up seismoscopes in his home, adding meteorological instruments a year later. He also studied telluric currents and well-water levels. He was a founding member of the Italian Seismological Society. [Ferrari (2003)]

Gzovsky, Michael Vladimirivich (1919–1971)

Russian geophysicist who studied the mechanical properties of rocks using the polarization-optical method of stress investigation. Gzovsky graduated from Moscow Geological Prospecting Institute in 1939 and taught there until 1950, serving as a military geologist during World War II. He then joined the Geophysical Institute of USSR Academy of Sciences. [Chapter 79.45]

H

Haeno, Seizô (1906–1942)

Japanese developer of radio-transmitted recording methods and seismic prospecting techniques. Haeno was born in Kagoshima and graduated from the Seismological Institute of Tokyo Imperial University in 1928, where he served as an assistant. Moving to the Geological Survey of Japan in 1937, he was appointed chief of its Fourth Division (geophysical exploration) in 1940 and was concurrently a lecturer in the Seismological Institute of Tokyo Imperial University. [Miyamura (2003)]

Hagiwara, Takahiro (1908–1999)

Japanese earthquake prediction researcher, designer of a variety of instruments (including a sea-floor seismometer and a tiltmeter), and author or editor of nine books on seismology. Hagiwara was born in Tokyo and graduated from the Institute of Seismology of Tokyo Imperial University in 1932. At the Earthquake Research Institute of Tokyo University, he rose from assistant (1933) to assistant professor (1941) to professor (1944) to director (1965–1967). He was chairman of the Coordinating Committee for Earthquake Prediction from 1969 to 1981, and chairman of the IUGG Committee for Earthquake Prediction from 1967 to 1971. [Miyamura (2003)]

Hakobyan, Solak G. (1916–1977)

Armenian geophysicist whose research was on magnetism, paleomagnetism, and the crustal structure of Armenia. Hakobyan worked at the Armenian Academy of Sciences from 1946 to 1961 in the Institute of Geological Sciences and thereafter in the Institute of Geophysics and Engineering Seismology, which he helped to found. From 1973 to 1977 he planned and built the Gamy Geophysical Observatory. He was editor and one of the authors of *Geology of Armenia*. He was a member of the Scientific Council of the Presidium of the USSR Academy of Sciences. [Chapter 79.4]

Hamilton, William Rowan (1805–1865)

Irish mathematician, astronomer, and physicist. He is best known for the “Hamilton’s equations,” a set of equations describing the positions and momenta of a collection of particles, and the “Hamilton’s principle.” Born in Dublin, Hamilton was a child prodigy and entered Trinity College at 18. Four years later he was appointed professor of astronomy at Trinity College and Astronomer Royal for Ireland, positions he held for life. [Bell (1937); James and James (1976); Britannica (2003)]

Haskell, Norman Abraham (1905–1970)

American geophysicist whose contributions were in the theory, dispersion, and energy of surface waves, computer programming, and elasto-dynamic theory of failure. His career included Reiber Laboratory (1936–1937), Western Geophysical Company (1937–1941), Columbia University (1941–1942), California Institute of Technology (1942–1945), US Smelting and Refining Co. (1946–1948), and the Air Force Cambridge Center (1948–1970). He was president of the Seismological Society of America from 1969 to 1970. [Ben-Menahem (1990)]

Hattori, Ichizô (1851–1929)

Early Japanese seismologist and scientific leader. He was the first president of the Seismological Society of Japan in 1880. He supervised recovery from the 1896 Sanriku tsunami and proposed a plan of reconstruction of Tokyo after the 1923 earthquake. Born inYamaguchi, Hattori studied abroad and graduated from Rutgers College in 1875. Hattori served the Ministry of Education of Japan in a variety of positions, including vice-dean of the Faculty of Science and Letters, dean of the Faculty of Law, and president of Tokyo University. He was governer of Hyogo Prefecture from 1900 to 1916, then named by the Emperor to the Upper House. [Miyamura (2003)]

Heck, Nicholas Hunter (1882–1953)

American geophysicist who studied earthquakes and tsunamis and wrote a book, *Earthquakes*. He worked on locating peaks in the sea floor, velocity of sound in seawater, and radio-acoustic location of vessels. Heck was born at Heckton Mills, Pennsylvania, and earned A.B. (1903) and C.E. (1904) degrees at Lehigh University. He worked for the US Coast and Geodetic Survey from 1904 to 1944, serving as chief of the Division of Geomagnetism and Seismology from 1922 to 1942. He was president of the Seismological Society of America in 1937–1939 and with Hodgson and Macelwane founded the Eastern Section of SSA in 1946. He was president of the International Association of Seismology from 1936 to 1940. He was the American Geophysical Union’s Bowie Medalist in 1942. [*NCAB* **45**: 217]

Hecker, Oskar August Ernst (1864–1938)

German seismologist who worked on Earth tide observations in Potsdam and built seismometers (horizontal pendula), and in particular experimented with damping. Hecker was director of the Central Bureau of the International Seismological Association (ISA) in Strasbourg and of the German main station earthquake research in Strasbourg (both as successor of Gerland). After World War I he became director of the *Reichsanstalt für Erdbebenforschung* in Jena. In 1922, he initiated, together with Wiechert, the foundation of the German Geophysical Society. [Schweitzer (2003)]

Heiskanen, Weikko Aleksanteri (1895–1971)

Finnish geodesist best known for his work in physical geodesy and the gravimetric method. Heiskanen was born in Kangaslampi and received his doctoral degree from the University of Helsinki. He was professor of geodesy at the Technical University of Finland from 1929 to 1949. He then divided his time between Finland (as director of the Finnish Geodetic Institute) and the United States (as professor of geodesy of Ohio State University). He wrote two standard texts (Heiskanen and Vening Meinesz, 1958; Heiskanen and Moritz, 1967). [McGraw-Hill (1980)]

Herglotz, Gustav (1881–1953)

Austrian and German mathematician, professor in Göttingen, Vienna, Leipzig, and again Göttingen. The inversion of measured seismic travel times into a velocity-depth function (Benndorf's problem) leads to an integral equation. Herglotz's main contribution to seismology was to solve this equation by converting it to an Abel's integral equation. After Herglotz's solution was available in 1907, Geiger, Gutenberg, and Wiechert applied it to their travel-time curves and calculated accurate 1D seismological Earth models. [Tietze (1954); Kertz (2002); Schweitzer (2003)]

Hess, Harry Hammond (1906–1969)

American geologist who among other accomplishments developed the seafloor spreading hypothesis in the early 1960s. He studied marine gravity anomalies and the geology of the West Indies, mineralogy and petrology, especially serpentines and peridotites, discovered (and named) guyots while mapping seafloor topography, developed the theory of seafloor spreading, and was one of the originators of Project Mohole. Hess was born in New York City and received the B.S. at Yale University in 1927 and the M.A. (1931) and Ph.D. (1933) at Princeton University in geology. He was an instructor at Rutgers University in 1932–1933, worked at the Carnegie Institution Geophysical Laboratory in 1933–1934, then from 1934 to 1969 rose to instructor, assistant and associate professor, and professor at Princeton University, including head of the geology department from 1950 to 1966. He was president of the Mineralogical Society of America in 1953 and president of the American Geophysical Union's Geodesy Section in 1953–1954 and the Tectonophysics Section in 1976–1978. He was president of the Geological Society of America (GSA) in 1963 and the society's Penrose Medalist in 1966. For the US National Academy of Sciences he was chairman of its geology section in 1960–1963, its Committee on Disposal of Radioactive Waste, and the Space Science Board. Hess developed the seafloor spreading hypothesis in his paper, "History of ocean basins" (in "Petrologic Studies," the Buddington Volume, pp. 599–620, Geological Society of America, 1962; see also Chapter 6 of this Handbook). [*NAS Biog. Mem.* **43**:109–128, 1973]

Hisada, Toshihiko (1914–1988)

Japanese anti-seismic structural designer. He conducted research on structural design, particularly that of wooden structures, and contributed to the anti-seismic safety plan for nuclear power plants. He helped to establish the Japanese Building Standard Law and its enforcement regulations. Born in Tokyo, Hisada graduated in architecture from Tokyo Imperial University in 1938 and joined the Japan Ministry of Finance. He moved to the Building Research Institute in 1948 and served as its director from 1966 to 1969. He was director of Kajima Technical Research Institute from 1969 to 1981. [Miyamura (2003)]

Hobbs, William Herbert (1864–1953)

American geologist. Hobbs conducted research in structural and dynamic geology, glaciology, seismology, geology of southwestern Massachusetts and Connecticut, the Earth's fracture system. He is noted for his paper, "The earthquake of 1872 in the Owens Valley, California" (*Gerl. Beitr. zur Geophys.* **10**, 352–385, 1910). Hobbs was instructor and professor at University of Wisconsin (1889–1906), Assistant Geologist of the US Geological Survey (1896–1906), and Professor at the University of Michigan (1906–1953). He participated in the Greenland expeditions (1906–1930) and is the author of the book *Earthquakes, an Introduction to Seismic Geology* (Appleton, New York, 1907). [*Proc. GSA for 1953*, pp. 131–139].

Hodgson, Ernest Atkinson (1886–1975)

American-born Canadian seismologist. He conducted field studies of Canadian earthquakes and increased the Canadian network of seismic stations from three to ten. He studied microseisms, rock bursts, the Earth's structure, and the seismicity of Canada. Hodgson was born in Utica, New York, and received the B.A. (1912) and M.A. (1913) from the University of Toronto and the Ph.D. (1932) from St. Louis University. He was employed by the Dominion Observatory in Ottawa from 1914 until 1951. He was chief of its seismological division starting in 1932 and assistant director of the observatory from 1947 to 1951. He was president of the Seismological Society of America from 1941 to 1943. Together with Heck and Macelwane he founded the Eastern Section of SSA, and they started the Bibliography of Seismology which was initially published by Dominion Observatory. [*BSSA* **65**:1893–1895, 1975]

Hodgson, John Humphrey (1913–)

Canadian geophysicist. He was born in Toronto, Ontario, and received the B.A. (1940), M.A. (1946), and Ph.D. (1952, in geophysics) degrees from University of Toronto. He was a field engineer with Schlumberger Well Surveying Co. from 1940 to 1944, assistant professor of geophysics at University of Toronto from 1945 to 1949, and worked for the Dominion Observatory of Ottawa, Canada, starting in 1949. He succeeded his farther,

E. Hodgson, in 1952 as Chief Seismologist of the Observatory and its successor, the Earth Physics Branch, retiring in 1973 to become project manager of the Regional Seismology Program for Southeast Asia of UNESCO. He conducted research on the structure of the Earth's crust, on fault-plane solutions, on the mechanics of earthquakes, and on rockbursts. He was president of IASPEI from 1963 to 1967. He was a member of the Royal Canadian Academy of Sciences.

Hoff, Karl Ernst Adolf von (1771–1837)

German scientist best known for his general worldwide catalog of earthquakes (Hoff, 1840), and twelve annual earthquake catalogs for the years 1821–1832 (Hoff, 1826–1835). Hoff was born in Gotha and was educated at the University of Jena and of Göttingen. He became counselor of chancellery in 1813 and president of the Superior Consistory, Gotha, in 1828. He was named the director of science and art collections, Gotha, 1822. See Chapter 45 (p. 760) of this Handbook. [Debus (1968)]

Holmes, Arthur (1890–1965)

English geophysicist best known for his works in radiometric dates and convection currents in the mantle. Holmes was born at Hebburn and studied geology, physics, and mathematics at Imperial College, London. He taught geology at Imperial College from 1912 to 1920 and became the professor of a new department of geology at Durham University in 1924. He was appointed Regius Professor of Geology and Mineralogy at the University of Edinburgh in 1943 and retired in 1956. He wrote a popular and influential textbook on physical geology in 1944. [McGraw-Hill (1980)]

Homma, Shôsaku (1913–1953)

Japanese seismologist and administrator. He studied the theory of propagation of surface and boundary waves and the effects of bottom topography on tsunami waves. As first director of the Matushiro Observatory in 1948, he established its seismological program. He also researched the patterns of Japanese earthquakes. Homma was born in Tokyo and graduated from the Seismological Institute of Tokyo Imperial University in 1936, then joined the Central Meteorological Observatory. In 1948 he became acting director of the newly established Matushiro Observatory and developed a comprehensive seismological program. [Miyamura (2003)]

Honda, Hirokichi (1906–1982)

Japanese seismologist and Earth modeler. He concluded from the 1930 Idu earthquake that it involved two reverse moments applied at the focus, the mechanism now known as the double-couple mechanism. His research involved the mechanisms of earthquakes including deep-focus events, observations of the frequency spectra of earthquakes, and tidal motions. From the partition of energy between reflected and transmitted waves, he concluded that the outer core of the Earth must not be rigid. Honda was born in Tottori-ken, Japan, and in 1929 graduated with a major in physics from Tokyo Imperial University. He joined the Central Meteorological Observatory of Japan and in 1940 became first chief of the Seismology section. In 1948 he became director of the Sendai Meteorological Observatory, and in 1951 he was appointed professor at Tôhoku University. In 1960 he moved to Tokyo University and in 1966 was appointed to the principal professorship of seismology formerly held by Takeo Matuzawa. Some of Honda's important publications include: (1) Honda, H. (1932). On the mechanism and the types of the seismograms of shallow earthquakes. Geophys. Mag. 5:69–88. (2) Honda, H. (1932). On the types of seismograms and the mechanism of deep earthquakes. Geophys. Mag. 5:301–326. (3) Honda, H. (1962). The earthquake mechanism and seismic waves. J. Phys. Earth 10:1–97. [Miyamura (2003)]

Hron, Franta (1937–1998)

Czechoslovakian-born Canadian geophysicist. Hron made numerous significant contributions to geophysics. He is noted for fundamental work with the seismic modeling method known as asymptotic ray theory (ART), which originated in the former Soviet Union and Czechoslovakia. Hron first brought this method to the attention of his colleagues in the West, and he and his students proceeded to make it one of the most powerful seisimic modeling methods in existence today.

Hron was born in Stary Klicov (present Czech Republic). He completed high school in two years and then studied mathematics and physics at Charles University in Prague. After earning his Ph.D., he taught at Charles University for a number of years. He left Prague in 1968 and emigrated to Canada, where he eventually found an academic position at the University of Alberta in Edmonton. He was named a full professor in 1983. [*The Leading Edge* **18**(1):138–139, 1999]

Hubbert, Marion King (1903–1989)

American geophysicist and petroleum geologist. Though best known for his estimates of petroleum and mineral resources, he also did important work on the theory of scale models and the mechanics of fracture of rocks, on ground water, and the entrapment of petroleum and natural gas. Hubbert was born in San Saba, Texas, and received the B.S. (1926), M.S. (1928), and Ph.D. (1937) in geology and physics at the University of Chicago. He was instructor of geophysics at Columbia University from 1931 to 1940, part of the Board of Economic Warfare from 1942 to 1943, with Shell Oil Company from 1943 to 1963, and professor of geology and geophysics at Stanford University from 1963 to 1968. At the same time (1964–1976) he worked for the US Geological Survey. He was president of the Geological Society of America in 1962 and received the GSA's Day

Medal in 1954, Penrose Medal in 1973, and Vetlesen Prize in 1981. He was a member of the National Academy of Science and chairman of the National Research Council's Division of Earth Sciences in 1963–1965. [*The Leading Edge*, **2**(2):16–24, 1983; *GSA Memorial* **24**:39–46]

Hudson, Donald Ellis (1916–1999)

American earthquake engineer. During his long career, Hudson published 81 papers and co-authored two textbooks. His widely used monograph, *Reading and Interpreting Strong Motion Accelergrams* (Earthquake Engineering Research Institute, 1979), is archived on the attached Handbook CD. Through his teaching and supervision of Ph.D. candidates he exerted a very significant influence on earthquake engineering worldwide. His interest in studying destructive earthquake motions led him to develop a low-cost seismoscope in the 1950s, which essentially measured the intensity of groundshaking; because this instrument did not require maintenance, it was widely deployed in seismic countries. He was instrumental in developing the first modern commercially available strong motion accelerograph, of which about ten thousand have been manufactured and installed in almost all seismic countries of the world by the Kinemetrics Company in Pasadena. He and his coworkers developed the first modern, multi-unit vibration generator for shaking buildings, bridges, dams, etc., and measuring dynamic properties.

Born in Alma, Michigan, Hudson obtained the BSME from the California Institute of Technology (Caltech) in 1938, and obtained the PhD degree, also from Caltech, in 1942. As this was in the midst of World War II, he worked at the Naval Ordnance Test Station in Pasadena studying the hydrodynamic performance of aircraft-launched torpedoes. At the end of the war he was appointed assistant professor (and later professor) of Mechanical Engineering at Caltech, remaining there until 1981. From 1981 to 1984 he served as Fred Champion Professor and Chairman of the Department of Civil Engineering at the University of Southern California. He then returned to Caltech as Professor Emeritus from 1984 to 1999 and continued his activities until ill health forced him to retire in 1997. [Jennings (2003)].

I

Iida, Kumiji (1909–2000)

Japanese geophysicist. At Nagoya University, he helped establish several new seismic observatories, assisted local governments on natural disaster problems, and conducted research on rock and soil mechanics and on tsunamis. Iida was born in Iiyama and graduated from the Seismological Institute of Tokyo Imperial University in 1934. He received the D.Sc. degree in 1946. After serving as an assistant at the Earthquake Research Institute, he moved in 1940 to the Geophysical Research Division of the Japanese Geological Survey, where he worked on geophysical surveying for oil and minerals. In 1954 he became professor of geophysics at Nagoya University, where he worked until 1972, then became professor at Aichi Institute of Technology. [Miyamura (2003)]

Imamura, Akitsune (1870–1948)

Japanese seismologist. He observed the 1923 Tokyo earthquake and promoted measurements of crustal motion for purposes of forecasting earthquakes. He served on a succession of earthquake committees and revived the Seismological Society of Japan in 1929. Born in Kagoshima, Imamura graduated in physics from Tokyo Imperial University in 1894 and received a D.Sc. degree in 1905 for his teleseismic study. In 1913 he joined the Imperial Earthquake Investigation Committee, served on its successor, the Council for Earthquake Disaster Prevention, starting in 1925, and founded the private Association for Earthquake Disaster Prevention in 1941 after the Council was discontinued. Based on his study of historical earthquakes, he warned of an impending earthquake in the Tokyo area before the 1923 Kwanto earthquake occurred. While Professor Omori was abroad and Tokyo was on fire, Imamura kept the Tokyo Imperial University observatory operating to study this earthquake and its numerous aftershocks. He succeeded Omori as professor when Omori died shortly after his return. In 1925 Imamura was elected to the Imperial Academy of Japan. He organized the new Seismological Society of Japan in 1929. To investigate crustal movements as forerunners of big earthquakes, he had the Military Land Survey repeat geodetic surveys throughout Japan. He also started mareographic recordings and earthquake observations along the Pacific coast of Kii Peninsula and Shikoku for the purpose of predicting the next Tokaido and Nankaido earthquakes. After retiring in 1931 he continued his observations at his own expense. Regrettably, World War II curtailed his program before two great earthquakes occurred in 1944 (Tonankai) and 1946 (Nankaido). Among his many publications are: (1) A diary on the great earthquake. Bull. Seism. Soc. Am., **14**:1(1924), 1–5, (2) Preliminary note on the great earthquake of southeastern Japan, on Sept. 1. 1923. Bull. Seism. Soc. Am., **14**:2(1924), 136–149, and (3) *Theoretical and Applied Seismology*, Maruzen, 1937, 358 pp. [Miyamura (2003)]

Ingram, Richard E., S.J. (1916–1967)

Irish seismologist. His published works include papers in mathematics and seismology, particularly as concerns the theory of the mechanism of earthquakes. Ingram was born in Belfast and studied at Belvedere College and at the University of Dublin, where he obtained a B.S. and M.S. in mathematics in 1939. In 1934 he began a term as Director of the the Jesuit-run Seismological Observatory at Rathfarnham Castle. In 1946 he continued his studies in mathematics at Johns Hopkins University, where he obtained his Ph. D. in 1948. He spent the following year working in mathematics and seismology at the California

Institute of Technology. In 1949 he returned to Rathfarnham Castle and resumed his role as Director of the Observatory in addition to teaching mathematics at the University College of Dublin. In 1961–1962 he was Professor of Mathematics at Georgetown University. He also worked with the US Coast and Geodetic Survey and the Dominion Observatory of Canada. Ingram was a member of the Council of the Royal Irish Academy. [Stauder (1999a)]

Inouye, Win (1905–2000)

Hawaiian-born Japanese seismologist who published extensively on local earthquakes in Japan and China and on the T-phase. Inouye was born in Hilo, Hawaii, and graduated from the Seismological Institute of Tokyo Imperial University in 1927. His first appointment was at Tsukuba Observatory as an assistant at the Earthquake Research Institute. From 1937 to 1942 he was a professor at Hsinking Technology Institute in Changchun, China, and conducted research for the Geological Survey of Manchuria. In 1943 he returned to Tokyo as head of the seismological section of the Central Meteorological Observatory, then became chief of the seismological division of the Meteorological Research Institute. From 1965 to 1975 he was a professor at Nihon University. [Miyamura (2003)]

Ishimoto, Mishio (1893–1940)

Japanese geophysicist and earthquake researcher. He studied the statistical relations between earthquake frequencies and recorded maximum amplitudes, the relation between intensity and acceleration, and the predominant periods in ground vibrations. He used seismic surveying to study the site of the Honshu-Kyushu tunnel. In 1933 he received the Imperial Academy Prize for developing high-sensitivity seismographs. Ishimoto was born in Tokyo and graduated in experimental physics from Tokyo Imperial University in 1917. He studied ship vibrations at Tokyo University Department of Naval Architecture until 1919,when he moved to Mitsubishi Research Institute. In 1925 he joined the new Earthquake Research Institute of Tokyo University as associate professor, and in 1928 he received the D.Sc. degree and was promoted to professor. He was the institute's director from 1931 to 1939. He was the author of five books. [Miyamura (2003)]

J

Jacobsen, Lydik Siegumfeldt (1897–1976)

Danish-born American engineer who was a pioneer in studying the dynamic characteristics of buildings and other structures, especially their response to earthquake vibrations. Jacobsen was born in Nyborg, Denmark, and received the A.B. (1921) and Ph.D. (1927) degrees in physics at Stanford University. He started work at Stanford in 1921 and became professor of mechanical engineering and director of the Vibration Laboratory and head of the mechanical engineering department. He coauthored (with R. S. Ayre) a prominent textbook, *Engineering Vibrations*. He was the first president of the Earthquake Engineering Research Institute, a co-founder of Agbabian-Jacobsen engineering consulting firm, and president of the Seismological Society of America in 1953–1955. [*BSSA* **67**:1239–1241, 1977]

Jaeger, John Conrad (1907–1979)

Australian mathematical physicist and geophysicist. Jaeger was born in Stanmore and earned a B.Sc in 1928 at Sydney University. In 1928 Jaeger entered Trinity College, Cambridge, and remained there until 1935, carrying out research in quantum theoretical topics and continuing his interests in pure and applied mathematics. In 1936 he became lecturer in mathematics at the University of Tasmania, where with H. S. Carslaw he coauthored two major works, *Operational Methods in Applied Mathematics* (Oxford, 1941), and *Conduction of Heat in Solids* (Oxford, 1947). He also wrote *An Introduction to the Laplace Transformation* (Methuen, 1949) and *An Introduction to Applied Mathematics* (Oxford, 1951), and more research papers on heat conduction. In 1952 he became chair of geophysics at the then-new Australian National University, the first chair in geophysics at an Australian university. He built what became an outstanding research center in the solid earth sciences. It eventually led to the establishment of the Research School of Earth Sciences, formalized shortly after his retirement in 1972. Jaeger's main areas of research at ANU were in terrestrial heat flow and in rock mechanics. He was active as a member of the International Heat Flow Commission (see Chapter 81.4 of this Handbook). Rock mechanics dominated the last 15 years of his active research, with particular concentration on the fracture of rocks and the friction at sliding rock surfaces. He published (with N. G. W. Cook) *Fundamentals of Rock Mechanics* (Chapman and Hall, 1969, and later editions). This book followed an earlier successful Methuen Monograph on *Elasticity, Fracture and Flow* (1956). [*Biogr. Mem. Fell. Roy. Soc. Lond.*, **28**:163–303, 1982; Paterson, 2000]

Janczewski, Edward Walery (1887–1959)

Polish geophysicist who began geophysical surveying in Poland. Janczewski was born in Cracow, entered the University of Cracow in 1906, and moved to Switzerland in 1909 where he studied at the Universities of Fribourg, Zürich, Lausanne, and Neuchâtel. He received a licentiate in mathematics and nature in 1912 and a doctorate in 1914. From 1916 to 1920 he studied Alpine seismicity. From 1921 to 1939 he worked for the Polish State Geological Institute in Warsaw. There he initiated geophysical surveying in Poland, first gravity then seismic refraction (1926–1928). He studied microseismicity in Poland in 1932. After World War II, he moved to Cracow where he was employed at the State

Geological Institute and lectured on geophysics at the Mining Academy. He organized and was appointed director of the Chair in Applied Geophysics there in 1948. He studied mining shocks in Upper Silesia. [Chapter 79.43]

Jeffreys, Harold (1891–1989)

English geophysicist, astronomer, and mathematician. Jeffreys' work in diverse areas of science had mathematical applications as their link. In geophysics he contributed greatly to seismology and physics of the Earth's interior, and the circulation of the atmosphere. In astronomy he studied the outer planets and the origin of the solar system. Jeferys was born near Durham, and in 1907 he went to Armstrong College, a part of Durham University, where he graduated in 1910 with distinction in mathematics. He continued his study at St. John's College, Cambridge, and became a fellow in 1914. Jeffreys worked in the Cavendish Laboratories on war-related work from 1915 to 1917. He joined the Meteorological Office from 1917 to 1922, working on hydrodynamical problems. He then returned to Cambridge and remained there. He first lectured in mathematics until 1932. From 1932 to 1946 he taught geophysics, then he became Plumian Professor of Astronomy. Jeffreys was president of IASPEI from 1957 to 1960. A photo of Sir Harold is included in Figure 2 of Chapter 81.3. Among Jeffreys's books are *The Earth: Its Origin, History and Physical Constitution* (1924, and many later editions), *Earthquakes and Mountains* (1935), *Theory of Probability* (1939), and written jointly with B.S. Jeffreys, *Methods of Mathematical Physics* (1946). [*Biogr. Mem. Fell. Roy. Soc. Lond.* **36**, 301–333, 1976; Jeffreys (1973).]

Jensen, Henry (1915–1974)

Danish seismologist. In 1953 he took over the seismology department from Lehmann. His special field was seismic noise, microseisms. He was much involved in the International Geophysical Year (IGY) in 1957–1958, and the establishment of a new seismograph station in northern Greenland, Station Nord (NOR), for IGY. Together with the new seismologist Joergen Hjelme, he established (1962–1964) and continued operation of four stations of the World Wide Standardized Seismograph Network: Copenhagen (COP) in Denmark, Nord (NOR) in northern Greenland, Kap Tobin (KTG), which replaced the Scoresbysund seismograph in Greenland, and Godhavn (GDH) on Disko Island in western Greenland, where Harboe had his station in the beginning of the century. In this time of build-up Erik Hjortenberg joined Danish seismology, still housed in the state institution, Danish Geodetic Institute. The fields of investigation of the three seismologists were mainly explosion studies of crustal structure and microseisms. [Chapter 79.18]

Joyner, William B. (1929–2001)

American seismologist. He was a scientific leader in the modeling and prediction of strong ground motion and also an effective liaison between the seismological and engineering communities, serving on numerous interdisciplinary committees focused on reducing future earthquake losses, especially through improved building codes. Joyner was born in Casper, Wyoming, and was educated in geology and geophysics at Harvard University. After a few years in petroleum exploration, in 1964 he joined the US Geological Survey, where he spent nearly his entire career. [*SRL* **72**:511–513, 2001]

K

Kanasewich, Ernest Raymond (1931–1998)

Canadian geophysicist who pioneered the development of seismology as a tool in hydrocarbon exploration. Kanasewich was born in rural Saskatchewan and received his bachelor's (1952) and master's in physics (1960) from the University of Alberta, and his Ph.D. (1962) from the University of British Columbia. In 1963 he was appointed assistant professor at the University of Alberta, becoming professor in 1971 and chair of the Department of Physics (1991–1996). He was also director of the Institute of Geophysics, Meteorology and Space Physics from 1991 to 1993. In the early 1980s, he and six other geophysicists, geologists, and geochemists were instrumental in launching the Lithoprobe project. He carried out wide-ranging scientific investigations in diverse fields in geophysics, and authored 2 books and over 118 papers. As the first Distinguished Lecturer for the Canadian Geophysical Union, 1990–1991, he lectured throughout Canada. In 1975, Kanasewich was elected as a fellow of the Royal Society of Canada. [*Am. Men & Women of Sci.*, 1995]

Karapetyan, Nadezhda K. (1927–1992)

Armenian seismologist. Karapetyan was a candidate of physical and mathematical sciences. She joined the Institute of Geophysics and Engineering Seismology, Armenia, in 1961 as a researcher. Her scientific researches were devoted to different problems of seismology, engineering seismology, and acoustics. She is author or co-author of many monographs on studies of seismic oscillations frequency spectrum, hodographs of seismic waves, earthquake nucleation, and earthquake repeatability. Karapetyan published over 100 scientific articles and 10 monographs. Many of her works were presented to European Seismological Commission's meetings. [Chapter 79.4]

Kárník, Vít (1926–1994)

Czech seismologist. Born in Prague, Kárník graduated with a degree in geodesy from the Technical University of Prague in 1950, then earned the Ph.D. there in 1958 and the D.Sc. in 1967. His main interests were the structure of the central European crust, history and monitoring of seismicity in Europe and the Mediterranean, classification and statistics of earthquakes, seismic zoning, and seismic hazard assessment and mitigation. Kárník was chairman of the subcommission for the seismicity of Europe in 1959–1990 and the vice-president of IASPEI from 1963 to 1967. He served as the UNESCO expert at the International Institute of Seismology and Earthquake Engineering in Tokyo in 1963–1964. In 1967 he was admitted to the Yugoslav Academy of Sciences and Arts in Zagreb. He carried out important work as the advisor of the Greek and Yugoslav governments on disaster response after strong earthquakes in 1978 and 1979. He was on the council (1981–1990) of the United Nations Disaster Relief Organization and a member of the scientific committee for the International Decade for Natural Disaster Reduction. In 1993 he was given the prestigious Sasakawa–DHA Disaster Prevention Award. He published a wide range of scientific literature concerning seismological problems, most notably, *Seismicity of the European Area* (2 volumes, D. Reidel Publ. Co., Dordrecht, Holland, 1969 and 1971). He was the co-author of *Erdbebenkatalog der Tschechoslowakei* (1957), of the "Seismic intensity scale version MSK 1964" (1965), and the *Manual of Seismological Observatory Practice* (1969). [Chapter 79.17]

Kawasumi, Hiroshi (1904–1972)

Japanese seismologist who devised the Kawasumi macroseismic magnitude scale Mk and laid the seismological foundations of earthquake engineering practice. He devoted his life to the study of earthquake hazards in metropolitan areas and gave impetus for local governments to develop earthquake disaster prevention policies. Kawasumi was born in Suwa and graduated from the Seismological Institute of Tokyo Imperial University in 1928. Upon earning his Dr. Sci. on propagation of seismic waves in 1937, he became an assistant professor of the Seismological Institute in 1941, professor of the Earthquake Research Institute in 1944, and director in 1963–1965. His early research was on problems of seismic wave propagation and distribution of initial motions. He then introduced a macroseismic magnitude, now known as Kawasumi magnitude, Mk, which was correlated to Gutenberg-Richter magnitude. He compiled magnitude catalogues of Japanese earthquakes for 1885–1950. From these catalogues and the intensity decay relation, Kawasumi published expected maximum intensities for Japan in 1951, which were used for many years for earthquake engineering purposes. Two of his most important papers are (1) Intensity and magnitude of shallow earthquake. Publ. BCIS, Ser. A, Trav. Sci., **19** (1954), 99–114, and (2) Proofs of 69 years periodicity and imminence of destructive earthquake in southern Kwanto district and problems in the countermeasures thereof. Chigaku Zasshi (J. of Geography), **79** (1970), 115–138 (in Japanese). [Miyamura (2003)]

Keimatsu, Mitsuo (1907–1976)

Japanese historian of Chinese earthquakes. He published many papers related to historical earthquakes of China, starting at the Central Meteorological Observatory in 1941. Keimatsu was born in Tokyo and graduated in history from Kyoto Imperial University in 1934. In 1962 he received a Dr. Sci. from Tôhoku University and a Dr. of Literature from Kyoto University. He became professor at Kanazawa University in 1963 and retired in 1972. [Miyamura (2003)]

Kharin, Dmitry A. (1902–1991)

Russian engineering seismologist who designed narrow-frequency-band seismometers and vibration recorders for engineering purposes. He also conducted research on oscillations of structures and seismic-station network organization, and he conducted seismic prospecting, especially in Siberia. Kharin graduated from Leningrad State University and from 1931 to 1946 worked as scientific secretary of the Seismological Institute of the USSR Academy of Sciences. In 1970 he received the Doctor of Engineering Science degree. In 1946 he was made chief of the Institute of Seismology's network of seismic stations in central Asia. He was awarded a Labor Red Banner and several medals for his work. [Chapter 79.45]

Kikuchi, Dairoku (1855–1917)

Early Japanese seismologist. His 1904 paper "Recent Seismological Investigation in Japan" (Publication of the IEIC in Foreign Languages, No. 19 (1904), pp. 1–120) described the development of seismology in Japan since the Meiji Restoration. Kikuchi was born in Yedo (Tokyo) and graduated in 1877 from Cambridge University. He became professor of mathematics of the University of Tokyo and received his D.Sc. degree in 1888. After the 1891 Mino-Owari earthquake he endeavored to establish the Imperial Earthquake Investigation Committee. In 1898–1901 he became president of the Tokyo Imperial University. At the time of his death he was president of the Imperial Academy of Japan. [Miyamura (2003)]

Kikumnik, Leonid Shmaevich (1937–1994)

Russian structural engineer. He helped to develop the USSR Seismic Building Code, working particularly in design of earthquake-resistant features such as isolation and steel-frame structural systems. He used the time-history of acceleration of strong earthquakes. He worked in the Department of Earthquake Engineering of the Central Research Institute of Structures in Moscow. He surveyed and analyzed the effects of many large earthquakes in the USSR. [Rzhevsky (2002)]

Kirchhoff, Gustav Robert (1824–1887)

German physicist. In seismology Kirchhoff is best known for originating the surface integral method to solve the Cauchy problem: finding a solution of the wave equation for given initial conditions on the wave function and its time derivative defined for the whole space. The solution was expressed as a surface integral of functions given as initial conditions properly time-shifted and weighted over a spherical surface centered at the observation point. It can be generalized to an integral over a non-spherical surface enclosing the region where the homogeneous wave equation is valid. It can be used, for example, to represent scattered waves from a rough surface as a surface integral. The integral reduces to ray-theoretical solution for the planar surface. See Chapter 9 of this Handbook. Kirchhoff was born in Königsberg (now Kaliningrad, Russia) and graduated from the Albertus University of Königsberg in 1847. He had already derived Kirchhoff's law, which allows calculation of the currents, voltages, and resistances of electrical networks. After teaching in Berlin and Breslau, he became professor of physics at the University of Heidelberg in 1854, where he collaborated with Robert Bunsen in founding the discipline of spectrum analysis. In 1875 he became professor of mathematical physics at the University of Berlin. [Britannica (2003)]

Kirnos, Dmitry Petrovich (1905–1995)

Russian geophysicist. Beginning in 1945 he developed a wide variety of observatory seismometers for more than 150 stations in USSR and abroad. Kirnos was a graduate of Leningrad State University. Starting in 1926 he worked in the Seismic Department of the USSR Academy of Sciences and in its Seismological Institute on seismic prospecting and engineering seismology. In 1936 he received the degree of Candidate of Science in physics and mathematics, in 1943 the degree of Senior Scientific Researcher, and in 1952 the Doctorate. He was deputy head of the seismology department of the Geophysical Institute and was appointed professor in 1971. He was awarded two Orders of Red Banner and three medals. He published many papers and two monographs: *Some Questions of Instrumental Seismology* in 1955 and (with E. F. Savarensky) *Elements of Seismology and Seismometry* in 1949 and 1955. [Chapter 79.45]

Kiss, Zoltán (1930–1989)

Hungarian seismologist. Kiss's research included seismicity of Hungary and vicinity, crustal thickness beneath Hungary, surface wave dispersion, and the seismic waves from blasting. Born in Szécsény, he received the M.S. in geophysics from the University of Sopron in 1953 and the Ph. D. in technical sciences from the Hungarian Academy of Sciences in 1971. From 1954 to 1970 he was a Research Fellow at Eötvös Loránd University, and from 1971 to 1989 a Research Fellow at Budapest Seismological Observatory. [Chapter 79.26]

Knott, Cargill Gilston (1856–1922)

Scottish physicist. He was born at Pericuik and studied under P. G. Tait at University of Edinburgh, serving as his assistant from 1879 to 1883, at which time he was appointed professor of physics at Tokyo Imperial College, Japan. He returned to Edinburgh in 1891, where he held appointments as Lecturer and Reader and received the D.Sc. degree. He conducted research on magnetism, including a magnetic survey of Japan, on the splitting of seismic energy at a boundary into reflected and refracted waves, explaining how this resulted in the complexity of a seismogram, and on early travel-time curves. In 1908 he wrote a handbook, *The Physics of Earthquake Phenomena*. He was awarded the LL.D. degree by University of St. Andrews. He became a member of the Royal Society in 1920. [*Proc. Roy. Soc. Lond. A.*, **102**, xxvii–xxviii]

Kogan, Sarra D. (1924–1983)

Russian geophysicist. At the Institute of Physics of the Earth of USSR Academy of Sciences from 1946 to 1983, Kogan conducted research on deep-focus earthquakes, comparison of seismic waves from explosions and earthquakes, and mantle and core structure. Kogan earned the M.S. in physics and mathematics at Moscow University in 1947, Ph.D. at Geophysical Institute of USSR Academy of Sciences in 1953, and D.Sc. at Institute of Physics of the Earth in 1979. [Chapter 79.45]

Kogan, Sarra Ya. (1919–1989)

Russian geophysicist who investigated seismic waves from explosions, wave absorption, energy of earthquakes and explosions, and frequency spectra of seismic waves. Kogan earned the M.S. in physics and mathematics at Tbilisi University in 1941, Ph.D. at Moscow University in 1946, and D.Sc. at Institute of Physics of the Earth of USSR Academy of Sciences in 1972. Employed as head of calculations and senior research fellow at Moscow Mechanical Institute from 1945 to 1956 and senior research fellow at Institute of Physics of the Earth, USSR Academy of Sciences 1956 to 1989. [Chapter 79.45]

Kolderup, Carl Fredrik (1869–1942)

Norwegian geologist. He was professor of geology at the Bergen Museum (later the University of Bergen). In 1899 he assumed responsibility for the collection of data on Norwegian earthquakes. He installed the first seismograph in Norway at the Bergen Museum in 1905. He collected data on Norwegian earthquakes, but his main research interests were geologic mapping and petrology. [Chapter 79.41]

Korchinsky, Ioseph Lutsianovich (1905–1993)

Russian engineer who became interested in earthquake-engineering problems in 1948 as a result of a study of the Achgabad earthquake. He studied the relations of seismic intensity, shaking duration, and spectral characteristics as they relate to the energy-absorption properties of buildings and damage by earthquakes. He was particularly concerned with building material types (wood, masonry, steel, reinforced concrete) as these related to damage under vibratory loads and as affected by damping, fatigue, and plastic behavior. He analyzed ground-acceleration records and advocated full-scale testing of structures. He was coauthor in 1961 of the book, *Fundamentals of Seismic Design*. He received the D.Sc. and worked as a professor at the Central Research Institute of Structures in Moscow. [Rzhevsky (2002)]

Kosminskaya, Irina Petrovna (1916–1996)

Russian geophysicist who used the deep seismic sounding method to study structure of sedimentary basins, roots of mountains and crustal transition zones, and the fine structure of the Mohorovicic boundary. She was particularly noted for her ability to organize large international observational projects. Kosminskaya graduated from Moscow Geological Prospecting Institute in 1941 and went to the Institute of Theoretical Geophysics (later the Institute of Physics of the Earth of the USSR Academy of Sciences), where she received her doctorate in 1965. She researched refracted-wave prospecting and deep seismic sounding (DSS) methods. She was a member of the Russian national committee on the Third International Geophysical Year (1957–1959), chairman of the Passive Continental Margins section of the Upper Mantle Project (1960–1970), convenor of DSS and island-arc sessions at IUGG Assemblies (1960–1979), and chairman and co-chairman of its sessions on controlled sources (1968–1993), head of the Passive Margins section of the International Geodetic Project (1970–1980), USSR representative to the International Deep Drilling Project (1974–1978), and scientific leader of that project's DSS study of the Baltic Shield (1979–1992), for which she was awarded the Medal of Helsinki University. She was a member of the editorial board of the Journal of Geophysics from 1976 to 1986. She published more than 120 scientific papers and in 1968 the monograph, *The Method of Deep Seismic Sounding of the Earth's Crust and Upper Mantle*. [Chapter 79.45]

Kostrov, Boris Vikrorovick (1933–1998)

Russian geophysicist whose work focused on the dynamics of elasticity and failure mechanics, on theory of the earthquake source and the propagation of cracks. Kostrov graduated in physics and mathematics from the Crimea State Pedagogical Institute in 1955 and started work in 1958 at the Yalta geophysical station of the Institute of Physics of the Earth, becoming head of the station in 1960. In 1964 he was appointed head of the Department of Wave Fields of the Institute of Physics of the Earth. One of Kostrov's important publications is *Principles of Earthquake Source Mechanics* (with S. Das), Cambridge University Press, Cambridge, 1988. [Chapter 79.45]

Koto, Bunjiro (1856–1935)

Japanese geologist whose report on the Mino-Owari earthquake of 1891 focused attention on the fault origin of earthquakes. Koto was born in Tsuwano and graduated in geology from Tokyo University in 1879. After studying in Germany, he became professor of geology at Tokyo University. He studied various Japanese earthquakes. [Miyamura (2003)]

Kovesligethy, Rudolf (1862–1934)

Hungarian seismologist who organized the Hungarian Seismological Service, originated a "geometric" theory of earthquake phenomena, and studied focal depths (using intensity distribution), travel times, and earthquake prediction. Koveslígethy was born in Verona, Italy, and received the Ph.D. degree from the University of Vienna in 1884. He was an assistant at Ogyalla Astronomical Observatory (OAS) from 1883 to 1887, an assistant at the Meteorological and Geomagnetic Institute in Budapest from 1887 to 1888, an assistant at the Physics Institute of the University of Sciences at Budapest from 1888 to 1893, appointed assistant professor of cosmology there in 1893 and professor in 1904. He was vice-director of OAS from 1899 to 1903, a member of the Permanent Seismological Commission in 1904, and director of the Seismological Observatory and Calculation Institute at Budapest from 1905 to 1934. He was elected to the Hungarian Academy of Sciences in 1909 and was general secretary of the International Association of Seismology from 1907 to 1922. A photo of Koveslígethy is included in Chapter 81.3 of this Handbook. [Chapter 79.26]

Kramers, Hendrik Anthony (1894–1952)

Dutch physicist best known by theoretical seismologists as the "K" in the WKBJ method, which was originally developed for quantum mechanics. Kramers was born in Rotterdam and was educated at the University of Leiden. He made important contributions in physics, including the derivation (with Ralph de Laer Kronig) of the equations relating the absorption to the dispersion of light, and the study of inelastic scattering of light. See Chapter 8 (p. 90) of this Handbook. [Britannica (2003)]

Krishna, Jai (1912–1999)

Indian engineer who formulated the Indian Standards for Earthquake-Resistant Design and designed many important structures in India incorporating these features, including dams, bridges, and nuclear power plants, with particular attention to brick buildings and concrete containers. He designed and installed strong motion and vibration-measuring seismic instruments. He carried out field investigations of earthquake damage, including the 1963 Skopje and 1966 Debar Yugoslavian earthquakes and the 1967 Koyna and 1970 Broach Indian earthquakes. He was also noted as a mentor of many students. Krishna was born in Musaffarnagar, Uttar Pradesh, and received the B.S. degree with honors in 1935 from Thomason College of Civil Engineering (now the University of Roorkee). As a student he received the Thomason Prize, Cautley Gold Medal, and Calcoff Reilly Gold Medal. He was awarded the D.Sc. in civil engineering from the University of London in 1954. He worked first for the State Public Works Department. He was appointed lecturer in 1939 at Thomason College, where he introduced new courses in soil mechanics (1948) and structural dynamics (1958). He became a professor at the University of Roorkee in 1960 and established the School of Research and Training in Earthquake Engineering there in 1967, serving as its director until 1977. He was sent by UNESCO as a visiting professor to the Institute of Earthquake Engineering and Seismology at Skopje, Yugoslavia, in 1967–1968. He was appointed vice chancellor of the University of Roorkee in 1971. Krishna was a fellow of the Indian National Science Academy, a founder and president of the Indian National Academy of Engineering and the Indian Society of Earthquake Technology, the Third World Academy of Science and the International Association of Earthquake Engineers, of which he was president from 1977 to 1980. He was awarded honorary degrees by Agra University and the University of Roorkee, the Bhatnagar Award (1966), the Indian National Design Award of the Indian Institution of Engineers (1971), the Moudgil Award of the Indian Standards Institution (1972), and the International Award of the Japan Society of Disaster Prevention (1988). He was conferred Padma Bhushan for meritorious service to the nation by the president of India in 1971. He published two books: *Plain and Reinforced Concrete Engineering* in 1951 and *Elements of Earthquake Engineering* in 1949, and many papers. [Chapters 79.28 and 80.9]

Krishnan, Maharajapuram Sitaram (1898–)

Indian geologist. He investigated the 1930 North Bengal, 1934 North Bihar, and 1967 Koyna earthquakes. Born at Maharajapuram, Krishnan received the B.A. degree in 1919 and the M.A. in 1921 from Madras University and the Ph.D. from London University in 1924. He then started work as assistant superintendent of the Geological Survey of India, working primarily on mineral deposits, advancing to assistant director in 1938. In 1951 he became director of the Indian School of Mines and Applied Geology, the first Indian to hold that office. In 1943 he wrote a textbook on the geology of India and Burma. He was foundation fellow of the Geological Society of India and its vice president since its inception in 1959. [Narula (1998)]

Krylov, Sergey V. (1931–1997)

Russian geophysicist. His seismic research largely involved deep seismic sounding of the Earth's lithosphere in Siberia, China, and Antarctica using both *P* and *S* waves and seismic tomography. He developed methods of deep seismic sounding, studied heterogeneities of the Earth's crust, developed methods of seismic tomography, zoning of seismoactive regions especially in the Baikal rift zone and the Altai-Sayany folded area. Krylov graduated from the Leningrad Mining Institute in 1955, and after a short period at the Buguruslan Geophysical Office he returned to Leningrad for graduate study from 1957 to 1960. He then joined the Siberian Department of the Institute of Geology and Geophysics, rising over the years to director (1961–1997). He also served as senior lecturer and professor at Novosibirsk University from 1965 to 1996. He was a Corresponding Member of the USSR Academy of Sciences. [Chapter 79.45]

Kubo, Keizaburo (1922–1995)

Japanese earthquake engineer. He worked on developing earthquake-resistant design of bridges, buried pipes, and other civil infrastructures. He surveyed the lifeline damage caused by the 1964 Niigata earthquake and the 1971 San Fernando earthquake. Kubo was born in Tokyo and graduated in civil engineering from Tokyo Imperial University in 1945. He stayed at that institution until 1983, first as lecturer in the Department of Civil Engineering in 1946, then as assistant professor in 1948, and professor at the Institute of Industrial Sciences in 1963. He moved to Saitama University in 1983 and to Tokai University from 1987 to 1993. He was Vice-President of the International Association of Earthquake Engineers from 1984 to 1988 and organized the World Conference on Earthquake Engineering in Japan in 1988. [Miyamura (2003)]

Kusakabe, Shirôta (1875–1924)

Japanese geophysicist. He studied the frequency of aftershocks and potential methods of earthquake prediction, and received the Imperial Academy Prize in 1914 for his studies of the mechanical properties of rocks. Kusakabe was born in Yamagata and graduated in physics from Tokyo Imperial University in 1900, receiving the D. Sc. degree there in 1906 for a study of elasticity and seismic-wave velocities of rocks. After three years of study in Europe, he took a professorship at the College of Science at the newly established Tôhoku University, where he initiated his seismological research. [Miyamura (2003)]

Kuznetsov, Vasili Petrovich (1907–)

Azerbaijani seismologist who focused on the seismicity and seismotectonics of Azerbaijan and on seismic microzonation at power station sites. Kuznetsov was born in Yaroslavl, Russia, and graduated in 1930 in physics from Yaroslavl University. In 1936 he started work in the Institute of Mechanics and Mathematics of the Academy of Sciences, defending his candidacy thesis in 1946 on characteristics of the atmosphere in Baku. Starting in 1937, he worked in the Geology Institute where he was head of the Earth Physics Department and the Laboratory of Seismic Microzonation. [Chapter 79.7]

L

LaCoste, Lucien J. B. (1908–1995)

American instrument engineer. LaCoste is noted for his invention of the zero-length spring in his paper, "A new type long period seismograph" (*Physics*, **5**, 178–180, 1934). The seismograph he and Arnold Romberg built was superior to any instrument then available. Although the majority of long-period vertical seismometers built since then employed the LaCoste suspension, neither LaCoste nor Romberg attempted to build them commercially. Instead they turned their attention to gravity instrumentation, in which the zero-length spring is also a key element. Their underwater, air/sea, and borehole gravity meters have enjoyed unparalleled success.

LaCoste attended the University of Texas at Austin, receiving his doctorate in 1933. In 1939 LaCoste and Romberg founded the LaCoste and Romberg Company, and began building gravity meters. He did not allow his company to get too large and never cut back on development. While disclaiming managerial ability and business acumen, LaCoste inspired a loyal group of highly skilled employees. He financed his development work out of the company's earnings. He turned down offers of government funding more than once, so that he had complete control over what he did. "I'll build it and, if they like it, they'll buy it," he said. [*Eos*, December 12, 1995, p. 516; *The Leading Edge,* **3**(12) 24–29, 1984]

Lamb, Horace (1849–1934)

English mathematician and geophysicist. Born in Stockport, Lamb was educated at Cambridge University and graduated in 1872. He became, in 1875, the first professor of mathematics at the University of Adelaide. Ten years later, he returned to England as the chair of mathematics at Manchester University, and retired in 1920. He is most noted for his contribution in fluid mechanics. His classic book, *Hydrodynamics*, was first published in 1895 and went through several editions. In 1904, Lamb published "On the propagation of tremors over the surface of an elastic solid" (*Phil. Trans. Roy. Soc. London A*, **203**:1–42), one of the fundamental contributions to theoretical seismology. He discussed the motion due to an origin at some depth, both for the case of simple harmonic motion and for that of a single impulse. Lamb showed that, at any place on the surface, considerable displacement would be set up on the arrival of two waves traveling over the surface from the internal source with the velocities of the compressional and distortional waves respectively. In addition, a greater displacement would be set on the arrival of the waves traveling over the surface from the part nearest the source, with the velocity of the type of waves discovered by Rayleigh. [*Obit. Nat. Fell. Roy. Soc. Lond.* **1**:375–391, 1935.]

Laplace, Pierre-Simon (1749–1827)

French mathematician, astronomer, and physicist best known for his work on celestial mechanics, probability, and for the "Laplace equation" and "Laplace transform." Born in Beaumont-en-Auge, Laplace entered Caen University at 16 to study theology, but left two years later for Paris. Laplace was soon appointed professor of mathematics at the Ecole Militaire and began to publish papers in mathematics. In 1773 he was elected to the Académie des Sciences. He was appointed examiner at the Royal Artillery Corps in 1775, and in this role in 1785, he examined and passed the young Napoleon Bonaparte. Laplace was a member of the committee of the Académie des Sciences to standardize weights and measures in May 1790. This led to the SI system of measurements. He helped found the Bureau des Longitudes in 1795 and became its leader as well as that of the Paris Observatory. Laplace's most important work, *Traité du Mécanique Céleste*, was published in 5 volumes. [Bell(1937); James and James (1976); Britannica (2003)]

Lasarev, Petr Petrovich (1878–1942)

Russian geophysicist who compiled a map of magnetic variations in the Soviet Union. Lasarev graduated first in medicine (1901) and then in theoretical physics (1903) from Moscow University. He worked as an assistant at Moscow University from 1907 to 1911, then held a professorship at Moscow High Technical College until 1925 while simultaneously heading the Physical Laboratory of the Academy of Sciences. From 1918 to 1920 he headed an Academy commission investigating the Kursk magnetic anomaly. He then became head of the Institute of Physics and Biophysics, then director of the Geophysical Institute and professor at Moscow Geological Prospecting Institute, and eventually a department head in the Institute of Theoretical Geophysics of the USSR Academy of Sciences. Besides work in physics, biophysics, and photochemistry, he worked on the theory of amorphous matter, the Earth's climate, geothermal heat, geomagnetism, and the history of science. [Chapter 79.45]

Láska, Václav (1862–?)

Czech astronomer and geophysicist. Born in Prague in 1862, Láska studied mathematics and physics at Charles University to become an astronomer, but soon his interests broadened into geomagnetism and meteorology. He soon habilitated in geodesy and moved to the Technical University of Lvov (Lemburg) in Austro-Hungarian Galicia in 1896, where he stayed for 16 years. There, besides his outstanding activities in mine geodesy, astronomy, and associated fields including petroleum geology, he performed fundamental research into seismology and geophysics. His contributions concerning the interpretation of seismograms and the study of microseisms drew considerable response among specialists throughout the world. Of special importance is Láska's rule for rapid estimation of epicentral distance. He became the reporter of the Vienna Academy of Sciences for research of earthquakes in Austro-Hungary. In 1901, at the first International Seismological Conference in Strasbourg, Láska presented a paper dealing with "pendulum unrest," or in today's terminology, meteorological microseisms, a monograph that remains of interest. After his return to Prague in 1911, Láska became professor of applied mathematics at Charles University. The end of the war and the declaration of the independent Czechoslovakia provided Láska with a new impulse to establish geophysics in the young republic. As the chairman of a commission of experts, he founded the Institute of Geophysics and prepared its statute and scientific program. [Chapter 79.17]

Lawson, Andrew Cowper (1861–1952)

Scottish-born American geologist. He published widely on the geology of California and other regions and on isostasy. He was chairman of the State Earthquake Investigation Commission appointed to investigate the 1906 San Francisco earthquake, and a principal author and editor of its report published by the Carnegie Institution of Washington. Lawson was born at Anstruther, Scotland, and received the A.B. (1883) and A.M. (1885) degrees at the University of Toronto and the Ph.D. (1888) at Johns Hopkins University, the LL.D. (1935) from the University of California, Berkeley, and the D.Sc. (1936) from Harvard University. After early field work in Canada, he accepted a professorship at Berkeley in 1890 and was head of the Department of Geological Sciences from 1901 to 1925 and dean of the College of Mines from 1914 to 1917. He was the second president of the Seismological Society of America in 1909–1910. He was vice-president of the Geological Society of America in 1908, president in 1926, and received the GSA's Penrose Medal in 1938. [Vaughan (1970); Britannica (2003)]

Lehmann, Inge (1888–1993)

Danish seismologist and discoverer of the Earth's inner core in her paper, "*P*'" (*Publ. Bur. Cent. Seismol. Internat, Trav. Sci. A*, **14**, 87–115, 1936). Lehmann was born in Osterbro and in 1907 began studying mathematics at the University of Copenhagen, graduating after some years' interruption in 1920. In 1925 she became assistant to Professor N.E. Norlund, who was also director of "Gradmaalingen" and was planning to install seismographic stations near Copenhagen and in Ivigtut and Scoresbysund in Greenland. She became involved with seismology early. She was immediately sent to seismological key points in Europe: to Gutenberg in Göttingen and Jeffreys in Cambridge. For a quarter of a century she was the seismology institution in Denmark, with technical help of army non-commissioned officer H. Rasmussen and since 1946 of Erik Moeller, and only occasional academic help. At her disposal she had the Copenhagen (COP) seismograph station with Wiechert and Galitzin-Wilip seismographs, and from 1936 a Benioff seismograph. In Greenland she had Wiechert seimographs in Ivigtut (IVI) from 1927 to 1953, and various replacements, which never came to function properly, 1955–1960. The other seismograph station in Greenland was placed in Scoresbysund on the east coast from 1928. Lehmann was much concerned with travel-time studies, where she advocated the personal collection and reading of seismograms from various stations rather than the collection of many readings done by various analysts. She is best known for her discovery that the Earth has an inner core, but she also studied the structure of the upper mantle. At a depth of 220 km she proposed a discontinuity, which is often referred as the Lehmann discontinuity. She also investigated the Danish earthquakes occurring during her time and she made a comprehensive overview of Danish earthquakes. A photo of Lehmann is included in Figure 4 of Chapter 81.3 of this Handbook. [Chapter 79.18; *Biogr. Mem. Fell. Roy. Soc. Lond.* **43**:285–301, 1997]

Levitski, Grygory Vasylyevich (1852–1917)

Russian astronomer who conducted seismic investigations and served in scientific offices. Levitski graduated from Petersburg University in 1874, from 1876 to 1879 worked at Pulkovo Observatory, and from 1879 to 1894 was a professor of astronomy at Charkov Universty, where he directed the construction of Charkov Observatory. From 1894 to 1908 he was a professor at Tartu University, serving as rector until 1905 and as director of the Tartu Observatory. From 1908 to 1914 he worked at Vilno and Warsaw. From 1915 to 1918 he taught at the Feminine Pedagological Institute at Petersburg University. He conducted research on sunspots, double stars, longitude measurements, gravity observations, and tidal oscillations, and he investigated the Achalkalaksk earthquake. He was a founding member of the Petersburg Imperial Academy of Sciences Permanent Seismic Commission, represented Russia at the founding of the International Seismological Union, and was a member of its Permanent Commission from 1905 until he died. [Heinloo and All (2003); Chapter 79.45]

Li, Shanbang (1902–1980)

Chinese geophysicist. Li was born in Xingning and graduated from Southeast University, Nanking, in 1930. He installed the Chiufeng seismic station in Peiping, supervised its operation, and compiled its Monthly Seismic Bulletin. He worked in geophysical exploration starting in 1937, discovering the Panzhihua iron deposit in 1943. He designed and built a seismometer and installed the Beipei seismic station near Chongqing. Starting in 1949, he worked on seismic instrumentation and observational seismology, pioneering the Chinese National Seismograph Network. He worked on the history of earthquakes in China.

His book *Earthquakes in China* was published in 1981. [Chapter 79.14]

Linehan, Daniel, S.J. (1904–1987)

American seismologist noted for his studies concerning the identification of the T-phase, the thickness of glacier deposits in New England, and (with Joseph Lynch) a seismic refraction study under St. Peter's Basilica in Rome for archeological purposes. Linehan was born in Beverly, Massachusetts, in 1904. He obtained an M.S. in physics from Boston College in 1931 and an M.S. in geology from Harvard in 1939. In 1935 he began working at the Jesuit-run Weston Observatory, where he was named director in 1950, a post he held until 1982. In 1948 he created the Department of Geophysics at Boston College. He held the position of director of that department until 1963, and of Professor of Geophysics until 1972. One of his first seismological studies concerned the identification of the T phase, representing a part water part solid earth path of propagation from marine epicenter earthquakes. He concentrated much of his seismological research on exploration geophysics, especially on the study of the thickness of glacier deposits in New England. In 1951, together with Joseph Lynch, he conducted a seismic refraction study under St. Peter's in Rome for archeological purposes. Between the years 1954 and 1958 he participated in three expeditions to Antarctica. From 1961 to 1964 he participated in various UNESCO seismological missions to Africa, Asia, and South America. He received honorary doctorate degrees from Le Moyne College, Holy Cross University, and Lowell Technical Institute. In 1958 he received the Distinguished Public Service Award from the US Navy. [Stauder (1999a)]

Liu, Hsi-Ping (1944–2001)

Chinese-born American seismologist. Working in the US Geological Survey, he made important contributions to experimental seismology, particularly the measurement of shear-wave velocities for ground-motion response studies. Liu was born in Chungking, China, and was educated at Tunghai University, Taiwan, and at Dartmouth College and the California Institute of Technology. [*SRL* **73**:875–876, 2002]

Liu, Huixian (1912–1992)

Chinese engineer who initiated earthquake-engineering studies and practice in China. He established the National Strong-Motion Instrument Network, designed China's largest shaking table, created the Chinese seismic intensity scale, and updated its seismic zoning map. He developed methods of dynamic analysis of buildings and dams, and established a national journal of earthquake engineering. Liu was born in Jiangxi, graduated from the Tangshan Institute of Railway Engineering, and in 1937 earned a Ph.D. from Cornell University. In 1954 he established the Institute of Engineering Mechanics and was its director and honorary director until his death. He led the compilation of *The Great Tangshan Earthquake* (1976). He was chairman of the Chinese Association of Earthquake Engineering and president of the China Association of Natural Disaster Prevention. He was elected to the Chinese Academy of Sciences in 1980. [Chapter 79.14]

Louderback, George Davis (1874–1957)

American engineering geologist who contributed to California seismic networks and was a key supporter of the Seismological Society of America (SSA). Louderback was born in San Francisco and received the A.B. degree in 1896 and the Ph.D. in 1899 at the University of California, Berkeley. After a brief stint at the University of Nevada (1900–1906), he worked at Berkeley for the rest of his career (1906–1944). He was dean of the College of Letters and Sciences from 1920 to 1922 and from 1930 to 1939. As head of the Department of Geological Sciences (1925–1939), he expanded the Berkeley Seismic Station network of stations and improved the instrumentation. He was the first secretary of SSA from 1906 to 1910, its president from 1929 to 1935, and editor of the SSA Bulletin from 1935 to 1957. He published on the geology of the Basin and Range province of the western United States, the Coast Ranges of California, and dam sites. He went on geological expeditions to China and the Philippines. [*Proc. GSA for 1969*, pp. 137–142]

Love, Augustus Edward Hough (1863–1940)

British mathematician and geophysicist. In 1904 Love discovered the Love waves in his paper, "The propagation of wave motion in an isotropic elastic solid medium" (Proc. London Math. Soc. (Series 2), 1, 291–344). Love was born in Weston-super-Mare, England, graduated from Cambridge, and held the Sedleian chair of natural philosophy at Oxford from 1899. He also published two influential books, *Mathematical Theory of Elasticity* (1892–1893; reprinted by Dover Publications, New York, 1944), and *Some Problems of Geodynamics* (Cambridge University Press, Cambridge, 1926). [*Obit. Not. Fell. Roy. Soc, III (1939–41)*, pp. 467–482; Britannica (2003)]

Lubimova, Elena Alexandrovna (1925–1985)

Russian geophysicist specializing in geothermics. Lubimova was born in Moscow and in 1949 graduated in physics from Moscow State University. Academician A.N. Tikhonov recommended her as a talented student to Prof. O. Yu. Schmidt for further work. In 1949 she joined the Geophysical Institute (later the Institute of Physics of the Earth) of the USSR Academy of Sciences, where she worked all her life. In 1969 she became head of the Laboratory of Geothermics. In 1955 she defended her Ph.D. thesis "The Thermal Field of the Earth," in 1966 she became Doctor in geophysics, and in 1972 professor. Lubimova is the author of 4 monographs and more than 130 scientific publications in different fields of geophysics and geothermics.

Her monograph, *The Thermics of the Earth and the Moon* (1968), became one of the principal books in theoretical and practical geothermics. She was an active member of several geophysical organizations. She was a founder of the International Heat Flow Commission of the IASPEI and served as its chairman from 1967 to 1979. She was also the vice-president of the Scientific Committee of Geothermal Research in Russia and coordinated a number of international projects. [Chapters 79.45 and 81.4]

Lyell, Charles (1797–1875)

Scottish geologist. Born in Kinnordy, Lyell attended Oxford University at age 19. He became professor of geology at King's College, London (1831–1833), but gave it up for geological studies throughout Europe. His three-volume *Principles of Geology* (1830–1833) went through 11 editions in his lifetime. In this enormously influential book, Lyell established the doctrine of uniformitarianism and discussed earthquake phenomena based on actual observations. He was knighted for scientific accomplishment in 1848. [*Obit., Proc. Roy. Soc. Lond.* **25**:xi–xiv, 1876]

Lynch, John Joseph, S.J. (1894–1987)

English-American seismologist. Lynch was born in London and immigrated with his family to Philadelphia. In 1920 he initiated his teaching career in physics at Fordham University. From 1923 to 1927, while in England for further studies, he spent the summers as an assistant in the seismological observatory at Oxford University. He returned to Fordham in 1928, eventually receiving his doctorate in physics there in 1939. He became director of the Jesuit-run seismological observatory at Fordham in 1928 and served in that role until 1977. In 1946, at the request of the government of the Dominican Republic, he investigated the damage to structures and landforms from the earthquake of August 4 of that year. In 1950, together with D. Linehan, he conducted a seismic refraction survey in the Vatican to examine the archeological ruins under St. Peter's Basilica. He served a term as Chairman of the Eastern Section of the Seismological Society of America, and from 1957 to 1970 he was president of the Jesuit Seismological Association. [Stauder (1999a)]

M

Ma, Xingyuan (1919–2001)

Chinese geologist who organized the national atlas of lithospheric dynamics of China as well as the national program for the Global Geoscience Transect in China. He also studied the Precambrian crust of China and its evolution. Ma was born in Changchun, Jilin, and earned a B.Sc. in geology at the Southwestern United University, China (1941), and a Ph.D. in geology at the University of Edinburgh (1948). He taught at Peking University (1949–1952), then became a professor at the China University of Geosciences (1952–1978). For the rest of his career he was vice director of the State Seismological Bureau and director of its Institute of Geology. He was admitted to the Chinese Academy of Sciences in 1980. [Ma (2002)]

Macelwane, James Bernard, S.J. (1883–1956)

American seismologist and geophysicist. Born in Port Clinton, Ohio, Macelwane received an M.A. in 1911 and an M.S. in 1912 from Saint Louis University, and in that same year he took charge of the seismographic station while continuing studies in physics. In 1920 he entered the University of California, Berkeley. He graduated with a Ph.D. degree in physics in 1923 with a thesis on the dispersion of seismic surface waves, a study he conducted under the direction of A.C. Lawson. He remained at Berkeley for two years as a professor of geology before returning to Saint Louis University in 1925. There he created the Department of Geophysics, remaining director of the department until his death. He reorganized the Jesuit Seismological Association, a network of seismographic stations distributed at some fifteen Jesuit colleges and universities throughout the United States. In 1936 he produced the first textbook on seismology to be published in America. Over his lifetime he worked in a variety of research problems in seismology, including travel times of seismic waves, the constitution of the Earth's interior, and the nature of microseisms and their relation to atmospheric storms. In 1944 he was elected a member of the National Academy of Science. He served as president of the Seismological Society of America in 1928 and as a member of the Board of Directors from 1925 to 1956. He was president of the American Geophysical Union from 1954 to 1956. Throughout his career Macelwane took an active part in the committees and commissions of these societies, as well as in the activities of the National Research Council and the International Union of Geodesy and Geophysics. In 1962, in recognition of his interest in young scientists, the American Geophysical Union instituted a medal in his honor to be awarded each year to a young geophysicist of outstanding ability. [Stauder (1999a)]

Magnitsky, Vladimir A. (1915–)

Russian geophysicist. He studied the figure of the Earth, its internal structure, its temperature structure, the nature of the upper-mantle low-velocity zone, crustal movements, mantle deformation, and the relation of these to seismology. Magnitsky graduated from Moscow Institute of Geodesy in 1940, earned the Ph.D. in 1944, and the D.Sc. in 1948 for a "Study of the Russian Platform Based on Geodetic and Gravimetric Data." He was for many years head of the institute's Geophysical Department. He was elected a corresponding member of the USSR Academy of Sciences in 1965 and became an academician in 1979. He was chairman of its Scientific Council on Physics of the Earth's Interior. He received the Academy's O. Yu. Schmidt Prize in 1969 and the Demidoff Prize in 1995. He served as president of the IASPEI from 1971 to 1975, and for many years was editor-in-chief of the journal Physics of the Solid Earth. A photo of Magnitsky is included in Figure 2 of Chapter 81.3 of this Handbook. [Chapter 79.45]

Mainka, Karl (1873–1943)

German seismologist who build a widely used horizontal seismometer (Mainka's bifilar cone pendulum). Mainka worked at the Central Buereau of the International Seismological Association in Strasbourg, analyzed the data of the first seismic station on Spitsbergen, and was founder and director of a seismic station in Ratibor (today Racibórz, Poland) to monitor the induced seismicity of this mining area (coal). [Anonymous (1944)]

Mallet, Robert (1810–1881)

Irish engineer and seismologist. Born in Dublin, Mallet graduated from Trinity College in 1830 and for the next 30 years worked at his father's factory as an engineer. In 1861 he became a consulting engineer in London. His scientific curiosity led him to publish papers on earthquake phenomena, starting in 1846 in the Philosophical Magazine. In 1850 he investigated the speed of seismic waves using small explosives, and in 1857 he investigated the Naples earthquake. His major publications include "On the dynamics of earthquakes; being an attempt to reduce their observed phenomena to the known laws of wave motion in solids and fluids," *Trans. Royal Irish Academy*, **21**, 51–105, 1848; "On the objects, construction, and use of certain new instruments for self-registration of the passage of earthquake shocks," *Trans. Royal Irish Academy*, **21**, 107–113, 1848; *The Great Neapolitan Earthquake of 1857: The first principles of observational seismology* (2 volumes, Chapman and Hall, London, 1862). He also compiled an extensive catalogue of recorded earthquakes and published the "Map of the global incidence of earthquakes in 1858" (see Color Plate 1 of Part A of this Handbook. [Davison, 1927]

Mandjgaladze, Peter (1944–1993)

Georgian geophysicist who studied electromagnetic and acoustic phenomena which accompanied earthquakes and made laboratory studies of the influence of electromagnetic fields on processes of destruction. He was chief geophysicist of the Experimental Methodology Geophysical Expedition of the Georgian Academy of Sciences and a professor at Rome University. While in Tokyo, he conducted experiments on stick-slip deformation of rocks. [Chapter 79.23]

Martel, Romeo Raoul (1890–1965)

Romeo Raoul Martel was a professor of structural engineering and faculty member at the California Institute of Technology for 42 years. A native of Iberville, Quebec, R.R. Martel graduated from Brown University in 1912 and taught civil engineering at colleges before coming west to work for the Atchison, Topeka & Santa Fe Railroad in Amarillo, Texas. In 1918 he joined the Caltech faculty and became a full professor of civil engineering in 1930.

R.R. Martel was a pioneer in the earthquake-resistant design of structures and in earthquake engineering research. He was largely responsible for the earthquake provisions in the first issue of the Uniform Building Code in 1927, and he was active on building code committees for the City of Pasadena, the California State Chamber of Commerce, the State Division of Architecture, and the American Standards Association.

In 1926 he was a delegate to the Council on Earthquake Protection at the Third Pan-Pacific Science Congress in Tokyo, and attended the World Engineering Congress in Tokyo in 1929 in a similar capacity. He was a member of the Advisory Committee on Engineering Seismology for the United States Coast and Geodetic Survey, and was one of the founding members of the Earthquake Engineering Research Institute. He was instrumental in organizing the first meetings of structural engineers in Southern California. These developed from a group of 12 men into the Structural Engineers Association of Southern California and later, into the Structural Engineers Association of California. These associations have had a profound influence on the development of structural engineering in California.

R.R. Martel served as consultant on the design and construction of many important and novel structures, including the Mt. Palomar telescope and the pumps for the Metropolitan Aqueduct, which were unusual in size and capacity. [Jennings, 2003].

Martemyanov, Askold I. (1927–1994)

Russian engineer who studied earthquakes in the field, evaluated their intensity based on building damage, and developed methods of restoring and strengthening damaged buildings and designing new buildings, particularly in rural areas. Martemyanov received B.S. and M.S. degrees in civil engineering at the Central Asian Polytechnical Institute in Tashkent and the D.Sci. in 1989 in structural engineering at the Central Russian Institute of Architectural Engineering in Moscow. He worked as laboratory head at the Institute of Mechanics, Academy of Sciences of Uzbekistan from 1964 to 1972, as department head at the Ministry of Rural Construction in Moscow from 1972 to 1978, and as department head at the State Committee on Construction in Moscow from 1978 to 1990. He was professor of industrial and civil engineering at the Engineering-Construction Institute at Moscow from 1990 to 1994. [Shirin, 2002]

Matthiesen, R.B. "Fritz" (1926–1981)

American engineer who was an effective advocate and leader within numerous earthquake and engineering professional organizations. Matthiesen was born in Oakland, California, and was educated at the University of Washington, the University of Illinois, Urbana, and the University of California, Berkeley, in civil engineering and structural mechanics. He was a leader in the dynamic testing of full-scale structures. After a decade of teaching and research at the University of California, Los Angeles, he assumed leadership of the federal government's effort to record strong earthquake shaking on the ground and in structures. [*BSSA* **72**:1046–1048, 1982]

Matuzawa, Takeo (1902–1989)

Japanese seismologist who studied individual Japanese earthquakes and the heat generated by earthquakes. Matuzawa was born in Saitama and graduated in physics from Tokyo Imperial University in 1923, becoming assistant professor of the Seismological Institute there in 1924. He received the D.Sc. degree in 1929. He became a professor in 1936 and led the explosion seismology research group in the 1950s and 1960s. He moved to Hokkaido University in 1960 and to Tokyo Kasei-Gakuin University in 1963. [Miyamura (2003)]

Maxwell, James Clerk (1831–1879)

Scottish-born British physicist. He derived Maxwell's equations, four partial differential equations that described the electromagnetic field. He also contributed to many areas in physics, including the Maxwell-Boltzmann kinetic theory of gases. Maxwell was born in Edinburgh and educated at the University of Edinburgh (1847–1850) and Cambridge University (1850–1854). He was professor at Marishal College, Aberdeen (1856–1860), King's College, London (1861–1865). In 1871 he was appointed the first Cavendish professor of experimental physics at Cambridge. [Weaver (1987); Britannica (2003)]

McAdie, Alexander George (1863–1943)

American meteorologist who helped organize the Seismological Society of America. His career encompassed the US Signal Corps (1882–1889), the US Weather Bureau (1890–1913), and Harvard University, where he was a professor and director of Blue Hill Observatory 1913–1931. He witnessed the 1906 San Francisco earthquake and was president of the Seismological Society of America in 1915. [*NCAB* **35**:107–108]

McEvilly, Thomas Vincent (1934–2002)

American seismologist who was crucial in the development of the Incorporated Research Institutions for Seismology (IRIS). His diverse research interests included seismicity, crustal structure, the seismic source, fault properties, and earthquake prediction, all of which he addressed in a long series of contributions (1966–2002) on the earthquakes and tectonics of the San Andreas fault. McEvilly graduated from St. Louis University in geophysics with a B.S. in 1956 and a Ph.D. in 1964. He was associated with Sprengnether Instrument Company from 1961 to 1968 and was a leading expert in seismic instrumentation. He joined the Department of Geology and Geophysics at the University of California, Berkeley in 1964, served at its chairman (1976–1980), and was director of the Earth Sciences Division of the National Lawrence Berkeley Laboratory (1982–1993). He served on many scientific advisory committees, notably for UNESCO, the National Science Foundation, the US Geological Survey, and the Department of Energy. He was editor of the Bulletin of the Seismological Society of America from 1976 to 1985. [Bakun (2003)]

McKeown, Francis A. (1920–1996)

American geologist. In his lifelong career with the US Geological Survey, he worked on predicting the nature of faulting induced by underground nuclear explosions and played a central role in framing and conducting multidisciplinary studies on the origin of seismic activity in the central Mississippi Valley region of the United States. McKeown was born in Camden, New Jersey, and was educated at the University of Pennsylvania and Johns Hopkins University in geology and chemistry. [R.A. Page, private communication]

Medvedev, Sergey Vasilyevich (1910–1977)

Russian engineer who studied the effects of strong seismic vibrations and developed construction guidelines for buildings. He created a scale for the measurement of earthquake strength and contributed to seismic and microseismic zonation. Medvedev graduated from Moscow Engineer-Construction Institute in 1935, and worked in construction engineering until 1944, when he joined the Seismological Institute of the USSR Academy of Sciences. After this organization became the Institute of the Physics of the Earth he became head of its Department of Engineering Seismology. [Chapter 79.45]

Menard, William Henry "Bill" (1920–1986)

American marine geologist who was the first to recognize and map the seafloor fracture zones of the eastern Pacific. He also mapped the magnetic lineations of the seafloor, a key element in the emergence of plate tectonics. Born in Fresno, California, Menard earned a B.S. (1942) and M.S. from California Insititute of Technology and a Ph.D. from Harvard University in 1949. He worked at the US Naval Electronics Laboratory under Dietz (1949–1955), then spent the rest of his career at Scripps Institute of Oceanography, where he became a professor in 1961. He also served as a technical advisor to the US Office of Science and Technology (1965–1966) and director of the US Geological Survey (1978–1981). He was the American Geophysical Union's Bowie Medalist in 1965. He made many other contributions related to the physiography and mineral resources of the seafloor, turbidity currents, and tectonics. He was the author of six books. [*NAS Biog. Mem.* **64**:267–276, 1994]

Mercalli, Giuseppe (1850–1914)

Italian geophysicist whose most important contribution to seismology was the development of the twelve-level Mercalli scale of intensity, an expansion of the ten-level Rossi-Forel scale (see Chapter 49). Mercalli was born in Milan and ordained as a priest in 1871. He graduated in natural sciences from the Polytechnic of Milan in 1874. He taught natural sciences in various schools in Italy until 1911, including the Universities of Catania and Naples. In 1883 he published *Vulcani e Fenomeni Vulcanici d'Italia*, which included maps of seismicity and data on the frequency and intensity of earthquakes. In addition to his studies of volcanism, he engaged in field studies of earthquakes in Italy and Spain and published two monographs on the regional seismicity of Liguria and Piedmont in 1891. He divided Italy into thirty seismic districts, in which earthquakes had more or less similar characteristics. [Ferrari (2003)]

Meyer, Robert Paul (1924–1997)

American seismologist whose principal research was on seismic measurements of crustal structure in the United States and abroad. He was born in Milwaukee, Wisconsin, and earned his B.S. (1948), M.S. (1950), and Ph.D. (1957) in geology and physics from the University of Wisconsin. He was a professor at the University of Wisconsin from 1957 to 1996. He co-authored (with J. S. Steinhart) *Explosion Studies of Continental Structure*, published by Carnegie Institution of Washington. He helped to design portable digital seismographs, was active in developing IRIS (Incorporated Research Institutions for Seismology), and worked on underground nuclear test monitoring. [*SRL* **68**:718–719, 1997]

Michell, John (1724–1793)

British astronomer, geologist, and seismologist. Davison (1927) considered Michell the first founder of seismology, because he based his research on evidence of modern observers. Michell graduated from Cambridge University in divinity. In 1762 he became Woodward professor of geology at Cambridge, and in 1764 became a village rector for the rest of his life. His major publication is "Conjectures concerning the cause, and observations upon the phenomena of earthquakes" (*Phil. Trans. Royal Soc. London*, **51**, 1760). As an astronomer, Michell discovered double stars and estimated stellar distances. [Davison (1927)]

Middlemiss, Charles Stewart (1859–1945)

British geologist working in India who investigated the 1885 Bengal, 1903 Kangra, and 1906 Calcutta earthquakes. Middlemiss was born in Hull and received the B.A. degree from Cambridge in 1881. He joined the Geological Survey of India in 1883, was promoted to deputy superintendent in 1889 and superintendent in 1895, and served as director from 1914 to 1916. He worked on the geology and mineral resources of a wide variety of areas in India and Burma. He was a fellow of the Asiatic Society of Bengal, the Geological Society of London, and the Royal Society. [Narula (1998)]

Miki, Haruo (1923–2000)

Japanese geophysicist. He was a pioneer in high-temperature, high-pressure studies in Japan to investigate the physics of the Earth's mantle and core. He also studied microseisms and explosion-produced seismic waves. Miki was born in Hyogo and graduated in geophysics from Kyoto University, where he spent his career, in 1945. He worked as assistant, lecturer, and associate professor and was promoted to professor in 1959. He was director of the university's Abuyama Seismological Observatory. [Miyamura (2003)]

Milne, John (1850–1913)

English mining engineer and seismologist, one of a handful of people who may truly be called a founder of seismology. Born in Liverpool and educated at King's College, London, Milne studied geology and mineralogy under Warrington Smyth at the Royal School of Mines. After some practical mining work in England, he studied mineralogy at the University of Freiberg, visiting the principal German mining distracts, then spent two years investigating the mineral resources of Newfoundland and Labrador. In 1874 he joined the Royal Geographical Society expedition to the Sinai. Milne was appointed professor of geology and mining in the Imperial College of the Public Works Department, Tokyo, in 1875 and spent several months reaching Japan by way of Siberia, Mongolia, and China. He soon turned to studying Japanese earthquakes; as he remarked, "they had earthquakes for breakfast, dinner, and supper, and to sleep on." After a destructive earthquake in 1880, he arranged a public meeting, and the Seismological Society of Japan (SSJ) was formed with Milne as secretary—the world's first professional society of seismology. For the next 15 years he edited and contributed many articles to the society's Transactions. Milne developed two seismographs that bear his name. The first was designed (by Thomas Gray and Milne originally, and improved later by Omori) for recording local earthquakes and is known as the Gray Milne, Gray Milne Ewing, or Ordinary seismograph. A bracket pendulum of 2 kg mass and 3 sec period was used for the horizontal component and a seismoscope triggered a rotating drum when an earthquake was detected. Static magnification was about 5 for the horizontal component and about 10 for the vertical. About 40 of these were installed at weather stations throughout Japan, being gradually replaced by more sensitive seismographs after the 1923 Kanto earthquake, and mostly discontinued in the mid-1940s. The second (Milne or, later, Milne-Shaw) seismograph was designed to record large earthquakes anywhere in the world; Milne published a detailed description in 1897. About 60 Milne seismographs were deployed around the world, forming the first worldwide seismograph network.

In 1895 Milne returned to England and settled at Shide, on the Isle of Wight. With a small grant from the British Association for the Advancement of Science (BAAS) and a few other small donations, Milne began his ambitious worldwide seismograph network and published observation registers and results in the "Reports of the BAAS Seismological Committee" and the "Shide Circulars." Milne's expenditures far exceeded the grants he received, and he simply spent his own money on what he believed. He also left a bequest of 1,000 pounds to continue his efforts, which later became the "International Seismological Summary." Milne is among the few seismologists who had devoted all their time and personal wealth to seismology, and is truly a founder of seismology. Many biographies have been written about John Milne (e.g., Perry, 1913–1914; Davison, 1927), and a lively description of Milne's observatory at Shide was given by Hoover (1912). On the attached Handbook CD#3, we have archived as computer-readable files all John Milne's reports published in the BAAS Annual Reports from 1895 to 1913, and the complete set of 27 "Shide Circulars." See Chapters 1 and 4 of this Handbook for Milne's contribution to seismology, and to international seismology, respectively. [*Obit. Proc. Roy. Soc. Lond. A*, **89**:xxii–xxv, 1913; Hoover (1912); Davison (1927)]

Minai, Ryô'ichirô (1930–1991)

Japanese engineer. His research was on the structural safety of extremely stressed buildings and earthquake-resistant design, leading to the Revised Japanese Building Code. Minai was born in Moriyama and graduated in 1955 from Kyoto University, where he became associate professor in 1961 and professor in 1966. [Miyamura (2003)]

Minakami, Takeshi (1909–1985)

Japanese volcanologist and seismologist. His research was largely on volcanism but also on the mechanisms of shallow earthquakes and on aftershocks. He predicted the eruption of Asama Volcano on the basis of earthquakes. Minakami was born in Takaoka and graduated in seismology from Tokyo Imperial University in 1934. He worked for the Earthquake Research Institute from 1934 to 1970, retiring with the rank of professor. He received the Japan Academy Prize in 1956 for his studies of volcanoes and was elected to the academy in 1981. He was Vice-President of the International Association of Volcanism and Chemistry of the Earth's Interior from 1971 to 1975. [Miyamura (2003)]

Minami, Kazuo (1907–1984)

American-born Japanese geotechnical engineer. He conducted research of soil liquefaction caused by earthquakes and other soil-foundation problems and was the author of four books. Minami was born in Seattle, Washington, and earned a Master's degree in civil engineering from Massachusetts Institute of Technology in 1932. He studied architecture at Wasada University in Japan, receiving the Ph.D. for a study of hollow-shell foundation design as related to earthquake resistance. He was a lecturer, assistant professor, and professor (1952) at the School of Science and Engineering of Wasada University. He was Secretary General of the International Association of Earthquake Engineering from 1963 to 1977. [Miyamura (2003)]

Minami, Tadao (1940–1999)

Japanese earthquake engineer. He conducted reconnaisance on devastating earthquakes, studied disaster mitigation, particularly of rural houses, stiffness deterioration of steel-reinforced concrete buildings, and elastic-plastic response spectra. Minami was born in Tokyo and graduated in architecture from the University of Tokyo in 1964, then earned an M.Sc. in engineering there in 1966. He started as a research associate at the Earthquake Research Institute (ERI) at Tokyo; then from 1977 to 1984 he was an associate professor at the University of Tsukuba, returning to ERI as professor in 1988. [Miyamura (2003)]

Mintrop, Ludger (1880–1956)

German mining surveyor and geophysicist, who in 1919 discovered the seismic headwave. He built small portable seismometers, and with his worldwide-active company established geophysics as a prospecting tool, in particular by applying seismic methods. [Schleusener (1956); Kertz (1991; 2002)]

Mohorovičić, Andrija (1857–1936)

Croatian geophysicist and a pioneer in the world seismological community for whom the Moho is named. Mohorovičić was born at Volosko, on the Istria peninsula south of Trieste in what was a part of the Austro-Hungarian empire. He had two careers, the first in meteorology and the second in seismology. In 1879 he graduated in mathematics and physics at Carl IV University in Prague, then returned to Croatia to teach in schools. At Bakar he established a meteorological station. In 1892 he became director of the Zagreb Meteorological Observatory. In 1893 he earned a Ph.D. at Zagreb University for a thesis on "Observations of clouds of daily and yearly period at Bakar." There he served first as private docent, becoming associate professor in 1910 and teaching geophysics and astronomy. In 1893 he became an associate member and in 1898 a member of the Yugoslav Academy of Sciences. In 1901 he was appointed head of the Meteorological Service of Croatia and Slavonia. Also in 1901 he set out to establish a seismic station at Zagreb, and beginning in 1906 his observatory reports included recorded earthquakes. By 1909 he had installed several Wiechert seismometers in the observatory and began having good seismograms to study. Following the 8 October 1909 earthquake near Zagreb, his research interests shifted definitively to seismology. Using data on the 1909 earthquake from 29 stations, he plotted a travel-time chart on which he recognized two sets of arrivals: direct and refracted P and

S pulses. He interpreted the refracted pulse as being due to a velocity increase at what is now known as the Mohorovičić discontinuity between the crust and mantle, and he calculated the seismic velocities in the two layers. The classic paper with his results was published in the Zagreb Meteorological Observatory Report for 1909. He also reported on the epicenters of Croatian and Slovenian earthquakes, and on the destructive effects of the 1908 Messina earthquake. He published a series of papers showing hodographs of recorded wave motions, and studied the velocity increase with depth in the Earth, discussing the problem of getting good time data, which was necessary to correlate observations from station to station. A photo of Mohorovičić is included in Figure 4 of Chapter 81.3 of this Handbook. [Chapter 79.16]

Mononobe, Nagaho (1888–1941)

Japanese engineer who advanced the science of earthquake-resistant construction. Mononobe was born in Akita-ken, graduated in civil engineering from Tokyo Imperial University in 1911, and received a doctorate there in 1920. In 1911 he was appointed Senior Engineer of the Japan Ministry of Inner Affairs. He was honored by the Imperial Academy in 1925 for his research on vibrations and on construction of earthquake-resistant towers. [Miyamura (2003)]

Montessus de Ballore, Fernand de (1851–1924)

French seismologist. His early catalog of world earthquakes eventually had 171,434 entries (see Chapter 45 in Part A of this Handbook). He was the first to report that nearly all of the world's earthquakes occur in two great-circle belts: one surrounding the Pacific Ocean and the other running from New Zealand thru the Sunda arc, the Himalayas, the Mediterranean and to the West Indies. He noted that these two belts are associated with Mesozoic geosynclines and rising mountain ranges. Montessus was born at Dompierre Sous Souvignes and educated at the Ecole Polytechnique. In 1881 he was sent on a military mission to El Salvador, where he became interested in earthquakes and began a lifertime career of studying seismicity. On his return to France he was appointed director of the Ecole Polytechnique but spent most of his time assembling data for the first of a series of monographs, including *La Geographie Sismologique* (A. Collin, Paris, 1905), and *La Geologie Sismologique* (A. Collin, Paris, 1924). In 1907 he moved to Chile as Director of its Seismological Service and established a central observatory at Santiago with 33 branch stations. He also lectured on earthquake seismology to architectural and engineering students. [*BSSA* **14**:177–180, 1924; Davison (1927)]

Mueller, Stephan (1930–1997)

German-born Swiss geophysicist. His principal research interest was the structure of the European lithosphere-asthenosphere. He worked on the structure and evolution of rifts, the propagation of shear and surface waves, reflection and refraction surveying, deep seismic soundings, the roots of the Alps, free Earth oscillations, Earth tides, and the lithospheric low-velocity zone. He modernized the Swiss seismic-station network, founded the Black Forest Observatory, and published a seismic hazard map of Switzerland. Mueller was born in Marktredwitz, Germany, and received the diploma in physics at the University of Stuttgart, Germany in 1957, the M.Sc. in electrical engineering at Columbia University, USA, in 1959, and the Ph.D. in geophysics at Stuttgart in 1962. He was professor of geophysics and head of the Geophysical Institute of the University of Karlsruhe, Germany, from 1964 to 1971, professor of geophysics and director of the Swiss Seismological Service at the Institute of Geophysics of the Federal Institute of Technology at Zürich from 1971 to 1995, and professor of geophysics at the University of Zürich from 1977 to 1995. He was president of the European Seismological Commission from 1972 to 1976, president of the governing council of the International Seismological Center from 1975 to 1985, president of the European Geophysical Society from 1978 to 1980, and president of IASPEI from 1987 to 1991. A photo of Mueller is included in Figure 3 of Chapter 81.3. [Chapter 79.41]

Müller, Gerhard (1940–2002)

German geophysicist, who died July 9, 2002. His research interests included synthetic seismograms, structure of the core and mantle, sources with volume change, generalized Maxwell bodies and postglacial uplift, seismic migration, the search for non-Newtonian gravitation, waves in random media, and crack experiments. Müller was the first who could estimate using body waves, *S* velocities of the Earth's inner core and who defined the seismic moment for a volume-changing source (explosion, implosion). Müller was born in Schwäbisch Gmünd and received a diploma in geophysics from the University of Mainz, 1965, a Ph.D. in geophysics from the Technical University of Clausthal, 1967, and a D.Sc. from the University of Karlsruhe, 1974. He was a research scientist (1969–1971) and assistant professor (1972–1979) at the University of Karlsruhe, then became a professor of mathematical geophysics at the University of Frankfurt in 1979. For many years Müller was editor of the *Zeitschrift für Geophysik/Journal of Geophysics* published by the German Geophysical Society (DGG) and the main DGG editor for the *Geophysical Journal International*. He was honoured for his scientific work with the Emil-Wiechert Medal of the DGG, honour mebership of the DGG, and with elections as Fellow of the American Geophysical Unions and as Associate of the Royal Astronomical Society. [Müller (2000); Kennett (2002); Zürn (2002)]

Murphy, Leonard Maurice (1916–1974)

American seismologist who established earthquake and tsunami warning systems and the WWSSN. Murphy was born in

Whitehall, New York, and earned B.S. (1938) and M.S. (1940) degrees from Fordham College. He worked for the US Coast and Geodetic Survey Seismology Division starting in 1942 and became its chief in 1957, then director of the NOAA Seismological Investigation Group. He originated the Coast and Geodetic Survey's earthquake location and reporting services in 1946 and helped to originate its tsunami warning system. He worked on seismic detection of underground nuclear blasts. He was in charge of setting up the World-Wide Standardized Seismograph Network in the 1960s. He was secretary of the IUGG Tsunami Research Committee. [BSSA **65**:807–808, 1975]

Murusidze, Georgi (1915–1988)

Georgian geophysicist. He started his career in seismic prospecting and worked on determining wave velocities using supercritical reflections. He established a network of seismic stations to study induced seismicity around Enguri dam. He mapped faults and seismogenic zones in western Georgia and studied the structure of the crust and upper mantle in Georgia and vicinity. Murusidze was head of the Department of Applied Seismology of the Institute of Geophysics of the Georgia Academy of Sciences. [Chapter 79.23]

Musha, Kinkichi (1891–1962)

Japanese historian of earthquakes. Musha was born in Tokyo and graduated in literature from Waseda University in 1913. He taught English, history, and geography at various schools from 1916 to 1948. Beginning in 1929 he also studied historical Japanese earthquakes and related luminous phenomena at the Earthquake Research Institute of Tokyo Imperial University and the Central Meteorological Observatory. In 1929 he received the First Science Grant of Mainichi-Shimbun. From 1949 to 1960 he worked for the Geological Survey Branch, Army Engineers Office of the Allied Occupation Force. [Miyamura (2003)]

Mushketov, Ivan Vasilyevich (1850–1902)

Russian geologist who investigated the 1878 and 1899 Achalkalaksk earthquakes and designed a model questionnaire for gathering data on earthquakes. He supplemented and updated Orlov's Catalog of the Earthquakes of the Russian Empire. In 1899 he started the establishment of seismic observatories in Russia. Mushketov graduated from Petersburg Mining Institute in 1872, became a professor there in 1882, and was a senior geologist of its Geological Committee. He was also a member of the Permanent Seismic Commission of the Petersburg Academy of Sciences. He conducted field studies in the Urals, in Tien-Shan, the Kuldzha and Caucusus coal fields, Buchara, Gissar, Amu-Darya and Kyzyl-Kum desert and the Seravshan, Elbrus, and Kazbek glaciers. [Chapter 79.45]

Muto, Kiyoshi (1903–1989)

Japanese architect. His research was on earthquake-resistant design of buildings, especially high-rise structures. Muto was born in Toride and graduated in architecture from Tokyo Imperial University in 1925. He was appointed assistant professor there in 1927 and professor in 1935, retiring in 1963 to become vice-president of Kajima Construction Company. He was elected the first President of the International Association for Earthquake Engineering in 1963. In 1969 he founded Muto Structural Dynamics Research Institute. Muto received the Imperial Award of the Japan Academy in 1964 for research on anti-seismic structures. He became a member of the academy in 1975 and was awarded the Order of Cultural Merit in 1983. [Chapter 80.10; Miyamura (2003)]

Myachkin, Victor Iosyfovich (1930–1980)

Russian geophysicist. He graduated in physics from Moscow State University in 1953, and at the Institute of Physics of the Earth of the USSR Academy of Sciences he studied rock pressure in mines in the USSR and Czechoslovakia. From 1966 to 1976 he organized and headed an expedition to survey the oceanic focal zone along the Kamchatka Peninsula. He also organized a study of the site of the Toctogul reservoir as it was filled and the Ashkabad testing area. He studied the seismicity of the Kopet-Dag seismic zone and avalanches in seismic zones. [Chapter 79.45]

N

Naito, Tachû (1886–1970)

Japanese researcher on earthquake-resistant construction of buildings, particularly iron-frame buildings. Naito was born in Yamanashi and graduated in architecture from Tokyo Imperial University in 1910. He then became lecturer at Waseda University and was promoted to professor in 1912. He received the D.Sc. degree in 1924 and was appointed to the Imperial Earthquake Investigation Committee after the 1923 Tokyo earthquake. He was elected to the Japan Academy in 1960 and appointed a Person of Cultural Merit in 1962. [Miyamura (2003)]

Nakagawa, Kyôji (1912–1992)

Japanese engineer, developed a large shaking table to test building construction methods and designed strong motion seismometers. He studied high-rise-building design using analog computers. Nakagawa was born in Tokyo, studied shipbuilding, and entered the navy in 1934. In 1949 he was employed by the Building Research Institute and in 1962 joined its International Institute of Seismology and Earthquake Engineering. From 1969 to 1982 he worked for Obayashi-gumi Construction Company, leading its Technological Research Institute. [Miyamura (2003)]

Nakamura, Kazuaki (1932–1987)

Japanese volcanologist and tectonicist. He elucidated the relationship between tectonic stress and volcanic eruption and developed a method to detect stress trajectory based on geological evidence. He established the tephrochronology of Izu-Oshima volcano and first described base-surge deposits at Taal volcano in the Philippines. In tectonics, he studied the relationship between ground cracks and underlying fault and demonstrated the stress trajectory within island arcs related to plate subduction. Born in Tokyo, Nakamura earned his B.Sc. (1955) and M.Sc. (1957) from Tokyo Imperial University, and D.Sc. (1964) in geology from the University of Tokyo. He spent his career at the University of Tokyo as an instructor (1964), associate professor (1965), and professor (1985) at the Earthquake Research Institute. [Fujii (2002a); Miyamura (2003)]

Nakamura, Saemon-Taro (1891–1974)

Japanese geophysicist. He studied earthquake prediction, geomagnetism of volcanic and seismically active regions, and dynamics of earthquakes. He was author of six books on earthquakes. Nakamura was born in Tokyo and graduated in physics from Tokyo Imperial University in 1914, joining the Central Meteorological Observatory. He received the D.Sc. degree in 1920. He was professor at Tôhoku Imperial University from 1924 to 1951 and at Kumamoto University from 1951 to 1956. [Miyamura (2003)]

Nakano, Hiroshi (1894–1929)

Japanese seismologist, who studied the generation of elastic waves and pointed out that Raleigh waves do not appear as such in the seismogram until they have traveled a significant distance from the epicenter. Nakano was born in Nagoya, graduated in physics from Tokyo Imperial University in 1919, and joined the staff of the Central Meteorological Observatory. [Miyamura (2003)]

Napetvaridze, Shio German (1914–1987)

Georgian engineer who studied seismostability, focusing on natural vibrations and intensity of seismic stress, particularly for dams. He worked on microzonation and stability of slopes and foundations under seismic stress. Napetvaridze graduated from the Georgian Polytechnical Institute in 1939. Starting in 1946 he worked in the Institute of Structural Mechanics and Earthquake Engineering of the Georgian Academy of Sciences, becoming head of the Department of Theory of Seismic Resistance of Structures and Engineering Seismology in 1960 and director in 1983. He received the degree of Doctor of Technics in 1958, became a professor in 1966 and a corresponding member of the Academy in 1974. [Chapter 80.8]

Nasu, Nobuji (1899–1983)

Japanese seismologist and earthquake engineer. He studied the 1927 Tango, 1930 North Idu, 1934 Bihar-Nepal, and 1948 Fukui earthquakes. He was one of the first to use seismic surveying in civil engineering studies, investigating the seafloor at the site of the Honshu-Kyushu tunnel. He studied earthquake-resistant design of nuclear power plants from 1958 to 1965 as a member of the Japanese Government Nuclear Power Committee. He was part of the Organizing Committee of the second World Conference on Earthquake Engineering in Tokyo in 1960 and helped establish the Japanese International Institute of Seismology and Earthquake Engineering in 1962. Nasu was born in Osaka and graduated in physics from Tokyo Imperial University (TIU) in 1924. He was first employed by the Imperial Earthquake Investigating Committee, then in 1926 became an assistant at TIU's Seismological Institute. In 1930 he moved to the Earthquake Research Institute. He received the D.Sc. degree in 1935 for research on the mechanics of earthquake generation based on aftershocks of the 1927 Tango earthquake. In 1943 he was appointed professor, and from 1953 to 1960 he was director of the Earthquake Research Institute. From 1960 to 1970 he was a professor at Waseda University's Engineering Research Laboratory. [Miyamura (2003)]

Navarro-Neumann, Manuel Sanchez, S.J. (1867–1941)

Spanish medical doctor and seismologist. Born in Malaga, Spain, in 1867, Navarro-Neumann obtained his doctorate in Medicine and Surgery in 1893. He entered the Society of Jesus in 1900, and in 1908 took charge of the seismological department of the Jesuit-run Observatory of Cartuja, Granada. In 1915 he was named director of the observatory, a position he held until the Jesuits were expelled from Spain in 1932. After the Spanish Civil War he was reinstated as director in 1938 and remained in that position until 1940. During his directorship he contributed greatly to the development of the observatory. He directed his research to the study of the seismicity of Spain, cataloging the earthquake activity of the Iberian peninsula from the year 1917 to 1929. He published over 300 articles on specific earthquakes and other topics in seismology. In 1916 he published the first seismology text in Spanish. [Stauder (1999a)]

Nazarov, Armen G. (1908–1983)

Armenian seismic engineer who conducted research on the dynamics of seismic resistance, seismic forces and spectra, and dispersion of seismic energy. Nazarov graduated from Tbilisi Industrial Institute, receiving the degree of Candidate in Engineering Science in 1937. He defended a doctoral thesis in 1945 on "Experience of Creating of the Theory of Buildings Seismic Resistance." He was then elected a corresponding member of the Armenian Academy of Sciences. In 1958 he was recognized as an Honored Man of Science and Engineering, and in 1960 he was elected to the Council of Seismology at the Presidium of the USSR Academy of Sciences. He served as president of the Committee of Engineering Seismology of the Interdisciplinary Council of Seismology and Seismic Resistant Building of this academy. In 1961 he founded the Giumry Institute of Geophysics and Engineering Seismology of the Armenian Academy of Sciences. [Chapter 79.4]

Nersesov, Igor Leonovich (1919–1995)

Russian geophysicist who studied seismicity, deep structure of the Earth, and earthquake hazard evaluation and prediction. Between 1960 and 1962 he directed a 3500-km seismic profile using fifty mobile seismic stations in the Pamir-Hindukush-Tien-Shan region, mapping the Earth's structure to 1000 km depth. He worked on discrimination of seismic waves from nuclear explosions and earthquakes. After graduation in geophysics from Moscow State University in 1943, Nersesov spent four years on aeromagnetic surveys in the Arctic. In 1949 he moved to the Geophysical Institute of the USSR Academy of Sciences, where he held many positions such as head of the Laboratory of Regional Seismicity, deputy director of the Institute of Physics of the Earth, and head of the Complex Seismological Expedition. He helped found and develop seismological organizations in former USSR republics such as Talgar in Kazakhstan and Garm in Tadzhikistan. [Chapter 79.45]

Neumann, Frank (1892–1964)

American seismologist. Neumann was born in Baltimore and worked as a geophysicist for the US Coast and Geodetic Survey from 1911 to 1925, then served as chief of its seismology branch from 1925 to 1953. He was a seismologist at the University of Washington in Seattle from 1953 to 1964. He worked with H.O. Wood on the revision of the Mercalli intensity scale, and studied individual earthquakes, surface wave velocities, and strong motion data. He was president of the Seismological Society of America in 1949–1951.

Newmark, Nathan Mortimore (1910–1981)

American civil and earthquake engineer, coauthor (with Rosenblueth) of *Fundamentals of Earthquake Engineering* (Prentice-Hall, Englewood Cliffs, 1971). He was noted for his work on design of earthquake-resistant structures such as the Alaskan pipeline and tall buildings. He developed methods of analysis of structural components under a variety of conditions. Newmark was born in Plainfield, New Jersey, and earned the B.S. in 1930 from Rutgers University and the M.S. (1932) and Ph.D. (1934) from the University of Illinois, all in civil engineering. He joined the University of Illinois in 1934, rising to professor, department head of civil engineering, and director of the Structural Engineering Laboratory. He received the President's Certification of Merit in 1948, the Croes, Mossieff, Norman, and Howard Medals of the Society of Civil Engineers, and the Wason Medal of the International Bridge and Structural Engineers. He was a member of the National Academy of Science and a founding member of the National Academy of Engineering. [*BSSA* **71**:1385–1386, 1981; *NAS Biogr. Mem.* **60**:168–181, 1991]

Newton, Isaac (1642–1727)

English physicist, mathematician, and astronomer whose advances underlie all of modern science, starting with his invention of the calculus (independently also by Gottfried Leibniz) and formulation of his famous laws of motion. Newton was born in Woolsthorpe and studied in Cambridge University. In 1669 he was appointed the Lucasian Professor at Cambridge. In 1687 his Principia was published under the patronage of Halley, setting forth his most significant advances. He was also the first to recognize the nature of white light as a mixture of wavelengths, and he invented the reflecting telescope. After 1693 he retired from research and became Master of the Mint. He was president of the Royal Society from 1703 to 1727. The SI unit of force, the newton, is named for him. [Bell (1937); James and James (1976); Britannica (2003)]

Nikiforov, Pavel M. (1881–1944)

Russian designer of seismic instruments and organizer of seismic networks. Nikiforov graduated in mathematics from St. Petersburg University in 1908 and became an assistant to Galitzin at the Imperial Academy of Sciences. In 1909 he became secretary and in 1916 head of its Seismic Commission. During 1918 to 1921 he was its only employee. In 1921 he became head of the Seismological Department of the USSR Academy of Sciences and in 1928 the first director of its newly organized Seismological Institute. Much of his work was on instrument design, especially for seismic prospecting. He was active in various regional surveys. He reorganized the Russian seismic-station network in the 1970s and promoted the idea of regional networks. [Chapter 79.45]

Nikolaenko, Nikolay Alexandrovich (1929–1988)

Russian engineer. He was professor and D.Sci. working at the Department of Earthquake Engineering of the Central Research Institute of Structures in Moscow begining in 1958. He developed mathematical and physical three-dimensional linear and nonlinear models for analysis of structural systems subject to earthquake forces. He was especially concerned with problems involving suspended objects and those containing fluids. [Rzhevsky (2002)]

Nishimura, Eiichi (1907–1964)

Japanese geophysicist. He studied Earth tidal motions and tilting of the ground before large earthquakes such as the 1943 Tottori M7.2 earthquake. Nishimura was born in Kyoto, graduated in geophysics from Kyoto Imperial University in 1937, and received the D.Sc. degree in 1942. He became an associate professor at Kyoto Imperial University in 1945 and professor in 1951. [Miyamura (2003)]

Nuttli, Otto William (1926–1988)

American seismologist who worked on the seismicity of the central and eastern United States (most notably the 1811–1812 New Madrid earthquakes) and developed a method of calculating magnitude from the amplitude of Lg phases of earthquake and blast seismograms. He also studied Earth structure and seismic wave attenuation. His career was largely spent as a professor at St. Louis University. He was editor of the Seismological Society of America Bulletin (1971–1975), president of the Seismological Society of America (1976–1977), and the SSA medalist in 1987. A national earthquake hazard mitigation award bears his name. [*BSSA* **77**:227–231, 1987; **78**:1387–1389, 1988]

Oddone, Emilio (1864–1940)

Italian seismologist who did research on seismometer design and rock properties. Oddone was born in Baldissero Canavese and graduated in physics from the University of Turin in 1886. From 1890 to 1892 he was an assistant at Rocca di Papa Geodynamical Observatory. In 1892 he became director of the Geodynamical Observatory at Pavia and also taught experimental physics. In 1904 he moved to the Central Office for Meteorology and Geodynamics, of which he was director from 1933 to 1935. He was a founding member of the Italian Seismological Society and served as vice-president (1922–1930) and president (1930–1936) of the International Association of Seismology. A photo of Oddone is included in Figure 1 of Chapter 81.3 of this Handbook. [Ferrari (2003)]

Odenbach, Frederick L., S.J. (1857–1933)

American seismologist who began the Jesuit Seismological Service, the first continental-scale seismic network using uniform instruments and a major contributor to the world seismic network. Odenbach was born in Rochester, New York, in 1857. He spent his career as a Professor of Physics at John Carroll University in Cleveland, where he was known for his innovative research in meteorology. In 1900 he became interested in seismology, and built his own seismograph. After the establishment of the International Seismological Center in Strasbourg in 1906 he conceived the idea of creating a network of Jesuit seismographic stations which could contribute significant data to this international enterprise. The idea came to fruition in the formation in 1909 of the Jesuit Seismological Service (later the Jesuit Seismological Association), composed of sixteen stations distributed at cooperating Jesuit colleges and universities spread throughout the United States and Canada, all equipped with identical two-component horizontal Wiechert seismographs. Data were reported to the Central Station in Cleveland. There the data were collated and forwarded to the ISC in Strasbourrg. Odenbach was also a founding member of the Seismological Society of America. [Stauder (1999a)]

Ohsaki, Yorihiko (1921–1999)

Japanese earthquake engineer and author of five books on the subject. Ohsaki was born in Kyoto, graduated in aeronautics from Tokyo Imperial University in 1943, and after the war entered the Technical Research Institute of the War Damage Restoration Bureau. He received the D. Eng. degree from University of Tokyo in 1958 for studies of building construction. He was director of the International Institute of Seismology and Earthquake Engineering. From 1971 to 1982 he was professor of architecture at the University of Tokyo. In 1982 he joined Shimizu Corporation and established the Ohsaki Research Institute. [Chapter 80.10; Miyamura (2003)]

Oldham, Richard Dixon (1858–1936)

Irish-born British geologist and seismologist. Born in Dublin and educated at Rugby and the Royal School of Mines, Oldham joined the Geological Survey of India in 1879 as assistant superintendant. He studied the geology of the Himalaya Mountains and other parts of India, the Andaman Islands, and Burma. He prepared the second edition of the Manual of Geology of India and compiled the Bibliography of Indian Geology. He cooperated with his father, T. Oldham, to publish the "Catalog of Indian Earthquakes from the Earliest Time to the End of A.D. 1869." In 1897 he interrupted his geological studies to investigate the Assam earthquake, authoring the major memoir describing its effects, including the first evidence that earthquakes sometimes produced accelerations greater than gravity. His studies of the seismograms of this earthquake led him to conclude that the "vorlaufer" (forerunners) preceding the largest amplitudes on seismograms were actually compressional and shear waves. He observed that shear waves were not observed beyond 130 degrees distance from the epicenter and proposed that the Earth has a core with a radius of 0.4 that of the Earth, composed of material with different physical properties than the rest of the Earth. He favored recrystallization as the cause of deep earthquakes. Oldham retired to England in 1904 and studied the physiography of the Rhône delta in France. He received the Lyell Medal of the Geological Society in 1908. He was elected a fellow of the Royal Society in 1911 and served on its council in 1920–1921. He was also a fellow of the Royal Geographic Society and an honorary fellow of the Imperial College of Science. He was president of the Geological Society of London in 1921–1922. [*Obit. Not. Fell. Roy. Soc. Lond.* **2**:111–113, 1936; Narula (1998)]

Oldham, Thomas (1816–1878)

Irish-born British geologist who conducted important early research in India. Oldham was born in Dublin and earned a B.A. degree from Trinity College there in 1838. He worked on mapping the geology of Ireland from 1839 to 1844, was appointed assistant professor at Trinity College in 1844 and professor in 1845. He became a Fellow of the Royal Society of London in 1846. He moved to India in 1850, where he organized the Geological Society of India and supervised its work, especially the mapping of the coal fields of India and Burma. He pointed out the association of petroleum with anticlines in 1855. He began the systematic mapping of the geology of India. He coordinated the field studies of the 1869 Cachar earthquake and with his son, R.D. Oldham, he prepared a catalog of Indian earthquakes. [Davison (1927); Narula (1998)]

O'Leary, William J., S.J. (1869–1939)

Irish seismologist and founder of seismic observatories. O'Leary was born in Dublin, studied mathematics, physics, and astronomy at the University of Louvain, and thereafter served as a professor of mathematics and physics at various Irish colleges. In 1909 he founded an observatory at the Jesuit-run Mungret College, Limerick, then in 1915 established the Rathfarnham Observatory in Dublin, where he constructed a seismograph of 1.5 tons in mass. In 1929 he went to Australia as the director of Riverview Observatory. In addition to observations in seismology he worked in collaboration with the Bosscha Observatory in Lembang, Java, to observe variable stars. He was a fellow of the Royal Astronomical Society and a member of the Royal Irish Academy and the Australian National Committee on Astronomy. [Stauder (1999a)]

Omori, Fusakichi (1868–1923)

Japanese seismologist and a seminal figure in his science. In 1899 he determined the linear relation between duration time of preliminary microtremor and hypocentral distance known as the Omori formula. He also discovered the hyperbolic decay of aftershock numbers, now known as Omori's law. He devised the Omori horizontal pendulum seismograph and other instruments. He published field investigation reports of many Japanese earthquakes and eruptions as well as foreign earthquakes such as the 1905 Kangra and the 1908 Messina earthquakes. Omori completed a catalogue of Japanese earthquakes in 1904 and propounded the first version of today's seismic gap hypothesis. In the interest of avoiding panic, he opposed Imamura's prediction of an earthquake near Tokyo. However, on Sept. 1, 1923, while visiting the Riverview Observatory in Sydney, Australia, he watched a seismograph recording the Tokyo earthquake and returned home, falling ill during the voyage. He met Imamura in the hospital of the ruined Tokyo and entrusted the future of seismology to him. Omori was born in Fukui and graduated in physics from Tokyo Imperial University in 1890. In 1892 he became a founding member of the Imperial Earthquake Investigation Committee. He visited Germany and Italy in 1894 and the United States in 1906 after the San Francisco earthquake. He became a professor of Tokyo Imperial University in 1897, received his D.Sc. degree in 1898, and was named to the Imperial Academy of Japan in 1906. His contribution as a founder of seismology covers very wide fields and his publications reach 103 in Japanese and 111 in English. A photo of Omori is included in Figure 4 of Chapter 81.3 of this Handbook. [Davison (1924, 1927); Miyamura (2003)]

Omote, Syun'ichirô (1912–2002)

Japanese seismologist. Starting in the late 1940s he studied the relationship between underground structure and earthquake damage, and his investigations in the Nagoya and Yokohama-Kawasaki areas are well known. He was prominent in the Research Group of Explosion Seismology starting in 1950, when a 50-ton civil engineering explosion at the Isibuti damsite was planned to be observed by seismologists. He played a leading role in organizing the Research Group of Small Explosion Experiments for seismic wave studies. He visited many developing countries to investigate damaging earthquakes and to give technical advice. He was involved in nuclear power plant safety, and his seismotectonic map of Japan has been used to estimate the design earthquake motions for nuclear plants. Omote was born in Tobata and graduated from the Seismological Institute of Tokyo Imperial University in 1936. He joined the Earthquake Research Institute as a research fellow, then rose to assistant in 1937, associate professor in 1944, and professor in 1961. He earned a D.Sc. degree in 1947 for his research on coda waves. In 1962 he moved to the Building Research Institute in the Ministry of Construction as the first director of the International Institute of Seismology and Earthquake Engineering (IISEE). After retiring from IISEE in 1971 he was a professor at Kyushu University and Kyushu Sangyo University. [Miyamura (2003)].

Orlov, Alexander Yakovlevich (1880–1954)

Russian astronomer who researched earth tides, gravity, polar motions, and latitude changes. Orlov graduated from Petersburg University in 1902, later becoming an assistant at Tartu Observatory, then astronomer-observer and from 1909 to 1913 head of the observatory and lecturer at Tartu University. From 1913 to 1934 he was director of Odessa Observatory and professor at Novorossian University. He also helped organize Poltava Gravimetric Observatory. From 1934 to 1938 he worked at the State

Astronomic and Geodesy Institutes in Moscow. From 1939 to 1941 he was director of Karpatskaya Astronomic Observatory. He helped organize the Far-eastern Latitude Station. From 1944 to 1951 he was director of the Kiev Astronomic Observatory. He became a member-correspondent of the USSR Academy of Sciences in 1927 and an Academician of the Ukraine Academy of Sciences in 1939. He created the Commission on Latitudes of the Astronomic Council of the USSR Academy of Sciences and headed it until 1952. [Chapter 79.45]

Osawa, Yutaka (1927–1991)

Japanese earthquake engineer. His research was on the resistance of buildings to earthquakes, in particular the elastic-plastic behavior of buildings, soil–structure interactions, strong motion measurements, core-wall problems, and motions of tall buildings. Osawa was born in Tokyo and graduated in architecture from the University of Tokyo in 1950, where he also earned the D. Eng degree in 1957. At the Earthquake Research Institute, he began as a research associate in 1955, becoming an associate professor in 1961 and professor in 1965. He was acting director of the institute in 1971 and director from 1975 to 1977. He was secretary-general of the International Association of Earthquake Engineering from 1977 to 1988. [Miyamura (2003)]

Ostrovsky, Aleksey Emelyonovich (1907–1988)

Russian geophysicist who worked in seismic prospecting, measurement of seismic-wave velocities in boreholes, crustal structure, earthquake precursors, and the relation of surface contours to floods. Ostrovsky attended Leningrad Pedologic Institute and the Central-Scientific-Research-Geological-Prospecting Institute. Starting in 1934, he worked for the Seismological Institute of the USSR Academy of Sciences, rising to head-of-laboratory in 1947 and professor in 1970. [Chapter 79.45]

Otsuka, Yanosuke (1903–1950)

Japanese geologist who made significant contributions in characteristics of surface faults in the circum-Pacific area, effects of tsunamis, tectonics of crustal movements, molluscan taxonomy, paleoclimatology, mapping of Cenozoic strata, and the Tertiary history of the Japanese islands. Otsuka was born in Tokyo and graduated from Tokyo Imperial University in 1929. He entered the Earthquake Research Institute as an assistant in 1930, becoming an associate professor in 1939 and a professor in 1943. He published valuable papers on Tertiary crustal deformation in Japan, for which he obtained a D.Sc. degree from Tokyo Imperial University in 1939. He died prematurely at the age of 47. [Information submitted by Arata Sugimura to Miyamura (2003)]

P

Pakiser, Louis C., Jr. (1919–2001)

American geophysicist and a leader in government earthquake research programs. Pakiser was born in Denver, Colorado, and was educated at the Colorado School of Mines in geological engineering. After a few years in the petroleum industry he joined the US Geological Survey in Washington, D.C., and applied potential-field techniques to study the structure of the Earth's crust. In the 1950s, he transferred to the Survey's Denver office where he expanded the scope of crustal studies by incorporating active seismic techniques. In the early 1960s he led a large geophysical effort to resolve the crustal structure of the western United States, as a basis for improving the ability to identify underground nuclear explosions using seismic waves. After the great 1964 Alaska earthquake, he relocated with several colleagues to the Survey's offices in Menlo Park, California, where he established and led the first organizational unit in the Survey focused on earthquakes, the Office of Earthquake Research and Crustal Studies (internationally recognized as the National Center for Earthquake Research). He worked with Frank Press and others in laying the groundwork for the National Earthquake Hazards Reduction Act of 1977, which established the multi-agency National Earthquake Hazards Reduction Program. A visionary and forceful scientific leader, Pakiser was in a very real sense the father of the earthquake program within the USGS. He was a passionate supporter of young scientists and worked vigorously to attract minorities to Earth science careers. [*SRL* **73**:459–460, 2002]

Palazzo, Luigi (1861–1933)

Italian geophysicist who studied terrestrial magnetism and prepared charts of magnetic declination and inclination in Italy. Palazzo was born in Turin and graduated in physics from the University of Turin in 1884. He joined the Central Office of Meteorology and Geodynamics in Rome in 1888, becoming its director in 1901. He taught geophysics and meteorology at the University of Rome starting in 1897. In 1903 he became a director of the Italian Seismological Society. He served on the commission studying the 1905 Calabria earthquake. He was a member of the Accademia Nazionale dei Lincei, the Pontificia Academica dei Scienze Naturali of Catania, and vice president of the Italian Meteorological Society. From 1905 to 1907 he was president of the International Association of Seismology. A photo of Palazzo is included in Figure 1 of Chapter 81.3 of this Handbook. [Ferrari (2003)]

Palmieri, Luigi (1807–1896)

Italian physicist. In his work at the Vesuvian Observatory, he recognized the connection between volcanic eruptions and seismic activity, distinguishing between small earthquakes and harmonic tremor. He designed a sensitive three-component seismoscope that recorded the occurrence of unfelt earthquakes. Palmieri was born at Faicchio and graduated in physics, mathematics, and philosophy from the University of Naples. In 1847 he was appointed to the chair of logic and mathematics at Naples University and in 1860 to a newly created chair of terrestrial physics. He pursued research on telluric currents and atmospheric electricity all his life, inventing an electrometer. Beginning in 1850 he included earthquakes in his studies. In 1852 he began work at the Vesuvian Observatory, being appointed head of it in 1854. He was a member of the Accademia Nazionale dei Lincei. He was also known for his "sismografo elettromagnetico," a collection of seismoscopes. [Debus (1968), Ferrari (2003)]

Papastamatiou, Dimitris (1941–2000)

Greek seismic engineer whose research interests were seismic hazard, strong motion data acquisition and analysis, site effects, response of structures to earthquakes, and sesimotectonic studies. He also conducted field studies of earthquakes. Papastamatiou was born in Athens and graduated in 1964 in engineering from the National Technical University of Athens (NTUA). In 1971 he earned a Ph.D. in engineering seismology from Imperial College (London) and worked for six years for Dames and Moore Co. in London as a seismic engineer. From 1980 to 1981 he was director of Geognosis Ltd. of London. He also established his own consulting company, Delta Pi Associates, and served as a UNESCO consultant to the Institute of Earthquake Engineering Seismology at Scopje, Yugoslavia. In 1988 he was appointed assistant professor of Engineering Seismology at NTUA, advancing to associate professor in 1994 and professor in the year of his death. [Chapter 79.25]

Pasechnik, Ivan P. (1910–1988)

Russian seismologist. He worked on development of seismographs, Antarctic seismology (coauthor of the first Soviet map of Antarctic seismology), structure of the Earth, particularly of the core–mantle boundary, and absorption of seismic waves. He led research on distinguishing explosions from earthquakes, including the problem of earthquakes triggered by explosions. Pasechnik started his studies at Kiev Institute for Mining and Geology, receiving the M.S. degree at Dnepropetrovsk Mining Institute in 1935, the Ph.D. in physics and mathematics in 1951 at the Geophysical Institute of the USSR Academy of Sciences, and the D.Sc. at the academy's Institute of Physics of the Earth in 1970. He became an engineer for the All-Soviet-Union Bureau of Geophysical Prospecting in 1937, director of the Institute of Physics of the Earth in 1939, and head of its Special Seismology Laboratory from 1954 to 1988. [Chapter 79.45]

Pekeris, Chaim Leib (1908–1993)

Lithuanian applied mathematician and geophysicist. Pekeris contributed to several areas in geophysics, including free oscillations of the Earth, tides in the oceans, and international constitutions of the Earth. Pekeris and his associates calculated the free vibrations of some realistic Earth models in a paper, "Oscillations of the Earth" (*Proc. Roy. Soc.* A, **252**, 80–95, 1959, by Z.S. Alterman, H. Jarosch, and C.L. Pekeris). Following the 1960 Chilean earthquake, three teams reported the Earth's free oscillations, using the gravimeter, the strain seismometer, and long-period seismometers. The measured periods in all cases were close to Pekeris's predicted values (Slichter, 1967).

Born in Alytus, Lithuania, Pekeris immigrated to the United States and began his studies at the Massachusetts Institute of Technology (MIT). After receiving his D.Sc. degree in 1934, he stayed at MIT as a Rockefeller Fellow (1934–1936) and then as a faculty member (1937–1941). He was a member of the War Research Group at Columbia University during World War II. Pekeris was a member of the Institute for Advanced Study in Princeton (1946–1948), and in 1949 he settled in Israel as professor and Head of the Department of Applied Mathematics which he established at the Weizmann Institute of Science. During 1950–1970 he oversaw the production of the Institute's digital electronic computers, the WEIZAC and the GOLEM (among the most advanced in the world at that time), for computation needs of the Institute. [Chapter 79.31.2]

Perrey, Alexis (1807–1882)

French mathematician and natural scientist who compiled earthquake catalogs for many years. Perrey trained for holy orders at the college of Langres starting in 1823, but abandoned the clerical life in 1830 and maintained himself first as a tutor, then as a lecturer in mathematics at several royal colleges. In 1838 he was named deputy professor of pure mathematics at Dijon College, then professor of applied mathematics in 1847. Most of his work was on meteorology, astronomy, and mathematics. However, for 29 years (1843–1871) he compiled and published annual lists of earthquakes. He also studied the periodicity of earthquakes. [Davison, 1927]

Petrushevsky, Boris Abramovich (1908–1986)

Russian geologist. He graduated from Moscow State University in 1930 and worked for various geological organizations at Moscow, including 35 years at the Institute of Physics of the Earth of the USSR Academy of Sciences. His specialty was the relation of geology to seismotectonics, including the seismicity of intraplate areas. He studied how geosynclinal areas are transformed into platforms, particularly the evolution and tectonic structure of eastern Asia. [Chapter 79.45]

Poisson, Siméon-Denis (1781–1840)

French mathematician. Poisson's name is attached to a wide variety of ideas, e.g., Poisson's integral, Poisson's equation in potential theory, Poisson distribution (a probability distribution that characterizes discrete events occurring independently of one another in time; see Chapter 82 of this Handbook), and Poisson's ratio in elasticity (the absolute ratio of the transverse strain to the longitudinal strain; see Chapter 83, and Eq. (1.6) in Appendix 2 of this Handbook). Poisson was completely dedicated to mathematics, and was reported to have said: "Life is good for only two things: to study mathematics and to teach it." Poisson is best known in seismology for having predicted theoretically the existence of longitudinal and transverse elastic waves in 1828. Poisson was born in Pithiviers and began to study mathematics in 1798 at the Ecole Polytechnique, and was named deputy professor at the Ecole Polytechnique in 1802 and a professor in 1806. In 1808 Poisson also became an astronomer at the Bureau des Longitudes. In 1809 he added another appointment, the chair of mechanics in the newly opened Faculté des Sciences. His two important papers are "Memoire sur lequilibre et le mouvement des corps elastique" (*Mem. Acad. Sci. Paris*, **8**, 357–570, 1929), and "Memoire sur la propagation due mouvement dans les milieux elastiques" (*Mem. Acad. Sci. Paris*, **10**, 578–605, 1931). [James and James (1976); Britannica (2003)]

Polyakov, Svyatoslav Vasilyevitch (1918–1992)

Russian engineer whose work focused on anti-seismic design, especially to large precast reinforced concrete panel buildings. He was a coauthor of *Fundamentals of Seismic Design* and four other books on earthquake engineering. He helped to create a large shaking table to test structures and to design the USSR Seismic Building Code. He participated in the field study and analysis of large USSR earthquakes. He was a professor and a D.Sci. working at the Central Research Institute of Structures in Moscow. [Rzhevsky (2002)]

Popov, Egor P. (1913–2001)

American structural and earthquake engineer. Egor Popov was a leading researcher on the seismic response of both reinforced concrete and steel structures and the author of a famous textbook, *Mechanics of Materials*. Born in Kiev, Russia, Professor Popov came to the United States during the Russian revolution and studied engineering at the University of California, Berkeley, the Massachusetts Institute of Technology, the California Institute of Technology, and Stanford. He joined the faculty at the University of California, Berkeley, in 1946 and was active in teaching and research for over 50 years. He is credited with many of the recent innovations in the seismic design of steel structures. His work saw application in the San Francisco–Oakland Bay Bridge, the San Francisco Museum of Modern Art, the Alaska Pipeline, and in many braced steel-frame structures constructed in seismic areas. He was also well known as a gifted and devoted teacher. He was widely honored for his accomplishments in both teaching and research, including, in 1977, election to the National Academy of Engineering. [*EERI Newsletter*, **35**(6), 4, 2001.]

R

Rahimov, Shamo Suleiman (1931–1965)

Azerbaijani seismologist. Rahimov was born in Baku, graduated in physics from Moscow State University in 1953, and started graduate study at the Institute of Physics of the Earth. He defended his candidacy thesis in 1958 on dispersion of Rayleigh waves. He developed a method of determining direction to the epicenter using surface waves and identified a new series of Rayleigh waves. He used seismic data from large Azerbiajan earthquakes to study the earth's core. He classified earthquakes as global and local. [Chapter 79.7]

Ramirez, Jesus Emilio, S.J. (1904–1983)

Colombian seismologist whose main research topic was the seismicity and seismotectonics of Colombia. Ramirez was born in Yolombo and obtained an M.S. degree from Boston College in 1931 and a Ph.D. in geophysics from Saint Louis University in 1939. His dissertation research, conducted under the direction of Macelwane, concerned a method to locate hurricanes using microseisms detected across a tripartite array of seismic stations. On returning to Colombia in 1941 he became the founder of the Jesuit-run Geophysical Institute of the Colombian Andes. He published several monographic studies of the more important earthquakes in Colombia. He was a member of the Colombian Academy of Exact Sciences, a corresponding member of the Royal Academy of Exact, Physical, and Natural Sciences of Madrid, and a member of many other scientific societies. [Stauder (1999a)]

Rayleigh, John William Strutt, Lord (1842–1919)

British physicist. In seismology, Lord Rayleigh is best known for his discovery of the third kind of seismic waves that bears his name in his paper, "On waves propagated along the plane surface of an elastic solid" (*Proc. London Math. Soc. (Series 1)*, **17**: 4–11, 1885). He contributed to many areas in physics, including acoustics, optics, and the discovery of argon, for which he won the Nobel prize in physics in 1904. In addition to the Rayleigh wave, he is also associated with, e.g., Rayleigh number in fluid mechanics, and Rayleigh scattering of electromagnetic radiation by particles. Rayleigh was born near Maldon, England, and graduated in mathematics from Cambridge University in 1873. He largely worked in his own laboratory, being a titled landowner for whom class norms weighed against a scientific career. In 1879 he succeeded Maxwell as the Cavendish professor of experimental physics at Cambridge, but only stayed for 5 years. [*Obit. Proc. Roy. Soc. Lond. A*, **98**:i–xix, 1921; Britannica (2003)]

Rebeur-Paschwitz, Ernst von (1861–1895)

German astronomer and geophysicist. In 1885 he built a tiltmeter, based on Zöllner's horizontal pendulum, that is considered the first instrument to actually record a teleseism (in 1889). He recognized that the ground motion he recorded consisted of two types of waves: body waves traveling through the deep parts of the Earth and surface waves traveling along the Earth's surface. In 1895 he proposed the foundation of an international network of earthquake observatories, and a central institute to collect data on seismic observations, but died prematurely at the age of 34. Rebeur-Paschwitz obtained his doctor's degree at the University of Berlin in 1893 and became a privat-dozent at the University of Halle. [Gerland (1898); Davison (1895); Davison (1927); Schweitzer (2003)]

Reid, Harry Fielding (1859–1944)

American geophysicist who originated the elastic rebound theory of earthquake occurrence. Reid was born in Baltimore and received the A.B. (1880) and Ph.D. in physics from Johns Hopkins University. He was professor of mathematics at Case University from 1886 to 1889, then moved in 1890 to Johns Hopkins University, where he became professor of geological physics in 1901. He was in charge of earthquake records for the US Geological Survey. He represented the United States at International Seismological Association meetings starting in 1906. He was a member of Lawson's California State Earthquake Investigation Commission and author of Volume 2 of its report on the 1906 San Francisco earthquake. He was also an authority on glaciers. He published on the theory of isostasy and the theory of the seismograph. He was president of the Seismological Society of America from 1912 to 1914 and of the American Geophysical Union from 1924 to 1926. [*NAS Biog. Mem.* **25**:1–12, 1947]

Repetti, William C., S.J. (1884–1966)

American seismologist who focused on deep Earth studies and Philippine earthquakes. Born in Washington, D.C., Repetti taught physics and mathematics at Fordham University and was director of the Jesuit-run seismographic station there from 1914 to 1918. In 1926 he studied geophysics at St. Louis University under Macelwane, receiving his Ph.D. in 1928. His dissertation concerned a study of the interior of the Earth from the travel times of body waves. He inferred the existence of several discontinuities, including a prominent one at a depth of 600 km. In 1928 he joined the staff of Manila Observatory in the Philippines, where he undertook a systematic study of the epicenters of Philippine earthquakes. Unfortunately most of the records of the observatory were destroyed during the war, but Repetti retrieved the listing of Philippine earthquakes from 1589 to 1899 (*Bull. Seism. Soc. Am.*, **36**:133–322, 1946). [Stauder (1999a)]

Rethy, Antal (1879–1975)

Hungarian geophysicist who published lists of earthquakes of Hungary and the Carpathian Basin, produced a map of Hungarian epicenters, and studied earthquake, meteorological, and climatological disasters of Hungary. Rethy was born in Budapest, receiving the Ph.D. degree in seismology from Ferenc Jozsef University of Science in 1912 and the D.Sc. in technical sciences from the Hungarian Academy of Sciences in 1927. He was a research fellow at the National Institute of Meteorology and Geomagnetism (NIMGM) in Budapest from 1900 to 1903, a privat-docent at the Budapest Economic University in 1923, director of the Turkish Meteorological Service from 1925 to 1927, deputy director of NIMGM from 1933 to 1935 and director from 1935 to 1948. He was also a professor at the Technical University, Budapest, in 1943. He served as general secretary of the Hungarian Meteorological Society in 1935. [Chapter 79.26]

Riccó, Annibale (1844–1919)

Italian geophysicist who studied volcanic activity at Stromboli and Etna. Riccó graduated in 1868 in natural sciences from the University of Modena and received a diploma in engineering in Milan. He taught physics, geodesy, and mineralogy at Modena. In 1878 he started teaching physics at the Engineering School of Naples. In 1870 he moved to Catania where he taught astrophysics and organized and managed the observatories of Catania and Etna. He was director of the geodynamical service of Sicily and the Sicilian Islands. He served on the commissions studying the 1894 and 1905 Calabrian earthquakes. He was a member of the Accademia Nazionale dei Lincei. [Ferrari (2003)]

Richter, Charles Francis (1900–1985)

American seismologist. He is noted for his studies of seismicity, especially that of southern California. He originated the magnitude scale for measuring the size of earthquakes, and he wrote the widely used textbook, *Elementary Seismology*, and (with Gutenberg) the reference work, *Seismicity of the Earth*. Richter was born in Butler County, Ohio, and received the A.B. degree from Stanford University in 1920 and the Ph.D. from the California Institute of Technology in 1928 in physics. He was hired in 1927 by the Carnegie Institution as one of the first employees of what became in 1937 the Caltech Seismological Laboratory. He was appointed to the Caltech faculty in 1937, becoming professor in 1952. Using Anderson's and Benioff's short-period seismometers, he (with Gutenberg) greatly improved the accuracy of earthquake time measurements, leading to sharper delineation of their locations and of the structure and properties of the Earth's interior. He published many studies of individual earthquakes and blasts. [*BSSA* **67**:1243–1244, 1977; *BSSA* **77**: 234–237, 1987; Goodstein (1991)]

Riznichenko, Yuri Vladimirovich (1911–1981)

Russian seismologist who made contributions in basic and applied seismology. He studied the internal structure of the Earth, seismicity, and physics of the earthquake source. He developed concepts and methods of determination of seismic activity, maximum possible earthquakes, shakability, and prediction of long-term seismic hazard. Riznichenko served in the army from 1935 to 1938, then studied at the Institute of Theoretical Geophysics under Gamburtsev. From 1941 to 1943 he worked in the Bashkirian oil expedition. From 1944 to 1950 he was a lecturer in seismic prospecting in the Moscow Geological Prospecting Institute, becoming professor in 1947. In 1955 he joined the Institute of Physics of the Earth, USSR Academy of Science, serving as head of the Department of Physics of Earthquakes until 1969, then head of the institute's laboratory. He became a corresponding member of the USSR Academy of Science in 1958. In 1969 he organized and headed the academy's Special Commission of the Joint Council on Seismology and Earthquake-proof Construction. This work led to the book, *Seismic Shakability of the USSR Territory*, and an atlas of maps of seismic shakability in 1979. Beginning in the 1970s he worked on the theory of seismic flow of rocks. He was a principal author of two monographs, *Selected Papers on Problems of Seismology* and *Seismology Prospecting in Layered Media* (1985). In 1961–1963 he took part in the Geneva meetings on prohibition of nuclear tests; from 1957 to 1961 he was first Soviet scientist, vice-president of IASPEI, and participated in the Pugwash movement of scientists for peace. From 1962 to 1981 he was the main editor of Izvestia Akademii Nauk. Fizika Zemli. He received two Orders of Red Banner of Labor. [Chapter 79.45]

Rosenblueth, Emilio (1926–1994)

Mexican engineer whose principal interest was earthquake engineering, especially hazard estimation. He was also one of the authors of the Mexico City Building Code. He worked on problems of tall buildings, dams, nuclear power plants, and soil dynamics. He had a special concern with the problems of ethics of legal regulations in relation to construction costs. Rosenblueth was born in Mexico City, received a B.Eng. from the National University of Mexico (UNAM) in 1948 and a Ph.D. from the University of Illinois, in 1951. He was a Professor at UNAM from 1955 to 1994, also serving as director of UNAM's Institute of Engineering from 1959 to 1966, and dean of research from 1966 to 1970. He was Undersecretary of Education of Mexico from 1977 to 1982. He was a founding member and a president of the International Association for Earthquake Engineering and an advisor to UNESCO and the Organization of American States. He authored, jointly with N. Newmark, *Fundamentals of Earthquake Engineering*. He was honored with the revered Mexican title of National Researcher. [Chapter 79.38]

Rosenthal, Elmar (1873–1919)

Estonian seismologist who worked for many years at the Central Bureau of the International Seismological Association (ISA) in Strasbourg. Here he was for many years the editor of the bulletins for instrumental earthquake observations published by ISA [Chapter 88; Schweitzer (2003)]

Rosova, Evdokiya Alexandrovna (1899–1971)

Russian seismologist. She graduated in theoretical physics from Leningrad State University in 1924 and started as a probationer at the Physics-Mathematics Institute of the USSR Academy of Sciences. In 1926 she was sent to Sverdlosk to head its seismic station and in 1928 organized the seismic station at Andizhan. Beginning in 1930 she worked for the Seismological Institute of the USSR Academy of Sciences, investigating the physical properties and structure of the Earth's crust based on analysis of seismograms. [Chapter 79.45]

Rothé, Edmond (1873–1942)

French physicist and seismologist. Born in Paris, Edmond Rothé earned a bachelor's degree in mathematics in 1895, another in physics in 1896, and the Ph.D. in 1899, all at the Sorbonne. He was appointed assistant professor at Grenoble in 1903 and moved to Nancy in 1905, where he taught physics and worked on a variety of projects including radio transmission and weather forecasting. In 1919 he was appointed professor of geophysics at Strasbourg, took charge of the Institut de Physique du Globe, and became head of the Central Seismological Bureau of France. In 1922 he was appointed director of the International Bureau of Seismology and secretary general of the Seismological Section of the IUGG. In 1929 he became dean of Strasbourg's science faculty. Between 1929 and 1942 he published 218 notes, scientific papers, and books. [Cara (2003)]

Rothé, Jean-Pierre (1906–1991)

French seismologist. Jean-Pierre Rothé was born in Nancy and received a diploma in physical sciences at Strasbourg in 1928 and the Ph.D. at Paris in 1937. He was appointed assistant professor at Strasbourg in 1928 in charge of weather forecasting and climatological studies. He spent 1932–1933 in Greenland on a geological expedition and 1934–1935 at the Institut de Physique du Globe in Paris. During this time his research was mainly on geomagnetism, meteorology, and the distribution of earthquakes. In 1942, on the death of his father, Edmond Rothé, he took over as secretary general of the International Association of Seismology, earning the position outright in 1948. In 1945, he became professor of sciences at Strasbourg and director of the Institut de Physique du Globe there. He was also director of the French Central Bureau of Seismology. In 1969 he published *Seismicity of the Earth, 1953–1965* (Unesco, Paris), a continuation of Gutenberg and Richter's work on the subject. In it he gave particular attention to the distribution of epicenters on the mid-ocean ridges. He also studied a 1954 deep earthquake south of Spain and worked on induced earthquakes. In 1981 he published "Fifty Years of History of the International Association of Seismology" (*Bull. Seism. Soc. Am.* **71**:905–923, 1981). [Cara (2003)]

Rubey, William Walden (1898–1974)

American geologist noted for his work on chemical differentiation of the Earth, origin of sea water and the atmosphere, induced seismicity, mechanics of faulting, and (with Hubbert) the role of fluid pressure on the fracture of rocks. Born in Moberley, Missouri, Rubey earned an A.B. from the University of Missouri in 1920. He was associated with the US Geological Survey throughout his life, starting in 1922. He was also an instructor of geology at Yale University (1922–1924), professor of geology and geophysics at the University of California, Los Angeles (1960–1966), and the first director of the Lunar Sciences Institute (1968–1971). He was president of the Geological Society of America in 1949–1950 and its Penrose Medalist in 1963, chairman of the National Research Council's Division of Geology and Geography in 1943–1946, and council chairman in 1951–1954. He served the National Science Foundation in the 1950s and 1960s and was a member of the US National Academy of Science. [*NAS Biog. Mem.* **49**:204–223, 1977; Rubey (1974)]

Rudolph, Emil (1854–1915)

German geographer and seismologist. Rudolph was co-worker with Gerland at the University and at the ISA Central Bureau, both in Strasbourg. There he edited the proceedings of the first two international conferences on seismology, worked on earthquake geography, and compiled and published in parallel with Milne the earliest regional and global earthquake catalogues with instrumental and macroseismic observations. A photo of Rudolf is included in Figure 3 of Chapter 81.3 of this Handbook. [Sapper (1915); Schweitzer (2003)]

Rudzki, Maurycy Pius (1862–1916)

Polish geophysicist and astronomer whose theoretical research was on mathematical models, elastic wave propagation, seismic-wave anisotropy, focal depth, reducing gravity measurements, figure of the Earth, thermal history of the Earth, and age of the Earth. Rudzki was born in Uhrynkowice (then Austria) and attended college in Warsaw, Kamieniec Podolski, the University of Lvov, and Kharkov, Ukraine. He received the Ph.D. in geology from the University of Vienna in 1886, then started in 1890 as a private docent at the University of Odessa. In 1895 he was appointed to the newly created chair of mathematical geophysics and meteorology at the Jagiellonian University in Cracow. In 1902 he became director of its astronomical observatory, to which he added a seismic station. In 1902 he became full professor and a member of the Academy of Sciences. He wrote the books *Physics of the Earth* (published in both Polish and German) and *Theoretical Astronomy* (in Polish), and many papers. [Chapter 79.43]

Rykunov, Lev Nikolaevich (1928–1999)

Russian seismologist of wide-ranging interests whose research included P waves diffracted at the Earth's core, microseisms, physical modeling, rigidity of the core, surface-wave propagation and heterogeneity, and deep sounding. He designed portable autonomous seismographs for mobile research sites. He studied the nodal points of Earth stress and correlated the patterns of global asymmetry and surface deformation with deep structures reaching to the Earth's core. He is noted for the large number of students that he mentored. Rykunov graduated in physics from Moscow State University in 1951 and received the doctorate in 1966. He remained at the university all his life, rising to professor and head of the Cathedra of Moscow State University. He became a Corresponding Member of the USSR Academy of Sciences in 1984 and an Academician in 1998. [Chapter 79.45]

S

Saderra-Masó, Miguel, S.J. (1865–1939)

Spanish seismologist who published the first research on Philippine seismology. Saderra-Masó arrived in Manila in 1890 and took charge of the Seismological Department of the Jesuit-run Manila Observatory in the following year. He dedicated himself to the study of the seismicity of the Philippines, interpreting it in terms of seismotectonic lines. In 1895 he published the first work on Philippine seismology (*La Sismología en Filipinas*, Imp. Ramirez, Manila). Apart from a brief return to Spain in 1896 to 1901, he held his post until he returned to Spain for health reasons in the 1930s. [Stauder (1999a)]

Sadovsky, Michail Aleksandrovich (1904–1994)

Russian geophysicist who studied the mechanics of explosions, seismic zoning, and earthquake prediction. Sadovsky graduated in physics and mechanics from the Leningrad Polytechnical Institute in 1928 and undertook gravity research at the Institute of Applied Geophysics of the USSR Academy of Sciences. He soon shifted his interests to seismology at the Academy's Institute of Seismology, in 1930 beginning research on the mechanics of explosions. He developed safety rules for construction and mining explosive work and methods of modeling explosions to control rock mass movements. After World War II he worked in the

Special Sector of the Institute of Chemical Physics of the USSR Academy of Sciences on nuclear explosions and monitoring. He was elected a director of the Institute of Physics of the Earth and after 1989 its honorary director. He was head of the Soviet experts at the Geneva talks on prohibition of nuclear weapons. He also worked on seismic zoning and earthquake prediction. He developed a block-heirarchial model of the Earth based on dissipation of energy. He received awards as Hero of Soviet Labor, five Orders of Lenin, the Order of the October Revolution, three Orders of the Red Banner of Labor, the Badge of Honor, many medals, and a Lenin prize and four State prizes for his work. [Chapter 79.45]

Šalamoun, Bedřich (1880–?)

Czech cartographer and geophysicist. Šalamoun was born in Prague and studied mathematics and projective geometry at Charles University and Technical University in Prague, habilitating in cartography at Charles University in 1922. In 1931 he was named professor of cartography with a duty to read also geophysics. His scientific papers were devoted to problems of cartography, geodesy, and general geophysics. His contributions to graphical methods of statistical evaluation of observations appeared to be very effective and were many times used for processing sets of geophysical data. [Chapter 79.17]

Sano, Toshikata (1880–1956)

Japanese earthquake engineer who formulated the structural-engineering system used in Japan. He developed a seismic coefficient (the Sano Seismic Scale) to evaluate the earthquake resistance of structures. Sano was born in Yamagata, graduated in architecture from Tokyo Imperial University in 1903, and received the D. Eng. degree in 1915 for a thesis on earthquake-proof construction. He started as a lecturer at Tokyo Imperial University, rising to assistant professor in 1906 and professor in 1918. After the 1923 Kwanto (Tokyo) earthquake he served as building division chief of the Tokyo City Building Bureau. From 1929 to 1939 he was a professor at Nihon University and faculty dean while also serving as lecturer at Tokyo Institute of Technology. He was elected a member of the Japan Academy in 1950. [Miyamura (2003)]

Santo, Tetsuo (1919–1997)

Japanese seismologist. He studied filter design for seismic recording systems and researched many aspects of surface waves, including their dispersion and use to determine crustal structure. He worked on the relation of microseisms to meteorological conditions and on global seismicity. He was born Tetsuo Akima in Saitama, changing his last name to Santo in 1952. He graduated in physics from Hokkaido Imperial University in 1944 and in 1946 joined the Earthquake Research Institute. In 1963 Santo moved to the Japanese International Institute of Seismology and Earthquake Engineering, serving as its director from 1971 to 1976. He was professor at Kobe University from 1976 to 1983. [Miyamura (2003)]

Sassa, Kenzo (1900–1981)

Japanese seismologist and geologist. He studied the distribution of deep-focus earthquakes, used microtremors to predict the eruption of Aso volcano, studied the tilting and deformation of the ground and crustal movements before earthquakes, and conducted research on landslides. Sassa was born in Ichinomiya, graduated in geophysics from Kyoto Imperial University in 1925. He became a lecturer there the following year, associate professor in 1929, and professor in 1945. He was dean of the faculty of science from 1957 to 1959 and director of the Disaster Prevention Research Institute from 1961 to 1963. From 1963 to 1974 he was president of Osaka Institute of Technology. [Miyamura (2003)]

Satô, Yasuo (1918–1996)

Japanese seismologist. He related observations of the seismic waves from the 1960 Chilean earthquake to oscillations of an elastic sphere, and studied the attenuation and dispersion of surface waves. He published *Theory of Elastic Waves* in 1978. He studied the distribution of intensity from large earthquakes using mailed questionaires, developing a twelve-level intensity scale. Satô was born in Dalian, China. He graduated in seismology from Tokyo Imperial University in 1941 and became an assistant at its Seismological Institute. In 1945 he moved to the Earthquake Research Institute, where he became associate professor in 1948 and professor in 1961. He received the D.Sc. degree for mathematical studies of surface waves using Fourier transforms in 1956. He was professor at Kagoshima University from 1975 to 1984. [Miyamura (2003)]

Savarensky, Evgeny Fedorovich (1911–1980)

Russian seismologist. He carried out research on the physical properties of the Earth, deep-focus earthquakes, surface waves and forecasting of earthquakes and tsunamis. Savarensky graduated in theoretical physics from Moscow University in 1930 and started work at the Moscow Institute of Geological Prospecting. In 1934 he moved to the Seismological Institute of the USSR Academy of Sciences, where he founded the Moscow Central Seismic Station in 1936. From 1948 to 1956 he worked for the Geophysical Institute of the Academy. In 1945 he started lecturing on seismology at Moscow University, and he completed his doctorate in 1949 on the angle of emergence of seismic waves. He became professor in 1950. In 1956 he became chairman of the Council of Seismology of the Academy of Sciences and in 1958 vice-chairman of the European Seismological Commission. In 1963 he became director of the General Seismology Department of the Institute of Physics of the Earth, editor of the journal "Physics of the Earth," and vice-president of the Tsunami Committee of IUGG. He was a member of the Highest Certification Commission of the USSR from 1965 to 1975. He became a corresponding member of the USSR Academy of Sciences in 1966. He was elected chairman of the European Seismological Commission in 1970, chairman of the IASPEI Commission on Earthquake Prediction in 1971, and vice-president of the European Assembly on Engineering Seismology in 1978. He also became director of the Seismological and Geoacoustical Section of the Geological Department of Moscow University. He edited or wrote such books as *Elements of Seismology and Seismometry, Manual on Making and Processing of Observations on Seismic Stations of the USSR, Earthquakes of the USSR, An Atlas of Earthquakes of the USSR*, and *Seismic Waves*. [Chapter 79.45]

Savinov, Oleg Alexandrovich (1910–1992)

Russian engineer. Throughout his career his interests were in soil dynamics, foundations, vibrotechnics, earthquake engineering of hydrotechnical structures, and seismic isolation. After graduating from the Leningrad Institute of Civil Engineering, Savinov conducted research on vibrations at the Leningrad Institute of Structures. From 1939 to 1942 he managed the construction of military and industrial facilities in Leningrad. From 1942 to 1944 he worked at evacuating factories to the Urals. From 1944 to 1946 he worked on construction of port facilities at Leningrad, becoming head of the Institute of Foundation Designing and lecturing at the Institute of Civil Engineering and the Institute of Means of Communication. In 1946 he became a laboratory head at the Leningrad State Research Institute of Bases and Underground Structures, working on foundation problems and vibration techniques. In 1955 he received the Ph.D. degree for a thesis entitled "Foundations under Dynamic Loads." In 1961 he organized and headed the Department of Dynamics and Earthquake Engineering of the All-Union Research Institute of Hydrotechnic, where he studied dams and nuclear power plants. In 1976 he was elected chairman of the Commission on Earthquake Resistance in Hydroconstructions. He was head of the Soil and Foundation Department of the Leningard Research Society of Building Industry and the Engineering Department of the Inter-Departmental Board on Seismology and Earthquake Engineering at the Presidium of the USSR Academy of Sciences. [Uzdin (2003)]

Schuster, Arthur (1851–1934)

German-born British physicist and seismologist who studied earthquake periodicity. Born in Frankfurt, Germany, in 1851, Schuster earned a doctorate from the University of Heidelberg in 1873 and shortly thereafter became a British citizen. In 1882 he joined the University of Manchester as a professor of applied mathematics, then professor of physics from 1889 to 1907. From 1907 to 1911 he was president of the International Association of Seismology. A photo of Schuster is included in Figure 1 of Chapter 81.3 of this Handbook. [*Obit. Not. Fell. Roy. Soc. Lond.* **1**:409–423, 1935]

Seed, H. Bolton (1922–1989)

Amercian soil mechanist and earthquake engineer. Harry Bolton Seed was born in Bolton, England, on August 19, 1922. He completed his undergraduate studies at the University of London, where he received a B.Sc. in civil engineering in 1944 and a Ph.D. in structural engineering in 1947. Following two years as assistant lecturer at Kings College, he came to the United States to study soil mechanics at Harvard University under the tutelage of Karl Terzaghi and Arthur Casagrande, receiving his S.M. degree in 1948. In 1950 Professor Seed joined the civil engineering faculty at the University of California, where he spent the remainder of his career as an engineering educator, researcher in geotechnical engineering, and consultant to numerous companies and public agencies.

A leading and highly influential figure in the important new area of geotechnical earthquake engineering, he contributed extensively to the understanding of soil liquefaction, the dynamic response of earth dams, soil–structure interaction, the response of piles and site effects on earthquake motions. Through his extensive consulting practice, he was involved in the design of over 100 dams, 20 nuclear power plants, and many other important structures. As a professor, he had 50 Ph.D. students who continue to influence broadly the field of soil mechanics. He received many national and international honors for his work. [*US Nat. Acad. Eng., Mem. Tributes*, **5**, 247–252, 1992].

Sekhniashvili, Emil Aleksey (1924–1992)

Georgian engineer whose work focused on the theory of vibrations, including the natural frequencies of beams and the stability of buildings under seismic stress. He was particularly concerned with reinforced concrete structures. Sekhniashvili graduated from Tbilisi Institute of Railway Transport Engineers (TIRTE) in 1939. He worked for the Institute of Structural Mechanics and Earthquake Engineering (ISMEE) of the Georgian Academy of Sciences from 1950 to 1952, on government structures from 1952 to 1956, then for TIRTE from 1956 to 1968, serving as head of the Department of Institute Building Engineers from 1961 to 1968. From 1968 to 1972 he was assistant director of the Tbitzniep Institute. From 1972 to 1988 he again worked on government structures. From 1988 to 1990 he was director of the ISMEE. He became a professor and a corresponding member of the Georgian Academy of Sciences in 1974. [Chapter 80.8]

Sekiya, Seikei (1854–1896)

Early Japanese seismologist. He became the first professor of seismology at Tokyo Imperial University in 1886. He conducted field investigations of the 1888 Bandai-san Volcano and the 1889 Kumamoto earthquake, setting up seismometers in the field to measure ground vibrations including aftershocks. He greatly expanded the number of seismic stations in Japan, and developed a four-level earthquake intensity scale. He was one of the first to demonstrate the complex three-dimensional ground motions of earthquakes, and showed that ground motions at the surface often exceed motions at depth. He worked with Omori on compiling the first catalog of historic Japanese earthquakes. Sekiya was born in Ogaki. He studied at the Daigaku-Nanko and Tokyo Kaisei-gakko from 1870 to 1876, and at University College in London, England, from 1876 to 1877. In 1880 he joined the Seismological Observatory of Tokyo University while serving concurrently as chief of the Seismological Section of the Geographical Bureau of the Japanese Ministry of Internal Affairs. He was secretary of the Earthquake Investigation Committee from 1893 to 1896. [Miyamura (2003)]

Serpieri, Alessandro (1823–1885)

Italian physicist who studied the 1873 Urbino, 1875 Rimini, and 1883 Ischia earthquakes. He circulated questionnaires to gather information on the shocks, investigated the ability of animals to anticipate them and electric and magnetic phenomena that preceded them. Serpieri was born in San Giovannia Marignano and received a religious education. He became a teacher and director of the meteorological observatory at Urbino. He worked in pure physics, astronomy, meteorology, and seismology. [Ferrari (2003)]

Sezawa, Katsutada (1895–1944)

Japanese seismologist. He conducted research on scattering and on the effect of viscosity on absorption. His book, *Theory of Vibrations*, was published in 1932. He received the Imperial Award of the Imperial Academy in 1931 for studies of the generation and propagation of seismic waves. He noted that Rayleigh waves could be either retrograde or prograde. Sezawa was born in Yamaguchi, graduated in naval architecture from Tokyo Imperial University in 1921, and was appointed assistant professor there in 1922. He received the D.Eng. degree in 1924 and the D.Sc. in 1939. He joined the Earthquake Research Institute in 1925, was appointed professor there in 1928, and was its director from 1942 to 1944. He was admitted to the Japan Academy in 1943. [Miyamura (2003)]

Shebalin, Nikolai Vissarionovich (1927–1996)

Russian seismologist whose research was on strong earthquakes in the USSR and the Balkan area, with particular concern for their recurrence and on engineering seismology. Shebalin graduated in physics from Moscow State University in 1950. While still a student he started work at Moscow Seismic Station and the Institute of Physics of the Earth. He helped to install many seismic stations in the USSR and worked on interpreting seismograms. He completed his Ph.D. in 1956. He created the Laboratory of Strong Earthquakes in 1969. He co-edited the *New Catalog of Strong Earthquakes in the USSR Territory from Ancient Times to 1975*. He was for many years chairman of the Seismology Section, National Geophysical Committee of the USSR, and was a vice-president of IUGG. He was a titular member of the Russian Academy of Natural Sciences. [Chapter 79.45]

Shida, Toshi (1876–1936)

Japanese seismologist. He was first to point out (in 1909) the quadrature pattern of earthquake first motions. He conducted research on deep earthquakes and the physics of the Earth's deep interior. He received the Imperial Award of the Imperial Academy in 1929 for a series of papers on Earth tides and the rigidity of the Earth. Shida was born in Sakura and graduated in physics from Tokyo Imperial University in 1901. He was appointed assistant professor at Kyoto University in 1909 and professor in 1918. He established the University's Kamigamo Observatory and the Beppu Geophysical Institute. He also created the Aso Volcanic Observatory in 1928 and the Abuyama Seismological Observatory in 1931. [Miyamura (2003)]

Sieberg, August (1875–1945)

German seismologist who mainly worked on earthquake catalogues, geographical distribution of earthquakes, tectonics, and analysis of macroseismic data. In 1912, he introduced a 12-grade intensity scale for macroseismic observations and in 1939 he published the first catalogue for historical earthquakes in Germany and surrounding areas. [Krumbach (1949); Schweitzer (2003)]

Silvestri, Orazio (1835–1890)

Italian volcanologist. He was born in Florence, graduated from the University of Pisa, and then taught natural history at a local grammar school. In 1863 he moved to the University of Catania and began studying volcanology and seismicity. In 1874 he moved to the Industrial Museum of Turin, where he served as professor of applied chemistry. In 1878 he returned to Catania as professor of mineralogy, geology, geophysics, and volcanology. He reorganized the university's volcanological institute. He was founding director of the Geodynamic Service of Etna. He was general secretary of the Accademia Gioenia di Scienze Naturali of Catania, a member of the Council of the Italian Meteorological Society, and served on the Royal Geological Committee. [Ferrari (2003)]

Slichter, Louis Byrne (1896–1978)

American geophysicist who developed tiltmeters and portable seismographs and studied the solid-earth tides, the gravity field, and free oscillations; he was also involved in Antarctic research. Slichter was born in Madison, Wisconsin, and earned a B.A. (1917), A.M. (1920), and Ph.D. (1922) in physics from the University of Wisconsin. He worked for the Submarine Signal Co. from 1922 to 1924, then in geophysical prospecting for ores from 1924 to 1931. He was a professor at Massachusetts Institute of Technology from 1931 to 1945, moving to the University of Wisconsin for a year, then in 1947 to the University of California at Los Angeles, where he was professor of geophysics and the first director of the Institute of Geophysics. He developed methods of electromagnetic prospecting for ores. In his studies on the thermal history of the Earth, he was one of the first to take into account the heat generated by radioactivity. He received the President's Certificate of Merit in 1947, the Jackling Award of the Institure of Mining, Metallurgical and Petroleum Engineers in 1960, and was Bowie Medalist of the American Geophysical Union in 1966. He received an honorary D.Sc. from the University of Wisconsin in 1967 and an LL.D. from the University of California at Los Angeles in 1969. He was a member of the National Academy of Science. [*BSSA* **69**:655–657, 1979]

Solovi'ev, Sergey Leonidovich (1930–1994)

Russian geophysicist who made significant contributions to the science of tsunamis. Solovi'ev graduated from Leningrad State University in 1953, and received the candidacy degree in 1956 and the doctorate in physics and mathematics in 1971 from the Geophysical Institute of the USSR Academy of Sciences. From 1956 to 1961 he worked at the Institute of Physics of the Earth and was secretary of its Council on Seismology. In 1957 he coauthored the *Atlas of Earthquakes of the USSR*. From 1961 to 1968 he was head of the seismology department in the Sakhalin Complex Scientific Research Institute, where from 1965 on he was scientific deputy director and from 1971 to 1977 director. He helped to create the tsunami warning system for the Pacific coast and headed this for fifteen years. He worked on tsunami prediction, especially their height, and helped to compile catalogs of Pacific tsunamis as well as a map of their sources and heights. He developed a fleet of vessels for studying the ocean floor along Russia's Pacific coast. He studied the seismicity and geologic structure of Sakhalin Island and the adjoining sea. In 1977–1978 he was chairman of the Joint Council on Seismology and Earthquake-proof Construction of the USSR Academy of Sciences. From 1978 to 1984 he was head of the seismology laboratory of the Institute of Oceanography of the Russian Academy of Sciences. He helped to develop ocean-bottom seismometers and studied the seismicity of oceanic areas. From 1967 to 1979 he was vice-chairman and chairman of the IUGG Committee on Tsunamis. He authored over 400 scientific papers and 12 monographs. He was elected a corresponding member of the USSR Academy of Sciences in 1972 and a member of the Russian Academy in 1991. [Chapter 79.45]

Somigliana, Carlo (1860–1955)

Italian physicist. He worked on statics, elastic dynamics and potential theory, and derived the fundamental equations of elastostatics known as Somigliana's tensor (see Chapter 8). He worked on the theory of propagation of seismic waves and on gravimetry. Somigliana graduated from the Scuola Normale Superiore of Pisa in 1881. He held the chair of mathematical physics at Turin from 1905 to 1935. He was a member of the Accademia Nazionale dei Lincei. [Ferrari (2003)]

Somville, Oscar (1880–1980)

Belgian seismologist. Somville earned a Ph.D. in physical and mathematical sciences in 1902 at the Catholic University of Louvain, and at the Royal Observatory of Belgium was appointed an assistant in 1903, an adjunct astronomer in 1905, and astronomer in 1909. He was head of the Observatory's Seismological Division until his retirement in 1945, and secretary of the National Committee for Geology and Geophysics from 1926 to 1945. He was responsible for maintenance of the Observatory's instruments, analyzing the seismograms, and printing of the bulletins. He published articles on the characteristics of the seismometers and on individual earthquakes. [Chapter 79.8]

Spencer, Carl P. (1955–1999)

English geophysicist best known for his work on ray tracing techniques (particularly the ray bending method), migration, inverse methods (particularly the partitioned matrix method for the joint inversion for earthquake location and velocity structure), extensions of ray theory to the Kirchhoff surface integral method, and extensions of migration methods using the Radon transform. He also worked extensively with his colleagues at the Geological Survey of Canada planning, running, acquiring, processing, and interpreting LITHOPROBE reflection and refraction seismic experiments. Spencer was born in Beddington and educated at the Imperial College of London and the University of Cambridge (Ph.D., 1981 in geophysics). After a post-doctoral fellowship at the Victoria University of Wellington, New Zealand, Carl served the Geological Survey of Canada from 1983 to 1991. He then served as a senior research scientist at the Schumberger Cambridge Research, Cambridge, England, until his premature death. [Chapman (2002)]

Stauder, William, S.J. (1922–2002)

American seismologist. Stauder's research interests included use of *S* waves in determining focal mechanisms, the nature of subduction plates, the seismicity of the New Madrid seismic zone, and aftershocks. He is best remembered for his review, "The focal mechanism of earthquakes" (*Adv. Geophysics*, **9**, 1–76, 1962).

Born in New Rochelle, New York, Stauder joined the Society of Jesus in 1939, being ordained in 1952. He graduated with a major in physics from St. Louis University, receiving the B.S. degree in 1943 and the M.S. degree in 1948. He received the Ph.D. degree with a major in geophysics from University of California at Berkeley in 1959. He then was appointed to the Faculty of the Department of Earth and Atmospheric Sciences of St. Louis University, becoming Chairman of the Department in 1971. In 1974 he became first Acting Dean and then Dean of the Graduate School. In 1989 he was appointed Academic Vice-President of the University and in 1998 was promoted to Associate Provost. He retired in 1999, although he continued to work part-time in the Office of Research Services.

Stauder was a member of the Geophysics Advisory Panel, Air Force Office of Scientific Research from 1961–1971, Panel on Seismology of the National Research Council Committee on the Alaskan Earthquake 1964–1972, Advisory Panel, National Center for Earthquake Research 1966–1976, Atomic Energy Committee on Triggering of Earthquakes 1969–1972, Vice-President and President of the Seismological Society of America 1964–1966. [Stauder (1999b)]

Stegena, Lajos (1921–1997)

Hungarian geophysicist who contributed to geochemical oil prospecting, geothermal map of Eastern Europe, geothermics of basins, velocity structure and geothermics of the Earth's crust along the European geotraverse, and historical earthquakes in Central Europe. Stegena was born in Keszegfalva and educated at the Technical University in Budapest (B.S. chemistry, 1942; Cand. Sci., geophysics, 1957; Dr. Sci., geophysics, 1964). He spent his career at Eötvös University in Budapest as geophysicist (1942–1963) and professor (1963–1997); he was chief of the cartography department from 1966 to 1987. In 1995 he was honored "Laureatus academiae" by the Hungarian Academy of Sciences. He was president of the International Heat Flow Commission of the IASPEI from 1979 to 1983. [Chapter 81.4]

Steinbrugge, Karl V. (1919–2001)

American earthquake engineer. Karl Steinbrugge was a Professor of Structural Design at the University of California at Berkeley, College of Environmental Design, 1950 to 1978, and a licensed civil and structural engineer in the State of California. Concurrently, he was Manager of the Earthquake Department of the Pacific Fire Rating Bureau in San Francisco (later the Insurance Services Office). His work included engineering investigations of potential earthquake damage to structures as well as many field studies of earthquakes and their effects. He was the first chairman of the California State Seismic Safety Commission (1975–1980) and president of the Earthquake Engineering Research Institute (1968–1970). On two occasions he served in the Executive Office of the President in Washington, D.C., helping to formulate earthquake hazard reduction policy. He served as consultant to many local, state, and federal agencies, including the Federal Emergency Management Agency, US Geological Survey, and the California Division of Mines and Geology. He has published more than 100 articles; one of his best-known works is Earthquakes, Volcanoes, and Tsunamis: An Anatomy of Hazards. In retirement, he served as a consultant to government and the private sector on earthquake hazard evaluation. [Jennings, 2003]

Stewart, Samuel W. (1933–1996)

American geophysicist who pioneered the application of computer methods and technology to seismological problems, most notably to the automatic detection and location of earthquakes recorded by a regional seismograph network. Stewart was born in Washington, D.C., and was educated in geology, mathematics, and geophysics at Princeton University, the University of Utah, and Saint Louis University. His entire career was with the US Geological Survey. [*SRL* **68**:488, 1997]

Stoneley, Robert (1894–1976)

British geophysicist who demonstrated the existence of wave trapped at a plane interface of two elastic media in his paper, "Elastic waves at the surface of separation of two solids" (*Proc. Roy. Soc. Lond. A*, **106**, 416–428, 1924). These are now called Stoneley waves. In 1912 Stoneley entered St. John's College, Cambridge, to study mathematics. He became a fellow of the Royal Society in 1935 and was active in the International Association of Seismology, serving as president from 1940 to 1951. A photo of Stoneley is included in Figure 1 of Chapter 81.3. [*Biogr. Mem. Fell. Roy. Soc. Lond.* **22**:555–564, 1976]

Suyehiro, Kyôji (1877–1932)

Japanese engineering seismologist. Suyehiro was born in Tokyo and graduated in naval architecture in 1900 from Tokyo Imperial University. He was appointed an assistant professor there in 1902 and earned a D.Eng. degree in 1909. He was then appointed a professor of applied mechanics and naval architecture, a position he held until his death. In 1927 he was elected to the Imperial Academy. When the University's Earthquake Research Institute was established in 1925 to replace the old Imperial Earthquake Investigation Committee, Suyehiro was appointed to be its first Director. His most notable publication is his lectures, "Engineering Seismology: Notes on American Lectures" (*Proc. Am. Soc. Civil Eng.* **58**(4), part II, 1932). [Miyamura (2003)]

Suzuki, Ziro (1923–1997)

Japanese seismologist and geophysicist. Suzuki was born in Tokyo, graduated from the Geophysical Institute of Tokyo Imperial University, and was appointed an assistant there in 1945. He moved to Tôhoku University in 1951 as associate professor, was appointed professor in 1961, and served as faculty dean from 1971 to 1974. He moved to Kogakuin University in 1986.

He became vice-president of IASPEI in 1979, was president from 1983 to 1987, and remained on the executive committee until 1991. He was a leader of the Japan Group for Explosion Seismology. His book, *General Geophysics*, was published in 1974. A photo of Suzuki is included in Figure 3 of Chapter 81.3 of this Handbook. [Miyamura (2003)]

T

Tacchini, Pietro (1838–1905)

Italian astronomer who developed an effective meteorological and seismological observatory system in Italy, establishing the Rocca di Papa, Casamiccola, Pavia, and Catania Observatories and stations at Etna, Mt. Cimone, Monte Rosa, Modena, Catanzaro, and Salerno. He founded the Italian Seismological Society in 1895 and started a museum of seismic instruments at the Central Office of Meteorology and Geodynamics. Tacchini graduated from the University of Modena in 1857 and carried out astronomical research at the observatories at Modena, Padua, and Palermo until 1879, at which time he moved to Rome. He was first appointed director of the Collegio Romano, then of the Central Office of Meteorology, which he renamed and reorganized as the Central Office of Meteorology and Geodynamics. He was a member of the Accademia Nazionale dei Lincei. [Ferrari (2003)]

Takahasi, Ryutarô (1904–1993)

Japanese geophysicist. He conducted research on tsunamis and storm surges, on seismic-wave generation and hypocenter determination, seismic-wave velocities, and on variation in magnetic field intensity during volcanic eruptions. Takahasi was born in Tokyo, graduated in physics from Tokyo Imperial University in 1927, and became an assistant there in the Earthquake Research Institute, associate professor in 1930, and professor in 1941. He received the D.Sc. degree in 1929 for research on crustal tilting due to tidal loading. He was director of the Earthquake Research Institute from 1960 to 1963. He moved to Chuo University in 1965. [Miyamura (2003)]

Takeyama, Kenzaburô (1908–1986)

Japanese earthquake engineer noted for helping develop safety standards for nuclear power plants and techniques for the construction of super-high-rise buildings. Takeyama was born in Tokyo, graduated in architecture from Tokyo Imperial University in 1932, and worked as an engineer in the Ministry of Finance until 1948. He received the D.Eng. degree in 1945 for research on the structure of wooden buildings. In 1948 he moved to the Building Research Institute of the Ministry of Construction, serving as its director from 1955 to 1962 and helping establish its Institute of Seismology and Earthquake Engineering. From 1963 to 1976 he was director of the Technological Research Institute of Kajima Corporation. [Miyamura (2003)]

Tams, Ernst (1882–1963)

German seismologist who had studied geophysics and in particular seismology in Göttingen at Wiechert and in Strasbourg at Gerland, before he spent most of his time working in Hamburg, where he discovered in 1920, independently from Angenheister, the principal difference between continental and oceanic crust, by analyzing surface-wave propagation. [Hiller (1964); Schweitzer (2003)]

Tanabashi, Ryô (1907–1974)

Japanese structural engineer. He conducted research on anti-seismic theory and design, torsional vibrations of buildings, material ductility, and nonlinear vibrations. Tanabashi was born in Shizuoka and graduated in architecture from Kyoto Imperial University in 1929. He was appointed lecturer there in 1931, assistant professor in 1933, and professor in 1945. He was director of the university's Disaster Prevention Research Institute from 1951 to 1953 and 1959 to 1961. He moved to Kinki University in 1970. [Chapter 80.10; Miyamura (2003)]

Taniguchi, Tadashi (1900–1995)

Japanese architect. He studied the damage produced by the 1923 Kwanto (Tokyo), 1975 Tajima, 1927 Tango, 1930 Izu, and 1935 Hsinchu-Taichung, Taiwan, earthquakes. He was awarded an Architecture Institute of Japan prize in 1939, for a study of damped vibrations of framed structures, and the institute's Great Prize in 1984 for studies and activities in the public welfare. Taniguchi was born in Oita, graduated in architecture from the Higher School of Engineering in 1921, and became an assistant in the architecture department of Tokyo Imperial University, where he studied earthquake-resistant design. Starting in 1925 he held a concurrent appointment in the Earthquake Research Institute. In 1928 he received the D.Sc. degree for a study of earthquake damage. He was appointed assistant professor at Tokyo Institute of Technology in 1929 and professor in 1936. He was professor of architecture at Kanagawa University from 1961 to 1978. [Miyamura (2003)]

Taramelli, Torquato (1845–1922)

Italian geologist. Most of his research was on geologic mapping and hydrogeology, but he was a founding member of the Italian Seismological Society and conducted detailed field studies of a number of earthquakes, often with Mercalli, including chronologies of forerunning phenomena and aftershocks. He also prepared a map of the Italian seismic areas. Taramelli studied at Pavia and graduated in natural sciences at Palermo. He taught science at the Technical Institute of Udine, fought under

Garabaldi, and starting in 1875 taught geology and paleontology at the University of Pavia. He was a member of the Accademia Nazionale dei Lincei. [Ferrari (2003)]

Tatsuno, Kingo (1854–1919)

Japanese engineer. He was a member of the Imperial Earthquake Investigating Committee and noted for his design of such buildings as the Bank of Japan and the Tokyo railroad station. Tatsuno was born in Karatsu, graduated in engineering from what was to become the Imperial University in 1879, and was dean of engineering at Tokyo Imperial University from 1898 to 1902.

Taylor, Frank Bursley (1860–1938)

American geologist who was an early advocate of continental-drift theory before the work of Wegener. He also made contributions to Pleistocene and recent geology. He was employed by the US Geological Survey 1900–1916. [*Proc. GSA for 1938*, pp. 191–200.]

Terada, Torahiko (1878–1935)

Japanese geologist who conducted laboratory model studies of fracture of rocks, researched the role of continental drift in the origin of Japan and the Sea of Japan, and conducted crustal studies. He was awarded the Imperial Award of the Imperial Academy in 1917 for studies of x-ray crystallography and was admitted to the Academy in 1925. Terada was born in Tokyo, graduated in physics from Tokyo Imperial University in 1903, and received the D.Sc. degree in 1908. He was appointed associate professor at the University in 1909. He conducted research in the University's physics department, Earthquake Research Institute, and Physical and Chemical Research Institute. [Miyamura (2003)]

Terazawa, Kwan'ichi (1882–1969)

Japanese mathematician and administrator. His book, *Introduction to Mathematics for Natural Scientists* (in Japanese), first published from Iwanami-shoten in 1931, went through several editions until 1954, and a companion book on applications was published in 1960. Terazawa was born in Yonezawa, graduated in physics from Tokyo Imperial University in 1908, and received a D.Sc. degree in 1917. He was a professor in physics at the Tokyo Imperial University from 1918 to his retirement in 1949. He concurrently served as professor in the Aeronautical Research Institute from 1921 to 1942 (and as director in 1942–1943) and in the Earthquake Research Institute from 1936 to 1942 (and as director in 1939–1942), and during 1938–1943 was also dean of the Faculty of Science. In 1951 he was admitted to the Japan Academy. [Maruyama (2002); Miyamura (2003)]

Tocher, Don (1926–1979)

American engineering seismologist who studied individual western US earthquakes, crustal structure, and seismic hazard, particularly as related to siting of critical facilities. He implemented the first FM-telemetric seismograph network (at Berkeley) and studied creep on the San Andreas and other faults. Tocher was born in Hollister, California, and earned A.B. (1945), M.A. (1952), and Ph.D. (1955) degrees at the University of California, Berkeley. He was a research seismologist at Berkeley from 1956 to 1964. He established and was director of the NOAA (later US Geological Survey) Earthquake Mechanisms Laboratory from 1964 to 1974, and worked for Woodward-Clyde Consultants from 1975 to 1979. He was editor of BSSA from 1956 to 1961, secretary of the Seismological Society of America from 1966 to1971, and president from 1973 to 1974. [*BSSA* **70**:400–402, 1980]

Tokarev, Pavel Ivanovich (1923–1993)

Russian geophysicist who conducted research on the relation of earthquakes to volcanism, eruption prediction, volcanic tremor, and the energies of volcanic earthquakes. He helped to create two new seismic stations in the Kamchatka area and the Karymsky seismic-volcanic observatory. Tokarev graduated in physics from Moscow State University in 1948 and started graduate study at the Institute of Physics of the Earth of the USSR Academy of Sciences. In 1957 he moved to the Klyuci, Kamchatka Volcanic Station. He worked in the USSR Academy of Sciences Laboratory of Volcanology from 1957 to 1961 and the Russian Academy's Institute of Volcanology from 1962 to 1993. From 1979 to 1989 he was chairman of the International Working Group on Earthquakes and Earth Surface Deformation in Volcanic Regions of the International Association of Volcanism and Geochemistry of the Earth's Interior. [Chapter 79.45]

Townley, Sidney Dean (1867–1946)

American astronomer and seismologist who authored (with M. Allen) "Descriptive Catalog of Earthquakes of the Pacific Coast of the United States 1769 to 1928" (*Bull. Seismol. Soc. Am.*, **29** 1–297, 1939). Townley earned a degree from the University of Wisconsin in 1890 and worked as an instructor at the University of California, Berkeley (1898–1903), then took charge of the International Latitude Observatory at Ukiah, California (1903–1907). He was part of the Lawson commission investigating the 1906 San Francisco earthquake. He was professor of mathematics, astronomy, and geodesy at Stanford University 1907–1932. For the Seismological Society of America he was secretary-treasurer (1911–1929), president (1935–1936), and editor of its Bulletin (1911–1935). [*BSSA* **36**:322–325]

Tskhakaia, Alexander (1902–1970)

Georgian seismologist. Tskhakaia was a candidate of physical-mathematical sciences. He headed the department of regional seismology of the Institute of Geophysics, Georgian Academics of Sciences, for many years. His scientific interests included macroseismic investigations of strong earthquakes, seismicity of the territory of Georgia and Transcaucasus, seismic zoning, etc. He was one of the authors of seismic zoning maps of Georgia till 1970. Sixteen seismic stations of regional type were installed and operated in Georgia under his direct guidance during 1933–1970. Tskhakaia authored more than 60 published works, including *Development of Seismology in Georgia*, Tbilisi, 1950, and *Seismic Conditions of the Caucasus*, Tbilisi, 1973. [Chapter 79.23]

Tsuboi, Chuji (1902–1982)

Japanese geophysicist. He conducted research on crustal movements, crustal strain, earthquake energy release, magnitudes and frequency-magnitude variations in different areas of Japan, and gravity surveys. Tsuboi was born in Tokyo, graduated in physics from Tokyo Imperial University in 1926, and remained there throughout his career. He began as an assistant in the Earthquake Research Institute, rising to associate professor in 1928 and professor in 1941. He received the D.Sc. degree in 1934. He moved to the newly established Geophysical Institute in 1943 and served as dean of the faculty of science of the University of Tokyo from 1961 to 1963. He received the Japan Academy Prize in 1952, the Asahi Cultural Prize in 1955, and membership in the Academy in 1960. [Miyamura (2003)]

Tsuboi, Yoshikatsu (1907–1990)

Japanese architect who designed many large big-shell buildings in Japan. Tsuboi was born in Tokyo and graduated in architecture from Tokyo Imperial University in 1932. He worked as an engineer for the Wakayama prefectural government until 1937, when he became a lecturer at Kyushu Imperial University, and in 1940 an assistant professor. In 1942 he moved to Tokyo Imperial University as professor of engineering, and from 1949 to 1968 was a professor at the University's Institute of Industrial Science. In 1940 he received a prize from the Architectural Institute of Japan for research on square plates. In 1987 he received the Japan Academy Prize for research on curved structures and their application to large buildings. [Miyamura (2003)]

Tsubokawa, Ietsune (1918–1994)

Japanese geodesist. His main contribution to seismology was his study of crustal deformation through geodetic measurements. In 1964 he found an example of anomalous crustal deformation preceding the 1965 Niigata Earthquake. In 1969 he posited a relation between duration of anomalous crustal deformation and the magnitude of subsequent earthquakes. Tsubokawa was born in Fukui and graduated in astronomy from Tokyo Imperial University in 1942. He served in the Military Land Survey and its successor agency, the Geographical Survey Institute in the Ministry of Construction. He received his D.Sc. degree in 1956 and became chief of the geodetic division in 1961. He moved to the University of Tokyo's Earthquake Research Institute in 1965 as a professor, and was its director from 1973 to 1975. After his retirement in 1976 he became director of International Latitude Observatory Mizusawa until he died. [Okubo (2002); Miyamura (2003)]

Tsuya, Hiromichi (1902–1988)

Japanese geologist. He established methods to describe active volcanic eruptions and described major eruptions in Japan during the period 1930–1960. His petrological study included pioneering work on Asama, Sakura-jima, Miyake-jima, Izu-ooshima, and Fuji volcanoes. He studied the geology of major earthquakes, in particular the earthquake faults of the 1943 Tottori and 1945 Mikawa earthquakes. He edited the National Report of the 1948 Fukui earthquake. Tsuya was born in Gifu, graduated in geology from Tokyo Imperial University in 1926, and entered the Earthquake Research Institute as an assistant to Seitaro Tsuboi. He was appointed associate professor in 1930 and professor in 1941, served as director for 1944–1953, and retired in 1963. [Fujii (2002b); Miyamura (2003)]

Turner, Herbert Hall (1861–1930)

English astronomer and seismologist. Born in England, Turner was first an astronomer, becoming chief assistant at the Royal Observatory, Greenwich, in 1884. In 1893 he became the Savilian professor of astronomy and director of the University Observatory, Oxford. His later career focused on seismology. He carried on the work of John Milne after Milne's death in 1913, and supervised the publication of the International Seismological Summary (ISS) from 1918 until his unexpected death in 1930 (see Chapter 88 of this Handbook). While working on the ISS, he discovered deep focus earthquakes (independently of Wadati). Turner served as the president of the International Association of Seismology from 1922 to 1930. A photo of Turner is included in Figure 1 of Chapter 81.3 of this Handbook. [*Obit. Proc. Roy. Soc. Lond. A* **133**:i–ix, 1931; Britannica (2003)]

Tuve, Merle Anthony (1901–1982)

American physicist. With Tatal he undertook refraction measurements of continental structure, and he was director of the Carnegie Institution of Washington's Department of Terrestrial Magnetism. Tuve was born in Canton, South Dakota, and received the B.S. (1922) and A.M. (1923) degrees from the University of Minnesota and the Ph.D. (1926) from Johns Hopkins University, as well as many honorary degrees. He was also a pioneer in radio astronomy. With Briet, he studied the layering of the ionosphere using pulsed radio waves. He helped to develop the proximity fuse during World War II. With Hofstad and Haydenberg, he made the first definitive measurements of the proton–proton force at nuclear distances. He was the first chairman of the Geophysics Research Board of the National Academy of Science and the American Geophysical Union's Bowie Medalist in 1963. [*NAS Biog. Mem.* **70**:406–422, 1996]

Tvaltvadze, Guri (1907–1970)

Georgian geophysicist. Tvaltvadze was a doctor of physical-mathematical sciences, professor, and head of the department of seismic prospecting of the Institute of Geophysics, Georgian Academy of Sciences. He organized this department in 1933, and methods of seismic prospecting and studies of deep structure of the Earth's crust of Caucasus were developed under his guidance. He introduced a new method of compilation of theoretical hodograph systems for numerous structures of the Earth's crust, and investigated crustal structures by using observational data of near earthquakes. Tvaltvadze authored 70 published works, including *Structure of the Earth's crust in Georgia and construction of theoretical hodograph system*, Tbilisi, 1960. [Chapter 79.23]

U

Uchida, Yoshikazu (1885–1972)

Japanese earthquake engineer. Uchida was born in Tokyo, graduated in architecture from Tokyo Imperial University in 1907, and worked as an engineer for Mitsubishi Company from 1907 to 1910. He returned to Tokyo Imperial University in 1910, was appointed assistant professor in 1918 on receiving the D.Eng. degree for research on anti-seismic building structures, became professor in 1921, and was university president from 1943 to 1945. He received membership in the Japan Academy in 1957 and the Cultural Merit award in 1972. [Miyamura (2003)]

Ulrich, Franklin P. (1891–1952)

American seismologist who studied strong (near-source) earthquake motions. He worked for the New York State Engineering Department in 1913–1914, then joined the US Coast and Geodetic Survey (1914–1952), earning a Meritorious Service medal on his retirement. As chief of the Seismological Field Survey group (1933–1952), he investigated the 1935 Helena (Montana) earthquake and the 1940 Imperial Valley (California) earthquake.

Umemura, Hajime (1918–1995)

Japanese earthquake engineer. He conducted research on anti-seismic construction of high-rise and nuclear-power-plant buildings, and he worked to improve regulations for building construction and retrofitting. Umemura was born in Shizuoka and graduated in architecture from Tokyo Imperial University in 1941. He was appointed associate professor there in 1943, received the D.Eng. degree in 1949, and became professor in 1963. He was a visiting professor at the University of California at Berkeley in 1966 and 1972. In 1978–1988 he was a professor at Shibaura Technical University. He was president of the International Association of Earthquake Engineers from 1984 to 1988. [Chapter 80.10; Miyamura (2003)]

V

Valle, Paolo Emilio (1913–1970)

Italian geophysicist. He worked on the propagation of Love waves and the measurement of their group velocity, and studied the internal constitution of the Earth and other planets. His interests included melting temperatures and low-velocity layers in the Earth, the elastic constants of solids, the equilibrium of self-gravitating bodies, and the modeling of seismic sequences. Valle was born in Rome and graduated in 1940 in physics from the University of Rome. He joined the National Institute of Geophysics in 1940 and became a visiting professor of geophysics there in 1951. He received the Vercelli Prize from the Accademia dei Lincei for studies of the temperature gradient in the Earth. [Ferrari (2003)]

Van Gils, Jean-Marie (1918–1989)

Belgian seismologist who published the first seismic zoning map of Belgium and neighboring countries, including an assessment of the impact of earthquakes, especially as related to the nuclear industry, and a catalog of European earthquakes. Van Gils received a master's degree in physical sciences from the Brussels Free University in 1952. He worked as an assistant at the Royal Observatory of Belgium from 1945 to 1953, becoming acting head of the Seismological Division from 1953 to 1983. He was secretary of the European Seismological Commission and editor and co-editor of its proceedings from 1970 to 1986. He installed four new seismic stations in Belgium and helped install the first station in Luxembourg. [Chapter 79.8]

Vening Meinesz, Felix Andries (1887–1966)

Dutch geophysicist whose gravimetric data stimulated the idea of convection currents in the mantle. Vening Meinesz was born in The Hague and studied civil engineering at the Technical University at Delft (1904–1910). He became a professor of cartography at the University of Utrecht in 1927 and professor of geophysics in 1935. He was the director of the Royal Meteorological and Geophysical Institute at De Bilt from 1945 to 1951. Vening Meinesz invented the pendulum method for determining gravity at sea. He co-authored a well-known geophysical text (Heiskanen and Vening Meinesz, 1958). [McGraw-Hill (1980)]

Vicentini, Guiseppe (1860–1944)

Italian physicist noted for his design and construction of seismographs, which included improved damping systems. Vicentini was born at Ala, graduated in physics at Padua in 1882, and was appointed as an assistant at the Technical Institute at Turin. In 1885 he became professor of instrumental physics at University of Cagliari, and in 1889 was appointed director of the physics laboratory and of the attached meteorological and geodynamic observatory at the University of Siena. In 1894 he moved to the University of Padua. He is also noted for his studies of electrical resistance and radioactivity. He was a member of the Accademia Nazionale dei Lincei. [Ferrari (2003)]

Volarovich, Michael Pavlovich (1900–1987)

Russian geophysicist. His research was on the properties of rocks and minerals under high pressure and temperature. He measured the viscosities of melted glasses, rocks, slags, and meteorites. He was an early user of electron microscopy and electrongraphy, and he used nuclear radiation to study mineral material. Volarovich graduated in physics from Moscow University in 1926, worked from 1926 to 1935 at the Institute of Physics and Biophysics, and after 1950 was head of its Mineral Resources Laboratory. In 1950 he became head of the High-Pressure Laboratory of the Institute of Physics of the Earth of the USSR Academy of Sciences. He also worked in colloid chemistry. [Chapter 79.45]

Vvedenskaya, Anna Victorovna (1923–1997)

Russian seismologist who investigated the physics of source mechanisms of earthquakes. She favored the double-dipole source of finite dimensions and used stereographic projections to elaborate the earthquake mechanism. Vvedenskaya graduated from Irkutsk State University in 1945 then continued her studies at the Geophysical Institute of the USSR Academy of Sciences, completing dissertations in 1950 and 1968 on problems of seismic sources. [Chapter 79.45]

W

Wadati, Kiyoo (1902–1995)

Japanese seismologist. He discovered deep focus earthquakes (independently of H.H. Turner), for which he received the D.Sc. degree in 1932, and confirmed the Mohorovičić discontinuity around Japan. The alignment of deep subduction-related earthquakes that marks a descending crustal slab is today called the Wadati-Benioff zone. The linear relation he established between S–P and P times dependent on V_P/V_S or Poisson's ratio of the medium, a useful tool to find the origin time and to check P and S readings, today known as the Wadati diagram. Wadati was born in Nagoya, graduated in physics from Tokyo Imperial University in 1925, and joined the Central Meteorological Observatory (CMO). In the late 1930s he determined that ground subsidence in Osaka is mainly caused by extraction of groundwater, perhaps the earliest example of applying geophysical techniques to a human activity. He was the director of the Central Meteorological Observatory and its successor, the Japan Meteorological Agency, from 1947 to 1963. He became a member of the Japan Academy in 1949. He served as director of the National Research Center for Disaster Prevention in 1963–1966, president of Saitama University in 1966–1973, and president of the Japan Academy for 1974–1980, and he served on many scientific and governmental committees throughout his life. He was IASPEI president in 1967–1971 and received the Medal of Seismological Society of America in 1980. In 1985 he was awarded the Order of Cultural Merit. A photo of Wadati is included in Figure 2 of Chapter 81.3 of this Handbook. [Aki (1981); Wadati (1989); Miyamura (2003)]

Wanner, Ernst (1900–1955)

Swiss seismologist. Wanner earned a Ph.D. in mathematics in 1926 at the Swiss Federal Institute of Technology in Zürich. In 1928 he was appointed head of the Swiss Seismological Service, a position he held until his death. From 1946 to 1955 he was also vice-director of the Swiss Central Meteorological Agency. He worked on the seismicity of Switzerland, published a catalog of earthquakes from 1858 to 1879, and a map of Swiss epicenters. He studied the 1946 Valais earthquake and its aftershocks and used these to calculate a crustal model beneath Zürich. [Chapter 79.41]

Watanabe, Hikaru (1934–2000)

Japanese seismologist. His research was primarily on microearthquakes and methods of determining magnitude at Abuyama Seismological Observatory. He sorted microseisms into high-frequency and low-frequency classes. He studied the occurence of microearthquakes as foreshocks. Watanabe was born at Kokura and graduated in geophysics from Kyoto University in 1957, where he was appointed assistant in 1961, associate professor in 1975, and professor in 1992. [Miyamura (2003)]

Wegener, Alfred Lothar (1880–1930)

German meteorologist, geophysicist, and polar scientist who proposed the theory of continental motion in 1912. The first issue of his monograph *Die Enstehung der Kontinente und Ozeane* (translated as *Origin of Continents and Oceans*, 4th edition, Dover, 1929) was published in 1915. Wegener was born in Berlin and obtained his doctorate in astronomy from the University of Berlin in 1905. He was appointed a lecturer in astronomy and meteorology at the University of Marburg, and moved to a special chair of meteorology and geophysics at the University of Graz, Austria, after World War I. [Benndorf (1931); Daintith and Gjertsen (1999); Kertz (1980; 2002)]

Wentzel, Gregor (1898–1978)

German physicist best known by theoretical seismologists as the "W" in the WKBJ method, which was originally developed for quantum mechanics. Wentzel was born in Dusseldorf and was educated at the Universities of Freiburg, Greifswald, and Munich. In 1928 he succeeded Schrödinger at the University of Zürich. He joined the faculty of the University of Chicago in 1948 and retired in 1970. Wentzel made contributions to many fields in physics, including quantum mechanics, quantum field theory, high-energy physics, and solid-state physics. See Chapter 8 (p. 90) of this Handbook. [Holmes (1990)]

West, William Dixon (1901–1994)

British geologist. He was born in Bournemouth, England, and entered Cambridge University in 1920, where he earned the B.A. degree and later the D.Sc. degree. He joined the Geological Survey of India in 1923, where he worked on the geology of the central provinces and the Himalayas, studied the petrology of the Deccan lavas, and helped to develop coal mines in Afghanistan. He investigated the Baluchistan earthquakes of 1931 and 1935 and showed that the fifteen shocks of this region were related to the Quetta reentrant folding. He prepared an isoseismal map of India. He was director of the Geological Survey of India from 1945 to 1951. He was appointed professor of geology and dean of engineering and technology at University of Saugar and eventually University vice chancellor. West was a founding fellow of the India National Science Academy and received its Wadia Medal in 1983; he also was a recipient of the Companion Order of the Indian Empire, a fellow of the Geological Society of Great Britain, and the first president of the Indian Association of Hydrogeologists. [Narula (1998)]

Wiechert, Emil (1861–1928)

German physicist and geophysicist who made fundamental contributions to seismology. Wiechert was born in Tilsit, East Prussia (now part of Russia), and studied physics and mathematics at the University of Königsberg, completing the Ph.D. in 1889. Wiechert was first a professor at Königsberg University and then at Göttingen, where he founded the Institute of Geophysics. At Königsberg he discovered that cathode rays consist of particles. In 1896 he predicted that the Earth had a core that he suggested consisted of iron. In 1898 he installed his first seismograph at Göttingen. In 1899 he presumed that the body waves of earthquakes consisted of both longitudinal and transverse pulses. He was the first to identify seismic pulses reflected at the Earth's surface. His large-mass inverted-pendulum seismographs, developed in the early 1900s, were installed at over 100 locations worldwide. Because they included damping, increased magnification, and improved paper recording, they constituted a significant advance in recording earthquakes. In 1907, he and Zoeppritz published travel-time charts for P, S, multiply reflected body waves (PP, SS, etc.), and surface waves. He and Geiger were the first to calculate accurately the variations in velocity of seismic waves with depth in the mantle, developing what is today called the Wiechert-Herglotz inversion method. He studied the seismic waves from quarry blasts and wave propagation in the atmosphere. He was noted for mentoring many students, some of whom became leading seismologists. A photo of Wiechert is included in Figure 4 of Chapter 81.3 of this Handbook. [Angenheister (1928); Gutenberg (1928); Kertz (2002); Hennings and Ritter (2003); Ritter and Schweitzer (2003b); Schweitzer (2003)]

Willis, Bailey (1857–1949)

American structural geologist. His seismic contributions included a fault map of California (with H. O. Wood) and articles on earthquake risk and damage to buildings and the seismicity of various areas. Born in Idlewild-on-the-Hudson, New York, Willis was employed during his career by the Northern Pacific Railroad, the US Geological Survey (where he became Chief Geologist), the Carnegie Institution of Washington, and Stanford University (where he was a professor and head of the geology department 1915–1922). He performed some of the earliest model experiments on mountain formation and wrote the textbook *Geological Structures* (1929) and several other books on his worldwide travels, which included China, Chile, the Himalayas, and the Levant. He was president of the Seismological Society of America (1921–1927). [*NAS Biog. Mem.* **35**:332–350, 1961]

Willmore, Patrick L. (1921–1994)

British seismologist. He worked for a time at Dominion Observatory in Ottawa, Canada, on seismic networks. After returning to the UK, Willmore became the first Director of the International Seismological Centre in Edinburgh. Founded and led the Edinburgh Seismology Research Group in the early 1960s, which subsequently became a unit of the British Geological Survey. He was editor of the *Manual of Seismological Observatory Practice*, 1979 (see Chapter 81.5). Willmore's main interest was seismometer design; the Willmore MK III instrument was a portable high-frequency instrument that was widely deployed in the 1980s. [Modified from R. Musson's e-mail communication, 2003].

Wilson, James Tinley (1914–1978)

American seismologist who was the first to use dispersion of surface waves to delineate crustal structure in America. He was born in Claremont, California, and received the A.B. (1935) and Ph.D. (1939) degrees at the University of California, Berkeley. He started at the University of Michigan in 1939, rising to professor in 1955, head of the geology department in 1956, and associate director of the Institute of Science and Technology in 1960. He was president of the Seismological Society of America in 1960–1961 and a member of the National Academy of Science. He was chairman of the National Research Council's Committee on Seismology and active in many other government committees related to seismology. He was a member of the IUGG Committee for Standardized Seismology and Seismographs. [*BSSA* **68**:1553–1554, 1978]

Wilson, John Tuzo (1908–1993)

Canadian geologist and geophysicist. In his paper, "A new class of fault and its bearing in continental drift" (*Nature*, **207**, 343–347, 1965), he introduced "transform fault," which links trenches (where the plates collide) and rifts (where the plates pull apart). (see Appendix 1 for expanded definition). In the second half of the 1960s, he championed the theory of plate tectonics, e.g., his paper, "A revolution in earth science" (*Geotimes*, **13**(10), 10–16, 1968). He also introduced the idea of the Wilson cycle, in his paper "Did the Atlantic close and then re-open again?" (*Nature*, **211**, 676–681, 1968). See Chapter 6 of this Handbook. Wilson was born in Ottawa and studied geophysics at the University of Toronto, the first student in Canada to take such a course. He did graduate work at Cambridge and Princeton, receiving a doctorate in geology in 1936. He worked for the Geological Survey of Canada, spent seven years in the army, and taught at the University of Toronto from 1946 to 1974. In 1974 became the director of the Ontario Science Center. Wilson was an inspiring teacher and a superb expositor of science to the public. He had more friends throughout the world than any geophysicist we know, and he served as the president of the International Union of Geodesy and Geophysics from 1957 to 1960, overseeing the immensely successful International Geophysical Year in 1957. [*Biogr. Mem. Fell. Roy. Soc. Lond.* **41**:533–552, 1995]

Wong, Wenhao (1889–1971)

Chinese geologist who introduced the study of earthquakes to the Geological Survey of China and studied the seismotectonics of China, the mountains of China, and the history of Chinese earthquakes. He directed the exploitation of the Yumen oilfield. Wong was born in Yinxian and earned a Ph.D. in 1912 from Louvain University in Belgium. He started work in the Institute of Geology of the China Ministry of Agriculture and Mines in 1913, and worked for its Geological Survey starting in 1918. He was president of the Geological Society of China in 1941 and elected to the Academia Sinica in 1948. He published two books: *Geology* in 1914, and *Mineral Resources in China* in 1919. [Chapter 79.14]

Wood, Harry Oscar (1879–1958)

American seismologist. While instructor in mineralogy and geology at the University of California, Berkeley, in 1904–1912, he taught the first course on seismology in the United States. Following the 1906 earthquake, Lawson assigned him the task of investigating the extent and nature of the earthquake damage in San Francisco (he wrote pp. 220–245 of Lawson's report). He prepared the first three issues of the Berkeley Seismic Bulletins. From 1912 to 1917 he was a research associate at the Hawaiian Volcano Observatory, operating the Hawaiian seismograph station, and reporting on Hawaiian earthquakes. In 1916 he published an analysis of the need for seismic studies of the southern California region. During World War I he worked at the National Bureau of Standards in Washington, where he developed a peizoelectric seismograph to record cannon detonations. In 1919–1920, he was acting associate secretary of the National Research Council and secretary of the newly formed American Geophysical Union. While in Washington, he interested the Carnegie Institution in southern California earthquakes. Carnegie hired him to establish what would become the Caltech seismic observatory. He studied earthquakes and blasts in southern California, helped develop the Wood-Anderson torsion seismometer, with Neumann modernized the Mercalli Intensity scale of earthquake effects, with Willis published the Fault Map of California, and with Heck and Allen wrote the first edition of the *Earthquake History of the United States*. [*Proc. GSA for 1958*, pp. 219–224; Goodstein (1991)]

Woodworth, Jay Backus (1865–1925)

American geologist who worked on the glacial and structural geology of New England and the geology of South America. He spent his career at Harvard University, rising from instructor in 1890 to professor. Notable for seismology, he organized the Harvard seismograph station. He also served as chairman of the National Research Council Committee on Use of Seismograph in War 1917–1918 and was president of the Seismological Society of America 1916–1918. [*BSSA* **16**:43–44, 1926]

Z

Zaborovsky, Alexander Ignatyevich (1894–1976)

Russian geophysicist who published works on the theory of the Earth's magnetism, refraction seismologic surveying, and well logging. Zaborovsky graduated from Petersburg University and became head of the Survey of the Kursk Magnetic Anomaly in 1917. Beginning in 1926 he conducted a magnetic survey of the Baltic region. He helped organize geophysical exploration for oil in Russia. Starting in 1954 he worked at the Institute of Physics of the Earth of the USSR Academy of Sciences, rising to be head of the Institute of Electricity. He taught geophysics at Moscow Geologic Institute and at Moscow State University. [Chapter 79.45]

Zapolsky, Konstantin Konstantinovich (1916–1992)

Russian geophysicist. Zapolsky graduated in geography from Moscow Geologic Prospecting Institute in 1939. While still a student, he started work at the Seismologic Institute of the USSR Academy of Sciences. He worked on seismic prospecting and well logging, specializing in construction of apparatus and study of the spectra of seismic waves. He studied the relation of earthquake magnitude, moment, and energy to the size of fault ruptures and the differences in the frequency spectra between main shocks, foreshocks, and aftershocks. [Chapter 79.45]

Zátopek, Alois (1907–1985)

Czech geophysicist, was born in Zašova in Moravia. After finishing his studies at Charles University in Prague in 1932, he joined the Czechoslovak Institute of Geophysics in 1934. In 1946 he was appointed associate professor of geophysics as a result of his investigations in the propagation of the East Alpine earthquakes in the Bohemian Massif. From 1965 to 1970 he was president of the Czechoslovak National Geophysical and Geodetic Committee. After World War II he carried out important work to organize Czechoslovak geophysics in general. His major interest always focused on research into the seismicity of Czechoslovakia, the classification of earthquake intensities, studies of magnitude quantity, and European meteorological microseisms. An excellent teacher, he educated a number of Czech geophysicists. During a sojourn in Tokyo, he held the post of senior councilor of a UNESCO development project. He served as a chairman of the European Seismological Council and twice as a member of the executive committee of the International Association of Seismology. He was a member of the Czechoslovak Academy of Sciences. [Chapter 79.17]

Zavriev, Kiriak Samsonovich (1891–1978)

Georgian engineer who worked on the dynamic theory of earthquake resistance, on anti-seismic belts of buildings, and on critical loading. Zavriev graduated from the Petersburg Institute of Engineers of Communications in 1914, worked there briefly, and from 1921 to 1930 worked at the Tbilisi State Polytechnical Institute, then at the Tbilisi Institute of the Engineers of Railway Transport from 1931 to 1959, and after 1959 at the Lenin Georgian Polytechnical Institute. From 1941 to 1943 he was chairman of the Bureau of Antiseismic Construction at the Presidium of the Georgian Academy of Sciences. From 1947 until his death he was director of the Institute of Structural Mechanics and Earthquake Engineering. He was a deputy of the Supreme Council of the Georgian Soviet Socialist Republic and decorated with two Lenin and four other orders. [Chapter 80.8]

Zhang, Heng (78–139)

Ancient Chinese astronomer who invented the "Houfeng Didong Yi," the world's first seismoscope. It identified the azimuth of the epicenter of the thousand-kilometer distant Gansu earthquake in 138. He worked on the lunar calendar and invented a tricycle. [Chapter 79.14]

Zharov, Aleksandr Matveevich (1935–1991)

Russian engineer. Starting in 1960 he worked in the Department of Earthquake Engineering of the Central Research Institute of Structures in Moscow on the relation of seismic ground movements to building damage. He developed mathematical models to relate the time-history of ground acceleration, including duration of shaking, to earthquake damage. He was concerned with problems of repeated shaking, frequency spectra, and intensity scales. He conducted field inspections of damage in the USSR and in Romania. [Rzhevsky (2002)]

Zoeppritz, Karl (1881–1908)

German geologist and geophysicist. His Ph.D. thesis in 1906 was on Alpine geology, but starting in 1906 he worked with Wiechert at Göttingen in developing early, accurate travel-time charts for body and surface waves, and determining seismic velocity variations with depth in the Earth. In 1907, they recognized surface-reflected pulses in seismograms. He was the first to include surface-reflected pulses in travel-time plots. Working with Geiger, he also calculated the variation in Poisson's ratio with depth in the mantle. He was one of the first if not the first to note that some earthquakes produce very weak surface waves, which is now known to be due to great depth of focus. In an article published posthumously in 1919 he derived the reflection and transmission coefficients for a plane wave at a boundary. [Ritter and Schweitzer (2003c)]

Acknowledgements

We wish to thank about one hundred people for their kindness in writing biographies of their deceased colleagues. Their names are given in the citations of this chapter and on the attached Handbook CDs.

References

Aki, K. (1981). Citation for the fifth award of the Medal of the Seismological Society of America, with letter of acceptance by Wadati, *Bull. Seismol. Soc. Am.*, **71**, 1373–1377.

Angenheister, G. H. (1928). Emil Wiechert. *Zeitschrift für Geophysik* **4**, 113–117.

Anonymous (1870–1871). Obituary notices of Fellows deceased. *Proc. Roy. Soc. London, A*, **19**, i–ix.

Anonymous (1944). Prof. Dr. Carl Mainka. *Beiträge zur angewamdten Geophysik* **11**, 1–2.

Bakun, W. H. (2002). Personal communication in an e-mail, archived as Bakun.pdf on Handbook CD#3, under the directory of \89Howell.

Ben-Menahem, A. (1990). "Vincit Veritas—A Portrait of the Life and Work of Norman Abraham Haskell." American Geophysical Union, Washington, D.C.

Ben-Menahem, A. (1995). A concise history of mainstream seismology: origins, legacy, and perspectives. *Bull. Seismol. Soc. Am.*, **85**, 1202–1225.

Bell, E. T. (1938). "Men of Mathematics." Simon and Schuster, New York.

Benndorf, H. (1931). Alfred Wegener. *Gerlands Beiträge zur Geophysik* **31**, 337–377.

Birch, F. (1979). Reminiscences and digressions. *Ann. Rev. Earth Planet. Sci.*, **7**, 1–9.

Blume, J. A. (1994). "John A. Blume," Connections: The EERI Oral History Series, Earthquake Research Institute, Oakland, CA.

Bock, Günter (2002). Biographical sketch submitted on April 13, 2002, and archived as Bock.pdf on Handbook CD#3, under the directory of \89Howell.

Britannica (2003). "Encyclopedia Britannica," 32 volumes. Encyclopedia Britannica, Inc., Chicago.

Bühler, W. K. (1981). "Gauss: A biographical study." Springer, Berlin, Heidelberg, New York 1981.

Bullard, E. C. (1974). The emergence of plate tectonics: a personal view. *Ann. Rev. Earth Planet. Sci.*, **3**, 1–30.

Cara, M. (2003). Personal communication in an e-mail, archived as Cara.pdf on Handbook CD#3, under the directory of \89Howell.

Cannell, D. M. (2001). "George Green: Mathematician and Physicist 1793–1841: The Background to His Life and Work. Second edition, Society for Industrial and Applied Mathematics, Philadelphia.

Daintith, J., and D. Gjertsen, (Eds.) (1999). "Oxford Dictioary of Scientists." Oxford University Press, Oxford.

Davison, C. (1895). Dr. E. von Rebeur-Paschwitz. *Nature* **52**, 599–600; See Chapter 79.24.3 on Handbook CD #2, under the directory of \7924Germany.

Davison, C. (1924). Fusakichi Omori and his work on earthquakes. *Bull. Seismol. Soc. Am.*,**14**, 240–255.

Davison, C. (1927). "The Founders of Seismology," Cambridge University Press, Cambridge.

Debus, A. G. (Ed.) (1968). "World Who's Who in Science." Marquis, Chicago.

Degenkolb, H. J. (1994). "Henry J. Degenkolb," Connections: The EERI Oral History Series, Earthquake Research Institute, Oakland, CA.

Dietz, R. S. (1994). Earth, sea, and sky: life and times of a journeyman geologist. *Ann. Rev. Earth Planet. Sci.*, **22**, 1–32.

Engdahl, E. R. (1998). Personal communication in an e-mail, archived as Engdahl.pdf on Handbook CD#3, under the directory of \89Howell.

Ferrari, G. (2003). Personal communication in an e-mail, archived as Ferrari.pdf on Handbook CD#3, under the directory of \89Howell.

Fujii, T. (2002a). A biography of Kazuaki Nakamura, archived as Fujii 2002a.pdf on Handbook CD#3, under the directory of \89Howell.

Fujii, T. (2002b). A biography of Hiromich Tsuya, archived as Fujii 2002b.pdf on Handbook CD#3, under the directory of \89Howell.

Gerland, G. (1898). Ernst Ludwig August v. Reuber-Paschwitz. *Beiträge zur Geophysik* **3**, 16–18.

Gerland, G. (1899). Dr. Reinhold Ehlert. *Beiträge zur Geophysik* **4**, 105–107.

Goodstein, J. R. (1991). "Millikan's School: A History of the California Institute of Technology". Norton, New York.

Gutenberg, B. (1928). Geh. Reg.-Rat Prof. Dr. E. Wiechert . Meteorologische Zeitschrift **45**, Kleine Mitteilungen, 183–185; see Chapter 79.24.3 on Handbook CD #2, under the directory of \7924Germany.

Hader, F. (1962). Viktor Conrad (26. VIII. 1876–1926. IV. 1962). *Wetter und Leben* **14**, 93–94.

Heinloo, O., and T. All (2003). Prof. G. V. Levitski (1852–1918) and Yuryev (Tartu) Seismological Station in 1896–1912. See Appendix 1 of Chapter 79.20 on Handbook CD #2, under the directory of \7920Estonia.

Hennings, R., and J. Ritter (2003). Bibliography of Emil Wiechert. See Chapter 79.24.3 on Handbook CD #2, under the directory of \7924Germany.

Hiller, W. (1964). In Memoriam Ernst Tams. *Zeitschrift für Geophysik* **30**, 49–50.

Holmes, F. L. (editor in chief) (1990). "Dictionary of Scientific Biography." Charles Scribner's Sons, New York.

Hoover, M. L. (1912). John Milne, Seismologist. *Bull. Seismol. Soc. Am.*, **2**, 2–7 (with 2 plates).

Husebye, E. (2002). A biography of Markus Båth, archived as Husebye.pdf on Handbook CD#3, under the directory of \89Howell.

James G., and R. C. James (1976). "Mathematics Dictionary." Fourth edition, van Nostrand and Reinhold, New York.

Jeffreys, H. (1973). Developments in geophysics. *Ann. Rev. Earth Planet. Sci.*, **1**, 1–13.

Jennings, P. C. (2003). Biographical Notes, archived as Jennings.pdf on Handbook CD#3, under the directory of \89Howell.

Kennett, B. L. N. (2002). Prof. Gerhard Müller 1940–2000. *Astronomy & Geophysics* **43**(6), 35.

Kertz, W. (1980). Alfred Wegener—Reformator der Geowissenschaften. *Physikalische Blätter* **36**, 347–353; See Chapter 79.24.3 on Handbook CD #2, under the directory of \7924Germany.

Kertz, W. (1991). Ludger Mintrop, der die angewandte Geophysik zum Erfolg brachte. Deutsche Geophysikalische Gesellschaft e.V., *Mitteilungen* **3/1991**, 2–16.

Kertz, W. (2002). Biographisches Lexikon zur Geschichte der Geophysik. Edited by K.-H. Glaßmeier & R. Kertz. Braunschweigische Wissenschaftliche Gesellschaft, Brauschweig 2002, 384 pp.

Klotz, O. (1917). Prince Boris Galitzin. *Bull. Seismol. Soc. Amer.*, **7**, 49–50.

Knopoff, L. (2000). Personal communication in an e-mail, archived as Knopoff.pdf on Handbook CD#3, under the directory of \89Howell. [Editor's Note: Professor Knopoff kindly supplied us with a one-page summary of his biography of Beno Gutenberg (*U.S. Nat. Acad. Sci. Biogr. Memoirs*, **76**:114–147, 1999). It is reproduced here with very minor editing.]

Kulhanek, L. (2002). A biography of Markus Båth, archived as Kulhanek.pdf on Handbook CD#3, under the directory of \89Howell.

Maruyama, T. (2002). A biography of Kwan-ichi Terazawa, archived as Maruyama.pdf on Handbook CD#3, under the directory of \89Howell.

McGraw-Hill (1980). "Modern Scientists and Engineers," McGraw-Hill, New York.

Miyamura, S. (2003). Biographies of 62 Japanese scientists and engineers, archived as a part of the Japan National Report on Handbook CD#2, under the directory of \79.33Japan.

Müller, Gerhard (2000). Biographical sketch submitted July 2, 2000, archived as Mueller.pdf on Handbook CD#3, under the directory of \89Howell.

Narula, P. L. (1998). Biographies of J. B. Auden, M. S. Krishnan, C. S. Middlemiss, R. D. Oldham, T. Oldham, and W. D. West, archived as Narula.pdf on Handbook CD, under the directory of \89Howell.

Okubo, S. (2002). A biography of Ietsune Tsubokawa, archived as Okubo.pdf on Handbook CD#3, under the directory of \89Howell.

Pyne, S. J. (1989). "Grove Karl Gilbert, a great engine of research." Univ. of Texas Press.

Reyners, M. E. 1992. Death of George Eiby. *Bull. New Zealand Nat. Soc. Earthq. Engin.*, **25**, 73.

Richter, C. F. (1958). "Elementary Seismology." W. H. Freeman & Co., San Francisco.

Ritter, J. R. R., and J. Schweitzer (2003a). Ludwig Carl Geiger (1882–1966). See Chapter 79.24.3 on Handbook CD #2, under the directory of \7924Germany.

Ritter, J. R. R., and J. Schweitzer (2003b). Emil Wiechert (1861–1928). See Chapter 79.24.3 on Handbook CD #2, under the directory of \7924Germany.

Ritter, J. R. R., and J. Schweitzer (2003c). Karl Zoeppritz (1881–1908). See Chapter 79.24.3 on Handbook CD #2, under the directory of \7924Germany.

Ritter, J. R. R., R. Meyer, and J. Schweitzer (2003). Gustav H. Angenheister (1878–1945) as Seismologist. See Chapter 79.24.3 on Handbook CD #2, under the directory of \7924Germany.

Rubey, W. W. (1974). Fifty years of Earth sciences—a renaissance. *Ann. Rev. Earth Planet. Sci.*, **2**, 1–24.

Rzhevsky, V. A. (2002). Biographies of V. A. Bychovsky, I. V. Goldenblat, L. S. Kilimnik, I. L. Korchinsky, N. A. Nikolaenko, S. V. Polyakov, and A. M. Zharov, archived as Rzhevsky.pdf on Handbook CD, under the directory of \89Howell.

Sapper, K. (1915). Emil Rudolph†. *Gerlands Beiträge zur Geophysik* **14**, without page numbers; See Chapter 79.24.3 on Handbook CD #2, under the directory of \7924Germany.

Schleusener, A. (1956). In Memoriam Prof. Dr. Dr.h.c. Ludger Mintrop. *Zeitschrift für Geophysik 22*, 58–64; See Chapter 79.24.3 on Handbook CD #2, under the directory of \7924Germany.

Schweidler, E. v. (1941). Hans Benndorf zum 70. Geburtstag. *Gerlands Beiträege zur Geophysik* **57**.

Schweitzer, J. (2003). Early German contributions to modern seismology. Chapter 79.24.1 on Handbook CD #2; and Biographical Notes in Chapter 79.24.3 on Handbook CD #2, under the directory of \7924Germany.

Shirin, V. V. (2002). A Biography of Askold I. Martemyanov by V.V. Shirin, archived as Shirin.pdf on Handbook CD#3, under the directory of \89Howell.

Slichter, L. B. (1967). Earth. free oscillations of . In: "International Dictionary of Geophysics" (K. Runcorn, general editor), pp. 331–343, Pergamon Press, Oxford.

Smith, W. D. 1992. A tribute to George Eiby. *Bull. New Zealand Nat. Soc. Earthq. Engin.*, **25**, 73.

Snelders, H. A. M. (1989). Berlage [j.], Hendrik Petrus (1896–1969). In: "Biografisch Woordenboek van Nederland 3," Den Haag.

Stauder, W. (1999a). Biographies of Jesuit seismologists, archived as Stauder1999a.pdf on Handbook CD#3, under the directory of \89Howell.

Stauder, W. (1999b). Biographical sketch submitted on September 10, 1999, and archived as Stauder1999b.pdf on Handbook CD#3, under the directory of \89Howell.

Tietze, H. (1954). G. Herglotz. *Jahrbuch der Bayerischen Akademie der Wissenschaften* **1954**, 188–194.

Uzdin, A. M. (2003). A Biography of Oleg Alexandrovich Savinov, archived as Uzdin.pdf on Handbook CD#3, under the directory of \89Howell.

Wadati, K. (1989). Born in a country of earthquakes. *Ann. Rev. Earth and Planet. Sci.*, **17**, 1–12.

Zürn, W. (2002). Nachruf auf Professor Dr. Gerhard Müller. Deutsche Geophysikalische Gesellschaft e.V., *Mitteilungen* **3/2002**, 23–26.

90

Digital Imagery of Faults and Volcanoes

Michael J. Rymer
US Geological Survey, Menlo Park, California, USA
David E. Wieprecht
US Geological Survey, Vancouver, Washington, USA
John C. Dohrenwend
Teasdale, Utah, USA

1. Introduction

Faults and volcanoes are results of Earth's dynamic crust and mantle and are found on each of the continents and in each of the oceans. This chapter presents photographs of a small sampling of faults and volcanoes from around the world; the photographs are also placed as computer-readable files on the attached Handbook CD under the directory \90Rymer. Photographs are images created from 35-mm slides scanned on a Kodak PIW film scanner or from high-resolution prints; large-format prints were scanned on a flatbed Linotype-Hell scanner. Satellite images are also included that were composed to enhance the locations of faults and volcanoes. These pictures and satellite images are but a small fraction of millions taken by scientists and others who study the Earth.

Photographs were selected for inclusion in this chapter to portray the grandeur of faults and volcanoes and to illustrate part of the range of Earth-surface phenomena. References that describe the general tectonic or volcanic setting are included for each image on the CD.

We present nine images in Color Plates 31 and 32 for illustrative purposes. Full citations of the respective photographers and their institutional affiliations are included in the full version of Chapter 90 on the attached Handbook CD under the directory \90Rymer.

2. Images Shown in Color Plate 31

Picture (a) in Color Plate 31 is a Landsat Thematic Mapper satellite image of Death Valley and associated faults, southeastern California. The Furnace Creek fault zone trends diagonally across the top of the image, and the Death Valley fault zone extends approximately north–south along the margins of the valley. Light blue color represents playa deposits.

Picture (b) shows a vigorous eruption column rising over the summit of 1252-m-high Augustine volcano, Alaska. At the time of this photograph (31 March 1986), a new lava dome was growing at the summit.

Mount Taranaki (Egmont) shown in Picture (c) is the second largest and most southerly volcano in New Zealand. This stratovolcano reaches a height of 2518 m. A parasitic cone, Fanthams Peak, is in the foreground.

Picture (d) shows a view of the scarp along the Johnson Valley segment of rupture associated with the 1992 Landers, California, earthquake. Vertical component of displacement at this site was about 1 m up to the west; corresponding right-lateral component was about 0.6 m. The total horizontal slip across fault ruptures both east and west of this site was about 3 m.

3. Images Shown in Color Plate 32

Picture (a) in Color Plate 32 shows the surface rupture on the Nojima fault, Awaji Island, associated with the 1995 Hyogo-ken Nambu (Kobe), Japan, earthquake.

Left-lateral offsets preserved in rice patties by the 1990 Luzon, Philippines, earthquake are shown in Picture (b).

Picture (c) is an oblique aerial view of the Pearce scarp at mouth of Miller Basin associated with the 1915 Pleasant Valley, Nevada, earthquake.

INTERNATIONAL HANDBOOK OF EARTHQUAKE AND ENGINEERING SEISMOLOGY, VOLUME 81B ISBN: 0-12-440658-0

Tall, faceted spurs and preserved mole track associated with the 1932 Chang Ma earthquake, Gansu Province, China, are shown in Picture (d). View is to the west. South-dipping facets imply normal component of slip along a south-dipping fault.

Picture (e) shows the rupture along the Bogd fault associated with the 1957 Gobi Altai earthquake, Mongolia. Measured left-lateral offsets in this part of the fault ranged from about 5 to 7 m. View is to the west; note for scale the three people walking along fault and a fourth sitting on an outcrop to the left in midground.

Acknowledgments

We thank the individual photographers for use of images in this chapter. Photographers whose images are included in Plates 31 and 32 are M. E. Yount, 31(b); D. L. Homer, 31(c); M. J. Rymer, 31(d); Takashi Nakata, 32(a) and 32(b); R. E. Wallace, 32(c); M. N. Machette, 32(d); and Carol Prentice, 32(e); see Chapter 90 on the attached Handbook CD for full citations of photographers.

Appendix 1

Glossary of Interest to Earthquake and Engineering Seismologists

Keiiti Aki
Observatoire Volcanologique du Piton de la Fournaise, La Réunion, France
W. H. K. Lee
US Geological Survey, Menlo Park, California, USA (retired)

The following glossary includes about 1500 specialized terms (including some abbreviations and acronyms) that readers may encounter in the literature in earthquake and engineering seismology and related fields. For each term, a brief definition is provided (if possible), with citation to the original paper (if known to us). Citations for further reading are also provided in many cases at the end of each term. We attempt to give commonly used definitions, but readers should realize that (1) a term might have a different definition in some literature, (2) the usage of a term might have changed over time, and (3) it is often not possible to define a specialized term with simple words. The purpose of this glossary is to provide a quick introduction and a starting point for readers to learn more from the cited references. In addition, we avoid introducing mathematical formulas in this glossary (except for a small number of cases). However, key formulas in earthquake seismology are given by Yehuda Ben-Zion in Appendix 2 of this Handbook.

This glossary was prepared from glossary terms submitted by many of the Handbook chapter authors and from consultation of some standard reference sources, including Aki and Richards (2002), Bormann (2002), Clark (1966), EERI (1989), Hancock and Skinner (2000), Jackson (1997), James and James (1976), Lee and Stewart (1981), Lide (2002), Meyers (2002), Naeim (2001), Richter (1958), Runcorn (1967), Sigurdsson (2000), USGS Photo Glossary (2002), and Yeats *et al.* (1997).

Additional terms and/or suggestions for revisions were provided by Robin Adams, John Anderson, Joe Andrews, Michael Asten, Yehuda Ben-Zion, M.G. Bonilla, Dave Boore, Ken Campbell, Chris Chapman, C.B. Crouse, Andrew Curtis, Alan Douglas, Luis Esteva, Bill Hall, Dave Hill, Bob Hutt, Porter Irwin, Raymond Jeanloz, Paul Jennings, Carl Kisslinger, Ota Kulhanek, Art McGarr, Claus Prodehl, Haruo Sato, Jim Savage, Arthur Snoke, Kiyoshi Suyehiro, Bob Tilling, Seiya Uyeda, and Felix Waldhauser. Yehuda Ben-Zion also provided cross-references to his "Key Formulas in Earthquake Seismology," Appendix 2 of this Handbook.

Alphabetization is strictly letter-by-letter of the glossary terms (in boldface), ignoring hyphenation, comma, and space between words. Words included in parentheses are also ignored in the alphabetization.

Although this glossary is extensive, it is by no means complete. If a desired term is not found below or if the description is found to be inadequate, please consult the index of this Handbook and some general references, especially Jackson (1997), Lide (2002), and Meyers (2002). Acronyms for some selected institutions are also included in this glossary (their Web site addresses are given, if known to us). An effective way to find out more about an institution is to visit its Web site. Readers interested in finding out more about a nation or an institution may find the information in Chapters 79, 80, or 81, provided that such a report was submitted to either IASPEI or IAEE.

A

aa: A common type of lava flow composed of a jumble of lava blocks, with a rough, jagged, spiny, and generally clinkery or slaggy surface. Of Hawaiian origin, the term aa is applied to similar lava flows at other volcanoes. See also clinker.

AAL: Average annualized loss, the average economic loss expected per year for a specific property, portfolio of properties, or region as a result of one or more damaging earthquakes.

INTERNATIONAL HANDBOOK OF EARTHQUAKE AND ENGINEERING SEISMOLOGY, VOLUME 81B ISBN: 0-12-440658-0

acausal signal: An output signal that also exists for times prior to the application of an impulsive input.

acceleration: The rate of change of particle velocity per unit time.

accelerogram: A record (or time history) of acceleration produced by an accelerograph.

accelerograph: A seismograph designed to record acceleration, especially for strong ground shaking caused by large earthquakes nearby. Incorporating analog or digital recording, a traditional strong-motion accelerograph begins recording when the motion exceeds a certain specified trigger level. At present, some seismic networks record high-dynamic-range acceleration continuously at free-field sites or in building structures. See also strong-motion accelerograph.

accelerometer: An acceleration sensor, or transducer, that converts ground acceleration to an electrical signal, typically voltage, proportional to the acceleration. In early analog accelerographs, the accelerometer converted acceleration to movements of a light beam. Accelerometers may be within an accelerograph, but for studies of structural systems, they are typically located remotely and their signals transmitted (usually by cables) to a central recorder.

acceptable risk: The probability associated with a social or economic consequence of an earthquake that is low enough (in comparison with other risks) to be judged acceptable by appropriate authorities. It is often used to represent a realistic basis for determining design requirements for engineered structures or for taking certain social or economic actions. See Chapter 75.

accretionary prism: A generally wedge-shaped mass of trench sediments and volcanics added to the margin of an overriding plate during the subduction of an oceanic plate. See Von Huene and Scholl (1991) and Duff (1993).

accretionary wedge: See accretionary prism.

ACH method: ACH stands for Aki, Christoffersson, and Husebye (1977), who developed a method for determining the 3D structure of the Earth using the teleseismic data from the 2D seismic array. The method was extended to the local earthquake data by Aki and Lee (1976). It opened the way to the research area now called seismic tomography, although it was invented independently of the medical tomography.

acoustic emission: A term in rock mechanics indicating seismic radiation from dynamic formation of a microcrack. See Chapter 32 (p. 509).

active fault: A fault that has moved in historic (e.g., past 10,000 years) or recent geological time (e.g., past 500,000 years). Although faults that move in earthquakes today are active, not all active faults generate earthquakes—some are capable of moving aseismically (see also silent earthquake and slow earthquake). More precise attempts (usually unsatisfactory) have been made to define active faults for legal or regulatory purposes. See Yeats *et al.* (1997, p. 449).

active margin: A continental margin that is also a plate boundary with earthquakes and/or volcanic activity.

active tectonic regime: A term that refers to regions where tectonic deformation is relatively large and earthquakes are relatively frequent, usually near plate boundaries.

active tectonics: The tectonic movements that are expected to occur or that have occurred within a time span of concern to society.

active volcano: By a widely used but poor definition, one that is erupting or that has erupted one or more times in recorded history (see Decker and Decker, 1998). Active and potentially active (i.e., presently dormant) volcanoes occur in narrow belts that collectively comprise less than 1% of the Earth's total surface (Simkin *et al.*, 1994).

ADC: Analog-to-digital converter. It is an electronic device that converts a voltage produced by, for example, an accelerometer to digital values, at a digitization rate of typically 100 samples or more per second.

adopting: A term in hazard mitigation referring to the initial commitment of resources in taking necessary precautions. See Chapter 75.

aelotropy: A term proposed by Kelvin (1904) for anisotropy. See Chapter 53 (p. 875).

aftershock: An earthquake occurring as a consequence of a larger earthquake in roughly the same location. The sequence of such earthquakes following a larger one generally shows a regular decrease in the rate of occurrence, first discovered by Omori (1894), indicating a stress relaxation process as the rocks accommodate to their new postearthquake state. See Chapter 43 and Eq. (7.4) in Appendix 2 of this Handbook.

afterslip: The increase in displacement on a fault by creep following a sudden coseismic slip.

AGSO: Australian Geological Survey Organisation (now Geoscience Australia; URL:http://www.agso.gov.au). See Chapter 79.05.

AGU: American Geophysical Union (URL:http://www.agu.org).

air-coupled surface wave: Despite the great density contrast between air and ground, atmospheric pressure disturbances traveling over the Earth's surface can produce surface waves if the phase velocity is equal to the acoustic velocity in the air. Such a coupling with air has been observed for flexural waves in ice sheets floating on the ocean and for Rayleigh waves in the ground with low-velocity surface layers. A simultaneous arrival of airwaves and tidal disturbances was observed after the well-known explosion of the volcano Krakatoa in 1883.

airgun: A nonexplosive seismic source normally used at sea. Compressed air (about 15 MPa) is fed into air chambers of variable size (1–30 L) and is repetitively released into water to generate sound waves to be emitted. See Chapter 27.

airwave: The audible sounds are sometimes generated by earthquakes; a local earthquake may sound like distant thunder.

Instrumental measurements show that these sounds arrive simultaneously with the first P waves (Hill *et al.*, 1976). Long-period (minutes to hours) acoustic-gravity waves are also excited by great earthquakes as well as by volcanic explosions, meteorite falls, and nuclear blasts in the atmosphere.

Airy phase: The portion of dispersed wave trains associated with the maxima or minima of the group velocity as a function of frequency. The Airy function can be used for an approximate calculation of the waveform. Examples are continental Rayleigh waves at periods around 15 seconds, mantle Rayleigh waves at periods of 200–250 seconds, and surface waves of period of the order of 10 seconds that travel across an ocean at a velocity near 1 km/sec. See Chapter 21 (p. 345) and Aki and Richards (2002, p. 256).

aleatory uncertainty: The uncertainty in seismic hazard analysis due to inherent random variability of the quantity being measured. Aleatory uncertainties cannot be reduced by refining modeling or analytical techniques. Cf.: epistemic uncertainty.

aliasing: A serious error in the analog-digital conversion, pointed out by Blackman and Tukey (1958), that can occur when the sampling rate is less than twice the highest frequency contained in the signal to be digitized. A long-period ghost that does not exist in the original signal may appear after digitization. See Aki and Richards (1980, pp. 576-579).

alluvium: The loose gravel, sand, silt, or clay deposited by streams after the last ice age (see also Holocene).

Alpide belt: A mountain belt with frequent earthquakes that extends from the Mediterranean region eastward through Turkey, Iran, and northern India.

altitude of ambiguity: The topographic relief (or error) required to create one fringe in an interferogram. The height of ambiguity is expressed in units of meters. See Chapter 37.

ampere (A): A base unit in the international system of units (SI) for electric current. The official definition of the ampere: "The ampere is that constant current which, if maintained in two straight parallel conductors of infinite length, of negligible circular-cross section, and placed 1 meter apart in vacuum, would produce between these conductors a force equal to 2×10^{-7} Newton per meter of length." See NBS (1981).

amplification (by recording site): The term also used for describing the increase in amplitude of seismic waves due to the recording site's condition. Most seismographs are installed at a site on or near a surface with irregular topography and lithologic structures of heterogeneous material created by eons of weathering, erosion, deposition, and other geological processes. These complex near-surface structures tend to amplify the amplitude of incident seismic waves. See also site effect and site response.

amplification (by seismograph): The magnification of the ground motion for visual recording. Modern seismographs are able to magnify the ground motion a million times or even more (i.e., they are able to resolve ground-motion amplitudes as small as the diameter of molecules or even atoms).

amplitude: The size of the wiggles on an earthquake recording, or the deflection of the recorded trace from the zero level to its peak. It is often measured from peak or trough to the zero level. It is called double or peak-to-peak amplitude when measured from peak to trough.

amplitude response: The modulus of the complex frequency response (Fourier transform of the output divided by that of input) of a system.

anelastic attenuation: See intrinsic attenuation.

angular coherence function: The measure of coherency of two transmitted wave fields as a function of incident angles. See Chapter 13.

anisotropic: A description of a medium whose physical properties depend on the direction. See also seismic anisotropy.

annual probability of exceedance: The probability that a given level of seismic hazard (typically some measure of ground motions, e.g., seismic magnitude or intensity) or seismic risk (typically economic loss or casualties) can be equaled or surpassed within an exposure time of 1 year.

ansätz: A German word used by mathematicians to describe a starting or trial solution.

anthropogenic earthquake: An earthquake that is a by-product of human activities. See Chapter 40.

anticline: A fold, generally convex upward, whose core contains the stratigraphically older rocks.

antiplane strain: The assumed 2D symmetry, used to simplify mathematical analysis or numerical calculation, in which displacement depends on only two spatial coordinates and whose direction is perpendicular to the plane of those coordinates. Strike-slip displacement that does not vary along a strike is an example of antiplane deformation. Examples are a Love wave from a line source and mode 3 crack. Cf.: plane strain. See also mode 3 crack, Love wave, and Love (1944).

apparent polar wonder (APW) path: In the 1950s, the tracks of the paleomagnetic poles from each continent introduced as the apparent polar wonder paths. See Chapter 6, Merrill *et al.* (1998), and Hancock and Skinner (2000).

apparent stress: The product of the rigidity and the ratio of the seismic energy to the seismic moment for an earthquake, as defined by Wyss and Brune (1968). It is the fictitious stress (radiation resistance) that, acting through the actual earthquake slip, would do the work equivalent to the seismic energy radiated in the faulting. More simply, it is the product of the seismic efficiency and the average absolute stress as explained in Aki and Richards (2002, p. 55). See Chapter 35 (pp. 583-584) and Eq. (6.4) in Appendix 2 of this Handbook.

apparent velocity: If T (in sec) and Δ (in km) denote the travel times and epicentral distances for a seismic phase arriving at points of the Earth's surface, then the apparent velocity is defined as $d\Delta/dT$. The apparent velocity is the reciprocal of the seismic ray parameter or horizontal slowness.

APW: See apparent polar wonder (APW) path.

aquifer: A permeable rock body sufficiently saturated to yield an economical quantity of water, as, for example, from a well drilled into it. See Chapter 39.

arc: See volcanic arc.

Archaean: An eon in geologic time, from 4 to 2.5 billion years ago. Archaean is from the Greek and means "beginning." See also geologic time.

Arias Intensity: A broadband measure of the strength of strong ground motion observed in an accelerogram, defined by the integral over all natural frequencies of the energy input to a damped single degree of freedom oscillator in response to an accelerogram. In practice, it is computed by $A = \frac{\cos^{-1}\zeta}{\sqrt{1-\zeta^2}} \int_0^\infty a^2(t)dt$, in which ζ is the fraction of critical damping (viscous) and $a(t)$ is the accelerogram. The coefficient of the integral is a weak function of the damping factor. See Lange (1968) and Arias (1970).

Arkhanes (Crete): A temple that collapsed in 1700 BC exactly while the sacrifice of a youth was ongoing. See Chapter 46.

Armageddon (Megiddo): The site of the retroactively prophesied battle of the apocalypse and the earthquake that will happen during this battle, as described in the Book of Revelation, and also the site of ancient Megiddo in northern Israel, which has been subject to many historical battles and many earthquakes. See Chapter 46.

array: See seismic array.

array aperture: The size of an array measured by the maximum separation between pairs of seismometers in an array. Traditionally arrays are divided into small aperture (maximum separation around 3 km), medium aperture (maximum separation 10–25 km), and large aperture (maximum separation 100–200 km), depending on their size.

array beam: The combination of the output from each element of an array (say, by delay and sum processing) to enhance signals with a given apparent surface speed and azimuth of approach. By choosing the speed and azimuth expected for a hypocenter of the source, this process can enhance the beam of seismic waves coming from the particular source location. See Chapter 23.

arrival: The appearance of a particular seismic phase (e.g., *P* wave) on a seismogram.

arrival time: The time of the onset of a seismic wave (e.g., *P* wave) as first recorded by a seismograph. Arrival times are the basic data for locating earthquake hypocenters and for determining seismic velocity structures of the Earth.

aseismic: In seismology, an adjective for an area where few or no earthquakes have been observed. In earthquake engineering, an adjective for structures that are designed and built to withstand earthquakes.

ash: See volcanic ash.

ashfall: See fallout.

asperity: A site on a fault surface of higher strength than the surroundings, at which stress concentrates prior to fault rupture. In seismology, this term (and also the term barrier) is often used to describe the heterogeneity of a fault. It was first introduced by Lay and Kanamori (1981) to describe regions of earthquake ruptures where relatively high release of seismic moment occurs. Cf.: barrier. See Scholz (2002).

ASRO: Abbreviated Seismic Research Observatories seismograph network, also known as the Modified High-Gain Long-Period seismograph network. See Chapter 20.

association (of arrivals): See signal association.

asthenosphere: The weak lower portion of the near-surface thermal boundary layer and underlying mantle that undergoes significant ductile deformation under long-term loads. See Chapter 51 (p. 836).

asymptotic ray theory: The mathematical theory used to describe high-frequency seismic rays. The ray ansatz is constructed as a series in inverse powers of frequency in the frequency domain, or integrals of the source impulse in the time domain, with amplitude coefficients and travel-time-only functions of position along the ray. See Chapter 9.

attenuation: A decrease in wave amplitude with propagation, caused by intrinsic absorption and/or scattering. See Chapter13 (pp. 195-197), Chapter 51 (pp. 850-851), and Eq. (3.9) in Appendix 2 of this Handbook.

attenuation relationship: A mathematical expression that relates a ground-motion parameter, such as the peak ground acceleration, to the source and propagation path parameters of an earthquake such as the magnitude, source-to-site distance, or fault type. Its coefficients are usually derived from statistical analysis of earthquake records. It is a common engineering term for a ground-motion relation. See Chapter 60.

attribute (of an arrival): A quantitative measure of a seismic arrival, such as onset time, direction of first motion, period, and amplitude.

A-type earthquake: An earthquake with clear *P* waves and *S* waves occurring beneath volcanoes. It is indistinguishable from normal shallow tectonic earthquakes. In the original definition by Minakami (1961), it was contrasted to the B-type earthquake attributed to an extremely shallow focal depth. In recent years, the contrast has been attributed to the source process, and it is more often called a volcano-tectonic (VT) or high-frequency (HF) earthquake. See Chapter 25.

azimuth: In seismology, the direction from a seismic source to a seismic station recording this event. It is usually measured in degrees clockwise from the north. In radar terminology, it is the along-track component of the vector between the ground and the satellite. The azimuth direction is parallel to the trajectory of the satellite. See Chapter 37.

azimuthal anisotropy: A term used when the seismic-wave properties depend on the azimuth of propagation in the horizontal plane. For example, *Pn* waves in the neighborhood of an oceanic spreading center exhibit azimuthally anisotropic velocity. See Chapter 53.

B

back arc: The region adjacent to a volcanic arc and on the opposite side of the oceanic trench and subducting plate. See also volcanic arc.

back-arc basin: A basin floored by oceanic crust behind a volcanic arc. See Duff (1993) and Uyeda (1982). See also back arc.

back-azimuth: The direction from the seismic station toward a seismic source. It is usually measured in degrees clockwise from the north. Cf.: azimuth.

background noise: See microseism.

background seismicity: The seismicity that occurs at a more or less steady rate in the absence of an earthquake sequence, such as a swarm or foreshock-mainshock-aftershock sequence. In seismic hazard analysis, it is the seismicity that cannot be attributed to a specific fault or source zone.

ballistic projectile: A collective term for fragmental volcanic products (tephra) of diverse shapes greater than 64 mm in size. Such projectiles can exit the volcanic vent at speeds of tens to hundreds of meters per second on trajectories that are little affected by the eruption dynamics or the prevailing wind direction; thus, they are typically restricted to within 5 km of vents. See also volcanic block, volcanic bomb, and cinder.

bandpass filter: A filter that removes low- and high-frequency portions of the input signal outside of the passband. See Chapter 22.

barrier: A site on a fault surface of higher strength than the surroundings that is capable of terminating rupture or across which rupture may jump. In seismology, this term (and also the term asperity) is often used to describe heterogeneity of a fault. It was first introduced by Das and Aki (1977) to construct a two-dimensional fault model. Cf.: asperity. See Scholz (2002).

basalt: A general term for mafic igneous rocks (or corresponding melts) produced by partial melting of the upper mantle and comprising most of the crust beneath the oceans. See Jackson (1997) for a petrologic definition.

base isolation: A technique to reduce the earthquake forces in a structure by the installation of horizontally flexible devices at the foundation level. Such devices greatly increase the lowest natural periods of the structure for horizontal motions and thereby lower the accelerations experienced by the structure. The most common applications of base isolation are to old historic buildings, hospitals, and bridges. See Chapter 67.

baseline: In triangulation networks, the scalar distance between two benchmarks that determines the scale of the network. In GPS, the same term means the vector difference in position between two benchmarks. In INSAR, it means the (vector) separation or (scalar) distance between two orbital trajectories. See Chapter 37.

baseline correction: A term used in the processing of accelerograms. It corrects both a recorded signal for the bias in the zero-acceleration value and any long-period drift in the zero level that may arise from instrumental noise (in the case of digital accelerographs) or from the digitization of analog accelerograms. See Chapter 58.

basement: In geology, the igneous and metamorphic rocks that underlie the sedimentary deposits and extend downward to the base of the crust.

base moment: The moment, in the plane of response, experienced at the base of the structure during earthquake response or calculated during the earthquake-resistant design process. The maximum base moment is an important response parameter, particularly for structural elements such as exterior columns at the lower levels of a building. It is sometimes called overturning moment. See Chapter 67.

base shear: The horizontal shearing force at the base of the structure. The maximum base shear, typically in the form of a fraction of the weight of the structure, is an important parameter in earthquake response studies and in earthquake-resistant design. See Chapter 67.

base surge: A highly turbulent, dilute cloud of volcanic ash, air, and steam that expands radially away from the base of an eruption column at high velocity. It commonly forms when an eruption starts in a lake or coastal environment and involves the explosive interaction of hot magma with water.

basin-induced surface wave: The surface wave generated at the edges of a basin or at a strong lateral discontinuity inside the basin by incident body (*S* or *P*) waves. Their amplitude and waveform depend on the basin structure along the ray path from the edge (or the discontinuity) to the site. If the site lies in the central part of a large sedimentary basin, the basin-induced surface waves appear much later than the direct and reverberated *S* waves. The term "secondary surface waves" sometimes used for them is not appropriate because their amplitudes can be of primary importance. See Chapter 61.

basin-transduced surface wave: The surface wave transduced at edges of a basin or at a strong lateral discontinuity inside the basin from incident surface waves. They are distinguished from basin-induced surface waves by the difference in incident waves. These surface waves are observed if the earthquake source is shallow and distant. Refraction and mode conversion of both Love and Rayleigh waves and transformation among them occur at the edge of a basin. See Chapter 61.

bay: In a building structure, the horizontal space between adjacent planar frames, walls, or lines of columns. See Chapter 69 and Park and Paulay (1975).

Bayesian method: A method accepting the interpretation of a probability not only as the limit of an experimental histogram but also as the representation of subjective information. See Chapter 16 and the biography of Thomas Bayes in Chapter 89.

BCIS: Bureau Central International de Séismologie. See Chapters 4, 41, and 88.

beam: See array beam.

bedrock: A relatively hard, solid rock that commonly underlies soil or other unconsolidated materials.

bending-moment fault: A fault formed due to bending of a flexed layer during folding. Normal faults characterize the convex side, placed in tension, and reverse faults characterize the concave side, placed in compression.

Benioff zone: See subduction zone.

BGS: British Geological Survey (URL: http://www.bgs.ac.uk). See Chapter 79.54.

Big Benioff: A short-period seismometer designed by Hugo Benioff with a 107.5-kilogram mass, used in the WWSSN program. Total weight of this seismometer is 206 kilograms (454 pounds). See Chapter 20 and the biography of Hugo Benioff in Chapter 89.

blind fault: A fault that does not extend upward to the Earth's surface. It usually terminates upward in the axial region of an anticline. If its dip is $< 45°$, it is a blind thrust. The Northridge thrust fault responsible for the 1994 Northridge, California, earthquake is an example of a blind fault. See Yeats *et al.* (1997).

block-and-ash flow: A small-volume pyroclastic flow characterized by a large fraction of dense to moderately vesicular juvenile blocks in a matrix of the same composition. Such flows are typically associated with collapse of vulcanian eruption columns or are produced from partial or complete collapse of unstable, viscous lava domes as they oversteepen at their fronts during active growth (e.g., as during the eruption of Unzen Volcano, Japan, in 1991). See Freundt *et al.* (2000).

block (or blocky) lava: A term sometimes applied to all lava flows with fractured surfaces that are covered with angular fragments (including aa). However, as emphasized by Macdonald, 1972, the term is more appropriately restricted to those lava flows that, even though similar in overall structure to aa, lack the jaggedly spinose features of aa and have fragments characterized by more regular forms (often approaching cubelike) and smoother surfaces.

block Toeplitz matrix: A matrix where all terms along each diagonal are the same. Given the entries in the leftmost column and the top row, a block Toeplitz matrix is fully specified. See Chapter 50.

body wave: Any wave that propagates through an unbounded continuum as opposed to surface waves that propagate along the boundary surface. There are two basic types of body waves: *P* wave and *S* wave. See also *P* wave, *S* wave, and Eq. (1.9) in Appendix 2 of this Handbook.

body-wave magnitude (m_B and m_b): An earthquake magnitude calculated from amplitude/period ratios of body waves. m_B and m_b are based on data recorded by relatively broadband and short-period seismographs, respectively. The original body-wave magnitude introduced by Gutenberg (1945) is denoted by m_B. Body-wave magnitudes assigned by USGS and ISC are m_b (ISC uses the notation M_b). See Chapter 44.

bog-burst: The result of when large deposits of peat on a hillside soak up large amounts of rainfall like a sponge and the weight of the absorbed water becomes too great for the strength of the bog holding it so that the peat deposit suddenly loses its cohesion, releasing an avalanche of water, mud, and organic material. In early records, the word "earthquake" is often used to describe bog-bursts and landslides as well as genuine earthquakes. See Chapter 48.5.

borehole sensor: A sensor installed below the Earth's surface in a borehole.

Born approximation: A perturbation method that iteratively solves an inhomogeneous wave equation, where waves are decomposed into primary waves, which satisfy the homogeneous wave equation and scattered waves of small amplitude. See Chapter 13 and the biography of Max Born in Chapter 89.

branch (of travel-time curve): A term used in seismology for a segment of the travel-time curves that is related to particular ray paths of the same type of seismic waves. An example is the receding branch in triplication. See also travel time and triplication.

breccia: A coarse-grained clastic rock composed of angular broken rock fragments in a fine-grained matrix.

brittle: An inability to accommodate inelastic strain without loss of strength. See Chapter 32 (p. 507).

brittle-ductile boundary: See brittle-ductile transition.

brittle-ductile transition: A zone within the Earth's crust that separates brittle rocks above from ductile (plastic or quasi-plastic) rocks below. Its depth is usually identified as the maximum focal depth of local earthquakes and occurs at depths around 15 km at which the temperature lies between 200° to 400°. See also brittle-plastic transition and Ito (1990).

brittle-failure (BF) earthquake: A term sometimes used in volcano seismology to distinguish an earthquake dominated by brittle processes (frictional slip or formation of a tensional crack) from those in which fluids are thought to play an active role (such as LP earthquakes) or harmonic tremor. The designation "BF earthquake" is generally synonymous with high-frequency, volcano tectonic, or A-type earthquake.

brittle-plastic transition: A simplified, more precise model of the brittle-ductile transition in which the plastic fracture is considered as the deformation mechanism in the ductile part.

broadband body-wave magnitude (m_B): See body-wave magnitude.

broadband seismogram: A seismogram recorded by a broadband seismograph. See Chapter 23.

broadband seismograph: A type of seismograph that can record seismic signals over a broad frequency range. To avoid the strong ambient noise caused by ocean waves (microseisms), two types of seismographs have traditionally been used to record seismic signals: one for periods longer than about 10 seconds, the other for those shorter than about 1 second. A broadband seismograph can record faithfully seismic signals in the period range roughly from 0.1 to 100 seconds, thanks to the improved linearity range of the seismometer and the dynamic range of the recorder. See Chapter 18.

broadband seismometer: A seismic sensor used in a broadband seismograph.

Brune model: The omega-squared circular-crack model proposed by Brune (1970). The physical basis of this model was refined by Madariaga (1976), who solved the mathematical problem of seismic radiation from an expanding circular crack that stops. See also omega-squared model, Chapter 12, and Aki and Richards (2002, pp. 560-565).

***b* slope:** See *b* value.

B-type earthquake: An earthquake occurring under volcanoes with emergent onset of *P* waves, no distinct *S* waves, and low-frequency content as compared with the usual tectonic earthquakes of the same magnitude. In the original definition by Minakami (1961), its character difference from the A-type earthquake was attributed to the extremely shallow focal depth (<1 km). More recently the difference has been attributed to the source process and is now more often called long-period (LP) or low-frequency earthquake. See Chapter 25.

bulk density: The mass or weight of a material divided by its volume, including the volume of its pore spaces.

bulk modulus: The modulus of volume elasticity that relates a change in volume to a change in hydrostatic pressure. It is the reciprocal of compressibility. See Birch (1966) for an introduction and a compilation of values, and Eq. (1.6) in Appendix 2 of this Handbook.

Burridge and Knopoff model: A spring-and-block model proposed by Burridge and Knopoff (1967) for earthquakes. See Chapter 12 (pp. 176-177).

***b* value:** A coefficient in the frequency-magnitude relation, $\log N(M) = a - bM$, obtained by Gutenberg and Richter (1941, 1949), where M is the earthquake magnitude and $N(M)$ is the number of earthquakes with magnitude greater than or equal to M. Estimated b values for most seismic zones fall between 0.6 and 1.1 but may be higher in volcanic regions. See Chapter 43 (p. 723) and Eq. (7.2) in Appendix 2 of this Handbook. See also the biographies of Beno Gutenberg and Charles Francis Richter in Chapter 89.

Byerlee's law: The common observation from laboratory friction experiments that the coefficients of friction for nearly all rocks that comprise the lithosphere fall in the range 0.6–1. See Chapters 32 and 40 and Eq. (5.2) in Appendix 2 of this Handbook.

C

***c*:** A symbol used to indicate the reflection at the core-mantle boundary for waves incident from the mantle. For example, an *S* arrival at the surface from an *S* wave reflected at the core-mantle boundary is referred to as *ScS*.

caldera: A circular or ovoid depression, generally 5–20 km in diameter, formed by collapse or subsidence of the central part of a volcano, often associated with large explosive eruptions.

CALEX: A computer program for determining parameters of the transfer function of a seismometer. See also Chapter 85.18 and transfer function.

calibration: The process of determining the response function and sensitivity of a seismic instrument. See Chapter 19.

calibration pulse: An electronic signal used to calibrate seismic instruments. See also calibration.

calorie (cal): A unit of energy originally defined as the heat required to raise the temperature of 1 gram of water by 1° Celsius. Different usage of the word "calorie" has been confusing (e.g., the calorie in nutrition is 1000 times the calorie in physics). The thermodynamic calorie is now defined as 4.184 joules. See Lide (2002).

Caltech: California Institute of Technology in Pasadena, California, USA (URL: http://www.caltech.edu).

capable fault: A mapped fault that is deemed a possible site for a future earthquake with magnitude greater than some specified threshold. In nuclear reactor siting, a fault capable of surface rupture but which may or may not generate earthquakes (American Nuclear Society, 1982).

cascade model of earthquakes: An extended version of the characteristic earthquake model in which several consecutive fault segments can rupture in a variety of combinations. The slip on each segment is assumed to be characteristic to the segment. See Chapter 5.

cataclasite: A cohesive fault rock comprising mineral fragments derived predominantly from brittle cataclastic fragmentation with textures ranging from essentially random-fabric to foliated. The cataclasite series (protocataclasite—cataclasite—ultracataclasite) reflects progressive reduction in grain size. See Chapter 29 (p. 458).

catalog of earthquakes: A chronological listing of earthquakes. Early catalogs were purely descriptive (i.e., they gave the date of each earthquake and some description of its effects). Modern catalogs are usually quantitative (i.e., earthquakes are listed as a set of numerical parameters describing origin time, hypocenter location, magnitude, moment tensor, etc.). See Chapter 41 for an instrumental-determined earthquake catalog (1900–1999), Chapter 42 for a catalog of deadly earthquakes (1500–2000), and Chapter 45 (pp. 759-761) for a discussion of historical and modern earthquake catalogs.

causal signal: An output signal that does not exist for times prior to the application of an impulsive input. In other words, an impulse response h(t) that vanishes for $t < 0$. See also impulse response, linear system, and convolution.

caustic point: A point on the Earth's surface where the rate of change of the epicentral distance traveled by a seismic ray with the change of its takeoff angle changes its sign. A large concentration of arriving seismic energy will generally be observed at the caustic point. See Chapter 21 (p. 340).

C-band: A radar frequency around 5 GHz with wavelength around 6 cm. See Chapter 37.

CDP: Crustal Dynamics Project, a NASA research program in the 1980s. See Chapter 37.

CDSN: China Digital Seismograph Network. See Chapter 20.

Cenozoic: An era in geologic time, from 66.4 million years ago to the present. See also geologic time.

central eruption: See summit eruption.

CERESIS: Centro Regional de Sismología para América del Sur, the South American regional seismological center based in Lima, Peru. See Chapter 81.9.

CGPS: Continuously operating GPS receivers and networks. See Chapter 37.

cgs system of units: An obsolete system of units based on (1) centimeter for length, (2) gram for mass, and (3) second for time. Some derived cgs units with special names are: (1) force in dyne (cm g s^{-2}), (2) energy in erg (cm^2 g s^{-2}), and (3) acceleration of free fall in gal (cm s^{-2}). See also International System of Units.

channel: In observational seismology, the signal output of a component of a seismic sensor. Modern seismic recorders are able to record simultaneously the signals from many channels. See also data logger.

chaos: A term that has been used since antiquity to describe various forms of randomness. It is now a commonly accepted technical term referring to the irregular, unpredictable, and apparently random behavior of deterministic dynamic systems. Solutions to deterministic equations are chaotic if adjacent solutions diverge exponentially in phase space. See Socolar (2002) or Wolfram (2002) for a general discussion of chaos and Chapter 14 for an application in seismology.

characteristic earthquake model: A fault-specific earthquake model in which a given fault segment generates an earthquake of a size and mechanism determined from the geometry of the segment. At a specific location along a fault, the displacement (slip) is the same in successive characteristic earthquakes. Other earthquakes occurring on the fault are much smaller than these. In its application to seismic hazard analysis, it refers to an earthquake of a specific size that is known or inferred to recur at the same location, usually at a rate greater than that extrapolated from the frequency-magnitude relation for smaller earthquakes in the area. See Chapter 5 (pp. 44-45), Chapter 30 (pp. 482-484), and Eq. (7.3) and Figure 15 in Appendix 2 of this Handbook.

charge: A basic property of matter that is conserved. Electric charge flows in electric currents or accumulates on nonmetallic surfaces. The unit of electric charge is measured in Coulombs. One electron has a negative charge of 1.60×10^{-19} Coulombs. See Chapter 38.

Christoffel matrix: A symmetric real-valued matrix, of which eigenvalues give the seismic velocities and eigenvectors give the polarization directions of the waves. See Chapters 9 and 53 and the biography of Elwin Brun Christoffel in Chapter 89.

cinder: A common fragmental volcanic product (tephra), classified as ranging between 2 and 64 mm in size.

cinder cone: A relatively small cone-shaped hill or mountain, rarely exceeding 1 km in height, that nearly always has a truncated top in which is a bowl-shaped crater (see Macdonald, 1972). As its name implies, this volcanic edifice is built entirely of tephra ejected during moderately explosive eruptions. See also cinder.

circular fault: A fault-plane geometry in which the rupture front starts from a point on the fault and spreads as a circle. See also Brune model and rupture front.

Circum-Pacific belt: The zone surrounding the Pacific Ocean that is characterized by frequent and strong earthquakes and many volcanoes as well as high tsunami hazard. See also Ring of Fire.

clinker: An angular, irregularly shaped, jagged fragment of lava, ranging widely in size from a few centimeters to many meters, that makes up the surface of aa lava flows (Macdonald, 1972). The term "slag" is sometimes also used for clinker.

CLVD: Compensated linear-vector dipole. It is the force system representing a crack opening or closing under the constraint of no volume change. The corresponding moment tensor has zero trace (purely deviatoric), traditionally called cone-type mechanism in Japan. See also Jost and Herrmann (1989), Julian *et al.* (1998), and non-double-couple earthquake.

CMG-3TB: A broadband-borehole-deployable seismometer manufactured by Guralp Systems Limited, Reading, England. See Chapter 20.

CMGLNM: Low-noise model for the Guralp Systems CMG-3T and CMG-3B broadband seismometers. See also CMG-3TB and Chapter 20.

CNES: Centre National d'Études Spatiales, the French space agency. See Chapter 37.

coda: The part of the seismogram that follows the arrival of a well-defined wave. For example, the teleseismic *P* coda follows the arrival of teleseismic *P* waves. See Chapter 21 (pp. 333-334).

coda attenuation: See coda *Q*.

coda normalization method: A method to use the power of local earthquake *S* coda at a given lapse time as a measure of earthquake radiation energy, which is used for measurements of site amplification factors, attenuation per unit travel distance, and source energy. This method is based on the assumption of a uniform spatial distribution of coda energy density of a local earthquake at a long lapse time measured from the earthquake origin time. See Chapter 13 (pp. 200-201).

coda *Q*: A parameter characterizing the amplitude decay of *S* coda of a local earthquake with the lapse time defined by Aki and Chouet (1975), assuming single *S* to *S* scattering. The coda lasts longer for higher coda *Q*.

coda wave (of a local earthquake): The seismograms of a local earthquake usually show some vibrations long after the passage of body waves and surface waves. This portion of the seismogram to its end is called the coda. These vibrations are believed to be back-scattering waves due to lateral inhomogeneity distributed throughout the Earth's lithosphere. Coda has been extensively used to obtain source spectra as well as to measure seismic attenuation and site amplification factors by the coda normalization method. See Chapter 13 (pp. 198-201).

coefficient of friction (μ_f): The coefficient of the linear relationship between shear stress (τ) and normal stress (σ_n) on a fault or joint that is slipping: $\tau = S_o + \mu_f \sigma_n$. See Chapter 32 (p. 506) and section 5 in Appendix 2 of this Handbook.

coefficient of internal friction (μ_i): The coefficient of the linear relationship between shear stress (τ) and normal stress (σ_n) resolved onto an incipient rupture plane: $\tau = S_o + \mu_i \sigma_n$. See Chapter 32 (p. 505) and section 5 in Appendix 2 of this Handbook.

coherence: The degree to which two wave fields are in phase. Mathematically, it is the normalized cross-power spectrum of the two wave fields. It is the frequency-domain counterpart of the correlation in the time domain. See Chapter 13.

cohesion (S_o): The inherent shear strength of a fault or joint surface, or shear strength at zero normal stress in the equation: $\tau = S_o + \mu_i \sigma_n$. See Chapter 32 (p. 506).

cohesionless: The condition of a sediment whose shear strength depends only on friction because there is no bonding between the grains. This condition is typical of clay-free sandy deposits. See also cohesion.

cohesive force: The force that acts in the cohesive zone of a crack located between the freely slipping crack surface and the intact elastic body ahead of the crack tip. The existence of a cohesive zone removes the stress singularity at the crack tip. See Aki and Richards (2002, pp. 548-552).

collision zone: A zone where two continents collide as a result of continued subduction of oceanic plate. Continental collision is the cause of mountain building or orogeny. See Chapter 6 and Duff (1993).

colluvial wedge: A prism-shaped deposit of fallen and washed material at the base of (and formed by erosion from) a fault scarp or other slope, commonly taken as evidence in outcrop of a scarp-forming event.

colluvium: The loose soil or rock fragments on or at the base of gentle slopes or hillsides, deposited by or moving under the influence of rain wash or downhill creep.

column collapse: A condition that occurs when an explosive eruption column, composed of volcanic gases, tephra, and air, becomes denser than the ambient atmosphere and collapses to ground level.

comparative subductology: The comparative study of subduction zones emphasizing the differences between the Mariana-type and Chilean-type subduction zones, caused by difference in the degree of plate coupling. See Chapter 6 and Uyeda (1982).

complex site effect: A dynamic amplification effect (e.g., that arising from a resonance condition, generated by propagation of earthquake waves in 2D/3D near-surface geological configurations, such as sedimentary valleys, and topographic irregularities). See Chapter 62.

composite volcano: See stratovolcano.

compressibility: The reciprocal of bulk modulus. See also bulk modulus.

compressional tectonic setting: A region undergoing lateral contraction and for which the vertical principal stress is the minimum. See Chapter 40.

compressional wave: See *P* wave.

compressive strength: The maximum compressive stress that can be applied to a material before failure occurs. See also strength.

compressive stress: The stress component normal to a given surface, such as a fault plane, that tends to push the materials together. Cf.: tensile stress. See also normal stress and Figure 1 in Appendix 2 of this Handbook.

COMPSYN: A complete synthetic software package for evaluating the representation theorem integrals on a fault surface. See Chapter 85.14.

confidence ellipsoid: In hypocenter location calculations, an ellipsoid surrounding the computed hypocenter within which the true hypocenter is (formally) located at some specified level of confidence.

conical wave: See head wave.

Conrad discontinuity: The seismic boundary between the upper and middle crust that is usually defined by an increase in seismic velocity from 6.2–6.4 km/sec to about 6.6–6.8 km/sec. The term has fallen into disuse in recent years due to the lack of universality of such a discontinuity. See Chapter 54 and the biography of V. Conrad in Chapter 89.

continental deformation: A term usually used to emphasize the contrast between deforming zones in the oceans and on the continents. See also continental tectonics and Chapter 31 in this Handbook.

continental drift: A theory that continents have displaced thousands of kilometers in the last few hundred million years. See Chapter 6.

continental tectonics: A term used to include the large-scale motions, interactions, and deformations of the continental lithosphere. It is often used in contrast to plate tectonics. Whereas deforming zones in the oceanic plates are usually narrow and confined, on continents they are often spread over wide areas, requiring a different approach to their description and analysis. See Chapter 31.

contractional jog: A step-over between en-échelon fault strands, where slip transfer involves contraction, and increased mean compressive stress within the step-over. See Chapter 29.

contractive soil: The granular soil that decreases or tends to decrease in volume during large shear deformation. The tendency to contract increases pore water pressures in undrained saturated soils during shear. Slopes and embankments underlain by contractive soils may suffer catastrophic strength loss and flow failure during earthquake shaking. See Chapter 70.

controlled-source seismology: The type of seismic investigation that utilizes man-made seismic sources, such as chemical explosions detonated in boreholes. See Chapter 54.

convergent margin: A plate boundary zone where two plates are converging to one another. Either plate may subduct under, collide into, or obduct over the other plate. See Chapter 27.

converted wave: The conversion of P to S waves and S to P waves that occurs at a discontinuity for non-normal incidence. These converted waves sometimes show distinct arrivals on the seismogram between the P and S arrivals and may be used to determine the location of the discontinuity.

convolution: The convolution of two functions $x(t)$ and $h(t)$ that is often written as $x(t) * h(t)$ and that is defined mathematically as $y(t) = x(t) * h(t) = \int x(\tau)h(t - \tau)d\tau$. In signal processing, $y(t)$ represents an output of a stationary linear system in terms of the input $x(t)$ and impulse response $h(t)$ of the system. See also impulse response and linear system.

core: The central part of the Earth's interior with upper boundary at a depth of about 2900 km. It represents about 16% of the Earth's volume and is divided into an inner solid core, and an outer fluid core. See Chapters 51 and 56.

core shadow zone: A gap in the emergence of P-waves that have traveled directly through the Earth extending from epicentral distances of about 103° to 144°. This gap or shadow of the core is caused by the sharp decrease in the P-wave velocity as the waves pass from the bottom of the mantle into the core. See Chapter 21 (p. 340).

corner frequency: In earthquake source studies, a parameter characterizing the far-field body-wave displacement spectrum. See also omega-squared model and Figure 7 and Eq. (6.1) in Appendix 2 of this Handbook.

corrected acceleration: An acceleration time history that has been "corrected" from the raw data recorded by an accelerograph. The correction typically involves removing drift, spikes, and any distorting effects created by the instrument or digitizing process. See Chapter 58.

corrected penetration resistance: A measure of an *in situ* property of soil based on a penetration test, such as standard penetration test resistance, corrected for the influence of overburden pressure, hammer energy, rod length, borehole diameter, and sampler type, or cone penetration resistance, corrected for the influence of overburden pressure. See Chapter 70.

coseismic: A term meaning occurring at the same time as an earthquake.

Coulomb friction: In a constitutive law that is governed by Coulomb friction, a slip occurs when shear stress exceeds strength defined by a coefficient of friction (a constant) times the normal stress. In contrast, the rate-and-state dependent law involves the coefficient of friction, which is not a constant but depends on the sliding speed (rate) and contacts (state) on the fault plane. See Chapter 73 and Eq. (5.1) in Appendix 2 of this Handbook. See also the biography of Charles Augstin de Coulomb in Chapter 89.

CPU: Central processing unit of a computer.

Crary wave: A train of sinusoidal waves with nearly constant frequency observed on a floating ice sheet (Crary, 1955). They are multireflected SV waves with horizontal phase velocity near the speed of compressional waves in ice.

crater: A bowl- or funnel-shaped depression, generally in the top of a volcanic cone, often the major vent for eruptions. The distinction between the terms "crater" and "caldera" is blurred, but generally caldera tends to be applied to depressions with diameters >5 km.

creep: The release of shear strain accomplished without significant radiated energy. See Chapter 32 (p. 516).

creep event: An episodic slip across a fault trace observed at the surface over minutes to days. See Chapter 36.

creep meter: An instrument for measuring displacement across a fault trace on the Earth's surface, usually with a baseline length around 10 meters. See Chapter 36.

creep wave: A regional strain wave propagating along active faults, as suggested by Savage (1971). However, these creep waves have not yet been directly observed. See Chapter 36.

Cretaceous pulse of spreading: A phenomenon that is interpreted as caused by mantle superplume activity. During the Cretaceous period (124–83 Ma) when there was no geomagnetic reversal, sea-floor spreading was 50–75% faster than in other periods. See Chapter 6 and Larson (1991).

critical damping: In vibration theory, critical damping occurs when the dashpot coefficient has a value equal to twice the mass times the oscillator's natural frequency expressed in radians per second. The fraction of critical damping (ζ) is defined by the ratio of the actual dashpot coefficient to the dashpot coefficient giving critical damping. In other words, critical damping is the value of the dashpot coefficient that marks the transition of free vibration of the oscillator from the underdamped case of exponentially decaying harmonic motion to a-periodic motion that decays without changing algebraic sign. This can be demonstrated by solving the homogeneous version of Eq. (1) of Chapter 67 of this Handbook, with ζ less than unity, equal to unity and larger than unity. Critical damping is important in seismometer design (see Richter, 1958, p.215-219). See also damping, and Eq. (3.10) in Appendix 2 of this Handbook.

critical facility: A man-made structure whose ongoing performance during an emergency is required or whose failure could threaten many lives. This may include (1) structures such as nuclear-power reactors or large dams whose failure might be catastrophic; (2) major communication, utility, and transportation systems; (3) high-occupancy buildings such as schools or offices; and (4) emergency facilities such as hospitals, police and fire stations, and disaster-response centers.

critically stressed crust: A state of stress in the Earth's brittle crust balanced by the frictional strength of the Earth's crust. See Chapter 34.

cross-axis sensitivity: The susceptibility of a sensor to produce a spurious signal in response to motion perpendicular to the sensing direction of the sensor. It is often due to limitations in the design or manufacture of the mechanical suspension in the sensor or misalignment of the components within the sensor. See Chapter 58.

crossover: The distance from an event where two different phases arrive at the same time, allowing constructive interference that sometimes enhances the signal amplitudes.

cross-talk: The appearance of a signal on a channel due to electrical leakage from a signal on an adjacent channel. It is often due to electrical problems with shielding, grounding, etc. See Chapter 58.

cross-track: The component of motion of a satellite perpendicular to its trajectory. See Chapter 37.

CRR: Cyclic resistance ratio.

crust: The outmost layer of the Earth above the Moho. The crust in continental regions is about 25–75 km thick, and that in oceanic areas is about 5–10 km thick. It represents less than 0.1% of the Earth's volume, with rocks that are chemically different from those in the mantle. See also mantle and Chapters 54 and 55.

CSB: China Seismological Bureau, formerly known as the State Seismological Bureau (SSB). See Chapters 79.14 and 80.6.

CSR: Cyclic stress ratio.

CUBE: Caltech-US Geological Survey Broadcast of Earthquakes, a program to develop and distribute real-time earthquake information in southern California. See Chapter 78.

cultural noise: The seismic noise generated by various human activities such as traffic, construction works, and industries. See Chapter 19.

cumulative seismic moment: The sum of the seismic moments of earthquakes that have occurred in a region from a fixed starting time until time t, as a function of t. It can be related to the cumulative tectonic displacement across a plate boundary lying in the region or to the cumulative tectonic strain within the region. See Chapter 5.

curtain of fire: See lava fountain.

CWB: Central Weather Bureau in Taipei, Taiwan (URL: http://www.cwb.gov.tw/V3.0e/index-e.htm).

cyclic mobility: A condition in which liquefaction occurs and flow deformation commences in response to static or dynamic loads but deformation is arrested by dilation at moderate to large shear strains. See Chapter 70.

cyclic resistance ratio (CRR): It is the capacity of a granular soil layer to resist liquefaction, expressed as a ratio of the dynamic stress required to initiate liquefaction to the static effective overburden pressure. See Chapter 70.

cyclic stress ratio (CSR): It is the seismic load placed on a granular soil layer, expressed as the ratio of the average horizontal dynamic stress generated by an earthquake to the static effective overburden pressure. See Chapter 70.

D

D″: A layer surrounding the fluid core of the Earth occupying the lowermost 200 km of the mantle, which is believed to be a region of strong thermal and chemical heterogeneity. See Chapter 51.

damage probability matrix: A matrix giving the probabilities of sustaining a range of losses for various levels of a ground-motion parameter.

damage scenario: A representation of the possible damage caused by an earthquake to the built environment in an area, in terms of parameters useful for economical and engineering assessment or postearthquake emergency management. See Chapter 62.

damping: In seismometry, a specific device (symbolized as a dashpot) that is added in the design of a seismometer to diminish the effects of resonance and of the transient free oscillation. In vibration analysis, it indicates the mechanism for the dissipation of the energy of motion. Viscous damping, which is proportional to the velocity of motion and is described by linear equations, is used to define different levels of response spectra and is commonly used to approximate the energy dissipation in the lower levels of earthquake response. See also critical damping and Chapter 67.

damping constant: A parameter quantifying the effect of damping.

DARPA: [United States] Defense Advanced Research Projects Agency (URL: http://www/darpa.mil).

data logger: A digital data acquisition unit, usually for multichannel recordings.

dB: See decibel.

debris avalanche: A flowing mixture of debris, rock, and moisture that moves downslope under the influence of gravity. It principally differs from debris flows in that it is not water-saturated. Debris avalanches are almost always associated with sector collapse at volcanoes. Cf.: debris flow. See Vallance (2000).

debris flow: A water-saturated mixture of rock debris and water having large sediment concentration that moves downslope in a laminar fashion under the influence of gravity. Depending on the ratio of debris, sediment, and water, debris flows can grade into ordinary water floods. Cf.: debris avalanche. See Vallance (2000).

decibel (dB): A logarithmic measure of relative signal amplitude, defined as $20 \log(A/A_o)$, where A_o is the signal amplitude at some reference level, typically the minimum signal resolvable by the recorder.

declustering: The procedure for removing dependent earthquakes, such as foreshocks and aftershocks. A declustered earthquake catalog contains only mainshocks and isolated earthquakes. The largest shock in each earthquake sequence is considered as the mainshock and remains in the catalog. See Chapter 43.

deconvolution: A procedure that removes the unwanted effects of convolution on signal. Cf.: convolution.

delay-and-sum processing: A procedure for forming an array beam. Time delays are applied to the output of each of the seismometers of an array to bring the desired signal into phase, followed by summing of the outputs with appropriate weights. See Chapter 23.

delay firing: See ripple firing.

DEM: Digital elevation model, an array of digitized elevations or topographic values. See Chapter 37.

density: Two definitions: (1) the quantity of something per unit measure, such as unit length, area, volume, or frequency (see also power spectral density); (2) the mass per unit volume of a substance under specified conditions of pressure and temperature.

depth phases (*pP, pwP, pS, sP, sS*): The depth phase symbol *pP* that is used for *P* waves propagated upward from the hypocenter, turned into downward propagating *P* waves by reflection at the free surface (*pwP* at the ocean surface), and observed at teleseismic distances. *sS*, *sP*, and *pS* have analogous meanings. For example, *sP* corresponds to a phase that ascends from the focus to the surface as an *S* wave and then, after reflection, travels as a *P* wave to the recording station. These phases are useful for an accurate determination of focal depth. See Chapters 21 and 41.

design ground motion: A level of ground motion used in structural design. It is usually specified by one or more specific strong-motion parameters or by one or more time series. The structure is designed to resist this motion at a specified level of response, for example, within a given ductility level. See Chapter 68.

design spectrum: The specification of the required strength or capacity of the structure plotted as a function of the natural period or frequency of the structure and of the damping appropriate to earthquake response at the required level. Design spectra are often composed of straight-line segments and/or simple curves (e.g., as in most building codes), but they can also be constructed from statistics of response spectra of a suite of ground motions appropriate to the design earthquake(s). To be implemented, the requirements of a design spectrum are associated with allowable levels of stresses, ductilities, displacements, or other measures of response. Cf.: response spectrum. See Housner (1970).

detection: The identification of arrival of a seismic signal with amplitudes above and/or with waveform and spectral contents different from ambient noise.

deterministic earthquake scenario: A representation, in terms of useful descriptive parameters, of an earthquake of specified size postulated to occur at a specified location (typically an active fault), and of its effects. See Chapter 62.

deterministic system: A dynamical system whose equations and initial conditions are fully specified, not stochastic or random. See Chapter 14.

deviatoric moment tensor: A moment tensor with no isotropic (explosive or implosive) component. The sum of its eigenvalues is zero.

diaphragm: In structural engineering, a planar structural element, such as a floor slab, that is very resistant to inplane deformations. In earthquake-response studies, floor diaphragms are often considered to respond to horizontal motion as rigid bodies. See Chapter 69.

differential pressure gauge: A device for measuring long-period (0.1–2000 seconds) pressure fluctuations on the seafloor. See Chapter 19.

differential travel time: The difference between the arrival time of one seismic phase and that of another phase on the same seismogram from the same source. Differential travel times are often used to eliminate the uncertainty in the event origin time and to reduce the uncertainties in the structure near the source or receiver. See Chapter 56.

diffracted *P*: The P wave that follows a ray path from a surface focus that grazes the Earth's core, is diffracted along the core-mantle boundary, and emerges at an epicentral distance of about 100°. Although geometrical optics predicts no direct arrivals of P waves in the shadow zone beyond this distance, P waves (especially long-period) are observed in the shadow zone up to distances of 130°. See Aki and Richards (2002, pp. 456-457).

diffraction: A wave phenomenon, which cannot be explained by geometrical optics, that commonly occurs when the curvature of the surface of an obstacle is large compared with the curvature of the incident wave front. See Chapter 21.

diffusion: A process describing transport of mass or heat caused by differences in chemical potential, pressure, or temperature. It can also describe seismic energy transport in randomly heterogeneous elastic media. See Chapters 13 and 36.

digital accelerograph or seismograph: An accelerograph or seismograph in which the analog output signal from an accelerometer or seismometer is first converted to digital samples and then stored as time series data in digital form. This contrasts with an analog accelerograph or seismograph, which records the analog output signal on paper, or optically on film, or on magnetic tape.

digital filter: A mathematical tool designed to modulate the frequency characteristics of a discrete time series by means of numerical calculations on a computer. See Chapters 22 and 63.

digitization: The conversion of an analog waveform (e.g., an accelerogram recorded optically on film) to a series of discrete x-y values corresponding to time-acceleration pairs, for use in subsequent computer processing. See Chapter 58.

dilatancy: The inelastic increase of a volume of rock caused by the formation of cracks. Dilatancy may occur in a rock mass prior to an earthquake with some observable effects, such as changes in elevation and effects of temporary hardening against fault rupture.

dilational jog: A step-over between en-échelon fault strands where slip transfer involves dilatation and reduced mean compressional stress within the step-over. See Chapter 29.

dilative: A type of granular soil that increases or tends to increase in volume during shear. See Chapter 70.

dilatometer: A borehole strainmeter measuring volumetric strain. See Chapter 36.

dip: The inclination of a planar geologic surface from the horizontal (measured in degrees). See also fault dip and Figure 13 in Appendix 2 of this Handbook.

dip slip: The component of the slip parallel with the dip of the fault.

dip-slip fault: A fault in which the slip is predominantly in the direction of the dip of the fault.

directed blast: See lateral blast.

directivity focusing: The variation in wave amplitude as a function of azimuth relative to strike or dip of the fault source caused by the propagation of the rupture front. See Chapter 60 (p. 1006) and Figure 6 in Appendix 2 of this Handbook.

directivity pulse: A concentrated pulselike ground motion generated by constructive interference of S waves traveling ahead of the tip of a rupturing fault. See Chapter 60 (p. 1006) and Figure 6 in Appendix 2 of this Handbook.

disaster: An accidental or uncontrollable event, actual or threatened, in which individuals or society undergoes severe danger or damage that disrupts the social structure of a society. See Chapter 75.

discrete wavenumber method: A simple and accurate method for calculating the complete Green's function for a variety of problems in elastodynamics. See Bouchon (2003) for the latest review.

discriminant: A characteristic feature of a seismic signal that can be used for discrimination. See also discrimination.

discrimination: The work of identifying different types of events, and specifically of distinguishing both between earthquakes and underground explosions and between underground nuclear and chemical explosions. See Chapter 24.

dislocation: In seismology, the displacement discontinuity (slip) associated with faulting on an internal surface. In physics, dislocation generally refers to the line bounding an internal surface subject to uniform slip, in which case the segments of the dislocation may be described as edge, screw, or mixed depending on whether the slip is perpendicular, parallel, or oblique to the line. See Chapters 1, 12, 15, and 36 and Savage (1980). See also dislocation modeling.

dislocation modeling: The determination of the dislocation distribution that would produce the observed surface deformation of an earthquake. See also dislocation.

dislocation superlattice: A regular distribution of dislocations treated as a reference state. See Chapter 15.

DISPCAL: A computer program for determining the generator constant of a seismometer. See Chapter 85.18.

dispersion: The spreading of wave duration with propagation distance due to frequency-dependent velocity. See Chapter 21.

dispersion relations: See Kramers-Kronig relations.

displacement: A position vector pointing to a moving material particle at a given time from some reference point.

In earthquake geology, it is a general term for the movement of one side of a fault relative to the other; the amount of displacement may be measured in any chosen direction, usually along the strike and the dip of the fault. In strong-motion seismology, it is the displacement of a point during earthquake shaking relative to the point at rest, typically obtained by doubly integrating the acceleration and applying certain corrective procedures. In geodesy, it is movement of a point on the Earth's surface, traditionally defined in a local (east, north, up) coordinate system centered at a reference point fixed to the Earth. In space geodesy, the reference point can be attached to the inertial frame. See Chapters 37 and 58.

dome: See lava dome.

Doppler effect: A shift in the frequency of a signal due to relative motion between the signal source and the sensor measuring the signal. The observed frequency is shifted from the source frequency depending on the direction with respect to the moving direction. See Chapter 37, and the biography of Christian Doppler in Chapter 89.

DORIS: Détermination d'Orbite et Radiopositionnement Intégré par Satellite, a Doppler satellite navigation system developed by the French Space Agency. See Chapter 37.

dormant volcano: By a widely used but inadequate definition, volcano that is not presently erupting but that is considered likely to do so in the future (see Decker and Decker, 1998). Volcanoes can remain dormant for hundreds, thousands, and, in rare cases, millions of years before reactivating to erupt again. In general, the longer a volcano remains dormant, the greater the likelihood that its next eruption will be large (Simkin and Siebert, 1994).

double couple: A force system consisting of two orthogonal couples with the same scalar moment and opposite sign. Its equivalence to a dynamic fault slip was first shown by Maruyama (1963) for an isotropic elastic medium. The force system equivalent to fault slip for an anisotropic medium was obtained by Burridge and Knopoff (1964). See Aki and Richards (2002, pp. 42-48).

double-difference: Three variations: (1) In earthquake seismology, the double-difference algorithm for earthquake location that optimally locates earthquakes in the presence of measurement errors and earth model uncertainty. The algorithm minimizes the residuals between observed and predicted travel-time differences (or double-difference) for pairs of earthquakes observed at common stations by iteratively adjusting the vector connecting their hypocenters. By linking many earthquakes together through a chain of near neighbors, it is possible to obtain high-resolution hypocenter locations over large distances without the use of station corrections. See Waldhauser and Ellsworth (2000) and references therein. (2) In INSAR, this term denotes the difference of two interferograms, each of which is the difference of two radar images. (3) In GPS, this term describes a linear combination of four ray paths involving two satellites and two receivers. See Chapter 37.

double-frequency microseism: A term for microseismic noise on the Earth generated by ocean waves with the main spectral peak at half the wave period (hence the seismic noise is at double frequency), as explained by Longuet-Higgins (1950). See Chapter 19.

drift (ratio): In earthquake engineering design and analysis, the ratio of the lateral interfloor deflection to the height between the two floors involved. The allowable drift ratio under design loading is often prescribed in building codes. See Chapter 67.

DSDP: Deep Sea Drilling Project (1968–1983). See Chapter 27.

DSP: Digital signal processor. The processor implements mathematical operations, such as a Fast Fourier transform, numerically. Recent electronic analog-to-digital converters use DSPs to attain high resolution. See Chapter 63.

DTED: Digital Terrain Elevation Data. See Chapter 37.

ductile: The ability to accommodate inelastic strain without loss of strength. See Chapter 32 (p. 516).

ductility: The property of a structure or a structural component that allows it to continue to have significant strength after it has yielded or begun to fail. Typically, a well-designed ductile structure or component will show, up to a point, increasing strength as its deflection increases beyond yielding, or cracking in the case of reinforced concrete or masonry. See Chapter 67.

ductility ratio: The ratio of the maximum deflection (or rotation) of a structure or structural component to the deflection (or rotation) at first yielding or cracking.

duration magnitude: The logarithm of the duration of a local earthquake seismogram that is generally proportional to earthquake magnitude calculated from direct-wave amplitudes. Its physical basis is given by the back-scattering theory of coda waves, after which the duration must be measured from the earthquake origin time. More often it is measured from the onset of P waves, requiring an additional minor, usually neglected, correction for the epicentral distance. See Lee and Stewart (1981, pp. 155-157) and Aki (1987).

DWWSSN: Digital World Wide Standardized Seismograph Network. See Chapter 20.

dynamic range: The amplitude ratio between the smallest and the largest signal that can be faithfully recorded by a system, usually expressed in decibels. See Chapter 18.

dynamic ray equation: An equation that governs the properties of a paraxial ray, which is defined as a ray generated by perturbing the source position or takeoff direction of a central ray. It is expressed as an ordinary differential equation for certain characteristics of a wave field and is related to the amplitude and curvature of wave front in the vicinity of the central ray as a function of the travel time or the ray-path length. See Chapter 9.

dynamics: A term concerned with the stresses or forces responsible for generating seismic motion. Cf.: kinematics. See Aki and Richards (2002, Box 5.3, p. 129).

dynamic stress drop: A term in earthquake source physics referring to the space-time history of stress drop while the fault is slipping, affected by the detailed processes of friction sliding and seismic wave radiation. It is more difficult to estimate from observations than static stress drop. Cf.: static stress drop. See Chapter 12, Chapter 33 (pp. 550-551), and Figure 14 in Appendix 2 of this Handbook.

dyne (dyn): A cgs unit for force; 1 dyne $= 10^{-5}$ Newton. Although it is an obsolete unit, it is still commonly used in seismology. See also Newton.

E

Earth: The third planet from the Sun in the solar system. It has a mean radius of 6371 km, a surface area of 5.1×10^8 km^2, a volume of 1.08×10^{12} km^3, and a mass of 5.98×10^{24} kg. Its internal structure consists of the crust, the mantle, and the core.

earthquake: A shaking of the Earth that is tectonic or volcanic in origin. A tectonic earthquake is caused by fault slip.

earthquake catalog: See catalog of earthquakes.

earthquake energy change: The total change in energy associated with an earthquake. This includes changes in elastic energy, gravitational potential, energy radiated in seismic waves, frictional dissipation, and surface energy of newly created surfaces. See Chapter 35 (pp. 571-572), Eq. (6.3) in Appendix 2 of this Handbook, and Scholz (2002).

earthquake engineering: The field that is defined as encompassing man's efforts to cope with the harmful effects of earthquakes (Housner, 1967). It may be subdivided into three parts: (1) a study of those aspects of seismology and geology that are pertinent to the problem, (2) an analysis of the dynamic behavior of structures under the action of earthquakes, and (3) the development and application of appropriate methods of planning, designing, and constructing. It overlaps with earth sciences on one hand and with social scientists, architects, planners, industry, and government on the other hand.

earthquake focal mechanism: A description of the orientation and sense of slip on the causative fault plane derived seismologically from the radiation pattern of seismic waves (traditionally by the sense of the first P-wave motion and polarization of S-wave motion, and now the moment tensor inversion of waveforms). From the seismic radiation pattern alone, it is not possible to distinguish between the fault plane and the auxiliary plane orthogonal to the slip direction. See Chapter 34 (p. 562).

earthquake hazard: See seismic hazard.

earthquake intensity: See seismic intensity.

earthquake precursor: An anomalous phenomenon preceding an earthquake. See Chapter 72.

earthquake prediction: The statement, in advance of the event, of the time, location, and magnitude of a future earthquake. Earthquake prediction programs have been promoted in Japan, China, the United States, the former Soviet Union, and other countries. See Chapter 72.

earthquake risk: See seismic risk.

earthquake sequence: A series of earthquakes originating in the same locality. See Chapter 43.

earthquake source parameters: The parameters that are specified for an earthquake source, depending on the applied model. They are origin time, hypocenter location, magnitude, focal mechanism, and moment tensor for a point-source model. They include fault geometry, rupture velocity, stress drop, and slip distribution for a finite-fault model. See section 6 in Appendix 2 of this Handbook.

earthquake storm: A sequence of large earthquakes occurring over a period of a few years to tens of years over distances of hundreds of kilometers (e.g., along the North Anatolian fault between 1939 to 1999, the Mojave region between 1950 and the present, and the eastern Mediterranean from about 340 AD to 380 AD). See Chapter 46.

earthquake stress drop: See dynamic stress drop and static stress drop. See also Figure 14 in Appendix 2 of this Handbook.

earthquake swarm: A series of earthquakes occurring in a limited area and time period in which there is not a clearly identified main shock of a much larger magnitude than the rest. See Chapter 43.

earthquake thermodynamics: A branch of physics of earthquakes that deals with consequences of the second law of thermodynamics on rock fracture and earthquake faulting. See Chapter 15.

earth tides: The periodic strains primarily at diurnal and semidiurnal periods generated in the solid Earth by the gravitational attraction of the Sun and Moon.

ECR: A computer program for synthetic seismogram calculations using the reflectivity method. See Chapter 85.10.

edge dislocation: See dislocation.

edge effect: A special amplification effect of ground motion found near the edge of a basin. In a 2D or 3D model of basin, basin-induced diffracted waves and surface waves generated at the edge by an incident body wave meet together in phase at some point causing constructive interference. As a result, the amplitude of ground motion there becomes much larger than in the case of a simple 1D model. The edge effect contributed importantly to the damage concentration in Kobe during the Hyogo-ken Nanbu earthquake of 1995. See Chapter 61.

edifice collapse: See sector collapse.

EDM: Electro-optical distance meter, the instrument and technique usually used for trilateration. See Chapter 37.

EERI: Earthquake Engineering Research Institute, Oakland, California, USA (URL: http://www.eeri.org). See Chapter 80.18.

effective elastic thickness: The conceptual thickness of the layer within the lithosphere that can support significant deviatoric stresses over geological time scales. See Chapter 31.

effective normal stress: The normal stress minus the pore pressure. See Nur and Byerlee (1971) and Chapter 40.

effective stress: In soil mechanics, the sum of the interparticle contact forces acting across a plane through a soil element divided by the area of the plane. On a horizontal plane, the effective stress is defined as the total overburden pressure minus the instantaneous pore water pressure. In rock mechanics, it also means the stress above the pore pressure. See Chapters 32 and 70.

effusive eruption: A nonexplosive extrusion of lava at the surface that typically produces lava flows and lava domes.

eikonal: From the Greek word *eikon*, which means "image." It was introduced in 1895 by H. Bruns to describe the phase function (or travel time) in wave solutions that can usefully be written $A(x)\exp[i\psi(x)]$. Constant values of the eikonal represent surfaces of constant phase or wave fronts. The eikonal equation is a partial differential equation for the eikonal. See also eikonal equation.

eikonal equation: The equation that, when considering a wave front originating from an impulsive source at a point and defining the travel time in space as its time of arrival, equates the magnitude of gradient vector of the travel time to the reciprocal of the local wave speed. The term is also used in relation to Hamilton's equations in analytical dynamics, which have some similarity with equations in ray theory. See Chapter 9 and the biography of William Rowan Hamilton in Chapter 89.

elasticity: The property of a body to recover its shape and size after being deformed. See Eq. (1.4) in Appendix 2 of this Handbook.

elastic rebound model: A model of the overall earthquake genesis process first put forth by H.F. Reid in 1910. It says the energy that is released in a tectonic earthquake accumulates slowly in the vicinity of the fault in the form of elastic strain energy in the rocks, until the stresses on the fault exceed its strength and sudden rupture occurs. The rock masses on the two sides of the fault move in a direction to reduce the strain in them. At the time this model was proposed, the source of the accumulating strain was not known.

elastodynamic equation: The basic equation governing the dynamic deformation of a solid body. It is derived from Newton's law on force and acceleration, the generalized Hooke's law on stress and strain, and the assumption of the existence of the strain energy function. For an infinitesimally small displacement, it can be reduced to a form equating density times the second time-derivative of the displacement to the sum of the body force and the divergence of the stress tensor (written in terms of the space-derivatives of the displacement). See Aki and Richards (2002, pp. 11-36), Chapter 15, and Eqs. (1.7) and (1.8) in Appendix 2 of this Handbook. See also the biographies of Isaac Newton and Robert Hooke in Chapter 89.

electrokinetic: A term referring to the electric and magnetic fields generated by fluid flow in a porous medium. See Chapter 38.

electromagnetic: The coupled electric- and magnetic-field effects. See Chapter 38.

elliptic crack: The fault-plane geometry in which the rupture front starts at a point on the fault and spreads as an ellipse. See also rupture front and Aki and Richards (2002, pp. 509-510, 552-560).

empirical Green's function method: A method for calculating strong ground motion for a large earthquake using actual records of small earthquakes originating on or near the fault plane of the large one, with Green's function representing the propagation-path effect.

end of the Bronze Age: A catastrophic period in 1200 BC $\pm$ 25 years during which numerous cities in the eastern Mediterranean and the Near East inexplicably collapsed politically and physically. See Chapter 46.

en-echelon: An overlapping or staggered arrangement, in a zone, of geologic features that are oriented obliquely to the orientation of the zone as a whole. The individual features are short relative to the zone. See Dennis (1967).

envelope broadening: The phenomenon of a wave envelope, which has been impulsive since the time of radiation from the source, broadening with increasing travel distance because of diffraction and scattering by distributed heterogeneities. See Chapter 13.

epicenter: A point on the Earth's surface immediately above the earthquake focus, called a hypocenter. See Chapter 21 and Lee and Stewart (1981, pp. 130-139).

epicentral distance: The distance from a site (usually a recording seismograph station) to the epicenter of an earthquake. It is commonly given in kilometers for local earthquakes and in degrees (1 degree is about 111 km) for teleseismic events.

epistemic uncertainty: The uncertainty in seismic hazard analysis due to imperfect knowledge in model parametrization and other limitations of the methods employed. Epistemic uncertainty can be reduced by improvements in the modeling and analysis. Cf.: aleatory uncertainty. See Chapter 56.

equivalent linearization (linearisation): The process of selecting a linear system and its parameters so that it approximates the response of a nonlinear system (e.g., selecting the stiffness and damping of a linear single degree of freedom structure so it can approximate the earthquake response of a similar structure with a yielding force-displacement relation).

erg: A cgs unit for energy; 1 erg = 10^{-7} joule. Although it is an obsolete unit, it is still commonly used in seismology. See also joule.

ERI: Earthquake Research Institute, University of Tokyo, Tokyo, Japan (URL: http://www.eri.u-tokyo.ac.jp/). See Chapter 79.33.

ERS-1: European Remote Sensing satellite 1. It carries a C-band SAR and was launched in 1991. See Chapter 37.

ERS-2: The twin of ERS-1, launched in 1995. See Chapter 37.

eruption: A general term applied loosely to any sudden expulsion—explosive or nonexplosive—of material (solid, liquid, or gas) from volcanic vents. Some volcanologists, however, prefer that the term be restricted to ejections of lava and/or fragmental solid debris, thereby distinguishing them from aqueous or gaseous emissions devoid of particulate matter. Such a distinction is important in the assessment of volcano hazards and in the communication of volcano hazard information during a volcanic emergency.

eruption column: A vertical plume of tephra and gases mixed with air that forms above a volcanic vent during an explosive volcanic eruption. It may rise to stratospheric heights (tens of kilometers) and can last for many hours.

ESA: European Space Agency. See Chapter 37.

ESP: Expanded spread profile.

ETHZ: Swiss Federal Institute of Technology, Zurich, Switzerland (URL: http://www.ethz.ch).

Euler pole: See Euler vector.

Euler vector: The rotation vector describing the relative motion between two tectonic plates. The magnitude of the Euler vector is the rotation rate, and its intersection with the Earth's surface is the Euler pole. The linear velocity at any point on the plate boundary is the vector product of the Euler vector and the radius vector to that point. See Chapter 7 and the biography of Leonhard Euler in Chapter 89.

event: A general term used to represent an earthquake or any similar seismic disturbance.

event horizon: A term in paleoseismology indicating the ground surface at the time of an earthquake in the geologic past. See Chapter 30.

exceedance probability: The probability that a specified value of a strong-motion parameter is exceeded by a future occurrence within a specified period of time. See Chapter 60.

experimental design: See optimal survey design.

explosion earthquake: A term in volcanology referring to a seismic event associated with explosive volcanic activities. In addition to the usual seismic waves such as *P* waves and *S* waves, it often shows airwaves that travel through the air as acoustic waves and that are transmitted back into the ground in the vicinity of the seismometer. See Chapter 25.

exposure time: The time period used in the specification of probabilistic seismic hazard and seismic risk. It is usually chosen to represent the design or economic life of a structure.

extensional tectonic setting: A region in which the crust is undergoing lateral expansion and for which the maximum principal stress is vertical. See Chapter 40.

extinct volcano: By a widely used but inadequate definition, volcano that is not erupting and is not expected to do so in the future (see Decker and Decker, 1998). However, volcanoes can hold surprises; for example, Mount Vesuvius was considered "extinct" before its explosive eruption in AD 79 that devastated Pompeii.

F

failure envelope: The boundary of stress conditions associated with rock failure, as defined in shear-stress versus normal-stress space. Typically, the slope of the failure envelope $d\sigma/d\sigma_n$ is steepest at low-normal stress (σ_n). The slope is also referred to as the pressure dependence of failure strength. See Chapter 32 (p. 508).

fallout: The settling and deposition of tephra from the atmospheric plume of an explosive eruption.

far-field: A displacement field from a seismic source observed at large distances as compared to the wavelength and/or the source dimension. See Eq. (3.4) and related material in Appendix 2 of this Handbook.

Fast Fourier transform (FFT): A computer algorithm that transforms a time-domain sample sequence to a frequency-domain sequence that describes the spectral content of the signal. This efficient algorithm for computing Fourier transform was first introduced by Cooley and Tukey (1965). See Brigham (1988) and Chapters 20 and 22.

fault: A fracture or fracture zone in the Earth along which the two sides have been displaced relative to one another parallel to the fracture. The accumulated displacement may range from a fraction of a meter to many kilometers. The type of fault is specified according to the direction of this slip. See also normal fault, reverse fault, and strike-slip fault; Bonilla (1970); and Figure 13 in Appendix 2 of this Handbook.

fault breccia: A fault rock made up of angular rock fragments derived from brittle fragmentation in a fine-grained or hydrothermal matrix. See Chapter 29.

fault creep: The quasi-static slip on a fault, a slip that occurs so slowly that radiation of seismic waves is negligible. See also creep event.

fault dip: The inclination of a fault plane relative to the horizontal in degrees. A dip of 0 degree is horizontal; 90 degrees is vertical. See Aki and Richards (2002, Fig. 4.13, p. 101) and Figure 13 in Appendix 2 of this Handbook.

fault-fracture mesh: A mesh structure of interlinked minor faults (shear fractures) and extension fractures occupying a substantial volume of the rock mass. See Chapter 29.

fault gouge: A clayey, soft material formed when the rocks in a fault zone are pulverized during slippage. It is predominantly made up of mineral fragments derived from brittle cataclastic fragmentation. Textures range from essentially random-fabric to foliated. See Chapters 29 and 39.

fault model: A model of an earthquake source in which a slip across a fault plane buried in an Earth model causes the earthquake ground motion. See Chapters 5 and 12, Kostrov and Das (1988), and Scholoz (2002). See also Brune model, Burridge and Knopoff model, Haskell model, Knopoff model, Kostrov model, omega-squared model, and specific barrier model.

fault-plane solution: See earthquake focal mechanism.

fault rake: A parameter describing the slip direction on a fault, measured as an angle in the fault plane between the fault strike and the slip vector. Slip in the direction of the maximum slope of the fault plane is a rake of 90°, positive for reverse faults, negative for normal faults. See Aki and Richards (2002, Fig. 4.13, p. 101) and Figure 13 in Appendix 2 of this Handbook.

fault rock: A rock whose textural/microstructural characteristics (usually involving grain-size reduction from the protolith) developed through deformation within a fault zone at depth. See Chapter 29.

fault sag: A narrow tectonic depression common in strike-slip fault zones. Fault sags are generally closed depressions a few hundred feet wide and approximately parallel to the fault zone; those that contain water are called sag ponds (Bonilla, 1970; after Gilbert, 1908, and Sharp, 1954).

fault scarp: A slope formed by the offset of the Earth's surface by a fault.

fault slip: The relative displacement of points on opposite sides of a fault, measured on the fault surface. See Aki and Richards (2002, Fig. 4.13, p. 101) and Figure 13 in Appendix 2 of this Handbook.

fault-slip rate: The rate of displacement on a fault averaged over a time period encompassing several earthquakes.

fault strike: The azimuthal direction of the horizontal in the fault plane. See Aki and Richards (2002, Fig. 4.13, p. 101) and Figure 13 in Appendix 2 of this Handbook.

fault trace: The intersection of a fault with the Earth's surface or with a horizontal plane.

fault-valve action: The fluid redistribution arising from a fault rupture transecting a superhydrostatic fluid-pressure gradient, leading to cyclic changes in fluid pressure and fault strength coupled to the earthquake stress cycle. See Chapter 29.

fault zone: The fault plane of the mathematical model of an earthquake is usually a planar surface in an elastic medium. The actual fault is a zone in which a nonelastic process occurs during the earthquake, sometimes called a breakdown zone or cohesive zone. It includes the fault gouge that has been studied geologically and petrologically. See Chapters 29 and 30.

fault-zone head wave: A P and S seismic wave propagating along a material interface in a fault-zone structure with the velocity and motion polarity of the corresponding body waves on the faster side of the fault. See also head wave.

fault-zone trapped mode: A term referring to seismic guided waves trapped in the low-velocity fault zone that are useful for studying the 3D structure of the fault zone. They have been used also to find the temporal change in the physical characteristics of the fault zone.

FDSN: Federation of Digital Seismograph Networks, a global organization (URL: http://www.fdsn.org). Its membership is comprised of groups responsible for the installation and maintenance of broadband seismographs either within their geographic borders or globally. Membership in the FDSN is open to all organizations that operate more than one broadband station. See Chapter 20.

feedback: The addition of part of the output signal of a linear system like an amplifier to the input signal in order to modify the response, mostly used to stabilize the system (negative feedback). See Chapter 18.

felt area: The area of the Earth's surface over which the effects of an earthquake are humanly perceptible. See Chapter 49.

Fennoscandia: The northern European region comprising the Caledonids and the Baltic shield of Finland, Scandinavia, and the Kola Peninsula of northwestern Russia.

FFT: Fast Fourier transform, an algorithm for fast computation of Fourier transforms.

filter: An electronic device or algorithm that acts on an input signal to produce an output signal with a desired modification in spectral characteristics. See also digital filter and Chapter 22.

finite element: A geometrical subdivision, generally small, of a larger structure or medium. Finite elements are commonly used in numerical studies of the response of complex structures or media to earthquake excitation. Typical finite elements include triangles and rectangles for analysis of 2D problems and tetrahedrons for 3D problems. In applications, a structure is divided into many finite elements, and field quantities such as density and displacement, for example, are assumed to vary over each element in a prescribed manner (e.g., linearly varying displacement) so the displacement and associated stresses and strains within each element are determined by the displacements at its corners or nodes. The displacements of the nodes of the finite elements then become the unknown vector in a matrix equation that is solved numerically. See Chapter 69.

fire fountain: See lava fountain.

first motion: The first half-cycle of a body wave, usually the direct P wave. The direction (polarity) of first motion, which for the P wave is either away from (called compression) or toward (called dilatation) the source, is assumed to be conserved along the ray path from the focus to a given station, given that corrections have been made for any polarity changes due to reflection of the wave at any boundary between source and receiver. First motions of P arrivals are often used for determining earthquake focal mechanism.

fissure eruption: A volcanic eruption fed by one or more fissure vents rather than by a central (point-source) vent or a tight cluster of central vents.

f-k analysis: The frequency (f) versus wavenumber (k) analysis that maps the power of seismic waves observed at an array as function of azimuth and slowness for a given frequency band. See also wavenumber filtering.

flank eruption: An eruption that takes place away from the summit area of the volcano (e.g., on rift zones or flanks of the volcano), synonymous with lateral eruption.

flank failure: See sector collapse.

flexural-slip fault: A bedding fault formed by layer parallel slip during folding.

flexural wave: A normal mode in an infinite plate in vacuum with motion antisymmetric with respect to the median plane of the plate. Examples in nature are the waves in floating ice. For short wavelengths, the period equation for a normal mode reduces to that for flexural waves in a plate modified slightly by the presence of water. For long wavelengths, however, the gravity term in the period equation dominates, and the mode approaches that of gravity waves in water.

fling step: The permanent displacement of the ground at or near a ruptured fault.

flow deformation: The shear deformation in soil that occurs with little resistance within a liquefied layer or element. Deformations may be of limited excursion, as occurs in dilative soils, or unlimited excursion, as occurs in contractive soils. See Chapter 70.

flow failure: The liquefaction-induced failure of a slope or embankment underlain by contractive soils leading to large, rapid ground displacements. These failures are characterized by large loss of shear strength, large lateral displacements (several meters or more), and severe disturbance to the liquefied and overlying soil layers. See Chapter 70.

fluid overpressure: A fluid pressure exceeding the hydrostatic pressure value appropriate to the depth. See Chapter 29.

fmax: Above a limiting frequency, fmax, the observed spectrum of a typical local earthquake decays more steeply with increasing frequency than is expected from the standard omega-squared model. This was first recognized by Hanks (1982) and has been attributed to the earthquake source effect (e.g., Kinoshita, 1992) as well as to the recording site effect (e.g., Anderson and Hough, 1984). See also omega-squared model.

focal depth: The conceptual depth of an earthquake focus. If determined from the first-motion arrival-time data, this represents the depth of rupture initiation (the hypocenter depth). If determined from waveforms that are long compared with the fault dimension, the depth represents some weighted average of the moment release in the earthquake (the centroid depth). See Chapter 11 (p. 153) and Lee and Stewart (1981, pp. 130-139).

focal mechanism: See earthquake focal mechanism.

focal sphere: An imaginary sphere with small radius surrounding the hypocenter, onto which the polarity of first motions is plotted in focal mechanism studies.

focus: See hypocenter.

focusing effect: The special amplification of ground motion found just above the corner of the basin bottom or layer interfaces. In a 2D or 3D basin with strong lateral variation, rays of an incident body wave are warped at these interfaces, and constructive (or destructive) interference takes place at some point on the surface. See Chapter 61.

footwall: The underlying side of a nonvertical fault surface. Cf.: hanging wall. See Figure 13 in Appendix 2 of this Handbook.

force-balance seismometer: A seismometer that compensates the inertial force acting on its suspended mass with a negative feedback force and thereby keeps the mass at rest relative to the ground. See Chapter 18.

fore arc: The region between the subduction thrust fault and a volcanic arc.

forensic seismology: The seismology applied to problems involving legal issues, such as the verification of compliance with any treaty limiting the testing of nuclear explosions. See Chapters 23 and 24.

foreshock: A smaller earthquake preceding the largest earthquake in an earthquake sequence. See Chapter 43.

forward problem: A typical problem in physical science of the modeling of natural phenomena. Through such a modeling, and given some parameters describing the system under study, one may predict the outcome of some possible observables. Solving a forward problem may be summarized as model parameters → model → prediction of observed data. Cf.: inverse problem. See Chapter 16 and Menke (1984).

Fourier transform (spectrum): An integral transform mapping signals defined in the time domain $f(t)$ into the complex frequency domain $F(\omega)$ by multiplying $\exp(i\omega t)$ and integrating over t from minus infinity to plus infinity. See Chapter 22 and the biography of Jean Baptiste Joseph Fourier in Chapter 89.

FR: Frictional regime.

fractal: A geometrical structure or object that is self-similar on all scales (i.e., scale invariance) and is quantified by a fractional (instead of an integer) dimension. See Mandelbrot and Frame (2002) for a general review of fractals.

fractal dimension: The fractional dimension of a fractal and the power in the power-law fractal scaling. See also fractal.
fractal statistics: A statistical distribution in which the number of objects has a power-law dependence on their size. See Chapters 14 and 82. See also fractal and fractal dimension.
fractionation: The separation of a mixture in successive stages, such as by differential solubility in a water-solvent mixture. See Chapter 39.
fracture: A general term for discontinuities in rock including faults, joints, and other breaks. See section 4 in Appendix 2 of this Handbook.
fracture toughness: A measure of the energy expended in an increment of fracture. It is more difficult to propagate a fracture in a material with high fracture toughness than for low fracture toughness. See Chapters 15 and 32 and Eq. (4.9a) in Appendix 2 of this Handbook.
fracture zone: An elongated zone of unusually irregular topography that separates regions of different water depths in the ocean floor. It is commonly found to displace the magnetic lineations and the axis of the midocean ridge, and it is interpreted as a transform fault. See Wilson (1965) and Uyeda (1978).
free-field motion: A strong ground motion that is not modified by the earthquake-caused motions of nearby buildings or other structures or geologic features. In analyses, it is often defined as the motion that would occur at the interface of the structure and the foundation if the structure were not present. A fully instrumented building site typically includes one or more accelerographs located some distance from the structure to obtain a better approximation of the free-field motion. See Chapter 67.
free oscillation: See normal mode.
frequency: The number of cycles of a periodic motion per unit time, or the reciprocal of period. In the context of the Fourier transform, frequency is also used for the angular frequency, which is 2π times the frequency defined above. See Chapters 18 and 60.
frequency response: The Fourier transform of the output signal of a linear system divided by the Fourier transform of the input signal. The frequency response of a linear system equals the Fourier transform of its impulse response. See also impulse response.
friction: The ratio of shear stress to normal stress on a planar discontinuity such as a fault or joint surface. See Chapter 23 (p. 506) and section 5 in Appendix 2 of this Handbook.
frictional instability: A condition that arises during quasistatic fault slip when the reduction of fault strength with displacement exceeds the elastic stiffness of the region surrounding the fault. Under these conditions, the slip rate accelerates rapidly. See Chapter 32 (p. 517) and material below Eq. (5.4f) in Appendix 2 of this Handbook.
Frictional regime (FR): It is that portion of the crust or lithosphere where fault motion is dominated by unstable (velocity-weakening) pressure-sensitive frictional sliding. See Chapter 29.
friction melting: The melting as a consequence of high temperature generated by friction during rapid localized fault slip. See Chapter 29.
frozen wave: The type of wave attributed to cracking open of the ground at the crest of the wave, sometimes with the emission of sand and water. In the epicentral area of a great earthquake, walls, embankments, and the like are sometimes left in the form of a wave. Frozen waves, concentric with very large impact craters, are also seen on the surface of the Moon.
fully decoupled explosion: An underground explosion, carried out within a cavity large enough that the walls of the cavity are not stressed beyond their elastic limit. See Chapter 24.

G

g: A commonly used unit for the acceleration due to the gravity of the Earth.
GA: Genetic algorithm.
Ga: An abbreviation for 1 billion years ago.
gal: A unit in acceleration (1 gal = 1 cm/s^2, or approximately one-thousandth of 1 g).
gas: See volcanic gas.
GDSN: Global Digital Seismic Network. See Chapter 20.
Geiger's method: A commonly used method to locate and determine the origin time of an earthquake using the first arrival times of seismic waves, assuming a travel-time curve or function of the Earth. Geiger (1912) applied the Gauss-Newton method (an optimization procedure by least squares) to solve the earthquake location problem, but his method was not practical until the advance of computers in the late 1950s. See Chapters 85.7, 85.8, and 85.17; Lee and Stewart (1981, pp. 132-139); and the biographies of Ludwig Geiger, Karl Friedrich Gauss, and Isaac Newton in Chapter 89.
genetic algorithm (GA): It is a learning algorithm used to maximize the adaptability of a system to its environment. It is often used as a computer algorithm in function optimization or object-selection problems. See McClean (2002) and Chapter 85.15.
geochronology: The science of dating Earth materials, geologic surfaces, and geologic processes. See Chapter 30.
geodesy: The study of the shape of the Earth and the dynamical processes and forces leading to its observed shape.
geodetic moment: The geodetic estimate of seismic moment. See Chapter 37 (pp. 616-617).
geodynamics: The study of the dynamics of the Earth's interior, including heat transfer and convection current.
geodynamo: The magneto-hydrodynamic convection in the Earth's core that is responsible for the generation of the Earth's main magnetic field. See Chapter 56.

geoid: The surface of constant potential energy, representing the balance between gravitational and rotational forces, corresponding to an idealization of the Earth's actual surface. In practice, deviations of the geoid from a reference ellipsoid are presented as the geoid.

geologic time: The time since the Earth was formed 4.6 billion years ago (Ga), encompassing the entire geologic history of the Earth. It is usually divided chronologically into the following hierarchies of time units: eon, era, period, and epoch. The eons are: (1) Hadean (4.6–4 Ga), (2) Archaean (4–2.5 Ga), (3) Proterozoic (2.5–0.545 Ga), and (4) Phanerozoic (545 million years ago–present). The Phanerozoic eon is divided into the following eras: (1) Paleozoic (545–245 Ma), (2) Mesozoic (245–66.4 Ma), and (3) Cenozoic (66.4–0 Ma). The Cenozoic era is divided into following periods: (1) Tertiary (66.4–1.6 Ma) and (2) Quaternary (1.6 Ma–present). The Quaternary period is divided into the following epochs: (1) Pleistocene (1.6 Ma–10,000 years ago) and (2) Holocene (10,000 years ago–present). For a detailed discussion of geologic time, see Hancock and Skinner (2000) and Eicher (2002).

geomagnetic polarity epoch: A period of geologic time during which the Earth's magnetic field was essentially or entirely of one polarity. A short period of opposite polarity within a polarity epoch is called an event. See Cox (1973) and Merrill *et al.* (1998).

geomagnetic pole: The point where the axis of the calculated best-fit dipole cuts the Earth's surface in the northern or southern hemisphere. These poles lie opposite each other and for the epoch 1990 are calculated to lie at 79.2° N, 288.9° E and 79.2° S, 108.9° E, respectively. See Merrill *et al.* (1998).

geomagnetic reversal: A change of the Earth's magnetic field between normal polarity and reversed polarity. See Chapter 6 in this Handbook, Cox (1973), and Merrill *et al.* (1998).

geomagnetism: The study of the properties, history, and origin of the Earth's magnetic field. See Merrill and McFadden (2002).

geometrical spreading: The decrease of wave energy per unit area of wave front, due to the expansion of the wave front with increasing distance.

geophone: A term commonly used for a relatively inexpensive electromagnetic seismometer that is lightweight and responds to high-frequency (>1 Hz) ground motions.

GEOSCOPE: French Global Digital Network. See Chapter 20.

geosyncline: A crustal down-warp that is several hundreds or thousands of kilometers long and contains an exceptional thickness of accumulated rocks, on the order of 10 to 15 km. See Billings (1972).

geotechnical engineering: The branch of engineering that deals with the analysis, design, and construction of foundations and retaining structures on, within, and with soil and rock. Common examples include structural foundations and earth or rock-fill dams.

geotechnical zonation: A mapping that portrays specified ground characteristics or mechanical properties in an area, typically showing zones that correspond to predefined geotechnical units. For example, a map of shear-wave velocities derived from soil-profile classifications. See Chapter 62.

geotherm: A profile of temperature as a function of depth inside the Earth. Cf.: geothermal gradient.

geothermal gradient: The rate of increase in temperature with depth in the Earth. By convention, this quantity is positive when increasing downward.

geyser: A fountain of hot water and steam ejected intermittently from a source heated by magma or hot rocks. See Chapter 39.

GFZ: GeoForschungsZentrum, Potsdam, Germany (URL: http://www.gfz-potsdam.de).

glass: See volcanic glass.

global plate-motion model: A set of Euler vectors specifying relative plate motions. Such models can be derived for time spans of millions of years using rates estimated from sea-floor magnetic anomalies, directions of motion from the orientations of transform faults, and the slip vectors of earthquakes on transforms and at subduction zones. They can also be derived for time spans of a few years using space-based geodesy. See Chapter 7.

glowing avalanche: An incandescent and turbulent mixture of pyroclasts, volcanic gases, and air that flows under gravity down the flanks of a volcano; essentially synonymous with nuée ardente, pyroclastic flow, and pyroclastic surge.

Gondwanaland: The protocontinent of the southern hemisphere in the late Paleozoic. It included Australia, South America, and Antarctica, as well as India, before they were drifted away from one another. It was named by the Austrian geologist Eduard Suess after Gondwana, a historic region in central India. See Chapter 6 in this Handbook, Duff (1993), and Hancock and Skinner (2000).

gouge: See fault gouge.

GPS: Global Positioning System. A dual-frequency L-band satellite navigation system operated by the US Department of Defense. This system allows extremely accurate position and time determinations anywhere on the Earth's surface. See Chapters 20 and 37.

graben: A down-dropped block of the Earth's crust resulting from extension, or pulling, of the crust. See also horst.

gravity anomaly: The difference between the observed local gravity and the theoretically calculated value based on a regional model.

gravity load: In earthquake engineering, the vertical load created in the elements of a structure by the force of gravity. Gravity loads are sometimes separated into dead load (the weight of the structure) and live load (the weight of furnishings and other contents). See Chapter 69.

gravity wave: The normal mode in a surface layer with very low shear velocity, such as unconsolidated sediments, that may be affected significantly by gravity at long periods such as tsunami in the ocean. Waves similar to the gravity waves in a fluid layer are possible in addition to the shortening of wavelength of normal modes by gravity. So-called visible waves with large amplitude and relatively long periods observed in the epicentral area of a great earthquake have been suggested to be gravity waves. See Chapter 28 (pp. 442-443).

great earthquake: An earthquake with magnitude greater than about 7 $^3/_4$ (or 8). See also magnitude.

Green's function: In seismology, the vector displacement field generated by an impulsive force applied at a point in the Earth. When combined with the source function describing the discontinuities in displacement and traction across an internal surface, it can represent, by compact formulas, seismic displacements caused by earthquake faults and buried explosions. Green's function was first introduced by Green (1828) as a solution to an inhomogeneous hyperbolic equation for a point source in space and time. See Aki and Richards (2002, pp. 27-28), Rickayzen (2002), Chapter 8 (p. 84) of this Handbook, and Eqs. (1.10)–(1.12), (3.1), (3.2), and (3.6) in Appendix 2 of this Handbook. See also the biography of George Green and ray Green's function in Chapter 89.

ground failure: A permanent deformation of the ground (e.g., liquefaction, fault displacement, and landslides), typically resulting from an earthquake capable of causing damage to engineered structures. See Chapter 70.

ground motion: The vibration of the ground primarily due to earthquakes. It is measured by a seismograph that records acceleration, velocity, or displacement. In engineering seismology, it is usually given in terms of a time series (an accelerogram), a response spectrum, or a Fourier spectrum. See Chapters 57, 60, and 74.

ground-motion parameter: A parameter characterizing ground motion, such as peak acceleration, peak velocity, and peak displacement (peak parameters), or ordinates of response spectra and Fourier spectra (spectral parameters). See Chapter 60.

ground-motion relation: See attenuation relationship.

ground oscillation: A geotechnical term specifying a ground condition in which liquefaction of subsurface layers decouples overlying nonliquefied layers from the underlying firm ground, allowing large transient ground motions or ground waves to develop within the liquefied and overlying layers. See Chapter 70.

ground roll: A term used in exploration seismology to refer to surface waves generated from explosions. They are characterized by low velocity, low frequency, and high amplitude, and they are observed in regions where the near-surface layering consists of poorly consolidated low-velocity sediments overlying more competent beds with higher velocity.

ground settlement: A permanent vertical displacement of the ground surface due to compaction or consolidation of underlying soil layers. See Chapter 70.

ground-shaking scenario: A representation for a site or region depicting the possible ground-shaking level or levels due to earthquake in terms of useful descriptive parameters. See Chapter 62.

ground truth: A jargon in nuclear testing detection seismology referring to information about the location of a seismic event that is derived from data not usually available to an analyst studying the event with a typical monitoring network. The ground truth can come from satellite observations, from fault traces that break out at the Earth's surface, or from operation of a local network. Ground-truth locations are sometimes associated with a number (e.g., GT5, meaning a location believed to be accurate to within 5 km, obtained from ground truth). See Chapter 24.

group velocity: For dispersive waves with frequency-dependent phase velocity $c(\omega)$, the wave packet or energy propagates with the group velocity given by $d\omega/dk$, where k is the wave number given by $\omega/c(\omega)$. See Aki and Richards (2002, pp. 253-254).

GS-21: A short-period borehole-deployable seismometer manufactured by Teledyne Geotech, Garland, Texas. See Chapter 20.

GSHAP: An acronym for the Global Seismic Hazard Assessment Program. See Chapter 74.

GSN: Global Seismograph Network, a network of 128 high-quality broadband seismograph stations distributed uniformly over the Earth's surface. See Chapter 20.

guided wave: A wave that is trapped in a wave guide by total reflections or bending of rays at the top and bottom boundaries. An outstanding example is the acoustic waves in the SOFAR channel, a low-velocity channel in the ocean. Since the absorption coefficient for sound in seawater is quite small for frequencies on the order of a few hundred cycles per second, transoceanic transmission is easily achieved. If the Earth's surface is considered as the top of a wave guide, surface waves, such as Rayleigh and Love (and their higher modes), are guided waves. The fault-zone trapped mode is a guided wave in the low-velocity fault zone. Where they can exist, guided waves may propagate to considerable distances because they are effectively spreading in only two spatial dimensions.

Gutenberg discontinuity: The seismic velocity discontinuity marking the core-mantle boundary (CMB) at which the velocity of P waves drops from about 13.7 km/sec to about 8.0 km/sec and that of S waves from about 7.3 km/sec to 0 km/sec. The CMB reflects the change from the solid mantle material to the fluid outer core. See Chapter 51.

Gutenberg-Richter's relation: An empirical relation expressing the frequency distribution of magnitudes of earthquakes occurring in a given area and time interval. It is given by $\log N(M) = a - bM$, where M is the earthquake magnitude, $N(M)$ is the number of earthquakes with magnitude greater than or equal to M, and a and b are constants (see also b value). It was obtained by Gutenberg and Richter (1941, 1949, 1954) and is equivalent to the power-law distribution of earthquake energies or moments. See Chapter 43 (p. 723) and Eqs. (7.1)–(7.2) in Appendix 2 of this Handbook. See also Ishimoto-Iida relation and the biographies of Beno Gutenberg and Charles Francis Richter in Chapter 89.

***G*-wave (*Gn*):** Another name for long-period Love waves. Because the group velocity of Love waves in the Earth is nearly constant (4.4 km/sec) over the period range from about 40 to 300 seconds, their waveform is rather impulsive, and they have received this additional name. They are called G waves after Gutenberg. It takes about 2.5 hours for G waves to make a round trip of the Earth. After a large earthquake, a sequence of G waves may be observed. They are named $G1, G2, \ldots, G$n, according to the arrival time. The odd numbers refer to G waves traveling in the direction from epicenter to station, and the even numbers to those leaving the epicenter in the opposite direction and approaching the station from the antipode of the epicenter. See Chapter 21 (p. 346).

H

Hadean: An eon in geologic time, from 4.6 to 4 billion years ago. It is named after Hades, the underworld of Greek mythology. See also geologic time.

Hales discontinuity: See N discontinuity.

half-life: The time required for a mass of a radioactive isotope to decay to one-half of its initial amount. See Chapter 39.

half-space: A mathematical model of a medium bounded by a planar surface but otherwise infinite. Properties within the model are commonly assumed to be homogeneous and isotropic, unlike the Earth itself, which is heterogeneous and anisotropic. Nevertheless, the half-space model is frequently used to perform some theoretical calculations in seismology.

hanging wall: The overlying side of a nonvertical fault surface. Cf.: footwall. See Figure 13 in Appendix 2 of this Handbook.

harmonic tremor: An oscillatory ground motion sustained for minutes to days or longer commonly associated with volcanic eruptions and often observed during the precursory phase of eruptive activity. The spectral content of harmonic tremor is commonly dominated by one or two spectral peaks in the frequency range 1–5 Hz. Cf.: volcanic tremor. See Chapter 25.

Haskell model: A representation of the earthquake source by a ramp-function slip propagating unilaterally at a uniform velocity along the long dimension of a narrow rectangular fault surface (Haskell, 1964). It is specified by five parameters (i.e., fault length, fault width, final slip, rupture velocity, and rise time). See Aki and Richards (2002, pp. 498-503), and Purcaru and Berckhemer (1982) for a compilation of earthquakes with these parameters determined by various authors. See also rectangular fault, Chapter 8 (pp. 99-100), and Eq. (3.8) in Appendix 2 of this Handbook.

Hawaiian eruption: An eruption of highly fluid (low-viscosity) magma that commonly produces lava fountains and forms thin and widespread lava flows, and generally very minor pyroclastic deposits around the vent.

hazard: See seismic hazard and volcano hazard.

hazard master model: A type of master model of earthquakes in which the probabilistic seismic hazard analysis (PSHA) is used for unifying multidisciplinary data regarding earthquakes in a given region. The integration of the multidisciplinary data is done in the end product, in contrast to a physical master model. See Chapter 5.

head wave: An interface wave observed in a half-space that is in welded contact with another half-space with higher velocity when the seismic source is located in the lower-velocity medium. The wave enters the higher-velocity medium at the critical angle of incidence and, after traveling along the interface at the higher velocity, reenters the lower-velocity medium at the critical angle. The wave front in the lower-velocity half-space is a part of the surface of an expanding cone. For this reason, head waves are sometimes called conical waves. See Aki and Richards (2002, pp. 203-209).

heat flow: In geophysics, a term that usually means the rate of heat flowing from the Earth's interior per unit area of the surface. It is measured near the Earth's surface as the product of thermal conductivity and vertical temperature gradient. The average heat flow of the Earth is estimated to be about 2 μcal cm^{-2} sec^{-1}. See also HFU and thermal conductivity, Chapter 81.4, and Morgan (2002).

heat flow unit (HFU): Terrestrial heat flow values are often given in cgs units of μcal/cm^2 sec, 10^{-6}cal/cm^2 sec, or μcal cm^{-2} sec^{-1}. In the SI units, heat flow is commonly given in mW m^{-2}, where mW is milli-watt. The conversion is 1 μcal cm^{-2} sec^{-1} = 41.84 mW m^{-2}. Occasionally some authors use the heat flow unit or HFU for μcal cm^{-2} sec^{-1}. See also heat flow.

helioseismology: The study of the Sun's free oscillations (both acoustic and gravity modes), typically through spatially resolved doppler spectroscopy.

hertz (Hz): An SI-derived unit for frequency, expressed in cycles per second (s^{-1}). It is named after Heinrich Rudolph Hertz (1857–1894), who discovered radio waves.

heterogeneity: A characteristic of medium whose physical properties change along the space coordinates. A critical parameter affecting seismic phenomena is the scale of heterogeneities as compared with the seismic wavelengths. For a relatively large wavelength, for example, an intrinsically isotropic medium with oriented heterogeneities may behave as a homogeneous anisotropic medium. See Chapter 53.

hexagonal symmetry: The symmetry by rotation of 60° around one axis, the 6-fold symmetry axis. Hexagonal symmetry and cylindrical symmetry are equivalent for the elastic properties. There are five independent elastic parameters in hexagonal symmetry systems. See Chapter 53.

HFU: Heat flow unit.

HGLP: High-gain long-period seismograph network. See Chapter 20.

high-frequency (HF) earthquake: In volcano seismology, a term generally synonymous with a volcano-tectonic, brittle-failure, or A-type earthquake. See Chapter 25.

high-pass filter: A filter that removes the low-frequency portion of the input signal. See Chapter 22.

high-resolution beam-forming: See maximum-likelihood method beam-forming.

hodograph: In the early era of seismology, the term sometimes used for the time of arrival of various seismic phases as a function of epicentral distance. See Chapter 1.

Holocene: An epoch in geologic time from 10,000 years ago to the present. See also geologic time.

horizontal-to-horizontal spectral ratio: The spectral ratio of horizontal components at a target site to those at a reference site. If a reference site is virtually site-effect free, this corresponds to the true site amplification of the target site. Care must be taken to use a deep-borehole station as a reference site because seismic motions at a depth are modified by the presence of the free surface differently from those at the surface. See Chapter 61.

horst: An upthrown block lying between two steep-angled fault blocks.

hot spot: A long-lived volcanic center that is thought to be the surface expression of a persistent rising plume of hot mantle material. See Chapter 6 (pp. 59-61) and Olson (2002). See also melting anomaly.

HPOT: Higher-Order Perturbation Theory, a 3D synthetic seismogram software package. See Chapters 10 and 85.16.

hummock: A term in volcanology referring to characteristic topographic features for debris-avalanche deposits (e.g., hummocky topography), as well demonstrated during the May 18, 1980, eruption of Mount St. Helens, Washington State. Their occurrence provides telltale evidence for one or more sector collapses of a volcanic edifice in its eruptive history.

H/V (horizintal-to-vertical) spectral ratio: The ratio of the Fourier amplitude spectra of horizontal and vertical components of ambient noise, or of earthquake ground motions, recorded at a site, typically used to identify the presence of site-specific dominant frequencies in such motions. If this technique is applied to microtremor data, it is sometimes called Nakamura's method. If applied to the *P* and *S* portions of seismic data from distant sources to invert for a 1D structure, it is called the receiver function method. See receiver function and Chapter 62.

hyaloclastic rock: A general term for a wide variety of glassy volcanic rocks formed by the fragmentation of magma or lava in the presence of water. The adjective hyaloclastic has Greek roots (*hualos*, "glass"; *klastos*, "broken").

hybrid earthquake: In volcano seismology, an earthquake with characteristics of both high-frequency and long-period earthquakes, often applied to earthquakes that have an impulsive high-frequency onset with distinct *P* and *S* waves followed by an extended low-frequency coda. Usage is variable.

hydraulic permeability: The rate of fluid flow through rock per unit pressure gradient. See Chapters 36 and 38.

hydroacoustic: A term pertaining to compressional (sound) waves in water, in particular in the ocean. Hydroacoustic waves may be generated by submarine explosions, volcanic eruptions, or earthquakes. See also SOFAR channel and *T* phase.

hydrology: The science that deals with water on and under the ground surface, including its properties, circulation, and distribution. See Chapter 39.

hydrophone: A device for measuring pressure fluctuations in the ocean using a piezoelectric sensor.

hydrostatic pressure: A term with two variations on its usage: in one usage, a stress state that is isotropic (i.e., shear stresses are zero); in the other usage, hydrostatic pressure as a special case of lithostatic pressure, where the density of the overlying column is chosen to be water density. See Chapter 33.

HYPO71: A computer program for earthquake location. See Chapter 85.17.

hypocenter: A point in the Earth where the rupture of the rocks initiates during an earthquake. Its position, in practice, is determined from arrival times of the onsets of *P* and *S* waves and is also called earthquake focus. The point on the Earth's surface vertically above the hypocenter is called the epicenter. See Lee and Stewart (1981, pp. 130-139).

hypocentral distance: The distance from an observation point to the hypocenter of an earthquake.

HYPOELLIPSE: A computer program for earthquake location. See Chapter 85.7.

HYPOINVERSE2000: A computer program for earthquake location. See Chapter 85.8.

I

***I* (*i*):** The symbol *I* that is used to indicate the part of a ray path that has traversed the Earth's inner core as a *P* wave.

For example, *PKIKP* refers to *P* waves that have penetrated to the interior of the inner core and returned to the surface without conversion to *S* waves throughout the entire path. On the other hand, *i* is used to indicate reflection at the boundary between the outer and the inner core (e.g., *PKiKP*) in the same manner that c is used for reflection at the core-mantle boundary.

IAEE: International Association for Earthquake Engineering. See Chapter 81.2

IASPEI: International Association of Seismology and Physics of the Earth's Interior (URL: http://www.iaspei.org). See Chapters 4 and 81.3.

IDA: International Deployment of Accelerometers (Agnew *et al.*, 1976). See Chapters 17 (p. 271) and 20.

IDNDR: [United Nations] International Decade for Natural Disaster Reduction, originally proposed by Frank Press. By resolution 44/236 of December 22, 1989, the United Nations General Assembly designated the 1990s as the International Decade for Natural Disaster Reduction. See Chapter 4 (pp. 35-36).

igneous rock: A rock that has formed from molten or partly molten material. See Chapter 83 and Press and Siever (1994).

IGS: International GPS Service, an organization responsible for worldwide coordination of continuous GPS measurements. See Chapter 37.

IISEE: International Institute of Seismology and Earthquake Enginnering, Tsukuba, Japan (URL: http://iisee.kenken.go.jp). See Chapter 79.33.

impulse response: The output signal of a stationary linear system to an impulsive (delta function) input.

incident wave: The wave that propagates toward an interface.

inclusion: See lithics and xenolith.

induced earthquake: An earthquake that results from changes in crustal stress and/or strength due to man-made sources (e.g., underground mining and filling of a high dam) or natural sources (e.g., the fault slip of a major earthquake). As defined less rigorously, the term "induced" is used interchangeably with "triggered" and applies to any earthquake associated with a stress change, large or small. Cf.: triggered earthquake. See Chapter 40.

infragravity wave: A wave that is a component of ocean waves at periods longer than 25 seconds. Infragravity waves control long-period (>30 seconds) vertical-component seismic noise levels at the seafloor and at coastal sites on land. Infragravity waves are trapped along shorelines as edge waves or surf beat and are generated by nonlinear mechanisms by the interaction of shorter-period (wind-driven) ocean waves along coastlines. See Chapter 19 (p. 307).

infrasonic: See infrasound.

infrasound: An acoustic wave at a frequency approximately between the seismic and audible range, typical of sound waves that propagate over great distances through the atmosphere. See Chapter 24 (p. 370).

INGV: Istituto Nazionale di Geofisica e Vulcanologia, Rome, Italy (URL: http://www.ingv.it).

inhomogeneous plane wave: A plane wave with an amplitude varying in a direction different from the direction of propagation. The velocity of propagation is lower than that of regular plane waves. For example, Rayleigh waves are composed of inhomogeneous *P* and *SV* waves propagating horizontally with amplitude decaying with depth. They are also called evanescent waves. See Aki and Richards (2002, pp. 149-157).

inner core: The central part of the Earth's core. It extends from a depth of about 5200 km to the center. It has nonzero rigidity, in contrast to the outer core and was discovered in 1936 by Inge Lehmann.

inner-core anisotropy: A term referring to elastic anisotropy of the inner core. The *P* velocity along the north–south direction in the inner core is about 3% faster than along east–west directions, although the amount appears to vary spatially. See Chapter 56.

inner-core rotation: The geodynamo is expected to drive the inner core to rotate relative to the mantle. The observational evidence for such a differential inner-core rotation has been reported, but it is still under debate. See Chapter 56.

INSAR: Interferometric (analysis of) synthetic aperture radar (images). See Chapter 37.

instrumental noise: The unwanted part of data originating from the instrument used for the measurement.

instrument response: A response that can be broken down into two stages: (1) the transformation of ground motion to an electrical energy and (2) the transformation of that electrical energy into an output signal that can produce a permanent record (generally photographic or digital stream recorded on a computer storage unit). In order to recover the ground motion at a recording site, one must deconvolve the contribution of the recording instrumentation. See Chapters 18 and 22 and Eq. (3.10) in Appendix 2 of this Handbook.

Intensity: When capitalized, a term that typically refers to the Modified Mercalli Intensity or one of the other common noninstrumental measures of the strength of earthquake motion. The values in an intensity scale are pegged to observed physical damage or perceived motions, etc. See also seismic intensity.

interface: In a continuum, a surface of discontinuity that separates two different media.

interferometry, 2-pass differential: An approach to calculating interferograms that uses two radar images and a digital elevation model, also called DEM-elimination. See Chapter 37.

interferometry, 3-pass differential: An approach to calculating interferograms, that uses three radar images but no elevation model, also called double-differencing. See Chapter 37.

International System of Units (SI): It is an internationally agreed-upon system of units, which is abbreviated as SI for "Système International d'Unités," recommended for use in all scientific and technical fields (NBS, 1981). The international system of units is comprised of 7 base units: (1) length in meter (m), (2) mass in kilogram (kg), (3) time in second (s), (4) electric current in ampere (A), (5) thermodynamic temperature in kelvin (K), (6) amount of substance in mole (mol), and (7) luminous intensity in candela (cd). It also has a number of derived and supplementary units. Some derived units with special names are (1) frequency in hertz (Hz), (2) force in newton (N), (3) pressure or stress in pascal (Pa), (4) energy, work, or quantity of heat in joule (J), (5) power in watt (W), (5) electric charge in coulomb (C), and (6) electric potential in volt (V). See Lide (2002) for a detailed review of SI and conversion factors.

Internet: A worldwide electronic network that consists of mutually connected computers. Seismic data or program files can be transferred on the Internet.

interplate earthquake: An earthquake that has its source in the interface between two lithospheric plates, such as in a subduction zone or on a transform fault.

intraplate earthquake: An earthquake that has its source in the interior of a lithospheric plate. Often the reason for the specific location of such an earthquake or its relation to an identifiable fault is not evident.

intrinsic attenuation: The attenuation of seismic waves due to absorption in the medium. See Chapter 13 (p. 196) and Eq. (3.9) in Appendix 2 of this Handbook.

inverse problem: A problem in which the intent is to infer the values of the parameters characterizing a system, given some observed data. Solving an inverse problem may be summarized as observed data → model → estimates of model parameters. An inverse problem is the opposite of a forward problem; it can be formulated as a problem of data fitting or, more generally, as a problem of probabilistic inference. Cf.: forward problem. See also inversion, Chapters 16 and 52, and Menke (1984).

inverse refraction diagram: A refraction diagram for tsunami drawn from coastal observation points. Iso-time contours correspond to the travel time encompassed by the ray path to each station from the source region on the map. See Chapter 27 (p. 443).

inversion: Given a model hypothesized for explaining a set of observed data, a procedure for determining the model parameters from the observed data. Various issues involved in this procedure, such as the evaluation of the resolution and error in the estimated parameters, constitute the inverse problem. In contrast, the prediction of observables for a model with assumed parameters is called the forward problem. See also inverse problem and Chapters 16 and 52.

IPGP: Institut de physique du globe de Paris, France (URL: http://www.ipgp.jussieu.fr). See Chapter 79.22.

IPOD: International Phase of Ocean Drilling (1975–1983).

IRIG: Inter-Range Instrumentation Group, which was responsible for standardization of time-code format and timing techniques in the White Sands Missile Range, New Mexico, in the 1960s.

IRIS: Incorporated Research Institutions for Seismology, later changed to the IRIS Consortium (URL: http://www.iris.washington.edu). It is a consortium of US universities that have research programs in seismology. The purpose of IRIS is to develop and operate the infrastructure needed for the acquisition and distribution of high-quality seismic data. See Chapter 20.

ISC: International Seismological Centre. See Chapter 81.6.

Ishimoto-Iida's relation: An empirical relation expressing the frequency of occurrence of the maximum amplitudes of earthquakes recorded at a seismograph station obtained by Ishimoto and Iida (1939). Although the concept of magnitude was not known to them, the power law found for the amplitude distribution is reducible to the power law for earthquake energy implied in Gutenberg-Richter's frequency-magnitude relation. They recognized the significance of its departure from the Boltzmann exponential law for statistical energy distribution. See Chapter 43, and Eqs. (7.1) and (7.2) in Appendix 2 of this Handbook.

isochron: In seismology, a line on a map or a chart connecting points of equal time or time interval (e.g., arrival times of the onset of seismic waves from a point impulsive source). In geochronology, it is a line connecting points of equal age. In marine geophysics, it is the magnetic lineation with equal age, often abbreviated as chron. See Cox (1973), Uyeda (1978), and Duff (1993).

isopach: As used in volcanology, a line joining points of equal thickness in a pyroclastic deposit.

isopleth: As used in volcanology, a line joining points where the sizes of the largest pyroclasts are the same.

isoseismal: A closed curve bounding the area within which the intensity from a particular earthquake was predominantly equal to or higher than a given value. See Chapter 49.

isostasy: The state of the Earth's structure in which the mass of any column of rock above a specified level is everywhere the same. When dynamic processes destroy this balance, it is restored because of the hydrostatic condition hypothesized below the level.

ISOSYN: An isochrone-integration synthetic software. See Chapter 85.14.

isotope: The various species of atoms of a chemical element with the same atomic number but different atomic weights.

ISS: International Seismological Summary, a series of publications (1918–1963) summarizing the seismic arrival times of earthquakes worldwide and providing their origin times and locations. See Chapters 4, 41, and 88.

ITRF: International terrestrial reference frame, the geodetic reference frame defined by a combination of VLBI, SLR, DORIS, and GPS currently used to represent absolute coordinates for subcentimeter geodetic measurements. See Chapter 37.

IUGG: International Union of Geodesy and Geophysics (URL: http://www.iugg.org). See Chapter 81.2.

J

J: The symbol used to indicate that part of a ray path has traversed the Earth's inner core as an *S*wave. Unambiguous observations of waves such as *PKJKP* and *SKJKP* have not yet been achieved, but they may be possible with suitable sources and instrument responses.

Jericho: A site of ancient buildings situated on the Dead Sea transform fault. The rebuilding of many of the over 20 levels of destruction at this site can be attributed to historical and prehistorical earthquakes. See Chapter 46.

JERS-1: Japanese Earth Resource Satellite 1. See Chapter 37.

JHD: Joint hypocentral determination. See Chapter 85.9.

JMA: Japan Meteorological Agency (URL: http://www.jma.go.jp).

JMA magnitude (M_J): The magnitude for earthquakes in Japan and its vicinity, assigned by JMA using data from the JMA network. M_J is fairly close to M_S for shallow earthquakes and m_B for intermediate and deep earthquakes. See Chapter 44.

JMA seismic intensity scale: The scale used by JMA to measure seismic intensity, modified four times since it started in 1884. Since 1996, the scale has 10 grades designated as 0, 1, 2, 3. 4, 5-, 5+, 6-, 6+, and 7. See Chapter 44.

joint transverse-angular coherence function: A measure of coherency between two transmitted wave fields as a function of incident angles and the transverse separation of receivers. See Chapter 13 (p. 202).

joule (J): An SI-derived unit for energy, work, or quantity of heat, expressed in meter^2 kilogram per second per second (m^2 kg s^{-2}) or Newton meter (N m). It is named after James Joule (1818–1889), who established the mechanical equivalence of heat (1 calorie = 4.184 joules).

JPL: Jet Propulsion Laboratory at Pasadena, California. See Chapter 37.

Julian day: As used in seismology, the number of a day during the year, counting continuously from January 1 as Day 1. For example, February 2 is Julian day 33 in this method of counting.

juvenile: An adjective commonly used to describe eruptive materials produced from the magma involved in the eruption, in contrast to lithic materials derived from preexisting rocks (volcanic or nonvolcanic). Cf.: lithics.

K

K: A term for a *P* wave in the outer core (the German word for core is *kern*). For example, *S* waves traveling steeply downward in the mantle, converted to *P* waves at the core boundary, propagated through the outer core as *P* waves, and converted back to *S*waves at reentry to the mantle are designated as *SKS*. Just as *PP*, *PPP*, etc., are used to designate surface reflections, *KK*, *KKK*, etc., are used for *P* waves in the core reflected at the core-mantle interface from below.

kelvin (K): A base unit of thermodynamic temperature in the international system of units (SI). The official definition: "The kelvin, unit of thermodynamic temperature, is the fraction 1/273.16 of the thermodynamic temperature of the triple point of water." See NBS (1981).

kilogram (kg): A base unit of mass in the international system of units (SI). The official definition: "The kilogram is the unit of mass; it is equal to the mass of the international prototype of mass of the kilogram." See NBS (1981).

kiloton: An energy unit, now defined as a trillion (10^{12}) calories. The term originates from an estimate of the energy released by the explosion of a thousand tons of TNT. See Chapter 24.

kinematic ray equation: The ordinary differential equation governing the geometry of a ray path. The independent variable is typically the travel time or ray-path length, and the dependent variables are the ray position and direction. See Chapter 9.

kinematics: A description of the deformation or motions within a seismic source region in geometrical terms such as fault slip or transformational strain, without reference to the stresses or forces that are responsible. Cf.: dynamics. See Aki and Richards (2002, Box 5.3, p. 129).

Kirchhoff surface integral method: The method that originates from Kirchhoff's solution of the Cauchy problem (i.e., finding a solution of the wave equation for given initial conditions on the wave function and its time derivative defined for the whole space). The solution was expressed as a surface integral of functions given as initial conditions properly time-shifted and weighted over a spherical surface centered at the observation point. It can be generalized to an integral over a nonspherical surface enclosing the region where the homogeneous wave equation is valid. It can be used, for example, to represent scattered waves from a rough surface as a surface integral. The integral reduces to ray-theoretical solution for the planar surface. See Chapter 9 and the biographies of Augustin Louis Cauchy and Gustav Kirchhoff in Chapter 89.

K-Net: Kyoshin Net, Japan (URL: http://www.k-net.bosai.go.jp). See Chapter 63.

Knopoff model: A shear crack model for earthquakes investigated by Knopoff (1958). See Chapter 12 and Scholz (2002). See also fault model.

Kostrov model: A self-similar circular rupture model for earthquakes introduced by Kostrov (1966). See Chapter 12 and Kostrov and Das (1988). See also fault model.

Kramers-Kronig relations: The relationships between the amplitude and phase of the frequency response resulting from the requirement that the corresponding impulse response is causal, also called dispersion relations. See also causal signal, frequency response, and impulse response and Aki and Richards (2002, pp. 167-169).

KS-36000: A broadband borehole-deployable seismometer manufactured by Geotech Instruments (formerly Teledyne Geotech), Garland, Texas. See Chapter 20.

KS-54000: A broadband borehole-deployable seismometer manufactured by Geotech Instruments (formerly Teledyne Geotech), Garland, Texas. See Chapter 20.

L

$L(LQ, LR)$: The symbol that is used to designate long-period surface waves. When the type of surface wave is known, *LQ* and *LR* are used for Love waves and Rayleigh waves, respectively. See Chapter 21 (pp. 344-346).

lahar: A widely used Indonesian term for volcanic debris flow. See also mud flow.

Landsat: A series of optical imaging satellites. See Chapter 37.

landslide: The downslope movement of soil and/or rock.

lapilli: Of Italian origin (meaning "little stones"), a term given to tephra ranging in size between 2 and 64 mm.

Laplace transform: An integral transform mapping signals $r(t)$ defined in the positive time domain ($r(t) = 0$ for $t < 0$) into the complex domain by multiplying by $\exp(-st)$ and integrating over t from 0 to infinity. See Chapter 22 (p. 351) and the biography of Pierre Simon de Laplace in Chapter 89.

lapse time: The time measured beginning from the earthquake origin time. See Chapter 13.

LASA: Large Aperture Seismic Array in Montana. LASA was a large-aperture (200-km diameter) seismic array consisting of 21 subarrays, each with 25 seismometers. It was in operation from 1965 to 1978. See Chapter 23.

lateral blast: A phenomenon sharing the characteristics of pyroclastic flows and surges, with an initial low-angle component of explosive energy release. It is often associated with sudden decompression of a highly pressurized magmatic system upon sector collapse, as was well documented for the May 18, 1980, eruption of Mount St. Helens, Washington State.

lateral-force coefficient: A term in earthquake engineering referring to a numerical coefficient, specified in terms of a fraction or percentage of the weight of a structure, used in earthquake-resistant design to specify the horizontal seismic forces to be resisted by the structure. See Chapter 68.

lateral spread and flow: Two terms referring to landslides that commonly form on gentle slopes and that have rapid fluid-like flow movement, like water.

lateral spread in ground failure: The lateral displacement of the ground down gentle slopes or toward a free face, as a consequence of liquefaction, accompanied by flow within subsurface granular layers. See Chapter 70.

Laurasia: The protocontinent of the northern hemisphere in the late Paleozoic time. It included most of North America, Greenland, and most of Eurasia (excluding India) before they drifted away from one another. See Duff (1993) and Hancock and Skinner (2000).

lava: An all-inclusive term for any magma once it is erupted from a volcano. Rocks and deposits formed from lava can vary widely in texture and appearance, depending on chemical composition, gas and/or crystal content, eruption mode (effusive or explosive), and depositional environment.

lava cone: A relatively small volcanic edifice built largely of lava flows around a central vent, but distinguished from a lava shield by having a slightly concave (upward) profile becoming steeper toward the top, similar to those of cinder cones or stratovolcanoes. Depending on the viscosity of the lava flows involved, there can be a continuum in characteristics between lava cones and lava shields.

lava delta: A wide fan-shaped area of new land created by prolonged lava entry (commonly via a lava-tube system) into the ocean, as well demonstrated by the 1983–present eruption at Kīlauea Volcano in Hawaii. Such new land is usually built on sloping layers of loose lava fragments and flows.

lava dome: A roughly symmetrical and hemispherical mound of lava formed during extrusion of magma that is so viscous that it accumulates over the vent.

lava flow: A stream of molten rock that pours from a volcanic vent during an effusive eruption. Lava flows can exhibit wide differences in size, fluidity, textures, and physical aspects, depending on lava viscosity and eruption rates.

lava fountain: A pillarlike uprushing jet of incandescent pyroclasts that falls to the ground near the vent. Lava fountains can also occur as sheetlike jets from one or more fissure vents (see also curtain of fire). Lava fountains can attain heights of hundreds of meters. See also Hawaiian eruption.

lava lake (or pond): An accumulation of molten lava, usually basaltic, contained in a vent, crater, or broad depression. It describes both lava lakes that are molten and those that are partly or completely solidified. If one or more vents sustain a lava lake, it can remain active for many months to years.

lava shield: A broadly convex, gently sloping volcanic landform built by multiple thin flows of fluid lava from a volcanic vent or a cluster of closely spaced vents. While the term applies to such landforms of any size, the larger lava shields tend to be called shield volcanoes.

lava tube: A natural conduit through which lava travels beneath the surface. Lava tubes form by the crusting over of lava channels and flows. A broad lava-flow field often consists of a main lava tube and a series of smaller tubes that supply lava to the front of one or more separate flows.

Layer 2: The upper part of the oceanic crust, characterized by velocities lower than about 6.5 km/sec and by high-velocity gradients (typically 0.5–1.0 km/sec per kilometer). See Chapter 55.

Layer 2A: A layer of particularly low velocities (typically 2–4 km/sec) found in the vicinity of ridge axes, bounded by a sharp velocity increase at its base, and commonly interpreted as an extrusive basalt layer. See Chapter 55.

Layer 3: The lower part of the oceanic crust, characterized by velocities normally in the range 6.5–7.2 km/sec and by low-velocity gradients (typically 0.1–0.2 km/sec per kilometer). See Chapter 55.

L-band: A radar frequency around 1.2 GHz, with wavelength around 25 cm. See Chapter 37.

LDEO: Lamont-Doherty Earth Observatory of Columbia University (located in Palisades, New York; URL: http://www.ldeo.columbia.edu). See Chapter 79.55.

LDGO: Lamont-Doherty Geological Observatory of Columbia University (located in Palisades, New York; URL: http://www.ldeo.columbia.edu). It was recently renamed the Lamont-Doherty Earth Observatory. See Chapter 79.55.

leaking mode: Normal modes in a layered half-space, in general, have cutoff frequencies below which the phase velocity exceeds the P and/or S velocities of the half-space, and the energy leaks through the half-space as body waves. Because of the leakage, the amplitude of leaky modes attenuates exponentially with distance. See Aki and Richards (2002, pp. 312-324).

left-lateral fault: A strike-slip fault across which a viewer would see the block on the other side moving to the left. Cf.: right-lateral fault.

Lenz's law: The law that states an induced electric current in a circuit always flows in a direction so as to oppose the change that causes it. It was deduced by the physicist Heinrich Friedrich Emil Lenz (1804–1865) in 1834. See Chapter 18 (p. 287) and Maxwell (1891).

***Lg* wave:** A short-period (1–6 seconds) large amplitude arrival with predominantly transverse motion (Ewing *et al.*, 1957). *Lg* waves propagate along the surface with velocities close to the average shear velocity in the upper part of the continental crust. The waves are observed only when the wave path is entirely continental. As little as 2° of intervening ocean is sufficient to eliminate the waves. When *Lg* waves arrive in two distinct groups, they are called *Lg*1 and *Lg*2. See Chapter 21 (pp. 336-337).

lifeline: A man-made structure that is important or critical for a community to function, such as roadways, pipelines, power lines, sewers, communications, and port facilities.

linear analysis: A calculation or theoretical study that assumes material properties stay within their linear ranges and that the deflections are small in comparison to characteristic lengths. These assumptions permit the response to be described by linear differential or matrix equations. See Chapter 67.

linearity: In mathematics, a function F is linear if $F(ax) = aF(x)$, and $F(x + y) = F(x) + F(y)$, where x and y are variables and a is a constant. A system is linear if its output can be expressed as the convolution of the input and the impulse response. When the input is a function of time in a linear system, the output can be obtained from it using its frequency response alone. See also convolution and frequency response. See Chapter 18 (p. 297) for testing the linearity of a seismometer.

linearize: To linearize an equation means to use a linear approximation to that equation, often found by keeping only the first-order terms in the Taylor expansion, or, in equivalent linearization, by choosing the parameters of a linear system to minimize the difference between some measure of the responses of a nonlinear system and its linear approximation.

linear system: A system that is completely determined by its frequency response. When the frequency response is time-invariant, it is called a stationary linear system. See also frequency response and linearity.

liquefaction: The transformation of a granular material from a solid state into a liquefied state as a consequence of increased pore water pressures and reduced effective stress. In engineering seismology, it refers to the loss of soil strength as a result of an increase in pore pressure due to ground motion. See Chapter 70.

lithics: A term referring to nonjuvenile fragments in pyroclastic deposits that represent preexisting solid volcanic or nonvolcanic rock incorporated and ejected during explosive eruptions.

lithosphere: The rigid upper portion of the near-surface thermal boundary layer of the Earth at temperatures below about 650°C that can store elastic energy under long-term loads. This definition may or may not agree with the seismic lithosphere, which is composed of the crust and the uppermost mantle lid above the low-velocity layer. See Chapters 51 and 54.

lithostatic stress: Two variations on the usage of this term: (1) usage (also called the lithostatic load or lithostatic pressure) that gives the vertical normal stress as a function of depth due to the gravitational force of the overlying rock column, (2) usage (also called the lithostatic stress state) that combines the lithostatic load with assumptions about the strain history and rheology of the rock column to produce the complete stress tensor. See Chapter 33.

***Li* wave:** A wave similar to *Lg* waves, but its existence is not as widely accepted as that of *Lg*. The velocity of *Li* waves is 3.8 km/sec (as compared to 3.5 km/sec for *Lg*) and may be associated with the lower continental crust (Bath, 1954).

local magnitude (M_L): A magnitude scale introduced by Richter (1935) for earthquakes in southern California. M_L was originally defined as the logarithm of the maximum amplitude of seismic waves on a seismogram written by the Wood-Anderson seismograph (Anderson and Wood, 1925) at a distance of 100 km from the epicenter. In practice, measurements are reduced to the standard distance of 100 km by a calibrating function established empirically. Because Wood-Anderson seismographs have been out of use since the 1970s, M_L is now computed with simulated Wood-Anderson records or by more practical methods. See Chapter 44 and the biography of Charles Francis Richter in Chapter 89.

local site conditions: A qualitative or quantitative description of the topography, geology, and soil profile at a site that affect ground motions during an earthquake. See Chapter 60.

locked fault: A fault that is not slipping because frictional resistance on the fault is greater than the shear stress across the fault. Such faults may store strain for extended periods, which is eventually released in an earthquake when frictional resistance is overcome. In contrast, a fault segment that slips sometimes without earthquakes is called creeping segment. See also creep event.

log-periodic behavior: A term for power-law (fractal) behavior when the power is not real but complex. See Chapter 14.

longitudinal wave: Another name for *P* wave. See Eq. (1.9a) in Appendix 2 of this Handbook.

long-period earthquake: A seismic event occurring under volcanoes with emergent onset of *P* waves, no distinct *S* waves, and a low-frequency content (1–5 Hz) as compared with the usual tectonic earthquake of the same magnitude, also called B-type or low-frequency earthquake. The observed features are attributed to the involvement of fluid such as magma and/or water in the source process. See Chouet (1996) for the review of its observation and interpretation. See also Chapter 25.

long-period seismograph: A type of seigmograph. In traditional seismometry with limited dynamic range, seismographs were naturally divided into two types: long-period seismographs for periods above the spectral peak (periods around 5 seconds) of the microseismic noise generated by ocean waves, and short-period seismographs for periods below it. A long-period seismograph response usually has a bandwidth about 10–100 seconds. See Chapter 18.

long (oceanic) wave: A kind of oceanic gravity wave. When the wavelength is much larger than the water depth, it is called long wave or shallow-water wave. Most tsunamis can be treated as long waves. See Chapter 28.

look vector: A vector parallel to the line of sight between a radar and its target on the ground. See Chapter 37.

loss: In earthquake damage, an adverse economic or social effect (or cumulative effects) of an earthquake (or earthquakes), usually specified as a monetary value or as a fraction or percentage of the total value of a property or portfolio of properties.

loss function: A mathematical expression or graphical relationship between a specified loss and a specified ground-motion parameter (often the Modified Mercalli intensity) for a given structure or class of structures.

Love-Rayleigh wave discrepancy: Observed phase velocities of Love waves and Rayleigh waves require a transverse isotropy in the crust and the uppermost mantle. Love waves are faster than predicted for isotropic models that explain Rayleigh waves. This discrepancy was first discovered by Aki and Kaminuma (1963) for Japan and has been found universally, in both tectonically active and stable regions. See Chapter 53.

Love wave: An *SH* wave trapped near the surface of the Earth and propagating along it. The existence of Love waves was first predicted by Love (1911) for a homogeneous layer overlying half-space with an *S* velocity greater than that of the layer. They can exist, in general, in a vertically heterogeneous media, but not in a homogeneous half-space with a planar surface. See Chapter 21 and the biography of Augustus Edward Hough Love in Chapter 89.

low-frequency earthquake: See long-period earthquake.

low-pass filter: A filter that removes the high-frequency portion of the input signal. See Chapter 22.

low-velocity zone: The region in the upper mantle in which seismic velocities are lower than the overlying lid (i.e., the uppermost layer of mantle), commonly associated with the presence of partial melt and with ductile flow in the asthenosphere. Also, it is any layer in which the seismic velocity of concern is lower than in the overlying layer. See Chapter 51.

LP: Long-period.

LSB: Least significant bit. It is the smallest change in value that can be represented in the output of an analog-to-digital converter (digitizer). See Chapter 20.

Lyapunov exponent: Solutions to deterministic equations are chaotic if adjacent solutions diverge exponentially in phase space; the exponent is known as the Lyapunov exponent. Solutions are chaotic if the exponent is positive. See also chaos.

M

Ma: An abbreviation for 1 million years ago (Megannum).

macroscopic precursor: An earthquake precursor detected by human sense organs, for example, anomalous animal behavior, gush of well water, earthquake lights, and rumbling sounds. See Rikitake and Hamada (2002).

macroseismic: The term pertaining to the observed (felt) effects of earthquakes. See Chapter 49.

Macroseismology: The study of any effects of earthquakes that are observable without instruments, such as felt by people, landslides, fissures, knocked-down chimneys.

magma: The molten rock that forms below the Earth's surface that contains variable amounts of dissolved gases (principally water vapor, carbon dioxide, and sulfur dioxide) and crystals (principally silicates and oxides).

magmatism: The production of molten igneous rock called magma. The magma either erupts at the Earth's surface as lava or ash (volcanism) or is trapped in subterranean chambers within the crust or upper mantle as rock bodies, termed "intrusions" (plutonism).

magnetic anomaly: The difference between the observed value of the magnetic field and the calculated value based on a standard model of the Earth's magnetic field. See Chapter 6 (pp. 55-56).

magnetic lineation: The linear-striped, patterned magnetic anomalies observed in oceanic areas. They are generally 10–20 km wide and many hundred kilometers long, with amplitude of many hundred nT (nano Tesla). Their origin is now explained by the Vine-Matthews-Morley hypothesis (Vine and Matthews, 1963). See Chapter 6, Raff and Mason (1961), Cox (1973), Uyeda (1978), Duff (1993), and Merrill *et al.* (1998).

magnetic permeability: A coefficient relating magnetic flux density to magnetic field intensity. See Chapter 38.

magnetic pole: A point on the Earth's surface where the magnetic inclination is observed to be +90° or −90°. The North Pole and the South Pole are not exactly opposite each other. For the 1990 epoch, the North Pole was located at 78.1° N, 256.3° E, and the South Pole was located at 64.9° S, 138.9° E. See Merrill *et al.* (1998). See also geomagnetic pole.

magnetic quiet zone: An area of the seafloor without magnetic lineations because the Earth's magnetic field did not reverse its polarity for a long interval (e.g., during the long Cretaceous normal magnetic epoch). See Chapter 6 and Larson and Pitman (1972).

magnetization: A property of magnetized matter such as amount of magnetism or magnetic dipole moment per unit volume. See Chapter 38.

magnetogram: A graphic record of the variation of some parameter of the Earth's magnetic field, produced by a magnetometer.

magnetometer: A device for measuring magnetic fields and/or recording the variation of some parameter of the Earth's magnetic field with time. See Chapter 38.

magnification curve: A diagram showing the amplification (e.g. of the seismic ground motion by a seismograph) as a function of frequency. See also amplification.

magnitude: In seismology, a quantity intended to measure the size of an earthquake and is independent of the place of observation. Richter magnitude or local magnitude (M_L) was originally defined in Richter (1935) as the logarithm of the maximum amplitude in micrometers of seismic waves in a seismogram written by a standard Wood-Anderson seismograph at a distance of 100 km from the epicenter. Empirical tables were constructed to reduce measurements to the standard distance of 100 km, and the zero of the scale was fixed arbitrarily to fit the smallest earthquake then recorded. The concept was extended later to construct magnitude scales based on other data, resulting in many types of magnitudes, such as body-wave magnitude (m_b), surface-wave magnitude (M_S), and moment magnitude (M_W). In some cases, magnitudes are estimated from seismic intensity data, tsunami data, or duration of coda waves. The word "magnitude" or the symbol M, without a subscript, is sometimes used when the specific type of magnitude is clear from the context or is not really important. Earthquakes are often classified by magnitude (M) as major if $M \geq 7$, as moderate if M ranges from 5 to 7, as small if M ranges from 3 to 5, and as micro if $M < 3$. An earthquake with magnitude greater than about 7 3/4 is often called great. See also body-wave magnitude, duration magnitude, local magnitude, moment magnitude, and surface-wave magnitude. See Chapter 44 and Lee and Stewart (1981, pp. 153-157).

mainshock: The largest shock in an earthquake sequence.

major earthquake: An earthquake with magnitude equal to or greater than 7. See also magnitude.

mantle: The zone of the Earth's interior below the crust and above the core. The mantle represents about 84% of the Earth's volume and is divided into the upper mantle and the lower mantle, with a transition zone between. See Chapter 51 and Jeanloz (2002).

mantle convection: The slow overturning of the solid crystalline mantle by subsolidus creep, driven by internal buoyancy forces. The term "subsolidus creep" refers to the fact that mantle convection occurs in a crystalline solid. The hot interior of the mantle deforms like an elastic solid on short timescales but like a viscous fluid on geologic timescales. The primary surface expression of mantle convection is plate tectonics and continental drift. See Chapter 6 and Olson (2002).

mantle plume: A concentrated mantle upwelling that forms a hot spot in the Earth's surface. See also melting anomaly, Chapter 6, and Olson (2002).

mantle Rayleigh wave: Another name for long-period Rayleigh waves, just as long-period Love waves are given another name, G waves.

marine microseism: An ever-present, nearly periodic seismic signal originating from ocean waves and surf. Its amplitude is typically between 0.1 and 10 microns and its period between 4 and 8 seconds. The marine microseisms divide the spectrum of seismic signals into a short-period and a long-period band. See also double-frequency microseism and microseism.

Maslov asymptotic ray theory: The generalization of ray theory by using an ansatz, which integrates over neighboring rays with appropriate phase delays. At caustics, ray theory breaks down as the rays are focused and the amplitude predicted is singular. Maslov asymptotic ray theory reduces to ray theory when the latter is valid, but it remains valid at caustics. See Chapter 9.

master model of earthquakes: A concept developed at the Southern California Earthquake Center to use as a framework for integrating multidisciplinary earth science information regarding earthquakes in southern California for the purpose of transmitting it to people living there. A first-generation master model for southern California is described in WGCEP (1995). See Chapter 5.

maximum credible earthquake (MCE): The maximum earthquake, compatible with the known tectonic framework, that appears capable of occurring in a given area. See Chapter 62.

maximum-likelihood method beam-forming: An array beam-forming method introduced by Capon (1969). See Liu *et al.* (2000) for a recent example of application.

maximum probable earthquake (MPE): The maximum earthquake that could strike a given area with a significant probability of occurrence. See Chapter 62.

Maxwell material: A theoretical viscoelastic material in which the short-time response to an applied force is that of an elastic solid, while the long-time response is that of a viscous fluid. See also viscoelasticity.

Maxwell's equations: A set of four partial differential equations that describe the large-scale electromagnetic phenomena by the behavior of the electric and magnetic vectors and how the electric and magnetic fields are related to each other. These equations were formulated by Maxwell (1864) based on the experimental results of Ampere, Coulomb, Faraday, and Lenz. See Maxwell (1891), Stratton (1941), and the biography of James Clerk Maxwell in Chapter 89.

MCE: Maximum credible earthquake.

MCS: Multichannel seismic profiling. See Chapter 27.

melting anomaly: Another name for hot spot. It is a region on Earth (often in an intraplate tectonic setting) of anomalously high rates of volcanism over many millions of years, presumably sustained by a persistent source of hot mantle material. The Hawaiian melting anomaly (or hot spot) is arguably the best example.

Mesozoic: An era in geologic time from 245 to 66.4 million years ago. See also geologic time.

metamorphic facies: A set of metamorphic rocks characterized by particular mineral associations, indicating origin under a specific pressure-temperature range. See Chapter 83.

metamorphic rock: A rock derived from preexisting rocks by mineralogical, chemical, and/or structural changes, essentially in the solid state, in response to changes in pressure, temperature, shearing stress, and chemical environment. See Chapter 83.

meter (m): A base unit of length in the international system of units (SI). The official definition: "The meter is the length of the path traveled by light in vacuum during a time interval of 1/299792458 of a second." See Lide (2002).

Metropolis method: A general method for sampling a probability, including high-dimensional spaces. See Chapter 16.

microcontinent: An isolated, relatively small mass of sialic crust surrounded by oceanic crust. It is commonly plastered onto continents at a convergent margin where it becomes a part of the orogenic belt identifiable as an exotic terrane or suspect terrane.

microcrack: An ensemble of minute cracks that can join to form a crack. See Chapter 15.

microearthquake: An earthquake with magnitude smaller than 3. See also magnitude.

microseism: A continuous ground motion constituting the background noise for any seismic measurement. Microseisms with frequencies higher than about 1 Hz are usually caused by artificial sources, such as traffic and machinery, and in post-1980 literature they are more commonly called microtremors, to be distinguished from longer-period microseisms due to natural disturbances. At a typical station in the interior of a continent, the microseisms have predominant periods of about 6 seconds. They are caused by surf pounding on steep coastlines and the pressure from standing ocean waves, which may be formed by waves traveling in opposite directions in the source region of a storm or near the coast (Longuet-Higgins, 1950). They are also called double-frequency microseisms and marine microseisms. See Chapters 19 and 21 and Aki and Richards (2002, pp. 616-617).

microtremor: See microseism.

microzonation: The identification and mapping at local or site scales of areas having different potentials for hazardous earthquake effects, such as ground-shaking intensity, liquefaction, or landslide potential.

midocean ridge: An undersea mountain range extending through the North and South Atlantic Oceans, the Indian Ocean, and the South Pacific Ocean that consists of many small and slightly offset segments. Each segment is characterized by a central rift where two tectonic plates are being pulled apart and new oceanic lithosphere is being created. See Chapter 6 of this Handbook, Cox (1973), Uyeda (1978), Duff (1993), and Hancock and Skinner (2000).

millisecond delay initiation: See ripple firing.

mitigation: The proactive use of information, land use, technology, and societies' resources to prevent or minimize the effects of disasters on society. See Chapter 75.

mixed dislocation: See dislocation.

MM: Modified Mercalli. See also Modified Mercalli seismic intensity scale.

MMI: Modified Mercalli intensity. See also Modified Mercalli seismic intensity scale.

modal: A term referring to natural modes of vibration or their properties (e.g., modal damping). See Chapter 67.

mode 1 crack: An opening crack in which displacement is directed normal to the crack surface. See section 4 and Figure 9 in Appendix 2 of this Handbook.

mode 2 crack: A shear crack in the plane-strain mode. See also plane strain and section 4 and Figure 9 in Appendix 2 of this Handbook.

mode 3 crack: A shear crack in the antiplane-strain mode. See antiplane strain and section 4 and Figures 9 and 10 in Appendix 2 of this Handbook.

moderate earthquake: An earthquake with magnitude ranges from 5 to 7. See also magnitude.

mode shape: The shape that a structure assumes when it oscillates solely at one of its natural frequencies. Mathematically, a mode shape is an eigenvector of the eigenvalue problem posed by linear free vibration of a structure. See Chapter 67.

Modified Mercalli seismic intensity scale: A 12-grade seismic-intensity scale (I to XII) published by H.O. Wood and F. Neumann in 1931 (Wood and Neumann, 1931). It is used in many countries in and outside of North America. See Richter (1958, pp. 137-139).

Moho: A short word for "Mohorovičić discontinuity," the seismic boundary between the crust and mantle, named after A. Mohorovičić who discovered it from the travel-time data in Europe (Mohorovičić, 1910). The velocity contrast across the boundary is such that the lower crust typically has a compressional-wave velocity of 6.5–7.4 km/sec, while the uppermost mantle a velocity greater than 7.6 km/sec with an average value of 8.1 km/sec. See Chapters 54 and 55 and the biography of Andrija Mohorovičić in Chapter 89.

Mohorovičić discontinuity: See Moho.

Mohr diagram: A 2D diagram of a particular plane (Mohr plane) of the 3D stress state. The Mohr plane contains two of the principal stresses $\sigma_i > \sigma_j$, and it depicts the shear and normal stresses on planes of all possible orientation whose normals lie in the Mohr plane. The shear and normal stresses define a circle (Mohr circle) of radius $(\sigma_i - \sigma_j)/2$, ordinate $(\sigma_i - \sigma_j)/\sin(2\beta)/2$, and abcissa $(\sigma_i + \sigma_j)/2 + (\sigma_i - \sigma_j)/\cos(2\beta)$. β is the angle between σ_i and the normal to the plane of interest. Details on use and construction of Mohr diagrams can be found in Jaeger and Cook (1971). See Chapter 32 of this Handbook (p. 508).

moment connection: A connection between elements of a structure that is capable of transmitting moment as well as shear force. In a moment frame, the joints between the beams and columns are typically designed to take the full moment developed in the beam. See Chapter 69 of this Handbook, Dowrick (1977), and Salmon and Johnson (1996).

moment frame: An unbraced planar frame. Moment frames resist lateral forces by flexure of the comprising beams and columns. See Chapter 69 of this Handbook, Park and Paulay (1975), Dowrick (1977), and Salmon and Johnson (1996).

moment magnitude (M_W): A magnitude computed using the scalar-seismic moment (M_O). It was introduced by Kanamori (1977) via the Gutenberg-Richter magnitude-energy relation in the form $\log M_O$ (in dyne cm) $= 1.5M_W + 16.1$, where the subscript letter W stands for the work at an earthquake fault. It was more directly related to the seismic moment by Hanks and Kanamori (1979) using the same formula, except that the constant term was 16.05 instead of 16.1 as pointed out in Chapter 44 (p. 736) of this Handbook. This minor difference was not important in the 1970s but is becoming a concern with the improved accuracy in moment measurement. See Chapter 44.

moment tensor: A symmetric second-order tensor that characterizes an internal seismic point source completely. For a finite source, it represents a point-source approximation and can be determined from the analysis of seismic waves whose wavelengths are much greater than the source dimensions. It was first introduced by Gilbert (1970). See Jost and Herrmann (1989), Aki and Richards (2002, pp. 49-58), Chapter 8 (pp. 97-98), Chapter 50, and section 2 in Appendix 2 of this Handbook.

moment-tensor inversion: A procedure to calculate the elements of a moment tensor of an earthquake. There are several methods available in either the time or frequency domain, using various seismic waves. See also moment tensor, Jost and Herrmann (1989), and Chapters 50 and 85.11, of this Handbook.

monitoring system: A system for monitoring earthquakes, volcanic eruptions, tsunami, and/or other phenomena, usually consisting of a network of seismic stations and/or arrays, sometimes complemented by other types of sensors.

monogenetic volcano: A small volcano that is of a type that generally has a single eruption, or short-lived eruptive sequence, during its lifetime.

Monte Carlo method: Any mathematical method using at its core a random (or, more frequently, pseudo-random) generation of numbers. Monte Carlo methods are typically used to randomly explore high-dimensional spaces, where a systematic exploration is unfeasible. See Chapter 16.

MPE: Maximum probable earthquake.

MSK seismic intensity scale: A 12-grade intensity scale (I to XII), proposed by S. Medvedev, W. Sponheuer, and V. Kárník, aiming at worldwide use (Sponheuer and Kárník, 1964) by further modifying the Modified Mercalli seismic intensity scale. See Chapter 49 (pp. 810-811).

mud flow: A term essentially synonymous with debris flow and lahar.

multi-degree of freedom: A term applied to a structure or other dynamical system that possess more than a single degree of freedom. The number of degrees of freedom of a complex system is equal to the number of independent variables needed to fully describe the spatial distribution of its response at any instant. See Chapter 67.

multiple lapse-time window analysis: A method to measure scattering loss and intrinsic absorption from the analysis of whole S-seismogram envelopes. This method is based on the radiative transfer theory for *S*-wave propagation through scattering media. See Chapter 13 and Sato and Fehler (1998, p. 189).

mylonite: A cohesive, penetratively foliated (and usually lineated) fault rock (L-S tectonite) with at least one of the major mineral constituents (usually quartz) having undergone grain-size reduction through dynamic recrystallization accompanying crystal plastic flow. The mylonite series (protomylonite–mylonite–ultramylonite) reflects progressive reduction in grain size, generally under greenschist facies metamorphic conditions. See Chapter 29 (p. 458).

mylonitic gneiss: A comparatively coarse-grained, banded mylonitic fault rock (L-S tectonite) developed in a high-strain ductile shear zone under amphibolite facies metamorphic conditions, in which most of the mineral constituents have undergone dynamic recrystallization. See Chapter 29.

N

NA-Sampler Package: A software implementation of the Neighborhood algorithm for a multidimensional parameter space search. See Chapter 85.15.

natural frequency: The reciprocal of natural period.

natural period: The time for one cycle of an oscillator or structure during free vibration, typically measured in seconds.

N discontinuity: A seismic boundary within the continental lithosphere at a depth of 60–120 km reported particularly in Eurasia, also called Hales discontinuity. See Chapter 54 (p. 898).

near field: A term for the area near the causative fault of an earthquake, often taken as extending a distance from the fault equal to the length of fault rupture. It is also used to specify a distance to a seismic source comparable to or shorter than the wavelength concerned. See Eqs. (3.6)–(3.7) and related material in Appendix 2 of this Handbook.

NEHRP: National Earthquake Hazards Reduction Program of the United States. See Chapter 79.55.

NEIC: [United States] National Earthquake Information Center in Golden, Colorado, operated by the USGS (http://neic.usgs.gov). It is also the site for the World Data Center for Seismology, Denver, Colorado. See Chapter 20.

neotectonics: The study of the post-Miocene structures and structural history of the Earth's crust. The Miocene ended about 5 million years ago.

newton (N): An SI-derived unit for force. It is defined as the force that produces an acceleration of 1 meter per second per second when it acts on a mass of 1 kilogram (i.e., $1\ \mathrm{N} = 1\ \mathrm{m\ kg\ s^{-2}}$). It is named after Isaac Newton (1642–1727), who formulated the three laws of motion.

NHNM: New high-noise model used by the USGS as a criterion for testing the instrumental noise of borehole seismometers to be installed at high-noise sites such as islands. See Chapter 20.

NLNM: New low-noise model used by the USGA as a criterion for testing the instrumental noise of borehole seismometers to be installed at low-noise sites. See Chapter 20.

noble gas: A gas in group 0 of the periodic table of the elements (neon, argon, krypton, xenon, and radon). It is monatomic and chemically inert. See Chapter 39.

node: A reference point, typically a corner, in finite-element analysis. Because the field quantities of interest, such as displacement, strain, and velocity, are assumed to vary within finite elements in specified ways, their values become functions of values of key variables at the nodes. See Chapter 69.

noise: The unwanted part of data superimposed on a wanted signal. Noise is often modeled as a stationary random process but may also contain signal-dependent components. One man's noise can be the other man's signal. See Chapter 18.

NOISECON: A computer program for comparing noise specifications. See Chapter 85.18.

non-double-couple earthquake: An earthquake produced by a source that includes displacement components other than simple shear displacements along a fault plane (a double-couple earthquake). Common examples include the abrupt extensional opening of a crack or a compensated linear vector dipole (CLVD). See Julian *et al.* (1998).

nonlinear analysis: In earthquake engineering, a calculation or theoretical development that includes material changes, such as yielding or cracking, or the nonlinear effects of large displacements, such as the increased moment at the base of a structure caused by lateral displacements of floor masses. These effects cannot be considered within the framework of linear equations. See Chapter 67.

nonstructural: The parts of a building system that do not support loads (e.g., partitions, exterior claddings, and lighting fixtures).

normal fault: A fault that involves lateral extension, where one block (the hanging wall) moves over another (the footwall) down the dip of the fault plane (i.e., the movement of the hanging wall is downward relative to the footwall). The maximum compressive stress is vertical; the least compressive stress is horizontal. Cf.: reverse fault. See Chapter 31 and Figure 13 of Appendix 2 of this Handbook.

normal mode: The linear, free vibrations of a system with a finite number of degrees of freedom, such as a finite number of mass particles connected by massless springs. Each mode is a simple harmonic vibration at a certain frequency, called an eigenfrequency or natural frequency. There are as many independent modes as the number of degrees of freedom. An arbitrary motion of the system can be expressed as a superposition of normal modes. Free vibrations of a finite continuum body, such as the Earth, are also called normal modes. In this case, there are an infinite number of normal modes, and an arbitrary motion of the body can be expressed by their superposition. The concept of normal modes has been extended to wave guides in which free waves with a certain phase velocity can exist without external force. Examples are Rayleigh waves in a half-space and Love waves in a layered half-space. In these cases, however, one cannot express an arbitrary motion by superposing normal modes. See Chapter 8 (pp. 90-92), Chapter 10, and Aki and Richards (2002, Chapters 7 and 8).

normal stress: The force per unit area exerted upon an infinitesimal area in the direction of the normal to that area. In most disciplines, tension across the area is reckoned positive, but in geology, the opposite convention may be used. Cf.: shear stress. See Chapter 33 and Figure 1 in Appendix 2 of this Handbook.

NORSAR: Norwegian Seismic Array in Norway, which became a permanent research institution (URL: http://www.norsar.no). See Chapters 23 and 79.41.

NSF: [US] National Science Foundation, the primary funding agency for research in science and engineering in the United States (URL: http://www.nsf.gov).

nuée ardente: A French term ("glowing cloud") that was first used to describe the deadly 1902 eruption of Montagne Pelée (Island of Martinique) that devastated the port city of St. Pierre (Lacroix, 1904). A nuée ardente is a density current consisting of hot pyroclastic fragments, volcanic gases, and air flowing at high speed from the crater region of a volcano—generally associated with the collapse of an eruption column or unstable lava dome. See also glowing avalanche, pyroclastic flow, and pyroclastic surge.

number of deaths: The number of people killed by an earthquake, given only very approximately. Widely different figures are often found in different reports of a given historical earthquake. Even in recent years, such inconsistencies may arise due to omission, double counting, differences in the definition of an earthquake-related death, and inclusion or exclusion of missing people. See Chapter 42.

O

oblique slip: A combination of strike slip and normal or reverse slip.

OBS: Ocean bottom seismometer. See also OBS/OBH and Chapter 27.

OBS/OBH: Ocean bottom seismometer/hydrophone. It is a receiver for seismic signals that is located at or near the seabed. An OBH is commonly moored a few meters above the seabed and has a single hydrophone sensor. An OBS commonly rests on the seabed and, in addition to a hydrophone, has a three-component seismograph, which may be inside the recording package or deployed as a separate package for better coupling. See Chapter 55.

oceanic spreading ridge: A fracture zone along the ocean bottom that accommodates upwelling of mantle material to the surface, thus creating new crust. A line of ridges, formed as molten rock, reaches the ocean bottom, solidifies, and topographically marks this fracture.

oceanic trench: In marine geology, a narrow and elongated deep depression of the seafloor associated with a subduction zone. See Uyeda (1978) and Duff (1993).

ODP: Ocean Drilling Program (1983–2003). See Chapter 27.

Okada model: A closed-form analytic solution model for the static displacement field due to shear and tensile faults in a homogeneous half-space. It has become a standard model used for calculating the displacement field for earthquake fault slips and pressure sources beneath volcanoes. See Okada (1985) and Chapter 37 of this Handbook (pp. 612-613). A computer program is included on the attached Handbook CD#1, under the directory \37Feigel.

omega-squared model: A widely used model of earthquake source associated with the far-field body wave characterized by the displacement amplitude spectrum having the flat low-frequency part and the high-frequency power-law decay (with the power of –2) part separated by the corner frequency. The height of the flat part is proportional to the seismic moment, and the corner frequency is inversely proportional to the linear dimension of the source. See Hanks (1979) and Aki and Richards (2002, Chapters 10 and 11) for its observational and physical bases. The observed seismic spectra tend to be lower than predicted by this model at frequencies higher than a limit called fmax. A practical procedure for computing the time history of strong ground motion, including the fmax effect, was given by Boore (1983) based on this model. See also Brune model, corner frequency, fmax, seismic moment, and Eq. (6.1) in Appendix 2 of this Handbook.

Omori's relation (Omori's law): A relation expressing the temporal decay of the aftershock frequency, occasionally called hyperbolic law. The modified Omori relation generalizes it to a power-law dependence on time with the power allowed to deviate from –1. See Chapter 43 and Eq. (7.4) in Appendix 2 of this Handbook. See also p value and the biography of Fusakichi Omori in Chapter 89.

1D amplification of seismic waves: In site effect studies, of a 1D structure means a structure varying only in the vertical direction. It is also called a horizontally layered structure. The amplification for incident body waves can be theoretically calculated by considering multiple transmission and reflection at each interface of horizontal layers, including the free surface. There is no amplification for surface waves incident on such a structure because it is assumed to be homogeneous in the horizontal directions. See Chapter 61.

onset: The first appearance of a seismic wavelet on a record.

ophiolite: A suite of mainly ultramafic and mafic rocks formed at midocean spreading centers and along zones of convergence of oceanic crustal plates. On-land exposures of tectonically accreted sections of ophiolite are a primary means of studying the rocks that make up the oceanic crust and uppermost mantle. See Chapter 55 and Dickinson (2002).

optimal survey design: The optimal design of experiment or survey based on the inverse theory. See Chapter 52 (pp. 868-870).

organized noise: Any seismic noise whose power spectrum is not uniformly distributed but is concentrated at particular wave numbers and frequencies. See Chapter 23.

origin time: The instant of time when an earthquake begins at the hypocenter. It is usually determined from arrival times of seismic waves by Geiger's method.

orogenesis: See orogeny.

orogeny: The process of crustal uplift, folding, and faulting by which systems of mountains are formed. See Chapter 6 (pp. 62-64).

orthorhombic symmetry: The symmetry with respect to three mutually orthogonal planes. There are nine independent elastic parameters for an orthorhombic material. Olivine crystals are orthorhombic. See Chapter 53.

oscillator: In earthquake engineering, an idealized damped mass-spring system used as a model of the response of a structure to earthquake ground motion. A seismograph is also an oscillator of this type. See also single-degree-of-freedom system.

outer rise: The upward flexure of the oceanic plate before subducting beneath the trench.

oversampling: A sampling technique in which the analog input signal is first sampled at a sampling rate much higher than the final sampling rate and subsequently decimated digitally. See Chapter 22.

P

$\bar{P}$(*Pg*): Travel-time curves at short distances (up to a few hundred km) for seismic sources in the Earth's crust usually consist of two intersecting straight lines, one with velocity about 6 km/sec at shorter distances and the other about 8 km/sec at greater distances. The former is attributed to direct *P* waves propagating through the crust and is designated as $\bar{P}$ or *Pg*, which stands for granitic layer. The latter is *Pn*. See Chapter 21 (pp. 335-338).

***P**:** A designation for *P* waves refracted through an intermediate layer in the Earth's crust with a velocity near 6.5 km/sec. The upper boundary of this layer has been called the Conrad discontinuity. See Chapter 21.

***P′*:** Another symbol for *PKP*.

pahoehoe: A common type of fluid lava flow produced by Hawaiian eruptions. Pahoehoe in solidified form is characterized by a continuous, smooth, billowy, or ropy surface. Like aa, the term is also of Hawaiian origin; both aa and pahoehoe have become universally accepted to describe lava flows of similar appearance and characteristics the world over.

paleoearthquake: An ancient earthquake.

paleomagnetic pole: An estimate of the geographical pole of the Earth at a certain time in the past by averaging the paleomagnetic field over a sufficient long period of time (e.g., 10^4–10^5 years) from rocks that acquired the paleomagnetic field at that time. See Chapter 6 and Merrill *et al.* (1998).

Paleomagnetics: The study of the geomagnetic field existing at the time of formation of a rock, based on the remanent magnetization that was induced in the rock at the time of its formation and is still preserved today. See Chapter 6 (pp. 52-54), and Merrill and McFadden (2002).

paleoseismic event: A paleoearthquake that caused surface faulting, displacements in young sediments, or other near-surface phenomena such as liquefaction. See Chapter 30.

paleoseismology: The study of paleoearthquakes decades, centuries, or millennia after their occurrence. See Chapter 30 and McCalpin and Nelson (1996).

Paleozoic: An era in geologic time from 545 to 245 million years ago. See also geologic time.

Pangea: A supercontinent (including most of the continental crust of the Earth) that was postulated by Alfred Wegener in 1912 to have existed from about 300 to 200 million years ago. Pangea is believed to have split into two protocontinents, Gondwana on the south and Laurasia on the north, from which the modern continents are derived by continental drift. See Chapter 6, Wegener (1924), Uyeda (1978), and Duff (1993).

parabolic approximation: An approximation to the wave equation in which the second derivative with respect to the global ray direction is neglected. It can be justified when the wavelength is much smaller than the characteristic length of the heterogeneity. The wave equation then becomes parabolic in type. This approximation includes diffraction and forward scattering but neglects backscattering in the formulation. See Chapter 13.

parameter: See attribute.

paraxial ray: A term originally defined for an optical system with a symmetrical axis of rotation. A ray of light near this axis that has a small inclination to the axis is called a paraxial ray. In seismic ray theory, it is defined as a ray generated by perturbing the source position or takeoff direction of a central ray. The dynamic ray equation governs the properties of the paraxial ray. See Chapter 9.

participation factor: A factor that apportions scaled versions of the input to a multidegree-of-freedom system to the various modes of the system. The factor is not unique and depends on how the mode shapes are normalized. In one common normalization scheme, the participation factors add to unity; in this case, the participation factor for a mode is the fraction of excitation that mode receives. See Chapter 67.

pascal (Pa): An SI-derived unit for pressure or stress. It is defined as a force of one newton per square meter (N m^{-2} or m^{-1} kg s^{-2}). In seismology, 1 MPa (i.e., 10^6 Pa) is frequently used, and 1 MPa = 10 bar in cgs units. It is named after Blaise Pascal (1623–1662) for his pioneering work on hydrostatics.

passband: The portion of the frequency spectrum that a filter passes with less than 3-decibel attenuation. See Ballou (1987).

passive margin: A continental margin that is not a plate boundary, formed, for example, as seafloor spreading carries the continental block away from the spreading center. If earthquakes occur on such a margin, they are intraplate events. Cf.: active margin.

passive-source seismology: A seismic investigation that utilizes data from naturally occurring seismic events. See Chapter 54.

path effect: The effect of the propagation path on seismic ground motions. It is implicitly assumed that the source, path, and site effects on ground motions are separable.

***P* coda:** The portion of *P* waves after the arrival of the primary waves. They may be due to *P*- to *S*-wave conversions at interfaces or to multiple reflections in layers or to scattering by 3D inhomogeneities.

$P_d(P_u,P_r)$: The travel-time curve for *P* waves near $\Delta = 20°$ shows a triplication due to a sharp velocity increase in the upper mantle below the low-velocity layer. Three branches are designated in the order of decreasing $dt/d\Delta$ as P_d (direct), P_u (upper), and P_r (refracted).

p-delta effect: A term in earthquake engineering for the destabilizing moment of gravity created when the masses of a vertical structure, such as a building, experience lateral displacements. See Dowrick (1977) and Salmon and Johnson (1996).

$P_dP(PdP)$: A wave like the surface reflection *PP*, except that the reflection occurs at an interface at depth *d*, expressed in kilometers (e.g., $P_{600}P$), instead of at the surface.

PDZ: Principal displacement zone.

peak acceleration: See peak ground acceleration.

peak ground acceleration (PGA): The maximum acceleration amplitude measured (or expected) in a strong-motion accelerogram of an earthquake. See Chapter 57.

Peléan eruption: A type of eruption named after the 1902 eruption of Montagne Pelée, Martinique, associated with relatively viscous magmas (e.g., andesitic to rhyolitic compositions). The most violent and destructive activity generally occurs early in the eruption and commonly involves nuées ardentes, followed by emplacement of lava domes and thick lava flows.

pendulum: A mechanical sensing element of a seismometer, consisting of a mass free to pivot about some point. In the simple or common pendulum, the mass is below the pivot and is constrained to move in a vertical plane; in the inverted pendulum, the mass is directly over the pivot; and in the horizontal pendulum, the mass is to one side of the pivot and is constrained to move in a nearly horizontal circle. See Chapter 18 (pp. 284-285).

percolation cluster: A term applied to a grid of sites in two or three dimensions. The probability that a site is permeable is specified, and there is a sudden onset of flow through the grid at a critical value of this probability. See Chapter 14.

peridotite: The rock comprising the uppermost mantle, composed mainly of olivine [(Mg, Fe)2SiO_4] and pyroxene [(Mg, Fe, Ca)SiO_3]. Other minerals, such as garnet, can also be present.

period: The length of time required to complete one cycle or a single oscillation of a periodic process. See also natural period.

period of a signal: The period of the dominant sinusoidal signal. Most seismic signals are aperiodic but can be mathematically represented as a superposition of periodic sinusoidal signals.

permittivity: A constant relating the force between two charges separated by a distance. See Chapter 38.

perovskite: A dense, high-pressure phase of pyroxene [[(Mg, Fe, Ca)SiO_3]. It is only stable at high pressures (above ~20 GPa) but is thought to be the predominant mineral of the Earth's mantle. See Jeanloz (2002).

Phanerozoic: An eon in geologic time from 545 million years ago to the present. It is from Greek and means "apparent life." See also geologic time.

phase response: The phase of the complex frequency response. See also frequency response.

phreatic: An adjective of Greek origin (from *phrear*, "well") relating to groundwater. It is commonly applied to explosive eruptions triggered by interaction of water with magma or with solidified but still hot volcanic rock that ejects steam and fragments of preexisting solid rock (lithics), but not magma. Phreatic is synonymous with steam blast.

phreatomagmatic eruption: A volcanic eruption that results partly from the interaction of external water and magma, with explosive conversion of water to steam and disruption of the erupted lava.

physical master model: A type of master model of earthquakes in which universally applicable laws of physics and chemistry are applied to geologic processes to integrate earth science information regarding earthquakes in a given region. Its construction has been an ultimate goal beyond that of the hazard master model at the Southern California Earthquake Center. See Chapter 5.

piercing line: A term in paleoseismology indicating a feature that crosses a fault. The point where a piercing line intersects a fault is called piercing point. See Chapter 30.

piezomagnetism: Any change in rock magnetism resulting from stress loading. See Chapter 38.

pillow lava: A type of submarine lava flow resulting from the underwater volcanic eruptions or from the entry of lava erupted on land into the ocean, with formation of bulbous and tubular masses with glassy surfaces upon chilling in water.

pinned connection: A connection that can transmit shear but not moment, like a hinge. Historically, actual pins were used in such connections, particularly in bridges, but in modern construction most such "pinned joints" are welded or bolted joints not designed to transmit significant moments. See Dowrick (1977) and Salmon and Johnson (1996).

PKIKP: A *P* wave transmitted through the outer and inner core of the Earth. See Chapter 21 (p. 342) and Chapter 53 (pp. 880-881).

PKJKP: A *P* wave that travels as a converted shear wave through the solid inner core (known as a *J* wave). The *PKJKP* phase, though ardently sought, has yet to be reliably observed. See Chapter 56.

***PKP* precursor:** The precursor to *PKP* phases, the first geometric arrivals beyond the core shadow. Sometimes on a short-period seismogram at distance range of about 128–140 degrees, clearly visible waves arrive a few to 10 seconds prior to the *PKP* phases. The *PKP* precursors are believed to be scattering from the heterogeneous lowermost mantle or the whole mantle. See Chapter 21 (pp. 341-342) and Chapter 56.

***PKP* triplication:** In a travel-time curve, a triplication, which has three distinct branches of seismic phases, is the result of wave propagation through a zone of a steep velocity increase with depth. The *PKP* triplication is associated with the marked *P* velocity increase from the bottom of fluid outer core to the top of the solid inner core. See Chapter 56.

planar frame: An assemblage of vertical columns and horizontal beams, with or without inclined braces, where all members lie in a single plane. See Park and Paulay (1975).

plane strain: Also called in-plane strain, in contrast to antiplane strain, used to simplify mathematical analysis of elastic problem in which displacement depends on only two spatial coordinates and its direction lies in the plane of those coordinates. Dip-slip displacement that does not vary along strike is an example of plane-strain deformation. Another example is a Rayleigh wave from a line source. Cf.: antiplane strain. See Love (1944).

plane wave: A wave for which the constant-phase surfaces are planes perpendicular to the direction of propagation. Cf.: spherical wave. See also inhomogeneous plane wave, Aki and Richards (2002, pp. 119-187), and section 3 of Appendix 2 in this Handbook.

plastic hinge: A term given to a point on a structural element, typically a beam or column, where the element has yielded to the extent that increased curvature occurs without an increase in the resisting moment. See Chapter 59.

plastic strain rate: The irreversible strain rate expressed by a tensor. See Chapter 15.

plate: A synonym for tectonic plate. In tectonics, a plate is an internally rigid segment of the Earth's lithosphere that is in motion relative to other plates and to deeper mantle beneath the lithosphere. See plate motion, plate tectonics, and Chapters 6 and 7.

plate boundary zone: The zone of diffuse deformation, often marked by a distribution of seismicity, active faulting, and rough topography, within which relative plate motion is accommodated. Plate boundary zones are typically narrow under ocean and broad under continent. See Chapter 7.

plate driving force: A force that drives the motion of Earth's lithospheric plates. Terms such as "ridge push," "slab pull," and "basal shear" have been used to describe it. See Chapter 34 (pp. 564-565).

plate motion: Three fundamental kinds of relative plate motions that combine geometrically to produce the observed patterns on Earth: Pairs of plates in contact at mutual boundaries move apart (plate divergence), move toward one another (plate convergence), or slip sideways past one another along structural ruptures (transform faults). The topology of plate boundaries on a sphere also dictates that three plates locally meet at common junctures (triple junctions), which may involve any of the three kinds of plate boundaries in any combination. The present Earth may be divided into six major plates and six smaller plates, with various additional microplates that jostle between the larger plates along some plate boundaries. See also Euler vector, Chapters 6 and 7, and Dickinson (2002).

plate tectonics: A single coherent theory that reconciles continental drift, seafloor spreading, and deep-focus earthquakes and provided a new paradigm for earth sciences. The lithosphere of the Earth is broken into six major and six smaller plates that are in relative motion with respect to each other, over the underlying hotter and more fluid asthenosphere. See Chapters 6 and 7 and Dickinson (2002).

plate wave: The period equation for normal modes in an infinite plate in a vacuum can be split into two. One of the equations governs the mode with motion symmetric with respect to the median of the plate, and the other governs the mode with antisymmetric motion. The former is sometimes called the $M1$ wave, and the latter $M2$. For example, $M11$ and $M12$ are the fundamental and first higher modes of the $M1$ wave, respectively. For very short waves, both $M11$ and $M21$ approach Rayleigh waves in an elastic half-space made of the plate material. For wavelengths that are long compared with plate thickness, $M21$ are called flexural waves. They are dispersive, with phase velocities decreasing to zero with increasing wavelength.

Pleistocene: An epoch in geologic time from 1.6 million years ago to 10,000 years ago. See also geologic time.

Plinian eruption: An explosive eruption that generates a high and sustained eruption column, producing fallout of pumice and ash.

plume: See mantle plume.

***PL* wave:** A train of long-period (30–50 seconds) waves observed in the interval between P waves and S waves for distances less than about 30 degrees. They show normal dispersion (longer periods arriving earlier). They are explained as a leaking mode of the crust-mantle wave guide. See Aki and Richards (2002, p. 323).

PML: Probable maximum loss.

P_n (*Pn*): Beyond a certain critical distance, generally in the range from 100 to 200 km, the first arrival from seismic sources in the crust corresponding to waves refracted from the top of the mantle. Called P_n, these waves are relatively small, with long-period motion followed by larger and sharper waves of shorter period called $\bar{P}$, which are propagated through the crust. The P_n wave has long been interpreted as a head (conical) wave along the interface of two homogeneous media, namely, crust and mantle. The observed amplitude, however, is usually greater than that predicted for head waves, implying that the velocity change is not exactly step-like but has a finite gradient at or below the transition zone. The designation P_n has been applied to short-period P waves that propagate over considerable distances (even up to 20 degrees) with horizontal phase velocities in the range 7.8–8.3 km/sec. An interpretation in terms of head waves at the Moho is unsatisfactory (although the horizontal velocity and travel times would be explained) because head waves must decay rapidly with distance. More likely is an explanation in terms of guided waves, within a high-Q layer several tens of kilometers in thickness at the top of the mantle. Cf.: S_n. See Chapter 21 (pp. 335-338) and Chapter 54.

point process: A series of events distributed according to a probabilistic rule. See Chapter 43.

Poisson distribution: A probability distribution that characterizes discrete events occurring independently of one another in time. See Chapter 82.

Poisson process: A point process in which events are statistical independently and uniformly distributed with respect to the independent variable (e.g., time). See Chapter 43 and the biography of Simeon Denis Poisson in Chapter 89.

Poisson's ratio: The absolute ratio of the transverse strain to the longitudinal strain. An elastic body such as a metal bar under uniaxial tension shows elongation along the tension axis and shrinkage in the transverse direction. Poisson's ratios for rocks are typically around 0.25. See Chapter 83, the biography of Simeon Denis Poisson in Chapter 89, and Eq. (1.6) in Appendix 2 of this Handbook.

polarity: See first motion.

polarity reversal: In seismology, the occurrence of waveforms that are mirror images of their related initial phases (e.g., the waveforms of depth phases with respect to their primary P or S waves).

polarization: The direction of particle motion relative to the direction of wave propagation. See Christoffel matrix, earthquake focal mechanism, seismic anisotropy, and shear-wave splitting.

polarization anisotropy: A term (in contrast to azimuthal anisotropy) used to refer to an anisotropic medium in which the seismic-wave properties depend apparently on the polarization of the waves, not on the azimuth of their propagation direction. SH/SV differences of velocity and Love/Rayleigh-wave discrepancies are often referred to as polarization anisotropy, while azimuthal variations of velocity would be due to azimuthal anisotropy. See Chapter 53.

polar wandering: The movement of geographic or paleomagnetic poles over geologic time. See Chapter 6 of this Handbook, Uyeda (1978), and Duff (1993).

poloidal: See spheroidal.

polygenetic volcano: A long-lived volcano that has erupted repeatedly throughout its lifetime, often in an episodic manner.

pore-fluid factor: The ratio of fluid pressure to overburden pressure ($\lambda_v = P_f/\sigma_v$) employed as a measure of the degree of overpressuring. For typical rock densities, $\lambda_v \sim 0.4$ and $\lambda_v = 1.0$ denote hydrostatic and lithostatic fluid-pressure conditions, respectively. See Chapter 29.

pore pressure: In soil mechanics, the pressure within the fluid that fills voids between soil particles. In rock mechanics, it is the pressure due to liquids in formations that offset the strengthening effect of normal stresses acting on a fault. At a given depth, the ambient pore pressure is often observed to be nearly equivalent to that of a column of water extending upward to the ground surface, a pore-pressure regime termed the hydrostatic. See Chapters 36, 40, and 70.

porosity: The volume of voids per unit volume of a porous material, usually expressed as a percentage.

poststack seismic data: A term in reflection seismology when the seismic data set is reduced by the stacking process such that each surface location is represented by a single stacked seismic trace.

potency: See seismic potency.

power spectral density: The power spectral density of a stationary time series is defined as the Fourier transform of its autocorrelation function. The power spectral density integrated over the whole frequency range is equal to the mean square of the dependent variable the time series, hence the name "spectral density." There is a factor of 2 difference depending on whether the frequency range is taken from 0 to ∞ or $-\infty$ to $+\infty$. See Aki and Richards (2002, pp. 610-611).

precursor time: The time span between the onset of an anomalous phenomenon and the occurrence of the main shock.

pre-event memory: In a triggered recording by a digital seismograph, the buffered signal memory from which the ground motion can be retrieved for a time segment prior to the time the instrument was triggered.

preliminary tremor: In early seismological studies, the term used for the small motion at the start of an earthquake record. For local earthquakes, this corresponds to the waves in the interval between the first *P* arrival and the main *S* arrival in modern terms. Omori found that its duration is proportional to the epicentral distance with the coefficient of about 8 km/sec. See Chapter 1 and the biography of Fusakichi Omori in Chapter 89.

pressure: If a material element has zero shear stress, then the principal stresses all share the same value; this value is the negative pressure. Any rotation of the coordinate system produces the same value for the normal stresses; hence a stress state with zero shear stress is isotropic, and the stress tensor can be written as the product of the scalar quantity pressure and the identity matrix. Even if shear stresses are present, the isotropic component of the stress state can be extracted from the general stress tensor. See Chapter 33.

pressure solution: A water-assisted ductile deformation mechanism whereby minerals are sequentially dissolved from highly stressed regions, transported through fluid by diffusion or advection, and precipitated in low-stress regions. See Chapter 32 (p. 528).

prestack seismic data: A term in reflection seismology referring to the entire seismic data set before reduction by the stacking process.

principal axes: The three orthogonal axes at which the rotated tensor has only diagonal elements. For example, in the case of the seismic moment tensor corresponding to a double-couple source, the principal axes are called the compression axis, tension axis, and null axis. See Chapter 50 and Figure 1 in Appendix 2 of this Handbook.

principal displacement zone (PDZ): The zone of concentrated slip (sometimes referred to as the fault core) accommodating most of the displacement within the more distributed deformation (damage zone) defining a major fault zone. See Chapter 29.

principal part: In early seismological studies, the term used for the largest motions seen on the seismogram from an earthquake. In modern terms, the principal part is mostly surface waves for teleseismic events and *S* waves for local earthquakes. See Chapter 1.

principal plane: The most favorably oriented plane for shear fracture. Given planes of equal shear strength and of all possible orientations, the principal plane is that which minimizes the work done in shear. The orientation is given by $\tan 2\beta = 1/\mu_*$, where β is the angle between the greatest principal stress and the normal to the plane of interest, and μ_* is the coefficient of friction for a fault or the coefficient of internal friction for an intact rock. See Chapter 32 (p. 508).

principal stresses: The three normal stresses in the particular coordinate frame (principal axes) oriented so that the stress tensor reduces to diagonal form (i.e., no shear stresses). See Chapters 33 and 34 and Figure 1 in Appendix 2 of this Handbook. See also normal stress, principal axes, and shear stress.

prior information: The information one may have on the parameters characterizing a system prior to any actual measurement. See Chapter 16.

probabilistic earthquake scenario: A representation, in terms of useful descriptive parameters, of earthquake effects with a specified probability of exceedance during a prescribed period in an area. See Chapter 62.

probabilistic seismic hazard analysis (PSHA): Available information on earthquake sources in a given region is combined with theoretical and empirical relations among earthquake magnitude, distance from the source, and local site conditions to evaluate the exceedance probability of a certain ground-motion parameter, such as the peak acceleration, at a given site during a prescribed period. See Chapters 5 and 65.

probability density: A function of a continuous statistical variable whose integral over a given interval gives the probability that the variable will fall within that interval. See Chapter 82.

probability of exceedance: The probability that, in a given area or site, an earthquake ground motion will be greater than a given value during some time period. See Chapter 74.

probable maximum loss (PML): A probable upper limit of the losses that are expected to occur as a result of a damaging earthquake, normally defined as the largest monetary loss associated with one or more earthquakes proposed to occur on specific faults or within specific source zones.

Proterozoic: An eon in geologic time from 2.5 billion years ago to 545 million years ago. It is from Greek and means "first life." See also geologic time.

PSD: Power spectral density.

pseudotachylyte: A black aphanitic fault rock retaining evidence of a melt origin (e.g., relic glass, devitrification, or quench textures) attributed to heat generated by localized seismic slip, also developed in association with impact structures or at the base of some landslides. See Chapter 29.

PSHA: Probabilistic seismic hazard analysis.

pull-apart basin: A topographic depression produced by extensional bends or step-overs along a strike-slip fault.

pumice: A low-density fragment of silicate glass (generally of dacitic to rhyolitic composition) containing few crystals but a large fraction of gas bubbles (voids) due to exsolution of gas from magma during eruption. See also scoria.

pushover analysis: An analysis used to estimate the capacity of a structure to resist collapse from strong ground motion. Typically used in earthquake-resistant design studies, the analysis employs a selected force profile that is increased in intensity until the analysis indicates a structural element will fail or yield. That member is replaced by its yielding resistance, or removed if failed, and the force profile increased until another member yields or fails. The process is repeated until the overall structure has a constant or declining resistance under increased load. The deflection profile at this final stage approximates the structure's ultimate nonlinear earthquake resistance. See Chapter 59.

***p* value:** The exponent of the time factor in the modified Omori relation showing the rate of decay of aftershock frequency. A higher p value indicates more rapid decay. Estimated p values for most aftershock sequences fall between 0.9 and 1.5. In the original Omori relation, the p value is fixed as 1. See Chapter 43.

***P* wave:** A compressional elastic wave in seismology, P standing for primary. In a homogeneous isotropic body, the velocity of P waves is equal to $\sqrt{[(\kappa+4/3\mu)/\rho]}$, where κ, μ, and ρ are bulk modulus, rigidity, and density, respectively, and the particle displacement associated with P waves at far-field is parallel to the direction of wave propagation. For this reason, P waves are sometimes called longitudinal waves. See Aki and Richards (2002, p. 122).

pyroclast: A general term for any fragment of lava ejected during an explosive volcanic eruption. By definition, pyroclastic deposits are composed of pyroclasts.

pyroclastic: An adjective of Greek derivation (from *pyro*, "fire"; *klastos*, "broken") relating to fragmental materials formed by the shredding of molten or semimolten lava and/or the shattering of preexisting solid rock (volcanic and other) during explosive eruptions, and to the volcanic deposits they form.

pyroclastic flow: Ground-hugging mixture of hot, generally dry pyroclastic debris and volcanic gases that sweeps along the ground surface at extremely high velocities (<250 km/hr). A continuum exists between pyroclastic flow and pyroclastic surge, depending on the ratio of solid materials to gases. Because of its relatively high aggregate density, the movement of pyroclastic flows tends to be more controlled by topography, mostly restricted to valley floors.

pyroclastic surge: Part of a continuum with pyroclastic flow. Pyroclastic surges have relatively lower ratios of solid materials to gases (lower aggregate density). In contrast to pyroclastic flows, the less dense, more mobile surges are less controlled by topography and can affect areas high on valley walls and even overtop ridges to enter adjacent valleys.

Q

***Q*:** A measure of the dissipative characteristics of a system or a material, proportional to the fractional loss of energy per cycle of oscillation. See also Q inverse and Eq. (3.9) in appendix 2 of this Handbook.

***Q* inverse (Q^{-1}):** The reciprocal of Q value, the fractional loss of wave energy per cycle divided by 2π. It is called spatial Q^{-1} if measured from the amplitude attenuation with distance and temporal Q^{-1} if measured from that with time. They can be different for dispersive waves. See Aki and Richards (2002, Box 5.7, pp. 162-163).

quasi-isotropic ray theory: A theory that describes rays in weakly anisotropic media and includes the coupling between quasi-shear rays. The quasi-isotropic ray ansatz depends on the signal period and the anisotropic part of the model parameters. See Chapter 9.

quasi-plastic (QP) regime: That portion of the crust or lithosphere where fault motion is governed by localized ductile shearing in which temperature-sensitive crystal plastic and/or diffusional flow mechanisms dominate. See Chapter 29.

Quaternary: A period in geologic time from 1.6 million years ago to the present. See also geologic time.

quiescence: A significant decrease in the rate of background seismicity from the long-time average rate.

R

radar: Radio detection and ranging. See Chapter 37.

RADARSAT: Multimode C-band radar satellite, launched by Canada in 1995. See Chapter 37.

radial anisotropy: A term sometimes used to denote hexagonal symmetry with a vertical symmetry axis, traditionally called transverse isotropy. See Chapter 53.

radiation damping: The dissipation of energy through the transmission of vibratory energy that is not reflected or confined. Radiation damping occurs, for example, in the idealized problem of a vibrating structure imbedded in a linear elastic half-space, even when the material of the half-space is not dissipative. See Chapter 67.

radiation pattern: The dependence of the amplitudes of seismic P and S waves on the direction and takeoff angle under which their seismic rays have left the seismic source. It is controlled by the type of source mechanism (e.g., the orientation of the earthquake fault plane and slip direction in space). See also earthquake focal mechanism and Eq. (3.7b) in Appendix 2 of this Handbook.

radiative transfer theory: A phenomenological theory that describes a multiple scattering process based on causality, geometrical spreading, and the energy conservation law, which neglects the interference between wave packets. It is often applied to model the propagation of high-frequency seismic-wave energy in heterogeneous Earth media. See Chapter 13 and Sato and Fehler (1998, p. 173).

radioactivity: The emission of radiation or energetic particles during the decay of an unstable atomic nucleus. See Chapter 39.

radiocarbon dating: The use of radioactive ^{14}C in geochronology. See Chapter 30.

radiogenic heat production: The rate of heat generated in a unit volume of rock by spontaneous radioactive decay of natural unstable isotopes in the rock. The primary heat-producing isotopes in the Earth are ^{232}Th, ^{238}U, ^{40}K, and ^{235}U. See Morgan (2002).

radiometric: A term pertaining to the measurement of geologic time by the analysis of certain radioisotopes in rocks and their known rates of decay.

random medium: A mathematical model for a medium whose spatial variation in parameters is described by random functions. Their autocorrelation functions or power-spectral-density functions give their statistical properties. See Chapter 13.

random vibration theory (RVT): A theory of the response of systems to probabilistic defined inputs. In seismology, a convenient way of estimating time domain properties, such as peak amplitudes and frequency, from the Fourier spectrum of ground motion and an estimate of the duration of the motion. For applications, see the review by Boore (2003).

random walk: The typical path followed by a molecule in Brownian motion, and by extension, the path followed in a parameter space when using a Monte Carlo method. See Chapter 16 and the biography of Robert Brown in Chapter 89.

range: In satellite geodesy, the distance along the line of sight between the satellite and the ground. See Chapter 37.

rate- and state-dependent friction: A class of constitutive relationships that describes sliding friction over a wide range of conditions and time scales, developed by Ruina (1983) and Rice and Ruina (1983) as a generalization of equations of Dieterich (1978, 1979). These relations consist of a base frictional resistance and depend on the sliding speed (rate) and contacts (state) on the fault plane. See Chapter 32 (p. 521), Chapter 73, and Eqs. (5.4a)–(5.4f) and Figure 12 in Appendix 2 of this Handbook.

rate strengthening: A term used to describe a positive dependence of fault shear strength on sliding velocity, which always produces aseismic fault slip. See Chapter 32 (p. 521).

rate weakening: A term used to describe a negative dependence of shear strength on sliding velocity, which can lead to frictional instability (seismic slip) under some circumstances. See Chapter 32 (p. 521) and the material below Eq. (5.4f) in Appendix 2 of this Handbook.

ray: A trajectory along which the constituent components of a wave (e.g., phase and energy) propagate. See Chapter 9 and Bleistein (2002).

ray expansion: The complete seismic response that can be expanded into a summation of terms associated with rays. Normally the response can be approximated by a limited number of rays in the time window and amplitude range of interest. See Chapter 9.

ray Green's function: The approximate Green's function obtained using asymptotic ray theory. See also Green's function, Chapter 9, and the biography of George Green in Chapter 89.

Rayleigh damping: A form of damping used in the matrix analysis of structures in which the damping matrix is a constant times the mass matrix plus another constant times the stiffness matrix. Rayleigh damping produces uncoupled modal vibrations with viscous damping factors in the modes that are dependent on the natural frequencies of the modes. The constants are chosen to produce desired damping factors over the important modes of response. See Chapter 67 and the biography of Lord Rayleigh in Chapter 89.

Rayleigh wave: The coupled *P* and *SV* wave trapped near the surface of the Earth and propagating along it. Their existence was first predicted by Rayleigh (1887) for a homogeneous half-space, for which the velocity of propagation is 0.88 to 0.95 times the shear velocity, depending on the Poisson's ratio, and their particle motion is retrograde elliptic near the surface. They can exist, in general, in a vertically heterogeneous media bounded by a free surface. See Aki and Richards (2002, pp. 155-157), Chapter 8 (pp. 87-88), Chapter 21 (pp. 334-335), and the biography of Lord Rayleigh in Chapter 89.

ray theory: A theory in which waves are treated like particles moving along rays. It is a valid approximation when the medium property, such as the wave velocity, varies smoothly within a wavelength. See Chapter 9.

receiver function: The spectral ratio of the horizontal component of *S* waves to the vertical component of *P* waves recorded at a single station from a teleseismic event. Assuming a horizontally layered structure beneath the station, it gives an estimate of the seismic velocity structure, particularly the nature of discontinuities, of the crust and the uppermost mantle.

record section: A display of seismograms recorded from a source (e.g., an explosion) according to their distance from that source in a time-distance graph.

rectangular fault: A fault-plane geometry with a rectangular shape. The long and short side is called, respectively, fault length and fault width. The rupture front is a line segment parallel to the width. It starts at a point on the fault and propagates along the fault length. See also Haskell model and rupture front.

recurrence interval: The average time interval between consecutive events that occur repeatedly. See Chapter 30.

recursive technique: A computational technique based on a formula that relates certain coefficients (e.g., for a filter) at the nth step with those at the $(n-1)$th step. Desired coefficients are obtained at the nth step by an iterative use of the formula starting with the initial values at the 0th step. See Chapter 50.

reduced travel time (T_r): The observed travel time T minus the distance from a source X divided by a suitable reduction velocity (V_r).

reduction velocity (V_r): A suitable reduction velocity that is mostly used for controlled-source P-wave crustal studies of 6 km/sec and for upper-mantle studies, 8 km/sec. For S-wave studies, the reduction velocity is usually that for P-wave studies divided by $\sqrt{3}$.

reflection coefficient: The ratio of amplitude of the reflected wave relative to the incident wave at an interface. In seismology, the coefficients are found by imposing boundary conditions that the seismic displacement and the traction acting on the discontinuity surface are continuous across the surface. These conditions physically mean a welded contact and absence of extra-seismic sources at the discontinuity. Coefficients have been defined for different normalizations and signs of the component waves (e.g., energy or displacement). See Chapter 9 (pp.110-112) and Aki and Richards (2002, pp. 128-149).

reflection seismology: The study of the Earth's structure by the use of seismic waves originating from a near-surface source and reflected back to the surface from subsurface discontinuities.

refraction diagram: A chart indicating wave fronts of a tsunami from the source. It is used to predict tsunami travel times from a given source area. See Chapter 28.

refraction seismology: The study of the Earth's structure by the use of seismic waves from a near-surface source refracted back to the surface after propagation along deep interfaces.

regional strain: A description of the overall deformation of a region. See Chapter 31.

regional wave: A seismic wave that propagates laterally in the crust and uppermost mantle. Examples are *Pg*, *Pn*, *Sn*, *Lg*, and *Rg*. These waves are distinguished from teleseismic waves, which propagate in the deep interior of the Earth. Regional waves typically propagate over epicentral distances up to 1500 km (somewhat farther for *Lg*), and their travel speed can be very different for different regions. For example, *Pn* wave, which travels along the top of mantle, can arrive 8 seconds earlier in some regions and 5 seconds later in other regions as compared to the global average arrival time for the same distance. See Chapter 24.

regression analysis: A statistical technique applied to data to determine the functional relationship between a dependent variable with one or more independent variables. See Draper and Smith (1998).

renewal process: A point process in which the statistical distribution of the intervals between consecutive events is statistically independent and stationary. See Chapter 43.

residual strength: The shear resistance within cohesive soils at very large shear deformations or within contractive granular soils in the liquefied state. See Chapter 70.

resistivity: The electrical resistance between the opposite faces of a unit cube of material. See Chapter 38.

response-spectral ordinate: A term in engineering seismology indicating the amplitude of a response spectrum at a specified value of undamped natural period or frequency. See Chapter 60.

response spectrum: The maximum response to a specified acceleration time series of a set of single-degree-of-freedom oscillators with chosen levels of viscous damping, plotted as a function of the undamped natural period or undamped natural frequency of the system. The response spectrum is used for the prediction of the earthquake response of buildings or other structures. Cf.: design spectrum. See Chapter 57.

return period: The average time between exceedance of a specified level of ground motion at a specific location, equal to the inverse of the annual probability of exceedance.

reverse fault: A fault that involves lateral shortening, where one block (the hanging wall) moves over another (the footwall) up the dip of the fault (i.e., the movement of the hanging wall is upward relative to the footwall). The maximum compressive stress is horizontal, and the least compressive stress is vertical. Reverse faults that dip less than about 30 degrees are often called thrust fault. Cf.: normal fault. See Chapter 31.

***Rg*:** A symbol for the fundamental-mode Rayleigh waves observed for continental paths in the period range 8–12 seconds (Ewing *et al.*, 1957), now often applied to such waves with periods down to about 1 second. See Chapter 21 (p. 337).

Richter magnitude: The magnitude of an earthquake based on the Richter magnitude scale (Richter, 1935). In popular use, the term is often mistaken to be any magnitude. See also magnitude.

Richter scale: Also called Richter magnitude scale, or local magnitude (M_L), introduced by Richter (1935). See also magnitude.

rift (or rift valley): An extended feature marked by a fault-caused trough that is created in a zone of divergent crustal deformation. Examples are the East African Rift Valley and the rift that often exists along the crest of an undersea ridge.

rift zone: An elongate system of structural weakness, defined by eruptive fissures, craters, volcanic vents, and geologic faults, that has undergone extension (ground has spread apart) by movement, storage, and rise of magma.

right-lateral fault: A strike-slip fault across which a viewer would see the block on the other side moving to the right. Cf.: left-lateral fault.

Ring of Fire: Another name for the Circum-Pacific belt within which about 90% of the world's earthquake energy release occurs. The next most seismic region (5–6 %) is the Alpide belt. See also Alpide belt and Circum-Pacific belt.

ripple firing, delay-firing, millisecond delay initiation: A term used to characterize quarry or mine-blasting activity in which a particular shot is carried out as a series of separate charges, fired in a sequence instead of all together. Ripple firing is the term commonly used, but experts usually refer to delay firing (where a common delay is 8 milliseconds between charges), and in some cases to millisecond delay initiation (where a specifically timed pattern of charges is executed that may entail delays that are designed for a particular blasting objective, such as fragmenting rock to a predetermined size and moving the fragments a prespecified distance). See Chapter 24.

rise time: In earthquake source studies, the time required for the completion of slip at a point on the fault plane. In the Haskell model, it is the parameter of ramp function used for describing the slip time history. See also Haskell model and Eqs. (3.5b) and (3.8) in Appendix 2 of this Handbook.

root mean square: The square root of the mean value of a set of squared values.

root n improvement: More precisely, "square root of n" improvement in signal-to-noise ratio, as expected from delay-and-sum processing of n independent recordings with noise that has uniform power. See Chapter 23.

ROV: Remotely operated vehicle. See Chapter 27.

RSAM: Real-time seismic amplitude measurement. Widely used by many volcano observatories, the RSAM system enables the continuous monitoring of total seismic energy release during rapidly escalating volcanic unrest, when conventional seismic-monitoring systems are saturated. See Endo and Murray (1991) for details. The RSAM and SSAM systems, together with associated PC-based software for data collection and analysis, form the core of an integrated mobile volcano observatory, which can be rapidly deployed at previously unmonitored or little monitored volcanoes (Murray *et al.*, 1996).

run-up: The maximum height above a reference sea level of the water brought onshore by a tsunami. See also tsunami run-up height.

rupture front: The instantaneous boundary between the slipping and locked parts of a fault during an earthquake. Rupture-front propagation along the fault length of a rectangular fault from one end to the other is referred to as unilateral. If it starts in the middle of the fault and propagates toward its two ends, it is called bilateral. Rupture may expand outward in a circular or an elliptic form from its center. See also circular fault, elliptic crack, rectangular fault, and Figures 9 and 10 in Appendix 2 of this Handbook.

rupture velocity: The velocity at which a rupture front moves along the surface of the fault during an earthquake.

RVT: Random vibration theory.

Rytov method: A method to find an approximate solution for the logarithm of the wave field (complex phase function) for waves in a smoothly varying inhomogeneous media. In Rytov approximation, the second-order term of the gradient of the complex phase function is often neglected. See Chapter 13.

S

$\bar{S}$ (*Sg*): An S wave propagating through the crust, like $\bar{P}$ or Pg. These S waves are seen with simple impulsive onsets at short distances (up to a few tens of kilometers). At greater distances, onset may not be impulsive (due to multiple paths all trapped in the crust), and Sg then is an alternate name for the wave also called Lg.

SAC: Seismic analysis code. See Chapter 85.5.

sag pond: A fault sag that contains water. See also fault sag.

sand boil: A term for sand and water ejected to the ground surface during strong earthquake shaking as a result of liquefaction at shallow depth. The conical sediment deposit remains as evidence of liquefaction. See Chapter 71.

SAR: Synthetic aperture radar. See Chapter 37.

saturation: Above a certain magnitude level, the magnitude determined from the amplitudes or amplitude/period ratios of specified seismic waves recorded by a specific seismograph increases only slowly or does not increase as the physical size of the earthquake as measured by the moment magnitude (M_W) increases. This behavior, sometimes called saturation, is due to the shape of the displacement spectrum of a typical seismic signal, which is characterized by a corner frequency above which the displacement amplitude decreases rapidly. As the size of the earthquake increases, the corner frequency moves to lower frequencies. When the frequency at which a specific type of magnitude is calculated falls above this corner frequency, that magnitude scale saturates. Saturation sets in at smaller magnitudes for scales based on seismograms from short-period seismographs. Magnitudes m_b, M_L, m_B, and M_S saturate for earthquakes with M_W larger than about 6.0, 6.5, 7.0, and 8.0, respectively. Earthquakes with magnitude $m_b > 6.5$, $M_L > 7.0$, $m_B > 7.5$, or $M_S > 8.5$ have rarely been found due to almost complete saturation. No saturation occurs for M_W and M_t because these scales are based on waves with very low frequencies that are always below the corner frequency. See Chapter 44.

ocean waves, and short-period seismographs for periods below it. A short-period seismograph response usually has a bandwidth from 0.1 second or less to about 1 second. See Chapter 18.

shutter ridge (also shutterridge): A ridge that has been displaced by strike-slip or oblique-slip faulting, thereby shutting off the drainage channels of streams crossing the fault (after Buwalda 1937 and Sharp 1954).

SI: International System of Units (in French, Système International d'Unités). See International System of Units.

Sigma-Delta modulation: A technique for analog-to-digital conversion commonly used in seismic acquisition systems. See Chapter 22.

signal: The wanted part of given data. The remaining part is called noise.

signal association: The term for the difficult work in detection of seismic events using a seismic network of sorting out lists of arrival times from each station and correctly identifying the sources of overlapping events. This work is also called event association. It is particularly challenging for global networks producing reliable seismic bulletins, that commonly locate more than 100 events per day. See Chapter 24.

signal detection: The detection of weak signals imposed upon noise that can be accomplished by noting significant changes in amplitude or frequency content. If the signal to be detected has a waveform that is known beforehand (e.g., from a prior event in the same location and recorded at the same station), then correlation methods or matched filtering can be applied for detection. See Chapter 24.

signal-to-noise improvement: A measure of how well a wanted signal is enhanced by processing, relative to the ambient noise and unwanted signals. Assuming that the processing leaves the signal unchanged, the signal-to-noise improvement is often expressed as the square root of the ratio of the noise power (or its average over channels for an array) before processing to the power after processing. See Chapter 23.

signal-to-noise ratio: The ratio of the power (variance) of the signal to that of the noise for a given frequency band, often measured in decibels (dB). See Chapters 19 and 27.

silent earthquake: A fault movement not accompanied by seismic radiation. See Chapter 36.

simulated annealing: A trick used to solve optimization problems, taking its roots in a thermodynamical analogy. Simulated annealing is based on the Metropolis algorithm. See Chapter 16.

SINFIT: A computer program for fitting sine wave to two signals. See Chapter 85.18.

single-degree-of-freedom (SDOF) system: An oscillator or structure composed of a single mass attached to a spring and dashpot. This system has a single mode or period of vibration whose response is described by a single variable. See Chapter 60.

single-frequency microseism: A secondary peak near 14-second period in seismic noise related to the action of ocean waves that is usually 20–40 dB smaller than the main peak due to the double-frequency microseism. It is called the single-frequency peak because the seismic noise is at the same frequency as the ocean waves. See Chapter 19.

SIO: Scripps Institution of Oceanography, University of California at San Diego, located in La Jolla, California (URL: http://www.sio.ucsd.edu).

SIR-C: Shuttle imaging radar C. See Chapter 37.

site category: The category of site geologic conditions affecting earthquake ground motions based on descriptions of the geology, measurements of the S-wave velocity standard penetration test, shear strength, or other properties of the subsurface. For example, in the NEHRP seismic provisions, the site geologic condition is classified into categories ranging from A (hard rock) to F (very soft soil), and different amplification factors are assigned for them. See Chapter 61.

site classification: The process of assigning a site category to a site by means of geologic properties (e.g., crystalline rock or Quaternary deposits) or by means of a geotechnical characterization of the soil profile (e.g., standard penetration test and S-wave velocity). See Chapter 62.

site effect: The effect of local geologic and topographic conditions at a recording site on ground motions. It is implicitly assumed that the source, path, and site effects on ground motions are separable. See Chapter 61.

site response: The modification of earthquake ground motion in the time or frequency domain caused by local site conditions. See Chapters 60 and 61.

Skempton's coefficient: A poroelastic constant B relating changes in mean stress σ_m to pore fluid pressure p, $B = dp/d\sigma_m$. See Chapter 32 (p. 526).

***SKS*:** A mantle shear wave transmitted through the outer core as a P wave. See Chapter 21 (pp. 341-343), Chapter 53, and Aki and Richards (2002, Box 5.9, pp. 181-182).

slab: A subducted oceanic plate, typically a dipping tabular region of relatively high seismic velocity, high Q, and low temperature. Deep earthquakes only occur within subducted slabs, and such seismicity extends to depths down to 700 km. See Chapter 51.

slab pull: The force of gravity causing the cooler and denser oceanic slab to sink into the hotter and less dense mantle material. The down-dip component of this force leads to down-dip extensional stress in the slab and may produce earthquakes within the subducted slab. Slab pull may also contribute to stress on the subduction thrust fault if the fault is locked.

slag: See clinker.

SLC: Single-look complex. Radar images include both phase and amplitude information, after processing by the synthetic aperture resolution reconstruction process. See Chapter 37.

slickenside: A polished and smoothly striated surface that results from slip along a fault surface. The striations themselves are slickenlines.

slip: See slip vector.

slip model: A kinematic model that describes the amount, distribution, and timing of a slip associated with an earthquake.

slip moment: A geological term for seismic moment. See Chapter 36.

slip rate: See fault slip rate.

slip vector: The direction and amount of slip in a fault plane, showing the relative motion between the two sides of the fault. See Chapter 31, Aki and Richards (2002, Fig. 4.13, p. 101), and Figure 13 in Appendix 2 of this Handbook.

slip wave: See creep wave.

slope-intercept method: A graphical technique used to infer seismic velocities and layer thicknesses from seismic travel-time data, based on the assumption that the crust is built up of uniform layers bounded by horizontal discontinuities. See Chapter 55.

slow earthquake: A fault failure occurring without significant seismic radiation. See Chapter 36.

slowness: The inverse of velocity.

SLR: Satellite laser ranging. See Chapter 37.

small earthquake: An earthquake with magnitude ranges from 3 to 5. See also magnitude.

SMSIM: A software package for simulating ground motions based on the stochastic method. See Chapter 85.13.

S_n(or *Sn*): Quite commonly, the term now applied to a prominent arrival of short-period shear waves that may be observed (with a straight-line travel-time curve) at epicentral distances as great as 40 degrees. Early use of the designation S_n was in reference to short-period *S* waves that were presumed to propagate as head waves along the top of the mantle. See Chapter 21 (pp. 335-336).

Snell's law: In seismology, the law describing the angles of seismic rays reflected and transmitted (i.e., refracted) at an interface in the Earth. It is derived from the conservation of wave slowness parallel to the interface. This law was originally discovered in optics by the Dutch physicist Willebrord Snell in about 1621. See Chapter 9 and Bleistein (2002).

SNR: Signal-to-noise ratio.

S/N ratio: A commonly used term for signal-to-noise ratio. See also SNR.

SOFAR channel: A depth region of low acoustic velocity under the ocean. It is a very efficient wave guide for the seismic T phase from earthquake sources. See Chapter 21.

soil: Two definitions: (1) In geotechnical engineering, all unconsolidable material above the bedrock; (2) in soil science, naturally occurring layers of mineral and/or organic constituents that differ from the underlying parent material in their physical, chemical, mineralogical, and morphological character because of pedogenic processes.

soil profile: The vertical arrangement of soil horizons down to the parent material or to bedrock.

soil-structure interaction: A term applied to the consequences of the deformation and forces induced into the soil by the movement of a structure. The common fixed-base assumption in the analysis of structures implies no soil-structure interaction.

sound wave: See airwave.

source depth: The depth of an earthquake hypocenter, a buried explosion, a mining collapse, or any other type of source that generates seismic waves. See also focal depth.

source-directivity effect: The effect of the earthquake source on the amplitude, frequency content, and duration of the seismic waves propagating in different directions, resulting from the propagation of fault rupture in one or more directions with finite speeds. Directivity effects are comparable to, but not exactly the same as, the Doppler effect for a moving oscillator. See Chapter 62, the material below Eq. (3.8), and Figure 6 in Appendix 2 of this Handbook.

source effect: The effect of the earthquake source on seismic motions. It is implicitly assumed that the source, path, and site effects on ground motions are separable. See Chapter 5 and Eq. (3.5) in Appendix 2 of this Handbook.

source function: A compact space-time history of the earthquake source process that will give the observed displacement waveforms as its convolution with the Green function for the wave propagation in the Earth and the instrument response of the recording instrument. See Kostrov and Das (1988).

source-to-site distance: The shortest distance between an observation point and the source of an earthquake that is represented as either a point (point-source distance measure) or a ruptured area (finite-source distance measure). See Chapter 60.

source zone: An area in which an earthquake is expected to originate; more specifically, an area considered to have a uniform rate of seismicity or a single probability distribution for purposes of a seismic hazard or seismic risk analysis.

SP: Short-period.

SPAC: Spatial autocorrelation method.

space-based geodesy: A space-based technology to measure the positions of geodetic monuments on the Earth. A fundamental departure from the traditional geodesy is its reference to an inertial reference frame rather than those fixed to the Earth. Current accuracy of better than a centimeter for sites thousands of kilometers apart is good enough for measuring the present-day plate motion. See Chapters 7 and 37.

Sparta earthquake: An earthquake in 469 BC, the occurrence of which triggered an uprising by indigenous people and neighboring enemies. See Chapter 46.

spasmodic burst: A rapid-fire sequence of high-frequency earthquakes often observed in volcanic or geothermal areas, presumably due to a cascading brittle-failure sequence along subjacent faults in a fracture mesh driven by transient increases in local fluid pressures. See Hill *et al.* (1990).

spatial autocorrelation method (SPAC): An array-processing method introduced by Aki (1957, 1965) that uses azimuthal averaging of the wave field as observed with a circular array. It has most application to the study of microtremors where the wave field is multidirectional or omnidirectional but the wave scalar velocity is approximately single-valued at each frequency. See Okada (2003) for a recent review.

specific barrier model: A fault model consisting of a rectangular fault plane filled with circular-crack subevents, developed by Papageorgiou and Aki (1983) for interpreting the observed power spectra of strong ground-motion records. It is a hybrid of deterministic and stochastic fault models in which the subevent ruptures are statistically independent from each other as the rupture front sweeps along the fault length. See Papageorgiou (2003) for its latest review.

spectrum intensity: It is a broadband measure of the intensity of strong ground motion defined by the area under a response spectrum between two selected natural periods or frequencies. It was first introduced by Housner (1952) as the area under the 20% damped-response spectrum between periods of 0.1 and 2.5 seconds.

spherical surface harmonics: The analog of the 2D Fourier series for the spherical surface. It combines Fourier harmonics and Legendre polynomials and is used to describe the Earth's normal modes of oscillation (among many other phenomena). See Aki and Richards (2002, Box 8.1, pp. 334-339).

spherical wave: A wave for which the constant-phase surfaces are spheres perpendicular to the direction of propagation. Cf.: plane wave. See Aki and Richards (2002, pp. 190-217).

spheroidal: A mode of vector field **u** in spherical coordinates for which the radial component of **rot u** is zero. Cf.: toroidal. See Aki and Richards (2002, p. 341).

SPOT: Satellite Pour l'Observation de la Terre, a series of optical imaging satellites with a resolution of 10 or 20 m for Earth observation. See Chapter 37.

spread: The layout of seismometer/geophone groups from which data from a single shot (the explosive charge) or vibrator sweep are recorded simultaneously.

spread footing: A support structure for a wall, column, or pier that is essentially a horizontal mat, usually of reinforced concrete, lacking piles or caissons. The footing works by spreading the load over an area wide enough to reduce the bearing stresses to permissible levels. See Park and T. Paulay (1975).

SRA: Seismic risk analysis.

SRO: Seismic Research Observatories seismograph network. See Chapter 20.

SRSS: Square root of the sum of squares. This technique is used in earthquake engineering analysis and design (introduced by Rosenblueth, 1951), wherein the maximum contributions of different modes to a variable of interest, determined from a response spectrum or spectra, are combined by squaring them, adding, and taking the square root of the sum. The result gives an estimate of how the modal responses, whose time of maximum occurrence is not known, combine to produce the maximum total response. See the biography of Emilio Rosenblueth in Chapter 89.

SRTM: Shuttle Radar Topographic Mission. See Chapter 37.

SSA: Seismological Society of America, El Cerrito, California (URL: http://www.seismosoc.org). See Chapter 79.55.

SSAM: Real-time seismic spectral amplitude measurement. The SSAM system aids in the rapid identification of precursory long-period (LP) volcanic seismicity during rapidly escalating volcanic unrest (see Stephens *et al.*, 1994, for details).

stable tectonic regime: A term that refers to a region where tectonic deformation is relatively low and earthquakes are relatively infrequent, usually far from plate boundaries.

stacking process: In reflection seismology, the process of summing several seismic traces that have been corrected for normal moveout. See Sheriff and Geldart (1995).

state of tectonic stress: The magnitude and orientation of principal stresses in the Earth. Principal stresses are usually aligned with horizontal and vertical directions. See Chapter 34.

static fatigue: A laboratory-observed phenomenon in mechanical engineering, material science, and rock mechanics consisting of an inherent time delay between the application of a stress increase and the occurrence of brittle failure (stress drop) induced by the stress change. See Chapter 32 (p. 523).

static stress drop: A theoretically well-defined quantity that compares the before and after stress-strain states of the elastic volume that surrounds the fault. The definition of the overall static stress drop reduces to a weighted integral over just the fault portion that slips during the earthquake. Cf.: dynamic stress drop. See Chapter 33 (pp. 548-550), Chapter 35 (pp. 580-581), and Figure 14 in Appendix 2 of this Handbook.

station: In seismology, the site where seismic instruments (e.g., seismographs) have been installed for permanent or temporary observations. Stations can be either single sites or arrays.

stationary linear system: See linear system.

stationary random process: An ensemble of time series for which the ensemble-averaged statistical properties such as mean and variance do not vary with time. If the statistical averages taken over time for a single member of the ensemble are the same as the corresponding averages taken across the ensemble, the stationary random process is also ergodic.

station calibration: In a global monitoring of seismic events, used to obtain the characteristics of waves arriving at a particular seismic station (e.g., the azimuthal variation and distance dependence of travel times for different seismic phases). More broadly, the term applies to the bias in magnitude estimation from amplitudes of various types of waves observed at a station with respect to the network average. See Chapter 24.

steady-state strength: The shear resistance of contractive soil in the liquefied state. With respect to liquefaction, the steady-state strength and the residual strength are used interchangeably. See Chapter 70.

steam blast: See phreatic.

steepest-ascent method: A method for finding a local maximum of a function that is based in small steps following the local direction of maximum ascent. See Chapter 16.

steepest-descent method: A method for finding a local minimum of a function that is based in small steps following the local direction of maximum descent. See Chapter 16.

step-over: The region where one fault ends and another en-echelon fault of the same orientation begins, described as being either right or left, depending on whether the bend or step is to the right or left as one progresses along the fault.

stick-slip: A periodic or quasi-periodic faulting cycle consisting of a relatively long period of static, predominately elastic loading followed by rapid fault slip seen on laboratory fault surfaces loaded at a constant rate of shear stressing. Stick-slip is considered the laboratory timescale equivalent of periodic earthquake recurrence. See Chapter 32 (p. 518).

stochastic process: A physical process with some random or statistical element in its structure. See Chapter 14.

Stoneley wave: A wave trapped at a plane interface of two elastic media. It is always possible at a solid-fluid interface with the phase velocity lower than the compressional velocity of the fluid. It can exist at a solid-solid interface only in restrictive cases. See Chapter 19, the biography of Robert Stoneley in Chapter 89, and Aki and Richards (2002, pp. 156-157).

strain: A change in shape and/or size of an infinitesimal material element per unit length or volume, usually referred to a reference state in which no forces are applied. For small deformation, the complete description of strain at a point in three dimensions requires the specification of the extensions (change in length/length) in the directions of each of the reference axes and the change in the angle between lines that were parallel to each pair of reference axes in the reference state. See Chapter 36.

strain event: An episodic strain transient recorded by strainmeters. See Chapter 36.

strain hardening: The property of a material to resist postyield strains with increasing stress, as opposed to the constant postyield stress of a perfectly plastic material.

strainmeter: Another basic type of seismic sensor, along with extensometers. They measure the motion of one point of the ground relative to another, in contrast to inertial seismometers that measure the ground motion relative to an inertial reference such as a suspended mass.

strain rate: The rate of strain accumulation in tectonic deformation. A typical value is 10^{-14} per second.

strain step: A steplike change in strain recorded by strainmeters. See Chapter 36.

stratovolcano: A volcanic landform produced by deposits from explosive and nonexplosive eruptions, resulting in a volcanic edifice built of interbedded layers of lavas and pyroclastic materials.

strength: The ability of rock to withstand an applied stress before failure. It is measured in units of stress by means of the uniaxial and triaxial compression test. See Handin (1966) and Scholz (2002).

stress: The force, resolved into normal and tangential components, exerted per unit area on an infinitesimal surface. A complete description of stress (i.e., stress tensor) at a point in a 3D reference frame requires the specification of the three components of stress on each of three surface elements orthogonal to one of the reference axes. Hence there are nine components of stress, though the typical symmetry conditions reduce the stress description to six independent components. In contrast, traction is a force vector defined for a particular surface. See Chapters 33, 34, and 36 and Figure 1 in Appendix 2 of this Handbook.

stress corrosion: A stress degradation process that is related to the action of some chemical agents. See Chapter 15.

stress drop: See dynamic stress drop, static stress drop, and Figure 14 in Appendix 2 of this Handbook.

stress glut: A term (in earthquake source studies) introduced by Backus and Mulcahy (1976). It is used to define the moment-density tensor with the dimensions of moment per unit volume. See also moment tensor, Aki and Richards (2002, p. 57), and Eq. (2.1b) in Appendix 2 of this Handbook.

stress map: A map showing the orientation and relative magnitude of horizontal principal stress orientations. See Chapter 34.

stress shadow: A region where earthquakes are temporarily inhibited because the ambient stress has been decreased by a nearby earthquake. A fault can remain in a stress shadow until stress accumulation recovers the pre-earthquake stress state. See Chapter 73.

stress tensor: See stress.

strike: The trend or bearing, relative to north, of the line defined by the intersection of a planar geologic surface (e.g., a fault or a bed) and a horizontal surface such as the ground. See also fault strike and Figure 13 in Appendix 2 of this Handbook.

strike-slip fault: A fault on which the movement is parallel to the strike of the fault (e.g., the San Andreas Fault). The maximum and minimum principal stresses are both horizontal. See Yeats *et al.* (1997, pp. 167-248).

Strombolian eruption: Slightly more violent than Hawaiian eruptions, the type of activity characterized by the intermittent explosion or fountaining of lava from a single vent or crater. Such eruptions typically occur every few minutes or so, sometimes rhythmically and sometimes irregularly.

strong ground motion: A ground motion having the potential to cause significant risk to a structure's architectural or structural components, or to its contents. One common practical designation of strong ground motion is a peak ground acceleration of 0.05g or larger. See Chapters 57, 59, and 60.

strong motion: See strong ground motion.

strong-motion accelerograph: An accelerograph designed to record accurately the strong ground motion generated by an earthquake. It was originally developed by earthquake engineers because the traditional seismographs designed by seismologists to record weak ground motions from earthquakes were too fragile and did not have sufficient dynamic range to record strong ground motions. A typical strong-motion accelerograph has a tri-axial accelerometer, which records acceleration up to 2 g on scale. See Chapters 57 and 63.

strong-motion instrument: See strong-motion accelerograph.

strong-motion parameter: A parameter characterizing the amplitude of strong ground motion in the time domain (time-domain parameter) or frequency domain (frequency-domain parameter). See Chapter 60.

strong-motion seismograph: See strong-motion accelerograph.

strong-motion sensor: In TriNet, the sensor that has a linear frequency response from DC to an upper corner at 50 or 80 Hz. USGS/Caltech sensors clip at 2 g, and the CDMG sensors clip at 4 g. A strong-motion sensor is often called an accelerometer. See Chapter 78.

strong-motion station: In TriNet, the station that has a triaxial set of strong-motion sensors, a data logger with GPS timing, and real-time or near-real-time communication capability. These strong-motion stations are deployed in free-field locations to serve as reference sites for important engineered structures. See Chapter 78.

structural geology: The study of geologic structures and their formation processes. Applied to earthquakes, it includes the relation between faults, folds, and deformation of rocks on all scales. See Chapter 31.

STS-1: A broadband vault-type seismometer manufactured by G. Streckeisen AG, Messgeraete, Switzerland. It uses the modern astatic leaf-spring suspension for vertical component invented by Wielandt (1975) in place of the classic zero-initial-length spring. See Chapters 18 and 20.

STS-2: A portable and less expensive version of the STS-1 broadband seismometer. See Chapter 20.

subduction thrust fault: The fault that accommodates the differential motion between the downgoing oceanic crustal plate and the continental plate. This fault is the contact between the top of the oceanic plate and the bottom of the newly formed continental accretionary wedge. It is also alternately referred to as the plate-boundary thrust fault, the thrust interface fault, or the megathrust fault.

subduction zone: A zone of convergence of two lithospheric plates characterized by thrusting of one plate into the Earth's mantle beneath the other. Processes within the subduction zone bring about melt generation in the mantle wedge and cause buildup of the overlying volcanic arc. Subduction zones (where most of the world's greatest earthquakes occur) are recognized from the systematic distribution of hypocenters of deep earthquakes called Benioff or Wadati-Benioff zones. See Chapter 6 (pp. 62-64) and the biographies of Hugo Benioff and Kiyoo Wadati in Chapter 89.

subplinian eruption: A small-scale Plinian eruption—intermediate in size and energy release between Strombolian and Plinian activity—that is characterized by pumice and ash deposits covering less than 500 km^2.

substructure: Generally, that portion of a structure that lies below the ground, for example, the basement parking in a tall building. In analyses, the substructure can include a portion of the surrounding ground as well as all or part of the structural foundation. See Chapter 69.

summit eruption: A general term referring to an eruption that takes place from one or more vents (fissural or cylindrical) located at or near the summit, or center, of the volcanic edifice, in contrast to eruptive outbreaks on the flanks of a volcano's downslope from the summit area. See also flank eruption.

superstructure: That portion of a structure that is above the surface of the ground or other reference elevation at which a significant change in the structure occurs. See Chapter 69.

surface faulting: The faulting that reaches the Earth's surface. The faulting is almost always coseismic. See Bonilla (1970).

surface *P* wave: The ray path of surface P that consists of two segments: an S-wave path from the source to the free surface with an apparent horizontal velocity equal to the P-wave velocity, and a P-wave path along the free surface to the receiver. The surface P waves appear at the critical distance and can be a sharper arrival than the direct S waves, although they attenuate very rapidly with distance. In some respects, they behave like head waves.

surface reflection (*PP, SS, SP, PS, PPP, SSS*): A wave that has undergone one or more reflections at the surface before arriving at the recording station. For example, *PP* is a *P* wave that initially leaves the hypocenter in the downward direction (in contrast to *pP*, which leaves in an upward direction) and is reflected once from the surface. Those reflected twice at the surface are denoted as *PPP*. Likewise, *PS* is a once-reflected wave arriving at the station as an *S* wave after conversion by reflection from *P* waves.

surface wave: A seismic wave that propagates on the surface of an elastic half-space or a layered elastic half-space with appropriate seismic velocities. See also Love wave, Rayleigh wave, and Chapters 11 and 21.

surface-wave magnitude (M_S): The earthquake magnitude determined from surface-wave records. The one introduced by Gutenberg (1945) uses maximum amplitudes of surface waves with periods around 20 seconds. This magnitude is computed only for events shallower than 50 or 60 km and saturates for earthquakes larger than magnitude 8. The surface-wave magnitudes assigned by USGS and ISC are calculated from a formula called the IASPEI formula, or Prague formula, developed by Vanek *et al.* (1962) using amplitude/period ratios. See Chapters 41 and 44.

***SV*:** See *SH*.

swarm: A sequence of earthquakes occurring closely clustered in space and time with no dominant main shock. See Chapters 25 and 43.

swath bathymetry: A seafloor bathymetry obtained by a multinarrow-beam echo sounder. One sounding sequence can provide bathymetry information over a width of more than 7 times the water depth. See Chapter 27.

***S* wave:** An elastic shear wave in seismology, *S* standing for "secondary." In a homogeneous isotropic body, the velocity of *S* waves is equal to $\sqrt{(\mu/\rho)}$, where μ and ρ are the rigidity and density, respectively. The particle displacement associated with *S* waves is perpendicular to the direction of wave propagation if the medium is isotropic. For this reason, waves are sometimes called transverse waves. See Aki and Richards (2002, p. 122) and Eq. (1.9b) in Appendix 2 of this Handbook.

synthetic ground motion: The time history of a strong ground motion, calculated for engineering purposes by a deterministic or stochastic simulation. See Boore (2003) for the latest review on simulation of ground motion using the stochastic model. See also Chapter 62.

synthetic seismogram: A computer-generated seismogram for models of seismic source and the Earth's structure. Various methods are used for different purposes, encompassing the whole field of theoretical seismology. Those for the vertically heterogeneous 1D Earth model have been well developed and described in many text books. Those for the 2D and 3D models are still under development. See Chapters 8 through 13 and 55.

T

tamped explosion: A single-fired chemical explosion in which the explosive is buried under the ground firmly to avoid venting into the atmosphere. See Chapter 24.

Taylor expansion: A mathematical technique commonly used to linearize an equation. It is based on Taylor's theorem, which describes approximating polynomials for rather general functions and provides estimates for errors. It is named after the English mathematician and philosopher Brook Tayler (1685–1731). See James and James (1976).

TDMT_INV: A software package for time-domain seismic moment-tensor inversion. See also moment-tensor inversion and Chapter 85.11.

tectonic: A term designating the rock structure and external forms resulting from deep-seated crustal and subcrustal forces in the Earth.

tectonic earthquake: An earthquake caused by the release of strain that has accumulated in the Earth as a result of broad-scale deformations (as contrasted with volcanic earthquakes and impact earthquakes).

tectonic plate: A large, relatively rigid plate of the lithosphere that moves relative to another one on the outer surface of the Earth.

tectonics: The study of the broad-scale movements, deformations, and the resulting structures of the Earth's crust and lithosphere through geologic time.

tectonoelectric: A term referring to electric fields generated by nonseismic crustal stress changes. See Chapter 38.

tectonomagnetic: A term referring to magnetic fields generated by nonseismic crustal stress changes. See Chapter 38.

tectonophysics: A branch of geophysics that deals with the deformation of the Earth.

teleseism: An earthquake at an epicentral distance larger than about 20 degrees or 2000 km from the place of observation. See Chapter 21.

teleseismic: A term pertaining to a seismic source at distances greater than 20 degrees (or about 2000 km) from the observation site.

tensile stress: The stress component normal to a given internal surface, such as a fault plane, that tends to pull the materials apart. Cf.: compressive stress. See also normal stress and Figure 1 in Appendix 2 of this Handbook.

tensor strainmeter: A borehole strainmeter measuring three horizontal strain components from which the principal strain components can be derived. See Chapter 36.

tephra: A collective term for airborne pyroclastic material, regardless of fragment size or shape, produced by explosive volcanic eruptions. Tephra can be of juvenile or lithic origin. See also juvenile and lithics.

TERRAscope: A seismic network of ultra-broadband (360 or 120 seconds on the low end and 8 or 50 Hz on the high end) instruments with data loggers that have onsite storage,

GPS timing, and real-time transmission capability. The TERRAscope network of 24 stations has been integrated into TriNet. See Chapter 78.

Tertiary: A period in geologic time from 66.4 to 1.6 million years ago. See also geologic time.

thermal boundary layer: The layer at the top and bottom of a convecting fluid where heat advection and heat conduction are in balance.

thermal conductivity: A proportionality constant that is based on the rate of heat flow in rock per unit area of the internal surface normal to the temperature gradient vector that is proportional to the magnitude of the temperature gradient.

thermal fluid pressurization: The transient thermal pressurization of fault zone fluids as a consequence of heat generated during seismic slip. See Chapter 29.

3D site amplification: Traditionally, site amplification of ground motion has been modeled by means of a depth-dependent 1D structure. 3D site amplification considers a structure varying in all directions due to features such as irregular topography and sub-surface interfaces. The computational load increases dramatically as one proceeds from 1D to 3D through the intermediate 2D approximation. See Chapter 61.

thrust fault: A reverse fault that dips at shallow angles, typically less than about 30 degrees. See Chapter 31.

TILTCAL: A computer program for determining the generator constant of a horizontal seismeter from a tilt experiment. See Chapter 85.18.

tiltmeter: An instrument for measuring the horizontal gradient of vertical displacement or tilt of the Earth's surface. See Chapter 36.

tilt noise: A noise on a horizontal-component seismogram created by rotation (local tilt) around a horizontal axis. Slight tilts produce large apparent accelerations as a component of the acceleration of gravity is rotated into the horizontal components. See Chapter 19.

time history: An engineering term for a seismogram or a time-dependent response. Examples include an accelerogram and the displacement of point in a structure. See Chapter 56.

tomogram: An image of (usually a slice through) the interior of a body (in this case, the Earth) formed using tomography. See Chapter 52.

tomography: The science and technological art of creating images of the interior of a body. See also seismic tomography and Chapter 52.

topographic site effect: A site effect caused by surface irregularities such as canyons or mountains. In practice, it is difficult to separate topographic effects from effects caused by subsurface layering. See Chapter 61.

tornillo: In volcano seismology, a term sometimes applied to long-period (LP) volcanic earthquakes with an especially monochromatic coda reminiscent of the profile of a screw (from the Spanish for "screw"). See Chapter 25 (Figure 4).

toroidal: A mode of vector field **u** in spherical coordinates for which both the radial component of **u** and div **u** vanish. Cf.: spheroidal. See Aki and Richards (2002, p. 341).

total overburden stress: The vertical stress in a soil layer due to the weight of overlying soil, water, and surface loads. See Chapter 70.

***T* phase, *T* wave:** A tertiary wave in seismology, introduced by Linehan (1940), next to *P* (primary) and *S* (secondary) waves. It is generated by an earthquake in or near the oceans, propagated in the oceans as an acoustic wave guided in the SOFAR channel and converted back to a seismic wave at the ocean-land boundary near the recording site. See Chapter 21.

trace: In observational seismology, a plot of the time history of seismic waveforms recorded by a channel of a seismograph. An analog signal, usually recorded on film or smoked paper by an accelerometer, is often referred to as a trace. See Chapter 58.

traction: The force exerted per unit area of a particular internal or external surface by the body on one side (containing the normal vector) of the surface to that on the other. This is a vector, in contrast to the stress tensor. Cf.: stress.

transducer: A device that converts a signal in one physical form to that in another. For example, a seismic transducer converts seismic motion into electrical current.

transfer function: The Laplace transform of the impulse response of a linear system (filter), or the Laplace transform of the output signal divided by that of the input signal. See Chapters 18 and 22.

transform fault: A strike-slip fault that connects segments of convergent or divergent plate boundary features, such as ridge-to-ridge, ridge-to-trench, etc. This type of fault was proposed by Wilson (1965), and confirmed by Sykes (1967). See Chapter 6 (p. 57), and the biography of J. Tuzo Wilson in Chapter 89.

transition zone: A term in the study of the Earth's interior referring to the depth range from 410 to about 800 km, in which solid-state phase transitions produce strong seismic velocity gradients and sharp discontinuities, sometimes used only for the depth range 410–660 km between the major discontinuities at those depths. See Chapter 51.

transmission coefficient: The ratio of amplitude of the transmitted wave relative to the incident wave at a subsurface discontinuity. The coefficient is found by imposing boundary conditions that the seismic displacement and the traction acting on the discontinuity surface are continuous across the surface. These conditions physically mean a welded contact and absence of extra-seismic sources at the discontinuity. Coefficients have been defined for different normalizations and signs of the component waves (e.g. energy or displacement). See Chapter 9 (pp. 110-112) and Aki and Richards (2002, pp. 128-149).

transmitting boundary: In analytical studies, a boundary to a region of interest that approximates the effects of media outside the boundary. Transmitting boundaries or boundary elements are often used in finite element studies to include the effects of far extents of solid or fluid domains.

transport equation: The ordinary differential equation describing the variation of the amplitude coefficients along a ray. See Chapter 9.

transverse coherence function: A measure of coherency of transmitted wave field as a function of the transverse separation of receivers. See Chapter 13.

transverse isotropy: A term introduced by Love (1927) to describe a medium with one axis of cylindrical symmetry. In the geophysical literature, this axis is often implicitly considered as vertical, so that the velocity and polarization are independent of the azimuth of propagation in the horizontal plane. See Chapter 53.

travel time: The time taken for the seismic waves to propagate from one point to another along the ray. High-frequency seismic waves in a smoothly varying medium propagate like a particle approximately without dispersion along a ray. See Chapter 9.

travel-time curve: A graph of travel time plotted against epicentral distance. See also epicentral distance, hodograph, and travel time.

trench: In global scale, an oceanic trench. In earthquake geology, it means a trench excavated across a fault trace for investigation of the fault-zone structure.

trench log: A map of an excavation or trench wall that exposes geologic formations or structures for detailed study. Trench logs constitute primary data records for many paleoseismic investigations. See Chapter 30.

triaxial seismometer: A set of three seismic sensors whose sensitive axes are perpendicular to each other. See Chapter 18.

triggered earthquake: An earthquake that results from stress changes (e.g., those from a distant earthquake) that are considered small compared to ambient stress levels at the earthquake source. Cf.: induced earthquake. See Chapter 40.

triggering: In seismometry, the turning on of an instrument, especially an accelerograph, by ground motion of a prescribed level. In soil mechanics, it is the onset of a liquefied condition. See Chapter 70.

TriNet: A word referring to the three agencies that built and operate a modern ground-motion network in southern California: the California Institute of Technology, the California Division of Mines and Geology, and the US Geological Survey. TriNet also refers to the seismic network itself. See Chapter 78.

triplication: The behavior of the travel-time curve showing three arrivals for the same range of epicentral distance caused by a sharp increase of velocity with depth. This happens when a part of ray paths refracted from below the zone of velocity increase appears at shorter distances than that from above the zone. These two travel-time branches are connected by the so-called receding branch associated with ray paths through the zone. See Aki and Richards (2002, Fig. 9.19, p. 421).

tromometer: Originally, a simple-pendulum seismoscope, observed with a microscope, to measure background vibrations. It was used through about 1900 as another term for seismometer. See Chapter 1.

tsunami: A train of gravity waves set up on the surface of the sea by a disturbance in the seabed such as a submarine earthquake, landslide, or volcanic explosion. Tsunami amplitudes are typically small in the open ocean but can reach damaging amplitudes near shore or in shallow or confined waters. They are also called seismic sea waves. See Chapter 26.

tsunami earthquake: An earthquake that generates much larger tsunamis than expected from its magnitude. See Chapter 26.

tsunamigenic earthquake: An earthquake that causes a tsunami. Most large, shallow earthquakes under the sea are tsunamigenic. See Chapter 26.

tsunami magnitude (M_t): The magnitude of an earthquake determined from the height of a tsunami at a given travel distance, representing the strength of a tsunami source. The tsunami magnitude M_t, as referenced to moment magnitude (M_W), was introduced by Abe (1979). See Chapters 26 and 44.

tsunami run-up height: The tsunami height on land above a reference sea level measured at the maximum inundation distance from the coast. See Chapter 26.

tube wave: In an empty cylindrical hole, a kind of surface wave that can propagate along the axis of the hole with energy confined to the vicinity of the hole. Tube waves exhibit dispersion with phase velocity increasing with the wavelength. At wavelengths much shorter than the hole radius, they approach Rayleigh waves. The phase velocity reaches the shear velocity at wavelengths of about 3 times the radius. Beyond this cutoff wavelength, they attenuate quickly by radiating S waves. In a fluid-filled cylindrical hole, in addition to a series of multireflected conical waves propagating in the fluid, tube waves exist without a cutoff for the entire period range. At short wavelengths, they approach Stoneley waves for the plane liquid-solid interface. For wavelengths longer than about 10 times the hole radius, the velocity of tube waves becomes constant, given in terms of the bulk modulus κ of the fluid and the rigidity μ of the solid, by $\upsilon = c/\sqrt{(1 + \kappa/\mu)}$, where c is the acoustic velocity in the fluid.

turbidite: A sea-bottom deposit formed by massive slope failures of large sedimentary deposits. These slopes fail in response to earthquake shaking or excessive sedimentation load.

2D site amplification: A site effect calculated for a geologic structure varying in the vertical direction and in only one horizontal direction. In a purely 2D case, incident waves are assumed to be homogeneous along the same horizontal direction. The case in which the incident wave field is allowed to be 3D, such as spherical waves, is sometimes called 2.5D. See Chapter 61.

U

Ubeidiya: Arguably, the oldest human-made structure on earth—a pebble floor in northern Israel from about 1.2 million years ago. Originally horizontal, it is now tilted at 60 degrees or so due to fault motion on the Dead Sea transform. See Chapter 46.

ULF: Ultra-low-frequency electromagnetic field. See Chapter 38.

ULP (ultra-long-period) earthquake: In volcano seismology, a long-period earthquake with waveforms having dominant periods of 100 seconds or longer. See Chouet (1996).

unconformity: The surface of erosion or nondeposition that separates younger strata from older rocks.

underplating: The addition of igneous material to the base of the crust, a process thought to have occurred at volcanic rifted margins and beneath some ocean islands, or the addition of material (including sediments) to the base of an accretionary prism by thrusting at subduction zones. See Chapter 55.

UNICROSP: A computer program for estimating seismic and instrumental noise. See Chapter 85.18.

uniform hazard (response) spectrum: In probabilistic hazard analysis, a response spectrum with ordinates having equal probability of being exceeded. See Chapter 60.

uniform shear beam: A beam with constant properties but arbitrary cross-section that can deform only in shear and whose response is described by a 1D wave equation. A cantilevered shear beam is often used as a simplified model of a building. See Chapter 67.

unreinforced masonry: The type of construction that employs brick, stone, clay tile, or similar materials but that lacks steel bars or other strengthening elements. See Chapter 67.

URL: Uniform resource locator. It is the address for locating a Web site and consists of the protocol and the Internet address.

USCGS: United States Coast and Geodetic Survey.

USGS: United States Geological Survey (URL: http://www.usgs.gov). See Chapter 79.55.

UTC: Universal Time Coordinated, a timescale defined by the International Time Bureau and agreed upon by international convention.

V

VBB: Very broadband, in reference to the 0.1-second to 360-second band of seismic signals. See Chapter 20.

VEI: Volcanic Explosivity Index.

velocity: The rate of change of particle position with respect to time. It is a vector quantity whose magnitude is expressed in units of distance over time and whose direction is the direction of the displacement vector.

velocity field: A term in geodesy describing the deformation within a region given by the velocities of various points relative to some reference frame. See Chapter 31.

velocity structure: In seismology, a generalized local, regional, or global model of the Earth that represents its structure in terms of the velocities of *P* and/or *S* waves. See Chapters 51, 54, and 55.

vent: See volcanic vent.

VLP (very-long-period) earthquake: In volcano seismology, a long-period earthquake with waveforms having dominant periods of 10s of seconds. See Chouet (1996).

vesicle: A cavity formed in a volcanic rock as a result of exsolution and expansion of gas phase when the rock was still in a molten state, either below the surface as magma or during the slow cooling and solidification of thick lava flows or lava lakes.

viscoelasticity: The property of a material that causes its response to an applied force to involve both elastic and viscous components. Several models of viscoelastic materials have been developed that depend on the initial and long-time response to a step in stress, such as Maxwell and Voigt models. The short-time response of a Maxwell material is that of an elastic solid, while the long-time response is that of a viscous fluid.

viscous damping: The most commonly used type of damping for analytical or numerical studies of earthquake response of a structure or a seismometer. A viscous damper generates a force proportional to the relative velocity of its two attachment points in a direction that opposes the motion. Viscous damping is one of the few forms of damping that can be described by linear equations. See Chapter 67.

visible earthquake wave: A slow wave with long-period and short-period wavelengths reported by eyewitnesses in the epicentral area of a great earthquake.

VLBI: Very-long-baseline interferometry. It uses radar frequency around 9 GHz with wavelength around 3 cm. See Chapter 37.

VLF: Very-low-frequency band of electromagnetic waves between 9 kHz and 30 kHz, representing wavelengths from 33 km to 10 km. See Chapter 20.

VLP: Very-long-period, typically in reference to the 27-second to 600-second band of seismic signals recorded by a modern GSN station. See Chapter 20.

volatile: A chemical species or compound that is dissolved in magma at high pressure and that appears as low-density gas at low pressure, the most common ones in magmas are water, carbon dioxide, and sulfur dioxide.

volcanic arc: A long arcuate chain of volcanoes that are created above subduction zones and that are currently active or were operative in the geologic past. The volcanic arc can form on land as volcanic mountains or in the sea as volcanic islands.

volcanic ash: A term for pyroclastic fragments of rocks, minerals, and volcanic glass smaller than 2 mm in size produced during explosive eruptions or the pyroclastic deposit formed by the accumulation of such fragments.

volcanic chain: The roughly linear alignment of a series of volcanoes (extinct, dormant, or active). Most volcanic chains are located along or near the boundaries of tectonic plates, but some can be of intraplate origin, the best-known example of which is the Hawaiian Ridge-Emperor Seamount Chain in the interior of the Pacific Plate.

volcanic degassing: A collective term for the processes by which volatiles and volcanic gases escape from magma or lava. Degassing can occur continuously or discontinuously at highly variable rates before, during, and after eruptions.

Volcanic Explosivity Index (VEI): First proposed by Newhall and Self (1982), it is a now widely used open-ended semi-quantitative classification scheme—based principally on the height of the eruption column and the erupted volume or mass—to describe the size of explosive eruptions. The largest historical eruption (Tambora, Indonesia, in 1815) is assigned VEI = 7; for comparison, the May 18, 1980, eruption of Mount St. Helens was ranked as VEI = 5. However, much larger explosive eruptions in the geologic past (e.g., voluminous caldera-forming events at Yellowstone volcanic system, Wyoming) can be assigned VEI = 8. Nonexplosive eruptions are VEI = 0 regardless of size. As with earthquakes, small to moderate-size explosive volcanic events (<VEI = 5) occur much more frequently than large events (>VEI = 5). Simkin and Siebert (1994) have assigned VEI estimates for all of the world's known eruptions during the last 10,000 years.

In terms of the energy involved, the energy of the 1980 Mount St. Helens (VEI = 5) eruption has been estimated to be about 2×10^{23} ergs and is comparable to the seismic energy of an $M_W = 8$ earthquake. The energy of the 1815 Tambora eruption (VEI = 7) has been estimated to be about 1×10^{26} ergs and is comparable to the seismic energy of the 1960 Chilean earthquake ($M_W = 9.5$), the largest instrumentally recorded earthquake so far.

volcanic gas: The gas produced from exsolution of volatiles, which were originally dissolved in magma at great pressure (depth), during magma rise into, and storage within, lower-pressure regions. Such gases are released from the magmatic system either passively (when the volcano is quiescent) or abruptly (upon rapid expansion and eruption). Explosively expanding volcanic gases initiate and propel eruptions.

volcanic glass: The quenched lava that contains no visible (or only a few submicroscopic) crystals.

volcanic plume: Synonymous with eruption column, the term also sometimes used to describe the quiescent degassing of water vapor and other gases above a volcanic vent.

volcanic tremor: Also called harmonic tremor, a term for continuous vibrations from volcanoes. The seismic signals generated by volcanic activity are highly variable in character, ranging from those indistinguishable from tectonic earthquakes to continuous vibrations with relatively low frequencies (0.5–5 Hz). Volcanic tremors have a very uniform appearance, whereas spasmodic tremors are pulsating and consist of higher frequencies with a more irregular appearance. Cf.: harmonic tremor. See Chapter 25.

volcanic unrest: A general term that applies to a volcano's behavior when it departs from its usual or background level of activity, as evidenced by visible or instrumentally measurable changes—seismic, geodetic, or geochemical—in the state of the volcano from volcano-monitoring studies. However, not all episodes of volcanic unrest, which can last from a few days to a few decades, culminate in eruptive activity.

volcanic vent: A surface opening at and through which volcanic materials are erupted, explosively or nonexplosively. Such openings can be fissures or point sources and vary widely in form and complexity.

volcano: A mountain, hill, or plateau formed by the accumulation of materials erupted—effusively or explosively—through one or more openings (volcanic vents) in the Earth's surface. It also refers to the vents themselves. However, some volcanoes, such as calderas and craters, have a negative topographic expression.

volcano hazard: A potentially damaging volcano-related process or product that occurs during or following eruptions; in quantitative hazard assessments, the probability of a given area being affected by potentially destructive phenomena within a given period of time. At the start of the 21st century, more than half a billion of the world's population are at risk from volcano hazards. For general discussions of volcano hazards and their monitoring and mitigation, see Tilling (1989, 2002) and Scarpa and Tilling (1996).

volcano monitoring: The general term applied to the systematic surveillance of the physical and chemical changes in the state of a volcano and its associated hydrothermal system before, during, and after volcanic unrest.

volcano risk: A term relating to the adverse impacts of volcano hazards, generally involving the consideration of the general relation: risk = volcano hazard x vulnerability (to the given hazard) x value (what is at risk). In probabilistic risk assessments, it is the probability or likely magnitude of human and economic loss, calculated from the same relation.

volcano seismology: The seismological study of the structures and processes beneath volcanoes. See Chapter 25 and Chouet (2003) for its latest reviews.

VT (volcano-tectonic) earthquake: A term often used to refer to ordinary (high-frequency or brittle-failure) earthquakes in a volcanic environment. This usage reflects an ambiguity as to whether the stresses leading to the earthquake are a result of regional tectonic processes or local magmatic processes, such as dike intrusion, or some combination thereof. Cf.: A-type earthquake.

Volume (Vol. n): A defined stage of the operations of processing strong motion data, typically consisting of Vol. 1 (raw acceleration data), Vol. 2 (corrected acceleration data), Vol. 3 (response spectra), and Vol. 4 (Fourier spectra). See Chapter 58.

vulcanian eruption: A term taken from descriptions of eruptions on the island of Vulcano (the home of the Roman god Vulcan), Italy, in the late 19th century. Characterized by discrete violent explosions lasting second to minutes in duration, vulcanian activity is generally considered to be more energetic than Strombolian activity.

W

Wadati-Benioff zone: See subduction zone.

watt (W): An SI-derived unit for power, or radiant flux, expressed in meter^2 kilogram second^{-3} (m^2 kg s^{-3}), or joule per second (J/s). It is named after James Watt (1736–1819), who invented the modern steam engine.

waveform: The complete analog or digital representation (without aliasing) of a wavelet or a whole trace in a seismogram, instead of the parameter data such as onset time and maximum amplitude.

waveform inversion: A term used in both Earth structure and earthquake source studies indicating the use of the whole waveform data instead of their particular characteristics, such as arrival time of a certain phase, sense of the first motion, or maximum amplitude. Various inversion procedures can be used for estimating the model parameters of Earth structure and earthquake source by minimizing the misfit between the observed and synthetic seismograms based on trial parameters. See also moment-tensor inversion and Chapter 31.

wavefront: A surface in 3D-space of points with the same travel time for a common source point. High-frequency seismic signals in a smoothly varying medium propagate like a particle (approximately without dispersion) along a ray. See also travel time.

wavelength: The spatial distance between adjacent points of equal phase of wave motions (e.g., crest to crest, or trough to trough).

wavenumber filtering: A technique for processing recordings from an array of seismometers to enhance signals with a given range of frequency ω and wavenumber vector $\mathbf{k}$. The wave field is assumed to be composed of plane harmonic waves with frequency ω propagating along the surface with horizontal slowness vector $\mathbf{k}/\omega$. The reciprocal of the absolute value of $\mathbf{k}/\omega$ is the apparent velocity. See Chapter 23.

Web: See World Wide Web.

WEGENER: An SLR geodetic network around the Mediterranean. See Chapter 37 and the biography of Afred Wegener in Chapter 89.

WGS84: World Geodetic System, 1984. It is a system of coordinates conventionally used for (coarse) GPS coordinates and includes an ellipsoid with inverse flattening $1/f = 298.25722$ and semi-major axis $= 6378.137$ km (DMA, 1987). The flattening f is defined as the difference between the equatorial and polar radius of the Earth divided by the former. See Chapter 37.

wide-angle seismic method: A modern term for seismic experiments in which the source and receiver are widely separated compared to the target depth. This term is preferred over the traditional seismic refraction method because of the recognition that much of the information on crustal structure in modern experiments comes from reflected phases. See Chapter 55.

Wilson cycle: A successive recurrence of opening and closing of ocean basins due to plate motions on a timescale of about 100 million years. It was proposed by Wilson (1968). See Chapter 6 (p. 63) and Dickinson (2002).

WKBJ approximation: A method for approximate solution of wave propagation through an inhomogeneous medium in which the velocity variations are relatively weak and occur over large distances compared with the wavelengths of interest. This method is named after Gregor Wentzel, Hendrik Kramers, Leon Brillouin, and Harlod Jeffreys. See Aki and Richards (2002, Box 9.6, pp. 434-437) and the biographies of Wentzel, Kramers, Brillouin, and Jeffreys in Chapter 89 of this Handbook.

WWW: World Wide Web, a mechanism for making information (stored in different formats at various computers) available through a computer network (e.g., the Internet). The Web was first created by Tim Berners-Lee in 1990. See Jackson (2002).

WWSSN: World-Wide Standardized Seismograph Network. See Chapters 17 and 20.

WWV: The American radio station that continuously broadcasts standard time of day and other radio and navigation information. WWV is located at Fort Collins, Colorado (Latitude: 40°40′ 49.0″ N, longitude: 105° 02′ 27.0″ W). See Chapter 20.

X

X band: The radar frequency around 9 GHz with wavelength around 3 cm. See Chapter 37.

xenolith: A piece of preexisting solid mantle or crustal rock that is picked up and incorporated into flowing magma; also sometimes called inclusion. It is a direct sample of the uppermost mantle and the crust.

Z

zero-initial-length spring: A helical spring that is prestressed so that the force it exerts in extension is proportional to the distance between the points of attachment rather than the difference between that distance and the initial length of the spring. Originally developed by L. LaCoste for use in gravimeters, it found important applications in the design of long-period vertical seismographs that can theoretically achieve an infinite period without instability. See Aki and Richards (2002, pp. 602-603).

z transform: A discrete equivalent of the Laplace transform. See Chapter 22 and Claerbout (1976, pp. 1-8).

References

Abe, K. (1979). Size of great earthquakes of 1837-1974 inferred from tsunami data. *J. Geophys. Res.*, **84**, 1561–1568.

Agnew, D., J. Berger, J. R. Buland, R. W. Farrell, and F. Gilbert (1976). International Deployment of Accelerometers: A network for very long period seismology. *EOS, Trans. Am. Geophys. Un.*, **57**, 180–188.

Aki, K. (1957). Space and time spectra of stationary stochastic waves, with special reference to microtremors. *Bull. Earthq. Res. Inst., Tokyo Univ.*, **35**, 415–456.

Aki, K. (1965). A note on the use of microseisms in determining the shallow structures of the earth's crust. *Geophysics*, **30**, 665–666.

Aki, K. (1966). Generation and propagation of G waves from the Niigata earthquake of June 16, 1964: Part 2. Estimation of earthquake moment, released energy and stress drop from the G wave spectra. *Bull. Earthq. Res. Inst.,Tokyo Univ.*, **44**, 73–88.

Aki, K. (1987). Magnitude-frequency relation for small earthquakes; a clue to the origin of fmax of large earthquakes. *J. Geophys. Res.*, **92**, 1349–1355.

Aki, K., A. Christoffersson, and E. S. Husebye (1977). Determination of the three-dimensional structure of the lithosphere. *J. Geophys. Res.*, **82**, 277–296.

Aki, K., and B. Chouet (1975). Origin of coda waves: Source, attenuation and scattering effects. *J. Geophys. Res.*, **80**, 3322–3342.

Aki, K., and K. Kaminuma (1963). Phase velocity of Love waves in Japan, Part I, Love waves from the Aleutian earthquake shock of March 9, 1957. *Bull. Earthq. Res. Inst., Tokyo Univ.*, **41**, 243–259.

Aki, K., and W. H. K. Lee (1976). Determination of three-dimensional velocity anomalies under a seismic array using first *P* arrival times from local earthquakes, Part I, A homogeneous initial model. *J. Geophys. Res.*, **81**, 4381–4399.

Aki, K., and P. G. Richards (1980). *Quantitative Seismology: Theory and Methods.* 2 volumes, W. H. Freeman and Company, San Francisco.

Aki, K., and P. G. Richards (2002). *Quantitative Seismology.* Second Edition, University Science Books, Sausalito, California.

American Nuclear Society (1982). Criteria and guidelines for assessing capability for surface faulting at nuclear power plant sites. ANSI/ANS 2.7, 8 pp.

Anderson, J. A., and H. O. Wood (1925). Description and theory of the torsion seismometer. *Bull. Seismol. Soc. Am.*, **15**, 1–72.

Anderson, J. G., and S. E. Hough (1984). A model for the shape of the Fourier amplitude spectrum of acceleration at high frequencies. *Bull. Seismol. Soc. Am.*, **74**, 1969–1993.

Arias, A. (1970). A measure of earthquake intensity. In: *Seismic Design for Nuclear Power Plants.* Edited by R. J. Hansen, p. 438–483, The M.I.T. Press, Cambridge.

Backus, G., and M. Mulcahy (1976). Moment tensors and other phenomenological descriptions of seismic sources. I. Continuous displacements. *Geophys. J. Roy. Astron. Soc.*, 46, 341–361.

Bak, P. (1996). *How Nature Works–The Science of Self-Organized Criticality.* Springer, New York.

Ballou, G. (1987). Filters and equalizers. In: *Handbook for Sound Engineers.* Edited by G. Ballou, p. 569–612, Howard W. Sams, Indianapolis.

Banzhaf, W. (2002). Self-organizing systems. In: *Encyclopedia of Science and Technology.* Third Edition, Volume 14, p. 589–598, Academic Press, San Diego.

Bath, M. (1954). The elastic waves *Lg* and *Rg* along Euroasiatic paths. *Arkiv för Geofysik*, **2**(13), 295–342.

Billings, M. P. (1972). *Structural Geology.* Third Edition, Prentice-Hall, Englewood Cliffs.

Birch, F. (1966). Compressibility; elastic constants. In: *Handbook of Physical Constants.* Edited by S. P. Clark, Revised edition, p. 97–173, Geological Society of America, New York.

Blackman, R. B., and J. W. Tukey (1958). *The Measurement of Power Spectra.* Dover Publications, New York.

Bleistein, N. (2002). Wave phenomena. In: *Encyclopedia of Science and Technology.* Third Edition, Volume 17, p. 789–804, Academic Press, San Diego.

Bonilla, M. G. (1970). Surface faulting and related effects. In: *Earthquake Engineering.* Edited by R. L. Wiegel, p. 47–74, Prentice-Hall, Englewood Cliffs.

Boore, D. M. (1983). Stochastic simulation of high-frequency ground motions based on seismological models of the radiated spectra. *Bull. Seismol. Soc. Am.*, **73**, 1865–1894.

Boore, D. M. (2003). Simulation of ground motion using the stochastic method. *Pure Appl. Geophys.*, **160**, 635–676.

Bormann, P. (Editor) (2002). *New Manual of Seismological Observatory Practice (NMSOP).* 2 volumes, GeoForschungsZentrum, Potsdam, Germany. [Computer readable files of NMSOP are archived on the attached Handbook CD #2.]

Bouchon, M. (2003). A review of the Discrete Wavenumber Method. *Pure. Appl. Geophys.*, **160**, 445–465.

Brigham, E. O. (1988). *The Fast Fourier Transform and its Applications.* Prentice Hall, Englewood Cliffs.

Brune, J. N. (1970). Tectonic stress and the spectra of seismic shear waves from earthquakes. *J. Geophys. Res.*, **75**, 4997–5009.

Burridge, R., and L. Knopoff (1964). Body force equivalents for seismic dislocations. *Bull. Seismol. Soc. Am.*, **54**, 1875–1888.

Burridge, R., and L. Knopoff (1967). Model and theoretical seismicity. *Bull. Seismol. Soc. Am.*, **67**, 341–371.

Buwalda, J. P. (1937). Shutterridges, characteristic physiographic features of active faults (abstract). *Geol. Soc. Am. Proc. 1936*, p. 307.

Capon, J. (1969). High-resolution frequency-wavenumber spectrum analysis. *Proc. IEEE*, **57**, 1408–1418.

Chouet, B. A. (1996). Long-period volcano seismicity: Its source and use in eruption forecasting. *Nature*, **380**, 309–316.

Chouet, B. A. (2003). Volcano seismology. *Pure Appl. Geophys.*, **160**, 739–788.

Claerbout, J. F. (1976). *Fundamentals of Geophysical Data Processing.* McGraw-Hill, New York.

Clark, S. P. (Editor) (1966). *Handbook of Physical Constants.* Revised edition, Geological Society of America, New York.

Cooley, J. W., and J. W. Tukey (1965). An algorithm for the machine calculation of complex Fourier series. *Math. Computation*, **19**, 297–301.

Cox, A. (Editor) (1973). *Plate Tectonics and Geomagnetic Reversals.* W. H. Freeman, San Francisco.

Crary, A. P. (1955). A brief study of ice tremors. *Bull. Seismol. Soc. Am.*, **45**, 1–10.

Das, S., and K. Aki. (1977). Fault plan with barriers: A versatile earthquake model. *J. Geophys. Res.*, **82**, 5658–5670.

Decker, R., and B. Decker (1998). *Volcanoes.* Third Edition, W. H. Freeman, New York.

Dennis, J. G. (1967). *International Tectonic Dictionary. Am. Assoc. Petroleum Geologists Memoir*, **7**, 196 pp.

Dickinson, W. R. (2002). Plate tectonics. In: *Encyclopedia of Science and Technology.* Third Edition, Volume 12, p. 475–489, Academic Press, San Diego.

Dieterich, J. H. (1978). Time-dependent friction and the mechanics of stick slip. *Pure Appl. Geophys.*, **116**, 790–806.

Dieterich, J. H. (1979). Modeling of rock friction 1. Experimental results and constitutive equations. *J. Geophys. Res.*, **84**, 2161–2168.

Dietz, R. (1961). Continent and ocean evolution by spreading of the sea floor. *Nature*, **190**, 854–857.

DMA (1987). Supplement to Department of Defense World Geodetic System 1984 Technical Report: Part II—Parameters, Formulas and Graphics for the Practical Application of WGS84. *DMA TR 8350.2-B*, Defense Mapping Agency.

Dowrick, D. (1977). *Earthquake Resistant Design for Engineers and Architects.* Second Edition, Wiley, New York.

Draper, N. R., and H. Smith (1998). *Applied Regression Analysis.* Third Edition, Wiley, New York.

Duff, P. M. D. (Editor) (1993). *'Holmes' Principles of Physical Geology.'* Fourth Edition, Chapman and Hall, London.

EERI (1989). Glossary by the EERI Committee on seismic risk. *Earthq. Spectra*, **5**, 700–702.

Eicher, D. L. (2002). Geologic time. In: *Encyclopedia of Science and Technology.* Third Edition, Volume 6, p. 623–648, Academic Press, San Diego.

Endo, E. T., T. L. Murray (1991). Real-time seismic amplitude measurement (RSAM): A volcano monitoring and prediction tool. *Bull. Volcan.*, **53**, 533–545.

Ewing, W., W. Jardetzky, and F. Press (1957). *Elastic Waves in Layered Media.* McGraw-Hill, New York.

Fedotov, S. A. (1965). Regularities of the distribution of strong earthquakes in Kamtchatka, the Kurile Islands, and north-east Japan. *Tr. Inst. Fiz. Zemli Akad. Nauk SSSR*, **36**, 66–93.

Freundt, A., C. J. N. Wilson, and S. N. Carey (2000). Ignimbrites and block-and-ash flow deposits. In: *Encyclopedia of Volcanoes,* p. 581–599, Academic Press, San Diego.

Geiger, L. (1912). Probability method for the determination of earthquake epicenters from the arrival time only (translation of the 1911 article). *Bull. St. Louis Univ.*, **8**, 60–71.

Gilbert, F. (1970). Excitation of the normal modes of the Earth by earthquake sources. *Geophys. J. Roy. Astron. Soc.*, **22**, 223–226.

Gilbert, G. K. (1908). Characteristics of the rift. In: *The California earthquake of April 18, 1906,* Report of the State Earthquake Investigation Commission (A. C. Lawson, Chairman), 1908, *Carnegie Institution of Washington Publication 87*, **1**, 30–35.

Green, G. (1828). An essay on the application of mathematical analysis to the theories of electricity and magnetism. Private publication, Nottingham.

Gutenberg, B. (1945). Amplitude of *P, PP* and *S* and magnitude of shallow earthquakes. *Bull. Seismol. Soc. Am.*, **35**, 57–69.

Gutenberg, B., and C. F. Richter (1941). Seismicity of the Earth. *Geol. Soc. Am. Special Paper*, No. 34, 131 pp.

Gutenberg, B., and C. F. Richter (1949; 1954). *Seismicity of the Earth and Associated Phenomena.* First Edition, 1949, Second Edition, 1954, Princeton University Press, Princeton.

Hancock, P. L., and B. J. Skinner (Editor) (2000). *The Oxford Companion to the Earth.* Oxford University Press, Oxford.

Handin, J. (1966). Strength and ductility. In: *Handbook of Physical Constants.* Edited by S. P. Clark, Revised edition, p. 223–289, Geological Society of America, New York.

Hanks, T. C. (1979). *b*-values and $\omega^{-\gamma}$ seismic source models: Implications for tectonic stress variations along active crustal fault zones and the estimation of high frequency strong ground motion. *J. Geophys. Res.*, **84**, 2235–2241.

Hanks, T. C. (1982). Fmax. *Bull. Seismol. Soc. Am.*, **72**, 1867–1879.

Hanks, T. C., and H. Kanamori (1979). Moment magnitude. *J. Geophys. Res.*, **84**, 2348–2350.

Haskell, N. A. (1964). Total energy and energy spectral density of elastic wave propagation from propagating faults. *Bull. Seismol. Soc. Am.*, **54**, 1811–1841.

Hess, H. H. (1962). History of ocean basins. In: *Petrologic Studies.* Buddington Volume, p. 599–620, Geological Society of America, New York.

Hill, D. P., F. G. Fisher, K. M. Lahr, and J. M. Coakley (1976). Earthquake sounds generated by body wave ground motion. *Bull. Seismol. Soc. Am.*, **66**, 1159–1172.

Hill, D. P., W. L. Ellsworth, M. J. S. Johnston, J. O. Langbein, D. H. Oppenheimer, A. M. Pitt, P. A. Reasenberg, M. L. Sorey, and S. R. McNutt (1990). The 1989 earthquake swarm beneath Mammoth Mountain, California: An initial look at the 4 May through 30 September activity. *Bull. Seismol. Soc. Am.*, **80**, 325–339.

Housner, G. W. (1952). Spectral intensity of strong motion earthquakes. Proc. Symp. Blast and Earthquake Effects on Structure, Earthquake Engineering Research Institute. [A computer readable file of this symposium proceeding is placed on the attached Handbook CD #2.]

Housner, G. W. (1967). Earthquake engineering. In: *International Dictionary of Geophysics.* Edited by S. K. Runcorn, p. 392–404, Pergamon Press, Oxford.

Housner, G. W. (1970). Design spectrum. Chapter 5 of *Earthquake Engineering.* Edited by Robert W. Weigel, p. 93–106, Prentice-Hall, Englewood Cliffs.

Ishimoto, M., and K. Iida (1939). Observations sur les seismes enregistres par le microseismographe construit dernierement. *Earthq. Res. Inst. Bull., Tokyo Univ.*, **17**, 443–478.

Ito, K. (1990). Regional variations of the cutoff depth of seismicity in the crust and their relation to heat flow and large earthquakes. *J. Phys. Earth*, **38**, 223–250.

Iyer, H. M., and K. Hirahara (Editors) (1993). *Seismic Tomography: Theory and practice.* Chapman and Hall, London.

Jackson, J. A. (Editor) (1997). *Glossary of Geology.* Fourth Edition, American Geological Institute, Alexandria, Virginia.

Jackson, M. (2002). WWW (World Wide Web). In: *Encyclopedia of Science and Technology.* Third Edition, Volume 17, p. 875–885, Academic Press, San Diego.

Jaeger, J. C., and N. G. W. Cook (1971). *Fundamental of Rock Mechanics.* Chapman and Hall, London.

James, G., and R. C. James (1976). *Mathematics Dictionary.* Fourth Edition, Van Nostrand Reinhold, New York.

Jeanloz, R. (2002). Earth's mantle. In: *Encyclopedia of Science and Technology.* Third Edition, Volume 4, p. 783–799, Academic Press, San Diego.

Jost, M. L., and R. B. Herrmann (1989). A student guide to and review of moment tensors. *Seismol. Res. Lett.,* **60**, 37–57.

Julian, B. R., A. D. Miller, and G. R. Foulger (1998). Non-double-couple earthquakes I. Theory. *Rev. Geophys.*, **36**, 525–549.

Kanamori, H. (1977). The energy release in great earthquakes. *J. Geophys. Res.*, **82**, 2876–2981.

Kelvin, Lord (W. Thomson) (1904). *Baltimore Lectures on Molecular Dynamics and the Wave Theory of Light.* Cambridge University Press, London (course delivered in 1884).

Kinoshita, S. (1992). Local characteristics of the fmax of bedrock motion in Tokyo metropolitan area. *J. Phys. Earth*, **40**, 487–515.

Knopoff, L. (1958). Energy released in earthquakes. *Geophys. J. Roy. Astron. Soc.*, **1**, 44–52.

Kostrov, B. V. (1966). Unsteady propagation of longitudinal shear cracks. *J. Appl. Math. Mech.*, **30**, 1241–1248.

Kostrov, B. V., and S. Das (1988). *Principles of Earthquake Source Mechanics.* Cambridge University Press, Cambridge.

Lacroix, A. (1904). La Montagne Pelée et ses éruptions. Masson et Cie., Paris.

Lange, J. G. (1968). Una Medida de Intensidad Sismica. Departmento de Obras Civiles, Univesidad de Chile, Santiago.

Larson, R. L. (1991). Geological consequences of superplumes. *Geology*, **19**, 963–966.

Larson, R. L., and W. C. Pitman (1972). World-wide correlation of Mesozoic magnetic anomalies and its implications. *Geol. Soc. Am. Bull.*, **83**, 3627–3644.

Lay, T., and H. Kanamori (1981). An asperity model of great earthquake sequences. In: *Earthquake Prediction: An International Review.* Edited by D. Simpson and P. Richards, p. 579–592, American Geophysical Union, Washington, D.C.

Lee, W. H. K., and S. W. Stewart (1981). *Principles and Applications of Microearthquake Networks.* Academic Press, New York. [A computer readable file of this book is placed on the attached Handbook CD #1.]

Lide, D. R. (2002). *CRC Handbook of Chemistry and Physics.* 83rd Edition, CRC Press, Boca Raton.

Linehan, D. (1940). Earthquakes in the West Indian region. *Trans. Am. Geophys. Un.*, **21**, 229–232.

Liu, H. P., D. M. Boore, W. B. Joyner, D. H. Oppenheimer, R. E. Warrick, W. Zhang, J. C. Hamilton, and L. T. Brown (2000). Comparison of phase velocities from array measurements of Rayleigh waves associated with microtremor and results calculated from borehole shear-wave velocity profiles. *Bull. Seismol. Soc. Am.,* **90**, 666–678.

Longuet-Higgins, M. S. (1950). A theory of the origin of microseisms. *Phil. Trans. Roy. Soc. London,* **A243**, 485–496.

Love, A. E. H. (1911). *Problems of Geodynamics.* Cambridge University Press, Cambridge.

Love, A. E. H. (1927; 1944). *A Treatise on the Mathematical Theory of Elasticity.* Fourth Edition, Cambridge University Press, Cambridge. Reprinted by Dover Publications, New York, in 1944.

Macdonald, G. A. (1972). *Volcanoes.* Prentice-Hall, Englewood Cliffs.

Madariaga, R. (1976). Dynamics of an expanding circular fault. *Bull. Seismol. Soc. Am.*, **66**, 639–667.

Mandelbrot B. B., and M. Frame (2002). Fractals. In: *Encyclopedia of Science and Technology.* Third Edition, Volume 6, p. 185–207, Academic Press, San Diego.

Maruyama, T. (1963). On the force equivalents of dynamic elastic dislocations with reference to the earthquake mechanism. *Bull. Earthq. Res. Inst., Tokyo Univ*., **41**, 467–486.

Maxwell, J. C. (1864). A dynamical theory of the electromagnetic field. [A reprint of this article was published in: *The World of Physics.* Edited by J. H. Weaver, Volume 1, p. 846–858, Simon and Schuster, New York, 1987.]

Maxwell, J. C. (1891). *A Treatise on Electricity and Magnetism.* Third Edition, Clarendon Press. Reprinted by Dover Publications, New York, in 1954.

McCalpin, J. P., and Nelson, A. R. (1996). Introduction to paleoseismology. In: *Paleoseismology.* Edited by J. P. McCalpin, p. 1–32, Academic Press, San Diego.

McClean, S. I. (2002). Data mining and knowledge recovery. In: *Encyclopedia of Science and Technology.* Third Edition, Volume 4, p. 229–246, Academic Press, San Diego.

Menke, W. (1984). *Geophysical Data Analysis: Discrete Inverse Theory.* Academic Press, Orlando.

Merrill, R. T., M. W. McElhinny, and P. L. McFadden (1998). *The Magnetic Field of the Earth: Paleomagnetism, the Core, and the Deep Mantle.* Academic Press, San Diego.

Merrill, R. T., and P. L. McFadden (2002). Geomagnetism. In: *Encyclopedia of Science and Technology.* Third Edition, Volume 6, p. 663–674, Academic Press, San Diego.

Meyers, R. A. (Editor-in-chief) (2002). *Encyclopedia of Science and Technology.* Third Edition, 18 volumes, Academic Press, San Diego.

Minakami, T. (1961). Study of eruptions and earthquakes originating from volcanos, I. *Int. Geol. Rev.*, **3**, 712–719.

Mohorovicic (1910). Earthquake of 8 October 1909 (Potres od 8. X 1909; Das Beben vom 8. X. 1909.). Yearly report of the Zagreb Meteorological Observatory for the year 1909, 63 pp.

Morgan, P. (2002). Heat flow. In: *Encyclopedia of Science and Technology.* Third Edition, Volume 7, p. 265–278, Academic Press, San Diego.

Murray, T. L., J. W. Ewert, A. B. Lockhart, and R. G. LaHusen (1996). The integrated mobile volcano-monitoring system used by the Volcano Disaster Assistance Program (VDAP). In: *Monitoring and Mitigation of Volcano Hazards.* Edited by R. Scarpa and R. I. Tilling, p. 315–362, Springer-Verlag, Heidelberg.

Naeim, F. (Editor) (2001). *The Seismic Design Handbook.* Second Edition, Kluwer Academic Publishers, Boston.

NBS (1981). The International National System of Units (SI). *National Bureau of Standards Special Publication 330,* U.S. Government Printing Office, Washington, D.C.

Newhall, C. G., and S. Self (1982). The volcanic explosivity index (VEI): An estimate of explosive magnitude for historical volcanism. *J. Geophys. Res.*, **87**, 1231–1238.

Nur, A., and J. D. Byerlee (1971). An exact effective stress law for elastic deformation of rock with fluids. *J. Geophys. Res.*, **76**, 6414–6419.

Okada, H. (2003). The Microseismic Survey Method. Society of Exploration Geophysicists of Japan. [Translated by Koya Suto, as Geophysical Monograph Series No. 12, Society of Exploration Geophysicists, Tulsa Oklahoma, USA, in press.]

Okada, Y. (1985). Surface deformation due to shear and tensile faults in a half-space. *Bull. Seismol. Soc. Am.*, **75**, 1135–1154.

Olson, P. (2002). Mantle convection and plumes. In: *Encyclopedia of Science and Technology.* Third Edition, Volume 9, p. 77–94, Academic Press, San Diego.

Omori, F. (1894). On the aftershocks of earthquakes. *Tokyo Univ. Coll. Sci. Imp. Jour.*, **7**, 111–200.

Papageorgiou, A. (2003). The barrier model and strong ground motion. *Pure. Appl. Geophys.*, **160**, 603–634.

Papageorgiou, A., and K. Aki (1983). A specific barrier model for the quantitative description of inhomogeneous faulting and the prediction of strong motion, Part I, Description of the model. *Bull. Seismol. Soc. Am.*, **73**, 693–722; Part II, Application of the model. *Bull. Seismol. Soc. Am.*, **73**, 955–978.

Park, R., and T. Paulay (1975). *Reinforced Concrete Structures.* Wiley, New York.

Press, F., and R. Siever (1994). *Understanding Earth.* W. H. Freeman, New York (and many later editions).

Purcaru, G., and H. Berckhemer (1982). Quantitative relations of seismic source parameters and a classification of earthquakes. *Tectonophysics*, **84**, 57–128.

Raff, A., and R. Mason (1961). Magnetic surveys off the west coast of North America, 40° N latitude to 50° N latitude. *Bull. Geol. Soc. Am.*, **72**, 1267–1270.

Rayleigh, Lord (J. W. S.) (1887). On waves propagated along the plane surface of an elastic solid. *Proc. London Math. Soc.*, **17**, 4–11.

Reiter, L. (1990). *Earthquake Hazard Analysis: Issues and Insights.* Columbia University Press, New York.

Rice, J. R., and A. L. Ruina (1983). Stability of steady frictional slipping. *J. Appl. Mech.*, **50**, 343–349.

Richter, C. F. (1935). An instrumental earthquake magnitude scale. *Bull. Seismol. Soc. Am.*, **25**, 1–32. [A computer readable file of this paper is placed on the attached Handbook CD #3.]

Richter, C. F. (1958). *Elementary Seismology.* W. H. Freeman, San Francisco.

Rickayzen, G. (2002). Green's functions. In: *Encyclopedia of Science and Technology.* Third Edition, Volume 7, p. 129–135, Academic Press, San Diego.

Rikitake, T., and K. Hamada (2002). Earthquake prediction. In: *Encyclopedia of Science and Technology.* Third Edition, Volume 4, p. 743–760, Academic Press, San Diego.

Rosenblueth, E. (1951). "A Basis for Aseismic Design of Structures." Doctoral thesis, University of Illinois, Urbana, Illinois.

Ruina, A. L. (1983). Slip instability and state variable friction laws. *J. Geophys. Res.*, **88**, 10,359–10,370.

Runcorn, S. K. (Editor) (1967). *International Dictionary of Geophysics.* 2 volumes, Pergamon Press, Oxford.

Salmon, C. G., and J. E. Johnson (1996). *Steel Structures—Design and Behavior.* Fourth Edition, Harper Collins, New York.

Sato, H., and M. Fehler (1998). *Seismic Wave Propagation and Scattering in the Heterogeneous Earth.* AIP Press/ Springer Verlag, New York.

Savage, J. C. (1971). A theory of creep waves propagating along a transform fault. *J. Geophys. Res.*, **76**, 1954–1966.

Savage, J. C. (1980). Dislocations in seismology. In: *Dislocations in Solids.* Edited by F. R. N. Nabarro, Volume 3, p. 251–339, North Holland Publishing Co., Amsterdam.

Savage, J. C., and M. D. Wood (1971). The relation between apparent stress and stress drop. *Bull. Seismol. Soc. Am.*, **61**, 1381–1388.

Scarpa, R., and R. I. Tilling (Editors) (1996). *Monitoring and Mitigation of Volcano Hazards.* Springer-Verlag, Heidelberg.

Scholz, C. H. (2002). *The Mechanics of Earthquakes and Faulting.* 2nd Edition, Cambridge University Press, Cambridge.

Sharp, R. P. (1954). Physiographic features of faulting in Southern California. *California Div. Mines and Geology Bull.*, **170**, Chapter 5, pt. 3, p. 21–28.

Sheriff, R. E., and L. P. Geldart (1995). *Exploration Seismology.* Second Edition, Cambridge University Press, Cambridge.

Sigurdsson, H. (Editor-in-Chief) (2000). *Encyclopedia of Volcanoes.* Academic Press, San Diego.

Simkin, T., and L. Siebert (1994). *Volcanoes of the World: A Regional Directory, Gazetteer, and Chronology of Volcanism During the Last 10,000 Years.* Second Edition, Geoscience Press, Inc., Tucson, Arizona (in association with the Smithsonian Institution, Washington, D.C.).

Simkin, T., J. D. Unger, R. I. Tilling, P. R. Vogt, and H. Spall (1994). Second Edition of *This Dynamic Planet: World Map of Volcanoes, Earthquakes, Impact Craters, and Plate Tectonics.* U.S. Geological Survey Special Map (Mercator projection: scale 1: 30,000,000 at the equator).

Socolar, J. (2002). Chaos. In: *Encyclopedia of Science and Technology.* Third Edition, Volume 2, p. 637–665, Academic Press, San Diego.

Sponhever, W. and V. Karnik (1964). Neve seismiche skala, In: Proc. 7th Symposium of the ESC 9.0 (Sponhever, W., ed.), Jena, 24–30 Sept. 1962, Veröff. Inst. F. Bodendyn. U. Erbebenforsch. Jena D. Deutschen Akad. D. Wiss, **77**, 69–76.

Stephens, C. D., B. A. Chouet, R. A. Page, J. C. Lahr, and J. A. Power (1994). Seismological aspects of the 1989-1990 eruptions at Redoubt Volcano, Alaska: The SSAM perspective. *J. Volcan. Geotherm. Res.*, **62**, 153–182.

Stratton, J. A. (1941). *The Electromagnetic Theory.* McGraw-Hill, New York.

Sykes, L. R. (1967). Mechanism of earthquakes and nature of faulting on the mid-oceanic ridges. *J. Geophys. Res.*, **75**, 5041–5055.

Tilling, R. I. (1989). Volcanic hazards and their mitigation: Progress and problems. *Rev. Geophys.*, **27**, 237–269.

Tilling, R. I. (2002). Volcanic hazards. In: *Encyclopedia of Science and Technology.* Third Edition, Volume 17, p. 559–577, Academic Press, San Diego.

USGS Photo Glossary (2002). Photo glossary of volcanic terms. In: Website of the U.S. Geological Survey Volcano Hazards Program; provides images and definitions of some common volcano terms: <http://volcanoes.usgs.gov/Products/Pglossary/pglossary.html>.

Uyeda, S. (1978). *The New View of the Earth.* W. H. Freeman, San Francisco.

Uyeda, S. (1982). Subduction zones: An introduction to comparative subductology. *Tectonophysics*, **81**, 133–159.

Vallance, J. W. (2000). Lahars. In: *Encyclopedia of Volcanoes.* p. 601–616, Academic Press, San Diego.

Vanek, J., A. Zapotek, V. Karnik, N. V. Kondorskaya, Y. V. Riznichenko, E. F. Savarensky, S. L. Solov'yov, and N. V. Shebalin (1962). Standardization of magnitude scales. *Izvestiya Akad. SSSR, Ser. Geofiz.*, **2**, 153–158.

Vine, F., and D. Matthews (1963). Magnetic anomalies over oceanic ridges. *Nature*, **199**, 947–949.

Von Huene, R., and D. Scholl (1991). Observation at convergent margins concerning sediment subduction, subduction erosion, and the growth of continental crust. *Rev. Geophys. Space Phys.*, **29**, 279–316.

Waldhauser, F., and W. L. Ellsworth (2000). A double-difference earthquake location algorithm: Method and application to the northern Hayward fault. *Bull. Seismol. Soc. Am.*, **90**, 1353–1368.

Wegener, A. (1924). *The Origin of Oceans and Continents.* Methuen, London (Dover, paperback ed., 1966).

WGCEP (Working Group on California Earthquake Probabilities) (1995). Seismic hazards in southern California: probable earthquakes, 1994-2014. *Bull. Seismol. Soc. Am.*, **85**, 379–439.

Wielandt, E. (1975). Ein astasiertes Vertikalpendel mit tragender Blattfeder. *J. Geophys.*, **41**, 545–547.

Wilson, J. T. (1965). A new class of fault and their bearing on continental drift. *Nature*, **207**, 343–347.

Wilson, J. T. (1968). Did the Atlantic close and then re-open again? *Nature*, **211**, 676–681.

Wolfram, S. (2002). *A New Kind of Science.* Wolfram Media, Champaign.

Wood, H. O., and F. Neumann (1931). Modified Mercalli intensity scale of 1931. *Bull. Seismol. Soc. Am.*, **21**, 277–283.

Wyss, M., and J. N. Brune (1968). Seismic moment, stress and source dimensions for earthquakes in California-Nevada region. *J. Geophys. Res.*, **73**, 4681–4694.

Yeats, R. S., K. Sieh, and C. R. Allen (1997). *The Geology of Earthquakes.* Oxford University Press, New York.

Appendix 2

Key Formulas in Earthquake Seismology

Yehuda Ben-Zion
University of Southern California, Los Angeles, California, USA

Introduction

The material below contains selective key formulas for different branches of earthquake seismology, with an emphasis on source processes, accompanied by brief explanations. The goal is to provide a concise collection of useful expressions in one place without elaborate background information and details. The formulas are organized in seven brief sections: 1. *Thermodynamics and elastodynamics*; 2. *Seismic source and representation integrals*; 3. *Elastodynamic waves*; 4. *Fracture*; 5. *Friction*; 6. *Earthquake source parameters and scaling relations*; and 7. *Seismicity patterns.* The choice and ordering of material is to some extent a matter of taste (and of course knowledge). Most formulas are followed only by definitions and basic clarifications, but in some places there is also a brief discussion of properties and implications of results. Further details on the different subjects, and many additional important formulas, can be found in the references given by numbers in the parentheses after the title of each section. Since most expressions are well-known, explicit references are usually not provided for each separate result; however, in some cases this is done for easy identification of sources. An excellent general reference for the material of sections 1, 2, 3, and 6 is the book *Quantitative Seismology* by Aki and Richards [1]. Good general references for sections 4 and 5 in relation to earthquakes are Rice [42] and Scholz [50]. Useful general references for section 6 are Brune and Thatcher [8] and Kanamori [25]. A good general reference for section 7 is Utsu [59].

The following notations are adopted throughout. Variables and parameters are written in italics (e.g., x), vectors are denoted with boldface (e.g., $\boldsymbol{x}$), and unit vectors are marked with the circumflex symbol (e.g., $\hat{\boldsymbol{n}}$). Cartesian components of vectors and tensors are denoted with subscripts (e.g., x_i, c_{ijkl}). The summation convention for repeating subscripts is used (e.g., $c_{ijkl}\varepsilon_{kl} = \sum_{k,l} c_{ijkl}\varepsilon_{kl}$). Overdots indicate time derivatives (e.g., $\ddot{u}_i = \partial^2 u_i/\partial t^2$), and a comma after a subscript implies a spatial derivative (e.g., $\sigma_{ij,j} = \sum_j \partial\sigma_{ij}/\partial x_j$). Additional notations are defined where needed.

1. Thermodynamics and Elastodynamics ([1], [37], [43], [55], [57])

The first law of thermodynamics implies that the energy balance during evolving deformation can be expressed as

$$d\Omega/dt + dQ/dt = d(K + U)/dt \tag{1.1}$$

where Ω, Q, K, and U are, respectively, mechanical work, heat, kinetic energy, and internal energy, all per unit mass or volume. The internal energy is an intrinsic potential such as gravity, magnetic field, and elastic strain energy. If the internal energy is known as a function of strain ε and entropy $\mathcal{S}$, it forms a complete equation of state, and all other equilibrium properties of a deforming system, such as specific heat and stress–strain relation, can be obtained from it by differentiation. For some applications, it is preferable to use the strain and temperature T as independent variables. In that case, the complete equation of state is given by the Helmholtz free energy F defined as

$$F = U - T\mathcal{S} \tag{1.2}$$

where F, U, and $\mathcal{S}$ are per unit mass or volume.

The second law of thermodynamics requires that the entropy production rate during any process in a closed system be non-negative

$$d\mathcal{S} \geq 0 \tag{1.3}$$

INTERNATIONAL HANDBOOK OF EARTHQUAKE AND ENGINEERING SEISMOLOGY, VOLUME 81B ISBN: 0-12-440658-0

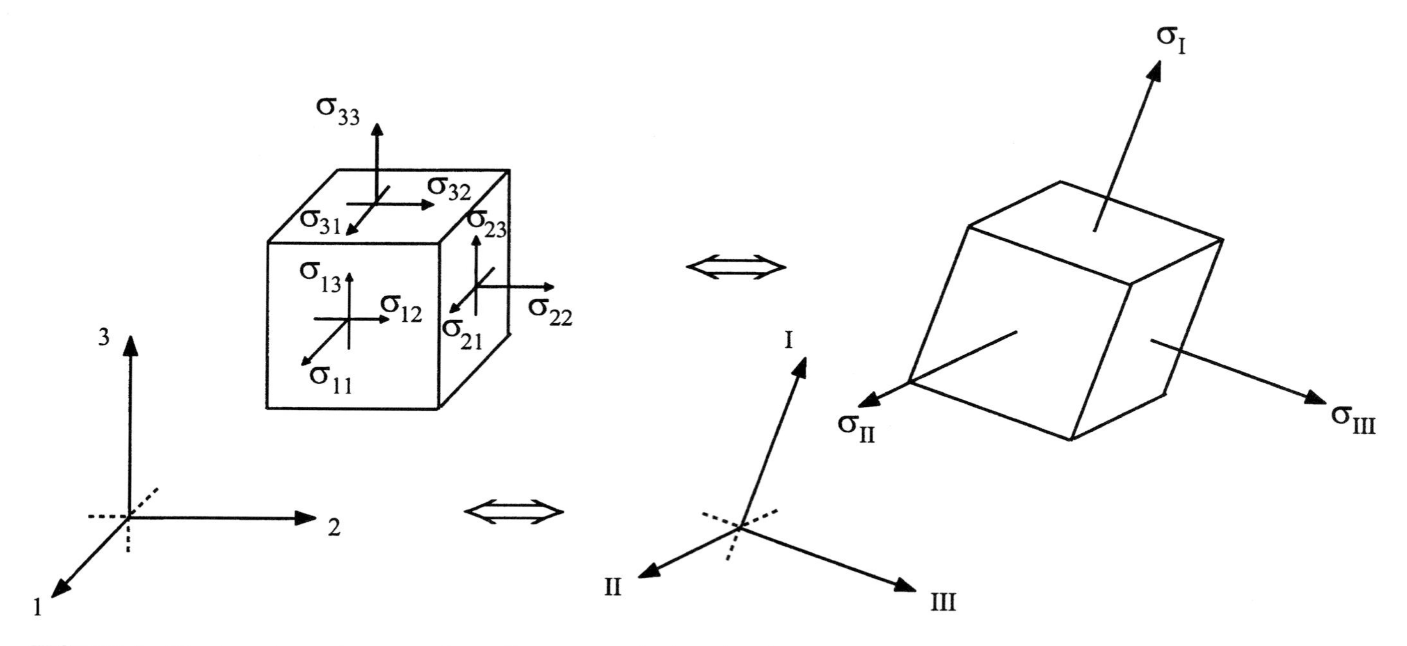

FIGURE 1 The stress tensor in a general coordinate system having nine shear and normal components (left) and the principal coordinate system having only three normal components (right).

with the equality characterizing reversible processes (e.g., purely elastic deformation) and the ">" sign characterizing irreversible ones (e.g., fracture and friction).

The stress tensor σ_{ij} and moduli tensor c_{ijkl} of elastic deformation can be obtained from the strain energy density function W and strain tensor ε_{ij} by the following derivatives

$$\sigma_{ij} = \rho' \partial W / \partial \varepsilon_{ij} \quad (1.4a)$$

and

$$c_{ijkl} = \partial^2 W / \partial \varepsilon_{ij} \partial \varepsilon_{kl} \quad (1.4b)$$

where ρ' is mass density if W is energy density per unit mass or 1 if it is energy density per unit volume. For deformation under adiabatic conditions, the appropriate choice for W is U, while for isothermal deformation it is F.

The stress–strain relation of a linear elastic solid is

$$\sigma_{ij} = c_{ijkl}\varepsilon_{kl}. \quad (1.4c)$$

The stress, strain, and elastic moduli tensors have the following symmetry properties

$$\begin{aligned} \sigma_{ij} &= \sigma_{ji}, \\ \varepsilon_{ij} &= \varepsilon_{ji}, \\ c_{ijkl} &= c_{jikl} = c_{ijlk} = c_{klij}. \end{aligned} \quad (1.4d)$$

With the above definitions and properties, the strain energy density per unit volume of a linear elastic solid is

$$W = \frac{1}{2} c_{ijkl}\varepsilon_{ij}\varepsilon_{kl} = \frac{1}{2}\sigma_{ij}\varepsilon_{ij}. \quad (1.4e)$$

Since the stress and strain tensors are symmetric and real-valued, they can always be diagonalized by transformation (Fig. 1) to a coordinate system consisting of three orthogonal directions (called the principal axes) normal to planes subjected only to normal stress and strain components (called the principal stress and strain values).

The stress–strain relation for a linear isotropic elastic solid is

$$\sigma_{ij} = \lambda\delta_{ij}\varepsilon_{kk} + 2\mu\varepsilon_{ij} \quad (1.5)$$

where λ and μ are the Lamé constants and δ_{ij} is the Kronecker delta function. The Lamé constants and bulk modulus K can be written in terms of Young modulus E and Poisson ratio ν as

$$\begin{aligned} \mu &= E/[2(1+\nu)], \\ \lambda &= \nu E/[(1+\nu)(1-2\nu)], \\ K &= E/[3(1-2\nu)] = \lambda + 2\mu/3. \end{aligned} \quad (1.6)$$

For a uniaxial linear elastic deformation of homogeneous isotropic bar under tensile stress σ_{11}, the axial strain components are $\varepsilon_{11} = \sigma_{11}/E$, $\varepsilon_{22} = \varepsilon_{33} = -\nu\varepsilon_{11}$ and the shear strain components are zero. During infinitesimal deformation of homogeneous isotropic elastic solid under pressure $-p$, the normal stress components are $\sigma_{11} = \sigma_{22} = \sigma_{33} = -p$ and the fractional reduction of volume is $\Delta V/V = (\varepsilon_{11} + \varepsilon_{22} + \varepsilon_{33}) = -p/K$.

The Cauchy equation of motion for a continuum in terms of stress and displacement $\boldsymbol{u}$ is

$$\sigma_{ij,j} + f_i = \rho \ddot{u}_i \quad (1.7)$$

where f_i is the i component of body force per unit volume.

Putting (1.5) into (1.7), and using the definition of infinitesimal strain in terms of displacement gradients $\varepsilon_{ij} = \frac{1}{2}(u_{i,j}+u_{j,i})$, gives the Navier equation of motion for a linear homogeneous isotropic elastic solid in terms of infinitesimal displacement

$$(\lambda + \mu)u_{k,ki} + \mu u_{i,kk} + f_i = \rho \ddot{u}_i. \tag{1.8}$$

Equation (1.8) has solutions in a homogeneous medium in terms of two types of plane body waves $\boldsymbol{u}(\boldsymbol{x}, t) = \hat{\boldsymbol{p}} u(\hat{\boldsymbol{n}} \cdot \boldsymbol{x} - ct)$, consisting of an arbitrary pulse shape u propagating in direction $\hat{\boldsymbol{n}}$ with velocity c and particle motion polarization $\hat{\boldsymbol{p}}$. One solution has velocity and polarization

$$c = v_P \equiv \sqrt{(\lambda + 2\mu)/\rho} \quad \text{and} \quad \hat{\boldsymbol{p}} \times \hat{\boldsymbol{n}} = 0 \tag{1.9a}$$

and the other

$$c = v_S \equiv \sqrt{\mu/\rho} \quad \text{and} \quad \hat{\boldsymbol{p}} \cdot \hat{\boldsymbol{n}} = 0. \tag{1.9b}$$

Solutions (1.9a) and (1.9b) describe longitudinal dilatational P waves and transverse shear S waves, respectively. More general wavefields can be represented as superpositions of plane waves (see section 3).

The strain energy density per unit volume of a plane S or P wave in a linear isotropic elastic solid is

$$W = \frac{1}{2}\sigma_{ij}\varepsilon_{ij} = \frac{1}{2}\rho\dot{u}^2 = K \tag{1.9c}$$

where K is the kinetic energy density per unit volume. The flux rate of energy transmission (i.e., energy per unit time across a unit area normal to the propagation direction) associated with a plane wave of elastic disturbance is $v_S \rho \dot{u}^2$ for S waves and $v_P \rho \dot{u}^2$ for P waves.

The displacement field generated by a distribution of body forces and surface tractions can be synthesized using the elastodynamic Green function $G_{ij}(\boldsymbol{x}, t; \boldsymbol{x}', t')$, giving the i component of displacement at $(\boldsymbol{x}, t)$ due to a localized unit body force operating at $(\boldsymbol{x}', t')$ in the j direction. The elastodynamic Green function satisfies the Navier equation of motion for a linear elastic solid

$$\rho \frac{\partial^2}{\partial t^2} G_{ij} = \delta_{ij}\delta(\boldsymbol{x} - \boldsymbol{x}')\delta(t - t') + \frac{\partial}{\partial x_n}\left(c_{inkl}\frac{\partial}{\partial x_l} G_{kj}\right) \tag{1.10}$$

where $\delta(\ )$ is the Dirac delta function. A complete determination of G_{ij} requires meeting initial conditions (taken usually to be $G = \partial G/\partial t = 0$ for $t \leq t'$ and $\boldsymbol{x} \neq \boldsymbol{x}'$) and specified boundary conditions on the surface of the medium.

If G_{ij} satisfies homogeneous boundary conditions (i.e., zero traction or zero displacement) on S, it has the following spatiotemporal reciprocity properties

$$G_{ij}(\boldsymbol{x}, t; \boldsymbol{x}', t') = G_{ji}(\boldsymbol{x}', -t'; \boldsymbol{x}, -t). \tag{1.11}$$

The displacement field in a solid with volume V and surface S subjected to body force $\boldsymbol{f}$ and surface traction $\boldsymbol{T}$ can be written using a Green function response that satisfies stress-free boundary conditions on S as

$$u_i(\boldsymbol{x}, t) = \int_{-\infty}^{t} dt' \int_V G_{ij}(\boldsymbol{x}, t; \boldsymbol{x}', t')\, f_j(\boldsymbol{x}', t')\, d^3\boldsymbol{x}' + \int_{-\infty}^{t} dt' \int_S G_{ij}(\boldsymbol{x}, t; \boldsymbol{x}', t')\, T_j(\boldsymbol{x}', t')\, d^2\boldsymbol{x}'. \tag{1.12}$$

2. Seismic Source and Representation Integrals

([1], [3], [4], [12], [21], [28], [34], [42], [57])

The total strain at $(\boldsymbol{x}, t)$ may be written as a sum of elastic and inelastic contributions (Fig. 2). The inelastic strain tensor in the faulting region, also called transformational strain, defines the seismic potency density tensor per unit volume,

$$\varepsilon_{ij}^P(\boldsymbol{x}, t) \equiv \text{seismic potency density tensor}. \tag{2.1a}$$

The corresponding transformational stress, also called stress glut, defines the seismic moment density tensor per unit volume,

$$c_{ijkl}(\boldsymbol{x})\varepsilon_{kl}^P(\boldsymbol{x}, t) \equiv m_{ij}(\boldsymbol{x}, t) \equiv \text{seismic moment density tensor}. \tag{2.1b}$$

Relation (2.1b) assumes that the tensor of elastic moduli is time-invariant. Since this does not hold during faulting, c_{ijkl} should be interpreted as a tensor of "effective" elastic constants with actual employed values depending on the application. If ε_{kl}^P is identified as the tensor of elastic strain drop during faulting, m_{ij} may be interpreted as the corresponding stress drop tensor $\Delta\sigma_{ij}$,

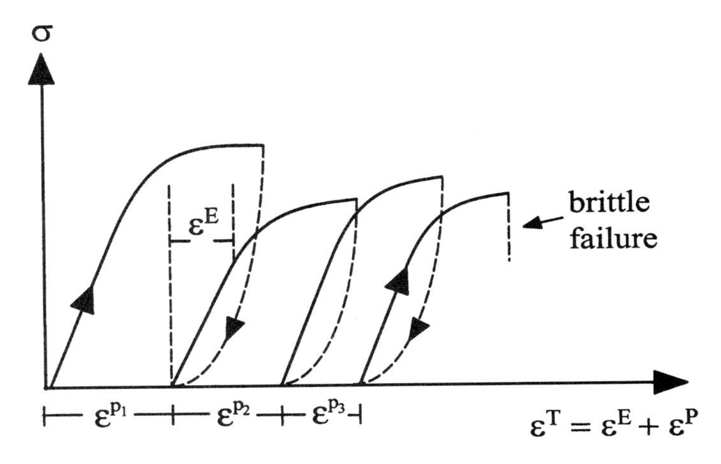

FIGURE 2 A schematic 1-D stress–strain diagram for large deformation with total strain ε^{T} having elastic ε^{E} and inelastic ε^{P} contributions.

and c_{ijkl} as effective elastic constants relating the elastic strain drop tensor to the stress drop tensor.

The seismic potency tensor is

$$P_{ij}(t) = \int_V \varepsilon_{ij}^P \, dV \quad (2.2a)$$

where the integral covers the inelastically deforming volume in the earthquake source region. Similarly, the seismic moment tensor is

$$M_{ij}(t) = \int_V c_{ijkl} \, \varepsilon_{kl}^P \, dV = \int_V \Delta\sigma_{ij} \, dV. \quad (2.2b)$$

For a displacement discontinuity $\Delta\boldsymbol{u}$ across a surface S with a unit normal $\hat{\boldsymbol{n}}$

$$P_{ij}(t) = \frac{1}{2} \int_S (\Delta u_i n_j + \Delta u_j n_i) \, dS \quad (2.3a)$$

and

$$M_{ij}(t) = \int_S c_{ijkl} \Delta u_k n_l \, dS. \quad (2.3b)$$

For isotropic material

$$M_{ij}(t) = \int_S [\lambda \delta_{ij} \Delta u_k n_k + \mu(\Delta u_j n_i + \Delta u_i n_j)] \, dS. \quad (2.3c)$$

The scalar seismic potency of shear faulting on a planar surface is the integral of slip over the rupture area or, equivalently, the product of the spatial average of the final slip distribution and the failure area A

$$P_0 = \langle \Delta u \rangle \, A. \quad (2.4a)$$

The corresponding scalar seismic moment is

$$M_0 = \mu P_0 = \mu \, \langle \Delta u \rangle \, A \quad (2.4b)$$

with μ being an effective rigidity in the source area.

The displacement field at $(\boldsymbol{x}, t)$ due to a distribution of m_{jk} $(= c_{pqjk} n_q \Delta u_p)$ along an internal surface S is

$$u_i(\boldsymbol{x}, t) = \int_{-\infty}^{t} dt' \int_S c_{pqjk}(\boldsymbol{x}') n_q(\boldsymbol{x}') \Delta u_p(\boldsymbol{x}', t') \times [\partial G_{ij}(\boldsymbol{x}, t; \boldsymbol{x}', t')/\partial x'_k] \, d^2x'. \quad (2.5a)$$

The displacement field at $(\boldsymbol{x}, t)$ due to a distribution of displacement discontinuities along an internal surface S can be written as

$$u_i(\boldsymbol{x}, t) = \int_{-\infty}^{t} dt' \int_S \Delta u_j(\boldsymbol{x}', t') B_{ij}(\boldsymbol{x}, t; \boldsymbol{x}', t', \hat{\boldsymbol{n}}) \, d^2x' \quad (2.5b)$$

where B_{ij} gives the i component of displacement at $(\boldsymbol{x}, t)$ due to a unit point dislocation in the j direction at $(\boldsymbol{x}', t')$ across a surface S with a unit normal $\hat{\boldsymbol{n}}(\boldsymbol{x}')$. The situation $\hat{\boldsymbol{e}}_j \cdot \hat{\boldsymbol{n}} = 0$ with $\hat{\boldsymbol{e}}_j$ being a unit vector in the j direction corresponds to a shear crack, while $\hat{\boldsymbol{e}}_j \times \hat{\boldsymbol{n}} = 0$ represents a tensile crack. For shear faulting in isotropic material

$$B_{ij}(\boldsymbol{x}, t; \boldsymbol{x}', t', \hat{\boldsymbol{n}}) = n_k \mu [\partial G_{ij}(\boldsymbol{x}, t; \boldsymbol{x}', t')/\partial x'_k + \partial G_{ik}(\boldsymbol{x}, t; \boldsymbol{x}', t')/\partial x'_j]. \quad (2.5c)$$

For 2-D static crack problems, the stress drop and spatial derivative of slip are proportional to the Hilbert transforms of each other. For example, the stress drop sustained by a crack on the plane $x_2 = 0$ in a region $[-L, L]$ along the x_1 axis due to a slip distribution in that region is

$$\Delta\sigma_{2j}(x_1, x_2 = 0) = \frac{\mu}{2\pi(1-\nu)} \times \int_{-L}^{L} \frac{d[\Delta u_j(x'_1) - \nu \delta_{j3} \Delta u_3(x'_1)]/dx'_1}{x_1 - x'_1} \, dx'_1 \quad (2.6)$$

where $j = 1$ and 3 correspond to in-plane and antiplane shear, respectively, and $j = 2$ corresponds to opening motion. Relation (2.6) stems from the fact that a line dislocation at $x_1 = x'_1$ produces stresses that decay with distance r from the source like $1/r$.

3. Elastodynamic Waves

([1], [10], [12], [30], [33], [42], [49], [51], [57], [61], [64])

As mentioned in section 1, the Navier equation of motion for a linear elastic solid has solutions in terms of P and S plane body waves and elastodynamic Green functions. Stress-free surfaces, interfaces separating solids with different elastic properties, and other heterogeneities produce reflected/transmitted/converted waves and additional seismic phases. A solid with a free surface, taken to be horizontal, is referred to as a half space. Shear waves with horizontal polarization in an isotropic half space or horizontally layered structures, called *SH* waves, excite on horizontal planes a single stress component (e.g., σ_{zy} for a coordinate system with z normal to horizontal planes and y in the polarization direction). Corresponding S waves with vertical polarization, called *SV* waves, excite on horizontal planes the other two components of stress (e.g., σ_{zx} and σ_{zz}). Incident *SH* waves produce upon interaction with horizontal interfaces only *SH* reflected/transmitted waves, while *SV* (and also P) waves interacting with horizontal interfaces produce in general both *SV* and P reflected/transmitted waves.

Rayleigh waves consist of combined P and SV elastic disturbances propagating along a free surface of elastic solid. For a Poissonian half space, having $\lambda = \mu$ and $\nu = 0.25$, the Rayleigh wave velocity is $v_R \approx 0.92 v_S$. In a half space with increasing v_S and v_P with depth, or a corresponding spherical situation, Rayleigh waves are dispersive. In such cases, the longer wavelength components are affected by deeper structure and, hence, propagate faster. Love waves are dispersive SH waves that travel, again with faster propagation for longer wavelength components, within a horizontal surface layer or the top section of a half space having increasing v_S with depth (or corresponding situations with spherical geometry). Stoneley waves are combined P and SV disturbances propagating along an interface between pairs of solids with elastic properties in a certain range or between a solid and a fluid.

Plane waves form a complete set of basis functions and, hence, can be used to represent any wavefield. In practice, they are useful in problems with geometrical elements that are naturally characterized by a Cartesian geometry (e.g., planar surfaces on which boundary conditions should be satisfied), and in high-frequency seismology with propagation distance much larger than the wavelength. Other complete sets of basis functions that are commonly used in wave propagation seismology are Bessel/Hankel functions and spherical harmonics. Bessel and Hankel functions are useful in problems with elements that have a cylindrical or spherical geometry (e.g., 2-D problems with a line source, 3-D problems with a point source, tunnel or borehole in a 3-D solid). Spherical harmonics provide a natural representation for the angular dependency of normal modes (free oscillations) of the Earth and are useful in other problems of low-frequency seismology (e.g., surface waves). The radial dependency of normal modes is described by spherical Bessel functions. In a spherically symmetric solid, there are two distinct types of normal modes, toroidal and spheroidal. Toroidal modes are analogous to SH and Love waves and are associated only with an angular motion. Spheroidal modes are analogous to P/SV and Rayleigh waves and are associated with both radial and angular motion.

The elastodynamic Green function in an unbounded homogeneous isotropic solid has P and S wave components

$$G_{ij}(\mathbf{x}, \mathbf{x}', t) = G_{ij}^P + G_{ij}^S \tag{3.1a}$$

where

$$G_{ij}^P = -(\partial^2/\partial x_i \partial x_j)\, h(r, t; v_P) \tag{3.1b}$$

and

$$G_{ij}^S = -(\delta_{ij}\nabla^2 - \partial^2/\partial x_i \partial x_j)\, h(r, t; v_S) \tag{3.1c}$$

with (Fig. 3)

$$h(r, t; c) = -\frac{1}{4\pi\rho r}(t - r/c)\, H(t - r/c). \tag{3.1d}$$

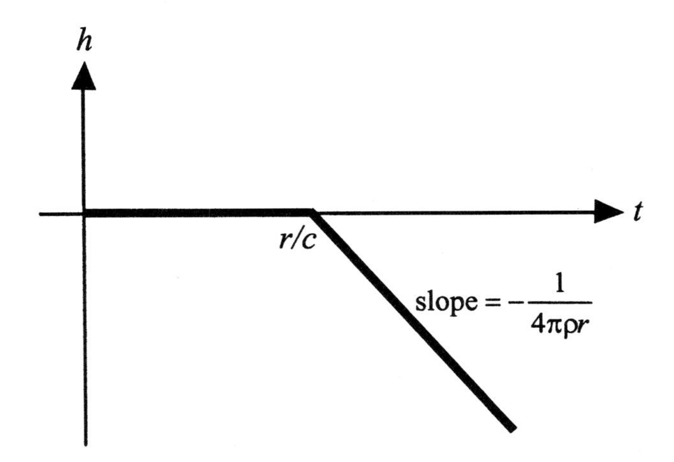

FIGURE 3 The function $h(r, t; c)$ at space-time distances (r, t) from a source in a solid with wave velocity c.

In (3.1b)–(3.1d), the source operates at zero time, r is the source-receiver distance, v_P and v_S are P and S wave velocities, and H is the Heaviside unit step function.

The elastodynamic Green functions for 2- and 3-D homogeneous isotropic solids can be written in the frequency domain [61] as

$$G_{ij}(\mathbf{x}, \mathbf{x}', \omega) = \frac{1}{4\pi\rho\omega^2}\left\{\delta_{ij}k_S^2 g(k_S r) - \frac{\partial}{\partial x_i}\frac{\partial}{\partial x_j}[g(k_P r) - g(k_S r)]\right\} \tag{3.2}$$

where ω is angular frequency, $k_S = \omega/v_S$, $k_P = \omega/v_P$, $g(kr) = \exp(-ikr)/r$ for a 3-D solid, and $g(kr) = i\pi H_0(kr)$ for a 2-D case with H_0 being the Hankel function of order zero.

The displacement field in an unbounded homogeneous isotropic material generated by a distribution of m_{jk} can be written from (2.5) and (3.1) as

$$u_i(\mathbf{x}, t) = u_i^P + u_i^S \tag{3.3a}$$

where

$$u_i^{P or S}(\mathbf{x}, t) = O_{ijk}^{P or S} \int_V \int_{-\infty}^{t} h(r, t - t'; v_{P or S}) \times m_{jk}(\mathbf{x}', t')\, dt'\, d^3x' \tag{3.3b}$$

with

$$O_{ijk}^P = \partial^3/\partial x_i \partial x_j \partial x_k \tag{3.3c}$$

and

$$O_{ijk}^S = \frac{1}{2}(\delta_{ij}\partial/\partial x_k + \delta_{ik}\partial/\partial x_j)\nabla^2 - \partial^3/\partial x_i \partial x_j \partial x_k. \tag{3.3d}$$

The far-field approximation of (3.3a)–(3.3d), valid for $r \gg a$ and $r \gg \lambda$ with a and λ being source dimension and wavelength of interest, is

$$u_i^{PorS}(\boldsymbol{x}, t) = \frac{R_{ijk}^{PorS}}{4\pi\rho r_0 v_{PorS}^3} \int_V \dot{m}_{jk}\left(\boldsymbol{x}', t - \frac{r_0 - \gamma_i x_i'}{v_{PorS}}\right) d^3x' \tag{3.4a}$$

where r_0 is a representative source-receiver distance, $\gamma_i = \partial r/\partial x_i = x_i/r$ are components of the unit vector in the source-receiver direction, the integral term is referred to as source effect, and the receiver plus radiation pattern functions R in front of the integral are

$$R_{ijk}^{P} = \gamma_i \gamma_j \gamma_k \tag{3.4b}$$

and

$$R_{ijk}^{S} = \frac{1}{2}(\delta_{ij}\gamma_k + \delta_{ik}\gamma_j) - \gamma_i\gamma_j\gamma_k. \tag{3.4c}$$

The γ_i terms in (3.4b) and (3.4c) are associated with the variable of interest at the receiver, while the other terms are associated with radiation patterns for P and S waves.

For a unidirectional slip in the x_1 direction on a fault surface in the x_1-x_3 plane (Fig. 4), the corresponding far-field displacement field is

$$u_i^{PorS}(\boldsymbol{x}, t) = \frac{R_{i12}^{PorS} v_S^2}{2\pi r_0 v_{PorS}^3} \int_S \Delta\dot{u}_1[x_1', x_3', t + (\gamma_1 x_1' + \gamma_3 x_3' - r_0)/v_{PorS}]\, d^2x'. \tag{3.5a}$$

The factor $1/r_0$ in (3.4a) and (3.5a) accounts for the geometric attenuation of body waves in the far-field. The ratio of the maximum amplitudes of far-field displacement S waves to P waves is essentially v_P^3/v_S^3, which is about 5 for a Poissonian solid. Denoting the source effect (integral term) in (3.5a) by $\Omega(\boldsymbol{x}', t)$, the zero-frequency asymptote of the far-field displacement source spectrum is

$$\Omega(\boldsymbol{x}', \omega = 0) = \int_S d^2x' \int_{-\infty}^{\infty} \Delta\dot{u}_1\, dt = \int_S \Delta u_1(\boldsymbol{x}', t = T_r)\, d^2x' = \langle \Delta u_1 \rangle A = P_0 \tag{3.5b}$$

where T_r is the time at the completion of the dynamic slip process (called the rise time), $\langle \Delta u_1 \rangle$ is the spatial average of slip over the fault, A is the rupture area, and P_0 is the scalar seismic potency (see section 2). The zero-frequency asymptote of the far-field spectrum of any plausible description of the source process is equal to the scalar seismic potency or moment divided by rigidity (see, e.g., equation (3.8) and Fig. 7). Details of different source processes are contained in the high-frequency portions of the spectra.

The far-field condition $r \gg \lambda$ is equivalent to a high-frequency criterion $|\omega| \gg v_{SorP}/r$. Thus (3.4) and (3.5) do not represent properly the low-frequency components of the fields radiated from distributions of slip and moment density tensors. In particular, these results do not include the final static components of the response. Equations (3.4) and (3.5) have approximate descriptions of space-time variations of the source process (viewed from a constant source-receiver angle and without terms that attenuate faster than $1/r$). In contrast, point-source and low-frequency approximations of the far-field radiation from a distribution of m_{jk} are associated with more limiting conditions at the source. The point-source approximation is valid for $\lambda \gg a$ and $r \gg \lambda$, and the corresponding results do not include spatial variations at the source. The low-frequency approximation is valid for $|\omega| T_r \ll 1$ and $r \gg \lambda$, and the corresponding results do not include space or time variations of the source process.

The complete displacement field in an unbounded homogeneous isotropic material due to a spatially localized body force with a source-time function $\boldsymbol{F}(t)$ is

$$u_i(\boldsymbol{x}, t) = \frac{1}{4\pi\rho}(3\gamma_i\gamma_j - \delta_{ij})\frac{1}{r^3} \int_{r/v_P}^{r/v_S} \tau F_j(t - \tau)\, d\tau + \frac{1}{4\pi\rho v_P^2}\gamma_i\gamma_j \frac{1}{r} F_j(t - r/v_P) - \frac{1}{4\pi\rho v_S^2}(\gamma_i\gamma_j - \delta_{ij})\frac{1}{r} F_j(t - r/v_S). \tag{3.6}$$

The first term with unseparated P and S waves decays with distance like $1/r^2$ for sources that are nonzero for times shorter than the interval $(r/v_S - r/v_P)$. Since this term is dominant for very short source-receiver distances ($r \to 0$), it is called the near-field term. The second and third terms with separate P and S wave contributions proportional to $1/r$ are dominant at large distances ($r \to \infty$) and are called far-field terms.

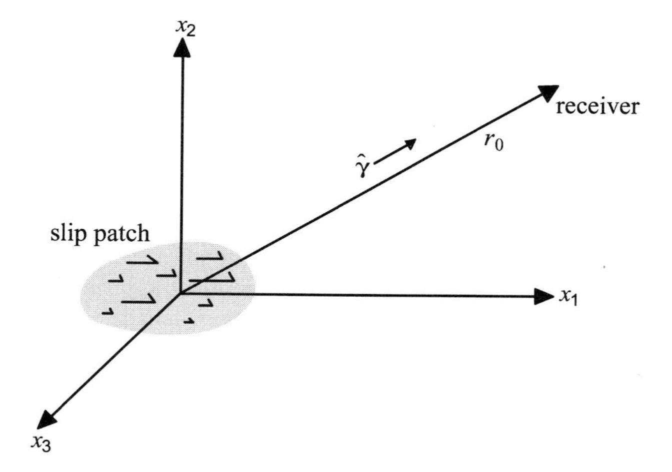

FIGURE 4 A coordinate system for analysis of far-field radiation at receiver with distance r_0 and angle $\hat{\gamma}$ from a fault with unidirectional slip $\Delta\boldsymbol{u} = [\Delta u_1(x_1, x_3, t), 0, 0]$.

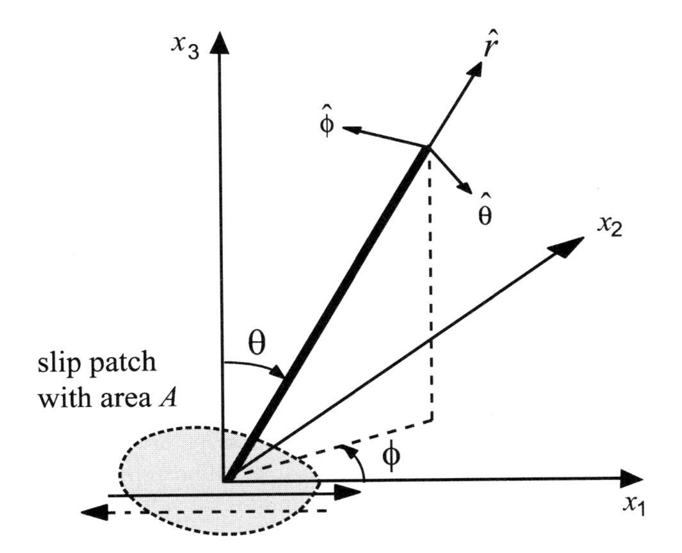

FIGURE 5 Cartesian and polar coordinate systems for analysis of radiation by a slip patch with area A and average slip $\langle \Delta u(t) \rangle$.

The corresponding displacement field generated by a time-dependent scalar seismic moment $M_0(t) = \mu \langle \Delta u(t) \rangle A$, associated with a shear dislocation (Fig. 5) parallel to a fault surface with an area A, is

$$\begin{aligned} \boldsymbol{u}(\boldsymbol{x}, t) = {} & \frac{1}{4\pi\rho} \boldsymbol{A}^N \frac{1}{r^4} \int_{r/v_P}^{r/v_S} \tau M_0(t-\tau)\, d\tau \\ & + \frac{1}{4\pi\rho v_P^2} \boldsymbol{A}^{IP} \frac{1}{r^2} M_0(t - r/v_P) \\ & + \frac{1}{4\pi\rho v_S^2} \boldsymbol{A}^{IS} \frac{1}{r^2} M_0(t - r/v_S) \\ & + \frac{1}{4\pi\rho v_P^3} \boldsymbol{A}^{FP} \frac{1}{r} \dot{M}_0(t - r/v_P) \\ & + \frac{1}{4\pi\rho v_S^3} \boldsymbol{A}^{FS} \frac{1}{r} \dot{M}_0(t - r/v_S). \end{aligned} \tag{3.7a}$$

The first term in (3.7a) with unseparated P and S waves gives the near-field response, the second and third terms with separate P- and S-wave contributions proportional to $1/r^2$ are intermediate-field terms, and the fourth and fifth terms with separate P and S waves proportional to $1/r$ are the far-field terms. The functions $\boldsymbol{A}$ in front of each term are radiation patterns for the various terms given by

$$\begin{aligned} \boldsymbol{A}^N &= 9 \sin 2\theta \cos\phi \hat{\boldsymbol{r}} - 6(\cos 2\theta \cos\phi \hat{\boldsymbol{\theta}} - \cos\theta \sin\phi \hat{\boldsymbol{\phi}}), \\ \boldsymbol{A}^{IP} &= 4 \sin 2\theta \cos\phi \hat{\boldsymbol{r}} - 2(\cos 2\theta \cos\phi \hat{\boldsymbol{\theta}} - \cos\theta \sin\phi \hat{\boldsymbol{\phi}}), \\ \boldsymbol{A}^{IS} &= -3 \sin 2\theta \cos\phi \hat{\boldsymbol{r}} + 3(\cos 2\theta \cos\phi \hat{\boldsymbol{\theta}} - \cos\theta \sin\phi \hat{\boldsymbol{\phi}}), \\ \boldsymbol{A}^{FP} &= \sin 2\theta \cos\phi \hat{\boldsymbol{r}}, \\ \boldsymbol{A}^{FS} &= \cos 2\theta \cos\phi \hat{\boldsymbol{\theta}} - \cos\theta \sin\phi \hat{\boldsymbol{\phi}}, \end{aligned} \tag{3.7b}$$

where $\hat{\boldsymbol{r}}$, $\hat{\boldsymbol{\theta}}$, and $\hat{\boldsymbol{\phi}}$ are unit direction vectors in a polar coordinate system for the source-receiver geometry (Fig. 5). The far-field radiation patterns provide the basis for routine derivation of fault-plane solutions from observed spatial distributions of first-motion polarities.

The largest amplitude arrivals at teleseismic distances are surface Rayleigh and Love waves, which attenuate with distance like $r^{-1/2}$ in contrast to the r^{-1} attenuation of body waves. For this reason, surface waves are used extensively to derive information from observed seismograms on low-frequency properties of earthquake sources and the velocity structure of the crust and upper mantle.

The far-field displacement amplitude spectrum from a unidirectional rupture on a rectangular fault with length L, width $W \ll L$, rupture velocity V_r in the positive length direction, and ramp source-time function with final slip D and rise time T_r is

$$|\Omega(\boldsymbol{x}, \omega)| = WLD \frac{\sin X}{X} \left| \frac{1 - e^{i\omega T}}{\omega T_r} \right| \tag{3.8}$$

where $X = (\omega L/2)[1/V_r - (\cos\Psi)/v_{PorS}]$ and Ψ is the angle between the rupture direction and receiver. The function $\sin X/X$, called the finiteness factor or directivity function, accounts for the finite apparent rupture duration at different directions and produces oscillations with nodes at $X = \pi, 2\pi, \ldots$ The term in the square bracket of X times L gives the apparent rupture duration at receiver direction Ψ (Fig. 6). The zero-frequency asymptote of (3.8), WLD, is the scalar seismic potency (see also (3.5b) and section 2), while at high frequencies (3.8) is a decaying oscillatory function with a high-frequency asymptote proportional to ω^{-2}.

In general, the low-frequency asymptote of the far-field displacement source spectrum, being independent of internal variations of the source process, is a stable constant value equal to the scalar potency or moment divided by rigidity. In contrast, the high-frequency spectral source behavior depends strongly on model assumptions and parameters. Kinematic rupture models that start from a point and spread with a uniform velocity to prescribed boundaries have far-field displacement spectra with high-frequency asymptotes proportional to ω^{-2}–ω^{-3} for most model assumptions and source-receiver directions (Fig. 7). The frequency at the intersection of the low- and high-frequency asymptotes is called the corner frequency ω_c. For source models with uniform rupture propagation over a continuous fault in a full space, ω_c is inversely proportional to the rupture dimension. However, the precise value of ω_c depends strongly on details of the source process, source-receiver direction, and wave type (P or S).

The attenuation of elastic strain energy during a stress cycle with angular frequency ω may be characterized in terms of a quality factor $Q(\omega)$ as

$$\frac{1}{Q(\omega)} = -\frac{\Delta E}{2\pi E} \tag{3.9a}$$

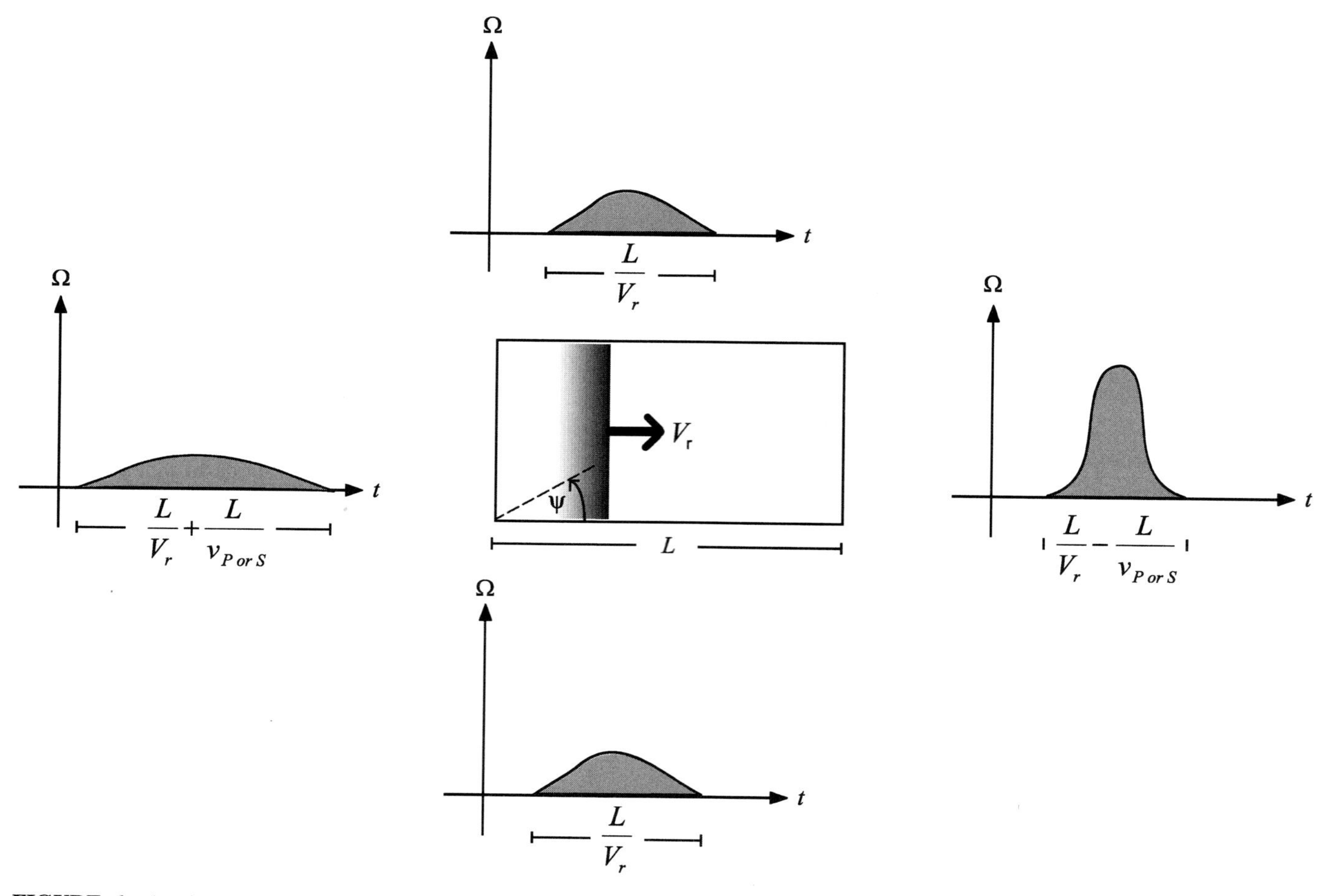

FIGURE 6 A schematic representation of displacement pulses $\Omega(t)$ radiated at different directions from a propagating rupture. The areas under the curves are the same, but the amplitudes and durations are different.

where E and ΔE are the peak and change of elastic strain energy during the cycle. An effective way of incorporating attenuation into solutions of elastic wave propagation problems, when Q is approximately constant over a wide range of frequencies, is to multiply the elastic body-wave velocities by the function

$$[1 + \ln(\omega/\omega_0)/\pi Q - i/2Q] \tag{3.9b}$$

where ω_0 is a reference angular frequency (e.g., 2π) for which the body-wave phase velocities are known and i is the imaginary unit.

The amplitude response $|X(\omega)|$ and phase delay $\phi(\omega)$ of a pendulum seismometer with mass M, spring constant k, and dashpot constant D are given by

$$|X(\omega)| = \frac{\omega^2}{\sqrt{\left(\omega^2 - \omega_s^2\right)^2 + 4\varepsilon^2\omega^2}} \tag{3.10a}$$

and

$$\phi(\omega) = \pi - \tan^{-1}\frac{2\varepsilon\omega}{\omega^2 - \omega_s^2} \tag{3.10b}$$

where $2\varepsilon = D/M, \omega_s = (k/M)^{1/2}$ and the dimensionless damping constant ε/ω_s is equal to half the reciprocal of the Q value of a damped oscillator.

4. Fracture ([5], [7], [15], [17], [28], [29], [31], [32], [42], [44], [46], [47], [50], [54])

Holes, notches, cracks, inclusions, and other flaws in solids amplify or concentrate stress near their boundaries. For example, the tensile stress on the boundary of a circular hole around the origin of the x_1-x_2 plane with remote loading σ^∞ in the x_2 direction is

$$\sigma_{\theta\theta} = \sigma^\infty(1 + 2\cos 2\theta) \tag{4.1}$$

where $\theta = 0$ on the positive x_1 axis. Equation (4.1) indicates a stress concentration by a factor of 3 at the intersections of the hole with the x_1 axis.

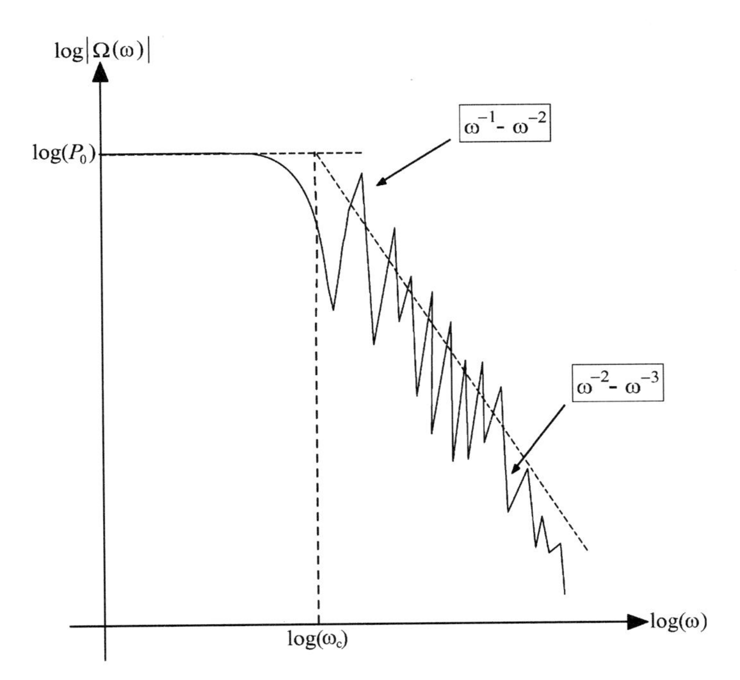

FIGURE 7 A schematic representation of radiated far-field displacement source spectrum with scalar seismic potency P_0 and corner frequency ω_c.

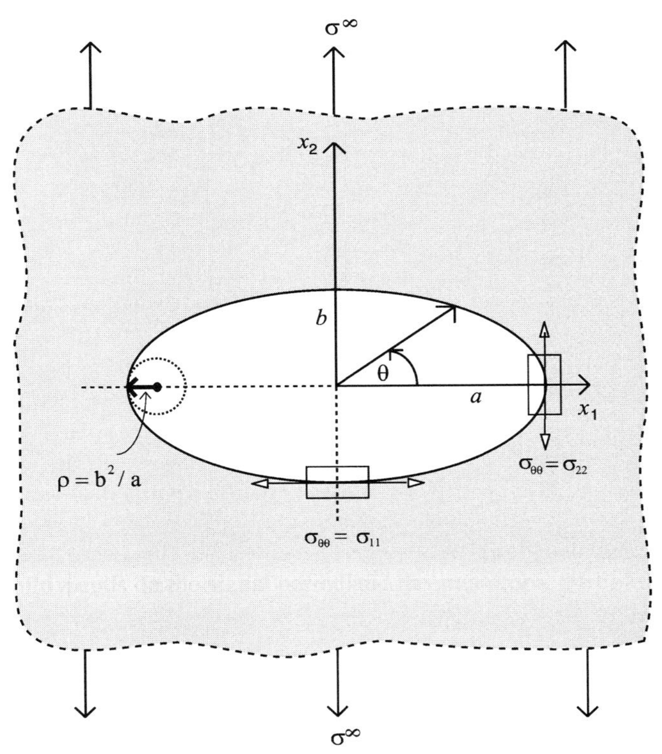

FIGURE 8 An elliptical hole with a radius of curvature ρ in a plate under remote tensile loading σ^∞.

The tensile stress at the x_1 intersections of an elliptical hole (Fig. 8) around the origin of the x_1-x_2 plane with major and minor axes a and b under remote loading σ^∞ in the x_2 direction is

$$\sigma_{\theta\theta}(x_1 = \pm a, x_2 = 0) = \sigma^\infty(1 + 2a/b) = \sigma^\infty(1 + 2\sqrt{a/\rho}) \tag{4.2}$$

where $\rho = (b^2/a)$ is the radius of curvature of the ellipse at $x_1 = \pm a$. The limit $b \to 0$ corresponds to a flat Griffith crack. In that limit, modeled as a sharp mathematical cut, the stress concentration at $x_1 = \pm a$ becomes unbounded. This is a general feature of all problems in Linear Elastic Fracture Mechanics (LEFM), where the breakdown processes at the crack tips are ignored. The region of validity of LEFM is at distances from the crack tip (Fig. 9) larger than the inelastic process zone and much smaller than any macroscopic dimension such as crack length or distances from boundaries. Nonlinear fracture mechanics models incorporate constitutive laws for the inelastic deformation in the process zone that eliminate the crack tip singularities (see, e.g., Fig. 11 for slip-weakening friction behind the rupture front).

For a flat Griffith crack of length $2a$ on the x_1 axis under remote tensile loading σ^∞ in the x_2 direction, the tensile stress along the x_1 axis and slip Δu_2 are

$$\sigma_{22}(|x_1| > a, x_2 = 0) = \sigma^\infty\,|x_1|/\sqrt{x_1^2 - a^2} \tag{4.3a}$$

and

$$\Delta u_2(|x_1| < a) = (4\sigma^\infty/E')\sqrt{a^2 - x_1^2} \tag{4.3b}$$

where $E' = E$ for plane stress and $E' = E/(1-\nu^2)$ for plane strain with E and ν being Young modulus and Poisson ratio, respectively. The $1/\sqrt{r}$ and $\sqrt{r'}$ functional dependencies in (4.3a) and (4.3b), with r and r' denoting the distances from the crack tip to points in the unbroken and broken material, respectively, are common to all 2D, 3D, static, dynamic, isotropic, and anisotropic problems in LEFM (Fig. 10).

A useful way of quantifying the strength of the singular crack tip field in LEFM problems is through the limit

$$K = \lim_{r\to 0}\left[\sigma\sqrt{2\pi r}\right] \tag{4.4}$$

where K is called the stress intensity factor.

The singular stress and displacement discontinuity fields along dynamic planar cracks under general mixed loading have the following general forms

$$\{\sigma_{22}, \sigma_{21}, \sigma_{23}\} = \frac{1}{\sqrt{2\pi r}}\{K_{\mathrm{I}}, K_{\mathrm{II}}, K_{\mathrm{III}}\} \tag{4.5a}$$

and

$$\{\Delta u_2, \Delta u_1, \Delta u_3\} = \frac{4(1-\nu)}{\mu} \times \sqrt{\frac{r'}{2\pi}}\{K_{\mathrm{I}} f_{\mathrm{I}}, K_{\mathrm{II}} f_{\mathrm{II}}, K_{\mathrm{III}} f_{\mathrm{III}}/(1-\nu)\}. \tag{4.5b}$$

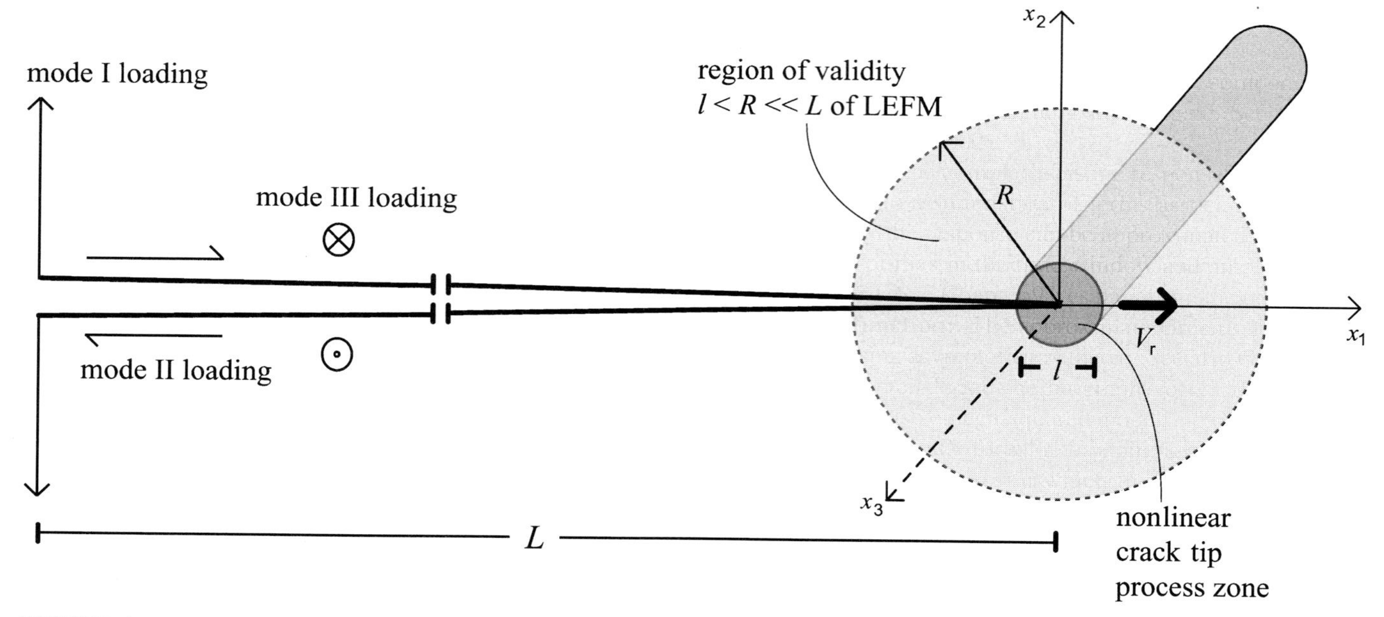

FIGURE 9 A schematic diagram of a 2-D crack illustrating the region of validity of Linear Elastic Fracture Mechanics and modes I, II, and III of rupture extending along the x_1 axis with velocity V_r.

The functions K_J are stress intensity factors for modes J = I, II, and III. A pure mode I corresponds to an "opening" tensile fracture with particle motion normal to the crack surface (Fig. 9). A pure mode II corresponds to a "sliding" in-plane shear fracture with particle motion in the crack plane and parallel to the direction of crack extension. A pure mode III corresponds to a "tear" antiplane shear fracture with particle motion in the crack plane and normal to the direction of crack extension (Fig. 10). For cracks that grow from an initial "nucleation" size with continuing slip behind the propagating front, the functions K_J decrease monotonically with the crack speed to zero: $K_{\mathrm{I}}(V_r \to v_R) \to 0$, $K_{\mathrm{II}}(V_r \to v_R) \to 0$, $K_{\mathrm{III}}(V_r \to v_S) \to 0$ with v_R and v_S being the Rayleigh and shear wave speeds, respectively. For cracks expanding from the origin at a uniform speed V_r, $K_{\mathrm{J}} \propto \Delta\sigma_{\mathrm{J}}\sqrt{V_r t}$ where $\Delta\sigma_{\mathrm{J}} = \sigma_{\mathrm{J}}^{\infty} - \sigma_{\mathrm{J}}^{crack}$ are stress drop components associated with the different modes. A static crack of length $2a$ in an infinite plate has $K_{\mathrm{J}} = \sigma_{\mathrm{J}}^{\infty}\sqrt{\pi a}$. The functions f_{J} in (4.5b) increase monotonically with the crack speed such that $f_{\mathrm{J}}(V_r = 0) = 1$ for all modes, and $f_{\mathrm{I}}(V_r \to v_R) \to \infty$, $f_{\mathrm{II}}(V_r \to v_R) \to \infty$, $f_{\mathrm{III}}(V_r \to v_S) \to \infty$. The products $K_{\mathrm{J}} f_{\mathrm{J}}$ governing the amplitudes of the displacement discontinuity fields remain finite. Explicitly,

$$f_{\mathrm{I}} = \frac{\alpha_P V_r^2}{(1-\nu)v_S^2 R}, \quad f_{\mathrm{II}} = \frac{\alpha_S V_r^2}{(1-\nu)v_S^2 R}, \quad f_{\mathrm{III}} = 1/\alpha_S \quad (4.5c)$$

where $\alpha_P = \sqrt{1 - V_r^2/v_P^2}$, $\alpha_S = \sqrt{1 - V_r^2/v_S^2}$, and $R = 4\alpha_P\alpha_S - (1+\alpha_S^2)^2$. The Rayleigh function R vanishes when $V_r = v_R$.

Shear fracture in mode II on a frictional interface can propagate at intersonic velocities [46] larger than the above speed limits for singular cracks (v_R for modes I and II; v_S for mode III) and smaller than the dilatational wave speed v_P. The singular solution, not just along a planar extension of the crack as in (4.5a) and (4.5b) but for all positions, includes an angular component that depends on V_r. The full singular solution predicts, for a homogeneous solid, rotations of maximum failure stress components away from the continuation of a planar crack and, hence, branching [15] at certain values of V_r. Branching is expected to be suppressed by high compressive normal stress, and for rupture in a weak layer (e.g., of a damaged fault zone rock) surrounded by a stronger material or along the interface between such a layer and the host solid (or other dynamically weak interfaces).

Inferred values of earthquake rupture velocity averaged over the failure area are usually about 75% of the estimated v_S of the host rock [25, 53]. Such V_r values may exceed the S-wave velocity of the (damaged) fault zone material, which can be considerably lower (e.g., by 25% or more) than that of the host rock. In some cases, there is evidence ([46] and references therein) that V_r exceeds along part of the rupture surface the v_R and v_S velocities of the host rock. Very low values of V_r (e.g., 1% v_S of the host rock or less) are sometimes inferred in association with so-called slow earthquakes (e.g., [22] and references therein).

During a self-similar growth of a circular crack with uniform V_r and $\Delta\sigma$, the slip is

$$\Delta u = [A(V_r)\Delta\sigma/\mu]\sqrt{V_r^2 t^2 - X^2}, \quad (4.6a)$$

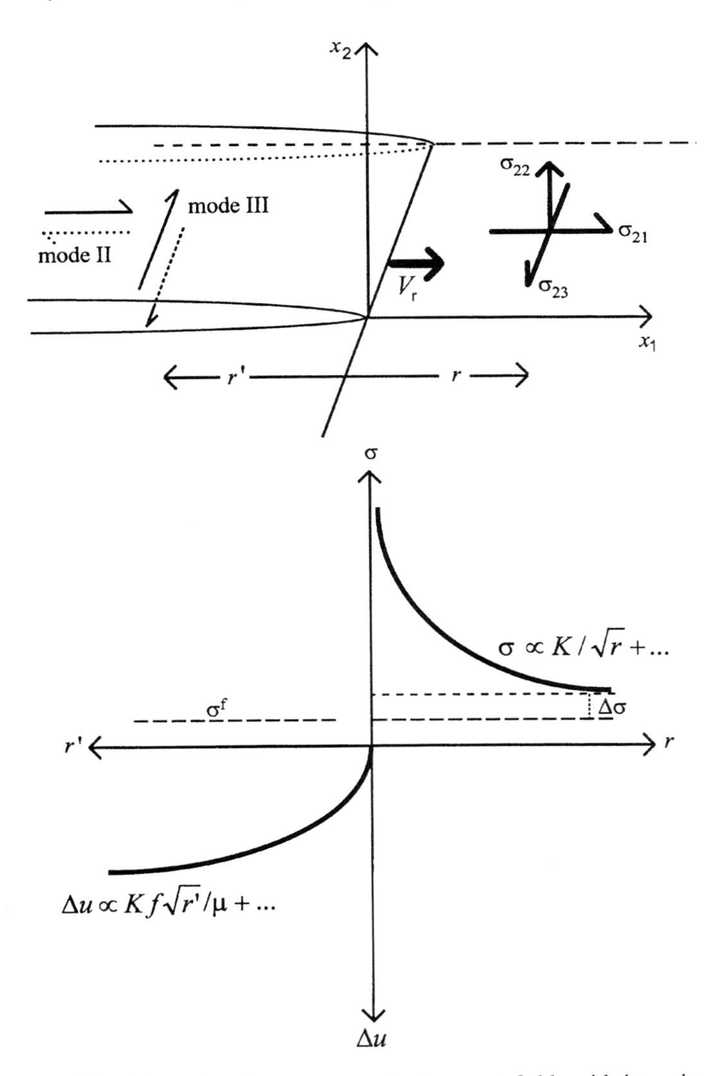

FIGURE 10 Singular stress and displacement fields with intensity factors K and Kf, respectively, for mode II or III shear crack having frictional stress σ^f and stress drop $\Delta\sigma$.

where $A(V_r)$ is of order 1 for $0 \leq V_r \leq v_R$ and X is the distance from the center of the crack.

The corresponding radiated far-field displacement is proportional to the stress drop

$$u \propto 2\pi A(V_r)V_r^3 t^2 \Delta\sigma/\mu. \tag{4.6b}$$

Equation (4.6b) can be used to estimate the dynamic stress drop from the initial pulse of seismic radiation.

For a crack growing in its plane, the energy release from the crack tip per new unit surface area is

$$G = \{(1-\nu)(K_{\rm I}^2 f_{\rm I} + K_{\rm II}^2 f_{\rm II}) + K_{\rm III}^2 f_{\rm III}\}/2\mu. \tag{4.7}$$

The different mode-components of G decrease monotonically with V_r in the subsonic regime, from maximum values at $V_r = 0$ to zero at the limiting speeds.

During a short-time crack extension of distance a with nonuniform velocity $V_r(t)$, prior to arrival back at the advancing tip of wave reflections from the other crack tip or boundaries, the stress intensity factor of a given mode can be written as the product

$$K = K^*(a)K[V_r(t)] \tag{4.8a}$$

where $K^*(a)$ is the "rest" stress intensity factor equal to that of a semi-infinite static crack that advanced the same distance a into the same initial stress field and the non-dimensional function $K[V_r(t)]$ decreases monotonically from unity at $V_r(t) = 0$ to zero at the appropriate limiting speed. Similarly, the corresponding energy release rate can be written as

$$G = G^*(a)G[V_r(t)] \tag{4.8b}$$

with similar interpretations of the terms.

The failure fracture criterion for initiation of rupture in a given mode J = I, II, III is

$$K_{\rm J} = K_{\rm Jc} \tag{4.9a}$$

where $K_{\rm J}$ is the appropriate stress intensity factor and $K_{\rm Jc}$ is a corresponding material property called fracture toughness.

The failure fracture criterion for continuation of dynamic rupture propagation is

$$G = G_c \tag{4.9b}$$

where G_c is a material property specifying the amount of energy required for the creation of a unit new surface area. For a purely brittle fracture process associated with the creation of a new surface by a reversible separation of atoms, $G_c = 2\Gamma_{\rm S}$ with $\Gamma_{\rm S}$ being the specific Griffith surface energy.

For a 2-D quasi-static crack extending in the x_1 direction, the integral

$$J = \int_{\rm C} (Wn_1 - n_i\sigma_{ij}\partial u_j/\partial x_1)\, dS \tag{4.10}$$

is path-independent for all contours C that begin and end on the crack. In (4.10), n_i are components of the outer unit normal to C and W is the strain energy density $W(\varepsilon) = \int_0^{\varepsilon} \sigma_{ij}\, d\varepsilon_{ij}$.

For frictionless cracks $J = G$, while for cracks that support a frictional stress σ_{2j}^f, $J_Q = G + \sigma_{2j}^f(\Delta u_j)_Q$ with Q being any point on the crack.

5. Friction ([5], [13], [31], [32], [35], [38], [41], [42], [45], [50])

The resistance to initial macroscopic tangential motion along a sliding surface is often described by the Coulomb friction

$$\tau_f = c + f_s\sigma_n \tag{5.1}$$

where c is a cohesion term representing resistance due to joined portions of the surface, f_s is the static coefficient of friction, and σ_n is normal stress. In cases where pore fluids are present, σ_n should be replaced by $\sigma_n^{effective} = (\sigma_n - p)$ with p being the pore pressure. The same holds for σ_n in all other expressions of this section. In friction experiments with many rock types (not including some clay materials), the data can be fitted by the lines

$$(c = 0,\ f_s = 0.85) \quad \text{for} \quad \sigma_n < 200 \text{ MPa} \tag{5.2a}$$

and

$$(c = 50 \text{ MPa},\ f_s = 0.6) \quad \text{for} \quad 200 \text{ MPa} < \sigma_n < 1700 \text{ MPa} \tag{5.2b}$$

with some scatter representing, at least in part, dependency on surface conditions and rock type. Equation (5.1) with coefficients given by (5.2) is referred to as Byerlee friction.

Equation (5.1) also describes the failure envelope on a Mohr diagram of rock fracturing experiments with $\sigma_n < 1000$ MPa. In that context, f_s is called the coefficient of internal friction, and inferred values of f_s are, as in Byerlee friction, about 0.7. However, values of c in shear fracture experiments are considerably larger than those associated with frictional sliding.

Frictional resistance in which the friction coefficient decreases with slip Δu is referred to as slip-weakening friction (Fig. 11). A simple triangular form often used in numerical calculations is

$$f = f_s - (f_s - f_d)\Delta u / D_c \quad \text{for} \quad \Delta u < D_c \tag{5.3a}$$

and

$$f = f_d \quad \text{for} \quad \Delta u \geq D_c \tag{5.3b}$$

where f_s and f_d are static and dynamic friction coefficients and D_c is a characteristic slip-weakening distance. In the context of the slip-weakening friction (5.3), the fracture energy density per unit fault area associated with the breakdown processes leading to strength degradation (see e.g., Fig. 14a) is

$$G_c = \sigma_n(f_s - f_d)D_c/2. \tag{5.3c}$$

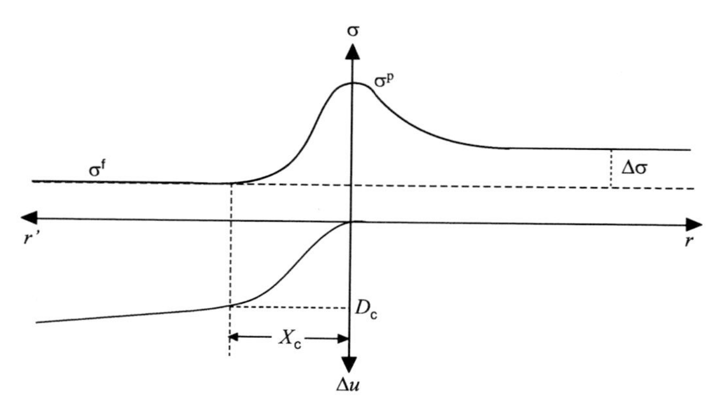

FIGURE 11 A schematic diagram of slip-weakening friction with strength $\sigma(\Delta u)$ decaying from a peak value σ^p to a residual level σ^f over a characteristic slip distance D_c. The sizes of the breakdown zone and stress drop are X_c and $\Delta\sigma$, respectively.

Values of G_c based on shear laboratory fracture data [31, 41] with various pressure-temperature conditions and rock types, interpreted with slip-weakening friction, are in the range 10^4–10^5 J/m^2. For comparisons, estimates of G_c based on initiation and stopping of earthquakes and modeling of seismic data are in the range 10^6–10^8 J/m^2, estimates of G_c in tensile lab fracture experiments with granite are in the range 3–50 J/m^2, and estimates of the specific Griffith surface energy for various materials are in the range 1–2 J/m^2.

The spatial region behind the tip of the sliding material where the strength degradation occurs and slip achieves a value of D_c (see Fig. 11) is called the process zone, degradation zone, or breakdown zone. The linear size of this region is

$$X_c = c\mu D_c / [\sigma_n(f_s - f_d)] \tag{5.3d}$$

where c is a dimensionless constant of order 2–3 and μ is rigidity. Laboratory values of D_c depend on the roughness of the sliding surface and possible existence of gouge. In experiments done so far [31, 41], it is in the range $(10^{-6} - 5 \times 10^{-4})$ m. Using $f_s - f_d \approx 0.05$ and a representative ratio for the seismogenic zone $\mu/\sigma_n \approx 300$ implies values of X_c in the range $(10^{-2}$–$10)$ m. Implications of the critical weakening distance to nucleation of slip instabilities are discussed below.

Rate- and state-dependent friction laws characterize the dependency of the friction coefficient on slip, slip velocity, history (represented by state variables), and normal stress. In a "standard" form of rate- and state-dependent friction (Fig. 12), with a single state variable θ and no dependency of the friction coefficient on normal stress, the friction coefficient can be written as

$$f = f_0 + a\ln(v/v_0) + b\ln(v_0\theta/L) \tag{5.4a}$$

where f_0 is a nominal friction coefficient (about 0.7 for most rocks as indicated in (5.2)), v and v_0 are current and reference values of sliding velocity, a is the amplitude of the initial response to a velocity jump, and b is the amplitude of gradual strength alteration over a characteristic slip distance L (also denoted by D_c) following a velocity jump. There are two common versions that describe the evolution of the state variable. In the "slowness" version, the state variable satisfies

$$d\theta/dt = 1 - v\theta/L \tag{5.4b}$$

whereas in the "slip" version

$$d\theta/dt = -(v\theta/L)\ln(v\theta/L). \tag{5.4c}$$

In both cases, during steady-state sliding with constant velocity and slip distance larger than L, $\theta = L/V$ and the steady-state dependency of the friction coefficient on the sliding velocity is

$$f_{ss} = f_0 + (a - b)\ln(v/v_0). \tag{5.4d}$$

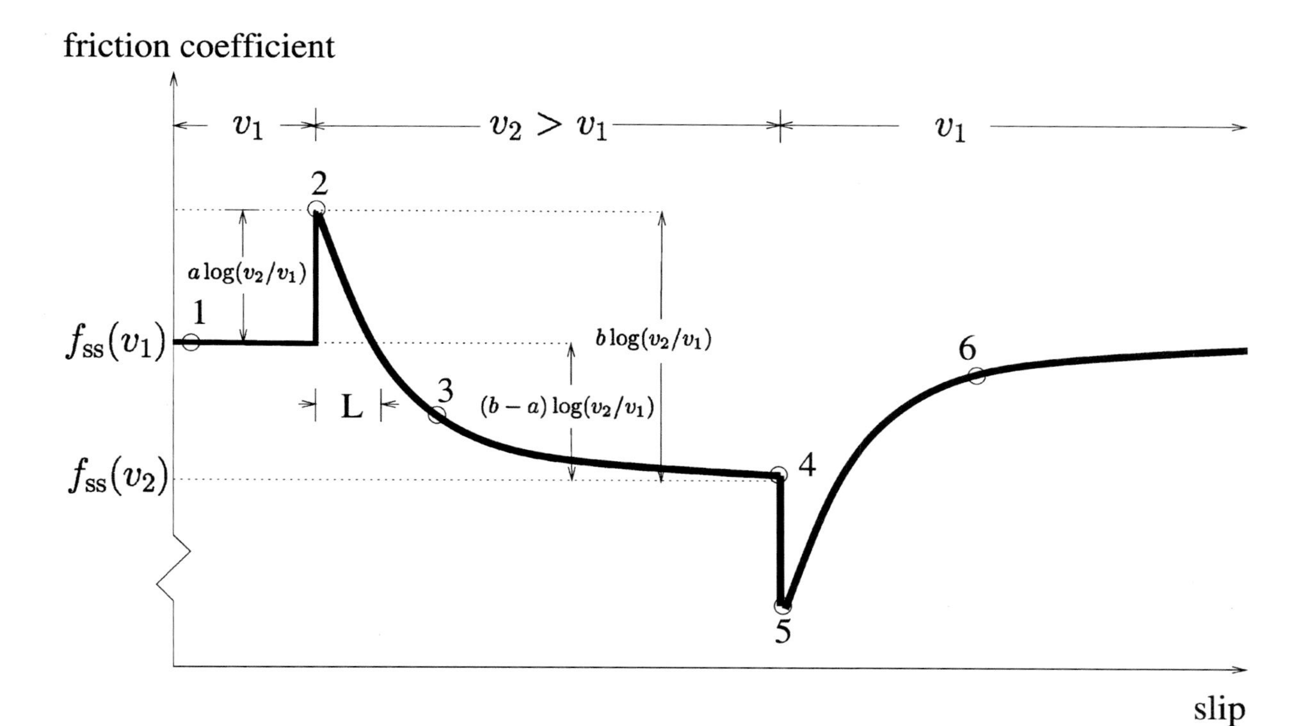

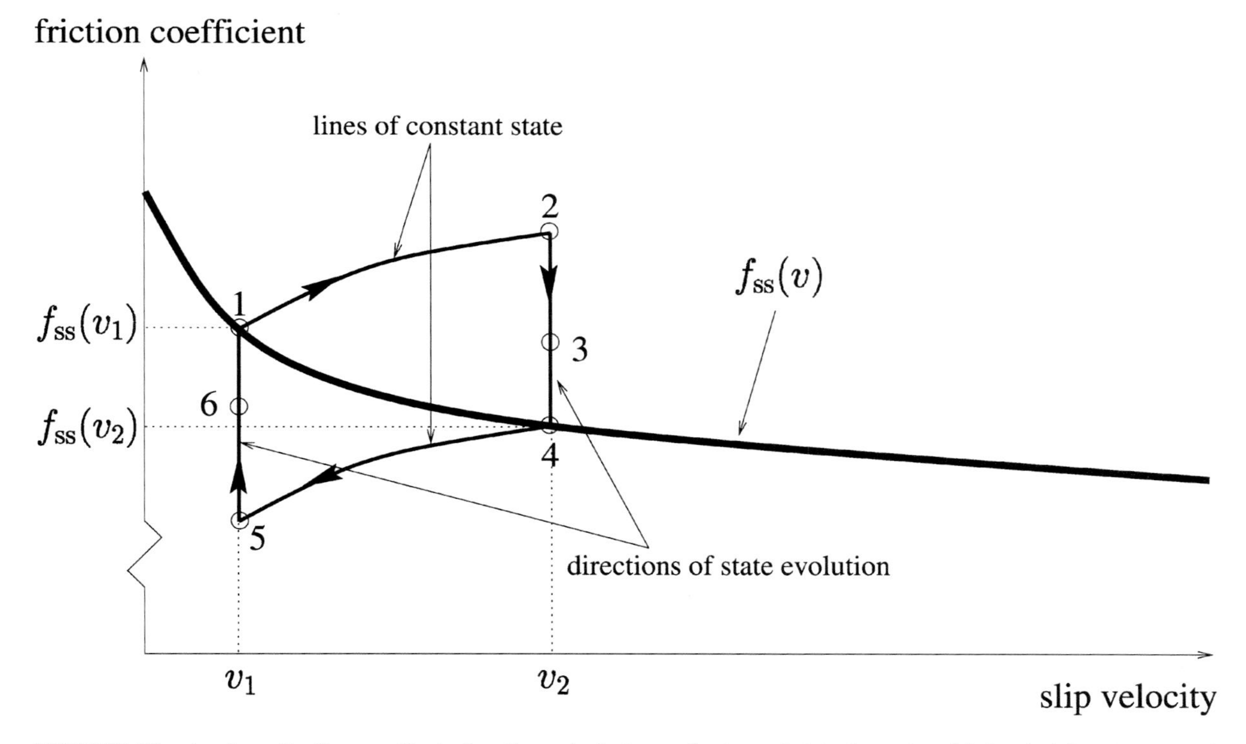

FIGURE 12 A schematic diagram illustrating the main features of rate- and state-dependent friction in laboratory experiments with sliding velocities in the range 10^{-6}–10^{-3} m/s. The top panel shows the response to velocity jumps. The bottom panel shows the steady-state frictional behavior in a velocity-weakening regime and evolution around the steady-state behavior corresponding to the velocity jumps in the top panel.

For the "slip" version (5.4c), the friction coefficient following a step change of sliding velocity from v_1 to v_2 can be written in the slip-weakening form

$$f = f_{ss}(v_2) + \exp(-\delta/L)[f_{ss}(v_1) - f_{ss}(v_2) + a\ln(v_2/v_1)] \tag{5.4e}$$

where $\delta = v_2 t$ is measured from the time of the velocity jump.

Expression (5.4a) as written is not appropriate for very low and very high values of sliding velocities. A simple regularization near $v = 0$, motivated by Arrhenius thermal activation of creep at asperity contacts, is to invert the equation to an exponential form for v and then replace $\exp(f/a)$ by $2\sinh(f/a)$. This leads to

$$f = a\sinh^{-1}\{(v/2v_0)\exp[(f_0 + b\ln(v_0\theta/L))/a]\}. \tag{5.4f}$$

In addition, v should be replaced by $|v|$ in (5.4b) and (5.4c) to allow velocity of either sign, and there should be an upper limit cutoff at some high slip velocity.

If $a > b$, the overall change of f with increasing sliding velocity is positive, the friction is velocity-strengthening, and only stable sliding is possible. On the other hand, if $a < b$, the overall change is negative, the friction is velocity-weakening, and dynamic instabilities can occur. The situation $a < b$ provides a necessary but not a sufficient condition for instability. The occurrence of instability requires a rate of weakening that is larger than the rate of stress reduction (stiffness of the system) on a slipping patch. For a given material under fixed pressure, temperature, surface roughness, and other relevant conditions, the rate of weakening is constant and is given approximately by $(b - a)\sigma_n$. The rate of stress reduction depends on the size of the zone that is slipping coherently. Infinitesimally small slip patches are infinitely stiff and are always stable, but as a slipping zone grows larger, its stiffness decreases and it can turn unstable. This happens when the slip patch reaches a critical "nucleation" size for which the rate of stress reduction is first equal (from above) to the rate of weakening. The nucleation zone size for a failure process governed by rate- and state-dependent friction is

$$h^* = C\mu L/[(b - a)\sigma_n] \tag{5.4g}$$

where C is a dimensionless constant of order 1. The nucleation size, at which there is a transition from aseismic slip to dynamic rupture, can be used to obtain an estimate for a minimum earthquake size. Observed values of L with sliding velocities in the range (10^{-8}–10^{-1}) m/s are generally somewhat smaller [13, 38] than those associated with D_c values of slip-weakening experiments. Observed values of $(b - a)$ in pressure-temperature conditions corresponding to the brittle seismogenic zone are about 0.02. The scalar potency release associated with the nucleation process can be calculated roughly by using $\mu/\sigma_n \approx 300$ to estimate h^* and multiplying the area of a circular patch with the obtained h^* by observed values of L (or D_c). Converting the result to an earthquake magnitude using empirical moment-magnitude or potency-magnitude scaling relations (see section 6) gives a range of minimum earthquake magnitude centered around -3.

6. Earthquake Source Parameters and Scaling Relations

([1], [2], [6], [8], [11], [14], [16], [19], [20], [22], [25], [26], [31], [35], [36], [39], [48], [50], [52], [53])

Faulting is associated with nonlinear inelastic deformation, intricate energy partition, evolving material properties, and other complexities. In general, faulting under natural conditions is inaccessible for direct observations, and earthquake source parameters are typically estimated from inversions of seismic and geodetic data in the far-field. This necessarily parameterizes the source process in terms of equivalent deformation in a linear elastic solid surrounding the inelastically deforming regions. Geological observations, lab studies, and measurements in mines provide limited direct information on the faulting process proper, although typically for conditions far removed from those operating in natural tectonic faults at seismogenic depths (e.g., 7.5 km for continental strike-slip faults).

The most common form of earthquake data consists of seismic catalogs that typically list the time, location, and magnitude M of earthquakes in a given space-time domain. Instead of (or in addition to) M, some catalogs list the scalar seismic moment M_0, which gives (as does the scalar potency P_0) a better physical characterization for the overall size of an earthquake source. The scalar moment and potency are typically derived from the zero-frequency asymptotes of far-field displacement spectra (see equations (3.5b) and (3.8), Fig. 7, and related results in sections 2 and 3). Additional important parameters that augment the information contained in the scalar moment and potency are radiated seismic energy E_R, stress drop $\Delta\sigma$, fracture energy G_c, rupture velocity V_r, and directivity.

With a proper distribution of stations around the fault, it is possible to derive fault-plane solutions from observed earthquake seismograms, and this has been done for many thousands of earthquakes (see equation (3.7b) and related material). Fault-plane solutions provide information on the strike, dip, and slip angles of the earthquake rupture (Fig. 13) and directions of maximum (P) and minimum (T) compressive principal stresses, with an ambiguity of an auxiliary set of quantities. Independent constraints from field observations, aftershock locations, and other information can be used, when available, to separate the earthquake fault plane and associated set of quantities from the auxiliary set. In cases of well-recorded earthquakes, seismic data can be inverted (usually with dislocation-based models) to provide detailed images of earthquake slip histories (typically so far with a resolution of about 3 km). At present, such slip models have been derived for several tens of earthquakes [36, 53].

Observed far-field displacement spectra can be fitted, after corrections to remove propagation and recording-site effects, by

$$\Omega(\omega) = P_0/[1 + (\omega/\omega_0)^{-\gamma}] \tag{6.1}$$

where P_0 is the scalar seismic potency, ω_0 is corner frequency, and observed values of the exponent γ typically fall in the range 1–3. A finite total radiated energy implies that for high-enough frequency $\gamma > 1.5$. The symbol denoting the corner frequency here is different from that used in the context of equation (3.8) because estimated values of the corner frequency obtained by fitting spectra to (6.1) differ in general from values obtained by the intersection of the low- and high-frequency spectral asymptotes. This illustrates the more general point that inferred values of seismological parameters often depend strongly on the estimation procedure.

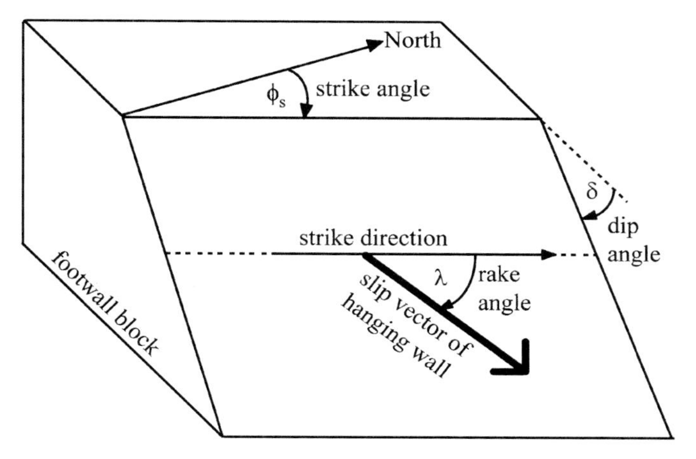

FIGURE 13 Fault and slip parameters. The strike $0 \leq \phi_s < 2\pi$ is the horizontal azimuth of the fault measured clockwise from the north. The dip $0 \leq \delta \leq \pi/2$ is the angle from the horizonal to the fault surface. The rake $-\pi \leq \lambda \leq \pi$ is the angle between the strike and slip direction of the hanging wall.

The static stress drop is defined as

$$\Delta\sigma_{static} = \sigma^0 - \sigma^f \tag{6.2a}$$

where σ^0 and σ^f are the initial and final stress values before and after the earthquake. Similarly, the dynamic stress drop may be defined as

$$\Delta\sigma_{dynamic} = \sigma^0 - \sigma^{dyn} \tag{6.2b}$$

where σ^{dyn} is a representative value of stress, such as average or minimum, during the active portion of dynamic slip (Fig. 14).

The average stress operating on the fault by the surrounding medium during an earthquake is

$$\bar{\sigma} = (\sigma^0 + \sigma^f)/2. \tag{6.2c}$$

Expression (6.2c) is usually used to denote the average stress on a fault area rather than at a point, in which case σ^0 and σ^f should be interpreted as spatial averages along the failure surface.

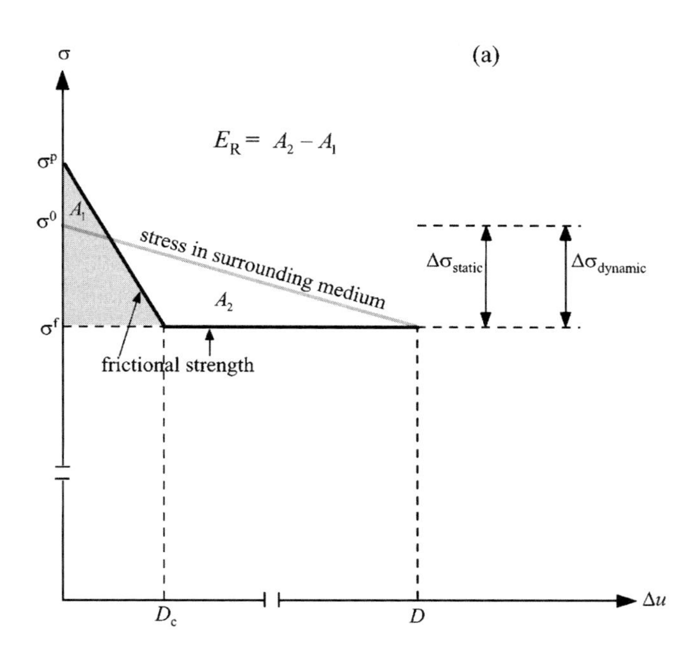

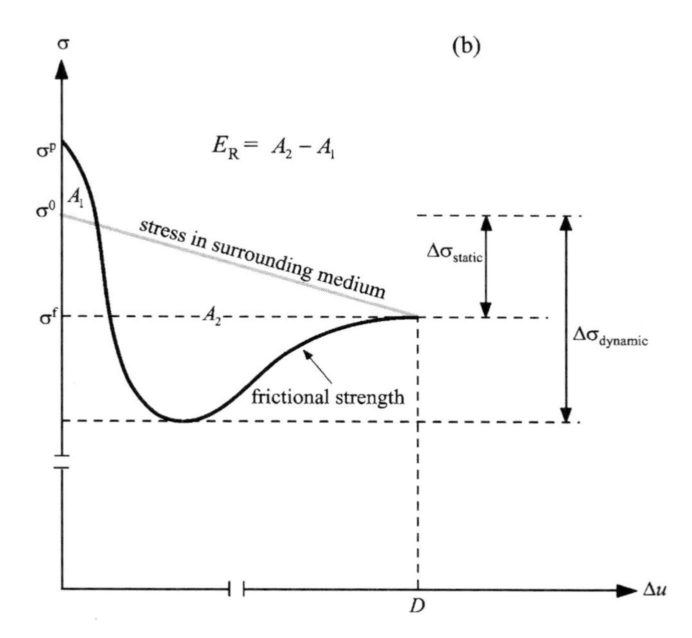

FIGURE 14 Components of energy changes associated with motion on a fault surface during an earthquake without (a) and with (b) a dynamic stress overshoot. The initial, peak, and final stress levels are denoted with σ^0, σ^p, and σ^f, respectively. The final slip is marked with D. The gray line represents the stress operating on the fault by the surrounding medium, and the black curve represents the frictional strength during slip. The radiated seismic energy E_R is the difference $A_2 - A_1$ between the areas below the gray and black lines. The area under the black curve gives the frictional heat plus fracture energy. The shaded area and D_c in (a) correspond to fracture energy G_c and critical slip-weakening distance, respectively.

The change of energy generated by an earthquake slip is

$$-\Delta E = \bar{\sigma}\,\langle \Delta u \rangle\, S = \bar{\sigma}(M_0/\mu) = \bar{\sigma}\,P_0 \quad (6.3)$$

where $\langle \Delta u \rangle$ and S are average slip and rupture area, respectively. The energy reduction in (6.3) involves changes of elastic strain energy, gravitational energy, and rotational energy of the Earth [11] and is partitioned among heat, fracture energy (which is a form of latent heat), and seismic radiation (see Fig. 14). The radiated seismic energy E_R can be estimated by integrating velocity seismograms with proper corrections for radiation pattern, attenuation, and other propagation and recording-site effects.

The apparent stress is defined as

$$\tau_a = E_R/(M_0/\mu) = E_R/P_0. \quad (6.4)$$

The seismic efficiency η is defined as

$$\eta = E_R/|\Delta E| = \tau_a/\bar{\sigma}. \quad (6.5)$$

The radiated seismic energy and surface magnitude of earthquakes in the magnitude range $5 \lesssim M_S \lesssim 8$ are related via the empirical Gutenberg-Richter relation

$$\log_{10} E_R = 1.5 M_S + 11.8 \quad (6.6)$$

where E_R is in erg (= dyne cm = 10^{-7} J).

The scalar seismic moment and magnitude of earthquakes with $M \gtrsim 3.5$ are related via the empirical relation

$$\log_{10} M_0 = 1.5M + 16.1 \quad (6.7a)$$

where M_0 is in dyne cm (10^{-7} J). Empirical scaling relations between moment or potency and magnitude over a broad magnitude range with single smooth lines require a quadratic term [6, 20]. For example, the scalar seismic potency and local magnitude of California earthquakes in the range $1.0 \lesssim M_L \lesssim 7.0$ are related via the empirical quadratic relation

$$\log_{10} P_0 = 0.06 M_L^2 + 0.98 M_L - 4.87 \quad (6.7b)$$

where P_0 is in km^2 cm.

For a classical crack sustaining a uniform stress drop over a failure area S

$$M_0 = c\,\Delta\sigma_{static} S^{3/2} \quad (6.8)$$

where c is a dimensionless constant that depends on the failure geometry and elastic properties. For example, $c = 16/(7\pi^{3/2}) \approx 0.41$ for a circular crack in an infinite Poissonian solid.

For a fractal-like failure in a rough stress field [16]

$$M_0 \propto S. \quad (6.9)$$

Equation (6.8) is used often to estimate the static stress drop from inferred values of M_0 and S. The obtained values, and their physical interpretation, depend on the methods used to estimate M_0 and S. The moment is typically derived from seismograms, but sometimes it is obtained from field values or geodetic data. The area is inferred (typically with large uncertainties) from directivity effects, pulse duration, and corner frequency in seismograms (see Fig. 7, equation (6.1), and related material), as well as from aftershock locations, geological observations, and geodetic data. The corner frequency can be estimated in principle from most seismograms (although the obtained values depend strongly, as mentioned before, on the data type and estimation procedure) and, hence, is used frequently. The large uncertainties in inferred values of S produce large uncertainties in estimates of $\Delta\sigma_{static}$.

Measurements of root-mean-square acceleration in seismograms and assumptions on a source model can be used to obtain a type of dynamic stress drop that may be referred to as $\Delta\sigma_{a_{rms}}$. Another type of dynamic stress drop may be obtained from comparing the initial slope of velocity seismograms to analytical results like equation (4.6b).

Inferred values of static and dynamic stress drops averaged over the failure area are usually in the range 10^{-2}–10^2 MPa [8, 25]. Inferred average values of apparent stress typically fall in the same range, and estimates of average seismic efficiency are about 0.06 or less [39].

7. Seismicity Patterns

([9], [16], [18], [19], [23], [24], [27], [40], [56], [58], [59], [60], [62], [63])

Seismicity exhibits a wide variety of fluctuations and patterns in space, time, and energy (or magnitude) domains. At present, analysis of seismicity is largely phenomenological with little theoretical foundation, and many studies of seismicity are essentially descriptive without quantitative formulation. Examples of reported patterns include foreshocks, aftershocks, time intervals of quiescence and accelerated seismic release, changes in b values of frequency-magnitude statistics, migration of seismicity along and between faults, switching of activity on a given fault between different modes of response, and other types of spatiotemporal clustering, periodicities, and gaps. Many functions have been employed to describe statistical aspects of seismicity including power law, exponential, normal, lognormal, Gamma, Weibull, Pareto, and Cauchy distributions. Formulas and properties of these functions can be found in *http://mathworld.wolfram.com* (see also [23] and [59]).

It is important to distinguish between regional seismicity patterns characterizing large spatial domains with many faults and patterns characterizing individual fault systems. Since instrumental earthquake catalogs exist only for short duration (e.g., 50–100 y) compared to recurrence times of large earthquakes (e.g., 100–5000 y), most observational studies of seismicity have focused on regional patterns for which more data are available. Recently some works examined earthquake patterns on individual faults by combining instrumental and geological data (see, e.g., [19] and [63]). In general, regional seismicity appears to be dominated by various forms of spatiotemporal clustering, while patterns associated with large individual faults (or fault segments) may include spatiotemporal periodicities such as quasi-periodic occurrence of system-size events and spatial regularity of microearthquake locations.

The frequency-moment statistics of regional earthquakes follow the power-law probability density function

$$n(M_0) \propto M_0^{-1-\beta} \tag{7.1a}$$

and corresponding power-law survivor function

$$N(M_0) \propto M_0^{-\beta} \tag{7.1b}$$

where $N(M_0) = \int_{M_0}^{\infty} n(M_0')\, dM_0'$.

Similar power-law relations hold for frequency-energy statistics of regional earthquakes. A maximum event size can be incorporated into (7.1a) and (7.1b) by multiplying the right sides of the equations with the exponential tapering function $\exp(-M_0/\hat{M}_0)$ having a corner moment $\hat{M}_0$ that characterizes the finite size effects. Frequency-moment distributions consisting of an initial power law for small and intermediate size events and exponential taper for large ones have been derived using several theoretical frameworks, including critical branching process, critical phase transition, and maximum entropy arguments.

Using the moment-magnitude relation (6.7a) in $n(M_0)dM_0$ of (7.1a) and $N(M_0)$ of (7.1b) leads to the discrete Gutenberg-Richter frequency-magnitude statistics

$$\log n(M) = a - bM \tag{7.2a}$$

and corresponding cumulative distribution

$$\log N(M) = A - bM \tag{7.2b}$$

where $b = 1.5\beta$ and observed b values of regional seismicity typically fall in the range 0.7–1.3. Observed frequency-moment and frequency-magnitude statistics of regional earthquakes are also analyzed with a tapered Pareto and other distributions [24, 58].

The frequency-magnitude statistics of earthquakes in large individual fault-systems, occupying narrow and long spatial domains, often consist of a Gutenberg-Richter type distribution of small events combined with enhanced statistics around a larger "characteristic" earthquake. This may be described empirically by a superposition (Fig. 15) of two separate populations, one following Gutenberg-Richter statistics over the magnitude interval $M_c < M < M_1$, and the other a Gaussian distribution centered on event size $M_2 > M_1$. Such statistics can be written as

$$\begin{aligned}\log[n(M)+1] = {} & (a - bM)H(M - M_c)H(M_1 - M) \\ & + c\exp[-(M - M_2)^2/2\sigma^2] \\ & \times H(\Delta M - |M - M_2|)\end{aligned} \tag{7.3}$$

where H is the unit step function and c is a normalization factor. If $|M_2 - M_1| > \Delta M$ as in Fig. 15, equation (7.3) has no events in the magnitude gap $M_1 < M < M_2 - \Delta M$. The distribution is assumed to describe data collected over sufficient duration so that $n + 1 \approx n$ for $M_c < M < M_1$ and $|M - M_2| < \Delta M$. In applications to observed data, c should be determined from the number of events associated with the "characteristic bump" around M_2 before estimating the other parameters of the Gaussian distribution. In some cases it may be possible to fix the value of M_2 based on independent geological data. It is also possible to replace the abrupt truncations in (7.3) with smooth tapering and to replace the Gaussian distribution with other peaked functions. One example is the symmetric rescaled Beta function $c(M + \Delta M - M_2)^{\alpha-1}(M_2 + \Delta M - M)^{\alpha-1}$ with $\alpha > 1$.

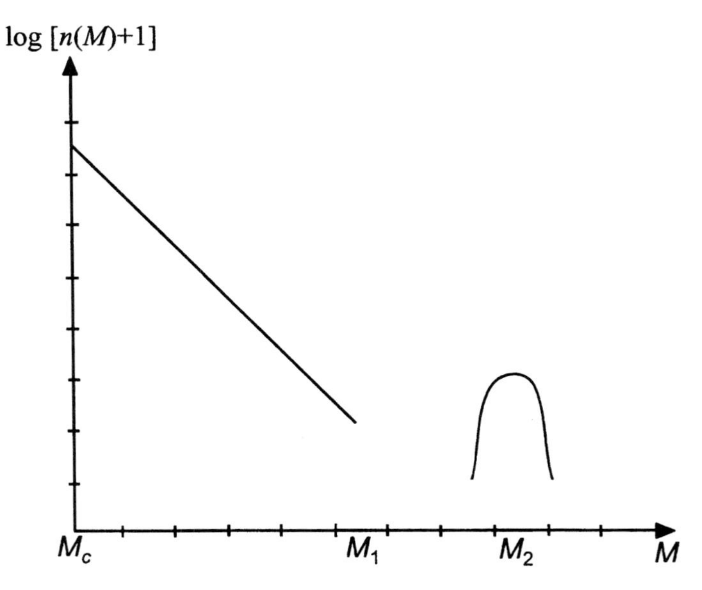

FIGURE 15 A schematic representation of discrete frequency-magnitude statistics of earthquakes on individual fault systems with two separate populations: a Gutenberg-Richter distribution of small events and a peaked Gaussian-like distribution around a large characteristic event size.

Aftershock decay rates can usually be described by the modified Omori law

$$\Delta N/\Delta t = K(t + c)^{-p} \tag{7.4}$$

where N is cumulative number of events, t is time after the mainshock, and observed values of the exponent p typically fall in the range 0.7–1.5. A finite total number of events implies that for large enough time $p > 1$.

The Epidemic-type Aftershock-sequences (ETAS) model combines the modified Omori law with the Gutenberg-Richter frequency-magnitude relation for a history-dependent occurrence rate of a point process in the form

$$\lambda(t|H_t) = \mu + \sum_{t_i < t} \frac{K_0 \exp[\alpha(M_i - M_c)]}{(t - t_i + c)^p} \tag{7.5}$$

where μ is a constant background rate, M_i is the magnitude of earthquake at time t_i, M_c is a lower magnitude cutoff, H_t denotes the history, and the factor $K_0 \exp[\alpha(M_i - M_c)]$ gives the number of events triggered by a parent earthquake with magnitude M_i.

Large earthquakes are sometimes preceded by a period of accelerated seismic activity in a broad surrounding region of dimension that generally scales with that of the large event. During such activation periods, several functions of various seismicity parameters (e.g., number N and moment M_0) can be fitted by a number of functional forms. One example is the power law time-to-failure relation of cumulative Benioff strain

$$\sum_{t_i<t} M_0^{1/2}(t_i) = A + B(t_f - t)^m \tag{7.6}$$

where t is time, t_f is failure time of the large event terminating the phase of accelerated seismic release, and observed values of the exponent m typically fall in the range 0.2–0.4.

Acknowledgments

The manuscript benefited greatly from many useful comments by Ralph Archuleta, Rafael Benites, Matthias Holschneider, Vladimir Lyakhovsky, Art McGarr, Takeshi Mikumo, Jim Rice, Paul Richards, Jim Savage, David Vere-Jones, Gert Zöller, and editor Willie Lee. I thank Zhigang Peng and Ory Dor for help with the text and figures preparation.

References

[1] Aki, K., and P.G. Richards (2002). *Quantitative Seismology* (second edition). University Science Books.
[2] Anderson, J.G. (2002). Strong-Motion Seismology. Chapter 57 of this Handbook.
[3] Ben-Menahem, A., and S.J. Singh (1981). *Seismic Waves and Sources*. Springer-Verlag.
[4] Ben-Zion, Y. (1989). The response of two joined quarter spaces to SH line sources located at the material discontinuity interface. *Geophys. J. Int.* **98**, 213–222.
[5] Ben-Zion, Y. (2001). Dynamic rupture in recent models of earthquake faults. *J. Mech. Phys. Solids* **49**, 2209–2244.
[6] Ben-Zion Y., and L. Zhu (2002). Potency-magnitude scaling relations for southern California earthquakes with $1.0 < M_L < 7.0$. *Geophys. J. Int.* **148**, F1–F5.
[7] Broberg, K.B. (1999). *Cracks and Fracture*. Academic Press.
[8] Brune, J.N., and W. Thatcher (2002). Strength and Energetics of Active Fault Zones. Chapter 35 of this Handbook.
[9] Bufe, C.G., and D.J. Varnes (1993). Predictive modeling of the seismic cycle of the greater San Francisco bay region. *J. Geophys. Res.* **98**, 9871–9883.
[10] Chapman, C.H. (2002). Seismic Ray Theory and Finite Frequency Extensions. Chapter 9 of this Handbook.
[11] Dahlen, F.A. (1977). The balance of energy in earthquake faulting. *Geophys. J. R. Astr. Soc* **48**, 239–261.
[12] Dahlen, F.A., and J. Tromp (1998). *Theoretical Global Seismology*. Princeton University Press.
[13] Dieterich, J.H., and B. Kilgore (1996). Implications of fault constitutive properties for earthquake prediction. *Proc. Natl. Acad. Sci. U.S.A.* **93**, 3787–3794.
[14] Feigl, K.L. (2002). Estimating Earthquake Source Parameters from Geodetic Measurements. Chapter 37 of this Handbook.
[15] Fineberg, J., and M. Marder (1999). Instability in dynamic fracture. *Phys. Reports* **313**, 1–108.
[16] Fisher, D.S., K. Dahmen, S. Ramanathan, and Y. Ben-Zion (1997). Statistics of earthquakes in simple models of heterogeneous faults. *Phys. Rev. Lett.* **78**, 4885–4888.
[17] Freund, L.B. (1990). *Dynamic Fracture Mechanics*. Cambridge University Press.
[18] Frohlich, C., and S.D. Davis (1993). Teleseismic b values; or, much ado about 1.0. *J. Geophys. Res.* **98**, 631–644.
[19] Grant, L.B. (2002). Paleoseismology. Chapter 30 of this Handbook.
[20] Hanks, T.C., and D.M. Boore (1984). Moment-magnitude relations in theory and practice. *J. Geophys. Res.* **89**, 6229–6235.
[21] Heaton, H.T., and R.E. Heaton, (1989). Static deformation from point forces and force couples located in welded elastic Poissonian half-spaces: Implications for seismic moment tensors. *Bull. Seism. Soc. Am.* **79**, 813–841.
[22] Johnston, M.J.S., and A.T. Linde (2002). Implications of Crustal Strains during Conventional, Slow, and Silent Earthquakes. Chapter 36 of this Handbook.
[23] Kagan, Y.Y. (1994). Observational evidence for earthquakes as a nonlinear dynamic process. *Physica D.* **77**, 160–192.
[24] Kagan, Y.Y. (2002). Seismic moment distribution revisited: I. Statistical results. *Geophys. J. Int.* **148**, 520–541.
[25] Kanamori, H. (1994). Mechanics of earthquakes. *Annu. Rev. Earth Planet. Sci.* **22**, 207–237.
[26] Kasahara, K. (1981). *Earthquake Mechanics*. Cambridge University Press.
[27] Kisslinger, C. (1996). Aftershocks and fault-zone properties. *Adv. Geophys.* **38**, 1–36.
[28] Kostrov, B.V., and S. Das (1988). *Principles of Earthquake Source Mechanics*. Cambridge University Press.
[29] Lawn, B. (1993). *Fracture of Brittle Solids* (2nd edition). Cambridge University Press.
[30] Lay, T., and T.C. Wallace (1995). *Modern Global Seismology*. Academic University Press.
[31] Li, V.C. (1987). Mechanics of Shear Rupture Applied to Earthquake Zones. In: *Fracture Mechanics of Rock* (B.K. Atkinson, Ed.), pp. 351–428. Academic Press.
[32] Lockner, D.A., and N.M. Beeler (2002). Rock Failure and Earthquakes. Chapter 31 of this Handbook.
[33] Lognonné, P., and E. Clévédé (2002). Normal Modes of the Earth and Planets. Chapter 10 of this Handbook.
[34] Madariaga, R. (1979). On the relation between seismic moment and stress drop in the presence of stress and strength heterogeneity. *J. Geophys. Res.* **84**, 2243–2250.
[35] Madariaga, R., and K.B. Olsen (2002). Earthquake Dynamics. Chapter 12 of this Handbook.
[36] Mai, P.M., and G.C. Beroza (2002). A spatial random field model to characterize complexity in earthquake slip. *J. Geophys. Res.*, **107**, 2308, doi:10.1029/2001/JB000588.
[37] Malvern, L.E. (1969). *Introduction to the Mechanics of a Continuous Medium*. Prentice Hall, Inc.
[38] Marone, C. (1998). Laboratory-derived friction laws and their application to seismic faulting. *Annu. Rev. Earth Planet. Sci.* **26**, 643–649.

[39] McGarr, A. (1999). On relating apparent stress to the stress causing earthquake fault slip. *J. Geophys. Res.* **104**, 3003–3011.
[40] Ogata, Y. (1999). Seismicity analysis through point-process modeling: A review. *Pure Appl. Geophys.* **155**, 471–507.
[41] Ohnaka, M. (2003). A constitutive scaling law and a unified comprehension for frictional slip failure, shear fracture of intact rock, and earthquake rupture. *J. Geophys. Res.* 108(B2), 2080, doi:10.1029/2000JB000123.
[42] Rice, J.R. (1980). The Mechanics of Earthquake Rupture. In: *Physics of the Earth's Interior* (A.M. Dziewonski and E. Boschi, Eds.), pp. 555–649. Italian Physical Society/North Holland, Amsterdam.
[43] Rice, J.R. (1993). Mechanics of Solids, section of the article "Mechanics." In: *Encyclopeadia Britannica*, 1993 printing of the 15th edition, vol. **23**, pp. 734–747 and 773.
[44] Rice, J.R. (2001). New Perspectives on Crack and Fault Dynamics. In: *Mechanics for a New Millennium* (H. Aref and J.W. Phillips, Eds.), pp. 1–23, Kluwer Academic Publications.
[45] Rice, J.R., N. Lapusta, and K. Ranjith (2001). Rate and state dependent friction and the stability of sliding between elastically deformable solids. *J. Mech. Phys. Solids* **49**, 1865–1898.
[46] Rosakis, A. (2002). Intersonic shear cracks and fault ruptures. *Advances Phys.* **51**, 1189–1257.
[47] Rudnicki, J.W. (1980). Fracture mechanics applied to the earth's crust. *Ann. Rev. Earth Planet. Sci.* **8**, 489–525.
[48] Ruff, L.J. (2002). State of Stress Within the Earth. Chapter 33 of this Handbook.
[49] Sato, H., M. Fehler, and R.-S. Wu (2002). Scattering and Attenuation of Seismic Waves in the Lithosphere. Chapter 13 of this Handbook.
[50] Scholz, C.H. (2002). *The Mechanics of Earthquakes and Faulting*. Cambridge University Press.
[51] Shearer, P. (1999). *Introduction to Seismology*. Cambridge University Press.
[52] Sibson, R.H. (2002). Geology of the Crustal Earthquake Source. Chapter 29 of this Handbook.
[53] Somerville, P.G., K. Irikura, R. Graves, S. Sawada, D.J. Wald, N. Abrahmson, Y. Iwasaki, T. Kagawa, N. Smith, and A. Kowada (1999). Characterizing crustal earthquake slip models for the prediction of strong ground motion. *Seismol. Res. Lett.* **70**, 59–80.
[54] Tada, H., P.C. Paris, and G.R. Irwin (1985). *The Stress Analysis of Cracks Handbook*, 2nd edition. Paris Productions Incorporated (and Del Research Corporation).
[55] Teisseyre, R., and E. Majewski (2002). Physics of Earthquakes. Chapter 15 of this Handbook.
[56] Turcotte, D.L., and B.D. Malamud (2002). Earthquakes as a Complex System. Chapter 14 of this Handbook.
[57] Udias, A. (2002). Theoretical Seismology: An Introduction. Chapter 8 of this Handbook.
[58] Utsu, T. (1999). Representation and analysis of the earthquake size distribution: A historical review and some new approaches. *Pure Appl. Geophys.* **155**, 509–535.
[59] Utsu, T. (2002). Statistical Features of Seismicity. Chapter 43 of this Handbook.
[60] Utsu T., Y. Ogata, and R.S. Matsu'ura (1995). The centenary of the Omori Formula for a decay law of aftershock activity. *J. Phys. Earth* **43**, 1–33.
[61] Varatharajulu, V., and Y.-H. Pao (1976). Scattering matrix for elastic waves. I. Theory. *J. Acoust. Soc. Am.* **60**, 556–566.
[62] Vere-Jones, D. (1994). Statistical Models for Earthquake Occurrence: Clusters, Cycles and Characteristic Earthquakes. In: *Proc. first US/Japan conference on the frontiers of statistical modeling: An informational approach*, pp. 105–136. Kluwer.
[63] Wesnousky, S.G. (1994). The Gutenberg-Richter or characteristic earthquake distribution, which is it? *Bull. Seismol. Soc. Amer.* **84**, 1940–1959.
[64] Wieland, E. (2002). Seismometry. Chapter 18 of this Handbook.

Appendix 3

Name Index of the Contributors of Biographical Sketches

compiled by William H. K. Lee

This Index contains the names of over 2100 contributors of biographical sketches. These biographical sketches were submitted by the contributors in their national or institutional reports (see Chapters 79 and 80 of this Handbook), or directly to the editors, over a period of about four years. Therefore, most of them contain information as of the end of the 20th century. The submitted biographical sketches were formatted to a more uniform style, and were condensed, if necessary, to fit within one printed page. Each biography usually includes some personal information, education, major positions held, research interests or contributions, and a list of selected publications. The edited biographical sketches are archived in a computer-searchable file on the attached Handbook CD#3, under the directory of \BioSketches. An index of the contributors' names is given below in alphabetic order by their family names.

INTERNATIONAL HANDBOOK OF EARTHQUAKE AND ENGINEERING SEISMOLOGY, VOLUME 81B ISBN: 0-12-440658-0

B

C

D

J

K

L

N

O

P

Q

R

S

T

U

V

Z

Appendix 4

User's Manual to the Attached CD-ROMs

Michael F. Diggles
U.S. Geological Survey, Menlo Park, California, USA

1. Introduction

The *International Handbook of Earthquake and Engineering Seismology*, in two parts, contains three CD-ROMs. The printed book, Part A, contains the disc IASPEI_Handbook_CD1. The printed book, Part B (this volume), contains two discs, IASPEI_Handbook_CD2 and IASPEI_Handbook_CD3.

The materials on IASPEI_Handbook_CD1 contain supplemental information for Chapters 1–56, and are divided into six categories: (1) full manuscripts of which only abridged versions are included in Part A due to space limitations, (2) large data sets for further data processing and analysis, (3) full references of which only abridged versions are included in Part A, (4) appendices and figures, (5) out-of-print books presented in computer-readable files, and (6) graphics files used to create the Handbook Part A, archived as ZIP files in the Graphics_Archive folder. However, some chapters do not have materials for archiving.

The materials on IASPEI_Handbook_CD2 and IASPEI_Handbook_CD3 contain supplemental information for Chapters 60–90 (please note that some chapters do not have materials for archiving) and Appendix 3. As in IASPEI_Handbook_CD1, the graphics used to create Handbook Part B are provided in the Graphics_Archive2 folder.

The folders are given in the order of the Handbook chapter in the form of: cAn, where c is the Chapter number, and A is the first author. If an author has more than one chapter, then n is numbered sequentially from 1; otherwise no number is given. Exceptions to the folder names are 79 and 80, which use country names, 81, which uses organization acronyms, and 85, which is software. Some folder and file names may differ from what is cited in the paper volume (e.g., substitution of underscores for dots and spaces; shorter filenames) in order to ensure cross-platform compatibility.

2. Portable Document Format (PDF) Files

The CD-ROMs make extensive use of Adobe Acrobat Reader. Using Adobe Acrobat Reader, you can view, search, and print the PDF files on this CD-ROM. Many PDFs open with displayed "bookmarks" on the left side of the screen that allow you to choose from among the sections of the text and the figures.

The Acrobat folder on IASPEI_Handbook_CD1 contains installers for Adobe Acrobat Reader 5.0.5 for both Windows (PC folder) and Macintosh (Mac folder). You can use the installers provided on that disc, or download the latest version (5.1 at publication time) of Adobe Acrobat Reader at no cost via the World Wide Web from the Adobe home page (http://www.adobe.com/products/acrobat/readstep.html).

Each of the three CD-ROMs contains a full-text index (index.pdx and associated files in the index folder) that is used for searching for words or sets of words in the PDF files on the discs, using the Search tool in Acrobat Reader. To make best use of the CD-ROMs, you will need to develop some familiarity with Acrobat Reader; an on-line Acrobat Reader User's Guide is available in Acrobat Reader under the "Help" menu.

Tip: If your copy of Acrobat Reader 5 has "Open Cross-Document Links In Same Window" ("Edit" menu under "Preferences," "General...", "Options") selected, you should **deselect** it. This will keep the main document open while you open and close other PDF files. In older versions of Acrobat, this menu item is in "File" "Preferences" "General..."

INTERNATIONAL HANDBOOK OF EARTHQUAKE AND ENGINEERING SEISMOLOGY, VOLUME 81B *ISBN: 0-12-440658-0*

3. Explanation of the Graphics_Archive and Graphics_Archive2 Folders

If a chapter in Part A or Part B of the Handbook has figure(s), then the figure file(s) are included in the Graphics_Archive folders in their respective CD-ROMs in a chapter folder as a ZIP file, AAnnFigs.zip, where AA is the senior author's name, nn is a sequential number if the author has more than one chapter (otherwise, it is omitted). The figure files are given in various formats as submitted by the authors or edited by the editors. The zip files also contain the figure captions files in PDF format. Color plates are also provided in their own archive folder, i.e., \ColorPlates.

4. System Requirements

The CD-ROMs were produced in accordance with the ISO 9660 Level 2 standard and Apple Computer's hierarchical file system standard. The data and text on this CD-ROM require either a UNIX system-based or Linux workstation, Macintosh or compatible computer, or an IBM or compatible personal computer, all equipped with a CD-ROM drive and a color monitor that can display 256 colors (16.7 million colors recommended).

4.1 PC Platform

The PC should have

1. Intel Pentium or equivalent processor-based personal computer
2. Microsoft Windows 95 OSR 2.0, Windows 98 SE, Windows Millennium, Windows NT 4.0 with Service Pack 5, Windows 2000, or Windows XP
3. 64 MB of RAM.

4.2 Macintosh Platform

The Macintosh should have

1. PowerPC processor
2. Mac OS software version 8.6, 9.0.4, 9.1, 9.2, or OS X; some features of Acrobat 5.0.5 may not be available for OS 8.6 and OS X due to OS limitations.
3. 64 MB of RAM.

4.3 UNIX Platform

Almost any UNIX system-based or Linux workstation can read these files.

4.4 All Platforms

Adobe Acrobat Reader 5.0 or higher. The PDF files on this CD-ROM can be read with Acrobat 4 but without the re-wrap feature or tags that provide accessibility to sight-disabled people through the use of assistive technology such as screen readers.

5. To Get Started

On a PC system with Windows, IASPEI_Handbook_CD1 should open automatically. If not, double-click on My Computer, double-click on the CD-ROM icon, and double-click on the file 00_Start_Here.pdf to open it with Acrobat Reader.

On a Macintosh, double-click on the CD-ROM icon and double-click on the file 00_Start_Here.pdf to open it with Acrobat Reader.

On a UNIX or Linux workstation, mount the CD-ROM. If you are unsure about the process of mounting CD-ROMs on the workstation, see your system administrator for instructions. Once the CD-ROM is mounted, use Acrobat Reader to open the file 00_Start_Here.pdf.

6. Directories on Handbook CD#1

IASPEI_Handbook_CD1 has the following files and directories at the root level (please note that some chapters do not have materials for archiving on the CD):

00_README.TXT
00_Start_Here.pdf
01Agnew
04Adams
10Lognnone
13Sato
15Teisseyre
16Mosegaard
17Lee1
18Wielandt
21Kulhanek
22Scherbaum
25McNutt
30Grant
31Jackson
32Lockner
33Ruff
35Brune
36Johnston1
37Feigl
38Johnston2
40McGarr
41Engdahl
42Utsu1
43Utsu2
44Utsu3
47Guidoboni
48_1Lee2
48_2Toppozada
48_3Satyabala
48_4Usami
48_5Musson1
49Musson2
50Sipkin
51Lay
54Mooney
55Minshull
56Song
Graphics_Archive

7. Directories on Handbook CD#2

IASPEI_Handbook_CD2 has the following files and directories at the root level (please note that some chapters do not have materials for archiving on the CD):

00_README.TXT
00_Start_Here.pdf
60Campbell
68Borcherdt
74Giardini
7902Albania
7903Argentina
7904Armenia
7905Australia
7906Austria
7907Azerbaijan
7908Belgium
7909Bolivia
7910Brazil
7912Canada
7913Chile
7914China(Beijing)
7915China(Taipei)
7916Croatia
7917CzechRepublic
7919ElSalvador
7920Estonia
7921Ethiopia
7922France
7923Georgia
7924Germany
7925Greece
7926Hungary
7928India
7929Indonesia
7930Iran
7931Israel
7932Italy
7933Japan
7934Kazakhstan
7935Kyrgyzstan
7936Luxembourg
7937Macedonia
7938Mexico
7939Moldova
7940NewZealand
7941Norway
7942Philippines
7943Poland
7944Portugal
7945Russia
7946Slovakia
7947Slovenia
7948SouthAfrica
7949Spain
7950Sweden
7951Switzerland
7952Turkey
7953Ukraine
7954UnitedKingdom
7955USA
7956Uzbekistan
7957Vietnam
8002Austria
8003Bulgaria
8005Chile
8006China
8007Croatia
8008Georgia
8009India
8010Japan
8011Kazakhstan
8012Kyrgrzstan
8013Macedonia
8015Russia
8016USA
8017Uzbekistan
8104IHFC
8105COE
8107JWG
8108CERESIS
82VereJones
Graphics_Archive2

8. Directories on Handbook CD#3

IASPEI_Handbook_CD3 has the following files and directories at the root level (please note that some chapters do not have materials for archiving on the CD):

00_README.TXT
00_Start_Here.pdf
84Yang
85Software
87Lahr
88Schweitzer
89Howell
90Rymer
Appendix3

Index for Part A and Part B

Locators in **bold** indicate main discussion or chapter pages by an author; those in *italic* indicate table or figure. Locators followed by 'p' refer to a color plate, e.g., 13p refers to color plate 13. Materials archived on the attached CD-ROMs are shown by CD number and directory name. Titles of archived books or monographs on the attached CD-ROMs are shown in *italic*.

A

B

D

G

I

J

K

L

M

O

P

S

T

U

Y

Z

Contents of Part A

I. History and Prefatory Essays

II. Theoretical Seismology

III. Observational Seismology

IV. Earthquake Geology and Mechanics

V. Seismicity of the Earth

VI. Earth's Structure

International Geophysics Series

EDITED BY

RENATA DMOWSKA
Division of Applied Science
Harvard University
Cambridge, Massachusetts

JAMES R. HOLTON
Department of Atmospheric Sciences
University of Washington
Seattle, Washington

H. THOMAS ROSSBY
Graduate School of Oceanography
University of Rhode Island
Narragansett, Rhode Island

Volume 1 BENO GUTENBERG. Physics of the Earth's Interior. 1959*

Volume 2 JOSEPH W. CHAMBERLAIN. Physics of the Aurora and Airglow. 1961*

Volume 3 S. K. RUNCORN (ed.). Continental Drift. 1962*

Volume 4 C. E. JUNGE. Air Chemistry and Radioactivity. 1963*

Volume 5 ROBERT G. FLEAGLE AND JOOST A. BUSINGER. An Introduction to Atmospheric Physics. 1963*

Volume 6 L. DEFOUR AND R. DEFAY. Thermodynamics of Clouds. 1963*

Volume 7 H. U. ROLL. Physics of the Marine Atmosphere. 1965*

Volume 8 RICHARD A. CRAIG. The Upper Atmosphere: Meteorology and Physics. 1965*

Volume 9 WILLIS L. WEBB. Structure of the Stratosphere and Mesosphere. 1966*

Volume 10 MICHELE CAPUTO. The Gravity Field of the Earth from Classical and Modern Methods. 1967*

Volume 11 S. MATSUSHITA AND WALLACE H. CAMPBELL (eds.). Physics of Geomagnetic Phenomena (In two Volumes). 1967*

Volume 12 K. YA KONDRATYEV. Radiation in the Atmosphere. 1969*

Volume 13 E. PALMÅN AND C. W. NEWTON. Atmospheric Circulation Systems: Their Structure and Physical Interpretation. 1969*

Volume 14 HENRY RISHBETH AND OWEN K. GARRIOTT. Introduction to Ionospheric Physics. 1969*

Volume 15 C. S. RAMAGE. Monsoon Meteorology. 1971*

Volume 16 JAMES R. HOLTON. An Introduction to Dynamic Meteorology. 1972*

Volume 17 K. C. YEH AND C. H. LIU. Theory of Ionospheric Waves. 1972*

Volume 18 M. I. BUDYKO. Climate and Life. 1974*

*Out of print.
[NYP]Not yet published.

Volume 19 MELVIN E. STERN. Ocean Circulation Physics. 1975*

Volume 20 J. A. JACOBS. The Earth's Core. 1975*

Volume 21 DAVID H. MILLER. Water at the Surface of the Earth: An Introduction to Ecosystem Hydrodynamics. 1977

Volume 22 JOSEPH W. CHAMBERLAIN. Theory of Planetary Atmospheres: An Introduction to Their Physics and Chemistry 1978*

Volume 23 JAMES R. HOLTON. An Introduction to Dynamic Meteorology, Second Edition. 1979*

Volume 24 ARNETT S. DENNIS. Weather Modification by Cloud Seeding. 1980*

Volume 25 ROBERT G. FLEAGLE AND JOOST A. BUSINGER. An Introduction to Atmospheric Physics, Second Edition. 1980

Volume 26 KUO-NAN LIOU. An Introduction to Atmospheric Radiation. 1980*

Volume 27 DAVID H. MILLER. Energy at the Surface of the Earth: An Introduction to the Energetics of Ecosystems. 1981

Volume 28 HELMUT G. LANDBERG. The Urban Climate. 1991

Volume 29 M. I. BUDKYO. The Earth's Climate: Past and Future. 1982*

Volume 30 ADRIAN E. GILL. Atmosphere-Ocean Dynamics. 1982

Volume 31 PAOLO LANZANO. Deformations of an Elastic Earth. 1982*

Volume 32 RONALD T. MERRILL AND MICHAEL W. MCELHINNY. The Earth's Magnetic Field: Its History, Origin, and Planetary Perspective. 1983*

Volume 33 JOHN S. LEWIS AND RONALD G. PRINN. Planets and Their Atmospheres: Origin and Evolution. 1983

Volume 34 ROLF MEISSNER. The Continental Crust: A Geophysical Approach. 1986

Volume 35 M. U. SAGITOV, B. BODKI, V. S. NAZARENKO, AND KH. G. TADZHIDINOV. Lunar Gravimetry. 1986

Volume 36 JOSEPH W. CHAMERLAIN AND DONALD M. HUNTEN. Theory of Planetary Atmospheres, 2nd Edition. 1989

Volume 37 J. A. JACOBS. The Earth's Core, 2nd Edition. 1987*

Volume 38 J. R. APEL. Principles of Ocean Physics. 1987

Volume 39 MARTIN A. UMAN. The Lightning Discharge. 1987*

Volume 40 DAVID G. ANDREWS, JAMES R. HOLTON, AND CONWAY B. LEOVY. Middle Atmosphere Dynamics. 1987

Volume 41 PETER WARNECK. Chemistry of the Natural Atmosphere. 1988*

Volume 42 S. PAL ARYA. Introduction to Micrometeorology. 1988*

Volume 43 MICHAEL C. KELLEY. The Earth's Ionosphere. 1989*

Volume 44 WILLIAM R. COTTON AND RICHARD A. ANTHES. Storm and Cloud Dynamics. 1989

Volume 45 WILLIAM MENKE. Geophysical Data Analysis: Discrete Inverse Theory, Revised Edition. 1989

Volume 46 S. GEORGE PHILANDER. El Niño, La Niña, and the Southern Oscillation. 1990

Volume 47 ROBERT A. BROWN. Fluid Mechanics of the Atmosphere. 1991

Volume 48 JAMES R. HOLTON. An Introduction to Dynamic Meteorology, Third Edition. 1992

Volume 49 ALEXANDER A. KAUFMAN. Geophysical Field Theory and Method.
Part A: Gravitational, Electric, and Magnetic Fields. 1992*
Part B: Electromagnetic Fields I. 1994*
Part C: Electromagnetic Fields II. 1994*

Volume 50 SAMUEL S. BUTCHER, GORDON H. ORIANS, ROBERT J. CHARLSON, AND GORDON V. WOLFE. Global Biogeochemical Cycles. 1992*

Volume 51 BRIAN EVANS AND TENG-FONG WONG. Fault Mechanics and Transport Properties of Rocks. 1992

Volume 52 ROBERT E. HUFFMAN. Atmospheric Ultraviolet Remote Sensing. 1992

Volume 53 ROBERT A. HOUZE, JR. Cloud Dynamics. 1993

Volume 54 PETER V. HOBBS. Aerosol-Cloud-Climate Interactions.1993

Volume 55 S. J. GIBOWICZ AND A. KIJKO. An Introduction to Mining Seismology. 1993

Volume 56 DENNIS L. HARTMANN. Global Physical Climatology. 1994

*Out of print.
[NYP]Not yet published.

Volume 57 Michael P. Ryan. Magmatic Systems. 1994

Volume 58 Thorne Lay and Terry C. Wallace. Modern Global Seismology. 1995

Volume 59 Daniel S. Wilks. Statistical Methods in the Atmospheric Sciences. 1995

Volume 60 Frederik Nebeker. Calculating the Weather. 1995

Volume 61 Murry L. Salby. Fundamentals of Atmospheric Physics. 1996

Volume 62 James P. McCalpin. Paleoseismology. 1996

Volume 63 Ronald T. Merrill, Michael W. McElhinny, and Phillip L. McFadden. The Magnetic Field of the Earth: Paleomagnetism, the Core, and the Deep Mantle. 1996

Volume 64 Neil D. Opdyke and James E. T. Channell. Magnetic Stratigraphy. 1996

Volume 65 Judith A. Curry and Peter J. Webster. Thermodynamics of Atmospheres and Oceans. 1998

Volume 66 Lakshmi H. Kantha and Carol Anne Clayson. Numerical Models of Oceans and Oceanic Processes. 2000

Volume 67 Lakshmi H. Kantha and Carol Anne Clayson. Small Scale Processes in Geophysical Fluid Flows. 2000

Volume 68 Raymond S. Bradley. Paleoclimatology, Second Edition. 1999

Volume 69 Lee-Lueng Fu and Anny Cazanave. Satellite Altimetry and Earth Sciences: A Handbook of Techniques and Applications. 2000

Volume 70 David A. Randall. General Circulation Model Development: Past, Present, and Future. 2000

Volume 71 Peter Warneck. Chemistry of the Natural Atmosphere, Second Edition. 2000

Volume 72 Michael C. Jacobson, Robert J. Charleson, Henning Rodhe, and Gordon H. Orians. Earth System Science: From Biogeochemical Cycles to Global Change. 2000

Volume 73 Michael W. McElhinny and Phillip L. McFadden. Paleomagnetism: Continents and Oceans. 2000

Volume 74 Andrew E. Dessler. The Chemistry and Physics of Stratospheric Ozone. 2000

Volume 75 Bruce Douglas, Michael Kearney, and Stephen Leatherman. Sea Level Rise: History and Consequences. 2000

Volume 76 Roman Teisseyre and Eugeniusz Majewski. Earthquake Thermodynamics and Phase Transformations in the Interior. 2001

Volume 77 Gerold Siedler, John Church, and John Gould. Ocean Circulation and Climate: Observing and Modelling The Global Ocean. 2001

Volume 78 Roger A. Pielke Sr. Mesoscale Meteorological Modeling, 2nd Edition. 2001

Volume 79 S. Pal Arya. Introduction to Micrometeorology. 2001

Volume 80 Barry Saltzman. Dynamical Paleoclimatology: Generalized Theory of Global Climate Change. 2002

Volume 81A William H. K. Lee, Hiroo Kanamori, Paul Jennings, and Carl Kisslinger. International Handbook of Earthquake and Engineering Seismology, Part A. 2002

Volume 81B William H. K. Lee, Hiroo Kanamori, Paul Jennings, and Carl Kisslinger. International Handbook of Earthquake and Engineering Seismology, Part B. 2003

Volume 82 Gordon G. Shepherd. Spectral Imaging of the Atmosphere. 2002

Volume 83 Robert P. Pearce. Meteorology at the Millennium. 2001

Volume 84 Kuo-Nan Liou. An Introduction to Atmospheric Radiation, 2nd Ed. 2002

Volume 85 Carmen J. Nappo. An Introduction to Atmospheric Gravity Waves. 2002

Volume 86 Michael E. Evans and Friedrich Heller. Environmental Magnetism: Principles and Applications of Enviromagnetics. 2003

*Out of print.
[NYP]Not yet published.